Algebra
for College Students

Algebra for College Students

Sixth Edition

Robert Blitzer
Miami Dade College

PEARSON
Prentice Hall

Upper Saddle River, New Jersey 07458

Library of Congress Cataloging-in-Publication Data

Blitzer, Robert.
Algebra for college students / Robert Blitzer.-6th ed.
p. cm.
Includes index.
ISBN 0-13-601974-9
1. Algebra—Textbook. I. Title.

QA154.3.B583 2009
512.9—dc22 2007052294

President: *Greg Tobin*
Editor in Chief: *Paul Murphy*
Editorial Director: *Christine Hoag*
Editorial Project Manager: *Dawn Nuttall*
Assistant Editor: *Christine Whitlock*
Editorial Assistant: *Georgina Brown*
Production Project Manager: *Barbara Mack*
Associate Managing Editor: *Bayani Mendoza de Leon*
Senior Managing Editor: *Linda Mihatov Behrens*
Operations Specialist: *Ilene Kahn*
Senior Operations Supervisor: *Diane Peirano*
Executive Director of Development: *Carol Trueheart*
Media Producer: *Audra J. Walsh*
Media Project Manager: *Richard Bretan*
Software Development: *Michael Duguay and Mary Durnwald*
Director of Marketing: *Amy Cronin*
Executive Marketing Manager: *Kate Valentine*
Senior Marketing Manager: *Michella Renda*
Marketing Manager: *Marlana Voerster*
Marketing Assistants: *Jill Kapinus and Nathaniel Koven*
Senior Art Director: *Juan R. López*
Interior Designer: *Juan R. López*
Cover Art Director: *Maureen Eide*
Cover Designer: *John Christiana*
AV Project Manager: *Thomas Benfatti*
Art Manuscript Coordinator: *Dan Missildine*
Cover Photo: *Shutterstock Images*
Manager, Cover Visual Research and Permissions: *Karen Sanatar*
Director, Image Resource Center: *Melinda Patelli*
Manager, Rights and Permissions: *Zina Arabia*
Manager, Visual Research: *Beth Brenzel*
Image Permission Coordinator: *Debbie Hewitson*
Photo Researchers: *Diane Austin; Rachel Lucas*
Art Studios: *Scientific Illustrators/Laserwords*
Compositor: *Prepare, Inc.*

Pearson Prentice Hall
Pearson Education, Inc.
Upper Saddle River, New Jersey 07458

Printed in the United States of America
10 9 8 7 6 5 4 3 2

ISBN-13: 978-0-13-601974-9
ISBN-10: 0-13-601974-9

Pearson Education Ltd., *London*
Pearson Education Australia PTY, Limited, *Sydney*
Pearson Education Singapore, Pte. Ltd
Pearson Education North Asia Ltd, *Hong Kong*
Pearson Education Canada, Ltd., *Toronto*
Pearson Educación de Mexico, S.A. de C.V.
Pearson Education—Japan, *Tokyo*
Pearson Education Malaysia, Pte. Ltd

CONTENTS

CHAPTER 12 Sequences, Induction, and Probability 843

PREFACE

Algebra for College Students, Sixth Edition, is written for students who have had the equivalent of one year of high school algebra. The content of the book is drawn from both intermediate algebra and college algebra, and provides comprehensive coverage of the topics required in a strong one-term course in intermediate algebra or a one-term algebra for college students course. I wrote the book to help diverse students, with different backgrounds and career plans, to succeed in these courses. *Algebra for College Students,* Sixth Edition, has two primary goals:

1. To help students acquire a solid foundation in the skills and concepts of algebra.
2. To show students how algebra can model and solve authentic real-world problems.

One major obstacle in the way of achieving these goals is the fact that very few students actually read their textbook. This has been a regular source of frustration for me and for my colleagues in the classroom. Anecdotal evidence gathered over years highlights two basic reasons why students do not take advantage of their textbook:

- "I'll never use this information."
- "I can't follow the explanations."

I've written every page of the Sixth Edition with the intent of eliminating these two objections. The ideas and tools I've used to do so are described in the features that follow. These features and their benefits are highlighted for the student in "A Brief Guide to Getting the Most from This Book," which appears inside the front cover.

What's New in the Sixth Edition?

- **New Applications and Real World Data.** I'm on a constant search for real-world data that can be used to illustrate unique algebraic applications. I researched hundreds of books, magazines, newspapers, almanacs, and online sites to prepare the Sixth Edition. With 244 worked-out examples and application exercises based on new data sets, the Sixth Edition contains a greater array of applications than any previous revision of this book.
- **"Make Sense?" Classroom Discussion Exercises.** Each exercise set contains four Critical Thinking exercises intended for classroom discussion in order to engage participation in the learning process. These items test conceptual understanding by asking students to determine whether statements are sensible, and to explain why or why not. Although sample answers are provided, students have skills and perspectives that frequently differ from those of math teachers, so answers and explanations may vary. The important part of this new feature is to let you ask students what they think about selected statements, determine whether they understand the concepts, and give them feedback to clarify any misunderstandings.
- **New Directions for the True/False Critical Thinking Exercises.** The Sixth Edition asks students to determine whether each statement in an itemized list is true or false. If the statement is false, students are then asked to make the necessary change or changes to produce a true statement.
- **Preview Exercises.** Each exercise set concludes with three problems to help students prepare for the next section. Some of these problems review previously covered material that students will need to be successful in the forthcoming section. Other problems are designed to get students thinking about concepts they will soon encounter.
- **More Detailed Directions When Comparing Mathematical Models with Actual Data.** The Sixth Edition asks students if values obtained from mathematical models underestimate or overestimate data displayed by graphs, and, if so, by how much.
- **Increased Study Tip Boxes.** The book's Study Tip boxes offer suggestions for problem solving, point out common errors to avoid, and provide informal hints and suggestions. These invaluable hints appear in greater abundance in the Sixth Edition.

- **New Chapter-Opening and Section-Opening Scenarios.** Every chapter and every section open with a scenario based on an application, the majority of which are unique to the Sixth Edition. These scenarios are revisited in the course of the chapter or section in one of the book's new examples, exercises, or discussions. The often-humorous tone of these openers is intended to help fearful and reluctant students overcome their negative perceptions about math.
- **819 New Examples and Exercises.** The Sixth Edition contains 29 detailed worked-out examples involving new data, 215 new application exercises, 280 "make sense" discussion exercises, 210 preview exercises, and 85 new exercises that appear in the various other categories of the exercise sets.

What Content and Organizational Changes Have Been Made to the Sixth Edition?

- **Section 2.1** (Introduction to Functions) and **Section 2.2** (Graphs of Functions) introduce functions over two sections, rather than only one section, as in the previous edition. This gradual approach allows for a discussion of functions represented by tables in Section 2.1.
- **Section 2.5** (The Point-Slope Form of the Equation of a Line) contains a more thoroughly developed example on writing equations of lines perpendicular to a given line.
- **Section 4.3** (Equations and Inequalities Involving Absolute Value) uses boundary points to show students how solving inequalities involving absolute value is connected to the graph of $f(x) = |x|$. Instructors are given the option of solving absolute value inequalities using boundary points or by the more traditional method of rewriting as equivalent compound inequalities.
- **Section 9.5** (Exponential and Logarithmic Equations) has been reorganized into four categories:

  Solving exponential equations using like bases
  Solving exponential equations using logarithms and logarithmic properties
  Solving logarithmic equations using the definition of a logarithm
  Solving logarithmic equations using the one-to-one property of logarithms.

  New examples appear throughout the section to ensure adequate coverage of each category.
- **Section 11.2** (Zeros of Polynomial Functions) contains a new discussion on the various kinds of zeros and a new example on finding these zeros.

What Familiar Features Have Been Retained in the Sixth Edition?

- **Detailed Worked-Out Examples.** Each worked example is titled, making clear the purpose of the example. Examples are clearly written and provide students with detailed step-by-step solutions. No steps are omitted and key steps are thoroughly explained to the right of the mathematics.
- **Explanatory Voice Balloons.** Voice balloons are used in a variety of ways to demystify mathematics. They translate algebraic ideas into everyday English, help clarify problem-solving procedures, present alternative ways of understanding concepts, and connect problem solving to concepts students have already learned.
- **Check Point Examples.** Each example is followed by a similar matched problem, called a Check Point, offering students the opportunity to test their understanding of the example by working a similar exercise. The answers to the Check Points are provided in the answer section.
- **Extensive and Varied Exercise Sets.** An abundant collection of exercises is included in an exercise set at the end of each section. Exercises are organized within eight category types: Practice Exercises, Practice Plus Exercises, Application Exercises, Writing in Mathematics, Technology Exercises, Critical Thinking Exercises,

Review Exercises, and Preview Exercises. This format makes it easy to create well-rounded homework assignments. The order of the practice exercises is exactly the same as the order of the section's worked examples. This parallel order enables students to refer to the titled examples and their detailed explanations to achieve success working the practice exercises.

- **Practice Plus Problems.** This category of exercises contains more challenging practice problems that often require students to combine several skills or concepts. With an average of ten practice plus problems per exercise set, instructors are provided with the option of creating assignments that take practice exercises to a more challenging level.
- **Mid-Chapter Check Points.** At approximately the midway point in each chapter, an integrated set of review exercises allows students to review and assimilate the skills and concepts they learned separately over several sections.
- **Graphing and Functions.** Graphing is introduced in Chapter 1 and functions are introduced in Chapter 2, with an integrated graphing functional approach emphasized throughout the book. Graphs and functions that model data appear in nearly every section and exercise set. Examples and exercises use graphs of functions to explore relationships between data and to provide ways of visualizing a problem's solution. Because functions are the core of this course, students are repeatedly shown how functions relate to equations and graphs.
- **Section Objectives.** Learning objectives are clearly stated at the beginning of each section. These objectives help students recognize and focus on the section's most important ideas. The objectives are restated in the margin at their point of use.
- **Integration of Technology Using Graphical and Numerical Approaches to Problems.** Side-by-side features in the technology boxes connect algebraic solutions to graphical and numerical approaches to problems. Although the use of graphing utilities is optional, students can use the explanatory voice balloons to understand different approaches to problems even if they are not using a graphing utility in the course.
- **Chapter Review Grids.** Each chapter contains a review chart that summarizes the definitions and concepts in every section of the chapter. Examples that illustrate these key concepts are also included in the chart.
- **End-of-Chapter Materials.** A comprehensive collection of review exercises for each of the chapter's sections follows the review grid. This is followed by a chapter test that enables students to test their understanding of the material covered in the chapter. Beginning with Chapter 2, each chapter concludes with a comprehensive collection of mixed cumulative review exercises.
- **Chapter Test Prep Video CD.** Packaged at the front of the text, this video CD provides students with step-by-step solutions for each of the exercises in the book's chapter tests.
- **Blitzer Bonuses.** These enrichment essays provide historical, interdisciplinary, and otherwise interesting connections to the algebra under study, showing students that math is an interesting and dynamic discipline.
- **Discovery.** Discover for Yourself boxes, found throughout the text, encourage students to further explore algebraic concepts. These explorations are optional and their omission does not interfere with the continuity of the topic under consideration.
- **Chapter Projects.** At the end of each chapter is a collaborative activity that gives students the opportunity to work cooperatively as they think and talk about mathematics. Additional group projects can be found in the *Instructor's Resource Manual.* Many of these exercises should result in interesting group discussions.

I hope that my passion for teaching, as well as my respect for the diversity of students I have taught and learned from over the years, is apparent throughout this new edition. By connecting algebra to the whole spectrum of learning, it is my intent to show students that their world is profoundly mathematical, and indeed, π is in the sky.

Robert Blitzer

Resources for the Sixth Edition

FOR STUDENTS

Student Solutions Manual Fully worked solutions to the odd-numbered section exercises plus all Check Points, Review/Preview Exercises, Mid-Chapter Check Points, Chapter Reviews, Chapter Tests, and Cumulative Reviews.

Worksheets Provide a ready-to-use lesson and exercise set for every section of the text with ample student work space.

CD Lecture Series A comprehensive set of CD-ROMS, tied to the textbook, containing short video clips of an instructor working key text examples/exercises. (Also available separately on DVD.)

MathXL® Tutorials on CD This interactive tutorial CD-ROM provides algorithmically generated practice exercises that are correlated at the objective level to the exercises in the textbook. Every practice exercise is accompanied by an example and a guided solution designed to involve students in the solution process.

Chapter Test Prep Video CD Provides step-by-step video solutions to each problem in each Chapter Test in the textbook. Packaged with a new text, inside the front cover.

FOR INSTRUCTORS

Instructor Resource Distribution Most instructor resources can be downloaded from the Web site, *www.prenhall.com*. Select "Browse our catalog," then click on "mathematics," select your course and choose your text. Under "resources," on the left side, select "instructor" and choose the supplement you need to download. You will be required to run through a one time registration before you can complete this process.

Math Adjunct Support Center The Pearson Math Adjunct Support Center is staffed by qualified mathematics instructors with over 50 years of combined experience at both the community college and university level. Assistance is provided for faculty in the following areas:

- Suggested syllabus consultation
- Tips on using materials packed with your book
- Book-specific content assistance
- Teaching suggestions including advice on classroom strategies

Instructor's Solutions Manual Fully worked solutions to every exercise in the text.

Instructor's Resource Manual with Tests Includes a Mini-Lecture, Skill Builder, and Additional Exercises for every section of the text; two short group Activities per chapter, several chapter test forms, both free-response and multiple-choice, as well as cumulative tests and final exams. Answers to all items also included.

TestGen® Easily create tests from section objectives. Questions are algorithmically generated allowing for unlimited versions. Edit problems or create your own. There's a chapter test file for each Chapter Test in the text.

Annotated Instructor's Edition Answers to exercises are printed on the same text page with graphing answers in a special Graphing Answer Section in the back of the text.

FOR BOTH

MathXL®

MathXL® is a powerful online homework, tutorial, and assessment system that accompanies this textbook. Instructors can create, edit, and assign online homework and tests using algorithmically generated exercises correlated at the objective level to the textbook. Student work is tracked in an online gradebook. Students can take chapter tests and receive personalized study plans based on their results. The study plan diagnoses weaknesses and links students directly to tutorial exercises for objectives they need to study. Students can also access video clips directly from selected exercises. MathXL is available to qualified adopters. For more information, visit our website at *www.mathxl.com* or contact your Prentice Hall sales representative.

MyMathLab®

MyMathLab® is a series of text-specific, customizable online courses for your textbooks. Powered by CourseCompass™ (Pearson Education's online teaching and learning environment) and MathXL® (our online homework, tutorial, and assessment system), MyMathLab gives you the tools you need to deliver all or a portion of your course online, whether students are in a lab setting or working from home. MyMathLab provides a rich and flexible set of course materials, featuring free-response exercises that are algorithmically generated for unlimited practice. Students can also use online tools, such as video lectures and a multimedia textbook, to improve their performance. Instructors can use MyMathLab's homework and test managers to select and assign online exercises correlated to the textbook, and can import TestGen tests for added flexibility. The online gradebook—designed specifically for mathematics—automatically tracks students' homework and test results and gives the instructor control over how to calculate final grades. MyMathLab also includes access to **Pearson's Tutor Center**, which provides students with tutoring via toll-free phone, fax, email, and interactive Web sessions. MyMathLab is available to qualified adopters. For more information, visit our website at *www.mymathlab.com* or contact your Pearson sales representative.

Acknowledgments

An enormous benefit of authoring a successful series is the broad-based feedback I receive from the students, dedicated users, and reviewers. Every change to this edition is the result of their thoughtful comments and suggestions. I would like to express my appreciation to all the reviewers, whose collective insights form the backbone of this revision. In particular, I would like to thank the following people for reviewing *Algebra for College Students*.

Gwen P. Aldridge	*Northwest Mississippi Community College*
Howard Anderson	*Skagit Valley College*
John Anderson	*Illinois Valley Community College*
Michael H. Andreoli	*Miami Dade College–North Campus*
Jan Archibald	*Ventura College*
Donna Beatty	*Ventura College*
Michael S. Bowen	*Ventura College*
Gale Brewer	*Amarillo College*
Hien Bui	*Hillsborough Community College*
Warren J. Burch	*Brevard Community College*
Alice Burstein	*Middlesex Community College*
Edie Carter	*Amarillo College*
Thomas B. Clark	*Trident Technical College*
Sandra Pryor Clarkson	*Hunter College*
Bettyann Daley	*University of Delaware*
Robert A. Davies	*Cuyahoga Community College*
Paige Davis	*Lurleen B. Wallace Community College*
Ben Divers, Jr.	*Ferrum College*
Irene Doo	*Austin Community College*
Charles C. Edgar	*Onondaga Community College*
Rhoderick Fleming	*Wake Technical Community College*
Susan Forman	*Bronx Community College*
Donna Gerken	*Miami-Dade College*
Marion K. Glasby	*Anne Arundel Community College*
Sue Glascoe	*Mesa Community College*
Jay Graening	*University of Arkansas*
Robert B. Hafer	*Brevard Community College*
Mary Lou Hammond	*Spokane Community College*
Donald Herrick	*Northern Illinois University*
Beth Hooper	*Golden West College*
Tracy Hoy	*College of Lake County*
Judy Kasabian	*Lansing Community College*
Gary Kersting	*North Central Michigan College*
Gary Knippenberg	*Lansing Community College*
Mary Kochler	*Cuyahoga Community College*
Mary A. Koehler	*Cuyahoga Community College*
Kristi Laird	*Jackson Community College*
Jennifer Lempke	*North Central Michigan College*
Sandy Lofstock	*St. Petersburg College*
Hank Martel	*Broward Community College*
Diana Martelly	*Miami-Dade College*

John Robert Martin	*Tarrant County Junior College*
Mikal McDowell	*Cedar Valley College*
Irwin Metviner	*State University of New York at Old Westbury*
Terri Moser	*Austin Community College*
Robert Musselman	*California State University, Fresno*
Kamilia Nemri	*Spokane Community College*
Allen R. Newhart	*Parkersburg Community College*
Steve O'Donnell	*Rogue Community College*
Jeff Parent	*Oakland Community College*
Tian Ren	*Queensborough Community College*
Kate Rozsa	*Mesa Community College*
Scott W. Satake	*Eastern Washington University*
Mike Schramm	*Indian River Community College*
Terri Seiver	*San Jacinto College*
Kathy Shepard	*Monroe County Community College*
Gayle Smith	*Lane Community College*
Linda Smoke	*Central Michigan University*
Dick Spangler	*Tacoma Community College*
Janette Summers	*University of Arkansas*
Kory Swart	*Kirkwood Community College*
Robert Thornton	*Loyola University*
Lucy C. Thrower	*Francis Marion College*
Andrew Walker	*North Seattle Community College*
Kathryn Wetzel	*Amarillo College*
Margaret Williamson	*Milwaukee Area Technical College*
Roberta Yellott	*McNeese State University*
Marilyn Zopp	*McHenry County College*

Additional acknowledgments are extended to Dan Miller and Kelly Barber, for preparing the solutions manuals, Brad Davis, for preparing the answer section and serving as accuracy checker, the Preparè, Inc. formatting team, for the book's brilliant paging, Aaron Darnall at Scientific Illustrators, for superbly illustrating the book, Diane Austin and Rachel Lucas, photo researchers, for obtaining the book's new photographs, and Barbara Mack, whose talents as production editor kept every aspect of this complex project moving through its many stages.

I would like to thank my editors at Prentice Hall, Paul Murphy and Chris Hoag, and Project Manager, Dawn Nuttall, who guided and coordinated the book from manuscript through production. Thanks to Maureen Eide and John Christiana for the beautiful cover, Juan López for the wonderful interior design, Kate Valentine and Patrice Jones for their innovative marketing efforts, and to the entire Pearson Education sales force for their confidence and enthusiasm about the book.

Robert Blitzer

TO THE STUDENT

I have written this book so that you can learn about the power of algebra and how it relates directly to your life outside the classroom. All concepts are carefully explained, important definitions and procedures are set off in boxes, and worked-out examples that present solutions in a step-by-step manner appear in every section. Each example is followed by a similar matched problem, called a Check Point, for you to try so that you can actively participate in the learning process as you read the book. (Answers to all Check Points appear in the back of the book.) Study Tips offer hints and suggestions and often point out common errors to avoid. A great deal of attention has been given to applying algebra to your life to make your learning experience both interesting and relevant.

As you begin your studies, I would like to offer some specific suggestions for using this book and for being successful in this course:

- **Attend all lectures.** No book is intended to be a substitute for valuable insights and interactions that occur in the classroom. In addition to arriving for lecture on time and being prepared, you will find it useful to read the section before it is covered in lecture. This will give you a clear idea of the new material that will be discussed.
- **Read the book.** Read each section with pen (or pencil) in hand. Move through the worked-out examples with great care. These examples provide a model for doing exercises in the exercise sets. As you proceed through the reading, do not give up if you do not understand every single word. Things will become clearer as you read on and see how various procedures are applied to specific worked-out examples.
- **Work problems every day and check your answers.** The way to learn mathematics is by doing mathematics, which means working the Check Points and assigned exercises in the exercise sets. The more exercises you work, the better you will understand the material.
- **Review for quizzes and tests.** After completing a chapter, study the chapter review chart, work the exercises in the Chapter Review, and work the exercises in the Chapter Test. Answers to all these exercises are given in the back of the book.

  The methods that I've used to help you read the book, work the problems, and review for tests are described in "A Brief Guide to Getting the Most from This Book" that appears inside the front cover. Spend a few minutes reviewing the guide to familiarize yourself with the book's features and their benefits.

- **Use the resources available with this book.** Additional resources to aid your study are described on page xiv. These resources include a Solutions Manual; a Chapter Test Prep Video CD; MyMathLab®, an online version of the book with links to multimedia resources; MathXL®, an online homework, tutorial, and assessment system of the text; and tutorial support at the Pearson Tutor Services.

I wrote this book in Point Reyes National Seashore, 40 miles north of San Francisco. The park consists of 75,000 acres with miles of pristine surf-washed beaches, forested ridges, and bays bordered by white cliffs. It was my hope to convey the beauty and excitement of mathematics using nature's unspoiled beauty as a source of inspiration and creativity. Enjoy the pages that follow as you empower yourself with the algebra needed to succeed in college, your career, and in your life.

Regards,

Bob

Robert Blitzer

Bob and his horse Jerid

ABOUT THE AUTHOR

Bob Blitzer is a native of Manhattan and received a Bachelor of Arts degree with dual majors in mathematics and psychology (minor: English literature) from the City College of New York. His unusual combination of academic interests led him toward a Master of Arts in mathematics from the University of Miami and a doctorate in behavioral sciences from Nova University. Bob's love for teaching mathematics was nourished for nearly 30 years at Miami Dade College, where he received numerous teaching awards, including Innovator of the Year from the League for Innovations in the Community College and an endowed chair based on excellence in the classroom. In addition to *Algebra for College Students,* Bob has written textbooks covering intoductory algebra, intermediate algebra, college algebra, algebra and trigonometry, precalculus, and liberal arts mathematics, all published by Prentice Hall. When not secluded in his Northern California writer's cabin, Bob can be found hiking the beaches and trails of Point Reyes National Seashore, and tending to the chores required by his beloved entourage of horses, chickens, and irritable roosters.

CHAPTER 1

Algebra, Mathematical Models, and Problem Solving

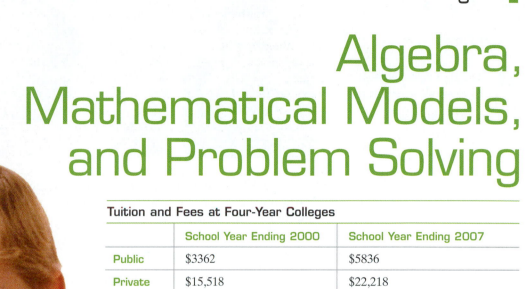

Tuition and Fees at Four-Year Colleges

	School Year Ending 2000	School Year Ending 2007
Public	\$3362	\$5836
Private	\$15,518	\$22,218

Source: The College Board

The cost of a college education is skyrocketing. If these trends continue, what can we expect by the 2010s and beyond? We can answer this question by representing the data mathematically. With such representations, called *mathematical models,* we can gain insights and predict what might occur in the future on a variety of issues, ranging from college costs to a possible Social Security doomsday, and even the changes that occur as we age.

Mathematical models involving college costs appear as Exercises 67–68 in Exercise Set 1.4. The insecurities of Social Security are explored in Exercise 75 in the Review Exercises. Some surprising changes that occur with aging appear as Example 2 in Section 1.1, Exercises 75–78 in Exercise Set 1.1, and Exercises 53–56 in Exercise Set 1.3.

SECTION 1.1

Algebraic Expressions and Real Numbers

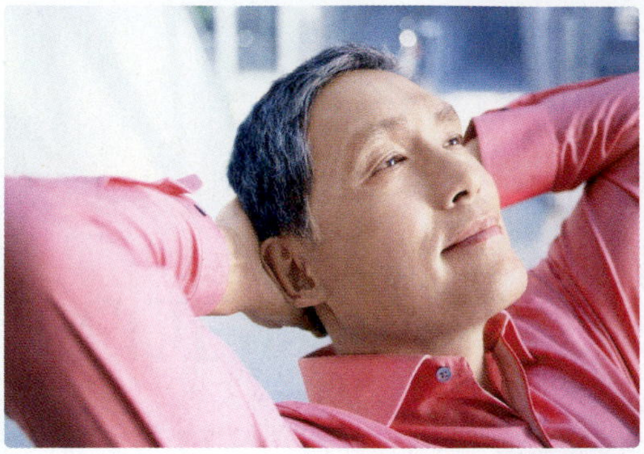

Objectives

1. Translate English phrases into algebraic expressions.
2. Evaluate algebraic expressions.
3. Use mathematical models.
4. Recognize the sets that make up the real numbers.
5. Use set-builder notation.
6. Use the symbols $\in$ and $\notin$.
7. Use inequality symbols.

As we get older, do we mellow out or become more neurotic? In this section, you will learn how the special language of algebra describes your world, including our improving emotional health with age.

Algebraic Expressions

Algebra uses letters, such as x and y, to represent numbers. If a letter is used to represent various numbers, it is called a **variable**. For example, imagine that you are basking in the sun on the beach. We can let x represent the number of minutes that you can stay in the sun without burning with no sunscreen. With a number 6 sunscreen, exposure time without burning is six times as long, or 6 times x. This can be written $6 \cdot x$, but it is usually expressed as $6x$. Placing a number and a letter next to one another indicates multiplication.

Notice that $6x$ combines the number 6 and the variable x using the operation of multiplication. A combination of variables and numbers using the operations of addition, subtraction, multiplication, or division, as well as powers or roots, is called an **algebraic expression**. Here are some examples of algebraic expressions:

$$x + 6, \quad x - 6, \quad 6x, \quad \frac{x}{6}, \quad 3x + 5, \quad x^2 - 3, \quad \sqrt{x} + 7.$$

Is every letter in algebra a variable? No. Some letters stand for a particular number. Such a letter is called a **constant**. For example, let d = the number of days in a week. The letter d represents just one number, namely 7, and is a constant.

1 Translate English phrases into algebraic expressions.

Translating English Phrases into Algebraic Expressions

Problem solving in algebra involves translating English phrases into algebraic expressions. Here is a list of words and phrases for the four basic operations:

Addition	Subtraction	Multiplication	Division
sum	difference	product	quotient
plus	minus	times	divide
increased by	decreased by	of (used with fractions)	per
more than	less than	twice	ratio

EXAMPLE 1 Translating English Phrases into Algebraic Expressions

Write each English phrase as an algebraic expression. Let x represent the number.

a. Nine less than six times a number

b. The quotient of five and a number, increased by twice the number

Solution

a. Nine less than six times a number

$$6x - 9$$

b. The quotient of five and a number, increased by twice the number

$$\frac{5}{x} + 2x$$

✓ CHECK POINT 1 Write each English phrase as an algebraic expression. Let x represent the number.

a. Five more than 8 times a number

b. The quotient of a number and 7, decreased by twice the number

Study Tip

After working each Check Point, check your answer in the answer section before continuing your reading.

2 Evaluate algebraic expressions.

Evaluating Algebraic Expressions

Evaluating an algebraic expression means to find the value of the expression for a given value of the variable.

EXAMPLE 2 Evaluating an Algebraic Expression

A test measuring neurotic traits, such as anxiety and hostility, indicates that people may become less neurotic as they get older. **Figure 1.1** shows the average level of neuroticism, on a scale of 0 to 50, for persons at various ages.

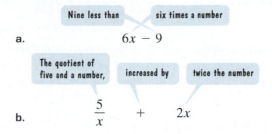

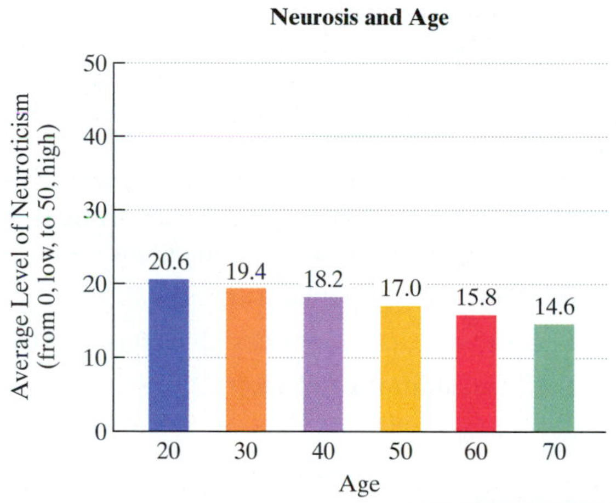

FIGURE 1.1

Source: L. M. Williams, "The Mellow Years? Neural Basis of Improving Emotional Stability over Age," *The Journal of Neuroscience*, June 14, 2006.

The algebraic expression $23 - 0.12x$ describes the average neurotic level for people who are x years old. Evaluate the expression for $x = 80$. Describe what the answer means in practical terms.

Solution We begin by substituting 80 for x. Because $x = 80$, we will be finding the average neurotic level at age 80.

$$23 - 0.12x$$

Replace x with 80.

$$= 23 - 0.12(80) = 23 - 9.6 = 13.4$$

Thus, at age 80, the average level of neuroticism on a scale of 0 to 50 is 13.4.

✓ CHECK POINT 2 Evaluate the expression from Example 2, $23 - 0.12x$, for $x = 10$. Describe what the answer means in practical terms.

Many algebraic expressions involve *exponents*. For example, the algebraic expression

$$233x^2 + 2296x + 16{,}197$$

approximates the number of registered lobbyists in Washington x years after 2000. The expression x^2 means $x \cdot x$, and is read "x to the second power" or "x squared." The exponent, 2, indicates that the base, x, appears as a factor two times.

Exponential Notation

If n is a counting number (1, 2, 3, and so on),

Exponent or Power

$$b^n = \underbrace{b \cdot b \cdot b \cdot \cdots \cdot b}_{\text{b appears as a factor n times.}}$$

Base

b^n is read "the nth power of b" or "b to the nth power." Thus, the nth power of b is defined as the product of n factors of b. The expression b^n is called an **exponential expression**. Furthermore, $b^1 = b$.

Using Technology

You can use a calculator to evaluate exponential expressions. For example, to evaluate 2^4, press the following keys:

Many Scientific Calculators

2 $\boxed{y^x}$ 4 $\boxed{=}$

Many Graphing Calculators

2 $\boxed{\wedge}$ 4 $\boxed{\text{ENTER}}$

Although calculators have special keys to evaluate powers of ten and to square bases, you can always use one of the sequences shown here.

For example,

$$8^2 = 8 \cdot 8 = 64, \quad 5^3 = 5 \cdot 5 \cdot 5 = 125, \quad \text{and} \quad 2^4 = 2 \cdot 2 \cdot 2 \cdot 2 = 16.$$

Many algebraic expressions involve more than one operation. Evaluating an algebraic expression without a calculator involves carefully applying the following order of operations agreement:

The Order of Operations Agreement

1. Perform operations within the innermost parentheses and work outward. If the algebraic expression involves a fraction, treat the numerator and the denominator as if they were each enclosed in parentheses.
2. Evaluate all exponential expressions.
3. Perform multiplications and divisions as they occur, working from left to right.
4. Perform additions and subtractions as they occur, working from left to right.

EXAMPLE 3 Evaluating an Algebraic Expression

Evaluate $7 + 5(x - 4)^3$ for $x = 6$.

Solution

$7 + 5(x - 4)^3 = 7 + 5(6 - 4)^3$ — Replace x with 6.

$= 7 + 5(2)^3$ — First work inside parentheses: $6 - 4 = 2$.

$= 7 + 5(8)$ — Evaluate the exponential expression: $2^3 = 2 \cdot 2 \cdot 2 = 8$.

$= 7 + 40$ — Multiply: $5(8) = 40$.

$= 47$ — Add.

✓ **CHECK POINT 3** Evaluate $8 + 6(x - 3)^2$ for $x = 13$.

3 Use mathematical models.

Formulas and Mathematical Models

An **equation** is formed when an equal sign is placed between two algebraic expressions. One aim of algebra is to provide a compact, symbolic description of the world. These descriptions involve the use of *formulas*. A **formula** is an equation that uses variables to express a relationship between two or more quantities. Here is an example of a formula:

$$C = \frac{5}{9}(F - 32).$$

Celsius temperature | is | $\frac{5}{9}$ of | the difference between Fahrenheit temperature and 32°.

The process of finding formulas to describe real-world phenomena is called **mathematical modeling**. Such formulas, together with the meaning assigned to the variables, are called **mathematical models**. We often say that these formulas model, or describe, the relationships among the variables.

EXAMPLE 4 Modeling the Number of Lobbyists in Washington

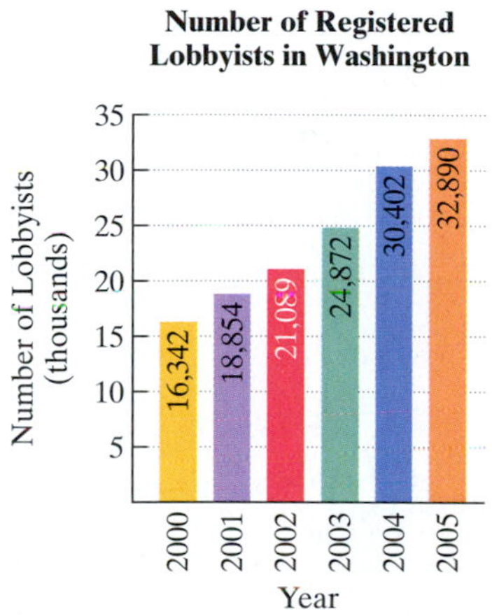

FIGURE 1.2

Source: Senate Office of Public Records

In 2006, the ease with which conniving lobbyists bought members of Congress put pressure on lawmakers to mend a broken system. The formula

$$L = 233x^2 + 2296x + 16{,}197$$

models the number of registered lobbyists, L, in Washington, x years after 2000.

a. Use the formula to find the number of lobbyists in Washington in 2004.

b. By how much is the model value for 2004 greater than or less than the actual data value shown in **Figure 1.2**?

Solution

a. Because 2004 is 4 years after 2000, we substitute 4 for x in the given formula. Then we use the order of operations to find L, the number of lobbyists in 2004.

$L = 233x^2 + 2296x + 16{,}197$	This is the given mathematical model.
$L = 233(4)^2 + 2296(4) + 16{,}197$	Replace each occurrence of x with 4.
$L = 233(16) + 2296(4) + 16{,}197$	Evaluate the exponential expression: $4^2 = 4 \cdot 4 = 16$.
$L = 3728 + 9184 + 16{,}197$	Multiply from left to right: $233(16) = 3728$ and $2296(4) = 9184$.
$L = 29{,}109$	Add.

The formula indicates that in 2004, there were 29,109 lobbyists in Washington.

b. The number of lobbyists for 2004 given in **Figure 1.2** is 30,402. The model value, 29,109, is less than the actual data value by 30,402 − 29,109, or by 1293 lobbyists.

✓ CHECK POINT 4

a. Use the formula in Example 4 to find the number of lobbyists in Washington in 2005.

b. By how much is the model value for 2005 greater than or less than the actual data value shown in **Figure 1.2**?

Sometimes a mathematical model gives an estimate that is not a good approximation or is extended to include values of the variable that do not make sense. In these cases, we say that **model breakdown** has occurred. Models that accurately describe data for the past ten years might not serve as reliable predictions for what can reasonably be expected to occur in the future. Model breakdown can occur when formulas are extended too far into the future.

4 Recognize the sets that make up the real numbers.

The Set of Real Numbers

Before we describe the set of real numbers, let's be sure you are familiar with some basic ideas about sets. A **set** is a collection of objects whose contents can be clearly determined. The objects in a set are called the **elements** of the set. For example, the set of numbers used for counting can be represented by

$$\{1, 2, 3, 4, 5, \dots\}.$$

Study Tip

Grouping symbols such as parentheses, ( ), and square brackets, [ ], are not used to represent sets. Only commas are used to separate the elements of a set. Separators such as colons or semicolons are not used.

The braces, { }, indicate that we are representing a set. This form of representation, called the **roster method**, uses commas to separate the elements of the set. The three dots after the 5, called an *ellipsis*, indicate that there is no final element and that the listing goes on forever.

Three common sets of numbers are the *natural numbers*, the *whole numbers*, and the *integers*.

Natural Numbers, Whole Numbers, and Integers

The Set of Natural Numbers

$$\{1, 2, 3, 4, 5, \dots\}$$

These are the numbers that we use for counting.

The Set of Whole Numbers

$$\{0, 1, 2, 3, 4, 5, \dots\}$$

The set of whole numbers includes 0 and the natural numbers.

The Set of Integers

$$\{\dots, -5, -4, -3, -2, -1, 0, 1, 2, 3, 4, 5, \dots\}$$

The set of integers includes the negatives of the natural numbers and the whole numbers.

5 Use set-builder notation.

A set can also be written in **set-builder notation**. In this notation, the elements of the set are described, but not listed. Here is an example:

$$\{x \mid x \text{ is a natural number less than } 6\}.$$

The set of all x | such that | x is a natural number less than 6.

The same set is written using the roster method as

$$\{1, 2, 3, 4, 5\}.$$

6 Use the symbols $\in$ and $\notin$.

The symbol $\in$ is used to indicate that a number or object is in a particular set. The symbol $\in$ is read "is an element of." Here is an example:

$$7 \in \{1, 2, 3, 4, 5, \dots\}.$$

7 | is an element of | the set of natural numbers.

The symbol $\notin$ is used to indicate that a number or object is not in a particular set. The symbol $\notin$ is read "is not an element of." Here is an example:

$$\frac{1}{2} \notin \{1, 2, 3, 4, 5, \ldots\}.$$

$\frac{1}{2}$ | is not an element of | the set of natural numbers.

EXAMPLE 5 Using the Symbols $\in$ and $\notin$

Determine whether each statement is true or false:

a. $100 \in \{x \mid x \text{ is an integer}\}$ **b.** $20 \notin \{5, 10, 15\}$.

Solution

a. Because 100 is an integer, the statement

$$100 \in \{x \mid x \text{ is an integer}\}$$

is true. The number 100 is an element of the set of integers.

b. Because 20 is not an element of $\{5, 10, 15\}$, the statement $20 \notin \{5, 10, 15\}$ is true.

✓ CHECK POINT 5 Determine whether each statement is true or false:

a. $13 \in \{x \mid x \text{ is an integer}\}$ **b.** $6 \notin \{7, 8, 9, 10\}$.

Another common set is the set of *rational numbers*. Each of these numbers can be expressed as an integer divided by a nonzero integer.

Rational Numbers

The set of **rational numbers** is the set of all numbers that can be expressed as a quotient of two integers, with the denominator not 0.

$$\left\{\frac{a}{b} \,\middle|\, a \text{ and } b \text{ are integers and } b \neq 0\right\}$$

This means that b is not equal to zero.

Three examples of rational numbers are

$$\frac{1}{4}, \quad \frac{-2}{3}, \quad \text{and } 5 = \frac{5}{1}.$$

$a = 1$, $b = 4$ | $a = -2$, $b = 3$ | $a = 5$, $b = 1$

Can you see that integers are also rational numbers because they can be written in terms of division by 1?

Rational numbers can be expressed in fraction or decimal notation. To express the fraction $\frac{a}{b}$ as a decimal, divide the denominator, b, into the numerator, a. In decimal notation, rational numbers either terminate (stop) or have a digit, or block of digits, that repeats. For example,

$$\frac{3}{8} = 3 \div 8 = 0.375 \quad \text{and} \quad \frac{7}{11} = 7 \div 11 = 0.6363\ldots = 0.\overline{63}.$$

The decimal stops: it is a terminating decimal. | This is a repeating decimal. The bar is written over the repeating part.

Some numbers cannot be expressed as terminating or repeating decimals. An example of such a number is $\sqrt{2}$, the square root of 2. The number $\sqrt{2}$ is a number that can be squared to give 2. No terminating or repeating decimal can be squared to get 2. However, some approximations have squares that come close to 2. We use the symbol $\approx$, which means "is approximately equal to."

- $\sqrt{2} \approx 1.4$ because $(1.4)^2 = (1.4)(1.4) = 1.96$.
- $\sqrt{2} \approx 1.41$ because $(1.41)^2 = (1.41)(1.41) = 1.9881$.
- $\sqrt{2} \approx 1.4142$ because $(1.4142)^2 = (1.4142)(1.4142) = 1.99996164$.

$\sqrt{2}$ is an example of an *irrational number*.

Irrational Numbers

The set of **irrational numbers** is the set of numbers whose decimal representations neither terminate nor repeat. Irrational numbers cannot be expressed as quotients of integers.

Examples of irrational numbers include

$$\sqrt{3} \approx 1.73205 \qquad \text{and} \qquad \pi(\text{pi}) \approx 3.141593.$$

Not all square roots are irrational. For example, $\sqrt{25} = 5$ because $5^2 = 5 \cdot 5 = 25$. Thus, $\sqrt{25}$ is a natural number, a whole number, an integer, and a rational number $\left(\sqrt{25} = \frac{5}{1}\right)$.

The set of *real numbers* is formed by combining the sets of rational numbers and irrational numbers. Thus, every real number is either rational or irrational, as shown in **Figure 1.3**.

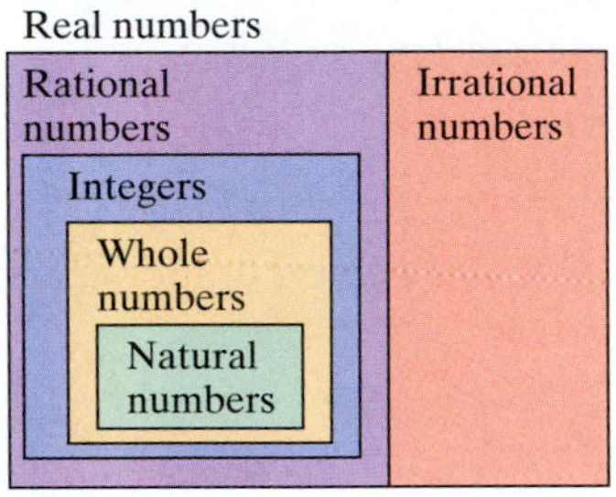

FIGURE 1.3 Every real number is either rational or irrational.

Real Numbers

The set of **real numbers** is the set of numbers that are either rational or irrational:

$$\{x \mid x \text{ is rational or } x \text{ is irrational}\}.$$

The Real Number Line

The **real number line** is a graph used to represent the set of real numbers. An arbitrary point, called the **origin**, is labeled 0. Select a point to the right of 0 and label it 1. The distance from 0 to 1 is called the **unit distance**. Numbers to the right of the origin are **positive** and numbers to the left of the origin are **negative**. The real number line is shown in **Figure 1.4**.

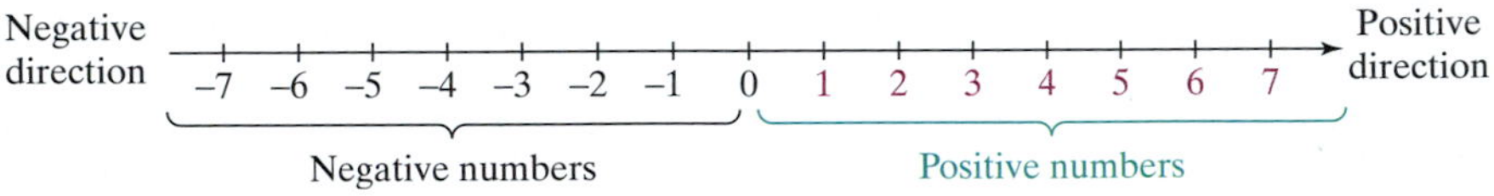

FIGURE 1.4 The real number line

Real numbers are **graphed** on a number line by placing a dot at the correct location for each number. The integers are easiest to locate. In **Figure 1.5**, we've graphed six rational numbers and three irrational numbers on a real number line.

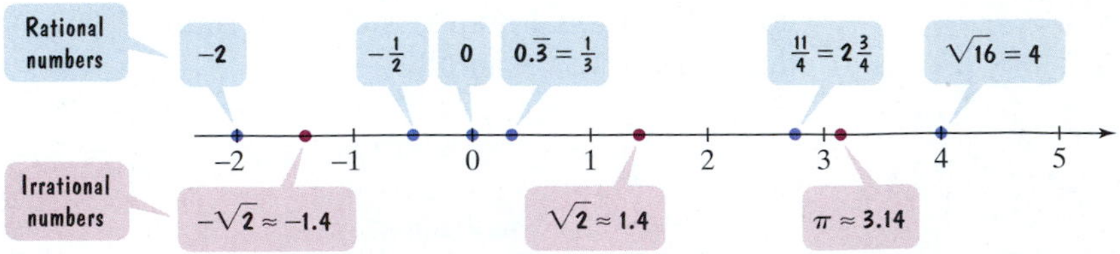

FIGURE 1.5 Graphing numbers on a real number line

Every real number corresponds to a point on the number line and every point on the number line corresponds to a real number. We say that there is a **one-to-one correspondence** between all the real numbers and all points on a real number line.

Study Tip

Wondering how we located $\sqrt{2}$ as a precise point on the number line in **Figure 1.5**? We used a right triangle with two legs of length 1. The remaining side has a length measuring $\sqrt{2}$.

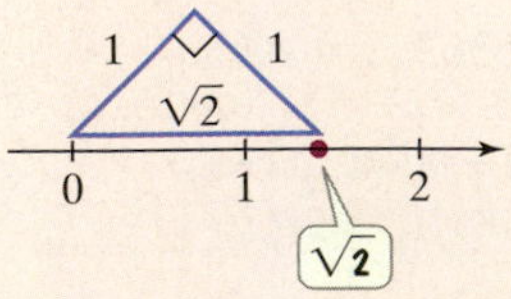

We'll have lots more to say about right triangles later in the book.

7 Use inequality symbols.

Ordering the Real Numbers

On the real number line, the real numbers increase from left to right. The lesser of two real numbers is the one farther to the left on a number line. The greater of two real numbers is the one farther to the right on a number line.

Look at the number line in **Figure 1.6**. The integers -4 and -1 are graphed.

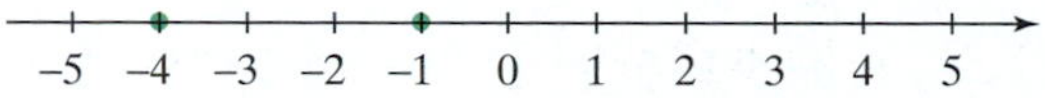

FIGURE 1.6

Observe that -4 is to the left of -1 on the number line. This means that -4 is less than -1.

$$-4 < -1$$

-4 is less than -1 because -4 is to the **left** of -1 on the number line.

In **Figure 1.6**, we can also observe that -1 is to the right of -4 on the number line. This means that -1 is greater than -4.

$$-1 > -4$$

-1 is greater than -4 because -1 is to the **right** of -4 on the number line.

The symbols $<$ and $>$ are called **inequality symbols**. These symbols always point to the lesser of the two real numbers when the inequality statement is true.

-4 is less than -1. $\quad -4 < -1 \quad$ The symbol points to -4, the lesser number.

-1 is greater than -4. $\quad -1 > -4 \quad$ The symbol still points to -4, the lesser number.

The symbols $<$ and $>$ may be combined with an equal sign, as shown in the following table:

Symbol	Meaning	Examples	Explanation
$a \le b$	a is less than or equal to b.	$2 \le 9$ $9 \le 9$	Because $2 < 9$ Because $9 = 9$
$b \ge a$	b is greater than or equal to a.	$9 \ge 2$ $2 \ge 2$	Because $9 > 2$ Because $2 = 2$

This inequality is true if either the $<$ part or the $=$ part is true.

This inequality is true if either the $>$ part or the $=$ part is true.

EXAMPLE 6 Using Inequality Symbols

Write out the meaning of each inequality. Then determine whether the inequality is true or false.

a. $-5 < -1$ **b.** $6 > -2$ **c.** $-6 \le 3$ **d.** $10 \ge 10$ **e.** $-9 \ge 6$

Solution The solution is illustrated by the number line in **Figure 1.7**.

FIGURE 1.7

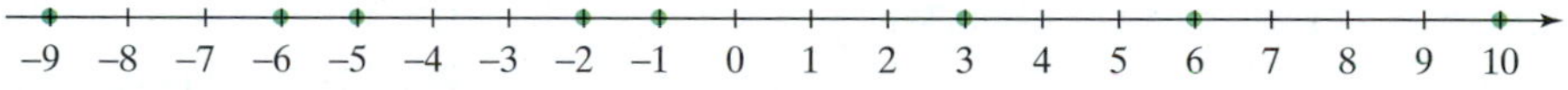

Inequality	Meaning
a. $-5 < -1$	"-5 is less than -1." Because -5 is to the left of -1 on the number line, the inequality is true.
b. $6 > -2$	"6 is greater than -2." Because 6 is to the right of -2 on the number line, the inequality is true.
c. $-6 \le 3$	"-6 is less than or equal to 3." Because $-6 < 3$ is true (-6 is to the left of 3 on the number line), the inequality is true.
d. $10 \ge 10$	"10 is greater than or equal to 10." Because $10 = 10$ is true, the inequality is true.
e. $-9 \ge 6$	"-9 is greater than or equal to 6." Because neither $-9 > 6$ nor $-9 = 6$ is true, the inequality is false.

Study Tip

Here are three similar English phrases that have very different translations:

- 7 minus a number: $7 - x$
- 7 less than a number: $x - 7$
- 7 is less than a number: $7 < x$

Think carefully about what is expressed in English before you translate into the language of algebra.

✓ CHECK POINT 6 Write out the meaning of each inequality. Then determine whether the inequality is true or false.

a. $-8 < -2$

b. $7 > -3$

c. $-1 \leq -4$

d. $5 \geq 5$

e. $2 \geq -14$

1.1 EXERCISE SET

MyMathLab · MathXL PRACTICE · WATCH · DOWNLOAD · READ · REVIEW

Study Tip

Researchers say the mind can be strengthened, just like your muscles, with regular training and rigorous practice. Think of the book's exercise sets as brain calisthenics. If you're feeling a bit sluggish before any of your mental workouts, try this warmup:

In the list below, say the color the word is printed in, not the word itself. Once you can do this in 15 seconds without an error, the warmup is over and it's time to move on to the assigned exercises.

Blue Yellow Red Green Yellow Green Blue Red Yellow Red

Practice Exercises

In Exercises 1–14, write each English phrase as an algebraic expression. Let x represent the number.

1. Five more than a number

2. A number increased by six

3. Four less than a number

4. Nine less than a number

5. Four times a number

6. Twice a number

7. Ten more than twice a number

8. Four more than five times a number

9. The difference of six and half of a number

10. The difference of three and half of a number

11. Two less than the quotient of four and a number

12. Three less than the quotient of five and a number

13. The quotient of three and the difference of five and a number

14. The quotient of six and the difference of ten and a number

In Exercises 15–26, evaluate each algebraic expression for the given value or values of the variable(s).

15. $7 + 5x$, for $x = 10$

16. $8 + 6x$, for $x = 5$

17. $6x - y$, for $x = 3$ and $y = 8$

18. $8x - y$, for $x = 3$ and $y = 4$

19. $x^2 + 3x$, for $x = \frac{1}{3}$

20. $x^2 + 2x$ for $x = \frac{1}{2}$

21. $x^2 - 6x + 3$, for $x = 7$

22. $x^2 - 7x + 4$, for $x = 8$

23. $4 + 5(x - 7)^3$, for $x = 9$

24. $6 + 5(x - 6)^3$, for $x = 8$

25. $x^2 - 3(x - y)$, for $x = 8$ and $y = 2$

26. $x^2 - 4(x - y)$, for $x = 8$ and $y = 3$

In Exercises 27–34, use the roster method to list the elements in each set.

27. $\{x | x$ is a natural number less than $5\}$

28. $\{x | x$ is a natural number less than $4\}$

29. $\{x | x$ is an integer between -8 and $-3\}$

30. $\{x | x$ is an integer between -7 and $-2\}$

31. $\{x | x$ is a natural number greater than $7\}$

32. $\{x | x$ is a natural number greater than $9\}$

33. $\{x | x$ is an odd whole number less than $11\}$

34. $\{x | x$ is an odd whole number less than $9\}$

In Exercises 35–48, use the meaning of the symbols $\in$ and $\notin$ to determine whether each statement is true or false.

35. $7 \in \{x | x$ is an integer$\}$

36. $9 \in \{x | x$ is an integer$\}$

37. $7 \in \{x | x$ is a rational number$\}$

38. $9 \in \{x | x$ is a rational number$\}$

39. $7 \in \{x | x \text{ is an irrational number}\}$

40. $9 \in \{x | x \text{ is an irrational number}\}$

41. $3 \notin \{x | x \text{ is an irrational number}\}$

42. $5 \notin \{x | x \text{ is an irrational number}\}$

43. $\frac{1}{2} \notin \{x | x \text{ is a rational number}\}$

44. $\frac{1}{4} \notin \{x | x \text{ is a rational number}\}$

45. $\sqrt{2} \notin \{x | x \text{ is a rational number}\}$

46. $\pi \notin \{x | x \text{ is a rational number}\}$

47. $\sqrt{2} \notin \{x | x \text{ is a real number}\}$

48. $\pi \notin \{x | x \text{ is a real number}\}$

In Exercises 49–64, write out the meaning of each inequality. Then determine whether the inequality is true or false.

49. $-6 < -2$

50. $-7 < -3$

51. $5 > -7$

52. $3 > -8$

53. $0 < -4$

54. $0 < -5$

55. $-4 \leq 1$

56. $-5 \leq 1$

57. $-2 \leq -6$

58. $-3 \leq -7$

59. $-2 \leq -2$

60. $-3 \leq -3$

61. $-2 \geq -2$

62. $-3 \geq -3$

63. $2 \leq -\frac{1}{2}$

64. $4 \leq -\frac{1}{2}$

Practice PLUS

By definition, an "and" statement is true only when the statements before and after the "and" connective are both true. Use this definition to determine whether each statement in Exercises 65–74 is true or false.

65. $0.\overline{3} > 0.3$ and $-10 < 4 + 6$.

66. $0.6 < 0.\overline{6}$ and $2 \cdot 5 \leq 4 + 6$.

67. $12 \in \{1, 2, 3, \dots\}$ and $\{3\} \in \{1, 2, 3, 4\}$.

68. $17 \in \{1, 2, 3, \dots\}$ and $\{4\} \in \{1, 2, 3, 4, 5\}$.

69. $\left(\frac{2}{5} + \frac{3}{5}\right) \in \{x | x \text{ is a natural number}\}$ and the value of $9x^2(x + 11) - 9(x + 11)x^2$, for $x = 100$, is 0.

70. $\left(\frac{14}{19} + \frac{5}{19}\right) \in \{x | x \text{ is a natural number}\}$ and the value of $12x^2(x + 10) - 12(x + 10)x^2$, for $x = 50$, is 0.

71. $\{x | x \text{ is an integer between } -3 \text{ and } 0\} = \{-3, -2, -1, 0\}$ and $-\pi > -3.5$.

72. $\{x | x \text{ is an integer between } -4 \text{ and } 0\} = \{-4, -3, -2, -1, 0\}$ and $-\frac{\pi}{2} > -2.3$.

73. Twice the sum of a number and three is represented by $2x + 3$ and $-1{,}100{,}000 \in \{x | x \text{ is an integer}\}$.

74. Three times the sum of a number and five is represented by $3x + 5$ and $-4{,}500{,}000 \in \{x | x \text{ is an integer}\}$.

Application Exercises

Why do people become less neurotic as they get older? One theory is that key centers of the brain tend to create less resistance to feelings of happiness as we age. The graph shows the average resistance to happiness, on a scale of 0 (no resistance) to 8 (completely resistant), for persons at various ages.

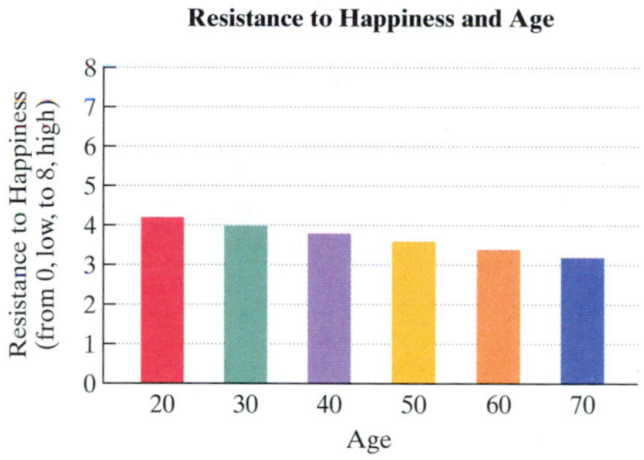

Source: L. M. Williams, *Ibid.*

The data in the graph can be modeled by the formula

$$R = 4.6 - 0.02x,$$

where R represents the average resistance to happiness, on a scale of 0 to 8, for a person who is x years old. Use this formula to solve Exercises 75–78.

75. According to the formula, what is the average resistance to happiness at age 20?

76. According to the formula, what is the average resistance to happiness at age 30?

77. What is the difference between the average resistance to happiness at age 30 and at age 50?

78. What is the difference between the average resistance to happiness at age 20 and at age 70?

In 2005, the average CEO was paid 821 times as much per hour as a full-time minimum-wage earner, who earned \$5.15 per hour. The graph shows the ratio of CEO compensation to minimum-wage salary from 2001 through 2005.

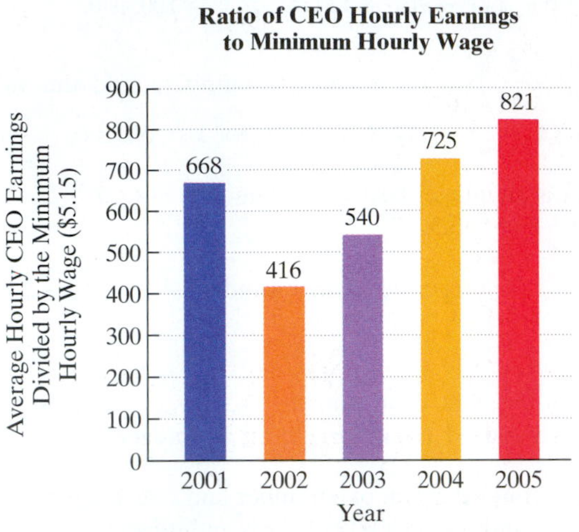

Source: Economic Policy Institute

The data in the graph can be modeled by the formula

$$R = 54x^2 - 263x + 828,$$

where R represents the ratio of CEO compensation to minimum-wage salary (\$5.15 per hour) x years after 2000. Use the formula to solve Exercises 79–80.

79. According to the formula, what was the ratio of CEO compensation to minimum-wage salary in 2005? Does the formula underestimate or overestimate the actual ratio shown by the bar graph? By how much?

80. According to the formula, what was the ratio of CEO compensation to minimum-wage salary in 2003? Does the formula underestimate or overestimate the actual ratio shown by the bar graph? By how much?

The formula

$$C = \frac{5}{9}(F - 32)$$

expresses the relationship between Fahrenheit temperature, F, and Celsius temperature, C. In Exercises 81–82, use the formula to convert the given Fahrenheit temperature to its equivalent temperature on the Celsius scale.

81. 50°F

82. 86°F

A football is kicked vertically upward from a height of 4 feet with an initial speed of 60 feet per second. The formula

$$h = 4 + 60t - 16t^2$$

describes the ball's height above the ground, h, in feet, t seconds after it was kicked. Use this formula to solve Exercises 83–84.

83. What is the ball's height 2 seconds after it was kicked?

84. What is the ball's height 3 seconds after it was kicked?

Writing in Mathematics

Writing about mathematics will help you to learn mathematics. For all writing exercises in this book, use complete sentences to respond to the question. Some writing exercises can be answered in a sentence. Others require a paragraph or two. You can decide how much you need to write as long as your writing clearly and directly answers the question in the exercise. Standard references such as a dictionary and a thesaurus should be helpful.

85. What is a variable?

86. What is an algebraic expression? Give an example with your explanation.

87. If n is a natural number, what does b^n mean? Give an example with your explanation.

88. What does it mean when we say that a formula models real-world phenomena?

89. What is model breakdown?

90. What is a set?

91. Describe the roster method for representing a set.

92. What are the natural numbers?

93. What are the whole numbers?

94. What are the integers?

95. Describe the rational numbers.

96. Describe the difference between a rational number and an irrational number.

97. What are the real numbers?

98. What is set-builder notation?

99. Describe the meanings of the symbols $\in$ and $\notin$. Provide an example showing the correct use of each symbol.

100. What is the real number line?

101. If you are given two real numbers, explain how to determine which one is the lesser.

Critical Thinking Exercises

Make Sense? *In Exercises 102–105, determine whether each statement "makes sense" or "does not make sense" and explain your reasoning.*

102. My mathematical model describes the data for the past ten years extremely well, so it will serve as an accurate prediction for what will occur in 2050.

103. My calculator will not display the value of 13^{1500}, so the algebraic expression $4x^{1500} - 3x + 7$ cannot be evaluated for $x = 13$ even without a calculator.

104. Regardless of what real numbers I substitute for x and y, I will always obtain zero when evaluating $2x^2y - 2yx^2$.

105. A model that describes the number of lobbyists x years after 2000 cannot be used to estimate the number in 2000.

In Exercises 106–109, determine whether each statement is true or false. If the statement is false, make the necessary change(s) to produce a true statement.

106. Every rational number is an integer.

107. Some whole numbers are not integers.

108. Some rational numbers are not positive.

109. Some irrational numbers are negative.

110. A bird lover visited a pet shop where there were twice 4 and 20 parrots. The bird lover purchased $\frac{1}{7}$ of the birds. English, of course, can be ambiguous and "twice 4 and 20" can mean $2(4 + 20)$ or $2 \cdot 4 + 20$. Explain how the conditions of the situation determine if "twice 4 and 20" means $2(4 + 20)$ or $2 \cdot 4 + 20$.

In Exercises 111–112, insert parentheses to make each statement true.

111. $2 \cdot 3 + 3 \cdot 5 = 45$

112. $8 + 2 \cdot 4 - 3 = 10$

113. Between which two consecutive integers is $-\sqrt{26}$? Do not use a calculator.

Preview Exercises

Exercises 114–116 will help you prepare for the material covered in the next section.

114. There are two real numbers whose distance is five units from zero on a real number line. What are these numbers?

115. Simplify: $\dfrac{16 + 3(2)^4}{12 - (10 - 6)}$.

116. Evaluate $2(3x + 5)$ and $6x + 10$ for $x = 4$.

SECTION 1.2

Operations with Real Numbers and Simplifying Algebraic Expressions

Objectives

1. Find a number's absolute value.
2. Add real numbers.
3. Find opposites.
4. Subtract real numbers.
5. Multiply real numbers.
6. Evaluate exponential expressions.
7. Divide real numbers.
8. Use the order of operations.
9. Use commutative, associative, and distributive properties.
10. Simplify algebraic expressions.

From both the political left and the right, watchdogs of the federal budget are warning of fiscal trouble. David Walker, Comptroller General of the United States and the nation's top auditor, admits to being "terrified" about the budget deficit in coming decades. To hear him tell it, the United States can be likened to Rome before the fall of the empire. America's financial condition is "worse than advertised," he says. It has a "broken business model. It faces deficits in its budget, its balance of payments, and its savings." In this section, we review operations with real numbers so that you can make numerical sense of the nation's finances.

Absolute Value

Absolute value is used to describe how to operate with positive and negative numbers.

Geometric Meaning of Absolute Value

The **absolute value** of a real number a, denoted by $|a|$, is the distance from 0 to a on the number line. This distance is always taken to be nonnegative.

1 Find a number's absolute value.

EXAMPLE 1 Finding Absolute Value

Find the absolute value:

a. $|-4|$ **b.** $|3.5|$ **c.** $|0|$.

Solution The solution is illustrated in **Figure 1.8**.

a. $|-4| = 4$ The absolute value of -4 is 4 because -4 is 4 units from 0.

b. $|3.5| = 3.5$ The absolute value of 3.5 is 3.5 because 3.5 is 3.5 units from 0.

c. $|0| = 0$ The absolute value of 0 is 0 because 0 is 0 units from itself.

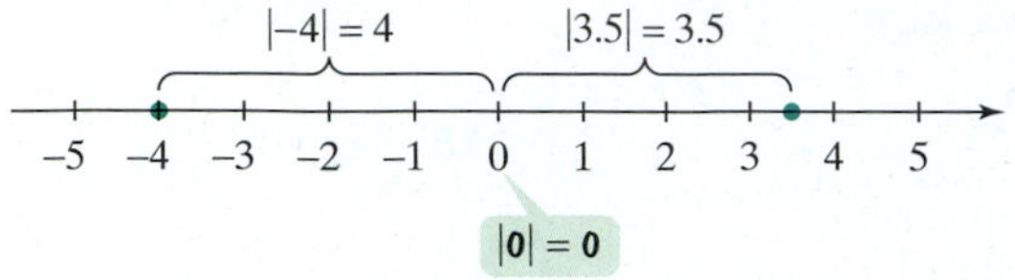

FIGURE 1.8

Can you see that the absolute value of a real number is either positive or zero? Zero is the only real number whose absolute value is 0:

$$|0| = 0.$$

The absolute value of a real number is never negative.

✓ **CHECK POINT 1** Find the absolute value:

a. $|-6|$ **b.** $|4.5|$ **c.** $|0|$.

2 Add real numbers.

Adding Real Numbers

Table 1.1 reviews how to add real numbers.

Table 1.1 Adding Real Numbers

Rule	Example
To add two real numbers with the same sign, add their absolute values. Use the common sign as the sign of the sum.	$(-7) + (-4) = -(\|-7\| + \|\ 4\|)$ $= -(7 + 4)$ $= -11$
To add two real numbers with different signs, subtract the smaller absolute value from the greater absolute value. Use the sign of the number with the greater absolute value as the sign of the sum.	$7 + (-15) = -(\|-15\| - \|7\|)$ $= -(15 - 7)$ $= -8$

EXAMPLE 2 Adding Real Numbers

Add: **a.** $-12 + (-14)$ **b.** $-0.3 + 0.7$ **c.** $-\frac{3}{4} + \frac{1}{2}$.

Solution

a. $-12 + (-14) = -26$

Add absolute values: $12 + 14 = 26$.

Use the common sign.

b. $-0.3 + 0.7 = 0.4$

Subtract absolute values: $0.7 - 0.3 = 0.4$.

Use the sign of the number with the greater absolute value. The sign of the sum is assumed to be positive.

c. $-\frac{3}{4} + \frac{1}{2} = -\frac{1}{4}$

Subtract absolute values: $\frac{3}{4} - \frac{1}{2} = \frac{3}{4} - \frac{2}{4} = \frac{1}{4}$.

Use the sign of the number with the greater absolute value.

✓ **CHECK POINT 2** Add:

a. $-10 + (-18)$ **b.** $-0.2 + 0.9$ **c.** $-\frac{3}{5} + \frac{1}{2}$.

If one of two numbers being added is zero, the sum is the other number. For example,

$$-3 + 0 = -3 \qquad \text{and} \qquad 0 + 2 = 2.$$

In general,

$$a + 0 = a \qquad \text{and} \qquad 0 + a = a.$$

We call 0 the **identity element of addition** or the **additive identity**. Thus, the additive identity can be deleted from a sum.

3 Find opposites.

Numbers with different signs but the same absolute value are called **opposites** or **additive inverses**. For example, 3 and -3 are additive inverses. When additive inverses are added, their sum is 0. For example,

$$3 + (-3) = 0 \qquad \text{and} \qquad -3 + 3 = 0.$$

Inverse Property of Addition

The sum of a real number and its additive inverse is 0, the additive identity.

$$a + (-a) = 0 \qquad \text{and} \qquad (-a) + a = 0$$

The symbol "$-$" is used to name the opposite, or additive inverse, of a. When a is a negative number, $-a$, its opposite, is positive. For example, if a is -4, its opposite is 4. Thus,

$$-(-4) = 4.$$

The opposite of -4 is 4.

In general, if a is any real number,

$$-(-a) = a.$$

EXAMPLE 3 Finding Opposites

Find $-x$ if **a.** $x = -6$ **b.** $x = \frac{1}{2}$.

Solution

a. If $x = -6$, then $-x = -(-6) = 6$. The opposite of -6 is 6.

b. If $x = \frac{1}{2}$, then $-x = -\frac{1}{2}$. The opposite of $\frac{1}{2}$ is $-\frac{1}{2}$.

✓ CHECK POINT 3 Find $-x$ if

a. $x = -8$ **b.** $x = \frac{1}{3}$.

We can define the absolute value of the real number a using opposites, without referring to a number line. The algebraic definition of the absolute value of a is given as follows:

Definition of Absolute Value

$$|a| = \begin{cases} a & \text{if } a \geq 0 \\ -a & \text{if } a < 0 \end{cases}$$

If a is nonnegative (that is, $a \geq 0$), the absolute value of a is the number itself: $|a| = a$. For example,

$$|5| = 5 \qquad |\pi| = \pi \qquad \left|\frac{1}{3}\right| = \frac{1}{3} \qquad |0| = 0$$

Zero is the only number whose absolute value is 0.

If a is a negative number (that is, $a < 0$), the absolute value of a is the opposite of a: $|a| = -a$. This makes the absolute value positive. For example,

$$|-3| = -(-3) = 3 \qquad |-\pi| = -(-\pi) = \pi \qquad \left|-\frac{1}{3}\right| = -\left(-\frac{1}{3}\right) = \frac{1}{3}.$$

This middle step is usually omitted.

4 Subtract real numbers.

Subtracting Real Numbers

Subtraction of real numbers is defined in terms of addition.

Definition of Subtraction

If a and b are real numbers,

$$a - b = a + (-b).$$

To subtract a real number, add its opposite or additive inverse.

Thus, to subtract real numbers,

1. Change the subtraction to addition.
2. Change the sign of the number being subtracted.
3. Add, using one of the rules for adding numbers with the same sign or different signs.

EXAMPLE 4 Subtracting Real Numbers

Subtract: **a.** $6 - 13$ **b.** $5.1 - (-4.2)$ **c.** $-\frac{11}{3} - \left(-\frac{4}{3}\right)$.

Solution

a. $6 - 13 = 6 + (-13) = -7$

Change the subtraction to addition. Replace 13 with its opposite.

b. $5.1 - (-4.2) = 5.1 + 4.2 = 9.3$

Change the subtraction to addition. Replace -4.2 with its opposite.

c. $-\frac{11}{3} - \left(-\frac{4}{3}\right) = -\frac{11}{3} + \frac{4}{3} = -\frac{7}{3}$

Change the subtraction to addition. Replace $-\frac{4}{3}$ with its opposite.

✓ CHECK POINT 4 Subtract:

a. $7 - 10$ **b.** $4.3 - (-6.2)$ **c.** $-\frac{4}{5} - \left(-\frac{1}{5}\right)$.

5 Multiply real numbers.

Multiplying Real Numbers

You can think of multiplication as repeated addition or subtraction that starts at 0. For example,

$$3(-4) = 0 + (-4) + (-4) + (-4) = -12$$

The numbers have different signs and the product is negative.

and

$$(-3)(-4) = 0 - (-4) - (-4) - (-4) = 0 + 4 + 4 + 4 = 12$$

The numbers have the same sign and the product is positive.

Table 1.2 reviews how to multiply real numbers.

Table 1.2 Multiplying Real Numbers

Rule	Example
The product of two real numbers with different signs is found by multiplying their absolute values. The product is negative.	$7(-5) = -35$
The product of two real numbers with the same sign is found by multiplying their absolute values. The product is positive.	$(-6)(-11) = 66$
The product of 0 and any real number is 0: $a \cdot 0 = 0$ and $0 \cdot a = 0$.	$-17(0) = 0$
If no number is 0, a product with an odd number of negative factors is found by multiplying absolute values. The product is negative.	$-2(-3)(-5) = -30$ Three (odd) negative factors
If no number is 0, a product with an even number of negative factors is found by multiplying absolute values. The product is positive.	$-2(3)(-5) = 30$ Two (even) negative factors

6 Evaluate exponential expressions.

Because exponents indicate repeated multiplication, rules for multiplying real numbers can be used to evaluate exponential expressions.

EXAMPLE 5 Evaluating Exponential Expressions

Evaluate: **a.** $(-6)^2$ **b.** -6^2 **c.** $(-5)^3$ **d.** $\left(-\frac{2}{3}\right)^4$.

Solution

a. $(-6)^2 = (-6)(-6) = 36$

Base is −6. Same signs give positive product.

b. $-6^2 = -(6 \cdot 6) = -36$

Base is 6. The negative is not inside parentheses and is not taken to the second power.

c. $(-5)^3 = (-5)(-5)(-5) = -125$

An odd number of negative factors gives a negative product.

d. $\left(-\frac{2}{3}\right)^4 = \left(-\frac{2}{3}\right)\left(-\frac{2}{3}\right)\left(-\frac{2}{3}\right)\left(-\frac{2}{3}\right) = \frac{16}{81}$

Base is $-\frac{2}{3}$.

An even number of negative factors gives a positive product.

CHECK POINT 5 Evaluate:

a. $(-5)^2$ **b.** -5^2 **c.** $(-4)^3$ **d.** $\left(-\frac{3}{5}\right)^4$.

7 Divide real numbers.

Dividing Real Numbers

If a and b are real numbers and b is not 0, then the quotient of a and b is defined as follows:

$$a \div b = a \cdot \frac{1}{b} \quad \text{or} \quad \frac{a}{b} = a \cdot \frac{1}{b}.$$

Thus, to find the quotient of a and b, we can divide by b or multiply by $\frac{1}{b}$. The nonzero real numbers b and $\frac{1}{b}$ are called **reciprocals**, or **multiplicative inverses**, of one another. When reciprocals are multiplied, their product is 1:

$$b \cdot \frac{1}{b} = 1.$$

Because division is defined in terms of multiplication, the sign rules for dividing numbers are the same as the sign rules for multiplying them.

Dividing Real Numbers

The quotient of two numbers with different signs is negative. The quotient of two numbers with the same sign is positive. The quotient is found by dividing absolute values.

EXAMPLE 6 Dividing Real Numbers

Divide: **a.** $\frac{20}{-5}$ **b.** $-\frac{3}{4} \div \left(-\frac{5}{9}\right)$.

Solution

a. $\frac{20}{-5} = -4$

Divide absolute values: $\frac{20}{5} = 4$.

Different signs: negative quotient

b. $-\frac{3}{4} \div \left(-\frac{5}{9}\right) = \frac{27}{20}$

Divide absolute values: $\frac{3}{4} \div \frac{5}{9} = \frac{3}{4} \cdot \frac{9}{5} = \frac{27}{20}$.

Same sign: positive quotient

CHECK POINT 6 Divide:

a. $\frac{32}{-4}$ **b.** $-\frac{2}{3} \div \left(-\frac{5}{4}\right)$.

We must be careful with division when 0 is involved. Zero divided by any nonzero real number is 0. For example,

$$\frac{0}{-5} = 0.$$

Can you see why $\frac{0}{-5}$ must be 0? The definition of division tells us that

$$\frac{0}{-5} = 0 \cdot \left(-\frac{1}{5}\right)$$

and the product of 0 and any real number is 0. By contrast, what happens if we divide -5 by 0. The answer must be a number that, when multiplied by 0, gives -5. However, any number multiplied by 0 is 0. Thus, we cannot divide -5, or any other real number, by 0.

Division by Zero

Division by zero is not allowed; it is undefined. A real number can never have a denominator of 0.

8 Use the order of operations.

Order of Operations

The rules for order of operations can be applied to positive and negative real numbers. Recall that if no grouping symbols are present, we

- Evaluate exponential expressions.
- Multiply and divide, from left to right.
- Add and subtract, from left to right.

EXAMPLE 7 Using the Order of Operations

Simplify: $4 - 7^2 + 8 \div 2(-3)^2$.

Solution

$4 - 7^2 + 8 \div 2(-3)^2$

$= 4 - 49 + 8 \div 2(9)$ Evaluate exponential expressions: $7^2 = 7 \cdot 7 = 49$ and $(-3)^2 = (-3)(-3) = 9$.

$= 4 - 49 + 4(9)$ Divide: $8 \div 2 = 4$.

$= 4 - 49 + 36$ Multiply: $4(9) = 36$.

$= -45 + 36$ Subtract: $4 - 49 = 4 + (-49) = -45$.

$= -9$ Add.

✓ **CHECK POINT 7** Simplify: $3 - 5^2 + 12 \div 2(-4)^2$.

If an expression contains grouping symbols, we perform operations within these symbols first. Common grouping symbols are parentheses, brackets, and braces. Other grouping symbols include fraction bars, absolute value symbols, and radical symbols, such as square root signs ($\sqrt{\ }$).

EXAMPLE 8 Using the Order of Operations

Simplify: $\dfrac{13 - 3(-2)^4}{3 - (6 - 10)}$.

Solution Simplify the numerator and the denominator separately. Then divide.

$$\frac{13 - 3(-2)^4}{3 - (6 - 10)}$$

$$= \frac{13 - 3(16)}{3 - (-4)}$$ Evaluate the exponential expression in the numerator: $(-2)^4 = (-2)(-2)(-2)(-2) = 16$. Subtract inside parentheses in the denominator: $6 - 10 = 6 + (-10) = -4$.

$$= \frac{13 - 48}{7}$$ Multiply in the numerator: $3(16) = 48$. Subtract in the denominator: $3 - (-4) = 3 + 4 = 7$.

$$= \frac{-35}{7}$$ Subtract in the numerator: $13 - 48 = 13 + (-48) = -35$.

$$= -5$$ Divide.

✓ **CHECK POINT 8** Simplify: $\frac{4 + 3(-2)^3}{2 - (6 - 9)}$.

9 Use commutative, associative, and distributive properties.

The Commutative, Associative, and Distributive Properties

Basic algebraic properties enable us to write *equivalent algebraic expressions*. Two algebraic expressions that have the same value for all replacements are called **equivalent algebraic expressions**. In Section 1.4, you will use such expressions to solve equations.

In arithmetic, when two numbers are added or multiplied, the order in which the numbers are written does not affect the answer. These facts are called **commutative properties**.

The Commutative Properties

Let a and b represent real numbers, variables, or algebraic expressions.

Addition: $a + b = b + a$

Multiplication: $ab = ba$

Changing order when adding or multiplying does not affect a sum or product.

EXAMPLE 9 Using the Commutative Properties

Write an algebraic expression equivalent to $3x + 7$ using each of the commutative properties.

Solution

Commutative of Addition

$3x + 7 = 7 + 3x$

Change the order of the addition.

Commutative of Multiplication

$3x + 7 = x \cdot 3 + 7$

Change the order of the multiplication.

✓ **CHECK POINT 9** Write an algebraic expression equivalent to $4x + 9$ using each of the commutative properties.

The **associative properties** enable us to form equivalent expressions by regrouping.

The Associative Properties

Let a, b, and c represent real numbers, variables, or algebraic expressions.

Addition: $(a + b) + c = a + (b + c)$

Multiplication: $(ab)c = a(bc)$

Changing grouping when adding or multiplying does not affect a sum or product.

Blitzer Bonus

The Associative Property and the English Language

In the English language, phrases can take on different meanings depending on the way the words are associated with commas. Here are three examples.

- Woman, without her man, is nothing.
  Woman, without her, man is nothing.
- What's the latest dope?
  What's the latest, dope?
- Population of Amsterdam broken down by age and sex
  Population of Amsterdam, broken down by age and sex

EXAMPLE 10 Using the Associative Properties

Use an associative property to write an equivalent expression and simplify:

a. $7 + (3 + x)$ **b.** $-6(5x)$.

Solution

a. $7 + (3 + x) = (7 + 3) + x = 10 + x$

b. $-6(5x) = (-6 \cdot 5)x = -30x$

✓ CHECK POINT 10 Use an associative property to write an equivalent expression and simplify:

a. $6 + (12 + x)$ **b.** $-7(4x)$.

The **distributive property** allows us to rewrite the product of a number and a sum as the sum of two products.

The Distributive Property

Let a, b, and c represent real numbers, variables, or algebraic expressions.

$$a(b + c) = ab + ac$$

Multiplication distributes over addition.

EXAMPLE 11 Using the Distributive Property

Use the distributive property to write an equivalent expression:

$$-2(3x + 5).$$

Solution

$$-2(3x + 5) = -2 \cdot 3x + (-2) \cdot 5 = -6x + (-10) = -6x - 10$$

✓ CHECK POINT 11 Use the distributive property to write an equivalent expression: $-4(7x + 2)$.

Table 1.3 shows a number of other forms of the distributive property.

Study Tip

When using a distributive property to remove parentheses, be sure to multiply *each term* inside the parentheses by the factor outside.

Incorrect

~~$5(3y + 7) = 5 \cdot 3y + 7$~~
~~$= 15y + 7$~~

Table 1.3 Other Forms of the Distributive Property

Property	Meaning	Example
$a(b - c) = ab - ac$	Multiplication distributes over subtraction.	$6(4x - 5) = 6 \cdot 4x - 6 \cdot 5$ $= 24x - 30$
$a(b + c + d) = ab + ac + ad$	Multiplication distributes over three or more terms in parentheses.	$5(x + 4 + 7y)$ $= 5x + 5 \cdot 4 + 5 \cdot 7y$ $= 5x + 20 + 35y$
$(b + c)a = ba + ca$	Multiplication on the right distributes over addition (or subtraction).	$(x + 10)8 = x \cdot 8 + 10 \cdot 8$ $= 8x + 80$

10 Simplify algebraic expressions.

Combining Like Terms and Simplifying Algebraic Expressions

The **terms** of an algebraic expression are those parts that are separated by addition. For example, consider the algebraic expression

$$7x - 9y + z - 3,$$

which can be expressed as

$$7x + (-9y) + z + (-3).$$

This expression contains four terms, namely $7x$, $-9y$, z, and -3.

The numerical part of a term is called its **coefficient**. In the term $7x$, the 7 is the coefficient. If a term containing one or more variables is written without a coefficient, the coefficient is understood to be 1. Thus, z means $1z$. If a term is a constant, its coefficient is that constant. Thus, the coefficient of the constant term -3 is -3.

$$7x + (-9y) + z + (-3)$$

Coefficient is 7.
Coefficient is −9.
Coefficient is 1; z means $1z$.
Coefficient is −3.

The parts of each term that are multiplied are called the **factors** of the term. The factors of the term $7x$ are 7 and x.

Like terms are terms that have exactly the same variable factors. For example, $3x$ and $7x$ are like terms. The distributive property in the form

$$ba + ca = (b + c)a$$

enables us to add or subtract like terms. For example,

$$3x + 7x = (3 + 7)x = 10x$$
$$7y^2 - y^2 = 7y^2 - 1y^2 = (7 - 1)y^2 = 6y^2.$$

This process is called **combining like terms**.

An algebraic expression is **simplified** when grouping symbols have been removed and like terms have been combined.

Study Tip

To combine like terms mentally, add or subtract the coefficients of the terms. Use this result as the coefficient of the terms' variable factor(s).

EXAMPLE 12 Simplifying an Algebraic Expression

Simplify: $7x + 12x^2 + 3x + x^2$.

Solution

$$7x + 12x^2 + 3x + x^2$$

$$= (7x + 3x) + (12x^2 + x^2)$$ Rearrange terms and group like terms using commutative and associative properties. This step is often done mentally.

$x^2 = 1x^2$

$$= (7 + 3)x + (12 + 1)x^2$$ Apply the distributive property.

$$= 10x + 13x^2$$ Simplify. Because $10x$ and $13x^2$ are not like terms, this is the final answer.

Using the commutative property of addition, we can write this simplified expression as $13x^2 + 10x$.

✓ **CHECK POINT 12** Simplify: $3x + 14x^2 + 11x + x^2$.

EXAMPLE 13 Simplifying an Algebraic Expression

Simplify: $4(7x - 3) - 10x$.

Solution

$4(7x - 3) - 10x$

$= 4 \cdot 7x - 4 \cdot 3 - 10x$ Use the distributive property to remove the parentheses.

$= 28x - 12 - 10x$ Multiply.

$= (28x - 10x) - 12$ Group like terms.

$= (28 - 10)x - 12$ Apply the distributive property.

$= 18x - 12$ Simplify.

CHECK POINT 13 Simplify: $8(2x - 5) - 4x$.

It is not uncommon to see algebraic expressions with parentheses preceded by a negative sign or subtraction. An expression of the form $-(b + c)$ can be simplified as follows:

$$-(b + c) = -1(b + c) = (-1)b + (-1)c = -b + (-c) = -b - c.$$

Do you see a fast way to obtain the simplified expression on the right? **If a negative sign or a subtraction symbol appears outside parentheses, drop the parentheses and change the sign of every term within the parentheses.** For example,

$$-(3x^2 - 7x - 4) = -3x^2 + 7x + 4.$$

EXAMPLE 14 Simplifying an Algebraic Expression

Simplify: $8x + 2[5 - (x - 3)]$.

Solution

$8x + 2[5 - (x - 3)]$

$= 8x + 2[5 - x + 3]$ Drop parentheses and change the sign of each term in parentheses: $-(x - 3) = -x + 3$.

$= 8x + 2[8 - x]$ Simplify inside brackets: $5 + 3 = 8$.

$= 8x + 16 - 2x$ Apply the distributive property: $2[8 - x] = 2 \cdot 8 - 2x = 16 - 2x$.

$= (8x - 2x) + 16$ Group like terms.

$= (8 - 2)x + 16$ Apply the distributive property.

$= 6x + 16$ Simplify.

CHECK POINT 14 Simplify: $6 + 4[7 - (x - 2)]$.

1.2 EXERCISE SET

MyMathLab MathXL PRACTICE WATCH DOWNLOAD READ REVIEW

Practice Exercises

In Exercises 1–12, find each absolute value.

1. $|-7|$
2. $|-10|$
3. $|4|$
4. $|13|$
5. $|-7.6|$
6. $|-8.3|$
7. $\left|\frac{\pi}{2}\right|$
8. $\left|\frac{\pi}{3}\right|$
9. $|-\sqrt{2}|$
10. $|-\sqrt{3}|$
11. $-\left|-\frac{2}{5}\right|$
12. $-\left|-\frac{7}{10}\right|$

In Exercises 13–28, add as indicated.

13. $-3 + (-8)$
14. $-5 + (-10)$
15. $-14 + 10$
16. $-15 + 6$
17. $-6.8 + 2.3$
18. $-7.9 + 2.4$
19. $\frac{11}{15} + \left(-\frac{3}{5}\right)$
20. $\frac{7}{10} + \left(-\frac{4}{5}\right)$
21. $-\frac{2}{9} - \frac{3}{4}$
22. $-\frac{3}{5} - \frac{4}{7}$
23. $-3.7 + (-4.5)$
24. $-6.2 + (-5.9)$
25. $0 + (-12.4)$
26. $0 + (-15.3)$
27. $12.4 + (-12.4)$
28. $15.3 + (-15.3)$

In Exercises 29–34, find $-x$ for the given value of x.

29. $x = 11$
30. $x = 13$
31. $x = -5$
32. $x = -9$
33. $x = 0$
34. $x = -\sqrt{2}$

In Exercises 35–46, subtract as indicated.

35. $3 - 15$
36. $4 - 20$
37. $8 - (-10)$
38. $7 - (-13)$
39. $-20 - (-5)$
40. $-30 - (-10)$
41. $\frac{1}{4} - \frac{1}{2}$
42. $\frac{1}{10} - \frac{2}{5}$
43. $-2.3 - (-7.8)$
44. $-4.3 - (-8.7)$
45. $0 - (-\sqrt{2})$
46. $0 - (-\sqrt{3})$

In Exercises 47–58, multiply as indicated.

47. $9(-10)$
48. $8(-10)$
49. $(-3)(-11)$
50. $(-7)(-11)$
51. $\frac{15}{13}(-1)$
52. $\frac{11}{13}(-1)$
53. $-\sqrt{2}\cdot 0$
54. $-\sqrt{3}\cdot 0$
55. $(-4)(-2)(-1)$
56. $(-5)(-3)(-2)$
57. $2(-3)(-1)(-2)(-4)$
58. $3(-2)(-1)(-5)(-3)$

In Exercises 59–70, evaluate each exponential expression.

59. $(-10)^2$
60. $(-8)^2$
61. -10^2
62. -8^2
63. $(-2)^3$
64. $(-3)^3$
65. $(-1)^4$
66. $(-4)^4$
67. $(-1)^{33}$
68. $(-1)^{35}$
69. $-\left(-\frac{1}{2}\right)^3$
70. $-\left(-\frac{1}{4}\right)^3$

In Exercises 71–82, divide as indicated or state that the division is undefined.

71. $\frac{12}{-4}$
72. $\frac{30}{-5}$
73. $\frac{-90}{-2}$
74. $\frac{-55}{-5}$
75. $\frac{0}{-4.6}$
76. $\frac{0}{-5.3}$
77. $-\frac{4.6}{0}$
78. $-\frac{5.3}{0}$
79. $-\frac{1}{2} \div \left(-\frac{7}{9}\right)$
80. $-\frac{1}{2} \div \left(-\frac{3}{5}\right)$
81. $6 \div \left(-\frac{2}{5}\right)$
82. $8 \div \left(-\frac{2}{9}\right)$

In Exercises 83–100, use the order of operations to simplify each expression.

83. $4(-5) - 6(-3)$
84. $8(-3) - 5(-6)$
85. $3(-2)^2 - 4(-3)^2$
86. $5(-3)^2 - 2(-2)^2$
87. $8^2 - 16 \div 2^2 \cdot 4 - 3$
88. $10^2 - 100 \div 5^2 \cdot 2 - 3$
89. $\frac{5\cdot 2 - 3^2}{[3^2 - (-2)]^2}$
90. $\frac{10 \div 2 + 3\cdot 4}{(12 - 3\cdot 2)^2}$
91. $8 - 3[-2(2 - 5) - 4(8 - 6)]$
92. $8 - 3[-2(5 - 7) - 5(4 - 2)]$
93. $\frac{2(-2) - 4(-3)}{5 - 8}$
94. $\frac{6(-4) - 5(-3)}{9 - 10}$
95. $\frac{(5 - 6)^2 - 2|3 - 7|}{89 - 3\cdot 5^2}$
96. $\frac{12 \div 3\cdot 5|2^2 + 3^2|}{7 + 3 - 6^2}$
97. $15 - \sqrt{3 - (-1)} + 12 \div 2\cdot 3$
98. $17 - |5 - (-2)| + 12 \div 2\cdot 3$
99. $20 + 1 - \sqrt{10^2 - (5 + 1)^2}\,(-2)$
100. $24 \div \sqrt{3\cdot(5 - 2)} \div [-1 - (-3)]^2$

In Exercises 101–104, write an algebraic expression equivalent to the given expression using each of the commutative properties.

101. $4x + 10$

102. $5x + 30$

103. $7x - 5$

104. $3x - 7$

In Exercises 105–110, use an associative property to write an algebraic expression equivalent to each expression and simplify.

105. $4 + (6 + x)$
106. $12 + (3 + x)$
107. $-7(3x)$
108. $-10(5x)$
109. $-\frac{1}{3}(-3y)$
110. $-\frac{1}{4}(-4y)$

In Exercises 111–116, use the distributive property to write an equivalent expression.

111. $3(2x + 5)$
112. $5(4x + 7)$
113. $-7(2x + 3)$
114. $-9(3x + 2)$
115. $-(3x - 6)$
116. $-(6x - 3)$

In Exercises 117–130, simplify each algebraic expression.

117. $7x + 5x$

118. $8x + 10x$

119. $6x^2 - x^2$

120. $9x^2 - x^2$

121. $6x + 10x^2 + 4x + 2x^2$

122. $9x + 5x^2 + 3x + 4x^2$

123. $8(3x - 5) - 6x$

124. $7(4x - 5) - 8x$

125. $5(3y - 2) - (7y + 2)$

126. $4(5y - 3) - (6y + 3)$

127. $7 - 4[3 - (4y - 5)]$

128. $6 - 5[8 - (2y - 4)]$

129. $18x^2 + 4 - [6(x^2 - 2) + 5]$

130. $14x^2 + 5 - [7(x^2 - 2) + 4]$

Practice PLUS

In Exercises 131–138, write each English phrase as an algebraic expression. Then simplify the expression. Let x represent the number.

131. A number decreased by the sum of the number and four

132. A number decreased by the difference between eight and the number

133. Six times the product of negative five and a number

134. Ten times the product of negative four and a number

135. The difference between the product of five times a number and twice the number

136. The difference between the product of six and a number and negative two times the number

137. The difference between eight times a number and six more than three times the number

138. Eight decreased by three times the sum of a number and six

Application Exercises

The bar graph shows the U.S. trade balance in goods and services, in billions of dollars, from 2000 through 2005. The most complete scorecard of the U.S. international trade performance deteriorated to a record \$805 billion deficit in 2005. Use the information shown by the graph to solve Exercises 139–142. Express answers in billions of dollars.

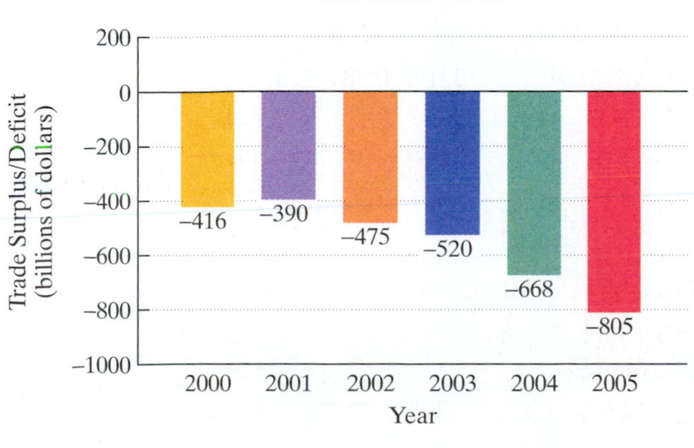

Source: Bureau of Economic Analysis

139. Find the difference between the 2000 trade deficit and the 2005 trade deficit.

140. Find the difference between the 2001 trade deficit and the 2005 trade deficit.

141. By how much did the 2005 deficit exceed twice the 2001 deficit?

142. Find the average trade deficit for 2003 and 2004 combined. By how much did the 2005 deficit exceed this average?

The bar graph shows the amount of money, in billions of dollars, collected and spent by the U.S. government from 2001 through 2006. Use the information from the graph to solve Exercises 143–146. Express answers in billions of dollars.

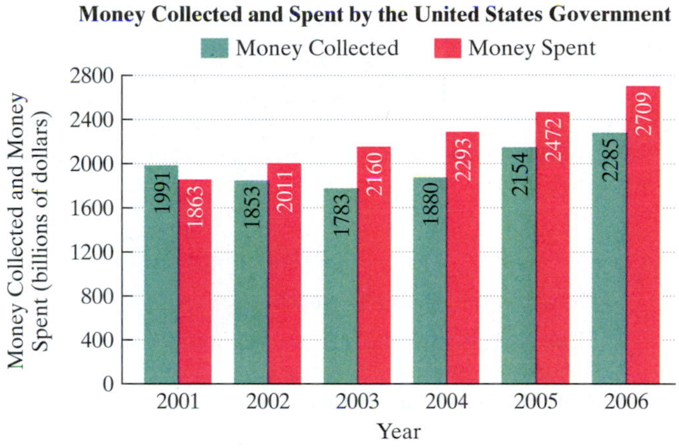

Source: Budget of the U.S. Government

143. In 2003, what was the difference between the amount of money collected and the amount spent? Was there a budget surplus or deficit in 2003?

144. In 2004, what was the difference between the amount of money collected and the amount spent? Was there a budget surplus or deficit in 2004?

145. What is the difference between the 2001 surplus and the 2006 deficit?

146. What is the difference between the 2001 surplus and the 2005 deficit?

The bar graph shows the percentage of U.S. adults who have been tested for the HIV virus, by age.

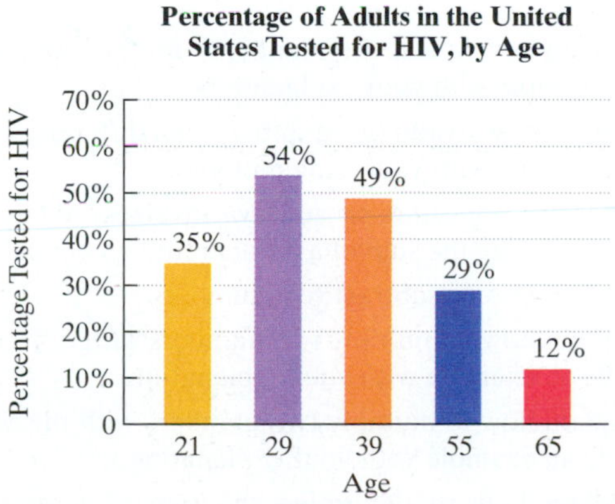

Source: National Center for Health Statistics

The data in the graph at the bottom of the previous page can be modeled by the formula

$$P = -0.05x^2 + 3.6x - 15,$$

where P represents the percentage of U.S. adults tested for HIV at age x. Use this formula to solve Exercises 147–148.

147. According to the formula, what percentage of U.S. adults who are 21 years old have been tested for HIV? Does the model underestimate or overestimate the percent displayed by the bar graph? By how much?

148. According to the formula, what percentage of U.S. adults who are 39 years old have been tested for HIV? Does the model underestimate or overestimate the percent displayed by the bar graph? By how much?

149. You had \$10,000 to invest. You put x dollars in a safe, government-insured certificate of deposit paying 5% per year. You invested the remainder of the money in noninsured corporate bonds paying 12% per year. Your total interest earned at the end of the year is given by the algebraic expression

$$0.05x + 0.12(10{,}000 - x).$$

a. Simplify the algebraic expression.

b. Use each form of the algebraic expression to determine your total interest earned at the end of the year if you invested \$6000 in the safe, government-insured certificate of deposit.

150. It takes you 50 minutes to get to campus. You spend t minutes walking to the bus stop and the rest of the time riding the bus. Your walking rate is 0.06 mile per minute and the bus travels at a rate of 0.5 mile per minute. The total distance walking and traveling by bus is given by the algebraic expression

$$0.06t + 0.5(50 - t).$$

a. Simplify the algebraic expression.

b. Use each form of the algebraic expression to determine the total distance that you travel if you spend 20 minutes walking to the bus stop.

Writing in Mathematics

151. What is the meaning of $|a|$ in terms of a number line?

152. Explain how to add two numbers with the same sign. Give an example with your explanation.

153. Explain how to add two numbers with different signs. Give an example with your explanation.

154. What are opposites, or additive inverses? What happens when finding the sum of a number and its opposite?

155. Explain how to subtract real numbers.

156. Explain how to multiply two numbers with different signs. Give an example with your explanation.

157. Explain how to multiply two numbers with the same sign. Give an example with your explanation.

158. Explain how to determine the sign of a product that involves more than two numbers.

159. Explain how to divide real numbers.

160. Why is $\frac{0}{4} = 0$, although $\frac{4}{0}$ is undefined?

161. What are equivalent algebraic expressions?

162. State a commutative property and give an example of how it is used to write equivalent algebraic expressions.

163. State an associative property and give an example of how it is used to write equivalent algebraic expressions.

164. State a distributive property and give an example of how it is used to write equivalent algebraic expressions.

165. What are the terms of an algebraic expression? How can you tell if terms are like terms?

166. What does it mean to simplify an algebraic expression?

167. If a negative sign appears outside parentheses, explain how to simplify the expression. Give an example.

168. What explanations can you offer for the trend in the percentage of U.S. adults tested for HIV, by age, shown in the bar graph in Exercises 147–148?

Critical Thinking Exercises

Make Sense? *In Exercises 169–172, determine whether each statement "makes sense" or "does not make sense" and explain your reasoning.*

169. My mathematical model, although it contains an algebraic expression that is not simplified, describes the data perfectly well, so it will describe the data equally well when simplified.

170. Subtraction actually means the addition of an additive inverse.

171. The terms $13x^2$ and $10x$ both contain the variable x, so I can combine them to obtain $23x^3$.

172. There is no number in front of the term x, so this means that the term has no coefficient.

In Exercises 173–177, determine whether each statement is true or false. If the statement is false, make the necessary change(s) to produce a true statement.

173. $16 \div 4 \cdot 2 = 16 \div 8 = 2$

174. $6 - 2(4 + 3) = 4(4 + 3) = 4(7) = 28$

175. $5 + 3(x - 4) = 8(x - 4) = 8x - 32$

176. $-x - x = -x + (-x) = 0$

177. $x - 0.02(x + 200) = 0.98x - 4$

In Exercises 178–179, insert parentheses to make each statement true.

178. $8 - 2 \cdot 3 - 4 = 14$

179. $2 \cdot 5 - \frac{1}{2} \cdot 10 \cdot 9 = 45$

180. Simplify: $\dfrac{9[4 - (1 + 6)] - (3 - 9)^2}{5 + \dfrac{12}{5 - \dfrac{6}{2 + 1}}}$.

Review Exercises

From here on, each exercise set will contain three review exercises. It is important to review previously covered topics to improve your understanding of the topics and to help maintain your mastery of the material. If you are not certain how to solve a review exercise, turn to the section and the worked-out example given in parentheses at the end of each exercise.

181. Write the following English phrase as an algebraic expression: "The quotient of ten and a number, decreased by four times the number." Let x represent the number. (Section 1.1, Example 1)

182. Evaluate $10 + 2(x - 5)^4$ for $x = 7$. (Section 1.1, Example 3)

183. Determine whether the following statement is true or false: $\frac{1}{2} \notin \{x | x \text{ is an irrational number}\}$. (Section 1.1, Example 5)

Preview Exercises

Exercises 184–186 will help you prepare for the material covered in the next section.

184. If $y = 4 - x^2$, find the value of y that corresponds to values of x for each integer starting with -3 and ending with 3.

185. If $y = 1 - x^2$, find the value of y that corresponds to values of x for each integer starting with -3 and ending with 3.

186. If $y = |x + 1|$, find the value of y that corresponds to values of x for each integer starting with -4 and ending with 2.

SECTION 1.3

Graphing Equations

Objectives

1. Plot points in the rectangular coordinate system.
2. Graph equations in the rectangular coordinate system.
3. Use the rectangular coordinate system to visualize relationships between variables.
4. Interpret information about a graphing utility's viewing rectangle or table.

The beginning of the seventeenth century was a time of innovative ideas and enormous intellectual progress in Europe. English theatergoers enjoyed a succession of exciting new plays by Shakespeare. William Harvey proposed the radical notion that the heart was a pump for blood rather than the center of emotion. Galileo, with his new-fangled invention called the telescope, supported the theory of Polish astronomer Copernicus that the sun, not the Earth, was the center of the solar system. Monteverdi was writing the world's first grand operas. French mathematicians Pascal and Fermat invented a new field of mathematics called probability theory.

Into this arena of intellectual electricity stepped French aristocrat René Descartes (1596–1650). Descartes (pronounced "day cart"), propelled by the creativity surrounding him, developed a new branch of mathematics that brought together algebra and geometry in a unified way—a way that visualized numbers as points on a graph, equations as geometric figures, and geometric figures as equations. This new branch of mathematics, called *analytic geometry*, established Descartes as one of the founders of modern thought and among the most original mathematicians and philosophers of any age. We begin this section by looking at Descartes's deceptively simple idea, called the **rectangular coordinate system** or (in his honor) the **Cartesian coordinate system**.

1 Plot points in the rectangular coordinate system.

Points and Ordered Pairs

Descartes used two number lines that intersect at right angles at their zero points, as shown in **Figure 1.9**. The horizontal number line is the ***x*-axis**. The vertical number line is the ***y*-axis**. The point of intersection of these axes is their zero points, called the **origin**. Positive numbers are shown to the right and above the origin. Negative numbers are shown to the left and below the origin. The axes divide the plane into four quarters, called **quadrants**. The points located on the axes are not in any quadrant.

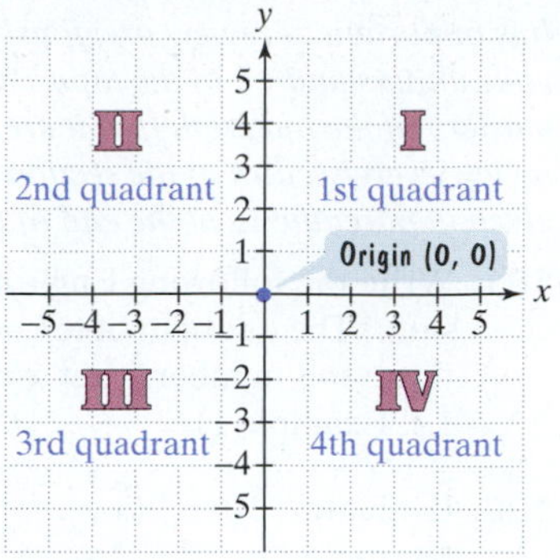

FIGURE 1.9 The rectangular coordinate system

Each point in the rectangular coordinate system corresponds to an **ordered pair** of real numbers, (x, y). Examples of such pairs are $(-5, 3)$ and $(3, -5)$. The first number in each pair, called the ***x*-coordinate**, denotes the distance and direction from the origin along the x-axis. The second number, called the ***y*-coordinate**, denotes vertical distance and direction along a line parallel to the y-axis or along the y-axis itself.

Figure 1.10 shows how we **plot**, or locate, the points corresponding to the ordered pairs $(-5, 3)$ and $(3, -5)$. We plot $(-5, 3)$ by going 5 units from 0 to the left along the x-axis. Then we go 3 units up parallel to the y-axis. We plot $(3, -5)$ by going 3 units from 0 to the right along the x-axis and 5 units down parallel to the y-axis. The phrase "the points corresponding to the ordered pairs $(-5, 3)$ and $(3, -5)$" are often abbreviated as "the points $(-5, 3)$ and $(3, -5)$."

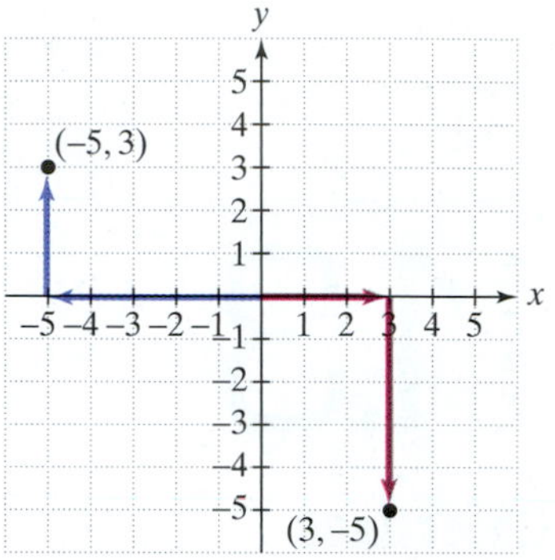

FIGURE 1.10 Plotting (−5, 3) and (3, −5)

Study Tip

The phrase *ordered pair* is used because order is important. The order in which coordinates appear makes a difference in a point's location. This is illustrated in **Figure 1.10**.

EXAMPLE 1 Plotting Points in the Rectangular Coordinate System

Plot the points: $A(-4, 5)$, $B(3, -4)$, $C(-5, 0)$, $D(-4, -2)$, $E(0, 3.5)$, and $F(0, 0)$.

Solution See **Figure 1.11**. We move from the origin and plot the points in the following way:

$A(-4, 5)$: 4 units left, 5 units up

$B(3, -4)$: 3 units right, 4 units down

$C(-5, 0)$: 5 units left, 0 units up or down

$D(-4, -2)$: 4 units left, 2 units down

$E(0, 3.5)$: 0 units right or left, 3.5 units up

$F(0, 0)$: 0 units right or left, 0 units up or down

Notice that the origin is represented by (0, 0).

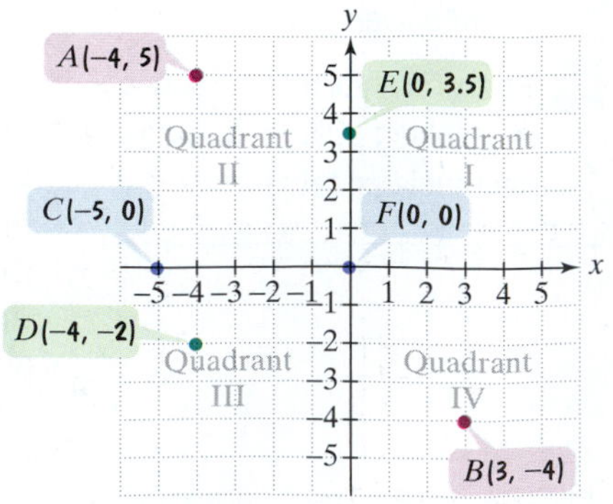

FIGURE 1.11 Plotting points

✓ **CHECK POINT 1** Plot the points:

$A(2, 5)$, $B(-1, 3)$, $C(-1.5, -4.5)$, and $D(0, -2)$.

2 Graph equations in the rectangular coordinate system.

Graphs of Equations

A relationship between two quantities can be expressed as an **equation in two variables**, such as

$$y = 4 - x^2.$$

A **solution of an equation in two variables**, x and y, is an ordered pair of real numbers with the following property: When the x-coordinate is substituted for x and the y-coordinate is substituted for y in the equation, we obtain a true statement. For example, consider the equation $y = 4 - x^2$ and the ordered pair $(3, -5)$. When 3 is substituted for x and -5 is substituted for y, we obtain the statement $-5 = 4 - 3^2$, or $-5 = 4 - 9$, or $-5 = -5$. Because this statement is true, the ordered pair $(3, -5)$ is a solution of the equation $y = 4 - x^2$. We also say that $(3, -5)$ **satisfies** the equation.

We can generate as many ordered-pair solutions as desired to $y = 4 - x^2$ by substituting numbers for x and then finding the corresponding values for y. For example, suppose we let $x = 3$:

Start with x. Compute y. Form the ordered pair (x, y).

x	$y = 4 - x^2$	Ordered Pair (x, y)
3	$y = 4 - 3^2 = 4 - 9 = -5$	$(3, -5)$

Let $x = 3$. $(3, -5)$ is a solution of $y = 4 - x^2$.

The **graph of an equation in two variables** is the set of all points whose coordinates satisfy the equation. One method for graphing such equations is the **point-plotting method**. First, we find several ordered pairs that are solutions of the equation. Next, we plot these ordered pairs as points in the rectangular coordinate system. Finally, we connect the points with a smooth curve or line. This often gives us a picture of all ordered pairs that satisfy the equation.

EXAMPLE 2 Graphing an Equation Using the Point-Plotting Method

Graph $y = 4 - x^2$. Select integers for x, starting with -3 and ending with 3.

Solution For each value of x, we find the corresponding value for y.

Start with x. Compute y. Form the ordered pair (x, y).

x	$y = 4 - x^2$	Ordered Pair (x, y)
-3	$y = 4-(-3)^2 = 4 - 9 = -5$	$(-3, -5)$
-2	$y = 4-(-2)^2 = 4 - 4 = 0$	$(-2, 0)$
-1	$y = 4-(-1)^2 = 4 - 1 = 3$	$(-1, 3)$
0	$y = 4 - 0^2 = 4 - 0 = 4$	$(0, 4)$
1	$y = 4 - 1^2 = 4 - 1 = 3$	$(1, 3)$
2	$y = 4 - 2^2 = 4 - 4 = 0$	$(2, 0)$
3	$y = 4 - 3^2 = 4 - 9 = -5$	$(3, -5)$

We selected integers from -3 to 3, inclusive, to include three negative numbers, 0, and three positive numbers. We also wanted to keep the resulting computations for y relatively simple.

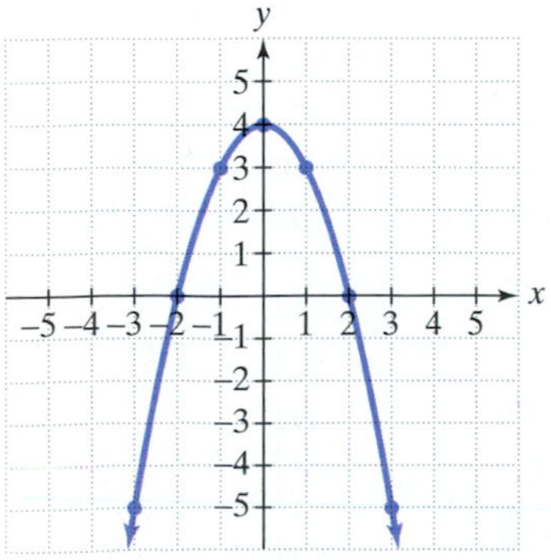

FIGURE 1.12 The graph of $y = 4 - x^2$

Now we plot the seven points and join them with a smooth curve, as shown in **Figure 1.12**. The graph of $y = 4 - x^2$ is a curve where the part of the graph to the right of the y-axis is a reflection of the part to the left of it and vice versa. The arrows on the left and the right of the curve indicate that it extends indefinitely in both directions. ■

✓ **CHECK POINT 2** Graph $y = 1 - x^2$. Select integers for x, starting with -3 and ending with 3.

EXAMPLE 3 Graphing an Equation Using the Point-Plotting Method

Graph $y = |x|$. Select integers for x, starting with -3 and ending with 3.

Solution For each value of x, we find the corresponding value for y.

x	$y = \lvert x \rvert$	Ordered Pair (x, y)
-3	$y = \lvert -3 \rvert = 3$	$(-3, 3)$
-2	$y = \lvert -2 \rvert = 3$	$(-2, 2)$
-1	$y = \lvert -1 \rvert = 3$	$(-1, 1)$
0	$y = \lvert 0 \rvert = 0$	$(0, 0)$
1	$y = \lvert 1 \rvert = 1$	$(1, 1)$
3	$y = \lvert 2 \rvert = 2$	$(2, 2)$
3	$y = \lvert 3 \rvert = 3$	$(3, 3)$

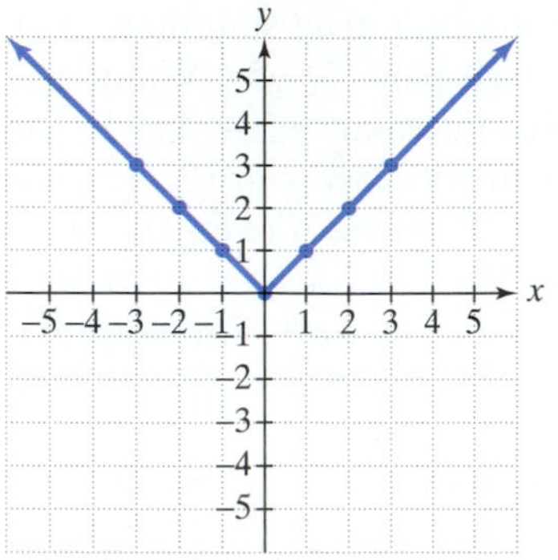

FIGURE 1.13 The graph of $y = |x|$

We plot the points and connect them, resulting in the graph shown in **Figure 1.13**. The graph is V-shaped and centered at the origin. For every point (x, y) on the graph, the point $(-x, y)$ is also on the graph. This shows that the absolute value of a positive number is the same as the absolute value of its opposite.

✓ **CHECK POINT 3** Graph $y = |x + 1|$. Select integers for x, starting with -4 and ending with 2.

3 Use the rectangular coordinate system to visualize relationships between variables.

Applications

The rectangular coordinate system allows us to visualize relationships between two variables by associating any equation in two variables with a graph. Graphs in the rectangular coordinate system can also be used to tell a story.

EXAMPLE 4 Telling a Story with a Graph

Too late for that flu shot now! It's only 8 A.M. and you're feeling lousy. Fascinated by the way that algebra models the world (your author is projecting a bit here), you construct a graph showing your body temperature from 8 A.M. through 3 P.M. You decide to let x represent the number of hours after 8 A.M. and y your body temperature at time x. The graph is shown in **Figure 1.14**. The symbol ⌇ on the y-axis shows that there is a break in values between 0 and 98. Thus, the first tick mark on the y-axis represents a temperature of 98°F.

a. What is your temperature at 8 A.M.?

b. During which period of time is your temperature decreasing?

c. Estimate your minimum temperature during the time period shown. How many hours after 8 A.M. does this occur? At what time does this occur?

d. During which period of time is your temperature increasing?

e. Part of the graph is shown as a horizontal line segment. What does this mean about your temperature and when does this occur?

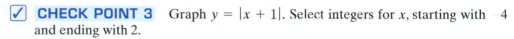

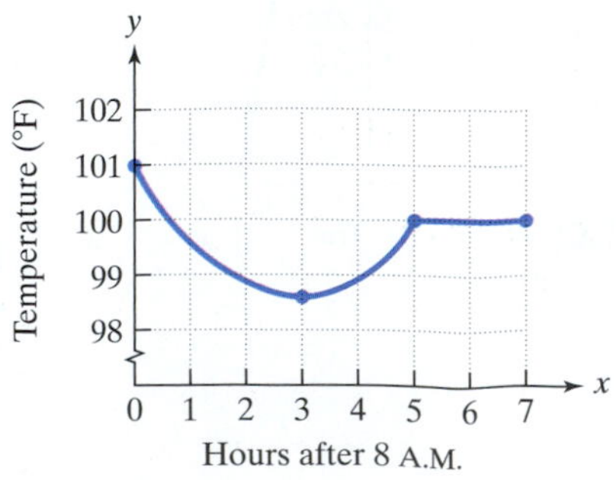

FIGURE 1.14 Body temperature from 8 A.M. through 3 P.M.

Solution

a. Because x is the number of hours after 8 A.M., your temperature at 8 A.M. corresponds to $x = 0$. Locate 0 on the horizontal axis and look at the point on the graph above 0. The first figure on the right shows that your temperature at 8 A.M. is 101°F.

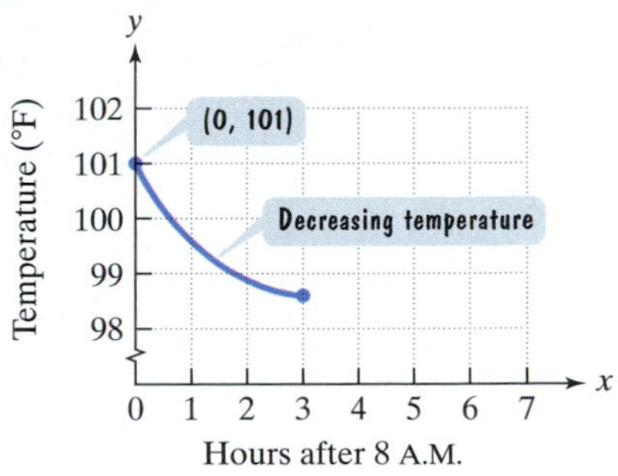

b. Your temperature is decreasing when the graph falls from left to right. This occurs between $x = 0$ and $x = 3$, also shown in the first figure on the right. Because x represents the number of hours after 8 A.M., your temperature is decreasing between 8 A.M. and 11 A.M.

c. Your minimum temperature can be found by locating the lowest point on the graph. This point lies above 3 on the horizontal axis, shown in the second figure on the right. The y-coordinate of this point falls more than midway between 98 and 99, at approximately 98.6. The lowest point on the graph, (3, 98.6), shows that your minimum temperature, 98.6°F, occurs 3 hours after 8 A.M., at 11 A.M.

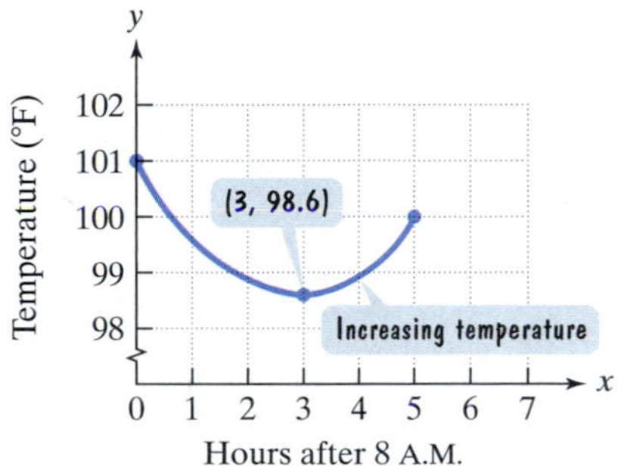

d. Your temperature is increasing when the graph rises from left to right. This occurs between $x = 3$ and $x = 5$, shown in the second figure. Because x represents the number of hours after 8 A.M., your temperature is increasing between 11 A.M. and 1 P.M.

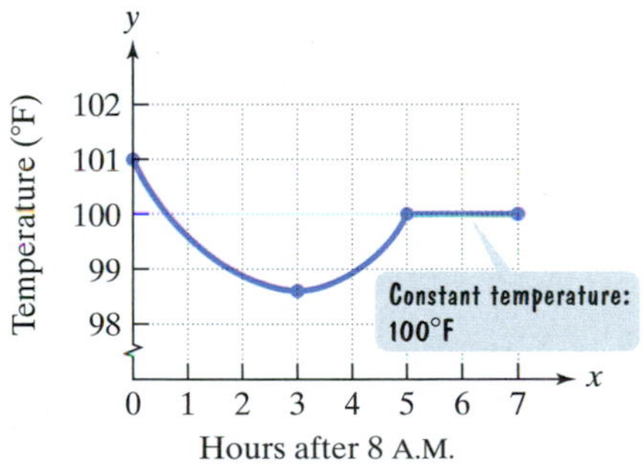

e. The horizontal line segment shown in the figure on the right indicates that your temperature is neither increasing nor decreasing. Your temperature remains the same, 100°F, between $x = 5$ and $x = 7$. Thus, your temperature is at a constant 100°F between 1 P.M. and 3 P.M.

✓ **CHECK POINT 4** When a person receives a drug injected into a muscle, the concentration of the drug in the body, measured in milligrams per 100 milliliters, depends on the time elapsed after the injection, measured in hours. The figure shows the graph of drug concentration over time, where x represents hours after the injection and y represents the drug concentration at time x.

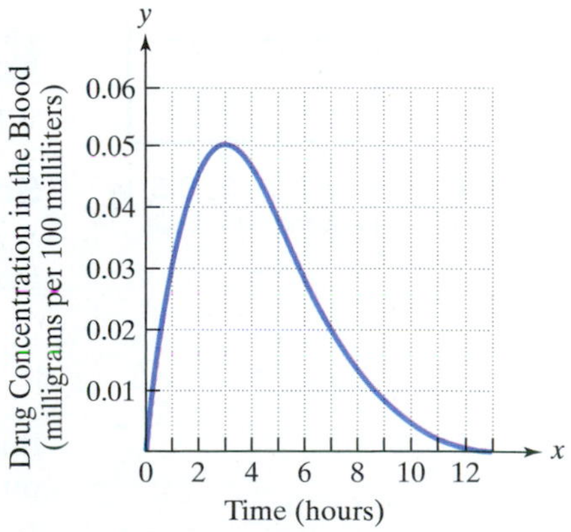

a. During which period of time is the drug concentration increasing?

b. During which period of time is the drug concentration decreasing?

c. What is the drug's maximum concentration and when does this occur?

d. What happens by the end of 13 hours?

4 Interpret information about a graphing utility's viewing rectangle or table.

Graphing Equations and Creating Tables Using a Graphing Utility

Graphing calculators and graphing software packages for computers are referred to as **graphing utilities** or graphers. A graphing utility is a powerful tool that quickly generates the graph of an equation in two variables. **Figures 1.15(a)** and **1.15(b)** show two such graphs for the equations in Examples 2 and 3.

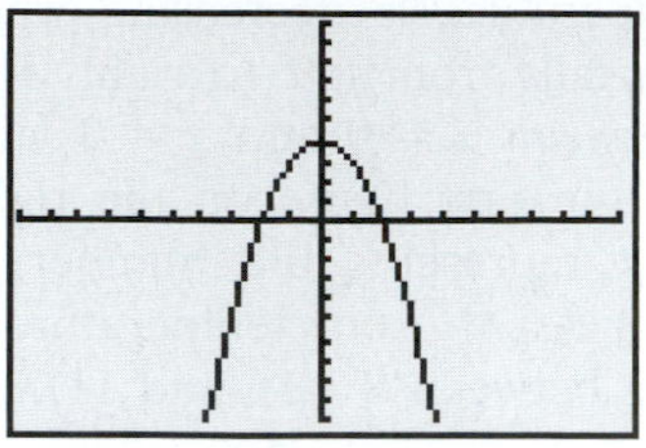

FIGURE 1.15(a) The graph of $y = 4 - x^2$

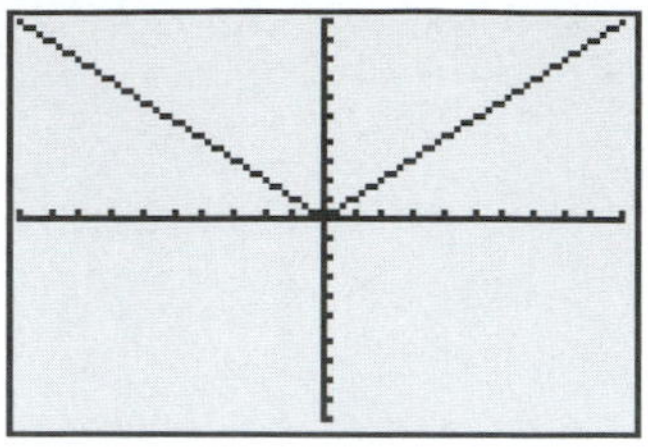

FIGURE 1.15(b) The graph of $y = |x|$

Study Tip

Even if you are not using a graphing utility in the course, read this part of the section. Knowing about viewing rectangles will enable you to understand the graphs that we display in the technology boxes throughout the book.

What differences do you notice between these graphs and the graphs that we drew by hand? They do seem a bit "jittery." Arrows do not appear on the left and right ends of the graphs. Furthermore, numbers are not given along the axes. For both graphs in **Figure 1.15**, the x-axis extends from -10 to 10 and the y-axis also extends from -10 to 10. The distance represented by each consecutive tick mark is one unit. We say that the **viewing rectangle**, or the **viewing window**, is $[-10, 10, 1]$ by $[-10, 10, 1]$.

$[-10,$	$10,$	$1]$	by	$[-10,$	$10,$	$1]$
The minimum x-value along the x-axis is -10.	The maximum x-value along the x-axis is 10.	Distance between consecutive tick marks on the x-axis is one unit.		The minimum y-value along the y-axis is -10.	The maximum y-value along the y-axis is 10.	Distance between consecutive tick marks on the y-axis is one unit.

To graph an equation in x and y using a graphing utility, enter the equation and specify the size of the viewing rectangle. The size of the viewing rectangle sets minimum and maximum values for both the x- and y-axes. Enter these values, as well as the values between consecutive tick marks, on the respective axes. The $[-10, 10, 1]$ by $[-10, 10, 1]$ viewing rectangle used in **Figure 1.15** is called the **standard viewing rectangle**.

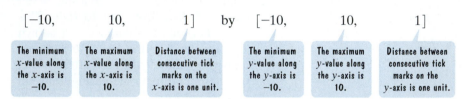

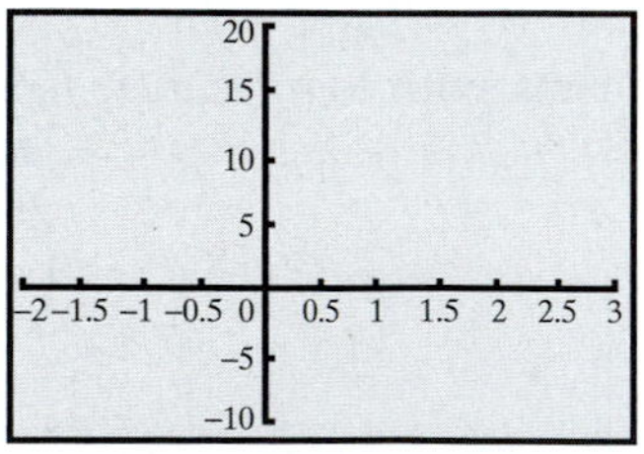

FIGURE 1.16 A $[-2, 3, 0.5]$ by $[-10, 20, 5]$ viewing rectangle

EXAMPLE 5 Understanding the Viewing Rectangle

What is the meaning of a $[-2, 3, 0.5]$ by $[-10, 20, 5]$ viewing rectangle?

Solution We begin with $[-2, 3, 0.5]$, which describes the x-axis. The minimum x-value is -2 and the maximum x-value is 3. The distance between consecutive tick marks is 0.5.

Next, consider $[-10, 20, 5]$, which describes the y-axis. The minimum y-value is -10 and the maximum y-value is 20. The distance between consecutive tick marks is 5.

Figure 1.16 illustrates a $[-2, 3, 0.5]$ by $[-10, 20, 5]$ viewing rectangle. To make things clearer, we've placed numbers by each tick mark. These numbers do not appear on the axes when you use a graphing utility to graph an equation.

✓ CHECK POINT 5 What is the meaning of a $[-100, 100, 50]$ by $[-100, 100, 10]$ viewing rectangle? Create a figure like the one in **Figure 1.16** that illustrates this viewing rectangle.

On most graphing utilities, the display screen is two-thirds as high as it is wide. By using a square setting, you can equally space the x and y tick marks. (This does not occur in the standard viewing rectangle.) Graphing utilities can also *zoom in* and *zoom out*. When you zoom in, you see a smaller portion of the graph, but you do so in greater detail. When you zoom out, you see a larger portion of the graph. Thus, zooming out may help you to develop a better understanding of the overall character of the graph. With practice, you will become more comfortable with graphing equations in two variables using your graphing utility. You will also develop a better sense of the size of the viewing rectangle that will reveal needed information about a particular graph.

Graphing utilities can also be used to create tables showing solutions of equations in two variables. Use the Table Setup function to choose the starting value of x and to input the increment, or change, between the consecutive x-values. The corresponding y-values are calculated based on the equation(s) in two variables in the [Y=] screen. In **Figure 1.17**, we used a TI-84 Plus to create a table for $y = 4 - x^2$ and $y = |x|$, the equations in Examples 2 and 3.

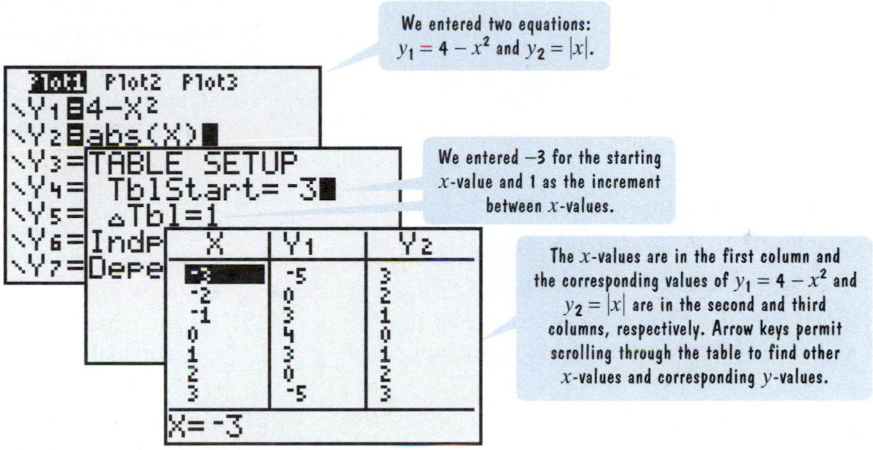
We entered two equations: $y_1 = 4 - x^2$ and $y_2 = |x|$.
Plot1 Plot2 Plot3
\Y1 = 4-X²
\Y2 = abs(X)
TABLE SETUP
TblStart=-3
ΔTbl=1
We entered −3 for the starting x-value and 1 as the increment between x-values.
X Y1 Y2
X=-3
The x-values are in the first column and the corresponding values of $y_1 = 4 - x^2$ and $y_2 = |x|$ are in the second and third columns, respectively. Arrow keys permit scrolling through the table to find other x-values and corresponding y-values.

FIGURE 1.17 Creating a table for $y_1 = 4 - x^2$ and $y_2 = |x|$

1.3 EXERCISE SET

MyMathLab PRACTICE WATCH DOWNLOAD READ REVIEW

Practice Exercises

In Exercises 1–10, plot the given point in a rectangular coordinate system.

1. $(1, 4)$
2. $(2, 5)$
3. $(-2, 3)$
4. $(-1, 4)$
5. $(-3, -5)$
6. $(-4, -2)$
7. $(4, -1)$
8. $(3, -2)$
9. $(-4, 0)$
10. $(0, -3)$

Graph each equation in Exercises 11–26. Let $x = -3, -2, -1, 0, 1, 2,$ and 3.

11. $y = x^2 - 4$
12. $y = x^2 - 9$
13. $y = x - 2$
14. $y = x + 2$
15. $y = 2x + 1$
16. $y = 2x - 4$
17. $y = -\frac{1}{2}x$
18. $y = -\frac{1}{2}x + 2$
19. $y = |x| + 1$
20. $y = |x| - 1$
21. $y = 2|x|$
22. $y = -2|x|$
23. $y = -x^2$
24. $y = -\frac{1}{2}x^2$
25. $y = x^3$
26. $y = x^3 - 1$

In Exercises 27–30, match the viewing rectangle with the correct figure. Then label the tick marks in the figure to illustrate this viewing rectangle.

27. $[-5, 5, 1]$ by $[-5, 5, 1]$
28. $[-10, 10, 2]$ by $[-4, 4, 2]$
29. $[-20, 80, 10]$ by $[-30, 70, 10]$
30. $[-40, 40, 20]$ by $[-1000, 1000, 100]$

a.
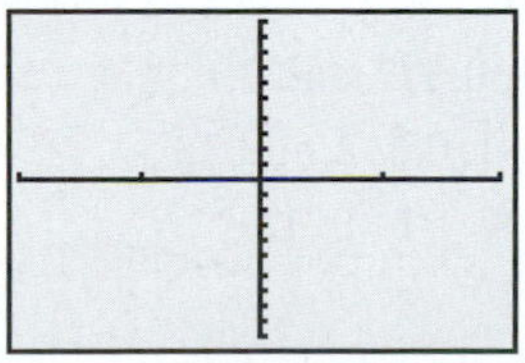

b.
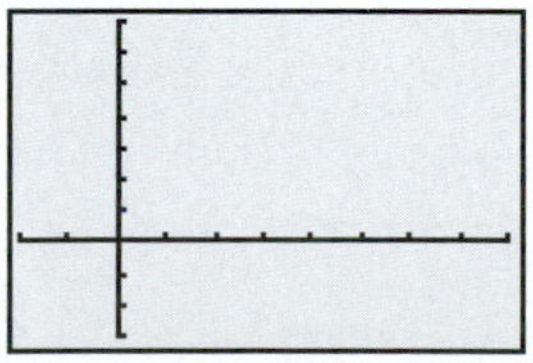

c.
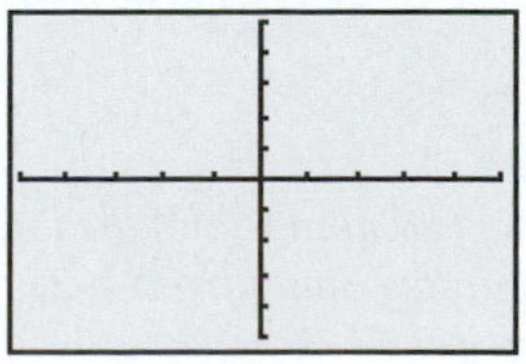

d.
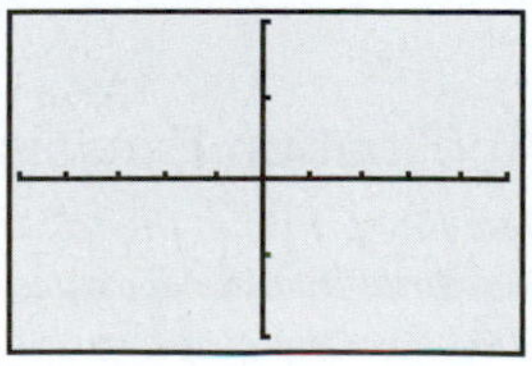

The table of values was generated by a graphing utility with a TABLE feature. Use the table to solve Exercises 31–38.

X	Y1	Y2
-3	9	5
-2	4	4
-1	1	3
0	0	2
1	1	1
2	4	0
3	9	-1

X= -3

31. Which equation corresponds to Y_2 in the table?

a. $y_2 = x + 8$

b. $y_2 = x - 2$

c. $y_2 = 2 - x$

d. $y_2 = 1 - 2x$

32. Which equation corresponds to Y_1 in the table?

a. $y_1 = -3x$

b. $y_1 = x^2$

c. $y_1 = -x^2$

d. $y_1 = 2 - x$

33. Does the graph of Y_2 pass through the origin?

34. Does the graph of Y_1 pass through the origin?

35. At which point does the graph of Y_2 cross the x-axis?

36. At which point does the graph of Y_2 cross the y-axis?

37. At which points do the graphs of Y_1 and Y_2 intersect?

38. For which values of x is $Y_1 = Y_2$?

Practice PLUS

In Exercises 39–42, write each English sentence as an equation in two variables. Then graph the equation.

39. The y-value is four more than twice the x-value.

40. The y-value is the difference between four and twice the x-value.

41. The y-value is three decreased by the square of the x-value.

42. The y-value is two more than the square of the x-value.

In Exercises 43–46, graph each equation.

43. $y = 5$ (Let $x = -3, -2, -1, 0, 1, 2,$ and 3.)

44. $y = -1$ (Let $x = -3, -2, -1, 0, 1, 2,$ and 3.)

45. $y = \frac{1}{x}$ (Let $x = -2, -1, -\frac{1}{2}, -\frac{1}{3}, \frac{1}{3}, \frac{1}{2}, 1,$ and 2.)

46. $y = -\frac{1}{x}$ (Let $x = -2, -1, -\frac{1}{2}, -\frac{1}{3}, \frac{1}{3}, \frac{1}{2}, 1,$ and 2.)

Application Exercises

The line graph at the top of the next column shows the top marginal income tax rates in the United States from 1925 through 2005. Use the graph to solve Exercises 47–52.

Top United States Marginal Tax Rates, 1925–2005

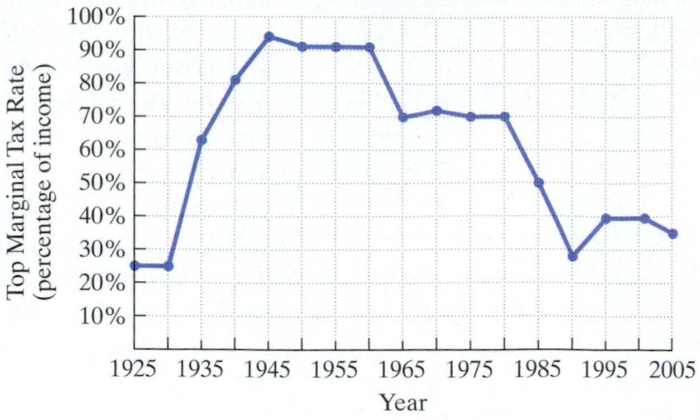

Source: National Taxpayers Union

47. Estimate the top marginal tax rate in 2005.

48. Estimate the top marginal tax rate in 1925.

49. For the period shown, during which year did the United States have the highest marginal tax rate? Estimate, to the nearest percent, the tax rate for that year.

50. For the period from 1950 through 2005, during which year did the United States have the lowest marginal tax rate? Estimate, to the nearest percent, the tax rate for that year.

51. For the period shown, during which ten-year period did the top marginal tax rate remain constant? Estimate, to the nearest percent, the tax rate for that period.

52. For the period shown, during which five-year period did the top marginal tax rate increase most rapidly? Estimate, to the nearest percent, the increase in the top tax rate for that period.

Contrary to popular belief, older people do not need less sleep than younger adults. However, the line graphs show that they awaken more often during the night. The numerous awakenings are one reason why some elderly individuals report that sleep is less restful than it had been in the past. Use the line graphs to solve Exercises 53–56.

Average Number of Awakenings During the Night, by Age and Gender

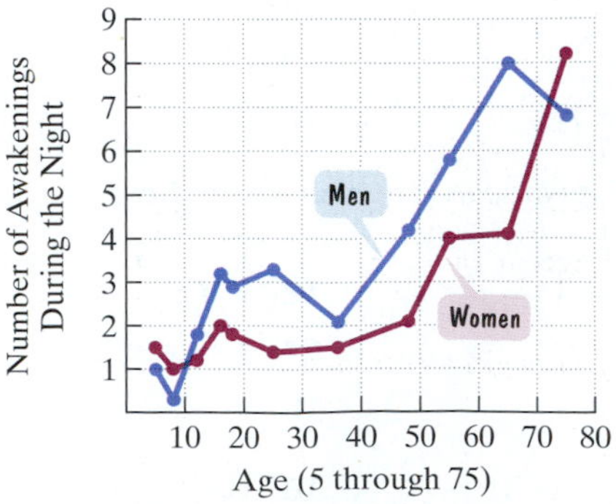

Source: Stephen Davis and Joseph Palladino, *Psychology,* 5th Edition, Prentice Hall, 2007

53. At which age, estimated to the nearest year, do women have the least number of awakenings during the night? What is the average number of awakenings at that age?

54. At which age do men have the greatest number of awakenings during the night? What is the average number of awakenings at that age?

55. Estimate, to the nearest tenth, the difference between the average number of awakenings during the night between 25-year-old men and 25-year-old women.

56. Estimate, to the nearest tenth, the difference between the average number of awakenings during the night between 18-year-old men and 18-year-old women.

In Exercises 57–60, match the story with the correct figure. The figures are labeled (a), (b), (c), and (d).

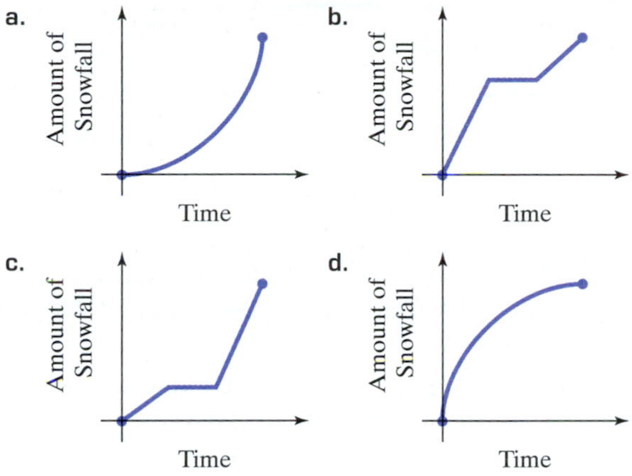

57. As the blizzard got worse, the snow fell harder and harder.

58. The snow fell more and more softly.

59. It snowed hard, but then it stopped. After a short time, the snow started falling softly.

60. It snowed softly, and then it stopped. After a short time, the snow started falling hard.

In Exercises 61–64, select the graph that best illustrates each story.

61. An airplane flew from Miami to San Francisco.

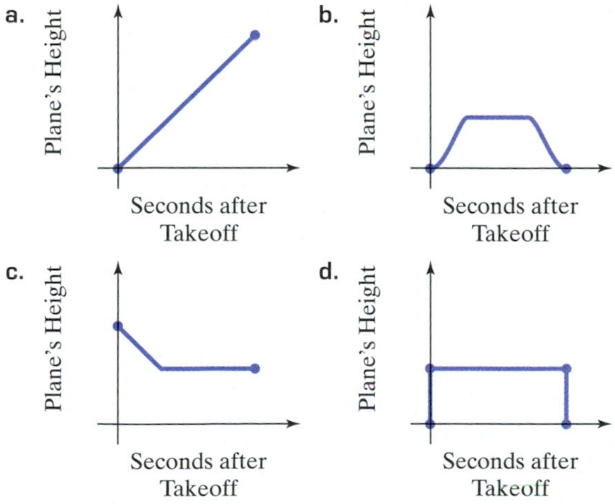

62. At noon, you begin to breathe in.

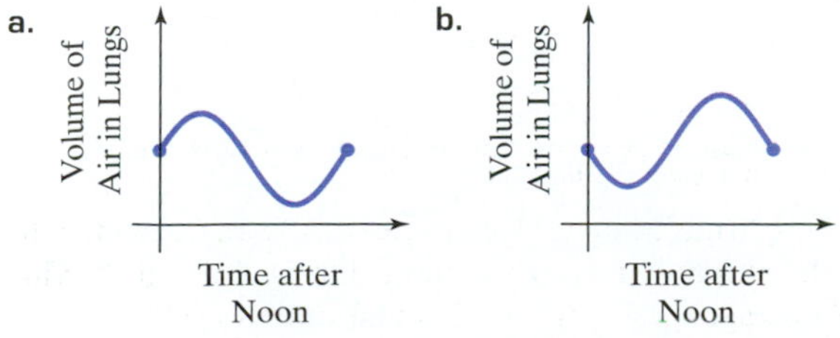

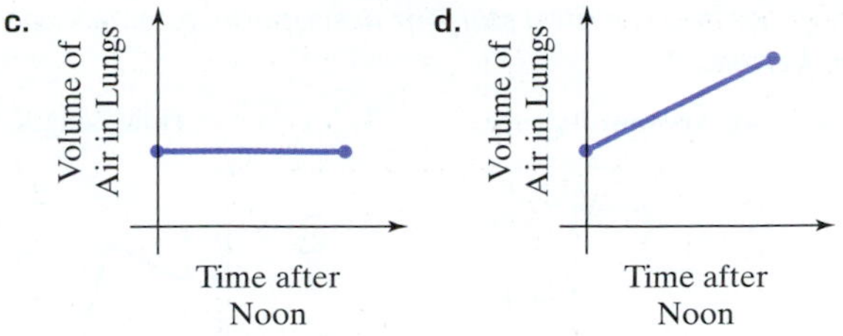

63. Measurements are taken of a person's height from birth to age 100.

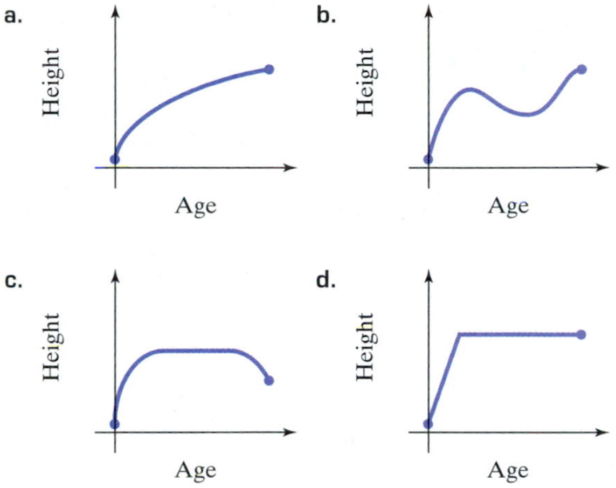

64. You begin your bike ride by riding down a hill. Then you ride up another hill. Finally, you ride along a level surface before coming to a stop.

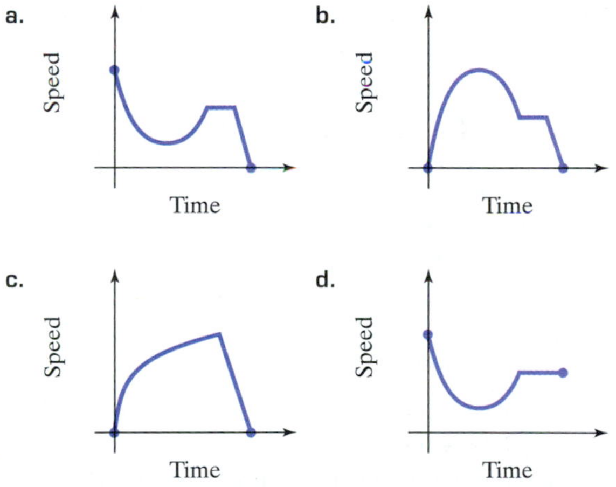

Writing in Mathematics

65. What is the rectangular coordinate system?

66. Explain how to plot a point in the rectangular coordinate system. Give an example with your explanation.

67. Explain why $(5, -2)$ and $(-2, 5)$ do not represent the same point.

68. Explain how to graph an equation in the rectangular coordinate system.

69. What does a $[-20, 2, 1]$ by $[-4, 5, 0.5]$ viewing rectangle mean?

In Exercises 70–71, write a story, or description, to match each title and graph.

70. **Checking Account Balance**

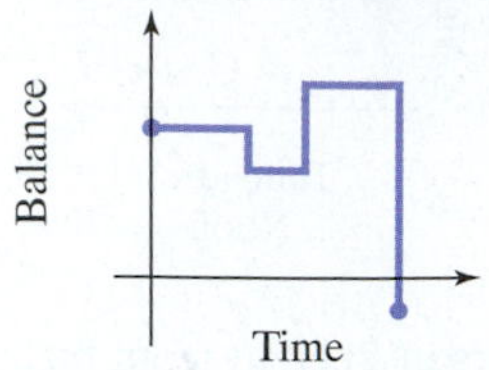

71. **Hair Length**

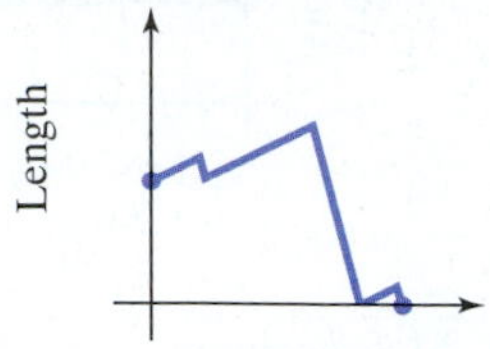

Technology Exercise

72. Use a graphing utility to verify each of your hand-drawn graphs in Exercises 11–26. Experiment with the viewing rectangle to make the graph displayed by the graphing utility resemble your hand-drawn graph as much as possible.

Critical Thinking Exercises

Make Sense? *In Exercises 73–76, determine whether each statement "makes sense" or "does not make sense" and explain your reasoning.*

73. The rectangular coordinate system provides a geometric picture of what an equation in two variables looks like.

74. There is something wrong with my graphing utility because it is not displaying numbers along the x- and y-axes.

75. A horizontal line is not a graph that tells the story of the number of calories that I burn throughout the day.

76. I told my story with a graph, so I can be confident that there is a mathematical model that perfectly describes the graph's data.

In Exercises 77–80, determine whether each statement is true or false. If the statement is false, make the necessary change(s) to produce a true statement.

77. If the product of a point's coordinates is positive, the point must be in quadrant I.

78. If a point is on the x-axis, it is neither up nor down, so $x = 0$.

79. If a point is on the y-axis, its x-coordinate must be 0.

80. The ordered pair $(2, 5)$ satisfies $3y - 2x = -4$.

The graph shows the costs at a parking garage that allows cars to be parked for up to ten hours per day. Closed dots indicate that points belong to the graph and open dots indicate that points are not part of the graph. Use the graph to solve Exercises 81–82.

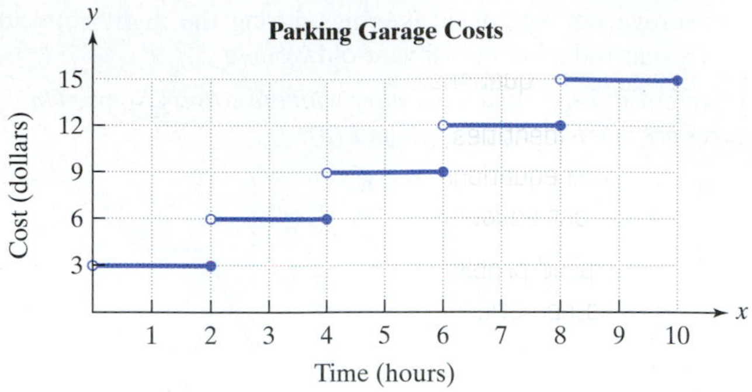

81. You park your car at the garage for four hours on Tuesday and five hours on Wednesday. What are the total parking garage costs for the two days?

82. On Thursday, you paid $12 for parking at the garage. Describe how long your car was parked.

Review Exercises

83. Find the absolute value: $|-14.3|$. (Section 1.2, Example 1)

84. Simplify: $[12 - (13 - 17)] - [9 - (6 - 10)]$. (Section 1.2, Examples 7 and 8)

85. Simplify: $6x - 5(4x + 3) - 10$. (Section 1.2, Example 13)

Preview Exercises

Exercises 86–88 will help you prepare for the material covered in the next section.

86. If -9 is substituted for x in the equation $4x - 3 = 5x + 6$, is the resulting statement true or false?

87. Simplify: $13 - 3(x + 2)$.

88. Simplify: $10\left(\frac{3x + 1}{2}\right)$.

SECTION 1.4

Solving Linear Equations

Objectives

1. Solve linear equations.
2. Recognize identities, conditional equations, and inconsistent equations.
3. Solve applied problems using mathematical models.

Teutul Time: Jr. and Sr. in a rare peaceful moment

Talk about a booming business cycle: TV's "American Chopper" is the hottest thing since *Easy Rider* and motorcycles are selling faster than they have in a generation. The 10.5 million weekly viewers of "American Chopper" are drawn as much by the titanic tussels between superstar bike builders Paul Senior and Paul Junior Teutul as by the craftsmanship in their elaborate theme bikes.

Although "American Chopper" is bringing more people into showrooms, the bar graph in **Figure 1.18** indicates that motorcycle sales were rising long before the Teutuls showed up. The rising sales from 1998 through 2005 can be modeled by the formula

$$N = 0.12x + 0.4,$$

where N represents the number of new motorcycles sold in the United States, in millions, x years after 1998. In 2004, cycle sales reached 1.1 million, topping 1 million for the first time since the post–*Easy Rider* days of the early 1970s. So, when will motorcycle sales double the 2004 level and reach 2.2 million? Substitute 2.2 for N in the formula $N = 0.12x + 0.4$:

$$2.2 = 0.12x + 0.4.$$

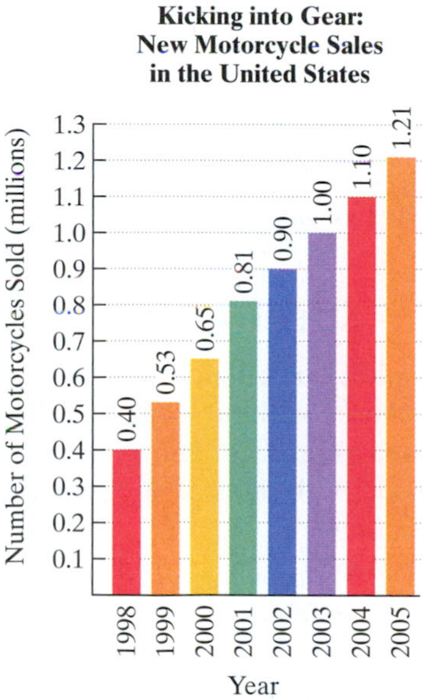

FIGURE 1.18

Source: Motorcycle Industry Council

Our goal is to determine the value of x, the number of years after 1998, when sales will reach 2.2 million. Notice that the exponent on the variable in this equation is 1. In this section, we will study how to determine the value of x in such equations. With this skill, you will be able to use certain mathematical models, such as the model for motorcycle sales, to project what might occur in the future.

1 Solve linear equations.

Solving Linear Equations in One Variable

We begin with a general definition of a linear equation in one variable.

Definition of a Linear Equation

A **linear equation in one variable x** is an equation that can be written in the form

$$ax + b = 0,$$

where a and b are real numbers, and $a \neq 0$ (a is not equal to 0).

An example of a linear equation in one variable is

$$4x + 12 = 0.$$

Solving an equation in x involves determining all values of x that result in a true statement when substituted into the equation. Such values are **solutions**, or **roots**, of the equation. For example, substitute -3 for x in $4x + 12 = 0$. We obtain

$$4(-3) + 12 = 0, \quad \text{or} \quad -12 + 12 = 0.$$

This simplifies to the true statement $0 = 0$. Thus, -3 is a solution of the equation $4x + 12 = 0$. We also say that -3 **satisfies** the equation $4x + 12 = 0$, because when we substitute -3 for x, a true statement results. The set of all such solutions is called the equation's **solution set**. For example, the solution set of the equation $4x + 12 = 0$ is $\{-3\}$.

Two or more equations that have the same solution set are called **equivalent equations**. For example, the equations

$$4x + 12 = 0 \quad \text{and} \quad 4x = -12 \quad \text{and} \quad x = -3$$

are equivalent equations because the solution set for each is $\{-3\}$. To solve a linear equation in x, we transform the equation into an equivalent equation one or more times. Our final equivalent equation should be of the form

$$x = \text{a number}.$$

The solution set of this equation is the set consisting of the number.

To generate equivalent equations, we will use the following properties:

The Addition and Multiplication Properties of Equality

The Addition Property of Equality

The same real number or algebraic expression may be added to both sides of an equation without changing the equation's solution set.

$$a = b \text{ and } a + c = b + c \text{ are equivalent equations.}$$

The Multiplication Property of Equality

The same nonzero real number may multiply both sides of an equation without changing the equation's solution set.

$$a = b \text{ and } ac = bc \text{ are equivalent equations as long as } c \neq 0.$$

Because subtraction is defined in terms of addition, the addition property also lets us subtract the same number from both sides of an equation without changing the equation's solution set. Similarly, because division is defined in terms of multiplication, the multiplication property of equality can be used to divide both sides of an equation by the same nonzero number to obtain an equivalent equation.

Table 1.4 illustrates how these properties are used to isolate x to obtain an equation of the form $x = $ a number.

Study Tip

Your final equivalent equation should not be of the form

$$-x = \text{a number}.$$

We're not finished. A negative sign should not precede the variable.

Isolate x by multiplying or dividing both sides of this equation by -1.

Table 1.4 Using Properties of Equality to Solve Linear Equations

	Equation	How to Isolate x	Solving the Equation	The Equation's Solution Set
These equations are solved using the Addition Property of Equality.	$x - 3 = 8$	Add 3 to both sides.	$x - 3 + 3 = 8 + 3$ $x = 11$	$\{11\}$
	$x + 7 = -15$	Subtract 7 from both sides.	$x + 7 - 7 = -15 - 7$ $x = -22$	$\{-22\}$
These equations are solved using the Multiplication Property of Equality.	$6x = 30$	Divide both sides by 6 (or multiply both sides by $\frac{1}{6}$).	$\frac{6x}{6} = \frac{30}{6}$ $x = 5$	$\{5\}$
	$\frac{x}{5} = 9$	Multiply both sides by 5.	$5 \cdot \frac{x}{5} = 5 \cdot 9$ $x = 45$	$\{45\}$

EXAMPLE 1 Solving a Linear Equation

Solve and check: $2x + 3 = 17$.

Solution Our goal is to obtain an equivalent equation with x isolated on one side and a number on the other side.

$$2x + 3 = 17$$ This is the given equation.

$$2x + 3 - 3 = 17 - 3$$ Subtract 3 from both sides.

$$2x = 14$$ Simplify.

$$\frac{2x}{2} = \frac{14}{2}$$ Divide both sides by 2.

$$x = 7$$ Simplify.

Now we check the proposed solution, 7, by replacing x with 7 in the original equation.

$$2x + 3 = 17$$ This is the original equation.

$$2 \cdot 7 + 3 \stackrel{?}{=} 17$$ Substitute 7 for x. The question mark indicates that we do not yet know if the two sides are equal.

$$14 + 3 \stackrel{?}{=} 17$$ Multiply: $2 \cdot 7 = 14$.

This statement is true.

$$17 = 17$$ Add: $14 + 3 = 17$.

Because the check results in a true statement, we conclude that the solution of the given equation is 7, or the solution set is $\{7\}$.

✓ **CHECK POINT 1** Solve and check: $4x + 5 = 29$.

Study Tip

We simplify algebraic expressions. We solve algebraic equations. Notice the differences between the procedures:

Simplifying an Algebraic Expression

Simplify: $3(x - 7) - (5x - 11)$.

This is not an equation. There is no equal sign.

Solution
$$\begin{aligned} &3(x - 7) - (5x - 11) \\ &= 3x - 21 - 5x + 11 \\ &= (3x - 5x) + (-21 + 11) \\ &= -2x + (-10) \\ &= -2x - 10 \end{aligned}$$

Stop! Further simplification is not possible. Avoid the common error of setting $-2x - 10$ equal to 0.

Solving an Algebraic Equation

Solve: $3(x - 7) - (5x - 11) = 14$.

This is an equation. There is an equal sign.

Solution
$$\begin{aligned} 3(x - 7) - (5x - 11) &= 14 \\ 3x - 21 - 5x + 11 &= 14 \\ -2x - 10 &= 14 \\ -2x - 10 + 10 &= 14 + 10 \\ -2x &= 24 \\ \frac{-2x}{-2} &= \frac{24}{-2} \\ x &= -12 \end{aligned}$$

Add 10 to both sides.

Divide both sides by −2.

The solution set is $\{-12\}$.

Here is a step-by-step procedure for solving a linear equation in one variable. Not all of these steps are necessary to solve every equation.

Solving a Linear Equation

1. Simplify the algebraic expression on each side by removing grouping symbols and combining like terms.
2. Collect all the variable terms on one side and all the numbers, or constant terms, on the other side.
3. Isolate the variable and solve.
4. Check the proposed solution in the original equation.

EXAMPLE 2 Solving a Linear Equation

Solve and check: $2x - 7 + x = 3x + 1 + 2x$.

Solution

Step 1. Simplify the algebraic expression on each side.

$$2x - 7 + x = 3x + 1 + 2x \quad \text{This is the given equation.}$$

$$3x - 7 = 5x + 1 \quad \text{Combine like terms: } 2x + x = 3x \text{ and } 3x + 2x = 5x.$$

Discover for Yourself

Solve the equation in Example 2 by collecting terms with the variable on the right and constant terms on the left. What do you observe?

Step 2. Collect variable terms on one side and constant terms on the other side. We will collect variable terms on the left by subtracting $5x$ from both sides. We will collect the numbers on the right by adding 7 to both sides.

$$3x - 5x - 7 = 5x - 5x + 1 \quad \text{Subtract } 5x \text{ from both sides.}$$

$$-2x - 7 = 1 \quad \text{Simplify.}$$

$$-2x - 7 + 7 = 1 + 7 \quad \text{Add 7 to both sides.}$$

$$-2x = 8 \quad \text{Simplify.}$$

Step 3. Isolate the variable and solve. We isolate x by dividing both sides by -2.

$$\frac{-2x}{-2} = \frac{8}{-2} \quad \text{Divide both sides by } -2.$$

$$x = -4 \quad \text{Simplify.}$$

Step 4. Check the proposed solution in the original equation. Substitute -4 for x in the original equation.

$$2x - 7 + x = 3x + 1 + 2x \quad \text{This is the original equation.}$$

$$2(-4) - 7 + (-4) \stackrel{?}{=} 3(-4) + 1 + 2(-4) \quad \text{Substitute } -4 \text{ for } x.$$

$$-8 - 7 + (-4) \stackrel{?}{=} -12 + 1 + (-8) \quad \text{Multiply: } 2(-4) = -8,\ 3(-4) = -12, \text{ and } 2(-4) = -8.$$

$$-15 + (-4) \stackrel{?}{=} -11 + (-8) \quad \text{Add or subtract from left to right: } -8 - 7 = -15 \text{ and } -12 + 1 = -11.$$

$$-19 = -19 \quad \text{Add.}$$

The true statement $-19 = -19$ verifies that -4 is the solution, or the solution set is $\{-4\}$.

✓ CHECK POINT 2 Solve and check: $2x - 12 + x = 6x - 4 + 5x$.

EXAMPLE 3 Solving a Linear Equation

Solve and check: $4(2x + 1) - 29 = 3(2x - 5)$.

Solution

Step 1. Simplify the algebraic expression on each side.

$$4(2x + 1) - 29 = 3(2x - 5)$$ This is the given equation.

$$8x + 4 - 29 = 6x - 15$$ Use the distributive property.

$$8x - 25 = 6x - 15$$ Simplify.

Step 2. Collect variable terms on one side and constant terms on the other side. We will collect the variable terms on the left by subtracting $6x$ from both sides. We will collect the numbers on the right by adding 25 to both sides.

$$8x - 6x - 25 = 6x - 6x - 15$$ Subtract $6x$ from both sides.

$$2x - 25 = -15$$ Simplify.

$$2x - 25 + 25 = -15 + 25$$ Add 25 to both sides.

$$2x = 10$$ Simplify.

Step 3. Isolate the variable and solve. We isolate x by dividing both sides by 2.

$$\frac{2x}{2} = \frac{10}{2}$$ Divide both sides by 2.

$$x = 5$$ Simplify.

(*The example continues on page 42.*)

Using Technology

Numeric and Graphic Connections

In many algebraic situations, technology provides numeric and visual insights into problem solving. For example, you can use a graphing utility to check the solution of a linear equation, giving numeric and geometric meaning to the solution. Enter each side of the equation separately under y_1 and y_2. Then use the table or the graphs to locate the x-value for which the y-values are the same. This x-value is the solution.

Let's verify our work in Example 3 and show that 5 is the solution of

$$4(2x + 1) - 29 = 3(2x - 5).$$

Enter $y_1 = 4(2x + 1) - 29$ in the $\boxed{y=}$ screen.

Enter $y_2 = 3(2x - 5)$ in the $\boxed{y=}$ screen.

Numeric Check

Display a table for y_1 and y_2.

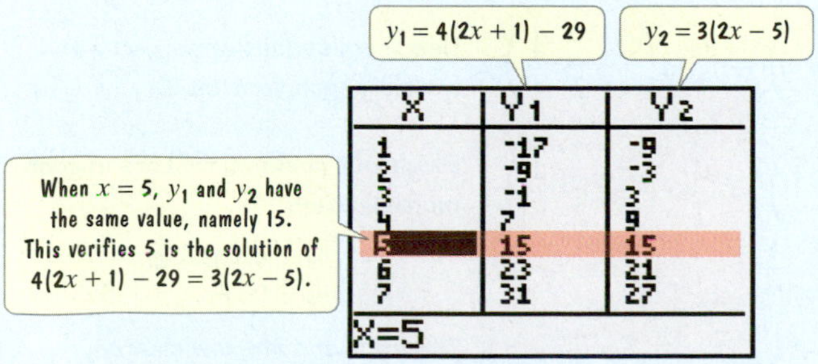
y1 = 4(2x + 1) − 29
y2 = 3(2x − 5)
When x = 5, y1 and y2 have the same value, namely 15. This verifies 5 is the solution of 4(2x + 1) − 29 = 3(2x − 5).
X
Y1
Y2
1 2 3 4 5 6 7
-17 -9 -1 7 15 23 31
-9 -3 3 9 15 21 27
X=5

Graphic Check

Display graphs for y_1 and y_2 and use the intersection feature. The solution is the x-coordinate of the intersection point.

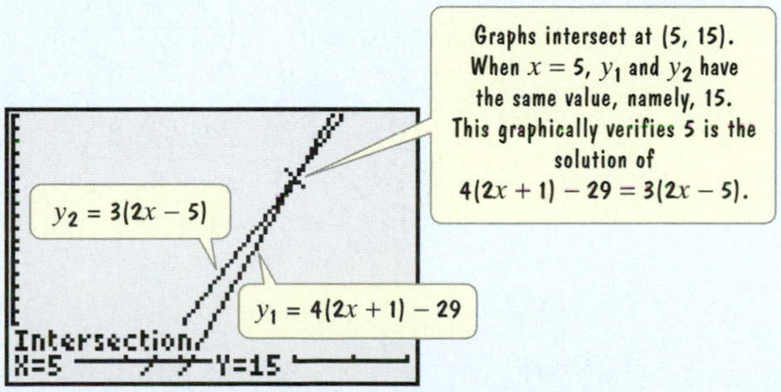
Graphs intersect at (5, 15). When x = 5, y1 and y2 have the same value, namely, 15. This graphically verifies 5 is the solution of 4(2x + 1) − 29 = 3(2x − 5).
y2 = 3(2x − 5)
y1 = 4(2x + 1) − 29
Intersection
X=5
Y=15

[0, 7, 1] by [−1, 20, 1]

Choose a large enough viewing rectangle so that you can see the intersection point.

Step 4. Check the proposed solution in the original equation. Substitute 5 for x in the original equation.

$$4(2x + 1) - 29 = 3(2x - 5)$$ This is the original equation.

$$4(2 \cdot 5 + 1) - 29 \stackrel{?}{=} 3(2 \cdot 5 - 5)$$ Substitute 5 for x.

$$4(11) - 29 \stackrel{?}{=} 3(5)$$ Simplify inside parentheses: $2 \cdot 5 + 1 = 10 + 1 = 11$ and $2 \cdot 5 - 5 = 10 - 5 = 5$.

$$44 - 29 \stackrel{?}{=} 15$$ Multiply: $4(11) = 44$ and $3(5) = 15$.

$$15 = 15$$ Subtract.

The true statement $15 = 15$ verifies that 5 is the solution, or the solution set is $\{5\}$.

CHECK POINT 3 Solve and check: $2(x - 3) - 17 = 13 - 3(x + 2)$.

Linear Equations with Fractions

Equations are easier to solve when they do not contain fractions. How do we remove fractions from an equation? We begin by multiplying both sides of the equation by the least common denominator (LCD) of any fractions in the equation. The least common denominator is the smallest number that all denominators will divide into. Multiplying every term on both sides of the equation by the least common denominator will eliminate the fractions in the equation. Example 4 shows how we "clear an equation of fractions."

EXAMPLE 4 Solving a Linear Equation Involving Fractions

Solve: $\frac{2x + 5}{5} + \frac{x - 7}{2} = \frac{3x + 1}{2}$.

Solution The denominators are 5, 2, and 2. The smallest number that is divisible by 5, 2, and 2 is 10. We begin by multiplying both sides of the equation by 10, the least common denominator.

$$\frac{2x + 5}{5} + \frac{x - 7}{2} = \frac{3x + 1}{2}$$ This is the given equation.

$$10\left(\frac{2x + 5}{5} + \frac{x - 7}{2}\right) = 10\left(\frac{3x + 1}{2}\right)$$ Multiply both sides by 10.

$$\frac{10}{1} \cdot \left(\frac{2x + 5}{5}\right) + \frac{10}{1} \cdot \left(\frac{x - 7}{2}\right) = \frac{10}{1} \cdot \left(\frac{3x + 1}{2}\right)$$ Use the distributive property and multiply each term by 10.

$$\frac{\overset{2}{\cancel{10}}}{1} \cdot \left(\frac{2x + 5}{\underset{1}{\cancel{5}}}\right) + \frac{\overset{5}{\cancel{10}}}{1} \cdot \left(\frac{x - 7}{\underset{1}{\cancel{2}}}\right) = \frac{\overset{5}{\cancel{10}}}{1} \cdot \left(\frac{3x + 1}{\underset{1}{\cancel{2}}}\right)$$ Divide out common factors in each multiplication.

$$2(2x + 5) + 5(x - 7) = 5(3x + 1)$$ The fractions are now cleared.

At this point, we have an equation similar to those we have previously solved. Use the distributive property to begin simplifying each side.

$$4x + 10 + 5x - 35 = 15x + 5$$ Use the distributive property.

$$9x - 25 = 15x + 5$$ Combine like terms on the left side: $4x + 5x = 9x$ and $10 - 35 = -25$.

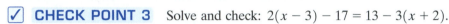

For variety, let's collect variable terms on the right and constant terms on the left.

$$9x - 9x - 25 = 15x - 9x + 5 \quad \text{Subtract } 9x \text{ from both sides.}$$

$$-25 = 6x + 5 \quad \text{Simplify.}$$

$$-25 - 5 = 6x + 5 - 5 \quad \text{Subtract 5 from both sides.}$$

$$-30 = 6x \quad \text{Simplify.}$$

Isolate x on the right side by dividing both sides by 6.

$$\frac{-30}{6} = \frac{6x}{6} \quad \text{Divide both sides by 6.}$$

$$-5 = x \quad \text{Simplify.}$$

Check the proposed solution in the original equation. Substitute -5 for x in the original equation. You should obtain $-7 = -7$. This true statement verifies that -5 is the solution, or the solution set is $\{-5\}$.

✓ CHECK POINT 4 Solve: $\frac{x+5}{7} + \frac{x-3}{4} = \frac{5}{14}$.

2 Recognize identities, conditional equations, and inconsistent equations.

Types of Equations

Equations can be placed into categories that depend on their solution sets.

An equation that is true for all real numbers for which both sides are defined is called an **identity**. An example of an identity is

$$x + 3 = x + 2 + 1.$$

Every number plus 3 is equal to that number plus 2 plus 1. Therefore, the solution set to this equation is the set of all real numbers. This set is written either as

$$\{x|x \text{ is a real number}\} \quad \text{or} \quad \mathbb{R}.$$

An equation that is not an identity, but that is true for at least one real number, is called a **conditional equation**. The equation $2x + 3 = 17$ is an example of a conditional equation. The equation is not an identity and is true only if x is 7.

An **inconsistent equation** is an equation that is not true for even one real number. An example of an inconsistent equation is

$$x = x + 7.$$

There is no number that is equal to itself plus 7. The equation $x = x + 7$ has no solution. Its solution set is written either as

$$\{\ \} \quad \text{or} \quad \varnothing.$$

These symbols stand for the empty set, a set with no elements.

Study Tip

If you are concerned by the vocabulary of equation types, keep in mind that there are three possible situations. We can state these situations informally as follows:

1. $x =$ a real number — Conditional equation
2. $x =$ all real numbers — Identity
3. $x =$ no real numbers. — Inconsistent equation

If you attempt to solve an identity or an inconsistent equation, you will eliminate the variable. A true statement such as $6 = 6$ or a false statement such as $2 = 3$ will be the result. **If a true statement results, the equation is an identity that is true for all real numbers. If a false statement results, the equation is an inconsistent equation with no solution.**

EXAMPLE 5 Categorizing an Equation

Solve and determine whether the equation

$$2(x + 1) = 2x + 3$$

is an identity, a conditional equation, or an inconsistent equation.

Solution Begin by applying the distributive property on the left side. We obtain

$$2x + 2 = 2x + 3.$$

Does something look strange about $2x + 2 = 2x + 3$? Can doubling a number and increasing the product by 2 give the same result as doubling the same number and increasing the product by 3? No. Let's continue solving the equation by subtracting $2x$ from both sides of $2x + 2 = 2x + 3$.

$$2x - 2x + 2 = 2x - 2x + 3$$

Keep reading. $2 = 3$ is not the solution.

$$2 = 3$$

The original equation is equivalent to the statement $2 = 3$, which is false for every value of x. The equation is inconsistent and has no solution. You can express this by writing "no solution" or using one of the symbols for the empty set, $\{\ \}$ or $\varnothing$.

Using Technology

Graphic Connections

How can technology visually reinforce the fact that the equation

$$2(x + 1) = 2x + 3$$

has no solution? Enter $y_1 = 2(x + 1)$ and $y_2 = 2x + 3$. The graphs of y_1 and y_2 appear to be parallel lines with no intersection point. This supports our conclusion that $2(x + 1) = 2x + 3$ is an inconsistent equation with no solution.

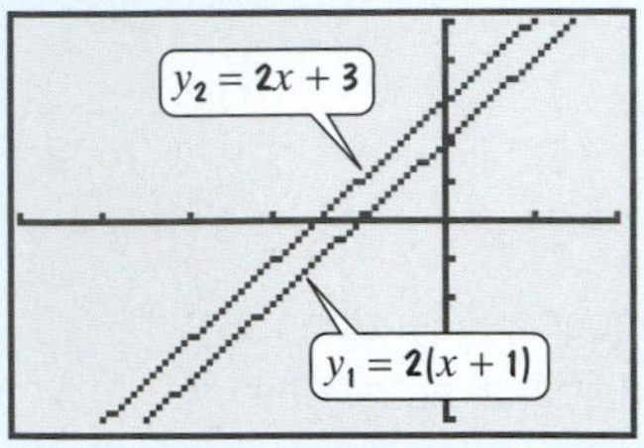

[−5, 2, 1] by [−5, 5, 1]

CHECK POINT 5 Solve and determine whether the equation

$$4x - 7 = 4(x - 1) + 3$$

is an identity, a conditional equation, or an inconsistent equation.

EXAMPLE 6 Categorizing an Equation

Solve and determine whether the equation

$$4x + 6 = 6(x + 1) - 2x$$

is an identity, a conditional equation, or an inconsistent equation.

Solution

$4x + 6 = 6(x + 1) - 2x$ — This is the given equation.

$4x + 6 = 6x + 6 - 2x$ — Apply the distributive property on the right side.

$4x + 6 = 4x + 6$ — Combine like terms on the right side: $6x - 2x = 4x$.

Can you see that the equation $4x + 6 = 4x + 6$ is true for every value of x? Let's continue solving the equation by subtracting $4x$ from both sides.

$$4x - 4x + 6 = 4x - 4x + 6$$

Keep reading. $6 = 6$ is not the solution.

$$6 = 6$$

The original equation is equivalent to the statement $6 = 6$, which is true for every value of x. The equation is an identity, and all real numbers are solutions. You can express this by writing "all real numbers" or using one of the following notations:

$$\{x \mid x \text{ is a real number}\} \quad \text{or} \quad \mathbb{R}.$$

Using Technology

Numeric Connections

A graphing utility's TABLE feature can be used to numerically verify that the solution set of

$$4x + 6 = 6(x + 1) - 2x$$

is the set of all real numbers.

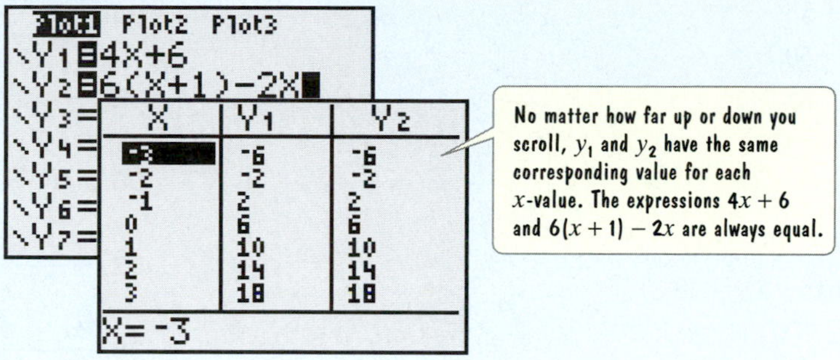

✓ **CHECK POINT 6** Solve and determine whether the equation

$$7x + 9 = 9(x + 1) - 2x$$

is an identity, a conditional equation, or an inconsistent equation.

3 Solve applied problems using mathematical models.

Applications

Our next example shows how the procedure for solving linear equations can be used to find the value of a variable in a mathematical model.

EXAMPLE 7 Motorcycle Sales in the United States

The formula

$$N = 0.12x + 0.4$$

models the number of new motorcycles sold in the United States, N, in millions, x years after 1998. When will new motorcycle sales reach 2.2 million?

Solution We are interested in when sales will reach 2.2 million, so substitute 2.2 for N in the formula and solve for x, the number of years after 1998.

$N = 0.12x + 0.4$	This is the given formula.
$2.2 = 0.12x + 0.4$	Replace N with 2.2.
$2.2 - 0.4 = 0.12x + 0.4 - 0.4$	Subtract 0.4 from both sides.
$1.8 = 0.12x$	Simplify.
$\dfrac{1.8}{0.12} = \dfrac{0.12x}{0.12}$	Divide both sides by 0.12.
$15 = x$	Simplify.

The model indicates that 15 years after 1998, or in 2013, new motorcycle sales will reach 2.2 million.

✓ **CHECK POINT 7** Use the formula in Example 7 to find when new motorcycle sales reached 1.6 million.

1.4 EXERCISE SET

MyMathLab — Math XL PRACTICE · WATCH · DOWNLOAD · READ · REVIEW

Practice Exercises

In Exercises 1–24, solve and check each linear equation.

1. $5x + 3 = 18$

2. $3x + 8 = 50$

3. $6x - 3 = 63$

4. $5x - 8 = 72$

5. $14 - 5x = -41$

6. $25 - 6x = -83$

7. $11x - (6x - 5) = 40$

8. $5x - (2x - 8) = 35$

9. $2x - 7 = 6 + x$

10. $3x + 5 = 2x + 13$

11. $7x + 4 = x + 16$

12. $8x + 1 = x + 43$

13. $8y - 3 = 11y + 9$

14. $5y - 2 = 9y + 2$

15. $3(x - 2) + 7 = 2(x + 5)$

16. $2(x - 1) + 3 = x - 3(x + 1)$

17. $3(x - 4) - 4(x - 3) = x + 3 - (x - 2)$

18. $2 - (7x + 5) = 13 - 3x$

19. $16 = 3(x - 1) - (x - 7)$

20. $5x - (2x + 2) = x + (3x - 5)$

21. $7(x + 1) = 4[x - (3 - x)]$

22. $2[3x - (4x - 6)] = 5(x - 6)$

23. $\frac{1}{2}(4z + 8) - 16 = -\frac{2}{3}(9z - 12)$

24. $\frac{3}{4}(24 - 8z) - 16 = -\frac{2}{3}(6z - 9)$

In Exercises 25–38, solve each equation.

25. $\frac{x}{3} = \frac{x}{2} - 2$

26. $\frac{x}{5} = \frac{x}{6} + 1$

27. $20 - \frac{x}{3} = \frac{x}{2}$

28. $\frac{x}{5} - \frac{1}{2} = \frac{x}{6}$

29. $\frac{3x}{5} = \frac{2x}{3} + 1$

30. $\frac{x}{2} = \frac{3x}{4} + 5$

31. $\frac{3x}{5} - x = \frac{x}{10} - \frac{5}{2}$

32. $2x - \frac{2x}{7} = \frac{x}{2} + \frac{17}{2}$

33. $\frac{x+3}{6} = \frac{2}{3} + \frac{x-5}{4}$

34. $\frac{x+1}{4} = \frac{1}{6} + \frac{2-x}{3}$

35. $\frac{x}{4} = 2 + \frac{x-3}{3}$

36. $5 + \frac{x-2}{3} = \frac{x+3}{8}$

37. $\frac{x+1}{3} = 5 - \frac{x+2}{7}$

38. $\frac{3x}{5} - \frac{x-3}{2} = \frac{x+2}{3}$

In Exercises 39–50, solve each equation. Then state whether the equation is an identity, a conditional equation, or an inconsistent equation.

39. $5x + 9 = 9(x + 1) - 4x$

40. $4x + 7 = 7(x + 1) - 3x$

41. $3(y + 2) = 7 + 3y$

42. $4(y + 5) = 21 + 4y$

43. $10x + 3 = 8x + 3$

44. $5x + 7 = 2x + 7$

45. $\frac{1}{2}(6z + 20) - 8 = 2(z - 4)$

46. $\frac{1}{3}(6z + 12) = \frac{1}{5}(20z + 30) - 8$

47. $-4x - 3(2 - 2x) = 7 + 2x$

48. $3x - 3(2 - x) = 6(x - 1)$

49. $y + 3(4y + 2) = 6(y + 1) + 5y$

50. $9y - 3(6 - 5y) = y - 2(3y + 9)$

In Exercises 51–54, use the [Y=] *screen to write the equation being solved. Then use the table to solve the equation.*

51.

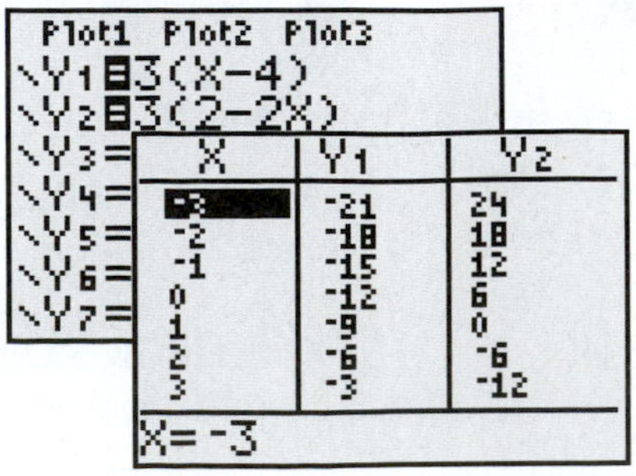

52. 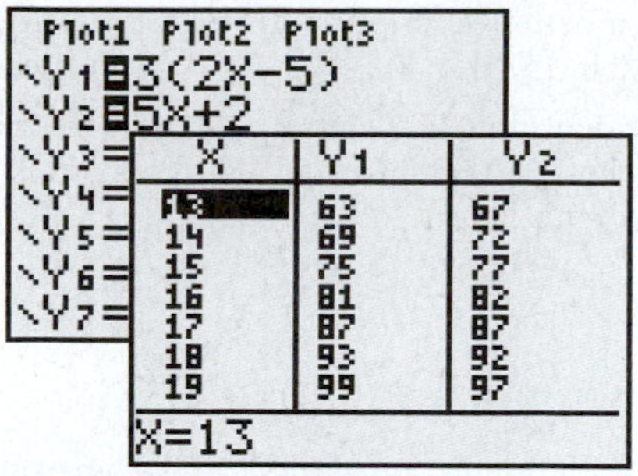

53. 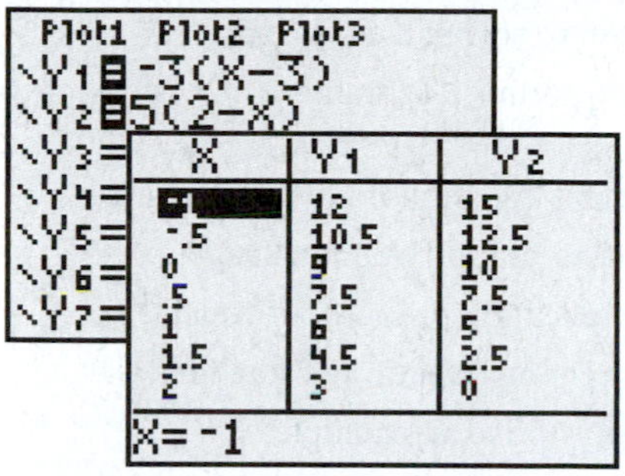

54. 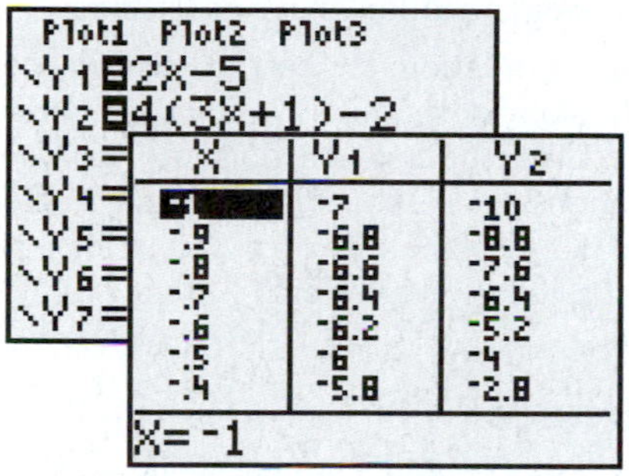

Practice PLUS

55. Evaluate $x^2 - x$ for the value of x satisfying $4(x - 2) + 2 = 4x - 2(2 - x)$.

56. Evaluate $x^2 - x$ for the value of x satisfying $2(x - 6) = 3x + 2(2x - 1)$.

57. Evaluate $x^2 - (xy - y)$ for x satisfying $\frac{3(x + 3)}{5} = 2x + 6$ and y satisfying $-2y - 10 = 5y + 18$.

58. Evaluate $x^2 - (xy - y)$ for x satisfying $\frac{13x - 6}{4} = 5x + 2$ and y satisfying $5 - y = 7(y + 4) + 1$.

In Exercises 59–66, solve each equation.

59. $[(3 + 6)^2 \div 3] \cdot 4 = -54x$

60. $2^3 - [4(5 - 3)^3] = -8x$

61. $5 - 12x = 8 - 7x - [6 \div 3(2 + 5^3) + 5x]$

62. $2(5x + 58) = 10x + 4(21 \div 3.5 - 11)$

63. $0.7x + 0.4(20) = 0.5(x + 20)$

64. $0.5(x + 2) = 0.1 + 3(0.1x + 0.3)$

65. $4x + 13 - \{2x - [4(x - 3) - 5]\} = 2(x - 6)$

66. $-2\{7 - [4 - 2(1 - x) + 3]\} = 10 - [4x - 2(x - 3)]$

Application Exercises

67. The bar graph shows the average cost of tuition and fees at private four-year colleges in the United States.

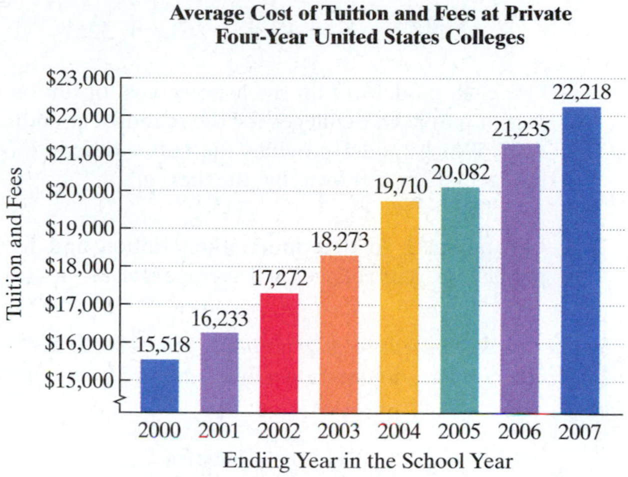

Source: The College Board

Here are two mathematical models for the data shown in the graph. In each formula, T represents the average cost of tuition and fees at private U.S. colleges for the school year ending x years after 2000.

Model 1 $T = 974x + 15{,}410$

Model 2 $T = -2.1x^2 + 988x + 15{,}395$

a. Use each model to find the average cost, to the nearest dollar, of tuition and fees at private U.S. colleges for the school year ending in 2007. By how much does each model underestimate or overestimate the actual cost shown for the school year ending in 2007?

b. Use model 1 to determine when tuition and fees at private four-year colleges will average $27,098.

68. The bar graph shows the average cost of tuition and fees at public four-year colleges in the United States.

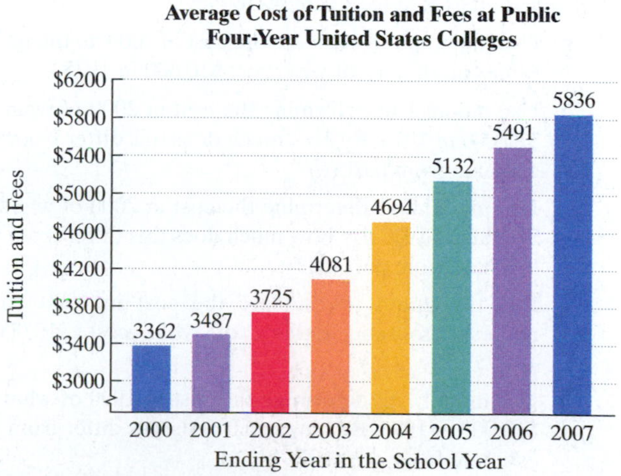

Source: The College Board

At the top of the next page are two mathematical models for the data shown in the graph. In each formula, T represents the average cost of tuition and fees at public U.S. colleges for the school year ending x years after 2000.

Model 1 $T = 383x + 3136$

Model 2 $T = 17x^2 + 261x + 3257$

a. Use each model to find the average cost of tuition and fees at public U.S. colleges for the school year ending in 2007. Which model provides the better description for the actual cost shown by the bar graph?

b. Use model 1 to determine when tuition and fees at public four-year colleges will average \$8498.

The line graph shows the cost of inflation. What cost \$10,000 in 1975 would cost the amount shown by the graph in subsequent years.

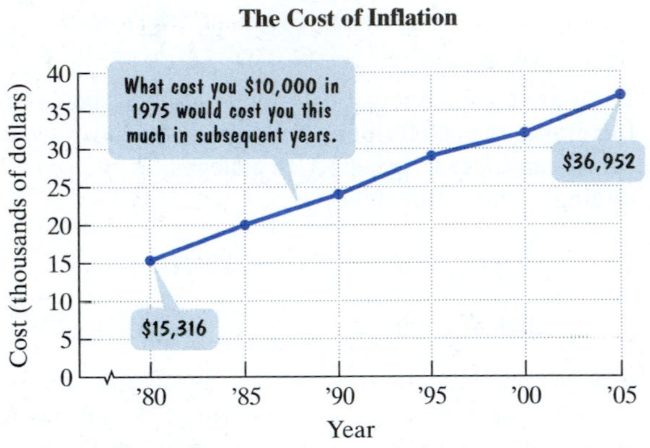

Source: Bureau of Labor Statistics

Here are two mathematical models for the data shown in the graph. In each formula, C represents the cost x years after 1980 of what cost \$10,000 in 1975.

Model 1 $C = 865x + 15{,}316$

Model 2 $C = -2x^2 + 900x + 15{,}397$

Use these models to solve Exercises 69–73.

69. a. Use the graph to estimate the cost in 2000, to the nearest thousand dollars, of what cost \$10,000 in 1975.

b. Use model 1 to determine the cost in 2000 of what cost \$10,000 in 1975. By how much does this differ from your estimate from part (a)?

c. Use model 2 to determine the cost in 2000 of what cost \$10,000 in 1975. By how much does this differ from your estimate from part (a)?

70. a. Use the graph to estimate the cost in 1990, to the nearest thousand dollars, of what cost \$10,000 in 1975.

b. Use model 1 to determine the cost in 1990 of what cost \$10,000 in 1975. By how much does this differ from your estimate from part (a)?

c. Use model 2 to determine the cost in 1990 of what cost \$10,000 in 1975. By how much does this differ from your estimate from part (a)?

71. Which model is a better description for the cost in 2005 of what cost \$10,000 in 1975?

72. Use model 1 to determine in which year the cost will be \$43,861 for what cost \$10,000 in 1975.

73. Use model 1 to determine in which year the cost will be \$54,241 for what cost \$10,000 in 1975.

Writing in Mathematics

74. What is a linear equation in one variable? Give an example of this type of equation.

75. What does it mean to solve an equation?

76. How do you determine if a number is a solution of an equation?

77. What are equivalent equations? Give an example.

78. What is the addition property of equality?

79. What is the multiplication property of equality?

80. Explain how to clear an equation of fractions.

81. What is an identity? Give an example.

82. What is a conditional equation? Give an example.

83. What is an inconsistent equation? Give an example.

84. Despite low rates of inflation, the cost of a college education continues to skyrocket. This is a departure from the trend during the 1970s: In constant dollars (which negate the effect of inflation), the cost of college actually decreased several times. What explanations can you offer for the increasing cost of a college education?

Technology Exercises

In Exercises 85–88, use your graphing utility to enter each side of the equation separately under y_1 and y_2. Then use the utility's TABLE *or* GRAPH *feature to solve the equation.*

85. $5x + 2(x - 1) = 3x + 10$

86. $2x + 3(x - 4) = 4x - 7$

87. $3(2x + 11) = 3(5 + x)$

88. $\frac{2x - 1}{3} - \frac{x - 5}{6} = \frac{x - 3}{4}$

Critical Thinking Exercises

Make Sense? *In Exercises 89–92, determine whether each statement "makes sense" or "does not make sense" and explain your reasoning*

89. Because $x = x + 5$ is an inconsistent equation, the graphs of $y = x$ and $y = x + 5$ should not intersect.

90. Because subtraction is defined in terms of addition, it's not necessary to state a separate subtraction property of equality to generate equivalent equations.

91. The number 3 satisfies the equation $7x + 9 = 9(x + 1) - 2x$, so $\{3\}$ is the equation's solution set.

92. I can solve $-2x = 10$ using the addition property of equality.

In Exercises 93–96, determine whether each statement is true or false. If the statement is false, make the necessary change(s) to produce a true statement.

93. The equation $-7x = x$ has no solution.

94. The equations $\frac{x}{x-4} = \frac{4}{x-4}$ and $x = 4$ are equivalent.

95. The equations $3y - 1 = 11$ and $3y - 7 = 5$ are equivalent.

96. If a and b are any real numbers, then $ax + b = 0$ always has only one number in its solution set.

97. Solve for x: $ax + b = c$.

98. Write three equations that are equivalent to $x = 5$.

99. If x represents a number, write an English sentence about the number that results in an inconsistent equation.

100. Find b such that $\frac{7x+4}{b} + 13 = x$ will have a solution set given by $\{-6\}$.

Review Exercises

In Exercises 101–102, perform the indicated operations.

101. $-\frac{1}{5} - \left(-\frac{1}{2}\right)$ (Section 1.2, Example 4)

102. $4(-3)(-1)(-5)$ (Section 1.2, Examples in **Table 1.2**)

103. Graph $y = x^2 - 4$. Let $x = -3, -2, -1, 0, 1, 2,$ and 3. (Section 1.3, Example 2)

Preview Exercises

Exercises 104–106 will help you prepare for the material covered in the next section.

104. Let x represent a number.

a. Write an equation in x that describes the following conditions:

Four less than three times the number is 32.

b. Solve the equation and determine the number.

105. Let x represent the number of countries in the world that are not free. The number of free countries exceeds the number of not-free countries by 44. Write an algebraic expression that represents the number of free countries.

106. You purchase a new car for \$20,000. Each year the value of the car decreases by \$2500. Write an algebraic expression that represents the car's value, in dollars, after x years.

MID-CHAPTER CHECK POINT Section 1.1–Section 1.4

What You Know: We reviewed a number of topics from introductory algebra, including the real numbers and their representations on number lines. We performed operations with real numbers and applied the order-of-operations agreement to expressions containing more than one operation. We used commutative, associative, and distributive properties to simplify algebraic expressions. We used the rectangular coordinate system to represent ordered pairs of real numbers and graph equations in two variables. Finally, we solved linear equations, including equations with fractions. We saw that some equations have no solution, whereas others have all real numbers as solutions.

In Exercises 1–14, simplify the expression or solve the equation, whichever is appropriate.

1. $-5 + 3(x + 5)$

2. $-5 + 3(x + 5) = 2(3x - 4)$

3. $3[7 - 4(5 - 2)]$

4. $\frac{x-3}{5} - 1 = \frac{x-5}{4}$

5. $\frac{-2^4 + (-2)^2}{-4 - (2 - 2)}$

6. $7x - [8 - 3(2x - 5)]$

7. $3(2x - 5) - 2(4x + 1) = -5(x + 3) - 2$

8. $3(2x - 5) - 2(4x + 1) - 5(x + 3) - 2$

9. $-4^2 \div 2 + (-3)(-5)$

10. $3x + 1 - (x - 5) = 2x - 4$

11. $\frac{3x}{4} - \frac{x}{3} + 1 = \frac{4x}{5} - \frac{3}{20}$

12. $(6 - 9)(8 - 12) \div \frac{5^2 + 4 \div 2}{8^2 - 9^2 + 8}$

13. $4x - 2(1 - x) = 3(2x + 1) - 5$

14. $\frac{3[4 - 3(-2)^2]}{2^2 - 2^4}$

In Exercises 15–17, graph each equation in a rectangular coordinate system.

15. $y = 2x - 1$ **16.** $y = 1 - |x|$ **17.** $y = x^2 + 2$

In Exercises 18–21, determine whether each statement is true or false.

18. $-\left|-\frac{\sqrt{3}}{5}\right| = -\frac{\sqrt{3}}{5}$

19. $\{x \mid x \text{ is a negative integer greater than } -4\} = \{-4, -3, -2, -1\}$

20. $-17 \notin \{x \mid x \text{ is a rational number}\}$

21. $-128 \div (2 \cdot 4) > (-128 \div 2) \cdot 4$

SECTION 1.5

Problem Solving and Using Formulas

Objectives

1. Solve algebraic word problems using linear equations.
2. Solve a formula for a variable.

The human race is undeniably becoming a faster race. Since the beginning of the past century, track-and-field records have fallen in everything from sprints to miles to marathons. The performance arc is clearly rising, but no one knows how much higher it can climb. At some point, even the best-trained body simply has to up and quit. The question is, just where is that point, and is it possible for athletes, trainers, and genetic engineers to push it higher? In this section, you will learn a problem-solving strategy that uses linear equations to determine if anyone will ever run a 3-minute mile.

1 Solve algebraic word problems using linear equations.

Problem Solving with Linear Equations

We have seen that a model is a mathematical representation of a real-world situation. In this section, we will be solving problems that are presented in English. This means that we must obtain models by translating from the ordinary language of English into the language of algebraic equations. To translate, however, we must understand the English prose and be familiar with the forms of algebraic language. Here are some general steps we will follow in solving word problems:

Study Tip

When solving word problems, particularly problems involving geometric figures, drawing a picture of the situation is often helpful. Label x on your drawing and, where appropriate, label other parts of the drawing in terms of x.

Strategy for Solving Word Problems

Step 1. Read the problem carefully. Attempt to state the problem in your own words and state what the problem is looking for. Let x (or any variable) represent one of the unknown quantities in the problem.

Step 2. If necessary, write expressions for any other unknown quantities in the problem in terms of x.

Step 3. Write an equation in x that models the verbal conditions of the problem.

Step 4. Solve the equation and answer the problem's question.

Step 5. Check the solution *in the original wording* of the problem, not in the equation obtained from the words.

World's Countries by Status of Freedom

FIGURE 1.19

Source: Larry Berman and Bruce Murphy, *Approaching Democracy*, 5th Edition, Prentice Hall, 2007

EXAMPLE 1 Status of Freedom in the World

The circle graph in **Figure 1.19** represents the 2006 breakdown of the world's 192 countries by free, partly free, or not free. The number of free countries exceeds the number of not-free countries by 44. The number of partly free countries exceeds the number of not-free countries by 13. Determine the number of countries in the world that fall into each of the categories in the circle graph.

Solution We must determine the numbers of countries that are free, partly free, and not free.

Step 1. Let x represent one of the quantities. We know something about free countries and partly free countries: The numbers exceed the number of not-free countries by 44 and 13, respectively. We will let

$$x = \text{the number of not-free countries.}$$

Step 2. Represent other unknown quantities in terms of x. Because the number of free countries exceeds the number of not-free countries by 44, let

$$x + 44 = \text{the number of free countries.}$$

Because the number of partly free countries exceeds the number of not-free countries by 13, let

$$x + 13 = \text{the number of partly free countries.}$$

Step 3. Write an equation in x that models the conditions. **Figure 1.19** categorizes the world's 192 countries by free, partly free, or not free.

The number of not-free countries	plus	the number of free countries	plus	the number of partly free countries	equals	192 countries.
x	$+$	$(x + 44)$	$+$	$(x + 13)$	$=$	192

Step 4. Solve the equation and answer the question.

$$x + (x + 44) + (x + 13) = 192$$ This is the equation that models the problem's conditions.

$$3x + 57 = 192$$ Remove parentheses, regroup, and combine like terms.

$$3x = 135$$ Subtract 57 from both sides.

$$x = 45$$ Divide both sides by 3.

Thus,

the number of not-free countries $= x = 45$,

the number of free countries $= x + 44 = 45 + 44 = 89$,

and the number of partly free countries $= x + 13 = 45 + 13 = 58$.

There are 45 countries that are not free, 89 countries that are free, and 58 countries that are partly free.

Step 5. Check the proposed solution in the original wording of the problem. The problem states that we are categorizing the world's 192 countries by status of freedom. By adding 45, 89, and 58, the numbers that we found in each category, we obtain

$$45 + 89 + 58 = 192,$$

as specified by the problem's conditions.

Study Tip

Modeling with the word *exceeds* can be a bit tricky. It's helpful to identify the smaller quantity. Then add to this quantity to represent the larger quantity. For example, suppose that Tim's height exceeds Tom's height by a inches. Tom is the shorter person. If Tom's height is represented by x, then Tim's height is represented by $x + a$.

✓ CHECK POINT 1 There are 46 countries in the world with Muslim majorities. Of these 46 countries, the number that are not free exceeds the number that are free by 20. The number that are partly free exceeds the number that are free by 17. Determine the numbers of countries with Muslim majorities that are free, not free, and partly free. (*Source*: 2006 data from Larry Berman and Bruce Murphy, *Ibid.*)

Mile Records

1886	4:12.3	1958	3:54.5
1923	4:10.4	1966	3:51.3
1933	4:07.6	1979	3:48.9
1945	4:01.3	1985	3:46.3
1954	3:59.4	1999	3:43.1

Source: U.S.A. Track and Field

EXAMPLE 2 Will Anyone Ever Run a Three-Minute Mile?

One yardstick for measuring how steadily—if slowly—athletic performance has improved is the mile run. In 1923, the record for the mile was a comparatively sleepy 4 minutes, 10.4 seconds. In 1954, Roger Bannister of Britain cracked the 4-minute mark, coming in at 3 minutes, 59.4 seconds. In the half-century since, about 0.3 second per year has been shaved off Bannister's record. If this trend continues, by which year will someone run a 3-minute mile?

Solution In solving this problem, we will express time for the mile run in seconds. Our interest is in a time of 3 minutes, or 180 seconds.

Step 1. Let x represent one of the quantities. Here is the critical information in the problem:

- In 1954, the record was 3 minutes, 59.4 seconds, or 239.4 seconds.
- The record has decreased by 0.3 second per year since then.

We are interested in when the record will be 180 seconds. Let

x = the number of years after 1954 when someone will run a 3-minute mile.

Step 2. Represent other unknown quantities in terms of x. There are no other unknown quantities to find, so we can skip this step.

Step 3. Write an equation in x that models the conditions.

The 1954 record time	decreased by	0.3 second per year for x years	equals	the 3-minute, or 180-second, mile.
239.4	$-$	$0.3x$	$=$	180

Step 4. Solve the equation and answer the question.

$$239.4 - 0.3x = 180$$ This is the equation that models the problem's conditions.

$$239.4 - 239.4 - 0.3x = 180 - 239.4$$ Subtract 239.4 from both sides.

$$-0.3x = -59.4$$ Simplify.

$$\frac{-0.3x}{-0.3} = \frac{-59.4}{-0.3}$$ Divide both sides by -0.3.

$$x = 198$$ Simplify.

Using current trends, by 198 years (gasp!) after 1954, or in 2152, someone will run a 3-minute mile.

Step 5. Check the proposed solution in the original wording of the problem. The problem states that the record time should be 180 seconds. Do we obtain 180 seconds if we decrease the 1954 record time, 239.4 seconds, by 0.3 second per year for 198 years, our proposed solution?

$$239.4 - 0.3(198) = 239.4 - 59.4 = 180$$

This verifies that, using current trends, the 3-minute mile will be run 198 years after 1954.

✓ **CHECK POINT 2** Cars in the United States are being driven longer. In 2005, the average age of a U.S. automobile was 9 years. Between 1990 and 2005, this average age in operation had increased by 0.17 year per year. If this trend continues, by which year will automobiles in the United States have an average age of 10.7 years? (*Source*: Bureau of Transportation Statistics)

EXAMPLE 3 Selecting a Long-Distance Carrier

You are choosing between two long-distance telephone plans. Plan A has a monthly fee of \$20 with a charge of \$0.05 per minute for all long-distance calls. Plan B has a monthly fee of \$5 with a charge of \$0.10 per minute for all long-distance calls. For how many minutes of long-distance calls will the costs for the two plans be the same?

Solution

Step 1. Let x represent one of the quantities. Let

x = the number of minutes of long-distance calls for which the two plans cost the same.

Step 2. Represent other unknown quantities in terms of x. There are no other unknown quantities, so we can skip this step.

Step 3. Write an equation in x that models the conditions. The monthly cost for plan A is the monthly fee, \$20, plus the per minute charge, \$0.05, times the number of minutes of long-distance calls, x. The monthly cost for plan B is the monthly fee, \$5, plus the per-minute charge, \$0.10, times the number of minutes of long-distance calls, x.

The monthly cost for plan A | must equal | the monthly cost for plan B.

$$20 + 0.05x = 5 + 0.10x$$

Step 4. Solve the equation and answer the question.

$$20 + 0.05x = 5 + 0.10x$$ This is the equation that models the problem's conditions.

$$20 = 5 + 0.05x$$ Subtract 0.05x from both sides.

$$15 = 0.05x$$ Subtract 5 from both sides.

$$\frac{15}{0.05} = \frac{0.05x}{0.05}$$ Divide both sides by 0.05.

$$300 = x$$ Simplify.

Because x represents the number of minutes of long-distance calls for which the two plans cost the same, the costs will be the same for 300 minutes of long-distance calls.

Step 5. Check the proposed solution in the original wording of the problem. The problem states that the costs for the two plans should be the same. Let's see if they are with 300 minutes of long-distance calls:

$$\text{Cost for plan A} = \$20 + \$0.05(300) = \$20 + \$15 = \$35$$

Monthly fee | Per-minute charge

$$\text{Cost for plan B} = \$5 + \$0.10(300) = \$5 + \$30 = \$35.$$

With 300 minutes, or 5 hours, of long-distance chatting, both plans cost \$35 for the month. Thus, the proposed solution, 300 minutes, satisfies the problem's conditions.

Using Technology

Numeric and Graphic Connections

We can use a graphing utility to numerically or graphically verify our work in Example 3.

The monthly cost for plan A | must equal | the monthly cost for plan B.

$$20 + 0.05x = 5 + 0.10x$$

Enter $y_1 = 20 + .05x$. Enter $y_2 = 5 + .10x$.

Numeric Check

Display a table for y_1 and y_2.

X	Y1	Y2
100	25	15
150	27.5	20
200	30	25
250	32.5	30
300	35	35
350	37.5	40
400	40	45

X=300

When $x = 300$, y_1 and y_2 have the same value, 35. With 300 minutes of calls, costs are the same, \$35, for both plans.

Graphic Check

Display graphs for y_1 and y_2. Use the intersection feature.

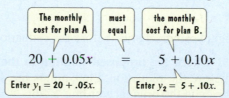

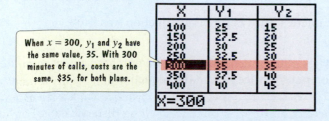

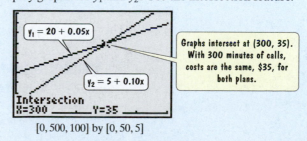

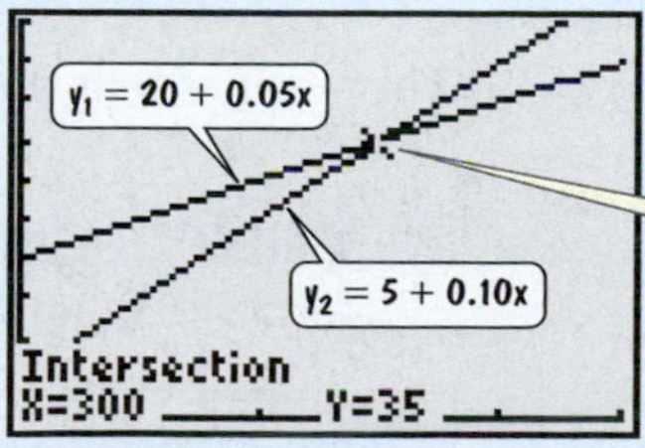

Graphs intersect at (300, 35). With 300 minutes of calls, costs are the same, \$35, for both plans.

[0, 500, 100] by [0, 50, 5]

✓ **CHECK POINT 3** You are choosing between two long-distance telephone plans. Plan A has a monthly fee of \$15 with a charge of \$0.08 per minute for all long-distance calls. Plan B has a monthly fee of \$3 with a charge of \$0.12 per minute for all long-distance calls. For how many minutes of long-distance calls will the costs for the two plans be the same?

EXAMPLE 4 A Price Reduction on a Digital Camera

Your local computer store is having a terrific sale on digital cameras. After a 40% price reduction, you purchase a digital camera for \$276. What was the camera's price before the reduction?

Solution

Step 1. Let x represent one of the quantities. We will let

$$x = \text{the original price of the digital camera prior to the reduction.}$$

Step 2. Represent other unknown quantities in terms of x. There are no other unknown quantities to find, so we can skip this step.

Step 3. Write an equation in x that models the conditions. The camera's original price minus the 40% reduction is the reduced price, \$276.

Study Tip

Observe that the original price, x, reduced by 40% is $x - 0.4x$ and *not* $x - 0.4$.

Original price	minus	the reduction (40% of the original price)	is	the reduced price, \$276.
x	$-$	$0.4x$	$=$	276

Step 4. Solve the equation and answer the question.

$$x - 0.4x = 276$$ This is the equation that models the problem's conditions.

$$0.6x = 276$$ Combine like terms: $x - 0.4x = 1x - 0.4x = 0.6x$.

$$\frac{0.6x}{0.6} = \frac{276}{0.6}$$ Divide both sides by 0.6.

$$x = 460$$ Simplify: $\begin{array}{r} 460. \\ 0.6\overline{)276.0} \end{array}$

The digital camera's price before the reduction was \$460.

Step 5. Check the proposed solution in the original wording of the problem. The price before the reduction, \$460, minus the 40% reduction should equal the reduced price given in the original wording, \$276:

$$460 - 40\% \text{ of } 460 = 460 - 0.4(460) = 460 - 184 = 276.$$

This verifies that the digital camera's price before the reduction was \$460.

✓ **CHECK POINT 4** After a 30% price reduction, you purchase a new computer for \$840. What was the computer's price before the reduction?

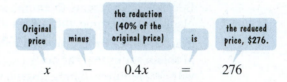

Solving geometry problems usually requires a knowledge of basic geometric ideas and formulas. Formulas for area, perimeter, and volume are given in **Table 1.5**.

Table 1.5 Common Formulas for Area, Perimeter, and Volume

Square
$A = s^2$
$P = 4s$

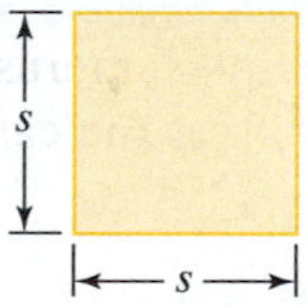

Rectangle
$A = lw$
$P = 2l + 2w$

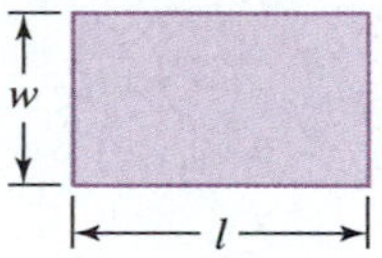

Circle
$A = \pi r^2$
$C = 2\pi r$

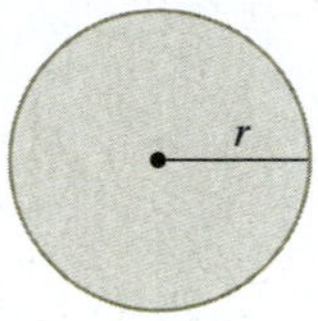

Triangle
$A = \frac{1}{2}bh$

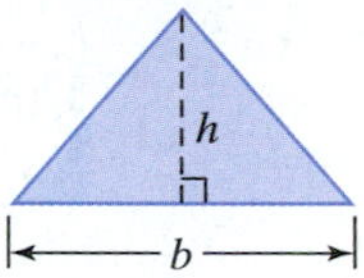

Trapezoid
$A = \frac{1}{2}h(a + b)$

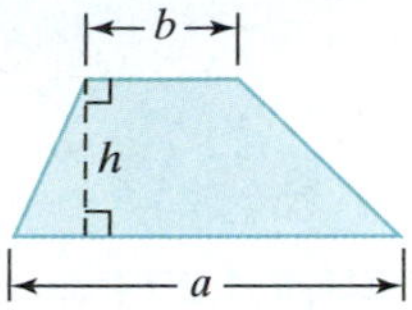

Cube
$V = s^3$

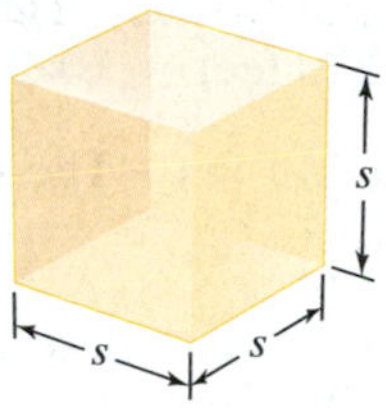

Rectangular Solid
$V = lwh$

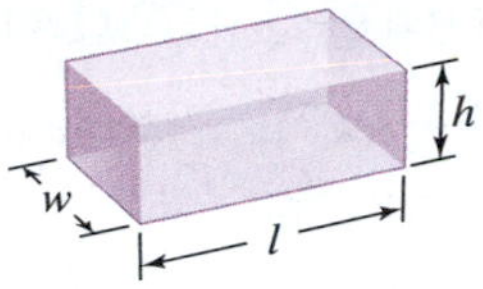

Circular Cylinder
$V = \pi r^2 h$

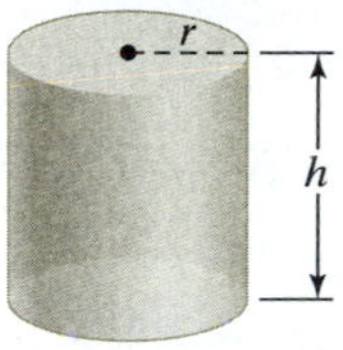

Sphere
$V = \frac{4}{3}\pi r^3$

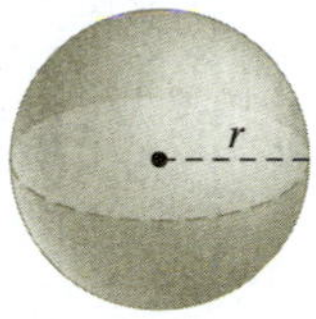

Cone
$V = \frac{1}{3}\pi r^2 h$

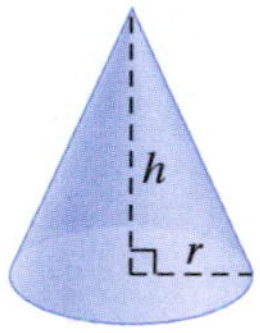

We will be using the formula for the perimeter of a rectangle, $P = 2l + 2w$, in our next example. The formula states that a rectangle's perimeter is the sum of twice its length and twice its width.

EXAMPLE 5 Finding the Dimensions of an American Football Field

The length of an American football field is 200 feet more than the width. If the perimeter of the field is 1040 feet, what are its dimensions?

Solution

Step 1. Let x represent one of the quantities. We know something about the length; the length is 200 feet more than the width. We will let

$$x = \text{the width.}$$

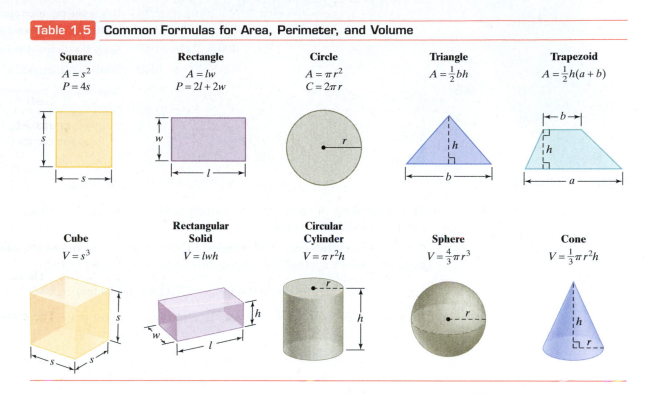

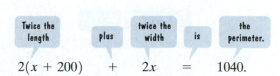

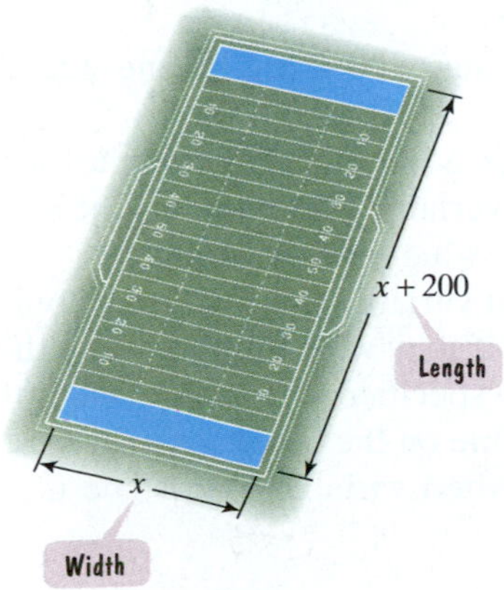

FIGURE 1.20 An American football field

Step 2. Represent other unknown quantities in terms of x. Because the length is 200 feet more than the width, let

$$x + 200 = \text{the length.}$$

Figure 1.20 illustrates an American football field and its dimensions.

Step 3. Write an equation in x that models the conditions. Because the perimeter of the field is 1040 feet,

Twice the length | plus | twice the width | is | the perimeter.

$$2(x + 200) \quad + \quad 2x \quad = \quad 1040.$$

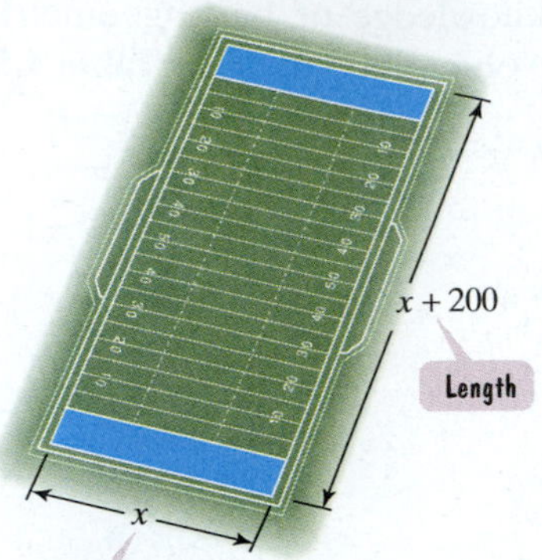

FIGURE 1.20 (repeated)

Step 4. Solve the equation and answer the question.

$2(x + 200) + 2x = 1040$	This is the equation that models the problem's conditions.
$2x + 400 + 2x = 1040$	Apply the distributive property.
$4x + 400 = 1040$	Combine like terms: $2x + 2x = 4x$.
$4x = 640$	Subtract 400 from both sides.
$x = 160$	Divide both sides by 4.

Thus,

$$\text{width} = x = 160.$$

$$\text{length} = x + 200 = 160 + 200 = 360.$$

The dimensions of an American football field are 160 feet by 360 feet. (The 360-foot length is usually described as 120 yards.)

Step 5. Check the proposed solution in the original wording of the problem. The perimeter of the football field using the dimensions that we found is

$$2(360 \text{ feet}) + 2(160 \text{ feet}) = 720 \text{ feet} + 320 \text{ feet} = 1040 \text{ feet}.$$

Because the problem's wording tells us that the perimeter is 1040 feet, our dimensions are correct.

CHECK POINT 5 The length of a rectangular basketball court is 44 feet more than the width. If the perimeter of the basketball court is 288 feet, what are its dimensions?

2 Solve a formula for a variable.

Solving a Formula for One of its Variables

We know that solving an equation is the process of finding the number (or numbers) that make the equation a true statement. All of the equations we have solved contained only one letter, x.

By contrast, formulas contain two or more letters, representing two or more variables. An example is the formula for the perimeter of a rectangle:

$$2l + 2w = P.$$

We say that this formula is solved for the variable P because P is alone on one side of the equation and the other side does not contain a P.

Solving a formula for a variable means using the addition and multiplication properties of equality to rewrite the formula so that the variable is isolated on one side of the equation. It does not mean obtaining a numerical value for that variable.

To solve a formula for one of its variables, treat that variable as if it were the only variable in the equation. Think of the other variables as if they were numbers. Use the addition property of equality to isolate all terms with the specified variable on one side of the equation and all terms without the specified variable on the other side. Then use the multiplication property of equality to get the specified variable alone. The next example shows how to do this.

EXAMPLE 6 Solving a Formula for a Variable

Solve the formula $2l + 2w = P$ for l.

Solution First, isolate $2l$ on the left by subtracting $2w$ from both sides. Then solve for l by dividing both sides by 2.

We need to isolate l.

$$2l + 2w = P$$ This is the given formula.

$$2l + 2w - 2w = P - 2w$$ Isolate $2l$ by subtracting $2w$ from both sides.

$$2l = P - 2w$$ Simplify.

$$\frac{2l}{2} = \frac{P - 2w}{2}$$ Solve for l by dividing both sides by 2.

$$l = \frac{P - 2w}{2}$$ Simplify.

✓ CHECK POINT 6 Solve the formula $2l + 2w = P$ for w.

EXAMPLE 7 Solving a Formula for a Variable

Circular Cylinder
$V = \pi r^2 h$

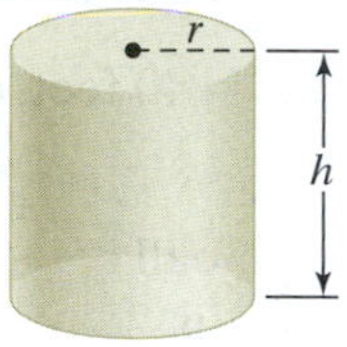

Table 1.5 on page 55 shows that the volume of a circular cylinder is given by the formula

$$V = \pi r^2 h,$$

where r is the radius of the circle at either end and h is the height. Solve this formula for h.

Solution Our goal is to get h by itself on one side of the formula. There is only one term with h, $\pi r^2 h$, and it is already isolated on the right side. We isolate h on the right by dividing both sides by πr^2.

We need to isolate h.

$$V = \pi r^2 h$$ This is the given formula.

$$\frac{V}{\pi r^2} = \frac{\pi r^2 h}{\pi r^2}$$ Isolate h by dividing both sides by πr^2.

$$\frac{V}{\pi r^2} = h$$ Simplify: $\frac{\pi r^2 h}{\pi r^2} = \frac{\pi r^2}{\pi r^2} \cdot h = 1h = h$.

Equivalently,

$$h = \frac{V}{\pi r^2}.$$

✓ CHECK POINT 7 The volume of a rectangular solid is the product of its length, width, and height:

$$V = lwh.$$

Solve this formula for h.

The Celsius scale is on the left and the Fahrenheit scale is on the right.

You'll be leaving the cold of winter for a vacation to Hawaii. CNN International reports a temperature in Hawaii of 30°C. Should you pack a winter coat? You can convert from Celsius temperature, C, to Fahrenheit temperature, F, using the formula

$$F = \frac{9}{5}C + 32.$$

A temperature of 30°C corresponds to a Fahrenheit temperature of

$$F = \frac{9}{5} \cdot 30 + 32 = \frac{9}{\cancel{5}_1} \cdot \frac{\cancel{30}^6}{1} + 32 = 54 + 32 = 86,$$

or a balmy 86°F. (Don't pack the coat.)

Visitors to the United States are more likely to be familiar with the Celsius temperature scale. For them, a useful formula is one that can be used to convert from Fahrenheit to Celsius. In Example 8, you will see how to obtain such a formula.

EXAMPLE 8 Solving a Formula for a Variable

Solve the formula

$$F = \frac{9}{5}C + 32$$

for C.

Solution We begin by multiplying both sides of the formula by 5 to clear the fraction. Then we isolate the variable C.

$$F = \frac{9}{5}C + 32 \quad \text{This is the given formula.}$$

$$5F = 5\left(\frac{9}{5}C + 32\right) \quad \text{Multiply both sides by 5.}$$

$$5F = 5\cdot\frac{9}{5}C + 5\cdot 32 \quad \text{Apply the distributive property.}$$

We need to isolate C.

$$5F = 9C + 160 \quad \text{Simplify.}$$

$$5F - 160 = 9C + 160 - 160 \quad \text{Subtract 160 from both sides.}$$

$$5F - 160 = 9C \quad \text{Simplify.}$$

$$\frac{5F - 160}{9} = \frac{9C}{9} \quad \text{Divide both sides by 9.}$$

$$\frac{5F - 160}{9} = C \quad \text{Simplify.}$$

Using the distributive property, we can express $5F - 160$ as $5(F - 32)$. Thus,

$$C = \frac{5F - 160}{9} = \frac{5(F - 32)}{9}.$$

This formula, used to convert from Fahrenheit to Celsius, is usually given as

$$C = \frac{5}{9}(F - 32).$$

✓ CHECK POINT 8 The formula

$$\frac{W}{2} - 3H = 53$$

models the recommended weight, W, in pounds, for a male, where H represents his height, in inches, over 5 feet. Solve this formula for W.

EXAMPLE 9 Solving a Formula for a Variable That Occurs Twice

The formula

$$A = P + Prt$$

describes the amount, A, that a principal of P dollars is worth after t years when invested at a simple annual interest rate, r. Solve this formula for P.

Solution Notice that all the terms with P already occur on the right side of the formula.

We need to isolate P.

$$A = P + Prt$$

We can use the distributive property in the form $ab + ac = a(b + c)$ to convert the two occurrences of P into one.

$$A = P + Prt \quad \text{This is the given formula.}$$

$$A = P(1 + rt) \quad \text{Use the distributive property to obtain a single occurrence of } P.$$

$$\frac{A}{1 + rt} = \frac{P(1 + rt)}{1 + rt} \quad \text{Divide both sides by } 1 + rt.$$

$$\frac{A}{1 + rt} = P \quad \text{Simplify: } \frac{P(1 + rt)}{1(1 + rt)} = \frac{P}{1} = P.$$

Equivalently,

$$P = \frac{A}{1 + rt}.$$

Study Tip

You cannot solve $A = P + Prt$ for P by subtracting Prt from both sides and writing

$$A - Prt = P.$$

When a formula is solved for a specified variable, that variable must be isolated on one side. The variable P occurs on both sides of

$$A - Prt = P.$$

✓ CHECK POINT 9 Solve the formula $P = C + MC$ for C.

Blitzer Bonus

Einstein's Famous Formula: $E = mc^2$

One of the most famous formulas in the world is $E = mc^2$, formulated by Albert Einstein. Einstein showed that any form of energy has mass and that mass itself is a form of energy. In this formula, E represents energy, in ergs, m represents mass, in grams, and c represents the speed of light. Because light travels at 30 billion centimeters per second, the formula indicates that 1 gram of mass will produce 900 billion billion ergs of energy.

Einstein's formula implies that the mass of a golf ball could provide the daily energy needs of the metropolitan Boston area. Mass and energy are equivalent, and the transformation of even a tiny amount of mass releases an enormous amount of energy. If this energy is released suddenly, a destructive force is unleashed, as in an atom bomb. When the release is gradual and controlled, the energy can be used to generate power.

The theoretical results implied by Einstein's formula $E = mc^2$ have not been realized because scientists have not yet developed a way of converting a mass completely to energy.

1.5 EXERCISE SET

MyMathLab MathXL PRACTICE WATCH DOWNLOAD READ REVIEW

Practice Exercises

Use the five-step strategy for solving word problems to find the number or numbers described in Exercises 1–10.

1. When five times a number is decreased by 4, the result is 26. What is the number?
2. When two times a number is decreased by 3, the result is 11. What is the number?
3. When a number is decreased by 20% of itself, the result is 20. What is the number?
4. When a number is decreased by 30% of itself, the result is 28. What is the number?
5. When 60% of a number is added to the number, the result is 192. What is the number?
6. When 80% of a number is added to the number, the result is 252. What is the number?
7. 70% of what number is 224?
8. 70% of what number is 252?
9. One number exceeds another by 26. The sum of the numbers is 64. What are the numbers?
10. One number exceeds another by 24. The sum of the numbers is 58. What are the numbers?

Practice PLUS

In Exercises 11–16, find all values of x satisfying the given conditions.

11. $y_1 = 13x - 4$, $y_2 = 5x + 10$, and y_1 exceeds y_2 by 2.

12. $y_1 = 10x + 6$, $y_2 = 12x - 7$, and y_1 exceeds y_2 by 3.

13. $y_1 = 10(2x - 1)$, $y_2 = 2x + 1$, and y_1 is 14 more than 8 times y_2.

14. $y_1 = 9(3x - 5)$, $y_2 = 3x - 1$, and y_1 is 51 less than 12 times y_2.

15. $y_1 = 2x + 6$, $y_2 = x + 8$, $y_3 = x$, and the difference between 3 times y_1 and 5 times y_2 is 22 less than y_3.

16. $y_1 = 2.5$, $y_2 = 2x + 1$, $y_3 = x$, and the difference between 2 times y_1 and 3 times y_2 is 8 less than 4 times y_3.

Application Exercises

In Exercises 17–48, use the five-step strategy for solving word problems.

17. The bar graph represents the average credit card debt for U.S. college students. The average credit card debt for juniors exceeds the debt for sophomores by \$421, and the average credit card debt for seniors exceeds the debt for sophomores by \$1265. The combined credit card debt for a sophomore, a junior, and a senior is \$6429. Determine the average credit card debt for a sophomore, a junior, and a senior.

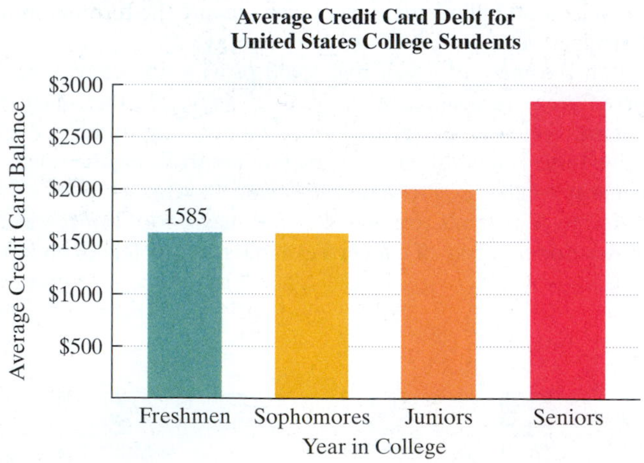

Source: Nellie Mae

18. The bar graph at the top of the next column represents the millions of barrels of oil consumed each day by the countries with the greatest oil consumption. Oil consumption in China exceeds Japan's by 0.8 million barrels per day, and oil consumption in the United States exceeds Japan's by 15 million barrels per day. Of the 82 million barrels of oil used by the world every day, the combined consumption for the United States, China, and Japan is 32.3 million barrels. Determine the daily oil consumption, in millions of barrels, for the United States, China, and Japan.

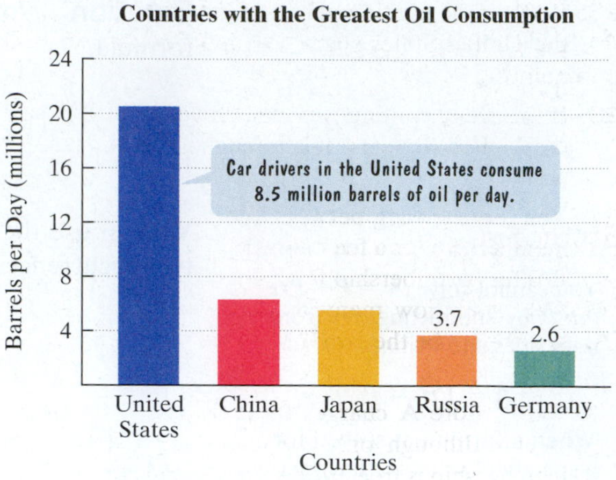

Source: U.S. Energy Information Administration

Solve Exercises 19–22 using the fact that the sum of the measures of the three angles of a triangle is 180°.

19. In a triangle, the measure of the first angle is twice the measure of the second angle. The measure of the third angle is 8° less than the measure of the second angle. What is the measure of each angle?

20. In a triangle, the measure of the first angle is three times the measure of the second angle. The measure of the third angle is 35° less than the measure of the second angle. What is the measure of each angle?

21. In a triangle, the measures of the three angles are consecutive integers. What is the measure of each angle?

22. In a triangle, the measures of the three angles are consecutive even integers. What is the measure of each angle?

According to one mathematical model, the average life expectancy for American men born in 1900 was 55 years. Life expectancy has increased by about 0.2 year for each birth year after 1900. Use this information to solve Exercises 23–24.

23. If this trend continues, for which birth year will the average life expectancy be 85 years?

24. If this trend continues, for which birth year will the average life expectancy be 91 years?

The line graph indicates that in 2005, 19.4% of people in the United States spoke a language other than English at home. For the period from 2000 through 2005, this had been increasing by approximately 0.4% per year. Use this information to solve Exercises 25–26.

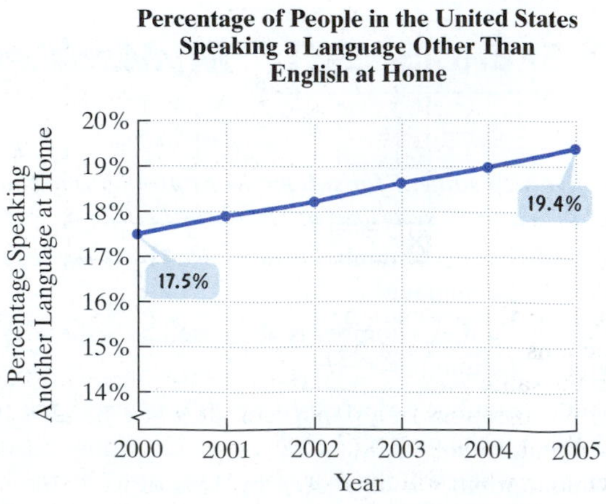

Source: U.S. Census Bureau

25. If this trend continues, by which year will 23% of people in the United States speak a language other than English at home?

26. If this trend continues, by which year will 25% of people in the United States speak a language other than English at home?

27. You are choosing between two health clubs. Club A offers membership for a fee of \$40 plus a monthly fee of \$25. Club B offers membership for a fee of \$15 plus a monthly fee of \$30. After how many months will the total cost at each health club be the same? What will be the total cost for each club?

28. Video Store A charges \$9 to rent a video game for one week. Although only members can rent from the store, membership is free. Video Store B charges only \$4 to rent a video game for one week. Only members can rent from the store and membership is \$50 per year. After how many video-game rentals will the total amount spent at each store be the same? What will be the total amount spent at each store?

29. The bus fare in a city is \$1.25. People who use the bus have the option of purchasing a monthly coupon book for \$15.00. With the coupon book, the fare is reduced to \$0.75. Determine the number of times in a month the bus must be used so that the total monthly cost without the coupon book is the same as the total monthly cost with the coupon book.

30. A coupon book for a bridge costs \$30 per month. The toll for the bridge is normally \$5.00, but it is reduced to \$3.50 for people who have purchased the coupon book. Determine the number of times in a month the bridge must be crossed so that the total monthly cost without the coupon book is the same as the total monthly cost with the coupon book.

31. In 2005, there were 13,300 students at college A, with a projected enrollment increase of 1000 students per year. In the same year, there were 26,800 students at college B, with a projected enrollment decline of 500 students per year.

a. According to these projections, when will the colleges have the same enrollment? What will be the enrollment in each college at that time?

b. Use the following table to numerically check your work in part (a). What equations were entered for y_1 and y_2 to obtain this table?

X	Y1	Y2
7	20300	23300
8	21300	22800
9	22300	22300
10	23300	21800
11	24300	21300
12	25300	20800
13	26300	20300

X=7

32. In 2000, the population of Greece was 10,600,000, with projections of a population decrease of 28,000 people per year. In the same year, the population of Belgium was 10,200,000, with projections of a population decrease of 12,000 people per year. (*Source:* United Nations) According to these projections, when will the two countries have the same population? What will be the population at that time?

33. After a 20% reduction, you purchase a television for \$336. What was the television's price before the reduction?

34. After a 30% reduction, you purchase a dictionary for \$30.80. What was the dictionary's price before the reduction?

35. Including 8% sales tax, an inn charges \$162 per night. Find the inn's nightly cost before the tax is added.

36. Including 5% sales tax, an inn charges \$252 per night. Find the inn's nightly cost before the tax is added.

The graph shows average yearly earnings in the United States by highest educational attainment. Use the relevant information shown in the graph to solve Exercises 37–38.

Average Earnings of Full-Time Workers in the United States, by Highest Educational Attainment

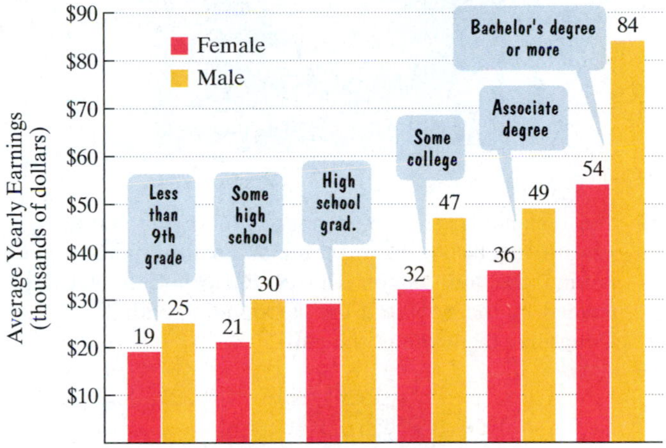

Source: U.S. Census Bureau

37. The annual salary for men with some college is an increase of 20.5% over the annual salary for men whose highest educational attainment is a high school degree. What is the annual salary, to the nearest thousand dollars, for men whose highest educational attainment is a high school degree?

38. The annual salary for women with an associate degree is an increase of 24.1% over the annual salary for women whose highest educational attainment is a high school degree. What is the annual salary, to the nearest thousand dollars, for women whose highest educational attainment is a high school degree?

Exercises 39–40 involve markup, the amount added to the dealer's cost of an item to arrive at the selling price of that item.

39. The selling price of a refrigerator is \$584. If the markup is 25% of the dealer's cost, what is the dealer's cost of the refrigerator?

40. The selling price of a scientific calculator is \$15. If the markup is 25% of the dealer's cost, what is the dealer's cost of the calculator?

41. A rectangular soccer field is twice as long as it is wide. If the perimeter of the soccer field is 300 yards, what are its dimensions?

42. A rectangular swimming pool is three times as long as it is wide. If the perimeter of the pool is 320 feet, what are its dimensions?

43. The length of the rectangular tennis court at Wimbledon is 6 feet longer than twice the width. If the court's perimeter is 228 feet, what are the court's dimensions?

44. The length of a rectangular pool is 6 meters less than twice the width. If the pool's perimeter is 126 meters, what are its dimensions?

45. The rectangular painting in the figure shown measures 12 inches by 16 inches and contains a frame of uniform width around the four edges. The perimeter of the rectangle formed by the painting and its frame is 72 inches. Determine the width of the frame.

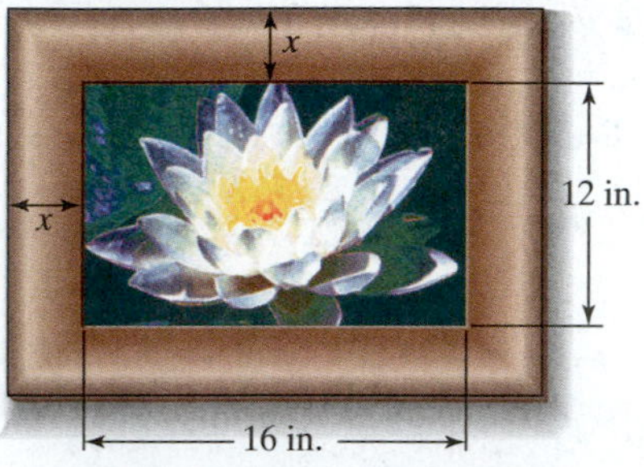

46. The rectangular swimming pool in the figure shown measures 40 feet by 60 feet and contains a path of uniform width around the four edges. The perimeter of the rectangle formed by the pool and the surrounding path is 248 feet. Determine the width of the path.

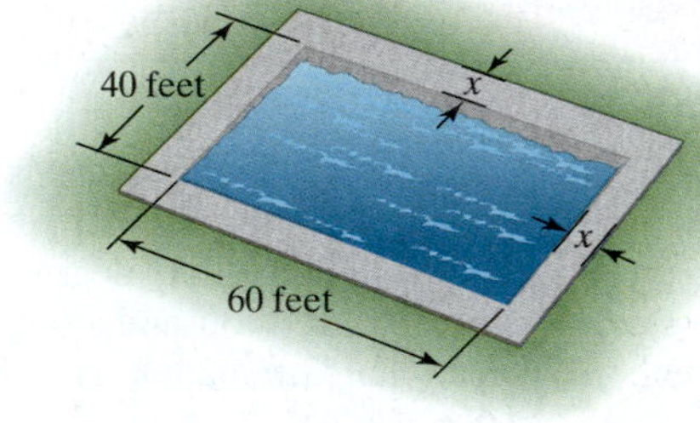

47. For a long-distance person-to-person telephone call, a telephone company charges \$0.43 for the first minute, \$0.32 for each additional minute, and a \$2.10 service charge. If the cost of a call is \$5.73, how long did the person talk?

48. A job pays an annual salary of \$33,150, which includes a holiday bonus of \$750. If paychecks are issued twice a month, what is the gross amount for each paycheck?

In Exercises 49–74, solve each formula for the specified variable.

49. $A = lw$ for l

50. $A = lw$ for w

51. $A = \frac{1}{2}bh$ for b

52. $A = \frac{1}{2}bh$ for h

53. $I = Prt$ for P

54. $I = Prt$ for t

55. $T = D + pm$ for p

56. $P = C + MC$ for M

57. $A = \frac{1}{2}h(a + b)$ for a

58. $A = \frac{1}{2}h(a + b)$ for b

59. $V = \frac{1}{3}\pi r^2 h$ for h

60. $V = \frac{1}{3}\pi r^2 h$ for r^2

61. $y - y_1 = m(x - x_1)$ for m

62. $y_2 - y_1 = m(x_2 - x_1)$ for m

63. $V = \dfrac{d_1 - d_2}{t}$ for d_1

64. $z = \dfrac{x - u}{s}$ for x

65. $Ax + By = C$ for x

66. $Ax + By = C$ for y

67. $s = \frac{1}{2}at^2 + vt$ for v

68. $s = \frac{1}{2}at^2 + vt$ for a

69. $L = a + (n - 1)d$ for n

70. $L = a + (n - 1)d$ for d

71. $A = 2lw + 2lh + 2wh$ for l

72. $A = 2lw + 2lh + 2wh$ for h

73. $IR + Ir = E$ for I

74. $A = \dfrac{x_1 + x_2 + x_3}{n}$ for n

Writing in Mathematics

75. In your own words, describe a step-by-step approach for solving algebraic word problems.

76. Write an original word problem that can be solved using a linear equation. Then solve the problem.

77. Explain what it means to solve a formula for a variable.

78. Did you have difficulties solving some of the problems that were assigned in this exercise set? Discuss what you did if this happened to you. Did your course of action enhance your ability to solve algebraic word problems?

79. The mile records in Example 2 on page 52 are a yardstick for measuring how athletes are getting better and better. Do you think that there is a limit to human performance? Explain your answer. If so, when might we reach it?

80. The bar graph in Exercises 37–38 shows average earnings of U.S. men and women, by highest educational attainment. Describe the trend shown by the graph. Discuss any aspects of the data that surprise you.

Technology Exercises

81. Use a graphing utility to numerically or graphically verify your work in any one exercise from Exercises 27–30. For assistance on how to do this, refer to the Using Technology box on page 53.

82. The formula $y = 55 + 0.2x$ models the average life expectancy, y, of American men born x years after 1900. Graph the formula in a [0, 200, 20] by [0, 100, 10] viewing rectangle. Then use the TRACE or ZOOM feature to verify your answer in Exercise 23 or 24.

83. In Exercises 25–26, we saw that in 2005, 19.4% of people in the United States spoke a language other than English at home, increasing by approximately 0.4% per year.

a. Write a formula that models the percentage of people in the United States speaking a language other than English at home, y, x years after 2005.

b. Enter the formula from part (a) as y_1 in your graphing utility. Then use either a table for y_1 or a graph of y_1 to numerically or graphically verify your answer to Exercise 25 or 26.

Critical Thinking Exercises

Make Sense? *In Exercises 84–87, determine whether each statement "makes sense" or "does not make sense" and explain your reasoning.*

84. I solved the formula for one of its variables, so now I have a numerical value for that variable.

85. Reasoning through a word problem often increases problem-solving skills in general.

86. The hardest part in solving a word problem is writing the equation that models the verbal conditions.

87. When traveling in Europe, the most useful form of the two Celsius-Fahrenheit conversion formulas is the formula used to convert from Fahrenheit to Celsius.

In Exercises 88–91, determine whether each statement is true or false. If the statement is false, make the necessary change(s) to produce a true statement.

88. If $I = prt$, then $t = I - pr$.

89. If $y = \frac{kx}{z}$, then $z = \frac{kx}{y}$.

90. It is not necessary to use the distributive property to solve $P = C + MC$ for C.

91. An item's price, x, reduced by $\frac{1}{3}$ is modeled by $x - \frac{1}{3}$.

92. The price of a dress is reduced by 40%. When the dress still does not sell, it is reduced by 40% of the reduced price. If the price of the dress after both reductions is \$72, what was the original price?

93. Suppose that we agree to pay you 8¢ for every problem in this chapter that you solve correctly and fine you 5¢ for every problem done incorrectly. If at the end of 26 problems we do not owe each other any money, how many problems did you solve correctly?

94. It was wartime when the Ricardos found out Mrs. Ricardo was pregnant. Ricky Ricardo was drafted and made out a will, deciding that \$14,000 in a savings account was to be divided between his wife and his child-to-be. Rather strangely, and certainly with gender bias, Ricky stipulated that if the child were a boy, he would get twice the amount of the mother's portion. If it were a girl, the mother would get twice the amount the girl was to receive. We'll never know what Ricky was thinking, for (as fate would have it) he did not return from war. Mrs. Ricardo gave birth to twins—a boy and a girl. How was the money divided?

95. A thief steals a number of rare plants from a nursery. On the way out, the thief meets three security guards, one after another. To each security guard, the thief is forced to give one-half the plants that he still has, plus 2 more. Finally, the thief leaves the nursery with 1 lone palm. How many plants were originally stolen?

96. Solve for C: $V = C - \frac{C - S}{L}N$.

Review Exercises

97. What does $-6 \leq -6$ mean? Is the statement true or false? (Section 1.1, Example 6)

98. Simplify: $\frac{(2 + 4)^2 + (-1)^5}{12 \div 2 \cdot 3 - 3}$. (Section 1.2, Example 8)

99. Solve: $\frac{2x}{3} - \frac{8}{3} = x$. (Section 1.4, Example 4)

Preview Exercises

Exercises 100–102 will help you prepare for the material covered in the next section.

100. In parts (a) and (b), complete each statement.

a. $b^4 \cdot b^3 = (b \cdot b \cdot b \cdot b)(b \cdot b \cdot b) = b^?$

b. $b^5 \cdot b^5 = (b \cdot b \cdot b \cdot b \cdot b)(b \cdot b \cdot b \cdot b \cdot b) = b^?$

c. Generalizing from parts (a) and (b), what should be done with the exponents when multiplying exponential expressions with the same base?

101. In parts (a) and (b), complete each statement.

a. $\frac{b^7}{b^3} = \frac{\cancel{b} \cdot \cancel{b} \cdot \cancel{b} \cdot b \cdot b \cdot b \cdot b}{\cancel{b} \cdot \cancel{b} \cdot \cancel{b}} = b^?$

b. $\frac{b^8}{b^2} = \frac{\cancel{b} \cdot \cancel{b} \cdot b \cdot b \cdot b \cdot b \cdot b \cdot b}{\cancel{b} \cdot \cancel{b}} = b^?$

c. Generalizing from parts (a) and (b), what should be done with the exponents when dividing exponential expressions with the same base?

102. Simplify: $\frac{1}{\left(-\frac{1}{2}\right)^3}$.

SECTION 1.6

Properties of Integral Exponents

Objectives

1. Use the product rule.
2. Use the quotient rule.
3. Use the zero-exponent rule.
4. Use the negative-exponent rule.
5. Use the power rule.
6. Find the power of a product.
7. Find the power of a quotient.
8. Simplify exponential expressions.

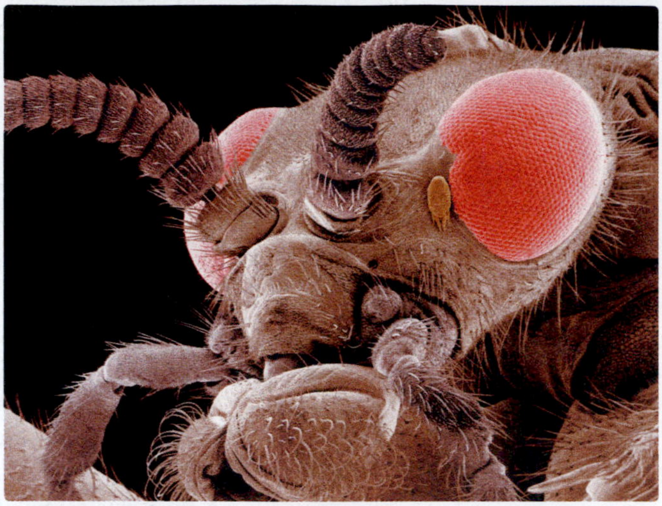

Our opening photo shows the head of a fly as seen under an electronic microscope. Some electronic microscopes can view objects that are less than 10^{-4} meter, or 0.0001 meter, in size. In this section, we'll make sense of the negative exponent in 10^{-4}, as we turn to integral exponents and their properties.

The Product and Quotient Rules

1 Use the product rule.

We have seen that exponents are used to indicate repeated multiplication. Now consider the multiplication of two exponential expressions, such as $b^4 \cdot b^3$. We are multiplying 4 factors of b and 3 factors of b. We have a total of 7 factors of b:

4 factors of b — 3 factors of b

$$b^4 \cdot b^3 = (b \cdot b \cdot b \cdot b)(b \cdot b \cdot b) = b^7.$$

Total: 7 factors of b

The product is exactly the same if we add the exponents:

$$b^4 \cdot b^3 = b^{4+3} = b^7.$$

This suggests the following rule:

The Product Rule

$$b^m \cdot b^n = b^{m+n}$$

When multiplying exponential expressions with the same base, add the exponents. Use this sum as the exponent of the common base.

EXAMPLE 1 Using the Product Rule

Multiply each expression using the product rule:

a. $b^8 \cdot b^{10}$ **b.** $(6x^4y^3)(5x^2y^7)$.

Solution

a. $b^8 \cdot b^{10} = b^{8+10} = b^{18}$

b. $(6x^4y^3)(5x^2y^7)$

$= 6 \cdot 5 \cdot x^4 \cdot x^2 \cdot y^3 \cdot y^7$ Use the associative and commutative properties. This step can be done mentally.

$= 30x^{4+2}y^{3+7}$

$= 30x^6y^{10}$

CHECK POINT 1 Multiply each expression using the product rule:

a. $b^6 \cdot b^5$ **b.** $(4x^3y^4)(10x^2y^6)$.

2 Use the quotient rule.

Now, consider the division of two exponential expressions, such as the quotient of b^7 and b^3. We are dividing 7 factors of b by 3 factors of b.

$$\frac{b^7}{b^3} = \frac{b \cdot b \cdot b \cdot b \cdot b \cdot b \cdot b}{b \cdot b \cdot b} = \boxed{\frac{b \cdot b \cdot b}{b \cdot b \cdot b}} \cdot b \cdot b \cdot b \cdot b = 1 \cdot b \cdot b \cdot b \cdot b = b^4$$

This factor is equal to 1.

The quotient is exactly the same if we subtract the exponents:

$$\frac{b^7}{b^3} = b^{7-3} = b^4.$$

This suggests the following rule:

The Quotient Rule

$$\frac{b^m}{b^n} = b^{m-n}, \quad b \neq 0$$

When dividing exponential expressions with the same nonzero base, subtract the exponent in the denominator from the exponent in the numerator. Use this difference as the exponent of the common base.

EXAMPLE 2 Using the Quotient Rule

Divide each expression using the quotient rule:

a. $\frac{(-2)^7}{(-2)^4}$ **b.** $\frac{30x^{12}y^9}{5x^3y^7}$.

Solution

a. $\frac{(-2)^7}{(-2)^4} = (-2)^{7-4} = (-2)^3$ or -8 $(-2)^3 = (-2)(-2)(-2) = -8$

b. $\frac{30x^{12}y^9}{5x^3y^7} = \frac{30}{5} \cdot \frac{x^{12}}{x^3} \cdot \frac{y^9}{y^7} = 6x^{12-3}y^{9-7} = 6x^9y^2$

CHECK POINT 2 Divide each expression using the quotient rule:

a. $\frac{(-3)^6}{(-3)^3}$ **b.** $\frac{27x^{14}y^8}{3x^3y^5}$.

3 Use the zero-exponent rule.

Zero as an Exponent

A nonzero base can be raised to the 0 power. The quotient rule can be used to help determine what zero as an exponent should mean. Consider the quotient of b^4 and b^4, where b is not zero. We can determine this quotient in two ways.

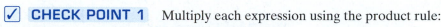

$$\frac{b^4}{b^4} = 1 \qquad\qquad \frac{b^4}{b^4} = b^{4-4} = b^0$$

Any nonzero expression divided by itself is 1.

Use the quotient rule and subtract exponents.

This means that b^0 must equal 1.

The Zero-Exponent Rule

If b is any real number other than 0,

$$b^0 = 1.$$

EXAMPLE 3 Using the Zero-Exponent Rule

Use the zero-exponent rule to simplify each expression:

a. 8^0 **b.** $(-6)^0$ **c.** -6^0 **d.** $5x^0$ **e.** $(5x)^0$.

Solution

a. $8^0 = 1$ *Any nonzero number raised to the 0 power is 1.*

b. $(-6)^0 = 1$

c. $-6^0 = -(6^0) = -1$

Only 6 is raised to the 0 power.

d. $5x^0 = 5 \cdot 1 = 5$ *Only x is raised to the 0 power.*

e. $(5x)^0 = 1$ *The entire expression, 5x, is raised to the 0 power.*

✓ CHECK POINT 3 Use the zero-exponent rule to simplify each expression:

a. 7^0 **b.** $(-5)^0$ **c.** -5^0 **d.** $10x^0$ **e.** $(10x)^0$.

4 Use the negative-exponent rule.

Negative Integers as Exponents

A nonzero base can be raised to a negative power. The quotient rule can be used to help determine what a negative integer as an exponent should mean. Consider the quotient of b^3 and b^5, where b is not zero. We can determine this quotient in two ways.

$$\frac{b^3}{b^5} = \frac{\cancel{b} \cdot \cancel{b} \cdot \cancel{b}}{\cancel{b} \cdot \cancel{b} \cdot \cancel{b} \cdot b \cdot b} = \frac{1}{b^2} \qquad\qquad \frac{b^3}{b^5} = b^{3-5} = b^{-2}$$

After dividing common factors, we have two factors of b in the denominator.

Use the quotient rule and subtract exponents.

Notice that $\frac{b^3}{b^5}$ equals both b^{-2} and $\frac{1}{b^2}$. This means that b^{-2} must equal $\frac{1}{b^2}$. This example is a special case of the **negative-exponent rule**.

The Negative-Exponent Rule

If b is any real number other than 0 and n is a natural number, then

$$b^{-n} = \frac{1}{b^n}.$$

EXAMPLE 4 Using the Negative-Exponent Rule

Use the negative-exponent rule to write each expression with a positive exponent. Simplify, if possible:

a. 9^{-2} **b.** $(-2)^{-5}$ **c.** $\frac{1}{6^{-2}}$ **d.** $7x^{-5}y^2$.

Solution

a. $9^{-2} = \frac{1}{9^2} = \frac{1}{81}$

b. $(-2)^{-5} = \frac{1}{(-2)^5} = \frac{1}{(-2)(-2)(-2)(-2)(-2)} = \frac{1}{-32} = -\frac{1}{32}$

Only the sign of the exponent, −5, changes. The base, −2, does not change sign.

c. $\frac{1}{6^{-2}} = \frac{1}{\frac{1}{6^2}} = 1 \cdot \frac{6^2}{1} = 6^2 = 36$

d. $7x^{-5}y^2 = 7 \cdot \frac{1}{x^5} \cdot y^2 = \frac{7y^2}{x^5}$

✓ CHECK POINT 4 Use the negative-exponent rule to write each expression with a positive exponent. Simplify, if possible:

a. 5^{-2} **b.** $(-3)^{-3}$ **c.** $\frac{1}{4^{-2}}$ **d.** $3x^{-6}y^4$.

In Example 4 and Check Point 4, did you notice that

$$\frac{1}{6^{-2}} = 6^2 \quad \text{and} \quad \frac{1}{4^{-2}} = 4^2?$$

In general, if a negative exponent appears in a denominator, an expression can be written with a positive exponent using

$$\frac{1}{b^{-n}} = b^n.$$

Negative Exponents in Numerators and Denominators

If b is any real number other than 0 and n is a natural number, then

$$b^{-n} = \frac{1}{b^n} \quad \text{and} \quad \frac{1}{b^{-n}} = b^n.$$

When a negative number appears as an exponent, switch the position of the base (from numerator to denominator or from denominator to numerator) and make the exponent positive. The sign of the base does not change.

EXAMPLE 5 Using Negative Exponents

Write each expression with positive exponents only. Then simplify, if possible:

a. $\frac{5^{-3}}{4^{-2}}$ **b.** $\frac{1}{6x^{-4}}$.

Solution

a. $\frac{5^{-3}}{4^{-2}} = \frac{4^2}{5^3} = \frac{4 \cdot 4}{5 \cdot 5 \cdot 5} = \frac{16}{125}$

Switch the position of each base to the other side of the fraction bar and change the sign of the exponent.

b. $\frac{1}{6x^{-4}} = \frac{x^4}{6}$ Switch the position of x to the other side of the fraction bar and change −4 to 4.

Don't switch the position of 6. It is not affected by a negative exponent.

CHECK POINT 5 Write each expression with positive exponents only. Then simplify, if possible:

a. $\frac{7^{-2}}{4^{-3}}$ b. $\frac{1}{5x^{-2}}$.

5 Use the power rule.

The Power Rule for Exponents (Powers to Powers)

The next property of exponents applies when an exponential expression is raised to a power. Here is an example:

$$(b^2)^4.$$

The exponential expression b^2 is raised to the fourth power.

There are 4 factors of b^2. Thus,

$$(b^2)^4 = b^2 \cdot b^2 \cdot b^2 \cdot b^2 = b^{2+2+2+2} = b^8.$$

Add exponents when multiplying with the same base.

We can obtain the answer, b^8, by multiplying the exponents:

$$(b^2)^4 = b^{2\cdot 4} = b^8.$$

This suggests the following rule:

The Power Rule (Powers to Powers)

$$(b^m)^n = b^{mn}$$

When an exponential expression is raised to a power, multiply the exponents. Place the product of the exponents on the base and remove the parentheses.

EXAMPLE 6 Using the Power Rule (Powers to Powers)

Simplify each expression using the power rule:

a. $(x^6)^4$ b. $(y^5)^{-3}$ c. $(b^{-4})^{-2}$.

Solution

a. $(x^6)^4 = x^{6\cdot 4} = x^{24}$

b. $(y^5)^{-3} = y^{5(-3)} = y^{-15} = \frac{1}{y^{15}}$

c. $(b^{-4})^{-2} = b^{(-4)(-2)} = b^8$

CHECK POINT 6 Simplify each expression using the power rule:

a. $(x^5)^3$ b. $(y^7)^{-2}$ c. $(b^{-3})^{-4}$.

6 Find the power of a product.

The Products-to-Powers Rule for Exponents

The next property of exponents applies when we are raising a product to a power. Here is an example:

$$(2x)^4.$$

The product $2x$ is raised to the fourth power.

There are four factors of $2x$. Thus,

$$(2x)^4 = 2x \cdot 2x \cdot 2x \cdot 2x = 2 \cdot 2 \cdot 2 \cdot 2 \cdot x \cdot x \cdot x \cdot x = 2^4x^4.$$

We can obtain the answer, 2^4x^4, by raising each factor within the parentheses to the fourth power:

$$(2x)^4 = 2^4x^4.$$

This suggests the following rule:

Products to Powers

$$(ab)^n = a^nb^n$$

When a product is raised to a power, raise each factor to that power.

EXAMPLE 7 Using the Products-to-Powers Rule

Simplify each expression using the products-to-powers rule:

a. $(6x)^3$ **b.** $(-2y^2)^4$ **c.** $(-3x^{-1}y^3)^{-2}$.

Solution

a. $(6x)^3 = 6^3x^3$ — Raise each factor to the third power.

$= 216x^3$ — Simplify: $6^3 = 6 \cdot 6 \cdot 6 = 216$.

b. $(-2y^2)^4 = (-2)^4(y^2)^4$ — Raise each factor to the fourth power.

$= (-2)^4y^{2 \cdot 4}$ — To raise an exponential expression to a power, multiply exponents: $(b^m)^n = b^{mn}$.

$= 16y^8$ — Simplify: $(-2)^4 = (-2)(-2)(-2)(-2) = 16$.

c. $(-3x^{-1}y^3)^{-2} = (-3)^{-2}(x^{-1})^{-2}(y^3)^{-2}$ — Raise each factor to the -2 power.

$= (-3)^{-2}x^{(-1)(-2)}y^{3(-2)}$ — Use $(b^m)^n = b^{mn}$ on the second and third factors.

$= (-3)^{-2}x^2y^{-6}$ — Simplify.

$= \dfrac{1}{(-3)^2} \cdot x^2 \cdot \dfrac{1}{y^6}$ — Apply $b^{-n} = \dfrac{1}{b^n}$ to the first and last factors.

$= \dfrac{x^2}{9y^6}$ — Simplify: $(-3)^2 = (-3)(-3) = 9$.

✓ CHECK POINT 7 Simplify each expression using the products-to-powers rule:

a. $(2x)^4$ **b.** $(-3y^2)^3$ **c.** $(-4x^5y^{-1})^{-2}$.

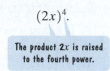

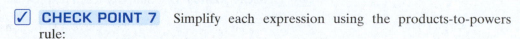

7 Find the power of a quotient.

The Quotients-to-Powers Rule for Exponents

The following rule is used to raise a quotient to a power:

Quotients to Powers

If b is a nonzero real number, then

$$\left(\frac{a}{b}\right)^n = \frac{a^n}{b^n}.$$

When a quotient is raised to a power, raise the numerator to that power and divide by the denominator to that power.

EXAMPLE 8 Using the Quotients-to-Powers Rule

Simplify each expression using the quotients-to-powers rule:

a. $\left(\frac{x^2}{4}\right)^3$ **b.** $\left(\frac{2x^3}{y^{-4}}\right)^5$ **c.** $\left(\frac{x^3}{y^2}\right)^{-4}$.

Solution

a. $\left(\frac{x^2}{4}\right)^3 = \frac{(x^2)^3}{4^3} = \frac{x^{2\cdot 3}}{4\cdot 4\cdot 4} = \frac{x^6}{64}$ Cube the numerator and the denominator.

b. $\left(\frac{2x^3}{y^{-4}}\right)^5 = \frac{(2x^3)^5}{(y^{-4})^5}$ Raise the numerator and the denominator to the fifth power.

$= \frac{2^5(x^3)^5}{(y^{-4})^5}$ Raise each factor in the numerator to the fifth power.

$= \frac{2^5 \cdot x^{3\cdot 5}}{y^{(-4)(5)}}$ Multiply exponents in both powers-to-powers expressions: $(b^m)^n = b^{mn}$.

$= \frac{32x^{15}}{y^{-20}}$ Simplify.

$= 32x^{15}y^{20}$ Move y to the other side of the fraction bar and change -20 to 20: $\frac{1}{b^{-n}} = b^n$.

c. $\left(\frac{x^3}{y^2}\right)^{-4} = \frac{(x^3)^{-4}}{(y^2)^{-4}}$ Raise the numerator and the denominator to the -4 power.

$= \frac{x^{3(-4)}}{y^{2(-4)}}$ Multiply exponents in both powers-to-powers expressions: $(b^m)^n = b^{mn}$.

$= \frac{x^{-12}}{y^{-8}}$ Simplify.

$= \frac{y^8}{x^{12}}$ Move each base to the other side of the fraction bar and make each exponent positive.

Study Tip

When simplifying exponential expressions, the first step should be to simplify inside parentheses. In Example 8, the expressions inside parentheses have different bases and cannot be simplified. This is why we begin with the quotients-to-powers rule.

✓ CHECK POINT 8 Simplify each expression using the quotients-to-powers rule:

a. $\left(\frac{x^5}{4}\right)^3$ **b.** $\left(\frac{2x^{-3}}{y^2}\right)^4$ **c.** $\left(\frac{x^{-3}}{y^4}\right)^{-5}$.

8 Simplify exponential expressions.

Simplifying Exponential Expressions

Properties of exponents are used to simplify exponential expressions. An exponential expression is **simplified** when

- No parentheses appear.
- No powers are raised to powers.
- Each base occurs only once.
- No negative or zero exponents appear.

Simplifying Exponential Expressions

	Example
1. If necessary, remove parentheses by using $(ab)^n = a^n b^n$ or $\left(\frac{a}{b}\right)^n = \frac{a^n}{b^n}$.	$(xy)^3 = x^3y^3$
2. If necessary, simplify powers to powers by using $(b^m)^n = b^{mn}$.	$(x^4)^3 = x^{4\cdot 3} = x^{12}$
3. If necessary, be sure that each base appears only once by using $b^m \cdot b^n = b^{m+n}$ or $\frac{b^m}{b^n} = b^{m-n}$.	$x^4 \cdot x^3 = x^{4+3} = x^7$
4. If necessary, rewrite exponential expressions with zero powers as 1 $(b^0 = 1)$. Furthermore, write the answer with positive exponents by using $b^{-n} = \frac{1}{b^n}$ or $\frac{1}{b^{-n}} = b^n$.	$\frac{x^5}{x^8} = x^{5-8} = x^{-3} = \frac{1}{x^3}$

The following example shows how to simplify exponential expressions. Throughout the example, assume that no variable in a denominator is equal to zero.

EXAMPLE 9 Simplifying Exponential Expressions

Simplify:

a. $(-2xy^{-14})(-3x^4y^5)^3$ **b.** $\left(\frac{25x^2y^4}{-5x^6y^{-8}}\right)^2$ **c.** $\left(\frac{x^{-4}y^7}{2}\right)^{-5}$.

Solution

a. $(-2xy^{-14})(-3x^4y^5)^3$

$= (-2xy^{-14})(-3)^3(x^4)^3(y^5)^3$ Cube each factor in the second parentheses.

$= (-2xy^{-14})(-27)x^{12}y^{15}$ Multiply the exponents when raising a power to a power: $(x^4)^3 = x^{4\cdot 3} = x^{12}$ and $(y^5)^3 = y^{5\cdot 3} = y^{15}$.

$= (-2)(-27)x^{1+12}y^{-14+15}$ Mentally rearrange factors and multiply like bases by adding the exponents.

$= 54x^{13}y$ Simplify.

b. $\left(\dfrac{25x^2y^4}{-5x^6y^{-8}}\right)^2$ The expression inside parentheses contains the same bases and can be simplified. Begin with this simplification.

$= (-5x^{2-6}y^{4-(-8)})^2$ Simplify inside the parentheses. Subtract the exponents when dividing.

$= (-5x^{-4}y^{12})^2$ Simplify.

$= (-5)^2(x^{-4})^2(y^{12})^2$ Square each factor in parentheses.

$= 25x^{-8}y^{24}$ Multiply the exponents when raising a power to a power: $(x^{-4})^2 = x^{-4(2)} = x^{-8}$ and $(y^{12})^2 = y^{12\cdot 2} = y^{24}$.

$= \dfrac{25y^{24}}{x^8}$ Simplify x^{-8} using $b^{-n} = \dfrac{1}{b^n}$.

c. $\left(\dfrac{x^{-4}y^7}{2}\right)^{-5}$

$= \dfrac{(x^{-4}y^7)^{-5}}{2^{-5}}$ Raise the numerator and the denominator to the -5 power.

$= \dfrac{(x^{-4})^{-5}(y^7)^{-5}}{2^{-5}}$ Raise each factor in the numerator to the -5 power.

$= \dfrac{x^{20}y^{-35}}{2^{-5}}$ Multiply the exponents when raising a power to a power: $(x^{-4})^{-5} = x^{-4(-5)} = x^{20}$ and $(y^7)^{-5} = y^{7(-5)} = y^{-35}$.

$= \dfrac{2^5x^{20}}{y^{35}}$ Move each base with a negative exponent to the other side of the fraction bar and make each negative exponent positive.

$= \dfrac{32x^{20}}{y^{35}}$ Simplify: $2^5 = 2\cdot 2\cdot 2\cdot 2\cdot 2 = 32$.

✓ CHECK POINT 9 Simplify:

a. $(-3x^{-6}y)(-2x^3y^4)^2$ **b.** $\left(\dfrac{10x^3y^5}{5x^6y^{-2}}\right)^2$ **c.** $\left(\dfrac{x^3y^5}{4}\right)^{-3}$.

Study Tip

Try to avoid the following common errors that can occur when simplifying exponential expressions.

Correct	Incorrect	Description of Error
$b^3\cdot b^4 = b^7$	~~$b^3\cdot b^4 = b^{12}$~~	The exponents should be added, not multiplied.
$3^2\cdot 3^4 = 3^6$	~~$3^2\cdot 3^4 = 9^6$~~	The common base should be retained, not multiplied.
$\dfrac{5^{16}}{5^4} = 5^{12}$	~~$\dfrac{5^{16}}{5^4} = 5^4$~~	The exponents should be subtracted, not divided.
$(4a)^3 = 64a^3$	~~$(4a)^3 = 4a^3$~~	Both factors should be cubed.
$b^{-n} = \dfrac{1}{b^n}$	~~$b^{-n} = -\dfrac{1}{b^n}$~~	Only the exponent should change sign.
$(a+b)^{-1} = \dfrac{1}{a+b}$	~~$(a+b)^{-1} = \dfrac{1}{a} + \dfrac{1}{b}$~~	The exponent applies to the entire expression $a + b$.

1.6 EXERCISE SET

MyMathLab — MathXL PRACTICE · WATCH · DOWNLOAD · READ · REVIEW

Practice Exercises

In Exercises 1–14, multiply using the product rule.

1. $b^4 \cdot b^7$
2. $b^5 \cdot b^9$
3. $x \cdot x^3$
4. $x \cdot x^4$
5. $2^3 \cdot 2^2$
6. $2^4 \cdot 2^2$
7. $3x^4 \cdot 2x^2$
8. $5x^3 \cdot 3x^2$
9. $(-2y^{10})(-10y^2)$
10. $(-4y^8)(-8y^4)$
11. $(5x^3y^4)(20x^7y^8)$
12. $(4x^5y^6)(20x^7y^4)$
13. $(-3x^4y^0z)(-7xyz^3)$
14. $(-9x^3yz^4)(-5xy^0z^2)$

In Exercises 15–24, divide using the quotient rule.

15. $\dfrac{b^{12}}{b^3}$
16. $\dfrac{b^{25}}{b^5}$
17. $\dfrac{15x^9}{3x^4}$
18. $\dfrac{18x^{11}}{3x^4}$
19. $\dfrac{x^9y^7}{x^4y^2}$
20. $\dfrac{x^9y^{12}}{x^2y^6}$
21. $\dfrac{50x^2y^7}{5xy^4}$
22. $\dfrac{36x^{12}y^4}{4xy^2}$
23. $\dfrac{-56a^{12}b^{10}c^8}{7ab^2c^4}$
24. $\dfrac{-66a^9b^7c^6}{6a^3bc^2}$

In Exercises 25–34, use the zero-exponent rule to simplify each expression.

25. 6^0
26. 9^0
27. $(-4)^0$
28. $(-2)^0$
29. -4^0
30. -2^0
31. $13y^0$
32. $17y^0$
33. $(13y)^0$
34. $(17y)^0$

In Exercises 35–52, write each expression with positive exponents only. Then simplify, if possible.

35. 3^{-2}
36. 4^{-2}
37. $(-5)^{-2}$
38. $(-7)^{-2}$
39. -5^{-2}
40. -7^{-2}
41. x^2y^{-3}
42. x^3y^{-4}
43. $8x^{-7}y^3$
44. $9x^{-8}y^4$
45. $\dfrac{1}{5^{-3}}$
46. $\dfrac{1}{2^{-5}}$
47. $\dfrac{1}{(-3)^{-4}}$
48. $\dfrac{1}{(-2)^{-4}}$
49. $\dfrac{x^{-2}}{y^{-5}}$
50. $\dfrac{x^{-3}}{y^{-7}}$
51. $\dfrac{a^{-4}b^7}{c^{-3}}$
52. $\dfrac{a^{-3}b^8}{c^{-2}}$

In Exercises 53–58, simplify each expression using the power rule.

53. $(x^6)^{10}$
54. $(x^3)^2$
55. $(b^4)^{-3}$
56. $(b^8)^{-3}$
57. $(7^{-4})^{-5}$
58. $(9^{-4})^{-5}$

In Exercises 59–72, simplify each expression using the products-to-powers rule.

59. $(4x)^3$
60. $(2x)^5$
61. $(-3x^7)^2$
62. $(-4x^9)^2$
63. $(2xy^2)^3$
64. $(3x^2y)^4$
65. $(-3x^2y^5)^2$
66. $(-3x^4y^6)^2$
67. $(-3x^{-2})^{-3}$
68. $(-2x^{-4})^{-3}$
69. $(5x^3y^{-4})^{-2}$
70. $(7x^2y^{-5})^{-2}$
71. $(-2x^{-5}y^4z^2)^{-4}$
72. $(-2x^{-4}y^5z^3)^{-4}$

In Exercises 73–84, simplify each expression using the quotients-to-powers rule.

73. $\left(\dfrac{2}{x}\right)^4$
74. $\left(\dfrac{y}{2}\right)^5$
75. $\left(\dfrac{x^3}{5}\right)^2$
76. $\left(\dfrac{x^4}{6}\right)^2$
77. $\left(-\dfrac{3x}{y}\right)^4$
78. $\left(-\dfrac{2x}{y}\right)^5$
79. $\left(\dfrac{x^4}{y^2}\right)^6$
80. $\left(\dfrac{x^5}{y^3}\right)^6$
81. $\left(\dfrac{x^3}{y^{-4}}\right)^3$
82. $\left(\dfrac{x^4}{y^{-2}}\right)^3$
83. $\left(\dfrac{a^{-2}}{b^3}\right)^{-4}$
84. $\left(\dfrac{a^{-3}}{b^5}\right)^{-4}$

In Exercises 85–116, simplify each exponential expression.

85. $\dfrac{x^3}{x^9}$
86. $\dfrac{x^6}{x^{10}}$
87. $\dfrac{20x^3}{-5x^4}$
88. $\dfrac{10x^5}{-2x^6}$
89. $\dfrac{16x^3}{8x^{10}}$
90. $\dfrac{15x^2}{3x^{11}}$

91. $\dfrac{20a^3b^8}{2ab^{13}}$

92. $\dfrac{72a^5b^{11}}{9ab^{17}}$

93. $x^3 \cdot x^{-12}$

94. $x^4 \cdot x^{-12}$

95. $(2a^5)(-3a^{-7})$

96. $(4a^2)(-2a^{-5})$

97. $\left(-\frac{1}{4}x^{-4}y^5z^{-1}\right)(-12x^{-3}y^{-1}z^4)$

98. $\left(-\frac{1}{3}x^{-5}y^4z^6\right)(-18x^{-2}y^{-1}z^{-7})$

99. $\dfrac{6x^2}{2x^{-8}}$

100. $\dfrac{12x^5}{3x^{-10}}$

101. $\dfrac{x^{-7}}{x^3}$

102. $\dfrac{x^{-10}}{x^4}$

103. $\dfrac{30x^2y^5}{-6x^8y^{-3}}$

104. $\dfrac{24x^2y^{13}}{-2x^5y^{-2}}$

105. $\dfrac{-24a^3b^{-5}c^5}{-3a^{-6}b^{-4}c^{-7}}$

106. $\dfrac{-24a^2b^{-2}c^8}{-8a^{-5}b^{-1}c^{-3}}$

107. $\left(\dfrac{x^3}{x^{-5}}\right)^2$

108. $\left(\dfrac{x^4}{x^{-11}}\right)^3$

109. $\left(\dfrac{-15a^4b^2}{5a^{10}b^{-3}}\right)^3$

110. $\left(\dfrac{-30a^{14}b^8}{10a^{17}b^{-2}}\right)^3$

111. $\left(\dfrac{3a^{-5}b^2}{12a^3b^{-4}}\right)^0$

112. $\left(\dfrac{4a^{-5}b^3}{12a^3b^{-5}}\right)^0$

113. $\left(\dfrac{x^{-5}y^8}{3}\right)^{-4}$

114. $\left(\dfrac{x^6y^{-7}}{2}\right)^{-3}$

115. $\left(\dfrac{20a^{-3}b^4c^5}{-2a^{-5}b^{-2}c}\right)^{-2}$

116. $\left(\dfrac{-2a^{-4}b^3c^{-1}}{3a^{-2}b^{-5}c^{-2}}\right)^{-4}$

Practice PLUS

In Exercises 117–124, simplify each exponential expression.

117. $\dfrac{9y^4}{x^{-2}} + \left(\dfrac{x^{-1}}{y^2}\right)^{-2}$

118. $\dfrac{7x^3}{y^{-9}} + \left(\dfrac{x^{-1}}{y^3}\right)^{-3}$

119. $\left(\dfrac{3x^4}{y^{-4}}\right)^{-1}\left(\dfrac{2x}{y^2}\right)^3$

120. $\left(\dfrac{2^{-1}x^{-2}y}{x^4y^{-1}}\right)^{-2}\left(\dfrac{xy^{-3}}{x^{-3}y}\right)^3$

121. $(-4x^3y^{-5})^{-2}(2x^{-8}y^{-5})$

122. $(-4x^{-4}y^5)^{-2}(-2x^5y^{-6})$

123. $\dfrac{(2x^2y^4)^{-1}(4xy^3)^{-3}}{(x^2y)^{-5}(x^3y^2)^4}$

124. $\dfrac{(3x^3y^2)^{-1}(2x^2y)^{-2}}{(xy^2)^{-5}(x^2y^3)^3}$

Application Exercises

The formula

$$A = 1000 \cdot 2^t$$

models the population, A, of aphids in a field of potato plants after t weeks. Use this formula to solve Exercises 125–126.

125. **a.** What is the present aphid population?

b. What will the aphid population be in 4 weeks?

c. What was the aphid population 3 weeks ago?

126. **a.** What is the present aphid population?

b. What will the aphid population be in 3 weeks?

c. What was the aphid population 2 weeks ago?

A rumor about algebra CDs that you can listen to as you sleep, allowing you to awaken refreshed and algebraically empowered, is spreading among the students in your math class. The formula

$$N = \frac{25}{1 + 24 \cdot 2^{-t}}$$

models the number of people in the class, N, who have heard the rumor after t minutes. Use this formula to solve Exercises 127–128.

127. **a.** How many people in the class started the rumor?

b. How many people in the class have heard the rumor after 4 minutes?

128. **a.** How many people in the class started the rumor?

b. How many people in the class, rounded to the nearest whole number, have heard the rumor after 6 minutes?

Use the graph of the rumor model to solve Exercises 129–132.

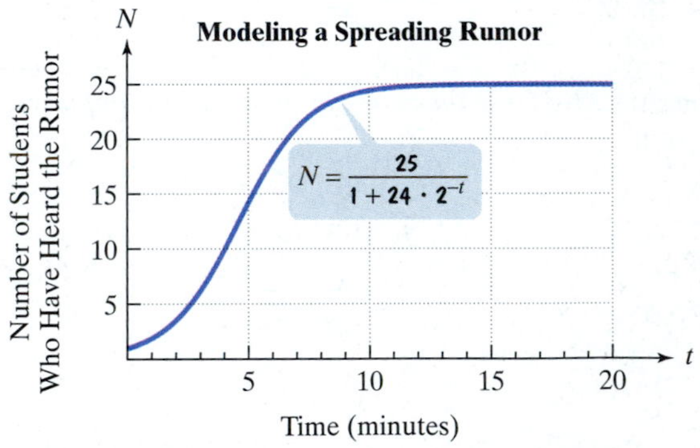

129. Identify your answers to Exercise 127, parts (a) and (b), as points on the graph.

130. Identify your answers to Exercise 128, parts (a) and (b), as points on the graph.

131. Which one of the following best describes the rate of growth of the rumor as shown by the graph?

a. The number of people in the class who heard the rumor grew steadily over time.

b. The number of people in the class who heard the rumor remained constant over time.

c. The number of people in the class who heard the rumor increased slowly at the beginning, but this rate of increase continued to escalate over time.

d. The number of people in the class who heard the rumor increased quite rapidly at the beginning, but this rate of increase eventually slowed down, ultimately limited by the number of students in the class.

132. Use the graph to determine how many people in the class eventually heard the rumor.

The astronomical unit (AU) is often used to measure distances within the solar system. One AU is equal to the average distance between Earth and the sun, or 92,955,630 miles. The distance, d, of the nth planet from the sun is modeled by the formula

$$d = \frac{3(2^{n-2}) + 4}{10},$$

where d is measured in astronomical units. Use this formula to solve Exercises 133–136.

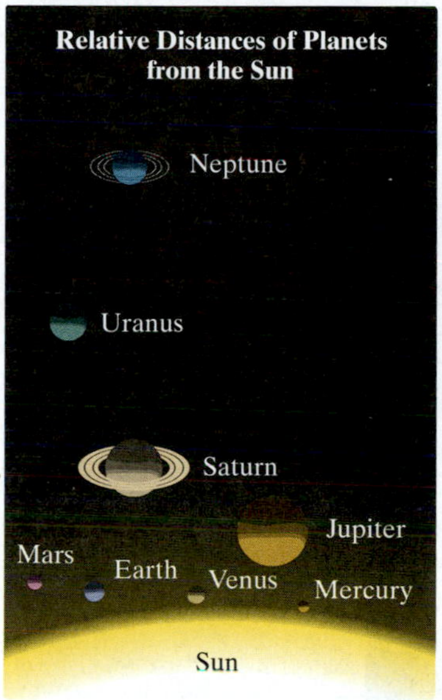

133. Substitute 1 for n and find the distance between Mercury and the sun.

134. Substitute 2 for n and find the distance between Venus and the sun.

135. How much farther from the sun is Jupiter than Earth?

136. How much farther from the sun is Uranus than Earth?

Writing in Mathematics

137. Explain the product rule for exponents. Use $b^2 \cdot b^3$ in your explanation.

138. Explain the quotient rule for exponents. Use $\frac{b^8}{b^2}$ in your explanation.

139. Explain how to find any nonzero number to the 0 power.

140. Explain the negative-exponent rule and give an example.

141. Explain the power rule for exponents. Use $(b^2)^3$ in your explanation.

142. Explain how to simplify an expression that involves a product raised to a power. Give an example.

143. Explain how to simplify an expression that involves a quotient raised to a power. Give an example.

144. How do you know if an exponential expression is simplified?

Technology Exercise

145. Enter the rumor formula

$$N = \frac{25}{1 + 24 \cdot 2^{-t}}$$

in your graphing utility as

$y_1 = 25$ [÷] [(] 1 [+] 24 [×] 2 [∧] [(−)] [x] [)].

Then use a table for y_1 to numerically verify your answers to Exercise 127 or 128.

Critical Thinking Exercises

Make Sense? *In Exercises 146–149, determine whether each statement "makes sense" or "does not make sense" and explain your reasoning.*

146. The properties $(ab)^n = a^n b^n$ and $\left(\frac{a}{b}\right)^n = \frac{a^n}{b^n}$ are like distributive properties of powers over multiplication and division.

147. If 7^{-2} is raised to the third power, the result is a number between 0 and 1.

148. There are many exponential expressions that are equal to $25x^{12}$, such as $(5x^6)^2$, $(5x^3)(5x^9)$, $25(x^3)^9$, and $5^2(x^2)^6$.

149. The expression $\frac{a^n}{b^0}$ is undefined because division by 0 is undefined.

In Exercises 150–157, determine whether each statement is true or false. If the statement is false, make the necessary change(s) to produce a true statement.

150. $2^2 \cdot 2^4 = 2^8$

151. $5^6 \cdot 5^2 = 25^8$

152. $2^3 \cdot 3^2 = 6^5$

153. $\frac{1}{(-2)^3} = 2^{-3}$

154. $\frac{2^8}{2^{-3}} = 2^5$

155. $2^4 + 2^5 = 2^9$

156. $2000.002 = (2 \times 10^3) + (2 \times 10^{-3})$

157. $40{,}000.04 = (4 \times 10^4) + (4 \times 10^{-2})$

In Exercises 158–161, simplify the expression. Assume that all variables used as exponents represent integers and that all other variables represent nonzero real numbers.

158. $x^{n-1} \cdot x^{3n+4}$

159. $(x^{-4n} \cdot x^n)^{-3}$

160. $\left(\frac{x^{3-n}}{x^{6-n}}\right)^{-2}$

161. $\left(\frac{x^n y^{3n+1}}{y^n}\right)^3$

Review Exercises

162. Graph $y = 2x - 1$ in a rectangular coordinate system. Let $x = -3, -2, -1, 0, 1, 2,$ and 3. (Section 1.3, Example 2)

163. Solve $Ax + By = C$ for y. (Section 1.5, Example 6)

164. The length of a rectangular playing field is 5 meters less than twice its width. If 230 meters of fencing enclose the field, what are its dimensions? (Section 1.5, Example 5)

Preview Exercises

Exercises 165–167 will help you prepare for the material covered in the next section.

165. If 6.2 is multiplied by 10^3, what does this multiplication do to the decimal point in 6.2?

166. If 8.5 is multiplied by 10^{-2}, what does this multiplication do to the decimal point in 8.5?

167. Write each computation as a single power of 10. Then evaluate this exponential expression.

a. $10^9 \times 10^{-4}$ **b.** $\dfrac{10^4}{10^{-2}}$

SECTION 1.7

Scientific Notation

Objectives

1. Convert from scientific to decimal notation.
2. Convert from decimal to scientific notation.
3. Perform computations with scientific notation.
4. Use scientific notation to solve problems.

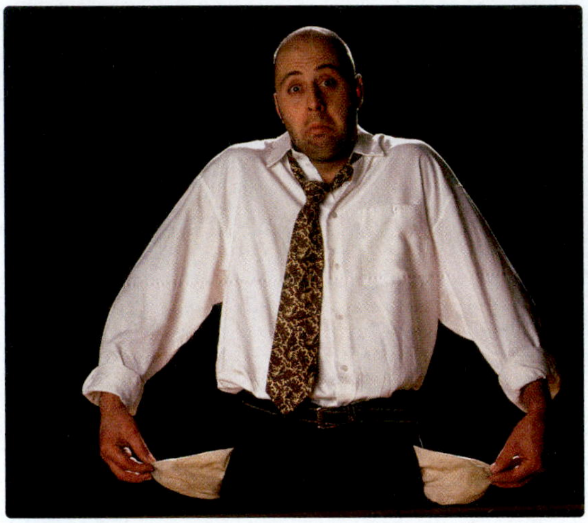

People who complain about paying their income tax can be divided into two types: men and women. Perhaps we can quantify the complaining by examining the data in **Figure 1.21**. The bar graphs show the U.S. population, in millions, and the total amount we paid in federal taxes, in trillions of dollars, for nine selected years.

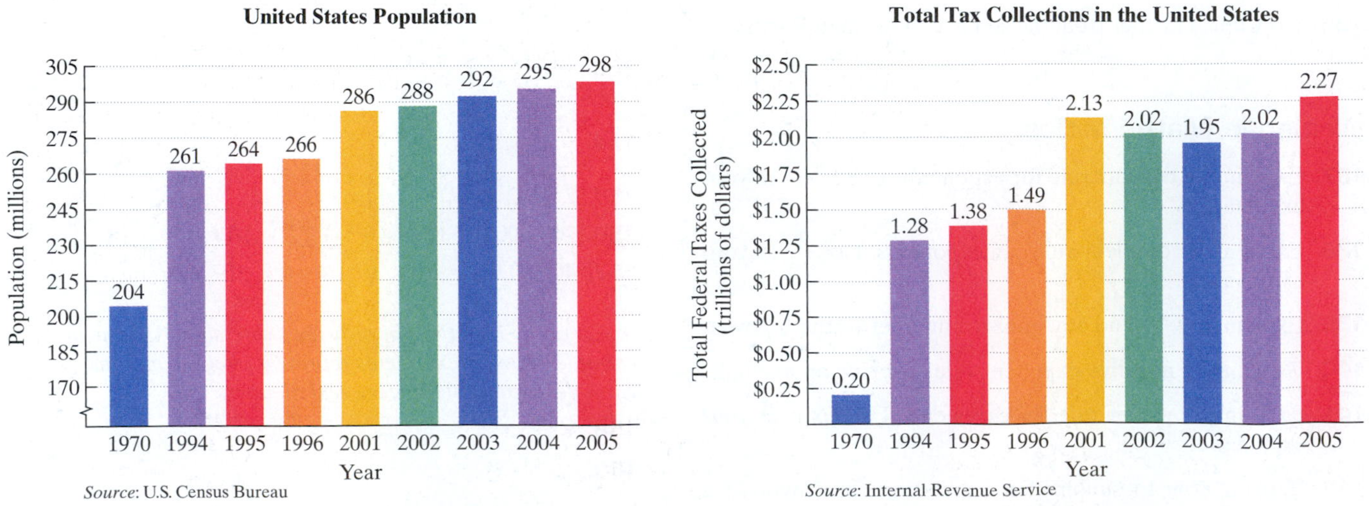

FIGURE 1.21 Population and total tax collections in the United States

The bar graph on the right shows that in 2005 total tax collections were \$2.27 trillion. How can we place this amount in the proper perspective? If the total tax collections were evenly divided among all Americans, how much would each citizen pay in taxes?

In this section, you will learn to use exponents to provide a way of putting large and small numbers in perspective. Using this skill, we will explore the per capita tax for some of the years shown in **Figure 1.21**.

1 Convert from scientific to decimal notation.

Scientific Notation

We have seen that in 2005 total tax collections were \$2.27 trillion. Because a trillion is 10^{12} (see **Table 1.6**), this amount can be expressed as

$$2.27 \times 10^{12}.$$

The number 2.27×10^{12} is written in a form called *scientific notation*.

Table 1.6 Names of Large Numbers

10^2	hundred
10^3	thousand
10^6	million
10^9	billion
10^{12}	trillion
10^{15}	quadrillion
10^{18}	quintillion
10^{21}	sextillion
10^{24}	septillion
10^{27}	octillion
10^{30}	nonillion
10^{100}	googol

Scientific Notation

A number is written in **scientific notation** when it is expressed in the form

$$a \times 10^n,$$

where the absolute value of a is greater than or equal to 1 and less than 10 $(1 \le |a| < 10)$, and n is an integer.

It is customary to use the multiplication symbol, ×, rather than a dot, when writing a number in scientific notation.

Converting from Scientific to Decimal Notation

Here are two examples of numbers in scientific notation:

$$6.4 \times 10^5 \quad \text{means} \quad 640{,}000.$$
$$2.17 \times 10^{-3} \quad \text{means} \quad 0.00217.$$

Do you see that the number with the positive exponent is relatively large and the number with the negative exponent is relatively small?

We can use n, the exponent on the 10 in $a \times 10^n$, to change a number in scientific notation to decimal notation. If n is **positive**, move the decimal point in a to the **right** n places. If n is **negative**, move the decimal point in a to the **left** $|n|$ places.

EXAMPLE 1 Converting from Scientific to Decimal Notation

Write each number in decimal notation:

a. 6.2×10^7 **b.** -6.2×10^7 **c.** 2.019×10^{-3} **d.** -2.019×10^{-3}.

Solution In each case, we use the exponent on the 10 to move the decimal point. In parts (a) and (b), the exponent is positive, so we move the decimal point to the right. In parts (c) and (d), the exponent is negative, so we move the decimal point to the left.

a. $6.2 \times 10^7 = 62{,}000{,}000$

$n = 7$ — Move the decimal point 7 places to the right.

b. $-6.2 \times 10^7 = -62{,}000{,}000$

$n = 7$ — Move the decimal point 7 places to the right.

c. $2.019 \times 10^{-3} = 0.002019$

$n = -3$ — Move the decimal point $|-3|$ places, or 3 places, to the left.

d. $-2.019 \times 10^{-3} = -0.002019$

$n = -3$ — Move the decimal point $|-3|$ places, or 3 places, to the left.

✓ CHECK POINT 1 Write each number in decimal notation:

a. -2.6×10^9 **b.** 3.017×10^{-6}.

2 Convert from decimal to scientific notation.

Converting from Decimal to Scientific Notation

To convert from decimal notation to scientific notation, we reverse the procedure of Example 1.

Converting from Decimal to Scientific Notation

Write the number in the form $a \times 10^n$.

- Determine a, the numerical factor. Move the decimal point in the given number to obtain a number whose absolute value is between 1 and 10, including 1.
- Determine n, the exponent on 10^n. The absolute value of n is the number of places the decimal point was moved. The exponent n is positive if the decimal point was moved to the left, negative if the decimal point was moved to the right, and 0 if the decimal point was not moved.

Using Technology

You can use your calculator's EE (enter exponent) or EXP key to convert from decimal to scientific notation. Here is how it's done for 0.0000000000802.

Many Scientific Calculators

Keystrokes

.0000000000802 EE =

Display

8.02 − 11

Many Graphing Calculators

Use the mode setting for scientific notation.

Keystrokes

.0000000000802 ENTER

Display

8.02E − 11

EXAMPLE 2 Converting from Decimal Notation to Scientific Notation

Write each number in scientific notation:

a. 34,970,000,000,000 **b.** −34,970,000,000,000
c. 0.0000000000802 **d.** −0.0000000000802.

Solution

a. $34{,}970{,}000{,}000{,}000 = 3.497 \times 10^{13}$

Move the decimal point to get a number whose absolute value is between 1 and 10.

The decimal point was moved 13 places to the left, so $n = 13$.

b. $-34{,}970{,}000{,}000{,}000 = -3.497 \times 10^{13}$

c. $0.0000000000802 = 8.02 \times 10^{-11}$

Move the decimal point to get a number whose absolute value is between 1 and 10.

The decimal point was moved 11 places to the right, so $n = -11$.

d. $-0.0000000000802 = -8.02 \times 10^{-11}$

Study Tip

If the absolute value of a number is greater than 10, it will have a positive exponent in scientific notation. If the absolute value of a number is less than 1, it will have a negative exponent in scientific notation.

✓ CHECK POINT 2 Write each number in scientific notation:

a. 5,210,000,000 **b.** −0.00000006893.

EXAMPLE 3 Expressing the Number of Cellphone Spam Messages in Scientific Notation

As feature-rich cellphones function more like PCs, digital intruders are targeting them with viruses, spam, and phishing scams. The bar graph in **Figure 1.22** on the next page shows the number of cellphone spam messages, in millions, from 2002 through 2005. Express the number of spam messages in 2005 in scientific notation.

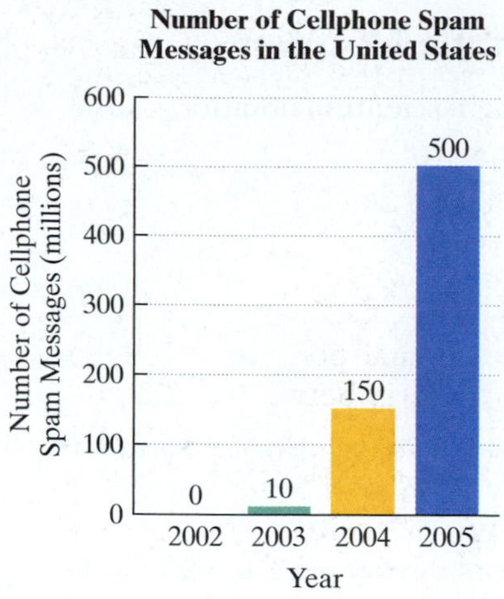

FIGURE 1.22

Source: Ferris Research

Solution Because a million is 10^6, the number of cellphone spam messages in 2005 can be expressed as

$$500 \times 10^6.$$

This factor is not between 1 and 10, so the number is not in scientific notation.

The voice balloon indicates that we need to convert 500 to scientific notation.

$$500 \times 10^6 = (5 \times 10^2) \times 10^6 = 5 \times (10^2 \times 10^6) = 5 \times 10^{2+6} = 5 \times 10^8$$

There were 5×10^8 cellphone spam messages in 2005.

Study Tip

Many of the large numbers we encounter in newspapers, magazines, and almanacs are expressed in millions (10^6), billions (10^9), and trillions (10^{12}). We can use exponential properties to describe the number of cellphone spam messages in 2005 using millions or billions.

$$500 \times 10^6 = 5 \times 10^8 = 0.5 \times 10^9$$

There were 500 million messages.

This expresses the number of messages in scientific notation.

There were half a billion messages.

CHECK POINT 3 In 2005, the federal cost of social security was \$519 billion. Express this amount in scientific notation.

3 Perform computations with scientific notation.

Computations with Scientific Notation

Properties of exponents are used to perform computations with numbers that are expressed in scientific notation.

Computations with Numbers in Scientific Notation

Multiplication

$$(a \times 10^n)(b \times 10^m) = (a \times b) \times 10^{n+m}$$

Add the exponents on 10 and multiply the other parts of the numbers separately.

Division

$$\frac{a \times 10^n}{b \times 10^m} = \left(\frac{a}{b}\right) \times 10^{n-m}$$

Subtract the exponents on 10 and divide the other parts of the numbers separately.

After the computation is completed, the answer may require an adjustment before it is back in scientific notation.

EXAMPLE 4 Computations with Scientific Notation

Perform the indicated computations, writing the answers in scientific notation:

a. $(6.1 \times 10^5)(4 \times 10^{-9})$ **b.** $\dfrac{1.8 \times 10^4}{3 \times 10^{-2}}$.

Solution

a. $(6.1 \times 10^5)(4 \times 10^{-9})$

$= (6.1 \times 4) \times (10^5 \times 10^{-9})$ Regroup factors.

$= 24.4 \times 10^{5+(-9)}$ Add the exponents on 10 and multiply the other parts.

$= 24.4 \times 10^{-4}$ Simplify.

$= (2.44 \times 10^1) \times 10^{-4}$ Convert 24.4 to scientific notation: $24.4 = 2.44 \times 10^1$.

$= 2.44 \times 10^{-3}$ $10^1 \times 10^{-4} = 10^{1+(-4)} = 10^{-3}$

b. $\dfrac{1.8 \times 10^4}{3 \times 10^{-2}} = \left(\dfrac{1.8}{3}\right) \times \left(\dfrac{10^4}{10^{-2}}\right)$ Regroup factors.

$= 0.6 \times 10^{4-(-2)}$ Subtract the exponents on 10 and divide the other parts.

$= 0.6 \times 10^6$ Simplify: $4 - (-2) = 4 + 2 = 6$.

$= (6 \times 10^{-1}) \times 10^6$ Convert 0.6 to scientific notation: $0.6 = 6 \times 10^{-1}$.

$= 6 \times 10^5$ $10^{-1} \times 10^6 = 10^{-1+6} = 10^5$

Using Technology

$(6.1 \times 10^5)(4 \times 10^{-9})$
on a Calculator:

Many Scientific Calculators

6.1 [EE] 5 [×] 4 [EE] 9 [+/−] [=]

Display

2.44 − 03

Many Graphing Calculators

6.1 [EE] 5 [×] 4 [EE] [(−)] 9 [ENTER]

Display (in scientific notation mode)

2.44 E − 3

✓ CHECK POINT 4 Perform the indicated computations, writing the answers in scientific notation:

a. $(7.1 \times 10^5)(5 \times 10^{-7})$ **b.** $\dfrac{1.2 \times 10^6}{3 \times 10^{-3}}$.

4 Use scientific notation to solve problems.

Applications: Putting Numbers in Perspective

We have seen that in 2005 the U.S. government collected \$2.27 trillion in taxes. Example 5 shows how we can use scientific notation to comprehend the meaning of a number such as 2.27 trillion.

EXAMPLE 5 Tax per Capita

In 2005, the U.S. government collected 2.27×10^{12} dollars in taxes. At that time, the U.S. population was approximately 298 million, or 2.98×10^8. If the total tax collections were evenly divided among all Americans, how much would each citizen pay? Express the answer in decimal notation, rounded to the nearest dollar.

Solution The amount that we would each pay, or the tax per capita, is the total amount collected, 2.27×10^{12}, divided by the number of Americans, 2.98×10^{8}.

$$\frac{2.27 \times 10^{12}}{2.98 \times 10^{8}} = \left(\frac{2.27}{2.98}\right) \times \left(\frac{10^{12}}{10^{8}}\right) \approx 0.7617 \times 10^{12-8} = 0.7617 \times 10^{4} = 7617$$

To obtain an answer in decimal notation, it is not necessary to express this number in scientific notation.

Move the decimal point 4 places to the right.

If total tax collections were evenly divided, we would each pay approximately \$7617 in taxes.

CHECK POINT 5 In 2004, the U. S. government collected 2.02×10^{12} dollars in taxes. At that time, the U.S. population was approximately 295 million or 2.95×10^{8}. Find the per capita tax, rounded to the nearest dollar, in 2004.

Many problems in algebra involve motion. Suppose that you ride your bike at an average speed of 12 miles per hour. What distance do you cover in 2 hours? Your distance is the product of your speed and the time that you travel:

$$\frac{12 \text{ miles}}{\cancel{\text{hour}}} \times 2\ \cancel{\text{hours}} = 24 \text{ miles}.$$

Your distance is 24 miles. Notice how the hour units cancel. The distance is expressed in miles.

In general, the distance covered by any moving body is the product of its average speed, or rate, and its time in motion.

A Formula for Motion

$$d = rt$$

Distance equals rate times time.

EXAMPLE 6 Using the Motion Formula

Light travels at a rate of approximately 1.86×10^{5} miles per second. It takes light 5×10^{2} seconds to travel from the sun to Earth. What is the distance between Earth and the sun?

Solution

$d = rt$	Use the motion formula.
$d = (1.86 \times 10^{5}) \times (5 \times 10^{2})$	Substitute the given values.
$d = (1.86 \times 5) \times (10^{5} \times 10^{2})$	Rearrange factors.
$d = 9.3 \times 10^{7}$	Add the exponents on 10 and multiply the other parts.

The distance between Earth and the sun is approximately 9.3×10^{7} miles, or 93 million miles.

CHECK POINT 6 A futuristic spacecraft traveling at 1.55×10^{3} miles per hour takes 20,000 hours (about 833 days) to travel from Venus to Mercury. What is the distance from Venus to Mercury?

1.7 EXERCISE SET

MyMathLab | MathXL PRACTICE | WATCH | DOWNLOAD | READ | REVIEW

Practice Exercises

In Exercises 1–14, write each number in decimal notation without the use of exponents.

1. 3.8×10^2

2. 9.2×10^2

3. 6×10^{-4}

4. 7×10^{-5}

5. -7.16×10^6

6. -8.17×10^6

7. 1.4×10^0

8. 2.4×10^0

9. 7.9×10^{-1}

10. 6.8×10^{-1}

11. -4.15×10^{-3}

12. -3.14×10^{-3}

13. -6.00001×10^{10}

14. -7.00001×10^{10}

In Exercises 15–30, write each number in scientific notation.

15. 32,000

16. 64,000

17. 638,000,000,000,000,000

18. 579,000,000,000,000,000

19. −317

20. −326

21. −5716

22. −3829

23. 0.0027

24. 0.0083

25. −0.00000000504

26. −0.00000000405

27. 0.007

28. 0.005

29. 3.14159

30. 2.71828

In Exercises 31–50, perform the indicated computations. Write the answers in scientific notation. If necessary, round the decimal factor in your scientific notation answer to two decimal places.

31. $(3 \times 10^4)(2.1 \times 10^3)$

32. $(2 \times 10^4)(4.1 \times 10^3)$

33. $(1.6 \times 10^{15})(4 \times 10^{-11})$

34. $(1.4 \times 10^{15})(3 \times 10^{-11})$

35. $(6.1 \times 10^{-8})(2 \times 10^{-4})$

36. $(5.1 \times 10^{-8})(3 \times 10^{-4})$

37. $(4.3 \times 10^8)(6.2 \times 10^4)$

38. $(8.2 \times 10^8)(4.6 \times 10^4)$

39. $\dfrac{8.4 \times 10^8}{4 \times 10^5}$

40. $\dfrac{6.9 \times 10^8}{3 \times 10^5}$

41. $\dfrac{3.6 \times 10^4}{9 \times 10^{-2}}$

42. $\dfrac{1.2 \times 10^4}{2 \times 10^{-2}}$

43. $\dfrac{4.8 \times 10^{-2}}{2.4 \times 10^6}$

44. $\dfrac{7.5 \times 10^{-2}}{2.5 \times 10^6}$

45. $\dfrac{2.4 \times 10^{-2}}{4.8 \times 10^{-6}}$

46. $\dfrac{1.5 \times 10^{-2}}{3 \times 10^{-6}}$

47. $\dfrac{480{,}000{,}000{,}000}{0.00012}$

48. $\dfrac{282{,}000{,}000{,}000}{0.00141}$

49. $\dfrac{0.00072 \times 0.003}{0.00024}$

50. $\dfrac{66{,}000 \times 0.001}{0.003 \times 0.002}$

Practice PLUS

In Exercises 51–58, solve each equation. Express the solution in scientific notation.

51. $(2 \times 10^{-5})x = 1.2 \times 10^9$

52. $(3 \times 10^{-2})x = 1.2 \times 10^4$

53. $\dfrac{x}{2 \times 10^8} = -3.1 \times 10^{-5}$

54. $\dfrac{x}{5 \times 10^{11}} = -2.9 \times 10^{-3}$

55. $x - (7.2 \times 10^{18}) = 9.1 \times 10^{18}$

56. $x - (5.3 \times 10^{-16}) = 8.4 \times 10^{-16}$

57. $(-1.2 \times 10^{-3})x = (1.8 \times 10^{-4})(2.4 \times 10^6)$

58. $(-7.8 \times 10^{-4})x = (3.9 \times 10^{-7})(6.8 \times 10^5)$

Application Exercises

The graph shows the net worth, in billions of dollars, of the five richest Americans. Use 10^9 for one billion and the figures shown to solve Exercises 59–62. Express all answers in scientific notation.

Net Worth of the Five Richest Americans

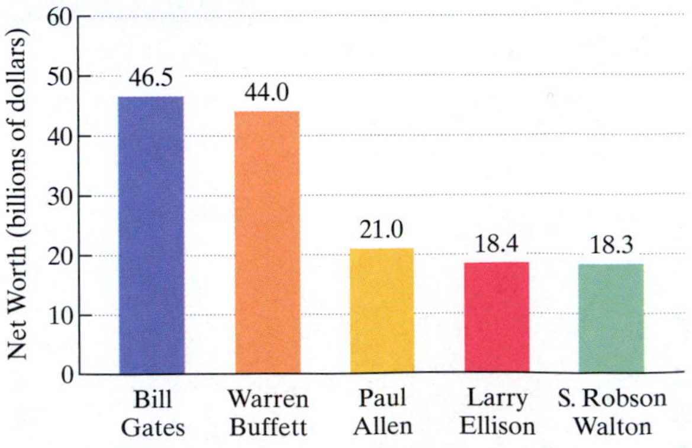

Source: Forbes Billionaires List, 2005

59. How much is Bill Gates worth?

60. How much is Warren Buffett worth?

61. By how much does Larry Ellison's worth exceed that of S. Robson Walton?

62. If each person doubled his net worth, by how much would Larry Ellison's worth exceed that of S. Robson Walton?

Our ancient ancestors hunted for their meat and expended a great deal of energy chasing it down. Today, our animal protein is raised in cages and on feedlots, delivered in great abundance nearly to our door. Use the numbers shown below to solve Exercises 63–66. Use 10^6 for one million and 10^9 for one billion.

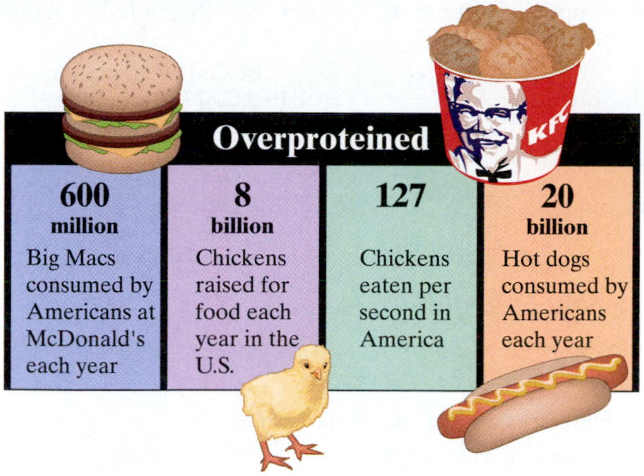

Source: *Time*, October 20, 2003

In Exercises 63–64, use 300 million, or 3×10^8, for the U.S. population. Express answers in decimal notation, rounded, if necessary, to the nearest whole number.

63. Find the number of hot dogs consumed by each American in a year.

64. If the consumption of Big Macs was divided evenly among all Americans, how many Big Macs would we each consume in a year?

In Exercises 65–66, use the fact that there are approximately 3.2×10^7 seconds in a year.

65. How many chickens are raised for food each second in the United States? Express the answer in scientific and decimal notations.

66. How many chickens are eaten per year in the United States? Express the answer in scientific notation.

The graph shows the cost, in billions of dollars, and the enrollment, in millions of people, for various federal social programs in 2005. Use the numbers shown to solve Exercises 67–69.

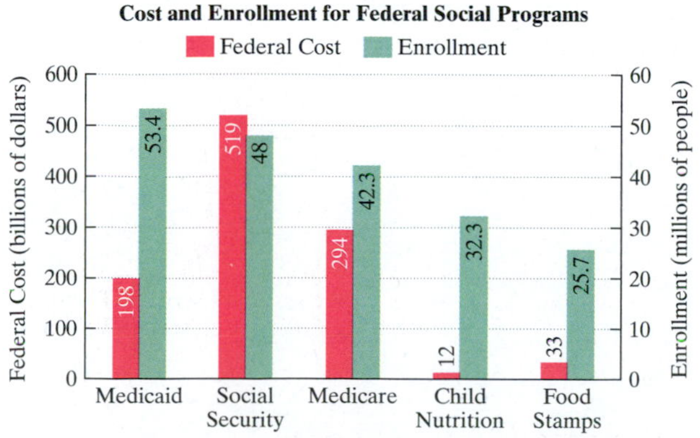

Source: Office of Management and Budget

67. a. What was the average per person benefit for Social Security? Express the answer in scientific notation and in decimal notation, rounded to the nearest dollar.

b. What was the average monthly per person benefit, rounded to the nearest dollar, for Social Security?

68. a. What was the average per person benefit for the food stamps program? Express the answer in scientific notation and in decimal notation, rounded to the nearest dollar.

b. What was the average monthly per person benefit, rounded to the nearest dollar, for the food stamps program?

69. Medicaid provides health insurance for the poor. Medicare provides health insurance for people 65 and older, as well as younger people who are disabled. Which program provides the greater per person benefit? By how much, rounded to the nearest dollar?

70. The area of Alaska is approximately 3.66×10^8 acres. The state was purchased in 1867 from Russia for \$7.2 million. What price per acre, to the nearest cent, did the United States pay Russia?

71. The mass of one oxygen molecule is 5.3×10^{-23} gram. Find the mass of 20,000 molecules of oxygen. Express the answer in scientific notation.

72. The mass of one hydrogen atom is 1.67×10^{-24} gram. Find the mass of 80,000 hydrogen atoms. Express the answer in scientific notation.

73. In Exercises 65–66, we used 3.2×10^7 as an approximation for the number of seconds in a year. Convert 365 days (one year) to hours, to minutes, and, finally, to seconds, to determine precisely how many seconds there are in a year. Express the answer in scientific notation.

Writing in Mathematics

74. How do you know if a number is written in scientific notation?

75. Explain how to convert from scientific to decimal notation and give an example.

76. Explain how to convert from decimal to scientific notation and give an example.

77. Describe one advantage of expressing a number in scientific notation over decimal notation.

Technology Exercises

78. Use a calculator to check any three of your answers in Exercises 1–14.

79. Use a calculator to check any three of your answers in Exercises 15–30.

80. Use a calculator with an EE or EXP key to check any four of your computations in Exercises 31–50. Display the result of the computation in scientific notation.

Critical Thinking Exercises

Make Sense? *In Exercises 81–84, determine whether each statement "makes sense" or "does not make sense" and explain your reasoning.*

81. For a recent year, total tax collections in the United States were $\$2.02 \times 10^7$.

82. I just finished reading a book that contained approximately 1.04×10^5 words.

83. If numbers in the form $a \times 10^n$ are listed from least to greatest, values of a need not appear from least to greatest.

84. When expressed in scientific notation, 58 million and 58 millionths have exponents on 10 with the same absolute value.

In Exercises 85–89, determine whether each statement is true or false. If the statement is false, make the necessary change(s) to produce a true statement.

85. $534.7 = 5.347 \times 10^3$

86. $\frac{8 \times 10^{30}}{4 \times 10^{-5}} = 2 \times 10^{25}$

87. $(7 \times 10^5) + (2 \times 10^{-3}) = 9 \times 10^2$

88. $(4 \times 10^3) + (3 \times 10^2) = 43 \times 10^2$

89. The numbers 8.7×10^{25}, 1.0×10^{26}, 5.7×10^{26}, and 3.7×10^{27} are listed from least to greatest.

In Exercises 90–91, perform the indicated additions. Write the answers in scientific notation.

90. $5.6 \times 10^{13} + 3.1 \times 10^{13}$

91. $8.2 \times 10^{-16} + 4.3 \times 10^{-16}$

92. Our hearts beat approximately 70 times per minute. Express in scientific notation how many times the heart beats over a lifetime of 80 years. Round the decimal factor in your scientific notation answer to two decimal places.

93. Give an example of a number where there is no advantage in using scientific notation over decimal notation.

Review Exercises

94. Simplify: $9(10x - 4) - (5x - 10)$. (Section 1.2, Example 14)

95. Solve: $\frac{4x - 1}{10} = \frac{5x + 2}{4} - 4$. (Section 1.4, Example 4)

96. Simplify: $(8x^4y^{-3})^{-2}$. (Section 1.6, Example 7)

Preview Exercises

Exercises 97–99 will help you prepare for the material covered in the first section of the next chapter.

97. Here are two sets of ordered pairs:

set 1: $\{(1, 5), (2, 5)\}$

set 2: $\{(5, 1), (5, 2)\}$

In which set is each x-coordinate paired with only one y-coordinate?

98. Evaluate $r^3 - 2r^2 + 5$ for $r = -5$.

99. Evaluate $5x + 7$ for $x = a + h$.

GROUP PROJECT

CHAPTER 1

One of the best ways to learn how to *solve* a word problem in algebra is to *design* word problems of your own. Creating a word problem makes you very aware of precisely how much information is needed to solve the problem. You must also focus on the best way to present information to a reader and on how much information to give. As you write your problem, you gain skills that will help you solve problems created by others.

The group should design five different word problems that can be solved using an algebraic equation. All of the problems should be on different topics. For example, the group should not have more than one problem on a price reduction. The group should turn in both the problems and their algebraic solutions.

Chapter 1 Summary

Definitions and Concepts / Examples

Section 1.1 Algebraic Expressions and Real Numbers

Letters that represent numbers are called variables.
An algebraic expression is a combination of variables, numbers, and operation symbols. English phrases can be translated into algebraic expressions:

- Addition: sum, plus, increased by, more than
- Subtraction: difference, minus, decreased by, less than
- Multiplication: product, times, of, twice
- Division: quotient, divide, per, ratio

Translate: Six less than the product of a number and five.

6 less than — the product of a number and 5

$$5x - 6$$

or $x \cdot 5 - 6$

Many algebraic expressions contain exponents. If b is a natural number, b^n, the nth power of b, is the product of n factors of b. Furthermore, $b^1 = b$.
Evaluating an algebraic expression means to find the value of the expression for a given value of the variable.

Evaluate $6 + 5(x - 10)^3$ for $x = 12$.

$$\begin{aligned} &6 + 5(12 - 10)^3 \\ &= 6 + 5 \cdot 2^3 \\ &= 6 + 5 \cdot 8 \\ &= 6 + 40 = 46 \end{aligned}$$

An equation is a statement that two expressions are equal. Formulas are equations that express relationships among two or more variables. Mathematical modeling is the process of finding formulas to describe real-world phenomena. Such formulas, together with the meaning assigned to the variables, are called mathematical models. The formulas are said to model, or describe, the relationships among the variables.

The formula

$$h = -16t^2 + 200t + 4$$

models the height, h, in feet, of fireworks t seconds after launch. What is the height after 2 seconds?

$$\begin{aligned} h &= -16(2)^2 + 200(2) + 4 \\ &= -16(4) + 200(2) + 4 \\ &= -64 + 400 + 4 = 340 \end{aligned}$$

The height is 340 feet.

A set is a collection of objects, called elements, enclosed in braces. The roster method uses commas to separate the elements of the set. Set-builder notation describes the elements of a set, but does not list them. The symbol $\in$ means that a number or object is in a set; $\notin$ means that a number or object is not in a set. The set of real numbers is the set of all numbers that can be represented by points on the number line. Sets that make up the real numbers include

Natural numbers: $\{1, 2, 3, 4, \ldots\}$

Whole numbers: $\{0, 1, 2, 3, 4, \ldots\}$

Integers: $\{\ldots, -4, -3, -2, -1, 0, 1, 2, 3, 4, \ldots\}$

Rational numbers: $\left\{\frac{a}{b} \middle| a \text{ and } b \text{ are integers and } b \neq 0\right\}$

Irrational numbers:
$\{x | x \text{ is a real number and } x \text{ is not a rational number}\}$.

In decimal form, rational numbers terminate or repeat.
In decimal form, irrational numbers do neither.

- Use the roster method to list the elements of

$$\{x | x \text{ is a natural number less than } 6\}.$$

Solution

$$\{1, 2, 3, 4, 5\}$$

- True or false:

$$\sqrt{2} \notin \{x | x \text{ is a rational number}\}.$$

Solution

The statement is true:

$\sqrt{2}$ is not a rational number.

The decimal form of $\sqrt{2}$ neither terminates nor repeats. Thus, $\sqrt{2}$ is an irrational number.

For any two real numbers, a and b, a is less than b if a is to the left of b on the number line.

Inequality Symbols

$<$: is less than
$>$: is greater than
$\leq$: is less than or equal to
$\geq$: is greater than or equal to

- $-1 < 5$, or -1 is less than 5, is true because -1 is to the left of 5 on a number line.
- $-3 \geq 7$, -3 is greater than or equal to 7, is false. Neither $-3 > 7$ nor $-3 = 7$ is true.

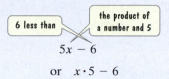

Definitions and Concepts	Examples

Section 1.2 Operations with Real Numbers and Simplifying Algebraic Expressions

Absolute Value

$$|a| = \begin{cases} a & \text{if } a \geq 0 \\ -a & \text{if } a < 0 \end{cases}$$

The opposite, or additive inverse, of a is $-a$. When a is a negative number, $-a$ is positive.

- $|6.03| = 6.03$
- $|0| = 0$
- $|-4.9| = -(-4.9) = 4.9$

Adding Real Numbers

To add two numbers with the same sign, add their absolute values and use their common sign. To add two numbers with different signs, subtract the smaller absolute value from the greater absolute value and use the sign of the number with the greater absolute value.

- $-4.1 + (-6.2) = -10.3$
- $-30 + 25 = -5$
- $12 + (-8) = 4$

Subtracting Real Numbers

$$a - b = a + (-b)$$

$$-\frac{3}{4} - \left(-\frac{1}{2}\right) = -\frac{3}{4} + \frac{1}{2} = -\frac{3}{4} + \frac{2}{4} = -\frac{1}{4}$$

Multiplying and Dividing Real Numbers

The product or quotient of two numbers with the same sign is positive and with different signs is negative. If no number is 0, a product with an even number of negative factors is positive and a product with an odd number of negative factors is negative. Division by 0 is undefined.

- $2(-6)(-1)(-5) = -60$

  Three (odd) negative factors give a negative product.
- $(-2)^3 = (-2)(-2)(-2) = -8$
- $-\frac{1}{3}\left(-\frac{2}{5}\right) = \frac{2}{15}$
- $\frac{-14}{2} = -7$

Order of Operations

1. Perform operations within grouping symbols, starting with the innermost grouping symbols. Grouping symbols include parentheses, brackets, fraction bars, absolute value symbols, and square root signs.
2. Evaluate exponential expressions.
3. Multiply and divide from left to right.
4. Add and subtract from left to right.

Simplify: $\frac{6(8-10)^3 + (-2)}{(-5)^2(-2)}$.

$$= \frac{6(-2)^3 + (-2)}{(-5)^2(-2)} = \frac{6(-8) + (-2)}{25(-2)}$$

$$= \frac{-48 + (-2)}{-50} = \frac{-50}{-50} = 1$$

Basic Algebraic Properties

Commutative: $a + b = b + a$

$ab = ba$

Associative: $(a + b) + c = a + (b + c)$

$(ab)c = a(bc)$

Distributive: $a(b + c) = ab + ac$

$a(b - c) = ab - ac$

$(b + c)a = ba + ca$

- Commutative

  $3x + 5 = 5 + 3x = 5 + x \cdot 3$
- Associative

  $-4(6x) = (-4 \cdot 6)x = -24x$
- Distributive

  $$-4(9x + 3) = -4(9x) + (-4) \cdot 3$$
  $$= -36x + (-12)$$
  $$= -36x - 12$$

Definitions and Concepts / Examples

Section 1.2 Operations with Real Numbers and Simplifying Algebraic Expressions (continued)

Simplifying Algebraic Expressions

Terms are separated by addition. Like terms have the same variable factors and are combined using the distributive property. An algebraic expression is simplified when grouping symbols have been removed and like terms have been combined.

Simplify: $7(3x - 4) - (10x - 5)$.

$$= 21x - 28 - 10x + 5$$
$$= 21x - 10x - 28 + 5$$
$$= 11x - 23$$

Section 1.3 Graphing Equations

The rectangular coordinate system consists of a horizontal number line, the x-axis, and a vertical number line, the y-axis, intersecting at their zero points, the origin. Each point in the system corresponds to an ordered pair of real numbers (x, y). The first number in the pair is the x-coordinate; the second number is the y-coordinate.

Plot: $(4, 2), (-3, 4), (-5, -4)$, and $(4, -3)$.

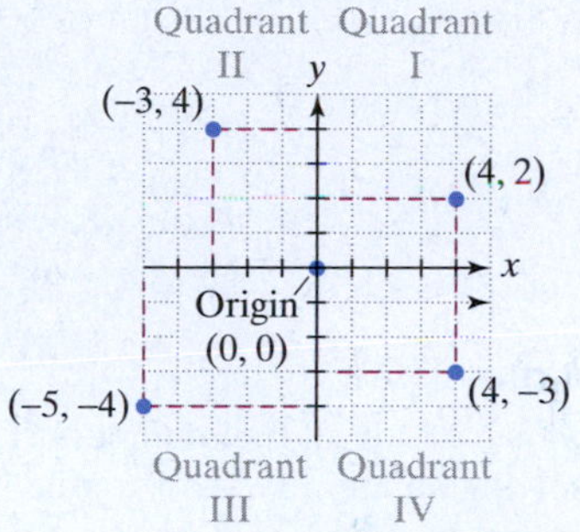

An ordered pair is a solution of an equation in two variables if replacing the variables by the corresponding coordinates results in a true statement. The ordered pair is said to satisfy the equation. The graph of the equation is the set of all points whose coordinates satisfy the equation. One method for graphing an equation is to plot ordered-pair solutions and connect them with a smooth curve or line.

Graph: $y = x^2 - 1$.

x	$y = x^2 - 1$
-2	$(-2)^2 - 1 = 3$
-1	$(-1)^2 - 1 = 0$
0	$0^2 - 1 = -1$
1	$1^2 - 1 = 0$
2	$2^2 - 1 = 3$

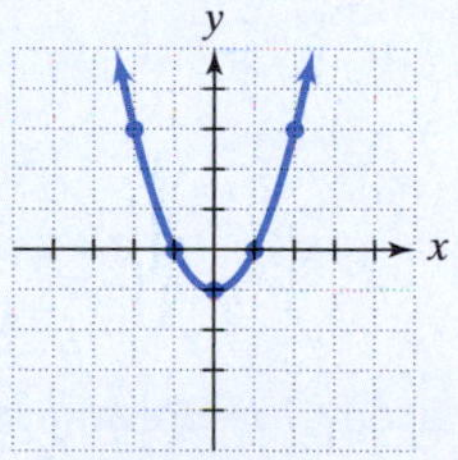

Section 1.4 Solving Linear Equations

A linear equation in one variable can be written in the form $ax + b = 0, a \neq 0$. A solution is a value of the variable that makes the equation a true statement. The set of all such solutions is the equation's solution set. Equivalent equations have the same solution set. To solve a linear equation,

1. Simplify each side.
2. Collect variable terms on one side and constant terms on the other side.
3. Isolate the variable and solve.
4. Check the proposed solution in the original equation.

Solve: $4(x - 5) = 2x - 14$.

$$4x - 20 = 2x - 14$$
$$4x - 2x - 20 = 2x - 2x - 14$$
$$2x - 20 = -14$$
$$2x - 20 + 20 = -14 + 20$$
$$2x = 6$$
$$\frac{2x}{2} = \frac{6}{2}$$
$$x = 3$$

Checking gives $-8 = -8$, so 3 is the solution, or $\{3\}$ is the solution set.

Definitions and Concepts	Examples

Section 1.4 Solving Linear Equations (continued)

Equations Containing Fractions

Multiply both sides (all terms) by the least common denominator. This clears the equation of fractions.

Solve: $\frac{x-2}{5} + \frac{x+2}{2} = \frac{x+4}{3}$.

$$30\left(\frac{x-2}{5} + \frac{x+2}{2}\right) = 30\left(\frac{x+4}{3}\right)$$

$$6(x-2) + 15(x+2) = 10(x+4)$$

$$6x - 12 + 15x + 30 = 10x + 40$$

$$21x + 18 = 10x + 40$$

$$11x = 22$$

$$x = 2$$

Checking gives $2 = 2$, so 2 is the solution, or $\{2\}$ is the solution set.

Types of Equations

An equation that is true for all real numbers, $\mathbb{R}$, is called an identity. When solving an identity, the variable is eliminated and a true statement, such as $3 = 3$, results. An equation that is not true for even one real number is called an inconsistent equation. A false statement, such as $3 = 7$, results when solving such an equation, whose solution set is $\varnothing$, the empty set. A conditional equation is not an identity, but is true for at least one real number.

Solve: $4x + 5 = 4(x + 2)$.

$$4x + 5 = 4x + 8$$

$$5 = 8, \quad \text{false}$$

The inconsistent equation has no solution: $\varnothing$.

Solve: $5x - 4 = 5(x + 1) - 9$.

$$5x - 4 = 5x + 5 - 9$$

$$5x - 4 = 5x - 4$$

$$-4 = -4, \quad \text{true}$$

All real numbers satisfy the identity: $\mathbb{R}$.

Section 1.5 Problem Solving and Using Formulas

Strategy for Solving Algebraic Word Problems

1. Let x represent one of the quantities.
2. Represent other unknown quantities in terms of x.
3. Write an equation that models the conditions.
4. Solve the equation and answer the question.
5. Check the proposed solution in the original wording of the problem.

After a 60% reduction, a suit sold for \$32. What was the original price?

Let $x =$ the original price.

Original price	minus	60% reduction	=	reduced price
x	$-$	$0.6x$	$=$	32

$$0.4x = 32$$

$$\frac{0.4x}{0.4} = \frac{32}{0.4}$$

$$x = 80$$

The original price was \$80. Check this amount using the first sentence in the problem's conditions.

To solve a formula for a variable, use the steps for solving a linear equation and isolate that variable on one side of the equation.

Solve for r: $E = I(R + r)$.

$$E = IR + Ir$$ We need to isolate r.

$$E - IR = Ir$$

$$\frac{E - IR}{I} = r$$

Definitions and Concepts	Examples
Section 1.6 Properties of Integral Exponents	
The Product Rule $b^m \cdot b^n = b^{m+n}$	$(-3x^{10})(5x^{20}) = -3 \cdot 5x^{10+20} = -15x^{30}$
The Quotient Rule $\frac{b^m}{b^n} = b^{m-n}, b \neq 0$	$\frac{5x^{20}}{10x^{10}} = \frac{5}{10} \cdot x^{20-10} = \frac{x^{10}}{2}$
Zero and Negative Exponents $b^0 = 1, b \neq 0$; $b^{-n} = \frac{1}{b^n}$ and $\frac{1}{b^{-n}} = b^n$	• $(3x)^0 = 1$ • $3x^0 = 3 \cdot 1 = 3$ • $\frac{2^{-3}}{4^{-2}} = \frac{4^2}{2^3} = \frac{16}{8} = 2$
Power Rule $(b^m)^n = b^{mn}$	$(x^5)^{-4} = x^{5(-4)} = x^{-20} = \frac{1}{x^{20}}$
Products to Powers $(ab)^n = a^n b^n$	$(5x^3y^{-4})^{-2} = 5^{-2} \cdot (x^3)^{-2} \cdot (y^{-4})^{-2} = 5^{-2}x^{-6}y^8 = \frac{y^8}{5^2x^6} = \frac{y^8}{25x^6}$
Quotients to Powers $\left(\frac{a}{b}\right)^n = \frac{a^n}{b^n}$	$\left(\frac{2}{x^3}\right)^{-4} = \frac{2^{-4}}{(x^3)^{-4}} = \frac{2^{-4}}{x^{-12}} = \frac{x^{12}}{2^4} = \frac{x^{12}}{16}$
An exponential expression is simplified when • No parentheses appear. • No powers are raised to powers. • Each base occurs only once. • No negative or zero exponents appear.	Simplify: $\frac{-5x^{-3}y^2}{-20x^2y^{-6}}$. $= \frac{-5}{-20} \cdot x^{-3-2} \cdot y^{2-(-6)}$ $= \frac{1}{4}x^{-5}y^8 = \frac{y^8}{4x^5}$
Section 1.7 Scientific Notation	
A number in scientific notation is expressed in the form $a \times 10^n$, where $\lvert a \rvert$ is greater than or equal to 1 and less than 10, and n is an integer.	Write in decimal notation: 3.8×10^{-3}. $3.8 \times 10^{-3} = .0038 = 0.0038$ Write in scientific notation: 26,000. $26{,}000 = 2.6 \times 10^4$
Computations with Numbers in Scientific Notation $(a \times 10^n)(b \times 10^m) = (a \times b) \times 10^{n+m}$ $\frac{a \times 10^n}{b \times 10^m} = \left(\frac{a}{b}\right) \times 10^{n-m}$	$(8 \times 10^3)(5 \times 10^{-8})$ $= 8 \cdot 5 \times 10^{3+(-8)}$ $= 40 \times 10^{-5}$ $= (4 \times 10^1) \times 10^{-5} = 4 \times 10^{-4}$

CHAPTER 1 REVIEW EXERCISES

1.1 *In Exercises 1–3, write each English phrase as an algebraic expression. Let x represent the number.*

1. Ten less than twice a number
2. Four more than the product of six and a number
3. The quotient of nine and a number, increased by half of the number

In Exercises 4–6, evaluate each algebraic expression for the given value or values of the variable.

4. $x^2 - 7x + 4$, for $x = 10$
5. $6 + 2(x - 8)^3$, for $x = 11$
6. $x^4 - (x - y)$, for $x = 2$ and $y = 1$

In Exercises 7–8, use the roster method to list the elements in each set.

7. $\{x|x$ is a natural number less than $3\}$
8. $\{x|x$ is an integer greater than -4 and less than $2\}$

In Exercises 9–11, determine whether each statement is true or false.

9. $0 \in \{x|x$ is a natural number$\}$
10. $-2 \in \{x|x$ is a rational number$\}$
11. $\frac{1}{3} \notin \{x|x$ is an irrational number$\}$

In Exercises 12–14, write out the meaning of each inequality. Then determine whether the inequality is true or false.

12. $-5 < 2$
13. $-7 \geq -3$
14. $-7 \leq -7$
15. You are riding along an expressway traveling x miles per hour. The formula

$$S = 0.015x^2 + x + 10$$

models the recommended safe distance, S, in feet, between your car and other cars on the expressway. What is the recommended safe distance when your speed is 60 miles per hour?

1.2 *In Exercises 16–18, find each absolute value.*

16. $|-9.7|$
17. $|5.003|$
18. $|0|$

In Exercises 19–30, perform the indicated operation.

19. $-2.4 + (-5.2)$
20. $-6.8 + 2.4$
21. $-7 - (-20)$
22. $(-3)(-20)$
23. $-\frac{3}{5} - \left(-\frac{1}{2}\right)$
24. $\left(\frac{2}{7}\right)\left(-\frac{3}{10}\right)$
25. $4(-3)(-2)(-10)$
26. $(-2)^4$
27. -2^5
28. $-\frac{2}{3} \div \frac{8}{5}$
29. $\frac{-35}{-5}$
30. $\frac{54.6}{-6}$
31. Find $-x$ if $x = -7$.

In Exercises 32–38, simplify each expression.

32. $-11 - [-17 + (-3)]$
33. $\left(-\frac{1}{2}\right)^3 \cdot 2^4$
34. $-3[4 - (6 - 8)]$
35. $8^2 - 36 \div 3^2 \cdot 4 - (-7)$
36. $\frac{(-2)^4 + (-3)^2}{2^2 - (-21)}$
37. $\frac{(7 - 9)^3 - (-4)^2}{2 + 2(8) \div 4}$
38. $4 - (3 - 8)^2 + 3 \div 6 \cdot 4^2$

In Exercises 39–43, simplify each algebraic expression.

39. $5(2x - 3) + 7x$
40. $5x + 7x^2 - 4x + 2x^2$
41. $3(4y - 5) - (7y + 2)$
42. $8 - 2[3 - (5x - 1)]$
43. $6(2x - 3) - 5(3x - 2)$

1.3 *In Exercises 44–46, plot the given point in a rectangular coordinate system.*

44. $(-1, 3)$
45. $(2, -5)$
46. $(0, -6)$

In Exercises 47–50, graph each equation. Let $x = -3, -2, -1, 0, 1, 2,$ and 3.

47. $y = 2x - 2$
48. $y = x^2 - 3$
49. $y = x$
50. $y = |x| - 2$
51. What does a $[-20, 40, 10]$ by $[-5, 5, 1]$ viewing rectangle mean? Draw axes with tick marks and label the tick marks to illustrate this viewing rectangle.

The caseload of Alzheimer's disease in the United States is expected to explode as baby boomers head into their later years. The graph shows the percentage of Americans with the disease, by age. Use the graph to solve Exercises 52–54.

Alzheimer's Prevalence in the United States, by Age

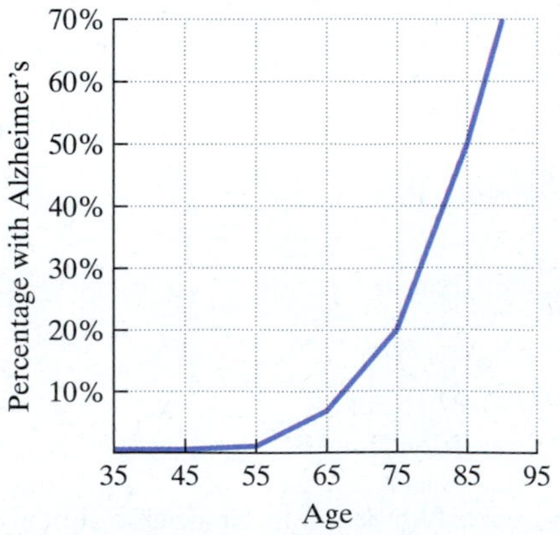

Source: Centers for Disease Control

52. What percentage of Americans who are 75 have Alzheimer's disease?
53. What age represents 50% prevalence of Alzheimer's disease?

54. Describe the trend shown by the graph.

55. Select the graph that best illustrates the following description: A train pulls into a station and lets off its passengers.

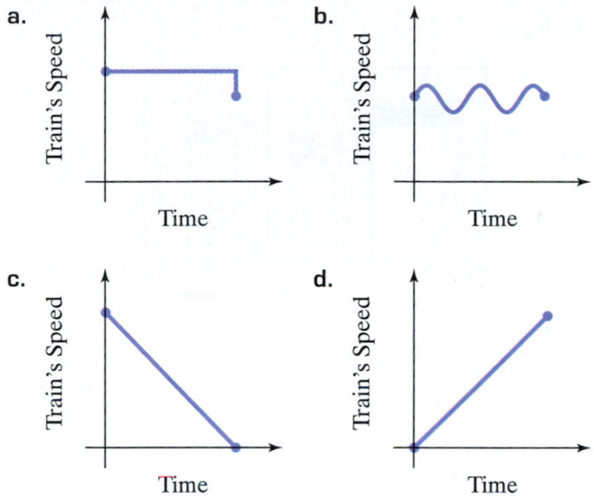

1.4 *In Exercises 56–61, solve and check each linear equation.*

56. $2x - 5 = 7$

57. $5x + 20 = 3x$

58. $7(x - 4) = x + 2$

59. $1 - 2(6 - x) = 3x + 2$

60. $2(x - 4) + 3(x + 5) = 2x - 2$

61. $2x - 4(5x + 1) = 3x + 17$

In Exercises 62–66, solve each equation.

62. $\frac{2x}{3} = \frac{x}{6} + 1$

63. $\frac{x}{2} - \frac{1}{10} = \frac{x}{5} + \frac{1}{2}$

64. $\frac{2x}{3} = 6 - \frac{x}{4}$

65. $\frac{x}{4} = 2 + \frac{x - 3}{3}$

66. $\frac{3x + 1}{3} - \frac{13}{2} = \frac{1 - x}{4}$

In Exercises 67–71, solve each equation. Then state whether the equation is an identity, a conditional equation, or an inconsistent equation.

67. $7x + 5 = 5(x + 3) + 2x$

68. $7x + 13 = 4x - 10 + 3x + 23$

69. $7x + 13 = 3x - 10 + 2x + 23$

70. $4(x - 3) + 5 = x + 5(x - 2)$

71. $(2x - 3)2 - 3(x + 1) = (x - 2)4 - 3(x + 5)$

72. The bar graph shows the number of corporations that owned the majority of the media industry in the United States for selected years. Through mergers and buyouts, by 2004, a majority of American newspapers, magazines, TV and radio stations, book publishers, and movie studios were owned by just five corporations: Time Warner, Disney, News Corp., Bertelsmann, and Viacom.

Number of Corporations Owning the Majority of United States Media

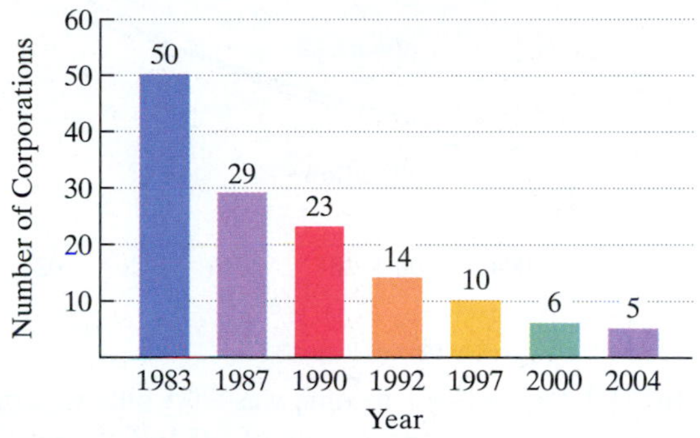

Source: Jonathan Teller-Elsberg et al., *Field Guide to the U.S. Economy*, The New Press, 2006

The data can be modeled by the formula $N = -2x + 40$, in which N represents the number of corporations owning the majority of U.S. media x years after 1983.

a. According to the model, in which year were there 12 corporations that owned the majority of media available to Americans?

b. Does the information provided by the model in part (a) overestimate or underestimate the number of corporations indicated by the graph? By how much?

1.5 *In Exercises 73–79, use the five-step strategy for solving word problems.*

73. The bar graph represents money earned, in millions of dollars, by each of the top five concert tours in 2005. Gross earnings by U2 exceeded earnings by The Eagles by \$143 million. Gross earnings by The Rolling Stones exceeded The Eagles' earnings by \$24 million. Combined, these three groups earned \$518 million on their concert tours. Determine the gross earnings on concert tours, in millions of dollars, for each of the three groups.

Top United States Concert Tours, 2005

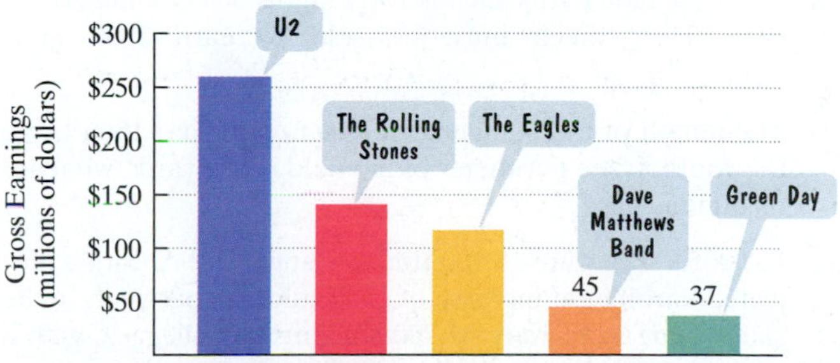

Source: Billboard magazine

74. One angle of a triangle measures 10° more than the second angle. The measure of the third angle is twice the sum of the measures of the first two angles. Determine the measure of each angle.

75. Without changes, the graphs show projections for the amount being paid in Social Security benefits and the amount going into the system. All data are expressed in billions of dollars.

Social Insecurity: Projected Income and Outflow of the Social Security System

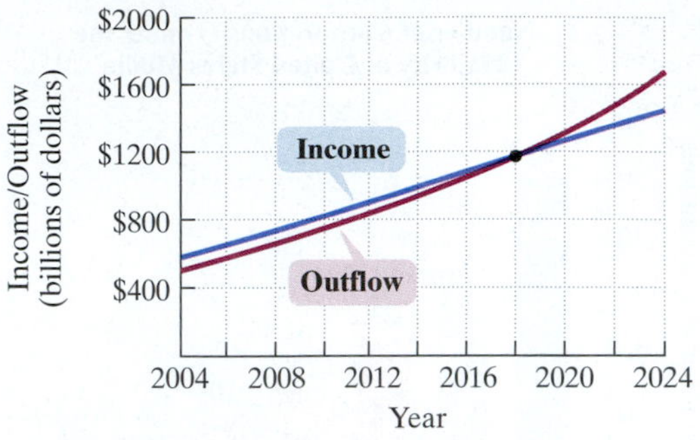

Source: 2004 Social Security Trustees Report

a. In 2004, the system's income was \$575 billion, projected to increase at an average rate of \$43 billion per year. In which year will the system's income be \$1177 billion?

b. The data for the system's outflow can be modeled by the formula

$$B = 0.07x^2 + 47.4x + 500,$$

where B represents the amount paid in benefits, in billions of dollars, x years after 2004. According to this model, what will be the amount paid in benefits for the year you determined in part (a)? Round to the nearest billion dollar.

c. How are your answers to parts (a) and (b) shown by the graphs?

76. You are choosing between two long-distance telephone plans. One plan has a monthly fee of \$15 with a charge of \$0.05 per minute. The other plan has a monthly fee of \$5 with a charge of \$0.07 per minute. For how many minutes of long-distance calls will the costs for the two plans be the same?

77. After a 20% price reduction, a cordless phone sold for \$48. What was the phone's price before the reduction?

78. A salesperson earns \$300 per week plus 5% commission of sales. How much must be sold to earn \$800 in a week?

79. The length of a rectangular field is 6 yards less than triple the width. If the perimeter of the field is 340 yards, what are its dimensions?

80. In 2005, there were 14,100 students at college A, with a projected enrollment increase of 1500 students per year. In the same year, there were 41,700 students at college B, with a projected enrollment decline of 800 students per year.

a. Let x represent the number of years after 2005. Write, but do not solve, an equation that can be used to find how many years after 2005 the colleges will have the same enrollment.

b. The following table is based on your equation in part (a). Y_1 represents one side of the equation and Y_2 represents the other side of the equation. Use the table to answer these questions: In which year will the colleges have the same enrollment? What will be the enrollment in each college at that time?

X	Y1	Y2
7	24600	36100
8	26100	35300
9	27600	34500
10	29100	33700
11	30600	32900
12	32100	32100
13	33600	31300

X=7

In Exercises 81–86, solve each formula for the specified variable.

81. $V = \frac{1}{3}Bh$ for h

82. $y - y_1 = m(x - x_1)$ for x

83. $E = I(R + r)$ for R

84. $C = \frac{5F - 160}{9}$ for F

85. $s = vt + gt^2$ for g

86. $T = gr + gvt$ for g

1.6 *In Exercises 87–101, simplify each exponential expression. Assume that no denominators are 0.*

87. $(-3x^7)(-5x^6)$

88. x^2y^{-5}

89. $\frac{3^{-2}x^4}{y^{-7}}$

90. $(x^3)^{-6}$

91. $(7x^3y)^2$

92. $\frac{16y^3}{-2y^{10}}$

93. $(-3x^4)(4x^{-11})$

94. $\frac{12x^7}{4x^{-3}}$

95. $\frac{-10a^5b^6}{20a^{-3}b^{11}}$

96. $(-3xy^4)(2x^2)^3$

97. $2^{-2} + \frac{1}{2}x^0$

98. $(5x^2y^{-4})^{-3}$

99. $(3x^4y^{-2})(-2x^5y^{-3})$

100. $\left(\frac{3xy^3}{5x^{-3}y^{-4}}\right)^2$

101. $\left(\frac{-20x^{-2}y^3}{10x^5y^{-6}}\right)^{-3}$

1.7 *In Exercises 102–103, write each number in decimal notation.*

102. 7.16×10^6

103. 1.07×10^{-4}

In Exercises 104–105, write each number in scientific notation.

104. −41,000,000,000,000

105. 0.00809

In Exercises 106–107, perform the indicated computations. Write the answers in scientific notation.

106. $(4.2 \times 10^{13})(3 \times 10^{-6})$

107. $\dfrac{5 \times 10^{-6}}{20 \times 10^{-8}}$

108. The human body contains approximately 3.2×10^4 microliters of blood for every pound of body weight. Each microliter of blood contains approximately 5×10^6 red blood cells. Express in scientific notation the approximate number of red blood cells in the body of a 180-pound person.

CHAPTER 1 TEST

Remember to use your Chapter Test Prep Video CD to see the worked-out solutions to the test questions you want to review.

1. Write the following English phrase as an algebraic expression:

Five less than the product of a number and four.

Let x represent the number.

2. Evaluate $8 + 2(x - 7)^4$, for $x = 10$.

3. Use the roster method to list the elements in the set:

$$\{x \mid x \text{ is a negative integer greater than } -5\}.$$

4. Determine whether the following statement is true or false:

$$\frac{1}{4} \notin \{x \mid x \text{ is a natural number}\}.$$

5. Write out the meaning of the inequality $-3 > -1$. Then determine whether the inequality is true or false.

6. The bar graph shows the number of billionaires in the United States from 2000 through 2004.

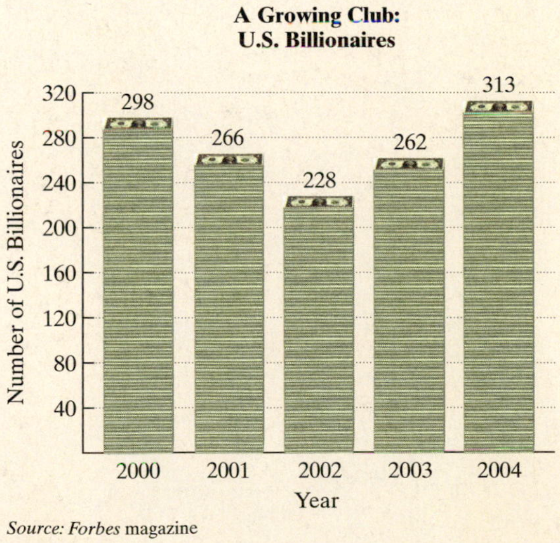

Source: Forbes magazine

The formula

$$N = 17x^2 - 65.4x + 302.2$$

models the number of billionaires, N, in the United States, x years after 2000. According to the formula, how many U.S. billionaires, to the nearest whole number, were there in 2003? Does the formula overestimate or underestimate the actual number shown by the bar graph? By how much?

7. Find the absolute value: $|-17.9|$.

In Exercises 8–12, perform the indicated operation.

8. $-10.8 + 3.2$

9. $-\frac{1}{4} - \left(-\frac{1}{2}\right)$

10. $2(-3)(-1)(-10)$

11. $-\frac{1}{4}\left(-\frac{1}{2}\right)$

12. $\dfrac{-27.9}{-9}$

In Exercises 13–18, simplify each expression.

13. $24 - 36 \div 4 \cdot 3$

14. $(5^2 - 2^4) + [9 \div (-3)]$

15. $\dfrac{(8 - 10)^3 - (-4)^2}{2 + 8(2) \div 4}$

16. $7x - 4(3x + 2) - 10$

17. $5(2y - 6) - (4y - 3)$

18. $9x - [10 - 4(2x - 3)]$

19. Plot $(-2, -4)$ in a rectangular coordinate system.

20. Graph $y = x^2 - 4$ in a rectangular coordinate system.

In Exercises 21–23, solve each equation. If the solution set is ∅ or ℝ, classify the equation as an inconsistent equation or an identity.

21. $3(2x - 4) = 9 - 3(x + 1)$

22. $\dfrac{2x - 3}{4} = \dfrac{x - 4}{2} - \dfrac{x + 1}{4}$

23. $3(x - 4) + x = 2(6 + 2x)$

In Exercises 24–28, use the five-step strategy for solving word problems.

24. Find two numbers such that the second number is 3 more than twice the first number and the sum of the two numbers is 72.

25. You bought a new car for $13,805. Its value is decreasing by $1820 per year. After how many years will its value be $4705?

26. Photo Shop A charges $1.60 to develop a roll of film plus $0.11 for each print. Photo Shop B charges $1.20 to develop a roll of film plus $0.13 per print. For how many prints will the amount spent at each photo shop be the same? What will be that amount?

27. After a 60% reduction, a jacket sold for \$20. What was the jacket's price before the reduction?

28. The length of a rectangular field exceeds the width by 260 yards. If the perimeter of the field is 1000 yards, what are its dimensions?

In Exercises 29–30, solve each formula for the specified variable.

29. $V = \frac{1}{3}lwh$ for h

30. $Ax + By = C$ for y

In Exercises 31–35, simplify each exponential expression.

31. $(-2x^5)(7x^{-10})$

32. $(-8x^{-5}y^{-3})(-5x^2y^{-5})$

33. $\frac{-10x^4y^3}{-40x^{-2}y^6}$

34. $(4x^{-5}y^2)^{-3}$

35. $\left(\frac{-6x^{-5}y}{2x^3y^{-4}}\right)^{-2}$

36. Write in decimal notation: 3.8×10^{-6}.

37. Write in scientific notation: 407,000,000,000.

38. Divide and write the answer in scientific notation:

$$\frac{4 \times 10^{-3}}{8 \times 10^{-7}}.$$

39. In 2006, world population was approximately 6.5×10^9. By some projections, world population will double by 2080. Express the population at that time in scientific notation.

CHAPTER 2

Functions and Linear Functions

A vast expanse of open water at the top of our world was once covered with ice. The melting of the Arctic ice caps has forced polar bears to swim as far as 40 miles, causing them to drown in significant numbers. Such deaths were rare in the past.

There is strong scientific consensus that human activities are changing the Earth's climate. Scientists now believe that there is a striking correlation between atmospheric carbon dioxide concentration and global temperature. As both of these variables increase at significant rates, there are warnings of a planetary emergency that threatens to condemn coming generations to a catastrophically diminished future.*

In this chapter, you'll learn to approach our climate crisis mathematically by creating formulas, called functions, that model data for average global temperature and carbon dioxide concentration over time. Understanding the concept of a function will give you a new perspective on many situations, ranging from global warming to using mathematics in a way that is similar to making a movie.

**Sources: Al Gore, An Inconvenient Truth, Rodale, 2006; Time, April 3, 2006*

Mathematical models involving global warming are developed in Exercises 69–70 in Exercise Set 2.5. Using mathematics in a way that is similar to making a movie is discussed in the Blitzer Bonus on page 138.

SECTION 2.1

Introduction to Functions

Objectives

1. Find the domain and range of a relation.
2. Determine whether a relation is a function.
3. Evaluate a function.

Actors Tommy Lee Jones and Will Smith

Top U.S. Last Names

Name	% of All Names
Smith	1.006%
Johnson	0.810%
Williams	0.699%
Brown	0.621%
Jones	0.621%

Source: Russell Ash, *The Top 10 of Everything 2006*

The top five U.S. last names shown above account for nearly 4% of the entire population. The table indicates a correspondence between a last name and the percentage of Americans who share that name. We can write this correspondence using a set of ordered pairs:

{(Smith, 1.006%), (Johnson, 0.810%), (Williams, 0.699%), (Brown, 0.621%), (Jones, 0.621%)}.

These braces indicate we are representing a set.

The mathematical term for a set of ordered pairs is a *relation*.

Definition of a Relation

A **relation** is any set of ordered pairs. The set of all first components of the ordered pairs is called the **domain** of the relation and the set of all second components is called the **range** of the relation.

1 Find the domain and range of a relation.

EXAMPLE 1 Finding the Domain and Range of a Relation

Find the domain and range of the relation:

{(Smith, 1.006%), (Johnson, 0.810%), (Williams, 0.699%), (Brown, 0.621%), (Jones, 0.621%)}.

Solution The domain is the set of all first components. Thus, the domain is

{Smith, Johnson, Williams, Brown, Jones}.

The range is the set of all second components. Thus, the range is

{1.006%, 0.810%, 0.699%, 0.621%}.

Although Brown and Jones are both shared by 0.621% of the U.S. population, it is not necessary to list 0.621% twice.

✓ **CHECK POINT 1** Find the domain and the range of the relation:

$$\{(0, 9.1), (10, 6.7), (20, 10.7), (30, 13.2), (34, 15.5)\}.$$

As you worked Check Point 1, did you wonder if there was a rule that assigned the "inputs" in the domain to the "outputs" in the range? For example, for the ordered pair (30, 13.2), how does the output 13.2 depend on the input 30? The ordered pair is based on the data in **Figure 2.1(a)**, which shows the percentage of first-year U.S. college students claiming no religious affiliation.

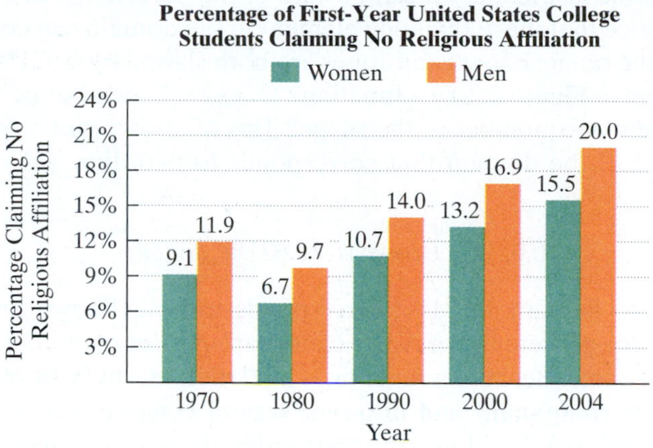

FIGURE 2.1(a)   Data for women and men

Source: John Macionis, *Sociology, 11th Edition*, Prentice Hall, 2007

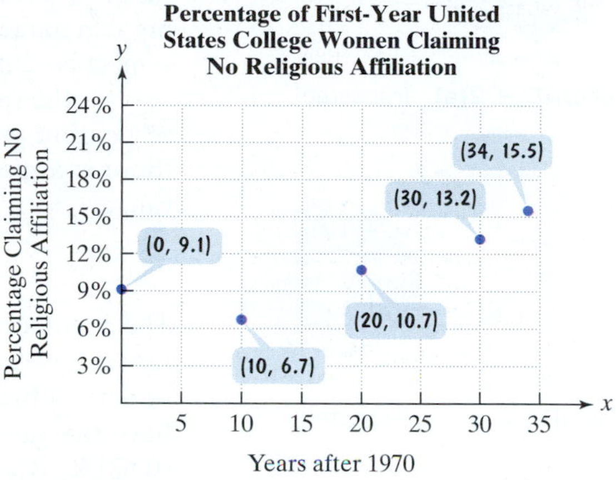

FIGURE 2.1(b)   Visually representing the relation for women's data

In **Figure 2.1(b)**, we used the data for college women to create the following ordered pairs:

$$\left(\text{years after 1970,} \quad \begin{array}{l} \text{percentage of first-year college} \\ \text{women claiming no religious} \\ \text{affiliation} \end{array} \right).$$

Consider, for example, the ordered pair (30, 13.2).

(30, 13.2)

30 years after 1970, or in 2000,   13.2% of first-year college women claimed no religious affiliation.

The five points in **Figure 2.1(b)** visually represent the relation formed from the women's data. Another way to visually represent the relation is as follows:

Domain		Range
0	→	9.1
10	→	6.7
20	→	10.7
30	→	13.2
34	→	15.5

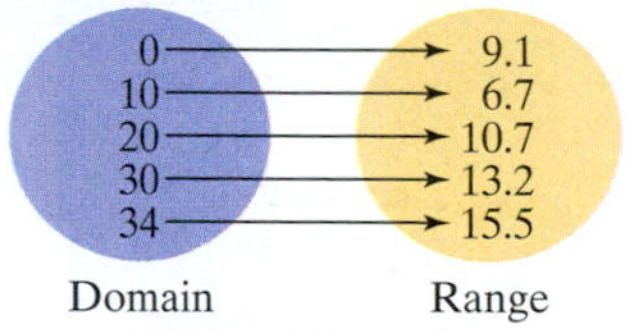

2   Determine whether a relation is a function.

Functions

Shown, again, in the margin are the top five U.S. last names and the percentage of Americans who share those names. We've used this information to define two relations. **Figure 2.2(a)** shows a correspondence between last names and percents sharing those names. **Figure 2.2(b)** shows a correspondence between percents sharing last names and those last names.

Top U.S. Last Names

Name	% of All Names
Smith	1.006%
Johnson	0.810%
Williams	0.699%
Brown	0.621%
Jones	0.621%

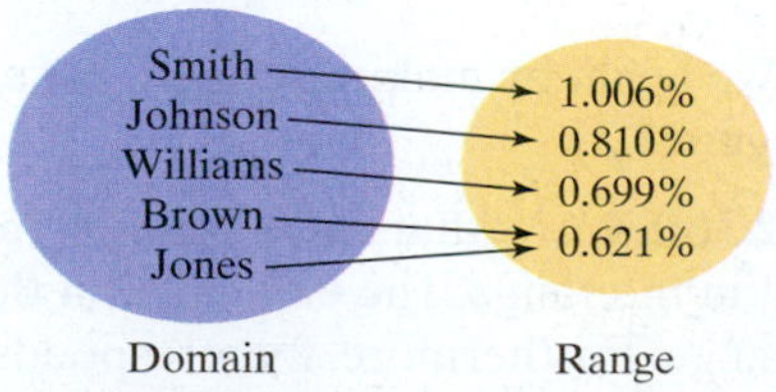

FIGURE 2.2(a)   Names correspond to percents.

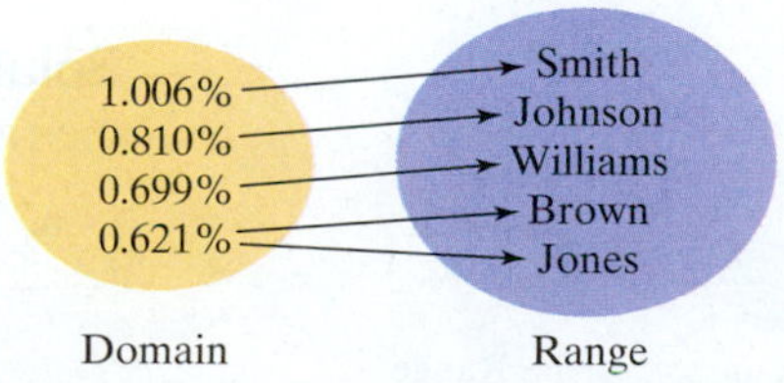

FIGURE 2.2(b)   Percents correspond to names.

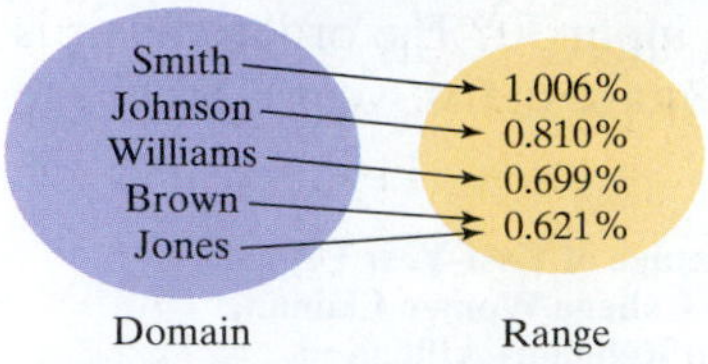

FIGURE 2.2(a) (repeated)

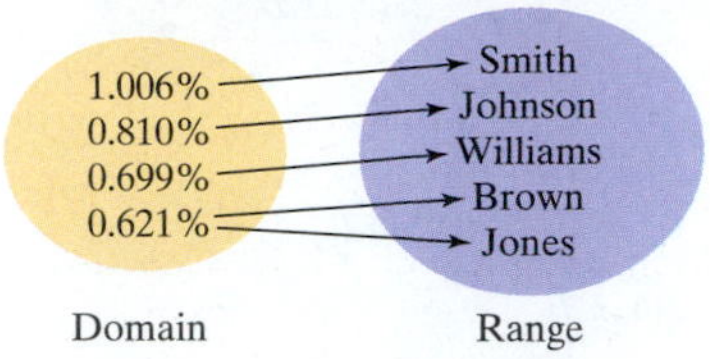

FIGURE 2.2(b) (repeated)

A relation in which each member of the domain corresponds to exactly one member of the range is a **function**. Can you see that the relation in **Figure 2.2(a)** is a function? Each last name in the domain corresponds to exactly one percent in the range. If we know the last name, we can be sure of the percentage of Americans sharing that name. Notice that more than one element in the domain can correspond to the same element in the range: Brown and Jones are both shared by 0.621% of Americans.

Is the relation in **Figure 2.2(b)** a function? Does each member of the domain correspond to precisely one member of the range? This relation is not a function because there is a member of the domain that corresponds to two different members of the range:

$$(0.621\%, \text{Brown}) \quad (0.621\%, \text{Jones}).$$

The member of the domain, 0.621%, corresponds to both Brown and Jones in the range. If we know the percentage of Americans sharing a last name, 0.621%, we cannot be sure of that last name. Because **a function is a relation in which no two ordered pairs have the same first component and different second components**, the ordered pairs (0.621%, Brown) and (0.621%, Jones) are not ordered pairs of a function.

Same first component

$$(0.621\%, \text{Brown}) \quad (0.621\%, \text{Jones})$$

Different second components

Definition of a Function

A **function** is a correspondence from a first set, called the **domain**, to a second set, called the **range**, such that each element in the domain corresponds to *exactly one* element in the range.

In Check Point 1, we considered a relation that gave a correspondence between years after 1970 and the percentage of first-year college women claiming no religious affiliation. Can you see that this relation is a function?

Each element in the domain

$$\{(0, 9.1), (10, 6.7), (20, 10.7), (30, 13.2), (34, 15.5)\}$$

corresponds to exactly one element in the range.

However, Example 2 illustrates that not every correspondence between sets is a function.

EXAMPLE 2 Determining Whether a Relation Is a Function

Determine whether each relation is a function:

a. $\{(1, 5), (2, 5), (3, 7), (4, 8)\}$ **b.** $\{(5, 1), (5, 2), (7, 3), (8, 4)\}$.

Solution We begin by making a figure for each relation that shows the domain and the range (**Figure 2.3**).

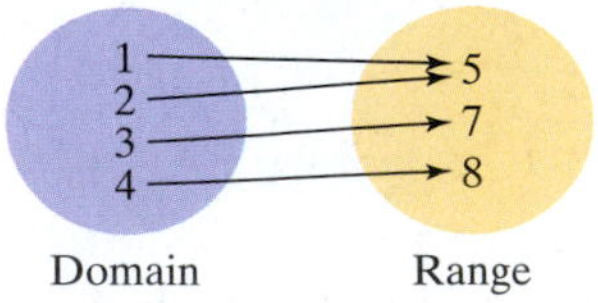

FIGURE 2.3(a)

a. **Figure 2.3(a)** shows that every element in the domain corresponds to exactly one element in the range. The element 1 in the domain corresponds to the element 5 in the range. Furthermore, 2 corresponds to 5, 3 corresponds to 7, and 4 corresponds to 8. No two ordered pairs in the given relation have the same first component and different second components. Thus, the relation is a function.

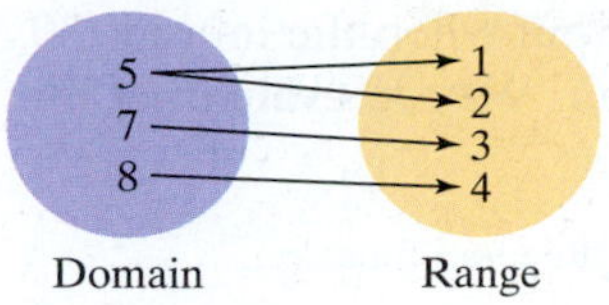

FIGURE 2.3(b)

b. **Figure 2.3(b)** shows that 5 corresponds to both 1 and 2. If any element in the domain corresponds to more than one element in the range, the relation is not a function. This relation is not a function because two ordered pairs have the same first component and different second components.

Same first component

(5, 1) (5, 2)

Different second components

Study Tip

If a relation is a function, reversing the components in each of its ordered pairs may result in a relation that is not a function.

Look at **Figure 2.3(a)** again. The fact that 1 and 2 in the domain correspond to the same number, 5, in the range does not violate the definition of a function. **A function can have two different first components with the same second component.** By contrast, a relation is not a function when two different ordered pairs have the same first component and different second components. Thus, the relation in Example 2(b) is not a function.

✓ CHECK POINT 2 Determine whether each relation is a function:

a. $\{(1, 2), (3, 4), (5, 6), (5, 7)\}$

b. $\{(1, 2), (3, 4), (6, 5), (7, 5)\}$.

3 Evaluate a function.

Functions as Equations and Function Notation

Functions are usually given in terms of equations rather than as sets of ordered pairs. For example, here is an equation that models the percentage of first-year college women claiming no religious affiliation as a function of time:

$$y = 0.012x^2 - 0.19x + 8.7.$$

The variable x represents the number of years after 1970. The variable y represents the percentage of first-year college women claiming no religious affiliation. The variable y is a function of the variable x. For each value of x, there is one and only one value of y. The variable x is called the **independent variable** because it can be assigned any value from the domain. Thus, x can be assigned any nonnegative integer representing the number of years after 1970. The variable y is called the **dependent variable** because its value depends on x. The percentage claiming no religious affiliation depends on the number of years after 1970. The value of the dependent variable, y, is calculated after selecting a value for the independent variable, x.

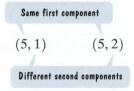

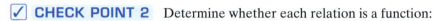

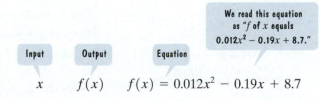

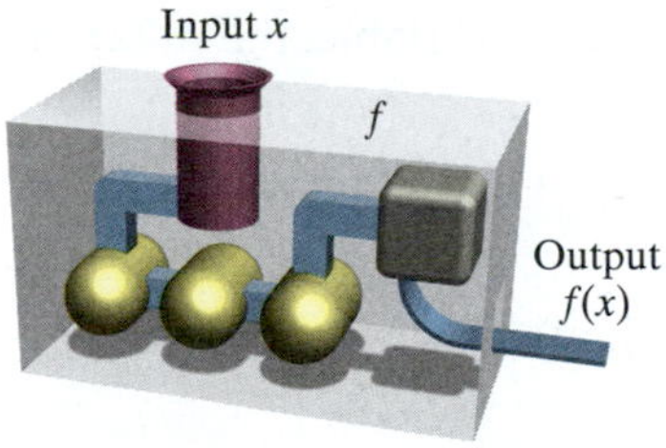

FIGURE 2.4 A "function machine" with inputs and outputs

If an equation in x and y gives one and only one value of y for each value of x, then the variable y is a function of the variable x. When an equation represents a function, the function is often named by a letter such as f, g, h, F, G, or H. Any letter can be used to name a function. Suppose that f names a function. Think of the domain as the set of the function's inputs and the range as the set of the function's outputs. As shown in **Figure 2.4**, the input is represented by x and the output by $f(x)$. The special notation $f(x)$, read "f of x" or "f at x," represents the **value of the function at the number x**.

Let's make this clearer by considering a specific example. We know that the equation

$$y = 0.012x^2 - 0.19x + 8.7$$

defines y as a function of x. We'll name the function f. Now, we can apply our new function notation.

Input: x Output: $f(x)$ Equation: $f(x) = 0.012x^2 - 0.19x + 8.7$

We read this equation as "f of x equals $0.012x^2 - 0.19x + 8.7$."

Study Tip

The notation $f(x)$ does *not* mean "f times x." The notation describes the value of the function at x.

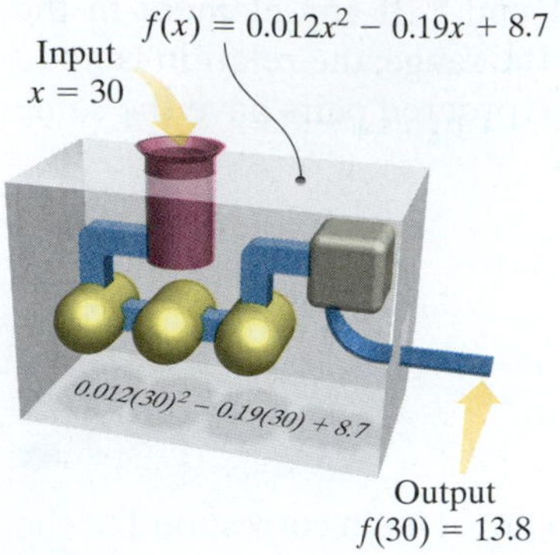

FIGURE 2.5 A function machine at work

Suppose we are interested in finding $f(30)$, the function's output when the input is 30. To find the value of the function at 30, we substitute 30 for x. We are **evaluating the function** at 30.

$f(x) = 0.012x^2 - 0.19x + 8.7$	This is the given function.
$f(30) = 0.012(30)^2 - 0.19(30) + 8.7$	Replace each occurrence of x with 30.
$= 0.012(900) - 0.19(30) + 8.7$	Evaluate the exponential expression: $30^2 = 30 \cdot 30 = 900$.
$= 10.8 - 5.7 + 8.7$	Perform the multiplications.
$f(30) = 13.8$	Subtract and add from left to right.

The statement $f(30) = 13.8$, read "f of 30 equals 13.8," tells us that the value of the function at 30 is 13.8. When the function's input is 30, its output is 13.8. **Figure 2.5** illustrates the input and output in terms of a function machine.

$$f(30) = 13.8$$

30 years after 1970, or in 2000, | 13.8% of first-year college women claimed no religious affiliation.

We have seen that in 2000, 13.2% actually claimed nonaffiliation, so our function that models the data slightly overestimates the percent for 2000.

Using Technology

Graphing utilities can be used to evaluate functions. The screens on the right show the evaluation of

$$f(x) = 0.012x^2 - 0.19x + 8.7$$

at 30 on a TI-84 Plus graphing calculator. The function f is named Y_1.

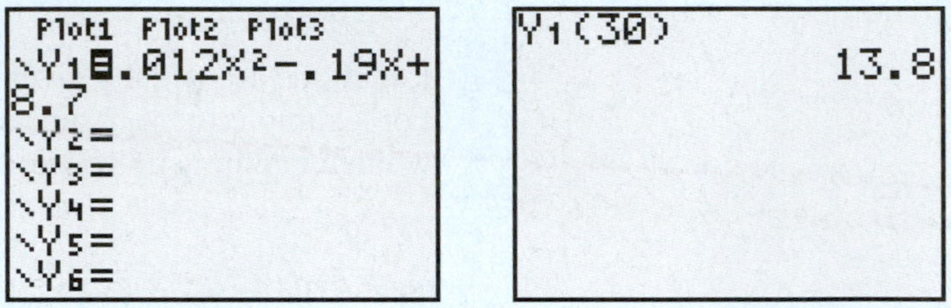

We used $f(x) = 0.012x^2 - 0.19x + 8.7$ to find $f(30)$. To find other function values, such as $f(40)$ or $f(55)$, substitute the specified input value, 40 or 55, for x in the function's equation.

If a function is named f and x represents the independent variable, the notation $f(x)$ corresponds to the y-value for a given x. Thus,

$$f(x) = 0.012x^2 - 0.19x + 8.7 \quad \text{and} \quad y = 0.012x^2 - 0.19x + 8.7$$

define the same function. This function may be written as

$$y = f(x) = 0.012x^2 - 0.19x + 8.7.$$

EXAMPLE 3 Using Function Notation

Find the indicated function value:

a. $f(4)$ for $f(x) = 2x + 3$

b. $g(-2)$ for $g(x) = 2x^2 - 1$

c. $h(-5)$ for $h(r) = r^3 - 2r^2 + 5$

d. $F(a + h)$ for $F(x) = 5x + 7$.

Solution

a.

$f(x) = 2x + 3$	This is the given function.
$f(4) = 2 \cdot 4 + 3$	To find f of 4, replace x with 4.
$= 8 + 3$	Multiply: $2 \cdot 4 = 8$.
$f(4) = 11$	Add.

f of 4 is 11.

b. $g(x) = 2x^2 - 1$ — This is the given function.

$g(-2) = 2(-2)^2 - 1$ — To find g of -2, replace x with -2.

$= 2(4) - 1$ — Evaluate the exponential expression: $(-2)^2 = 4$.

$= 8 - 1$ — Multiply: $2(4) = 8$.

$g(-2) = 7$ — Subtract.

g of −2 is 7.

c. $h(r) = r^3 - 2r^2 + 5$ — The function's name is h and r represents the independent variable.

$h(-5) = (-5)^3 - 2(-5)^2 + 5$ — To find h of -5, replace each occurrence of r with -5.

$= -125 - 2(25) + 5$ — Evaluate exponential expressions.

$= -125 - 50 + 5$ — Multiply.

$h(-5) = -170$ — $-125 - 50 = -175$ and $-175 + 5 = -170$.

h of −5 is −170.

d. $F(x) = 5x + 7$ — This is the given function.

$F(a + h) = 5(a + h) + 7$ — Replace x with $a + h$.

$F(a + h) = 5a + 5h + 7$ — Apply the distributive property.

F of $a + h$ is $5a + 5h + 7$.

✓ CHECK POINT 3 Find the indicated function value:

a. $f(6)$ for $f(x) = 4x + 5$

b. $g(-5)$ for $g(x) = 3x^2 - 10$

c. $h(-4)$ for $h(r) = r^2 - 7r + 2$

d. $F(a + h)$ for $F(x) = 6x + 9$.

Functions Represented by Tables and Function Notation

Function notation can be applied to functions that are represented by tables.

EXAMPLE 4 Using Function Notation

Function f is defined by the following table:

x	$f(x)$
−2	5
−1	0
0	3
1	1
2	4

a. Explain why the table defines a function.

b. Find the domain and the range of the function.

Find the indicated function value:

c. $f(-1)$

d. $f(0)$.

e. Find x such that $f(x) = 4$.

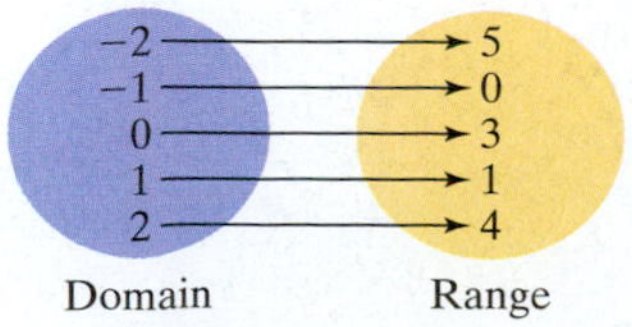

FIGURE 2.6

Solution

a. Values in the first column of the table make up the domain, or input values. Values in the second column of the table make up the range, or output values. We see that every element in the domain corresponds to exactly one element in the range, shown in **Figure 2.6**. Therefore, the relation given by the table is a function.

The voice balloons pointing to appropriate parts of the table illustrate the solution to parts (b)-(e).

x	$f(x)$
-2	5
-1	0
0	3
1	1
2	4

c. $f(-1) = 0$: When the input is -1, the output is 0.

d. $f(0) = 3$: When the input is 0, the output is 3.

e. $f(x) = 4$ when $x = 2$: The output, $f(x)$, is 4 when the input, x, is 2.

b. The domain is the set of inputs: $\{-2, -1, 0, 1, 2\}$.

b. The range is the set of outputs: $\{5, 0, 3, 1, 4\}$.

✓ CHECK POINT 4 Function g is defined by the following table:

x	$g(x)$
0	3
1	0
2	1
3	2
4	3

a. Explain why the table defines a function.

b. Find the domain and the range of the function.

Find the indicated function value:

c. $g(1)$

d. $g(3)$

e. Find x such that $g(x) = 3$.

2.1 EXERCISE SET

Practice Exercises

In Exercises 1–8, determine whether each relation is a function. Give the domain and range for each relation.

1. $\{(1, 2), (3, 4), (5, 5)\}$
2. $\{(4, 5), (6, 7), (8, 8)\}$
3. $\{(3, 4), (3, 5), (4, 4), (4, 5)\}$
4. $\{(5, 6), (5, 7), (6, 6), (6, 7)\}$
5. $\{(-3, -3), (-2, -2), (-1, -1), (0, 0)\}$
6. $\{(-7, -7), (-5, -5), (-3, -3), (0, 0)\}$
7. $\{(1, 4), (1, 5), (1, 6)\}$
8. $\{(4, 1), (5, 1), (6, 1)\}$

In Exercises 9–22, find the indicated function values.

9. $f(x) = x + 1$
   a. $f(0)$ b. $f(5)$ c. $f(-8)$
   d. $f(2a)$ e. $f(a + 2)$
10. $f(x) = x + 3$
   a. $f(0)$ b. $f(5)$ c. $f(-8)$
   d. $f(2a)$ e. $f(a + 2)$
11. $g(x) = 3x - 2$
   a. $g(0)$ b. $g(-5)$ c. $g\left(\frac{2}{3}\right)$
   d. $g(4b)$ e. $g(b + 4)$
12. $g(x) = 4x - 3$
   a. $g(0)$ b. $g(-5)$ c. $g\left(\frac{3}{4}\right)$
   d. $g(5b)$ e. $g(b + 5)$
13. $h(x) = 3x^2 + 5$
   a. $h(0)$ b. $h(-1)$ c. $h(4)$
   d. $h(-3)$ e. $h(4b)$
14. $h(x) = 2x^2 - 4$
   a. $h(0)$ b. $h(-1)$ c. $h(5)$
   d. $h(-3)$ e. $h(5b)$
15. $f(x) = 2x^2 + 3x - 1$
   a. $f(0)$ b. $f(3)$ c. $f(-4)$
   d. $f(b)$
   e. $f(5a)$

16. $f(x) = 3x^2 + 4x - 2$

a. $f(0)$ b. $f(3)$ c. $f(-5)$

d. $f(b)$

e. $f(5a)$

17. $f(x) = \dfrac{2x - 3}{x - 4}$

a. $f(0)$ b. $f(3)$ c. $f(-4)$

d. $f(-5)$ e. $f(a + h)$

f. Why must 4 be excluded from the domain of f?

18. $f(x) = \dfrac{3x - 1}{x - 5}$

a. $f(0)$ b. $f(3)$ c. $f(-3)$

d. $f(10)$ e. $f(a + h)$

f. Why must 5 be excluded from the domain of f?

19.

x	$f(x)$
−4	3
−2	6
0	9
2	12
4	15

a. $f(-2)$

b. $f(2)$

c. For what value of x is $f(x) = 9$?

20.

x	$f(x)$
−5	4
−3	8
0	12
3	16
5	20

a. $f(-3)$

b. $f(3)$

c. For what value of x is $f(x) = 12$?

21.

x	$h(x)$
−2	2
−1	1
0	0
1	1
2	2

a. $h(-2)$

b. $h(1)$

c. For what values of x is $h(x) = 1$?

22.

x	$h(x)$
−2	−2
−1	−1
0	0
1	−1
2	−2

a. $h(-2)$

b. $h(1)$

c. For what values of x is $h(x) = -1$?

Practice PLUS

In Exercises 23–24, let $f(x) = x^2 - x + 4$ *and* $g(x) = 3x - 5$.

23. Find $g(1)$ and $f(g(1))$.

24. Find $g(-1)$ and $f(g(-1))$.

In Exercises 25–26, let f and g be defined by the following table:

x	$f(x)$	$g(x)$
−2	6	0
−1	3	4
0	−1	1
1	−4	−3
2	0	−6

25. Find $\sqrt{f(-1) - f(0)} - [g(2)]^2 + f(-2) \div g(2) \cdot g(-1)$.

26. Find $|f(1) - f(0)| - [g(1)]^2 + g(1) \div f(-1) \cdot g(2)$.

In Exercises 27–28, find $f(-x) - f(x)$ *for the given function f. Then simplify the expression.*

27. $f(x) = x^3 + x - 5$

28. $f(x) = x^2 - 3x + 7$

In Exercises 29–30, each function is defined by two equations. The equation in the first row gives the output for negative numbers in the domain. The equation in the second row gives the output for non-negative numbers in the domain. Find the indicated function values.

29. $f(x) = \begin{cases} 3x + 5 & \text{if } x < 0 \\ 4x + 7 & \text{if } x \geq 0 \end{cases}$

a. $f(-2)$ b. $f(0)$

c. $f(3)$ d. $f(-100) + f(100)$

30. $f(x) = \begin{cases} 6x - 1 & \text{if } x < 0 \\ 7x + 3 & \text{if } x \geq 0 \end{cases}$

a. $f(-3)$ b. $f(0)$

c. $f(4)$ d. $f(-100) + f(100)$

Application Exercises

31. The bar graph shows the breakdown of political ideologies in the United States.

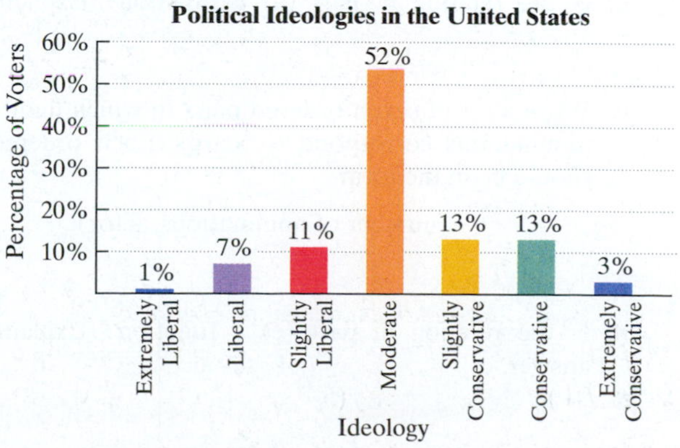

Source: Center for Political Studies, University of Michigan

(Refer to the graph at the bottom of page 103 as you work this exercise.)

a. Write a set of seven ordered pairs in which political ideologies correspond to percentages. Each ordered pair should be in the form

(ideology, percent).

Use EL, L, SL, M, SC, C, and EC to represent the respective ideologies from left to right.

b. Is the relation in part (a) a function? Explain your answer.

c. Write a set of seven ordered pairs in which percentages correspond to political ideologies. Each ordered pair should be in the form

(percent, ideology).

d. Is the relation in part (c) a function? Explain your answer.

32. Actors with the most Oscar nominations include Jack Nicholson, Laurence Olivier, Paul Newman, Spencer Tracy, Marlon Brando, Jack Lemmon, and Al Pacino. The bar graph shows the number of nominations for each of these men.

Actors with the Most Oscar Nominations

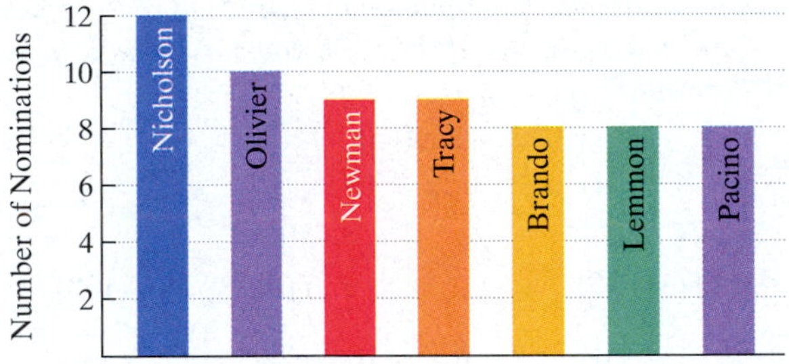

Source: Russell Ash, *The Top 10 of Everything 2007*

a. Write a set of seven ordered pairs in which actors correspond to numbers of Oscar nominations. Each ordered pair should be in the form

(actor, number of nominations).

b. Is the relation in part (a) a function? Explain your answer.

c. Write a set of seven ordered pairs in which numbers of nominations correspond to actors. Each ordered pair should be in the form

(number of nominations, actor).

d. Is the relation in part (c) a function? Explain your answer.

Writing in Mathematics

33. What is a relation? Describe what is meant by its domain and its range.

34. Explain how to determine whether a relation is a function. What is a function?

35. Does $f(x)$ mean f times x when referring to function f? If not, what does $f(x)$ mean? Provide an example with your explanation.

36. For people filing a single return, federal income tax is a function of adjusted gross income because for each value of adjusted gross income there is a specific tax to be paid. By contrast, the price of a house is not a function of the lot size on which the house sits because houses on same-sized lots can sell for many different prices.

a. Describe an everyday situation between variables that is a function.

b. Describe an everyday situation between variables that is not a function.

Critical Thinking Exercises

Make Sense? *In Exercises 37–40, determine whether each statement "makes sense" or "does not make sense" and explain your reasoning.*

37. Today's temperature is a function of the time of day.

38. My height is a function of my age.

39. Although I presented my function as a set of ordered pairs, I could have shown the correspondences using a table or using points plotted in a rectangular coordinate system.

40. My function models how the chance of divorce depends on the number of years of marriage, so the range is $\{x \mid x \text{ is the number of years of marriage}\}$.

In Exercises 41–46, determine whether each statement is true or false. If the statement is false, make the necessary change(s) to produce a true statement.

41. All relations are functions.

42. No two ordered pairs of a function can have the same second components and different first components.

Using the tables that define f and g, determine whether each statement in Exercises 43–46 is true or false.

x	$f(x)$
−4	−1
−3	−2
−2	−3
−1	−4

x	$g(x)$
−1	−4
−2	−3
−3	−2
−4	−1

43. The domain of f = the range of f

44. The range of f = the domain of g

45. $f(-4) - f(-2) = 2$

46. $g(-4) + f(-4) = 0$

47. If $f(x) = 3x + 7$, find $\frac{f(a+h) - f(a)}{h}$.

48. Give an example of a relation with the following characteristics: The relation is a function containing two ordered pairs. Reversing the components in each ordered pair results in a relation that is not a function.

49. If $f(x + y) = f(x) + f(y)$ and $f(1) = 3$, find $f(2)$, $f(3)$, and $f(4)$. Is $f(x + y) = f(x) + f(y)$ for all functions?

Review Exercises

50. Simplify: $24 \div 4[2 - (5 - 2)]^2 - 6$. (Section 1.2, Example 7)

51. Simplify: $\left(\frac{3x^2y^{-2}}{y^3}\right)^{-2}$. (Section 1.6, Example 9)

52. Solve: $\frac{x}{3} = \frac{3x}{5} + 4$. (Section 1.4, Example 4)

Preview Exercises

Exercises 53–55 will help you prepare for the material covered in the next section.

53. Graph $y = 2x$. Select integers for x, starting with -2 and ending with 2.

54. Graph $y = 2x + 4$. Select integers for x, starting with -2 and ending with 2.

55. Use the following graph to solve this exercise.

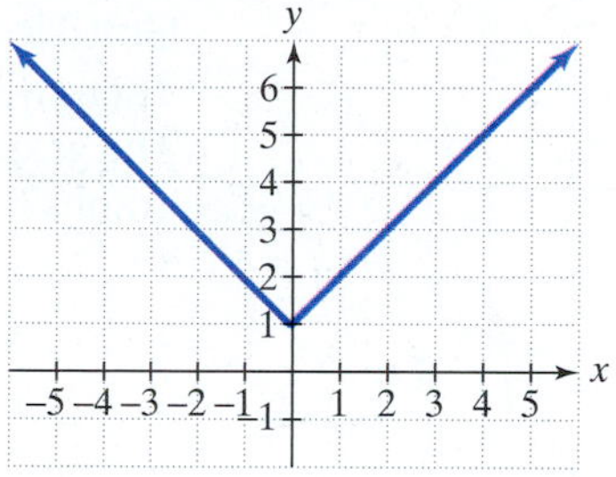

a. What is the y-coordinate when the x-coordinate is 2?

b. What are the x-coordinates when the y-coordinate is 4?

c. Describe the x-coordinates of all points on the graph.

d. Describe the y-coordinates of all points on the graph.

SECTION 2.2

Graphs of Functions

Objectives

1. Graph functions by plotting points.
2. Use the vertical line test to identify functions.
3. Obtain information about a function from its graph.
4. Identify the domain and range of a function from its graph.

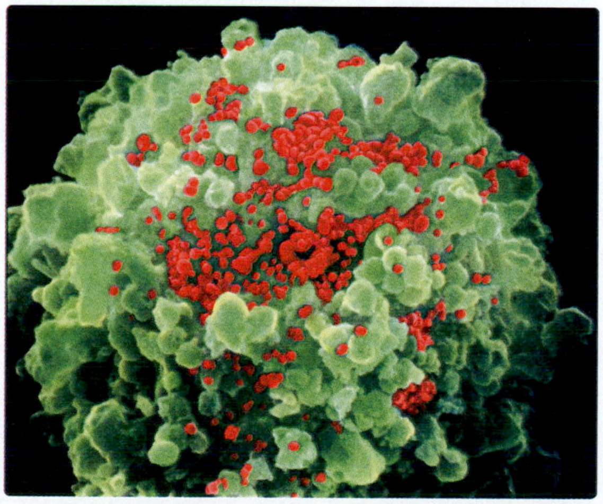

Magnified 6000 times, this color-scanned image shows a T-lymphocyte blood cell (green) infected with the HIV virus (red). Depletion of the number of T cells causes destruction of the immune system.

The number of T cells in a person with HIV is a function of time after infection. In this section, we'll analyze the graph of this function, using the rectangular coordinate system to visualize what functions look like.

1 Graph functions by plotting points.

Graphs of Functions

The **graph of a function** is the graph of its ordered pairs. For example, the graph of $f(x) = 2x$ is the set of points (x, y) in the rectangular coordinate system satisfying $y = 2x$. Similarly, the graph of $g(x) = 2x + 4$ is the set of points (x, y) in the rectangular coordinate system satisfying the equation $y = 2x + 4$. In the next example, we graph both of these functions in the same rectangular coordinate system.

EXAMPLE 1 Graphing Functions

Graph the functions $f(x) = 2x$ and $g(x) = 2x + 4$ in the same rectangular coordinate system. Select integers for x, starting with -2 and ending with 2.

Solution We begin by setting up a partial table of coordinates for each function. Then, we plot the five points in each table and connect them, as shown in **Figure 2.7**. The graph of each function is a straight line. Do you see a relationship between the two graphs? The graph of g is the graph of f shifted vertically up by 4 units.

x	$f(x) = 2x$	(x, y) or $(x, f(x))$
-2	$f(-2) = 2(-2) = -4$	$(-2, -4)$
-1	$f(-1) = 2(-1) = -2$	$(-1, -2)$
0	$f(0) = 2 \cdot 0 = 0$	$(0, 0)$
1	$f(1) = 2 \cdot 1 = 2$	$(1, 2)$
2	$f(2) = 2 \cdot 2 = 4$	$(2, 4)$

Choose x. Compute $f(x)$ by evaluating f at x. Form the ordered pair.

x	$g(x) = 2x + 4$	(x, y) or $(x, g(x))$
-2	$g(-2) = 2(-2) + 4 = 0$	$(-2, 0)$
-1	$g(-1) = 2(-1) + 4 = 2$	$(-1, 2)$
0	$g(0) = 2 \cdot 0 + 4 = 4$	$(0, 4)$
1	$g(1) = 2 \cdot 1 + 4 = 6$	$(1, 6)$
2	$g(2) = 2 \cdot 2 + 4 = 8$	$(2, 8)$

Choose x. Compute $g(x)$ by evaluating g at x. Form the ordered pair.

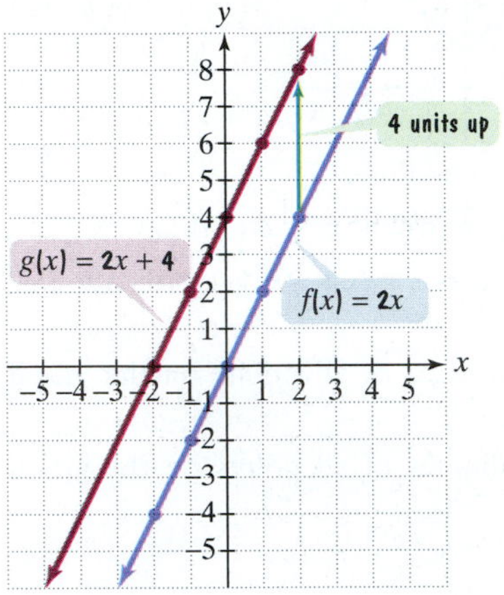

FIGURE 2.7

The graphs in Example 1 are straight lines. All functions with equations of the form $f(x) = mx + b$ graph as straight lines. Such functions, called **linear functions**, will be discussed in detail in Sections 2.4–2.5.

Using Technology

We can use a graphing utility to check the tables and the graphs in Example 1 for the functions

$$f(x) = 2x \quad \text{and} \quad g(x) = 2x + 4.$$

Enter $y_1 = 2x$ in the $\boxed{y=}$ screen.

Enter $y_2 = 2x + 4$ in the $\boxed{y=}$ screen.

Checking Tables

We entered -2 for the starting x-value and 1 as an increment between x-values to check our tables in Example 1.

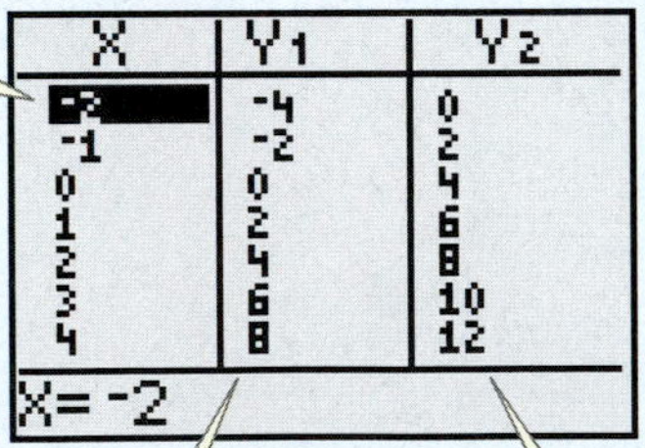

Use the first five ordered pairs (x, y_1) to check the first table.

Use the first five ordered pairs (x, y_2) to check the second table.

Checking Graphs

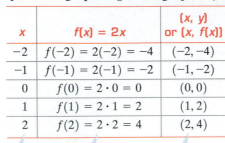

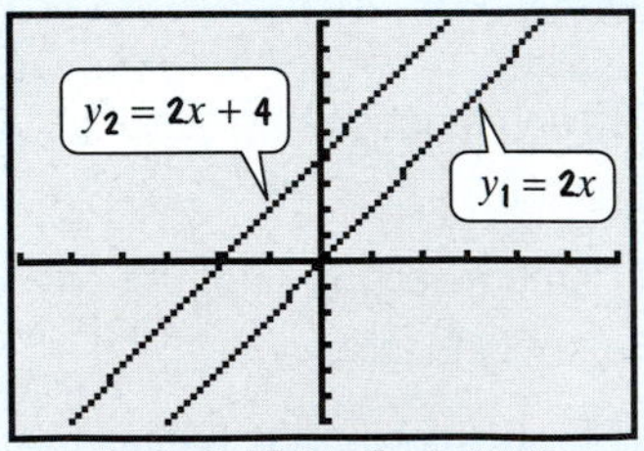

$[-6, 6, 1]$ by $[-6, 9, 1]$

We selected this viewing rectangle, or window, to match **Figure 2.7.**

CHECK POINT 1 Graph the functions $f(x) = 2x$ and $g(x) = 2x - 3$ in the same rectangular coordinate system. Select integers for x, starting with -2 and ending with 2. How is the graph of g related to the graph of f?

2 Use the vertical line test to identify functions.

The Vertical Line Test

Not every graph in the rectangular coordinate system is the graph of a function. The definition of a function specifies that no value of x can be paired with two or more different values of y. Consequently, if a graph contains two or more different points with the same first coordinate, the graph cannot represent a function. This is illustrated in **Figure 2.8**. Observe that points sharing a common first coordinate are vertically above or below each other.

This observation is the basis of a useful test for determining whether a graph defines y as a function of x. The test is called the **vertical line test**.

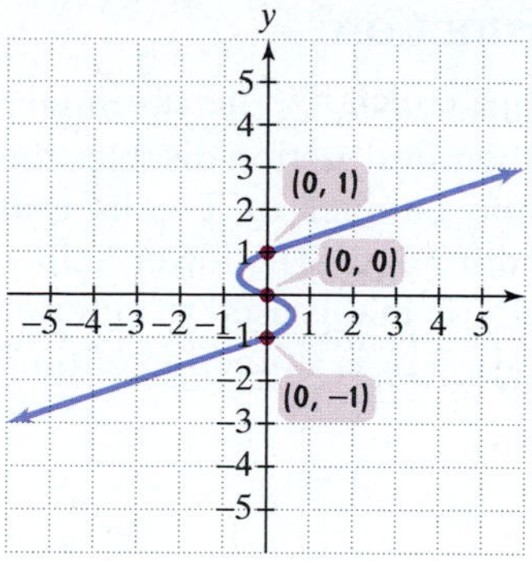

FIGURE 2.8 y is not a function of x because 0 is paired with three values of y, namely, 1, 0, and −1.

The Vertical Line Test for Functions

If any vertical line intersects a graph in more than one point, the graph does not define y as a function of x.

EXAMPLE 2 Using the Vertical Line Test

Use the vertical line test to identify graphs in which y is a function of x.

a.
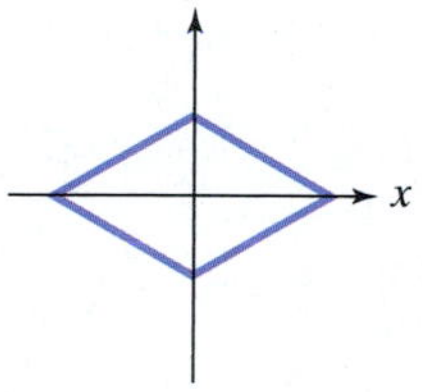

b.
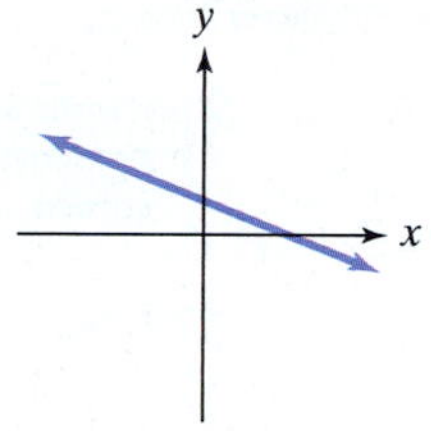

c.
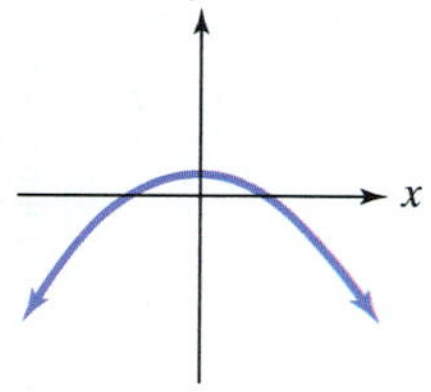

d.
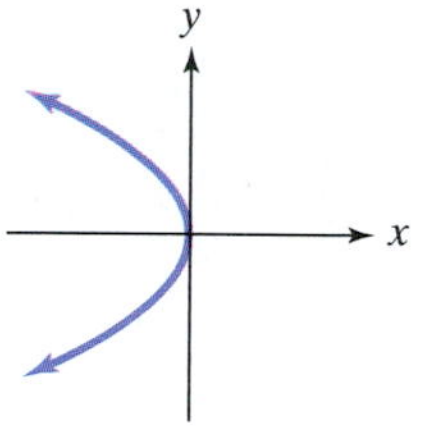

Solution y is a function of x for the graphs in (b) and (c).

a.
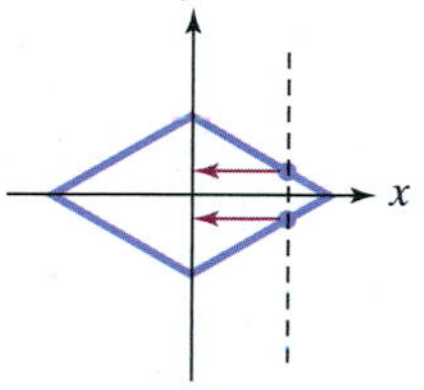

y **is not a function** of x. Two values of y correspond to one x-value.

b.
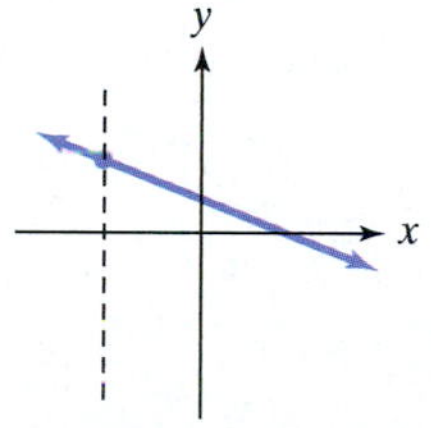

y **is a function** of x.

c.
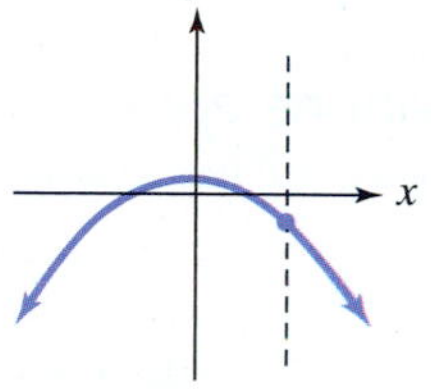

y **is a function** of x.

d.
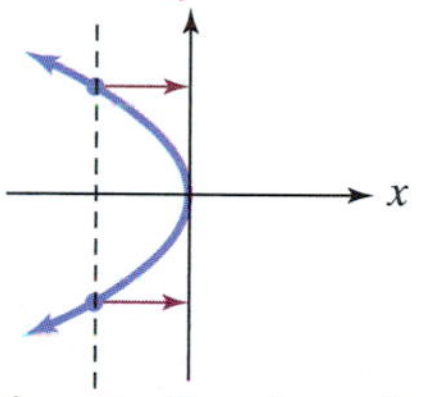

y **is not a function** of x. Two values of y correspond to one x-value.

✓ **CHECK POINT 2** Use the vertical line test to identify graphs in which y is a function of x.

a.
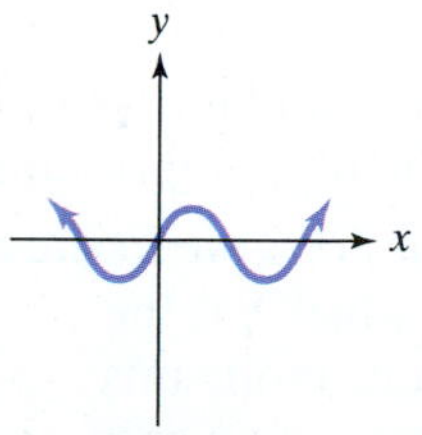

b.
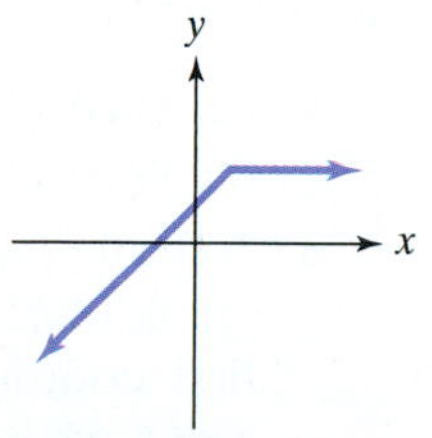

c.
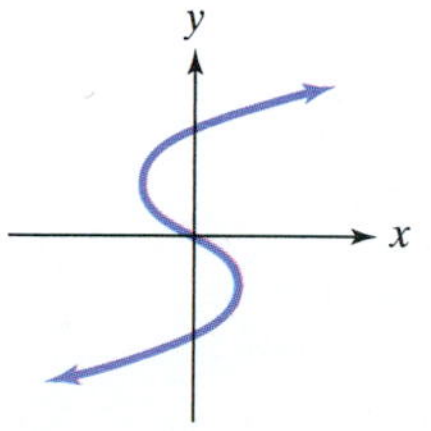

3 Obtain information about a function from its graph.

Obtaining Information from Graphs

You can obtain information about a function from its graph. At the right or left of a graph, you will often find closed dots, open dots, or arrows.

- A closed dot indicates that the graph does not extend beyond this point and the point belongs to the graph.
- An open dot indicates that the graph does not extend beyond this point and the point does not belong to the graph.
- An arrow indicates that the graph extends indefinitely in the direction in which the arrow points.

EXAMPLE 3 Analyzing the Graph of a Function

The human immunodeficiency virus, or HIV, infects and kills helper T cells. Because T cells stimulate the immune system to produce antibodies, their destruction disables the body's defenses against other pathogens. By counting the number of T cells that remain active in the body, the progression of HIV can be monitored. The fewer helper T cells, the more advanced the disease. **Figure 2.9** shows a graph that is used to monitor the average progression of the disease. The number of T cells, $f(x)$, is a function of time after infection, x.

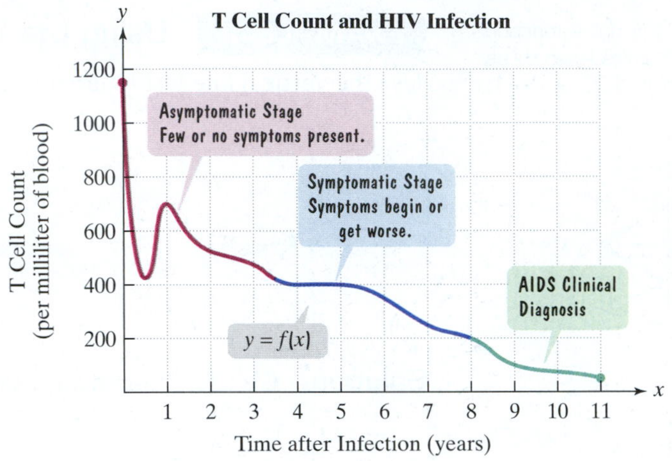

FIGURE 2.9

Source: B.E. Pruitt et al., *Human Sexuality*, Prentice Hall, 2007

a. Explain why f represents the graph of a function.

b. Use the graph to find $f(8)$.

c. For what value of x is $f(x) = 350$?

d. Describe the general trend shown by the graph.

Solution

a. No vertical line can be drawn that intersects the graph of f more than once. By the vertical line test, f represents the graph of a function.

b. To find $f(8)$, or f of 8, we locate 8 on the x-axis. **Figure 2.10** shows the point on the graph of f for which 8 is the first coordinate. From this point, we look to the y-axis to find the corresponding y-coordinate. We see that the y-coordinate is 200. Thus,

$$f(8) = 200.$$

When the time after infection is 8 years, the T cell count is 200 cells per milliliter of blood. (AIDS clinical diagnosis is given at a T cell count of 200 or below.)

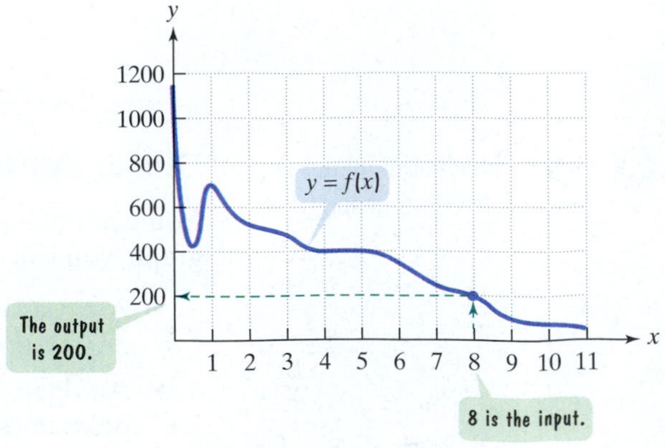

FIGURE 2.10 Finding $f(8)$

c. To find the value of x for which $f(x) = 350$, we approximately locate 350 on the y-axis. **Figure 2.11** shows that there is one point on the graph of f for which 350 is the second coordinate. From this point, we look to the x-axis to find the corresponding x-coordinate. We see that the x-coordinate is 6. Thus,

$$f(x) = 350 \text{ for } x = 6.$$

A T cell count of 350 occurs 6 years after infection.

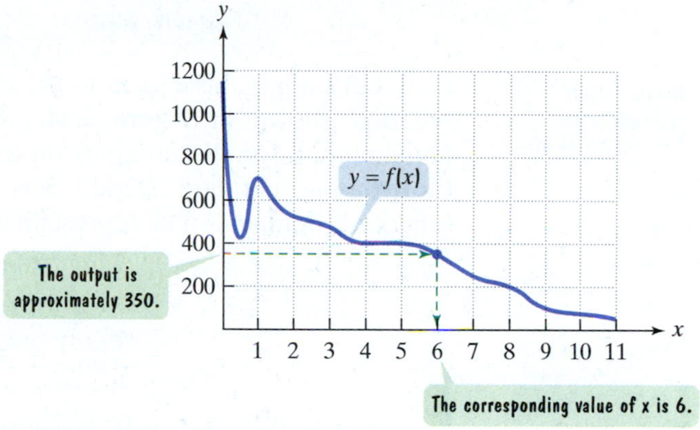

FIGURE 2.11 Finding x for which $f(x) = 350$

d. **Figure 2.12** uses voice balloons to describe the general trend shown by the graph.

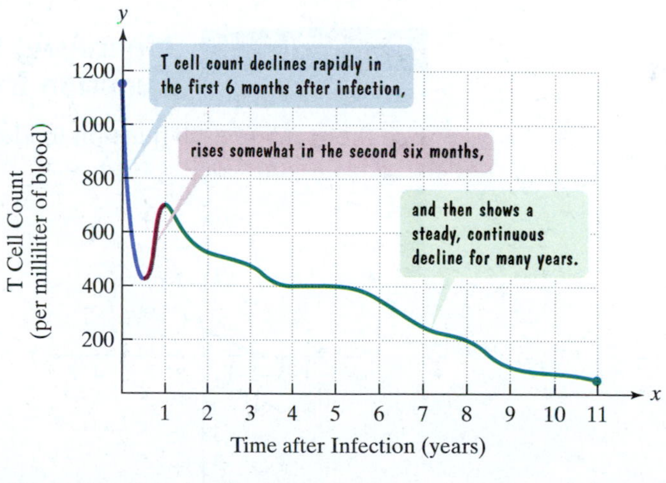

FIGURE 2.12 Describing changing T cell count over time in a person infected with HIV

✓ CHECK POINT 3

a. Use the graph of f in **Figure 2.9** on page 108 to find $f(5)$.

b. For what value of x is $f(x) = 100$?

c. Estimate the minimum T cell count during the asymptomatic stage.

4 Identify the domain and range of a function from its graph.

Identifying Domain and Range from a Function's Graph

Figure 2.13 illustrates how the graph of a function is used to determine the function's domain and its range.

Domain: set of inputs

Found on the x-axis

Range: set of outputs

Found on the y-axis

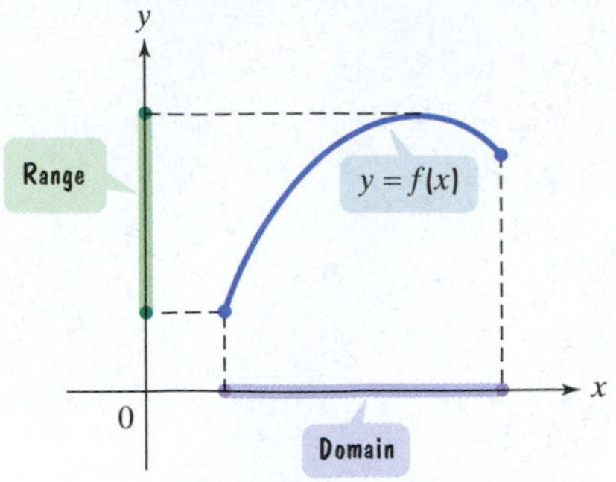

FIGURE 2.13 Domain and range of f

Let's apply these ideas to the graph of the function shown in **Figure 2.14.** To find the domain, look for all the inputs on the x-axis that correspond to points on the graph. Can you see that they extend from -4 to 2, inclusive? Using set-builder notation, the function's domain can be represented as follows:

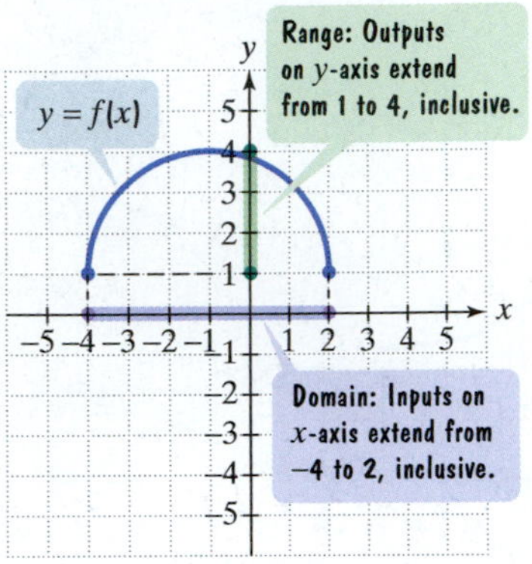

FIGURE 2.14 Domain and range of f

$$\{ x \mid -4 \le x \le 2 \}.$$

The set of all x — such that — x is greater than or equal to -4 and less than or equal to 2.

To find the range, look for all the outputs on the y-axis that correspond to points on the graph. They extend from 1 to 4, inclusive. Using set-builder notation, the function's range can be represented as follows:

$$\{ y \mid 1 \le y \le 4 \}$$

The set of all y — such that — y is greater than or equal to 1 and less than or equal to 4.

EXAMPLE 4 Identifying the Domain and Range of a Function from Its Graph

Use the graph of each function to identify its domain and its range.

a.

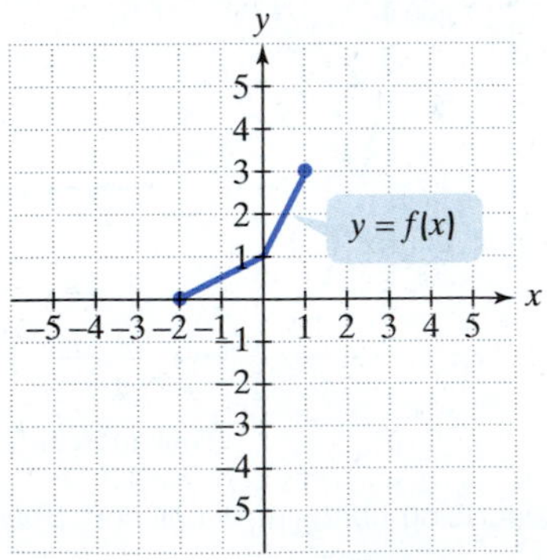

b.

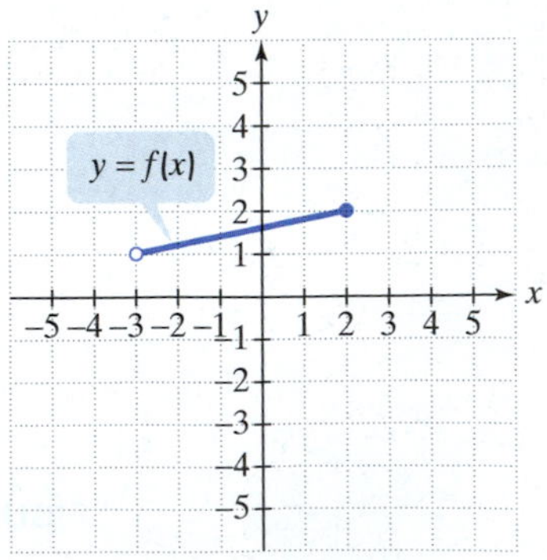

c.

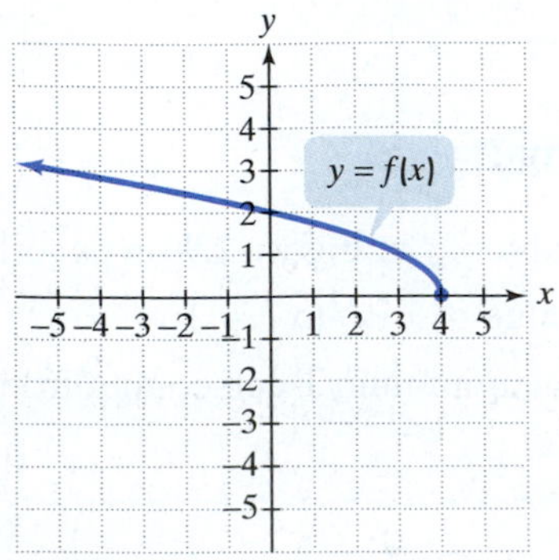

d.

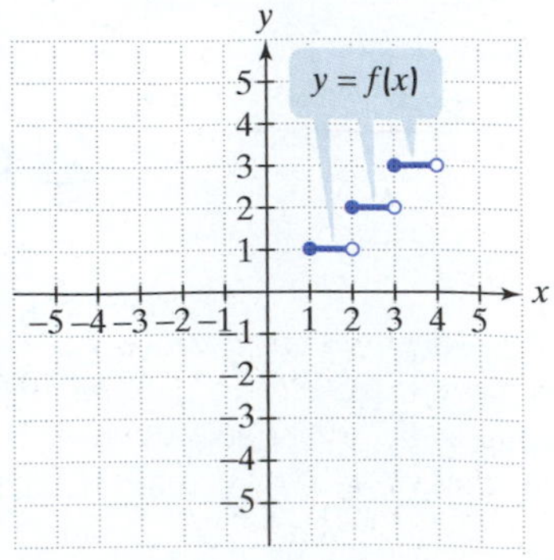

Solution For the graph of each function, the domain is highlighted in purple on the x-axis and the range is highlighted in green on the y-axis.

a.

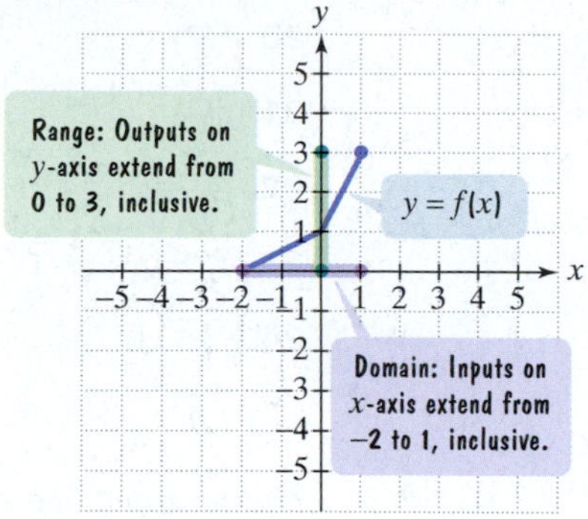

$$\text{Domain} = \{x \mid -2 \le x \le 1\}$$
$$\text{Range} = \{y \mid 0 \le y \le 3\}$$

b.

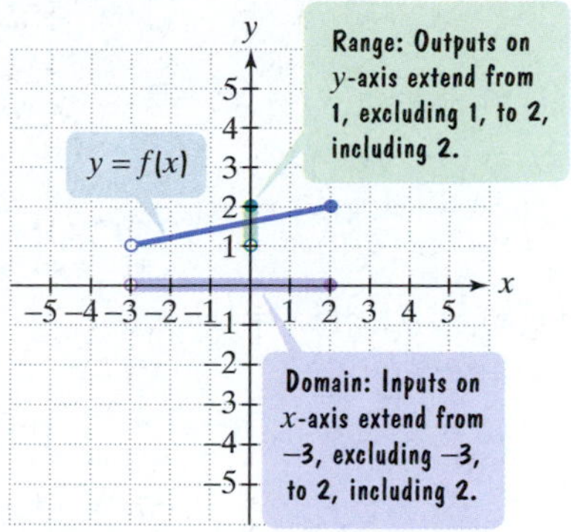

$$\text{Domain} = \{x \mid -3 < x \le 2\}$$
$$\text{Range} = \{y \mid 1 < y \le 2\}$$

c.

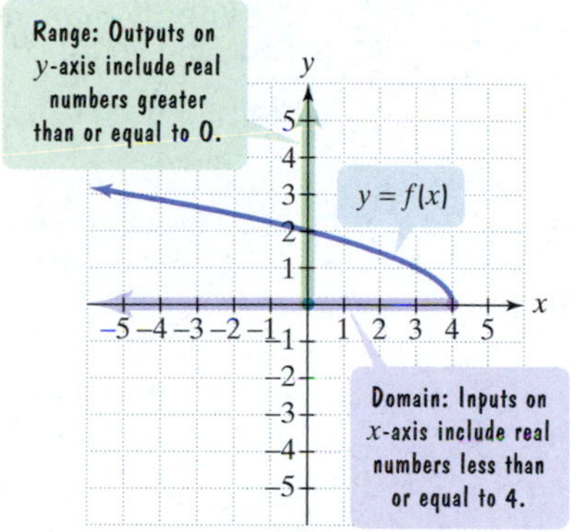

$$\text{Domain} = \{x \mid x \le 4\}$$
$$\text{Range} = \{y \mid y \ge 0\}$$

d.

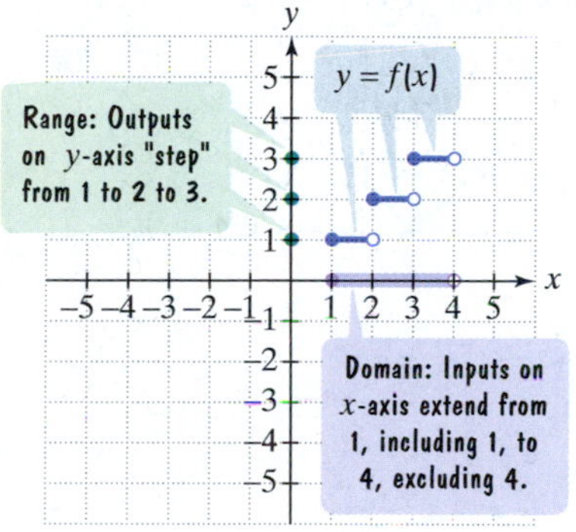

$$\text{Domain} = \{x \mid 1 \le x < 4\}$$
$$\text{Range} = \{y \mid y = 1, 2, 3\}$$

✓ CHECK POINT 4 Use the graph of each function to identify its domain and its range.

a.

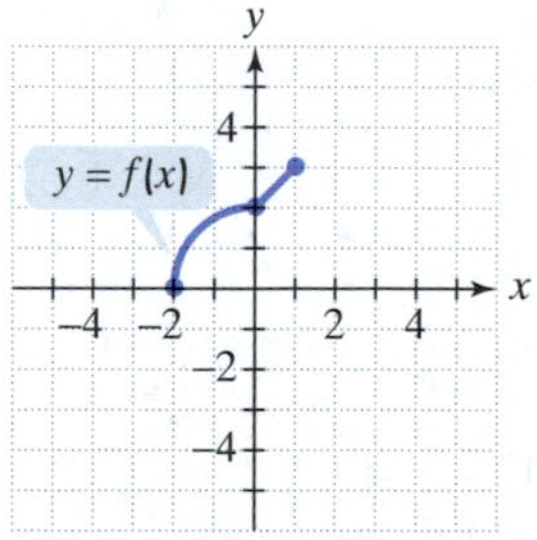

b.

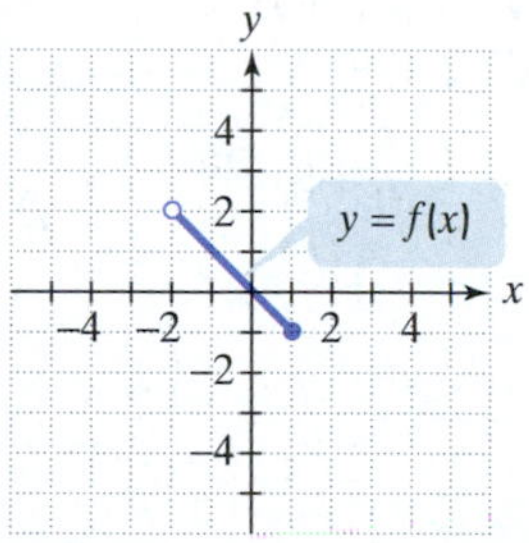

c.

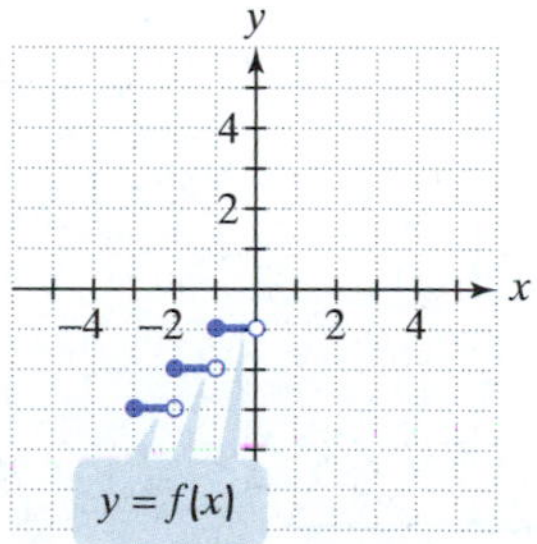

2.2 EXERCISE SET

MyMathLab · MathXL PRACTICE · WATCH · DOWNLOAD · READ · REVIEW

Practice Exercises

In Exercises 1–10, graph the given functions, f and g, in the same rectangular coordinate system. Select integers for x, starting with −2 and ending with 2. Once you have obtained your graphs, describe how the graph of g is related to the graph of f.

1. $f(x) = x,\ g(x) = x + 3$

2. $f(x) = x,\ g(x) = x - 4$

3. $f(x) = -2x,\ g(x) = -2x - 1$

4. $f(x) = -2x,\ g(x) = -2x + 3$

5. $f(x) = x^2,\ g(x) = x^2 + 1$

6. $f(x) = x^2,\ g(x) = x^2 - 2$

7. $f(x) = |x|,\ g(x) = |x| - 2$

8. $f(x) = |x|,\ g(x) = |x| + 1$

9. $f(x) = x^3,\ g(x) = x^3 + 2$

10. $f(x) = x^3,\ g(x) = x^3 - 1$

In Exercises 11–18, use the vertical line test to identify graphs in which y is a function of x.

11.

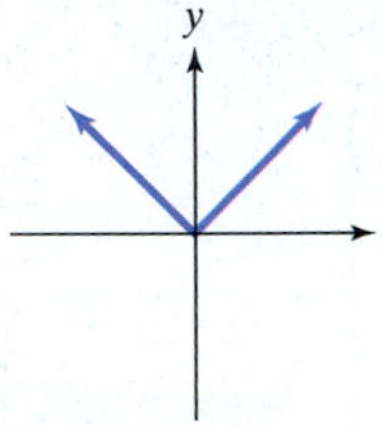

12.

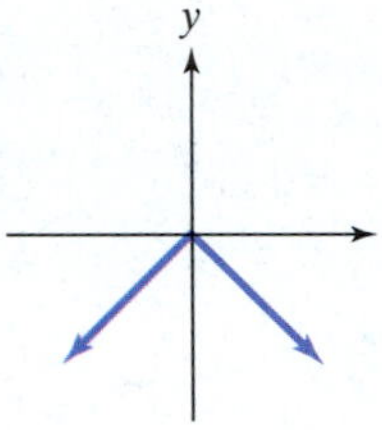

13.

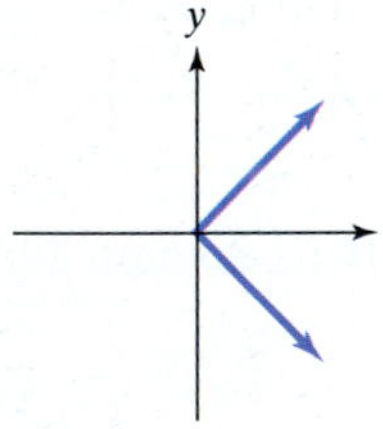

14.

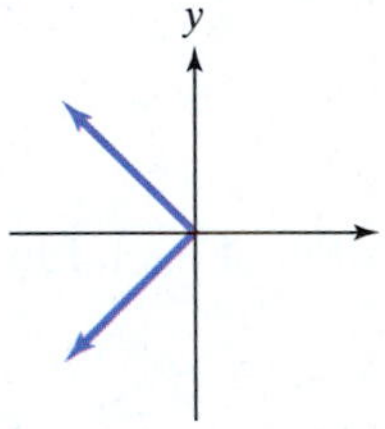

15.

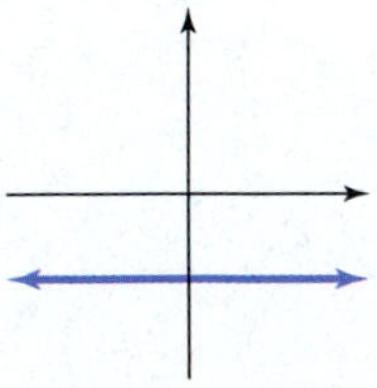

16.

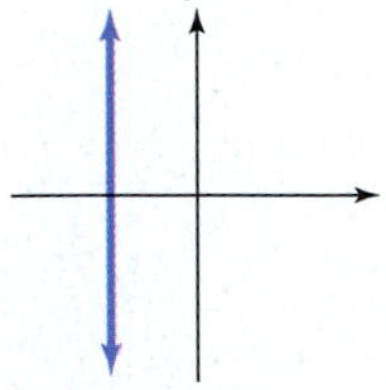

17.

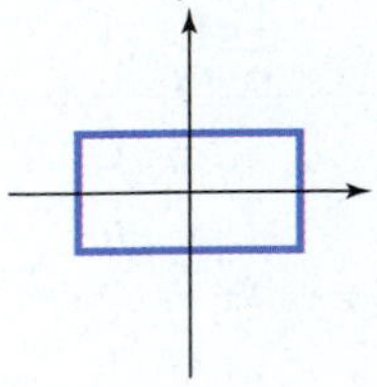

18.

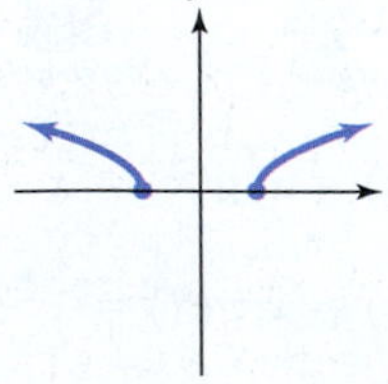

In Exercises 19–24, use the graph of f to find each indicated function value.

19. $f(-2)$

20. $f(2)$

21. $f(4)$

22. $f(-4)$

23. $f(-3)$

24. $f(-1)$

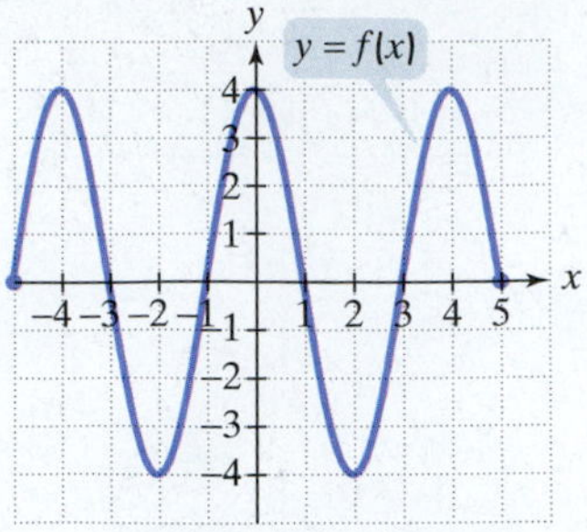

Use the graph of g to solve Exercises 25–30.

25. Find $g(-4)$.

26. Find $g(2)$.

27. Find $g(-10)$.

28. Find $g(10)$.

29. For what value of x is $g(x) = 1$?

30. For what value of x is $g(x) = -1$?

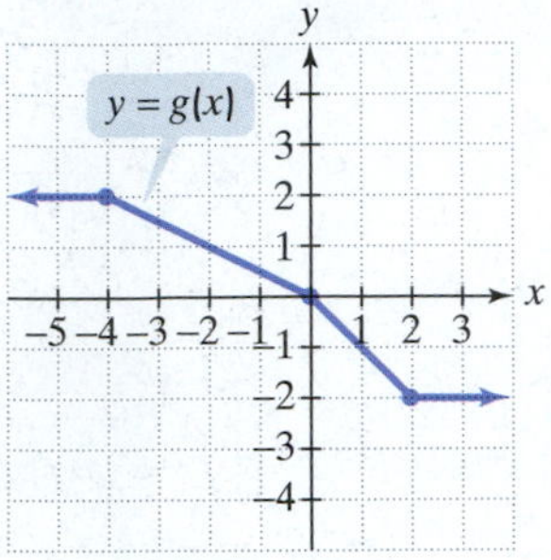

In Exercises 31–40, use the graph of each function to identify its domain and its range.

31.

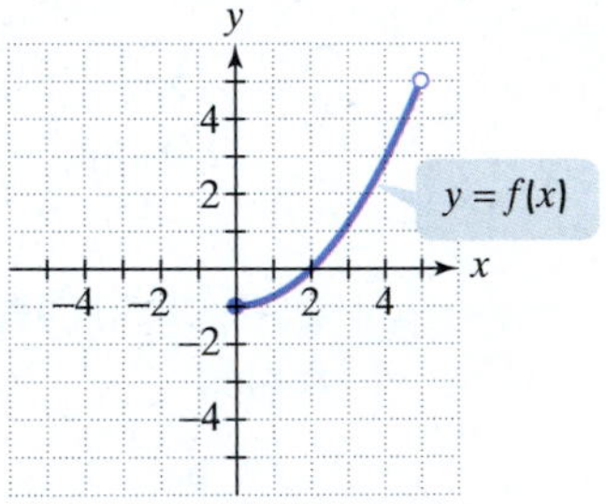

32.

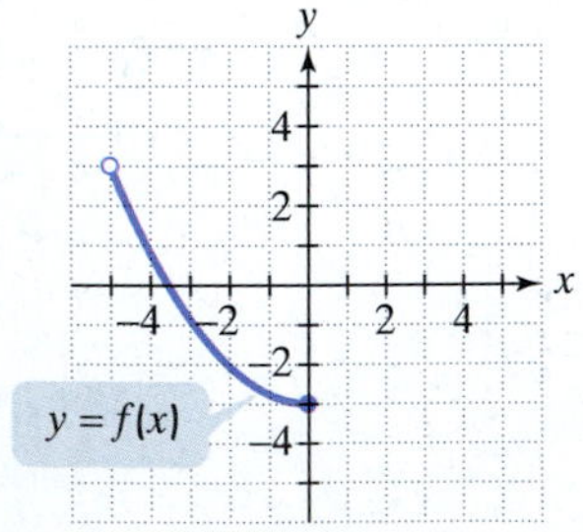

33.

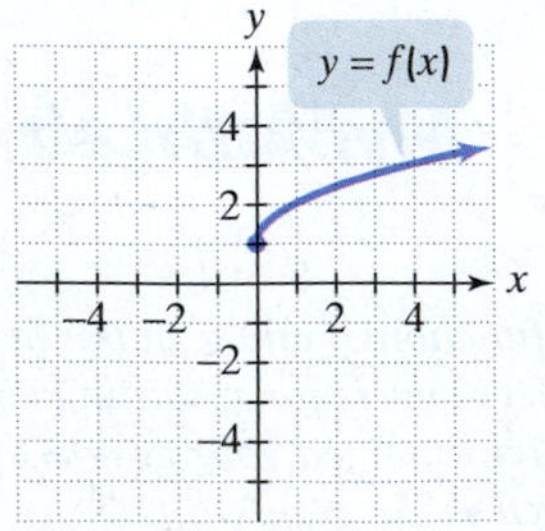

34.

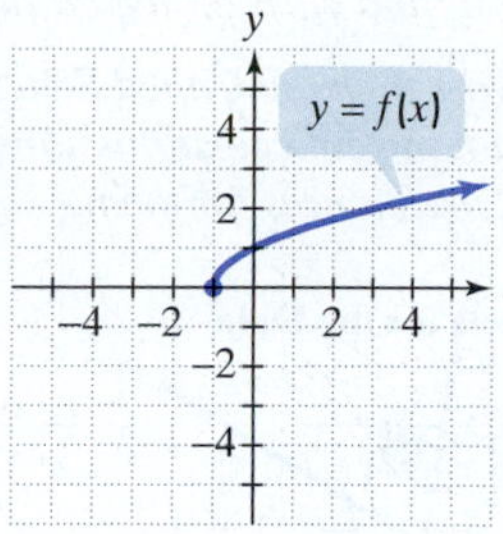

35.

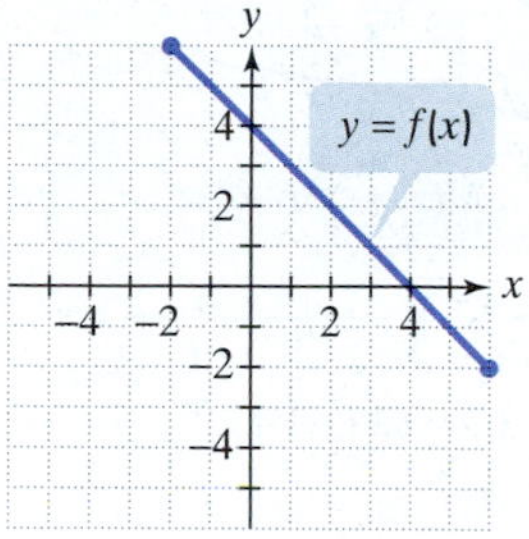

36.

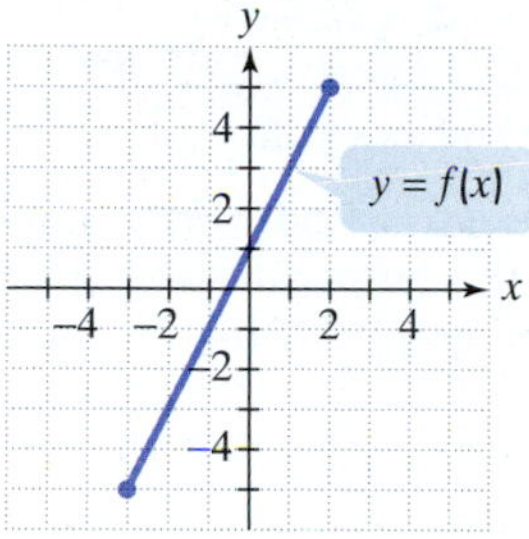

37.

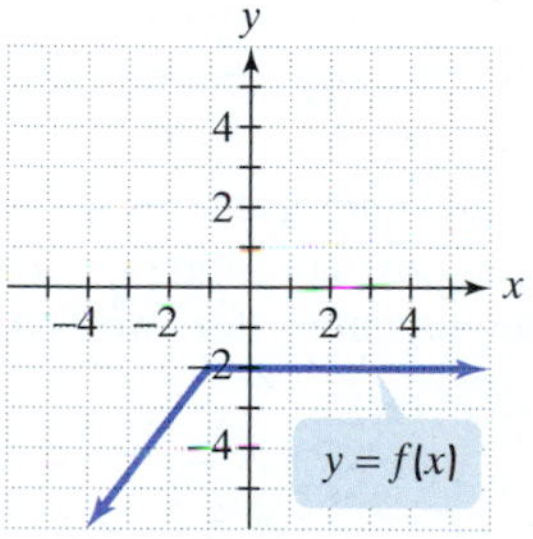

38.

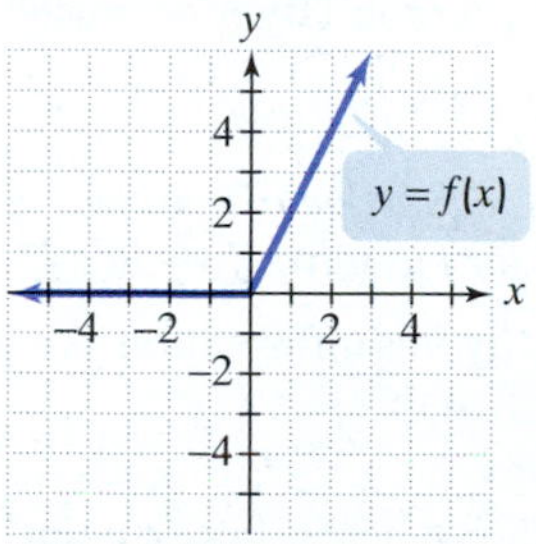

39.

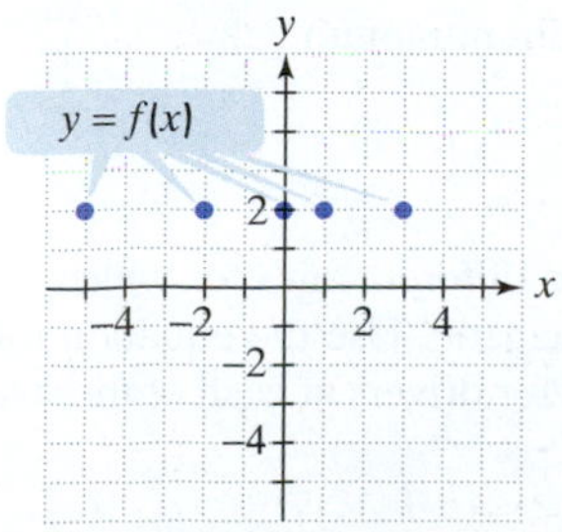

40.

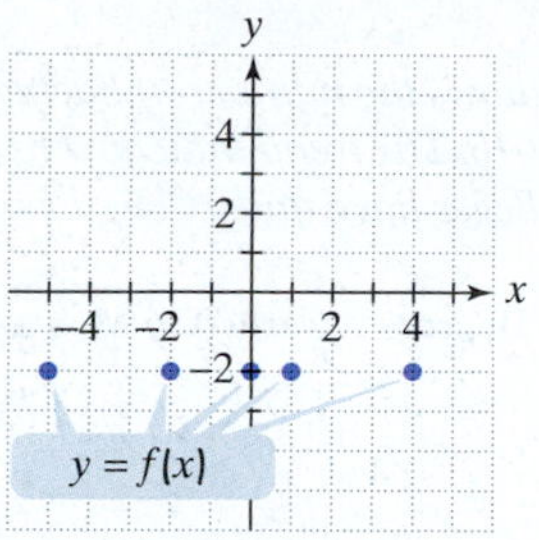

Practice PLUS

41. Use the graph of f to determine each of the following. Where applicable, use set-builder notation.

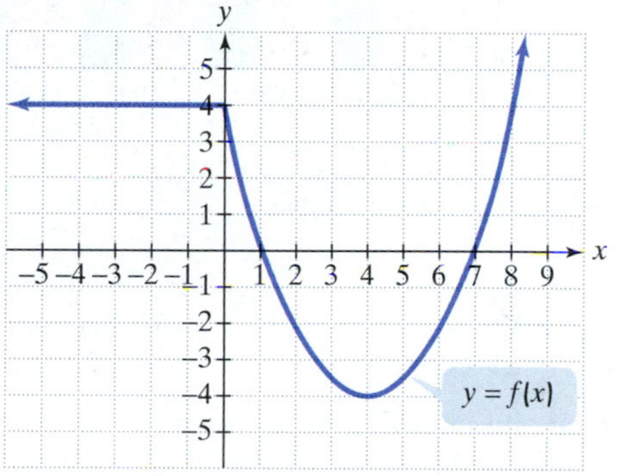

a. the domain of f

b. the range of f

c. $f(-3)$

d. the values of x for which $f(x) = -2$

e. the points where the graph of f crosses the x-axis

f. the point where the graph of f crosses the y-axis

g. values of x for which $f(x) < 0$

h. Is $f(-8)$ positive or negative?

42. Use the graph of f to determine each of the following. Where applicable, use set-builder notation.

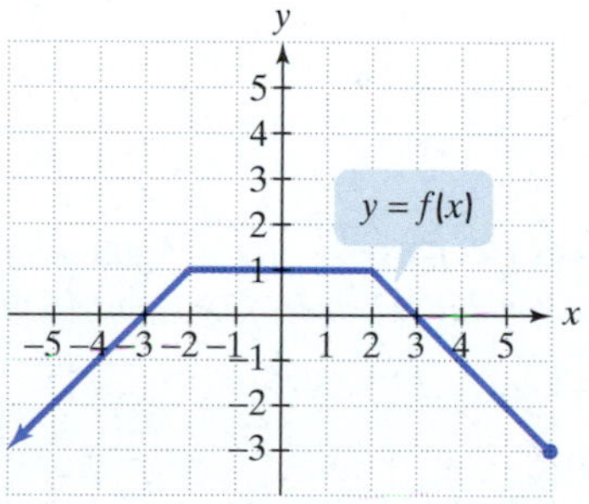

a. the domain of f

b. the range of f

c. $f(-4)$

d. the values of x for which $f(x) = -3$

e. the points where the graph of f crosses the x-axis

f. the point where the graph of f crosses the y-axis

g. values of x for which $f(x) > 0$

h. Is $f(-2)$ positive or negative?

Application Exercises

The male minority? The bar graph shows the number of bachelor's degrees, in thousands, awarded to men and women in the United States for selected years, with projections for 2010. The trend indicated by the graphs is among the hottest topics of debate among college-admissions officers. Some private liberal arts colleges have quietly begun special efforts to recruit men–including admissions preferences for them.

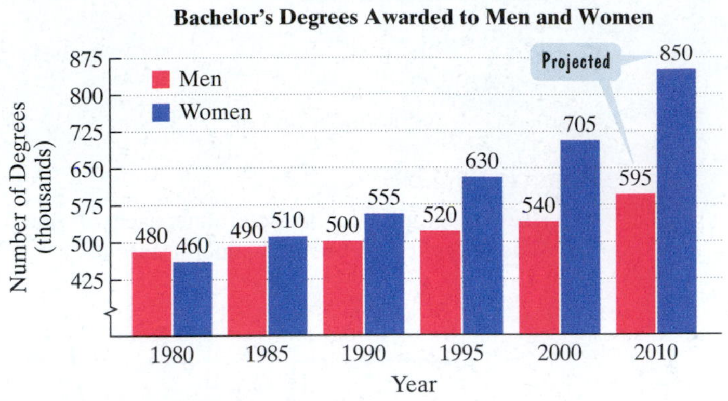

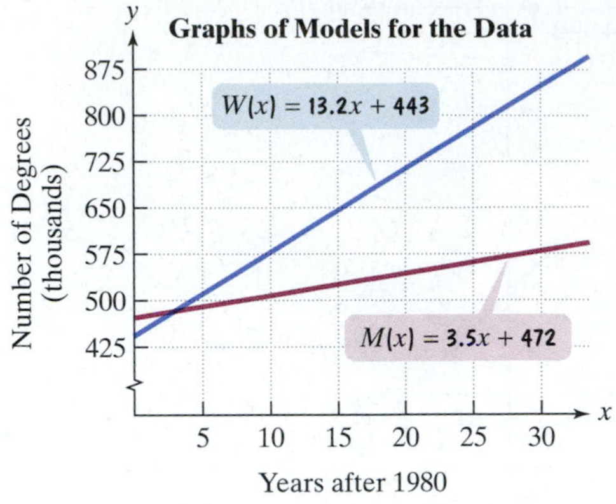

Source: Department of Education

The function $M(x) = 3.5x + 472$ *models the number of bachelor's degrees, in thousands, awarded to men x years after 1980. The function* $W(x) = 13.2x + 443$ *models the number of bachelor's degrees, in thousands, awarded to women x years after 1980. The graphs of functions M and W are shown to the right of the actual data. Use this information to solve Exercises 43–46.*

43. **a.** Find and interpret $W(20)$. Identify this information as a point on the graph of the function for women.

b. Does $W(20)$ overestimate or underestimate the actual data shown by the bar graph? By how much?

44. **a.** Find and interpret $M(20)$. Identify this information as a point on the graph of the function for men.

b. Does $M(20)$ overestimate or underestimate the actual data shown by the bar graph? By how much?

45. **a.** Use the two functions to find and interpret $W(10) - M(10)$. Identify this information as an appropriate distance between the graphs of the functions for women and men.

b. Does $W(10) - M(10)$ overestimate or underestimate the actual data shown by the bar graph? By how much?

46. **a.** Use the two functions to find and interpret $W(5) - M(5)$. Identify this information as an appropriate distance between the graphs of the functions for women and men.

b. Does $W(5) - M(5)$ overestimate or underestimate the actual data shown by the bar graph? By how much?

The function $f(x) = 0.4x^2 - 36x + 1000$ *models the number of accidents, f(x), per 50 million miles driven as a function of a driver's age, x, in years, where x includes drivers from ages 16 through 74, inclusive. The graph of f is shown. Use the equation for f to solve Exercises 47–50.*

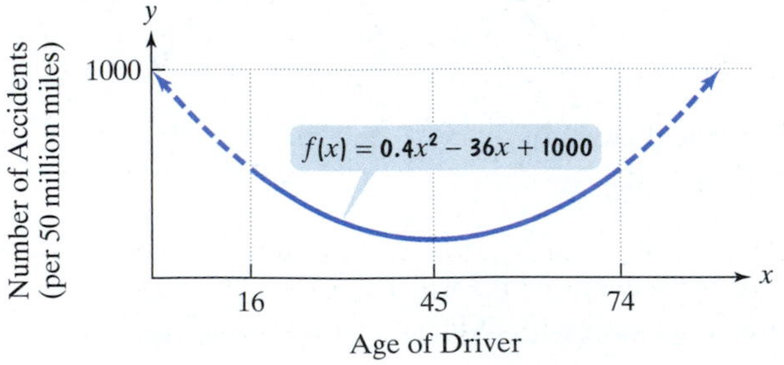

47. Find and interpret $f(20)$. Identify this information as a point on the graph of f.

48. Find and interpret $f(50)$. Identify this information as a point on the graph of f.

49. For what value of x does the graph reach its lowest point? Use the equation for f to find the minimum value of y. Describe the practical significance of this minimum value.

50. Use the graph to identify two different ages for which drivers have the same number of accidents. Use the equation for f to find the number of accidents for drivers at each of these ages.

The figure shows the percentage of Jewish Americans in the U.S. population, $f(x)$, x years after 1900. Use the graph to solve Exercises 51–58.

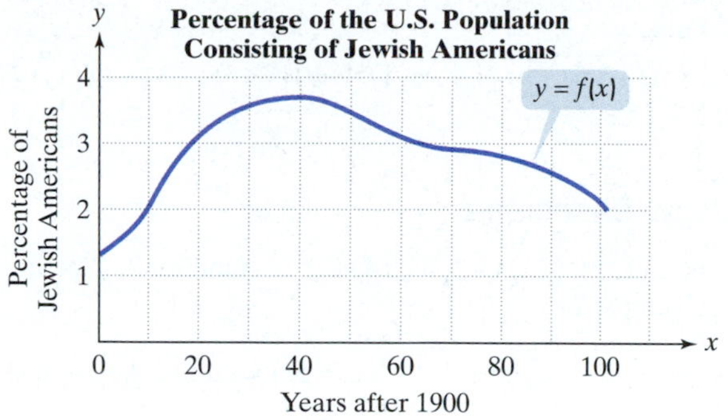

Source: American Jewish Yearbook

51. Use the graph to find a reasonable estimate of $f(60)$. What does this mean in terms of the variables in this situation?

52. Use the graph to find a reasonable estimate of $f(100)$. What does this mean in terms of the variables in this situation?

53. For what value or values of x is $f(x) = 3$? Round to the nearest year. What does this mean in terms of the variables in this situation?

54. For what value or values of x is $f(x) = 2.5$? Round to the nearest year. What does this mean in terms of the variables in this situation?

55. In which year did the percentage of Jewish Americans in the U.S. population reach a maximum? What is a reasonable estimate of the percentage for that year?

56. In which year was the percentage of Jewish Americans in the U.S. population at a minimum? What is a reasonable estimate of the percentage for that year?

57. Explain why f represents the graph of a function.

58. Describe the general trend shown by the graph.

The figure shows the cost of mailing a first-class letter, $f(x)$, as a function of its weight, x, in ounces, for weights not exceeding 3.5 ounces. Use the graph to solve Exercises 59–62.

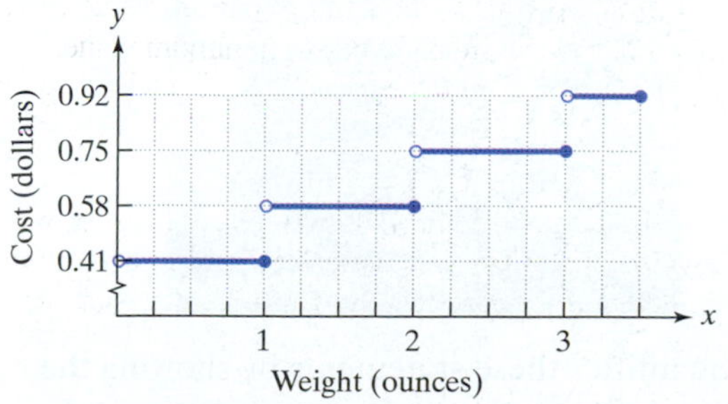

59. Find $f(3)$. What does this mean in terms of the variables in this situation?

60. Find $f(3.5)$. What does this mean in terms of the variables in this situation?

61. What is the cost of mailing a letter that weighs 1.5 ounces?

62. What is the cost of mailing a letter that weighs 1.8 ounces?

Writing in Mathematics

63. What is the graph of a function?

64. Explain how the vertical line test is used to determine whether a graph represents a function.

65. Explain how to identify the domain and range of a function from its graph.

66. Do you believe that the trend shown by the graphs for Exercises 43–46 should be reversed by providing admissions preferences for men? Explain your position on this issue.

Technology Exercises

67. Use a graphing utility to verify the pairs of graphs that you drew by hand in Exercises 1–10.

68. The function

$$f(x) = -0.00002x^3 + 0.008x^2 - 0.3x + 6.95$$

models the number of annual physician visits, $f(x)$, by a person of age x. Graph the function in a [0, 100, 5] by [0, 40, 2] viewing rectangle. What does the shape of the graph indicate about the relationship between one's age and the number of annual physician visits? Use the TRACE or minimum function capability to find the coordinates of the minimum point on the graph of the function. What does this mean?

Critical Thinking Exercises

Make Sense? *In Exercises 69–72, determine whether each statement "makes sense" or "does not make sense" and explain your reasoning.*

69. I knew how to use point plotting to graph the equation $y = x^2 - 1$, so there was really nothing new to learn when I used the same technique to graph the function $f(x) = x^2 - 1$.

70. The graph of my function revealed aspects of its behavior that were not obvious by just looking at its equation.

71. I graphed a function showing how paid vacation days depend on the number of years a person works for a company. The domain was the number of paid vacation days.

72. I graphed a function showing how the number of annual physician visits depends on a person's age. The domain was the number of annual physician visits.

In Exercises 73–78, determine whether each statement is true or false. If the statement is false, make the necessary change(s) to produce a true statement.

73. The graph of every line is a function.

74. A horizontal line can intersect the graph of a function in more than one point.

Use the graph of f to determine whether each statement in Exercises 75–78 is true or false.

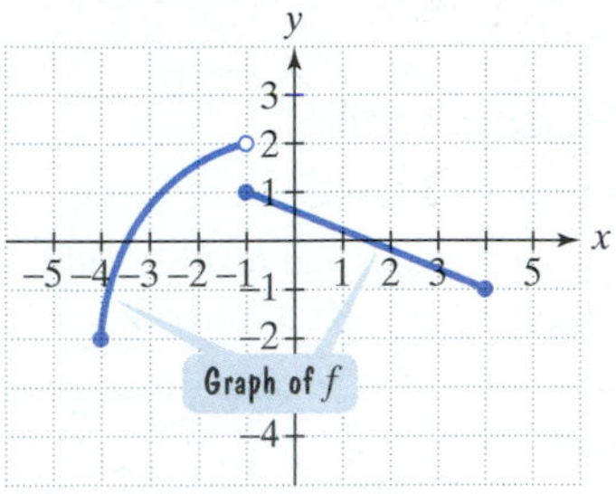

75. The domain of f is $\{x|-4 \le x \le 4\}$.

76. The range of f is $\{y|-2 \le y \le 2\}$.

77. $f(-1) - f(4) = 2$

78. $f(0) = 2.1$

In Exercises 79–80, let f be defined by the following graph:

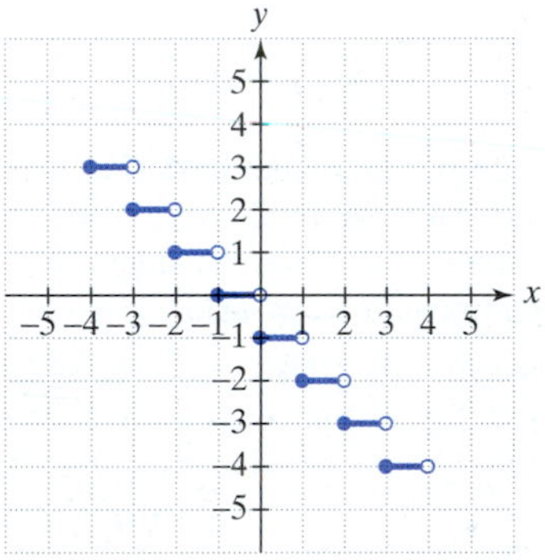

79. Find

$$\sqrt{f(-1.5) + f(-0.9)} - [f(\pi)]^2 + f(-3) \div f(1) \cdot f(-\pi).$$

80. Find

$$\sqrt{f(-2.5) - f(1.9)} - [f(-\pi)]^2 + f(-3) \div f(1) \cdot f(\pi).$$

Review Exercises

81. Is $\{(1, 1), (2, 2), (3, 3), (4, 4)\}$ a function? (Section 2.1, Example 2)

82. Solve: $12 - 2(3x + 1) = 4x - 5$. (Section 1.4, Example 3)

83. The length of a rectangle exceeds 3 times the width by 8 yards. If the perimeter of the rectangle is 624 yards, what are its dimensions? (Section 1.5, Example 5)

Preview Exercises

Exercises 84–86 will help you prepare for the material covered in the next section.

84. If $f(x) = \dfrac{4}{x - 3}$, why must 3 be excluded from the domain of f?

85. If $f(x) = x^2 + x$ and $g(x) = x - 5$, find $f(4) + g(4)$.

86. Simplify: $7.4x^2 - 15x + 4046 - (-3.5x^2 + 20x + 2405)$.

SECTION 2.3

The Algebra of Functions

Objectives

1 Find the domain of a function.

2 Use the algebra of functions to combine functions and determine domains.

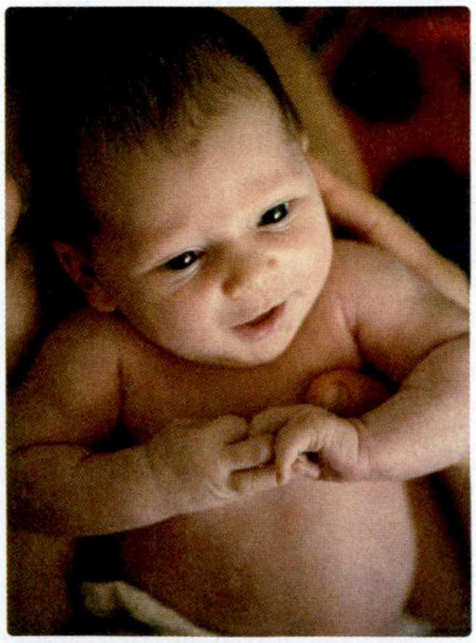

We're born. We die. **Figure 2.15** quantifies these statements by showing the number of births and deaths in the United States for six selected years.

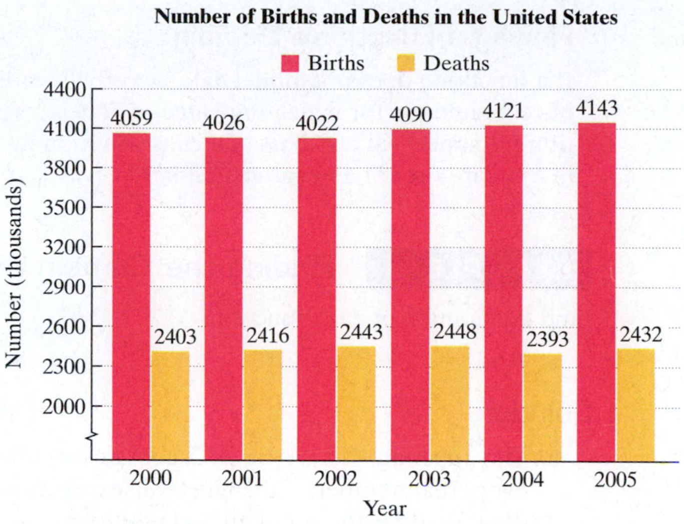

FIGURE 2.15

Source: U.S. Department of Health and Human Services

In this section, we look at these data from the perspective of functions. By considering the yearly change in the U.S. population, you will see that functions can be subtracted using procedures that will remind you of combining algebraic expressions.

1 Find the domain of a function.

The Domain of a Function

We begin with two functions that model the data in **Figure 2.15**.

$$B(x) = 7.4x^2 - 15x + 4046 \qquad D(x) = -3.5x^2 + 20x + 2405$$

Number of births, $B(x)$, in thousands, x years after 2000

Number of deaths, $D(x)$, in thousands, x years after 2000

The years in **Figure 2.15** extend from 2000 through 2005. Because x represents the number of years after 2000,

$$\text{Domain of } B = \{x \mid x = 0, 1, 2, 3, 4, 5\}$$

and

$$\text{Domain of } D = \{x \mid x = 0, 1, 2, 3, 4, 5\}.$$

Functions that model data often have their domains explicitly given with the function's equation. However, for most functions, only an equation is given and the domain is not specified. In cases like this, the domain of a function f is the largest set of real numbers for which the value of $f(x)$ is a real number. For example, consider the function

$$f(x) = \frac{1}{x-3}.$$

Because division by 0 is undefined (and not a real number), the denominator, $x - 3$, cannot be 0. Thus, x cannot equal 3. The domain of the function consists of all real numbers other than 3, represented by

$$\text{Domain of } f = \{x \mid x \text{ is a real number and } x \neq 3\}.$$

In Chapter 7, we will be studying square root functions such as

$$g(x) = \sqrt{x}.$$

The equation tells us to take the square root of x. Because only nonnegative numbers have square roots that are real numbers, the expression under the square root sign, x, must be nonnegative. Thus,

$$\text{Domain of } g = \{x \mid x \text{ is a nonnegative real number}\}.$$

Equivalently,

$$\text{Domain of } g = \{x \mid x \geq 0\}.$$

Using Technology

Graphic Connections

You can graph a function and visually determine its domain. Look for inputs on the x-axis that correspond to points on the graph. For example, consider

$$g(x) = \sqrt{x}, \text{ or } y = \sqrt{x}.$$

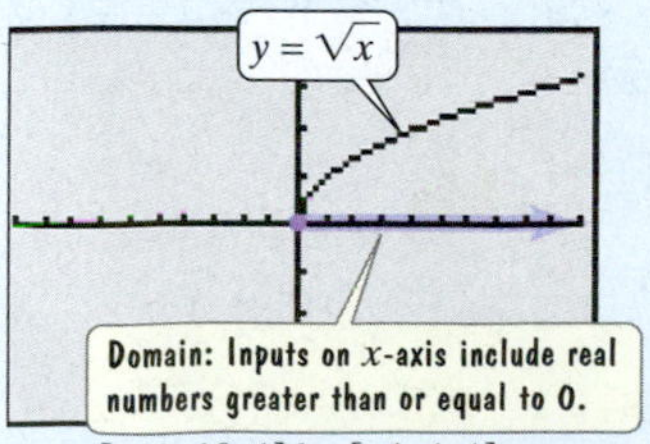

[−10, 10, 1] by [−4, 4, 1]

This verifies that $\{x \mid x \geq 0\}$ is the domain.

Finding a Function's Domain

If a function f does not model data or verbal conditions, its domain is the largest set of real numbers for which the value of $f(x)$ is a real number. Exclude from a function's domain real numbers that cause division by zero and real numbers that result in a square root of a negative number.

EXAMPLE 1 Finding the Domain of a Function

Find the domain of each function:

a. $f(x) = 3x + 2$ **b.** $g(x) = \dfrac{3x + 2}{x + 1}$.

Solution

a. The function $f(x) = 3x + 2$ contains neither division nor a square root. For every real number, x, the algebraic expression $3x + 2$ is a real number. Thus, the domain of f is the set of all real numbers.

$$\text{Domain of } f = \{x \mid x \text{ is a real number}\}$$

b. The function $g(x) = \dfrac{3x + 2}{x + 1}$ contains division. Because division by 0 is undefined, we must exclude from the domain the value of x that causes $x + 1$ to be 0. Thus, x cannot equal -1.

$$\text{Domain of } g = \{x \mid x \text{ is a real number and } x \neq -1\}$$

✓ **CHECK POINT 1** Find the domain of each function:

a. $f(x) = \dfrac{1}{2}x + 3$

b. $g(x) = \dfrac{7x + 4}{x + 5}$.

2 Use the algebra of functions to combine functions and determine domains.

The Algebra of Functions

We can combine functions using addition, subtraction, multiplication, and division by performing operations with the algebraic expressions that appear on the right side of the equations. For example, the functions $f(x) = 2x$ and $g(x) = x - 1$ can be combined to form the sum, difference, product, and quotient of f and g. Here's how it's done:

For each function, $f(x) = 2x$ and $g(x) = x - 1$.

Sum: $f + g$
$$(f + g)(x) = f(x) + g(x) = 2x + (x - 1) = 3x - 1$$

Difference: $f - g$
$$(f - g)(x) = f(x) - g(x) = 2x - (x - 1) = 2x - x + 1 = x + 1$$

Product: fg
$$(fg)(x) = f(x) \cdot g(x) = 2x(x - 1) = 2x^2 - 2x$$

Quotient: $\dfrac{f}{g}$
$$\left(\frac{f}{g}\right)(x) = \frac{f(x)}{g(x)} = \frac{2x}{x - 1}, x \neq 1.$$

The domain for each of these functions consists of all real numbers that are common to the domains of f and g. In the case of the quotient function $\dfrac{f(x)}{g(x)}$, we must remember not to divide by 0, so we add the further restriction that $g(x) \neq 0$.

The Algebra of Functions: Sum, Difference, Product, and Quotient of Functions

Let f and g be two functions. The **sum** $f + g$, the **difference** $f - g$, the **product** fg, and the **quotient** $\frac{f}{g}$ are functions whose domains are the set of all real numbers common to the domains of f and g, defined as follows:

1. Sum: $(f + g)(x) = f(x) + g(x)$
2. Difference: $(f - g)(x) = f(x) - g(x)$
3. Product: $(fg)(x) = f(x) \cdot g(x)$
4. Quotient: $\left(\frac{f}{g}\right)(x) = \frac{f(x)}{g(x)}$, provided $g(x) \neq 0$.

EXAMPLE 2 Using the Algebra of Functions

Let $f(x) = x^2 - 3$ and $g(x) = 4x + 5$. Find each of the following:

a. $(f + g)(x)$ b. $(f + g)(3)$.

Solution

a. $(f + g)(x) = f(x) + g(x) = (x^2 - 3) + (4x + 5) = x^2 + 4x + 2$
Thus,

$$(f + g)(x) = x^2 + 4x + 2.$$

b. We find $(f + g)(3)$ by substituting 3 for x in the equation for $f + g$.

$$(f + g)(x) = x^2 + 4x + 2$$ This is the equation for $f + g$.

Substitute 3 for x.

$$(f + g)(3) = 3^2 + 4 \cdot 3 + 2 = 9 + 12 + 2 = 23$$

✓ **CHECK POINT 2** Let $f(x) = 3x^2 + 4x - 1$ and $g(x) = 2x + 7$. Find each of the following:

a. $(f + g)(x)$ b. $(f + g)(4)$.

EXAMPLE 3 Using the Algebra of Functions

Let $f(x) = \frac{4}{x}$ and $g(x) = \frac{3}{x + 2}$. Find each of the following:

a. $(f - g)(x)$ b. the domain of $f - g$.

Solution

a. $(f - g)(x) = f(x) - g(x) = \frac{4}{x} - \frac{3}{x + 2}$

(In Chapter 6, we will discuss how to perform the subtraction with these algebraic fractions. Perhaps you remember how to do so from your work in introductory algebra. For now, we will leave these fractions in the form shown.)

b. The domain of $f - g$ is the set of all real numbers that are common to the domain of f and the domain of g. Thus, we must find the domains of f and g. We will do so for f first.

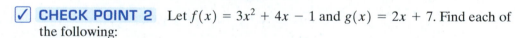

Note that $f(x) = \frac{4}{x}$ is a function involving division. Because division by 0 is undefined, x cannot equal 0.

$$\text{Domain of } f = \{x \mid x \text{ is a real number and } x \neq 0\}$$

The function $g(x) = \frac{3}{x+2}$ is also a function involving division. Because division by 0 is undefined, x cannot equal -2.

$$\text{Domain of } g = \{x \mid x \text{ is a real number and } x \neq -2\}$$

To find $f(x) - g(x)$, x must be in both domains listed. Thus,

$$\text{Domain of } f - g = \{x \mid x \text{ is a real number and } x \neq 0 \text{ and } x \neq -2\}.$$

✓ CHECK POINT 3 Let $f(x) = \frac{5}{x}$ and $g(x) = \frac{7}{x-8}$. Find each of the following:

a. $(f - g)(x)$

b. the domain of $f - g$.

EXAMPLE 4 Using the Algebra of Functions

Let $f(x) = x^2 + x$ and $g(x) = x - 5$. Find each of the following:

a. $(f + g)(4)$

b. $(f - g)(x)$ and $(f - g)(-3)$

c. $\left(\frac{f}{g}\right)(x)$ and $\left(\frac{f}{g}\right)(7)$

d. $(fg)(-2)$.

Solution

a. We can find $(f + g)(4)$ using $f(4)$ and $g(4)$.

$$f(x) = x^2 + x \qquad g(x) = x - 5$$
$$f(4) = 4^2 + 4 = 20 \qquad g(4) = 4 - 5 = -1$$

Thus,

$$(f + g)(4) = f(4) + g(4) = 20 + (-1) = 19.$$

We can also find $(f + g)(4)$ by first finding $(f + g)(x)$ and then substituting 4 for x:

$$(f + g)(x) = f(x) + g(x)$$ This is the definition of the sum $f + g$.

$$= (x^2 + x) + (x - 5)$$ Substitute the given functions.

$$= x^2 + 2x - 5.$$ Simplify.

Using $(f + g)(x) = x^2 + 2x - 5$, we have

$$(f + g)(4) = 4^2 + 2 \cdot 4 - 5 = 16 + 8 - 5 = 19.$$

b. $$(f - g)(x) = f(x) - g(x)$$ This is the definition of the difference $f - g$.

$$= (x^2 + x) - (x - 5)$$ Substitute the given functions.

$$= x^2 + x - x + 5$$ Remove parentheses and change the sign of each term in the second set of parentheses.

$$= x^2 + 5$$ Simplify.

Using $(f - g)(x) = x^2 + 5$, we have

$$(f - g)(-3) = (-3)^2 + 5 = 9 + 5 = 14.$$

c. $\left(\frac{f}{g}\right)(x) = \frac{f(x)}{g(x)}$ This is the definition of the quotient $\frac{f}{g}$.

$= \frac{x^2 + x}{x - 5}$ Substitute the given functions.

Using $\left(\frac{f}{g}\right)(x) = \frac{x^2 + x}{x - 5}$, we have

$$\left(\frac{f}{g}\right)(7) = \frac{7^2 + 7}{7 - 5} = \frac{56}{2} = 28.$$

d. We can find $(fg)(-2)$ using the fact that

$$(fg)(-2) = f(-2) \cdot g(-2).$$

$$f(x) = x^2 + x \qquad g(x) = x - 5$$

$$f(-2) = (-2)^2 + (-2) = 4 - 2 = 2 \qquad g(-2) = -2 - 5 = -7$$

Thus,

$$(fg)(-2) = f(-2) \cdot g(-2) = 2(-7) = -14.$$

We could also have found $(fg)(-2)$ by multiplying $f(x) \cdot g(x)$ and then substituting -2 into the product. We will discuss how to multiply expressions such as $x^2 + x$ and $x - 5$ in Chapter 5.

CHECK POINT 4 Let $f(x) = x^2 - 2x$ and $g(x) = x + 3$. Find each of the following:

a. $(f + g)(5)$

b. $(f - g)(x)$ and $(f - g)(-1)$

c. $\left(\frac{f}{g}\right)(x)$ and $\left(\frac{f}{g}\right)(7)$

d. $(fg)(-4)$.

EXAMPLE 5 Applying the Algebra of Functions

We opened the section with functions that model the number of births and deaths in the United States from 2000 through 2005:

$$B(x) = 7.4x^2 - 15x + 4046 \qquad D(x) = -3.5x^2 + 20x + 2405.$$

Number of births, $B(x)$, in thousands, x years after 2000

Number of deaths, $D(x)$, in thousands, x years after 2000

a. Write a function that models the change in U.S. population for each year from 2000 through 2005.

b. Use the function from part (a) to find the change in U.S. population in 2003.

c. Does the result in part (b) overestimate or underestimate the actual population change in 2003 obtained from the data in **Figure 2.15** on page 117? By how much?

Solution

a. The change in population is the number of births minus the number of deaths. Thus, we will find the difference function, $B - D$.

$(B - D)(x)$

$= B(x) - D(x)$

$= (7.4x^2 - 15x + 4046) - (-3.5x^2 + 20x + 2405)$ Substitute the given functions.

$= 7.4x^2 - 15x + 4046 + 3.5x^2 - 20x - 2405$ Remove parentheses and change the sign of each term in the second set of parentheses.

$= (7.4x^2 + 3.5x^2) + (-15x - 20x) + (4046 - 2405)$ Group like terms.

$= 10.9x^2 - 35x + 1641$ Combine like terms.

The function

$$(B - D)(x) = 10.9x^2 - 35x + 1641$$

models the change in U.S. population, in thousands, x years after 2000.

b. Because 2003 is 3 years after 2000, we substitute 3 for x in the difference function $(B - D)(x)$.

$$(B - D)(x) = 10.9x^2 - 35x + 1641 \quad \text{Use the difference function } B - D.$$
$$(B - D)(3) = 10.9(3)^2 - 35(3) + 1641 \quad \text{Substitute 3 for } x.$$
$$= 10.9(9) - 35(3) + 1641 \quad \text{Evaluate the exponential expression: } 3^2 = 9.$$
$$= 98.1 - 105 + 1641 \quad \text{Perform the multiplications.}$$
$$= 1634.1 \quad \text{Subtract and add from left to right.}$$

We see that $(B - D)(3) = 1634.1$. The model indicates that there was a population increase of 1634.1 thousand, or approximately 1,634,000 people, in 2003.

c. The data for 2003 in **Figure 2.15** on page 117 show 4090 thousand births and 2448 thousand deaths.

$$\text{population change} = \text{births} - \text{deaths}$$
$$= 4090 - 2448 = 1642$$

The actual population increase was 1642 thousand, or 1,642,000. Our model gave us an increase of 1634.1 thousand. Thus, the model underestimates the actual increase by $1642 - 1634.1$, or 7.9 thousand people.

CHECK POINT 5 Use the birth and death models from Example 5.

a. Write a function that models the total number of births and deaths in the United States for each year from 2000 through 2005.

b. Use the function from part (a) to find the total number of births and deaths in the United States in 2005.

c. Does the result in part (b) overestimate or underestimate the actual number of total births and deaths in 2005 obtained from the data in **Figure 2.15** on page 117? By how much?

2.3 EXERCISE SET

MyMathLab MathXL PRACTICE WATCH DOWNLOAD READ REVIEW

Practice Exercises

In Exercises 1–10, find the domain of each function.

1. $f(x) = 3x + 5$
2. $f(x) = 4x + 7$
3. $g(x) = \dfrac{1}{x + 4}$
4. $g(x) = \dfrac{1}{x + 5}$
5. $f(x) = \dfrac{2x}{x - 3}$
6. $f(x) = \dfrac{4x}{x - 2}$
7. $g(x) = x + \dfrac{3}{5 - x}$
8. $g(x) = x + \dfrac{7}{6 - x}$
9. $f(x) = \dfrac{1}{x + 7} + \dfrac{3}{x - 9}$
10. $f(x) = \dfrac{1}{x + 8} + \dfrac{3}{x - 10}$

In Exercises 11–16, find $(f + g)(x)$ and $(f + g)(5)$.

11. $f(x) = 3x + 1, g(x) = 2x - 6$
12. $f(x) = 4x + 2, g(x) = 2x - 9$

13. $f(x) = x - 5, g(x) = 3x^2$

14. $f(x) = x - 6, g(x) = 2x^2$

15. $f(x) = 2x^2 - x - 3, g(x) = x + 1$

16. $f(x) = 4x^2 - x - 3, g(x) = x + 1$

In Exercises 17–28, for each pair of functions, f and g, determine the domain of f + g.

17. $f(x) = 3x + 7, g(x) = 9x + 10$

18. $f(x) = 7x + 4, g(x) = 5x - 2$

19. $f(x) = 3x + 7, g(x) = \dfrac{2}{x - 5}$

20. $f(x) = 7x + 4, g(x) = \dfrac{2}{x - 6}$

21. $f(x) = \dfrac{1}{x}, g(x) = \dfrac{2}{x - 5}$

22. $f(x) = \dfrac{1}{x}, g(x) = \dfrac{2}{x - 6}$

23. $f(x) = \dfrac{8x}{x - 2}, g(x) = \dfrac{6}{x + 3}$

24. $f(x) = \dfrac{9x}{x - 4}, g(x) = \dfrac{7}{x + 8}$

25. $f(x) = \dfrac{8x}{x - 2}, g(x) = \dfrac{6}{2 - x}$

26. $f(x) = \dfrac{9x}{x - 4}, g(x) = \dfrac{7}{4 - x}$

27. $f(x) = x^2, g(x) = x^3$

28. $f(x) = x^2 + 1, g(x) = x^3 - 1$

In Exercises 29–48, let

$$f(x) = x^2 + 4x \quad \text{and} \quad g(x) = 2 - x.$$

Find each of the following.

29. $(f + g)(x)$ and $(f + g)(3)$

30. $(f + g)(x)$ and $(f + g)(4)$

31. $f(-2) + g(-2)$

32. $f(-3) + g(-3)$

33. $(f - g)(x)$ and $(f - g)(5)$

34. $(f - g)(x)$ and $(f - g)(6)$

35. $f(-2) - g(-2)$

36. $f(-3) - g(-3)$

37. $(fg)(-2)$

38. $(fg)(-3)$

39. $(fg)(5)$

40. $(fg)(6)$

41. $\left(\dfrac{f}{g}\right)(x)$ and $\left(\dfrac{f}{g}\right)(1)$

42. $\left(\dfrac{f}{g}\right)(x)$ and $\left(\dfrac{f}{g}\right)(3)$

43. $\left(\dfrac{f}{g}\right)(-1)$

44. $\left(\dfrac{f}{g}\right)(0)$

45. The domain of $f + g$

46. The domain of $f - g$

47. The domain of $\dfrac{f}{g}$

48. The domain of fg

Practice PLUS

Use the graphs of f and g to solve Exercises 49–56.

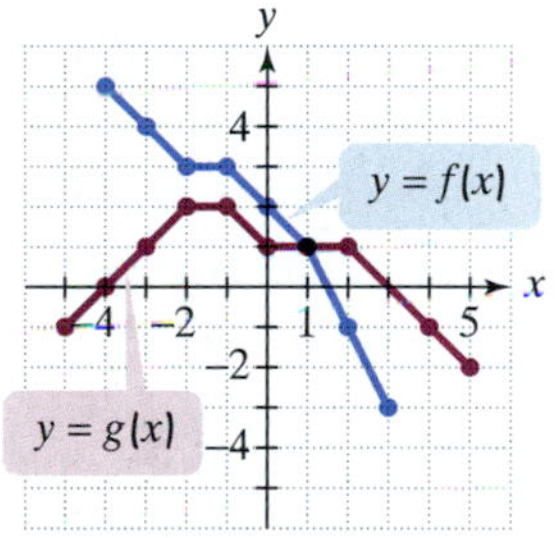

49. Find $(f + g)(-3)$.

50. Find $(g - f)(-2)$.

51. Find $(fg)(2)$.

52. Find $\left(\dfrac{g}{f}\right)(3)$.

53. Find the domain of $f + g$.

54. Find the domain of $\dfrac{f}{g}$.

55. Graph $f + g$.

56. Graph $f - g$.

Use the table defining f and g to solve Exercises 57–60.

x	$f(x)$	$g(x)$
−2	5	0
−1	3	−2
0	−2	4
1	−6	−3
2	0	1

57. Find $(f + g)(1) - (g - f)(-1)$.

58. Find $(f + g)(-1) - (g - f)(0)$.

59. Find $(fg)(-2) - \left[\left(\dfrac{f}{g}\right)(1)\right]^2$.

60. Find $(fg)(2) - \left[\left(\dfrac{g}{f}\right)(0)\right]^2$.

Application Exercises

The bar graph shows the population of the United States, in millions, for five selected years.

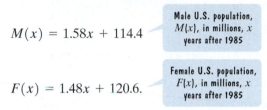

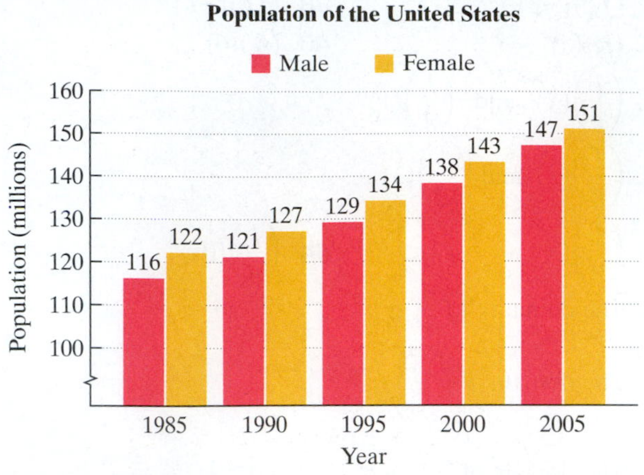

Source: U.S. Census Bureau

Here are two functions that model the data:

$$M(x) = 1.58x + 114.4$$ Male U.S. population, $M(x)$, in millions, x years after 1985

$$F(x) = 1.48x + 120.6.$$ Female U.S. population, $F(x)$, in millions, x years after 1985

Use these functions to solve Exercises 61–63.

61. a. Write a function that models the total U.S. population for the years shown in the bar graph.

b. Use the function from part (a) to find the total U.S. population in 2005.

c. Does the result in part (b) overestimate or underestimate the actual total U.S. population in 2005 shown by the bar graph? By how much?

62. a. Write a function that models the difference between the female U.S. population and the male U.S. population for the years shown in the bar graph.

b. Use the function from part (a) to find how many more women than men there were in the U.S. population in 2005.

c. Does the result in part (b) overestimate or underestimate the actual difference between the female and male population in 2005 shown by the bar graph? By how much?

63. a. Write a function that models the ratio of men to women in the U.S. population for the years shown in the bar graph.

b. Use the function from part (a) to find the ratio of men to women, correct to three decimal places, in 2000.

c. Does the result in part (b) overestimate or underestimate the actual ratio of men to women in 2000 shown by the bar graph? By how much?

64. A company that sells radios has yearly fixed costs of \$600,000. It costs the company \$45 to produce each radio. Each radio will sell for \$65. The company's costs and revenue are modeled by the following functions:

$$C(x) = 600{,}000 + 45x$$ This function models the company's costs.

$$R(x) = 65x$$ This function models the company's revenue.

Find and interpret $(R - C)(20{,}000)$, $(R - C)(30{,}000)$, and $(R - C)(40{,}000)$.

Writing in Mathematics

65. If a function is defined by an equation, explain how to find its domain.

66. If equations for functions f and g are given, explain how to find $f + g$.

67. If the equations of two functions are given, explain how to obtain the quotient function and its domain.

68. If equations for functions f and g are given, describe two ways to find $(f - g)(3)$.

Technology Exercises

In Exercises 69–72, graph each of the three functions in the same $[-10, 10, 1]$ *by* $[-10, 10, 1]$ *viewing rectangle.*

69. $y_1 = 2x + 3$
$y_2 = 2 - 2x$
$y_3 = y_1 + y_2$

70. $y_1 = x - 4$
$y_2 = 2x$
$y_3 = y_1 - y_2$

71. $y_1 = x$
$y_2 = x - 4$
$y_3 = y_1 \cdot y_2$

72. $y_1 = x^2 - 2x$
$y_2 = x$
$y_3 = \dfrac{y_1}{y_2}$

73. In Exercise 72, use the TRACE feature to trace along y_3. What happens at $x = 0$? Explain why this occurs.

Critical Thinking Exercises

Make Sense? *In Exercises 74–77, determine whether each statement "makes sense" or "does not make sense" and explain your reasoning.*

74. There is an endless list of real numbers that cannot be included in the domain of $f(x) = \sqrt{x}$.

75. I used a function to model data from 1980 through 2005. The independent variable in my model represented the number of years after 1980, so the function's domain was $\{x \mid x = 0, 1, 2, 3, \ldots, 25\}$.

76. If I have equations for functions f and g, and 3 is in both domains, then there are always two ways to determine $(f + g)(3)$.

77. I have two functions. Function f models total world population x years after 2000 and function g models population of the world's more-developed regions x years after 2000. I can use $f - g$ to determine the population of the world's less-developed regions for the years in both function's domains.

In Exercises 78–81, determine whether each statement is true or false. If the statement is false, make the necessary change(s) to produce a true statement.

78. If $(f + g)(a) = 0$, then $f(a)$ and $g(a)$ must be opposites, or additive inverses.

79. If $(f - g)(a) = 0$, then $f(a)$ and $g(a)$ must be equal.

80. If $\left(\frac{f}{g}\right)(a) = 0$, then $f(a)$ must be 0.

81. If $(fg)(a) = 0$, then $f(a)$ must be 0.

Review Exercises

82. Solve for b: $R = 3(a + b)$. (Section 1.5, Example 6)

83. Solve: $3(6 - x) = 3 - 2(x - 4)$. (Section 1.4, Example 3)

84. If $f(x) = 6x - 4$, find $f(b + 2)$. (Section 2.1, Example 3)

Preview Exercises

Exercises 85–87 will help you prepare for the material covered in the next section.

85. Consider $4x - 3y = 6$.

a. What is the value of x when $y = 0$?

b. What is the value of y when $x = 0$?

86. a. Graph $y = 2x + 4$. Select integers for x from -3 to 1, inclusive.

b. At what point does the graph cross the x-axis?

c. At what point does the graph cross the y-axis?

87. Solve for y: $5x + 3y = -12$.

MID-CHAPTER CHECK POINT Section 2.1–Section 2.3

What You Know: We learned that a function is a relation in which no two ordered pairs have the same first component and different second components. We represented functions as equations and used function notation. We graphed functions and applied the vertical line test to identify graphs of functions. We determined the domain and range of a function from its graph, using inputs on the x-axis for the domain and outputs on the y-axis for the range. Finally, we developed an algebra of functions to combine functions and determine their domains.

In Exercises 1–6, determine whether each relation is a function. Give the domain and range for each relation.

1. $\{(2, 6), (1, 4), (2, -6)\}$

2. $\{(0, 1), (2, 1), (3, 4)\}$

3.

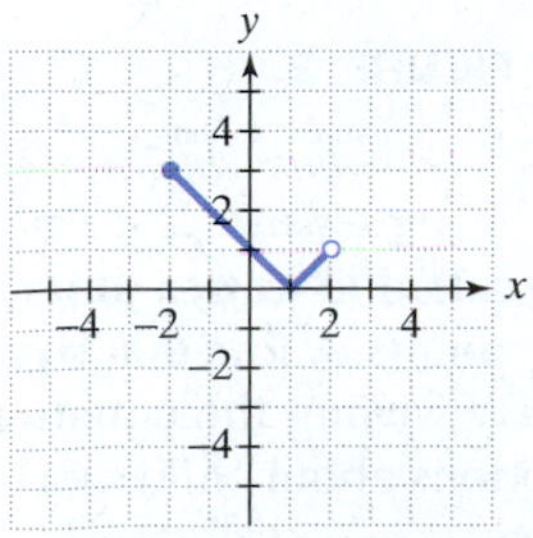

4.

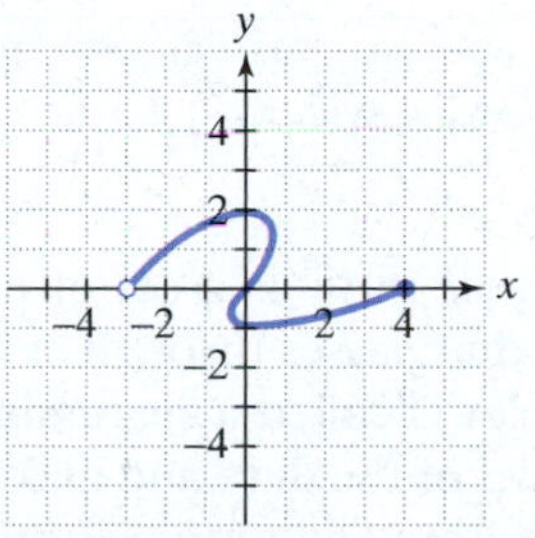

5.

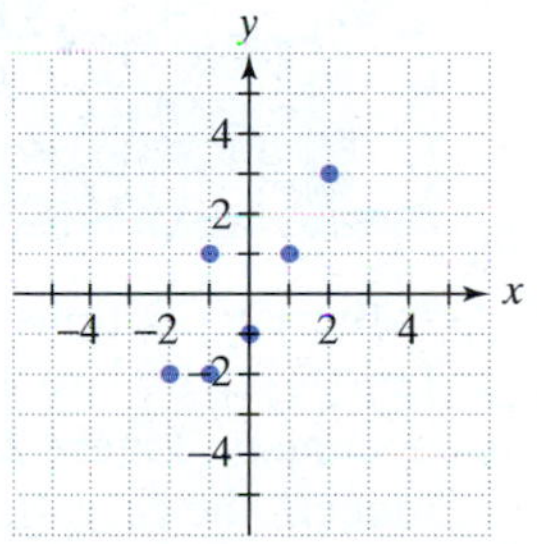

6.

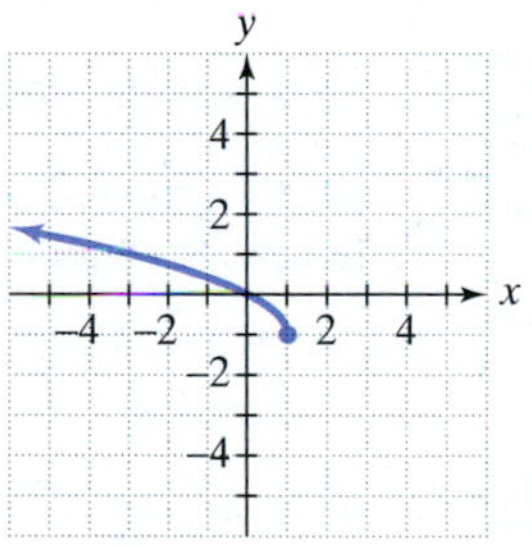

Use the graph of f to solve Exercises 7–12.

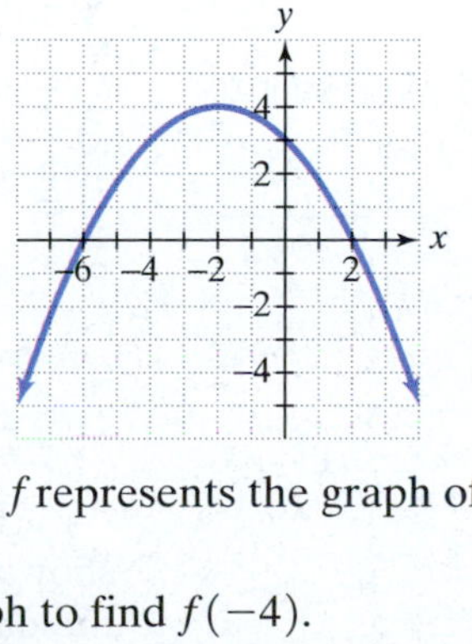

7. Explain why f represents the graph of a function.

8. Use the graph to find $f(-4)$.

9. For what value or values of x is $f(x) = 4$?

10. For what value or values of x is $f(x) = 0$?

11. Find the domain of f.

12. Find the range of f.

In Exercises 13–14, find the domain of each function.

13. $f(x) = (x + 2)(x - 2)$

14. $g(x) = \dfrac{1}{(x + 2)(x - 2)}$

In Exercises 15–22, let

$$f(x) = x^2 - 3x + 8 \text{ and } g(x) = -2x - 5.$$

Find each of the following.

15. $f(0) + g(-10)$

16. $f(-1) - g(3)$

17. $f(a) + g(a + 3)$

18. $(f + g)(x)$ and $(f + g)(-2)$

19. $(f - g)(x)$ and $(f - g)(5)$

20. $(fg)(-1)$

21. $\left(\dfrac{f}{g}\right)(x)$ and $\left(\dfrac{f}{g}\right)(-4)$

22. The domain of $\dfrac{f}{g}$

SECTION 2.4

Linear Functions and Slope

Objectives

1. Use intercepts to graph a linear function in standard form.
2. Compute a line's slope.
3. Find a line's slope and y-intercept from its equation.
4. Graph linear functions in slope-intercept form.
5. Graph horizontal or vertical lines.
6. Interpret slope as rate of change.
7. Find a function's average rate of change.
8. Use slope and y-intercept to model data.

Is there a relationship between literacy and child mortality? As the percentage of adult females who are literate increases, does the mortality of children under age five decrease? **Figure 2.16**, based on data from the United Nations, indicates that this is, indeed, the case. Each point in the figure represents one country.

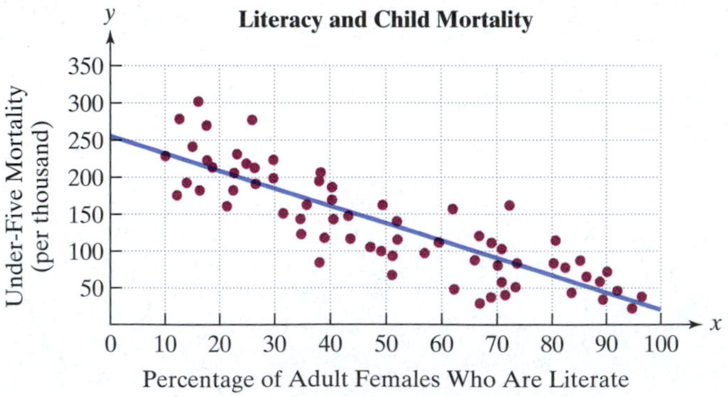

FIGURE 2.16
Source: United Nations

Data presented in a visual form as a set of points is called a **scatter plot**. Also shown in **Figure 2.16** is a line that passes through or near the points. A line that best fits the data points in a scatter plot is called a **regression line**. By writing the equation of this line, we can obtain a model of the data and make predictions about child mortality based on the percentage of literate adult females in a country.

Data often fall on or near a line. In the remainder of this chapter, we will use equations to model such data and make predictions. We begin with a discussion of graphing linear functions using intercepts.

1 Use intercepts to graph a linear function in standard form.

Graphing Using Intercepts

The equation of the regression line in **Figure 2.16** is

$$y = -2.39x + 254.47.$$

The variable x represents the percentage of adult females in a country who are literate. The variable y represents child mortality, per thousand, for children under five in that country. Using function notation, we can rewrite the equation as

$$f(x) = -2.39x + 254.47.$$

A function such as this, whose graph is a straight line, is called a **linear function**. There is another way that we can write the function's equation

$$y = -2.39x + 254.47.$$

We will collect the x- and y-terms on the left side. This is done by adding $2.39x$ to both sides:

$$2.39x + y = 254.47.$$

The form of this equation is $Ax + By = C$.

$$2.39x \quad + \quad y = 254.47$$

A, the coefficient of x, is 2.39. | B, the coefficient of y, is 1. | C, the constant on the right, is 254.47.

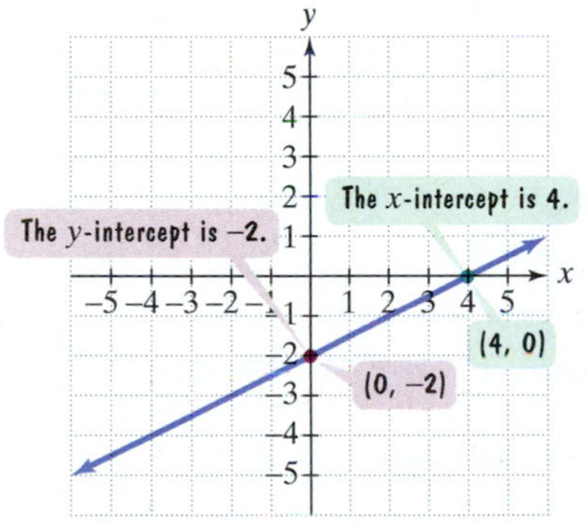

FIGURE 2.17 The graph of $2x - 4y = 8$

All equations of the form $Ax + By = C$ are straight lines when graphed, as long as A and B are not both zero. Such an equation is called the **standard form of the equation of a line**. To graph equations of this form, we will use two important points: the **intercepts**.

An ***x*-intercept** of a graph is the x-coordinate of a point where the graph intersects the x-axis. For example, look at the graph of $2x - 4y = 8$ in **Figure 2.17**. The graph crosses the x-axis at $(4, 0)$. Thus, the x-intercept is 4. **The *y*-coordinate corresponding to an *x*-intercept is always zero.**

A ***y*-intercept** of a graph is the y-coordinate of a point where the graph intersects the y-axis. The graph of $2x - 4y = 8$ in **Figure 2.17** shows that the graph crosses the y-axis at $(0, -2)$. Thus, the y-intercept is -2. **The *x*-coordinate corresponding to a *y*-intercept is always zero.**

Study Tip

Mathematicians tend to use two ways to describe intercepts. Did you notice that we are using single numbers? If a graph's x-intercept is a, it passes through the point $(a, 0)$. If a graph's y-intercept is b, it passes through the point $(0, b)$.

Some books state that the x-intercept is the *point* $(a, 0)$ and the x-intercept is *at a* on the x-axis. Similarly, the y-intercept is the *point* $(0, b)$ and the y-intercept is *at b* on the y-axis. In these descriptions, the intercepts are the actual points where a graph crosses the axes.

Although we'll describe intercepts as single numbers, we'll immediately state the point on the x- or y-axis that the graph passes through. Here's the important thing to keep in mind:

x-intercept: The corresponding y-coordinate is 0.

y-intercept: The corresponding x-coordinate is 0.

When graphing using intercepts, it is a good idea to use a third point, a check-point, before drawing the line. A checkpoint can be obtained by selecting a value for either variable, other than 0, and finding the corresponding value for the other variable. The checkpoint should lie on the same line as the x- and y-intercepts. If it does not, recheck your work and find the error.

Using Intercepts to Graph $Ax + By = C$

1. Find the x-intercept. Let $y = 0$ and solve for x.
2. Find the y-intercept. Let $x = 0$ and solve for y.
3. Find a checkpoint, a third ordered-pair solution.
4. Graph the equation by drawing a line through the three points.

EXAMPLE 1 Using Intercepts to Graph a Linear Equation

Graph: $4x - 3y = 6$.

Solution

Step 1. Find the x-intercept. Let $y = 0$ and solve for x.

$$4x - 3 \cdot 0 = 6 \quad \text{Replace } y \text{ with 0 in } 4x - 3y = 6.$$
$$4x = 6 \quad \text{Simplify.}$$
$$x = \frac{6}{4} = \frac{3}{2} \quad \text{Divide both sides by 4.}$$

The x-intercept is $\frac{3}{2}$, so the line passes through $\left(\frac{3}{2}, 0\right)$ or $(1.5, 0)$.

Step 2. Find the y-intercept. Let $x = 0$ and solve for y.

$$4 \cdot 0 - 3y = 6 \quad \text{Replace } x \text{ with 0 in } 4x - 3y = 6.$$
$$-3y = 6 \quad \text{Simplify.}$$
$$y = -2 \quad \text{Divide both sides by } -3.$$

The y-intercept is -2, so the line passes through $(0, -2)$.

Step 3. Find a checkpoint, a third ordered-pair solution. For our checkpoint, we will let $x = 1$ and find the corresponding value for y.

$$4x - 3y = 6 \quad \text{This is the given equation.}$$
$$4 \cdot 1 - 3y = 6 \quad \text{Substitute 1 for } x.$$
$$4 - 3y = 6 \quad \text{Simplify.}$$
$$-3y = 2 \quad \text{Subtract 4 from both sides.}$$
$$y = -\frac{2}{3} \quad \text{Divide both sides by } -3.$$

The checkpoint is the ordered pair $\left(1, -\frac{2}{3}\right)$.

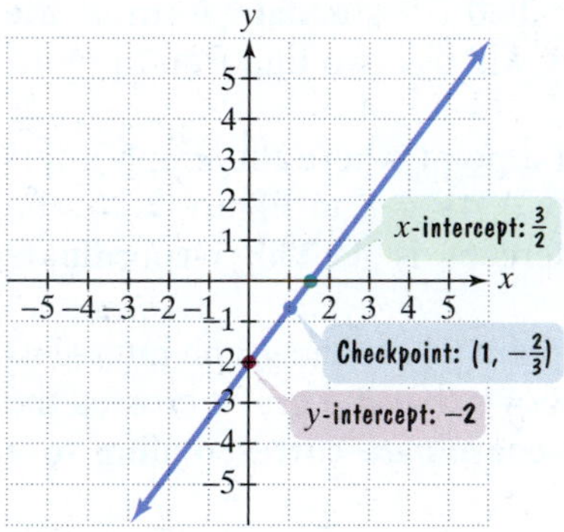

FIGURE 2.18 The graph of $4x - 3y = 6$

Step 4. Graph the equation by drawing a line through the three points. The three points in **Figure 2.18** lie along the same line. Drawing a line through the three points results in the graph of $4x - 3y = 6$.

Using Technology

You can use a graphing utility to graph equations of the form $Ax + By = C$. Begin by solving the equation for y. For example, to graph $4x - 3y = 6$, solve the equation for y.

$$4x - 3y = 6 \quad \text{This is the equation to be graphed.}$$
$$4x - 4x - 3y = -4x + 6 \quad \text{Add } -4x \text{ to both sides.}$$
$$-3y = -4x + 6 \quad \text{Simplify.}$$
$$\frac{-3y}{-3} = \frac{-4x + 6}{-3} \quad \text{Divide both sides by } -3.$$
$$y = \frac{4}{3}x - 2 \quad \text{Simplify.}$$

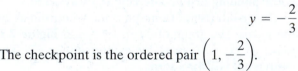

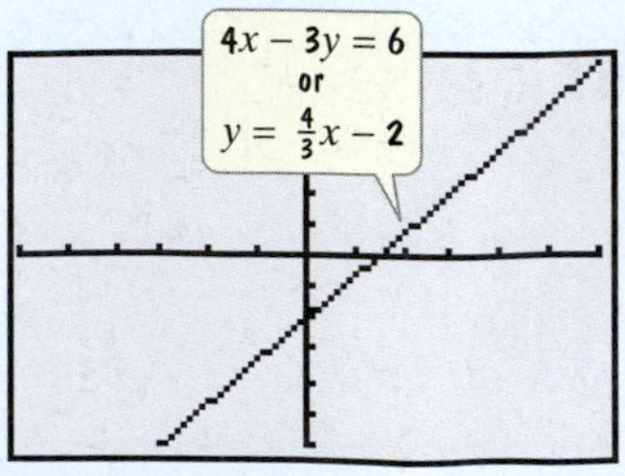

This is the equation to enter in your graphing utility. The graph of $y = \frac{4}{3}x - 2$, or, equivalently, $4x - 3y = 6$ is shown above in a $[-6, 6, 1]$ by $[-6, 6, 1]$ viewing rectangle.

✓ **CHECK POINT 1** Graph: $3x - 2y = 6$.

2 Compute a line's slope.

The Slope of a Line

Mathematicians have developed a useful measure of the steepness of a line, called the **slope** of the line. Slope compares the vertical change (the **rise**) to the horizontal change (the **run**) when moving from one fixed point to another along the line. To calculate the slope of a line, we use a ratio that compares the change in y (the rise) to the change in x (the run).

Definition of Slope

The **slope** of the line through the distinct points (x_1, y_1) and (x_2, y_2) is

$$\frac{\text{Change in } y}{\text{Change in } x} = \frac{\text{Rise}}{\text{Run}} = \frac{y_2 - y_1}{x_2 - x_1},$$

Vertical change (Rise); Horizontal change (Run)

where $x_2 - x_1 \neq 0$.

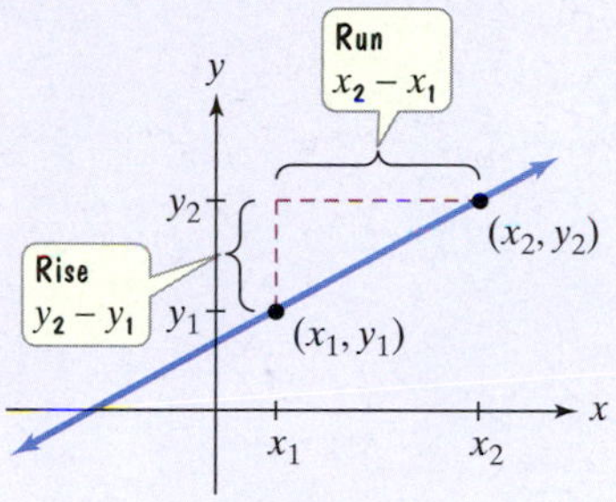

It is common notation to let the letter m represent the slope of a line. The letter m is used because it is the first letter of the French verb *monter*, meaning "to rise," or "to ascend."

EXAMPLE 2 Using the Definition of Slope

Find the slope of the line passing through each pair of points:

a. $(-3, -4)$ and $(-1, 6)$ **b.** $(-1, 3)$ and $(-4, 5)$.

Solution

a. Let $(x_1, y_1) = (-3, -4)$ and $(x_2, y_2) = (-1, 6)$. The slope is obtained as follows:

$$m = \frac{\text{Change in } y}{\text{Change in } x} = \frac{y_2 - y_1}{x_2 - x_1} = \frac{6 - (-4)}{-1 - (-3)} = \frac{6 + 4}{-1 + 3} = \frac{10}{2} = 5.$$

The situation is illustrated in **Figure 2.19**. The slope of the line is 5, or $\frac{10}{2}$. For every vertical change, or rise, of 10 units, there is a corresponding horizontal change, or run, of 2 units. The slope is positive and the line rises from left to right.

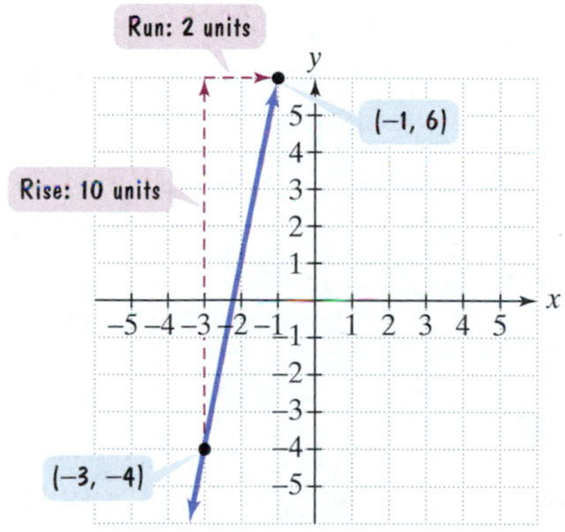

FIGURE 2.19 Visualizing a slope of 5

Study Tip

When computing slope, it makes no difference which point you call (x_1, y_1) and which point you call (x_2, y_2). If we let $(x_1, y_1) = (-1, 6)$ and $(x_2, y_2) = (-3, -4)$, the slope is still 5:

$$m = \frac{y_2 - y_1}{x_2 - x_1} = \frac{-4 - 6}{-3 - (-1)} = \frac{-10}{-2} = 5.$$

However, you should not subtract in one order in the numerator $(y_2 - y_1)$ and then in a different order in the denominator $(x_1 - x_2)$.

$$\frac{-4 - 6}{-1 - (-3)} = \frac{-10}{2} = -5.$$ Incorrect! The slope is not −5.

b. To find the slope of the line passing through $(-1, 3)$ and $(-4, 5)$, we can let $(x_1, y_1) = (-1, 3)$ and $(x_2, y_2) = (-4, 5)$. The slope is computed as follows:

$$m = \frac{\text{Change in } y}{\text{Change in } x} = \frac{y_2 - y_1}{x_2 - x_1} = \frac{5 - 3}{-4 - (-1)} = \frac{2}{-3} = -\frac{2}{3}.$$

The situation is illustrated in **Figure 2.20**. The slope of the line is $-\frac{2}{3}$. For every vertical change of -2 units (2 units down), there is a corresponding horizontal change of 3 units. The slope is negative and the line falls from left to right.

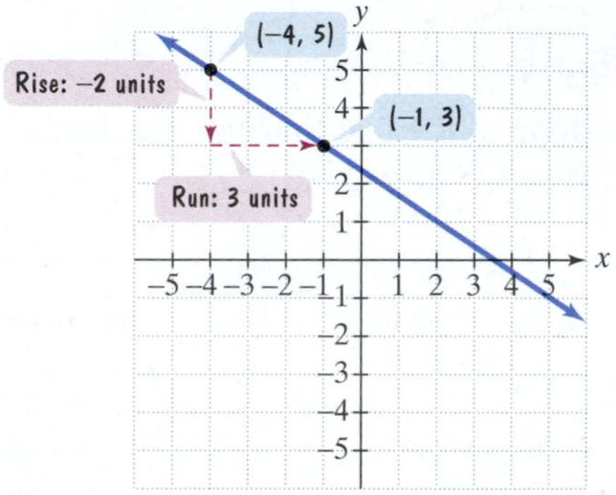

FIGURE 2.20 Visualizing a slope of $-\frac{2}{3}$

✓ CHECK POINT 2 Find the slope of the line passing through each pair of points:

a. $(-3, 4)$ and $(-4, -2)$ **b.** $(4, -2)$ and $(-1, 5)$.

Example 2 illustrates that a line with a positive slope is rising from left to right and a line with a negative slope is falling from left to right. By contrast, a horizontal line neither rises nor falls and has a slope of zero. A vertical line has no horizontal change, so $x_2 - x_1 = 0$ in the formula for slope. Because we cannot divide by zero, the slope of a vertical line is undefined. This discussion is summarized in **Table 2.1.**

Table 2.1 Possibilities for a Line's Slope

Positive Slope	Negative Slope	Zero Slope	Undefined Slope
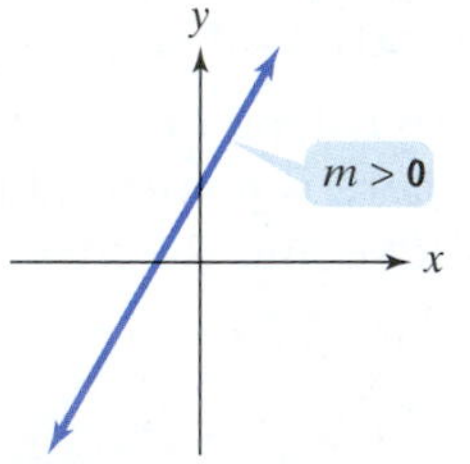	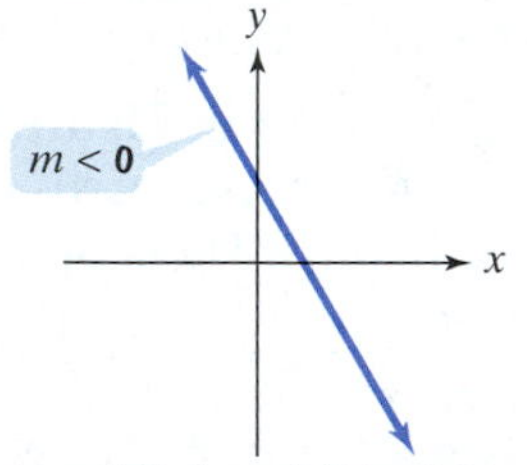	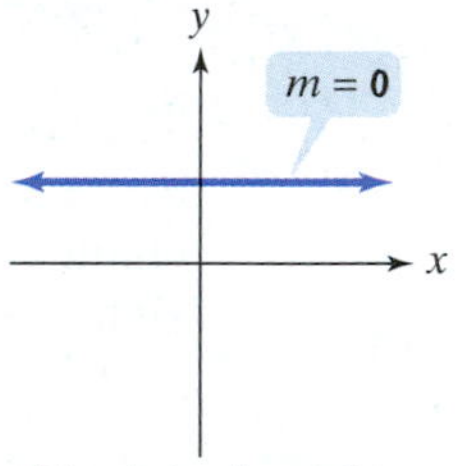	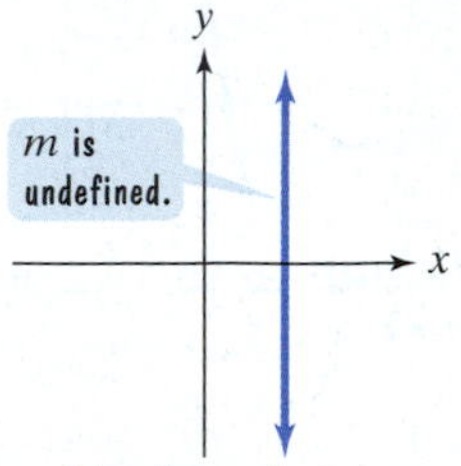
Line rises from left to right.	Line falls from left to right.	Line is horizontal.	Line is vertical.

Study Tip

Always be clear in the way you use language, especially in mathematics. For example, it's not a good idea to say that a line has "no slope." This could mean that the slope is zero or that the slope is undefined.

The Slope-Intercept Form of the Equation of a Line

We opened this section with a linear function that modeled child mortality as a function of literacy. The function's equation can be expressed as

$$y = -2.39x + 254.47 \quad \text{or} \quad f(x) = -2.39x + 254.47.$$

What is the significance of −2.39, the x-coefficient, or of 254.47, the constant term? To answer this question, let's look at an equation in the same form with simpler numbers. In particular, consider the equation $y = 2x + 4$.

Figure 2.21 shows the graph of $y = 2x + 4$. Verify that the x-intercept is −2 by setting y equal to 0 and solving for x. Similarly, verify that the y-intercept is 4 by setting x equal to 0 and solving for y.

Now that we have two points on the line, we can calculate the slope of the graph of $y = 2x + 4$.

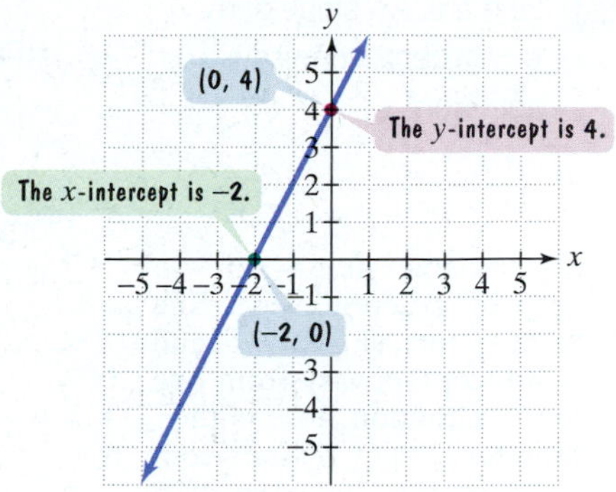

FIGURE 2.21 The graph of $y = 2x + 4$

$$\text{Slope} = \frac{\text{Change in } y}{\text{Change in } x}$$
$$= \frac{4 - 0}{0 - (-2)} = \frac{4}{2} = 2$$

We see that the slope of the line is 2, the same as the coefficient of x in the equation $y = 2x + 4$. The y-intercept is 4, the same as the constant in the equation $y = 2x + 4$.

$$y = 2x + 4$$

The slope is 2. The y-intercept is 4.

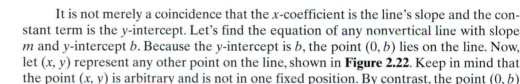

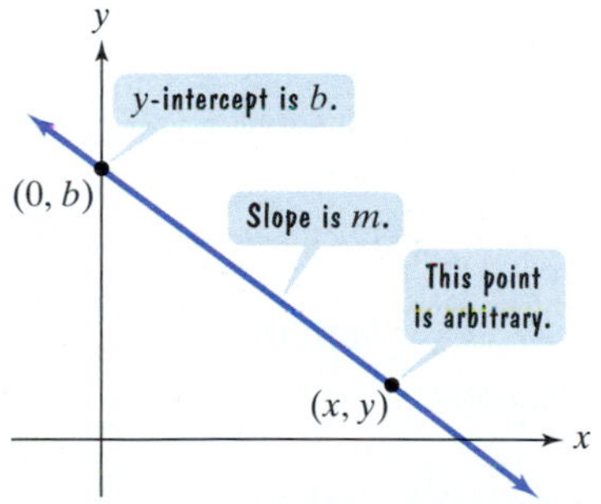

FIGURE 2.22 A line with slope m and y-intercept b

It is not merely a coincidence that the x-coefficient is the line's slope and the constant term is the y-intercept. Let's find the equation of any nonvertical line with slope m and y-intercept b. Because the y-intercept is b, the point $(0, b)$ lies on the line. Now, let (x, y) represent any other point on the line, shown in **Figure 2.22**. Keep in mind that the point (x, y) is arbitrary and is not in one fixed position. By contrast, the point $(0, b)$ is fixed.

Regardless of where the point (x, y) is located, the steepness of the line in **Figure 2.22** remains the same. Thus, the ratio for slope stays a constant m. This means that for all points along the line,

$$m = \frac{\text{Change in } y}{\text{Change in } x} = \frac{y - b}{x - 0} = \frac{y - b}{x}.$$

We can clear the fraction by multiplying both sides by x, the least common denominator.

$$m = \frac{y - b}{x}$$ This is the slope of the line in Figure 2.22.

$$mx = \frac{y - b}{x} \cdot x$$ Multiply both sides by x.

$$mx = y - b$$ Simplify: $\frac{y - b}{\cancel{x}} \cdot \cancel{x} = y - b$.

$$mx + b = y - b + b$$ Add b to both sides and solve for y.

$$mx + b = y$$ Simplify.

Now, if we reverse the two sides, we obtain the slope-intercept form of the equation of a line.

Slope-Intercept Form of the Equation of a Line

The **slope-intercept form of the equation** of a nonvertical line with slope m and y-intercept b is

$$y = mx + b.$$

3 Find a line's slope and y-intercept from its equation.

The slope-intercept form of a line's equation, $y = mx + b$, can be expressed in function notation by replacing y with $f(x)$:

$$f(x) = mx + b.$$

We have seen that functions in this form are called **linear functions**. Thus, in the equation of a linear function, the x-coefficient is the line's slope and the constant term is the y-intercept. Here are two examples:

$$y = 2x - 4 \qquad f(x) = \frac{1}{2}x + 2.$$

The slope is 2. The y-intercept is -4. The slope is $\frac{1}{2}$. The y-intercept is 2.

Study Tip

The variables in $y = mx + b$ vary in different ways. The values for slope, m, and y-intercept, b, vary from one line's equation to another. However, they remain constant in the equation of a single line. By contrast, the variables x and y represent the infinitely many points, (x, y), on a single line. Thus, these variables vary in both the equation of a single line, as well as from one equation to another.

If a linear function's equation is in standard form, $Ax + By = C$, do you see how we can identify the line's slope and y-intercept? Solve the equation for y and convert to slope-intercept form.

EXAMPLE 3 Converting from Standard Form to Slope-Intercept Form

Give the slope and the y-intercept for the line whose equation is

$$5x + 3y = -12.$$

Solution We convert $5x + 3y = -12$ to slope-intercept form by solving the equation for y. In this form, the coefficient of x is the line's slope and the constant term is the y-intercept.

$$5x + 3y = -12$$ This is the given equation in standard form, $Ax + By = C$.

Our goal is to isolate y.

$$5x - 5x + 3y = -5x - 12$$ Add $-5x$ to both sides.

$$3y = -5x - 12$$ Simplify.

$$\frac{3y}{3} = \frac{-5x - 12}{3}$$ Divide both sides by 3.

$$y = -\frac{5}{3}x - 4$$ Divide each term in the numerator by 3.

The slope is $-\frac{5}{3}$. The y-intercept is -4.

CHECK POINT 3 Give the slope and the y-intercept for the line whose equation is $8x - 4y = 20$.

---

4 Graph linear functions in slope-intercept form.

If a linear function's equation is in slope-intercept form, we can use the y-intercept and the slope to obtain its graph.

Graphing $y = mx + b$ Using the Slope and y-Intercept

1. Plot the point containing the y-intercept on the y-axis. This is the point $(0, b)$.
2. Obtain a second point using the slope, m. Write m as a fraction, and use rise over run, starting at the point containing the y-intercept, to plot this point.
3. Use a straightedge to draw a line through the two points. Draw arrowheads at the ends of the line to show that the line continues indefinitely in both directions.

Study Tip

Writing the slope, m, as a fraction allows you to identify the rise (the fraction's numerator) and the run (the fraction's denominator).

EXAMPLE 4 Graphing by Using the Slope and *y*-Intercept

Graph the line whose equation is $y = 3x - 4$.

Solution The equation $y = 3x - 4$ is in the form $y = mx + b$. The slope, m, is the coefficient of x. The y-intercept, b, is the constant term.

$$y = 3x + (-4)$$

The slope is 3. The y-intercept is -4.

Now that we have identified the slope and the y-intercept, we use the three-step procedure to graph the equation.

Step 1. Plot the point containing the *y*-intercept on the *y*-axis. The y-intercept is -4. We plot the point $(0, -4)$, shown in **Figure 2.23(a)**.

Step 2. Obtain a second point using the slope, *m*. Write *m* as a fraction, and use rise over run, starting at the point containing the *y*-intercept, to plot this point. We express the slope, 3, as a fraction.

$$m = \frac{3}{1} = \frac{\text{Rise}}{\text{Run}}$$

We plot the second point on the line by starting at $(0, -4)$, the first point. Based on the slope, we move 3 units *up* (the rise) and 1 unit to the *right* (the run). This puts us at a second point on the line, $(1, -1)$, shown in **Figure 2.23(b)**.

Step 3. Use a straightedge to draw a line through the two points. The graph of $y = 3x - 4$ is shown in **Figure 2.23(c)**.

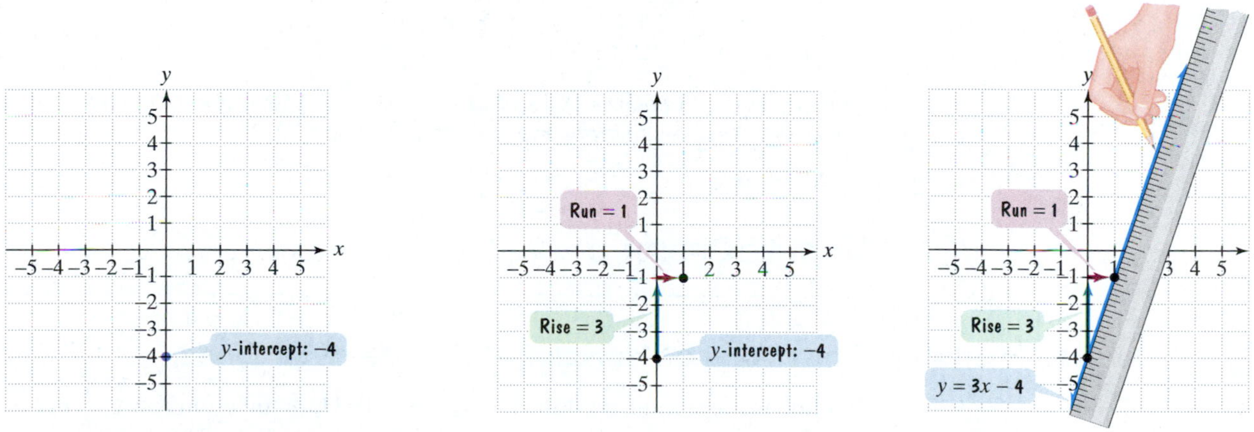

(a) The y-intercept is -4, so $(0, -4)$ is a point on the line.

(b) The slope is 3.

(c) The graph of $y = 3x - 4$.

FIGURE 2.23 Graphing $y = 3x - 4$ using the y-intercept and slope

✓ **CHECK POINT 4** Graph the line whose equation is $y = 4x - 3$.

EXAMPLE 5 Graphing by Using the Slope and *y*-Intercept

Graph the linear function: $f(x) = -\frac{3}{2}x + 2$.

Solution The equation of the line is in the form $f(x) = mx + b$. We can find the slope, m, by identifying the coefficient of x. We can find the y-intercept, b, by identifying the constant term.

$$f(x) = -\frac{3}{2}x + 2$$

The slope is $-\frac{3}{2}$. The y-intercept is 2.

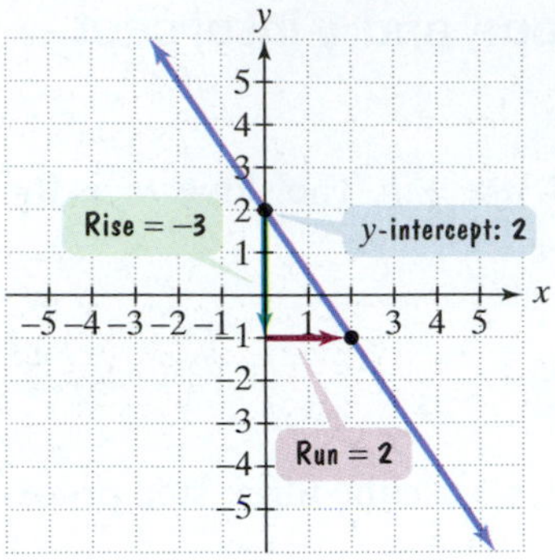

FIGURE 2.24 The graph of $f(x) = -\frac{3}{2}x + 2$

Now that we have identified the slope $-\frac{3}{2}$, and the y-intercept, 2, we use the three-step procedure to graph the equation.

Step 1. Plot the point containing the y-intercept on the y-axis. The y-intercept is 2. We plot (0, 2), shown in **Figure 2.24**.

Step 2. Obtain a second point using the slope, m. Write m as a fraction, and use rise over run, starting at the point containing the y-intercept, to plot this point. The slope, $-\frac{3}{2}$, is already written as a fraction.

$$m = -\frac{3}{2} = \frac{-3}{2} = \frac{\text{Rise}}{\text{Run}}$$

We plot the second point on the line by starting at (0, 2), the first point. Based on the slope, we move 3 units *down* (the rise) and 2 units to the *right* (the run). This puts us at a second point on the line, (2, −1), shown in **Figure 2.24**.

Step 3. Use a straightedge to draw a line through the two points. The graph of the linear function $f(x) = -\frac{3}{2}x + 2$ is shown as a blue line in **Figure 2.24**.

Discover for Yourself

Obtain a second point in Example 5 by writing the slope as follows:

$$m = \frac{3}{-2} = \frac{\text{Rise}}{\text{Run}}.$$

$-\frac{3}{2}$ can be expressed as $\frac{-3}{2}$ or $\frac{3}{-2}$.

Obtain a second point in **Figure 2.24** by moving *up* 3 units and to the *left* 2 units, starting at (0, 2). What do you observe once you draw the line?

✓ CHECK POINT 5 Graph the linear function: $f(x) = -\frac{2}{3}x$.

5 Graph horizontal or vertical lines.

Equations of Horizontal and Vertical Lines

If a line is horizontal, its slope is zero: $m = 0$. Thus, the equation $y = mx + b$ becomes $y = b$, where b is the y-intercept. All horizontal lines have equations of the form $y = b$.

EXAMPLE 6 Graphing a Horizontal Line

Graph $y = -4$ in the rectangular coordinate system.

Solution All ordered pairs that are solutions of $y = -4$ have a value of y that is always −4. Any value can be used for x. In the table on the right, we have selected three of the possible values for x −2, 0, and 3. The table shows that three ordered pairs that are solutions of $y = -4$ are (−2, −4), (0, −4), and (3, −4). Drawing a line that passes through the three points gives the horizontal line shown in **Figure 2.25**.

x	$y = -4$	(x, y)
−2	−4	(−2, −4)
0	−4	(0, −4)
3	−4	(3, −4)

For all choices of x, y is a constant −4.

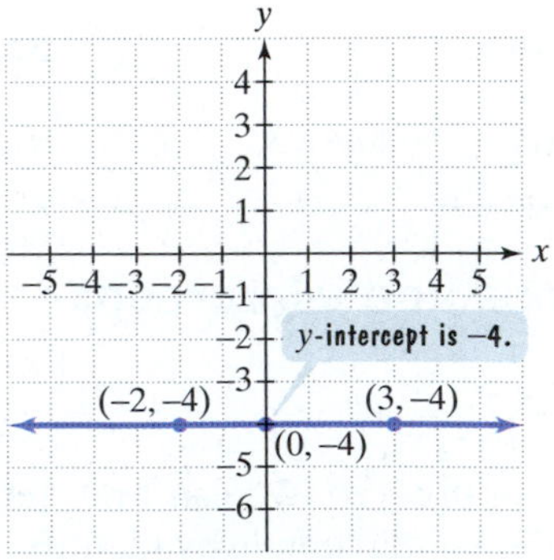

FIGURE 2.25 The graph of $y = -4$ or $f(x) = -4$

✓ CHECK POINT 6 Graph $y = 3$ in the rectangular coordinate system.

Equation of a Horizontal Line

A horizontal line is given by an equation of the form

$$y = b,$$

where b is the y-intercept.

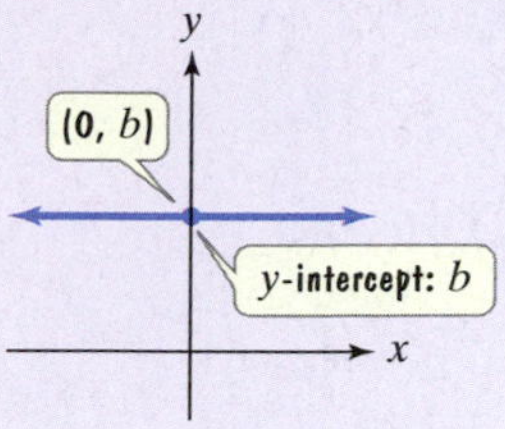

Because any vertical line can intersect the graph of a horizontal line $y = b$ only once, a horizontal line is the graph of a function. Thus, we can express the equation $y = b$ as $f(x) = b$. This linear function is often called a **constant function**.

Next, let's see what we can discover about the graph of an equation of the form $x = a$ by looking at an example.

EXAMPLE 7 Graphing a Vertical Line

Graph the linear equation: $x = 2$.

Solution All ordered pairs that are solutions of $x = 2$ have a value of x that is always 2. Any value can be used for y. In the table on the right, we have selected three of the possible values for y: -2, 0, and 3. The table shows that three ordered pairs that are solutions of $x = 2$ are $(2, -2)$, $(2, 0)$, and $(2, 3)$. Drawing a line that passes through the three points gives the vertical line shown in **Figure 2.26**.

For all choices of y,

$x = 2$	y	(x, y)
2	-2	$(2, -2)$
2	0	$(2, 0)$
2	3	$(2, 3)$

x is always 2.

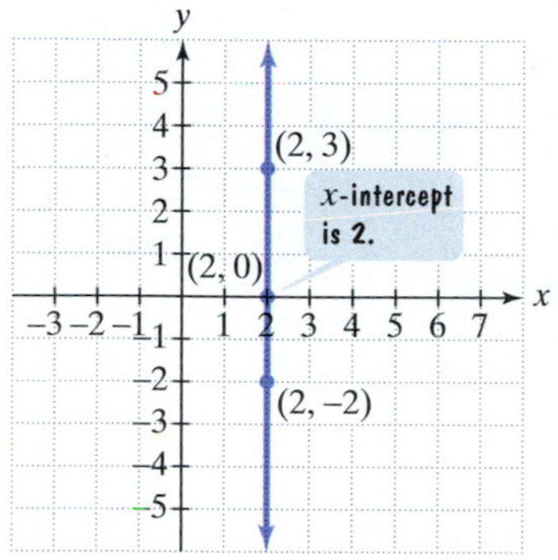

FIGURE 2.26 The graph of $x = 2$

Equation of a Vertical Line

A vertical line is given by an equation of the form

$$x = a,$$

where a is the x-intercept.

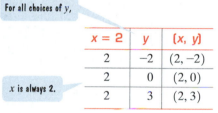

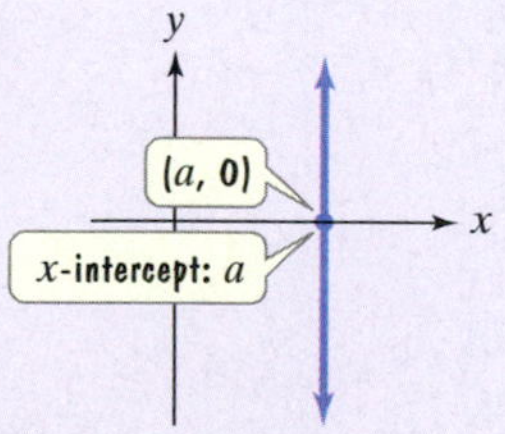

Does a vertical line represent the graph of a linear function? No. Look at the graph of $x = 2$ in **Figure 2.26**. A vertical line drawn through $(2, 0)$ intersects the graph infinitely many times. This shows that infinitely many outputs are associated with the input 2. **No vertical line is a linear function.**

✓ **CHECK POINT 7** Graph the linear equation: $x = -3$.

Study Tip

The linear equations in Examples 6 and 7, $y = -4$ and $x = 2$, each show only one variable. However, these are equations in two variables in the standard form $Ax + By = C$:

- $y = -4$ means $0x + 1y = -4$.
- $x = 2$ means $1x + 0y = 2$.

6 Interpret slope as rate of change.

Slope as Rate of Change

Slope is defined as the ratio of a change in y to a corresponding change in x. It describes how fast y is changing with respect to x. For a linear function, slope may be interpreted as the rate of change of the dependent variable per unit change in the independent variable.

Our next example shows how slope can be interpreted as a rate of change in an applied situation. When calculating slope in applied problems, keep track of the units in the numerator and the denominator.

EXAMPLE 8 Slope as a Rate of Change

The line graphs in **Figure 2.27** show the percentage of Americans in two age groups who reported using illegal drugs in the previous month. Find the slope of the line segment for the 12–17 age group. Describe what this slope represents.

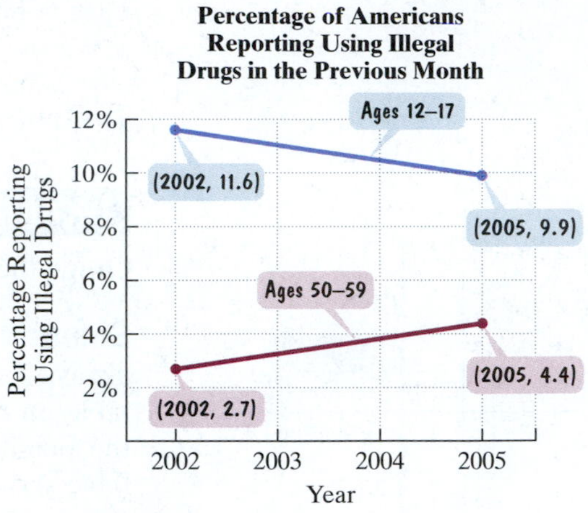

FIGURE 2.27

Source: Substance Abuse and Mental Health Services Administration

Solution We will let x represent a year and y the percentage reporting illegal drug use in the previous month for that year. The two points shown on the line segment for the 12–17 age group have the following coordinates:

$$(2002, 11.6) \quad \text{and} \quad (2005, 9.9).$$

In 2002, 11.6% reported using illegal drugs in the previous month.

In 2005, 9.9% reported using illegal drugs in the previous month.

Now we compute the slope:

$$m = \frac{\text{Change in } y}{\text{Change in } x} = \frac{9.9 - 11.6}{2005 - 2002}$$

The unit in the numerator is *percent*.

The unit in the denominator is *year*.

$$= \frac{-1.7}{3} \approx -0.57$$

The slope indicates that for the 12–17 age group, the percentage reporting using illegal drugs in the previous month decreased by approximately 0.57% each year. The rate of change is a decrease of approximately 0.57% per year.

✓ CHECK POINT 8 Use the graph in **Figure 2.27** to find the slope of the line segment for the 50–59 age group. Express the slope correct to two decimal places and describe what it represents.

7 Find a function's average rate of change.

The Average Rate of Change of a Function

If the graph of a function is not a straight line, the **average rate of change** between any two points is the slope of the line containing the two points. For example, **Figure 2.28** shows the graph of a particular man's height, in inches, as a function of his age, in years. Two points on the graph are labeled (13, 57) and (18, 76). At age 13, this man was 57 inches tall, and at age 18, he was 76 inches tall.

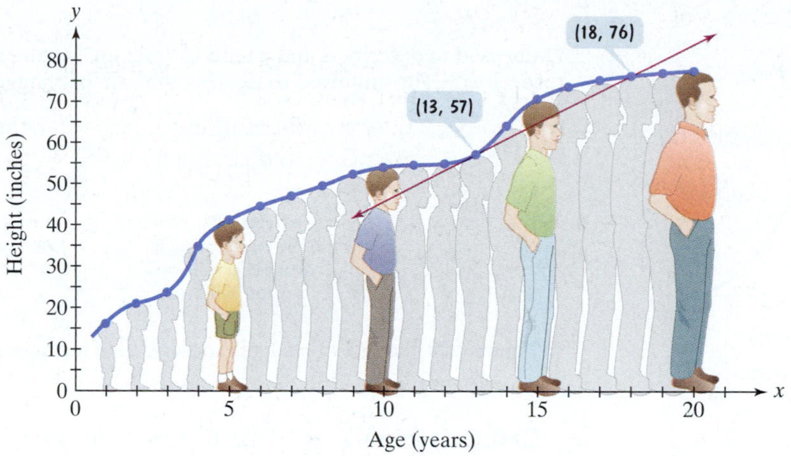

FIGURE 2.28 Height as a function of age

The man's average growth rate between ages 13 and 18 is the slope of the line containing (13, 57) and (18, 76):

$$m = \frac{\text{Change in } y}{\text{Change in } x} = \frac{76 - 57}{18 - 13} = \frac{19}{5} = 3\frac{4}{5}.$$

This man's average rate of change, or average growth rate, from age 13 to age 18 was $3\frac{4}{5}$, or 3.8, inches per year.

For any function, $y = f(x)$, the slope of the line between any two points is the **average change in y per unit change in x.**

EXAMPLE 9 Finding the Average Rate of Change

When a person receives a drug injected into a muscle, the concentration of the drug in the body, measured in milligrams per 100 milliliters, is a function of the time elapsed after the injection, measured in hours. **Figure 2.29** shows the graph of such a function, where x represents hours after the injection and $f(x)$ is the drug's concentration at time x. Find the average rate of change in the drug's concentration between 3 and 7 hours.

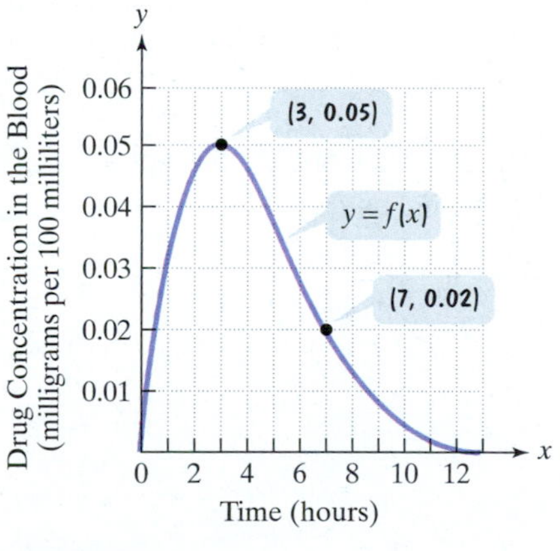

FIGURE 2.29 Concentration of a drug as a function of time

Solution At 3 hours, the drug's concentration is 0.05 and at 7 hours, the concentration is 0.02. The average rate of change in its concentration between 3 and 7 hours is the slope of the line connecting the points (3, 0.05) and (7, 0.02).

$$m = \frac{\text{Change in } y}{\text{Change in } x} = \frac{0.02 - 0.05}{7 - 3} = \frac{-0.03}{4} = -0.0075$$

The average rate of change is -0.0075. This means that the drug's concentration is decreasing at an average rate of 0.0075 milligram per 100 milliliters per hour.

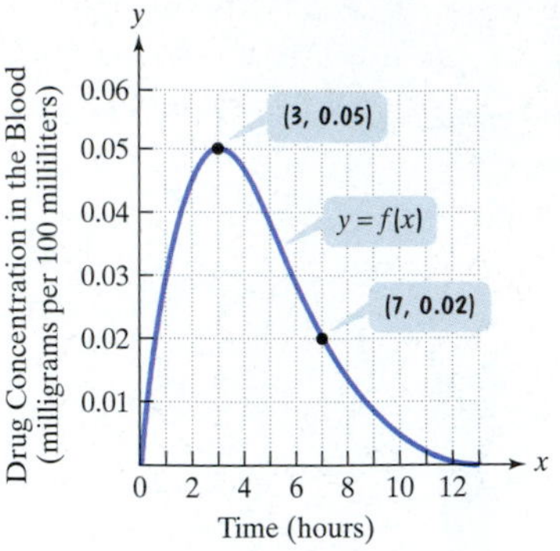

FIGURE 2.29 (repeated)

Study Tip

Units used to describe x and y tend to "pile up" when expressing the rate of change of y with respect to x. The unit used to express the rate of change of y with respect to x is

the unit used to describe y **per** the unit used to describe x.

In Figure 2.29, y, or drug concentration, is described in milligrams per 100 milliliters.

In Figure 2.29, x, or time, is described in hours.

In **Figure 2.29**, the rate of change is described in terms of milligrams per 100 milliliters per hour.

CHECK POINT 9 Use **Figure 2.29** to find the average rate of change in the drug's concentration between 1 hour and 3 hours.

Blitzer Bonus

How Calculus Studies Change

Take a rapid sequence of still photographs of a moving scene and project them onto a screen at thirty shots a second or faster. Our eyes see the results as continuous motion. The small difference between one frame and the next cannot be detected by the human visual system. The idea of calculus likewise regards continuous motion as made up of a sequence of still configurations. Calculus masters the mystery of movement by "freezing the frame" of a continuous changing process, instant by instant. For example, **Figure 2.30** shows a male's changing height over intervals of time. Over the period of time from P to D, his average rate of growth is his change in height—that is, his height at time D minus his height at time P—divided by the change in time from P to D. This is the slope of line PD.

The lines PD, PC, PB, and PA shown in **Figure 2.30** have slopes that show average growth rates for successively shorter periods of time. Calculus makes these time frames so small that they approach a single point—that is, a single instant in time. This point is shown as point P in **Figure 2.30**. The slope of the line that touches the graph at P gives the male's growth rate at one instant in time, P.

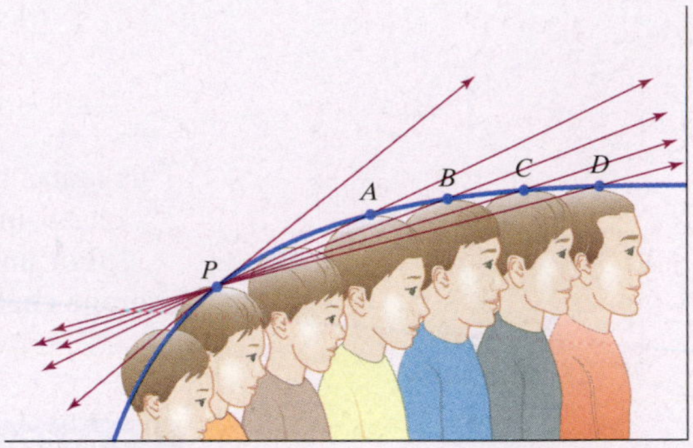

FIGURE 2.30 Analyzing continuous growth over intervals of time and at an instant in time

8 Use slope and y-intercept to model data.

Modeling Data with the Slope-Intercept Form of the Equation of a Line

Linear functions are useful for modeling data that fall on or near a line. For example, the bar graph in **Figure 2.31(a)** gives the average ticket price for a U.S. movie in the indicated year. The data are displayed as a set of four points in the scatter plot in **Figure 2.31(b)**.

Average Ticket Price for a Movie in the United States

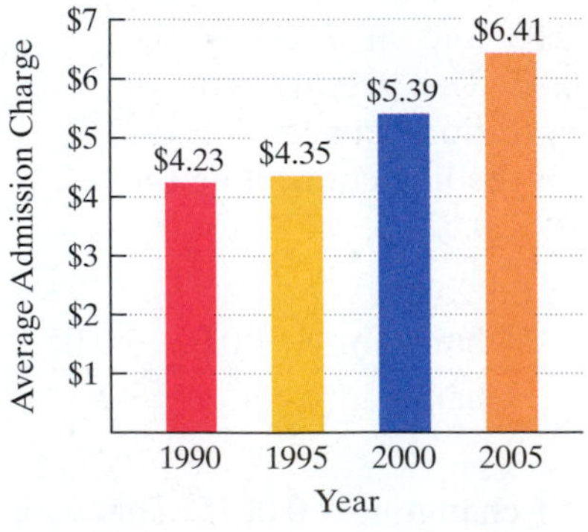

FIGURE 2.31(a)

Source: National Association of Theatre Owners

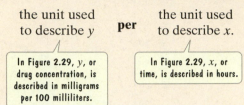

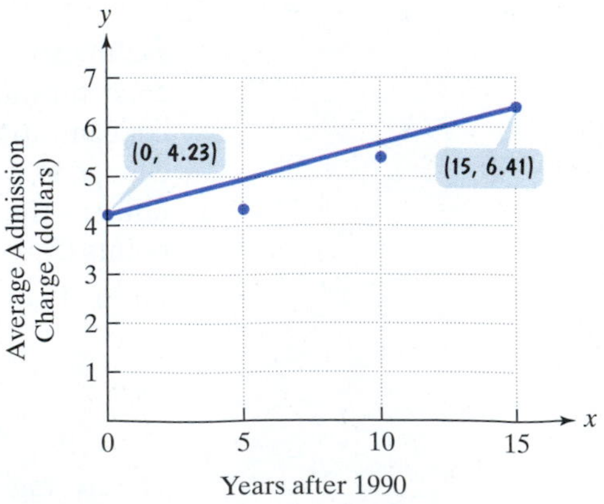

FIGURE 2.31(b)

Also shown on the scatter plot in **Figure 2.31(b)** is a line that passes through or near the four points. Example 10 illustrates how we can use the equation $y = mx + b$ to obtain a model for the data and make predictions about what you can expect to pay for a movie ticket in the future.

EXAMPLE 10 Modeling with the Slope-Intercept Form of the Equation

a. Use the scatter plot in **Figure 2.31(b)** to find a function in the form $T(x) = mx + b$ that models the average ticket price for a movie, $T(x)$, x years after 1990.

b. Use the model to predict the average ticket price in 2010.

Solution

a. We will use the line segment passing through the points $(0, 4.23)$ and $(15, 6.41)$ to obtain a model. We need values for m, the slope, and b, the y-intercept.

$$T(x) = mx + b$$

$m = \frac{\text{Change in } y}{\text{Change in } x}$

$m = \frac{6.41 - 4.23}{15 - 0} \approx 0.15$

The point $(0, 4.23)$ lies on the line segment, so the y-intercept is 4.23: $b = 4.23$.

The average ticket price for a movie, $T(x)$, x years after 1990 can be modeled by the linear function

$$T(x) = 0.15x + 4.23.$$

The slope, approximately 0.15, indicates an increase in ticket price of about \$0.15 per year from 1990 through 2005.

b. Now let's use this function to predict the average ticket price in 2010. Because 2010 is 20 years after 1990, substitute 20 for x in $T(x) = 0.15x + 4.23$ and evaluate the function at 20.

$$T(20) = 0.15(20) + 4.23 = 7.23$$

Our model predicts an average ticket price of \$7.23 in 2010.

CHECK POINT 10 The table shows the median age of first marriage for U.S. women. (The median age is the age in the middle when all the ages of first-married women are arranged from youngest to oldest.) **Figure 2.32** shows a scatter plot based on the data, as well as a line that passes through or near the four points.

Women's Median Age of First Marriage

Year	Median Age
1990	23.9
1995	24.5
2000	25.1
2003	25.3

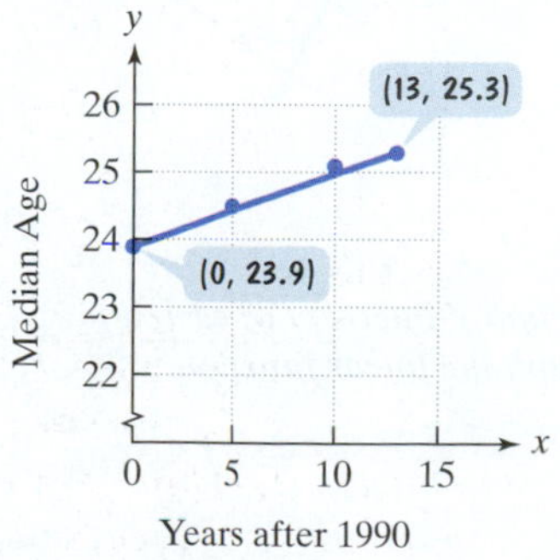

FIGURE 2.32
Source: U.S. Census Bureau

a. Use **Figure 2.32** on page 139 to find a function in the form $A(x) = mx + b$ that models the median age of first marriage for U.S. women, $A(x)$, x years after 1990.

b. Use the model to predict the median age of first marriage for U.S. women in 2030.

2.4 EXERCISE SET

MyMathLab MathXL PRACTICE WATCH DOWNLOAD READ REVIEW

Practice Exercises

In Exercises 1–14, use intercepts and a checkpoint to graph each linear function.

1. $x + y = 4$
2. $x + y = 2$
3. $x + 3y = 6$
4. $2x + y = 4$
5. $6x - 2y = 12$
6. $6x - 9y = 18$
7. $3x - y = 6$
8. $x - 4y = 8$
9. $x - 3y = 9$
10. $2x - y = 5$
11. $2x = 3y + 6$
12. $3x = 5y - 15$
13. $6x - 3y = 15$
14. $8x - 2y = 12$

In Exercises 15–24, find the slope of the line passing through each pair of points or state that the slope is undefined. Then indicate whether the line through the points rises, falls, is horizontal, or is vertical.

15. $(2, 4)$ and $(3, 8)$
16. $(3, 1)$ and $(5, 4)$
17. $(-1, 4)$ and $(2, 5)$
18. $(-3, -2)$ and $(2, 5)$
19. $(2, 5)$ and $(-1, 5)$
20. $(-6, -3)$ and $(4, -3)$
21. $(-7, 1)$ and $(-4, -3)$
22. $(2, -1)$ and $(-6, 3)$
23. $\left(\frac{7}{2}, -2\right)$ and $\left(\frac{7}{2}, \frac{1}{4}\right)$
24. $\left(\frac{3}{2}, -6\right)$ and $\left(\frac{3}{2}, \frac{1}{6}\right)$

In Exercises 25–26, find the slope of each line.

25.

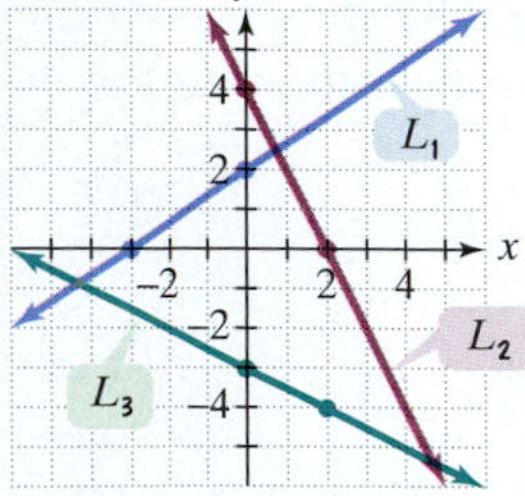

26.

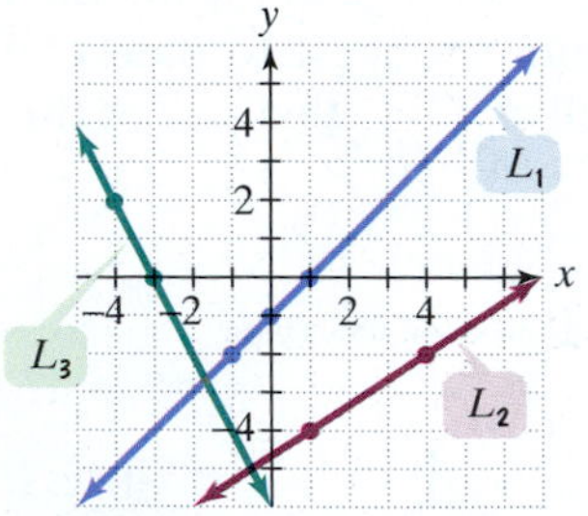

In Exercises 27–38, give the slope and y-intercept of each line whose equation is given. Then graph the linear function.

27. $y = 2x + 1$
28. $y = 3x + 2$
29. $y = -2x + 1$
30. $y = -3x + 2$
31. $f(x) = \frac{3}{4}x - 2$
32. $f(x) = \frac{3}{4}x - 3$
33. $f(x) = -\frac{3}{5}x + 7$
34. $f(x) = -\frac{2}{5}x + 6$
35. $y = -\frac{1}{2}x$
36. $y = -\frac{1}{3}x$
37. $y = -\frac{1}{2}$
38. $y = -\frac{1}{3}$

In Exercises 39–46,

a. *Rewrite the given equation in slope-intercept form by solving for y.*

b. *Give the slope and y-intercept.*

c. *Use the slope and y-intercept to graph the linear function.*

39. $2x + y = 0$
40. $3x + y = 0$
41. $5y = 4x$
42. $4y = 3x$
43. $3x + y = 2$
44. $2x + y = 4$
45. $5x + 3y = 15$
46. $7x + 2y = 14$

In Exercises 47–60, graph each equation in a rectangular coordinate system.

47. $y = 3$
48. $y = 5$
49. $f(x) = -2$
50. $f(x) = -4$
51. $3y = 18$
52. $5y = -30$
53. $f(x) = 2$
54. $f(x) = 1$
55. $x = 5$
56. $x = 4$
57. $3x = -12$
58. $4x = -12$
59. $x = 0$
60. $y = 0$

Practice PLUS

In Exercises 61–64, find the slope of the line passing through each pair of points or state that the slope is undefined. Assume that all variables represent positive real numbers. Then indicate whether the line through the points rises, falls, is horizontal, or is vertical.

61. $(0, a)$ and $(b, 0)$

62. $(-a, 0)$ and $(0, -b)$

63. (a, b) and $(a, b + c)$

64. $(a - b, c)$ and $(a, a + c)$

In Exercises 65–66, give the slope and y-intercept of each line whose equation is given. Assume that $B \neq 0$.

65. $Ax + By = C$

66. $Ax = By - C$

In Exercises 67–68, find the value of y if the line through the two given points is to have the indicated slope.

67. $(3, y)$ and $(1, 4)$, $m = -3$

68. $(-2, y)$ and $(4, -4)$, $m = \frac{1}{3}$

In Exercises 69–70, graph each linear function.

69. $3x - 4f(x) = 6$

70. $6x - 5f(x) = 20$

71. If one point on a line is $(3, -1)$ and the line's slope is -2, find the y-intercept.

72. If one point on a line is $(2, -6)$ and the line's slope is $-\frac{3}{2}$, find the y-intercept.

Use the figure to make the lists in Exercises 73–74.

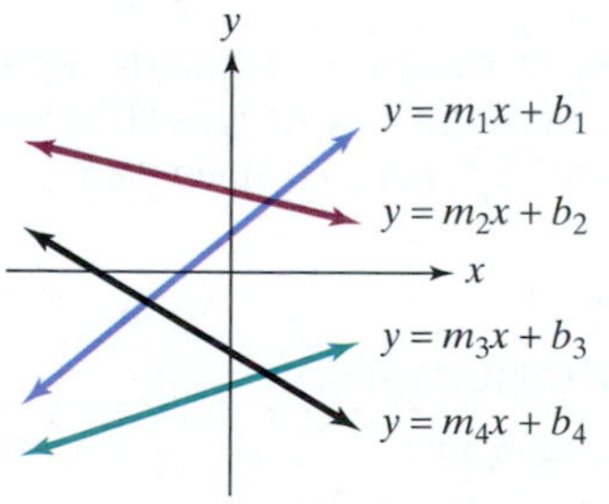

73. List the slopes m_1, m_2, m_3, and m_4 in order of decreasing size.

74. List the y-intercepts b_1, b_2, b_3, and b_4 in order of decreasing size.

Application Exercises

In Exercises 75–78, a linear function that models data is described. Find the slope of each model. Then describe what this means in terms of the rate of change of the dependent variable per unit change in the independent variable.

75. The linear function $f(x) = 0.01x + 57.7$ models the global average temperature of Earth, $f(x)$, in degrees Fahrenheit, x years after 1995.

76. The linear function $f(x) = 2x + 10$ models the amount, $f(x)$, in billions of dollars, that the drug industry spent on marketing information about drugs to doctors x years after 2000. (*Source:* IMS Health)

77. The linear function $f(x) = -0.52x + 24.7$ models the percentage of U.S. adults who smoked cigarettes, $f(x)$, x years after 1997. (*Source:* National Center for Health Statistics)

78. The linear function $f(x) = -0.28x + 1.7$ models the percentage of U.S. taxpayers who were audited by the IRS, $f(x)$, x years after 1996. (*Source:* IRS)

Divorce rates are typically higher for couples who marry in their teens. The graph shows the percentage of marriages ending in divorce by wife's age at marriage. Use the information shown to solve Exercises 79–80.

Percentage of Marriages Ending in Divorce by Wife's Age at Marriage

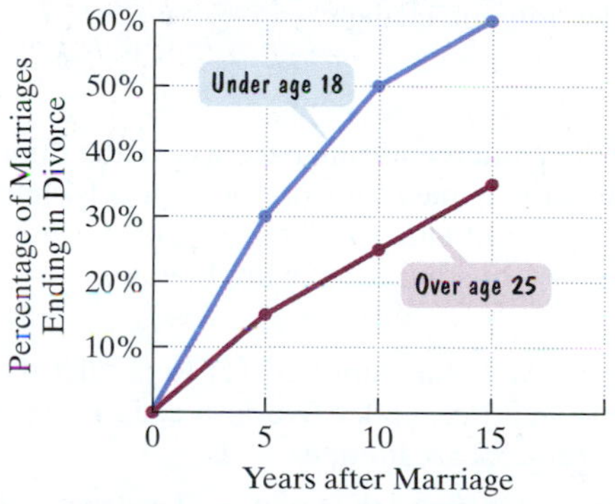

Source: B. E. Pruitt et al., *Human Sexuality*, Prentice Hall, 2007

79. a. What percentage of marriages in which the wife is under 18 when she marries end in divorce within the first five years?

b. What percentage of marriages in which the wife is under 18 when she marries end in divorce within the first ten years?

c. Find the average rate of change in the percentage of marriages ending in divorce between five and ten years of marriage in which the wife is under 18 when she marries.

80. a. What percentage of marriages in which the wife is over age 25 when she marries end in divorce within the first five years?

b. What percentage of marriages in which the wife is over age 25 when she marries end in divorce within the first ten years?

c. Find the average rate of change in the percentage of marriages ending in divorce between five and ten years of marriage in which the wife is over age 25 when she marries.

81. Shown, again, is the scatter plot that indicates a relationship between the percentage of adult females in a country who are literate and the mortality of children under five. Also shown is a line that passes through or near the points.

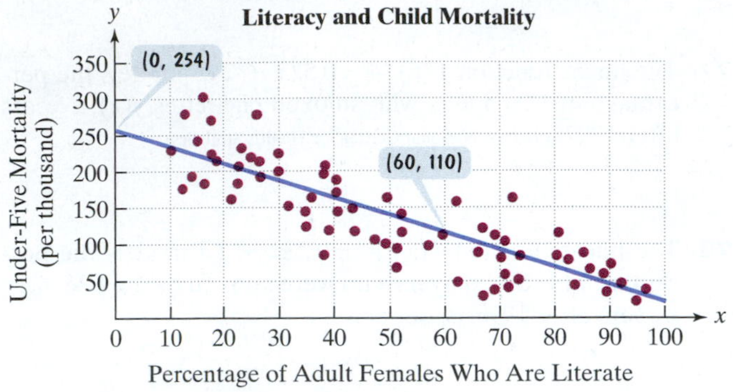

Source: United Nations

a. According to the graph, what is the y-intercept of the line? Describe what this represents in this situation.

b. Use the coordinates of the two points shown to compute the slope of the line. Describe what this means in terms of the rate of change.

c. Use the y-intercept from part (a) and the slope from part (b) to write a linear function that models child mortality, $f(x)$, per thousand, for children under five in a country where x% of adult women are literate.

d. Use the function from part (c) to predict the mortality rate of children under five in a country where 50% of adult females are literate.

82. The scatter plot shows the number of college students in the United States, in thousands, enrolled exclusively in online education from 2002 through 2007. Also shown is a line that passes through or near the six data points.

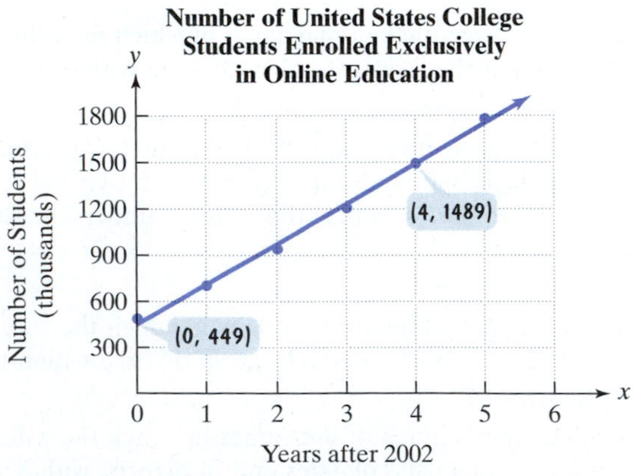

Source: U.S. Distance Learning Association

a. Use the coordinates of the two points shown to compute the slope of the line. Describe what this means in terms of the rate of change.

b. Use the y-intercept shown and the slope from part (a) to write a linear function that models the number of college students enrolled exclusively in online education, $L(x)$, in thousands, x years after 2002.

c. Use the function from part (b) to predict the number of college students who will be enrolled exclusively in online education in 2010.

The bar graph shows that as online news has grown, traditional news media have slipped.

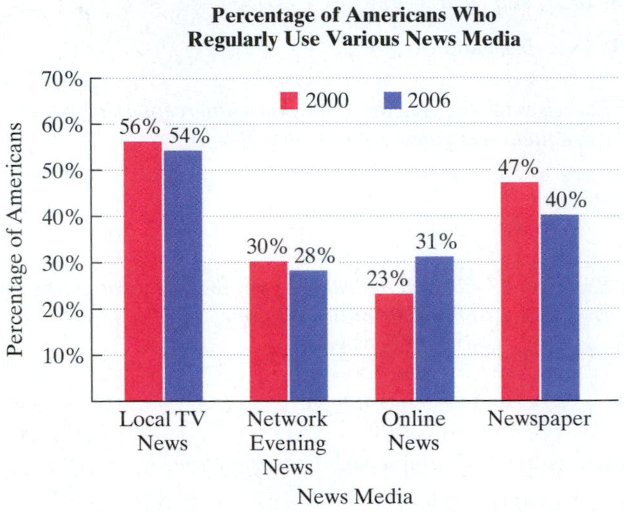

Source: Pew Research Center

In Exercises 83–84, find a linear function in slope-intercept form that models the given description. Each function should model the percentage of Americans, $P(x)$, who regularly used the news outlet x years after 2000.

83. In 2000, 47% of Americans regularly used newspapers for getting news and this has decreased at an average rate of approximately 1.2% per year since then.

84. In 2000, 23% of Americans regularly used online news for getting news and this has increased at an average rate of approximately 1.3% per year since then.

Writing in Mathematics

85. What is a scatter plot?

86. What is a regression line?

87. What is the standard form of the equation of a line?

88. What is an x-intercept of a graph?

89. What is a y-intercept of a graph?

90. If you are given the standard form of the equation of a line, explain how to find the x-intercept.

91. If you are given the standard form of the equation of a line, explain how to find the y-intercept.

92. What is the slope of a line?

93. Describe how to calculate the slope of a line passing through two points.

94. What does it mean if the slope of a line is zero?

95. What does it mean if the slope of a line is undefined?

96. Describe how to find the slope of a line whose equation is given.

97. Describe how to graph a line using the slope and y-intercept. Provide an original example with your description.

98. Describe the graph of $y = b$.

99. Describe the graph of $x = a$.

100. If the graph of a function is not a straight line, explain how to find the average rate of change between two points.

101. Take another look at the scatter plot in Exercise 81. Although there is a relationship between literacy and child mortality, we cannot conclude that increased literacy causes child mortality to decrease. Offer two or more possible explanations for the data in the scatter plot.

Technology Exercises

102. Use a graphing utility to verify any three of your hand-drawn graphs in Exercises 1–14. Solve the equation for y before entering it.

In Exercises 103–106, use a graphing utility to graph each linear function. Then use the TRACE *feature to trace along the line and find the coordinates of two points. Use these points to compute the line's slope. Check your result by using the coefficient of x in the line's equation.*

103. $y = 2x + 4$

104. $y = -3x + 6$

105. $f(x) = -\frac{1}{2}x - 5$

106. $f(x) = \frac{3}{4}x - 2$

Critical Thinking Exercises

Make Sense? *In Exercises 107–110, determine whether each statement "makes sense" or "does not make sense" and explain your reasoning.*

107. The graph of my linear function at first rose from left to right, reached a maximum point, and then fell from left to right.

108. A linear function that models tuition and fees at public four-year colleges from 2000 through 2006 has negative slope.

109. The function $S(x) = 49{,}100x + 1700$ models the average salary for a college professor, $S(x)$, x years after 2000.

110. The federal minimum wage was \$5.15 per hour from 1997 through 2006, so $f(x) = 5.15$ models the minimum wage, $f(x)$, in dollars, for the domain $\{1997, 1998, 1999, \ldots, 2006\}$.

In Exercises 111–114, determine whether each statement is true or false. If the statement is false, make the necessary change(s) to produce a true statement.

111. A linear function with nonnegative slope has a graph that rises from left to right.

112. Every line in the rectangular coordinate system has an equation that can be expressed in slope-intercept form.

113. The graph of the linear function $5x + 6y = 30$ is a line passing through the point $(6, 0)$ with slope $-\frac{5}{6}$.

114. The graph of $x = 7$ in the rectangular coordinate system is the single point $(7, 0)$.

In Exercises 115–116, find the coefficients that must be placed in each shaded area so that the function's graph will be a line satisfying the specified conditions.

115. $\square x + \square y = 12$; x-intercept $= -2$; y-intercept $= 4$

116. $\square x + \square y = 12$; y-intercept $= -6$; slope $= \frac{1}{2}$

117. For the linear function

$$f(x) = mx + b,$$

a. Find $f(x_1 + x_2)$.

b. Find $f(x_1) + f(x_2)$.

c. Is $f(x_1 + x_2) = f(x_1) + f(x_2)$?

Review Exercises

118. Simplify: $\left(\frac{4x^2}{y^{-3}}\right)^2$. (Section 1.6, Example 9)

119. Multiply and write the answer in scientific notation:

$$(8 \times 10^{-7})(4 \times 10^3).$$

(Section 1.7, Example 4)

120. Simplify: $5 - [3(x - 4) - 6x]$. (Section 1.2, Example 14)

Preview Exercises

Exercises 121–123 will help you prepare for the material covered in the next section.

121. Write the equation $y - 5 = 7(x + 4)$ in slope-intercept form.

122. Write the equation $y + 3 = -\frac{7}{3}(x - 1)$ in slope-intercept form.

123. The equation of a line is $x + 4y - 8 = 0$.

a. Write the equation in slope-intercept form and determine the slope.

b. The product of the line's slope in part (a) and the slope of a second line is -1. What is the slope of the second line?

SECTION 2.5

The Point-Slope Form of the Equation of a Line

Objectives

1. Use the point-slope form to write equations of a line.
2. Model data with linear functions and make predictions.
3. Find slopes and equations of parallel and perpendicular lines.

If present trends continue, is it possible that our descendants could live to be 200 years of age? To answer this question, we need to develop a function that models life expectancy by birth year. In this section, you will learn to use another form of a line's equation to obtain functions that model data.

Point-Slope Form

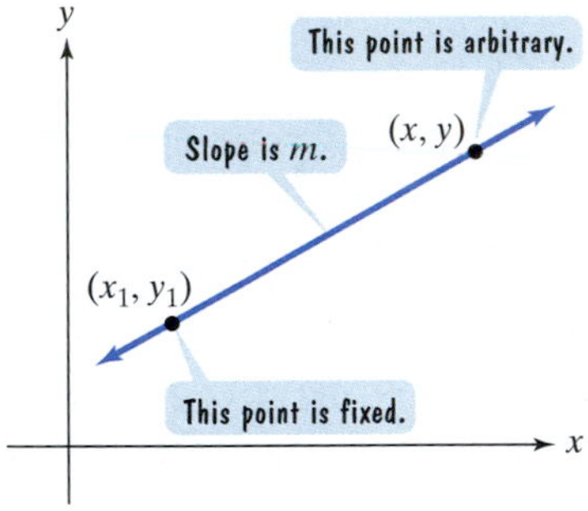

FIGURE 2.33 A line passing through (x_1, y_1) with slope m

We can use the slope of a line to obtain another useful form of the line's equation. Consider a nonvertical line that has slope m and contains the point (x_1, y_1). Now, let (x, y) represent any other point on the line, shown in **Figure 2.33**. Keep in mind that the point (x, y) is arbitrary and is not in one fixed position. By contrast, the point (x_1, y_1) is fixed.

Regardless of where the point (x, y) is located, the steepness of the line in **Figure 2.33** remains the same. Thus, the ratio for slope stays a constant m. This means that for all points (x, y) along the line

$$m = \frac{\text{Change in } y}{\text{Change in } x} = \frac{y - y_1}{x - x_1}.$$

We can clear the fraction by multiplying both sides by $x - x_1$, the least common denominator, where $x - x_1 \neq 0$.

$$m = \frac{y - y_1}{x - x_1} \quad \text{This is the slope of the line in Figure 2.33.}$$

$$m(x - x_1) = \frac{y - y_1}{x - x_1} \cdot (x - x_1) \quad \text{Multiply both sides by } x - x_1.$$

$$m(x - x_1) = y - y_1 \quad \text{Simplify: } \frac{y - y_1}{\cancel{x - x_1}} \cdot \cancel{(x - x_1)} = y - y_1.$$

Now, if we reverse the two sides, we obtain the **point-slope form** of the equation of a line.

Study Tip

When writing the point-slope form of a line's equation, you will never substitute numbers for x and y. You will substitute values for x_1, y_1, and m.

Point-Slope Form of the Equation of a Line

The **point-slope form of the equation** of a nonvertical line with slope m that passes through the point (x_1, y_1) is

$$y - y_1 = m(x - x_1).$$

For example, the point-slope form of the equation of the line passing through $(1, 5)$ with slope 2 $(m = 2)$ is

$$y - 5 = 2(x - 1).$$

1 Use the point-slope form to write equations of a line.

Using the Point-Slope Form to Write a Line's Equation

If we know the slope of a line and a point not containing the y-intercept through which the line passes, the point-slope form is the equation that we should use. Once we have obtained this equation, it is customary to solve for y and write the equation in slope-intercept form. Examples 1 and 2 illustrate these ideas.

EXAMPLE 1 Writing the Point-Slope Form and the Slope-Intercept Form

Write the point-slope form and the slope-intercept form of the equation of the line with slope 7 that passes through the point $(-4, 5)$.

Solution We begin with the point-slope form of the equation of a line with $m = 7$, $x_1 = -4$, and $y_1 = 5$.

$$y - y_1 = m(x - x_1) \quad \text{This is the point-slope form of the equation.}$$
$$y - 5 = 7[x - (-4)] \quad \text{Substitute the given values.}$$
$$y - 5 = 7(x + 4) \quad \text{We now have the point-slope form of the equation of the given line.}$$

Now we solve this equation for y and write an equivalent equation in slope-intercept form $(y = mx + b)$.

We need to isolate y.

$$y - 5 = 7(x + 4) \quad \text{This is the point-slope form of the equation.}$$
$$y - 5 = 7x + 28 \quad \text{Use the distributive property.}$$
$$y = 7x + 33 \quad \text{Add 5 to both sides.}$$

The slope-intercept form of the line's equation is $y = 7x + 33$. Using function notation, the equation is $f(x) = 7x + 33$.

✓ **CHECK POINT 1** Write the point-slope form and the slope-intercept form of the equation of the line with slope -2 that passes through the point $(4, -3)$.

EXAMPLE 2 Writing the Point-Slope Form and the Slope-Intercept Form

A line passes through the points $(1, -3)$ and $(-2, 4)$. (See **Figure 2.34**.) Find an equation of the line

a. in point-slope form. **b.** in slope-intercept form.

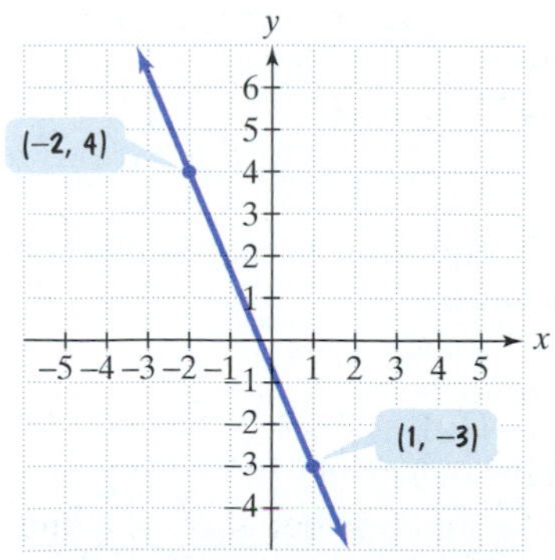

FIGURE 2.34 Line passing through $(1, -3)$ and $(-2, 4)$

Solution

a. To use the point-slope form, we need to find the slope. The slope is the change in the y-coordinates divided by the corresponding change in the x-coordinates.

$$m = \frac{4 - (-3)}{-2 - 1} = \frac{7}{-3} = -\frac{7}{3} \quad \text{This is the definition of slope using } (1, -3) \text{ and } (-2, 4).$$

We can take either point on the line to be (x_1, y_1). Let's use $(x_1, y_1) = (1, -3)$. Now, we are ready to write the point-slope form of the equation.

$$y - y_1 = m(x - x_1) \quad \text{This is the point-slope form of the equation.}$$
$$y - (-3) = -\frac{7}{3}(x - 1) \quad \text{Substitute: } (x_1, y_1) = (1, -3) \text{ and } m = -\frac{7}{3}.$$
$$y + 3 = -\frac{7}{3}(x - 1) \quad \text{Simplify.}$$

This equation is the point-slope form of the equation of the line shown in **Figure 2.34**.

b. Now, we solve this equation for y and write an equivalent equation in slope-intercept form $(y = mx + b)$.

We need to isolate y.

$$y + 3 = -\frac{7}{3}(x - 1)$$ This is the point-slope form of the equation.

$$y + 3 = -\frac{7}{3}x + \frac{7}{3}$$ Use the distributive property.

$$y = -\frac{7}{3}x - \frac{2}{3}$$ Subtract 3 from both sides: $\frac{7}{3} - 3 = \frac{7}{3} - \frac{9}{3} = -\frac{2}{3}$.

This equation is the slope-intercept form of the equation of the line shown in **Figure 2.34** on the previous page. Using function notation, the equation is $f(x) = -\frac{7}{3}x - \frac{2}{3}$.

Discover for Yourself

If you are given two points on a line, you can use either point for (x_1, y_1) when you write the point-slope form of its equation. Rework Example 2 using $(-2, 4)$ for (x_1, y_1). Once you solve for y, you should obtain the same slope-intercept form of the equation as the one shown in the last line of the solution to Example 2.

✓ CHECK POINT 2 A line passes through the points $(6, -3)$ and $(2, 5)$. Find an equation of the line

a. in point-slope form.

b. in slope-intercept form.

Here is a summary of the various forms for equations of lines:

Equations of Lines

1. Standard form: $Ax + By = C$
2. Slope-intercept form: $y = mx + b$ or $f(x) = mx + b$
3. Horizontal line: $y = b$
4. Vertical line: $x = a$
5. Point-slope form: $y - y_1 = m(x - x_1)$

In Examples 1 and 2, we eventually wrote a line's equation in slope-intercept form, or in function notation. But where do we start our work?

Starting with $y = mx + b$	Starting with $y - y_1 = m(x - x_1)$
Begin with the slope-intercept form if you know	Begin with the point-slope form if you know
• The slope of the line and the y-intercept.	• The slope of the line and a point on the line not containing the y-intercept
or	or
• Two points on the line, one of which contains the y-intercept.	• Two points on the line, neither of which contains the y-intercept.

2 Model data with linear functions and make predictions.

Applications

We have seen that linear functions are useful for modeling data that fall on or near a line. For example, the bar graph in **Figure 2.35(a)** gives the life expectancy for American men and women born in the indicated year. The data for the men are displayed as a set of six points in the scatter plot in **Figure 2.35(b)**.

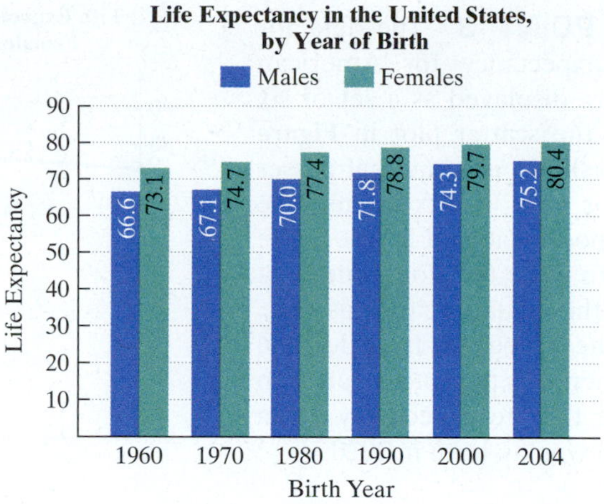

FIGURE 2.35(a) Data for men and women

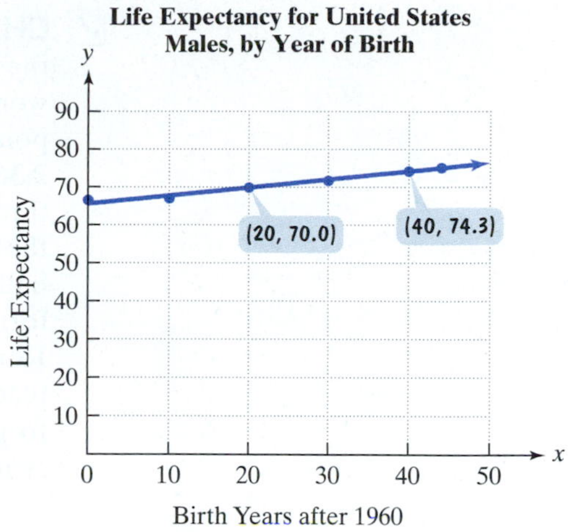

FIGURE 2.35(b) A scatter plot for the men's data

Source: National Center for Health Statistics

Also shown on the scatter plot in **Figure 2.35(b)** is a line that passes through or near the six points. By writing the equation of this line, we can obtain a model for life expectancy and make predictions about how long American men will live in the future.

EXAMPLE 3 Modeling Life Expectancy

Write the slope-intercept form of the equation of the line shown in **Figure 2.35(b)**. Use the equation to predict the life expectancy of an American man born in 2020.

Solution The line in **Figure 2.35(b)** passes through (20, 70.0) and (40, 74.3). We start by finding its slope.

$$m = \frac{\text{Change in } y}{\text{Change in } x} = \frac{74.3 - 70.0}{40 - 20} = \frac{4.3}{20} = 0.215$$

The slope indicates that for each subsequent birth year, a man's life expectancy is increasing by 0.215 years.

Now we write the line's equation in slope-intercept form.

$$y - y_1 = m(x - x_1)$$ Begin with the point-slope form.

$$y - 70.0 = 0.215(x - 20)$$ Either ordered pair can be (x_1, y_1). Let $(x_1, y_1) = (20, 70.0)$. From above, $m = 0.215$.

$$y - 70.0 = 0.215x - 4.3$$ Apply the distributive property: $0.215(20) = 4.3$.

$$y = 0.215x + 65.7$$ Add 70 to both sides and solve for y.

A linear function that models life expectancy, $f(x)$, for American men born x years after 1960 is

$$f(x) = 0.215x + 65.7.$$

Now let's use this function to predict the life expectancy of an American man born in 2020. Because 2020 is 60 years after 1960, substitute 60 for x and evaluate the function at 60.

$$f(60) = 0.215(60) + 65.7 = 78.6$$

Our model predicts that American men born in 2020 will have a life expectancy of 78.6 years.

Using Technology

You can use a graphing utility to obtain a model for a scatter plot in which the data points fall on or near a straight line. After entering the data in **Figure 2.35(b)**, a graphing utility displays a scatter plot of the data and the regression line, that is, the line that best fits the data.

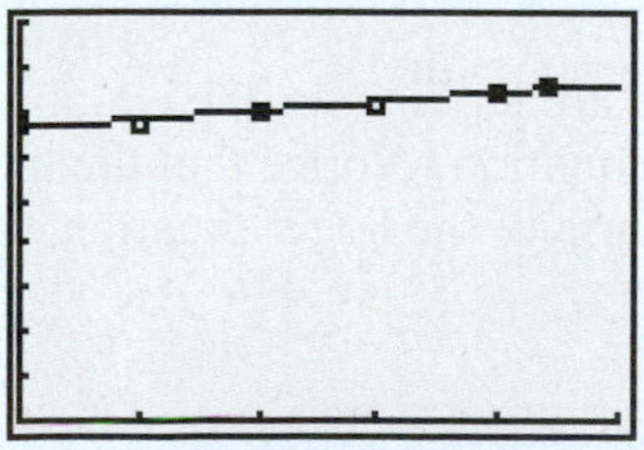

[0, 50, 10] by [0, 90, 10]

Also displayed is the regression line's equation.

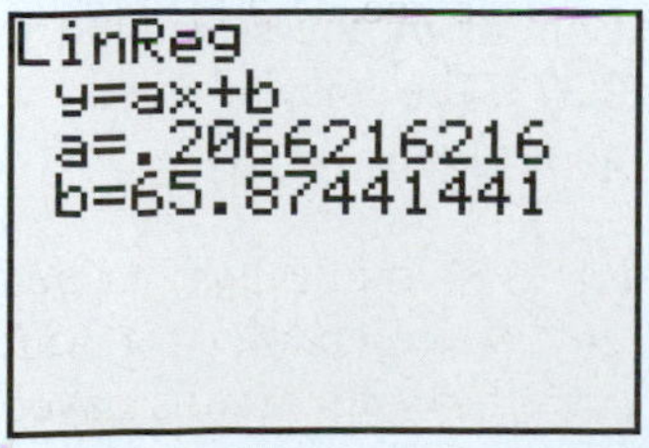

✓ CHECK POINT 3 The data for the life expectancy for American women are displayed as a set of six points in the scatter plot in **Figure 2.36**. Also shown is a line that passes through or near the six points. Use the data points labeled by the voice balloons to write the slope-intercept form of the equation of this line. Round the slope to two decimal places. Then use the linear function to predict the life expectancy of an American woman born in 2020.

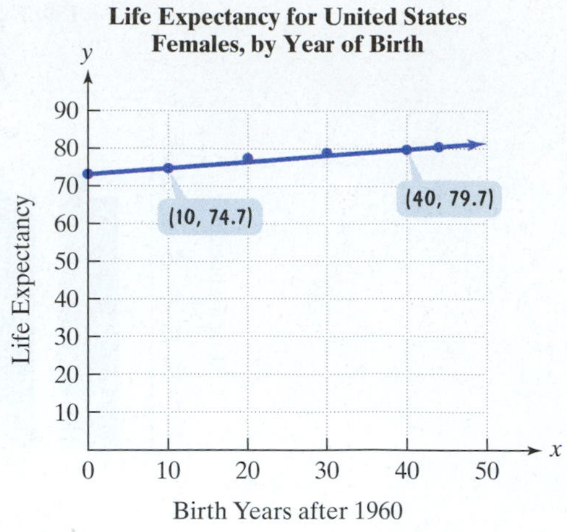

FIGURE 2.36

3 Find slopes and equations of parallel and perpendicular lines.

Parallel and Perpendicular Lines

Two nonintersecting lines that lie in the same plane are **parallel**. If two lines do not intersect, the ratio of the vertical change to the horizontal change is the same for each line. Because two parallel lines have the same "steepness," they must have the same slope.

Slope and Parallel Lines

1. If two nonvertical lines are parallel, then they have the same slope.
2. If two distinct nonvertical lines have the same slope, then they are parallel.
3. Two distinct vertical lines, both with undefined slopes, are parallel.

EXAMPLE 4 Writing Equations of a Line Parallel to a Given Line

Write an equation of the line passing through $(-3, 1)$ and parallel to the line whose equation is $y = 2x + 1$. Express the equation in point-slope form and slope-intercept form.

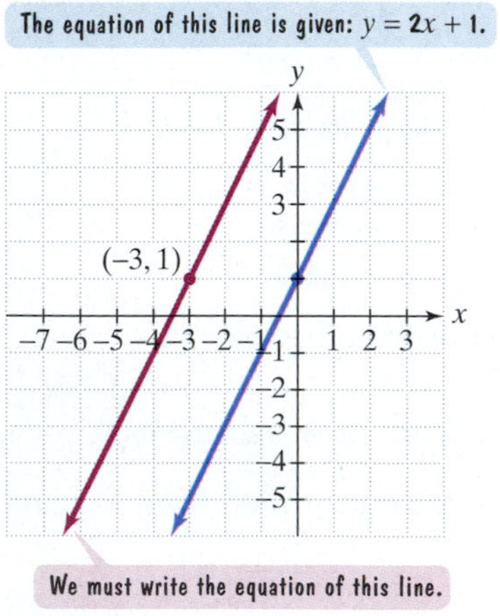

FIGURE 2.37

Solution The situation is illustrated in **Figure 2.37**. We are looking for the equation of the red line shown on the left. How do we obtain this equation? Notice that the line passes through the point $(-3, 1)$. Using the point-slope form of the line's equation, we have $x_1 = -3$ and $y_1 = 1$.

$$y - y_1 = m(x - x_1)$$

$y_1 = 1$ $\quad$ $x_1 = -3$

Now the only thing missing from the equation of the red line is m, the slope. Do we know anything about the slope of either line in **Figure 2.37**? The answer is yes; we know the slope of the blue line on the right, whose equation is given.

$$y = 2x + 1$$

The slope of the blue line on the right in Figure 2.37 is 2.

Parallel lines have the same slope. Because the slope of the blue line is 2, the slope of the red line, the line whose equation we must write, is also 2: $m = 2$. We now have values for x_1, y_1, and m for the red line.

$$y - y_1 = m(x - x_1)$$

$y_1 = 1$ $m = 2$ $x_1 = -3$

The point-slope form of the red line's equation is

$$y - 1 = 2[x - (-3)] \text{ or}$$
$$y - 1 = 2(x + 3).$$

Solving for y, we obtain the slope-intercept form of the equation.

$y - 1 = 2x + 6$ Apply the distributive property.

$y = 2x + 7$ Add 1 to both sides. This is the slope-intercept form, $y = mx + b$, of the equation. Using function notation, the equation is $f(x) = 2x + 7$.

✓ **CHECK POINT 4** Write an equation of the line passing through $(-2, 5)$ and parallel to the line whose equation is $y = 3x + 1$. Express the equation in point-slope form and slope-intercept form.

Two lines that intersect at a right angle (90°) are said to be **perpendicular**, shown in **Figure 2.38**. The relationship between the slopes of perpendicular lines is not as obvious as the relationship between parallel lines. **Figure 2.38** shows line AB, with slope $\frac{c}{d}$. Rotate line AB through 90° counterclockwise to obtain line $A'B'$, perpendicular to line AB. The figure indicates that the rise and the run of the new line are reversed from the original line, but the run is now negative. This means that the slope of the new line is $-\frac{d}{c}$. Notice that the product of the slopes of the two perpendicular lines is -1:

$$\left(\frac{c}{d}\right)\left(-\frac{d}{c}\right) = -1.$$

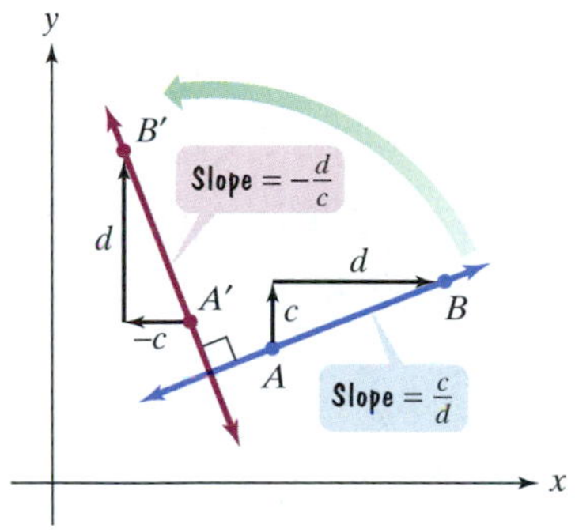

FIGURE 2.38
Slopes of perpendicular lines

This relationship holds for all nonvertical perpendicular lines and is summarized in the following box:

Slope and Perpendicular Lines

1. If two nonvertical lines are perpendicular, then the product of their slopes is -1.
2. If the product of the slopes of two lines is -1, then the lines are perpendicular.
3. A horizontal line having zero slope is perpendicular to a vertical line having undefined slope.

An equivalent way of stating this relationship is to say that **one line is perpendicular to another line if its slope is the *negative reciprocal* of the slope of the other line**. For example, if a line has slope 5, any line having slope $-\frac{1}{5}$ is perpendicular to it. Similarly, if a line has slope $-\frac{3}{4}$, any line having slope $\frac{4}{3}$ is perpendicular to it.

EXAMPLE 5 Writing Equations of a Line Perpendicular to a Given Line

a. Find the slope of any line that is perpendicular to the line whose equation is $x + 4y = 8$.

b. Write the equation of the line passing through $(3, -5)$ and perpendicular to the line whose equation is $x + 4y = 8$. Express the equation in point-slope form and slope-intercept form.

Solution

a. We begin by writing the equation of the given line, $x + 4y = 8$, in slope-intercept form. Solve for y.

$$x + 4y = 8 \quad \text{This is the given equation.}$$
$$4y = -x + 8 \quad \text{To isolate the } y\text{-term, subtract } x \text{ from both sides.}$$
$$y = -\frac{1}{4}x + 2 \quad \text{Divide both sides by 4.}$$

Slope is $-\frac{1}{4}$.

The given line has slope $-\frac{1}{4}$. Any line perpendicular to this line has a slope that is the negative reciprocal of $-\frac{1}{4}$. Thus, the slope of any perpendicular line is 4.

b. Let's begin by writing the point-slope form of the perpendicular line's equation. Because the line passes through the point $(3, -5)$, we have $x_1 = 3$ and $y_1 = -5$. In part (a), we determined that the slope of any line perpendicular to $x + 4y = 8$ is 4, so the slope of this particular perpendicular line must also be 4: $m = 4$.

$$y - y_1 = m(x - x_1)$$

$y_1 = -5$ $m = 4$ $x_1 = 3$

The point-slope form of the perpendicular line's equation is

$$y - (-5) = 4(x - 3) \text{ or}$$
$$y + 5 = 4(x - 3).$$

How can we express this equation in slope-intercept form, $y = mx + b$? We need to solve for y.

$$y + 5 = 4(x - 3) \quad \text{This is the point-slope form of the line's equation.}$$
$$y + 5 = 4x - 12 \quad \text{Apply the distributive property.}$$
$$y = 4x - 17 \quad \text{Subtract 5 from both sides of the equation and solve for } y.$$

The point-slope form of the perpendicular line's equation is

$$y = 4x - 17 \quad \text{or} \quad f(x) = 4x - 17.$$

✓ CHECK POINT 5

a. Find the slope of any line that is perpendicular to the line whose equation is $x + 3y = 12$.

b. Write the equation of the line passing through $(-2, -6)$ and perpendicular to the line whose equation is $x + 3y = 12$. Express the equation in point-slope form and slope-intercept form.

2.5 EXERCISE SET

MyMathLab | MathXL PRACTICE | WATCH | DOWNLOAD | READ | REVIEW

Practice Exercises

Write the point-slope form of the line's equation satisfying each of the conditions in Exercises 1–28. Then use the point-slope form of the equation to write the slope-intercept form of the equation in function notation.

1. Slope = 3, passing through $(2, 5)$
2. Slope = 4, passing through $(3, 1)$
3. Slope = 5, passing through $(-2, 6)$
4. Slope = 8, passing through $(-4, 1)$
5. Slope = -4, passing through $(-3, -2)$
6. Slope = -6, passing through $(-2, -4)$
7. Slope = -5, passing through $(-2, 0)$

8. Slope $= -4$, passing through $(0, -3)$

9. Slope $= -1$, passing through $\left(-2, -\frac{1}{2}\right)$

10. Slope $= -1$, passing through $\left(-\frac{1}{4}, -4\right)$

11. Slope $= \frac{1}{4}$, passing through the origin

12. Slope $= \frac{1}{5}$, passing through the origin

13. Slope $= -\frac{2}{3}$, passing through $(6, -4)$

14. Slope $= -\frac{2}{5}$, passing through $(15, -4)$

15. Passing through $(6, 3)$ and $(5, 2)$

16. Passing through $(1, 3)$ and $(2, 4)$

17. Passing through $(-2, 0)$ and $(0, 4)$

18. Passing through $(2, 0)$ and $(0, -1)$

19. Passing through $(-6, 13)$ and $(-2, 5)$

20. Passing through $(-3, 2)$ and $(2, -8)$

21. Passing through $(1, 9)$ and $(4, -2)$

22. Passing through $(4, -8)$ and $(8, -3)$

23. Passing through $(-2, -5)$ and $(3, -5)$

24. Passing through $(-1, -4)$ and $(3, -4)$

25. Passing through $(7, 8)$ with x-intercept $= 3$

26. Passing through $(-4, 5)$ with y-intercept $= -3$

27. x-intercept $= 2$ and y-intercept $= -1$

28. x-intercept $= -2$ and y-intercept $= 4$

In Exercises 29–44, the equation of a line is given. Find the slope of a line that is **a.** *parallel to the line with the given equation; and* **b.** *perpendicular to the line with the given equation.*

29. $y = 5x$

30. $y = 3x$

31. $y = -7x$

32. $y = -9x$

33. $y = \frac{1}{2}x + 3$

34. $y = \frac{1}{4}x - 5$

35. $y = -\frac{2}{5}x - 1$

36. $y = -\frac{3}{7}x - 2$

37. $4x + y = 7$

38. $8x + y = 11$

39. $2x + 4y = 8$

40. $3x + 2y = 6$

41. $2x - 3y = 5$

42. $3x - 4y = -7$

43. $x = 6$

44. $y = 9$

In Exercises 45–48, write an equation for line L in point-slope form and slope-intercept form.

45.

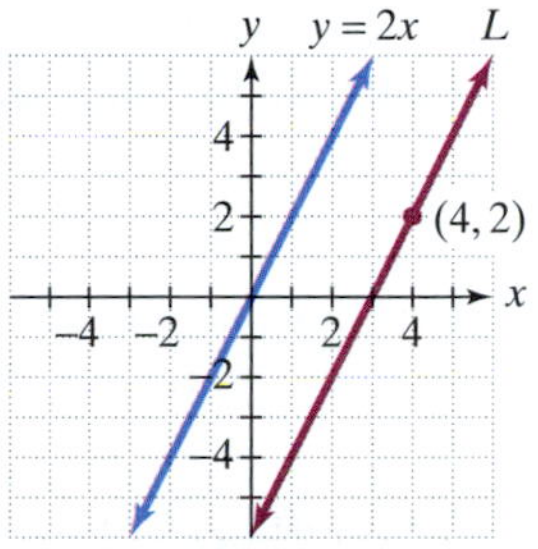

L is parallel to y = 2x.

46.

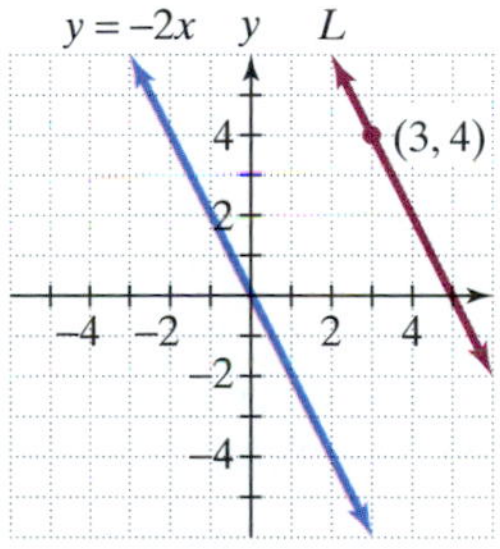

L is parallel to y = −2x.

47.

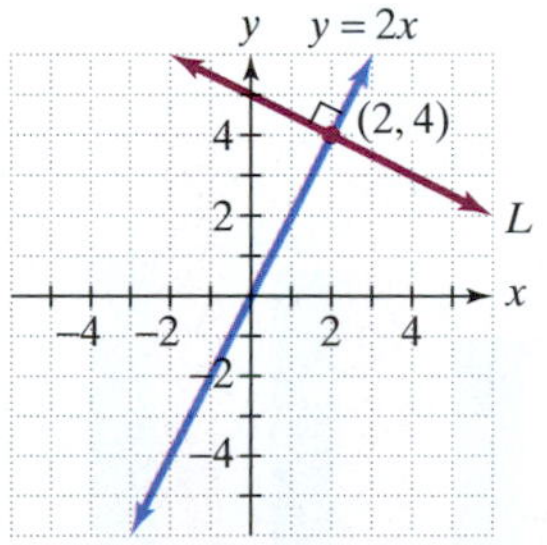

L is perpendicular to y = 2x.

48.

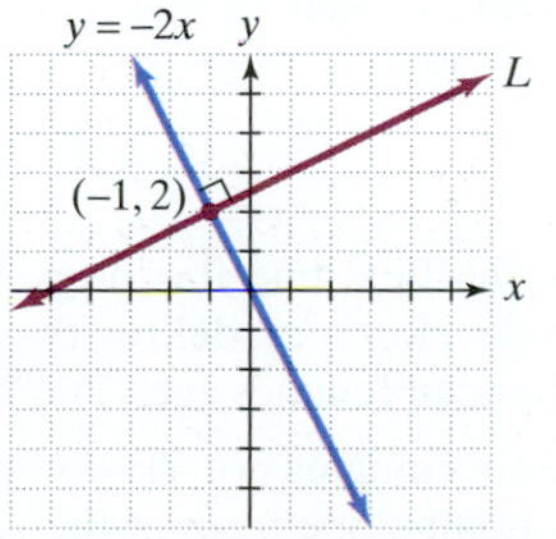

L is perpendicular to y = −2x.

In Exercises 49–56, use the given conditions to write an equation for each line in point-slope form and slope-intercept form.

49. Passing through $(-8, -10)$ and parallel to the line whose equation is $y = -4x + 3$

50. Passing through $(-2, -7)$ and parallel to the line whose equation is $y = -5x + 4$

51. Passing through $(2, -3)$ and perpendicular to the line whose equation is $y = \frac{1}{5}x + 6$

52. Passing through $(-4, 2)$ and perpendicular to the line whose equation is $y = \frac{1}{3}x + 7$

53. Passing through $(-2, 2)$ and parallel to the line whose equation is $2x - 3y = 7$

54. Passing through $(-1, 3)$ and parallel to the line whose equation is $3x - 2y = 5$

55. Passing through $(4, -7)$ and perpendicular to the line whose equation is $x - 2y = 3$

56. Passing through $(5, -9)$ and perpendicular to the line whose equation is $x + 7y = 12$

Practice PLUS

In Exercises 57–64, write the slope-intercept form of the equation of a function f whose graph satisfies the given conditions.

57. The graph of f passes through $(-1, 5)$ and is perpendicular to the line whose equation is $x = 6$.

58. The graph of f passes through $(-2, 6)$ and is perpendicular to the line whose equation is $x = -4$.

59. The graph of f passes through $(-6, 4)$ and is perpendicular to the line that has an x-intercept of 2 and a y-intercept of -4.

60. The graph of f passes through $(-5, 6)$ and is perpendicular to the line that has an x-intercept of 3 and a y-intercept of -9.

61. The graph of f is perpendicular to the line whose equation is $3x - 2y = 4$ and has the same y-intercept as this line.

62. The graph of f is perpendicular to the line whose equation is $4x - y = 6$ and has the same y-intercept as this line.

63. The graph of f is the graph of $g(x) = 4x - 3$ shifted down 2 units.

64. The graph of f is the graph of $g(x) = 2x - 5$ shifted up 3 units.

65. What is the slope of a line that is parallel to the line whose equation is $Ax + By = C, B \neq 0$?

66. What is the slope of a line that is perpendicular to the line whose equation is $Ax + By = C, A \neq 0$ and $B \neq 0$?

Application Exercises

Americans are getting married later in life, or not getting married at all. In 2006, nearly half of Americans ages 25 through 29 were unmarried. The bar graph shows the percentage of never-married men and women in this age group. The data are displayed as two sets of four points each, one scatter plot for the percentage of never-married American men and one for the percentage of never-married American women. Also shown for each scatter plot is a line that passes through or near the four points. Use these lines to solve Exercises 67–68.

Percentage of United States Population Never Married, Ages 25–29

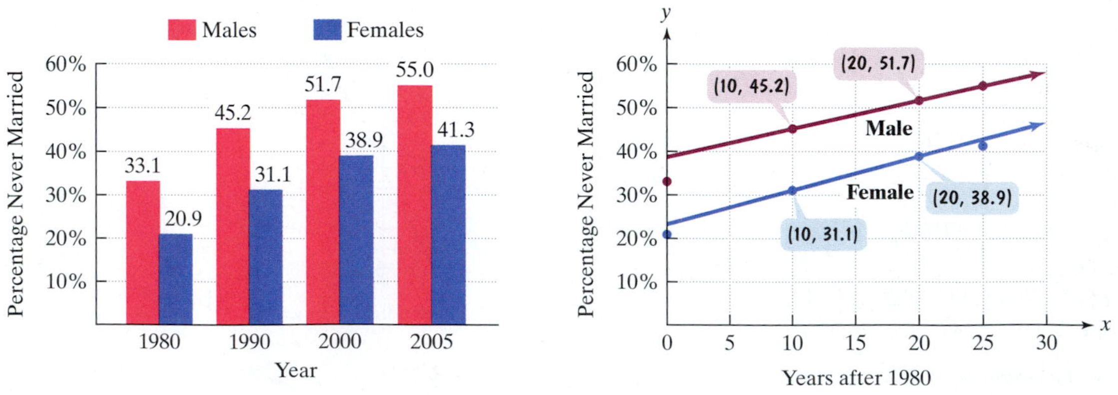

Source: U.S. Census Bureau

67. In this exercise, you will use the blue line for the women shown on the scatter plot to develop a model for the percentage of never-married American females ages 25–29.

a. Use the two points whose coordinates are shown by the voice balloons to find the point-slope form of the equation of the line that models the percentage of never-married American females ages 25–29, y, x years after 1980.

b. Write the equation from part (a) in slope-intercept form. Use function notation.

c. Use the linear function to predict the percentage of never-married American females, ages 25–29, in 2020.

68. In this exercise, you will use the red line for the men shown on the scatter plot on page 152 to develop a model for the percentage of never-married American males ages 25–29.

a. Use the two points whose coordinates are shown by the voice balloons to find the point-slope form of the equation of the line that models the percentage of never-married American males ages 25–29, y, x years after 1980.

b. Write the equation from part (a) in slope-intercept form. Use function notation.

c. Use the linear function to predict the percentage of never-married American males, ages 25–29, in 2015.

The amount of carbon dioxide in the atmosphere, measured in parts per million, has been increasing as a result of the burning of oil and coal. The buildup of gases and particles traps heat and raises the planet's temperature, a phenomenon called the greenhouse effect. In Exercises 69–70, you will develop linear models involving variables related to global warming.

69. The bar graph shows the average global temperature, in degrees Fahrenheit, for seven selected years.

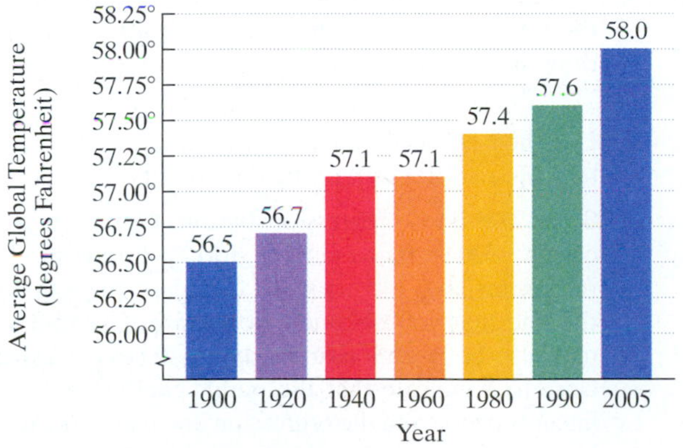

Source: National Oceanic and Atmospheric Administration

a. Let x represent the number of years after 1900 and let y represent the average global temperature. Create a scatter plot that displays the data as a set of seven points in a rectangular coordinate system.

b. Draw a line through the two points that show the average global temperatures for 1940 and 1990. Use the coordinates of these points to write the line's equation in point-slope form and slope-intercept form.

c. Use the slope-intercept form of the equation from part (b) to predict the average global temperature in 2050.

70. The pre-industrial concentration of atmospheric carbon dioxide was 280 parts per million. The bar graph at the top of the next column shows the average atmospheric concentration of carbon dioxide, in parts per million, for seven selected years.

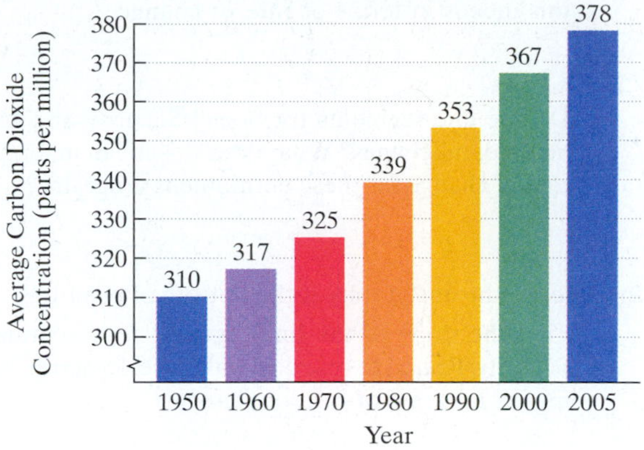

Source: National Oceanic and Atmospheric Administration

a. Let x represent the number of years after 1950 and let y represent the average atmospheric concentration of carbon dioxide. Create a scatter plot that displays the data as a set of seven points in a rectangular coordinate system.

b. Draw a line through the two points that show the average atmospheric concentration of carbon dioxide for 1960 and 2000. Use the coordinates of these points to write the line's equation in point-slope form and slope-intercept form.

c. Use the slope-intercept form of the equation from part (b) to predict the average atmospheric concentration of carbon dioxide in 2050.

In 2007, the U.S. government faced the prospect of paying out more and more in Social Security, Medicare, and Medicaid benefits. The line graphs show the projected costs of these entitlement programs, in billions of dollars, from 2007 through 2016. Use this information to solve Exercises 71–72.

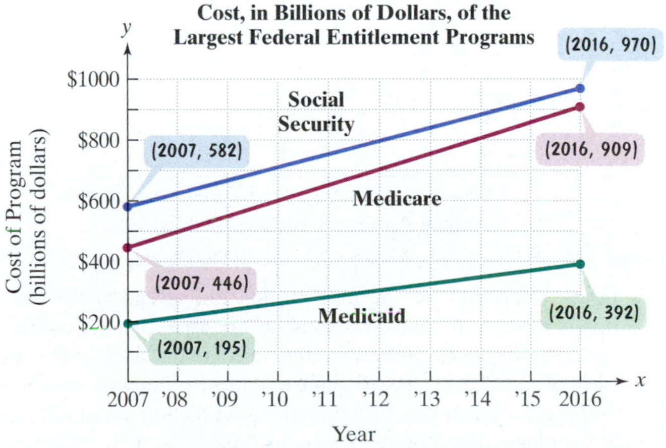

Source: Congressional Budget Office

71. a. Find the slope of the line segment representing Social Security. Round to one decimal place. Describe what this means in terms of rate of change.

b. Find the slope of the line segment representing Medicare. Round to one decimal place. Describe what this means in terms of rate of change.

c. Do the line segments for Social Security and Medicare lie on parallel lines? What does this mean in terms of the rate of change for these entitlement programs?

72. Refer to the line graphs at the bottom of page 153.

a. Find the slope of the line segment representing Social Security. Round to one decimal place. Describe what this means in terms of rate of change.

b. Find the slope of the line segment representing Medicaid. Round to one decimal place. Describe what this means in terms of rate of change.

c. Do the line segments for Social Security and Medicaid lie on parallel lines? What does this mean in terms of the rate of change for these entitlement programs?

73. Just as money doesn't buy happiness for individuals, the two don't necessarily go together for countries either. However, the scatter plot does show a relationship between a country's annual per capita income and the percentage of people in that country who call themselves "happy."

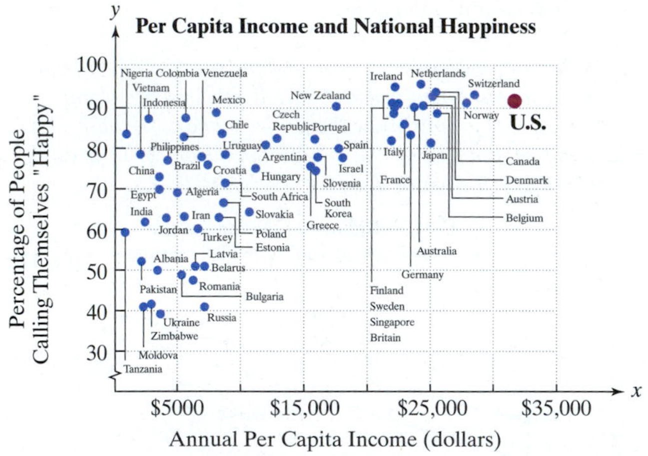

Source: Richard Layard, *Happiness: Lessons from a New Science*, Penguin, 2005

Draw a line that fits the data so that the spread of the data points around the line is as small as possible. Use the coordinates of two points along your line to write the slope-intercept form of its equation. Express the equation in function notation and use the linear function to make a prediction about national happiness based on per capita income.

Writing in Mathematics

74. Describe how to write the equation of a line if its slope and a point along the line are known.

75. Describe how to write the equation of a line if two points along the line are known.

76. If two lines are parallel, describe the relationship between their slopes.

77. If two lines are perpendicular, describe the relationship between their slopes.

78. If you know a point on a line and you know the equation of a line parallel to this line, explain how to write the line's equation.

79. In Example 3 on page 147, we developed a model that predicted American men born in 2020 will have a life expectancy of 78.6 years. Describe something that might occur that would make this prediction inaccurate.

Technology Exercises

80. The lines whose equations are $y = \frac{1}{3}x + 1$ and $y = -3x - 2$ are perpendicular because the product of their slopes, $\frac{1}{3}$ and -3, respectively, is -1.

a. Use a graphing utility to graph the equations in a $[-10, 10, 1]$ by $[-10, 10, 1]$ viewing rectangle. Do the lines appear to be perpendicular?

b. Now use the zoom square feature of your utility. Describe what happens to the graphs. Explain why this is so.

81. a. Use the statistical menu of your graphing utility to enter the seven data points shown in the scatter plot that you drew in Exercise 69(a).

b. Use the DRAW menu and the scatter plot capability to draw a scatter plot of the data points.

c. Select the linear regression option. Use your utility to obtain values for a and b for the equation of the regression line, $y = ax + b$. Compare this equation to the one that you obtained by hand in Exercise 69. You may also be given a **correlation coefficient**, r. Values of r close to 1 indicate that the points can be modeled by a linear function and the regression line has a positive slope. Values of r close to -1 indicate that the points can be modeled by a linear function and the regression line has a negative slope. Values of r close to 0 indicate no linear relationship between the variables. In this case, a linear model does not accurately describe the data.

d. Use the appropriate sequence (consult your manual) to graph the regression equation on top of the points in the scatter plot.

82. Repeat Exercise 81 using the seven data points shown in the scatter plot that you drew in Exercise 70(a).

Critical Thinking Exercises

Make Sense? *In Exercises 83–86, determine whether each statement "makes sense" or "does not make sense" and explain your reasoning.*

83. I can use any two points in a scatter plot to write the point-slope form of the equation of the line through those points. However, the other data points in the scatter plot might not fall on, or even near, this line.

84. I have linear functions that model changes for men and women over the same time period. The functions have the same slope, so their graphs are parallel lines, indicating that the rate of change for men is the same as the rate of change for women.

85. Some of the steel girders in this photo of the Eiffel Tower appear to be perpendicular. I can verify my observation by determining that their slopes are negative reciprocals.

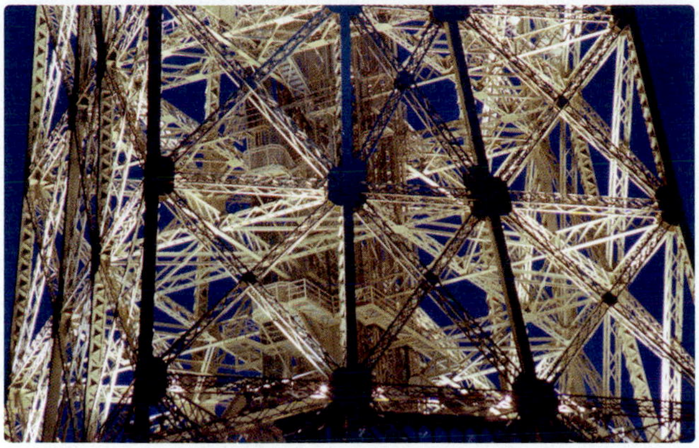

86. When writing equations of lines, it's always easiest to begin by writing the point-slope form of the equation.

In Exercises 87–90, determine whether each statement is true or false. If the statement is false, make the necessary change(s) to produce a true statement.

87. The standard form of the equation of a line passing through $(-3, -1)$ and perpendicular to the line whose equation is $y = -\frac{2}{5}x - 4$ is $5x - 2y = -13$.

88. If I change the subtraction signs to addition signs in $y - 12 = 8(x - 2)$, the y-intercept of the corresponding graph will change from -4 to 4.

89. $y - 5 = 2(x - 1)$ is an equation of a line passing through $(4, 11)$.

90. The function $\{(-1, 4), (3, 6), (5, 7), (11, 10)\}$ can be described using $y - 7 = \frac{1}{2}(x - 5)$ with a domain of $\{-1, 3, 5, 11\}$.

91. Determine the value of B so that the line whose equation is $By = 8x - 1$ has slope -2.

92. Determine the value of A so that the line whose equation is $Ax + y = 2$ is perpendicular to the line containing the points $(1, -3)$ and $(-2, 4)$.

93. Consider a line whose x-intercept is -3 and whose y-intercept is -6. Provide the missing coordinate for the following two points that lie on this line: $(-40, \quad)$ and $(\quad, -200)$.

94. Prove that the equation of a line passing through $(a, 0)$ and $(0, b)(a \neq 0, b \neq 0)$ can be written in the form $\frac{x}{a} + \frac{y}{b} = 1$. Why is this called the *intercept form* of a line?

Review Exercises

95. If $f(x) = 3x^2 - 8x + 5$, find $f(-2)$. (Section 2.1, Example 3)

96. If $f(x) = x^2 - 3x + 4$ and $g(x) = 2x - 5$, find $(fg)(-1)$. (Section 2.3, Example 4)

97. The sum of the angles of a triangle is 180°. Find the three angles of a triangle if one angle is 20° greater than the smallest angle and the third angle is twice the smallest angle. (Section 1.5, Example 1)

Preview Exercises

Exercises 98–100 will help you prepare for the material covered in the first section of the next chapter.

98. a. Does $(-5, -6)$ satisfy $2x - y = -4$?
    b. Does $(-5, -6)$ satisfy $3x - 5y = 15$?

99. Graph $y = -x - 1$ and $4x - 3y = 24$ in the same rectangular coordinate system. At what point do the graphs intersect?

100. Solve: $7x - 2(-2x + 4) = 3$.

GROUP PROJECT

CHAPTER 2

In Example 3 on page 147, we used the data in **Figure 2.35** to develop a linear function that modeled life expectancy. For this group exercise, you might find it helpful to pattern your work after **Figure 2.35** and the solution to Example 3. Group members should begin by consulting an almanac, newspaper, magazine, or the Internet to find data that appear to lie approximately on or near a line. Working by hand or using a graphing utility, group members should construct scatter plots for the data that were collected. If working by hand, draw a line that approximately fits the data in each scatter plot and then write its equation as a function in slope-intercept form. If using a graphing utility, obtain the equation of each regression line. Then use each linear function's equation to make predictions about what might occur in the future. Are there circumstances that might affect the accuracy of the prediction? List some of these circumstances.

Chapter 2 Summary

Definitions and Concepts	Examples

Section 2.1 Introduction to Functions

A relation is any set of ordered pairs. The set of first components of the ordered pairs is the domain and the set of second components is the range. A function is a relation in which each member of the domain corresponds to exactly one member of the range. No two ordered pairs of a function can have the same first component and different second components.

The domain of the relation $\{(1, 2), (3, 4), (3, 7)\}$ is $\{1, 3\}$. The range is $\{2, 4, 7\}$. The relation is not a function: 3, in the domain, corresponds to both 4 and 7 in the range.

If a function is defined by an equation, the notation $f(x)$, read "f of x" or "f at x," describes the value of the function at the number, or input, x.

If $f(x) = 7x - 5$, then

$$\begin{aligned} f(a+2) &= 7(a+2) - 5 \\ &= 7a + 14 - 5 \\ &= 7a + 9. \end{aligned}$$

Section 2.2 Graphs of Functions

The graph of a function is the graph of its ordered pairs.

The Vertical Line Test for Functions

If any vertical line intersects a graph in more than one point, the graph does not define y as a function of x.

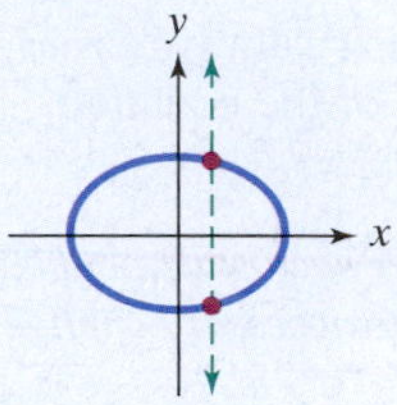

Not the graph of a function

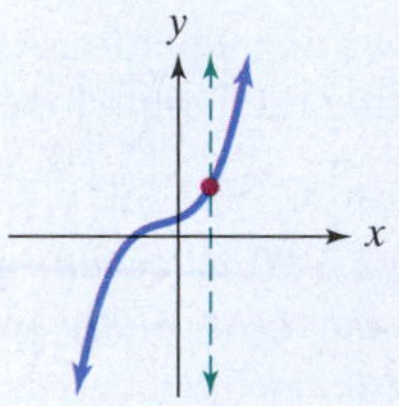

The graph of a function

At the left or right of a function's graph, you will often find closed dots, open dots, or arrows. A closed dot shows that the graph ends and the point belongs to the graph. An open dot shows that the graph ends and the point does not belong to the graph. An arrow indicates that the graph extends indefinitely.

The graph of a function can be used to determine the function's domain and its range. To find the domain, look for all the inputs on the x-axis that correspond to points on the graph. To find the range, look for all the outputs on the y-axis that correspond to points on the graph.

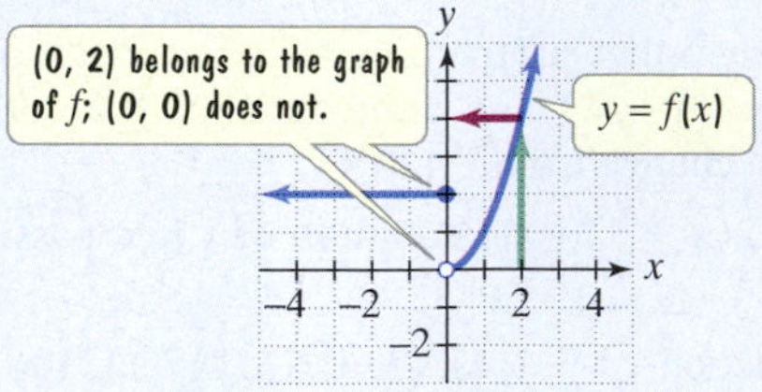

To find $f(2)$, locate 2 on the x-axis. The graph shows $f(2) = 4$.

Domain of $f = \{x \mid x \text{ is a real number}\}$

Range of $f = \{y \mid y > 0\}$

Section 2.3 The Algebra of Functions

A Function's Domain

If a function f does not model data or verbal conditions, its domain is the largest set of real numbers for which the value of $f(x)$ is a real number. Exclude from a function's domain real numbers that cause division by zero and real numbers that result in a square root of a negative number.

$$f(x) = 7x + 13$$

Domain of $f = \{x \mid x \text{ is a real number}\}$

$$g(x) = \frac{7x}{12 - x}$$

Domain of $g = \{x \mid x \text{ is a real number and } x \neq 12\}$

Definitions and Concepts	Examples

Section 2.3 The Algebra of Functions (continued)

The Algebra of Functions

Let f and g be two functions. The sum $f + g$, the difference $f - g$, the product fg, and the quotient $\frac{f}{g}$ are functions whose domains are the set of all real numbers common to the domains of f and g, defined as follows:

1. Sum: $(f + g)(x) = f(x) + g(x)$
2. Difference: $(f - g)(x) = f(x) - g(x)$
3. Product: $(fg)(x) = f(x) \cdot g(x)$
4. Quotient: $\left(\frac{f}{g}\right)(x) = \frac{f(x)}{g(x)}, g(x) \neq 0.$

Let $f(x) = x^2 + 2x$ and $g(x) = 4 - x$.

- $(f + g)(x) = (x^2 + 2x) + (4 - x) = x^2 + x + 4$

  $(f + g)(-2) = (-2)^2 + (-2) + 4 = 4 - 2 + 4 = 6$
- $(f - g)(x) = (x^2 + 2x) - (4 - x) = x^2 + 2x - 4 + x$

  $= x^2 + 3x - 4$

  $(f - g)(5) = 5^2 + 3 \cdot 5 - 4 = 25 + 15 - 4 = 36$
- $(fg)(1) = f(1) \cdot g(1) = (1^2 + 2 \cdot 1)(4 - 1)$

  $= 3(3) = 9$
- $\left(\frac{f}{g}\right)(x) = \frac{x^2 + 2x}{4 - x}, x \neq 4$

  $\left(\frac{f}{g}\right)(3) = \frac{3^2 + 2 \cdot 3}{4 - 3} = \frac{9 + 6}{1} = 15$

Section 2.4 Linear Functions and Slope

Data presented in a visual form as a set of points is called a scatter plot. A line that best fits the data points is called a regression line.

A function whose graph is a straight line is called a linear function. All linear functions can be written in the form $f(x) = mx + b$.

$f(x) = 3x + 10$ is a linear function.

$g(x) = 3x^2 + 10$ is not a linear function.

If a graph intersects the x-axis at $(a, 0)$, then a is an x-intercept. If a graph intersects the y-axis at $(0, b)$, then b is a y-intercept. The standard form of the equation of a line,

$$Ax + By = C,$$

can be graphed using intercepts and a checkpoint.

Graph using intercepts: $4x + 3y = 12$.

x-intercept: (Set $y = 0$.) $4x = 12$, $x = 3$ — Line passes through (3, 0).

y-intercept: (Set $x = 0$.) $3y = 12$, $y = 4$ — Line passes through (0, 4).

Checkpoint: Let $x = 2$.

$$4 \cdot 2 + 3y = 12$$
$$8 + 3y = 12$$
$$3y = 4$$
$$y = \frac{4}{3}$$

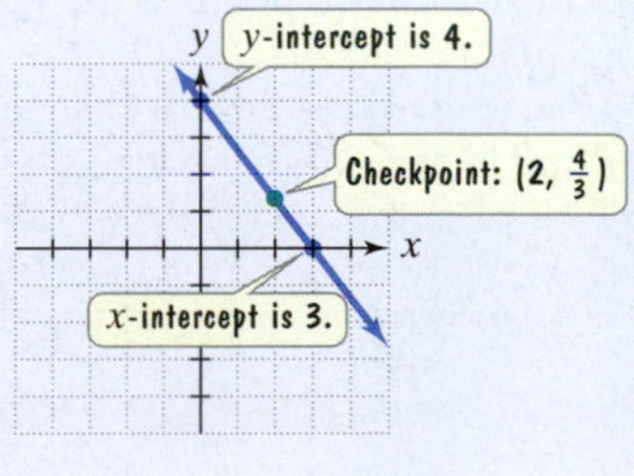

Definitions and Concepts | Examples

Section 2.4 Linear Functions and Slope (continued)

The slope, m, of the line through the points (x_1, y_1) and (x_2, y_2) is

$$m = \frac{y_2 - y_1}{x_2 - x_1}, \quad x_2 - x_1 \neq 0.$$

If the slope is positive, the line rises from left to right. If the slope is negative, the line falls from left to right. The slope of a horizontal line is 0. The slope of a vertical line is undefined.

For points $(-7, 2)$ and $(3, -4)$, the slope of the line through the points is

$$m = \frac{\text{Change in } y}{\text{Change in } x} = \frac{-4 - 2}{3 - (-7)} = \frac{-6}{10} = -\frac{3}{5}.$$

The slope is negative, so the line falls.

For points $(2, -5)$ and $(2, 16)$, the slope of the line through the points is:

$$m = \frac{\text{Change in } y}{\text{Change in } x} = \frac{16 - (-5)}{2 - 2} = \frac{21}{0}.$$

undefined

The slope is undefined, so the line is vertical.

The slope-intercept form of the equation of a nonvertical line with slope m and y-intercept b is

$$y = mx + b.$$

Using function notation, the equation is

$$f(x) = mx + b.$$

Graph: $f(x) = -\frac{3}{4}x + 1.$

Slope is $-\frac{3}{4}$.

y-intercept is 1.

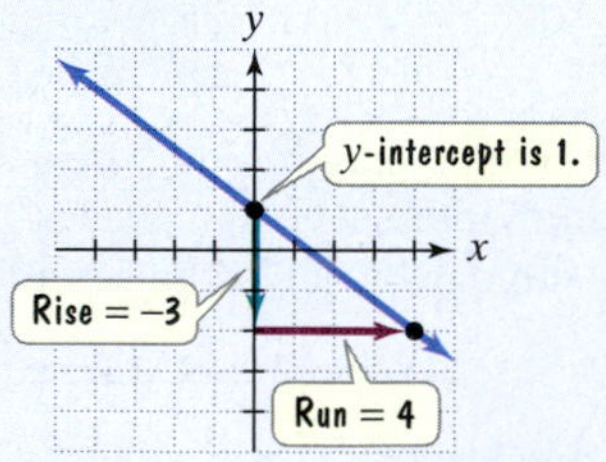

Horizontal and Vertical Lines

The graph of $y = b$, or $f(x) = b$, is a horizontal line. The y-intercept is b. The linear function $f(x) = b$ is called a constant function.
The graph of $x = a$ is a vertical line. The x-intercept is a. A vertical line is not a linear function.

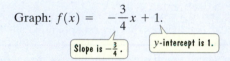

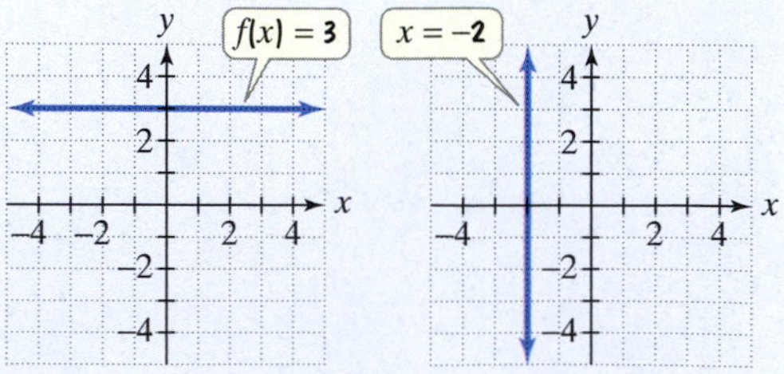

If the graph of a function is not a straight line, the average rate of change between any two points is the slope of the line containing the two points.
For a linear function, slope is the rate of change of the dependent variable per unit change of the independent variable.

The function

$$p(t) = -0.59t + 80.75$$

slope

models the percentage, $p(t)$, of Americans smoking cigarettes t years after 1900. The slope, -0.59, shows that the percentage of smokers is decreasing by 0.59% per year.

Definitions and Concepts | Examples

Section 2.5 The Point-Slope Form of the Equation of a Line

The point-slope form of the equation of a nonvertical line with slope m that passes through the point (x_1, y_1) is

$$y - y_1 = m(x - x_1).$$

Slope $= -4$, passing through $(-1, 5)$

$m = -4$ $x_1 = -1$ $y_1 = 5$

The point-slope form of the line's equation is

$$y - 5 = -4[x - (-1)].$$

Simplify:

$$y - 5 = -4(x + 1).$$

To write the point-slope form of the line passing through two points, begin by using the points to compute the slope, m. Use either given point as (x_1, y_1) and write the point-slope equation:

$$y - y_1 = m(x - x_1).$$

Solving this equation for y gives the slope-intercept form of the line's equation.

Write equations in point-slope form and in slope-intercept form of the line passing through $(4, 1)$ and $(3, -2)$.

$$m = \frac{-2 - 1}{3 - 4} = \frac{-3}{-1} = 3$$

Using $(4, 1)$ as (x_1, y_1), the point-slope form of the equation is

$$y - 1 = 3(x - 4).$$

Solve for y to obtain the slope-intercept form.

$$y - 1 = 3x - 12$$
$$y = 3x - 11$$

In function notation,

$$f(x) = 3x - 11.$$

Nonvertical parallel lines have the same slope. If the product of the slopes of two lines is -1, then the lines are perpendicular. One line is perpendicular to another line if its slope is the negative reciprocal of the slope of the other. A horizontal line having zero slope is perpendicular to a vertical line having undefined slope.

Write equations in point-slope form and in slope-intercept form of the line passing through $(2, -1)$

x_1 y_1

and perpendicular to $y = -\frac{1}{5}x + 6$.

slope

The slope, m, of the perpendicular line is 5, the negative reciprocal of $-\frac{1}{5}$.

$$y - (-1) = 5(x - 2)$$

Point-slope form of the equation

$$y + 1 = 5(x - 2)$$
$$y + 1 = 5x - 10$$
$$y = 5x - 11 \text{ or } f(x) = 5x - 11$$

Slope-intercept form of the equation

CHAPTER 2 REVIEW EXERCISES

2.1 *In Exercises 1–3, determine whether each relation is a function. Give the domain and range for each relation.*

1. $\{(3, 10), (4, 10), (5, 10)\}$

2. $\{(1, 12), (2, 100), (3, \pi), (4, -6)\}$

3. $\{(13, 14), (15, 16), (13, 17)\}$

In Exercises 4–5, find the indicated function values.

4. $f(x) = 7x - 5$
   a. $f(0)$ b. $f(3)$ c. $f(-10)$
   d. $f(2a)$ e. $f(a + 2)$
5. $g(x) = 3x^2 - 5x + 2$
   a. $g(0)$ b. $g(5)$ c. $g(-4)$
   d. $g(b)$ e. $g(4a)$

2.2 *In Exercises 6–7, graph the given functions, f and g, in the same rectangular coordinate system. Select integers for x, starting with −2 and ending with 2. Once you have obtained your graphs, describe how the graph of g is related to the graph of f.*

6. $f(x) = x^2, \quad g(x) = x^2 - 1$

7. $f(x) = |x|, \quad g(x) = |x| + 2$

In Exercises 8–13, use the vertical line test to identify graphs in which y is a function of x.

8.

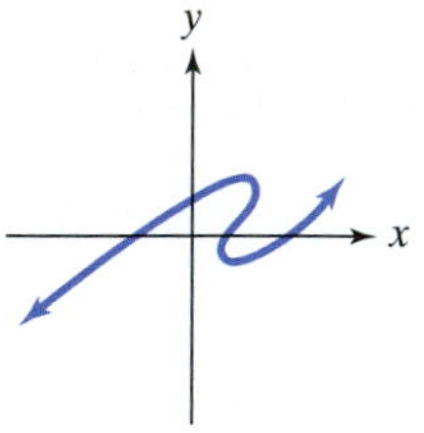

9.

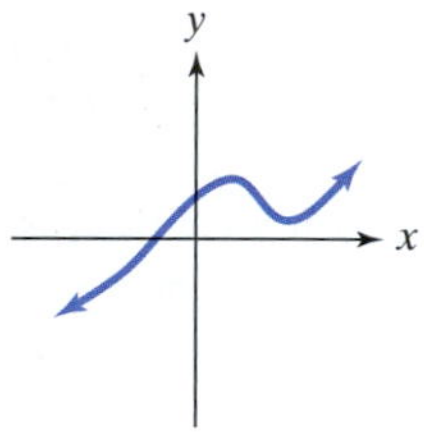

10.

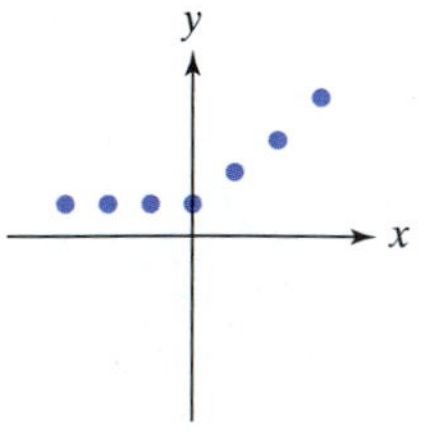

11.

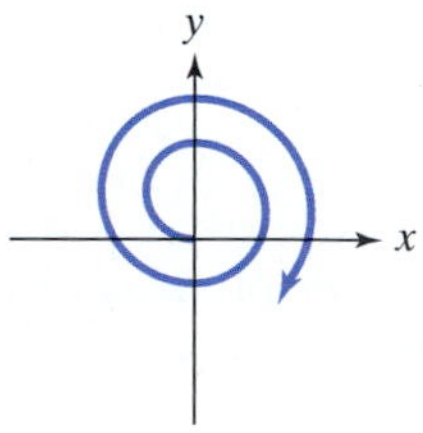

12.

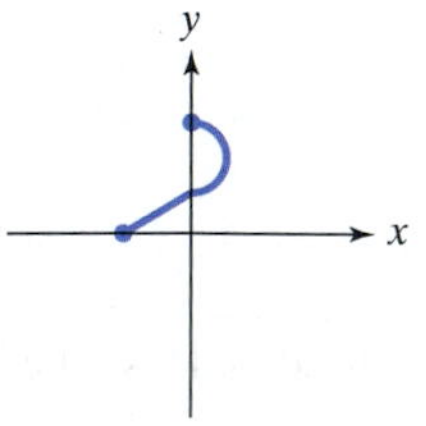

13.

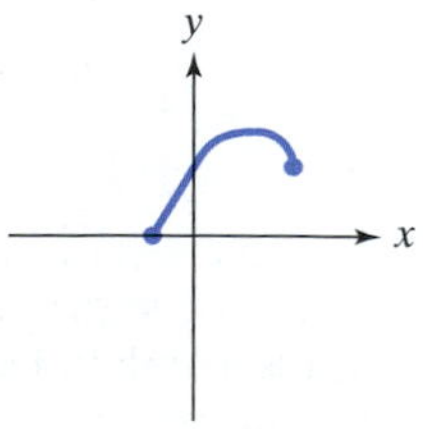

Use the graph of f to solve Exercises 14–18.

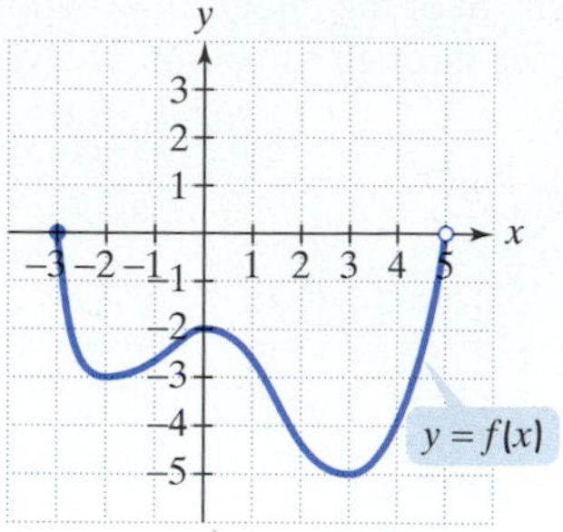

14. Find $f(-2)$.
15. Find $f(0)$.
16. For what value of x is $f(x) = -5$?
17. Find the domain of f.
18. Find the range of f.
19. The graph shows the height, in meters, of an eagle in terms of its time, in seconds, in flight.

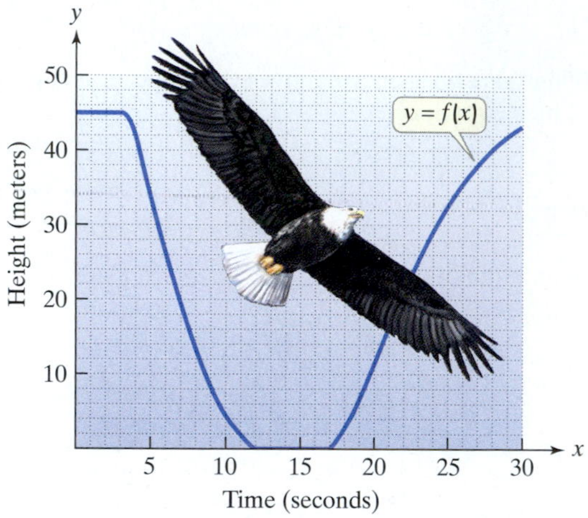

   a. Use the graph to explain why the eagle's height is a function of its time in flight.
   b. Find $f(15)$. Describe what this means in practical terms.
   c. What is a reasonable estimate of the eagle's maximum height?
   d. For what values of x is $f(x) = 20$? Describe what this means in practical terms.
   e. Use the graph of the function to write a description of the eagle's flight.

2.3 *In Exercises 20–22, find the domain of each function.*

20. $f(x) = 7x - 3$

21. $g(x) = \dfrac{1}{x + 8}$

22. $f(x) = x + \dfrac{3x}{x - 5}$

In Exercises 23–24, find **a.** $(f + g)(x)$ *and* **b.** $(f + g)(3)$.

23. $f(x) = 4x - 5,\ g(x) = 2x + 1$

24. $f(x) = 5x^2 - x + 4,\ g(x) = x - 3$

In Exercises 25–26, for each pair of functions, f and g, determine the domain of f + g.

25. $f(x) = 3x + 4,\ g(x) = \dfrac{5}{4 - x}$

26. $f(x) = \dfrac{7x}{x + 6},\ g(x) = \dfrac{4}{x + 1}$

In Exercises 27–34, let

$$f(x) = x^2 - 2x \quad \text{and} \quad g(x) = x - 5.$$

Find each of the following.

27. $(f + g)(x)$ and $(f + g)(-2)$

28. $f(3) + g(3)$

29. $(f - g)(x)$ and $(f - g)(1)$

30. $f(4) - g(4)$

31. $(fg)(-3)$

32. $\left(\dfrac{f}{g}\right)(x)$ and $\left(\dfrac{f}{g}\right)(4)$

33. The domain of $f - g$

34. The domain of $\dfrac{f}{g}$

2.4 *In Exercises 35–37, use intercepts and a checkpoint to graph each linear function.*

35. $x + 2y = 4$

36. $2x - 3y = 12$

37. $4x = 8 - 2y$

In Exercises 38–41, find the slope of the line passing through each pair of points or state that the slope is undefined. Then indicate whether the line through the points rises, falls, is horizontal, or is vertical.

38. $(5, 2)$ and $(2, -4)$

39. $(-2, 3)$ and $(7, -3)$

40. $(3, 2)$ and $(3, -1)$

41. $(-3, 4)$ and $(-1, 4)$

In Exercises 42–44, give the slope and y-intercept of each line whose equation is given. Then graph the linear function.

42. $y = 2x - 1$

43. $f(x) = -\dfrac{1}{2}x + 4$

44. $y = \dfrac{2}{3}x$

In Exercises 45–47, rewrite the equation in slope-intercept form. Give the slope and y-intercept.

45. $2x + y = 4$

46. $-3y = 5x$

47. $5x + 3y = 6$

In Exercises 48–52, graph each equation in a rectangular coordinate system.

48. $y = 2$

49. $7y = -21$

50. $f(x) = -4$

51. $x = 3$

52. $2x = -10$

53. The function $f(t) = -0.27t + 70.45$ models record time, $f(t)$, in seconds, for the women's 400-meter run t years after 1900. What is the slope of this model? Describe what this means in terms of rate of change.

54. The stated intent of the 1994 "don't ask, don't tell" policy was to reduce the number of discharges of gay men and lesbians from the military. The line graph shows the number of active-duty gay servicemembers discharged from the military for homosexuality under the policy.

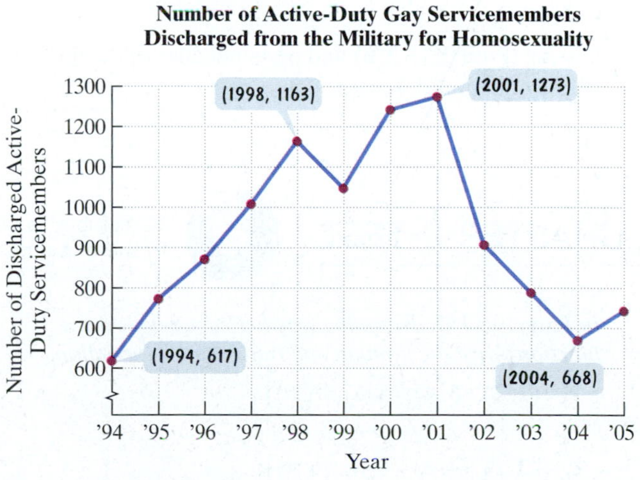

Source: General Accountability Office

a. Find the average rate of change, rounded to the nearest whole number, from 1994 through 1998. Describe what this means.

b. Find the average rate of change, rounded to the nearest whole number, from 2001 through 2004. Describe what this means.

55. The graph shows that a linear function describes the relationship between Fahrenheit temperature, F, and Celsius temperature, C.

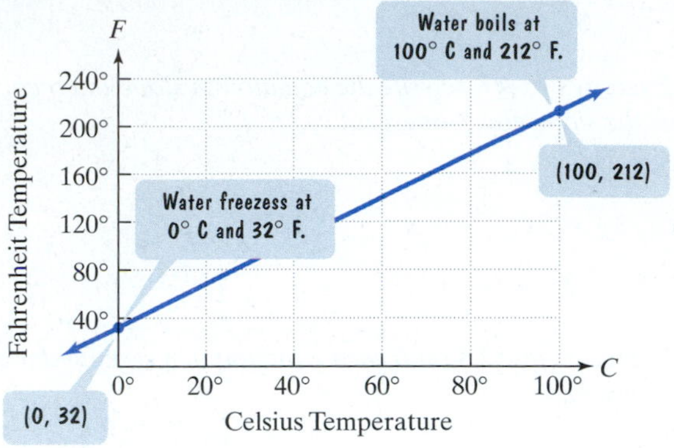

a. Use the points labeled by the voice balloons to find a function in the form $F = mC + b$ that expresses Fahrenheit temperature, F, in terms of Celsius temperature, C.

b. Use the function from part (a) to find the Fahrenheit temperature when the Celsius temperature is 30°.

2.5 *In Exercises 56–59, use the given conditions to write an equation for each line in point-slope form and in slope-intercept form.*

56. Pasing through $(-3, 2)$ with slope -6

57. Passing through $(1, 6)$ and $(-1, 2)$

58. Passing through $(4, -7)$ and parallel to the line whose equation is $3x + y - 9 = 0$

59. Passing through $(-2, 6)$ and perpendicular to the line whose equation is $y = \frac{1}{3}x + 4$

60. The bar graph shows the number of Americans, in millions, living below the poverty level from 2001 through 2005. The data are displayed as five points in a scatter plot. Also shown is a line that passes through or near the points.

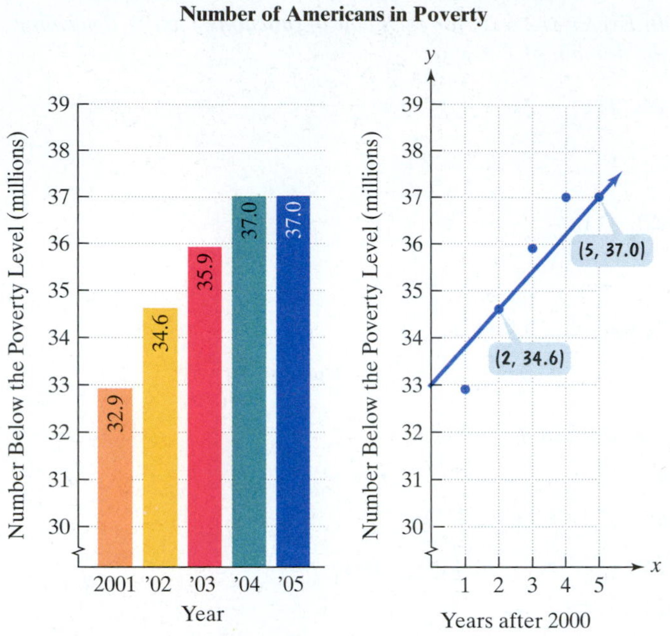

Source: U.S. Census Bureau

a. Use the two points whose coordinates are shown by the voice balloons to find the point-slope form of the equation of the line that models the number of Americans, y, in millions, living below the poverty level x years after 2000.

b. Write the equation from part (a) in slope-intercept form. Use function notation.

c. If trends shown from 2001 through 2005 continue, use the linear function to predict the number of Americans who will be living below the poverty level in 2010.

CHAPTER 2 TEST

Remember to use your Chapter Test Prep Video CD to see the worked-out solutions to the test questions you want to review.

In Exercises 1–2, determine whether each relation is a function. Give the domain and range for each relation.

1. $\{(1, 2), (3, 4), (5, 6), (6, 6)\}$

2. $\{(2, 1), (4, 3), (6, 5), (6, 6)\}$

3. If $f(x) = 3x - 2$, find $f(a + 4)$.

4. If $f(x) = 4x^2 - 3x + 6$, find $f(-2)$.

5. Graph $f(x) = x^2 - 1$ and $g(x) = x^2 + 1$ in the same rectangular coordinate system. Select integers for x, starting with -2 and ending with 2. Once you have obtained your graphs, describe how the graph of g is related to the graph of f.

In Exercises 6–7, identify the graph or graphs in which y is a function of x.

6.

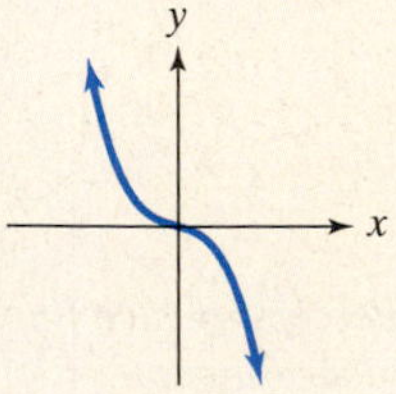

7.

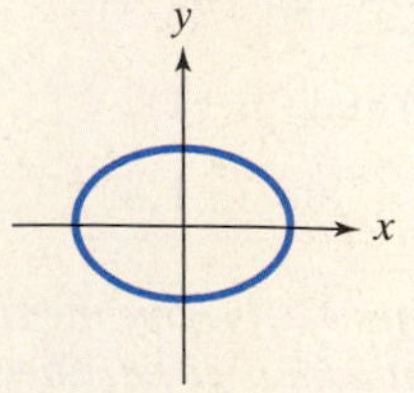

Use the graph of f to solve Exercises 8–11.

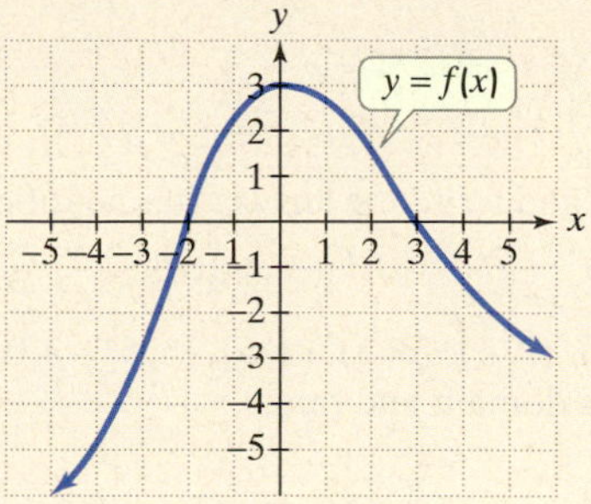

8. Find $f(6)$.
9. List two values of x for which $f(x) = 0$.
10. Find the domain of f.
11. Find the range of f.
12. Find the domain of $f(x) = \frac{6}{10 - x}$.

In Exercises 13–17, let

$$f(x) = x^2 + 4x \quad \text{and} \quad g(x) = x + 2.$$

Find each of the following.

13. $(f + g)(x)$ and $(f + g)(3)$
14. $(f - g)(x)$ and $(f - g)(-1)$
15. $(fg)(-5)$
16. $\left(\frac{f}{g}\right)(x)$ and $\left(\frac{f}{g}\right)(2)$
17. The domain of $\frac{f}{g}$

In Exercises 18–20, graph each linear function.

18. $4x - 3y = 12$
19. $f(x) = -\frac{1}{3}x + 2$
20. $f(x) = 4$

In Exercises 21–22, find the slope of the line passing through each pair of points or state that the slope is undefined. Then indicate whether the line through the points rises, falls, is horizontal, or is vertical.

21. $(5, 2)$ and $(1, 4)$
22. $(4, 5)$ and $(4, -5)$

The function $V(t) = 3.6t + 140$ models the number of Super Bowl viewers, $V(t)$, in millions, t years after 1995. Use the model to solve Exercises 23–24.

23. Find $V(10)$. Describe what this means in terms of the variables in the model.
24. What is the slope of this model? Describe what this means in terms of rate of change.

In Exercises 25–27, use the given conditions to write an equation for each line in point-slope form and slope-intercept form.

25. Passing through $(-1, -3)$ and $(4, 2)$
26. Passing through $(-2, 3)$ and perpendicular to the line whose equation is $y = -\frac{1}{2}x - 4$
27. Passing through $(6, -4)$ and parallel to the line whose equation is $x + 2y = 5$
28. The scatter plot shows the number of sentenced inmates in the United States per 100,000 residents from 2001 through 2005. Also shown is a line that passes through or near the data points.

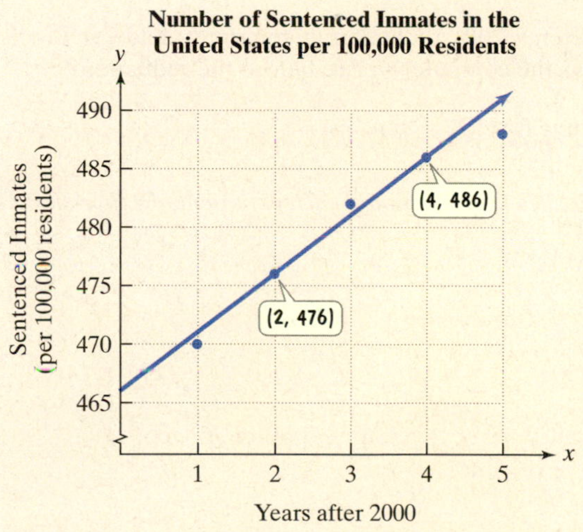

Source: U.S. Justice Department

a. Use the two points whose coordinates are shown by the voice balloons to find the point-slope form of the equation of the line that models the number of inmates per 100,000 residents, y, x years after 2000.

b. Write the equation from part (a) in slope-intercept form. Use function notation.

c. Use the linear function to predict the number of sentenced inmates in the United States per 100,000 residents in 2010.

CUMULATIVE REVIEW EXERCISES (CHAPTERS 1-2)

1. Use the roster method to list the elements in the set:
$$\{x | x \text{ is a whole number less than } 4\}.$$
2. Determine whether the following statement is true or false:
$$\pi \notin \{x | x \text{ is an irrational number}\}.$$

In Exercises 3–4, use the order of operations to simplify each expression.

3. $\dfrac{8 - 3^2 \div 9}{|-5| - [5 - (18 \div 6)]^2}$
4. $4 - (2 - 9)^0 + 3^2 \div 1 + 3$
5. Simplify: $3 - [2(x - 2) - 5x]$.

In Exercises 6–8, solve each equation. If the solution set is $\varnothing$ or $\mathbb{R}$, classify the equation as an inconsistent equation or an identity.

6. $2 + 3x - 4 = 2(x - 3)$
7. $4x + 12 - 8x = -6(x - 2) + 2x$
8. $\dfrac{x - 2}{4} = \dfrac{2x + 6}{3}$
9. After a 20% reduction, a computer sold for \$1800. What was the computer's price before the reduction?
10. Solve for t: $A = p + prt$.

In Exercises 11–12, simplify each exponential expression.

11. $(3x^4y^{-5})^{-2}$
12. $\left(\dfrac{3x^2y^{-4}}{x^{-3}y^2}\right)^2$
13. Multiply and write the answer in scientific notation:
$$(7 \times 10^{-8})(3 \times 10^2).$$
14. Is $\{(1, 5), (2, 5), (3, 5), (4, 5), (6, 5)\}$ a function? Give the relation's domain and range.
15. Graph $f(x) = |x| - 1$ and $g(x) = |x| + 2$ in the same rectangular coordinate system. Select integers for x, starting with -2 and ending with 2. Once you have obtained your graphs, describe how the graph of g is related to the graph of f.
16. Find the domain of $f(x) = \dfrac{1}{15 - x}$.
17. If $f(x) = 3x^2 - 4x + 2$ and $g(x) = x^2 - 5x - 3$, find $(f - g)(x)$ and $(f - g)(-1)$.

In Exercises 18-19, graph each linear function.

18. $f(x) = -2x + 4$.
19. $x - 2y = 6$
20. Write equations in point-slope form and slope-intercept form for the line passing through $(3, -5)$ and parallel to the line whose equation is $y = 4x + 7$.

CHAPTER 3

Systems of Linear Equations

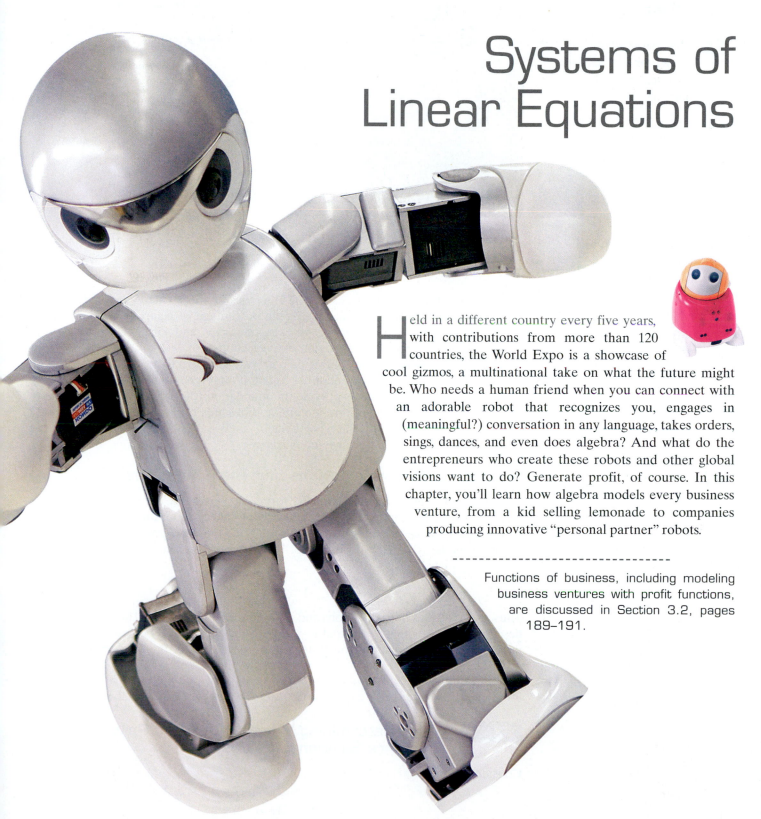

Held in a different country every five years, with contributions from more than 120 countries, the World Expo is a showcase of cool gizmos, a multinational take on what the future might be. Who needs a human friend when you can connect with an adorable robot that recognizes you, engages in (meaningful?) conversation in any language, takes orders, sings, dances, and even does algebra? And what do the entrepreneurs who create these robots and other global visions want to do? Generate profit, of course. In this chapter, you'll learn how algebra models every business venture, from a kid selling lemonade to companies producing innovative "personal partner" robots.

Functions of business, including modeling business ventures with profit functions, are discussed in Section 3.2, pages 189–191.

SECTION 3.1

Systems of Linear Equations in Two Variables

Objectives

1. Determine whether an ordered pair is a solution of a system of linear equations.
2. Solve systems of linear equations by graphing.
3. Solve systems of linear equations by substitution.
4. Solve systems of linear equations by addition.
5. Select the most efficient method for solving a system of linear equations.
6. Identify systems that do not have exactly one ordered-pair solution.

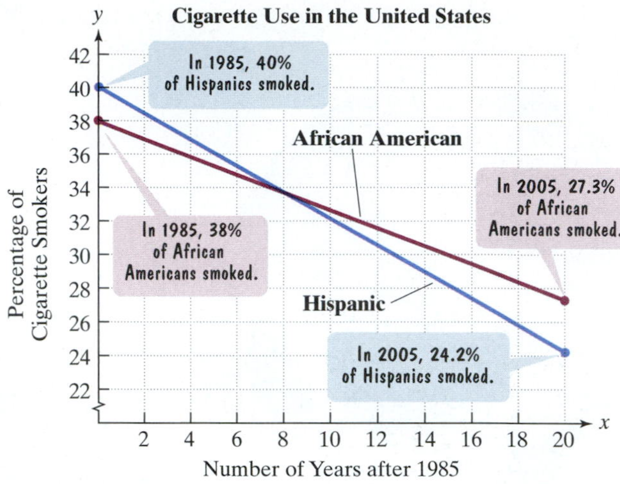

FIGURE 3.1

Source: Department of Health and Human Services

Although we still see celebrities smoking in movies, in music videos, and on television, there has been a remarkable decline in the percentage of cigarette smokers in the United States. The decline among African Americans and Hispanics, illustrated in **Figure 3.1**, can be analyzed using a pair of linear models in two variables.

In the first three sections of this chapter, you will learn to model your world with two equations in two variables and three equations in three variables. The methods you learn for solving these systems provide the foundation for solving complex problems involving thousands of equations containing thousands of variables. In this section's exercise set, you will apply these methods to analyze the decrease in cigarette use among whites, African Americans, and Hispanics.

1 Determine whether an ordered pair is a solution of a system of linear equations.

Systems of Linear Equations and Their Solutions

We have seen that all equations in the form $Ax + By = C$ are straight lines when graphed. Two such equations are called a **system of linear equations**, or a **linear system**. A **solution of a system of linear equations** is an ordered pair that satisfies both equations in the system. For example, $(3, 4)$ satisfies the system

$$x + y = 7 \quad (3 + 4 \text{ is, indeed, } 7.)$$
$$x - y = -1. \quad (3 - 4 \text{ is, indeed, } -1.)$$

Thus, $(3, 4)$ satisfies both equations and is a solution of the system. The solution can be described by saying that $x = 3$ and $y = 4$. The solution can also be described using set notation. The solution set of the system is $\{(3, 4)\}$—that is, the set consisting of the ordered pair $(3, 4)$.

A system of linear equations can have exactly one solution, no solution, or infinitely many solutions. We begin with systems with exactly one solution.

EXAMPLE 1 Determining Whether Ordered Pairs Are Solutions of a System

Consider the system:

$$x + 2y = -7$$
$$2x - 3y = 0.$$

Determine if each ordered pair is a solution of the system:

a. $(-3, -2)$ **b.** $(1, -4)$.

Solution

a. We begin by determining whether $(-3, -2)$ is a solution. Because -3 is the x-coordinate and -2 is the y-coordinate of $(-3, -2)$, we replace x with -3 and y with -2.

$$\begin{aligned} x + 2y &= -7 \\ -3 + 2(-2) &\stackrel{?}{=} -7 \\ -3 + (-4) &\stackrel{?}{=} -7 \\ -7 &= -7, \quad \text{true} \end{aligned} \qquad \begin{aligned} 2x - 3y &= 0 \\ 2(-3) - 3(-2) &\stackrel{?}{=} 0 \\ -6 - (-6) &\stackrel{?}{=} 0 \\ -6 + 6 &\stackrel{?}{=} 0 \\ 0 &= 0, \quad \text{true} \end{aligned}$$

The pair $(-3, -2)$ satisfies both equations: It makes each equation true. Thus, the ordered pair is a solution, of the system.

b. To determine whether $(1, -4)$ is a solution, we replace x with 1 and y with -4.

$$\begin{aligned} x + 2y &= -7 \\ 1 + 2(-4) &\stackrel{?}{=} -7 \\ 1 + (-8) &\stackrel{?}{=} -7 \\ -7 &= -7, \quad \text{true} \end{aligned} \qquad \begin{aligned} 2x - 3y &= 0 \\ 2\cdot 1 - 3(-4) &\stackrel{?}{=} 0 \\ 2 - (-12) &\stackrel{?}{=} 0 \\ 2 + 12 &\stackrel{?}{=} 0 \\ 14 &= 0, \quad \text{false} \end{aligned}$$

The pair $(1, -4)$ fails to satisfy *both* equations: It does not make both equations true. Thus, the ordered pair is not a solution of the system.

✓ CHECK POINT 1 Consider the system:

$$\begin{aligned} 2x + 5y &= -24 \\ 3x - 5y &= 14. \end{aligned}$$

Determine if each ordered pair is a solution of the system:

a. $(-7, -2)$ **b.** $(-2, -4)$.

2 Solve systems of linear equations by graphing.

Solving Linear Systems by Graphing

The solution of a system of two linear equations in two variables can be found by graphing both of the equations in the same rectangular coordinate system. For a system with one solution, **the coordinates of the point of intersection give the system's solution**.

Study Tip

When solving linear systems by graphing, neatly drawn graphs are essential for determining points of intersection.

- Use rectangular coordinate graph paper.
- Use a ruler or straightedge.
- Use a pencil with a sharp point.

Solving Systems of Two Linear Equations in Two Variables, x and y, by Graphing

1. Graph the first equation.
2. Graph the second equation on the same set of axes.
3. If the lines intersect at a point, determine the coordinates of this point of intersection. The ordered pair is the solution to the system.
4. Check the solution in both equations.

EXAMPLE 2 Solving a Linear System by Graphing

Solve by graphing:

$$y = -x - 1$$
$$4x - 3y = 24.$$

Solution

Step 1. Graph the first equation. We use the y-intercept and the slope to graph $y = -x - 1$.

$$y = -x - 1$$

The slope is −1. The y-intercept is −1.

The graph of the linear function is shown as a blue line in **Figure 3.2**.

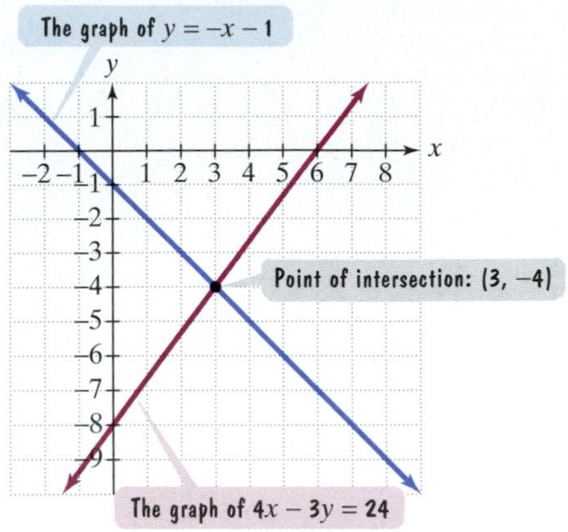

FIGURE 3.2

Step 2. Graph the second equation on the same axes. We use intercepts to graph $4x - 3y = 24$.

x-intercept (Set $y = 0$.)	y-intercept (Set $x = 0$.)
$4x - 3 \cdot 0 = 24$	$4 \cdot 0 - 3y = 24$
$4x = 24$	$-3y = 24$
$x = 6$	$y = -8$

The x-intercept is 6, so the line passes through $(6, 0)$. The y-intercept is −8, so the line passes through $(0, -8)$. The graph of $4x - 3y = 24$ is shown as a red line in **Figure 3.2**.

Step 3. Determine the coordinates of the intersection point. This ordered pair is the system's solution. Using **Figure 3.2**, it appears that the lines intersect at $(3, -4)$. The "apparent" solution of the system is $(3, -4)$.

Step 4. Check the solution in both equations.

Check $(3, -4)$ in $y = -x - 1$:

$$-4 \stackrel{?}{=} -3 - 1$$
$$-4 = -4, \quad \text{true}$$

Check $(3, -4)$ in $4x - 3y = 24$:

$$4(3) - 3(-4) \stackrel{?}{=} 24$$
$$12 + 12 \stackrel{?}{=} 24$$
$$24 = 24, \quad \text{true}$$

Because both equations are satisfied, $(3, -4)$ is the solution and $\{(3, -4)\}$ is the solution set.

Using Technology

A graphing utility can be used to solve the system in Example 2. Solve each equation for y, graph the equations, and use the intersection feature. The utility displays the solution $(3, -4)$ as $x = 3$, $y = -4$.

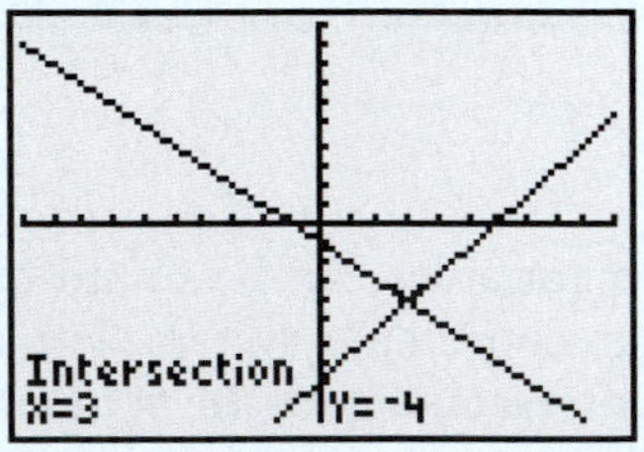

[−10, 10, 1] by [−10, 10, 1]

✓ **CHECK POINT 2** Solve by graphing:

$$y = -2x + 6$$
$$2x - y = -2.$$

3 Solve systems of linear equations by substitution.

Eliminating a Variable Using the Substitution Method

Finding the solution to a linear system by graphing equations may not be easy to do. For example, a solution of $\left(-\frac{2}{3}, \frac{157}{29}\right)$ would be difficult to "see" as an intersection point on a graph.

Let's consider a method that does not depend on finding a system's solution visually: the substitution method. This method involves converting the system to one equation in one variable by an appropriate substitution.

Study Tip

In step 1, you can choose which variable to isolate in which equation. If possible, solve for a variable whose coefficient is 1 or −1 to avoid working with fractions.

Solving Linear Systems by Substitution

1. Solve either of the equations for one variable in terms of the other. (If one of the equations is already in this form, you can skip this step.)
2. Substitute the expression found in step 1 into the other equation. This will result in an equation in one variable.
3. Solve the equation containing one variable.
4. Back-substitute the value found in step 3 into one of the original equations. Simplify and find the value of the remaining variable.
5. Check the proposed solution in both of the system's given equations.

EXAMPLE 3 Solving a System by Substitution

Solve by the substitution method:

$$y = -2x + 4$$
$$7x - 2y = 3.$$

Solution

Step 1. Solve either of the equations for one variable in terms of the other. This step has already been done for us. The first equation, $y = -2x + 4$, is solved for y in terms of x.

Step 2. Substitute the expression from step 1 into the other equation. We substitute the expression $-2x + 4$ for y in the second equation:

$$y = \boxed{-2x + 4} \qquad 7x - 2\boxed{y} = 3. \qquad \text{Substitute } -2x + 4 \text{ for } y.$$

This gives us an equation in one variable, namely

$$7x - 2(-2x + 4) = 3.$$

The variable y has been eliminated.

Step 3. Solve the resulting equation containing one variable.

$$7x - 2(-2x + 4) = 3 \quad \text{This is the equation containing one variable.}$$
$$7x + 4x - 8 = 3 \quad \text{Apply the distributive property.}$$
$$11x - 8 = 3 \quad \text{Combine like terms.}$$
$$11x = 11 \quad \text{Add 8 to both sides.}$$
$$x = 1 \quad \text{Divide both sides by 11.}$$

Step 4. Back-substitute the obtained value into one of the original equations. We now know that the x-coordinate of the solution is 1. To find the y-coordinate, we back-substitute the x-value into either original equation. We will use

$$y = -2x + 4.$$

Substitute 1 for x.

$$y = -2 \cdot 1 + 4 = -2 + 4 = 2$$

With $x = 1$ and $y = 2$, the proposed solution is $(1, 2)$.

Step 5. Check the proposed solution in both of the system's given equations. Replace x with 1 and y with 2.

$$\begin{aligned} y &= -2x + 4 \\ 2 &\stackrel{?}{=} -2 \cdot 1 + 4 \\ 2 &\stackrel{?}{=} -2 + 4 \\ 2 &= 2, \quad \text{true} \end{aligned} \qquad \begin{aligned} 7x - 2y &= 3 \\ 7(1) - 2(2) &\stackrel{?}{=} 3 \\ 7 - 4 &\stackrel{?}{=} 3 \\ 3 &= 3, \quad \text{true} \end{aligned}$$

The pair $(1, 2)$ satisfies both equations. The solution is $(1, 2)$ and the system's solution set is $\{(1, 2)\}$.

✓ **CHECK POINT 3** Solve by the substitution method:

$$\begin{aligned} y &= 3x - 7 \\ 5x - 2y &= 8. \end{aligned}$$

EXAMPLE 4 Solving a System by Substitution

Solve by the substitution method:

$$\begin{aligned} 5x + 2y &= 1 \\ x - 3y &= 7. \end{aligned}$$

Solution

Step 1. Solve either of the equations for one variable in terms of the other. We begin by isolating one of the variables in either of the equations. By solving for x in the second equation, which has a coefficient of 1, we can avoid fractions.

$x - 3y = 7$ This is the second equation in the given system.

$x = 3y + 7$ Solve for x by adding $3y$ to both sides.

Step 2. Substitute the expression from step 1 into the other equation. We substitute $3y + 7$ for x in the first equation.

$$x = \boxed{3y + 7} \qquad 5\boxed{x} + 2y = 1$$

This gives us an equation in one variable, namely

$$5(3y + 7) + 2y = 1.$$

The variable x has been eliminated.

Step 3. Solve the resulting equation containing one variable.

$5(3y + 7) + 2y = 1$ This is the equation containing one variable.

$15y + 35 + 2y = 1$ Apply the distributive property.

$17y + 35 = 1$ Combine like terms.

$17y = -34$ Subtract 35 from both sides.

$y = -2$ Divide both sides by 17.

Step 4. Back-substitute the obtained value into one of the original equations. We back-substitute -2 for y into one of the original equations to find x. Let's use both equations to show that we obtain the same value for x in either case.

Using the first equation:

$$\begin{aligned} 5x + 2y &= 1 \\ 5x + 2(-2) &= 1 \\ 5x - 4 &= 1 \\ 5x &= 5 \\ x &= 1 \end{aligned}$$

Using the second equation:

$$\begin{aligned} x - 3y &= 7 \\ x - 3(-2) &= 7 \\ x + 6 &= 7 \\ x &= 1 \end{aligned}$$

With $x = 1$ and $y = -2$, the proposed solution is $(1, -2)$.

Study Tip

The equation from step 1, in which one variable is expressed in terms of the other, is equivalent to one of the original equations. It is often easiest to back-substitute an obtained value into this equation to find the value of the other variable. After obtaining both values, get into the habit of checking the ordered-pair solution in *both* equations of the system.

Step 5. Check. Take a moment to show that $(1, -2)$ satisfies both given equations. The solution is $(1, -2)$ and the solution set is $\{(1, -2)\}$.

✓ **CHECK POINT 4** Solve by the substitution method:

$$\begin{aligned} 3x + 2y &= 4 \\ 2x + y &= 1. \end{aligned}$$

4 Solve systems of linear equations by addition.

Eliminating a Variable Using the Addition Method

The substitution method is most useful if one of the given equations has an isolated variable. A third method for solving a linear system is the addition method. Like the substitution method, the addition method involves eliminating a variable and ultimately solving an equation containing only one variable. However, this time we eliminate a variable by adding the equations.

For example, consider the following system of linear equations:

$$\begin{aligned} 3x - 4y &= 11 \\ -3x + 2y &= -7. \end{aligned}$$

When we add these two equations, the x-terms are eliminated. This occurs because the coefficients of the x-terms, 3 and -3, are opposites (additive inverses) of each other:

$$\begin{aligned} 3x - 4y &= 11 \\ \underline{-3x + 2y} &= \underline{-7} \\ \text{Add: } 0x - 2y &= 4 \\ -2y &= 4 \\ y &= -2. \end{aligned}$$

The sum is an equation in one variable.

Divide both sides by -2 and solve for y.

Now we can back-substitute -2 for y into one of the original equations to find x. It does not matter which equation you use; you will obtain the same value for x in either case. If we use either equation, we can show that $x = 1$ and the solution $(1, -2)$ satisfies both equations in the system.

When we use the addition method, we want to obtain two equations whose sum is an equation containing only one variable. The key step is to **obtain, for one of the variables, coefficients that differ only in sign.** To do this, we may need to multiply one or both equations by some nonzero number so that the coefficients of one of the variables, x or y, become opposites. Then when the two equations are added, this variable will be eliminated.

EXAMPLE 5 Solving a System by the Addition Method

Solve by the addition method:

$$\begin{aligned} 3x + 4y &= -10 \\ 5x - 2y &= 18. \end{aligned}$$

Study Tip

Although the addition method is also known as the elimination method, variables are eliminated when using both the substitution and addition methods. The name *addition method* specifically tells us that the elimination of a variable is accomplished by adding two equations.

Solution We must rewrite one or both equations in equivalent forms so that the coefficients of the same variable (either x or y) are opposites of each other. Consider the terms in y in each equation, that is, $4y$ and $-2y$. To eliminate y, we can multiply each term of the second equation by 2 and then add equations.

$$3x + 4y = -10 \xrightarrow{\text{No change}} 3x + 4y = -10$$

$$5x - 2y = 18 \xrightarrow{\text{Multiply by 2.}} \underline{10x - 4y} = \underline{36}$$

$$\begin{aligned} \text{Add: } 13x + 0y &= 26 \\ 13x &= 26 \\ x &= 2 \end{aligned}$$

Divide both sides by 13 and solve for x.

Thus, $x = 2$. We back-substitute this value into either one of the given equations. We'll use the first one.

$$3x + 4y = -10 \quad \text{This is the first equation in the given system.}$$
$$3(2) + 4y = -10 \quad \text{Substitute 2 for } x.$$
$$6 + 4y = -10 \quad \text{Multiply.}$$
$$4y = -16 \quad \text{Subtract 6 from both sides.}$$
$$y = -4 \quad \text{Divide both sides by 4.}$$

We see that $x = 2$ and $y = -4$. The ordered pair $(2, -4)$ can be shown to satisfy both equations in the system. Consequently, the solution is $(2, -4)$ and the solution set is $\{(2, -4)\}$.

Solving Linear Systems by Addition

1. If necessary, rewrite both equations in the form $Ax + By = C$.
2. If necessary, multiply either equation or both equations by appropriate nonzero numbers so that the sum of the x-coefficients or the sum of the y-coefficients is 0.
3. Add the equations in step 2. The sum will be an equation in one variable.
4. Solve the equation in one variable.
5. Back-substitute the value obtained in step 4 into either of the given equations and solve for the other variable.
6. Check the solution in both of the original equations.

✓ **CHECK POINT 5** Solve by the addition method:

$$4x - 7y = -16$$
$$2x + 5y = 9.$$

EXAMPLE 6 Solving a System by the Addition Method

Solve by the addition method:

$$7x = 5 - 2y$$
$$3y = 16 - 2x.$$

Solution

Step 1. Rewrite both equations in the form $Ax + By = C$. We first arrange the system so that variable terms appear on the left and constants appear on the right. We obtain

$$7x + 2y = 5 \quad \text{Add } 2y \text{ to both sides of the first equation.}$$
$$2x + 3y = 16. \quad \text{Add } 2x \text{ to both sides of the second equation.}$$

Step 2. If necessary, multiply either equation or both equations by appropriate numbers so that the sum of the x-coefficients or the sum of the y-coefficients is 0. We can eliminate x or y. Let's eliminate y by multiplying the first equation by 3 and the second equation by -2.

$$7x + 2y = 5 \xrightarrow{\text{Multiply by 3.}} 21x + 6y = 15$$
$$2x + 3y = 16 \xrightarrow{\text{Multiply by } -2.} \underline{-4x - 6y = -32}$$

Step 3. Add the equations.

$$\text{Add:} \quad 17x + 0y = -17$$
$$17x = -17$$

Step 4. Solve the equation in one variable. We solve $17x = -17$ by dividing both sides by 17.

$$\frac{17x}{17} = \frac{-17}{17} \qquad \text{Divide both sides by 17.}$$

$$x = -1 \qquad \text{Simplify.}$$

Step 5. Back-substitute and find the value of the other variable. We can back-substitute -1 for x into either one of the given equations. We'll use the second one.

$$3y = 16 - 2x \qquad \text{This is the second equation in the given system.}$$

$$3y = 16 - 2(-1) \qquad \text{Substitute } -1 \text{ for } x.$$

$$3y = 16 + 2 \qquad \text{Multiply.}$$

$$3y = 18 \qquad \text{Add.}$$

$$y = 6 \qquad \text{Divide both sides by 3.}$$

We found that $x = -1$ and $y = 6$. The proposed solution is $(-1, 6)$.

Step 6. Check. Take a moment to show that $(-1, 6)$ satisfies both given equations. The solution is $(-1, 6)$ and the solution set is $\{(-1, 6)\}$.

✓ **CHECK POINT 6** Solve by the addition method:

$$3x = 2 - 4y$$
$$5y = -1 - 2x.$$

Some linear systems have solutions that are not integers. If the value of one variable turns out to be a "messy" fraction, back-substitution might lead to cumbersome arithmetic. If this happens, you can return to the original system and use the addition method a second time to find the value of the other variable.

EXAMPLE 7 Solving a System by the Addition Method

Solve by the addition method:

$$\frac{x}{2} - 5y = 32$$
$$\frac{3x}{2} - 7y = 45.$$

Solution

Step 1. Rewrite both equations in the form $Ax + By = C$. Although each equation is already in this form, the coefficients of x are not integers. There is less chance for error if the coefficients for x and y in $Ax + By = C$ are integers. Consequently, we begin by clearing fractions. Multiply both sides of each equation by 2.

$$\frac{x}{2} - 5y = 32 \xrightarrow{\text{Multiply by 2.}} x - 10y = 64$$

$$\frac{3x}{2} - 7y = 45 \xrightarrow{\text{Multiply by 2.}} 3x - 14y = 90$$

Step 2. If necessary, multiply either equation or both equations by appropriate numbers so that the sum of the x-coefficients or the sum of the y-coefficients is 0. We will eliminate x. Multiply the first equation with integral coefficients by -3 and leave the second equation unchanged.

$$x - 10y = 64 \xrightarrow{\text{Multiply by } -3.} -3x + 30y = -192$$

$$3x - 14y = 90 \xrightarrow{\text{No change}} \underline{3x - 14y = 90}$$

Step 3. Add the equations.

$$\text{Add: } 0x + 16y = -102$$

$$16y = -102$$

Step 4. Solve the equation in one variable. We solve $16y = -102$ by dividing both sides by 16.

$$\frac{16y}{16} = \frac{-102}{16} \quad \text{Divide both sides by 16.}$$

$$y = -\frac{102}{16} = -\frac{51}{8} \quad \text{Simplify.}$$

Step 5. Back-substitute and find the value of the other variable. Back-substitution of $-\frac{51}{8}$ for y into either of the given equations results in cumbersome arithmetic. Instead, let's use the addition method on the system with integral coefficients from step 1 to find the value of x. Thus, we eliminate y by multiplying the first equation by -7 and the second equation by 5.

$$\begin{array}{rcl} x - 10y = 64 & \xrightarrow{\text{Multiply by } -7.} & -7x + 70y = -448 \\ 3x - 14y = 90 & \xrightarrow{\text{Multiply by } 5.} & \underline{15x - 70y = 450} \\ & \text{Add:} & 8x \qquad = 2 \\ & & x = \frac{2}{8} = \frac{1}{4} \end{array}$$

We found that $x = \frac{1}{4}$ and $y = -\frac{51}{8}$. The proposed solution is $\left(\frac{1}{4}, -\frac{51}{8}\right)$.

Step 6. Check. For this system, a calculator is helpful in showing that $\left(\frac{1}{4}, -\frac{51}{8}\right)$ satisfies both of the original equations of the system. The solution is $\left(\frac{1}{4}, -\frac{51}{8}\right)$ and the solution set is $\left\{\left(\frac{1}{4}, -\frac{51}{8}\right)\right\}$.

✓ CHECK POINT 7 Solve by the addition method:

$$\frac{3x}{2} - 2y = \frac{5}{2}$$

$$x - \frac{5y}{2} = -\frac{3}{2}.$$

5 Select the most efficient method for solving a system of linear equations.

Comparing the Three Solution Methods

The following chart compares the graphing, substitution, and addition methods for solving systems of linear equations in two variables. With increased practice, you will find it easier to select the best method for solving a particular linear system.

Comparing Solution Methods

Method	Advantages	Disadvantages
Graphing	You can see the solutions.	If the solutions do not involve integers or are too large to be seen on the graph, it's impossible to tell exactly what the solutions are.
Substitution	Gives exact solutions. Easy to use if a variable is on one side by itself.	Solutions cannot be seen. Introduces extensive work with fractions when no variable has a coefficient of 1 or -1.
Addition	Gives exact solutions. Easy to use if no variable has a coefficient of 1 or -1.	Solutions cannot be seen.

6 Identify systems that do not have exactly one ordered-pair solution.

Linear Systems Having No Solution or Infinitely Many Solutions

We have seen that a system of linear equations in two variables represents a pair of lines. The lines either intersect at one point, are parallel, or are identical. Thus, there are three possibilities for the number of solutions to a system of two linear equations.

The Number of Solutions to a System of Two Linear Equations

The number of solutions to a system of two linear equations in two variables is given by one of the following. (See **Figure 3.3**.)

Number of Solutions	What This Means Graphically
Exactly one ordered-pair solution	The two lines intersect at one point.
No solution	The two lines are parallel.
Infinitely many solutions	The two lines are identical.

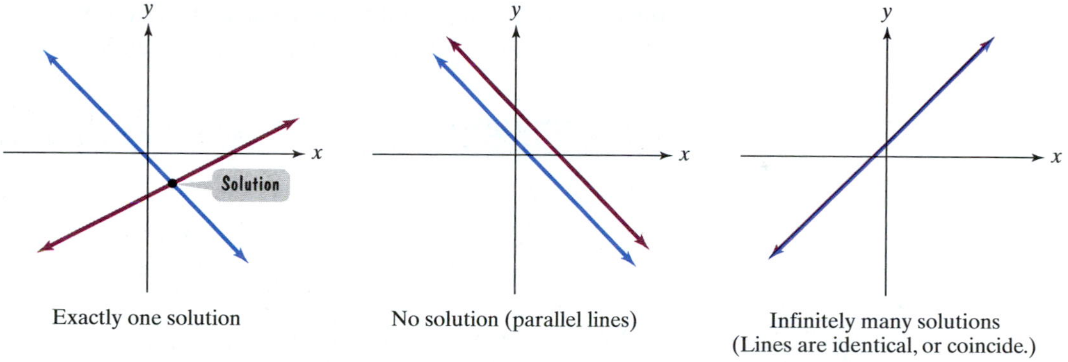

FIGURE 3.3 Possible graphs for a system of two linear equations in two variables

A linear system with no solution is called an **inconsistent system**. If you attempt to solve such a system by substitution or addition, you will eliminate both variables. A false statement, such as $0 = 6$, will be the result.

EXAMPLE 8 A System with No Solution

Solve the system:

$$3x - 2y = 6$$
$$6x - 4y = 18.$$

Solution Because no variable is isolated, we will use the addition method. To obtain coefficients of x that differ only in sign, we multiply the first equation by -2.

$$3x - 2y = 6 \xrightarrow{\text{Multiply by } -2.} -6x + 4y = -12$$
$$6x - 4y = 18 \xrightarrow{\text{No change}} 6x - 4y = 18$$
$$\text{Add: } 0 = 6$$

There are no values of x and y for which $0 = 6$. No values of x and y satisfy $0x + 0y = 6$.

The false statement $0 = 6$ indicates that the system is inconsistent and has no solution. The solution set is the empty set, $\varnothing$.

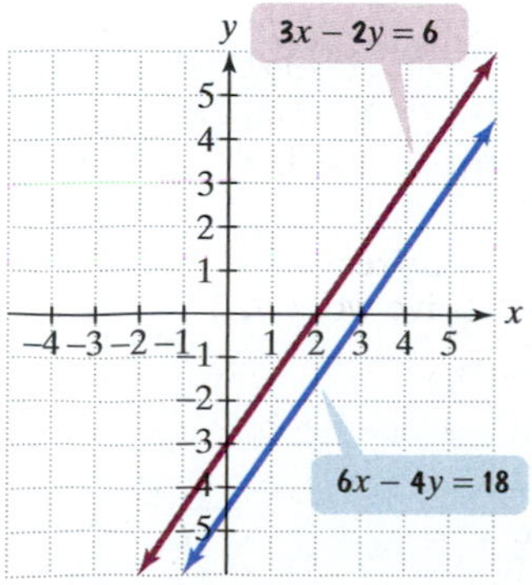

FIGURE 3.4 The graph of an inconsistent system

The lines corresponding to the two equations in Example 8 are shown in **Figure 3.4**. The lines are parallel and have no point of intersection.

Discover for Yourself

Show that the graphs of $3x - 2y = 6$ and $6x - 4y = 18$ must be parallel lines by solving each equation for y. What is the slope and the y-intercept for each line? What does this mean? If a linear system is inconsistent, what must be true about the slopes and the y-intercepts for the system's graphs?

✓ CHECK POINT 8 Solve the system:

$$5x - 2y = 4$$
$$-10x + 4y = 7$$

A linear system that has at least one solution is called a **consistent system**. Lines that intersect and lines that coincide both represent consistent systems. If the lines coincide, then the consistent system has infinitely many solutions, represented by every point on the coinciding lines.

The equations in a linear system with infinitely many solutions are called **dependent**. If you attempt to solve such a system by substitution or addition, you will eliminate both variables. However, a true statement, such as $10 = 10$, will be the result.

EXAMPLE 9 A System with Infinitely Many Solutions

Solve the system:

$$y = 3x - 2$$
$$15x - 5y = 10.$$

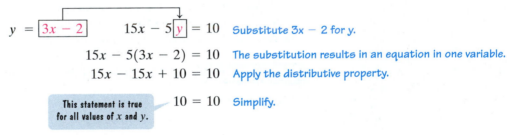

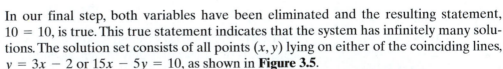

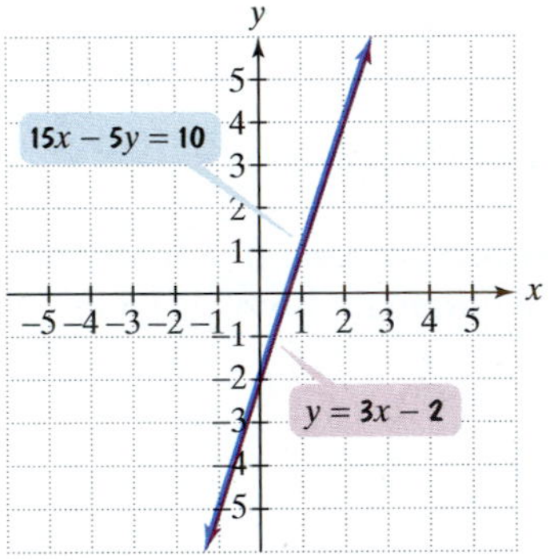

FIGURE 3.5 The graph of a system with infinitely many solutions

Study Tip

Although the system in Example 9 has infinitely many solutions, this does not mean that any ordered pair of numbers you can form will be a solution. The ordered pair (x, y) must satisfy one of the system's equations, $y = 3x - 2$ or $15x - 5y = 10$, and there are infinitely many such ordered pairs. Because the graphs are coinciding lines, the ordered pairs that are solutions of one of the equations are also solutions of the other equation.

Solution Because the variable y is isolated in $y = 3x - 2$, the first equation, we can use the substitution method. We substitute the expression for y into the second equation.

$y = \boxed{3x - 2}$ $\qquad 15x - 5\boxed{y} = 10$ Substitute $3x - 2$ for y.

$15x - 5(3x - 2) = 10$ The substitution results in an equation in one variable.

$15x - 15x + 10 = 10$ Apply the distributive property.

$10 = 10$ Simplify.

This statement is true for all values of x and y.

In our final step, both variables have been eliminated and the resulting statement, $10 = 10$, is true. This true statement indicates that the system has infinitely many solutions. The solution set consists of all points (x, y) lying on either of the coinciding lines, $y = 3x - 2$ or $15x - 5y = 10$, as shown in **Figure 3.5**.

We express the solution set for the system in one of two equivalent ways:

$$\{(x, y) \mid y = 3x - 2\} \quad \text{or} \quad \{(x, y) \mid 15x - 5y = 10\}.$$

The set of all ordered pairs (x, y) such that $y = 3x - 2$

The set of all ordered pairs (x, y) such that $15x - 5y = 10$

✓ CHECK POINT 9 Solve the system:

$$x = 4y - 8$$
$$5x - 20y = -40.$$

3.1 EXERCISE SET

MyMathLab — Math XL PRACTICE — WATCH — DOWNLOAD — READ — REVIEW

Practice Exercises

In Exercises 1–6, determine whether the given ordered pair is a solution of the system.

1. $(7, -5)$
$x - y = 12$
$x + y = 2$

2. $(-3, 1)$
$x - y = -4$
$2x + 10y = 4$

3. $(2, -1)$
$3x + 4y = 2$
$2x + 5y = 1$

4. $(4, 2)$
$2x - 5y = -2$
$3x + 4y = 18$

5. $(5, -3)$
$y = 2x - 13$
$4x + 9y = -7$

6. $(-3, -4)$
$y = 3x + 5$
$5x - 2y = -7$

In Exercises 7–24, solve each system by graphing. Identify systems with no solution and systems with infinitely many solutions, using set notation to express their solution sets.

7. $x + y = 4$
$x - y = 2$

8. $x + y = 6$
$x - y = -4$

9. $2x + y = 4$
$y = 4x + 1$

10. $x + 2y = 4$
$y = -2x - 1$

11. $3x - 2y = 6$
$x - 4y = -8$

12. $4x + y = 4$
$3x - y = 3$

13. $2x + 3y = 6$
$4x = -6y + 12$

14. $3x - 3y = 6$
$2x = 2y + 4$

15. $y = 2x - 2$
$y = -5x + 5$

16. $y = -x + 1$
$y = 3x + 5$

17. $3x - y = 4$
$6x - 2y = 4$

18. $2x - y = -4$
$4x - 2y = 6$

19. $2x + y = 4$
$4x + 3y = 10$

20. $4x - y = 9$
$x - 3y = 16$

21. $x - y = 2$
$y = 1$

22. $x + 2y = 1$
$x = 3$

23. $3x + y = 3$
$6x + 2y = 12$

24. $2x - 3y = 6$
$4x - 6y = 24$

In Exercises 25–42, solve each system by the substitution method. Identify inconsistent systems and systems with dependent equations, using set notation to express their solution sets.

25. $x + y = 6$
$y = 2x$

26. $x + y = 10$
$y = 4x$

27. $2x + 3y = 9$
$x = y + 2$

28. $3x - 4y = 18$
$y = 1 - 2x$

29. $y = -3x + 7$
$5x - 2y = 8$

30. $x = 3y + 8$
$2x - y = 6$

31. $4x + y = 5$
$2x - 3y = 13$

32. $x - 3y = 3$
$3x + 5y = -19$

33. $x - 2y = 4$
$2x - 4y = 5$

34. $x - 3y = 6$
$2x - 6y = 5$

35. $2x + 5y = -4$
$3x - y = 11$

36. $2x + 5y = 1$
$-x + 6y = 8$

37. $2(x - 1) - y = -3$
$y = 2x + 3$

38. $x + y - 1 = 2(y - x)$
$y = 3x - 1$

39. $\frac{x}{4} - \frac{y}{4} = -1$
$x + 4y = -9$

40. $\frac{x}{6} - \frac{y}{2} = \frac{1}{3}$
$x + 2y = -3$

41. $y = \frac{2}{5}x - 2$
$2x - 5y = 10$

42. $y = \frac{1}{3}x + 4$
$3y = x + 12$

In Exercises 43–58, solve each system by the addition method. Identify inconsistent systems and systems with dependent equations, using set notation to express their solution sets.

43. $x + y = 7$
$x - y = 3$

44. $2x + y = 3$
$x - y = 3$

45. $12x + 3y = 15$
$2x - 3y = 13$

46. $4x + 2y = 12$
$3x - 2y = 16$

47. $x + 3y = 2$
$4x + 5y = 1$

48. $x + 2y = -1$
$2x - y = 3$

49. $6x - y = -5$
$4x - 2y = 6$

50. $x - 2y = 5$
$5x - y = -2$

51. $3x - 5y = 11$
$2x - 6y = 2$

52. $4x - 3y = 12$
$3x - 4y = 2$

53. $2x - 5y = 13$
$5x + 3y = 17$

54. $4x + 5y = -9$
$6x - 3y = -3$

55. $2x + 6y = 8$
$3x + 9y = 12$

56. $x - 3y = -6$
$3x - 9y = 9$

57. $2x - 3y = 4$
$4x + 5y = 3$

58. $4x - 3y = 8$
$2x - 5y = -14$

In Exercises 59–82, solve each system by the method of your choice. Identify inconsistent systems and systems with dependent equations, using set notation to express solution sets.

59. $3x - 7y = 1$
$2x - 3y = -1$

60. $2x - 3y = 2$
$5x + 4y = 51$

61. $x = y + 4$
$3x + 7y = -18$

62. $y = 3x + 5$
$5x - 2y = -7$

63. $9x + \frac{4y}{3} = 5$
$4x - \frac{y}{3} = 5$

64. $\frac{x}{6} - \frac{y}{5} = -4$
$\frac{x}{4} - \frac{y}{6} = -2$

65. $\frac{1}{4}x - \frac{1}{9}y = \frac{2}{3}$
$\frac{1}{2}x - \frac{1}{3}y = 1$

66. $\frac{1}{16}x - \frac{3}{4}y = -1$
$\frac{3}{4}x + \frac{5}{2}y = 11$

67. $x = 3y - 1$
$2x - 6y = -2$

68. $x = 4y - 1$
$2x - 8y = -2$

69. $y = 2x + 1$
$y = 2x - 3$

70. $y = 2x + 4$
$y = 2x - 1$

71. $0.4x + 0.3y = 2.3$
$0.2x - 0.5y = 0.5$

72. $0.2x - y = -1.4$
$0.7x - 0.2y = -1.6$

73. $5x - 40 = 6y$
$2y = 8 - 3x$

74. $4x - 24 = 3y$
$9y = 3x - 1$

75. $3(x + y) = 6$
$3(x - y) = -36$

76. $4(x - y) = -12$
$4(x + y) = -20$

77. $3(x - 3) - 2y = 0$
$2(x - y) = -x - 3$

78. $5x + 2y = -5$
$4(x + y) = 6(2 - x)$

79. $x + 2y - 3 = 0$
$12 = 8y + 4x$

80. $2x - y - 5 = 0$
$10 = 4x - 2y$

81. $3x + 4y = 0$
$7x = 3y$

82. $5x + 8y = 20$
$4y = -5x$

Practice PLUS

In Exercises 83–84, solve each system by the method of your choice.

83. $\frac{x+2}{2} - \frac{y+4}{3} = 3$
$\frac{x+y}{5} = \frac{x-y}{2} - \frac{5}{2}$

84. $\frac{x-y}{3} = \frac{x+y}{2} - \frac{1}{2}$
$\frac{x+2}{2} - 4 = \frac{y+4}{3}$

In Exercises 85–86, solve each system for x and y, expressing either value in terms of a or b, if necessary. Assume that $a \neq 0$ and $b \neq 0$.

85. $5ax + 4y = 17$
$ax + 7y = 22$

86. $4ax + by = 3$
$6ax + 5by = 8$

87. For the linear function $f(x) = mx + b$, $f(-2) = 11$ and $f(3) = -9$. Find m and b.

88. For the linear function $f(x) = mx + b$, $f(-3) = 23$ and $f(2) = -7$. Find m and b.

Use the graphs of the linear functions to solve Exercises 89–90.

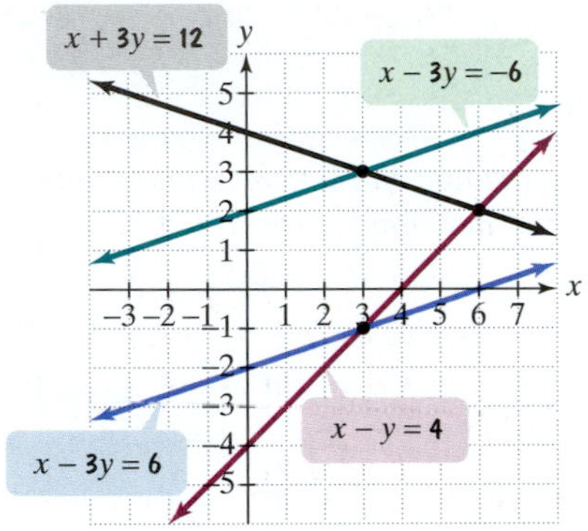

89. Write the linear system whose solution set is $\{(6, 2)\}$. Express each equation in the system in slope-intercept form.

90. Write the linear system whose solution set is $\varnothing$. Express each equation in the system in slope-intercept form.

Application Exercises

91. In 1915, the average U.S. household contained more than four people. In 2007, the average was 2.5. Large families are increasingly rare. The graphs below illustrate this trend.

Smaller United States Families

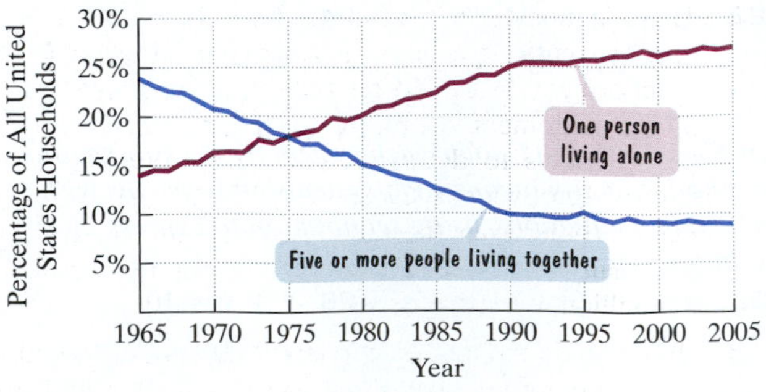

Source: U.S. Census Bureau

a. Use the graphs to estimate the point of intersection. In what year was the percentage of all U.S. households consisting of five or more people the same as the percentage of all U.S. households consisting of one person living alone? What percentage of all U.S. households did each of these groups comprise?

b. The function $0.6x + y = 24$ models the percentage, y, of all U.S. households consisting of five or more people x years after 1965. The function $y = 0.4x + 14$ models the percentage, y, of all U.S. households consisting of one person living alone x years after 1965. Use these models to determine when the percentage of all U.S. households consisting of five or more people was the same as the percentage of all U.S. households consisting of one person living alone. According to the models, what percentage of all U.S. households did each of these groups comprise?

c. How well do the models in part (b) describe the point of intersection of the graphs that you estimated in part (a)?

92. The graph shows that from 2000 through 2006, Americans unplugged land lines and switched to cellphones.

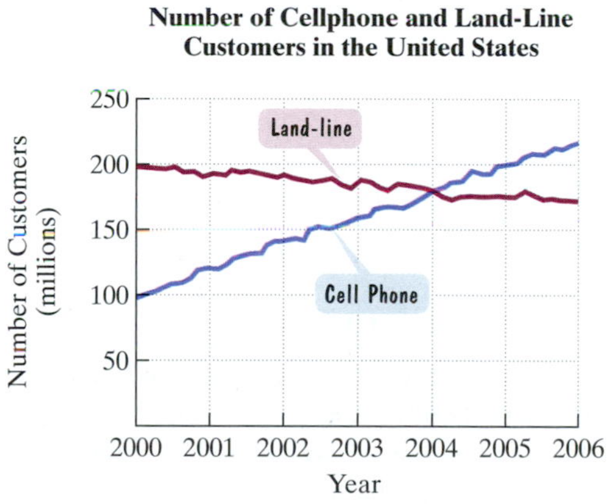

Source: Federal Communications Commission

a. Use the graphs to estimate the point of intersection. In what year was the number of cellphone and land-line customers the same? How many millions of customers were there for each?

b. The function $4.3x + y = 198$ models the number of land-line customers, in millions, x years after 2000. The function $y = 19.8x + 98$ models the number of cellphone customers, in millions, x years after 2000. Use these models to determine the year, rounded to the nearest year, when the number of cellphone and land-line customers was the same. According to the models, how many millions of customers, rounded to the nearest ten million, were there for each?

c. How well do the models in part (b) describe the point of intersection of the graphs that you estimated in part (a)?

93. Although Social Security is a problem, some projections indicate that there's a much bigger time bomb ticking in the federal budget, and that's Medicare. In 2000, the cost of Social Security was 5.48% of the gross domestic product, increasing by 0.04% of the GDP per year. In 2000, the cost of Medicare was 1.84% of the gross domestic product, increasing by 0.17% of the GDP per year. (*Source:* Congressional Budget Office)

a. Write a function that models the cost of Social Security as a percentage of the GDP x years after 2000.

b. Write a function that models the cost of Medicare as a percentage of the GDP x years after 2000.

c. In which year will the cost of Medicare and Social Security be the same? For that year, what will be the cost of each program as a percentage of the GDP? Which program will have the greater cost after that year?

94. The graph indicates that in 1984, there were 72 meals per person at take-out restaurants. For the period shown, this number increased by an average of 2.25 meals per person per year. In 1984, there were 94 meals per person at on-premise dining facilities and this number decreased by an average of 0.55 meal per person per year.

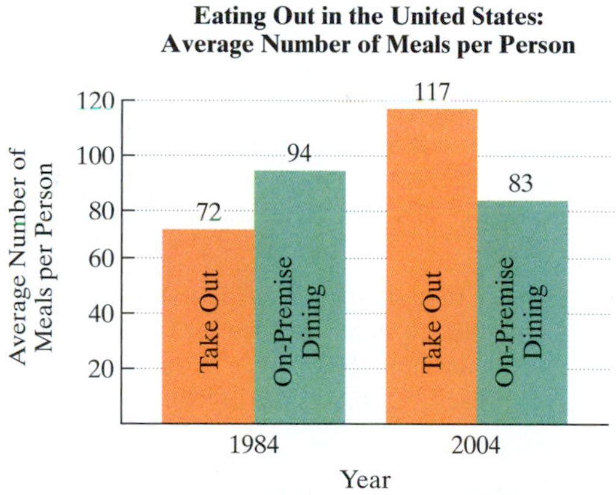

Source: The NPD Group

a. Write a function that models the average number of meals per person at take-out restaurants x years after 1984.

b. Write a function that models the average number of meals per person at on-premise dining facilities x years after 1984.

c. In which year, to the nearest whole year, was the average number of meals per person for take-out and on-premise restaurants the same? For that year, how many meals per person, to the nearest whole number, were there for each kind of restaurant? Which kind of restaurant had the greater number of meals per person after that year?

The bar graph shows the percentage of Americans who used cigarettes, by ethnicity, in 1985 and 2005. For each of the groups shown, cigarette use has been linearly decreasing. Use this information to solve Exercises 95–96.

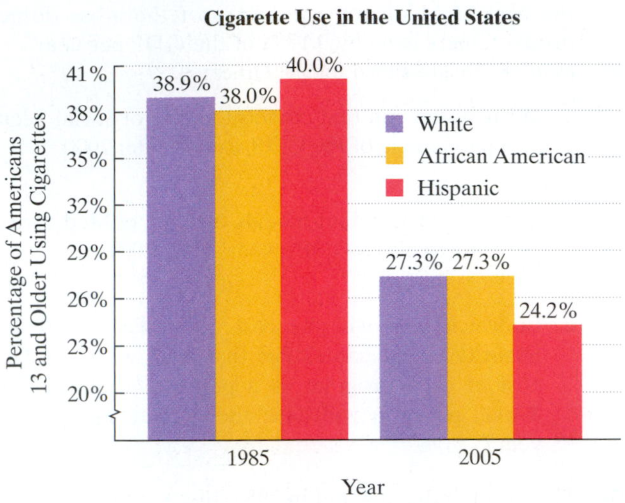

Source: Department of Health and Human Services

95. In this exercise, let x represent the number of years after 1985 and let y represent the percentage of Americans in one of the groups shown who used cigarettes.

a. Use the data points (0, 38) and (20, 27.3) to find the slope-intercept equation of the line that models the percentage of African Americans who used cigarettes, y, x years after 1985. Round the value of the slope m to two decimal places.

b. Use the data points (0, 40) and (20, 24.2) to find the slope-intercept equation of the line that models the percentage of Hispanics who used cigarettes, y, x years after 1985.

c. Use the models from parts (a) and (b) to find the year during which cigarette use was the same for African Americans and Hispanics. What percentage of each group used cigarettes during that year?

96. In this exercise, let x represent the number of years after 1985 and let y represent the percentage of Americans in one of the groups shown who used cigarettes.

a. Use the data points (0, 38.9) and (20, 27.3) to find the slope-intercept equation of the line that models the percentage of whites who used cigarettes, y, x years after 1985.

b. Use the data points (0, 40) and (20, 24.2) to find the slope-intercept equation of the line that models the percentage of Hispanics who used cigarettes, y, x years after 1985.

c. Use the models from parts (a) and (b) to find the year, to the nearest whole year, during which cigarette use was the same for whites and Hispanics. What percentage of each group, to the nearest percent, used cigarettes during that year?

An important application of systems of equations arises in connection with supply and demand. As the price of a product increases, the demand for that product decreases. However, at higher prices, suppliers are willing to produce greater quantities of the product. Exercises 97–98 involve supply and demand.

97. A chain of electronics stores sells hand-held color televisions. The weekly demand and supply models are given as follows:

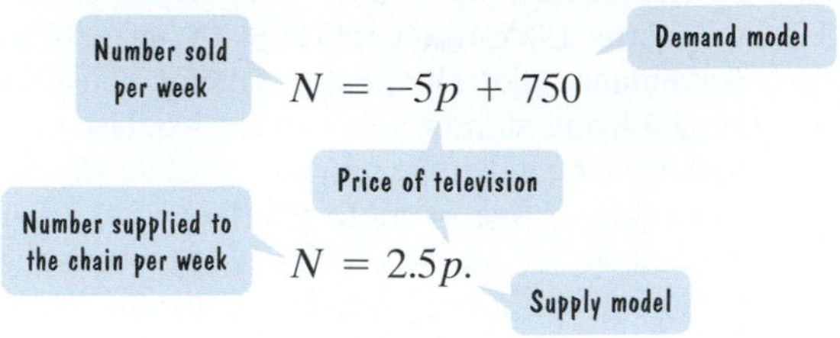

a. How many hand-held color televisions can be sold and supplied at \$120 per television?

b. Find the price at which supply and demand are equal. At this price, how many televisions can be supplied and sold each week?

98. At a price of p dollars per ticket, the number of tickets to a rock concert that can be sold is given by the demand model $N = -25p + 7800$. At a price of p dollars per ticket, the number of tickets that the concert's promoters are willing to make available is given by the supply model $N = 5p + 6000$.

a. How many tickets can be sold and supplied for \$50 per ticket?

b. Find the ticket price at which supply and demand are equal. At this price, how many tickets will be supplied and sold?

Writing in Mathematics

99. What is a system of linear equations? Provide an example with your description.

100. What is a solution of a system of linear equations?

101. Explain how to determine if an ordered pair is a solution of a system of linear equations.

102. Explain how to solve a system of linear equations by graphing.

103. Explain how to solve a system of equations using the substitution method. Use $y = 3 - 3x$ and $3x + 4y = 6$ to illustrate your explanation.

104. Explain how to solve a system of equations using the addition method. Use $5x + 8y = -1$ and $3x + y = 7$ to illustrate your explanation.

105. When is it easier to use the addition method rather than the substitution method to solve a system of equations?

106. When using the addition or substitution method, how can you tell if a system of linear equations has no solution? What is the relationship between the graphs of the two equations?

107. When using the addition or substitution method, how can you tell if a system of linear equations has infinitely many solutions? What is the relationship between the graphs of the two equations?

Technology Exercise

108. Verify your solutions to any five exercises from Exercises 7–24 by using a graphing utility to graph the two equations in the system in the same viewing rectangle. Then use the intersection feature to display the solution.

Critical Thinking Exercises

Make Sense? *In Exercises 109–112, determine whether each statement "makes sense" or "does not make sense" and explain your reasoning.*

109. Even if a linear system has a solution set involving fractions, such as $\{(\frac{8}{11}, \frac{43}{11})\}$, I can use graphs to determine if the solution set is reasonable.

110. If I add the equations on the right and solve the resulting equation for x, I will obtain the x-coordinate of the intersection point of the lines represented by the equations on the left.

$$\begin{aligned} 4x - 6y &= 1 \longrightarrow 20x - 30y = 5 \\ 3x + 5y &= -8 \longrightarrow 18x + 30y = -8 \end{aligned}$$

111. In the previous chapter, we developed models for life expectancy, y, for U.S. men and women born x years after 1960:

$$y = 0.22x + 65.7 \quad \text{Men}$$
$$y = 0.17x + 72.9. \quad \text{Women}$$

The system indicates that life expectancy for men is increasing at a faster rate than for women, so if these trends continue, life expectancies for men and women will be the same for some future birth year.

112. Here are two models that describe winning times for the Olympic 400-meter run, y, in seconds, x years after 1968:

$$y = -0.02433x + 44.43 \quad \text{Men}$$
$$y = -0.08883x + 50.86. \quad \text{Women}$$

The system indicates that winning times have been decreasing more rapidly for women than for men, so if these trends continue, there will be a year when winning times for men and women are the same.

In Exercises 113–116, determine whether each statement is true or false. If the statement is false, make the necessary change(s) to produce a true statement.

113. The addition method cannot be used to eliminate either variable in a system of two equations in two variables.

114. The solution set of the system

$$\begin{aligned} 5x - y &= 1 \\ 10x - 2y &= 2 \end{aligned}$$

is $\{(2, 9)\}$.

115. A system of linear equations can have a solution set consisting of precisely two ordered pairs.

116. The solution set of the system

$$\begin{aligned} y &= 4x - 3 \\ y &= 4x + 5 \end{aligned}$$

is the empty set.

117. Determine a and b so that $(2, 1)$ is a solution of this system:

$$\begin{aligned} ax - by &= 4 \\ bx + ay &= 7. \end{aligned}$$

118. Write a system of equations having $\{(-2, 7)\}$ as a solution set. (More than one system is possible.)

119. Solve the system for x and y in terms of a_1, b_1, c_1, a_2, b_2, and c_2:

$$a_1x + b_1y = c_1$$
$$a_2x + b_2y = c_2.$$

Review Exercises

120. Solve: $6x = 10 + 5(3x - 4)$. (Section 1.4, Example 3)

121. Simplify: $(4x^2y^4)^2(-2x^5y^0)^3$. (Section 1.6, Example 9)

122. If $f(x) = x^2 - 3x + 7$, find $f(-1)$. (Section 2.1, Example 3)

Preview Exercises

Exercises 123–125 will help you prepare for the material covered in the next section.

123. The formula $I = Pr$ is used to find the simple interest, I, earned for one year when the principal, P, is invested at an annual interest rate, r. Write an expression for the total interest earned on a principal of x dollars at a rate of 15% $(r = 0.15)$ and a principal of y dollars at a rate of 7% $(r = 0.07)$.

124. A chemist working on a flu vaccine needs to obtain 50 milliliters of a 30% sodium-iodine solution. How many milliliters of sodium-iodine are needed in the solution?

125. A company that manufactures running shoes sells them at \$80 per pair. Write an expression for the revenue that is generated by selling x pairs of shoes.

SECTION 3.2

Problem Solving and Business Applications Using Systems of Equations

Objectives

1. Solve problems using systems of equations.
2. Use functions to model revenue, cost, and profit, and perform a break-even analysis.

Driving through your neighborhood, you see kids selling lemonade. Would it surprise you to know that this activity can be analyzed using functions and systems of equations? By doing so, you will view profit and loss in the business world in a new way. In this section, we use systems of equations to solve problems and model business ventures.

1 Solve problems using systems of equations.

A Strategy for Solving Word Problems Using Systems of Equations

When we solved problems in Chapter 1, we let x represent a quantity that was unknown. Problems in this section involve two unknown quantities. We will let x and y represent these quantities. We then translate from the verbal conditions of the problem into a *system* of linear equations.

EXAMPLE 1 Solving a Problem Involving Energy Efficiency of Building Materials

A heat-loss survey by an electric company indicated that a wall of a house containing 40 square feet of glass and 60 square feet of plaster lost 1920 Btu (British thermal units) of heat. A second wall containing 10 square feet of glass and 100 square feet of plaster lost 1160 Btu of heat. Determine the heat lost per square foot for the glass and for the plaster.

Study Tip

There is great value in reasoning through a word problem. This value comes from the problem-solving skills that are attained and is often more important than the specific problem or its solution.

Solution

Step 1. Use variables to represent unknown quantities.

Let x = the heat lost per square foot for the glass.

Let y = the heat lost per square foot for the plaster.

Step 2. Write a system of equations that models the problem's conditions. The heat loss for each wall is the heat lost by the glass plus the heat lost by the plaster. One wall containing 40 square feet of glass and 60 square feet of plaster lost 1920 Btu of heat.

Heat lost by the glass + heat lost by the plaster = total heat lost.

(Number of ft²) · (heat lost per ft²) + (number of ft²) · (heat lost per ft²) = total heat lost.

$$40 \cdot x + 60 \cdot y = 1920$$

A second wall containing 10 square feet of glass and 100 square feet of plaster lost 1160 Btu of heat.

Heat lost by the glass + heat lost by the plaster = total heat lost.

(Number of ft²) · (heat lost per ft²) + (number of ft²) · (heat lost per ft²) = total heat lost.

$$10 \cdot x + 100 \cdot y = 1160$$

Step 3. Solve the system and answer the problem's question. The system

$$40x + 60y = 1920$$
$$10x + 100y = 1160$$

can be solved by addition. We'll multiply the second equation by -4 and then add equations to eliminate x.

$$40x + 60y = 1920 \xrightarrow{\text{No change}} 40x + 60y = 1920$$
$$10x + 100y = 1160 \xrightarrow{\text{Multiply by } -4.} -40x - 400y = -4640$$
$$\text{Add: } -340y = -2720$$
$$y = \frac{-2720}{-340} = 8$$

Now we can find the value of x by back-substituting 8 for y in either of the system's equations.

$10x + 100y = 1160$ We'll use the second equation.

$10x + 100(8) = 1160$ Back-substitute 8 for y.

$10x + 800 = 1160$ Multiply.

$10x = 360$ Subtract 800 from both sides.

$x = 36$ Divide both sides by 10.

We see that $x = 36$ and $y = 8$. Because x represents heat lost per square foot for the glass and y for the plaster, the glass lost 36 Btu of heat per square foot and the plaster lost 8 Btu per square foot.

Step 4. Check the proposed solution in the original wording of the problem. The problem states that the wall with 40 square feet of glass and 60 square feet of plaster lost 1920 Btu.

$$40(36) + 60(8) = 1440 + 480 = 1920 \text{ Btu of heat}$$

Proposed solution is 36 Btu per ft² for glass and 8 Btu per ft² for plaster.

Our proposed solution checks with the first statement. The problem also states that the wall with 10 square feet of glass and 100 square feet of plaster lost 1160 Btu.

$$10(36) + 100(8) = 360 + 800 = 1160 \text{ Btu of heat}$$

Our proposed solution also checks with the second statement.

✓ **CHECK POINT 1** University of Arkansas researchers discovered that we underestimate the number of calories in restaurant meals. The next time you eat out, take the number of calories you think you ate and double it. The researchers concluded that this number should be a more accurate estimate. The actual number of calories in one portion of hamburger and fries and two portions of fettuccine Alfredo is 4240. The actual number of calories in two portions of hamburger and fries and one portion of fettuccine Alfredo is 3980. Find the actual number of calories in each of these dishes. (*Source*: *Consumer Reports*, January/February, 2007)

Test Your Calorie I.Q.

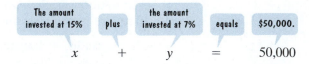

Hamburger and Fries
Average guess:
777 calories

Fettuccine Alfredo
Average guess:
704 calories

Next, we will solve problems involving investments, mixtures, and motion with systems of equations. We will continue using our four-step problem-solving strategy. We will also use tables to help organize the information in the problems.

Dual Investments with Simple Interest

Simple interest involves interest calculated only on the amount of money that we invest, called the **principal**. The formula $I = Pr$ is used to find the simple interest, I, earned for one year when the principal, P, is invested at an annual interest rate, r. Dual investment problems involve different amounts of money in two or more investments, each paying a different rate.

EXAMPLE 2 Solving a Dual Investment Problem

Your grandmother needs your help. She has $50,000 to invest. Part of this money is to be invested in noninsured bonds paying 15% annual interest. The rest of this money is to be invested in a government-insured certificate of deposit paying 7% annual interest. She told you that she requires $6000 per year in extra income from both of these investments. How much money should be placed in each investment?

Solution

Step 1. Use variables to represent unknown quantities.

Let x = the amount invested in the 15% noninsured bonds.

Let y = the amount invested in the 7% certificate of deposit.

Step 2. Write a system of equations that models the problem's conditions. Because Grandma has $50,000 to invest,

The amount invested at 15% | plus | the amount invested at 7% | equals | $50,000.

$$x + y = 50{,}000$$

Furthermore, Grandma requires $6000 in total interest. We can use a table to organize the information in the problem and obtain a second equation.

	Principal (amount invested)	×	Interest rate	=	Interest earned
15% Investment	x		0.15		$0.15x$
7% Investment	y		0.07		$0.07y$

The interest for the two investments combined must be $6000.

Interest from the 15% investment	plus	interest from the 7% investment	is	$6000.
$0.15x$	$+$	$0.07y$	$=$	6000

Step 3. Solve the system and answer the problem's question. The system

$$x + y = 50{,}000$$
$$0.15x + 0.07y = 6000$$

can be solved by substitution or addition. Substitution works well because both variables in the first equation have coefficients of 1. Addition also works well; if we multiply the first equation by -0.15 or -0.07, adding equations will eliminate a variable. We will use addition.

$$x + y = 50{,}000 \xrightarrow{\text{Multiply by } -0.07.} -0.07x - 0.07y = -3500$$
$$0.15x + 0.07y = 6000 \xrightarrow{\text{No change}} \underline{0.15x + 0.07y = 6000}$$
$$\text{Add:} \quad 0.08x = 2500$$
$$x = \frac{2500}{0.08}$$
$$x = 31{,}250$$

Because x represents the amount that should be invested at 15%, Grandma should place $31,250 in 15% noninsured bonds. Now we can find y, the amount that she should place in the 7% certificate of deposit. We do so by back-substituting 31,250 for x in either of the system's equations.

$x + y = 50{,}000$ — We'll use the first equation.

$31{,}250 + y = 50{,}000$ — Back-substitute 31,250 for x.

$y = 18{,}750$ — Subtract 31,250 from both sides.

Because $x = 31{,}250$ and $y = 18{,}750$, Grandma should invest $31,250 at 15% and $18,750 at 7%.

Step 4. Check the proposed answers in the original wording of the problem. Has Grandma invested $50,000?

$$\$31{,}250 + \$18{,}750 = \$50{,}000$$

Yes, all her money was placed in the dual investments. Can she count on $6000 interest? The interest earned on $31,250 at 15% is ($31,250)(0.15), or $4687.50. The interest earned on $18,750 at 7% is ($18,750)(0.07), or $1312.50. The total interest is $4687.50 + $1312.50, or $6000, exactly as it should be. You've made your grandmother happy. (Now if you would just visit her more often . . .)

✓ **CHECK POINT 2** You inherited $5000 with the stipulation that for the first year the money had to be invested in two funds paying 9% and 11% annual interest. How much did you invest at each rate if the total interest earned for the year was $487?

Problems Involving Mixtures

Chemists and pharmacists often have to change the concentration of solutions and other mixtures. In these situations, the amount of a particular ingredient in the solution or mixture is expressed as a percentage of the total.

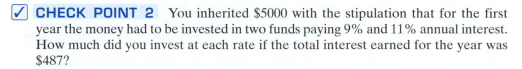

EXAMPLE 3 Solving a Mixture Problem

A chemist working on a flu vaccine needs to mix a 10% sodium-iodine solution with a 60% sodium-iodine solution to obtain 50 milliliters of a 30% sodium-iodine solution. How many milliliters of the 10% solution and of the 60% solution should be mixed?

Solution

Step 1. Use variables to represent unknown quantities.

Let $x =$ the number of milliliters of the 10% solution to be used in the mixture.

Let $y =$ the number of milliliters of the 60% solution to be used in the mixture.

Step 2. Write a system of equations that models the problem's conditions. The situation is illustrated in **Figure 3.6**. The chemist needs 50 milliliters of a 30% sodium-iodine solution. We form a table that shows the amount of sodium-iodine in each of the three solutions.

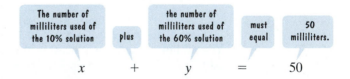

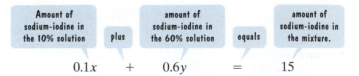

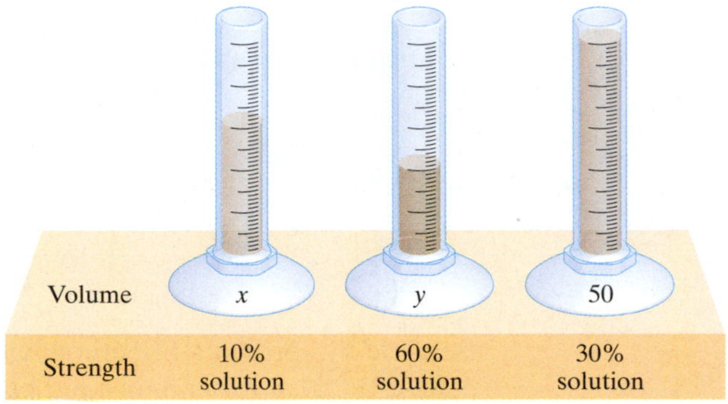

FIGURE 3.6

Solution	Number of Milliliters	×	Percent of Sodium-Iodine	=	Amount of Sodium-Iodine
10% Solution	x		$10\% = 0.1$		$0.1x$
60% Solution	y		$60\% = 0.6$		$0.6y$
30% Mixture	50		$30\% = 0.3$		$0.3(50) = 15$

The chemist needs to obtain a 50-milliliter mixture.

The number of milliliters used of the 10% solution	plus	the number of milliliters used of the 60% solution	must equal	50 milliliters.
x	$+$	y	$=$	50

The 50-milliliter mixture must be 30% sodium-iodine. The amount of sodium-iodine must be 30% of 50, or $(0.3)(50) = 15$ milliliters.

Amount of sodium-iodine in the 10% solution	plus	amount of sodium-iodine in the 60% solution	equals	amount of sodium-iodine in the mixture.
$0.1x$	$+$	$0.6y$	$=$	15

Step 3. Solve the system and answer the problem's question. The system

$$x + y = 50$$
$$0.1x + 0.6y = 15$$

can be solved by substitution or addition. Let's use substitution. The first equation can easily be solved for x or y. Solving for y, we obtain $y = 50 - x$.

$$y = \boxed{50 - x} \qquad 0.1x + 0.6\boxed{y} = 15$$

We substitute $50 - x$ for y in the second equation. This gives us an equation in one variable.

$$0.1x + 0.6(50 - x) = 15 \quad \text{This equation contains one variable, } x.$$
$$0.1x + 30 - 0.6x = 15 \quad \text{Apply the distributive property.}$$
$$-0.5x + 30 = 15 \quad \text{Combine like terms.}$$
$$-0.5x = -15 \quad \text{Subtract 30 from both sides.}$$
$$x = \frac{-15}{-0.5} = 30 \quad \text{Divide both sides by } -0.5.$$

Back-substituting 30 for x in either of the system's equations ($x + y = 50$ is easier to use) gives $y = 20$. Because x represents the number of milliliters of the 10% solution and y the number of milliliters of the 60% solution, the chemist should mix 30 milliliters of the 10% solution with 20 milliliters of the 60% solution.

Step 4. Check the proposed solution in the original wording of the problem. The problem states that the chemist needs 50 milliliters of a 30% sodium-iodine solution. The amount of sodium-iodine in this mixture is 0.3(50), or 15 milliliters. The amount of sodium-iodine in 30 milliliters of the 10% solution is 0.1(30), or 3 milliliters. The amount of sodium-iodine in 20 milliliters of the 60% solution is $0.6(20) = 12$ milliliters. The amount of sodium-iodine in the two solutions used in the mixture is 3 milliliters + 12 milliliters, or 15 milliliters, exactly as it should be.

✓ **CHECK POINT 3** A chemist needs to mix a 12% acid solution with a 20% acid solution to obtain 160 ounces of a 15% acid solution. How many ounces of each of the acid solutions must be used?

Study Tip

Problems involving dual investments and problems involving mixtures are both based on the same idea: The total amount times the rate gives the amount.

Dual Investment Problems: principal · rate = interest

(principal: total money amount; rate: rate of interest; interest: amount of interest)

Percents are expressed as decimals in these equations.

Mixture Problems: solution · concentration = ingredient

(solution: total solution amount; concentration: percent of ingredient, or rate of concentration; ingredient: amount of ingredient)

Our dual investment problem involved mixing two investments. Our mixture problem involved mixing two liquids. The equations in these problems are obtained from similar conditions:

Dual Investment Problems

Interest from investment 1 + interest from investment 2 = amount of interest from mixed investments.

Mixture Problems

Ingredient amount in solution 1 + ingredient amount in solution 2 = amount of ingredient in mixture.

Being aware of the similarities between dual investment and mixture problems should make you a better problem solver in a variety of situations that involve mixtures.

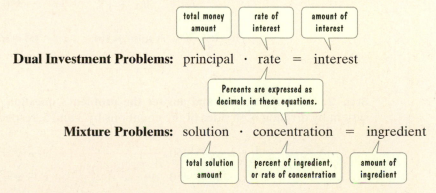

Problems Involving Motion

We have seen that the distance, d, covered by any moving body is the product of its average rate, r, and its time in motion, t:

$$d = rt. \quad \text{Distance equals rate times time.}$$

Wind and water current have the effect of increasing or decreasing a traveler's rate.

Study Tip

It is not always necessary to use x and y to represent a problem's variables. Select letters that help you remember what the variables represent. For example, in Example 4, you may prefer using p and w rather than x and y:

p = plane's average rate in still air

w = wind's average rate.

EXAMPLE 4 Solving a Motion Problem

When a small airplane flies with the wind, it can travel 450 miles in 3 hours. When the same airplane flies in the opposite direction against the wind, it takes 5 hours to fly the same distance. Find the average rate of the plane in still air and the average rate of the wind.

Solution

Step 1. Use variables to represent unknown quantities.

Let x = the average rate of the plane in still air.

Let y = the average rate of the wind.

Step 2. Write a system of equations that models the problem's conditions. As it travels with the wind, the plane's rate is increased. The net rate is its rate in still air, x, plus the rate of the wind, y, given by the expression $x + y$. As it travels against the wind, the plane's rate is decreased. The net rate is its rate in still air, x, minus the rate of the wind, y, given by the expression $x - y$. Here is a chart that summarizes the problem's information and includes the increased and decreased rates.

	Rate	×	Time	=	Distance
Trip with the Wind	$x + y$		3		$3(x + y)$
Trip against the Wind	$x - y$		5		$5(x - y)$

The problem states that the distance in each direction is 450 miles. We use this information to write our system of equations.

The distance of the trip with the wind is 450 miles.

$$3(x + y) = 450$$

The distance of the trip against the wind is 450 miles.

$$5(x - y) = 450$$

Step 3. Solve the system and answer the problem's question. We can simplify the system by dividing both sides of the equations by 3 and 5, respectively.

$$3(x + y) = 450 \xrightarrow{\text{Divide by 3.}} x + y = 150$$
$$5(x - y) = 450 \xrightarrow{\text{Divide by 5.}} x - y = 90$$

Solve the system on the right by the addition method.

$$\begin{aligned} x + y &= 150 \\ x - y &= 90 \\ \text{Add: } 2x &= 240 \\ x &= 120 \quad \text{Divide both sides by 2.} \end{aligned}$$

Back-substituting 120 for x in either of the system's equations gives $y = 30$. Because $x = 120$ and $y = 30$, the average rate of the plane in still air is 120 miles per hour and the average rate of the wind is 30 miles per hour.

Step 4. Check the proposed solution in the original wording of the problem. The problem states that the distance in each direction is 450 miles. The average rate of the plane with the wind is $120 + 30 = 150$ miles per hour. In 3 hours, it travels $150 \cdot 3$, or 450 miles, which checks with the stated condition. Furthermore, the average rate of the plane against the wind is $120 - 30 = 90$ miles per hour. In 5 hours, it travels $90 \cdot 5 = 450$ miles, which is the stated distance.

✓ **CHECK POINT 4** With the current, a motorboat can travel 84 miles in 2 hours. Against the current, the same trip takes 3 hours. Find the average rate of the boat in still water and the average rate of the current.

---

2 Use functions to model revenue, cost, and profit, and perform a break-even analysis.

Functions of Business: Break-Even Analysis

Suppose that a company produces and sells x units of a product. Its *revenue* is the money generated by selling x units of the product. Its *cost* is the cost of producing x units of the product.

Revenue and Cost Functions

A company produces and sells x units of a product.

Revenue Function

$$R(x) = (\text{price per unit sold})x$$

Cost Function

$$C(x) = \text{fixed cost} + (\text{cost per unit produced})x$$

The point of intersection of the graphs of the revenue and cost functions is called the **break-even point**. The x-coordinate of the point reveals the number of units that a company must produce and sell so that money coming in, the revenue, is equal to money going out, the cost. The y-coordinate of the break-even point gives the amount of money coming in and going out. Example 5 illustrates the use of the substitution method in determining a company's break-even point.

EXAMPLE 5 Finding a Break-Even Point

Technology is now promising to bring light, fast, and beautiful wheelchairs to millions of disabled people. A company is planning to manufacture these radically different wheelchairs. Fixed cost will be \$500,000 and it will cost \$400 to produce each wheelchair. Each wheelchair will be sold for \$600.

a. Write the cost function, C, of producing x wheelchairs.

b. Write the revenue function, R, from the sale of x wheelchairs.

c. Determine the break-even point. Describe what this means.

Solution

a. The cost function is the sum of the fixed cost and variable cost.

Fixed cost of \$500,000 — plus — Variable cost: \$400 for each chair produced

$$C(x) = 500{,}000 + 400x$$

b. The revenue function is the money generated from the sale of x wheelchairs.

Revenue per chair, \$600, times — the number of chairs sold

$$R(x) = 600x$$

c. The break-even point occurs where the graphs of C and R intersect. Thus, we find this point by solving the system

$$\begin{aligned} C(x) &= 500{,}000 + 400x \\ R(x) &= 600x \end{aligned} \quad \text{or} \quad \begin{aligned} y &= 500{,}000 + 400x \\ y &= 600x. \end{aligned}$$

Using substitution, we can substitute $600x$ for y in the first equation.

$$\begin{aligned} 600x &= 500{,}000 + 400x && \text{Substitute } 600x \text{ for } y \text{ in } y = 500{,}000 + 400x. \\ 200x &= 500{,}000 && \text{Subtract } 400x \text{ from both sides.} \\ x &= 2500 && \text{Divide both sides by 200.} \end{aligned}$$

Back-substituting 2500 for x in either of the system's equations (or functions), we obtain

$$R(2500) = 600(2500) = 1{,}500{,}000.$$

We used $R(x) = 600x$.

The break-even point is (2500, 1,500,000). This means that the company will break even if it produces and sells 2500 wheelchairs. At this level, the money coming in is equal to the money going out: \$1,500,000.

Figure 3.7 shows the graphs of the revenue and cost functions for the wheelchair business. Similar graphs and models apply no matter how small or large a business venture may be.

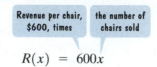

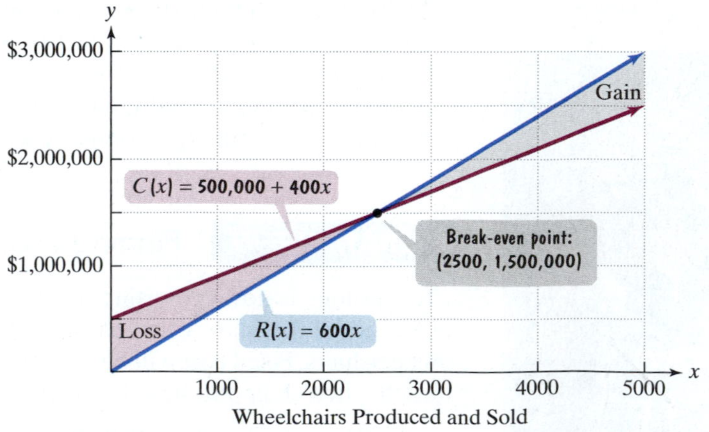

FIGURE 3.7

The intersection point confirms that the company breaks even by producing and selling 2500 wheelchairs. Can you see what happens for $x < 2500$? The red cost graph lies above the blue revenue graph. The cost is greater than the revenue and the business is losing money. Thus, if they sell fewer than 2500 wheelchairs, the result is a *loss*. By contrast, look at what happens for $x > 2500$. The blue revenue graph lies above the red cost graph. The revenue is greater than the cost and the business is making money. Thus, if they sell more than 2500 wheelchairs, the result is a *gain*.

✓ CHECK POINT 5 A company that manufactures running shoes has a fixed cost of \$300,000. Additionally, it costs \$30 to produce each pair of shoes. The shoes are sold at \$80 per pair.

a. Write the cost function, C, of producing x pairs of running shoes.

b. Write the revenue function, R, from the sale of x pairs of running shoes.

c. Determine the break-even point. Describe what this means.

What does every entrepreneur, from a kid selling lemonade to Donald Trump, want to do? Generate profit, of course. The *profit* made is the money taken in, or the revenue, minus the money spent, or the cost. This relationship between revenue and cost allows us to define the *profit function*, $P(x)$.

The Profit Function

The profit, $P(x)$, generated after producing and selling x units of a product is given by the **profit function**

$$P(x) = R(x) - C(x),$$

where R and C are the revenue and cost functions, respectively.

EXAMPLE 6 Writing a Profit Function

Use the revenue and cost functions for the wheelchair business in Example 5,

$$R(x) = 600x \quad \text{and} \quad C(x) = 500{,}000 + 400x,$$

to write the profit function for producing and selling x wheelchairs.

Solution The profit function is the difference between the revenue function and the cost function.

$$\begin{aligned} P(x) &= R(x) - C(x) && \text{This is the definition of the profit function.} \\ &= 600x - (500{,}000 + 400x) && \text{Substitute the given functions.} \\ &= 600x - 500{,}000 - 400x && \text{Distribute } -1 \text{ to each term in parentheses.} \\ &= 200x - 500{,}000 && \text{Simplify: } 600x - 400x = 200x. \end{aligned}$$

The profit function is $P(x) = 200x - 500{,}000$. ■

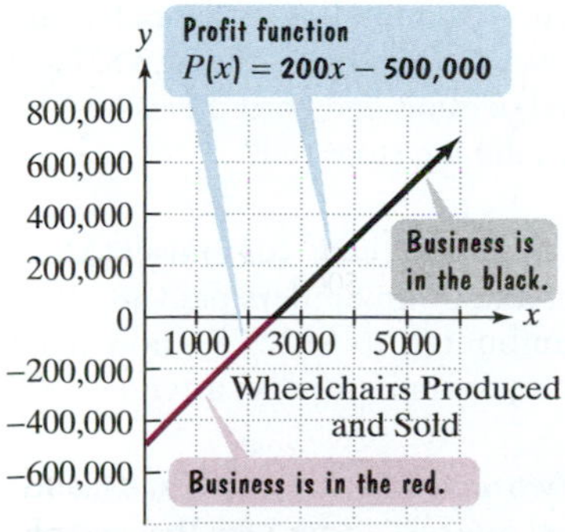

FIGURE 3.8

The graph of the profit function for the wheelchair business, $P(x) = 200x - 500{,}000$, is shown in **Figure 3.8**. The red portion lies below the x-axis and shows a loss when fewer than 2500 wheelchairs are sold. The business is "in the red." The black portion lies above the x-axis and shows a gain when more than 2500 wheelchairs are sold. The wheelchair business is "in the black."

✓ CHECK POINT 6 Use the revenue and cost functions that you obtained in Check Point 5 to write the profit function for producing and selling x pairs of running shoes. $P(x) = 50x - 300{,}000$

3.2 EXERCISE SET

MyMathLab Math XL PRACTICE WATCH DOWNLOAD READ REVIEW

Practice Exercises

In Exercises 1–4, let x represent one number and let y represent the other number. Use the given conditions to write a system of equations. Solve the system and find the numbers.

1. The sum of two numbers is 7. If one number is subtracted from the other, the result is -1. Find the numbers.
2. The sum of two numbers is 2. If one number is subtracted from the other, the result is 8. Find the numbers.
3. Three times a first number decreased by a second number is 1. The first number increased by twice the second number is 12. Find the numbers.
4. The sum of three times a first number and twice a second number is 8. If the second number is subtracted from twice the first number, the result is 3. Find the numbers.

In Exercises 5–8, cost and revenue functions for producing and selling x units of a product are given. Cost and revenue are expressed in dollars.

a. *Find the number of units that must be produced and sold to break even. At this level, what is the dollar amount coming in and going out?*
b. *Write the profit function from producing and selling x units of the product.*

5. $C(x) = 25{,}500 + 15x$
   $R(x) = 32x$
6. $C(x) = 15{,}000 + 12x$
   $R(x) = 32x$
7. $C(x) = 105x + 70{,}000$
   $R(x) = 245x$
8. $C(x) = 1.2x + 1500$
   $R(x) = 1.7x$

Application Exercises

In Exercises 9–40, use the four-step strategy to solve each problem.

9. At some point, it's time to kick, or gently ease, kids off the parental gravy train. The circle graph shows the percentage of parents who think significant financial support should end at various milestones.

Percentage of Parents Ending a Child's Financial Support at Various Milestones

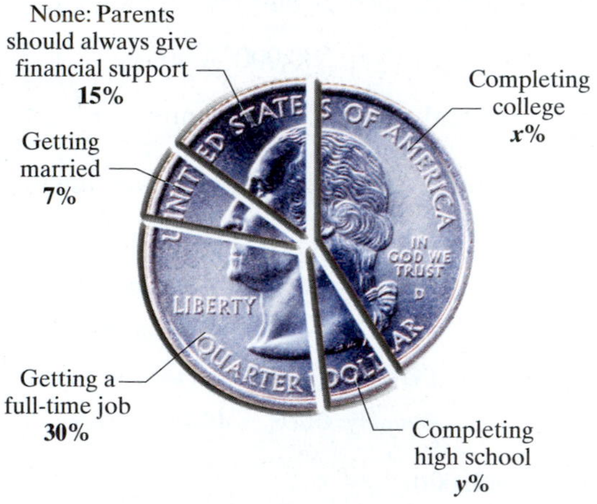
None: Parents should always give financial support
15%
Completing college
x%
Getting married
7%
Getting a full-time job
30%
Completing high school
y%

Source: Consumer Reports Money Adviser, July 2006

A total of 48% of parents would end financial support after completing education. The difference in the percentage who would end this support after completing college and after completing high school is 34%. Find the percentage of parents who would end financial support after a child completes college and the percentage who would end financial support after a child completes high school.

10. In 2007, there were approximately 730,000 homeless people in the United States. The circle graph shows the breakdown of the nation's homeless population.

The United States Homeless Population

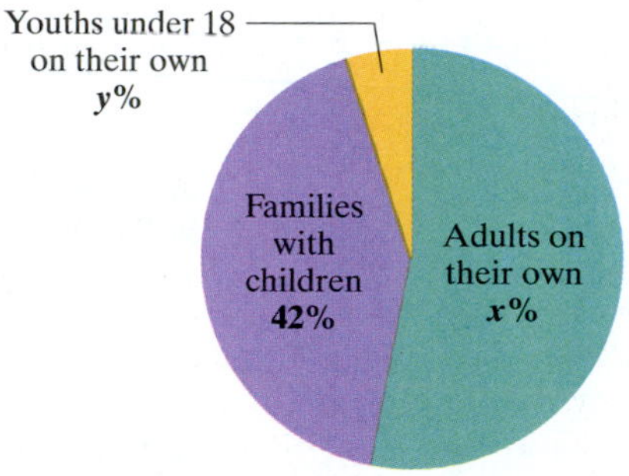
Youths under 18 on their own
y%
Families with children
42%
Adults on their own
x%

Source: U.S. Department of Housing and Urban Development

A total of 58% of the homeless consist of people on their own. The difference in the percentage of the population consisting of adults on their own and youths on their own is 48%. Find the percentage of the U.S. homeless population consisting of adults on their own and the percentage consisting of youths under 18 on their own.

11. One week a computer store sold a total of 36 computers and external hard drives. The revenue from these sales was $27,710. If computers sold for $1180 per unit and hard drives for $125 per unit, how many of each did the store sell?
12. There were 180 people at a civic club fundraiser. Members paid $4.50 per ticket and nonmembers paid $8.25 per ticket. If total receipts amounted to $1222.50, how many members and how many nonmembers attended the fundraiser?
13. You invested $7000 in two accounts paying 6% and 8% annual interest. If the total interest earned for the year was $520, how much was invested at each rate?

14. You invested $11,000 in stocks and bonds, paying 5% and 8% annual interest. If the total interest earned for the year was $730, how much was invested in stocks and how much was invested in bonds?

15. You invested money in two funds. Last year, the first fund paid a dividend of 9% and the second a dividend of 3%, and you received a total of $900. This year, the first fund paid a 10% dividend and the second only 1%, and you received a total of $860. How much money did you invest in each fund?

16. You invested money in two funds. Last year, the first fund paid a dividend of 8% and the second a dividend of 5%, and you received a total of $1330. This year, the first fund paid a 12% dividend and the second only 2%, and you received a total of $1500. How much money did you invest in each fund?

17. Things did not go quite as planned. You invested $20,000, part of it in a stock that paid 12% annual interest. However, the rest of the money suffered a 5% loss. If the total annual income from both investments was $1890, how much was invested at each rate?

18. Things did not go quite as planned. You invested $30,000, part of it in a stock that paid 14% annual interest. However, the rest of the money suffered a 6% loss. If the total annual income from both investments was $200, how much was invested at each rate?

19. A wine company needs to blend a California wine with a 5% alcohol content and a French wine with a 9% alcohol content to obtain 200 gallons of wine with a 7% alcohol content. How many gallons of each kind of wine must be used?

20. A jeweler needs to mix an alloy with a 16% gold content and an alloy with a 28% gold content to obtain 32 ounces of a new alloy with a 25% gold content. How many ounces of each of the original alloys must be used?

21. For thousands of years, gold has been considered one of Earth's most precious metals. One hundred percent pure gold is 24-karat gold, which is too soft to be made into jewelry. In the United States, most gold jewelry is 14-karat gold, approximately 58% gold. If 18-karat gold is 75% gold and 12-karat gold is 50% gold, how much of each should be used to make a 14-karat gold bracelet weighing 300 grams?

23. The manager of a candystand at a large multiplex cinema has a popular candy that sells for $1.60 per pound. The manager notices a different candy worth $2.10 per pound that is not selling well. The manager decides to form a mixture of both types of candy to help clear the inventory of the more expensive type. How many pounds of each kind of candy should be used to create a 75-pound mixture selling for $1.90 per pound?

24. A grocer needs to mix raisins at $2.00 per pound with granola at $3.25 per pound to obtain 10 pounds of a mixture that costs $2.50 per pound. How many pounds of raisins and how many pounds of granola must be used?

25. A coin purse contains a mixture of 15 coins in nickels and dimes. The coins have a total value of $1.10. Determine the number of nickels and the number of dimes in the purse.

26. A coin purse contains a mixture of 15 coins in dimes and quarters. The coins have a total value of $3.30. Determine the number of dimes and the number of quarters in the purse.

27. When a small plane flies with the wind, it can travel 800 miles in 5 hours. When the plane flies in the opposite direction, against the wind, it takes 8 hours to fly the same distance. Find the rate of the plane in still air and the rate of the wind.

28. When a plane flies with the wind, it can travel 4200 miles in 6 hours. When the plane flies in the opposite direction, against the wind, it takes 7 hours to fly the same distance. Find the rate of the plane in still air and the rate of the wind.

29. A boat's crew rowed 16 kilometers downstream, with the current, in 2 hours. The return trip upstream, against the current, covered the same distance, but took 4 hours. Find the crew's rowing rate in still water and the rate of the current.

30. A motorboat traveled 36 miles downstream, with the current, in 1.5 hours. The return trip upstream, against the current, covered the same distance, but took 2 hours. Find the boat's rate in still water and the rate of the current.

22. In the "Peanuts" cartoon shown, solve the problem that is sending Peppermint Patty into an agitated state. How much cream and how much milk, to the nearest hundredth of a gallon, must be mixed together to obtain 50 gallons of cream that contains 12.5% butterfat?

PEANUTS © United Feature Syndicate, Inc.

31. With the current, you can canoe 24 miles in 4 hours. Against the same current, you can canoe only $\frac{3}{4}$ of this distance in 6 hours. Find your rate in still water and the rate of the current.

32. With the current, you can row 24 miles in 3 hours. Against the same current, you can row only $\frac{2}{3}$ of this distance in 4 hours. Find your rowing rate in still water and the rate of the current.

33. A student has two test scores. The difference between the scores is 12 and the mean, or average, of the scores is 80. What are the two test scores?

34. A student has two test scores. The difference between the scores is 8 and the mean, or average, of the scores is 88. What are the two test scores?

In Exercises 35–36, an isosceles triangle containing two angles with equal measure is shown. The degree measure of each triangle's three interior angles and an exterior angle is represented with variables. Find the measure of the three interior angles.

35.

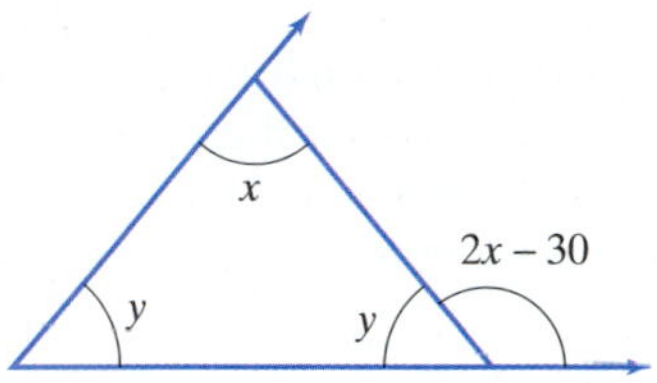

36.

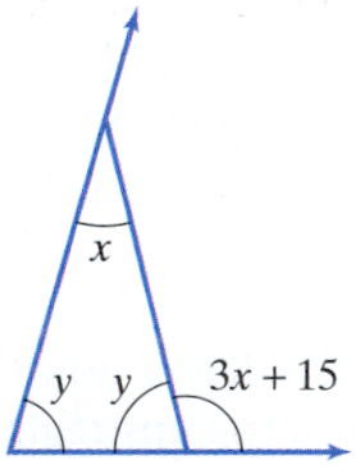

37. A rectangular lot whose perimeter is 220 feet is fenced along three sides. An expensive fencing along the lot's length costs \$20 per foot, and an inexpensive fencing along the two side widths costs only \$8 per foot. The total cost of the fencing along the three sides comes to \$2040. What are the lot's dimensions?

38. A rectangular lot whose perimeter is 260 feet is fenced along three sides. An expensive fencing along the lot's length costs \$16 per foot, and an inexpensive fencing along the two side widths costs only \$5 per foot. The total cost of the fencing along the three sides comes to \$1780. What are the lot's dimensions?

39. A new restaurant is to contain two-seat tables and four-seat tables. Fire codes limit the restaurant's maximum occupancy to 56 customers. If the owners have hired enough servers to handle 17 tables of customers, how many of each kind of table should they purchase?

40. A hotel has 200 rooms. Those with kitchen facilities rent for \$100 per night and those without kitchen facilities rent for \$80 per night. On a night when the hotel was completely occupied, revenues were \$17,000. How many of each type of room does the hotel have?

The figure shows the graphs of the cost and revenue functions for a company that manufactures and sells small radios. Use the information in the figure to solve Exercises 41–46.

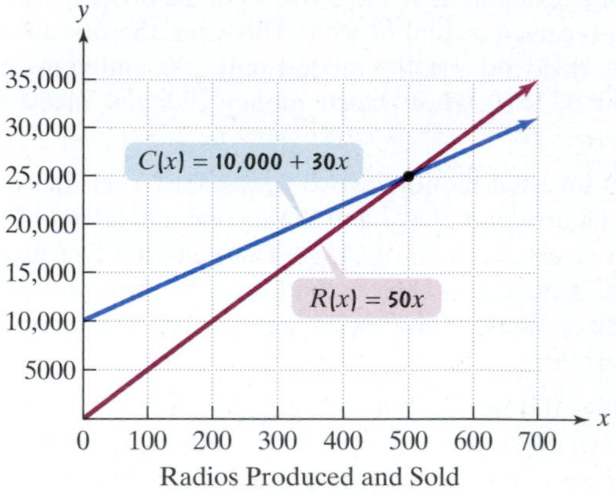

41. How many radios must be produced and sold for the company to break even?

42. More than how many radios must be produced and sold for the company to have a profit?

43. Use the formulas shown in the voice balloons to find $R(200) - C(200)$. Describe what this means for the company.

44. Use the formulas shown in the voice balloons to find $R(300) - C(300)$. Describe what this means for the company.

45. a. Use the formulas shown in the voice balloons to write the company's profit function, P, from producing and selling x radios.

b. Find the company's profit if 10,000 radios are produced and sold.

46. a. Use the formulas shown in the voice balloons to write the company's profit function, P, from producing and selling x radios.

b. Find the company's profit if 20,000 radios are produced and sold.

Exercises 47–50 describe a number of business ventures. For each exercise,

a. *Write the cost function, C.*

b. *Write the revenue function, R.*

c. *Determine the break-even point. Describe what this means.*

47. A company that manufactures small canoes has a fixed cost of \$18,000. It costs \$20 to produce each canoe. The selling price is \$80 per canoe. (In solving this exercise, let x represent the number of canoes produced and sold.)

48. A company that manufactures bicycles has a fixed cost of \$100,000. It costs \$100 to produce each bicycle. The selling price is \$300 per bike. (In solving this exercise, let x represent the number of bicycles produced and sold.)

49. You invest in a new play. The cost includes an overhead of \$30,000, plus production costs of \$2500 per performance. A sold-out performance brings in \$3125. (In solving this exercise, let x represent the number of sold-out performances.)

50. You invested \$30,000 and started a business writing greeting cards. Supplies cost 2¢ per card and you are selling each card for 50¢. (In solving this exercise, let x represent the number of cards produced and sold.)

Writing in Mathematics

51. Describe the conditions in a problem that enable it to be solved using a system of linear equations.

52. Write a word problem that can be solved by translating to a system of linear equations. Then solve the problem.

53. Describe a revenue function for a business venture.

54. Describe a cost function for a business venture. What are the two kinds of costs that are modeled by this function?

55. What is the profit function for a business venture and how is it determined?

56. Describe the break-even point for a business.

57. The law of supply and demand states that, in a free market economy, a commodity tends to be sold at its equilibrium price. At this price, the amount that the seller will supply is the same amount that the consumer will buy. Explain how graphs can be used to determine the equilibrium price.

58. Many students hate mixture problems and decide to ignore them, stating, "I'll just skip that one on the test." If you share this opinion, describe what you find particularly unappealing about this kind of problem.

Technology Exercises

In Exercises 59–60, graph the revenue and cost functions in the same viewing rectangle. Then use the intersection feature to determine the break-even point.

59. $R(x) = 50x, \quad C(x) = 20x + 180$

60. $R(x) = 92.5x, \quad C(x) = 52x + 1782$

61. Use the procedure in Exercises 59–60 to verify your work for any one of the break-even points that you found in Exercises 47–50.

Critical Thinking Exercises

Make Sense? *In Exercises 62–65, determine whether each statement "makes sense" or "does not make sense" and explain your reasoning.*

62. A system of linear equations can be used to model and compare the fees charged by two different taxicab companies.

63. I should mix 6 liters of a 50% acid solution with 4 liters of a 25% acid solution to obtain 10 liters of a 75% acid solution.

64. If I know the perimeter of this rectangle and triangle, each in the same unit of measure, I can use a system of linear equations to determine values for x and y.

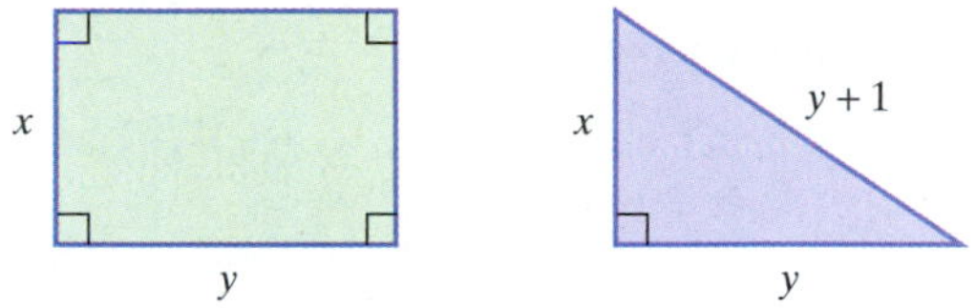

65. You told me that you flew against the wind from Miami to Seattle, 2800 miles, in 7 hours and, at the same time, your friend flew with the wind from Seattle to Miami in only 5.6 hours. You have not given me enough information to determine the average rate of the wind.

66. The radiator in your car contains 4 gallons of antifreeze and water. The mixture is 45% antifreeze. How much of this mixture should be drained and replaced with pure antifreeze in order to have a 60% antifreeze solution? Round to the nearest tenth of a gallon.

67. A marching band has 52 members, and there are 24 in the pom-pom squad. They wish to form several hexagons and squares like those diagrammed below. Can it be done with no people left over?

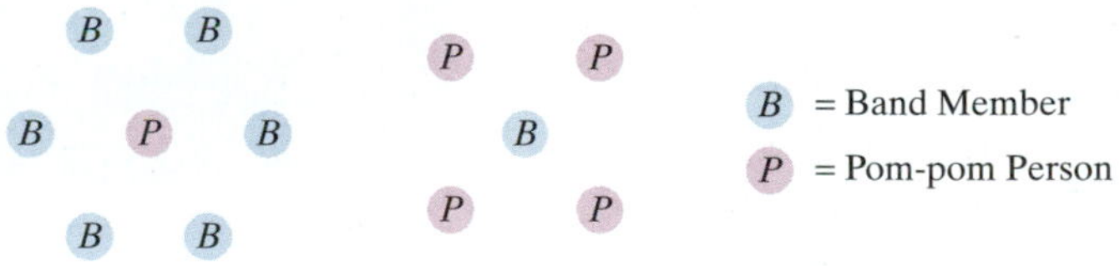

68. A boy has as many brothers as he has sisters. Each of his sisters has twice as many brothers as she has sisters. How many boys and girls are in this family?

69. When entering your test score into a computer, your professor accidently reversed the two digits. This error reduced your score by 36 points. Your professor told you that the sum of the digits of your actual score was 14, corrected the error, and agreed to give you extra credit if you could determine the actual score without looking back at the test. What was your actual test score? (*Hint*: Let t = the tens-place digit of your actual score and let u = the units-place digit of your actual score. Thus, $10t + u$ represents your actual test score.)

70. A dealer paid a total of \$67 for mangos and avocados. The mangos were sold at a profit of 20% on the dealer's cost, but the avocados started to spoil, resulting in a selling price of a 2% loss on the dealer's cost. The dealer made a profit of \$8.56 on the total transaction. How much did the dealer pay for the mangos and for the avocados?

Review Exercises

In Exercises 71–72, use the given conditions to write an equation for each line in point-slope form and slope-intercept form.

71. Passing through $(-2, 5)$ and $(-6, 13)$

(Section 2.5, Example 2)

72. Passing through $(-3, 0)$ and parallel to the line whose equation is $-x + y = 7$

(Section 2.5, Example 4)

73. Find the domain of $g(x) = \frac{x-2}{3-x}$.

(Section 2.3, Example 1)

Preview Exercises

Exercises 74–76 will help you prepare for the material covered in the next section.

74. If $x = 3, y = 2,$ and $z = -3$, does the ordered triple (x, y, z) satisfy the equation $2x - y + 4z = -8$?

75. Consider the following equations:

$$5x - 2y - 4z = 3 \quad \text{Equation 1}$$
$$3x + 3y + 2z = -3. \quad \text{Equation 2}$$

Use these equations to eliminate z. Copy Equation 1 and multiply Equation 2 by 2. Then add the equations.

76. Write an equation involving a, b, and c based on the following description:

When the value of x in $y = ax^2 + bx + c$ is 4, the value of y is 1682.

SECTION 3.3

Systems of Linear Equations in Three Variables

Objectives

1. Verify the solution of a system of linear equations in three variables.
2. Solve systems of linear equations in three variables.
3. Identify inconsistent and dependent systems.
4. Solve problems using systems in three variables.

All animals sleep, but the length of time they sleep varies widely: Cattle sleep for only a few minutes at a time. We humans seem to need more sleep than other animals, up to eight hours a day. Without enough sleep, we have difficulty concentrating, make mistakes in routine tasks, lose energy, and feel bad-tempered. There is a relationship between hours of sleep and death rate per year per 100,000 people. How many hours of sleep will put you in the group with the minimum death rate? In this section, we will answer this question by solving a system of linear equations with more than two variables.

1 Verify the solution of a system of linear equations in three variables.

Systems of Linear Equations in Three Variables and Their Solutions

An equation such as $x + 2y - 3z = 9$ is called a *linear equation in three variables*. In general, any equation of the form

$$Ax + By + Cz = D,$$

where A, B, C, and D are real numbers such that A, B, and C are not all 0, is a **linear equation in three variables: x, y, and z.** The graph of this linear equation in three variables is a plane in three-dimensional space.

The process of solving a system of three linear equations in three variables is geometrically equivalent to finding the point of intersection (assuming that there is

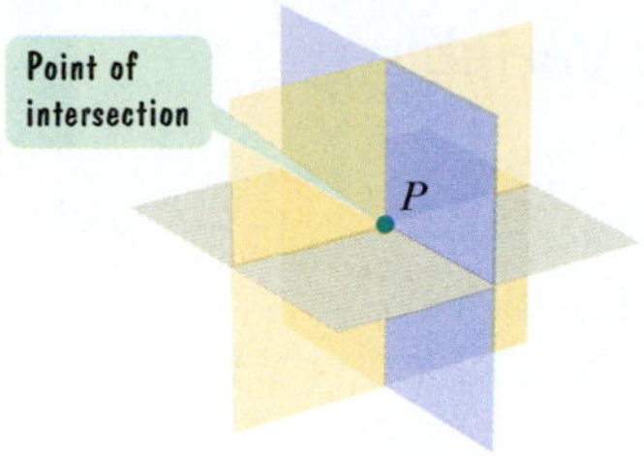

FIGURE 3.9

one) of three planes in space (see **Figure 3.9**). A **solution** of a system of linear equations in three variables is an ordered triple of real numbers that satisfies all equations in the system. The **solution set** of the system is the set of all its solutions.

EXAMPLE 1 Determining Whether an Ordered Triple Satisfies a System

Show that the ordered triple $(-1, 2, -2)$ is a solution of the system:

$$\begin{aligned} x + 2y - 3z &= 9 \\ 2x - y + 2z &= -8 \\ -x + 3y - 4z &= 15. \end{aligned}$$

Solution Because -1 is the x-coordinate, 2 is the y-coordinate, and -2 is the z-coordinate of $(-1, 2, -2)$, we replace x with -1, y with 2, and z with -2 in each of the three equations.

$$\begin{aligned} x + 2y - 3z &= 9 \\ -1 + 2(2) - 3(-2) &\stackrel{?}{=} 9 \\ -1 + 4 + 6 &\stackrel{?}{=} 9 \\ 9 &= 9, \quad \text{true} \end{aligned}$$

$$\begin{aligned} 2x - y + 2z &= -8 \\ 2(-1) - 2 + 2(-2) &\stackrel{?}{=} -8 \\ -2 - 2 - 4 &\stackrel{?}{=} -8 \\ -8 &= -8, \quad \text{true} \end{aligned}$$

$$\begin{aligned} -x + 3y - 4z &= 15 \\ -(-1) + 3(2) - 4(-2) &\stackrel{?}{=} 15 \\ 1 + 6 + 8 &\stackrel{?}{=} 15 \\ 15 &= 15, \quad \text{true} \end{aligned}$$

The ordered triple $(-1, 2, -2)$ satisfies the three equations: It makes each equation true. Thus, the ordered triple is a solution of the system.

✓ CHECK POINT 1 Show that the ordered triple $(-1, -4, 5)$ is a solution of the system:

$$\begin{aligned} x - 2y + 3z &= 22 \\ 2x - 3y - z &= 5 \\ 3x + y - 5z &= -32. \end{aligned}$$

2 Solve systems of linear equations in three variables.

Solving Systems of Linear Equations in Three Variables by Eliminating Variables

The method for solving a system of linear equations in three variables is similar to that used on systems of linear equations in two variables. We use addition to eliminate any variable, reducing the system to two equations in two variables. Once we obtain a system of two equations in two variables, we use addition or substitution to eliminate a variable. The result is a single equation in one variable. We solve this equation to get the value of the remaining variable. Other variable values are found by back-substitution.

Study Tip

It does not matter which variable you eliminate, as long as you eliminate the same variable in two different pairs of equations.

Solving Linear Systems in Three Variables by Eliminating Variables

1. Reduce the system to two equations in two variables. This is usually accomplished by taking two different pairs of equations and using the addition method to eliminate the same variable from both pairs.
2. Solve the resulting system of two equations in two variables using addition or substitution. The result is an equation in one variable that gives the value of that variable.
3. Back-substitute the value of the variable found in step 2 into either of the equations in two variables to find the value of the second variable.
4. Use the values of the two variables from steps 2 and 3 to find the value of the third variable by back-substituting into one of the original equations.
5. Check the proposed solution in each of the original equations.

EXAMPLE 2 Solving a System in Three Variables

Solve the system:

$$\begin{aligned} 5x - 2y - 4z &= 3 && \text{Equation 1} \\ 3x + 3y + 2z &= -3 && \text{Equation 2} \\ -2x + 5y + 3z &= 3. && \text{Equation 3} \end{aligned}$$

Solution There are many ways to proceed. Because our initial goal is to reduce the system to two equations in two variables, **the central idea is to take two different pairs of equations and eliminate the same variable from both pairs.**

Step 1. Reduce the system to two equations in two variables. We choose any two equations and use the addition method to eliminate a variable. Let's eliminate z using Equations 1 and 2. We do so by multiplying Equation 2 by 2. Then we add equations.

$$\begin{aligned} &\text{(Equation 1)}\ 5x - 2y - 4z = 3 \xrightarrow{\text{No change}} & 5x - 2y - 4z &= 3 \\ &\text{(Equation 2)}\ 3x + 3y + 2z = -3 \xrightarrow{\text{Multiply by 2.}} & 6x + 6y + 4z &= -6 \\ & & \text{Add: } 11x + 4y \qquad &= -3 \quad \text{Equation 4} \end{aligned}$$

Now we must eliminate the *same* variable from another pair of equations. We can eliminate z using Equations 2 and 3. First, we multiply Equation 2 by -3. Next, we multiply Equation 3 by 2. Finally, we add equations.

$$\begin{aligned} &\text{(Equation 2)}\ 3x + 3y + 2z = -3 \xrightarrow{\text{Multiply by } -3.} & -9x - 9y - 6z &= 9 \\ &\text{(Equation 3)}\ -2x + 5y + 3z = 3 \xrightarrow{\text{Multiply by 2.}} & -4x + 10y + 6z &= 6 \\ & & \text{Add: } -13x + y \qquad &= 15 \quad \text{Equation 5} \end{aligned}$$

Equations 4 and 5 give us a system of two equations in two variables.

Step 2. Solve the resulting system of two equations in two variables. We will use the addition method to solve Equations 4 and 5 for x and y. To do so, we multiply Equation 5 by -4 and add this to Equation 4.

$$\begin{aligned} &\text{(Equation 4)}\ 11x + 4y = -3 \xrightarrow{\text{No change}} & 11x + 4y &= -3 \\ &\text{(Equation 5)}\ -13x + y = 15 \xrightarrow{\text{Multiply by } -4.} & 52x - 4y &= -60 \\ & & \text{Add: } 63x \qquad &= -63 \\ & & x \qquad &= -1 \quad \text{Divide both sides by 63.} \end{aligned}$$

Step 3. Use back-substitution in one of the equations in two variables to find the value of the second variable. We back-substitute -1 for x in either Equation 4 or 5 to find the value of y. We will use Equation 5.

$$\begin{aligned} -13x + y &= 15 && \text{Equation 5} \\ -13(-1) + y &= 15 && \text{Substitute } -1 \text{ for } x. \\ 13 + y &= 15 && \text{Multiply.} \\ y &= 2 && \text{Subtract 13 from both sides.} \end{aligned}$$

Step 4. Back-substitute the values found for two variables into one of the original equations to find the value of the third variable. We can now use any one of the original equations and back-substitute the values of x and y to find the value for z. We will use Equation 2.

$$\begin{aligned} 3x + 3y + 2z &= -3 && \text{Equation 2} \\ 3(-1) + 3(2) + 2z &= -3 && \text{Substitute } -1 \text{ for } x \text{ and 2 for } y. \\ 3 + 2z &= -3 && \text{Multiply and then add: } 3(-1) + 3(2) = -3 + 6 = 3. \\ 2z &= -6 && \text{Subtract 3 from both sides.} \\ z &= -3 && \text{Divide both sides by 2.} \end{aligned}$$

With $x = -1$, $y = 2$, and $z = -3$, the proposed solution is the ordered triple $(-1, 2, -3)$.

Step 5. Check. Check the proposed solution, $(-1, 2, -3)$, by substituting the values for x, y, and z into each of the three original equations. These substitutions yield three true statements. Thus, the solution is $(-1, 2, -3)$ and the solution set is $\{(-1, 2, -3)\}$.

✓ **CHECK POINT 2** Solve the system:

$$\begin{aligned} x + 4y - z &= 20 \\ 3x + 2y + z &= 8 \\ 2x - 3y + 2z &= -16. \end{aligned}$$

In some examples, one of the variables is missing from a given equation. In this case, the missing variable should be eliminated from the other two equations, thereby making it possible to omit one of the elimination steps. We illustrate this idea in Example 3.

EXAMPLE 3 Solving a System of Equations with a Missing Term

Solve the system:

$$\begin{aligned} x + z &= 8 && \text{Equation 1} \\ x + y + 2z &= 17 && \text{Equation 2} \\ x + 2y + z &= 16. && \text{Equation 3} \end{aligned}$$

Solution

Step 1. Reduce the system to two equations in two variables. Because Equation 1 contains only x and z, we can omit one of the elimination steps by eliminating y using Equations 2 and 3. This will give us two equations in x and z. To eliminate y using Equations 2 and 3, we multiply Equation 2 by -2 and add Equation 3.

$$\begin{aligned} &\text{(Equation 2)} \quad x + y + 2z = 17 \xrightarrow{\text{Multiply by } -2.} && -2x - 2y - 4z = -34 \\ &\text{(Equation 3)} \quad x + 2y + z = 16 \xrightarrow{\text{No change}} && \underline{x + 2y + z = 16} \\ & && \text{Add: } -x \qquad - 3z = -18 \quad \text{Equation 4} \end{aligned}$$

Equation 4 and the given Equation 1 provide us with a system of two equations in two variables:

$$\begin{aligned} x + z &= 8 && \text{Equation 1} \\ -x - 3z &= -18. && \text{Equation 4} \end{aligned}$$

Step 2. Solve the resulting system of two equations in two variables. We will solve Equations 1 and 4 for x and z.

$$\begin{aligned} x + z &= 8 && \text{Equation 1} \\ \underline{-x - 3z} &= \underline{-18} && \text{Equation 4} \\ \text{Add: } -2z &= -10 \\ z &= 5 && \text{Divide both sides by } -2. \end{aligned}$$

Step 3. Use back-substitution in one of the equations in two variables to find the value of the second variable. To find x, we back-substitute 5 for z in either Equation 1 or 4. We will use Equation 1.

$$\begin{aligned} x + z &= 8 && \text{Equation 1} \\ x + 5 &= 8 && \text{Substitute 5 for } z. \\ x &= 3 && \text{Subtract 5 from both sides.} \end{aligned}$$

Step 4. Back-substitute the values found for two variables into one of the original equations to find the value of the third variable. To find y, we back-substitute 3 for x and 5 for z into Equation 2, $x + y + 2z = 17$, or Equation 3, $x + 2y + z = 16$. We can't use Equation 1, $x + z = 8$, because y is missing in this equation. We will use Equation 2.

$$x + y + 2z = 17 \quad \text{Equation 2}$$
$$3 + y + 2(5) = 17 \quad \text{Substitute 3 for x and 5 for z.}$$
$$y + 13 = 17 \quad \text{Multiply and add.}$$
$$y = 4 \quad \text{Subtract 13 from both sides.}$$

We found that $z = 5$, $x = 3$, and $y = 4$. Thus, the proposed solution is the ordered triple $(3, 4, 5)$.

Step 5. Check. Substituting 3 for x, 4 for y, and 5 for z into each of the three original equations yields three true statements. Consequently, the solution is $(3, 4, 5)$ and the solution set is $\{(3, 4, 5)\}$.

✓ CHECK POINT 3 Solve the system:

$$\begin{aligned} 2y - z &= 7 \\ x + 2y + z &= 17 \\ 2x - 3y + 2z &= -1. \end{aligned}$$

3 Identify inconsistent and dependent systems.

Inconsistent and Dependent Systems

A system of three linear equations in three variables represents three planes. The three planes need not intersect at one point. The planes may have no common point of intersection and represent an **inconsistent system** with no solution. **Figure 3.10** illustrates some of the geometric possibilities for inconsistent systems.

Three planes are parallel with no common intersection point.

Two planes are parallel with no common intersection point.

Planes intersect two at a time. There is no intersection point common to all three planes.

FIGURE 3.10 Three planes may have no common point of intersection.

If you attempt to solve an inconsistent system algebraically, at some point in the solution process you will eliminate all three variables. A false statement, such as $0 = -10$, will be the result. For example, consider the system

$$\begin{aligned} 2x + 5y + z &= 12 \quad \text{Equation 1} \\ x - 2y + 4z &= -10 \quad \text{Equation 2} \\ -3x + 6y - 12z &= 20. \quad \text{Equation 3} \end{aligned}$$

Suppose we reduce the system to two equations in two variables by eliminating x. To eliminate x using Equations 2 and 3, we multiply Equation 2 by 3 and add Equation 3:

$$x - 2y + 4z = -10 \xrightarrow{\text{Multiply by 3.}} 3x - 6y + 12z = -30$$
$$-3x + 6y - 12z = 20 \xrightarrow{\text{No change}} \underline{-3x + 6y - 12z = 20}$$
$$\text{Add:} \quad 0 = -10$$

There are no values of x, y, and z for which $0 = -10$. The false statement $0 = -10$ indicates that the system is inconsistent and has no solution. The solution set is the empty set, $\varnothing$.

We have seen that a linear system that has at least one solution is called a **consistent system**. Planes that intersect at one point and planes that intersect at infinitely many points both represent consistent systems. **Figure 3.11** illustrates two different cases of three planes that intersect at infinitely many points. The equations in these linear systems with infinitely many solutions are called **dependent**.

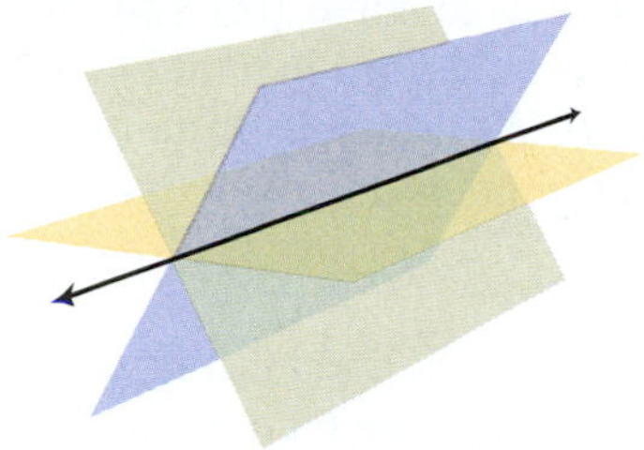

The planes intersect along a common line.

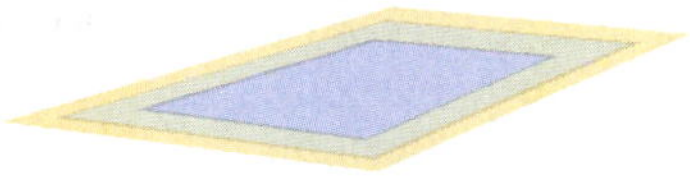

The planes are identical, or coincide.

FIGURE 3.11 Three planes may intersect at infinitely many points.

If you attempt to solve a system with dependent equations algebraically, at some point in the solution process you will eliminate all three variables. A true statement, such as $0 = 0$, will be the result. If this occurs as you are solving a linear system, simply state that the equations are dependent.

4 Solve problems using systems in three variables.

Applications

Systems of equations may allow us to find models for data without using a graphing utility. Three data points that do not lie on or near a line determine the graph of a function of the form

$$y = ax^2 + bx + c, a \neq 0.$$

Such a function is called a **quadratic function**. If $a > 0$, its graph is shaped like a bowl, making it ideal for modeling situations in which values of y are decreasing and then increasing. In Chapter 8, we'll have lots of interesting things to tell you about quadratic functions and their graphs.

The process of determining a function whose graph contains given points is called **curve fitting**. In our next example, we fit the curve whose equation is $y = ax^2 + bx + c$ to three data points. Using a system of equations, we find values for a, b, and c.

EXAMPLE 4 Modeling Data Relating Sleep and Death Rate

In a study relating sleep and death rate, the following data were obtained. Use the function $y = ax^2 + bx + c$ to model the data.

x (Average Number of Hours of Sleep)	y (Death Rate per Year per 100,000 Males)
4	1682
7	626
9	967

Average hours of sleep — Yearly death rates per 100,000 males

x	y
4	1682
7	626
9	967

Using Technology

The graph of

$$y = 104.5x^2 - 1501.5x + 6016$$

is displayed in a [3, 12, 1] by [500, 2000, 100] viewing rectangle. The minimum function feature shows that the lowest point on the graph is approximately (7.2, 622.5). Men who average 7.2 hours of sleep are in the group with the lowest death rate, approximately 622.5 deaths per 100,000 males.

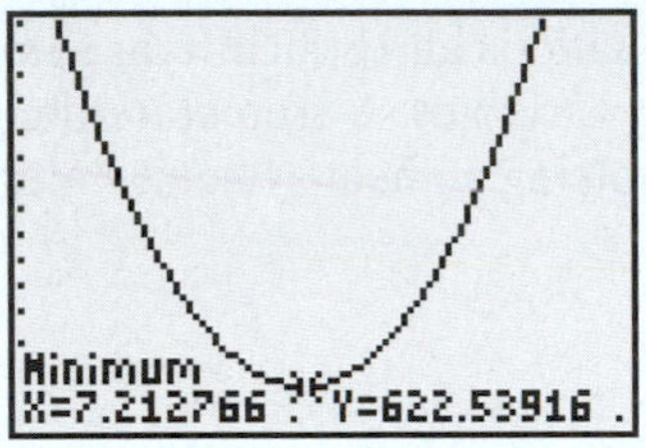

Solution We need to find values for a, b, and c in $y = ax^2 + bx + c$. We can do so by solving a system of three linear equations in a, b, and c. We obtain the three equations by using the values of x and y from the data, repeated in the margin, as follows:

$\mathbf{y = ax^2 + bx + c}$ Use the quadratic function to model the data.

When $x = 4$, $y = 1682$: $1682 = a \cdot 4^2 + b \cdot 4 + c$ or $16a + 4b + c = 1682$

When $x = 7$, $y = 626$: $626 = a \cdot 7^2 + b \cdot 7 + c$ or $49a + 7b + c = 626$

When $x = 9$, $y = 967$: $967 = a \cdot 9^2 + b \cdot 9 + c$ or $81a + 9b + c = 967.$

The easiest way to solve this system is to eliminate c from two pairs of equations, obtaining two equations in a and b. Solving this system gives $a = 104.5$, $b = -1501.5$, and $c = 6016$. We now substitute the values for a, b, and c into $y = ax^2 + bx + c$. The function that models the given data is

$$y = 104.5x^2 - 1501.5x + 6016.$$

We can use the model that we obtained in Example 4 to find the death rate of males who average, say, 6 hours of sleep. First, write the model in function notation:

$$f(x) = 104.5x^2 - 1501.5x + 6016.$$

Substitute 6 for x:

$$f(6) = 104.5(6)^2 - 1501.5(6) + 6016 = 769.$$

According to the model, the death rate for males who average 6 hours of sleep is 769 deaths per 100,000 males.

✓ CHECK POINT 4 Find the quadratic function $y = ax^2 + bx + c$ whose graph passes through the points $(1, 4)$, $(2, 1)$, and $(3, 4)$.

Problems involving three unknowns can be solved using the same strategy for solving problems with two unknown quantities. You can let x, y, and z represent the unknown quantities. We then translate from the verbal conditions of the problem to a system of three equations in three variables. Problems of this type are included in the exercise set that follows.

3.3 EXERCISE SET

MyMathLab — PRACTICE WATCH DOWNLOAD READ REVIEW

Practice Exercises

In Exercises 1–4, determine if the given ordered triple is a solution of the system.

1. $(2, -1, 3)$

$$\begin{aligned} x + y + z &= 4 \\ x - 2y - z &= 1 \\ 2x - y - z &= -1 \end{aligned}$$

2. $(5, -3, -2)$

$$\begin{aligned} x + y + z &= 0 \\ x + 2y - 3z &= 5 \\ 3x + 4y + 2z &= -1 \end{aligned}$$

3. $(4, 1, 2)$

$$\begin{aligned} x - 2y &= 2 \\ 2x + 3y &= 11 \\ y - 4z &= -7 \end{aligned}$$

4. $(-1, 3, 2)$

$$\begin{aligned} x - 2z &= -5 \\ y - 3z &= -3 \\ 2x - z &= -4 \end{aligned}$$

Solve each system in Exercises 5–22. If there is no solution or if there are infinitely many solutions and a system's equations are dependent, so state.

5.
$$\begin{aligned} x + y + 2z &= 11 \\ x + y + 3z &= 14 \\ x + 2y - z &= 5 \end{aligned}$$

6.
$$\begin{aligned} 2x + y - 2z &= -1 \\ 3x - 3y - z &= 5 \\ x - 2y + 3z &= 6 \end{aligned}$$

7. $$\begin{aligned} 4x - y + 2z &= 11 \\ x + 2y - z &= -1 \\ 2x + 2y - 3z &= -1 \end{aligned}$$

8. $$\begin{aligned} x - y + 3z &= 8 \\ 3x + y - 2z &= -2 \\ 2x + 4y + z &= 0 \end{aligned}$$

9. $$\begin{aligned} 3x + 2y - 3z &= -2 \\ 2x - 5y + 2z &= -2 \\ 4x - 3y + 4z &= 10 \end{aligned}$$

10. $$\begin{aligned} 2x + 3y + 7z &= 13 \\ 3x + 2y - 5z &= -22 \\ 5x + 7y - 3z &= -28 \end{aligned}$$

11. $$\begin{aligned} 2x - 4y + 3z &= 17 \\ x + 2y - z &= 0 \\ 4x - y - z &= 6 \end{aligned}$$

12. $$\begin{aligned} x + z &= 3 \\ x + 2y - z &= 1 \\ 2x - y + z &= 3 \end{aligned}$$

13. $$\begin{aligned} 2x + y &= 2 \\ x + y - z &= 4 \\ 3x + 2y + z &= 0 \end{aligned}$$

14. $$\begin{aligned} x + 3y + 5z &= 20 \\ y - 4z &= -16 \\ 3x - 2y + 9z &= 36 \end{aligned}$$

15. $$\begin{aligned} x + y &= -4 \\ y - z &= 1 \\ 2x + y + 3z &= -21 \end{aligned}$$

16. $$\begin{aligned} x + y &= 4 \\ x + z &= 4 \\ y + z &= 4 \end{aligned}$$

17. $$\begin{aligned} 2x + y + 2z &= 1 \\ 3x - y + z &= 2 \\ x - 2y - z &= 0 \end{aligned}$$

18. $$\begin{aligned} 3x + 4y + 5z &= 8 \\ x - 2y + 3z &= -6 \\ 2x - 4y + 6z &= 8 \end{aligned}$$

19. $$\begin{aligned} 5x - 2y - 5z &= 1 \\ 10x - 4y - 10z &= 2 \\ 15x - 6y - 15z &= 3 \end{aligned}$$

20. $$\begin{aligned} x + 2y + z &= 4 \\ 3x - 4y + z &= 4 \\ 6x - 8y + 2z &= 8 \end{aligned}$$

21. $$\begin{aligned} 3(2x + y) + 5z &= -1 \\ 2(x - 3y + 4z) &= -9 \\ 4(1 + x) &= -3(z - 3y) \end{aligned}$$

22. $$\begin{aligned} 7z - 3 &= 2(x - 3y) \\ 5y + 3z - 7 &= 4x \\ 4 + 5z &= 3(2x - y) \end{aligned}$$

In Exercises 23–26, find the quadratic function $y = ax^2 + bx + c$ whose graph passes through the given points.

23. $(-1, 6), (1, 4), (2, 9)$

24. $(-2, 7), (1, -2), (2, 3)$

25. $(-1, -4), (1, -2), (2, 5)$

26. $(1, 3), (3, -1), (4, 0)$

In Exercises 27–28, let x represent the first number, y the second number, and z the third number. Use the given conditions to write a system of equations. Solve the system and find the numbers.

27. The sum of three numbers is 16. The sum of twice the first number, 3 times the second number, and 4 times the third number is 46. The difference between 5 times the first number and the second number is 31. Find the three numbers.

28. The following is known about three numbers: Three times the first number plus the second number plus twice the third number is 5. If 3 times the second number is subtracted from the sum of the first number and 3 times the third number, the result is 2. If the third number is subtracted from the sum of 2 times the first number and 3 times the second number, the result is 1. Find the numbers.

Practice PLUS

Solve each system in Exercises 29–30.

29. $$\begin{aligned} \frac{x+2}{6} - \frac{y+4}{3} + \frac{z}{2} &= 0 \\ \frac{x+1}{2} + \frac{y-1}{2} - \frac{z}{4} &= \frac{9}{2} \\ \frac{x-5}{4} + \frac{y+1}{3} + \frac{z-2}{2} &= \frac{19}{4} \end{aligned}$$

30. $$\begin{aligned} \frac{x+3}{2} - \frac{y-1}{2} + \frac{z+2}{4} &= \frac{3}{2} \\ \frac{x-5}{2} + \frac{y+1}{3} - \frac{z}{4} &= -\frac{25}{6} \\ \frac{x-3}{4} - \frac{y+1}{2} + \frac{z-3}{2} &= -\frac{5}{2} \end{aligned}$$

In Exercises 31–32, find the equation of the quadratic function $y = ax^2 + bx + c$ whose graph is shown. Select three points whose coordinates appear to be integers.

31.

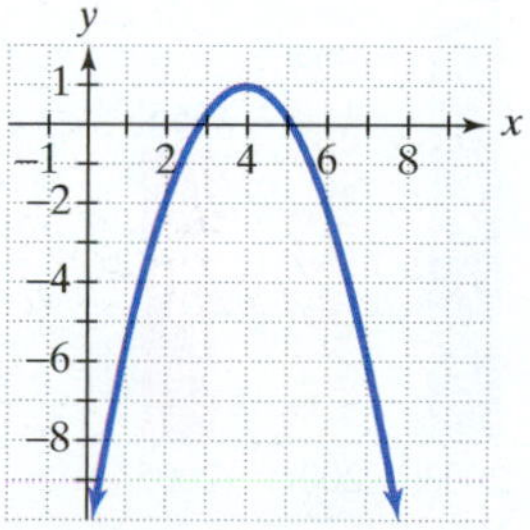

32.

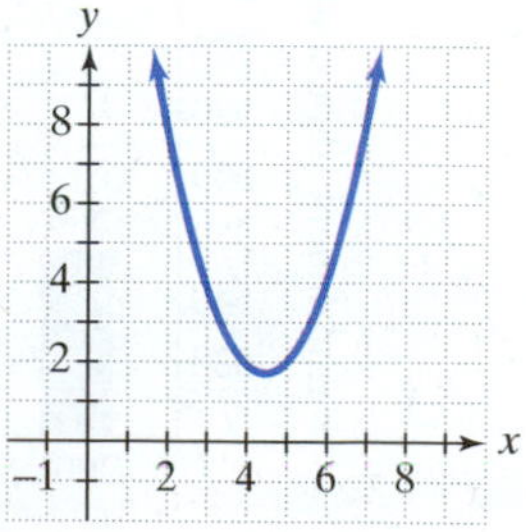

In Exercises 33–34, solve each system for (x, y, z) in terms of the nonzero constants a, b, and c.

33. $ax - by - 2cz = 21$
$ax + by + cz = 0$
$2ax - by + cz = 14$

34. $ax - by + 2cz = -4$
$ax + 3by - cz = 1$
$2ax + by + 3cz = 2$

Application Exercises

35. U.S. spending on energy efficiency fell with power industry deregulation in the mid-1990s, but has nearly doubled since 1999. The bar graph shows the nationwide annual spending, in billions of dollars, on energy-efficiency programs.

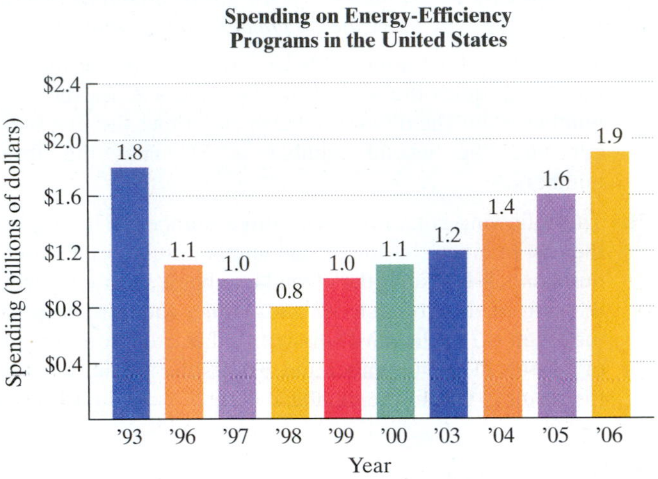

Source: American Council for an Energy-Efficient Economy

a. Write the data for 1993, 1998, and 2006 as ordered pairs (x, y), where x is the number of years after 1993 and y is the spending on energy-efficiency programs, in billions of dollars, in that year.

b. The three data points in part (a) can be modeled by the quadratic function $y = ax^2 + bx + c$, where $a > 0$. Substitute each ordered pair into this function, one ordered pair at a time, and write a system of linear equations in three variables that can be used to find values for a, b, and c. It is not necessary to solve the system.

36. The bar graph shows foreign student enrollment, in thousands, in U.S. colleges and universities.

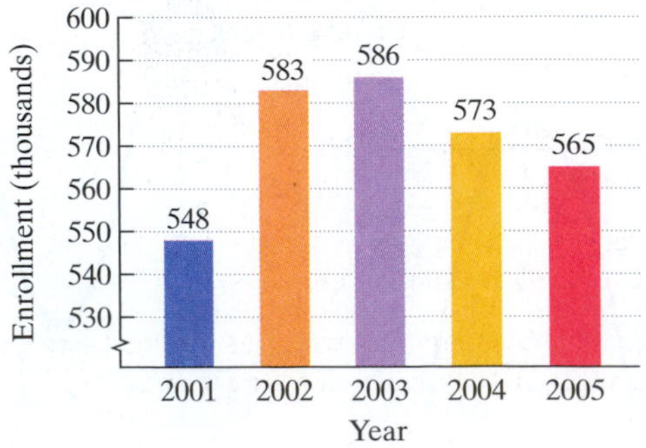

Source: Institute of International Education

a. Write the data for 2001, 2003, and 2005 as ordered pairs (x, y), where x is the number of years after 2001 and y is that year's foreign student enrollment, in thousands.

b. The three data points in part (a) can be modeled by the quadratic function $y = ax^2 + bx + c$, where $a < 0$. Substitute each ordered pair into this function, one ordered pair at a time, and write a system of linear equations in three variables that can be used to find values for a, b, and c. It is not necessary to solve the system.

37. You throw a ball straight up from a rooftop. The ball misses the rooftop on its way down and eventually strikes the ground. A mathematical model can be used to describe the ball's height above the ground, y, after x seconds. Consider the following data.

x, seconds after the ball is thrown	y, ball's height, in feet, above the ground
1	224
3	176
4	104

a. Find the quadratic function $y = ax^2 + bx + c$ whose graph passes through the given points.

b. Use the function in part (a) to find the value for y when $x = 5$. Describe what this means.

38. A mathematical model can be used to describe the relationship between the number of feet a car travels once the brakes are applied, y, and the number of seconds the car is in motion after the brakes are applied, x. A research firm collects the data shown below.

x, seconds in motion after brakes are applied	y, feet car travels once the brakes are applied
1	46
2	84
3	114

a. Find the quadratic function $y = ax^2 + bx + c$ whose graph passes through the given points.

b. Use the function in part (a) to find the value for y when $x = 6$. Describe what this means.

In Exercises 39–46, use the four-step strategy to solve each problem. Use x, y, and z to represent unknown quantities. Then translate from the verbal conditions of the problem to a system of three equations in three variables.

39. The bar graph at the top of the next column shows the average annual U.S. household spending on selected items. The combined spending on rent, cars, and books is \$12,691. The difference between spending on rent and spending on cars is \$6204. The difference between spending on cars and spending on books is $2573. Find the average annual household spending on rent, cars, and books.

Average Annual United States Household Spending on Selected Items

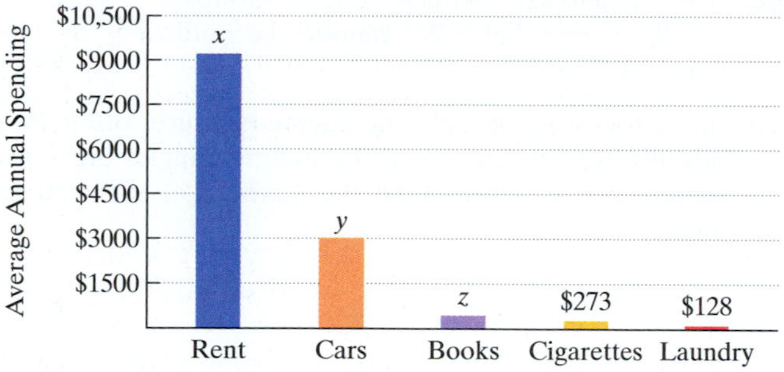

Source: Ori Heffetz, *Consumption and the Visibility of Consumer Expenditures*, Princeton University

40. The bar graph indicates that George Washington, Franklin Roosevelt, and William Howard Taft were the three U.S. presidents with the most Supreme Court appointments. Combined, they appointed 26 judges to the Supreme Court. The difference between the number of appointments by Washington and Roosevelt was 2. The difference between the number of appointments by Roosevelt and Taft was 3. Find the number of judges appointed to the Supreme Court by Washington, Roosevelt, and Taft.

United States Presidents with the Most Supreme Court Appointments

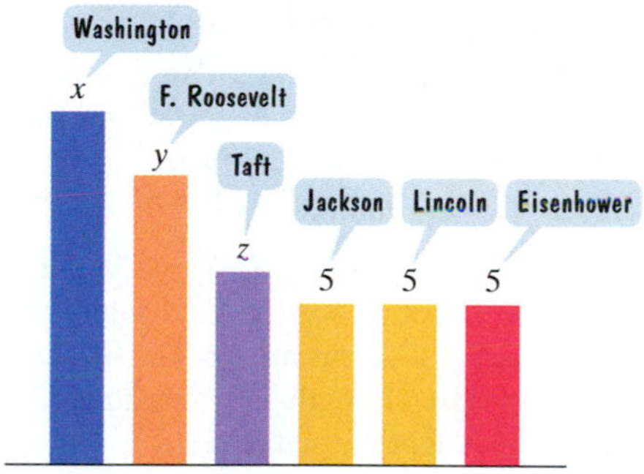

Source: The Associated Press

41. A person invested $6700 for one year, part at 8%, part at 10%, and the remainder at 12%. The total annual income from these investments was $716. The amount of money invested at 12% was $300 more than the amounts invested at 8% and 10% combined. Find the amount invested at each rate.

42. A person invested $17,000 for one year, part at 10%, part at 12%, and the remainder at 15%. The total annual income from these investments was $2110. The amount of money invested at 12% was $1000 less than the amounts invested at 10% and 15% combined. Find the amount invested at each rate.

43. At a college production of *Streetcar Named Desire*, 400 tickets were sold. The ticket prices were $8, $10, and $12, and the total income from ticket sales was $3700. How many tickets of each type were sold if the combined number of $8 and $10 tickets sold was 7 times the number of $12 tickets sold?

44. A certain brand of razor blades comes in packages of 6, 12, and 24 blades, costing $2, $3, and $4 per package, respectively. A store sold 12 packages containing a total of 162 razor blades and took in $35. How many packages of each type were sold?

45. Three foods have the following nutritional content per ounce.

	Calories	Protein (in grams)	Vitamin C (in milligrams)
Food A	40	5	30
Food B	200	2	10
Food C	400	4	300

If a meal consisting of the three foods allows exactly 660 calories, 25 grams of protein, and 425 milligrams of vitamin C, how many ounces of each kind of food should be used?

46. A furniture company produces three types of desks: a children's model, an office model, and a deluxe model. Each desk is manufactured in three stages: cutting, construction, and finishing. The time requirements for each model and manufacturing stage are given in the following table.

	Children's model	Office model	Deluxe model
Cutting	2 hr	3 hr	2 hr
Construction	2 hr	1 hr	3 hr
Finishing	1 hr	1 hr	2 hr

Each week the company has available a maximum of 100 hours for cutting, 100 hours for construction, and 65 hours for finishing. If all available time must be used, how many of each type of desk should be produced each week?

Writing in Mathematics

47. What is a system of linear equations in three variables?

48. How do you determine whether a given ordered triple is a solution of a system of linear equations in three variables?

49. Describe in general terms how to solve a system in three variables.

50. Describe what happens when using algebraic methods to solve an inconsistent system.

51. Describe what happens when using algebraic methods to solve a system with dependent equations.

52. AIDS is taking a deadly toll on southern Africa. Describe how to use the techniques that you learned in this section to obtain a model for African life span using projections with AIDS. Let x represent the number of years after 1985 and let y represent African life span in that year.

African Life Span

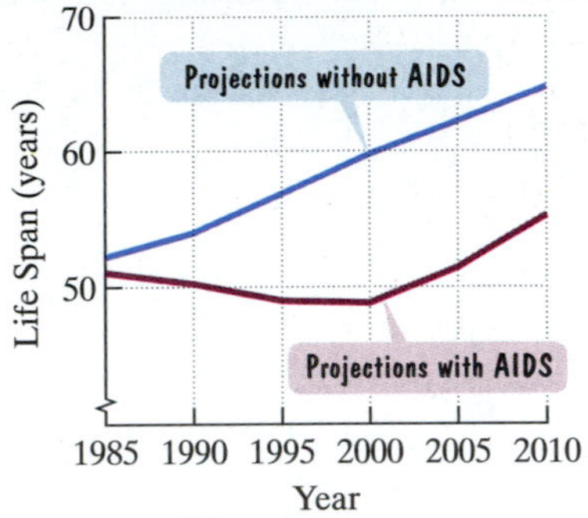

Source: United Nations

Technology Exercises

53. Does your graphing utility have a feature that allows you to solve linear systems by entering coefficients and constant terms? If so, use this feature to verify the solutions to any five exercises that you worked by hand from Exercises 5–16.

54. Verify your results in Exercises 23–26 by using a graphing utility to graph the quadratic function. Trace along the curve and convince yourself that the three points given in the exercise lie on the function's graph.

Critical Thinking Exercises

Make Sense? *In Exercises 55–58, determine whether each statement "makes sense" or "does not make sense" and explain your reasoning.*

55. Solving a system in three variables, I found that $x = 3$ and $y = -1$. Because z represents a third variable, z cannot equal 3 or -1.

56. A system of linear equations in three variables, x, y, and z, cannot contain an equation in the form $y = mx + b$.

57. I'm solving a three-variable system in which one of the given equations has a missing term, so it will not be necessary to use any of the original equations twice when I reduce the system to two equations in two variables.

58. Because the percentage of the U.S. population that was foreign-born decreased from 1910 through 1970 and then increased after that, a quadratic function of the form $f(x) = ax^2 + bx + c$, rather than a linear function of the form $f(x) = mx + b$, should be used to model the data.

In Exercises 59–62, determine whether each statement is true or false. If the statement is false, make the necessary change(s) to produce a true statement.

59. The ordered triple (2, 15, 14) is the only solution of the equation $x + y - z = 3$.

60. The equation $x - y - z = -6$ is satisfied by $(2, -3, 5)$.

61. If two equations in a system are $x + y - z = 5$ and $x + y - z = 6$, then the system must be inconsistent.

62. An equation with four variables, such as $x + 2y - 3z + 5w = 2$, cannot be satisfied by real numbers.

63. In the following triangle, the degree measures of the three interior angles and two of the exterior angles are represented with variables. Find the measure of each interior angle.

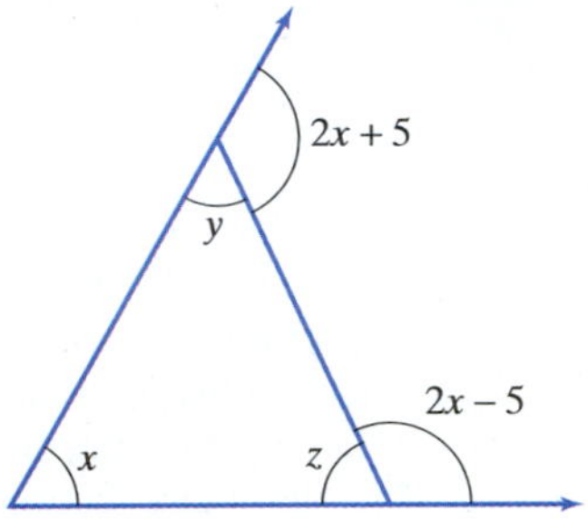

64. A modernistic painting consists of triangles, rectangles, and pentagons, all drawn so as to not overlap or share sides. Within each rectangle are drawn 2 red roses, and each pentagon contains 5 carnations. How many triangles, rectangles, and pentagons appear in the painting if the painting contains a total of 40 geometric figures, 153 sides of geometric figures, and 72 flowers?

65. Two blocks of wood having the same length and width are placed on the top and bottom of a table, as shown in (a). Length A measures 32 centimeters. The blocks are rearranged as shown in (b). Length B measures 28 centimeters. Determine the height of the table.

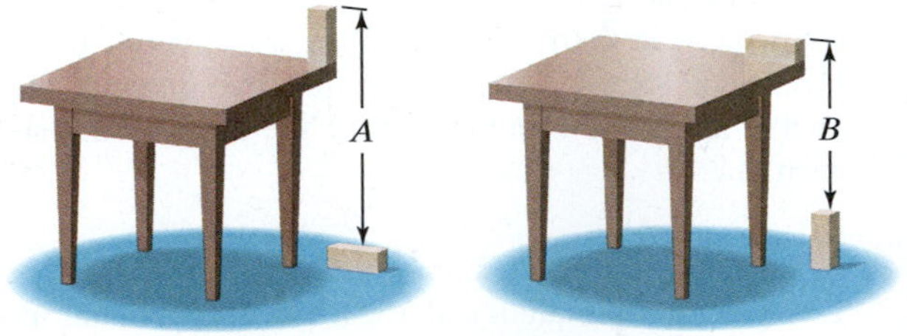

(a) (b)

Review Exercises

In Exercises 66–68, graph each linear function.

66. $f(x) = -\frac{3}{4}x + 3$ (Section 2.4, Example 5)

67. $-2x + y = 6$ (Section 2.4, Example 1)

68. $f(x) = -5$ (Section 2.4, Example 6)

Preview Exercises

Exercises 69–71 will help you prepare for the material covered in the next section.

69. Solve the system:

$$\begin{aligned} x + 2y &= -1 \\ y &= 1. \end{aligned}$$

What makes it fairly easy to find the solution?

70. Solve the system:

$$\begin{aligned} x + y + 2z &= 19 \\ y + 2z &= 13 \\ z &= 5. \end{aligned}$$

What makes it fairly easy to find the solution?

71. Consider the following array of numbers:

$$\begin{bmatrix} 1 & 2 & -1 \\ 4 & -3 & -15 \end{bmatrix}.$$

Rewrite the array as follows: Multiply each number in the top row by -4 and add this product to the corresponding number in the bottom row. Do not change the numbers in the top row.

MID-CHAPTER CHECK POINT Section 3.1–Section 3.3

What You Know: We learned to solve systems of linear equations. We solved systems in two variables by graphing, by the substitution method, and by the addition method. We solved systems in three variables by eliminating a variable, reducing the system to two equations in two variables. We saw that some systems, called inconsistent systems, have no solution, whereas other systems, called dependent systems, have infinitely many solutions. We used systems of linear equations to solve a variety of applied problems, including dual investment problems, mixture problems, motion problems, and business problems.

In Exercises 1–8, solve each system by the method of your choice.

1. $x = 3y - 7$
$4x + 3y = 2$

2. $3x + 4y = -5$
$2x - 3y = 8$

3. $\dfrac{2x}{3} + \dfrac{y}{5} = 6$
$\dfrac{x}{6} - \dfrac{y}{2} = -4$

4. $y = 4x - 5$
$8x - 2y = 10$

5. $2x + 5y = 3$
$3x - 2y = 1$

6. $\dfrac{x}{12} - y = \dfrac{1}{4}$
$4x - 48y = 16$

7. $2x - y + 2z = -8$
$x + 2y - 3z = 9$
$3x - y - 4z = 3$

8. $x - 3z = -5$
$2x - y + 2z = 16$
$7x - 3y - 5z = 19$

In Exercises 9–10, solve each system by graphing.

9. $2x - y = 4$
$x + y = 5$

10. $y = x + 3$
$y = -\dfrac{1}{2}x$

11. A company is planning to manufacture PDAs (personal digital assistants). The fixed cost will be $400,000 and it will cost $20 to produce each PDA. Each PDA will be sold for $100.

a. Write the cost function, C, of producing x PDAs.

b. Write the revenue function, R, from the sale of x PDAs.

c. Write the profit function, P, from producing and selling x PDAs.

d. Determine the break-even point. Describe what this means.

In Exercises 12–18, solve each problem.

12. Roses sell for $3 each and carnations for $1.50 each. If a mixed bouquet of 20 flowers consisting of roses and carnations costs $39, how many of each type of flower is in the bouquet?

13. You invested $15,000 in two funds paying 5% and 6% annual interest. At the end of the year, the total interest from these investments was $837. How much was invested at each rate?

14. The manager of a gardening center needs to mix a plant food that is 13% nitrogen with one that is 18% nitrogen to obtain 50 gallons of a plant food that is 16% nitrogen. How many gallons of each of the plant foods must be used?

15. With the current, you can row 9 miles in 2 hours. Against the current, your return trip takes 6 hours. Find your average rowing rate in still water and the average rate of the current.

16. You invested $8000 in two funds paying 2% and 5% annual interest. At the end of the year, the interest from the 5% investment exceeded the interest from the 2% investment by $85. How much money was invested at each rate?

17. Find the quadratic function $y = ax^2 + bx + c$ whose graph passes through the points $(-1, 0)$, $(1, 4)$, and $(2, 3)$.

18. A coin collection contains a mixture of 26 coins in nickels, dimes, and quarters. The coins have a total value of \$4.00. The number of quarters is 2 less than the number of nickels and dimes combined. Determine the number of nickels, the number of dimes, and the number of quarters in the collection.

SECTION 3.4

Matrix Solutions to Linear Systems

Objectives

1. Write the augmented matrix for a linear system.
2. Perform matrix row operations.
3. Use matrices to solve linear systems in two variables.
4. Use matrices to solve linear systems in three variables.
5. Use matrices to identify inconsistent and dependent systems.

The data below show that we spend a lot of time sprucing up.

Average Number of Minutes per Day Americans Spend on Grooming

	Ages 15–19	Ages 20–24	Ages 45–54	Ages 65+	Married	Single
Men	37	37	34	28	31	34
Women	59	49	46	46	44	50

Source: Bureau of Labor Statistics' American Time-Use Survey

The 12 numbers inside the brackets are arranged in two rows and six columns. This rectangular array of 12 numbers, arranged in rows and columns and placed in brackets, is an example of a **matrix** (plural: **matrices**). The numbers inside the brackets are called **elements** of the matrix. Matrices are used to display information and to solve systems of linear equations.

1 Write the augmented matrix for a linear system.

Augmented Matrices

A matrix gives us a shortened way of writing a system of equations. The first step in solving a system of linear equations using matrices is to write the *augmented matrix*. An **augmented matrix** has a vertical bar separating the columns of the matrix into two groups. The coefficients of each variable are placed to the left of the vertical line and the constants are placed to the right. If any variable is missing, its coefficient is 0. On the next page are two examples.

System of Linear Equations	Augmented Matrix
$\begin{aligned} x + 3y &= 5 \\ 2x - y &= -4 \end{aligned}$	$\left[\begin{array}{cc\|c} 1 & 3 & 5 \\ 2 & -1 & -4 \end{array}\right]$
$\begin{aligned} 3x + 4y &= 19 \\ 2y + 3z &= 8 \\ 4x - 5z &= 7 \end{aligned}$	$\left[\begin{array}{ccc\|c} 3 & 4 & 0 & 19 \\ 0 & 2 & 3 & 8 \\ 4 & 0 & -5 & 7 \end{array}\right]$.

Our goal in solving a linear system using matrices is to produce a matrix with 1s down the diagonal from upper left to lower right on the left side of the vertical bar, called the **main diagonal**, and 0s below the 1s. In general, the matrix will be one of the following forms.

This is the desired form for systems with two equations. $\left[\begin{array}{cc|c} 1 & a & b \\ 0 & 1 & c \end{array}\right]$

$\left[\begin{array}{ccc|c} 1 & a & b & c \\ 0 & 1 & d & e \\ 0 & 0 & 1 & f \end{array}\right]$ This is the desired form for systems with three equations.

The last row of these matrices gives us the value of one variable. The values of the other variables can then be found by back-substitution.

2 Perform matrix row operations.

Matrix Row Operations

A matrix with 1s down the main diagonal and 0s below the 1s is said to be in **row-echelon form**. How do we produce a matrix in this form? We use **row operations** on the augmented matrix. These row operations are just like what you did when solving a linear system by the addition method. The difference is that we no longer write the variables, usually represented by x, y, and z.

Matrix Row Operations

The following row operations produce matrices that represent systems with the same solution set:

1. Two rows of a matrix may be interchanged. This is the same as interchanging two equations in a linear system.
2. The elements in any row may be multiplied by a nonzero number. This is the same as multiplying both sides of an equation by a nonzero number.
3. The elements in any row may be multiplied by a nonzero number, and these products may be added to the corresponding elements in any other row. This is the same as multiplying an equation by a nonzero number and then adding equations to eliminate a variable.

Two matrices are **row equivalent** if one can be obtained from the other by a sequence of row operations.

Study Tip

When performing the row operation

$$kR_i + R_j$$

you use row i to find the products. However, **elements in row i do not change. It is the elements in row j that change:** Add k times the elements in row i to the corresponding elements in row j. Replace elements in row j by these sums.

Each matrix row operation in the preceding box can be expressed symbolically as follows:

1. Interchange the elements in the ith and jth rows: $R_i \leftrightarrow R_j$.
2. Multiply each element in the ith row by k: kR_i.
3. Add k times the elements in row i to the corresponding elements in row j: $kR_i + R_j$.

EXAMPLE 1 Performing Matrix Row Operations

Use the matrix

$$\left[\begin{array}{rrr|r} 3 & 18 & -12 & 21 \\ 1 & 2 & -3 & 5 \\ -2 & -3 & 4 & -6 \end{array}\right]$$

and perform each indicated row operation:

a. $R_1 \leftrightarrow R_2$ **b.** $\frac{1}{3}R_1$ **c.** $2R_2 + R_3$.

Solution

a. The notation $R_1 \leftrightarrow R_2$ means to interchange the elements in row 1 and row 2. This results in the row-equivalent matrix

$$\left[\begin{array}{rrr|r} 1 & 2 & -3 & 5 \\ 3 & 18 & -12 & 21 \\ -2 & -3 & 4 & -6 \end{array}\right].$$

This was row 2; now it's row 1.

This was row 1; now it's row 2.

b. The notation $\frac{1}{3}R_1$ means to multiply each element in row 1 by $\frac{1}{3}$. This results in the row-equivalent matrix

$$\left[\begin{array}{rrr|r} \frac{1}{3}(3) & \frac{1}{3}(18) & \frac{1}{3}(-12) & \frac{1}{3}(21) \\ 1 & 2 & -3 & 5 \\ -2 & -3 & 4 & -6 \end{array}\right] = \left[\begin{array}{rrr|r} 1 & 6 & -4 & 7 \\ 1 & 2 & -3 & 5 \\ -2 & -3 & 4 & -6 \end{array}\right].$$

c. The notation $2R_2 + R_3$ means to add 2 times the elements in row 2 to the corresponding elements in row 3. Replace the elements in row 3 by these sums. First, we find 2 times the elements in row 2, namely, 1, 2, -3 and 5:

$$2(1) \text{ or } 2, \quad 2(2) \text{ or } 4, \quad 2(-3) \text{ or } -6, \quad 2(5) \text{ or } 10.$$

Now we add these products to the corresponding elements in row 3. Although we use row 2 to find the products, row 2 does not change. It is the elements in row 3 that change, resulting in the row-equivalent matrix

Replace row 3 by the sum of itself and 2 times row 2.

$$\left[\begin{array}{rrr|r} 3 & 18 & -12 & 21 \\ 1 & 2 & -3 & 5 \\ -2+2=0 & -3+4=1 & 4+(-6)=-2 & -6+10=4 \end{array}\right] = \left[\begin{array}{rrr|r} 3 & 18 & -12 & 21 \\ 1 & 2 & -3 & 5 \\ 0 & 1 & -2 & 4 \end{array}\right].$$

✓ CHECK POINT 1 Use the matrix

$$\left[\begin{array}{rrr|r} 4 & 12 & -20 & 8 \\ 1 & 6 & -3 & 7 \\ -3 & -2 & 1 & -9 \end{array}\right]$$

and perform each indicated row operation:

a. $R_1 \leftrightarrow R_2$

b. $\frac{1}{4}R_1$

c. $3R_2 + R_3$.

3 Use matrices to solve linear systems in two variables.

Solving Linear Systems in Two Variables Using Matrices

The process that we use to solve linear systems using matrix row operations is often called **Gaussian elimination**, after the German mathematician Carl Friedrich Gauss (1777–1855). Here are the steps used in solving linear systems in two variables with matrices:

Solving Linear Systems in Two Variables Using Matrices

1. Write the augmented matrix for the system.
2. Use matrix row operations to simplify the matrix to a row-equivalent matrix in row-echelon form, with 1s down the main diagonal from upper left to lower right, and a 0 below the 1 in the first column.

$$\left[\begin{array}{cc|c} 1 & * & * \\ * & * & * \end{array}\right] \rightarrow \left[\begin{array}{cc|c} 1 & * & * \\ 0 & * & * \end{array}\right] \rightarrow \left[\begin{array}{cc|c} 1 & * & * \\ 0 & 1 & * \end{array}\right]$$

Get 1 in the upper left-hand corner.

Use the 1 in the first column to get 0 below it.

Get 1 in the second row, second column position.

3. Write the system of linear equations corresponding to the matrix from step 2 and use back-substitution to find the system's solution.

EXAMPLE 2 Using Matrices to Solve a Linear System

Use matrices to solve the system:

$$\begin{aligned} 4x - 3y &= -15 \\ x + 2y &= -1. \end{aligned}$$

Solution

Step 1. Write the augmented matrix for the system.

Linear System	Augmented Matrix
$\begin{aligned} 4x - 3y &= -15 \\ x + 2y &= -1 \end{aligned}$	$\left[\begin{array}{cc|c} 4 & -3 & -15 \\ 1 & 2 & -1 \end{array}\right]$

Step 2. Use matrix row operations to simplify the matrix to row-echelon form, with 1s down the main diagonal from upper left to lower right, and a 0 below the 1 in the first column. Our first step in achieving this goal is to get 1 in the top position of the first column.

We want 1 in this position.

$$\left[\begin{array}{cc|c} 4 & -3 & -15 \\ 1 & 2 & -1 \end{array}\right]$$

To get 1 in this position, we interchange row 1 and row 2: $R_1 \leftrightarrow R_2$.

$$\left[\begin{array}{cc|c} 1 & 2 & -1 \\ 4 & -3 & -15 \end{array}\right]$$

This was row 2; now it's row 1.

This was row 1; now it's row 2.

Now we want a 0 below the 1 in the first column.

We want 0 in this position.

$$\left[\begin{array}{cc|c} 1 & 2 & -1 \\ 4 & -3 & -15 \end{array}\right]$$

We want 0 in this position.
$$\left[\begin{array}{cc|c} 1 & 2 & -1 \\ 4 & -3 & -15 \end{array}\right]$$

Let's get a 0 where there is now a 4. If we multiply the top row of numbers by -4 and add these products to the second row of numbers, we will get 0 in this position: $-4R_1 + R_2$. *We change only row 2.*

Replace row 2 by $-4R_1 + R_2$.
$$\left[\begin{array}{cc|c} 1 & 2 & -1 \\ -4(1)+4 & -4(2)+(-3) & -4(-1)+(-15) \end{array}\right] = \left[\begin{array}{cc|c} 1 & 2 & -1 \\ 0 & -11 & -11 \end{array}\right]$$

We move on to the second column. We want 1 in the second row, second column.

We want 1 in this position.
$$\left[\begin{array}{cc|c} 1 & 2 & -1 \\ 0 & -11 & -11 \end{array}\right]$$

To get 1 in the desired position, we multiply -11 by its multiplicative inverse, or reciprocal, $-\frac{1}{11}$. Therefore, we multiply all the numbers in the second row by $-\frac{1}{11}$: $-\frac{1}{11}R_2$.

$-\frac{1}{11}R_2$
$$\left[\begin{array}{cc|c} 1 & 2 & -1 \\ -\frac{1}{11}(0) & -\frac{1}{11}(-11) & -\frac{1}{11}(-11) \end{array}\right] = \left[\begin{array}{cc|c} 1 & 2 & -1 \\ 0 & 1 & 1 \end{array}\right]$$

We now have the desired matrix in row-echelon form, with 1s down the main diagonal and a 0 below the 1 in the first column.

Step 3. Write the system of linear equations corresponding to the matrix from step 2 and use back-substitution to find the system's solution. The system represented by the matrix from step 2 is

$$\left[\begin{array}{cc|c} 1 & 2 & -1 \\ 0 & 1 & 1 \end{array}\right] \rightarrow \begin{aligned} 1x + 2y &= -1 \\ 0x + 1y &= 1 \end{aligned} \quad \text{or} \quad \begin{aligned} x + 2y &= -1 \quad (1) \\ y &= 1. \quad (2) \end{aligned}$$

We immediately see from Equation (2) that the value for y is 1. To find x, we back-substitute 1 for y in Equation (1).

$$\begin{aligned} x + 2y &= -1 && \text{Equation (1)} \\ x + 2\cdot 1 &= -1 && \text{Substitute 1 for } y. \\ x + 2 &= -1 && \text{Multiply.} \\ x &= -3 && \text{Subtract 2 from both sides.} \end{aligned}$$

With $x = -3$ and $y = 1$, the proposed solution is $(-3, 1)$. Take a moment to show that $(-3, 1)$ satisfies both equations. The solution is $(-3, 1)$ and the solution set is $\{(-3, 1)\}$.

✓ CHECK POINT 2 Use matrices to solve the system:

$$\begin{aligned} 2x - y &= -4 \\ x + 3y &= 5. \end{aligned}$$

4 Use matrices to solve linear systems in three variables.

Solving Linear Systems in Three Variables Using Matrices

Gaussian elimination is also used to solve a system of linear equations in three variables. Most of the work involves using matrix row operations to obtain a matrix with 1s down the main diagonal and 0s below the 1s in the first and second columns.

Solving Linear Systems in Three Variables Using Matrices

1. Write the augmented matrix for the system.
2. Use matrix row operations to simplify the matrix to a row-equivalent matrix in row-echelon form, with 1s down the main diagonal from upper left to lower right, and 0s below the 1s in the first and second columns.

$$\left[\begin{array}{ccc|c} 1 & * & * & * \\ * & * & * & * \\ * & * & * & * \end{array}\right] \rightarrow \left[\begin{array}{ccc|c} 1 & * & * & * \\ 0 & * & * & * \\ 0 & * & * & * \end{array}\right] \rightarrow \left[\begin{array}{ccc|c} 1 & * & * & * \\ 0 & 1 & * & * \\ 0 & * & * & * \end{array}\right] \rightarrow \left[\begin{array}{ccc|c} 1 & * & * & * \\ 0 & 1 & * & * \\ 0 & 0 & * & * \end{array}\right] \rightarrow \left[\begin{array}{ccc|c} 1 & * & * & * \\ 0 & 1 & * & * \\ 0 & 0 & 1 & * \end{array}\right]$$

Get 1 in the upper left-hand corner.

Use the 1 in the first column to get 0s below it.

Get 1 in the second row, second column position.

Use the 1 in the second column to get 0 below it.

Get 1 in the third row, third column position.

3. Write the system of linear equations corresponding to the matrix from step 2 and use back-substitution to find the system's solution.

EXAMPLE 3 Using Matrices to Solve a Linear System

Use matrices to solve the system:

$$\begin{aligned} 3x + y + 2z &= 31 \\ x + y + 2z &= 19 \\ x + 3y + 2z &= 25. \end{aligned}$$

Solution

Step 1. Write the augmented matrix for the system.

Linear System	Augmented Matrix
$$\begin{aligned} 3x + y + 2z &= 31 \\ x + y + 2z &= 19 \\ x + 3y + 2z &= 25 \end{aligned}$$	$$\left[\begin{array}{ccc\|c} 3 & 1 & 2 & 31 \\ 1 & 1 & 2 & 19 \\ 1 & 3 & 2 & 25 \end{array}\right]$$

Step 2. Use matrix row operations to simplify the matrix to row-echelon form, with 1s down the main diagonal from upper left to lower right, and 0s below the 1s in the first and second columns. Our first step in achieving this goal is to get 1 in the top position of the first column.

We want 1 in this position.

$$\left[\begin{array}{ccc|c} 3 & 1 & 2 & 31 \\ 1 & 1 & 2 & 19 \\ 1 & 3 & 2 & 25 \end{array}\right]$$

To get 1 in this position, we interchange row 1 and row 2: $R_1 \leftrightarrow R_2$. (We could also interchange row 1 and row 3 to attain our goal.)

$$\left[\begin{array}{ccc|c} 1 & 1 & 2 & 19 \\ 3 & 1 & 2 & 31 \\ 1 & 3 & 2 & 25 \end{array}\right]$$

This was row 2; now it's row 1.

This was row 1; now it's row 2.

Now we want to get 0s below the 1 in the first column.

We want 0 in these positions.

$$\left[\begin{array}{ccc|c} 1 & 1 & 2 & 19 \\ 3 & 1 & 2 & 31 \\ 1 & 3 & 2 & 25 \end{array}\right]$$

We want 0 in these positions.

$$\left[\begin{array}{ccc|c} 1 & 1 & 2 & 19 \\ 3 & 1 & 2 & 31 \\ 1 & 3 & 2 & 25 \end{array}\right]$$

To get a 0 where there is now a 3, multiply the top row of numbers by -3 and add these products to the second row of numbers: $-3R_1 + R_2$. To get a 0 in the bottom of the first column where there is now a 1, multiply the top row of numbers by -1 and add these products to the third row of numbers: $-1R_1 + R_3$. Although we are using row 1 to find the products, the numbers in row 1 do not change.

Replace row 2 by $-3R_1 + R_2$.

Replace row 3 by $-1R_1 + R_3$.

$$\left[\begin{array}{ccc|c} 1 & 1 & 2 & 19 \\ -3(1)+3 & -3(1)+1 & -3(2)+2 & -3(19)+31 \\ -1(1)+1 & -1(1)+3 & -1(2)+2 & -1(19)+25 \end{array}\right] = \left[\begin{array}{ccc|c} 1 & 1 & 2 & 19 \\ 0 & -2 & -4 & -26 \\ 0 & 2 & 0 & 6 \end{array}\right]$$

We want 1 in this position.

We move on to the second column. To get 1 in the desired position, we multiply -2 by its reciprocal, $-\frac{1}{2}$. Therefore, we multiply all the numbers in the second row by $-\frac{1}{2}$: $-\frac{1}{2}R_2$.

$-\frac{1}{2}R_2$

$$\left[\begin{array}{ccc|c} 1 & 1 & 2 & 19 \\ -\frac{1}{2}(0) & -\frac{1}{2}(-2) & -\frac{1}{2}(-4) & -\frac{1}{2}(-26) \\ 0 & 2 & 0 & 6 \end{array}\right] = \left[\begin{array}{ccc|c} 1 & 1 & 2 & 19 \\ 0 & 1 & 2 & 13 \\ 0 & 2 & 0 & 6 \end{array}\right].$$

We want 0 in this position.

We are not yet done with the second column. The voice balloon shows that we want to get a 0 where there is now a 2. If we multiply the second row of numbers by -2 and add these products to the third row of numbers, we will get 0 in this position: $-2R_2 + R_3$. Although we are using the numbers in row 2 to find the products, the numbers in row 2 do not change.

Replace row 3 by $-2R_2 + R_3$.

$$\left[\begin{array}{ccc|c} 1 & 1 & 2 & 19 \\ 0 & 1 & 2 & 13 \\ -2(0)+0 & -2(1)+2 & -2(2)+0 & -2(13)+6 \end{array}\right] = \left[\begin{array}{ccc|c} 1 & 1 & 2 & 19 \\ 0 & 1 & 2 & 13 \\ 0 & 0 & -4 & -20 \end{array}\right]$$

We want 1 in this position.

We move on to the third column. To get 1 in the desired position, we multiply -4 by its reciprocal, $-\frac{1}{4}$. Therefore, we multiply all the numbers in the third row by $-\frac{1}{4}$: $-\frac{1}{4}R_3$.

$-\frac{1}{4}R_3$

$$\left[\begin{array}{ccc|c} 1 & 1 & 2 & 19 \\ 0 & 1 & 2 & 13 \\ -\frac{1}{4}(0) & -\frac{1}{4}(0) & -\frac{1}{4}(-4) & -\frac{1}{4}(-20) \end{array}\right] = \left[\begin{array}{ccc|c} 1 & 1 & 2 & 19 \\ 0 & 1 & 2 & 13 \\ 0 & 0 & 1 & 5 \end{array}\right].$$

We now have the desired matrix in row-echelon form, with 1s down the main diagonal and 0s below the 1s in the first and second columns.

Step 3. Write the system of linear equations corresponding to the matrix from step 2 and use back-substitution to find the system's solution. The system represented by the matrix from step 2 is

$$\left[\begin{array}{ccc|c} 1 & 1 & 2 & 19 \\ 0 & 1 & 2 & 13 \\ 0 & 0 & 1 & 5 \end{array}\right] \rightarrow \begin{aligned} 1x + 1y + 2z &= 19 \\ 0x + 1y + 2z &= 13 \\ 0x + 0y + 1z &= 5 \end{aligned} \quad \text{or} \quad \begin{aligned} x + y + 2z &= 19 \quad (1) \\ y + 2z &= 13. \quad (2) \\ z &= 5 \quad (3) \end{aligned}$$

Using Technology

Most graphing utilities can convert an augmented matrix to row-echelon form, with 1s down the main diagonal and 0s below the 1s. However, row-echelon form is not unique. Your graphing utility might give a row-echelon form different from the one you obtained by hand. However, all row-echelon forms for a given system's augmented matrix produce the same solution to the system. Enter the augmented matrix and name it A. Then use the REF (row-echelon form) command on matrix A.

We immediately see from equation (3) that the value for z is 5. To find y, we back-substitute 5 for z in the second equation.

$$y + 2z = 13 \quad \text{Equation (2)}$$
$$y + 2(5) = 13 \quad \text{Substitute 5 for } z.$$
$$y = 3 \quad \text{Solve for } y.$$

Finally, back-substitute 3 for y and 5 for z in the first equation.

$$x + y + 2z = 19 \quad \text{Equation (1)}$$
$$x + 3 + 2(5) = 19 \quad \text{Substitute 3 for } y \text{ and 5 for } z.$$
$$x + 13 = 19 \quad \text{Multiply and add.}$$
$$x = 6 \quad \text{Subtract 13 from both sides.}$$

The solution of the original system is $(6, 3, 5)$ and the solution set is $\{(6, 3, 5)\}$. Check to see that the solution satisfies all three equations in the given system. ■

✓ CHECK POINT 3 Use matrices to solve the system:

$$\begin{aligned} 2x + y + 2z &= 18 \\ x - y + 2z &= 9 \\ x + 2y - z &= 6. \end{aligned}$$

Modern supercomputers are capable of solving systems with more than 600,000 variables. The augmented matrices for such systems are huge, but the solution using matrices is exactly like what we did in Example 3. Work with the augmented matrix, one column at a time. First, get 1s down the main diagonal from upper left to lower right. Then get 0s below the 1s.

5 Use matrices to identify inconsistent and dependent systems.

Inconsistent Systems and Systems with Dependent Equations

When solving a system using matrices, you might obtain a matrix with a row in which the numbers to the left of the vertical bar are all zeros, but a nonzero number appears on the right. In such a case, the system is inconsistent and has no solution. For example, a system of equations that yields the following matrix is an inconsistent system:

$$\left[\begin{array}{cc|c} 1 & -2 & 3 \\ 0 & 0 & -4 \end{array}\right].$$

The second row of the matrix represents the equation $0x + 0y = -4$, which is false for all values of x and y.

If you obtain a matrix in which a 0 appears across an entire row, the system contains dependent equations and has infinitely many solutions. This row of zeros represents $0x + 0y = 0$ or $0x + 0y + 0z = 0$. These equations are satisfied by infinitely many ordered pairs or triples.

3.4 EXERCISE SET

MyMathLab | MathXL PRACTICE | WATCH | DOWNLOAD | READ | REVIEW

Practice Exercises

In Exercises 1–14, perform each matrix row operation and write the new matrix.

1. $\left[\begin{array}{cc|c} 2 & 2 & 5 \\ 1 & -\frac{3}{2} & 5 \end{array}\right] R_1 \leftrightarrow R_2$

2. $\left[\begin{array}{cc|c} -6 & 9 & 4 \\ 1 & -\frac{3}{2} & 4 \end{array}\right] R_1 \leftrightarrow R_2$

3. $\left[\begin{array}{cc|c} -6 & 8 & -12 \\ 3 & 5 & -2 \end{array}\right] -\frac{1}{6}R_1$

4. $\left[\begin{array}{cc|c} -2 & 3 & -10 \\ 4 & 2 & 5 \end{array}\right] -\frac{1}{2}R_1$

5. $\left[\begin{array}{cc|c} 1 & -3 & 5 \\ 2 & 6 & 4 \end{array}\right] -2R_1 + R_2$

6. $\left[\begin{array}{cc|c} 1 & -3 & 1 \\ 2 & 1 & -5 \end{array}\right] -2R_1 + R_2$

7. $\left[\begin{array}{cc|c} 1 & -\frac{3}{2} & \frac{7}{2} \\ 3 & 4 & 2 \end{array}\right] -3R_1 + R_2$

8. $\left[\begin{array}{cc|c} 1 & -\frac{2}{5} & \frac{3}{4} \\ 4 & 2 & -1 \end{array}\right] -4R_1 + R_2$

9. $\left[\begin{array}{ccc|c} 2 & -6 & 4 & 10 \\ 1 & 5 & -5 & 0 \\ 3 & 0 & 4 & 7 \end{array}\right] \frac{1}{2}R_1$

10. $\left[\begin{array}{ccc|c} 3 & -12 & 6 & 9 \\ 1 & -4 & 4 & 0 \\ 2 & 0 & 7 & 4 \end{array}\right] \frac{1}{3}R_1$

11. $\left[\begin{array}{ccc|c} 1 & -3 & 2 & 0 \\ 3 & 1 & -1 & 7 \\ 2 & -2 & 1 & 3 \end{array}\right] -3R_1 + R_2$

12. $\left[\begin{array}{ccc|c} 1 & -1 & 5 & -6 \\ 3 & 3 & -1 & 10 \\ 1 & 3 & 2 & 5 \end{array}\right] -3R_1 + R_2$

13. $\left[\begin{array}{ccc|c} 1 & 1 & -1 & 6 \\ 2 & -1 & 1 & -3 \\ 3 & -1 & -1 & 4 \end{array}\right] \begin{array}{l} -2R_1 + R_2 \\ -3R_1 + R_3 \end{array}$

14. $\left[\begin{array}{ccc|c} 1 & 2 & 1 & 2 \\ -2 & -1 & 2 & 5 \\ 1 & 3 & -2 & -8 \end{array}\right] \begin{array}{l} 2R_1 + R_2 \\ -1R_1 + R_3 \end{array}$

In Exercises 15–38, solve each system using matrices. If there is no solution or if there are infinitely many solutions and a system's equations are dependent, so state.

15. $\begin{aligned} x + y &= 6 \\ x - y &= 2 \end{aligned}$

16. $\begin{aligned} x + 2y &= 11 \\ x - y &= -1 \end{aligned}$

17. $\begin{aligned} 2x + y &= 3 \\ x - 3y &= 12 \end{aligned}$

18. $\begin{aligned} 3x - 5y &= 7 \\ x - y &= 1 \end{aligned}$

19. $\begin{aligned} 5x + 7y &= -25 \\ 11x + 6y &= -8 \end{aligned}$

20. $\begin{aligned} 3x - 5y &= 22 \\ 4x - 2y &= 20 \end{aligned}$

21. $\begin{aligned} 4x - 2y &= 5 \\ -2x + y &= 6 \end{aligned}$

22. $\begin{aligned} -3x + 4y &= 12 \\ 6x - 8y &= 16 \end{aligned}$

23. $\begin{aligned} x - 2y &= 1 \\ -2x + 4y &= -2 \end{aligned}$

24. $\begin{aligned} 3x - 6y &= 1 \\ 2x - 4y &= \frac{2}{3} \end{aligned}$

25. $\begin{aligned} x + y - z &= -2 \\ 2x - y + z &= 5 \\ -x + 2y + 2z &= 1 \end{aligned}$

26. $\begin{aligned} x - 2y - z &= 2 \\ 2x - y + z &= 4 \\ -x + y - 2z &= -4 \end{aligned}$

27. $\begin{aligned} x + 3y &= 0 \\ x + y + z &= 1 \\ 3x - y - z &= 11 \end{aligned}$

28. $\begin{aligned} 3y - z &= -1 \\ x + 5y - z &= -4 \\ -3x + 6y + 2z &= 11 \end{aligned}$

29. $\begin{aligned} 2x + 2y + 7z &= -1 \\ 2x + y + 2z &= 2 \\ 4x + 6y + z &= 15 \end{aligned}$

30. $\begin{aligned} 3x + 2y + 3z &= 3 \\ 4x - 5y + 7z &= 1 \\ 2x + 3y - 2z &= 6 \end{aligned}$

31. $\begin{aligned} x + y + z &= 6 \\ x - z &= -2 \\ y + 3z &= 11 \end{aligned}$

32. $\begin{aligned} x + y + z &= 3 \\ -y + 2z &= 1 \\ -x + z &= 0 \end{aligned}$

33. $\begin{aligned} x - y + 3z &= 4 \\ 2x - 2y + 6z &= 7 \\ 3x - y + 5z &= 14 \end{aligned}$

34. $\begin{aligned} 3x - y + 2z &= 4 \\ -6x + 2y - 4z &= 1 \\ 5x - 3y + 8z &= 0 \end{aligned}$

35. $\begin{aligned} x - 2y + z &= 4 \\ 5x - 10y + 5y &= 20 \\ -2x + 4y - 2z &= -8 \end{aligned}$

36. $\begin{aligned} x - 3y + z &= 2 \\ 4x - 12y + 4z &= 8 \\ -2x + 6y - 2z &= -4 \end{aligned}$

37. $\begin{aligned} x + y &= 1 \\ y + 2z &= -2 \\ 2x - z &= 0 \end{aligned}$

38. $\begin{aligned} x + 3y &= 3 \\ y + 2z &= -8 \\ x - z &= 7 \end{aligned}$

Practice PLUS

In Exercises 39–40, write the system of linear equations represented by the augmented matrix. Use w, x, y, and z, for the variables. Once the system is written, use back-substitution to find its solution set, $\{(w, x, y, z)\}$.

39. $\left[\begin{array}{cccc|c} 1 & -1 & 1 & 1 & 3 \\ 0 & 1 & -2 & -1 & 0 \\ 0 & 0 & 1 & 6 & 17 \\ 0 & 0 & 0 & 1 & 3 \end{array}\right]$

40. $\left[\begin{array}{cccc|c} 1 & 2 & -1 & 0 & 2 \\ 0 & 1 & 1 & -2 & -3 \\ 0 & 0 & 1 & -1 & -2 \\ 0 & 0 & 0 & 1 & 3 \end{array}\right]$

In Exercises 41–42, perform each matrix row operation and write the new matrix.

41. $\left[\begin{array}{cccc|c} 1 & -1 & 1 & 1 & 3 \\ 0 & 1 & -2 & -1 & 0 \\ 2 & 0 & 3 & 4 & 11 \\ 5 & 1 & 2 & 4 & 6 \end{array}\right] \begin{array}{l} \\ \\ -2R_1 + R_3 \\ -5R_1 + R_4 \end{array}$

42. $\left[\begin{array}{cccc|c} 1 & -5 & 2 & -2 & 4 \\ 0 & 1 & -3 & -1 & 0 \\ 3 & 0 & 2 & -1 & 6 \\ -4 & 1 & 4 & 2 & -3 \end{array}\right] \begin{array}{l} \\ \\ -3R_1 + R_3 \\ 4R_1 + R_4 \end{array}$

In Exercises 43–44, solve each system using matrices. You will need to use matrix row operations to obtain matrices like those in Exercises 39 and 40, with 1s down the main diagonal and 0s below the 1s. Express the solution set as $\{(w, x, y, z)\}$.

43. $\begin{aligned} w + x + y + z &= 4 \\ 2w + x - 2y - z &= 0 \\ w - 2x - y - 2z &= -2 \\ 3w + 2x + y + 3z &= 4 \end{aligned}$

44. $\begin{aligned} w + x + y + z &= 5 \\ w + 2x - y - 2z &= -1 \\ w - 3x - 3y - z &= -1 \\ 2w - x + 2y - z &= -2 \end{aligned}$

Application Exercises

45. A ball is thrown straight upward. The graph shows the ball's height, $s(t)$, in feet, after t seconds.

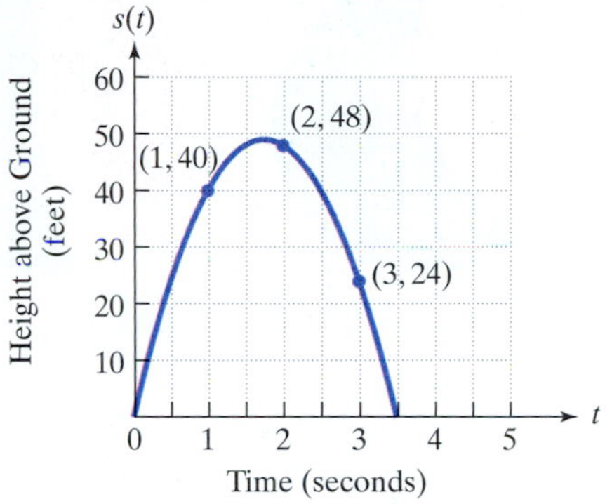

a. Find the quadratic function

$$s(t) = at^2 + bt + c$$

whose graph passes through the three points labeled on the graph. Solve the system of linear equations involving a, b, and c using matrices.

b. Find and interpret $s(3.5)$. Identify your solution as a point on the graph shown.

46. A football is kicked straight upward. The graph shows the football's height, $s(t)$, in feet, after t seconds.

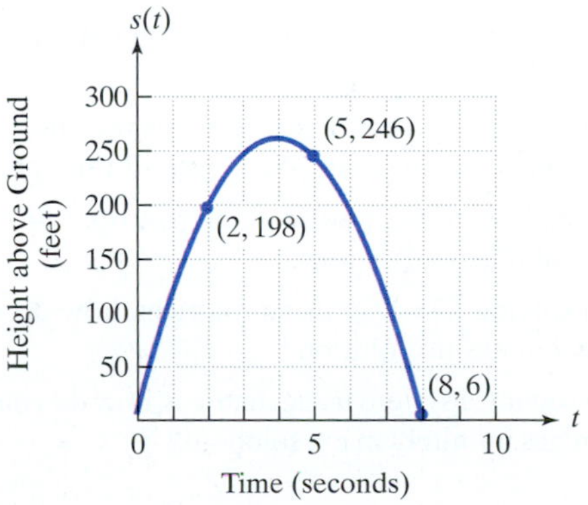

a. Find the quadratic function

$$s(t) = at^2 + bt + c$$

whose graph passes through the three points labeled on the graph. Solve the system of linear equations involving a, b, and c using matrices.

b. Find and interpret $s(7)$. Identify your solution as a point on the graph shown.

Write a system of linear equations in three variables to solve Exercises 47–48. Then use matrices to solve the system. Exercises 47–48 are based on a Time/CNN telephone poll that included never-married single women between the ages of 18 and 49 and never-married single men between the ages of 18 and 49. The circle graphs show the results for one of the questions in the poll.

If You Couldn't Find the Perfect Mate, Would You Marry Someone Else?

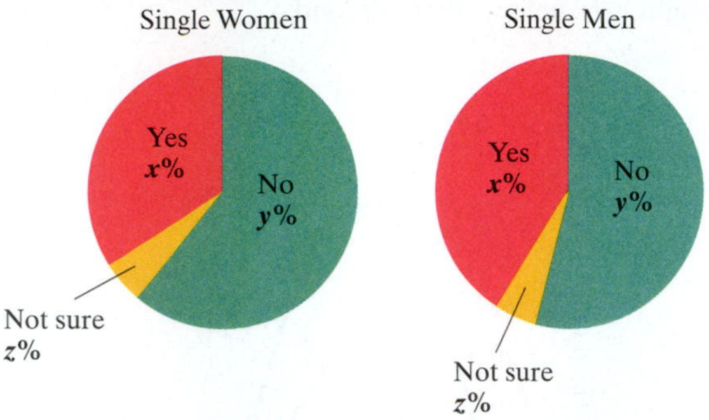

47. For single women in the poll, the percentage who said no exceeded the combined percentages for those who said yes and those who said not sure by 22%. If the percentage who said yes is doubled, it is 7% more than the percentage who said no. Find the percentage of single women who responded yes, no, and not sure.

48. For single men in the poll, the percentage who said no exceeded the combined percentages for those who said yes and those who said not sure by 8%. If the percentage who said yes is doubled, it is 28% more than the percentage who said no. Find the percentage of single men who responded yes, no, and not sure.

Writing in Mathematics

49. What is a matrix?

50. Describe what is meant by the augmented matrix of a system of linear equations.

51. In your own words, describe each of the three matrix row operations. Give an example of each of the operations.

52. Describe how to use matrices and row operations to solve a system of linear equations.

53. When solving a system using matrices, how do you know if the system has no solution?

54. When solving a system using matrices, how do you know if the system has infinitely many solutions?

Technology Exercises

55. Most graphing utilities can perform row operations on matrices. Consult the owner's manual for your graphing utility to learn proper keystroke sequences for performing these operations. Then duplicate the row operations of any three exercises that you solved from Exercises 3–12.

56. If your graphing utility has a REF (row-echelon form) command, use this feature to verify your work with any five systems from Exercises 15–38.

57. A matrix with 1s down the main diagonal and 0s in every position *above and below* each 1 is said to be in **reduced row-echelon form**.

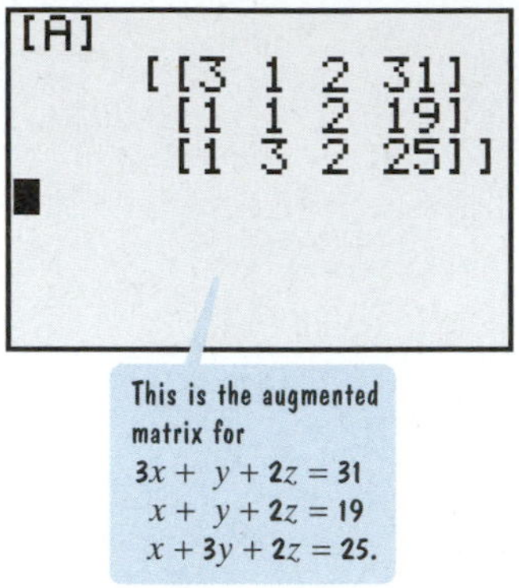

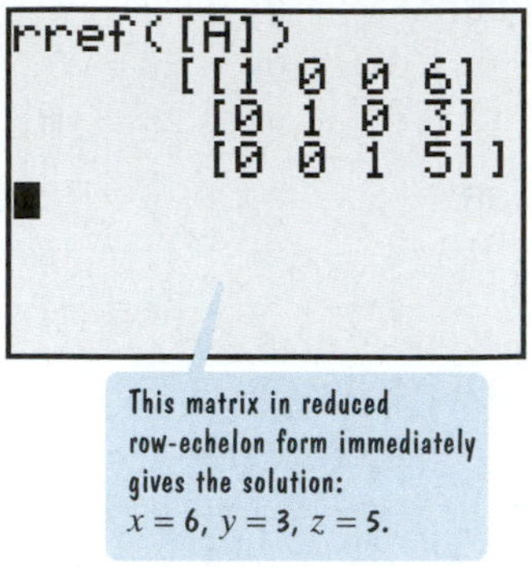

If your graphing utility has a RREF (reduced row-echelon form) command, use this feature to verify your work with any five systems from Exercises 15–38.

Critical Thinking Exercises

Make Sense? *In Exercises 58–61, determine whether each statement "makes sense" or "does not make sense" and explain your reasoning.*

58. Matrix row operations remind me of what I did when solving a linear system by the addition method, although I no longer write the variables.

59. When I use matrices to solve linear systems, the only arithmetic involves multiplication or a combination of multiplication and addition.

60. When I use matrices to solve linear systems, I spend most of my time using row operations to express the system's augmented matrix in row-echelon form.

61. Using row operations on an augmented matrix, I obtain a row in which 0s appear to the left of the vertical bar, but 6 appears on the right, so the system I'm working with has infinitely many solutions.

In Exercises 62–65, determine whether each statement is true or false. If the statement is false, make the necessary change(s) to produce a true statement.

62. A matrix row operation such as $-\frac{4}{5}R_1 + R_2$ is not permitted because of the negative fraction.

63. The augmented matrix for the system

$$\begin{aligned} x - 3y &= 5 \\ y - 2z &= 7 \\ 2x + z &= 4 \end{aligned} \quad \text{is} \quad \left[\begin{array}{cc|c} 1 & -3 & 5 \\ 1 & -2 & 7 \\ 2 & 1 & 4 \end{array}\right].$$

64. In solving a linear system of three equations in three variables, we begin with the augmented matrix and use row operations to obtain a row-equivalent matrix with 0s down the diagonal from left to right and 1s below each 0.

65. The row operation $kR_i + R_j$ indicates that it is the elements in row i that change.

66. The vitamin content per ounce for three foods is given in the following table.

	Milligrams per Ounce		
	Thiamin	Riboflavin	Niacin
Food A	3	7	1
Food B	1	5	3
Food C	3	8	2

a. Use matrices to show that no combination of these foods can provide exactly 14 milligrams of thiamin, 32 milligrams of riboflavin, and 9 milligrams of niacin.

b. Use matrices to describe in practical terms what happens if the riboflavin requirement is increased by 5 milligrams and the other requirements stay the same.

Review Exercises

67. If $f(x) = -3x + 10$, find $f(2a - 1)$. (Section 2.1, Example 3)

68. If $f(x) = 3x$ and $g(x) = 2x - 3$, find $(fg)(-1)$. (Section 2.3, Example 4)

69. Simplify: $\dfrac{-4x^8y^{-12}}{12x^{-3}y^{24}}$. (Section 1.6, Example 9)

Preview Exercises

Exercises 70–72 will help you prepare for the material covered in the next section. Simplify the expression in each exercise.

70. $2(-5) - (-3)(4)$

71. $\dfrac{2(-5) - 1(-4)}{5(-5) - 6(-4)}$

72. $2(-30 - (-3)) - 3(6 - 9) + (-1)(1 - 15)$

SECTION 3.5 Determinants and Cramer's Rule

Objectives

1. Evaluate a second-order determinant.
2. Solve a system of linear equations in two variables using Cramer's rule.
3. Evaluate a third-order determinant.
4. Solve a system of linear equations in three variables using Cramer's rule.
5. Use determinants to identify inconsistent and dependent systems.

A portion of Charles Babbage's unrealized Difference Engine

As cyberspace absorbs more and more of our work, play, shopping, and socializing, where will it all end? Which activities will still be offline in 2025?

Our technologically transformed lives can be traced back to the English inventor Charles Babbage (1791–1871). Babbage knew of a method for solving linear systems called *Cramer's rule*, in honor of the Swiss geometer Gabriel Cramer (1704–1752). Cramer's rule was simple, but involved numerous multiplications for large systems. Babbage designed a machine, called the "difference engine," that consisted of toothed wheels on shafts for performing these multiplications. Despite the fact that only one-seventh of the functions ever worked, Babbage's invention demonstrated how complex calculations could be handled mechanically. In 1944, scientists at IBM used the lessons of the difference engine to create the world's first computer.

Those who invented computers hoped to relegate the drudgery of repeated computation to a machine. In this section, we look at a method for solving linear systems that played a critical role in this process. The method uses real numbers, called *determinants*, that are associated with arrays of numbers. As with matrix methods, solutions are obtained by writing down the coefficients and constants of a linear system and performing operations with them.

1 Evaluate a second-order determinant.

The Determinant of a 2 × 2 Matrix

A matrix of **order** $m \times n$ has m rows and n columns. If $m = n$, a matrix has the same number of rows as columns and is called a **square matrix**. Associated with every square matrix is a real number, called its **determinant**. The determinant for a 2×2 square matrix is defined as follows:

Study Tip

To evaluate a second-order determinant, find the difference of the product of the two diagonals.

$$\begin{vmatrix} a_1 & b_1 \\ a_2 & b_2 \end{vmatrix} = a_1b_2 - a_2b_1$$

Definition of the Determinant of a 2 × 2 Matrix

The determinant of the matrix $\begin{bmatrix} a_1 & b_1 \\ a_2 & b_2 \end{bmatrix}$ is denoted by $\begin{vmatrix} a_1 & b_1 \\ a_2 & b_2 \end{vmatrix}$ and is defined by

$$\begin{vmatrix} a_1 & b_1 \\ a_2 & b_2 \end{vmatrix} = a_1b_2 - a_2b_1.$$

We also say that the **value** of the **second-order determinant** $\begin{vmatrix} a_1 & b_1 \\ a_2 & b_2 \end{vmatrix}$ is $a_1b_2 - a_2b_1$.

Example 1 illustrates that the determinant of a matrix may be positive or negative. The determinant can also have 0 as its value.

EXAMPLE 1 Evaluating the Determinant of a 2 × 2 Matrix

Evaluate the determinant of each of the following matrices:

a. $\begin{bmatrix} 5 & 6 \\ 7 & 3 \end{bmatrix}$

b. $\begin{bmatrix} 2 & 4 \\ -3 & -5 \end{bmatrix}$.

Discover for Yourself

Write and then evaluate three determinants, one whose value is positive, one whose value is negative, and one whose value is 0.

Solution We multiply and subtract as indicated.

a. $\begin{vmatrix} 5 & 6 \\ 7 & 3 \end{vmatrix} = 5 \cdot 3 - 7 \cdot 6 = 15 - 42 = -27$ The value of the second-order determinant is −27.

b. $\begin{vmatrix} 2 & 4 \\ -3 & -5 \end{vmatrix} = 2(-5) - (-3)(4) = -10 + 12 = 2$ The value of the second-order determinant is 2.

✓ **CHECK POINT 1** Evaluate the determinant of each of the following matrices:

a. $\begin{bmatrix} 10 & 9 \\ 6 & 5 \end{bmatrix}$ **b.** $\begin{bmatrix} 4 & 3 \\ -5 & -8 \end{bmatrix}$.

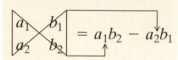

2 Solve a system of linear equations in two variables using Cramer's rule.

Solving Systems of Linear Equations in Two Variables Using Determinants

Determinants can be used to solve a linear system in two variables. In general, such a system appears as

$$a_1x + b_1y = c_1$$
$$a_2x + b_2y = c_2.$$

Let's first solve this system for x using the addition method. We can solve for x by eliminating y from the equations. Multiply the first equation by b_2 and the second equation by $-b_1$. Then add the two equations:

$$\begin{array}{lcrcl} a_1x + b_1y = c_1 & \xrightarrow{\text{Multiply by } b_2.} & a_1b_2x + b_1b_2y & = & c_1b_2 \\ a_2x + b_2y = c_2 & \xrightarrow{\text{Multiply by } -b_1.} & -a_2b_1x - b_1b_2y & = & -c_2b_1 \\ & \text{Add:} & (a_1b_2 - a_2b_1)x & = & c_1b_2 - c_2b_1 \\ & & x & = & \dfrac{c_1b_2 - c_2b_1}{a_1b_2 - a_2b_1} \end{array}$$

Because

$$\begin{vmatrix} c_1 & b_1 \\ c_2 & b_2 \end{vmatrix} = c_1b_2 - c_2b_1 \quad \text{and} \quad \begin{vmatrix} a_1 & b_1 \\ a_2 & b_2 \end{vmatrix} = a_1b_2 - a_2b_1,$$

we can express our answer for x as the quotient of two determinants:

$$x = \frac{\begin{vmatrix} c_1 & b_1 \\ c_2 & b_2 \end{vmatrix}}{\begin{vmatrix} a_1 & b_1 \\ a_2 & b_2 \end{vmatrix}}.$$

Similarly, we could use the addition method to solve our system for y, again expressing y as the quotient of two determinants. This method of using determinants to solve the linear system, called **Cramer's rule**, is summarized in the box.

Solving a Linear System in Two Variables Using Determinants

Cramer's Rule

If

$$a_1x + b_1y = c_1$$
$$a_2x + b_2y = c_2$$

then

$$x = \frac{\begin{vmatrix} c_1 & b_1 \\ c_2 & b_2 \end{vmatrix}}{\begin{vmatrix} a_1 & b_1 \\ a_2 & b_2 \end{vmatrix}} \quad \text{and} \quad y = \frac{\begin{vmatrix} a_1 & c_1 \\ a_2 & c_2 \end{vmatrix}}{\begin{vmatrix} a_1 & b_1 \\ a_2 & b_2 \end{vmatrix}},$$

where

$$\begin{vmatrix} a_1 & b_1 \\ a_2 & b_2 \end{vmatrix} \neq 0.$$

Here are some helpful tips when solving

$$a_1x + b_1y = c_1$$
$$a_2x + b_2y = c_2$$

using determinants:

1. Three different determinants are used to find x and y. The determinants in the denominators for x and y are identical. The determinants in the numerators for x and y differ. In abbreviated notation, we write

$$x = \frac{D_x}{D} \quad \text{and} \quad y = \frac{D_y}{D}, \text{ where } D \neq 0.$$

2. The elements of D, the determinant in the denominator, are the coefficients of the variables in the system.

$$D = \begin{vmatrix} a_1 & b_1 \\ a_2 & b_2 \end{vmatrix}$$

3. D_x, the determinant in the numerator of x, is obtained by replacing the x-coefficients, in D, a_1 and a_2, with the constants on the right sides of the equations, c_1 and c_2.

$$D = \begin{vmatrix} a_1 & b_1 \\ a_2 & b_2 \end{vmatrix} \quad \text{and} \quad D_x = \begin{vmatrix} c_1 & b_1 \\ c_2 & b_2 \end{vmatrix}$$

Replace the column with a_1 and a_2 with the constants c_1 and c_2 to get D_x.

4. D_y, the determinant in the numerator for y, is obtained by replacing the y-coefficients, in D, b_1 and b_2, with the constants on the right sides of the equations, c_1 and c_2.

$$D = \begin{vmatrix} a_1 & b_1 \\ a_2 & b_2 \end{vmatrix} \quad \text{and} \quad D_y = \begin{vmatrix} a_1 & c_1 \\ a_2 & c_2 \end{vmatrix}$$

Replace the column with b_1 and b_2 with the constants c_1 and c_2 to get D_y.

EXAMPLE 2 Using Cramer's Rule to Solve a Linear System

Use Cramer's rule to solve the system:

$$5x - 4y = 2$$
$$6x - 5y = 1.$$

Solution Because

$$x = \frac{D_x}{D} \quad \text{and} \quad y = \frac{D_y}{D},$$

we will set up and evaluate the three determinants D, D_x, and D_y.

1. D, the determinant in both denominators, consists of the x- and y-coefficients.

$$D = \begin{vmatrix} 5 & -4 \\ 6 & -5 \end{vmatrix} = (5)(-5) - (6)(-4) = -25 + 24 = -1$$

Because this determinant is not zero, we continue to use Cramer's rule to solve the system.

2. D_x, the determinant in the numerator for x, is obtained by replacing the x-coefficients in D, 5 and 6, by the constants on the right sides of the equations, 2 and 1.

$$D_x = \begin{vmatrix} 2 & -4 \\ 1 & -5 \end{vmatrix} = (2)(-5) - (1)(-4) = -10 + 4 = -6$$

3. D_y, the determinant in the numerator for y, is obtained by replacing the y-coefficients in D, -4 and -5, by the constants on the right sides of the equations, 2 and 1.

$$D_y = \begin{vmatrix} 5 & 2 \\ 6 & 1 \end{vmatrix} = (5)(1) - (6)(2) = 5 - 12 = -7$$

4. Thus,

$$x = \frac{D_x}{D} = \frac{-6}{-1} = 6 \quad \text{and} \quad y = \frac{D_y}{D} = \frac{-7}{-1} = 7.$$

As always, the ordered pair $(6, 7)$ should be checked by substituting these values into the original equations. The solution is $(6, 7)$ and the solution set is $\{(6, 7)\}$. ▬

✓ **CHECK POINT 2**   Use Cramer's rule to solve the system:

$$5x + 4y = 12$$
$$3x - 6y = 24.$$

3 Evaluate a third-order determinant.

The Determinant of a 3 × 3 Matrix

The determinant for a 3×3 matrix is defined in terms of second-order determinants:

Definition of the Determinant of a 3 × 3 Matrix

A third-order determinant is defined by

$$\begin{vmatrix} a_1 & b_1 & c_1 \\ a_2 & b_2 & c_2 \\ a_3 & b_3 & c_3 \end{vmatrix} = a_1 \begin{vmatrix} b_2 & c_2 \\ b_3 & c_3 \end{vmatrix} \overset{\text{Subtract.}}{-} a_2 \begin{vmatrix} b_1 & c_1 \\ b_3 & c_3 \end{vmatrix} \overset{\text{Add.}}{+} a_3 \begin{vmatrix} b_1 & c_1 \\ b_2 & c_2 \end{vmatrix}.$$

Each a on the right comes from the first column.

Here are some tips that should be helpful when evaluating the determinant of a 3×3 matrix:

Evaluating the Determinant of a 3 × 3 Matrix

1. Each of the three terms in the definition above contains two factors—a numerical factor and a second-order determinant.

2. The numerical factor in each term is an element from the first column of the third-order determinant.

3. The minus sign precedes the second term.

4. The second-order determinant that appears in each term is obtained by crossing out the row and the column containing the numerical factor.

$$a_1 \begin{vmatrix} b_2 & c_2 \\ b_3 & c_3 \end{vmatrix} - a_2 \begin{vmatrix} b_1 & c_1 \\ b_3 & c_3 \end{vmatrix} + a_3 \begin{vmatrix} b_1 & c_1 \\ b_2 & c_2 \end{vmatrix}$$

$$\downarrow \qquad\qquad \downarrow \qquad\qquad \downarrow$$

$$\begin{vmatrix} a_1 & b_1 & c_1 \\ a_2 & b_2 & c_2 \\ a_3 & b_3 & c_3 \end{vmatrix} \quad \begin{vmatrix} a_1 & b_1 & c_1 \\ a_2 & b_2 & c_2 \\ a_3 & b_3 & c_3 \end{vmatrix} \quad \begin{vmatrix} a_1 & b_1 & c_1 \\ a_2 & b_2 & c_2 \\ a_3 & b_3 & c_3 \end{vmatrix}$$

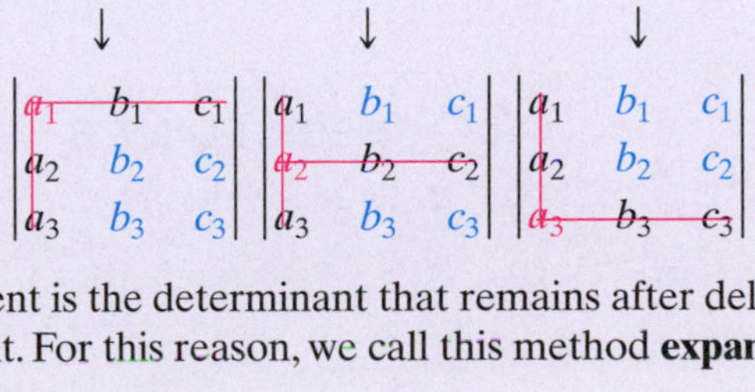

The **minor** of an element is the determinant that remains after deleting the row and column of that element. For this reason, we call this method **expansion by minors**.

EXAMPLE 3 Evaluating the Determinant of a 3 × 3 Matrix

Evaluate the determinant of the following matrix:

$$\begin{bmatrix} 4 & 1 & 0 \\ -9 & 3 & 4 \\ -3 & 8 & 1 \end{bmatrix}.$$

Solution We know that each of the three terms in the determinant contains a numerical factor and a second-order determinant. The numerical factors are from the first column of the given matrix. They are shown in red in the following matrix:

$$\begin{bmatrix} 4 & 1 & 0 \\ -9 & 3 & 4 \\ -3 & 8 & 1 \end{bmatrix}.$$

We find the minor for each numerical factor by deleting the row and column of that element:

$$\begin{bmatrix} 4 & 1 & 0 \\ -9 & 3 & 4 \\ -3 & 8 & 1 \end{bmatrix} \quad \begin{bmatrix} 4 & 1 & 0 \\ -9 & 3 & 4 \\ -3 & 8 & 1 \end{bmatrix} \quad \begin{bmatrix} 4 & 1 & 0 \\ -9 & 3 & 4 \\ -3 & 8 & 1 \end{bmatrix}$$

The minor for 4 is $\begin{vmatrix} 3 & 4 \\ 8 & 1 \end{vmatrix}$.

The minor for -9 is $\begin{vmatrix} 1 & 0 \\ 8 & 1 \end{vmatrix}$.

The minor for -3 is $\begin{vmatrix} 1 & 0 \\ 3 & 4 \end{vmatrix}$.

Now we have three numerical factors, 4, −9, and −3, and three second-order determinants. We multiply each numerical factor by its second-order determinant to find the three terms of the third-order determinant:

$$4\begin{vmatrix} 3 & 4 \\ 8 & 1 \end{vmatrix}, \quad -9\begin{vmatrix} 1 & 0 \\ 8 & 1 \end{vmatrix}, \quad -3\begin{vmatrix} 1 & 0 \\ 3 & 4 \end{vmatrix}.$$

Based on the preceding definition, we subtract the second term from the first term and add the third term:

Don't forget to supply the minus sign.

$$\begin{vmatrix} 4 & 1 & 0 \\ -9 & 3 & 4 \\ -3 & 8 & 1 \end{vmatrix} = 4\begin{vmatrix} 3 & 4 \\ 8 & 1 \end{vmatrix} - (-9)\begin{vmatrix} 1 & 0 \\ 8 & 1 \end{vmatrix} - 3\begin{vmatrix} 1 & 0 \\ 3 & 4 \end{vmatrix}$$ Begin by evaluating the three second-order determinants.

$= 4(3 \cdot 1 - 8 \cdot 4) + 9(1 \cdot 1 - 8 \cdot 0) - 3(1 \cdot 4 - 3 \cdot 0)$

$= 4(3 - 32) + 9(1 - 0) - 3(4 - 0)$ Multiply within parentheses.

$= 4(-29) + 9(1) - 3(4)$ Subtract within parentheses.

$= -116 + 9 - 12$ Multiply.

$= -119$ Add and subtract as indicated.

Using Technology

A graphing utility can be used to evaluate the determinant of a matrix. Enter the matrix and call it A. Then use the determinant command. The screen below verifies our result in Example 3.

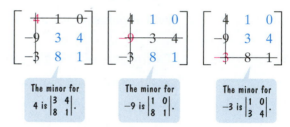

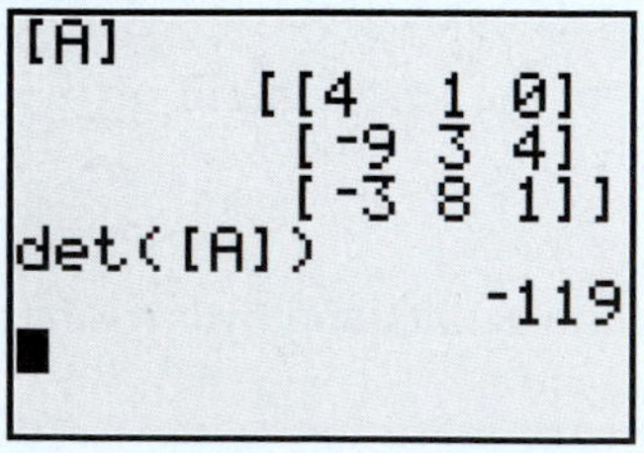

✓ CHECK POINT 3 Evaluate the determinant of the following matrix:

$$\begin{bmatrix} 2 & 1 & 7 \\ -5 & 6 & 0 \\ -4 & 3 & 1 \end{bmatrix}.$$

4 Solve a system of linear equations in three variables using Cramer's rule.

Solving Systems of Linear Equations in Three Variables Using Determinants

Cramer's rule can be applied to solving systems of linear equations in three variables. The determinants in the numerator and denominator of the quotients determining each variable are third-order determinants.

Solving Three Equations in Three Variables Using Determinants

Cramer's Rule

If

$$\begin{aligned} a_1x + b_1y + c_1z &= d_1 \\ a_2x + b_2y + c_2z &= d_2 \\ a_3x + b_3y + c_3z &= d_3 \end{aligned}$$

then,

$$x = \frac{D_x}{D}, \quad y = \frac{D_y}{D}, \quad \text{and} \quad z = \frac{D_z}{D}.$$

These four third-order determinants are given by

$$D = \begin{vmatrix} a_1 & b_1 & c_1 \\ a_2 & b_2 & c_2 \\ a_3 & b_3 & c_3 \end{vmatrix}$$

These are the coefficients of the variables x, y, and z. $D \neq 0$

$$D_x = \begin{vmatrix} d_1 & b_1 & c_1 \\ d_2 & b_2 & c_2 \\ d_3 & b_3 & c_3 \end{vmatrix}$$

Replace x-coefficients in D with the constants on the right of the three equations.

$$D_y = \begin{vmatrix} a_1 & d_1 & c_1 \\ a_2 & d_2 & c_2 \\ a_3 & d_3 & c_3 \end{vmatrix}$$

Replace y-coefficients in D with the constants on the right of the three equations.

$$D_z = \begin{vmatrix} a_1 & b_1 & d_1 \\ a_2 & b_2 & d_2 \\ a_3 & b_3 & d_3 \end{vmatrix}.$$

Replace z-coefficients in D with the constants on the right of the three equations.

EXAMPLE 4 Using Cramer's Rule to Solve a Linear System in Three Variables

Use Cramer's rule to solve:

$$\begin{aligned} x + 2y - z &= -4 \\ x + 4y - 2z &= -6 \\ 2x + 3y + z &= 3. \end{aligned}$$

Solution Because

$$x = \frac{D_x}{D}, \quad y = \frac{D_y}{D}, \quad \text{and} \quad z = \frac{D_z}{D},$$

we need to set up and evaluate four determinants.

Step 1. Set up the determinants.

1. D, the determinant in all three denominators, consists of the x-, y-, and z-coefficients.

$$D = \begin{vmatrix} 1 & 2 & -1 \\ 1 & 4 & -2 \\ 2 & 3 & 1 \end{vmatrix}$$

$$\begin{aligned} x + 2y - z &= -4 \\ x + 4y - 2z &= -6 \\ 2x + 3y + z &= 3 \end{aligned}$$

The system we are solving (repeated)

2. D_x, the determinant in the numerator for x, is obtained by replacing the x-coefficients in D, 1, 1, and 2, with the constants on the right sides of the equations, -4, -6, and 3.

$$D_x = \begin{vmatrix} -4 & 2 & -1 \\ -6 & 4 & -2 \\ 3 & 3 & 1 \end{vmatrix}$$

3. D_y, the determinant in the numerator for y, is obtained by replacing the y-coefficients in D, 2, 4, and 3, with the constants on the right sides of the equations, -4, -6, and 3.

$$D_y = \begin{vmatrix} 1 & -4 & -1 \\ 1 & -6 & -2 \\ 2 & 3 & 1 \end{vmatrix}$$

4. D_z, the determinant in the numerator for z, is obtained by replacing the z-coefficients in D, -1, -2, and 1, with the constants on the right sides of the equations, -4, -6, and 3.

$$D_z = \begin{vmatrix} 1 & 2 & -4 \\ 1 & 4 & -6 \\ 2 & 3 & 3 \end{vmatrix}$$

Step 2. Evaluate the four determinants.

$$D = \begin{vmatrix} 1 & 2 & -1 \\ 1 & 4 & -2 \\ 2 & 3 & 1 \end{vmatrix} = 1\begin{vmatrix} 4 & -2 \\ 3 & 1 \end{vmatrix} - 1\begin{vmatrix} 2 & -1 \\ 3 & 1 \end{vmatrix} + 2\begin{vmatrix} 2 & -1 \\ 4 & -2 \end{vmatrix}$$

$$= 1(4 + 6) - 1(2 + 3) + 2(-4 + 4)$$

$$= 1(10) - 1(5) + 2(0) = 5$$

Study Tip

To find D_x, D_y, and D_z, you'll need to apply the evaluation process for a 3×3 determinant three times. The values of D_x, D_y, and D_z cannot be obtained from the numbers that occur in the computation of D.

Using the same technique to evaluate each determinant, we obtain

$$D_x = -10, \quad D_y = 5, \quad \text{and} \quad D_z = 20.$$

Step 3. Substitute these four values and solve the system.

$$x = \frac{D_x}{D} = \frac{-10}{5} = -2$$

$$y = \frac{D_y}{D} = \frac{5}{5} = 1$$

$$z = \frac{D_z}{D} = \frac{20}{5} = 4$$

The ordered triple $(-2, 1, 4)$ can be checked by substitution into the original three equations. The solution is $(-2, 1, 4)$ and the solution set is $\{(-2, 1, 4)\}$.

✓ CHECK POINT 4 Use Cramer's rule to solve the system:

$$\begin{aligned} 3x - 2y + z &= 16 \\ 2x + 3y - z &= -9 \\ x + 4y + 3z &= 2. \end{aligned}$$

5 Use determinants to identify inconsistent and dependent systems.

Cramer's Rule with Inconsistent and Dependent Systems

If D, the determinant in the denominator, is 0, the variables described by the quotient of determinants are not real numbers. However, when $D = 0$, this indicates that the system is inconsistent or contains dependent equations. This gives rise to the following two situations:

Determinants: Inconsistent and Dependent Systems

1. If $D = 0$ and at least one of the determinants in the numerator is not 0, then the system is inconsistent. The solution set is $\varnothing$.
2. If $D = 0$ and all the determinants in the numerators are 0, then the equations in the system are dependent. The system has infinitely many solutions.

Although we have focused on applying determinants to solve linear systems, they have other applications, some of which we consider in the exercise set that follows.

3.5 EXERCISE SET

MyMathLab · MathXL PRACTICE · WATCH · DOWNLOAD · READ · REVIEW

Practice Exercises

Evaluate each determinant in Exercises 1–10.

1. $\begin{vmatrix} 5 & 7 \\ 2 & 3 \end{vmatrix}$

2. $\begin{vmatrix} 4 & 8 \\ 5 & 6 \end{vmatrix}$

3. $\begin{vmatrix} -4 & 1 \\ 5 & 6 \end{vmatrix}$

4. $\begin{vmatrix} 7 & 9 \\ -2 & -5 \end{vmatrix}$

5. $\begin{vmatrix} -7 & 14 \\ 2 & -4 \end{vmatrix}$

6. $\begin{vmatrix} 1 & -3 \\ -8 & 2 \end{vmatrix}$

7. $\begin{vmatrix} -5 & -1 \\ -2 & -7 \end{vmatrix}$

8. $\begin{vmatrix} \frac{1}{5} & \frac{1}{6} \\ -6 & 5 \end{vmatrix}$

9. $\begin{vmatrix} \frac{1}{2} & \frac{1}{2} \\ \frac{1}{8} & -\frac{3}{4} \end{vmatrix}$

10. $\begin{vmatrix} \frac{2}{3} & \frac{1}{3} \\ -\frac{1}{2} & \frac{3}{4} \end{vmatrix}$

For Exercises 11–26, use Cramer's rule to solve each system or to determine that the system is inconsistent or contains dependent equations.

11. $x + y = 7$
$x - y = 3$

12. $2x + y = 3$
$x - y = 3$

13. $12x + 3y = 15$
$2x - 3y = 13$

14. $x - 2y = 5$
$5x - y = -2$

15. $4x - 5y = 17$
$2x + 3y = 3$

16. $3x + 2y = 2$
$2x + 2y = 3$

17. $x - 3y = 4$
$3x - 4y = 12$

18. $2x - 9y = 5$
$3x - 3y = 11$

19. $3x - 4y = 4$
$2x + 2y = 12$

20. $3x = 7y + 1$
$2x = 3y - 1$

21. $2x = 3y + 2$
$5x = 51 - 4y$

22. $y = -4x + 2$
$2x = 3y + 8$

23. $3x = 2 - 3y$
$2y = 3 - 2x$

24. $x + 2y - 3 = 0$
$12 = 8y + 4x$

25. $4y = 16 - 3x$
$6x = 32 - 8y$

26. $2x = 7 + 3y$
$4x - 6y = 3$

Evaluate each determinant in Exercises 27–32.

27. $\begin{vmatrix} 3 & 0 & 0 \\ 2 & 1 & -5 \\ 2 & 5 & -1 \end{vmatrix}$

28. $\begin{vmatrix} 4 & 0 & 0 \\ 3 & -1 & 4 \\ 2 & -3 & 5 \end{vmatrix}$

29. $\begin{vmatrix} 3 & 1 & 0 \\ -3 & 4 & 0 \\ -1 & 3 & -5 \end{vmatrix}$

30. $\begin{vmatrix} 2 & -4 & 2 \\ -1 & 0 & 5 \\ 3 & 0 & 4 \end{vmatrix}$

31. $\begin{vmatrix} 1 & 1 & 1 \\ 2 & 2 & 2 \\ -3 & 4 & -5 \end{vmatrix}$

32. $\begin{vmatrix} 1 & 2 & 3 \\ 2 & 2 & -3 \\ 3 & 2 & 1 \end{vmatrix}$

In Exercises 33–40, use Cramer's rule to solve each system.

33. $x + y + z = 0$
$2x - y + z = -1$
$-x + 3y - z = -8$

34. $x - y + 2z = 3$
$2x + 3y + z = 9$
$-x - y + 3z = 11$

35. $4x - 5y - 6z = -1$
$x - 2y - 5z = -12$
$2x - y = 7$

36. $x - 3y + z = -2$
$x + 2y = 8$
$2x - y = 1$

37. $\begin{aligned} x + y + z &= 4 \\ x - 2y + z &= 7 \\ x + 3y + 2z &= 4 \end{aligned}$

38. $\begin{aligned} 2x + 2y + 3z &= 10 \\ 4x - y + z &= -5 \\ 5x - 2y + 6z &= 1 \end{aligned}$

39. $\begin{aligned} x + 2z &= 4 \\ 2y - z &= 5 \\ 2x + 3y &= 13 \end{aligned}$

40. $\begin{aligned} 3x + 2z &= 4 \\ 5x - y &= -4 \\ 4y + 3z &= 22 \end{aligned}$

Practice PLUS

In Exercises 41–42, evaluate each determinant.

41. $\begin{vmatrix} \begin{vmatrix} 3 & 1 \\ -2 & 3 \end{vmatrix} & \begin{vmatrix} 7 & 0 \\ 1 & 5 \end{vmatrix} \\ \begin{vmatrix} 3 & 0 \\ 0 & 7 \end{vmatrix} & \begin{vmatrix} 9 & -6 \\ 3 & 5 \end{vmatrix} \end{vmatrix}$

42. $\begin{vmatrix} \begin{vmatrix} 5 & 0 \\ 4 & -3 \end{vmatrix} & \begin{vmatrix} -1 & 0 \\ 0 & -1 \end{vmatrix} \\ \begin{vmatrix} 7 & -5 \\ 4 & 6 \end{vmatrix} & \begin{vmatrix} 4 & 1 \\ -3 & 5 \end{vmatrix} \end{vmatrix}$

In Exercises 43–44, write the system of linear equations for which Cramer's rule yields the given determinants.

43. $D = \begin{vmatrix} 2 & -4 \\ 3 & 5 \end{vmatrix}, \quad D_x = \begin{vmatrix} 8 & -4 \\ -10 & 5 \end{vmatrix}$

44. $D = \begin{vmatrix} 2 & -3 \\ 5 & 6 \end{vmatrix}, \quad D_x = \begin{vmatrix} 8 & -3 \\ 11 & 6 \end{vmatrix}$

In Exercises 45–48, solve each equation for x.

45. $\begin{vmatrix} -2 & x \\ 4 & 6 \end{vmatrix} = 32$

46. $\begin{vmatrix} x + 3 & -6 \\ x - 2 & -4 \end{vmatrix} = 28$

47. $\begin{vmatrix} 1 & x & -2 \\ 3 & 1 & 1 \\ 0 & -2 & 2 \end{vmatrix} = -8$

48. $\begin{vmatrix} 2 & x & 1 \\ -3 & 1 & 0 \\ 2 & 1 & 4 \end{vmatrix} = 39$

Application Exercises

Determinants are used to find the area of a triangle whose vertices are given by three points in a rectangular coordinate system. The area of a triangle with vertices (x_1, y_1), (x_2, y_2), and (x_3, y_3) is

$$\text{Area} = \pm\frac{1}{2}\begin{vmatrix} x_1 & y_1 & 1 \\ x_2 & y_2 & 1 \\ x_3 & y_3 & 1 \end{vmatrix},$$

where the ± symbol indicates that the appropriate sign should be chosen to yield a positive area. Use this information to work Exercises 49–50.

49. Use determinants to find the area of the triangle whose vertices are $(3, -5)$, $(2, 6)$, and $(-3, 5)$.

50. Use determinants to find the area of the triangle whose vertices are $(1, 1)$, $(-2, -3)$, and $(11, -3)$.

Determinants are used to show that three points lie on the same line (are collinear). If

$$\begin{vmatrix} x_1 & y_1 & 1 \\ x_2 & y_2 & 1 \\ x_3 & y_3 & 1 \end{vmatrix} = 0,$$

then the points (x_1, y_1), (x_2, y_2), and (x_3, y_3) are collinear. If the determinant does not equal 0, then the points are not collinear. Use this information to work Exercises 51–52.

51. Are the points $(3, -1)$, $(0, -3)$, and $(12, 5)$ collinear?

52. Are the points $(-4, -6)$, $(1, 0)$, and $(11, 12)$ collinear?

Determinants are used to write an equation of a line passing through two points. An equation of the line passing through the distinct points (x_1, y_1) and (x_2, y_2) is given by

$$\begin{vmatrix} x & y & 1 \\ x_1 & y_1 & 1 \\ x_2 & y_2 & 1 \end{vmatrix} = 0.$$

Use this information to work Exercises 53–54.

53. Use the determinant to write an equation for the line passing through $(3, -5)$ and $(-2, 6)$. Then expand the determinant, expressing the line's equation in slope-intercept form.

54. Use the determinant to write an equation for the line passing through $(-1, 3)$ and $(2, 4)$. Then expand the determinant, expressing the line's equation in slope-intercept form.

Writing in Mathematics

55. Explain how to evaluate a second-order determinant.

56. Describe the determinants D, D_x, and D_y in terms of the coefficients and constants in a system of two equations in two variables.

57. Explain how to evaluate a third-order determinant.

58. When expanding a determinant by minors, when is it necessary to supply a minus sign?

59. Without going into too much detail, describe how to solve a linear system in three variables using Cramer's rule.

60. In applying Cramer's rule, what does it mean if $D = 0$?

61. The process of solving a linear system in three variables using Cramer's rule can involve tedious computation. Is there a way of speeding up this process, perhaps using Cramer's rule to find the value for only one of the variables? Describe how this process might work, presenting a specific example with your description. Remember that your goal is still to find the value for each variable in the system.

62. If you could use only one method to solve linear systems in three variables, which method would you select? Explain why this is so.

Technology Exercises

63. Use the feature of your graphing utility that evaluates the determinant of a square matrix to verify any five of the determinants that you evaluated by hand in Exercises 1–10 or 27–32.

64. What is the fastest method for solving a linear system with your graphing utility?

Critical Thinking Exercises

Make Sense? *In Exercises 65–68, determine whether each statement "makes sense" or "does not make sense" and explain your reasoning.*

65. I'm solving a linear system using a determinant that contains two rows and three columns.

66. I can speed up the tedious computations required by Cramer's rule by using the value of D to determine the value of D_x.

67. When using Cramer's rule to solve a linear system, the number of determinants that I set up and evaluate is the same as the number of variables in the system.

68. Using Cramer's rule to solve a linear system, I found the value of D to be zero, so the system is inconsistent.

In Exercises 69–72, determine whether each statement is true or false. If the statement is false, make the necessary change(s) to produce a true statement.

69. Only one 2×2 determinant is needed to evaluate

$$\begin{vmatrix} 2 & 3 & -2 \\ 0 & 1 & 3 \\ 0 & 4 & -1 \end{vmatrix}.$$

70. If $D = 0$, then every variable has a value of 0.

71. Because there are different determinants in the numerators of x and y, if a system is solved using Cramer's rule, x and y cannot have the same value.

72. Using Cramer's rule, we use $\frac{D}{D_y}$ to get the value of y.

73. What happens to the value of a second-order determinant if the two columns are interchanged?

74. Consider the system

$$a_1x + b_1y = c_1$$
$$a_2x + b_2y = c_2.$$

Use Cramer's rule to prove that if the first equation of the system is replaced by the sum of the two equations, the resulting system has the same solution as the original system.

75. Show that the equation of a line through (x_1, y_1) and (x_2, y_2) is given by the determinant on the right.

$$\begin{vmatrix} x & y & 1 \\ x_1 & y_1 & 1 \\ x_2 & y_2 & 1 \end{vmatrix} = 0$$

Review Exercises

76. Solve: $6x - 4 = 2 + 6(x - 1)$. (Section 1.4, Example 6)

77. Solve for y: $-2x + 3y = 7$. (Section 1.5, Example 6)

78. Solve: $\frac{4x + 1}{3} = \frac{x - 3}{6} + \frac{x + 5}{6}$. (Section 1.4, Example 4)

Preview Exercises

Exercises 79–81 will help you prepare for the material covered in the first section of the next chapter.

79. Solve: $\frac{x + 3}{4} = \frac{x - 2}{3} + \frac{1}{4}$.

80. Solve: $-2x - 4 = x + 5$.

81. Use set notation to describe values of x for which $2(x + 4)$ is greater than $2x + 3$.

GROUP PROJECT

CHAPTER 3

The group is going into business for the next year. Your first task is to determine the product that you plan to manufacture and sell. Choose something unique, but realistic, reflecting the abilities and interests of group members. How much will it cost to produce each unit of your product? What will be the selling price for each unit? What is a reasonable estimate of your fixed cost for the entire year? Be sure to include utilities, labor, materials, marketing, and anything else that is relevant to your business venture. Once you have determined the product and these three figures,

a. Write the cost function, C, of producing x units of your product.

b. Write the revenue function, R, from the sale of x units of your product.

c. Determine the break-even point. Does this seem realistic in terms of your product and its target market?

d. Graph the cost function and the revenue function in the same rectangular coordinate system. Label the figure just like **Figure 3.7** on page 190. Indicate the regions that show the loss and gain for your business venture.

e. Write the profit function, P, from producing and selling x units of your product. Graph the profit function, in the same style as **Figure 3.8** on page 191. Indicate where the business is in the red and where it is in the black.

f. MTV's *The Real World* has offered to pay the fixed cost for the business venture, videotaping all group interactions for its forthcoming *The Real World: Profit and Loss*. Group members need to determine whether or not to accept the offer.

Chapter 3 Summary

Definitions and Concepts	Examples

Section 3.1 Systems of Linear Equations in Two Variables

A system of linear equations in two variables, x and y, consists of two equations that can be written in the form $Ax + By = C$. The solution set is the set of all ordered pairs that satisfy both equations. Using the graphing method, a solution of a linear system is a point common to the graphs of both equations in the system.

Solve by graphing: $2x - y = 6$
$x + y = 6.$

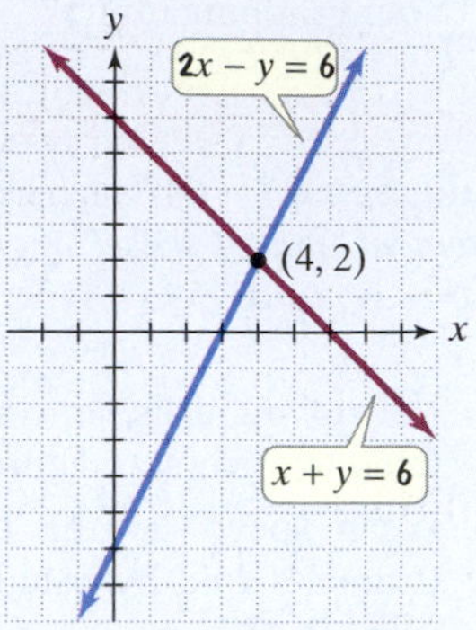

The intersection point gives the solution: (4, 2). The solution set is $\{(4, 2)\}$.

To solve a linear system by the substitution method,

1. Solve either equation for one variable in terms of the other.
2. Substitute the expression for that variable into the other equation.
3. Solve the equation in one variable.
4. Back-substitute the value of the variable into one of the original equations and find the value of the other variable.
5. Check the proposed solution in both equations.

Solve by the substitution method:

$$y = 2x - 3$$
$$4x - 3y = 5.$$

Substitute $2x - 3$ for y in the second equation.

$$4x - 3(2x - 3) = 5$$
$$4x - 6x + 9 = 5$$
$$-2x + 9 = 5$$
$$-2x = -4$$
$$x = 2$$

Find y. Substitute 2 for x in $y = 2x - 3$.

$$y = 2(2) - 3 = 4 - 3 = 1$$

The ordered pair (2, 1) checks. The solution is (2, 1) and $\{(2, 1)\}$ is the solution set.

To solve a linear system by the addition method,

1. Write equations in $Ax + By = C$ form.
2. Multiply one or both equations by nonzero numbers so that coefficients of one variable are opposites.
3. Add equations.
4. Solve the resulting equation for the variable.
5. Back-substitute the value of the variable into either original equation and find the value of the remaining variable.
6. Check the proposed solution in both equations.

Solve by the addition method:

$$2x + y = 10$$
$$3x + 4y = 25.$$

Eliminate y. Multiply the first equation by -4.

$$-8x - 4y = -40$$
$$3x + 4y = 25$$
$$\text{Add: } -5x = -15$$
$$x = 3$$

Find y. Back-substitute 3 for x. Use the first equation, $2x + y = 10$.

$$2(3) + y = 10$$
$$6 + y = 10$$
$$y = 4$$

The ordered pair (3, 4) checks. The solution is (3, 4) and $\{(3, 4)\}$ is the solution set.

Definitions and Concepts

Examples

Section 3.1 Systems of Linear Equations in Two Variables (continued)

A linear system with at least one solution is a consistent system. A system that has no solution, with ∅ as its solution set, is an inconsistent system. A linear system with infinitely many solutions has dependent equations. Solving inconsistent systems by substitution or addition leads to a false statement, such as $0 = 3$. Solving systems with dependent equations leads to a true statement, such as $7 = 7$.

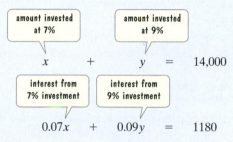

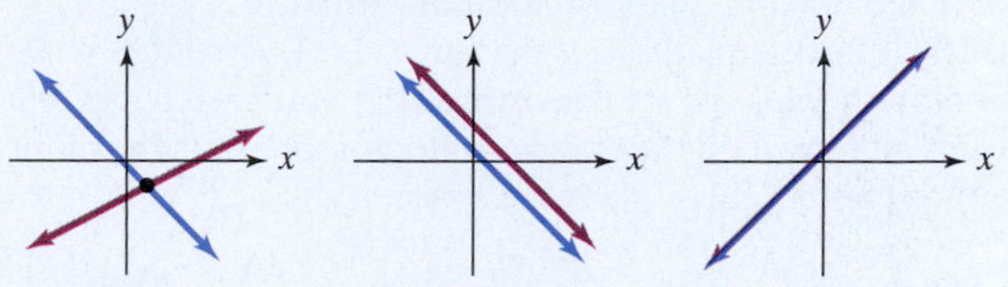

One solution: consistent | No solution: inconsistent | Infinitely many solutions: dependent and consistent

Section 3.2 Problem Solving and Business Applications Using Systems of Equations

A Problem-Solving Strategy

1. Use variables, usually x and y, to represent unknown quantities.
2. Write a system of equations describing the problem's conditions.
3. Solve the system and answer the problem's question.
4. Check proposed answers in the problem's wording.

You invested \$14,000 in two stocks paying 7% and 9% interest. Total year-end interest was \$1180. How much was invested at each rate?

Let x = amount invested at 7% and
y = amount invested at 9%.

(amount invested at 7%) (amount invested at 9%)

$$x + y = 14{,}000$$

(interest from 7% investment) (interest from 9% investment)

$$0.07x + 0.09y = 1180$$

Solving by substitution or addition, $x = 4000$ and $y = 10{,}000$. Thus, \$4000 was invested at 7% and \$10,000 at 9%.

Functions of Business

A company produces and sells x units of a product.

Revenue Function

$$R(x) = (\text{price per unit sold})x$$

Cost Function

$$C(x) = \text{fixed cost} + (\text{cost per unit produced})x$$

Profit Function

$$P(x) = R(x) - C(x)$$

The point of intersection of the graphs of R and C is the break even point. The x-coordinate of the point reveals the number of units that a company must produce and sell so that the money coming in, the revenue, is equal to the money going out, the cost. The y-coordinate gives the amount of money coming in and going out.

A company that manufactures lamps has a fixed cost of \$80,000 and it costs \$20 to produce each lamp. Lamps are sold for \$70.

a. Write the cost function.

$$C(x) = 80{,}000 + 20x$$

(Fixed cost) (Variable cost: \$20 per lamp.)

b. Write the revenue function.

$$R(x) = 70x$$

(Revenue per lamp, \$70, times number of lamps sold)

c. Find the break-even point.

Solve

$$y = 80{,}000 + 20x$$
$$y = 70x$$

by substitution. Solving

$$70x = 80{,}000 + 20x$$

yields $x = 1600$. Back-substituting, $y = 112{,}000$.
The break-even point is (1600, 112,000): The company breaks even if it sells 1600 lamps. At this level, money coming in equals money going out: \$112,000.

Definitions and Concepts	Examples

Section 3.3 Systems of Linear Equations in Three Variables

A system of linear equations in three variables, x, y, and z, consists of three equations of the form $Ax + By + Cz = D$. The solution set is the set consisting of the ordered triple that satisfies all three equations. The solution represents the point of intersection of three planes in space.

Is $(2, -1, 3)$ a solution of

$$\begin{aligned} 3x + 5y - 2z &= -5 \\ 2x + 3y - z &= -2 \\ 2x + 4y + 6z &= 18? \end{aligned}$$

Replace x with 2, y with -1, and z with 3. Using the first equation, we obtain:

$$\begin{aligned} 3 \cdot 2 + 5(-1) - 2(3) &\stackrel{?}{=} -5 \\ 6 - 5 - 6 &\stackrel{?}{=} -5 \\ -5 &= -5, \quad \text{true} \end{aligned}$$

The ordered triple $(2, -1, 3)$ satisfies the first equation. In a similar manner, it satisfies the other two equations and is a solution.

To solve a linear system in three variables by eliminating variables,

1. Reduce the system to two equations in two variables.
2. Solve the resulting system of two equations in two variables.
3. Use back-substitution in one of the equations in two variables to find the value of the second variable.
4. Back-substitute the values for two variables into one of the original equations to find the value of the third variable.
5. Check.

If all variables are eliminated and a false statement results, the system is inconsistent and has no solution. If a true statement results, the system contains dependent equations and has infinitely many solutions.

Solve:

$$\begin{aligned} 2x + 3y - 2z &= 0 && \text{Equation 1} \\ x + 2y - z &= 1 && \text{Equation 2} \\ 3x - y + z &= -15. && \text{Equation 3} \end{aligned}$$

Add Equations 2 and 3 to eliminate z.

$$4x + y = -14 \quad \text{Equation 4}$$

Eliminate z again. Multiply Equation 3 by 2 and add to Equation 1.

$$8x + y = -30 \quad \text{Equation 5}$$

Multiply Equation 4 by -1 and add to Equation 5.

$$\begin{aligned} -4x - y &= 14 \\ \underline{8x + y} &= \underline{-30} \\ \text{Add: } 4x \quad &= -16 \\ x &= -4 \end{aligned}$$

Substitute -4 for x in Equation 4.

$$\begin{aligned} 4(-4) + y &= -14 \\ y &= 2 \end{aligned}$$

Substitute -4 for x and 2 for y in Equation 3.

$$\begin{aligned} 3(-4) - 2 + z &= -15 \\ -14 + z &= -15 \\ z &= -1 \end{aligned}$$

Checking verifies that $(-4, 2, -1)$ is the solution and $\{(-4, 2, -1)\}$ is the solution set.

Definitions and Concepts	Examples

Section 3.3 Systems of Linear Equations in Three Variables (continued)

Curve Fitting

Curve fitting is determining a function whose graph contains given points. Three points that do not lie on a line determine the graph of a quadratic function

$$y = ax^2 + bx + c.$$

Use the three given points to create a system of three equations. Solve the system to find a, b, and c.

Find the quadratic function $y = ax^2 + bx + c$ whose graph passes through the points $(-1, 2)$, $(1, 8)$, and $(2, 14)$.
Use $y = ax^2 + bx + c$.

When $x = -1, y = 2$: $\quad 2 = a(-1)^2 + b(-1) + c$

When $x = 1, y = 8$: $\quad 8 = a \cdot 1^2 + b \cdot 1 + c$

When $x = 2, y = 14$: $\quad 14 = a \cdot 2^2 + b \cdot 2 + c$

Solving,

$$\begin{aligned} a - b + c &= 2 \\ a + b + c &= 8 \\ 4a + 2b + c &= 14, \end{aligned}$$

$a = 1, b = 3$, and $c = 4$. The quadratic function, $y = ax^2 + bx + c$, is $y = x^2 + 3x + 4$.

Section 3.4 Matrix Solutions to Linear Systems

A matrix is a rectangular array of numbers. The augmented matrix of a linear system is obtained by writing the coefficients of each variable, a vertical bar, and the constants of the system.

$$\begin{aligned} x + 4y &= 9 \\ 3x + y &= 5 \end{aligned}$$

The augmented matrix is $\left[\begin{array}{cc|c} 1 & 4 & 9 \\ 3 & 1 & 5 \end{array}\right]$.

The following row operations produce matrices that represent systems with the same solution. Two matrices are row equivalent if one can be obtained from the other by a sequence of these row operations.

1. Interchange the elements in the ith and jth rows: $R_i \leftrightarrow R_j$.
2. Multiply each element in the ith row by k: kR_i.
3. Add k times the elements in row i to the corresponding elements in row j: $kR_i + R_j$.

Find the result of the row operation $-4R_1 + R_2$:

$$\left[\begin{array}{ccc|c} 1 & 0 & -2 & 5 \\ 4 & -1 & 2 & 6 \\ 3 & -7 & 9 & 10 \end{array}\right].$$

Add -4 times the elements in row 1 to the corresponding elements in row 2.

$$\left[\begin{array}{ccc|c} 1 & 0 & -2 & 5 \\ -4(1) + 4 & -4(0) + (-1) & -4(-2) + 2 & -4(5) + 6 \\ 3 & -7 & 9 & 10 \end{array}\right]$$

$$= \left[\begin{array}{ccc|c} 1 & 0 & -2 & 5 \\ 0 & -1 & 10 & -14 \\ 3 & -7 & 9 & 10 \end{array}\right]$$

Solving Linear Systems Using Matrices

1. Write the augmented matrix for the system.
2. Use matrix row operations to simplify the matrix to row-eschelon form, with 1s down the main diagonal from upper left to lower right, and 0s below the 1s.
3. Write the system of linear equations corresponding to the matrix from step 2 and use back-substitution to find the system's solution.

If you obtain a matrix with a row containing 0s to the left of the vertical bar and a nonzero number on the right, the system is inconsistent. If 0s appear across an entire row, the system contains dependent equations.

Solve using matrices:

$$\begin{aligned} 3x + y &= 5 \\ x + 4y &= 9. \end{aligned}$$

$$\left[\begin{array}{cc|c} 3 & 1 & 5 \\ 1 & 4 & 9 \end{array}\right] \xrightarrow{R_1 \leftrightarrow R_2} \left[\begin{array}{cc|c} 1 & 4 & 9 \\ 3 & 1 & 5 \end{array}\right]$$

$$\xrightarrow{-3R_1 + R_2} \left[\begin{array}{cc|c} 1 & 4 & 9 \\ 0 & -11 & -22 \end{array}\right] \xrightarrow{-\frac{1}{11}R_2} \left[\begin{array}{cc|c} 1 & 4 & 9 \\ 0 & 1 & 2 \end{array}\right]$$

$$\rightarrow x + 4y = 9$$
$$y = 2.$$

When $y = 2$, $x + 4 \cdot 2 = 9$, so $x = 1$. The solution is $(1, 2)$ and the solution set is $\{(1, 2)\}$.

Definitions and Concepts	Examples

Section 3.5 Determinants and Cramer's Rule

A square matrix has the same number of rows as columns. A determinant is a real number associated with a square matrix. The determinant is denoted by placing vertical bars about the array of numbers. The value of a second-order determinant is

$$\begin{vmatrix} a_1 & b_1 \\ a_2 & b_2 \end{vmatrix} = a_1b_2 - a_2b_1.$$

Evaluate:

$$\begin{vmatrix} 2 & -1 \\ 3 & 4 \end{vmatrix} = 2(4) - 3(-1) = 8 + 3 = 11.$$

Cramer's Rule for Two Linear Equations in Two Variables

If

$$a_1x + b_1y = c_1$$
$$a_2x + b_2y = c_2,$$

then

$$x = \frac{\begin{vmatrix} c_1 & b_1 \\ c_2 & b_2 \end{vmatrix}}{\begin{vmatrix} a_1 & b_1 \\ a_2 & b_2 \end{vmatrix}} = \frac{D_x}{D} \quad \text{and} \quad y = \frac{\begin{vmatrix} a_1 & c_1 \\ a_2 & c_2 \end{vmatrix}}{\begin{vmatrix} a_1 & b_1 \\ a_2 & b_2 \end{vmatrix}} = \frac{D_y}{D}, \quad D \neq 0.$$

If $D = 0$ and any numerator is not zero, the system is inconsistent and has no solution. If all determinants are 0, the system contains dependent equations and has infinitely many solutions.

Solve by Cramer's rule:

$$5x + 3y = 7$$
$$-x + 2y = 9.$$

$$D = \begin{vmatrix} 5 & 3 \\ -1 & 2 \end{vmatrix} = 5(2) - (-1)(3) = 10 + 3 = 13$$

$$D_x = \begin{vmatrix} 7 & 3 \\ 9 & 2 \end{vmatrix} = 7 \cdot 2 - 9 \cdot 3 = 14 - 27 = -13$$

$$D_y = \begin{vmatrix} 5 & 7 \\ -1 & 9 \end{vmatrix} = 5(9) - (-1)(7) = 45 + 7 = 52$$

$$x = \frac{D_x}{D} = \frac{-13}{13} = -1, \; y = \frac{D_y}{D} = \frac{52}{13} = 4$$

The solution is $(-1, 4)$ and the solution set is $\{(-1, 4)\}$.

The value of a third-order determinant is

$$\begin{vmatrix} a_1 & b_1 & c_1 \\ a_2 & b_2 & c_2 \\ a_3 & b_3 & c_3 \end{vmatrix}$$

$$= a_1\begin{vmatrix} b_2 & c_2 \\ b_3 & c_3 \end{vmatrix} - a_2\begin{vmatrix} b_1 & c_1 \\ b_3 & c_3 \end{vmatrix} + a_3\begin{vmatrix} b_1 & c_1 \\ b_2 & c_2 \end{vmatrix}.$$

Each second-order determinant is called a minor.

Evaluate:

$$\begin{vmatrix} 1 & -2 & 1 \\ 3 & 1 & -2 \\ 5 & 5 & 3 \end{vmatrix}$$

$$= 1\begin{vmatrix} 1 & -2 \\ 5 & 3 \end{vmatrix} - 3\begin{vmatrix} -2 & 1 \\ 5 & 3 \end{vmatrix} + 5\begin{vmatrix} -2 & 1 \\ 1 & -2 \end{vmatrix}$$

$$= 1(3 - (-10)) - 3(-6 - 5) + 5(4 - 1)$$

$$= 1(13) - 3(-11) + 5(3)$$

$$= 13 + 33 + 15 = 61.$$

Definitions and Concepts	Examples

Section 3.5 Determinants and Cramer's Rule (continued)

Cramer's Rule for Three Linear Equations in Three Variables

If

$$a_1x + b_1y + c_1z = d_1$$
$$a_2x + b_2y + c_2z = d_2$$
$$a_3x + b_3y + c_3z = d_3,$$

then

$$x = \frac{D_x}{D}, \quad y = \frac{D_y}{D}, \quad z = \frac{D_z}{D}.$$

$$D = \begin{vmatrix} a_1 & b_1 & c_1 \\ a_2 & b_2 & c_2 \\ a_3 & b_3 & c_3 \end{vmatrix} \neq 0, \quad D_x = \begin{vmatrix} d_1 & b_1 & c_1 \\ d_2 & b_2 & c_2 \\ d_3 & b_3 & c_3 \end{vmatrix}$$

$$D_y = \begin{vmatrix} a_1 & d_1 & c_1 \\ a_2 & d_2 & c_2 \\ a_3 & d_3 & c_3 \end{vmatrix}, \quad D_z = \begin{vmatrix} a_1 & b_1 & d_1 \\ a_2 & b_2 & d_2 \\ a_3 & b_3 & d_3 \end{vmatrix}$$

If $D = 0$ and any numerator is not zero, the system is inconsistent. If all determinants are 0, the system contains dependent equations.

Solve by Cramer's rule:

$$x - 2y + z = 4$$
$$3x + y - 2z = 3$$
$$5x + 5y + 3z = -8.$$

$$D = \begin{vmatrix} 1 & -2 & 1 \\ 3 & 1 & -2 \\ 5 & 5 & 3 \end{vmatrix} = 61$$

This evaluation is shown on the bottom of page 234.

$$D_x = \begin{vmatrix} 4 & -2 & 1 \\ 3 & 1 & -2 \\ -8 & 5 & 3 \end{vmatrix} = 61$$

$$D_y = \begin{vmatrix} 1 & 4 & 1 \\ 3 & 3 & -2 \\ 5 & -8 & 3 \end{vmatrix} = -122$$

$$D_z = \begin{vmatrix} 1 & -2 & 4 \\ 3 & 1 & 3 \\ 5 & 5 & -8 \end{vmatrix} = -61$$

$$x = \frac{D_x}{D} = \frac{61}{61} = 1, \quad y = \frac{D_y}{D} = \frac{-122}{61} = -2,$$

$$z = \frac{D_z}{D} = \frac{-61}{61} = -1.$$

The solution is $(1, -2, -1)$ and the solution set is $\{(1, -2, -1)\}$.

CHAPTER 3 REVIEW EXERCISES

3.1 *In Exercises 1–2, determine whether the given ordered pair is a solution of the system.*

1. $(4, 2)$

$$2x - 5y = -2$$
$$3x + 4y = 4$$

2. $(-5, 3)$

$$-x + 2y = 11$$
$$y = -\frac{x}{3} + \frac{4}{3}$$

In Exercises 3–6, solve each system by graphing. Identify systems with no solution and systems with infinitely many solutions, using set notation to express their solution sets.

3. $x + y = 5$
$3x - y = 3$

4. $3x - 2y = 6$
$6x - 4y = 12$

5. $y = \frac{3}{5}x - 3$
$2x - y = -4$

6. $y = -x + 4$
$3x + 3y = -6$

In Exercises 7–13, solve each system by the substitution method or the addition method. Identify systems with no solution and systems with infinitely many solutions, using set notation to express their solution sets.

7. $2x - y = 2$
$x + 2y = 11$

8. $y = -2x + 3$
$3x + 2y = -17$

9. $3x + 2y = -8$
$2x + 5y = 2$

10. $5x - 2y = 14$
$3x + 4y = 11$

11. $y = 4 - x$
$3x + 3y = 12$

12. $\frac{x}{8} + \frac{3y}{4} = \frac{19}{8}$
$-\frac{x}{2} + \frac{3y}{4} = \frac{1}{2}$

13. $x - 2y + 3 = 0$
$2x - 4y + 7 = 0$

3.2

In Exercises 14–18, use the four-step strategy to solve each problem.

14. An appliance store is having a sale on small TVs and stereos. One day a salesperson sells 3 of the TVs and 4 stereos for \$2530. The next day the salesperson sells 4 of the same TVs and 3 of the same stereos for \$2510. What are the prices of a TV and a stereo?

15. You invested \$9000 in two funds paying 4% and 7% annual interest. At the end of the year, the total interest from these investments was \$555. How much was invested at each rate?

16. A chemist needs to mix a solution that is 34% silver nitrate with one that is 4% silver nitrate to obtain 100 milliliters of a mixture that is 7% silver nitrate. How many milliliters of each of the solutions must be used?

17. When a plane flies with the wind, it can travel 2160 miles in 3 hours. When the plane flies in the opposite direction, against the wind, it takes 4 hours to fly the same distance. Find the rate of the plane in still air and the rate of the wind.

18. The perimeter of a rectangular table top is 34 feet. The difference between 4 times the length and 3 times the width is 33 feet. Find the dimensions.

The cost and revenue functions for producing and selling x units of a new graphing calculator are

$$C(x) = 22{,}500 + 40x \quad \text{and} \quad R(x) = 85x.$$

Use these functions to solve Exercises 19–21.

19. Find the loss or the gain from selling 400 graphing calculators.

20. Determine the break-even point. Describe what this means.

21. Write the profit function, P, from producing and selling x graphing calculators.

22. A company is planning to manufacture computer desks. The fixed cost will be \$60,000 and it will cost \$200 to produce each desk. Each desk will be sold for \$450.

a. Write the cost function, C, of producing x desks.

b. Write the revenue function, R, from the sale of x desks.

c. Determine the break-even point. Describe what this means.

3.3

23. Is $(-3, -2, 5)$ a solution of the system

$$\begin{aligned} x + y + z &= 0 \\ 2x - 3y + z &= 5 \\ 4x + 2y + 4z &= 3? \end{aligned}$$

Solve each system in Exercises 24–26 by eliminating variables using the addition method. If there is no solution or if there are infinitely many solutions and a system's equations are dependent, so state.

24.
$$\begin{aligned} 2x - y + z &= 1 \\ 3x - 3y + 4z &= 5 \\ 4x - 2y + 3z &= 4 \end{aligned}$$

25.
$$\begin{aligned} x + 2y - z &= 5 \\ 2x - y + 3z &= 0 \\ 2y + z &= 1 \end{aligned}$$

26.
$$\begin{aligned} 3x - 4y + 4z &= 7 \\ x - y - 2z &= 2 \\ 2x - 3y + 6z &= 5 \end{aligned}$$

27. Find the quadratic function $y = ax^2 + bx + c$ whose graph passes through the points $(1, 4)$, $(3, 20)$, and $(-2, 25)$.

28. The bar graph shows the average debt in the United States, not including real estate mortgages, by age group. The difference between the average debt for the 30–39 age group and the 18–29 age group is \$8100. The difference between the average debt for the 40–49 age group and the 30–39 age group is \$3100. The combined average debt for these three age groups is \$44,200. Find the average debt for each of these age groups.

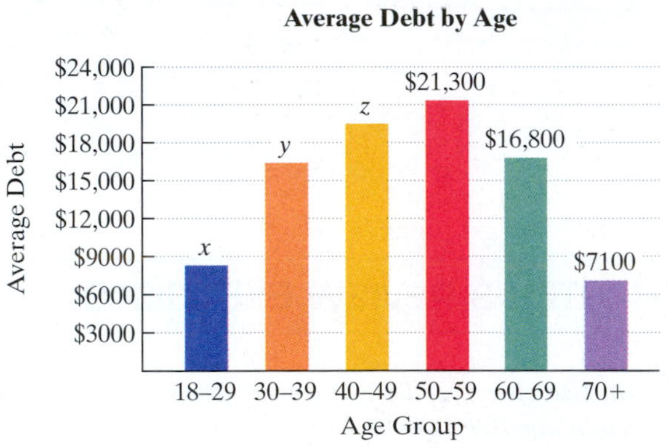

Source: Experian

3.4

In Exercises 29–32, perform each matrix row operation and write the new matrix.

29. $\left[\begin{array}{cc|c} 1 & -8 & 3 \\ 0 & 7 & -14 \end{array}\right] \; \frac{1}{7}R_2$

30. $\left[\begin{array}{cc|c} 1 & -3 & 1 \\ 2 & 1 & -5 \end{array}\right] \; -2R_1 + R_2$

31. $\left[\begin{array}{ccc|c} 2 & -2 & 1 & -1 \\ 1 & 2 & -1 & 2 \\ 6 & 4 & 3 & 5 \end{array}\right] \; \frac{1}{2}R_1$

32. $\left[\begin{array}{ccc|c} 1 & 2 & 2 & 2 \\ 0 & 1 & -1 & 2 \\ 0 & 5 & 4 & 1 \end{array}\right] \; -5R_2 + R_3$

In Exercises 33–36, solve each system using matrices. If there is no solution or if a system's equations are dependent, so state.

33. $x + 4y = 7$
$3x + 5y = 0$

34. $2x - 3y = 8$
$-6x + 9y = 4$

35. $x + 2y + 3z = -5$
$2x + y + z = 1$
$x + y - z = 8$

36. $x - 2y + z = 0$
$y - 3z = -1$
$2y + 5z = -2$

3.5 *In Exercises 37–40, evaluate each determinant.*

37. $\begin{vmatrix} 3 & 2 \\ -1 & 5 \end{vmatrix}$

38. $\begin{vmatrix} -2 & -3 \\ -4 & -8 \end{vmatrix}$

39. $\begin{vmatrix} 2 & 4 & -3 \\ 1 & -1 & 5 \\ -2 & 4 & 0 \end{vmatrix}$

40. $\begin{vmatrix} 4 & 7 & 0 \\ -5 & 6 & 0 \\ 3 & 2 & -4 \end{vmatrix}$

In Exercises 41–44, use Cramer's rule to solve each system. If there is no solution or if a system's equations are dependent, so state.

41. $x - 2y = 8$
$3x + 2y = -1$

42. $7x + 2y = 0$
$2x + y = -3$

43. $x + 2y + 2z = 5$
$2x + 4y + 7z = 19$
$-2x - 5y - 2z = 8$

44. $2x + y = -4$
$y - 2z = 0$
$3x - 2z = -11$

45. Use the quadratic function $y = ax^2 + bx + c$ to model the following data:

x (Age of a Driver)	y (Average Number of Automobile Accidents per Day in the United States)
20	400
40	150
60	400

Use Cramer's rule to determine values for a, b, and c. Then use the model to write a statement about the average number of automobile accidents in which 30-year-old drivers and 50-year-old drivers are involved daily.

CHAPTER 3 TEST

Remember to use your Chapter Test Prep Video CD to see the worked-out solutions to the test questions you want to review.

1. Solve by graphing

$$x + y = 6$$
$$4x - y = 4.$$

In Exercises 2–4, solve each system by the substitution method or the addition method. Identify systems with no solution and systems with infinitely many solutions, using set notation to express their solution sets.

2. $5x + 4y = 10$
$3x + 5y = -7$

3. $x = y + 4$
$3x + 7y = -18$

4. $4x = 2y + 6$
$y = 2x - 3$

In Exercises 5–8, solve each problem.

5. In a new development, 50 one- and two-bedroom condominiums were sold. Each one-bedroom condominium sold for \$120 thousand and each two-bedroom condominium sold for \$150 thousand. If sales totaled \$7050 thousand, how many of each type of unit was sold?

6. You invested \$9000 in two funds paying 6% and 7% annual interest. At the end of the year, the total interest from these investments was \$610. How much was invested at each rate?

7. You need to mix a 6% peroxide solution with a 9% peroxide solution to obtain 36 ounces of an 8% peroxide solution. How many ounces of each of the solutions must be used?

8. A paddleboat on the Mississippi River travels 48 miles downstream, with the current, in 3 hours. The return trip, against the current, takes the paddleboat 4 hours. Find the boat's rate in still water and the rate of the current.

Use this information to solve Exercises 9–11: A company is planning to produce and sell a new line of computers. The fixed cost will be \$360,000 and it will cost \$850 to produce each computer. Each computer will be sold for \$1150.

9. Write the cost function, C, of producing x computers.

10. Write the revenue function, R, from the sale of x computers.

11. Determine the break-even point. Describe what this means.

12. The cost and revenue functions for producing and selling x units of a toaster oven are

$$C(x) = 40x + 350{,}000 \quad \text{and} \quad R(x) = 125x.$$

Write the profit function, P, from producing and selling x toaster ovens.

13. Solve by eliminating variables using the addition method:

$$\begin{aligned} x + y + z &= 6 \\ 3x + 4y - 7z &= 1 \\ 2x - y + 3z &= 5. \end{aligned}$$

14. Perform the indicated matrix row operation and write the new matrix.

$$\left[\begin{array}{rrr|r} 1 & 0 & -4 & 5 \\ 6 & -1 & 2 & 10 \\ 2 & -1 & 4 & -3 \end{array}\right] -6R_1 + R_2$$

In Exercises 15–16, solve each system using matrices.

15. $$\begin{aligned} 2x + y &= 6 \\ 3x - 2y &= 16 \end{aligned}$$

16. $$\begin{aligned} x - 4y + 4z &= -1 \\ 2x - y + 5z &= 6 \\ -x + 3y - z &= 5 \end{aligned}$$

In Exercises 17–18, evaluate each determinant.

17. $\begin{vmatrix} -1 & -3 \\ 7 & 4 \end{vmatrix}$

18. $\begin{vmatrix} 3 & 4 & 0 \\ -1 & 0 & -3 \\ 4 & 2 & 5 \end{vmatrix}$

In Exercises 19–20, use Cramer's rule to solve each system.

19. $$\begin{aligned} 4x - 3y &= 14 \\ 3x - y &= 3 \end{aligned}$$

20. $$\begin{aligned} 2x + 3y + z &= 2 \\ 3x + 3y - z &= 0 \\ x - 2y - 3z &= 1 \end{aligned}$$

CUMULATIVE REVIEW EXERCISES (CHAPTERS 1–3)

1. Simplify: $\dfrac{6(8-10)^3 + (-2)}{(-5)^2(-2)}$.

2. Simplify: $7x - [5 - 2(4x - 1)]$.

In Exercises 3–5, solve each equation.

3. $5 - 2(3 - x) = 2(2x + 5) + 1$

4. $\dfrac{3x}{5} + 4 = \dfrac{x}{3}$

5. $3x - 4 = 2(3x + 2) - 3x$

6. For a summer sales job, you are choosing between two pay arrangements: a weekly salary of \$200 plus 5% commission on sales, or a straight 15% commission. For how many dollars of sales will the earnings be the same regardless of the pay arrangement?

7. Simplify: $\dfrac{-5x^6y^{-10}}{20x^{-2}y^{20}}$.

8. If $f(x) = -4x + 5$, find $f(a + 2)$.

9. Find the domain of $f(x) = \dfrac{4}{x + 3}$.

10. If $f(x) = 2x^2 - 5x + 2$ and $g(x) = x^2 - 2x + 3$, find $(f - g)(x)$ and $(f - g)(3)$.

In Exercises 11–12, graph each linear function.

11. $f(x) = -\frac{2}{3}x + 2$

12. $2x - y = 6$

In Exercises 13–14, use the given conditions to write an equation for each line in point-slope form and slope-intercept form.

13. Passing through $(2, 4)$ and $(4, -2)$

14. Passing through $(-1, 0)$ and parallel to the line whose equation is $3x + y = 6$

In Exercises 15–16, solve each system by eliminating variables using the addition method.

15. $$\begin{aligned} 3x + 12y &= 25 \\ 2x - 6y &= 12 \end{aligned}$$

16. $$\begin{aligned} x + 3y - z &= 5 \\ -x + 2y + 3z &= 13 \\ 2x - 5y - z &= -8 \end{aligned}$$

17. If two pads of paper and 19 pens are sold for \$5.40 and 7 of the same pads and 4 of the same pens sell for \$6.40, find the cost of one pad and one pen.

18. Evaluate:

$$\begin{vmatrix} 0 & 1 & -2 \\ -7 & 0 & -4 \\ 3 & 0 & 5 \end{vmatrix}.$$

19. Solve using matrices:

$$\begin{aligned} 2x + 3y - z &= -1 \\ x + 2y + 3z &= 2 \\ 3x + 5y - 2z &= -3. \end{aligned}$$

20. Solve using Cramer's rule (determinants):

$$\begin{aligned} 3x + 4y &= -1 \\ -2x + y &= 8. \end{aligned}$$

CHAPTER 4

Inequalities and Problem Solving

You are in Yosemite National Park in California, surrounded by evergreen forests, alpine meadows, and sheer walls of granite. The beauty of soaring cliffs, plunging waterfalls, gigantic trees, rugged canyons, mountains, and valleys is overwhelming. This is so different from where you live and attend college, a region in which grasslands predominate. What variables affect whether regions are forests, grasslands, or deserts, and what kinds of mathematical models are used to describe the incredibly diverse land that forms the surface of our planet?

The role that temperature and precipitation play in determining whether regions are forests, grasslands, or deserts can be modeled using inequalities in two variables. You will use these models and their graphs in Example 4 of Section 4.4.

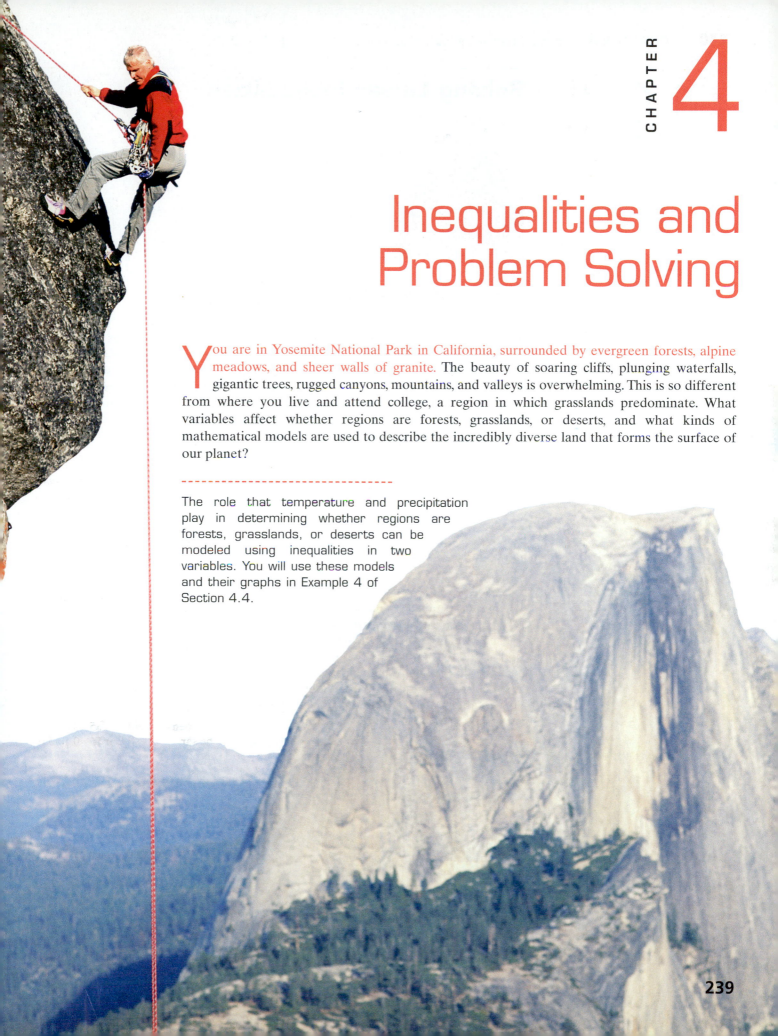

SECTION 4.1

Solving Linear Inequalities

Objectives

1. Use interval notation.
2. Solve linear inequalities.
3. Recognize inequalities with no solution or all real numbers as solutions.
4. Solve applied problems using linear inequalities.

You can go online and obtain a list of telecommunication companies that provide residential long-distance phone service. The list contains the monthly fee, the monthly minimum, and the rate per minute for each service provider. You've chosen a plan that has a monthly fee of $15 with a charge of $0.08 per minute for all long-distance calls. Suppose you are limited by how much money you can spend for the month: You can spend at most $35. If we let x represent the number of minutes of long-distance calls in a month, we can write an inequality that describes the given conditions:

The monthly fee of $15	plus	the charge of $0.08 per minute for x minutes	must be less than or equal to	$35.
15	+	$0.08x$	$\le$	35.

Using the commutative property of addition, we can express this inequality as

$$0.08x + 15 \le 35.$$

Placing an inequality symbol between a linear expression $(mx + b)$ and a constant results in a *linear inequality in one variable*. In this section, we will study how to solve linear inequalities such as the one shown above. **Solving an inequality** is the process of finding the set of numbers that make the inequality a true statement. These numbers are called the **solutions** of the inequality and we say that they **satisfy** the inequality. The set of all solutions is called the **solution set** of the inequality. Set-builder notation and a new notation, called *interval notation*, are used to represent solution sets. We begin this section by looking at interval notation.

1 Use interval notation.

Interval Notation

Some sets of real numbers can be represented using **interval notation**. Suppose that a and b are two real numbers such that $a < b$.

Interval Notation	Graph
The **open interval** (a, b) represents the set of real numbers between, but not including, a and b. $(a, b) = \{x \mid a < x < b\}$ x is greater than a $(a < x)$ and x is less than b $(x < b)$.	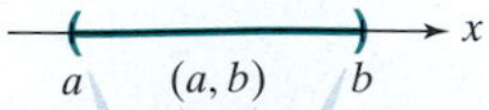 The parentheses in the graph and in interval notation indicate that a and b, the endpoints, are excluded from the interval.

(continued)

Interval Notation	Graph
The **closed interval** $[a, b]$ represents the set of real numbers between, and including, a and b. $$[a, b] = \{x \mid a \leq x \leq b\}$$ x is greater than or equal to a $(a \leq x)$ and x is less than or equal to b $(x \leq b)$.	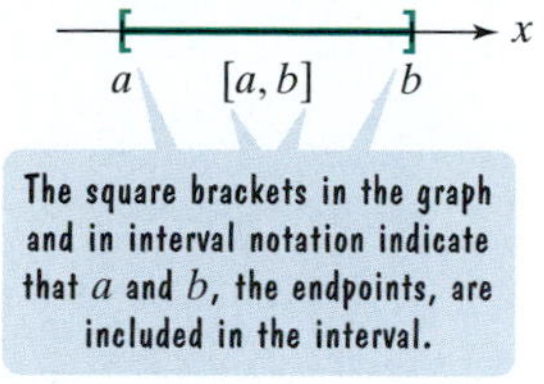
The **infinite interval** (a, ∞) represents the set of real numbers that are greater than a. $$(a, \infty) = \{x \mid x > a\}$$ The infinity symbol does not represent a real number. It indicates that the interval extends indefinitely to the right.	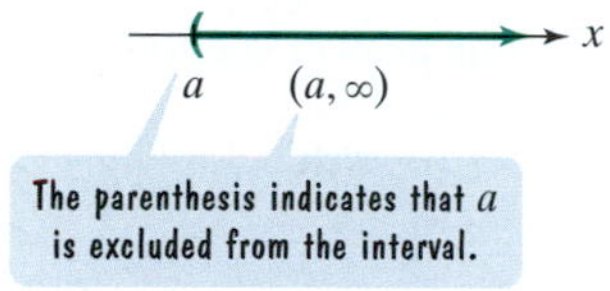
The **infinite interval** $(-\infty, b]$ represents the set of real numbers that are less than or equal to b. $$(-\infty, b] = \{x \mid x \leq b\}$$ The negative infinity symbol indicates that the interval extends indefinitely to the left.	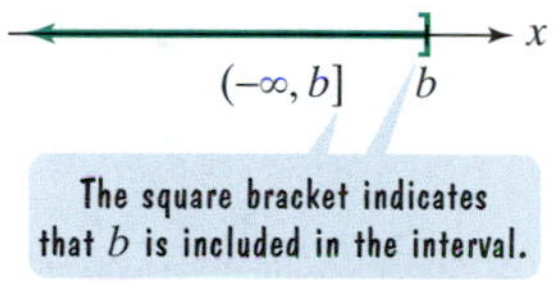

Parentheses and Brackets in Interval Notation

Parentheses indicate endpoints that are not included in an interval. Square brackets indicate endpoints that are included in an interval. Parentheses are always used with ∞ or $-\infty$.

Table 4.1 lists nine possible types of intervals used to describe sets of real numbers.

Table 4.1 Intervals on the Real Number Line

Let a and b be real numbers such that $a < b$.

Interval Notation	Set-Builder Notation	Graph
(a, b)	$\{x \mid a < x < b\}$	
$[a, b]$	$\{x \mid a \leq x \leq b\}$	
$[a, b)$	$\{x \mid a \leq x < b\}$	
$(a, b]$	$\{x \mid a < x \leq b\}$	
(a, ∞)	$\{x \mid x > a\}$	
$[a, \infty)$	$\{x \mid x \geq a\}$	
$(-\infty, b)$	$\{x \mid x < b\}$	
$(-\infty, b]$	$\{x \mid x \leq b\}$	
$(-\infty, \infty)$	$\{x \mid x \text{ is a real number}\}$ or $\mathbb{R}$ (set of all real numbers)	

EXAMPLE 1 Interpreting Interval Notation

Express each interval in set-builder notation and graph:

a. $(-1, 4]$ **b.** $[2.5, 4]$ **c.** $(-4, \infty)$.

Solution

a. $(-1, 4] = \{x \mid -1 < x \leq 4\}$

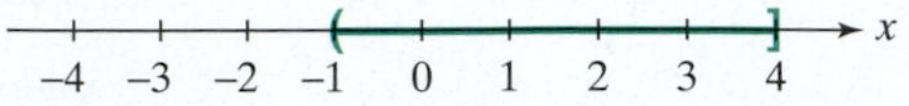

b. $[2.5, 4] = \{x \mid 2.5 \leq x \leq 4\}$

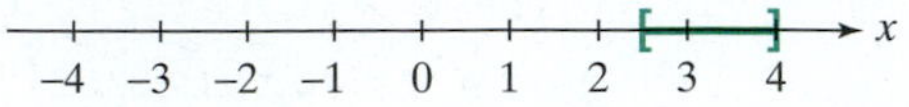

c. $(-4, \infty) = \{x \mid x > -4\}$

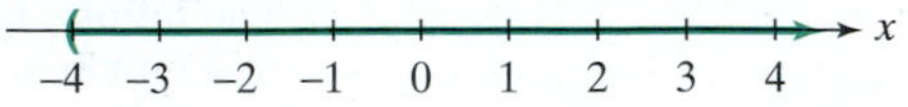

See graphing answer section.

✓ CHECK POINT 1 Express each interval in set-builder notation and graph:

a. $[-2, 5)$ **b.** $[1, 3.5]$ **c.** $(-\infty, -1)$.

---

2 Solve linear inequalities.

Solving Linear Inequalities in One Variable

We know that a linear equation in x can be expressed as $ax + b = 0$. A **linear inequality in x** can be written in one of the following forms: $ax + b < 0$, $ax + b \leq 0$, $ax + b > 0$, $ax + b \geq 0$. In each form, $a \neq 0$.

Back to our question that opened this section: How many minutes of long-distance calls can you make in a month if you can spend at most \$35? We answer the question by solving the linear inequality

$$0.08x + 15 \leq 35$$

for x. The solution procedure is nearly identical to that for solving the equation

$$0.08x + 15 = 35.$$

Our goal is to get x by itself on the left side. We do this by first subtracting 15 from both sides to isolate $0.08x$:

$$0.08x + 15 \leq 35 \quad \text{This is the given inequality.}$$
$$0.08x + 15 - 15 \leq 35 - 15 \quad \text{Subtract 15 from both sides.}$$
$$0.08x \leq 20. \quad \text{Simplify.}$$

Finally, we isolate x from $0.08x$ by dividing both sides of the inequality by 0.08:

$$\frac{0.08x}{0.08} \leq \frac{20}{0.08} \quad \text{Divide both sides by 0.08.}$$
$$x \leq 250. \quad \text{Simplify.}$$

With at most \$35 per month to spend, you can make no more than 250 minutes of long-distance calls each month.

We started with the inequality $0.08x + 15 \leq 35$ and obtained the inequality $x \leq 250$ in the final step. Both of these inequalities have the same solution set, namely $\{x \mid x \leq 250\}$. Inequalities such as these, with the same solution set, are said to be **equivalent**.

We isolated x from $0.08x$ by dividing both sides of $0.08x \leq 20$ by 0.08, a positive number. Let's see what happens if we divide both sides of an inequality by a negative number. Consider the inequality $10 < 14$. Divide 10 and 14 by -2:

$$\frac{10}{-2} = -5 \quad \text{and} \quad \frac{14}{-2} = -7.$$

Because -5 lies to the right of -7 on the number line, -5 is greater than -7:

$$-5 > -7.$$

Notice that the direction of the inequality symbol is reversed:

$$\begin{array}{c} 10 < 14 \\ \downarrow \\ -5 > -7. \end{array}$$

Dividing by -2 changes the direction of the inequality symbol.

In general, **when we multiply or divide both sides of an inequality by a negative number, the direction of the inequality symbol is reversed**. When we reverse the direction of the inequality symbol, we say that we change the *sense* of the inequality.

We can isolate a variable in a linear inequality the same way we can isolate a variable in a linear equation. The following properties are used to create equivalent inequalities:

Properties of Inequalities

Property	The Property in Words	Example
The Addition Property of Inequality If $a < b$, then $a + c < b + c$. If $a < b$, then $a - c < b - c$.	If the same quantity is added to or subtracted from both sides of an inequality, the resulting inequality is equivalent to the original one.	$2x + 3 < 7$ Subtract 3: $2x + 3 - 3 < 7 - 3.$ Simplify: $2x < 4.$
The Positive Multiplication Property of Inequality If $a < b$ and c is positive, then $ac < bc$. If $a < b$ and c is positive, then $\frac{a}{c} < \frac{b}{c}$.	If we multiply or divide both sides of an inequality by the same positive quantity, the resulting inequality is equivalent to the original one.	$2x < 4$ Divide by 2: $\frac{2x}{2} < \frac{4}{2}.$ Simplify: $x < 2.$
The Negative Multiplication Property of Inequality If $a < b$ and c is negative, then $ac > bc$. If $a < b$ and c is negative, then $\frac{a}{c} > \frac{b}{c}$.	If we multiply or divide both sides of an inequality by the same negative quantity and reverse the direction of the inequality symbol, the resulting inequality is equivalent to the original one.	$-4x < 20$ Divide by -4 and change the sense of the inequality: $\frac{-4x}{-4} > \frac{20}{-4}.$ Simplify: $x > -5.$

If an inequality does not contain fractions, it can be solved using the following procedure. (In Example 4, we will see how to clear fractions.) Notice, again, how similar this procedure is to the procedure for solving a linear equation.

Solving a Linear Inequality

1. Simplify the algebraic expression on each side.
2. Use the addition property of inequality to collect all the variable terms on one side and all the constant terms on the other side.
3. Use the multiplication property of inequality to isolate the variable and solve. Change the sense of the inequality when multiplying or dividing both sides by a negative number.
4. Express the solution set in set-builder or interval notation and graph the solution set on a number line.

EXAMPLE 2 Solving a Linear Inequality

Solve and graph the solution set on a number line:

$$3x - 5 > -17.$$

Solution

Step 1. Simplify each side. Because each side is already simplified, we can skip this step.

Step 2. Collect variable terms on one side and constant terms on the other side. The variable term, $3x$, is already on the left side of $3x - 5 > -17$. We will collect constant terms on the right side by adding 5 to both sides.

$$3x - 5 > -17 \quad \text{This is the given inequality.}$$
$$3x - 5 + 5 > -17 + 5 \quad \text{Add 5 to both sides.}$$
$$3x > -12 \quad \text{Simplify.}$$

Step 3. Isolate the variable and solve. We isolate the variable, x, by dividing both sides by 3. Because we are dividing by a positive number, we do not reverse the direction of the inequality symbol.

$$\frac{3x}{3} > \frac{-12}{3} \quad \text{Divide both sides by 3.}$$
$$x > -4 \quad \text{Simplify.}$$

Step 4. Express the solution set in set-builder or interval notation and graph the set on a number line. The solution set consists of all real numbers that are greater than -4, expressed as $\{x|x > -4\}$ in set-builder notation. The interval notation for this solution set is $(-4, \infty)$. The graph of the solution set is shown as follows:

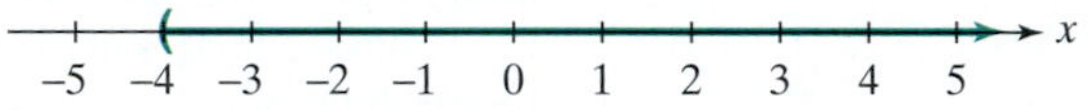

CHECK POINT 2 Solve and graph the solution set on a number line:

$$4x - 3 > -23.$$

Discover for Yourself

As a partial check, select one number from the solution set of $3x - 5 > -17$, the inequality in Example 2. Substitute that number into the original inequality. Perform the resulting computations. You should obtain a true statement.

Is it possible to perform a partial check using a number that is not in the solution set? What should happen in this case? Try doing this.

EXAMPLE 3 Solving a Linear Inequality

Solve and graph the solution set on a number line:

$$-2x - 4 > x + 5.$$

Solution

Step 1. Simplify each side. Because each side is already simplified, we can skip this step.

Step 2. Collect variable terms on one side and constant terms on the other side. We will collect variable terms on the left and constant terms on the right.

$$-2x - 4 > x + 5 \quad \text{This is the given inequality.}$$
$$-2x - 4 - x > x + 5 - x \quad \text{Subtract } x \text{ from both sides.}$$
$$-3x - 4 > 5 \quad \text{Simplify.}$$
$$-3x - 4 + 4 > 5 + 4 \quad \text{Add 4 to both sides.}$$
$$-3x > 9 \quad \text{Simplify.}$$

Step 3. Isolate the variable and solve. We isolate the variable, x, by dividing both sides by -3. Because we are dividing by a negative number, we must reverse the direction of the inequality symbol.

$$\frac{-3x}{-3} < \frac{9}{-3} \quad \text{Divide both sides by } -3 \text{ and change the sense of the inequality.}$$
$$x < -3 \quad \text{Simplify.}$$

Study Tip

You can solve

$$-2x - 4 > x + 5$$

by isolating x on the right side. Add $2x$ to both sides.

$$-2x - 4 + 2x > x + 5 + 2x$$
$$-4 > 3x + 5$$

Now subtract 5 from both sides.

$$-4 - 5 > 3x + 5 - 5$$
$$-9 > 3x$$

Finally, divide both sides by 3.

$$\frac{-9}{3} > \frac{3x}{3}$$
$$-3 > x$$

This last inequality means the same thing as $x < -3$.

Step 4. Express the solution set in set-builder or interval notation and graph the set on a number line. The solution set consists of all real numbers that are less than -3, expressed in set-builder notation as $\{x|x < -3\}$. The interval notation for this solution set is $(-\infty, -3)$. The graph of the solution set is shown as follows:

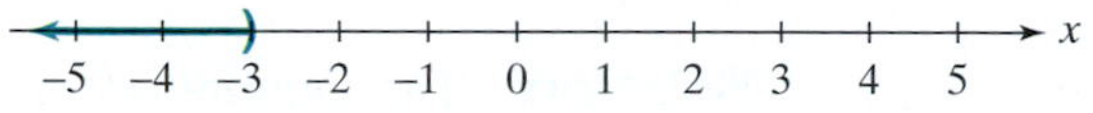

x
−5 −4 −3 −2 −1 0 1 2 3 4 5

Using Technology

Numeric and Graphic Connections

You can use a graphing utility to check the solution set of a linear inequality. Enter each side of the inequality separately under y_1 and y_2. Then use the table or the graphs. To use the table, first locate the x-value for which the y-values are the same. Then scroll up or down to locate x values for which y_1 is greater than y_2 or for which y_1 is less than y_2. To use the graphs, locate the intersection point and then find the x-values for which the graph of y_1 lies above the graph of y_2 ($y_1 > y_2$) or for which the graph of y_1 lies below the graph of y_2 ($y_1 < y_2$).

Let's verify our work in Example 3 and show that $(-\infty, -3)$ is the solution set of

$$-2x - 4 > x + 5.$$

Enter $y_1 = -2x - 4$ in the $\boxed{y=}$ screen.

Enter $y_2 = x + 5$ in the $\boxed{y=}$ screen.

We are looking for values of x for which y_1 is greater than y_2.

Numeric Check

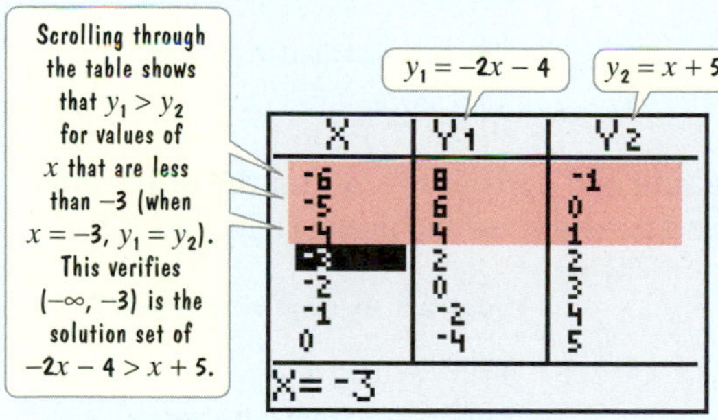

Scrolling through the table shows that $y_1 > y_2$ for values of x that are less than −3 (when $x = -3$, $y_1 = y_2$). This verifies $(-\infty, -3)$ is the solution set of $-2x - 4 > x + 5$.
$y_1 = -2x - 4$
$y_2 = x + 5$
X Y1 Y2
-6 8 -1
-5 6 0
-4 4 1
-3 2 2
-2 0 3
-1 -2 4
0 -4 5
X=-3

Graphic Check

Display the graphs for y_1 and y_2. Use the intersection feature. The solution set is the set of x-values for which the graph of y_1 lies above the graph of y_2.

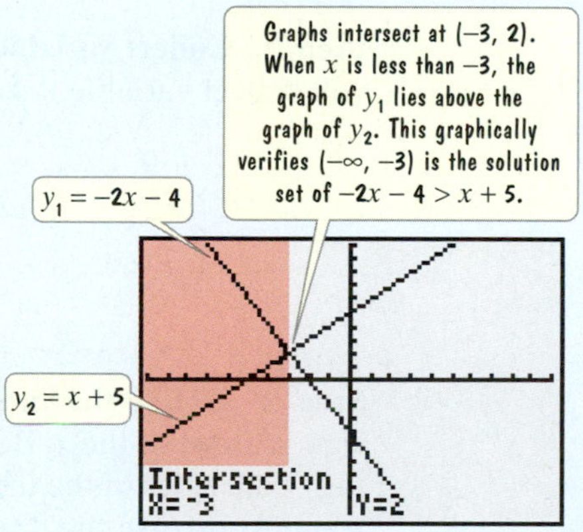

Graphs intersect at (−3, 2). When x is less than −3, the graph of y_1 lies above the graph of y_2. This graphically verifies $(-\infty, -3)$ is the solution set of $-2x - 4 > x + 5$.
$y_1 = -2x - 4$
$y_2 = x + 5$
Intersection
X=-3
Y=2

[−10, 10, 1] by [−10, 10, 1]

✓ **CHECK POINT 3** Solve and graph the solution set: $3x + 1 > 7x - 15$.

If an inequality contains fractions, begin by multiplying both sides by the least common denominator. This will clear the inequality of fractions.

EXAMPLE 4 Solving a Linear Inequality Containing Fractions

Solve and graph the solution set on a number line:

$$\frac{x+3}{4} \geq \frac{x-2}{3} + \frac{1}{4}.$$

Solution The denominators are 4, 3, and 4. The least common denominator is 12. We begin by multiplying both sides of the inequality by 12.

$$\frac{x+3}{4} \geq \frac{x-2}{3} + \frac{1}{4}$$

This is the given inequality.

$$12\left(\frac{x+3}{4}\right) \geq 12\left(\frac{x-2}{3} + \frac{1}{4}\right)$$

Multiply both sides by 12. Multiplying by a positive number preserves the sense of the inequality.

$$\frac{12}{1}\cdot\frac{x+3}{4} \geq \frac{12}{1}\cdot\frac{x-2}{3} + \frac{12}{1}\cdot\frac{1}{4}$$

Multiply each term by 12. Use the distributive property on the right side.

$$\frac{\overset{3}{\cancel{12}}}{1}\cdot\frac{x+3}{\underset{1}{\cancel{4}}} \geq \frac{\overset{4}{\cancel{12}}}{1}\cdot\frac{x-2}{\underset{1}{\cancel{3}}} + \frac{\overset{3}{\cancel{12}}}{1}\cdot\frac{1}{\underset{1}{\cancel{4}}}$$

Divide out common factors in each multiplication.

$$3(x+3) \geq 4(x-2) + 3$$

The fractions are now cleared.

Now that the fractions have been cleared, we follow the four steps that we used in the previous examples.

Step 1. Simplify each side.

$$3(x+3) \geq 4(x-2) + 3$$

This is the inequality with the fractions cleared.

$$3x + 9 \geq 4x - 8 + 3$$

Use the distributive property.

$$3x + 9 \geq 4x - 5$$

Simplify.

Step 2. Collect variable terms on one side and constant terms on the other side. We will collect variable terms on the left and constant terms on the right.

$$3x + 9 - 4x \geq 4x - 5 - 4x$$

Subtract 4x from both sides.

$$-x + 9 \geq -5$$

Simplify.

$$-x + 9 - 9 \geq -5 - 9$$

Subtract 9 from both sides.

$$-x \geq -14$$

Simplify.

Step 3. Isolate the variable and solve. To isolate x, we must eliminate the negative sign in front of the x. Because $-x$ means $-1x$, we can do this by multiplying (or dividing) both sides of the inequality by -1. We are multiplying by a negative number. Thus, we must reverse the direction of the inequality symbol.

$$(-1)(-x) \leq (-1)(-14)$$

Multiply both sides by -1 and change the sense of the inequality.

$$x \leq 14$$

Simplify.

Step 4. Express the solution set in set-builder or interval notation and graph the set on a number line. The solution set consists of all real numbers that are less than or equal to 14, expressed in set-builder notation as $\{x \mid x \leq 14\}$. The interval notation for this solution set is $(-\infty, 14]$. The graph of the solution set is shown as follows:

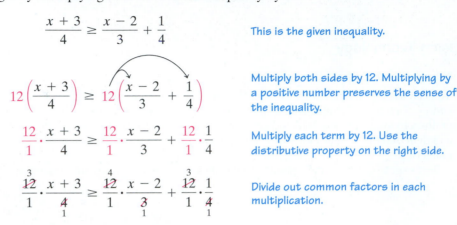

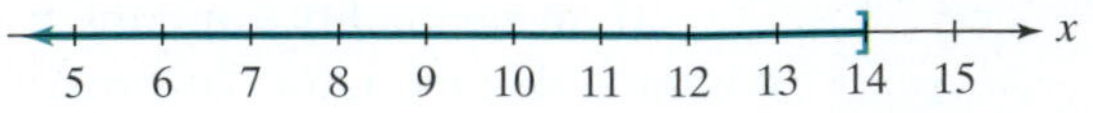

✓ **CHECK POINT 4** Solve and graph the solution set on a number line:

$$\frac{x-4}{2} \geq \frac{x-2}{3} + \frac{5}{6}.$$

3 Recognize inequalities with no solution or all real numbers as solutions.

Inequalities with Unusual Solution Sets

We have seen that some equations have no solution. This is also true for some inequalities. An example of such an inequality is

$$x > x + 1.$$

There is no number that is greater than itself plus 1. This inequality has no solution and its solution set is ∅, the empty set.

By contrast, some inequalities are true for all real numbers. An example of such an inequality is

$$x < x + 1.$$

Every real number is less than itself plus 1. The solution set is $\{x | x \text{ is a real number}\}$ or ℝ. In interval notation, the solution set is $(-\infty, \infty)$.

If you attempt to solve an inequality that has no solution, you will eliminate the variable and obtain a false statement, such as $0 > 1$. If you attempt to solve an inequality that is true for all real numbers, you will eliminate the variable and obtain a true statement, such as $0 < 1$.

Using Technology

Graphic Connections

The graphs of

$y_1 = 2(x + 4)$ and $y_2 = 2x + 3$

are parallel lines. The graph of y_1 is always above the graph of y_2. Every value of x satisfies the inequality $y_1 > y_2$. Thus, the solution set of the inequality

$$2(x + 4) > 2x + 3$$

is $(-\infty, \infty)$.

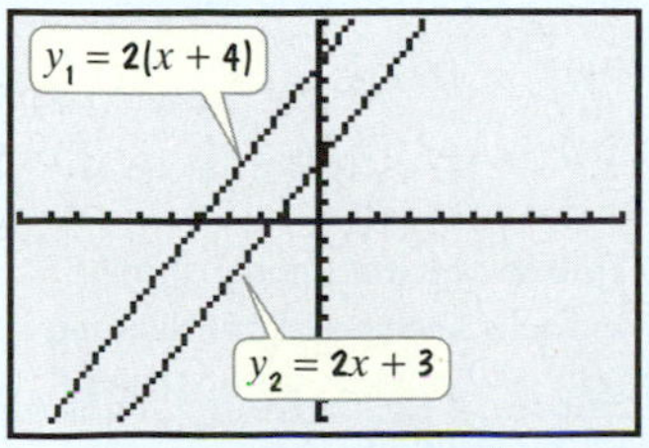

[−10, 10, 1] by [−10, 10, 1]

EXAMPLE 5 Solving Linear Inequalities

Solve each inequality:

a. $2(x + 4) > 2x + 3$ **b.** $x + 7 \leq x - 2$.

Solution

a.

$2(x + 4) > 2x + 3$	*This is the given inequality.*
$2x + 8 > 2x + 3$	*Apply the distributive property.*
$2x + 8 - 2x > 2x + 3 - 2x$	*Subtract 2x from both sides.*
$8 > 3$	*Simplify. The statement $8 > 3$ is true.*

The inequality $8 > 3$ is true for all values of x. Because this inequality is equivalent to the original inequality, the original inequality is true for all real numbers. The solution set is

$$\{x | x \text{ is a real number}\} \text{ or } \mathbb{R} \text{ or } (-\infty, \infty).$$

b.

$x + 7 \leq x - 2$	*This is the given inequality.*
$x + 7 - x \leq x - 2 - x$	*Subtract x from both sides.*
$7 \leq -2$	*Simplify. The statement $7 \leq -2$ is false.*

The inequality $7 \leq -2$ is false for all values of x. Because this inequality is equivalent to the original inequality, the original inequality has no solution. The solution set is ∅.

✓ **CHECK POINT 5** Solve each inequality:

a. $3(x + 1) > 3x + 2$

b. $x + 1 \leq x - 1$.

4 Solve applied problems using linear inequalities.

Applications

Commonly used English phrases such as "at least" and "at most" indicate inequalities. **Table 4.2** lists sentences containing these phrases and their algebraic translations into inequalities.

Table 4.2 English Sentences and Inequalities

English Sentence	Inequality
x is at least 5.	$x \geq 5$
x is at most 5.	$x \leq 5$
x is between 5 and 7.	$5 < x < 7$
x is no more than 5.	$x \leq 5$
x is no less than 5.	$x \geq 5$

Our next example shows how to use an inequality to select the better deal when considering two pricing options. We use our strategy for solving word problems, translating from the verbal conditions of the problem to a linear inequality.

EXAMPLE 6 Selecting the Better Deal

Acme Car rental agency charges \$4 a day plus \$0.15 per mile, whereas Interstate rental agency charges \$20 a day and \$0.05 per mile. How many miles must be driven to make the daily cost of an Acme rental a better deal than an Interstate rental?

Solution

Step 1. Let x represent one of the unknown quantities. We are looking for the number of miles that must be driven in a day to make Acme the better deal. Thus,

let x = the number of miles driven in a day.

Step 2. Represent other unknown quantities in terms of x. We are not asked to find another quantity, so we can skip this step.

Step 3. Write an inequality in x that models the conditions. Acme is a better deal than Interstate if the daily cost of Acme is less than the daily cost of Interstate.

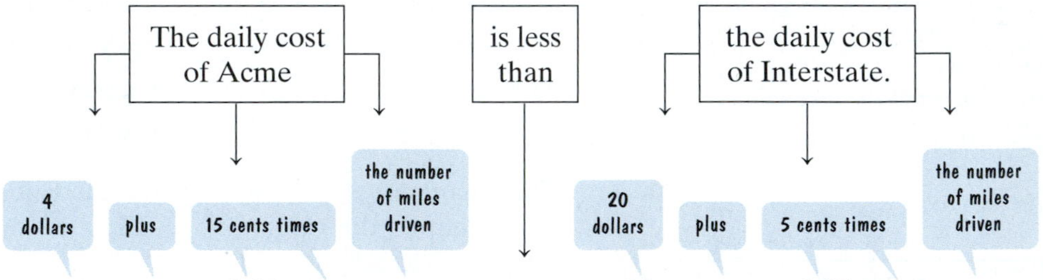

$$4 + 0.15 \cdot x < 20 + 0.05 \cdot x$$

Step 4. Solve the inequality and answer the question.

$$4 + 0.15x < 20 + 0.05x$$ This is the inequality that models the verbal conditions.

$$4 + 0.15x - 0.05x < 20 + 0.05x - 0.05x$$ Subtract 0.05x from both sides.

$$4 + 0.1x < 20$$ Simplify.

$$4 + 0.1x - 4 < 20 - 4$$ Subtract 4 from both sides.

$$0.1x < 16$$ Simplify.

$$\frac{0.1x}{0.1} < \frac{16}{0.1}$$ Divide both sides by 0.1.

$$x < 160$$ Simplify.

Thus, driving fewer than 160 miles per day makes Acme the better deal.

Using Technology

Graphic Connections

The graphs of the daily cost models for the car rental agencies

$$y_1 = 4 + 0.15x$$

and $y_2 = 20 + 0.05x$

are shown in a [0, 300, 10] by [0, 40, 4] viewing rectangle. The graphs intersect at (160, 28). To the left of $x = 160$, the graph of Acme's daily cost lies below that of Interstate's daily cost. This shows that for fewer than 160 miles per day, Acme offers the better deal.

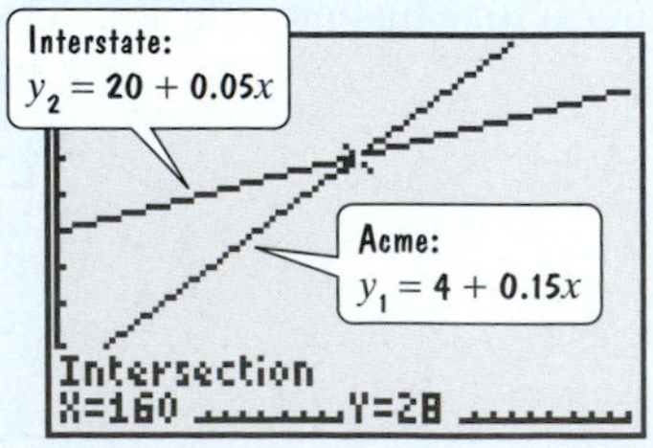

Step 5. Check the proposed solution in the original wording of the problem. One way to do this is to take a mileage less than 160 miles per day to see if Acme is the better deal. Suppose that 150 miles are driven in a day.

$$\text{Cost for Acme} = 4 + 0.15(150) = 26.50$$
$$\text{Cost for Interstate} = 20 + 0.05(150) = 27.50$$

Acme has a lower daily cost, making Acme the better deal.

CHECK POINT 6 A car can be rented from Basic Rental for \$260 per week with no extra charge for mileage. Continental charges \$80 per week plus 25 cents for each mile driven to rent the same car. How many miles must be driven in a week to make the rental cost for Basic Rental a better deal than Continental's?

4.1 EXERCISE SET

MyMathLab — MathXL PRACTICE — WATCH — DOWNLOAD — READ — REVIEW

Practice Exercises

In Exercises 1–14, express each interval in set-builder notation and graph the interval on a number line.

1. $(1, 6]$
2. $(-2, 4]$
3. $[-5, 2)$
4. $[-4, 3)$
5. $[-3, 1]$
6. $[-2, 5]$
7. $(2, \infty)$
8. $(3, \infty)$
9. $[-3, \infty)$
10. $[-5, \infty)$
11. $(-\infty, 3)$
12. $(-\infty, 2)$
13. $(-\infty, 5.5)$
14. $(-\infty, 3.5]$

In Exercises 15–46, solve each linear inequality. Other than ∅, graph the solution set on a number line.

15. $5x + 11 < 26$
16. $2x + 5 < 17$
17. $3x - 8 \geq 13$
18. $8x - 2 \geq 14$
19. $-9x \geq 36$
20. $-5x \leq 30$
21. $8x - 11 \leq 3x - 13$
22. $18x + 45 \leq 12x - 8$
23. $4(x + 1) + 2 \geq 3x + 6$
24. $8x + 3 > 3(2x + 1) + x + 5$
25. $2x - 11 < -3(x + 2)$
26. $-4(x + 2) > 3x + 20$
27. $1 - (x + 3) \geq 4 - 2x$
28. $5(3 - x) \leq 3x - 1$
29. $\frac{x}{4} - \frac{1}{2} \leq \frac{x}{2} + 1$
30. $\frac{3x}{10} + 1 \geq \frac{1}{5} - \frac{x}{10}$
31. $1 - \frac{x}{2} > 4$
32. $7 - \frac{4}{5}x < \frac{3}{5}$
33. $\frac{x - 4}{6} \geq \frac{x - 2}{9} + \frac{5}{18}$
34. $\frac{4x - 3}{6} + 2 \geq \frac{2x - 1}{12}$
35. $4(3x - 2) - 3x < 3(1 + 3x) - 7$
36. $3(x - 8) - 2(10 - x) < 5(x - 1)$
37. $8(x + 1) \leq 7(x + 5) + x$
38. $4(x - 1) \geq 3(x - 2) + x$
39. $3x < 3(x - 2)$
40. $5x < 5(x - 3)$
41. $7(x + 4) - 13 < 12 + 13(3 + x)$
42. $-3[7x - (2x - 3)] > -2(x + 1)$
43. $6 - \frac{2}{3}(3x - 12) \leq \frac{2}{5}(10x + 50)$
44. $\frac{2}{7}(7 - 21x) - 4 > 10 - \frac{3}{11}(11x - 11)$
45. $3[3(x + 5) + 8x + 7] + 5[3(x - 6) - 2(3x - 5)] < 2(4x + 3)$

46. $5[3(2 - 3x) - 2(5 - x)] - 6[5(x - 2) - 2(4x - 3)] < 3x + 19$

47. Let $f(x) = 3x + 2$ and $g(x) = 5x - 8$. Find all values of x for which $f(x) > g(x)$.

48. Let $f(x) = 2x - 9$ and $g(x) = 5x + 4$. Find all values of x for which $f(x) > g(x)$.

49. Let $f(x) = \frac{2}{5}(10x + 15)$ and $g(x) = \frac{1}{4}(8 - 12x)$. Find all values of x for which $g(x) \le f(x)$.

50. Let $f(x) = \frac{3}{5}(10x - 15) + 9$ and $g(x) = \frac{3}{8}(16 - 8x) - 7$. Find all values of x for which $g(x) \le f(x)$.

51. Let $f(x) = 1 - (x + 3) + 2x$. Find all values of x for which $f(x)$ is at least 4.

52. Let $f(x) = 2x - 11 + 3(x + 2)$. Find all values of x for which $f(x)$ is at most 0.

Practice PLUS

In Exercises 53–54, solve each linear inequality and graph the solution set on a number line.

53. $2(x + 3) > 6 - \{4[x - (3x - 4) - x] + 4\}$

54. $3(4x - 6) < 4 - \{5x - [6x - (4x - (3x + 2))]\}$

In Exercises 55–56, write an inequality with x isolated on the left side that is equivalent to the given inequality.

55. $ax + b > c$; Assume $a < 0$.

56. $\frac{ax + b}{c} > b$; Assume $a > 0$ and $c < 0$.

In Exercises 57–58, use the graphs of y_1 and y_2 to solve each inequality.

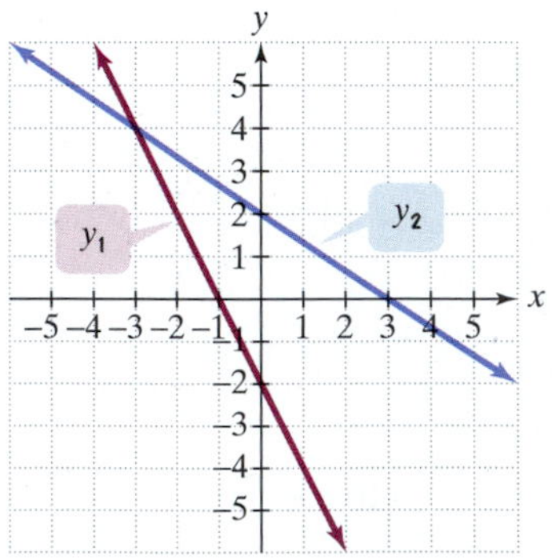

57. $y_1 \ge y_2$

58. $y_1 \le y_2$

In Exercises 59–60, use the table of values for the linear functions y_1 and y_2 to solve each inequality.

X	Y1	Y2
-1.8	9.4	7.4
-1.7	9.1	7.6
-1.6	8.8	7.8
-1.5	8.5	8
-1.4	8.2	8.2
-1.3	7.9	8.4
-1.2	7.6	8.6

X=-1.8

59. $y_1 < y_2$

60. $y_1 > y_2$

Application Exercises

The graphs show that the three components of love, namely passion, intimacy, and commitment, progress differently over time. Passion peaks early in a relationship and then declines. By contrast, intimacy and commitment build gradually. Use the graphs to solve Exercises 61–68.

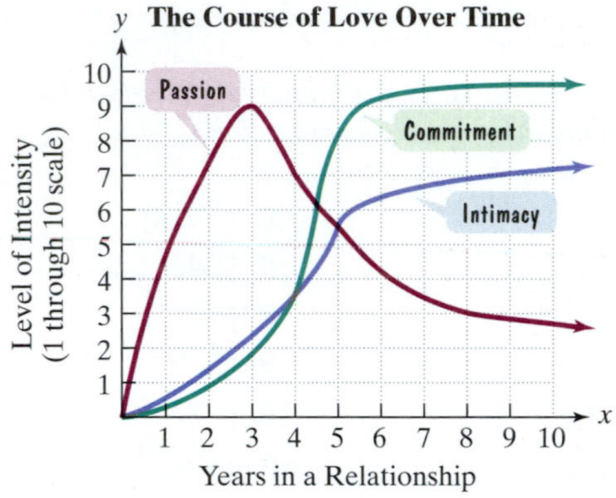

Source: R. J. Sternberg, A Triangular Theory of Love, *Psychological Review, 93, 119–135.*

61. Use interval notation to write an inequality that expresses for which years in a relationship intimacy is greater than commitment.

62. Use interval notation to write an inequality that expresses for which years in a relationship passion is greater than or equal to intimacy.

63. What is the relationship between passion and intimacy on the interval $[5, 7)$?

64. What is the relationship between intimacy and commitment on the interval $[4, 7)$?

65. What is the relationship between passion and commitment for $\{x | 6 < x < 8\}$?

66. What is the relationship between passion and commitment for $\{x | 7 < x < 9\}$?

67. What is the maximum level of intensity for passion? After how many years in a relationship does this occur?

68. After approximately how many years do levels of intensity for commitment exceed the maximum level of intensity for passion?

69. The percentage, P, of U.S. voters who use electronic voting systems, such as optical scans, in national elections can be modeled by the formula

$$P = 3.1x + 25.8,$$

where x is the number of years after 1994. In which years will more than 63% of U.S. voters use electronic systems?

70. The percentage, P, of U.S. voters who use punch cards or lever machines in national elections can be modeled by the formula

$$P = -2.5x + 63.1,$$

where x is the number of years after 1994. In which years will fewer than 38.1% of U.S. voters use punch cards or lever machines?

The Olympic 500-meter speed skating times have generally been decreasing over time. The formulas

$$W = -0.19t + 57 \quad \text{and} \quad M = -0.15t + 50$$

model the winning times, in seconds, for women, W, and men, M, t years after 1900. Use these models to solve Exercises 71–72.

71. Find values of t such that $W < M$. Describe what this means in terms of winning times.

72. Find values of t such that $W > M$. Describe what this means in terms of winning times.

In Exercises 73–80, use the strategy for solving word problems, translating from the verbal conditions of the problem to a linear inequality.

73. A truck can be rented from Basic Rental for \$50 a day plus \$0.20 per mile. Continental charges \$20 per day plus \$0.50 per mile to rent the same truck. How many miles must be driven in a day to make the rental cost for Basic Rental a better deal than Continental's?

74. You are choosing between two long-distance telephone plans. Plan A has a monthly fee of \$15 with a charge of \$0.08 per minute for all long-distance calls. Plan B has a monthly fee of \$3 with a charge of \$0.12 per minute for all long-distance calls. How many minutes of long-distance calls in a month make plan A the better deal?

75. A city commission has proposed two tax bills. The first bill requires that a homeowner pay \$1800 plus 3% of the assessed home value in taxes. The second bill requires taxes of \$200 plus 8% of the assessed home value. What price range of home assessment would make the first bill a better deal for the homeowner?

76. A local bank charges \$8 per month plus 5¢ per check. The credit union charges \$2 per month plus 8¢ per check. How many checks should be written each month to make the credit union a better deal?

77. A company manufactures and sells blank audiocassette tapes. The weekly fixed cost is \$10,000 and it costs \$0.40 to produce each tape. The selling price is \$2.00 per tape. How many tapes must be produced and sold each week for the company to have a profit?

78. A company manufactures and sells personalized stationery. The weekly fixed cost is \$3000 and it costs \$3.00 to produce each package of stationery. The selling price is \$5.50 per package. How many packages of stationery must be produced and sold each week for the company to have a profit?

79. An elevator at a construction site has a maximum capacity of 3000 pounds. If the elevator operator weighs 200 pounds and each cement bag weighs 70 pounds, how many bags of cement can be safely lifted on the elevator in one trip?

80. An elevator at a construction site has a maximum capacity of 2500 pounds. If the elevator operator weighs 160 pounds and each cement bag weighs 60 pounds, how many bags of cement can be safely lifted on the elevator in one trip?

Writing in Mathematics

81. When graphing the solutions of an inequality, what does a parenthesis signify? What does a bracket signify?

82. When solving an inequality, when is it necessary to change the sense of the inequality? Give an example.

83. Describe ways in which solving a linear inequality is similar to solving a linear equation.

84. Describe ways in which solving a linear inequality is different from solving a linear equation.

85. When solving a linear inequality, describe what happens if the solution set is $(-\infty, \infty)$.

86. When solving a linear inequality, describe what happens if the solution set is $\varnothing$.

87. What is the slope of each model in Exercises 69–70? What does this mean in terms of the percentage of U.S. voters using electronic voting systems and more traditional methods, such as punch cards or lever machines? What explanations can you offer for these changes in vote-counting systems?

Technology Exercises

In Exercises 88–89, solve each inequality using a graphing utility. Graph each side separately. Then determine the values of x for which the graph on the left side lies above the graph on the right side.

88. $-3(x - 6) > 2x - 2$

89. $-2(x + 4) > 6x + 16$

90. Use a graphing utility's TABLE feature to verify your work in Exercises 88–89.

Use the same technique employed in Exercises 88–89 to solve each inequality in Exercises 91–92. In each case, what conclusion can you draw? What happens if you try solving the inequalities algebraically?

91. $12x - 10 > 2(x - 4) + 10x$

92. $2x + 3 > 3(2x - 4) - 4x$

93. A bank offers two checking account plans. Plan A has a base service charge of \$4.00 per month plus 10¢ per check. Plan B charges a base service charge of \$2.00 per month plus 15¢ per check.

a. Write models for the total monthly costs for each plan if x checks are written.

b. Use a graphing utility to graph the models in the same [0, 50, 1] by [0, 10, 1] viewing rectangle.

c. Use the graphs (and the intersection feature) to determine for what number of checks per month plan A will be better than plan B.

d. Verify the result of part (c) algebraically by solving an inequality.

Critical Thinking Exercises

Make Sense? *In Exercises 94–97, determine whether each statement "makes sense" or "does not make sense" and explain your reasoning.*

94. I began the solution of $5 - 3(x + 2) > 10x$ by simplifying the left side, obtaining $2x + 4 > 10x$.

95. I have trouble remembering when to reverse the direction of an inequality symbol, so I avoid this difficulty by collecting variable terms on an appropriate side.

96. If you tell me that three times a number is less than two times that number, it's obvious that no number statisfies this condition, and there is no need for me to write and solve an inequality.

97. Whenever I solve a linear inequality in which the coefficients of the variable on each side are the same, the solution set is $\varnothing$ or $(-\infty, \infty)$.

In Exercises 98–101, determine whether each statement is true or false. If the statement is false, make the necessary change(s) to produce a true statement.

98. The inequality $3x > 6$ is equivalent to $2 > x$.

99. The smallest real number in the solution set of $2x > 6$ is 4.

100. If x is at least 7, then $x > 7$.

101. The inequality $-3x > 6$ is equivalent to $-2 > x$.

102. Find a so that the solution set of $ax + 4 \le -12$ is $[8, \infty)$.

103. What's wrong with this argument? Suppose x and y represent two real numbers, where $x > y$.

$2 > 1$	This is a true statement.
$2(y - x) > 1(y - x)$	Multiply both sides by $y - x$.
$2y - 2x > y - x$	Use the distributive property.
$y - 2x > -x$	Subtract y from both sides.
$y > x$	Add $2x$ to both sides.

The final inequality, $y > x$, is impossible because we were initially given $x > y$.

Review Exercises

104. If $f(x) = x^2 - 2x + 5$, find $f(-4)$. (Section 2.1, Example 3)

105. Solve the system:

$$\begin{aligned} 2x - y - z &= -3 \\ 3x - 2y - 2z &= -5 \\ -x + y + 2z &= 4. \end{aligned}$$

(Section 3.3, Example 2)

106. Simplify: $\left(\frac{2x^4y^{-2}}{4xy^3}\right)^3$. (Section 1.6, Example 9)

Preview Exercises

Exercises 107–109 will help you prepare for the material covered in the next section.

107. Consider the sets $A = \{1, 2, 3, 4\}$ and $B = \{3, 4, 5, 6, 7\}$.

a. Write the set consisting of elements common to both set A and set B.

b. Write the set consisting of elements that are members of set A or of set B or of both sets.

108. a. Solve: $x - 3 < 5$.

b. Solve: $2x + 4 < 14$.

c. Give an example of a number that satisfies the inequality in part (a) and the inequality in part (b).

d. Give an example of a number that satisfies the inequality in part (a), but not the inequality in part (b).

109. a. Solve: $2x - 6 \ge -4$.

b. Solve: $5x + 2 \ge 17$.

c. Give an example of a number that satisfies the inequality in part (a) and the inequality in part (b).

d. Give an example of a number that satisfies the inequality in part (a), but not the inequality in part (b).

SECTION 4.2

Objectives

1. Find the intersection of two sets.
2. Solve compound inequalities involving *and*.
3. Find the union of two sets.
4. Solve compound inequalities involving *or*.

Compound Inequalities

Sixty-six percent of U.S. adults are overweight or obese (30 or more pounds over a healthy weight). In this section's exercise set, you'll use *compound inequalities* to analyze data that show we are becoming a weightier nation.

A **compound inequality** is formed by joining two inequalities with the word *and* or the word *or*.

Examples of Compound Inequalities

- $x - 3 < 5$ and $2x + 4 < 14$
- $3x - 5 \leq 13$ or $5x + 2 > -3$

Compound inequalities illustrate the importance of the words *and* and *or* in mathematics, as well as in everyday English.

1 Find the intersection of two sets.

Compound Inequalities Involving *And*

If A and B are sets, we can form a new set consisting of all elements that are in both A and B. This set is called the *intersection* of the two sets.

Definition of the Intersection of Sets

The **intersection** of sets A and B, written $A \cap B$, is the set of elements common to both set A **and** set B. This definition can be expressed in set-builder notation as follows:

$$A \cap B = \{x \mid x \in A \text{ AND } x \in B\}.$$

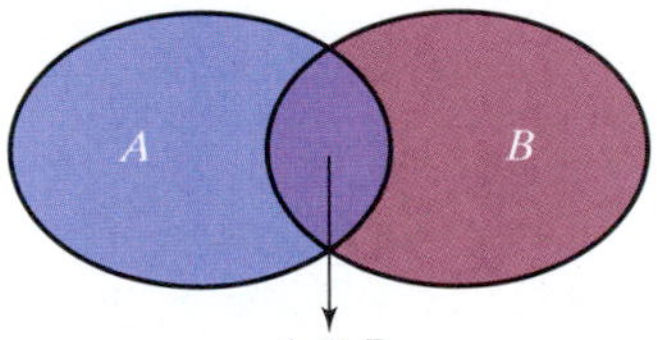

FIGURE 4.1 Picturing the intersection of two sets

Figure 4.1 shows a useful way of picturing the intersection of sets A and B. The figure indicates that $A \cap B$ contains those elements that belong to both A and B at the same time.

EXAMPLE 1 Finding the Intersection of Two Sets

Find the intersection: $\{7, 8, 9, 10, 11\} \cap \{6, 8, 10, 12\}$.

Solution The elements common to $\{7, 8, 9, 10, 11\}$ and $\{6, 8, 10, 12\}$ are 8 and 10. Thus,

$$\{7, 8, 9, 10, 11\} \cap \{6, 8, 10, 12\} = \{8, 10\}.$$

✓ **CHECK POINT 1** Find the intersection: $\{3, 4, 5, 6, 7\} \cap \{3, 7, 8, 9\}$.

2 Solve compound inequalities involving *and.*

A number is a **solution of a compound inequality formed by the word *and*** if it is a solution of both inequalities. For example, the solution set of the compound inequality

$$x \leq 6 \quad \text{and} \quad x \geq 2$$

is the set of values of x that satisfy both $x \leq 6$ and $x \geq 2$. Thus, the solution set is the intersection of the solution sets of the two inequalities.

What are the numbers that satisfy both $x \leq 6$ and $x \geq 2$? These numbers are easier to see if we graph the solution set to each inequality on a number line. These graphs are shown in **Figure 4.2.** The intersection is shown in the third graph.

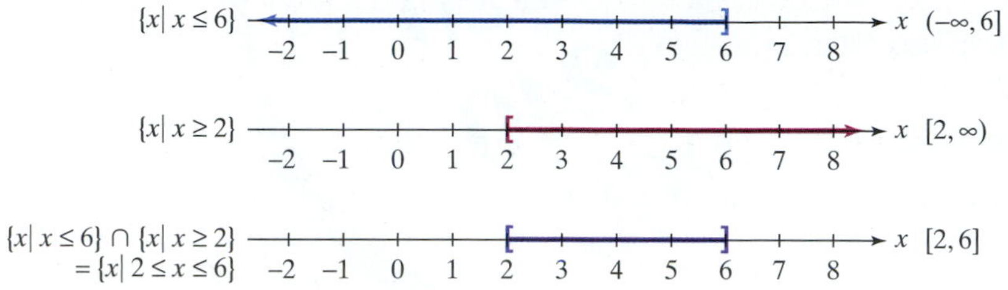

FIGURE 4.2 Numbers satisfying both $x \leq 6$ and $x \geq 2$

The numbers common to both sets are those that are less than or equal to 6 and greater than or equal to 2. This set is $\{x|2 \leq x \leq 6\}$, or, in interval notation, [2, 6].

Here is a procedure for finding the solution set of a compound inequality containing the word *and.*

Solving Compound Inequalities Involving *AND*

1. Solve each inequality separately.
2. Graph the solution set to each inequality on a number line and take the intersection of these solution sets. This intersection appears as the portion of the number line that the two graphs have in common.

EXAMPLE 2 Solving a Compound Inequality with *And*

Solve: $x - 3 < 5$ and $2x + 4 < 14$.

Solution

Step 1. Solve each inequality separately.

$$\begin{aligned} x - 3 &< 5 \quad \text{and} \quad & 2x + 4 &< 14 \\ x &< 8 & 2x &< 10 \\ & & x &< 5 \end{aligned}$$

Step 2. Take the intersection of the solution sets of the two inequalities. We graph the solution sets of $x < 8$ and $x < 5$. The intersection is shown in the third graph.

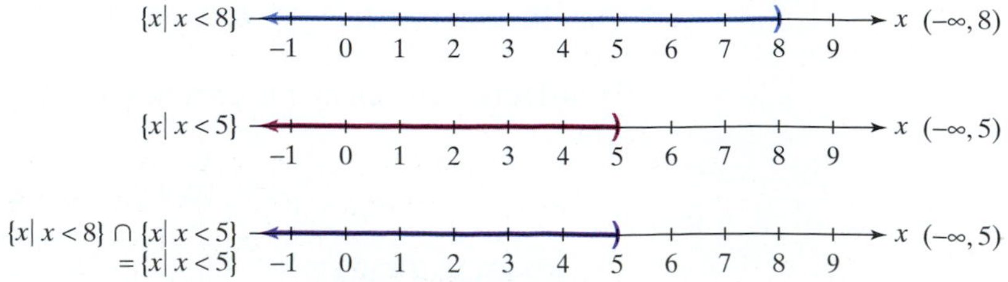

The numbers common to both sets are those that are less than 5. The solution set is $\{x|x < 5\}$, or, in interval notation, $(-\infty, 5)$. Take a moment to check that any number in $(-\infty, 5)$ satisfies both of the original inequalities.

CHECK POINT 2 Solve: $x + 2 < 5$ and $2x - 4 < -2$.

EXAMPLE 3 Solving a Compound Inequality with *And*

Solve: $2x - 7 > 3$ and $5x - 4 < 6$.

Solution

Step 1. Solve each inequality separately.

$$\begin{aligned} 2x - 7 &> 3 \quad \text{and} \quad & 5x - 4 &< 6 \\ 2x &> 10 & 5x &< 10 \\ x &> 5 & x &< 2 \end{aligned}$$

Step 2. Take the intersection of the solution sets of the two inequalities. We graph the solution sets of $x > 5$ and $x < 2$. We use these graphs to find their intersection.

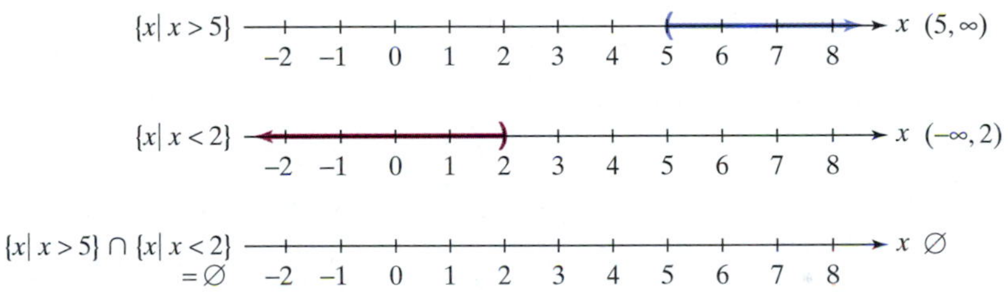

There is no number that is both greater than 5 and at the same time less than 2. Thus, the solution set is the empty set, $\varnothing$.

CHECK POINT 3 Solve: $4x - 5 > 7$ and $5x - 2 < 3$.

If $a < b$, the compound inequality

$$a < x \text{ and } x < b$$

can be written in the shorter form

$$a < x < b.$$

For example, the compound inequality

$$-3 < 2x + 1 \text{ and } 2x + 1 < 3$$

can be abbreviated

$$-3 < 2x + 1 < 3.$$

The word *and* does not appear when the inequality is written in the shorter form, although it is implied. The shorter form enables us to solve both inequalities at once. By performing the same operations on all three parts of the inequality, our goal is to **isolate x in the middle.**

EXAMPLE 4 Solving a Compound Inequality

Solve and graph the solution set:

$$-3 < 2x + 1 \le 3.$$

Solution We would like to isolate x in the middle. We can do this by first subtracting 1 from all three parts of the compound inequality. Then we isolate x from $2x$ by dividing all three parts of the inequality by 2.

$$-3 < 2x + 1 \le 3 \quad \text{This is the given inequality.}$$

$$-3 - 1 < 2x + 1 - 1 \le 3 - 1 \quad \text{Subtract 1 from all three parts.}$$

$$-4 < 2x \le 2 \quad \text{Simplify.}$$

$$\frac{-4}{2} < \frac{2x}{2} \le \frac{2}{2} \quad \text{Divide each part by 2.}$$

$$-2 < x \le 1 \quad \text{Simplify.}$$

The solution set consists of all real numbers greater than -2 and less than or equal to 1, represented by $\{x|-2 < x \le 1\}$ in set-builder notation and $(-2, 1]$ in interval notation. The graph is shown as follows:

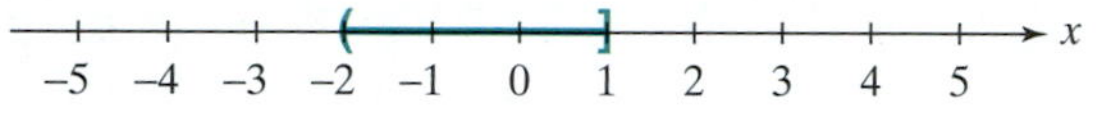

Using Technology

Numeric and Graphic Connections

Let's verify our work in Example 4 and show that $(-2, 1]$ is the solution set of $-3 < 2x + 1 \le 3$.

Numeric Check

To check numerically, enter $y_1 = 2x + 1$.

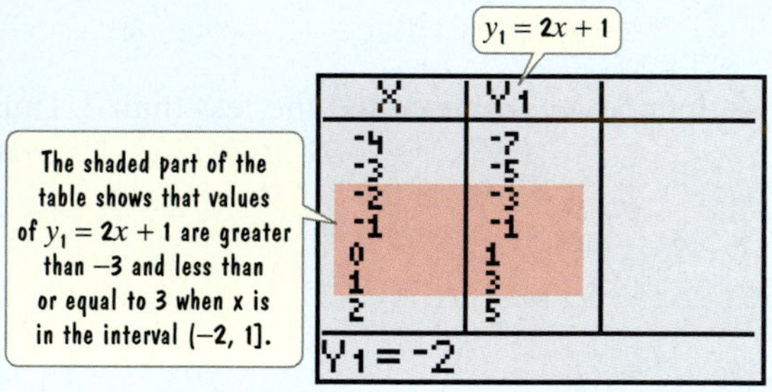

Graphic Check

To check graphically, graph each part of

$$-3 < 2x + 1 \le 3.$$

Enter $y_1 = -3$. Enter $y_2 = 2x + 1$. Enter $y_3 = 3$.

The figure shows that the graph of $y_2 = 2x + 1$ lies above the graph of $y_1 = -3$ and on or below the graph of $y_3 = 3$ when x is in the interval $(-2, 1]$.

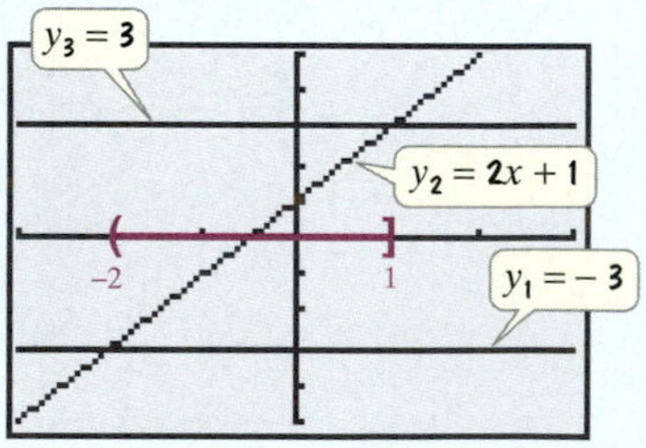

[−3, 3, 1] by [−5, 5, 1]

✓ CHECK POINT 4 Solve and graph the solution set: $1 \le 2x + 3 < 11$.

---

3 Find the union of two sets.

Compound Inequalities Involving *Or*

Another set that we can form from sets A and B consists of elements that are in A or B or in both sets. This set is called the *union* of the two sets.

Definition of the Union of Sets

The **union** of sets A and B, written $A \cup B$, is the set of elements that are members of set A **or** of set B or of both sets. This definition can be expressed in set-builder notation as follows:

$$A \cup B = \{x|x \in A \text{ OR } x \in B\}.$$

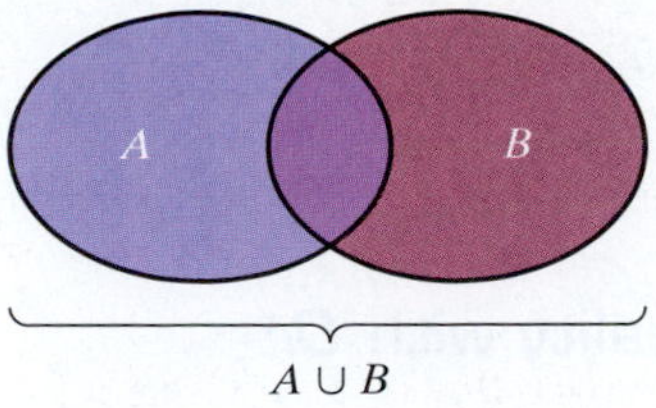

FIGURE 4.3 Picturing the union of two sets

Figure 4.3 shows a useful way of picturing the union of sets A and B. The figure indicates that $A \cup B$ is formed by joining the sets together.

We can find the union of set A and set B by listing the elements of set A. Then, we include any elements of set B that have not already been listed. Enclose all elements that are listed with braces. This shows that the union of two sets is also a set.

EXAMPLE 5 Finding the Union of Two Sets

Find the union: $\{7, 8, 9, 10, 11\} \cup \{6, 8, 10, 12\}$.

Solution To find $\{7, 8, 9, 10, 11\} \cup \{6, 8, 10, 12\}$, start by listing all the elements from the first set, namely 7, 8, 9, 10, and 11. Now list all the elements from the second set that are not in the first set, namely 6 and 12. The union is the set consisting of all these elements. Thus,

$$\{7, 8, 9, 10, 11\} \cup \{6, 8, 10, 12\} = \{6, 7, 8, 9, 10, 11, 12\}.$$

Although 8 and 10 appear in both sets, do not list 8 and 10 twice.

CHECK POINT 5 Find the union: $\{3, 4, 5, 6, 7\} \cup \{3, 7, 8, 9\}$.

Study Tip

The words *union* and *intersection* are helpful in distinguishing these two operations. Union, as in a marriage union, suggests joining things, or uniting them. Intersection, as in the intersection of two crossing streets, brings to mind the area common to both, suggesting things that overlap.

4 Solve compound inequalities involving *or*.

A number is a **solution of a compound inequality formed by the word *or*** if it is a solution of either inequality. Thus, the solution set of a compound inequality formed by the word *or* is the union of the solution sets of the two inequalities.

Solving Compound Inequalities Involving *OR*

1. Solve each inequality separately.
2. Graph the solution set to each inequality on a number line and take the union of these solution sets. This union appears as the portion of the number line representing the total collection of numbers in the two graphs.

EXAMPLE 6 Solving a Compound Inequality with *Or*

Solve: $2x - 3 < 7$ or $35 - 4x \le 3$.

Solution

Step 1. Solve each inequality separately.

$$\begin{aligned} 2x - 3 &< 7 \quad \text{or} \quad & 35 - 4x &\le 3 \\ 2x &< 10 & -4x &\le -32 \\ x &< 5 & x &\ge 8 \end{aligned}$$

Step 2. Take the union of the solution sets of the two inequalities. We graph the solution sets of $x < 5$ and $x \ge 8$. We use these graphs to find their union.

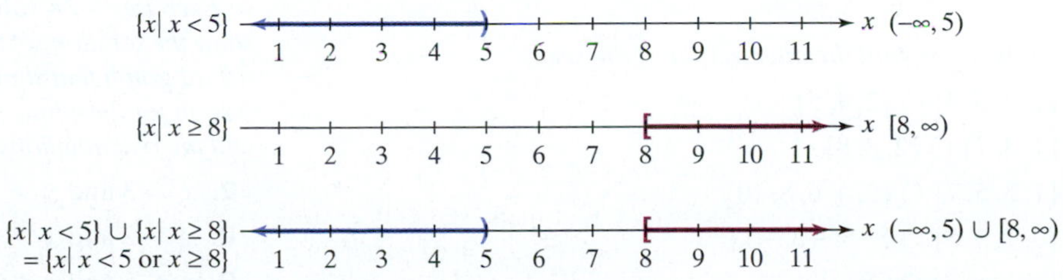

The solution set consists of all numbers that are less than 5 or greater than or equal to 8. The solution set is $\{x | x < 5 \text{ or } x \ge 8\}$, or, in interval notation, $(-\infty, 5) \cup [8, \infty)$. There is no shortcut way to express this union when interval notation is used.

CHECK POINT 6 Solve: $3x - 5 \le -2$ or $10 - 2x < 4$.

EXAMPLE 7 Solving a Compound Inequality with *Or*

Solve: $3x - 5 \le 13$ or $5x + 2 > -3$.

Solution

Step 1. Solve each inequality separately.

$$\begin{aligned} 3x - 5 &\le 13 \quad \text{or} \quad & 5x + 2 &> -3 \\ 3x &\le 18 & 5x &> -5 \\ x &\le 6 & x &> -1 \end{aligned}$$

Step 2. Take the union of the solution sets of the two inequalities. We graph the solution sets of $x \le 6$ and $x > -1$. We use these graphs to find their union.

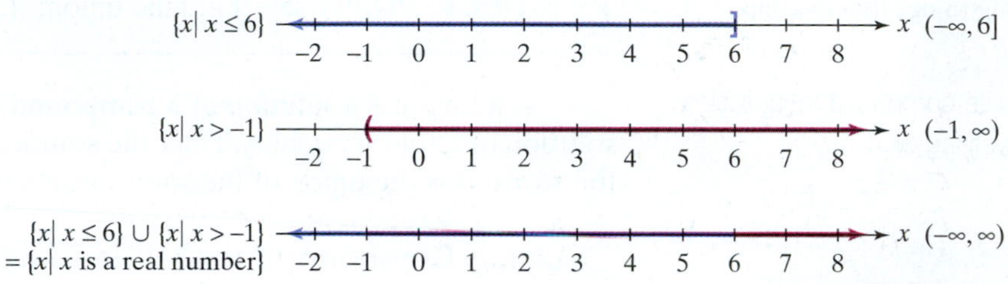

Because all real numbers are either less than or equal to 6 or greater than -1 or both, the union of the two sets fills the entire number line. Thus, the solution set is $\{x | x \text{ is a real number}\}$, or $\mathbb{R}$. The solution set in interval notation is $(-\infty, \infty)$. Any real number that you select will satisfy at least one of the original inequalities.

CHECK POINT 7 Solve: $2x + 5 \ge 3$ or $2x + 3 < 3$.

4.2 EXERCISE SET

MyMathLab — Math XL PRACTICE — WATCH — DOWNLOAD — READ — REVIEW

Practice Exercises

In Exercises 1–6, find the intersection of the sets.

1. $\{1, 2, 3, 4\} \cap \{2, 4, 5\}$
2. $\{1, 3, 7\} \cap \{2, 3, 8\}$
3. $\{1, 3, 5, 7\} \cap \{2, 4, 6, 8, 10\}$
4. $\{0, 1, 3, 5\} \cap \{-5, -3, -1\}$
5. $\{a, b, c, d\} \cap \varnothing$
6. $\{w, y, z\} \cap \varnothing$

In Exercises 7–24, solve each compound inequality. Use graphs to show the solution set to each of the two given inequalities, as well as a third graph that shows the solution set of the compound inequality. Except for the empty set, express the solution set in both set-builder and interval notations.

7. $x > 3$ and $x > 6$
8. $x > 2$ and $x > 4$
9. $x \le 5$ and $x \le 1$
10. $x \le 6$ and $x \le 2$

11. $x < 2$ and $x \geq -1$

12. $x < 3$ and $x \geq -1$

13. $x > 2$ and $x < -1$

14. $x > 3$ and $x < -1$

15. $5x < -20$ and $3x > -18$

16. $3x \leq 15$ and $2x > -6$

17. $x - 4 \leq 2$ and $3x + 1 > -8$

18. $3x + 2 > -4$ and $2x - 1 < 5$

19. $2x > 5x - 15$ and $7x > 2x + 10$

20. $6 - 5x > 1 - 3x$ and $4x - 3 > x - 9$

21. $4(1 - x) < -6$ and $\frac{x - 7}{5} \leq -2$

22. $5(x - 2) > 15$ and $\frac{x - 6}{4} \leq -2$

23. $x - 1 \leq 7x - 1$ and $4x - 7 < 3 - x$

24. $2x + 1 > 4x - 3$ and $x - 1 \geq 3x + 5$

In Exercises 25–32, solve each inequality and graph the solution set on a number line. Express the solution set in both set-builder and interval notations.

25. $6 < x + 3 < 8$

26. $7 < x + 5 < 11$

27. $-3 \leq x - 2 < 1$

28. $-6 < x - 4 \leq 1$

29. $-11 < 2x - 1 \leq -5$

30. $3 \leq 4x - 3 < 19$

31. $-3 \leq \frac{2x}{3} - 5 < -1$

32. $-6 \leq \frac{x}{2} - 4 < -3$

In Exercises 33–38, find the union of the sets.

33. $\{1, 2, 3, 4\} \cup \{2, 4, 5\}$

34. $\{1, 3, 7, 8\} \cup \{2, 3, 8\}$

35. $\{1, 3, 5, 7\} \cup \{2, 4, 6, 8, 10\}$

36. $\{0, 1, 3, 5\} \cup \{2, 4, 6\}$

37. $\{a, e, i, o, u\} \cup \varnothing$

38. $\{e, m, p, t, y\} \cup \varnothing$

In Exercises 39–54, solve each compound inequality. Use graphs to show the solution set to each of the two given inequalities, as well as a third graph that shows the solution set of the compound inequality. Express the solution set in both set-builder and interval notations.

39. $x > 3$ or $x > 6$

40. $x > 2$ or $x > 4$

41. $x \leq 5$ or $x \leq 1$

42. $x \leq 6$ or $x \leq 2$

43. $x < 2$ or $x \geq -1$

44. $x < 3$ or $x \geq -1$

45. $x \geq 2$ or $x < -1$

46. $x \geq 3$ or $x < -1$

47. $3x > 12$ or $2x < -6$

48. $3x < 3$ or $2x > 10$

49. $3x + 2 \leq 5$ or $5x - 7 \geq 8$

50. $2x - 5 \leq -11$ or $5x + 1 \geq 6$

51. $4x + 3 < -1$ or $2x - 3 \geq -11$

52. $2x + 1 < 15$ or $3x - 4 \geq -1$

53. $-2x + 5 > 7$ or $-3x + 10 > 2x$

54. $16 - 3x \geq -8$ or $13 - x > 4x + 3$

55. Let $f(x) = 2x + 3$ and $g(x) = 3x - 1$. Find all values of x for which $f(x) \geq 5$ and $g(x) > 11$.

56. Let $f(x) = 4x + 5$ and $g(x) = 3x - 4$. Find all values of x for which $f(x) \geq 5$ and $g(x) \leq 2$.

57. Let $f(x) = 3x - 1$ and $g(x) = 4 - x$. Find all values of x for which $f(x) < -1$ or $g(x) < -2$.

58. Let $f(x) = 2x - 5$ and $g(x) = 3 - x$. Find all values of x for which $f(x) \geq 3$ or $g(x) < 0$.

Practice PLUS

In Exercises 59–60, write an inequality with x isolated in the middle that is equivalent to the given inequality. Assume $a > 0, b > 0$, and $c > 0$.

59. $-c < ax - b < c$

60. $-2 < \frac{ax - b}{c} < 2$

In Exercises 61–62, use the graphs of y_1, y_2, and y_3 to solve each compound inequality.

61. $-3 \leq 2x - 1 \leq 5$

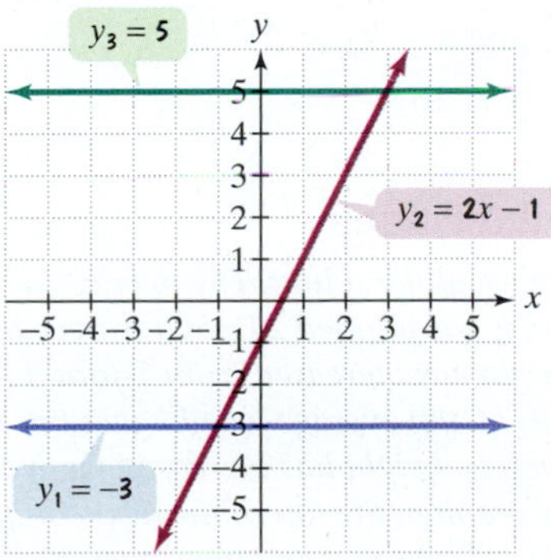

62. $x - 2 < 2x - 1 < x + 2$

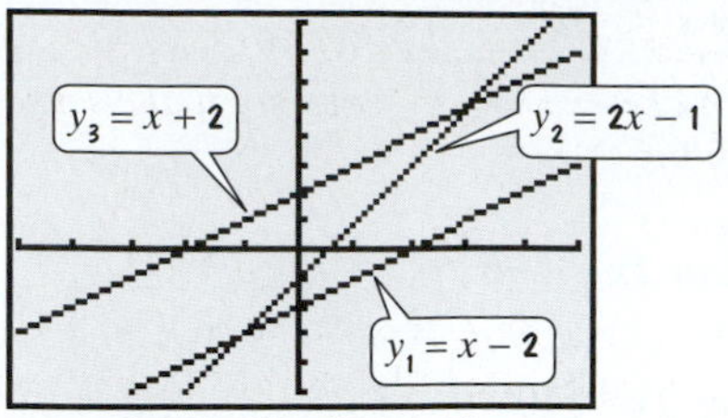

[−5, 5, 1] by [−5, 8, 1]

63. Solve $x - 2 < 2x - 1 < x + 2$, the inequality in Exercise 62, using algebraic methods. (*Hint:* Rewrite the inequality as $2x - 1 > x - 2$ and $2x - 1 < x + 2$.)

64. Use the hint given in Exercise 63 to solve $x \leq 3x - 10 \leq 2x$.

In Exercises 65–66, use the table to solve each inequality.

65. $-2 \leq 5x + 3 < 13$

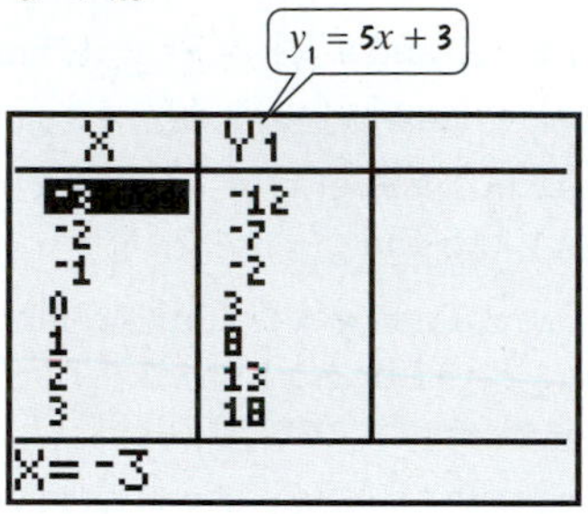

66. $-3 < 2x - 5 \leq 3$

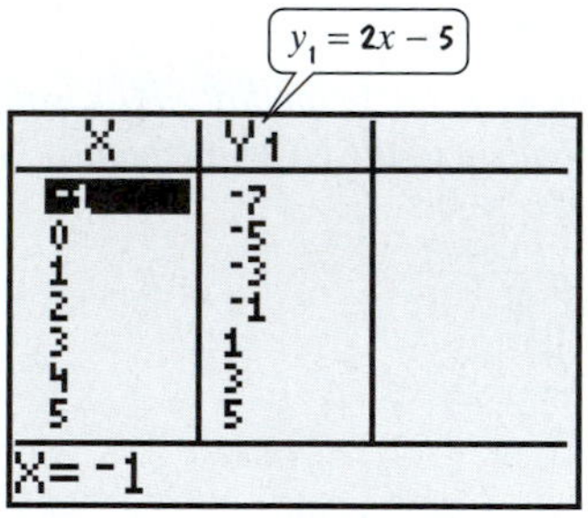

In Exercises 67–68, use the roster method to find the set of negative integers that are solutions of each inequality.

67. $5 - 4x \geq 1$ and $3 - 7x < 31$

68. $-5 < 3x + 4 \leq 16$

Application Exercises

We are becoming weightier adults. The average weight of U.S. women, ages 20–74, has jumped 24 pounds over four decades, while average height has increased from 5-foot-3 to 5-foot-4. For U.S. men, ages 20–74, the average weight gain has been 25 pounds, while average height has increased from 5-foot-8 to 5-foot-9. The bar graph at the top of the next column shows the average weight of U.S. women and men, ages 20–74, for five selected years over four decades.

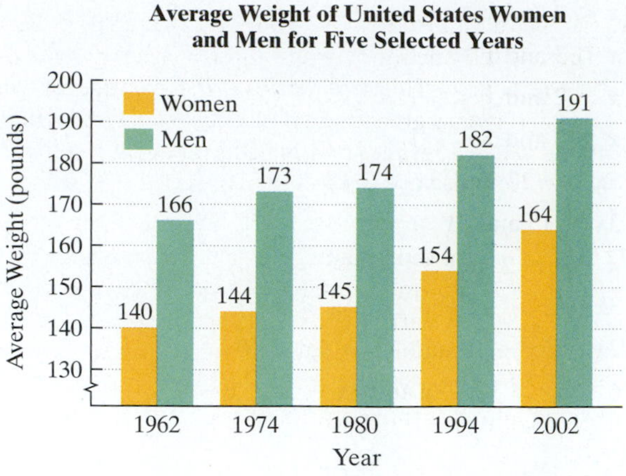

Source: Centers for Disease Control and Prevention

In Exercises 69–78, use the years shown in the graph to find each set.

69. $\{x|x$ is a year for which women's weight $\geq 144\} \cap \{x|x$ is a year for which men's weight $\leq 182\}$

70. $\{x|x$ is a year for which women's weight $\geq 144\} \cap \{x|x$ is a year for which men's weight $< 182\}$

71. $\{x|x$ is a year for which women's weight $\geq 144\} \cup \{x|x$ is a year for which men's weight $\leq 182\}$

72. $\{x|x$ is a year for which women's weight $\geq 144\} \cup \{x|x$ is a year for which men's weight $< 182\}$

73. $\{x|x$ is a year for which women's weight $< 144\} \cup \{x|x$ is a year for which men's weight $> 182\}$

74. $\{x|x$ is a year for which women's weight $< 144\} \cup \{x|x$ is a year for which men's weight $\geq 182\}$

75. $\{x|x$ is a year for which women's weight $< 144\} \cap \{x|x$ is a year for which men's weight $> 182\}$

76. $\{x|x$ is a year for which women's weight $< 144\} \cap \{x|x$ is a year for which men's weight $\geq 182\}$

77. $\{x|x$ is a year for which $166 \leq$ men's weight $< 174\}$

78. $\{x|x$ is a year for which $145 <$ women's weight $\leq 164\}$

79. A basic cellular phone plan costs \$20 per month for 60 calling minutes. Additional time costs \$0.40 per minute. The formula

$$C = 20 + 0.40(x - 60)$$

gives the monthly cost for this plan, C, for x calling minutes, where $x > 60$. How many calling minutes are possible for a monthly cost of at least \$28 and at most \$40?

80. The formula for converting Fahrenheit temperature, F, to Celsius temperature, C, is

$$C = \frac{5}{9}(F - 32).$$

If Celsius temperature ranges from 15° to 35°, inclusive, what is the range for the Fahrenheit temperature? Use interval notation to express this range.

81. On the first of four exams, your grades are 70, 75, 87, and 92. There is still one more exam, and you are hoping to earn a B in the course. This will occur if the average of your five exam grades is greater than or equal to 80 and less than 90. What range of grades on the fifth exam will result in earning a B? Use interval notation to express this range.

82. On the first of four exams, your grades are 82, 75, 80, and 90. There is still a final exam, and it counts as two grades. You are hoping to earn a B in the course: This will occur if the average of your six exam grades is greater than or equal to 80 and less than 90. What range of grades on the final exam will result in earning a B? Use interval notation to express this range.

83. The toll to a bridge is \$3.00. A three-month pass costs \$7.50 and reduces the toll to \$0.50. A six-month pass costs \$30 and permits crossing the bridge for no additional fee. How many crossing per three-month period does it take for the three-month pass to be the best deal?

84. Parts for an automobile repair cost \$175. The mechanic charges \$34 per hour. If you receive an estimate for at least \$226 and at most \$294 for fixing the car, what is the time interval that the mechanic will be working on the job?

Writing in Mathematics

85. Describe what is meant by the intersection of two sets. Give an example.

86. Explain how to solve a compound inequality involving *and*.

87. Why is $1 < 2x + 3 < 9$ a compound inequality? What are the two inequalities and what is the word that joins them?

88. Explain how to solve $1 < 2x + 3 < 9$.

89. Describe what is meant by the union of two sets. Give an example.

90. Explain how to solve a compound inequality involving *or*.

Technology Exercises

In Exercises 91–94, solve each inequality using a graphing utility. Graph each of the three parts of the inequality separately in the same viewing rectangle. The solution set consists of all values of x for which the graph of the linear function in the middle lies between the graphs of the constant functions on the left and the right.

91. $1 < x + 3 < 9$

92. $-1 < \frac{x+4}{2} < 3$

93. $1 \le 4x - 7 \le 3$

94. $2 \le 4 - x \le 7$

95. Use a graphing utility's TABLE feature to verify your work in Exercises 91–94.

Critical Thinking Exercises

Make Sense? *In Exercises 96–99, determine whether each statement "makes sense" or "does not make sense" and explain your reasoning.*

96. I've noticed that when solving some compound inequalities with *or*, there is no way to express the solution set using a single interval, but this does not happen with *and* compound inequalities.

97. Compound inequalities with *and* have solutions that satisfy both inequalities, whereas compound inequalities with *or* have solutions that satisfy at least one of the inequalities.

98. I'm considering the compound inequality $x < 8$ and $x > a$, and I'm certain that there are no values of a that make the solution set $\varnothing$.

99. I'm considering the compound inequality $x < 8$ and $x > a$, and I'm certain that there are no values of a that make the solution set $(-\infty, \infty)$.

In Exercises 100–103, determine whether each statement is true or false. If the statement is false, make the necessary change(s) to produce a true statement.

100. $(-\infty, -1] \cap [-4, \infty) = [-4, -1]$

101. $(-\infty, 3) \cup (-\infty, -2) = (-\infty, -2)$

102. The union of two sets can never give the same result as the intersection of those same two sets.

103. The solution set of the compound inequality $x < a$ and $x > a$ is the set of all real numbers excluding a.

104. Solve and express the solution set in interval notation: $-7 \le 8 - 3x \le 20$ and $-7 < 6x - 1 < 41$.

The graphs of $f(x) = \sqrt{4 - x}$ *and* $g(x) = \sqrt{x + 1}$ *are shown in a* $[-3, 10, 1]$ *by* $[-2, 5, 1]$ *viewing rectangle.*

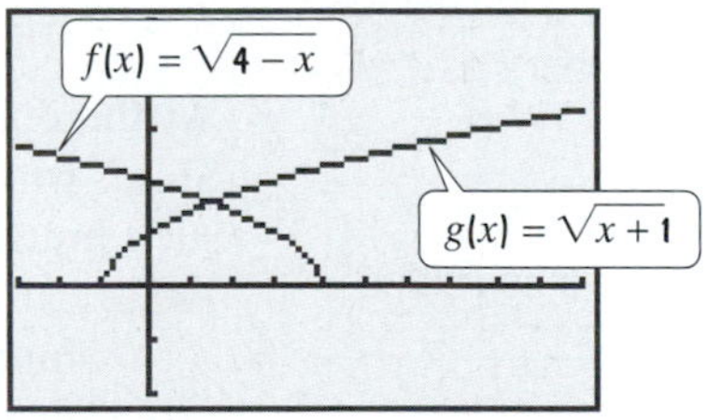

In Exercises 105–108, use the graphs and interval notation to express the domain of the given function.

105. The domain of f

106. The domain of g

107. The domain of $f + g$

108. The domain of $\frac{f}{g}$

109. At the end of the day, the change machine at a laundrette contained at least \$3.20 and at most \$5.45 in nickels, dimes, and quarters. There were 3 fewer dimes than twice the number of nickels and 2 more quarters than twice the number of nickels. What was the least possible number and the greatest possible number of nickels?

Review Exercises

110. If $f(x) = x^2 - 3x + 4$ and $g(x) = 2x - 5$, find $(g - f)(x)$ and $(g - f)(-1)$. (Section 2.3, Example 4)

111. Use function notation to write the equation of the line passing through (4, 2) and perpendicular to the line whose equation is $4x - 2y = 8$. (Section 2.5, Example 5)

112. Simplify: $4 - [2(x - 4) - 5]$. (Section 1.2, Example 14)

Preview Exercises

Exercises 113–115 will help you prepare for the material covered in the next section.

113. Find all values of x satisfying $1 - 4x = 3$ or $1 - 4x = -3$.

114. Find all values of x satisfying $3x - 1 = x + 5$ or $3x - 1 = -(x + 5)$.

115. **a.** Substitute -5 for x and determine whether -5 satisfies $|2x + 3| \geq 5$.

b. Does 0 satisfy $|2x + 3| \geq 5$?

SECTION 4.3

Equations and Inequalities Involving Absolute Value

*M*A*S*H was set in the early 1950s during the Korean War. By the final episode, the show had lasted four times as long as the Korean War.*

Objectives

1. Solve absolute value equations.
2. Use boundary points to solve absolute value inequalities.
3. Use equivalent compound inequalities to solve absolute value inequalities.
4. Recognize absolute value inequalities with no solution or all real numbers as solutions.
5. Solve problems using absolute value inequalities.

At the end of the twentieth century, there were 94 million households in the United States with television sets. The television program viewed by the greatest percentage of such households in that century was the final episode of $M{*}A{*}S{*}H$. Over 50 million American households watched this program.

Numerical information, such as the number of households watching a television program, is often given with a margin of error. Inequalities involving absolute value are used to describe errors in polling, as well as errors of measurement in manufacturing, engineering, science, and other fields. In this section, you will learn to solve equations and inequalities containing absolute value. With these skills, you will be able to analyze the percentage of households that watched the final episode of $M{*}A{*}S{*}H$.

Equations Involving Absolute Value

1 Solve absolute value equations.

We have seen that the absolute value of a, denoted $|a|$, is the distance from 0 to a on a number line. Now consider **absolute value equations**, such as

$$|x| = 2.$$

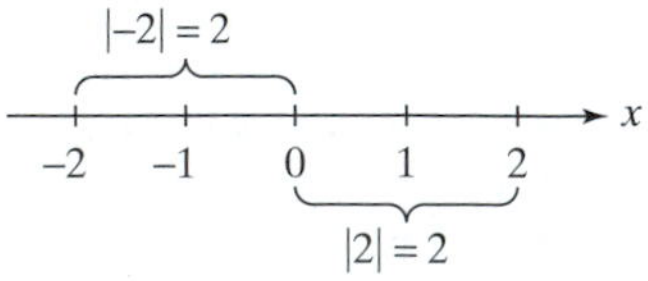

FIGURE 4.4 If $|x| = 2$, then $x = 2$ or $x = -2$.

This means that we must determine real numbers whose distance from the origin on a number line is 2. **Figure 4.4** shows that there are two numbers such that $|x| = 2$, namely, 2 and -2. We write $x = 2$ or $x = -2$. This observation can be generalized as follows:

Rewriting an Absolute Value Equation Without Absolute Value Bars

If c is a positive real number and X represents any algebraic expression, then $|X| = c$ is equivalent to $X = c$ or $X = -c$.

Using Technology

Graphic Connections

You can use a graphing utility to verify the solution set of an absolute value equation. Consider, for example,

$$|2x - 3| = 11.$$

Graph $y_1 = |2x - 3|$ and $y_2 = 11$. The graphs are shown in a $[-10, 10, 1]$ by $[-1, 15, 1]$ viewing rectangle. The x-coordinates of the intersection points are -4 and 7, verifying that $\{-4, 7\}$ is the solution set.

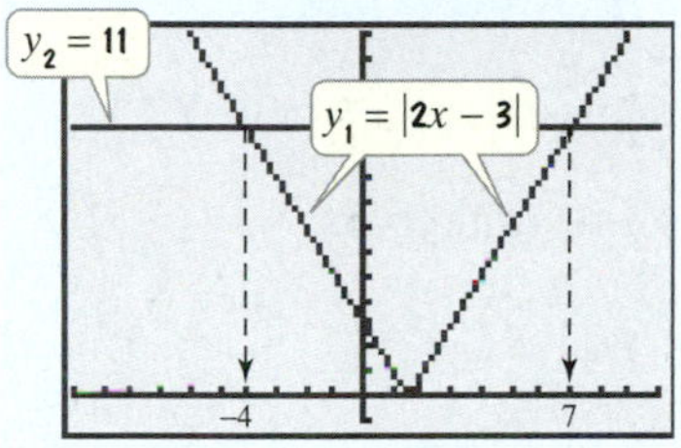

[−10, 10, 1] by [−1, 15, 1]

EXAMPLE 1 Solving an Equation Involving Absolute Value

Solve: $|2x - 3| = 11$.

Solution

$|2x - 3| = 11$			This is the given equation.
$2x - 3 = 11$	or	$2x - 3 = -11$	Rewrite the equation without absolute value bars.
$2x = 14$		$2x = -8$	Add 3 to both sides of each equation.
$x = 7$		$x = -4$	Divide both sides of each equation by 2.

Check 7:	Check −4:	
$|2x - 3| = 11$	$|2x - 3| = 11$	This is the original equation.
$|2(7) - 3| \stackrel{?}{=} 11$	$|2(-4) - 3| \stackrel{?}{=} 11$	Substitute the proposed solutions.
$|14 - 3| \stackrel{?}{=} 11$	$|-8 - 3| \stackrel{?}{=} 11$	Perform operations inside the absolute value bars.
$|11| \stackrel{?}{=} 11$	$|-11| \stackrel{?}{=} 11$	
$11 = 11$, true	$11 = 11$, true	These true statements indicate that 7 and −4 are solutions.

The solutions are -4 and 7. We can also say that the solution set is $\{-4, 7\}$.

✓ CHECK POINT 1 Solve: $|2x - 1| = 5$.

EXAMPLE 2 Solving an Equation Involving Absolute Value

Solve: $5|1 - 4x| - 15 = 0$.

Solution

$5|1 - 4x| - 15 = 0$			This is the given equation.

We need to isolate $|1 - 4x|$, the absolute value expression.

$5|1 - 4x| = 15$			Add 15 to both sides.
$|1 - 4x| = 3$			Divide both sides by 5.
$1 - 4x = 3$	or	$1 - 4x = -3$	Rewrite $|X| = c$ as $X = c$ or $X = -c$.
$-4x = 2$		$-4x = -4$	Subtract 1 from both sides of each equation.
$x = -\frac{1}{2}$		$x = 1$	Divide both sides of each equation by −4.

Take a moment to check $-\frac{1}{2}$ and 1, the proposed solutions, in the original equation, $5|1 - 4x| - 15 = 0$. In each case, you should obtain the true statement $0 = 0$. The solutions are $-\frac{1}{2}$ and 1, and the solution set is $\left\{-\frac{1}{2}, 1\right\}$.

✓ CHECK POINT 2 Solve: $2|1 - 3x| - 28 = 0$.

The absolute value of a number is never negative. Thus, if X is an algebraic expression and c is a negative number, then $|X| = c$ has no solution. For example, the equation $|3x - 6| = -2$ has no solution because $|3x - 6|$ cannot be negative. The solution set is $\varnothing$, the empty set.

The absolute value of 0 is 0. Thus, if X is an algebraic expression and $|X| = 0$, the solution is found by solving $X = 0$. For example, the solution of $|x - 2| = 0$ is obtained by solving $x - 2 = 0$. The solution is 2 and the solution set is $\{2\}$.

Some equations have two absolute value expressions, such as

$$|3x - 1| = |x + 5|.$$

These absolute value expressions are equal when the expressions inside the absolute value bars are equal to or opposites of each other.

> **Rewriting an Absolute Value Equation with Two Absolute Values Without Absolute Value Bars**
>
> If $|X_1| = |X_2|$, then $X_1 = X_2$ or $X_1 = -X_2$.

EXAMPLE 3 **Solving an Absolute Value Equation with Two Absolute Values**

Solve: $|3x - 1| = |x + 5|$.

Solution We rewrite the equation without absolute value bars.

$$|X_1| = |X_2| \quad \text{means} \quad X_1 = X_2 \quad \text{or} \quad X_1 = -X_2$$

$$|3x - 1| = |x + 5| \quad \text{means} \quad 3x - 1 = x + 5 \quad \text{or} \quad 3x - 1 = -(x + 5).$$

We now solve the two equations that do not contain absolute value bars.

$$
\begin{aligned}
3x - 1 &= x + 5 \quad & \text{or} \quad 3x - 1 &= -(x + 5) \\
2x - 1 &= 5 \quad & 3x - 1 &= -x - 5 \\
2x &= 6 \quad & 4x - 1 &= -5 \\
x &= 3 \quad & 4x &= -4 \\
& & x &= -1
\end{aligned}
$$

Take a moment to complete the solution process by checking the two proposed solutions in the original equation. The solutions are -1 and 3, and the solution set is $\{-1, 3\}$. ∎

✓ **CHECK POINT 3** Solve: $|2x - 7| = |x + 3|$.

❷ Use boundary points to solve absolute value inequalities.

Solving Inequalities Involving Absolute Value Using Boundary Points

Graphs can help us visualize the solution sets of equations and inequalities involving absolute value. Let's first consider the equation

$$|x| = 2.$$

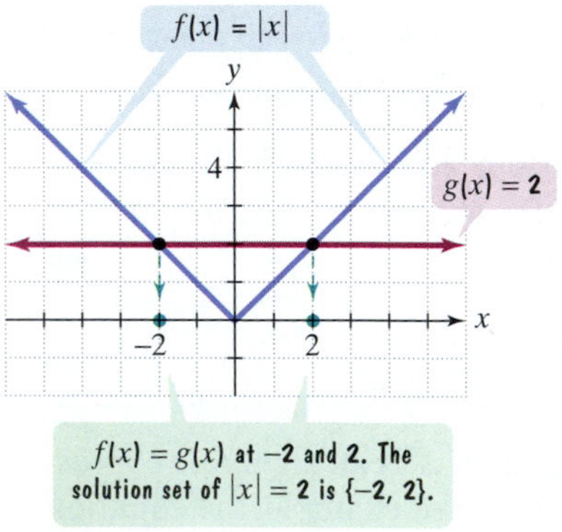

Figure 4.5 shows the graphs of $f(x) = |x|$ and $g(x) = 2$. The x-coordinates of the intersection points illustrate that $\{-2, 2\}$ is the solution set of $|x| = 2$.

Now let's see what the graphs of $f(x) = |x|$ and $g(x) = 2$ can tell us about the solution sets of the following inequalities involving absolute value:

$$|x| < 2 \qquad \text{and} \qquad |x| > 2.$$

Look for values of x where the graph of $f(x) = |x|$ lies *below* the graph of $g(x) = 2$.

Look for values of x where the graph of $f(x) = |x|$ lies *above* the graph of $g(x) = 2$.

$f(x) = g(x)$ at **−2** and **2**. The solution set of $|x| = 2$ is $\{-2, 2\}$.

FIGURE 4.5

Figure 4.6(a) illustrates the solution set of $|x| < 2$. **Figure 4.6(b)** illustrates the solution set of $|x| > 2$.

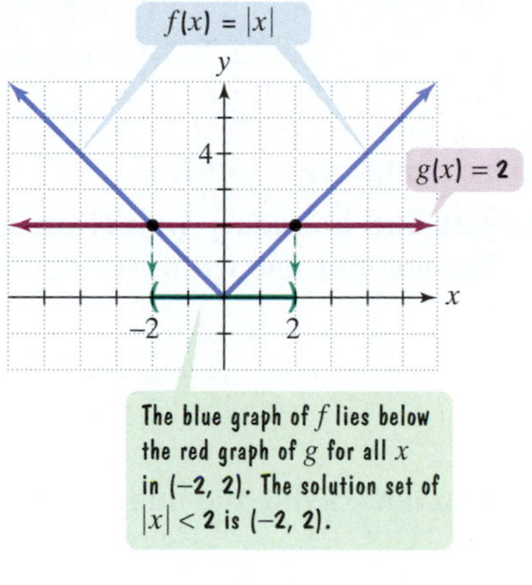

FIGURE 4.6(a)

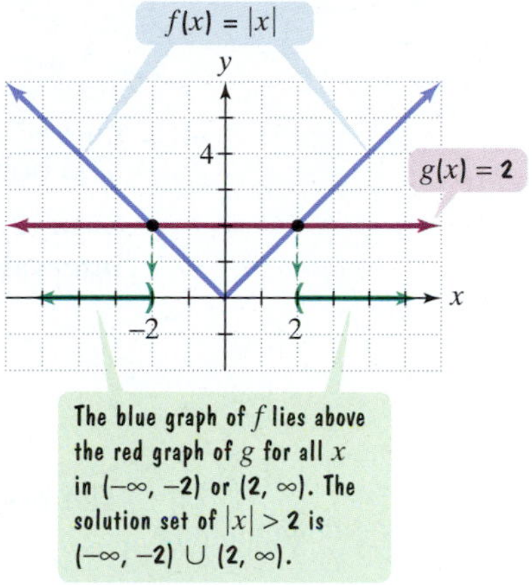

FIGURE 4.6(b)

Can you see how -2 and 2, the solutions of $|x| = 2$, serve as boundary points that divide the x-axis into intervals? On each interval, the blue graph of f is either below the red graph of g or above the red graph of g.

Boundary points play a fundamental role in solving absolute value inequalities in the form $|X| < c$ or $|X| > c$, where X is an algebraic expression and c is a positive number. The boundary points are found by solving the equation $|X| = c$.

Using Boundary Points to Solve Absolute Value Inequalities

If X is an algebraic expression and c is a positive number, the inequalities $|X| < c$ and $|X| > c$ can be solved by the following procedure.

1. Solve the equation $|X| = c$. The solutions are the **boundary points**.
2. Locate these boundary points on a number line, thereby dividing the number line into intervals.
3. Choose one representative number, called a **test value**, within each interval and substitute that number into the given inequality.
   a. If a true statement results, then all numbers, x, in the interval satisfy the given inequality.
   b. If a false statement results, then no numbers, x, in the interval satisfy the given inequality.
4. Write the solution set, selecting the interval or intervals that satisfy the given inequality.

This procedure is valid if $<$ is replaced by $\leq$, or $>$ is replaced by $\geq$. However, if the inequality involves $\leq$ or $\geq$, include the boundary points (the solutions of $|X| = c$) in the solution set.

Study Tip

Each test value must be chosen from the *interior* of an interval. Test values should not be endpoints of intervals.

EXAMPLE 4 Solving an Absolute Value Inequality Using Boundary Points

Solve and graph the solution set on a number line:

$$|x - 4| < 3.$$

Solution

Step 1. Solve the equation $|X| = c$. We find the boundary points for $|x - 4| < 3$ by solving $|x - 4| = 3$.

$$|x - 4| = 3 \quad \text{This is the equation needed to find the boundary points.}$$

$$x - 4 = 3 \quad \text{or} \quad x - 4 = -3 \quad \text{Rewrite } |X| = c \text{ as } X = c \text{ or } X = -c.$$

$$x = 7 \qquad\qquad x = 1 \quad \text{Add 4 to both sides of each equation and solve for } x.$$

The boundary points are 1 and 7.

Step 2. Locate the boundary points on a number line and separate the line into intervals. The number line with the boundary points is shown as follows:

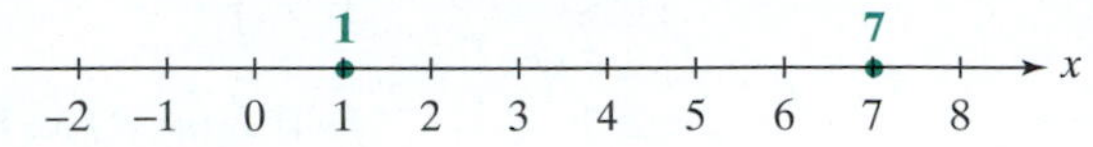

The boundary points divide the number line into three intervals:

$$(-\infty, 1) \qquad (1, 7) \qquad (7, \infty).$$

Step 3. Choose one test value within each interval and substitute that value into the given inequality.

Interval	Test Value	Substitute into $\|x - 4\| < 3$	Conclusion
$(-\infty, 1)$	0	$\|0 - 4\| \overset{?}{<} 3$ $\|-4\| \overset{?}{<} 3$ $4 < 3$, false	$(-\infty, 1)$ does not belong to the solution set.
$(1, 7)$	2	$\|2 - 4\| \overset{?}{<} 3$ $\|-2\| \overset{?}{<} 3$ $2 < 3$, true	$(1, 7)$ belongs to the solution set.
$(7, \infty)$	8	$\|8 - 4\| \overset{?}{<} 3$ $\|-4\| \overset{?}{<} 3$ $4 < 3$, false	$(7, \infty)$ does not belong to the solution set.

Step 4. Write the solution set, selecting the interval or intervals that satisfy the given inequality. Based on our work in step 3, we see that the solution set of the given inequality, $|x - 4| < 3$, is $(1, 7)$. The solution set can be expressed in set-builder notation as $\{x | 1 < x < 7\}$. The graph of the solution set is shown as follows:

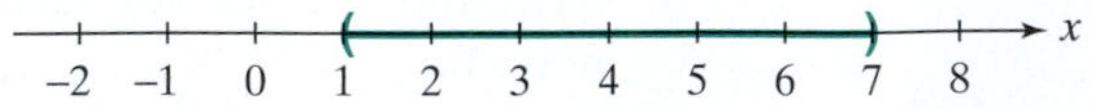

We can use the rectangular coordinate system to visualize the solution set of

$$|x - 4| < 3.$$

Figure 4.7 shows the graphs of $f(x) = |x - 4|$ and $g(x) = 3$. The solution set of $|x - 4| < 3$ consists of all values of x for which the blue graph of f lies below the red graph of g. These x-values make up the interval $(1, 7)$, which is the solution set.

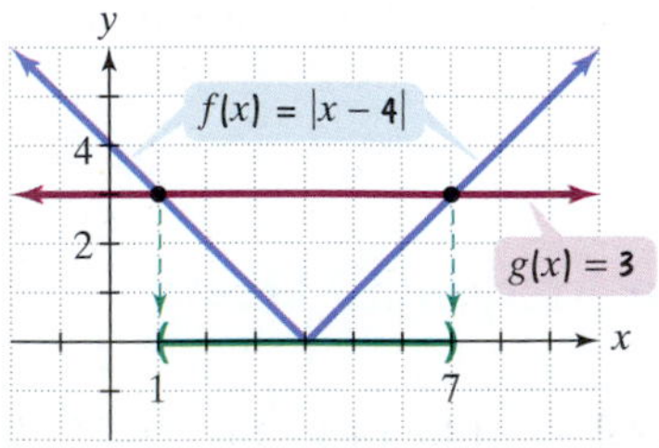

FIGURE 4.7 The solution set of $|x - 4| < 3$ is $(1, 7)$.

✓ CHECK POINT 4 Solve and graph the solution set on a number line: $|x - 2| < 5$.

EXAMPLE 5 Solving an Absolute Value Inequality Using Boundary Points

Solve and graph the solution set on a number line:

$$|2x + 3| \geq 5.$$

Solution

Step 1. Solve the equation $|X| = c$. We find the boundary points by solving $|2x + 3| = 5$.

$|2x + 3| = 5$			This is the equation needed to find the boundary points.
$2x + 3 = 5$	or	$2x + 3 = -5$	Rewrite $|X| = c$ as $X = c$ or $X = -c$.
$2x = 2$		$2x = -8$	Subtract 3 from both sides of each equation.
$x = 1$		$x = -4$	Divide both sides of each equation by 2 and solve for x.

The boundary points are -4 and 1.

Step 2. Locate the boundary points on a number line and separate the line into intervals. The number line with the boundary points is shown as follows:

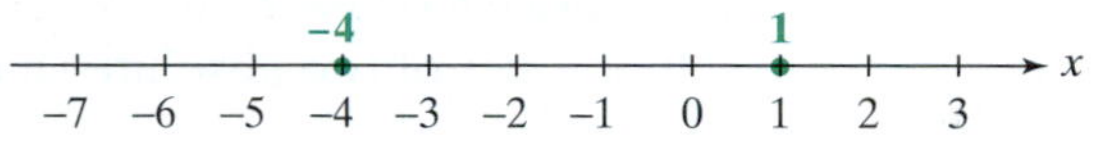

The boundary points divide the number line into three intervals:

$$(-\infty, -4) \quad (-4, 1) \quad (1, \infty).$$

Step 3. Choose one test value within each interval and substitute that value into the given inequality.

Interval	Test Value	Substitute into $|2x + 3| \geq 5$	Conclusion
$(-\infty, -4)$	-5	$|2(-5) + 3| \overset{?}{\geq} 5$ $|-10 + 3| \overset{?}{\geq} 5$ $|-7| \overset{?}{\geq} 5$ $7 \geq 5$, true	$(-\infty, -4)$ belongs to the solution set.
$(-4, 1)$	0	$|2 \cdot 0 + 3| \overset{?}{\geq} 5$ $|0 + 3| \overset{?}{\geq} 5$ $|3| \overset{?}{\geq} 5$ $3 \geq 5$, false	$(-4, 1)$ does not belong to the solution set.
$(1, \infty)$	2	$|2 \cdot 2 + 3| \overset{?}{\geq} 5$ $|4 + 3| \overset{?}{\geq} 5$ $|7| \overset{?}{\geq} 5$ $7 \geq 5$, true	$(1, \infty)$ belongs to the solution set.

Step 4. Write the solution set, selecting the interval or intervals that satisfy the given inequality. Based on our work in step 3, we see that all x in $(-\infty, -4)$ or $(1, \infty)$ belong to the solution set of $|2x + 3| \geq 5$. However, because the inequality

involves $\geq$ (greater than or *equal to*), we must also include the solutions of $|2x + 3| = 5$, namely the boundary points -4 and 1, in the solution set. Thus, the solution set of $|2x + 3| \geq 5$ is

$$(-\infty, -4] \cup [1, \infty)$$
$$\text{or} \quad \{x | x \leq -4 \quad \text{or} \quad x \geq 1\}.$$

The graph of the solution set on a number line is shown as follows:

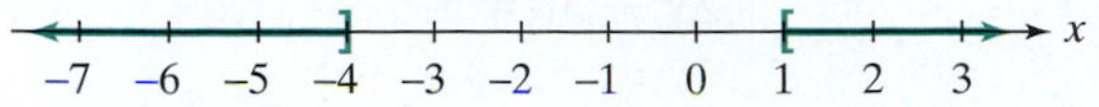

CHECK POINT 5 Solve and graph the solution set on a number line: $|2x - 5| \geq 3$.

---

3 Use equivalent compound inequalities to solve absolute value inequalities.

Solving Inequalities Involving Absolute Value Using Equivalent Compound Inequalities

In **Figure 4.6** on page 265, we used graphs of $f(x) = |x|$ and $g(x) = 2$ to visualize the solution sets of two absolute value inequalities:

- The solution set of $|x| < 2$ is $(-2, 2)$ or $\{x | -2 < x < 2\}$.
- The solution set of $|x| > 2$ is $(-\infty, -2) \cup (2, \infty)$ or $\{x | x < -2 \text{ or } x > 2\}$.

We can verify these results by interpreting $|x|$ as the distance from 0 to x on a number line.

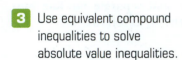

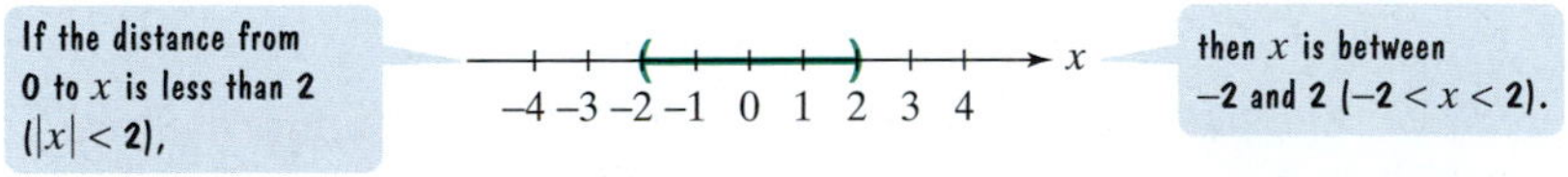

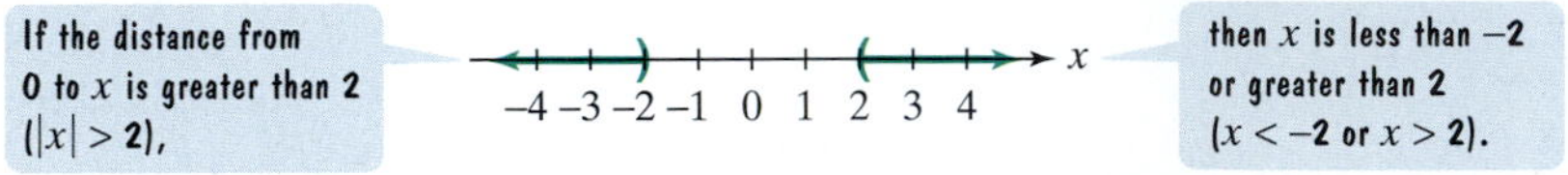

Generalizing from these observations gives us a second method for solving inequalities with absolute value. This method involves rewriting the given inequality without absolute value bars.

Using Equivalent Compound Inequalities to Solve Absolute Value Inequalities

If X is an algebraic expression and c is a positive number,

1. The solutions of $|X| < c$ are the numbers that satisfy $-c < X < c$.
2. The solutions of $|X| > c$ are the numbers that satisfy $X < -c$ or $X > c$.

These rules are valid if $<$ is replaced by $\leq$ and $>$ is replaced by $\geq$.

Let's rework Examples 4 and 5 by converting to equivalent compound inequalities. Although this method is faster than using boundary points, it's easy to make an error when rewriting absolute value inequalities without absolute value bars. Ask your instructor if one method is preferred over another.

EXAMPLE 6 Solving Absolute Value Inequalities Using Equivalent Compound Inequalities

Solve each inequality:

a. $|x - 4| < 3$ **b.** $|2x + 3| \geq 5$.

Solution

a. We rewrite the inequality without absolute value bars.

$|X| < c$ **means** $-c < X < c.$

$$|x - 4| < 3 \quad \text{means} \quad -3 < x - 4 < 3.$$

We solve the compound inequality by adding 4 to all three parts.

$$-3 < x - 4 < 3$$
$$-3 + 4 < x - 4 + 4 < 3 + 4$$
$$1 < x < 7$$

The solution set is all real numbers greater than 1 and less than 7, denoted by $\{x|1 < x < 7\}$ or $(1, 7)$.

b. We rewrite the inequality without absolute value bars.

$|X| \geq c$ **means** $X \leq -c$ **or** $X \geq c.$

$$|2x + 3| \geq 5 \quad \text{means} \quad 2x + 3 \leq -5 \quad \text{or} \quad 2x + 3 \geq 5.$$

We solve this compound inequality by solving each of these inequalities separately. Then we take the union of their solution sets.

$2x + 3 \leq -5$ or $2x + 3 \geq 5$ *These are the inequalities without absolute value bars.*

$2x \leq -8$ $\quad 2x \geq 2$ *Subtract 3 from both sides.*

$x \leq -4$ $\quad x \geq 1$ *Divide both sides by 2.*

The solution set consists of all numbers that are less than or equal to -4 or greater than or equal to 1. The solution set is $\{x|x \leq -4 \text{ or } x \geq 1\}$, or, in interval notation, $(-\infty, -4] \cup [1, \infty)$.

Study Tip

If X is a linear expression, the graph of the solution set for $|X| > c$ will be divided into two intervals whose union cannot be represented as a single interval. The graph of the solution set for $|X| < c$ will be a single interval. Avoid the common error of rewriting $|X| > c$ as $-c < X > c$.

✓ CHECK POINT 6 Solve each inequality using equivalent compound inequalities. (These are the inequalities that you solved in Check Points 4 and 5 using boundary points.)

a. $|x - 2| < 5$ **b.** $|2x - 5| \geq 3$

EXAMPLE 7 Solving an Absolute Value Inequality Using an Equivalent Compound Inequality

Solve and graph the solution set on a number line: $-2|3x + 5| + 7 \geq -13$.

Study Tip

If you use the boundary point method, you'll still need to first isolate the absolute value expression. You can then use boundary points to solve

$$|3x + 5| \leq 10.$$

Solution

$$-2|3x + 5| + 7 \geq -13$$ This is the given inequality.

We need to isolate $|3x + 5|$, the absolute value expression.

$$-2|3x + 5| + 7 - 7 \geq -13 - 7$$ Subtract 7 from both sides.

$$-2|3x + 5| \geq -20$$ Simplify.

$$\frac{-2|3x + 5|}{-2} \leq \frac{-20}{-2}$$ Divide both sides by −2 and change the sense of the inequality.

$$|3x + 5| \leq 10$$ Simplify.

$$-10 \leq 3x + 5 \leq 10$$ Rewrite without absolute value bars: $|X| \leq c$ means $-c \leq X \leq c$.

Now we need to isolate x in the middle.

$$-10 - 5 \leq 3x + 5 - 5 \leq 10 - 5$$ Subtract 5 from all three parts.

$$-15 \leq 3x \leq 5$$ Simplify.

$$\frac{-15}{3} \leq \frac{3x}{3} \leq \frac{5}{3}$$ Divide each part by 3.

$$-5 \leq x \leq \frac{5}{3}$$ Simplify.

The solution set is $\{x | -5 \leq x \leq \frac{5}{3}\}$ in set-builder notation and $\left[-5, \frac{5}{3}\right]$ in interval notation. The graph is shown as follows:

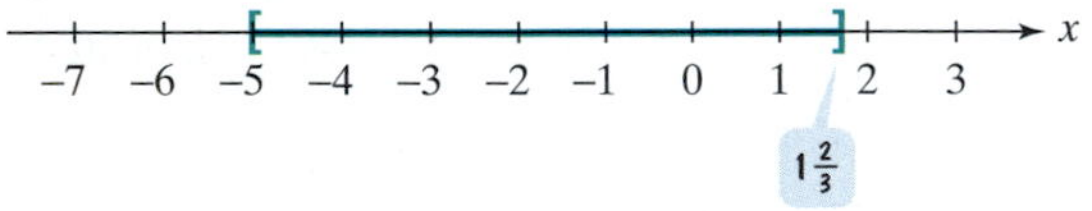

CHECK POINT 7 Solve and graph the solution set on a number line: $-3|5x - 2| + 20 \geq -19$.

4 Recognize absolute value inequalities with no solution or all real numbers as solutions.

Absolute Value Inequalities with Unusual Solution Sets

We have been working with $|X| < c$ and $|X| > c$, where c is a positive number. Now let's see what happens to these inequalities if c is a negative number. Consider, for example, $|x| < -2$. Because $|x|$ always has a value that is greater than or equal to 0, there is no number whose absolute value is less than −2. The inequality $|x| < -2$ has no solution. The solution set is $\varnothing$.

Now consider the inequality $|x| > -2$. Because $|x|$ is never negative, all numbers have an absolute value greater than −2. All real numbers satisfy the inequality $|x| > -2$. The solution set is $(-\infty, \infty)$.

Absolute Value Inequalities with Unusual Solution Sets

If X is an algebraic expression and c is a negative number,

1. The inequality $|X| < c$ has no solution.
2. The inequality $|X| > c$ is true for all real numbers for which X is defined.

5 Solve problems using absolute value inequalities.

Applications

When you were between the ages of 6 and 14, how would you have responded to this question:

What is bad about being a kid?

Table 4.3 What Is Bad about Being a Kid?

Kids Say	
Getting bossed around	17%
School, homework	15%
Can't do everything I want	11%
Chores	9%
Being grounded	9%

Source: Penn, Schoen, and Berland using 1172 interviews with children ages 6 to 14 from May 14 to June 1, 1999, Margin of error: ±2.9 %

Note the margin of error.

In a random sample of 1172 children ages 6 to 14, 17% of the children responded, "Getting bossed around." The problem is that this is a single random sample. Do 17% of kids in the entire population of children ages 6 to 14 think that getting bossed around is a bad thing?

If you look at the results of a poll like the one in **Table 4.3**, you will observe that a **margin of error** is reported. The margin of error is ±2.9%. This means that the actual percentage of children who feel getting bossed around is a bad thing is at most 2.9% greater than or less than 17%. If x represents the percentage of children in the population who think that getting bossed around is a bad thing, then the poll's margin of error can be expressed as an absolute value inequality:

$$|x - 17| \le 2.9.$$

EXAMPLE 8 Analyzing a Poll's Margin of Error

The inequality

$$|x - 9| \le 2.9$$

describes the percentage of children in the population who think that being grounded is a bad thing about being a kid. (See **Table 4.3**.) Solve the inequality and interpret the solution.

Solution We can solve the inequality using boundary points or an equivalent compound inequality. Let's use an equivalent compound inequality and rewrite without absolute value bars.

$$|X| \le c \quad \text{means} \quad -c \le X \le c.$$

$$|x - 9| \le 2.9 \quad \text{means} \quad -2.9 \le x - 9 \le 2.9.$$

We solve the compound inequality by adding 9 to all three parts.

$$-2.9 \le x - 9 \le 2.9$$

$$-2.9 + 9 \le x - 9 + 9 \le 2.9 + 9$$

$$6.1 \le x \le 11.9$$

The percentage of children in the population who think that being grounded is a bad thing is somewhere between a low of 6.1% and a high of 11.9%. Notice that these percents are 2.9% above and below the given 9%, and that 2.9% is the poll's margin of error.

✓ **CHECK POINT 8** Solve the inequality:

$$|x - 11| \le 2.9.$$

Interpret the solution in terms of the information in **Table 4.3**.

4.3 EXERCISE SET

MyMathLab · MathXL PRACTICE · WATCH · DOWNLOAD · READ · REVIEW

Practice Exercises

In Exercises 1–38, find the solution set for each equation.

1. $|x| = 8$
2. $|x| = 6$
3. $|x - 2| = 7$
4. $|x + 1| = 5$
5. $|2x - 1| = 7$
6. $|2x - 3| = 11$
7. $\left|\frac{4x - 2}{3}\right| = 2$
8. $\left|\frac{3x - 1}{5}\right| = 1$
9. $|x| = -8$
10. $|x| = -6$
11. $|x + 3| = 0$
12. $|x + 2| = 0$
13. $2|y + 6| = 10$
14. $3|y + 5| = 12$
15. $3|2x - 1| = 21$
16. $2|3x - 2| = 14$
17. $|6y - 2| + 4 = 32$
18. $|3y - 1| + 10 = 25$
19. $7|5x| + 2 = 16$
20. $7|3x| + 2 = 16$
21. $|x + 1| + 5 = 3$
22. $|x + 1| + 6 = 2$
23. $|4y + 1| + 10 = 4$
24. $|3y - 2| + 8 = 1$
25. $|2x - 1| + 3 = 3$
26. $|3x - 2| + 4 = 4$
27. $|5x - 8| = |3x + 2|$
28. $|4x - 9| = |2x + 1|$
29. $|2x - 4| = |x - 1|$
30. $|6x| = |3x - 9|$
31. $|2x - 5| = |2x + 5|$
32. $|3x - 5| = |3x + 5|$
33. $|x - 3| = |5 - x|$
34. $|x - 3| = |6 - x|$
35. $|2y - 6| = |10 - 2y|$
36. $|4y + 3| = |4y + 5|$
37. $\left|\frac{2x}{3} - 2\right| = \left|\frac{x}{3} + 3\right|$
38. $\left|\frac{x}{2} - 2\right| = \left|x - \frac{1}{2}\right|$

In Exercises 39–74, solve and graph the solution set on a number line. Use either boundary points or an equivalent compound inequality, or the method specified by your instructor.

39. $|x| < 3$
40. $|x| < 5$
41. $|x - 2| < 1$
42. $|x - 1| < 5$
43. $|x + 2| \le 1$
44. $|x + 1| \le 5$
45. $|2x - 6| < 8$
46. $|3x + 5| < 17$
47. $|x| > 3$
48. $|x| > 5$
49. $|x + 3| > 1$
50. $|x - 2| > 5$
51. $|x - 4| \ge 2$
52. $|x - 3| \ge 4$
53. $|3x - 8| > 7$
54. $|5x - 2| > 13$
55. $|2(x - 1) + 4| \le 8$
56. $|3(x - 1) + 2| \le 20$
57. $\left|\frac{2x + 6}{3}\right| < 2$
58. $\left|\frac{3x - 3}{4}\right| < 6$
59. $\left|\frac{2x + 2}{4}\right| \ge 2$
60. $\left|\frac{3x - 3}{9}\right| \ge 1$
61. $\left|3 - \frac{2x}{3}\right| > 5$
62. $\left|3 - \frac{3x}{4}\right| > 9$
63. $|x - 2| < -1$
64. $|x - 3| < -2$
65. $|x + 6| > -10$
66. $|x + 4| > -12$
67. $|x + 2| + 9 \le 16$
68. $|x - 2| + 4 \le 5$
69. $2|2x - 3| + 10 > 12$
70. $3|2x - 1| + 2 > 8$
71. $-4|1 - x| < -16$
72. $-2|5 - x| < -6$
73. $3 \le |2x - 1|$

74. $9 \le |4x + 7|$

75. Let $f(x) = |5 - 4x|$. Find all values of x for which $f(x) = 11$.

76. Let $f(x) = |2 - 3x|$. Find all values of x for which $f(x) = 13$.

77. Let $f(x) = |3 - x|$ and $g(x) = |3x + 11|$. Find all values of x for which $f(x) = g(x)$.

78. Let $f(x) = |3x + 1|$ and $g(x) = |6x - 2|$. Find all values of x for which $f(x) = g(x)$.

79. Let $g(x) = |-1 + 3(x + 1)|$. Find all values of x for which $g(x) \le 5$.

80. Let $g(x) = |-3 + 4(x + 1)|$. Find all values of x for which $g(x) \le 3$.

81. Let $h(x) = |2x - 3| + 1$. Find all values of x for which $h(x) > 6$.

82. Let $h(x) = |2x - 4| - 6$. Find all values of x for which $h(x) > 18$.

Practice PLUS

83. When 3 times a number is subtracted from 4, the absolute value of the difference is at least 5. Use interval notation to express the set of all real numbers that satisfy this condition.

84. When 4 times a number is subtracted from 5, the absolute value of the difference is at most 13. Use interval notation to express the set of all real numbers that satisfy this condition.

In Exercises 85–86, solve each inequality. Assume that $a > 0$ and $c > 0$. Use set-builder notation to express each solution set.

85. $|ax + b| < c$

86. $|ax + b| \ge c$

In Exercises 87–88, use the graph of $f(x) = |4 - x|$ to solve each equation or inequality.

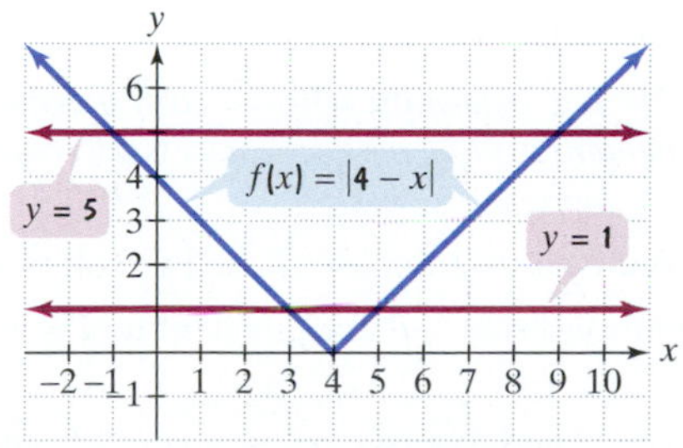

87. $|4 - x| = 1$

88. $|4 - x| < 5$

In Exercises 89–90, use the table to solve each inequality.

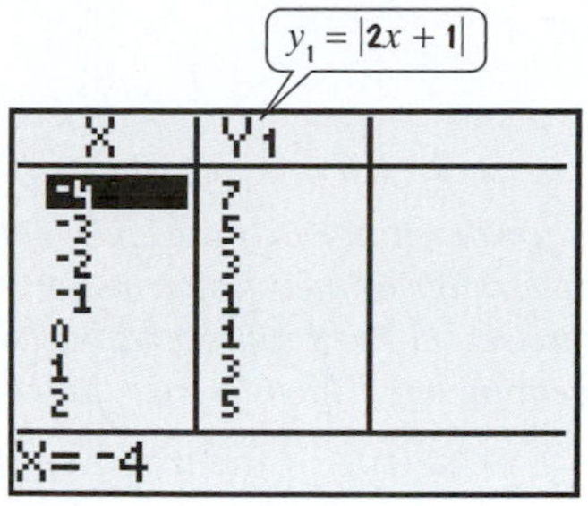

89. $|2x + 1| \le 3$

90. $|2x + 1| \ge 3$

Application Exercises

The three television programs viewed by the greatest percentage of U.S. households in the twentieth century are shown in the table. The data are from a random survey of 4000 TV households by Nielsen Media Research. In Exercises 91–92, let x represent the actual viewing percentage in the U.S. population.

TV Programs with the Greatest U.S. Audience Viewing Percentage of the Twentieth Century

Program	Viewing Percentage in Survey
1 *M*A*S*H* Feb. 28, 1983	60.2%
2 *Dallas* Nov. 21, 1980	53.3%
3 *Roots Part 8* Jan. 30, 1977	51.1%

Source: Nielsen Media Research

91. Solve the inequality: $|x - 60.2| \le 1.6$. Interpret the solution in terms of the information in the table. What is the margin of error?

92. Solve the inequality: $|x - 51.1| \le 1.6$. Interpret the solution in terms of the information in the table. What is the margin of error?

93. The inequality $|T - 57| \le 7$ describes the range of monthly average temperature, T, in degrees Fahrenheit, for San Francisco, California. Solve the inequality and interpret the solution.

94. The inequality $|T - 50| \le 22$ describes the range of monthly average temperature, T, in degrees Fahrenheit, for Albany, New York. Solve the inequality and interpret the solution.

The specifications for machine parts are given with tolerance limits that describe a range of measurements for which the part is acceptable. In Exercises 95–96, x represents the length of a machine part, in centimeters. The tolerance limit is 0.01 centimeter.

95. Solve: $|x - 8.6| \le 0.01$. If the length of the machine part is supposed to be 8.6 centimeters, interpret the solution.

96. Solve: $|x - 9.4| \le 0.01$. If the length of the machine part is supposed to be 9.4 centimeters, interpret the solution.

97. If a coin is tossed 100 times, we would expect approximately 50 of the outcomes to be heads. It can be demonstrated that a coin is unfair if h, the number of outcomes that result in heads, satisfies $\left|\frac{h - 50}{5}\right| \ge 1.645$. Describe the number of outcomes that result in heads that determine an unfair coin that is tossed 100 times.

Writing in Mathematics

98. Explain how to solve an equation containing one absolute value expression.

99. Explain why the procedure that you described in Exercise 98 does not apply to the equation $|x - 5| = -3$. What is the solution set of this equation?

100. Describe how to solve an absolute value equation with two absolute values.

101. Describe one method for solving an absolute value inequality.

102. Explain why the procedure that you described in Exercise 101 does not apply to the inequality $|x - 5| < -3$. What is the solution set of this inequality?

103. Explain why the procedure that you described in Exercise 101 does not apply to the inequality $|x - 5| > -3$. What is the solution set of this inequality?

104. The final episode of $M * A * S * H$ was viewed by more than 58% of U.S. television households. Is it likely that a popular television series in the twenty-first century will achieve a 58% market share? Explain your answer.

Technology Exercises

In Exercises 105–107, solve each equation using a graphing utility. Graph each side separately in the same viewing rectangle. The solutions are the x-coordinates of the intersection points.

105. $|x + 1| = 5$

106. $|3(x + 4)| = 12$

107. $|2x - 3| = |9 - 4x|$

In Exercises 108–110, solve each inequality using a graphing utility. Graph each side separately in the same viewing rectangle. The solution set consists of all values of x for which the graph of the left side lies below the graph of the right side.

108. $|2x + 3| < 5$

109. $\left|\frac{2x - 1}{3}\right| < \frac{5}{3}$

110. $|x + 4| < -1$

In Exercises 111–113, solve each inequality using a graphing utility. Graph each side separately in the same viewing rectangle. The solution set consists of all values of x for which the graph of the left side lies above the graph of the right side.

111. $|2x - 1| > 7$

112. $|0.1x - 0.4| + 0.4 > 0.6$

113. $|x + 4| > -1$

114. Use a graphing utility to verify the solution sets for any five equations or inequalities that you solved by hand in Exercises 1–74.

Critical Thinking Exercises

Make Sense? *In Exercises 115–118, determine whether each statement "makes sense" or "does not make sense" and explain your reasoning.*

115. I have problems setting up the appropriate compound inequality for an absolute value inequality, so I use the boundary point method.

116. I noticed that the graph of $f(x) = |x - 3|$ lies below the graph of $g(x) = 4$ in $(-1, 7)$, so the solution set of $|x - 3| > 4$ must be $(-\infty, -1) \cup (7, \infty)$.

117. Because the absolute value of any expression is never less than a negative number, I can immediately conclude that the inequality $|2x - 5| - 9 < -4$ has no solution.

118. I'll win the contest if I can complete the crossword puzzle in 20 minutes plus or minus 5 minutes, so my winning time, x, is modeled by $|x - 20| \le 5$.

In Exercises 119–122, determine whether each statement is true or false. If the statement is false, make the necessary change(s) to produce a true statement.

119. All absolute value equations have two solutions.

120. The equation $|x| = -6$ is equivalent to $x = 6$ or $x = -6$.

121. Values of -5 and 5 satisfy $|x| = 5$, $|x| \le 5$, and $|x| \ge -5$.

122. The absolute value of any linear expression is greater than 0 for all real numbers except the number for which the expression is equal to 0.

123. Write an absolute value inequality for which the interval shown is the solution.

a.
b.

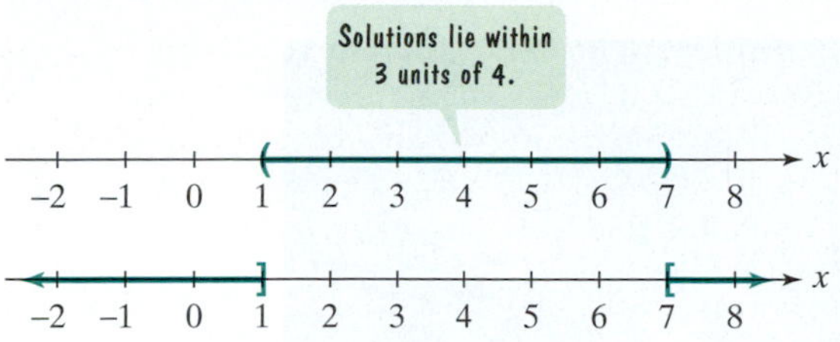

124. The percentage, p, of defective products manufactured by a company is given by $|p - 0.3\%| \le 0.2\%$. If 100,000 products are manufactured and the company offers a \$5 refund for each defective product, describe the company's cost for refunds.

125. Solve: $|2x + 5| = 3x + 4$.

Review and Preview Exercises

Exercises 126–128 will enable you to review graphing linear functions. In addition, they will help you prepare for the material covered in the next section. In each exercise, graph the linear function.

126. $3x - 5y = 15$ (Section 2.4, Example 1)

127. $f(x) = -\frac{2}{3}x$ (Section 2.4, Example 5)

128. $f(x) = -2$ (Section 2.4, Example 6)

MID-CHAPTER CHECK POINT Section 4.1–Section 4.3

What You Know: We learned to solve linear inequalities, expressing solution sets in set-builder and interval notations. We know that it is necessary to change the sense of an inequality when multiplying or dividing both sides by a negative number. We solved compound inequalities with *and* by finding the intersection of solution sets and with *or* by finding the union of solution sets. Finally, we solved equations and inequalities involving absolute value by carefully rewriting the given equation or inequality without absolute value bars, or, in the case of inequalities, using boundary points. For positive values of c, we wrote $|X| = c$ as $X = c$ or $X = -c$. Without using boundary points, we wrote $|X| < c$ as $-c < X < c$, and we wrote $|X| > c$ as $X < -c$ or $X > c$.

In Exercises 1–18, solve each inequality or equation.

1. $4 - 3x \ge 12 - x$
2. $5 \le 2x - 1 < 9$
3. $|4x - 7| = 5$
4. $-10 - 3(2x + 1) > 8x + 1$
5. $2x + 7 < -11$ or $-3x - 2 < 13$
6. $|3x - 2| \le 4$
7. $|x + 5| = |5x - 8|$
8. $5 - 2x \ge 9$ and $5x + 3 > -17$
9. $3x - 2 > -8$ or $2x + 1 < 9$
10. $\frac{x}{2} + 3 \le \frac{x}{3} + \frac{5}{2}$
11. $\frac{2}{3}(6x - 9) + 4 > 5x + 1$
12. $|5x + 3| > 2$
13. $7 - \left|\frac{x}{2} + 2\right| \le 4$
14. $5(x - 2) - 3(x + 4) \ge 2x - 20$
15. $\frac{x + 3}{4} < \frac{1}{3}$
16. $5x + 1 \ge 4x - 2$ and $2x - 3 > 5$
17. $3 - |2x - 5| = -6$
18. $3 + |2x - 5| = -6$

In Exercises 19–22, solve each problem.

19. A car rental agency rents a certain car for \$40 per day with unlimited mileage or \$24 per day plus \$0.20 per mile. How far can a customer drive this car per day for the \$24 option to cost no more than the unlimited mileage option?
20. To receive a B in a course, you must have an average of at least 80% but less than 90% on five exams. Your grades on the first four exams were 95%, 79%, 91%, and 86%. What range of grades on the fifth exam will result in a B for the course?
21. A retiree requires an annual income of at least \$9000 from an investment paying 7.5% annual interest. How much should the retiree invest to achieve the desired return?
22. A company that manufactures compact discs has fixed monthly overhead costs of \$60,000. Each disc costs \$0.18 to produce and sells for \$0.30. How many discs should be produced and sold each month for the company to have a profit of at least \$30,000?

SECTION 4.4

Linear Inequalities in Two Variables

Objectives

1. Graph a linear inequality in two variables.
2. Use mathematical models involving linear inequalities.
3. Graph a system of linear inequalities.

This book was written in Point Reyes National Seashore, 40 miles north of San Francisco. The park consists of 75,000 acres with miles of pristine surf-washed beaches, forested ridges, and bays bordered by white cliffs.

Like your author, many people are kept inspired and energized surrounded by nature's unspoiled beauty. In this section, you will see how systems of inequalities model whether a region's natural beauty manifests itself in forests, grasslands, or deserts.

Linear Inequalities in Two Variables and Their Solutions

We have seen that equations in the form $Ax + By = C$ are straight lines when graphed. If we change the symbol $=$ to $>$, $<$, $\geq$, or $\leq$, we obtain a **linear inequality in two variables**. Some examples of linear inequalities in two variables are $x + y > 2$, $3x - 5y \leq 15$, and $2x - y < 4$.

A **solution of an inequality in two variables**, x and y, is an ordered pair of real numbers with the following property: When the x-coordinate is substituted for x and the y-coordinate is substituted for y in the inequality, we obtain a true statement. For example, (3, 2) is a solution of the inequality $x + y > 1$. When 3 is substituted for x and 2 is substituted for y, we obtain the true statement $3 + 2 > 1$, or $5 > 1$. Because there are infinitely many pairs of numbers that have a sum greater than 1, the inequality $x + y > 1$ has infinitely many solutions. Each ordered-pair solution is said to **satisfy** the inequality. Thus, (3, 2) satisfies the inequality $x + y > 1$.

1 Graph a linear inequality in two variables.

The Graph of a Linear Inequality in Two Variables

We know that the graph of an equation in two variables is the set of all points whose coordinates satisfy the equation. Similarly, the **graph of an inequality in two variables** is the set of all points whose coordinates satisfy the inequality.

Let's use **Figure 4.8** to get an idea of what the graph of a linear inequality in two variables looks like. Part of the figure shows the graph of the linear equation $x + y = 2$. The line divides the points in the rectangular coordinate system into three sets. First, there is the set of points along the line, satisfying $x + y = 2$. Next, there is the set of points in the green region above the line. Points in the green region satisfy the linear inequality $x + y > 2$. Finally, there is the set of points in the purple region below the line. Points in the purple region satisfy the linear inequality $x + y < 2$.

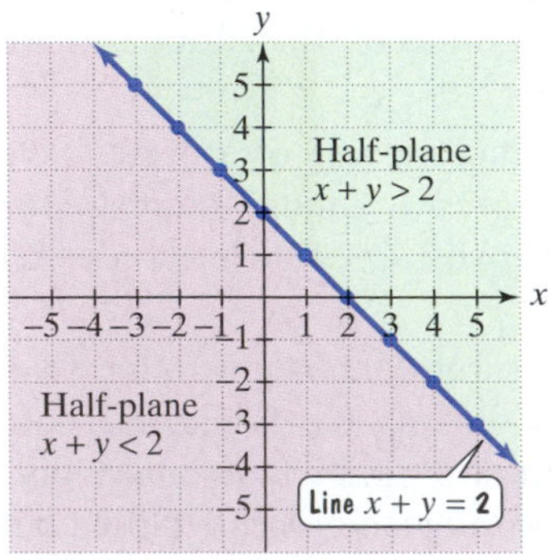

FIGURE 4.8

A **half-plane** is the set of all the points on one side of a line. In **Figure 4.8**, the green region is a half-plane. The purple region is also a half-plane. A half-plane is the graph of a linear inequality that involves $>$ or $<$. The graph of a linear inequality that involves $\geq$ or $\leq$ is a half-plane and a line. A solid line is used to show that a line is part of a graph. A dashed line is used to show that a line is not part of a graph.

Graphing a Linear Inequality in Two Variables

1. Replace the inequality symbol with an equal sign and graph the corresponding linear equation. Draw a solid line if the original inequality contains a $\leq$ or $\geq$ symbol. Draw a dashed line if the original inequality contains a $<$ or $>$ symbol.
2. Choose a test point from one of the half-planes. (Do not choose a point on the line.) Substitute the coordinates of the test point into the inequality.
3. If a true statement results, shade the half-plane containing this test point. If a false statement results, shade the half-plane not containing this test point.

EXAMPLE 1 Graphing a Linear Inequality in Two Variables

Graph: $2x - 3y \geq 6$.

Solution

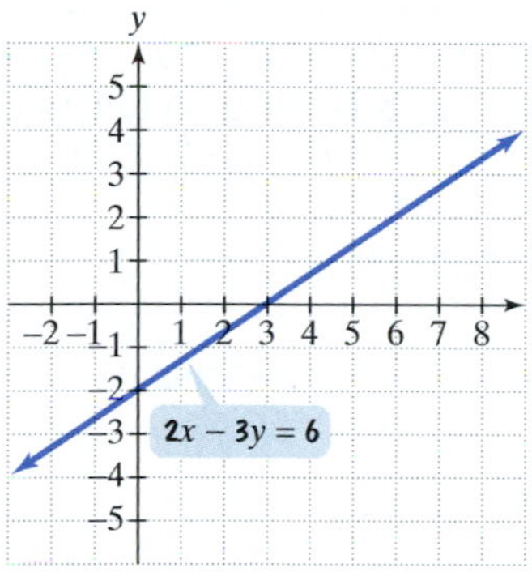

FIGURE 4.9 Preparing to graph $2x - 3y \geq 6$

Step 1. Replace the inequality symbol by = and graph the linear equation. We need to graph $2x - 3y = 6$. We can use intercepts to graph this line.

We set $y = 0$ to find the x-intercept:

$$2x - 3y = 6$$
$$2x - 3 \cdot 0 = 6$$
$$2x = 6$$
$$x = 3.$$

We set $x = 0$ to find the y-intercept:

$$2x - 3y = 6$$
$$2 \cdot 0 - 3y = 6$$
$$-3y = 6$$
$$y = -2.$$

The x-intercept is 3, so the line passes through (3, 0). The y-intercept is −2, so the line passes through (0, −2). Using the intercepts, the line is shown in **Figure 4.9** as a solid line. This is because the inequality $2x - 3y \geq 6$ contains a $\geq$ symbol, in which equality is included.

Step 2. Choose a test point from one of the half-planes and not from the line. Substitute its coordinates into the inequality. The line $2x - 3y = 6$ divides the plane into three parts—the line itself and two half-planes. The points in one half-plane satisfy $2x - 3y > 6$. The points in the other half-plane satisfy $2x - 3y < 6$. We need to find which half-plane belongs to the solution of $2x - 3y \geq 6$. To do so, we test a point from either half-plane. The origin, (0, 0), is the easiest point to test.

$$2x - 3y \geq 6 \quad \text{This is the given inequality.}$$
$$2 \cdot 0 - 3 \cdot 0 \overset{?}{\geq} 6 \quad \text{Test (0, 0) by substituting 0 for x and 0 for y.}$$
$$0 - 0 \overset{?}{\geq} 6 \quad \text{Multiply.}$$
$$0 \geq 6 \quad \text{This statement is false.}$$

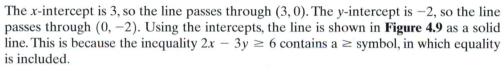

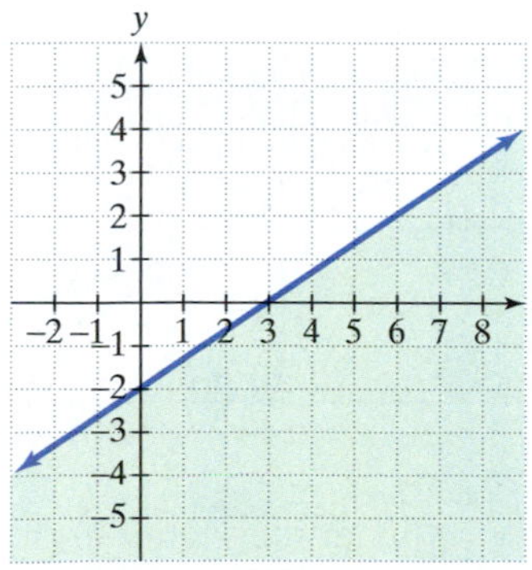

FIGURE 4.10 The graph of $2x - 3y \geq 6$

Step 3. If a false statement results, shade the half-plane not containing the test point. Because 0 is not greater than or equal to 6, the test point, (0, 0), is not part of the solution set. Thus, the half-plane below the solid line $2x - 3y = 6$ is part of the solution set. The solution set is the line and the half-plane that does not contain the point (0, 0), indicated by shading this half-plane. The graph is shown using green shading and a blue line in **Figure 4.10**.

✓ **CHECK POINT 1** Graph: $4x - 2y \geq 8$.

When graphing a linear inequality, choose a test point that lies in one of the half-planes and *not on the line dividing the half-planes*. The test point (0, 0) is convenient because it is easy to calculate when 0 is substituted for each variable. However, if (0, 0) lies on the dividing line and not in a half-plane, a different test point must be selected.

EXAMPLE 2 Graphing a Linear Inequality in Two Variables

Graph: $y > -\frac{2}{3}x$.

Solution

Step 1. Replace the inequality symbol by = and graph the linear equation. Because we are interested in graphing $y > -\frac{2}{3}x$, we begin by graphing $y = -\frac{2}{3}x$. We can use the slope and the y-intercept to graph this linear function.

$$y = -\frac{2}{3}x + 0$$

Slope $= \frac{-2}{3} = \frac{\text{rise}}{\text{run}}$ y-intercept = 0

The y-intercept is 0, so the line passes through (0, 0). Using the y-intercept and the slope, the line is shown in **Figure 4.11** as a dashed line. This is because the inequality $y > -\frac{2}{3}x$ contains a > symbol, in which equality is not included.

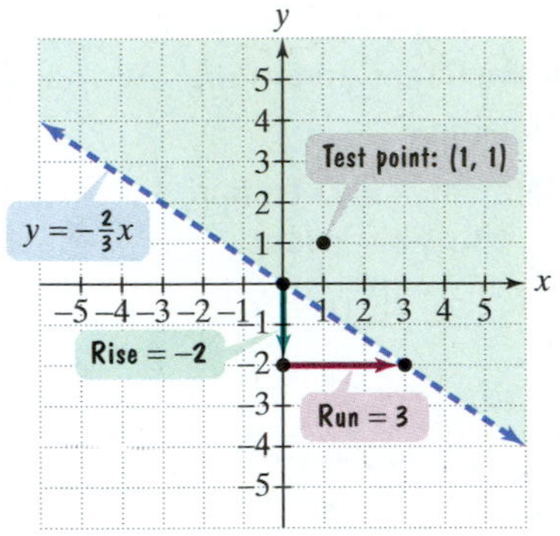

FIGURE 4.11 The graph of $y > -\frac{2}{3}x$

Step 2. Choose a test point from one of the half-planes and not from the line. Substitute its coordinates into the inequality. We cannot use (0, 0) as a test point because it lies on the line and not in a half-plane. Let's use (1, 1), which lies in the half-plane above the line.

$$y > -\frac{2}{3}x \quad \text{This is the given inequality.}$$

$$1 \overset{?}{>} -\frac{2}{3}\cdot 1 \quad \text{Test (1, 1) by substituting 1 for } x \text{ and 1 for } y.$$

$$1 > -\frac{2}{3} \quad \text{This statement is true.}$$

Step 3. If a true statement results, shade the half-plane containing the test point. Because 1 is greater than $-\frac{2}{3}$, the test point (1, 1) is part of the solution set. All the points on the same side of the line $y = -\frac{2}{3}x$ as the point (1, 1) are members of the solution set. The solution set is the half-plane that contains the point (1, 1), indicated by shading this half-plane. The graph is shown using green shading and a dashed blue line in **Figure 4.11**.

Using Technology

Most graphing utilities can graph inequalities in two variables with the SHADE feature. The procedure varies by model, so consult your manual. For most graphing utilities, you must first solve for y if it is not already isolated. The figure shows the graph of $y > -\frac{2}{3}x$. Most displays do not distinguish between dashed and solid boundary lines.

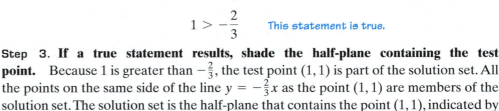

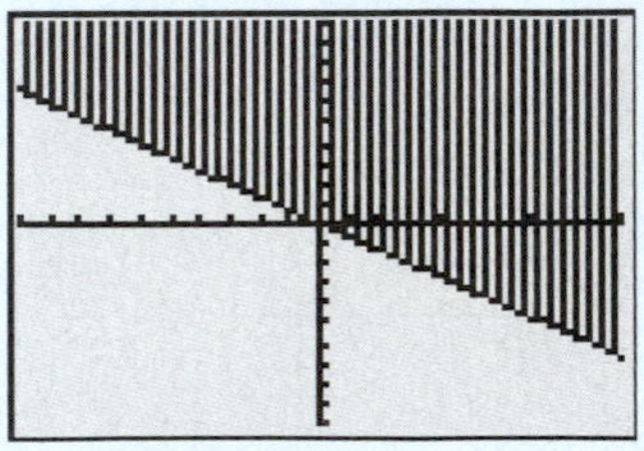

✓ CHECK POINT 2 Graph: $y > -\frac{3}{4}x$.

Graphing Linear Inequalities without Using Test Points

You can graph inequalities in the form $y > mx + b$ or $y < mx + b$ without using test points. The inequality symbol indicates which half-plane to shade.

- If $y > mx + b$, shade the half-plane above the line $y = mx + b$.
- If $y < mx + b$, shade the half-plane below the line $y = mx + b$.

Observe how this is illustrated in **Figure 4.11**. The graph of $y > -\frac{2}{3}x$ is the half-plane above the line $y = -\frac{2}{3}x$.

Study Tip

Continue using test points to graph inequalities in the form $Ax + By > C$ or $Ax + By < C$. The graph of $Ax + By > C$ can lie above or below the line given by $Ax + By = C$, depending on the value of B. The same comment applies to the graph of $Ax + By < C$.

It is also not necessary to use test points when graphing inequalities involving half-planes on one side of a vertical or a horizontal line.

For the Vertical Line $x = a$:	For the Horizontal Line $y = b$:
• If $x > a$, shade the half-plane to the right of $x = a$.	• If $y > b$, shade the half-plane above $y = b$.
• If $x < a$, shade the half-plane to the left of $x = a$.	• If $y < b$, shade the half-plane below $y = b$.

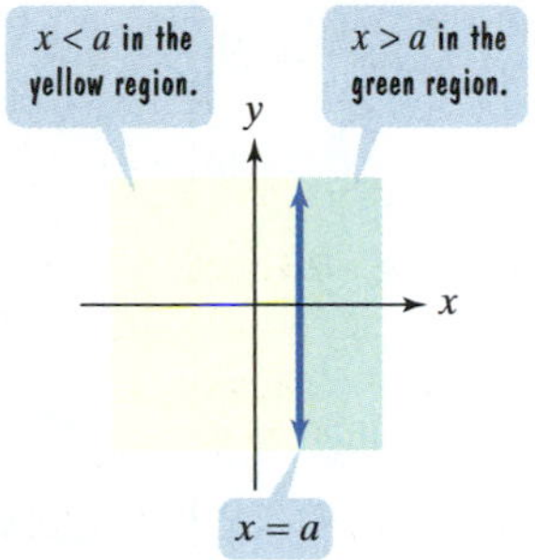
$x < a$ in the yellow region.
$x > a$ in the green region.
y
x
$x = a$

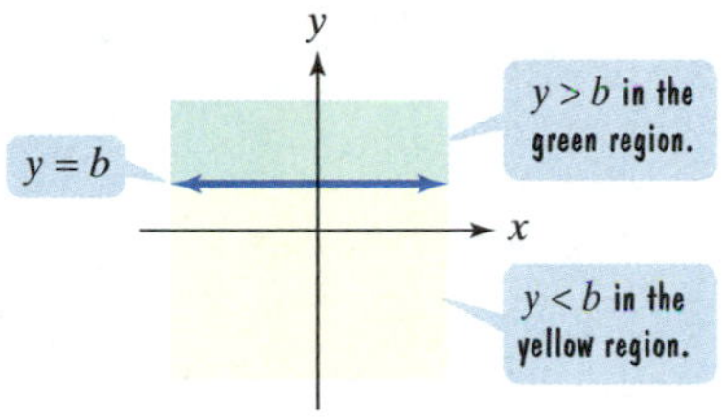
y
$y = b$
$y > b$ in the green region.
x
$y < b$ in the yellow region.

EXAMPLE 3 Graphing Inequalities without Using Test Points

Graph each inequality in a rectangular coordinate system:

a. $y \le -3$ **b.** $x > 2$.

Solution

a. $y \le -3$

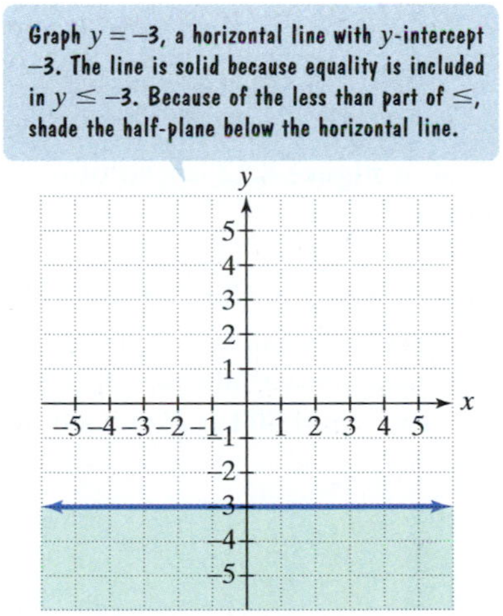
Graph $y = -3$, a horizontal line with y-intercept -3. The line is solid because equality is included in $y \le -3$. Because of the less than part of $\le$, shade the half-plane below the horizontal line.
y
x
5 4 3 2 1 −1 −2 −3 −4 −5
−5 −4 −3 −2 −1 1 2 3 4 5

b. $x > 2$

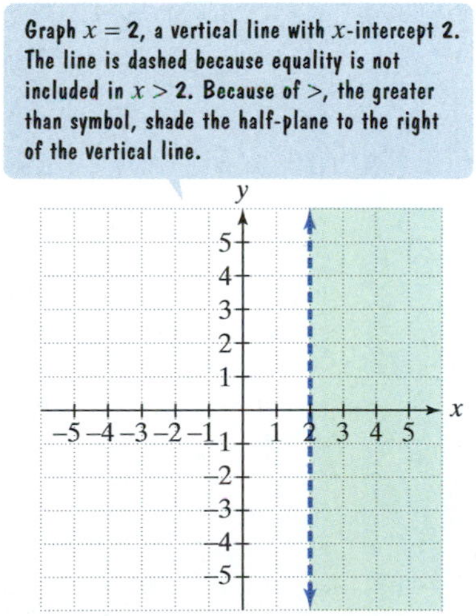
Graph $x = 2$, a vertical line with x-intercept 2. The line is dashed because equality is not included in $x > 2$. Because of $>$, the greater than symbol, shade the half-plane to the right of the vertical line.
y
x
5 4 3 2 1 −1 −2 −3 −4 −5
−5 −4 −3 −2 −1 1 2 3 4 5

✓ **CHECK POINT 3** Graph each inequality in a rectangular coordinate system:

a. $y > 1$ **b.** $x \le -2$.

2 Use mathematical models involving linear inequalities.

Modeling with Systems of Linear Inequalities

Just as two or more linear equations make up a system of linear equations, two or more linear inequalities make up a **system of linear inequalities**. A **solution of a system of linear inequalities** in two variables is an ordered pair that satisfies each inequality in the system.

EXAMPLE 4 Forests, Grasslands, Deserts, and Systems of Inequalities

Temperature and precipitation affect whether or not trees and forests can grow. At certain levels of precipitation and temperature, only grasslands and deserts will exist. **Figure 4.12** shows three kinds of regions—deserts, grasslands, and forests—that result from various ranges of temperature, T, and precipitation, P.

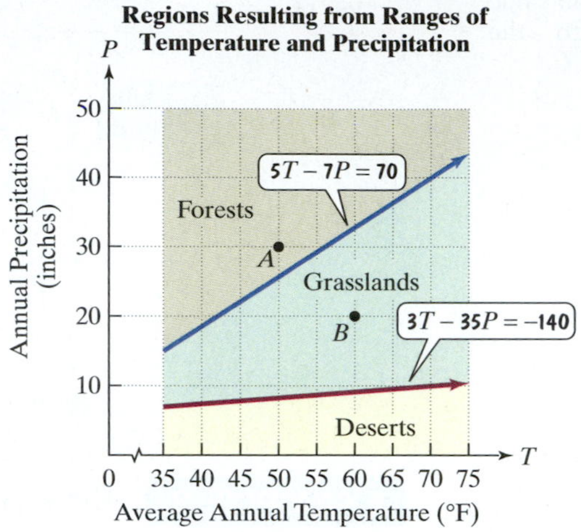

FIGURE 4.12

Source: A. Miller and J. Thompson, *Elements of Meteorology*

Systems of inequalities can be used to model where forests, grasslands, and deserts occur. Because these regions occur when the average annual temperature, T, is 35°F or greater, each system contains the inequality $T \geq 35$.

Forests occur if	Grasslands occur if	Deserts occur if
$T \geq 35$ $5T - 7P < 70.$	$T \geq 35$ $5T - 7P \geq 70$ $3T - 35P \leq -140.$	$T \geq 35$ $3T - 35P > -140.$

Show that point A in **Figure 4.12** is a solution of the system of inequalities that models where forests occur.

Solution Point A has coordinates (50, 30). This means that if a region has an average annual temperature of 50°F and an average annual precipitation of 30 inches, a forest occurs. We can show that (50, 30) satisfies the system of inequalities for forests by substituting 50 for T and 30 for P in each inequality in the system.

$$T \geq 35 \qquad\qquad 5T - 7P < 70$$

$$50 \geq 35, \quad \text{true} \qquad\qquad 5 \cdot 50 - 7 \cdot 30 \stackrel{?}{<} 70$$

$$250 - 210 \stackrel{?}{<} 70$$

$$40 < 70, \quad \text{true}$$

The coordinates (50, 30) make each inequality true. Thus, (50, 30) satisfies the system for forests.

✓ **CHECK POINT 4** Show that point B in **Figure 4.12** is a solution of the system of inequalities that models where grasslands occur.

3 Graph a system of linear inequalities.

Graphing Systems of Linear Inequalities

The **solution set of a system of linear inequalities in two variables** is the set of all ordered pairs that satisfy each inequality in the system. Thus, to graph a system of inequalities in two variables, begin by graphing each individual inequality in the same rectangular coordinate system. Then find the region, if there is one, that is common to every graph in the system. This region of intersection gives a picture of the system's solution set.

EXAMPLE 5 Graphing a System of Linear Inequalities

Graph the solution set of the system:

$$x - y < 1$$
$$2x + 3y \geq 12.$$

Solution Replacing each inequality symbol with an equal sign indicates that we need to graph $x - y = 1$ and $2x + 3y = 12$. We can use intercepts to graph these lines.

$x - y = 1$		$2x + 3y = 12$

x-intercept: $x - 0 = 1$ Set $y = 0$ in each equation. x-intercept: $2x + 3 \cdot 0 = 12$

$\quad\quad\quad\quad\quad x = 1$ $\quad\quad\quad\quad\quad\quad\quad\quad\quad\quad\quad\quad 2x = 12$

The line passes through $(1, 0)$. $\quad\quad\quad\quad\quad\quad\quad\quad\quad\quad x = 6$

$\quad\quad\quad\quad\quad\quad\quad\quad\quad\quad\quad\quad\quad\quad\quad\quad$ The line passes through $(6, 0)$.

y-intercept: $0 - y = 1$ Set $x = 0$ in each equation. y-intercept: $2 \cdot 0 + 3y = 12$

$\quad\quad\quad\quad\quad -y = 1$ $\quad\quad\quad\quad\quad\quad\quad\quad\quad\quad\quad\quad 3y = 12$

$\quad\quad\quad\quad\quad\quad y = -1$ $\quad\quad\quad\quad\quad\quad\quad\quad\quad\quad\quad y = 4$

The line passes through $(0, -1)$ $\quad\quad$ The line passes through $(0, 4)$.

Now we are ready to graph the solution set of the system of linear inequalities.

Graph $x - y < 1$. The blue line, $x - y = 1$, is dashed: Equality is not included in $x - y < 1$. Because $(0, 0)$ makes the inequality true $(0 - 0 < 1$, or $0 < 1$, is true), shade the half-plane containing $(0, 0)$ in yellow.

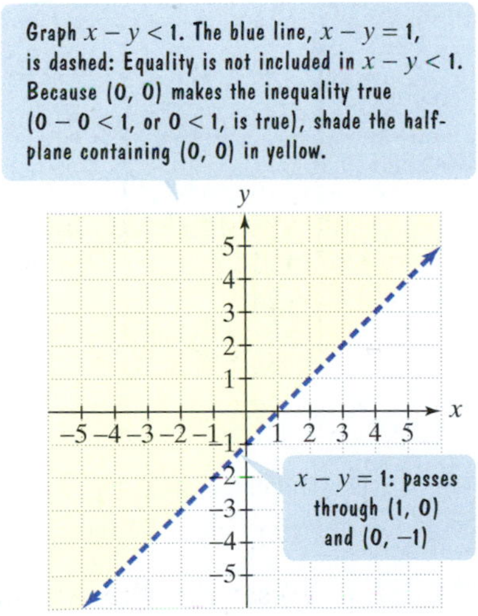

The graph of $x - y < 1$

Add the graph of $2x + 3y \geq 12$. The red line, $2x + 3y = 12$, is solid: Equality is included in $2x + 3y \geq 12$. Because $(0, 0)$ makes the inequality false $(2 \cdot 0 + 3 \cdot 0 \geq 12$, or $0 \geq 12$, is false), shade the half-plane not containing $(0, 0)$ using green vertical shading.

$2x + 3y = 12$: passes through $(6, 0)$ and $(0, 4)$

$x - y = 1$

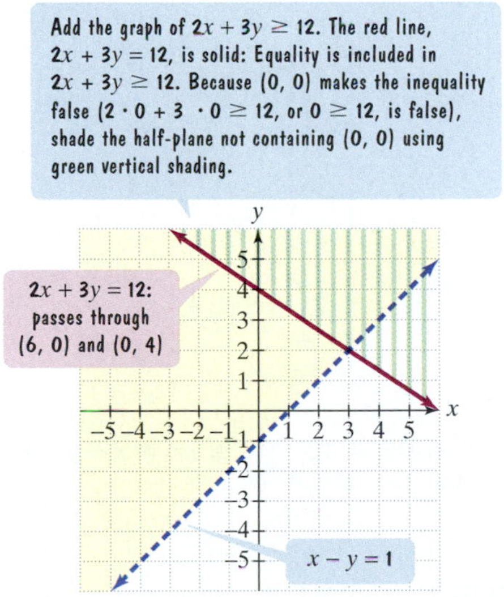

Adding the graph of $2x + 3y \geq 12$

The solution set of the system is graphed as the intersection (the overlap) of the two half-planes. This is the region in which the yellow shading and the green vertical shading overlap.

This open dot shows $(3, 2)$ is not in the solution set. It does not satisfy $x - y < 1$.

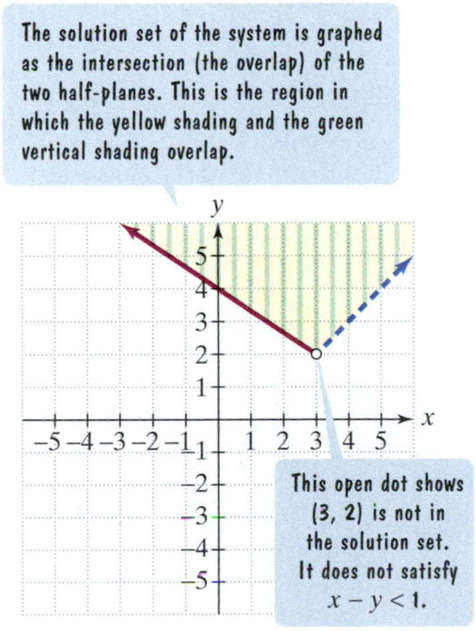

The graph of $x - y < 1$ and $2x + 3y \geq 12$

☑ **CHECK POINT 5** Graph the solution set of the system:

$$x - 3y < 6$$
$$2x + 3y \geq -6.$$

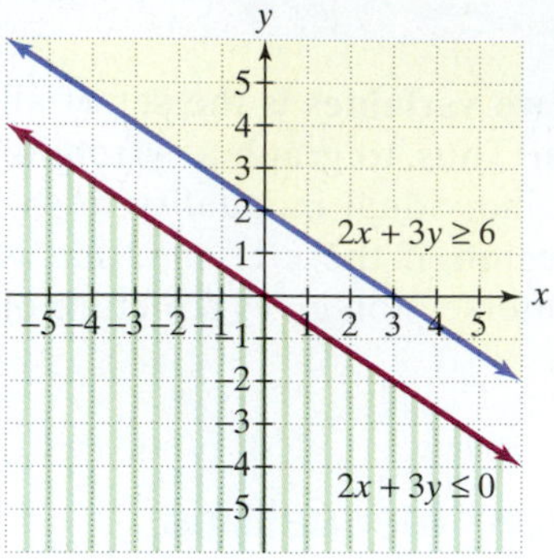

FIGURE 4.13 A system of inequalities with no solution

A system of inequalities has no solution if there are no points in the rectangular coordinate system that simultaneously satisfy each inequality in the system. For example, the system

$$2x + 3y \geq 6$$
$$2x + 3y \leq 0,$$

whose separate graphs are shown in **Figure 4.13**, has no overlapping region. Thus, the system has no solution. The solution set is ∅, the empty set.

EXAMPLE 6 Graphing a System of Inequalities

Graph the solution set of the system:

$$x - y < 2$$
$$-2 \leq x < 4$$
$$y < 3.$$

Solution We begin by graphing $x - y < 2$, the first given inequality. The line $x - y = 2$ has an x-intercept of 2 and a y-intercept of -2. The test point $(0, 0)$ makes the inequality $x - y < 2$ true. The graph of $x - y < 2$ is shown in **Figure 4.14**.

Now, let's consider the second given inequality, $-2 \leq x < 4$. Replacing the inequality symbols by =, we obtain $x = -2$ and $x = 4$, graphed as red vertical lines in **Figure 4.15**. The line of $x = 4$ is not included. Because x is between -2 and 4, we shade the region between the vertical lines. We must intersect this region with the yellow region in **Figure 4.14**. The resulting region is shown in yellow and green vertical shading in **Figure 4.15**.

Finally, let's consider the third given inequality, $y < 3$. Replacing the inequality symbol by =, we obtain $y = 3$, which graphs as a horizontal line. Because of the less than symbol in $y < 3$, the graph consists of the half-plane below the line $y = 3$. We must intersect this half-plane with the region in **Figure 4.15**. The resulting region is shown in yellow and green vertical shading in **Figure 4.16**. This region represents the graph of the solution set of the given system.

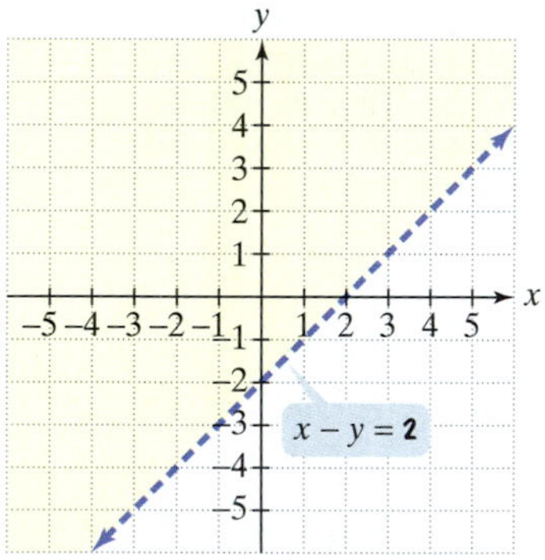

FIGURE 4.14 The graph of $x - y < 2$

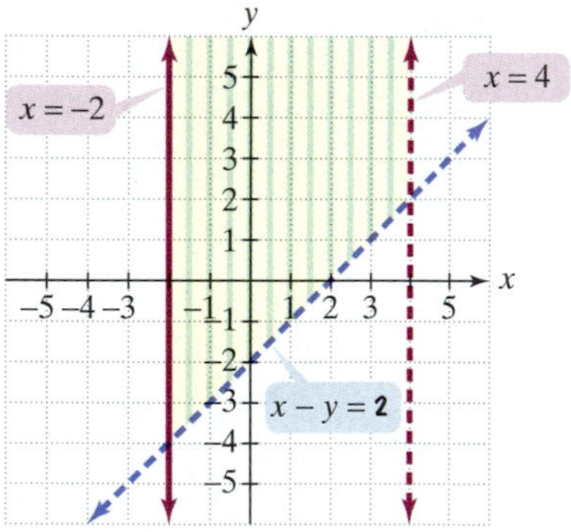

FIGURE 4.15 The graph of $x - y < 2$ and $-2 \leq x < 4$

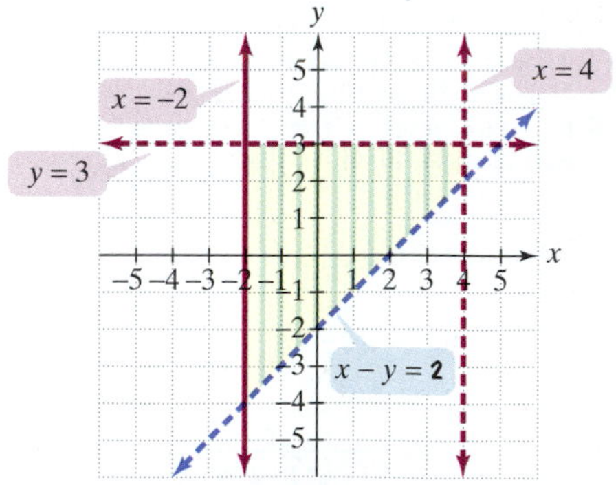

FIGURE 4.16 The graph of $x - y < 2$ and $-2 \leq x < 4$ and $y < 3$

CHECK POINT 6 Graph the solution set of the system:

$$x + y < 2$$
$$-2 \leq x < 1$$
$$y > -3.$$

4.4 EXERCISE SET

MyMathLab · MathXL PRACTICE · WATCH · DOWNLOAD · READ · REVIEW

Practice Exercises

In Exercises 1–22, graph each inequality.

1. $x + y \geq 3$
2. $x + y \geq 2$
3. $x - y < 5$
4. $x - y < 6$
5. $x + 2y > 4$
6. $2x + y > 6$
7. $3x - y \leq 6$
8. $x - 3y \leq 6$
9. $\frac{x}{2} + \frac{y}{3} < 1$
10. $\frac{x}{4} + \frac{y}{2} < 1$
11. $y > \frac{1}{3}x$
12. $y > \frac{1}{4}x$
13. $y \leq 3x + 2$
14. $y \leq 2x - 1$
15. $y < -\frac{1}{4}x$
16. $y < -\frac{1}{3}x$
17. $x \leq 2$
18. $x \leq -4$
19. $y > -4$
20. $y > -2$
21. $y \geq 0$
22. $x \leq 0$

In Exercises 23–46, graph the solution set of each system of inequalities or indicate that the system has no solution.

23. $3x + 6y \leq 6$, $2x + y \leq 8$
24. $x - y \geq 4$, $x + y \leq 6$
25. $2x - 5y \leq 10$, $3x - 2y > 6$
26. $2x - y \leq 4$, $3x + 2y > -6$
27. $y > 2x - 3$, $y < -x + 6$
28. $y < -2x + 4$, $y < x - 4$
29. $x + 2y \leq 4$, $y \geq x - 3$
30. $x + y \leq 4$, $y \geq 2x - 4$
31. $x \leq 2$, $y \geq -1$
32. $x \leq 3$, $y \leq -1$
33. $-2 \leq x < 5$
34. $-2 < y \leq 5$
35. $x - y \leq 1$, $x \geq 2$
36. $4x - 5y \geq -20$, $x \geq -3$
37. $x + y > 4$, $x + y < -1$
38. $x + y > 3$, $x + y < -2$
39. $x + y > 4$, $x + y > -1$
40. $x + y > 3$, $x + y > -2$
41. $x - y \leq 2$, $x \geq -2$, $y \leq 3$
42. $3x + y \leq 6$, $x \geq -2$, $y \leq 4$
43. $x \geq 0$, $y \geq 0$, $2x + 5y \leq 10$, $3x + 4y \leq 12$
44. $x \geq 0$, $y \geq 0$, $2x + y \leq 4$, $2x - 3y \leq 6$
45. $3x + y \leq 6$, $2x - y \leq -1$, $x \geq -2$, $y \leq 4$
46. $2x + y \leq 6$, $x + y \geq 2$, $1 \leq x \leq 2$, $y \leq 3$

Practice PLUS

In Exercises 47–48, write each sentence as a linear inequality in two variables. Then graph the inequality.

47. The y-variable is at least 4 more than the product of -2 and the x-variable.
48. The y-variable is at least 2 more than the product of -3 and the x-variable.

In Exercises 49–50, write the given sentences as a system of linear inequalities in two variables. Then graph the system.

49. The sum of the x-variable and the y-variable is at most 4. The y-variable added to the product of 3 and the x-variable does not exceed 6.
50. The sum of the x-variable and the y-variable is at most 3. The y-variable added to the product of 4 and the x-variable does not exceed 6.

In Exercises 51–52, rewrite each inequality in the system without absolute value bars. Then graph the rewritten system in rectangular coordinates.

51. $|x| \leq 2$, $|y| \leq 3$
52. $|x| \leq 1$, $|y| \leq 2$

*The graphs of solution sets of systems of inequalities involve finding the intersection of the solution sets of two or more inequalities. By contrast, in Exercises 53–54 you will be graphing the **union** of the solution sets of two inequalities.*

53. Graph the union of $y > \frac{3}{2}x - 2$ and $y < 4$.
54. Graph the union of $x - y \geq -1$ and $5x - 2y \leq 10$.

Without graphing, in Exercises 55–58, determine if each system has no solution or infinitely many solutions.

55. $3x + y < 9$, $3x + y > 9$
56. $6x - y \leq 24$, $6x - y > 24$
57. $3x + y \leq 9$, $3x + y \geq 9$
58. $6x - y \leq 24$, $6x - y \geq 24$

Application Exercises

Maximum heart rate, H, in beats per minute is a function of age, a, modeled by the formula

$$H = 220 - a,$$

where $10 \le a \le 70$. The bar graph shows the target heart rate ranges for four types of exercise goals in terms of maximum heart rate.

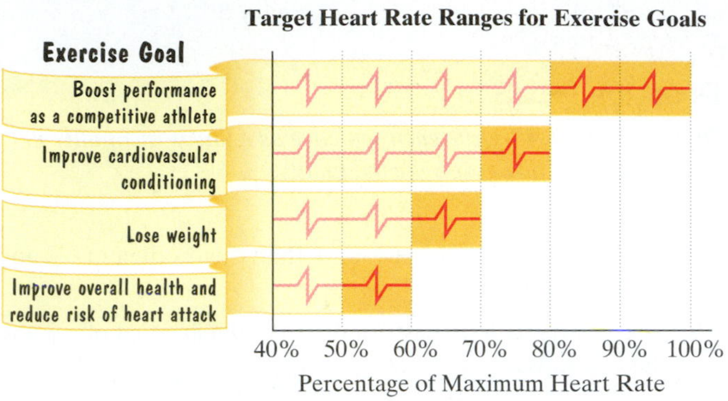

Source: Vitality

In Exercises 59–62, systems of inequalities will be used to model three of the target heart rate ranges shown in the bar graph. We begin with the target heart rate range for cardiovascular conditioning, modeled by the following system of inequalities:

$10 \le a \le 70$ — Heart rate ranges apply to ages 10 through 70, inclusive.

$H \ge 0.7(220 - a)$ — Target heart rate range is greater than or equal to 70% of maximum heart rate

$H \le 0.8(220 - a).$ — and less than or equal to 80% of maximum heart rate.

The graph of this system is shown in the figure. Use the graph to solve Exercises 59–60.

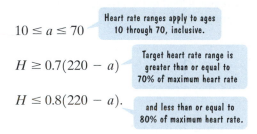

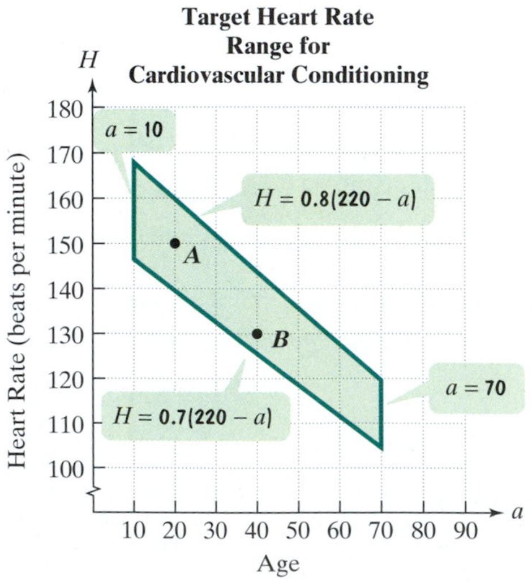

59. **a.** What are the coordinates of point A and what does this mean in terms of age and heart rate?

b. Show that point A is a solution of the system of inequalities.

60. **a.** What are the coordinates of point B and what does this mean in terms of age and heart rate?

b. Show that point B is a solution of the system of inequalities.

61. Write a system of inequalities that models the target heart rate range for the goal of losing weight.

62. Write a system of inequalities that models the target heart rate range for improving overall health.

63. On your next vacation, you will divide lodging between large resorts and small inns. Let x represent the number of nights spent in large resorts. Let y represent the number of nights spent in small inns.

a. Write a system of inequalities that models the following conditions:

> You want to stay at least 5 nights. At least one night should be spent at a large resort. Large resorts average \$200 per night and small inns average \$100 per night. Your budget permits no more than \$700 for lodging.

b. Graph the solution set of the system of inequalities in part (a).

c. Based on your graph in part (b), how many nights could you spend at a large resort and still stay within your budget?

64. **a.** An elevator can hold no more than 2000 pounds. If children average 80 pounds and adults average 160 pounds, write a system of inequalities that models when the elevator holding x children and y adults is overloaded.

b. Graph the solution set of the system of inequalities in part (a).

Writing in Mathematics

65. What is a linear inequality in two variables? Provide an example with your description.

66. How do you determine if an ordered pair is a solution of an inequality in two variables, x and y?

67. What is a half-plane?

68. What does a solid line mean in the graph of an inequality?

69. What does a dashed line mean in the graph of an inequality?

70. Explain how to graph $x - 2y < 4$.

71. What is a system of linear inequalities?

72. What is a solution of a system of linear inequalities?

73. Explain how to graph the solution set of a system of inequalities.

74. What does it mean if a system of linear inequalities has no solution?

Technology Exercises

Graphing utilities can be used to shade regions in the rectangular coordinate system, thereby graphing an inequality in two variables. Read the section of the user's manual for your graphing utility that describes how to shade a region. Then use your graphing utility to graph the inequalities in Exercises 75–78.

75. $y \le 4x + 4$

76. $y \ge \frac{2}{3}x - 2$

77. $2x + y \le 6$

78. $3x - 2y \ge 6$

79. Does your graphing utility have any limitations in terms of graphing inequalities? If so, what are they?

80. Use a graphing utility with a SHADE feature to verify any five of the graphs that you drew by hand in Exercises 1–22.

81. Use a graphing utility with a SHADE feature to verify any five of the graphs that you drew by hand for the systems in Exercises 23–46.

Critical Thinking Exercises

Make Sense? *In Exercises 82–85, determine whether each statement "makes sense" or "does not make sense" and explain your reasoning.*

82. When graphing a linear inequality, I should always use (0, 0) as a test point because it's easy to perform the calculations when 0 is substituted for each variable.

83. If you want me to graph $x < 3$, you need to tell me whether to use a number line or a rectangular coordinate system.

84. When graphing $3x - 4y < 12$, it's not necessary for me to graph the linear equation $3x - 4y = 12$ because the inequality contains a $<$ symbol, in which equality is not included.

85. Linear inequalities can model situations in which I'm interested in purchasing two items at different costs, I can spend no more than a specified amount on both items, and I want to know how many of each item I can purchase.

In Exercises 86–89, determine whether each statement is true or false. If the statement is false, make the necessary change(s) to produce a true statement.

86. The graph of $3x - 5y < 10$ consists of a dashed line and a shaded half-plane below the line.

87. The graph of $y \ge -x + 1$ consists of a solid line that rises from left to right and a shaded half-plane above the line.

88. The ordered pair $(-2, 40)$ satisfies the following system:

$$y \ge 9x + 11$$
$$13x + y > 14.$$

89. For the graph of $y < x - 3$, the points $(0, -3)$ and $(8, 5)$ lie on the graph of the corresponding linear equation, but neither point is a solution of the inequality.

90. Write a linear inequality in two variables whose graph is shown.

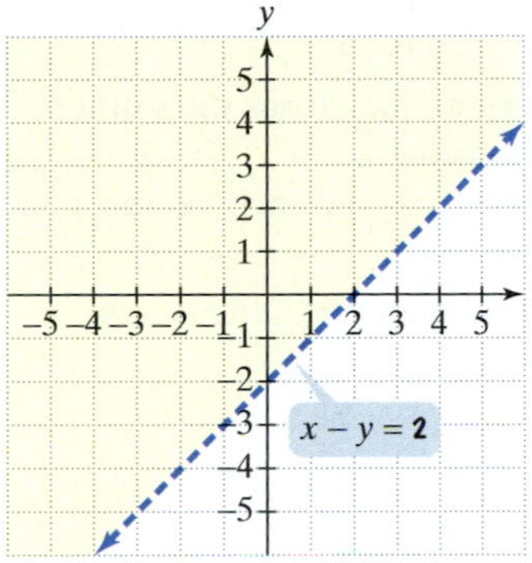

In Exercises 91–92, write a system of inequalities for each graph.

91.

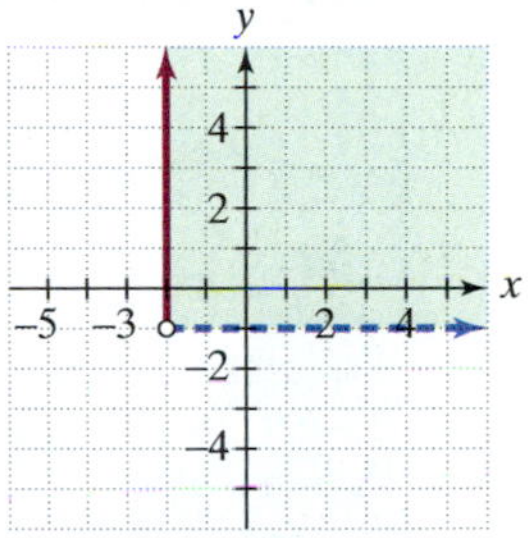

92.

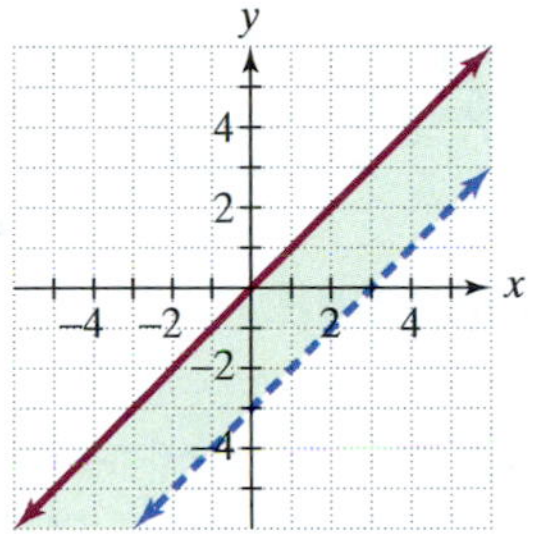

93. Write a linear inequality in two variables satisfying the following conditions: The points $(-3, -8)$ and $(4, 6)$ lie on the graph of the corresponding linear equation and each point is a solution of the inequality. The point $(1, 1)$ is also a solution.

94. Write a system of inequalities whose solution set includes every point in the rectangular coordinate system.

95. Sketch the graph of the solution set for the following system of inequalities:

$$y \ge nx + b\ (n < 0, b > 0)$$
$$y \le mx + b\ (m > 0, b > 0).$$

Review Exercises

96. Solve using matrices:

$$3x - y = 8$$
$$x - 5y = -2.$$

(Section 3.4, Example 2)

97. Solve by graphing:

$$y = 3x - 2$$
$$y = -2x + 8.$$

(Section 3.1, Example 2)

98. Evaluate:

$$\begin{vmatrix} 8 & 2 & -1 \\ 3 & 0 & 5 \\ 6 & -3 & 4 \end{vmatrix}.$$

(Section 3.5, Example 3)

Preview Exercises

Exercises 99–101 will help you prepare for the material covered in the next section.

99. **a.** Graph the solution set of the system:

$$\begin{aligned} x + y &\geq 6 \\ x &\leq 8 \\ y &\leq 5. \end{aligned}$$

b. List the points that form the corners of the graphed region in part (a).

c. Evaluate $3x + 2y$ at each of the points obtained in part (b).

100. **a.** Graph the solution set of the system:

$$\begin{aligned} x &\geq 0 \\ y &\geq 0 \\ 3x - 2y &\leq 6 \\ y &\leq -x + 7. \end{aligned}$$

b. List the points that form the corners of the graphed region in part (a).

c. Evaluate $2x + 5y$ at each of the points obtained in part (b).

101. Bottled water and medical supplies are to be shipped to survivors of an earthquake by plane. The bottled water weighs 20 pounds per container and medical kits weigh 10 pounds per kit. Each plane can carry no more than 80,000 pounds. If x represents the number of bottles of water to be shipped per plane and y represents the number of medical kits per plane, write an inequality that models each plane's 80,000 pound weight restriction.

SECTION 4.5

Linear Programming

Objectives

1. Write an objective function modeling a quantity that must be maximized or minimized.
2. Use inequalities to model limitations in a situation.
3. Use linear programming to solve problems.

West Berlin children at Tempelhof airport watch fleets of U.S. airplanes bringing in supplies to circumvent the Soviet blockade. The airlift began June 28, 1948 and continued for 15 months.

The Berlin Airlift (1948–1949) was an operation by the United States and Great Britain in response to military action by the former Soviet Union: Soviet troops closed all roads and rail lines between West Germany and Berlin, cutting off supply routes to the city. The Allies used a mathematical technique developed during World War II to maximize the quantities of supplies transported. During the 15-month airlift, 278,228 flights provided basic necessities to blockaded Berlin, saving one of the world's great cities.

In this section, we will look at an important application of systems of linear inequalities. Such systems arise in **linear programming**, a method for solving problems in which a particular quantity that must be maximized or minimized is limited by other factors. Linear programming is one of the most widely used tools in management science. It helps businesses allocate resources to manufacture products in a way that will maxmize profit. Linear programming accounts for more than 50% and perhaps as much as 90% of all computing time used for management decisions in business. The Allies used linear programming to save Berlin.

1 Write an objective function modeling a quantity that must be maximized or minimized.

Objective Functions in Linear Programming

Many problems involve quantities that must be maximized or minimized. Businesses are interested in maximizing profit. A relief operation in which bottled water and medical kits are shipped to earthquake survivors needs to maximize the number of survivors helped by this shipment. An **objective function** is an algebraic expression in two or more variables describing a quantity that must be maximized or minimized.

EXAMPLE 1 Writing an Objective Function

Bottled water and medical supplies are to be shipped to survivors of an earthquake by plane. Each container of bottled water will serve 10 people and each medical kit will aid 6 people. If x represents the number of bottles of water to be shipped and y represents the number of medical kits, write the objective function that models the number of people that can be helped.

Solution Because each bottle of water serves 10 people and each medical kit aids 6 people, we have

The number of people helped | is | 10 times the number of bottles of water | plus | 6 times the number of medical kits.

$$= \quad 10x \quad + \quad 6y.$$

Using z to represent the number of people helped, the objective function is

$$z = 10x + 6y.$$

Unlike the functions that we have seen so far, the objective function is an equation in three variables. For a value of x and a value of y, there is one and only one value of z. Thus, z is a function of x and y.

✓ **CHECK POINT 1** A company manufactures bookshelves and desks for computers. Let x represent the number of bookshelves manufactured daily and y the number of desks manufactured daily. The company's profits are \$25 per bookshelf and \$55 per desk. Write the objective function that models the company's total daily profit, z, from x bookshelves and y desks. (Check Points 2 through 4 are related to this situation, so keep track of your answers.)

2 Use inequalities to model limitations in a situation.

Constraints in Linear Programming

Ideally, the number of earthquake survivors helped in Example 1 should increase without restriction so that every survivor receives water and medical supplies. However, the planes that ship these supplies are subject to weight and volume restrictions. In linear programming problems, such restrictions are called **constraints**. Each constraint is expressed as a linear inequality. The list of constraints forms a system of linear inequalities.

EXAMPLE 2 Writing a Constraint

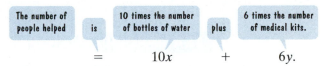

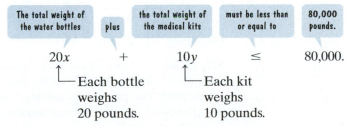

Each plane can carry no more than 80,000 pounds. The bottled water weighs 20 pounds per container and each medical kit weighs 10 pounds. Let x represent the number of bottles of water to be shipped and y the number of medical kits. Write an inequality that models this constraint.

Solution Because each plane can carry no more than 80,000 pounds, we have

The total weight of the water bottles | plus | the total weight of the medical kits | must be less than or equal to | 80,000 pounds.

$$20x \quad + \quad 10y \quad \leq \quad 80{,}000.$$

Each bottle weighs 20 pounds. (under $20x$)

Each kit weighs 10 pounds. (under $10y$)

The plane's weight constraint is modeled by the inequality

$$20x + 10y \leq 80{,}000.$$

✓ **CHECK POINT 2** To maintain high quality, the company in Check Point 1 should not manufacture more than a total of 80 bookshelves and desks per day. Write an inequality that models this constraint.

In addition to a weight constraint on its cargo, each plane has a limited amount of space in which to carry supplies. Example 3 demonstrates how to express this constraint.

EXAMPLE 3 Writing a Constraint

Each plane can carry a total volume of supplies that does not exceed 6000 cubic feet. Each water bottle is 1 cubic foot and each medical kit also has a volume of 1 cubic foot. With x still representing the number of water bottles and y the number of medical kits, write an inequality that models this second constraint.

Solution Because each plane can carry a volume of supplies that does not exceed 6000 cubic feet, we have

The total volume of the water	plus	the total volume of the medical kits	must be less than or equal to	6000 cubic feet.
$1x$	$+$	$1y$	$\leq$	$6000.$
Each bottle is 1 cubic foot.		Each kit is 1 cubic foot.		

The plane's volume constraint is modeled by the inequality $x + y \leq 6000.$

In summary, here's what we have described so far in this aid-to-earthquake-survivors situation:

$$z = 10x + 6y$$

This is the objective function modeling the number of people helped with x bottles of water and y medical kits.

$$20x + 10y \leq 80{,}000$$
$$x + y \leq 6000.$$

These are the constraints based on each plane's weight and volume limitations.

✓ **CHECK POINT 3** To meet customer demand, the company in Check Point 1 must manufacture between 30 and 80 bookshelves per day, inclusive. Furthermore, the company must manufacture at least 10 and no more than 30 desks per day. Write an inequality that models each of these sentences. Then summarize what you have described about this company by writing the objective function for its profits and the three constraints.

---

3 Use linear programming to solve problems.

Solving Problems with Linear Programming

The problem in the earthquake situation described previously is to maximize the number of survivors who can be helped, subject to each plane's weight and volume constraints. The process of solving this problem is called *linear programming*, based on a theorem that was proven during World War II.

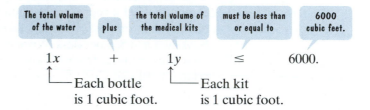

Solving a Linear Programming Problem

Let $z = ax + by$ be an objective function that depends on x and y. Furthermore, z is subject to a number of linear constraints on x and y. If a maximum or minimum value of z exists, it can be determined as follows:

1. Graph the system of inequalities representing the constraints.
2. Find the value of the objective function at each corner, or **vertex**, of the graphed region. The maximum and minimum of the objective function occur at one or more of the corner points.

EXAMPLE 4 Solving a Linear Programming Problem

Determine how many bottles of water and how many medical kits should be sent on each plane to maximize the number of earthquake survivors who can be helped.

Solution We must maximize $z = 10x + 6y$ subject to the following constraints:

$$20x + 10y \le 80{,}000$$
$$x + y \le 6000.$$

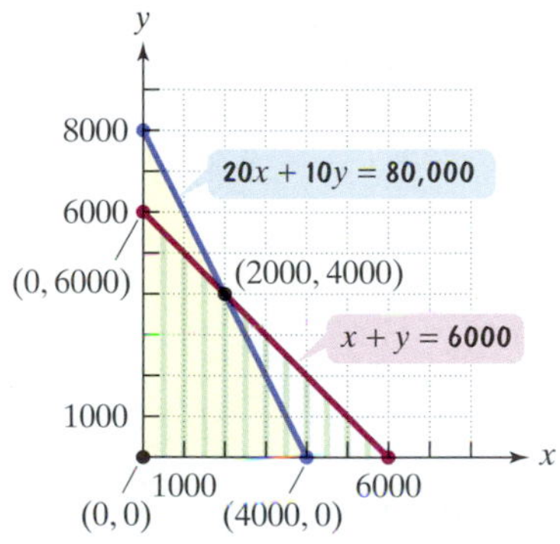

FIGURE 4.17 The region in quadrant I representing the constraints $20x + 10y \le 80{,}000$ and $x + y \le 6000$

Step 1. Graph the system of inequalities representing the constraints. Because x (the number of bottles of water per plane) and y (the number of medical kits per plane) must be nonnegative, we need to graph the system of inequalities in quadrant I and its boundary only.

To graph the inequality $20x + 10y \le 80{,}000$, we graph the equation $20x + 10y = 80{,}000$ as a solid blue line (**Figure 4.17**). Setting $y = 0$, the x-intercept is 4000 and setting $x = 0$, the y-intercept is 8000. Using $(0, 0)$ as a test point, the inequality is satisfied, so we shade below the blue line, as shown in yellow in **Figure 4.17**.

Now we graph $x + y \le 6000$ by first graphing $x + y = 6000$ as a solid red line. Setting $y = 0$, the x-intercept is 6000. Setting $x = 0$, the y-intercept is 6000. Using $(0, 0)$ as a test point, the inequality is satisfied, so we shade below the red line, as shown using green vertical shading in **Figure 4.17**.

We use the addition method to find where the lines $20x + 10y = 80{,}000$ and $x + y = 6000$ intersect.

$$\begin{array}{rcl} 20x + 10y = 80{,}000 & \xrightarrow{\text{No change}} & 20x + 10y = 80{,}000 \\ x + y = 6000 & \xrightarrow{\text{Multiply by } -10.} & \underline{-10x - 10y = -60{,}000} \\ & \text{Add:} & 10x = 20{,}000 \\ & & x = 2000 \end{array}$$

Back-substituting 2000 for x in $x + y = 6000$, we find $y = 4000$, so the intersection point is $(2000, 4000)$.

The system of inequalities representing the constraints is shown by the region in which the yellow shading and the green vertical shading overlap in **Figure 4.17**. The graph of the system of inequalities is shown again in **Figure 4.18**. The red and blue line segments are included in the graph.

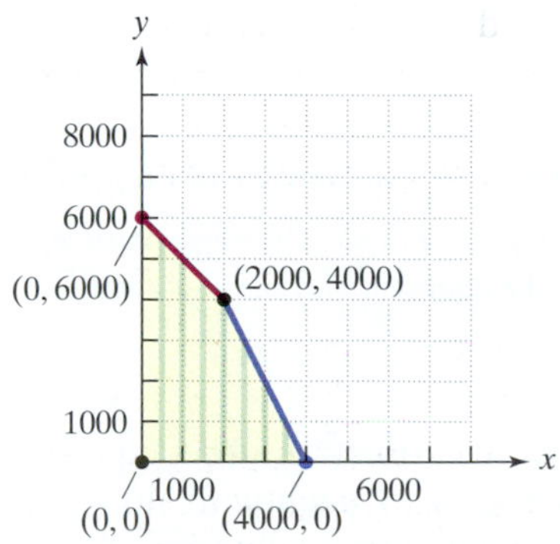

FIGURE 4.18

Step 2. Find the value of the objective function at each corner of the graphed region. The maximum and minimum of the objective function occur at one or more of the corner points. We must evaluate the objective function, $z = 10x + 6y$, at the four corners, or vertices, of the region in **Figure 4.18**.

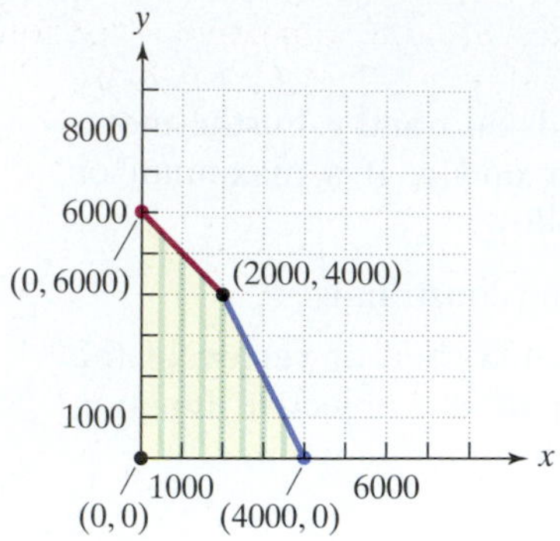

FIGURE 4.18 (repeated)

Corner (x, y)	Objective Function $z = 10x + 6y$
(0, 0)	$z = 10(0) + 6(0) = 0$
(4000, 0)	$z = 10(4000) + 6(0) = 40{,}000$
(2000, 4000)	$z = 10(2000) + 6(4000) = 44{,}000$ ← maximum
(0, 6000)	$z = 10(0) + 6(6000) = 36{,}000$

Thus, the maximum value of z is 44,000 and this occurs when $x = 2000$ and $y = 4000$. In practical terms, this means that the maximum number of earthquake survivors who can be helped with each plane shipment is 44,000. This can be accomplished by sending 2000 water bottles and 4000 medical kits per plane.

CHECK POINT 4 For the company in Check Points 1–3, how many bookshelves and how many desks should be manufactured per day to obtain maximum profit? What is the maximum daily profit?

EXAMPLE 5 Solving a Linear Programming Problem

Find the maximum value of the objective function

$$z = 2x + y$$

subject to the following constraints:

$$x \geq 0,\ y \geq 0$$
$$x + 2y \leq 5$$
$$x - y \leq 2.$$

Solution We begin by graphing the region in quadrant I ($x \geq 0$, $y \geq 0$) formed by the constraints. The graph is shown in **Figure 4.19**.

Now we evaluate the objective function at the four vertices of this region.

Objective function: $z = 2x + y$

At (0, 0): $z = 2 \cdot 0 + 0 = 0$
At (2, 0): $z = 2 \cdot 2 + 0 = 4$
At (3, 1): $z = 2 \cdot 3 + 1 = 7$ — Maximum value of z
At (0, 2.5): $z = 2 \cdot 0 + 2.5 = 2.5$

Thus, the maximum value of z is 7, and this occurs when $x = 3$ and $y = 1$.

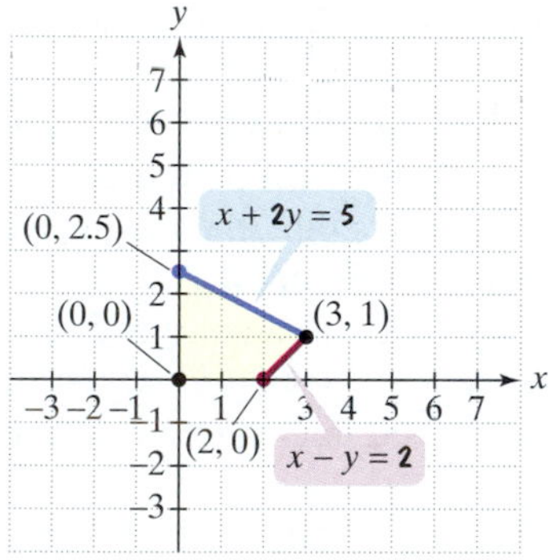

FIGURE 4.19 The graph of $x + 2y \leq 5$ and $x - y \leq 2$ in quadrant I

We can see why the objective function in Example 5 has a maximum value that occurs at a vertex by solving the equation for y.

$$z = 2x + y \qquad \text{This is the objective function of Example 5.}$$
$$y = -2x + z \qquad \text{Solve for } y. \text{ Recall that the slope-intercept form of the equation of a line is } y = mx + b.$$

Slope $= -2$; y-intercept $= z$

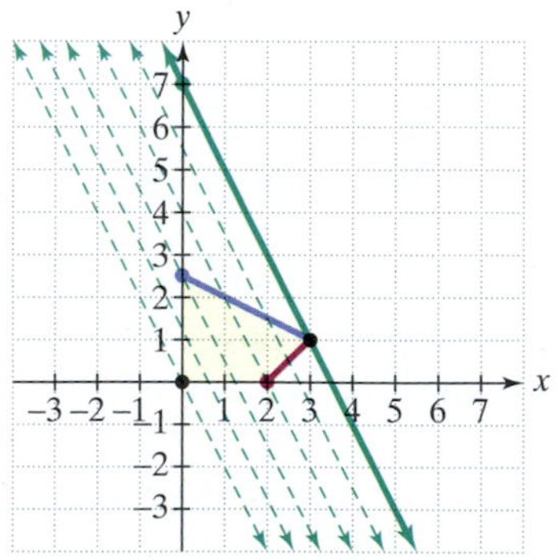

FIGURE 4.20 The line with slope −2 with the greatest y-intercept that intersects the shaded region passes through one of the vertices of the region.

In this form, z represents the y-intercept of the objective function. The equation describes infinitely many parallel lines (one for each value of z), each with slope −2. The process in linear programming involves finding the maximum z-value for all lines that intersect the region determined by the constraints. Of all the lines whose slope is −2, we're looking for the one with the greatest y-intercept that intersects the given region. As we see in **Figure 4.20**, such a line will pass through one (or possibly more) of the vertices of the region.

✓ **CHECK POINT 5** Find the maximum value of the objective function $z = 3x + 5y$ subject to the constraints $x \geq 0, y \geq 0, x + y \geq 1, x + y \leq 6$.

4.5 EXERCISE SET

MyMathLab | MathXL PRACTICE | WATCH | DOWNLOAD | READ | REVIEW

Practice Exercises

In Exercises 1–4, find the value of the objective function at each corner of the graphed region. What is the maximum value of the objective function? What is the minimum value of the objective function?

1. Objective Function
$z = 5x + 6y$

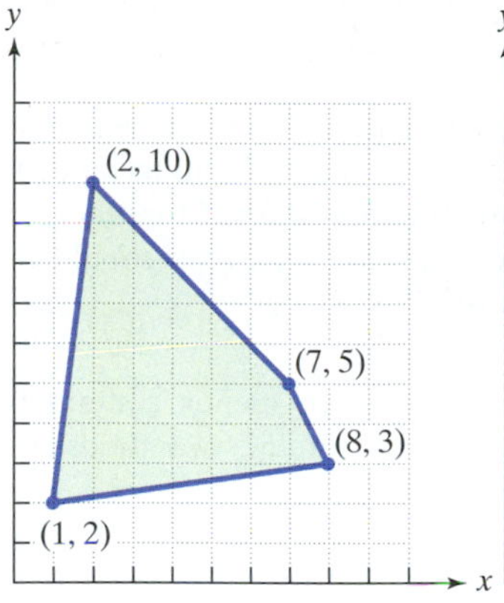

2. Objective Function
$z = 3x + 2y$

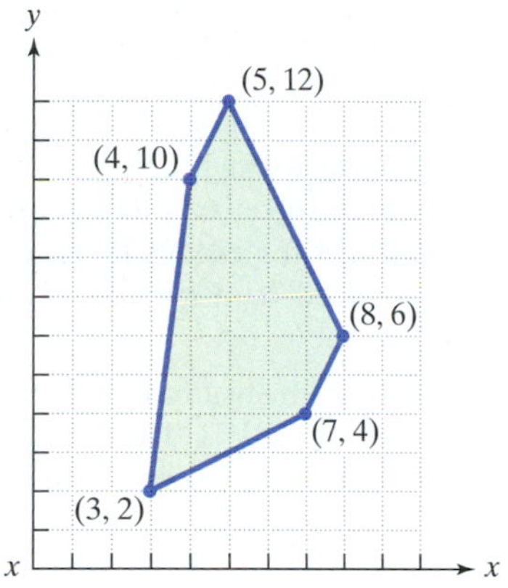

3. Objective Function
$z = 40x + 50y$

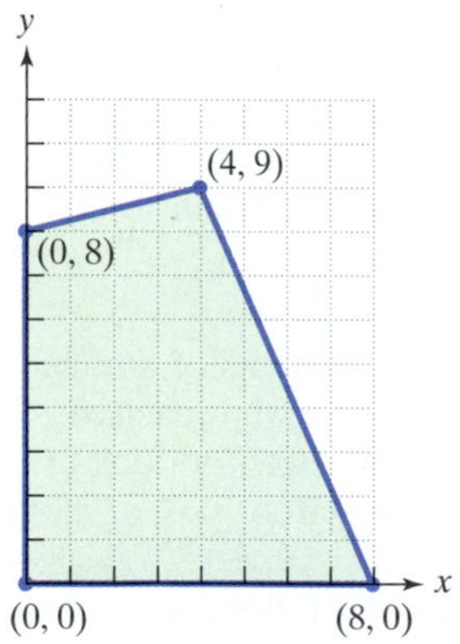

4. Objective Function
$z = 30x + 45y$

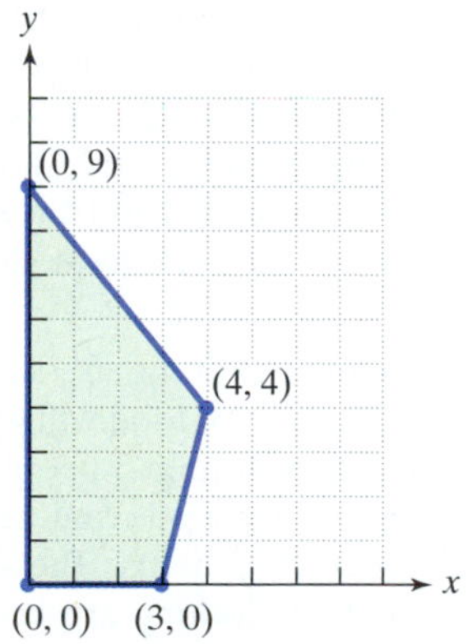

In Exercises 5–14, an objective function and a system of linear inequalities representing constraints are given.

a. *Graph the system of inequalities representing the constraints.*

b. *Find the value of the objective function at each corner of the graphed region.*

c. *Use the values in part (b) to determine the maximum value of the objective function and the values of x and y for which the maximum occurs.*

5. Objective Function $z = 3x + 2y$
Constraints $x \geq 0, y \geq 0$
$2x + y \leq 8$
$x + y \geq 4$

6. Objective Function $z = 2x + 3y$
Constraints $x \geq 0, y \geq 0$
$2x + y \leq 8$
$2x + 3y \leq 12$

7. Objective Function $z = 4x + y$
Constraints $x \geq 0, y \geq 0$
$2x + 3y \leq 12$
$x + y \geq 3$

8. Objective Function $z = x + 6y$
Constraints $x \geq 0, y \geq 0$
$2x + y \leq 10$
$x - 2y \geq -10$

9. Objective Function $z = 3x - 2y$
Constraints $1 \leq x \leq 5$
$y \geq 2$
$x - y \geq -3$

10. Objective Function $z = 5x - 2y$
Constraints $0 \leq x \leq 5$
$0 \leq y \leq 3$
$x + y \geq 2$

11. Objective Function $z = 4x + 2y$
Constraints $x \geq 0, y \geq 0$
$2x + 3y \leq 12$
$3x + 2y \leq 12$
$x + y \geq 2$

12. Objective Function $z = 2x + 4y$
Constraints $x \geq 0, y \geq 0$
$x + 3y \geq 6$
$x + y \geq 3$
$x + y \leq 9$

13. Objective Function $z = 10x + 12y$
Constraints $x \geq 0, y \geq 0$
$2x + y \leq 10$
$2x + 3y \leq 18$

14. Objective Function $z = 5x + 6y$

Constraints
$$\begin{aligned} x \geq 0,\ y &\geq 0 \\ 2x + y &\geq 10 \\ x + 2y &\geq 10 \\ x + y &\leq 10 \end{aligned}$$

Application Exercises

15. A television manufacturer makes rear-projection and plasma televisions. The profit per unit is \$125 for the rear-projection televisions and \$200 for the plasma televisions.

a. Let x = the number of rear-projection televisions manufactured in a month and y = the number of plasma televisions manufactured in a month. Write the objective function that models the total monthly profit.

b. The manufacturer is bound by the following constraints:
- Equipment in the factory allows for making at most 450 rear-projection televisions in one month.
- Equipment in the factory allows for making at most 200 plasma televisions in one month.
- The cost to the manufacturer per unit is \$600 for the rear-projection televisions and \$900 for the plasma televisions. Total monthly costs cannot exceed \$360,000.

Write a system of three inequalities that models these constraints.

c. Graph the system of inequalities in part (b). Use only the first quadrant and its boundary, because x and y must both be nonnegative.

d. Evaluate the objective function for total monthly profit at each of the five vertices of the graphed region. [The vertices should occur at (0, 0), (0, 200), (300, 200), (450, 100), and (450, 0).]

e. Complete the missing portions of this statement: The television manufacturer will make the greatest profit by manufacturing ____ rear-projection televisions each month and ____ plasma televisions each month. The maximum monthly profit is \$______.

16. a. A student earns \$10 per hour for tutoring and \$7 per hour as a teacher's aid. Let x = the number of hours each week spent tutoring and y = the number of hours each week spent as a teacher's aid. Write the objective function that models total weekly earnings.

b. The student is bound by the following constraints:
- To have enough time for studies, the student can work no more than 20 hours a week.
- The tutoring center requires that each tutor spend at least three hours a week tutoring.
- The tutoring center requires that each tutor spend no more than eight hours a week tutoring.

Write a system of three inequalities that models these constraints.

c. Graph the system of inequalities in part (b). Use only the first quadrant and its boundary, because x and y are nonnegative.

d. Evaluate the objective function for total weekly earnings at each of the four vertices of the graphed region. [The vertices should occur at (3, 0), (8, 0), (3, 17), and (8, 12).]

e. Complete the missing portions of this statement: The student can earn the maximum amount per week by tutoring for __ hours per week and working as a teacher's aid for ___ hours per week. The maximum amount that the student can earn each week is \$____.

Use the two steps for solving a linear programming problem, given in the box on page 289, to solve the problems in Exercises 17–23.

17. A manufacturer produces two models of mountain bicycles. The times (in hours) required for assembling and painting each model are given in the following table:

	Model A	Model B
Assembling	5	4
Painting	2	3

The maximum total weekly hours available in the assembly department and the paint department are 200 hours and 108 hours, respectively. The profits per unit are \$25 for model A and \$15 for model B. How many of each type should be produced to maximize profit?

18. A large institution is preparing lunch menus containing foods A and B. The specifications for the two foods are given in the following table:

Food	Units of Fat per Ounce	Units of Carbohydrates per Ounce	Units of Protein per Ounce
A	1	2	1
B	1	1	1

Each lunch must provide at least 6 units of fat per serving, no more than 7 units of protein, and at least 10 units of carbohydrates. The institution can purchase food A for \$0.12 per ounce and food B for \$0.08 per ounce. How many ounces of each food should a serving contain to meet the dietary requirement at the least cost?

19. Food and clothing are shipped to survivors of a hurricane. Each carton of food will feed 12 people, while each carton of clothing will help 5 people. Each 20-cubic-foot box of food weighs 50 pounds and each 10-cubic-foot box of clothing weighs 20 pounds. The commercial carriers transporting food and clothing are bound by the following constraints:
- The total weight per carrier cannot exceed 19,000 pounds.
- The total volume must be less than 8000 cubic feet.

How many cartons of food and how many cartons of clothing should be sent with each plane shipment to maximize the number of people who can be helped?

20. On June 24, 1948, the former Soviet Union blocked all land and water routes through East Germany to Berlin. A gigantic airlift was organized using American and British planes to bring food, clothing, and other supplies to the more than 2 million people in West Berlin. The cargo capacity was 30,000 cubic feet for an American plane and 20,000 cubic feet for a British plane. To break the Soviet blockade, the Western Allies had to maximize cargo capacity, but were subject to the following restrictions:

- No more than 44 planes could be used.
- The larger American planes required 16 personnel per flight, double that of the requirement for the British planes. The total number of personnel available could not exceed 512.
- The cost of an American flight was \$9000 and the cost of a British flight was \$5000. Total weekly costs could not exceed \$300,000.

Find the number of American planes and the number of British planes that were used to maximize cargo capacity.

21. A theater is presenting a program on drinking and driving for students and their parents. The proceeds will be donated to a local alcohol information center. Admission is \$2.00 for parents and \$1.00 for students. However, the situation has two constraints: The theater can hold no more than 150 people and every two parents must bring at least one student. How many parents and students should attend to raise the maximum amount of money?

22. You are about to take a test that contains computation problems worth 6 points each and word problems worth 10 points each. You can do a computation problem in 2 minutes and a word problem in 4 minutes. You have 40 minutes to take the test and may answer no more than 12 problems. Assuming you answer all the problems attempted correctly, how many of each type of problem must you do to maximize your score? What is the maximum score?

23. In 1978, a ruling by the Civil Aeronautics Board allowed Federal Express to purchase larger aircraft. Federal Express's options included 20 Boeing 727s that United Airlines was retiring and/or the French-built Dassault Fanjet Falcon 20. To aid in their decision, executives at Federal Express analyzed the following data:

	Boeing 727	Falcon 20
Direct Operating Cost	\$1400 per hour	\$500 per hour
Payload	42,000 pounds	6000 pounds

Federal Express was faced with the following constraints:

- Hourly operating cost was limited to \$35,000.
- Total payload had to be at least 672,000 pounds.
- Only twenty 727s were available.

Given the constraints, how many of each kind of aircraft should Federal Express have purchased to maximize the number of aircraft?

Writing in Mathematics

24. What kinds of problems are solved using the linear programming method?

25. What is an objective function in a linear programming problem?

26. What is a constraint in a linear programming problem? How is a constraint represented?

27. In your own words, describe how to solve a linear programming problem.

28. Describe a situation in your life in which you would like to maximize something, but are limited by at least two constraints. Can linear programming be used in this situation? Explain your answer.

Critical Thinking Exercises

Make Sense? *In Exercises 29–32, determine whether each statement "makes sense" or "does not make sense" and explain your reasoning.*

29. In order to solve a linear programming problem, I use the graph representing the constraints and the graph of the objective function.

30. I use the coordinates of each vertex from my graph representing the constraints to find the values that maximize or minimize an objective function.

31. I need to be able to graph systems of linear inequalities in order to solve linear programming problems.

32. An important application of linear programming for businesses involves maximizing profit.

33. Suppose that you inherit \$10,000. The will states how you must invest the money. Some (or all) of the money must be invested in stocks and bonds. The requirements are that at least \$3000 be invested in bonds, with expected returns of \$0.08 per dollar, and at least \$2000 be invested in stocks, with expected returns of \$0.12 per dollar. Because the stocks are medium risk, the final stipulation requires that the investment in bonds should never be less than the investment in stocks. How should the money be invested so as to maximize your expected returns?

34. Consider the objective function $z = Ax + By$ ($A > 0$ and $B > 0$) subject to the following constraints: $2x + 3y \leq 9$, $x - y \leq 2$, $x \geq 0$, and $y \geq 0$. Prove that the objective function will have the same maximum value at the vertices $(3, 1)$ and $(0, 3)$ if $A = \frac{2}{3}B$.

Review Exercises

35. Simplify: $(2x^4y^3)(3xy^4)^3$. (Section 1.6, Example 9)

36. Solve for L: $3P = \dfrac{2L - W}{4}$. (Section 1.5, Example 8)

37. If $f(x) = x^3 + 2x^2 - 5x + 4$, find $f(-1)$. (Section 2.1, Example 3)

Preview Exercises

Exercises 38–40 will help you prepare for the material covered in the first section of the next chapter.

In Exercises 38–39, simplify each algebraic expression.

38. $(-9x^3 + 7x^2 - 5x + 3) + (13x^3 + 2x^2 - 8x - 6)$

39. $(7x^3 - 8x^2 + 9x - 6) - (2x^3 - 6x^2 - 3x + 9)$

40. The figures show the graphs of two functions.

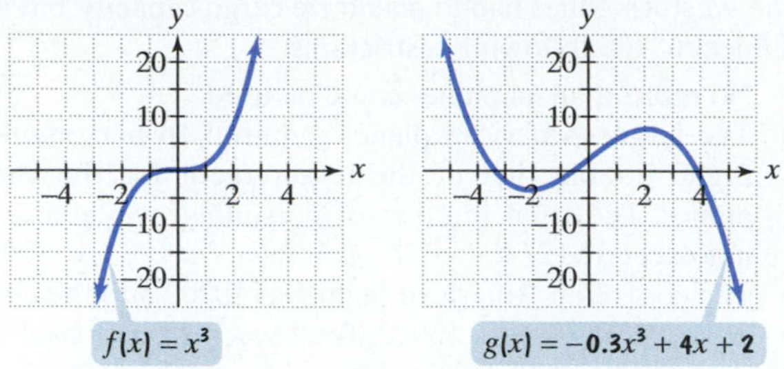

a. Which function, f or g, has a graph that rises to the left and falls to the right?

b. Which function, f or g, has a graph that falls to the left and rises to the right?

GROUP PROJECT

CHAPTER 4

Each group member should research one situation that provides two different pricing options. These can involve areas such as public transportation options (with or without coupon books) or long-distance telephone plans or anything of interest. Be sure to bring in all the details for each option. At the group meeting, select the two pricing situations that are most interesting and relevant. Using each situation, write a word problem about selecting the better of the two options. The word problem should be one that can be solved using a linear inequality. The group should turn in the two problems and their solutions.

Chapter 4 Summary

Definitions and Concepts	Examples

Section 4.1 Solving Linear Inequalities

A linear inequality in one variable can be written in the form $ax + b < 0$, $ax + b \leq 0$, $ax + b > 0$, or $ax + b \geq 0$. The set of all numbers that make the inequality a true statement is its solution set. Graphs of solution sets are shown on a number line by shading all points representing numbers that are solutions. Parentheses indicate endpoints that are not solutions. Square brackets indicate endpoints that are solutions.

- $(-2, 1] = \{x | -2 < x \leq 1\}$

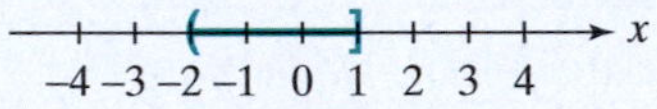

- $[-2, \infty) = \{x | x \geq -2\}$

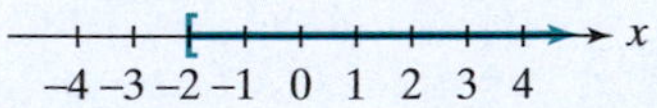

Solving a Linear Inequality

1. Simplify each side.
2. Collect variable terms on one side and constant terms on the other side.
3. Isolate the variable and solve.

If an inequality is multiplied or divided by a negative number, the direction of the inequality symbol must be reversed.

Solve: $2(x + 3) - 5x \leq 15$.

$$2x + 6 - 5x \leq 15$$
$$-3x + 6 \leq 15$$
$$-3x \leq 9$$
$$\frac{-3x}{-3} \geq \frac{9}{-3}$$
$$x \geq -3$$

Solution set: $\{x | x \geq -3\}$ or $[-3, \infty)$

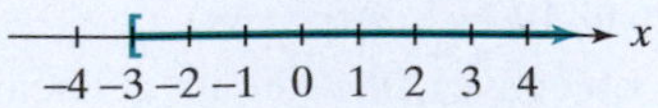

Definitions and Concepts | Examples

Section 4.2 Compound Inequalities

Intersection ($\cap$) and Union ($\cup$)

$A \cap B$ is the set of elements common to both set A and set B.
$A \cup B$ is the set of elements that are members of set A or set B or of both sets.

$\{1, 3, 5, 7\} \cap \{5, 7, 9, 11\} = \{5, 7\}$
$\{1, 3, 5, 7\} \cup \{5, 7, 9, 11\} = \{1, 3, 5, 7, 9, 11\}$

A compound inequality is formed by joining two inequalities with the word *and* or *or*.

When the connecting word is *and*, graph each inequality separately and take the intersection of their solution sets.

Solve: $x + 1 > 3$ and $x + 4 \le 8$.

$x > 2$ and $x \le 4$

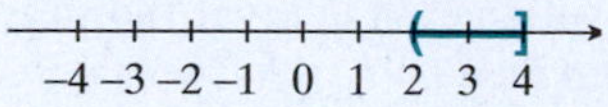

Solution set: $\{x|2 < x \le 4\}$ or $(2, 4]$

The compound inequality $a < x < b$ means $a < x$ and $x < b$. Solve by isolating the variable in the middle.

Solve: $-1 < \dfrac{2x + 1}{3} \le 2$.

$-3 < 2x + 1 \le 6$ Multiply by 3.

$-4 < 2x \le 5$ Subtract 1.

$-2 < x \le \dfrac{5}{2}$ Divide by 2.

Solution set: $\left\{x \middle| -2 < x \le \dfrac{5}{2}\right\}$ or $\left(-2, \dfrac{5}{2}\right]$

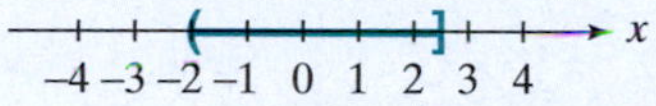

When the connecting word in a compound inequality is *or*, graph each inequality separately and take the union of their solution sets.

Solve: $x - 2 > -3$ or $2x \le -6$.

$x > -1$ or $x \le -3$

Solution set: $\{x|x \le -3 \text{ or } x > -1\}$ or $(-\infty, -3] \cup (-1, \infty)$

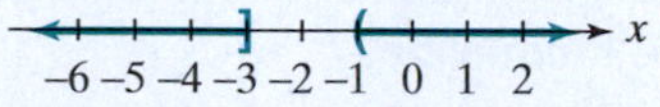

Section 4.3 Equations and Inequalities Involving Absolute Value

Absolute Value Equations

1. If $c > 0$, then $|X| = c$ means $X = c$ or $X = -c$.
2. If $c < 0$, then $|X| = c$ has no solution.
3. If $c = 0$, then $|X| = 0$ means $X = 0$.

Solve: $|2x - 7| = 3$.

$2x - 7 = 3$ or $2x - 7 = -3$

$2x = 10$ $\qquad$ $2x = 4$

$x = 5$ $\qquad$ $x = 2$

The solution set is $\{2, 5\}$.

Absolute Value Equations with Two Absolute Value Bars

If $|X_1| = |X_2|$, then $X_1 = X_2$ or $X_1 = -X_2$.

Solve: $|x - 6| = |2x + 1|$.

$x - 6 = 2x + 1$ or $x - 6 = -(2x + 1)$

$-x - 6 = 1$ $\qquad$ $x - 6 = -2x - 1$

$-x = 7$ $\qquad$ $3x - 6 = -1$

$x = -7$ $\qquad$ $3x = 5$

$x = \dfrac{5}{3}$

The solutions are -7 and $\dfrac{5}{3}$, and the solution set is $\left\{-7, \dfrac{5}{3}\right\}$.

Definitions and Concepts	Examples

Section 4.3 Equations and Inequalities Involving Absolute Value (continued)

Using Boundary Points to Solve Absolute Value Inequalities

If c is a positive number, solve $|X| < c$ and $|X| > c$ using the following procedure.

1. Solve $|X| = c$. The solutions are the boundary points.
2. Locate these boundary points on a number line, thereby dividing the number line into intervals.
3. Choose a test value within each interval and substitute that number into the given inequality. If a true statement results, all numbers in the interval satisfy the inequality.
4. Write the solution set, selecting the interval(s) that satisfy the given inequality.

This procedure is valid if $<$ is replaced by $\le$ or if $>$ is replaced by $\ge$. In these cases, include the boundary points in the solution set.

Solve: $|3x + 6| > 12$.
First solve $|3x + 6| = 12$.

$$3x + 6 = 12 \quad \text{or} \quad 3x + 6 = -12$$
$$3x = 6 \qquad\qquad 3x = -18$$
$$x = 2 \qquad\qquad x = -6$$

Boundary points are -6 and 2.

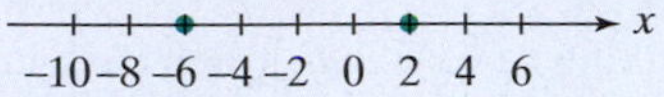

Intervals are $(-\infty, -6)$ $(-6, 2)$ $(2, \infty)$

Test value: −7 | Test value: 0 | Test value: 3

Test each value in $|3x + 6| > 12$.

Test −7: $|3(-7) + 6| \overset{?}{>} 12$
$15 > 12$, true
$(-\infty, -6)$ is in the solution set.

Test 0: $|3 \cdot 0 + 6| \overset{?}{>} 12$
$6 > 12$, false
$(-6, 2)$ is not in the solution set.

Test 3: $|3 \cdot 3 + 6| \overset{?}{>} 12$
$15 > 12$, true
$(2, \infty)$ is in the solution set.

The solution set is $(-\infty, -6) \cup (2, \infty)$ or $\{x | x < -6 \text{ or } x > 2\}$.

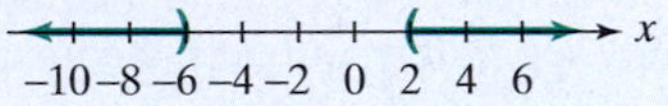

Using Equivalent Compound Inequalities to Solve Absolute Value Inequalities

If c is a positive number,

1. The solutions of $|X| < c$ are the numbers that satisfy $-c < X < c$.
2. The solutions of $|X| > c$ are the numbers that satisfy $X < -c$ or $X > c$.

In each case, the absolute value inequality is rewritten as an equivalent compound inequality without absolute value bars.

Solve: $|x - 4| < 3$.

$$-3 < x - 4 < 3$$
$$1 < x < 7 \quad \text{Add 4.}$$

The solution set is $\{x | 1 < x < 7\}$ or $(1, 7)$.

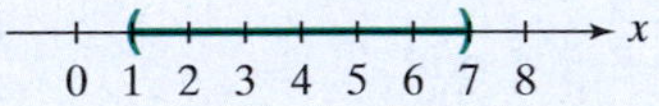

Solve: $\left|\frac{x}{3} - 1\right| \ge 2$.

$$\frac{x}{3} - 1 \le -2 \quad \text{or} \quad \frac{x}{3} - 1 \ge 2.$$
$$x - 3 \le -6 \quad \text{or} \quad x - 3 \ge 6 \quad \text{Multiply by 3.}$$
$$x \le -3 \quad \text{or} \quad x \ge 9 \quad \text{Add 3.}$$

The solution set is $\{x | x \le -3 \text{ or } x \ge 9\}$ or $(-\infty, -3] \cup [9, \infty)$.

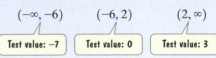

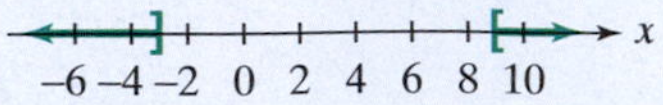

Definitions and Concepts	Examples

Section 4.3 Equations and Inequalities Involving Absolute Value (continued)

Absolute Value Inequalities with Unusual Solution Sets

If c is a negative number,

1. $|X| < c$ has no solution.
2. $|X| > c$ is true for all real numbers for which X is defined.

- $|x - 4| < -3$ has no solution. The solution set is $\varnothing$.
- $|3x + 6| > -12$ is true for all real numbers. The solution set is $(-\infty, \infty)$.

Section 4.4 Linear Inequalities in Two Variables

If the equal sign in $Ax + By = C$ is replaced with an inequality symbol, the result is a linear inequality in two variables. Its graph is the set of all points whose coordinates satisfy the inequality. To obtain the graph,

1. Replace the inequality symbol with an equal sign and graph the boundary line. Use a solid line for $\le$ or $\ge$ and a dashed line for $<$ or $>$.
2. Choose a test point not on the line and substitute its coordinates into the inequality.
3. If a true statement results, shade the half-plane containing the test point. If a false statement results, shade the half-plane not containing the test point.

Graph: $x - 2y \le 4$.

1. Graph $x - 2y = 4$. Use a solid line because the inequality symbol is $\le$.
2. Test $(0, 0)$.

$$x - 2y \le 4$$
$$0 - 2 \cdot 0 \overset{?}{\le} 4$$
$$0 \le 4, \quad \text{true}$$

3. The inequality is true. Shade the half-plane containing $(0, 0)$.

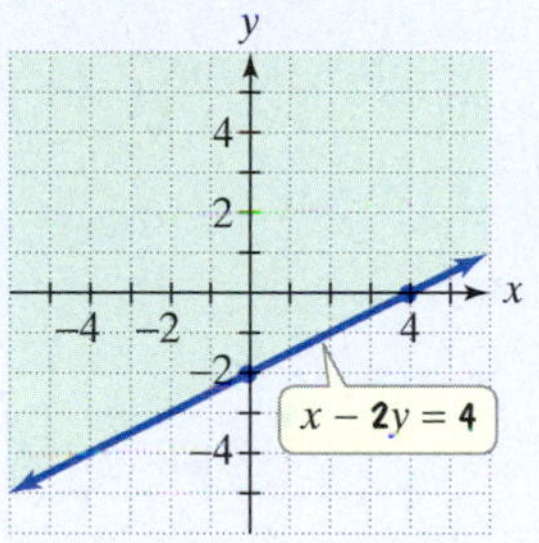

Two or more linear inequalities make up a system of linear inequalities. A solution is an ordered pair satisfying all inequalities in the system. To graph a system of inequalities, graph each inequality in the system. The overlapping region, if there is one, represents the solutions of the system. If there is no overlapping region, the system has no solution.

Graph the solutions of the system:

$$y \le -2x$$
$$x - y \ge 3.$$

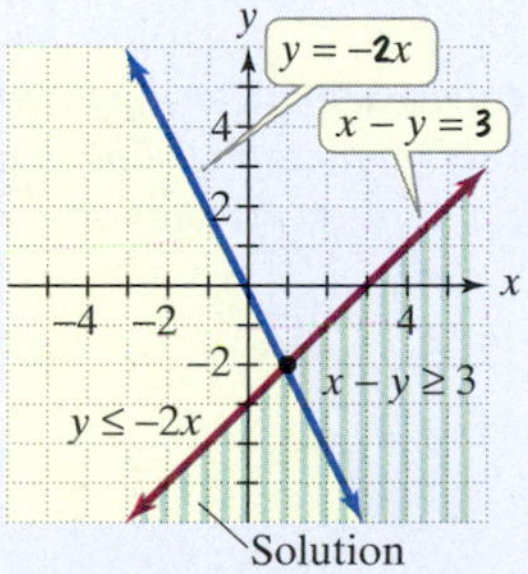

Definitions and Concepts	Examples

Section 4.5 Linear Programming

Linear programming is a method for solving problems in which a particular quantity that must be maximized or minimized is limited. An objective function is an algebraic expression in three variables modeling a quantity that must be maximized or minimized. Constraints are restrictions, expressed as linear inequalities.

Solving a Linear Programming Problem

1. Graph the system of inequalities representing the constraints.
2. Find the value of the objective function at each corner, or vertex, of the graphed region. The maximum and minimum of the objective function occur at one or more vertices.

Find the maximum value of the objective function $z = 3x + 2y$ subject to the following constraints: $x \geq 0, y \geq 0, 2x + 3y \leq 18, 2x + y \leq 10$.

1. Graph the system of inequalities representing the constraints.

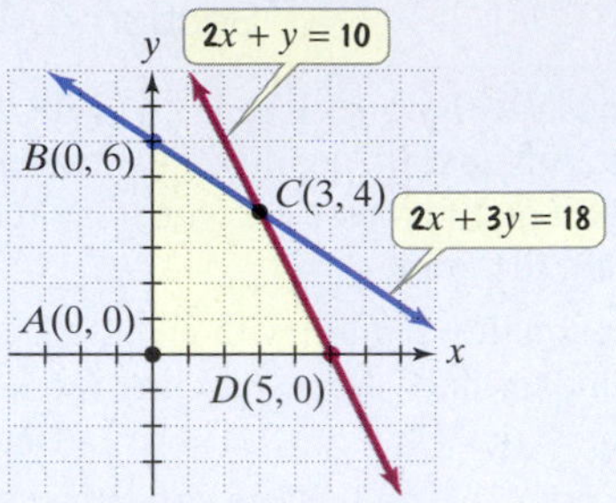

2. Evaluate the objective function at each vertex.

Vertex	$z = 3x + 2y$
$A(0, 0)$	$z = 3(0) + 2(0) = 0$
$B(0, 6)$	$z = 3(0) + 2(6) = 12$
$C(3, 4)$	$z = 3(3) + 2(4) = 17$
$D(5, 0)$	$z = 3(5) + 2(0) = 15$

The maximum value of the objective function is 17.

CHAPTER 4 REVIEW EXERCISES

4.1 *In Exercises 1–3, express each interval in set-builder notation and graph the interval on a number line.*

1. $(-2, 3]$
2. $[-1.5, 2]$
3. $(-1, \infty)$

In Exercises 4–9, solve each linear inequality. Other than $\varnothing$, graph the solution set on a number line. Express the solution set in both set-builder and interval notations.

4. $-6x + 3 \leq 15$
5. $6x - 9 \geq -4x - 3$
6. $\frac{x}{3} - \frac{3}{4} - 1 > \frac{x}{2}$
7. $6x + 5 > -2(x - 3) - 25$
8. $3(2x - 1) - 2(x - 4) \geq 7 + 2(3 + 4x)$
9. $2x + 7 \leq 5x - 6 - 3x$
10. A person can choose between two charges on a checking account. The first method involves a fixed cost of \$11 per month plus 6¢ for each check written. The second method involves a fixed cost of \$4 per month plus 20¢ for each check written. How many checks should be written to make the first method a better deal?
11. A salesperson earns \$500 per month plus a commission of 20% of sales. Describe the sales needed to receive a total income that exceeds \$3200 per month.

4.2 *In Exercises 12–15, let $A = \{a, b, c\}$, $B = \{a, c, d, e\}$, and $C = \{a, d, f, g\}$. Find the indicated set.*

12. $A \cap B$
13. $A \cap C$
14. $A \cup B$
15. $A \cup C$

In Exercises 16–26, solve each compound inequality. Except for the empty set, express the solution set in both set-builder and interval notations. Graph the solution set on a number line.

16. $x \le 3$ and $x < 6$

17. $x \le 3$ or $x < 6$

18. $-2x < -12$ and $x - 3 < 5$

19. $5x + 3 \le 18$ and $2x - 7 \le -5$

20. $2x - 5 > -1$ and $3x < 3$

21. $2x - 5 > -1$ or $3x < 3$

22. $x + 1 \le -3$ or $-4x + 3 < -5$

23. $5x - 2 \le -22$ or $-3x - 2 > 4$

24. $5x + 4 \ge -11$ or $1 - 4x \ge 9$

25. $-3 < x + 2 \le 4$

26. $-1 \le 4x + 2 \le 6$

27. To receive a B in a course, you must have an average of at least 80% but less than 90% on five exams. Your grades on the first four exams were 95%, 79%, 91%, and 86%. What range of grades on the fifth exam will result in a B for the course? Use interval notation to express this range.

4.3

In Exercises 28–31, find the solution set for each equation.

28. $|2x + 1| = 7$

29. $|3x + 2| = -5$

30. $2|x - 3| - 7 = 10$

31. $|4x - 3| = |7x + 9|$

In Exercises 32–36, solve and graph the solution set on a number line. Except for the empty set, express the solution set in both set-builder and interval notations.

32. $|2x + 3| \le 15$

33. $\left|\frac{2x + 6}{3}\right| > 2$

34. $|2x + 5| - 7 < -6$

35. $-4|x + 2| + 5 \le -7$

36. $|2x - 3| + 4 \le -10$

37. Approximately 90% of the population sleeps h hours daily, where h is modeled by the inequality $|h - 6.5| \le 1$. Write a sentence describing the range for the number of hours that most people sleep. Do *not* use the phrase "absolute value" in your description.

4.4

In Exercises 38–43, graph each inequality in a rectangular coordinate system.

38. $3x - 4y > 12$

39. $x - 3y \le 6$

40. $y \le -\frac{1}{2}x + 2$

41. $y > \frac{3}{5}x$

42. $x \le 2$

43. $y > -3$

In Exercises 44–52, graph the solution set of each system of inequalities or indicate that the system has no solution.

44. $2x - y \le 4$
$x + y \ge 5$

45. $y < -x + 4$
$y > x - 4$

46. $-3 \le x < 5$

47. $-2 < y \le 6$

48. $x \ge 3$
$y \le 0$

49. $2x - y > -4$
$x \ge 0$

50. $x + y \le 6$
$y \ge 2x - 3$

51. $3x + 2y \ge 4$
$x - y \le 3$
$x \ge 0,\ y \ge 0$

52. $2x - y > 2$
$2x - y < -2$

4.5

53. Find the value of the objective function $z = 2x + 3y$ at each corner of the graphed region shown. What is the maximum value of the objective function? What is the minimum value of the objective function?

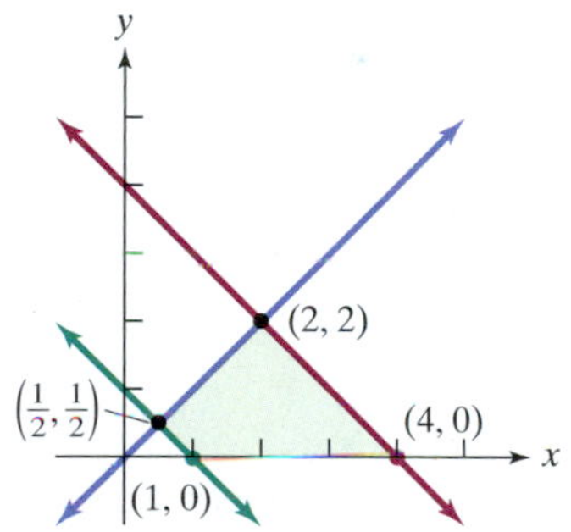

In Exercises 54–56, graph the region determined by the constraints. Then find the maximum value of the given objective function, subject to the constraints.

54. Objective Function $z = 2x + 3y$
Constraints $x \ge 0,\ y \ge 0$
$x + y \le 8$
$3x + 2y \ge 6$

55. Objective Function $z = x + 4y$
Constraints $0 \le x \le 5,\ 0 \le y \le 7$
$x + y \ge 3$

56. Objective Function $z = 5x + 6y$
Constraints $x \ge 0,\ y \ge 0$
$y \le x$
$2x + y \le 12$
$2x + 3y \ge 6$

57. A paper manufacturing company converts wood pulp to writing paper and newsprint. The profit on a unit of writing paper is \$500 and the profit on a unit of newsprint is \$350.

a. Let x represent the number of units of writing paper produced daily. Let y represent the number of units of newsprint produced daily. Write the objective function that models total daily profit.

b. The manufacturer is bound by the following constraints:

- Equipment in the factory allows for making at most 200 units of paper (writing paper and newsprint) in a day.
- Regular customers require at least 10 units of writing paper and at least 80 units of newsprint daily.

Write a system of inequalities that models these constraints.

c. Graph the inequalities in part (b). Use only the first quadrant, because x and y must both be positive. (*Suggestion*: Let each unit along the x- and y-axes represent 20.)

d. Evaluate the objective function at each of the three vertices of the graphed region.

e. Complete the missing portions of this statement: The company will make the greatest profit by producing ____ units of writing paper and ___ units of newsprint each day. The maximum daily profit is $______.

58. A manufacturer of lightweight tents makes two models whose specifications are given in the following table.

	Cutting Time per Tent	Assembly Time per Tent
Model A	0.9 hour	0.8 hour
Model B	1.8 hours	1.2 hours

Each month, the manufacturer has no more than 864 hours of labor available in the cutting department and at most 672 hours in the assembly division. The profits come to $25 per tent for model A and $40 per tent for model B. How many of each should be manufactured monthly to maximize the profit?

CHAPTER 4 TEST

Remember to use your Chapter Test Prep Video CD to see the worked-out solutions to the test questions you want to review.

In Exercises 1–2, express each interval in set-builder notation and graph the interval on a number line.

1. $[-3, 2)$

2. $(-\infty, -1]$

In Exercises 3–4, solve and graph the solution set on a number line. Express the solution set in both set-builder and interval notations.

3. $3(x + 4) \geq 5x - 12$

4. $\frac{x}{6} + \frac{1}{8} \leq \frac{x}{2} - \frac{3}{4}$

5. You are choosing between two telephone plans for local calls. Plan A charges $25 per month for unlimited calls. Plan B has a monthly fee of $13 with a charge of $0.06 per local call. How many local telephone calls in a month make plan A the better deal?

6. Find the intersection: $\{2, 4, 6, 8, 10\} \cap \{4, 6, 12, 14\}$.

7. Find the union: $\{2, 4, 6, 8, 10\} \cup \{4, 6, 12, 14\}$.

In Exercises 8–12, solve each compound inequality. Except for the empty set, express the solution set in both set-builder and interval notations. Graph the solution set on a number line.

8. $2x + 4 < 2$ and $x - 3 > -5$

9. $x + 6 \geq 4$ and $2x + 3 \geq -2$

10. $2x - 3 < 5$ or $3x - 6 \leq 4$

11. $x + 3 \leq -1$ or $-4x + 3 < -5$

12. $-3 \leq \frac{2x + 5}{3} < 6$

In Exercises 13–14, find the solution set for each equation.

13. $|5x + 3| = 7$

14. $|6x + 1| = |4x + 15|$

In Exercises 15–16, solve and graph the solution set on a number line. Express the solution set in both set-builder and interval notations.

15. $|2x - 1| < 7$

16. $|2x - 3| \geq 5$

17. The inequality $|b - 98.6| > 8$ describes a person's body temperature, b, in degrees Fahrenheit, when hyperthermia (extremely high body temperature) or hypothermia (extremely low body temperature) occurs. Solve the inequality and interpret the solution.

In Exercises 18–20, graph each inequality in a rectangular coordinate system.

18. $3x - 2y < 6$ **19.** $y \geq \frac{1}{2}x - 1$ **20.** $y \leq -1$

In Exercises 21–23, graph the solution set of each system of inequalities.

21. $x + y \geq 2$
$x - y \geq 4$

22. $3x + y \leq 9$
$2x + 3y \geq 6$
$x \geq 0, y \geq 0$

23. $-2 < x \leq 4$

24. Find the maximum value of the objective function $z = 3x + 5y$ subject to the following constraints: $x \geq 0, y \geq 0, x + y \leq 6, x \geq 2$.

25. A manufacturer makes two types of jet skis, regular and deluxe. The profit on a regular jet ski is \$200 and the profit on the deluxe model is \$250. To meet customer demand, the company must manufacture at least 50 regular jet skis per week and at least 75 deluxe models. To maintain high quality, the total number of both models of jet skis manufactured by the company should not exceed 150 per week. How many jet skis of each type should be manufactured per week to obtain maximum profit? What is the maximum weekly profit?

CUMULATIVE REVIEW EXERCISES (CHAPTERS 1–4)

In Exercises 1–2, solve each equation.

1. $5(x + 1) + 2 = x - 3(2x + 1)$

2. $\dfrac{2(x + 6)}{3} = 1 + \dfrac{4x - 7}{3}$

3. Simplify: $\dfrac{-10x^2y^4}{15x^7y^{-3}}$.

4. If $f(x) = x^2 - 3x + 4$, find $f(-3)$ and $f(2a)$.

5. If $f(x) = 3x^2 - 4x + 1$ and $g(x) = x^2 - 5x - 1$, find $(f - g)(x)$ and $(f - g)(2)$.

6. Use function notation to write the equation of the line passing through (2, 3) and perpendicular to the line whose equation is $y = 2x - 3$.

In Exercises 7–10, graph each equation or inequality in a rectangular coordinate system.

7. $f(x) = 2x + 1$

8. $y > 2x$

9. $2x - y \geq 6$

10. $f(x) = -1$

11. Solve the system:

$$\begin{aligned} 3x - y + z &= -15 \\ x + 2y - z &= 1 \\ 2x + 3y - 2z &= 0 \end{aligned}$$

12. Solve using matrices:

$$\begin{aligned} 2x - y &= -4 \\ x + 3y &= 5. \end{aligned}$$

13. Evaluate: $\begin{vmatrix} 4 & 3 \\ -1 & -5 \end{vmatrix}$.

14. A motel with 60 rooms charges \$90 per night for rooms with kitchen facilities and \$80 per night for rooms without kitchen facilities. When all rooms are occupied, the nightly revenue is \$5260. How many rooms of each kind are there?

15. Which of the following are functions?

a.

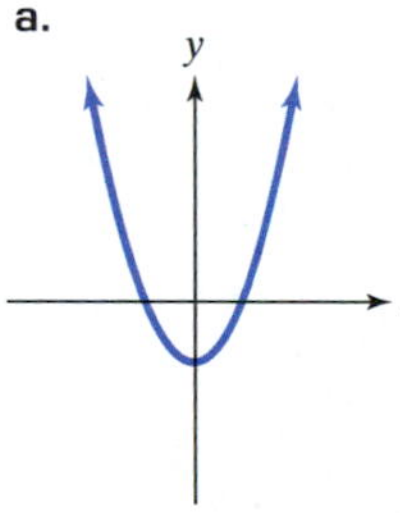

b.

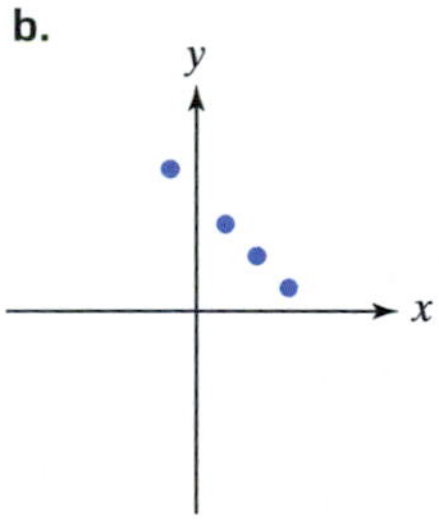

c.

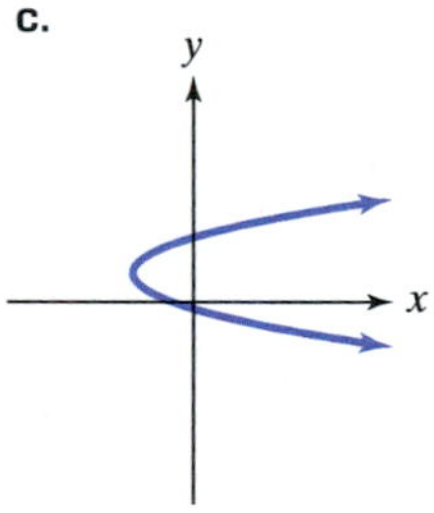

In Exercises 16–20, solve and graph the solution set on a number line. Express the solution set in both set-builder and interval notations.

16. $\dfrac{x}{4} - \dfrac{3}{4} - 1 \leq \dfrac{x}{2}$

17. $2x + 5 \leq 11$ and $-3x > 18$

18. $x - 4 \geq 1$ or $-3x + 1 \geq -5 - x$

19. $|2x + 3| \leq 17$

20. $|3x - 8| > 7$

CHAPTER 5

Polynomials, Polynomial Functions, and Factoring

New carry-on restrictions are wreaking havoc at the airport. You were made paranoid by news reports of lost, damaged, delayed, and pilfered checked luggage, so you overpacked a carry-on bag. The good news is that everything in your bag is permitted under the new restrictions. The bad news is that your bag is too large to carry on. The airline informed you that the sum of the length, width, and depth of a piece of carry-on luggage cannot exceed 40 inches. This bit of unexpected bad news, propelled by the airport's overall chaos, is enough to throw you into a real hissy fit. However, here's something that might help you get a grip on your emotions, if not on your luggage: your carry-on debacle has been modeled by a function, with an accompanying graph no less, for your reading pleasure on the plane. Have a good flight!

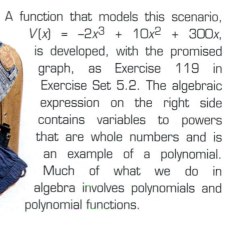

A function that models this scenario, $V(x) = -2x^3 + 10x^2 + 300x$, is developed, with the promised graph, as Exercise 119 in Exercise Set 5.2. The algebraic expression on the right side contains variables to powers that are whole numbers and is an example of a polynomial. Much of what we do in algebra involves polynomials and polynomial functions.

SECTION 5.1

Introduction to Polynomials and Polynomial Functions

Objectives

1. Use the vocabulary of polynomials.
2. Evaluate polynomial functions.
3. Determine end behavior.
4. Add polynomials.
5. Subtract polynomials.

In 1980, U.S. doctors diagnosed 41 cases of a rare form of cancer, Kaposi's sarcoma, that involved skin lesions, pneumonia, and severe immunological deficiencies. All cases involved gay men ranging in age from 26 to 51. By the end of 2003, approximately 930,000 Americans, straight and gay, male and female, old and young, were infected with the HIV virus.

Modeling AIDS-related data and making predictions about the epidemic's havoc is serious business. **Figure 5.1** shows the number of AIDS cases diagnosed in the United States from 1983 through 2003.

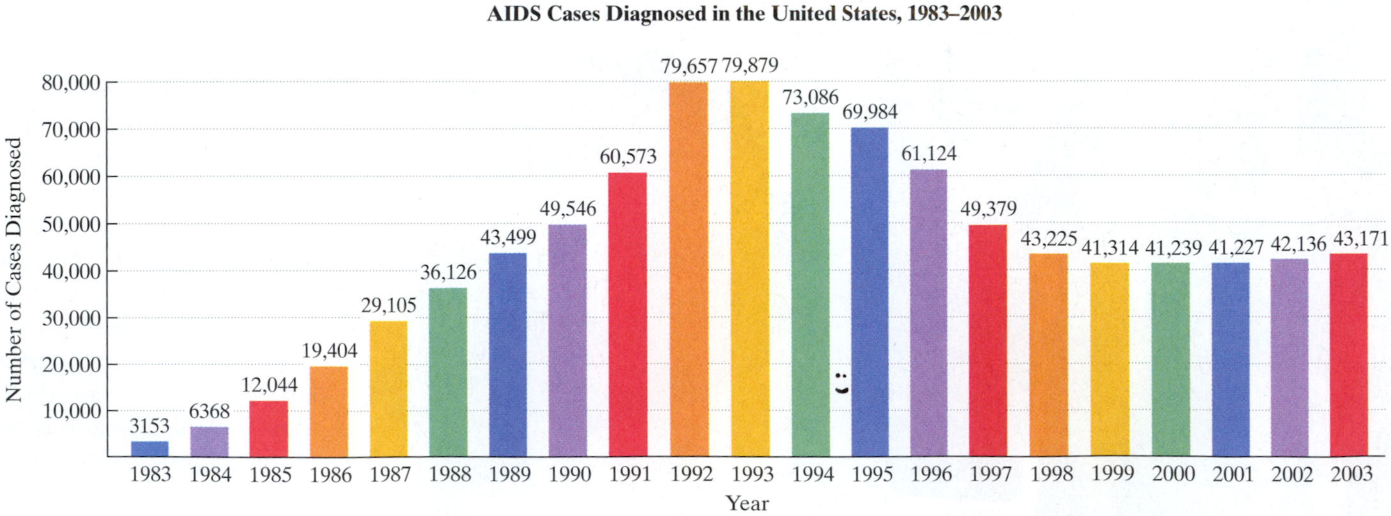

FIGURE 5.1

Source: Department of Health and Human Services

Changing circumstances and unforeseen events can result in models for AIDS-related data that are not particularly useful over long periods of time. For example, the function

$$f(x) = -49x^3 + 806x^2 + 3776x + 2503$$

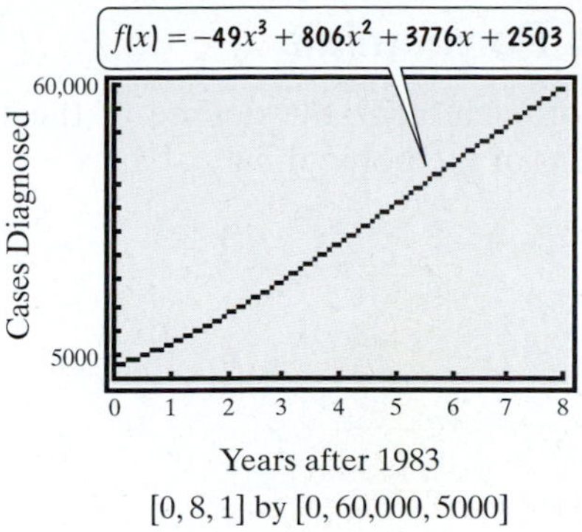

FIGURE 5.2 The graph of a function modeling the number of AIDS cases from 1983 through 1991

models the number of AIDS cases diagnosed in the United States x years after 1983. The model was obtained using a portion of the data shown in **Figure 5.1**, namely cases diagnosed from 1983 through 1991, inclusive. **Figure 5.2** shows the graph of f from 1983 through 1991.

The voice balloon in **Figure 5.2** displays the algebraic expression used to define f. This algebraic expression is an example of a *polynomial*. A **polynomial** is a single term or the sum of two or more terms containing variables with whole-number exponents. Functions containing polynomials are used in such diverse areas as science, business, medicine, psychology, and sociology. In this section, we present basic ideas about polynomials and polynomial functions. We will use our knowledge of combining like terms to find sums and differences of polynomials.

How We Describe Polynomials

1 Use the vocabulary of polynomials.

Consider the polynomial

$$-49x^3 + 806x^2 + 3776x + 2503.$$

This polynomial contains four terms. It is customary to write the terms in the order of descending powers of the variable. This is the **standard form** of a polynomial.

Some polynomials contain only one variable. Each term of such a polynomial in x is of the form ax^n. If $a \neq 0$, the **degree** of ax^n is n. For example, the degree of the term $-49x^3$ is 3.

Study Tip

We can express 0 in many ways, including $0x$, $0x^2$, and $0x^3$. It is impossible to assign a single exponent on the variable. This is why 0 has no defined degree.

The Degree of ax^n

If $a \neq 0$ and n is a whole number, the degree of ax^n is n. The degree of a nonzero constant is 0. The constant 0 has no defined degree.

Here is the polynomial modeling AIDS cases and the degree of each of its four terms:

$$-49x^3 + 806x^2 + 3776x + 2503$$

degree 3 | degree 2 | degree 1 | degree of nonzero constant: 0

Notice that the exponent on x for the term $3776x$ is understood to be 1: $3776x^1$. For this reason, the degree of $3776x$ is 1. You can think of 2503 as $2503x^0$; thus, its degree is 0.

A polynomial is simplified when it contains no grouping symbols and no like terms. A simplified polynomial that has exactly one term is called a **monomial**. A **binomial** is a simplified polynomial that has two terms. A **trinomial** is a simplified polynomial with three terms. Simplified polynomials with four or more terms have no special names.

Some polynomials contain two or more variables. Here is an example of a polynomial in two variables, x and y:

$$7x^2y^3 - 17x^4y^2 + xy - 6y^2 + 9.$$

A polynomial in two variables, x and y, contains the sum of one or more monomials of the form ax^ny^m. The constant a is the **coefficient**. The exponents, n and m, represent whole numbers. The **degree of the term** ax^ny^m is the sum of the exponents of the variables, $n + m$.

The **degree of a polynomial** is the greatest degree of any term of the polynomial. If there is precisely one term of the greatest degree, it is called the **leading term**. Its coefficient is called the **leading coefficient**.

EXAMPLE 1 Using the Vocabulary of Polynomials

Determine the coefficient of each term, the degree of each term, the degree of the polynomial, the leading term, and the leading coefficient of the polynomial

$$7x^2y^3 - 17x^4y^2 + xy - 6y^2 + 9.$$

Solution

Term	Coefficient	Degree (Sum of Exponents on the Variables)
$7x^2y^3$	7	$2 + 3 = 5$
$-17x^4y^2$	-17	$4 + 2 = 6$
xy	1	$1 + 1 = 2$
$-6y^2$	-6	$0 + 2 = 2$
9	9	$0 + 0 = 0$

Think of xy as $1x^1y^1$.

Think of $-6y^2$ as $-6x^0y^2$.

Think of 9 as $9x^0y^0$.

The degree of the polynomial is the greatest degree of any term of the polynomial, which is 6. The leading term is the term of the greatest degree, which is $-17x^4y^2$. Its coefficient, -17, is the leading coefficient.

✓ CHECK POINT 1 Determine the coefficient of each term, the degree of each term, the degree of the polynomial, the leading term, and the leading coefficient of the polynomial

$$8x^4y^5 - 7x^3y^2 - x^2y - 5x + 11.$$

If a polynomial contains three or more variables, the degree of a term is the sum of the exponents of all the variables. Here is an example of a polynomial in three variables, x, y, and z:

The coefficients are $\frac{1}{4}$, -2, 6 and 5.

$$\frac{1}{4}xy^2z^4 - 2xyz + 6x^2 + 5.$$

Degree: $1 + 2 + 4 = 7$ | Degree: $1 + 1 + 1 = 3$ | Degree: 2 | Degree: 0

The degree of this polynomial is the greatest degree of any term of the polynomial, which is 7.

2 Evaluate polynomial functions.

Polynomial Functions

The expression $4x^3 - 5x^2 + 3$ is a polynomial. If we write

$$f(x) = 4x^3 - 5x^2 + 3,$$

then we have a **polynomial function**. In a polynomial function, the expression that defines the function is a polynomial. How do we evaluate a polynomial function? Use substitution, just as we did to evaluate other functions in Chapter 2.

EXAMPLE 2 Evaluating a Polynomial Function

The polynomial function

$$f(x) = -49x^3 + 806x^2 + 3776x + 2503$$

models the number of AIDS cases diagnosed in the United States, $f(x)$, x years after 1983, where $0 \le x \le 8$. Find $f(6)$ and describe what this means in practical terms.

Solution To find $f(6)$, or f of 6, we replace each occurrence of x in the function's formula with 6.

$$f(x) = -49x^3 + 806x^2 + 3776x + 2503$$ This is the given function.

$$f(6) = -49(6)^3 + 806(6)^2 + 3776(6) + 2503$$ Replace each occurrence of x with 6.

$$= -49(216) + 806(36) + 3776(6) + 2503$$ Evaluate exponential expressions.

$$= -10{,}584 + 29{,}016 + 22{,}656 + 2503$$ Multiply.

$$= 43{,}591$$ Add.

Thus, $f(6) = 43{,}591$. According to the model, this means that 6 years after 1983, in 1989, there were 43,591 AIDS cases diagnosed in the United States. (The actual number, shown in **Figure 5.1**, is 43,499.)

Using Technology

Once each occurrence of x in $f(x) = -49x^3 + 806x^2 + 3776x + 2503$ is replaced with 6, the resulting computation can be performed using a scientific calculator or a graphing calculator.

$$-49(6)^3 + 806(6)^2 + 3776(6) + 2503$$

Many Scientific Calculators

49 [+/−] [×] 6 [y^x] 3 [+] 806 [×] 6 [y^x] 2 [+] 3776 [×] 6 [+] 2503 [=]

Many Graphing Calculators

[(−)] 49 [×] 6 [^] 3 [+] 806 [×] 6 [^] 2 [+] 3776 [×] 6 [+] 2503 [ENTER]

The display should be 43591. This number can also be obtained by using a graphing utility's feature that evaluates a function or by using its table feature.

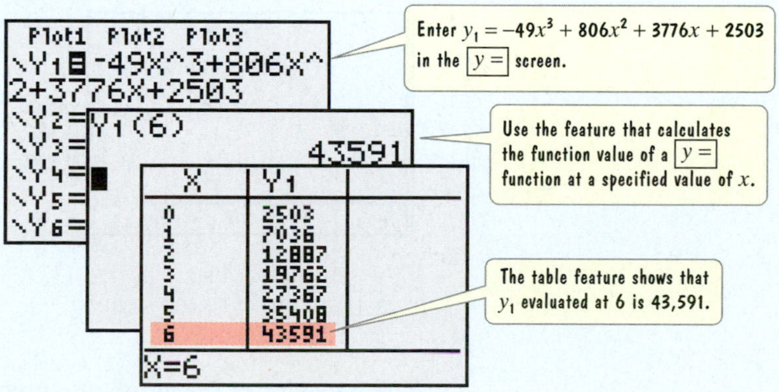
Plot1 Plot2 Plot3
\Y1 = -49X^3+806X^2+3776X+2503
\Y2=
\Y3=
\Y4=
\Y5=
\Y6=
Y1(6)
43591
X
Y1
0 2503
1 7036
2 12887
3 19762
4 27367
5 35408
6 43591
X=6
Enter $y_1 = -49x^3 + 806x^2 + 3776x + 2503$ in the $y =$ screen.
Use the feature that calculates the function value of a $y =$ function at a specified value of x.
The table feature shows that y_1 evaluated at 6 is 43,591.

CHECK POINT 2 For the polynomial function

$$f(x) = 4x^3 - 3x^2 - 5x + 6,$$

find $f(2)$.

Smooth, Continuous Graphs

Polynomial functions of degree 2 or higher have graphs that are *smooth* and *continuous*. By **smooth**, we mean that the graph contains only rounded curves with no sharp corners. By **continuous**, we mean that the graph has no breaks and can be drawn without lifting your pencil from the rectangular coordinate system. These ideas are illustrated in **Figure 5.3**.

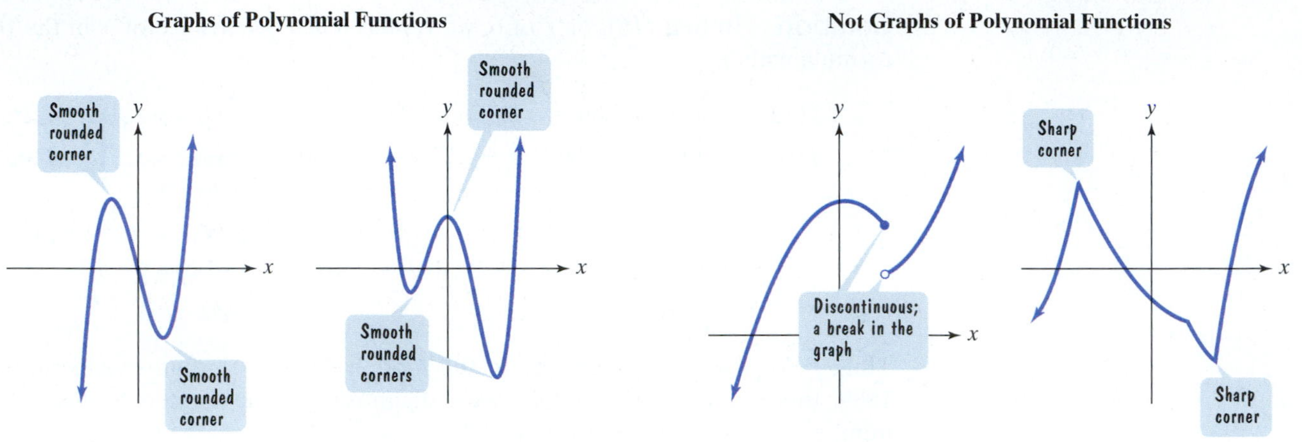

FIGURE 5.3 Recognizing graphs of polynomial functions

3 Determine end behavior.

End Behavior of Polynomial Functions

Figure 5.4 shows the graph of the function

$$f(x) = -49x^3 + 806x^2 + 3776x + 2503,$$

which models U.S. AIDS cases from 1983 through 1991. Look what happens to the graph when we extend the year up through 2005. By year 21 (2004), the values of y are negative and the function no longer models AIDS cases. We've added an arrow to the graph at the far right to emphasize that it continues to decrease without bound. This far-right *end behavior* of the graph is one reason that this function is inappropriate for modeling AIDS cases into the future.

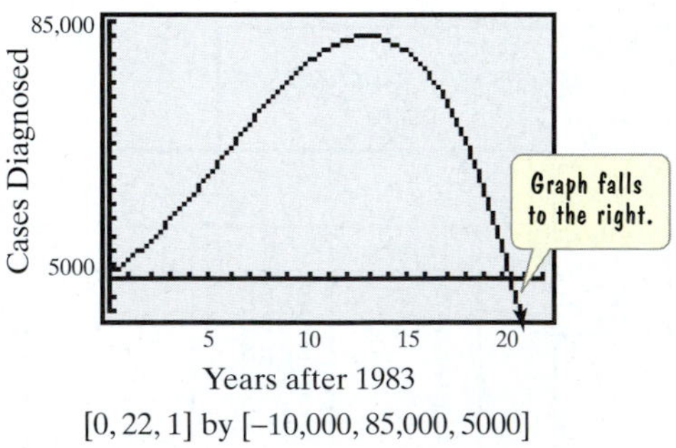

FIGURE 5.4 By extending the viewing rectangle, we see that y is eventually negative and the function no longer models the number of AIDS cases.

The behavior of the graph of a function to the far left or the far right is called its **end behavior**. Although the graph of a polynomial function may have intervals where it increases and intervals where it decreases, the graph will eventually rise or fall without bound as it moves far to the left or far to the right.

How can you determine whether the graph of a polynomial function goes up or down at each end? **The end behavior depends upon the leading term.** In particular, the sign of the leading coefficient and the degree of the polynomial reveal the graph's end behavior. With regard to end behavior, only the leading term—that is, the term of the greatest degree—counts, as summarized by the **Leading Coefficient Test**.

The Leading Coefficient Test

As x increases or decreases without bound, the graph of a polynomial function eventually rises or falls. In particular,

1. For odd-degree polynomials:

If the leading coefficient is positive, the graph falls to the left and rises to the right. (↙, ↗)

If the leading coefficient is negative, the graph rises to the left and falls to the right. (↖, ↘)

2. For even-degree polynomials:

If the leading coefficient is positive, the graph rises to the left and rises to the right. (↖, ↗)

If the leading coefficient is negative, the graph falls to the left and falls to the right. (↙, ↘)

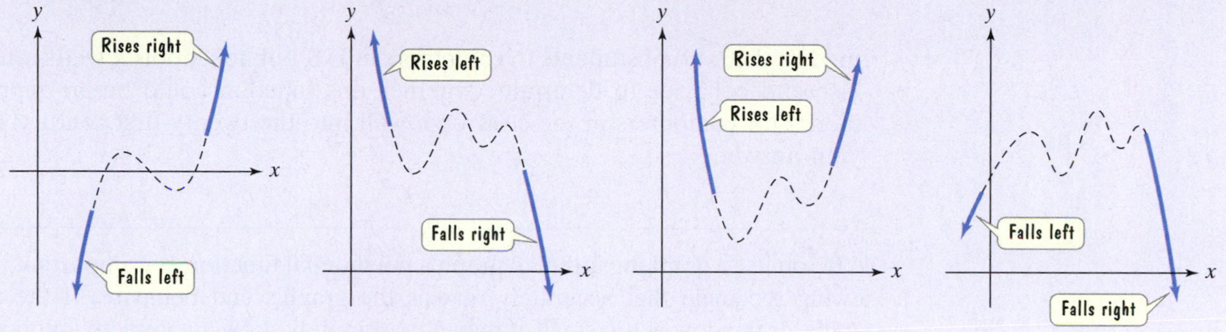

Study Tip

Odd-degree polynomial functions have graphs with opposite behavior at each end. Even-degree polynomial functions have graphs with the same behavior at each end.

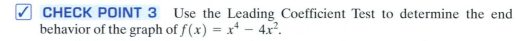

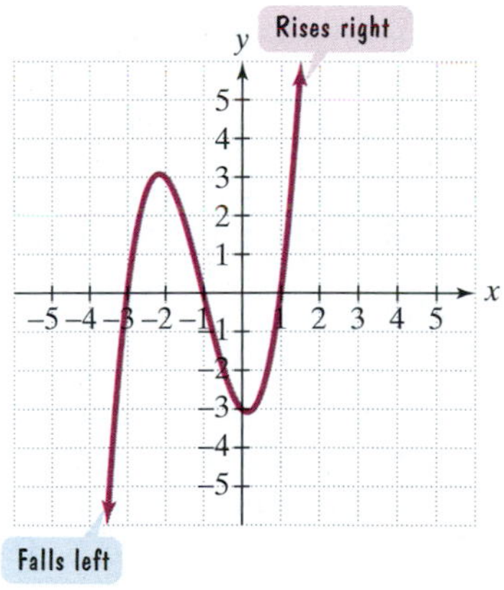

FIGURE 5.5 The graph of $f(x) = x^3 + 3x^2 - x - 3$

EXAMPLE 3 Using the Leading Coefficient Test

Use the Leading Coefficient Test to determine the end behavior of the graph of

$$f(x) = x^3 + 3x^2 - x - 3.$$

Solution We begin by identifying the sign of the leading coefficient and the degree of the polynomial.

$$f(x) = x^3 + 3x^2 - x - 3$$

The leading coefficient, 1, is positive. The degree of the polynomial, 3, is odd.

The degree of the function f is 3, which is odd. Odd-degree polynomial functions have graphs with opposite behavior at each end. The leading coefficient, 1, is positive. Thus, the graph falls to the left and rises to the right (↙, ↗). The graph of f is shown in **Figure 5.5**.

✓ **CHECK POINT 3** Use the Leading Coefficient Test to determine the end behavior of the graph of $f(x) = x^4 - 4x^2$.

EXAMPLE 4 Using the Leading Coefficient Test

Use end behavior to explain why

$$f(x) = -49x^3 + 806x^2 + 3776x + 2503$$

is only an appropriate model for AIDS cases for a limited time period.

Solution We begin by identifying the sign of the leading coefficient and the degree of the polynomial.

$$f(x) = -49x^3 + 806x^2 + 3776x + 2503$$

The leading coefficient, −49, is negative. The degree of the polynomial, 3, is odd.

The degree of f in $f(x) = -49x^3 + 806x^2 + 3776x + 2053$ is 3, which is odd. Odd-degree polynomial functions have graphs with opposite behavior at each end. The leading coefficient, -49, is negative. Thus, the graph rises to the left and falls to the right (↖, ↘). The fact that the graph falls to the right indicates that at some point the number of AIDS cases will be negative, an impossibility. If a function has a graph that decreases without bound over time, it will not be capable of modeling nonnegative phenomena over long time periods. Model breakdown will eventually occur.

✓ CHECK POINT 4 The polynomial function

$$f(x) = -0.27x^3 + 9.2x^2 - 102.9x + 400$$

models the ratio of students to computers in U.S. public schools x years after 1980. Use end behavior to determine whether this function could be an appropriate model for computers in the classroom well into the twenty-first century. Explain your answer.

If you use a graphing utility to graph a polynomial function, it is important to select a viewing rectangle that accurately reveals the graph's end behavior. If the viewing rectangle, or window, is too small, it may not accurately show a complete graph with the appropriate end behavior.

EXAMPLE 5 Using the Leading Coefficient Test

The graph of $f(x) = -x^4 + 8x^3 + 4x^2 + 2$ was obtained with a graphing utility using a $[-8, 8, 1]$ by $[-10, 10, 1]$ viewing rectangle. The graph is shown in **Figure 5.6**. Is this a complete graph that shows the end behavior of the function?

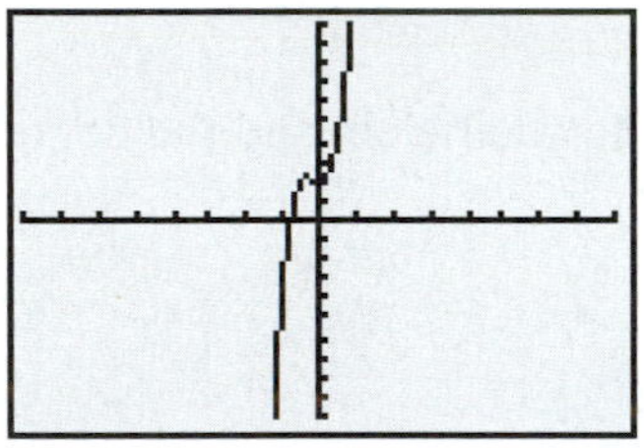

[−8, 8, 1] by [−10, 10, 1]

FIGURE 5.6

Solution We begin by identifying the sign of the leading coefficient and the degree of the polynomial.

$$f(x) = -x^4 + 8x^3 + 4x^2 + 2$$

The leading coefficient, −1, is negative.

The degree of the polynomial, 4, is even.

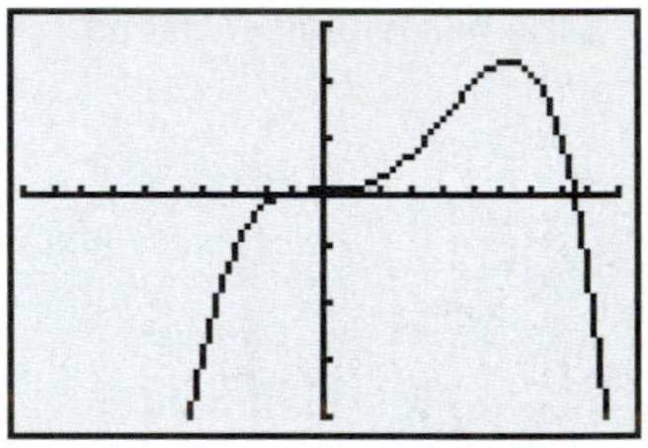

[−10, 10, 1] by [−1000, 750, 250]

FIGURE 5.7

The degree of f is 4, which is even. Even-degree polynomial functions have graphs with the same behavior at each end. The leading coefficient, -1, is negative. Thus, the graph should fall to the left and fall to the right (↙, ↘). The graph in **Figure 5.6** is falling to the left, but it is not falling to the right. Therefore, the graph is not complete enough to show end behavior. A more complete graph of the function is shown in a larger viewing rectangle in **Figure 5.7**.

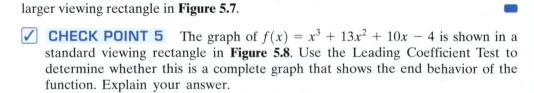

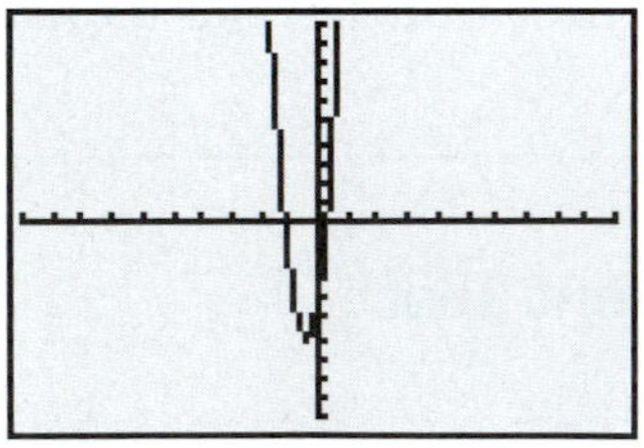

FIGURE 5.8

✓ CHECK POINT 5 The graph of $f(x) = x^3 + 13x^2 + 10x - 4$ is shown in a standard viewing rectangle in **Figure 5.8**. Use the Leading Coefficient Test to determine whether this is a complete graph that shows the end behavior of the function. Explain your answer.

---

4 Add polynomials.

Adding Polynomials

Polynomials are added by combining like terms. Here are two examples that illustrate the use of the distributive property in adding monomials and combining like terms:

$$-9x^3 + 13x^3 = (-9 + 13)x^3 = 4x^3$$

Add coefficients and keep the same variable factor(s).

$$-7x^3y^2 + 4x^3y^2 = (-7 + 4)x^3y^2 = -3x^3y^2.$$

EXAMPLE 6 Adding Polynomials

Add: $(-6x^3 + 5x^2 + 4) + (2x^3 + 7x^2 - 10)$.

Solution

$$(-6x^3 + 5x^2 + 4) + (2x^3 + 7x^2 - 10)$$

$$= -6x^3 + 5x^2 + 4 + 2x^3 + 7x^2 - 10$$ Remove the parentheses. Like terms are shown in the same color.

$$= \underbrace{-6x^3 + 2x^3} + \underbrace{5x^2 + 7x^2} + \underbrace{4 - 10}$$ Rearrange the terms so that like terms are adjacent.

$$= \quad -4x^3 \quad + 12x^2 \quad - 6$$ Combine like terms.

$$= -4x^3 + 12x^2 - 6$$ This is the same sum as above, written more concisely.

✓ **CHECK POINT 6** Add: $(-7x^3 + 4x^2 + 3) + (4x^3 + 6x^2 - 13)$.

EXAMPLE 7 Adding Polynomials

Add: $(5x^3y - 4x^2y - 7y) + (2x^3y + 6x^2y - 4y - 5)$.

Solution

$$(5x^3y - 4x^2y - 7y) + (2x^3y + 6x^2y - 4y - 5)$$ The given problem involves adding polynomials in two variables.

$$= 5x^3y - 4x^2y - 7y + 2x^3y + 6x^2y - 4y - 5$$ Remove the parentheses. Like terms are shown in the same color.

$$= \underbrace{5x^3y + 2x^3y} - \underbrace{4x^2y + 6x^2y} - \underbrace{7y - 4y} \underbrace{- 5}$$ Rearrange the terms so that like terms are adjacent.

$$= \quad 7x^3y \quad + 2x^2y \quad - 11y \quad - 5$$ Combine like terms.

Polynomials can be added by arranging like terms in columns. Then combine like terms, column by column. Here's the solution to Example 7 using columns and a vertical format:

$$\begin{array}{r} 5x^3y - 4x^2y - 7y \phantom{{}-5} \\ 2x^3y + 6x^2y - 4y - 5 \\ \hline 7x^3y + 2x^2y - 11y - 5 \end{array}$$

✓ **CHECK POINT 7** Add: $(7xy^3 - 5xy^2 - 3y) + (2xy^3 + 8xy^2 - 12y - 9)$.

---

5 Subtract polynomials.

Subtracting Polynomials

We subtract real numbers by adding the opposite, or additive inverse, of the number being subtracted. For example,

$$8 - 3 = 8 + (-3) = 5.$$

Similarly, we subtract one polynomial from another by adding the opposite of the polynomial being subtracted.

Subtracting Polynomials

To subtract two polynomials, change the sign of every term of the second polynomial. Add this result to the first polynomial.

Study Tip

You can also subtract polynomials using a vertical format. Here's the solution to Example 8 using a vertical format. Notice that you still distribute the negative sign, thereby adding the opposite.

$$\begin{array}{r} 7x^3 - 8x^2 + 9x - 6 \\ -(2x^3 - 6x^2 - 3x + 9) \\ \hline 7x^3 - 8x^2 + 9x - 6 \\ + -2x^3 + 6x^2 + 3x - 9 \\ \hline 5x^3 - 2x^2 + 12x - 15 \end{array}$$

EXAMPLE 8 Subtracting Polynomials

Subtract: $(7x^3 - 8x^2 + 9x - 6) - (2x^3 - 6x^2 - 3x + 9)$.

Solution

$$(7x^3 - 8x^2 + 9x - 6) - (2x^3 - 6x^2 - 3x + 9)$$

$$= (7x^3 - 8x^2 + 9x - 6) + (-2x^3 + 6x^2 + 3x - 9)$$

Change the sign of each term of the second polynomial and add the two polynomials. Like terms are shown in the same color.

$$= \underbrace{7x^3 - 2x^3}_{5x^3} \underbrace{- 8x^2 + 6x^2}_{-2x^2} \underbrace{+ 9x + 3x}_{+12x} \underbrace{- 6 - 9}_{-15}$$

Rearrange terms.

$$= 5x^3 - 2x^2 + 12x - 15$$

Combine like terms.

CHECK POINT 8 Subtract: $(14x^3 - 5x^2 + x - 9) - (4x^3 - 3x^2 - 7x + 1)$.

Study Tip

Be careful of the order in Example 9. For example, subtracting 2 from 5 is equivalent to $5 - 2$. In general, subtracting B from A means $A - B$. The order of the resulting algebraic expression is not the same as the order in English.

EXAMPLE 9 Subtracting Polynomials

Subtract $-2x^5y^2 - 3x^3y + 7$ from $3x^5y^2 - 4x^3y - 3$.

Solution

$$(3x^5y^2 - 4x^3y - 3) - (-2x^5y^2 - 3x^3y + 7)$$

$$= 3x^5y^2 - 4x^3y - 3 + 2x^5y^2 + 3x^3y - 7$$

Change subtraction to addition and change the sign of every term of the second polynomial. Like terms are shown in the same color.

$$= \underbrace{3x^5y^2 + 2x^5y^2}_{5x^5y^2} \underbrace{- 4x^3y + 3x^3y}_{-x^3y} \underbrace{- 3 - 7}_{-10}$$

Rearrange terms.

$$= 5x^5y^2 - x^3y - 10$$

Combine like terms.

CHECK POINT 9 Subtract $-7x^2y^5 - 4xy^3 + 2$ from $6x^2y^5 - 2xy^3 - 8$.

5.1 EXERCISE SET

MyMathLab MathXL PRACTICE WATCH DOWNLOAD READ REVIEW

Practice Exercises

In Exercises 1–10, determine the coefficient of each term, the degree of each term, the degree of the polynomial, the leading term, and the leading coefficient of the polynomial.

1. $-x^4 + x^2$
2. $x^3 - 4x^2$
3. $5x^3 + 7x^2 - x + 9$
4. $11x^3 - 6x^2 + x + 3$

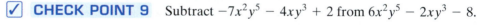

5. $3x^2 - 7x^4 - x + 6$

6. $2x^2 - 9x^4 - x + 5$

7. $x^3y^2 - 5x^2y^7 + 6y^2 - 3$

8. $12x^4y - 5x^3y^7 - x^2 + 4$

9. $x^5 + 3x^2y^4 + 7xy + 9x - 2$

10. $3x^6 + 4x^4y^4 - x^3y + 4x^2 - 5$

In Exercises 11–20, let

$$f(x) = x^2 - 5x + 6 \quad \text{and} \quad g(x) = 2x^3 - x^2 + 4x - 1.$$

Find the indicated function values.

11. $f(3)$ **12.** $f(4)$ **13.** $f(-1)$

14. $f(-2)$ **15.** $g(3)$ **16.** $g(2)$

17. $g(-2)$ **18.** $g(-3)$ **19.** $g(0)$

20. $f(0)$

In Exercises 21–24, identify which graphs are not those of polynomial functions.

21.

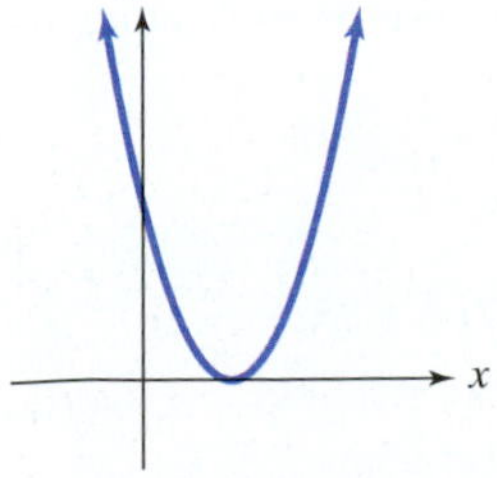

22.

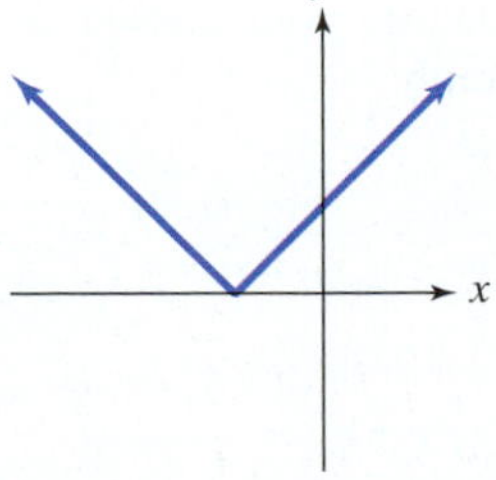

23.

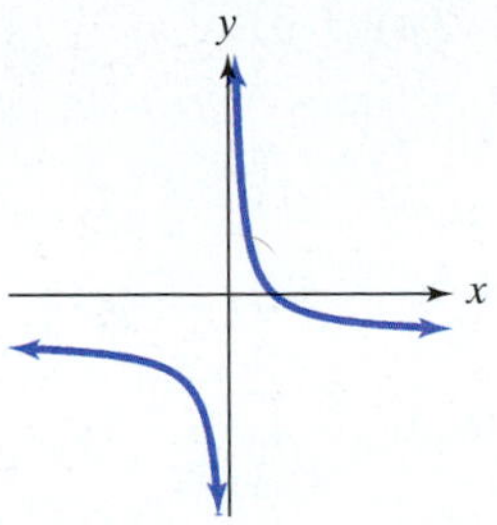

24.

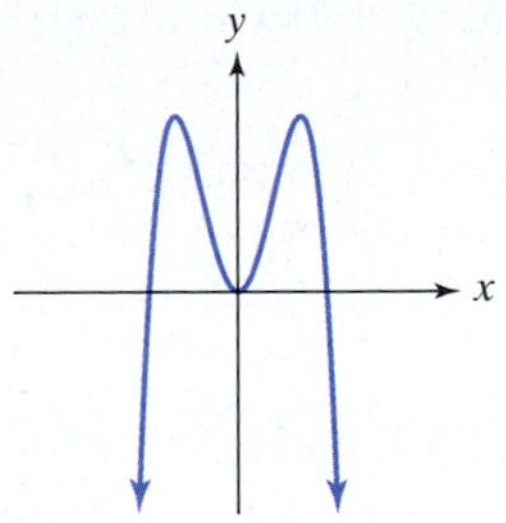

In Exercises 25–28, use the Leading Coefficient Test to determine the end behavior of the graph of the given polynomial function. Then use this end behavior to match the polynomial function with its graph. [The graphs are labeled (a) through (d).]

25. $f(x) = -x^4 + x^2$

26. $f(x) = x^3 - 4x^2$

27. $f(x) = x^2 - 6x + 9$

28. $f(x) = -x^3 - x^2 + 5x - 3$

a.

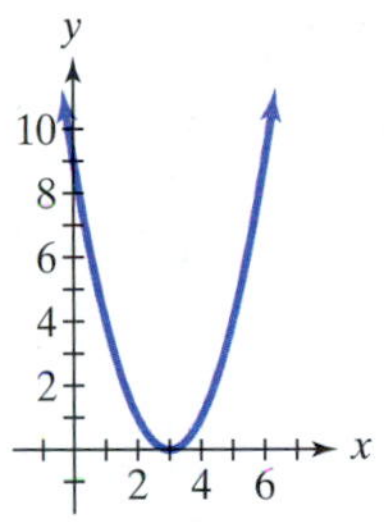

b.

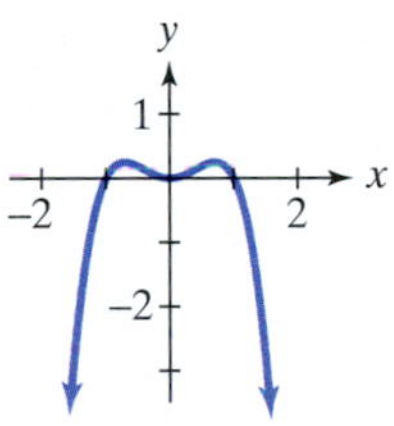

c.

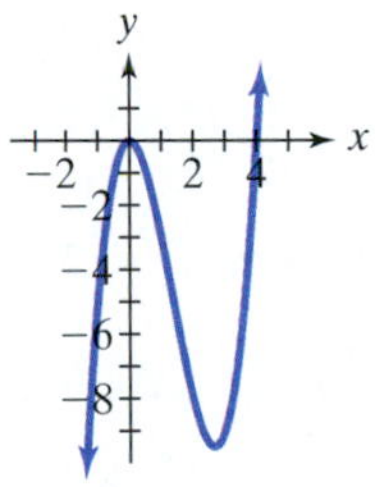

d.

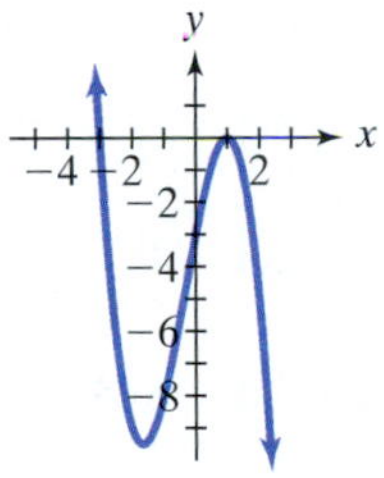

In Exercises 29–40, add the polynomials. Assume that all variable exponents represent whole numbers.

29. $(-6x^3 + 5x^2 - 8x + 9) + (17x^3 + 2x^2 - 4x - 13)$

30. $(-7x^3 + 6x^2 - 11x + 13) + (19x^3 - 11x^2 + 7x - 17)$

31. $\left(\frac{2}{5}x^4 + \frac{2}{3}x^3 + \frac{5}{8}x^2 + 7\right) + \left(-\frac{4}{5}x^4 + \frac{1}{3}x^3 - \frac{1}{4}x^2 - 7\right)$

32. $\left(\frac{1}{5}x^4 + \frac{1}{3}x^3 + \frac{3}{8}x^2 + 6\right) + \left(-\frac{3}{5}x^4 + \frac{2}{3}x^3 - \frac{1}{2}x^2 - 6\right)$

33. $(7x^2y - 5xy) + (2x^2y - xy)$

34. $(-4x^2y + xy) + (7x^2y + 8xy)$

35. $(5x^2y + 9xy + 12) + (-3x^2y + 6xy + 3)$

36. $(8x^2y + 12xy + 14) + (-2x^2y + 7xy + 4)$

37. $(9x^4y^2 - 6x^2y^2 + 3xy) + (-18x^4y^2 - 5x^2y - xy)$

38. $(10x^4y^2 - 3x^2y^2 + 2xy) + (-16x^4y^2 - 4x^2y - xy)$

39. $(x^{2n} + 5x^n - 8) + (4x^{2n} - 7x^n + 2)$

40. $(6y^{2n} + y^n + 5) + (3y^{2n} - 4y^n - 15)$

In Exercises 41–50, subtract the polynomials. Assume that all variable exponents represent whole numbers.

41. $(17x^3 - 5x^2 + 4x - 3) - (5x^3 - 9x^2 - 8x + 11)$

42. $(18x^3 - 2x^2 - 7x + 8) - (9x^3 - 6x^2 - 5x + 7)$

43. $(13y^5 + 9y^4 - 5y^2 + 3y + 6) - (-9y^5 - 7y^3 + 8y^2 + 11)$

44. $(12y^5 + 7y^4 - 3y^2 + 6y + 7) - (-10y^5 - 8y^3 + 3y^2 + 14)$

45. $(x^3 + 7xy - 5y^2) - (6x^3 - xy + 4y^2)$

46. $(x^4 - 7xy - 5y^3) - (6x^4 - 3xy + 4y^3)$

47. $(3x^4y^2 + 5x^3y - 3y) - (2x^4y^2 - 3x^3y - 4y + 6x)$

48. $(5x^4y^2 + 6x^3y - 7y) - (3x^4y^2 - 5x^3y - 6y + 8x)$

49. $(7y^{2n} + y^n - 4) - (6y^{2n} - y^n - 1)$

50. $(8x^{2n} + x^n - 4) - (9x^{2n} - x^n - 2)$

51. Subtract $-5a^2b^4 - 8ab^2 - ab$ from $3a^2b^4 - 5ab^2 + 7ab$.

52. Subtract $-7a^2b^4 - 8ab^2 - ab$ from $13a^2b^4 - 17ab^2 + ab$.

53. Subtract $-4x^3 - x^2y + xy^2 + 3y^3$ from $x^3 + 2x^2y - y^3$.

54. Subtract $-6x^3 + x^2y - xy^2 + 2y^3$ from $x^3 + 2xy^2 - y^3$.

Practice PLUS

55. Add $6x^4 - 5x^3 + 2x$ to the difference between $4x^3 + 3x^2 - 1$ and $x^4 - 2x^2 + 7x - 3$.

56. Add $5x^4 - 2x^3 + 7x$ to the difference between $2x^3 + 5x^2 - 3$ and $-x^4 - x^2 - x - 1$.

57. Subtract $9x^2y^2 - 3x^2 - 5$ from the sum of $-6x^2y^2 - x^2 - 1$ and $5x^2y^2 + 2x^2 - 1$.

58. Subtract $6x^2y^3 - 2x^2 - 7$ from the sum of $-5x^2y^3 + 3x^2 - 4$ and $4x^2y^3 - 2x^2 - 6$.

In Exercises 59–64, let

$$f(x) = -3x^3 - 2x^2 - x + 4$$
$$g(x) = x^3 - x^2 - 5x - 4$$
$$h(x) = -2x^3 + 5x^2 - 4x + 1.$$

Find the indicated function, function value, or polynomial.

59. $(f - g)(x)$ and $(f - g)(-1)$

60. $(g - h)(x)$ and $(g - h)(-1)$

61. $(f + g - h)(x)$ and $(f + g - h)(-2)$

62. $(g + h - f)(x)$ and $(g + h - f)(-2)$

63. $2f(x) - 3g(x)$

64. $-2g(x) - 3h(x)$

Application Exercises

The polynomial function

$$f(x) = -1844x^2 + 54{,}923x + 111{,}568$$

models the cumulative number of deaths from AIDS in the United States, $f(x)$, x years after 1990. Use this function to solve Exercises 65–66.

65. Find and interpret $f(10)$.

66. Find and interpret $f(8)$.

The graph of the polynomial function in Exercises 65–66 is shown in the figure. Use the graph to solve Exercises 67–68.

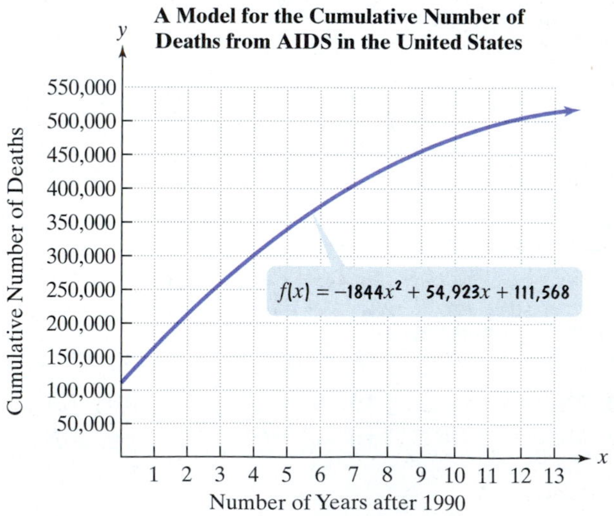

67. Identify your answer from Exercise 65 as a point on the graph.

68. Identify your answer from Exercise 66 as a point on the graph.

The bar graph shows the actual data for the cumulative number of deaths from AIDS in the United States from 1990 through 2003.

Cumulative Number of Deaths from AIDS in the United States

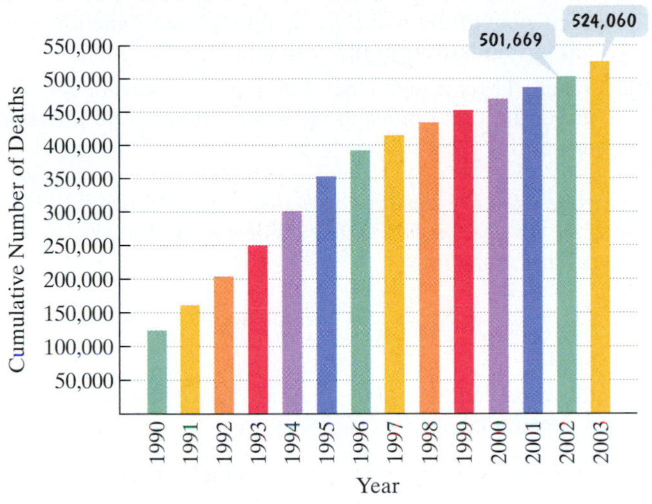

Source: Centers for Disease Control

The data in the bar graph can be modeled by the following second- and third-degree polynomial functions:

Cumulative number of AIDS deaths x years after 1990

$$f(x) = -1844x^2 + 54{,}923x + 111{,}568$$

$$g(x) = 11x^3 - 2066x^2 + 56{,}036x + 110{,}590$$

Use these functions to solve Exercises 69–71.

69. Use both functions to find the cumulative number of AIDS deaths in 2003. Which function provides a better description for the actual number shown in the bar graph?

70. Use both functions to find the cumulative number of AIDS deaths in 2002. Which function provides a better description for the actual number shown in the bar graph?

71. Use the Leading Coefficient Test to determine the end behavior to the right for the graph of f. Will this function be useful in modeling the cumulative number of AIDS deaths over an extended period of time? Explain your answer.

72. The common cold is caused by a rhinovirus. After x days of invasion by the viral particles, the number of particles in our bodies, $f(x)$, in billions, can be modeled by the polynomial function

$$f(x) = -0.75x^4 + 3x^3 + 5.$$

Use the Leading Coefficient Test to determine the graph's end behavior to the right. What does this mean about the number of viral particles in our bodies over time?

73. The polynomial function

$$f(x) = -0.87x^3 + 0.35x^2 + 81.62x + 7684.94$$

models the number of thefts, $f(x)$, in thousands, in the United States x years after 1987. Will this function be useful in modeling the number of thefts over an extended period of time? Explain your answer.

74. A herd of 100 elk is introduced to a small island. The number of elk, $f(x)$, after x years is modeled by the polynomial function

$$f(x) = -x^4 + 21x^2 + 100.$$

Use the Leading Coefficient Test to determine the graph's end behavior to the right. What does this mean about what will eventually happen to the elk population?

Writing in Mathematics

75. What is a polynomial?

76. Explain how to determine the degree of each term of a polynomial.

77. Explain how to determine the degree of a polynomial.

78. Explain how to determine the leading coefficient of a polynomial.

79. What is a polynomial function?

80. What do we mean when we describe the graph of a polynomial function as smooth and continuous?

81. What is meant by the end behavior of a polynomial function?

82. Explain how to use the Leading Coefficient Test to determine the end behavior of a polynomial function.

83. Why is a polynomial function of degree 3 with a negative leading coefficient not appropriate for modeling nonnegative real-world phenomena over a long period of time?

84. Explain how to add polynomials.

85. Explain how to subtract polynomials.

86. In a favorable habitat and without natural predators, a population of reindeer is introduced to an island preserve. The reindeer population t years after their introduction is modeled by the polynomial function

$$f(t) = -0.125t^5 + 3.125t^4 + 4000.$$

Discuss the growth and decline of the reindeer population. Describe the factors that might contribute to this population model.

Technology Exercises

Write a polynomial function that imitates the end behavior of each graph in Exercises 87–90. The dashed portions of the graphs indicate that you should focus only on imitating the left and right end behavior of the graph. You can be flexible about what occurs between the left and right ends. Then use your graphing utility to graph the polynomial function and verify that you imitated the end behavior shown in the given graph.

87.

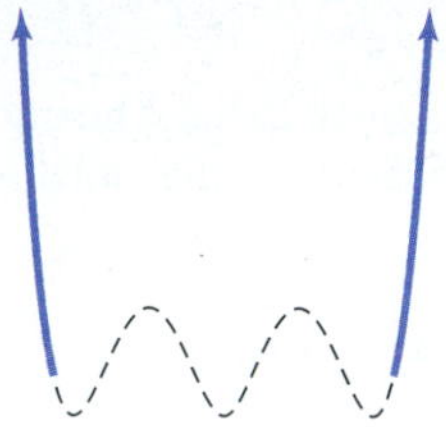

88.

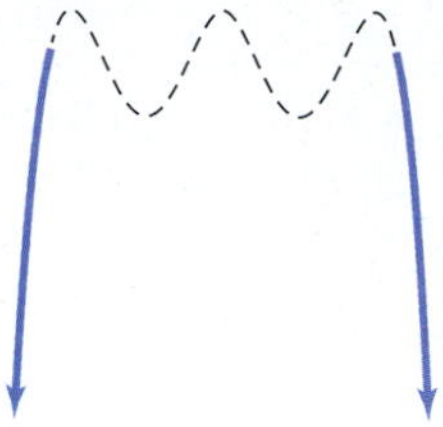

89.

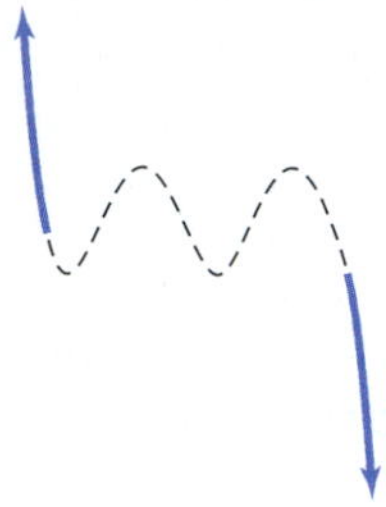

90.

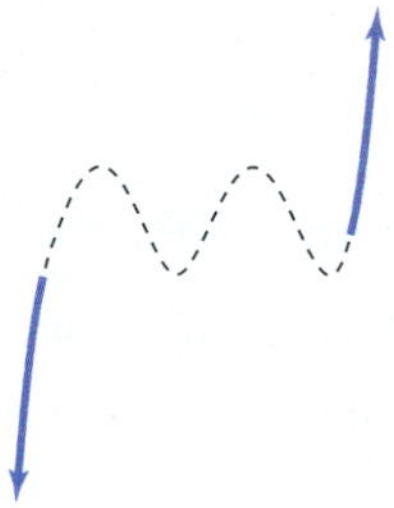

In Exercises 91–94, use a graphing utility with a viewing rectangle large enough to show end behavior to graph each polynomial function.

91. $f(x) = x^3 + 13x^2 + 10x - 4$

92. $f(x) = -2x^3 + 6x^2 + 3x - 1$

93. $f(x) = -x^4 + 8x^3 + 4x^2 + 2$

94. $f(x) = -x^5 + 5x^4 - 6x^3 + 2x + 20$

In Exercises 95–96, use a graphing utility to graph f and g in the same viewing rectangle. Then use the ZOOM OUT *feature to show that f and g have identical end behavior.*

95. $f(x) = x^3 - 6x + 1, \quad g(x) = x^3$

96. $f(x) = -x^4 + 2x^3 - 6x, \quad g(x) = -x^4$

Critical Thinking Exercises

Make Sense? *In Exercises 97–100, determine whether each statement "makes sense" or "does not make sense" and explain your reasoning.*

97. Many English words have prefixes with meanings similar to those used to describe polynomials, such as *monologue*, *binocular*, and *tricuspid*.

98. I can determine a polynomial's leading coefficient by inspecting the coefficient of the first term.

99. When I'm trying to determine end behavior, it's the coefficient of the first term of a polynomial function written in standard form that I should inspect.

100. When I rearrange the terms of a polynomial, it's important that I move the sign in front of a term with that term.

In Exercises 101–104, determine whether each statement is true or false. If the statement is false, make the necessary change(s) to produce a true statement.

101. $4x^3 + 7x^2 - 5x + \frac{2}{x}$ is a polynomial containing four terms.

102. If two polynomials of degree 2 are added, the sum must be a polynomial of degree 2.

103. $(x^2 - 7x) - (x^2 - 4x) = -11x$ for all values of x.

104. All terms of a polynomial are monomials.

In Exercises 105–106, perform the indicated operations. Assume that exponents represent whole numbers.

105. $(x^{2n} - 3x^n + 5) + (4x^{2n} - 3x^n - 4) - (2x^{2n} - 5x^n - 3)$

106. $(y^{3n} - 7y^{2n} + 3) - (-3y^{3n} - 2y^{2n} - 1) + (6y^{3n} - y^{2n} + 1)$

107. From what polynomial must $4x^2 + 2x - 3$ be subtracted to obtain $5x^2 - 5x + 8$?

Review Exercises

108. Solve: $9(x - 1) = 1 + 3(x - 2)$. (Section 1.4, Example 3)

109. Graph: $2x - 3y < -6$. (Section 4.4, Example 1)

110. Write the point-slope form and slope-intercept form of equations of a line passing through the point $(-2, 5)$ and parallel to the line whose equation is $3x - y = 9$. (Section 2.5, Example 4)

Preview Exercises

Exercises 111–113 will help you prepare for the material covered in the next section.

111. Multiply: $(2x^3y^2)(5x^4y^7)$.

112. Use the distributive property to multiply: $2x^4(8x^4 + 3x)$.

113. Simplify and express the polynomial in standard form:

$$3x(x^2 + 4x + 5) + 7(x^2 + 4x + 5).$$

SECTION 5.2

Multiplication of Polynomials

Old Dog... New Chicks

Objectives

1. Multiply monomials.
2. Multiply a monomial and a polynomial.
3. Multiply polynomials when neither is a monomial.
4. Use FOIL in polynomial multiplication.
5. Square binomials.
6. Multiply the sum and difference of two terms.
7. Find the product of functions.
8. Use polynomial multiplication to evaluate functions.

Can that be Axl, your author's yellow lab, sharing a special moment with a baby chick? And if it is (it is), what possible relevance can this have to multiplying polynomials? An answer is promised before you reach the exercise set. For now, let's begin by reviewing how to multiply monomials, a skill that you will apply in every polynomial multiplication problem.

Multiplying Monomials

1 Multiply monomials.

To multiply monomials, begin by multiplying the coefficients. Then multiply the variables. Use the product rule for exponents to multiply the variables: Retain the variable and add the exponents.

EXAMPLE 1 Multiplying Monomials

Multiply:

a. $(5x^3y^4)(-6x^7y^8)$ **b.** $(4x^3y^2z^5)(2x^5y^2z^4)$.

Solution

a.
$$(5x^3y^4)(-6x^7y^8) = 5(-6)x^3 \cdot x^7 \cdot y^4 \cdot y^8$$
Rearrange factors. This step is usually done mentally.
$$= -30x^{3+7}y^{4+8}$$
Multiply coefficients and add exponents.
$$= -30x^{10}y^{12}$$
Simplify.

b.
$$(4x^3y^2z^5)(2x^5y^2z^4) = 4 \cdot 2 \cdot x^3 \cdot x^5 \cdot y^2 \cdot y^2 \cdot z^5 \cdot z^4$$
Rearrange factors.
$$= 8x^{3+5}y^{2+2}z^{5+4}$$
Multiply coefficients and add exponents.
$$= 8x^8y^4z^9$$
Simplify.

✓ **CHECK POINT 1** Multiply:

a. $(6x^5y^7)(-3x^2y^4)$ **b.** $(10x^4y^3z^6)(3x^6y^3z^2)$.

2 Multiply a monomial and a polynomial.

Multiplying a Monomial and a Polynomial That Is Not a Monomial

We use the distributive property to multiply a monomial and a polynomial that is not a monomial. For example,

$$3x^2(2x^3 + 5x) = 3x^2 \cdot 2x^3 + 3x^2 \cdot 5x = 3 \cdot 2x^{2+3} + 3 \cdot 5x^{2+1} = 6x^5 + 15x^3.$$

Monomial   Binomial   Multiply coefficients and add exponents.

To multiply a monomial and a polynomial, multiply each term of the polynomial by the monomial. Once the monomial factor is distributed, we multiply the resulting monomials using the procedure shown in Example 1.

EXAMPLE 2 Multiplying a Monomial and a Trinomial

Multiply:

a. $4x^3(6x^5 - 2x^2 + 3)$ **b.** $5x^3y^4(2x^7y - 6x^4y^3 - 3)$.

Solution

a. $4x^3(6x^5 - 2x^2 + 3) = 4x^3 \cdot 6x^5 - 4x^3 \cdot 2x^2 + 4x^3 \cdot 3$ Use the distributive property.

$= 24x^8 - 8x^5 + 12x^3$ Multiply coefficients and add exponents.

b. $5x^3y^4(2x^7y - 6x^4y^3 - 3)$

$= 5x^3y^4 \cdot 2x^7y - 5x^3y^4 \cdot 6x^4y^3 - 5x^3y^4 \cdot 3$ Use the distributive property.

$= 10x^{10}y^5 - 30x^7y^7 - 15x^3y^4$ Multiply coefficients and add exponents.

✓ CHECK POINT 2 Multiply:

a. $6x^4(2x^5 - 3x^2 + 4)$

b. $2x^4y^3(5xy^6 - 4x^3y^4 - 5)$.

---

3 Multiply polynomials when neither is a monomial.

Multiplying Polynomials when Neither Is a Monomial

How do we multiply two polynomials if neither is a monomial? For example, consider

$$(3x + 7)(x^2 + 4x + 5).$$

Binomial   Trinomial

One way to perform this multiplication is to distribute $3x$ throughout the trinomial

$$3x(x^2 + 4x + 5)$$

and 7 throughout the trinomial

$$7(x^2 + 4x + 5).$$

Then combine the like terms that result.

Multiplying Polynomials when Neither Is a Monomial

Multiply each term of one polynomial by each term of the other polynomial. Then combine like terms.

EXAMPLE 3 Multiplying a Binomial and a Trinomial

Multiply: $(3x + 7)(x^2 + 4x + 5)$.

Solution

$(3x + 7)(x^2 + 4x + 5)$

$= 3x(x^2 + 4x + 5) + 7(x^2 + 4x + 5)$ — Multiply the trinomial by each term of the binomial.

$= 3x \cdot x^2 + 3x \cdot 4x + 3x \cdot 5 + 7x^2 + 7 \cdot 4x + 7 \cdot 5$ — Use the distributive property.

$= 3x^3 + 12x^2 + 15x + 7x^2 + 28x + 35$ — Multiply monomials: Multiply coefficients and add exponents.

$= 3x^3 + 19x^2 + 43x + 35$ — Combine like terms: $12x^2 + 7x^2 = 19x^2$ and $15x + 28x = 43x$.

CHECK POINT 3 Multiply: $(3x + 2)(2x^2 - 2x + 1)$.

Another method for solving Example 3 is to use a vertical format similar to that used for multiplying whole numbers.

$$\begin{array}{rrrr} & x^2 & +\ 4x & +\ 5 \\ & & 3x & +\ 7 \\ \hline & 7x^2 & +\ 28x & +\ 35 \\ 3x^3 & +\ 12x^2 & +\ 15x & \\ \hline 3x^3 & +\ 19x^2 & +\ 43x & +\ 35 \end{array}$$

Write like terms in the same column.

$7(x^2 + 4x + 5)$

$3x(x^2 + 4x + 5)$

Combine like terms.

EXAMPLE 4 Multiplying a Binomial and a Trinomial

Multiply: $(2x^2y + 3y)(5x^4y - 4x^2y + y)$.

Solution

$(2x^2y + 3y)(5x^4y - 4x^2y + y)$

$= 2x^2y(5x^4y - 4x^2y + y) + 3y(5x^4y - 4x^2y + y)$ — Multiply the trinomial by each term of the binomial.

$= 2x^2y \cdot 5x^4y - 2x^2y \cdot 4x^2y + 2x^2y \cdot y + 3y \cdot 5x^4y - 3y \cdot 4x^2y + 3y \cdot y$ — Use the distributive property.

$= 10x^6y^2 - 8x^4y^2 + 2x^2y^2 + 15x^4y^2 - 12x^2y^2 + 3y^2$ — Multiply coefficients and add exponents.

$= 10x^6y^2 + 7x^4y^2 - 10x^2y^2 + 3y^2$ — Combine like terms: $-8x^4y^2 + 15x^4y^2 = 7x^4y^2$ and $2x^2y^2 - 12x^2y^2 = -10x^2y^2$.

CHECK POINT 4 Multiply: $(4xy^2 + 2y)(3xy^4 - 2xy^2 + y)$.

4 Use FOIL in polynomial multiplication.

The Product of Two Binomials: FOIL

Frequently we need to find the product of two binomials. One way to perform this multiplication is to distribute each term in the first binomial throughout the second binomial. For example, we can find the product of the binomials $7x + 2$ and $4x + 5$ as follows:

$$\begin{aligned}(7x + 2)(4x + 5) &= 7x(4x + 5) + 2(4x + 5)\\ &= 7x(4x) + 7x(5) + 2(4x) + 2(5)\\ &= 28x^2 + 35x + 8x + 10.\end{aligned}$$

Distribute $7x$ over $4x + 5$.

Distribute 2 over $4x + 5$.

We'll combine these like terms later. For now, our interest is in how to obtain *each* of these four terms.

We can also find the product of $7x + 2$ and $4x + 5$ using a method called FOIL, which is based on our work shown above. Any two binomials can be quickly multiplied using the FOIL method, in which **F** represents the product of the **first** terms in each binomial, **O** represents the product of the **outside** terms, **I** represents the product of the two **inside** terms, and **L** represents the product of the **last**, or second, terms in each binomial. For example, we can use the FOIL method to find the product of the binomials $7x + 2$ and $4x + 5$ as follows:

last, first, inside, outside

$$\begin{aligned}(7x + 2)(4x + 5) &= \overset{F}{28x^2} + \overset{O}{35x} + \overset{I}{8x} + \overset{L}{10}\\ &= 28x^2 + 43x + 10\end{aligned}$$

Product of First terms | Product of Outside terms | Product of Inside terms | Product of Last terms

Combine like terms.

In general, here's how to use the FOIL method to find the product of $ax + b$ and $cx + d$:

Using the FOIL Method to Multiply Binomials

last, first, inside, outside

$$(ax + b)(cx + d) = \overset{F}{ax \cdot cx} + \overset{O}{ax \cdot d} + \overset{I}{b \cdot cx} + \overset{L}{b \cdot d}$$

Product of First terms | Product of Outside terms | Product of Inside terms | Product of Last terms

EXAMPLE 5 Using the FOIL Method

Multiply:

a. $(x + 3)(x + 2)$ **b.** $(3x + 5y)(x - 2y)$ **c.** $(5x^3 - 6)(4x^3 - x)$.

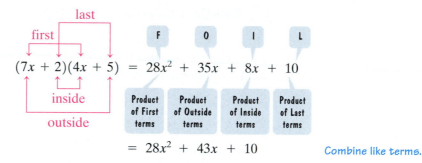

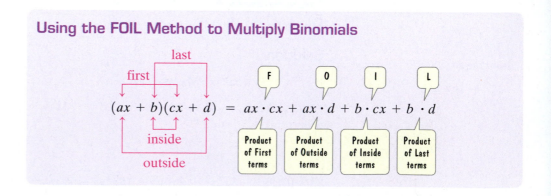

Solution

a. (first, outside, inside, last — F O I L)

$$(x+3)(x+2) = x \cdot x + x \cdot 2 + 3 \cdot x + 3 \cdot 2$$
$$= x^2 + 2x + 3x + 6$$
$$= x^2 + 5x + 6 \quad \text{Combine like terms.}$$

b. (first, outside, inside, last — F O I L)

$$(3x+5y)(x-2y) = 3x \cdot x + 3x(-2y) + 5y \cdot x + 5y(-2y)$$
$$= 3x^2 - 6xy + 5xy - 10y^2$$
$$= 3x^2 - xy - 10y^2 \quad \text{Combine like terms.}$$

c. (first, outside, inside, last — F O I L)

$$(5x^3-6)(4x^3-x) = 5x^3 \cdot 4x^3 + 5x^3(-x) + (-6)(4x^3) + (-6)(-x)$$
$$= 20x^6 - 5x^4 - 24x^3 + 6x \quad \text{There are no like terms to combine.}$$

CHECK POINT 5 Multiply:

a. $(x+5)(x+3)$ **b.** $(7x+4y)(2x-y)$ **c.** $(4x^3-5)(x^3-3x)$.

5 Square binomials.

The Square of a Binomial

Let us find $(A+B)^2$, the square of a binomial sum. To do so, we begin with the FOIL method and look for a general rule.

(F O I L)

$$(A+B)^2 = (A+B)(A+B) = A \cdot A + A \cdot B + A \cdot B + B \cdot B$$
$$= A^2 + 2AB + B^2$$

This result implies the following rule, which is often called a **special-product formula**:

The Square of a Binomial Sum

$$(A+B)^2 = A^2 + 2AB + B^2$$

The square of a binomial sum is first term squared plus 2 times the product of the terms plus last term squared.

Study Tip

Caution! The square of a sum is *not* the sum of the squares.

$(A+B)^2 \neq A^2 + B^2$

The middle term $2AB$ is missing.

$(x+5)^2 \neq x^2 + 25$

Incorrect

Show that $(x+5)^2$ and x^2+25 are not equal by substituting 3 for x in each expression and simplifying.

EXAMPLE 6 Finding the Square of a Binomial Sum

Multiply:

a. $(x + 5)^2$ b. $(3x + 2y)^2$.

Solution Use the special-product formula shown.

$(A + B)^2 = \quad A^2 \quad + \quad 2AB \quad + \quad B^2$

	(First Term)^2	+	2 · Product of the Terms	+	(Last Term)^2	= Product
a. $(x + 5)^2 =$	x^2	+	$2 \cdot x \cdot 5$	+	5^2	$= x^2 + 10x + 25$
b. $(3x + 2y)^2 =$	$(3x)^2$	+	$2 \cdot 3x \cdot 2y$	+	$(2y)^2$	$= 9x^2 + 12xy + 4y^2$

✓ CHECK POINT 6 Multiply:

a. $(x + 8)^2$ b. $(4x + 5y)^2$.

The formula for the square of a binomial sum can be interpreted geometrically by analyzing the areas in **Figure 5.9**.

Area of the large square: $(A + B)^2$

Sum of the areas of the four smaller rectangles inside the large square: $A^2 + AB + AB + B^2$
$= A^2 + 2AB + B^2$

Conclusion:
$(A + B)^2 = A^2 + 2AB + B^2$

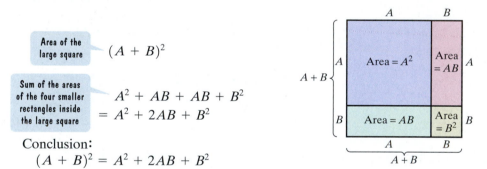

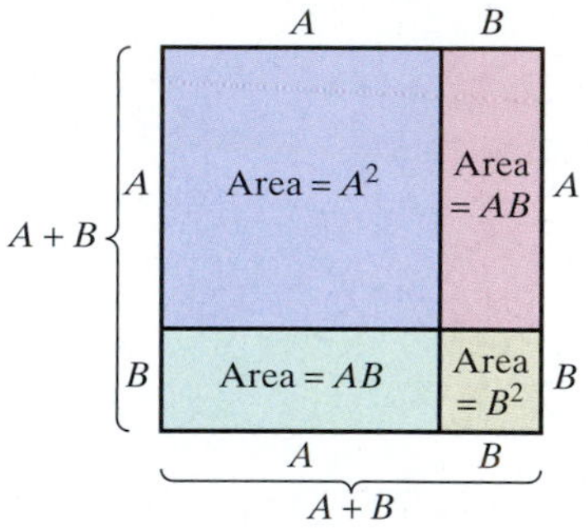

FIGURE 5.9

A similar pattern occurs for $(A - B)^2$, the square of a binomial difference. Using the FOIL method on $(A - B)^2$, we obtain the following rule:

The Square of a Binomial Difference

$$(A - B)^2 = A^2 - 2AB + B^2$$

The square of a binomial difference | is | first term squared | minus | 2 times the product of the terms | plus | last term squared.

EXAMPLE 7 Finding the Square of a Binomial Difference

Multiply:

a. $(x - 8)^2$ b. $\left(\frac{1}{2}x - 4y^3\right)^2$.

Solution Use the special-product formula shown.

$$(A - B)^2 = A^2 - 2AB + B^2$$

	(First Term)^2	−	2 · Product of the Terms	+	(Last Term)^2	= Product
a. $(x - 8)^2 =$	x^2	−	$2 \cdot x \cdot 8$	+	8^2	$= x^2 - 16x + 64$
b. $\left(\frac{1}{2}x - 4y^3\right)^2 =$	$\left(\frac{1}{2}x\right)^2$	−	$2 \cdot \frac{1}{2}x \cdot 4y^3$	+	$(4y^3)^2$	$= \frac{1}{4}x^2 - 4xy^3 + 16y^6$

✓ CHECK POINT 7 Multiply:

a. $(x - 5)^2$ **b.** $(2x - 6y^4)^2$.

6 Multiply the sum and difference of two terms.

Multiplying the Sum and Difference of Two Terms

We can use the FOIL method to multiply $A + B$ and $A - B$ as follows:

F O I L

$$(A + B)(A - B) = A^2 - AB + AB - B^2 = A^2 - B^2.$$

Notice that the outside and inside products have a sum of 0 and the terms cancel. The FOIL multiplication provides us with a quick rule for multiplying the sum and difference of two terms, which is another example of a special-product formula.

The Product of the Sum and Difference of Two Terms

$$(A + B)(A - B) = A^2 - B^2$$

The product of the sum and the difference of the same two terms | is | the square of the first term minus the square of the second term.

EXAMPLE 8 **Finding the Product of the Sum and Difference of Two Terms**

Multiply:

a. $(x + 8)(x - 8)$ **b.** $(9x + 5y)(9x - 5y)$ **c.** $(6a^2b - 3b)(6a^2b + 3b)$.

Solution Use the special-product formula shown.

$$(A + B)(A - B) = A^2 - B^2$$

First term squared − Second term squared = Product

a. $(x + 8)(x - 8) = x^2 - 8^2 = x^2 - 64$

b. $(9x + 5y)(9x - 5y) = (9x)^2 - (5y)^2 = 81x^2 - 25y^2$

c. $(6a^2b - 3b)(6a^2b + 3b) = (6a^2b)^2 - (3b)^2 = 36a^4b^2 - 9b^2$

✓ CHECK POINT 8 Multiply:

a. $(x + 3)(x - 3)$
b. $(5x + 7y)(5x - 7y)$
c. $(5ab^2 - 4a)(5ab^2 + 4a)$.

Special products can sometimes be used to find the products of certain trinomials, as illustrated in Example 9.

EXAMPLE 9 Using the Special Products

Multiply:

a. $(7x + 5 + 4y)(7x + 5 - 4y)$ b. $(3x + y + 1)^2$.

Solution

a. By grouping the first two terms within each set of parentheses, we can find the product using the form for the sum and difference of two terms.

$$(A + B) \cdot (A - B) = A^2 - B^2$$

$$[(7x + 5) + 4y] \cdot [(7x + 5) - 4y] = (7x + 5)^2 - (4y)^2$$
$$= (7x)^2 + 2 \cdot 7x \cdot 5 + 5^2 - (4y)^2$$
$$= 49x^2 + 70x + 25 - 16y^2$$

b. We can group the terms so that the formula for the square of a binomial can be applied.

Discover for Yourself

Group $(3x + y + 1)^2$ as $[3x + (y + 1)]^2$. Verify that you get the same product as we obtained in Example 9(b).

$$(A + B)^2 = A^2 + 2 \cdot A \cdot B + B^2$$

$$[(3x + y) + 1]^2 = (3x + y)^2 + 2 \cdot (3x + y) \cdot 1 + 1^2$$
$$= 9x^2 + 6xy + y^2 + 6x + 2y + 1$$

✓ CHECK POINT 9 Multiply:

a. $(3x + 2 + 5y)(3x + 2 - 5y)$
b. $(2x + y + 3)^2$.

7 Find the product of functions.

Multiplication of Polynomial Functions

In Chapter 2, we developed an algebra of functions, defining the product of functions f and g as follows:

$$(fg)(x) = f(x) \cdot g(x).$$

Now that we know how to multiply polynomials, we can find the product of functions.

EXAMPLE 10 Using the Algebra of Functions

Let $f(x) = x - 5$ and $g(x) = x - 2$. Find:

a. $(fg)(x)$ b. $(fg)(1)$.

Solution

a. $(fg)(x) = f(x) \cdot g(x)$ — This is the definition of the product function, fg.

$= (x-5)(x-2)$ — Substitute the given functions.

F O I L

$= x^2 - 2x - 5x + 10$ — Multiply by the FOIL method.

$= x^2 - 7x + 10$ — Combine like terms.

Thus,

$$(fg)(x) = x^2 - 7x + 10.$$

b. We use the product function to find $(fg)(1)$—that is, the value of the function fg at 1. Replace x with 1.

$$(fg)(1) = 1^2 - 7 \cdot 1 + 10 = 4$$

Example 10 involved linear and quadratic functions.

$$f(x) = x - 5 \qquad g(x) = x - 2 \qquad (fg)(x) = x^2 - 7x + 10$$

These are linear functions of the form $f(x) = mx + b$.

This is a quadratic function of the form $f(x) = ax^2 + bx + c$.

All three of these functions are polynomial functions. A linear function is a first-degree polynomial function. A quadratic function is a second-degree polynomial function.

✓ CHECK POINT 10 Let $f(x) = x - 3$ and $g(x) = x - 7$. Find:

a. $(fg)(x)$

b. $(fg)(2)$.

---

8 Use polynomial multiplication to evaluate functions.

If you are given a function, f, calculus can reveal how it is changing at any instant in time. The algebraic expression

$$\frac{f(a+h) - f(a)}{h}$$

plays an important role in this process. Our work with polynomial multiplication can be used to evaluate the numerator of this expression.

EXAMPLE 11 Using Polynomial Multiplication to Evaluate Functions

Given $f(x) = x^2 - 7x + 3$, find and simplify each of the following:

a. $f(a+4)$ b. $f(a+h) - f(a)$.

Solution

a. We find $f(a+4)$, read "f at a plus 4," by replacing x with $a + 4$ each time that x appears in the polynomial.

$$f(x) = x^2 - 7x + 3$$

Replace x with $a + 4$. Replace x with $a + 4$. Replace x with $a + 4$. Copy the 3. There is no x in this term.

$$f(a+4) = (a+4)^2 - 7(a+4) + 3$$

$= a^2 + 8a + 16 - 7a - 28 + 3$ — Multiply as indicated.

$= a^2 + a - 9$ — Combine like terms: $8a - 7a = a$ and $16 - 28 + 3 = -9$.

b. To find $f(a+h) - f(a)$, we first replace each occurrence of x in $f(x) = x^2 - 7x + 3$ with $a + h$ and then replace each occurrence of x with a. Then we perform the resulting operations and simplify.

This is $f(a+h)$. Use $f(x) = x^2 - 7x + 3$ and replace x with $a + h$.

This is $f(a)$. Use $f(x) = x^2 - 7x + 3$ and replace x with a.

$$f(a+h) - f(a) = \boxed{(a+h)^2 - 7(a+h) + 3} - (a^2 - 7a + 3)$$

$$= (a^2 + 2ah + h^2 - 7a - 7h + 3) - (a^2 - 7a + 3)$$ Perform the multiplications required by $f(a + h)$.

$$= (a^2 + 2ah + h^2 - 7a - 7h + 3) + (-a^2 + 7a - 3)$$ Change the sign of each term of the second polynomial and add the two polynomials. Like terms are shown in the same color.

$$= a^2 - a^2 - 7a + 7a + 3 - 3 + 2ah + h^2 - 7h$$ Group like terms.

$$= 2ah + h^2 - 7h$$ Simplify. Observe that $a^2 - a^2 = 0$, $-7a + 7a = 0$, and $3 - 3 = 0$.

✓ CHECK POINT 11 Given $f(x) = x^2 - 5x + 4$, find and simplify each of the following:

a. $f(a + 3)$

b. $f(a + h) - f(a)$.

Blitzer Bonus

Labrador Retrievers and Polynomial Multiplication

The color of a Labrador retriever is determined by its pair of genes. A single gene is inherited at random from each parent. The black-fur gene, B, is dominant. The yellow-fur gene, Y, is recessive. This means that labs with at least one black-fur gene (BB or BY) have black coats. Only labs with two yellow-fur genes (YY) have yellow coats.

Axl, your author's yellow lab, inherited his genetic makeup from two black BY parents.

Second BY parent, a black lab with a recessive yellow-fur gene

First BY parent, a black lab with a recessive yellow-fur gene

	B	Y
B	BB	BY
Y	BY	YY

The table shows the four possible combinations of color genes that BY parents can pass to their offspring.

Because YY is one of four possible outcomes, the probability that a yellow lab like Axl will be the offspring of these black parents is $\frac{1}{4}$.

The probabilities suggested by the table can be modeled by the expression $\left(\frac{1}{2}B + \frac{1}{2}Y\right)^2$.

$$\left(\frac{1}{2}B + \frac{1}{2}Y\right)^2 = \left(\frac{1}{2}B\right)^2 + 2\left(\frac{1}{2}B\right)\left(\frac{1}{2}Y\right) + \left(\frac{1}{2}Y\right)^2$$

$$= \frac{1}{4}BB + \frac{1}{2}BY + \frac{1}{4}YY$$

The probability of a black lab with two dominant black genes is $\frac{1}{4}$.

The probability of a black lab with a recessive yellow gene is $\frac{1}{2}$.

The probability of a yellow lab with two recessive yellow genes is $\frac{1}{4}$.

5.2 EXERCISE SET

MyMathLab MathXL PRACTICE WATCH DOWNLOAD READ REVIEW

Practice Exercises

Throughout the practice exercises, assume that any variable exponents represent whole numbers.

In Exercises 1–8, multiply the monomials.

1. $(3x^2)(5x^4)$
2. $(4x^2)(6x^4)$
3. $(3x^2y^4)(5xy^7)$
4. $(6x^4y^2)(3x^7y)$
5. $(-3xy^2z^5)(2xy^7z^4)$
6. $(11x^2yz^4)(-3xy^5z^6)$
7. $(-8x^{2n}y^{n-5})\left(-\frac{1}{4}x^ny^3\right)$
8. $(-9x^{3n}y^{n-3})\left(-\frac{1}{3}x^ny^2\right)$

In Exercises 9–22, multiply the monomial and the polynomial.

9. $4x^2(3x + 2)$
10. $5x^2(6x + 7)$
11. $2y(y^2 - 5y)$
12. $3y(y^2 - 4y)$
13. $5x^3(2x^5 - 4x^2 + 9)$
14. $6x^3(3x^5 - 5x^2 + 7)$
15. $4xy(7x + 3y)$
16. $5xy(8x + 3y)$
17. $3ab^2(6a^2b^3 + 5ab)$
18. $5ab^2(10a^2b^3 + 7ab)$
19. $-4x^2y(3x^4y^2 - 7xy^3 + 6)$
20. $-3x^2y(10x^2y^4 - 2xy^3 + 7)$
21. $-4x^n\left(3x^{2n} - 5x^n + \frac{1}{2}x\right)$
22. $-10x^n\left(4x^{2n} - 3x^n + \frac{1}{5}x\right)$

In Exercises 23–34, find each product using either a horizontal or a vertical format.

23. $(x - 3)(x^2 + 2x + 5)$
24. $(x + 4)(x^2 - 5x + 8)$
25. $(x - 1)(x^2 + x + 1)$
26. $(x - 2)(x^2 + 2x + 4)$
27. $(a - b)(a^2 + ab + b^2)$
28. $(a + b)(a^2 - ab + b^2)$
29. $(x^2 + 2x - 1)(x^2 + 3x - 4)$
30. $(x^2 - 2x + 3)(x^2 + x + 1)$
31. $(x - y)(x^2 - 3xy + y^2)$
32. $(x - y)(x^2 - 4xy + y^2)$
33. $(xy + 2)(x^2y^2 - 2xy + 4)$
34. $(xy + 3)(x^2y^2 - 2xy + 5)$

In Exercises 35–54, use the FOIL method to multiply the binomials.

35. $(x + 4)(x + 7)$
36. $(x + 5)(x + 8)$
37. $(y + 5)(y - 6)$
38. $(y + 5)(y - 8)$
39. $(5x + 3)(2x + 1)$
40. $(4x + 3)(5x + 1)$
41. $(3y - 4)(2y - 1)$
42. $(5y - 2)(3y - 1)$
43. $(3x - 2)(5x - 4)$
44. $(2x - 3)(4x - 5)$
45. $(x - 3y)(2x + 7y)$
46. $(3x - y)(2x + 5y)$
47. $(7xy + 1)(2xy - 3)$
48. $(3xy - 1)(5xy + 2)$
49. $(x - 4)(x^2 - 5)$
50. $(x - 5)(x^2 - 3)$
51. $(8x^3 + 3)(x^2 - 5)$
52. $(7x^3 + 5)(x^2 - 2)$
53. $(3x^n - y^n)(x^n + 2y^n)$
54. $(5x^n - y^n)(x^n + 4y^n)$

In Exercises 55–68, multiply using one of the rules for the square of a binomial.

55. $(x + 3)^2$
56. $(x + 4)^2$
57. $(y - 5)^2$
58. $(y - 6)^2$
59. $(2x + y)^2$
60. $(4x + y)^2$
61. $(5x - 3y)^2$
62. $(3x - 4y)^2$
63. $(2x^2 + 3y)^2$
64. $(4x^2 + 5y)^2$
65. $(4xy^2 - xy)^2$
66. $(5xy^2 - xy)^2$
67. $(a^n + 4b^n)^2$
68. $(3a^n - b^n)^2$

In Exercises 69–82, multiply using the rule for the product of the sum and difference of two terms.

69. $(x + 4)(x - 4)$
70. $(x + 5)(x - 5)$
71. $(5x + 3)(5x - 3)$
72. $(3x + 2)(3x - 2)$

73. $(4x + 7y)(4x - 7y)$

74. $(8x + 7y)(8x - 7y)$

75. $(y^3 + 2)(y^3 - 2)$

76. $(y^3 + 3)(y^3 - 3)$

77. $(1 - y^5)(1 + y^5)$

78. $(2 - y^5)(2 + y^5)$

79. $(7xy^2 - 10y)(7xy^2 + 10y)$

80. $(3xy^2 - 4y)(3xy^2 + 4y)$

81. $(5a^n - 7)(5a^n + 7)$

82. $(10b^n - 3)(10b^n + 3)$

In Exercises 83–94, find each product.

83. $[(2x + 3) + 4y][(2x + 3) - 4y]$

84. $[(3x + 2) + 5y][(3x + 2) - 5y]$

85. $(x + y + 3)(x + y - 3)$

86. $(x + y + 4)(x + y - 4)$

87. $(5x + 7y - 2)(5x + 7y + 2)$

88. $(7x + 5y - 2)(7x + 5y + 2)$

89. $[5y + (2x + 3)][5y - (2x + 3)]$

90. $[8y + (3x + 2)][8y - (3x + 2)]$

91. $(x + y + 1)^2$

92. $(x + y + 2)^2$

93. $(x + 1)(x - 1)(x^2 + 1)$

94. $(x + 2)(x - 2)(x^2 + 4)$

95. Let $f(x) = x - 2$ and $g(x) = x + 6$. Find each of the following.

- **a.** $(fg)(x)$
- **b.** $(fg)(-1)$
- **c.** $(fg)(0)$

96. Let $f(x) = x - 4$ and $g(x) = x + 10$. Find each of the following.

- **a.** $(fg)(x)$
- **b.** $(fg)(-1)$
- **c.** $(fg)(0)$

97. Let $f(x) = x - 3$ and $g(x) = x^2 + 3x + 9$. Find each of the following.

- **a.** $(fg)(x)$
- **b.** $(fg)(-2)$
- **c.** $(fg)(0)$

98. Let $f(x) = x + 3$ and $g(x) = x^2 - 3x + 9$. Find each of the following.

- **a.** $(fg)(x)$
- **b.** $(fg)(-2)$
- **c.** $(fg)(0)$

In Exercises 99–102, find each of the following and simplify:

a. $f(a + 2)$ **b.** $f(a + h) - f(a)$.

99. $f(x) = x^2 - 3x + 7$

100. $f(x) = x^2 - 4x + 9$

101. $f(x) = 3x^2 + 2x - 1$

102. $f(x) = 4x^2 + 5x - 1$

Practice PLUS

In Exercises 103–112, perform the indicated operation or operations.

103. $(3x + 4y)^2 - (3x - 4y)^2$

104. $(5x + 2y)^2 - (5x - 2y)^2$

105. $(5x - 7)(3x - 2) - (4x - 5)(6x - 1)$

106. $(3x + 5)(2x - 9) - (7x - 2)(x - 1)$

107. $(2x + 5)(2x - 5)(4x^2 + 25)$

108. $(3x + 4)(3x - 4)(9x^2 + 16)$

109. $(x - 1)^3$

110. $(x - 2)^3$

111. $\dfrac{(2x - 7)^5}{(2x - 7)^3}$

112. $\dfrac{(5x - 3)^6}{(5x - 3)^4}$

Application Exercises

In Exercises 113–114, find the area of the large rectangle in two ways:

- **a.** *Find the sum of the areas of the four smaller rectangles.*
- **b.** *Multiply the length and the width of the large rectangle using the FOIL method. Compare this product with your answer to part (a).*

113.

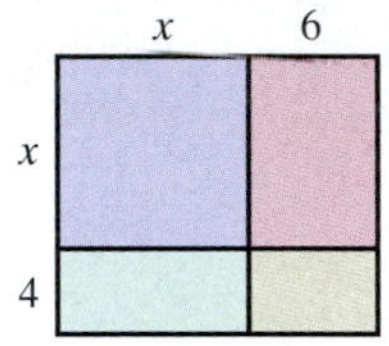

114.

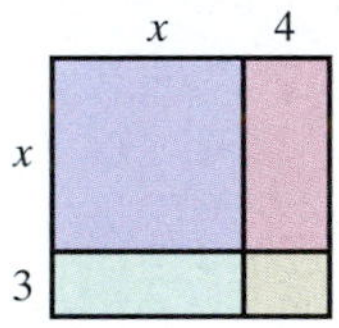

In Exercises 115–116, express each polynomial in standard form—that is, in descending powers of x.

- **a.** *Write a polynomial that represents the area of the large rectangle.*
- **b.** *Write a polynomial that represents the area of the small, unshaded rectangle.*
- **c.** *Write a polynomial that represents the area of the shaded region.*

115.

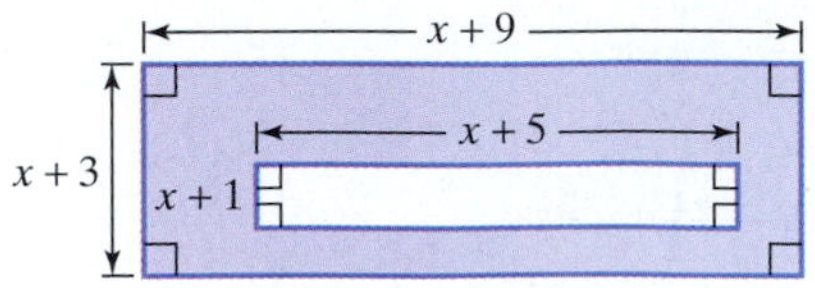

116.

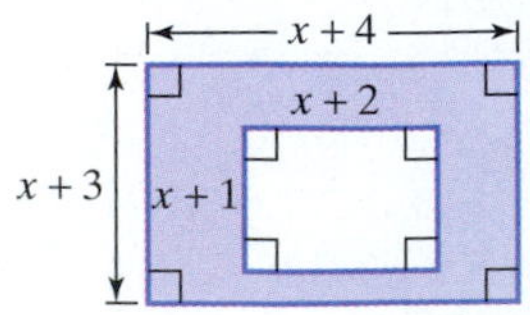

In Exercises 117–118, express each polynomial in standard form.

a. *Write a polynomial that represents the area of the rectangular base of the open box.*

b. *Write a polynomial that represents the volume of the open box.*

117.

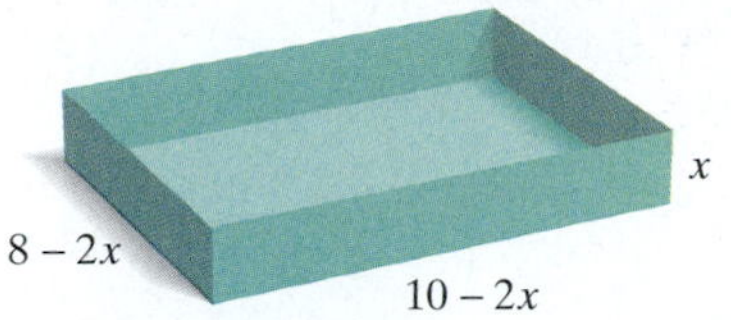

118.

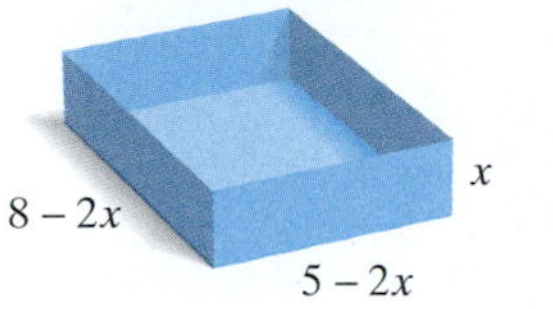

119. A popular model of carry-on luggage has a length that is 10 inches greater than its depth. Airline regulations require that the sum of the length, width, and depth cannot exceed 40 inches. These conditions, with the assumption that this sum *is* 40 inches, can be modeled by a function that gives the volume of the luggage, V, in cubic inches, in terms of its depth, x, in inches.

Volume = depth · length · width: 40 − (depth + length)

$$V(x) = x \cdot (x+10) \cdot [40-(x+x+10)]$$

$$V(x) = x(x+10)(30-2x)$$

a. Perform the multiplications in the formula for $V(x)$ and express the formula in standard form.

b. Use the function's formula from part (a) and the Leading Coefficient Test to determine the end behavior of its graph.

c. Does the end behavior to the right make this function useful in modeling the volume of carry-on luggage as its depth continues to increase?

d. Use the formula from part (a) to find $V(10)$. Describe what this means in practical terms.

e. The graph of the function modeling the volume of carry-on luggage is shown below. Identify your answer from part (d) as a point on the graph.

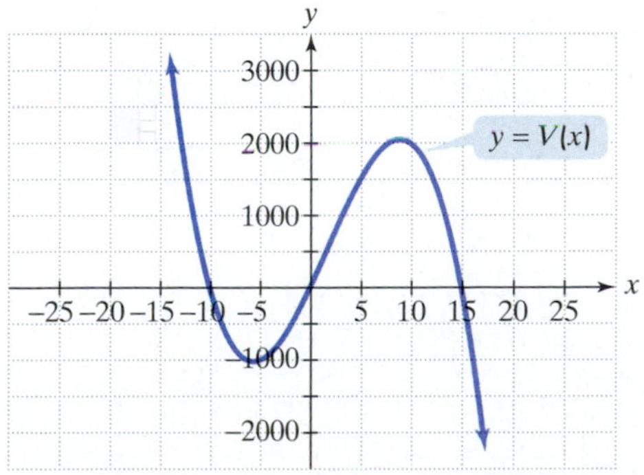

f. Use the graph to describe a realistic domain, x, for the volume function, where x represents the depth of the carry-on luggage. Use set-builder notation or interval notation to express this realistic domain.

120. Before working this exercise, be sure that you have read the Blitzer Bonus on page 326. The table shows the four combinations of color genes that a YY yellow lab and a BY black lab can pass to their offspring.

	B	Y
Y	BY	YY
Y	BY	YY

a. How many combinations result in a yellow lab with two recessive yellow genes? What is the probability of a yellow lab?

b. How many combinations result in a black lab with a recessive yellow gene? What is the probability of a black lab?

c. Find the product of Y and $\frac{1}{2}B + \frac{1}{2}Y$. How does this product model the probabilities that you determined in parts (a) and (b)?

Writing in Mathematics

121. Explain how to multiply monomials. Give an example.

122. Explain how to multiply a monomial and a polynomial that is not a monomial. Give an example.

123. Explain how to multiply a binomial and a trinomial.

124. What is the FOIL method and when is it used? Give an example of the method.

125. Explain how to square a binomial sum. Give an example.

126. Explain how to square a binomial difference. Give an example.

127. Explain how to find the product of the sum and difference of two terms. Give an example with your explanation.

128. How can the graph of function fg be obtained from the graphs of functions f and g?

129. Explain how to find $f(a+h) - f(a)$ for a given function f.

Technology Exercises

In Exercises 130–133, use a graphing utility to graph the functions y_1 and y_2. Select a viewing rectangle that is large enough to show the end behavior of y_2. What can you conclude? Verify your conclusions using polynomial multiplication.

130. $y_1 = (x-2)^2$
$y_2 = x^2 - 4x + 4$

131. $y_1 = (x-4)(x^2-3x+2)$
$y_2 = x^3 - 7x^2 + 14x - 8$

132. $y_1 = (x - 1)(x^2 + x + 1)$
$y_2 = x^3 - 1$

133. $y_1 = (x + 1.5)(x - 1.5)$
$y_2 = x^2 - 2.25$

134. Graph $f(x) = x + 4$, $g(x) = x - 2$, and the product function, fg, in a $[-6, 6, 1]$ by $[-10, 10, 1]$ viewing rectangle. Trace along the curves and show that $(fg)(1) = f(1) \cdot g(1)$.

Critical Thinking Exercises

Make Sense? *In Exercises 135–138, determine whether each statement "makes sense" or "does not make sense" and explain your reasoning.*

135. Knowing the difference between factors and terms is important: In $(3x^2y)^2$, I can distribute the exponent 2 on each factor, but in $(3x^2 + y)^2$, I cannot do the same thing on each term.

136. I used the FOIL method to find the product of $x + 5$ and $x^2 + 2x + 1$.

137. Instead of using the formula for the square of a binomial sum, I prefer to write the binomial sum twice and then apply the FOIL method.

138. Special-product formulas have patterns that make their multiplications quicker than using the FOIL method.

In Exercises 139–142, determine whether each statement is true or false. If the statement is false, make the necessary change(s) to produce a true statement.

139. If f is a polynomial function, then
$$f(a + h) - f(a) = f(a) + f(h) - f(a) = f(h).$$

140. $(x - 5)^2 = x^2 - 5x + 25$

141. $(x + 1)^2 = x^2 + 1$

142. Suppose a square garden has an area represented by $9x^2$ square feet. If one side is made 7 feet longer and the other side is made 2 feet shorter, then the trinomial that represents the area of the larger garden is $9x^2 + 15x - 14$ square feet.

143. Express the area of the plane figure shown as a polynomial in standard form.

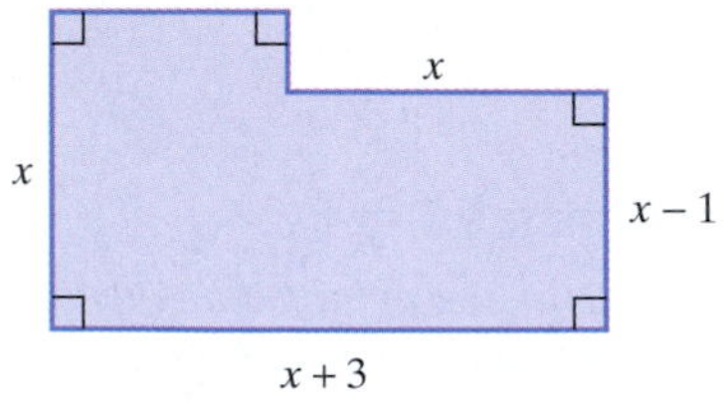

In Exercises 144–145, represent the volume of each figure as a polynomial in standard form.

144.

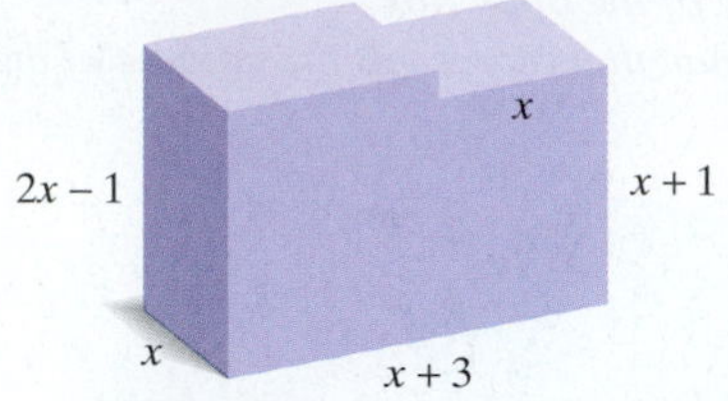

145.

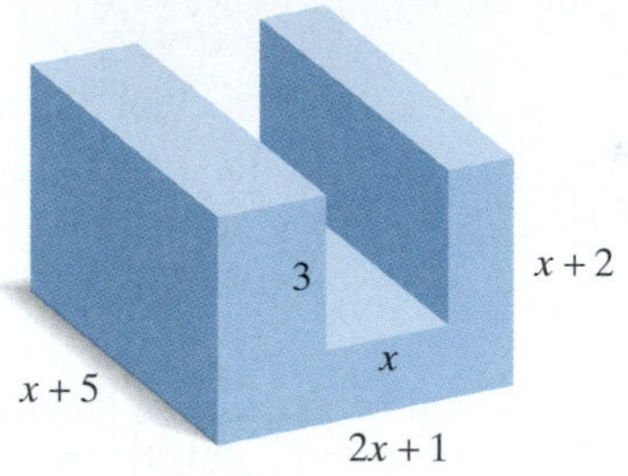

146. Simplify: $(y^n + 2)(y^n - 2) - (y^n - 3)^2$.

147. The product of two consecutive odd integers is 22 less than the square of the greater integer. Find the integers.

Review Exercises

148. Solve: $|3x + 4| \geq 10$. (Section 4.3, Example 5 or Example 6(b))

149. Solve: $2 - 6x \leq 20$. (Section 4.1, Example 2)

150. Write in scientific notation: 8,034,000,000. (Section 1.7, Example 2)

Preview Exercises

Exercises 151–153 will help you prepare for the material covered in the next section.

151. Replace each boxed question mark with a polynomial that results in the given product.

a. $3x^3 \cdot \boxed{?} = 9x^5$

b. $2x^3y^2 \cdot \boxed{?} = 12x^5y^4$

In Exercises 152–153, a polynomial is given in factored form. Use multipication to find the product of the factors.

152. $(x - 5)(x^2 + 3)$

153. $(x + 4)(3x - 2y)$

SECTION 5.3

Greatest Common Factors and Factoring By Grouping

Objectives

1. Factor out the greatest common factor of a polynomial.
2. Factor out a common factor with a negative coefficient.
3. Factor by grouping.

The inability to understand numbers and their meanings is called innumeracy *by mathematics professor John Allen Paulos. Paulos has written a book about mathematical illiteracy. Entitled* Innumeracy, *the book seeks to explain why so many people are numerically inept and to show how the problem can be corrected.*

Jasper Johns *0 Through 9*, 1961. Oil on canvas, 137 × 105 cm. The Saatchi Collection, Courtesy of the Leo Castelli Gallery. © Jasper Johns/Licensed by VAGA, New York, NY.

Did you know that one of the most common ways that you are given numerical information is with percents? Unfortunately, many people are innumerate when it comes to this topic. For example, a computer whose price has been reduced by 40% and then another 40% is not selling at 20% of its original price.

To cure this bout of innumeracy (see Exercise 81 in Exercise Set 5.3), we turn to a process that reverses polynomial multiplication. For example, we can multiply polynomials and show that

$$7x(3x + 4) = 21x^2 + 28x.$$

We can also reverse this process and express the resulting polynomial as

$$21x^2 + 28x = 7x(3x + 4).$$

Factoring a polynomial consisting of the sum of monomials means finding an equivalent expression that is a product.

Factoring $21x^2 + 28x$

Sum of monomials | Equivalent expression that is a product

$$21x^2 + 28x = 7x(3x + 4)$$

The factors of $21x^2 + 28x$ are $7x$ and $3x + 4$.

In this chapter, we will be factoring over the set of integers, meaning that the coefficients in the factors are integers. Polynomials that cannot be factored using integer coefficients are called **prime polynomials over the set of integers**.

1 Factor out the greatest common factor of a polynomial.

Factoring Out the Greatest Common Factor

In any factoring problem, the first step is to look for the *greatest common factor*. The **greatest common factor**, abbreviated GCF, is an expression with the greatest coefficient and of the highest degree that divides each term of the polynomial. Can you see that $7x$ is the greatest common factor of $21x^2 + 28x$? 7 is the greatest integer that divides both 21 and 28. Furthermore, x is the greatest power of x that divides x^2 and x.

The variable part of the greatest common factor always contains the *smallest* power of a variable that appears in all terms of the polynomial. For example, consider the polynomial

$$21x^2 + 28x.$$

x^1, or x, is the variable raised to the smallest exponent.

We see that x is the variable part of the greatest common factor, $7x$.

When factoring a monomial from a polynomial, determine the greatest common factor of all terms in the polynomial. Sometimes there may not be a GCF other than 1. When a GCF other than 1 exists, we use the following procedure:

Factoring a Monomial From a Polynomial

1. Determine the greatest common factor of all terms in the polynomial.
2. Express each term as the product of the GCF and its other factor.
3. Use the distributive property to factor out the GCF.

EXAMPLE 1 Factoring Out the Greatest Common Factor

Factor: $21x^2 + 28x$.

Solution The GCF of the two terms of the polynomial is $7x$.

$$21x^2 + 28x$$

$$= 7x(3x) + 7x(4)$$ Express each term as the product of the GCF and its other factor.

$$= 7x(3x + 4)$$ Factor out the GCF.

We can check this factorization by multiplying $7x$ and $3x + 4$, obtaining the original polynomial as the answer.

✓ **CHECK POINT 1** Factor: $20x^2 + 30x$.

EXAMPLE 2 Factoring Out the Greatest Common Factor

Factor:

a. $9x^5 + 15x^3$ b. $16x^2y^3 - 24x^3y^4$ c. $12x^5y^4 - 4x^4y^3 + 2x^3y^2$.

Solution

a. First, determine the greatest common factor.

3 is the greatest integer that divides 9 and 15.

$$9x^5 + 15x^3$$

x^3 is the variable raised to the smallest exponent.

The GCF of the two terms of the polynomial is $3x^3$.

$$9x^5 + 15x^3$$
$$= 3x^3 \cdot 3x^2 + 3x^3 \cdot 5 \quad \text{Express each term as the product of the GCF and its other factor.}$$
$$= 3x^3(3x^2 + 5) \quad \text{Factor out the GCF.}$$

b. Begin by determining the greatest common factor.

8 is the greatest integer that divides 16 and 24.

$$16x^2y^3 - 24x^3y^4$$

The variables raised to the smallest exponents are x^2 and y^3.

The GCF of the two terms of the polynomial is $8x^2y^3$.

$$16x^2y^3 - 24x^3y^4$$
$$= 8x^2y^3 \cdot 2 - 8x^2y^3 \cdot 3xy \quad \text{Express each term as the product of the GCF and its other factor.}$$
$$= 8x^2y^3(2 - 3xy) \quad \text{Factor out the GCF.}$$

c. First, determine the greatest common factor of the three terms.

2 is the greatest integer that divides 12, 4, and 2.

$$12x^5y^4 - 4x^4y^3 + 2x^3y^2$$

The variables raised to the smallest exponents are x^3 and y^2.

The GCF is $2x^3y^2$.

$$12x^5y^4 - 4x^4y^3 + 2x^3y^2$$
$$= 2x^3y^2 \cdot 6x^2y^2 - 2x^3y^2 \cdot 2xy + 2x^3y^2 \cdot 1 \quad \text{Express each term as the product of the GCF and its other factor.}$$

You can obtain the factors shown in black by dividing each term of the given polynomial by $2x^3y^2$, the GCF.

$$\frac{12x^5y^4}{2x^3y^2} = 6x^2y^2 \qquad \frac{4x^4y^3}{2x^3y^2} = 2xy \qquad \frac{2x^3y^2}{2x^3y^2} = 1$$

$$= 2x^3y^2(6x^2y^2 - 2xy + 1) \quad \text{Factor out the GCF.}$$

Because factoring reverses the process of multiplication, all factorizations can be checked by multiplying. Take a few minutes to check each of the three factorizations in Example 2. Use the distributive property to multiply the factors. This should give the original polynomial.

✓ CHECK POINT 2 Factor:

a. $9x^4 + 21x^2$

b. $15x^3y^2 - 25x^4y^3$

c. $16x^4y^5 - 8x^3y^4 + 4x^2y^3$.

---

2 Factor out a common factor with a negative coefficient.

When the leading coefficient of a polynomial is negative, it is often desirable to factor out a common factor with a negative coefficient. The common factor is the GCF preceded by a negative sign.

EXAMPLE 3 Using a Common Factor with a Negative Coefficient

Factor: $-3x^3 + 12x^2 - 15x$.

Solution The GCF is $3x$. Because the leading coefficient, -3, is negative, we factor out a common factor with a negative coefficient. We will factor out the opposite of the GCF, or $-3x$.

$$-3x^3 + 12x^2 - 15x$$

$$= -3x(x^2) - 3x(-4x) - 3x(5)$$ Express each term as the product of the common factor and its other factor.

$$= -3x(x^2 - 4x + 5)$$ Factor out the opposite of the GCF.

CHECK POINT 3 Factor out a common factor with a negative coefficient: $-2x^3 + 10x^2 - 6x$.

Factoring by Grouping

Up to now, we have factored a monomial from a polynomial. By contrast, in our next example, the greatest common factor of the polynomial is a binomial.

EXAMPLE 4 Factoring Out the Greatest Common Binomial Factor

Factor:

a. $2(x - 7) + 9a(x - 7)$ **b.** $5y(a - b) - (a - b)$.

Solution Let's identify the common binomial factor in each part of the problem.

$2(x - 7) + 9a(x - 7)$ — The GCF, a binomial, is $x - 7$.

$5y(a - b) - (a - b)$ — The GCF, a binomial, is $a - b$.

We factor out each common binomial factor as follows.

a. $2(x - 7) + 9a(x - 7)$

$= (x - 7)2 + (x - 7)9a$ This step, usually omitted, shows each term as the product of the GCF and its other factor, in that order.

$= (x - 7)(2 + 9a)$ Factor out the GCF.

b. $5y(a - b) - (a - b)$

$= 5y(a - b) - 1(a - b)$ Write $-(a - b$ as $-1(a - b)$ to aid in the factoring.

$= (a - b)(5y - 1)$ Factor out the GCF.

CHECK POINT 4 Factor:

a. $3(x - 4) + 7a(x - 4)$

b. $7x(a + b) - (a + b)$.

3 Factor by grouping.

Some polynomials have only a greatest common factor of 1. However, by a suitable grouping of the terms, it still may be possible to factor. This process, called **factoring by grouping**, is illustrated in Example 5.

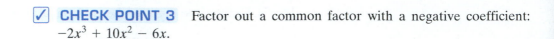

EXAMPLE 5 Factoring by Grouping

Factor: $x^3 - 5x^2 + 3x - 15$.

Solution There is no factor other than 1 common to all four terms. However, we can group terms that have a common factor:

$$\boxed{x^3 - 5x^2} + \boxed{3x - 15}$$

Common factor is x^2. Common factor is 3.

Discover for Yourself

In Example 5, group the terms as follows:

$$x^2 + 3x - 5x^2 - 15.$$

Factor out the common factor from each group and complete the factoring process. Describe what happens. What can you conclude?

We now factor the given polynomial as follows:

$$x^3 - 5x^2 + 3x - 15$$

$= (x^3 - 5x^2) + (3x - 15)$ Group terms with common factors.

$= x^2(x - 5) + 3(x - 5)$ Factor out the greatest common factor from the grouped terms. The remaining two terms have $x - 5$ as a common binomial factor.

$= (x - 5)(x^2 + 3).$ Factor out the GCF.

Thus, $x^3 - 5x^2 + 3x - 15 = (x - 5)(x^2 + 3)$. Check the factorization by multiplying the right side of the equation using the FOIL method. Because the factorization is correct, you should obtain the original polynomial.

✓ **CHECK POINT 5** Factor: $x^3 - 4x^2 + 5x - 20$.

Factoring by Grouping

1. Group terms that have a common monomial factor. There will usually be two groups. Sometimes the terms must be rearranged.
2. Factor out the common monomial factor from each group.
3. Factor out the remaining common binomial factor (if one exists).

EXAMPLE 6 Factoring by Grouping

Factor: $3x^2 + 12x - 2xy - 8y$.

Solution There is no factor other than 1 common to all four terms. However, we can group terms that have a common factor:

$$\boxed{3x^2 + 12x} + \boxed{-2xy - 8y}.$$

Common factor is $3x$: $3x^2 + 12x = 3x(x + 4)$.

Use $-2y$, rather than $2y$, as the common factor: $-2xy - 8y = -2y(x + 4)$. In this way, the common binomial factor, $x + 4$, appears.

The voice balloons illustrate that it is sometimes necessary to use a factor with a negative coefficient to obtain a common binomial factor for the two groupings. We now factor the given polynomial as follows:

$$3x^2 + 12x - 2xy - 8y$$

$= (3x^2 + 12x) + (-2xy - 8y)$ Group terms with common factors.

$= 3x(x + 4) - 2y(x + 4)$ Factor out the common factors from the grouped terms.

$= (x + 4)(3x - 2y).$ Factor out the GCF.

Thus, $3x^2 + 12x - 2xy - 8y = (x + 4)(3x - 2y)$. Using the commutative property of multiplication, the factorization can also be expressed as $(3x - 2y)(x + 4)$. Verify the factorization by showing that, regardless of the order, FOIL multiplication gives the original polynomial.

CHECK POINT 6 Factor: $4x^2 + 20x - 3xy - 15y$.

Factoring by grouping sometimes requires that the terms be rearranged before the groupings are made. For example, consider the polynomial

$$3x^2 - 8y + 12x - 2xy.$$

The first two terms have no common factor other than 1. We must rearrange the terms and try a different grouping. Example 6 showed one such rearrangement of two groupings.

5.3 EXERCISE SET

MyMathLab MathXL PRACTICE WATCH DOWNLOAD READ REVIEW

Practice Exercises

Throughout the practice exercises, assume that any variable exponents represent whole numbers.

In Exercises 1–22, factor the greatest common factor from each polynomial.

1. $10x^2 + 4x$
2. $12x^2 + 9x$
3. $y^2 - 4y$
4. $y^2 - 7y$
5. $x^3 + 5x^2$
6. $x^3 + 7x^2$
7. $12x^4 - 8x^2$
8. $20x^4 - 8x^2$
9. $32x^4 + 2x^3 + 8x^2$
10. $9x^4 + 18x^3 + 6x^2$
11. $4x^2y^3 + 6xy$
12. $6x^3y^2 + 9xy$
13. $30x^2y^3 - 10xy^2$
14. $27x^2y^3 - 18xy^2$
15. $12xy - 6xz + 4xw$
16. $14xy - 10xz + 8xw$
17. $15x^3y^6 - 9x^4y^4 + 12x^2y^5$
18. $15x^4y^6 - 3x^3y^5 + 12x^4y^4$
19. $25x^3y^6z^2 - 15x^4y^4z^4 + 25x^2y^5z^3$
20. $49x^4y^3z^5 - 70x^3y^5z^4 + 35x^4y^4z^3$
21. $15x^{2n} - 25x^n$
22. $12x^{3n} - 9x^{2n}$

In Exercises 23–34, factor out the negative of the greatest common factor.

23. $-4x + 12$
24. $-5x + 20$
25. $-8x - 48$
26. $-7x - 63$
27. $-2x^2 + 6x - 14$
28. $-2x^2 + 8x - 12$
29. $-5y^2 + 40x$
30. $-9y^2 + 45x$
31. $-4x^3 + 32x^2 - 20x$
32. $-5x^3 + 50x^2 - 10x$
33. $-x^2 - 7x + 5$
34. $-x^2 - 8x + 8$

In Exercises 35–44, factor the greatest common binomial factor from each polynomial.

35. $4(x + 3) + a(x + 3)$
36. $5(x + 4) + a(x + 4)$
37. $x(y - 6) - 7(y - 6)$
38. $x(y - 9) - 5(y - 9)$
39. $3x(x + y) - (x + y)$
40. $7x(x + y) - (x + y)$
41. $4x^2(3x - 1) + 3x - 1$
42. $6x^2(5x - 1) + 5x - 1$
43. $(x + 2)(x + 3) + (x - 1)(x + 3)$
44. $(x + 4)(x + 5) + (x - 1)(x + 5)$

In Exercises 45–68, factor by grouping.

45. $x^2 + 3x + 5x + 15$
46. $x^2 + 2x + 4x + 8$
47. $x^2 + 7x - 4x - 28$
48. $x^2 + 3x - 5x - 15$
49. $x^3 - 3x^2 + 4x - 12$
50. $x^3 - 2x^2 + 5x - 10$
51. $xy - 6x + 2y - 12$

52. $xy - 5x + 9y - 45$

53. $xy + x - 7y - 7$

54. $xy + x - 5y - 5$

55. $10x^2 - 12xy + 35xy - 42y^2$

56. $3x^2 - 6xy + 5xy - 10y^2$

57. $4x^3 - x^2 - 12x + 3$

58. $3x^3 - 2x^2 - 6x + 4$

59. $x^2 - ax - bx + ab$

60. $x^2 + ax - bx - ab$

61. $x^3 - 12 - 3x^2 + 4x$

62. $2x^3 - 10 + 4x^2 - 5x$

63. $ay - by + bx - ax$

64. $cx - dx + dy - cy$

65. $ay^2 + 2by^2 - 3ax - 6bx$

66. $3a^2x + 6a^2y - 2bx - 4by$

67. $x^ny^n + 3x^n + y^n + 3$

68. $x^ny^n - x^n + 2y^n - 2$

Practice PLUS

In Exercises 69–78, factor each polynomial.

69. $ab - c - ac + b$

70. $ab - 3c - ac + 3b$

71. $x^3 - 5 + 4x^3y - 20y$

72. $x^3 - 2 + 3x^3y - 6y$

73. $2y^7(3x - 1)^5 - 7y^6(3x - 1)^4$

74. $3y^9(3x - 2)^7 - 5y^8(3x - 2)^6$

75. $ax^2 + 5ax - 2a + bx^2 + 5bx - 2b$

76. $ax^2 + 3ax - 11a + bx^2 + 3bx - 11b$

77. $ax + ay + az - bx - by - bz + cx + cy + cz$

78. $ax^2 + ay^2 - az^2 + bx^2 + by^2 - bz^2 + cx^2 + cy^2 - cz^2$

Application Exercises

79. A ball is thrown straight upward. The function

$$f(t) = -16t^2 + 40t$$

describes the ball's height above the ground, $f(t)$, in feet, t seconds after it is thrown.

a. Find and interpret $f(2)$.

b. Find and interpret $f(2.5)$.

c. Factor the polynomial $-16t^2 + 40t$ and write the function in factored form.

d. Use the factored form of the function to find $f(2)$ and $f(2.5)$. Do you get the same answers as you did in parts (a) and (b)? If so, does this prove that your factorization is correct? Explain.

80. An explosion causes debris to rise vertically. The function

$$f(t) = -16t^2 + 72t$$

describes the height of the debris above the ground, $f(t)$, in feet, t seconds after the explosion.

a. Find and interpret $f(2)$.

b. Find and interpret $f(4.5)$.

c. Factor the polynomial $-16t^2 + 72t$ and write the function in factored form.

d. Use the factored form of the function to find $f(2)$ and $f(4.5)$. Do you get the same answers as you did in parts (a) and (b)? If so, does this prove that your factorization is correct? Explain.

81. Your computer store is having an incredible sale. The price on one model is reduced by 40%. Then the sale price is reduced by another 40%. If x is the computer's original price, the sale price can be represented by

$$(x - 0.4x) - 0.4(x - 0.4x).$$

a. Factor out $(x - 0.4x)$ from each term. Then simplify the resulting expression.

b. Use the simplified expression from part (a) to answer these questions. With a 40% reduction followed by a 40% reduction, is the computer selling at 20% of its original price? If not, at what percentage of the original price is it selling?

82. Your local electronics store is having an end-of-the-year sale. The price on a plasma television had been reduced by 30%. Now the sale price is reduced by another 30%. If x is the television's original price, the sale price can be represented by

$$(x - 0.3x) - 0.3(x - 0.3x).$$

a. Factor out $(x - 0.3x)$ from each term. Then simplify the resulting expression.

b. Use the simplified expression from part (a) to answer these questions. With a 30% reduction followed by a 30% reduction, is the television selling at 40% of its original price? If not, at what percentage of the original price is it selling?

Exercises 83–84 involve compound interest. **Compound interest** *is interest computed on your original savings as well as on any accumulated interest.*

83. After 2 years, the balance, A, in an account with principal P and interest rate r compounded annually is given by the formula

$$A = P + Pr + (P + Pr)r.$$

Use factoring by grouping to express the formula as $A = P(1 + r)^2$.

84. After 3 years, the balance, A, in an account with principal P and interest rate r compounded annually is given by the formula

$$A = P(1 + r)^2 + P(1 + r)^2 r.$$

Use factoring by grouping to express the formula as $A = P(1 + r)^3$.

85. The area of the skating rink with semicircular ends shown is $A = \pi r^2 + 2rl$. Express the area, A, in factored form.

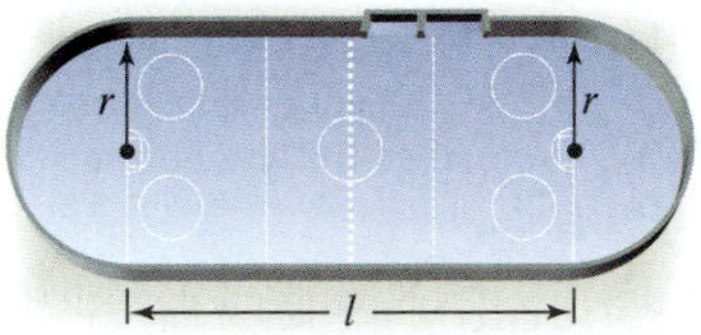

86. The amount of sheet metal needed to manufacture a cylindrical tin can, that is, its surface area, S, is $S = 2\pi r^2 + 2\pi rh$. Express the surface area, S, in factored form.

Writing in Mathematics

87. What is factoring?

88. If a polynomial has a greatest common factor other than 1, explain how to find its GCF.

89. Using an example, explain how to factor out the greatest common factor of a polynomial.

90. Suppose that a polynomial contains four terms and can be factored by grouping. Explain how to obtain the factorization.

91. Use two different groupings to factor

$$ac - ad + bd - bc$$

in two ways. Then explain why the two factorizations are the same.

92. Write a sentence that uses the word *factor* as a noun. Then write a sentence that uses the word *factor* as a verb.

Technology Exercises

In Exercises 93–96, use a graphing utility to graph the function on each side of the equation in the same viewing rectangle. Use end behavior to show a complete picture of the polynomial function on the left side. Do the graphs coincide? If so, this means that the polynomial on the left side has been factored correctly. If not, factor the polynomial correctly and then use your graphing utility to verify the factorization.

93. $x^2 - 4x = x(x - 4)$

94. $x^2 - 2x + 5x - 10 = (x - 2)(x - 5)$

95. $x^2 + 2x + x + 2 = x(x + 2) + 1$

96. $x^3 - 3x^2 + 4x - 12 = (x^2 + 4)(x - 3)$

Critical Thinking Exercises

Make Sense? *In Exercises 97–100, determine whether each statement "makes sense" or "does not make sense" and explain your reasoning.*

97. After I've factored a polynomial, my answer cannot always be checked by multiplication.

98. The word *greatest* in greatest common factor is helpful because it tells me to look for the greatest power of a variable appearing in all terms.

99. Although $20x^3$ appears in both $20x^3 + 8x^2$ and $20x^3 + 10x$, I'll need to factor $20x^3$ in different ways to obtain each polynomial's factorization.

100. You grouped the polynomial's terms using different groupings than I did, yet we both obtained the same factorization.

In Exercises 101–104, determine whether each statement is true or false. If the statement is false, make the necessary change(s) to produce a true statement.

101. Because the GCF of $9x^3 + 6x^2 + 3x$ is $3x$, it is not necessary to write the 1 when $3x$ is factored from the last term.

102. Some polynomials with four terms, such as $x^3 + x^2 + 4x - 4$, cannot be factored by grouping.

103. The polynomial $28x^3 - 7x^2 + 36x - 9$ can be factored by grouping terms as follows:

$$(28x^3 + 36x) + (-7x^2 - 9).$$

104. $x^2 - 2$ is a factor of $2 - 50x - x^2 + 25x^3$.

In Exercises 105–107, factor each polynomial. Assume that all variable exponents represent whole numbers.

105. $x^{4n} + x^{2n} + x^{3n}$

106. $3x^{3m}y^m - 6x^{2m}y^{2m}$

107. $8y^{2n+4} + 16y^{2n+3} - 12y^{2n}$

In Exercises 108–109, write a polynomial that fits the given description. Do not use a polynomial that appeared in this section or in the exercise set.

108. The polynomial has three terms and can be factored using a greatest common factor that has both a negative coefficient and a variable.

109. The polynomial has four terms and can be factored by grouping.

Review Exercises

110. Solve by Cramer's rule:

$$3x - 2y = 8$$
$$2x - 5y = 10.$$

(Section 3.5, Example 2)

111. Determine whether each relation is a function.
  a. $\{(0, 5), (3, -5), (5, 5), (7, -5)\}$
  b. $\{(1, 2), (3, 4), (5, 5), (5, 6)\}$ (Section 2.1, Example 2)

112. The length of a rectangle is 2 feet greater than twice its width. If the rectangle's perimeter is 22 feet, find the length and width. (Section 1.5, Example 5)

Preview Exercises

Exercises 113–115 will help you prepare for the material covered in the next section. In each exercise, replace the boxed question mark with an integer that results in the given product. Some trial and error may be necessary.

113. $(x + 3)(x + \boxed{?}) = x^2 + 7x + 12$

114. $(x - \boxed{?})(x - 12) = x^2 - 14x + 24$

115. $(x + 3y)(x - \boxed{?}y) = x^2 - 4xy - 21y^2$

SECTION 5.4

Factoring Trinomials

Objectives

1. Factor a trinomial whose leading coefficient is 1.
2. Factor using a substitution.
3. Factor a trinomial whose leading coefficient is not 1.
4. Factor trinomials by grouping.

A great deal of trial and error is involved in finding your way out of this maze. Trial and error play an important role in problem solving and can be helpful in leading to correct solutions. In this section, you will use trial and error to factor trinomials following a problem-solving process that is not very different from learning to traverse the maze.

1 Factor a trinomial whose leading coefficient is 1.

Factoring a Trinomial Whose Leading Coefficient Is 1

In Section 5.2, we used the FOIL method to multiply two binomials. The product was often a trinomial. The following are some examples:

Factored Form		F O I L		Trinomial Form
$(x + 3)(x + 4)$	$=$	$x^2 + 4x + 3x + 12$	$=$	$x^2 + 7x + 12$
$(x - 3)(x - 4)$	$=$	$x^2 - 4x - 3x + 12$	$=$	$x^2 - 7x + 12$
$(x + 3)(x - 5)$	$=$	$x^2 - 5x + 3x - 15$	$=$	$x^2 - 2x - 15$

Observe that each trinomial is of the form $x^2 + bx + c$, where the coefficient of the squared term is 1. Our goal in the first part of this section is to start with the trinomial form and, assuming that it is factorable, return to the factored form.

The first FOIL multiplication shown in our list indicates that

$$(x + 3)(x + 4) = x^2 + 7x + 12.$$

Let's reverse the sides of this equation:

$$x^2 + 7x + 12 = (x + 3)(x + 4).$$

We can use $x^2 + 7x + 12 = (x + 3)(x + 4)$ to make several important observations about the factors on the right side.

The first term of each factor is x. The product of the First terms is $x \cdot x = x^2$.

3 and 4 are factors of 12. The product of the Last terms is $3 \cdot 4 = 12$.

The sum of the Outside and Inside products is $4x + 3x = 7x$.

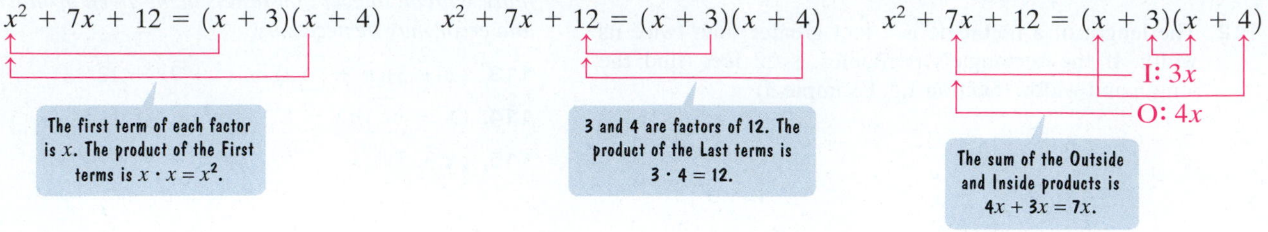

These observations provide us with a procedure for factoring $x^2 + bx + c$.

A Strategy for Factoring $x^2 + bx + c$

1. Enter x as the first term of each factor.

$$(x \quad)(x \quad) = x^2 + bx + c$$

2. List pairs of factors of the constant c.

3. Try various combinations of these factors. Select the combination in which the sum of the Outside and Inside products is equal to bx.

$$(x + \square)(x + \square) = x^2 + bx + c$$

I
O
Sum of O + I

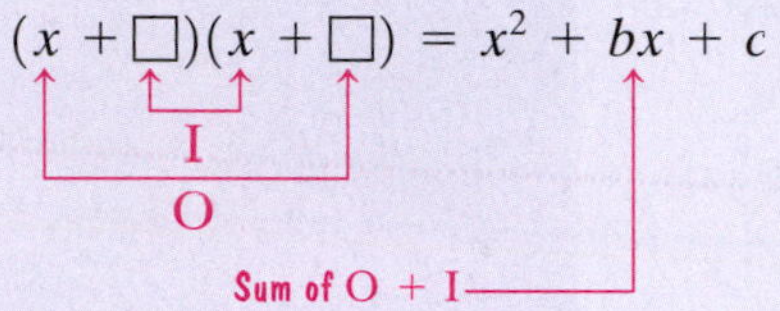

4. Check your work by multiplying the factors using the FOIL method. You should obtain the original trinomial.

If none of the possible combinations yield an Outside product and an Inside product whose sum is equal to bx, the trinomial cannot be factored using integers and is called **prime** over the set of integers.

EXAMPLE 1   **Factoring a Trinomial Whose Leading Coefficient Is 1**

Factor: $x^2 + 5x + 6$.

Solution

Step 1. Enter x as the first term of each factor.

$$x^2 + 5x + 6 = (x \quad)(x \quad)$$

Step 2. List pairs of factors of the constant, 6.

Factors of 6	6, 1	3, 2	−6, −1	−3, −2

Step 3. Try various combinations of these factors. The correct factorization of $x^2 + 5x + 6$ is the one in which the sum of the Outside and Inside products is equal to $5x$. At the top of the next page is a list of the possible factorizations.

Possible Factorizations of $x^2 + 5x + 6$	Sum of Outside and Inside Products (Should Equal $5x$)
$(x + 6)(x + 1)$	$x + 6x = 7x$
$(x + 3)(x + 2)$	$2x + 3x = 5x$
$(x - 6)(x - 1)$	$-x - 6x = -7x$
$(x - 3)(x - 2)$	$-2x - 3x = -5x$

This is the required middle term.

Thus,

$$x^2 + 5x + 6 = (x + 3)(x + 2).$$

Check this result by multiplying the right side using the FOIL method. You should obtain the original trinomial. Because of the commutative property, the factorization can also be expressed as

$$x^2 + 5x + 6 = (x + 2)(x + 3).$$

In factoring a trinomial of the form $x^2 + bx + c$, you can speed things up by listing the factors of c and then finding their sums. We are interested in a sum of b. For example, in factoring $x^2 + 5x + 6$, we are interested in the factors of 6 whose sum is 5.

Factors of 6	6, 1	3, 2	−6, −1	−3, −2
Sum of Factors	7	5	−7	−5

This is the desired sum.

Thus, $x^2 + 5x + 6 = (x + 3)(x + 2)$.

Using Technology

Numeric and Graphic Connections

If a polynomial contains one variable, a graphing utility can be used to check its factorization. For example, the factorization in Example 1 can be checked graphically or numerically.

Given polynomial — The factorization we are checking

$$x^2 + 5x + 6 = (x + 3)(x + 2)$$

Enter $y_1 = x^2 + 5x + 6$ in the [y=] screen.

Enter $y_2 = (x + 3)(x + 2)$ in the [y=] screen.

Numeric Check

Use the [TABLE] feature.

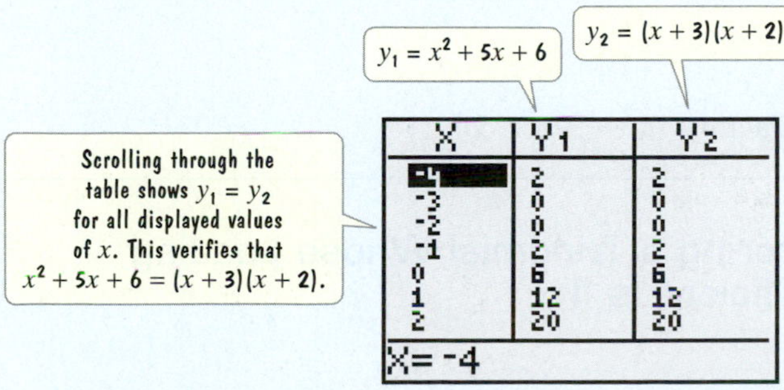

Graphic Check

Use the [GRAPH] feature to display graphs for y_1 and y_2.

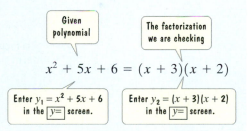

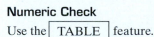

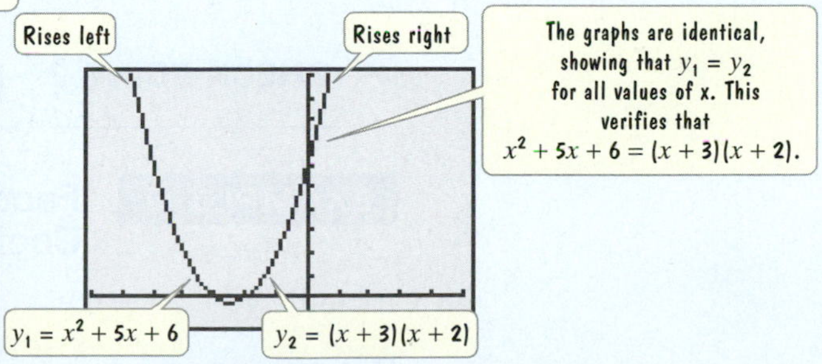

[−7, 5, 1] by [−1, 10, 1]

Notice that the graph of the quadratic function is shaped like a bowl. The graph of the even-degree quadratic function exhibits the same behavior at each end, rising to the left and rising to the right (↖, ↗).

✓ **CHECK POINT 1** Factor: $x^2 + 6x + 8$.

EXAMPLE 2 Factoring a Trinomial Whose Leading Coefficient Is 1

Factor: $x^2 - 14x + 24$.

Solution

Step 1. Enter x as the first term of each factor.

$$x^2 - 14x + 24 = (x \quad)(x \quad)$$

To find the second term of each factor, we must find two integers whose product is 24 and whose sum is −14.

Step 2. List pairs of factors of the constant, 24. Because the desired sum, −14, is negative, we will list only the negative pairs of factors of 24.

Negative Factors of 24	−24, −1	−12, −2	−8, −3	−6, −4

Step 3. Try various combinations of these factors. We are interested in the factors whose sum is −14.

Negative Factors of 24	−24, −1	−12, −2	−8, −3	−6, −4
Sum of Factors	−25	−14	−11	−10

This is the desired sum.

Thus, $x^2 - 14x + 24 = (x - 12)(x - 2)$.

Study Tip

To factor $x^2 + bx + c$ when c is positive, find two numbers with the same sign as the middle term.

$$x^2 + 5x + 6 = (x + 3)(x + 2)$$

Same signs

$$x^2 - 14x + 24 = (x - 12)(x - 2)$$

Same signs

✓ **CHECK POINT 2** Factor: $x^2 - 9x + 20$.

EXAMPLE 3 Factoring a Trinomial Whose Leading Coefficient Is 1

Factor: $y^2 + 7y - 60$.

Solution

Step 1. Enter y as the first term of each factor.

$$y^2 + 7y - 60 = (y \quad)(y \quad)$$

To find the second term of each factor, we must find two integers whose product is −60 and whose sum is 7.

Steps 2 and 3. List pairs of factors of the constant, −60, and try various combinations of these factors. Because the desired sum, 7, is positive, the positive factor of −60 must be farther from 0 than the negative factor is. Thus, we will only list pairs of factors of −60 in which the positive factor has the larger absolute value.

Some Factors of −60	60, −1	30, −2	20, −3	15, −4	12, −5	10, −6
Sum of Factors	59	28	17	11	7	4

This is the desired sum.

Thus, $y^2 + 7y - 60 = (y + 12)(y - 5)$.

Study Tip

To factor $x^2 + bx + c$ when c is negative, find two numbers with opposite signs whose sum is the coefficient of the middle term.

$$y^2 + 7y - 60 = (y + 12)(y - 5)$$

Negative

Opposite signs

CHECK POINT 3 Factor: $y^2 + 19y - 66$.

EXAMPLE 4 Factoring a Trinomial in Two Variables

Factor: $x^2 - 4xy - 21y^2$.

Solution

Step 1. Enter x as the first term of each factor. Because the last term of the trinomial contains y^2, the second term of each factor must contain y.

$$x^2 - 4xy - 21y^2 = (x \quad ?y)(x \quad ?y)$$

The question marks indicate that we are looking for the coefficients of y in each factor. To find these coefficients, we must find two integers whose product is −21 and whose sum is −4.

Steps 2 and 3. List pairs of factors of the coefficient of the last term, −21, and try various combinations of these factors. We are interested in the factors whose sum is −4.

Factors of −21	1, −21	3, −7	−1, 21	−3, 7
Sum of Factors	−20	−4	20	4

This is the desired sum.

Thus, $x^2 - 4xy - 21y^2 = (x + 3y)(x - 7y)$ or $(x - 7y)(x + 3y)$.

Step 4. Verify the factorization using the FOIL method.

$$(x + 3y)(x - 7y) = x^2 - 7xy + 3xy - 21y^2 = x^2 - 4xy - 21y^2$$

Because the product of the factors is the original polynomial, the factorization is correct.

CHECK POINT 4 Factor: $x^2 - 5xy + 6y^2$.

Can every trinomial be factored? The answer is no. For example, consider

$$x^2 + x - 5 = (x \quad)(x \quad).$$

To find the second term of each factor, we must find two integers whose product is -5 and whose sum is 1. Because no such integers exist, $x^2 + x - 5$ cannot be factored. This trinomial is prime.

To factor some polynomials, more than one technique must be used. **Always begin by trying to factor out the greatest common factor.** A polynomial is **factored completely** when it is written as the product of prime polynomials.

Study Tip

The technology box on page 341 shows how to use a graphing utility to check a polynomial's factorization graphically or numerically. The graphing utility verifies that the factorization is equivalent to the original polynomial. However, a graphing utility cannot verify that the factorization is complete.

EXAMPLE 5 Factoring Completely

Factor: $8x^3 - 40x^2 - 48x$.

Solution The GCF of the three terms of the polynomial is $8x$. We begin by factoring out $8x$. Then we factor the remaining trinomial.

$$8x^3 - 40x^2 - 48x$$

$$= 8x(x^2 - 5x - 6) \quad \text{Factor out the GCF.}$$

$$= 8x(x \quad)(x \quad) \quad \text{Begin factoring } x^2 - 5x - 6. \text{ Find two integers whose product is } -6 \text{ and whose sum is } -5.$$

$$= 8x(x - 6)(x + 1) \quad \text{The integers are } -6 \text{ and } 1.$$

Thus,

$$8x^3 - 40x^2 - 48x = 8x(x - 6)(x + 1).$$

Be sure to include the GCF in the factorization.

You can check this factorization by multiplying the binomials using the FOIL method. Then use the distributive property and multiply each term in this product by $8x$. Try doing this now. Because the factorization is correct, you should obtain the original polynomial.

✓ CHECK POINT 5 Factor: $3x^3 - 15x^2 - 42x$.

Some trinomials, such as $-x^2 + 5x + 6$, have a leading coefficient of -1. Because it is easier to factor a trinomial with a positive leading coefficient, begin by factoring out -1. For example,

$$-x^2 + 5x + 6 = -1(x^2 - 5x - 6) = -(x - 6)(x + 1).$$

2 Factor using a substitution.

In some trinomials, the highest power is greater than 2, and the exponent in one of the terms is half that of the other term. By letting u equal the variable to the smaller power, the trinomial can be written in a form that makes its possible factorization more obvious. Here are some examples:

Given Trinomial	Substitution	New Trinomial
$x^6 - 8x^3 + 15$ or $(x^3)^2 - 8x^3 + 15$	$u = x^3$	$u^2 - 8u + 15$
$x^4 - 8x^2 - 9$ or $(x^2)^2 - 8x^2 - 9$	$u = x^2$	$u^2 - 8u - 9$

In each case, we factor the given trinomial by working with the new trinomial on the right. If a factorization is found, we replace all occurrences of u in the factorization with the substitution shown in the middle column.

EXAMPLE 6 Factoring by Substitution

Factor: $x^6 - 8x^3 + 15$.

Solution Notice that the exponent on x^3 is half that of the exponent on x^6. We will let u equal the variable to the power that is half of 6. Thus, let $u = x^3$.

$$(x^3)^2 - 8x^3 + 15$$ This is the given polynomial, with x^6 written as $(x^3)^2$.

$$= u^2 - 8u + 15$$ Let $u = x^3$. Rewrite the trinomial in terms of u.

$$= (u-5)(u-3)$$ Factor.

$$= (x^3-5)(x^3-3)$$ Now substitute x^3 for u.

Thus, the given trinomial can be factored as

$$x^6 - 8x^3 + 15 = (x^3-5)(x^3-3).$$

Check this result using FOIL multiplication on the right.

✓ CHECK POINT 6 Factor: $x^6 - 7x^3 + 10$.

3 Factor a trinomial whose leading coefficient is not 1.

Factoring a Trinomial Whose Leading Coefficient Is Not 1

How do we factor a trinomial such as $5x^2 - 14x + 8$? Notice that the leading coefficient is 5. We must find two binomials whose product is $5x^2 - 14x + 8$. The product of the First terms must be $5x^2$:

$$(5x \quad)(x \quad).$$

From this point on, the factoring strategy is exactly the same as the one we use to factor a trinomial whose leading coefficient is 1.

A Strategy for Factoring $ax^2 + bx + c$

Assume, for the moment, that there is no greatest common factor.

1. Find two **F**irst terms whose product is ax^2:

$$(\square x + \quad)(\square x + \quad) = ax^2 + bx + c.$$

2. Find two **L**ast terms whose product is c:

$$(\square x + \square)(\square x + \square) = ax^2 + bx + c.$$

3. By trial and error, perform steps 1 and 2 until the sum of the **O**utside product and **I**nside product is bx:

$$(\square x + \square)(\square x + \square) = ax^2 + bx + c.$$

I

O

Sum of O + I

If no such combinations exist, the polynomial is prime.

Study Tip

The *error* part of the factoring strategy plays an important role in the process. If you do not get the correct factorization the first time, this is not a bad thing. This error is often helpful in leading you to the correct factorization.

EXAMPLE 7 Factoring a Trinomial Whose Leading Coefficient Is Not 1

Factor: $5x^2 - 14x + 8$.

Solution

Step 1. Find two First terms whose product is $5x^2$.

$$5x^2 - 14x + 8 = (5x \quad)(x \quad)$$

Step 2. Find two Last terms whose product is 8. The number 8 has pairs of factors that are either both positive or both negative. Because the middle term, $-14x$, is negative, both factors must be negative. The negative factorizations of 8 are $-1(-8)$ and $-2(-4)$.

Step 3. Try various combinations of these factors. The correct factorization of $5x^2 - 14x + 8$ is the one in which the sum of the Outside and Inside products is equal to $-14x$. Here is a list of the possible factorizations:

Possible Factorizations of $5x^2 - 14x + 8$	Sum of Outside and Inside Products (Should Equal $-14x$)
$(5x - 1)(x - 8)$	$-40x - x = -41x$
$(5x - 8)(x - 1)$	$-5x - 8x = -13x$
$(5x - 2)(x - 4)$	$-20x - 2x = -22x$
$(5x - 4)(x - 2)$	$-10x - 4x = -14x$

This is the required middle term.

Thus,

$$5x^2 - 14x + 8 = (5x - 4)(x - 2).$$

Show that this factorization is correct by multiplying the factors using the FOIL method. You should obtain the original trinomial.

Study Tip

With practice, you will find that you don't have to list every possible factorization of the trinomial. As you practice factoring, you will be able to narrow down the list of possible factorizations to just a few. When it comes to factoring, practice makes perfect.

✓ CHECK POINT 7 Factor: $3x^2 - 20x + 28$.

EXAMPLE 8 Factoring a Trinomial Whose Leading Coefficient Is Not 1

Factor: $8x^6 - 10x^5 - 3x^4$.

Solution The GCF of the three terms of the polynomial is x^4. We begin by factoring out x^4.

$$8x^6 - 10x^5 - 3x^4 = x^4(8x^2 - 10x - 3)$$

Now we factor the remaining trinomial, $8x^2 - 10x - 3$.

Step 1. Find two First terms whose product is $8x^2$.

$$8x^2 - 10x - 3 \stackrel{?}{=} (8x \quad)(x \quad)$$

$$8x^2 - 10x - 3 \stackrel{?}{=} (4x \quad)(2x \quad)$$

Step 2. Find two Last terms whose product is -3. The possible factorizations are $1(-3)$ and $-1(3)$.

Step 3. Try various combinations of these factors. The correct factorization of $8x^2 - 10x - 3$ is the one in which the sum of the Outside and Inside products is equal to $-10x$. At the top of the next page is a list of the possible factorizations.

Study Tip

Here are some suggestions for reducing the list of possible factorizations for $ax^2 + bx + c$.

1. If b is relatively small, avoid the larger factors of a.
2. If c is positive, the signs in both binomial factors must match the sign of b.
3. If the trinomial has no common factor, no binomial factor can have a common factor.
4. Reversing the signs in the binomial factors reverses the sign of bx, the middle term.

Possible Factorizations of $8x^2 - 10x - 3$	Sum of Outside and Inside Products (Should Equal $-10x$)
$(8x + 1)(x - 3)$	$-24x + x = -23x$
$(8x - 3)(x + 1)$	$8x - 3x = 5x$
$(8x - 1)(x + 3)$	$24x - x = 23x$
$(8x + 3)(x - 1)$	$-8x + 3x = -5x$
$(4x + 1)(2x - 3)$	$-12x + 2x = -10x$
$(4x - 3)(2x + 1)$	$4x - 6x = -2x$
$(4x - 1)(2x + 3)$	$12x - 2x = 10x$
$(4x + 3)(2x - 1)$	$-4x + 6x = 2x$

This is the required middle term.

The factorization of $8x^2 - 10x - 3$ is $(4x + 1)(2x - 3)$. Now we include the GCF in the complete factorization of the given polynomial. Thus,

$$8x^6 - 10x^5 - 3x^4 = x^4(8x^2 - 10x - 3) = x^4(4x + 1)(2x - 3).$$

This is the complete factorization with the GCF, x^4, included.

CHECK POINT 8 Factor: $6x^6 + 19x^5 - 7x^4$.

We have seen that not every trinomial can be factored. For example, consider

$$6x^2 + 14x + 7 = (6x + \square)(x + \square)$$
$$6x^2 + 14x + 7 = (3x + \square)(2x + \square).$$

The possible factors for the last term are 1 and 7. However, regardless of how these factors are placed in the boxes shown, the sum of the Outside and Inside products is not equal to $14x$. Thus, the trinomial $6x^2 + 14x + 7$ cannot be factored and is prime.

EXAMPLE 9 Factoring a Trinomial in Two Variables

Factor: $3x^2 - 13xy + 4y^2$.

Solution

Step 1. Find two First terms whose product is $3x^2$.

$$3x^2 - 13xy + 4y^2 = (3x \quad ?y)(x \quad ?y)$$

The question marks indicate that we are looking for the coefficients of y in each factor.

Steps 2 and 3. List pairs of factors of the coefficient of the last term, 4, and try various combinations of these factors. The correct factorization is the one in which the sum of the Outside and Inside products is equal to $-13xy$. Because of the negative coefficient, -13, we will consider only the negative pairs of factors of 4. The possible factorizations are $-1(-4)$ and $-2(-2)$.

Possible Factorizations of $3x^2 - 13xy + 4y^2$	Sum of Outside and Inside Products (Should Equal $-13xy$)
$(3x - y)(x - 4y)$	$-12xy - xy = -13xy$
$(3x - 4y)(x - y)$	$-3xy - 4xy = -7xy$
$(3x - 2y)(x - 2y)$	$-6xy - 2xy = -8xy$

This is the required middle term.

Thus,

$$3x^2 - 13xy + 4y^2 = (3x - y)(x - 4y).$$

✓ **CHECK POINT 9** Factor: $2x^2 - 7xy + 3y^2$.

EXAMPLE 10 Factoring by Substitution

Factor: $6y^4 + 13y^2 + 6$.

Solution Notice that the exponent on y^2 is half that of the exponent on y^4. We will let u equal the variable to the smaller power. Thus, let $u = y^2$.

$6(y^2)^2 + 13y^2 + 6$	This is the given polynomial, with y^4 written as $(y^2)^2$.
$= 6u^2 + 13u + 6$	Let $u = y^2$. Rewrite the trinomial in terms of u.
$= (3u + 2)(2u + 3)$	Factor the trinomial.
$= (3y^2 + 2)(2y^2 + 3)$	Now substitute y^2 for u.

Therefore, $6y^4 + 13y^2 + 6 = (3y^2 + 2)(2y^2 + 3)$. Check using FOIL multiplication.

✓ **CHECK POINT 10** Factor: $3y^4 + 10y^2 - 8$.

4 Factor trinomials by grouping.

Factoring Trinomials by Grouping

A second method for factoring $ax^2 + bx + c$, $a \neq 1$, is called the **grouping method**. This method involves both trial and error, as well as grouping. The trial and error in factoring $ax^2 + bx + c$ depends upon finding two numbers, p and q, for which $p + q = b$. Then we factor $ax^2 + px + qx + c$ using grouping.

Let's see how this works by looking at a particular factorization:

$$15x^2 - 7x - 2 = (3x - 2)(5x + 1).$$

If we multiply using FOIL on the right, we obtain

$$(3x - 2)(5x + 1) = 15x^2 + 3x - 10x - 2.$$

In this case, the desired numbers, p and q, are $p = 3$ and $q = -10$. Compare these numbers to ac and b in the given polynomial.

$a = 15$ $b = -7$ $c = -2$

$$15x^2 - 7x - 2$$

$ac = 15(-2) = -30$

Can you see that p and q, 3 and -10, are factors of ac, or -30? Furthermore, p and q have a sum of b, namely -7. By expressing the middle term, $-7x$, in terms of p and q, we can factor by grouping as follows:

$15x^2 - 7x - 2$	
$= 15x^2 + (3x - 10x) - 2$	Rewrite $-7x$ as $3x - 10x$.
$= (15x^2 + 3x) + (-10x - 2)$	Group terms.
$= 3x(5x + 1) - 2(5x + 1)$	Factor from each group.
$= (5x + 1)(3x - 2)$.	Factor out $5x + 1$, the common binomial factor.

Factoring $ax^2 + bx + c$ Using Grouping $(a \neq 1)$

1. Multiply the leading coefficient, a, and the constant, c.
2. Find the factors of ac whose sum is b.
3. Rewrite the middle term, bx, as a sum or difference using the factors from step 2.
4. Factor by grouping.

EXAMPLE 11 Factoring a Trinomial by Grouping

Factor by grouping: $12x^2 - 5x - 2$.

Solution The trinomial is of the form $ax^2 + bx + c$.

$$12x^2 - 5x - 2$$

$a = 12$ $b = -5$ $c = -2$

Step 1. Multiply the leading coefficient, a, and the constant, c. Using $a = 12$ and $c = -2$,

$$ac = 12(-2) = -24.$$

Step 2. Find the factors of ac whose sum is b. We want the factors of -24 whose sum is b, or -5. The factors of -24 whose sum is -5 are -8 and 3.

Step 3. Rewrite the middle term, $-5x$, as a sum or difference using the factors from step 2, -8 and 3.

$$12x^2 - 5x - 2 = 12x^2 - 8x + 3x - 2$$

Step 4. Factor by grouping.

$$= (12x^2 - 8x) + (3x - 2) \quad \text{Group terms.}$$
$$= 4x(3x - 2) + 1(3x - 2) \quad \text{Factor from each group.}$$
$$= (3x - 2)(4x + 1) \quad \text{Factor out } 3x - 2\text{, the common binomial factor.}$$

Thus,

$$12x^2 - 5x - 2 = (3x - 2)(4x + 1).$$

Discover for Yourself

In step 2, we found that the desired numbers were -8 and 3. We wrote $-5x$ as $-8x + 3x$. What happens if we write $-5x$ as $3x - 8x$? Use factoring by grouping on

$$12x^2 - 5x - 2$$
$$= 12x^2 + 3x - 8x - 12.$$

Describe what happens.

✓ CHECK POINT 11 Factor by grouping: $8x^2 - 22x + 5$.

5.4 EXERCISE SET

MyMathLab MathXL PRACTICE WATCH DOWNLOAD READ REVIEW

Practice Exercises

In Exercises 1–30, factor each trinomial, or state that the trinomial is prime. Check each factorization using FOIL multiplication.

1. $x^2 + 5x + 6$
2. $x^2 + 10x + 9$
3. $x^2 + 8x + 12$
4. $x^2 + 8x + 15$
5. $x^2 + 9x + 20$
6. $x^2 + 11x + 24$
7. $y^2 + 10y + 16$
8. $y^2 + 9y + 18$
9. $x^2 - 8x + 15$
10. $x^2 - 5x + 6$
11. $y^2 - 12y + 20$
12. $y^2 - 25y + 24$
13. $a^2 + 5a - 14$
14. $a^2 + a - 12$
15. $x^2 + x - 30$

16. $x^2 + 14x - 32$

17. $x^2 - 3x - 28$

18. $x^2 - 4x - 21$

19. $y^2 - 5y - 36$

20. $y^2 - 3y - 40$

21. $x^2 - x + 7$

22. $x^2 + 3x + 8$

23. $x^2 - 9xy + 14y^2$

24. $x^2 - 8xy + 15y^2$

25. $x^2 - xy - 30y^2$

26. $x^2 - 3xy - 18y^2$

27. $x^2 + xy + y^2$

28. $x^2 - xy + y^2$

29. $a^2 - 18ab + 80b^2$

30. $a^2 - 18ab + 45b^2$

In Exercises 31–38, factor completely.

31. $3x^2 + 3x - 18$

32. $4x^2 - 4x - 8$

33. $2x^3 - 14x^2 + 24x$

34. $2x^3 + 6x^2 + 4x$

35. $3y^3 - 15y^2 + 18y$

36. $4y^3 + 12y^2 - 72y$

37. $2x^4 - 26x^3 - 96x^2$

38. $3x^4 + 54x^3 + 135x^2$

In Exercises 39–44, factor by introducing an appropriate substitution.

39. $x^6 - x^3 - 6$

40. $x^6 + x^3 - 6$

41. $x^4 - 5x^2 - 6$

42. $x^4 - 4x^2 - 5$

43. $(x + 1)^2 + 6(x + 1) + 5$ (Let $u = x + 1$.)

44. $(x + 1)^2 + 8(x + 1) + 7$ (Let $u = x + 1$.)

In Exercises 45–68, use the method of your choice to factor each trinomial, or state that the trinomial is prime. Check each factorization using FOIL multiplication.

45. $3x^2 + 8x + 5$

46. $2x^2 + 9x + 7$

47. $5x^2 + 56x + 11$

48. $5x^2 - 16x + 3$

49. $3y^2 + 22y - 16$

50. $5y^2 + 33y - 14$

51. $4y^2 + 9y + 2$

52. $8y^2 + 10y + 3$

53. $10x^2 + 19x + 6$

54. $6x^2 + 19x + 15$

55. $8x^2 - 18x + 9$

56. $4x^2 - 27x + 18$

57. $6y^2 - 23y + 15$

58. $16y^2 - 6y - 27$

59. $6y^2 + 14y + 3$

60. $4y^2 + 22y - 5$

61. $3x^2 + 4xy + y^2$

62. $2x^2 + 3xy + y^2$

63. $6x^2 - 7xy - 5y^2$

64. $6x^2 - 5xy - 6y^2$

65. $15x^2 - 31xy + 10y^2$

66. $15x^2 + 11xy - 14y^2$

67. $3a^2 - ab - 14b^2$

68. $15a^2 - ab - 6b^2$

In Exercises 69–82, factor completely.

69. $15x^3 - 25x^2 + 10x$

70. $10x^3 + 24x^2 + 14x$

71. $24x^4 + 10x^3 - 4x^2$

72. $15x^4 - 39x^3 + 18x^2$

73. $15y^5 - 2y^4 - y^3$

74. $10y^5 - 17y^4 + 3y^3$

75. $24x^2 + 3xy - 27y^2$

76. $12x^2 + 10xy - 8y^2$

77. $6a^2b - 2ab - 60b$

78. $8a^2b + 34ab - 84b$

79. $12x^2y - 34xy^2 + 14y^3$

80. $12x^2y - 46xy^2 + 14y^3$

81. $13x^3y^3 + 39x^3y^2 - 52x^3y$

82. $4x^3y^5 + 24x^2y^5 - 64xy^5$

In Exercises 83–92, factor by introducing an appropriate substitution.

83. $2x^4 - x^2 - 3$

84. $5x^4 + 2x^2 - 3$

85. $2x^6 + 11x^3 + 15$

86. $2x^6 + 13x^3 + 15$

87. $2y^{10} + 7y^5 + 3$

88. $5y^{10} + 29y^5 - 42$

89. $5(x + 1)^2 + 12(x + 1) + 7$ (Let $u = x + 1$.)

90. $3(x + 1)^2 - 5(x + 1) + 2$ (Let $u = x + 1$.)

91. $2(x - 3)^2 - 5(x - 3) - 7$

92. $3(x - 2)^2 - 5(x - 2) - 2$

Practice PLUS

In Exercises 93–100, factor completely.

93. $x^2 - 0.5x + 0.06$

94. $x^2 + 0.3x - 0.04$

95. $x^2 - \frac{3}{49} + \frac{2}{7}x$

96. $x^2 - \frac{6}{25} + \frac{1}{5}x$

97. $acx^2 - bcx + adx - bd$

98. $acx^2 - bcx - adx + bd$

99. $-4x^5y^2 + 7x^4y^3 - 3x^3y^4$

100. $-5x^4y^3 + 7x^3y^4 - 2x^2y^5$

101. If $(fg)(x) = 3x^2 - 22x + 39$, find f and g.

102. If $(fg)(x) = 4x^2 - x - 5$, find f and g.

In Exercises 103–104, a large rectangle formed by a number of smaller rectangles is shown. Factor the sum of the areas of the smaller rectangles to determine the dimensions of the large rectangle.

103.

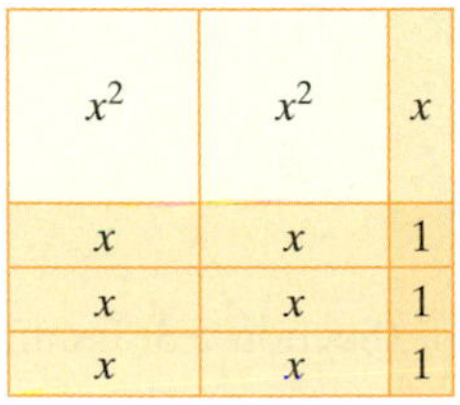

x^2	x^2	x
x	x	1
x	x	1
x	x	1

104.

x^2	x	x	x	x
x	1	1	1	1

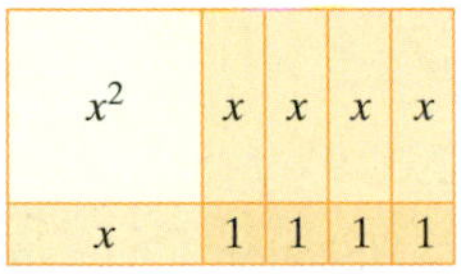

Application Exercises

105. A diver jumps directly upward from a board that is 32 feet high. The function

$$f(t) = -16t^2 + 16t + 32$$

describes the driver's height above the water, $f(t)$, in feet, after t seconds.
   a. Find and interpret $f(1)$.

   b. Find and interpret $f(2)$.

   c. Factor the expression for $f(t)$ and write the function in completely factored form.
   d. Use the factored form of the function to find $f(1)$ and $f(2)$.

106. The function $V(x) = 3x^3 - 2x^2 - 8x$ describes the volume, $V(x)$, in cubic inches, of the box shown whose height is x inches.

   a. Find and interpret $V(4)$.

   b. Factor the expression for $V(x)$ and write the function in completely factored form.
   c. Use the factored form of the function to find $V(4)$ and $V(5)$.

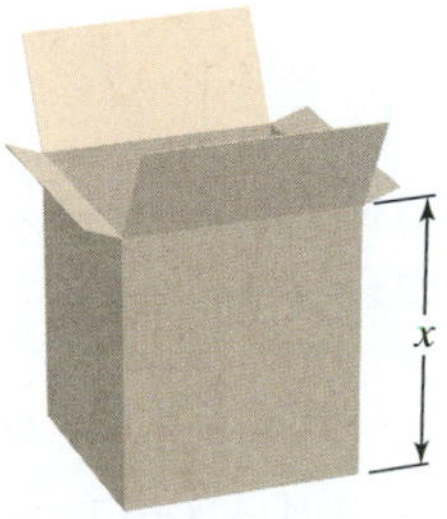

107. Find the area of the large rectangle shown below in two ways.

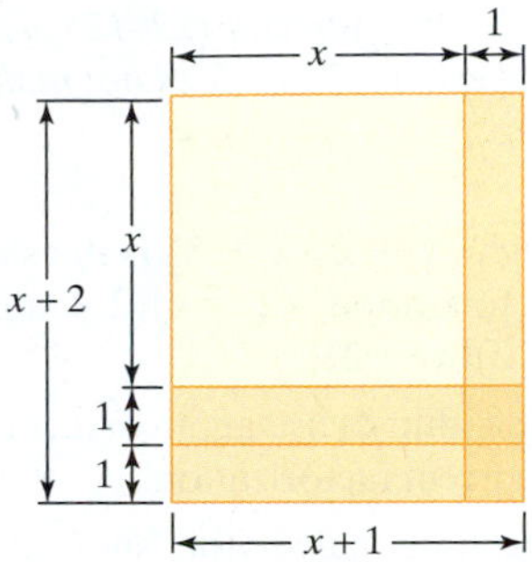

a. Find the sum of the areas of the six smaller rectangles and squares.
   b. Express the area of the large rectangle as the product of its length and width.
   c. Explain how the figure serves as a geometric model for the factorization of the sum that you wrote in part (a).

108. If x represents a positive integer, factor $x^3 + 3x^2 + 2x$ to show that the trinomial represents the product of three consecutive integers.

Writing in Mathematics

109. Explain how to factor $x^2 + 8x + 15$.

110. Give two helpful suggestions for factoring $x^2 - 5x + 6$.

111. In factoring $x^2 + bx + c$, describe how the last term in each binomial factor is related to b and to c.

112. Describe the first thing that you should try doing when factoring a polynomial.

113. What does it mean to factor completely?

114. Explain how to factor $x^6 - 7x^3 + 10$ by substitution.

115. Is it possible to factor $x^6 - 7x^3 + 10$ without using substitution? How might this be done?

116. Explain how to factor $2x^2 - x - 1$.

Technology Exercises

In Exercises 117–120, use a graphing utility to graph the function on each side of the equation in the same viewing rectangle. Use end behavior to show a complete picture of the polynomial function on the left side. Do the graphs coincide? If so, this means that the polynomial on the left side has been factored correctly. If not, factor the polynomial correctly and then use your graphing utility to verify the factorization.

117. $x^2 + 7x + 12 = (x + 4)(x + 3)$

118. $x^2 - 7x + 6 = (x - 2)(x - 3)$

119. $6x^3 + 5x^2 - 4x = x(3x + 4)(2x - 1)$

120. $x^4 - x^2 - 20 = (x^2 + 5)(x^2 - 4)$

121. Use the ⎡TABLE⎤ feature of a graphing utility to verify any two of your factorizations in Exercises 39–44.

Critical Thinking Exercises

Make Sense? *In Exercises 122–125, determine whether each statement "makes sense" or "does not make sense" and explain your reasoning.*

122. Although $(x + 2)(x - 5)$ is the same as $(x - 5)(x + 2)$, the factorization $(2 - x)(2 + x)$ is not the same as $-(x - 2)(x + 2)$.

123. I'm often able to use an incorrect factorization to lead me to the correct factorization.

124. My graphing calculator showed the same graph for $y_1 = 20x^3 - 70x^2 + 60x$ and $y_2 = 10x(2x^2 - 7x + 6)$, so I can conclude that the complete factorization of $20x^3 - 70x^2 + 60x$ is $10x(2x^2 - 7x + 6)$.

125. First factoring out the greatest common factor makes it easier for me to determine how to factor the remaining factor, assuming that it is not prime.

In Exercises 126–129, determine whether each statement is true or false. If the statement is false, make the necessary change(s) to produce a true statement.

126. Once a GCF is factored from $6y^6 - 19y^5 + 10y^4$, the remaining trinomial factor is prime.

127. One factor of $8y^2 - 51y + 18$ is $8y - 3$.

128. We can immediately tell that $6x^2 - 11xy - 10y^2$ is prime because 11 is a prime number and the polynomial contains two variables.

129. A factor of $12x^2 - 19xy + 5y^2$ is $4x - y$.

In Exercises 130–131, find all integers b so that the trinomial can be factored.

130. $4x^2 + bx - 1$

131. $3x^2 + bx + 5$

In Exercises 132–137, factor each polynomial. Assume that all variable exponents represent whole numbers.

132. $9x^{2n} + x^n - 8$

133. $4x^{2n} - 9x^n + 5$

134. $a^{2n+2} - a^{n+2} - 6a^2$

135. $b^{2n+2} + 3b^{n+2} - 10b^2$

136. $3c^{n+2} - 10c^{n+1} + 3c^n$

137. $2d^{n+2} - 5d^{n+1} + 3d^n$

Review Exercises

138. Solve: $-2x \le 6$ and $-2x + 3 < -7$. (Section 4.2, Example 2)

139. Solve the system:

$$\begin{aligned} 2x - y - 2z &= -1 \\ x - 2y - z &= 1 \\ x + y + z &= 4. \end{aligned}$$

(Section 3.3, Example 2)

140. Factor: $4x^3 + 8x^2 - 5x - 10$. (Section 5.3, Example 5)

Preview Exercises

Exercises 141–143 will help you prepare for the material covered in the next section. In each exercise, factor the polynomial. (You'll soon be learning techniques that will shorten the factoring process.)

141. $x^2 + 14x + 49$

142. $x^2 - 8x + 16$

143. $x^2 - 25$ (or $x^2 + 0x - 25$)

MID-CHAPTER CHECK POINT Section 5.1–Section 5.4

What You Know: We learned the vocabulary of polynomials and observed the smooth, continuous graphs of polynomial functions. We used the Leading Coefficient Test to describe the end behavior of these graphs. We learned to add, subtract, and multiply polynomials. We used a number of fast methods for finding products of polynomials, including the FOIL method for multiplying binomials, special-product formulas for squaring binomials $[(A + B)^2 = A^2 + 2AB + B^2; (A - B)^2 = A^2 - 2AB + B^2]$, and a special-product formula for the product of the sum and difference of two terms $[(A + B)(A - B) = A^2 - B^2]$. We learned to factor out a polynomial's greatest common factor and to use grouping to factor polynomials with more than three terms. We factored polynomials with three terms, beginning with trinomials with leading coefficient 1 and moving on to $ax^2 + bx + c$, with $a \neq 1$. We saw that the factoring process should begin by looking for a GCF and, if there is one, factoring it out first.

In Exercises 1–18, perform the indicated operations.

1. $(-8x^3 + 6x^2 - x + 5) - (-7x^3 + 2x^2 - 7x - 12)$

2. $(6x^2yz^4)\left(-\frac{1}{3}x^5y^2z\right)$

3. $5x^2y\left(6x^3y^2 - 7xy - \frac{2}{5}\right)$

4. $(3x - 5)(x^2 + 3x - 8)$

5. $(x^2 - 2x + 1)(2x^2 + 3x - 4)$

6. $(x^2 - 2x + 1) - (2x^2 + 3x - 4)$

7. $(6x^3y - 11x^2y - 4y) + (-11x^3y + 5x^2y - y - 6) - (-x^3y + 2y - 1)$

8. $(2x + 5)(4x - 1)$

9. $(2xy - 3)(5xy + 2)$

10. $(3x - 2y)(3x + 2y)$

11. $(3xy + 1)(2x^2 - 3y)$

12. $(7x^3y + 5x)(7x^3y - 5x)$
13. $3(x + h)^2 - 2(x + h) + 5 - (3x^2 - 2x + 5)$
14. $(x^2 - 3)^2$
15. $(x^2 - 3)(x^3 + 5x + 2)$
16. $(2x + 5y)^2$
17. $(x + 6 + 3y)(x + 6 - 3y)$
18. $(x + y + 5)^2$

In Exercises 19–30, factor completely, or state that the polynomial is prime.

19. $x^2 - 5x - 24$
20. $15xy + 5x + 6y + 2$
21. $5x^2 + 8x - 4$
22. $35x^2 + 10x - 50$
23. $9x^2 - 9x - 18$
24. $10x^3y^2 - 20x^2y^2 + 35x^2y$
25. $18x^2 + 21x + 5$
26. $12x^2 - 9xy - 16x + 12y$
27. $9x^2 - 15x + 4$
28. $3x^6 + 11x^3 + 10$
29. $25x^3 + 25x^2 - 14x$
30. $2x^4 - 6x - x^3y + 3y$

SECTION 5.5

Factoring Special Forms

Objectives

1. Factor the difference of two squares.
2. Factor perfect square trinomials.
3. Use grouping to obtain the difference of two squares.
4. Factor the sum or difference of two cubes.

Bees use honeycombs to store honey and house larvae. They construct honey storage cells from wax. Each cell has the shape of a six-sided figure whose sides are all the same length and whose angles all have the same measure, called a regular hexagon. The cells fit together perfectly, preventing dirt or predators from entering. Squares or equilateral triangles would fit equally well, but regular hexagons provide the largest storage space for the amount of wax used.

In this section, we develop factoring techniques by reversing the formulas for special products discussed in Section 5.2. Like the construction of honeycombs, these factorizations can be visualized by perfectly fitting together "cells" of squares and rectangles to form larger rectangles.

1 Factor the difference of two squares.

Factoring the Difference of Two Squares

A method for factoring the difference of two squares is obtained by reversing the special product for the sum and difference of two terms.

The Difference of Two Squares

If A and B are real numbers, variables, or algebraic expressions, then

$$A^2 - B^2 = (A + B)(A - B).$$

In words: The difference of the squares of two terms factors as the product of a sum and a difference of those terms.

EXAMPLE 1 Factoring the Difference of Two Squares

Factor:

a. $9x^2 - 100$ **b.** $36y^6 - 49x^4$.

Solution We must express each term as the square of some monomial. Then we use the formula for factoring $A^2 - B^2$.

a. $9x^2 - 100 = (3x)^2 - 10^2 = (3x + 10)(3x - 10)$

$$A^2 - B^2 = (A + B)(A - B)$$

b. $36y^6 - 49x^4 = (6y^3)^2 - (7x^2)^2 = (6y^3 + 7x^2)(6y^3 - 7x^2)$

In order to apply the factoring formula for $A^2 - B^2$, each term must be the square of an integer or a polynomial.

- A number that is the square of an integer is called a **perfect square**. For example, 100 is a perfect square because $100 = 10^2$.
- Any exponential expression involving a perfect-square coefficient and variables to even powers is a perfect square. For example, $100y^6$ is a perfect square because $100y^6 = (10y^3)^2$.

Study Tip

It's helpful to recognize perfect squares. Here are 16 perfect squares, each printed in boldface.

$\mathbf{1} = 1^2$	$\mathbf{25} = 5^2$	$\mathbf{81} = 9^2$	$\mathbf{169} = 13^2$
$\mathbf{4} = 2^2$	$\mathbf{36} = 6^2$	$\mathbf{100} = 10^2$	$\mathbf{196} = 14^2$
$\mathbf{9} = 3^2$	$\mathbf{49} = 7^2$	$\mathbf{121} = 11^2$	$\mathbf{225} = 15^2$
$\mathbf{16} = 4^2$	$\mathbf{64} = 8^2$	$\mathbf{144} = 12^2$	$\mathbf{256} = 16^2$

✓ CHECK POINT 1 Factor:

a. $16x^2 - 25$ **b.** $100y^6 - 9x^4$.

Be careful when determining whether or not to apply the factoring formula for the difference of two squares.

Prime Over the Integers

- $x^2 - 5$ — 5 is not a perfect square.
- $x^7 - 25$ — 7 is an odd power. x^7 is not the square of any integer power of x.

Factorable

- $1 - x^6y^4$ — Perfect square: $1 = 1^2$. Even powers. Perfect square: $x^6y^4 = (x^3y^2)^2$

When factoring, always check first for common factors. If there are common factors, factor out the GCF and then factor the resulting polynomial.

EXAMPLE 2 Factoring Out the GCF and Then Factoring the Difference of Two Squares

Factor: $3y - 3x^6y^5$.

Solution The GCF of the two terms of the polynomial is $3y$. We begin by factoring out $3y$.

$$3y - 3x^6y^5 = 3y(1 - x^6y^4) = 3y[1^2 - (x^3y^2)^2] = 3y(1 + x^3y^2)(1 - x^3y^2)$$

Factor out the GCF. $A^2 - B^2 = (A + B)(A - B)$

CHECK POINT 2 Factor: $6y - 6x^2y^7$.

We have seen that a polynomial is factored completely when it is written as the product of prime polynomials. To be sure that you have factored completely, check to see whether any factors with more than one term in the factored polynomial can be factored further. If so, continue factoring.

EXAMPLE 3 A Repeated Factorization

Factor completely: $81x^4 - 16$.

Solution

$81x^4 - 16 = (9x^2)^2 - 4^2$ — Express as the difference of two squares.

$= (9x^2 + 4)(9x^2 - 4)$ — The factors are the sum and difference of the expressions being squared.

$= (9x^2 + 4)[(3x)^2 - 2^2]$ — The factor $9x^2 - 4$ is the difference of two squares and can be factored.

$= (9x^2 + 4)(3x + 2)(3x - 2)$ — The factors of $9x^2 - 4$ are the sum and difference of the expressions being squared.

Study Tip

Factoring $81x^4 - 16$ as

$$(9x^2 + 4)(9x^2 - 4)$$

is not a complete factorization. The second factor, $9x^2 - 4$, is itself a difference of two squares and can be factored.

Are you tempted to further factor $9x^2 + 4$, the sum of two squares, in Example 3? Resist the temptation! **The sum of two squares, $A^2 + B^2$, with no common factor other than 1 is a prime polynomial.**

CHECK POINT 3 Factor completely: $16x^4 - 81$.

In our next example, we begin with factoring by grouping. We can then factor further using the difference of two squares.

EXAMPLE 4 Factoring Completely

Factor completely: $x^3 + 5x^2 - 9x - 45$.

Solution

$x^3 + 5x^2 - 9x - 45$

$= (x^3 + 5x^2) + (-9x - 45)$ — Group terms with common factors.

$= x^2(x + 5) - 9(x + 5)$ — Factor out the common factor from each group.

$= (x + 5)(x^2 - 9)$ — Factor out $x + 5$, the common binomial factor, from both terms.

$= (x + 5)(x + 3)(x - 3)$ — Factor $x^2 - 3^2$, the difference of two squares.

CHECK POINT 4 Factor completely: $x^3 + 7x^2 - 4x - 28$.

In Examples 1–4, we used the formula for factoring the difference of two squares. Although we obtained the formula by reversing the special product for the sum and difference of two terms, it can also be obtained geometrically.

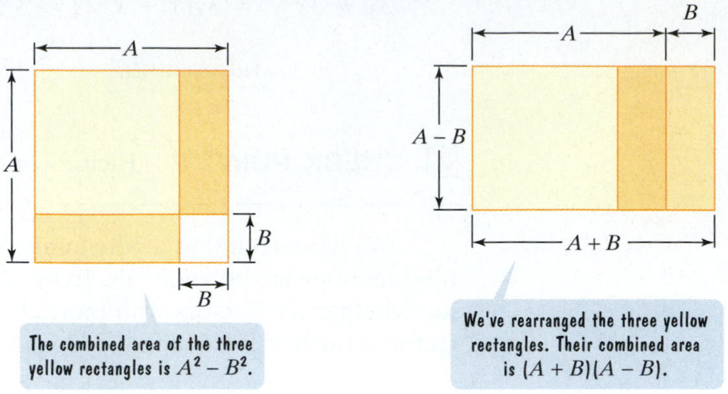
A
A
B
B
The combined area of the three yellow rectangles is $A^2 - B^2$.
B
A
A − B
A + B
We've rearranged the three yellow rectangles. Their combined area is $(A + B)(A - B)$.

Because the three yellow rectangles make up the same combined area in both figures,

$$A^2 - B^2 = (A + B)(A - B).$$

2 Factor perfect square trinomials.

Factoring Perfect Square Trinomials

Our next factoring technique is obtained by reversing the special products for squaring binomials. The trinomials that are factored using this technique are called **perfect square trinomials**.

Factoring Perfect Square Trinomials

Let A and B be real numbers, variables, or algebraic expressions.

1. $A^2 + 2AB + B^2 = (A + B)^2$ (Same sign)
2. $A^2 - 2AB + B^2 = (A - B)^2$ (Same sign)

The two items in the box show that perfect square trinomials, $A^2 + 2AB + B^2$ and $A^2 - 2AB + B^2$, come in two forms: one in which the coefficient of the middle term is positive and one in which the coefficient of the middle term is negative. Here's how to recognize a perfect square trinomial:

1. The first and last terms are squares of monomials or integers.
2. The middle term is twice the product of the expressions being squared in the first and last terms.

EXAMPLE 5 Factoring Perfect Square Trinomials

Factor:

a. $x^2 + 14x + 49$ **b.** $4x^2 + 12xy + 9y^2$ **c.** $9y^4 - 12y^2 + 4$.

Solution

a. $x^2 + 14x + 49 = x^2 + 2 \cdot x \cdot 7 + 7^2 = (x + 7)^2$

$A^2 + 2AB + B^2 = (A + B)^2$

The middle term has a positive sign.

b. We suspect that $4x^2 + 12xy + 9y^2$ is a perfect square trinomial because $4x^2 = (2x)^2$ and $9y^2 = (3y)^2$. The middle term can be expressed as twice the product of $2x$ and $3y$.

$$4x^2 + 12xy + 9y^2 = (2x)^2 + 2 \cdot 2x \cdot 3y + (3y)^2 = (2x + 3y)^2$$

$$A^2 + 2AB + B^2 = (A + B)^2$$

c. $9y^4 - 12y^2 + 4 = (3y^2)^2 - 2 \cdot 3y^2 \cdot 2 + 2^2 = (3y^2 - 2)^2$

$$A^2 - 2AB + B^2 = (A - B)^2$$

The middle term has a negative sign.

✓ CHECK POINT 5 Factor:

a. $x^2 + 6x + 9$

b. $16x^2 + 40xy + 25y^2$

c. $4y^4 - 20y^2 + 25.$

---

3 Use grouping to obtain the difference of two squares.

Using Special Forms When Factoring by Grouping

If a polynomial contains four terms, try factoring by grouping. In the next example, we group the terms to obtain the difference of two squares. One of the squares is a perfect square trinomial.

EXAMPLE 6 Using Grouping to Obtain the Difference of Two Squares

Factor: $x^2 - 8x + 16 - y^2.$

Solution

$x^2 - 8x + 16 - y^2$

$= (x^2 - 8x + 16) - y^2$ — Group as a perfect square trinomial minus y^2 to obtain a difference of two squares.

$= (x - 4)^2 - y^2$ — Factor the perfect square trinomial.

$= (x - 4 + y)(x - 4 - y)$ — Factor the difference of two squares. The factors are the sum and difference of the expressions being squared.

✓ CHECK POINT 6 Factor: $x^2 + 10x + 25 - y^2.$

EXAMPLE 7 Using Grouping to Obtain the Difference of Two Squares

Factor: $a^2 - b^2 + 10b - 25.$

Solution Grouping into two groups of two terms does not result in a common binomial factor. Let's look for a perfect square trinomial. Can you see that the perfect square trinomial is the expression being subtracted from a^2?

$a^2 - b^2 + 10b - 25$

$= a^2 - (b^2 - 10b + 25)$ — Factor out −1 and group as a^2 − (perfect square trinomial) to obtain a difference of two squares.

$= a^2 - (b - 5)^2$ — Factor the perfect square trinomial.

$= [a + (b - 5)][a - (b - 5)]$ — Factor the difference of squares. The factors are the sum and difference of the expressions being squared.

$= (a + b - 5)(a - b + 5)$ — Simplify.

✓ CHECK POINT 7 Factor: $a^2 - b^2 + 4b - 4$.

4 Factor the sum or difference of two cubes.

Factoring the Sum or Difference of Two Cubes

Here are two multiplications that lead to factoring formulas for the sum of two cubes and the difference of two cubes:

$$\begin{aligned}(A + B)(A^2 - AB + B^2) &= A(A^2 - AB + B^2) + B(A^2 - AB + B^2)\\ &= A^3 - A^2B + AB^2 + A^2B - AB^2 + B^3\\ &= A^3 + B^3\end{aligned}$$

The product results in the sum of two cubes.

Combine like terms: $-A^2B + A^2B = 0$ and $AB^2 - AB^2 = 0$.

and

$$\begin{aligned}(A - B)(A^2 + AB + B^2) &= A(A^2 + AB + B^2) - B(A^2 + AB + B^2)\\ &= A^3 + A^2B + AB^2 - A^2B - AB^2 - B^3\\ &= A^3 - B^3.\end{aligned}$$

The product results in the difference of two cubes.

Combine like terms: $A^2B - A^2B = 0$ and $AB^2 - AB^2 = 0$.

By reversing the two sides of these equations, we obtain formulas that allow us to factor a sum or difference of two cubes. These formulas should be memorized.

Factoring the Sum or Difference of Two Cubes

1. Factoring the Sum of Two Cubes

$$A^3 + B^3 = (A + B)(A^2 - AB + B^2)$$

Same signs Opposite signs

2. Factoring the Difference of Two Cubes

$$A^3 - B^3 = (A - B)(A^2 + AB + B^2)$$

Same signs Opposite signs

EXAMPLE 8 Factoring the Sum of Two Cubes

Factor:

a. $x^3 + 125$ b. $x^6 + 64y^3$.

Solution We must express each term as the cube of some monomial. Then we use the formula for factoring $A^3 + B^3$.

a. $x^3 + 125 = x^3 + 5^3 = (x + 5)(x^2 - x \cdot 5 + 5^2) = (x + 5)(x^2 - 5x + 25)$

$A^3 + B^3 = (A + B)(A^2 - AB + B^2)$

b. $x^6 + 64y^3 = (x^2)^3 + (4y)^3 = (x^2 + 4y)[(x^2)^2 - x^2 \cdot 4y + (4y)^2]$

$A^3 + B^3 = (A + B)(A^2 - AB + B^2)$

$= (x^2 + 4y)(x^4 - 4x^2y + 16y^2)$

Study Tip

When factoring the sum or difference of cubes, it is helpful to recognize the following cubes:

$$1 = 1^3$$
$$8 = 2^3$$
$$27 = 3^3$$
$$64 = 4^3$$
$$125 = 5^3$$
$$216 = 6^3$$
$$1000 = 10^3.$$

CHECK POINT 8 Factor:

a. $x^3 + 27$

b. $x^6 + 1000y^3$.

EXAMPLE 9 Factoring the Difference of Two Cubes

Factor:

a. $x^3 - 216$ b. $8 - 125x^3y^3$.

Solution We must express each term as the cube of some monomial. Then we use the formula for factoring $A^3 - B^3$.

a. $x^3 - 216 = x^3 - 6^3 = (x - 6)(x^2 + x \cdot 6 + 6^2) = (x - 6)(x^2 + 6x + 36)$

$A^3 - B^3 = (A - B)(A^2 + AB + B^2)$

b. $8 - 125x^3y^3 = 2^3 - (5xy)^3 = (2 - 5xy)[2^2 + 2 \cdot 5xy + (5xy)^2]$

$A^3 - B^3 = (A - B)(A^2 + AB + B^2)$

$= (2 - 5xy)(4 + 10xy + 25x^2y^2)$

CHECK POINT 9 Factor:

a. $x^3 - 8$

b. $1 - 27x^3y^3$.

5.5 EXERCISE SET

MyMathLab MathXL PRACTICE WATCH DOWNLOAD READ REVIEW

Practice Exercises

In Exercises 1–22, factor each difference of two squares. Assume that any variable exponents represent whole numbers.

1. $x^2 - 4$
2. $x^2 - 16$
3. $9x^2 - 25$
4. $4x^2 - 9$
5. $9 - 25y^2$
6. $16 - 49y^2$
7. $36x^2 - 49y^2$
8. $64x^2 - 25y^2$
9. $x^2y^2 - 1$
10. $x^2y^2 - 100$
11. $9x^4 - 25y^6$
12. $25x^4 - 9y^6$
13. $x^{14} - y^4$
14. $x^4 - y^{10}$
15. $(x - 3)^2 - y^2$
16. $(x - 6)^2 - y^2$
17. $a^2 - (b - 2)^2$
18. $a^2 - (b - 3)^2$
19. $x^{2n} - 25$
20. $x^{2n} - 36$
21. $1 - a^{2n}$
22. $4 - b^{2n}$

In Exercises 23–48, factor completely, or state that the polynomial is prime.

23. $2x^3 - 8x$
24. $2x^3 - 72x$
25. $50 - 2y^2$
26. $72 - 2y^2$
27. $8x^2 - 8y^2$
28. $6x^2 - 6y^2$
29. $2x^3y - 18xy$

30. $2x^3y - 32xy$
31. $a^3b^2 - 49ac^2$
32. $4a^3c^2 - 16ax^2y^2$
33. $5y - 5x^2y^7$
34. $2y - 2x^6y^3$
35. $8x^2 + 8y^2$
36. $6x^2 + 6y^2$
37. $x^2 + 25y^2$
38. $x^2 + 36y^2$
39. $x^4 - 16$
40. $x^4 - 1$
41. $81x^4 - 1$
42. $1 - 81x^4$
43. $2x^5 - 2xy^4$
44. $3x^5 - 3xy^4$
45. $x^3 + 3x^2 - 4x - 12$
46. $x^3 + 3x^2 - 9x - 27$
47. $x^3 - 7x^2 - x + 7$
48. $x^3 - 6x^2 - x + 6$

In Exercises 49–64, factor any perfect square trinomials, or state that the polynomial is prime.

49. $x^2 + 4x + 4$
50. $x^2 + 2x + 1$
51. $x^2 - 10x + 25$
52. $x^2 - 14x + 49$
53. $x^4 - 4x^2 + 4$
54. $x^4 - 6x^2 + 9$
55. $9y^2 + 6y + 1$
56. $4y^2 + 4y + 1$
57. $64y^2 - 16y + 1$
58. $25y^2 - 10y + 1$
59. $x^2 - 12xy + 36y^2$
60. $x^2 + 16xy + 64y^2$
61. $x^2 - 8xy + 64y^2$
62. $x^2 - 9xy + 81y^2$
63. $9x^2 + 48xy + 64y^2$
64. $16x^2 - 40xy + 25y^2$

In Exercises 65–74, factor by grouping to obtain the difference of two squares.

65. $x^2 - 6x + 9 - y^2$
66. $x^2 - 12x + 36 - y^2$
67. $x^2 + 20x + 100 - x^4$
68. $x^2 + 16x + 64 - x^4$
69. $9x^2 - 30x + 25 - 36y^2$
70. $25x^2 - 20x + 4 - 81y^2$
71. $x^4 - x^2 - 2x - 1$
72. $x^4 - x^2 - 6x - 9$
73. $z^2 - x^2 + 4xy - 4y^2$
74. $z^2 - x^2 + 10xy - 25y^2$

In Exercises 75–94, factor using the formula for the sum or difference of two cubes.

75. $x^3 + 64$
76. $x^3 + 1$
77. $x^3 - 27$
78. $x^3 - 1000$
79. $8y^3 + 1$
80. $27y^3 + 1$
81. $125x^3 - 8$
82. $27x^3 - 8$
83. $x^3y^3 + 27$
84. $x^3y^3 + 64$
85. $64x - x^4$
86. $216x - x^4$
87. $x^6 + 27y^3$
88. $x^6 + 8y^3$
89. $125x^6 - 64y^6$
90. $125x^6 - y^6$
91. $x^9 + 1$
92. $x^9 - 1$
93. $(x - y)^3 - y^3$
94. $x^3 + (x + y)^3$

Practice PLUS

In Exercises 95–104, factor completely.

95. $0.04x^2 + 0.12x + 0.09$
96. $0.09x^2 - 0.12x + 0.04$
97. $8x^4 - \frac{x}{8}$
98. $27x^4 + \frac{x}{27}$
99. $x^6 - 9x^3 + 8$
100. $x^6 + 9x^3 + 8$
101. $x^8 - 15x^4 - 16$
102. $x^8 + 15x^4 - 16$
103. $x^5 - x^3 - 8x^2 + 8$
104. $x^5 - x^3 + 27x^2 - 27$

105. The figure shows four yellow rectangles that fit together to form a large square.

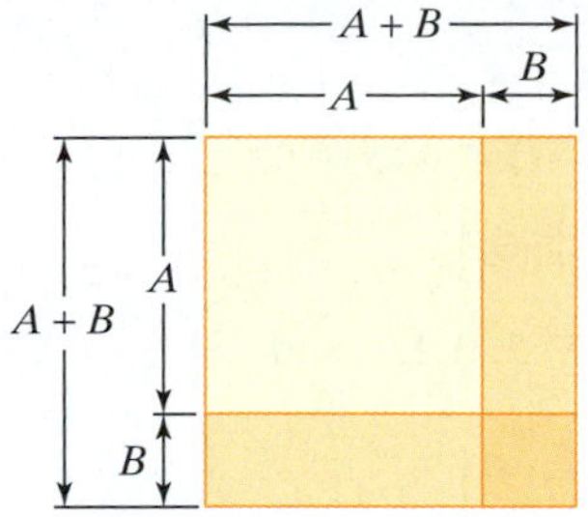

a. Express the area of the large square in terms of one of its sides, $A + B$.

b. Write an expression for the area of each of the four rectangles that form the large square.

c. Use the sum of the areas from part (b) to write a second expression for the area of the large square.

d. Set the expression from part (c) equal to the expression from part (a). What factoring technique have you established?

Application Exercises

In Exercises 106–109, find the formula for the area of the shaded region and express it in factored form.

106.

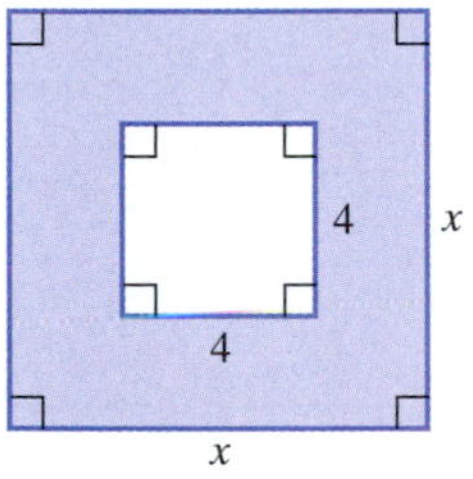

107.

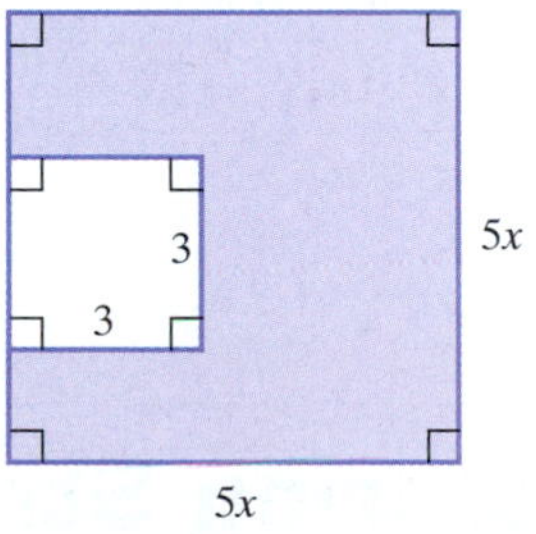

108.

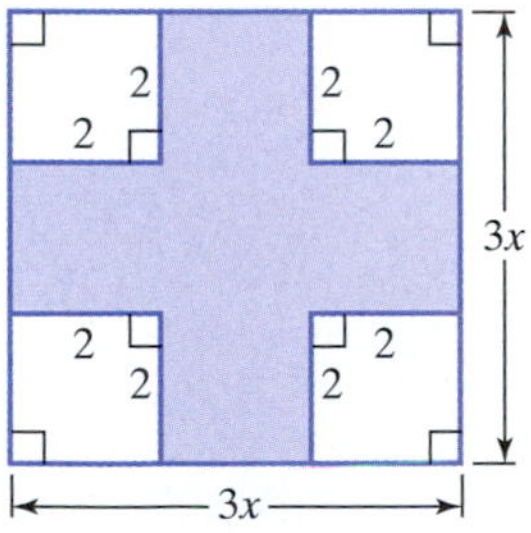

109.

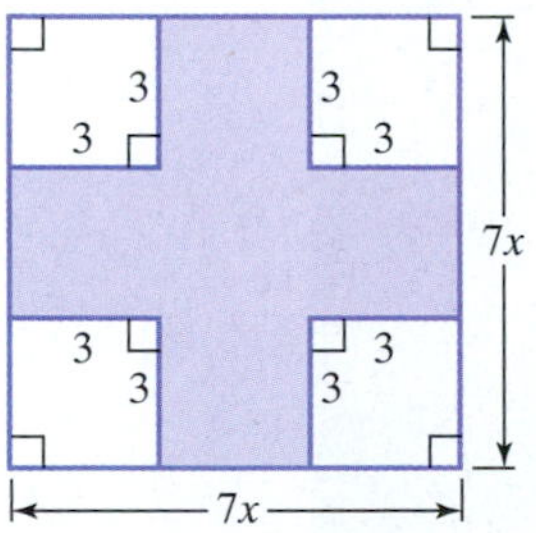

In Exercises 110–111, find the formula for the volume of the region outside the smaller rectangular solid and inside the larger rectangular solid. Then express the volume in factored form.

110.

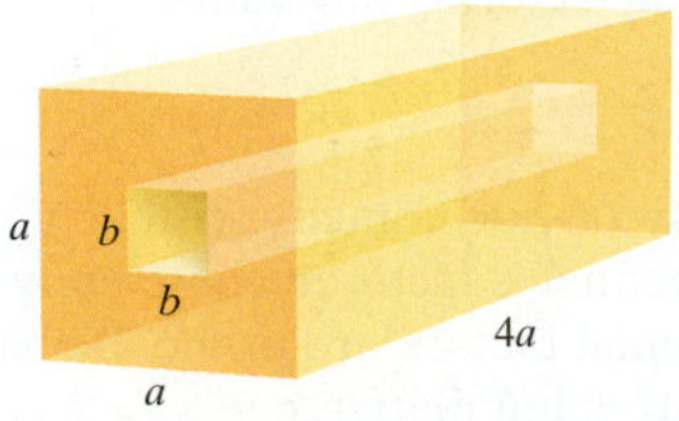

111.

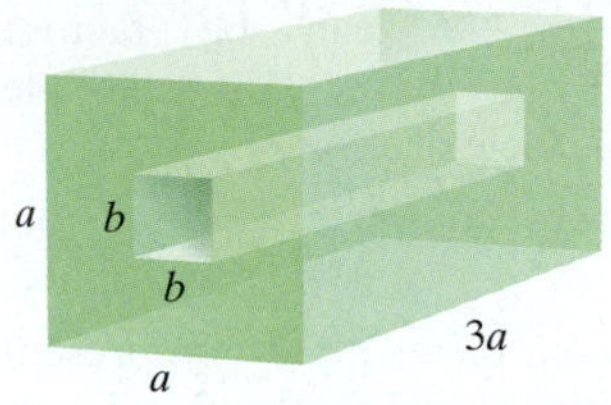

Writing in Mathematics

112. Explain how to factor the difference of two squares. Provide an example with your explanation.

113. What is a perfect square trinomial and how is it factored?

114. Explain how to factor $x^2 - y^2 + 8x - 16$. Should the expression be grouped into two groups of two terms? If not, why not, and what sort of grouping should be used?

115. Explain how to factor $x^3 + 1$.

Technology Exercises

In Exercises 116–123, use a graphing utility to graph the function on each side of the equation in the same viewing rectangle. Use end behavior to show a complete picture of the polynomial function on the left side. Do the graphs coincide? If so, this means that the polynomial on the left side has been factored correctly. If not, factor the polynomial correctly and then use your graphing utility to verify the factorization.

116. $9x^2 - 4 = (3x + 2)(3x - 2)$

117. $x^2 + 4x + 4 = (x + 4)^2$

118. $9x^2 + 12x + 4 = (3x + 2)^2$

119. $25 - (x^2 + 4x + 4) = (x + 7)(x - 3)$

120. $(2x + 3)^2 - 9 = 4x(x + 3)$

121. $(x - 3)^2 + 8(x - 3) + 16 = (x - 1)^2$

122. $x^3 - 1 = (x - 1)(x^2 - x + 1)$

123. $(x + 1)^3 + 1 = (x + 1)(x^2 + x + 1)$

124. Use the [TABLE] feature of a graphing utility to verify any two of your factorizations in Exercises 67–68 or 77–78.

Critical Thinking Exercises

Make Sense? *In Exercises 125–128, determine whether each statement "makes sense" or "does not make sense" and explain your reasoning.*

125. Although I can factor the difference of squares and perfect square trinomials using trial-and-error associated with FOIL, recognizing these special forms shortens the process.

126. Although $x^3 + 2x^2 - 5x - 6$ can be factored as $(x + 1)(x + 3)(x - 2)$, I have not yet learned techniques to obtain this factorization.

127. I factored $4x^2 - 100$ completely and obtained $(2x + 10)(2x - 10)$.

128. You told me that the area of a square is represented by $9x^2 + 12x + 4$ square inches, so I factored and concluded that the length of one side must be $3x + 2$ inches.

In Exercises 129–132, determine whether each statement is true or false. If the statement is false, make the necessary change(s) to produce a true statement.

129. $9x^2 + 15x + 25 = (3x + 5)^2$

130. $x^3 - 27 = (x - 3)(x^2 + 6x + 9)$

131. $x^3 - 64 = (x - 4)^3$

132. $4x^2 - 121 = (2x - 11)^2$

In Exercises 133–136, factor each polynomial completely. Assume that any variable exponents represent whole numbers.

133. $y^3 + x + x^3 + y$

134. $36x^{2n} - y^{2n}$

135. $x^{3n} + y^{12n}$

136. $4x^{2n} + 20x^n y^m + 25y^{2m}$

137. Factor $x^6 - y^6$ first as the difference of squares and then as the difference of cubes. From these two factorizations, determine a factorization for $x^4 + x^2y^2 + y^4$.

In Exercises 138–139, find all integers k so that the trinomial is a perfect square trinomial.

138. $kx^2 + 8xy + y^2$

139. $64x^2 - 16x + k$

Review Exercises

140. Solve: $2x + 2 \geq 12$ and $\frac{2x - 1}{3} \leq 7$. (Section 4.2, Example 2)

141. Solve using matrices:

$$\begin{aligned} 3x - 2y &= -8 \\ x + 6y &= 4. \end{aligned}$$

(Section 3.4, Example 2)

142. Factor: $3x^2 + 21x - xy - 7y$. (Section 5.3, Example 6)

Preview Exercises

Exercises 143–145 will help you prepare for the material covered in the next section. In each exercise, factor completely.

143. $2x^3 + 8x^2 + 8x$

144. $5x^3 - 40x^2y + 35xy^2$

145. $9b^2x + 9b^2y - 16x - 16y$

SECTION 5.6

A General Factoring Strategy

Objectives

1. Use the appropriate method for factoring a polynomial.
2. Use a general strategy for factoring polynomials.

Successful problem solving involves understanding the problem, devising a plan for solving it, and then carrying out the plan. In this section, you will learn a step-by-step strategy that provides a plan and direction for solving factoring problems.

1 Use the appropriate method for factoring a polynomial.

A Strategy for Factoring Polynomials

It is important to practice factoring a wide variety of polynomials so that you can quickly select the appropriate technique. The polynomial is factored completely when all its polynomial factors, except possibly for monomial factors, are prime. Because of the commutative property, the order of the factors does not matter.

2 Use a general strategy for factoring polynomials.

Here is a general strategy for factoring polynomials:

A Strategy for Factoring a Polynomial

1. If there is a common factor, factor out the GCF or factor out a common factor with a negative coefficient.
2. Determine the number of terms in the polynomial and try factoring as follows:
   a. If there are two terms, can the binomial be factored by using one of the following special forms?

      Difference of two squares: $A^2 - B^2 = (A + B)(A - B)$

      Sum of two cubes: $A^3 + B^3 = (A + B)(A^2 - AB + B^2)$

      Difference of two cubes: $A^3 - B^3 = (A - B)(A^2 + AB + B^2)$

   b. If there are three terms, is the trinomial a perfect square trinomial? If so, factor by using one of the following special forms:

      $$A^2 + 2AB + B^2 = (A + B)^2$$
      $$A^2 - 2AB + B^2 = (A - B)^2.$$

      If the trinomial is not a perfect square trinomial, try factoring by trial and error or grouping.

   c. If there are four or more terms, try factoring by grouping.
3. Check to see if any factors with more than one term in the factored polynomial can be factored further. If so, factor completely.

Remember to check the factored form by multiplying or by using the TABLE or GRAPH feature of a graphing utility.

The following examples and those in the exercise set are similar to the previous factoring problems. However, these factorizations are not all of the same type. They are intentionally mixed to promote the development of a general factoring strategy.

Using Technology

Graphic Connections

The polynomial functions $y_1 = 2x^3 + 8x^2 + 8x$ and $y_2 = 2x(x + 2)^2$ have identical graphs. This verifies that

$$2x^3 + 8x^2 + 8x = 2x(x + 2)^2.$$

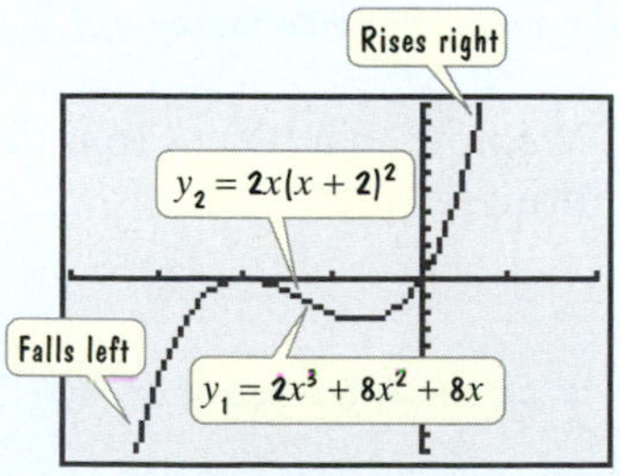

$[-4, 2, 1]$ by $[-10, 10, 1]$

The degree of y_1 is 3, which is odd. Odd-degree polynomial functions have graphs with opposite behavior at each end. The leading coefficient, 2, is positive. The graph should fall to the left and rise to the right (↙, ↗). The viewing rectangle used is complete enough to show this end behavior.

EXAMPLE 1 Factoring a Polynomial

Factor: $2x^3 + 8x^2 + 8x$.

Solution

Step 1. If there is a common factor, factor out the GCF. Because $2x$ is common to all terms, we factor it out.

$$2x^3 + 8x^2 + 8x = 2x(x^2 + 4x + 4) \quad \text{Factor out the GCF.}$$

Step 2. Determine the number of terms and factor accordingly. The factor $x^2 + 4x + 4$ has three terms and is a perfect square trinomial. We factor using $A^2 + 2AB + B^2 = (A + B)^2$.

$$2x^3 + 8x^2 + 8x = 2x(x^2 + 4x + 4)$$
$$= 2x(x^2 + 2 \cdot x \cdot 2 + 2^2)$$

$A^2 \; + \; 2AB \; + \; B^2$

$$= 2x(x + 2)^2 \quad A^2 + 2AB + B^2 = (A + B)^2$$

Step 3. Check to see if factors can be factored further. In this problem, they cannot. Thus,

$$2x^3 + 8x^2 + 8x = 2x(x + 2)^2.$$

✓ **CHECK POINT 1** Factor: $3x^3 - 30x^2 + 75x$.

EXAMPLE 2 Factoring a Polynomial

Factor: $4x^2y - 16xy - 20y$.

Solution

Step 1. If there is a common factor, factor out the GCF. Because $4y$ is common to all terms, we factor it out.

$$4x^2y - 16xy - 20y = 4y(x^2 - 4x - 5) \quad \text{Factor out the GCF.}$$

Step 2. Determine the number of terms and factor accordingly. The factor $x^2 - 4x - 5$ has three terms, but it is not a perfect square trinomial. We factor it using trial and error.

$$4x^2y - 16xy - 20y = 4y(x^2 - 4x - 5) = 4y(x + 1)(x - 5)$$

Step 3. Check to see if factors can be factored further. In this case, they cannot, so we have factored completely.

✓ **CHECK POINT 2** Factor: $3x^2y - 12xy - 36y$.

EXAMPLE 3 Factoring a Polynomial

Factor: $9b^2x - 16y - 16x + 9b^2y$.

Solution

Step 1. If there is a common factor, factor out the GCF. Other than 1 or -1, there is no common factor.

Step 2. Determine the number of terms and factor accordingly. There are four terms. We try factoring by grouping. Notice that the first and last terms have a common factor of $9b^2$ and the two middle terms have a common factor of -16. Thus, we begin by rearranging the terms.

$$9b^2x - 16y - 16x + 9b^2y$$

$$= (9b^2x + 9b^2y) + (-16x - 16y) \quad \text{Rearrange terms and group terms with common factors.}$$

$$= 9b^2(x + y) - 16(x + y) \quad \text{Factor from each group.}$$

$$= (x + y)(9b^2 - 16) \quad \text{Factor out the common binomial factor, } x + y.$$

Step 3. Check to see if factors can be factored further. We note that $9b^2 - 16$ is the difference of two squares, $(3b)^2 - 4^2$, so we continue factoring.

$$9b^2x - 16y - 16x + 9b^2y$$

$$= (x + y)[(3b)^2 - 4^2] \quad \text{Express } 9b^2 - 16 \text{ as the difference of squares.}$$

$$= (x + y)(3b + 4)(3b - 4) \quad \text{The factors of } 9b^2 - 16 \text{ are the sum and difference of the expressions being squared.}$$

✓ **CHECK POINT 3** Factor: $16a^2x - 25y - 25x + 16a^2y$.

EXAMPLE 4 Factoring a Polynomial

Factor: $x^2 - 25a^2 + 8x + 16$.

Solution

Step 1. If there is a common factor, factor out the GCF. Other than 1 or -1, there is no common factor.

Step 2. Determine the number of terms and factor accordingly. There are four terms. We try factoring by grouping. Grouping into two groups of two terms does not result in a common binomial factor. Let's try grouping as a difference of squares.

$$x^2 - 25a^2 + 8x + 16$$

$$= (x^2 + 8x + 16) - 25a^2$$ Rearrange terms and group as a perfect square trinomial minus $25a^2$ to obtain a difference of squares.

$$= (x + 4)^2 - (5a)^2$$ Factor the perfect square trinomial.

$$= (x + 4 + 5a)(x + 4 - 5a)$$ Factor the difference of squares. The factors are the sum and difference of the expressions being squared.

Step 3. Check to see if factors can be factored further. In this case, they cannot, so we have factored completely.

CHECK POINT 4 Factor: $x^2 - 36a^2 + 20x + 100$.

EXAMPLE 5 Factoring a Polynomial

Factor: $3x^{10} + 3x$.

Solution

Step 1. If there is a common factor, factor out the GCF. Because $3x$ is common to both terms, we factor it out.

$$3x^{10} + 3x = 3x(x^9 + 1)$$ Factor out the GCF.

Step 2. Determine the number of terms and factor accordingly. The factor $x^9 + 1$ has two terms. This binomial can be expressed as $(x^3)^3 + 1^3$, so it can be factored as the sum of two cubes.

$$3x^{10} + 3x = 3x(x^9 + 1)$$

$$= 3x[(x^3)^3 + 1^3] = 3x(x^3 + 1)[(x^3)^2 - x^3 \cdot 1 + 1^2]$$

$$A^3 + B^3 = (A + B)(A^2 - AB + B^2)$$

$$= 3x(x^3 + 1)(x^6 - x^3 + 1)$$ Simplify.

Step 3. Check to see if factors can be factored further. We note that $x^3 + 1$ is the sum of two cubes, $x^3 + 1^3$, so we continue factoring.

$$3x^{10} + 3x$$

$$= 3x(x^3 + 1)(x^6 - x^3 + 1)$$ This is our factorization in the previous step.

$$A^3 + B^3 = (A + B)(A^2 - AB + B^2)$$

$$= 3x(x + 1)(x^2 - x + 1)(x^6 - x^3 + 1)$$ Factor completely by factoring $x^3 + 1^3$, the sum of cubes.

CHECK POINT 5 Factor: $x^{10} + 512x$. *Hint*: $512 = 8^3$.

5.6 EXERCISE SET

MyMathLab — MathXL PRACTICE · WATCH · DOWNLOAD · READ · REVIEW

Practice Exercises

In Exercises 1–68, factor completely, or state that the polynomial is prime.

1. $x^3 - 16x$
2. $x^3 - x$
3. $3x^2 + 18x + 27$
4. $8x^2 + 40x + 50$
5. $81x^3 - 3$
6. $24x^3 - 3$
7. $x^2y - 16y + 32 - 2x^2$
8. $12x^2y - 27y - 4x^2 + 9$
9. $4a^2b - 2ab - 30b$
10. $32y^2 - 48y + 18$
11. $ay^2 - 4a - 4y^2 + 16$
12. $ax^2 - 16a - 2x^2 + 32$
13. $11x^5 - 11xy^2$
14. $4x^9 - 400x$
15. $4x^5 - 64x$
16. $7x^5 - 7x$
17. $x^3 - 4x^2 - 9x + 36$
18. $x^3 - 5x^2 - 4x + 20$
19. $2x^5 + 54x^2$
20. $3x^5 + 24x^2$
21. $3x^4y - 48y^5$
22. $32x^4y - 2y^5$
23. $12x^3 + 36x^2y + 27xy^2$
24. $18x^3 + 48x^2y + 32xy^2$
25. $x^2 - 12x + 36 - 49y^2$
26. $x^2 - 10x + 25 - 36y^2$
27. $4x^2 + 25y^2$
28. $16x^2 + 49y^2$
29. $12x^3y - 12xy^3$
30. $9x^2y^2 - 36y^2$
31. $6bx^2 + 6by^2$
32. $6x^2 - 66$
33. $x^4 - xy^3 + x^3y - y^4$
34. $x^3 - xy^2 + x^2y - y^3$
35. $x^2 - 4a^2 + 12x + 36$
36. $x^2 - 49a^2 + 14x + 49$
37. $5x^3 + x^6 - 14$
38. $6x^3 + x^6 - 16$
39. $4x - 14 + 2x^3 - 7x^2$
40. $3x^3 + 8x + 9x^2 + 24$
41. $54x^3 - 16y^3$
42. $54x^3 - 250y^3$
43. $x^2 + 10x - y^2 + 25$
44. $x^2 + 6x - y^2 + 9$
45. $x^8 - y^8$
46. $x^8 - 1$
47. $x^3y - 16xy^3$
48. $x^3y - 100xy^3$
49. $x + 8x^4$
50. $x + 27x^4$
51. $16y^2 - 4y - 2$
52. $32y^2 + 4y - 6$
53. $14y^3 + 7y^2 - 10y$
54. $5y^3 - 45y^2 + 70y$
55. $27x^2 + 36xy + 12y^2$
56. $125x^2 + 50xy + 5y^2$
57. $12x^3 + 3xy^2$
58. $3x^4 + 27x^2$
59. $x^6y^6 - x^3y^3$
60. $x^3 - 2x^2 - x + 2$
61. $(x + 5)(x - 3) + (x + 5)(x - 7)$
62. $(x + 4)(x - 9) + (x + 4)(2x - 3)$
63. $a^2(x - y) + 4(y - x)$
64. $b^2(x - 3) + c^2(3 - x)$
65. $(c + d)^4 - 1$
66. $(c + d)^4 - 16$
67. $p^3 - pq^2 + p^2q - q^3$
68. $p^3 - pq^2 - p^2q + q^3$

Practice PLUS

In Exercises 69–80, factor completely.

69. $x^4 - 5x^2y^2 + 4y^4$
70. $x^4 - 10x^2y^2 + 9y^4$
71. $(x + y)^2 + 6(x + y) + 9$
72. $(x - y)^2 - 8(x - y) + 16$
73. $(x - y)^4 - 4(x - y)^2$
74. $(x + y)^4 - 100(x + y)^2$
75. $2x^2 - 7xy^2 + 3y^4$
76. $3x^2 + 5xy^2 + 2y^4$
77. $x^3 - y^3 - x + y$
78. $x^3 + y^3 + x^2 - y^2$
79. $x^6y^3 + x^3 - 8x^3y^3 - 8$
80. $x^6y^3 - x^3 + x^3y^3 - 1$

Application Exercises

In Exercises 81–86,

a. *Write an expression for the area of the shaded region.*

b. *Write the expression in factored form.*

81.

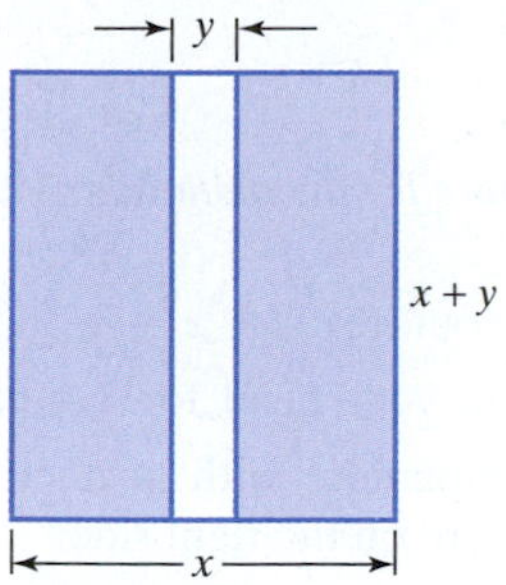

82.

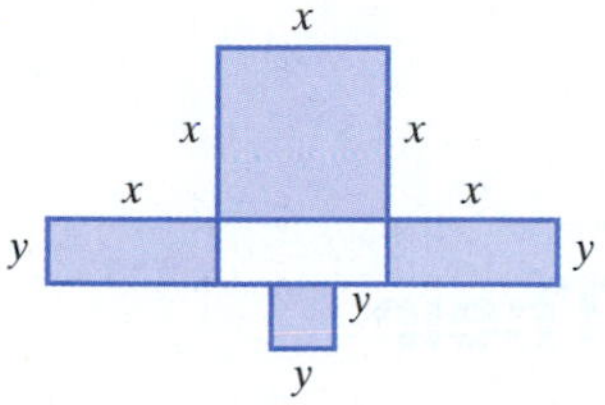

83.

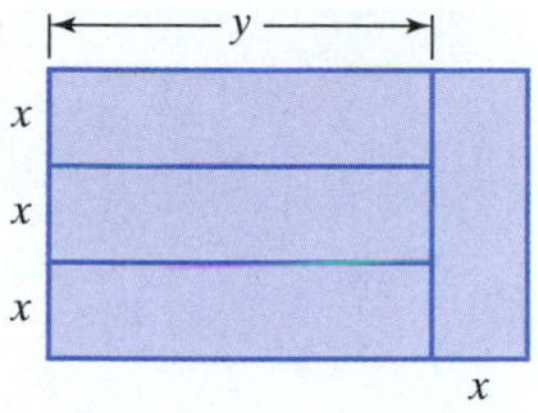

84.

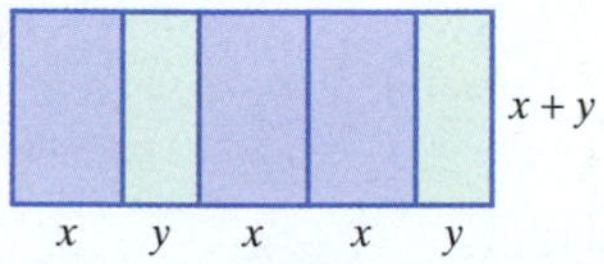

85.

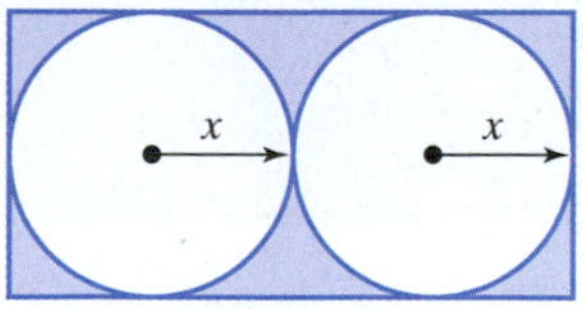

86.

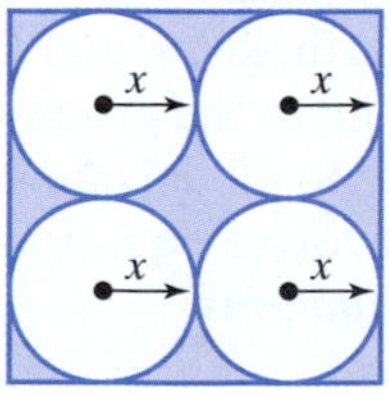

Writing in Mathematics

87. Describe a strategy that can be used to factor polynomials.

88. Describe some of the difficulties in factoring polynomials. What suggestions can you offer to overcome these difficulties?

Technology Exercises

In Exercises 89–92, use a graphing utility to graph the function on each side of the equation in the same viewing rectangle. Use end behavior to show a complete picture of the polynomial on the left side. Do the graphs coincide? If so, the factorization is correct. If not, factor correctly and then use your graphing utility to verify the factorization.

89. $4x^2 - 12x + 9 = (4x - 3)^2$

90. $2x^3 + 10x^2 - 2x - 10 = 2(x + 5)(x^2 + 1)$

91. $x^4 - 16 = (x^2 + 4)(x + 2)(x - 2)$

92. $x^3 + 1 = (x + 1)^3$

93. Use the TABLE feature of a graphing utility to verify any two of your complete factorizations in Exercises 15–20.

Critical Thinking Exercises

Make Sense? *In Exercises 94–97, determine whether each statement "makes sense" or "does not make sense" and explain your reasoning.*

94. It takes a great deal of practice to get good at factoring a wide variety of polynomials.

95. Multiplying polynomials is relatively mechanical, but factoring often requires a great deal of thought.

96. The factorable trinomial $4x^2 + 8x + 3$ and the prime trinomial $4x^2 + 8x + 1$ are in the form $ax^2 + bx + c$, but $b^2 - 4ac$ is a perfect square only in the case of the factorable trinomial.

97. You told me that the volume of a rectangular solid is represented by $5x^3 + 30x^2 + 40x$ cubic inches, so I factored completely and concluded that the dimensions are $5x$ inches, $x + 2$ inches, and $x + 5$ inches.

In Exercises 98–101, determine whether each statement is true or false. If the statement is false, make the necessary change(s) to produce a true statement.

98. $x^4 - 16$ is factored completely as $(x^2 + 4)(x^2 - 4)$.

99. The trinomial $x^2 - 4x - 4$ is a prime polynomial.

100. $x^2 + 36 = (x + 6)^2$

101. $x^3 - 64 = (x + 4)(x^2 + 4x - 16)$

In Exercises 102–104, factor completely. Assume that variable exponents represent whole numbers.

102. $x^{2n+3} - 10x^{n+3} + 25x^3$

103. $3x^{n+2} - 13x^{n+1} + 4x^n$

104. $x^{4n+1} - xy^{4n}$

105. In certain circumstances, the sum of two perfect squares can be factored by adding and subtracting the same perfect square. For example,

$$x^4 + 4 = x^4 + 4x^2 + 4 - 4x^2.$$ Add and subtract $4x^2$.

Use this first step to factor $x^4 + 4$.

106. Express $x^3 + x + 2x^4 + 4x^2 + 2$ as the product of two polynomials of degree 2.

Review Exercises

107. Solve: $\frac{3x-1}{5} + \frac{x+2}{2} = -\frac{3}{10}$.

(Section 1.4, Example 4)

108. Simplify: $(4x^3y^{-1})^2(2x^{-3}y)^{-1}$.

(Section 1.6, Example 9)

109. Evaluate: $\begin{vmatrix} 0 & -3 & 2 \\ 1 & 5 & 3 \\ -2 & 1 & 4 \end{vmatrix}$.

(Section 3.5, Example 3)

Preview Exercises

Exercises 110–112 will help you prepare for the material covered in the next section.

110. Evaluate $(2x + 3)(x - 4)$ in your head for $x = 4$.

111. Evaluate $-16(t - 6)(t + 4)$ in your head for $t = 6$.

112. Express as an equivalent equation with a factored trinomial on the left side and zero on the right side:

$$x^2 + (x + 7)^2 = (x + 8)^2.$$

SECTION 5.7

Polynomial Equations and Their Applications

Objectives

1. Solve quadratic equations by factoring.
2. Solve higher-degree polynomial equations by factoring.
3. Solve problems using polynomial equations.

Motion and change are the very essence of life. Moving air brushes against our faces; rain falls on our heads; birds fly past us; plants spring from the earth, grow, and then die; and rocks thrown upward reach a maximum height before falling to the ground. In this section, you will use quadratic functions and factoring strategies to model and visualize motion. Analyzing the where and when of moving objects involves equations in which the highest exponent on the variable is 2, called *quadratic equations*.

The Standard Form of a Quadratic Equation

We begin by defining a quadratic equation.

Definition of a Quadratic Equation

A **quadratic equation** in x is an equation that can be written in the **standard form**

$$ax^2 + bx + c = 0,$$

where a, b, and c are real numbers, with $a \neq 0$. A quadratic equation in x is also called a **second-degree polynomial equation** in x.

Here is an example of a quadratic equation in standard form:

$$x^2 - 12x + 27 = 0.$$

$a = 1$ $b = -12$ $c = 27$

1 Solve quadratic equations by factoring.

Solving Quadratic Equations by Factoring

We can factor the left side of the quadratic equation $x^2 - 12x + 27 = 0$. We obtain $(x - 3)(x - 9) = 0$. If a quadratic equation has zero on one side and a factored expression on the other side, it can be solved using the **zero-product principle**.

The Zero-Product Principle

If the product of two algebraic expressions is zero, then at least one of the factors is equal to zero.

$$\text{If } AB = 0, \text{ then } A = 0 \text{ or } B = 0.$$

For example, consider the equation $(x - 3)(x - 9) = 0$. According to the zero-product principle, this product can be zero only if at least one of the factors is zero. We set each individual factor equal to zero and solve the resulting equations for x.

$$(x - 3)(x - 9) = 0$$

$$x - 3 = 0 \quad \text{or} \quad x - 9 = 0$$

$$x = 3 \qquad\qquad x = 9$$

The solutions of the original quadratic equation, $x^2 - 12x + 27 = 0$, are 3 and 9. The solution set is $\{3, 9\}$.

Solving a Quadratic Equation by Factoring

1. If necessary, rewrite the equation in the standard form $ax^2 + bx + c = 0$, moving all terms to one side, thereby obtaining zero on the other side.
2. Factor completely.
3. Apply the zero-product principle, setting each factor containing a variable equal to zero.
4. Solve the equations in step 3.
5. Check the solutions in the original equation.

EXAMPLE 1 Solving a Quadratic Equation by Factoring

Solve: $2x^2 - 5x = 12$.

Solution

Step 1. Move all terms to one side and obtain zero on the other side. Subtract 12 from both sides and write the equation in standard form.

$$2x^2 - 5x - 12 = 12 - 12$$

$$2x^2 - 5x - 12 = 0$$

Step 2. Factor.

$$(2x + 3)(x - 4) = 0$$

Steps 3 and 4. Set each factor equal to zero and solve the resulting equations.

$$2x + 3 = 0 \quad \text{or} \quad x - 4 = 0$$

$$2x = -3 \qquad\qquad x = 4$$

$$x = -\frac{3}{2}$$

Step 5. Check the solutions in the original equation.

Check $-\frac{3}{2}$:

$$2x^2 - 5x = 12$$
$$2\left(-\frac{3}{2}\right)^2 - 5\left(-\frac{3}{2}\right) \stackrel{?}{=} 12$$
$$2\left(\frac{9}{4}\right) - 5\left(-\frac{3}{2}\right) \stackrel{?}{=} 12$$
$$\frac{9}{2} + \frac{15}{2} \stackrel{?}{=} 12$$
$$\frac{24}{2} \stackrel{?}{=} 12$$
$$12 = 12, \quad \text{true}$$

Check 4:

$$2x^2 - 5x = 12$$
$$2(4)^2 - 5(4) \stackrel{?}{=} 12$$
$$2(16) - 5(4) \stackrel{?}{=} 12$$
$$32 - 20 \stackrel{?}{=} 12$$
$$12 = 12, \quad \text{true}$$

The solutions are $-\frac{3}{2}$ and 4, and the solution set is $\left\{-\frac{3}{2}, 4\right\}$.

✓ **CHECK POINT 1** Solve: $2x^2 - 9x = 5$.

Study Tip

Do not confuse factoring a polynomial with solving a quadratic equation by factoring.

Factoring a Polynomial

Factor: $2x^2 - 5x - 12$.

This is not an equation. There is no equal sign.

Solution: $(2x + 3)(x - 4)$

Stop! Avoid the common error of setting each factor equal to zero.

Solving a Quadratic Equation

Solve: $2x^2 - 5x - 12 = 0$.

This is an equation. There is an equal sign.

Solution: $(2x + 3)(x - 4) = 0$

$$2x + 3 = 0 \quad \text{or} \quad x - 4 = 0$$
$$x = -\frac{3}{2} \qquad\qquad x = 4$$

The solution set is $\left\{-\frac{3}{2}, 4\right\}$.

There is an important relationship between a quadratic equation in standard form, such as

$$2x^2 - 5x - 12 = 0$$

and a quadratic function, such as

$$y = 2x^2 - 5x - 12.$$

The solutions of $ax^2 + bx + c = 0$ correspond to the x-intercepts of the graph of the quadratic function $y = ax^2 + bx + c$. For example, you can visualize the solutions of $2x^2 - 5x - 12 = 0$ by looking at the x-intercepts of the graph of the quadratic function $y = 2x^2 - 5x - 12$. The graph, shaped like a bowl, is shown in **Figure 5.10**. The solutions of the equation $2x^2 - 5x - 12 = 0$, $-\frac{3}{2}$ and 4, appear as the graph's x-intercepts.

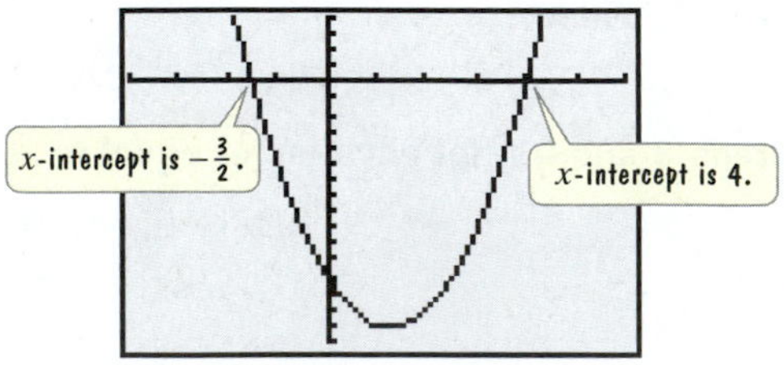

[−4, 6, 1] by [−16, 4, 1]

FIGURE 5.10

EXAMPLE 2 Solving Quadratic Equations by Factoring

Solve:

a. $5x^2 = 20x$ **b.** $x^2 + 4 = 8x - 12$ **c.** $(x - 7)(x + 5) = -20.$

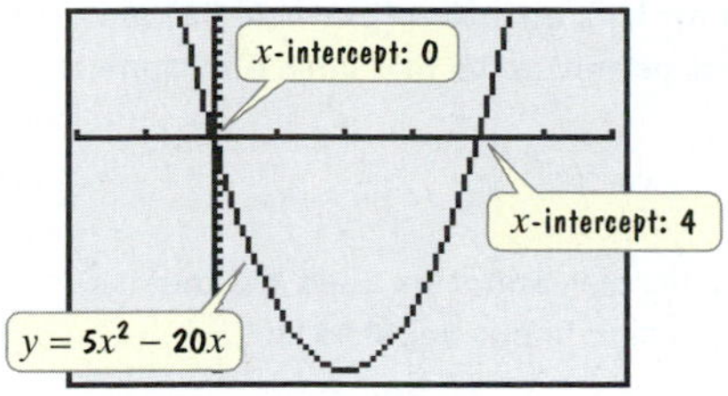

[−2, 6, 1] by [−20, 10, 1]

FIGURE 5.11 The solution set of $5x^2 = 20x$, or $5x^2 - 20x = 0$, is $\{0, 4\}$.

Solution

a.

$5x^2 = 20x$	This is the given equation.
$5x^2 - 20x = 0$	Subtract 20x from both sides and write the equation in standard form.
$5x(x - 4) = 0$	Factor.
$5x = 0$ or $x - 4 = 0$	Set each factor equal to 0.
$x = 0$ $\quad$ $x = 4$	Solve the resulting equations.

Check by substituting 0 and 4 into the given equation. The graph of $y = 5x^2 - 20x$, obtained with a graphing utility, is shown in **Figure 5.11**. The x-intercepts are 0 and 4. This verifies that the solutions are 0 and 4, and the solution set is $\{0, 4\}$.

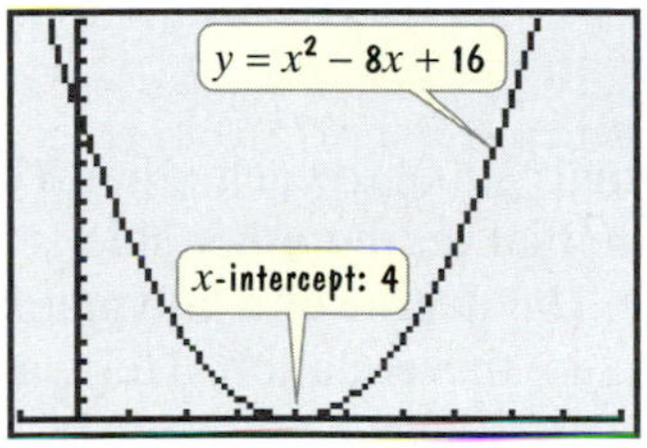

[−1, 10, 1] by [0, 20, 1]

FIGURE 5.12 The solution set of $x^2 + 4 = 8x - 12$, or $x^2 - 8x + 16 = 0$, is $\{4\}$.

b.

$x^2 + 4 = 8x - 12$	This is the given equation.
$x^2 - 8x + 16 = 0$	Write the equation in standard form by subtracting 8x and adding 12 on both sides.
$(x - 4)(x - 4) = 0$	Factor.
$x - 4 = 0$ or $x - 4 = 0$	Set each factor equal to 0.
$x = 4$ $\quad$ $x = 4$	Solve the resulting equations.

Notice that there is only one solution (or, if you prefer, a repeated solution.) The trinomial $x^2 - 8x + 16$ is a perfect square trinomial that could have been factored as $(x - 4)^2$. The graph of $y = x^2 - 8x + 16$, obtained with a graphing utility, is shown in **Figure 5.12**. The graph has only one x-intercept at 4. This verifies that the equation's solution is 4 and the solution set is $\{4\}$.

c. Be careful! Although the left side of $(x - 7)(x + 5) = -20$ is factored, we cannot use the zero-product principle. Why not? The right side of the equation is not 0. So we begin by multiplying the factors on the left side of the equation. Then we add 20 to both sides to obtain 0 on the right side.

$(x - 7)(x + 5) = -20$	This is the given equation.
$x^2 - 2x - 35 = -20$	Use the FOIL method to multiply on the left side.
$x^2 - 2x - 15 = 0$	Add 20 to both sides.
$(x + 3)(x - 5) = 0$	Factor.
$x + 3 = 0$ or $x - 5 = 0$	Set each factor equal to 0.
$x = -3$ $\quad$ $x = 5$	Solve the resulting equations.

Check by substituting −3 and 5 into the given equation. The graph of $y = x^2 - 2x - 15$, obtained with a graphing utility, is shown in **Figure 5.13**. The x-intercepts are −3 and 5. This verifies that the solutions are −3 and 5, and the solution set is $\{-3, 5\}$.

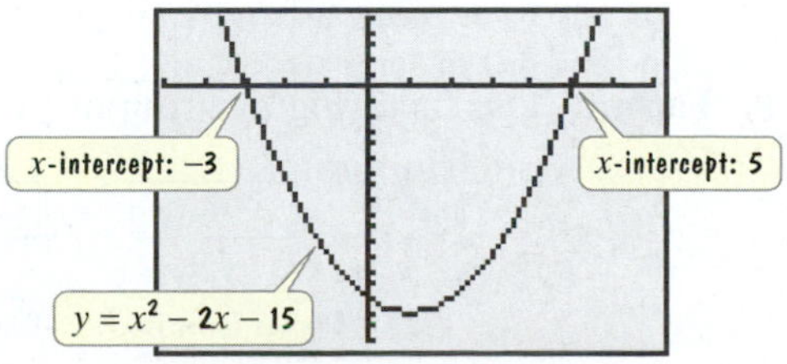

[−5, 7, 1] by [−18, 5, 1]

FIGURE 5.13 The solution set of $(x - 7)(x + 5) = -20$, or $x^2 - 2x - 15 = 0$, is $\{-3, 5\}$.

Study Tip

Avoid the following errors:

$$5x^2 = 20x$$

~~$\frac{5x^2}{x} = \frac{20x}{x}$~~

~~$5x = 20$~~

~~$x = 4$~~

Never divide both sides of an equation by x. Division by zero is undefined and x may be zero. Indeed, the solutions for this equation (Example 2a) are 0 and 4. Dividing both sides by x does not permit us to find both solutions.

$$(x - 7)(x + 5) = -20$$

~~$x - 7 = -20$ or $x + 5 = -20$~~

~~$x = -13$ or $x = -25$~~

The zero-product principle cannot be used because the right side of the equation is not equal to 0.

✓ CHECK POINT 2 Solve:

a. $3x^2 = 2x$

b. $x^2 + 7 = 10x - 18$

c. $(x - 2)(x + 3) = 6$.

---

2 Solve higher-degree polynomial equations by factoring.

Polynomial Equations

A **polynomial equation** is the result of setting two polynomials equal to each other. The equation is in **standard form** if one side is 0 and the polynomial on the other side is in standard form, that is, in descending powers of the variable. The **degree of a polynomial equation** is the same as the highest degree of any term in the equation. Here are examples of three polynomial equations:

$$3x + 5 = 14 \qquad 2x^2 + 7x = 4 \qquad x^3 + x^2 = 4x + 4.$$

This equation is of degree 1 because 1 is the highest degree.

This equation is of degree 2 because 2 is the highest degree.

This equation is of degree 3 because 3 is the highest degree.

Notice that a polynomial equation of degree 1 is a linear equation. A polynomial equation of degree 2 is a quadratic equation.

Some polynomial equations of degree 3 or higher can be solved by moving all terms to one side, thereby obtaining 0 on the other side. Once the equation is in standard form, factor and then set each factor equal to 0.

EXAMPLE 3 Solving a Polynomial Equation by Factoring

Solve by factoring: $x^3 + x^2 = 4x + 4$.

Solution

Step 1. Move all terms to one side and obtain zero on the other side. Subtract $4x$ and subtract 4 from both sides.

$$x^3 + x^2 - 4x - 4 = 4x + 4 - 4x - 4$$

$$x^3 + x^2 - 4x - 4 = 0$$

Step 2. Factor. Use factoring by grouping. Group terms that have a common factor.

$$\boxed{x^3 + x^2} + \boxed{-4x - 4} = 0$$

Common factor is x^2.

Common factor is -4.

Using Technology

Numeric Connections

A graphing utility's TABLE feature can be used to numerically verify that $\{-2, -1, 2\}$ is the solution set of

$$x^3 + x^2 = 4x + 4$$

Enter $y_1 = x^3 + x^2$. Enter $y_2 = 4x + 4$.

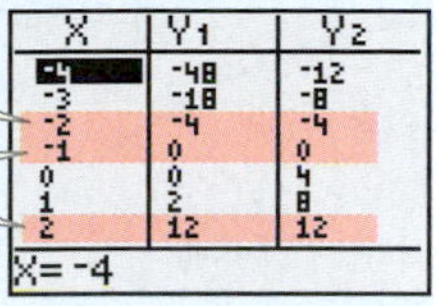

Discover for Yourself

Suggest a method involving intersecting graphs that can be used with a graphing utility to verify that $\{-2, -1, 2\}$ is the solution set of

$$x^3 + x^2 = 4x + 4.$$

Apply this method to verify the solution set.

$$x^2(x + 1) - 4(x + 1) = 0 \quad \text{Factor } x^2 \text{ from the first two terms and } -4 \text{ from the last two terms.}$$

$$(x + 1)(x^2 - 4) = 0 \quad \text{Factor out the common binomial, } x + 1, \text{ from each term.}$$

$$(x + 1)(x + 2)(x - 2) = 0 \quad \text{Factor completely by factoring } x^2 - 4 \text{ as the difference of two squares.}$$

Steps 3 and 4. Set each factor equal to zero and solve the resulting equations.

$$x + 1 = 0 \quad \text{or} \quad x + 2 = 0 \quad \text{or} \quad x - 2 = 0$$
$$x = -1 \qquad\qquad x = -2 \qquad\qquad x = 2$$

Step 5. Check the solutions in the original equation. Check the three solutions, -1, -2, and 2, by substituting them into the original equation. Can you verify that the solutions are -1, -2, and 2, and the solution set is $\{-2, -1, 2\}$?

Using Technology

Graphic Connections

You can use a graphing utility to check the solutions to $x^3 + x^2 - 4x - 4 = 0$. Graph $y = x^3 + x^2 - 4x - 4$, as shown on the right. Is the graph complete? Because the degree, 3, is odd, and the leading coefficient, 1, is positive, it should fall to the left and rise to the right (↙, ↗). The graph shows this end behavior and is therefore complete. The x-intercepts are $-2, -1$, and 2, corresponding to the equation's solutions.

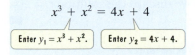

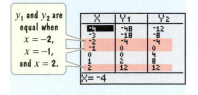

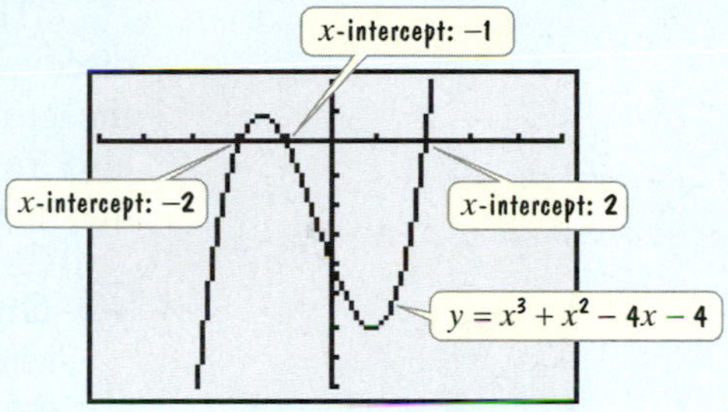

$[-5, 5, 1]$ by $[-8, 2, 1]$

CHECK POINT 3 Solve by factoring: $2x^3 + 3x^2 = 8x + 12$.

---

3 Solve problems using polynomial equations.

Applications of Polynomial Equations

Solving polynomial equations by factoring can be used to answer questions about variables contained in mathematical models.

EXAMPLE 4 Modeling Motion

You throw a ball straight up from a rooftop 384 feet high with an initial speed of 32 feet per second. The function

$$s(t) = -16t^2 + 32t + 384$$

describes the ball's height above the ground, $s(t)$, in feet, t seconds after you throw it. The ball misses the rooftop on its way down and eventually strikes the ground. How long will it take for the ball to hit the ground?

Solution The ball hits the ground when $s(t)$, its height above the ground, is 0 feet. Thus, we substitute 0 for $s(t)$ in the given function and solve for t.

$$s(t) = -16t^2 + 32t + 384 \quad \text{This is the function that models the ball's height.}$$

$$0 = -16t^2 + 32t + 384 \quad \text{Substitute 0 for } s(t).$$

$$0 = -16(t^2 - 2t - 24) \quad \text{Factor out } -16.$$

$$0 = -16(t-6)(t+4) \quad \text{Factor } t^2 - 2t - 24\text{, the trinomial.}$$

Do not set the constant, −16, equal to zero: $-16 \neq 0$.

$$t - 6 = 0 \quad \text{or} \quad t + 4 = 0 \quad \text{Set each variable factor equal to 0.}$$

$$t = 6 \qquad\qquad t = -4 \quad \text{Solve for } t.$$

Because we begin describing the ball's height at $t = 0$, we discard the solution $t = -4$. The ball hits the ground after 6 seconds.

Figure 5.14 shows the graph of the function $s(t) = -16t^2 + 32t + 384$. The horizontal axis is labeled t, for the ball's time in motion. The vertical axis is labeled $s(t)$, for the ball's height above the ground at time t. Because time and height are both positive, the function is graphed in quadrant I only.

The graph visually shows what we discovered algebraically: The ball hits the ground after 6 seconds. The graph also reveals that the ball reaches its maximum height, 400 feet, after 1 second. Then the ball begins to fall.

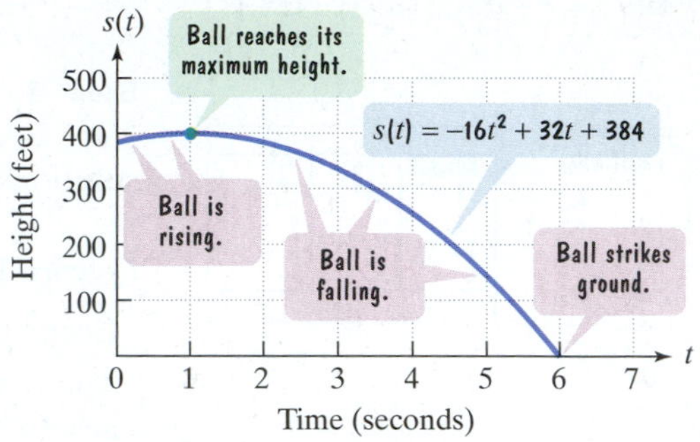

FIGURE 5.14

✓ **CHECK POINT 4** Use the function $s(t) = -16t^2 + 32t + 384$ to determine when the ball's height is 336 feet. Identify your meaningful solution as a point on the graph in **Figure 5.14**.

In our next example, we use our five-step strategy for solving word problems.

EXAMPLE 5 Solving a Problem Involving Landscape Design

A rectangular garden measures 80 feet by 60 feet. A large path of uniform width is to be added along both shorter sides and one longer side of the garden. The landscape designer doing the work wants to double the garden's area with the addition of this path. How wide should the path be?

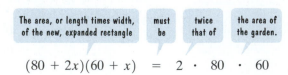

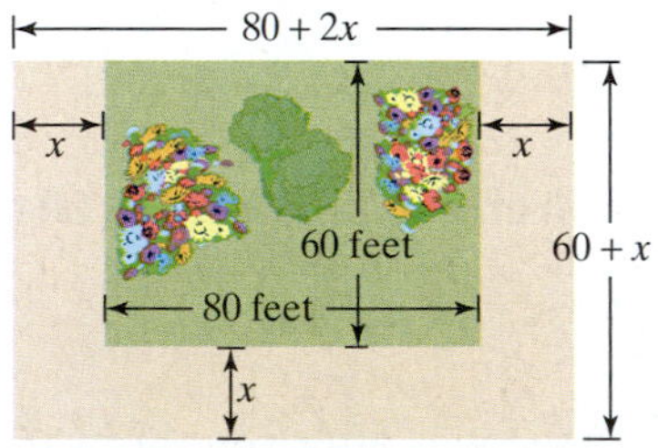

FIGURE 5.15 The garden's area is to be doubled by adding the path.

Solution

Step 1. Let x represent one of the unknown quantities. We will let

$$x = \text{the width of the path.}$$

The situation is illustrated in **Figure 5.15**. The figure shows the original 80-by-60 foot rectangular garden and the path of width x added along both shorter sides and one longer side.

Step 2. Represent other unknown quantities in terms of x. Because the path is added along both shorter sides and one longer side, **Figure 5.15** shows that

$$80 + 2x = \text{the length of the new, expanded rectangle}$$

$$60 + x = \text{the width of the new, expanded rectangle.}$$

Step 3. Write an equation that models the conditions. The area of the rectangle must be doubled by the addition of the path.

The area, or length times width, of the new, expanded rectangle | must be | twice that of | the area of the garden.

$$(80 + 2x)(60 + x) = 2 \cdot 80 \cdot 60$$

Step 4. Solve the equation and answer the question.

$$(80 + 2x)(60 + x) = 2 \cdot 80 \cdot 60$$ This is the equation that models the problem's conditions.

$$4800 + 200x + 2x^2 = 9600$$ Multiply. Use FOIL on the left side.

$$2x^2 + 200x - 4800 = 0$$ Subtract 9600 from both sides and write the equation in standard form.

$$2(x^2 + 100x - 2400) = 0$$ Factor out 2, the GCF.

$$2(x - 20)(x + 120) = 0$$ Factor the trinomial.

$$x - 20 = 0 \quad \text{or} \quad x + 120 = 0$$ Set each variable factor equal to 0.

$$x = 20 \quad \text{or} \quad x = -120$$ Solve for x.

The path cannot have a negative width. Because -120 is geometrically impossible, we use $x = 20$. The width of the path should be 20 feet.

Step 5. Check the proposed solution in the original wording of the problem. Has the landscape architect doubled the garden's area with the 20-foot-wide path? The area of the garden is 80 feet times 60 feet, or 4800 square feet. Because $80 + 2x$ and $60 + x$ represent the length and width of the expanded rectangle,

$$80 + 2x = 80 + 2 \cdot 20 = 120 \text{ feet is the expanded rectangle's length.}$$
$$60 + x = 60 + 20 = 80 \text{ feet is the expanded rectangle's width.}$$

The area of the expanded rectangle is 120 feet times 80 feet, or 9600 square feet. This is double the area of the garden, 4800 square feet, as specified by the problem's conditions.

✓ **CHECK POINT 5** A rectangular garden measures 16 feet by 12 feet. A path of uniform width is to be added so as to surround the entire garden. The landscape artist doing the work wants the garden and path to cover an area of 320 square feet. How wide should the path be?

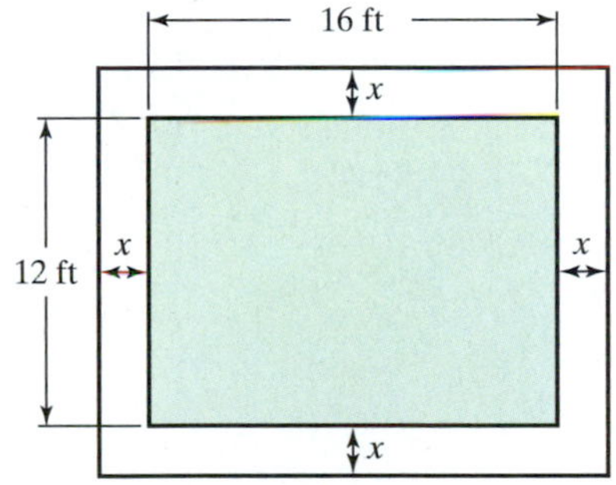

The solution to our next problem relies on knowing the **Pythagorean Theorem**. The theorem relates the lengths of the three sides of a **right triangle**, a triangle with one angle measuring 90°. The side opposite the 90° angle is called the **hypotenuse**. The other sides are called **legs**. The legs form the two sides of the right angle.

The Pythagorean Theorem

The sum of the squares of the lengths of the legs of a right triangle equals the square of the length of the hypotenuse.

If the legs have lengths a and b, and the hypotenuse has length c, then

$$a^2 + b^2 = c^2.$$

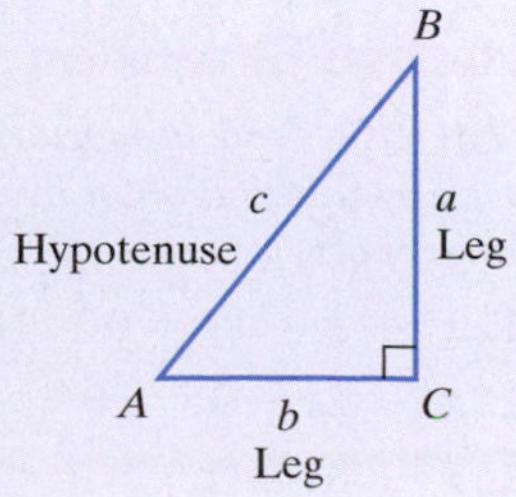

EXAMPLE 6 Using the Pythagorean Theorem to Obtain a Polynomial Equation

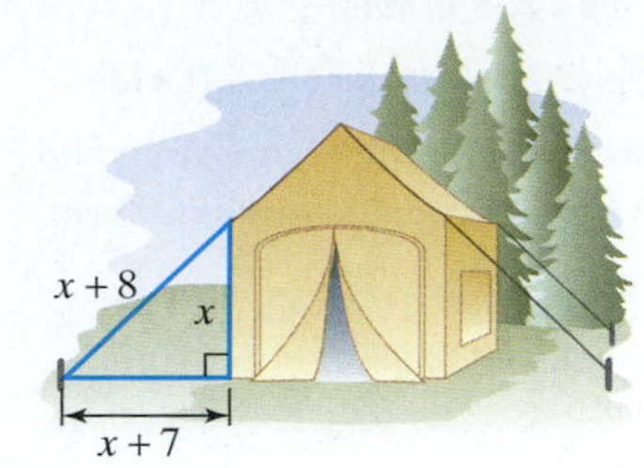

FIGURE 5.16

Figure 5.16 shows a tent with wires attached to help stabilize it. The length of each wire is 8 feet greater than the distance from the ground to where it is attached to the tent. The distance from the base of the tent to where the wire is anchored exceeds this height by 7 feet. Find the length of each wire used to stabilize the tent.

Solution **Figure 5.16** shows a right triangle. The length of the legs are x and $x + 7$. The length of the hypotenuse, $x + 8$, represents the length of the wire. We use the Pythagorean Theorem to find this length.

$$\text{leg}^2 + \text{leg}^2 = \text{hypotenuse}^2$$

$$x^2 + (x+7)^2 = (x+8)^2$$ This is the equation arising from the Pythagorean Theorem.

$$x^2 + x^2 + 14x + 49 = x^2 + 16x + 64$$ Square $x + 7$ and $x + 8$.

$$2x^2 + 14x + 49 = x^2 + 16x + 64$$ Combine like terms: $x^2 + x^2 = 2x^2$.

$$x^2 - 2x - 15 = 0$$ Subtract $x^2 + 16x + 64$ from both sides and write the quadratic equation in standard form.

$$(x-5)(x+3) = 0$$ Factor the trinomial.

$$x - 5 = 0 \quad \text{or} \quad x + 3 = 0$$ Set each factor equal to 0.

$$x = 5 \qquad\qquad x = -3$$ Solve for x.

Because x represents the distance from the ground to where the wire is attached, x cannot be negative. Thus, we only use $x = 5$. **Figure 5.16** shows that the length of the wire is $x + 8$ feet. The length of the wire is $5 + 8$ feet, or 13 feet.

We can check to see that the lengths of the three sides of the right triangle, x, $x + 7$, and $x + 8$, satisfy the Pythagorean Theorem when $x = 5$. The lengths are 5 feet, 12 feet, and 13 feet.

$$\text{leg}^2 + \text{leg}^2 = \text{hypotenuse}^2$$

$$5^2 + 12^2 \stackrel{?}{=} 13^2$$

$$25 + 144 \stackrel{?}{=} 169$$

$$169 = 169, \quad \text{true}$$

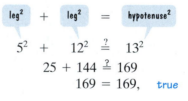

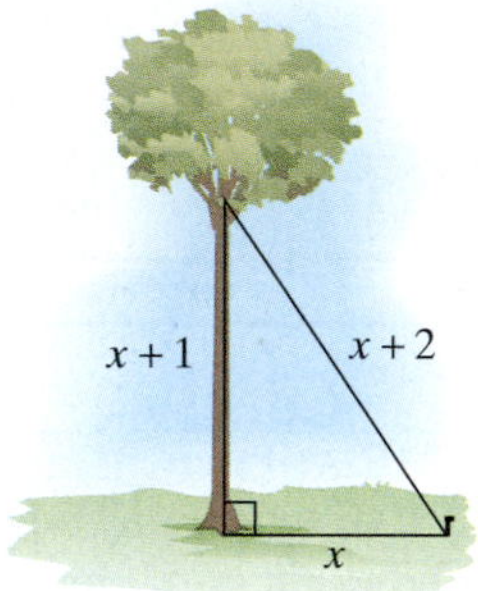

FIGURE 5.17

✓ CHECK POINT 6 A guy wire is attached to a tree to help it grow straight. The situation is illustrated in **Figure 5.17**. The length of the wire is 2 feet greater than the distance from the base of the tree to the stake. Find the length of the wire.

5.7 EXERCISE SET

MyMathLab — MathXL PRACTICE, WATCH, DOWNLOAD, READ, REVIEW

Practice Exercises

In Exercises 1–36, use factoring to solve each quadratic equation. Check by substitution or by using a graphing utility and identifying x-intercepts.

1. $x^2 + x - 12 = 0$
2. $x^2 - 2x - 15 = 0$
3. $x^2 + 6x = 7$
4. $x^2 - 4x = 45$
5. $3x^2 + 10x - 8 = 0$
6. $2x^2 - 5x - 3 = 0$
7. $5x^2 = 8x - 3$
8. $7x^2 = 30x - 8$
9. $3x^2 = 2 - 5x$
10. $5x^2 = 2 + 3x$
11. $x^2 = 8x$

12. $x^2 = 4x$

13. $3x^2 = 5x$

14. $2x^2 = 5x$

15. $x^2 + 4x + 4 = 0$

16. $x^2 + 6x + 9 = 0$

17. $x^2 = 14x - 49$

18. $x^2 = 12x - 36$

19. $9x^2 = 30x - 25$

20. $4x^2 = 12x - 9$

21. $x^2 - 25 = 0$

22. $x^2 - 49 = 0$

23. $9x^2 = 100$

24. $4x^2 = 25$

25. $x(x - 3) = 18$

26. $x(x - 4) = 21$

27. $(x - 3)(x + 8) = -30$

28. $(x - 1)(x + 4) = 14$

29. $x(x + 8) = 16(x - 1)$

30. $x(x + 9) = 4(2x + 5)$

31. $(x + 1)^2 - 5(x + 2) = 3x + 7$

32. $(x + 1)^2 = 2(x + 5)$

33. $x(8x + 1) = 3x^2 - 2x + 2$

34. $2x(x + 3) = -5x - 15$

35. $\frac{x^2}{18} + \frac{x}{2} + 1 = 0$

36. $\frac{x^2}{4} - \frac{5x}{2} + 6 = 0$

In Exercises 37–46, use factoring to solve each polynomial equation. Check by substitution or by using a graphing utility and identifying x-intercepts.

37. $x^3 + 4x^2 - 25x - 100 = 0$

38. $x^3 - 2x^2 - x + 2 = 0$

39. $x^3 - x^2 = 25x - 25$

40. $x^3 + 2x^2 = 16x + 32$

41. $3x^4 - 48x^2 = 0$

42. $5x^4 - 20x^2 = 0$

43. $x^4 - 4x^3 + 4x^2 = 0$

44. $x^4 - 6x^3 + 9x^2 = 0$

45. $2x^3 + 16x^2 + 30x = 0$

46. $3x^3 - 9x^2 - 30x = 0$

In Exercises 47–50, determine the x-intercepts of the graph of each quadratic function. Then match the function with its graph, labeled (a)–(d). Each graph is shown in a $[-10, 10, 1]$ by $[-10, 10, 1]$ viewing rectangle.

47. $y = x^2 - 6x + 8$

48. $y = x^2 - 2x - 8$

49. $y = x^2 + 6x + 8$

50. $y = x^2 + 2x - 8$

a.

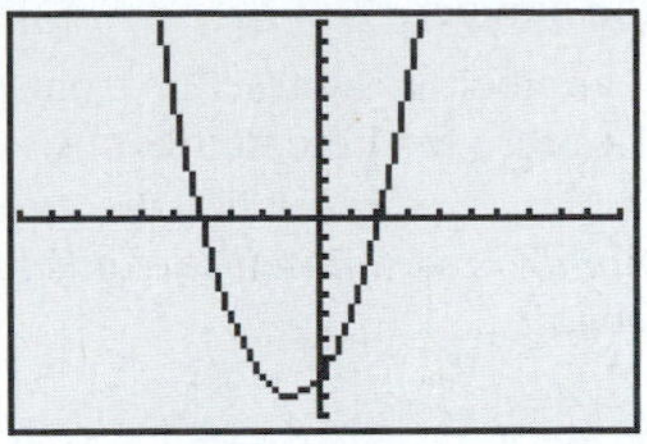

b.

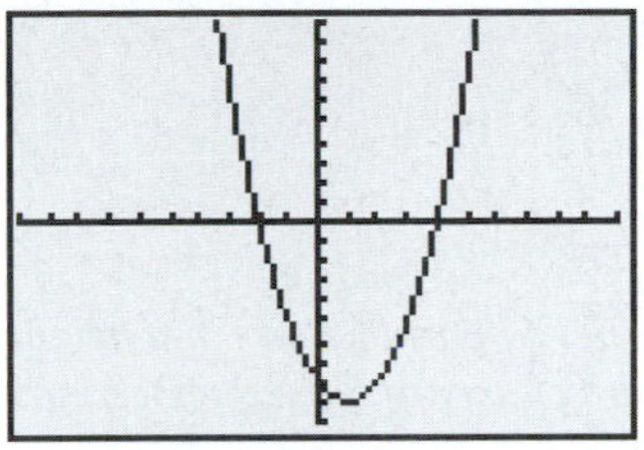

c.

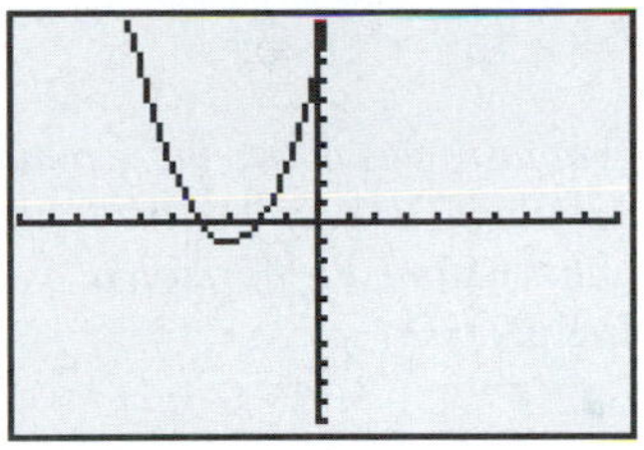

d.

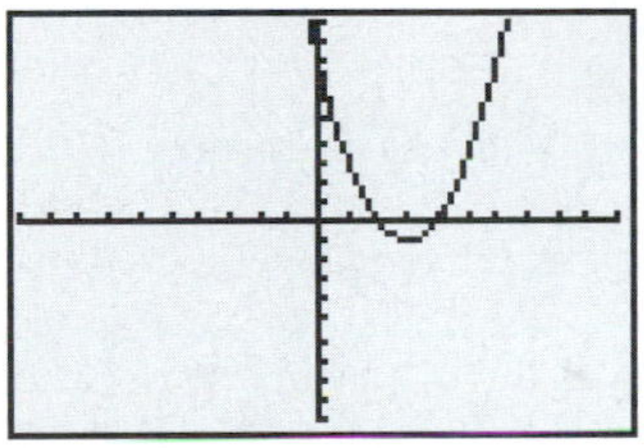

Practice PLUS

In Exercises 51–54, solve each polynomial equation.

51. $x(x + 1)^3 - 42(x + 1)^2 = 0$

52. $x(x - 2)^3 - 35(x - 2)^2 = 0$

53. $-4x[x(3x - 2) - 8](25x^2 - 40x + 16) = 0$

54. $-7x[x(2x - 5) - 12](9x^2 + 30x + 25) = 0$

In Exercises 55–58, find all values of c satisfying the given conditions.

55. $f(x) = x^2 - 4x - 27$ and $f(c) = 5$.

56. $f(x) = 5x^2 - 11x + 6$ and $f(c) = 4$.

57. $f(x) = 2x^3 + x^2 - 8x + 2$ and $f(c) = 6$.

58. $f(x) = x^3 + 4x^2 - x + 6$ and $f(c) = 10$.

In Exercises 59–62, find all numbers satisfying the given conditions.

59. The product of the number decreased by 1 and increased by 4 is 24.

60. The product of the number decreased by 6 and increased by 2 is 20.

61. If 5 is subtracted from 3 times the number, the result is the square of 1 less than the number.

62. If the square of the number is subtracted from 61, the result is the square of 1 more than the number.

In Exercises 63–64, list all numbers that must be excluded from the domain of the given function.

63. $f(x) = \dfrac{3}{x^2 + 4x - 45}$

64. $f(x) = \dfrac{7}{x^2 - 3x - 28}$

Application Exercises

A gymnast dismounts the uneven parallel bars at a height of 8 feet with an initial upward velocity of 8 feet per second. The function

$$s(t) = -16t^2 + 8t + 8$$

describes the height of the gymnast's feet above the ground, $s(t)$, in feet, t seconds after dismounting. The graph of the function is shown, with unlabeled tick marks along the horizontal axis. Use the function to solve Exercises 65–66.

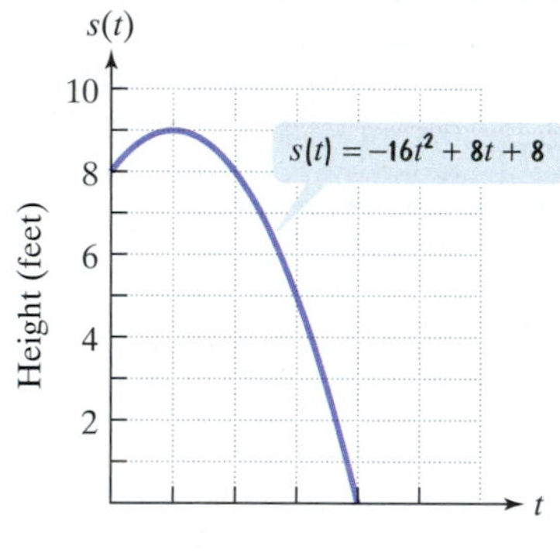

65. How long will it take the gymnast to reach the ground? Use this information to provide a number on each tick mark along the horizontal axis in the figure shown.

66. When will the gymnast be 8 feet above the ground? Identify the solution(s) as one or more points on the graph.

In a round-robin chess tournament, each player is paired with every other player once. The function

$$f(x) = \frac{x^2 - x}{2}$$

models the number of chess games, $f(x)$, that must be played in a round-robin tournament with x chess players. Use this function to solve Exercises 67–68.

67. In a round-robin chess tournament, 21 games were played. How many players were entered in the tournament?

68. In a round-robin chess tournament, 36 games were played. How many players were entered in the tournament?

The graph of the quadratic function in Exercises 67–68 is shown. Use the graph to solve Exercises 69–70.

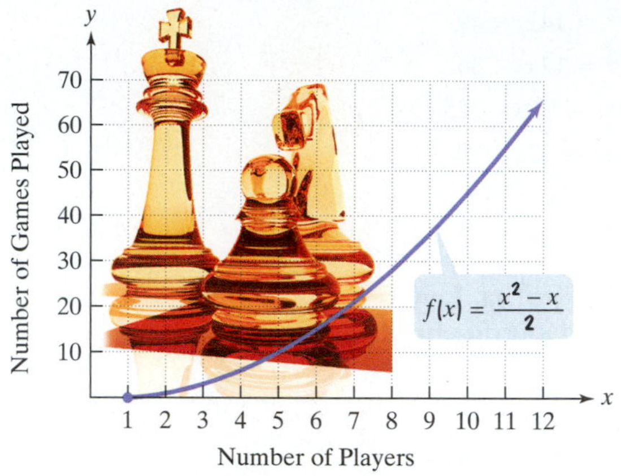

69. Identify your solution to Exercise 67 as a point on the graph.

70. Identify your solution to Exercise 68 as a point on the graph.

71. The length of a rectangular sign is 3 feet longer than the width. If the sign's area is 54 square feet, find its length and width.

72. A rectangular parking lot has a length that is 3 yards greater than the width. The area of the parking lot is 180 square yards. Find the length and the width.

73. Each side of a square is lengthened by 3 inches. The area of this new, larger square is 64 square inches. Find the length of a side of the original square.

74. Each side of a square is lengthened by 2 inches. The area of this new, larger square is 36 square inches. Find the length of a side of the original square.

75. A pool measuring 10 meters by 20 meters is surrounded by a path of uniform width, as shown in the figure. If the area of the pool and the path combined is 600 square meters, what is the width of the path?

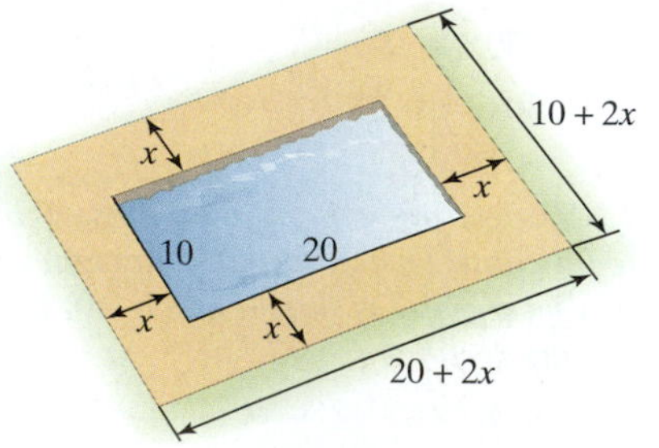

76. A vacant rectangular lot is being turned into a community vegetable garden measuring 15 meters by 12 meters. A path of uniform width is to surround the garden. If the area of the lot is 378 square meters, find the width of the path surrounding the garden.

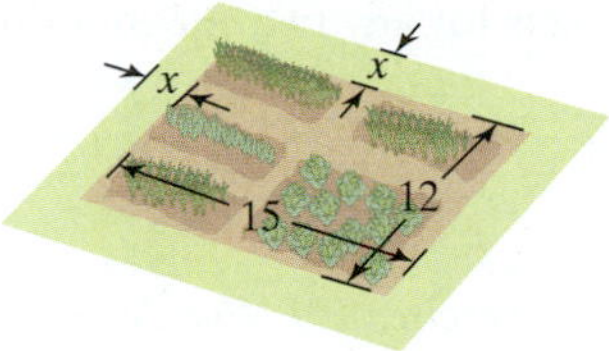

77. As part of a landscaping project, you put in a flower bed measuring 10 feet by 12 feet. You plan to surround the bed with a uniform border of low-growing plants.

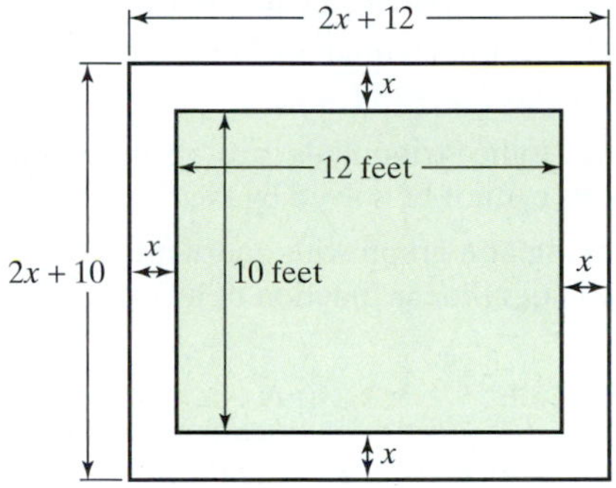

a. Write a polynomial that describes the area of the uniform border that surrounds your flower bed. *Hint:* The area of the border is the area of the large rectangle shown in the figure minus the area of the flower bed.

b. The low growing plants surrounding the flower bed require 1 square foot each when mature. If you have 168 of these plants, how wide a strip around the flower bed should you prepare for the border?

78. As part of a landscaping project, you put in a flower bed measuring 20 feet by 30 feet. To finish off the project, you are putting in a uniform border of pine bark around the outside of the rectangular garden. You have enough pine bark to cover 336 square feet. How wide should the border be?

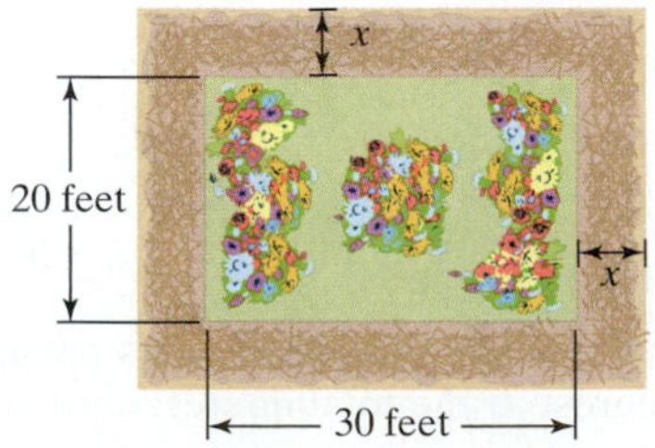

79. A machine produces open boxes using square sheets of metal. The figure illustrates that the machine cuts equal-sized squares measuring 2 inches on a side from the corners and then shapes the metal into an open box by turning up the sides. If each box must have a volume of 200 cubic inches, find the length and width of the open box.

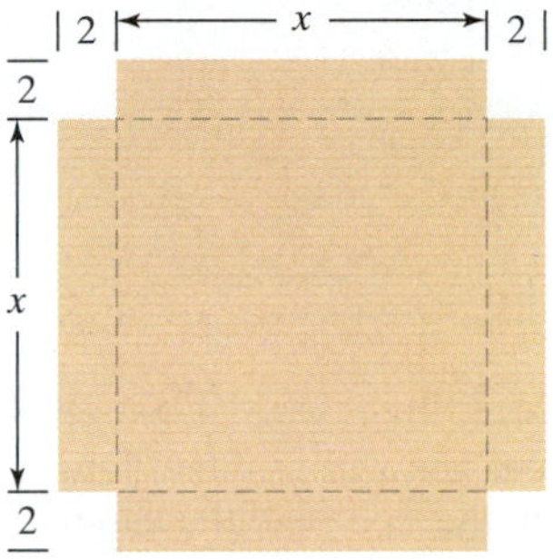

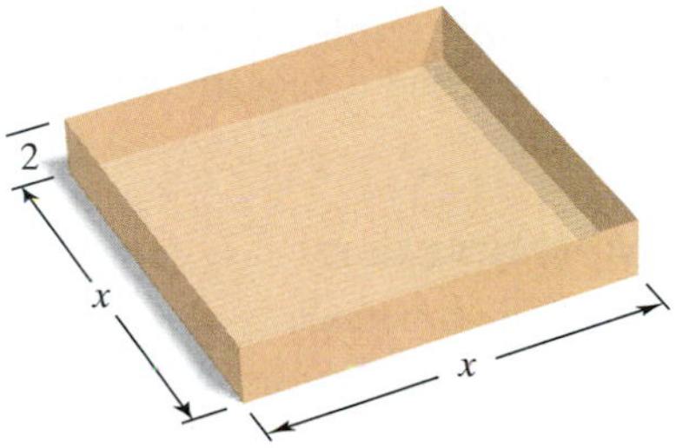

80. A machine produces open boxes using square sheets of metal. The machine cuts equal-sized squares measuring 3 inches on a side from the corners and then shapes the metal into an open box by turning up the sides. If each box must have a volume of 75 cubic inches, find the length and width of the open box.

81. The rectangular floor of a closet is divided into two right triangles by drawing a diagonal, as shown in the figure. One leg of the right triangle is 2 feet more than twice the other leg. The hypotenuse is 13 feet. Determine the closet's length and width.

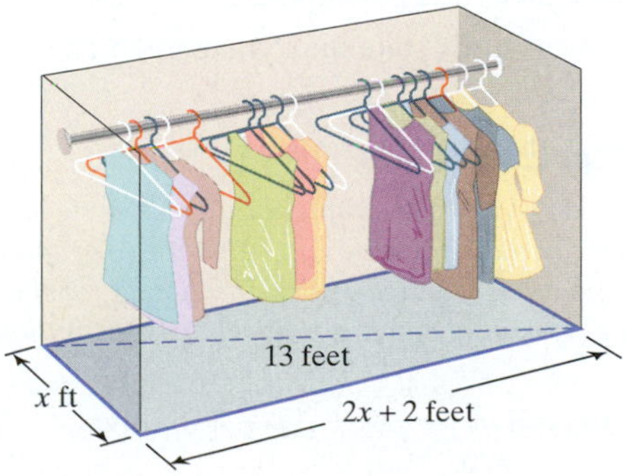

82. A piece of wire measuring 20 feet is attached to a telephone pole as a guy wire. The distance along the ground from the bottom of the pole to the end of the wire is 4 feet greater than the height where the wire is attached to the pole. How far up the pole does the guy wire reach?

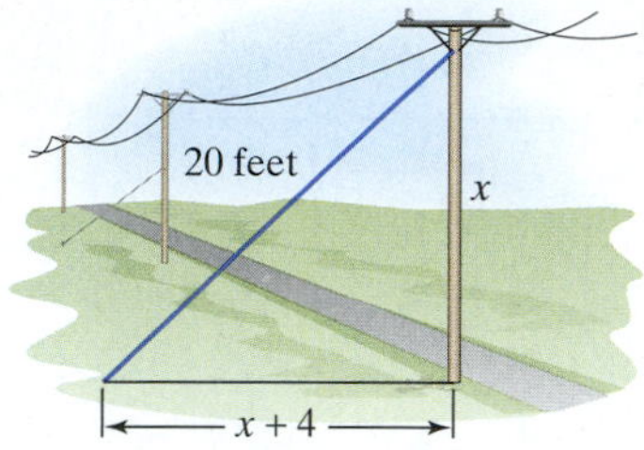

83. A tree is supported by a wire anchored in the ground 15 feet from its base. The wire is 4 feet longer than the height that it reaches on the tree. Find the length of the wire.

84. A tree is supported by a wire anchored in the ground 5 feet from its base. The wire is 1 foot longer than the height that it reaches on the tree. Find the length of the wire.

Writing in Mathematics

85. What is a quadratic equation?

86. What is the zero-product principle?

87. Explain how to solve $x^2 - x = 6$.

88. Describe the relationship between the solutions of a quadratic equation and the graph of the corresponding quadratic function.

89. What is a polynomial equation? When is it in standard form?

90. What is the degree of a polynomial equation? What are polynomial equations of degree 1 and degree 2, respectively, called?

91. Explain how to solve $x^3 + x^2 = x + 1$.

92. If something is thrown straight up, or possibly dropped, describe a situation in which it is important to know how long it will take the object to hit the ground or possibly the water.

93. A toy rocket is launched vertically upward. Using a quadratic equation, we find that the rocket will reach a height of 220 feet at 2.5 seconds and again at 5.5 seconds. How can this be?

94. Describe a situation in which a landscape designer might use polynomials and polynomial equations.

95. In your own words, state the Pythagorean Theorem.

Technology Exercises

In Exercises 96–99, use a graphing utility with a viewing rectangle large enough to show end behavior to graph each polynomial function. Then use the x-intercepts for the graph to solve the polynomial equation. Check by substitution.

96. Use the graph of $y = x^2 + 3x - 4$ to solve $x^2 + 3x - 4 = 0$.

97. Use the graph of $y = x^3 + 3x^2 - x - 3$ to solve $x^3 + 3x^2 - x - 3 = 0$.

98. Use the graph of $y = 2x^3 - 3x^2 - 11x + 6$ to solve $2x^3 - 3x^2 - 11x + 6 = 0$.

99. Use the graph of $y = -x^4 + 4x^3 - 4x^2$ to solve $-x^4 + 4x^3 - 4x^2 = 0$.

100. Use the [TABLE] feature of a graphing utility to verify the solution sets for any two equations in Exercises 31–32 or 39–40.

Critical Thinking Exercises

Make Sense? *In Exercises 101–104, determine whether each statement "makes sense" or "does not make sense" and explain your reasoning.*

101. I'm working with a quadratic function that describes the length of time a ball has been thrown into the air and its height above the ground, and I find the function's graph more meaningful than its equation.

102. I set the quadratic equation $2x^2 - 5x = 12$ equal to zero and obtained $2x^2 - 5x = 0$.

103. Because some trinomials are prime, some quadratic equations cannot be solved by factoring.

104. I'm looking at a graph with one x-intercept, so it must be the graph of a linear function or a vertical line.

In Exercises 105–108, determine whether each statement is true or false. If the statement is false, make the necessary change(s) to produce a true statement.

105. Quadratic equations solved by factoring always have two different solutions.

106. If $4x(x^2 + 49) = 0$, then

$$4x = 0 \quad \text{or} \quad x^2 + 49 = 0$$
$$x = 0 \quad \text{or} \quad x = 7 \quad \text{or} \quad x = -7.$$

107. If -4 is a solution of $7y^2 + (2k - 5)y - 20 = 0$, then k must equal 14.

108. Some quadratic equations have more than two solutions.

109. Write a quadratic equation in standard form whose solutions are -3 and 7.

110. Solve: $|x^2 + 2x - 36| = 12$.

Review Exercises

111. Solve: $|3x - 2| = 8$. (Section 4.3, Example 1)

112. Simplify: $3(5 - 7)^2 + \sqrt{16} + 12 \div (-3)$. (Section 1.2, Example 7)

113. You invested \$3000 in two accounts paying 5% and 8% annual interest. If the total interest earned for the year is \$189, how much was invested at each rate? (Section 3.2, Example 2)

Preview Exercises

Exercises 114–116 will help you prepare for the material covered in the first section of the next chapter.

114. If $f(x) = \frac{120x}{100 - x}$, find $f(20)$.

115. Find the domain of $f(x) = \frac{4}{x - 2}$.

116. Factor the numerator and the denominator. Then simplify by dividing out the common factor in the numerator and the denominator.

$$\frac{x^2 - 7x - 18}{2x^2 + 3x - 2}$$

GROUP PROJECT

CHAPTER 5

Divide the group in half. Without looking at any factoring problems in the book, each group should use polynomial multiplication to create five factoring problems. Make sure that some of your problems require at least two factoring strategies. Next, exchange problems with the other half of the group. Work to factor the five problems. After completing the factorizations, evaluate the factoring problems that you were given. Are they too easy? Too difficult? Can the polynomials really be factored? Share your responses with the half of the group that wrote the problems. Finally, grade each other's work in factoring the polynomials. Each factoring problem is worth 20 points. You may award partial credit. If you take off points, explain why points are deducted and how you decided to take off a particular number of points for the error(s) that you found.

Chapter 5 Summary

Definitions and Concepts	Examples

Section 5.1 Introduction to Polynomials and Polynomial Functions

A polynomial is a single term or the sum of two or more terms containing variables with whole-number exponents. A monomial is a polynomial with exactly one term; a binomial has exactly two terms; a trinomial has exactly three terms. If $a \neq 0$, the degree of ax^n is n and the degree of ax^ny^m is $n + m$. The degree of a nonzero constant is 0. The constant 0 has no defined degree. The degree of a polynomial is the greatest degree of any term. The leading term is the term of greatest degree. Its coefficient is called the leading coefficient.

Examples

$$7x^3y - 4x^5y^4 - 2x^4y$$

Degree is $3 + 1 = 4$. | Degree is $5 + 4 = 9$. | Degree is $4 + 1 = 5$.

The degree of the polynomial is 9. The leading term is $-4x^5y^4$. The leading coefficient is -4. This polynomial is a trinomial.

In a polynomial function, the expression that defines the function is a polynomial. Polynomial functions have graphs that are smooth and continuous. The behavior of the graph of a polynomial function to the far left or the far right is called its end behavior.

The Leading Coefficient Test

1. Odd-degree polynomial functions have graphs with opposite behavior at each end. If the leading coefficient is positive, the graph falls to the left and rises to the right. If the leading coefficient is negative, the graph rises to the left and falls to the right.
2. Even-degree polynomial functions have graphs with the same behavior at each end. If the leading coefficient is positive, the graph rises to the left and rises to the right. If the leading coefficient is negative, the graph falls to the left and falls to the right.

Examples

Describe the end behavior of the graph of each polynomial function:

- $f(x) = -2x^3 + 3x^2 + 11x - 6$

  The degree, 3, is odd. The leading coefficient, -2, is negative. The graph rises to the left and falls to the right.

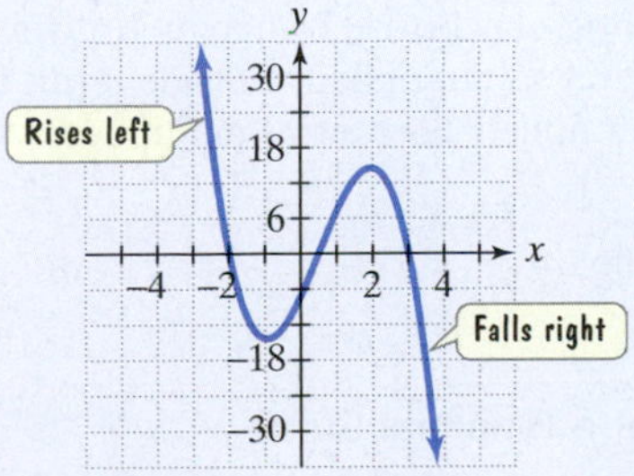

Definitions and Concepts	Examples

Section 5.1 Introduction to Polynomials and Polynomial Functions (continued)

- $f(x) = x^2 - 4$

  The degree, 2, is even. The leading coefficient, 1, is positive. The graph rises to the left and rises to the right.

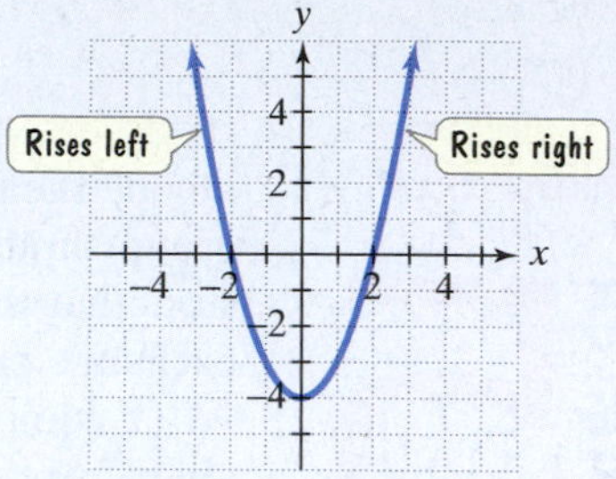

To add polynomials, add like terms.

$$(6x^3y + 5x^2y - 7y) + (-9x^3y + x^2y + 6y)$$
$$= (6x^3y - 9x^3y) + (5x^2y + x^2y) + (-7y + 6y)$$
$$= -3x^3y + 6x^2y - y$$

To subtract two polynomials, change the sign of every term of the second polynomial. Add this result to the first polynomial.

$$(5y^3 - 9y^2 - 4) - (3y^3 - 12y^2 - 5)$$
$$= (5y^3 - 9y^2 - 4) + (-3y^3 + 12y^2 + 5)$$
$$= (5y^3 - 3y^3) + (-9y^2 + 12y^2) + (-4 + 5)$$
$$= 2y^3 + 3y^2 + 1$$

Section 5.2 Multiplication of Polynomials

To multiply monomials, multiply coefficients and add exponents.

$$(-2x^2y^4)(-3x^3y)$$
$$= (-2)(-3)x^{2+3}y^{4+1} = 6x^5y^5$$

To multiply a monomial and a polynomial, multiply each term of the polynomial by the monomial.

$$7x^2y(4x^3y^5 - 2xy - 1)$$
$$= 7x^2y \cdot 4x^3y^5 - 7x^2y \cdot 2xy - 7x^2y \cdot 1$$
$$= 28x^5y^6 - 14x^3y^2 - 7x^2y$$

To multiply polynomials if neither is a monomial, multiply each term of one by each term of the other.

$$(x^3 + 2x)(5x^2 - 3x + 4)$$
$$= x^3(5x^2 - 3x + 4) + 2x(5x^2 - 3x + 4)$$
$$= 5x^5 - 3x^4 + 4x^3 + 10x^3 - 6x^2 + 8x$$
$$= 5x^5 - 3x^4 + 14x^3 - 6x^2 + 8x$$

The FOIL method may be used when multiplying two binomials: First terms multiplied. Outside terms multiplied. Inside terms multiplied. Last terms multiplied.

$$(5x - 3y)(2x + y) = \underbrace{5x \cdot 2x}_{F} + \underbrace{5x \cdot y}_{O} + \underbrace{(-3y) \cdot 2x}_{I} + \underbrace{(-3y) \cdot y}_{L}$$
$$= 10x^2 + 5xy - 6xy - 3y^2$$
$$= 10x^2 - xy - 3y^2$$

The Square of a Binomial Sum

$$(A + B)^2 = A^2 + 2AB + B^2$$

$$(x^2 + 6)^2 = (x^2)^2 + 2 \cdot x^2 \cdot 6 + 6^2$$
$$= x^4 + 12x^2 + 36$$

Definitions and Concepts	Examples

Section 5.2 Multiplication of Polynomials (continued)

The Square of a Binomial Difference

$$(A - B)^2 = A^2 - 2AB + B^2$$

$$(4x - 5)^2 = (4x)^2 - 2 \cdot 4x \cdot 5 + 5^2$$
$$= 16x^2 - 40x + 25$$

The Product of the Sum and Difference of Two Terms

$$(A + B)(A - B) = A^2 - B^2$$

- $(3x + 7y)(3x - 7y) = (3x)^2 - (7y)^2$
  $= 9x^2 - 49y^2$
- $[(x + 2) - 4y][(x + 2) + 4y]$
  $= (x + 2)^2 - (4y)^2 = x^2 + 4x + 4 - 16y^2$

Section 5.3 Greatest Common Factors and Factoring by Grouping

Factoring a polynomial consisting of the sum of monomials means finding an equivalent expression that is a product. Polynomials that cannot be factored using integer coefficients are called prime polynomials over the integers. The greatest common factor, GCF, is an expression that divides every term of the polynomial. The GCF is the product of the largest common numerical factor and the variable of lowest degree common to every term of the polynomial. To factor a monomial from a polynomial, express each term as the product of the GCF and its other factor. Then use the distributive property to factor out the GCF. When the leading coefficient of a polynomial is negative, it is often desirable to factor out a common factor with a negative coefficient.

Factor: $4x^4y^2 - 12x^2y^3 + 20xy^2$.
(GCF is $4xy^2$)
$$= 4xy^2 \cdot x^3 - 4xy^2 \cdot 3xy + 4xy^2 \cdot 5$$
$$= 4xy^2(x^3 - 3xy + 5)$$

Factor: $-25x^3 + 10x^2 - 15x$.
(Use $-5x$ as a common factor.)
$$= -5x(5x^2) - 5x(-2x) - 5x(3)$$
$$= -5x(5x^2 - 2x + 3)$$

To factor by grouping, factor out the GCF from each group. Then factor out the remaining factor.

$$xy + 7x - 2y - 14$$
$$= x(y + 7) - 2(y + 7)$$
$$= (y + 7)(x - 2)$$

Section 5.4 Factoring Trinomials

To factor a trinomial of the form $x^2 + bx + c$, find two numbers whose product is c and whose sum is b. The factorization is

$$(x + \text{one number})(x + \text{other number}).$$

Factor: $x^2 + 9x + 20$.
Find two numbers whose product is 20 and whose sum is 9. The numbers are 4 and 5.
$$x^2 + 9x + 20 = (x + 4)(x + 5)$$

In some trinomials, the highest power is greater than 2, and the exponent in one of the terms is half that of the other term. Factor by introducing a substitution. Let u equal the variable to the smaller power.

Factor: $x^6 - 7x^3 + 12$.
$$= (x^3)^2 - 7x^3 + 12$$
$$= u^2 - 7u + 12 \quad \text{Let } u = x^3.$$
$$= (u - 4)(u - 3)$$
$$= (x^3 - 4)(x^3 - 3)$$

To factor $ax^2 + bx + c$ by trial and error, try various combinations of factors of ax^2 and c until a middle term of bx is obtained for the sum of the outside and inside products.

Factor: $2x^2 + 7x - 15$.

Factors of $2x^2$: $2x, x$

Factors of -15: 1 and -15, -1 and 15, 3 and -5, -3 and 5

$$(2x - 3)(x + 5)$$

Sum of outside and inside products should equal $7x$.
$$10x - 3x = 7x$$
Thus, $2x^2 + 7x - 15 = (2x - 3)(x + 5)$.

Definitions and Concepts	Examples
Section 5.4 Factoring Trinomials (continued)	
To factor $ax^2 + bx + c$ by grouping, find the factors of ac whose sum is b. Write bx as a sum or difference using these factors. Then factor by grouping.	Factor: $2x^2 + 7x - 15$. Find the factors of $2(-15)$, or -30, whose sum is 7. They are 10 and -3. $2x^2 + 7x - 15$ $= 2x^2 + 10x - 3x - 15$ $= 2x(x + 5) - 3(x + 5) = (x + 5)(2x - 3)$
Section 5.5 Factoring Special Forms	
The Difference of Two Squares $A^2 - B^2 = (A + B)(A - B)$	$16x^2 - 9y^2$ $= (4x)^2 - (3y)^2 = (4x + 3y)(4x - 3y)$
Perfect Square Trinomials $A^2 + 2AB + B^2 = (A + B)^2$ $A^2 - 2AB + B^2 = (A - B)^2$	• $x^2 + 20x + 100 = x^2 + 2 \cdot x \cdot 10 + 10^2 = (x + 10)^2$ • $9x^2 - 30x + 25 = (3x)^2 - 2 \cdot 3x \cdot 5 + 5^2 = (3x - 5)^2$
Sum or Difference of Cubes $A^3 + B^3 = (A + B)(A^2 - AB + B^2)$ $A^3 - B^3 = (A - B)(A^2 + AB + B^2)$	$125x^3 - 8 = (5x)^3 - 2^3$ $= (5x - 2)[(5x)^2 + 5x \cdot 2 + 2^2]$ $= (5x - 2)(25x^2 + 10x + 4)$
When using factoring by grouping, terms can sometimes be grouped to obtain the difference of two squares. One of the squares is a perfect square trinomial.	$\underbrace{x^2 + 18x + 81} - 25y^2$ $= (x + 9)^2 - (5y)^2$ $= (x + 9 + 5y)(x + 9 - 5y)$
Section 5.6 A General Factoring Strategy	
A Factoring Strategy **1.** Factor out the GCF or a common factor with a negative coefficient. **2. a.** If two terms, try $A^2 - B^2 = (A + B)(A - B)$ $A^3 + B^3 = (A + B)(A^2 - AB + B^2)$ $A^3 - B^3 = (A - B)(A^2 + AB + B^2)$. **b.** If three terms, try $A^2 + 2AB + B^2 = (A + B)^2$ $A^2 - 2AB + B^2 = (A - B)^2$. If not a perfect square trinomial, try trial and error or grouping. **c.** If four terms, try factoring by grouping. **3.** See if any factors can be factored further.	Factor: $3x^4 + 12x^3 - 3x^2 - 12x$. The GCF is $3x$. $3x^4 + 12x^3 - 3x^2 - 12x$ $= 3x(x^3 + 4x^2 - x - 4)$ Four terms: Try grouping. $= 3x[x^2(x + 4) - 1(x + 4)]$ $= 3x(x + 4)(x^2 - 1)$ This can be factored further. $= 3x(x + 4)(x + 1)(x - 1)$

Definitions and Concepts	Examples

Section 5.7 Polynomial Equations and Their Applications

Definitions and Concepts

A quadratic equation in x can be written in the standard form

$$ax^2 + bx + c, a \neq 0.$$

A polynomial equation is the result of setting two polynomials equal to each other. The equation is in standard form if one side is 0 and the polynomial on the other side is in standard form, that is in descending powers of the variable. In standard form, its degree is the highest degree of any term in the equation. A polynomial equation of degree 1 is a linear equation and of degree 2 a quadratic equation. Some polynomial equations can be solved by writing the equation in standard form, factoring, and then using the zero-product principle: If a product is 0, then at least one of the factors is equal to 0.

Examples

Solve: $5x^2 + 7x = 6$.

$$5x^2 + 7x - 6 = 0$$
$$(5x - 3)(x + 2) = 0$$
$$5x - 3 = 0 \quad \text{or} \quad x + 2 = 0$$
$$5x = 3 \qquad\qquad x = -2$$
$$x = \frac{3}{5}$$

The solutions are -2 and $\frac{3}{5}$, and the solution set is $\left\{-2, \frac{3}{5}\right\}$. (The solutions are the x-intercepts of the graph of $y = 5x^2 + 7x - 6$.)

CHAPTER 5 REVIEW EXERCISES

5.1 *In Exercises 1–2, determine the coefficient of each term, the degree of each term, the degree of the polynomial, the leading term, and the leading coefficient of the polynomial.*

1. $-5x^3 + 7x^2 - x + 2$

2. $8x^4y^2 - 7xy^6 - x^3y$

3. If $f(x) = x^3 - 4x^2 + 3x - 1$, find $f(-2)$.

4. The bar graph shows the number of Americans living with AIDS from 1999 through 2003. The data can be modeled by the function

$$f(x) = 220x^3 - 986x^2 + 24{,}104x + 311{,}243,$$

where $f(x)$ is the number of Americans living with AIDS x years after 1999.

Number of People in the United States Living with AIDS

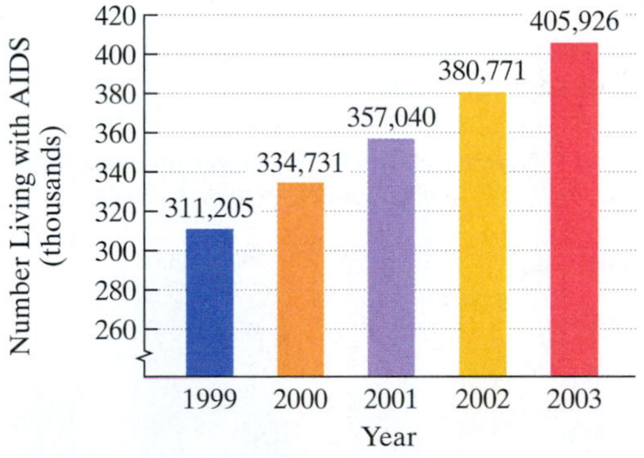

Source: Centers for Disease Control

a. Find and interpret $f(3)$.

b. Does your answer in part (a) underestimate or overestimate the actual number shown by the graph? By how much?

In Exercises 5–8, use the Leading Coefficient Test to determine the end behavior of the graph of the given polynomial function. Then use this end behavior to match the polynomial function with its graph. [The graphs are labeled (a) through (d).]

5. $f(x) = -x^3 + x^2 + 2x$

6. $f(x) = x^6 - 6x^4 + 9x^2$

7. $f(x) = x^5 - 5x^3 + 4x$

8. $f(x) = -x^4 + 1$

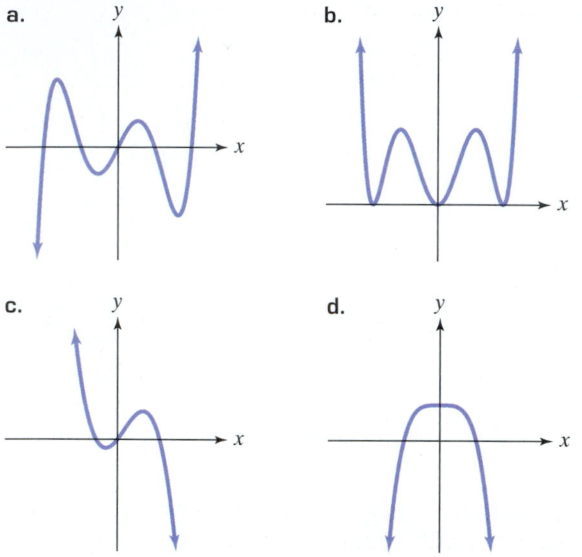

9. Tech firms might be rebounding from the dot-com bust, but enrollment in college computer programs keeps falling. The bar graph shows the number of newly-declared computer science and computer engineering majors for the fall term in U.S. and Canadian colleges from 1999 through 2003. The data can be modeled by the function

$$f(x) = -365x^4 + 2728x^3 - 7106x^2 + 7372x + 20{,}787,$$

where $f(x)$ is the number of newly-declared computer majors for the fall term x years after 1999.

Number of Newly-Declared Computer Majors in U.S. and Canadian Colleges

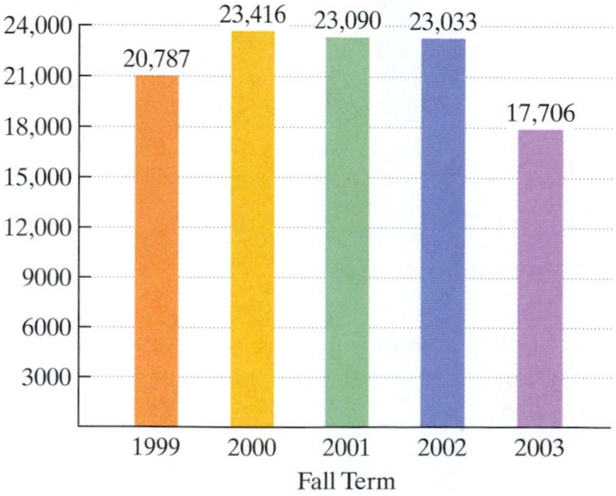

Source: Computing Research Association

a. Find $f(0)$ and $f(1)$. What do you observe when you compare these values with the appropriate numbers shown by the bar graph?

b. Use end behavior to explain why the function that models the data is valid only for a limited period of time.

In Exercises 10–11, add the polynomials.

10. $(-8x^3 + 5x^2 - 7x + 4) + (9x^3 - 11x^2 + 6x - 13)$

11. $(7x^3y - 13x^2y - 6y) + (5x^3y + 11x^2y - 8y - 17)$

In Exercises 12–13, subtract the polynomials.

12. $(7x^3 - 6x^2 + 5x - 11) - (-8x^3 + 4x^2 - 6x - 3)$

13. $(4x^3y^2 - 7x^3y - 4) - (6x^3y^2 - 3x^3y + 4)$

14. Subtract $-2x^3 - x^2y + xy^2 + 7y^3$ from $x^3 + 4x^2y - y^3$.

5.2 *In Exercises 15–27, multiply the polynomials.*

15. $(4x^2yz^5)(-3x^4yz^2)$

16. $6x^3\left(\frac{1}{3}x^5 - 4x^2 - 2\right)$

17. $7xy^2(3x^4y^2 - 5xy - 1)$

18. $(2x + 5)(3x^2 + 7x - 4)$

19. $(x^2 + x - 1)(x^2 + 3x + 2)$

20. $(4x - 1)(3x - 5)$

21. $(3xy - 2)(5xy + 4)$

22. $(3x + 7y)^2$

23. $(x^2 - 5y)^2$

24. $(2x + 7y)(2x - 7y)$

25. $(3xy^2 - 4x)(3xy^2 + 4x)$

26. $[(x + 3) + 5y][(x + 3) - 5y]$

27. $(x + y + 4)^2$

28. Let $f(x) = x - 3$ and $g(x) = 2x + 5$. Find $(fg)(x)$ and $(fg)(-4)$.

29. Let $f(x) = x^2 - 7x + 2$. Find each of the following and simplify:

a. $f(a - 1)$

b. $f(a + h) - f(a)$.

5.3 *In Exercises 30–33, factor the greatest common factor from each polynomial.*

30. $16x^3 + 24x^2$

31. $2x - 36x^2$

32. $21x^2y^2 - 14xy^2 + 7xy$

33. $18x^3y^2 - 27x^2y$

In Exercises 34–35, factor out a common factor with a negative coefficient.

34. $-12x^2 + 8x - 48$

35. $-x^2 - 11x + 14$

In Exercises 36–38, factor by grouping.

36. $x^3 - x^2 - 2x + 2$

37. $xy - 3x - 5y + 15$

38. $5ax - 15ay + 2bx - 6by$

5.4 *In Exercises 39–47, factor each trinomial completely, or state that the trinomial is prime.*

39. $x^2 + 8x + 15$

40. $x^2 + 16x - 80$

41. $x^2 + 16xy - 17y^2$

42. $3x^3 - 36x^2 + 33x$

43. $3x^2 + 22x + 7$

44. $6x^2 - 13x + 6$

45. $5x^2 - 6xy - 8y^2$

46. $6x^3 + 5x^2 - 4x$

47. $2x^2 + 11x + 15$

In Exercises 48–51, factor by introducing an appropriate substitution.

48. $x^6 + x^3 - 30$

49. $x^4 - 10x^2 - 39$

50. $(x + 5)^2 + 10(x + 5) + 24$

51. $5x^6 + 17x^3 + 6$

5.5 *In Exercises 52–55, factor each difference of two squares.*

52. $4x^2 - 25$

53. $1 - 81x^2y^2$

54. $x^8 - y^6$

55. $(x - 1)^2 - y^2$

In Exercises 56–60, factor any perfect square trinomials, or state that the polynomial is prime.

56. $x^2 + 16x + 64$

57. $9x^2 - 6x + 1$

58. $25x^2 + 20xy + 4y^2$

59. $49x^2 + 7x + 1$

60. $25x^2 - 40xy + 16y^2$

In Exercises 61–62, factor by grouping to obtain the difference of two squares.

61. $x^2 + 18x + 81 - y^2$

62. $z^2 - 25x^2 + 10x - 1$

In Exercises 63–65, factor using the formula for the sum or difference of two cubes.

63. $64x^3 + 27$

64. $125x^3 - 8$

65. $x^3y^3 + 1$

5.6 *In Exercises 66–90, factor completely, or state that the polynomial is prime.*

66. $15x^2 + 3x$

67. $12x^4 - 3x^2$

68. $20x^4 - 24x^3 + 28x^2 - 12x$

69. $x^3 - 15x^2 + 26x$

70. $-2y^4 + 24y^3 - 54y^2$

71. $9x^2 - 30x + 25$

72. $5x^2 - 45$

73. $2x^3 - x^2 - 18x + 9$

74. $6x^2 - 23xy + 7y^2$

75. $2y^3 + 12y^2 + 18y$

76. $x^2 + 6x + 9 - 4a^2$

77. $8x^3 - 27$

78. $x^5 - x$

79. $x^4 - 6x^2 + 9$

80. $x^2 + xy + y^2$

81. $4a^3 + 32$

82. $x^4 - 81$

83. $ax + 3bx - ay - 3by$

84. $27x^3 - 125y^3$

85. $10x^3y + 22x^2y - 24xy$

86. $6x^6 + 13x^3 - 5$

87. $2x + 10 + x^2y + 5xy$

88. $y^3 + 2y^2 - 25y - 50$

89. $a^8 - 1$

90. $9(x - 4) + y^2(4 - x)$

In Exercises 91–92,

a. *Write an expression for the area of the shaded region.*

b. *Write the expression in factored form.*

91.

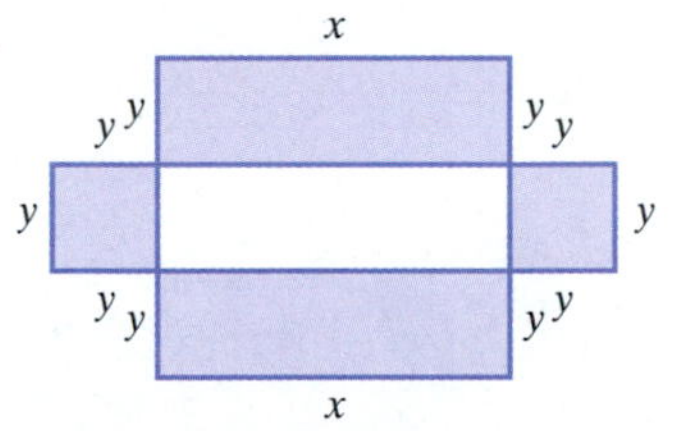

92.

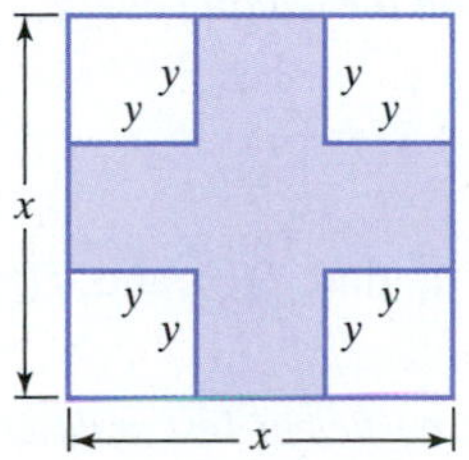

5.7 *In Exercises 93–97, use factoring to solve each polynomial equation.*

93. $x^2 + 6x + 5 = 0$

94. $3x^2 = 22x - 7$

95. $(x + 3)(x - 2) = 50$

96. $3x^2 = 12x$

97. $x^3 + 5x^2 = 9x + 45$

98. A model rocket is launched from the top of a cliff 144 feet above sea level. The function

$$s(t) = -16t^2 + 128t + 144$$

describes the rocket's height above the water, $s(t)$, in feet, t seconds after it is launched. The rocket misses the edge of the cliff on its way down and eventually lands in the ocean. How long will it take for the rocket to hit the water?

99. How much distance do you need to bring your car to a complete stop? A function used by those who study automobile safety is

$$d(x) = \frac{x^2}{20} + x,$$

where $d(x)$ is the stopping distance, in feet, for a car traveling at x miles per hour.

a. If it takes you 40 feet to come to a complete stop, how fast was your car traveling?

b. The graph of the quadratic function that models stopping distance is shown. Identify your solution from part (a) as a point on the graph.

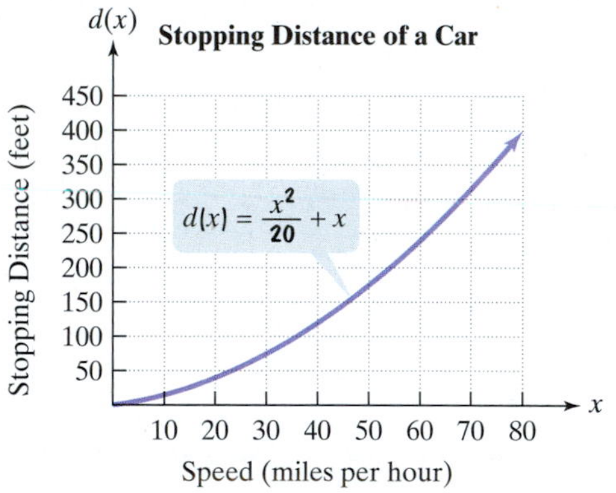

c. Describe the trend shown by the graph.

100. The length of a rectangular sign is 3 feet longer than the width. If the sign has space for 54 square feet of advertising, find its length and its width.

101. A painting measuring 10 inches by 16 inches is surrounded by a frame of uniform width. If the combined area of the painting and frame is 280 square inches, determine the width of the frame.

102. A lot is in the shape of a right triangle. The longer leg of the triangle is 20 yards longer than twice the length of the shorter leg. The hypotenuse is 30 yards longer than twice the length of the shorter leg. What are the lengths of the three sides?

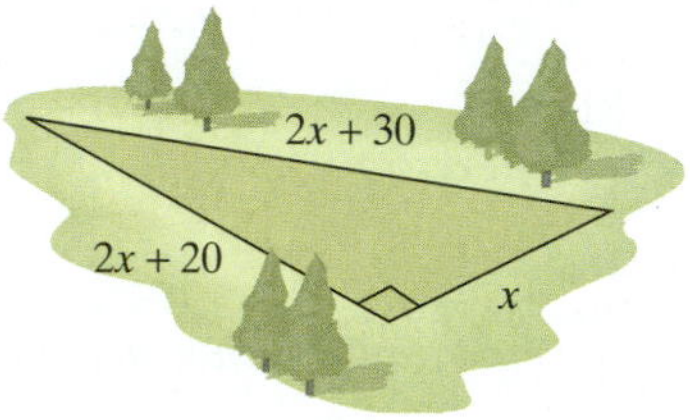

CHAPTER 5 TEST

Remember to use your Chapter Test Prep Video CD to see the worked-out solutions to the test questions you want to review.

In Exercises 1–2, give the degree and the leading coefficient of the polynomial.

1. $7x - 5 + x^2 - 6x^3$

2. $4xy^3 + 7x^4y^5 - 3xy^4$

3. If $f(x) = 3x^3 + 5x^2 - x + 6$, find $f(0)$ and $f(-2)$.

In Exercises 4–5, use the Leading Coefficient Test to describe the end behavior of the graph of the polynomial function.

4. $f(x) = -16x^2 + 160x$

5. $f(x) = 4x^3 + 12x^2 - x - 3$

In Exercises 6–13, perform the indicated operations.

6. $(4x^3y - 19x^2y - 7y) + (3x^3y + x^2y + 6y - 9)$

7. $(6x^2 - 7x - 9) - (-5x^2 + 6x - 3)$

8. $(-7x^3y)(-5x^4y^2)$

9. $(x - y)(x^2 - 3xy - y^2)$

10. $(7x - 9y)(3x + y)$

11. $(2x - 5y)(2x + 5y)$

12. $(4y - 7)^2$

13. $[(x + 2) + 3y][(x + 2) - 3y]$

14. Let $f(x) = x + 2$ and $g(x) = 3x - 5$. Find $(fg)(x)$ and $(fg)(-5)$.

15. Let $f(x) = x^2 - 5x + 3$. Find $f(a + h) - f(a)$ and simplify.

In Exercises 16–33, factor completely, or state that the polynomial is prime.

16. $14x^3 - 15x^2$

17. $81y^2 - 25$

18. $x^3 + 3x^2 - 25x - 75$

19. $25x^2 - 30x + 9$

20. $x^2 + 10x + 25 - 9y^2$

21. $x^4 + 1$

22. $y^2 - 16y - 36$

23. $14x^2 + 41x + 15$

24. $5x^3 - 5$

25. $12x^2 - 3y^2$

26. $12x^2 - 34x + 10$

27. $3x^4 - 3$

28. $x^8 - y^8$

29. $12x^2y^4 + 8x^3y^2 - 36x^2y$

30. $x^6 - 12x^3 - 28$

31. $x^4 - 2x^2 - 24$

32. $12x^2y - 27xy + 6y$

33. $y^4 - 3y^3 + 2y^2 - 6y$

In Exercises 34–37, solve each polynomial equation.

34. $3x^2 = 5x + 2$

35. $(5x + 4)(x - 1) = 2$

36. $15x^2 - 5x = 0$

37. $x^3 - 4x^2 - x + 4 = 0$

38. A baseball is thrown straight up from a rooftop 448 feet high. The function

$$s(t) = -16t^2 + 48t + 448$$

describes the ball's height above the ground, $s(t)$, in feet, t seconds after it is thrown. How long will it take for the ball to hit the ground?

39. An architect is allowed 15 square yards of floor space to add a small bedroom to a house. Because of the room's design in relationship to the existing structure, the width of the rectangular floor must be 7 yards less than two times the length. Find the length and width of the rectangular floor that the architect is permitted.

40. Find the lengths of the three sides of the right triangle in the figure shown.

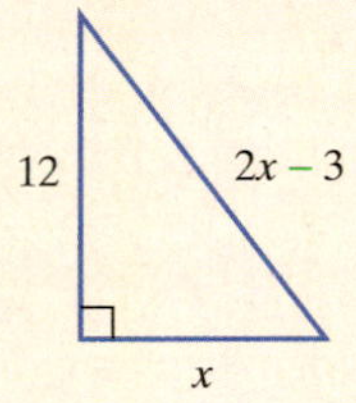

CUMULATIVE REVIEW EXERCISES (CHAPTERS 1–5)

In Exercises 1–7, solve each equation, inequality, or system of equations.

1. $8(x + 2) - 3(2 - x) = 4(2x + 6) - 2$

2. $2x + 4y = -6$
$x = 2y - 5$

3. $2x - y + 3z = 0$
$2y + z = 1$
$x + 2y - z = 5$

4. $2x + 4 < 10$ and $3x - 1 > 5$

5. $|2x - 5| \geq 9$

6. $2x^2 = 7x - 5$

7. $2x^3 + 6x^2 = 20x$

8. Solve for x: $x = \dfrac{ax + b}{c}$.

9. Use function notation to write the equation of the line passing through $(-2, -3)$ and $(2, 5)$.

10. In a campuswide election for student government president, 2800 votes were cast for the two candidates. If the winner had 160 more votes than the loser, how many votes were cast for each candidate?

In Exercises 11–13, graph each equation or inequality in a rectangular coordinate system.

11. $f(x) = -\frac{1}{3}x + 1$

12. $4x - 5y < 20$

13. $y \leq -1$

14. Simplify: $-\dfrac{8x^3y^6}{16x^9y^{-4}}$.

15. Write in scientific notation: 0.0000706.

In Exercises 16–17, perform the indicated operations.

16. $(3x^2 - y)^2$

17. $(3x^2 - y)(3x^2 + y)$

In Exercises 18–20, factor completely.

18. $x^3 - 3x^2 - 9x + 27$

19. $x^6 - x^2$

20. $14x^3y^2 - 28x^4y^2$

CHAPTER 6

Rational Expressions, Functions, and Equations

A few candy factoids:

- The average American spends \$84 per year on candy, consuming 23.9 pounds annually.
- 55% of candy sales are "impulse buys."
- The 1978 criminal defense of the man who murdered San Francisco's mayor was based on the perpetrator's temporary insanity due to overconsumption of candy and junk food.
- The word *candy* originated with the Sanskrit word *khanda*, meaning "pieces of crystallized sugar."
- A function consisting of the quotient of two rational expressions models what happens in your mouth after eating "pieces of crystallized sugar."

In this chapter, you will see how functions involving fractional expressions provide insights into phenomena as diverse as the aftereffects of candy on your mouth, the cost of environmental cleanup, the relationship between heart rate and life span, and even our ongoing processes of learning and forgetting.

Here's where you'll encounter these applications:

- Sugar and your mouth: Exercise Set 6.1, Exercises 109–112
- Environmental cleanup: Section 6.1, Example 1, and Section 6.6, Example 6
- Heart rate and life span: Exercise Set 6.8, Exercises 29–32
- Learning and forgetting: Exercise Set 6.6, Exercises 53–60.

SECTION 6.1

Rational Expressions and Functions: Multiplying and Dividing

Objectives

1. Evaluate rational functions.
2. Find the domain of a rational function.
3. Interpret information given by the graph of a rational function.
4. Simplify rational expressions.
5. Multiply rational expressions.
6. Divide rational expressions.

Environmental scientists and municipal planners often make decisions using **cost-benefit models**. These mathematical models estimate the cost of removing a pollutant from the atmosphere as a function of the percentage of pollutant removed. What kinds of functions describe the cost of reducing environmental pollution? In this section, we introduce this new category of functions, called *rational functions*.

Rational Expressions

A **rational expression** consists of a polynomial divided by a nonzero polynomial. Here are some examples of rational expressions:

$$\frac{120x}{100-x}, \quad \frac{2x+1}{2x^2-x-1}, \quad \frac{3x^2+12xy-15y^2}{6x^3-6xy^2}.$$

1 Evaluate rational functions.

Rational Functions

A **rational function** is a function defined by a formula that is a rational expression.

EXAMPLE 1 Using a Cost-Benefit Model

The rational function

$$f(x) = \frac{120x}{100-x}$$

models the cost, $f(x)$, in thousands of dollars, to remove $x\%$ of the pollutants that a city has discharged into a lake. Find and interpret each of the following:

a. $f(20)$ **b.** $f(80)$.

Solution We use substitution to evaluate a rational function, just as we did to evaluate other functions in Chapter 2.

a. $f(x) = \frac{120x}{100-x}$ This is the given rational function.

$f(20) = \frac{120(20)}{100-20}$ To find f of 20, replace x with 20.

$= \frac{2400}{80} = 30$ Perform the indicated operations.

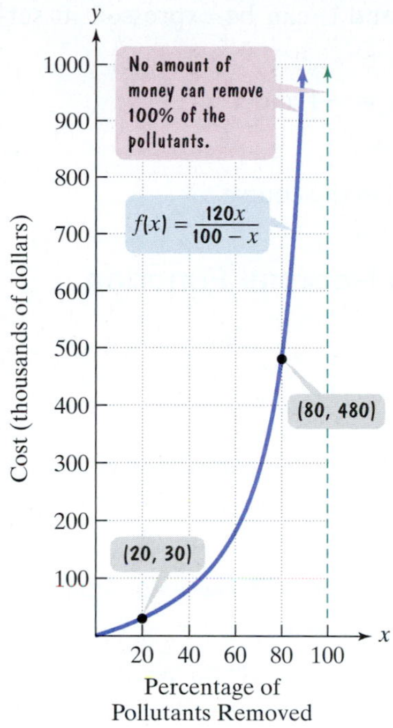

FIGURE 6.1

Thus, $f(20) = 30$. This means that the cost to remove 20% of the lake's pollutants is \$30 thousand. **Figure 6.1** illustrates the solution by the point (20, 30) on the graph of the rational function.

b. $f(x) = \frac{120x}{100 - x}$ This is the given rational function.

$f(80) = \frac{120(80)}{100 - 80} = \frac{9600}{20} = 480$ To find f of 80, replace x with 80.

Thus, $f(80) = 480$. This means that the cost to remove 80% of the lake's pollutants is \$480 thousand. **Figure 6.1** illustrates the solution by the point (80, 480) on the graph of the cost-benefit model. The graph illustrates that costs rise steeply as the percentage of pollutants removed increases.

CHECK POINT 1 Use the rational function in Example 1 to find and interpret: **a.** $f(40)$ **b.** $f(60)$. Identify each solution as a point on the graph of the rational function in **Figure 6.1**.

2 Find the domain of a rational function.

The Domain of a Rational Function

Does the cost-benefit model

$$f(x) = \frac{120x}{100 - x}$$

indicate that the city can clean up its lake completely? To do this, the city must remove 100% of the pollutants. The problem is that the rational function is undefined for $x = 100$.

$$f(x) = \frac{120x}{100 - x}$$

If $x = 100$, the value of the denominator is 0.

Notice how the graph of the rational function in **Figure 6.1** approaches, but never touches, the dashed green vertical line whose equation is $x = 100$. The graph continues to rise more and more steeply, visually showing the escalating costs. By never touching the dashed vertical line, the graph illustrates that no amount of money will be enough to remove all pollutants from the lake.

In Chapter 2, we learned to exclude from a function's domain real numbers that cause division by zero. Thus, the **domain of a rational function** is the set of all real numbers except those for which the denominator is zero. We can find the domain by determining when the denominator is zero. For the cost-benefit model, the denominator is zero when $x = 100$. Furthermore, for this model, negative values of x and values of x greater than 100 are not meaningful. The domain of the function is $[0, 100)$ and excludes 100.

Inspection can sometimes be used to find a rational function's domain. Here are two examples.

$$f(x) = \frac{4}{x - 2} \qquad g(x) = \frac{x - 3}{(x + 1)(x - 1)}$$

This numerator *can* be zero, so there is no need to exclude 3 from the domain.

This denominator would equal zero if $x = 2$.

This factor would equal zero if $x = -1$.

This factor would equal zero if $x = 1$.

The domain of f can be expressed in set-builder or interval notation:

$$\text{Domain of } f = \{x \mid x \text{ is a real number and } x \neq 2\}$$
$$\text{Domain of } f = (-\infty, 2) \cup (2, \infty).$$

$$g(x) = \frac{x - 3}{(x + 1)(x - 1)}$$

This factor would equal zero if $x = -1$.

This factor would equal zero if $x = 1$.

Function g (repeated)

Likewise, the domain of g, which excludes both -1 and 1, can be expressed in set-builder or interval notation:

$$\text{Domain of } g = \{x \mid x \text{ is a real number and } x \neq -1 \text{ and } x \neq 1\}$$
$$\text{Domain of } g = (-\infty, -1) \cup (-1, 1) \cup (1, \infty).$$

Ask your professor if a particular notation is preferred in the course.

EXAMPLE 2 Finding the Domain of a Rational Function

Find the domain of f if

$$f(x) = \frac{2x + 1}{2x^2 - x - 1}.$$

Solution The domain of f is the set of all real numbers except those for which the denominator is zero. We can identify such numbers by setting the denominator equal to zero and solving for x.

$$2x^2 - x - 1 = 0 \quad \text{Set the denominator equal to 0.}$$
$$(2x + 1)(x - 1) = 0 \quad \text{Factor.}$$
$$2x + 1 = 0 \quad \text{or} \quad x - 1 = 0 \quad \text{Set each factor equal to 0.}$$
$$2x = -1 \qquad x = 1 \quad \text{Solve the resulting equations.}$$
$$x = -\frac{1}{2}$$

Because $-\frac{1}{2}$ and 1 make the denominator zero, these are the values to exclude. Thus,

$$\text{Domain of } f = \left\{x \mid x \text{ is a real number and } x \neq -\frac{1}{2} \text{ and } x \neq 1\right\}$$

or

$$\text{Domain of } f = \left(-\infty, -\frac{1}{2}\right) \cup \left(-\frac{1}{2}, 1\right) \cup (1, \infty).$$

Using Technology

We can use the TABLE feature of a graphing utility (with $\Delta\text{Tbl} = .5$) to verify our work with

$$f(x) = \frac{2x + 1}{2x^2 - x - 1}.$$

Enter

$y_1 =$ (2 X + 1) ÷ (2 X ^ 2 − X − 1)

and press TABLE.

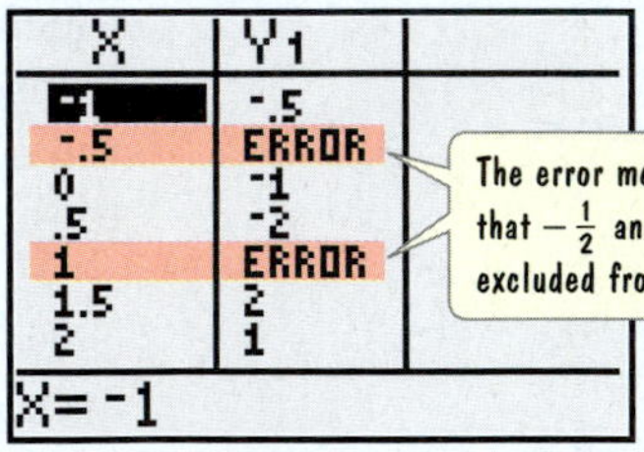

The error messages verify that $-\frac{1}{2}$ and 1 must be excluded from the domain.

CHECK POINT 2 Find the domain of f if

$$f(x) = \frac{x - 5}{2x^2 + 5x - 3}.$$

Does every rational function have values to exclude? The answer is no. For example, consider

$$f(x) = \frac{2}{x^2 + 1}.$$

No real-number values of x cause this denominator to equal zero.

The domain of f consists of the set of all real numbers. Recall that this set can be expressed in three different ways:

$$\text{Domain of } f = \{x | x \text{ is a real number}\} \text{ or } \mathbb{R} \text{ or } (-\infty, \infty).$$

3 Interpret information given by the graph of a rational function.

What Graphs of Rational Functions Look Like

In everyday speech, a continuous process is one that goes on without interruption and without abrupt changes. In mathematics, a continuous function has much the same meaning. The graph of a continuous function does not have interrupting breaks, such as holes, gaps, or jumps. Thus, the graph of a continuous function can be drawn without lifting a pencil off the paper.

Most rational functions are not continuous functions. For example, the graph of the rational function

$$f(x) = \frac{2x + 1}{2x^2 - x - 1}$$

is shown in **Figure 6.2**. In Example 2, we excluded $-\frac{1}{2}$ and 1 from this function's domain. Unlike the graph of a polynomial function, this graph has two breaks in it—one at each of the excluded values. At $-\frac{1}{2}$, a hole in the graph appears. The graph is composed of two distinct branches. Each branch approaches, but never touches, the dashed vertical line drawn at 1, the other excluded value.

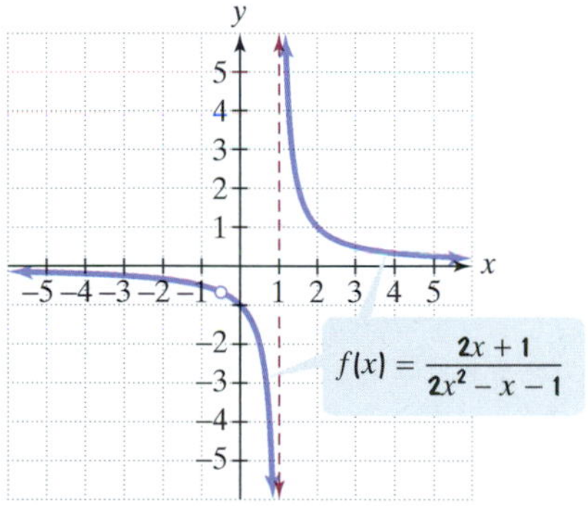

FIGURE 6.2

A vertical line that the graph of a function approaches, but does not touch, is said to be a **vertical asymptote** of the graph. In **Figure 6.2**, the line $x = 1$ is a vertical asymptote of the graph of f. A rational function may have no vertical asymptotes, one vertical asymptote, or several vertical asymptotes. The graph of a rational function never intersects a vertical asymptote.

Unlike the graph of a polynomial function, the graph of the rational function in **Figure 6.2** does not go up or down at each end. At the far left and the far right, the graph is getting closer to, but not actually reaching, the x-axis. This shows that as x increases or decreases without bound, the function values are approaching 0. The line $y = 0$ (that is, the x-axis) is a **horizontal asymptote** of the graph. Many, but not all, rational functions have horizontal asymptotes.

For the remainder of this section, we will focus on the rational expressions that define rational functions. Operations with these expressions should remind you of those performed in arithmetic with fractions.

Using Technology

When using a graphing utility to graph a rational function, you might not be pleased with the quality of the display. The graph of the rational function

$$y = \frac{2x + 1}{2x^2 - x - 1}$$

is shown using the DOT mode in a $[-6, 6, 1]$ by $[-6, 6, 1]$ viewing rectangle. In this mode, the utility displays unconnected points that have been calculated. Would you agree that the graph's behavior is better illustrated by the hand-drawn graph in **Figure 6.2**?

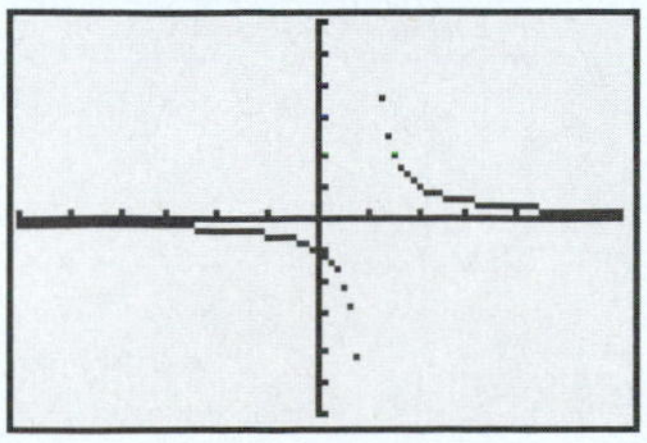

4 Simplify rational expressions.

Simplifying Rational Expressions

A rational expression is **simplified** if its numerator and denominator have no common factors other than 1 or -1. The following procedure can be used to simplify rational expressions:

Simplifying Rational Expressions

1. Factor the numerator and the denominator completely.
2. Divide both the numerator and the denominator by any common factors.

EXAMPLE 3 Simplifying a Rational Expression

Simplify: $\dfrac{x^2 + 4x + 3}{x + 1}$.

Solution

$$\frac{x^2 + 4x + 3}{x + 1} = \frac{(x + 1)(x + 3)}{1(x + 1)}$$ Factor the numerator and denominator.

$$= \frac{\cancel{(x + 1)}(x + 3)}{1\cancel{(x + 1)}}$$ Divide out the common factor, $x + 1$.

$$= x + 3$$

Simplifying a rational expression can change the numbers that make it undefined. For example, we just showed that

$$\frac{x^2 + 4x + 3}{x + 1} = x + 3.$$

This is undefined for $x = -1$. The simplified form is defined for all real numbers.

Thus, to equate the two expressions, we must restrict the values of x in the simplified expression to exclude -1. We can write

$$\frac{x^2 + 4x + 3}{x + 1} = x + 3, x \neq -1.$$

Without this restriction, the expressions are not equal. The slight difference between them is illustrated using the graphs of

$$f(x) = \frac{x^2 + 4x + 3}{x + 1} \quad \text{and} \quad g(x) = x + 3$$

in **Figure 6.3**.

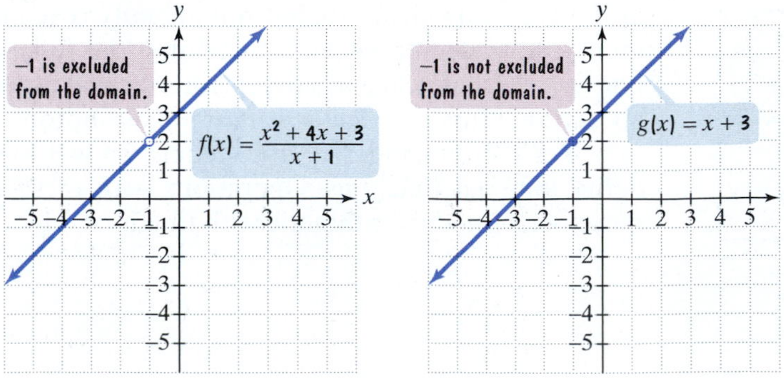

FIGURE 6.3 Visualizing the difference between $\dfrac{x^2 + 4x + 3}{x + 1}$ and $x + 3$

Hereafter, we will assume that a simplified rational expression is equal to the original rational expression for all real numbers except those for which either denominator is 0.

CHECK POINT 3 Simplify: $\dfrac{x^2 + 7x + 10}{x + 2}$.

EXAMPLE 4 Simplifying Rational Expressions

Simplify:

a. $\dfrac{x^2 - 7x - 18}{2x^2 + 3x - 2}$ b. $\dfrac{3x^2 + 12xy - 15y^2}{6x^3 - 6xy^2}$.

Solution

a. $\dfrac{x^2 - 7x - 18}{2x^2 + 3x - 2} = \dfrac{(x - 9)(x + 2)}{(2x - 1)(x + 2)}$ Factor the numerator and denominator.

$= \dfrac{(x - 9)\cancel{(x + 2)}}{(2x - 1)\cancel{(x + 2)}}$ Divide out the common factor, $x + 2$.

$= \dfrac{x - 9}{2x - 1}$

b. $\dfrac{3x^2 + 12xy - 15y^2}{6x^3 - 6xy^2} = \dfrac{3(x^2 + 4xy - 5y^2)}{6x(x^2 - y^2)}$ Factor out the GCF in the numerator and denominator.

$= \dfrac{3(x + 5y)(x - y)}{6x(x + y)(x - y)}$ Factor the numerator and denominator completely.

$= \dfrac{\overset{1}{\cancel{3}}(x + 5y)\cancel{(x - y)}}{\underset{2}{\cancel{6}}x(x + y)\cancel{(x - y)}}$ Divide out the common factor, $3(x - y)$.

$= \dfrac{x + 5y}{2x(x + y)}$ It is not necessary to carry out the multiplication in the denominator.

CHECK POINT 4 Simplify:

a. $\dfrac{x^2 - 2x - 15}{3x^2 + 8x - 3}$ b. $\dfrac{3x^2 + 9xy - 12y^2}{9x^3 - 9xy^2}$.

Study Tip

When simplifying rational expressions, you can only divide out, or cancel, factors common to the numerator and denominator. **It is incorrect to divide out common terms from the numerator and denominator.**

Incorrect!

$$\frac{x + 4}{x} = 4 \qquad \frac{x^2 - 4}{4} = x^2 - 1 \qquad \frac{\overset{x}{\cancel{x^2}} - \overset{3}{\cancel{9}}}{\underset{1}{\cancel{x}} - \underset{1}{\cancel{3}}} = x - 3$$

The first two expressions have no common factors in their numerators and denominators. Only when expressions are multiplied can they be factors. **If you can't factor, then don't try to cancel.** The third rational expression can be simplified as follows:

Correct

$$\frac{x^2 - 9}{x - 3} = \frac{(x + 3)\cancel{(x - 3)}}{1\cancel{(x - 3)}} = x + 3.$$

Divide out the common factor, $x - 3$.

5 Multiply rational expressions.

Multiplying Rational Expressions

The product of two rational expressions is the product of their numerators divided by the product of their denominators. For example,

$$\frac{x^2}{y+3} \cdot \frac{x+5}{y-7} = \frac{x^2(x+5)}{(y+3)(y-7)}.$$

Multiply numerators.

Multiply denominators.

Here is a step-by-step procedure for multiplying rational expressions. Before multiplying, divide out any factors common to both a numerator and a denominator.

Multiplying Rational Expressions

1. Factor all numerators and denominators completely.
2. Divide numerators and denominators by common factors.
3. Multiply the remaining factors in the numerators and multiply the remaining factors in the denominators.

EXAMPLE 5 Multiplying Rational Expressions

Multiply: $\frac{x+3}{x-4} \cdot \frac{x^2-2x-8}{x^2-9}$.

Solution

$$\frac{x+3}{x-4} \cdot \frac{x^2-2x-8}{x^2-9}$$

$$= \frac{1(x+3)}{1(x-4)} \cdot \frac{(x-4)(x+2)}{(x+3)(x-3)}$$ Factor all numerators and denominators completely.

$$= \frac{1\cancel{(x+3)}}{1\cancel{(x-4)}} \cdot \frac{\cancel{(x-4)}(x+2)}{\cancel{(x+3)}(x-3)}$$ Divide numerators and denominators by common factors.

$$= \frac{x+2}{x-3}$$ Multiply the remaining factors in the numerators and in the denominators.

✓ **CHECK POINT 5** Multiply: $\frac{x+4}{x-7} \cdot \frac{x^2-4x-21}{x^2-16}$.

Some rational expressions contain factors in the numerator and denominator that are opposites, or additive inverses. Here is an example of such an expression:

$$\frac{(2x+5)(2x-5)}{3(5-2x)}.$$

The factors $2x - 5$ and $5 - 2x$ are opposites. They differ only in sign.

Although you can factor out -1 from the numerator or the denominator and then divide out the common factor, there is an even faster way to simplify this rational expression.

Simplifying Rational Expressions with Opposite Factors in the Numerator and Denominator

The quotient of two polynomials that have opposite signs and are additive inverses is -1.

For example,

$$\frac{(2x+5)(2x-5)}{3(5-2x)} = \frac{(2x+5)\overset{(-1)}{\cancel{(2x-5)}}}{3\cancel{(5-2x)}} = \frac{-(2x+5)}{3} \text{ or } -\frac{2x+5}{3} \text{ or } \frac{-2x-5}{3}.$$

Factoring out -1 is done mentally: $\frac{(2x+5)(-1)\cancel{(5-2x)}}{3\cancel{(5-2x)}}$.

EXAMPLE 6 Multiplying Rational Expressions

Multiply: $\frac{5x+5}{7x-7x^2} \cdot \frac{2x^2+x-3}{4x^2-9}$.

Solution

$$\frac{5x+5}{7x-7x^2} \cdot \frac{2x^2+x-3}{4x^2-9}$$

$$= \frac{5(x+1)}{7x(1-x)} \cdot \frac{(2x+3)(x-1)}{(2x+3)(2x-3)}$$ Factor all numerators and denominators completely.

$$= \frac{5(x+1)}{7x\cancel{(1-x)}} \cdot \frac{\cancel{(2x+3)}\overset{(-1)}{\cancel{(x-1)}}}{\cancel{(2x+3)}(2x-3)}$$ Divide numerators and denominators by common factors. Because $1-x$ and $x-1$ are opposites, their quotient is -1.

$$= \frac{-5(x+1)}{7x(2x-3)} \text{ or } -\frac{5(x+1)}{7x(2x-3)}$$ Multiply the remaining factors in the numerators and in the denominators.

✓ **CHECK POINT 6** Multiply: $\frac{4x+8}{6x-3x^2} \cdot \frac{3x^2-4x-4}{9x^2-4}$.

6 Divide rational expressions.

Dividing Rational Expressions

The quotient of two rational expressions is the product of the first expression and the multiplicative inverse, or reciprocal, of the second expression. The reciprocal is found by interchanging the numerator and the denominator of the expression.

Dividing Rational Expressions

If P, Q, R, and S are polynomials, where $Q \neq 0$, $R \neq 0$, and $S \neq 0$, then

$$\frac{P}{Q} \div \frac{R}{S} = \frac{P}{Q} \cdot \frac{S}{R} = \frac{PS}{QR}.$$

Change division to multiplication.

Replace $\frac{R}{S}$ with its reciprocal by interchanging its numerator and denominator.

Thus, **we find the quotient of two rational expressions by inverting the divisor and multiplying**. For example,

$$\frac{x}{7} \div \frac{6}{y} = \frac{x}{7} \cdot \frac{y}{6} = \frac{xy}{42}.$$

Change the division to multiplication.

Replace $\frac{6}{y}$ with its reciprocal by interchanging its numerator and denominator.

EXAMPLE 7 Dividing Rational Expressions

Divide:

a. $(4x^2 - 25) \div \frac{2x + 5}{14}$ b. $\frac{x^2 + 3x - 10}{2x} \div \frac{x^2 - 5x + 6}{x^2 - 3x}$.

Study Tip

When performing operations with rational expressions, if a rational expression is written without a denominator, it is helpful to write the expression with a denominator of 1. In Example 7(a), we wrote

$4x^2 - 25$ as $\frac{4x^2 - 25}{1}$.

Solution

a. $(4x^2 - 25) \div \frac{2x + 5}{14}$

$= \frac{4x^2 - 25}{1} \div \frac{2x + 5}{14}$ Write $4x^2 - 25$ with a denominator of 1.

$= \frac{4x^2 - 25}{1} \cdot \frac{14}{2x + 5}$ Invert the divisor and multiply.

$= \frac{(2x + 5)(2x - 5)}{1} \cdot \frac{14}{1(2x + 5)}$ Factor.

$= \frac{\cancel{(2x + 5)}(2x - 5)}{1} \cdot \frac{14}{1\cancel{(2x + 5)}}$ Divide the numerator and denominator by the common factor, $2x + 5$.

$= 14(2x - 5)$ Multiply the remaining factors in the numerators and in the denominators.

b. $\frac{x^2 + 3x - 10}{2x} \div \frac{x^2 - 5x + 6}{x^2 - 3x}$

$= \frac{x^2 + 3x - 10}{2x} \cdot \frac{x^2 - 3x}{x^2 - 5x + 6}$ Invert the divisor and multiply.

$= \frac{(x + 5)(x - 2)}{2x} \cdot \frac{x(x - 3)}{(x - 3)(x - 2)}$ Factor.

$= \frac{(x + 5)\cancel{(x - 2)}}{2\cancel{x}} \cdot \frac{\cancel{x}\cancel{(x - 3)}}{\cancel{(x - 3)}\cancel{(x - 2)}}$ Divide numerators and denominators by common factors.

$= \frac{x + 5}{2}$ Multiply the remaining factors in the numerators and in the denominators.

CHECK POINT 7 Divide:

a. $(9x^2 - 49) \div \frac{3x - 7}{9}$

b. $\frac{x^2 - x - 12}{5x} \div \frac{x^2 - 10x + 24}{x^2 - 6x}$.

6.1 EXERCISE SET

MyMathLab · MathXL PRACTICE · WATCH · DOWNLOAD · READ · REVIEW

Practice Exercises

In Exercises 1–6, use the given rational function to find the indicated function values. If a function value does not exist, so state.

1. $f(x) = \frac{x^2 - 9}{x + 3}$; $f(-2), f(0), f(5)$
2. $f(x) = \frac{x^2 - 16}{x + 4}$; $f(-2), f(0), f(5)$
3. $f(x) = \frac{x^2 - 2x - 3}{4 - x}$; $f(-1), f(4), f(6)$
4. $f(x) = \frac{x^2 - 3x - 4}{3 - x}$; $f(-1), f(3), f(5)$
5. $g(t) = \frac{2t^3 - 5}{t^2 + 1}$; $g(-1), g(0), g(2)$
6. $g(t) = \frac{2t^3 - 1}{t^2 + 4}$; $g(-1), g(0), g(2)$

In Exercises 7–16, find the domain of the given rational function. Use the notation of your choice (set-builder or interval) or the notation requested by your professor.

7. $f(x) = \frac{x - 2}{x - 5}$
8. $f(x) = \frac{x - 3}{x - 6}$
9. $f(x) = \frac{x - 4}{(x - 1)(x + 3)}$
10. $f(x) = \frac{x - 5}{(x - 2)(x + 4)}$
11. $f(x) = \frac{2x}{(x + 5)^2}$
12. $f(x) = \frac{2x}{(x + 7)^2}$
13. $f(x) = \frac{3x}{x^2 - 8x + 15}$
14. $f(x) = \frac{3x}{x^2 - 13x + 36}$
15. $f(x) = \frac{(x - 1)^2}{3x^2 - 2x - 8}$
16. $f(x) = \frac{(x - 1)^2}{4x^2 - 13x + 3}$

The graph of a rational function, f, is shown in the figure. Use the graph to solve Exercises 17–26.

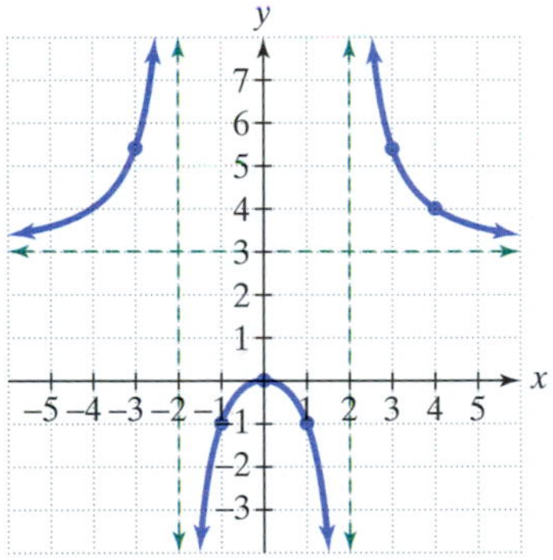

17. Find $f(4)$.
18. Find $f(1)$.
19. What is the domain of f? What is the range of f?
20. What are the equations of the vertical asymptotes of the graph of f?
21. Describe the end behavior of the graph at the far left. What is the equation of the horizontal asymptote?
22. Describe the end behavior of the graph at the far right. What is the equation of the horizontal asymptote?
23. Explain how the graph shows that $f(-2)$ does not exist.

(In Exercises 24–26, continue to use the graph on the previous page.)

24. Explain how the graph shows that $f(2)$ does not exist.

25. How can you tell that this is not the graph of a polynomial function?

26. List two real numbers that are not function values of f.

In Exercises 27–50, simplify each rational expression. If the rational expression cannot be simplified, so state.

27. $\frac{x^2 - 4}{x - 2}$

28. $\frac{x^2 - 25}{x - 5}$

29. $\frac{x + 2}{x^2 - x - 6}$

30. $\frac{x + 1}{x^2 - 2x - 3}$

31. $\frac{4x + 20}{x^2 + 5x}$

32. $\frac{5x + 30}{x^2 + 6x}$

33. $\frac{4y - 20}{y^2 - 25}$

34. $\frac{6y - 42}{y^2 - 49}$

35. $\frac{3x - 5}{25 - 9x^2}$

36. $\frac{5x - 2}{4 - 25x^2}$

37. $\frac{y^2 - 49}{y^2 - 14y + 49}$

38. $\frac{y^2 - 9}{y^2 - 6y + 9}$

39. $\frac{x^2 + 7x - 18}{x^2 - 3x + 2}$

40. $\frac{x^2 - 4x - 5}{x^2 + 5x + 4}$

41. $\frac{3x + 7}{3x + 10}$

42. $\frac{2x + 3}{2x + 5}$

43. $\frac{x^2 - x - 12}{16 - x^2}$

44. $\frac{x^2 - 7x + 12}{9 - x^2}$

45. $\frac{x^2 + 3xy - 10y^2}{3x^2 - 7xy + 2y^2}$

46. $\frac{x^2 + 2xy - 3y^2}{2x^2 + 5xy - 3y^2}$

47. $\frac{x^3 - 8}{x^2 - 4}$

48. $\frac{x^3 - 1}{x^2 - 1}$

49. $\frac{x^3 + 4x^2 - 3x - 12}{x + 4}$

50. $\frac{x^3 - 2x^2 + x - 2}{x - 2}$

In Exercises 51–72, multiply as indicated.

51. $\frac{x - 3}{x + 7} \cdot \frac{3x + 21}{2x - 6}$

52. $\frac{x - 2}{x + 3} \cdot \frac{2x + 6}{5x - 10}$

53. $\frac{x^2 - 49}{x^2 - 4x - 21} \cdot \frac{x + 3}{x}$

54. $\frac{x^2 - 25}{x^2 - 3x - 10} \cdot \frac{x + 2}{x}$

55. $\frac{x^2 - 9}{x^2 - x - 6} \cdot \frac{x^2 + 5x + 6}{x^2 + x - 6}$

56. $\frac{x^2 - 1}{x^2 - 4} \cdot \frac{x^2 - 5x + 6}{x^2 - 2x - 3}$

57. $\frac{x^2 + 4x + 4}{x^2 + 8x + 16} \cdot \frac{(x + 4)^3}{(x + 2)^3}$

58. $\frac{x^2 - 2x + 1}{x^2 - 4x + 4} \cdot \frac{(x - 2)^3}{(x - 1)^3}$

59. $\frac{8y + 2}{y^2 - 9} \cdot \frac{3 - y}{4y^2 + y}$

60. $\frac{6y + 2}{y^2 - 1} \cdot \frac{1 - y}{3y^2 + y}$

61. $\frac{y^3 - 8}{y^2 - 4} \cdot \frac{y + 2}{2y}$

62. $\frac{y^2 + 6y + 9}{y^3 + 27} \cdot \frac{1}{y + 3}$

63. $(x - 3) \cdot \frac{x^2 + x + 1}{x^2 - 5x + 6}$

64. $(x + 1) \cdot \frac{x + 2}{x^2 + 7x + 6}$

65. $\frac{x^2 + xy}{x^2 - y^2} \cdot \frac{4x - 4y}{x}$

66. $\frac{x^2 - y^2}{x} \cdot \frac{x^2 + xy}{x + y}$

67. $\frac{x^2 + 2xy + y^2}{x^2 - 2xy + y^2} \cdot \frac{4x - 4y}{3x + 3y}$

68. $\frac{2x^2 - 3xy - 2y^2}{3x^2 - 4xy + y^2} \cdot \frac{3x^2 - 2xy - y^2}{x^2 + xy - 6y^2}$

69. $\frac{4a^2 + 2ab + b^2}{2a + b} \cdot \frac{4a^2 - b^2}{8a^3 - b^3}$

70. $\frac{27a^3 - 8b^3}{b^2 - b - 6} \cdot \frac{bc - b - 3c + 3}{3ac - 2bc - 3a + 2b}$

71. $\frac{10z^2 + 13z - 3}{3z^2 - 8z + 5} \cdot \frac{2z^2 - 3z - 2z + 3}{25z^2 - 10z + 1} \cdot \frac{15z^2 - 28z + 5}{4z^2 - 9}$

72. $\frac{2z^2 - 2z - 12}{z^2 - 49} \cdot \frac{4z^2 - 1}{2z^2 + 5z + 2} \cdot \frac{2z^2 - 13z - 7}{2z^2 - 7z + 3}$

In Exercises 73–90, divide as indicated.

73. $\frac{x + 5}{7} \div \frac{4x + 20}{9}$

74. $\frac{x + 1}{3} \div \frac{3x + 3}{7}$

75. $\frac{4}{y - 6} \div \frac{40}{7y - 42}$

76. $\frac{7}{y - 5} \div \frac{28}{3y - 15}$

77. $\frac{x^2 - 2x}{15} \div \frac{x - 2}{5}$

78. $\frac{x^2 - x}{15} \div \frac{x - 1}{5}$

79. $\frac{y^2 - 25}{2y - 2} \div \frac{y^2 + 10y + 25}{y^2 + 4y - 5}$

80. $\dfrac{y^2+y}{y^2-4} \div \dfrac{y^2-1}{y^2+5y+6}$

81. $(x^2-16) \div \dfrac{x^2+3x-4}{x^2+4}$

82. $(x^2+4x-5) \div \dfrac{x^2-25}{x+7}$

83. $\dfrac{y^2-4y-21}{y^2-10y+25} \div \dfrac{y^2+2y-3}{y^2-6y+5}$

84. $\dfrac{y^2+4y-21}{y^2+3y-28} \div \dfrac{y^2+14y+48}{y^2+4y-32}$

85. $\dfrac{8x^3-1}{4x^2+2x+1} \div \dfrac{x-1}{(x-1)^2}$

86. $\dfrac{x^2-9}{x^3-27} \div \dfrac{x^2+6x+9}{x^2+3x+9}$

87. $\dfrac{x^2-4y^2}{x^2+3xy+2y^2} \div \dfrac{x^2-4xy+4y^2}{x+y}$

88. $\dfrac{xy-y^2}{x^2+2x+1} \div \dfrac{2x^2+xy-3y^2}{2x^2+5xy+3y^2}$

89. $\dfrac{x^4-y^8}{x^2+y^4} \div \dfrac{x^2-y^4}{3x^2}$

90. $\dfrac{(x-y)^3}{x^3-y^3} \div \dfrac{x^2-2xy+y^2}{x^2-y^2}$

Practice PLUS

In Exercises 91–98, perform the indicated operation or operations.

91. $\dfrac{x^3-4x^2+x-4}{2x^3-8x^2+x-4} \cdot \dfrac{2x^3+2x^2+x+1}{x^4-x^3+x^2-x}$

92. $\dfrac{y^3+y^2+yz^2+z^2}{y^3+y+y^2+1} \cdot \dfrac{y^3+y+y^2z+z}{2y^2+2yz-yz^2-z^3}$

93. $\dfrac{ax-ay+3x-3y}{x^3+y^3} \div \dfrac{ab+3b+ac+3c}{xy-x^2-y^2}$

94. $\dfrac{a^3+b^3}{ac-ad-bc+bd} \div \dfrac{ab-a^2-b^2}{ac-ad+bc-bd}$

95. $\dfrac{a^2b+b}{3a^2-4a-20} \cdot \dfrac{a^2+5a}{2a^2+11a+5} \div \dfrac{ab^2}{6a^2-17a-10}$

96. $\dfrac{a^2-8a+15}{2a^3-10a^2} \cdot \dfrac{2a^2+3a}{3a^3-27a} \div \dfrac{14a+21}{a^2-6a-27}$

97. $\dfrac{a-b}{4c} \div \left(\dfrac{b-a}{c} \div \dfrac{a-b}{c^2}\right)$

98. $\left(\dfrac{a-b}{4c} \div \dfrac{b-a}{c}\right) \div \dfrac{a-b}{c^2}$

In Exercises 99–102, find $\dfrac{f(a+h)-f(a)}{h}$ *and simplify.*

99. $f(x) = 7x - 4$

100. $f(x) = -3x + 5$

101. $f(x) = x^2 - 5x + 3$

102. $f(x) = 3x^2 - 4x + 7$

In Exercises 103–104, let

$$f(x) = \frac{(x+2)^2}{1-2x} \quad \text{and} \quad g(x) = \frac{x+2}{2x-1}.$$

103. Find $\left(\dfrac{f}{g}\right)(x)$ and the domain of $\dfrac{f}{g}$.

104. Find $\left(\dfrac{g}{f}\right)(x)$ and the domain of $\dfrac{g}{f}$.

Application Exercises

The rational function

$$f(x) = \frac{130x}{100-x}$$

models the cost, $f(x)$, in millions of dollars, to inoculate x% of the population against a particular strain of flu. The graph of the rational function is shown. Use the function's equation to solve Exercises 105–108.

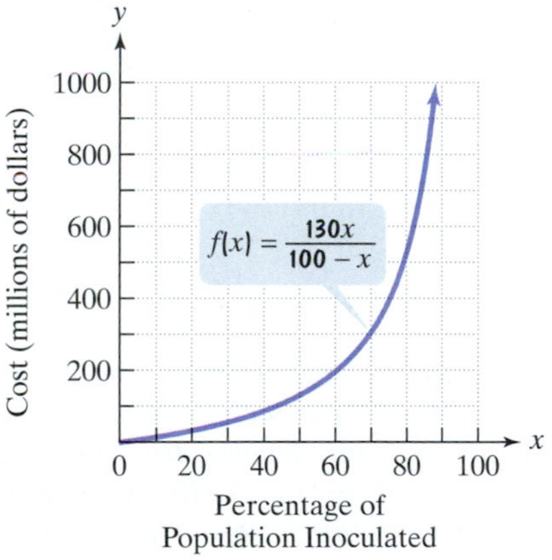

105. Find and interpret $f(60)$. Identify your solution as a point on the graph.

106. Find and interpret $f(80)$. Identify your solution as a point on the graph.

107. What value of x must be excluded from the rational function's domain? Does the cost model indicate that we can inoculate all of the population against the flu? Explain.

108. What happens to the cost as x approaches 100%? How is this shown by the graph? Explain what this means.

The function

$$f(x) = \frac{6.5x^2 - 20.4x + 234}{x^2 + 36}$$

models the pH level, $f(x)$, of the human mouth x minutes after a person eats food containing sugar. The graph of this function is shown in the figure. Use the function's equation or graph, as specified, to solve Exercises 109–112.

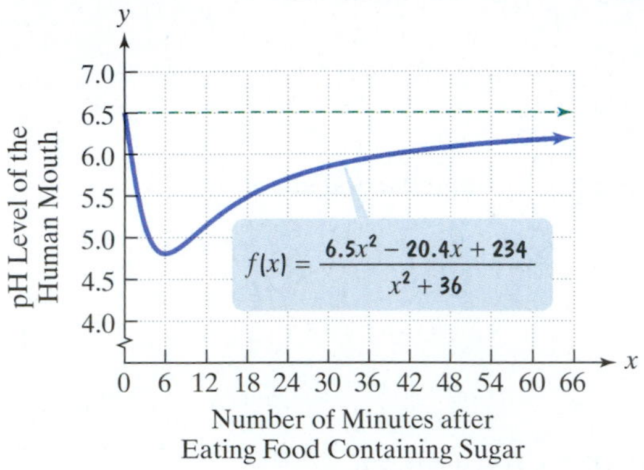

109. After eating sugar, when is the pH level the lowest? Use the function's equation to determine the pH level, to the nearest tenth, at this time.

110. Use the graph to obtain a reasonable estimate, to the nearest tenth, of the pH level of the human mouth 42 minutes after a person eats food containing sugar.

111. According to the graph, what is the normal pH level of the human mouth? What does the end behavior at the far right of the graph indicate in terms of the mouth's pH level over time?

112. Use the graph to describe what happens to the pH level during the first hour.

Among all deaths from a particular disease, the percentage that are smoking related (21–39 cigarettes per day) is a function of the disease's ***incidence ratio****. The incidence ratio describes the number of times more likely smokers are than nonsmokers to die from the disease. The following table shows the incidence ratios for heart disease and lung cancer for two age groups.*

Incidence Ratios

	Heart Disease	Lung Cancer
Ages 55–64	1.9	10
Ages 65–74	1.7	9

Source: Alexander M. Walker, *Observations and Inference*, Epidemiology Resources Inc., 1991.

For example, the incidence ratio of 9 in the table means that smokers between the ages of 65 and 74 are 9 times more likely than nonsmokers in the same age group to die from lung cancer.

The rational function

$$P(x) = \frac{100(x - 1)}{x}$$

models the percentage of smoking-related deaths among all deaths from a disease, $P(x)$, in terms of the disease's incidence ratio, x. The graph of the rational function is shown. Use this function to solve Exercises 113–116.

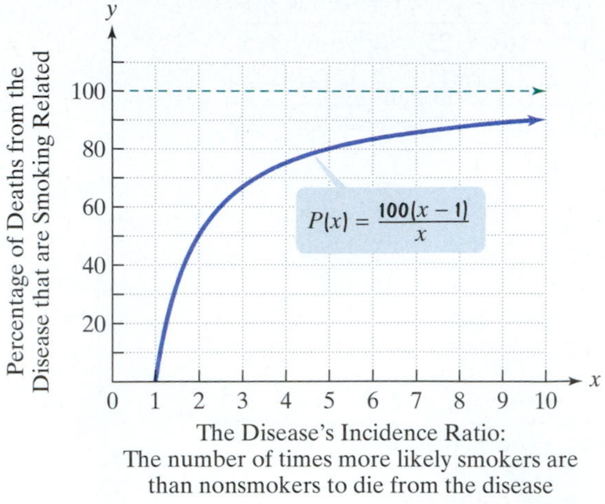

113. Find $P(10)$. Describe what this means in terms of the incidence ratio, 10, given in the table. Identify your solution as a point on the graph.

114. Find $P(9)$. Round to the nearest percent. Describe what this means in terms of the incidence ratio, 9, given in the table. Identify your solution as a point on the graph.

115. What is the horizontal asymptote of the graph? Describe what this means about the percentage of deaths caused by smoking with increasing incidence ratios.

116. According to the model and its graph, is there a disease for which all deaths are caused by smoking? Explain your answer.

Writing in Mathematics

117. What is a rational expression? Give an example with your explanation.

118. What is a rational function? Provide an example.

119. What is the domain of a rational function?

120. If you are given the equation of a rational function, explain how to determine its domain.

121. Describe two ways the graph of a rational function differs from the graph of a polynomial function.

122. What is a vertical asymptote?

123. What is a horizontal asymptote?

124. Explain how to simplify a rational expression.

125. Explain how to simplify a rational expression with opposite factors in the numerator and the denominator.

126. Explain how to multiply rational expressions.

127. Explain how to divide rational expressions.

128. Although your friend has a family history of heart disease, he smokes, on average, 25 cigarettes per day. He sees the table showing incidence ratios for heart disease (see Exercises 113–116) and feels comfortable that they are less than 2, compared to 9 and 10 for lung cancer. He claims that all family deaths have been from heart disease and decides not to give up smoking. Use the given function and its graph to describe some additional information not given in the table that might influence his decision.

Technology Exercises

In Exercises 129–132, determine if the multiplication or division has been performed correctly by graphing the function on each side of the equation in the same viewing rectangle. If the graphs do not coincide, correct the expression on the right side and then verify your correction using the graphing utility.

129. $\frac{x^2+x}{3x} \cdot \frac{6x}{x+1} = 2x$

130. $\frac{x^3-25x}{x^2-3x-10} \cdot \frac{x+2}{x} = x+5$

131. $\frac{x^2-9}{x+4} \div \frac{x-3}{x+4} = x-3$

132. $(x-5) \div \frac{2x^2-11x+5}{4x^2-1} = 2x-1$

133. Use the TABLE feature of a graphing utility to verify the domains that you determined for any two functions in Exercises 7–16.

134. **a.** Graph $f(x) = \frac{x^2-x-2}{x-2}$ and $g(x) = x+1$ in the same viewing rectangle. What do you observe?

b. Simplify the formula in the definition of function f. Do f and g represent exactly the same function? Explain.

c. Display the graphs of f and g separately in the utility's viewing rectangle. TRACE along each of the curves until you get to $x = 2$. What difference do you observe? What does this mean?

Critical Thinking Exercises

Make Sense? *In Exercises 135–138, determine whether each statement "makes sense" or "does not make sense" and explain your reasoning.*

135. I cannot simplify, multiply, or divide rational expressions without knowing how to factor polynomials.

136. I simplified $\frac{2(x+2)-5(x+1)}{(x+2)(x+1)}$ by dividing the numerator and the denominator by $(x+2)(x+1)$.

137. The values to exclude from the domain of $f(x) = \frac{x-3}{x-7}$ are 3 and 7.

138. When performing the division

$$\frac{3x}{x+2} \div \frac{(x+2)^2}{x-4},$$

I began by dividing the numerator and the denominator by the common factor, $x+2$.

In Exercises 139–142, determine whether each statement is true or false. If the statement is false, make the necessary change(s) to produce a true statement.

139. $\frac{x^2-25}{x-5} = x-5$

140. $\frac{x^2+7}{7} = x^2+1$

141. The domain of $f(x) = \frac{7}{x(x-3)+5(x-3)}$ is $(-\infty, 3) \cup (3, \infty)$.

142. The restrictions on the values of x when performing the division

$$\frac{f(x)}{g(x)} \div \frac{h(x)}{k(x)}$$

are $g(x) \neq 0$, $k(x) \neq 0$, and $h(x) \neq 0$.

143. Graph: $f(x) = \frac{x^2-x-2}{x-2}$.

144. Simplify: $\frac{6x^{3n}+6x^{2n}y^n}{x^{2n}-y^{2n}}$.

145. Divide: $\frac{y^{2n}-1}{y^{2n}+3y^n+2} \div \frac{y^{2n}+y^n-12}{y^{2n}-y^n-6}$.

146. Solve for x in terms of a and write the resulting rational expression in simplified form. Include any necessary restrictions on a.

$$a^2(x-1) = 4(x-1) + 5a + 10$$

Review Exercises

147. Graph: $4x - 5y \geq 20$. (Section 4.4, Example 1)

148. Multiply: $(2x-5)(x^2-3x-6)$. (Section 5.2, Example 3)

149. Simplify: $\left(\frac{ab^{-3}c^{-4}}{4a^5b^{10}c^{-3}}\right)^{-2}$. (Section 1.6, Example 9)

Preview Exercises

Exercises 150–152 will help you prepare for the material covered in the next section. In each exercise, perform the indicated operation. Where possible, reduce the answer to its lowest terms.

150. $\frac{7}{10} - \frac{3}{10}$ **151.** $\frac{1}{2} + \frac{2}{3}$ **152.** $\frac{7}{15} - \frac{3}{10}$

SECTION 6.2

Adding and Subtracting Rational Expressions

Objectives

1. Add rational expressions with the same denominator.
2. Subtract rational expressions with the same denominator.
3. Find the least common denominator.
4. Add and subtract rational expressions with different denominators.
5. Add and subtract rational expressions with opposite denominators.

Did you know that people in California are at far greater risk of injury or death from drunk drivers than from earthquakes? According to the U.S. Bureau of Justice statistics, half the arrests for driving under the influence of alcohol involve drivers ages 25 through 34. The rational function

$$f(x) = \frac{27{,}725(x - 14)}{x^2 + 9} - 5x$$

models the number of arrests for driving under the influence, $f(x)$, per 100,000 drivers, as a function of a driver's age, x.

The formula for function f involves the subtraction of two expressions. It is possible to perform this subtraction and express the function's formula as a single rational expression. In this section, we will draw on your experience from arithmetic to add and subtract rational expressions. We return to the driving-under-the-influence model and its graph in Exercise 83 in the section's exercise set.

1 Add rational expressions with the same denominator.

Addition and Subtraction when Denominators Are the Same

To add rational numbers having the same denominators, such as $\frac{2}{9}$ and $\frac{5}{9}$, we add the numerators and place the sum over the common denominator:

$$\frac{2}{9} + \frac{5}{9} = \frac{2+5}{9} = \frac{7}{9}.$$

We add rational expressions with the same denominator in an identical manner.

Adding Rational Expressions with Common Denominators

If $\frac{P}{R}$ and $\frac{Q}{R}$ are rational expressions, then

$$\frac{P}{R} + \frac{Q}{R} = \frac{P+Q}{R}.$$

To add rational expressions with the same denominator, add numerators and place the sum over the common denominator. If possible, factor and simplify the result.

EXAMPLE 1 Adding Rational Expressions when Denominators Are the Same

Add: $\frac{x^2 + 2x - 2}{x^2 + 3x - 10} + \frac{5x + 12}{x^2 + 3x - 10}$.

Solution

$$\frac{x^2 + 2x - 2}{x^2 + 3x - 10} + \frac{5x + 12}{x^2 + 3x - 10}$$

$$= \frac{x^2 + 2x - 2 + 5x + 12}{x^2 + 3x - 10}$$

Add numerators. Place this sum over the common denominator.

$$= \frac{x^2 + 7x + 10}{x^2 + 3x - 10}$$

Combine like terms: $2x + 5x = 7x$ and $-2 + 12 = 10$.

$$= \frac{(x + 2)\cancel{(x + 5)}}{(x - 2)\cancel{(x + 5)}}$$

Factor and simplify by dividing out the common factor, $x + 5$.

$$= \frac{x + 2}{x - 2}$$

✓ **CHECK POINT 1** Add: $\frac{x^2 - 5x - 15}{x^2 + 5x + 6} + \frac{2x + 5}{x^2 + 5x + 6}$.

The following box shows how to subtract rational expressions with the same denominator:

2 Subtract rational expressions with the same denominator.

Subtracting Rational Expressions with Common Denominators

If $\frac{P}{R}$ and $\frac{Q}{R}$ are rational expressions, then

$$\frac{P}{R} - \frac{Q}{R} = \frac{P - Q}{R}.$$

To subtract rational expressions with the same denominator, subtract numerators and place the difference over the common denominator. If possible, factor and simplify the result.

EXAMPLE 2 Subtracting Rational Expressions when Denominators Are the Same

Subtract: $\frac{3y^3 - 5x^3}{x^2 - y^2} - \frac{4y^3 - 6x^3}{x^2 - y^2}$.

Solution

$$\frac{3y^3 - 5x^3}{x^2 - y^2} - \frac{4y^3 - 6x^3}{x^2 - y^2} = \frac{3y^3 - 5x^3 - (4y^3 - 6x^3)}{x^2 - y^2}$$

Subtract numerators and include parentheses to indicate that both terms are subtracted. Place this difference over the common denominator.

$$= \frac{3y^3 - 5x^3 - 4y^3 + 6x^3}{x^2 - y^2}$$

Remove parentheses and then distribute the minus to change the sign of each term.

$$= \frac{x^3 - y^3}{x^2 - y^2}$$

Combine like terms: $-5x^3 + 6x^3 = x^3$ and $3y^3 - 4y^3 = -y^3$.

$$= \frac{\cancel{(x - y)}(x^2 + xy + y^2)}{(x + y)\cancel{(x - y)}}$$

Factor and simplify by dividing out the common factor, $x - y$.

$$= \frac{x^2 + xy + y^2}{x + y}$$

Study Tip

When a numerator is being subtracted, be sure to **subtract every term in that expression.**

The − sign applies to the entire numerator, $4y^3 - 6x^3$.

Insert parentheses to indicate this.

The sign of every term of $4y^3 - 6x^3$ changes.

$$\frac{3y^3 - 5x^3}{x^2 - y^2} - \frac{4y^3 - 6x^3}{x^2 - y^2} = \frac{3y^3 - 5x^3 - (4y^3 - 6x^3)}{x^2 - y^2} = \frac{3y^3 - 5x^3 - 4y^3 + 6x^3}{x^2 - y^2}$$

The entire numerator of the second rational expression must be subtracted. Avoid the common error of subtracting only the first term.

Incorrect!

$-6x^3$ must also be subtracted.

$$\frac{3y^3 - 5x^3}{x^2 - y^2} - \frac{4y^3 - 6x^3}{x^2 - y^2} = \frac{3y^3 - 5x^3 - 4y^3 - 6x^3}{x^2 - y^2}$$

✓ CHECK POINT 2 Subtract: $\frac{5x - y}{x^2 - y^2} - \frac{4x - 2y}{x^2 - y^2}$.

3 Find the least common denominator.

Finding the Least Common Denominator

We can gain insight into adding rational expressions with different denominators by looking closely at what we do when adding fractions with different denominators. For example, suppose that we want to add $\frac{1}{2}$ and $\frac{2}{3}$. We must first write the fractions with the same denominator. We look for the smallest number that contains both 2 and 3 as factors. This number, 6, is then used as the *least common denominator*, or LCD.

The **least common denominator** of several rational expressions is a polynomial consisting of the product of all prime factors in the denominators, with each factor raised to the greatest power of its occurrence in any denominator.

Finding the Least Common Denominator

1. Factor each denominator completely.
2. List the factors of the first denominator.
3. Add to the list in step 2 any factors of the second denominator that do not appear in the list.
4. Form the product of each different factor from the list in step 3. This product is the least common denominator.

EXAMPLE 3 Finding the Least Common Denominator

Find the LCD of

$$\frac{3}{10x^2} \text{ and } \frac{7}{15x}.$$

Solution

Step 1. Factor each denominator completely.

$$10x^2 = 5 \cdot 2x^2 \quad (\text{or } 5 \cdot 2 \cdot x \cdot x)$$
$$15x = 5 \cdot 3x$$

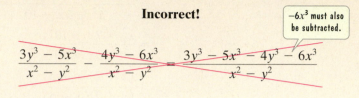

Step 2. List the factors of the first denominator.

$$5, 2, x^2 \quad (\text{or } 5, 2, x, x)$$

Step 3. Add any unlisted factors from the second denominator. Two factors from $5 \cdot 3x$ are already in our list. These factors include 5 and x. We add the unlisted factor, 3, to our list. We have

$$3, 5, 2, x^2.$$

Step 4. The least common denominator is the product of all factors in the final list. Thus,

$$3 \cdot 5 \cdot 2 \cdot x^2,$$

or $30x^2$, is the least common denominator.

Study Tip

After factoring denominators completely, the LCD can be determined by taking each factor to the highest power it appears in any factorization. The *Mathematics Teacher* magazine accused the LCD of keeping up with the Joneses. The LCD wants everything (all of the factors) the other denominators have.

✓ CHECK POINT 3 Find the LCD of

$$\frac{7}{6x^2} \text{ and } \frac{2}{9x}.$$

EXAMPLE 4 Finding the Least Common Denominator

Find the LCD of

$$\frac{9}{7x^2 + 28x} \text{ and } \frac{11}{x^2 + 8x + 16}.$$

Solution

Step 1. Factor each denominator completely.

$$7x^2 + 28x = 7x(x + 4)$$
$$x^2 + 8x + 16 = (x + 4)^2$$

Step 2. List the factors of the first denominator.

$$7, x, (x + 4)$$

Step 3. Add any unlisted factors from the second denominator. The second denominator is $(x + 4)^2$, or $(x + 4)(x + 4)$. One factor of $x + 4$ is already in our list, but the other factor is not. We add a second factor of $x + 4$ to the list. We have

$$7, x, (x + 4), (x + 4).$$

Step 4. The least common denominator is the product of all factors in the final list. Thus,

$$7x(x + 4)(x + 4), \text{ or } 7x(x + 4)^2,$$

is the least common denominator.

✓ CHECK POINT 4 Find the LCD of

$$\frac{7}{5x^2 + 15x} \text{ and } \frac{9}{x^2 + 6x + 9}.$$

---

4 Add and subtract rational expressions with different denominators.

Addition and Subtraction When Denominators Are Different

Finding the least common denominator for two (or more) rational expressions is the first step needed to add or subtract the expressions. For example, to add $\frac{1}{2}$ and $\frac{2}{3}$, we first determine that the LCD is 6. Then we write each fraction in terms of the LCD.

$$\frac{1}{2}+\frac{2}{3}=\frac{1}{2}\cdot\frac{3}{3}+\frac{2}{3}\cdot\frac{2}{2}$$

Multiply the numerator and denominator of each fraction by whatever extra factors are required to form 6, the LCD.

$\frac{3}{3}=1$ and $\frac{2}{2}=1$. Multiplying by 1 does not change a fraction's value.

$$=\frac{3}{6}+\frac{4}{6}$$ Perform the required multiplications.

$$=\frac{3+4}{6}$$ Add numerators. Place this sum over the LCD.

$$=\frac{7}{6}$$ Simplify.

We follow the same steps in adding or subtracting rational expressions with different denominators.

Adding and Subtracting Rational Expressions That Have Different Denominators

1. Find the LCD of the rational expressions.
2. Rewrite each rational expression as an equivalent expression whose denominator is the LCD. To do so, multiply the numerator and the denominator of each rational expression by any factor(s) needed to convert the denominator into the LCD.
3. Add or subtract numerators, placing the resulting expression over the LCD.
4. If possible, simplify the resulting rational expression.

EXAMPLE 5 Adding Rational Expressions with Different Denominators

Add: $\frac{3}{10x^2}+\frac{7}{15x}$.

Solution

Step 1. Find the least common denominator. In Example 3, we found that the LCD for these rational expressions is $30x^2$.

Step 2. Write equivalent expressions with the LCD as denominators. We must rewrite each rational expression with a denominator of $30x^2$.

$$\frac{3}{10x^2}\cdot\frac{3}{3}=\frac{9}{30x^2} \qquad \frac{7}{15x}\cdot\frac{2x}{2x}=\frac{14x}{30x^2}$$

Multiply the numerator and denominator by 3 to get $30x^2$, the LCD.

Multiply the numerator and denominator by $2x$ to get $30x^2$, the LCD.

Because $\frac{3}{3}=1$ and $\frac{2x}{2x}=1$, we are not changing the value of either rational expression, only its appearance. In summary, we have

$$\frac{3}{10x^2}+\frac{7}{15x}$$ The LCD is $30x^2$.

$$=\frac{3}{10x^2}\cdot\frac{3}{3}+\frac{7}{15x}\cdot\frac{2x}{2x}$$ Write equivalent expressions with the LCD.

$$=\frac{9}{30x^2}+\frac{14x}{30x^2}.$$ Perform the required multiplications.

Steps 3 and 4. Add numerators, putting this sum over the LCD. Simplify, if possible.

$$=\frac{9+14x}{30x^2} \text{ or } \frac{14x+9}{30x^2}$$ The numerator is prime and further simplification is not possible.

Study Tip

It is incorrect to add rational expressions by adding numerators and adding denominators. Avoid this common error.

Incorrect!

$$\frac{3}{10x^2}+\frac{7}{15x}=\frac{3+7}{10x^2+15x}=\frac{10}{10x^2+15x}$$

✓ **CHECK POINT 5** Add: $\frac{7}{6x^2} + \frac{2}{9x}$.

EXAMPLE 6 Adding Rational Expressions with Different Denominators

Add: $\frac{x}{x-3} + \frac{x-1}{x+3}$.

Solution

Step 1. Find the least common denominator. Begin by factoring the denominators.

$$x - 3 = 1(x - 3)$$
$$x + 3 = 1(x + 3)$$

The factors of the first denominator are 1 and $x - 3$. The only factor from the second denominator that is unlisted is $x + 3$. Thus, the least common denominator is $1(x - 3)(x + 3)$, or $(x - 3)(x + 3)$.

Step 2. Write equivalent expressions with the LCD as denominators.

$$\frac{x}{x-3} + \frac{x-1}{x+3}$$
$$= \frac{x(x+3)}{(x-3)(x+3)} + \frac{(x-1)(x-3)}{(x-3)(x+3)}$$

Multiply each numerator and denominator by the extra factor required to form $(x - 3)(x + 3)$, the LCD.

Steps 3 and 4. Add numerators, putting this sum over the LCD. Simplify, if possible.

$$= \frac{x(x+3) + (x-1)(x-3)}{(x-3)(x+3)}$$

$$= \frac{x^2 + 3x + x^2 - 4x + 3}{(x-3)(x+3)}$$

Perform the multiplications using the distributive property and FOIL.

$$= \frac{2x^2 - x + 3}{(x-3)(x+3)}$$

Combine like terms: $x^2 + x^2 = 2x^2$ and $3x - 4x = -x$.

The numerator is prime and further simplification is not possible.

✓ **CHECK POINT 6** Add: $\frac{x}{x-4} + \frac{x-2}{x+4}$.

EXAMPLE 7 Subtracting Rational Expressions with Different Denominators

Subtract: $\frac{x-1}{x^2+x-6} - \frac{x-2}{x^2+4x+3}$.

Solution

Step 1. Find the least common denominator. Begin by factoring the denominators.

$$x^2 + x - 6 = (x + 3)(x - 2)$$
$$x^2 + 4x + 3 = (x + 3)(x + 1)$$

The factors of the first denominator are $x + 3$ and $x - 2$. The only factor from the second denominator that is unlisted is $x + 1$. Thus, the least common denominator is $(x + 3)(x - 2)(x + 1)$.

Step 2. Write equivalent expressions with the LCD as denominators.

$$\frac{x-1}{x^2+x-6} - \frac{x-2}{x^2+4x+3}$$

$$= \frac{x-1}{(x+3)(x-2)} - \frac{x-2}{(x+3)(x+1)}$$ Factor denominators. The LCD is $(x+3)(x-2)(x+1)$.

$$= \frac{(x-1)(x+1)}{(x+3)(x-2)(x+1)} - \frac{(x-2)(x-2)}{(x+3)(x-2)(x+1)}$$ Multiply each numerator and denominator by the extra factor required to form $(x+3)(x-2)(x+1)$, the LCD.

Steps 3 and 4. Subtract numerators, putting this difference over the LCD. Simplify, if possible.

$$= \frac{(x-1)(x+1)-(x-2)(x-2)}{(x+3)(x-2)(x+1)}$$

$$= \frac{x^2-1-(x^2-4x+4)}{(x+3)(x-2)(x+1)}$$ Perform the multiplications in the numerator. Don't forget the parentheses.

$$= \frac{x^2-1-x^2+4x-4}{(x+3)(x-2)(x+1)}$$ Remove parentheses and change the sign of each term in parentheses.

$$= \frac{4x-5}{(x+3)(x-2)(x+1)}$$ Combine like terms.

The numerator is prime and further simplification is not possible.

✓ **CHECK POINT 7** Subtract: $\frac{2x-3}{x^2-5x+6} - \frac{x+4}{x^2-2x-3}$.

EXAMPLE 8 Adding and Subtracting Rational Expressions with Different Denominators

Perform the indicated operations:

$$\frac{3y+2}{y-5} + \frac{4}{3y+4} - \frac{7y^2+24y+28}{3y^2-11y-20}.$$

Solution

Step 1. Find the least common denominator. Begin by factoring the denominators.

$$y-5 = 1(y-5)$$
$$3y+4 = 1(3y+4)$$
$$3y^2-11y-20 = (3y+4)(y-5)$$

The factors of the first denominator are 1 and $y-5$. The only factor from the second denominator that is unlisted is $3y+4$. Adding this factor to our list, we have 1, $y-5$, and $3y+4$. We have listed all factors from the third denominator. Thus, the least common denominator is $1(y-5)(3y+4)$, or $(y-5)(3y+4)$.

Step 2. Write equivalent expressions with the LCD as denominators.

$$\frac{3y+2}{y-5} + \frac{4}{3y+4} - \frac{7y^2+24y+28}{3y^2-11y-20}$$

$$= \frac{3y+2}{y-5} + \frac{4}{3y+4} - \frac{7y^2+24y+28}{(3y+4)(y-5)}$$ Factor denominators. The LCD is $(y-5)(3y+4)$.

$$= \frac{(3y+2)(3y+4)}{(y-5)(3y+4)} + \frac{4(y-5)}{(y-5)(3y+4)} - \frac{7y^2+24y+28}{(3y+4)(y-5)}$$

Multiply the first two numerators and denominators by the extra factor required to form the LCD.

Step 3. Add and subtract numerators, putting this result over the LCD. Simplify, if possible.

$$= \frac{(3y+2)(3y+4) + 4(y-5) - (7y^2+24y+28)}{(y-5)(3y+4)}$$

$$= \frac{9y^2 + 18y + 8 + 4y - 20 - 7y^2 - 24y - 28}{(y-5)(3y+4)}$$

Perform multiplications. Remove parentheses and change the sign of each term.

$$= \frac{2y^2 - 2y - 40}{(y-5)(3y+4)}$$

Combine like terms: $9y^2 - 7y^2 = 2y^2$, $18y + 4y - 24y = -2y$, and $8 - 20 - 28 = -40$.

$$= \frac{2(y^2 - y - 20)}{(y-5)(3y+4)}$$

Factor out the GCF in the numerator.

$$= \frac{2(y+4)\cancel{(y-5)}}{\cancel{(y-5)}(3y+4)}$$

Factor completely and simplify.

$$= \frac{2(y+4)}{3y+4}$$

CHECK POINT 8 Perform the indicated operations:

$$\frac{y-1}{y-2} + \frac{y-6}{y^2-4} - \frac{y+1}{y+2}.$$

---

5 Add and subtract rational expressions with opposite denominators.

In some situations, we need to add or subtract rational expressions with denominators that are opposites, or additive inverses. Multiply the numerator and the denominator of either of the rational expressions by -1. Then they will have the same denominators.

EXAMPLE 9 Adding Rational Expressions when Denominators Are Opposites

Add: $\frac{4x-16y}{x-5y} + \frac{x-6y}{5y-x}$.

Solution

$$\frac{4x-16y}{x-5y} + \frac{x-6y}{5y-x}$$

The denominators, $x - 5y$ and $5y - x$, are opposites, or additive inverses.

$$= \frac{4x-16y}{x-5y} + \frac{(-1)}{(-1)} \cdot \frac{x-6y}{5y-x}$$

Multiply the numerator and denominator of the second rational expression by -1.

$$= \frac{4x-16y}{x-5y} + \frac{-x+6y}{-5y+x}$$

Perform the multiplications by -1.

$$= \frac{4x-16y}{x-5y} + \frac{-x+6y}{x-5y}$$

Rewrite $-5y + x$ as $x - 5y$. Both rational expressions have the same denominator.

$$= \frac{4x-16y-x+6y}{x-5y}$$

Add numerators. Place this sum over the common denominator.

$$= \frac{3x-10y}{x-5y}$$

Combine like terms.

CHECK POINT 9 Add: $\frac{4x-7y}{x-3y} + \frac{x-2y}{3y-x}$.

6.2 EXERCISE SET

MyMathLab MathXL PRACTICE WATCH DOWNLOAD READ REVIEW

Practice Exercises

In Exercises 1–16, perform the indicated operations. These exercises involve addition and subtraction when denominators are the same. Simplify the result, if possible.

1. $\frac{2}{9x} + \frac{4}{9x}$
2. $\frac{11}{6x} + \frac{4}{6x}$
3. $\frac{x}{x-5} + \frac{9x+3}{x-5}$
4. $\frac{x}{x-3} + \frac{11x+5}{x-3}$
5. $\frac{x^2-2x}{x^2+3x} + \frac{x^2+x}{x^2+3x}$
6. $\frac{x^2+7x}{x^2-5x} + \frac{x^2-4x}{x^2-5x}$
7. $\frac{y^2}{y^2-9} + \frac{9-6y}{y^2-9}$
8. $\frac{y^2}{y^2-25} + \frac{25-10y}{y^2-25}$
9. $\frac{3x}{4x-3} - \frac{2x-1}{4x-3}$
10. $\frac{3x}{7x-4} - \frac{2x-1}{7x-4}$
11. $\frac{x^2-2}{x^2+6x-7} - \frac{19-4x}{x^2+6x-7}$
12. $\frac{x^2+6x+2}{x^2+x-6} - \frac{2x-1}{x^2+x-6}$
13. $\frac{20y^2+5y+1}{6y^2+y-2} - \frac{8y^2-12y-5}{6y^2+y-2}$
14. $\frac{y^2+3y-6}{y^2-5y+4} - \frac{4y-4-2y^2}{y^2-5y+4}$
15. $\frac{2x^3-3y^3}{x^2-y^2} - \frac{x^3-2y^3}{x^2-y^2}$
16. $\frac{4y^3-3x^3}{y^2-x^2} - \frac{3y^3-2x^3}{y^2-x^2}$

In Exercises 17–28, find the least common denominator of the rational expressions.

17. $\frac{11}{25x^2}$ and $\frac{14}{35x}$
18. $\frac{7}{15x^2}$ and $\frac{9}{24x}$
19. $\frac{2}{x-5}$ and $\frac{3}{x^2-25}$
20. $\frac{2}{x+3}$ and $\frac{5}{x^2-9}$
21. $\frac{7}{y^2-100}$ and $\frac{13}{y(y-10)}$
22. $\frac{7}{y^2-4}$ and $\frac{15}{y(y+2)}$
23. $\frac{8}{x^2-16}$ and $\frac{x}{x^2-8x+16}$
24. $\frac{3}{x^2-25}$ and $\frac{x}{x^2-10x+25}$
25. $\frac{7}{y^2-5y-6}$ and $\frac{y}{y^2-4y-5}$
26. $\frac{3}{y^2-y-20}$ and $\frac{y}{2y^2+7y-4}$
27. $\frac{7y}{2y^2+7y+6}, \frac{3}{y^2-4}$, and $\frac{-7y}{2y^2-3y-2}$
28. $\frac{5y}{y^2-9}, \frac{8}{y^2+6y+9}$, and $\frac{-5y}{2y^2+5y-3}$

In Exercises 29–66, perform the indicated operations. These exercises involve addition and subtraction when denominators are different. Simplify the result, if possible.

29. $\frac{3}{5x^2} + \frac{10}{x}$
30. $\frac{7}{2x^2} + \frac{4}{x}$
31. $\frac{4}{x-2} + \frac{3}{x+1}$
32. $\frac{2}{x-3} + \frac{7}{x+2}$
33. $\frac{3x}{x^2+x-2} + \frac{2}{x^2-4x+3}$
34. $\frac{7x}{x^2+2x-8} + \frac{3}{x^2-3x+2}$
35. $\frac{x-6}{x+5} + \frac{x+5}{x-6}$
36. $\frac{x-2}{x+7} + \frac{x+7}{x-2}$
37. $\frac{3x}{x^2-25} - \frac{4}{x+5}$
38. $\frac{8x}{x^2-16} - \frac{5}{x+4}$
39. $\frac{3y+7}{y^2-5y+6} - \frac{3}{y-3}$
40. $\frac{2y+9}{y^2-7y+12} - \frac{2}{y-3}$
41. $\frac{x^2-6}{x^2+9x+18} - \frac{x-4}{x+6}$
42. $\frac{x^2-39}{x^2+3x-10} - \frac{x-7}{x-2}$
43. $\frac{4x+1}{x^2+7x+12} + \frac{2x+3}{x^2+5x+4}$
44. $\frac{3x-2}{x^2-x-6} + \frac{4x-3}{x^2-9}$

45. $\frac{x+4}{x^2-x-2}-\frac{2x+3}{x^2+2x-8}$

46. $\frac{2x+1}{x^2-7x+6}-\frac{x+3}{x^2-5x-6}$

47. $4+\frac{1}{x-3}$

48. $7+\frac{1}{x-5}$

49. $\frac{y-7}{y^2-16}+\frac{7-y}{16-y^2}$

50. $\frac{y-3}{y^2-25}+\frac{y-3}{25-y^2}$

51. $\frac{x+7}{3x+6}+\frac{x}{4-x^2}$

52. $\frac{x+5}{4x+12}+\frac{x}{9-x^2}$

53. $\frac{2x}{x-4}+\frac{64}{x^2-16}-\frac{2x}{x+4}$

54. $\frac{x}{x-3}+\frac{x+2}{x^2-2x-3}-\frac{4}{x+1}$

55. $\frac{5x}{x^2-y^2}-\frac{7}{y-x}$

56. $\frac{9x}{x^2-y^2}-\frac{10}{y-x}$

57. $\frac{3}{5x+6}-\frac{4}{x-2}+\frac{x^2-x}{5x^2-4x-12}$

58. $\frac{x-1}{x^2+2x+1}-\frac{3}{2x-2}+\frac{x}{x^2-1}$

59. $\frac{3x-y}{x^2-9xy+20y^2}+\frac{2y}{x^2-25y^2}$

60. $\frac{x+2y}{x^2+4xy+4y^2}-\frac{2x}{x^2-4y^2}$

61. $\frac{3x}{x^2-4}+\frac{5x}{x^2+x-2}-\frac{3}{x^2-4x+4}$

62. $\frac{1}{x}+\frac{4}{x^2-4}-\frac{2}{x^2-2x}$

63. $\frac{6a+5b}{6a^2+5ab-4b^2}-\frac{a+2b}{9a^2-16b^2}$

64. $\frac{5a-b}{a^2+ab-2b^2}-\frac{3a+2b}{a^2+5ab-6b^2}$

65. $\frac{1}{m^2+m-2}-\frac{3}{2m^2+3m-2}+\frac{2}{2m^2-3m+1}$

66. $\frac{5}{2m^2-5m-3}+\frac{3}{2m^2+5m+2}-\frac{1}{m^2-m-6}$

Practice PLUS

In Exercises 67–74, perform the indicated operations. Simplify the result, if possible.

67. $\left(\frac{2x+3}{x+1}\cdot\frac{x^2+4x-5}{2x^2+x-3}\right)-\frac{2}{x+2}$

68. $\frac{1}{x^2-2x-8}\div\left(\frac{1}{x-4}-\frac{1}{x+2}\right)$

69. $\left(2-\frac{6}{x+1}\right)\left(1+\frac{3}{x-2}\right)$

70. $\left(4-\frac{3}{x+2}\right)\left(1+\frac{5}{x-1}\right)$

71. $\left(\frac{1}{x+h}-\frac{1}{x}\right)\div h$

72. $\left(\frac{5}{x-5}-\frac{2}{x+3}\right)\div(3x+25)$

73. $\left(\frac{1}{a^3-b^3}\cdot\frac{ac+ad-bc-bd}{1}\right)-\frac{c-d}{a^2+ab+b^2}$

74. $\frac{ab}{a^2+ab+b^2}+\left(\frac{ac-ad-bc+bd}{ac-ad+bc-bd}\div\frac{a^3-b^3}{a^3+b^3}\right)$

75. If $f(x)=\frac{2x-3}{x+5}$ and $g(x)=\frac{x^2-4x-19}{x^2+8x+15}$, find $(f-g)(x)$ and the domain of $f-g$.

76. If $f(x)=\frac{2x-1}{x^2+x-6}$ and $g(x)=\frac{x+2}{x^2+5x+6}$, find $(f-g)(x)$ and the domain of $f-g$.

Application Exercises

You plan to drive from Miami, Florida, to Atlanta, Georgia. Your trip involves approximately 470 miles of travel in Florida and 250 miles in Georgia. The speed limit is 70 miles per hour in Florida and 65 miles per hour in Georgia. If you average x miles per hour over these speed limits, the total driving time, T(x), in hours, is given by the function

$$T(x)=\frac{470}{x+70}+\frac{250}{x+65}.$$

The graph of T is shown in the figure. Use the function's equation to solve Exercises 77–82.

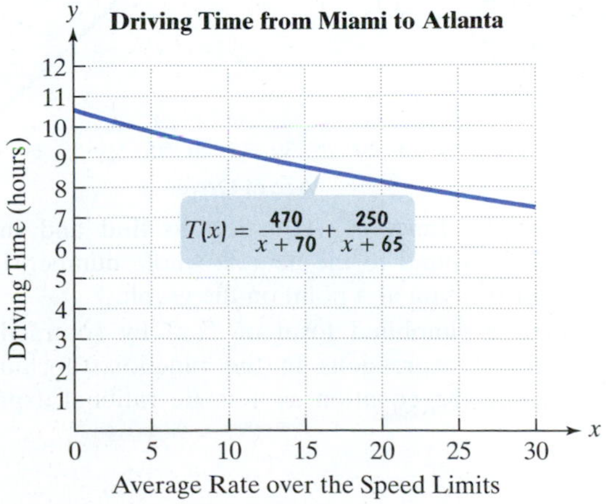

(Be sure to refer to the information at the bottom of the previous page as you solve Exercises 77–82.)

77. Find and interpret $T(0)$. Round to the nearest hour. Identify your solution as a point on the graph.

78. Find and interpret $T(5)$. Round to the nearest hour. Identify your solution as a point on the graph.

79. Find a simplified form of $T(x)$ by adding the rational expressions in the function's formula. Then use this form of the function to find $T(0)$.

80. Find a simplified form of $T(x)$ by adding the rational expressions in the function's formula. Then use this form of the function to find $T(5)$.

81. Use the graph to answer this question. If you want the driving time to be 9 hours, how much over the speed limits do you need to drive? Round to the nearest mile per hour. Does this seem like a realistic driving time for the trip? Explain your answer.

82. Use the graph to answer this question. If you want the driving time to be 8 hours, how much over the speed limits do you need to drive? Round to the nearest mile per hour. Does this seem like a realistic driving time for the trip? Explain your answer.

83. In the section opener, we saw that the rational function

$$f(x) = \frac{27{,}725(x - 14)}{x^2 + 9} - 5x$$

models the number of arrests, $f(x)$, per 100,000 drivers, for driving under the influence of alcohol as a function of a driver's age, x. The graph of f for an appropriate domain is shown in the figure.

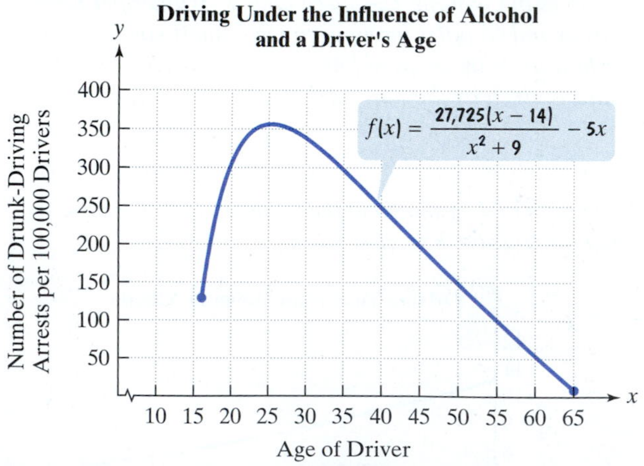

a. Use the function's equation to find and interpret $f(20)$. Round to the nearest whole number. Identify your solution as a point on the graph.

b. Find a simplified form of $f(x)$ by subtracting the rational expressions in the function's formula and writing the equation as a single rational expression.

c. Use the graph to determine the age, to the nearest five years, that corresponds to the greatest number of arrests. Then use the form of the function that you obtained in part (b) to determine the number of arrests, per 100,000 drivers, for this age group. Round to the nearest whole number.

In Exercises 84–85, express the perimeter of each rectangle as a single rational expression.

84.

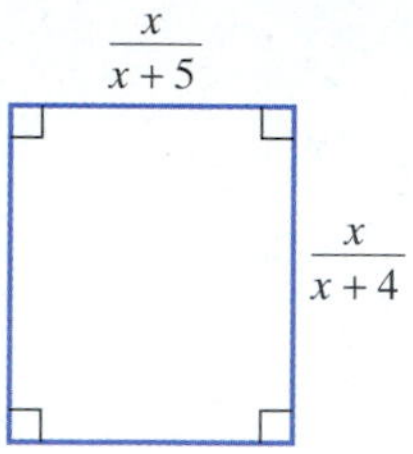

85.

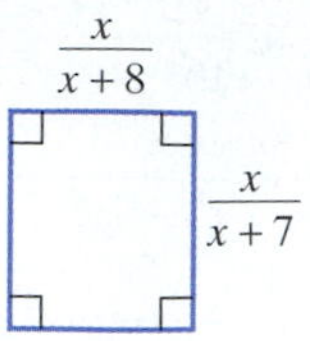

Writing in Mathematics

86. Explain how to add rational expressions when denominators are the same. Give an example with your explanation.

87. Explain how to subtract rational expressions when denominators are the same. Give an example with your explanation.

88. Explain how to find the least common denominator for denominators of $x^2 - 100$ and $x^2 - 20x + 100$.

89. Explain how to add rational expressions that have different denominators. Use $\frac{5}{x+1} + \frac{3}{x+4}$ in your explanation.

90. Explain how to add rational expressions when denominators are opposites. Use an example to support your explanation.

Explain the error in Exercises 91–92. Then rewrite the right side of the equation to correct the error that now exists.

91. $\frac{1}{a} + \frac{1}{b} = \frac{1}{a+b}$

92. $\frac{1}{x} + \frac{3}{7} = \frac{4}{x+7}$

Critical Thinking Exercises

Make Sense? *In Exercises 93–96, determine whether each statement "makes sense" or "does not make sense" and explain your reasoning.*

93. When a numerator is being subtracted, I find that inserting parentheses helps me to distribute the negative sign to every term.

94. The reason I can rewrite rational expressions with a common denominator is that 1 is the multiplicative identity.

95. The fastest way for me to add $\frac{5}{x-7} + \frac{3}{7-x}$ is by using $(x-7)(7-x)$ as the LCD.

96. Although $\dfrac{2x^3 + 11x^2}{x + 3} + \dfrac{5x^3 + 4x^2}{x + 3}$ looks more complicated than $\dfrac{2}{x + 3} + \dfrac{5}{x - 3}$, it takes me more steps to perform the less complicated-looking addition.

In Exercises 97–100, determine whether each statement is true or false. If the statement is false, make the necessary change(s) to produce a true statement.

97. $\dfrac{2}{x + 3} + \dfrac{3}{x + 4} = \dfrac{5}{2x + 7}$

98. $\dfrac{a}{b} + \dfrac{a}{c} = \dfrac{a}{b + c}$

99. $6 + \dfrac{1}{x} = \dfrac{7}{x}$

100. $\dfrac{1}{x + 3} + \dfrac{x + 3}{2} = \dfrac{1}{\cancel{(x + 3)}} + \dfrac{\cancel{(x + 3)}}{2} = 1 + \dfrac{1}{2} = \dfrac{3}{2}$

In Exercises 101–103, perform the indicated operations.

101. $\dfrac{1}{x^n - 1} - \dfrac{1}{x^n + 1} - \dfrac{1}{x^{2n} - 1}$

102. $\left(1 - \dfrac{1}{x}\right)\left(1 - \dfrac{1}{x + 1}\right)\left(1 - \dfrac{1}{x + 2}\right)\left(1 - \dfrac{1}{x + 3}\right)$

103. $(x - y)^{-1} + (x - y)^{-2}$

Review Exercises

104. Simplify: $\left(\dfrac{3x^2y^{-2}}{y^3}\right)^{-2}$. (Section 1.6, Example 9)

105. Solve: $|3x - 1| \le 14$. (Section 4.3, Example 4 or Example 6(a))

106. Factor completely: $50x^3 - 18x$. (Section 5.5, Example 2)

Preview Exercises

Exercises 107–109 will help you prepare for the material covered in the next section.

107. Multiply and simplify: $x^2y^2\left(\dfrac{1}{x} + \dfrac{y}{x^2}\right)$.

108. Multiply and simplify: $x(x + h)\left(\dfrac{1}{x + h} - \dfrac{1}{x}\right)$.

109. Divide: $\dfrac{x^2 - 1}{x^2} \div \dfrac{x^2 - 4x + 3}{x^2}$.

SECTION 6.3

Objectives

1. Simplify complex rational expressions by multiplying by 1.
2. Simplify complex rational expressions by dividing.

Complex Rational Expressions

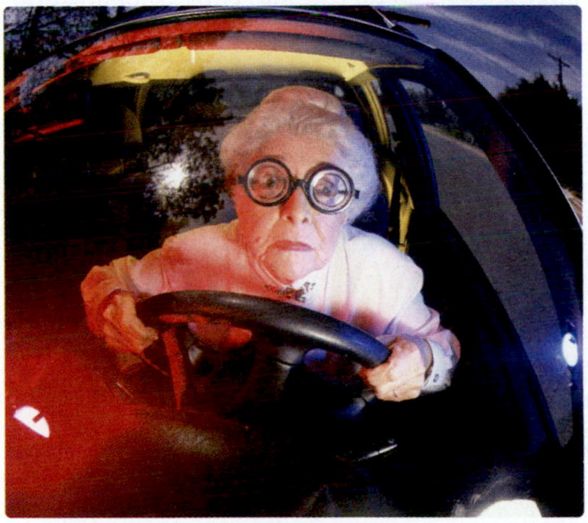

One area in finance of great interest to us ordinary folks when we buy a new car is the amount of each monthly payment. If P is the principal, or the amount borrowed, i is the monthly interest rate, and n is the number of monthly payments, then the amount, A, of each monthly payment is

$$A = \frac{Pi}{1 - \dfrac{1}{(1 + i)^n}}.$$

Do you notice anything unusual about the rational expression for the amount of each payment? It has a separate rational expression in its denominator.

$$\frac{Pi}{1 - \dfrac{1}{(1 + i)^n}}$$

Numerator

Main fraction bar

Denominator

A separate rational expression occurs in the denominator.

Complex rational expressions, also called **complex fractions**, have numerators or denominators containing one or more rational expressions. Here is another example of such an expression:

Numerator

$$\frac{\frac{1}{x}+\frac{y}{x^2}}{\frac{1}{y}+\frac{x}{y^2}}.$$

Main fraction bar

Separate rational expressions occur in the numerator and the denominator.

Denominator

In this section, we study two methods for simplifying complex rational expressions.

1 Simplify complex rational expressions by multiplying by 1.

Simplifying Complex Rational Expressions by Multiplying by 1

One method for simplifying a complex rational expression is to find the least common denominator of all the rational expressions in its numerator and denominator. Then multiply each term in its numerator and denominator by this least common denominator. Because we are multiplying by a form of 1, we will obtain an equivalent expression that does not contain fractions in the numerator or denominator.

Simplifying a Complex Rational Expression by Multiplying by 1 in the Form $\frac{\text{LCD}}{\text{LCD}}$

1. Find the LCD of all rational expressions within the complex rational expression.
2. Multiply both the numerator and the denominator of the complex rational expression by this LCD.
3. Use the distributive property and multiply each term in the numerator and denominator by this LCD. Simplify each term. No fractional expressions should remain within the numerator or denominator of the main fraction.
4. If possible, factor and simplify.

EXAMPLE 1 Simplifying a Complex Rational Expression

Simplify:

$$\frac{\frac{1}{x}+\frac{y}{x^2}}{\frac{1}{y}+\frac{x}{y^2}}.$$

Solution The denominators in the complex rational expression are x, x^2, y, and y^2. The LCD is x^2y^2. Multiply both the numerator and the denominator of the complex rational expression by x^2y^2.

$$\frac{\frac{1}{x}+\frac{y}{x^2}}{\frac{1}{y}+\frac{x}{y^2}}=\frac{x^2y^2}{x^2y^2}\cdot\frac{\left(\frac{1}{x}+\frac{y}{x^2}\right)}{\left(\frac{1}{y}+\frac{x}{y^2}\right)}$$

Multiply the numerator and the denominator by x^2y^2.

In all four rational expressions, we have divided numerators and denominators by common boxed factors.

$$= \frac{\boxed{x}^2y^2 \cdot \frac{1}{\boxed{x}} + \boxed{x^2}y^2 \cdot \frac{y}{\boxed{x^2}}}{x^2\boxed{y}^2 \cdot \frac{1}{\boxed{y}} + x^2\boxed{y^2} \cdot \frac{x}{\boxed{y^2}}}$$ Use the distributive property.

$$= \frac{xy^2 + y^3}{x^2y + x^3}$$ Simplify: $x^2y^2 \cdot \frac{1}{x} = xy^2$; $x^2y^2 \cdot \frac{y}{x^2} = y^3$; $x^2y^2 \cdot \frac{1}{y} = x^2y$; $x^2y^2 \cdot \frac{x}{y^2} = x^3$.

$$= \frac{y^2(x+y)}{x^2(y+x)}$$ Factor and simplify.

$$= \frac{y^2}{x^2}$$

CHECK POINT 1 Simplify:

$$\frac{\frac{x}{y} - 1}{\frac{x^2}{y^2} - 1}.$$

Study Tip

In Section 1.6, we introduced the negative-exponent rule:

$$b^{-n} = \frac{1}{b^n}, b \neq 0.$$

See pages 66–68 if you need to review negative integers as exponents.

Complex rational expressions are often written with negative exponents. For example,

$$\frac{x^{-1} + x^{-2}y}{y^{-1} + xy^{-2}} \quad \text{means} \quad \frac{\frac{1}{x} + \frac{y}{x^2}}{\frac{1}{y} + \frac{x}{y^2}}.$$

This is the expression that we simplified in Example 1. If an expression contains negative exponents, first rewrite it as an equivalent expression with positive exponents. Then simplify by multiplying the numerator and the denominator by the LCD.

EXAMPLE 2 Simplifying a Complex Rational Expression

Simplify:

$$\frac{\frac{1}{x+h} - \frac{1}{x}}{h}.$$

Solution The denominators in the complex rational expression are $x + h$ and x. The LCD is $x(x + h)$. Multiply both the numerator and the denominator of the complex rational expression by $x(x + h)$.

$$\frac{\frac{1}{x+h} - \frac{1}{x}}{h} = \frac{x(x+h)}{x(x+h)} \cdot \frac{\left(\frac{1}{x+h} - \frac{1}{x}\right)}{h}$$ Multiply the numerator and the denominator by $x(x + h)$.

$$= \frac{x(x+h) \cdot \frac{1}{x+h} - x(x+h) \cdot \frac{1}{x}}{x(x+h)h}$$ Use the distributive property in the numerator to multiply every term by the LCD.

$$= \frac{x - (x+h)}{x(x+h)h}$$ Simplify: $x(x+h) \cdot \frac{1}{(x+h)} = x$ and $x(x+h) \cdot \frac{1}{x} = x + h$.

$$= \frac{x - x - h}{x(x+h)h}$$ Remove parentheses and change the sign of each term.

$$= \frac{-h}{x(x+h)h} \quad \text{Simplify } \frac{x-x-h}{x(x+h)h}.$$

$$= \frac{-\cancel{h}}{x(x+h)\cancel{h}} \quad \text{Divide the numerator and the denominator by the common factor, } h.$$

$$= -\frac{1}{x(x+h)}$$

✓ **CHECK POINT 2** Simplify:

$$\frac{\frac{1}{x+7} - \frac{1}{x}}{7}.$$

---

2 Simplify complex rational expressions by dividing.

Simplifying Complex Rational Expressions by Dividing

A second method for simplifying a complex rational expression is to combine its numerator into a single rational expression and combine its denominator into a single rational expression. Then perform the division by inverting the denominator and multiplying.

Simplifying a Complex Rational Expression by Dividing

1. If necessary, add or subtract to get a single rational expression in the numerator.
2. If necessary, add or subtract to get a single rational expression in the denominator.
3. Perform the division indicated by the main fraction bar: Invert the denominator of the complex rational expression and multiply.
4. If possible, simplify.

EXAMPLE 3 Simplifying a Complex Rational Expression

Simplify:

$$\frac{\frac{x+1}{x} + \frac{x+1}{x-1}}{\frac{x+2}{x} - \frac{2}{x-1}}.$$

Solution

Step 1. Add to get a single rational expression in the numerator.

$$\frac{x+1}{x} + \frac{x+1}{x-1}$$

The LCD is $x(x-1)$.

$$= \frac{(x+1)(x-1)}{x(x-1)} + \frac{x(x+1)}{x(x-1)} = \frac{(x+1)(x-1)+x(x+1)}{x(x-1)} = \frac{x^2-1+x^2+x}{x(x-1)} = \frac{2x^2+x-1}{x(x-1)}$$

Step 2. Subtract to get a single rational expression in the denominator.

$$\frac{x+2}{x} - \frac{2}{x-1}$$

The LCD is $x(x-1)$.

$$= \frac{(x+2)(x-1)}{x(x-1)} - \frac{2x}{x(x-1)} = \frac{(x+2)(x-1)-2x}{x(x-1)} = \frac{x^2+x-2-2x}{x(x-1)} = \frac{x^2-x-2}{x(x-1)}$$

Steps 3 and 4. Perform the division indicated by the main fraction bar: Invert and multiply. If possible, simplify.

$$\frac{\frac{x+1}{x}+\frac{x+1}{x-1}}{\frac{x+2}{x}-\frac{2}{x-1}} = \frac{\frac{2x^2+x-1}{x(x-1)}}{\frac{x^2-x-2}{x(x-1)}}$$

These are the single rational expressions from steps 1 and 2.

$$= \frac{2x^2+x-1}{x(x-1)} \cdot \frac{x(x-1)}{x^2-x-2} = \frac{(2x-1)\cancel{(x+1)}}{\cancel{x(x-1)}} \cdot \frac{\cancel{x(x-1)}}{(x-2)\cancel{(x+1)}} = \frac{2x-1}{x-2}$$

Invert and multiply.

CHECK POINT 3 Simplify:

$$\frac{\frac{x+1}{x-1}-\frac{x-1}{x+1}}{\frac{x-1}{x+1}+\frac{x+1}{x-1}}.$$

Which of the two methods do you prefer? Let's try them both in Example 4.

EXAMPLE 4 Simplifying a Complex Rational Expression: Comparing Methods

Simplify:

$$\frac{1-x^{-2}}{1-4x^{-1}+3x^{-2}}.$$

Solution First rewrite the expression without negative exponents.

$$\frac{1-x^{-2}}{1-4x^{-1}+3x^{-2}} = \frac{1-\frac{1}{x^2}}{1-\frac{4}{x}+\frac{3}{x^2}}$$

The negative exponents affect only the variables and not the constants.

Method 1 Multiplying by 1

$$\frac{1-\frac{1}{x^2}}{1-\frac{4}{x}+\frac{3}{x^2}} = \frac{x^2}{x^2}\cdot\frac{\left(1-\frac{1}{x^2}\right)}{\left(1-\frac{4}{x}+\frac{3}{x^2}\right)}$$

Multiply the numerator and denominator by x^2, the LCD of all fractions.

$$= \frac{x^2\cdot 1 - x^2\cdot\frac{1}{x^2}}{x^2\cdot 1 - x^2\cdot\frac{4}{x}+x^2\cdot\frac{3}{x^2}}$$

Apply the distributive property.

$$= \frac{x^2-1}{x^2-4x+3}$$

Simplify.

$$= \frac{(x+1)\cancel{(x-1)}}{(x-3)\cancel{(x-1)}}$$

Factor and simplify.

$$= \frac{x+1}{x-3}$$

Method 2 Dividing

$$\frac{1-\frac{1}{x^2}}{1-\frac{4}{x}+\frac{3}{x^2}} = \frac{\frac{x^2}{x^2}-\frac{1}{x^2}}{\frac{x^2}{x^2}-\frac{4}{x}\cdot\frac{x}{x}+\frac{3}{x^2}}$$

Get a single rational expression in the numerator and in the denominator.

$$= \frac{\frac{x^2-1}{x^2}}{\frac{x^2-4x+3}{x^2}}$$

$$= \frac{x^2-1}{x^2}\cdot\frac{x^2}{x^2-4x+3}$$

Invert and multiply.

$$= \frac{(x+1)\cancel{(x-1)}}{\cancel{x^2}}\cdot\frac{\cancel{x^2}}{\cancel{(x-1)}(x-3)}$$

Factor and simplify.

$$= \frac{x+1}{x-3}$$

CHECK POINT 4 Simplify by the method of your choice:

$$\frac{1-4x^{-2}}{1-7x^{-1}+10x^{-2}}.$$

6.3 EXERCISE SET

MyMathLab MathXL PRACTICE WATCH DOWNLOAD READ REVIEW

Practice Exercises

In Exercises 1–40, simplify each complex rational expression by the method of your choice.

1. $\dfrac{4+\frac{2}{x}}{1-\frac{3}{x}}$

2. $\dfrac{5-\frac{2}{x}}{3+\frac{1}{x}}$

3. $\dfrac{\frac{3}{x}+\frac{x}{3}}{\frac{x}{3}-\frac{3}{x}}$

4. $\dfrac{\frac{x}{5}-\frac{5}{x}}{\frac{1}{5}+\frac{1}{x}}$

5. $\dfrac{\frac{1}{x}+\frac{1}{y}}{\frac{1}{x}-\frac{1}{y}}$

6. $\dfrac{\frac{x}{y}+\frac{1}{x}}{\frac{y}{x}+\frac{1}{x}}$

7. $\dfrac{8x^{-2}-2x^{-1}}{10x^{-1}-6x^{-2}}$

8. $\dfrac{12x^{-2}-3x^{-1}}{15x^{-1}-9x^{-2}}$

9. $\dfrac{\frac{1}{x-2}}{1-\frac{1}{x-2}}$

10. $\dfrac{\frac{1}{x+2}}{1+\frac{1}{x+2}}$

11. $\dfrac{\frac{1}{x+5}-\frac{1}{x}}{5}$

12. $\dfrac{\frac{1}{x+6}-\frac{1}{x}}{6}$

13. $\dfrac{\frac{4}{x+4}}{\frac{1}{x+4}-\frac{1}{x}}$

14. $\dfrac{\frac{7}{x+7}}{\frac{1}{x+7}-\frac{1}{x}}$

15. $\dfrac{\frac{1}{x-1}+1}{\frac{1}{x+1}-1}$

16. $\dfrac{\frac{1}{x+1}-1}{\frac{1}{x-1}+1}$

17. $\dfrac{x^{-1}+y^{-1}}{(x+y)^{-1}}$

18. $(x^{-1}+y^{-1})^{-1}$

19. $\dfrac{\frac{x+2}{x-2}-\frac{x-2}{x+2}}{\frac{x-2}{x+2}+\frac{x+2}{x-2}}$

20. $\dfrac{\frac{x+1}{x-1}-\frac{x-1}{x+1}}{\frac{x-1}{x+1}+\frac{x+2}{x-1}}$

21. $\dfrac{\frac{2}{x^3y}+\frac{5}{xy^4}}{\frac{5}{x^3y}-\frac{3}{xy}}$

22. $\dfrac{\frac{3}{xy^2}+\frac{2}{x^2y}}{\frac{1}{x^2y}+\frac{2}{xy^3}}$

23. $\dfrac{\frac{3}{x+2}-\frac{3}{x-2}}{\frac{5}{x^2-4}}$

24. $\dfrac{\frac{3}{x+1}-\frac{3}{x-1}}{\frac{5}{x^2-1}}$

25. $\dfrac{3a^{-1}+3b^{-1}}{4a^{-2}-9b^{-2}}$

26. $\dfrac{5a^{-1}-2b^{-1}}{25a^{-2}-4b^{-2}}$

27. $\dfrac{\frac{4x}{x^2-4}-\frac{5}{x-2}}{\frac{2}{x-2}+\frac{3}{x+2}}$

28. $\dfrac{\frac{2}{x+3}+\frac{5x}{x^2-9}}{\frac{4}{x+3}+\frac{2}{x-3}}$

29. $\dfrac{\frac{2y}{y^2+4y+3}}{\frac{1}{y+3}+\frac{2}{y+1}}$

30. $\dfrac{\frac{5y}{y^2-5y+6}}{\frac{3}{y-3}+\frac{2}{y-2}}$

31. $\dfrac{\frac{2}{a^2}-\frac{1}{ab}-\frac{1}{b^2}}{\frac{1}{a^2}-\frac{3}{ab}+\frac{2}{b^2}}$

32. $\dfrac{\frac{2}{b^2}-\frac{5}{ab}-\frac{3}{a^2}}{\frac{2}{b^2}+\frac{7}{ab}+\frac{3}{a^2}}$

33. $\dfrac{\frac{2x}{x^2-25}+\frac{1}{3x-15}}{\frac{5}{x-5}+\frac{3}{4x-20}}$

34. $\dfrac{\frac{7x}{2x-2}+\frac{x}{x^2-1}}{\frac{4}{x+1}-\frac{1}{3x+3}}$

35. $\dfrac{\frac{3}{x+2y}-\frac{2y}{x^2+2xy}}{\frac{3y}{x^2+2xy}+\frac{5}{x}}$

36. $\dfrac{\frac{1}{x^3-y^3}}{\frac{1}{x-y}-\frac{1}{x^2+xy+y^2}}$

37. $\dfrac{\dfrac{2}{m^2 - 3m + 2} + \dfrac{2}{m^2 - m - 2}}{\dfrac{2}{m^2 - 1} + \dfrac{2}{m^2 + 4m + 3}}$

38. $\dfrac{\dfrac{m}{m^2 - 9} - \dfrac{2}{m^2 - 4m + 4}}{\dfrac{3}{m^2 - 5m + 6} + \dfrac{m}{m^2 + m - 6}}$

39. $\dfrac{\dfrac{2}{a^2 + 2a - 8} + \dfrac{1}{a^2 + 5a + 4}}{\dfrac{1}{a^2 - 5a + 6} + \dfrac{2}{a^2 - a - 2}}$

40. $\dfrac{\dfrac{3}{a^2 + 10a + 25} - \dfrac{1}{a^2 - a - 2}}{\dfrac{4}{a^2 + 6a + 5} - \dfrac{2}{a^2 + 3a - 10}}$

Practice PLUS

In Exercises 41–46, perform the indicated operations. Simplify the result, if possible.

41. $\dfrac{\dfrac{x-1}{x^2-4}}{1 + \dfrac{1}{x-2}} - \dfrac{1}{x-2}$

42. $\dfrac{\dfrac{x-3}{x^2-16}}{1 + \dfrac{1}{x-4}} - \dfrac{1}{x-4}$

43. $\dfrac{3}{1 - \dfrac{3}{3+x}} - \dfrac{3}{\dfrac{3}{3-x} - 1}$

44. $\dfrac{5}{1 - \dfrac{5}{5+x}} - \dfrac{5}{\dfrac{5}{5-x} - 1}$

45. $\dfrac{x}{1 - \dfrac{1}{1 + \dfrac{1}{x}}}$

46. $\dfrac{\dfrac{1}{x+1}}{x - \dfrac{1}{x + \dfrac{1}{x}}}$

In Exercises 47–48, let $f(x) = \dfrac{1+x}{1-x}$.

47. Find $f\left(\dfrac{1}{x+3}\right)$ and simplify.

48. Find $f\left(\dfrac{1}{x-6}\right)$ and simplify.

In Exercises 49–50, use the given rational function to find and simplify

$$\frac{f(a+h) - f(a)}{h}.$$

49. $f(x) = \dfrac{3}{x}$

50. $f(x) = \dfrac{1}{x^2}$

Application Exercises

51. How much are your monthly payments on a loan? If P is the principal, or amount borrowed, i is the monthly interest rate (as a decimal), and n is the number of monthly payments, then the amount, A, of each monthly payment is

$$A = \frac{Pi}{1 - \dfrac{1}{(1+i)^n}}.$$

a. Simplify the complex rational expression for the amount of each payment.

b. You purchase a \$20,000 automobile at 1% monthly interest to be paid over 48 months. How much do you pay each month? Use the simplified rational expression from part (a) and a calculator. Round to the nearest dollar.

52. The average rate on a round-trip commute having a one-way distance d is given by the complex rational expression

$$\frac{2d}{\dfrac{d}{r_1} + \dfrac{d}{r_2}},$$

in which r_1 and r_2 are the rates on the outgoing and return trips, respectively.

a. Simplify the complex rational expression.

b. Find your average rate if you drive to campus averaging 30 miles per hour and return home on the same route averaging 40 miles per hour. Use the simplified rational expression from part (a).

53. If three resistors with resistances R_1, R_2, and R_3 are connected in parallel, their combined resistance, R, is given by the formula

$$R = \frac{1}{\dfrac{1}{R_1} + \dfrac{1}{R_2} + \dfrac{1}{R_3}}.$$

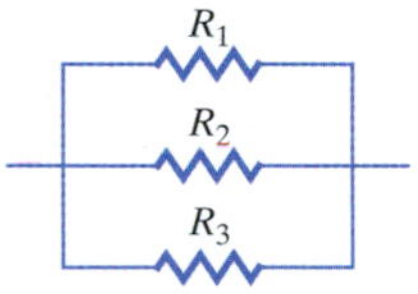

Simplify the complex rational expression on the right side of the formula. Then find R, to the nearest hundredth of an ohm, when R_1 is 4 ohms, R_2 is 8 ohms, and R_3 is 12 ohms.

54. A camera lens has a measurement called its focal length, f. When an object is in focus, its distance from the lens, p, and its image distance from the lens, q, satisfy the formula

$$f = \frac{1}{\dfrac{1}{p} + \dfrac{1}{q}}.$$

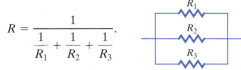

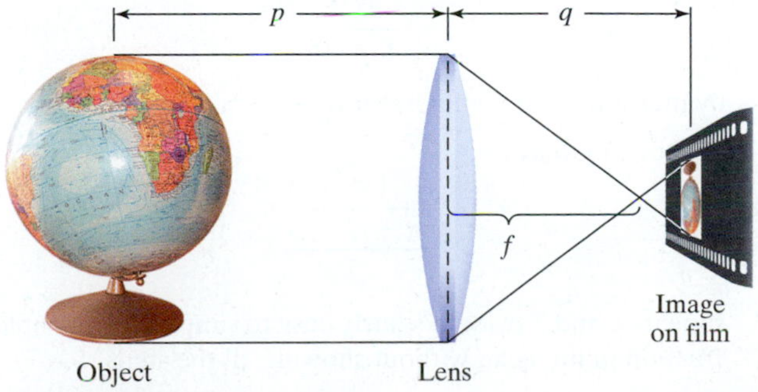

Simplify the complex rational expression on the right side of the formula.

Writing in Mathematics

55. What is a complex rational expression? Give an example with your explanation.

56. Describe two ways to simplify $\dfrac{\frac{2}{x}+\frac{2}{y}}{\frac{2}{x}-\frac{2}{y}}$.

57. Which method do you prefer for simplifying complex rational expressions? Why?

58. Of the four complex rational expressions in Exercises 51–54, which one do you find most useful? Explain how you might use this expression in a practical situation.

Technology Exercises

In Exercises 59–62, use a graphing utility to determine if the simplification is correct by graphing the function on each side of the equation in the same viewing rectangle. If the graphs do not coincide, correct the expression on the right side and then verify your correction using the graphing utility.

59. $\dfrac{x-\frac{1}{2x+1}}{1-\frac{x}{2x+1}}=2x-1$

60. $\dfrac{\frac{1}{x}+1}{\frac{1}{x}}=2$

61. $\dfrac{\frac{1}{x}+\frac{1}{3}}{\frac{1}{3x}}=x+\dfrac{1}{3}$

62. $\dfrac{\frac{x}{3}}{\frac{2}{x+1}}=\dfrac{3(x+1)}{2}$

Critical Thinking Exercises

Make Sense? *In Exercises 63–66, determine whether each statement "makes sense" or "does not make sense" and explain your reasoning.*

63. I simplified

$$\frac{\frac{1+3x}{xy}}{5+4y}$$

by multiplying the numerator by xy.

64. By noticing that

$$\frac{\frac{1}{x+7}-\frac{1}{x}}{7}$$

repeats x and 7 twice, it's fairly easy to simplify the complex fraction in my head without showing all the steps.

65. I simplified

$$\frac{3-\frac{6}{x+5}}{1+\frac{7}{x-4}}$$

by multiplying by 1 and obtained $\dfrac{3-6(x-4)}{1+7(x+5)}$.

66. Before simplifying $\dfrac{1-x^{-2}}{1-5x^{-3}}$, I wrote the complex fraction without negative exponents as

$$\frac{1-\frac{1}{x^2}}{1-\frac{1}{5x^3}}.$$

67. Simplify:

$$\frac{\frac{x+h}{x+h+1}-\frac{x}{x+1}}{h}.$$

68. Simplify:

$$x+\cfrac{1}{x+\cfrac{1}{x+\cfrac{1}{x}}}.$$

69. If $f(x)=\dfrac{1}{x+1}$, find $f(f(a))$ and simplify.

70. Let x represent the first of two consecutive integers. Find a simplified expression that represents the reciprocal of the sum of the reciprocals of the two integers.

Review Exercises

71. Solve: $x^2+27=12x$. (Section 5.7, Example 2)

72. Multiply: $(4x^2-y)^2$. (Section 5.2, Example 7)

73. Solve: $-4<3x-7<8$. (Section 4.2, Example 4)

Preview Exercises

Exercises 74–76 will help you prepare for the material covered in the next section.

74. Simplify: $\dfrac{8x^4y^5}{4x^3y^2}$.

75. Divide 737 by 21 without using a calculator. Write the answer as

$$\text{quotient}+\frac{\text{remainder}}{\text{divisor}}.$$

76. Simplify: $6x^2+3x-(6x^2-4x)$.

SECTION 6.4

Division of Polynomials

Objectives

1. Divide a polynomial by a monomial.
2. Use long division to divide by a polynomial containing more than one term.

During the 1980s, the controversial economist Arthur Laffer promoted the idea that tax *increases* lead to a *reduction* in government revenue. Called supply-side economics, the theory uses rational functions as models. One such function

$$f(x) = \frac{80x - 8000}{x - 110}, \quad 30 \le x \le 100$$

models the government tax revenue, $f(x)$, in tens of billions of dollars, as a function of the tax rate, x. The graph of the rational function is shown in **Figure 6.4**. The graph shows tax revenue decreasing quite dramatically as the tax rate increases. At a tax rate of (gasp) 100%, the government takes all our money and no one has an incentive to work. With no income earned, zero dollars in tax revenue is generated.

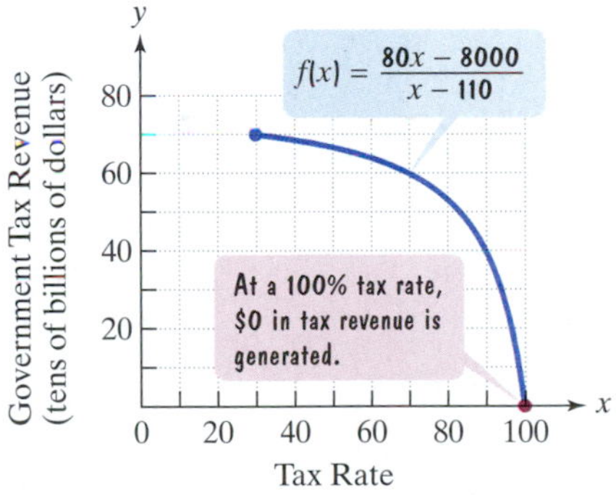

FIGURE 6.4

Like all rational functions, the Laffer model consists of the quotient of polynomials. Although the rational expression in the model cannot be simplified, it is possible to perform the division, thereby expressing the function in another form. In this section, you will learn to divide polynomials.

1 Divide a polynomial by a monomial.

Dividing a Polynomial by a Monomial

We have seen that to divide monomials, we divide the coefficients and subtract the exponents when bases are the same. For example,

$$\frac{25x^{12}}{5x^4} = \frac{25}{5}x^{12-4} = 5x^8 \quad \text{and} \quad \frac{60x^4y^3}{-30x^2y} = \frac{60}{-30}x^{4-2}y^{3-1} = -2x^2y^2.$$

How do we divide a polynomial that is not a monomial by a monomial? We divide each term of the polynomial by the monomial. For example,

$$\frac{10x^8 + 15x^6}{5x^3} = \frac{10x^8}{5x^3} + \frac{15x^6}{5x^3} = \frac{10}{5}x^{8-3} + \frac{15}{5}x^{6-3} = 2x^5 + 3x^3.$$

Polynomial dividend

Monomial divisor

Divide the first term by $5x^3$.

Divide the second term by $5x^3$.

Dividing a Polynomial That Is Not a Monomial by a Monomial

To divide a polynomial by a monomial, divide each term of the polynomial by the monomial.

EXAMPLE 1 Dividing a Polynomial by a Monomial

Divide: $(15x^3 - 5x^2 + x + 5) \div (5x)$.

Solution

$$\frac{15x^3 - 5x^2 + x + 5}{5x}$$ Rewrite the division in a vertical format.

$$= \frac{15x^3}{5x} - \frac{5x^2}{5x} + \frac{x}{5x} + \frac{5}{5x}$$ Divide each term of the polynomial by the monomial.

$$= 3x^2 - x + \frac{1}{5} + \frac{1}{x}$$ Simplify each quotient.

CHECK POINT 1 Divide: $(16x^3 - 32x^2 + 2x + 4) \div 4x$.

EXAMPLE 2 Dividing a Polynomial by a Monomial

Divide $8x^4y^5 - 10x^4y^3 + 12x^2y^3$ by $4x^3y^2$.

Solution

$$\frac{8x^4y^5 - 10x^4y^3 + 12x^2y^3}{4x^3y^2}$$ Express the division in a vertical format.

$$= \frac{8x^4y^5}{4x^3y^2} - \frac{10x^4y^3}{4x^3y^2} + \frac{12x^2y^3}{4x^3y^2}$$ Divide each term of the polynomial by the monomial.

$$= 2xy^3 - \frac{5}{2}xy + \frac{3y}{x}$$ Simplify each quotient.

CHECK POINT 2 Divide $15x^4y^5 - 5x^3y^4 + 10x^2y^2$ by $5x^2y^3$.

2 Use long division to divide by a polynomial containing more than one term.

Dividing by a Polynomial Containing More Than One Term

We now look at division by a polynomial containing more than one term, such as

$$x - 2\overline{)x^2 - 14x + 24}.$$

Divisor has two terms and is a binomial. The polynomial dividend has three terms and is a trinomial.

When a divisor has more than one term, the four steps used to divide whole numbers—**divide, multiply, subtract, bring down the next term**—form the repetitive procedure for polynomial long division.

EXAMPLE 3 Dividing a Polynomial by a Binomial

Divide $x^2 - 14x + 24$ by $x - 2$.

Solution The following steps illustrate how polynomial division is very similar to numerical division.

$$x - 2\overline{)x^2 - 14x + 24}$$ Arrange the terms of the dividend $(x^2 - 14x + 24)$ and the divisor $(x - 2)$ in descending powers of x.

$$\begin{array}{r} x \phantom{{}-14x+24} \\ x-2\overline{)x^2-14x+24} \end{array}$$

Divide x^2 (the first term in the dividend) by x (the first term in the divisor): $\frac{x^2}{x} = x$. Align like terms.

$x(x-2) = x^2 - 2x$

$$\begin{array}{r} x \phantom{{}-14x+24} \\ x-2\overline{)x^2-14x+24} \\ x^2-2x \phantom{{}+24} \end{array}$$

Multiply each term in the divisor $(x-2)$ by x, aligning terms of the product under like terms in the dividend.

Change signs of the polynomial being subtracted.

$$\begin{array}{r} x \phantom{{}-14x+24} \\ x-2\overline{)x^2-14x+24} \\ \ominus x^2 \oplus 2x \phantom{{}+24} \\ \hline -12x \phantom{{}+24} \end{array}$$

Subtract $x^2 - 2x$ from $x^2 - 14x$ by changing the sign of each term in the lower expression and adding.

$$\begin{array}{r} x \phantom{{}-14x+24} \\ x-2\overline{)x^2-14x+24} \\ x^2-2x \quad\downarrow \\ \hline -12x+24 \end{array}$$

Bring down 24 from the original dividend and add algebraically to form a new dividend.

$$\begin{array}{r} x-12 \\ x-2\overline{)x^2-14x+24} \\ x^2-2x \phantom{{}+24} \\ \hline -12x+24 \end{array}$$

Find the second term of the quotient. Divide the first term of $-12x + 24$ by x, the first term of the divisor: $\frac{-12x}{x} = -12$.

$-12(x-2) = -12x + 24$

$$\begin{array}{r} x-12 \\ x-2\overline{)x^2-14x+24} \\ x^2-2x \phantom{{}+24} \\ \hline -12x+24 \\ \oplus\!-12x \overset{\ominus}{+} 24 \\ \hline 0 \end{array}$$

Remainder

Multiply the divisor $(x-2)$ by -12, aligning under like terms in the new dividend. Then subtract to obtain the remainder of 0.

The quotient is $x - 12$. Because the remainder is 0, we can conclude that $x - 2$ is a factor of $x^2 - 14x + 24$ and

$$\frac{x^2 - 14x + 24}{x - 2} = x - 12.$$

After performing polynomial long division, the answer can be checked. Find the product of the divisor and the quotient, and add the remainder. If the result is the dividend, the answer to the division problem is correct. For example, let's check our work in Example 3.

$$\frac{x^2 - 14x + 24}{x - 2} = x - 12$$

Dividend: $x^2 - 14x + 24$. Divisor: $x - 2$. Quotient to be checked: $x - 12$.

Multiply the divisor and the quotient, and add the remainder, 0:

$$(x - 2)(x - 12) + 0 = x^2 - 12x - 2x + 24 + 0 = x^2 - 14x + 24.$$

Divisor: $(x-2)$. Quotient: $(x-12)$. Remainder: 0. This is the dividend: $x^2 - 14x + 24$.

Because we obtained the dividend, the quotient is correct.

☑ **CHECK POINT 3** Divide $3x^2 - 14x + 16$ by $x - 2$.

Before considering additional examples, let's summarize the general procedure for dividing by a polynomial that contains more than one term.

Long Division of Polynomials

1. Arrange the terms of both the dividend and the divisor in descending powers of any variable.
2. **Divide** the first term in the dividend by the first term in the divisor. The result is the first term of the quotient.
3. **Multiply** every term in the divisor by the first term in the quotient. Write the resulting product beneath the dividend with like terms lined up.
4. **Subtract** the product from the dividend.
5. **Bring down** the next term in the original dividend and write it next to the remainder to form a new dividend.
6. Use this new expression as the dividend and repeat this process until the remainder can no longer be divided. This will occur when the degree of the remainder (the highest exponent on a variable in the remainder) is less than the degree of the divisor.

In our next long division, we will obtain a nonzero remainder.

EXAMPLE 4 Long Division of Polynomials

Divide $4 - 5x - x^2 + 6x^3$ by $3x - 2$.

Solution We begin by writing the dividend in descending powers of x.

$$4 - 5x - x^2 + 6x^3 = 6x^3 - x^2 - 5x + 4$$

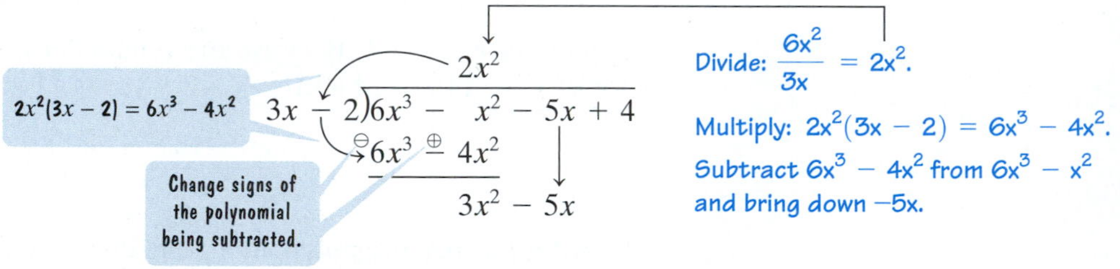

Now we divide $3x^2$ by $3x$ to obtain x, multiply x and the divisor, and subtract.

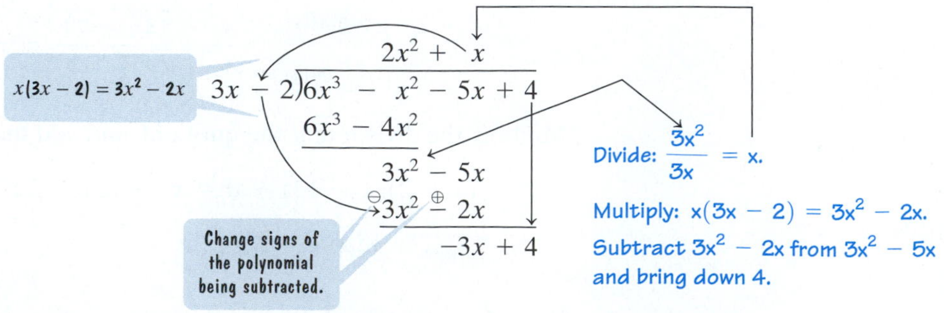

Now we divide $-3x$ by $3x$ to obtain -1, multiply -1 and the divisor, and subtract.

$$
\begin{array}{r}
2x^2 + x - 1 \\
3x - 2\overline{\smash{)}6x^3 - x^2 - 5x + 4} \\
\underline{6x^3 - 4x^2} \qquad\qquad \\
3x^2 - 5x \qquad \\
\underline{3x^2 - 2x} \qquad \\
-3x + 4 \\
\underline{\oplus{-3x} \ominus 2} \\
2
\end{array}
$$

$-1(3x - 2) = -3x + 2$

Change signs of the polynomial being subtracted.

Remainder

Divide: $\frac{-3x}{3x} = -1$.

Multiply: $-1(3x - 2) = -3x + 2$.

Subtract $-3x + 2$ from $-3x + 4$, leaving a remainder of 2.

The quotient is $2x^2 + x - 1$ and the remainder is 2. When there is a nonzero remainder, as in this example, list the quotient, plus the remainder above the divisor. Thus,

$$\frac{6x^3 - x^2 - 5x + 4}{3x - 2} = \underbrace{2x^2 + x - 1}_{\text{Quotient}} + \frac{2}{3x - 2}.$$

Remainder above divisor

Check this result by showing that the product of the divisor and the quotient,

$$(3x - 2)(2x^2 + x - 1),$$

plus the remainder, 2, is the dividend, $6x^3 - x^2 - 5x + 4$.

✓ CHECK POINT 4 Divide $-9 + 7x - 4x^2 + 4x^3$ by $2x - 1$.

If a power of x is missing in either a dividend or a divisor, add that power of x with a coefficient of 0 and then divide. In this way, like terms will be aligned as you carry out the long division.

EXAMPLE 5 Long Division of Polynomials

Divide $6x^4 + 5x^3 + 3x - 5$ by $3x^2 - 2x$.

Solution We write the dividend, $6x^4 + 5x^3 + 3x - 5$, as $6x^4 + 5x^3 + 0x^2 + 3x - 5$ to keep all like terms aligned.

Multiply.

$$
\begin{array}{r}
2x^2 + 3x + 2 \\
3x^2 - 2x\overline{\smash{)}6x^4 + 5x^3 + 0x^2 + 3x - 5} \\
\underline{\ominus 6x^4 \oplus 4x^3} \qquad\qquad\qquad\quad \\
9x^3 + 0x^2 \qquad\qquad \\
\underline{\ominus 9x^3 \oplus 6x^2} \qquad\qquad \\
6x^2 + 3x \qquad \\
\underline{\ominus 6x^2 \oplus 4x} \qquad \\
7x - 5
\end{array}
$$

$2x^2(3x^2 - 2x) = 6x^4 - 4x^3$

$3x(3x^2 - 2x) = 9x^3 - 6x^2$

$2(3x^2 - 2x) = 6x^2 - 4x$

Remainder

The division process is finished because the degree of $7x - 5$, which is 1, is less than the degree of the divisor $3x^2 - 2x$, which is 2.

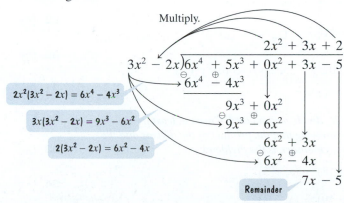

$$\begin{array}{r} 2x^2 + 3x + 2 \\ 3x^2 - 2x\overline{)6x^4 + 5x^3 + 0x^2 + 3x - 5} \\ \vdots \\ \overline{7x - 5} \end{array}$$

Remainder

Shown in the margin is the original division problem, with the quotient and the remainder. We see that the answer is

$$\frac{6x^4 + 5x^3 + 3x - 5}{3x^2 - 2x} = 2x^2 + 3x + 2 + \frac{7x - 5}{3x^2 - 2x}.$$

✓ **CHECK POINT 5** Divide $2x^4 + 3x^3 - 7x - 10$ by $x^2 - 2x$.

6.4 EXERCISE SET

MyMathLab — MathXL PRACTICE | WATCH | DOWNLOAD | READ | REVIEW

Practice Exercises

In Exercises 1–12, divide the polynomial by the monomial.

1. $\dfrac{25x^7 - 15x^5 + 10x^3}{5x^3}$
2. $\dfrac{49x^7 - 28x^5 + 14x^3}{7x^3}$
3. $\dfrac{18x^3 + 6x^2 - 9x - 6}{3x}$
4. $\dfrac{25x^3 + 50x^2 - 40x - 10}{5x}$
5. $(28x^3 - 7x^2 - 16x) \div (4x^2)$
6. $(70x^3 - 10x^2 - 14x) \div (7x^2)$
7. $(25x^8 - 50x^7 + 3x^6 - 40x^5) \div (-5x^5)$
8. $(18x^7 - 9x^6 + 20x^5 - 10x^4) \div (-2x^4)$
9. $(18a^3b^2 - 9a^2b - 27ab^2) \div (9ab)$
10. $(12a^2b^2 + 6a^2b - 15ab^2) \div (3ab)$
11. $(36x^4y^3 - 18x^3y^2 - 12x^2y) \div (6x^3y^3)$
12. $(40x^4y^3 - 20x^3y^2 - 50x^2y) \div (10x^3y^3)$

In Exercises 13–36, divide as indicated. Check at least five of your answers by showing that the product of the divisor and the quotient, plus the remainder, is the dividend.

13. $(x^2 + 8x + 15) \div (x + 5)$
14. $(x^2 + 3x - 10) \div (x - 2)$
15. $(x^3 - 2x^2 - 5x + 6) \div (x - 3)$
16. $(x^3 + 5x^2 + 7x + 2) \div (x + 2)$
17. $(x^2 - 7x + 12) \div (x - 5)$
18. $(2x^2 + x - 9) \div (x - 2)$
19. $(2x^2 + 13x + 5) \div (2x + 3)$
20. $(8x^2 + 6x - 25) \div (4x + 9)$
21. $(x^3 + 3x^2 + 5x + 4) \div (x + 1)$
22. $(x^3 + 6x^2 - 2x + 3) \div (x - 1)$
23. $(4y^3 + 12y^2 + 7y - 3) \div (2y + 3)$
24. $(6y^3 + 7y^2 + 12y - 5) \div (3y - 1)$
25. $(9x^3 - 3x^2 - 3x + 4) \div (3x + 2)$
26. $(2x^3 + 13x^2 + 9x - 6) \div (2x + 3)$
27. $(4x^3 - 6x - 11) \div (2x - 4)$
28. $(2x^3 + 6x - 4) \div (x + 4)$
29. $(4y^3 - 5y) \div (2y - 1)$
30. $(6y^3 - 5y) \div (2y - 1)$
31. $(4y^4 - 17y^2 + 14y - 3) \div (2y - 3)$
32. $(2y^4 - y^3 + 16y^2 - 4) \div (2y - 1)$
33. $(4x^4 + 3x^3 + 4x^2 + 9x - 6) \div (x^2 + 3)$
34. $(3x^5 - x^3 + 4x^2 - 12x - 8) \div (x^2 - 2)$
35. $(15x^4 + 3x^3 + 4x^2 + 4) \div (3x^2 - 1)$
36. $(18x^4 + 9x^3 + 3x^2) \div (3x^2 + 1)$

In Exercises 37–40, find a simplified expression for $\left(\frac{f}{g}\right)(x)$.

37. $f(x) = 8x^3 - 38x^2 + 49x - 10,$

$g(x) = 4x - 1$

38. $f(x) = 2x^3 - 9x^2 - 17x + 39,$

$g(x) = 2x - 3$

39. $f(x) = 2x^4 - 7x^3 + 7x^2 - 9x + 10,$

$g(x) = 2x - 5$

40. $f(x) = 4x^4 + 6x^3 + 3x - 1,$

$g(x) = 2x^2 + 1$

Practice PLUS

In Exercises 41–50, divide as indicated.

41. $\dfrac{x^4 + y^4}{x + y}$

42. $\dfrac{x^5 + y^5}{x + y}$

43. $\dfrac{3x^4 + 5x^3 + 7x^2 + 3x - 2}{x^2 + x + 2}$

44. $\dfrac{x^4 - x^3 - 7x^2 - 7x - 2}{x^2 - 3x - 2}$

45. $\dfrac{4x^3 - 3x^2 + x + 1}{x^2 + x + 1}$

46. $\dfrac{x^4 - x^2 + 1}{x^2 + x + 1}$

47. $\dfrac{x^5 - 1}{x^2 - x + 2}$

48. $\dfrac{5x^5 - 7x^4 + 3x^3 - 20x^2 + 28x - 12}{x^3 - 4}$

49. $\dfrac{4x^3 - 7x^2y - 16xy^2 + 3y^3}{x - 3y}$

50. $\dfrac{12x^3 - 19x^2y + 13xy^2 - 10y^3}{4x - 5y}$

In Exercises 51–52, find $\left(\frac{f-g}{h}\right)(x)$ *and the domain of* $\frac{f-g}{h}$.

51. $f(x) = 3x^3 + 4x^2 - x - 4,\ g(x) = -5x^3 + 22x^2 - 28x - 12,$

$h(x) = 4x + 1$

52. $f(x) = x^3 + 9x^2 - 6x + 25,\ g(x) = -3x^3 + 2x^2 - 14x + 5,$

$h(x) = 2x + 4$

In Exercises 53–54, solve each equation for x in terms of a and simplify.

53. $ax + 2x + 4 = 3a^3 + 10a^2 + 6a$

54. $ax - 3x + 6 = a^3 - 6a^2 + 11a$

Application Exercises

In the section opener, we saw that

$$f(x) = \frac{80x - 8000}{x - 110}, \quad 30 \le x \le 100$$

models the government tax revenue, $f(x)$, in tens of billions of dollars, as a function of the tax rate percentage, x. Use this function to solve Exercises 55–58. Round to the nearest ten billion dollars.

55. Find and interpret $f(30)$. Identify the solution as a point on the graph of the function in **Figure 6.4** on page 425.

56. Find and interpret $f(70)$. Identify the solution as a point on the graph of the function in **Figure 6.4** on page 425.

57. Rewrite the function by using long division to perform

$$(80x - 8000) \div (x - 110).$$

Then use this new form of the function to find $f(30)$. Do you obtain the same answer as you did in Exercise 55? Which form of the function do you find easier to use?

58. Rewrite the function by using long division to perform

$$(80x - 8000) \div (x - 110).$$

Then use this new form of the function to find $f(70)$. Do you obtain the same answer as you did in Exercise 56? Which form of the function do you find easier to use?

Writing in Mathematics

59. Explain how to divide a polynomial that is not a monomial by a monomial. Give an example.

60. In your own words, explain how to divide by a polynomial containing more than one term. Use $\frac{x^2 + 4}{x + 2}$ in your explanation.

61. When performing polynomial long division, explain when to stop dividing.

62. After performing polynomial long division, explain how to check the answer.

63. When performing polynomial long division with missing terms, explain the advantage of writing the missing terms with zero coefficients.

64. The idea of supply-side economics is that an increase in the tax rate may actually reduce government revenue. What explanation can you offer for this theory?

Technology Exercises

In Exercises 65–67, use a graphing utility to determine if the division has been performed correctly. Graph the function on each side of the equation in the same viewing rectangle. If the graphs do not coincide, correct the expression on the right side by using polynomial long division. Then verify your correction using the graphing utility.

65. $(6x^2 + 16x + 8) \div (3x + 2) = 2x + 4$

66. $(4x^3 + 7x^2 + 8x + 20) \div (2x + 4) = 2x^2 - \frac{1}{2}x + 3$

67. $(3x^4 + 4x^3 - 32x^2 - 5x - 20) \div (x + 4) = 3x^3 - 8x^2 + 5$

68. Use the TABLE feature of a graphing utility to verify any two division results that you obtained in Exercises 13–36.

Critical Thinking Exercises

Make Sense? *In Exercises 69–72, determine whether each statement "makes sense" or "does not make sense" and explain your reasoning.*

69. When performing the division

$$(2x^3 + 13x^2 + 9x - 6) \div (2x + 3),$$

I mentally cover up the +3, the second term of the binomial divisor, before dividing into the dividend.

70. When performing the division $(x^5 + 1) \div (x + 1)$, there's no need for me to follow all the steps involved in polynomial long division because I can work the problem in my head and see that the quotient must be $x^4 + 1$.

71. Because of exponential properties, the degree of the quotient must be the difference between the degree of the dividend and the degree of the divisor.

72. When performing the division $(x^3 + 1) \div (x + 2)$, the purpose of rewriting $x^3 + 1$ as $x^3 + 0x^2 + 0x + 1$ is to keep all like terms aligned.

In Exercises 73–76, determine whether each statement is true or false. If the statement is false, make the necessary change(s) to produce a true statement.

73. All long-division problems can be done by the alternative method of factoring the dividend and canceling identical factors in the dividend and the divisor.

74. Polynomial long division always shows that the answer is a polynomial.

75. The long division process should be continued until the degree of the remainder is the same as the degree of the divisor.

76. If a polynomial long-division problem results in a remainder that is zero, then the divisor is a factor of the dividend.

In Exercises 77–78, divide as indicated.

77. $(x^{3n} - 4x^{2n} - 2x^n - 12) \div (x^n - 5)$

78. $(x^{3n} + 1) \div (x^n + 1)$

79. When $2x^2 - 7x + 9$ is divided by a polynomial, the quotient is $2x - 3$ and the remainder is 3. Find the polynomial.

80. Find k so that the remainder is 0:
$(20x^3 + 23x^2 - 10x + k) \div (4x + 3)$.

Review Exercises

81. Solve: $|2x - 3| > 4$. (Section 4.3, Example 5 or Example 6(b))

82. Write 40,610,000 in scientific notation. (Section 1.7, Example 2)

83. Simplify: $2x - 4[x - 3(2x + 1)]$. (Section 1.2, Example 14)

Preview Exercises

Exercises 84–86 will help you prepare for the material covered in the next section.

84. a. Divide: $\dfrac{5x^3 + 6x + 8}{x + 2}$.

b. Find the sum of the numbers in each column, designated by □, in the following array of numbers. Then describe the relationship between the four numbers in the bottom row of the array and your answer to the division problem in part (a).

−2	5	0	6	8
		−10	20	−52
	5	□	□	□

85. a. Divide: $\dfrac{3x^3 - 4x^2 + 2x - 1}{x + 1}$.

b. Find the sum of the numbers in each column, designated by □, in the following array of numbers. Then describe the relationship between the four numbers in the bottom row of the array and your answer to the division problem in part (a).

−1	3	−4	2	−1
		−3	7	−9
	3	□	□	□

86. Divide $2x^3 - 3x^2 - 11x + 6$ by $x - 3$. Use your answer to factor $2x^3 - 3x^2 - 11x + 6$ completely.

MID-CHAPTER CHECK POINT Section 6.1–Section 6.4

What You Know: We learned that the domain of a rational function is the set of all real numbers except those for which the denominator is zero. We saw that graphs of rational functions have vertical asymptotes or breaks at these excluded values. We learned to simplify rational expressions by dividing the numerator and the denominator by common factors. We performed a variety of operations with rational expressions, including multiplication, division, addition, and subtraction. We used two methods (multiplying by 1 and dividing) to simplify complex rational expressions. Finally, we used long division when dividing by a polynomial with more than one term.

1. Simplify: $\dfrac{x^2 - x - 6}{x^2 + 3x - 18}$

In Exercises 2–19, perform the indicated operation(s) and, if possible, simplify.

2. $\dfrac{2x^2 - 8x - 11}{x^2 + 3x - 4} + \dfrac{x^2 + 14x - 13}{x^2 + 3x - 4}$

3. $\dfrac{x^3 - 27}{4x^2 - 4x} \cdot \dfrac{4x}{x - 3}$

4. $5 + \dfrac{7}{x - 2}$

5. $\dfrac{x - \dfrac{4}{x + 6}}{\dfrac{1}{x + 6} + x}$

6. $(2x^4 - 13x^3 + 17x^2 + 18x - 24) \div (x - 4)$

7. $\dfrac{x^3y - y^3x}{x^2y - xy^2}$

8. $(28x^8y^3 - 14x^6y^2 + 3x^2y^2) \div (7x^2y)$

9. $\dfrac{2x - 1}{x + 6} - \dfrac{x + 3}{x - 2}$

10. $\dfrac{3}{x - 2} - \dfrac{2}{x + 2} - \dfrac{x}{x^2 - 4}$

11. $\dfrac{3x^2 - 7x - 6}{3x^2 - 13x - 10} \div \dfrac{2x^2 - x - 1}{4x^2 - 18x - 10}$

12. $\dfrac{3}{7 - x} + \dfrac{x - 2}{x - 7}$

13. $(6x^4 - 3x^3 - 11x^2 + 2x + 4) \div (3x^2 - 1)$

14. $\dfrac{5 + \dfrac{2}{x}}{3 - \dfrac{1}{x}}$

15. $\dfrac{x}{x^2 - 7x + 6} - \dfrac{x}{x^2 - 2x - 24}$

16. $\dfrac{\dfrac{3}{x + 1} + \dfrac{4}{x}}{\dfrac{4}{x}}$

17. $\dfrac{x^2 - x - 6}{x + 1} \div \left(\dfrac{x^2 - 9}{x^2 - 1} \cdot \dfrac{x - 1}{x + 3}\right)$

18. $(64x^3 + 4) \div (4x + 2)$

19. $\dfrac{x + 1}{x^2 + x - 2} - \dfrac{1}{x^2 - 3x + 2} + \dfrac{2x}{x^2 - 4}$

20. Find the domain of $f(x) = \dfrac{5x - 10}{x^2 + 5x - 14}$. Then simplify the function's equation.

SECTION 6.5

Synthetic Division and the Remainder Theorem

Objectives

1. Divide polynomials using synthetic division.
2. Evaluate a polynomial function using the Remainder Theorem.
3. Show that a number is a solution of a polynomial equation using the Remainder Theorem.

A moth has moved into your closet. She appeared in your bedroom at night, but somehow her relatively stout body escaped your clutches. Within a few weeks, swarms of moths in your tattered wardrobe suggest that Mama Moth was in the family way. There must be at least 200 critters nesting in every crevice of your clothing.

Two hundred plus moth-tykes from one female moth—is this possible? Indeed it is. The number of eggs, $f(x)$, in a female moth is a function of her abdominal width, x, in millimeters, modeled by

$$f(x) = 14x^3 - 17x^2 - 16x + 34, \quad 1.5 \le x \le 3.5.$$

Because there are 200 moths feasting on your favorite sweaters, Mama's abdominal width can be estimated by finding the solutions of the polynomial equation

$$14x^3 - 17x^2 - 16x + 34 = 200.$$

With mathematics present even in your quickly disappearing attire, we move from rags to a shortcut for long division, called *synthetic division*. In the exercise set, you will use this shortcut to find Mama Moth's abdominal width.

1 Divide polynomials using synthetic division.

Dividing Polynomials Using Synthetic Division

We can use **synthetic division** to divide polynomials if the divisor is of the form $x - c$. This method provides a quotient more quickly than long division. Let's compare the two methods showing $x^3 + 4x^2 - 5x + 5$ divided by $x - 3$.

Long Division

$$\begin{array}{r}
x^2 + 7x + 16 \\
x - 3 \overline{)\, x^3 + 4x^2 - 5x + 5} \\
\ominus x^3 \oplus - 3x^2 \qquad\qquad \\
\hline
7x^2 - 5x \qquad \\
\ominus 7x^2 \oplus - 21x \qquad \\
\hline
16x + 5 \\
\ominus 16x \oplus - 48 \\
\hline
53
\end{array}$$

Quotient: $x^2 + 7x + 16$

Dividend: $x^3 + 4x^2 - 5x + 5$

Divisor: $x - c$; $c = 3$

Remainder: 53

Synthetic Division

$$\begin{array}{r|rrrr}
3 & 1 & 4 & -5 & 5 \\
 & & 3 & 21 & 48 \\
\hline
 & 1 & 7 & 16 & 53
\end{array}$$

Notice the relationship between the polynomials in the long division process and the numbers that appear in synthetic division.

These are the coefficients of the dividend $x^3 + 4x^2 - 5x + 5$.

The divisor is $x - 3$. This is 3, or c, in $x - c$.

$$\begin{array}{r|rrrr} 3 & 1 & 4 & -5 & 5 \\ & & 3 & 21 & 48 \\ \hline & 1 & 7 & 16 & 53 \end{array}$$

These are the coefficients of the quotient $x^2 + 7x + 16$.

This is the remainder.

Now let's look at the steps involved in synthetic division.

Synthetic Division

To divide a polynomial by $x - c$:

Example

1. Arrange polynomials in descending powers, with a 0 coefficient for any missing term.

$$x - 3\overline{)x^3 + 4x^2 - 5x + 5}$$

2. Write c for the divisor, $x - c$. To the right, write the coefficients of the dividend.

$$\begin{array}{r|rrrr} 3 & 1 & 4 & -5 & 5 \end{array}$$

3. Write the leading coefficient of the dividend on the bottom row.

$$\begin{array}{r|rrrr} 3 & 1 & 4 & -5 & 5 \\ \hline & 1 & & & \end{array}$$

↓Bring down 1.

4. Multiply c (in this case, 3) times the value just written on the bottom row. Write the product in the next column in the second row.

$$\begin{array}{r|rrrr} 3 & 1 & 4 & -5 & 5 \\ & & 3 & & \\ \hline & 1 & & & \end{array}$$

Multiply by 3: $3 \cdot 1 = 3$.

5. Add the values in this new column, writing the sum in the bottom row.

$$\begin{array}{r|rrrr} 3 & 1 & 4 & -5 & 5 \\ & & 3 & & \\ \hline & 1 & 7 & & \end{array}$$

Add.

6. Repeat this series of multiplications and additions until all columns are filled in.

$$\begin{array}{r|rrrr} 3 & 1 & 4 & -5 & 5 \\ & & 3 & 21 & \\ \hline & 1 & 7 & 16 & \end{array}$$

Add.

Multiply by 3: $3 \cdot 7 = 21$.

$$\begin{array}{r|rrrr} 3 & 1 & 4 & -5 & 5 \\ & & 3 & 21 & 48 \\ \hline & 1 & 7 & 16 & 53 \end{array}$$

Add.

Multiply by 3: $3 \cdot 16 = 48$.

7. Use the numbers in the last row to write the quotient, plus the remainder above the divisor. **The degree of the first term of the quotient is one less than the degree of the first term of the dividend.** The final value in this row is the remainder.

Written from 1 7 16 53 the last row of the synthetic division

$$x - 3\overline{)x^3 + 4x^2 - 5x + 5}\quad\text{quotient: } 1x^2 + 7x + 16 + \frac{53}{x - 3}$$

EXAMPLE 1 Using Synthetic Division

Use synthetic division to divide $5x^3 + 6x + 8$ by $x + 2$.

Solution The divisor must be in the form $x - c$. Thus, we write $x + 2$ as $x - (-2)$. This means that $c = -2$. Writing a 0 coefficient for the missing x^2-term in the dividend, we can express the division as follows:

$$x - (-2)\overline{)5x^3 + 0x^2 + 6x + 8}.$$

Now we are ready to set up the problem so that we can use synthetic division.

Use the coefficients of the dividend $5x^3 + 0x^2 + 6x + 8$ in descending powers of x.

This is c in $x - (-2)$.

$$\underline{-2|}\ \ 5\quad 0\quad 6\quad 8$$

We begin the synthetic division process by bringing down 5. This is followed by a series of multiplications and additions.

1. Bring down 5.

$$\begin{array}{r|rrrr} -2 & 5 & 0 & 6 & 8 \\ & & & & \\ \hline & 5 & & & \end{array}$$

2. Multiply: $-2(5) = -10$.

$$\begin{array}{r|rrrr} -2 & 5 & 0 & 6 & 8 \\ & & -10 & & \\ \hline & 5 & & & \end{array}$$

Multiply 5 by −2.

3. Add: $0 + (-10) = -10$.

$$\begin{array}{r|rrrr} -2 & 5 & 0 & 6 & 8 \\ & & -10 & & \\ \hline & 5 & -10 & & \end{array}$$

Add.

4. Multiply: $-2(-10) = 20$.

$$\begin{array}{r|rrrr} -2 & 5 & 0 & 6 & 8 \\ & & -10 & 20 & \\ \hline & 5 & -10 & & \end{array}$$

Multiply −10 by −2.

5. Add: $6 + 20 = 26$.

$$\begin{array}{r|rrrr} -2 & 5 & 0 & 6 & 8 \\ & & -10 & 20 & \\ \hline & 5 & -10 & 26 & \end{array}$$

Add.

6. Multiply: $-2(26) = -52$.

$$\begin{array}{r|rrrr} -2 & 5 & 0 & 6 & 8 \\ & & -10 & 20 & -52 \\ \hline & 5 & -10 & 26 & \end{array}$$

Multiply 26 by −2.

7. Add: $8 + (-52) = -44$.

$$\begin{array}{r|rrrr} -2 & 5 & 0 & 6 & 8 \\ & & -10 & 20 & -52 \\ \hline & 5 & -10 & 26 & -44 \end{array}$$

Add.

The numbers in the last row represent the coefficients of the quotient and the remainder. The degree of the first term of the quotient is one less than that of the dividend. Because the degree of the dividend, $5x^3 + 6x + 8$, is 3, the degree of the quotient is 2. This means that the 5 in the last row represents $5x^2$.

$$\begin{array}{r|rrrr} -2 & 5 & 0 & 6 & 8 \\ & & -10 & 20 & -52 \\ \hline & 5 & -10 & 26 & -44 \end{array}$$

The quotient is $5x^2 - 10x + 26$. *The remainder is −44.*

Thus,

$$\begin{array}{r} 5x^2 - 10x + 26 - \dfrac{44}{x+2} \\ x + 2\overline{)5x^3 + \ 6x + \ 8} \end{array}$$

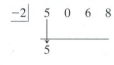

✓ **CHECK POINT 1** Use synthetic division to divide

$$x^3 - 7x - 6 \text{ by } x + 2.$$

2 Evaluate a polynomial function using the Remainder Theorem.

The Remainder Theorem

We have seen that the answer to a long division problem can be checked: Find the product of the divisor and the quotient and add the remainder. The result should be the dividend. If the divisor is $x - c$, we can express this idea symbolically:

$$f(x) = (x - c)q(x) + r.$$

Dividend | Divisor | Quotient | The remainder, r, is a constant when dividing by $x - c$.

Now let's evaluate f at c.

$$f(c) = (c - c)q(c) + r$$ Find $f(c)$ by letting $x = c$ in $f(x) = (x - c)q(x) + r$. This will give an expression for r.

$$f(c) = 0 \cdot q(c) + r$$ $c - c = 0$

$$f(c) = r$$ $0 \cdot q(c) = 0$ and $0 + r = r$.

What does this last equation mean? If a polynomial is divided by $x - c$, the remainder is the value of the polynomial at c. This result is called the **Remainder Theorem**.

The Remainder Theorem

If the polynomial $f(x)$ is divided by $x - c$, then the remainder is $f(c)$.

Example 2 shows how we can use the Remainder Theorem to evaluate a polynomial function at 2. Rather than substituting 2 for x, we divide the function by $x - 2$. The remainder is $f(2)$.

EXAMPLE 2 **Using the Remainder Theorem to Evaluate a Polynomial Function**

Given $f(x) = x^3 - 4x^2 + 5x + 3$, use the Remainder Theorem to find $f(2)$.

Solution By the Remainder Theorem, if $f(x)$ is divided by $x - 2$, then the remainder is $f(2)$. We'll use synthetic division to divide.

$$\begin{array}{r|rrrr} 2 & 1 & -4 & 5 & 3 \\ & & 2 & -4 & 2 \\ \hline & 1 & -2 & 1 & 5 \end{array}$$

Remainder

The remainder, 5, is the value of $f(2)$. Thus, $f(2) = 5$. We can verify that this is correct by evaluating $f(2)$ directly. Using $f(x) = x^3 - 4x^2 + 5x + 3$, we obtain

$$f(2) = 2^3 - 4 \cdot 2^2 + 5 \cdot 2 + 3 = 8 - 16 + 10 + 3 = 5.$$

✓ **CHECK POINT 2** Given $f(x) = 3x^3 + 4x^2 - 5x + 3$, use the Remainder Theorem to find $f(-4)$.

3 Show that a number is a solution of a polynomial equation using the Remainder Theorem.

If the polynomial $f(x)$ is divided by $x - c$ and the remainder is zero, then $f(c) = 0$. This means that c is a solution of the polynomial equation $f(x) = 0$.

EXAMPLE 3 Using the Remainder Theorem

Show that 3 is a solution of the equation

$$2x^3 - 3x^2 - 11x + 6 = 0.$$

Then solve the polynomial equation.

Solution One way to show that 3 is a solution is to substitute 3 for x in the equation and obtain 0. An easier way is to use synthetic division and the Remainder Theorem.

Proposed solution

$$\begin{array}{r|rrrr} 3 & 2 & -3 & -11 & 6 \\ & & 6 & 9 & -6 \\ \hline & 2 & 3 & -2 & 0 \end{array}$$

Remainder

$$\begin{array}{r} 2x^2 + 3x - 2 \\ x - 3\overline{)2x^3 - 3x^2 - 11x + 6} \end{array}$$

Equivalently,

$$2x^3 - 3x^2 - 11x + 6 = (x-3)(2x^2 + 3x - 2).$$

Because the remainder is 0, the polynomial has a value of 0 when $x = 3$. Thus, 3 is a solution of the given equation.

The synthetic division also shows that $x - 3$ divides the polynomial with a zero remainder. Thus, $x - 3$ is a factor of the polynomial, as shown to the right of the synthetic division. The other factor is the quotient found in the last row of the synthetic division. Now we can solve the polynomial equation.

$2x^3 - 3x^2 - 11x + 6 = 0$	This is the given equation.
$(x-3)(2x^2 + 3x - 2) = 0$	Factor using the result from the synthetic division.
$(x-3)(2x-1)(x+2) = 0$	Factor the trinomial.
$x - 3 = 0$ or $2x - 1 = 0$ or $x + 2 = 0$	Set each factor equal to 0.
$x = 3$ $\quad x = \frac{1}{2}$ $\quad x = -2$	Solve for x.

The solutions are $-2, \frac{1}{2}$, and 3, and the solution set is $\{-2, \frac{1}{2}, 3\}$.

Using Technology

Graphic Connections

Because the solution set of

$$2x^3 - 3x^2 - 11x + 6 = 0$$

is $\{-2, \frac{1}{2}, 3\}$, this implies that the polynomial function

$$f(x) = 2x^3 - 3x^2 - 11x + 6$$

has x-intercepts at $-2, \frac{1}{2}$, and 3. This is verified by the graph of f.

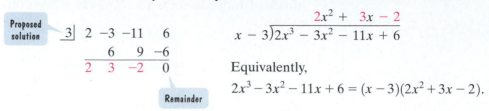

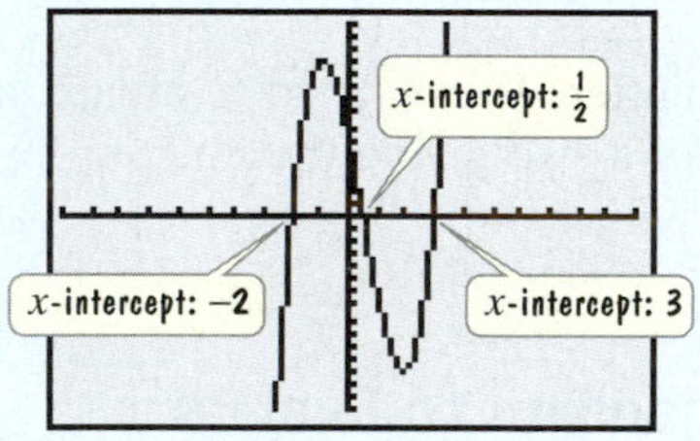

[−10, 10, 1] by [−15, 15, 1]

✓ **CHECK POINT 3** Use synthetic division to show that −1 is a solution of the equation

$$15x^3 + 14x^2 - 3x - 2 = 0.$$

Then solve the polynomial equation.

6.5 EXERCISE SET

MyMathLab — MathXL PRACTICE · WATCH · DOWNLOAD · READ · REVIEW

Practice Exercises

In Exercises 1–18, divide using synthetic division. In the first two exercises, begin the process as shown.

1. $(2x^2 + x - 10) \div (x - 2)$ $\quad 2\,|\ \ 2 \quad 1 \quad -10$
2. $(x^2 + x - 2) \div (x - 1)$ $\quad 1\,|\ \ 1 \quad 1 \quad -2$
3. $(3x^2 + 7x - 20) \div (x + 5)$
4. $(5x^2 - 12x - 8) \div (x + 3)$
5. $(4x^3 - 3x^2 + 3x - 1) \div (x - 1)$
6. $(5x^3 - 6x^2 + 3x + 11) \div (x - 2)$
7. $(6x^5 - 2x^3 + 4x^2 - 3x + 1) \div (x - 2)$
8. $(x^5 + 4x^4 - 3x^2 + 2x + 3) \div (x - 3)$
9. $(x^2 - 5x - 5x^3 + x^4) \div (5 + x)$

10. $(x^2 - 6x - 6x^3 + x^4) \div (6 + x)$

11. $(3x^3 + 2x^2 - 4x + 1) \div \left(x - \frac{1}{3}\right)$

12. $(2x^4 - x^3 + 2x^2 - 3x + 1) \div \left(x - \frac{1}{2}\right)$

13. $\dfrac{x^5 + x^3 - 2}{x - 1}$

14. $\dfrac{x^7 + x^5 - 10x^3 + 12}{x + 2}$

15. $\dfrac{x^4 - 256}{x - 4}$

16. $\dfrac{x^7 - 128}{x - 2}$

17. $\dfrac{2x^5 - 3x^4 + x^3 - x^2 + 2x - 1}{x + 2}$

18. $\dfrac{x^5 - 2x^4 - x^3 + 3x^2 - x + 1}{x - 2}$

In Exercises 19–26, use synthetic division and the Remainder Theorem to find the indicated function value.

19. $f(x) = 2x^3 - 11x^2 + 7x - 5;\ f(4)$

20. $f(x) = x^3 - 7x^2 + 5x - 6;\ f(3)$

21. $f(x) = 3x^3 - 7x^2 - 2x + 5;\ f(-3)$

22. $f(x) = 4x^3 + 5x^2 - 6x - 4;\ f(-2)$

23. $f(x) = x^4 + 5x^3 + 5x^2 - 5x - 6;\ f(3)$

24. $f(x) = x^4 - 5x^3 + 5x^2 + 5x - 6;\ f(2)$

25. $f(x) = 2x^4 - 5x^3 - x^2 + 3x + 2;\ f\left(-\frac{1}{2}\right)$

26. $f(x) = 6x^4 + 10x^3 + 5x^2 + x + 1;\ f\left(-\frac{2}{3}\right)$

In Exercises 27–32, use synthetic division to show that the number given to the right of each equation is a solution of the equation. Then solve the polynomial equation.

27. $x^3 - 4x^2 + x + 6 = 0;\ -1$

28. $x^3 - 2x^2 - x + 2 = 0;\ -1$

29. $2x^3 - 5x^2 + x + 2 = 0;\ 2$

30. $2x^3 - 3x^2 - 11x + 6 = 0;\ -2$

31. $6x^3 + 25x^2 - 24x + 5 = 0;\ -5$

32. $3x^3 + 7x^2 - 22x - 8 = 0;\ -4$

Practice PLUS

In Exercises 33–36, use the graph or the table to determine a solution of each equation. Use synthetic division to verify that this number is a solution of the equation. Then solve the polynomial equation.

33. $x^3 + 2x^2 - 5x - 6 = 0$

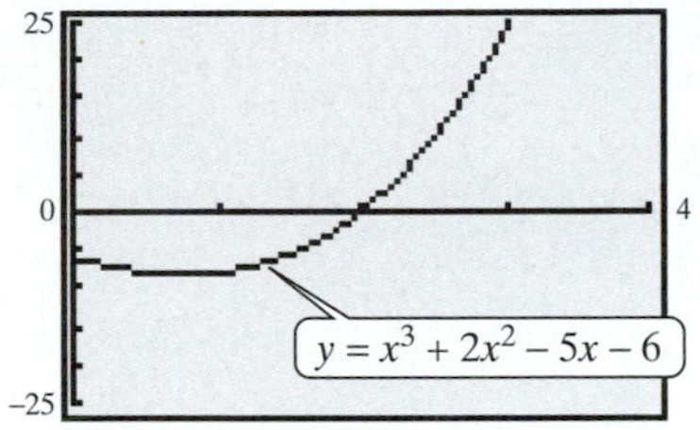

[0, 4, 1] by [−25, 25, 5]

34. $2x^3 + x^2 - 13x + 6 = 0$

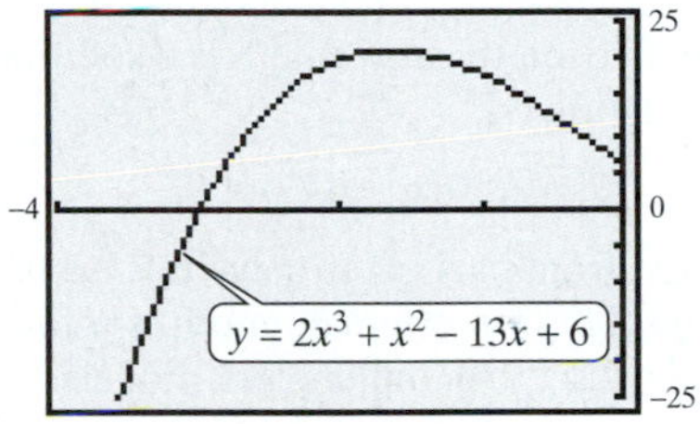

[−4, 0, 1] by [−25, 25, 5]

35. $6x^3 - 11x^2 + 6x - 1 = 0$

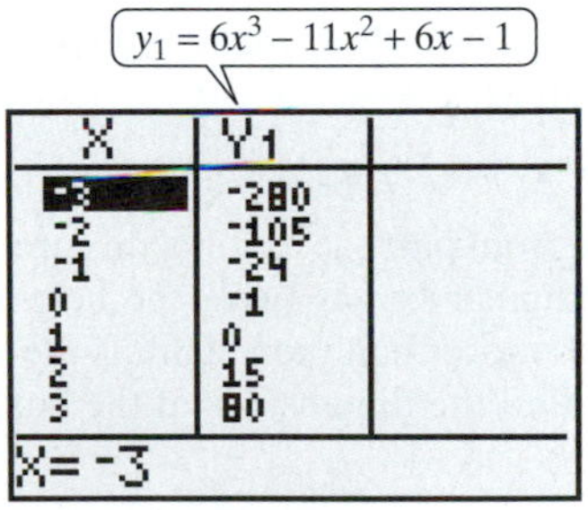

36. $2x^3 + 11x^2 - 7x - 6 = 0$

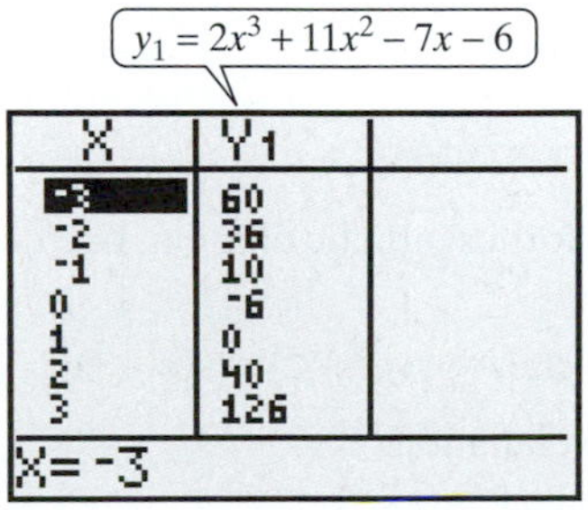

In Exercises 37–38, perform the given operations.

37. $(22x - 24 + 7x^3 - 29x^2 + 4x^4)(x + 4)^{-1}$

38. $(9 - x^2 + 6x + 2x^3)(x + 1)^{-1}$

In Exercises 39–40, write a polynomial that represents the length of each rectangle.

39.

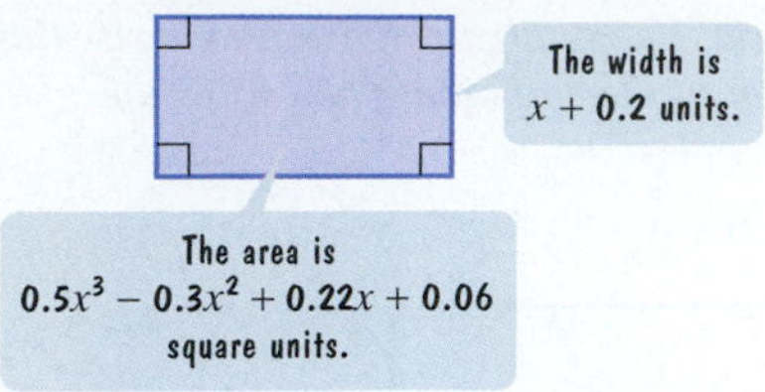
The width is
$x + 0.2$ units.
The area is
$0.5x^3 - 0.3x^2 + 0.22x + 0.06$
square units.

40.

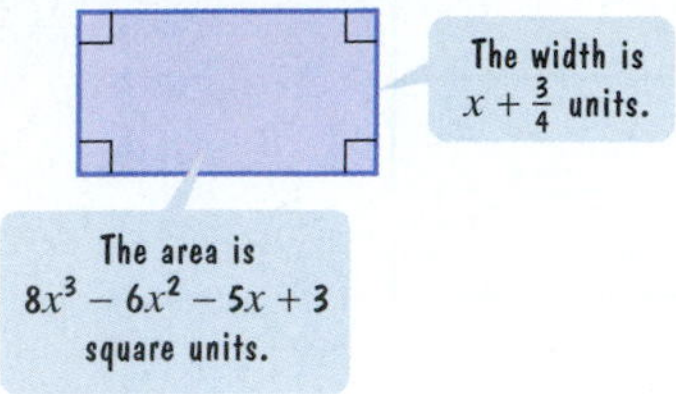
The width is
$x + \frac{3}{4}$ units.
The area is
$8x^3 - 6x^2 - 5x + 3$
square units.

Application Exercises

41. a. Use synthetic division to show that 3 is a solution of the polynomial equation

$$14x^3 - 17x^2 - 16x - 177 = 0.$$

b. Use the solution from part (a) to solve this problem. The number of eggs, $f(x)$, in a female moth is a function of her abdominal width, x, in millimeters, modeled by

$$f(x) = 14x^3 - 17x^2 - 16x + 34.$$

What is the abdominal width when there are 211 eggs?

42. a. Use synthetic division to show that 2 is a solution of the polynomial equation

$$2h^3 + 14h^2 - 72 = 0.$$

b. Use the solution from part (a) to solve this problem. The width of a rectangular box is twice the height and the length is 7 inches more than the height. If the volume is 72 cubic inches, find the dimensions of the box.

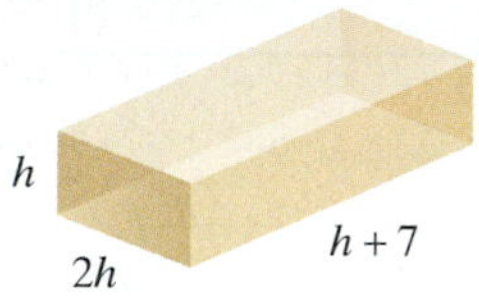
h
$2h$
$h + 7$

Writing in Mathematics

43. Explain how to perform synthetic division. Use the division problem

$$(2x^3 - 3x^2 - 11x + 7) \div (x - 3)$$

to support your explanation.

44. State the Remainder Theorem.

45. Explain how the Remainder Theorem can be used to find $f(-6)$ if $f(x) = x^4 + 7x^3 + 8x^2 + 11x + 5$. What advantage is there to using the Remainder Theorem in this situation rather than evaluating $f(-6)$ directly?

46. Explain how the Remainder Theorem and synthetic division can be used to determine whether -4 is a solution of the following equation:

$$5x^3 + 22x^2 + x - 28 = 0.$$

Technology Exercise

47. For each equation that you solved in Exercises 27–32, use a graphing utility to graph the polynomial function on the left side of the equation. Use end behavior to obtain a complete graph. Then use the graph's x-intercepts to verify your solutions.

Critical Thinking Exercises

Make Sense? *In Exercises 48–51, determine whether each statement "makes sense" or "does not make sense" and explain your reasoning.*

48. There are certain kinds of polynomial divisions that I can perform using long division, but not using synthetic division.

49. Every time I divide polynomials using synthetic division, I am using a highly condensed form of the long division procedure where omitting the variables and exponents does not involve the loss of any essential data.

50. The only nongraphic method that I have for evaluating a function at a given value is to substitute that value into the function's equation.

51. The Remainder Theorem gives me a method for factoring certain polynomials that I could not factor using the factoring strategies from the previous chapter.

52. Synthetic division is a process for dividing a polynomial by $x - c$. The coefficient of x is 1. How might synthetic division be used if you are dividing by $2x - 4$?

53. Use synthetic division to show that 5 is a solution of

$$x^4 - 4x^3 - 9x^2 + 16x + 20 = 0.$$

Then solve the polynomial equation. *Hint:* Use factoring by grouping when working with the quotient factor.

Review Exercises

54. Solve: $4x + 3 - 13x - 7 < 2(3 - 4x)$. (Section 4.1, Example 3)

55. Solve: $2x(x + 3) + 6(x - 3) = -28$. (Section 5.7, Example 2)

56. Solve by Cramer's rule:

$$\begin{aligned} 7x - 6y &= 17 \\ 3x + y &= 18. \end{aligned}$$

(Section 3.5, Example 2)

Preview Exercises

Exercises 57–59 will help you prepare for the material covered in the next section. In each exercise, find the LCD of the rational expressions.

57. $\frac{x + 4}{2x}, \frac{x + 20}{3x}$

58. $\frac{x}{3}, \frac{9}{x}, 4$

59. $\frac{2x}{x - 3}, \frac{6}{x + 3}, -\frac{28}{x^2 - 9}$

SECTION 6.6

Rational Equations

Objectives

1. Solve rational equations.
2. Solve problems involving rational functions that model applied situations.

The lake is in one of the city's favorite parks and the time has come to clean it up. Voters in the city have committed \$80 thousand for the cleanup. We know that

$$f(x) = \frac{120x}{100 - x}$$

models the cost, $f(x)$, in thousands of dollars, to remove x% of the lake's pollutants. What percentage of the pollutants can be removed for \$80 thousand?

To determine the percentage, we use the given cost-benefit model. Voters have committed \$80 thousand, so substitute 80 for $f(x)$:

$$80 = \frac{120x}{100 - x}.$$

This equation contains a rational expression.

Now we have to solve the equation and find the value for x. This variable represents the percentage of the pollutants that can be removed for \$80 thousand.

A **rational equation**, also called a **fractional equation**, is an equation containing one or more rational expressions. The equation shown above is an example of a rational equation. Do you see that there is a variable in a denominator? This is a characteristic of many rational equations. In this section, you will learn a procedure for solving such equations.

1 Solve rational equations.

Solving Rational Equations

We have seen that the LCD is used to add and subtract rational expressions. By contrast, when solving rational equations, **the LCD is used as a multiplier that clears an equation of fractions**.

EXAMPLE 1 Solving a Rational Equation

Solve: $\dfrac{x+4}{2x} + \dfrac{x+20}{3x} = 3.$

Solution Notice that the variable x appears in both denominators. We must avoid any values of the variable that make a denominator zero.

$$\frac{x+4}{2x} + \frac{x+20}{3x} = 3$$

This denominator would equal zero if $x = 0$. This denominator would equal zero if $x = 0$.

We see that x cannot equal zero.

The denominators are $2x$ and $3x$. The least common denominator is $6x$. We begin by multiplying both sides of the equation by $6x$. We will also write the restriction that x cannot equal zero to the right of the equation.

$$\frac{x+4}{2x}+\frac{x+20}{3x}=3,\ x\neq 0 \quad \text{This is the given equation.}$$

$$6x\left(\frac{x+4}{2x}+\frac{x+20}{3x}\right)=6x\cdot 3 \quad \text{Multiply both sides by } 6x\text{, the LCD.}$$

$$\frac{6x}{1}\cdot\frac{x+4}{2x}+\frac{6x}{1}\cdot\frac{x+20}{3x}=18x \quad \text{Use the distributive property.}$$

$$3(x+4)+2(x+20)=18x \quad \text{Divide out common factors in the multiplications.}$$

Observe that the equation is now cleared of fractions.

$$3x+12+2x+40=18x \quad \text{Use the distributive property.}$$

$$5x+52=18x \quad \text{Combine like terms.}$$

$$52=13x \quad \text{Subtract } 5x \text{ from both sides.}$$

$$4=x \quad \text{Divide both sides by 13.}$$

The proposed solution, 4, is not part of the restriction $x \neq 0$. It should check in the original equation.

Check 4:

$$\frac{x+4}{2x}+\frac{x+20}{3x}=3$$

$$\frac{4+4}{2\cdot 4}+\frac{4+20}{3\cdot 4}\stackrel{?}{=}3$$

$$\frac{8}{8}+\frac{24}{12}\stackrel{?}{=}3$$

$$1+2\stackrel{?}{=}3$$

$$3=3, \quad \text{true}$$

This true statement verifies that the solution is 4 and the solution set is $\{4\}$.

✓ **CHECK POINT 1** Solve: $\dfrac{x+6}{2x}+\dfrac{x+24}{5x}=2.$

The following steps may be used to solve a rational equation:

Solving Rational Equations

1. List restrictions on the variable. Avoid any values of the variable that make a denominator zero.
2. Clear the equation of fractions by multiplying both sides by the LCD of all rational expressions in the equation.
3. Solve the resulting equation.
4. Reject any proposed solution that is in the list of restrictions on the variable. Check other proposed solutions in the original equation.

EXAMPLE 2 Solving a Rational Equation

Solve: $\dfrac{x+1}{x+10}=\dfrac{x-2}{x+4}.$

Solution

Step 1. List restrictions on the variable.

This denominator would equal zero if $x = -10$.

$$\frac{x+1}{x+10} = \frac{x-2}{x+4}$$

This denominator would equal zero if $x = -4$.

The restrictions are $x \neq -10$ and $x \neq -4$.

Step 2. Multiply both sides by the LCD. The denominators are $x + 10$ and $x + 4$. Thus, the LCD is $(x + 10)(x + 4)$.

$$\frac{x+1}{x+10} = \frac{x-2}{x+4}, \quad x \neq -10, x \neq -4 \qquad \text{This is the given equation.}$$

$$(x+10)(x+4)\cdot\left(\frac{x+1}{x+10}\right) = (x+10)(x+4)\cdot\left(\frac{x-2}{x+4}\right) \qquad \text{Multiply both sides by the LCD.}$$

$$(x+4)(x+1) = (x+10)(x-2) \qquad \text{Simplify.}$$

Step 3. Solve the resulting equation.

$$(x+4)(x+1) = (x+10)(x-2) \qquad \text{This is the equation cleared of fractions.}$$
$$x^2 + 5x + 4 = x^2 + 8x - 20 \qquad \text{Use FOIL multiplication on each side.}$$
$$5x + 4 = 8x - 20 \qquad \text{Subtract } x^2 \text{ from both sides.}$$
$$-3x + 4 = -20 \qquad \text{Subtract } 8x \text{ from both sides.}$$
$$-3x = -24 \qquad \text{Subtract 4 from both sides.}$$
$$x = 8 \qquad \text{Divide both sides by } -3.$$

Step 4. Check the proposed solution in the original equation. The proposed solution, 8, is not part of the restriction that $x \neq -10$ and $x \neq -4$. Substitute 8 for x in the given equation. You should obtain the true statement $\frac{1}{2} = \frac{1}{2}$. The solution is 8 and the solution set is $\{8\}$.

✓ CHECK POINT 2 Solve: $\dfrac{x-3}{x+1} = \dfrac{x-2}{x+6}$.

Study Tip

We simplify rational expressions. We solve rational equations. Notice the differences between the procedures.

Simplifying a Rational Expression	Solving a Rational Equation
Simplify: $\frac{9}{4x} - \frac{5}{2x} - \frac{3}{4}$. This is not an equation. There is no equal sign.	Solve: $\frac{9}{4x} - \frac{5}{2x} = \frac{3}{4}$. This is an equation. There is an equal sign.
Solution The LCD is $4x$. Rewrite each expression with this LCD and retain the LCD. $\frac{9}{4x} - \frac{5}{2x} - \frac{3}{4}$ $= \frac{9}{4x} - \frac{5\cdot 2}{2x\cdot 2} - \frac{3\cdot x}{4\cdot x}$ $= \frac{9}{4x} - \frac{10}{4x} - \frac{3x}{4x}$ $= \frac{9-10-3x}{4x} = \frac{-1-3x}{4x}$	**Solution** The LCD is $4x$. Multiply both sides by this LCD and clear the fractions. $4x\left(\frac{9}{4x} - \frac{5}{2x}\right) = 4x\cdot\frac{3}{4}$ $\frac{4x}{1}\cdot\frac{9}{4x} - \frac{4x}{1}\cdot\frac{5}{2x} = 4x\cdot\frac{3}{4}$ $9 - 10 = 3x$ $-1 = 3x$ $-\frac{1}{3} = x$ The solution set is $\left\{-\frac{1}{3}\right\}$.

You only eliminate the denominators when solving a rational equation with an equal sign. You should never begin by eliminating the denominators when simplifying a rational expression involving addition or subtraction with no equal sign.

EXAMPLE 3 Solving a Rational Equation

Solve: $\frac{x}{x-3} = \frac{3}{x-3} + 9$.

Solution

Step 1. List restrictions on the variable.

$$\frac{x}{x-3} = \frac{3}{x-3} + 9$$

These denominators are zero if $x = 3$.

The restriction is $x \neq 3$.

Step 2. Multiply both sides by the LCD. The LCD is $x - 3$.

$$\frac{x}{x-3} = \frac{3}{x-3} + 9, \quad x \neq 3$$ This is the given equation.

$$(x-3)\cdot\frac{x}{x-3} = (x-3)\left(\frac{3}{x-3} + 9\right)$$ Multiply both sides by the LCD.

$$\cancel{(x-3)}\cdot\frac{x}{\cancel{x-3}} = \cancel{(x-3)}\cdot\frac{3}{\cancel{x-3}} + 9(x-3)$$ Use the distributive property on the right side.

$$x = 3 + 9(x-3)$$ Simplify.

Step 3. Solve the resulting equation.

$x = 3 + 9(x-3)$	This is the equation cleared of fractions.
$x = 3 + 9x - 27$	Use the distributive property on the right side.
$x = 9x - 24$	Combine numerical terms.
$-8x = -24$	Subtract 9x from both sides.
$x = 3$	Divide both sides by -8.

Step 4. Check proposed solutions. The proposed solution, 3, is *not* a solution because of the restriction that $x \neq 3$. Notice that 3 makes both of the denominators zero in the original equation. There is no solution for this equation. The solution set is $\varnothing$, the empty set.

Study Tip

Reject any proposed solution that causes any denominator in a rational equation to equal 0.

CHECK POINT 3 Solve: $\frac{8x}{x+1} = 4 - \frac{8}{x+1}$.

Examples 4 and 5 involve rational equations that become quadratic equations after clearing fractions.

EXAMPLE 4 Solving a Rational Equation

Solve: $\frac{x}{3} + \frac{9}{x} = 4$.

Solution

Step 1. List restrictions on the variable.

$$\frac{x}{3} + \frac{9}{x} = 4$$

This denominator would equal zero if $x = 0$.

The restriction is $x \neq 0$.

Step 2. Multiply both sides by the LCD. The denominators are 3 and x. Thus, the LCD is $3x$.

$$\frac{x}{3} + \frac{9}{x} = 4, \quad x \neq 0 \qquad \text{This is the given equation.}$$

$$3x\left(\frac{x}{3} + \frac{9}{x}\right) = 3x \cdot 4 \qquad \text{Multiply both sides by the LCD.}$$

$$3x \cdot \frac{x}{3} + 3x \cdot \frac{9}{x} = 12x \qquad \text{Use the distributive property on the left side.}$$

$$x^2 + 27 = 12x \qquad \text{Simplify.}$$

Step 3. Solve the resulting equation. Can you see that we have a quadratic equation? Write the equation in standard form and solve for x.

$$x^2 + 27 = 12x \qquad \text{This is the equation cleared of fractions.}$$

$$x^2 - 12x + 27 = 0 \qquad \text{Subtract 12x from both sides.}$$

$$(x - 9)(x - 3) = 0 \qquad \text{Factor.}$$

$$x - 9 = 0 \quad \text{or} \quad x - 3 = 0 \qquad \text{Set each factor equal to 0.}$$

$$x = 9 \qquad\qquad x = 3 \qquad \text{Solve the resulting equations.}$$

Step 4. Check proposed solutions in the original equation. The proposed solutions, 9 and 3, are not part of the restriction that $x \neq 0$. Substitute 9 for x, and then 3 for x, in the given equation. In each case, you should obtain the true statement $4 = 4$. The solutions are 3 and 9, and the solution set is $\{3, 9\}$.

Using Technology

Graphic Connections

The graphs of

$$y_1 = \frac{x}{3} + \frac{9}{x}$$

and

$$y_2 = 4$$

have two intersection points. The x-coordinates of the points are 3 and 9. This verifies that

$$\frac{x}{3} + \frac{9}{x} = 4$$

has both 3 and 9 as solutions.

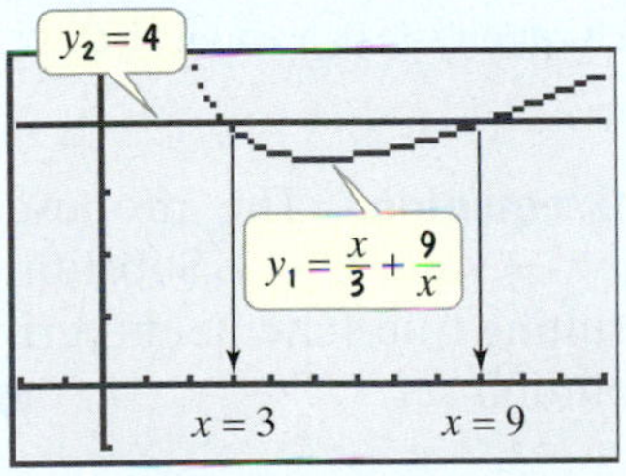

[−2, 12, 1] by [−1, 5, 1]

✓ **CHECK POINT 4** Solve: $\frac{x}{2} + \frac{12}{x} = 5$.

EXAMPLE 5 Solving a Rational Equation

Solve: $\frac{2x}{x-3} + \frac{6}{x+3} = -\frac{28}{x^2 - 9}$.

Solution

Step 1. List restrictions on the variable. By factoring denominators, it makes it easier to see values that make the denominators zero.

$$\frac{2x}{x-3} + \frac{6}{x+3} = -\frac{28}{(x+3)(x-3)}$$

This denominator is zero if $x = 3$.

This denominator is zero if $x = -3$.

This denominator is zero if $x = -3$ or $x = 3$.

The restrictions are $x \neq -3$ and $x \neq 3$.

Step 2. Multiply both sides by the LCD. The LCD is $(x+3)(x-3)$.

$$\frac{2x}{x-3}+\frac{6}{x+3}=-\frac{28}{(x+3)(x-3)}, \quad x \neq -3, x \neq 3$$ This is the given equation with a denominator factored.

$$(x+3)(x-3)\left(\frac{2x}{x-3}+\frac{6}{x+3}\right)=(x+3)(x-3)\left(-\frac{28}{(x+3)(x-3)}\right)$$ Multiply both sides by the LCD.

$$(x+3)\cancel{(x-3)}\cdot\frac{2x}{\cancel{x-3}}+\cancel{(x+3)}(x-3)\cdot\frac{6}{\cancel{x+3}}=\cancel{(x+3)(x-3)}\left(-\frac{28}{\cancel{(x+3)(x-3)}}\right)$$ Use the distributive property on the left side.

$$2x(x+3)+6(x-3)=-28$$ Simplify.

Step 3. Solve the resulting equation.

$2x(x+3)+6(x-3)=-28$	This is the equation cleared of fractions.
$2x^2+6x+6x-18=-28$	Use the distributive property twice on the left side.
$2x^2+12x-18=-28$	Combine like terms.
$2x^2+12x+10=0$	Add 28 to both sides and write the quadratic equation in standard form.
$2(x^2+6x+5)=0$	Factor out the GCF.
$2(x+5)(x+1)=0$	Factor the trinomial.
$x+5=0$ or $x+1=0$	Set each variable factor equal to 0.
$x=-5$ $\quad$ $x=-1$	Solve for x.

Step 4. Check the proposed solutions in the original equation. The proposed solutions, -5 and -1, are not part of the restriction that $x \neq -3$ and $x \neq 3$. Substitute -5 for x, and then -1 for x, in the given equation. The resulting true statements verify that -5 and -1 are the solutions, and $\{-5,-1\}$ is the solution set.

✓ **CHECK POINT 5** Solve: $\frac{3}{x-3}+\frac{5}{x-4}=\frac{x^2-20}{x^2-7x+12}$.

2 Solve problems involving rational functions that model applied situations.

Applications of Rational Equations

Solving rational equations can be used to answer questions about variables contained in rational functions.

EXAMPLE 6 Using a Cost-Benefit Model

The function

$$f(x)=\frac{120x}{100-x}$$

models the cost, $f(x)$, in thousands of dollars, to remove x% of a lake's pollutants. If voters commit \$80 thousand for this project, what percentage of the pollutants can be removed?

Solution Substitute 80, the cost in thousands of dollars, for $f(x)$ and solve the resulting rational equation for x.

$80 = \frac{120x}{100 - x}$	The LCD is 100 − x.
$(100 - x)80 = (100 - x) \cdot \frac{120x}{100 - x}$	Multiply both sides by the LCD.
$80(100 - x) = 120x$	Simplify.
$8000 - 80x = 120x$	Use the distributive property on the left side.
$8000 = 200x$	Add 80x to both sides.
$40 = x$	Divide both sides by 200.

If voters commit \$80 thousand, 40% of the lake's pollutants can be removed.

✓ CHECK POINT 6 Use the cost-benefit model in Example 6 to answer this question: If voters in the city commit \$120 thousand for the project, what percentage of the lake's pollutants can be removed?

6.6 EXERCISE SET

MyMathLab · MathXL PRACTICE · WATCH · DOWNLOAD · READ · REVIEW

Practice Exercises

In Exercises 1–34, solve each rational equation. If an equation has no solution, so state.

1. $\frac{1}{x} + 2 = \frac{3}{x}$
2. $\frac{1}{x} - 3 = \frac{4}{x}$
3. $\frac{5}{x} + \frac{1}{3} = \frac{6}{x}$
4. $\frac{4}{x} + \frac{1}{2} = \frac{5}{x}$
5. $\frac{x-2}{2x} + 1 = \frac{x+1}{x}$
6. $\frac{7x-4}{5x} = \frac{9}{5} - \frac{4}{x}$
7. $\frac{3}{x+1} = \frac{5}{x-1}$
8. $\frac{6}{x+3} = \frac{4}{x-3}$
9. $\frac{x-6}{x+5} = \frac{x-3}{x+1}$
10. $\frac{x+2}{x+10} = \frac{x-3}{x+4}$
11. $\frac{x+6}{x+3} = \frac{3}{x+3} + 2$
12. $\frac{3x+1}{x-4} = \frac{6x+5}{2x-7}$
13. $1 - \frac{4}{x+7} = \frac{5}{x+7}$
14. $5 - \frac{2}{x-5} = \frac{3}{x-5}$
15. $\frac{4x}{x+2} + \frac{2}{x-1} = 4$
16. $\frac{3x}{x+1} + \frac{4}{x-2} = 3$
17. $\frac{8}{x^2-9} + \frac{4}{x+3} = \frac{2}{x-3}$
18. $\frac{32}{x^2-25} = \frac{4}{x+5} + \frac{2}{x-5}$
19. $x + \frac{7}{x} = -8$
20. $x + \frac{6}{x} = -7$
21. $\frac{6}{x} - \frac{x}{3} = 1$
22. $\frac{x}{2} - \frac{12}{x} = 1$
23. $\frac{x+6}{3x-12} = \frac{5}{x-4} + \frac{2}{3}$
24. $\frac{1}{5x+5} = \frac{3}{x+1} - \frac{7}{5}$
25. $\frac{1}{x-1} + \frac{1}{x+1} = \frac{2}{x^2-1}$
26. $\frac{1}{x-2} + \frac{1}{x+2} = \frac{4}{x^2-4}$
27. $\frac{5}{x+4} + \frac{3}{x+3} = \frac{12x+19}{x^2+7x+12}$
28. $\frac{2x-1}{x^2+2x-8} + \frac{2}{x+4} = \frac{1}{x-2}$

29. $\frac{4x}{x+3} - \frac{12}{x-3} = \frac{4x^2+36}{x^2-9}$

30. $\frac{2}{x+3} - \frac{5}{x+1} = \frac{3x+5}{x^2+4x+3}$

31. $\frac{4}{x^2+3x-10} + \frac{1}{x^2+9x+20} = \frac{2}{x^2+2x-8}$

32. $\frac{4}{x^2+3x-10} - \frac{1}{x^2+x-6} = \frac{3}{x^2-x-12}$

33. $\frac{3y}{y^2+5y+6} + \frac{2}{y^2+y-2} = \frac{5y}{y^2+2y-3}$

34. $\frac{y-1}{y^2-4} + \frac{y}{y^2-y-2} = \frac{2y-1}{y^2+3y+2}$

In Exercises 35–38, a rational function g is given. Find all values of a for which g(a) is the indicated value.

35. $g(x) = \frac{x}{2} + \frac{20}{x}; g(a) = 7$

36. $g(x) = \frac{x}{4} + \frac{5}{x}; g(a) = 3$

37. $g(x) = \frac{5}{x+2} + \frac{25}{x^2+4x+4}; g(a) = 20$

38. $g(x) = \frac{3x-2}{x+1} + \frac{x+2}{x-1}; g(a) = 4$

Practice PLUS

In Exercises 39–46, solve or simplify, whichever is appropriate.

39. $\frac{x+2}{x^2-x} - \frac{6}{x^2-1}$

40. $\frac{x+3}{x^2-x} - \frac{8}{x^2-1}$

41. $\frac{x+2}{x^2-x} - \frac{6}{x^2-1} = 0$

42. $\frac{x+3}{x^2-x} - \frac{8}{x^2-1} = 0$

43. $\frac{1}{x^3-8} + \frac{3}{(x-2)(x^2+2x+4)} = \frac{2}{x^2+2x+4}$

44. $\frac{2}{x^3-1} + \frac{4}{(x-1)(x^2+x+1)} = -\frac{1}{x^2+x+1}$

45. $\frac{1}{x^3-8} + \frac{3}{(x-2)(x^2+2x+4)} - \frac{2}{x^2+2x+4}$

46. $\frac{2}{x^3-1} + \frac{4}{(x-1)(x^2+x+1)} + \frac{1}{x^2+x+1}$

In Exercises 47–48, find all values of a for which $f(a) = g(a) + 1$.

47. $f(x) = \frac{x+2}{x+3}, g(x) = \frac{x+1}{x^2+2x-3}$

48. $f(x) = \frac{4}{x-3}, g(x) = \frac{10}{x^2+x-12}$

In Exercises 49–50, find all values of a for which $(f+g)(a) = h(a)$.

49. $f(x) = \frac{5}{x-4}, g(x) = \frac{3}{x-3}, h(x) = \frac{x^2-20}{x^2-7x+12}$

50. $f(x) = \frac{6}{x+3}, g(x) = \frac{2x}{x-3}, h(x) = -\frac{28}{x^2-9}$

Application Exercises

The function

$$f(x) = \frac{250x}{100-x}$$

models the cost, f(x), in millions of dollars, to remove x% of a river's pollutants. Use this function to solve Exercises 51–52.

51. If the government commits $375 million for this project, what percentage of the pollutants can be removed?

52. If the government commits $750 million for this project, what percentage of the pollutants can be removed?

In an experiment about memory, students in a language class are asked to memorize 40 vocabulary words in Latin, a language with which they are not familiar. After studying the words for one day, students are tested each day thereafter to see how many words they remember. The class average is then found. The function

$$f(x) = \frac{5x+30}{x}$$

models the average number of Latin words remembered by the students, f(x), after x days. The graph of the rational function is shown. Use the function to solve Exercises 53–56.

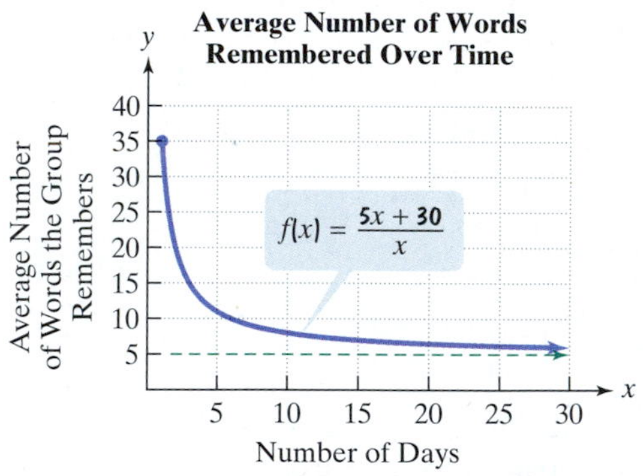

53. After how many days do the students remember 8 words? Identify your solution as a point on the graph.

54. After how many days do the students remember 7 words? Identify your solution as a point on the graph.

55. What is the horizontal asymptote of the graph? Describe what this means about the average number of Latin words remembered by the students over an extended period of time.

56. According to the graph, between which two days do students forget the most? Describe the trend shown by the memory function's graph.

Rational functions can be used to model learning. Many of these functions model the proportion of correct responses as a function of the number of trials of a particular task. One such model, called a learning curve, is

$$f(x) = \frac{0.9x - 0.4}{0.9x + 0.1},$$

where $f(x)$ is the proportion of correct responses after x trials. If $f(x) = 0$, there are no correct responses. If $f(x) = 1$, all responses are correct. The graph of the rational function is shown. Use the function to solve Exercises 57–60.

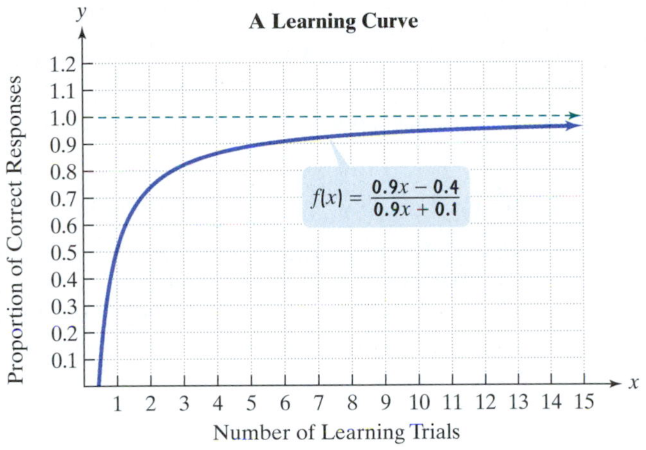

57. How many learning trials are necessary for 0.95 of the responses to be correct? Identify your solution as a point on the graph.

58. How many learning trials are necessary for 0.5 of the responses to be correct? Identify your solution as a point on the graph.

59. Describe the trend shown by the graph in terms of learning new tasks. What happens initially and what happens as time increases?

60. What is the horizontal asymptote of the graph? Once the performance level approaches peak efficiency, what effect does additional practice have on performance? Describe how this is shown by the graph.

61. A company wants to increase the 10% peroxide content of its product by adding pure peroxide (100% peroxide). If x liters of pure peroxide are added to 500 liters of its 10% solution, the concentration, $C(x)$, of the new mixture is given by

$$C(x) = \frac{x + 0.1(500)}{x + 500}.$$

How many liters of pure peroxide should be added to produce a new product that is 28% peroxide?

62. Suppose that x liters of pure acid are added to 200 liters of a 35% acid solution.

a. Write a function that gives the concentration, $C(x)$, of the new mixture. (*Hint*: See Exercise 61.)

b. How many liters of pure acid should be added to produce a new mixture that is 74% acid?

63. Bringing Up the Other Baby If you thought bringing up a child was expensive, take a look at pet expenditures. The bar graph shows retail sales, in billions of dollars, of nonfood pet supplies and nonfood baby-care supplies from 2001 through 2005.

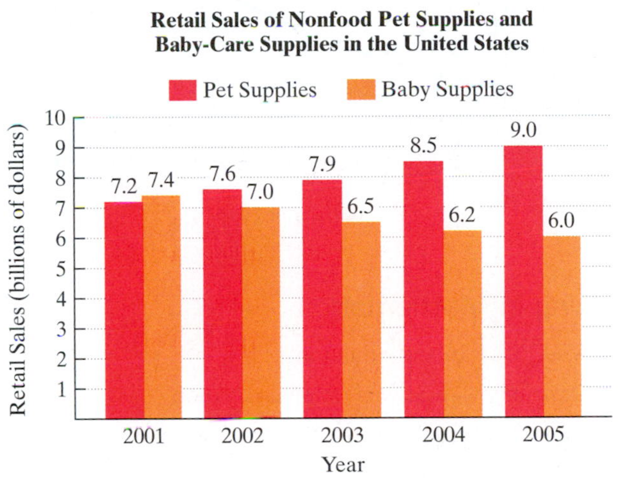

Source: Packaged Facts

a. In 2001, \$7.2 billion was spent on pet supplies and this has been increasing by approximately \$0.45 billion per year. Write a linear function that models the amount spent on pet supplies, $f(x)$, in billions of dollars, x years after 2001.

b. In 2001, \$7.4 billion was spent on baby supplies and this has been decreasing by approximately \$0.38 billion per year. Write a linear function that models the amount spent on baby supplies, $g(x)$, in billions of dollars, x years after 2001.

c. Use the functions that you wrote in parts (a) and (b) to write a rational function that models the ratio of the amount spent on pet supplies to the amount spent on baby supplies, $h(x)$, x years after 2001.

d. Use the rational function from part (c) to determine how many times more was spent on pet supplies than on baby supplies in 2005. Round to one decimal place.

e. Use the data shown in the bar graph to determine how many times more was spent on pet supplies than on baby supplies in 2005. Round to one decimal place. How well does the function in part (d) model this number? (*This exercise continues on the next page.*)

f. Use the rational function h from part (c) to determine in which year we will spend 3.25 times as much on pet supplies than on baby supplies.

g. The graph of the rational function h from part (c) is shown. Identify your solution from part (f) as a point on the graph.

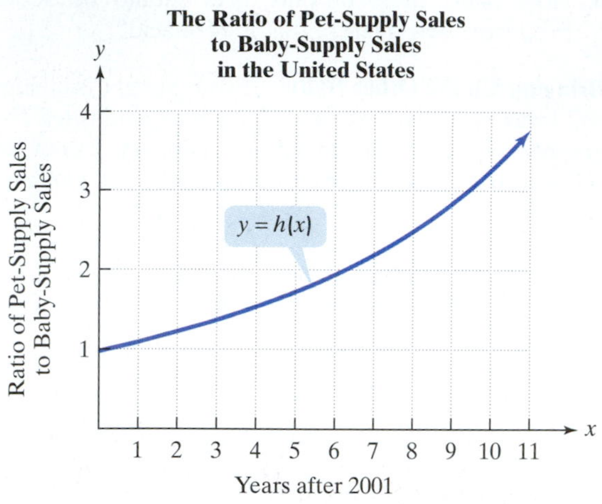

Writing in Mathematics

64. What is a rational equation?

65. Explain how to solve a rational equation.

66. Explain how to find restrictions on the variable in a rational equation.

67. Why should restrictions on the variable in a rational equation be listed before you begin solving the equation?

68. Describe similarities and differences between the procedures needed to solve the following problems:

$$\text{Add: } \frac{2}{x} + \frac{3}{4}. \qquad \text{Solve: } \frac{2}{x} + \frac{3}{4} = 1.$$

69. Rational functions model learning and forgetting. Use the graphs in Exercises 53–56 and in Exercises 57–60 to describe one similarity and one difference between learning and forgetting over time.

70. Does the graph of the learning curve shown in Exercises 57–60 indicate that practice makes perfect? Explain. Does this have anything to do with what psychologists call "the curse of perfection"?

Technology Exercises

In Exercises 71–75, use a graphing utility to solve each rational equation. Graph each side of the equation in the given viewing rectangle. The solution is the first coordinate of the point(s) of intersection. Check by direct substitution.

71. $\frac{x}{2} + \frac{x}{4} = 6$; $[-5, 10, 1]$ by $[-5, 10, 1]$

72. $\frac{50}{x} = 2x$; $[-10, 10, 1]$ by $[-20, 20, 2]$

73. $x + \frac{6}{x} = -5$; $[-10, 10, 1]$ by $[-7, 10, 1]$

74. $\frac{2}{x} = x + 1$; $[-5, 5, 1]$ by $[-5, 5, 1]$

75. $\frac{3}{x} - \frac{x + 21}{3x} = \frac{5}{3}$; $[-5, 5, 1]$ by $[-5, 5, 1]$

Critical Thinking Exercises

Make Sense? *In Exercises 76–79, determine whether each statement "makes sense" or "does not make sense" and explain your reasoning.*

76. I must have made an error if a rational equation produces no solution.

77. I added two rational expressions and found the solution set.

78. I can solve the equation $\frac{40}{x} = \frac{15}{x - 20}$ by multiplying both sides by the LCD or by setting the cross product of 40 and $x - 20$ equal to the cross product of 15 and x.

79. I'm solving a rational equation that became a quadratic equation, so my rational equation will have two solutions.

In Exercises 80–83, determine whether each statement is true or false. If the statement is false, make the necessary change(s) to produce a true statement.

80. Once a rational equation is cleared of fractions, all solutions of the resulting equation are also solutions of the rational equation.

81. We find

$$\frac{4}{x} - \frac{2}{x + 1}$$

by multiplying each term by the LCD, $x(x + 1)$, thereby clearing fractions.

82. All real numbers satisfy the equation $\frac{7}{x} - \frac{2}{x} = \frac{5}{x}$.

83. In order to find a number to add to the numerator and denominator of $\frac{3}{16}$ to result in $\frac{1}{2}$, we could solve the following rational equation:

$$\frac{3 + x}{16 + x} = \frac{1}{2}.$$

84. Solve: $\left(\frac{1}{x + 1} + \frac{x}{1 - x}\right) \div \left(\frac{x}{x + 1} - \frac{1}{x - 1}\right) = -1.$

85. Solve: $\left|\frac{x + 1}{x + 8}\right| = \frac{2}{3}.$

86. Write an original rational equation that has no solution.

87. Solve

$$\left(\frac{4}{x - 1}\right)^2 + 2\left(\frac{4}{x - 1}\right) + 1 = 0$$

by introducing the substitution $u = \frac{4}{x - 1}$.

Review Exercises

88. Graph the solution set:

$$x + 2y \geq 2$$
$$x - y \geq -4.$$

(Section 4.4, Example 5)

89. Solve:

$$\frac{x-4}{2} - \frac{1}{5} = \frac{7x+1}{20}.$$

(Section 1.4, Example 4)

90. Solve for F:

$$C = \frac{5F - 160}{9}.$$

(Section 1.5, Example 8)

Preview Exercises

Exercises 91–93 will help you prepare for the material covered in the next section.

91. Solve for p: $qf + pf = pq$.

92. Solve: $\frac{40}{x} + \frac{40}{x+30} = 2.$

93. A plane flies at an average rate of 450 miles per hour. It can travel 980 miles with the wind in the same amount of time as it travels 820 miles against the wind. Solve the equation

$$\frac{980}{450+x} = \frac{820}{450-x}$$

to find the average rate of the wind, x, in miles per hour.

SECTION 6.7 Formulas and Applications of Rational Equations

Objectives

1. Solve a formula with a rational expression for a variable.
2. Solve business problems involving average cost.
3. Solve problems involving time in motion.
4. Solve problems involving work.

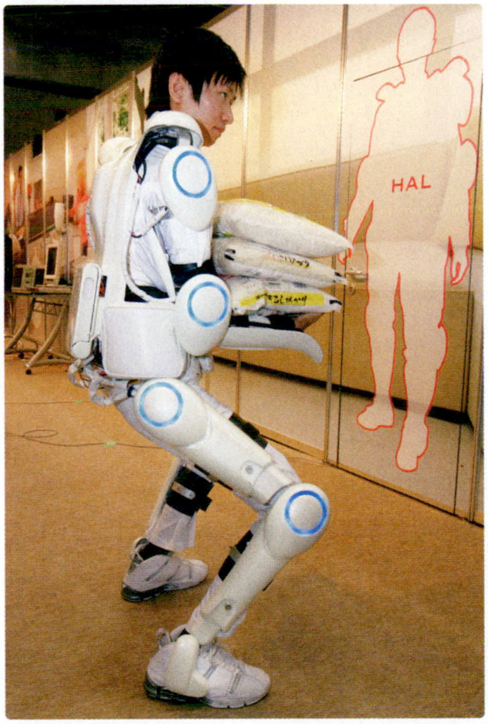

Your grandmother appears to be slowing down. Enter . . . Mecha–Grandma! Japanese researchers have developed the robotic exoskeleton shown here to help the elderly and disabled walk and even lift heavy objects like the three 22-pound bags of rice in the photo. It's called the Hybrid Assistive Limb, or HAL. (The inventor has obviously never seen *2001: A Space Odyssey*.) HAL's brain is a computer housed in a back-pack that learns to mimic the wearer's gait and posture. Bioelectric sensors pick up signals transmitted from the brain to the muscles, so it can anticipate movements the moment the wearer thinks of them. A commercial version is available at a hefty cost ranging between \$14,000 and \$20,000. (*Source*: sanlab.kz.tsukuba.ac.jp)

The cost of manufacturing robotic exoskeletons can be modeled by rational functions. In this section, you will see that high production levels of HAL can eventually make this amazing invention more affordable for the elderly and people with disabilities.

1 Solve a formula with a rational expression for a variable.

Solving a Formula for a Variable

Formulas and mathematical models frequently contain rational expressions. We solve for a specified variable using the procedure for solving rational equations. The goal is to get the specified variable alone on one side of the equation. To do so, collect all terms with this variable on one side and all other terms on the other side. It is sometimes necessary to factor out the variable you are solving for.

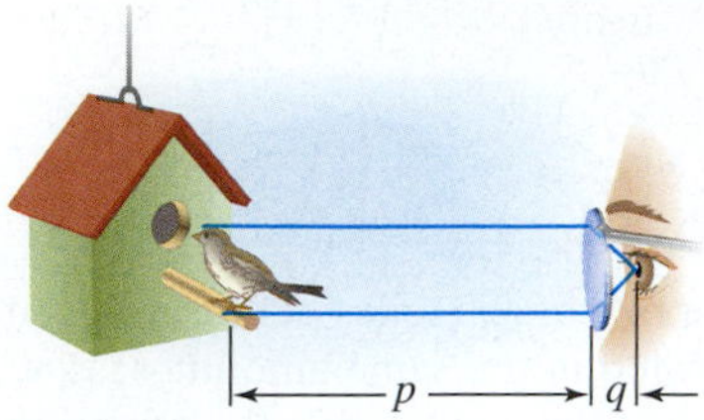

FIGURE 6.5

EXAMPLE 1 Solving for a Variable in a Formula

If you wear glasses, did you know that each lens has a measurement called its focal length, f? When an object is in focus, its distance from the lens, p, and the distance from the lens to your retina, q, satisfy the formula

$$\frac{1}{p} + \frac{1}{q} = \frac{1}{f}.$$

(See **Figure 6.5**.) Solve this formula for p.

Solution Our goal is to isolate the variable p. We begin by multiplying both sides by the least common denominator, pqf, to clear the equation of fractions.

$$\frac{1}{p} + \frac{1}{q} = \frac{1}{f}$$ This is the given formula.

$$pqf\left(\frac{1}{p} + \frac{1}{q}\right) = pqf\left(\frac{1}{f}\right)$$ Multiply both sides by pqf, the LCD.

$$\cancel{p}qf\left(\frac{1}{\cancel{p}}\right) + p\cancel{q}f\left(\frac{1}{\cancel{q}}\right) = pq\cancel{f}\left(\frac{1}{\cancel{f}}\right)$$ Use the distributive property on the left side.

$$qf + pf = pq$$ Simplify.

Observe that the formula is now cleared of fractions. Collect terms with p, the specified variable, on one side of the equation. To do so, subtract pf from both sides.

$$qf + pf = pq$$ This is the equation cleared of fractions.

$$qf = pq - pf$$ Subtract pf from both sides.

$$qf = p(q - f)$$ Factor out p, the specified variable.

$$\frac{qf}{q - f} = \frac{p(q - f)}{q - f}$$ Divide both sides by $q - f$ and solve for p.

$$\frac{qf}{q - f} = p$$ Simplify.

✓ CHECK POINT 1 Solve $\frac{1}{x} + \frac{1}{y} = \frac{1}{z}$ for x.

---

2 Solve business problems involving average cost.

Business Problems Involving Average Cost

We have seen that the cost function for a business is the sum of its fixed and variable costs:

$$C(x) = (\text{fixed cost}) + cx$$

Cost per unit times the number of units produced, x

The **average cost** per unit for a company to produce x units is the sum of its fixed and variable costs divided by the number of units produced. The **average cost function** is a rational function that is denoted by $\overline{C}$. Thus,

Cost of producing x units: fixed plus variable costs

$$\overline{C}(x) = \frac{(\text{fixed cost}) + cx}{x}.$$

Number of units produced

EXAMPLE 2 Average Cost for a Business

We return to the robotic exoskeleton described in the section opener. Suppose a company that manufactures this invention has a fixed monthly cost of \$1,000,000 and that it costs \$5000 to produce each robotic system.

a. Write the cost function, C, of producing x robotic systems.

b. Write the average cost function, $\overline{C}$, of producing x robotic systems.

c. How many robotic systems must be produced each month for the company to reach an average cost of \$5500 per system?

Solution

a. The cost function, C, is the sum of the fixed cost and and the variable costs.

$$C(x) = 1{,}000{,}000 + 5000x$$

Fixed cost is \$1,000,000.

Variable cost: \$5000 for each robotic system produced

b. The average cost function, $\overline{C}$, is the sum of fixed and variable costs divided by the number of virtual reality systems produced.

$$\overline{C}(x) = \frac{1{,}000{,}000 + 5000x}{x}$$

c. We are interested in the company's production level that results in an average cost of \$5500 per robotic system. Substitute 5500, the average cost, for $\overline{C}(x)$ and solve the resulting rational equation for x.

$$5500 = \frac{1{,}000{,}000 + 5000x}{x} \quad \text{Substitute 5500 for } \overline{C}(x).$$

$$5500x = 1{,}000{,}000 + 5000x \quad \text{Multiply both sides by the LCD, } x.$$

$$500x = 1{,}000{,}000 \quad \text{Subtract } 5000x \text{ from both sides.}$$

$$x = 2000 \quad \text{Divide both sides by 500.}$$

The company must produce 2000 robotic exoskeletons each month for an average cost of \$5500 per robotic system.

Figure 6.6 shows the graph of the average cost function in Example 2. As the production level increases, the average cost of producing each robotic exoskeleton decreases. The horizontal asymptote, $y = 5000$, is also shown in the figure. This means that the more robotic systems produced each month, the closer the average cost per system for the company comes to \$5000. The least possible cost per robotic exoskeleton is approaching \$5000. Competitively low prices take place with high production levels, posing a major problem for small businesses.

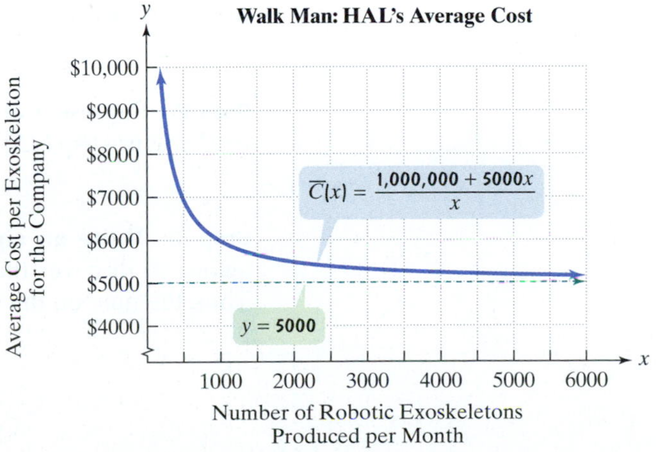

FIGURE 6.6

✓ **CHECK POINT 2** A company is planning to manufacture wheelchairs that are light, fast, and beautiful. Fixed monthly cost will be \$500,000 and it will cost \$400 to produce each radically innovative chair.

a. Write the cost function, C, of producing x wheelchairs.

b. Write the average cost function, $\overline{C}$, of producing x wheelchairs.

c. How many wheelchairs must be produced each month for the company to reach an average cost of \$450 per chair?

---

3 Solve problems involving time in motion.

Problems Involving Motion

We have seen that the distance, d, covered by any moving body is the product of its average rate, r, and its time in motion, t: $d = rt$. Rational expressions appear in motion problems when the conditions of the problem involve the time traveled. We can obtain an expression for t, the time traveled, by dividing both sides of $d = rt$ by r.

$$d = rt \quad \text{Distance equals rate times time.}$$

$$\frac{d}{r} = \frac{rt}{r} \quad \text{Divide both sides by } r.$$

$$\frac{d}{r} = t \quad \text{Simplify.}$$

Time in Motion

$$t = \frac{d}{r}$$

$$\text{Time traveled} = \frac{\text{Distance traveled}}{\text{Rate of travel}}$$

EXAMPLE 3 A Motion Problem Involving Time

You commute to work a distance of 40 miles and return on the same route at the end of the day. Your average rate on the return trip is 30 miles per hour faster than your average rate on the outgoing trip. If the round trip takes 2 hours, what is your average rate on the outgoing trip to work?

Solution

Step 1. Let x represent one of the quantities. Let

$$x = \text{the rate on the outgoing trip.}$$

Step 2. Represent other unknown quantities in terms of x. Because the average rate on the return trip is 30 miles per hour faster than the average rate on the outgoing trip, let

$$x + 30 = \text{the rate on the return trip.}$$

Step 3. Write an equation that models the conditions. By reading the problem again, we discover that the crucial idea is that the time for the round trip is 2 hours. Thus, the time on the outgoing trip plus the time on the return trip is 2 hours.

	Distance	Rate	Time = $\frac{\text{Distance}}{\text{Rate}}$
Outgoing Trip	40	x	$\frac{40}{x}$
Return Trip	40	$x + 30$	$\frac{40}{x+30}$

The sum of these times is 2 hours.

We are now ready to write an equation that models the problem's conditions.

Time on the outgoing trip | plus | time on the return trip | equals | 2 hours.

$$\frac{40}{x} + \frac{40}{x+30} = 2$$

Step 4. Solve the equation and answer the question.

$$\frac{40}{x} + \frac{40}{x+30} = 2 \quad \text{This is the equation that models the problem's conditions.}$$

$$x(x+30)\left(\frac{40}{x} + \frac{40}{x+30}\right) = 2x(x+30) \quad \text{Multiply both sides by the LCD, } x(x+30).$$

$$x(x+30)\cdot\frac{40}{x} + x(x+30)\cdot\frac{40}{x+30} = 2x^2 + 60x \quad \text{Use the distributive property on each side.}$$

$$40(x+30) + 40x = 2x^2 + 60x \quad \text{Simplify.}$$

$$40x + 1200 + 40x = 2x^2 + 60x \quad \text{Use the distributive property.}$$

$$80x + 1200 = 2x^2 + 60x \quad \text{Combine like terms: } 40x + 40x = 80x.$$

$$0 = 2x^2 - 20x - 1200 \quad \text{Subtract } 80x + 1200 \text{ from both sides.}$$

$$0 = 2(x^2 - 10x - 600) \quad \text{Factor out the GCF.}$$

$$0 = 2(x-30)(x+20) \quad \text{Factor completely.}$$

$$x - 30 = 0 \quad \text{or} \quad x + 20 = 0 \quad \text{Set each variable factor equal to 0.}$$

$$x = 30 \qquad x = -20 \quad \text{Solve for x.}$$

Because x represents the rate on the outgoing trip, we reject the negative value, -20. The rate on the outgoing trip is 30 miles per hour. At an outgoing rate of 30 miles per hour, the round trip should take 2 hours.

Step 5. Check the proposed solution in the original wording of the problem. Does the round trip take 2 hours? Because the rate on the return trip is 30 miles per hour faster than the rate on the outgoing trip, the rate on the return trip is $30 + 30$, or 60 miles per hour.

$$\text{Time on the outgoing trip} = \frac{\text{Distance}}{\text{Rate}} = \frac{40}{30} = \frac{4}{3} \text{ hours}$$

$$\text{Time on the return trip} = \frac{\text{Distance}}{\text{Rate}} = \frac{40}{60} = \frac{2}{3} \text{ hour}$$

The total time for the round trip is $\frac{4}{3} + \frac{2}{3} = \frac{6}{3}$, or 2 hours. This checks with the original conditions of the problem.

Using Technology

Graphic Connections

The graph of the rational function for the time for the round trip,

$$f(x) = \frac{40}{x} + \frac{40}{x+30},$$

is shown in a [0, 60, 30] by [0, 10, 1] viewing rectangle.

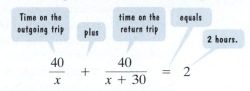

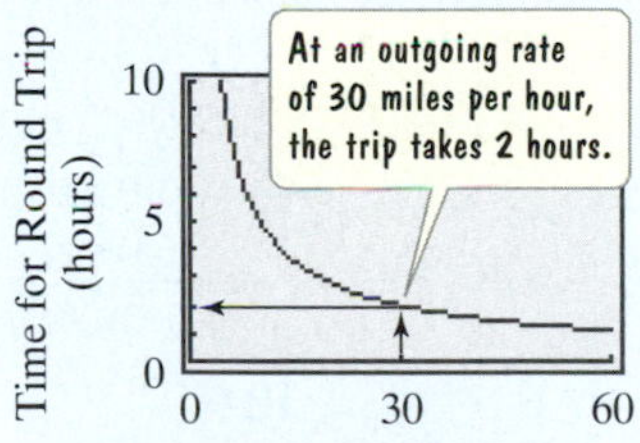

The graph is falling from left to right. This shows that the time for the round trip decreases as the rate increases.

✓ CHECK POINT 3 After riding at a steady speed for 40 miles, a bicyclist had a flat tire and walked 5 miles to a repair shop. The cycling rate was 4 times faster than the walking rate. If the time spent cycling and walking was 5 hours, at what rate was the cyclist riding?

4 Solve problems involving work.

Problems Involving Work

You are thinking of designing your own Web site. You estimate that it will take 15 hours to do the job. In 1 hour, $\frac{1}{15}$ of the job is completed. In 2 hours, $\frac{2}{15}$ of the job is completed. In 3 hours, the fractional part of the job done is $\frac{3}{15}$, or $\frac{1}{5}$. In x hours, the fractional part of the job that you can complete is $\frac{x}{15}$.

Your friend, who has experience developing Web sites, took 10 hours working on her own to design an impressive site. You wonder about the possibility of working together. How long would it take both of you to design your Web site?

Problems involving work usually have two (or more) people or machines working together to complete a job. The amount of time it takes each person to do the job working alone is frequently known. The question deals with how long it will take both people working together to complete the job.

In work problems, **the number 1 represents one whole job completed**. For example, the completion of your Web site is represented by 1. Equations in work problems are often based on the following condition:

Fractional part of the job done by the first person + fractional part of the job done by the second person = 1 (one whole job completed).

EXAMPLE 4 Solving a Problem Involving Work

You can design a Web site in 15 hours. Your friend can design the same site in 10 hours. How long will it take to design the Web site if you both work together?

Solution

Step 1. Let x represent one of the quantities. Let

$x =$ the time, in hours, for you and your friend working together to design the Web site.

Step 2. Represent other unknown quantities in terms of x. Because there are no other unknown quantities, we can skip this step.

Step 3. Write an equation that models the conditions. We construct a table to help find the fractional parts of the task completed by you and your friend in x hours.

	Fractional part of job completed in 1 hour	Time working together	Fractional part of job completed in x hours
You	$\frac{1}{15}$	x	$\frac{x}{15}$
Your friend	$\frac{1}{10}$	x	$\frac{x}{10}$

You can design the site in 15 hours.

Your friend can design the site in 10 hours.

Fractional part of the job done by you + fractional part of the job done by your friend = one whole job.

$$\frac{x}{15} + \frac{x}{10} = 1$$

Step 4. Solve the equation and answer the question.

$$\frac{x}{15} + \frac{x}{10} = 1$$ This is the equation that models the problem's conditions.

$$30\left(\frac{x}{15} + \frac{x}{10}\right) = 30 \cdot 1$$ Multiply both sides by 30, the LCD.

$$30 \cdot \frac{x}{15} + 30 \cdot \frac{x}{10} = 30$$ Use the distributive property on the left side.

$$2x + 3x = 30$$ Simplify: $\frac{\overset{2}{\cancel{30}}}{1} \cdot \frac{x}{\underset{1}{\cancel{15}}} = 2x$ and $\frac{\overset{3}{\cancel{30}}}{1} \cdot \frac{x}{\underset{1}{\cancel{10}}} = 3x$.

$$5x = 30$$ Combine like terms.

$$x = 6$$ Divide both sides by 5.

If you both work together, you can design your Web site in 6 hours.

Step 5. Check the proposed solution in the original wording of the problem. Will you both complete the job in 6 hours? Because you can design the site in 15 hours, in 6 hours, you can complete $\frac{6}{15}$, or $\frac{2}{5}$, of the job. Because your friend can design the site in 10 hours, in 6 hours, she can complete $\frac{6}{10}$, or $\frac{3}{5}$, of the job. Notice that $\frac{2}{5} + \frac{3}{5} = 1$, which represents the completion of the entire job, or one whole job.

Study Tip

Let

a = the time it takes person A to do a job working alone, and
b = the time it takes person B to do the same job working alone.

If x represents the time it takes for A and B to complete the entire job working together, then the situation can be modeled by the rational equation

$$\frac{x}{a} + \frac{x}{b} = 1.$$

✓ CHECK POINT 4 A new underwater tunnel is being built using tunnel-boring machines that begin at opposite ends of the tunnel. One tunnel-boring machine can complete the tunnel in 18 months. A faster machine can tunnel to the other side in 9 months. If both machines start at opposite ends and work at the same time, in how many months will the tunnel be finished?

EXAMPLE 5 Solving a Problem Involving Work

After designing your Web site, you and your friend decide to go into business setting up sites for others. With lots of practice, you can now work together and design a modest site in 4 hours. Your friend is still a faster worker. Working alone, you require 6 more hours than she does to design a site for a client. How many hours does it take your friend to design a Web site if she works alone?

Solution

Step 1. Let x represent one of the quantities. Let

x = the time, in hours, for your friend to design a Web site working alone.

Step 2. Represent other unknown quantities in terms of x. Because you require 6 more hours than your friend to do the job, let

$x + 6$ = the time, in hours, for you to design a Web site working alone.

Step 3. Write an equation that models the conditions. Working together, you and your friend can complete the job in 4 hours. We construct a table to find the fractional part of the task completed by you and your friend in 4 hours.

		Fractional part of job completed in 1 hour	Time working together	Fractional part of job completed in 4 hours
Your friend can design the site in x hours.	Your friend	$\frac{1}{x}$	4	$\frac{4}{x}$
You can design the site in $x + 6$ hours.	You	$\frac{1}{x+6}$	4	$\frac{4}{x+6}$

Because you can both complete the job in 4 hours,

Fractional part of the job done by your friend + fractional part of the job done by you = one whole job.

$$\frac{4}{x} + \frac{4}{x+6} = 1.$$

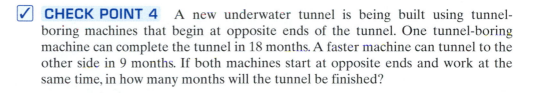

Step 4. Solve the equation and answer the question.

$$\frac{4}{x}+\frac{4}{x+6}=1$$ This is the equation that models the problem's conditions.

$$x(x+6)\left(\frac{4}{x}+\frac{4}{x+6}\right)=x(x+6)\cdot 1$$ Multiply both sides by $x(x+6)$, the LCD.

$$x(x+6)\cdot\frac{4}{x}+x(x+6)\cdot\frac{4}{(x+6)}=x^2+6x$$ Use the distributive property on each side.

$$4(x+6)+4x=x^2+6x$$ Simplify.

$$4x+24+4x=x^2+6x$$ Use the distributive property.

$$8x+24=x^2+6x$$ Combine like terms: $4x+4x=8x$.

$$0=x^2-2x-24$$ Subtract $8x+24$ from both sides and write the quadratic equation in standard form.

$$0=(x-6)(x+4)$$ Factor.

$$x-6=0 \quad \text{or} \quad x+4=0$$ Set each factor equal to 0.

$$x=6 \qquad\qquad x=-4$$ Solve for x.

Because x represents the time for your friend to design a Web site working alone, we reject the negative value, -4. Your friend can design a Web site in 6 hours.

Step 5. Check the proposed solution in the original wording of the problem. Will you both complete the job in 4 hours? Working alone, your friend takes 6 hours. Because you require 6 more hours than your friend, you require 12 hours to complete the job on your own.

In 4 hours, your friend completes $\frac{4}{6}=\frac{2}{3}$ of the job.

In 4 hours, you complete $\frac{4}{12}=\frac{1}{3}$ of the job.

Notice that $\frac{2}{3}+\frac{1}{3}=1$, which represents the completion of the entire job, or the design of one Web site.

✓ **CHECK POINT 5** An experienced carpenter can panel a room 3 times faster than an apprentice can. Working together, they can panel the room in 6 hours. How long would it take each person working alone to do the job?

6.7 EXERCISE SET

MyMathLab | Math XL PRACTICE | WATCH | DOWNLOAD | READ | REVIEW

Practice Exercises

In Exercises 1–14, solve each formula for the specified variable.

1. $\frac{V_1}{V_2}=\frac{P_2}{P_1}$ for P_1 (chemistry)

2. $\frac{V_1}{V_2}=\frac{P_2}{P_1}$ for V_2 (chemistry)

3. $\frac{1}{p}+\frac{1}{q}=\frac{1}{f}$ for f (optics)

4. $\frac{1}{p}+\frac{1}{q}=\frac{1}{f}$ for q (optics)

5. $P=\frac{A}{1+r}$ for r (investment)

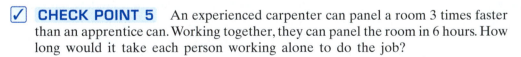

6. $S = \dfrac{a}{1-r}$ for r (mathematics)

7. $F = \dfrac{Gm_1m_2}{d^2}$ for m_1 (physics)

8. $F = \dfrac{Gm_1m_2}{d^2}$ for m_2 (physics)

9. $z = \dfrac{x - \bar{x}}{s}$ for x (statistics)

10. $z = \dfrac{x - \bar{x}}{s}$ for s (statistics)

11. $I = \dfrac{E}{R + r}$ for R (electronics)

12. $I = \dfrac{E}{R + r}$ for r (electronics)

13. $f = \dfrac{f_1f_2}{f_1 + f_2}$ for f_1 (optics)

14. $f = \dfrac{f_1f_2}{f_1 + f_2}$ for f_2 (optics)

Application Exercises

The figure shows the graph of the average cost function for the company described in Check Point 2 that manufactures wheelchairs. Use the graph to solve Exercises 15–18.

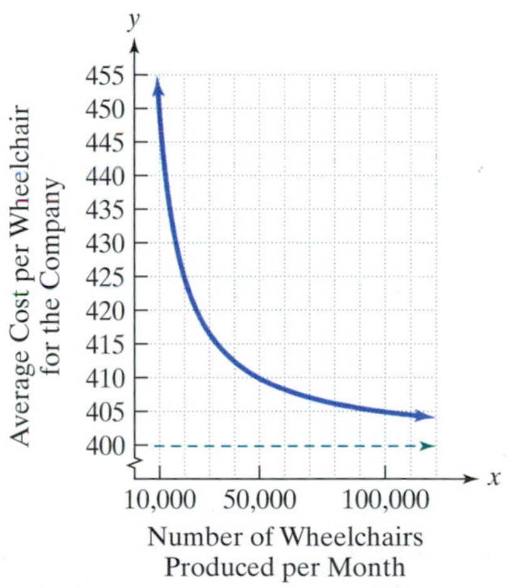

15. How many wheelchairs must be produced each month for the company to have an average cost of $410 per chair?

16. How many wheelchairs must be produced each month for the company to have an average cost of $425 per chair?

17. What is the equation of the horizontal asymptote shown by the dashed green line? What is the meaning of the horizontal asymptote as production level increases?

18. Describe the end behavior of the graph at the far right. Is there a production level that results in an average cost of $400 per chair? Explain your answer.

19. A company is planning to manufacture mountain bikes. Fixed monthly cost will be $100,000 and it will cost $100 to produce each bicycle.

a. Write the cost function, C, of producing x mountain bikes.

b. Write the average cost function, $\overline{C}$, of producing x mountain bikes.

c. How many mountain bikes must be produced each month for the company to have an average cost of $300 per bike?

20. A company is planning to manufacture small canoes. Fixed monthly cost will be $20,000 and it will cost $20 to produce each canoe.

a. Write the cost function, C, of producing x canoes.

b. Write the average cost function, $\overline{C}$, of producing x canoes.

c. How many canoes must be produced each month for the company to have an average cost of $40 per canoe?

It's vacation time. You drive 90 miles along a scenic highway and then take a 5-mile run along a hiking trail. Your driving rate is nine times that of your running rate. The graph shows the total time you spend driving and running, $f(x)$, as a function of your running rate, x. Use the graph of this rational function to solve Exercises 21–24.

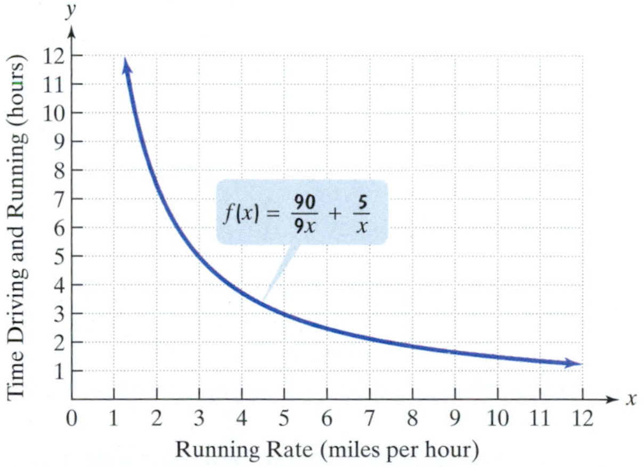

21. If the total time for driving and running is 3 hours, what is your running rate?

22. If the total time for driving and running is 5 hours, what is your running rate?

23. Describe the behavior of the graph as x approaches 0. What does this show about the time driving and running as your running rate is close to zero miles per hour (a stupefied crawl)?

24. The graph is falling from left to right. What does this show?

Use a rational equation to solve Exercises 25–36. Each exercise is a problem involving motion.

25. A car can travel 300 miles in the same amount of time it takes a bus to travel 180 miles. If the rate of the bus is 20 miles per hour slower than the rate of the car, find the average rate for each.

26. A passenger train can travel 240 miles in the same amount of time it takes a freight train to travel 160 miles. If the rate of the freight train is 20 miles per hour slower than the rate of the passenger train, find the average rate of each.

27. You ride your bike to campus a distance of 5 miles and return home on the same route. Going to campus, you ride mostly downhill and average 9 miles per hour faster than on your return trip home. If the round trip takes one hour and ten minutes—that is, $\frac{7}{6}$ hours—what is your average rate on the return trip?

28. An engine pulls a train 140 miles. Then a second engine, whose average rate is 5 miles per hour faster than the first engine, takes over and pulls the train 200 miles. The total time required for both engines is 9 hours. Find the average rate of each engine.

29. In still water, a boat averages 7 miles per hour. It takes the same amount of time to travel 20 miles downstream, with the current, as 8 miles upstream, against the current. What is the rate of the water's current?

	Distance	Rate	Time = $\frac{\text{Distance}}{\text{Rate}}$
With the Current	20	$7 + x$	
Against the Current	8	$7 - x$	

These times are equal.

30. In still water, a boat averages 8 miles per hour. It takes the same amount of time to travel 30 miles downstream, with the current, as 18 miles upstream, against the current. What is the rate of the water's current?

	Distance	Rate	Time = $\frac{\text{Distance}}{\text{Rate}}$
With the Current	30	$8 + x$	
Against the Current	18	$8 - x$	

These times are equal.

31. The rate of the jet stream is 100 miles per hour. Traveling with the jet stream, an airplane can fly 2400 miles in the same amount of time as it takes to fly 1600 miles against the jet stream. What is the airplane's average rate in calm air?

32. The wind is blowing at an average rate of 10 miles per hour. Riding with the wind, a bicyclist can cycle 75 miles in the same amount of time it takes to cycle 15 miles against the wind. What is the bicyclist's average rate in calm air?

33. A moving sidewalk at an airport glides at a rate of 1.8 feet per second. Walking on the moving sidewalk, you travel 100 feet forward in the same time it takes to travel 40 feet in the opposite direction. Find your walking speed on a nonmoving sidewalk.

34. A moving sidewalk at an airport glides at a rate of 1.8 feet per second. Walking on the moving sidewalk, you travel 105 feet forward in the same time it takes to travel 50 feet in the opposite direction. Find your walking speed on a nonmoving sidewalk. Round to the nearest tenth.

35. Two runners, one averaging 8 miles per hour and the other 6 miles per hour, start at the same place and run along the same trail. The slower runner arrives at the end of the trail a half hour after the faster runner. How far did each person run?

36. Two sailboats, one averaging 20 miles per hour and the other 18 miles per hour, start at the same place and follow the same course. The slower boat arrives at the end of the course $\frac{1}{6}$ of an hour after the faster sailboat. How far did each boat travel?

Use a rational equation to solve Exercises 37–48. Each exercise is a problem involving work.

37. You promised your parents that you would wash the family car. You have not started the job and they are due home in 20 minutes. You can wash the car in 45 minutes and your sister claims she can do it in 30 minutes. If you work together, how long will it take to do the job? Will this give you enough time before your parents return?

38. You must leave for campus in half an hour, or you will be late for class. Unfortunately, you are snowed in. You can shovel the driveway in 45 minutes and your brother claims he can do it in 36 minutes. If you shovel together, how long will it take to clear the driveway? Will this give you enough time before you have to leave?

39. A pool can be filled by one pipe in 6 hours and by a second pipe in 12 hours. How long will it take using both pipes to fill the pool?

40. A pond can be filled by one pipe in 8 hours and by a second pipe in 24 hours. How long will it take using both pipes to fill the pond?

41. Working with your cousin, you can refinish a table in 3 hours. Working alone, your cousin can complete the job in 4 hours. How long would it take you to refinish the table working alone?

42. Working with your cousin, you can split a cord of firewood in 5 hours. Working alone, your cousin can complete the job in 7 hours. How long would it take you to split the firewood working alone?

43. An earthquake strikes and an isolated area is without food or water. Three crews arrive. One can dispense needed supplies in 20 hours, a second in 30 hours, and a third in 60 hours. How long will it take all three crews working together to dispense food and water?

44. A hurricane strikes and a rural area is without food or water. Three crews arrive. One can dispense needed supplies in 10 hours, a second in 15 hours, and a third in 20 hours. How long will it take all three crews working together to dispense food and water?

45. An office has an old copying machine and a new one. Working together, it takes both machines 6 hours to make all the copies of the annual financial report. Working alone, it takes the old copying machine 5 hours longer than the new one to make all the copies of the report. How long would it take the new copying machine to make all the copies working alone?

46. A demolition company wants to build a brick wall to hide from public view the area where they store wrecked cars. Working together, an experienced bricklayer and an apprentice can build the wall in 12 hours. Working alone, it takes the apprentice 10 hours longer than the experienced bricklayer to do the job. How long would it take the experienced bricklayer to build the wall working alone?

47. A faucet can fill a sink in 5 minutes. It takes twice that long for the drain to empty the sink. How long will it take to fill the sink if the drain is open and the faucet is on?

48. A pool can be filled by a pipe in 3 hours. It takes 3 times as long for another pipe to empty the pool. How long will it take to fill the pool if both pipes are open?

Exercises 49–56 contain a variety of problems. Use a rational equation to solve each exercise.

49. What number multiplied by the numerator and added to the denominator of $\frac{4}{5}$ makes the resulting fraction equivalent to $\frac{3}{2}$?

50. What number multiplied by the numerator and subtracted from the denominator of $\frac{9}{11}$ makes the resulting fraction equivalent to $-\frac{12}{5}$?

51. The sum of 2 times a number and twice its reciprocal is $\frac{20}{3}$. Find the number(s).

52. If 2 times the reciprocal of a number is subtracted from 3 times the number, the difference is 1. Find the number(s).

53. You have 35 hits in 140 times at bat. Your batting average is $\frac{35}{140}$, or 0.25. How many consecutive hits must you get to increase your batting average to 0.30?

54. You have 30 hits in 120 times at bat. Your batting average is $\frac{30}{120}$, or 0.25. How many consecutive hits must you get to increase your batting average to 0.28?

55. If one pipe can fill a pool in a hours and a second pipe can fill the pool in b hours, write a formula for the time, x, in terms of a and b, for the number of hours it takes both pipes, working together, to fill the pool.

56. If one pipe can fill a pool in a hours and a second pipe can empty the pool in b hours, write a formula for the time, x, in terms of a and b, for the number of hours it takes to fill the pool with both of these pipes open.

Writing in Mathematics

57. Without showing the details, explain how to solve the formula

$$\frac{1}{R} = \frac{1}{R_1} + \frac{1}{R_2}$$

for R_1. (The formula is used in electronics.)

58. Explain how to find the average cost function for a business.

59. How does the average cost function illustrate a problem for small businesses?

60. What is the relationship among time traveled, distance traveled, and rate of travel?

61. If you know how many hours it takes for you to do a job, explain how to find the fractional part of the job you can complete in x hours.

62. If you can do a job in 6 hours and your friend can do the same job in 3 hours, explain how to find how long it takes to complete the job working together. It is not necessary to solve the problem.

63. When two people work together to complete a job, describe one factor that can result in more or less time than the time given by the rational equations we have been using.

Technology Exercises

64. For Exercises 19–20, use a graphing utility to graph the average cost function described by the problem's conditions. Then TRACE along the curve and find the point that visually shows the solution in part (c).

65. For Exercises 45–46, use a graphing utility to graph the function representing the sum of the fractional parts of the job done by the two machines or the two people. Then TRACE along the curve and find the point that visually shows the problem's solution.

66. A boat can travel 10 miles per hour in still water. The boat travels 24 miles upstream, against the current, and then 24 miles downstream, with the current.

a. Let x = the rate of the current. Write a function in terms of x that models the total time for the boat to travel upstream and downstream.

b. Use a graphing utility to graph the rational function in part (a).

c. TRACE along the curve and determine the current's rate if the trip's time is 5 hours. Then verify this result algebraically.

Critical Thinking Exercises

Make Sense? *Read this excerpt from an advertisement for bikes with aerodynamic coverings.*

Our high-performance bicycles have the aerodynamic design of custom racing bikes, but are practical for everyday riding. The aerodynamic covering will increase your average speed by 10 miles per hour. Cyclists using our bicycles, versus bikes without aerodynamic coverings, reduced time on a 75-mile test run by 2 hours.

Now you are interested in finding the average rate of the bikes with the aerodynamic coverings on the 75-mile test run. With this goal in mind, determine whether each statement in Exercises 67–70 "makes sense" or "does not make sense" and explain your reasoning.

67. I decided to organize the critical information from the advertisement in a table with the following entries:

	Distance	Rate	Time
With covering	75	$x + 10$	$\frac{75}{x+10}$
Without covering	75	x	$\frac{75}{x}$

68. The ad stated that bikes with coverings reduced time on the 75-mile test run by 2 hours, so I used my table from Exercise 67 and modeled this condition with the rational equation

$$\frac{75}{x} = \frac{75}{x+10} - 2.$$

69. The equation in x that modeled the conditions had a positive and a negative value for x, so I rejected the negative solution.

70. My professor verified that 15 is the correct value for x in the equation modeling the conditions, so I used my table from Exercise 67 and concluded that the average rate of the covered bikes on the 75-mile test run was 15 miles per hour.

In Exercises 71–74, determine whether each statement is true or false. If the statement is false, make the necessary change(s) to produce a true statement.

71. As production level increases, the average cost for a company to produce each unit of its product also increases.

72. To solve $qf + pf = pq$ for p, subtract qf from both sides and then divide by f.

73. If you plan a theater trip that costs $300 to rent a limousine and $25 per ticket, the cost per person, $f(x)$, for a group of x people is modeled by the rational function

$$f(x) = \frac{300 + 25x}{x}.$$

74. If you can clean the house in 3 hours and your sloppy friend can completely mess it up in 6 hours, then $\frac{x}{3} - \frac{x}{6} = 1$ can be used to find how long it takes to clean the house if you both "work" together.

75. Solve $\frac{1}{s} = f + \frac{1-f}{p}$ for f.

76. A new schedule for a train requires it to travel 351 miles in $\frac{1}{4}$ hour less time than before. To accomplish this, the rate of the train must be increased by 2 miles per hour. What should the average rate of the train be so that it can keep on the new schedule?

77. It takes Mr. Todd 4 hours longer to prepare an order of pies than it takes Mrs. Lovett. They bake together for 2 hours when Mrs. Lovett leaves. Mr. Todd takes 7 additional hours to complete the work. Working alone, how long does it take Mrs. Lovett to prepare the pies?

Review Exercises

78. Factor: $x^2 + 4x + 4 - 9y^2$. (Section 5.6, Example 4)

79. Solve using matrices:

$$2x + 5y = -5$$
$$x + 2y = -1.$$

(Section 3.4, Example 2)

80. Solve the system:

$$\begin{aligned} x + y + z &= 4 \\ 2x + 5y &= 1 \\ x - y - 2x &= 0. \end{aligned}$$

(Section 3.3, Example 3)

Preview Exercises

Exercises 81–83 will help you prepare for the material covered in the next section.

81. a. If $y = kx^2$, find the value of k using $x = 2$ and $y = 64$.

b. Substitute the value for k into $y = kx^2$ and write the resulting equation.

c. Use the equation from part (b) to find y when $x = 5$.

82. a. If $y = \frac{k}{x}$, find the value of k using $x = 8$ and $y = 12$.

b. Substitute the value for k into $y = \frac{k}{x}$ and write the resulting equation.

c. Use the equation from part (b) to find y when $x = 3$.

83. If $S = \frac{kA}{P}$, find the value of k using $A = 60{,}000$, $P = 40$, and $S = 12{,}000$.

SECTION 6.8

Modeling Using Variation

Objectives

1. Solve direct variation problems.
2. Solve inverse variation problems.
3. Solve combined variation problems.
4. Solve problems involving joint variation.

Have you ever wondered how telecommunication companies estimate the number of phone calls expected per day between two cities? The formula

$$C = \frac{0.02P_1P_2}{d^2}$$

shows that the daily number of phone calls, C, increases as the populations of the cities, P_1 and P_2, in thousands, increase, and decreases as the distance, d, between the cities increases.

Certain formulas occur so frequently in applied situations that they are given special names. Variation formulas show how one quantity changes in relation to other quantities. Quantities can vary *directly*, *inversely*, or *jointly*. In this section, we look at situations that can be modeled by each of these kinds of variation. And think of this. The next time you get one of those "all-circuits-are-busy" messages, you will be able to use a variation formula to estimate how many other callers you're competing with for those precious 5-cent minutes.

1 Solve direct variation problems.

Direct Variation

When you swim underwater, the pressure in your ears depends on the depth at which you are swimming. The formula

$$p = 0.43d$$

describes the water pressure, p, in pounds per square inch, at a depth of d feet. We can use this linear function to determine the pressure in your ears at various depths.

In each case, use $p = 0.43d$:

If $d = 20$, $p = 0.43(20) = 8.6$. At a depth of 20 feet, water pressure is 8.6 pounds per square inch.

Doubling the depth doubles the pressure.

If $d = 40$, $p = 0.43(40) = 17.2$. At a depth of 40 feet, water pressure is 17.2 pounds per square inch.

Doubling the depth doubles the pressure.

If $d = 80$, $p = 0.43(80) = 34.4$. At a depth of 80 feet, water pressure is 34.4 pounds per square inch.

The formula $p = 0.43d$ illustrates that water pressure is a constant multiple of your underwater depth. If your depth is doubled, the pressure is doubled; if your depth is tripled, the pressure is tripled; and so on. Because of this, the pressure in your ears is said to **vary directly** as your underwater depth. The **equation of variation** is

$$p = 0.43d.$$

Generalizing, we obtain the following statement:

Direct Variation

If a situation is described by an equation in the form

$$y = kx,$$

where k is a nonzero constant, we say that **y varies directly as x** or **y is directly proportional to x**. The number k is called the **constant of variation** or the **constant of proportionality**.

Can you see that **the direct variation equation, $y = kx$, is a special case of the linear function $y = mx + b$?** When $m = k$ and $b = 0$, $y = mx + b$ becomes $y = kx$. Thus, the slope of a direct variation equation is k, the constant of variation. Because b, the y-intercept, is 0, the graph of a direct variation equation is a line passing through the origin. This is illustrated in **Figure 6.7**, which shows the graph of $p = 0.43d$: Water pressure varies directly as depth.

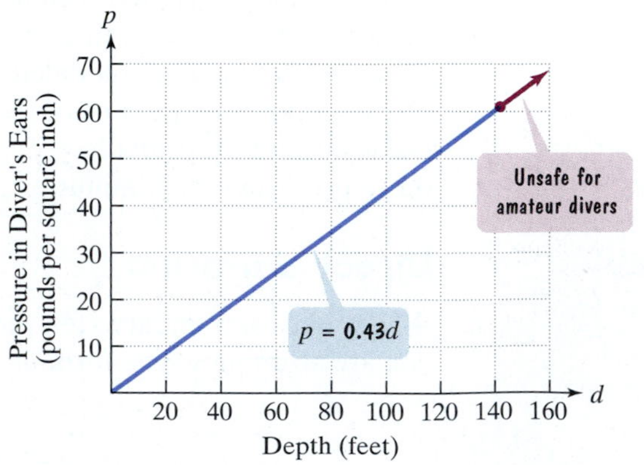

FIGURE 6.7 Water pressure at various depths

Problems involving direct variation can be solved using the following procedure. This procedure applies to direct variation problems, as well as to the other kinds of variation problems that we will discuss.

Solving Variation Problems

1. Write an equation that models the given English statement.
2. Substitute the given pair of values into the equation in step 1 and find the value of k, the constant of variation.
3. Substitute the value of k into the equation in step 1.
4. Use the equation from step 3 to answer the problem's question.

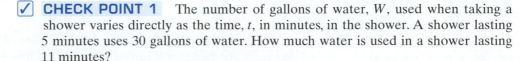

EXAMPLE 1 Solving a Direct Variation Problem

Many areas of Northern California depend on the snowpack of the Sierra Nevada mountain range for their water supply. The volume of water produced from melting snow varies directly as the volume of snow. Meteorologists have determined that 250 cubic centimeters of snow will melt to 28 cubic centimeters of water. How much water does 1200 cubic centimeters of melting snow produce?

Solution

Step 1. Write an equation. We know that *y varies directly as x* is expressed as

$$y = kx.$$

By changing letters, we can write an equation that models the following English statement: Volume of water, W, varies directly as volume of snow, S.

$$W = kS$$

Step 2. Use the given values to find k. We are told that 250 cubic centimeters of snow will melt to 28 cubic centimeters of water. Substitute 28 for W and 250 for S in the direct variation equation. Then solve for k.

$W = kS$	Volume of water varies directly as volume of melting snow.
$28 = k(250)$	250 cubic centimeters of snow melt to 28 cubic centimeters of water.
$\frac{28}{250} = \frac{k(250)}{250}$	Divide both sides by 250.
$0.112 = k$	Simplify.

Step 3. Substitute the value of k into the equation.

$W = kS$	This is the equation from step 1.
$W = 0.112S$	Replace k, the constant of variation, with 0.112.

Step 4. Answer the problem's question. How much water does 1200 cubic centimeters of melting snow produce? Substitute 1200 for S in $W = 0.112S$ and solve for W.

$W = 0.112S$	Use the equation from step 3.
$W = 0.112(1200)$	Substitute 1200 for S.
$W = 134.4$	Multiply.

A snowpack measuring 1200 cubic centimeters will produce 134.4 cubic centimeters of water.

✓ **CHECK POINT 1** The number of gallons of water, W, used when taking a shower varies directly as the time, t, in minutes, in the shower. A shower lasting 5 minutes uses 30 gallons of water. How much water is used in a shower lasting 11 minutes?

The direct variation equation $y = kx$ is a linear function. If $k > 0$, then the slope of the line is positive. Consequently, as x increases, y also increases.

A direct variation situation can involve variables to higher powers. For example, y can vary directly as x^2 ($y = kx^2$) or as x^3 ($y = kx^3$).

Direct Variation with Powers

y* varies directly as the *n*th power of *x if there exists some nonzero constant k such that

$$y = kx^n.$$

We also say that ***y* is directly proportional to the *n*th power of *x***.

Direct variation with exponents that are whole numbers is modeled by polynomial functions. In our next example, the graph of the variation equation is the graph of a quadratic function.

EXAMPLE 2 Solving a Direct Variation Problem

The distance, s, that a body falls from rest varies directly as the square of the time, t, of the fall. If skydivers fall 64 feet in 2 seconds, how far will they fall in 4.5 seconds?

Solution

Step 1. **Write an equation.** We know that y *varies directly as* the square of x is expressed as

$$y = kx^2.$$

By changing letters, we can write an equation that models the following English statement: Distance, s, varies directly as the square of time, t, of the fall.

$$s = kt^2$$

Step 2. **Use the given values to find k.** Skydivers fall 64 feet in 2 seconds. Substitute 64 for s and 2 for t in the direct variation equation. Then solve for k.

$s = kt^2$ — Distance varies directly as the square of time.

$64 = k \cdot 2^2$ — Skydivers fall 64 feet in 2 seconds.

$64 = 4k$ — Simplify: $2^2 = 4$.

$\frac{64}{4} = \frac{4k}{4}$ — Divide both sides by 4.

$16 = k$ — Simplify.

Step 3. **Substitute the value of k into the equation.**

$s = kt^2$ — Use the equation from step 1.

$s = 16t^2$ — Replace k, the constant of variation, with 16.

Step 4. **Answer the problem's question.** How far will the skydivers fall in 4.5 seconds? Substitute 4.5 for t in $s = 16t^2$ and solve for s.

$$s = 16(4.5)^2 = 16(20.25) = 324$$

Thus, in 4.5 seconds, the skydivers will fall 324 feet.

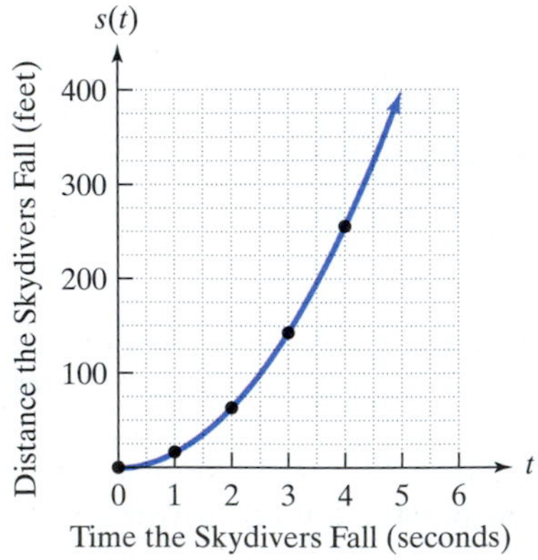

FIGURE 6.8 The graph of $s(t) = 16t^2$

We can express the variation equation from Example 2 in function notation, writing

$$s(t) = 16t^2.$$

The distance that a body falls from rest is a function of the time, t, of the fall. The graph of this quadratic function is shown in **Figure 6.8**. The graph increases rapidly from left to right, showing the effects of the acceleration of gravity.

✓ CHECK POINT 2 The distance required to stop a car varies directly as the square of its speed. If it requires 200 feet to stop a car traveling 60 miles per hour, how many feet are required to stop a car traveling 100 miles per hour?

2 Solve inverse variation problems.

Inverse Variation

The distance from San Francisco to Los Angeles is 420 miles. The time that it takes to drive from San Francisco to Los Angeles depends on the average rate at which one drives and is given by

$$\text{Time} = \frac{420}{\text{Rate}}.$$

For example, if you average 30 miles per hour, the time for the drive is

$$\text{Time} = \frac{420}{30} = 14,$$

or 14 hours. If you average 50 miles per hour, the time for the drive is

$$\text{Time} = \frac{420}{50} = 8.4,$$

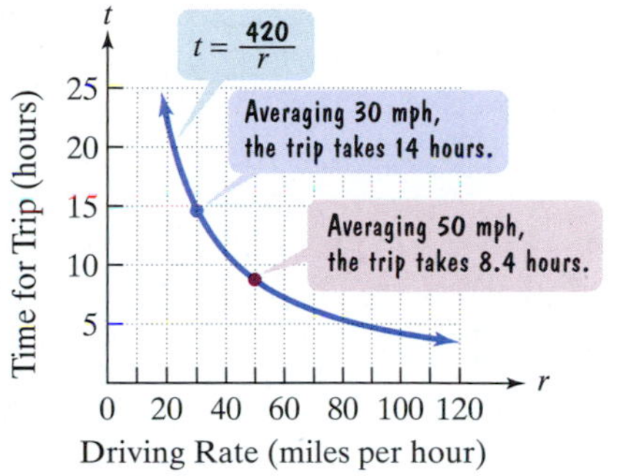

FIGURE 6.9

or 8.4 hours. As your rate (or speed) increases, the time for the trip decreases and vice versa. This is illustrated by the graph in **Figure 6.9**.

We can express the time for the San Francisco–Los Angeles trip using t for time and r for rate:

$$t = \frac{420}{r}.$$

This equation is an example of an **inverse variation** equation. Time, t, **varies inversely** as rate, r. When two quantities vary inversely, one quantity increases as the other decreases, and vice versa.

Generalizing, we obtain the following statement:

Inverse Variation

If a situation is described by an equation in the form

$$y = \frac{k}{x},$$

where k is a nonzero constant, we say that **y varies inversely as x** or **y is inversely proportional to x**. The number k is called the **constant of variation**.

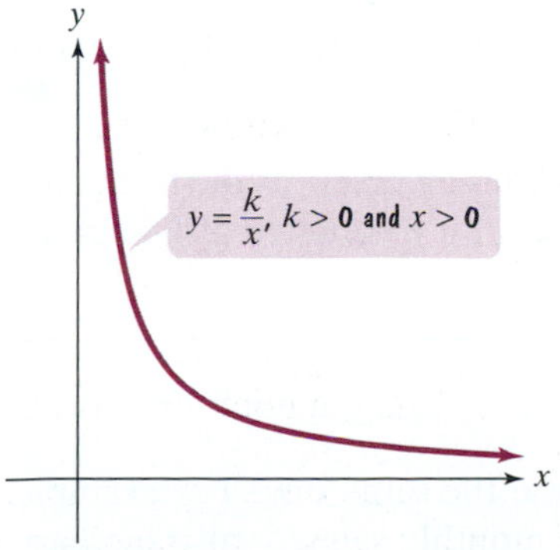

FIGURE 6.10 The graph of the inverse variation equation

Notice that **the inverse variation equation**

$$y = \frac{k}{x}, \quad \text{or} \quad f(x) = \frac{k}{x},$$

is a rational function. For $k > 0$ and $x > 0$, the graph of the function takes on the shape shown in **Figure 6.10**.

We use the same procedure to solve inverse variation problems as we did to solve direct variation problems. Example 3 illustrates this procedure.

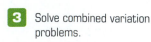

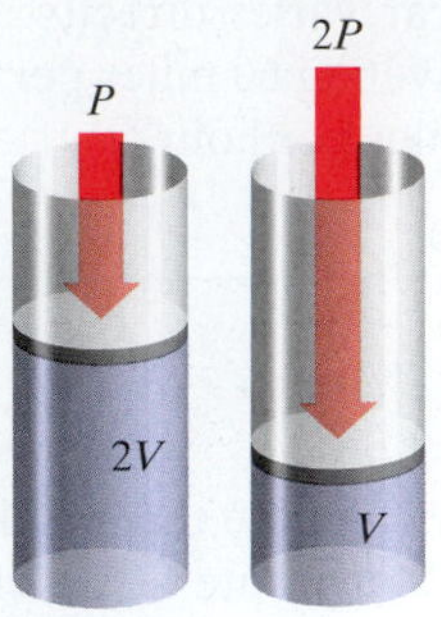

Doubling the pressure halves the volume.

EXAMPLE 3 Solving an Inverse Variation Problem

When you use a spray can and press the valve at the top, you decrease the pressure of the gas in the can. This decrease of pressure causes the volume of the gas in the can to increase. Because the gas needs more room than is provided in the can, it expands in spray form through the small hole near the valve. In general, if the temperature is constant, the pressure, P, of a gas in a container varies inversely as the volume, V, of the container. The pressure of a gas sample in a container whose volume is 8 cubic inches is 12 pounds per square inch. If the sample expands to a volume of 22 cubic inches, what is the new pressure of the gas?

Solution

Step 1. Write an equation. We know that *y varies inversely as x* is expressed as

$$y = \frac{k}{x}.$$

By changing letters, we can write an equation that models the following English statement: The pressure, P, of a gas in a container varies inversely as the volume, V.

$$P = \frac{k}{V}.$$

Step 2. Use the given values to find k. The pressure of a gas sample in a container whose volume is 8 cubic inches is 12 pounds per square inch. Substitute 12 for P and 8 for V in the inverse variation equation. Then solve for k.

$$P = \frac{k}{V}$$ Pressure varies inversely as volume.

$$12 = \frac{k}{8}$$ The pressure in an 8-cubic-inch container is 12 pounds per square inch.

$$12 \cdot 8 = \frac{k}{8} \cdot 8$$ Multiply both sides by 8.

$$96 = k$$ Simplify.

Step 3. Substitute the value of k into the equation.

$$P = \frac{k}{V}$$ Use the equation from step 1.

$$P = \frac{96}{V}$$ Replace k, the constant of variation, with 96.

Step 4. Answer the problem's question. We need to find the pressure when the volume expands to 22 cubic inches. Substitute 22 for V and solve for P.

$$P = \frac{96}{V} = \frac{96}{22} = 4\frac{4}{11}$$

When the volume is 22 cubic inches, the pressure of the gas is $4\frac{4}{11}$ pounds per square inch.

✓ **CHECK POINT 3** The length of a violin string varies inversely as the frequency of its vibrations. A violin string 8 inches long vibrates at a frequency of 640 cycles per second. What is the frequency of a 10-inch string?

3 Solve combined variation problems.

Combined Variation

In **combined variation**, direct and inverse variation occur at the same time. For example, as the advertising budget, A, of a company increases, its monthly sales, S, also increase. Monthly sales vary directly as the advertising budget:

$$S = kA.$$

By contrast, as the price of the company's product, P, increases, its monthly sales, S, decrease. Monthly sales vary inversely as the price of the product:

$$S = \frac{k}{P}.$$

We can combine these two variation equations into one equation:

$$S = \frac{kA}{P}.$$

Monthly sales, S, vary directly as the advertising budget, A, and inversely as the price of the product, P.

The following example illustrates an application of combined variation.

EXAMPLE 4 Solving a Combined Variation Problem

The owners of Rollerblades Now determine that the monthly sales, S, of its skates vary directly as its advertising budget, A, and inversely as the price of the skates, P. When \$60,000 is spent on advertising and the price of the skates is \$40, the monthly sales are 12,000 pairs of rollerblades.

a. Write an equation of variation that models this situation.

b. Determine monthly sales if the amount of the advertising budget is increased to \$70,000.

Solution

a. Write an equation.

$$S = \frac{kA}{P}.$$

Translate "sales vary directly as the advertising budget and inversely as the skates' price."

Use the given values to find k.

$$12{,}000 = \frac{k(60{,}000)}{40}$$ When \$60,000 is spent on advertising ($A = 60{,}000$) and the price is \$40 ($P = 40$), monthly sales are 12,000 units ($S = 12{,}000$).

$$12{,}000 = k \cdot 1500$$ Divide 60,000 by 40.

$$\frac{12{,}000}{1500} = \frac{k \cdot 1500}{1500}$$ Divide both sides of the equation by 1500.

$$8 = k$$ Simplify.

Therefore, the equation of variation that models monthly sales is

$$S = \frac{8A}{P}.$$ Substitute 8 for k in $S = \frac{kA}{P}$.

b. The advertising budget is increased to \$70,000, so $A = 70{,}000$. The skates' price is still \$40, so $P = 40$.

$$S = \frac{8A}{P}$$ This is the equation from part (a).

$$S = \frac{8(70{,}000)}{40}$$ Substitute 70,000 for A and 40 for P.

$$S = 14{,}000$$ Simplify.

With a \$70,000 advertising budget and \$40 price, the company can expect to sell 14,000 pairs of rollerblades in a month (up from 12,000).

✓ **CHECK POINT 4** The number of minutes needed to solve an exercise set of variation problems varies directly as the number of problems and inversely as the number of people working to solve the problems. It takes 4 people 32 minutes to solve 16 problems. How many minutes will it take 8 people to solve 24 problems?

4 Solve problems involving joint variation.

Joint Variation

Joint variation is a variation in which a variable varies directly as the product of two or more other variables. Thus, the equation $y = kxz$ is read "y varies jointly as x and z."

Joint variation plays a critical role in Isaac Newton's formula for gravitation:

$$F = G\frac{m_1 m_2}{d^2}.$$

The formula states that the force of gravitation, F, between two bodies varies jointly as the product of their masses, m_1 and m_2, and inversely as the square of the distance between them, d. (G is the gravitational constant.) The formula indicates that gravitational force exists between any two objects in the universe, increasing as the distance between the bodies decreases. One practical result is that the pull of the moon on the oceans is greater on the side of Earth closer to the moon. This gravitational imbalance is what produces tides.

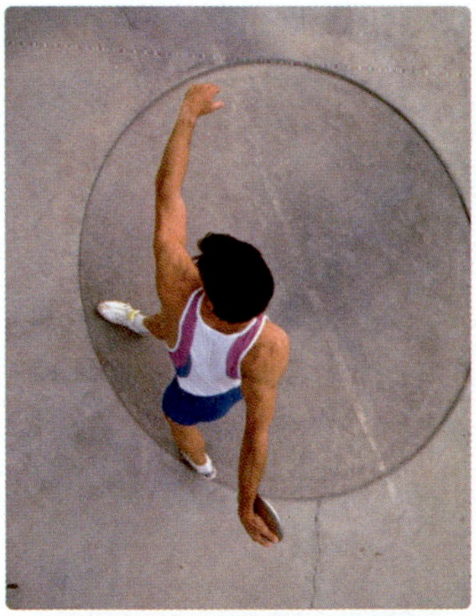

EXAMPLE 5 Modeling Centrifugal Force

The centrifugal force, C, of a body moving in a circle varies jointly with the radius of the circular path, r, and the body's mass, m, and inversely with the square of the time, t, it takes to move about one full circle. A 6-gram body moving in a circle with radius 100 centimeters at a rate of 1 revolution in 2 seconds has a centrifugal force of 6000 dynes. Find the centrifugal force of an 18-gram body moving in a circle with radius 100 centimeters at a rate of 1 revolution in 3 seconds.

Solution

$$C = \frac{krm}{t^2}$$

Translate "Centrifugal force, C, varies jointly with radius, r, and mass, m, and inversely with the square of time, t."

$$6000 = \frac{k(100)(6)}{2^2}$$

A 6-gram body ($m = 6$) moving in a circle with radius 100 centimeters ($r = 100$) at 1 revolution in 2 seconds ($t = 2$) has a centrifugal force of 6000 dynes ($C = 6000$).

$$6000 = 150k$$

Simplify: $\frac{100(6)}{2^2} = \frac{600}{4} = 150.$

$$40 = k$$

Divide both sides by 150 and solve for k.

$$C = \frac{40rm}{t^2}$$

Substitute 40 for k in the model for centrifugal force.

$$C = \frac{40(100)(18)}{3^2}$$

Find centrifugal force, C, of an 18-gram body ($m = 18$) moving in a circle with radius 100 centimeters ($r = 100$) at 1 revolution in 3 seconds ($t = 3$).

$$= 8000$$

Simplify.

The centrifugal force is 8000 dynes.

CHECK POINT 5 The volume of a cone, V, varies jointly as its height, h, and the square of its radius r. A cone with a radius measuring 6 feet and a height measuring 10 feet has a volume of 120π cubic feet. Find the volume of a cone having a radius of 12 feet and a height of 2 feet.

6.8 EXERCISE SET

MyMathLab MathXL PRACTICE WATCH DOWNLOAD READ REVIEW

Practice Exercises

Use the four-step procedure for solving variation problems given on page 465 to solve Exercises 1–10.

1. y varies directly as x. $y = 65$ when $x = 5$. Find y when $x = 12$.
2. y varies directly as x. $y = 45$ when $x = 5$. Find y when $x = 13$.
3. y varies inversely as x. $y = 12$ when $x = 5$. Find y when $x = 2$.
4. y varies inversely as x. $y = 6$ when $x = 3$. Find y when $x = 9$.
5. y varies directly as x and inversely as the square of z. $y = 20$ when $x = 50$ and $z = 5$. Find y when $x = 3$ and $z = 6$.
6. a varies directly as b and inversely as the square of c. $a = 7$ when $b = 9$ and $c = 6$. Find a when $b = 4$ and $c = 8$.
7. y varies jointly as x and z. $y = 25$ when $x = 2$ and $z = 5$. Find y when $x = 8$ and $z = 12$.
8. C varies jointly as A and T. $C = 175$ when $A = 2100$ and $T = 4$. Find C when $A = 2400$ and $T = 6$.
9. y varies jointly as a and b, and inversely as the square root of c. $y = 12$ when $a = 3, b = 2$, and $c = 25$. Find y when $a = 5, b = 3$, and $c = 9$.
10. y varies jointly as m and the square of n, and inversely as p. $y = 15$ when $m = 2, n = 1$, and $p = 6$. Find y when $m = 3, n = 4$, and $p = 10$.

Practice PLUS

In Exercises 11–20, write an equation that expresses each relationship. Then solve the equation for y.

11. x varies jointly as y and z.
12. x varies jointly as y and the square of z.
13. x varies directly as the cube of z and inversely as y.
14. x varies directly as the cube root of z and inversely as y.
15. x varies jointly as y and z and inversely as the square root of w.
16. x varies jointly as y and z and inversely as the square of w.
17. x varies jointly as z and the sum of y and w.
18. x varies jointly as z and the difference between y and w.
19. x varies directly as z and inversely as the difference between y and w.
20. x varies directly as z and inversely as the sum of y and w.

Application Exercises

Use the four-step procedure for solving variation problems given on page 465 to solve Exercises 21–28.

21. An alligator's tail length, T, varies directly as its body length, B. An alligator with a body length of 4 feet has a tail length of 3.6 feet. What is the tail length of an alligator whose body length is 6 feet?

22. An object's weight on the moon, M, varies directly as its weight on Earth, E. Neil Armstrong, the first person to step on the moon on July 20, 1969, weighed 360 pounds on Earth (with all of his equipment on) and 60 pounds on the moon. What is the moon weight of a person who weighs 186 pounds on Earth?
23. The height that a ball bounces varies directly as the height from which it was dropped. A tennis ball dropped from 12 inches bounces 8.4 inches. From what height was the tennis ball dropped if it bounces 56 inches?
24. The distance that a spring will stretch varies directly as the force applied to the spring. A force of 12 pounds is needed to stretch a spring 9 inches. What force is required to stretch the spring 15 inches?

25. If all men had identical body types, their weight would vary directly as the cube of their height. Shown below is Robert Wadlow, who reached a record height of 8 feet 11 inches (107 inches) before his death at age 22. If a man who is 5 feet 10 inches tall (70 inches) with the same body type as Mr. Wadlow weighs 170 pounds, what was Robert Wadlow's weight shortly before his death?

26. On a dry asphalt road, a car's stopping distance varies directly as the square of its speed. A car traveling at 45 miles per hour can stop in 67.5 feet. What is the stopping distance for a car traveling at 60 miles per hour?

27. The figure shows that a bicyclist tips the cycle when making a turn. The angle B, formed by the vertical direction and the bicycle, is called the banking angle. The banking angle varies inversely as the cycle's turning radius. When the turning radius is 4 feet, the banking angle is 28°. What is the banking angle when the turning radius is 3.5 feet?

28. The water temperature of the Pacific Ocean varies inversely as the water's depth. At a depth of 1000 meters, the water temperature is 4.4° Celsius. What is the water temperature at a depth of 5000 meters?

Heart rates and life spans of most mammals can be modeled using inverse variation. The bar graph shows the average heart rate and the average life span of five mammals. You will use the data to solve Exercises 29–30.

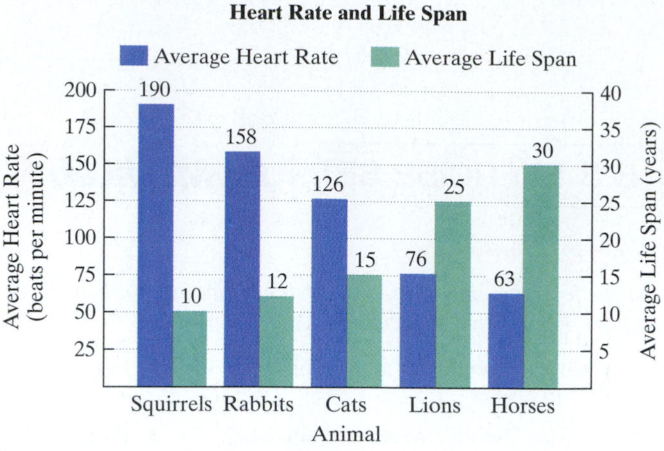

Source: *The Handy Science Answer Book*, Visible Ink Press, 2003

29. a. A mammal's average life span, L, in years, varies inversely as its average heart rate, R, in beats per minute. Use the data shown for horses to write the equation that models this relationship.

b. Is the inverse variation equation in part (a) an exact model or an approximate model for the data shown for lions?

c. Elephants have an average heart rate of 27 beats per minute. Determine their average life span.

30. a. A mammal's average life span, L, in years, varies inversely as its average heart rate, R, in beats per minute. Use the data shown for cats to write the equation that models this relationship.

b. Is the inverse variation equation in part (a) an exact model or an approximate model for the data shown for squirrels?

c. Mice have an average heart rate of 634 beats per minute. Determine their average life span, rounded to the nearest year.

The figure shows the graph of the inverse variation model that you wrote in Exercise 29(a) or Exercise 30(a). Use the graph of this rational function to solve Exercises 31–32.

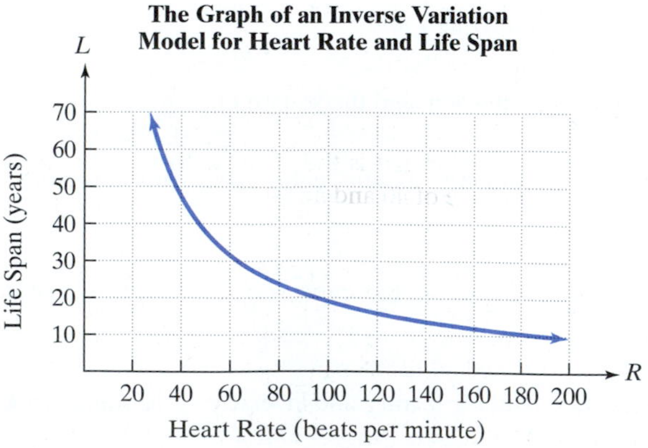

31. a. If a mammal has a life span of 20 years, use the graph to estimate its heart rate, rounded to the nearest 10 beats per minute.
b. Use the inverse variation equation that you wrote in Exercise 29(a) or Exercise 30(a) to determine the heart rate, rounded to the nearest beat per minute, for a mammal with a life span of 20 years.
c. The bar graph uses two bars to display the data for horses. How is this information shown on the graph of the inverse variation model?

32. a. If a mammal has a life span of 50 years, use the graph to estimate its heart rate, rounded to the nearest 10 beats per minute.
b. Use the inverse variation equation that you wrote in Exercise 29(a) or Exercise 30(a) to determine the heart rate, rounded to the nearest beat per minute, for a mammal with a life span of 50 years.
c. The bar graph uses two bars to display the data for lions. How is this information shown on the graph of the inverse variation model?

Continue to use the four-step procedure for solving variation problems given on page 465 to solve Exercises 33–40.

33. Radiation machines, used to treat tumors, produce an intensity of radiation that varies inversely as the square of the distance from the machine. At 3 meters, the radiation intensity is 62.5 milliroentgens per hour. What is the intensity at a distance of 2.5 meters?

34. The illumination provided by a car's headlight varies inversely as the square of the distance from the headlight. A car's headlight produces an illumination of 3.75 footcandles at a distance of 40 feet. What is the illumination when the distance is 50 feet?

35. Body-mass index, or BMI, takes both weight and height into account when assessing whether an individual is underweight or overweight. BMI varies directly as one's weight, in pounds, and inversely as the square of one's height, in inches. In adults, normal values for the BMI are between 20 and 25, inclusive. Values below 20 indicate that an individual is underweight and values above 30 indicate that an individual is obese. A person who weighs 180 pounds and is 5 feet, or 60 inches, tall has a BMI of 35.15. What is the BMI, to the nearest tenth, for a 170 pound person who is 5 feet 10 inches tall. Is this person overweight?

36. One's intelligence quotient, or IQ, varies directly as a person's mental age and inversely as that person's chronological age. A person with a mental age of 25 and a chronological age of 20 has an IQ of 125. What is the chronological age of a person with a mental age of 40 and an IQ of 80?

37. The heat loss of a glass window varies jointly as the window's area and the difference between the outside and inside temperatures. A window 3 feet wide by 6 feet long loses 1200 Btu per hour when the temperature outside is 20° colder than the temperature inside. Find the heat loss through a glass window that is 6 feet wide by 9 feet long when the temperature outside is 10° colder than the temperature inside.

38. Kinetic energy varies jointly as the mass and the square of the velocity. A mass of 8 grams and velocity of 3 centimeters per second has a kinetic energy of 36 ergs. Find the kinetic energy for a mass of 4 grams and velocity of 6 centimeters per second.

39. Sound intensity varies inversely as the square of the distance from the sound source. If you are in a movie theater and you change your seat to one that is twice as far from the speakers, how does the new sound intensity compare to that of your original seat?

40. Many people claim that as they get older, time seems to pass more quickly. Suppose that the perceived length of a period of time is inversely proportional to your age. How long will a year seem to be when you are three times as old as you are now?

41. The average number of daily phone calls, C, between two cities varies jointly as the product of their populations, P_1 and P_2, and inversely as the square of the distance, d, between them.
a. Write an equation that expresses this relationship.
b. The distance between San Francisco (population: 777,000) and Los Angeles (population: 3,695,000) is 420 miles. If the average number of daily phone calls between the cities is 326,000, find the value of k to two decimal places and write the equation of variation.
c. Memphis (population: 650,000) is 400 miles from New Orleans (population: 490,000). Find the average number of daily phone calls, to the nearest whole number, between these cities.

42. The force of wind blowing on a window positioned at a right angle to the direction of the wind varies jointly as the area of the window and the square of the wind's speed. It is known that a wind of 30 miles per hour blowing on a window measuring 4 feet by 5 feet exerts a force of 150 pounds. During a storm with winds of 60 miles per hour, should hurricane shutters be placed on a window that measures 3 feet by 4 feet and is capable of withstanding 300 pounds of force?

43. The table shows the values for the current, I, in an electric circuit and the resistance, R, of the circuit.

I (amperes)	0.5	1.0	1.5	2.0	2.5	3.0	4.0	5.0
R (ohms)	12	6.0	4.0	3.0	2.4	2.0	1.5	1.2

a. Graph the ordered pairs in the table of values, with values of I along the x-axis and values of R along the y-axis. Connect the eight points with a smooth curve.
b. Does current vary directly or inversely as resistance? Use your graph and explain how you arrived at your answer.
c. Write an equation of variation for I and R, using one of the ordered pairs in the table to find the constant of variation. Then use your variation equation to verify the other seven ordered pairs in the table.

Writing in Mathematics

44. What does it mean if two quantities vary directly?

45. In your own words, explain how to solve a variation problem.

46. What does it mean if two quantities vary inversely?

47. Explain what is meant by combined variation. Give an example with your explanation.

48. Explain what is meant by joint variation. Give an example with your explanation.

In Exercises 49–50, describe in words the variation shown by the given equation.

49. $z = \dfrac{k\sqrt{x}}{y^2}$

50. $z = kx^2\sqrt{y}$

51. We have seen that the daily number of phone calls between two cities varies jointly as their populations and inversely as the square of the distance between them. This model, used by telecommunication companies to estimate the line capacities needed among various cities, is called the *gravity model*. Compare the model to Newton's formula for gravitation on page 470 and describe why the name *gravity model* is appropriate.

Technology Exercise

52. Use a graphing utility to graph any three of the variation equations in Exercises 21–28. Then [TRACE] along each curve and identify the point that corresponds to the problem's solution.

Critical Thinking Exercises

Make Sense? *In Exercises 53–56, determine whether each statement "makes sense" or "does not make sense" and explain your reasoning.*

53. I'm using an inverse variation equation and I need to determine the value of the dependent variable when the independent variable is zero.

54. The graph of this direct variation equation has a positive constant of variation and shows one variable increasing as the other variable decreases.

55. When all is said and done, it seems to me that direct variation equations are special kinds of linear functions and inverse variation equations are special kinds of rational functions.

56. Using the language of variation, I can now state the formula for the area of a trapezoid, $A = \frac{1}{2}h(b_1 + b_2)$, as, "A trapezoid's area varies jointly with its height and the sum of its bases."

57. In a hurricane, the wind pressure varies directly as the square of the wind velocity. If wind pressure is a measure of a hurricane's destructive capacity, what happens to this destructive power when the wind speed doubles?

58. The heat generated by a stove element varies directly as the square of the voltage and inversely as the resistance. If the voltage remains constant, what needs to be done to triple the amount of heat generated?

59. Galileo's telescope brought about revolutionary changes in astronomy. A comparable leap in our ability to observe the universe took place as a result of the Hubble Space Telescope. The space telescope can see stars and galaxies whose brightness is $\frac{1}{50}$ of the faintest objects now observable using ground-based telescopes. Use the fact that the brightness of a point source, such as a star, varies inversely as the square of its distance from an observer to show that the space telescope can see about seven times farther than a ground-based telescope.

Review Exercises

60. Evaluate:

$$\begin{vmatrix} -1 & 2 \\ 3 & -4 \end{vmatrix}.$$

(Section 3.5, Example 1)

61. Factor completely:

$$x^2y - 9y - 3x^2 + 27.$$

(Section 5.6, Example 3)

62. Find the degree of

$$7xy + x^2y^2 - 5x^3 - 7.$$

(Section 5.1, Example 1)

Preview Exercises

Exercises 63–65 will help you prepare for the material covered in the first section of the next chapter.

63. If $f(x) = \sqrt{3x + 12}$, find $f(-1)$.

64. If $f(x) = \sqrt{3x + 12}$, find $f(8)$.

65. Use the graph of $f(x) = \sqrt{3x + 12}$ to identify the function's domain and its range. Express the domain and the range in interval notation.

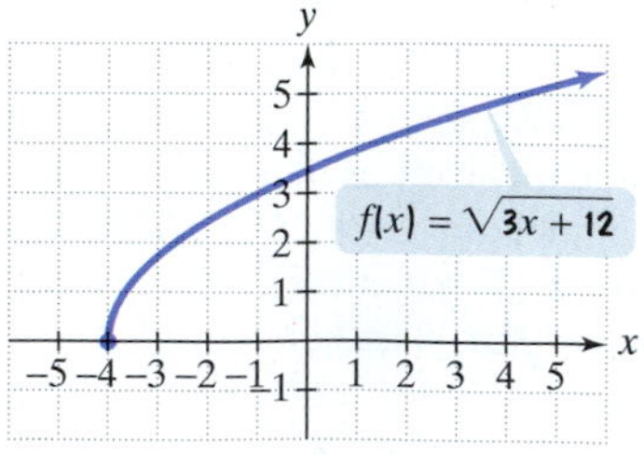

GROUP PROJECT

CHAPTER 6

Additional group projects can be found in the *Instructor's Resource Manual.*

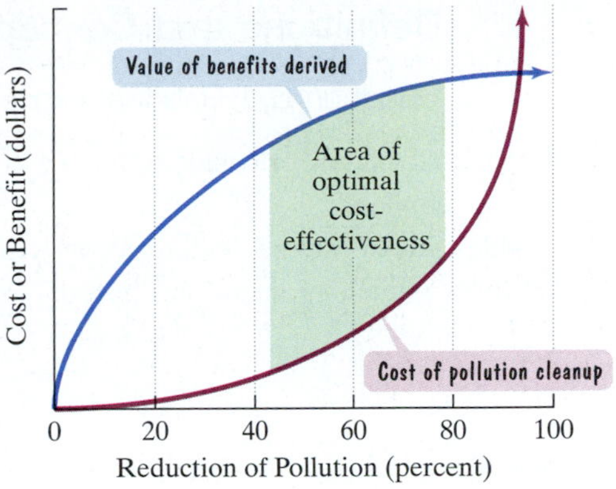

A cost-benefit analysis compares the estimated costs of a project with the benefits that will be achieved. Costs and benefits are given monetary values and compared using a benefit-cost ratio. As shown in the figure, a favorable ratio for a project means that the benefits outweigh the costs and the project is cost-effective. As a group, select an environmental project that was implemented in your area of the country. Research the cost and benefit graphs that resulted in the implementation of this project. How were the benefits converted into monetary terms? Is there an equation for either the cost model or the benefit model? Group members may need to interview members of environmental groups and businesses that were part of this project. You may wish to consult an environmental science textbook to find out more about cost-benefit analyses. After doing your research, the group should write or present a report explaining why the cost-benefit analysis resulted in the project's implementation.

Chapter 6 Summary

Definitions and Concepts / Examples

Section 6.1 Rational Expressions and Functions: Multiplying and Dividing

Definitions and Concepts

A rational expression consists of a polynomial divided by a nonzero polynomial. A rational function is defined by a formula that is a rational expression. The domain of a rational function is the set of all real numbers except those for which the denominator is zero. Graphs of rational functions often contain disconnected branches. The graphs often approach, but do not touch, vertical or horizontal lines, called vertical asymptotes and horizontal asymptotes, respectively.

Examples

Let $f(x) = \dfrac{x+2}{x^2+x-6}$. Find the domain of f.

$$x^2 + x - 6 = 0$$
$$(x+3)(x-2) = 0$$
$$x + 3 = 0 \quad \text{or} \quad x - 2 = 0$$
$$x = -3 \qquad\qquad x = 2$$

These values of x make the denominator zero. Exclude them from the domain.
Domain of $f = (-\infty, -3) \cup (-3, 2) \cup (2, \infty)$.
The branches of the graph of f break at -3 and 2.

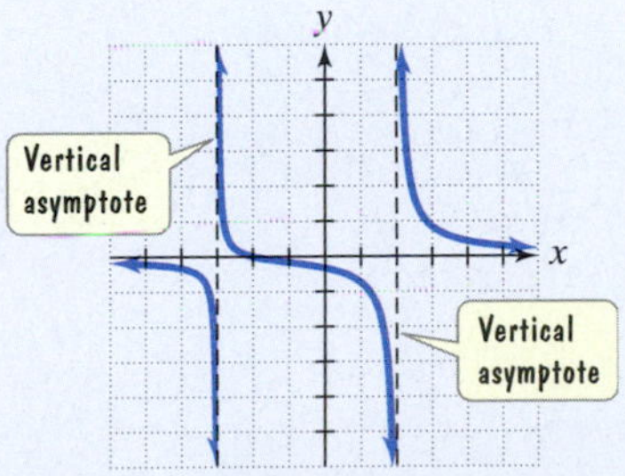

Simplifying Rational Expressions

1. Factor the numerator and the denominator completely.
2. Divide both the numerator and the denominator by any common factors.

Example

Simplify: $\dfrac{4x+28}{x^2-49}$.

$$\frac{4x+28}{x^2-49} = \frac{4\cancel{(x+7)}}{\cancel{(x+7)}(x-7)} = \frac{4}{x-7}$$

Definitions and Concepts	Examples

Section 6.1 Rational Expressions and Functions: Multiplying and Dividing (continued)

Multiplying Rational Expressions

1. Factor completely.
2. Divide numerators and denominators by common factors.
3. Multiply remaining factors in the numerators and in the denominators.

$$\frac{x^2+3x-18}{x^2-3x}\cdot\frac{x^2}{x^2-36}$$
$$=\frac{\cancel{(x+6)}\cancel{(x-3)}}{\cancel{x}\cancel{(x-3)}}\cdot\frac{\cancel{x}\cdot x}{\cancel{(x+6)}(x-6)}$$
$$=\frac{x}{x-6}$$

Dividing Rational Expressions

Invert the divisor and multiply.

$$\frac{5y+15}{(y+5)^2}\div\frac{y^2-9}{y+5}$$
$$=\frac{5y+15}{(y+5)^2}\cdot\frac{y+5}{y^2-9}=\frac{5\cancel{(y+3)}}{(y+5)\cancel{(y+5)}}\cdot\frac{\cancel{(y+5)}}{\cancel{(y+3)}(y-3)}$$
$$=\frac{5}{(y+5)(y-3)}$$

Section 6.2 Adding and Subtracting Rational Expressions

Adding or Subtracting Rational Expressions

If the denominators are the same, add or subtract the numerators. Place the result over the common denominator. If the denominators are different, write all rational expressions with the least common denominator. The LCD is a polynomial consisting of the product of all prime factors in the denominators, with each factor raised to the greatest power of its occurrence in any denominator. Once all rational expressions are written in terms of the LCD, add or subtract as described above. Simplify the result, if possible.

$$\frac{x+1}{2x-2}-\frac{2x}{x^2+2x-3}$$
$$=\frac{x+1}{2(x-1)}-\frac{2x}{(x-1)(x+3)}$$ LCD is $2(x-1)(x+3)$.
$$=\frac{(x+1)(x+3)}{2(x-1)(x+3)}-\frac{2x\cdot 2}{2(x-1)(x+3)}$$
$$=\frac{x^2+4x+3-4x}{2(x-1)(x+3)}=\frac{x^2+3}{2(x-1)(x+3)}$$

Section 6.3 Complex Rational Expressions

Complex rational expressions have numerators or denominators containing one or more rational expressions. Complex rational expressions can be simplified by multiplying the numerator and the denominator by the LCD. They can also be simplified by obtaining single expressions in the numerator and denominator and then dividing.

Simplify: $\dfrac{\frac{x+3}{x}}{x-\frac{9}{x}}$.

Multiplying by the LCD, x:

$$\frac{x}{x}\cdot\frac{\left(\frac{x+3}{x}\right)}{\left(x-\frac{9}{x}\right)}=\frac{x+3}{x^2-9}=\frac{\cancel{(x+3)}}{\cancel{(x+3)}(x-3)}=\frac{1}{x-3}$$

Simplifying by dividing:

$$\frac{\frac{x+3}{x}}{x-\frac{9}{x}}=\frac{\frac{x+3}{x}}{\frac{x}{1}\cdot\frac{x}{x}-\frac{9}{x}}=\frac{\frac{x+3}{x}}{\frac{x^2-9}{x}}=\frac{x+3}{x}\cdot\frac{x}{x^2-9}$$

LCD is x.

$$=\frac{\cancel{x+3}}{\cancel{x}}\cdot\frac{\cancel{x}}{\cancel{(x+3)}(x-3)}=\frac{1}{x-3}$$

Definitions and Concepts | Examples

Section 6.4 Division of Polynomials

To divide a polynomial by a monomial, divide each term of the polynomial by the monomial.

$$\frac{9x^3y^4 - 12x^3y^2 - 7x^2y^2}{3xy^2}$$

$$= \frac{9x^3y^4}{3xy^2} - \frac{12x^3y^2}{3xy^2} - \frac{7x^2y^2}{3xy^2} = 3x^2y^2 - 4x^2 - \frac{7x}{3}$$

To divide by a polynomial containing more than one term, use long division. If necessary, arrange the dividend in descending powers of the variable. If a power of a variable is missing, add that power with a coefficient of 0. Repeat the four steps of the long-division process—divide, multiply, subtract, bring down the next term—until the degree of the remainder is less than the degree of the divisor.

Divide: $(2x^3 - x^2 - 7) \div (x - 2)$.

$$\begin{array}{r} 2x^2 + 3x + 6 \\ x - 2\overline{)2x^3 - x^2 + 0x - 7} \\ \underline{2x^3 - 4x^2} \qquad\qquad \\ 3x^2 + 0x \qquad \\ \underline{3x^2 - 6x} \qquad \\ 6x - 7 \\ \underline{6x - 12} \\ 5 \end{array}$$

The answer is $2x^2 + 3x + 6 + \frac{5}{x - 2}$.

Section 6.5 Synthetic Division and the Remainder Theorem

A shortcut to long division, called synthetic division, can be used to divide a polynomial by a binomial of the form $x - c$.

Here is the division problem shown above using synthetic division.

$$\begin{array}{r|rrrr} 2 & 2 & -1 & 0 & -7 \\ & \downarrow & 4 & 6 & 12 \\ \hline & 2 & 3 & 6 & 5 \end{array}$$

Coefficients of the dividend, $2x^3 - x^2 + 0x - 7$

This is c in $x - c$. For $x - 2$, c is 2.

Coefficients of quotient

Remainder

The answer is $2x^2 + 3x + 6 + \frac{5}{x - 2}$.

The Remainder Theorem

If the polynomial $f(x)$ is divided by $x - c$, then the remainder is $f(c)$. This can be used to evaluate a polynomial function at c. Rather than substituting c for x, divide the function by $x - c$. The remainder is $f(c)$.

If $f(x) = x^4 - 3x^2 - 2x + 5$, use the Remainder Theorem to find $f(-2)$.

$$\begin{array}{r|rrrrr} -2 & 1 & 0 & -3 & -2 & 5 \\ & & -2 & 4 & -2 & 8 \\ \hline & 1 & -2 & 1 & -4 & 13 \end{array}$$

Remainder

Thus, $f(-2) = 13$.

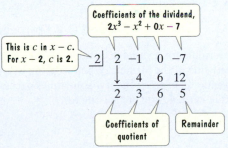

Definitions and Concepts	Examples

Section 6.6 Rational Equations

A rational equation is an equation containing one or more rational expressions.

Solving Rational Equations

1. List restrictions on the variable.
2. Clear fractions by multiplying both sides by the LCD.
3. Solve the resulting equation.
4. Reject any proposed solution in the list of restrictions. Check other proposed solutions in the original equation.

Solve: $\dfrac{7x}{x^2-4}+\dfrac{5}{x-2}=\dfrac{2x}{x^2-4}$.

$$\frac{7x}{(x+2)(x-2)}+\frac{5}{x-2}=\frac{2x}{(x+2)(x-2)}$$

Denominators would equal 0 if $x=-2$ or $x=2$. Restrictions: $x\neq-2$ and $x\neq 2$.

LCD is $(x+2)(x-2)$.

$$(x+2)(x-2)\left[\frac{7x}{(x+2)(x-2)}+\frac{5}{x-2}\right]=(x+2)(x-2)\cdot\frac{2x}{(x+2)(x-2)}$$

$$7x+5(x+2)=2x$$

$$7x+5x+10=2x$$

$$12x+10=2x$$

$$10=-10x$$

$$-1=x$$

The proposed solution, -1, is not part of the restriction $x\neq-2$ and $x\neq 2$. It checks. The solution is -1 and the solution set is $\{-1\}$.

Section 6.7 Formulas and Applications of Rational Equations

To solve a formula for a variable, get the specified variable alone on one side of the formula. When working with formulas containing rational expressions, it is sometimes necessary to factor out the variable you are solving for.

Solve: $\dfrac{e}{E}=\dfrac{r}{r+R}$ for r.

$$E(r+R)\cdot\frac{e}{E}=E(r+R)\cdot\frac{r}{r+R}\qquad\text{LCD is } E(r+R).$$

$$e(r+R)=Er$$

$$er+eR=Er$$

$$eR=Er-er$$

$$eR=(E-e)r$$

$$\frac{eR}{E-e}=r$$

The average cost function, $\overline{C}$, for a business is a rational function representing the average cost for the company to produce each unit of its product. The function consists of the sum of fixed and variable costs divided by the number of units produced. As production level increases, the average cost to produce each unit of a product decreases.

A company has a fixed cost of \$80,000 monthly and it costs \$20 to produce each unit of its product. Its cost function, C, of producing x units is

$$C(x)=80{,}000+20x.$$

Its average cost function, $\overline{C}$, is

$$\overline{C}(x)=\frac{80{,}000+20x}{x}.$$

Definitions and Concepts	Examples

Section 6.7 Formulas and Applications of Rational Equations (continued)

Problems involving time in motion and problems involving work translate into rational equations. Motion problems involving time are solved using

$$\text{Time traveled} = \frac{\text{Distance traveled}}{\text{Rate of travel}}.$$

Work problems are solved using the following condition:

Fraction of job done by first + fraction of job done by second = 1

It takes a cyclist who averages 16 miles per hour in still air the same time to travel 48 miles with the wind as 16 miles against the wind. What is the wind's rate?

$$x = \text{wind's rate}$$
$$16 + x = \text{cyclist's rate with wind}$$
$$16 - x = \text{cyclist's rate against wind}$$

	Distance	Rate	Time = $\frac{\text{Distance}}{\text{Rate}}$
With wind	48	$16 + x$	$\frac{48}{16 + x}$
Against wind	16	$16 - x$	$\frac{16}{16 - x}$

These times are equal.

$$(16 + x)(16 - x) \cdot \frac{48}{16 + x} = \frac{16}{16 - x} \cdot (16 + x)(16 - x)$$
$$48(16 - x) = 16(16 + x)$$

Solving this equation, $x = 8$. The wind's rate is 8 miles per hour.

Section 6.8 Modeling Using Variation

English Statement	Equation
y varies directly as x. y is directly proportional to x.	$y = kx$
y varies directly as x^n. y is directly proportional to x^n.	$y = kx^n$
y varies inversely as x. y is inversely proportional to x.	$y = \frac{k}{x}$
y varies inversely as x^n. y is inversely proportional to x^n.	$y = \frac{k}{x^n}$
y varies directly as x and inversely as z.	$y = \frac{kx}{z}$
y varies jointly as x and z.	$y = kxz$

Solving Variation Problems

1. Write an equation that models the given English statement.
2. Substitute the pair of values into the equation in step 1 and find k.
3. Substitute k into the equation in step 1.
4. Use the equation in step 3 to answer the problem's question.

The time that it takes you to drive a certain distance varies inversely as your driving rate. Averaging 40 miles per hour, it takes you 10 hours to drive the distance. How long would the trip take averaging 50 miles per hour?

1. $$t = \frac{k}{r}$$ Time, t, varies inversely as rate, r.
2. It takes 10 hours at 40 miles per hour.
$$10 = \frac{k}{40}$$
$$k = 10(40) = 400$$
3. $t = \frac{400}{r}$
4. How long at 50 miles per hour? Substitute 50 for r.
$$t = \frac{400}{50} = 8$$

It takes 8 hours at 50 miles per hour.

CHAPTER 6 REVIEW EXERCISES

6.1

1. If $f(x) = \frac{x^2 + 2x - 3}{x^2 - 4}$, find the following function values. If a function value does not exist, so state.

a. $f(4)$
b. $f(0)$
c. $f(-2)$
d. $f(-3)$

In Exercises 2–3, find the domain of the given rational function.

2. $f(x) = \frac{x - 6}{(x - 3)(x + 4)}$

3. $f(x) = \frac{x + 2}{x^2 + x - 2}$

In Exercises 4–8, simplify each rational expression. If the rational expression cannot be simplified, so state.

4. $\frac{5x^3 - 35x}{15x^2}$

5. $\frac{x^2 + 6x - 7}{x^2 - 49}$

6. $\frac{6x^2 + 7x + 2}{2x^2 - 9x - 5}$

7. $\frac{x^2 + 4}{x^2 - 4}$

8. $\frac{x^3 - 8}{x^2 - 4}$

In Exercises 9–15, multiply or divide as indicated.

9. $\frac{5x^2 - 5}{3x + 12} \cdot \frac{x + 4}{x - 1}$

10. $\frac{2x + 5}{4x^2 + 8x - 5} \cdot \frac{4x^2 - 4x + 1}{x + 1}$

11. $\frac{x^2 - 9x + 14}{x^3 + 2x^2} \cdot \frac{x^2 - 4}{x^2 - 4x + 4}$

12. $\frac{1}{x^2 + 8x + 15} \div \frac{3}{x + 5}$

13. $\frac{x^2 + 16x + 64}{2x^2 - 128} \div \frac{x^2 + 10x + 16}{x^2 - 6x - 16}$

14. $\frac{y^2 - 16}{y^3 - 64} \div \frac{y^2 - 3y - 18}{y^2 + 5y + 6}$

15. $\frac{x^2 - 4x + 4 - y^2}{2x^2 - 11x + 15} \cdot \frac{x^4y}{x - 2 + y} \div \frac{x^3y - 2x^2y - x^2y^2}{3x - 9}$

16. Deer are placed into a newly acquired habitat. The deer population over time is modeled by a rational function whose graph is shown in the figure. Use the graph in the next column to answer each of the following questions.

a. How many deer were introduced into the habitat?

b. What is the population after 10 years?

c. What is the equation of the horizontal asymptote shown in the figure? What does this mean in terms of the deer population?

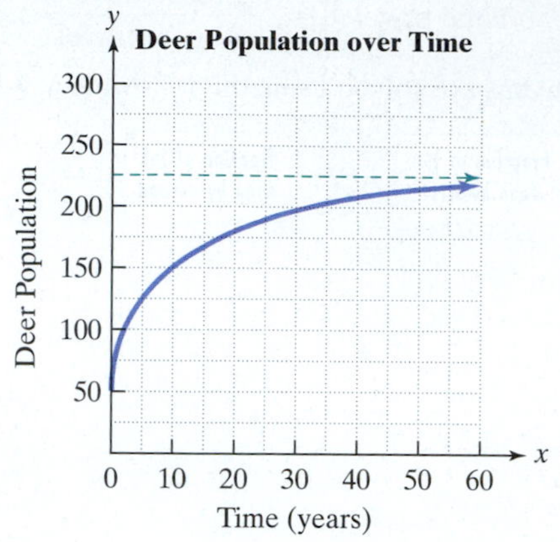

6.2

In Exercises 17–19, add or subtract as indicated. Simplify the result, if possible.

17. $\frac{4x + 1}{3x - 1} + \frac{8x - 5}{3x - 1}$

18. $\frac{2x - 7}{x^2 - 9} - \frac{x - 4}{x^2 - 9}$

19. $\frac{4x^2 - 11x + 4}{x - 3} - \frac{x^2 - 4x + 10}{x - 3}$

In Exercises 20–21, find the least common denominator of the rational expressions.

20. $\frac{7}{9x^3}$ and $\frac{5}{12x}$

21. $\frac{x + 7}{x^2 + 2x - 35}$ and $\frac{x}{x^2 + 9x + 14}$

In Exercises 22–28, perform the indicated operations. Simplify the result, if possible.

22. $\frac{1}{x} + \frac{2}{x - 5}$

23. $\frac{2}{x^2 - 5x + 6} + \frac{3}{x^2 - x - 6}$

24. $\frac{x - 3}{x^2 - 8x + 15} + \frac{x + 2}{x^2 - x - 6}$

25. $\frac{3x^2}{9x^2 - 16} - \frac{x}{3x + 4}$

26. $\frac{y}{y^2 + 5y + 6} - \frac{2}{y^2 + 3y + 2}$

27. $\frac{x}{x + 3} + \frac{x}{x - 3} - \frac{9}{x^2 - 9}$

28. $\frac{3x^2}{x - y} + \frac{3y^2}{y - x}$

6.3 *In Exercises 29–34, simplify each complex rational expression.*

29. $\dfrac{\frac{3}{x} - 3}{\frac{8}{x} - 8}$

30. $\dfrac{\frac{5}{x} + 1}{1 - \frac{25}{x^2}}$

31. $\dfrac{3 - \frac{1}{x+3}}{3 + \frac{1}{x+3}}$

32. $\dfrac{\frac{4}{x+3}}{\frac{2}{x-2} - \frac{1}{x^2 + x - 6}}$

33. $\dfrac{\frac{2}{x^2 - x - 6} + \frac{1}{x^2 - 4x + 3}}{\frac{3}{x^2 + x - 2} - \frac{2}{x^2 + 5x + 6}}$

34. $\dfrac{x^{-2} + x^{-1}}{x^{-2} - x^{-1}}$

6.4 *In Exercises 35–36, divide the polynomial by the monomial.*

35. $(15x^3 - 30x^2 + 10x - 2) \div (5x^2)$

36. $(36x^4y^3 + 12x^2y^3 - 60x^2y^2) \div (6xy^2)$

In Exercises 37–40, divide as indicated.

37. $(6x^2 - 5x + 5) \div (2x + 3)$

38. $(10x^3 - 26x^2 + 17x - 13) \div (5x - 3)$

39. $(x^6 + 3x^5 - 2x^4 + x^2 - 3x + 2) \div (x - 2)$

40. $(4x^4 + 6x^3 + 3x - 1) \div (2x^2 + 1)$

6.5 *In Exercises 41–43, divide using synthetic division.*

41. $(4x^3 - 3x^2 - 2x + 1) \div (x + 1)$

42. $(3x^4 - 2x^2 - 10x - 20) \div (x - 2)$

43. $(x^4 + 16) \div (x + 4)$

In Exercises 44–45, use synthetic division and the Remainder Theorem to find the indicated function value.

44. $f(x) = 2x^3 - 5x^2 + 4x - 1;\ f(2)$

45. $f(x) = 3x^4 + 7x^3 + 8x^2 + 2x + 4;\ f\left(-\frac{1}{3}\right)$

In Exercises 46–47, use synthetic division to determine whether or not the number given to the right of each equation is a solution of the equation.

46. $2x^3 - x^2 - 8x + 4 = 0;\ -2$

47. $x^4 - x^3 - 7x^2 + x + 6 = 0;\ 4$

48. Use synthetic division to show that $\frac{1}{2}$ is a solution of

$$6x^3 + x^2 - 4x + 1 = 0.$$

Then solve the polynomial equation.

6.6 *In Exercises 49–55, solve each rational equation. If an equation has no solution, so state.*

49. $\dfrac{3}{x} + \dfrac{1}{3} = \dfrac{5}{x}$

50. $\dfrac{5}{3x + 4} = \dfrac{3}{2x - 8}$

51. $\dfrac{1}{x - 5} - \dfrac{3}{x + 5} = \dfrac{6}{x^2 - 25}$

52. $\dfrac{x + 5}{x + 1} - \dfrac{x}{x + 2} = \dfrac{4x + 1}{x^2 + 3x + 2}$

53. $\dfrac{2}{3} - \dfrac{5}{3x} = \dfrac{1}{x^2}$

54. $\dfrac{2}{x - 1} = \dfrac{1}{4} + \dfrac{7}{x + 2}$

55. $\dfrac{2x + 7}{x + 5} - \dfrac{x - 8}{x - 4} = \dfrac{x + 18}{x^2 + x - 20}$

56. The function

$$f(x) = \frac{4x}{100 - x}$$

models the cost, $f(x)$, in millions of dollars, to remove $x\%$ of pollutants from a river due to pesticide runoff from area farms. What percentage of the pollutants can be removed for \$16 million?

6.7 *In Exercises 57–61, solve each formula for the specified variable.*

57. $P = \dfrac{R - C}{n}$ for C

58. $\dfrac{P_1V_1}{T_1} = \dfrac{P_2V_2}{T_2}$ for T_1

59. $T = \dfrac{A - P}{Pr}$ for P

60. $\dfrac{1}{R} = \dfrac{1}{R_1} + \dfrac{1}{R_2}$ for R

61. $I = \dfrac{nE}{R + nr}$ for n

62. A company is planning to manufacture affordable graphing calculators. Fixed monthly cost will be \$50,000, and it will cost \$25 to produce each calculator.

a. Write the cost function, C, of producing x graphing calculators.

b. Write the average cost function, $\overline{C}$, of producing x graphing calculators.

c. How many graphing calculators must be produced each month for the company to have an average cost of \$35 per graphing calculator?

63. After riding at a steady rate for 60 miles, a bicyclist had a flat tire and walked 8 miles to a repair shop. The cycling rate was 3 times faster than the walking rate. If the time spent cycling and walking was 7 hours, at what rate was the cyclist riding?

64. The current of a river moves at 3 miles per hour. It takes a boat a total of 3 hours to travel 12 miles upstream, against the current, and return the same distance traveling downstream, with the current. What is the boat's rate in still water?

65. Working alone, two people can clean their house in 3 hours and 6 hours, respectively. They have agreed to clean together so that they can finish in time to watch a TV program that begins in $1\frac{1}{2}$ hours. How long will it take them to clean the house working together? Can they finish before the program starts?

66. Working together, two crews can clear snow from the city's streets in 20 hours. Working alone, the faster crew requires 9 hours less time than the slower crew. How many hours would it take each crew to clear the streets working alone?

67. An inlet faucet can fill a small pond in 60 minutes. The pond can be emptied by an outlet pipe in 80 minutes. You begin filling the empty pond. By accident, the outlet pipe that empties the pond is left open. Under these conditions, how long will it take for the pond to fill?

6.8 *Solve the variation problems in Exercises 68–73.*

68. A company's profit varies directly as the number of products it sells. The company makes a profit of \$1175 on the sale of 25 products. What is the company's profit when it sells 105 products?

69. The distance that a body falls from rest varies directly as the square of the time of the fall. If skydivers fall 144 feet in 3 seconds, how far will they fall in 10 seconds?

70. The pitch of a musical tone varies inversely as its wavelength. A tone has a pitch of 660 vibrations per second and a wavelength of 1.6 feet. What is the pitch of a tone that has a wavelength of 2.4 feet?

71. The loudness of a stereo speaker, measured in decibels, varies inversely as the square of your distance from the speaker. When you are 8 feet from the speaker, the loudness is 28 decibels. What is the loudness when you are 4 feet from the speaker?

72. The time required to assemble computers varies directly as the number of computers assembled and inversely as the number of workers. If 30 computers can be assembled by 6 workers in 10 hours, how long would it take 5 workers to assemble 40 computers?

73. The volume of a pyramid varies jointly as its height and the area of its base. A pyramid with a height of 15 feet and a base with an area of 35 square feet has a volume of 175 cubic feet. Find the volume of a pyramid with a height of 20 feet and a base with an area of 120 square feet.

CHAPTER 6 TEST

Remember to use your Chapter Test Prep Video CD to see the worked-out solutions to the test questions you want to review.

1. Find the domain of $f(x) = \dfrac{x^2 - 2x}{x^2 - 7x + 10}$. Then simplify the right side of the function's equation.

In Exercises 2–11, perform the indicated operations. Simplify where possible.

2. $\dfrac{x^2}{x^2 - 16} \cdot \dfrac{x^2 + 7x + 12}{x^2 + 3x}$

3. $\dfrac{x^3 + 27}{x^2 - 1} \div \dfrac{x^2 - 3x + 9}{x^2 - 2x + 1}$

4. $\dfrac{x^2 + 3x - 10}{x^2 + 4x + 3} \cdot \dfrac{x^2 + x - 6}{x^2 + 10x + 25} \cdot \dfrac{x + 1}{x - 2}$

5. $\dfrac{x^2 - 6x - 16}{x^3 + 3x^2 + 2x} \cdot (x^2 - 3x - 4) \div \dfrac{x^2 - 7x + 12}{3x}$

6. $\dfrac{x^2 - 5x - 2}{6x^2 - 11x - 35} - \dfrac{x^2 - 7x + 5}{6x^2 - 11x - 35}$

7. $\dfrac{x}{x + 3} + \dfrac{5}{x - 3}$

8. $\dfrac{2}{x^2 - 4x + 3} + \dfrac{3x}{x^2 + x - 2}$

9. $\dfrac{5x}{x^2 - 4} - \dfrac{2}{x^2 + x - 2}$

10. $\dfrac{x - 4}{x - 5} - \dfrac{3}{x + 5} - \dfrac{10}{x^2 - 25}$

11. $\dfrac{1}{10 - x} + \dfrac{x - 1}{x - 10}$

In Exercises 12–13, simplify each rational expression.

12. $\dfrac{\dfrac{x}{4} - \dfrac{1}{x}}{1 + \dfrac{x + 4}{x}}$

13. $\dfrac{\dfrac{1}{x} - \dfrac{3}{x + 2}}{\dfrac{2}{x^2 + 2x}}$

In Exercises 14–16, divide as indicated.

14. $(12x^4y^3 + 16x^2y^3 - 10x^2y^2) \div (4x^2y)$

15. $(9x^3 - 3x^2 - 3x + 4) \div (3x + 2)$

16. $(3x^4 + 2x^3 - 8x + 6) \div (x^2 - 1)$

17. Divide using synthetic division:

$$(3x^4 + 11x^3 - 20x^2 + 7x + 35) \div (x + 5).$$

18. Given that

$$f(x) = x^4 - 2x^3 - 11x^2 + 5x + 34,$$

use synthetic division and the Remainder Theorem to find $f(-2)$.

19. Use synthetic division to decide whether -2 is a solution of $2x^3 - 3x^2 - 11x + 6 = 0$.

In Exercises 20–21, solve each rational equation.

20. $\dfrac{x}{x+4} = \dfrac{11}{x^2 - 16} + 2$

21. $\dfrac{x+1}{x^2 + 2x - 3} - \dfrac{1}{x+3} = \dfrac{1}{x-1}$

22. Park rangers introduce 50 elk into a wildlife preserve. The function

$$f(t) = \frac{250(3t + 5)}{t + 25}$$

models the elk population, $f(t)$, after t years. How many years will it take for the population to increase to 125 elk?

23. Solve for a: $R = \dfrac{as}{a + s}$.

24. A company is planning to manufacture portable satellite radio players. Fixed monthly cost will be $300,000 and it will cost $10 to produce each player.

a. Write the cost function, C, of producing x players.

b. Write the average cost function, $\overline{C}$, of producing x players.

c. How many portable satellite radio players must be produced each month for the company to have an average cost of $25 per player?

25. It takes one pipe 3 hours to fill a pool and a second pipe 4 hours to drain the pool. The pool is empty and the first pipe begins to fill it. The second pipe is accidently left open, so the water is also draining out of the pool. Under these conditions, how long will it take to fill the pool?

26. A motorboat averages 20 miles per hour in still water. It takes the boat the same amount of time to travel 3 miles with the current as it does to travel 2 miles against the current. What is the current's rate?

27. The intensity of light received at a source varies inversely as the square of the distance from the source. A particular light has an intensity of 20 foot-candles at 15 feet. What is the light's intensity at 10 feet?

CUMULATIVE REVIEW EXERCISES (CHAPTERS 1–6)

In Exercises 1–5, solve each equation, inequality, or system.

1. $2x + 5 \le 11$ and $-3x > 18$

2. $2x^2 = 7x + 4$

3.
$$\begin{aligned} 4x + 3y + 3z &= 4 \\ 3x \qquad + 2z &= 2 \\ 2x - 5y \qquad &= -4 \end{aligned}$$

4. $|3x - 4| \le 10$

5. $\dfrac{x}{x-8} + \dfrac{6}{x-2} = \dfrac{x^2}{x^2 - 10x + 16}$

6. Solve for s: $I = \dfrac{2R}{w + 2s}$.

7. Solve by graphing:
$$\begin{aligned} 2x - y &= 4 \\ x + y &= 5. \end{aligned}$$

8. Use function notation to write the equation of the line with slope -3 and passing through the point $(1, -5)$.

In Exercises 9–11, graph each equation, inequality, or system in a rectangular coordinate system.

9. $y = |x| + 2$

10. $y \ge 2x - 1$
$x \ge 1$

11. $2x - y < 4$

In Exercises 12–15, perform the indicated operations.

12. $[(x + 2) + 3y][(x + 2) - 3y]$

13. $\dfrac{2x^2 + x - 1}{2x^2 - 9x + 4} \div \dfrac{6x + 15}{3x^2 - 12x}$

14. $\dfrac{3x}{x^2 - 9x + 20} - \dfrac{5}{2x - 8}$

15. $(3x^2 + 10x + 10) \div (x + 2)$

In Exercises 16–17, factor completely.

16. $xy - 6x + 2y - 12$

17. $24x^3y + 16x^2y - 30xy$

18. A baseball is thrown straight up from a height of 64 feet. The function

$$s(t) = -16t^2 + 48t + 64$$

describes the ball's height above the ground, $s(t)$, in feet, t seconds after it is thrown. How long will it take for the ball to hit the ground?

19. The local cable television company offers two deals. Basic cable service with one movie channel costs \$35 per month. Basic service with two movie channels costs \$45 per month. Find the charge for the basic cable service and the charge for each movie channel.

20. A rectangular garden 10 feet wide and 12 feet long is surrounded by a rock border of uniform width. The area of the garden and rock border combined is 168 square feet. What is the width of the rock border?

CHAPTER 7

Radicals, Radical Functions, and Rational Exponents

Can mathematical models be created for events that appear to involve random behavior, such as stock market fluctuations or air turbulence? Chaos theory, a new frontier of mathematics, offers models and computer-generated images that reveal order and underlying patterns where only the erratic and the unpredictable had been observed. Because most behavior is chaotic, the computer has become a canvas that looks more like the real world than anything previously seen. Magnified portions of these computer images yield repetitions of the original structure, as well as new and unexpected patterns. The computer generates these visualizations of chaos by plotting large numbers of points for functions whose domains are nonreal numbers involving the square root of negative one.

We define $\sqrt{-1}$ in Section 7.7 and hint at chaotic possibilities in the Blitzer Bonus on page 551. If you are intrigued by how the operations of nonreal numbers in Section 7.7 reveal that the world is not random (rather, the underlying patterns are far more intricate than we had previously assumed), we suggest reading *Chaos* by James Gleick, published by Penguin Books.

SECTION 7.1

Radical Expressions and Functions

Objectives

1. Evaluate square roots.
2. Evaluate square root functions.
3. Find the domain of square root functions.
4. Use models that are square root functions.
5. Simplify expressions of the form $\sqrt{a^2}$.
6. Evaluate cube root functions.
7. Simplify expressions of the form $\sqrt[3]{a^3}$.
8. Find even and odd roots.
9. Simplify expressions of the form $\sqrt[n]{a^n}$.

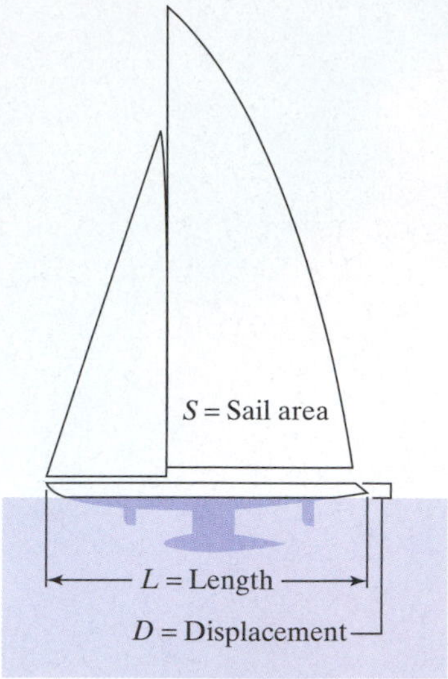

The America's Cup is the supreme event in ocean sailing. Competition is fierce and the costs are huge. Competitors look to mathematics to provide the critical innovation that can make the difference between winning and losing. The basic dimensions of competitors' yachts must satisfy an inequality containing square roots and cube roots:

$$L + 1.25\sqrt{S} - 9.8\sqrt[3]{D} \leq 16.296.$$

In the inequality, L is the yacht's length, in meters, S is its sail area, in square meters, and D is its displacement, in cubic meters.

In this section, we introduce a new category of expressions and functions that contain roots. You will see why square root functions are used to describe phenomena that are continuing to grow but whose growth is leveling off.

1 Evaluate square roots.

Square Roots

From our earlier work with exponents, we are aware that the square of 5 and the square of -5 are both 25:

$$5^2 = 25 \quad \text{and} \quad (-5)^2 = 25.$$

The reverse operation of squaring a number is finding the *square root* of the number. For example,

- One square root of 25 is 5 because $5^2 = 25$.
- Another square root of 25 is -5 because $(-5)^2 = 25$.

In general, **if $b^2 = a$, then b is a square root of a.**

The symbol $\sqrt{}$ is used to denote the *positive* or *principal square root* of a number. For example,

- $\sqrt{25} = 5$ because $5^2 = 25$ and 5 is positive.
- $\sqrt{100} = 10$ because $10^2 = 100$ and 10 is positive.

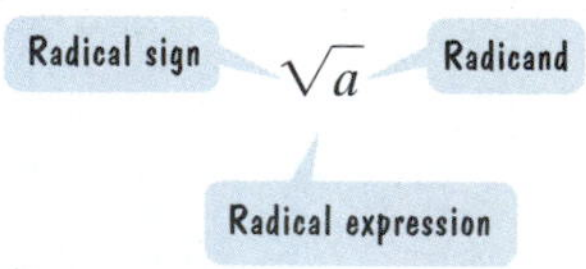

The symbol $\sqrt{}$ that we use to denote the principal square root is called a **radical sign**. The number under the radical sign is called the **radicand**. Together we refer to the radical sign and its radicand as a **radical expression**.

Definition of the Principal Square Root

If a is a nonnegative real number, the nonnegative number b such that $b^2 = a$, denoted by $b = \sqrt{a}$, is the **principal square root** of a.

The symbol $-\sqrt{\ }$ is used to denote the negative square root of a number. For example,

- $-\sqrt{25} = -5$ because $(-5)^2 = 25$ and -5 is negative.
- $-\sqrt{100} = -10$ because $(-10)^2 = 100$ and -10 is negative.

EXAMPLE 1 Evaluating Square Roots

Evaluate:

a. $\sqrt{81}$ **b.** $-\sqrt{9}$ **c.** $\sqrt{\frac{4}{49}}$

d. $\sqrt{0.0064}$ **e.** $\sqrt{36 + 64}$ **f.** $\sqrt{36} + \sqrt{64}$.

Solution

a. $\sqrt{81} = 9$ The principal square root of 81 is 9 because $9^2 = 81$.

b. $-\sqrt{9} = -3$ The negative square root of 9 is -3 because $(-3)^2 = 9$.

c. $\sqrt{\frac{4}{49}} = \frac{2}{7}$ The principal square root of $\frac{4}{49}$ is $\frac{2}{7}$ because $\left(\frac{2}{7}\right)^2 = \frac{4}{49}$.

d. $\sqrt{0.0064} = 0.08$ The principal square root of 0.0064 is 0.08 because $(0.08)^2 = (0.08)(0.08) = 0.0064$.

e. $\sqrt{36 + 64} = \sqrt{100}$ Simplify the radicand.

$= 10$ Take the principal square root of 100, which is 10.

f. $\sqrt{36} + \sqrt{64} = 6 + 8$ $\sqrt{36} = 6$ because $6^2 = 36$. $\sqrt{64} = 8$ because $8^2 = 64$.

$= 14$

Study Tip

In Example 1, parts (e) and (f), observe that $\sqrt{36 + 64}$ is not equal to $\sqrt{36} + \sqrt{64}$. In general,

$$\sqrt{a + b} \neq \sqrt{a} + \sqrt{b}$$

and

$$\sqrt{a - b} \neq \sqrt{a} - \sqrt{b}.$$

✓ CHECK POINT 1 Evaluate:

a. $\sqrt{64}$ **b.** $-\sqrt{49}$ **c.** $\sqrt{\frac{16}{25}}$

d. $\sqrt{0.0081}$ **e.** $\sqrt{9 + 16}$ **f.** $\sqrt{9} + \sqrt{16}$.

Let's see what happens to the radical expression $\sqrt{x}$ if x is a negative number. Is the square root of a negative number a real number? For example, consider $\sqrt{-25}$. Is there a real number whose square is -25? No. Thus, $\sqrt{-25}$ is not a real number. In general, **a square root of a negative number is not a real number**.

2 Evaluate square root functions.

Square Root Functions

Because each nonnegative real number, x, has precisely one principal square root, $\sqrt{x}$, there is a **square root function** defined by

$$f(x) = \sqrt{x}.$$

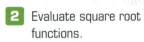

The domain of this function is $[0, \infty)$. We can graph $f(x) = \sqrt{x}$ by selecting nonnegative real numbers for x. It is easiest to choose perfect squares, numbers that have rational square roots. **Table 7.1** shows five such choices for x and the calculations for the corresponding outputs. We plot these ordered pairs as points in the rectangular coordinate system and connect the points with a smooth curve. The graph of $f(x) = \sqrt{x}$ is shown in **Figure 7.1**.

Table 7.1

x	$f(x) = \sqrt{x}$	(x, y) or $(x, f(x))$
0	$f(0) = \sqrt{0} = 0$	$(0, 0)$
1	$f(1) = \sqrt{1} = 1$	$(1, 1)$
4	$f(4) = \sqrt{4} = 2$	$(4, 2)$
9	$f(9) = \sqrt{9} = 3$	$(9, 3)$
16	$f(16) = \sqrt{16} = 4$	$(16, 4)$

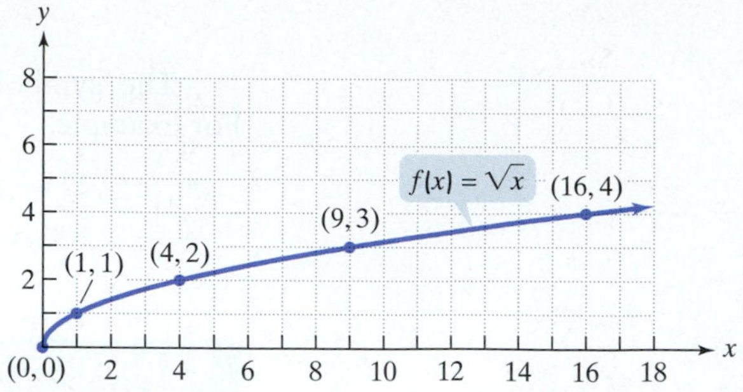

FIGURE 7.1 The graph of the square root function $f(x) = \sqrt{x}$

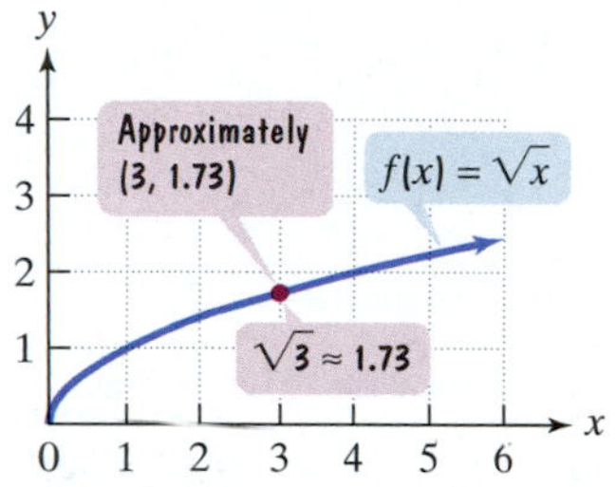

FIGURE 7.2 Visualizing $\sqrt{3}$ as a point on the graph of $f(x) = \sqrt{x}$

Is it possible to choose values of x for **Table 7.1** that are not squares of integers, or perfect squares? Yes. For example, we can let $x = 3$. Thus, $f(3) = \sqrt{3}$. Because 3 is not a perfect square, $\sqrt{3}$ is an irrational number, one that cannot be expressed as a quotient of integers. We can use a calculator to find a decimal approximation of $\sqrt{3}$.

Many Scientific Calculators	**Many Graphing Calculators**
3 [√]	[√] 3 [ENTER]

Rounding the displayed number to two decimal places, $\sqrt{3} \approx 1.73$. This information is shown visually as a point, approximately (3, 1.73), on the graph of $f(x) = \sqrt{x}$ in **Figure 7.2**.

To evaluate a square root function, we use substitution, just as we have done to evaluate other functions.

EXAMPLE 2 Evaluating Square Root Functions

For each function, find the indicated function value:

a. $f(x) = \sqrt{5x - 6};\ f(2)$ **b.** $g(x) = -\sqrt{64 - 8x};\ g(-3)$.

Solution

a. $f(2) = \sqrt{5 \cdot 2 - 6}$ — Substitute 2 for x in $f(x) = \sqrt{5x - 6}$.

$= \sqrt{4} = 2$ — Simplify the radicand and take the square root.

b. $g(-3) = -\sqrt{64 - 8(-3)}$ — Substitute −3 for x in $g(x) = -\sqrt{64 - 8x}$.

$= -\sqrt{88} \approx -9.38$ — Simplify the radicand: $64 - 8(-3) = 64 - (-24) = 64 + 24 = 88$. Then use a calculator to approximate $\sqrt{88}$.

✓ CHECK POINT 2 For each function, find the indicated function value:

a. $f(x) = \sqrt{12x - 20};\ f(3)$

b. $g(x) = -\sqrt{9 - 3x};\ g(-5)$.

3 Find the domain of square root functions.

We have seen that the domain of a function f is the largest set of real numbers for which the value of $f(x)$ is a real number. Because only nonnegative numbers have real square roots, the domain of a square root function is the set of real numbers for which the radicand is nonnegative.

EXAMPLE 3 Finding the Domain of a Square Root Function

Find the domain of

$$f(x) = \sqrt{3x + 12}.$$

Solution The domain is the set of real numbers, x, for which the radicand, $3x + 12$, is nonnegative. We set the radicand greater than or equal to 0 and solve the resulting inequality.

$$3x + 12 \geq 0$$
$$3x \geq -12$$
$$x \geq -4$$

The domain of f is $\{x|x \geq -4\}$ or $[-4, \infty)$.

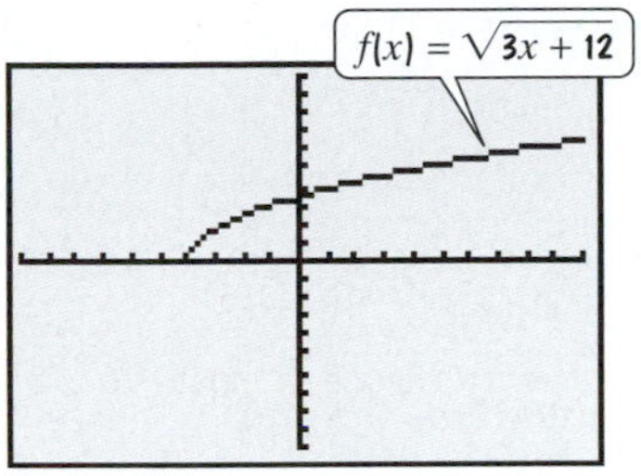

FIGURE 7.3

Figure 7.3 shows the graph of $f(x) = \sqrt{3x + 12}$ in a $[-10, 10, 1]$ by $[-10, 10, 1]$ viewing rectangle. The graph appears only for $x \geq -4$, verifying $[-4, \infty)$ as the domain. Can you see how the graph also illustrates this square root function's range? The graph only appears for nonnegative values of y. Thus, the range is $\{y|y \geq 0\}$ or $[0, \infty)$.

✓ **CHECK POINT 3** Find the domain of

$$f(x) = \sqrt{9x - 27}.$$

4 Use models that are square root functions.

The graph of the square root function $f(x) = \sqrt{x}$ is increasing from left to right. However, the rate of increase is slowing down as the graph moves to the right. This is why square root functions are often used to model growing phenomena with growth that is leveling off.

EXAMPLE 4 Modeling with a Square Root Function

By 2005, the amount of "clutter," including commercials and plugs for other shows, had increased to the point where an "hour-long" drama on cable TV was 45.4 minutes. The graph in **Figure 7.4** shows the average number of nonprogram minutes in an hour of prime-time cable television. Although the minutes of clutter grew from 1996 through 2005, the growth was leveling off. The data can be modeled by the function

$$M(x) = 0.7\sqrt{x} + 12.5,$$

where $M(x)$ is the average number of nonprogram minutes in an hour of prime-time cable x years after 1996. According to the model, in 2002, how many cluttered minutes disrupted cable TV action in an hour? Round to the nearest tenth of a minute. What is the difference between the actual data and the number of minutes that you obtained?

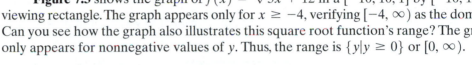

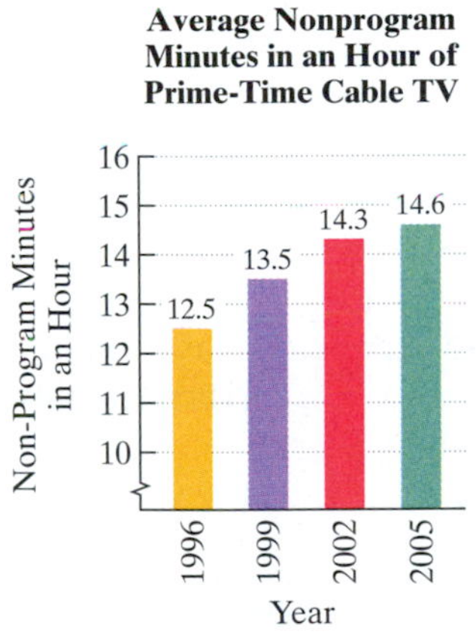

FIGURE 7.4

Source: Nielsen Monitor-Plus

Solution Because 2002 is 6 years after 1996, we substitute 6 for x and evaluate the function at 6.

$$M(x) = 0.7\sqrt{x} + 12.5 \quad \text{Use the given function.}$$
$$M(6) = 0.7\sqrt{6} + 12.5 \quad \text{Substitute 6 for x.}$$
$$\approx 14.2 \quad \text{Use a calculator.}$$

The model indicates that there were approximately 14.2 nonprogram minutes in an hour of prime-time cable in 2002. **Figure 7.4** shows 14.3 minutes, so the difference is $14.3 - 14.2$, or 0.1 minute.

✓ **CHECK POINT 4** If the trend from 1996 through 2005 continues, use the square root function in Example 4 to predict how many cluttered minutes, rounded to the nearest tenth, there will be in an hour in 2010.

5 Simplify expressions of the form $\sqrt{a^2}$.

Simplifying Expressions of the Form $\sqrt{a^2}$

You may think that $\sqrt{a^2} = a$. However, this is not necessarily true. Consider the following examples:

$$\sqrt{4^2} = \sqrt{16} = 4$$
$$\sqrt{(-4)^2} = \sqrt{16} = 4.$$

The result is not −4, but rather the absolute value of −4, or 4.

Here is a rule for simplifying expressions of the form $\sqrt{a^2}$:

Simplifying $\sqrt{a^2}$

For any real number a,

$$\sqrt{a^2} = |a|.$$

In words, the principal square root of a^2 is the absolute value of a.

Using Technology

Graphic Connections

The graphs of $f(x) = \sqrt{x^2}$ and $g(x) = |x|$ are shown in a $[-10, 10, 1]$ by $[-2, 10, 1]$ viewing rectangle. The graphs are the same. Thus,

$$\sqrt{x^2} = |x|.$$

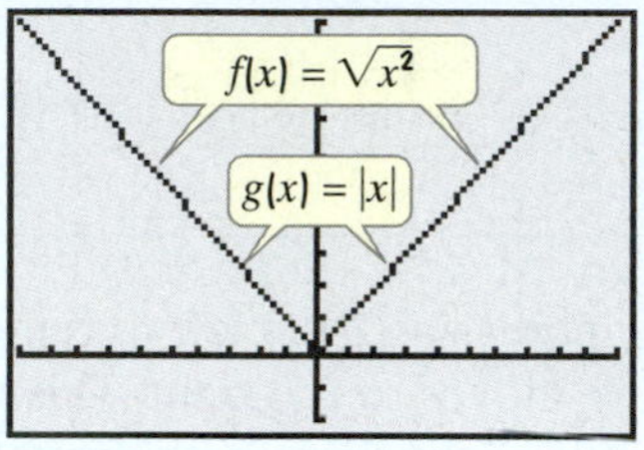

EXAMPLE 5 Simplifying Radical Expressions

Simplify each expression:

a. $\sqrt{(-6)^2}$ **b.** $\sqrt{(x+5)^2}$ **c.** $\sqrt{25x^6}$ **d.** $\sqrt{x^2 - 4x + 4}$.

Solution The principal square root of an expression squared is the absolute value of that expression. In parts (a) and (b), we are given squared radicands. In parts (c) and (d), it will first be necessary to express the radicand as an expression that is squared.

a. $\sqrt{(-6)^2} = |-6| = 6$

b. $\sqrt{(x+5)^2} = |x+5|$

c. To simplify $\sqrt{25x^6}$, first write $25x^6$ as an expression that is squared: $25x^6 = (5x^3)^2$. Then simplify.

$$\sqrt{25x^6} = \sqrt{(5x^3)^2} = |5x^3| \text{ or } 5|x^3|$$

d. To simplify $\sqrt{x^2 - 4x + 4}$, first write $x^2 - 4x + 4$ as an expression that is squared by factoring the perfect square trinomial: $x^2 - 4x + 4 = (x-2)^2$. Then simplify.

$$\sqrt{x^2 - 4x + 4} = \sqrt{(x-2)^2} = |x - 2|$$

✓ CHECK POINT 5 Simplify each expression:

a. $\sqrt{(-7)^2}$

b. $\sqrt{(x+8)^2}$

c. $\sqrt{49x^{10}}$

d. $\sqrt{x^2 - 6x + 9}$.

In some situations, we are told that no radicands involve negative quantities raised to even powers. When the expression being squared is nonnegative, it is not necessary to use absolute value when simplifying $\sqrt{a^2}$. For example, assuming that no radicands contain negative quantities that are squared,

$$\sqrt{x^6} = \sqrt{(x^3)^2} = x^3$$
$$\sqrt{25x^2 + 10x + 1} = \sqrt{(5x+1)^2} = 5x + 1.$$

6 Evaluate cube root
functions.

Cube Roots and Cube Root Functions

Finding the square root of a number reverses the process of squaring a number. Similarly, finding the cube root of a number reverses the process of cubing a number. For example, $2^3 = 8$, and so the cube root of 8 is 2. The notation that we use is $\sqrt[3]{8} = 2$.

> ### Definition of the Cube Root of a Number
> The **cube root** of a real number a is written $\sqrt[3]{a}$.
>
> $$\sqrt[3]{a} = b \quad \text{means that} \quad b^3 = a.$$

For example,

$$\sqrt[3]{64} = 4 \quad \text{because} \quad 4^3 = 64.$$
$$\sqrt[3]{-27} = -3 \quad \text{because} \quad (-3)^3 = -27.$$

In contrast to square roots, the cube root of a negative number is a real number. All real numbers have cube roots. The cube root of a positive number is positive. The cube root of a negative number is negative.

Because every real number, x, has precisely one cube root, $\sqrt[3]{x}$, there is a **cube root function** defined by

$$f(x) = \sqrt[3]{x}.$$

The domain of this function is the set of all real numbers. We can graph $f(x) = \sqrt[3]{x}$ by selecting perfect cubes, numbers that have rational cube roots, for x. **Table 7.2** shows five such choices for x and the calculations for the corresponding outputs. We plot these ordered pairs as points in the rectangular coordinate system and connect the points with a smooth curve. The graph of $f(x) = \sqrt[3]{x}$ is shown in **Figure 7.5**.

Study Tip

Some cube roots occur so frequently that you might want to memorize them.

$$\sqrt[3]{1} = 1$$
$$\sqrt[3]{8} = 2$$
$$\sqrt[3]{27} = 3$$
$$\sqrt[3]{64} = 4$$
$$\sqrt[3]{125} = 5$$
$$\sqrt[3]{216} = 6$$
$$\sqrt[3]{1000} = 10$$

Table 7.2

x	$f(x) = \sqrt[3]{x}$	(x, y) or $(x, f(x))$
-8	$f(-8) = \sqrt[3]{-8} = -2$	$(-8, -2)$
-1	$f(-1) = \sqrt[3]{-1} = -1$	$(-1, -1)$
0	$f(0) = \sqrt[3]{0} = 0$	$(0, 0)$
1	$f(1) = \sqrt[3]{1} = 1$	$(1, 1)$
8	$f(8) = \sqrt[3]{8} = 2$	$(8, 2)$

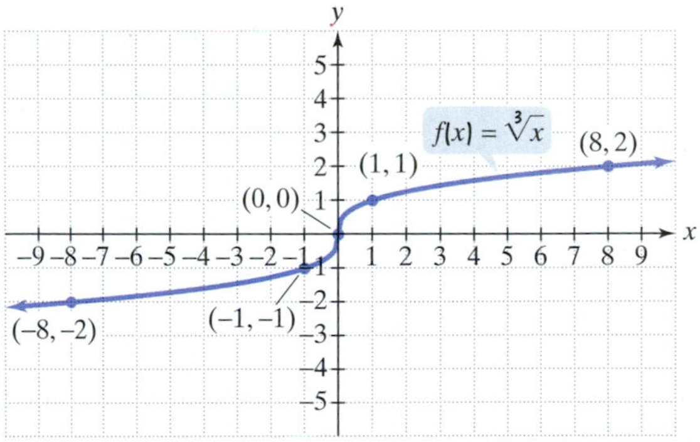

FIGURE 7.5 The graph of the cube root function $f(x) = \sqrt[3]{x}$

Notice that both the domain and the range of $f(x) = \sqrt[3]{x}$ are the set of all real numbers.

EXAMPLE 6   Evaluating Cube Root Functions

For each function, find the indicated function value:

a. $f(x) = \sqrt[3]{x - 2}; \quad f(127)$ **b.** $g(x) = \sqrt[3]{8x - 8}; \quad g(-7).$

Solution

a. $f(x) = \sqrt[3]{x-2}$ — This is the given function.

$f(127) = \sqrt[3]{127-2}$ — Substitute 127 for x.

$= \sqrt[3]{125}$ — Simplify the radicand.

$= 5$ — $\sqrt[3]{125} = 5$ because $5^3 = 125$.

b. $g(x) = \sqrt[3]{8x-8}$ — This is the given function.

$g(-7) = \sqrt[3]{8(-7)-8}$ — Substitute −7 for x.

$= \sqrt[3]{-64}$ — Simplify the radicand: $8(-7) - 8 = -56 - 8 = -64$.

$= -4$ — $\sqrt[3]{-64} = -4$ because $(-4)^3 = -64$.

CHECK POINT 6 For each function, find the indicated function value:

a. $f(x) = \sqrt[3]{x-6};\ f(33)$

b. $g(x) = \sqrt[3]{2x+2};\ g(-5)$.

7 Simplify expressions of the form $\sqrt[3]{a^3}$.

Because the cube root of a positive number is positive and the cube root of a negative number is negative, absolute value is not needed to simplify expressions of the form $\sqrt[3]{a^3}$.

Simplifying $\sqrt[3]{a^3}$

For any real number a,

$$\sqrt[3]{a^3} = a.$$

In words, the cube root of any expression cubed is that expression.

EXAMPLE 7 Simplifying a Cube Root

Simplify: $\sqrt[3]{-64x^3}$.

Solution Begin by expressing the radicand as an expression that is cubed: $-64x^3 = (-4x)^3$. Then simplify.

$$\sqrt[3]{-64x^3} = \sqrt[3]{(-4x)^3} = -4x$$

We can check our answer by cubing $-4x$:

$$(-4x)^3 = (-4)^3x^3 = -64x^3.$$

By obtaining the original radicand, we know that our simplification is correct.

CHECK POINT 7 Simplify: $\sqrt[3]{-27x^3}$.

8 Find even and odd roots.

Even and Odd *n*th Roots

Up to this point, we have focused on square roots and cube roots. Other radical expressions have different roots. For example, the fifth root of a, written $\sqrt[5]{a}$, is the number b for which $b^5 = a$. Thus,

$$\sqrt[5]{32} = 2 \qquad \text{because} \qquad 2^5 = 2\cdot2\cdot2\cdot2\cdot2 = 32.$$

The radical expression $\sqrt[n]{a}$ represents the ***n*th root** of a. The number n is called the **index**. An index of 2 represents a square root and is not written. An index of 3 represents a cube root.

If the index n in $\sqrt[n]{a}$ is an odd number, a root is said to be an **odd root**. A cube root is an odd root. Other odd roots have the same characteristics as cube roots.

- Every real number has exactly one real root when n is odd. An odd root of a positive number is positive and an odd root of a negative number is negative.

$$3^5 = 3 \cdot 3 \cdot 3 \cdot 3 \cdot 3 = 243, \text{ so the fifth root of 243 is 3.}$$
$$(-3)^5 = (-3)(-3)(-3)(-3)(-3) = -243, \text{ so the fifth root of } -243 \text{ is } -3.$$

- The (odd) nth root of a, $\sqrt[n]{a}$, is the number b for which $b^n = a$.

$$\sqrt[5]{243} = 3 \qquad \sqrt[5]{-243} = -3$$

$3^5 = 243$ $\qquad$ $(-3)^5 = -243$

Study Tip

Some higher even and odd roots occur so frequently that you might want to memorize them.

Fourth Roots	Fifth Roots
$\sqrt[4]{1} = 1$	$\sqrt[5]{1} = 1$
$\sqrt[4]{16} = 2$	$\sqrt[5]{32} = 2$
$\sqrt[4]{81} = 3$	$\sqrt[5]{243} = 3$
$\sqrt[4]{256} = 4$	
$\sqrt[4]{625} = 5$	

If the index n in $\sqrt[n]{a}$ is an even number, a root is said to be an **even root**. A square root is an even root. Other even roots have the same characteristics as square roots.

- Every positive real number has two real roots when n is even. One root is positive and one is negative.

$$2^4 = 2 \cdot 2 \cdot 2 \cdot 2 = 16 \quad \text{and} \quad (-2)^4 = (-2)(-2)(-2)(-2) = 16,$$

so both 2 and -2 are fourth roots of 16.

- The positive root, called the **principal *n*th root** and represented by $\sqrt[n]{a}$, is the nonnegative number b for which $b^n = a$. The symbol $-\sqrt[n]{a}$ is used to denote the negative nth root.

$$\sqrt[4]{16} = 2 \qquad -\sqrt[4]{16} = -2$$

$2^4 = 16$ $\qquad$ $(-2)^4 = 16$

- **An even root of a negative number is not a real number.**

$$\sqrt[4]{-16} \text{ is not a real number.}$$

EXAMPLE 8 Finding Even and Odd Roots

Find the indicated root, or state that the expression is not a real number:

a. $\sqrt[4]{81}$ **b.** $-\sqrt[4]{81}$ **c.** $\sqrt[4]{-81}$ **d.** $\sqrt[5]{-32}$.

Solution

a. $\sqrt[4]{81} = 3$ — The principal fourth root of 81 is 3 because $3^4 = 3 \cdot 3 \cdot 3 \cdot 3 = 81$.

b. $-\sqrt[4]{81} = -3$ — The negative fourth root of 81 is -3 because $(-3)^4 = (-3)(-3)(-3)(-3) = 81$.

c. $\sqrt[4]{-81}$ is not a real number because the index, 4, is even and the radicand, -81, is negative. No real number can be raised to the fourth power to give a negative result such as -81. Real numbers to even powers can only result in nonnegative numbers.

d. $\sqrt[5]{-32} = -2$ because $(-2)^5 = (-2)(-2)(-2)(-2)(-2) = -32$. An odd root of a negative real number is always negative.

✓ **CHECK POINT 8** Find the indicated root, or state that the expression is not a real number:

a. $\sqrt[4]{16}$ **b.** $-\sqrt[4]{16}$ **c.** $\sqrt[4]{-16}$ **d.** $\sqrt[5]{-1}$.

9 Simplify expressions of the form $\sqrt[n]{a^n}$.

Simplifying Expressions of the Form $\sqrt[n]{a^n}$

We have seen that

$$\sqrt{a^2} = |a| \quad \text{and} \quad \sqrt[3]{a^3} = a.$$

Expressions of the form $\sqrt[n]{a^n}$ can be simplified in the same manner. Unless a is known to be nonnegative, absolute value notation is needed when n is even. When the index is odd, absolute value bars are not used.

Simplifying $\sqrt[n]{a^n}$

For any real number a,

1. If n is even, $\sqrt[n]{a^n} = |a|$.
2. If n is odd, $\sqrt[n]{a^n} = a$.

EXAMPLE 9 Simplifying Radical Expressions

Simplify:

a. $\sqrt[4]{(x-3)^4}$ b. $\sqrt[5]{(2x+7)^5}$ c. $\sqrt[6]{(-5)^6}$.

Solution Each expression involves the nth root of a radicand raised to the nth power. Thus, each radical expression can be simplified. Absolute value bars are necessary in parts (a) and (c) because the index, n, is even.

a. $\sqrt[4]{(x-3)^4} = |x-3|$ $\sqrt[n]{a^n} = |a|$ if n is even.

b. $\sqrt[5]{(2x+7)^5} = 2x+7$ $\sqrt[n]{a^n} = a$ if n is odd.

c. $\sqrt[6]{(-5)^6} = |-5| = 5$ $\sqrt[n]{a^n} = |a|$ if n is even.

✓ CHECK POINT 9 Simplify:

a. $\sqrt[4]{(x+6)^4}$ b. $\sqrt[5]{(3x-2)^5}$ c. $\sqrt[6]{(-8)^6}$.

7.1 EXERCISE SET

MyMathLab MathXL PRACTICE WATCH DOWNLOAD READ REVIEW

Practice Exercises

In Exercises 1–20, evaluate each expression, or state that the expression is not a real number.

1. $\sqrt{36}$
2. $\sqrt{16}$
3. $-\sqrt{36}$
4. $-\sqrt{16}$
5. $\sqrt{-36}$
6. $\sqrt{-16}$
7. $\sqrt{\frac{1}{25}}$
8. $\sqrt{\frac{1}{49}}$
9. $-\sqrt{\frac{9}{16}}$
10. $-\sqrt{\frac{4}{25}}$
11. $\sqrt{0.81}$
12. $\sqrt{0.49}$
13. $-\sqrt{0.04}$
14. $-\sqrt{0.64}$
15. $\sqrt{25-16}$
16. $\sqrt{144+25}$
17. $\sqrt{25}-\sqrt{16}$
18. $\sqrt{144}+\sqrt{25}$
19. $\sqrt{16-25}$
20. $\sqrt{25-144}$

In Exercises 21–26, find the indicated function values for each function. If necessary, round to two decimal places. If the function value is not a real number and does not exist, so state.

21. $f(x) = \sqrt{x-2}$; $f(18), f(3), f(2), f(-2)$
22. $f(x) = \sqrt{x-3}$; $f(28), f(4), f(3), f(-1)$
23. $g(x) = -\sqrt{2x+3}$; $g(11), g(1), g(-1), g(-2)$
24. $g(x) = -\sqrt{2x+1}$; $g(4), g(1), g\left(-\frac{1}{2}\right), g(-1)$
25. $h(x) = \sqrt{(x-1)^2}$; $h(5), h(3), h(0), h(-5)$
26. $h(x) = \sqrt{(x-2)^2}$; $h(5), h(3), h(0), h(-5)$

In Exercises 27–32, find the domain of each square root function. Then use the domain to match the radical function with its graph. [The graphs are labeled (a) through (f) and are shown in [−10, 10, 1] by [−10, 10, 1] viewing rectangles on the next page.]

27. $f(x) = \sqrt{x-3}$
28. $f(x) = \sqrt{x+2}$
29. $f(x) = \sqrt{3x+15}$
30. $f(x) = \sqrt{3x-15}$

31. $f(x) = \sqrt{6 - 2x}$

32. $f(x) = \sqrt{8 - 2x}$

a.

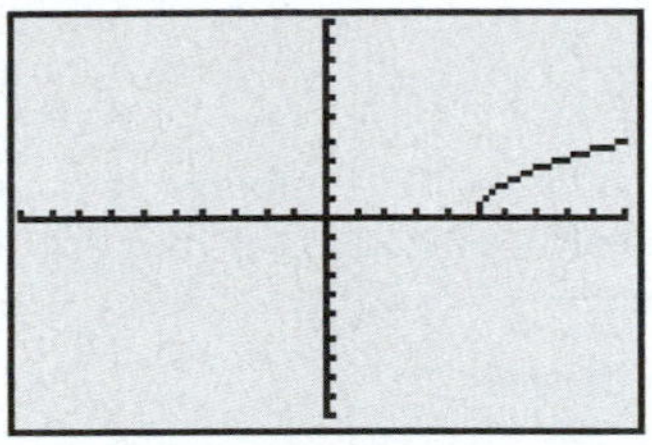

b.

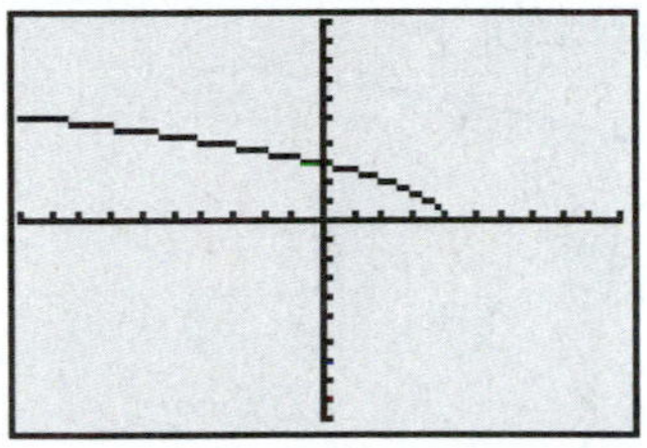

c.

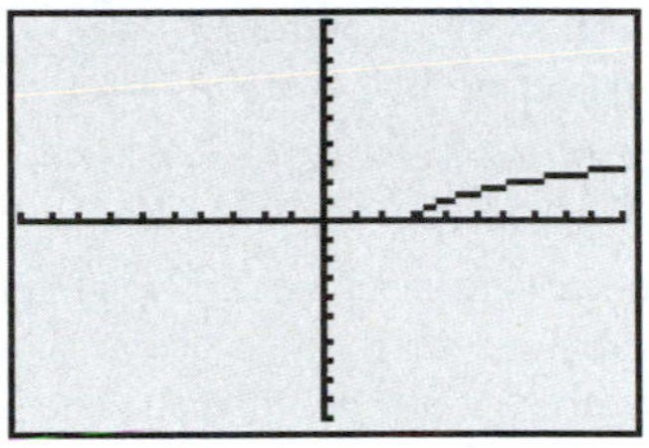

d.

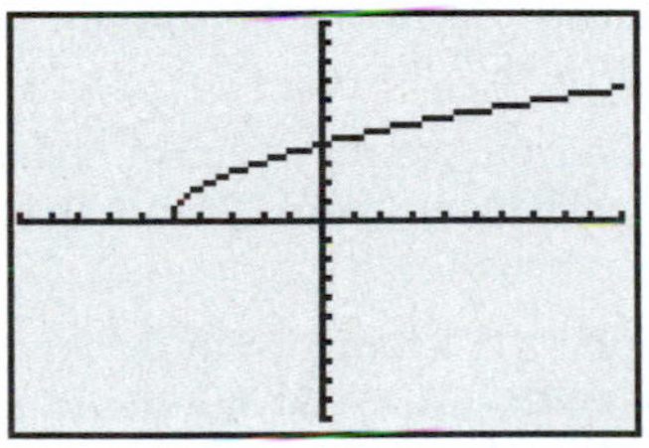

e.

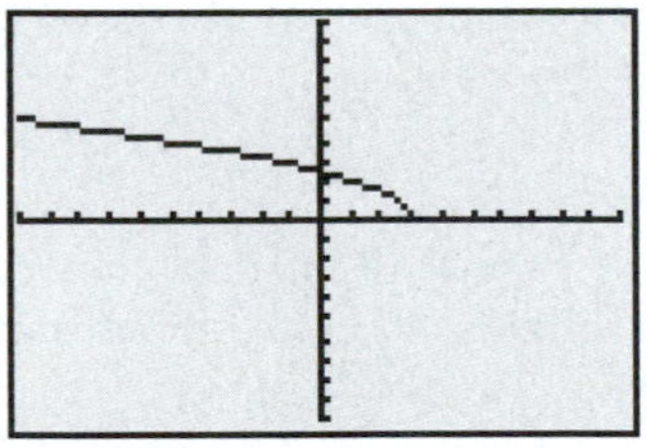

f.

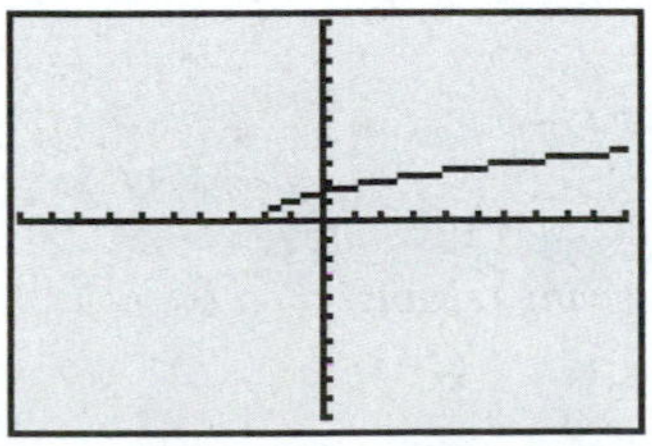

In Exercises 33–46, simplify each expression.

33. $\sqrt{5^2}$ **34.** $\sqrt{7^2}$

35. $\sqrt{(-4)^2}$ **36.** $\sqrt{(-10)^2}$

37. $\sqrt{(x - 1)^2}$ **38.** $\sqrt{(x - 2)^2}$

39. $\sqrt{36x^4}$ **40.** $\sqrt{81x^4}$

41. $-\sqrt{100x^6}$

42. $-\sqrt{49x^6}$

43. $\sqrt{x^2 + 12x + 36}$

44. $\sqrt{x^2 + 14x + 49}$

45. $-\sqrt{x^2 - 8x + 16}$

46. $-\sqrt{x^2 - 10x + 25}$

In Exercises 47–54, find each cube root.

47. $\sqrt[3]{27}$ **48.** $\sqrt[3]{64}$

49. $\sqrt[3]{-27}$ **50.** $\sqrt[3]{-64}$

51. $\sqrt[3]{\frac{1}{125}}$ **52.** $\sqrt[3]{\frac{1}{1000}}$

53. $\sqrt[3]{\frac{-27}{1000}}$ **54.** $\sqrt[3]{\frac{-8}{125}}$

In Exercises 55–58, find the indicated function values for each function.

55. $f(x) = \sqrt[3]{x - 1}$; $f(28), f(9), f(0), f(-63)$

56. $f(x) = \sqrt[3]{x - 3}$; $f(30), f(11), f(2), f(-122)$

57. $g(x) = -\sqrt[3]{8x - 8}$; $g(2), g(1), g(0)$

58. $g(x) = -\sqrt[3]{2x + 1}$; $g(13), g(0), g(-63)$

In Exercises 59–76, find the indicated root, or state that the expression is not a real number.

59. $\sqrt[4]{1}$ **60.** $\sqrt[5]{1}$

61. $\sqrt[4]{16}$ **62.** $\sqrt[4]{81}$

63. $-\sqrt[4]{16}$ **64.** $-\sqrt[4]{81}$

65. $\sqrt[4]{-16}$ **66.** $\sqrt[4]{-81}$

67. $\sqrt[5]{-1}$ **68.** $\sqrt[7]{-1}$

69. $\sqrt[6]{-1}$ **70.** $\sqrt[8]{-1}$

71. $-\sqrt[4]{256}$ **72.** $-\sqrt[4]{10{,}000}$

73. $\sqrt[6]{64}$ **74.** $\sqrt[5]{32}$

75. $-\sqrt[5]{32}$ **76.** $-\sqrt[6]{64}$

In Exercises 77–90, simplify each expression. Include absolute value bars where necessary.

77. $\sqrt[3]{x^3}$ **78.** $\sqrt[5]{x^5}$

79. $\sqrt[4]{y^4}$ **80.** $\sqrt[6]{y^6}$

81. $\sqrt[3]{-8x^3}$ **82.** $\sqrt[3]{-125x^3}$

83. $\sqrt[3]{(-5)^3}$ **84.** $\sqrt[3]{(-6)^3}$

85. $\sqrt[4]{(-5)^4}$ **86.** $\sqrt[6]{(-6)^6}$

87. $\sqrt[4]{(x + 3)^4}$ **88.** $\sqrt[4]{(x + 5)^4}$

89. $\sqrt[5]{-32(x - 1)^5}$

90. $\sqrt[5]{-32(x - 2)^5}$

Practice PLUS

In Exercises 91–94, complete each table and graph the given function. Identify the function's domain and range.

91. $f(x) = \sqrt{x} + 3$

x	$f(x) = \sqrt{x} + 3$
0	
1	
4	
9	

92. $f(x) = \sqrt{x} - 2$

x	$f(x) = \sqrt{x} - 2$
0	
1	
4	
9	

93. $f(x) = \sqrt{x - 3}$

x	$f(x) = \sqrt{x - 3}$
3	
4	
7	
12	

94. $f(x) = \sqrt{4 - x}$

x	$f(x) = \sqrt{4 - x}$
−5	
0	
3	
4	

In Exercises 95–98, find the domain of each function.

95. $f(x) = \dfrac{\sqrt[3]{x}}{\sqrt{30 - 2x}}$

96. $f(x) = \dfrac{\sqrt[3]{x}}{\sqrt{80 - 5x}}$

97. $f(x) = \dfrac{\sqrt{x - 1}}{\sqrt{3 - x}}$

98. $f(x) = \dfrac{\sqrt{x - 2}}{\sqrt{7 - x}}$

In Exercises 99–100, evaluate each expression.

99. $\sqrt[3]{\sqrt[4]{16} + \sqrt{625}}$

100. $\sqrt[3]{\sqrt{\sqrt{169} + \sqrt{9}} + \sqrt{\sqrt[3]{1000} + \sqrt[3]{216}}}$

Application Exercises

101. The function $f(x) = 2.9\sqrt{x} + 20.1$ models the median height, $f(x)$, in inches, of boys who are x months of age. The graph of f is shown.

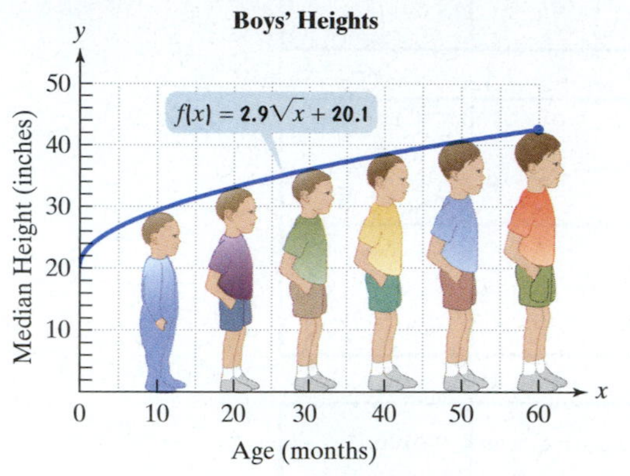

Source: Laura Walther Nathanson, *The Portable Pediatrician for Parents*

a. According to the model, what is the median height of boys who are 48 months, or four years, old? Use a calculator and round to the nearest tenth of an inch. The actual median height for boys at 48 months is 40.8 inches. Does the model overestimate or underestimate the actual height? By how much?

b. Use the model to find the average rate of change, in inches per month, between birth and 10 months. Round to the nearest tenth.

c. Use the model to find the average rate of change, in inches per month, between 50 and 60 months. Round to the nearest tenth. How does this compare with your answer in part (b)? How is this difference shown by the graph?

102. The function $f(x) = 3.1\sqrt{x} + 19$ models the median height, $f(x)$, in inches, of girls who are x months of age. The graph of f is shown.

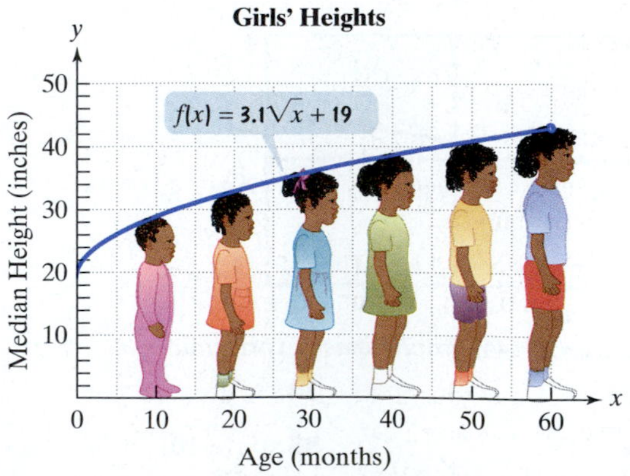

Source: Laura Walther Nathanson, *The Portable Pediatrician for Parents*

a. According to the model, what is the median height of girls who are 48 months, or four years, old? Use a calculator and round to the nearest tenth of an inch. The actual median height for girls at 48 months is 40.2 inches. Does the model overestimate or underestimate the actual height? By how much?

b. Use the model to find the average rate of change, in inches per month, between birth and 10 months. Round to the nearest tenth.

c. Use the model to find the average rate of change, in inches per month, between 50 and 60 months. Round to the nearest tenth. How does this compare with your answer in part (b)? How is this difference shown by the graph?

Police use the function $f(x) = \sqrt{20x}$ to estimate the speed of a car, $f(x)$, in miles per hour, based on the length, x, in feet, of its skid marks upon sudden braking on a dry asphalt road. Use the function to solve Exercises 103–104.

103. A motorist is involved in an accident. A police officer measures the car's skid marks to be 245 feet long. Estimate the speed at which the motorist was traveling before braking. If the posted speed limit is 50 miles per hour and the motorist tells the officer he was not speeding, should the officer believe him? Explain.

104. A motorist is involved in an accident. A police officer measures the car's skid marks to be 45 feet long. Estimate the speed at which the motorist was traveling before braking. If the posted speed limit is 35 miles per hour and the motorist tells the officer she was not speeding, should the officer believe her? Explain.

Writing in Mathematics

105. What are the square roots of 36? Explain why each of these numbers is a square root.

106. What does the symbol $\sqrt{}$ denote? Which of your answers in Exercise 105 is given by this symbol? Write the symbol needed to obtain the other answer.

107. Explain why $\sqrt{-1}$ is not a real number.

108. Explain how to find the domain of a square root function.

109. Explain how to simplify $\sqrt{a^2}$. Give an example with your explanation.

110. Explain why $\sqrt[3]{8}$ is 2. Then describe what is meant by the cube root of a real number.

111. Describe two differences between odd and even roots.

112. Explain how to simplify $\sqrt[n]{a^n}$ if n is even and if n is odd. Give examples with your explanations.

113. Explain the meaning of the words *radical*, *radicand*, and *index*. Give an example with your explanation.

114. Describe the trend in a boy's growth from birth through five years, shown in the graph for Exercise 101. Why is a square root function a useful model for the data?

Technology Exercises

115. Use a graphing utility to graph $y_1 = \sqrt{x}$, $y_2 = \sqrt{x+4}$, and $y_3 = \sqrt{x-3}$ in the same $[-5, 10, 1]$ by $[0, 6, 1]$ viewing rectangle. Describe one similarity and one difference that you observe among the graphs. Use the word *shift* in your response.

116. Use a graphing utility to graph $y = \sqrt{x}$, $y = \sqrt{x} + 4$, and $y = \sqrt{x} - 3$ in the same $[-1, 10, 1]$ by $[-10, 10, 1]$ viewing rectangle. Describe one similarity and one difference that you observe among the graphs.

117. Use a graphing utility to graph $f(x) = \sqrt{x}$, $g(x) = -\sqrt{x}$, $h(x) = \sqrt{-x}$, and $k(x) = -\sqrt{-x}$ in the same $[-10, 10, 1]$ by $[-4, 4, 1]$ viewing rectangle. Use the graphs to describe the domains and the ranges of functions f, g, h, and k.

118. Use a graphing utility to graph $y_1 = \sqrt{x^2}$ and $y_2 = -x$ in the same viewing rectangle.

a. For what values of x is $\sqrt{x^2} = -x$?

b. For what values of x is $\sqrt{x^2} \neq -x$?

Critical Thinking Exercises

Make Sense? *In Exercises 119–122, determine whether each statement "makes sense" or "does not make sense" and explain your reasoning.*

119. $\sqrt[4]{(-8)^4}$ cannot be positive 8 because the power and the index cancel each other.

120. If I am given any real number, that number has exactly one odd root and two even roots.

121. I need to restrict the domains of radical functions with even indices, but these restrictions are not necessary when indices are odd.

122. Using my calculator, I determined that $5^5 = 3125$, so 5 must be the fifth root of 3125.

In Exercises 123–126, determine whether each statement is true or false. If the statement is false, make the necessary change(s) to produce a true statement.

123. The domain of $f(x) = \sqrt[3]{x-4}$ is $[4, \infty)$.

124. If n is odd and b is negative, then $\sqrt[n]{b}$ is not a real number.

125. If $x = -2$, then $\sqrt{x^6} = x^3$.

126. The expression $\sqrt[n]{4}$ represents increasingly larger numbers for $n = 2, 3, 4, 5, 6$, and so on.

127. Write a function whose domain is $(-\infty, 5]$.

128. Let $f(x) = \sqrt{x-3}$ and $g(x) = \sqrt{x+1}$. Find the domain of $f + g$ and $\frac{f}{g}$.

129. Simplify: $\sqrt{(2x+3)^{10}}$.

In Exercises 130–131, graph each function by hand. Then describe the relationship between the function that you graphed and the graph of $f(x) = \sqrt{x}$.

130. $g(x) = \sqrt{x} + 2$

131. $h(x) = \sqrt{x+3}$

Review Exercises

132. Simplify: $3x - 2[x - 3(x + 5)]$. (Section 1.2, Example 14)

133. Simplify: $(-3x^{-4}y^3)^{-2}$. (Section 1.6, Example 7c)

134. Solve: $|3x - 4| > 11$. (Section 4.3, Example 5 or Example 6b)

Preview Exercises

Exercises 135–137 will help you prepare for the material covered in the next section. In each exercise, use properties of exponents to simplify the expression. Be sure that no negative exponents appear in your simplified expression. (If you have forgotten how to simplify an exponential expression, see the box on page 71.)

135. $(2^3x^5)(2^4x^{-6})$

136. $\frac{32x^2}{16x^5}$

137. $(x^{-2}y^3)^4$

SECTION 7.2 Rational Exponents

Objectives

1. Use the definition of $a^{\frac{1}{n}}$.
2. Use the definition of $a^{\frac{m}{n}}$.
3. Use the definition of $a^{-\frac{m}{n}}$.
4. Simplify expressions with rational exponents.
5. Simplify radical expressions using rational exponents.

Marine iguanas of the Galápagos Islands

The Galápagos Islands are a chain of volcanic islands lying 600 miles west of Ecuador. They are famed for over 5000 species of plants and animals, including a rare flightless cormorant, marine iguanas, and giant tortoises weighing more than 600 pounds. Early in 2001, the plants and wildlife that live in the Galápagos were at risk from a massive oil spill that flooded 150,000 gallons of toxic fuel into one of the world's most fragile ecosystems. The long-term danger of the accident is that fuel sinking to the ocean floor will destroy algae that is vital to the food chain. Any imbalance in the food chain could threaten the rare Galápagos plant and animal species that have evolved for thousands of years in isolation with little human intervention.

At risk on these ecologically vulnerable islands are unique flora and fauna that helped to inspire Charles Darwin's theory of evolution. Darwin made an enormous collection of the island's plant species. The function

$$f(x) = 29x^{\frac{1}{3}}$$

models the number of plant species, $f(x)$, on the various islands of the Galápagos in terms of the area, x, in square miles, of a particular island. But x to the *what* power? How can we interpret the information given by this function? In this section, we turn our attention to rational exponents such as $\frac{1}{3}$ and their relationship to roots of real numbers.

Defining Rational Exponents

We define rational exponents so that their properties are the same as the properties for integer exponents. For example, suppose that $x = 7^{\frac{1}{3}}$. We know that exponents are multiplied when an exponential expression is raised to a power. For this to be true,

$$x^3 = \left(7^{\frac{1}{3}}\right)^3 = 7^{\frac{1}{3} \cdot 3} = 7^1 = 7.$$

We see that $x^3 = 7$. This means that x is the number whose cube is 7. Thus, $x = \sqrt[3]{7}$. Remember that we began with $x = 7^{\frac{1}{3}}$. This means that

$$7^{\frac{1}{3}} = \sqrt[3]{7}.$$

We can generalize this idea with the following definition:

1 Use the definition of $a^{\frac{1}{n}}$.

The Definition of $a^{\frac{1}{n}}$

If $\sqrt[n]{a}$ represents a real number and $n \geq 2$ is an integer, then

$$a^{\frac{1}{n}} = \sqrt[n]{a}.$$

The denominator of the rational exponent is the radical's index.

If a is negative, n must be odd. If a is nonnegative, n can be any index.

EXAMPLE 1 Using the Definition of $a^{\frac{1}{n}}$

Use radical notation to rewrite each expression. Simplify, if possible:

a. $64^{\frac{1}{2}}$ **b.** $(-125)^{\frac{1}{3}}$ **c.** $(6x^2y)^{\frac{1}{5}}$.

Solution

a. $64^{\frac{1}{2}} = \sqrt{64} = 8$

The denominator is the index.

b. $(-125)^{\frac{1}{3}} = \sqrt[3]{-125} = -5$

c. $(6x^2y)^{\frac{1}{5}} = \sqrt[5]{6x^2y}$

✓ CHECK POINT 1 Use radical notation to rewrite each expression. Simplify, if possible:

a. $25^{\frac{1}{2}}$ **b.** $(-8)^{\frac{1}{3}}$ **c.** $(5xy^2)^{\frac{1}{4}}$.

In our next example, we begin with radical notation and rewrite the expression with rational exponents.

The radical's index becomes the exponent's denominator.

$$\sqrt[n]{a} = a^{\frac{1}{n}}$$

The radicand becomes the base.

EXAMPLE 2 Using the Definition of $a^{\frac{1}{n}}$

Rewrite with rational exponents:

a. $\sqrt[5]{13ab}$ b. $\sqrt[7]{\frac{xy^2}{17}}$.

Solution Parentheses are needed to show that the entire radicand becomes the base.

a. $\sqrt[5]{13ab} = (13ab)^{\frac{1}{5}}$

The index is the exponent's denominator.

b. $\sqrt[7]{\frac{xy^2}{17}} = \left(\frac{xy^2}{17}\right)^{\frac{1}{7}}$

✓ CHECK POINT 2 Rewrite with rational exponents:

a. $\sqrt[4]{5xy}$ b. $\sqrt[5]{\frac{a^3b}{2}}$.

2 Use the definition of $a^{\frac{m}{n}}$.

Can rational exponents have numerators other than 1? The answer is yes. If the numerator is some other integer, we still want to multiply exponents when raising a power to a power. For this reason,

$$a^{\frac{2}{3}} = (a^{\frac{1}{3}})^2 \text{ and } a^{\frac{2}{3}} = (a^2)^{\frac{1}{3}}.$$

This means $(\sqrt[3]{a})^2$. This means $\sqrt[3]{a^2}$.

Thus,

$$a^{\frac{2}{3}} = \left(\sqrt[3]{a}\right)^2 = \sqrt[3]{a^2}.$$

Do you see that the denominator, 3, of the rational exponent is the same as the index of the radical? The numerator, 2, of the rational exponent serves as an exponent in each of the two radical forms. We generalize these ideas with the following definition:

The Definition of $a^{\frac{m}{n}}$

If $\sqrt[n]{a}$ represents a real number, $\frac{m}{n}$ is a positive rational number reduced to lowest terms, and $n \geq 2$ is an integer, then

$$a^{\frac{m}{n}} = (\sqrt[n]{a})^m$$

First take the nth root of a.

and

$$a^{\frac{m}{n}} = \sqrt[n]{a^m}.$$

First raise a to the m power.

The first form of the definition shown in the box involves taking the root first. This form is often preferable because smaller numbers are involved. Notice that the rational exponent consists of two parts, indicated by the following voice balloons:

The numerator is the exponent.

$$a^{\frac{m}{n}} = (\sqrt[n]{a})^m.$$

The denominator is the radical's index.

EXAMPLE 3 Using the Definition of $a^{\frac{m}{n}}$

Use radical notation to rewrite each expression and simplify:

a. $1000^{\frac{2}{3}}$ b. $16^{\frac{3}{2}}$ c. $-32^{\frac{3}{5}}$.

Solution

a. $(1000)^{\frac{2}{3}} = (\sqrt[3]{1000})^2 = 10^2 = 100$

The denominator of $\frac{2}{3}$ is the root and the numerator is the exponent.

b. $16^{\frac{3}{2}} = (\sqrt{16})^3 = 4^3 = 64$

c. $-32^{\frac{3}{5}} = -(\sqrt[5]{32})^3 = -2^3 = -8$

The base is 32 and the negative sign is not affected by the exponent.

Using Technology

Here are the calculator keystroke sequences for $1000^{\frac{2}{3}}$:

Many Scientific Calculators

1000 [y^x] [( ] 2 [÷] 3 [ )] [=]

Many Graphing Calculators

1000 [∧] [( ] 2 [÷] 3 [ )] [ENTER]

✓ **CHECK POINT 3** Use radical notation to rewrite each expression and simplify:

a. $8^{\frac{4}{3}}$ b. $25^{\frac{3}{2}}$ c. $-81^{\frac{3}{4}}$.

In our next example, we begin with radical notation and rewrite the expression with rational exponents. When changing from radical form to exponential form, the index becomes the denominator of the rational exponent.

EXAMPLE 4 Using the Definition of $a^{\frac{m}{n}}$

Rewrite with rational exponents:

a. $\sqrt[3]{7^5}$ b. $(\sqrt[4]{13xy})^9$.

Solution

a. $\sqrt[3]{7^5} = 7^{\frac{5}{3}}$

The index is the exponent's denominator.

b. $(\sqrt[4]{13xy})^9 = (13xy)^{\frac{9}{4}}$

✓ **CHECK POINT 4** Rewrite with rational exponents:

a. $\sqrt[3]{6^4}$ b. $(\sqrt[5]{2xy})^7$.

3 Use the definition of $a^{-\frac{m}{n}}$.

Can a rational exponent be negative? Yes. The way that negative rational exponents are defined is similar to the way that negative integer exponents are defined.

The Definition of $a^{-\frac{m}{n}}$

If $a^{\frac{m}{n}}$ is a nonzero real number, then

$$a^{-\frac{m}{n}} = \frac{1}{a^{\frac{m}{n}}}.$$

EXAMPLE 5 Using the Definition of $a^{-\frac{m}{n}}$

Rewrite each expression with a positive exponent. Simplify, if possible:

a. $36^{-\frac{1}{2}}$ **b.** $125^{-\frac{1}{3}}$ **c.** $16^{-\frac{3}{4}}$ **d.** $(7xy)^{-\frac{4}{7}}$.

Using Technology

Here are the calculator keystroke sequences for $16^{-\frac{3}{4}}$:

Many Scientific Calculators

16 [y^x] [( ] 3 [+/−] [÷] 4 [ )] [=]

Many Graphing Calculators

16 [∧] [( ] [(−)] 3 [÷] 4 [ )] [ENTER]

Solution

a. $36^{-\frac{1}{2}} = \frac{1}{36^{\frac{1}{2}}} = \frac{1}{\sqrt{36}} = \frac{1}{6}$

b. $125^{-\frac{1}{3}} = \frac{1}{125^{\frac{1}{3}}} = \frac{1}{\sqrt[3]{125}} = \frac{1}{5}$

c. $16^{-\frac{3}{4}} = \frac{1}{16^{\frac{3}{4}}} = \frac{1}{\left(\sqrt[4]{16}\right)^3} = \frac{1}{2^3} = \frac{1}{8}$

d. $(7xy)^{-\frac{4}{7}} = \frac{1}{(7xy)^{\frac{4}{7}}}$

✓ CHECK POINT 5 Rewrite each expression with a positive exponent. Simplify, if possible:

a. $100^{-\frac{1}{2}}$

b. $8^{-\frac{1}{3}}$

c. $32^{-\frac{3}{5}}$

d. $(3xy)^{-\frac{5}{9}}$.

Properties of Rational Exponents

The same properties apply to rational exponents as to integer exponents. The following is a summary of these properties:

Properties of Rational Exponents

If m and n are rational exponents, and a and b are real numbers for which the following expressions are defined, then

1. $b^m \cdot b^n = b^{m+n}$ When multiplying exponential expressions with the same base, add the exponents. Use this sum as the exponent of the common base.
2. $\frac{b^m}{b^n} = b^{m-n}$ When dividing exponential expressions with the same base, subtract the exponents. Use this difference as the exponent of the common base.
3. $(b^m)^n = b^{mn}$ When an exponential expression is raised to a power, multiply the exponents. Place the product of the exponents on the base and remove the parentheses.
4. $(ab)^n = a^n b^n$ When a product is raised to a power, raise each factor to that power and multiply.
5. $\left(\frac{a}{b}\right)^n = \frac{a^n}{b^n}$ When a quotient is raised to a power, raise the numerator to that power and divide by the denominator to that power.

4 Simplify expressions with rational exponents.

We can use these properties to simplify exponential expressions with rational exponents. As with integer exponents, an expression with rational exponents is **simplified** when:

- No parentheses appear.
- No powers are raised to powers.
- Each base occurs only once.
- No negative or zero exponents appear.

EXAMPLE 6 Simplifying Expressions with Rational Exponents

Simplify:

a. $6^{\frac{1}{7}} \cdot 6^{\frac{4}{7}}$ b. $\dfrac{32x^{\frac{1}{2}}}{16x^{\frac{3}{4}}}$ c. $\left(8.3^{\frac{3}{4}}\right)^{\frac{2}{3}}$ d. $\left(x^{-\frac{2}{5}}y^{\frac{1}{3}}\right)^{\frac{1}{2}}$.

Study Tip

To simplify $6^{\frac{1}{7}} \cdot 6^{\frac{4}{7}}$, **do not multiply the numerical bases.**

Incorrect!

~~$6^{\frac{1}{7}} \cdot 6^{\frac{4}{7}} = 36^{\frac{1}{7}+\frac{4}{7}}$~~

Solution

a. $6^{\frac{1}{7}} \cdot 6^{\frac{4}{7}} = 6^{\frac{1}{7}+\frac{4}{7}}$ To multiply with the same base, add exponents.

$= 6^{\frac{5}{7}}$ Simplify: $\frac{1}{7} + \frac{4}{7} = \frac{5}{7}$.

b. $\dfrac{32x^{\frac{1}{2}}}{16x^{\frac{3}{4}}} = \dfrac{32}{16}x^{\frac{1}{2}-\frac{3}{4}}$ Divide coefficients. To divide with the same base, subtract exponents.

$= 2x^{\frac{2}{4}-\frac{3}{4}}$ Write exponents in terms of the LCD, 4.

$= 2x^{-\frac{1}{4}}$ Subtract: $\frac{2}{4} - \frac{3}{4} = -\frac{1}{4}$.

$= \dfrac{2}{x^{\frac{1}{4}}}$ Rewrite with a positive exponent: $a^{-\frac{m}{n}} = \dfrac{1}{a^{\frac{m}{n}}}$.

c. $\left(8.3^{\frac{3}{4}}\right)^{\frac{2}{3}} = 8.3^{\left(\frac{3}{4}\right)\left(\frac{2}{3}\right)}$ To raise a power to a power, multiply exponents.

$= 8.3^{\frac{1}{2}}$ Multiply: $\frac{3}{4} \cdot \frac{2}{3} = \frac{6}{12} = \frac{1}{2}$.

d. $\left(x^{-\frac{2}{5}}y^{\frac{1}{3}}\right)^{\frac{1}{2}} = \left(x^{-\frac{2}{5}}\right)^{\frac{1}{2}}\left(y^{\frac{1}{3}}\right)^{\frac{1}{2}}$ To raise a product to a power, raise each factor to the power.

$= x^{-\frac{1}{5}}y^{\frac{1}{6}}$ Multiply: $-\frac{2}{5} \cdot \frac{1}{2} = -\frac{1}{5}$ and $\frac{1}{3} \cdot \frac{1}{2} = \frac{1}{6}$.

$= \dfrac{y^{\frac{1}{6}}}{x^{\frac{1}{5}}}$ Rewrite with positive exponents.

✓ CHECK POINT 6 Simplify:

a. $7^{\frac{1}{2}} \cdot 7^{\frac{1}{3}}$ b. $\dfrac{50x^{\frac{1}{3}}}{10x^{\frac{4}{3}}}$ c. $\left(9.1^{\frac{2}{5}}\right)^{\frac{3}{4}}$ d. $\left(x^{-\frac{3}{5}}y^{\frac{1}{4}}\right)^{\frac{1}{3}}$.

5 Simplify radical expressions using rational exponents.

Using Rational Exponents to Simplify Radical Expressions

Some radical expressions can be simplified using rational exponents. We will use the following procedure:

Simplifying Radical Expressions Using Rational Exponents

1. Rewrite each radical expression as an exponential expression with a rational exponent.
2. Simplify using properties of rational exponents.
3. Rewrite in radical notation if rational exponents still appear.

EXAMPLE 7 Simplifying Radical Expressions Using Rational Exponents

Use rational exponents to simplify:

a. $\sqrt[10]{x^5}$ **b.** $\sqrt[3]{27a^{15}}$ **c.** $\sqrt[4]{x^6y^2}$ **d.** $\sqrt{x}\cdot\sqrt[3]{x}$ **e.** $\sqrt[3]{\sqrt{x}}$.

Solution

a.

$\sqrt[10]{x^5} = x^{\frac{5}{10}}$ Rewrite as an exponential expression.

$= x^{\frac{1}{2}}$ Simplify the exponent.

$= \sqrt{x}$ Rewrite in radical notation.

b.

$\sqrt[3]{27a^{15}} = (27a^{15})^{\frac{1}{3}}$ Rewrite as an exponential expression.

$= 27^{\frac{1}{3}}(a^{15})^{\frac{1}{3}}$ Raise each factor in parentheses to the $\frac{1}{3}$ power.

$= \sqrt[3]{27}\cdot a^{15\left(\frac{1}{3}\right)}$ To raise a power to a power, multiply exponents.

$= 3a^5$ $\sqrt[3]{27} = 3$. Multiply exponents: $15\cdot\frac{1}{3} = 5$.

c.

$\sqrt[4]{x^6y^2} = (x^6y^2)^{\frac{1}{4}}$ Rewrite as an exponential expression.

$= (x^6)^{\frac{1}{4}}(y^2)^{\frac{1}{4}}$ Raise each factor in parentheses to the $\frac{1}{4}$ power.

$= x^{\frac{6}{4}}y^{\frac{2}{4}}$ To raise powers to powers, multiply.

$= x^{\frac{3}{2}}y^{\frac{1}{2}}$ Simplify.

$= (x^3y)^{\frac{1}{2}}$ $a^nb^n = (ab)^n$

$= \sqrt{x^3y}$ Rewrite in radical notation.

d.

$\sqrt{x}\cdot\sqrt[3]{x} = x^{\frac{1}{2}}\cdot x^{\frac{1}{3}}$ Rewrite as exponential expressions.

$= x^{\frac{1}{2}+\frac{1}{3}}$ To multiply with the same base, add exponents.

$= x^{\frac{3}{6}+\frac{2}{6}}$ Write exponents in terms of the LCD, 6.

$= x^{\frac{5}{6}}$ Add: $\frac{3}{6}+\frac{2}{6}=\frac{5}{6}$.

$= \sqrt[6]{x^5}$ Rewrite in radical notation.

e.

$\sqrt[3]{\sqrt{x}} = \sqrt[3]{x^{\frac{1}{2}}}$ Write the radicand as an exponential expression.

$= \left(x^{\frac{1}{2}}\right)^{\frac{1}{3}}$ Write the entire expression in exponential form.

$= x^{\frac{1}{6}}$ To raise powers to powers, multiply: $\frac{1}{2}\cdot\frac{1}{3}=\frac{1}{6}$.

$= \sqrt[6]{x}$ Rewrite in radical notation.

✓ CHECK POINT 7 Use rational exponents to simplify:

a. $\sqrt[6]{x^3}$ **b.** $\sqrt[3]{8a^{12}}$ **c.** $\sqrt[8]{x^4y^2}$

d. $\dfrac{\sqrt{x}}{\sqrt[3]{x}}$ **e.** $\sqrt{\sqrt[3]{x}}$.

7.2 EXERCISE SET

MyMathLab — Math XL PRACTICE, WATCH, DOWNLOAD, READ, REVIEW

Practice Exercises

In Exercises 1–20, use radical notation to rewrite each expression. Simplify, if possible.

1. $49^{\frac{1}{2}}$ **2.** $100^{\frac{1}{2}}$

3. $(-27)^{\frac{1}{3}}$ **4.** $(-64)^{\frac{1}{3}}$

5. $-16^{\frac{1}{4}}$ **6.** $-81^{\frac{1}{4}}$

7. $(xy)^{\frac{1}{3}}$ **8.** $(xy)^{\frac{1}{4}}$

9. $(2xy^3)^{\frac{1}{5}}$ **10.** $(3xy^4)^{\frac{1}{5}}$

11. $81^{\frac{3}{2}}$ **12.** $25^{\frac{3}{2}}$

13. $125^{\frac{2}{3}}$ **14.** $1000^{\frac{2}{3}}$

15. $(-32)^{\frac{3}{5}}$ **16.** $(-27)^{\frac{2}{3}}$

17. $27^{\frac{2}{3}} + 16^{\frac{3}{4}}$

18. $4^{\frac{5}{2}} - 8^{\frac{2}{3}}$

19. $(xy)^{\frac{4}{7}}$

20. $(xy)^{\frac{4}{9}}$

In Exercises 21–38, rewrite each expression with rational exponents.

21. $\sqrt{7}$

22. $\sqrt{13}$

23. $\sqrt[3]{5}$

24. $\sqrt[3]{6}$

25. $\sqrt[5]{11x}$

26. $\sqrt[5]{13x}$

27. $\sqrt{x^3}$

28. $\sqrt{x^5}$

29. $\sqrt[5]{x^3}$

30. $\sqrt[7]{x^4}$

31. $\sqrt[5]{x^2y}$

32. $\sqrt[7]{xy^3}$

33. $(\sqrt{19xy})^3$

34. $(\sqrt{11xy})^3$

35. $(\sqrt[6]{7xy^2})^5$

36. $(\sqrt[6]{9x^2y})^5$

37. $2x\sqrt[3]{y^2}$

38. $4x\sqrt[5]{y^2}$

In Exercises 39–54, rewrite each expression with a positive rational exponent. Simplify, if possible.

39. $49^{-\frac{1}{2}}$

40. $9^{-\frac{1}{2}}$

41. $27^{-\frac{1}{3}}$

42. $125^{-\frac{1}{3}}$

43. $16^{-\frac{3}{4}}$

44. $81^{-\frac{5}{4}}$

45. $8^{-\frac{2}{3}}$

46. $32^{-\frac{4}{5}}$

47. $\left(\frac{8}{27}\right)^{-\frac{1}{3}}$

48. $\left(\frac{8}{125}\right)^{-\frac{1}{3}}$

49. $(-64)^{-\frac{2}{3}}$

50. $(-8)^{-\frac{2}{3}}$

51. $(2xy)^{-\frac{7}{10}}$

52. $(4xy)^{-\frac{4}{7}}$

53. $5xz^{-\frac{1}{3}}$

54. $7xz^{-\frac{1}{4}}$

In Exercises 55–78, use properties of rational exponents to simplify each expression. Assume that all variables represent positive numbers.

55. $3^{\frac{3}{4}} \cdot 3^{\frac{1}{4}}$

56. $5^{\frac{2}{3}} \cdot 5^{\frac{1}{3}}$

57. $\frac{16^{\frac{3}{4}}}{16^{\frac{1}{4}}}$

58. $\frac{100^{\frac{3}{4}}}{100^{\frac{1}{4}}}$

59. $x^{\frac{1}{2}} \cdot x^{\frac{1}{3}}$

60. $x^{\frac{1}{2}} \cdot x^{\frac{2}{3}}$

61. $\frac{x^{\frac{4}{5}}}{x^{\frac{1}{5}}}$

62. $\frac{x^{\frac{3}{7}}}{x^{\frac{1}{7}}}$

63. $\frac{x^{\frac{1}{3}}}{x^{\frac{3}{4}}}$

64. $\frac{x^{\frac{1}{4}}}{x^{\frac{3}{5}}}$

65. $\left(5^{\frac{2}{3}}\right)^3$

66. $\left(3^{\frac{4}{5}}\right)^5$

67. $\left(y^{-\frac{2}{3}}\right)^{\frac{1}{4}}$

68. $\left(y^{-\frac{3}{4}}\right)^{\frac{1}{6}}$

69. $\left(2x^{\frac{1}{5}}\right)^5$

70. $\left(2x^{\frac{1}{4}}\right)^4$

71. $(25x^4y^6)^{\frac{1}{2}}$

72. $(125x^9y^6)^{\frac{1}{3}}$

73. $\left(x^{\frac{1}{2}}y^{-\frac{3}{5}}\right)^{\frac{1}{2}}$

74. $\left(x^{\frac{1}{4}}y^{-\frac{2}{5}}\right)^{\frac{1}{3}}$

75. $\frac{3^{\frac{1}{2}} \cdot 3^{\frac{3}{4}}}{3^{\frac{1}{4}}}$

76. $\frac{5^{\frac{3}{4}} \cdot 5^{\frac{1}{2}}}{5^{\frac{1}{4}}}$

77. $\frac{\left(3y^{\frac{1}{4}}\right)^3}{y^{\frac{1}{12}}}$

78. $\frac{\left(2y^{\frac{1}{5}}\right)^4}{y^{\frac{3}{10}}}$

In Exercises 79–112, use rational exponents to simplify each expression. If rational exponents appear after simplifying, write the answer in radical notation. Assume that all variables represent positive numbers.

79. $\sqrt[8]{x^2}$

80. $\sqrt[10]{x^2}$

81. $\sqrt[3]{8a^6}$

82. $\sqrt[3]{27a^{12}}$

83. $\sqrt[5]{x^{10}y^{15}}$

84. $\sqrt[5]{x^{15}y^{20}}$

85. $(\sqrt[3]{xy})^{18}$

86. $(\sqrt[3]{xy})^{21}$

87. $\sqrt[10]{(3y)^2}$

88. $\sqrt[12]{(3y)^2}$

89. $(\sqrt[6]{2a})^4$

90. $(\sqrt[8]{2a})^6$

91. $\sqrt[9]{x^6y^3}$

92. $\sqrt[4]{x^2y^6}$

93. $\sqrt{2} \cdot \sqrt[3]{2}$

94. $\sqrt{3} \cdot \sqrt[3]{3}$

95. $\sqrt[5]{x^2} \cdot \sqrt{x}$

96. $\sqrt[7]{x^2} \cdot \sqrt{x}$

97. $\sqrt[4]{a^2b} \cdot \sqrt[3]{ab}$

98. $\sqrt[6]{ab^2} \cdot \sqrt[3]{a^2b}$

99. $\frac{\sqrt[4]{x}}{\sqrt[5]{x}}$

100. $\frac{\sqrt[3]{x}}{\sqrt[4]{x}}$

101. $\frac{\sqrt[3]{y^2}}{\sqrt[6]{y}}$

102. $\frac{\sqrt[5]{y^2}}{\sqrt[10]{y^3}}$

103. $\sqrt[4]{\sqrt{x}}$

104. $\sqrt[5]{\sqrt{x}}$

105. $\sqrt{\sqrt{x^2y}}$

106. $\sqrt{\sqrt{xy^2}}$

107. $\sqrt[4]{\sqrt[3]{2x}}$

108. $\sqrt[5]{\sqrt[3]{2x}}$

109. $(\sqrt[4]{x^3y^5})^{12}$

110. $(\sqrt[5]{x^4y^2})^{20}$

111. $\frac{\sqrt[4]{a^5b^5}}{\sqrt{ab}}$

112. $\frac{\sqrt[4]{a^3b^3}}{\sqrt{ab}}$

Practice PLUS

In Exercises 113–116, use the distributive property or the FOIL method to perform each multiplication.

113. $x^{\frac{1}{3}}\left(x^{\frac{1}{3}} - x^{\frac{2}{3}}\right)$

114. $x^{-\frac{1}{4}}\left(x^{\frac{9}{4}} - x^{\frac{1}{4}}\right)$

115. $\left(x^{\frac{1}{2}} - 3\right)\left(x^{\frac{1}{2}} + 5\right)$

116. $\left(x^{\frac{1}{3}} - 2\right)\left(x^{\frac{1}{3}} + 6\right)$

In Exercises 117–120, factor out the greatest common factor from each expression.

117. $6x^{\frac{1}{2}} + 2x^{\frac{3}{2}}$

118. $8x^{\frac{1}{4}} + 4x^{\frac{5}{4}}$

119. $15x^{\frac{1}{3}} - 60x$

120. $7x^{\frac{1}{3}} - 70x$

In Exercises 121–124, simplify each expression. Assume that all variables represent positive numbers.

121. $(49x^{-2}y^{4})^{-\frac{1}{2}}\left(xy^{\frac{1}{2}}\right)$

122. $(8x^{-6}y^{3})^{\frac{1}{3}}\left(x^{\frac{5}{6}}y^{-\frac{1}{3}}\right)^{6}$

123. $\left(\dfrac{x^{-\frac{5}{4}}y^{\frac{1}{3}}}{x^{-\frac{3}{4}}}\right)^{-6}$

124. $\left(\dfrac{x^{\frac{1}{2}}y^{-\frac{7}{4}}}{y^{-\frac{5}{4}}}\right)^{-4}$

Application Exercises

The Galápagos Islands, lying 600 miles west of Ecuador, are famed for their extraordinary wildlife. The function

$$f(x) = 29x^{\frac{1}{3}}$$

models the number of plant species, $f(x)$, on the various islands of the Galápagos chain in terms of the area, x, in square miles, of a particular island. Use the function to solve Exercises 125–126.

125. How many species of plants are on a Galápagos island that has an area of 8 square miles?

126. How many species of plants are on a Galápagos island that has an area of 27 square miles?

The function

$$f(x) = 70x^{\frac{3}{4}}$$

models the number of calories per day, $f(x)$, a person needs to maintain life in terms of that person's weight, x, in kilograms. (1 kilogram is approximately 2.2 pounds.) Use this model and a calculator to solve Exercises 127–128. Round answers to the nearest calorie.

127. How many calories per day does a person who weighs 80 kilograms (approximately 176 pounds) need to maintain life?

128. How many calories per day does a person who weighs 70 kilograms (approximately 154 pounds) need to maintain life?

The way that we perceive the temperature on a cold day depends on both air temperature and wind speed. The windchill is what the air temperature would have to be with no wind to achieve the same chilling effect on the skin. In 2002, the National Weather Service issued new windchill temperatures, shown in the table at the top of the next column. (One reason for this new windchill index is that the wind speed is now calculated at 5 feet, the average height of the human body's face, rather than 33 feet, the height of the standard anemometer, an instrument that calculates wind speed.) The windchill temperatures shown in the table can be calculated using

$$C = 35.74 + 0.6215t - 35.74\sqrt[25]{v^4} + 0.4275t\sqrt[25]{v^4},$$

in which C is the windchill, in degrees Fahrenheit, t is the air temperature, in degrees Fahrenheit, and v is the wind speed, in miles per hour. Use the formula to solve Exercises 129–132.

129. a. Rewrite the equation for calculating windchill temperatures using rational exponents.

b. Use the form of the equation in part (a) and a calculator to find the windchill temperature, to the nearest degree, when the air temperature is 25°F and the wind speed is 30 miles per hour.

New Windchill Temperature Index

Wind Speed (miles per hour)	Air Temperature (°F) 30	25	20	15	10	5	0	−5	−10	−15	−20	−25
5	25	19	13	7	1	−5	−11	−16	−22	−28	−34	−40
10	21	15	9	3	−4	−10	−16	−22	−28	−35	−41	−47
15	19	13	6	0	−7	−13	−19	−26	−32	−39	−45	−51
20	17	11	4	−2	−9	−15	−22	−29	−35	−42	−48	−55
25	16	9	3	−4	−11	−17	−24	−31	−37	−44	−51	−58
30	15	8	1	−5	−12	−19	−26	−33	−39	−46	−53	−60
35	14	7	0	−7	−14	−21	−27	−34	−41	−48	−55	−62
40	13	6	−1	−8	−15	−22	−29	−36	−43	−50	−57	−64
45	12	5	−2	−9	−16	−23	−30	−37	−44	−51	−58	−65
50	12	4	−3	−10	−17	−24	−31	−38	−45	−52	−60	−67
55	11	4	−3	−11	−18	−25	−32	−39	−46	−54	−61	−68
60	10	3	−4	−11	−19	−26	−33	−40	−48	−55	−62	−69

Frostbite occurs in 15 minutes or less.

Source: National Weather Service

130. a. Rewrite the equation for calculating windchill temperatures using rational exponents.

b. Use the form of the equation in part (a) and a calculator to find the windchill temperature, to the nearest degree, when the air temperature is 35°F and the wind speed is 15 miles per hour.

131. a. Substitute 0 for t in the equation with rational exponents from Exercise 129(a) and write a function $C(v)$ that gives the windchill temperature as a function of wind speed for an air temperature of 0°F.

b. Find and interpret $C(25)$. Use a calculator and round to the nearest degree.

c. Identify your solution to part (b) on the graph shown.

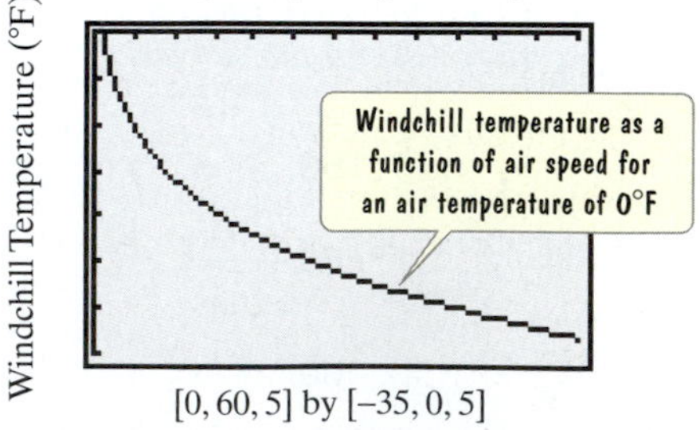

[0, 60, 5] by [−35, 0, 5]

132. a. Substitute 30 for t in the equation with rational exponents from Exercise 130(a) and write a function $C(v)$ that gives the windchill temperature as a function of wind speed for an air temperature of 30°F. Simplify the function's formula so that it contains exactly two terms.

b. Find and interpret $C(40)$. Use a calculator and round to the nearest degree.

c. Identify your solution to part (b) on the graph shown.

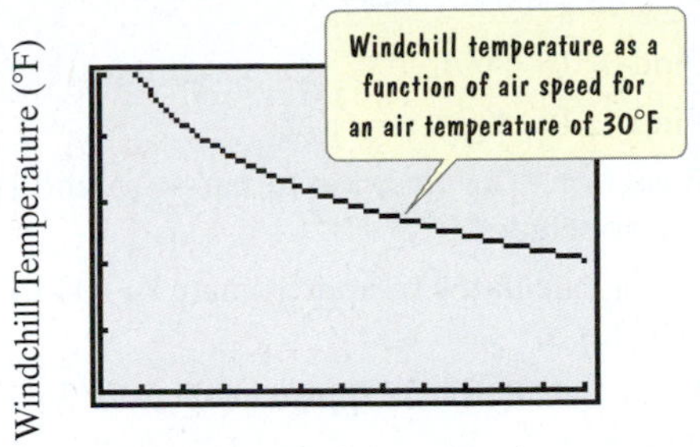

Air Speed (miles per hour)
[0, 60, 5] by [0, 25, 5]

Your job is to determine whether or not yachts are eligible for the America's Cup, the supreme event in ocean sailing. The basic dimensions of competitors' yachts must satisfy

$$L + 1.25\sqrt{S} - 9.8\sqrt[3]{D} \leq 16.296,$$

where L is the yacht's length, in meters, S is its sail area, in square meters, and D is its displacement, in cubic meters. Use this information to solve Exercises 133–134.

133. a. Rewrite the inequality using rational exponents.

b. Use your calculator to determine if a yacht with length 20.85 meters, sail area 276.4 square meters, and displacement 18.55 cubic meters is eligible for the America's Cup.

134. a. Rewrite the inequality using rational exponents.

b. Use your calculator to determine if a yacht with length 22.85 meters, sail area 312.5 square meters, and displacement 22.34 cubic meters is eligible for the America's Cup.

Writing in Mathematics

135. What is the meaning of $a^{\frac{1}{n}}$? Give an example to support your explanation.

136. What is the meaning of $a^{\frac{m}{n}}$? Give an example.

137. What is the meaning of $a^{-\frac{m}{n}}$? Give an example.

138. Explain why $a^{\frac{1}{n}}$ is negative when n is odd and a is negative. What happens if n is even and a is negative? Why?

139. In simplifying $36^{\frac{3}{2}}$, is it better to use $a^{\frac{m}{n}} = \sqrt[n]{a^m}$ or $a^{\frac{m}{n}} = (\sqrt[n]{a})^m$? Explain.

140. How can you tell if an expression with rational exponents is simplified?

141. Explain how to simplify $\sqrt[3]{x} \cdot \sqrt{x}$.

142. Explain how to simplify $\sqrt[3]{\sqrt{x}}$.

Technology Exercises

143. Use a scientific or graphing calculator to verify your results in Exercises 15–18.

144. Use a scientific or graphing calculator to verify your results in Exercises 45–50.

Exercises 145–147 show a number of simplifications, not all of which are correct. Enter the left side of each equation as y_1 and the right side as y_2. Then use your graphing utility's TABLE *feature to determine if the simplification is correct. If it is not, correct the right side and use the* TABLE *feature to verify your simplification.*

145. $x^{\frac{3}{5}} \cdot x^{-\frac{1}{10}} = x^{\frac{1}{2}}$

146. $\left(x^{-\frac{1}{2}} \cdot x^{\frac{3}{4}}\right)^{-2} = x^{\frac{1}{2}}$

147. $\dfrac{x^{\frac{1}{4}}}{x^{\frac{1}{2}} \cdot x^{-\frac{3}{4}}} = \dfrac{1}{x^{\frac{1}{2}}}$

Critical Thinking Exercises

Make Sense? *In Exercises 148–151, determine whether each statement "makes sense" or "does not make sense" and explain your reasoning.*

148. By adding the exponents, I simplified $7^{\frac{1}{2}} \cdot 7^{\frac{1}{2}}$ and obtained 49.

149. When I use the definition for $a^{\frac{m}{n}}$, I usually prefer to first raise a to the m power because smaller numbers are involved.

150. There's no question that $(-64)^{\frac{1}{3}} = -64^{\frac{1}{3}}$, so I can conclude that $(-64)^{\frac{1}{2}} = -64^{\frac{1}{2}}$.

151. I checked the following simplification and every step is correct:

$$\begin{aligned} 5\left(4 - 5^{\frac{1}{2}}\right) &= 5 \cdot 4 - 5 \cdot 5^{\frac{1}{2}} \\ &= 20 - 25^{\frac{1}{2}} \\ &= 20 - \sqrt{25} \\ &= 20 - 5 = 15. \end{aligned}$$

In Exercises 152–155, determine whether each statement is true or false. If the statement is false, make the necessary change(s) to produce a true statement.

152. If n is odd, then $(-b)^{\frac{1}{n}} = -b^{\frac{1}{n}}$.

153. $(a + b)^{\frac{1}{n}} = a^{\frac{1}{n}} + b^{\frac{1}{n}}$

154. $8^{-\frac{2}{3}} = -4$

155. $4^{-3.5} = \dfrac{1}{128}$

156. A mathematics professor recently purchased a birthday cake for her son with the inscription

$$\text{Happy } \left(2^{\frac{5}{2}} \cdot 2^{\frac{3}{4}} \div 2^{\frac{1}{4}}\right)\text{th Birthday.}$$

How old is the son?

157. The birthday boy in Exercise 156, excited by the inscription on the cake, tried to wolf down the whole thing. Professor Mom, concerned about the possible metamorphosis of her son into a blimp, exclaimed, "Hold on! It is your birthday, so why not take $\dfrac{8^{-\frac{4}{3}} + 2^{-2}}{16^{-\frac{3}{4}} + 2^{-1}}$ of the cake? I'll eat half of what's left over." How much of the cake did the professor eat?

158. Simplify: $\left[3 + \left(27^{\frac{2}{3}} + 32^{\frac{2}{5}}\right)\right]^{\frac{3}{2}} - 9^{\frac{1}{2}}$.

159. Find the domain of $f(x) = (x - 3)^{\frac{1}{2}}(x + 4)^{-\frac{1}{2}}$.

Review Exercises

160. Write the equation of the linear function whose graph passes through (5, 1) and (4, 3). (Section 2.5, Example 2)

161. Graph $y \le -\frac{3}{2}x + 3$. (Section 4.4, Example 2)

162. Solve by Cramer's rule:

$$5x - 3y = 3$$
$$7x + y = 25.$$

(Section 3.5, Example 2)

Preview Exercises

Exercises 163–165 will help you prepare for the material covered in the next section.

163. **a.** Find $\sqrt{16} \cdot \sqrt{4}$.

b. Find $\sqrt{16 \cdot 4}$.

c. Based on your answers to parts (a) and (b), what can you conclude?

164. **a.** Use a calculator to approximate $\sqrt{300}$ to two decimal places.

b. Use a calculator to approximate $10\sqrt{3}$ to two decimal places.

c. Based on your answers to parts (a) and (b), what can you conclude?

165. Simplify: **a.** $\sqrt[3]{x^{21}}$ **b.** $\sqrt[6]{y^{24}}$.

SECTION 7.3

Multiplying and Simplifying Radical Expressions

Objectives

1. Use the product rule to multiply radicals.
2. Use factoring and the product rule to simplify radicals.
3. Multiply radicals and then simplify.

George Tooker (b. 1920) "Mirror II" 1963, egg tempera on gesso panel, 20 × 20 in., 1968.4.

Gift of R. H. Donnelley Erdman (PA 1956). Addison Gallery of American Art, Phillips Academy, Andover, Massachusetts.

We opened this book with a model that described our improving emotional health as we age. Unfortunately, not everything gets better. The aging process is also accompanied by a number of physical transformations, including changes in vision that require glasses for reading, the onset of wrinkles and sagging skin, and a decrease in heart response. A change in heart response occurs fairly early; after 20, our hearts become less adept at accelerating in response to exercise. In this section's exercise set, you will see how a radical function models changes in heart function throughout the aging process, as we turn to multiplying and simplifying radical expressions.

1 Use the product rule to multiply radicals.

The Product Rule for Radicals

A rule for multiplying radicals can be generalized by comparing $\sqrt{25}\cdot\sqrt{4}$ and $\sqrt{25\cdot4}$. Notice that

$$\sqrt{25}\cdot\sqrt{4} = 5\cdot2 = 10 \quad\text{and}\quad \sqrt{25\cdot4} = \sqrt{100} = 10.$$

Because we obtain 10 in both situations, the original radical expressions must be equal. That is,

$$\sqrt{25}\cdot\sqrt{4} = \sqrt{25\cdot4}.$$

This result is a special case of the **product rule for radicals** that can be generalized as follows:

The Product Rule for Radicals

If $\sqrt[n]{a}$ and $\sqrt[n]{b}$ are real numbers, then

$$\sqrt[n]{a}\cdot\sqrt[n]{b} = \sqrt[n]{ab}.$$

The product of two nth roots is the nth root of the product of the radicands.

Study Tip

The product rule can be used only when the radicals have the same index. If indices differ, rational exponents can be used, as in $\sqrt{x}\cdot\sqrt[3]{x}$, which was Example 7(d) in the previous section.

EXAMPLE 1 Using the Product Rule for Radicals

Multiply:

a. $\sqrt{3}\cdot\sqrt{7}$ b. $\sqrt{x+7}\cdot\sqrt{x-7}$ c. $\sqrt[3]{7}\cdot\sqrt[3]{9}$ d. $\sqrt[8]{10x}\cdot\sqrt[8]{8x^4}$.

Solution In each problem, the indices are the same. Thus, we multiply by multiplying the radicands.

a. $\sqrt{3}\cdot\sqrt{7} = \sqrt{3\cdot7} = \sqrt{21}$

b. $\sqrt{x+7}\cdot\sqrt{x-7} = \sqrt{(x+7)(x-7)} = \sqrt{x^2-49}$

This is not equal to $\sqrt{x^2}-\sqrt{49}$.

c. $\sqrt[3]{7}\cdot\sqrt[3]{9} = \sqrt[3]{7\cdot9} = \sqrt[3]{63}$

d. $\sqrt[8]{10x}\cdot\sqrt[8]{8x^4} = \sqrt[8]{10x\cdot8x^4} = \sqrt[8]{80x^5}$

✓ CHECK POINT 1 Multiply:

a. $\sqrt{5}\cdot\sqrt{11}$ b. $\sqrt{x+4}\cdot\sqrt{x-4}$

c. $\sqrt[3]{6}\cdot\sqrt[3]{10}$ d. $\sqrt[7]{2x}\cdot\sqrt[7]{6x^3}$.

2 Use factoring and the product rule to simplify radicals.

Using Factoring and the Product Rule to Simplify Radicals

In Chapter 5, we saw that a number that is the square of an integer is a **perfect square**. For example, 100 is a perfect square because $100 = 10^2$. A number is a **perfect cube** if it is the cube of an integer. Thus, 125 is a perfect cube because $125 = 5^3$. In general, a number is a **perfect *n*th power** if it is the nth power of an integer. Thus, p is a perfect nth power if there is an integer q such that $p = q^n$.

A radical of index n is **simplified** when its radicand has no factors other than 1 that are perfect nth powers. For example, $\sqrt{300}$ is not simplified because it can be expressed as $\sqrt{100\cdot3}$ and 100 is a perfect square. We can use the product rule in the form

$$\sqrt[n]{ab} = \sqrt[n]{a}\cdot\sqrt[n]{b}$$

to simplify $\sqrt[n]{ab}$ when a or b is a perfect nth power. Consider $\sqrt{300}$. To simplify, we factor 300 so that one of its factors is the greatest perfect square possible.

Using Technology

You can use a calculator to provide numerical support that $\sqrt{300} = 10\sqrt{3}$. First find an approximation for $\sqrt{300}$:

300 [√] ≈ 17.32

or

[√] 300 [ENTER] ≈ 17.32.

Now find an aproximation for $10\sqrt{3}$:

10 [×] 3 [√] ≈ 17.32

or

10 [√] 3 [ENTER] ≈ 17.32.

Correct to two decimal places,

$\sqrt{300} \approx 17.32$ and $10\sqrt{3} \approx 17.32$.

This verifies that

$\sqrt{300} = 10\sqrt{3}$.

Use this technique to support the numerical results for the answers in this section. *Caution:* A simplified radical does not mean a decimal approximation.

$\sqrt{300} = \sqrt{100 \cdot 3}$ — Factor 300. 100 is the greatest perfect square factor.

$= \sqrt{100} \cdot \sqrt{3}$ — Use the product rule: $\sqrt[n]{ab} = \sqrt[n]{a} \cdot \sqrt[n]{b}$.

$= 10\sqrt{3}$ — Write $\sqrt{100}$ as 10. We read $10\sqrt{3}$ as "ten times the square root of three."

Simplifying Radical Expressions by Factoring

A radical expression whose index is n is **simplified** when its radicand has no factors that are perfect nth powers. To simplify, use the following procedure:

1. Write the radicand as the product of two factors, one of which is the greatest perfect nth power.
2. Use the product rule to take the nth root of each factor.
3. Find the nth root of the perfect nth power.

EXAMPLE 2 Simplifying Radicals by Factoring

Simplify by factoring:

a. $\sqrt{75}$ **b.** $\sqrt[3]{54}$ **c.** $\sqrt[5]{64}$ **d.** $\sqrt{500xy^2}$.

Solution

a. $\sqrt{75} = \sqrt{25 \cdot 3}$ — 25 is the greatest perfect square that is a factor of 75.

$= \sqrt{25} \cdot \sqrt{3}$ — Take the square root of each factor: $\sqrt[n]{ab} = \sqrt[n]{a} \cdot \sqrt[n]{b}$.

$= 5\sqrt{3}$ — Write $\sqrt{25}$ as 5.

b. $\sqrt[3]{54} = \sqrt[3]{27 \cdot 2}$ — 27 is the greatest perfect cube that is a factor of 54: $27 = 3^3$.

$= \sqrt[3]{27} \cdot \sqrt[3]{2}$ — Take the cube root of each factor: $\sqrt[n]{ab} = \sqrt[n]{a} \cdot \sqrt[n]{b}$.

$= 3\sqrt[3]{2}$ — Write $\sqrt[3]{27}$ as 3.

c. $\sqrt[5]{64} = \sqrt[5]{32 \cdot 2}$ — 32 is the greatest perfect fifth power that is a factor of 64: $32 = 2^5$.

$= \sqrt[5]{32} \cdot \sqrt[5]{2}$ — Take the fifth root of each factor: $\sqrt[n]{ab} = \sqrt[n]{a} \cdot \sqrt[n]{b}$.

$= 2\sqrt[5]{2}$ — Write $\sqrt[5]{32}$ as 2.

d. $\sqrt{500xy^2} = \sqrt{100y^2 \cdot 5x}$ — $100y^2$ is the greatest perfect square that is a factor of $500xy^2$: $100y^2 = (10y)^2$.

$= \sqrt{100y^2} \cdot \sqrt{5x}$ — Factor into two radicals.

$= 10|y|\sqrt{5x}$ — Take the square root of $100y^2$.

✓ CHECK POINT 2 Simplify by factoring:

a. $\sqrt{80}$ **b.** $\sqrt[3]{40}$

c. $\sqrt[4]{32}$ **d.** $\sqrt{200x^2y}$.

EXAMPLE 3 Simplifying a Radical Function

If

$$f(x) = \sqrt{2x^2 + 4x + 2},$$

express the function, f, in simplified form.

Solution Begin by factoring the radicand. The GCF is 2. Simplification is possible if we obtain a factor that is a perfect square.

$$f(x) = \sqrt{2x^2 + 4x + 2}$$ This is the given function.

$$= \sqrt{2(x^2 + 2x + 1)}$$ Factor out the GCF.

$$= \sqrt{2(x + 1)^2}$$ Factor the perfect square trinomial: $A^2 + 2AB + B^2 = (A + B)^2$.

$$= \sqrt{2} \cdot \sqrt{(x + 1)^2}$$ Take the square root of each factor. The factor $(x + 1)^2$ is a perfect square.

$$= \sqrt{2}|x + 1|$$ Take the square root of $(x + 1)^2$.

In simplified form,

$$f(x) = \sqrt{2}|x + 1|.$$

Using Technology

Graphic Connections

The graphs of

$$f(x) = \sqrt{2x^2 + 4x + 2}, \quad g(x) = \sqrt{2}|x + 1|, \quad \text{and} \quad h(x) = \sqrt{2}(x + 1)$$

are shown in **Figure 7.6** in three separate $[-5, 5, 1]$ by $[-5, 5, 1]$ viewing rectangles. The graphs in **Figure 7.6 (a)** and **(b)** are identical. This verifies that our simplification in Example 3 is correct: $\sqrt{2x^2 + 4x + 2} = \sqrt{2}|x + 1|$. Now compare the graphs in **Figure 7.6 (a)** and **(c)**. Can you see that they are not the same? This illustrates the importance of not leaving out absolute value bars:

$$\sqrt{2x^2 + 4x + 2} \neq \sqrt{2}(x + 1).$$

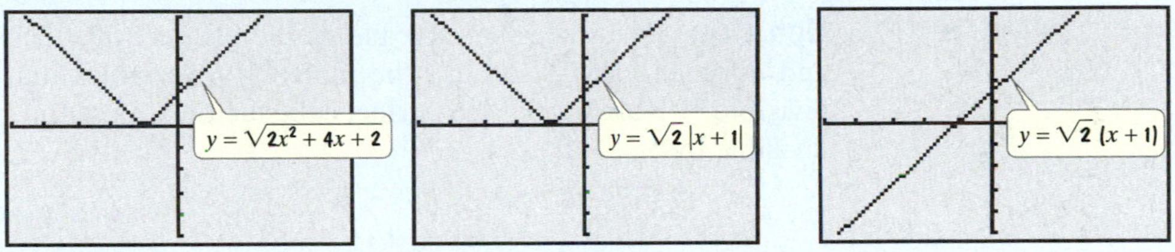

FIGURE 7.6 (a) (b) (c)

☑ **CHECK POINT 3** If $f(x) = \sqrt{3x^2 - 12x + 12}$, express the function, f, in simplified form.

For the remainder of this chapter, in situations that do not involve functions, we will **assume that no radicands involve negative quantities raised to even powers. Based upon this assumption, absolute value bars are not necessary when taking even roots.**

Simplifying When Variables to Even Powers in a Radicand Are Nonnegative Quantities

For any nonnegative real number a,

$$\sqrt[n]{a^n} = a.$$

In simplifying an nth root, how do we find variable factors in the radicand that are perfect nth powers? The **perfect nth powers have exponents that are divisible by n.** Simplification is possible by observation or by using rational exponents. Here are some examples:

- $\sqrt{x^6} = \sqrt{(x^3)^2} = x^3$ or $\sqrt{x^6} = (x^6)^{\frac{1}{2}} = x^3$

  6 is divisible by the index, 2. Thus, x^6 is a perfect square.

- $\sqrt[3]{y^{21}} = \sqrt[3]{(y^7)^3} = y^7$ or $\sqrt[3]{y^{21}} = (y^{21})^{\frac{1}{3}} = y^7$

  21 is divisible by the index, 3. Thus, y^{21} is a perfect cube.

- $\sqrt[6]{z^{24}} = \sqrt[6]{(z^4)^6} = z^4$ or $\sqrt[6]{z^{24}} = (z^{24})^{\frac{1}{6}} = z^4$.

  24 is divisible by the index, 6. Thus, z^{24} is a perfect 6th power.

EXAMPLE 4 Simplifying a Radical by Factoring

Simplify: $\sqrt{x^5y^{13}z^7}$.

Solution We write the radicand as the product of the greatest perfect square factor and another factor. Because the index is 2, variables that have exponents that are divisible by 2 are part of the perfect square factor. We use the greatest exponents that are divisible by 2.

Discover for Yourself

Square the answer in Example 4 and show that it is correct. If it is a square root, you should obtain the given radicand, $x^5y^{13}z^7$.

$$\sqrt{x^5y^{13}z^7} = \sqrt{x^4 \cdot x \cdot y^{12} \cdot y \cdot z^6 \cdot z}$$ Use the greatest even power of each variable.

$$= \sqrt{(x^4y^{12}z^6)(xyz)}$$ Group the perfect square factors.

$$= \sqrt{x^4y^{12}z^6} \cdot \sqrt{xyz}$$ Factor into two radicals.

$$= x^2y^6z^3\sqrt{xyz}$$ $\sqrt{x^4y^{12}z^6} = (x^4y^{12}z^6)^{\frac{1}{2}} = x^2y^6z^3$

CHECK POINT 4 Simplify: $\sqrt{x^9y^{11}z^3}$.

EXAMPLE 5 Simplifying a Radical by Factoring

Simplify: $\sqrt[3]{32x^8y^{16}}$.

Solution We write the radicand as the product of the greatest perfect cube factor and another factor. Because the index is 3, variables that have exponents that are divisible by 3 are part of the perfect cube factor. We use the greatest exponents that are divisible by 3.

$$\sqrt[3]{32x^8y^{16}} = \sqrt[3]{8 \cdot 4 \cdot x^6 \cdot x^2 \cdot y^{15} \cdot y}$$ Identify perfect cube factors.

$$= \sqrt[3]{(8x^6y^{15})(4x^2y)}$$ Group the perfect cube factors.

$$= \sqrt[3]{8x^6y^{15}} \cdot \sqrt[3]{4x^2y}$$ Factor into two radicals.

$$= 2x^2y^5\sqrt[3]{4x^2y}$$ $\sqrt[3]{8} = 2$ and $\sqrt[3]{x^6y^{15}} = (x^6y^{15})^{\frac{1}{3}} = x^2y^5$.

CHECK POINT 5 Simplify: $\sqrt[3]{40x^{10}y^{14}}$. $2x^3y^4\sqrt[3]{5xy^2}$

EXAMPLE 6 Simplifying a Radical by Factoring

Simplify: $\sqrt[5]{64x^3y^7z^{29}}$.

Solution We write the radicand as the product of the greatest perfect 5th power and another factor. Because the index is 5, variables that have exponents that are divisible by 5 are part of the perfect fifth factor. We use the greatest exponents that are divisible by 5.

$$\sqrt[5]{64x^3y^7z^{29}} = \sqrt[5]{32 \cdot 2 \cdot x^3 \cdot y^5 \cdot y^2 \cdot z^{25} \cdot z^4}$$ Identify perfect fifth factors.

$$= \sqrt[5]{(32y^5z^{25})(2x^3y^2z^4)}$$ Group the perfect fifth factors.

$$= \sqrt[5]{32y^5z^{25}} \cdot \sqrt[5]{2x^3y^2z^4}$$ Factor into two radicals.

$$= 2yz^5\sqrt[5]{2x^3y^2z^4}$$ $\sqrt[5]{32} = 2$ and $\sqrt[5]{y^5z^{25}} = (y^5z^{25})^{\frac{1}{5}} = yz^5$.

CHECK POINT 6 Simplify: $\sqrt[5]{32x^{12}y^2z^8}$.

3 Multiply radicals and then simplify.

Multiplying and Simplifying Radicals

We have seen how to use the product rule when multiplying radicals with the same index. Sometimes after multiplying, we can simplify the resulting radical.

EXAMPLE 7 Multiplying Radicals and Then Simplifying

Multiply and simplify:

a. $\sqrt{15}\cdot\sqrt{3}$ **b.** $7\sqrt[3]{4}\cdot 5\sqrt[3]{6}$ **c.** $\sqrt[4]{8x^3y^2}\cdot\sqrt[4]{8x^5y^3}$.

Solution

a. $\sqrt{15}\cdot\sqrt{3} = \sqrt{15\cdot 3}$ — Use the product rule.

$= \sqrt{45} = \sqrt{9\cdot 5}$ — 9 is the greatest perfect square factor of 45.

$= \sqrt{9}\cdot\sqrt{5} = 3\sqrt{5}$

b. $7\sqrt[3]{4}\cdot 5\sqrt[3]{6} = 35\sqrt[3]{4\cdot 6}$ — Use the product rule.

$= 35\sqrt[3]{24} = 35\sqrt[3]{8\cdot 3}$ — 8 is the greatest perfect cube factor of 24.

$= 35\sqrt[3]{8}\cdot\sqrt[3]{3} = 35\cdot 2\cdot\sqrt[3]{3}$

$= 70\sqrt[3]{3}$

c. $\sqrt[4]{8x^3y^2}\cdot\sqrt[4]{8x^5y^3} = \sqrt[4]{8x^3y^2\cdot 8x^5y^3}$ — Use the product rule.

$= \sqrt[4]{64x^8y^5}$ — Multiply.

$= \sqrt[4]{16\cdot 4\cdot x^8\cdot y^4\cdot y}$ — Identify perfect fourth factors.

$= \sqrt[4]{(16x^8y^4)(4y)}$ — Group the perfect fourth factors.

$= \sqrt[4]{16x^8y^4}\cdot\sqrt[4]{4y}$ — Factor into two radicals.

$= 2x^2y\sqrt[4]{4y}$ — $\sqrt[4]{16} = 2$ and $\sqrt[4]{x^8y^4} = (x^8y^4)^{\frac{1}{4}} = x^2y$.

Study Tip

Confused about when you should write an expression under one radical and when you should separate the radicals?

- Use $\sqrt[n]{a}\cdot\sqrt[n]{b} = \sqrt[n]{ab}$, writing under one radical, when *multiplying*.
- Use $\sqrt[n]{ab} = \sqrt[n]{a}\cdot\sqrt[n]{b}$, factoring into two radicals, when *simplifying*.

✓ CHECK POINT 7 Multiply and simplify:

a. $\sqrt{6}\cdot\sqrt{2}$

b. $10\sqrt[3]{16}\cdot 5\sqrt[3]{2}$

c. $\sqrt[4]{4x^2y}\cdot\sqrt[4]{8x^6y^3}$.

7.3 EXERCISE SET

MyMathLab MathXL PRACTICE WATCH DOWNLOAD READ REVIEW

Practice Exercises

In Exercises 1–20, use the product rule to multiply.

1. $\sqrt{3}\cdot\sqrt{5}$
2. $\sqrt{7}\cdot\sqrt{5}$
3. $\sqrt[3]{2}\cdot\sqrt[3]{9}$
4. $\sqrt[3]{5}\cdot\sqrt[3]{4}$
5. $\sqrt[4]{11}\cdot\sqrt[4]{3}$
6. $\sqrt[5]{9}\cdot\sqrt[5]{3}$
7. $\sqrt{3x}\cdot\sqrt{11y}$
8. $\sqrt{5x}\cdot\sqrt{11y}$
9. $\sqrt[5]{6x^3}\cdot\sqrt[5]{4x}$
10. $\sqrt[4]{6x^2}\cdot\sqrt[4]{3x}$
11. $\sqrt{x+3}\cdot\sqrt{x-3}$
12. $\sqrt{x+6}\cdot\sqrt{x-6}$
13. $\sqrt[6]{x-4}\cdot\sqrt[6]{(x-4)^4}$
14. $\sqrt[6]{x-5}\cdot\sqrt[6]{(x-5)^4}$
15. $\sqrt{\frac{2x}{3}}\cdot\sqrt{\frac{3}{2}}$
16. $\sqrt{\frac{2x}{5}}\cdot\sqrt{\frac{5}{2}}$
17. $\sqrt[4]{\frac{x}{7}}\cdot\sqrt[4]{\frac{3}{y}}$
18. $\sqrt[4]{\frac{x}{3}}\cdot\sqrt[4]{\frac{7}{y}}$
19. $\sqrt[7]{7x^2y}\cdot\sqrt[7]{11x^3y^2}$
20. $\sqrt[9]{12x^2y^3}\cdot\sqrt[9]{3x^3y^4}$

In Exercises 21–32, simplify by factoring.

21. $\sqrt{50}$
22. $\sqrt{27}$
23. $\sqrt{45}$
24. $\sqrt{28}$
25. $\sqrt{75x}$
26. $\sqrt{40x}$
27. $\sqrt[3]{16}$
28. $\sqrt[3]{54}$
29. $\sqrt[3]{27x^3}$
30. $\sqrt[3]{250x^3}$
31. $\sqrt[3]{-16x^2y^3}$
32. $\sqrt[3]{-32x^2y^3}$

In Exercises 33–38, express the function, f, in simplified form. Assume that x can be any real number.

33. $f(x) = \sqrt{36(x+2)^2}$
34. $f(x) = \sqrt{81(x-2)^2}$
35. $f(x) = \sqrt[3]{32(x+2)^3}$
36. $f(x) = \sqrt[3]{48(x-2)^3}$

37. $f(x) = \sqrt{3x^2 - 6x + 3}$

38. $f(x) = \sqrt{5x^2 - 10x + 5}$

In Exercises 39–60, simplify by factoring. Assume that all variables in a radicand represent positive real numbers and no radicands involve negative quantities raised to even powers.

39. $\sqrt{x^7}$

40. $\sqrt{x^5}$

41. $\sqrt{x^8y^9}$

42. $\sqrt{x^6y^7}$

43. $\sqrt{48x^3}$

44. $\sqrt{40x^3}$

45. $\sqrt[3]{y^8}$

46. $\sqrt[3]{y^{11}}$

47. $\sqrt[3]{x^{14}y^3z}$

48. $\sqrt[3]{x^3y^{17}z^2}$

49. $\sqrt[3]{81x^8y^6}$

50. $\sqrt[3]{32x^9y^{17}}$

51. $\sqrt[3]{(x + y)^5}$

52. $\sqrt[3]{(x + y)^4}$

53. $\sqrt[5]{y^{17}}$

54. $\sqrt[5]{y^{18}}$

55. $\sqrt[5]{64x^6y^{17}}$

56. $\sqrt[5]{64x^7y^{16}}$

57. $\sqrt[4]{80x^{10}}$

58. $\sqrt[4]{96x^{11}}$

59. $\sqrt[4]{(x - 3)^{10}}$

60. $\sqrt[4]{(x - 2)^{14}}$

In Exercises 61–82, multiply and simplify. Assume that all variables in a radicand represent positive real numbers and no radicands involve negative quantities raised to even powers.

61. $\sqrt{12} \cdot \sqrt{2}$

62. $\sqrt{3} \cdot \sqrt{6}$

63. $\sqrt{5x} \cdot \sqrt{10y}$

64. $\sqrt{8x} \cdot \sqrt{10y}$

65. $\sqrt{12x} \cdot \sqrt{3x}$

66. $\sqrt{20x} \cdot \sqrt{5x}$

67. $\sqrt{50xy} \cdot \sqrt{4xy^2}$

68. $\sqrt{5xy} \cdot \sqrt{10xy^2}$

69. $2\sqrt{5} \cdot 3\sqrt{40}$

70. $3\sqrt{15} \cdot 5\sqrt{6}$

71. $\sqrt[3]{12} \cdot \sqrt[3]{4}$

72. $\sqrt[4]{4} \cdot \sqrt[4]{8}$

73. $\sqrt{5x^3} \cdot \sqrt{8x^2}$

74. $\sqrt{2x^7} \cdot \sqrt{12x^4}$

75. $\sqrt[3]{25x^4y^2} \cdot \sqrt[3]{5xy^{12}}$

76. $\sqrt[3]{6x^7y} \cdot \sqrt[3]{9x^4y^{12}}$

77. $\sqrt[4]{8x^2y^3z^6} \cdot \sqrt[4]{2x^4yz}$

78. $\sqrt[4]{4x^2y^3z^3} \cdot \sqrt[4]{8x^4yz^6}$

79. $\sqrt[5]{8x^4y^6z^2} \cdot \sqrt[5]{8xy^7z^4}$

80. $\sqrt[5]{8x^4y^3z^3} \cdot \sqrt[5]{8xy^9z^8}$

81. $\sqrt[3]{x - y} \cdot \sqrt[3]{(x - y)^7}$

82. $\sqrt[3]{x - 6} \cdot \sqrt[3]{(x - 6)^7}$

Practice PLUS

In Exercises 83–90, simplify each expression. Assume that all variables in a radicand represent positive real numbers and no radicands involve negative quantities raised to even powers.

83. $-2x^2y(\sqrt[3]{54x^3y^7z^2})$

84. $\dfrac{-x^2y^7}{2}(\sqrt[3]{-32x^4y^9z^7})$

85. $-3y(\sqrt[5]{64x^3y^6})$

86. $-4x^2y^7(\sqrt[5]{-32x^{11}y^{17}})$

87. $(-2xy^2\sqrt{3x})(xy\sqrt{6x})$

88. $(-5x^2y^3z\sqrt{2xyz})(-x^4z\sqrt{10xz})$

89. $(2x^2y\sqrt[4]{8xy})(-3xy^2\sqrt[4]{2x^2y^3})$

90. $(5a^2b\sqrt[4]{8a^2b})(4ab\sqrt[4]{4a^3b^2})$

Application Exercises

The function

$$d(x) = \sqrt{\frac{3x}{2}}$$

models the distance, d(x), in miles, that a person h feet high can see to the horizon. Use this function to solve Exercises 91–92.

91. The pool deck on a cruise ship is 72 feet above the water. How far can passengers on the pool deck see? Write the answer in simplified radical form. Then use the simplified radical form and a calculator to express the answer to the nearest tenth of a mile.

92. The captain of a cruise ship is on the star deck, which is 120 feet above the water. How far can the captain see? Write the answer in simplified radical form. Then use the simplified radical form and a calculator to express the answer to the nearest tenth of a mile.

Paleontologists use the function

$$W(x) = 4\sqrt{2x}$$

to estimate the walking speed of a dinosaur, W(x), in feet per second, where x is the length, in feet, of the dinosaur's leg. The graph of W is shown in the figure. Use this information to solve Exercises 93–94.

Dinosaur Walking Speeds

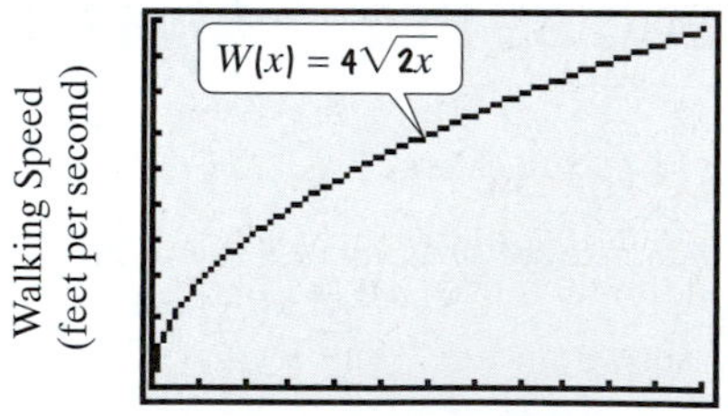

Leg Length (feet)
[0, 12, 1] by [0, 20, 2]

93. What is the walking speed of a dinosaur whose leg length is 6 feet? Use the function's equation and express the answer in simplified radical form. Then use the function's graph to estimate the answer to the nearest foot per second.

94. What is the walking speed of a dinosaur whose leg length is 10 feet? Use the function's equation and express the answer in simplified radical form. Then use the function's graph to estimate the answer to the nearest foot per second.

Your **cardiac index** *is your heart's output, in liters of blood per minute, divided by your body's surface area, in square meters. The cardiac index, $C(x)$, can be modeled by*

$$C(x) = \frac{7.644}{\sqrt[4]{x}}, \quad 10 \le x \le 80,$$

where x is an individual's age, in years. The graph of the function is shown. Use the function to solve Exercises 95–96.

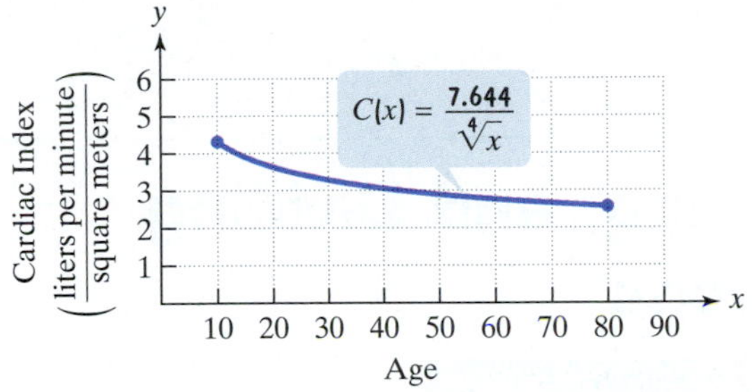

95. a. Find the cardiac index of a 32-year-old. Express the denominator in simplified radical form and reduce the fraction.

b. Use the form of the answer in part (a) and a calculator to express the cardiac index to the nearest hundredth. Identify your solution as a point on the graph.

96. a. Find the cardiac index of an 80-year-old. Express the denominator in simplified radical form and reduce the fraction.

b. Use the form of the answer in part (a) and a calculator to express the cardiac index to the nearest hundredth. Identify your solution as a point on the graph.

Writing in Mathematics

97. What is the product rule for radicals? Give an example to show how it is used.

98. Explain why $\sqrt{50}$ is not simplified. What do we mean when we say a radical expression is simplified?

99. In simplifying an *n*th root, explain how to find variable factors in the radicand that are perfect *n*th powers.

100. Without showing all the details, explain how to simplify $\sqrt[3]{16x^{14}}$.

101. As you get older, what would you expect to happen to your heart's output? Explain how this is shown in the graph for Exercises 95–96. Is this trend taking place progressively more rapidly or more slowly over the entire interval? What does this mean about this aspect of aging?

Technology Exercises

102. Use a calculator to provide numerical support for your simplifications in Exercises 21–24 and 27–28. In each case, find a decimal approximation for the given expression. Then find a decimal approximation for your simplified expression. The approximations should be the same.

In Exercises 103–106, determine if each simplification is correct by graphing the function on each side of the equation with your graphing utility. Use the given viewing rectangle. The graphs should be the same. If they are not, correct the right side of the equation and then use your graphing utility to verify the simplification.

103. $\sqrt{x^4} = x^2$; [0, 5, 1] by [0, 20, 1]

104. $\sqrt{8x^2} = 4x\sqrt{2}$; [−5, 5, 1] by [−5, 20, 1]

105. $\sqrt{3x^2 - 6x + 3} = (x - 1)\sqrt{3}$; [−5, 5, 1] by [−5, 5, 1]

106. $\sqrt[3]{2x} \cdot \sqrt[3]{4x^2} = 4x$; [−10, 10, 1] by [−10, 10, 1]

Critical Thinking Exercises

Make Sense? *In Exercises 107–110, determine whether each statement "makes sense" or "does not make sense" and explain your reasoning.*

107. Because the product rule for radicals applies when $\sqrt[n]{a}$ and $\sqrt[n]{b}$ are real numbers, I can use it to find $\sqrt[3]{16} \cdot \sqrt[3]{-4}$, but not to find $\sqrt{8} \cdot \sqrt{-2}$.

108. I multiply *n*th roots by taking the *n*th root of the product of the radicands.

109. I need to know how to factor a trinomial to simplify $\sqrt{x^2 - 10x + 25}$.

110. I know that I've simplified a radical expression when it contains a single radical.

In Exercises 111–114, determine whether each statement is true or false. If the statement is false, make the necessary change(s) to produce a true statement.

111. $2\sqrt{5} \cdot 6\sqrt{5} = 12\sqrt{5}$

112. $\sqrt[3]{4} \cdot \sqrt[3]{4} = 4$

113. $\sqrt{12} = 2\sqrt{6}$

114. $\sqrt[3]{3^{15}} = 243$

115. If a number is tripled, what happens to its square root?

116. What must be done to a number so that its cube root is tripled?

117. If $f(x) = \sqrt[3]{2x}$ and $(fg)(x) = 2x$, find $g(x)$.

118. Graph $f(x) = \sqrt{(x - 1)^2}$ by hand.

Review Exercises

119. Solve: $2x - 1 \le 21$ and $2x + 2 \ge 12$. (Section 4.2, Example 2)

120. Solve:

$$\begin{aligned} 5x + 2y &= 2 \\ 4x + 3y &= -4. \end{aligned}$$

(Section 3.1, Example 6)

121. Factor: $64x^3 - 27$. (Section 5.5, Example 9)

Preview Exercises

Exercises 122–124 will help you prepare for the material covered in the next section.

122. **a.** Simplify: $21x + 10x$.

b. Simplify: $21\sqrt{2} + 10\sqrt{2}$.

123. **a.** Simplify: $4x - 12x$.

b. Simplify: $4\sqrt[3]{2} - 12\sqrt[3]{2}$.

124. Simplify: $\dfrac{\sqrt[4]{7y^5}}{\sqrt[4]{x^{12}}}$.

SECTION 7.4

Adding, Subtracting, and Dividing Radical Expressions

Objectives

1. Add and subtract radical expressions.
2. Use the quotient rule to simplify radical expressions.
3. Use the quotient rule to divide radical expressions.

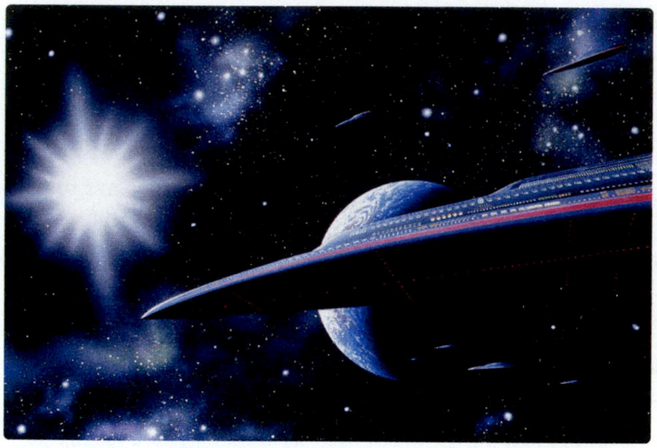

The future is now: You have the opportunity to explore the cosmos in a starship traveling near the speed of light. The experience will enable you to understand the mysteries of the universe first hand, transporting you to unimagined levels of knowing and being. The down side: According to Einstein's theory of relativity, close to the speed of light, your aging rate relative to friends on Earth is nearly zero. You will return from your two-year journey to an unknown futuristic world. In this section's exercise set, we provide an expression involving radical division that models your return to this unrecognizable world. To make sense of the model, we turn to various operations with radicals, including addition, subtraction, and division.

1 Add and subtract radical expressions.

Adding and Subtracting Radical Expressions

We know that like terms have exactly the same variable factors and can be combined. For example,

$$7x + 6x = (7 + 6)x = 13x.$$

Two or more radical expressions that have the same indices *and* the same radicands are called **like radicals**. Like radicals are combined in exactly the same way that we combine like terms. For example,

$$7\sqrt{11} + 6\sqrt{11} = (7 + 6)\sqrt{11} = 13\sqrt{11}.$$

7 square roots of 11 plus 6 square roots of 11 result in 13 square roots of 11.

EXAMPLE 1 Adding and Subtracting Like Radicals

Simplify (add or subtract) by combining like radical terms:

a. $7\sqrt{2} + 8\sqrt{2}$

b. $\sqrt[3]{5} - 4x\sqrt[3]{5} + 8\sqrt[3]{5}$

c. $8\sqrt[6]{5x} - 5\sqrt[6]{5x} + 4\sqrt[3]{5x}$.

Solution

a. $7\sqrt{2} + 8\sqrt{2}$

$= (7 + 8)\sqrt{2}$ — Apply the distributive property.

$= 15\sqrt{2}$ — Simplify.

b. $\sqrt[3]{5} - 4x\sqrt[3]{5} + 8\sqrt[3]{5}$

$= (1 - 4x + 8)\sqrt[3]{5}$ — Apply the distributive property.

$= (9 - 4x)\sqrt[3]{5}$ — Simplify.

c. $8\sqrt[6]{5x} - 5\sqrt[6]{5x} + 4\sqrt[3]{5x}$

$= (8 - 5)\sqrt[6]{5x} + 4\sqrt[3]{5x}$ — Apply the distributive property to the two terms with like radicals.

$= 3\sqrt[6]{5x} + 4\sqrt[3]{5x}$ — The indices, 6 and 3, differ. These are not like radicals and cannot be combined.

✓ CHECK POINT 1 Simplify by combining like radical terms:

a. $8\sqrt{13} + 2\sqrt{13}$

b. $9\sqrt[3]{7} - 6x\sqrt[3]{7} + 12\sqrt[3]{7}$

c. $7\sqrt[4]{3x} - 2\sqrt[4]{3x} + 2\sqrt[3]{3x}$.

In some cases, radical expressions can be combined once they have been simplified. For example, to add $\sqrt{2}$ and $\sqrt{8}$, we can write $\sqrt{8}$ as $\sqrt{4 \cdot 2}$ because 4 is a perfect square factor of 8.

$$\sqrt{2} + \sqrt{8} = \sqrt{2} + \sqrt{4 \cdot 2} = 1\sqrt{2} + 2\sqrt{2} = (1 + 2)\sqrt{2} = 3\sqrt{2}$$

Always begin by simplifying radical terms. This makes it possible to identify and combine any like radicals.

EXAMPLE 2 Combining Radicals That First Require Simplification

Simplify by combining like radical terms, if possible:

a. $7\sqrt{18} + 5\sqrt{8}$ **b.** $4\sqrt{27x} - 8\sqrt{12x}$ **c.** $7\sqrt{3} - 2\sqrt{5}$.

Solution

a. $7\sqrt{18} + 5\sqrt{8}$

$= 7\sqrt{9 \cdot 2} + 5\sqrt{4 \cdot 2}$ — Factor the radicands using the greatest perfect square factors.

$= 7\sqrt{9} \cdot \sqrt{2} + 5\sqrt{4} \cdot \sqrt{2}$ — Take the square root of each factor.

$= 7 \cdot 3 \cdot \sqrt{2} + 5 \cdot 2 \cdot \sqrt{2}$ — $\sqrt{9} = 3$ and $\sqrt{4} = 2$.

$= 21\sqrt{2} + 10\sqrt{2}$ — Multiply.

$= (21 + 10)\sqrt{2}$ — Apply the distributive property.

$= 31\sqrt{2}$ — Simplify.

b. $4\sqrt{27x} - 8\sqrt{12x}$

$= 4\sqrt{9\cdot 3x} - 8\sqrt{4\cdot 3x}$ Factor the radicands using the greatest perfect square factors.

$= 4\sqrt{9}\cdot\sqrt{3x} - 8\sqrt{4}\cdot\sqrt{3x}$ Take the square root of each factor.

$= 4\cdot 3\cdot\sqrt{3x} - 8\cdot 2\cdot\sqrt{3x}$ $\sqrt{9} = 3$ and $\sqrt{4} = 2$.

$= 12\sqrt{3x} - 16\sqrt{3x}$ Multiply.

$= (12 - 16)\sqrt{3x}$ Apply the distributive property.

$= -4\sqrt{3x}$ Simplify.

c. $7\sqrt{3} - 2\sqrt{5}$ cannot be simplified. The radical expressions have different radicands, namely 3 and 5, and are not like terms.

✓ CHECK POINT 2 Simplify by combining like radical terms, if possible:

a. $3\sqrt{20} + 5\sqrt{45}$

b. $3\sqrt{12x} - 6\sqrt{27x}$

c. $8\sqrt{5} - 6\sqrt{2}$.

EXAMPLE 3 Adding and Subtracting with Higher Indices

Simplify by combining like radical terms, if possible:

a. $2\sqrt[3]{16} - 4\sqrt[3]{54}$ **b.** $5\sqrt[3]{xy^2} + \sqrt[3]{8x^4y^5}$.

Solution

a. $2\sqrt[3]{16} - 4\sqrt[3]{54}$

$= 2\sqrt[3]{8\cdot 2} - 4\sqrt[3]{27\cdot 2}$ Factor the radicands using the greatest perfect cube factors.

$= 2\sqrt[3]{8}\cdot\sqrt[3]{2} - 4\sqrt[3]{27}\cdot\sqrt[3]{2}$ Take the cube root of each factor.

$= 2\cdot 2\cdot\sqrt[3]{2} - 4\cdot 3\cdot\sqrt[3]{2}$ $\sqrt[3]{8} = 2$ and $\sqrt[3]{27} = 3$.

$= 4\sqrt[3]{2} - 12\sqrt[3]{2}$ Multiply.

$= (4 - 12)\sqrt[3]{2}$ Apply the distributive property.

$= -8\sqrt[3]{2}$ Simplify.

b. $5\sqrt[3]{xy^2} + \sqrt[3]{8x^4y^5}$

$= 5\sqrt[3]{xy^2} + \sqrt[3]{(8x^3y^3)xy^2}$ Factor the second radicand using the greatest perfect cube factor.

$= 5\sqrt[3]{xy^2} + \sqrt[3]{8x^3y^3}\cdot\sqrt[3]{xy^2}$ Take the cube root of each factor.

$= 5\sqrt[3]{xy^2} + 2xy\sqrt[3]{xy^2}$ $\sqrt[3]{8} = 2$ and $\sqrt[3]{x^3y^3} = (x^3y^3)^{\frac{1}{3}} = xy$.

$= (5 + 2xy)\sqrt[3]{xy^2}$ Apply the distributive property.

✓ CHECK POINT 3 Simplify by combining like radical terms, if possible:

a. $3\sqrt[3]{24} - 5\sqrt[3]{81}$

b. $5\sqrt[3]{x^2y} + \sqrt[3]{27x^5y^4}$.

Dividing Radical Expressions

We have seen that the root of a product is the product of the roots. The root of a quotient can also be expressed as the quotient of roots. Here is an example:

$$\sqrt{\frac{64}{4}} = \sqrt{16} = 4 \quad \text{and} \quad \frac{\sqrt{64}}{\sqrt{4}} = \frac{8}{2} = 4.$$

This expression is the square root of a quotient.

This expression is the quotient of two square roots.

The two procedures produce the same result, 4. This is a special case of the **quotient rule for radicals**.

2 Use the quotient rule to simplify radical expressions.

The Quotient Rule for Radicals

If $\sqrt[n]{a}$ and $\sqrt[n]{b}$ are real numbers and $b \neq 0$, then

$$\sqrt[n]{\frac{a}{b}} = \frac{\sqrt[n]{a}}{\sqrt[n]{b}}.$$

The nth root of a quotient is the quotient of the nth roots of the numerator and denominator.

We know that a radical is simplified when its radicand has no factors other than 1 that are perfect nth powers. The quotient rule can be used to simplify some radicals. Keep in mind that all variables in radicands represent positive real numbers.

EXAMPLE 4 Using the Quotient Rule to Simplify Radicals

Simplify using the quotient rule:

a. $\sqrt[3]{\frac{16}{27}}$ **b.** $\sqrt{\frac{x^2}{25y^6}}$ **c.** $\sqrt[4]{\frac{7y^5}{x^{12}}}$.

Solution We simplify each expression by taking the roots of the numerator and the denominator. Then we use factoring to simplify the resulting radicals, if possible.

a. $$\sqrt[3]{\frac{16}{27}} = \frac{\sqrt[3]{16}}{\sqrt[3]{27}} = \frac{\sqrt[3]{8 \cdot 2}}{3} = \frac{\sqrt[3]{8} \cdot \sqrt[3]{2}}{3} = \frac{2\sqrt[3]{2}}{3}$$

b. $$\sqrt{\frac{x^2}{25y^6}} = \frac{\sqrt{x^2}}{\sqrt{25y^6}} = \frac{x}{5(y^6)^{\frac{1}{2}}} = \frac{x}{5y^3}$$

Try to do this step mentally.

c. $$\sqrt[4]{\frac{7y^5}{x^{12}}} = \frac{\sqrt[4]{7y^5}}{\sqrt[4]{x^{12}}} = \frac{\sqrt[4]{y^4 \cdot 7y}}{\sqrt[4]{x^{12}}} = \frac{y\sqrt[4]{7y}}{x^3}$$

✓ CHECK POINT 4 Simplify using the quotient rule:

a. $\sqrt[3]{\frac{24}{125}}$ **b.** $\sqrt{\frac{9x^3}{y^{10}}}$ **c.** $\sqrt[3]{\frac{8y^7}{x^{12}}}$.

By reversing the two sides of the quotient rule, we obtain a procedure for dividing radical expressions.

3 Use the quotient rule to divide radical expressions.

Dividing Radical Expressions

If $\sqrt[n]{a}$ and $\sqrt[n]{b}$ are real numbers and $b \neq 0$, then

$$\frac{\sqrt[n]{a}}{\sqrt[n]{b}} = \sqrt[n]{\frac{a}{b}}.$$

To divide two radical expressions with the same index, divide the radicands and retain the common index.

Study Tip

It can be confusing about when to write a quotient under a single radical and when to use separate radicals for a quotient's numerator and denominator.

- Use

$$\frac{\sqrt[n]{a}}{\sqrt[n]{b}} = \sqrt[n]{\frac{a}{b}},$$

writing under one radical (as in Example 5), when *dividing*.

- Use

$$\sqrt[n]{\frac{a}{b}} = \frac{\sqrt[n]{a}}{\sqrt[n]{b}},$$

with separate radicals for the numerator and the denominator (as in Example 4), when *simplifying* a quotient.

EXAMPLE 5 Dividing Radical Expressions

Divide and, if possible, simplify:

a. $\dfrac{\sqrt{48x^3}}{\sqrt{6x}}$ **b.** $\dfrac{\sqrt{45xy}}{2\sqrt{5}}$ **c.** $\dfrac{\sqrt[3]{16x^5y^2}}{\sqrt[3]{2xy^{-1}}}$.

Solution In each part of this problem, the indices in the numerator and the denominator are the same. Perform each division by dividing the radicands and retaining the common index.

a. $\dfrac{\sqrt{48x^3}}{\sqrt{6x}} = \sqrt{\dfrac{48x^3}{6x}} = \sqrt{8x^2} = \sqrt{4x^2 \cdot 2} = \sqrt{4x^2} \cdot \sqrt{2} = 2x\sqrt{2}$

b. $\dfrac{\sqrt{45xy}}{2\sqrt{5}} = \dfrac{1}{2} \cdot \sqrt{\dfrac{45xy}{5}} = \dfrac{1}{2} \cdot \sqrt{9xy} = \dfrac{1}{2} \cdot 3\sqrt{xy}$ or $\dfrac{3\sqrt{xy}}{2}$

c. $\dfrac{\sqrt[3]{16x^5y^2}}{\sqrt[3]{2xy^{-1}}} = \sqrt[3]{\dfrac{16x^5y^2}{2xy^{-1}}}$ Divide the radicands and retain the common index.

$= \sqrt[3]{8x^{5-1}y^{2-(-1)}}$ Divide factors in the radicand. Subtract exponents on common bases.

$= \sqrt[3]{8x^4y^3}$ Simplify.

$= \sqrt[3]{(8x^3y^3)x}$ Factor using the greatest perfect cube factor.

$= \sqrt[3]{8x^3y^3} \cdot \sqrt[3]{x}$ Factor into two radicals.

$= 2xy\sqrt[3]{x}$ Simplify.

✓ CHECK POINT 5 Divide and, if possible, simplify:

a. $\dfrac{\sqrt{40x^5}}{\sqrt{2x}}$ **b.** $\dfrac{\sqrt{50xy}}{2\sqrt{2}}$ **c.** $\dfrac{\sqrt[3]{48x^7y}}{\sqrt[3]{6xy^{-2}}}$.

7.4 EXERCISE SET

MyMathLab MathXL PRACTICE WATCH DOWNLOAD READ REVIEW

Practice Exercises

In this exercise set, assume that all variables represent positive real numbers.

In Exercises 1–10, add or subtract as indicated.

1. $8\sqrt{5} + 3\sqrt{5}$

2. $7\sqrt{3} + 2\sqrt{3}$

3. $9\sqrt[3]{6} - 2\sqrt[3]{6}$

4. $9\sqrt[3]{7} - 4\sqrt[3]{7}$

5. $4\sqrt[5]{2} + 3\sqrt[5]{2} - 5\sqrt[5]{2}$

6. $6\sqrt[5]{3} + 2\sqrt[5]{3} - 3\sqrt[5]{3}$

7. $3\sqrt{13} - 2\sqrt{5} - 2\sqrt{13} + 4\sqrt{5}$

8. $8\sqrt{17} - 5\sqrt{19} - 6\sqrt{17} + 4\sqrt{19}$

9. $3\sqrt{5} - \sqrt[3]{x} + 4\sqrt{5} + 3\sqrt[3]{x}$

10. $6\sqrt{7} - \sqrt[3]{x} + 2\sqrt{7} + 5\sqrt[3]{x}$

In Exercises 11–28, add or subtract as indicated. You will need to simplify terms to identify the like radicals.

11. $\sqrt{3} + \sqrt{27}$

12. $\sqrt{5} + \sqrt{20}$

13. $7\sqrt{12} + \sqrt{75}$

14. $5\sqrt{12} + \sqrt{75}$

15. $3\sqrt{32x} - 2\sqrt{18x}$

16. $5\sqrt{45x} - 2\sqrt{20x}$

17. $5\sqrt[3]{16} + \sqrt[3]{54}$

18. $3\sqrt[3]{24} + \sqrt[3]{81}$

19. $3\sqrt{45x^3} + \sqrt{5x}$

20. $8\sqrt{45x^3} + \sqrt{5x}$

21. $\sqrt[3]{54xy^3} + y\sqrt[3]{128x}$

22. $\sqrt[3]{24xy^3} + y\sqrt[3]{81x}$

23. $\sqrt[3]{54x^4} - \sqrt[3]{16x}$

24. $\sqrt[3]{81x^4} - \sqrt[3]{24x}$

25. $\sqrt{9x - 18} + \sqrt{x - 2}$

26. $\sqrt{4x - 12} + \sqrt{x - 3}$

27. $2\sqrt[3]{x^4y^2} + 3x\sqrt[3]{xy^2}$

28. $4\sqrt[3]{x^4y^2} + 5x\sqrt[3]{xy^2}$

In Exercises 29–44, simplify using the quotient rule.

29. $\sqrt{\dfrac{11}{4}}$

30. $\sqrt{\dfrac{19}{25}}$

31. $\sqrt[3]{\dfrac{19}{27}}$

32. $\sqrt[3]{\dfrac{11}{64}}$

33. $\sqrt{\dfrac{x^2}{36y^8}}$

34. $\sqrt{\dfrac{x^2}{144y^{12}}}$

35. $\sqrt{\dfrac{8x^3}{25y^6}}$

36. $\sqrt{\dfrac{50x^3}{81y^8}}$

37. $\sqrt[3]{\dfrac{x^4}{8y^3}}$

38. $\sqrt[3]{\dfrac{x^5}{125y^3}}$

39. $\sqrt[3]{\dfrac{50x^8}{27y^{12}}}$

40. $\sqrt[3]{\dfrac{81x^8}{8y^{15}}}$

41. $\sqrt[4]{\dfrac{9y^6}{x^8}}$

42. $\sqrt[4]{\dfrac{13y^7}{x^{12}}}$

43. $\sqrt[5]{\dfrac{64x^{13}}{y^{20}}}$

44. $\sqrt[5]{\dfrac{64x^{14}}{y^{15}}}$

In Exercises 45–66, divide and, if possible, simplify.

45. $\dfrac{\sqrt{40}}{\sqrt{5}}$

46. $\dfrac{\sqrt{200}}{\sqrt{10}}$

47. $\dfrac{\sqrt[3]{48}}{\sqrt[3]{6}}$

48. $\dfrac{\sqrt[3]{54}}{\sqrt[3]{2}}$

49. $\dfrac{\sqrt{54x^3}}{\sqrt{6x}}$

50. $\dfrac{\sqrt{72x^3}}{\sqrt{2x}}$

51. $\dfrac{\sqrt{x^5y^3}}{\sqrt{xy}}$

52. $\dfrac{\sqrt{x^7y^6}}{\sqrt{x^3y^2}}$

53. $\dfrac{\sqrt{200x^3}}{\sqrt{10x^{-1}}}$

54. $\dfrac{\sqrt{500x^3}}{\sqrt{10x^{-1}}}$

55. $\dfrac{\sqrt{48a^8b^7}}{\sqrt{3a^{-2}b^{-3}}}$

56. $\dfrac{\sqrt{54a^7b^{11}}}{\sqrt{3a^{-4}b^{-2}}}$

57. $\dfrac{\sqrt{72xy}}{2\sqrt{2}}$

58. $\dfrac{\sqrt{50xy}}{2\sqrt{2}}$

59. $\dfrac{\sqrt[3]{24x^3y^5}}{\sqrt[3]{3y^2}}$

60. $\dfrac{\sqrt[3]{250x^5y^3}}{\sqrt[3]{2x^3}}$

61. $\dfrac{\sqrt[4]{32x^{10}y^8}}{\sqrt[4]{2x^2y^{-2}}}$

62. $\dfrac{\sqrt[5]{96x^{12}y^{11}}}{\sqrt[5]{3x^2y^{-2}}}$

63. $\dfrac{\sqrt[3]{x^2 + 5x + 6}}{\sqrt[3]{x + 2}}$

64. $\dfrac{\sqrt[3]{x^2 + 7x + 12}}{\sqrt[3]{x + 3}}$

65. $\dfrac{\sqrt[3]{a^3 + b^3}}{\sqrt[3]{a + b}}$

66. $\dfrac{\sqrt[3]{a^3 - b^3}}{\sqrt[3]{a - b}}$

Practice PLUS

In Exercises 67–76, perform the indicated operations.

67. $\dfrac{\sqrt{32}}{5} + \dfrac{\sqrt{18}}{7}$

68. $\dfrac{\sqrt{27}}{2} + \dfrac{\sqrt{75}}{7}$

69. $3x\sqrt{8xy^2} - 5y\sqrt{32x^3} + \sqrt{18x^3y^2}$

70. $6x\sqrt{3xy^2} - 4x^2\sqrt{27xy} - 5\sqrt{75x^5y}$

71. $5\sqrt{2x^3} + \dfrac{30x^3\sqrt{24x^2}}{3x^2\sqrt{3x}}$

72. $7\sqrt{2x^3} + \dfrac{40x^3\sqrt{150x^2}}{5x^2\sqrt{3x}}$

73. $2x\sqrt{75xy} - \dfrac{\sqrt{81xy^2}}{\sqrt{3x^{-2}y}}$

74. $5\sqrt{8x^2y^3} - \dfrac{9x^2\sqrt{64y}}{3x\sqrt{2y^{-2}}}$

75. $\dfrac{15x^4\sqrt[3]{80x^3y^2}}{5x^3\sqrt[3]{2x^2y}} - \dfrac{75\sqrt[3]{5x^3y}}{25\sqrt[3]{x^{-1}}}$

76. $\dfrac{16x^4\sqrt[3]{48x^3y^2}}{8x^3\sqrt[3]{3x^2y}} - \dfrac{20\sqrt[3]{2x^3y}}{4\sqrt[3]{x^{-1}}}$

In Exercises 77–80, find $\left(\frac{f}{g}\right)(x)$ and the domain of $\frac{f}{g}$. Express each quotient function in simplified form.

77. $f(x) = \sqrt{48x^5},\quad g(x) = \sqrt{3x^2}$

78. $f(x) = \sqrt{x^2 - 25},\quad g(x) = \sqrt{x + 5}$

79. $f(x) = \sqrt[3]{32x^6},\ g(x) = \sqrt[3]{2x^2}$

80. $f(x) = \sqrt[3]{2x^6},\ g(x) = \sqrt[3]{16x}$

Application Exercises

Exercises 81–84 involve the perimeter and area of various geometric figures. Refer to ***Table 1.5*** *on page 55 if you've forgotten any of the formulas.*

In Exercises 81–82, find the perimeter and area of each rectangle. Express answers in simplified radical form.

81.

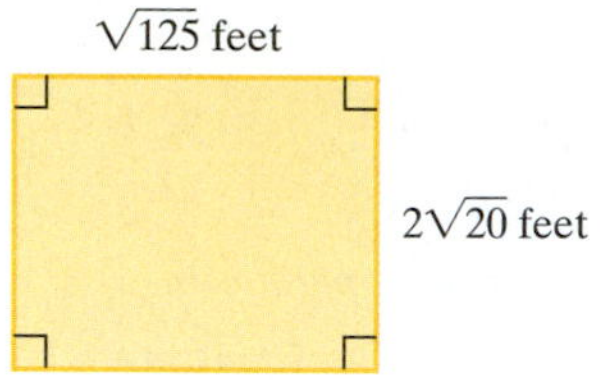

82.

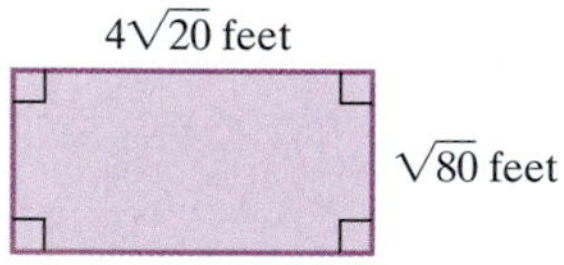

83. Find the perimeter of the triangle in simplified radical form.

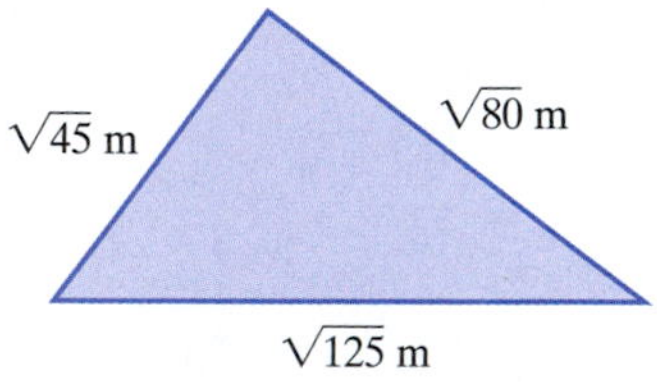

84. Find the area of the trapezoid in simplified radical form.

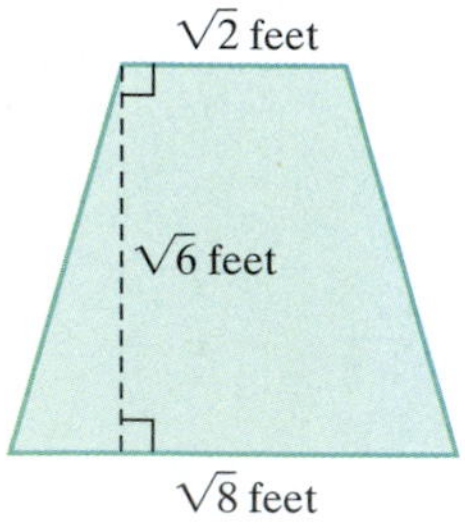

85. America is getting older. The graph shows the projected elderly U.S. population for ages 65–84 and for ages 85 and older.

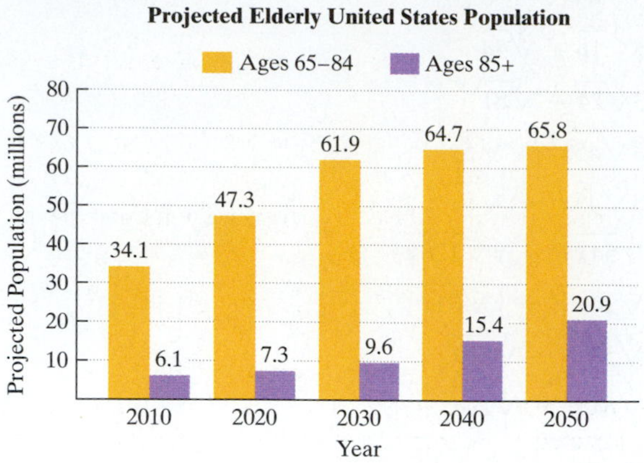

Source: U.S. Census Bureau

The function $f(x) = 5\sqrt{x} + 34.1$ models the projected number of Americans ages 65–84, $f(x)$, in millions, x years after 2010.

a. Use the function to find $f(40) - f(10)$. Express this difference in simplified radical form. What does this simplified radical represent?

b. Use a calculator and write your answer in part (a) to the nearest tenth. Does this rounded decimal overestimate or underestimate the difference in the projected data shown by the bar graph? By how much?

86. What does travel in space have to do with radicals? Imagine that in the future we will be able to travel in starships at velocities approaching the speed of light (approximately 186,000 miles per second). According to Einstein's theory of relativity, time would pass more quickly on Earth than it would in the moving starship. The radical expression

$$R_f\frac{\sqrt{c^2 - v^2}}{\sqrt{c^2}}$$

gives the aging rate of an astronaut relative to the aging rate of a friend, R_f, on Earth. In the expression, v is the astronaut's velocity and c is the speed of light.

a. Use the quotient rule and simplify the expression that shows your aging rate relative to a friend on Earth. Working in a step-by-step manner, express your aging rate as

$$R_f\sqrt{1 - \left(\frac{v}{c}\right)^2}.$$

b. You are moving at velocities approaching the speed of light. Substitute c, the speed of light, for v in the simplified expression from part (a). Simplify completely. Close to the speed of light, what is your aging rate relative to a friend on Earth? What does this mean?

Writing in Mathematics

87. What are like radicals? Give an example with your explanation.

88. Explain how to add like radicals. Give an example with your explanation.

89. If only like radicals can be combined, why is it possible to add $\sqrt{2}$ and $\sqrt{8}$?

90. Explain how to simplify a radical expression using the quotient rule. Provide an example.

91. Explain how to divide radical expressions with the same index.

92. In Exercise 85, use the data displayed by the bar graph to explain why we used a square root function to model the projected population for the 65–84 age group, but not for the 85+ group.

Technology Exercises

93. Use a calculator to provide numerical support for any four exercises that you worked from Exercises 1–66 that do not contain variables. Begin by finding a decimal approximation for the given expression. Then find a decimal approximation for your answer. The two decimal approximations should be the same.

In Exercises 94–96, determine if each operation is performed correctly by graphing the function on each side of the equation with your graphing utility. Use the given viewing rectangle. The graphs should be the same. If they are not, correct the right side of the equation and then use your graphing utility to verify the correction.

94. $\sqrt{4x} + \sqrt{9x} = 5\sqrt{x}$
[0, 5, 1] by [0, 10, 1]

95. $\sqrt{16x} - \sqrt{9x} = \sqrt{7x}$
[0, 5, 1] by [0, 5, 1]

96. $x\sqrt{8} + x\sqrt{2} = x\sqrt{10}$
[−5, 5, 1] by [−15, 15, 1]

Critical Thinking Exercises

Make Sense? *In Exercises 97–100, determine whether each statement "makes sense" or "does not make sense" and explain your reasoning.*

97. I divide *n*th roots by taking the *n*th root of the quotient of the radicands.

98. The unlike radicals $3\sqrt{2}$ and $5\sqrt{3}$ remind me of the unlike terms $3x$ and $5y$ that cannot be combined by addition or subtraction.

99. I simplified the terms of $3\sqrt[3]{81} + 2\sqrt[3]{54}$, and then I was able to identify and add the like radicals.

100. Without using a calculator, it's easier for me to estimate the decimal value of $\sqrt{72} + \sqrt{32} + \sqrt{18}$ by first simplifying.

In Exercises 101–104, determine whether each statement is true or false. If the statement is false, make the necessary change(s) to produce a true statement.

101. $\sqrt{5} + \sqrt{5} = \sqrt{10}$

102. $4\sqrt{3} + 5\sqrt{3} = 9\sqrt{6}$

103. If any two radical expressions are completely simplified, they can then be combined through addition or subtraction.

104. $\dfrac{\sqrt{-8}}{\sqrt{2}} = \sqrt{\dfrac{-8}{2}} = \sqrt{-4} = -2$

105. If an irrational number is decreased by $2\sqrt{18} - \sqrt{50}$, the result is $\sqrt{2}$. What is the number?

106. Simplify: $\dfrac{\sqrt{20}}{3} + \dfrac{\sqrt{45}}{4} - \sqrt{80}$.

107. Simplify: $\dfrac{6\sqrt{49xy}\,\sqrt{ab^2}}{7\sqrt{36x^{-3}y^{-5}}\,\sqrt{a^{-9}b^{-1}}}$.

Review Exercises

108. Solve: $2(3x - 1) - 4 = 2x - (6 - x)$.
(Section 1.4, Example 3)

109. Factor: $x^2 - 8xy + 12y^2$.
(Section 5.4, Example 4)

110. Add: $\dfrac{2}{x^2 + 5x + 6} + \dfrac{3x}{x^2 + 6x + 9}$.

(Section 6.2, Example 6)

Preview Exercises

Exercises 111–113 will help you prepare for the material covered in the next section.

111. a. Multiply: $7(x + 5)$.

b. Multiply: $\sqrt{7}(x + \sqrt{5})$.

112. a. Multiply: $(x + 5)(6x + 3)$.

b. Multiply: $(\sqrt{2} + 5)(6\sqrt{2} + 3)$.

113. Multiply and simplify:

$$\frac{10y}{\sqrt[5]{4x^3y}} \cdot \frac{\sqrt[5]{8x^2y^4}}{\sqrt[5]{8x^2y^4}}.$$

MID-CHAPTER CHECK POINT Section 7.1–Section 7.4

✓ **What You Know:** We learned to find roots of numbers. We saw that the domain of a square root function is the set of real numbers for which the radicand is nonnegative. We learned to simplify radical expressions, using $\sqrt[n]{a^n} = |a|$ if n is even and $\sqrt[n]{a^n} = a$ if n is odd. The definition $a^{\frac{m}{n}} = (\sqrt[n]{a})^m = \sqrt[n]{a^m}$ connected rational exponents and radicals. Finally, we performed various operations with radicals, including multiplication, addition, subtraction, and division.

In Exercises 1–23, simplify the given expression or perform the indicated operation(s) and, if possible, simplify. Assume that all variables represent positive real numbers.

1. $\sqrt{100} - \sqrt[3]{-27}$
2. $\sqrt{8x^5y^7}$
3. $3\sqrt[3]{4x^2} + 2\sqrt[3]{4x^2}$
4. $(3\sqrt[3]{4x^2})(2\sqrt[3]{4x^2})$
5. $27^{\frac{2}{3}} + (-32)^{\frac{3}{5}}$
6. $(64x^3y^{\frac{1}{4}})^{\frac{1}{3}}$
7. $5\sqrt{27} - 4\sqrt{48}$
8. $\sqrt{\dfrac{500x^3}{4y^4}}$
9. $\dfrac{x}{\sqrt[4]{x}}$
10. $\sqrt[3]{54x^5}$
11. $\dfrac{\sqrt[3]{160}}{\sqrt[3]{2}}$
12. $\sqrt[5]{\dfrac{x^{10}}{y^{20}}}$
13. $\dfrac{(x^{\frac{2}{3}})^2}{(x^{\frac{1}{4}})^3}$
14. $\sqrt[6]{x^6y^4}$
15. $\sqrt[7]{(x-2)^3}\cdot\sqrt[7]{(x-2)^6}$
16. $\sqrt[4]{32x^{11}y^{17}}$
17. $4\sqrt[3]{16} + 2\sqrt[3]{54}$
18. $\dfrac{\sqrt[7]{x^4y^9}}{\sqrt[7]{x^{-5}y^7}}$
19. $(-125)^{-\frac{2}{3}}$
20. $\sqrt{2}\cdot\sqrt[3]{2}$
21. $\sqrt[3]{\dfrac{32x}{y^4}}\cdot\sqrt[3]{\dfrac{2x^2}{y^2}}$
22. $\sqrt{32xy^2}\cdot\sqrt{2x^3y^5}$
23. $4x\sqrt{6x^4y^3} - 7y\sqrt{24x^6y}$

In Exercises 24–25, find the domain of each function.

24. $f(x) = \sqrt{30 - 5x}$
25. $g(x) = \sqrt[3]{3x - 15}$

SECTION 7.5

Multiplying with More Than One Term and Rationalizing Denominators

Objectives

1. Multiply radical expressions with more than one term.
2. Use polynomial special products to multiply radicals.
3. Rationalize denominators containing one term.
4. Rationalize denominators containing two terms.
5. Rationalize numerators.

PEANUTS © United Feature Syndicate, Inc.

The late Charles Schulz, creator of the "Peanuts" comic strip, transfixed 350 million readers worldwide with the joys and angst of his hapless Charlie Brown and Snoopy, a romantic self-deluded beagle. In 18,250 comic strips that spanned nearly 50 years, mathematics was often featured. Is the discussion of radicals shown above the real thing, or is it just an algebraic scam? By the time you complete this section on multiplying and dividing radicals, you will be able to decide.

1 Multiply radical expressions with more than one term.

Multiplying Radical Expressions with More Than One Term

Radical expressions with more than one term are multiplied in much the same way that polynomials with more than one term are multiplied. Example 1 uses the distributive property and the FOIL method to perform multiplications.

EXAMPLE 1 Multiplying Radicals

Multiply:

a. $\sqrt{7}(x+\sqrt{2})$ b. $\sqrt[3]{x}(\sqrt[3]{6}-\sqrt[3]{x^2})$ c. $(5\sqrt{2}+2\sqrt{3})(4\sqrt{2}-3\sqrt{3})$.

Solution

a. $\sqrt{7}(x+\sqrt{2})$

$=\sqrt{7}\cdot x+\sqrt{7}\cdot\sqrt{2}$ Use the distributive property.

$=x\sqrt{7}+\sqrt{14}$ Multiply the radicals.

b. $\sqrt[3]{x}(\sqrt[3]{6}-\sqrt[3]{x^2})$

$=\sqrt[3]{x}\cdot\sqrt[3]{6}-\sqrt[3]{x}\cdot\sqrt[3]{x^2}$ Use the distributive property.

$=\sqrt[3]{6x}-\sqrt[3]{x^3}$ Multiply the radicals: $\sqrt[n]{a}\cdot\sqrt[n]{b}=\sqrt[n]{ab}$.

$=\sqrt[3]{6x}-x$ Simplify: $\sqrt[3]{x^3}=x$.

c. first, last, inside, outside

$(5\sqrt{2}+2\sqrt{3})(4\sqrt{2}-3\sqrt{3})$ Use the FOIL method.

F O I L

$=(5\sqrt{2})(4\sqrt{2})+(5\sqrt{2})(-3\sqrt{3})+(2\sqrt{3})(4\sqrt{2})+(2\sqrt{3})(-3\sqrt{3})$

$=20\cdot 2-15\sqrt{6}+8\sqrt{6}-6\cdot 3$ Multiply. Note that $\sqrt{2}\cdot\sqrt{2}=\sqrt{4}=2$ and $\sqrt{3}\cdot\sqrt{3}=\sqrt{9}=3$.

$=40-15\sqrt{6}+8\sqrt{6}-18$ Complete the multiplications.

$=(40-18)+(-15\sqrt{6}+8\sqrt{6})$ Group like terms. Try to do this step mentally.

$=22-7\sqrt{6}$ Combine numerical terms and like radicals.

CHECK POINT 1 Multiply:

a. $\sqrt{6}(x+\sqrt{10})$

b. $\sqrt[3]{y}(\sqrt[3]{y^2}-\sqrt[3]{7})$

c. $(6\sqrt{5}+3\sqrt{2})(2\sqrt{5}-4\sqrt{2})$.

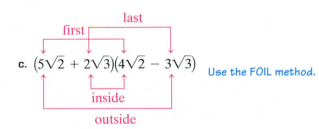

2 Use polynomial special products to multiply radicals.

Some radicals can be multiplied using the special products for multiplying polynomials.

EXAMPLE 2 Using Special Products to Multiply Radicals

Multiply:

a. $(\sqrt{3}+\sqrt{7})^2$ b. $(\sqrt{7}+\sqrt{3})(\sqrt{7}-\sqrt{3})$ c. $(\sqrt{a}-\sqrt{b})(\sqrt{a}+\sqrt{b})$.

Solution Use the special-product formulas shown.

$$(A+B)^2 = A^2 + 2\cdot A\cdot B + B^2$$

a. $(\sqrt{3}+\sqrt{7})^2 = (\sqrt{3})^2 + 2\cdot\sqrt{3}\cdot\sqrt{7} + (\sqrt{7})^2$ Use the special product for $(A+B)^2$.

$= 3 + 2\sqrt{21} + 7$ Multiply the radicals.

$= 10 + 2\sqrt{21}$ Simplify.

$$(A+B)\cdot(A-B) = A^2 - B^2$$

b. $(\sqrt{7}+\sqrt{3})(\sqrt{7}-\sqrt{3}) = (\sqrt{7})^2 - (\sqrt{3})^2$ Use the special product for $(A+B)(A-B)$.

$= 7 - 3$ Simplify: $(\sqrt{a})^2 = a$.

$= 4$

$$(A-B)\cdot(A+B) = A^2 - B^2$$

c. $(\sqrt{a}-\sqrt{b})(\sqrt{a}+\sqrt{b}) = (\sqrt{a})^2 - (\sqrt{b})^2 = a - b$

Radical expressions that involve the sum and difference of the same two terms are called **conjugates**. For example,

$$\sqrt{7}+\sqrt{3} \quad \text{and} \quad \sqrt{7}-\sqrt{3}$$

are conjugates of each other. Parts (b) and (c) of Example 2 illustrate that the product of two radical expressions need not be a radical expression:

$$(\sqrt{7}+\sqrt{3})(\sqrt{7}-\sqrt{3}) = 4$$
$$(\sqrt{a}-\sqrt{b})(\sqrt{a}+\sqrt{b}) = a - b.$$

The product of conjugates does not contain a radical.

Later in this section, we will use conjugates to simplify quotients.

CHECK POINT 2 Multiply:

a. $(\sqrt{5}+\sqrt{6})^2$ b. $(\sqrt{6}+\sqrt{5})(\sqrt{6}-\sqrt{5})$

c. $(\sqrt{a}-\sqrt{7})(\sqrt{a}+\sqrt{7})$.

3 Rationalize denominators containing one term.

Rationalizing Denominators Containing One Term

You can use a calculator to compare the approximate values for $\frac{1}{\sqrt{3}}$ and $\frac{\sqrt{3}}{3}$. The two approximations are the same. This is not a coincidence:

$$\frac{1}{\sqrt{3}} = \frac{1}{\sqrt{3}}\cdot\boxed{\frac{\sqrt{3}}{\sqrt{3}}} = \frac{\sqrt{3}}{\sqrt{9}} = \frac{\sqrt{3}}{3}.$$

Any number divided by itself is 1. Multiplication by 1 does not change the value of $\frac{1}{\sqrt{3}}$.

This process involves rewriting a radical expression as an equivalent expression in which the denominator no longer contains any radicals. The process is called **rationalizing the denominator**. When the denominator contains a single radical with an nth root, **multiply the numerator and the denominator by a radical of index n that produces a perfect nth power in the denominator's radicand**.

EXAMPLE 3 Rationalizing Denominators

Rationalize each denominator:

a. $\frac{\sqrt{5}}{\sqrt{6}}$ **b.** $\sqrt[3]{\frac{7}{25}}$.

Solution

a. If we multiply the numerator and the denominator of $\frac{\sqrt{5}}{\sqrt{6}}$ by $\sqrt{6}$, the denominator becomes $\sqrt{6}\cdot\sqrt{6} = \sqrt{36} = 6$. The denominator's radicand, 36, is a perfect square. The denominator no longer contains a radical. Therefore, we multiply by 1, choosing $\frac{\sqrt{6}}{\sqrt{6}}$ for 1.

$$\frac{\sqrt{5}}{\sqrt{6}} = \frac{\sqrt{5}}{\sqrt{6}}\cdot\frac{\sqrt{6}}{\sqrt{6}}$$ Multiply the numerator and denominator by $\sqrt{6}$ to remove the radical in the denominator.

$$= \frac{\sqrt{30}}{\sqrt{36}}$$ Multiply numerators and multiply denominators. The denominator's radicand, 36, is a perfect square.

$$= \frac{\sqrt{30}}{6}$$ Simplify: $\sqrt{36} = 6$.

b. Using the quotient rule, we can express $\sqrt[3]{\frac{7}{25}}$ as $\frac{\sqrt[3]{7}}{\sqrt[3]{25}}$. We have cube roots, so we want the denominator's radicand to be a perfect cube. Right now, the denominator's radicand is 25 or 5^2. We know that $\sqrt[3]{5^3} = 5$. If we multiply the numerator and the denominator of $\frac{\sqrt[3]{7}}{\sqrt[3]{25}}$ by $\sqrt[3]{5}$, the denominator becomes

$$\sqrt[3]{25}\cdot\sqrt[3]{5} = \sqrt[3]{5^2}\cdot\sqrt[3]{5} = \sqrt[3]{5^3} = 5.$$

The denominator's radicand, 5^3, is a perfect cube. The denominator no longer contains a radical. Therefore, we multiply by 1, choosing $\frac{\sqrt[3]{5}}{\sqrt[3]{5}}$ for 1.

$$\sqrt[3]{\frac{7}{25}} = \frac{\sqrt[3]{7}}{\sqrt[3]{25}}$$ Use the quotient rule and rewrite as the quotient of radicals.

$$= \frac{\sqrt[3]{7}}{\sqrt[3]{5^2}}$$ Write the denominator's radicand as an exponential expression.

$$= \frac{\sqrt[3]{7}}{\sqrt[3]{5^2}}\cdot\frac{\sqrt[3]{5}}{\sqrt[3]{5}}$$ Multiply the numerator and denominator by $\sqrt[3]{5}$ to remove the radical in the denominator.

$$= \frac{\sqrt[3]{35}}{\sqrt[3]{5^3}}$$ Multiply numerators and denominators. The denominator's radicand, 5^3, is a perfect cube.

$$= \frac{\sqrt[3]{35}}{5}$$ Simplify: $\sqrt[3]{5^3} = 5$.

Study Tip

Rationalizing a numerical denominator makes that denominator a rational number.

CHECK POINT 3 Rationalize each denominator:

a. $\frac{\sqrt{3}}{\sqrt{7}}$ **b.** $\sqrt[3]{\frac{2}{9}}$.

Example 3 showed that it is helpful to express the denominator's radicand using exponents. In this way, we can easily find the extra factor or factors needed to produce a perfect nth power. For example, suppose that $\sqrt[5]{8}$ appears in the denominator. We want a perfect fifth power. By expressing $\sqrt[5]{8}$ as $\sqrt[5]{2^3}$, we would multiply the numerator and denominator by $\sqrt[5]{2^2}$ because

$$\sqrt[5]{2^3} \cdot \sqrt[5]{2^2} = \sqrt[5]{2^5} = 2.$$

EXAMPLE 4 Rationalizing Denominators

Rationalize each denominator:

a. $\sqrt{\dfrac{3x}{5y}}$ **b.** $\dfrac{\sqrt[3]{x}}{\sqrt[3]{36y}}$ **c.** $\dfrac{10y}{\sqrt[5]{4x^3y}}$.

Solution By examining each denominator, you can determine how to multiply by 1. Let's first look at the denominators. For the square root, we must produce exponents of 2 in the radicand. For the cube root, we need exponents of 3, and for the fifth root, we want exponents of 5.

- $\sqrt{5y}$
  Multiply by $\sqrt{5y}$: $\sqrt{5y} \cdot \sqrt{5y} = \sqrt{25y^2} = 5y$.
- $\sqrt[3]{36y}$ or $\sqrt[3]{6^2y}$
  Multiply by $\sqrt[3]{6y^2}$: $\sqrt[3]{6^2y} \cdot \sqrt[3]{6y^2} = \sqrt[3]{6^3y^3} = 6y$.
- $\sqrt[5]{4x^3y}$ or $\sqrt[5]{2^2x^3y}$
  Multiply by $\sqrt[5]{2^3x^2y^4}$: $\sqrt[5]{2^2x^3y} \cdot \sqrt[5]{2^3x^2y^4} = \sqrt[5]{2^5x^5y^5} = 2xy$.

a. $$\sqrt{\frac{3x}{5y}} = \frac{\sqrt{3x}}{\sqrt{5y}} = \frac{\sqrt{3x}}{\sqrt{5y}} \cdot \frac{\sqrt{5y}}{\sqrt{5y}} = \frac{\sqrt{15xy}}{\sqrt{25y^2}} = \frac{\sqrt{15xy}}{5y}$$

Multiply by 1. $25y^2$ is a perfect square.

b. $$\frac{\sqrt[3]{x}}{\sqrt[3]{36y}} = \frac{\sqrt[3]{x}}{\sqrt[3]{6^2y}} = \frac{\sqrt[3]{x}}{\sqrt[3]{6^2y}} \cdot \frac{\sqrt[3]{6y^2}}{\sqrt[3]{6y^2}} = \frac{\sqrt[3]{6xy^2}}{\sqrt[3]{6^3y^3}} = \frac{\sqrt[3]{6xy^2}}{6y}$$

Multiply by 1. 6^3y^3 is a perfect cube.

c. $$\frac{10y}{\sqrt[5]{4x^3y}} = \frac{10y}{\sqrt[5]{2^2x^3y}} = \frac{10y}{\sqrt[5]{2^2x^3y}} \cdot \frac{\sqrt[5]{2^3x^2y^4}}{\sqrt[5]{2^3x^2y^4}}$$

Multiply by 1.

$$= \frac{10y\sqrt[5]{2^3x^2y^4}}{\sqrt[5]{2^5x^5y^5}} = \frac{10y\sqrt[5]{8x^2y^4}}{2xy} = \frac{5\sqrt[5]{8x^2y^4}}{x}$$

$2^5x^5y^5$ is a perfect 5th power. Simplify: Divide numerator and denominator by $2y$.

✓ CHECK POINT 4 Rationalize each denominator:

a. $\sqrt{\dfrac{2x}{7y}}$ **b.** $\dfrac{\sqrt[3]{x}}{\sqrt[3]{9y}}$ **c.** $\dfrac{6x}{\sqrt[5]{8x^2y^4}}$.

4 Rationalize denominators containing two terms.

Rationalizing Denominators Containing Two Terms

How can we rationalize a denominator if the denominator contains two terms with one or more square roots? **Multiply the numerator and the denominator by the conjugate of the denominator.** Here are three examples of such expressions:

- $\dfrac{8}{3\sqrt{2}+4}$ — The conjugate of the denominator is $3\sqrt{2}-4$.
- $\dfrac{2+\sqrt{5}}{\sqrt{6}-\sqrt{3}}$ — The conjugate of the denominator is $\sqrt{6}+\sqrt{3}$.
- $\dfrac{h}{\sqrt{x+h}-\sqrt{x}}$ — The conjugate of the denominator is $\sqrt{x+h}+\sqrt{x}$.

The product of the denominator and its conjugate is found using the formula

$$(A+B)(A-B)=A^2-B^2.$$

The simplified product will not contain a radical.

EXAMPLE 5 Rationalizing a Denominator Containing Two Terms

Rationalize the denominator: $\dfrac{8}{3\sqrt{2}+4}$.

Solution The conjugate of the denominator is $3\sqrt{2}-4$. If we multiply the numerator and the denominator by $3\sqrt{2}-4$, the simplified denominator will not contain a radical. Therefore, we multiply by 1, choosing $\dfrac{3\sqrt{2}-4}{3\sqrt{2}-4}$ for 1.

$$\frac{8}{3\sqrt{2}+4}=\frac{8}{3\sqrt{2}+4}\cdot\frac{3\sqrt{2}-4}{3\sqrt{2}-4}$$ Multiply by 1.

$$=\frac{8(3\sqrt{2}-4)}{(3\sqrt{2})^2-4^2}$$ $(A+B)(A-B)=A^2-B^2$

Leave the numerator in factored form. This helps simplify, if possible.

$$=\frac{8(3\sqrt{2}-4)}{18-16}$$ $(3\sqrt{2})^2=9\cdot 2=18$

$$=\frac{8(3\sqrt{2}-4)}{2}$$ This expression can still be simplified.

$$=\frac{\overset{4}{\cancel{8}}(3\sqrt{2}-4)}{\underset{1}{\cancel{2}}}$$ Divide the numerator and denominator by 2.

$$=4(3\sqrt{2}-4)\quad\text{or}\quad 12\sqrt{2}-16$$

✓ **CHECK POINT 5** Rationalize the denominator: $\dfrac{18}{2\sqrt{3}+3}$.

EXAMPLE 6 Rationalizing a Denominator Containing Two Terms

Rationalize the denominator: $\dfrac{2+\sqrt{5}}{\sqrt{6}-\sqrt{3}}$.

Solution The conjugate of the denominator of $\frac{2+\sqrt{5}}{\sqrt{6}-\sqrt{3}}$ is $\sqrt{6}+\sqrt{3}$. Multiplication of both the numerator and denominator by $\sqrt{6}+\sqrt{3}$ will rationalize the denominator. This will produce a rational number in the denominator.

$$\frac{2+\sqrt{5}}{\sqrt{6}-\sqrt{3}} = \frac{2+\sqrt{5}}{\sqrt{6}-\sqrt{3}} \cdot \frac{\sqrt{6}+\sqrt{3}}{\sqrt{6}+\sqrt{3}}$$

Multiply by 1.

F O I L

$$= \frac{2\sqrt{6}+2\sqrt{3}+\sqrt{5}\cdot\sqrt{6}+\sqrt{5}\cdot\sqrt{3}}{(\sqrt{6})^2-(\sqrt{3})^2}$$

Use FOIL in the numerator and $(A-B)(A+B)=A^2-B^2$ in the denominator.

$$= \frac{2\sqrt{6}+2\sqrt{3}+\sqrt{30}+\sqrt{15}}{6-3}$$

$$= \frac{2\sqrt{6}+2\sqrt{3}+\sqrt{30}+\sqrt{15}}{3}$$

Further simplification is not possible.

✓ **CHECK POINT 6** Rationalize the denominator: $\frac{3+\sqrt{7}}{\sqrt{5}-\sqrt{2}}$.

5 Rationalize numerators.

Rationalizing Numerators

We have seen that square root functions are often used to model growing phenomena with growth that is leveling off. **Figure 7.7** shows a male's height as a function of his age. The pattern of his growth suggests modeling with a square root function.

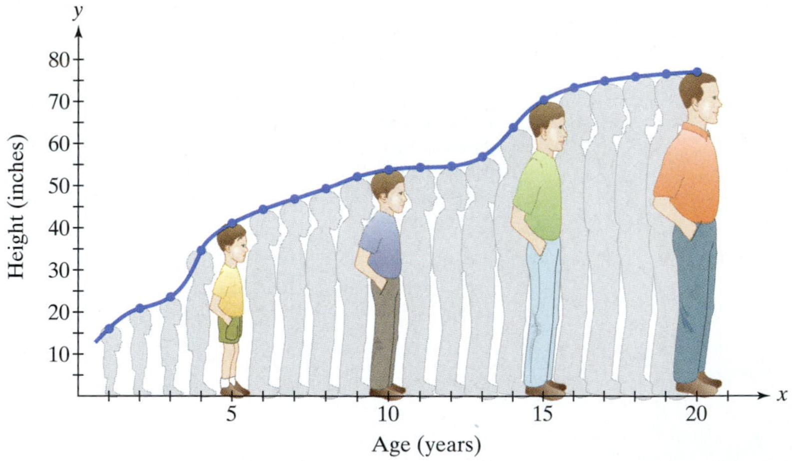

FIGURE 7.7

If we use $f(x)=\sqrt{x}$ to model height, $f(x)$, at age x, the expression

$$\frac{f(a+h)-f(a)}{h} = \frac{\sqrt{a+h}-\sqrt{a}}{h}$$

describes the man's average growth rate from age a to age $a+h$. Can you see that this expression is not defined if $h=0$? However, to explore the man's average growth rates for successively shorter periods of time, we need to know what happens to the expression as h takes on values that get closer and closer to 0.

What happens to growth near the instant in time that the man is age a? The question is answered in calculus by **rationalizing the numerator**. The procedure is similar to rationalizing the denominator. **To rationalize a numerator, multiply by 1 to eliminate the radical in the *numerator*.**

EXAMPLE 7 Rationalizing a Numerator

Rationalize the numerator:

$$\frac{\sqrt{a+h}-\sqrt{a}}{h}.$$

Solution The conjugate of the numerator is $\sqrt{a+h}+\sqrt{a}$. If we multiply the numerator and the denominator by $\sqrt{a+h}+\sqrt{a}$, the simplified numerator will not contain radicals. Therefore, we multiply by 1, choosing $\frac{\sqrt{a+h}+\sqrt{a}}{\sqrt{a+h}+\sqrt{a}}$ for 1.

$$\frac{\sqrt{a+h}-\sqrt{a}}{h}=\frac{\sqrt{a+h}-\sqrt{a}}{h}\cdot\frac{\sqrt{a+h}+\sqrt{a}}{\sqrt{a+h}+\sqrt{a}}$$ Multiply by 1.

$$=\frac{\left(\sqrt{a+h}\right)^2-\left(\sqrt{a}\right)^2}{h\left(\sqrt{a+h}+\sqrt{a}\right)}$$ $(A-B)(A+B)=A^2-B^2$. Leave the denominator in factored form.

$$=\frac{a+h-a}{h\left(\sqrt{a+h}+\sqrt{a}\right)}$$ $\left(\sqrt{a+h}\right)^2=a+h$ and $\left(\sqrt{a}\right)^2=a$.

$$=\frac{h}{h\left(\sqrt{a+h}+\sqrt{a}\right)}$$ Simplify the numerator.

$$=\frac{1}{\sqrt{a+h}+\sqrt{a}}$$ Simplify by dividing the numerator and the denominator by h.

✓ **CHECK POINT 7** Rationalize the numerator: $\frac{\sqrt{x+3}-\sqrt{x}}{3}$.

7.5 EXERCISE SET

MyMathLab | MathXL PRACTICE | WATCH | DOWNLOAD | READ | REVIEW

Practice Exercises

In this exercise set, assume that all variables represent positive real numbers.

In Exercises 1–38, multiply as indicated. If possible, simplify any radical expressions that appear in the product.

1. $\sqrt{2}\left(x+\sqrt{7}\right)$
2. $\sqrt{5}\left(x+\sqrt{3}\right)$
3. $\sqrt{6}\left(7-\sqrt{6}\right)$
4. $\sqrt{3}\left(5-\sqrt{3}\right)$
5. $\sqrt{3}\left(4\sqrt{6}-2\sqrt{3}\right)$
6. $\sqrt{6}\left(4\sqrt{6}-3\sqrt{2}\right)$
7. $\sqrt[3]{2}\left(\sqrt[3]{6}+4\sqrt[3]{5}\right)$
8. $\sqrt[3]{3}\left(\sqrt[3]{6}+7\sqrt[3]{4}\right)$
9. $\sqrt[3]{x}\left(\sqrt[3]{16x^2}-\sqrt[3]{x}\right)$
10. $\sqrt[3]{x}\left(\sqrt[3]{24x^2}-\sqrt[3]{x}\right)$
11. $\left(5+\sqrt{2}\right)\left(6+\sqrt{2}\right)$
12. $\left(7+\sqrt{2}\right)\left(8+\sqrt{2}\right)$
13. $\left(6+\sqrt{5}\right)\left(9-4\sqrt{5}\right)$
14. $\left(4+\sqrt{5}\right)\left(10-3\sqrt{5}\right)$
15. $\left(6-3\sqrt{7}\right)\left(2-5\sqrt{7}\right)$
16. $\left(7-2\sqrt{7}\right)\left(5-3\sqrt{7}\right)$
17. $\left(\sqrt{2}+\sqrt{7}\right)\left(\sqrt{3}+\sqrt{5}\right)$
18. $\left(\sqrt{3}+\sqrt{2}\right)\left(\sqrt{10}+\sqrt{11}\right)$
19. $\left(\sqrt{2}-\sqrt{7}\right)\left(\sqrt{3}-\sqrt{5}\right)$
20. $\left(\sqrt{3}-\sqrt{2}\right)\left(\sqrt{10}-\sqrt{11}\right)$
21. $\left(3\sqrt{2}-4\sqrt{3}\right)\left(2\sqrt{2}+5\sqrt{3}\right)$
22. $\left(3\sqrt{5}-2\sqrt{3}\right)\left(4\sqrt{5}+5\sqrt{3}\right)$
23. $\left(\sqrt{3}+\sqrt{5}\right)^2$
24. $\left(\sqrt{2}+\sqrt{7}\right)^2$
25. $\left(\sqrt{3x}-\sqrt{y}\right)^2$
26. $\left(\sqrt{2x}-\sqrt{y}\right)^2$
27. $\left(\sqrt{5}+7\right)\left(\sqrt{5}-7\right)$
28. $\left(\sqrt{6}+2\right)\left(\sqrt{6}-2\right)$
29. $\left(2-5\sqrt{3}\right)\left(2+5\sqrt{3}\right)$
30. $\left(3-5\sqrt{2}\right)\left(3+5\sqrt{2}\right)$
31. $\left(3\sqrt{2}+2\sqrt{3}\right)\left(3\sqrt{2}-2\sqrt{3}\right)$
32. $\left(4\sqrt{3}+3\sqrt{2}\right)\left(4\sqrt{3}-3\sqrt{2}\right)$
33. $\left(3-\sqrt{x}\right)\left(2-\sqrt{x}\right)$
34. $\left(4-\sqrt{x}\right)\left(3-\sqrt{x}\right)$

35. $(\sqrt[3]{x}-4)(\sqrt[3]{x}+5)$

36. $(\sqrt[3]{x}-3)(\sqrt[3]{x}+7)$

37. $(x+\sqrt[3]{y^2})(2x-\sqrt[3]{y^2})$

38. $(x-\sqrt[5]{y^3})(2x+\sqrt[5]{y^3})$

In Exercises 39–64, rationalize each denominator.

39. $\frac{\sqrt{2}}{\sqrt{5}}$

40. $\frac{\sqrt{7}}{\sqrt{3}}$

41. $\sqrt{\frac{11}{x}}$

42. $\sqrt{\frac{6}{x}}$

43. $\frac{9}{\sqrt{3y}}$

44. $\frac{12}{\sqrt{3y}}$

45. $\frac{1}{\sqrt[3]{2}}$

46. $\frac{1}{\sqrt[3]{3}}$

47. $\frac{6}{\sqrt[3]{4}}$

48. $\frac{10}{\sqrt[3]{5}}$

49. $\sqrt[3]{\frac{2}{3}}$

50. $\sqrt[3]{\frac{3}{4}}$

51. $\frac{4}{\sqrt[3]{x}}$

52. $\frac{7}{\sqrt[3]{x}}$

53. $\sqrt[3]{\frac{2}{y^2}}$

54. $\sqrt[3]{\frac{5}{y^2}}$

55. $\frac{7}{\sqrt[3]{2x^2}}$

56. $\frac{10}{\sqrt[3]{4x^2}}$

57. $\sqrt[3]{\frac{2}{xy^2}}$

58. $\sqrt[3]{\frac{3}{xy^2}}$

59. $\frac{3}{\sqrt[4]{x}}$

60. $\frac{5}{\sqrt[4]{x}}$

61. $\frac{6}{\sqrt[5]{8x^3}}$

62. $\frac{10}{\sqrt[5]{16x^2}}$

63. $\frac{2x^2y}{\sqrt[5]{4x^2y^4}}$

64. $\frac{3xy^2}{\sqrt[5]{8xy^3}}$

In Exercises 65–74, simplify each radical expression and then rationalize the denominator.

65. $\frac{9}{\sqrt{3x^2y}}$

66. $\frac{25}{\sqrt{5x^2y}}$

67. $-\sqrt{\frac{75a^5}{b^3}}$

68. $-\sqrt{\frac{150a^3}{b^5}}$

69. $\sqrt{\frac{7m^2n^3}{14m^3n^2}}$

70. $\sqrt{\frac{5m^4n^6}{15m^3n^4}}$

71. $\frac{3}{\sqrt[4]{x^5y^3}}$

72. $\frac{5}{\sqrt[4]{x^2y^7}}$

73. $\frac{12}{\sqrt[3]{-8x^5y^8}}$

74. $\frac{15}{\sqrt[3]{-27x^4y^{11}}}$

In Exercises 75–92, rationalize each denominator. Simplify, if possible.

75. $\frac{8}{\sqrt{5}+2}$

76. $\frac{15}{\sqrt{6}+1}$

77. $\frac{13}{\sqrt{11}-3}$

78. $\frac{17}{\sqrt{10}-2}$

79. $\frac{6}{\sqrt{5}+\sqrt{3}}$

80. $\frac{12}{\sqrt{7}+\sqrt{3}}$

81. $\frac{\sqrt{a}}{\sqrt{a}-\sqrt{b}}$

82. $\frac{\sqrt{b}}{\sqrt{a}-\sqrt{b}}$

83. $\frac{25}{5\sqrt{2}-3\sqrt{5}}$

84. $\frac{35}{5\sqrt{2}-3\sqrt{5}}$

85. $\frac{\sqrt{5}+\sqrt{3}}{\sqrt{5}-\sqrt{3}}$

86. $\frac{\sqrt{11}-\sqrt{5}}{\sqrt{11}+\sqrt{5}}$

87. $\frac{\sqrt{x}+1}{\sqrt{x}+3}$

88. $\frac{\sqrt{x}-2}{\sqrt{x}-5}$

89. $\frac{5\sqrt{3}-3\sqrt{2}}{3\sqrt{2}-2\sqrt{3}}$

90. $\frac{2\sqrt{6}+\sqrt{5}}{3\sqrt{6}-\sqrt{5}}$

91. $\frac{2\sqrt{x}+\sqrt{y}}{\sqrt{y}-2\sqrt{x}}$

92. $\frac{3\sqrt{x}+\sqrt{y}}{\sqrt{y}-3\sqrt{x}}$

In Exercises 93–104, rationalize each numerator. Simplify, if possible.

93. $\sqrt{\frac{3}{2}}$

94. $\sqrt{\frac{5}{3}}$

95. $\frac{\sqrt[3]{4x}}{\sqrt[3]{y}}$

96. $\frac{\sqrt[3]{2x}}{\sqrt[3]{y}}$

97. $\frac{\sqrt{x}+3}{\sqrt{x}}$

98. $\frac{\sqrt{x}+4}{\sqrt{x}}$

99. $\frac{\sqrt{a}+\sqrt{b}}{\sqrt{a}-\sqrt{b}}$

100. $\frac{\sqrt{a}-\sqrt{b}}{\sqrt{a}+\sqrt{b}}$

101. $\frac{\sqrt{x+5}-\sqrt{x}}{5}$

102. $\frac{\sqrt{x+7}-\sqrt{x}}{7}$

103. $\frac{\sqrt{x}+\sqrt{y}}{x^2-y^2}$

104. $\frac{\sqrt{x}-\sqrt{y}}{x^2-y^2}$

Practice PLUS

In Exercises 105–112, add or subtract as indicated. Begin by rationalizing denominators for all terms in which denominators contain radicals.

105. $\sqrt{2}+\frac{1}{\sqrt{2}}$

106. $\sqrt{5}+\frac{1}{\sqrt{5}}$

107. $\sqrt[3]{25}-\frac{15}{\sqrt[3]{5}}$

108. $\sqrt[4]{8}-\frac{20}{\sqrt[4]{2}}$

109. $\sqrt{6}-\sqrt{\frac{1}{6}}+\sqrt{\frac{2}{3}}$

110. $\sqrt{15} - \sqrt{\frac{5}{3}} + \sqrt{\frac{3}{5}}$

111. $\frac{2}{\sqrt{2} + \sqrt{3}} + \sqrt{75} - \sqrt{50}$

112. $\frac{5}{\sqrt{2} + \sqrt{7}} - 2\sqrt{32} + \sqrt{28}$

113. Let $f(x) = x^2 - 6x - 4$. Find $f(3 - \sqrt{13})$.

114. Let $f(x) = x^2 + 4x - 2$. Find $f(-2 + \sqrt{6})$.

115. Let $f(x) = \sqrt{9 + x}$. Find $f(3\sqrt{5}) \cdot f(-3\sqrt{5})$.

116. Let $f(x) = x^2$. Find $f(\sqrt{a + 1} - \sqrt{a - 1})$.

Application Exercises

117. The early Greeks believed that the most pleasing of all rectangles were **golden rectangles**, whose ratio of width to height is

$$\frac{w}{h} = \frac{2}{\sqrt{5} - 1}.$$

The Parthenon at Athens fits into a golden rectangle once the triangular pediment is reconstructed.

Rationalize the denominator of the golden ratio. Then use a calculator and find the ratio of width to height, correct to the nearest hundredth, in golden rectangles.

118. In the "Peanuts" cartoon shown in the section opener on page 524, Woodstock appears to be working steps mentally. Fill in the missing steps that show how to go from $\frac{7\sqrt{2 \cdot 2 \cdot 3}}{6}$ to $\frac{7}{3}\sqrt{3}$.

In Exercises 119–120, write expressions for the perimeter and area of each figure. Then simplify these expressions. Assume that all measures are given in inches.

119.

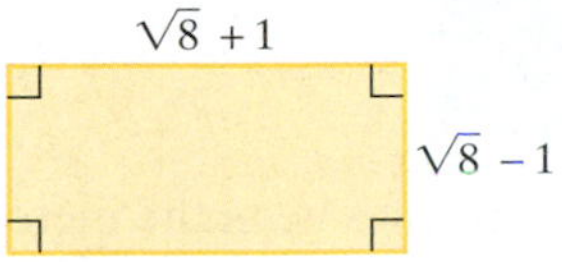

120.

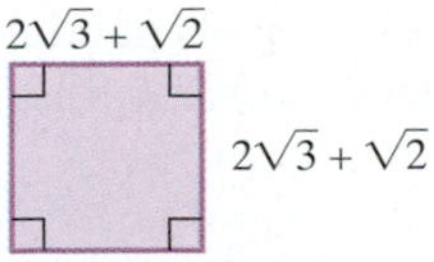

The Pythagorean Theorem for right triangles tells us that the length of the hypotenuse is the square root of the sum of the squares of the lengths of the legs. In Exercises 121–122, use the Pythagorean Theorem to find the length of each hypotenuse in simplified radical form. Assume that all measures are given in inches.

121.

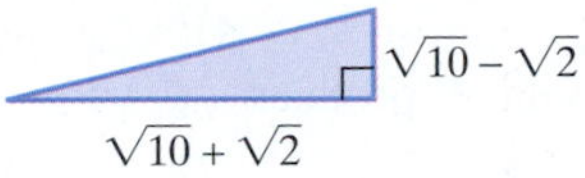

122.

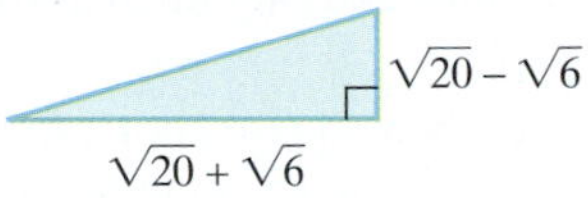

Writing in Mathematics

123. Explain how to perform this multiplication: $\sqrt{2}(\sqrt{7} + \sqrt{10})$.

124. Explain how to perform this multiplication: $(2 + \sqrt{3})(4 + \sqrt{3})$.

125. Explain how to perform this multiplication: $(2 + \sqrt{3})^2$.

126. What are conjugates? Give an example with your explanation.

127. Describe how to multiply conjugates.

128. Describe what it means to rationalize a denominator. Use both $\frac{1}{\sqrt{5}}$ and $\frac{1}{5 + \sqrt{5}}$ in your explanation.

129. When a radical expression has its denominator rationalized, we change the denominator so that it no longer contains any radicals. Doesn't this change the value of the radical expression? Explain.

130. Square the real number $\frac{2}{\sqrt{3}}$. Observe that the radical is eliminated from the denominator. Explain whether this process is equivalent to rationalizing the denominator.

Technology Exercises

In Exercises 131–134, determine if each operation is performed correctly by graphing the function on each side of the equation with your graphing utility. Use the given viewing rectangle. The graphs should be the same. If they are not, correct the right side of the equation and then use your graphing utility to verify the correction.

131. $(\sqrt{x} - 1)(\sqrt{x} - 1) = x + 1$
$[0, 5, 1]$ by $[-1, 2, 1]$

132. $(\sqrt{x} + 2)(\sqrt{x} - 2) = x^2 - 4$ for $x \geq 0$
$[0, 10, 1]$ by $[-10, 10, 1]$

133. $(\sqrt{x} + 1)^2 = x + 1$
$[0, 8, 1]$ by $[0, 15, 1]$

134. $\frac{3}{\sqrt{x + 3} - \sqrt{x}} = \sqrt{x + 3} + \sqrt{x}$
$[0, 8, 1]$ by $[0, 6, 1]$

Critical Thinking Exercises

Make Sense? *In Exercises 135–138, determine whether each statement "makes sense" or "does not make sense" and explain your reasoning.*

135. I use the same ideas to multiply $(\sqrt{2}+5)(\sqrt{2}+4)$ that I did to find the binomial product $(x+5)(x+4)$.

136. I used a special-product formula and simplified as follows: $(\sqrt{2}+\sqrt{5})^2 = 2 + 5 = 7$.

137. In some cases when I multiply a square root expression and its conjugate, the simplified product contains a radical.

138. I use the fact that 1 is the multiplicative identity to both rationalize denominators and rewrite rational expressions with a common denominator.

In Exercises 139–142, determine whether each statement is true or false. If the statement is false, make the necessary change(s) to produce a true statement.

139. $\dfrac{\sqrt{3}+7}{\sqrt{3}-2} = -\dfrac{7}{2}$

140. $\dfrac{4}{\sqrt{x+y}} = \dfrac{4\sqrt{x-y}}{x-y}$

141. $\dfrac{4\sqrt{x}}{\sqrt{x}-y} = \dfrac{4x+4y\sqrt{x}}{x-y^2}$

142. $(\sqrt{x}-7)^2 = x - 49$

143. Solve:

$$7[(2x-5)-(x+1)] = (\sqrt{7}+2)(\sqrt{7}-2).$$

144. Simplify: $(\sqrt{2+\sqrt{3}}+\sqrt{2-\sqrt{3}})^2$.

145. Rationalize the denominator: $\dfrac{1}{\sqrt{2}+\sqrt{3}+\sqrt{4}}$.

Review Exercises

146. Add: $\dfrac{2}{x-2} + \dfrac{3}{x^2-4}$.

(Section 6.2, Example 6)

147. Solve: $\dfrac{2}{x-2} + \dfrac{3}{x^2-4} = 0$.

(Section 6.6, Example 5)

148. If $f(x) = x^4 - 3x^2 - 2x + 5$, use synthetic division and the Remainder Theorem to find $f(-2)$. (Section 6.5, Example 2)

Preview Exercises

Exercises 149–151 will help you prepare for the material covered in the next section.

149. Multiply: $(\sqrt{x+4}+1)^2$.

150. Solve: $4x^2 - 16x + 16 = 4(x+4)$.

151. Solve: $26 - 11x = 16 - 8x + x^2$.

SECTION 7.6

Radical Equations

Objectives

1. Solve radical equations.
2. Use models that are radical functions to solve problems.

One of the most dramatic developments in the U.S. work force has been the increase in the number of women. In 1960, approximately 38% of women belonged to the labor force. By 2000, that number had increased to over 60%. With higher levels of education than ever before, most women now choose careers in the labor force rather than homemaking.

The function

$$f(x) = 3.5\sqrt{x} + 38$$

models the percentage of U.S. women in the labor force, $f(x)$, x years after 1960. How can we predict the year when, say, 70% of women will participate in the U.S. work force? Substitute 70 for $f(x)$ in $f(x) = 3.5\sqrt{x} + 38$ and solve for x:

$$70 = 3.5\sqrt{x} + 38.$$

The resulting equation contains a variable in the radicand and is called a *radical equation*. A **radical equation** is an equation in which the variable occurs in a square root, cube root, or any higher root. Some examples of radical equations are

$$\sqrt{2x+3} = 5, \quad \sqrt{3x+1} - \sqrt{x+4} = 1, \quad \text{and} \quad \sqrt[3]{3x-1} + 4 = 0.$$

Variables occur in radicands.

In this section, you will learn how to solve radical equations. Solving such equations will enable you to solve new kinds of problems using radical functions.

1 Solve radical equations.

Solving Radical Equations

Consider the following radical equation:

$$\sqrt{x} = 9.$$

We solve the equation by squaring both sides:

Squaring both sides eliminates the square root.

$$(\sqrt{x})^2 = 9^2$$
$$x = 81.$$

The proposed solution, 81, can be checked in the original equation, $\sqrt{x} = 9$. Because $\sqrt{81} = 9$, the solution is 81 and the solution set is $\{81\}$.

In general, we solve radical equations with square roots by squaring both sides of the equation. We solve radical equations with nth roots by raising both sides of the equation to the nth power. Unfortunately, if n is even, all the solutions of the equation raised to the even power may not be solutions of the original equation. Consider, for example, the equation

$$x = 4.$$

If we square both sides, we obtain

$$x^2 = 16.$$

$x^2 - 16 = 0$ — Subtract 16 from both sides and write the quadratic equation in standard form.

$(x+4)(x-4) = 0$ — Factor.

$x + 4 = 0$ or $x - 4 = 0$ — Set each factor equal to 0.

$x = -4$ $\quad$ $x = 4$ — Solve the resulting equations.

The equation $x^2 = 16$ has two solutions, -4 and 4. By contrast, only 4 is a solution of the original equation, $x = 4$. For this reason, **when raising both sides of an equation to an even power, always check proposed solutions in the original equation**.

Here is a general method for solving radical equations with nth roots:

Solving Radical Equations Containing *n*th Roots

1. If necessary, arrange terms so that one radical is isolated on one side of the equation.
2. Raise both sides of the equation to the nth power to eliminate the nth root.
3. Solve the resulting equation. If this equation still contains radicals, repeat steps 1 and 2.
4. Check all proposed solutions in the original equation.

EXAMPLE 1 Solving a Radical Equation

Solve: $\sqrt{2x + 3} = 5$.

Solution

Step 1. Isolate a radical on one side. The radical, $\sqrt{2x + 3}$, is already isolated on the left side of the equation, so we can skip this step.

Step 2. Raise both sides to the *n*th power. Because n, the index, is 2, we square both sides.

$$\sqrt{2x + 3} = 5 \quad \text{This is the given equation.}$$

$$\left(\sqrt{2x + 3}\right)^2 = 5^2 \quad \text{Square both sides to eliminate the radical.}$$

$$2x + 3 = 25 \quad \text{Simplify.}$$

Step 3. Solve the resulting equation.

$$2x + 3 = 25 \quad \text{The resulting equation is a linear equation.}$$

$$2x = 22 \quad \text{Subtract 3 from both sides.}$$

$$x = 11 \quad \text{Divide both sides by 2.}$$

Step 4. Check the proposed solution in the original equation. Because both sides were raised to an even power, this check is essential.

Check 11:

$$\sqrt{2x + 3} = 5$$

$$\sqrt{2 \cdot 11 + 3} \stackrel{?}{=} 5$$

$$\sqrt{25} \stackrel{?}{=} 5$$

$$5 = 5, \quad \text{true}$$

The solution is 11 and the solution set is $\{11\}$.

✓ CHECK POINT 1 Solve: $\sqrt{3x + 4} = 8$.

EXAMPLE 2 Solving a Radical Equation

Solve: $\sqrt{x - 3} + 6 = 5$.

Solution

Step 1. Isolate a radical on one side. The radical, $\sqrt{x - 3}$, can be isolated by subtracting 6 from both sides. We obtain

$$\sqrt{x - 3} = -1.$$

A principal square root cannot be negative. This equation has no solution. Let's continue the solution procedure to see what happens.

Step 2. Raise both sides to the *n*th power. Because n, the index, is 2, we square both sides.

$$\left(\sqrt{x - 3}\right)^2 = (-1)^2$$

$$x - 3 = 1 \quad \text{Simplify.}$$

Step 3. Solve the resulting equation.

$$x - 3 = 1 \quad \text{The resulting equation is a linear equation.}$$

$$x = 4 \quad \text{Add 3 to both sides.}$$

Step 4. Check the proposed solution in the original equation.

Check 4:

$$\sqrt{x-3}+6=5$$
$$\sqrt{4-3}+6\stackrel{?}{=}5$$
$$\sqrt{1}+6\stackrel{?}{=}5$$
$$1+6\stackrel{?}{=}5$$
$$7=5,\quad \text{false}$$

This false statement indicates that 4 is not a solution. Thus, the equation has no solution. The solution set is $\varnothing$, the empty set.

Example 2 illustrates that extra solutions may be introduced when you raise both sides of a radical equation to an even power. Such solutions, which are not solutions of the given equation, are called **extraneous solutions**. Thus, 4 is an extraneous solution of $\sqrt{x-3}+6=5$.

✓ **CHECK POINT 2** Solve: $\sqrt{x-1}+7=2$.

EXAMPLE 3 Solving a Radical Equation

Solve:

$$x+\sqrt{26-11x}=4.$$

Solution

Step 1. Isolate a radical on one side. We isolate the radical, $\sqrt{26-11x}$, by subtracting x from both sides.

$$x+\sqrt{26-11x}=4 \quad \text{This is the given equation.}$$
$$\sqrt{26-11x}=4-x \quad \text{Subtract } x \text{ from both sides.}$$

Step 2. Square both sides.

$$\left(\sqrt{26-11x}\right)^2=(4-x)^2$$
$$26-11x=16-8x+x^2$$

Simplify. Use the special-product formula $(A-B)^2=A^2-2AB+B^2$ to square $4-x$.

Step 3. Solve the resulting equation. Because of the x^2-term, the resulting equation is a quadratic equation. We need to write this quadratic equation in standard form. We can obtain zero on the left side by subtracting 26 and adding $11x$ on both sides.

$$26-26-11x+11x=16-26-8x+11x+x^2$$
$$0=x^2+3x-10 \quad \text{Simplify.}$$
$$0=(x+5)(x-2) \quad \text{Factor.}$$
$$x+5=0 \quad \text{or} \quad x-2=0 \quad \text{Set each factor equal to zero.}$$
$$x=-5 \qquad\qquad x=2 \quad \text{Solve for } x.$$

Step 4. Check the proposed solutions in the original equation.

Check −5:

$$x+\sqrt{26-11x}=4$$
$$-5+\sqrt{26-11(-5)}\stackrel{?}{=}4$$
$$-5+\sqrt{81}\stackrel{?}{=}4$$
$$-5+9\stackrel{?}{=}4$$
$$4=4,\quad \text{true}$$

Check 2:

$$x+\sqrt{26-11x}=4$$
$$2+\sqrt{26-11\cdot 2}\stackrel{?}{=}4$$
$$2+\sqrt{4}\stackrel{?}{=}4$$
$$2+2\stackrel{?}{=}4$$
$$4=4,\quad \text{true}$$

The solutions are −5 and 2, and the solution set is $\{-5, 2\}$.

Study Tip

Be sure to square *both sides* of an equation. Do *not* square each term.

Correct:

$$\left(\sqrt{26-11x}\right)^2=(4-x)^2$$

Incorrect!

$$~~\left(\sqrt{26-11x}\right)^2=4^2-x^2~~$$

Using Technology

Graphic Connections

You can use a graphing utility to provide a graphic check that $\{-5, 2\}$ is the solution set of $x + \sqrt{26 - 11x} = 4$.

Use the given equation

$$x + \sqrt{26 - 11x} = 4.$$

Enter $y_1 = x + \sqrt{26 - 11x}$ in the $\boxed{y=}$ screen.

Enter $y_2 = 4$ in the $\boxed{y=}$ screen.

Display graphs for y_1 and y_2. The solutions are the x-coordinates of the intersection points. These x-coordinates are -5 and 2. This verifies $\{-5, 2\}$ as the solution set of $x + \sqrt{26 - 11x} = 4$.

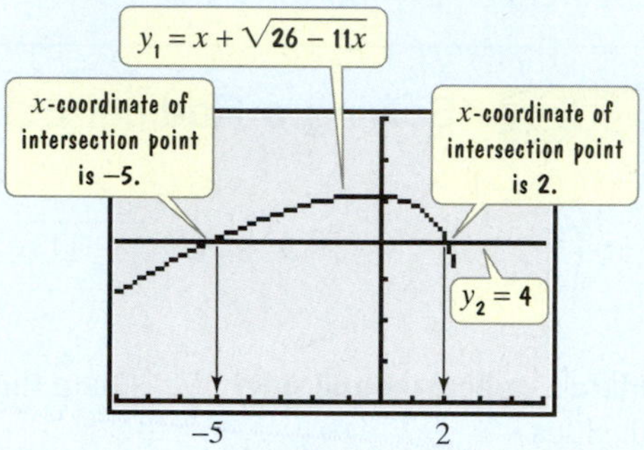

[−8, 5, 1] by [0, 7, 1]

Use the equivalent equation

$$x + \sqrt{26 - 11x} - 4 = 0.$$

Enter $y_1 = x + \sqrt{26 - 11x} - 4$ in the $\boxed{y=}$ screen.

Display the graph for y_1. The solutions are the x-intercepts. The x-intercepts are -5 and 2. This verifies $\{-5, 2\}$ as the solution set of $x + \sqrt{26 - 11x} - 4 = 0$.

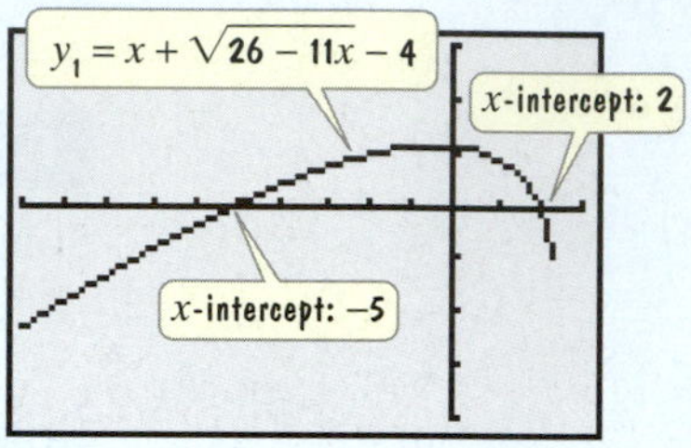

[−10, 3, 1] by [−4, 3, 1]

✓ CHECK POINT 3 Solve: $\sqrt{6x + 7} - x = 2$.

The solution of a radical equation with two or more square root expressions involves isolating a radical, squaring both sides, and then repeating this process. Let's consider an equation containing two square root expressions.

EXAMPLE 4 Solving an Equation That Has Two Radicals

Solve: $\sqrt{3x + 1} - \sqrt{x + 4} = 1$.

Solution

Step 1. Isolate a radical on one side. We can isolate the radical $\sqrt{3x + 1}$ by adding $\sqrt{x + 4}$ to both sides. We obtain

$$\sqrt{3x + 1} = \sqrt{x + 4} + 1.$$

Step 2. Square both sides.

$$\left(\sqrt{3x + 1}\right)^2 = \left(\sqrt{x + 4} + 1\right)^2$$

Squaring the expression on the right side of the equation can be a bit tricky. We have to use the formula

$$(A + B)^2 = A^2 + 2AB + B^2.$$

Focusing on just the right side, here is how the squaring is done:

$$(A + B)^2 = A^2 + 2 \cdot A \cdot B + B^2$$

$$(\sqrt{x + 4} + 1)^2 = (\sqrt{x + 4})^2 + 2 \cdot \sqrt{x + 4} \cdot 1 + 1^2 = x + 4 + 2\sqrt{x + 4} + 1.$$

Now let's return to squaring both sides.

$(\sqrt{3x+1})^2 = (\sqrt{x+4}+1)^2$	Square both sides of the equation with an isolated radical.
$3x + 1 = x + 4 + 2\sqrt{x+4} + 1$	$(\sqrt{3x+1})^2 = 3x + 1$; Square the right side using the formula for $(A + B)^2$.
$3x + 1 = x + 5 + 2\sqrt{x+4}$	Combine numerical terms on the right side: $4 + 1 = 5$.

Can you see that the resulting equation still contains a radical, namely $\sqrt{x+4}$? Thus, we need to repeat the first two steps.

Repeat Step 1. Isolate a radical on one side. We isolate $2\sqrt{x+4}$, the radical term, by subtracting $x + 5$ from both sides. We obtain

$3x + 1 = x + 5 + 2\sqrt{x+4}$	This is the equation from our last step.
$2x - 4 = 2\sqrt{x+4}.$	Subtract x and subtract 5 from both sides.

Although we can simplify the equation by dividing both sides by 2, this sort of simplification is not always helpful. Thus, we will work with the equation in this form.

Repeat Step 2. Square both sides.

Be careful in squaring both sides. Use $(A - B)^2 = A^2 - 2AB + B^2$ to square the left side. Use $(AB)^2 = A^2B^2$ to square the right side.

$(2x - 4)^2 = (2\sqrt{x+4})^2$	Square both sides.
$4x^2 - 16x + 16 = 4(x + 4)$	Square both 2 and $\sqrt{x+4}$ on the right side.

Step 3. Solve the resulting equation. We solve this quadratic equation by writing it in standard form.

$4x^2 - 16x + 16 = 4x + 16$	Use the distributive property.
$4x^2 - 20x = 0$	Subtract $4x + 16$ from both sides.
$4x(x - 5) = 0$	Factor.
$4x = 0$ or $x - 5 = 0$	Set each factor equal to zero.
$x = 0$ $\quad$ $x = 5$	Solve for x.

Step 4. Check the proposed solutions in the original equation.

Check 0:	Check 5:
$\sqrt{3x+1} - \sqrt{x+4} = 1$	$\sqrt{3x+1} - \sqrt{x+4} = 1$
$\sqrt{3 \cdot 0 + 1} - \sqrt{0+4} \stackrel{?}{=} 1$	$\sqrt{3 \cdot 5 + 1} - \sqrt{5+4} \stackrel{?}{=} 1$
$\sqrt{1} - \sqrt{4} \stackrel{?}{=} 1$	$\sqrt{16} - \sqrt{9} \stackrel{?}{=} 1$
$1 - 2 \stackrel{?}{=} 1$	$4 - 3 \stackrel{?}{=} 1$
$-1 = 1$, false	$1 = 1$, true

The check indicates that 0 is not a solution. It is an extraneous solution brought about by squaring each side of the equation. The only solution is 5 and the solution set is $\{5\}$.

CHECK POINT 4 Solve: $\sqrt{x+5} - \sqrt{x-3} = 2$.

EXAMPLE 5 Solving a Radical Equation

Solve: $(3x - 1)^{\frac{1}{3}} + 4 = 0.$

Solution Although we can rewrite the equation in radical form

$$\sqrt[3]{3x - 1} + 4 = 0,$$

it is not necessary to do so. Because the equation involves a cube root, we isolate the radical term—that is, the term with the rational exponent—and cube both sides.

$(3x - 1)^{\frac{1}{3}} + 4 = 0$ This is the given equation.

$(3x - 1)^{\frac{1}{3}} = -4$ Subtract 4 from both sides and isolate the term with the rational exponent.

$\left[(3x - 1)^{\frac{1}{3}}\right]^3 = (-4)^3$ Cube both sides.

$3x - 1 = -64$ Multiply exponents on the left side and simplify.

$3x = -63$ Add 1 to both sides.

$x = -21$ Divide both sides by 3.

Because both sides were raised to an odd power, it is not essential to check the proposed solution, -21. However, checking is always a good idea. Do so now and verify that -21 is the solution and the solution set is $\{-21\}$.

Example 5 illustrates that a radical equation with rational exponents can be solved by

1. isolating the expression with the rational exponent, and
2. raising both sides of the equation to a power that is the reciprocal of the rational exponent.

Keep in mind that it is essential to check proposed solutions when both sides have been raised to even powers. Thus, equations with rational exponents such as $\frac{1}{2}$ and $\frac{1}{4}$ must be checked.

✓ CHECK POINT 5 Solve: $(2x - 3)^{\frac{1}{3}} + 3 = 0.$

2 Use models that are radical functions to solve problems.

Applications of Radical Equations

Radical equations can be solved to answer questions about variables contained in radical functions.

EXAMPLE 6 Women in the Labor Force

The bar graph in **Figure 7.8** on the next page shows the percentage of U.S. women in the labor force from 1960 through 2000, with a projection for 2010. The function $f(x) = 3.5\sqrt{x} + 38$ models the percentage of U.S. women in the labor force, $f(x)$, x years after 1960. According to the model, when will 70% of U.S. women participate in the work force?

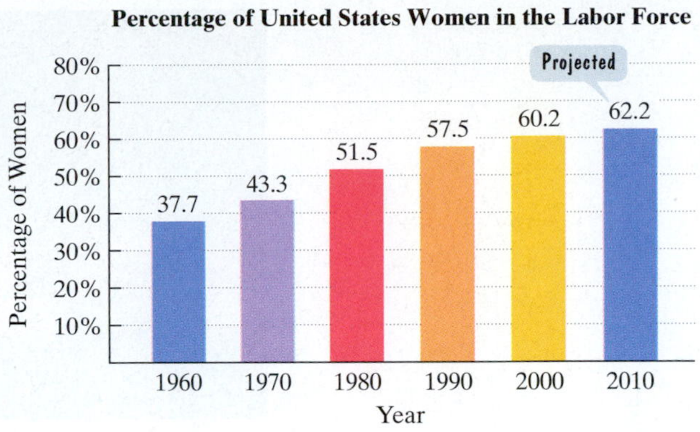

FIGURE 7.8

Source: U.S. Department of Labor

Solution To find when 70% of U.S. women will be in the work force, substitute 70 for $f(x)$ in the given function. Then solve for x, the number of years after 1960.

$$f(x) = 3.5\sqrt{x} + 38 \quad \text{This is the given function.}$$

$$70 = 3.5\sqrt{x} + 38 \quad \text{Substitute 70 for } f(x).$$

$$32 = 3.5\sqrt{x} \quad \text{Subtract 38 from both sides.}$$

$$\frac{32}{3.5} = \sqrt{x} \quad \text{Divide both sides by 3.5.}$$

$$\left(\frac{32}{3.5}\right)^2 = (\sqrt{x})^2 \quad \text{Square both sides.}$$

$$84 \approx x \quad \text{Use a calculator.}$$

The model indicates that 70% of U.S. women will be in the labor force approximately 84 years after 1960. Because $1960 + 84 = 2044$, this is projected to occur in 2044.

CHECK POINT 6 Use the function in Example 6 to project when 73% of U.S. women will participate in the work force.

7.6 EXERCISE SET

MyMathLab — PRACTICE WATCH DOWNLOAD READ REVIEW

Practice Exercises

In Exercises 1–38, solve each radical equation.

1. $\sqrt{3x - 2} = 4$
2. $\sqrt{5x - 1} = 8$
3. $\sqrt{5x - 4} - 9 = 0$
4. $\sqrt{3x - 2} - 5 = 0$
5. $\sqrt{3x + 7} + 10 = 4$
6. $\sqrt{2x + 5} + 11 = 6$
7. $x = \sqrt{7x + 8}$
8. $x = \sqrt{6x + 7}$
9. $\sqrt{5x + 1} = x + 1$
10. $\sqrt{2x + 1} = x - 7$
11. $x = \sqrt{2x - 2} + 1$
12. $x = \sqrt{3x + 7} - 3$
13. $x - 2\sqrt{x - 3} = 3$
14. $3x - \sqrt{3x + 7} = -5$
15. $\sqrt{2x - 5} = \sqrt{x + 4}$
16. $\sqrt{6x + 2} = \sqrt{5x + 3}$
17. $\sqrt[3]{2x + 11} = 3$
18. $\sqrt[3]{6x - 3} = 3$
19. $\sqrt[3]{2x - 6} - 4 = 0$
20. $\sqrt[3]{4x - 3} - 5 = 0$
21. $\sqrt{x - 7} = 7 - \sqrt{x}$
22. $\sqrt{x - 8} = \sqrt{x} - 2$
23. $\sqrt{x + 2} + \sqrt{x - 1} = 3$
24. $\sqrt{x - 4} + \sqrt{x + 4} = 4$
25. $2\sqrt{4x + 1} - 9 = x - 5$
26. $2\sqrt{x - 3} + 4 = x + 1$
27. $(2x + 3)^{\frac{1}{3}} + 4 = 6$
28. $(3x - 6)^{\frac{1}{3}} + 5 = 8$

29. $(3x + 1)^{\frac{1}{4}} + 7 = 9$

30. $(2x + 3)^{\frac{1}{4}} + 7 = 10$

31. $(x + 2)^{\frac{1}{2}} + 8 = 4$

32. $(x - 3)^{\frac{1}{2}} + 8 = 6$

33. $\sqrt{2x - 3} - \sqrt{x - 2} = 1$

34. $\sqrt{x + 2} + \sqrt{3x + 7} = 1$

35. $3x^{\frac{1}{3}} = (x^2 + 17x)^{\frac{1}{3}}$

36. $2(x - 1)^{\frac{1}{3}} = (x^2 + 2x)^{\frac{1}{3}}$

37. $(x + 8)^{\frac{1}{4}} = (2x)^{\frac{1}{4}}$

38. $(x - 2)^{\frac{1}{4}} = (3x - 8)^{\frac{1}{4}}$

Practice PLUS

39. If $f(x) = x + \sqrt{x + 5}$, find all values of x for which $f(x) = 7$.

40. If $f(x) = x - \sqrt{x - 2}$, find all values of x for which $f(x) = 4$.

41. If $f(x) = (5x + 16)^{\frac{1}{3}}$ and $g(x) = (x - 12)^{\frac{1}{3}}$, find all values of x for which $f(x) = g(x)$.

42. If $f(x) = (9x + 2)^{\frac{1}{4}}$ and $g(x) = (5x + 18)^{\frac{1}{4}}$, find all values of x for which $f(x) = g(x)$.

In Exercises 43–46, solve each formula for the specified variable.

43. $r = \sqrt{\frac{3V}{\pi h}}$ for V

44. $r = \sqrt{\frac{A}{4\pi}}$ for A

45. $t = 2\pi\sqrt{\frac{l}{32}}$ for l

46. $v = \sqrt{\frac{FR}{m}}$ for m

47. If 5 times a number is decreased by 4, the principal square root of this difference is 2 less than the number. Find the number(s).

48. If a number is decreased by 3, the principal square root of this difference is 5 less than the number. Find the number(s).

In Exercises 49–50, find the x-intercept(s) of the graph of each function without graphing the function.

49. $f(x) = \sqrt{x + 16} - \sqrt{x} - 2$

50. $f(x) = \sqrt{2x - 3} - \sqrt{2x} + 1$

Application Exercises

A basketball player's hang time is the time spent in the air when shooting a basket. The formula

$$t = \frac{\sqrt{d}}{2}$$

models hang time, t, in seconds, in terms of the vertical distance of a player's jump, d, in feet. Use this formula to solve Exercises 51–52.

51. When Michael Wilson of the Harlem Globetrotters slam-dunked a basketball 12 feet, his hang time for the shot was approximately 1.16 seconds. What was the vertical distance of his jump, rounded to the nearest tenth of a foot?

52. If hang time for a shot by a professional basketball player is 0.85 second, what is the vertical distance of the jump, rounded to the nearest tenth of a foot?

Use the graph of the formula for hang time to solve Exercises 53–54.

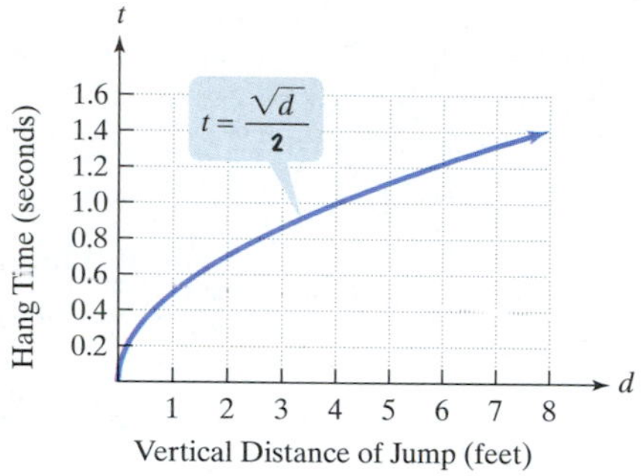

53. How is your answer to Exercise 51 shown on the graph?

54. How is your answer to Exercise 52 shown on the graph?

The graph shows the less income people have, the more likely they are to report that their health is fair or poor. The function

$$f(x) = -4.4\sqrt{x} + 38$$

models the percentage of Americans reporting fair or poor health, f(x), in terms of annual income, x, in thousands of dollars. Use this function to solve Exercises 55–56.

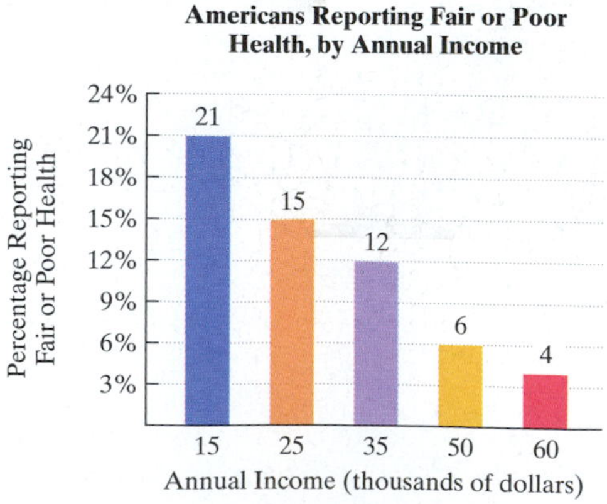

Source: William Kornblum and Joseph Julian, *Social Problems Twelfth Edition*, Prentice Hall, 2007

55. **a.** Find and interpret $f(25)$. Does this underestimate or overestimate the percent displayed by the graph? By how much?

b. According to the model, what annual income corresponds to 14% reporting fair or poor health? Round to the nearest thousand dollars.

56. **a.** Find and interpret $f(60)$. Round to one decimal place. Does this underestimate or overestimate the percent displayed by the graph? By how much?

b. According to the model, what annual income corresponds to 24% reporting fair or poor health? Round to the nearest thousand dollars.

The function

$$f(x) = 29x^{\frac{1}{3}}$$

models the number of plant species, $f(x)$, on the islands of the Galápagos in terms of the area, x, in square miles, of a particular island. Use the function to solve Exercises 57–58.

57. What is the area of a Galápagos island that has 87 species of plants?

58. What is the area of a Galápagos island that has 58 species of plants?

For each planet in our solar system, its year is the time it takes the planet to revolve once around the sun. The function

$$f(x) = 0.2x^{\frac{3}{2}}$$

models the number of Earth days in a planet's year, $f(x)$, where x is the average distance of the planet from the sun, in millions of kilometers. Use the function to solve Exercises 59–60.

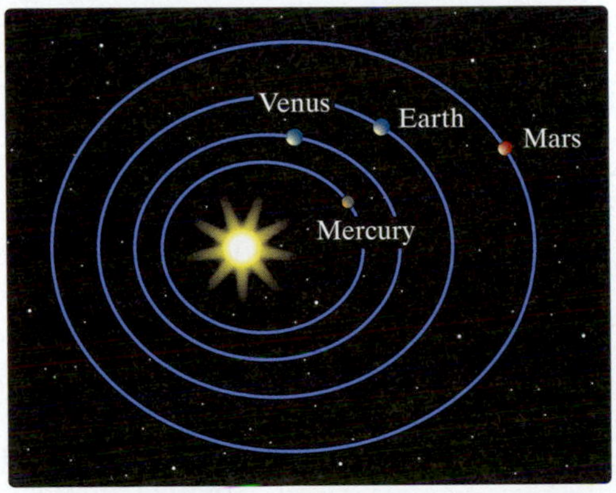

59. We, of course, have 365 Earth days in our year. What is the average distance of Earth from the sun? Use a calculator and round to the nearest million kilometers.

60. There are approximately 88 Earth days in the year of the planet Mercury. What is the average distance of Mercury from the sun? Use a calculator and round to the nearest million kilometers.

Writing in Mathematics

61. What is a radical equation?

62. In solving $\sqrt{2x - 1} + 2 = x$, why is it a good idea to isolate the radical term? What if we don't do this and simply square each side? Describe what happens.

63. What is an extraneous solution to a radical equation?

64. Explain why $\sqrt{x} = -1$ has no solution.

65. Explain how to solve a radical equation with rational exponents.

66. In Example 6 of the section, we used a square root function that modeled an increase in the percentage of U.S. women in the labor force, although the rate of increase in this percentage was leveling off. Describe an event that might occur in the future that could result in an ever-increasing rate in the percentage of women in the labor force. Would a square root function be appropriate for modeling this trend? Explain your answer.

67. The graph for Exercises 55–56 shows that the less income people have, the more likely they are to report fair or poor health. What explanations can you offer for this trend?

Technology Exercises

In Exercises 68–72, use a graphing utility to solve each radical equation. Graph each side of the equation in the given viewing rectangle. The equation's solution set is given by the x-coordinate(s) of the point(s) of intersection. Check by substitution.

68. $\sqrt{2x + 2} = \sqrt{3x - 5}$
$[-1, 10, 1]$ by $[-1, 5, 1]$

69. $\sqrt{x} + 3 = 5$
$[-1, 6, 1]$ by $[-1, 6, 1]$

70. $\sqrt{x^2 + 3} = x + 1$
$[-1, 6, 1]$ by $[-1, 6, 1]$

71. $4\sqrt{x} = x + 3$
$[-1, 10, 1]$ by $[-1, 14, 1]$

72. $\sqrt{x} + 4 = 2$
$[-2, 18, 1]$ by $[0, 10, 1]$

Critical Thinking Exercises

Make Sense? *In Exercises 73–76, determine whether each statement "makes sense" or "does not make sense" and explain your reasoning.*

73. When checking a radical equation's proposed solution, I can substitute into the original equation or any equation that is part of the solution process.

74. After squaring both sides of a radical equation, the only solution that I obtained was extraneous, so $\varnothing$ must be the solution set of the original equation.

75. When I raise both sides of an equation to any power, there's always the possibility of extraneous solutions.

76. Now that I know how to solve radical equations, I can use models that are radical functions to determine the value of the independent variable when a function value is known.

In Exercises 77–80, determine whether each statement is true or false. If the statement is false, make the necessary change(s) to produce a true statement.

77. The first step in solving $\sqrt{x+6} = x + 2$ is to square both sides, obtaining $x + 6 = x^2 + 4$.

78. The equations $\sqrt{x+4} = -5$ and $x + 4 = 25$ have the same solution set.

79. The equation $-\sqrt{x} = 9$ has no solution.

80. The equation $\sqrt{x^2 + 9x + 3} = -x$ has no solution because a principal square root is always nonnegative.

81. Find the length of the three sides of the right triangle shown in the figure.

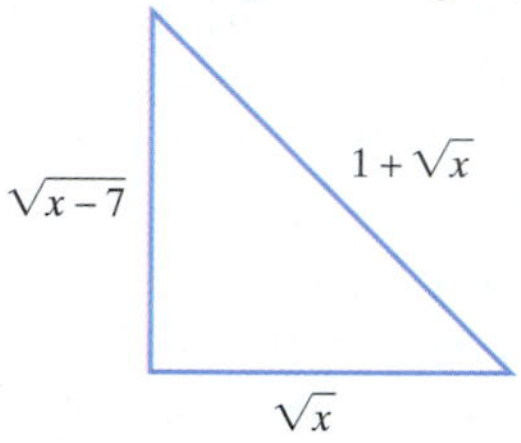

In Exercises 82–84, solve each equation.

82. $\sqrt[3]{x\sqrt{x}} = 9$

83. $\sqrt{\sqrt{x} + \sqrt{x+9}} = 3$

84. $(x - 4)^{\frac{2}{3}} = 25$

Review Exercises

85. Divide using synthetic division:
$$(4x^4 - 3x^3 + 2x^2 - x - 1) \div (x + 3).$$
(Section 6.5, Example 1)

86. Divide: $\dfrac{3x^2 - 12}{x^2 + 2x - 8} \div \dfrac{6x + 18}{x + 4}.$

(Section 6.1, Example 7)

87. Factor: $y^2 - 6y + 9 - 25x^2$.
(Section 5.5, Example 6)

Preview Exercises

Exercises 88–90 will help you prepare for the material covered in the next section.

88. Simplify: $(-5 + 7x) - (-11 - 6x)$.

89. Multiply: $(7 - 3x)(-2 - 5x)$.

90. Rationalize the denominator: $\dfrac{7 + 4\sqrt{2}}{2 - 5\sqrt{2}}$.

SECTION 7.7

Complex Numbers

Objectives

1. Express square roots of negative numbers in terms of *i*.
2. Add and subtract complex numbers.
3. Multiply complex numbers.
4. Divide complex numbers.
5. Simplify powers of *i*.

THE KID WHO LEARNED ABOUT MATH ON THE STREET
If you divide 6,973 by 0, you die.
Once, this guy tried to find the square root of -9, and his eyeballs turned black.
This girl my brother knows found out exactly what π equals, but she went nuts.
R. Chast

Who is this kid warning us about our eyeballs turning black if we attempt to find the square root of −9? Don't believe what you hear on the street. Although square roots of negative numbers are not real numbers, they do play a significant role in algebra. In this section, we move beyond the real numbers and discuss square roots with negative radicands.

1 Express square roots of negative numbers in terms of i.

The Imaginary Unit i

In Chapter 8, we will study equations whose solutions involve the square roots of negative numbers. Because the square of a real number is never negative, there is no real number x such that $x^2 = -1$. To provide a setting in which such equations have solutions, mathematicians invented an expanded system of numbers, the complex numbers. The *imaginary number i*, defined to be a solution of the equation $x^2 = -1$, is the basis of this new set.

> ### The Imaginary Unit i
> The **imaginary unit** i is defined as
> $$i = \sqrt{-1}, \quad \text{where} \quad i^2 = -1.$$

Using the imaginary unit i, we can express the square root of any negative number as a real multiple of i. For example,

$$\sqrt{-25} = \sqrt{25(-1)} = \sqrt{25}\sqrt{-1} = 5i.$$

We can check that $\sqrt{-25} = 5i$ by squaring $5i$ and obtaining -25.

$$(5i)^2 = 5^2 i^2 = 25(-1) = -25$$

> ### The Square Root of a Negative Number
> If b is a positive real number, then
> $$\sqrt{-b} = \sqrt{b(-1)} = \sqrt{b}\sqrt{-1} = \sqrt{b}i \quad \text{or} \quad i\sqrt{b}.$$

EXAMPLE 1 Expressing Square Roots of Negative Numbers as Multiples of i

Write as a multiple of i:

a. $\sqrt{-9}$ **b.** $\sqrt{-3}$ **c.** $\sqrt{-80}$.

Study Tip

We allow the use of the product rule $\sqrt{ab} = \sqrt{a}\sqrt{b}$ when a is positive and b is -1. However, you cannot use $\sqrt{ab} = \sqrt{a}\sqrt{b}$ when both a and b are negative.

Solution

a. $\sqrt{-9} = \sqrt{9(-1)} = \sqrt{9}\sqrt{-1} = 3i$

b. $\sqrt{-3} = \sqrt{3(-1)} = \sqrt{3}\sqrt{-1} = \sqrt{3}\,i$ *Be sure not to write i under the radical.*

c. $\sqrt{-80} = \sqrt{80(-1)} = \sqrt{80}\sqrt{-1} = \sqrt{16 \cdot 5}\sqrt{-1} = 4\sqrt{5}\,i$

In order to avoid writing i under a radical, let's agree to write i before any radical. Consequently, we express the multiple of i in part (b) as $i\sqrt{3}$ and the multiple of i in part (c) as $4i\sqrt{5}$. ■

☑ **CHECK POINT 1** Write as a multiple of i:

a. $\sqrt{-64}$ **b.** $\sqrt{-11}$ **c.** $\sqrt{-48}$.

A new system of numbers, called *complex numbers*, is based on adding multiples of i, such as $5i$, to the real numbers.

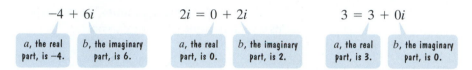

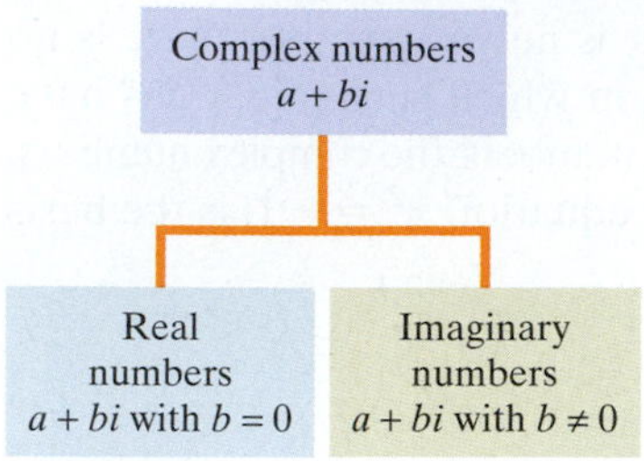

FIGURE 7.9
The complex number system

Complex Numbers and Imaginary Numbers

The set of all numbers in the form

$$a + bi,$$

with real numbers a and b, and i, the imaginary unit, is called the set of **complex numbers**. The real number a is called the **real part**, and the real number b is called the **imaginary part** of the complex number $a + bi$. If $b \neq 0$, then the complex number is called an **imaginary number** (**Figure 7.9**).

Here are some examples of complex numbers. Each number can be written in the form $a + bi$.

$$-4 + 6i \qquad 2i = 0 + 2i \qquad 3 = 3 + 0i$$

$-4 + 6i$: a, the real part, is −4. b, the imaginary part, is 6.

$2i = 0 + 2i$: a, the real part, is 0. b, the imaginary part, is 2.

$3 = 3 + 0i$: a, the real part, is 3. b, the imaginary part, is 0.

Can you see that b, the imaginary part, is not zero in the first two complex numbers? Because $b \neq 0$, these complex numbers are imaginary numbers. By contrast, the imaginary part of the complex number on the right is zero. This complex number is not an imaginary number. The number 3, or $3 + 0i$, is a real number.

2 Add and subtract complex numbers.

Adding and Subtracting Complex Numbers

The form of a complex number $a + bi$ is like the binomial $a + bx$. Consequently, we can add, subtract, and multiply complex numbers using the same methods we used for binomials, remembering that $i^2 = -1$.

Adding and Subtracting Complex Numbers

1. $(a + bi) + (c + di) = (a + c) + (b + d)i$
   In words, this says that you add complex numbers by adding their real parts, adding their imaginary parts, and expressing the sum as a complex number.
2. $(a + bi) - (c + di) = (a - c) + (b - d)i$
   In words, this says that you subtract complex numbers by subtracting their real parts, subtracting their imaginary parts, and expressing the difference as a complex number.

EXAMPLE 2 Adding and Subtracting Complex Numbers

Perform the indicated operations, writing the result in the form $a + bi$:

a. $(5 - 11i) + (7 + 4i)$ **b.** $(-5 + 7i) - (-11 - 6i)$.

Solution

a. $(5 - 11i) + (7 + 4i)$

$= 5 - 11i + 7 + 4i$ Remove the parentheses.

$= 5 + 7 - 11i + 4i$ Group real and imaginary terms.

$= (5 + 7) + (-11 + 4)i$ Add real parts and add imaginary parts.

$= 12 - 7i$ Simplify.

Study Tip

The following examples, using the same integers as in Example 2, show how operations with complex numbers are just like operations with polynomials.

a. $(5 - 11x) + (7 + 4x)$
$= 12 - 7x$

b. $(-5 + 7x) - (-11 - 6x)$
$= -5 + 7x + 11 + 6x$
$= 6 + 13x$

b. $(-5 + 7i) - (-11 - 6i)$

$= -5 + 7i + 11 + 6i$ — Remove the parentheses. Change signs of real and imaginary parts in the complex number being subtracted.

$= -5 + 11 + 7i + 6i$ — Group real and imaginary terms.

$= (-5 + 11) + (7 + 6)i$ — Add real parts and add imaginary parts.

$= 6 + 13i$ — Simplify.

CHECK POINT 2 Add or subtract as indicated:

a. $(5 - 2i) + (3 + 3i)$

b. $(2 + 6i) - (12 - 4i)$.

3 Multiply complex numbers.

Multiplying Complex Numbers

Multiplication of complex numbers is performed the same way as multiplication of polynomials, using the distributive property and the FOIL method. After completing the multiplication, we replace any occurrences of i^2 with -1. This idea is illustrated in the next example.

EXAMPLE 3 Multiplying Complex Numbers

Find the products:

a. $4i(3 - 5i)$ **b.** $(7 - 3i)(-2 - 5i)$.

Solution

a. $4i(3 - 5i)$

$= 4i \cdot 3 - 4i \cdot 5i$ — Distribute $4i$ throughout the parentheses.

$= 12i - 20i^2$ — Multiply.

$= 12i - 20(-1)$ — Replace i^2 with -1.

$= 20 + 12i$ — Simplify to $12i + 20$ and write in $a + bi$ form.

b. $(7 - 3i)(-2 - 5i)$

F O I L

$= -14 - 35i + 6i + 15i^2$ — Use the FOIL method.

$= -14 - 35i + 6i + 15(-1)$ — $i^2 = -1$

$= -14 - 15 - 35i + 6i$ — Group real and imaginary terms.

$= -29 - 29i$ — Combine real and imaginary terms.

CHECK POINT 3 Find the products:

a. $7i(2 - 9i)$ **b.** $(5 + 4i)(6 - 7i)$.

Consider the multiplication problem

$$5i \cdot 2i = 10i^2 = 10(-1) = -10.$$

The problem $5i \cdot 2i$ can also be given in terms of square roots of negative numbers:

$$\sqrt{-25} \cdot \sqrt{-4}.$$

Because the product rule for radicals only applies to real numbers, multiplying radicands is incorrect. **When performing operations with square roots of negative numbers, begin by expressing all square roots in terms of *i*.** Then perform the indicated operation.

CORRECT:

$$\sqrt{-25} \cdot \sqrt{-4} = \sqrt{25}\sqrt{-1} \cdot \sqrt{4}\sqrt{-1} = 5i \cdot 2i = 10i^2 = 10(-1) = -10$$

INCORRECT!

$$\sqrt{-25} \cdot \sqrt{-4} = \sqrt{(-25)(-4)} = \sqrt{100} = 10$$

EXAMPLE 4 Multiplying Square Roots of Negative Numbers

Multiply: $\sqrt{-3} \cdot \sqrt{-5}$.

Solution

$$\sqrt{-3} \cdot \sqrt{-5} = \sqrt{3}\sqrt{-1} \cdot \sqrt{5}\sqrt{-1}$$

$= i\sqrt{3} \cdot i\sqrt{5}$ Express square roots in terms of i.

$= i^2\sqrt{15}$ $\sqrt{3} \cdot \sqrt{5} = \sqrt{15}$ and $i \cdot i = i^2$.

$= (-1)\sqrt{15}$ $i^2 = -1$

$= -\sqrt{15}$

CHECK POINT 4 Multiply: $\sqrt{-5} \cdot \sqrt{-7}$.

4 Divide complex numbers.

Conjugates and Division

It is possible to multiply imaginary numbers and obtain a real number. Here is an example:

$$(4 + 7i)(4 - 7i) = 16 - 28i + 28i - 49i^2 = 16 - 49i^2 = 16 - 49(-1) = 65.$$

(F: 16, O: −28i, I: 28i, L: −49i²; Replace i^2 with −1.)

You can also perform $(4 + 7i)(4 - 7i)$ using the formula

$$(A + B)(A - B) = A^2 - B^2.$$

A real number is obtained even faster:

$$(4 + 7i)(4 - 7i) = 4^2 - (7i)^2 = 16 - 49i^2 = 16 - 49(-1) = 65.$$

The **conjugate** of the complex number $a + bi$ is $a - bi$. The **conjugate** of the complex number $a - bi$ is $a + bi$. The multiplication problem that we just performed involved conjugates. The multiplication of conjugates always results in a real number:

$$(a + bi)(a - bi) = a^2 - (bi)^2 = a^2 - b^2i^2 = a^2 - b^2(-1) = a^2 + b^2.$$

The product eliminates i.

Conjugates are used to divide complex numbers. The goal of the division procedure is to obtain a real number in the denominator. This real number becomes the denominator of a and b in the quotient $a + bi$. By multiplying the numerator and the denominator of the division by the conjugate of the denominator, you will obtain this real number in the denominator. Here are two examples of such divisions:

- $\dfrac{7+4i}{2-5i}$ — The conjugate of the denominator is $2+5i$.
- $\dfrac{5i-4}{3i}$ or $\dfrac{5i-4}{0+3i}$. — The conjugate of the denominator is $0-3i$, or $-3i$.

The procedure for dividing complex numbers, illustrated in Examples 5 and 6, should remind you of rationalizing denominators.

EXAMPLE 5 Using Conjugates to Divide Complex Numbers

Divide and simplify to the form $a + bi$:

$$\frac{7+4i}{2-5i}.$$

Solution The conjugate of the denominator is $2 + 5i$. Multiplication of both the numerator and the denominator by $2 + 5i$ will eliminate i from the denominator while maintaining the value of the expression.

$$\frac{7+4i}{2-5i} = \frac{7+4i}{2-5i}\cdot\frac{2+5i}{2+5i}$$ Multiply by 1.

F O I L

$$= \frac{14+35i+8i+20i^2}{2^2-(5i)^2}$$ Use FOIL in the numerator and $(A-B)(A+B) = A^2 - B^2$ in the denominator.

$$= \frac{14+43i+20i^2}{4-25i^2}$$ Simplify.

$$= \frac{14+43i+20(-1)}{4-25(-1)}$$ $i^2 = -1$

$$= \frac{14+43i-20}{4+25}$$ Perform the multiplications involving -1.

$$= \frac{-6+43i}{29}$$ Combine real terms in the numerator and denominator.

$$= -\frac{6}{29}+\frac{43}{29}i$$ Express the answer in the form $a + bi$.

✓ **CHECK POINT 5** Divide and simplify to the form $a + bi$:

$$\frac{6+2i}{4-3i}.$$

EXAMPLE 6 Using Conjugates to Divide Complex Numbers

Divide and simplify to the form $a + bi$:

$$\frac{5i-4}{3i}.$$

Solution The denominator of $\frac{5i-4}{3i}$ is $3i$, or $0+3i$. The conjugate of the denominator is $0-3i$. Multiplication of both the numerator and the denominator by $-3i$ will eliminate i from the denominator while maintaining the value of the expression.

$$\frac{5i-4}{3i} = \frac{5i-4}{3i}\cdot\frac{-3i}{-3i}$$ Multiply by 1.

$$= \frac{-15i^2+12i}{-9i^2}$$ Multiply. Use the distributive property in the numerator.

$$= \frac{-15(-1)+12i}{-9(-1)}$$ $i^2 = -1$

$$= \frac{15+12i}{9}$$ Perform the multiplications involving -1.

$$= \frac{15}{9}+\frac{12}{9}i$$ Express the division in the form $a+bi$.

$$= \frac{5}{3}+\frac{4}{3}i$$ Simplify real and imaginary parts.

CHECK POINT 6 Divide and simplify to the form $a+bi$:

$$\frac{3-2i}{4i}.$$

---

5 Simplify powers of i.

Powers of i

Using the fact that $i^2=-1$, any integral power of i greater than or equal to 2 can be simplified to either $-i$, i, -1, or 1. Here are some examples:

$$i^3 = i^2\cdot i = (-1)i = -i$$
$$i^4 = (i^2)^2 = (-1)^2 = 1$$
$$i^5 = i^4\cdot i = (i^2)^2\cdot i = (-1)^2\cdot i = i$$
$$i^6 = (i^2)^3 = (-1)^3 = -1$$

Here is a procedure for simplifying powers of i:

Simplifying Powers of i

1. Express the given power of i in terms of i^2.
2. Replace i^2 with -1 and simplify. Use the fact that -1 to an even power is 1 and -1 to an odd power is -1.

EXAMPLE 7 Simplifying Powers of i

Simplify:

a. i^{12} **b.** i^{39} **c.** i^{50}.

Solution

a. $i^{12} = (i^2)^6 = (-1)^6 = 1$

b. $i^{39} = i^{38}i = (i^2)^{19}i = (-1)^{19}i = (-1)i = -i$

c. $i^{50} = (i^2)^{25} = (-1)^{25} = -1$

CHECK POINT 7 Simplify:

a. i^{16} **b.** i^{25} **c.** i^{35}.

Blitzer Bonus

The Patterns of Chaos

One of the new frontiers of mathematics suggests that there is an underlying order in things that appear to be random, such as the hiss and crackle of background noises as you tune a radio. Irregularities in the heartbeat, some of them severe enough to cause a heart attack, or irregularities in our sleeping patterns, such as insomnia, are examples of chaotic behavior. Chaos in the mathematical sense does not mean a complete lack of form or arrangement. In mathematics, chaos is used to describe something that appears to be random, but actually contains underlying patterns that are far more intricate than previously assumed. The patterns of chaos appear in images like the one on the right and the one in the chapter opener, called the Mandelbrot set. Magnified portions of this image yield repetitions of the original structure, as well as new and unexpected patterns. The Mandelbrot set transforms the hidden structure of chaotic events into a source of wonder and inspiration.

The Mandelbrot set is made possible by opening up graphing to include complex numbers in the form $a + bi$. Each complex number is plotted like an ordered pair in a coordinate system consisting of a real axis and an imaginary axis. Plot certain complex numbers in this system, add color to the magnified boundary of the graph, and the patterns of chaos begin to appear.

R. F. Voss "29-Fold M-set Seahorse" computer-generated image

7.7 EXERCISE SET

MyMathLab MathXL PRACTICE WATCH DOWNLOAD READ REVIEW

Practice Exercises

In Exercises 1–16, express each number in terms of i and simplify, if possible.

1. $\sqrt{-100}$
2. $\sqrt{-49}$
3. $\sqrt{-23}$
4. $\sqrt{-21}$
5. $\sqrt{-18}$
6. $\sqrt{-125}$
7. $\sqrt{-63}$
8. $\sqrt{-28}$
9. $-\sqrt{-108}$
10. $-\sqrt{-300}$
11. $5 + \sqrt{-36}$
12. $7 + \sqrt{-4}$
13. $15 + \sqrt{-3}$
14. $20 + \sqrt{-5}$
15. $-2 - \sqrt{-18}$
16. $-3 - \sqrt{-27}$

In Exercises 17–32, add or subtract as indicated. Write the result in the form $a + bi$.

17. $(3 + 2i) + (5 + i)$
18. $(6 + 5i) + (4 + 3i)$
19. $(7 + 2i) + (1 - 4i)$
20. $(-2 + 6i) + (4 - i)$
21. $(10 + 7i) - (5 + 4i)$
22. $(11 + 8i) - (2 + 5i)$
23. $(9 - 4i) - (10 + 3i)$
24. $(8 - 5i) - (6 + 2i)$
25. $(3 + 2i) - (5 - 7i)$
26. $(-7 + 5i) - (9 - 11i)$
27. $(-5 + 4i) - (-13 - 11i)$
28. $(-9 + 2i) - (-17 - 6i)$
29. $8i - (14 - 9i)$
30. $15i - (12 - 11i)$
31. $(2 + i\sqrt{3}) + (7 + 4i\sqrt{3})$
32. $(4 + i\sqrt{5}) + (8 + 6i\sqrt{5})$

In Exercises 33–62, find each product. Write imaginary results in the form $a + bi$.

33. $2i(5 + 3i)$
34. $5i(4 + 7i)$
35. $3i(7i - 5)$
36. $8i(4i - 3)$
37. $-7i(2 - 5i)$
38. $-6i(3 - 5i)$
39. $(3 + i)(4 + 5i)$
40. $(4 + i)(5 + 6i)$
41. $(7 - 5i)(2 - 3i)$
42. $(8 - 4i)(3 - 2i)$
43. $(6 - 3i)(-2 + 5i)$
44. $(7 - 2i)(-3 + 6i)$
45. $(3 + 5i)(3 - 5i)$
46. $(2 + 7i)(2 - 7i)$
47. $(-5 + 3i)(-5 - 3i)$
48. $(-4 + 2i)(-4 - 2i)$
49. $(3 - i\sqrt{2})(3 + i\sqrt{2})$
50. $(5 - i\sqrt{3})(5 + i\sqrt{3})$
51. $(2 + 3i)^2$
52. $(3 + 2i)^2$
53. $(5 - 2i)^2$
54. $(5 - 3i)^2$
55. $\sqrt{-7} \cdot \sqrt{-2}$
56. $\sqrt{-7} \cdot \sqrt{-3}$
57. $\sqrt{-9} \cdot \sqrt{-4}$
58. $\sqrt{-16} \cdot \sqrt{-4}$
59. $\sqrt{-7} \cdot \sqrt{-25}$
60. $\sqrt{-3} \cdot \sqrt{-36}$
61. $\sqrt{-8} \cdot \sqrt{-3}$
62. $\sqrt{-9} \cdot \sqrt{-5}$

In Exercises 63–84, divide and simplify to the form $a + bi$.

63. $\dfrac{2}{3 + i}$
64. $\dfrac{3}{4 + i}$
65. $\dfrac{2i}{1 + i}$
66. $\dfrac{5i}{2 + i}$

67. $\frac{7}{4-3i}$

68. $\frac{9}{1-2i}$

69. $\frac{6i}{3-2i}$

70. $\frac{5i}{2-3i}$

71. $\frac{1+i}{1-i}$

72. $\frac{1-i}{1+i}$

73. $\frac{2-3i}{3+i}$

74. $\frac{2+3i}{3-i}$

75. $\frac{5-2i}{3+2i}$

76. $\frac{6-3i}{4+2i}$

77. $\frac{4+5i}{3-7i}$

78. $\frac{5-i}{3-2i}$

79. $\frac{7}{3i}$

80. $\frac{5}{7i}$

81. $\frac{8-5i}{2i}$

82. $\frac{3+4i}{5i}$

83. $\frac{4+7i}{-3i}$

84. $\frac{5+i}{-4i}$

In Exercises 85–100, simplify each expression.

85. i^{10}

86. i^{14}

87. i^{11}

88. i^{15}

89. i^{22}

90. i^{46}

91. i^{200}

92. i^{400}

93. i^{17}

94. i^{21}

95. $(-i)^4$

96. $(-i)^6$

97. $(-i)^9$

98. $(-i)^{13}$

99. $i^{24} + i^2$

100. $i^{28} + i^{30}$

Practice PLUS

In Exercises 101–108, perform the indicated operation(s) and write the result in the form a + bi.

101. $(2-3i)(1-i) - (3-i)(3+i)$

102. $(8+9i)(2-i) - (1-i)(1+i)$

103. $(2+i)^2 - (3-i)^2$

104. $(4-i)^2 - (1+2i)^2$

105. $5\sqrt{-16} + 3\sqrt{-81}$

106. $5\sqrt{-8} + 3\sqrt{-18}$

107. $\frac{i^4 + i^{12}}{i^8 - i^7}$

108. $\frac{i^8 + i^{40}}{i^4 + i^3}$

109. Let $f(x) = x^2 - 2x + 2$. Find $f(1+i)$.

110. Let $f(x) = x^2 - 2x + 5$. Find $f(1-2i)$.

In Exercises 111–114, simplify each evaluation to the form a + bi.

111. Let $f(x) = x - 3i$ and $g(x) = 4x + 2i$. Find $(fg)(-1)$.

112. Let $f(x) = 12x - i$ and $g(x) = 6x + 3i$. Find $(fg)\left(-\frac{1}{3}\right)$.

113. Let $f(x) = \frac{x^2 + 19}{2 - x}$. Find $f(3i)$.

114. Let $f(x) = \frac{x^2 + 11}{3 - x}$. Find $f(4i)$.

Application Exercises

Complex numbers are used in electronics to describe the current in an electric circuit. Ohm's law relates the current in a circuit, I, in amperes, the voltage of the circuit, E, in volts, and the resistance of the circuit, R, in ohms, by the formula E = IR. Use this formula to solve Exercises 115–116.

115. Find E, the voltage of a circuit, if $I = (4 - 5i)$ amperes and $R = (3 + 7i)$ ohms.

116. Find E, the voltage of a circuit, if $I = (2 - 3i)$ amperes and $R = (3 + 5i)$ ohms.

117. The mathematician Girolamo Cardano is credited with the first use (in 1545) of negative square roots in solving the now-famous problem, "Find two numbers whose sum is 10 and whose product is 40." Show that the complex numbers $5 + i\sqrt{15}$ and $5 - i\sqrt{15}$ satisfy the conditions of the problem. (Cardano did not use the symbolism $i\sqrt{15}$ or even $\sqrt{-15}$. He wrote R.m 15 for $\sqrt{-15}$, meaning "radix minus 15." He regarded the numbers 5 + R.m 15 and 5 − R.m 15 as "fictitious" or "ghost numbers," and considered the problem "manifestly impossible." But in a mathematically adventurous spirit, he exclaimed, "Nevertheless, we will operate.")

Writing in Mathematics

118. What is the imaginary unit i?

119. Explain how to write $\sqrt{-64}$ as a multiple of i.

120. What is a complex number? Explain when a complex number is a real number and when it is an imaginary number. Provide examples with your explanation.

121. Explain how to add complex numbers. Give an example.

122. Explain how to subtract complex numbers. Give an example.

123. Explain how to find the product of $2i$ and $5 + 3i$.

124. Explain how to find the product of $2i + 3$ and $5 + 3i$.

125. Explain how to find the product of $2i + 3$ and $2i - 3$.

126. Explain how to find the product of $\sqrt{-1}$ and $\sqrt{-4}$. Describe a common error in the multiplication that needs to be avoided.

127. What is the conjugate of $2 + 3i$? What happens when you multiply this complex number by its conjugate?

128. Explain how to divide complex numbers. Provide an example with your explanation.

129. Explain each of the three jokes in the cartoon on page 544.

130. A stand-up comedian uses algebra in some jokes, including one about a telephone recording that announces "You have just reached an imaginary number. Please multiply by i and dial again." Explain the joke.

Explain the error in Exercises 131–132.

131. $\sqrt{-9} + \sqrt{-16} = \sqrt{-25} = i\sqrt{25} = 5i$

132. $(\sqrt{-9})^2 = \sqrt{-9} \cdot \sqrt{-9} = \sqrt{81} = 9$

Critical Thinking Exercises

Make Sense? *In Exercises 133–136, determine whether each statement "makes sense" or "does not make sense" and explain your reasoning.*

133. The joke in the cartoon below is based on the math teacher not realizing that the average of complex real numbers is sometimes a complex imaginary number.

ROBOTMAN by Jim Meddick

Robotman reprinted by permission of Newspaper Enterprise Association Inc.

134. The word *imaginary* in imaginary numbers tells me that these numbers are undefined.

135. By writing the imaginary number $5i$, I can immediately see that 5 is the constant and i is the variable.

136. When I add or subtract complex numbers, I am basically combining like terms.

In Exercises 137–140, determine whether each statement is true or false. If the statement is false, make the necessary change(s) to produce a true statement.

137. Some irrational numbers are not complex numbers.

138. $(3 + 7i)(3 - 7i)$ is an imaginary number.

139. $\dfrac{7 + 3i}{5 + 3i} = \dfrac{7}{5}$

140. In the complex number system, $x^2 + y^2$ (the sum of two squares) can be factored as $(x + yi)(x - yi)$.

In Exercises 141–143, perform the indicated operations and write the result in the form $a + bi$.

141. $\dfrac{4}{(2 + i)(3 - i)}$

142. $\dfrac{1 + i}{1 + 2i} + \dfrac{1 - i}{1 - 2i}$

143. $\dfrac{8}{1 + \dfrac{2}{i}}$

Review Exercises

144. Simplify:

$$\frac{\dfrac{x}{y^2} + \dfrac{1}{y}}{\dfrac{y}{x^2} + \dfrac{1}{x}}.$$

(Section 6.3, Example 1)

145. Solve for x: $\dfrac{1}{x} + \dfrac{1}{y} = \dfrac{1}{z}$.

(Section 6.7, Example 1)

146. Solve:

$$2x - \frac{x - 3}{8} = \frac{1}{2} + \frac{x + 5}{2}.$$

(Section 1.4, Example 4)

Preview Exercises

Exercises 147–149 will help you prepare for the material covered in the first section of the next chapter.

147. Solve by factoring: $2x^2 + 7x - 4 = 0$.

148. Solve by factoring: $x^2 = 9$.

149. Use substitution to determine if $-\sqrt{6}$ is a solution of the quadratic equation $3x^2 = 18$.

GROUP PROJECT

CHAPTER 7

Group members should consult an almanac, newspaper, magazine, or the Internet and return to the group with as much data as possible that show phenomena that are continuing to grow over time, but whose growth is leveling off. Select the five data sets that you find most intriguing. Let x represent the number of years after the first year in each data set. Model the data by hand using

$$f(x) = a\sqrt{x} + b.$$

Use the first and last data points to find values for a and b. The first data point corresponds to $x = 0$. Its second coordinate gives the value of b. To find a, substitute the second data point into $f(x) = a\sqrt{x} + b$, with the value that you obtained for b. Now solve the equation and find a. Substitute a and b into $f(x) = a\sqrt{x} + b$ to obtain a square root function that models each data set. Then use the function to make predictions about what might occur in the future. Are there circumstances that might affect the accuracy of the predictions? List some of these circumstances.

Chapter 7 Summary

Definitions and Concepts	Examples
Section 7.1 Radical Expressions and Functions	
If $b^2 = a$, then b is a square root of a. The principal square root of a, designated $\sqrt{a}$, is the nonnegative number satisfying $b^2 = a$. The negative square root of a is written $-\sqrt{a}$. A square root of a negative number is not a real number. A radical function in x is a function defined by an expression containing a root of x. The domain of a square root function is the set of real numbers for which the radicand is nonnegative.	Let $f(x) = \sqrt{6 - 2x}$. $f(-15) = \sqrt{6 - 2(-15)} = \sqrt{6 + 30} = \sqrt{36} = 6$ $f(5) = \sqrt{6 - 2\cdot5} = \sqrt{6 - 10} = \sqrt{-4}$, not a real number Domain of f: Set the radicand greater than or equal to zero. $6 - 2x \geq 0$ $-2x \geq -6$ $x \leq 3$ Domain of $f = \{x \mid x \leq 3\}$ or $(-\infty, 3]$
The cube root of a real number a is written $\sqrt[3]{a}$. $\sqrt[3]{a} = b$ means that $b^3 = a$. The nth root of a real number a is written $\sqrt[n]{a}$. The number n is the index. Every real number has one root when n is odd. The odd nth root of a, $\sqrt[n]{a}$, is the number b for which $b^n = a$. Every positive real number has two real roots when n is even. An even root of a negative number is not a real number. If n is even, then $\sqrt[n]{a^n} = \lvert a \rvert$. If n is odd, then $\sqrt[n]{a^n} = a$.	• $\sqrt[3]{-8} = -2$ because $(-2)^3 = -8$. • $\sqrt[4]{-16}$ is not a real number. • $\sqrt{x^2 - 14x + 49} = \sqrt{(x - 7)^2} = \lvert x - 7 \rvert$ • $\sqrt[3]{125(x + 6)^3} = 5(x + 6)$
Section 7.2 Rational Exponents	
• $a^{\frac{1}{n}} = \sqrt[n]{a}$ • $a^{\frac{m}{n}} = (\sqrt[n]{a})^m$ or $\sqrt[n]{a^m}$ • $a^{-\frac{m}{n}} = \dfrac{1}{a^{\frac{m}{n}}}$	• $121^{\frac{1}{2}} = \sqrt{121} = 11$ • $64^{\frac{1}{3}} = \sqrt[3]{64} = 4$ • $27^{\frac{5}{3}} = (\sqrt[3]{27})^5 = 3^5 = 3\cdot3\cdot3\cdot3\cdot3 = 243$ • $16^{-\frac{3}{4}} = \dfrac{1}{16^{\frac{3}{4}}} = \dfrac{1}{(\sqrt[4]{16})^3} = \dfrac{1}{2^3} = \dfrac{1}{8}$ • $(\sqrt[3]{7xy})^4 = (7xy)^{\frac{4}{3}}$
Properties of integer exponents are true for rational exponents. An expression with rational exponents is simplified when no parentheses appear, no powers are raised to powers, each base occurs once, and no negative or zero exponents appear.	Simplify: $\left(8x^{\frac{1}{3}}y^{-\frac{1}{2}}\right)^{\frac{1}{3}}$. $= 8^{\frac{1}{3}}\left(x^{\frac{1}{3}}\right)^{\frac{1}{3}}\left(y^{-\frac{1}{2}}\right)^{\frac{1}{3}}$ $= 2x^{\frac{1}{9}}y^{-\frac{1}{6}} = \dfrac{2x^{\frac{1}{9}}}{y^{\frac{1}{6}}}$
Some radical expressions can be simplifed using rational exponents. Rewrite the expression using rational exponents, simplify, and rewrite in radical notation if rational exponents still appear.	• $\sqrt[9]{x^3} = x^{\frac{3}{9}} = x^{\frac{1}{3}} = \sqrt[3]{x}$ • $\sqrt[5]{x^2}\cdot\sqrt[4]{x} = x^{\frac{2}{5}}\cdot x^{\frac{1}{4}} = x^{\frac{2}{5}+\frac{1}{4}}$ $= x^{\frac{8}{20}+\frac{5}{20}} = x^{\frac{13}{20}} = \sqrt[20]{x^{13}}$

Definitions and Concepts	Examples

Section 7.3 Multiplying and Simplifying Radical Expressions

The product rule for radicals can be used to multiply radicals

$$\sqrt[n]{a}\cdot\sqrt[n]{b}=\sqrt[n]{ab}.$$

$$\sqrt[3]{7x}\cdot\sqrt[3]{10y^2}=\sqrt[3]{7x\cdot 10y^2}=\sqrt[3]{70xy^2}$$

The product rule for radicals can be used to simplify radicals:

$$\sqrt[n]{ab}=\sqrt[n]{a}\cdot\sqrt[n]{b}.$$

A radical expression with index n is simplified when its radicand has no factors that are perfect nth powers. To simplify, write the radicand as the product of two factors, one of which is the greatest perfect nth power. Then use the product rule to take the nth root of each factor. If all variables in a radicand are positive, then

$$\sqrt[n]{a^n}=a.$$

Some radicals can be simplified after multiplication is performed.

- Simplify: $\sqrt[3]{54x^7y^{11}}$.

$$=\sqrt[3]{27\cdot 2\cdot x^6\cdot x\cdot y^9\cdot y^2}$$

$$=\sqrt[3]{(27x^6y^9)(2xy^2)}$$

$$=\sqrt[3]{27x^6y^9}\cdot\sqrt[3]{2xy^2}=3x^2y^3\sqrt[3]{2xy^2}$$

- Assuming positive variables, multiply and simplify: $\sqrt[4]{4x^2y}\cdot\sqrt[4]{4xy^3}$.

$$=\sqrt[4]{4x^2y\cdot 4xy^3}=\sqrt[4]{16x^3y^4}$$

$$=\sqrt[4]{16y^4}\cdot\sqrt[4]{x^3}=2y\sqrt[4]{x^3}$$

Section 7.4 Adding, Subtracting, and Dividing Radical Expressions

Like radicals have the same indices and radicands. Like radicals can be added or subtracted using the distributive property. In some cases, radicals can be combined once they have been simplified.

$$4\sqrt{18}-6\sqrt{50}$$

$$=4\sqrt{9\cdot 2}-6\sqrt{25\cdot 2}=4\cdot 3\sqrt{2}-6\cdot 5\sqrt{2}$$

$$=12\sqrt{2}-30\sqrt{2}=-18\sqrt{2}$$

The quotient rule for radicals can be used to simplify radicals:

$$\sqrt[n]{\frac{a}{b}}=\frac{\sqrt[n]{a}}{\sqrt[n]{b}}.$$

$$\sqrt[3]{-\frac{8}{x^{12}}}=\frac{\sqrt[3]{-8}}{\sqrt[3]{x^{12}}}=-\frac{2}{x^4}$$

$\sqrt[3]{x^{12}}=(x^{12})^{\frac{1}{3}}=x^4$

The quotient rule for radicals can be used to divide radicals with the same indices:

$$\frac{\sqrt[n]{a}}{\sqrt[n]{b}}=\sqrt[n]{\frac{a}{b}}.$$

Some radicals can be simplified after the division is performed.

Assuming a positive variable, divide and simplify:

$$\frac{\sqrt[4]{64x^5}}{\sqrt[4]{2x^{-2}}}=\sqrt[4]{32x^{5-(-2)}}=\sqrt[4]{32x^7}$$

$$=\sqrt[4]{16\cdot 2\cdot x^4\cdot x^3}=\sqrt[4]{16x^4}\cdot\sqrt[4]{2x^3}$$

$$=2x\sqrt[4]{2x^3}.$$

Section 7.5 Multiplying with More Than One Term and Rationalizing Denominators

Radical expressions with more than one term are multiplied in much the same way that polynomials with more than one term are multiplied.

- $\sqrt{5}(2\sqrt{6}-\sqrt{3})=2\sqrt{30}-\sqrt{15}$
- $(4\sqrt{3}-2\sqrt{2})(\sqrt{3}+\sqrt{2})$

F O I L

$$=4\sqrt{3}\cdot\sqrt{3}+4\sqrt{3}\cdot\sqrt{2}-2\sqrt{2}\cdot\sqrt{3}-2\sqrt{2}\cdot\sqrt{2}$$

$$=4\cdot 3+4\sqrt{6}-2\sqrt{6}-2\cdot 2$$

$$=12+4\sqrt{6}-2\sqrt{6}-4=8+2\sqrt{6}$$

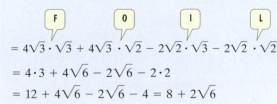

Definitions and Concepts	Examples

Section 7.5 Multiplying with More Than One Term and Rationalizing Denominators (continued)

Radical expressions that involve the sum and difference of the same two terms are called conjugates. Use

$$(A + B)(A - B) = A^2 - B^2$$

to multiply conjugates.

$$\begin{aligned}(8 + 2\sqrt{5})(8 - 2\sqrt{5}) &= 8^2 - (2\sqrt{5})^2 \\ &= 64 - 4\cdot 5 \\ &= 64 - 20 = 44\end{aligned}$$

The process of rewriting a radical expression as an equivalent expression without any radicals in the denominator is called rationalizing the denominator. When the denominator contains a single radical with an nth root, multiply the numerator and the denominator by a radical of index n that produces a perfect nth power in the denominator's radicand.

Rationalize the denominator: $\frac{7}{\sqrt[3]{2x}}$.

$$= \frac{7}{\sqrt[3]{2x}} \cdot \frac{\sqrt[3]{4x^2}}{\sqrt[3]{4x^2}} = \frac{7\sqrt[3]{4x^2}}{\sqrt[3]{8x^3}} = \frac{7\sqrt[3]{4x^2}}{2x}$$

If the denominator contains two terms, rationalize the denominator by multiplying the numerator and the denominator by the conjugate of the denominator.

$$\begin{aligned}\frac{13}{5 - \sqrt{3}} &= \frac{13}{5 - \sqrt{3}} \cdot \frac{5 + \sqrt{3}}{5 + \sqrt{3}} \\ &= \frac{13(5 + \sqrt{3})}{5^2 - (\sqrt{3})^2} \\ &= \frac{13(5 + \sqrt{3})}{25 - 3} = \frac{13(5 + \sqrt{3})}{22}\end{aligned}$$

Section 7.6 Radical Equations

A radical equation is an equation in which the variable occurs in a radicand.

Solving Radical Equations Containing *n*th Roots

1. Isolate one radical on one side of the equation.
2. Raise both sides to the nth power.
3. Solve the resulting equation.
4. Check proposed solutions in the original equation. Solutions of an equation to an even power that is radical-free, but not the original equation, are called extraneous solutions.

Solve: $\sqrt{6x + 13} - 2x = 1$.

$\sqrt{6x + 13} = 2x + 1$ — Isolate the radical.

$(\sqrt{6x + 13})^2 = (2x + 1)^2$ — Square both sides.

$6x + 13 = 4x^2 + 4x + 1$ — $(A + B)^2 = A^2 + 2AB + B^2$

$0 = 4x^2 - 2x - 12$ — Subtract $6x + 13$ from both sides.

$0 = 2(2x^2 - x - 6)$ — Factor out the GCF.

$0 = 2(2x + 3)(x - 2)$ — Factor completely.

$2x + 3 = 0$ or $x - 2 = 0$ — Set variable factors equal to zero.

$2x = -3$ $x = 2$ — Solve for x.

$x = -\frac{3}{2}$

Check both proposed solutions. 2 checks, but $-\frac{3}{2}$ is extraneous. The solution is 2 and the solution set is $\{2\}$.

Section 7.7 Complex Numbers

The imaginary unit i is defined as

$$i = \sqrt{-1}, \quad \text{where} \quad i^2 = -1.$$

The set of numbers in the form $a + bi$ is called the set of complex numbers; a is the real part and b is the imaginary part. If $b = 0$, the complex number is a real number. If $b \neq 0$, the complex number is an imaginary number.

- $\sqrt{-81} = \sqrt{81(-1)} = \sqrt{81}\sqrt{-1} = 9i$
- $\sqrt{-75} = \sqrt{75(-1)} = \sqrt{25\cdot 3}\sqrt{-1} = 5i\sqrt{3}$

Definitions and Concepts	Examples
Section 7.7 Complex Numbers (continued)	
To add or subtract complex numbers, add or subtract their real parts and add or subtract their imaginary parts.	$(2-4i)-(7-10i)$ $=2-4i-7+10i$ $=(2-7)+(-4+10)i=-5+6i$
To multiply complex numbers, multiply as if they were polynomials. After completing the multiplication, replace i^2 with -1. When performing operations with square roots of negative numbers, begin by expressing all square roots in terms of i. Then multiply.	F O I L • $(2-3i)(4+5i)=8+10i-12i-15i^2$ $=8+10i-12i-15(-1)$ $=23-2i$ • $\sqrt{-36}\cdot\sqrt{-100}=\sqrt{36(-1)}\cdot\sqrt{100(-1)}$ $=6i\cdot 10i=60i^2=60(-1)=-60$
The complex numbers $a+bi$ and $a-bi$ are conjugates. Conjugates can be multiplied using the formula $(A+B)(A-B)=A^2-B^2.$ The multiplication of conjugates results in a real number.	$(3+5i)(3-5i)=3^2-(5i)^2$ $=9-25i^2$ $=9-25(-1)=34$
To divide complex numbers, multiply the numerator and the denominator by the conjugate of the denominator in order to obtain a real number in the denominator. This real number becomes the denominator of a and b in the quotient $a+bi$.	$\frac{5+2i}{4-i}=\frac{5+2i}{4-i}\cdot\frac{4+i}{4+i}=\frac{20+5i+8i+2i^2}{16-i^2}$ $=\frac{20+13i+2(-1)}{16-(-1)}$ $=\frac{20+13i-2}{16+1}$ $=\frac{18+13i}{17}=\frac{18}{17}+\frac{13}{17}i$
To simplify powers of i, rewrite the expression in terms of i^2. Then replace i^2 with -1 and simplify.	Simplify: i^{27}. $i^{27}=i^{26}\cdot i=(i^2)^{13}i$ $=(-1)^{13}i=(-1)i=-i$

CHAPTER 7 REVIEW EXERCISES

7.1 *In Exercises 1–5, find the indicated root, or state that the expression is not a real number.*

1. $\sqrt{81}$

2. $-\sqrt{\frac{1}{100}}$

3. $\sqrt[3]{-27}$

4. $\sqrt[4]{-16}$

5. $\sqrt[5]{-32}$

In Exercises 6–7, find the indicated function values for each function. If necessary, round to two decimal places. If the function value is not a real number and does not exist, so state.

6. $f(x)=\sqrt{2x-5};\quad f(15), f(4), f\left(\frac{5}{2}\right), f(1)$

7. $g(x)=\sqrt[3]{4x-8};\quad g(4), g(0), g(-14)$

In Exercises 8–9, find the domain of each square root function.

8. $f(x)=\sqrt{x-2}$

9. $g(x)=\sqrt{100-4x}$

In Exercises 10–15, simplify each expression. Assume that each variable can represent any real number, so include absolute value bars where necessary.

10. $\sqrt{25x^2}$

11. $\sqrt{(x+14)^2}$

12. $\sqrt{x^2-8x+16}$

13. $\sqrt[3]{64x^3}$

14. $\sqrt[4]{16x^4}$

15. $\sqrt[5]{-32(x+7)^5}$

7.2 *In Exercises 16–18, use radical notation to rewrite each expression. Simplify, if possible.*

16. $(5xy)^{\frac{1}{3}}$

17. $16^{\frac{3}{2}}$

18. $32^{\frac{4}{5}}$

In Exercises 19–20, rewrite each expression with rational exponents.

19. $\sqrt{7x}$

20. $(\sqrt[3]{19xy})^5$

In Exercises 21–22, rewrite each expression with a positive rational exponent. Simplify, if possible.

21. $8^{-\frac{2}{3}}$

22. $3x(ab)^{-\frac{4}{5}}$

In Exercises 23–26, use properties of rational exponents to simplify each expression.

23. $x^{\frac{1}{3}} \cdot x^{\frac{1}{4}}$

24. $\dfrac{5^{\frac{1}{2}}}{5^{\frac{1}{3}}}$

25. $(8x^6y^3)^{\frac{1}{3}}$

25. $\left(x^{-\frac{2}{3}}y^{\frac{1}{4}}\right)^{\frac{1}{2}}$

In Exercises 27–31, use rational exponents to simplify each expression. If rational exponents appear after simplifying, write the answer in radical notation.

27. $\sqrt[3]{x^9y^{12}}$

28. $\sqrt[9]{x^3y^9}$

29. $\sqrt{x} \cdot \sqrt[3]{x}$

30. $\dfrac{\sqrt[3]{x^2}}{\sqrt[4]{x^2}}$

31. $\sqrt[5]{\sqrt[3]{x}}$

32. The function $f(x) = 350x^{\frac{2}{3}}$ models the expenditures, $f(x)$, in millions of dollars, for the U.S. National Park Service x years after 1985. According to this model, what will expenditures be in 2012?

7.3 *In Exercises 33–35, use the product rule to multiply.*

33. $\sqrt{3x} \cdot \sqrt{7y}$

34. $\sqrt[5]{7x^2} \cdot \sqrt[5]{11x}$

35. $\sqrt[6]{x-5} \cdot \sqrt[6]{(x-5)^4}$

36. If $f(x) = \sqrt{7x^2 - 14x + 7}$, express the function, f, in simplified form. Assume that x can be any real number.

In Exercises 37–39, simplify by factoring. Assume that all variables in a radicand represent positive real numbers.

37. $\sqrt{20x^3}$

38. $\sqrt[3]{54x^8y^6}$

39. $\sqrt[4]{32x^3y^{11}z^5}$

In Exercises 40–43, multiply and simplify, if possible. Assume that all variables in a radicand represent positive real numbers.

40. $\sqrt{6x^3} \cdot \sqrt{4x^2}$

41. $\sqrt[3]{4x^2y} \cdot \sqrt[3]{4xy^4}$

42. $\sqrt[5]{2x^4y^3z^4} \cdot \sqrt[5]{8xy^6z^7}$

43. $\sqrt{x+1} \cdot \sqrt{x-1}$

7.4 *Assume that all variables represent positive real numbers. In Exercises 44–47, add or subtract as indicated.*

44. $6\sqrt[3]{3} + 2\sqrt[3]{3}$

45. $5\sqrt{18} - 3\sqrt{8}$

46. $\sqrt[3]{27x^4} + \sqrt[3]{xy^6}$

47. $2\sqrt[3]{6} - 5\sqrt[3]{48}$

In Exercises 48–50, simplify using the quotient rule.

48. $\sqrt[3]{\dfrac{16}{125}}$

49. $\sqrt{\dfrac{x^3}{100y^4}}$

50. $\sqrt[4]{\dfrac{3y^5}{16x^{20}}}$

In Exercises 51–54, divide and, if possible, simplify.

51. $\dfrac{\sqrt{48}}{\sqrt{2}}$

52. $\dfrac{\sqrt[3]{32}}{\sqrt[3]{2}}$

53. $\dfrac{\sqrt[4]{64x^7}}{\sqrt[4]{2x^2}}$

54. $\dfrac{\sqrt{200x^3y^2}}{\sqrt{2x^{-2}y}}$

7.5 *Assume that all variables represent positive real numbers.*

In Exercises 55–62, multiply as indicated. If possible, simplify any radical expressions that appear in the product.

55. $\sqrt{3}(2\sqrt{6} + 4\sqrt{15})$

56. $\sqrt[3]{5}(\sqrt[3]{50} - \sqrt[3]{2})$

57. $(\sqrt{7} - 3\sqrt{5})(\sqrt{7} + 6\sqrt{5})$

58. $(\sqrt{x} - \sqrt{11})(\sqrt{y} - \sqrt{11})$

59. $(\sqrt{5} + \sqrt{8})^2$

60. $(2\sqrt{3} - \sqrt{10})^2$

61. $(\sqrt{7} + \sqrt{13})(\sqrt{7} - \sqrt{13})$

62. $(7 - 3\sqrt{5})(7 + 3\sqrt{5})$

In Exercises 63–75, rationalize each denominator. Simplify, if possible.

63. $\dfrac{4}{\sqrt{6}}$

64. $\sqrt{\dfrac{2}{7}}$

65. $\dfrac{12}{\sqrt[3]{9}}$

66. $\sqrt{\dfrac{2x}{5y}}$

67. $\dfrac{14}{\sqrt[3]{2x^2}}$

68. $\sqrt[4]{\dfrac{7}{3x}}$

69. $\dfrac{5}{\sqrt[5]{32x^4y}}$

70. $\dfrac{6}{\sqrt{3} - 1}$

71. $\dfrac{\sqrt{7}}{\sqrt{5} + \sqrt{3}}$

72. $\dfrac{10}{2\sqrt{5} - 3\sqrt{2}}$

73. $\dfrac{\sqrt{x} + 5}{\sqrt{x} - 3}$

74. $\dfrac{\sqrt{7} + \sqrt{3}}{\sqrt{7} - \sqrt{3}}$

75. $\dfrac{2\sqrt{3} + \sqrt{6}}{2\sqrt{6} + \sqrt{3}}$

In Exercises 76–79, rationalize each numerator. Simplify, if possible.

76. $\sqrt{\dfrac{2}{7}}$

77. $\dfrac{\sqrt[3]{3x}}{\sqrt[3]{y}}$

78. $\dfrac{\sqrt{7}}{\sqrt{5} + \sqrt{3}}$

79. $\dfrac{\sqrt{7} + \sqrt{3}}{\sqrt{7} - \sqrt{3}}$

7.6 *In Exercises 80–84, solve each radical equation.*

80. $\sqrt{2x+4} = 6$

81. $\sqrt{x-5} + 9 = 4$

82. $\sqrt{2x-3} + x = 3$

83. $\sqrt{x-4} + \sqrt{x+1} = 5$

84. $(x^2 + 6x)^{\frac{1}{3}} + 2 = 0$

85. In 2007, state tobacco taxes ranged from \$0.07 per pack in South Carolina to \$2.58 in New Jersey. The graph shows the average state cigarette tax per pack from 2001 through 2007.

Average State Cigarette Tax per Pack

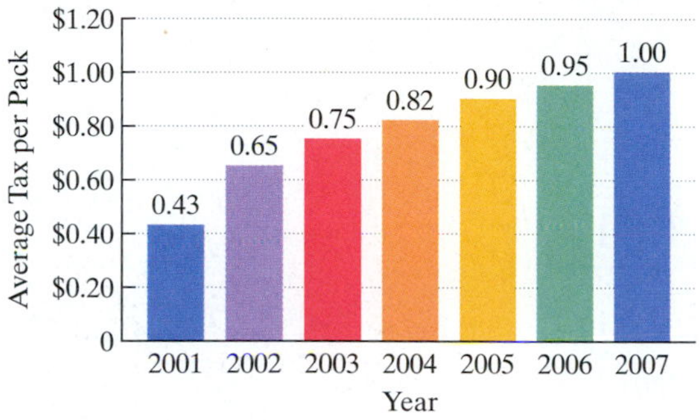

Source: Campaign for Tobacco-Free Kids

The function

$$f(x) = 0.23\sqrt{x} + 0.43$$

models the average state cigarette tax per pack, $f(x)$, x years after 2001.

a. Find and interpret $f(6)$. Round to two decimal places. Does this underestimate or overestimate the tax displayed by the graph? By how much?

b. According to the model, in which year will the average state cigarette tax be \$1.12 per pack?

86. Out of a group of 50,000 births, the number of people, $f(x)$, surviving to age x is modeled by the function

$$f(x) = 5000\sqrt{100 - x}.$$

To what age will 20,000 people in the group survive?

7.7 *In Exercises 87–89, express each number in terms of i and simplify, if possible.*

87. $\sqrt{-81}$

88. $\sqrt{-63}$

89. $-\sqrt{-8}$

In Exercises 90–99, perform the indicated operation. Write the result in the form a + bi.

90. $(7 + 12i) + (5 - 10i)$

91. $(8 - 3i) - (17 - 7i)$

92. $4i(3i - 2)$

93. $(7 - 5i)(2 + 3i)$

94. $(3 - 4i)^2$

95. $(7 + 8i)(7 - 8i)$

96. $\sqrt{-8} \cdot \sqrt{-3}$

97. $\dfrac{6}{5+i}$

98. $\dfrac{3+4i}{4-2i}$

99. $\dfrac{5+i}{3i}$

In Exercises 100–101, simplify each expression.

100. i^{16}

101. i^{23}

CHAPTER 7 TEST

Remember to use your Chapter Test Prep Video CD to see the worked-out solutions to the test questions you want to review.

1. Let $f(x) = \sqrt{8 - 2x}$.

a. Find $f(-14)$.

b. Find the domain of f.

2. Evaluate: $27^{-\frac{4}{3}}$.

3. Simplify: $\left(25x^{-\frac{1}{2}}y^{\frac{1}{4}}\right)^{\frac{1}{2}}$.

In Exercises 4–5, use rational exponents to simplify each expression. If rational exponents appear after simplifying, write the answer in radical notation.

4. $\sqrt[8]{x^4}$

5. $\sqrt[4]{x} \cdot \sqrt[5]{x}$

In Exercises 6–9, simplify each expression. Assume that each variable can represent any real number.

6. $\sqrt{75x^2}$

7. $\sqrt{x^2 - 10x + 25}$

8. $\sqrt[3]{16x^4y^8}$

9. $\sqrt[5]{-\dfrac{32}{x^{10}}}$

In Exercises 10–17, perform the indicated operation and, if possible, simplify. Assume that all variables represent positive real numbers.

10. $\sqrt[3]{5x^2} \cdot \sqrt[3]{10y}$

11. $\sqrt[4]{8x^3y} \cdot \sqrt[4]{4xy^2}$

12. $3\sqrt{18} - 4\sqrt{32}$

13. $\sqrt[3]{8x^4} + \sqrt[3]{xy^6}$

14. $\dfrac{\sqrt[3]{16x^8}}{\sqrt[3]{2x^4}}$

15. $\sqrt{3}(4\sqrt{6} - \sqrt{5})$

16. $(5\sqrt{6} - 2\sqrt{2})(\sqrt{6} + \sqrt{2})$

17. $(7 - \sqrt{3})^2$

In Exercises 18–20, rationalize each denominator. Simplify, if possible. Assume all variables represent positive real numbers.

18. $\sqrt{\dfrac{5}{x}}$

19. $\dfrac{5}{\sqrt[3]{5x^2}}$

20. $\dfrac{\sqrt{2} - \sqrt{3}}{\sqrt{2} + \sqrt{3}}$

In Exercises 21–23, solve each radical equation.

21. $3 + \sqrt{2x - 3} = x$

22. $\sqrt{x + 9} - \sqrt{x - 7} = 2$

23. $(11x + 6)^{\frac{1}{3}} + 3 = 0$

24. The function

$$f(x) = 2.9\sqrt{x} + 20.1$$

models the average height, $f(x)$, in inches, of boys who are x months of age, $0 \le x \le 60$. Find the age at which the average height of boys is 40.4 inches.

25. Express in terms of i and simplify: $\sqrt{-75}$.

In Exercises 26–29, perform the indicated operation. Write the result in the form $a + bi$.

26. $(5 - 3i) - (6 - 9i)$

27. $(3 - 4i)(2 + 5i)$

28. $\sqrt{-9} \cdot \sqrt{-4}$

29. $\dfrac{3 + i}{1 - 2i}$

30. Simplify: i^{35}.

CUMULATIVE REVIEW EXERCISES (CHAPTERS 1–7)

In Exercises 1–5, solve each equation, inequality, or system.

1.
$$\begin{aligned} 2x - y + z &= -5 \\ x - 2y - 3z &= 6 \\ x + y - 2z &= 1 \end{aligned}$$

2. $3x^2 - 11x = 4$

3. $2(x + 4) < 5x + 3(x + 2)$

4. $\dfrac{1}{x + 2} + \dfrac{15}{x^2 - 4} = \dfrac{5}{x - 2}$

5. $\sqrt{x + 2} - \sqrt{x + 1} = 1$

6. Graph the solution set of the system:

$$\begin{aligned} x + 2y &< 2 \\ 2y - x &> 4. \end{aligned}$$

In Exercises 7–15, perform the indicated operations.

7. $\dfrac{8x^2}{3x^2 - 12} \div \dfrac{40}{x - 2}$

8. $\dfrac{x + \dfrac{1}{y}}{y + \dfrac{1}{x}}$

9. $(2x - 3)(4x^2 - 5x - 2)$

10. $\dfrac{7x}{x^2 - 2x - 15} - \dfrac{2}{x - 5}$

11. $7(8 - 10)^3 - 7 + 3 \div (-3)$

12. $\sqrt{80x} - 5\sqrt{20x} + 2\sqrt{45x}$

13. $\dfrac{\sqrt{3} - 2}{2\sqrt{3} + 5}$

14. $(2x^3 - 3x^2 + 3x - 4) \div (x - 2)$

15. $\left(2\sqrt{3} + 5\sqrt{2}\right)\left(\sqrt{3} - 4\sqrt{2}\right)$

In Exercises 16–17, factor completely.

16. $24x^2 + 10x - 4$

17. $16x^4 - 1$

18. The amount of light provided by a light bulb varies inversely as the square of the distance from the bulb. The illumination provided is 120 lumens at a distance of 10 feet. How many lumens are provided at a distance of 15 feet?

19. You invested \$6000 in two accounts paying 7% and 9% annual interest. At the end of the year, the total interest from these investments was \$510. How much was invested at each rate?

20. Although there are 2332 students enrolled in the college, this is 12% fewer students than there were enrolled last year. How many students were enrolled last year?

CHAPTER 8

Quadratic Equations and Functions

We are surrounded by evidence that the world is profoundly mathematical. After turning a somersault, a diver's path can be modeled by a quadratic function, $f(x) = ax^2 + bx + c$, as can the path of a football tossed from quarterback to receiver or the path of a flipped coin. Even if you throw an object directly upward, although its path is straight and vertical, its changing height over time is described by a quadratic function. And tailgaters beware: whether you're driving a car, a motorcycle, or a truck on dry or wet roads, an array of quadratic functions that model your required stopping distances at various speeds are available to help you become a safer driver.

The quadratic functions surrounding our long history of throwing things appear throughout the chapter, including Example 6 in Section 8.3 and Example 5 in Section 8.5. Tailgaters should pay close attention to the Section 8.5 opener, Exercises 73–74 and 88–89 in Exercise Set 8.5, and Exercises 30–31 in the Chapter Review Exercises.

SECTION 8.1

The Square Root Property and Completing the Square

Objectives

1. Solve quadratic equations using the square root property.
2. Complete the square of a binomial.
3. Solve quadratic equations by completing the square.
4. Solve problems using the square root property.

"In the future there will be two kinds of corporations; those that go global, and those that go bankrupt."

C. Michael Armstrong, CEO, AT&T

For better or worse, ours is the era of the multinational corporation. New technology that the multinational corporations control is expanding their power. And their numbers are growing fast. There were approximately 300 multinationals in 1900. By 1970, there were close to 7000, and by 1990 the number had swelled to 30,000. In 2001, more than 65,000 multinational corporations enveloped the world.

In this section, you will learn two new methods for solving quadratic equations. These methods are called the *square root method* and *completing the square*. Using these techniques, you will explore the growth of multinational corporations and make predictions about the number of global corporations in the future.

Study Tip

Here is a summary of what we already know about quadratic equations and quadratic functions.

1. A **quadratic equation** in x can be written in the standard form

$$ax^2 + bx + c = 0, \quad a \neq 0.$$

2. Some quadratic equations can be solved by factoring.

Solve: $2x^2 + 7x - 4 = 0.$

$$(2x - 1)(x + 4) = 0$$

$$2x - 1 = 0 \quad \text{or} \quad x + 4 = 0$$

$$2x = 1 \qquad\qquad x = -4$$

$$x = \tfrac{1}{2}$$

The solutions are -4 and $\frac{1}{2}$, and the solution set is $\left\{-4, \frac{1}{2}\right\}$.

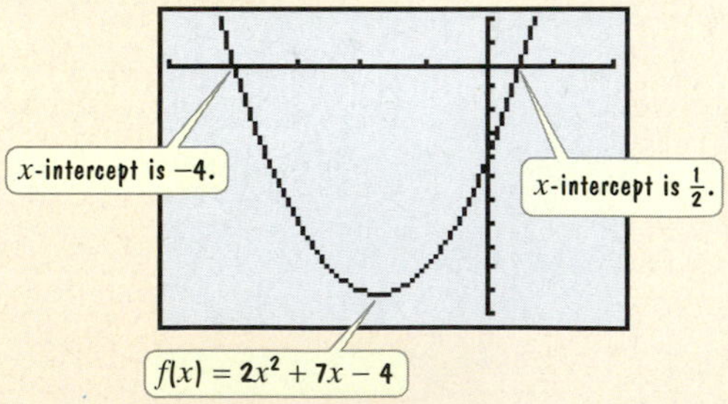

FIGURE 8.1

3. A polynomial function of the form

$$f(x) = ax^2 + bx + c, \quad a \neq 0$$

is a **quadratic function**. Graphs of quadratic functions are shaped like bowls or inverted bowls, with the same behavior at each end.

4. The real solutions of $ax^2 + bx + c = 0$ correspond to the x-intercepts for the graph of the quadratic function $f(x) = ax^2 + bx + c$. For example, the solutions of the equation $2x^2 + 7x - 4 = 0$ are -4 and $\frac{1}{2}$. **Figure 8.1** shows that the solutions appear as x-intercepts on the graph of the quadratic function $f(x) = 2x^2 + 7x - 4$.

Now that we've summarized what we know, let's look at where we go. How do we solve a quadratic equation, $ax^2 + bx + c = 0$, if the trinomial $ax^2 + bx + c$ cannot be factored? Methods other than factoring are needed. In this section, we look at other ways of solving quadratic equations.

1 Solve quadratic equations using the square root property.

The Square Root Property

Let's begin with a relatively simple quadratic equation:

$$x^2 = 9.$$

The value of x must be a number whose square is 9. There are two numbers whose square is 9:

$$x = \sqrt{9} = 3 \quad \text{or} \quad x = -\sqrt{9} = -3.$$

Thus, the solutions of $x^2 = 9$ are 3 and -3. This is an example of the **square root property**.

Discover for Yourself

Solve $x^2 = 9$, or

$$x^2 - 9 = 0,$$

by factoring. What is the advantage of using the square root property?

The Square Root Property

If u is an algebraic expression and d is a nonzero real number, then $u^2 = d$ is equivalent to $u = \sqrt{d}$ or $u = -\sqrt{d}$:

$$\text{If } u^2 = d, \quad \text{then } u = \sqrt{d} \text{ or } u = -\sqrt{d}.$$

Equivalently,

$$\text{If } u^2 = d, \quad \text{then } u = \pm\sqrt{d}.$$

Notice that $u = \pm\sqrt{d}$ is a shorthand notation to indicate that $u = \sqrt{d}$ or $u = -\sqrt{d}$. Although we usually read $u = \pm\sqrt{d}$ as "u equals plus or minus the square root of d," we actually mean that u is the positive square root of d or the negative square root of d.

EXAMPLE 1 Solving a Quadratic Equation by the Square Root Property

Solve: $3x^2 = 18$.

Solution To apply the square root property, we need a squared expression by itself on one side of the equation.

$$3x^2 = 18$$

We want x^2 by itself.

We can get x^2 by itself if we divide both sides by 3.

$$3x^2 = 18 \quad \text{This is the original equation.}$$

$$\frac{3x^2}{3} = \frac{18}{3} \quad \text{Divide both sides by 3.}$$

$$x^2 = 6 \quad \text{Simplify.}$$

$$x = \sqrt{6} \quad \text{or} \quad x = -\sqrt{6} \quad \text{Apply the square root property.}$$

Now let's check these proposed solutions in the original equation. Because the equation has an x^2-term and no x-term, we can check both values, $\pm\sqrt{6}$, at once.

Check $\sqrt{6}$ and $-\sqrt{6}$:

$$3x^2 = 18 \quad \text{This is the original equation.}$$
$$3(\pm\sqrt{6})^2 \stackrel{?}{=} 18 \quad \text{Substitute the proposed solutions.}$$
$$3 \cdot 6 \stackrel{?}{=} 18 \quad (\pm\sqrt{6})^2 = 6$$
$$18 = 18, \quad \text{true}$$

The solutions are $-\sqrt{6}$ and $\sqrt{6}$. The solution set is $\{-\sqrt{6}, \sqrt{6}\}$ or $\{\pm\sqrt{6}\}$.

CHECK POINT 1 Solve: $4x^2 = 28$.

In this section, we will express irrational solutions in simplified radical form, rationalizing denominators when possible.

EXAMPLE 2 Solving a Quadratic Equation by the Square Root Property

Solve: $2x^2 - 7 = 0$.

Solution To solve by the square root property, we isolate the squared expression on one side of the equation.

$$2x^2 - 7 = 0$$

We want x^2 by itself.

$$2x^2 - 7 = 0 \quad \text{This is the original equation.}$$
$$2x^2 = 7 \quad \text{Add 7 to both sides.}$$
$$x^2 = \frac{7}{2} \quad \text{Divide both sides by 2.}$$
$$x = \sqrt{\frac{7}{2}} \quad \text{or} \quad x = -\sqrt{\frac{7}{2}} \quad \text{Apply the square root property.}$$

Because the proposed solutions are opposites, we can rationalize both denominators at once:

$$\pm\sqrt{\frac{7}{2}} = \pm\frac{\sqrt{7}}{\sqrt{2}} \cdot \frac{\sqrt{2}}{\sqrt{2}} = \pm\frac{\sqrt{14}}{2}.$$

Substitute these values into the original equation and verify that the solutions are $-\frac{\sqrt{14}}{2}$ and $\frac{\sqrt{14}}{2}$. The solution set is $\left\{-\frac{\sqrt{14}}{2}, \frac{\sqrt{14}}{2}\right\}$ or $\left\{\pm\frac{\sqrt{14}}{2}\right\}$.

CHECK POINT 2 Solve: $3x^2 - 11 = 0$.

Some quadratic equations have solutions that are imaginary numbers.

EXAMPLE 3 Solving a Quadratic Equation by the Square Root Property

Solve: $9x^2 + 25 = 0$.

Solution We begin by isolating the squared expression on one side of the equation.

$$9x^2 + 25 = 0$$

We need to isolate x^2.

$9x^2 + 25 = 0$	This is the original equation.
$9x^2 = -25$	Subtract 25 from both sides.
$x^2 = -\frac{25}{9}$	Divide both sides by 9.
$x = \sqrt{-\frac{25}{9}}$ or $x = -\sqrt{-\frac{25}{9}}$	Apply the square root property.
$x = \sqrt{\frac{25}{9}}\sqrt{-1}$ $\quad$ $x = -\sqrt{\frac{25}{9}}\sqrt{-1}$	
$x = \frac{5}{3}i$ $\quad$ $x = -\frac{5}{3}i$	$\sqrt{-1} = i$

Using Technology

Graphic Connections

The graph of

$$f(x) = 9x^2 + 25$$

has no x-intercepts. This shows that

$$9x^2 + 25 = 0$$

has no real solutions. Example 3 algebraically establishes that the solutions are imaginary numbers.

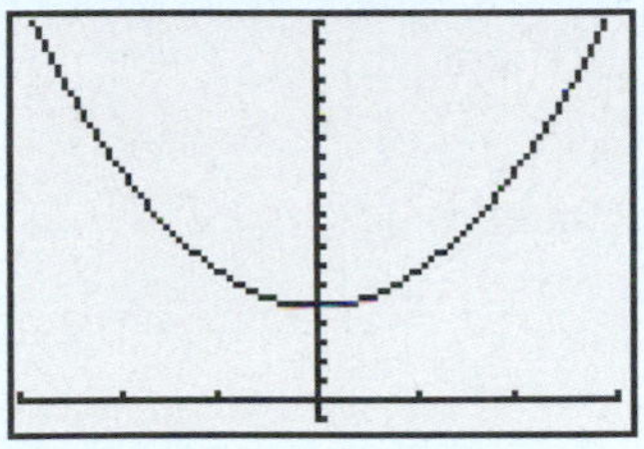

$[-3, 3, 1]$ by $[-5, 100, 5]$

Because the equation has an x^2-term and no x-term, we can check both proposed solutions, $\pm\frac{5}{3}i$, at once.

Check $\frac{5}{3}i$ and $-\frac{5}{3}i$:

$9x^2 + 25 = 0$	This is the original equation.
$9\left(\pm\frac{5}{3}i\right)^2 + 25 \stackrel{?}{=} 0$	Substitute the proposed solutions.
$9\left(\frac{25}{9}i^2\right) + 25 \stackrel{?}{=} 0$	$\left(\pm\frac{5}{3}i\right)^2 = \left(\pm\frac{5}{3}\right)^2 i^2 = \frac{25}{9}i^2$
$25i^2 + 25 \stackrel{?}{=} 0$ ($i^2 = -1$)	$9 \cdot \frac{25}{9} = 25$
$25(-1) + 25 \stackrel{?}{=} 0$	Replace i^2 with -1.
$0 = 0$,	true

The solutions are $-\frac{5}{3}i$ and $\frac{5}{3}i$. The solution set is $\left\{-\frac{5}{3}i, \frac{5}{3}i\right\}$ or $\left\{\pm\frac{5}{3}i\right\}$.

✓ **CHECK POINT 3** Solve: $4x^2 + 9 = 0$.

Can we solve an equation such as $(x - 1)^2 = 5$ using the square root property? Yes. The equation is in the form $u^2 = d$, where u^2, the squared expression, is by itself on the left side.

$$(x - 1)^2 = 5$$

This is u^2 in $u^2 = d$ with $u = x - 1$.

This is d in $u^2 = d$ with $d = 5$.

Discover for Yourself

Try solving

$$(x - 1)^2 = 5$$

by writing the equation in standard form and factoring. What problem do you encounter?

EXAMPLE 4 Solving a Quadratic Equation by the Square Root Property

Solve by the square root property: $(x - 1)^2 = 5$.

Solution

$(x - 1)^2 = 5$	This is the original equation.
$x - 1 = \sqrt{5}$ or $x - 1 = -\sqrt{5}$	Apply the square root property.
$x = 1 + \sqrt{5}$ $\quad$ $x = 1 - \sqrt{5}$	Add 1 to both sides in each equation.

Check $1 + \sqrt{5}$:

$$(x-1)^2 = 5$$
$$(1+\sqrt{5}-1)^2 \stackrel{?}{=} 5$$
$$(\sqrt{5})^2 \stackrel{?}{=} 5$$
$$5 = 5, \quad \text{true}$$

Check $1 - \sqrt{5}$:

$$(x-1)^2 = 5$$
$$(1-\sqrt{5}-1)^2 \stackrel{?}{=} 5$$
$$(-\sqrt{5})^2 \stackrel{?}{=} 5$$
$$5 = 5, \quad \text{true}$$

The solutions are $1 \pm \sqrt{5}$, and the solution set is $\{1+\sqrt{5}, 1-\sqrt{5}\}$ or $\{1 \pm \sqrt{5}\}$.

✓ CHECK POINT 4 Solve: $(x-3)^2 = 10$.

2 Complete the square of a binomial.

Completing the Square

We return to the question that opened this section: How do we solve a quadratic equation, $ax^2 + bx + c = 0$, if the trinomial $ax^2 + bx + c$ cannot be factored? We can convert the equation into an equivalent equation that can be solved using the square root property. This is accomplished by **completing the square**.

Completing the Square

If $x^2 + bx$ is a binomial, then by adding $\left(\frac{b}{2}\right)^2$, which is the square of half the coefficient of x, a perfect square trinomial will result.

The coefficient of x^2 must be 1 to complete the square.

$$x^2 + bx + \left(\frac{b}{2}\right)^2 = \left(x + \frac{b}{2}\right)^2.$$

EXAMPLE 5 Completing the Square

What term should be added to each binomial so that it becomes a perfect square trinomial? Write and factor the trinomial.

a. $x^2 + 8x$ **b.** $x^2 - 7x$ **c.** $x^2 + \frac{3}{5}x$

Solution To complete the square, we must add a term to each binomial. The term that should be added is the square of half the coefficient of x.

$x^2 + 8x$: Add $\left(\frac{8}{2}\right)^2 = 4^2$. Add 16 to complete the square.

$x^2 - 7x$: Add $\left(\frac{-7}{2}\right)^2$, or $\frac{49}{4}$, to complete the square.

$x^2 + \frac{3}{5}x$: Add $\left(\frac{1}{2} \cdot \frac{3}{5}\right)^2 = \left(\frac{3}{10}\right)^2$. Add $\frac{9}{100}$ to complete the square.

a. The coefficient of the x-term in $x^2 + 8x$ is 8. Half of 8 is 4, and $4^2 = 16$. Add 16. The result is a perfect square trinomial.

$$x^2 + 8x + 16 = (x+4)^2$$

b. The coefficient of the x-term in $x^2 - 7x$ is -7. Half of -7 is $-\frac{7}{2}$, and $\left(-\frac{7}{2}\right)^2 = \frac{49}{4}$. Add $\frac{49}{4}$. The result is a perfect square trinomial.

$$x^2 - 7x + \frac{49}{4} = \left(x - \frac{7}{2}\right)^2$$

c. The coefficient of the x-term in $x^2 + \frac{3}{5}x$ is $\frac{3}{5}$. Half of $\frac{3}{5}$ is $\frac{1}{2} \cdot \frac{3}{5}$, or $\frac{3}{10}$, and $\left(\frac{3}{10}\right)^2 = \frac{9}{100}$. Add $\frac{9}{100}$. The result is a perfect square trinomial.

$$x^2 + \frac{3}{5}x + \frac{9}{100} = \left(x + \frac{3}{10}\right)^2$$

Study Tip

You may not be accustomed to factoring perfect square trinomials in which fractions are involved. The constant in the factorization is always half the coefficient of x.

$$x^2 - 7x + \frac{49}{4} = \left(x - \frac{7}{2}\right)^2$$

Half the coefficient of x, -7, is $-\frac{7}{2}$.

$$x^2 + \frac{3}{5}x + \frac{9}{100} = \left(x + \frac{3}{10}\right)^2$$

Half the coefficient of x, $\frac{3}{5}$, is $\frac{3}{10}$.

CHECK POINT 5 What term should be added to each binomial so that it becomes a perfect square trinomial? Write and factor the trinomial.

a. $x^2 + 10x$

b. $x^2 - 3x$

c. $x^2 + \frac{3}{4}x$.

3 Solve quadratic equations by completing the square.

Solving Quadratic Equations by Completing the Square

We can solve *any* quadratic equation by completing the square. If the coefficient of the x^2-term is one, we add the square of half the coefficient of x to both sides of the equation. **When you add a constant term to one side of the equation to complete the square, be certain to add the same constant to the other side of the equation.** These ideas are illustrated in Example 6.

EXAMPLE 6 Solving a Quadratic Equation by Completing the Square

Solve by completing the square: $x^2 - 6x + 4 = 0$.

Solution We begin by subtracting 4 from both sides. This is done to isolate the binomial $x^2 - 6x$ so that we can complete the square.

$x^2 - 6x + 4 = 0$ This is the original equation.

$x^2 - 6x = -4$ Subtract 4 from both sides.

Next, we work with $x^2 - 6x = -4$ and complete the square. Find half the coefficient of the x-term and square it. The coefficient of the x-term is -6. Half of -6 is -3 and $(-3)^2 = 9$. Thus, we add 9 to both sides of the equation.

$x^2 - 6x + 9 = -4 + 9$ Add 9 to both sides to complete the square.

$(x - 3)^2 = 5$ Factor and simplify.

$x - 3 = \sqrt{5}$ or $x - 3 = -\sqrt{5}$ Apply the square root property.

$x = 3 + \sqrt{5}$ $\quad$ $x = 3 - \sqrt{5}$ Add 3 to both sides in each equation.

The solutions are $3 \pm \sqrt{5}$, and the solution set is $\{3 + \sqrt{5}, 3 - \sqrt{5}\}$ or $\{3 \pm \sqrt{5}\}$.

Study Tip

When you complete the square for the binomial expression $x^2 + bx$, you obtain a different polynomial. When you solve a quadratic equation by completing the square, you obtain an equation with the same solution set because the constant needed to complete the square is added to *both sides*.

If you solve a quadratic equation by completing the square and the solutions are rational numbers, the equation can also be solved by factoring. By contrast, quadratic equations with irrational solutions cannot be solved by factoring. However, all quadratic equations can be solved by completing the square.

✓ **CHECK POINT 6** Solve by completing the square: $x^2 + 4x - 1 = 0$.

We have seen that the leading coefficient must be 1 in order to complete the square. If the coefficient of the x^2-term in a quadratic equation is not 1, you must divide each side of the equation by this coefficient before completing the square. For example, to solve $9x^2 - 6x - 4 = 0$ by completing the square, first divide every term by 9:

$$\frac{9x^2}{9} - \frac{6x}{9} - \frac{4}{9} = \frac{0}{9}$$
$$x^2 - \frac{6}{9}x - \frac{4}{9} = 0$$
$$x^2 - \frac{2}{3}x - \frac{4}{9} = 0.$$

Now that the coefficient of the x^2-term is 1, we can solve by completing the square.

EXAMPLE 7 Solving a Quadratic Equation by Completing the Square

Solve by completing the square: $9x^2 - 6x - 4 = 0$.

Solution

$$9x^2 - 6x - 4 = 0$$ This is the original equation.

$$x^2 - \frac{2}{3}x - \frac{4}{9} = 0$$ Divide both sides by 9.

$$x^2 - \frac{2}{3}x = \frac{4}{9}$$ Add $\frac{4}{9}$ to both sides to isolate the binomial.

$$x^2 - \frac{2}{3}x + \frac{1}{9} = \frac{4}{9} + \frac{1}{9}$$ Complete the square: Half of $-\frac{2}{3}$ is $-\frac{2}{6}$, or $-\frac{1}{3}$, and $\left(-\frac{1}{3}\right)^2 = \frac{1}{9}$.

$$\left(x - \frac{1}{3}\right)^2 = \frac{5}{9}$$ Factor and simplify.

$$x - \frac{1}{3} = \sqrt{\frac{5}{9}} \quad \text{or} \quad x - \frac{1}{3} = -\sqrt{\frac{5}{9}}$$ Apply the square root property.

$$x - \frac{1}{3} = \frac{\sqrt{5}}{3} \qquad x - \frac{1}{3} = -\frac{\sqrt{5}}{3}$$ $\sqrt{\frac{5}{9}} = \frac{\sqrt{5}}{\sqrt{9}} = \frac{\sqrt{5}}{3}$

$$x = \frac{1}{3} + \frac{\sqrt{5}}{3} \qquad x = \frac{1}{3} - \frac{\sqrt{5}}{3}$$ Add $\frac{1}{3}$ to both sides and solve for x.

$$x = \frac{1 + \sqrt{5}}{3} \qquad x = \frac{1 - \sqrt{5}}{3}$$ Express solutions with a common denominator.

The solutions are $\frac{1 \pm \sqrt{5}}{3}$ and the solution set is $\left\{\frac{1 \pm \sqrt{5}}{3}\right\}$.

Using Technology

Graphic Connections

Obtain a decimal approximation for each solution of

$$9x^2 - 6x - 4 = 0,$$

the equation in Example 7.

$$\frac{1 + \sqrt{5}}{3} \approx 1.1$$
$$\frac{1 - \sqrt{5}}{3} \approx -0.4$$

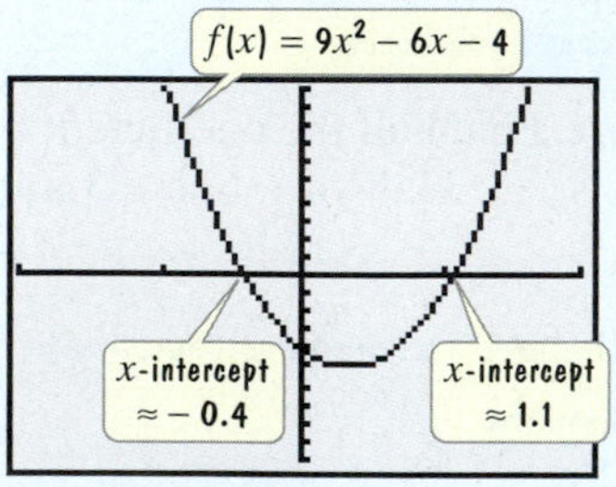

[−2, 2, 1] by [−10, 10, 1]

The x-intercepts of $f(x) = 9x^2 - 6x - 4$ verify the solutions.

✓ **CHECK POINT 7** Solve by completing the square: $2x^2 + 3x - 4 = 0$.

4 Solve problems using the square root property.

Applications

We all want a wonderful life with fulfilling work, good health, and loving relationships. And let's be honest: Financial security wouldn't hurt! Achieving this goal depends on understanding how money in a savings account grows in remarkable ways as a result of *compound interest*. **Compound interest** is interest computed on your original investment as well as on any accumulated interest. For example, suppose you deposit \$1000, the principal, in a savings account at a rate of 5%. **Table 8.1** shows how the investment grows if the interest earned is automatically added on to the principal.

Table 8.1 Compound Interest on \$1000

Year	Starting Balance	Interest Earned: $I = Pr$	New Balance
1	\$1000	$\$1000 \times 0.05 = \$50$	\$1050
2	\$1050	$\$1050 \times 0.05 = \$52.50$	\$1102.50
3	\$1102.50	$\$1102.50 \times 0.05 \approx \$55.13$	\$1157.63

A faster way to determine the amount, A, in an account subject to compound interest is to use the following formula:

A Formula for Compound Interest

Suppose that an amount of money, P, is invested at interest rate r, compounded annually. In t years, the amount, A, or balance, in the account is given by the formula

$$A = P(1 + r)^t.$$

Some compound interest problems can be solved using quadratic equations.

EXAMPLE 8 Solving a Compound Interest Problem

You invested \$1000 in an account whose interest is compounded annually. After 2 years, the amount, or balance, in the account is \$1210. Find the annual interest rate.

Solution We are given that

P (the amount invested) = \$1000

t (the time of the investment) = 2 years

A (the amount, or balance, in the account) = \$1210.

We are asked to find the annual interest rate, r. We substitute the three given values into the compound interest formula and solve for r.

$$A = P(1 + r)^t \quad \text{Use the compound interest formula.}$$

$$1210 = 1000(1 + r)^2 \quad \text{Substitute the given values.}$$

$$\frac{1210}{1000} = (1 + r)^2 \quad \text{Divide both sides by 1000.}$$

$$\frac{121}{100} = (1 + r)^2 \quad \text{Simplify the fraction.}$$

$$1 + r = \sqrt{\frac{121}{100}} \quad \text{or} \quad 1 + r = -\sqrt{\frac{121}{100}} \quad \text{Apply the square root property.}$$

$$1 + r = \frac{11}{10} \qquad 1 + r = -\frac{11}{10} \qquad \sqrt{\frac{121}{100}} = \frac{\sqrt{121}}{\sqrt{100}} = \frac{11}{10}$$

$$r = \frac{11}{10} - 1 \qquad r = -\frac{11}{10} - 1 \qquad \text{Subtract 1 from both sides and solve for } r.$$

$$r = \frac{1}{10} \qquad r = -\frac{21}{10} \qquad \frac{11}{10} - 1 = \frac{11}{10} - \frac{10}{10} = \frac{1}{10} \text{ and } -\frac{11}{10} - 1 = -\frac{11}{10} - \frac{10}{10} = -\frac{21}{10}.$$

Because the interest rate cannot be negative, we reject $-\frac{21}{10}$. Thus, the annual interest rate is $\frac{1}{10} = 0.10 = 10\%$.

We can check this answer using the formula $A = P(1 + r)^t$. If \$1000 is invested for 2 years at 10% interest, compounded annually, the balance in the account is

$$A = \$1000(1 + 0.10)^2 = \$1000(1.10)^2 = \$1210.$$

Because this is precisely the amount given by the problem's conditions, the annual interest rate is, indeed, 10% compounded annually.

CHECK POINT 8 You invested \$3000 in an account whose interest is compounded annually. After 2 years, the amount, or balance, in the account is \$4320. Find the annual interest rate.

In Chapter 5, we solved problems using the Pythagorean Theorem. Recall that in a right triangle, the side opposite the 90° angle is the hypotenuse and the other sides are legs. The Pythagorean Theorem states that the sum of the squares of the lengths of the legs equals the square of the length of the hypotenuse. Some problems that involve the Pythagorean Theorem can be solved using the square root property.

EXAMPLE 9 Using the Pythagorean Theorem and the Square Root Property

a. A wheelchair ramp with a length of 122 inches has a horizontal distance of 120 inches. What is the ramp's vertical distance?

b. Construction laws are very specific when it comes to access ramps for the disabled. Every vertical rise of 1 inch requires a horizontal run of 12 inches. Does this ramp satisfy the requirement?

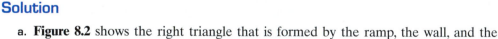

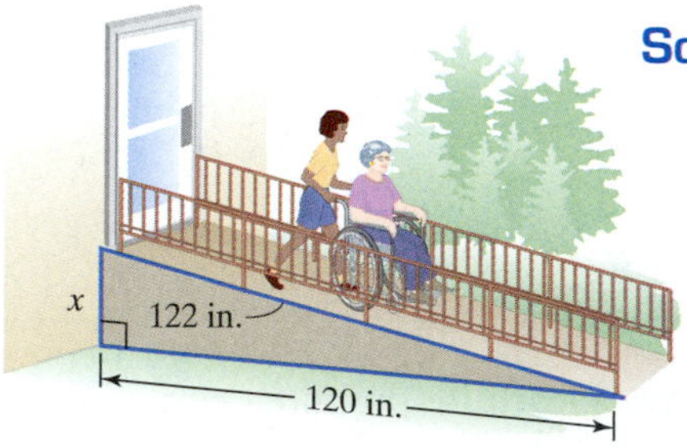

FIGURE 8.2

Solution

a. **Figure 8.2** shows the right triangle that is formed by the ramp, the wall, and the ground. We can find x, the ramp's vertical distance, using the Pythagorean Theorem.

(leg)²	plus	(leg)²	equals	(hypotenuse)²
x^2	$+$	120^2	$=$	122^2

We solve this equation using the square root property.

$x^2 + 120^2 = 122^2$ — This is the equation resulting from the Pythagorean Theorem.

$x^2 + 14{,}400 = 14{,}884$ — Square 120 and 122.

$x^2 = 484$ — Isolate x^2 by subtracting 14,400 from both sides.

$x = \sqrt{484}$ or $x = -\sqrt{484}$ — Apply the square root property.

$x = 22$ $\qquad$ $x = -22$

Because x represents the ramp's vertical distance, we reject the negative value. Thus, the ramp's vertical distance is 22 inches.

b. Every vertical rise of 1 inch requires a horizontal run of 12 inches. Because the ramp has a vertical distance of 22 inches, it requires a horizontal distance of 22(12) inches, or 264 inches. The horizontal distance is only 120 inches, so this ramp does not satisfy construction laws for access ramps for the disabled.

CHECK POINT 9 A 50-foot supporting wire is to be attached to an antenna. The wire is anchored 20 feet from the base of the antenna. How high up the antenna is the wire attached? Express the answer in simplified radical form. Then find a decimal approximation to the nearest tenth of a foot.

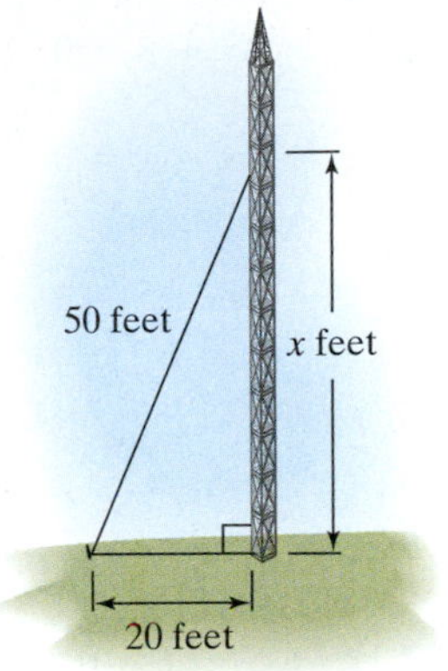

8.1 EXERCISE SET

MyMathLab MathXL PRACTICE WATCH DOWNLOAD READ REVIEW

Practice Exercises

In Exercises 1–22, solve each equation by the square root property. If possible, simplify radicals or rationalize denominators. Express imaginary solutions in the form a + bi.

1. $3x^2 = 75$
2. $5x^2 = 20$
3. $7x^2 = 42$
4. $8x^2 = 40$
5. $16x^2 = 25$
6. $4x^2 = 49$
7. $3x^2 - 2 = 0$
8. $3x^2 - 5 = 0$
9. $25x^2 + 16 = 0$
10. $4x^2 + 49 = 0$
11. $(x + 7)^2 = 9$
12. $(x + 3)^2 = 64$
13. $(x - 3)^2 = 5$
14. $(x - 4)^2 = 3$
15. $2(x + 2)^2 = 16$
16. $3(x + 2)^2 = 36$
17. $(x - 5)^2 = -9$
18. $(x - 5)^2 = -4$
19. $\left(x + \frac{3}{4}\right)^2 = \frac{11}{16}$
20. $\left(x + \frac{2}{5}\right)^2 = \frac{7}{25}$
21. $x^2 - 6x + 9 = 36$
22. $x^2 - 6x + 9 = 49$

In Exercises 23–34, determine the constant that should be added to the binomial so that it becomes a perfect square trinomial. Then write and factor the trinomial.

23. $x^2 + 2x$
24. $x^2 + 4x$
25. $x^2 - 14x$
26. $x^2 - 10x$
27. $x^2 + 7x$
28. $x^2 + 9x$
29. $x^2 - \frac{1}{2}x$
30. $x^2 - \frac{1}{3}x$
31. $x^2 + \frac{4}{3}x$
32. $x^2 + \frac{4}{5}x$
33. $x^2 - \frac{9}{4}x$
34. $x^2 - \frac{9}{5}x$

In Exercises 35–56, solve each quadratic equation by completing the square.

35. $x^2 + 4x = 32$
36. $x^2 + 6x = 7$
37. $x^2 + 6x = -2$

38. $x^2 + 2x = 5$

39. $x^2 - 8x + 1 = 0$

40. $x^2 + 8x - 5 = 0$

41. $x^2 + 2x + 2 = 0$

42. $x^2 - 4x + 8 = 0$

43. $x^2 + 3x - 1 = 0$

44. $x^2 - 3x - 5 = 0$

45. $x^2 + \frac{4}{7}x + \frac{3}{49} = 0$

46. $x^2 + \frac{6}{5}x + \frac{8}{25} = 0$

47. $x^2 + x - 1 = 0$

48. $x^2 - 7x + 3 = 0$

49. $2x^2 + 3x - 5 = 0$

50. $2x^2 + 5x - 3 = 0$

51. $3x^2 + 6x + 1 = 0$

52. $3x^2 - 6x + 2 = 0$

53. $3x^2 - 8x + 1 = 0$

54. $2x^2 + 3x - 4 = 0$

55. $8x^2 - 4x + 1 = 0$

56. $9x^2 - 6x + 5 = 0$

57. If $f(x) = (x - 1)^2$, find all values of x for which $f(x) = 36$.

58. If $f(x) = (x + 2)^2$, find all values of x for which $f(x) = 25$.

59. If $g(x) = \left(x - \frac{2}{5}\right)^2$, find all values of x for which $g(x) = \frac{9}{25}$.

60. If $g(x) = \left(x + \frac{1}{3}\right)^2$, find all values of x for which $g(x) = \frac{4}{9}$.

61. If $h(x) = 5(x + 2)^2$, find all values of x for which $h(x) = -125$.

62. If $h(x) = 3(x - 4)^2$, find all values of x for which $h(x) = -12$.

Practice PLUS

63. Three times the square of the difference between a number and 2 is -12. Find the number(s).

64. Three times the square of the difference between a number and 9 is -27. Find the number(s).

In Exercises 65–68, solve the formula for the specified variable. Because each variable is nonnegative, list only the principal square root. If possible, simplify radicals or rationalize denominators.

65. $h = \frac{v^2}{2g}$ for v

66. $s = \frac{kwd^2}{l}$ for d

67. $A = P(1 + r)^2$ for r

68. $C = \frac{kP_1P_2}{d^2}$ for d

In Exercises 69–72, solve each quadratic equation by completing the square.

69. $\frac{x^2}{3} + \frac{x}{9} - \frac{1}{6} = 0$

70. $\frac{x^2}{2} - \frac{x}{6} - \frac{3}{4} = 0$

71. $x^2 - bx = 2b^2$

72. $x^2 - bx = 6b^2$

73. The ancient Greeks used a geometric method for completing the square in which they literally transformed a figure into a square.

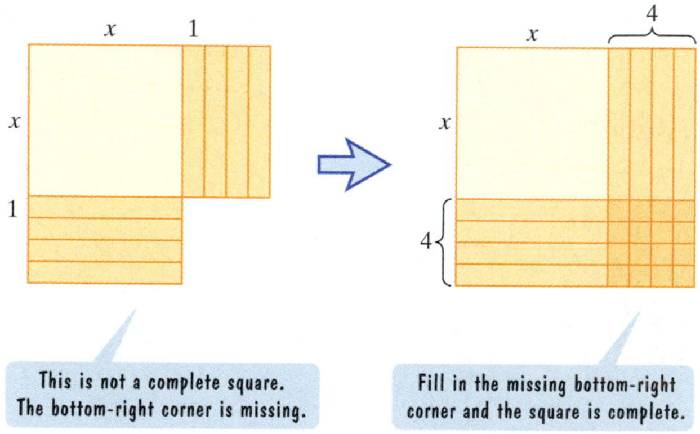
x
1
x
1
x
4
x
4
This is not a complete square. The bottom-right corner is missing.
Fill in the missing bottom-right corner and the square is complete.

a. Write a binomial in x that represents the combined area of the small square and the eight rectangular stripes that make up the incomplete square on the left.

b. What is the area of the region in the bottom-right corner that literally completes the square?

c. Write a trinomial in x that represents the combined area of the small square, the eight rectangular stripes, and the bottom-right corner that make up the complete square on the right.

d. Use the length of each side of the complete square on the right to express its area as the square of a binomial.

74. An **isosceles right triangle** has legs that are the same length and acute angles each measuring 45°.

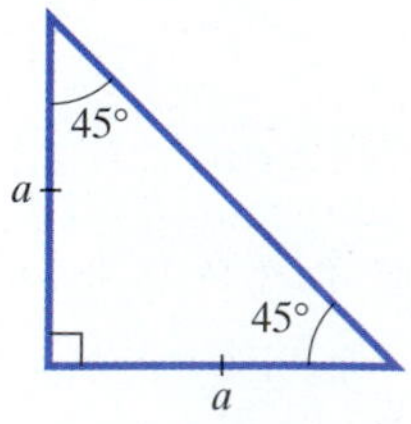
45°
45°
a
a

a. Write an expression in terms of a that represents the length of the hypotenuse.
b. Use your result from part (a) to write a sentence that describes the length of the hypotenuse of an isosceles right triangle in terms of the length of a leg.

Application Exercises

In Exercises 75–78, use the compound interest formula

$$A = P(1 + r)^t$$

to find the annual interest rate, r.

75. In 2 years, an investment of \$2000 grows to \$2880.

76. In 2 years, an investment of \$2000 grows to \$2420.

77. In 2 years, an investment of \$1280 grows to \$1445.

78. In 2 years, an investment of \$80,000 grows to \$101,250.

Of the one hundred largest economies in the world, 53 are multinational corporations. In 1970, there were approximately 7000 multinational corporations. By 2001, more than 65,000 corporations enveloped the world. The graph shows this rapid growth from 1970 through 2001, including the starting dates of some notable corporations.

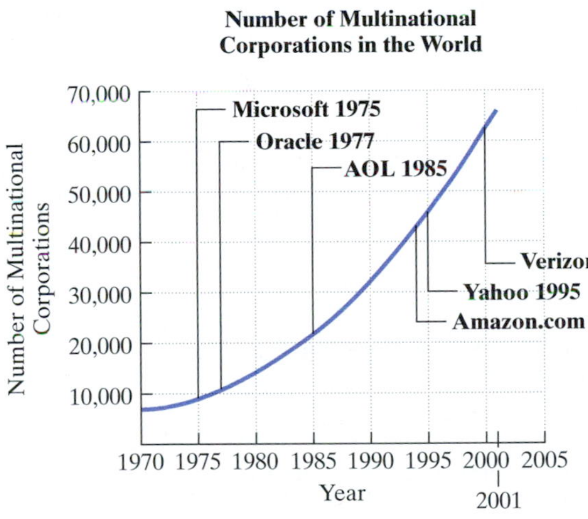

Source: Medard Gabel, *Global Inc*, The New Press, 2003

The data shown can be modeled by the function

$$f(x) = 62.2x^2 + 7000,$$

where $f(x)$ represents the number of multinational corporations in the world x years after 1970. Use this function and the square root property to solve Exercises 79–80.

79. In which year were there 32,000 multinational corporations? How well does the function model the actual number of corporations for that year shown in the graph?

80. In which year were there 46,000 multinational corporations? How well does the function model the actual number of corporations for that year shown in the graph?

The function $s(t) = 16t^2$ models the distance, $s(t)$, in feet, that an object falls in t seconds. Use this function and the square root property to solve Exercises 81–82. Express answers in simplified radical form. Then use your calculator to find a decimal approximation to the nearest tenth of a second.

81. A sky diver jumps from an airplane and falls for 4800 feet before opening a parachute. For how many seconds was the diver in a free fall?

82. A sky diver jumps from an airplane and falls for 3200 feet before opening a parachute. For how many seconds was the diver in a free fall?

Use the Pythagorean Theorem and the square root property to solve Exercises 83–88. Express answers in simplified radical form. Then find a decimal approximation to the nearest tenth.

83. A rectangular park is 6 miles long and 3 miles wide. How long is a pedestrian route that runs diagonally across the park?

84. A rectangular park is 4 miles long and 2 miles wide. How long is a pedestrian route that runs diagonally across the park?

85. The base of a 30-foot ladder is 10 feet from the building. If the ladder reaches the flat roof, how tall is the building?

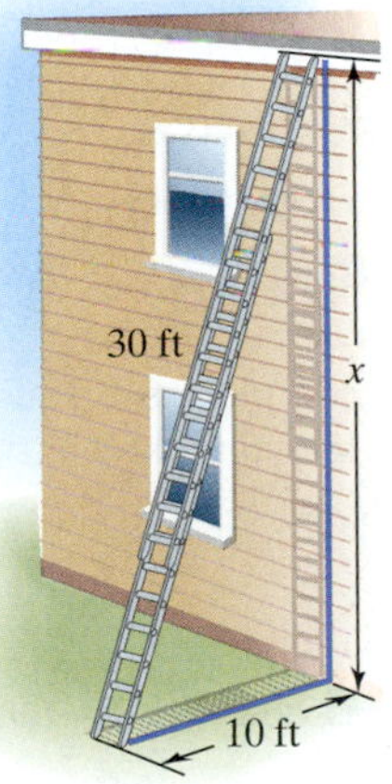

86. The doorway into a room is 4 feet wide and 8 feet high. What is the length of the longest rectangular panel that can be taken through this doorway diagonally?

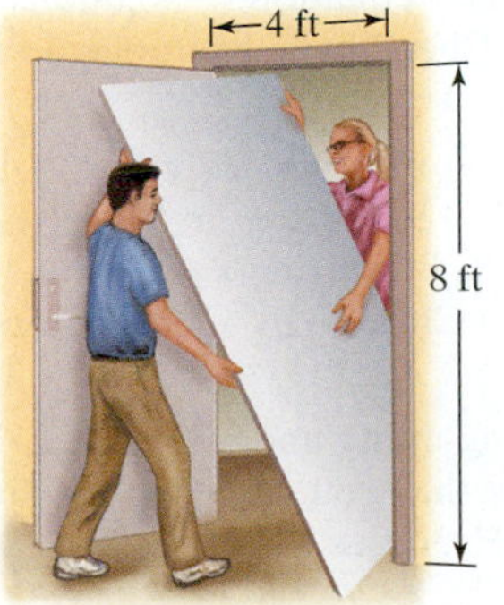

87. A supporting wire is to be attached to the top of a 50-foot antenna. If the wire must be anchored 50 feet from the base of the antenna, what length of wire is required?

88. A supporting wire is to be attached to the top of a 70-foot antenna. If the wire must be anchored 70 feet from the base of the antenna, what length of wire is required?

89. A square flower bed is to be enlarged by adding 2 meters on each side. If the larger square has an area of 196 square meters, what is the length of a side of the original square?

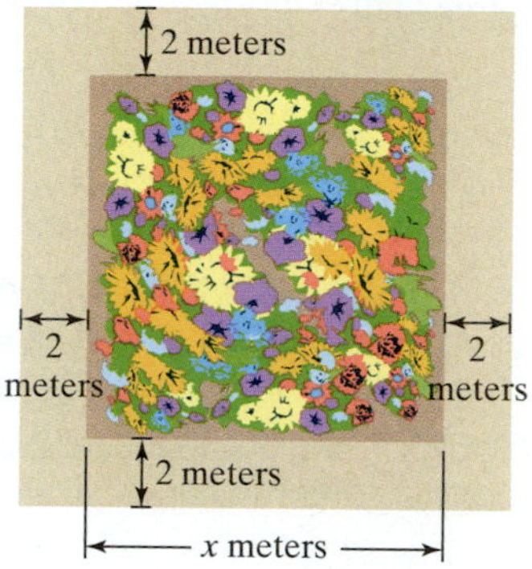

90. A square flower bed is to be enlarged by adding 4 feet on each side. If the larger square has an area of 225 square feet, what is the length of a side of the original square?

Writing in Mathematics

91. What is the square root property?

92. Explain how to solve $(x - 1)^2 = 16$ using the square root property.

93. Explain how to complete the square for a binomial. Use $x^2 + 6x$ to illustrate your explanation.

94. Explain how to solve $x^2 + 6x + 8 = 0$ by completing the square.

95. What is compound interest?

96. In your own words, describe the compound interest formula

$$A = P(1 + r)^t.$$

Technology Exercises

97. Use a graphing utility to solve $4 - (x + 1)^2 = 0$. Graph $y = 4 - (x + 1)^2$ in a $[-5, 5, 1]$ by $[-5, 5, 1]$ viewing rectangle. The equation's solutions are the graph's x-intercepts. Check by substitution in the given equation.

98. Use a graphing utility to solve $(x - 1)^2 - 9 = 0$. Graph $y = (x - 1)^2 - 9$ in a $[-5, 5, 1]$ by $[-9, 3, 1]$ viewing rectangle. The equation's solutions are the graph's x-intercepts. Check by substitution in the given equation.

99. Use a graphing utility and x-intercepts to verify any of the real solutions that you obtained for five of the quadratic equations in Exercises 35–56.

Critical Thinking Exercises

Make Sense? *In Exercises 100–103, determine whether each statement "makes sense" or "does not make sense" and explain your reasoning.*

100. When the coefficient of the x-term in a quadratic equation is negative and I'm solving by completing the square, I add a negative constant to each side of the equation.

101. When I complete the square for the binomial $x^2 + bx$, I obtain a different polynomial, but when I solve a quadratic equation by completing the square, I obtain an equation with the same solution set.

102. When I use the square root property to determine the length of a right triangle's side, I don't even bother to list the negative square root.

103. When I solved $4x^2 + 10x = 0$ by completing the square, I added 25 to both sides of the equation.

In Exercises 104–107, determine whether each statement is true or false. If the statement is false, make the necessary change(s) to produce a true statement.

104. The graph of $y = (x - 2)^2 + 3$ cannot have x-intercepts.

105. The equation $(x - 5)^2 = 12$ is equivalent to $x - 5 = 2\sqrt{3}$.

106. In completing the square for $2x^2 - 6x = 5$, we should add 9 to both sides.

107. Although not every quadratic equation can be solved by completing the square, they can all be solved by factoring.

108. Solve for y: $\frac{x^2}{a^2} + \frac{y^2}{b^2} = 1$.

109. Solve by completing the square:

$$x^2 + x + c = 0.$$

110. Solve by completing the square:

$$x^2 + bx + c = 0.$$

111. Solve: $x^4 - 8x^2 + 15 = 0$.

Review Exercises

112. Simplify: $4x - 2 - 3[4 - 2(3 - x)]$. (Section 1.2, Example 14)

113. Factor: $1 - 8x^3$. (Section 5.5, Example 9)

114. Divide: $(x^4 - 5x^3 + 2x^2 - 6) \div (x - 3)$. (Section 6.5, Example 1)

Preview Exercises

Exercises 115–117 will help you prepare for the material covered in the next section.

115. a. Solve by factoring: $8x^2 + 2x - 1 = 0$.

b. The quadratic equation in part (a) is in the standard form $ax^2 + bx + c = 0$. Compute $b^2 - 4ac$. Is $b^2 - 4ac$ a perfect square?

116. a. Solve by factoring: $9x^2 - 6x + 1 = 0$.

b. The quadratic equation in part (a) is in the standard form $ax^2 + bx + c = 0$. Compute $b^2 - 4ac$.

117. a. Clear fractions in the following equation and write in the form $ax^2 + bx + c = 0$:

$$3 + \frac{4}{x} = -\frac{2}{x^2}.$$

b. For the equation you wrote in part (a), compute $b^2 - 4ac$.

SECTION 8.2

The Quadratic Formula

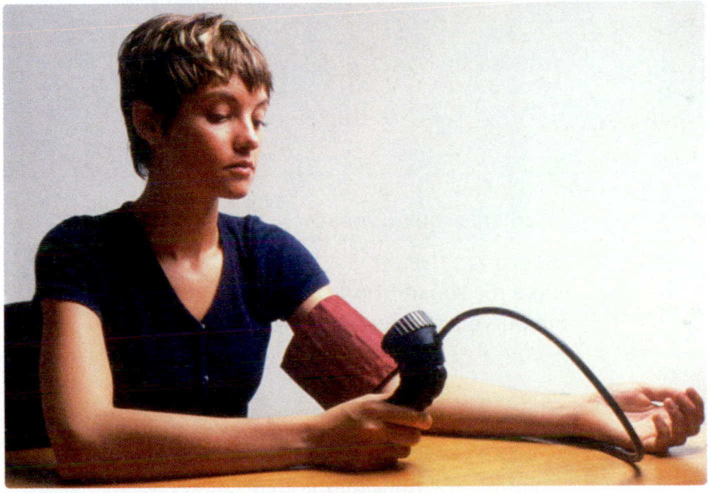

Objectives

1. Solve quadratic equations using the quadratic formula.
2. Use the discriminant to determine the number and type of solutions.
3. Determine the most efficient method to use when solving a quadratic equation.
4. Write quadratic equations from solutions.
5. Use the quadratic formula to solve problems.

Until fairly recently, many doctors believed that your blood pressure was theirs to know and yours to worry about. Today, however, people are encouraged to find out their blood pressure. That pumped-up cuff that squeezes against your upper arm measures blood pressure in millimeters (mm) of mercury (Hg). Blood pressure is given in two numbers: systolic pressure over diastolic pressure, such as 120 over 80. Systolic pressure is the pressure of blood against the artery walls when the heart contracts. Diastolic pressure is the pressure of blood against the artery walls when the heart is at rest.

In this section, we will derive a formula that will enable you to solve quadratic equations more quickly than the method of completing the square. Using this formula, we will work with functions that model changing systolic pressure for men and women with age.

1 Solve quadratic equations using the quadratic formula.

Solving Quadratic Equations Using the Quadratic Formula

We can use the method of completing the square to derive a formula that can be used to solve all quadratic equations. The derivation given on the next page also shows a particular quadratic equation, $3x^2 - 2x - 4 = 0$, to specifically illustrate each of the steps.

Deriving the Quadratic Formula

Standard Form of a Quadratic Equation	Comment	A Specific Example
$ax^2 + bx + c = 0, a > 0$	This is the given equation.	$3x^2 - 2x - 4 = 0$
$x^2 + \frac{b}{a}x + \frac{c}{a} = 0$	Divide both sides by the coefficient of x^2.	$x^2 - \frac{2}{3}x - \frac{4}{3} = 0$
$x^2 + \frac{b}{a}x = -\frac{c}{a}$	Isolate the binomial by adding $-\frac{c}{a}$ on both sides.	$x^2 - \frac{2}{3}x = \frac{4}{3}$
$x^2 + \frac{b}{a}x + \left(\frac{b}{2a}\right)^2 = -\frac{c}{a} + \left(\frac{b}{2a}\right)^2$ (half)^2	Complete the square. Add the square of half the coefficient of x to both sides.	$x^2 - \frac{2}{3}x + \left(-\frac{1}{3}\right)^2 = \frac{4}{3} + \left(-\frac{1}{3}\right)^2$ (half)^2
$x^2 + \frac{b}{a}x + \frac{b^2}{4a^2} = -\frac{c}{a} + \frac{b^2}{4a^2}$		$x^2 - \frac{2}{3}x + \frac{1}{9} = \frac{4}{3} + \frac{1}{9}$
$\left(x + \frac{b}{2a}\right)^2 = -\frac{c}{a} \cdot \frac{4a}{4a} + \frac{b^2}{4a^2}$	Factor on the left side and obtain a common denominator on the right side.	$\left(x - \frac{1}{3}\right)^2 = \frac{4}{3} \cdot \frac{3}{3} + \frac{1}{9}$
$\left(x + \frac{b}{2a}\right)^2 = \frac{-4ac + b^2}{4a^2}$ $\left(x + \frac{b}{2a}\right)^2 = \frac{b^2 - 4ac}{4a^2}$	Add fractions on the right side.	$\left(x - \frac{1}{3}\right)^2 = \frac{12 + 1}{9}$ $\left(x - \frac{1}{3}\right)^2 = \frac{13}{9}$
$x + \frac{b}{2a} = \pm\sqrt{\frac{b^2 - 4ac}{4a^2}}$	Apply the square root property.	$x - \frac{1}{3} = \pm\sqrt{\frac{13}{9}}$
$x + \frac{b}{2a} = \pm\frac{\sqrt{b^2 - 4ac}}{2a}$	Take the square root of the quotient, simplifying the denominator.	$x - \frac{1}{3} = \pm\frac{\sqrt{13}}{3}$
$x = \frac{-b}{2a} \pm \frac{\sqrt{b^2 - 4ac}}{2a}$	Solve for x by subtracting $\frac{b}{2a}$ from both sides.	$x = \frac{1}{3} \pm \frac{\sqrt{13}}{3}$
$x = \frac{-b \pm \sqrt{b^2 - 4ac}}{2a}$	Combine fractions on the right side.	$x = \frac{1 \pm \sqrt{13}}{3}$

The formula shown at the bottom of the left column is called the *quadratic formula*. A similar proof shows that the same formula can be used to solve quadratic equations if a, the coefficient of the x^2-term, is negative.

The Quadratic Formula

The solutions of a quadratic equation in standard form $ax^2 + bx + c = 0$, with $a \neq 0$, are given by the **quadratic formula**:

$$x = \frac{-b \pm \sqrt{b^2 - 4ac}}{2a}.$$

x equals negative b plus or minus the square root of $b^2 - 4ac$, all divided by $2a$.

To use the quadratic formula, write the quadratic equation in standard form if necessary. Then determine the numerical values for a (the coefficient of the x^2-term), b (the coefficient of the x-term), and c (the constant term). Substitute the values of a, b, and c into the quadratic formula and evaluate the expression. The $\pm$ sign indicates that there are two (not necessarily distinct) solutions of the equation.

EXAMPLE 1 Solving a Quadratic Equation Using the Quadratic Formula

Solve using the quadratic formula: $8x^2 + 2x - 1 = 0$.

Solution The given equation is in standard form. Begin by identifying the values for a, b, and c.

$$8x^2 + 2x - 1 = 0$$

$a = 8$ $b = 2$ $c = -1$

Substituting these values into the quadratic formula and simplifying gives the equation's solutions.

$$x = \frac{-b \pm \sqrt{b^2 - 4ac}}{2a}$$ Use the quadratic formula.

$$x = \frac{-2 \pm \sqrt{2^2 - 4(8)(-1)}}{2(8)}$$ Substitute the values for a, b, and c: $a = 8$, $b = 2$, and $c = -1$.

$$= \frac{-2 \pm \sqrt{4 - (-32)}}{16}$$ $2^2 - 4(8)(-1) = 4 - (-32)$

$$= \frac{-2 \pm \sqrt{36}}{16}$$ $4 - (-32) = 4 + 32 = 36$

$$= \frac{-2 \pm 6}{16}$$ $\sqrt{36} = 6$

Now we will evaluate this expression in two different ways to obtain the two solutions. On the left, we will *add* 6 to −2. On the right, we will *subtract* 6 from −2.

$$x = \frac{-2 + 6}{16} = \frac{4}{16} = \frac{1}{4} \quad \text{or} \quad x = \frac{-2 - 6}{16} = \frac{-8}{16} = -\frac{1}{2}$$

The solutions are $-\frac{1}{2}$ and $\frac{1}{4}$, and the solution set is $\left\{-\frac{1}{2}, \frac{1}{4}\right\}$.

Using Technology

Graphic Connections

The graph of the quadratic function

$$y = 8x^2 + 2x - 1$$

has x-intercepts at $-\frac{1}{2}$ and $\frac{1}{4}$. This verifies that $\left\{-\frac{1}{2}, \frac{1}{4}\right\}$ is the solution set of the quadratic equation

$$8x^2 + 2x - 1 = 0.$$

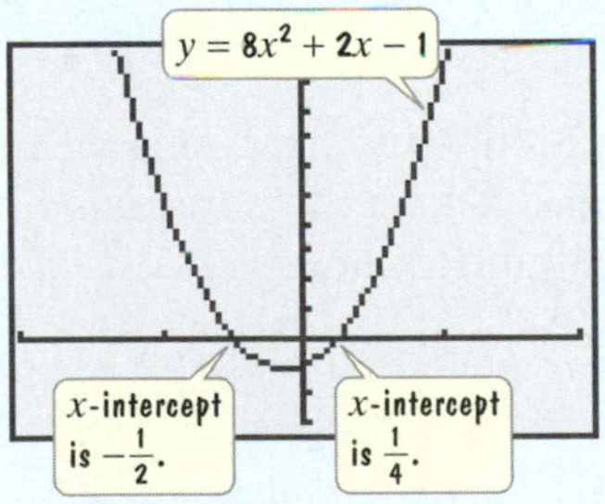

[−2, 2, 1] by [−3, 10, 1]

In Example 1, the solutions of $8x^2 + 2x - 1 = 0$ are rational numbers. This means that the equation can also be solved by factoring. The reason that the solutions are rational numbers is that $b^2 - 4ac$, the radicand in the quadratic formula, is 36, which is a perfect square. If a, b, and c are rational numbers, all quadratic equations for which $b^2 - 4ac$ is a perfect square have rational solutions.

✓ **CHECK POINT 1** Solve using the quadratic formula: $2x^2 + 9x - 5 = 0$.

EXAMPLE 2 Solving a Quadratic Equation Using the Quadratic Formula

Solve using the quadratic formula:

$$2x^2 = 4x + 1.$$

Solution The quadratic equation must be in standard form to identify the values for a, b, and c. To move all terms to one side and obtain zero on the right, we subtract $4x + 1$ from both sides. Then we can identify the values for a, b, and c.

$$2x^2 = 4x + 1$$ This is the given equation.

$$2x^2 - 4x - 1 = 0$$ Subtract $4x + 1$ from both sides.

$a = 2$ $b = -4$ $c = -1$

Substituting these values into the quadratic formula and simplifying gives the equation's solutions.

$$x = \frac{-b \pm \sqrt{b^2 - 4ac}}{2a}$$ Use the quadratic formula.

$$x = \frac{-(-4) \pm \sqrt{(-4)^2 - 4(2)(-1)}}{2(2)}$$ Substitute the values for a, b, and c: $a = 2$, $b = -4$, and $c = -1$.

$$= \frac{4 \pm \sqrt{16 - (-8)}}{4}$$ $(-4)^2 - 4(2)(-1) = 16 - (-8)$

$$= \frac{4 \pm \sqrt{24}}{4}$$ $16 - (-8) = 16 + 8 = 24$

$$= \frac{4 \pm 2\sqrt{6}}{4}$$ $\sqrt{24} = \sqrt{4 \cdot 6} = \sqrt{4} \cdot \sqrt{6} = 2\sqrt{6}$

$$= \frac{2(2 \pm \sqrt{6})}{4}$$ Factor out 2 from the numerator.

$$= \frac{2 \pm \sqrt{6}}{2}$$ Divide the numerator and denominator by 2.

The solutions are $\frac{2 \pm \sqrt{6}}{2}$, and the solution set is $\left\{\frac{2 + \sqrt{6}}{2}, \frac{2 - \sqrt{6}}{2}\right\}$ or $\left\{\frac{2 \pm \sqrt{6}}{2}\right\}$.

Using Technology

You can use a graphing utility to verify that the solutions of $2x^2 - 4x - 1 = 0$ are $\frac{2 \pm \sqrt{6}}{2}$. Begin by entering $y_1 = 2x^2 - 4x - 1$ in the Y= screen. Then evaluate this function at each of the proposed solutions.

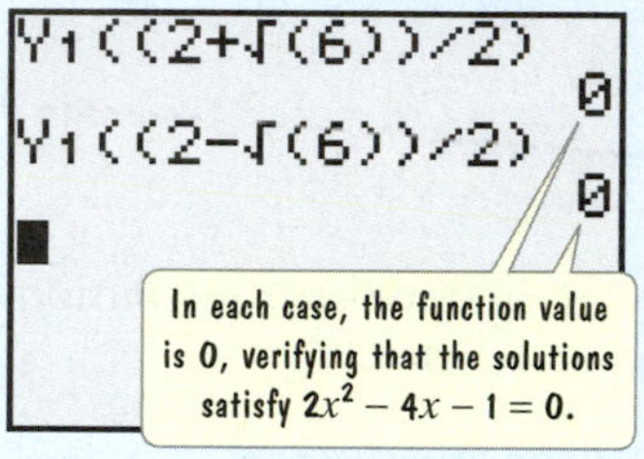

In each case, the function value is 0, verifying that the solutions satisfy $2x^2 - 4x - 1 = 0$.

In Example 2, the solutions of $2x^2 = 4x + 1$ are irrational numbers. This means that the equation cannot be solved by factoring. The reason that the solutions are irrational numbers is that $b^2 - 4ac$, the radicand in the quadratic formula, is 24, which is not a perfect square. Notice, too, that the solutions, $\frac{2 + \sqrt{6}}{2}$ and $\frac{2 - \sqrt{6}}{2}$, are conjugates.

Study Tip

Many students use the quadratic formula correctly until the last step, where they make an error in simplifying the solutions. Be sure to factor the numerator before dividing the numerator and the denominator by the greatest common factor.

$$\frac{4 \pm 2\sqrt{6}}{4} = \frac{2(2 \pm \sqrt{6})}{4} = \frac{\overset{1}{\cancel{2}}(2 \pm \sqrt{6})}{\underset{2}{\cancel{4}}} = \frac{2 \pm \sqrt{6}}{2}$$

You cannot divide just one term in the numerator and the denominator by their greatest common factor.

Incorrect!

$$\frac{\overset{1}{\cancel{4}} \pm 2\sqrt{6}}{\underset{1}{\cancel{4}}} = 1 \pm 2\sqrt{6} \qquad \frac{4 \pm \overset{1}{\cancel{2}}\sqrt{6}}{\underset{2}{\cancel{4}}} = \frac{4 \pm \sqrt{6}}{2}$$

Can all irrational solutions of quadratic equations be simplified? No. The following solutions cannot be simplified:

$$\frac{5 \pm 2\sqrt{7}}{2} \qquad \frac{-4 \pm 3\sqrt{7}}{2}.$$

Other than 1, terms in each numerator have no common factor.

CHECK POINT 2 Solve using the quadratic formula: $2x^2 = 6x - 1$.

EXAMPLE 3 Solving a Quadratic Equation Using the Quadratic Formula

Solve using the quadratic formula:

$$3x^2 + 2 = -4x.$$

Solution Begin by writing the quadratic equation in standard form.

$$3x^2 + 2 = -4x \qquad \text{This is the given equation.}$$

$$3x^2 + 4x + 2 = 0 \qquad \text{Add } 4x \text{ to both sides.}$$

$a = 3$ $\quad b = 4$ $\quad c = 2$

Substituting these values into the quadratic formula and simplifying gives the equation's solutions.

$$x = \frac{-b \pm \sqrt{b^2 - 4ac}}{2a} \qquad \text{Use the quadratic formula.}$$

$$x = \frac{-4 \pm \sqrt{4^2 - 4 \cdot 3 \cdot 2}}{2 \cdot 3} \qquad \text{Substitute the values for } a, b, \text{ and } c\text{: } a = 3, b = 4, \text{ and } c = 2.$$

$$= \frac{-4 \pm \sqrt{16 - 24}}{6} \qquad \text{Multiply under the radical.}$$

$$= \frac{-4 \pm \sqrt{-8}}{6} \qquad \text{Subtract under the radical.}$$

$$= \frac{-4 \pm 2i\sqrt{2}}{6} \qquad \sqrt{-8} = \sqrt{8(-1)} = \sqrt{8}\sqrt{-1} = i\sqrt{8} = i\sqrt{4 \cdot 2} = 2i\sqrt{2}$$

$$= \frac{2(-2 \pm i\sqrt{2})}{6} \qquad \text{Factor out 2 from the numerator.}$$

$$= \frac{-2 \pm i\sqrt{2}}{3} \qquad \text{Divide the numerator and denominator by 2.}$$

$$= -\frac{2}{3} \pm i\frac{\sqrt{2}}{3} \qquad \text{Express in the form } a + bi, \text{ writing } i \text{ before the square root.}$$

The solutions are $-\frac{2}{3} \pm i\frac{\sqrt{2}}{3}$, and the solution set is $\left\{-\frac{2}{3} + i\frac{\sqrt{2}}{3}, -\frac{2}{3} - i\frac{\sqrt{2}}{3}\right\}$ or $\left\{-\frac{2}{3} \pm i\frac{\sqrt{2}}{3}\right\}$.

Using Technology

Graphic Connections

The graph of the quadratic function

$$y = 3x^2 + 4x + 2$$

has no x-intercepts. This verifies that the equation in Example 3

$$3x^2 + 2 = -4x, \quad \text{or} \quad 3x^2 + 4x + 2 = 0$$

has imaginary solutions.

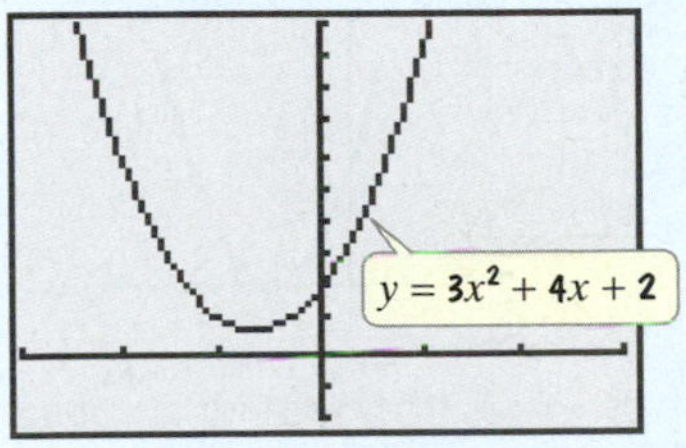

[−3, 3, 1] by [−2, 10, 1]

In Example 3, the solutions of $3x^2 + 2 = -4x$ are imaginary numbers. This means that the equation cannot be solved using factoring. The reason that the solutions are imaginary numbers is that $b^2 - 4ac$, the radicand in the quadratic formula, is -8, which is negative. Notice, too, that the solutions are complex conjugates.

✓ CHECK POINT 3 Solve using the quadratic formula: $3x^2 + 5 = -6x$.

Some rational equations can be solved using the quadratic formula. For example, consider the equation

$$3 + \frac{4}{x} = -\frac{2}{x^2}.$$

The denominators are x and x^2. The least common denominator is x^2. We clear fractions by multiplying both sides of the equation by x^2. Notice that x cannot equal zero.

$$x^2\left(3 + \frac{4}{x}\right) = x^2\left(-\frac{2}{x^2}\right),\quad x \neq 0$$

$$3x^2 + \frac{4}{x}\cdot x^2 = x^2\left(-\frac{2}{x^2}\right) \qquad \text{Use the distributive property.}$$

$$3x^2 + 4x = -2 \qquad \text{Simplify.}$$

By adding 2 to both sides of $3x^2 + 4x = -2$, we obtain the standard form of the quadratic equation:

$$3x^2 + 4x + 2 = 0.$$

This is the equation that we solved in Example 3. The two imaginary solutions are not part of the restriction that $x \neq 0$.

2 Use the discriminant to determine the number and type of solutions.

The Discriminant

The quantity $b^2 - 4ac$, which appears under the radical sign in the quadratic formula, is called the **discriminant**. **Table 8.2** shows how the discriminant of the quadratic equation $ax^2 + bx + c = 0$ determines the number and type of solutions.

Study Tip

Checking irrational and imaginary solutions can be time-consuming. The solutions given by the quadratic formula are always correct, unless you have made a careless error. Checking for computational errors or errors in simplification is sufficient.

Table 8.2 The Discriminant and the Kinds of Solutions to $ax^2 + bx + c = 0$

Discriminant $b^2 - 4ac$	Kinds of Solutions to $ax^2 + bx + c = 0$	Graph of $y = ax^2 + bx + c$
$b^2 - 4ac > 0$	**Two unequal real solutions:** If a, b, and c are rational numbers and the discriminant is a perfect square, the solutions are *rational*. If the discriminant is not a perfect square, the solutions are *irrational* conjugates.	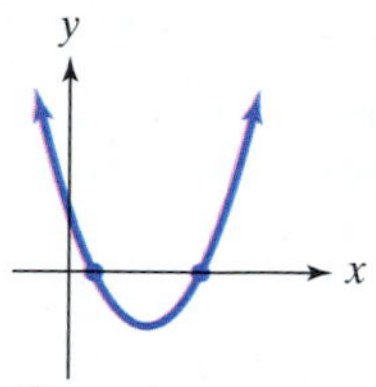 Two x-intercepts
$b^2 - 4ac = 0$	**One solution (a repeated solution) that is a real numbers:** If a, b, and c are rational numbers, the repeated solution is also a rational number.	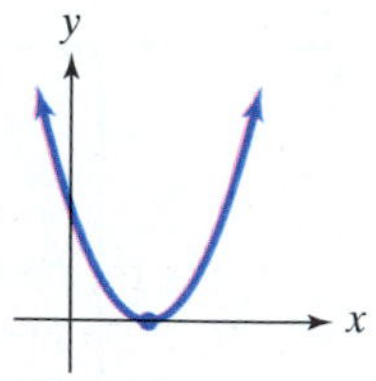 One x-intercept
$b^2 - 4ac < 0$	**No real solution; two imaginary solutions:** The solutions are complex conjugates.	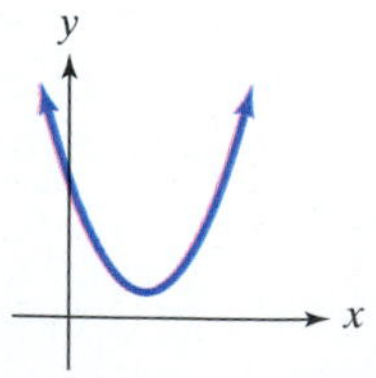 No x-intercepts

EXAMPLE 4 Using the Discriminant

For each equation, compute the discriminant. Then determine the number and type of solutions:

a. $3x^2 + 4x - 5 = 0$ b. $9x^2 - 6x + 1 = 0$ c. $3x^2 - 8x + 7 = 0.$

Study Tip

The discriminant is $b^2 - 4ac$. It is not $\sqrt{b^2 - 4ac}$, so do not give the discriminant as a radical.

Solution Begin by identifying the values for a, b, and c in each equation. Then compute $b^2 - 4ac$, the discriminant.

a. $3x^2 + 4x - 5 = 0$

$a = 3$ $b = 4$ $c = -5$

Substitute and compute the discriminant:

$$b^2 - 4ac = 4^2 - 4 \cdot 3(-5) = 16 - (-60) = 16 + 60 = 76.$$

The discriminant, 76, is a positive number that is not a perfect square. Thus, there are two real irrational solutions. (These solutions are conjugates of each other.)

b. $9x^2 - 6x + 1 = 0$

$a = 9$ $b = -6$ $c = 1$

Substitute and compute the discriminant:

$$b^2 - 4ac = (-6)^2 - 4 \cdot 9 \cdot 1 = 36 - 36 = 0.$$

The discriminant, 0, shows that there is only one real solution. This real solution is a rational number.

c. $3x^2 - 8x + 7 = 0$

$a = 3$ $b = -8$ $c = 7$

$$b^2 - 4ac = (-8)^2 - 4 \cdot 3 \cdot 7 = 64 - 84 = -20$$

The negative discriminant, -20, shows that there are two imaginary solutions. (These solutions are complex conjugates of each other.)

✓ CHECK POINT 4 For each equation, compute the discriminant. Then determine the number and type of solutions:

a. $x^2 + 6x + 9 = 0$
b. $2x^2 - 7x - 4 = 0$
c. $3x^2 - 2x + 4 = 0.$

3 Determine the most efficient method to use when solving a quadratic equation.

Determining Which Method to Use

All quadratic equations can be solved by the quadratic formula. However, if an equation is in the form $u^2 = d$, such as $x^2 = 5$ or $(2x + 3)^2 = 8$, it is faster to use the square root property, taking the square root of both sides. If the equation is not in the form $u^2 = d$, write the quadratic equation in standard form $(ax^2 + bx + c = 0)$. Try to solve the equation by factoring. If $ax^2 + bx + c$ cannot be factored, then solve the quadratic equation by the quadratic formula.

Because we used the method of completing the square to derive the quadratic formula, we no longer need it for solving quadratic equations. However, we will use completing the square in Chapter 10 to help graph certain kinds of equations.

Table 8.3 summarizes our observations about which technique to use when solving a quadratic equation.

Table 8.3 Determining the Most Efficient Technique to Use When Solving a Quadratic Equation

Description and Form of the Quadratic Equation	Most Efficient Solution Method	Example
$ax^2 + bx + c = 0$ and $ax^2 + bx + c$ can be factored easily.	Factor and use the zero-product principle.	$3x^2 + 5x - 2 = 0$ $(3x - 1)(x + 2) = 0$ $3x - 1 = 0$ or $x + 2 = 0$ $x = \frac{1}{3}$ $\quad x = -2$
$ax^2 + c = 0$ The quadratic equation has no x-term. $(b = 0)$	Solve for x^2 and apply the square root property.	$4x^2 - 7 = 0$ $4x^2 = 7$ $x^2 = \frac{7}{4}$ $x = \pm\frac{\sqrt{7}}{2}$
$u^2 = d$; u is a first-degree polynomial.	Use the square root property.	$(x + 4)^2 = 5$ $x + 4 = \pm\sqrt{5}$ $x = -4 \pm \sqrt{5}$
$ax^2 + bx + c = 0$ and $ax^2 + bx + c$ cannot be factored or the factoring is too difficult.	Use the quadratic formula: $x = \frac{-b \pm \sqrt{b^2 - 4ac}}{2a}$.	$x^2 - 2x - 6 = 0$ $a = 1$ $\quad b = -2$ $\quad c = -6$ $x = \frac{-(-2) \pm \sqrt{(-2)^2 - 4(1)(-6)}}{2(1)}$ $= \frac{2 \pm \sqrt{4 - 4(1)(-6)}}{2(1)}$ $= \frac{2 \pm \sqrt{28}}{2} = \frac{2 \pm \sqrt{4}\sqrt{7}}{2}$ $= \frac{2 \pm 2\sqrt{7}}{2} = \frac{2(1 \pm \sqrt{7})}{2}$ $= 1 \pm \sqrt{7}$

4 Write quadratic equations from solutions.

Writing Quadratic Equations from Solutions

Using the zero-product principle, the equation $(x - 3)(x + 5) = 0$ has two solutions, 3 and -5. By applying the zero-product principle in reverse, we can find a quadratic equation that has two given numbers as its solutions.

The Zero-Product Principle in Reverse

If $A = 0$ or $B = 0$, then $AB = 0$.

EXAMPLE 5 Writing Equations from Solutions

Write a quadratic equation with the given solution set:

a. $\left\{-\frac{5}{3}, \frac{1}{2}\right\}$ **b.** $\{-5i, 5i\}$.

Jasper Johns, "Zero." © Jasper Johns/Licensed by VAGA, New York, NY.

The special properties of zero make it possible to write a quadratic equation from its solutions.

Solution

a. Because the solution set is $\left\{-\frac{5}{3}, \frac{1}{2}\right\}$, then

$$x = -\frac{5}{3} \quad \text{or} \quad x = \frac{1}{2}.$$

$$x + \frac{5}{3} = 0 \quad \text{or} \quad x - \frac{1}{2} = 0$$ Obtain zero on one side of each equation.

$$3x + 5 = 0 \quad \text{or} \quad 2x - 1 = 0$$ Clear fractions, multiplying by 3 and 2, respectively.

$$(3x + 5)(2x - 1) = 0$$ Use the zero-product principle in reverse: If $A = 0$ or $B = 0$, then $AB = 0$.

$$6x^2 - 3x + 10x - 5 = 0$$ Use the FOIL method to multiply.

$$6x^2 + 7x - 5 = 0$$ Combine like terms.

Thus, one equation is $6x^2 + 7x - 5 = 0$. Many other quadratic equations have $\left\{-\frac{5}{3}, \frac{1}{2}\right\}$ for their solution sets. These equations can be obtained by multiplying both sides of $6x^2 + 7x - 5 = 0$ by any nonzero real number.

b. Because the solution set is $\{-5i, 5i\}$, then

$$x = -5i \quad \text{or} \quad x = 5i.$$

$$x + 5i = 0 \quad \text{or} \quad x - 5i = 0$$ Obtain zero on one side of each equation.

$$(x + 5i)(x - 5i) = 0$$ Use the zero-product principle in reverse: If $A = 0$ or $B = 0$, then $AB = 0$.

$$x^2 - (5i)^2 = 0$$ Multiply conjugates using $(A + B)(A - B) = A^2 - B^2$.

$$x^2 - 25i^2 = 0$$ $(5i)^2 = 5^2i^2 = 25i^2$

$$x^2 - 25(-1) = 0$$ $i^2 = -1$

$$x^2 + 25 = 0$$ This is the required equation.

✓ **CHECK POINT 5** Write a quadratic equation with the given solution set:

a. $\left\{-\frac{3}{5}, \frac{1}{4}\right\}$ **b.** $\{-7i, 7i\}$.

5 Use the quadratic formula to solve problems.

Applications

Quadratic equations can be solved to answer questions about variables contained in quadratic functions.

EXAMPLE 6 Blood Pressure and Age

The graphs in **Figure 8.3** illustrate that a person's normal systolic blood pressure, measured in millimeters of mercury (mm Hg), depends on his or her age. The function

$$P(A) = 0.006A^2 - 0.02A + 120$$

models a man's normal systolic pressure, $P(A)$, at age A.

a. Find the age, to the nearest year, of a man whose normal systolic blood pressure is 125 mm Hg.

b. Use the graphs in **Figure 8.3** to describe the differences between the normal systolic blood pressures of men and women as they age.

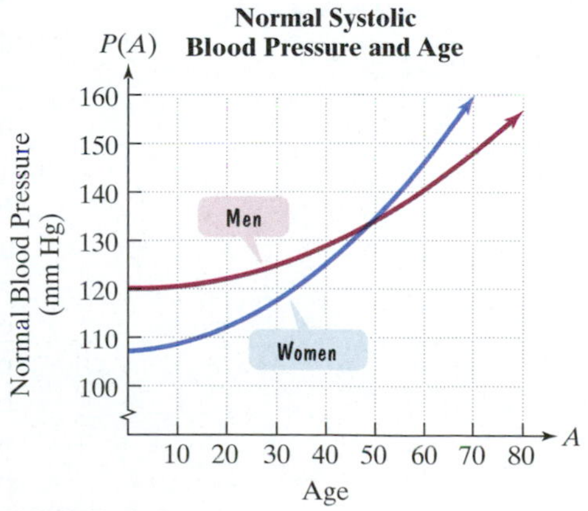

FIGURE 8.3

Solution

a. We are interested in the age of a man with a normal systolic blood pressure of 125 millimeters of mercury. Thus, we substitute 125 for $P(A)$ in the given function for men. Then we solve for A, the man's age.

$$P(A) = 0.006A^2 - 0.02A + 120 \quad \text{This is the given function for men.}$$

$$125 = 0.006A^2 - 0.02A + 120 \quad \text{Substitute 125 for } P(A).$$

$$0 = 0.006A^2 - 0.02A - 5 \quad \text{Subtract 125 from both sides and write the quadratic equation in standard form.}$$

$a = 0.006$ $b = -0.02$ $c = -5$

Because the trinomial on the right side of the equation is prime, we solve using the quadratic formula.

Notice that the variable is A, rather than the usual x.

$$A = \frac{-b \pm \sqrt{b^2 - 4ac}}{2a} \quad \text{Use the quadratic formula.}$$

$$= \frac{-(-0.02) \pm \sqrt{(-0.02)^2 - 4(0.006)(-5)}}{2(0.006)} \quad \text{Substitute the values for } a, b, \text{ and } c\text{: } a = 0.006,\ b = -0.02, \text{ and } c = -5.$$

$$= \frac{0.02 \pm \sqrt{0.1204}}{0.012} \quad \text{Use a calculator to simplify the expression under the square root.}$$

$$\approx \frac{0.02 \pm 0.347}{0.012} \quad \text{Use a calculator: } \sqrt{0.1204} \approx 0.347.$$

$$A \approx \frac{0.02 + 0.347}{0.012} \quad \text{or} \quad A \approx \frac{0.02 - 0.347}{0.012}$$

$$A \approx 31 \qquad\qquad A \approx -27 \quad \text{Use a calculator and round to the nearest integer.}$$

Reject this solution. Age cannot be negative.

The positive solution, $A \approx 31$, indicates that 31 is the approximate age of a man whose normal systolic blood pressure is 125 mm Hg. This is illustrated by the black lines with the arrows on the red graph representing men in **Figure 8.4**.

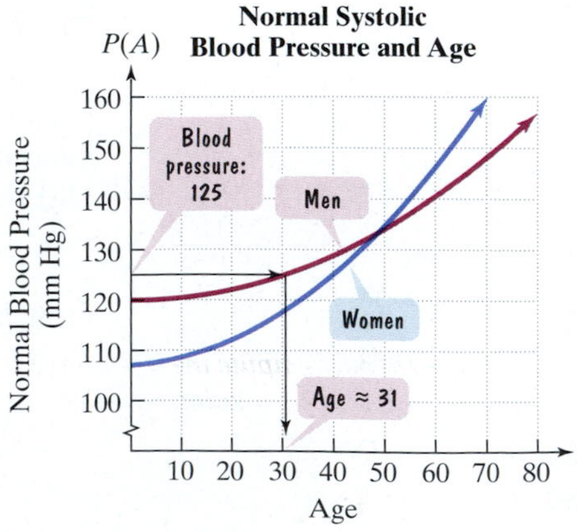

FIGURE 8.4

b. Take a second look at the graphs in **Figure 8.4**. Before approximately age 50, the blue graph representing women's normal systolic blood pressure lies below the red graph representing men's normal systolic blood pressure. Thus, up to age 50, women's normal systolic blood pressure is lower than men's, although it is increasing at a faster rate. After age 50, women's normal systolic blood pressure is higher than men's.

Using Technology

On most calculators, here is how to approximate

$$\frac{0.02 + \sqrt{0.1204}}{0.012}.$$

Many Scientific Calculators

[( ] .02 [+] .1204 [√] [ )] [÷] .012 [=]

Many Graphing Calculators

[( ] .02 [+] [√] .1204 [ )] [÷] .012 [ENTER]

If your calculator displays an open parenthesis after √, you'll need to enter another closed parenthesis here.

✓ CHECK POINT 6 The function $P(A) = 0.01A^2 + 0.05A + 107$ models a woman's normal systolic blood pressure, $P(A)$, at age A. Use this function to find the age, to the nearest year, of a woman whose normal systolic blood pressure is 115 mm Hg. Use the blue graph in **Figure 8.4** to verify your solution.

8.2 EXERCISE SET

MyMathLab MathXL PRACTICE WATCH DOWNLOAD READ REVIEW

Practice Exercises

In Exercises 1–16, solve each equation using the quadratic formula. Simplify solutions, if possible.

1. $x^2 + 8x + 12 = 0$
2. $x^2 + 8x + 15 = 0$
3. $2x^2 - 7x = -5$
4. $5x^2 + 8x = -3$
5. $x^2 + 3x - 20 = 0$
6. $x^2 + 5x - 10 = 0$
7. $3x^2 - 7x = 3$
8. $4x^2 + 3x = 2$
9. $6x^2 = 2x + 1$
10. $2x^2 = -4x + 5$
11. $4x^2 - 3x = -6$
12. $9x^2 + x = -2$
13. $x^2 - 4x + 8 = 0$
14. $x^2 + 6x + 13 = 0$
15. $3x^2 = 8x - 7$
16. $3x^2 = 4x - 6$
17. $2x(x - 2) = x + 12$
18. $2x(x + 4) = 3x - 3$

In Exercises 19–30, compute the discriminant. Then determine the number and type of solutions for the given equation.

19. $x^2 + 8x + 3 = 0$
20. $x^2 + 7x + 4 = 0$
21. $x^2 + 6x + 8 = 0$
22. $x^2 + 2x - 3 = 0$
23. $2x^2 + x + 3 = 0$
24. $2x^2 - 4x + 3 = 0$
25. $2x^2 + 6x = 0$
26. $3x^2 - 5x = 0$
27. $5x^2 + 3 = 0$
28. $5x^2 + 4 = 0$
29. $9x^2 = 12x - 4$
30. $4x^2 = 20x - 25$

In Exercises 31–46, solve each equation by the method of your choice. Simplify solutions, if possible.

31. $3x^2 - 4x = 4$
32. $2x^2 - x = 1$
33. $x^2 - 2x = 1$
34. $2x^2 + 3x = 1$
35. $(2x - 5)(x + 1) = 2$
36. $(2x + 3)(x + 4) = 1$
37. $(3x - 4)^2 = 16$
38. $(2x + 7)^2 = 25$
39. $\frac{x^2}{2} + 2x + \frac{2}{3} = 0$
40. $\frac{x^2}{3} - x - \frac{1}{6} = 0$
41. $(3x - 2)^2 = 10$
42. $(4x - 1)^2 = 15$
43. $\frac{1}{x} + \frac{1}{x + 2} = \frac{1}{3}$
44. $\frac{1}{x} + \frac{1}{x + 3} = \frac{1}{4}$
45. $(2x - 6)(x + 2) = 5(x - 1) - 12$
46. $7x(x - 2) = 3 - 2(x + 4)$

In Exercises 47–60, write a quadratic equation in standard form with the given solution set.

47. $\{-3, 5\}$
48. $\{-2, 6\}$
49. $\left\{-\frac{2}{3}, \frac{1}{4}\right\}$

50. $\left\{-\frac{5}{6}, \frac{1}{3}\right\}$

51. $\{-6i, 6i\}$

52. $\{-8i, 8i\}$

53. $\{-\sqrt{2}, \sqrt{2}\}$

54. $\{-\sqrt{3}, \sqrt{3}\}$

55. $\{-2\sqrt{5}, 2\sqrt{5}\}$

56. $\{-3\sqrt{5}, 3\sqrt{5}\}$

57. $\{1 + i, 1 - i\}$

58. $\{2 + i, 2 - i\}$

59. $\{1 + \sqrt{2}, 1 - \sqrt{2}\}$

60. $\{1 + \sqrt{3}, 1 - \sqrt{3}\}$

Practice PLUS

Exercises 61–64 describe quadratic equations. Match each description with the graph of the corresponding quadratic function. Each graph is shown in a $[-10, 10, 1]$ *by* $[-10, 10, 1]$ *viewing rectangle.*

61. A quadratic equation whose solution set contains imaginary numbers

62. A quadratic equation whose discriminant is 0

63. A quadratic equation whose solution set is $\{3 \pm \sqrt{2}\}$

64. A quadratic equation whose solution set contains integers

a.

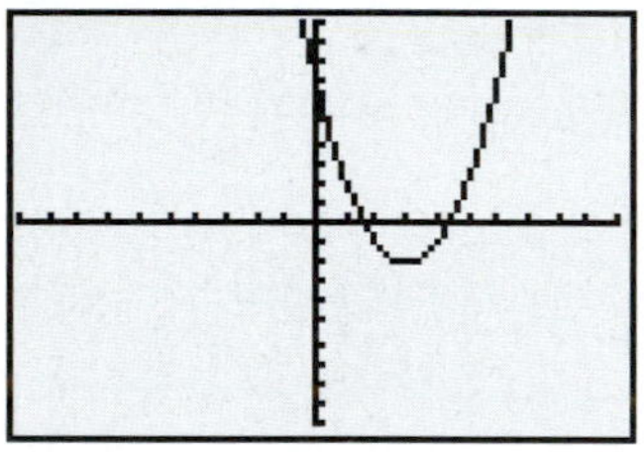

b.

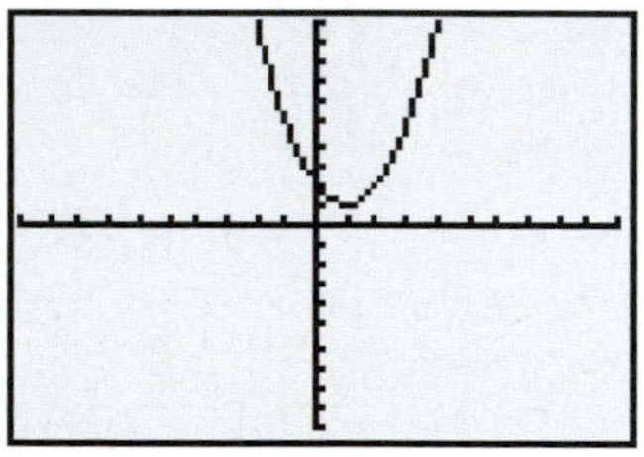

c.

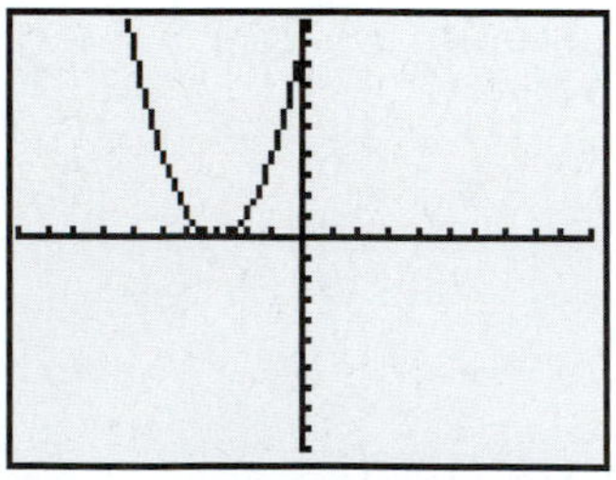

d.

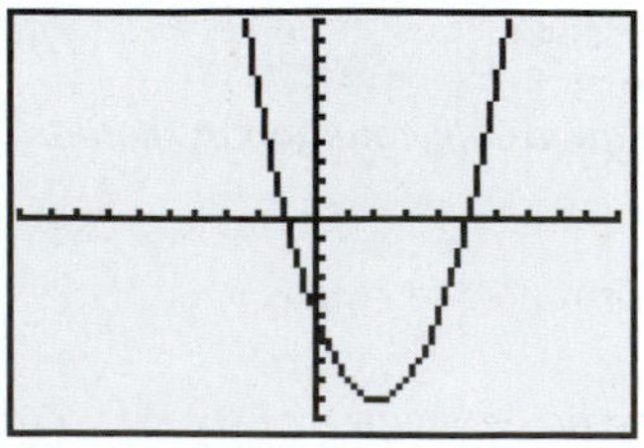

65. When the sum of 6 and twice a positive number is subtracted from the square of the number, 0 results. Find the number.

66. When the sum of 1 and twice a negative number is subtracted from twice the square of the number, 0 results. Find the number.

In Exercises 67–72, solve each equation by the method of your choice.

67. $\frac{1}{x^2 - 3x + 2} = \frac{1}{x + 2} + \frac{5}{x^2 - 4}$

68. $\frac{x - 1}{x - 2} + \frac{x}{x - 3} = \frac{1}{x^2 - 5x + 6}$

69. $\sqrt{2}x^2 + 3x - 2\sqrt{2} = 0$

70. $\sqrt{3}x^2 + 6x + 7\sqrt{3} = 0$

71. $|x^2 + 2x| = 3$

72. $|x^2 + 3x| = 2$

Application Exercises

A driver's age has something to do with his or her chance of getting into a fatal car crash. The bar graph shows the number of fatal vehicle crashes per 100 million miles driven for drivers of various age groups. For example, 25-year-old drivers are involved in 4.1 fatal crashes per 100 million miles driven. Thus, when a group of 25-year-old Americans have driven a total of 100 million miles, approximately 4 have been in accidents in which someone died.

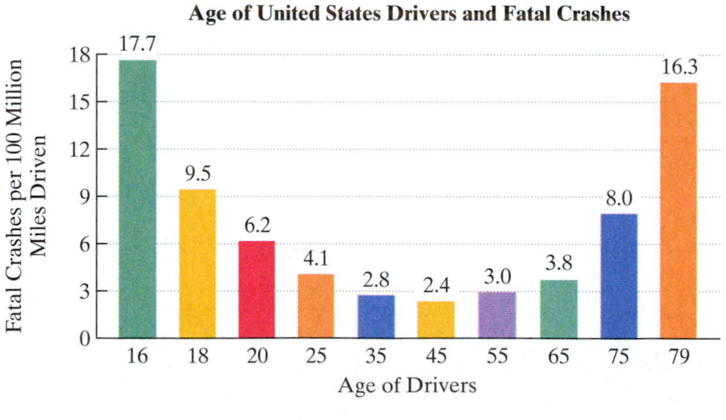

Source: Insurance Institute for Highway Safety

The number of fatal vehicle crashes per 100 million miles, $f(x)$*, for drivers of age* x *can be modeled by the quadratic function*

$$f(x) = 0.013x^2 - 1.19x + 28.24.$$

Use the function to solve Exercises 73–74.

73. What age groups are expected to be involved in 3 fatal crashes per 100 million miles driven? How well does the function model the trend in the actual data shown in the bar graph?

74. What age groups are expected to be involved in 10 fatal crashes per 100 million miles driven? How well does the function model the trend in the actual data shown in the bar graph?

Throwing events in track and field include the shot put, the discus throw, the hammer throw, and the javelin throw. The distance that an athlete can achieve depends on the initial velocity of the object thrown and the angle above the horizontal at which the object leaves the hand.

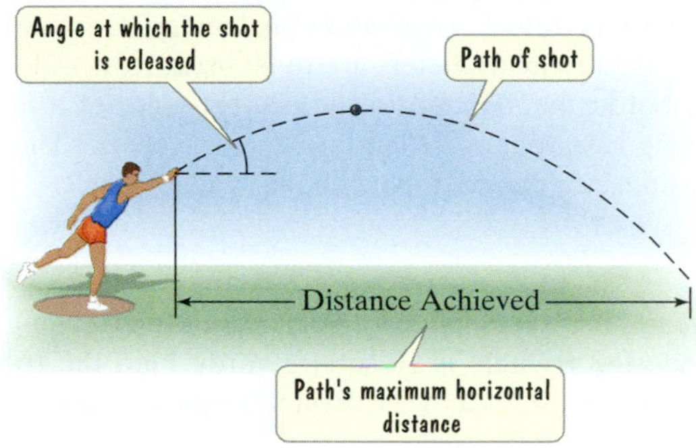

In Exercises 75–76, an athlete whose event is the shot put releases the shot with the same initial velocity, but at different angles.

75. When the shot is released at an angle of 35°, its path can be modeled by the function

$$f(x) = -0.01x^2 + 0.7x + 6.1,$$

in which x is the shot's horizontal distance, in feet, and $f(x)$ is its height, in feet. This function is shown by one of the graphs, (a) or (b), in the figure. Use the function to determine the shot's maximum distance. Use a calculator and round to the nearest tenth of a foot. Which graph, (a) or (b), shows the shot's path?

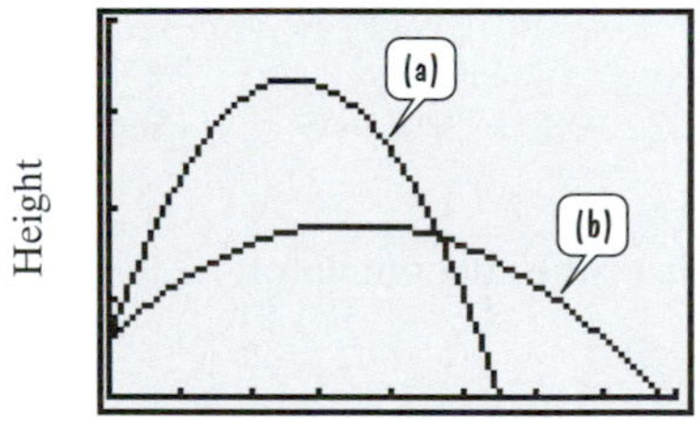

76. When the shot is released at an angle of 65°, its path can be modeled by the function

$$f(x) = -0.04x^2 + 2.1x + 6.1,$$

in which x is the shot's horizontal distance, in feet, and $f(x)$ is its height, in feet. This function is shown by one of the graphs, (a) or (b), in the figure above. Use the function to determine the shot's maximum distance. Use a calculator and round to the nearest tenth of a foot. Which graph, (a) or (b), shows the shot's path?

77. The length of a rectangle is 4 meters longer than the width. If the area is 8 square meters, find the rectangle's dimensions. Round to the nearest tenth of a meter.

78. The length of a rectangle exceeds twice its width by 3 inches. If the area is 10 square inches, find the rectangle's dimensions. Round to the nearest tenth of an inch.

79. The longer leg of a right triangle exceeds the shorter leg by 1 inch, and the hypotenuse exceeds the longer leg by 7 inches. Find the lengths of the legs. Round to the nearest tenth of a inch.

80. The hypotenuse of a right triangle is 6 feet long. One leg is 2 feet shorter than the other. Find the lengths of the legs. Round to the nearest tenth of a foot.

81. A rain gutter is made from sheets of aluminum that are 20 inches wide. As shown in the figure, the edges are turned up to form right angles. Determine the depth of the gutter that will allow a cross-sectional area of 13 square inches. Show that there are two different solutions to the problem. Round to the nearest tenth of an inch.

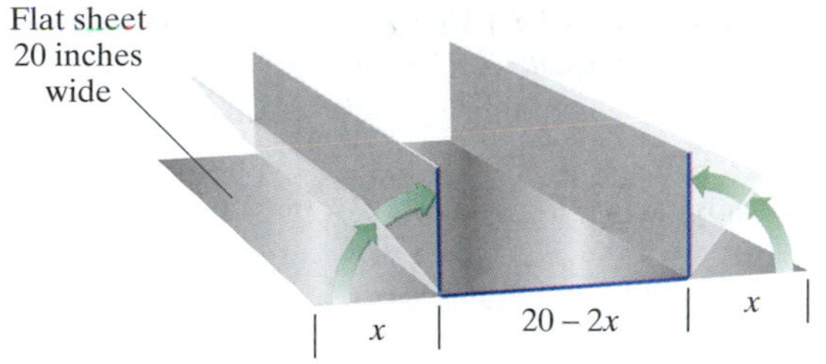

82. A piece of wire is 8 inches long. The wire is cut into two pieces and then each piece is bent into a square. Find the length of each piece if the sum of the areas of these squares is to be 2 square inches.

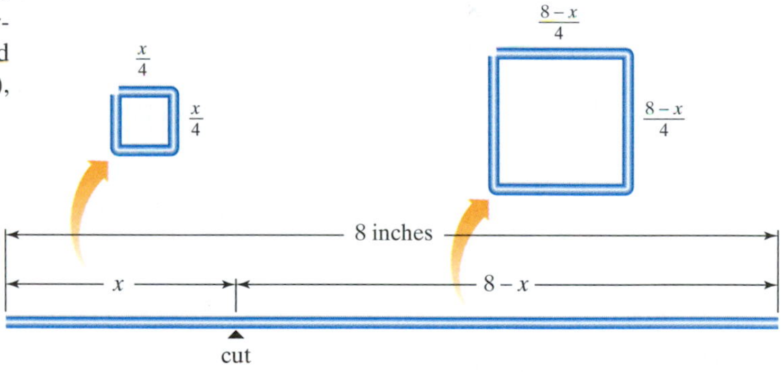

83. Working together, two people can mow a large lawn in 4 hours. One person can do the job alone 1 hour faster than the other person. How long does it take each person working alone to mow the lawn? Round to the nearest tenth of an hour.

84. A pool has an inlet pipe to fill it and an outlet pipe to empty it. It takes 2 hours longer to empty the pool than it does to fill it. The inlet pipe is turned on to fill the pool, but the outlet pipe is accidentally left open. Despite this, the pool fills in 8 hours. How long does it take the outlet pipe to empty the pool? Round to the nearest tenth of an hour.

Writing in Mathematics

85. What is the quadratic formula and why is it useful?

86. Without going into specific details for every step, describe how the quadratic formula is derived.

87. Explain how to solve $x^2 + 6x + 8 = 0$ using the quadratic formula.

88. If a quadratic equation has imaginary solutions, how is this shown on the graph of the corresponding quadratic function?

89. What is the discriminant and what information does it provide about a quadratic equation?

90. If you are given a quadratic equation, how do you determine which method to use to solve it?

91. Explain how to write a quadratic equation from its solution set. Give an example with your explanation.

Technology Exercises

92. Use a graphing utility to graph the quadratic function related to any five of the quadratic equations in Exercises 19–30. How does each graph illustrate what you determined algebraically using the discriminant?

93. Reread Exercise 81. The cross-sectional area of the gutter is given by the quadratic function

$$f(x) = x(20 - 2x).$$

Graph the function in a [0, 10, 1] by [0, 60, 5] viewing rectangle. Then TRACE along the curve or use the maximum function feature to determine the depth of the gutter that will maximize its cross-sectional area and allow the greatest amount of water to flow. What is the maximum area? Does the situation described in Exercise 81 take full advantage of the sheets of aluminum?

Critical Thinking Exercises

Makes Sense? *In Exercises 94–97, determine whether each statement "makes sense" or "does not make sense" and explain your reasoning.*

94. Because I want to solve $25x^2 - 169 = 0$ fairly quickly, I'll use the quadratic formula.

95. I simplified $\frac{3 + 2\sqrt{3}}{2}$ to $3 + \sqrt{3}$ because 2 is a factor of $2\sqrt{3}$.

96. I need to find a square root to determine the discriminant.

97. I obtained -17 for the discriminant, so there are two imaginary irrational solutions.

In Exercises 98–101, determine whether each statement is true or false. If the statement is false, make the necessary change(s) to produce a true statement.

98. Any quadratic equation that can be solved by completing the square can be solved by the quadratic formula.

99. The quadratic formula is developed by applying factoring and the zero-product principle to the quadratic equation $ax^2 + bx + c = 0$.

100. In using the quadratic formula to solve the quadratic equation $5x^2 = 2x - 7$, we have $a = 5, b = 2$, and $c = -7$.

101. The quadratic formula can be used to solve the equation $x^2 - 9 = 0$.

102. Solve for t: $s = -16t^2 + v_0t$.

103. A rectangular swimming pool is 12 meters long and 8 meters wide. A tile border of uniform width is to be built around the pool using 120 square meters of tile. The tile is from a discontinued stock (so no additional materials are available) and all 120 square meters are to be used. How wide should the border be? Round to the nearest tenth of a meter. If zoning laws require at least a 2-meter-wide border around the pool, can this be done with the available tile?

104. The area of the shaded region outside the rectangle and inside the triangle is 10 square yards. Find the triangle's height, represented by $2x$. Round to the nearest tenth of a yard.

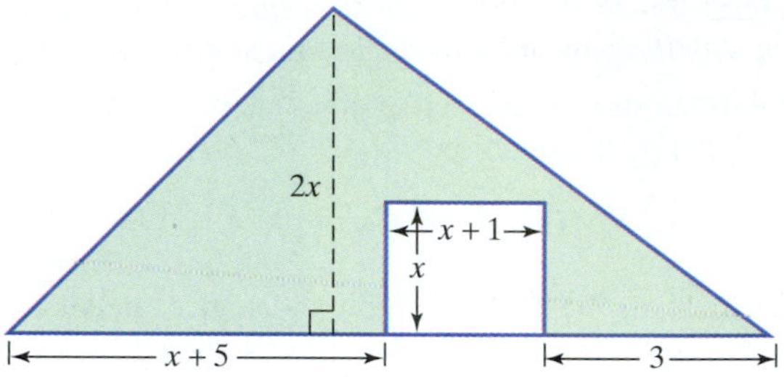

Review Exercises

105. Solve: $|5x + 2| = |4 - 3x|$. (Section 4.3, Example 3)

106. Solve: $\sqrt{2x - 5} - \sqrt{x - 3} = 1$. (Section 7.6, Example 4)

107. Rationalize the denominator: $\frac{5}{\sqrt{3} + x}$. (Section 7.5, Example 5)

Preview Exercises

Exercises 108–110 will help you prepare for the material covered in the next section.

108. Use point plotting to graph $f(x) = x^2$ and $g(x) = x^2 + 2$ in the same rectangular coordinate system.

109. Use point plotting to graph $f(x) = x^2$ and $g(x) = (x + 2)^2$ in the same rectangular coordinate system.

110. Find the x-intercepts for the graph of $f(x) = -2(x - 3)^2 + 8$.

SECTION 8.3

Quadratic Functions and Their Graphs

Objectives

1. Recognize characteristics of parabolas.
2. Graph parabolas in the form $f(x) = a(x - h)^2 + k$.
3. Graph parabolas in the form $f(x) = ax^2 + bx + c$.
4. Determine a quadratic function's minimum or maximum value.
5. Solve problems involving a quadratic function's minimum or maximum value.

We have a long history of throwing things. Before 400 B.C., the Greeks competed in games that included discus throwing. In the seventeenth century, English soldiers organized cannonball-throwing competitions. In 1827, a Yale University student, disappointed over failing an exam, took out his frustrations at the passing of a collection plate in chapel. Seizing the monetary tray, he flung it in the direction of a large open space on campus. Yale students see this act of frustration as the origin of the Frisbee.

In this section, we study quadratic functions and their graphs. By graphing functions that model the paths of the things we throw, you will be able to determine both the maximum height and the distance of these objects.

1 Recognize characteristics of parabolas.

Graphs of Quadratic Functions

The graph of any quadratic function

$$f(x) = ax^2 + bx + c, \quad a \neq 0,$$

is called a **parabola**. Parabolas are shaped like bowls or inverted bowls, as shown in **Figure 8.5**. If the coefficient of x^2 (the value of a in $ax^2 + bx + c$) is positive, the parabola opens upward. If the coefficient of x^2 is negative, the parabola opens downward. The **vertex** (or turning point) of the parabola is the lowest point on the graph when it opens upward and the highest point on the graph when it opens downward.

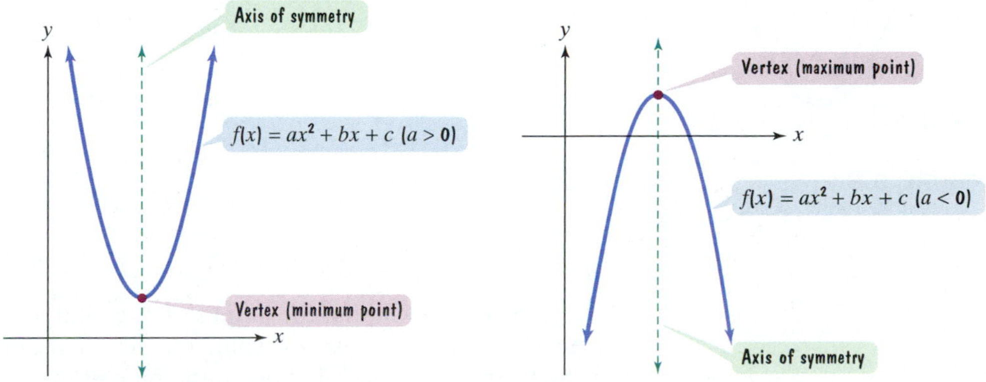

$a > 0$: Parabola opens upward. $a < 0$: Parabola opens downward.

FIGURE 8.5 Characteristics of graphs of quadratic functions

The two halves of a parabola are mirror images of each other. A "mirror line" through the vertex, called the **axis of symmetry**, divides the figure in half. If a parabola is folded along its axis of symmetry, the two halves match exactly.

2 Graph parabolas in the form $f(x) = a(x - h)^2 + k$.

Graphing Quadratic Functions in the Form $f(x) = a(x - h)^2 + k$

One way to obtain the graph of a quadratic function is to use point plotting. Let's begin by graphing the functions $f(x) = x^2$, $g(x) = 2x^2$, and $h(x) = \frac{1}{2}x^2$ in the same rectangular coordinate system. Select integers for x, starting with -3 and ending with 3. A partial table of coordinates for each function is shown below. The three parabolas are shown in **Figure 8.6**.

x	$f(x) = x^2$	(x, y) or $(x, f(x))$
-3	$f(-3) = (-3)^2 = 9$	$(-3, 9)$
-2	$f(-2) = (-2)^2 = 4$	$(-2, 4)$
-1	$f(-1) = (-1)^2 = 1$	$(-1, 1)$
0	$f(0) = 0^2 = 0$	$(0, 0)$
1	$f(1) = 1^2 = 1$	$(1, 1)$
2	$f(2) = 2^2 = 4$	$(2, 4)$
3	$f(3) = 3^2 = 9$	$(3, 9)$

x	$g(x) = 2x^2$	(x, y) or $(x, g(x))$
-3	$g(-3) = 2(-3)^2 = 18$	$(-3, 18)$
-2	$g(-2) = 2(-2)^2 = 8$	$(-2, 8)$
-1	$g(-1) = 2(-1)^2 = 2$	$(-1, 2)$
0	$g(0) = 2 \cdot 0^2 = 0$	$(0, 0)$
1	$g(1) = 2 \cdot 1^2 = 2$	$(1, 2)$
2	$g(2) = 2 \cdot 2^2 = 8$	$(2, 8)$
3	$g(3) = 2 \cdot 3^2 = 18$	$(3, 18)$

x	$h(x) = \frac{1}{2}x^2$	(x, y) or $(x, h(x))$
-3	$h(-3) = \frac{1}{2}(-3)^2 = \frac{9}{2}$	$\left(-3, \frac{9}{2}\right)$
-2	$h(-2) = \frac{1}{2}(-2)^2 = 2$	$(-2, 2)$
-1	$h(-1) = \frac{1}{2}(-1)^2 = \frac{1}{2}$	$\left(-1, \frac{1}{2}\right)$
0	$h(0) = \frac{1}{2} \cdot 0^2 = 0$	$(0, 0)$
1	$h(1) = \frac{1}{2} \cdot 1^2 = \frac{1}{2}$	$\left(1, \frac{1}{2}\right)$
2	$h(2) = \frac{1}{2} \cdot 2^2 = 2$	$(2, 2)$
3	$h(3) = \frac{1}{2} \cdot 3^2 = \frac{9}{2}$	$\left(3, \frac{9}{2}\right)$

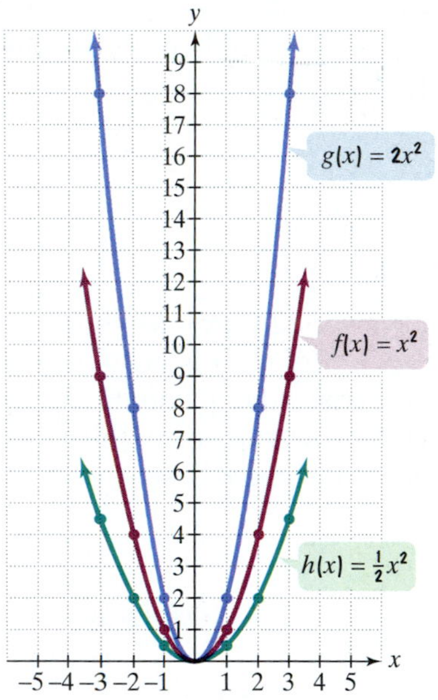

FIGURE 8.6

Can you see that the graphs of f, g, and h all have the same vertex, $(0, 0)$? They also have the same axis of symmetry, the y-axis, or $x = 0$. This is true for all graphs of the form $f(x) = ax^2$. However, the blue graph of $g(x) = 2x^2$ is a narrower parabola than the red graph of $f(x) = x^2$. By contrast, the green graph of $h(x) = \frac{1}{2}x^2$ is a flatter parabola than the red graph of $f(x) = x^2$.

Is there a more efficient method than point plotting to obtain the graph of a quadratic function? The answer is yes. The method is based on comparing graphs of the form $g(x) = a(x - h)^2 + k$ to those of the form $f(x) = ax^2$.

In **Figure 8.7(a)**, the graph of $f(x) = ax^2$ for $a > 0$ is shown in black. The parabola's vertex is (0, 0) and it opens upward. In **Figure 8.7(b)**, the graph of $f(x) = ax^2$ for $a < 0$ is shown in black. The parabola's vertex is (0, 0) and it opens downward.

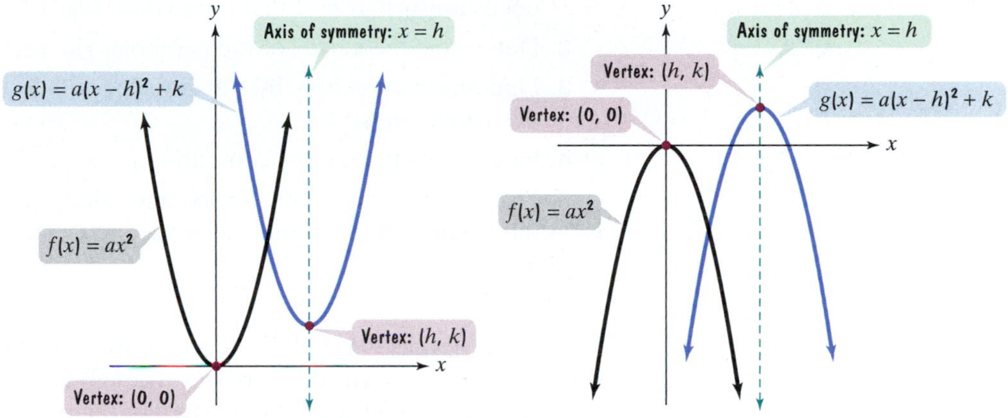

(a) $a > 0$: Parabola opens upward. **(b)** $a < 0$: Parabola opens downward.

FIGURE 8.7 Moving, or shifting, the graph of $f(x) = ax^2$

Figure 8.7(a) and **8.7(b)** also show the graph of $g(x) = a(x - h)^2 + k$ in blue. Compare these graphs to those of $f(x) = ax^2$. Observe that h determines a horizontal move, or shift, and k determines a vertical move, or shift, of the graph of $f(x) = ax^2$:

$$g(x) = a(x - h)^2 + k.$$

If $h > 0$, the graph of $f(x) = ax^2$ is shifted h units to the right.

If $k > 0$, the graph of $y = a(x - h)^2$ is shifted k units up.

Consequently, the vertex (0, 0) on the black graph of $f(x) = ax^2$ moves to the point (h, k) on the blue graph of $g(x) = a(x - h)^2 + k$. The axis of symmetry is the vertical line whose equation is $x = h$.

The form of the expression for g is convenient because it immediately identifies the vertex of the parabola as (h, k).

Quadratic Functions in the Form $f(x) = a(x - h)^2 + k$

The graph of

$$f(x) = a(x - h)^2 + k, \qquad a \neq 0$$

is a parabola whose vertex is the point (h, k). The parabola is symmetric with respect to the line $x = h$. If $a > 0$, the parabola opens upward; if $a < 0$, the parabola opens downward.

The sign of a in $f(x) = a(x - h)^2 + k$ determines whether the parabola opens upward or downward. Furthermore, if $|a|$ is small, the parabola opens more flatly than if $|a|$ is large. On the next page is a general procedure for graphing parabolas whose equations are in this form.

Graphing Quadratic Functions with Equations in the Form $f(x) = a(x - h)^2 + k$

To graph $f(x) = a(x - h)^2 + k$,

1. Determine whether the parabola opens upward or downward. If $a > 0$, it opens upward. If $a < 0$, it opens downward.
2. Determine the vertex of the parabola. The vertex is (h, k).
3. Find any x-intercepts by solving $f(x) = 0$. The equation's real solutions are the x-intercepts.
4. Find the y-intercept by computing $f(0)$.
5. Plot the intercepts, the vertex, and additional points as necessary. Connect these points with a smooth curve that is shaped like a bowl or an inverted bowl.

In the graphs that follow, we will show each axis of symmetry as a dashed vertical line. Because this vertical line passes through the vertex, (h, k), its equation is $x = h$. The line is dashed because it is not part of the parabola.

Study Tip

It's easy to make a sign error when finding h, the x-coordinate of the vertex. In

$$f(x) = a(x - h)^2 + k,$$

***h* is the number that follows the subtraction sign.**

- $f(x) = -2(x - 3)^2 + 8$

  The number *after* the subtraction is 3: $h = 3$.

- $f(x) = (x + 3)^2 + 1$
  $= (x - (-3))^2 + 1$

  The number *after* the subtraction is −3: $h = -3$.

EXAMPLE 1 Graphing a Quadratic Function in the Form $f(x) = a(x - h)^2 + k$

Graph the quadratic function $f(x) = -2(x - 3)^2 + 8$.

Solution We can graph this function by following the steps in the preceding box. We begin by identifying values for a, h, and k.

$$f(x) = a(x - h)^2 + k$$

$a = -2$ $h = 3$ $k = 8$

$$f(x) = -2(x - 3)^2 + 8$$

Step 1. Determine how the parabola opens. Note that a, the coefficient of x^2, is -2. Thus, $a < 0$; this negative value tells us that the parabola opens downward.

Step 2. Find the vertex. The vertex of the parabola is at (h, k). Because $h = 3$ and $k = 8$, the parabola has its vertex at $(3, 8)$.

Step 3. Find the x-intercepts by solving $f(x) = 0$. Replace $f(x)$ with 0 in $f(x) = -2(x - 3)^2 + 8$.

$$0 = -2(x - 3)^2 + 8$$ Find x-intercepts, setting f(x) equal to 0.

$$2(x - 3)^2 = 8$$ Solve for x. Add $2(x - 3)^2$ to both sides of the equation.

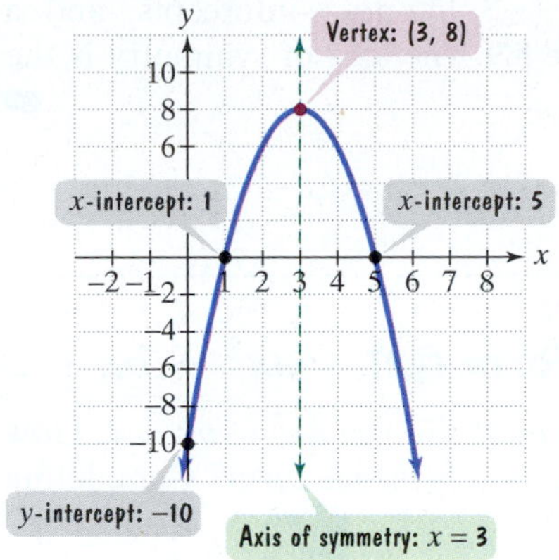

FIGURE 8.8 The graph of $f(x) = -2(x-3)^2 + 8$

$$(x-3)^2 = 4 \quad \text{Divide both sides by 2.}$$

$$x - 3 = \sqrt{4} \quad \text{or} \quad x - 3 = -\sqrt{4} \quad \text{Apply the square root property.}$$

$$x - 3 = 2 \qquad x - 3 = -2 \quad \sqrt{4} = 2$$

$$x = 5 \qquad x = 1 \quad \text{Add 3 to both sides in each equation.}$$

The x-intercepts are 5 and 1. The parabola passes through $(5, 0)$ and $(1, 0)$.

Step 4. Find the y-intercept by computing $f(0)$. Replace x with 0 in $f(x) = -2(x-3)^2 + 8$.

$$f(0) = -2(0-3)^2 + 8 = -2(-3)^2 + 8 = -2(9) + 8 = -10$$

The y-intercept is -10. The parabola passes through $(0, -10)$.

Step 5. Graph the parabola. With a vertex at $(3, 8)$, x-intercepts at 5 and 1, and a y-intercept at -10, the graph of f is shown in **Figure 8.8**. The axis of symmetry is the vertical line whose equation is $x = 3$.

✓ **CHECK POINT 1** Graph the quadratic function $f(x) = -(x-1)^2 + 4$.

EXAMPLE 2 Graphing a Quadratic Function in the Form $f(x) = a(x-h)^2 + k$

Graph the quadratic function $f(x) = (x+3)^2 + 1$.

Solution We begin by finding values for a, h, and k.

$$f(x) = a(x-h)^2 + k \quad \text{Form of quadratic function}$$

$$f(x) = (x+3)^2 + 1 \quad \text{Given function}$$

$$f(x) = 1(x-(-3))^2 + 1$$

$a = 1$ $h = -3$ $k = 1$

Step 1. Determine how the parabola opens. Note that a, the coefficient of x^2, is 1. Thus, $a > 0$; this positive value tells us that the parabola opens upward.

Step 2. Find the vertex. The vertex of the parabola is at (h, k). Because $h = -3$ and $k = 1$, the parabola has its vertex at $(-3, 1)$.

Step 3. Find the x-intercepts by solving $f(x) = 0$. Replace $f(x)$ with 0 in $f(x) = (x+3)^2 + 1$. Because the vertex is at $(-3, 1)$, which lies above the x-axis, and the parabola opens upward, it appears that this parabola has no x-intercepts. We can verify this observation algebraically.

$$0 = (x+3)^2 + 1 \quad \text{Find possible x-intercepts, setting } f(x) \text{ equal to 0.}$$

$$-1 = (x+3)^2 \quad \text{Solve for x. Subtract 1 from both sides.}$$

$$x + 3 = \sqrt{-1} \quad \text{or} \quad x + 3 = -\sqrt{-1} \quad \text{Apply the square root property.}$$

$$x + 3 = i \qquad x + 3 = -i \quad \sqrt{-1} = i$$

$$x = -3 + i \qquad x = -3 - i \quad \text{The solutions are } -3 \pm i.$$

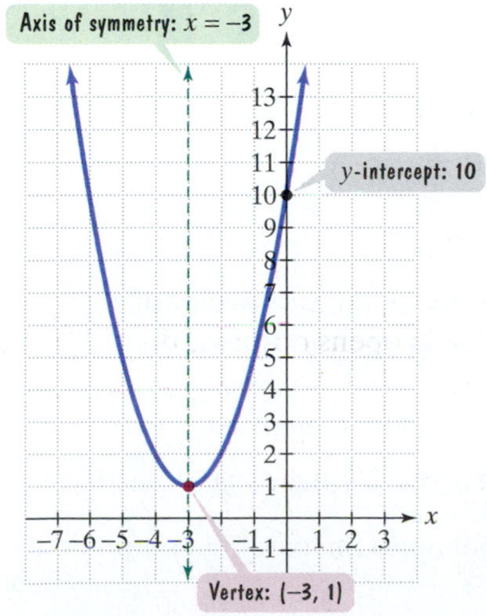

FIGURE 8.9 The graph of $f(x) = (x+3)^2 + 1$

Because this equation has no real solutions, the parabola has no x-intercepts.

Step 4. Find the y-intercept by computing $f(0)$. Replace x with 0 in $f(x) = (x+3)^2 + 1$.

$$f(0) = (0+3)^2 + 1 = 3^2 + 1 = 9 + 1 = 10$$

The y-intercept is 10. The parabola passes through $(0, 10)$.

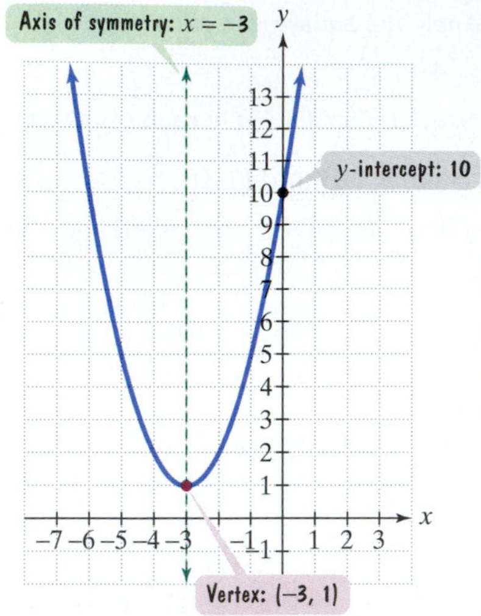

FIGURE 8.9 (repeated)

Step 5. Graph the parabola. With a vertex at $(-3, 1)$, no x-intercepts, and a y-intercept at 10, the graph of f is shown in **Figure 8.9**. The axis of symmetry is the vertical line whose equation is $x = -3$.

✓ CHECK POINT 2 Graph the quadratic function $f(x) = (x - 2)^2 + 1$.

3 Graph parabolas in the form $f(x) = ax^2 + bx + c$.

Graphing Quadratic Functions in the Form $f(x) = ax^2 + bx + c$

Quadratic functions are frequently expressed in the form $f(x) = ax^2 + bx + c$. How can we identify the vertex of a parabola whose equation is in this form? Completing the square provides the answer to this question.

$$f(x) = ax^2 + bx + c$$

$$= a\left(x^2 + \frac{b}{a}x\right) + c \qquad \text{Factor out } a \text{ from } ax^2 + bx.$$

$$= a\left(x^2 + \frac{b}{a}x + \frac{b^2}{4a^2}\right) + c - a\left(\frac{b^2}{4a^2}\right)$$

Complete the square by adding the square of half the coefficient of x.

By completing the square, we added $a \cdot \frac{b^2}{4a^2}$. To avoid changing the function's equation, we must subtract this term.

$$= a\left(x + \frac{b}{2a}\right)^2 + c - \frac{b^2}{4a} \qquad \text{Write the trinomial as the square of a binomial and simplify the constant term.}$$

Now let's compare the form of this equation with a quadratic function in the form $f(x) = a(x - h)^2 + k$.

The form we know how to graph: $f(x) = a(x - h)^2 + k$

$h = -\frac{b}{2a}$ $\quad$ $k = c - \frac{b^2}{4a}$

Equation under discussion: $f(x) = a\left(x - \left(-\frac{b}{2a}\right)\right)^2 + c - \frac{b^2}{4a}$

The important part of this observation is that h, the x-coordinate of the vertex, is $-\frac{b}{2a}$. The y-coordinate can be found by evaluating the function at $-\frac{b}{2a}$.

The Vertex of a Parabola Whose Equation Is $f(x) = ax^2 + bx + c$

Consider the parabola defined by the quadratic function $f(x) = ax^2 + bx + c$. The parabola's vertex is $\left(-\frac{b}{2a}, f\left(-\frac{b}{2a}\right)\right)$. The x-coordinate is $-\frac{b}{2a}$. The y-coordinate is found by substituting the x-coordinate into the parabola's equation and evaluating the function at this value of x.

EXAMPLE 3 Finding a Parabola's Vertex

Find the vertex for the parabola whose equation is $f(x) = 3x^2 + 12x + 8$.

Solution We know that the x-coordinate of the vertex is $x = -\frac{b}{2a}$. Let's identify the numbers a, b, and c in the given equation, which is in the form $f(x) = ax^2 + bx + c$.

$$f(x) = 3x^2 + 12x + 8$$

$a = 3$ $b = 12$ $c = 8$

Substitute the values of a and b into the equation for the x-coordinate:

$$x = -\frac{b}{2a} = -\frac{12}{2 \cdot 3} = -\frac{12}{6} = -2.$$

The x-coordinate of the vertex is -2. We substitute -2 for x into the equation of the function, $f(x) = 3x^2 + 12x + 8$, to find the y-coordinate:

$$f(-2) = 3(-2)^2 + 12(-2) + 8 = 3(4) + 12(-2) + 8 = 12 - 24 + 8 = -4.$$

The vertex is $(-2, -4)$.

✓ **CHECK POINT 3** Find the vertex for the parabola whose equation is $f(x) = 2x^2 + 8x - 1$.

We can apply our five-step procedure and graph parabolas in the form $f(x) = ax^2 + bx + c$.

Graphing Quadratic Functions with Equations in the Form $f(x) = ax^2 + bx + c$

To graph $f(x) = ax^2 + bx + c$,

1. Determine whether the parabola opens upward or downward. If $a > 0$, it opens upward. If $a < 0$, it opens downward.
2. Determine the vertex of the parabola. The vertex is $\left(-\frac{b}{2a}, f\left(-\frac{b}{2a}\right)\right)$.
3. Find any x-intercepts by solving $f(x) = 0$. The real solutions of $ax^2 + bx + c = 0$ are the x-intercepts.
4. Find the y-intercept by computing $f(0)$. Because $f(0) = c$ (the constant term in the function's equation), the y-intercept is c and the parabola passes through $(0, c)$.
5. Plot the intercepts, the vertex, and additional points as necessary. Connect these points with a smooth curve.

EXAMPLE 4 Graphing a Quadratic Function in the Form $f(x) = ax^2 + bx + c$

Graph the quadratic function $f(x) = -x^2 - 2x + 1$. Use the graph to identify the function's domain and its range.

Solution

Step 1. Determine how the parabola opens. Note that a, the coefficient of x^2, is -1. Thus, $a < 0$; this negative value tells us that the parabola opens downward.

Step 2. Find the vertex. We know that the x-coordinate of the vertex is $x = -\frac{b}{2a}$. We identify a, b, and c in $f(x) = ax^2 + bx + c$.

$$f(x) = -x^2 - 2x + 1$$

$a = -1$ $b = -2$ $c = 1$

Substitute the values of a and b into the equation for the x-coordinate:

$$x = -\frac{b}{2a} = -\frac{-2}{2(-1)} = -\left(\frac{-2}{-2}\right) = -1.$$

The x-coordinate of the vertex is -1. We substitute -1 for x into the equation of the function, $f(x) = -x^2 - 2x + 1$, to find the y-coordinate:

$$f(-1) = -(-1)^2 - 2(-1) + 1 = -1 + 2 + 1 = 2.$$

The vertex is $(-1, 2)$.

Step 3. Find the x-intercepts by solving $f(x) = 0$. Replace $f(x)$ with 0 in $f(x) = -x^2 - 2x + 1$. We obtain $0 = -x^2 - 2x + 1$. This equation cannot be solved by factoring. We will use the quadratic formula to solve it.

$$-x^2 - 2x + 1 = 0$$

$a = -1$ $b = -2$ $c = 1$

$$x = \frac{-b \pm \sqrt{b^2 - 4ac}}{2a} = \frac{-(-2) \pm \sqrt{(-2)^2 - 4(-1)(1)}}{2(-1)} = \frac{2 \pm \sqrt{4 - (-4)}}{-2}$$

To locate the x-intercepts, we need decimal approximations. Thus, there is no need to simplify the radical form of the solutions.

$$x = \frac{2 + \sqrt{8}}{-2} \approx -2.4 \quad \text{or} \quad x = \frac{2 - \sqrt{8}}{-2} \approx 0.4$$

The x-intercepts are approximately -2.4 and 0.4. The parabola passes through $(-2.4, 0)$ and $(0.4, 0)$.

Step 4. Find the y-intercept by computing $f(0)$. Replace x with 0 in $f(x) = -x^2 - 2x + 1$.

$$f(0) = -0^2 - 2 \cdot 0 + 1 = 1$$

The y-intercept is 1, which is the constant term in the function's equation. The parabola passes through $(0, 1)$.

Step 5. Graph the parabola. With a vertex at $(-1, 2)$, x-intercepts at -2.4 and 0.4, and a y-intercept at 1, the graph of f is shown in **Figure 8.10(a)**. The axis of symmetry is the vertical line whose equation is $x = -1$.

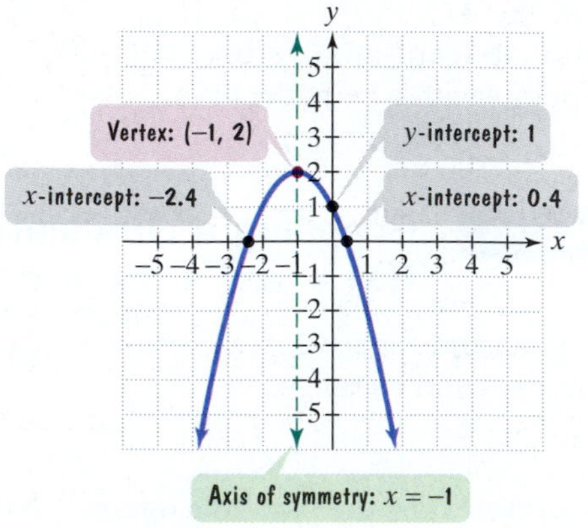

FIGURE 8.10(a) The graph of $f(x) = -x^2 - 2x + 1$

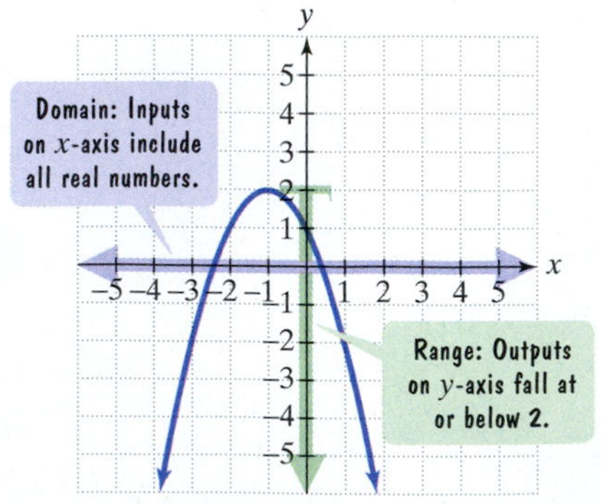

FIGURE 8.10(b) Determining the domain and range of $f(x) = -x^2 - 2x + 1$

Study Tip

The domain of any quadratic function includes all real numbers. If the vertex is the graph's highest point, the range includes all real numbers at or below the y-coordinate of the vertex. If the vertex is the graph's lowest point, the range includes all real numbers at or above the y-coordinate of the vertex.

Now we are ready to determine the domain and range of $f(x) = -x^2 - 2x + 1$. We can use the parabola, shown again in **Figure 8.10(b)**, to do so. To find the domain, look for all the inputs on the x-axis that correspond to points on the graph. As the graph widens and continues to fall at both ends, can you see that these inputs include all real numbers?

$$\text{Domain of } f \text{ is } \{x \mid x \text{ is a real number}\} \text{ or } (-\infty, \infty).$$

To find the range, look for all the outputs on the y-axis that correspond to points on the graph. **Figure 8.10(b)** shows that the parabola's vertex, $(-1, 2)$, is the highest point on the graph. Because the y-coordinate of the vertex is 2, outputs on the y-axis fall at or below 2.

$$\text{Range of } f \text{ is } \{y \mid y \leq 2\} \text{ or } (-\infty, 2].$$

✓ **CHECK POINT 4** Graph the quadratic function $f(x) = -x^2 + 4x + 1$. Use the graph to identify the function's domain and its range.

4 Determine a quadratic function's minimum or maximum value.

Minimum and Maximum Values of Quadratic Functions

Consider the quadratic function $f(x) = ax^2 + bx + c$. If $a > 0$, the parabola opens upward and the vertex is its lowest point. If $a < 0$, the parabola opens downward and the vertex is its highest point. The x-coordinate of the vertex is $-\frac{b}{2a}$. Thus, we can find the minimum or maximum value of f by evaluating the quadratic function at $x = -\frac{b}{2a}$.

Minimum and Maximum: Quadratic Functions

Consider the quadratic function $f(x) = ax^2 + bx + c$.

1. If $a > 0$, then f has a minimum that occurs at $x = -\frac{b}{2a}$. This minimum value is $f\left(-\frac{b}{2a}\right)$.
2. If $a < 0$, then f has a maximum that occurs at $x = -\frac{b}{2a}$. This maximum value is $f\left(-\frac{b}{2a}\right)$.

In each case, the value of x gives the location of the minimum or maximum value. The value of y, or $f\left(-\frac{b}{2a}\right)$, gives that minimum or maximum value.

EXAMPLE 5 **Obtaining Information about a Quadratic Function from Its Equation**

Consider the quadratic function $f(x) = -3x^2 + 6x - 13$.

a. Determine, without graphing, whether the function has a minimum value or a maximum value.
b. Find the minimum or maximum value and determine where it occurs.
c. Identify the function's domain and its range.

Solution We begin by identifying a, b, and c in the function's equation:

$$f(x) = -3x^2 + 6x - 13.$$

$a = -3$ $b = 6$ $c = -13$

$f(x) = -3x^2 + 6x - 13$

$a = -3$ $b = 6$ $c = -13$

The given function (repeated)

a. Because $a < 0$, the function has a maximum value.

b. The maximum value occurs at

$$x = -\frac{b}{2a} = -\frac{6}{2(-3)} = -\frac{6}{-6} = -(-1) = 1.$$

The maximum value occurs at $x = 1$ and the maximum value of $f(x) = -3x^2 + 6x - 13$ is

$$f(1) = -3 \cdot 1^2 + 6 \cdot 1 - 13 = -3 + 6 - 13 = -10.$$

We see that the maximum is -10 at $x = 1$.

c. Like all quadratic functions, the domain is $\{x|x \text{ is a real number}\}$ or $(-\infty, \infty)$. Because the function's maximum value is -10, the range includes all real numbers at or below -10. The range is $\{y|y \leq -10\}$ or $(-\infty, -10]$.

We can use the graph of $f(x) = -3x^2 + 6x - 13$ to visualize the results of Example 5. **Figure 8.11** shows the graph in a $[-6, 6, 1]$ by $[-50, 20, 10]$ viewing rectangle. The maximum function feature verifies that the function's maximum is -10 at $x = 1$. Notice that x gives the location of the maximum and y gives the maximum value. Notice, too, that the maximum value is -10 and not the ordered pair $(1, -10)$.

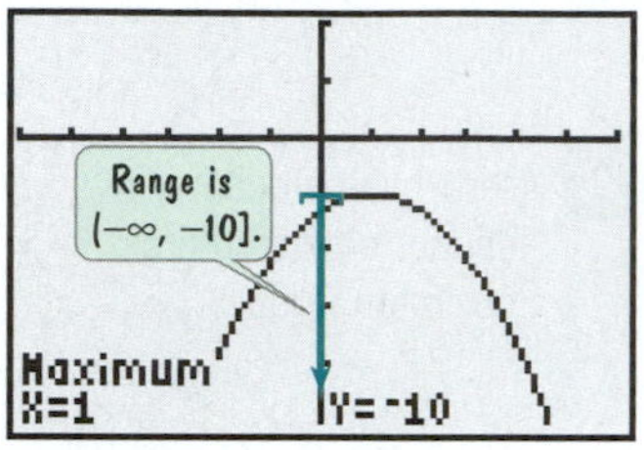

[−6, 6, 1] by [−50, 20, 10]

FIGURE 8.11

CHECK POINT 5 Repeat parts (a) through (c) of Example 5 using the quadratic function $f(x) = 4x^2 - 16x + 1000$.

---

5 Solve problems involving a quadratic function's minimum or maximum value.

Applications of Quadratic Functions

Many applied problems involve finding the maximum or minimum value of a quadratic function, as well as where this value occurs.

EXAMPLE 6 Parabolic Paths of a Shot Put

An athlete whose event is the shot put releases the shot with the same initial velocity, but at different angles. **Figure 8.12** shows the parabolic paths for shots released at angles of 35° and 65°.

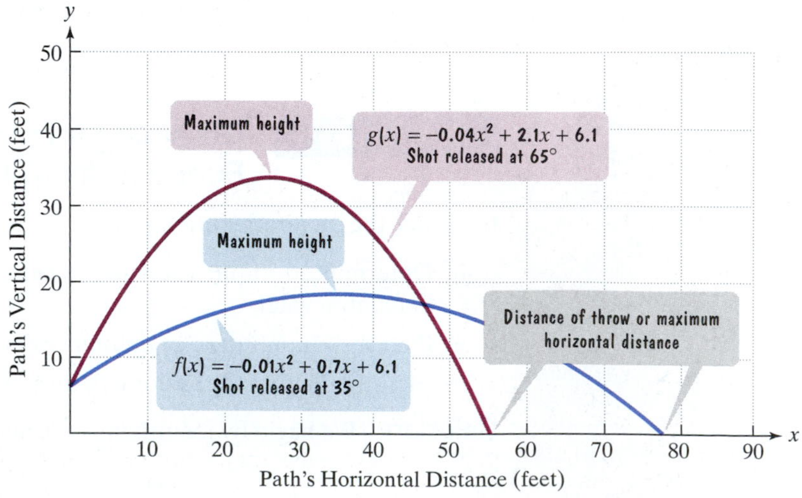

FIGURE 8.12 Two paths of a shot put

When the shot is released at an angle of 35°, its path can be modeled by the function

$$f(x) = -0.01x^2 + 0.7x + 6.1,$$

in which x is the shot's horizontal distance, in feet, and $f(x)$ is its height, in feet. What is the maximum height of this shot's path?

Solution The quadratic function is in the form $f(x) = ax^2 + bx + c$, with $a = -0.01$ and $b = 0.7$. Because $a < 0$, the function has a maximum that occurs at $x = -\frac{b}{2a}$.

$$x = -\frac{b}{2a} = -\frac{0.7}{2(-0.01)} = -(-35) = 35$$

This means that the shot's maximum height occurs when its horizontal distance is 35 feet. Can you see how this is shown by the blue graph of f in **Figure 8.12**? The maximum height of this path is

$$f(35) = -0.01(35)^2 + 0.7(35) + 6.1 = 18.35$$

or 18.35 feet.

✓ **CHECK POINT 6** Use function g, whose equation and graph are shown in **Figure 8.12**, to find the maximum height, to the nearest tenth of a foot, when the shot is released at an angle of 65°.

Quadratic functions can also be modeled from verbal conditions. Once we have obtained a quadratic function, we can then use the x-coordinate of the vertex to determine its maximum or minimum value. Here is a step-by-step strategy for solving these kinds of problems:

Strategy for Solving Problems Involving Maximizing or Minimizing Quadratic Functions

1. Read the problem carefully and decide which quantity is to be maximized or minimized.
2. Use the conditions of the problem to express the quantity as a function in one variable.
3. Rewrite the function in the form $f(x) = ax^2 + bx + c$.
4. Calculate $-\frac{b}{2a}$. If $a > 0$, f has a minimum at $x = -\frac{b}{2a}$. This minimum value is $f\left(-\frac{b}{2a}\right)$. If $a < 0$, f has a maximum at $x = -\frac{b}{2a}$. This maximum value is $f\left(-\frac{b}{2a}\right)$.
5. Answer the question posed in the problem.

EXAMPLE 7 Minimizing a Product

Among all pairs of numbers whose difference is 10, find a pair whose product is as small as possible. What is the minimum product?

Solution

Step 1. Decide what must be maximized or minimized. We must minimize the product of two numbers. Calling the numbers x and y, and calling the product P, we must minimize

$$P = xy.$$

Step 2. Express this quantity as a function in one variable. In the formula $P = xy$, P is expressed in terms of two variables, x and y. However, because the difference of the numbers is 10, we can write

$$x - y = 10.$$

We can solve this equation for y in terms of x (or vice versa), substitute the result into $P = xy$, and obtain P as a function of one variable.

$$-y = -x + 10 \quad \text{Subtract x from both sides of}$$
$$\qquad\qquad\qquad x - y = 10.$$
$$y = x - 10 \quad \text{Multiply both sides of the equation}$$
$$\qquad\qquad\qquad \text{by } -1 \text{ and solve for y.}$$

Now we substitute $x - 10$ for y in $P = xy$.

$$P = xy = x(x - 10).$$

Because P is now a function of x, we can write

$$P(x) = x(x - 10).$$

Step 3. Write the function in the form $f(x) = ax^2 + bx + c$. We apply the distributive property to obtain

$$P(x) = x(x - 10) = x^2 - 10x.$$

$$\boxed{a = 1} \qquad \boxed{b = -10}$$

Using Technology

Numeric Connections

The ⎡TABLE⎤ feature of a graphing utility can be used to verify our work in Example 7.

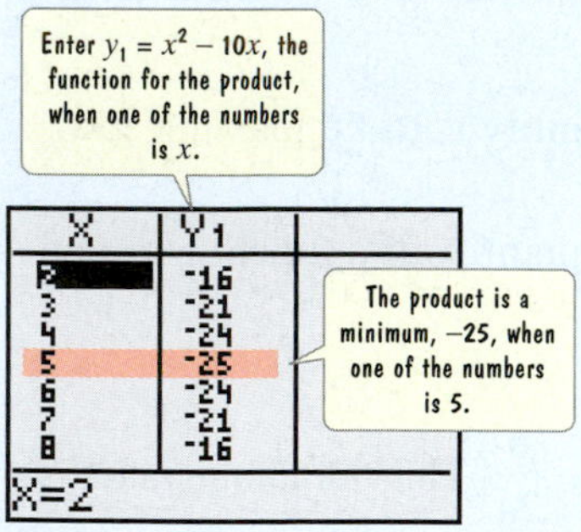

Enter $y_1 = x^2 - 10x$, the function for the product, when one of the numbers is x.

The product is a minimum, -25, when one of the numbers is 5.

Step 4. Calculate $-\dfrac{b}{2a}$. If $a > 0$, the function has a minimum at this value. The voice balloons show that $a = 1$ and $b = -10$.

$$x = -\frac{b}{2a} = -\frac{-10}{2(1)} = -(-5) = 5$$

This means that the product, P, of two numbers whose difference is 10 is a minimum when one of the numbers, x, is 5.

Step 5. Answer the question posed by the problem. The problem asks for the two numbers and the minimum product. We found that one of the numbers, x, is 5. Now we must find the second number, y.

$$y = x - 10 = 5 - 10 = -5.$$

The number pair whose difference is 10 and whose product is as small as possible is 5, -5. The minimum product is $5(-5)$, or -25. ∎

✓ **CHECK POINT 7** Among all pairs of numbers whose difference is 8, find a pair whose product is as small as possible. What is the minimum product?

EXAMPLE 8   Maximizing Area

You have 100 yards of fencing to enclose a rectangular region. Find the dimensions of the rectangle that maximize the enclosed area. What is the maximum area?

Solution

Step 1. Decide what must be maximized or minimized. We must maximize area. What we do not know are the rectangle's dimensions, x and y.

Step 2. Express this quantity as a function in one variable. Because we must maximize area, we have $A = xy$. We need to transform this into a function in which A is represented by one variable. Because you have 100 yards of fencing, the perimeter of the rectangle is 100 yards. This means that

$$2x + 2y = 100.$$

We can solve this equation for y in terms of x, substitute the result into $A = xy$, and obtain A as a function in one variable. We begin by solving for y.

$$2y = 100 - 2x \quad \text{Subtract } 2x \text{ from both sides.}$$

$$y = \frac{100 - 2x}{2} \quad \text{Divide both sides by 2.}$$

$$y = 50 - x \quad \text{Divide each term in the numerator by 2.}$$

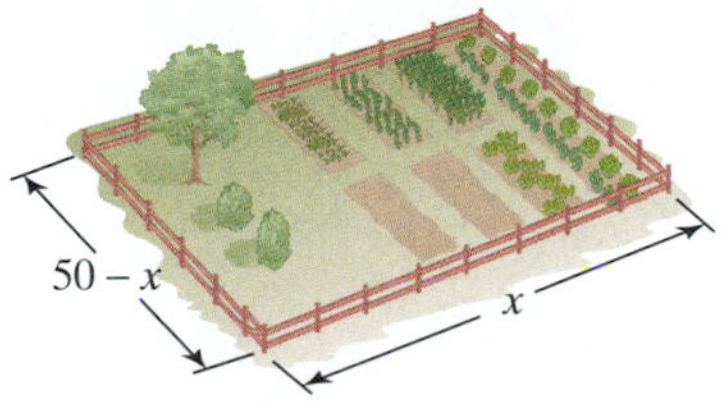

FIGURE 8.13 What value of x will maximize the rectangle's area?

Now we substitute $50 - x$ for y in $A = xy$.

$$A = xy = x(50 - x)$$

The rectangle and its dimensions are illustrated in **Figure 8.13**. Because A is now a function of x, we can write

$$A(x) = x(50 - x).$$

This function models the area, $A(x)$, of any rectangle whose perimeter is 100 yards in terms of one of its dimensions, x.

Step 3. Write the function in the form $f(x) = ax^2 + bx + c$. We apply the distributive property to obtain

$$A(x) = x(50 - x) = 50x - x^2 = -x^2 + 50x.$$

$a = -1$ $\quad$ $b = 50$

Using Technology

Graphic Connections

The graph of the area function

$$A(x) = x(50 - x)$$

was obtained with a graphing utility using a [0, 50, 2] by [0, 700, 25] viewing rectangle. The maximum function feature verifies that a maximum area of 625 square yards occurs when one of the dimensions is 25 yards.

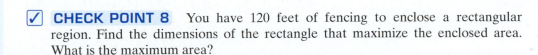

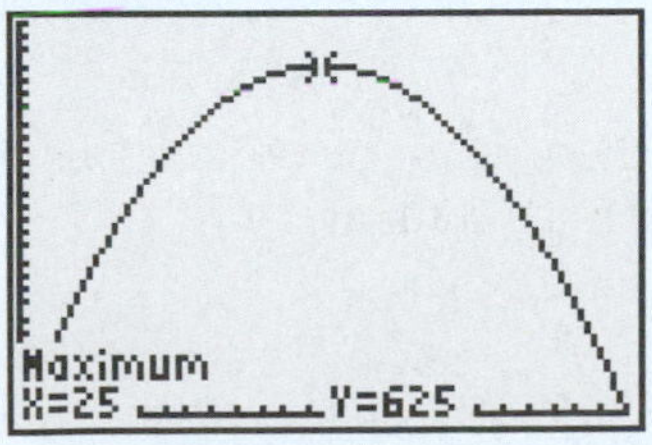

Step 4. Calculate $-\frac{b}{2a}$. If $a < 0$, the function has a maximum at this value. The voice balloons show that $a = -1$ and $b = 50$.

$$x = -\frac{b}{2a} = -\frac{50}{2(-1)} = 25$$

This means that the area, $A(x)$, of a rectangle with perimeter 100 yards is a maximum when one of the rectangle's dimensions, x, is 25 yards.

Step 5. Answer the question posed by the problem. We found that $x = 25$. **Figure 8.13** shows that the rectangle's other dimension is $50 - x = 50 - 25 = 25$. The dimensions of the rectangle that maximize the enclosed area are 25 yards by 25 yards. The rectangle that gives the maximum area is actually a square with an area of 25 yards · 25 yards, or 625 square yards.

✓ **CHECK POINT 8** You have 120 feet of fencing to enclose a rectangular region. Find the dimensions of the rectangle that maximize the enclosed area. What is the maximum area?

8.3 EXERCISE SET

MyMathLab MathXL PRACTICE WATCH DOWNLOAD READ REVIEW

Practice Exercises

In Exercises 1–4, the graph of a quadratic function is given. Write the function's equation, selecting from the following options:

$$f(x) = (x + 1)^2 - 1, g(x) = (x + 1)^2 + 1,$$
$$h(x) = (x - 1)^2 + 1, j(x) = (x - 1)^2 - 1.$$

1.

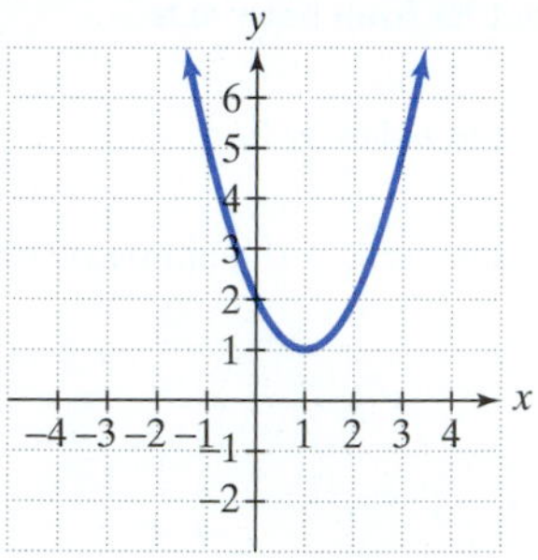

2.

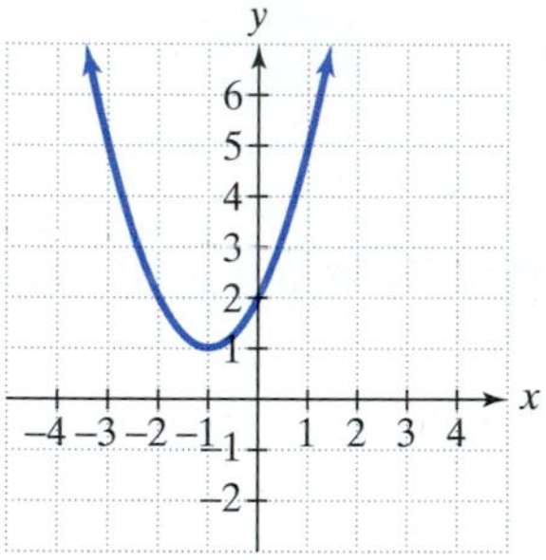

3.

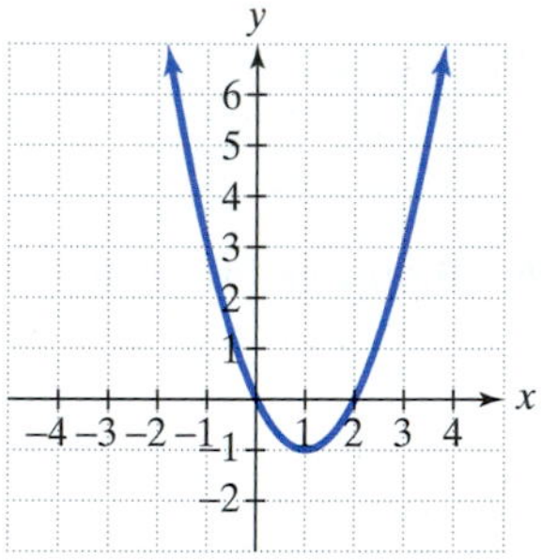

4.

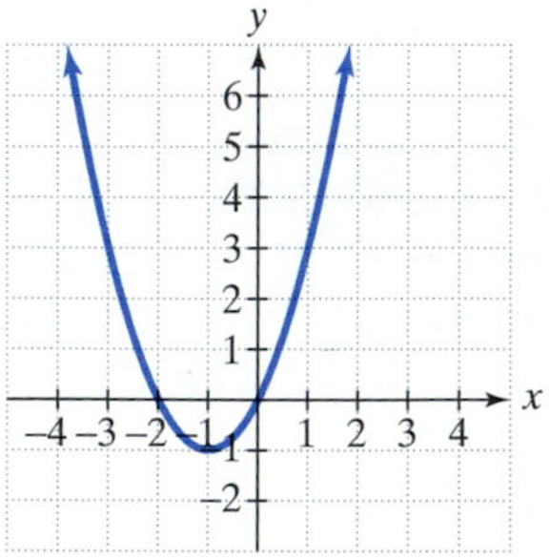

In Exercises 5–8, the graph of a quadratic function is given. Write the function's equation, selecting from the following options:

$$f(x) = x^2 + 2x + 1, g(x) = x^2 - 2x + 1,$$
$$h(x) = x^2 - 1, j(x) = -x^2 - 1.$$

5.

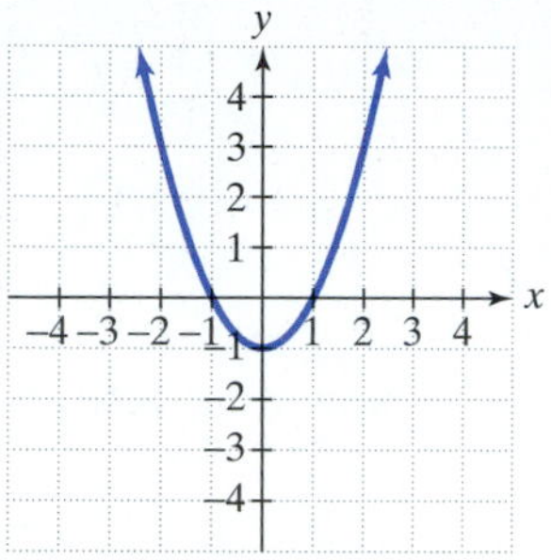

6.

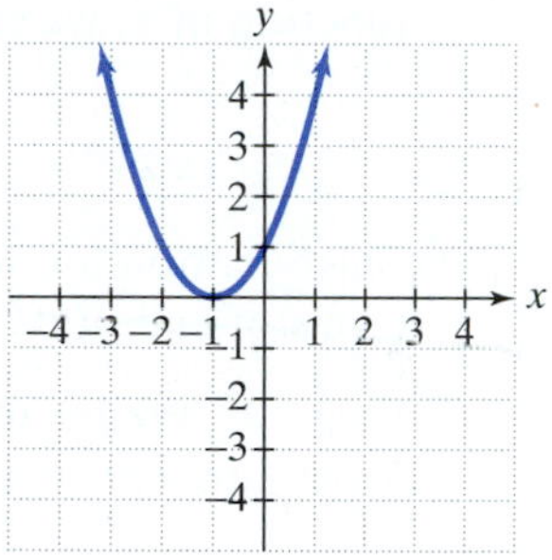

7.

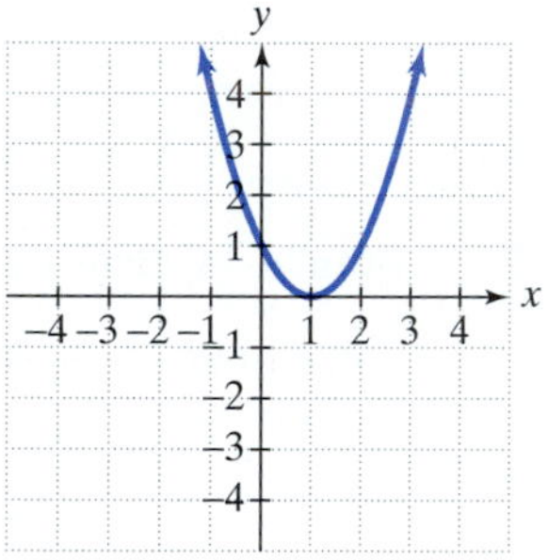

8.

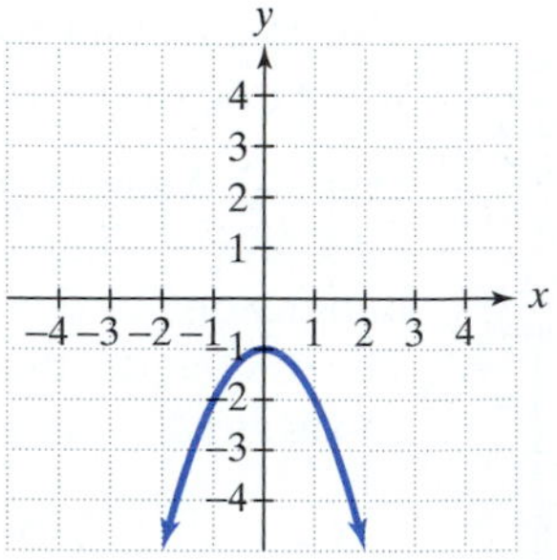

In Exercises 9–16, find the coordinates of the vertex for the parabola defined by the given quadratic function.

9. $f(x) = 2(x - 3)^2 + 1$

10. $f(x) = -3(x - 2)^2 + 12$

11. $f(x) = -2(x + 1)^2 + 5$

12. $f(x) = -2(x + 4)^2 - 8$

13. $f(x) = 2x^2 - 8x + 3$

14. $f(x) = 3x^2 - 12x + 1$

15. $f(x) = -x^2 - 2x + 8$

16. $f(x) = -2x^2 + 8x - 1$

In Exercises 17–38, use the vertex and intercepts to sketch the graph of each quadratic function. Use the graph to identify the function's range.

17. $f(x) = (x - 4)^2 - 1$

18. $f(x) = (x - 1)^2 - 2$

19. $f(x) = (x - 1)^2 + 2$

20. $f(x) = (x - 3)^2 + 2$

21. $y - 1 = (x - 3)^2$

22. $y - 3 = (x - 1)^2$

23. $f(x) = 2(x + 2)^2 - 1$

24. $f(x) = \frac{5}{4} - \left(x - \frac{1}{2}\right)^2$

25. $f(x) = 4 - (x - 1)^2$

26. $f(x) = 1 - (x - 3)^2$

27. $f(x) = x^2 - 2x - 3$

28. $f(x) = x^2 - 2x - 15$

29. $f(x) = x^2 + 3x - 10$

30. $f(x) = 2x^2 - 7x - 4$

31. $f(x) = 2x - x^2 + 3$

32. $f(x) = 5 - 4x - x^2$

33. $f(x) = x^2 + 6x + 3$

34. $f(x) = x^2 + 4x - 1$

35. $f(x) = 2x^2 + 4x - 3$

36. $f(x) = 3x^2 - 2x - 4$

37. $f(x) = 2x - x^2 - 2$

38. $f(x) = 6 - 4x + x^2$

In Exercises 39–44, an equation of a quadratic function is given.

a. *Determine, without graphing, whether the function has a minimum value or a maximum value.*

b. *Find the minimum or maximum value and determine where it occurs.*

c. *Identify the function's domain and its range.*

39. $f(x) = 3x^2 - 12x - 1$

40. $f(x) = 2x^2 - 8x - 3$

41. $f(x) = -4x^2 + 8x - 3$

42. $f(x) = -2x^2 - 12x + 3$

43. $f(x) = 5x^2 - 5x$

44. $f(x) = 6x^2 - 6x$

Practice PLUS

In Exercises 45–48, give the domain and the range of each quadratic function whose graph is described.

45. The vertex is $(-1, -2)$ and the parabola opens up.

46. The vertex is $(-3, -4)$ and the parabola opens down.

47. Maximum $= -6$ at $x = 10$

48. Minimum $= 18$ at $x = -6$

In Exercises 49–52, write an equation of the parabola that has the same shape as the graph of $f(x) = 2x^2$, but with the given point as the vertex.

49. $(5, 3)$

50. $(7, 4)$

51. $(-10, -5)$

52. $(-8, -6)$

In Exercises 53–56, write an equation of the parabola that has the same shape as the graph of $f(x) = 3x^2$ or $g(x) = -3x^2$, but with the given maximum or minimum.

53. Maximum $= 4$ at $x = -2$

54. Maximum $= -7$ at $x = 5$

55. Minimum $= 0$ at $x = 11$

56. Minimum $= 0$ at $x = 9$

Application Exercises

57. The graph shows U.S. adult wine consumption, in gallons per person, for selected years from 1980 through 2005.

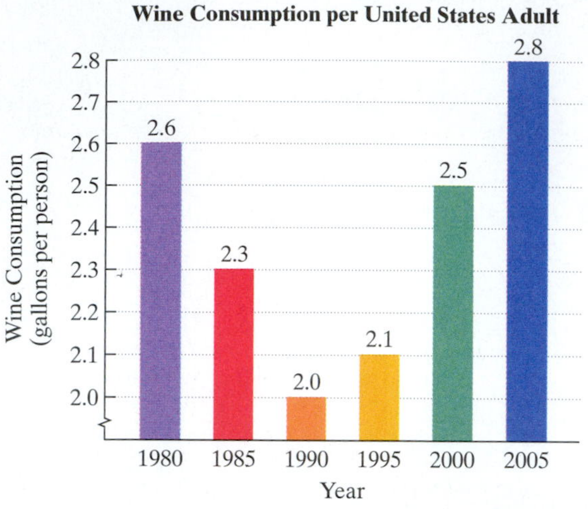

Source: Adams Business Media

The function

$$f(x) = 0.004x^2 - 0.094x + 2.6$$

models U.S. wine consumption, $f(x)$, in gallons per person, x years after 1980.

a. According to this function, what was U.S. adult wine consumption in 2005? Does this overestimate or underestimate the value shown by the graph? By how much?

b. According to this function, in which year was wine consumption at a minimum? Round to the nearest year. What does the function give for per capita consumption for that year? Does this seem reasonable in terms of the data shown by the graph or has model breakdown occurred?

58. The graph shows the number of movie tickets sold in the United States, in billions, from 2000 through 2005.

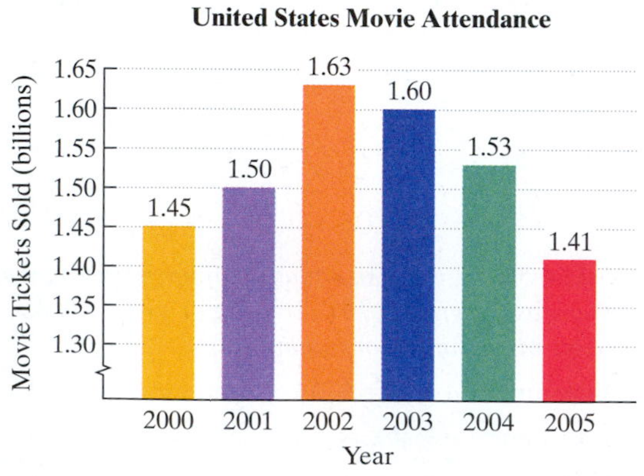

Source: National Association of Theater Owners

The function

$$f(x) = -0.03x^2 + 0.14x + 1.43$$

models U.S. movie attendance, $f(x)$, in billions of tickets sold, x years after 2000.

a. According to this function, how many billions of movie tickets were sold in 2005? Does this overestimate or underestimate the number shown by the graph? By how much?

b. According to this function, in which year was movie attendance at a maximum? Round to the nearest year. What does the function give for the billions of tickets sold for that year? By how much does this differ from the number shown by the graph?

59. A person standing close to the edge on the top of a 160-foot building throws a baseball vertically upward. The quadratic function

$$s(t) = -16t^2 + 64t + 160$$

models the ball's height above the ground, $s(t)$, in feet, t seconds after it was thrown.

a. After how many seconds does the ball reach its maximum height? What is the maximum height?

b. How many seconds does it take until the ball finally hits the ground? Round to the nearest tenth of a second.

c. Find $s(0)$ and describe what this means.

d. Use your results from parts (a) through (c) to graph the quadratic function. Begin the graph with $t = 0$ and end with the value of t for which the ball hits the ground.

60. A person standing close to the edge on the top of a 200-foot building throws a baseball vertically upward. The quadratic function

$$s(t) = -16t^2 + 64t + 200$$

models the ball's height above the ground, $s(t)$, in feet, t seconds after it was thrown.

a. After how many seconds does the ball reach its maximum height? What is the maximum height?

b. How many seconds does it take until the ball finally hits the ground? Round to the nearest tenth of a second.

c. Find $s(0)$ and describe what this means.

d. Use your results from parts (a) through (c) to graph the quadratic function. Begin the graph with $t = 0$ and end with the value of t for which the ball hits the ground.

61. Among all pairs of numbers whose sum is 16, find a pair whose product is as large as possible. What is the maximum product?

62. Among all pairs of numbers whose sum is 20, find a pair whose product is as large as possible. What is the maximum product?

63. Among all pairs of numbers whose difference is 16, find a pair whose product is as small as possible. What is the minimum product?

64. Among all pairs of numbers whose difference is 24, find a pair whose product is as small as possible. What is the minimum product?

65. You have 600 feet of fencing to enclose a rectangular plot that borders on a river. If you do not fence the side along the river, find the length and width of the plot that will maximize the area. What is the largest area that can be enclosed?

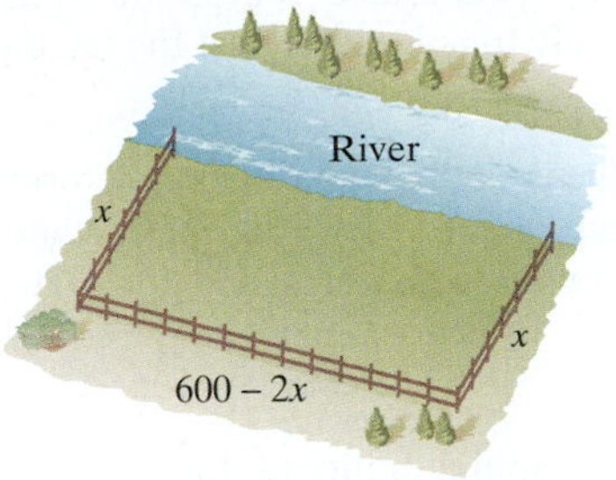

66. You have 200 feet of fencing to enclose a rectangular plot that borders on a river. If you do not fence the side along the river, find the length and width of the plot that will maximize the area. What is the largest area that can be enclosed?

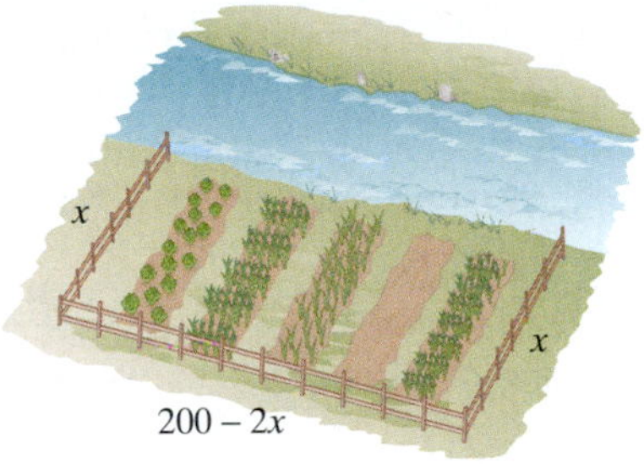

67. You have 50 yards of fencing to enclose a rectangular region. Find the dimensions of the rectangle that maximize the enclosed area. What is the maximum area?

68. You have 80 yards of fencing to enclose a rectangular region. Find the dimensions of the rectangle that maximize the enclosed area. What is the maximum area?

69. A rain gutter is made from sheets of aluminum that are 20 inches wide by turning up the edges to form right angles. Determine the depth of the gutter that will maximize its cross-sectional area and allow the greatest amount of water to flow. What is the maximum cross-sectional area?

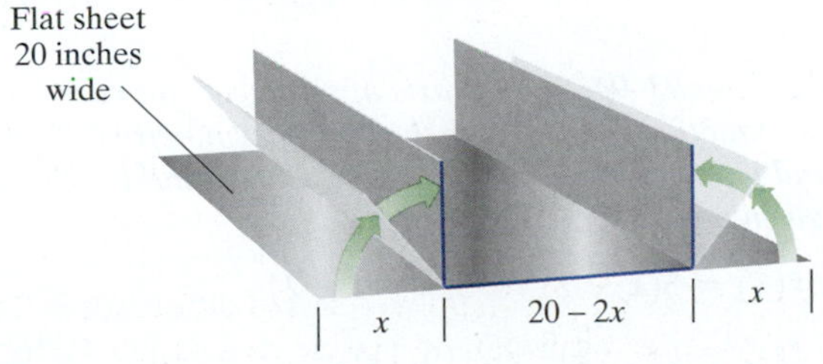

70. A rain gutter is made from sheets of aluminum that are 12 inches wide by turning up the edges to form right angles. Determine the depth of the gutter that will maximize its cross-sectional area and allow the greatest amount of water to flow. What is the maximum cross-sectional area?

In Chapter 3, we saw that the profit, $P(x)$, generated after producing and selling x units of a product is given by the function

$$P(x) = R(x) - C(x),$$

where R and C are the revenue and cost functions, respectively. Use these functions to solve Exercises 71–72.

71. Hunky Beef, a local sandwich store, has a fixed weekly cost of \$525.00, and variable costs for making a roast beef sandwich are \$0.55.

a. Let x represent the number of roast beef sandwiches made and sold each week. Write the weekly cost function, C, for Hunky Beef.

b. The function $R(x) = -0.001x^2 + 3x$ describes the money that Hunky Beef takes in each week from the sale of x roast beef sandwiches. Use this revenue function and the cost function from part (a) to write the store's weekly profit function, P.

c. Use the store's profit function to determine the number of roast beef sandwiches it should make and sell each week to maximize profit. What is the maximum weekly profit?

72. Virtual Fido is a company that makes electronic virtual pets. The fixed weekly cost is \$3000, and variable costs for each pet are \$20.

a. Let x represent the number of virtual pets made and sold each week. Write the weekly cost function, C, for Virtual Fido.

b. The function $R(x) = -x^2 + 1000x$ describes the money that Virtual Fido takes in each week from the sale of x virtual pets. Use this revenue function and the cost function from part (a) to write the weekly profit function, P.

c. Use the profit function to determine the number of virtual pets that should be made and sold each week to maximize profit. What is the maximum weekly profit?

Writing in Mathematics

73. What is a parabola? Describe its shape.

74. Explain how to decide whether a parabola opens upward or downward.

75. Describe how to find a parabola's vertex if its equation is in the form $f(x) = a(x - h)^2 + k$. Give an example.

76. Describe how to find a parabola's vertex if its equation is in the form $f(x) = ax^2 + bx + c$. Use $f(x) = x^2 - 6x + 8$ as an example.

77. A parabola that opens upward has its vertex at (1, 2). Describe as much as you can about the parabola based on this information. Include in your discussion the number of x-intercepts (if any) for the parabola.

Technology Exercises

78. Use a graphing utility to verify any five of your hand-drawn graphs in Exercises 17–38.

79. a. Use a graphing utility to graph $y = 2x^2 - 82x + 720$ in a standard viewing rectangle. What do you observe?

b. Find the coordinates of the vertex for the given quadratic function.

c. The answer to part (b) is $(20.5, -120.5)$. Because the leading coefficient, 2, of the given function is positive, the vertex is a minimum point on the graph. Use this fact to help find a viewing rectangle that will give a relatively complete picture of the parabola. With an axis of symmetry at $x = 20.5$, the setting for x should extend past this, so try Xmin = 0 and Xmax = 30. The setting for y should include (and probably go below) the y-coordinate of the graph's minimum point, so try Ymin = −130. Experiment with Ymax until your utility shows the parabola's major features.

d. In general, explain how knowing the coordinates of a parabola's vertex can help determine a reasonable viewing rectangle on a graphing utility for obtaining a complete picture of the parabola.

In Exercises 80–83, find the vertex for each parabola. Then determine a reasonable viewing rectangle on your graphing utility and use it to graph the parabola.

80. $y = -0.25x^2 + 40x$

81. $y = -4x^2 + 20x + 160$

82. $y = 5x^2 + 40x + 600$

83. $y = 0.01x^2 + 0.6x + 100$

84. The following data show fuel efficiency, in miles per gallon, for all U.S. automobiles in the indicated year.

x (Years after 1940)	y (Average Number of Miles per Gallon for U.S. Automobiles)
1940: 0	14.8
1950: 10	13.9
1960: 20	13.4
1970: 30	13.5
1980: 40	15.9
1990: 50	20.2
2000: 60	22.0

Source: U.S. Department of Transportation

a. Use a graphing utility to draw a scatter plot of the data. Explain why a quadratic function is appropriate for modeling these data.

b. Use the quadratic regression feature to find the quadratic function that best fits the data.

c. Use the equation in part (b) to determine the worst year for automobile fuel efficiency. What was the average number of miles per gallon for that year?

d. Use a graphing utility to draw a scatter plot of the data and graph the quadratic function of best fit on the scatter plot.

Critical Thinking Exercises

Make Sense? *In Exercises 85–88, determine whether each statement "makes sense" or "does not make sense" and explain your reasoning.*

85. Parabolas that open up appear to form smiles $(a > 0)$, while parabolas that open down frown $(a < 0)$.

86. I must have made an error when graphing this parabola because its axis of symmetry is the y-axis.

87. I like to think of a parabola's vertex as the point where it intersects its axis of symmetry.

88. I threw a baseball vertically upward and its path was a parabola.

In Exercises 89–92, determine whether each statement is true or false. If the statement is false, make the necessary change(s) to produce a true statement.

89. No quadratic functions have a range of $(-\infty, \infty)$.

90. The vertex of the parabola described by $f(x) = 2(x - 5)^2 - 1$ is at $(5, 1)$.

91. The graph of $f(x) = -2(x + 4)^2 - 8$ has one y-intercept and two x-intercepts.

92. The maximum value of y for the quadratic function $f(x) = -x^2 + x + 1$ is 1.

In Exercises 93–94, find the axis of symmetry for each parabola whose equation is given. Use the axis of symmetry to find a second point on the parabola whose y-coordinate is the same as the given point.

93. $f(x) = 3(x + 2)^2 - 5;\quad (-1, -2)$

94. $f(x) = (x - 3)^2 + 2; (6, 11)$

In Exercises 95–96, write the equation of each parabola in $f(x) = a(x - h)^2 + k$ form.

95. Vertex: $(-3, -4)$; The graph passes through the point $(1, 4)$.

96. Vertex: $(-3, -1)$; The graph passes through the point $(-2, -3)$.

97. A rancher has 1000 feet of fencing to construct six corrals, as shown in the figure. Find the dimensions that maximize the enclosed area. What is the maximum area?

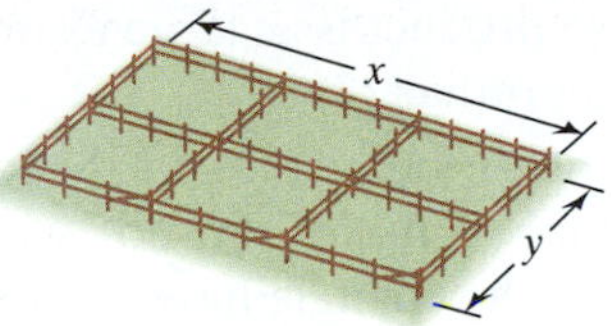

98. The annual yield per lemon tree is fairly constant at 320 pounds when the number of trees per acre is 50 or fewer. For each additional tree over 50, the annual yield per tree for all trees on the acre decreases by 4 pounds due to overcrowding. Find the number of trees that should be planted on an acre to produce the maximum yield. How many pounds is the maximum yield?

Review Exercises

99. Solve: $\frac{2}{x+5} + \frac{1}{x-5} = \frac{16}{x^2-25}$. (Section 6.6, Example 5)

100. Simplify: $\dfrac{1 + \frac{2}{x}}{1 - \frac{4}{x^2}}$. (Section 6.3, Example 1)

101. Solve using determinants (Cramer's Rule):

$$2x + 3y = 6$$
$$x - 4y = 14.$$

(Section 3.5, Example 2)

Preview Exercises

Exercises 102–104 will help you prepare for the material covered in the next section.

In Exercises 102–103, solve each quadratic equation for u.

102. $u^2 - 8u - 9 = 0$

103. $2u^2 - u - 10 = 0$

104. If $u = x^{\frac{1}{3}}$, rewrite $5x^{\frac{2}{3}} + 11x^{\frac{1}{3}} + 2 = 0$ as a quadratic equation in u. [*Hint*: $x^{\frac{2}{3}} = \left(x^{\frac{1}{3}}\right)^2$.]

MID-CHAPTER CHECK POINT Section 8.1–Section 8.3

What You Know: We saw that not all quadratic equations can be solved by factoring. We learned three new methods for solving these equations: the square root property, completing the square, and the quadratic formula. We saw that the discriminant of $ax^2 + bx + c = 0$, namely $b^2 - 4ac$, determines the number and type of the equation's solutions. We graphed quadratic functions using vertices, intercepts, and additional points, as necessary. We learned that the vertex of $f(x) = a(x - h)^2 + k$ is (h, k) and the vertex of $f(x) = ax^2 + bx + c$ is $\left(-\frac{b}{2a}, f\left(-\frac{b}{2a}\right)\right)$. We used the vertex to solve problems that involved minimizing or maximizing quadratic functions.

In Exercises 1–13, solve each equation by the method of your choice. Simplify solutions, if possible.

1. $(3x - 5)^2 = 36$

2. $5x^2 - 2x = 7$

3. $3x^2 - 6x - 2 = 0$

4. $x^2 + 6x = -2$

5. $5x^2 + 1 = 37$

6. $x^2 - 5x + 8 = 0$

7. $2x^2 + 26 = 0$

8. $(2x + 3)(x + 2) = 10$

9. $(x + 3)^2 = 24$

10. $\frac{1}{x^2} - \frac{4}{x} + 1 = 0$

11. $x(2x - 3) = -4$

12. $\frac{x^2}{3} + \frac{x}{2} = \frac{2}{3}$

13. $\frac{2x}{x^2 + 6x + 8} = \frac{x}{x+4} - \frac{2}{x+2}$

14. Solve by completing the square: $x^2 + 10x - 3 = 0$.

In Exercises 15–18, graph the given quadratic function. Give each function's domain and range.

15. $f(x) = (x - 3)^2 - 4$

16. $g(x) = 5 - (x + 2)^2$

17. $h(x) = -x^2 - 4x + 5$

18. $f(x) = 3x^2 - 6x + 1$

In Exercises 19–20, without solving the equation, determine the number and type of solutions.

19. $2x^2 + 5x + 4 = 0$

20. $10x(x + 4) = 15x - 15$

In Exercises 21–22, write a quadratic equation in standard form with the given solution set.

21. $\left\{-\frac{1}{2}, \frac{3}{4}\right\}$

22. $\{-2\sqrt{3}, 2\sqrt{3}\}$

23. A company manufactures and sells bath cabinets. The function

$$P(x) = -x^2 + 150x - 4425$$

models the company's daily profit, $P(x)$, when x cabinets are manufactured and sold per day. How many cabinets should be manufactured and sold per day to maximize the company's profit? What is the maximum daily profit?

24. Among all pairs of numbers whose sum is -18, find a pair whose product is as large as possible. What is the maximum product?

25. The base of a triangle measures 40 inches minus twice the measure of its height. For what measure of the height does the triangle have a maximum area? What is the maximum area?

SECTION 8.4

Equations Quadratic in Form

Objective

1 Solve equations that are quadratic in form.

"My husband asked me if we have any cheese puffs. Like he can't go and lift the couch cushion up himself."
—Roseanne Barr

How important is it for you to have a clean house? The percentage of people who find this to be quite important varies by age. In the exercise set, you will work with a function that models this phenomenon. Your work will be based on equations that are not quadratic, but that can be written as quadratic equations using an appropriate substitution. Here are some examples:

Given Equation	Substitution	New Equation
$x^4 - 10x^2 + 9 = 0$ or $(x^2)^2 - 10x^2 + 9 = 0$	$u = x^2$	$u^2 - 10u + 9 = 0$
$5x^{\frac{2}{3}} + 11x^{\frac{1}{3}} + 2 = 0$ or $5\left(x^{\frac{1}{3}}\right)^2 + 11x^{\frac{1}{3}} + 2 = 0$	$u = x^{\frac{1}{3}}$	$5u^2 + 11u + 2 = 0$

1 Solve equations that are quadratic in form.

An equation that is **quadratic in form** is one that can be expressed as a quadratic equation using an appropriate substitution. Both of the preceding given equations are quadratic in form.

In an equation that is quadratic in form, the variable factor in one term is the square of the variable factor in the other variable term. The third term is a constant. By letting u equal the variable factor that reappears squared, a quadratic equation in u will result. Now it's easy. Solve this quadratic equation for u. Finally, use your substitution to find the values for the variable in the given equation. Example 1 shows how this is done.

EXAMPLE 1 Solving an Equation Quadratic in Form

Solve: $x^4 - 8x^2 - 9 = 0.$

Solution Can you see that the variable factor in one term is the square of the variable factor in the other variable term?

$$x^4 - 8x^2 - 9 = 0$$

x^4 is the square of x^2: $(x^2)^2 = x^4$.

We will let u equal the variable factor that reappears squared. Thus,

$$\text{let } u = x^2.$$

Now we write the given equation as a quadratic equation in u and solve for u.

$x^4 - 8x^2 - 9 = 0$	This is the given equation.
$(x^2)^2 - 8x^2 - 9 = 0$	The given equation contains x^2 and x^2 squared.
$u^2 - 8u - 9 = 0$	Let $u = x^2$. Replace x^2 with u.
$(u - 9)(u + 1) = 0$	Factor.
$u - 9 = 0$ or $u + 1 = 0$	Apply the zero-product principle.
$u = 9$ $\quad u = -1$	Solve for u.

We're not done! Why not? We were asked to solve for x and we have values for u. We use the original substitution, $u = x^2$, to solve for x. Replace u with x^2 in each equation shown, namely $u = 9$ and $u = -1$.

$$x^2 = 9 \qquad x^2 = -1$$
$$x = \pm\sqrt{9} \qquad x = \pm\sqrt{-1} \quad \text{Apply the square root property.}$$
$$x = \pm 3 \qquad x = \pm i$$

Substitute these values into the given equation and verify that the solutions are $-3, 3, -i$, and i. The solution set is $\{-3, 3, -i, i\}$. The graph in the Using Technology box shows that only the real solutions, -3 and 3, appear as x-intercepts.

CHECK POINT 1 Solve: $x^4 - 5x^2 + 6 = 0.$

Using Technology

Graphic Connections

The graph of

$$y = x^4 - 8x^2 - 9$$

has x-intercepts at -3 and 3. This verifies that the real solutions of

$$x^4 - 8x^2 - 9 = 0$$

are -3 and 3. The imaginary solutions, $-i$ and i, are not shown as intercepts.

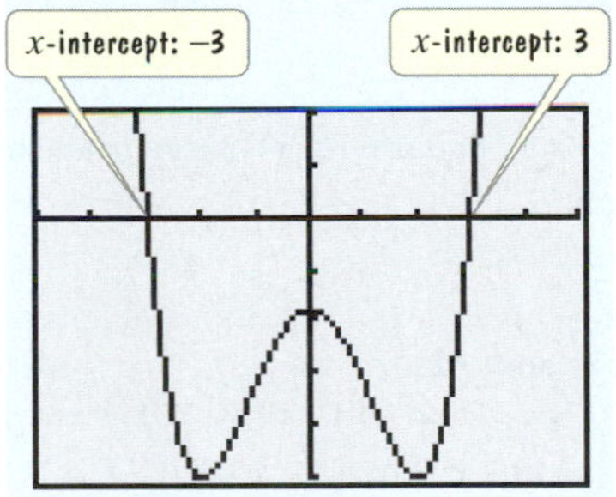

[−5, 5, 1] by [−25, 10, 5]

If checking proposed solutions is not overly cumbersome, you should do so either algebraically or with a graphing utility. The Using Technology box shows a check of the two real solutions in Example 1. Are there situations when solving equations quadratic in form where a check is essential? Yes. **If at any point in the solution process both sides of an equation are raised to an even power, a check is required.** Extraneous solutions that are not solutions of the given equation may have been introduced.

EXAMPLE 2 Solving an Equation Quadratic in Form

Solve: $2x - \sqrt{x} - 10 = 0.$

Solution To identify exponents on the terms, let's rewrite $\sqrt{x}$ as $x^{\frac{1}{2}}$. The equation can be expressed as

$$2x^1 - x^{\frac{1}{2}} - 10 = 0.$$

By expressing the equation as $2x^1 - x^{\frac{1}{2}} - 10 = 0$, we can see that the variable factor in one term is the square of the variable factor in the other variable term.

$$2x^1 - x^{\frac{1}{2}} - 10 = 0$$

x^1 is the square of $x^{\frac{1}{2}}$: $\left(x^{\frac{1}{2}}\right)^2 = x^1$.

We will let u equal the variable factor that reappears squared. Thus,

$$\text{let } u = x^{\frac{1}{2}}.$$

Now we write the given equation as a quadratic equation in u and solve for u.

$2x - \sqrt{x} - 10 = 0$	This is the given equation.
$2x^1 - x^{\frac{1}{2}} - 10 = 0$	This is the given equation in exponential form.
$2(x^{\frac{1}{2}})^2 - x^{\frac{1}{2}} - 10 = 0$	The equation contains $x^{\frac{1}{2}}$ and $x^{\frac{1}{2}}$ squared.
$2u^2 - u - 10 = 0$	Let $u = x^{\frac{1}{2}}$. Replace $x^{\frac{1}{2}}$ with u.
$(2u - 5)(u + 2) = 0$	Factor.
$2u - 5 = 0$ or $u + 2 = 0$	Set each factor equal to zero.
$u = \frac{5}{2}$ $\quad$ $u = -2$	Solve for u.

Use the original substitution, $u = x^{\frac{1}{2}}$, to solve for x. Replace u with $x^{\frac{1}{2}}$ in each of the preceding equations.

$x^{\frac{1}{2}} = \frac{5}{2}$ or $x^{\frac{1}{2}} = -2$	Replace u with $x^{\frac{1}{2}}$.
$\left(x^{\frac{1}{2}}\right)^2 = \left(\frac{5}{2}\right)^2$ $\quad$ $\left(x^{\frac{1}{2}}\right)^2 = (-2)^2$	Solve for x by squaring both sides of each equation.
$x = \frac{25}{4}$ $\quad$ $x = 4$	Square $\frac{5}{2}$ and -2.

Both sides are raised to even powers. We must check.

It is essential to check both proposed solutions in the original equation.

Check $\frac{25}{4}$:	**Check 4:**
$2x - \sqrt{x} - 10 = 0$	$2x - \sqrt{x} - 10 = 0$
$2 \cdot \frac{25}{4} - \sqrt{\frac{25}{4}} - 10 \stackrel{?}{=} 0$	$2 \cdot 4 - \sqrt{4} - 10 \stackrel{?}{=} 0$
$\frac{25}{2} - \frac{5}{2} - 10 \stackrel{?}{=} 0$	$8 - 2 - 10 \stackrel{?}{=} 0$
$\frac{20}{2} - 10 \stackrel{?}{=} 0$	$6 - 10 \stackrel{?}{=} 0$
$0 = 0$, true	$-4 = 0$, false

The check indicates that 4 is not a solution. It is an extraneous solution brought about by squaring each side of one of the equations. The only solution is $\frac{25}{4}$ and the solution set is $\left\{\frac{25}{4}\right\}$.

✓ CHECK POINT 2 Solve: $x - 2\sqrt{x} - 8 = 0$.

The equations in Examples 1 and 2 can be solved by methods other than using substitutions.

$$x^4 - 8x^2 - 9 = 0 \qquad 2x - \sqrt{x} - 10 = 0$$

This equation can be solved directly by factoring: $(x^2 - 9)(x^2 + 1) = 0$.

This equation can be solved by isolating the radical term: $2x - 10 = \sqrt{x}$. Then square both sides.

In the examples that follow, solving the equations by methods other than first introducing a substitution becomes increasingly difficult.

EXAMPLE 3 Solving an Equation Quadratic in Form

Solve: $(x^2 - 5)^2 + 3(x^2 - 5) - 10 = 0$.

Solution This equation contains $x^2 - 5$ and $x^2 - 5$ squared. We

$$\text{let } u = x^2 - 5.$$

$(x^2 - 5)^2 + 3(x^2 - 5) - 10 = 0$	This is the given equation.
$u^2 + 3u - 10 = 0$	Let $u = x^2 - 5$.
$(u + 5)(u - 2) = 0$	Factor.
$u + 5 = 0$ or $u - 2 = 0$	Set each factor equal to zero.
$u = -5$ $\quad u = 2$	Solve for u.

Use the original substitution, $u = x^2 - 5$, to solve for x. Replace u with $x^2 - 5$ in each of the preceding equations.

$x^2 - 5 = -5$ or $x^2 - 5 = 2$	Replace u with $x^2 - 5$.
$x^2 = 0$ $\quad x^2 = 7$	Solve for x by isolating x^2.
$x = 0$ $\quad x = \pm\sqrt{7}$	Apply the square root property.

Although we did not raise both sides of an equation to an even power, checking the three proposed solutions in the original equation is a good idea. Do this now and verify that the solutions are $-\sqrt{7}$, 0, and $\sqrt{7}$, and the solution set is $\{-\sqrt{7}, 0, \sqrt{7}\}$.

Study Tip

Solve Example 3 by first simplifying the given equation's left side. Then factor out x^2 and solve the resulting equation. Do you get the same solutions? Which method, substitution or first simplifying, is faster?

CHECK POINT 3 Solve: $(x^2 - 4)^2 + (x^2 - 4) - 6 = 0$.

EXAMPLE 4 Solving an Equation Quadratic in Form

Solve: $10x^{-2} + 7x^{-1} + 1 = 0$.

Solution The variable factor in one term is the square of the variable factor in the other variable term.

$$10x^{-2} + 7x^{-1} + 1 = 0$$

x^{-2} is the square of x^{-1}: $(x^{-1})^2 = x^{-2}$.

We will let u equal the variable factor that reappears squared. Thus,

$$\text{let } u = x^{-1}.$$

Now we write the given equation as a quadratic equation in u and solve for u.

$10x^{-2} + 7x^{-1} + 1 = 0$	This is the given equation.
$10(x^{-1})^2 + 7x^{-1} + 1 = 0$	The equation contains x^{-1} and x^{-1} squared.
$10u^2 + 7u + 1 = 0$	Let $u = x^{-1}$.
$(5u + 1)(2u + 1) = 0$	Factor.
$5u + 1 = 0$ or $2u + 1 = 0$	Set each factor equal to zero.
$5u = -1$ $\quad$ $2u = -1$	Solve each equation for u.
$u = -\frac{1}{5}$ $\quad$ $u = -\frac{1}{2}$	

Use the original substitution, $u = x^{-1}$, to solve for x. Replace u with x^{-1} in each of the preceding equations.

$x^{-1} = -\frac{1}{5}$ or $x^{-1} = -\frac{1}{2}$	Replace u with x^{-1}.
$(x^{-1})^{-1} = \left(-\frac{1}{5}\right)^{-1}$ $\quad$ $(x^{-1})^{-1} = \left(-\frac{1}{2}\right)^{-1}$	Solve for x by raising both sides of each equation to the -1 power.
$x = -5$ $\quad$ $x = -2$	

$\left(-\frac{1}{5}\right)^{-1} = \frac{1}{-\frac{1}{5}} = -5$ $\qquad$ $\left(-\frac{1}{2}\right)^{-1} = \frac{1}{-\frac{1}{2}} = -2$

We did not raise both sides of an equation to an even power. A check will show that both -5 and -2 are solutions of the original equation. The solution set is $\{-5, -2\}$.

✓ CHECK POINT 4 Solve: $2x^{-2} + x^{-1} - 1 = 0$.

EXAMPLE 5 Solving an Equation Quadratic in Form

Solve: $5x^{\frac{2}{3}} + 11x^{\frac{1}{3}} + 2 = 0$.

Solution The variable factor in one term is the square of the variable factor in the other variable term.

$$5x^{\frac{2}{3}} + 11x^{\frac{1}{3}} + 2 = 0$$

$x^{\frac{2}{3}}$ is the square of $x^{\frac{1}{3}}$: $\left(x^{\frac{1}{3}}\right)^2 = x^{\frac{2}{3}}$.

We will let u equal the variable factor that reappears squared. Thus,

$$\text{let } u = x^{\frac{1}{3}}.$$

Now we write the given equation as a quadratic equation in u and solve for u.

$5x^{\frac{2}{3}} + 11x^{\frac{1}{3}} + 2 = 0$	This is the given equation.
$5\left(x^{\frac{1}{3}}\right)^2 + 11\left(x^{\frac{1}{3}}\right) + 2 = 0$	The given equation contains $x^{\frac{1}{3}}$ and $x^{\frac{1}{3}}$ squared.
$5u^2 + 11u + 2 = 0$	Let $u = x^{\frac{1}{3}}$.
$(5u + 1)(u + 2) = 0$	Factor.
$5u + 1 = 0$ or $u + 2 = 0$	Set each factor equal to 0.
$u = -\frac{1}{5}$ $\quad$ $u = -2$	Solve for u.

Use the original substitution, $u = x^{\frac{1}{3}}$, to solve for x. Replace u with $x^{\frac{1}{3}}$ in each of the preceding equations.

$$x^{\frac{1}{3}} = -\frac{1}{5} \quad \text{or} \quad x^{\frac{1}{3}} = -2 \qquad \text{Replace } u \text{ with } x^{\frac{1}{3}} \text{ in } u = -\frac{1}{5} \text{ and } u = -2.$$

$$\left(x^{\frac{1}{3}}\right)^3 = \left(-\frac{1}{5}\right)^3 \qquad \left(x^{\frac{1}{3}}\right)^3 = (-2)^3 \qquad \text{Solve for } x \text{ by cubing both sides of each equation.}$$

$$x = -\frac{1}{125} \qquad x = -8$$

We did not raise both sides of an equation to an even power. A check will show that both -8 and $-\frac{1}{125}$ are solutions of the original equation. The solution set is $\left\{-8, -\frac{1}{125}\right\}$.

✓ **CHECK POINT 5** Solve: $3x^{\frac{2}{3}} - 11x^{\frac{1}{3}} - 4 = 0$.

8.4 EXERCISE SET

MyMathLab — MathXL PRACTICE · WATCH · DOWNLOAD · READ · REVIEW

Practice Exercises

In Exercises 1–32, solve each equation by making an appropriate substitution. If at any point in the solution process both sides of an equation are raised to an even power, a check is required.

1. $x^4 - 5x^2 + 4 = 0$
2. $x^4 - 13x^2 + 36 = 0$
3. $x^4 - 11x^2 + 18 = 0$
4. $x^4 - 9x^2 + 20 = 0$
5. $x^4 + 2x^2 = 8$
6. $x^4 + 4x^2 = 5$
7. $x + \sqrt{x} - 2 = 0$
8. $x + \sqrt{x} - 6 = 0$
9. $x - 4x^{\frac{1}{2}} - 21 = 0$
10. $x - 6x^{\frac{1}{2}} + 8 = 0$
11. $x - 13\sqrt{x} + 40 = 0$
12. $2x - 7\sqrt{x} - 30 = 0$
13. $(x - 5)^2 - 4(x - 5) - 21 = 0$
14. $(x + 3)^2 + 7(x + 3) - 18 = 0$
15. $(x^2 - 1)^2 - (x^2 - 1) = 2$
16. $(x^2 - 2)^2 - (x^2 - 2) = 6$
17. $(x^2 + 3x)^2 - 8(x^2 + 3x) - 20 = 0$
18. $(x^2 - 2x)^2 - 11(x^2 - 2x) + 24 = 0$
19. $x^{-2} - x^{-1} - 20 = 0$
20. $x^{-2} - x^{-1} - 6 = 0$
21. $2x^{-2} - 7x^{-1} + 3 = 0$
22. $20x^{-2} + 9x^{-1} + 1 = 0$
23. $x^{-2} - 4x^{-1} = 3$
24. $x^{-2} - 6x^{-1} = -4$
25. $x^{\frac{2}{3}} - x^{\frac{1}{3}} - 6 = 0$
26. $x^{\frac{2}{3}} + 2x^{\frac{1}{3}} - 3 = 0$
27. $x^{\frac{2}{5}} + x^{\frac{1}{5}} - 6 = 0$
28. $x^{\frac{2}{5}} + x^{\frac{1}{5}} - 2 = 0$
29. $2x^{\frac{1}{2}} - x^{\frac{1}{4}} = 1$
30. $2x^{\frac{1}{2}} - 5x^{\frac{1}{4}} = 3$
31. $\left(x - \frac{8}{x}\right)^2 + 5\left(x - \frac{8}{x}\right) - 14 = 0$
32. $\left(x - \frac{10}{x}\right)^2 + 6\left(x - \frac{10}{x}\right) - 27 = 0$

In Exercises 33–38, find the x-intercepts of the given function, f. Then use the x-intercepts to match each function with its graph. [The graphs are labeled (a) through (f).]

33. $f(x) = x^4 - 5x^2 + 4$

34. $f(x) = x^4 - 13x^2 + 36$

35. $f(x) = x^{\frac{1}{3}} + 2x^{\frac{1}{6}} - 3$

36. $f(x) = x^{-2} - x^{-1} - 6$

37. $f(x) = (x + 2)^2 - 9(x + 2) + 20$

38. $f(x) = 2(x + 2)^2 + 5(x + 2) - 3$

a.

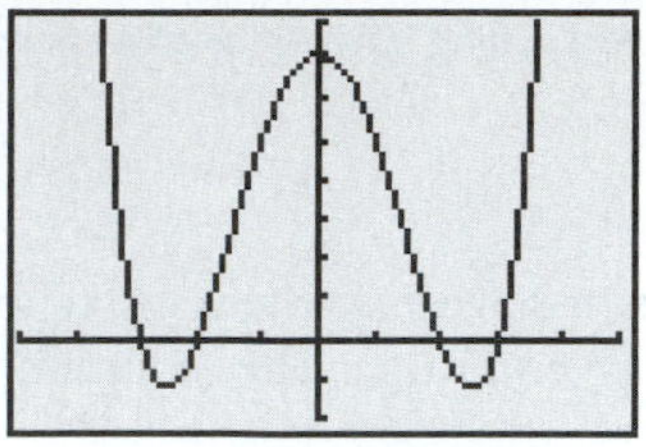

[−5, 5, 1] by [−10, 40, 5]

c.

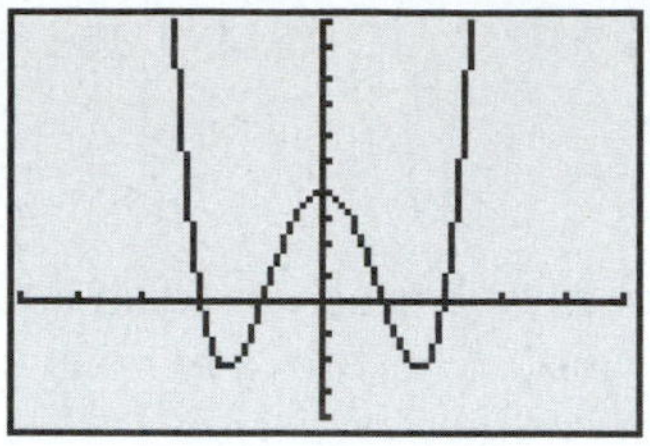

[−5, 5, 1] by [−4, 10, 1]

e.

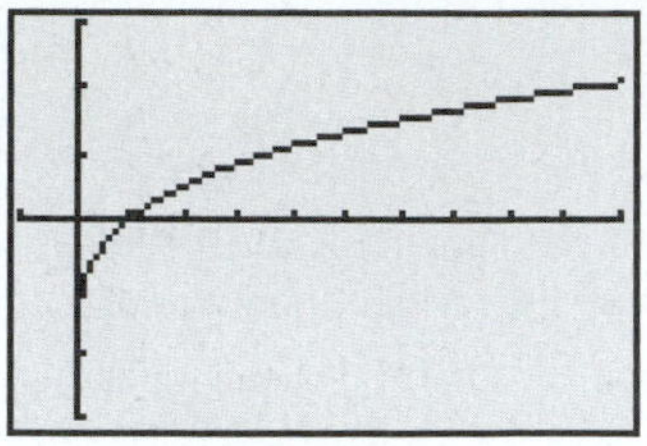

[−1, 10, 1] by [−3, 3, 1]

b.

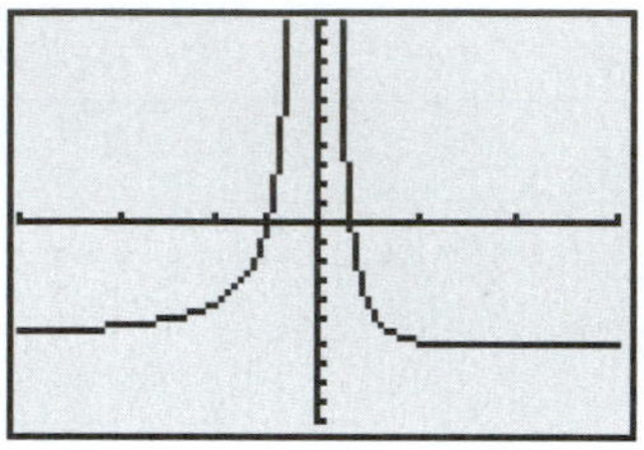

[−3, 3, 1] by [−10, 10, 1]

d.

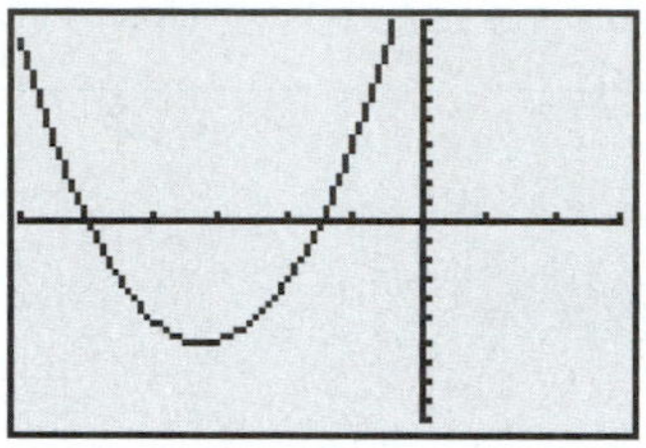

[−6, 3, 1] by [−10, 10, 1]

f.

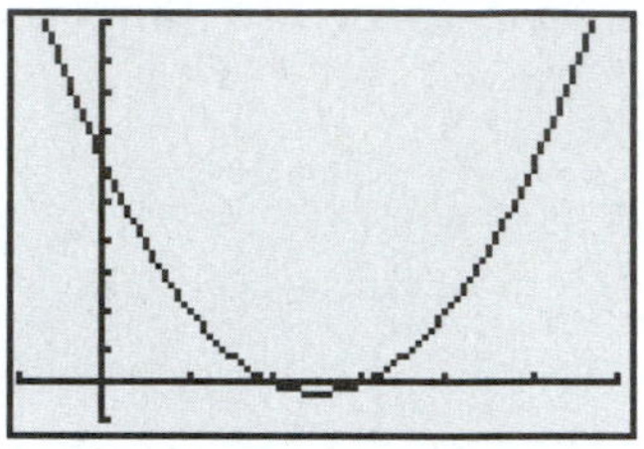

[−1, 6, 1] by [−1, 10, 1]

Practice PLUS

39. Let $f(x) = (x^2 + 3x - 2)^2 - 10(x^2 + 3x - 2)$. Find all x such that $f(x) = -16$.

40. Let $f(x) = (x^2 + 2x - 2)^2 - 7(x^2 + 2x - 2)$. Find all x such that $f(x) = -6$.

41. Let $f(x) = 3\left(\frac{1}{x} + 1\right)^2 + 5\left(\frac{1}{x} + 1\right)$. Find all x such that $f(x) = 2$.

42. Let $f(x) = 2x^{\frac{2}{3}} + 3x^{\frac{1}{3}}$. Find all x such that $f(x) = 2$.

43. Let $f(x) = \frac{x}{x - 4}$ and $g(x) = 13\sqrt{\frac{x}{x - 4}} - 36$. Find all x such that $f(x) = g(x)$.

44. Let $f(x) = \frac{x}{x - 2} + 10$ and $g(x) = -11\sqrt{\frac{x}{x - 2}}$. Find all x such that $f(x) = g(x)$.

45. Let $f(x) = 3(x - 4)^{-2}$ and $g(x) = 16(x - 4)^{-1}$. Find all x such that $f(x)$ exceeds $g(x)$ by 12.

46. Let $f(x) = 6\left(\frac{2x}{x - 3}\right)^2$ and $g(x) = 5\left(\frac{2x}{x - 3}\right)$. Find all x such that $f(x)$ exceeds $g(x)$ by 6.

Application Exercises

How important is it for you to have a clean house? The bar graph indicates that the percentage of people who find this to be quite important varies by age. The percentage, P(x), who find having a clean house very important can be modeled by the function

$$P(x) = 0.04(x + 40)^2 - 3(x + 40) + 104,$$

where x is the number of years a person's age is above or below 40. Thus, x is positive for people over 40 and negative for people under 40. Use the function to solve Exercises 47–48.

The Importance of Having a Clean House, by Age

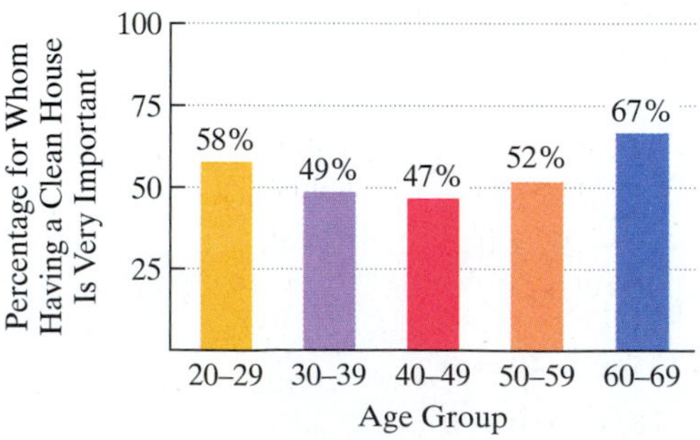

Source: Soap and Detergent Association

47. According to the model, at which ages do 60% of us feel that having a clean house is very important? Substitute 60 for $P(x)$ and solve the quadratic-in-form equation. How well does the function model the data shown in the bar graph?

48. According to the model, at which ages do 50% of us feel that having a clean house is very important? Substitute 50 for $P(x)$ and solve the quadratic-in-form equation. How well does the function model the data shown in the bar graph?

Writing in Mathematics

49. Explain how to recognize an equation that is quadratic in form. Provide two original examples with your explanation.

50. Describe two methods for solving this equation:

$$x - 5\sqrt{x} + 4 = 0.$$

Technology Exercises

51. Use a graphing utility to verify the solutions of any five equations in Exercises 1-32 that you solved algebraically. The real solutions should appear as x-intercepts on the graph of the function related to the given equation.

Use a graphing utility to solve the equations in Exercises 52–59. Check by direct substitution.

52. $x^6 - 7x^3 - 8 = 0$

53. $3(x-2)^{-2} - 4(x-2)^{-1} + 1 = 0$

54. $x^4 - 10x^2 + 9 = 0$

55. $2x + 6\sqrt{x} = 8$

56. $2(x+1)^2 = 5(x+1) + 3$

57. $(x^2 - 3x)^2 + 2(x^2 - 3x) - 24 = 0$

58. $x^{\frac{1}{2}} + 4x^{\frac{1}{4}} = 5$

59. $x^{\frac{2}{3}} - 3x^{\frac{1}{3}} + 2 = 0$

Critical Thinking Exercises

Make Sense? *In Exercises 60–63, determine whether each statement "makes sense" or "does not make sense" and explain your reasoning.*

60. When I solve an equation that is quadratic in form, it's important to write down the substitution that I am making.

61. Although I've rewritten an equation that is quadratic in form as $au^2 + bu + c = 0$ and solved for u, I'm not finished.

62. Checking is always a good idea, but it's never necessary when solving an equation that is quadratic in form.

63. The equation $5x^{\frac{2}{3}} + 11x^{\frac{1}{3}} + 2 = 0$ is quadratic in form, but when I reverse the variable terms and obtain $11x^{\frac{1}{3}} + 5x^{\frac{2}{3}} + 2 = 0$, the resulting equation is no longer quadratic in form.

In Exercises 64–67, determine whether each statement is true or false. If the statement is false, make the necessary change(s) to produce a true statement.

64. If an equation is quadratic in form, there is only one method that can be used to obtain its solution set.

65. An equation with three terms that is quadratic in form has a variable factor in one term that is the square of the variable factor in another term.

66. Because x^6 is the square of x^3, the equation $x^6 - 5x^3 + 6x = 0$ is quadratic in form.

67. To solve $x - 9\sqrt{x} + 14 = 0$, we let $\sqrt{u} = x$.

In Exercises 68–70, use a substitution to solve each equation.

68. $x^4 - 5x^2 - 2 = 0$

69. $5x^6 + x^3 = 18$

70. $\sqrt{\frac{x+4}{x-1}} + \sqrt{\frac{x-1}{x+4}} = \frac{5}{2}\left(\text{Let } u = \sqrt{\frac{x+4}{x-1}}.\right)$

Review Exercises

71. Simplify:

$$\frac{2x^2}{10x^3 - 2x^2}.$$

(Section 6.1, Example 4)

72. Divide: $\frac{2+i}{1-i}$. (Section 7.7, Example 5)

73. Solve using matrices:

$$2x + y = 6$$
$$x - 2y = 8.$$

(Section 3.4, Example 2)

Preview Exercises

Exercises 74–76 will help you prepare for the material covered in the next section.

74. Solve: $2x^2 + x = 15$.

75. Solve: $x^3 + x^2 = 4x + 4$.

76. Simplify: $\frac{x+1}{x+3} - 2$.

SECTION 8.5

Polynomial and Rational Inequalities

Objectives

1. Solve polynomial inequalities.
2. Solve rational inequalities.
3. Solve problems modeled by polynomial or rational inequalities.

Tailgaters beware: If your car is going 35 miles per hour on dry pavement, your required stopping distance is 160 feet, or the width of a football field. At 65 miles per hour, the distance required is 410 feet, or approximately the length of one and one-tenth football fields. **Figure 8.14** shows stopping distances for cars at various speeds on dry roads and on wet roads.

Using Technology

We used the statistical menu of a graphing utility and the quadratic regression program to obtain the quadratic function that models stopping distance on dry pavement. After entering the appropriate data from **Figure 8.14**, namely

(35, 160), (45, 225), (55, 310), (65, 410),

we obtained the results shown in the screen.

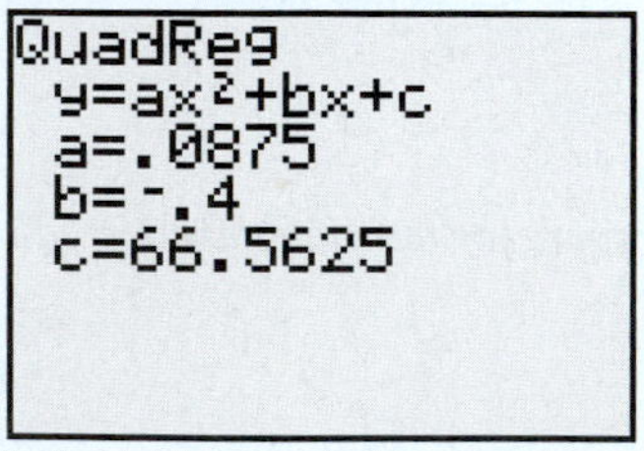

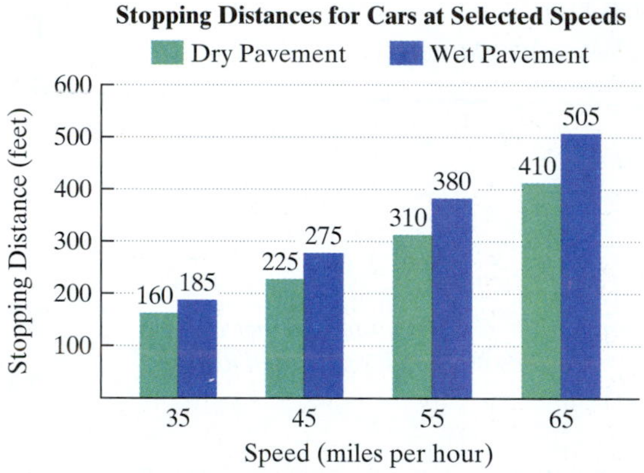

FIGURE 8.14

Source: National Highway Traffic Safety Administration

A car's required stopping distance, $f(x)$, in feet, on dry pavement traveling at x miles per hour can be modeled by the quadratic function

$$f(x) = 0.0875x^2 - 0.4x + 66.6.$$

How can we use this function to determine speeds on dry pavement requiring stopping distances that exceed the length of one and one-half football fields, or 540 feet? We must solve the inequality

$$0.0875x^2 - 0.4x + 66.6 > 540.$$

Required stopping distance — exceeds — 540 feet.

We begin by subtracting 540 from both sides. This will give us zero on the right:

$$0.0875x^2 - 0.4x + 66.6 - 540 > 540 - 540$$
$$0.0875x^2 - 0.4x - 473.4 > 0.$$

The form of this inequality is $ax^2 + bx + c > 0$. Such a quadratic inequality is called a *polynomial inequality*.

Definition of a Polynomial Inequality

A polynomial inequality is any inequality that can be put in one of the forms

$$f(x) < 0, \quad f(x) > 0, \quad f(x) \le 0, \quad \text{or} \quad f(x) \ge 0,$$

where f is a polynomial function.

In this section, we establish the basic techniques for solving polynomial inequalities. We will also use these techniques to solve inequalities involving rational functions.

1 Solve polynomial inequalities.

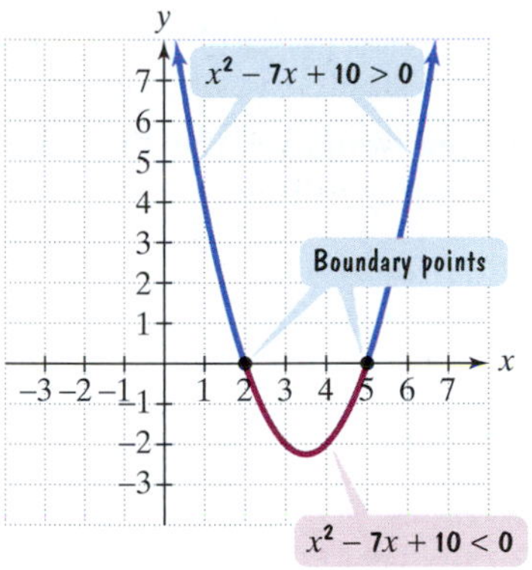

FIGURE 8.15

Solving Polynomial Inequalities

Graphs can help us visualize the solutions of polynomial inequalities. For example, the graph of $f(x) = x^2 - 7x + 10$ is shown in **Figure 8.15**. The x-intercepts, 2 and 5, are **boundary points** between where the graph lies above the x-axis, shown in blue, and where the graph lies below the x-axis, shown in red.

Locating the x-intercepts of a polynomial function, f, is an important step in finding the solution set for polynomial inequalities in the form $f(x) < 0$ or $f(x) > 0$. We use the x-intercepts of f as boundary points that divide the real number line into intervals. On each interval, the graph of f is either above the x-axis $[f(x) > 0]$ or below the x-axis $[f(x) < 0]$. For this reason, x-intercepts play a fundamental role in solving polynomial inequalities. The x-intercepts are found by solving the equation $f(x) = 0$.

Procedure for Solving Polynomial Inequalities

1. Express the inequality in the form
$$f(x) < 0 \quad \text{or} \quad f(x) > 0,$$
where f is a polynomial function.
2. Solve the equation $f(x) = 0$. The real solutions are the **boundary points**.
3. Locate these boundary points on a number line, thereby dividing the number line into intervals.
4. Choose one representative number, called a **test value**, within each interval and evaluate f at that number.
   a. If the value of f is positive, then $f(x) > 0$ for all numbers, x, in the interval.
   b. If the value of f is negative, then $f(x) < 0$ for all numbers, x, in the interval.
5. Write the solution set, selecting the interval or intervals that satisfy the given inequality.

This procedure is valid if $<$ is replaced by $\le$ or $>$ is replaced by $\ge$. However, if the inequality involves $\le$ or $\ge$, include the boundary points [the solutions of $f(x) = 0$] in the solution set.

EXAMPLE 1 Solving a Polynomial Inequality

Solve and graph the solution set on a real number line: $2x^2 + x > 15$.

Solution

Step 1. Express the inequality in the form $f(x) < 0$ or $f(x) > 0$. We begin by rewriting the inequality so that 0 is on the right side.

Using Technology

Graphic Connections

The solution set for

$$2x^2 + x > 15$$

or, equivalently,

$$2x^2 + x - 15 > 0$$

can be verified with a graphing utility. The graph of $f(x) = 2x^2 + x - 15$ was obtained using a $[-10, 10, 1]$ by $[-16, 6, 1]$ viewing rectangle. The graph lies above the x-axis, representing $>$, for all x in $(-\infty, -3)$ or $\left(\frac{5}{2}, \infty\right)$.

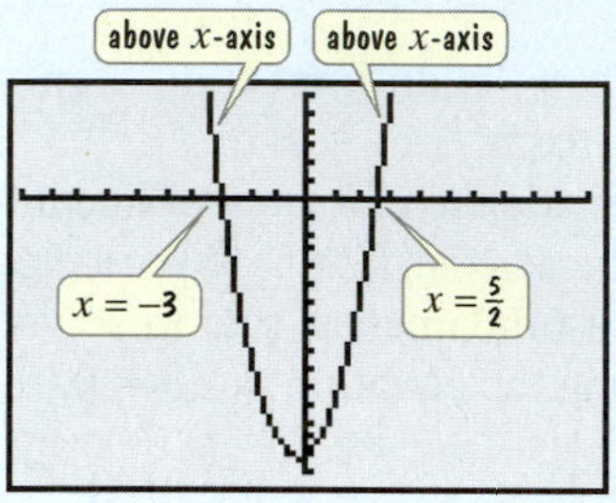

$$2x^2 + x > 15 \quad \text{This is the given inequality.}$$

$$2x^2 + x - 15 > 15 - 15 \quad \text{Subtract 15 from both sides.}$$

$$2x^2 + x - 15 > 0 \quad \text{Simplify.}$$

This inequality is equivalent to the one we wish to solve. It is in the form $f(x) > 0$, where $f(x) = 2x^2 + x - 15$.

Step 2. Solve the equation $f(x) = 0$. We find the x-intercepts of $f(x) = 2x^2 + x - 15$ by solving the equation $2x^2 + x - 15 = 0$.

$$2x^2 + x - 15 = 0 \quad \text{This polynomial equation is a quadratic equation.}$$

$$(2x - 5)(x + 3) = 0 \quad \text{Factor.}$$

$$2x - 5 = 0 \quad \text{or} \quad x + 3 = 0 \quad \text{Set each factor equal to 0.}$$

$$x = \frac{5}{2} \qquad x = -3 \quad \text{Solve for } x.$$

The x-intercepts of f are -3 and $\frac{5}{2}$. We will use these x-intercepts as boundary points on a number line.

Step 3. Locate the boundary points on a number line and separate the line into intervals. The number line with the boundary points is shown as follows:

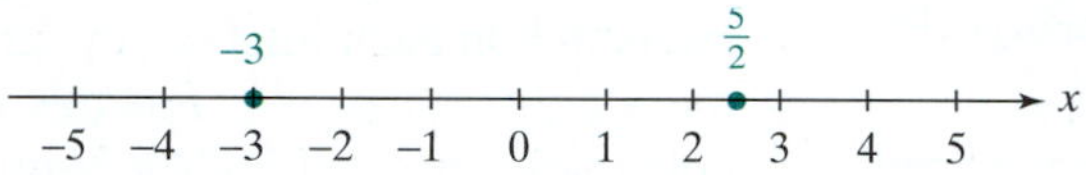

The boundary points divide the number line into three intervals:

$$(-\infty, -3) \quad \left(-3, \tfrac{5}{2}\right) \quad \left(\tfrac{5}{2}, \infty\right).$$

Step 4. Choose one test value within each interval and evaluate f at that number.

Interval	Test Value	Substitute into $f(x) = 2x^2 + x - 15$	Conclusion
$(-\infty, -3)$	-4	$f(-4) = 2(-4)^2 + (-4) - 15$ $= 13$, positive	$f(x) > 0$ for all x in $(-\infty, -3)$.
$\left(-3, \frac{5}{2}\right)$	0	$f(0) = 2 \cdot 0^2 + 0 - 15$ $= -15$, negative	$f(x) < 0$ for all x in $\left(-3, \frac{5}{2}\right)$.
$\left(\frac{5}{2}, \infty\right)$	3	$f(3) = 2 \cdot 3^2 + 3 - 15$ $= 6$, positive	$f(x) > 0$ for all x in $\left(\frac{5}{2}, \infty\right)$.

Step 5. Write the solution set, selecting the interval or intervals that satisfy the given inequality. We are interested in solving $f(x) > 0$, where $f(x) = 2x^2 + x - 15$. Based on our work in step 4, we see that $f(x) > 0$ for all x in $(-\infty, -3)$ or $\left(\frac{5}{2}, \infty\right)$. Thus, the solution set of the given inequality, $2x^2 + x > 15$, or, equivalently, $2x^2 + x - 15 > 0$, is

$$(-\infty, -3) \cup \left(\tfrac{5}{2}, \infty\right) \text{ or } \left\{x \mid x < -3 \text{ or } x > \tfrac{5}{2}\right\}.$$

The graph of the solution set on a number line is shown as follows:

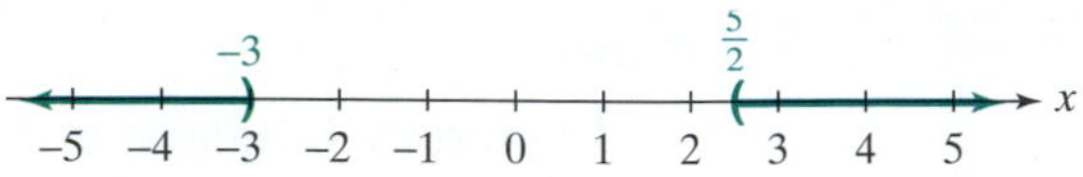

✓ **CHECK POINT 1** Solve and graph the solution set on a real number line: $x^2 - x > 20$.

EXAMPLE 2 Solving a Polynomial Inequality

Solve and graph the solution set on a real number line: $x^3 + x^2 \le 4x + 4$.

Solution

Step 1. Express the inequality in the form $f(x) \le 0$ or $f(x) \ge 0$. We begin by rewriting the inequality so that 0 is on the right side.

$$x^3 + x^2 \le 4x + 4 \qquad \text{This is the given inequality.}$$

$$x^3 + x^2 - 4x - 4 \le 4x + 4 - 4x - 4 \qquad \text{Subtract } 4x + 4 \text{ from both sides.}$$

$$x^3 + x^2 - 4x - 4 \le 0 \qquad \text{Simplify.}$$

This inequality is equivalent to the one we wish to solve. It is in the form $f(x) \le 0$, where $f(x) = x^3 + x^2 - 4x - 4$.

Step 2. Solve the equation $f(x) = 0$. We find the x-intercepts of $f(x) = x^3 + x^2 - 4x - 4$ by solving the equation $x^3 + x^2 - 4x - 4 = 0$.

$$x^3 + x^2 - 4x - 4 = 0 \qquad \text{This polynomial equation is of degree 3.}$$

$$x^2(x + 1) - 4(x + 1) = 0 \qquad \text{Factor } x^2 \text{ from the first two terms and } -4 \text{ from the last two terms.}$$

$$(x + 1)(x^2 - 4) = 0 \qquad \text{A common factor of } x + 1 \text{ is factored from the expression.}$$

$$(x + 1)(x + 2)(x - 2) = 0 \qquad \text{Factor completely.}$$

$$x + 1 = 0 \quad \text{or} \quad x + 2 = 0 \quad \text{or} \quad x - 2 = 0 \qquad \text{Set each factor equal to 0.}$$

$$x = -1 \qquad x = -2 \qquad x = 2 \qquad \text{Solve for } x.$$

The x-intercepts of f are -2, -1, and 2. We will use these x-intercepts as boundary points on a number line.

Step 3. Locate the boundary points on a number line and separate the line into intervals. The number line with the boundary points is shown as follows:

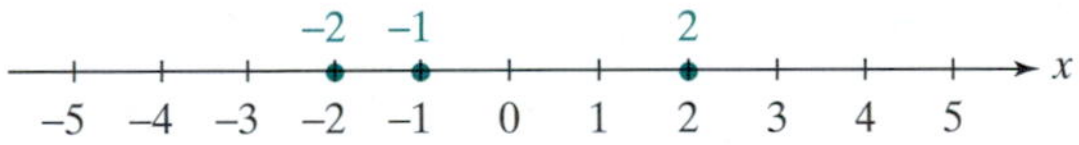

The boundary points divide the number line into four intervals:

$$(-\infty, -2) \quad (-2, -1) \quad (-1, 2) \quad (2, \infty).$$

Step 4. Choose one test value within each interval and evaluate f at that number.

Using Technology

Graphic Connections

The solution set for

$$x^3 + x^2 \le 4x + 4$$

or, equivalently,

$$x^3 + x^2 - 4x - 4 \le 0$$

can be verified with a graphing utility. The graph of $f(x) = x^3 + x^2 - 4x - 4$ lies on or below the x-axis, representing $\le$, for all x in $(-\infty, -2]$ or $[-1, 2]$.

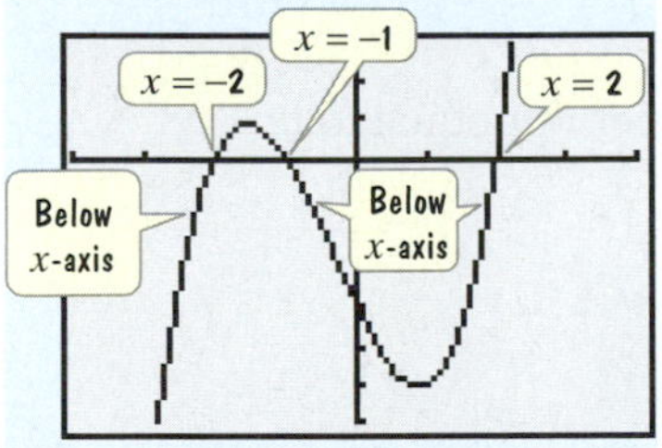

[−4, 4, 1] by [−7, 3, 1]

Interval	Test Value	Substitute into $f(x) = x^3 + x^2 - 4x - 4$	Conclusion
$(-\infty, -2)$	-3	$f(-3) = (-3)^3 + (-3)^2 - 4(-3) - 4$ $= -10$, negative	$f(x) < 0$ for all x in $(-\infty, -2)$.
$(-2, -1)$	-1.5	$f(-1.5) = (-1.5)^3 + (-1.5)^2 - 4(-1.5) - 4$ $= 0.875$, positive	$f(x) > 0$ for all x in $(-2, -1)$.
$(-1, 2)$	0	$f(0) = 0^3 + 0^2 - 4 \cdot 0 - 4$ $= -4$, negative	$f(x) < 0$ for all x in $(-1, 2)$.
$(2, \infty)$	3	$f(3) = 3^3 + 3^2 - 4 \cdot 3 - 4$ $= 20$, positive	$f(x) > 0$ for all x in $(2, \infty)$.

Interval	Conclusion
$(-\infty, -2)$	$f(x) < 0$
$(-2, -1)$	$f(x) > 0$
$(-1, 2)$	$f(x) < 0$
$(2, \infty)$	$f(x) > 0$

Summary of step 4 (repeated)

Step 5. Write the solution set, selecting the interval or intervals that satisfy the given inequality. We are interested in solving $f(x) \leq 0$, where $f(x) = x^3 + x^2 - 4x - 4$. Based on our work in step 4 that we've summarized in the margin, we see that $f(x) < 0$ for all x in $(-\infty, -2)$ or $(-1, 2)$. However, because the inequality involves $\leq$ (less than or *equal to*), we must also include the solutions of $x^3 + x^2 - 4x - 4 = 0$, namely -2, -1, and 2, in the solution set. Thus, the solution set of the given inequality, $x^3 + x^2 \leq 4x + 4$, or, equivalently, $x^3 + x^2 - 4x - 4 \leq 0$, is

$$(-\infty, -2] \cup [-1, 2]$$

$$\text{or} \quad \{x \mid x \leq -2 \text{ or } -1 \leq x \leq 2\}.$$

The graph of the solution set on a number line is shown as follows:

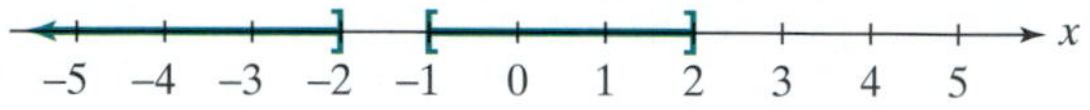

✓ **CHECK POINT 2** Solve and graph the solution set on a real number line: $x^3 + 3x^2 \leq x + 3$.

2 Solve rational inequalities.

Solving Rational Inequalities

A **rational inequality** is any inequality that can be put in one of the forms

$$f(x) < 0, \quad f(x) > 0, \quad f(x) \leq 0, \quad \text{or} \quad f(x) \geq 0,$$

where f is a rational function. An example of a rational inequality is

$$\frac{3x + 3}{2x + 4} > 0.$$

This inequality is in the form $f(x) > 0$, where f is the rational function given by

$$f(x) = \frac{3x + 3}{2x + 4}.$$

The graph of f is shown in **Figure 8.16**.

We can find the x-intercept of f by setting the numerator equal to 0:

$$3x + 3 = 0$$
$$3x = -3$$
$$x = -1.$$

f has an x-intercept at -1 and passes through $(-1, 0)$.

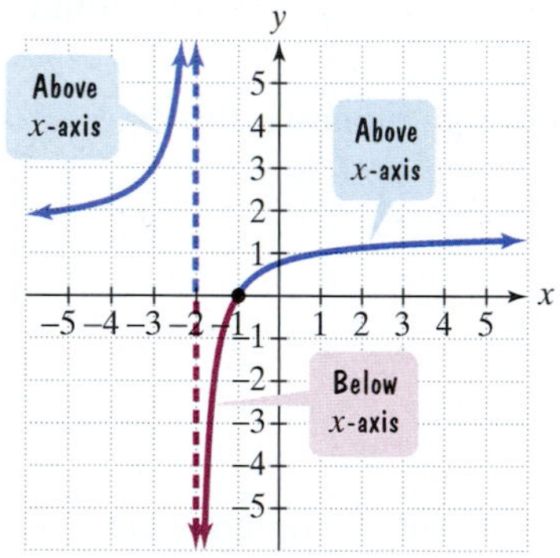

FIGURE 8.16 The graph of $f(x) = \dfrac{3x + 3}{2x + 4}$

We can determine where f is undefined by setting the denominator equal to 0:

$$2x + 4 = 0$$
$$2x = -4$$
$$x = -2.$$

f is undefined at -2. Figure 8.16 shows that the function's vertical asymptote is $x = -2$.

By setting both the numerator and the denominator of f equal to 0, we obtained the solutions -2 and -1. These numbers separate the x-axis into three intervals: $(-\infty, -2)$, $(-2, -1)$, and $(-1, \infty)$. On each interval, the graph of f is either above the x-axis $[f(x) > 0]$ or below the x-axis $[f(x) < 0]$.

Examine the graph in **Figure 8.16** carefully. Can you see that it is above the x-axis for all x in $(-\infty, -2)$ or $(-1, \infty)$, shown in blue? Thus, the solution set of $\dfrac{3x + 3}{2x + 4} > 0$ is $(-\infty, -2) \cup (-1, \infty)$. By contrast, the graph of f lies below the x-axis for all x in $(-2, -1)$, shown in red. Thus, the solution set of $\dfrac{3x + 3}{2x + 4} < 0$ is $(-2, -1)$.

The first step in solving a rational inequality is to bring all terms to one side, obtaining zero on the other side. Then express the rational function on the nonzero side as a single quotient. The second step is to set the numerator and the denominator of f equal to zero. The solutions of these equations serve as boundary points that separate the real number line into intervals. At this point, the procedure is the same as the one we used for solving polynomial inequalities.

EXAMPLE 3 Solving a Rational Inequality

Solve and graph the solution set: $\frac{x+3}{x-7} < 0.$

Solution

> **Study Tip**
>
> Many students want to solve
>
> $$\frac{x+3}{x-7} < 0$$
>
> by first multiplying both sides by $x - 7$ to clear fractions. This is incorrect. The problem is that $x - 7$ contains a variable and can be positive or negative, depending on the value of x. Thus, we do not know whether or not to reverse the sense of the inequality.

Step 1. Express the inequality so that one side is zero and the other side is a single quotient. The given inequality is already in this form. The form is $f(x) < 0$, where $f(x) = \frac{x+3}{x-7}$.

Step 2. Set the numerator and the denominator of f equal to zero. The real solutions are the boundary points.

$$x + 3 = 0 \qquad x - 7 = 0$$

Set the numerator and denominator equal to 0. These are the values that make the previous quotient zero or undefined.

$$x = -3 \qquad x = 7$$

Solve for x.

We will use these solutions as boundary points on a number line.

Step 3. Locate the boundary points on a number line and separate the line into intervals. The number line with the boundary points is shown as follows:

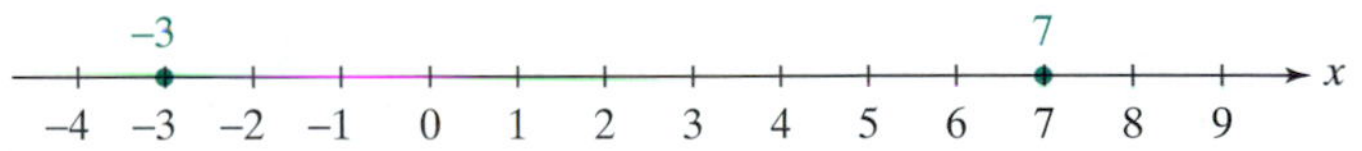

The boundary points divide the number line into three intervals:

$$(-\infty, -3) \quad (-3, 7) \quad (7, \infty).$$

Step 4. Choose one test value within each interval and evaluate f at that number.

Interval	Test Value	Substitute into $f(x) = \frac{x+3}{x-7}$	Conclusion
$(-\infty, -3)$	-4	$f(-4) = \frac{-4+3}{-4-7}$ $= \frac{-1}{-11} = \frac{1}{11}$, positive	$f(x) > 0$ for all x in $(-\infty, -3)$.
$(-3, 7)$	0	$f(0) = \frac{0+3}{0-7}$ $= -\frac{3}{7}$, negative	$f(x) < 0$ for all x in $(-3, 7)$.
$(7, \infty)$	8	$f(8) = \frac{8+3}{8-7}$ $= 11$, positive	$f(x) > 0$ for all x in $(7, \infty)$.

Step 5. Write the solution set selecting the interval or intervals that satisfy the given inequality. We are interested in solving $f(x) < 0$, where $f(x) = \frac{x+3}{x-7}$. Based on our work in step 4, we see that $f(x) < 0$ for all x in $(-3, 7)$.

Because $f(x) < 0$ for all x in $(-3, 7)$, the solution set of the given inequality, $\frac{x+3}{x-7} < 0$, is

$$(-3, 7) \quad \text{or} \quad \{x \mid -3 < x < 7\}.$$

The graph of the solution set on a number line is shown as follows:

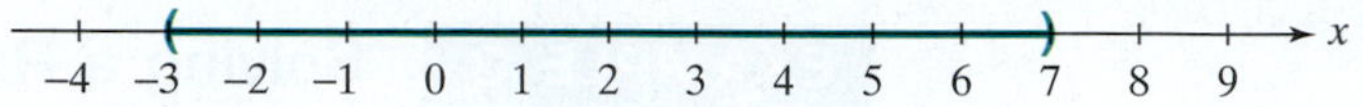

✓ **CHECK POINT 3** Solve and graph the solution set: $\frac{x-5}{x+2} < 0$.

EXAMPLE 4 Solving a Rational Inequality

Solve and graph the solution set: $\frac{x+1}{x+3} \geq 2$.

Study Tip

Do not begin solving

$$\frac{x+1}{x+3} \geq 2$$

by multiplying both sides by $x + 3$. We do not know if $x + 3$ is positive or negative. Thus, we do not know whether or not to reverse the sense of the inequality.

Solution

Step 1. Express the inequality so that one side is zero and the other side is a single quotient. We subtract 2 from both sides to obtain zero on the right.

$$\frac{x+1}{x+3} \geq 2 \qquad \text{This is the given inequality.}$$

$$\frac{x+1}{x+3} - 2 \geq 0 \qquad \text{Subtract 2 from both sides, obtaining 0 on the right.}$$

$$\frac{x+1}{x+3} - \frac{2(x+3)}{x+3} \geq 0 \qquad \text{The least common denominator is } x+3. \text{ Express 2 in terms of this denominator.}$$

$$\frac{x+1-2(x+3)}{x+3} \geq 0 \qquad \text{Subtract rational expressions.}$$

$$\frac{x+1-2x-6}{x+3} \geq 0 \qquad \text{Apply the distributive property.}$$

$$\frac{-x-5}{x+3} \geq 0 \qquad \text{Simplify.}$$

This inequality is equivalent to the one we wish to solve. It is in the form $f(x) \geq 0$, where $f(x) = \frac{-x-5}{x+3}$.

Study Tip

Never include the value that causes a rational function's denominator to equal zero in the solution set of a rational inequality. Division by zero is undefined.

Step 2. Set the numerator and the denominator of f equal to zero. The real solutions are the boundary points.

$$-x - 5 = 0 \qquad x + 3 = 0 \qquad \text{Set the numerator and denominator equal to 0. These are the values that make the previous quotient zero or undefined.}$$

$$x = -5 \qquad x = -3 \qquad \text{Solve for } x.$$

We will use these solutions as boundary points on a number line.

Step 3. Locate the boundary points on a number line and separate the line into intervals. The number line with the boundary points is shown as follows:

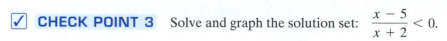

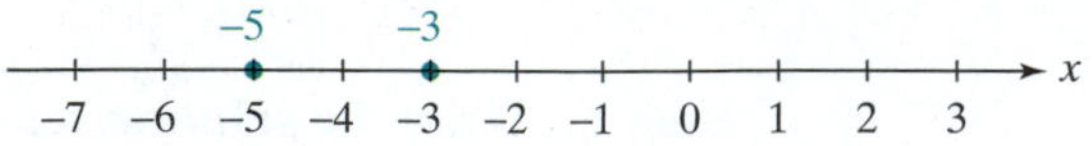

The boundary points divide the number line into three intervals:

$$(-\infty, -5) \quad (-5, -3) \quad (-3, \infty).$$

Step 4. Choose one test value within each interval and evaluate f at that number.

Interval	Test Value	Substitute into $f(x) = \frac{-x-5}{x+3}$	Conclusion
$(-\infty, -5)$	-6	$f(-6) = \frac{-(-6)-5}{-6+3}$ $= -\frac{1}{3}$, negative	$f(x) < 0$ for all x in $(-\infty, -5)$.
$(-5, -3)$	-4	$f(-4) = \frac{-(-4)-5}{-4+3}$ $= 1$, positive	$f(x) > 0$ for all x in $(-5, -3)$.
$(-3, \infty)$	0	$f(0) = \frac{-0-5}{0+3}$ $= -\frac{5}{3}$, negative	$f(x) < 0$ for all x in $(-3, \infty)$.

Step 5. Write the solution set, selecting the interval or intervals that satisfy the given inequality. We are interested in solving $f(x) \geq 0$, where $f(x) = \frac{-x-5}{x+3}$. Based on our work in step 4, we see that $f(x) > 0$ for all x in $(-5, -3)$. However, because the inequality involves $\geq$ (greater than or *equal to*), we must also include the solution of $f(x) = 0$, namely the value that we obtained when we set the numerator of f equal to zero. Thus, we must include -5 in the solution set. The solution set of the given inequality is

$$[-5, -3) \text{ or } \{x \mid -5 \leq x < -3\}.$$

The graph of the solution set on a number line is shown as follows:

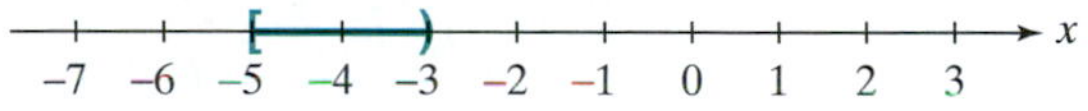

Using Technology

Graphic Connections

The solution set for

$$\frac{x+1}{x+3} \geq 2$$

or, equivalently,

$$\frac{-x-5}{x+3} \geq 0$$

can be verified with a graphing utility. The graph of $f(x) = \frac{-x-5}{x+3}$ lies on or above the x-axis, representing $\geq$, for all x in $[-5, -3)$.

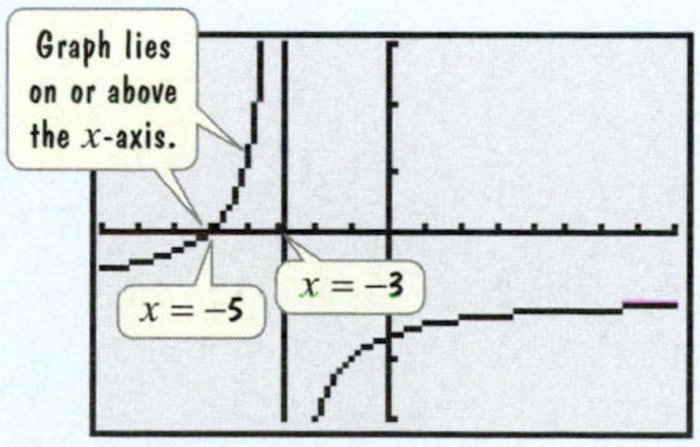

[−8, 8, 1] by [−3, 3, 1]

✓ **CHECK POINT 4** Solve and graph the solution set: $\frac{2x}{x+1} \geq 1$.

3 Solve problems modeled by polynomial or rational inequalities.

Applications

Polynomial inequalities can be solved to answer questions about variables contained in polynomial functions.

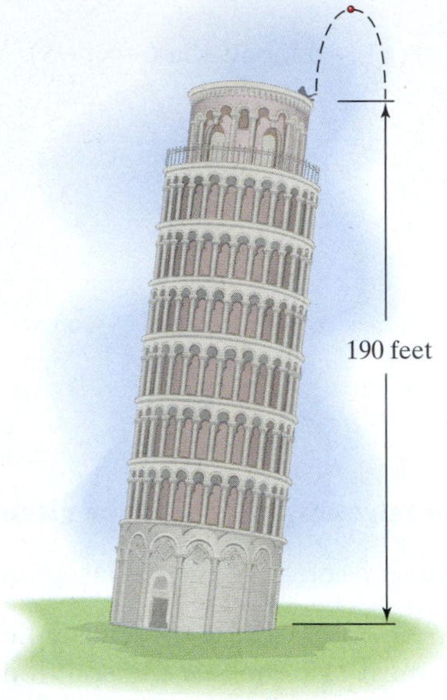

FIGURE 8.17 Throwing a ball from the top of the Leaning Tower of Pisa

EXAMPLE 5 Modeling the Position of a Free-Falling Object

A ball is thrown vertically upward from the top of the Leaning Tower of Pisa (190 feet high) with an initial velocity of 96 feet per second (**Figure 8.17**). The function

$$s(t) = -16t^2 + 96t + 190$$

models the ball's height above the ground, $s(t)$, in feet, t seconds after it was thrown. During which time period will the ball's height exceed that of the tower?

Solution Using the problem's question and the given model for the ball's height, $s(t) = -16t^2 + 96t + 190$, we obtain a polynomial inequality.

$$-16t^2 + 96t + 190 > 190$$

When will the ball's height | exceed that | of the tower?

$-16t^2 + 96t + 190 > 190$	This is the inequality that models the problem's question. We must find t.
$-16t^2 + 96t > 0$	Subtract 190 from both sides. This inequality is in the form $f(t) > 0$, where $f(t) = -16t^2 + 96t$.
$-16t^2 + 96t = 0$	Solve the equation $f(t) = 0$.
$-16t(t - 6) = 0$	Factor.
$-16t = 0$ or $t - 6 = 0$	Set each factor equal to 0.
$t = 0$ $\quad$ $t = 6$	Solve for t. The boundary points are 0 and 6.

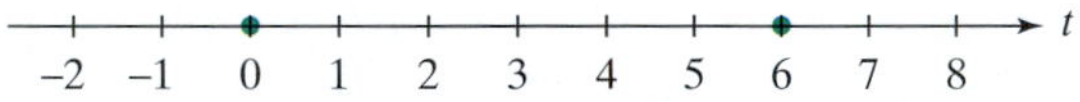

Locate these values on a number line, with $t \geq 0$.

The intervals are $(-\infty, 0)$, $(0, 6)$, and $(6, \infty)$. For our purposes, the mathematical model is useful only from $t = 0$ until the ball hits the ground. (By setting $-16t^2 + 96t + 190$ equal to zero, we find $t \approx 7.57$; the ball hits the ground after approximately 7.57 seconds.) Thus, we use $(0, 6)$ and $(6, 7.57)$ for our intervals.

Interval	Test Value	Substitute into $f(t) = -16t^2 + 96t$	Conclusion
$(0, 6)$	1	$f(1) = -16 \cdot 1^2 + 96 \cdot 1$ $= 80$, positive	$f(t) > 0$ for all t in $(0, 6)$.
$(6, 7.57)$	7	$f(7) = -16 \cdot 7^2 + 96 \cdot 7$ $= -112$, negative	$f(t) < 0$ for all t in $(6, 7.57)$.

We are interested in solving $f(t) > 0$, where $f(t) = -16t^2 + 96t$. We see that $f(t) > 0$ for all t in $(0, 6)$. This means that the ball's height exceeds that of the tower between 0 and 6 seconds.

Using Technology

Graphic Connections

The graphs of

$$y_1 = -16x^2 + 96x + 190$$

and

$$y_2 = 190$$

are shown in a

$$[0, 8, 1] \text{ by } [0, 360, 36]$$

seconds in motion — height, in feet

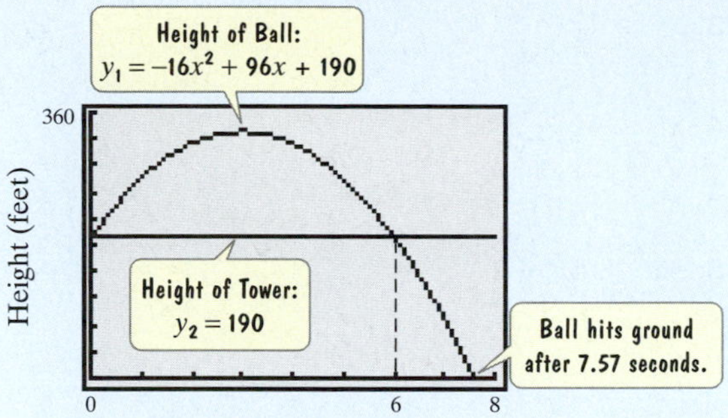

Seconds in Motion
[0, 8, 1] by [0, 360, 36]

viewing rectangle. The graphs show that the ball's height exceeds that of the tower between 0 and 6 seconds.

☑ **CHECK POINT 5** An object is propelled straight up from ground level with an initial velocity of 80 feet per second. Its height at time t is modeled by

$$s(t) = -16t^2 + 80t,$$

where the height, $s(t)$, is measured in feet and the time, t, is measured in seconds. In which time interval will the object be more than 64 feet above the ground?

8.5 EXERCISE SET

MyMathLab — MathXL PRACTICE — WATCH — DOWNLOAD — READ — REVIEW

Practice Exercises

Solve each polynomial inequality in Exercises 1–40 and graph the solution set on a real number line.

1. $(x - 4)(x + 2) > 0$

2. $(x + 3)(x - 5) > 0$

3. $(x - 7)(x + 3) \le 0$

4. $(x + 1)(x - 7) \le 0$

5. $x^2 - 5x + 4 > 0$

6. $x^2 - 4x + 3 < 0$

7. $x^2 + 5x + 4 > 0$

8. $x^2 + x - 6 > 0$

9. $x^2 - 6x + 8 \le 0$

10. $x^2 - 2x - 3 \ge 0$

11. $3x^2 + 10x - 8 \le 0$

12. $9x^2 + 3x - 2 \ge 0$

13. $2x^2 + x < 15$

14. $6x^2 + x > 1$

15. $4x^2 + 7x < -3$

16. $3x^2 + 16x < -5$

17. $x^2 - 4x \ge 0$

18. $x^2 + 2x < 0$

19. $2x^2 + 3x > 0$

20. $3x^2 - 5x \le 0$

21. $-x^2 + x \ge 0$

22. $-x^2 + 2x \geq 0$

23. $x^2 \leq 4x - 2$

24. $x^2 \leq 2x + 2$

25. $3x^2 > 4x + 2$

26. $3x^2 > 10x - 5$

27. $2x^2 - 5x \geq 1$

28. $5x^2 + 8x \geq 11$

29. $x^2 - 6x + 9 < 0$

30. $4x^2 - 4x + 1 \geq 0$

31. $(x - 1)(x - 2)(x - 3) \geq 0$

32. $(x + 1)(x + 2)(x + 3) \geq 0$

33. $x^3 + 2x^2 - x - 2 \geq 0$

34. $x^3 + 2x^2 - 4x - 8 \geq 0$

35. $x^3 - 3x^2 - 9x + 27 < 0$

36. $x^3 + 7x^2 - x - 7 < 0$

37. $x^3 + x^2 + 4x + 4 > 0$

38. $x^3 - x^2 + 9x - 9 > 0$

39. $x^3 \geq 9x^2$

40. $x^3 \leq 4x^2$

Solve each rational inequality in Exercises 41–56 and graph the solution set on a real number line.

41. $\dfrac{x - 4}{x + 3} > 0$

42. $\dfrac{x + 5}{x - 2} > 0$

43. $\dfrac{x + 3}{x + 4} < 0$

44. $\dfrac{x + 5}{x + 2} < 0$

45. $\dfrac{-x + 2}{x - 4} \geq 0$

46. $\dfrac{-x - 3}{x + 2} \leq 0$

47. $\dfrac{4 - 2x}{3x + 4} \leq 0$

48. $\dfrac{3x + 5}{6 - 2x} \geq 0$

49. $\dfrac{x}{x - 3} > 0$

50. $\dfrac{x + 4}{x} > 0$

51. $\dfrac{x + 1}{x + 3} < 2$

52. $\dfrac{x}{x - 1} > 2$

53. $\dfrac{x + 4}{2x - 1} \leq 3$

54. $\dfrac{1}{x - 3} < 1$

55. $\dfrac{x - 2}{x + 2} \leq 2$

56. $\dfrac{x}{x + 2} \geq 2$

In Exercises 57–60, use the given functions to find all values of x that satisfy the required inequality.

57. $f(x) = 2x^2,\ g(x) = 5x - 2;\ f(x) \geq g(x)$

58. $f(x) = 4x^2,\ g(x) = 9x - 2;\ f(x) < g(x)$

59. $f(x) = \dfrac{2x}{x + 1},\ g(x) = 1;\ f(x) < g(x)$

60. $f(x) = \dfrac{x}{2x - 1},\ g(x) = 1;\ f(x) \geq g(x)$

Practice PLUS

Solve each inequality in Exercises 61–66 and graph the solution set on a real number line.

61. $|x^2 + 2x - 36| > 12$

62. $|x^2 + 6x + 1| > 8$

63. $\dfrac{3}{x + 3} > \dfrac{3}{x - 2}$

64. $\frac{1}{x+1} > \frac{2}{x-1}$

65. $\frac{x^2 - x - 2}{x^2 - 4x + 3} > 0$

66. $\frac{x^2 - 3x + 2}{x^2 - 2x - 3} > 0$

In Exercises 67–68, use the graph of the polynomial function to solve each inequality.

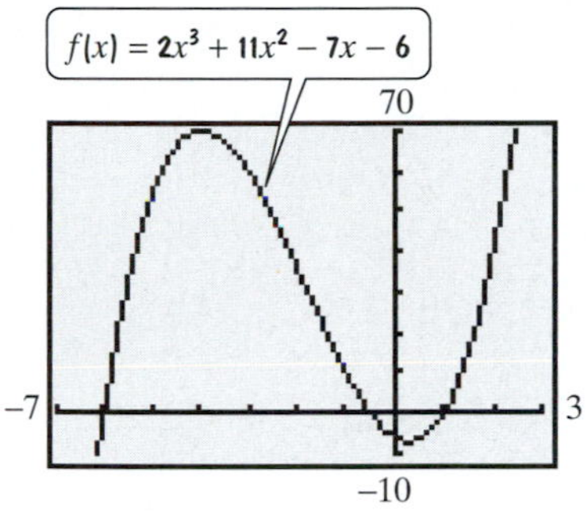

67. $2x^3 + 11x^2 \geq 7x + 6$

68. $2x^3 + 11x^2 < 7x + 6$

In Exercises 69–70, use the graph of the rational function to solve each inequality.

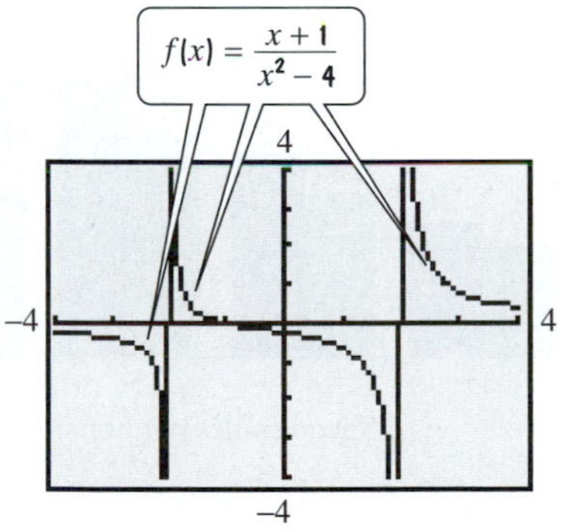

69. $\frac{1}{4(x+2)} \leq -\frac{3}{4(x-2)}$

70. $\frac{1}{4(x+2)} > -\frac{3}{4(x-2)}$

Application Exercises

71. You throw a ball straight up from a rooftop 160 feet high with an initial speed of 48 feet per second. The function

$$s(t) = -16t^2 + 48t + 160$$

models the ball's height above the ground, $s(t)$, in feet, t seconds after it was thrown. During which time period will the ball's height exceed that of the rooftop?

72. Divers in Acapulco, Mexico, dive headfirst from the top of a cliff 87 feet above the Pacific Ocean. The function

$$s(t) = -16t^2 + 8t + 87$$

models a diver's height above the ocean, $s(t)$, in feet, t seconds after leaping. During which time period will the diver's height exceed that of the cliff?

The functions

$$f(x) = 0.0875x^2 - 0.4x + 66.6$$

Dry pavement

and

Wet pavement

$$g(x) = 0.0875x^2 + 1.9x + 11.6$$

model a car's stopping distance, $f(x)$ or $g(x)$, in feet, traveling at x miles per hour. Function f models stopping distance on dry pavement and function g models stopping distance on wet pavement. The graphs of these functions are shown for $\{x|x \geq 30\}$. Notice that the figure does not specify which graph is the model for dry roads and which is the model for wet roads. Use this information to solve Exercises 73–74.

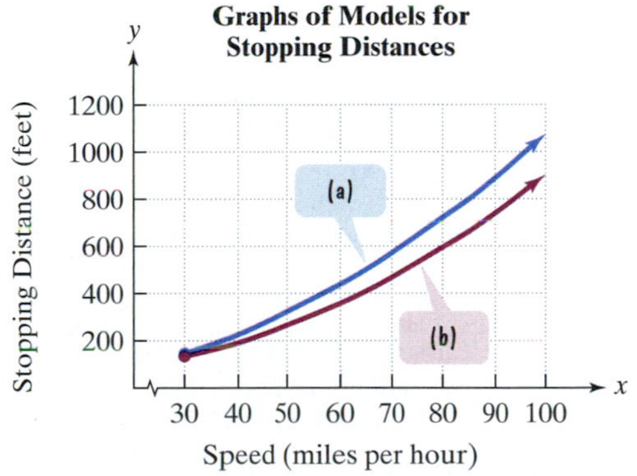

73. a. Use the given functions to find the stopping distance on dry pavement and the stopping distance on wet pavement for a car traveling at 35 miles per hour. Round to the nearest foot.

b. Based on your answers to part (a), which rectangular coordinate graph shows stopping distances on dry pavement and which shows stopping distances on wet pavement?

c. How well do your answers to part (a) model the actual stopping distances shown in **Figure 8.14** on page 616?

d. Determine speeds on dry pavement requiring stopping distances that exceed the length of one and one-half football fields, or 540 feet. Round to the nearest mile per hour. How is this shown on the appropriate graph of the models?

74. a. Use the given functions to find the stopping distance on dry pavement and the stopping distance on wet pavement for a car traveling at 55 miles per hour. Round to the nearest foot.

b. Based on your answers to part (a), which rectangular coordinate graph on page 627 shows stopping distances on dry pavement and which shows stopping distances on wet pavement?

c. How well do your answers to part (a) model the actual stopping distances shown in **Figure 8.14** on page 616?

d. Determine speeds on wet pavement requiring stopping distances that exceed the length of one and one-half football fields, or 540 feet. Round to the nearest mile per hour. How is this shown on the appropriate graph of the models on page 627?

A company manufactures wheelchairs. The average cost function, $\overline{C}$, of producing x wheelchairs per month is given by

$$\overline{C}(x) = \frac{500{,}000 + 400x}{x}.$$

The graph of the rational function is shown. Use the function to solve Exercises 75–76.

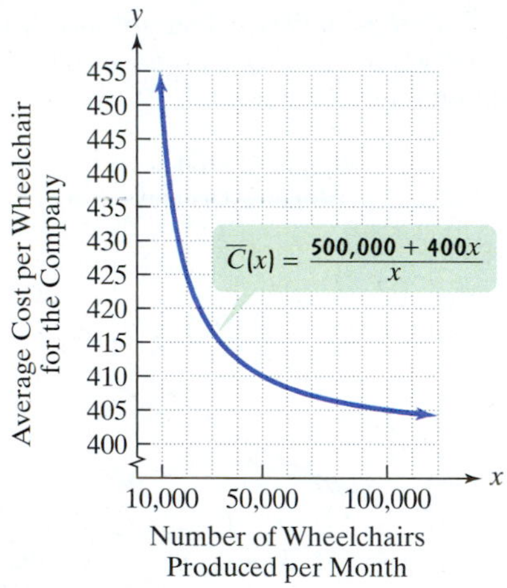

75. Describe the company's production level so that the average cost of producing each wheelchair does not exceed \$425. Use a rational inequality to solve the problem. Then explain how your solution is shown on the graph.

76. Describe the company's production level so that the average cost of producing each wheelchair does not exceed \$410. Use a rational inequality to solve the problem. Then explain how your solution is shown on the graph.

77. The perimeter of a rectangle is 50 feet. Describe the possible length of a side if the area of the rectangle is not to exceed 114 square feet.

78. The perimeter of a rectangle is 180 feet. Describe the possible lengths of a side if the area of the rectangle is not to exceed 800 square feet.

Writing in Mathematics

79. What is a polynomial inequality?

80. What is a rational inequality?

81. Describe similarities and differences between the solutions of

$$(x - 2)(x + 5) \ge 0 \quad \text{and} \quad \frac{x - 2}{x + 5} \ge 0.$$

Technology Exercises

Solve each inequality in Exercises 82–87 using a graphing utility.

82. $x^2 + 3x - 10 > 0$

83. $2x^2 + 5x - 3 \le 0$

84. $\frac{x - 4}{x - 1} \le 0$

85. $\frac{x + 2}{x - 3} \le 2$

86. $\frac{1}{x + 1} \le \frac{2}{x + 4}$

87. $x^3 + 2x^2 - 5x - 6 > 0$

The graph shows stopping distances for trucks at various speeds on dry roads and on wet roads. Use this information to solve Exercises 88–89.

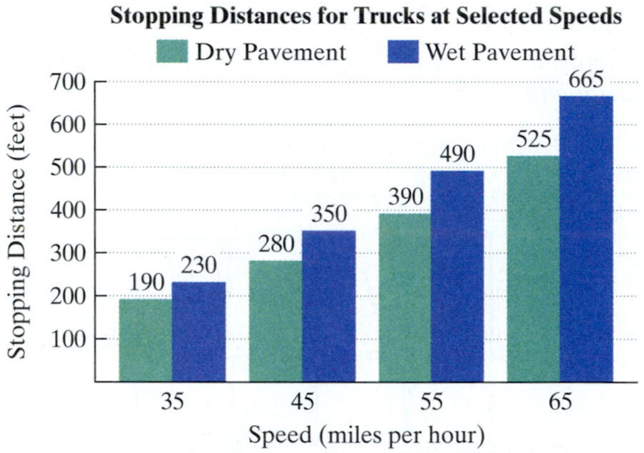

Source: National Highway Traffic Safety Administration

88. a. Use the statistical menu of your graphing utility and the quadratic regression program to obtain the quadratic function that models a truck's stopping distance, $f(x)$, in feet, on dry pavement traveling at x miles per hour. Round the x-coefficient and the constant term to one decimal place.

b. Use the function from part (a) to determine speeds on dry pavement requiring stopping distances that exceed 455 feet. Round to the nearest mile per hour.

89. a. Use the statistical menu of your graphing utility and the quadratic regression program to obtain the quadratic function that models a truck's stopping distance, $f(x)$, in feet, on wet pavement traveling at x miles per hour. Round the x-coefficient and the constant term to one decimal place.

b. Use the function from part (a) to determine speeds on wet pavement requiring stopping distances that exceed 446 feet.

Critical Thinking Exercises

Make Sense? *In Exercises 90–93, determine whether each statement "makes sense" or "does not make sense" and explain your reasoning.*

90. When solving $f(x) > 0$, where f is a polynomial function, I only pay attention to the sign of f at each test value and not the actual function value.

91. I'm solving a polynomial inequality that has a value for which the polynomial function is undefined.

92. Because it takes me longer to come to a stop on a wet road than on a dry road, graph (a) for Exercises 73–74 is the model for stopping distances on wet pavement and graph (b) is the model for stopping distances on dry pavement.

93. I began the solution of the rational inequality $\frac{x+1}{x+3} \geq 2$ by setting both $x + 1$ and $x + 3$ equal to zero.

In Exercises 94–97, determine whether each statement is true or false. If the statement is false, make the necessary change(s) to produce a true statement.

94. The solution set of $x^2 > 25$ is $(5, \infty)$.

95. The inequality $\frac{x-2}{x+3} < 2$ can be solved by multiplying both sides by $x + 3$, resulting in the equivalent inequality $x - 2 < 2(x + 3)$.

96. $(x + 3)(x - 1) \geq 0$ and $\frac{x+3}{x-1} \geq 0$ have the same solution set.

97. The inequality $\frac{x-2}{x+3} < 2$ can be solved by multiplying both sides by $(x + 3)^2$, $x \neq -3$, resulting in the equivalent inequality $(x - 2)(x + 3) < 2(x + 3)^2$.

98. Write a quadratic inequality whose solution set is $[-3, 5]$.

99. Write a rational inequality whose solution set is $(-\infty, -4) \cup [3, \infty)$.

In Exercises 100–103, use inspection to describe each inequality's solution set. Do not solve any of the inequalities.

100. $(x - 2)^2 > 0$

101. $(x - 2)^2 \leq 0$

102. $(x - 2)^2 < -1$

103. $\frac{1}{(x-2)^2} > 0$

104. The graphing calculator screen shows the graph of $y = 4x^2 - 8x + 7$.

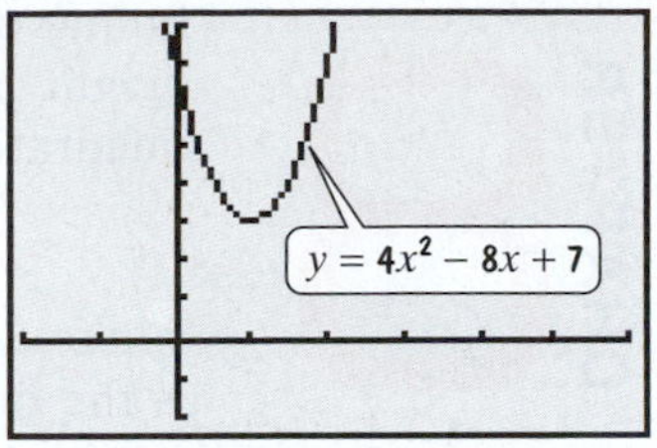

[−2, 6, 1] by [−2, 8, 1]

a. Use the graph to describe the solution set for $4x^2 - 8x + 7 > 0$.

b. Use the graph to describe the solution set for $4x^2 - 8x + 7 < 0$.

c. Use an algebraic approach to verify each of your descriptions in parts (a) and (b).

105. The graphing calculator screen shows the graph of $y = \sqrt{27 - 3x^2}$. Write and solve a quadratic inequality that explains why the graph only appears for $-3 \leq x \leq 3$.

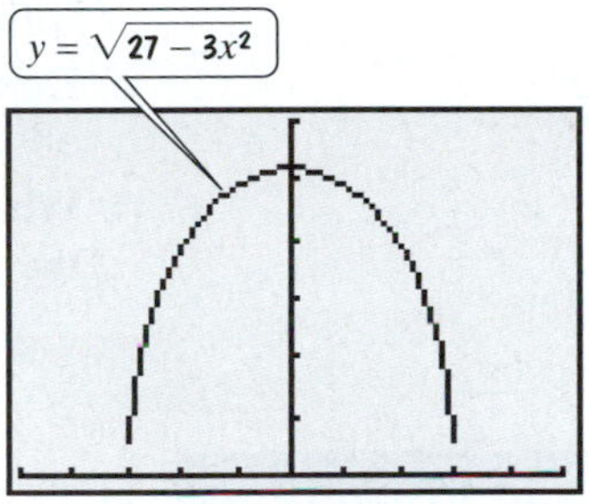

[−5, 5, 1] by [0, 6, 1]

Review Exercises

106. Solve: $\left|\frac{x-5}{3}\right| < 8$. (Section 4.3, Example 4 or Example 6(a))

107. Divide:

$$\frac{2x+6}{x^2+8x+16} \div \frac{x^2-9}{x^2+3x-4}.$$

(Section 6.1, Example 7)

108. Factor completely: $x^4 - 16y^4$. (Section 5.5, Example 3)

Preview Exercises

Exercises 109–111 will help you prepare for the material covered in the first section of the next chapter. In each exercise, use point plotting to graph the function. Begin by setting up a table of coordinates, selecting integers from −3 to 3, inclusive, for x.

109. $f(x) = 2^x$ **110.** $f(x) = 2^{-x}$ **111.** $f(x) = 2^x + 1$

GROUP PROJECT

CHAPTER 8

Throughout the chapter, we have considered functions that model the position of free-falling objects. Any object that is falling, or vertically projected into the air, has its height above the ground, $s(t)$, in feet, after t seconds in motion, modeled by the quadratic function

$$s(t) = -16t^2 + v_0t + s_0,$$

where v_0 is the original velocity (initial velocity) of the object, in feet per second, and s_0 is the original height (initial height) of the object, in feet, above the ground. In this exercise, group members will be working with this position function. The exercise is appropriate for groups of three to five people.

a. Drop a ball from a height of 3 feet, 6 feet, and 12 feet. Record the number of seconds it takes for the ball to hit the ground.
b. For each of the three initial positions, use the position function to determine the time required for the ball to hit the ground.
c. What factors might result in differences between the times that you recorded and the times indicated by the function?
d. What appears to be happening to the time required for a free-falling object to hit the ground as its initial height is doubled? Verify this observation algebraically and with a graphing utility.
e. Repeat part (a) using a sheet of paper rather than a ball. What differences do you observe? What factor seems to be ignored in the position function?
f. What is meant by the acceleration of gravity and how does this number appear in the position function for a free-falling object?

Chapter 8 Summary

Definitions and Concepts	Examples

Section 8.1 The Square Root Property and Completing the Square

The Square Root Property

If u is an algebraic expression and d is a real number, then

$$\text{If } u^2 = d, \quad \text{then} \quad u = \sqrt{d} \quad \text{or} \quad u = -\sqrt{d}.$$

Equivalently,

$$\text{If } u^2 = d, \quad \text{then} \quad u = \pm\sqrt{d}.$$

Solve:

$$(x-6)^2 = 50.$$
$$x - 6 = \pm\sqrt{50}$$
$$x - 6 = \pm\sqrt{25 \cdot 2}$$
$$x - 6 = \pm 5\sqrt{2}$$
$$x = 6 \pm 5\sqrt{2}$$

The solutions are $6 \pm 5\sqrt{2}$ and the solution set is $\{6 \pm 5\sqrt{2}\}$.

Completing the Square

If $x^2 + bx$ is a binomial, then by adding $\left(\frac{b}{2}\right)^2$, the square of half the coefficient of x, a perfect square trinomial will result. That is,

$$x^2 + bx + \left(\frac{b}{2}\right)^2 = \left(x + \frac{b}{2}\right)^2.$$

Complete the square:

$$x^2 + \frac{2}{7}x.$$

Half of $\frac{2}{7}$ is $\frac{1}{2} \cdot \frac{2}{7} = \frac{1}{7}$ and $\left(\frac{1}{7}\right)^2 = \frac{1}{49}$.

$$x^2 + \frac{2}{7}x + \frac{1}{49} = \left(x + \frac{1}{7}\right)^2$$

Definitions and Concepts	Examples

Section 8.1 The Square Root Property and Completing the Square (continued)

Solving Quadratic Equations by Completing the Square

1. If the coefficient of x^2 is not 1, divide both sides by this coefficient.
2. Isolate variable terms on one side.
3. Complete the square by adding the square of half the coefficient of x to both sides.
4. Factor the perfect square trinomial.
5. Solve by applying the square root property.

Solve by completing the square:

$$2x^2 + 16x - 6 = 0.$$

$$\frac{2x^2}{2} + \frac{16x}{2} - \frac{6}{2} = \frac{0}{2} \quad \text{Divide by 2.}$$

$$x^2 + 8x - 3 = 0 \quad \text{Simplify.}$$

$$x^2 + 8x = 3 \quad \text{Add 3.}$$

The coefficient of x is 8. Half of 8 is 4 and $4^2 = 16$. Add 16 to both sides.

$$x^2 + 8x + 16 = 3 + 16$$

$$(x + 4)^2 = 19$$

$$x + 4 = \pm\sqrt{19}$$

$$x = -4 \pm \sqrt{19}$$

Section 8.2 The Quadratic Formula

The solutions of a quadratic equation in standard form

$$ax^2 + bx + c = 0, \quad a \neq 0,$$

are given by the quadratic formula

$$x = \frac{-b \pm \sqrt{b^2 - 4ac}}{2a}.$$

Solve using the quadratic formula:

$$2x^2 = 6x - 3.$$

First write the equation in standard form by subtracting $6x$ and adding 3 on both sides.

$$2x^2 - 6x + 3 = 0$$

$a = 2$ $b = -6$ $c = 3$

$$x = \frac{-(-6) \pm \sqrt{(-6)^2 - 4\cdot2\cdot3}}{2\cdot2} = \frac{6 \pm \sqrt{36 - 24}}{4}$$

$$= \frac{6 \pm \sqrt{12}}{4} = \frac{6 \pm \sqrt{4\cdot3}}{4} = \frac{6 \pm 2\sqrt{3}}{4}$$

$$= \frac{\cancel{2}(3 \pm \sqrt{3})}{\cancel{2}\cdot2} = \frac{3 \pm \sqrt{3}}{2}$$

The Discriminant

The discriminant, $b^2 - 4ac$, of the quadratic equation $ax^2 + bx + c = 0$ determines the number and type of solutions.

Discriminant	Solutions
Positive perfect square, with a, b, and c rational numbers	2 real rational solutions
Positive and not a perfect square	2 real irrational solutions
Zero, with a, b, and c rational numbers	1 real rational solution
Negative	2 imaginary solutions

- $2x^2 - 7x - 4 = 0$

$a = 2$ $b = -7$ $c = -4$

$$b^2 - 4ac = (-7)^2 - 4(2)(-4)$$

$$= 49 - (-32) = 49 + 32 = 81$$

Positive perfect square

The equation has 2 real rational solutions.

Definitions and Concepts	Examples

Section 8.2 The Quadratic Formula (continued)

Writing Quadratic Equations from Solutions

The zero-product principle in reverse makes it possible to write a quadratic equation from solutions:

$$\text{If } A = 0 \quad \text{or} \quad B = 0, \quad \text{then} \quad AB = 0.$$

Write a quadratic equation with the solution set $\{-2\sqrt{3}, 2\sqrt{3}\}$.

$$x = -2\sqrt{3} \qquad\qquad x = 2\sqrt{3}$$

$$x + 2\sqrt{3} = 0 \quad \text{or} \quad x - 2\sqrt{3} = 0$$

$$(x + 2\sqrt{3})(x - 2\sqrt{3}) = 0$$

$$x^2 - (2\sqrt{3})^2 = 0$$

$$x^2 - 12 = 0$$

Section 8.3 Quadratic Functions and Their Graphs

The graph of the quadratic function

$$f(x) = a(x - h)^2 + k, \quad a \neq 0,$$

is called a parabola. The vertex, or turning point, is (h, k). The graph opens upward if a is positive and downward if a is negative. The axis of symmetry is a vertical line passing through the vertex. The graph can be obtained using the vertex, x-intercepts, if any, (set $f(x)$ equal to zero and solve), and the y-intercept (set $x = 0$).

Graph: $f(x) = -(x + 3)^2 + 1$.

$$f(x) = -1(x - (-3))^2 + 1$$

$a = -1$ $\quad h = -3$ $\quad k = 1$

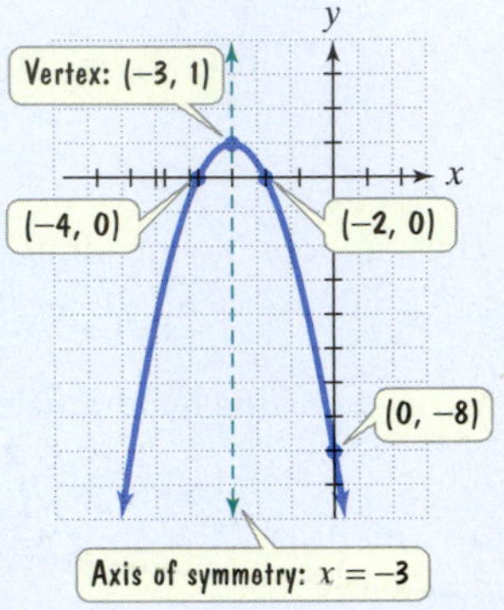

- Vertex: $(h, k) = (-3, 1)$
- Opens downward because $a < 0$
- x-intercepts: Set $f(x) = 0$.

$$0 = -(x + 3)^2 + 1$$

$$(x + 3)^2 = 1$$

$$x + 3 = \pm\sqrt{1}$$

$$x + 3 = 1 \quad \text{or} \quad x + 3 = -1$$

$$x = -2 \qquad\qquad x = -4$$

- y-intercept: Set $x = 0$.

$$f(0) = -(0 + 3)^2 + 1 = -9 + 1 = -8$$

A parabola whose equation is in the form

$$f(x) = ax^2 + bx + c, \quad a \neq 0,$$

has its vertex at

$$\left(-\frac{b}{2a}, f\left(-\frac{b}{2a}\right)\right).$$

The parabola is graphed as described in the left column above. The only difference is how we determine the vertex. If $a > 0$, then f has a minimum that occurs at $x = -\frac{b}{2a}$. This minimum value is $f\left(-\frac{b}{2a}\right)$. If $a < 0$, then f has a maximum that occurs at $x = -\frac{b}{2a}$. This maximum value is $f\left(-\frac{b}{2a}\right)$.

Graph:

$$f(x) = x^2 - 6x + 5.$$

$a = 1$ $\quad b = -6$ $\quad c = 5$

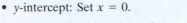

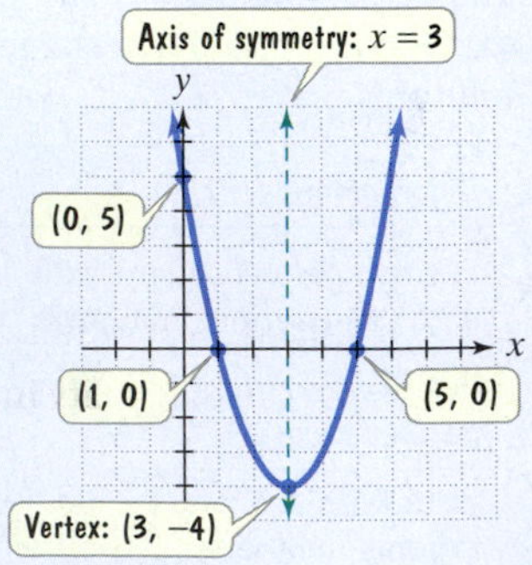

- Vertex:

$$x = -\frac{b}{2a} = -\frac{-6}{2 \cdot 1} = 3$$

$$f(3) = 3^2 - 6 \cdot 3 + 5 = -4$$

Vertex is at $(3, -4)$.

- Opens upward because $a > 0$.
- x-intercepts: Set $f(x) = 0$.

$$x^2 - 6x + 5 = 0$$

$$(x - 1)(x - 5) = 0$$

$$x = 1 \quad \text{or} \quad x = 5$$

- y-intercept: Set $x = 0$.

$$f(0) = 0^2 - 6 \cdot 0 + 5 = 5$$

Definitions and Concepts

Examples

Section 8.4 Equations Quadratic in Form

An equation that is quadratic in form is one that can be expressed as a quadratic equation using an appropriate substitution. In these equations, the variable factor in one term is the square of the variable factor in the other variable term. Let $u =$ the variable factor that reappears squared. If at any point in the solution process both sides of an equation are raised to an even power, a check is required.

Solve:

$$x^{\frac{2}{3}} - 3x^{\frac{1}{3}} + 2 = 0.$$

$$\left(x^{\frac{1}{3}}\right)^2 - 3x^{\frac{1}{3}} + 2 = 0$$

Let $u = x^{\frac{1}{3}}$.

$$u^2 - 3u + 2 = 0$$

$$(u-1)(u-2) = 0$$

$$u - 1 = 0 \quad \text{or} \quad u - 2 = 0$$

$$u = 1 \qquad\qquad u = 2$$

$$x^{\frac{1}{3}} = 1 \qquad\qquad x^{\frac{1}{3}} = 2$$

$$\left(x^{\frac{1}{3}}\right)^3 = 1^3 \qquad\qquad \left(x^{\frac{1}{3}}\right)^3 = 2^3$$

$$x = 1 \qquad\qquad x = 8$$

The solutions are 1 and 8, and the solution set is $\{1, 8\}$.

Section 8.5 Polynomial and Rational Inequalities

Solving Polynomial Inequalities

1. Express the inequality in the form

$$f(x) < 0 \quad \text{or} \quad f(x) > 0,$$

   where f is a polynomial function.
2. Solve the equation $f(x) = 0$. The real solutions are the boundary points.
3. Locate these boundary points on a number line, thereby dividing the number line into intervals.
4. Choose one representative number, called a test value, within each interval and evaluate f at that number.
   a. If the value of f is positive, then $f(x) > 0$ for all x in the interval.
   b. If the value of f is negative, then $f(x) < 0$ for all x in the interval.
5. Write the solution set, selecting the interval or intervals that satisfy the given inequality.

This procedure is valid if $<$ is replaced by $\leq$ and $>$ is replaced by $\geq$. In these cases, include the boundary points in the solution set.

Solve: $2x^2 + x - 6 > 0$.

The form of the inequality is $f(x) > 0$ with $f(x) = 2x^2 + x - 6$. Solve $f(x) = 0$.

$$2x^2 + x - 6 = 0$$

$$(2x - 3)(x + 2) = 0$$

$$2x - 3 = 0 \quad \text{or} \quad x + 2 = 0$$

$$x = \frac{3}{2} \qquad\qquad x = -2$$

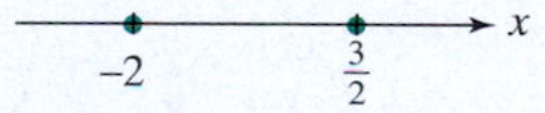

Use −3, 0, and 2 as test values.

$$f(-3) = 2(-3)^2 + (-3) - 6 = 9, \text{ positive}$$

$f(x) > 0$ for all x in $(-\infty, -2)$.

$$f(0) = 2 \cdot 0^2 + 0 - 6 = -6, \text{ negative}$$

$f(x) < 0$ for all x in $\left(-2, \frac{3}{2}\right)$.

$$f(2) = 2 \cdot 2^2 + 2 - 6 = 4, \text{ positive}$$

$f(x) > 0$ for all x in $\left(\frac{3}{2}, \infty\right)$.

The solution set is $\left\{x \mid x < -2 \text{ or } x > \frac{3}{2}\right\}$ or $(-\infty, -2) \cup \left(\frac{3}{2}, \infty\right)$.

Solving Rational Inequalities

1. Express the inequality in the form

$$f(x) < 0 \quad \text{or} \quad f(x) > 0,$$

   where f is a rational function.
2. Set the numerator and the denominator of f equal to zero. The real solutions are the boundary points.
3. Locate these boundary points on a number line, thereby dividing the number line into intervals.

Solve: $\dfrac{x}{x+4} \geq 2$.

$$\frac{x}{x+4} - \frac{2(x+4)}{x+4} \geq 0$$

$$\frac{-x-8}{x+4} \geq 0$$

The form of the inequality is $f(x) \geq 0$ with $f(x) = \dfrac{-x-8}{x+4}$.

Set the numerator and the denominator equal to zero.

Definitions and Concepts	Examples

Section 8.5 Polynomial and Rational Inequalities (continued)

Definitions and Concepts

4. Choose one representative number, called a test value, within each interval and evaluate f at that number.
   a. If the value of f is positive, then $f(x) > 0$ for all x in the interval.
   b. If the value of f is negative, then $f(x) < 0$ for all x in the interval.
5. Write the solution set, selecting the interval or intervals that satisfy the given inequality.

This procedure is valid if $<$ is replaced by $\le$ and $>$ is replaced by $\ge$. In these cases, include any values that make the numerator of f zero. Always exclude any values that make the denominator zero.

Examples

$$-x - 8 = 0 \qquad x + 4 = 0$$
$$-8 = x \qquad x = -4$$

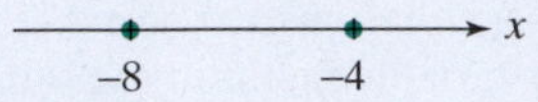

Use -9, -7, and -3 as test values.

$$f(-9) = \frac{-(-9) - 8}{-9 + 4} = \frac{1}{-5}, \text{ negative}$$
$f(x) < 0$ for all x in $(-\infty, -8)$.

$$f(-7) = \frac{-(-7) - 8}{-7 + 4} = \frac{-1}{-3} = \frac{1}{3}, \text{ positive}$$
$f(x) > 0$ for all x in $(-8, -4)$.

$$f(-3) = \frac{-(-3) - 8}{-3 + 4} = \frac{-5}{1} = -5, \text{ negative}$$
$f(x) < 0$ for all x in $(-4, \infty)$.

Because of $\ge$, include -8, the value that makes the numerator zero, in the solution set.
The solution set is $\{x | -8 \le x < -4\}$ or $[-8, -4)$.

CHAPTER 8 REVIEW EXERCISES

8.1 *In Exercises 1–5, solve each equation by the square root property. If possible, simplify radicals or rationalize denominators. Express imaginary solutions in the form $a + bi$.*

1. $2x^2 - 3 = 125$
2. $3x^2 - 150 = 0$
3. $3x^2 - 2 = 0$
4. $(x - 4)^2 = 18$
5. $(x + 7)^2 = -36$

In Exercises 6–7, determine the constant that should be added to the binomial so that it becomes a perfect square trinomial. Then write and factor the trinomial.

6. $x^2 + 20x$
7. $x^2 - 3x$

In Exercises 8–10, solve each quadratic equation by completing the square.

8. $x^2 - 12x + 27 = 0$
9. $x^2 - 7x - 1 = 0$
10. $2x^2 + 3x - 4 = 0$
11. In 2 years, an investment of $2500 grows to $2916. Use the compound interest formula
$$A = P(1 + r)^t$$
to find the annual interest rate, r.
12. The function $W(t) = 3t^2$ models the weight of a human fetus, $W(t)$, in grams, after t weeks, where $0 \le t \le 39$. After how many weeks does the fetus weigh 588 grams?
13. A building casts a shadow that is double the length of the building's height. If the distance from the end of the shadow to the top of the building is 300 meters, how high is the building? Express the answer in simplified radical form. Then find a decimal approximation to the nearest tenth of a meter.

8.2 *In Exercises 14–16, solve each equation using the quadratic formula. Simplify solutions, if possible.*

14. $x^2 = 2x + 4$
15. $x^2 - 2x + 19 = 0$
16. $2x^2 = 3 - 4x$

In Exercises 17–19, without solving the given quadratic equation, determine the number and type of solutions.

17. $x^2 - 4x + 13 = 0$
18. $9x^2 = 2 - 3x$
19. $2x^2 + 4x = 3$

In Exercises 20–26, solve each equation by the method of your choice. Simplify solutions, if possible.

20. $3x^2 - 10x - 8 = 0$
21. $(2x - 3)(x + 2) = x^2 - 2x + 4$

22. $5x^2 - x - 1 = 0$

23. $x^2 - 16 = 0$

24. $(x - 3)^2 - 8 = 0$

25. $3x^2 - x + 2 = 0$

26. $\frac{5}{x+1} + \frac{x-1}{4} = 2$

In Exercises 27–29, write a quadratic equation in standard form with the given solution set.

27. $\left\{-\frac{1}{3}, \frac{3}{5}\right\}$

28. $\{-9i, 9i\}$

29. $\{-4\sqrt{3}, 4\sqrt{3}\}$

30. The graph shows stopping distances for motorcycles at various speeds on dry roads and on wet roads.

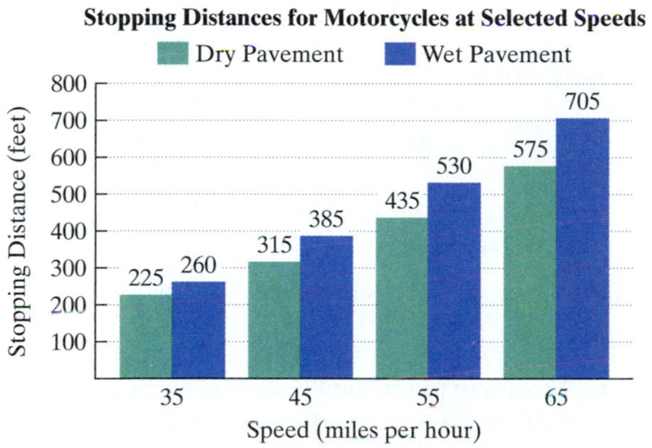

Source: National Highway Traffic Safety Administration

The functions

$$f(x) = 0.125x^2 - 0.8x + 99$$

Dry pavement

and

Wet pavement

$$g(x) = 0.125x^2 + 2.3x + 27$$

model a motorcycle's stopping distance, $f(x)$ or $g(x)$, in feet traveling at x miles per hour. Function f models stopping distance on dry pavement and function g models stopping distance on wet pavement.

a. Use function g to find the stopping distance on wet pavement for a motorcycle traveling at 35 miles per hour. Round to the nearest foot. Does your rounded answer overestimate or underestimate the stopping distance shown by the graph? By how many feet?

b. Use function f to determine a motorcycle's speed requiring a stopping distance on dry pavement of 267 feet.

31. The graphs of the functions in Exercise 30 are shown for $\{x | x \geq 30\}$.

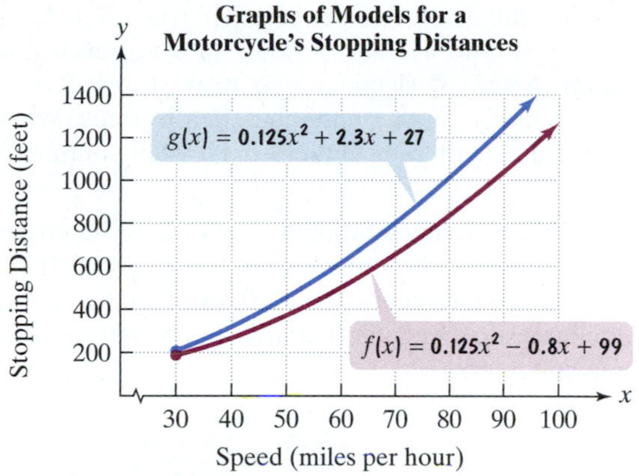

a. How is your answer to Exercise 30(a) shown on the graph of g?

b. How is your answer to Exercise 30(b) shown on the graph of f?

32. A baseball is hit by a batter. The function

$$s(t) = -16t^2 + 140t + 3$$

models the ball's height above the ground, $s(t)$, in feet, t seconds after it is hit. How long will it take for the ball to strike the ground? Round to the nearest tenth of a second.

8.3 *In Exercises 33–36, use the vertex and intercepts to sketch the graph of each quadratic function.*

33. $f(x) = -(x + 1)^2 + 4$

34. $f(x) = (x + 4)^2 - 2$

35. $f(x) = -x^2 + 2x + 3$

36. $f(x) = 2x^2 - 4x - 6$

37. The function

$$f(x) = -0.02x^2 + x + 1$$

models the yearly growth of a young redwood tree, $f(x)$, in inches, with x inches of rainfall per year. How many inches of rainfall per year result in maximum tree growth? What is the maximum yearly growth?

38. A model rocket is launched upward from a platform 40 feet above the ground. The quadratic function

$$s(t) = -16t^2 + 400t + 40$$

models the rocket's height above the ground, $s(t)$, in feet, t seconds after it was launched. After how many seconds does the rocket reach its maximum height? What is the maximum height?

39. The function

$$f(x) = 104.5x^2 - 1501.5x + 6016$$

models the death rate per year per 100,000 males, $f(x)$, for U.S. men who average x hours of sleep each night. How many hours of sleep, to the nearest tenth of an hour, corresponds to the minimum death rate? What is this minimum death rate, to the nearest whole number?

40. A field bordering a straight stream is to be enclosed. The side bordering the stream is not to be fenced. If 1000 yards of fencing material is to be used, what are the dimensions of the largest rectangular field that can be fenced? What is the maximum area?

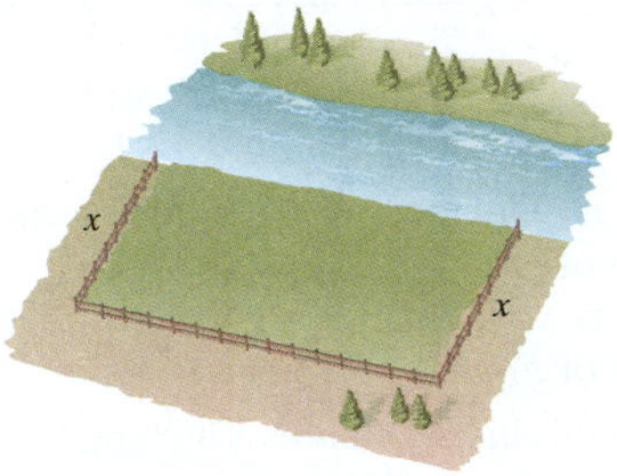

41. Among all pairs of numbers whose difference is 14, find a pair whose product is as small as possible. What is the minimum product?

8.4 *In Exercises 42–47, solve each equation by making an appropriate substitution. When necessary, check proposed solutions.*

42. $x^4 - 6x^2 + 8 = 0$

43. $x + 7\sqrt{x} - 8 = 0$

44. $(x^2 + 2x)^2 - 14(x^2 + 2x) = 15$

45. $x^{-2} + x^{-1} - 56 = 0$

46. $x^{\frac{2}{3}} - x^{\frac{1}{3}} - 12 = 0$

47. $x^{\frac{1}{2}} + 3x^{\frac{1}{4}} - 10 = 0$

8.5 *In Exercises 48–52, solve each inequality and graph the solution set on a real number line.*

48. $2x^2 + 5x - 3 < 0$

49. $2x^2 + 9x + 4 \geq 0$

50. $x^3 + 2x^2 > 3x$

51. $\dfrac{x - 6}{x + 2} > 0$

52. $\dfrac{x + 3}{x - 4} \leq 5$

53. A model rocket is launched from ground level. The function

$$s(t) = -16t^2 + 48t$$

models the rocket's height above the ground, $s(t)$, in feet, t seconds after it was launched. During which time period will the rocket's height exceed 32 feet?

54. The function

$$H(x) = \frac{15}{8}x^2 - 30x + 200$$

models heart rate, $H(x)$, in beats per minute, x minutes after a strenuous workout.

a. What is the heart rate immediately following the workout?

b. According to the model, during which intervals of time after the strenuous workout does the heart rate exceed 110 beats per minute? For which of these intervals has model breakdown occurred? Which interval provides a more realistic answer? How did you determine this?

CHAPTER 8 TEST

Remember to use your Chapter Test Prep Video CD to see the worked-out solutions to the test questions you want to review.

Express solutions to all equations in simplified form. Rationalize denominators, if possible.

In Exercises 1–2, solve each equation by the square root property.

1. $2x^2 - 5 = 0$

2. $(x - 3)^2 = 20$

In Exercises 3–4, determine the constant that should be added to the binomial so that it becomes a perfect square trinomial. Then write and factor the trinomial.

3. $x^2 - 16x$

4. $x^2 + \frac{2}{5}x$

5. Solve by completing the square: $x^2 - 6x + 7 = 0$.

6. Use the measurements determined by the surveyor to find the width of the pond. Express the answer in simplified radical form.

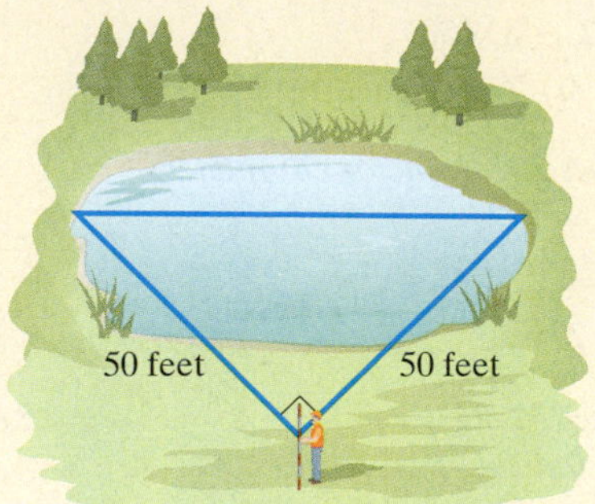

In Exercises 7–8, without solving the given quadratic equation, determine the number and type of solutions.

7. $3x^2 + 4x - 2 = 0$
8. $x^2 = 4x - 8$

In Exercises 9–12, solve each equation by the method of your choice.

9. $2x^2 + 9x = 5$
10. $x^2 + 8x + 5 = 0$
11. $(x + 2)^2 + 25 = 0$
12. $2x^2 - 6x + 5 = 0$

In Exercises 13–14, write a quadratic equation in standard form with the given solution set.

13. $\{-3, 7\}$
14. $\{-10i, 10i\}$
15. By 2007, cellphones with 2-gigabyte storage cards could hold 2000 songs. The bar graph shows the percentage of new cellphones that played music from 2005 through 2007, with projections for 2008 and 2009.

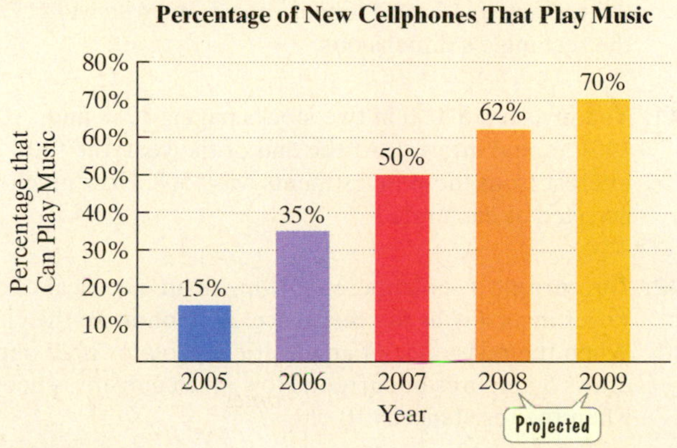

Source: Ovum

The function

$$f(x) = -1.8x^2 + 21x + 15$$

models the percentage of new cellphones that play music, $f(x)$, x years after 2005.

a. Use the function to find the percentage of new cellphones that will play music in 2009. Does this overestimate or underestimate the percentage shown by the graph? By how much?

b. Use the function to determine the first year in which 75% of new cellphones will play music.

In Exercises 16–17, use the vertex and intercepts to sketch the graph of each quadratic function.

16. $f(x) = (x + 1)^2 + 4$
17. $f(x) = x^2 - 2x - 3$

A baseball player hits a pop fly into the air. The function

$$s(t) = -16t^2 + 64t + 5$$

models the ball's height above the ground, $s(t)$, in feet, t seconds after it is hit. Use the function to solve Exercises 18–19.

18. When does the baseball reach its maximum height? What is that height?

19. After how many seconds does the baseball hit the ground? Round to the nearest tenth of a second.

20. The function $f(x) = -x^2 + 46x - 360$ models the daily profit, $f(x)$, in hundreds of dollars, for a company that manufactures x computers daily. How many computers should be manufactured each day to maximize profit? What is the maximum daily profit?

In Exercises 21–23, solve each equation by making an appropriate substitution. When necessary, check proposed solutions.

21. $(2x - 5)^2 + 4(2x - 5) + 3 = 0$
22. $x^4 - 13x^2 + 36 = 0$
23. $x^{\frac{2}{3}} - 9x^{\frac{1}{3}} + 8 = 0$

In Exercises 24–25, solve each inequality and graph the solution set on a real number line.

24. $x^2 - x - 12 < 0$
25. $\dfrac{2x + 1}{x - 3} \leq 3$

CUMULATIVE REVIEW EXERCISES (CHAPTERS 1–8)

In Exercises 1–13, solve each equation, inequality, or system.

1. $9(x - 1) = 1 + 3(x - 2)$
2. $3x + 4y = -7$
   $x - 2y = -9$
3. $x - y + 3z = -9$
   $2x + 3y - z = 16$
   $5x + 2y - z = 15$
4. $7x + 18 \le 9x - 2$
5. $4x - 3 < 13$ and $-3x - 4 \ge 8$
6. $2x + 4 > 8$ or $x - 7 \ge 3$
7. $|2x - 1| < 5$
8. $\left|\frac{2}{3}x - 4\right| = 2$
9. $\frac{4}{x - 3} - \frac{6}{x + 3} = \frac{24}{x^2 - 9}$
10. $\sqrt{x + 4} - \sqrt{x - 3} = 1$
11. $2x^2 = 5 - 4x$
12. $x^{\frac{2}{3}} - 5x^{\frac{1}{3}} + 6 = 0$
13. $2x^2 + x - 6 \le 0$

In Exercises 14–17, graph each function, equation, or inequality in a rectangular coordinate system.

14. $x - 3y = 6$
15. $f(x) = \frac{1}{2}x - 1$
16. $3x - 2y > -6$
17. $f(x) = -2(x - 3)^2 + 2$

In Exercises 18–28, perform the indicated operations, and simplify, if possible.

18. $4[2x - 6(x - y)]$
19. $(-5x^3y^2)(4x^4y^{-6})$
20. $(8x^2 - 9xy - 11y^2) - (7x^2 - 4xy + 5y^2)$
21. $(3x - 1)(2x + 5)$
22. $(3x^2 - 4y)^2$
23. $\frac{3x}{x + 5} - \frac{2}{x^2 + 7x + 10}$
24. $\frac{1 - \frac{9}{x^2}}{1 + \frac{3}{x}}$
25. $\frac{x^2 - 6x + 8}{3x + 9} \div \frac{x^2 - 4}{x + 3}$
26. $\sqrt{5xy} \cdot \sqrt{10x^2y}$
27. $4\sqrt{72} - 3\sqrt{50}$
28. $(5 + 3i)(7 - 3i)$

In Exercises 29–31, factor completely.

29. $81x^4 - 1$
30. $24x^3 - 22x^2 + 4x$
31. $x^3 + 27y^3$

In Exercises 32–34, let $f(x) = x^2 + 3x - 15$ and $g(x) = x - 2$. Find each indicated expression.

32. $(f - g)(x)$ and $(f - g)(5)$
33. $\left(\frac{f}{g}\right)(x)$ and the domain of $\frac{f}{g}$
34. $\frac{f(a + h) - f(a)}{h}$
35. Divide using synthetic division:

    $(3x^3 - x^2 + 4x + 8) \div (x + 2).$
36. Solve for R: $I = \frac{R}{R + r}$.
37. Write the slope-intercept form of the equation of the line through $(-2, 5)$ and parallel to the line whose equation is $3x + y = 9$.
38. Evaluate the determinant: $\begin{vmatrix} -2 & -4 \\ 5 & 7 \end{vmatrix}$.
39. The price of a computer is reduced by 30% to \$434. What was the original price?
40. The area of a rectangle is 52 square yards. The length of the rectangle is 1 yard longer than 3 times its width. Find the rectangle's dimensions.
41. You invested \$4000 in two stocks paying 12% and 14% annual interest. At the end of the year, the total interest from these investments was \$508. How much was invested at each rate?
42. The current, I, in amperes, flowing in an electrical circuit varies inversely as the resistance, R, in ohms, in the circuit. When the resistance of an electric percolator is 22 ohms, it draws 5 amperes of current. How much current is needed when the resistance is 10 ohms?

CHAPTER 9

Exponential and Logarithmic Functions

Can I put aside $25,000 when I'm 20 and wind up sitting on half a million dollars by my early fifties? Will population growth lead to a future without comfort or individual choice? Why did I feel I was walking too slowly on my visit to New York City? Are Californians at greater risk from drunk drivers than from earthquakes? What is the difference between earthquakes measuring 6 and 7 on the Richter scale? And what can possibly be causing merchants at our local shopping mall to grin from ear to ear as they watch the browsers?

The functions that you will be learning about in this chapter will provide you with the mathematics for answering these questions. You will see how these remarkable functions enable us to predict the future and rediscover the past.

You'll be sitting on $500,000 in Example 8 of Section 9.5. Here's where you'll find the other models related to our questions:

- World population growth: Section 9.6, Examples 4 and 5
- Population and walking speed: Section 9.6, Check Point 3, and Review Exercises, Exercise 84
- Alcohol and risk of a car accident: Section 9.5, Example 7
- Earthquake intensity: Section 9.3, Example 9

We open the chapter with those grinning merchants and the sound of ka-ching!

SECTION 9.1

Exponential Functions

Objectives

1. Evaluate exponential functions.
2. Graph exponential functions.
3. Evaluate functions with base e.
4. Use compound interest formulas.

Just browsing? Take your time. Researchers know, to the dollar, the average amount the typical consumer spends per minute at the shopping mall. And the longer you stay, the more you spend. So if you say you're just browsing, that's just fine with the mall merchants. Browsing is time and, as shown in **Figure 9.1**, time is money.

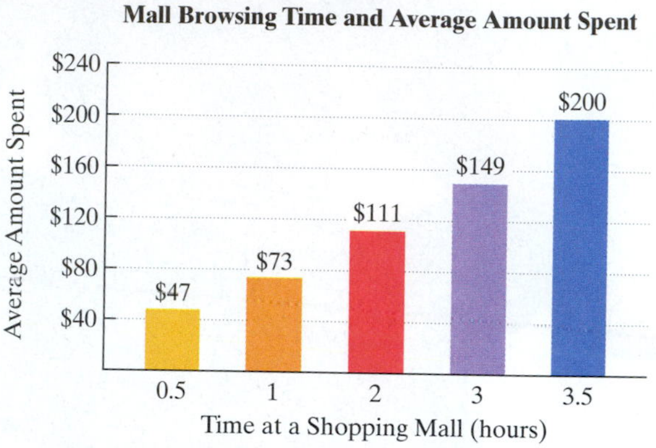

FIGURE 9.1

Source: International Council of Shopping Centers Research, 2006

The data in **Figure 9.1** can be modeled by the function

$$f(x) = 42.2(1.56)^x,$$

where $f(x)$ is the average amount spent, in dollars, at a shopping mall after x hours. Can you see how this function is different from polynomial functions? The variable x is in the exponent. Functions whose equations contain a variable in the exponent are called **exponential functions**. Many real-life situations, including population growth, growth of epidemics, radioactive decay, and other changes that involve rapid increase or decrease, can be described using exponential functions.

Definition of an Exponential Function

The **exponential function f with base b** is defined by

$$f(x) = b^x \quad \text{or} \quad y = b^x,$$

where b is a positive constant other than 1 ($b > 0$ and $b \neq 1$) and x is any real number.

Here are some examples of exponential functions:

$f(x) = 2^x$ — Base is 2.

$g(x) = 10^x$ — Base is 10.

$h(x) = 3^{x+1}$ — Base is 3.

$j(x) = \left(\frac{1}{2}\right)^{x-1}$. — Base is $\frac{1}{2}$.

Each of these functions has a constant base and a variable exponent.

By contrast, the following functions are not exponential functions:

$F(x) = x^2$ — Variable is the base and not the exponent.

$G(x) = 1^x$ — The base of an exponential function must be a positive constant other than 1.

$H(x) = (-1)^x$ — The base of an exponential function must be positive.

$J(x) = x^x$. — Variable is both the base and the exponent.

Why is $G(x) = 1^x$ not classified as an exponential function? The number 1 raised to any power is 1. Thus, the function G can be written as $G(x) = 1$, which is a constant function.

Why is $H(x) = (-1)^x$ not an exponential function? The base of an exponential function must be positive to avoid having to exclude many values of x from the domain that result in nonreal numbers in the range:

$$H(x) = (-1)^x \qquad H\left(\frac{1}{2}\right) = (-1)^{\frac{1}{2}} = \sqrt{-1} = i.$$

Not an exponential function

All values of x resulting in even roots of negative numbers produce nonreal numbers.

1 Evaluate exponential functions.

You will need a calculator to evaluate exponential expressions. Most scientific calculators have a $\boxed{y^x}$ key. Graphing calculators have a $\boxed{\wedge}$ key. To evaluate expressions of the form b^x, enter the base b, press $\boxed{y^x}$ or $\boxed{\wedge}$, enter the exponent x, and finally press $\boxed{=}$ or $\boxed{\text{ENTER}}$.

EXAMPLE 1 Evaluating an Exponential Function

The exponential function $f(x) = 42.2(1.56)^x$ models the average amount spent, $f(x)$, in dollars, at a shopping mall after x hours. What is the average amount spent, to the nearest dollar, after four hours?

Solution Because we are interested in the amount spent after four hours, substitute 4 for x and evaluate the function.

$$f(x) = 42.2(1.56)^x \quad \text{This is the given function.}$$
$$f(4) = 42.2(1.56)^4 \quad \text{Substitute 4 for x.}$$

Use a scientific or graphing calculator to evaluate $f(4)$. Press the following keys on your calculator to do this:

Scientific calculator: 42.2 $\boxed{\times}$ 1.56 $\boxed{y^x}$ 4 $\boxed{=}$

Graphing calculator: 42.2 $\boxed{\times}$ 1.56 $\boxed{\wedge}$ 4 $\boxed{\text{ENTER}}$.

The display should be approximately 249.92566.

$$f(4) = 42.2(1.56)^4 \approx 249.92566 \approx 250$$

Thus, the average amount spent after four hours at a mall is $250.

✓ CHECK POINT 1 Use the exponential function in Example 1 to find the average amount spent, to the nearest dollar, after three hours at a shopping mall. Does this rounded function value underestimate or overestimate the amount shown in **Figure 9.1**? By how much?

2 Graph exponential functions.

Graphing Exponential Functions

We are familiar with expressions involving b^x where x is a rational number. For example,

$$b^{1.7} = b^{\frac{17}{10}} = \sqrt[10]{b^{17}} \quad \text{and} \quad b^{1.73} = b^{\frac{173}{100}} = \sqrt[100]{b^{173}}.$$

However, note that the definition of $f(x) = b^x$ includes all real numbers for the domain x. You may wonder what b^x means when x is an irrational number, such as $b^{\sqrt{3}}$ or b^{π}. Using closer and closer approximations for $\sqrt{3}\left(\sqrt{3} \approx 1.73205\right)$, we can think of $b^{\sqrt{3}}$ as the value that has the successively closer approximations

$$b^{1.7}, b^{1.73}, b^{1.732}, b^{1.73205}, \ldots.$$

In this way, we can graph the exponential function with no holes, or points of discontinuity, at the irrational domain values.

EXAMPLE 2 Graphing an Exponential Function

Graph: $f(x) = 2^x$.

Solution We begin by setting up a table of coordinates.

x	$f(x) = 2^x$
-3	$f(-3) = 2^{-3} = \frac{1}{8}$
-2	$f(-2) = 2^{-2} = \frac{1}{4}$
-1	$f(-1) = 2^{-1} = \frac{1}{2}$
0	$f(0) = 2^0 = 1$
1	$f(1) = 2^1 = 2$
2	$f(2) = 2^2 = 4$
3	$f(3) = 2^3 = 8$

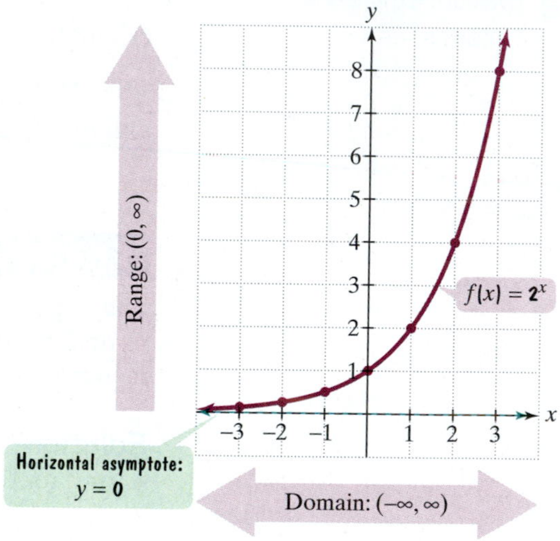

FIGURE 9.2 The graph of $f(x) = 2^x$

We plot these points, connecting them with a continuous curve. **Figure 9.2** shows the graph of $f(x) = 2^x$. Observe that the graph approaches, but never touches, the negative portion of the x-axis. Thus, the x-axis, or $y = 0$, is a horizontal asymptote. The range is the set of all positive real numbers. Although we used integers for x in our table of coordinates, you can use a calculator to find additional points. For example, $f(0.3) = 2^{0.3} \approx 1.231$ and $f(0.95) = 2^{0.95} \approx 1.932$. The points (0.3, 1.231) and (0.95, 1.932) approximately fit the graph.

☑ **CHECK POINT 2** Graph: $f(x) = 3^x$.

EXAMPLE 3 Graphing an Exponential Function

Graph: $g(x) = \left(\frac{1}{2}\right)^x$.

Solution We begin by setting up a table of coordinates. We compute the function values by noting that

$$g(x) = \left(\frac{1}{2}\right)^x = (2^{-1})^x = 2^{-x}.$$

x	$g(x) = \left(\frac{1}{2}\right)^x$ or 2^{-x}
-3	$g(-3) = 2^{-(-3)} = 2^3 = 8$
-2	$g(-2) = 2^{-(-2)} = 2^2 = 4$
-1	$g(-1) = 2^{-(-1)} = 2^1 = 2$
0	$g(0) = 2^{-0} = 1$
1	$g(1) = 2^{-1} = \frac{1}{2^1} = \frac{1}{2}$
2	$g(2) = 2^{-2} = \frac{1}{2^2} = \frac{1}{4}$
3	$g(3) = 2^{-3} = \frac{1}{2^3} = \frac{1}{8}$

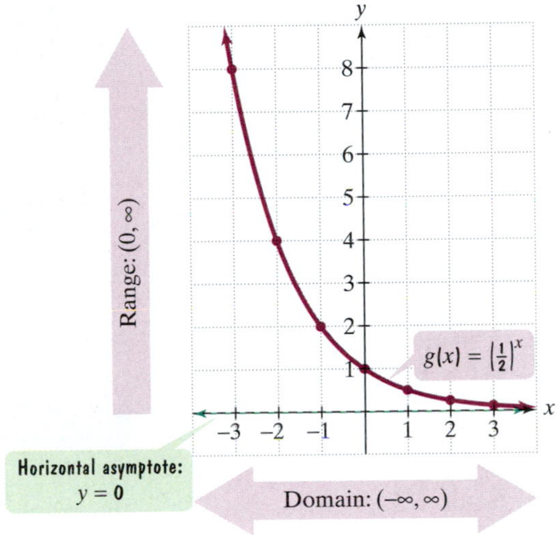

FIGURE 9.3 The graph of $g(x) = \left(\frac{1}{2}\right)^x$

We plot these points, connecting them with a continuous curve. **Figure 9.3** shows the graph of $g(x) = \left(\frac{1}{2}\right)^x$. This time the graph approaches, but never touches, the *positive* portion of the x-axis. Once again, the x-axis, or $y = 0$, is a horizontal asymptote. The range consists of all positive real numbers.

Do you notice a relationship between the graphs of $f(x) = 2^x$ and $g(x) = \left(\frac{1}{2}\right)^x$ in **Figures 9.2** and **9.3**? The graph of $g(x) = \left(\frac{1}{2}\right)^x$ is a mirror image, or reflection, of the graph of $f(x) = 2^x$ about the y-axis.

✓ CHECK POINT 3 Graph: $f(x) = \left(\frac{1}{3}\right)^x$. Note that $f(x) = \left(\frac{1}{3}\right)^x = (3^{-1})^x = 3^{-x}$.

Four exponential functions have been graphed in **Figure 9.4**. Compare the black and green graphs, where $b > 1$, to those in blue and red, where $b < 1$. When $b > 1$, the value of y increases as the value of x increases. When $b < 1$, the value of y decreases as the value of x increases. Notice that all four graphs pass through (0, 1). These graphs illustrate general characteristics of exponential functions, listed in the box on the next page.

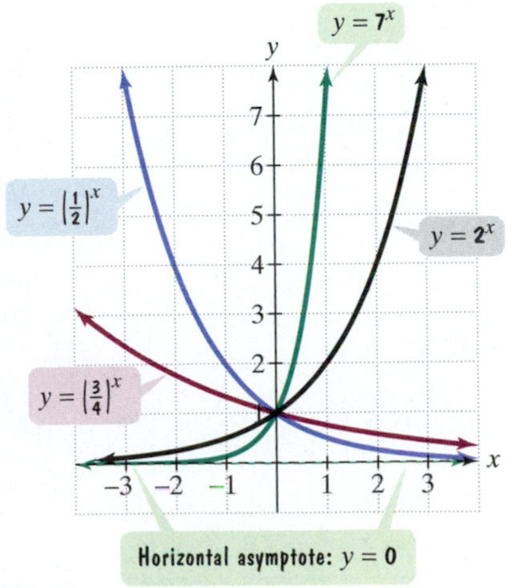

FIGURE 9.4 Graphs of four exponential functions

Characteristics of Exponential Functions of the Form $f(x) = b^x$

1. The domain of $f(x) = b^x$ consists of all real numbers: $(-\infty, \infty)$. The range of $f(x) = b^x$ consists of all positive real numbers: $(0, \infty)$.
2. The graphs of all exponential functions of the form $f(x) = b^x$ pass through the point $(0, 1)$ because $f(0) = b^0 = 1 (b \neq 0)$. The y-intercept is 1.
3. If $b > 1$, $f(x) = b^x$ has a graph that goes up to the right and is an increasing function. The greater the value of b, the steeper the increase.
4. If $0 < b < 1$, $f(x) = b^x$ has a graph that goes down to the right and is a decreasing function. The smaller the value of b, the steeper the decrease.
5. The graph of $f(x) = b^x$ approaches, but does not touch, the x-axis. The x-axis, or $y = 0$, is a horizontal asymptote.

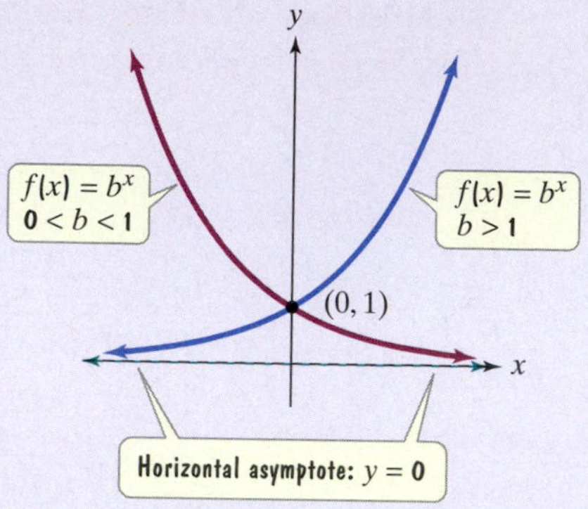

EXAMPLE 4 Graphing Exponential Functions

Graph $f(x) = 3^x$ and $g(x) = 3^{x+1}$ in the same rectangular coordinate system. How is the graph of g related to the graph of f?

Solution We begin by setting up a table showing some of the coordinates for f and g, selecting integers from -2 to 2 for x. Notice that $x + 1$ is the exponent for $g(x) = 3^{x+1}$.

x	$f(x) = 3^x$	$g(x) = 3^{x+1}$
-2	$f(-2) = 3^{-2} = \frac{1}{9}$	$g(-2) = 3^{-2+1} = 3^{-1} = \frac{1}{3}$
-1	$f(-1) = 3^{-1} = \frac{1}{3}$	$g(-1) = 3^{-1+1} = 3^0 = 1$
0	$f(0) = 3^0 = 1$	$g(0) = 3^{0+1} = 3^1 = 3$
1	$f(1) = 3^1 = 3$	$g(1) = 3^{1+1} = 3^2 = 9$
2	$f(2) = 3^2 = 9$	$g(2) = 3^{2+1} = 3^3 = 27$

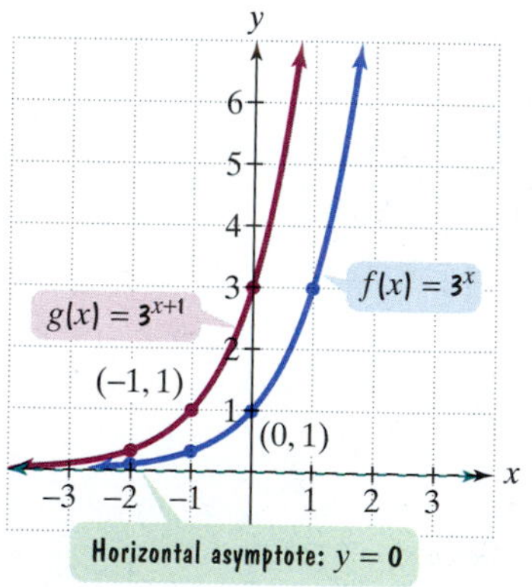

FIGURE 9.5

We plot the points for each function and connect them with a smooth curve. Because of the scale on the y-axis, the points on each function corresponding to $x = 2$ are not shown. **Figure 9.5** shows the graphs of $f(x) = 3^x$ and $g(x) = 3^{x+1}$. The graph of g is the graph of f shifted one unit to the left.

✓ **CHECK POINT 4** Graph $f(x) = 3^x$ and $g(x) = 3^{x-1}$ in the same rectangular coordinate system. Select integers from -2 to 2 for x. How is the graph of g related to the graph of f?

EXAMPLE 5 Graphing Exponential Functions

Graph $f(x) = 2^x$ and $g(x) = 2^x - 3$ in the same rectangular coordinate system. How is the graph of g related to the graph of f?

Solution We begin by setting up a table showing some of the coordinates for f and g, selecting integers from -2 to 2 for x.

x	$f(x) = 2^x$	$g(x) = 2^x - 3$
-2	$f(-2) = 2^{-2} = \frac{1}{4}$	$g(-2) = 2^{-2} - 3 = \frac{1}{4} - 3 = -2\frac{3}{4}$
-1	$f(-1) = 2^{-1} = \frac{1}{2}$	$g(-1) = 2^{-1} - 3 = \frac{1}{2} - 3 = -2\frac{1}{2}$
0	$f(0) = 2^0 = 1$	$g(0) = 2^0 - 3 = 1 - 3 = -2$
1	$f(1) = 2^1 = 2$	$g(1) = 2^1 - 3 = 2 - 3 = -1$
2	$f(2) = 2^2 = 4$	$g(2) = 2^2 - 3 = 4 - 3 = 1$

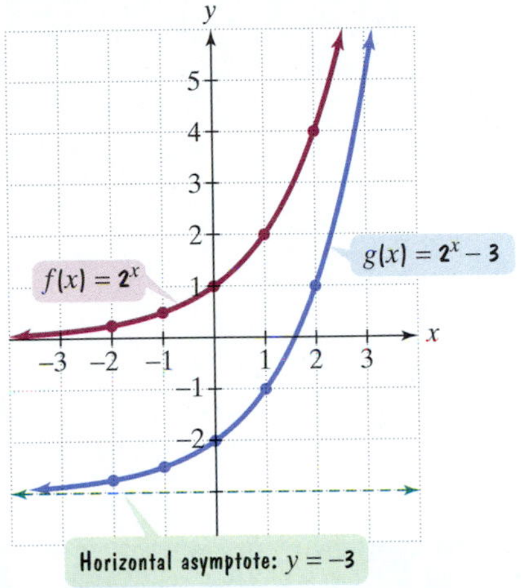

FIGURE 9.6

We plot the points for each function and connect them with a smooth curve. **Figure 9.6** shows the graphs of $f(x) = 2^x$ and $g(x) = 2^x - 3$. The graph of g is the graph of f shifted down three units. As a result, $y = -3$ is the horizontal asymptote for g.

✓ **CHECK POINT 5** Graph $f(x) = 2^x$ and $g(x) = 2^x + 3$ in the same rectangular coordinate system. Select integers from -2 to 2 for x. How is the graph of g related to the graph of f?

3 Evaluate functions with base e.

The Natural Base e

An irrational number, symbolized by the letter e, appears as the base in many applied exponential functions. The number e is defined as the value that $\left(1 + \frac{1}{n}\right)^n$ approaches as n gets larger and larger. **Table 9.1** shows values of $\left(1 + \frac{1}{n}\right)^n$ for increasingly large values of n. As n increases, the approximate value of e to nine decimal places is

$$e \approx 2.718281827.$$

The irrational number e, approximately 2.72, is called the **natural base**. The function $f(x) = e^x$ is called the **natural exponential function**.

Table 9.1

n	$\left(1 + \frac{1}{n}\right)^n$
1	2
2	2.25
5	2.48832
10	2.59374246
100	2.704813829
1000	2.716923932
10,000	2.718145927
100,000	2.718268237
1,000,000	2.718280469
1,000,000,000	2.718281827

As n takes on increasingly large values, the expression $\left(1 + \frac{1}{n}\right)^n$ approaches e.

Using Technology

Graphic Connections

As n increases, the graph of $y = \left(1 + \frac{1}{n}\right)^n$ approaches the graph of $y = e$.

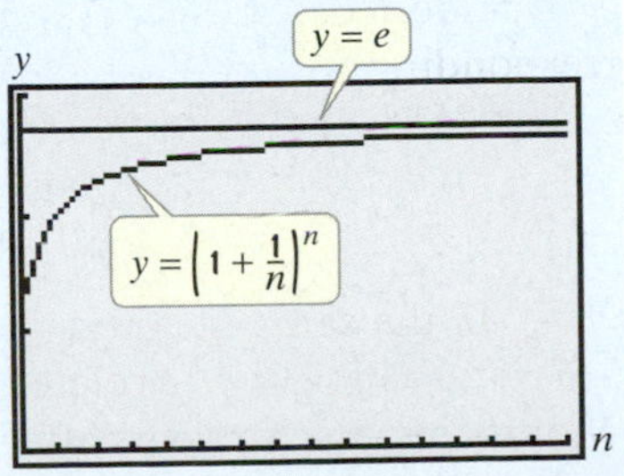

[0, 15, 1] by [0, 3, 1]

Use a scientific or graphing calculator with an $\boxed{e^x}$ key to evaluate e to various powers. For example, to find e^2, press the following keys on most calculators:

Scientific calculator: $2\ \boxed{e^x}$

Graphing calculator: $\boxed{e^x}\ 2\ \boxed{\text{ENTER}}$.

The display should be approximately 7.389.

$$e^2 \approx 7.389$$

The number e lies between 2 and 3. Because $2^2 = 4$ and $3^2 = 9$, it makes sense that e^2, approximately 7.389, lies between 4 and 9.

Because $2 < e < 3$, the graph of $y = e^x$ is between the graphs of $y = 2^x$ and $y = 3^x$, shown in **Figure 9.7**.

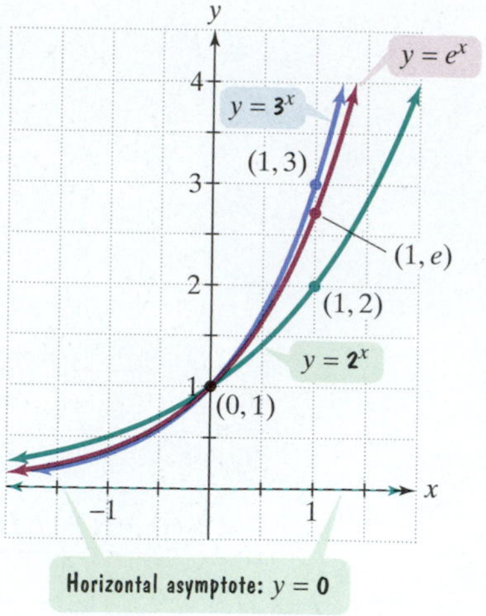

FIGURE 9.7 Graphs of three exponential functions

EXAMPLE 6 Gray Wolf Population

Insatiable killer. That's the reputation the gray wolf acquired in the United States in the nineteenth and early twentieth centuries. Although the label was undeserved, an estimated two million wolves were shot, trapped, or poisoned. By 1960, the population was reduced to 800 wolves. **Figure 9.8** shows the rebounding population in two recovery areas after the gray wolf was declared an endangered species and received federal protection.

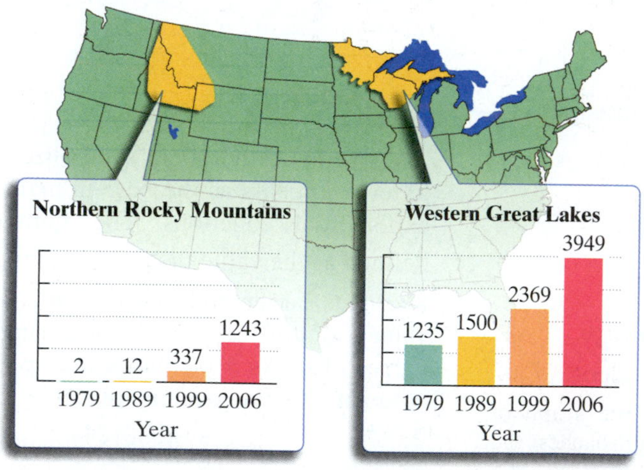

FIGURE 9.8

Source: U.S. Fish and Wildlife Service

The exponential function

$$f(x) = 1.26e^{0.247x}$$

models the gray wolf population of the Northern Rocky Mountains, $f(x)$, x years after 1978. If the wolf is not removed from the endangered species list and trends shown in **Figure 9.8** continue, project its population in the recovery area in 2010.

Solution Because 2010 is 32 years after 1978, we substitute 32 for x in the given function.

$$f(x) = 1.26e^{0.247x} \quad \text{This is the given function.}$$
$$f(32) = 1.26e^{0.247(32)} \quad \text{Substitute 32 for x.}$$

Perform this computation on your calculator.

Scientific calculator: 1.26 [×] [(] .247 [×] 32 [)] [e^x] [=]

Graphing calculator: 1.26 [×] [e^x] [(] .247 [×] 32 [)] [ENTER]

This parenthesis is given on some calculators.

The display should be approximately 3412.1973. Thus,

$$f(32) = 1.26e^{0.247(32)} \approx 3412.$$

This indicates that the gray wolf population of the Northern Rocky Mountains in the year 2010 is projected to be 3412.

CHECK POINT 6 The exponential function $f(x) = 1066e^{0.042x}$ models the gray wolf population of the Western Great Lakes, $f(x)$, x years after 1978. If trends shown in **Figure 9.8** continue, project the gray wolf's population in the recovery area in 2012.

4 Use compound interest formulas.

Compound Interest

In Chapter 8, we saw that the amount of money, A, that a principal, P, will be worth after t years at interest rate r, compounded annually, is given by the formula

$$A = P(1 + r)^t.$$

Most savings institutions have plans in which interest is paid more than once a year. If compound interest is paid twice a year, the compounding period is six months. We say that the interest is **compounded semiannually**. When compound interest is paid four times a year, the compounding period is three months and the interest is said to be **compounded quarterly**. Some plans allow for monthly compounding or daily compounding.

In general, when compound interest is paid n times a year, we say that there are **n compounding periods per year**. The formula $A = P(1 + r)^t$ can be adjusted to take into account the number of compounding periods in a year. If there are n compounding periods per year, the formula becomes

$$A = P\left(1 + \frac{r}{n}\right)^{nt}.$$

Some banks use **continuous compounding**, where the number of compounding periods increases infinitely (compounding interest every trillionth of a second, every quadrillionth of a second, etc.). As n, the number of compounding periods in a year, increases without bound, the expression $\left(1 + \frac{1}{n}\right)^n$ approaches e. As a result, the formula for continuous compounding is $A = Pe^{rt}$. Although continuous compounding sounds terrific, it yields only a fraction of a percent more interest over a year than daily compounding.

Formulas for Compound Interest

After t years, the balance, A, in an account with principal P and annual interest rate r (in decimal form) is given by the following formulas:

1. For n compoundings per year: $A = P\left(1 + \frac{r}{n}\right)^{nt}$
2. For continuous compounding: $A = Pe^{rt}$.

EXAMPLE 7 Choosing between Investments

You decide to invest \$8000 for 6 years and you have a choice between two accounts. The first pays 7% per year, compounded monthly. The second pays 6.85% per year, compounded continuously. Which is the better investment?

Solution The better investment is the one with the greater balance in the account after 6 years. Let's begin with the account with monthly compounding. We use the compound interest model with $P = 8000$, $r = 7\% = 0.07$, $n = 12$ (monthly compounding means 12 compoundings per year), and $t = 6$.

$$A = P\left(1 + \frac{r}{n}\right)^{nt} = 8000\left(1 + \frac{0.07}{12}\right)^{12\cdot 6} \approx 12{,}160.84$$

The balance in this account after 6 years is \$12,160.84.

For the second investment option, we use the model for continuous compounding with $P = 8000$, $r = 6.85\% = 0.0685$, and $t = 6$.

$$A = Pe^{rt} = 8000e^{0.0685(6)} \approx 12{,}066.60$$

The balance in this account after 6 years is \$12,066.60, slightly less than the previous amount. Thus, the better investment is the 7% monthly compounding option.

✓ CHECK POINT 7 A sum of \$10,000 is invested at an annual rate of 8%. Find the balance in the account after 5 years subject to **a.** quarterly compounding and **b.** continuous compounding.

9.1 EXERCISE SET

Practice Exercises

In Exercises 1–10, approximate each number using a calculator. Round your answer to three decimal places.

1. $2^{3.4}$ **2.** $3^{2.4}$

3. $3^{\sqrt{5}}$ **4.** $5^{\sqrt{3}}$

5. $4^{-1.5}$ **6.** $6^{-1.2}$

7. $e^{2.3}$ **8.** $e^{3.4}$

9. $e^{-0.95}$ **10.** $e^{-0.75}$

In Exercises 11–16, set up a table of coordinates for each function. Select integers from −2 to 2, inclusive, for x. Then use the table of coordinates to match the function with its graph. [The graphs are labeled (a) through (f).]

11. $f(x) = 3^x$ **12.** $f(x) = 3^{x-1}$

13. $f(x) = 3^x - 1$ **14.** $f(x) = -3^x$

15. $f(x) = 3^{-x}$ **16.** $f(x) = -3^{-x}$

a.

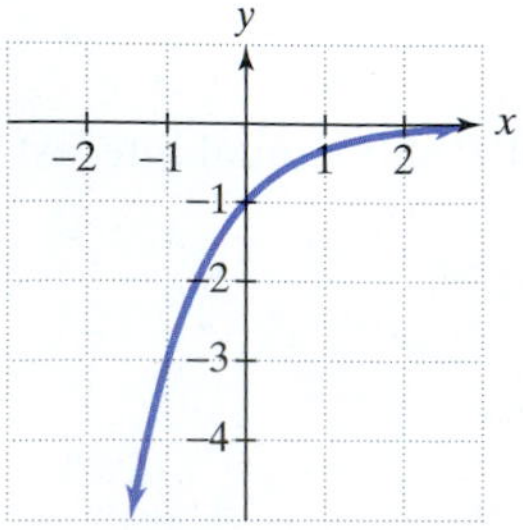

b.

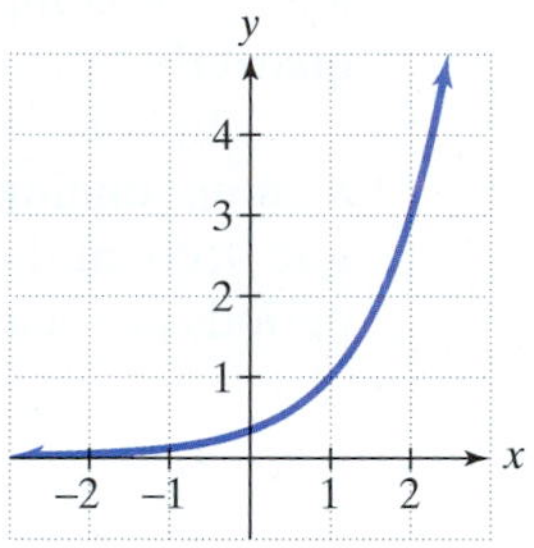

c.

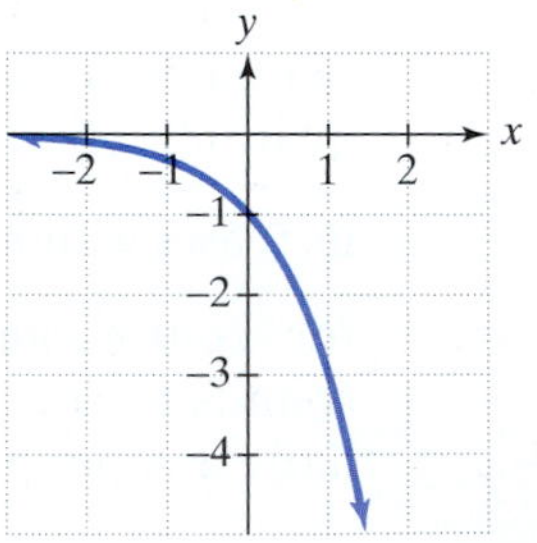

d.

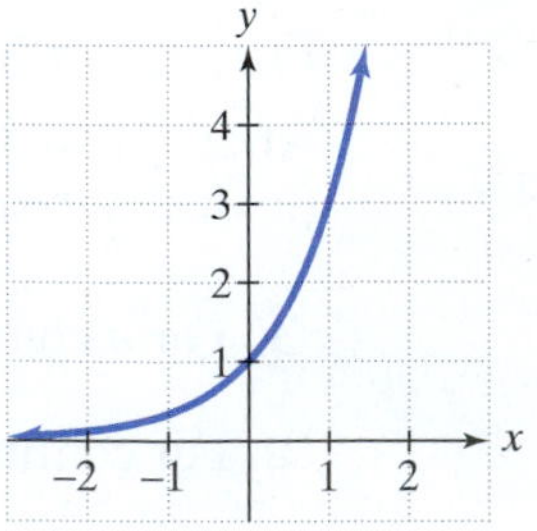

e.

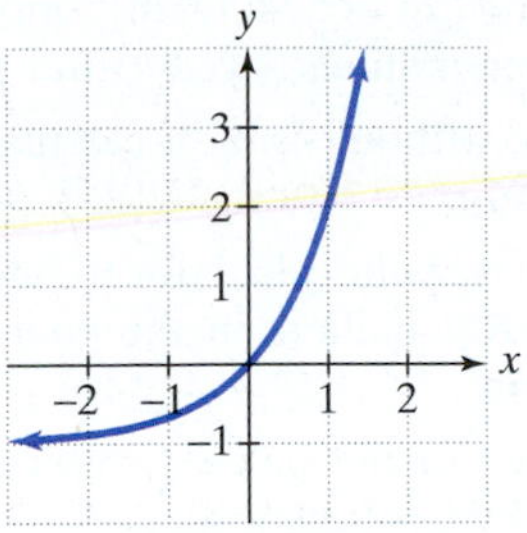

f.

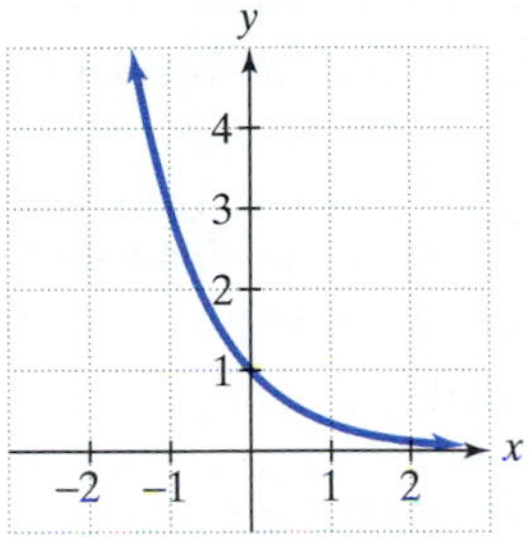

In Exercises 17–24, graph each function by making a table of coordinates. If applicable, use a graphing utility to confirm your hand-drawn graph.

17. $f(x) = 4^x$

18. $f(x) = 5^x$

19. $g(x) = \left(\frac{3}{2}\right)^x$

20. $g(x) = \left(\frac{4}{3}\right)^x$

21. $h(x) = \left(\frac{1}{2}\right)^x$

22. $h(x) = \left(\frac{1}{3}\right)^x$

23. $f(x) = (0.6)^x$

24. $f(x) = (0.8)^x$

In Exercises 25–38, graph functions f and g in the same rectangular coordinate system. Select integers from −2 to 2, inclusive, for x. Then describe how the graph of g is related to the graph of f. If applicable, use a graphing utility to confirm your hand-drawn graphs.

25. $f(x) = 2^x$ and $g(x) = 2^{x+1}$

26. $f(x) = 2^x$ and $g(x) = 2^{x+2}$

27. $f(x) = 2^x$ and $g(x) = 2^{x-2}$

28. $f(x) = 2^x$ and $g(x) = 2^{x-1}$

29. $f(x) = 2^x$ and $g(x) = 2^x + 1$

30. $f(x) = 2^x$ and $g(x) = 2^x + 2$

31. $f(x) = 2^x$ and $g(x) = 2^x - 2$

32. $f(x) = 2^x$ and $g(x) = 2^x - 1$

33. $f(x) = 3^x$ and $g(x) = -3^x$

34. $f(x) = 3^x$ and $g(x) = 3^{-x}$

35. $f(x) = 2^x$ and $g(x) = 2^{x+1} - 1$

36. $f(x) = 2^x$ and $g(x) = 2^{x+1} - 2$

37. $f(x) = 3^x$ and $g(x) = \frac{1}{3} \cdot 3^x$

38. $f(x) = 3^x$ and $g(x) = 3 \cdot 3^x$

Use the compound interest formulas, $A = P\left(1 + \frac{r}{n}\right)^{nt}$ *and* $A = Pe^{rt}$, *to solve Exercises 39–42. Round answers to the nearest cent.*

39. Find the accumulated value of an investment of \$10,000 for 5 years at an interest rate of 5.5% if the money is **a.** compounded semiannually; **b.** compounded monthly; **c.** compounded continuously.

40. Find the accumulated value of an investment of \$5000 for 10 years at an interest rate of 6.5% if the money is **a.** compounded semiannually; **b.** compounded monthly; **c.** compounded continuously.

41. Suppose that you have \$12,000 to invest. Which investment yields the greater return over 3 years: 7% compounded monthly or 6.85% compounded continuously?

42. Suppose that you have \$6000 to invest. Which investment yields the greater return over 4 years: 8.25% compounded quarterly or 8.3% compounded semiannually?

Practice PLUS

In Exercises 43–48, use each exponential function's graph to determine the function's domain and range.

43.

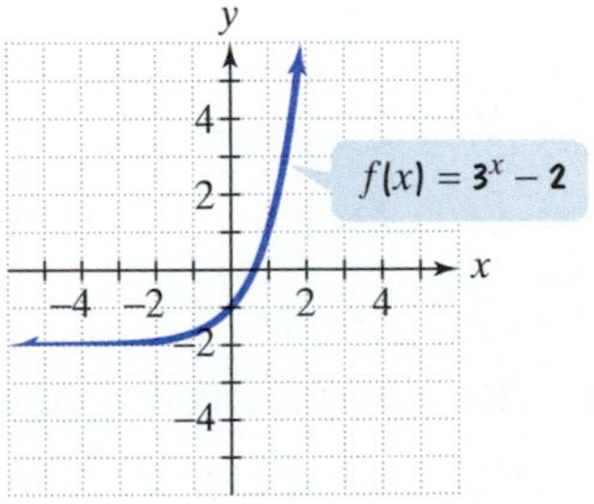

44.

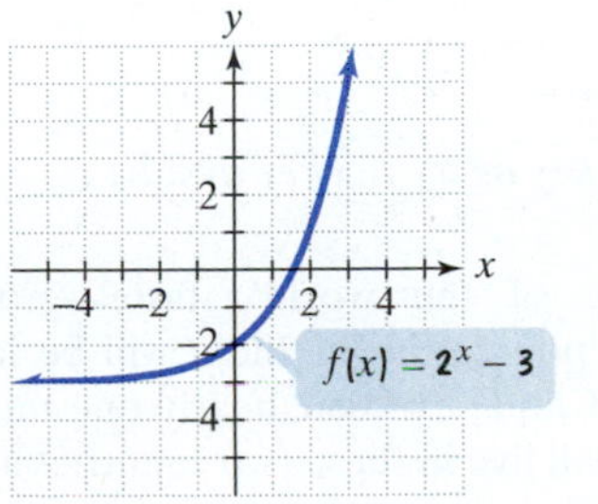

45.

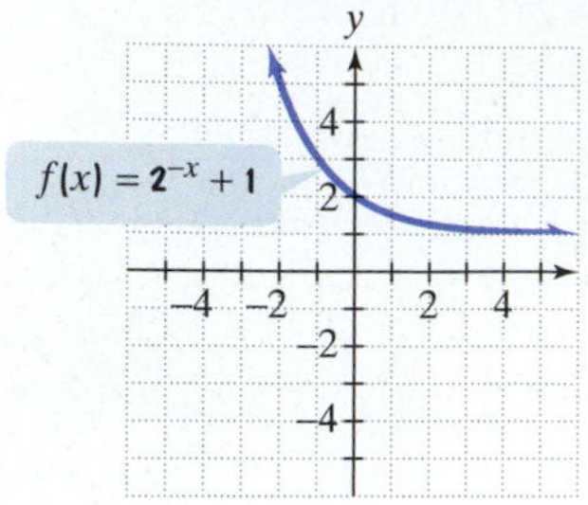

46.

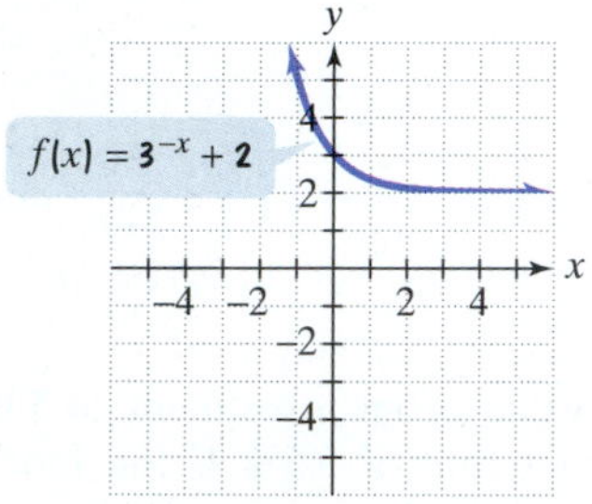

47.

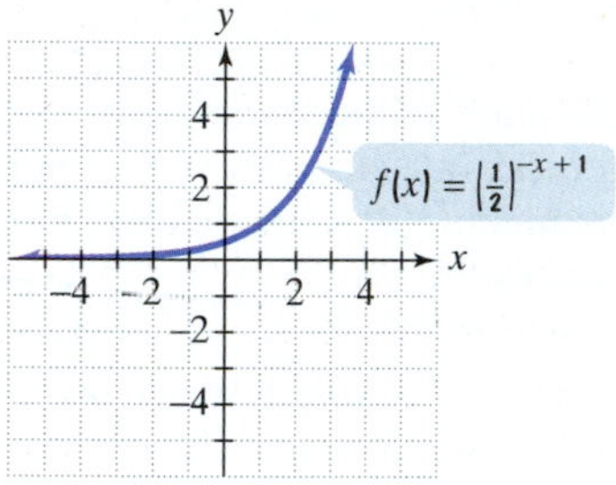

48.

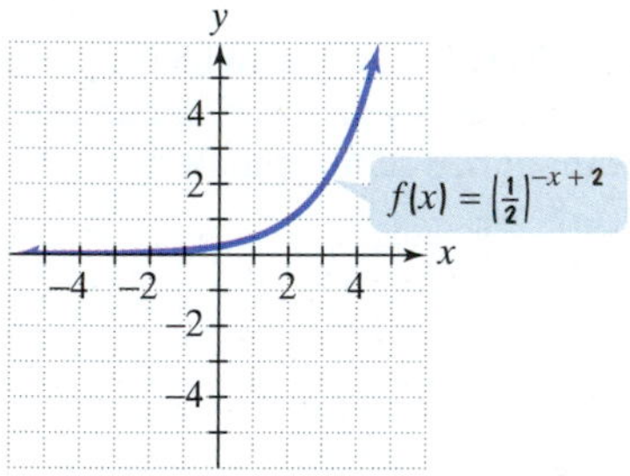

In Exercises 49–50, graph f and g in the same rectangular coordinate system. Then find the point of intersection of the two graphs.

49. $f(x) = 2^x$, $g(x) = 2^{-x}$

50. $f(x) = 2^{x+1}$, $g(x) = 2^{-x+1}$

51. Graph $y = 2^x$ and $x = 2^y$ in the same rectangular coordinate system.

52. Graph $y = 3^x$ and $x = 3^y$ in the same rectangular coordinate system.

Application Exercises

Use a calculator with a $\boxed{y^x}$ *key or a* $\boxed{\wedge}$ *key to solve Exercises 53–56.*

53. India is currently one of the world's fastest-growing countries. By 2040, the population of India will be larger than the population of China; by 2050, nearly one-third of the world's population will live in these two countries alone. The exponential function $f(x) = 574(1.026)^x$ models the population of India, $f(x)$, in millions, x years after 1974.

a. Substitute 0 for x and, without using a calculator, find India's population in 1974.

b. Substitute 27 for x and use your calculator to find India's population, to the nearest million, in the year 2001 as modeled by this function.

c. Find India's population, to the nearest million, in the year 2028 as predicted by this function.

d. Find India's population, to the nearest million, in the year 2055 as predicted by this function.

e. What appears to be happening to India's population every 27 years?

54. The 1986 explosion at the Chernobyl nuclear power plant in the former Soviet Union sent about 1000 kilograms of radioactive cesium-137 into the atmosphere. The function $f(x) = 1000(0.5)^{\frac{x}{30}}$ describes the amount, $f(x)$, in kilograms, of cesium-137 remaining in Chernobyl x years after 1986. If even 100 kilograms of cesium-137 remain in Chernobyl's atmosphere, the area is considered unsafe for human habitation. Find $f(80)$ and determine if Chernobyl will be safe for human habitation by 2066.

The formula $S = C(1 + r)^t$ models inflation, where C = the value today, r = the annual inflation rate, and S = the inflated value t years from now. Use this formula to solve Exercises 55–56. Round answers to the nearest dollar.

55. If the inflation rate is 6%, how much will a house now worth \$465,000 be worth in 10 years?

56. If the inflation rate is 3%, how much will a house now worth \$510,000 be worth in 5 years?

Use a calculator with an $\boxed{e^x}$ *key to solve Exercises 57–63.*

The graph shows the number of words, in millions, in the U.S. federal tax code for selected years from 1955 through 2005. The data can be modeled by

$$f(x) = 0.15x + 1.44 \quad \textit{and} \quad g(x) = 1.87e^{0.0344x}$$

in which $f(x)$ and $g(x)$ represent the number of words, in millions, in the federal tax code x years after 1955. Use these functions to solve Exercises 57–58. Round answers to one decimal place.

Number of Words, in Millions, in the Federal Tax Code

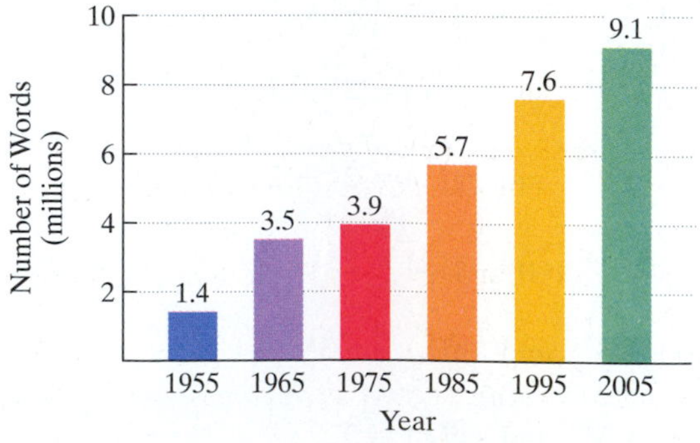

Source: The Tax Foundation

57. a. According to the linear model, how many millions of words were in the federal tax code in 2005?

b. According to the exponential model, how many millions of words were in the federal tax code in 2005?

c. Which function is a better model for the data in 2005?

58. a. According to the linear model, how many millions of words were in the federal tax code in 1975?

b. According to the exponential model, how many millions of words were in the federal tax code in 1975?

c. Which function is a better model for the data in 1975?

59. In college, we study large volumes of information—information that, unfortunately, we do not often retain for very long. The function

$$f(x) = 80e^{-0.5x} + 20$$

describes the percentage of information, $f(x)$, that a particular person remembers x weeks after learning the information.

a. Substitute 0 for x and, without using a calculator, find the percentage of information remembered at the moment it is first learned.

b. Substitute 1 for x and find the percentage of information that is remembered after 1 week.

c. Find the percentage of information that is remembered after 4 weeks.

d. Find the percentage of information that is remembered after one year (52 weeks).

60. In 1626, Peter Minuit persuaded the Wappinger Indians to sell him Manhattan Island for \$24. If the Native Americans had put the \$24 into a bank account paying 5% interest, how much would the investment have been worth in the year 2005 if interest were compounded

a. monthly?

b. continuously?

The function

$$f(x) = \frac{90}{1 + 270e^{-0.122x}}$$

models the percentage, $f(x)$, of people x years old with some coronary heart disease. Use this function to solve Exercises 61–62. Round answers to the nearest tenth of a percent.

61. Evaluate $f(30)$ and describe what this means in practical terms.

62. Evaluate $f(70)$ and describe what this means in practical terms.

63. The function

$$N(t) = \frac{30{,}000}{1 + 20e^{-1.5t}}$$

describes the number of people, $N(t)$, who become ill with influenza t weeks after its initial outbreak in a town with 30,000 inhabitants. The horizontal asymptote in the graph indicates that there is a limit to the epidemic's growth.

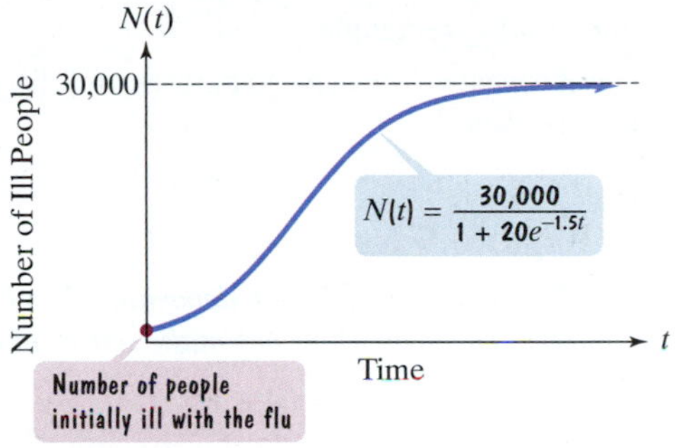

a. How many people became ill with the flu when the epidemic began? (When the epidemic began, $t = 0$.)

b. How many people were ill by the end of the third week?

c. Why can't the spread of an epidemic simply grow indefinitely? What does the horizontal asymptote shown in the graph indicate about the limiting size of the population that becomes ill?

Writing in Mathematics

64. What is an exponential function?

65. What is the natural exponential function?

66. Use a calculator to obtain an approximate value for e to as many decimal places as the display permits. Then use the calculator to evaluate $\left(1 + \frac{1}{x}\right)^x$ for $x = 10, 100, 1000, 10{,}000, 100{,}000$, and $1{,}000{,}000$. Describe what happens to the expression as x increases.

67. Write an example similar to Example 7 on page 648 in which continuous compounding at a slightly lower yearly interest rate is a better investment than compounding n times per year.

68. Describe how you could use the graph of $f(x) = 2^x$ to obtain a decimal approximation for $\sqrt{2}$.

Technology Exercises

69. You have \$10,000 to invest. One bank pays 5% interest compounded quarterly and the other pays 4.5% interest compounded monthly.

a. Use the formula for compound interest to write a function for the balance in each account at any time t in years.

b. Use a graphing utility to graph both functions in an appropriate viewing rectangle. Based on the graphs, which bank offers the better return on your money?

70. a. Graph $y = e^x$ and $y = 1 + x + \frac{x^2}{2}$ in the same viewing rectangle.

b. Graph $y = e^x$ and $y = 1 + x + \frac{x^2}{2} + \frac{x^3}{6}$ in the same viewing rectangle.

c. Graph $y = e^x$ and $y = 1 + x + \frac{x^2}{2} + \frac{x^3}{6} + \frac{x^4}{24}$ in the same viewing rectangle.

d. Describe what you observe in parts (a)–(c). Try generalizing this observation.

Critical Thinking Exercises

Make Sense? *In Exercises 71–74, determine whether each statement "makes sense" or "does not make sense" and explain your reasoning.*

71. My graph of $f(x) = 3 \cdot 2^x$ shows that the horizontal asymptote for f is $x = 3$.

72. I'm using a photocopier to reduce an image over and over by 50%, so the exponential function $f(x) = \left(\frac{1}{2}\right)^x$ models the new image size, where x is the number of reductions.

73. I'm looking at data that show yogurt sales in the United States, in billions of dollars, for selected years from 1978 through 2006, and a linear function appears to be a better choice than an exponential function for modeling sales during this period.

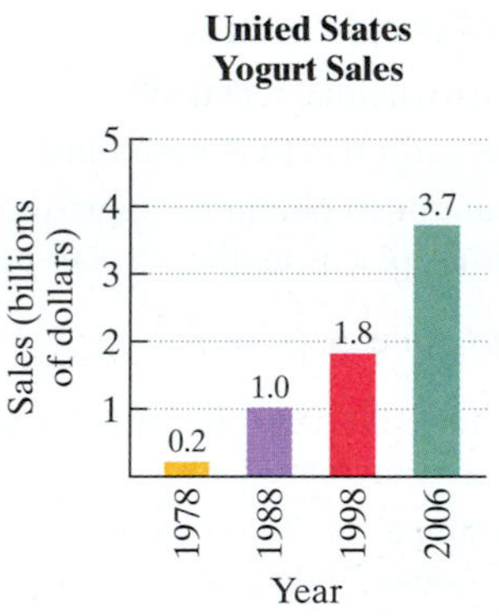

Source: AC Nielsen

74. I use the natural base e when determining how much money I'd have in a bank account that earns compound interest subject to continuous compounding.

In Exercises 75–78, determine whether each statement is true or false. If the statement is false, make the necessary change(s) to produce a true statement.

75. As the number of compounding periods increases on a fixed investment, the amount of money in the account over a fixed interval of time will increase without bound.

76. The functions $f(x) = 3^{-x}$ and $g(x) = -3^x$ have the same graph.

77. If $f(x) = 2^x$, then $f(a + b) = f(a) + f(b)$.

78. The functions $f(x) = \left(\frac{1}{3}\right)^x$ and $g(x) = 3^{-x}$ have the same graph.

79. The graphs labeled (a)–(d) in the figure represent $y = 3^x$, $y = 5^x$, $y = \left(\frac{1}{3}\right)^x$, and $y = \left(\frac{1}{5}\right)^x$, but not necessarily in that order. Which is which? Describe the process that enables you to make this decision.

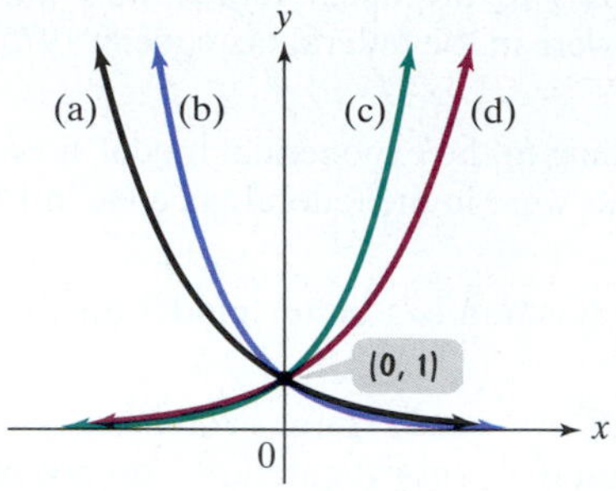

80. The hyperbolic cosine and hyperbolic sine functions are defined by

$$\cosh x = \frac{e^x + e^{-x}}{2} \quad \text{and} \quad \sinh x = \frac{e^x - e^{-x}}{2}.$$

Prove that $(\cosh x)^2 - (\sinh x)^2 = 1$.

Review Exercises

81. Solve for b: $D = \frac{ab}{a + b}$. (Section 6.7, Example 1)

82. Evaluate: $\begin{vmatrix} 3 & -2 \\ 7 & -5 \end{vmatrix}$. (Section 3.5, Example 1)

83. Solve: $x(x - 3) = 10$. (Section 5.7, Example 2)

Preview Exercises

Exercises 84–86 will help you prepare for the material covered in the next section.

84. Let $f(x) = 3x - 4$ and $g(x) = x^2 + 6$.

a. Find $f(5)$.

b. Find $g(f(5))$.

85. Simplify: $3\left(\frac{x - 2}{3}\right) + 2$.

86. Solve for y: $x = 7y - 5$.

SECTION 9.2

Composite and Inverse Functions

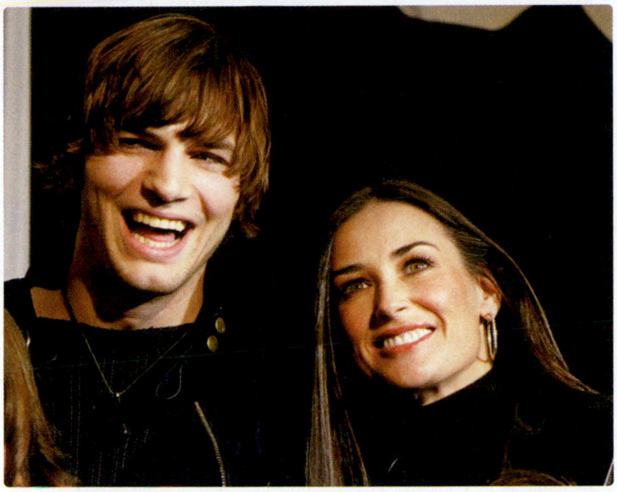

Objectives

1. Form composite functions.
2. Verify inverse functions.
3. Find the inverse of a function.
4. Use the horizontal line test to determine if a function has an inverse function.
5. Use the graph of a one-to-one function to graph its inverse function.

In most societies, women say they prefer to marry men who are older than themselves, whereas men say they prefer women who are younger. Evolutionary psychologists attribute these preferences to female concern with a partner's material resources and male concern with a partner's fertility (*Source:* David M. Buss, *Psychological Inquiry*, 6, 1–30). When the man is considerably older than the woman, people rarely comment. However, when the woman is older, as in the relationship between actors Ashton Kutcher and Demi Moore, people take notice.

Exercises 73–74 of this section's exercise set present ordered pairs that describe the preferred age in a mate in various countries. We then reverse the order of the components in each ordered pair. In the next section, we do the same thing to the ordered pairs of exponential functions. To understand what occurs when we interchange the components of ordered pairs, we turn to the topics of *composite* and *inverse functions*.

1 Form composite functions.

Composite Functions

In Chapter 2, we saw that functions could be combined using addition, subtraction, multiplication, and division. Now let's consider another way of combining two functions. To help understand this new combination, suppose that your local computer store is having a sale. The models that are on sale cost either \$300 less than the regular price or 85% of the regular price. If x represents the computer's regular price, the discounts can be described with the following functions:

$$f(x) = x - 300 \qquad g(x) = 0.85x.$$

The computer is on sale for \$300 less than its regular price.

The computer is on sale for 85% of its regular price.

At the store, you bargain with the salesperson. Eventually, she makes an offer you can't refuse. The sale price will be 85% of the regular price followed by a \$300 reduction:

$$0.85x - 300.$$

85% of the regular price

followed by a \$300 reduction

In terms of functions f and g, this offer can be obtained by taking the output of $g(x) = 0.85x$, namely $0.85x$, and using it as the input of f:

$$f(x) = x - 300$$

Replace x with $0.85x$, the output of $g(x) = 0.85x$.

$$f(0.85x) = 0.85x - 300.$$

Because $0.85x$ is $g(x)$, we can write this last equation as

$$f(g(x)) = 0.85x - 300.$$

We read this equation as "f of g of x is equal to $0.85x - 300$." We call $f(g(x))$ the **composition of the function f with g**, or a **composite function**. This composite function is written $f \circ g$. Thus,

$$(f \circ g)(x) = f(g(x)) = 0.85x - 300.$$

This can be read "f of g of x" or "f composed with g of x."

Like all functions, we can evaluate $f \circ g$ for a specified value of x in the function's domain. For example, here's how to find the value of the composite function describing the offer you cannot refuse at 1400:

$$(f \circ g)(x) = 0.85x - 300$$

Replace x with 1400.

$$(f \circ g)(1400) = 0.85(1400) - 300 = 1190 - 300 = 890.$$

This means that a computer that regularly sells for \$1400 is on sale for \$890 subject to both discounts. We can use a partial table of coordinates for each of the discount functions, g and f, to numerically verify this result.

x (Computer's regular price)	$g(x) = 0.85x$ (85% of the regular price)
1200	1020
1300	1105
1400	1190

x (85% of the regular price)	$f(x) = x - 300$ ($300 reduction)
1020	720
1105	805
1190	890

Using these tables, we can find $(f \circ g)(1400)$:

$$(f \circ g)(1400) = f(g(1400)) = f(1190) = 890.$$

The table for g shows that $g(1400) = 1190$.

The table for f shows that $f(1190) = 890$.

This verifies that a computer that regularly sells for \$1400 is on sale for \$890 subject to both discounts.

Before you run out to buy a computer, let's generalize our discussion of the computer's double discount and define the composition of any two functions.

The Composition of Functions

The **composition of the function f with g** is denoted by $f \circ g$ and is defined by the equation

$$(f \circ g)(x) = f(g(x)).$$

The **domain of the composite function $f \circ g$** is the set of all x such that

1. x is in the domain of g and
2. $g(x)$ is in the domain of f.

EXAMPLE 1 Forming Composite Functions

Given $f(x) = 3x - 4$ and $g(x) = x^2 + 6$, find each of the following composite functions:

a. $(f \circ g)(x)$ **b.** $(g \circ f)(x)$.

Solution

a. We begin with $(f \circ g)(x)$, the composition of f with g. Because $(f \circ g)(x)$ means $f(g(x))$, we must replace each occurrence of x in the equation for f with $g(x)$.

$f(x) = 3x - 4$ — This is the given equation for f.

Replace x with $g(x)$.

$(f \circ g)(x) = f(g(x)) = 3g(x) - 4$

$= 3(x^2 + 6) - 4$ — Because $g(x) = x^2 + 6$, replace $g(x)$ with $x^2 + 6$.

$= 3x^2 + 18 - 4$ — Use the distributive property.

$= 3x^2 + 14$ — Simplify.

Thus, $(f \circ g)(x) = 3x^2 + 14$.

b. Next, we find $(g \circ f)(x)$, the composition of g with f. Because $(g \circ f)(x)$ means $g(f(x))$, we must replace each occurrence of x in the equation for g with $f(x)$.

$g(x) = x^2 + 6$ — This is the given equation for g.

Replace x with $f(x)$.

$(g \circ f)(x) = g(f(x)) = (f(x))^2 + 6$

$= (3x - 4)^2 + 6$ — Because $f(x) = 3x - 4$, replace $f(x)$ with $3x - 4$.

$= 9x^2 - 24x + 16 + 6$ — Use $(A - B)^2 = A^2 - 2AB + B^2$ to square $3x - 4$.

$= 9x^2 - 24x + 22$ — Simplify.

Thus, $(g \circ f)(x) = 9x^2 - 24x + 22$.

Notice that $f \circ g$ is not the same function as $g \circ f$.

✓ CHECK POINT 1 Given $f(x) = 5x + 6$ and $g(x) = x^2 - 1$, find each of the following composite functions:

a. $(f \circ g)(x)$ **b.** $(g \circ f)(x)$.

Inverse Functions

Here are two functions that describe situations related to the price of a computer, x:

$$f(x) = x - 300 \qquad g(x) = x + 300.$$

Function f subtracts \$300 from the computer's price and function g adds \$300 to the computer's price. Let's see what $f(g(x))$ does. Put $g(x)$ into f:

$$f(x) = x - 300$$

This is the given equation for f.

Replace x with $g(x)$.

$$f(g(x)) = g(x) - 300$$
$$= x + 300 - 300$$
$$= x.$$

Because $g(x) = x + 300$, replace $g(x)$ with $x + 300$.

This is the computer's original price.

By putting $g(x)$ into f and finding $f(g(x))$, we see that the computer's price, x, went through two changes: the first, an increase; the second, a decrease:

$$x + 300 - 300.$$

The final price of the computer, x, is identical to its starting price, x.

In general, if the changes made to x by function g are undone by the changes made by function f, then

$$f(g(x)) = x.$$

Assume, also, that this "undoing" takes place in the other direction:

$$g(f(x)) = x.$$

Under these conditions, we say that each function is the *inverse function* of the other. The fact that g is the inverse of f is expressed by renaming g as f^{-1}, read "f-inverse." For example, the inverse functions

$$f(x) = x - 300 \qquad g(x) = x + 300$$

are usually named as follows:

$$f(x) = x - 300 \qquad f^{-1}(x) = x + 300.$$

We can use partial tables of coordinates for f and f^{-1} to gain numerical insight into the relationship between a function and its inverse function.

Computer's regular price — \$300 reduction

x	$f(x) = x - 300$
1200	900
1300	1000
1400	1100

Ordered pairs for f: (1200, 900), (1300, 1000), (1400, 1100)

Price with \$300 reduction — \$300 price increase

x	$f^{-1}(x) = x + 300$
900	1200
1000	1300
1100	1400

Ordered pairs for f^{-1}: (900, 1200), (1000, 1300), (1100, 1400)

The tables illustrate that if a function f is the set of ordered pairs (x, y), then the inverse, f^{-1}, is the set of ordered pairs (y, x). Using these tables, we can see how one function's changes to x are undone by the other function:

$$(f^{-1} \circ f)(1300) = f^{-1}(f(1300)) = f^{-1}(1000) = 1300.$$

The table for f shows that $f(1300) = 1000$.

The table for f^{-1} shows that $f^{-1}(1000) = 1300$.

The final price of the computer, \$1300, is identical to its starting price, \$1300.

With these ideas in mind, we present the formal definition of the inverse of a function:

Study Tip

The notation f^{-1} represents the inverse function of f. The -1 is *not* an exponent. The notation f^{-1} does *not* mean $\frac{1}{f}$:

$$f^{-1} \neq \frac{1}{f}.$$

Definition of the Inverse of a Function

Let f and g be two functions such that

$$f(g(x)) = x \qquad \text{for every } x \text{ in the domain of } g$$

and

$$g(f(x)) = x \qquad \text{for every } x \text{ in the domain of } f.$$

The function g is the **inverse of the function** $\boldsymbol{f}$, and is denoted by f^{-1} (read "f-inverse"). Thus, $f(f^{-1}(x)) = x$ and $f^{-1}(f(x)) = x$. The domain of f is equal to the range of f^{-1}, and vice versa.

2 Verify inverse functions.

EXAMPLE 2 Verifying Inverse Functions

Show that each function is the inverse of the other:

$$f(x) = 5x \qquad \text{and} \qquad g(x) = \frac{x}{5}.$$

Solution To show that f and g are inverses of each other, we must show that $f(g(x)) = x$ and $g(f(x)) = x$. We begin with $f(g(x))$.

$$f(x) = 5x \qquad \text{This is the given equation for } f.$$

Replace x with $g(x)$.

$$f(g(x)) = 5g(x) = 5\left(\frac{x}{5}\right) = x \qquad \text{Because } g(x) = \frac{x}{5}, \text{ replace } g(x) \text{ with } \frac{x}{5}. \text{ Then simplify.}$$

Next, we find $g(f(x))$.

$$g(x) = \frac{x}{5} \qquad \text{This is the given equation for } g.$$

Replace x with $f(x)$.

$$g(f(x)) = \frac{f(x)}{5} = \frac{5x}{5} = x \qquad \text{Because } f(x) = 5x, \text{ replace } g(x) \text{ with } 5x. \text{ Then simplify.}$$

Because g is the inverse of f (and vice versa), we can use inverse notation and write

$$f(x) = 5x \qquad \text{and} \qquad f^{-1}(x) = \frac{x}{5}.$$

Notice how f^{-1} undoes the change produced by f : f changes x by multiplying by 5 and f^{-1} undoes this change by dividing by 5. ■

✓ CHECK POINT 2 Show that each function is the inverse of the other:

$$f(x) = 7x \quad \text{and} \quad g(x) = \frac{x}{7}.$$

EXAMPLE 3 Verifying Inverse Functions

Show that each function is the inverse of the other:

$$f(x) = 3x + 2 \quad \text{and} \quad g(x) = \frac{x-2}{3}.$$

Solution To show that f and g are inverses of each other, we must show that $f(g(x)) = x$ and $g(f(x)) = x$. We begin with $f(g(x))$.

$$f(x) = 3x + 2$$ This is the equation for f.

Replace x with $g(x)$.

$$f(g(x)) = 3g(x) + 2 = 3\left(\frac{x-2}{3}\right) + 2 = (x-2) + 2 = x$$

$g(x) = \frac{x-2}{3}$

Next, we find $g(f(x))$.

$$g(x) = \frac{x-2}{3}$$ This is the equation for g.

Replace x with $f(x)$.

$$g(f(x)) = \frac{f(x) - 2}{3} = \frac{(3x+2) - 2}{3} = \frac{3x}{3} = x$$

$f(x) = 3x + 2$

Because g is the inverse of f (and vice versa), we can use inverse notation and write

$$f(x) = 3x + 2 \quad \text{and} \quad f^{-1}(x) = \frac{x-2}{3}.$$

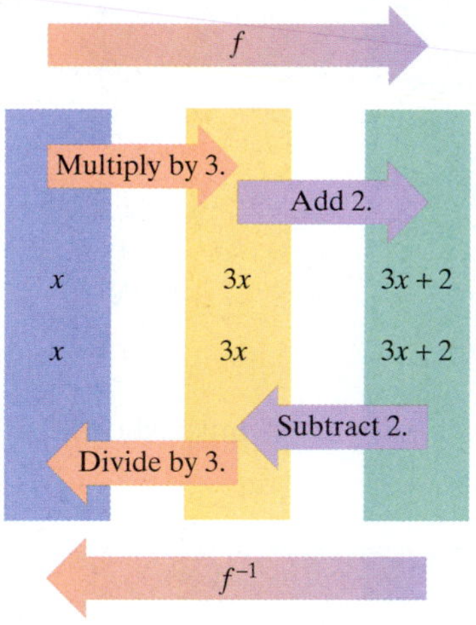

FIGURE 9.9 f^{-1} undoes the changes produced by f.

Notice how f^{-1} undoes the changes produced by f: f changes x by *multiplying* by 3 and *adding* 2, and f^{-1} undoes this by *subtracting* 2 and *dividing* by 3. This "undoing" process is illustrated in **Figure 9.9**.

✓ CHECK POINT 3 Show that each function is the inverse of the other:

$$f(x) = 4x - 7 \quad \text{and} \quad g(x) = \frac{x+7}{4}.$$

3 Find the inverse of a function.

Finding the Inverse of a Function

The definition of the inverse of a function tells us that the domain of f is equal to the range of f^{-1}, and vice versa. This means that if the function f is the set of ordered pairs (x, y), then the inverse of f is the set of ordered pairs (y, x). If a function is defined by an equation, we can obtain the equation for f^{-1}, the inverse of f, by interchanging the role of x and y in the equation for the function f.

Finding the Inverse of a Function

The equation for the inverse of a function f can be found as follows:

1. Replace $f(x)$ with y in the equation for $f(x)$.
2. Interchange x and y.
3. Solve for y. If this equation does not define y as a function of x, the function f does not have an inverse function and this procedure ends. If this equation does define y as a function of x, the function f has an inverse function.
4. If f has an inverse function, replace y in step 3 with $f^{-1}(x)$. We can verify our result by showing that $f(f^{-1}(x)) = x$ and $f^{-1}(f(x)) = x$.

The procedure for finding a function's inverse uses a *switch-and-solve* strategy. Switch x and y, then solve for y.

EXAMPLE 4 Finding the Inverse of a Function

Find the inverse of $f(x) = 7x - 5$.

Solution

Step 1. Replace $f(x)$ with y:

$$y = 7x - 5.$$

Step 2. Interchange x and y:

$$x = 7y - 5. \quad \text{This is the inverse function.}$$

Step 3. Solve for y:

$$x + 5 = 7y \quad \text{Add 5 to both sides.}$$

$$\frac{x+5}{7} = y. \quad \text{Divide both sides by 7.}$$

Step 4. Replace y with $f^{-1}(x)$:

$$f^{-1}(x) = \frac{x+5}{7}. \quad \text{The equation is written with } f^{-1} \text{ on the left.}$$

Thus, the inverse of $f(x) = 7x - 5$ is $f^{-1}(x) = \frac{x+5}{7}$. (Verify this result by showing that $f(f^{-1}(x)) = x$ and $f^{-1}(f(x)) = x$.)

The inverse function, f^{-1}, undoes the changes produced by f. f changes x by multiplying by 7 and subtracting 5. f^{-1} undoes this by adding 5 and dividing by 7.

✓ **CHECK POINT 4** Find the inverse of $f(x) = 2x + 7$.

EXAMPLE 5 Finding the Inverse of a Function

Find the inverse of $f(x) = x^3 + 1$.

Solution

Step 1. Replace $f(x)$ with y: $y = x^3 + 1$.

Step 2. Interchange x and y: $x = y^3 + 1$.

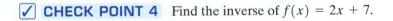

Step 3. Solve for *y*. We need to solve $x = y^3 + 1$ for y.

Our goal is to isolate y. Because $\sqrt[3]{y^3} = y$, we will take the cube root of both sides of the equation.

$$x - 1 = y^3 \quad \text{Subtract 1 from both sides of } x = y^3 + 1.$$

$$\sqrt[3]{x-1} = \sqrt[3]{y^3} \quad \text{Take the cube root on both sides.}$$

$$\sqrt[3]{x-1} = y \quad \text{Simplify.}$$

Step 4. Replace *y* with $f^{-1}(x)$: $f^{-1}(x) = \sqrt[3]{x-1}$.

Thus, the inverse of $f(x) = x^3 + 1$ is $f^{-1}(x) = \sqrt[3]{x-1}$.

✓ **CHECK POINT 5** Find the inverse of $f(x) = 4x^3 - 1$.

---

4 Use the horizontal line test to determine if a function has an inverse function.

The Horizontal Line Test and One-to-One Functions

Let's see what happens if we try to find the inverse of the quadratic function $f(x) = x^2$.

Step 1. Replace $f(x)$ with *y*: $y = x^2$.

Step 2. Interchange *x* and *y*: $x = y^2$.

Step 3. Solve for *y*: We apply the square root property to solve $y^2 = x$ for y. We obtain

$$y = \pm\sqrt{x}.$$

The $\pm$ in this last equation shows that for certain values of x (all positive real numbers), there are two values of y. Because this equation does not represent y as a function of x, the quadratic function $f(x) = x^2$ does not have an inverse function.

We can use a few of the solutions of $y = x^2$ to illustrate numerically that this function does not have an inverse:

Four solutions of $y = x^2$: $(-2, 4)$, $(-1, 1)$, $(1, 1)$, $(2, 4)$,

Interchange x and y in each ordered pair: $(4, -2)$, $(1, -1)$, $(1, 1)$, $(4, 2)$.

The input 1 is associated with two outputs, −1 and 1.

The input 4 is associated with two outputs, −2 and 2.

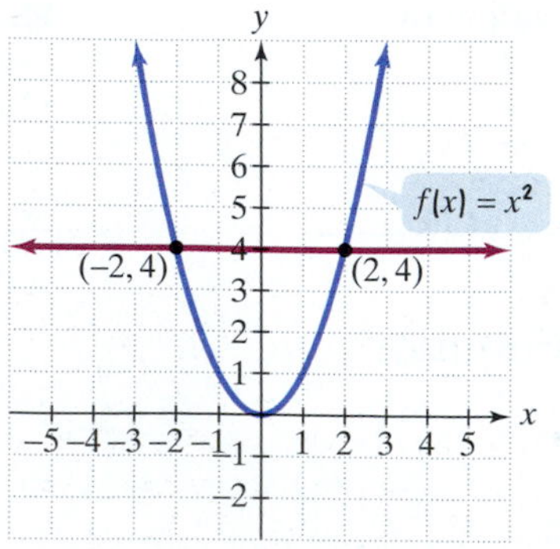

FIGURE 9.10 The horizontal line intersects the graph twice.

A function provides exactly one output for each input. Thus, the ordered pairs in the bottom row do not define a function.

Can we look at the graph of a function and tell if it represents a function with an inverse? Yes. The graph of the quadratic function $f(x) = x^2$ is shown in **Figure 9.10**. Four units above the x-axis, a horizontal line is drawn. This line intersects the graph at two of its points, $(-2, 4)$ and $(2, 4)$. Inverse functions have ordered pairs with the coordinates reversed. We just saw what happened when we interchanged x and y. We obtained $(4, -2)$ and $(4, 2)$, and these ordered pairs do not define a function.

If any horizontal line, such as the one in **Figure 9.10**, intersects a graph at two or more points, the set of these points will not define a function when their coordinates are reversed. This suggests the **horizontal line test** for inverse functions.

Discover for Yourself

How might you restrict the domain of $f(x) = x^2$, graphed in **Figure 9.10**, so that the remaining portion of the graph passes the horizontal line test?

The Horizontal Line Test for Inverse Functions

A function f has an inverse that is a function, f^{-1}, if there is no horizontal line that intersects the graph of the function f at more than one point.

EXAMPLE 6 Applying the Horizontal Line Test

Which of the following graphs represent functions that have inverse functions?

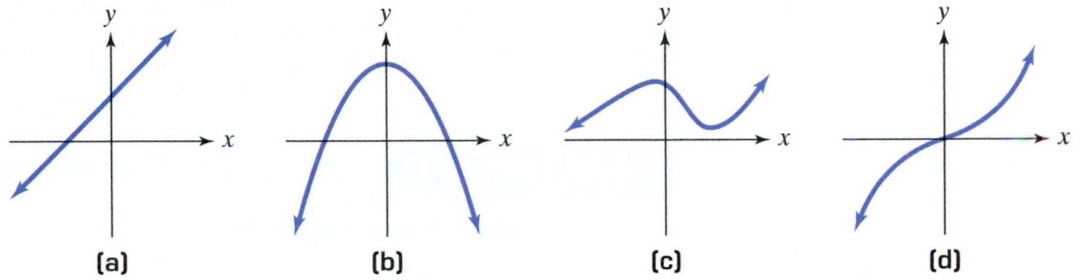

Solution Notice that horizontal lines can be drawn in graphs **(b)** and **(c)** that intersect the graphs more than once. These graphs do not pass the horizontal line test. These are not the graphs of functions with inverse functions. By contrast, no horizontal line can be drawn in graphs **(a)** and **(d)** that intersects the graphs more than once. These graphs pass the horizontal line test. Thus, the graphs in parts **(a)** and **(d)** represent functions that have inverse functions.

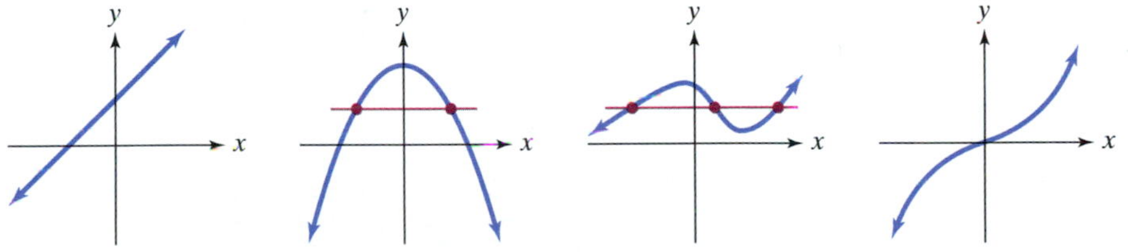

(a) Has an inverse function (b) No inverse function (c) No inverse function (d) Has an inverse function

✓ CHECK POINT 6 Which of the following graphs represent functions that have inverse functions?

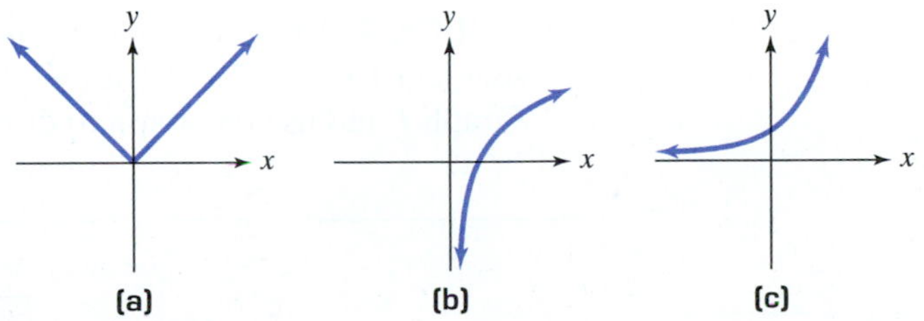

A function passes the horizontal line test when no two different ordered pairs have the same second component. This means that if $x_1 \neq x_2$, then $f(x_1) \neq f(x_2)$. Such a function is called a **one-to-one function**. Thus, **a one-to-one function is a function in which no two different ordered pairs have the same second component. Only one-to-one functions have inverse functions.** Any function that passes the horizontal line test is a one-to-one function. Any one-to-one function has a graph that passes the horizontal line test.

5 Use the graph of a one-to-one function to graph its inverse function.

Graphs of f and f^{-1}

There is a relationship between the graph of a one-to-one function, f, and its inverse, f^{-1}. Because inverse functions have ordered pairs with the coordinates reversed, if the point (a, b) is on the graph of f, then the point (b, a) is on the graph of f^{-1}. The points (a, b) and (b, a) are symmetric with respect to the line $y = x$. Thus, **the graph of f^{-1} is a reflection of the graph of f about the line $y = x$.** This is illustrated in **Figure 9.11**.

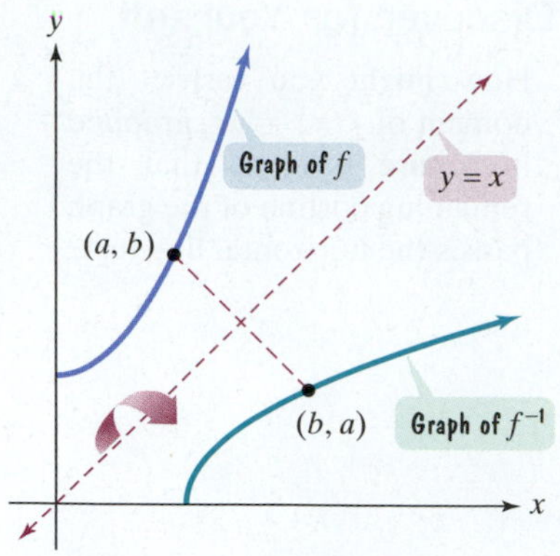

FIGURE 9.11 The graph of f^{-1} is a reflection of the graph of f about $y = x$.

EXAMPLE 7 Graphing the Inverse Function

Use the graph of f in **Figure 9.12** to draw the graph of its inverse function.

Solution We begin by noting that no horizontal line intersects the graph of f, shown again in blue in **Figure 9.13**, at more than one point, so f does have an inverse function. Because the points $(-3, -2)$, $(-1, 0)$, and $(4, 2)$ are on the graph of f, the graph of the inverse function, f^{-1}, has points with these ordered pairs reversed. Thus, $(-2, -3)$, $(0, -1)$, and $(2, 4)$ are on the graph of f^{-1}. We can use these points to graph f^{-1}. The graph of f^{-1} is shown in green in **Figure 9.13**. Note that the green graph of f^{-1} is the reflection of the blue graph of f about the line $y = x$.

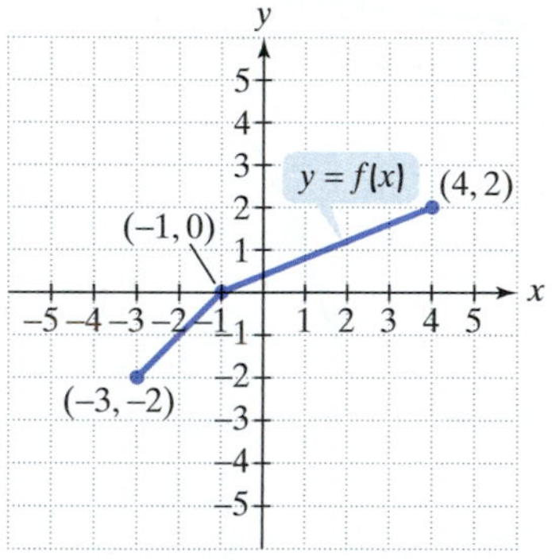

FIGURE 9.12

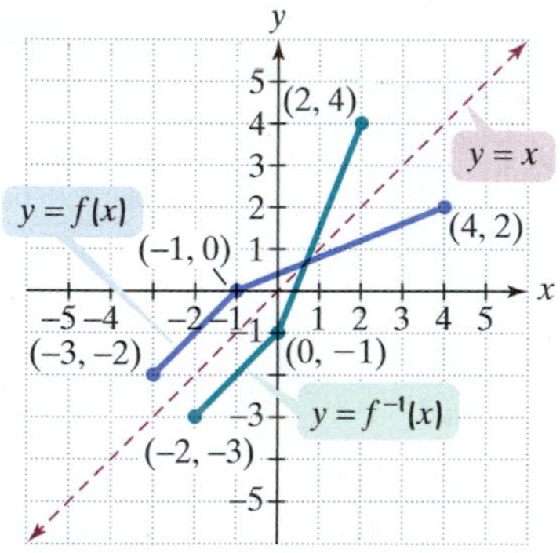

FIGURE 9.13 The graphs of f and f^{-1}

✓ CHECK POINT 7 The graph of function f consists of two line segments, one segment from $(-2, -2)$ to $(-1, 0)$ and a second segment from $(-1, 0)$ to $(1, 2)$. Graph f and use the graph to draw the graph of its inverse function.

9.2 EXERCISE SET

MyMathLab MathXL PRACTICE WATCH DOWNLOAD READ REVIEW

Practice Exercises

In Exercises 1–14, find

a. $(f \circ g)(x)$;

b. $(g \circ f)(x)$;

c. $(f \circ g)(2)$.

1. $f(x) = 2x, \quad g(x) = x + 7$

2. $f(x) = 3x, \quad g(x) = x - 5$

3. $f(x) = x + 4, \quad g(x) = 2x + 1$

4. $f(x) = 5x + 2, \quad g(x) = 3x - 4$

5. $f(x) = 4x - 3, \quad g(x) = 5x^2 - 2$

6. $f(x) = 7x + 1, \quad g(x) = 2x^2 - 9$

7. $f(x) = x^2 + 2, \quad g(x) = x^2 - 2$

8. $f(x) = x^2 + 1, \quad g(x) = x^2 - 3$

9. $f(x) = \sqrt{x}, \quad g(x) = x - 1$

10. $f(x) = \sqrt{x}, \quad g(x) = x + 2$

11. $f(x) = 2x - 3, \quad g(x) = \frac{x+3}{2}$

12. $f(x) = 6x - 3, \quad g(x) = \frac{x+3}{6}$

13. $f(x) = \frac{1}{x}, \quad g(x) = \frac{1}{x}$

14. $f(x) = \frac{1}{x}, \quad g(x) = \frac{2}{x}$

In Exercises 15–24, find $f(g(x))$ and $g(f(x))$ and determine whether each pair of functions f and g are inverses of each other.

15. $f(x) = 4x$ and $g(x) = \frac{x}{4}$

16. $f(x) = 6x$ and $g(x) = \frac{x}{6}$

17. $f(x) = 3x + 8$ and $g(x) = \frac{x-8}{3}$

18. $f(x) = 4x + 9$ and $g(x) = \frac{x-9}{4}$

19. $f(x) = 5x - 9$ and $g(x) = \frac{x+5}{9}$

20. $f(x) = 3x - 7$ and $g(x) = \frac{x+3}{7}$

21. $f(x) = \frac{3}{x-4}$ and $g(x) = \frac{3}{x} + 4$

22. $f(x) = \frac{2}{x-5}$ and $g(x) = \frac{2}{x} + 5$

23. $f(x) = -x$ and $g(x) = -x$

24. $f(x) = \sqrt[3]{x-4}$ and $g(x) = x^3 + 4$

The functions in Exercises 25–44 are all one-to-one. For each function,

a. *Find an equation for $f^{-1}(x)$, the inverse function.*

b. *Verify that your equation is correct by showing that $f(f^{-1}(x)) = x$ and $f^{-1}(f(x)) = x$.*

25. $f(x) = x + 3$

26. $f(x) = x + 5$

27. $f(x) = 2x$

28. $f(x) = 4x$

29. $f(x) = 2x + 3$

30. $f(x) = 3x - 1$

31. $f(x) = x^3 + 2$

32. $f(x) = x^3 - 1$

33. $f(x) = (x + 2)^3$

34. $f(x) = (x - 1)^3$

35. $f(x) = \frac{1}{x}$

36. $f(x) = \frac{2}{x}$

37. $f(x) = \sqrt{x}$

38. $f(x) = \sqrt[3]{x}$

39. $f(x) = x^2 + 1$, for $x \geq 0$

40. $f(x) = x^2 - 1$, for $x \geq 0$

41. $f(x) = \frac{2x + 1}{x - 3}$

42. $f(x) = \frac{2x - 3}{x + 1}$

43. $f(x) = \sqrt[3]{x - 4} + 3$

44. $f(x) = x^{\frac{3}{5}}$

Which graphs in Exercises 45–50 represent functions that have inverse functions?

45.

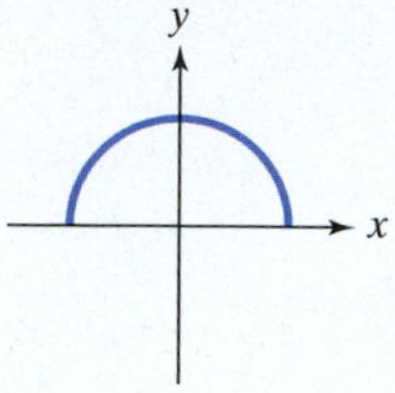

46.

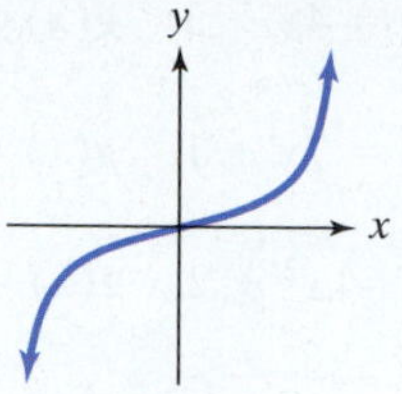

47.

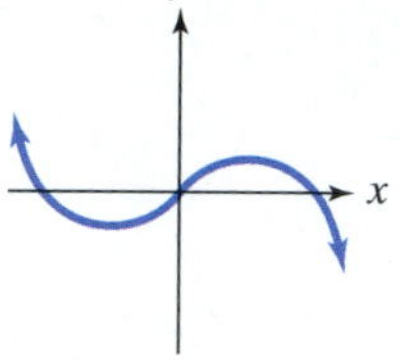

48.

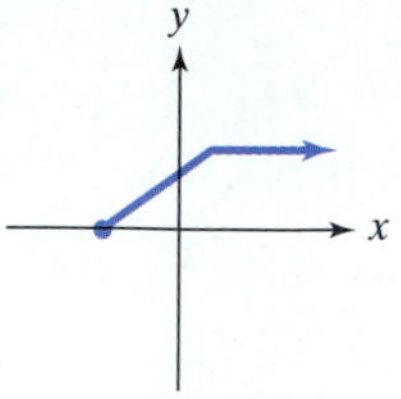

49.

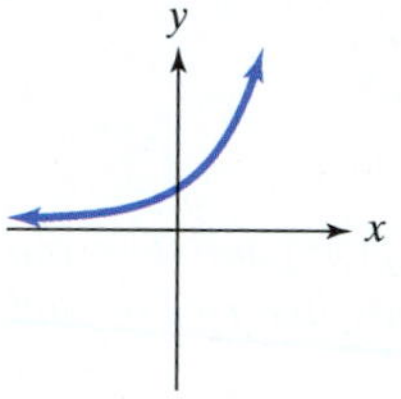

50.

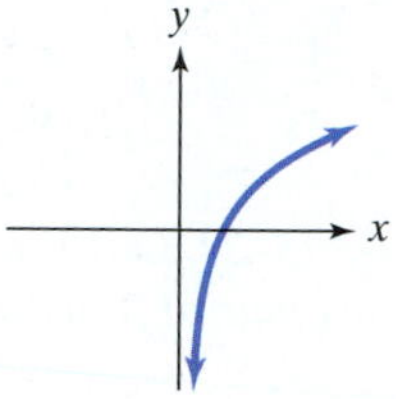

In Exercises 51–54, use the graph of f to draw the graph of its inverse function.

51.

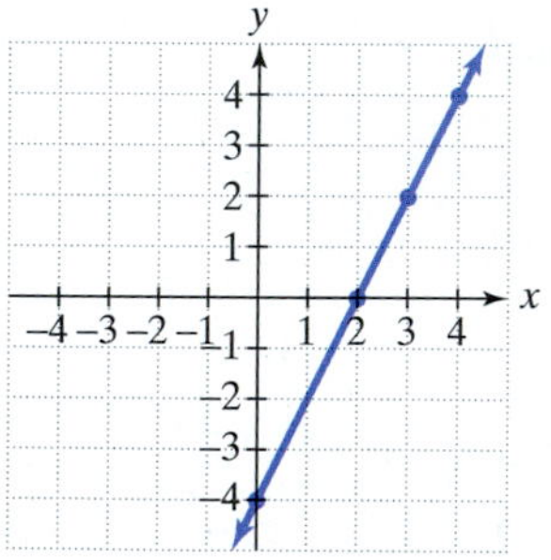

52.

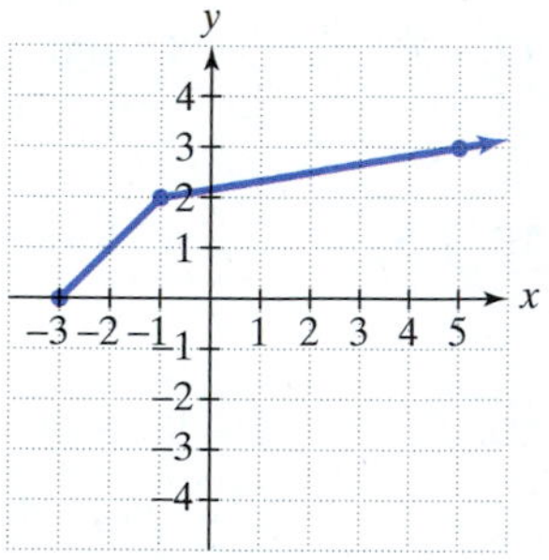

53.

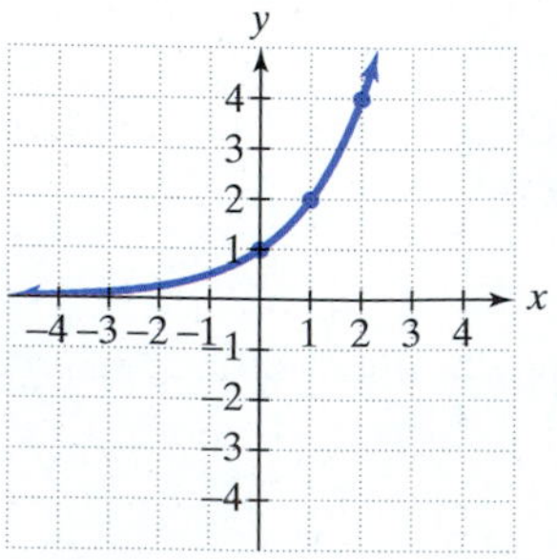

54.

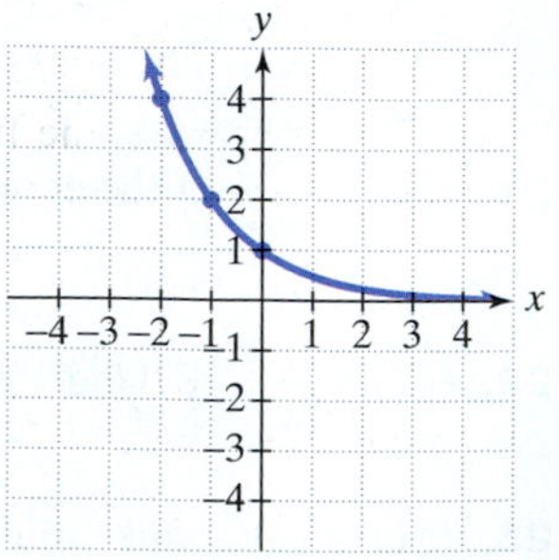

Practice PLUS

In Exercises 55–60, f and g are defined by the following tables. Use the tables to evaluate each composite function.

x	$f(x)$
−1	1
0	4
1	5
2	−1

x	$g(x)$
−1	0
1	1
4	2
10	−1

55. $f(g(1))$
56. $f(g(4))$
57. $(g \circ f)(-1)$
58. $(g \circ f)(0)$
59. $f^{-1}(g(10))$
60. $f^{-1}(g(1))$

In Exercises 61–64, use the graphs of f and g to evaluate each composite function.

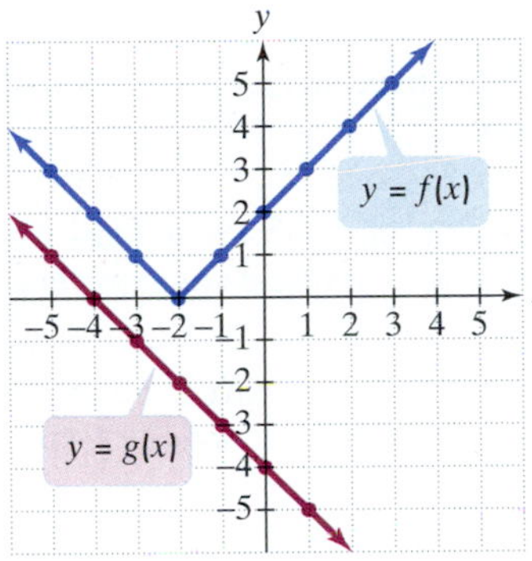

61. $(f \circ g)(-1)$
62. $(f \circ g)(1)$
63. $(g \circ f)(0)$
64. $(g \circ f)(-1)$

In Exercises 65–70, let

$$f(x) = 2x - 5$$
$$g(x) = 4x - 1$$
$$h(x) = x^2 + x + 2.$$

Evaluate the indicated function without finding an equation for the function.

65. $(f \circ g)(0)$
66. $(g \circ f)(0)$
67. $f^{-1}(1)$
68. $g^{-1}(7)$
69. $g(f[h(1)])$
70. $f(g[h(1)])$

Application Exercises

71. The regular price of a computer is x dollars. Let $f(x) = x - 400$ and $g(x) = 0.75x$.

a. Describe what the functions f and g model in terms of the price of the computer.

b. Find $(f \circ g)(x)$ and describe what this models in terms of the price of the computer.

c. Repeat part (b) for $(g \circ f)(x)$.

d. Which composite function models the greater discount on the computer, $f \circ g$ or $g \circ f$? Explain.

e. Find f^{-1} and describe what this models in terms of the price of the computer.

72. The regular price of a pair of jeans is x dollars. Let $f(x) = x - 5$ and $g(x) = 0.6x$.

a. Describe what functions f and g model in terms of the price of the jeans.

b. Find $(f \circ g)(x)$ and describe what this models in terms of the price of the jeans.

c. Repeat part (b) for $(g \circ f)(x)$.

d. Which composite function models the greater discount on the jeans, $f \circ g$ or $g \circ f$? Explain.

e. Find f^{-1} and describe what this models in terms of the price of the jeans.

In the section opener, we noted that in most societies, women say they prefer to marry men who are older than themselves, whereas men say they prefer women who are younger. The graph shows the preferred age in a mate in five selected countries. Use the information in the graph to solve Exercises 73–74.

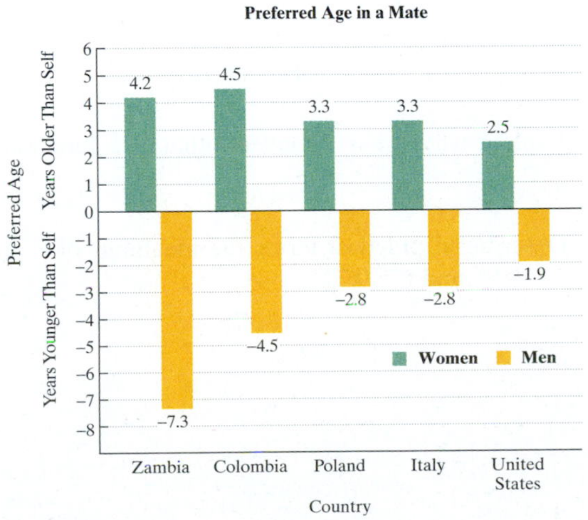

Source: Carole Wade and Carol Tavris, *Psychology*, 6th Edition, Prentice Hall, 2000

73. a. Consider a function, f, whose domain is the set of the five countries shown in the graph on page 665. Let the range be the set of the average number of years women in each of the respective countries prefer men who are older than themselves. Write function f as a set of ordered pairs.

b. Write the relation that is the inverse of f as a set of ordered pairs. Is this relation a function? Explain your answer.

74. a. Consider a function, f, whose domain is the set of the five countries shown in the graph on page 665. Let the range be the set of the average number of years men in each of the respective countries prefer women who are younger than themselves. Write f as a set of ordered pairs.

b. Write the relation that is the inverse of f as a set of ordered pairs. Is this relation a function? Explain your answer.

75. The graph represents the probability that two people in the same room share a birthday as a function of the number of people in the room. Call the function f.

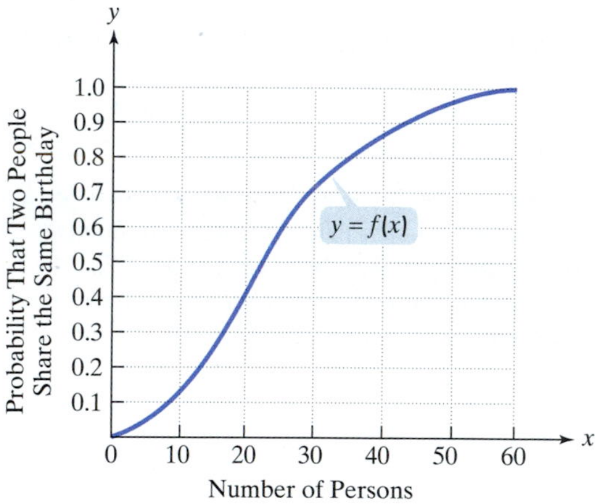

a. Explain why f has an inverse that is a function.

b. Describe in practical terms the meanings of $f^{-1}(0.25)$, $f^{-1}(0.5)$, and $f^{-1}(0.7)$.

76. A study of 900 working women in Texas showed that their feelings changed throughout the day. As the graph indicates, the women felt better as time passed, except for a blip at lunchtime.

Average Level of Happiness at Different Times of Day

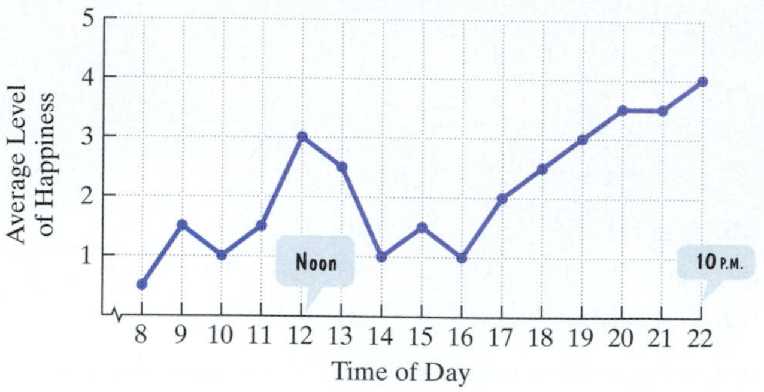

Source: D. Kahneman et al., "A Survey Method for Characterizing Daily Life Experience," *Science.*

a. Does the graph have an inverse that is a function? Explain your answer.

b. Identify two or more times of day when the average happiness level is 3. Express your answers as ordered pairs.

c. Do the ordered pairs in part (b) indicate that the graph represents a one-to-one function? Explain your answer.

77. The formula

$$y = f(x) = \frac{9}{5}x + 32$$

is used to convert from x degrees Celsius to y degrees Fahrenheit. The formula

$$y = g(x) = \frac{5}{9}(x - 32)$$

is used to convert from x degrees Fahrenheit to y degrees Celsius. Show that f and g are inverse functions.

Writing in Mathematics

78. Describe a procedure for finding $(f \circ g)(x)$.

79. Explain how to determine if two functions are inverses of each other.

80. Describe how to find the inverse of a one-to-one function.

81. What is the horizontal line test and what does it indicate?

82. Describe how to use the graph of a one-to-one function to draw the graph of its inverse function.

83. How can a graphing utility be used to visually determine if two functions are inverses of each other?

Technology Exercises

In Exercises 84–91, use a graphing utility to graph each function. Use the graph to determine whether the function has an inverse that is a function (that is, whether the function is one-to-one).

84. $f(x) = x^2 - 1$

85. $f(x) = \sqrt[3]{2 - x}$

86. $f(x) = \dfrac{x^3}{2}$

87. $f(x) = \dfrac{x^4}{4}$

88. $f(x) = |x - 2|$

89. $f(x) = (x - 1)^3$

90. $f(x) = -\sqrt{16 - x^2}$

91. $f(x) = x^3 + x + 1$

In Exercises 92–94, use a graphing utility to graph f and g in the same viewing rectangle. In addition, graph the line y = x and visually determine if f and g are inverses.

92. $f(x) = 4x + 4, \quad g(x) = 0.25x - 1$

93. $f(x) = \dfrac{1}{x} + 2, \quad g(x) = \dfrac{1}{x - 2}$

94. $f(x) = \sqrt[3]{x} - 2, \quad g(x) = (x + 2)^3$

Critical Thinking Exercises

Make Sense? *In Exercises 95–98, determine whether each statement "makes sense" or "does not make sense" and explain your reasoning.*

95. This diagram illustrates that $f(g(x)) = x^2 + 4$.

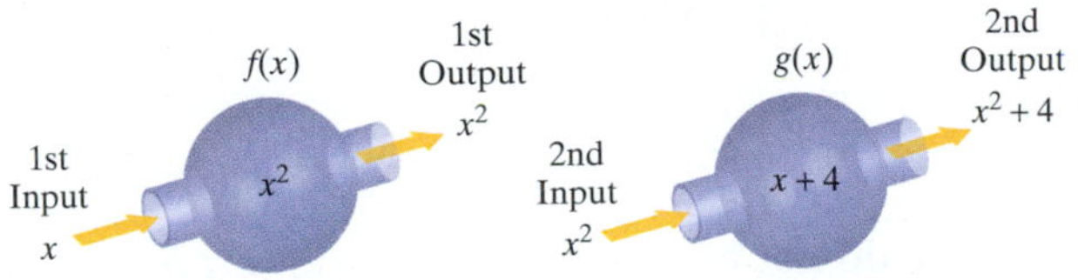

96. I must have made a mistake in finding the composite functions $f \circ g$ and $g \circ f$, because I notice that $f \circ g$ is the same function as $g \circ f$.

97. Regardless of what exponential function I graph, its shape indicates that it always has an inverse function.

98. I'm working with the linear function $f(x) = 3x + 5$ and I do not need to find f^{-1} in order to determine the value of $(f \circ f^{-1})(17)$.

In Exercises 99–102, determine whether each statement is true or false. If the statement is false, make the necessary change(s) to produce a true statement.

99. The inverse of $\{(1, 4), (2, 7)\}$ is $\{(2, 7), (1, 4)\}$.

100. The function $f(x) = 5$ is one-to-one.

101. If $f(x) = \sqrt{x}$ and $g(x) = 2x - 1$, then $(f \circ g)(5) = g(2)$.

102. If $f(x) = 3x$, then $f^{-1}(x) = \dfrac{1}{3x}$.

103. If $h(x) = \sqrt{3x^2 + 5}$, find functions f and g so that $h(x) = (f \circ g)(x)$.

104. If $f(x) = 3x$ and $g(x) = x + 5$, find $(f \circ g)^{-1}(x)$ and $(g^{-1} \circ f^{-1})(x)$.

105. Show that

$$f(x) = \frac{3x - 2}{5x - 3}$$

is its own inverse.

106. Consider the two functions defined by $f(x) = m_1x + b_1$ and $g(x) = m_2x + b_2$. Prove that the slope of the composite function of f with g is equal to the product of the slopes of the two functions.

Review Exercises

107. Divide and write the quotient in scientific notation:

$$\frac{4.3 \times 10^5}{8.6 \times 10^{-4}}.$$

(Section 1.7, Example 4)

108. Graph: $f(x) = x^2 - 4x + 3$.
(Section 8.3, Example 4)

109. Solve: $\sqrt{x + 4} - \sqrt{x - 1} = 1$.
(Section 7.6, Example 4)

Preview Exercises

Exercises 110–112 will help you prepare for the material covered in the next section.

110. What problem do you encounter when using the switch-and-solve strategy to find the inverse of $f(x) = 2^x$?

111. 25 to what power gives 5? $(25^? = 5)$

112. Solve: $(x - 3)^2 > 0$.

SECTION 9.3

Logarithmic Functions

Objectives

1. Change from logarithmic to exponential form.
2. Change from exponential to logarithmic form.
3. Evaluate logarithms.
4. Use basic logarithmic properties.
5. Graph logarithmic functions.
6. Find the domain of a logarithmic function.
7. Use common logarithms.
8. Use natural logarithms.

The earthquake that ripped through northern California on October 17, 1989, measured 7.1 on the Richter scale, killed more than 60 people, and injured more than 2400. Shown here is San Francisco's Marina district, where shock waves tossed houses off their foundations and into the street.

A higher measure on the Richter scale is more devastating than it seems because for each increase in one unit on the scale, there is a tenfold increase in the intensity of an earthquake. In this section, our focus is on the inverse of the exponential function, called the logarithmic function. The logarithmic function will help you to understand diverse phenomena, including earthquake intensity, human memory, and the pace of life in large cities.

The Definition of Logarithmic Functions

No horizontal line can be drawn that intersects the graph of an exponential function at more than one point. This means that the exponential function is one-to-one and has an inverse. Let's use our switch-and-solve strategy to find the inverse.

All exponential functions have inverse functions.

$$f(x) = b^x$$

Step 1. Replace $f(x)$ with y: $\quad y = b^x.$

Step 2. Interchange x and y: $\quad x = b^y.$

Step 3. Solve for y: $\quad ?$

The question mark indicates that we do not have a method for solving $b^y = x$ for y. To isolate the exponent y, a new notation, called *logarithmic notation*, is needed. This notation gives us a way to name the inverse of $f(x) = b^x$. **The inverse function of the exponential function with base b is called the *logarithmic function with base b*.**

Definition of the Logarithmic Function

For $x > 0$ and $b > 0, b \neq 1$,

$$y = \log_b x \text{ is equivalent to } b^y = x.$$

The function $f(x) = \log_b x$ is the **logarithmic function with base b.**

The equations

$$y = \log_b x \quad \text{and} \quad b^y = x$$

are different ways of expressing the same thing. The first equation is in **logarithmic form** and the second equivalent equation is in **exponential form**.

Notice that a **logarithm, y, is an exponent. Logarithmic form allows us to isolate this exponent.** You should learn the location of the base and exponent in each form.

Location of Base and Exponent in Exponential and Logarithmic Forms

Logarithmic Form: $y = \log_b x$ (Exponent: y; Base: b) Exponential Form: $b^y = x$ (Exponent: y; Base: b)

Study Tip

To change from logarithmic form to the more familiar exponential form, use this pattern:

$$y = \log_b x \quad \text{means} \quad b^y = x.$$

1 Change from logarithmic to exponential form.

EXAMPLE 1 Changing from Logarithmic to Exponential Form

Write each equation in its equivalent exponential form:

a. $2 = \log_5 x$ **b.** $3 = \log_b 64$ **c.** $\log_3 7 = y$.

Solution We use the fact that $y = \log_b x$ means $b^y = x$.

a. $2 = \log_5 x$ means $5^2 = x$. (Logarithms are exponents.)
b. $3 = \log_b 64$ means $b^3 = 64$. (Logarithms are exponents.)

c. $\log_3 7 = y$ or $y = \log_3 7$ means $3^y = 7$.

✓ CHECK POINT 1 Write each equation in its equivalent exponential form:

a. $3 = \log_7 x$ **b.** $2 = \log_b 25$ **c.** $\log_4 26 = y$.

2 Change from exponential to logarithmic form.

EXAMPLE 2 Changing from Exponential to Logarithmic Form

Write each equation in its equivalent logarithmic form:

a. $12^2 = x$ **b.** $b^3 = 8$ **c.** $e^y = 9$.

Solution We use the fact that $b^y = x$ means $y = \log_b x$. In logarithmic form, the exponent is isolated on one side of the equal sign.

a. $12^2 = x$ means $2 = \log_{12} x$. (Exponents are logarithms.)
b. $b^3 = 8$ means $3 = \log_b 8$. (Exponents are logarithms.)

c. $e^y = 9$ means $y = \log_e 9$.

✓ CHECK POINT 2 Write each equation in its equivalent logarithmic form:

a. $2^5 = x$ **b.** $b^3 = 27$ **c.** $e^y = 33$.

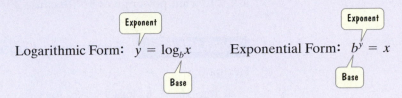

3 Evaluate logarithms.

Remembering that logarithms are exponents makes it possible to evaluate some logarithms by inspection. The logarithm of x with base b, $\log_b x$, is the exponent to which b must be raised to get x. For example, suppose we want to evaluate $\log_2 32$. We ask, 2 to what power gives 32? Because $2^5 = 32$, we can conclude that $\log_2 32 = 5$.

EXAMPLE 3 Evaluating Logarithms

Evaluate:

a. $\log_2 16$ **b.** $\log_3 9$ **c.** $\log_{25} 5$.

Solution

Logarithmic Expression	Question Needed for Evaluation	Logarithmic Expression Evaluated
a. $\log_2 16$	2 to what power gives 16?	$\log_2 16 = 4$ because $2^4 = 16$.
b. $\log_3 9$	3 to what power gives 9?	$\log_3 9 = 2$ because $3^2 = 9$.
c. $\log_{25} 5$	25 to what power gives 5?	$\log_{25} 5 = \frac{1}{2}$ because $25^{\frac{1}{2}} = \sqrt{25} = 5$.

CHECK POINT 3 Evaluate:

a. $\log_{10} 100$ **b.** $\log_3 3$ **c.** $\log_{36} 6$.

4 Use basic logarithmic properties.

Basic Logarithmic Properties

Because logarithms are exponents, they have properties that can be verified using the properties of exponents.

Basic Logarithmic Properties Involving 1

1. $\log_b b = 1$ because 1 is the exponent to which b must be raised to obtain b. $(b^1 = b)$
2. $\log_b 1 = 0$ because 0 is the exponent to which b must be raised to obtain 1. $(b^0 = 1)$

EXAMPLE 4 Using Properties of Logarithms

Evaluate:

a. $\log_7 7$ **b.** $\log_5 1$.

Solution

a. Because $\log_b b = 1$, we conclude $\log_7 7 = 1$.

b. Because $\log_b 1 = 0$, we conclude $\log_5 1 = 0$.

CHECK POINT 4 Evaluate:

a. $\log_9 9$ **b.** $\log_8 1$.

Now that we are familiar with logarithmic notation, let's resume and finish the switch-and-solve strategy for finding the inverse of $f(x) = b^x$.

Step 1. Replace $f(x)$ with y: $y = b^x$.

Step 2. Interchange x and y: $x = b^y$.

Step 3. Solve for y: $y = \log_b x$.

Step 4. Replace y with $f^{-1}(x)$: $f^{-1}(x) = \log_b x$.

The completed switch-and-solve strategy illustrates that if $f(x) = b^x$, then $f^{-1}(x) = \log_b x$. The inverse of an exponential function is the logarithmic function with the same base.

In Section 9.2, we saw how inverse functions "undo" one another. In particular,

$$f(f^{-1}(x)) = x \quad \text{and} \quad f^{-1}(f(x)) = x.$$

Applying these relationships to exponential and logarithmic functions, we obtain the following **inverse properties of logarithms:**

Inverse Properties of Logarithms

For $b > 0$ and $b \neq 1$,

$\log_b b^x = x$ — The logarithm with base b of b raised to a power equals that power.

$b^{\log_b x} = x$ — b raised to the logarithm with base b of a number equals that number.

EXAMPLE 5 Using Inverse Properties of Logarithms

Evaluate:

a. $\log_4 4^5$ **b.** $6^{\log_6 9}$.

Solution

a. Because $\log_b b^x = x$, we conclude $\log_4 4^5 = 5$.

b. Because $b^{\log_b x} = x$, we conclude $6^{\log_6 9} = 9$.

✓ CHECK POINT 5 Evaluate:

a. $\log_7 7^8$ **b.** $3^{\log_3 17}$.

5 Graph logarithmic functions.

Graphs of Logarithmic Functions

How do we graph logarithmic functions? We use the fact that a logarithmic function is the inverse of an exponential function. This means that the logarithmic function reverses the coordinates of the exponential function. It also means that the graph of the logarithmic function is a reflection of the graph of the exponential function about the line $y = x$.

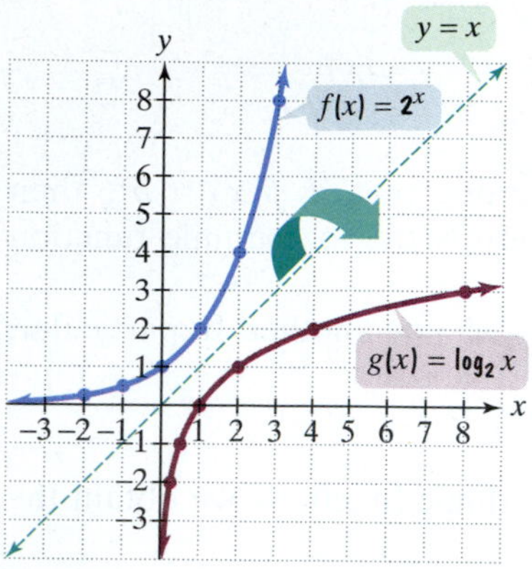

FIGURE 9.14 The graphs of $f(x) = 2^x$ and its inverse function

EXAMPLE 6 Graphs of Exponential and Logarithmic Functions

Graph $f(x) = 2^x$ and $g(x) = \log_2 x$ in the same rectangular coordinate system.

Solution We first set up a table of coordinates for $f(x) = 2^x$. Reversing these coordinates gives the coordinates for the inverse function $g(x) = \log_2 x$.

x	-2	-1	0	1	2	3
$f(x) = 2^x$	$\frac{1}{4}$	$\frac{1}{2}$	1	2	4	8

Reverse coordinates.

x	$\frac{1}{4}$	$\frac{1}{2}$	1	2	4	8
$g(x) = \log_2 x$	-2	-1	0	1	2	3

We now plot the ordered pairs from each table, connecting them with smooth curves. **Figure 9.14** shows the graphs of $f(x) = 2^x$ and its inverse function $g(x) = \log_2 x$. The graph of the inverse can also be drawn by reflecting the graph of $f(x) = 2^x$ about the line $y = x$.

Study Tip

You can obtain a partial table of coordinates for $g(x) = \log_2 x$ without having to obtain and reverse coordinates for $f(x) = 2^x$. Because $g(x) = \log_2 x$ means $2^{g(x)} = x$, we begin with values for $g(x)$ and compute corresponding values for x:

Use $x = 2^{g(x)}$ to compute x. For example, if $g(x) = -2$, $x = 2^{-2} = \frac{1}{2^2} = \frac{1}{4}$.

Start with values for $g(x)$.

x	$\frac{1}{4}$	$\frac{1}{2}$	1	2	4	8
$g(x) = \log_2 x$	-2	-1	0	1	2	3

CHECK POINT 6 Graph $f(x) = 3^x$ and $g(x) = \log_3 x$ in the same rectangular coordinate system.

Figure 9.15 illustrates the relationship between the graph of an exponential function, shown in blue, and its inverse, a logarithmic function, shown in red, for bases greater than 1 and for bases between 0 and 1. Also shown and labeled are the exponential function's horizontal asymptote $(y = 0)$ and the logarithmic function's vertical asymptote $(x = 0)$.

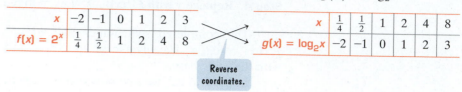

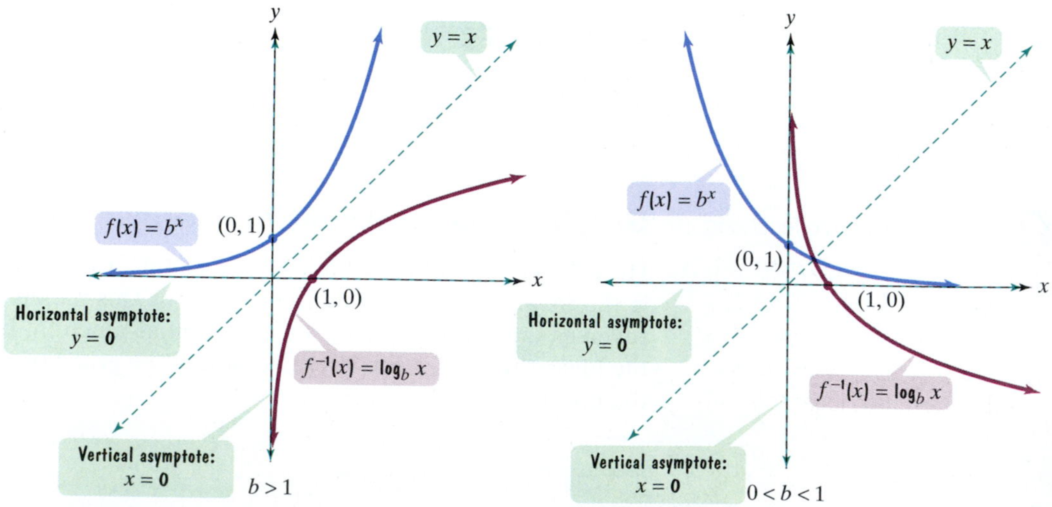

FIGURE 9.15 Graphs of exponential and logarithmic functions

The red graphs in **Figure 9.15** illustrate the following general characteristics of logarithmic functions:

Characteristics of Logarithmic Functions of the Form $f(x) = \log_b x$

1. The domain of $f(x) = \log_b x$ consists of all positive real numbers: $(0, \infty)$. The range of $f(x) = \log_b x$ consists of all real numbers: $(-\infty, \infty)$.
2. The graphs of all logarithmic functions of the form $f(x) = \log_b x$ pass through the point (1, 0) because $f(1) = \log_b 1 = 0$. The x-intercept is 1. There is no y-intercept.
3. If $b > 1$, $f(x) = \log_b x$ has a graph that goes up to the right and is an increasing function.
4. If $0 < b < 1$, $f(x) = \log_b x$ has a graph that goes down to the right and is a decreasing function.
5. The graph of $f(x) = \log_b x$ approaches, but does not touch, the y-axis. The y-axis, or $x = 0$, is a vertical asymptote.

6 Find the domain of a logarithmic function.

The Domain of a Logarithmic Function

In Section 9.1, we learned that the domain of an exponential function of the form $f(x) = b^x$ includes all real numbers and its range is the set of positive real numbers. Because the logarithmic function reverses the domain and the range of the exponential function, the **domain of a logarithmic function of the form $f(x) = \log_b x$ is the set of all positive real numbers**. Thus, $\log_2 8$ is defined because the value of x in the logarithmic expression, 8, is greater than zero and therefore is included in the domain of the logarithmic function $f(x) = \log_2 x$. However, $\log_2 0$ and $\log_2(-8)$ are not defined because 0 and -8 are not positive real numbers and therefore are excluded from the domain of the logarithmic function $f(x) = \log_2 x$. In general, **the domain of $f(x) = \log_b g(x)$ consists of all x for which $g(x) > 0$.**

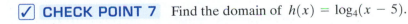

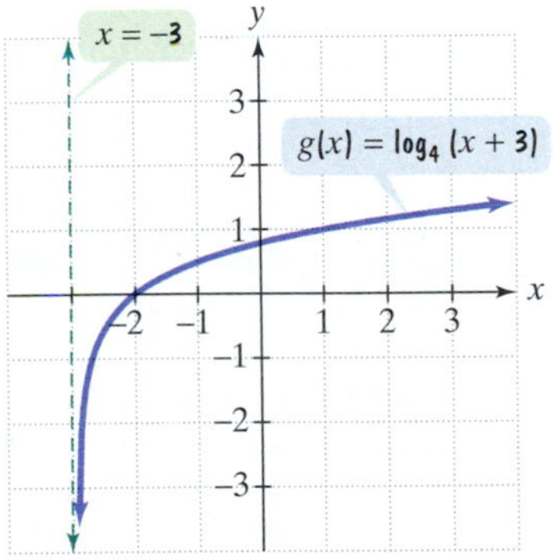

FIGURE 9.16 The domain of $g(x) = \log_4(x + 3)$ is $(-3, \infty)$.

EXAMPLE 7 Finding the Domain of a Logarithmic Function

Find the domain of $g(x) = \log_4(x + 3)$.

Solution The domain of g consists of all x for which $x + 3 > 0$. Solving this inequality for x, we obtain $x > -3$. Thus, the domain of g is $\{x | x > -3\}$ or $(-3, \infty)$. This is illustrated in **Figure 9.16**. The vertical asymptote is $x = -3$ and all points on the graph of g have x-coordinates that are greater than -3.

✓ **CHECK POINT 7** Find the domain of $h(x) = \log_4(x - 5)$.

7 Use common logarithms.

Common Logarithms

The logarithmic function with base 10 is called the **common logarithmic function**. The function $f(x) = \log_{10} x$ is usually expressed as $f(x) = \log x$. A calculator with a LOG key can be used to evaluate common logarithms. On the next page are some examples.

Logarithm	Most Scientific Calculator Keystrokes	Most Graphing Calculator Keystrokes	Display (or Approximate Display)
log 1000	1000 [LOG]	[LOG] 1000 [ENTER]	3
$\log\frac{5}{2}$	[(] 5 [÷] 2 [)] [LOG]	[LOG] [(] 5 [÷] 2 [)] [ENTER]	0.39794
$\frac{\log 5}{\log 2}$	5 [LOG] [÷] 2 [LOG] [=]	[LOG] 5 [÷] [LOG] 2 [ENTER]	2.32193
log(−3)	3 [+/−] [LOG]	[LOG] [(−)] 3 [ENTER]	[ERROR]

Some graphing calculators display an open parenthesis when the [LOG] key is pressed. In this case, remember to close the set of parentheses after entering the function's domain value: [LOG] 5 [)] [÷] [LOG] 2 [)] [ENTER].

The error message or [NONREAL ANS] message given by many calculators for log(−3) is a reminder that the domain of the common logarithmic function, $f(x) = \log x$, is the set of positive real numbers. In general, the domain of $f(x) = \log g(x)$ consists of all x for which $g(x) > 0$.

Many real-life phenomena start with rapid growth and then the growth begins to level off. This type of behavior can be modeled by logarithmic functions.

EXAMPLE 8 Modeling Height of Children

The percentage of adult height attained by a boy who is x years old can be modeled by

$$f(x) = 29 + 48.8 \log(x + 1),$$

where x represents the boy's age and $f(x)$ represents the percentage of his adult height. Approximately what percentage of his adult height has a boy attained at age eight?

Solution We substitute the boy's age, 8, for x and evaluate the function at 8.

$$f(x) = 29 + 48.8 \log(x + 1)$$ *This is the given function.*

$$f(8) = 29 + 48.8 \log(8 + 1)$$ *Substitute 8 for x.*

$$= 29 + 48.8 \log 9$$ *Graphing calculator keystrokes:* 29 [+] 48.8 [LOG] 9 [ENTER]

$$\approx 76$$

Thus, an 8-year-old boy has attained approximately 76% of his adult height.

✓ CHECK POINT 8 Use the function in Example 8 to answer this question: Approximately what percentage of his adult height has a boy attained at age ten?

The basic properties of logarithms that were listed earlier in this section can be applied to common logarithms.

Properties of Common Logarithms

General Properties	Common Logarithms
1. $\log_b 1 = 0$	1. $\log 1 = 0$
2. $\log_b b = 1$	2. $\log 10 = 1$
3. $\log_b b^x = x$	3. $\log 10^x = x$
4. $b^{\log_b x} = x$	4. $10^{\log x} = x$

Inverse properties

The property $\log 10^x = x$ can be used to evaluate common logarithms involving powers of 10. For example,

$$\log 100 = \log 10^2 = 2, \quad \log 1000 = \log 10^3 = 3, \quad \text{and} \quad \log 10^{7.1} = 7.1.$$

EXAMPLE 9 Earthquake Intensity

The magnitude, R, on the Richter scale of an earthquake of intensity I is given by

$$R = \log\frac{I}{I_0},$$

where I_0 is the intensity of a barely felt zero-level earthquake. The earthquake that destroyed San Francisco in 1906 was $10^{8.3}$ times as intense as a zero-level earthquake. What was its magnitude on the Richter scale?

Solution Because the earthquake was $10^{8.3}$ times as intense as a zero-level earthquake, the intensity, I, is $10^{8.3}I_0$.

$$R = \log\frac{I}{I_0}$$ This is the formula for magnitude on the Richter scale.

$$R = \log\frac{10^{8.3}I_0}{I_0}$$ Substitute $10^{8.3}I_0$ for I.

$$= \log 10^{8.3}$$ Simplify.

$$= 8.3$$ Use the property $\log 10^x = x$.

San Francisco's 1906 earthquake registered 8.3 on the Richter scale.

✓ **CHECK POINT 9** Use the formula in Example 9 to solve this problem. If an earthquake is 10,000 times as intense as a zero-level quake ($I = 10{,}000I_0$), what is its magnitude on the Richter scale?

8 Use natural logarithms.

Natural Logarithms

The logarithmic function with base e is called the **natural logarithmic function**. The function $f(x) = \log_e x$ is usually expressed as $f(x) = \ln x$, read "el en of x." A calculator with an LN key can be used to evaluate natural logarithms. Keystrokes are identical to those shown for common logarithmic evaluations on page 674.

Like the domain of all logarithmic functions, the domain of the natural logarithmic function $f(x) = \ln x$ is the set of all positive real numbers. Thus, the domain of $f(x) = \ln g(x)$ consists of all x for which $g(x) > 0$.

EXAMPLE 10 Finding Domains of Natural Logarithmic Functions

Find the domain of each function:

a. $f(x) = \ln(3 - x)$ **b.** $g(x) = \ln(x - 3)^2$.

Solution

a. The domain of $f(x) = \ln(3 - x)$ consists of all x for which $3 - x > 0$. Solving this inequality for x, we obtain $x < 3$. Thus, the domain of f is $\{x | x < 3\}$ or $(-\infty, 3)$. This is verified by the graph in **Figure 9.17**.

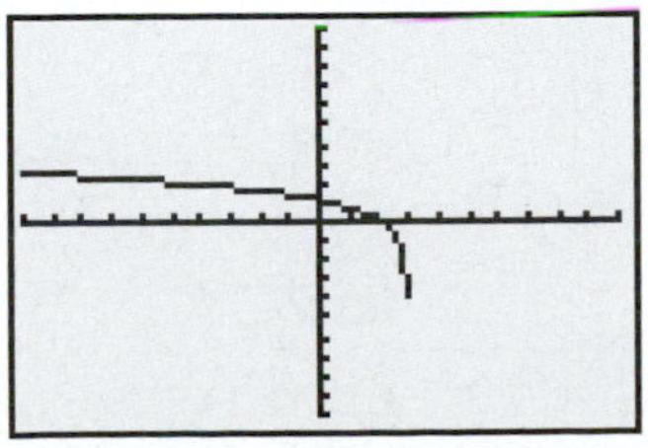

[−10, 10, 1] by [−10, 10, 1]

FIGURE 9.17 The domain of $f(x) = \ln(3 - x)$ is $(-\infty, 3)$.

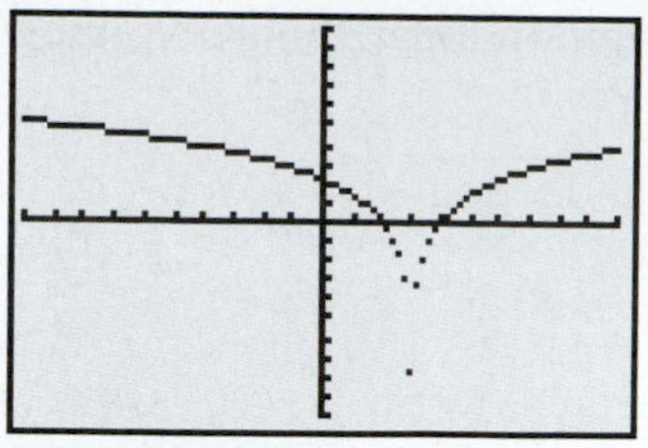

$[-10, 10, 1]$ by $[-10, 10, 1]$

FIGURE 9.18 3 is excluded from the domain of $g(x) = \ln(x-3)^2$.

b. The domain of $g(x) = \ln(x-3)^2$ consists of all x for which $(x-3)^2 > 0$. It follows that the domain of g is all real numbers except 3. Thus, the domain of g is $\{x \mid x \neq 3\}$ or $(-\infty, 3) \cup (3, \infty)$. This is shown by the graph in **Figure 9.18**. To make it more obvious that 3 is excluded from the domain, we used a DOT format.

CHECK POINT 10 Find the domain of each function:

a. $f(x) = \ln(4 - x)$

b. $g(x) = \ln x^2$.

The basic properties of logarithms that were listed earlier in this section can be applied to natural logarithms.

Properties of Natural Logarithms

General Properties		Natural Logarithms
1. $\log_b 1 = 0$		1. $\ln 1 = 0$
2. $\log_b b = 1$		2. $\ln e = 1$
3. $\log_b b^x = x$	Inverse properties	3. $\ln e^x = x$
4. $b^{\log_b x} = x$		4. $e^{\ln x} = x$

Examine the inverse properties, $\ln e^x = x$ and $e^{\ln x} = x$. Can you see how ln and e "undo" one another? For example,

$$\ln e^2 = 2, \quad \ln e^{7x^2} = 7x^2, \quad e^{\ln 2} = 2, \quad \text{and} \quad e^{\ln 7x^2} = 7x^2.$$

EXAMPLE 11 Dangerous Heat: Temperature in an Enclosed Vehicle

When the outside air temperature is anywhere from 72° to 96° Fahrenheit, the temperature in an enclosed vehicle climbs by 43° in the first hour. The bar graph in **Figure 9.19** shows the temperature increase throughout the hour. The function

$$f(x) = 13.4 \ln x - 11.6$$

models the temperature increase, $f(x)$, in degrees Fahrenheit, after x minutes. Use the function to find the temperature increase, to the nearest degree, after 50 minutes. How well does the function model the actual increase shown in **Figure 9.19**?

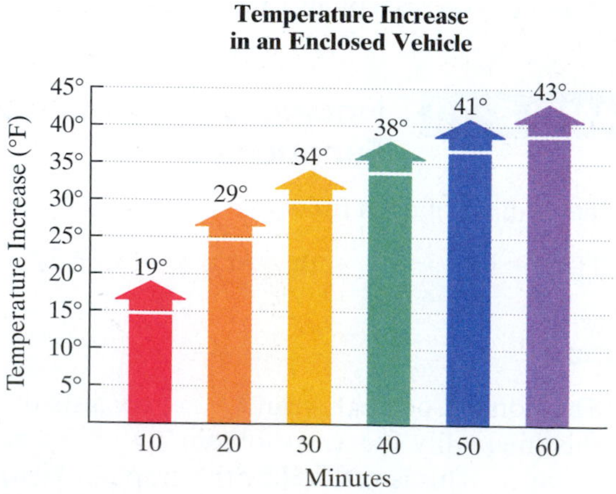

FIGURE 9.19

Source: Professor Jan Null, San Francisco State University

Solution We find the temperature increase after 50 minutes by substituting 50 for x and evaluating the function at 50.

$$f(x) = 13.4 \ln x - 11.6$$ This is the given function.

$$f(50) = 13.4 \ln 50 - 11.6$$ Substitute 50 for x.

$$\approx 41$$ Graphing calculator keystrokes: 13.4 [ln] 50 [−] 11.6 [ENTER]. On some calculators, a parenthesis is needed after 50.

According to the function, the temperature will increase by approximately 41° after 50 minutes. Because the increase shown in **Figure 9.19** is 41°, the function models the actual increase extremely well.

CHECK POINT 11 Use the function in Example 11 to find the temperature increase, to the nearest degree, after 30 minutes. How well does the function model the actual increase shown in **Figure 9.19**?

9.3 EXERCISE SET

MyMathLab — MathXL PRACTICE · WATCH · DOWNLOAD · READ · REVIEW

Practice Exercises

In Exercises 1–8, write each equation in its equivalent exponential form.

1. $4 = \log_2 16$

2. $6 = \log_2 64$

3. $2 = \log_3 x$

4. $2 = \log_9 x$

5. $5 = \log_b 32$

6. $3 = \log_b 27$

7. $\log_6 216 = y$

8. $\log_5 125 = y$

In Exercises 9–20, write each equation in its equivalent logarithmic form.

9. $2^3 = 8$

10. $5^4 = 625$

11. $2^{-4} = \frac{1}{16}$

12. $5^{-3} = \frac{1}{125}$

13. $\sqrt[3]{8} = 2$

14. $\sqrt[3]{64} = 4$

15. $13^2 = x$

16. $15^2 = x$

17. $b^3 = 1000$

18. $b^3 = 343$

19. $7^y = 200$

20. $8^y = 300$

In Exercises 21–42, evaluate each expression without using a calculator.

21. $\log_4 16$

22. $\log_7 49$

23. $\log_2 64$

24. $\log_3 27$

25. $\log_5 \frac{1}{5}$

26. $\log_6 \frac{1}{6}$

27. $\log_2 \frac{1}{8}$

28. $\log_3 \frac{1}{9}$

29. $\log_7 \sqrt{7}$

30. $\log_6 \sqrt{6}$

31. $\log_2 \frac{1}{\sqrt{2}}$

32. $\log_3 \frac{1}{\sqrt{3}}$

33. $\log_{64} 8$

34. $\log_{81} 9$

35. $\log_5 5$

36. $\log_{11} 11$

37. $\log_4 1$

38. $\log_6 1$

39. $\log_5 5^7$

40. $\log_4 4^6$

41. $8^{\log_8 19}$

42. $7^{\log_7 23}$

43. Graph $f(x) = 4^x$ and $g(x) = \log_4 x$ in the same rectangular coordinate system.

44. Graph $f(x) = 5^x$ and $g(x) = \log_5 x$ in the same rectangular coordinate system.

45. Graph $f(x) = \left(\frac{1}{2}\right)^x$ and $g(x) = \log_{\frac{1}{2}} x$ in the same rectangular coordinate system.

46. Graph $f(x) = \left(\frac{1}{4}\right)^x$ and $g(x) = \log_{\frac{1}{4}} x$ in the same rectangular coordinate system.

In Exercises 47–52, find the domain of each logarithmic function.

47. $f(x) = \log_5(x + 4)$

48. $f(x) = \log_5(x + 6)$

49. $f(x) = \log(2 - x)$

50. $f(x) = \log(7 - x)$

51. $f(x) = \ln(x - 2)^2$

52. $f(x) = \ln(x - 7)^2$

In Exercises 53–66, evaluate each expression without using a calculator.

53. $\log 100$

54. $\log 1000$

55. $\log 10^7$

56. $\log 10^8$

57. $10^{\log 33}$

58. $10^{\log 53}$

59. $\ln 1$

60. $\ln e$

61. $\ln e^6$

62. $\ln e^7$

63. $\ln \frac{1}{e^6}$

64. $\ln \frac{1}{e^7}$

65. $e^{\ln 125}$

66. $e^{\ln 300}$

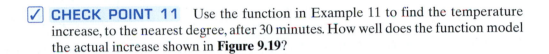

In Exercises 67–72, simplify each expression.

67. $\ln e^{9x}$

68. $\ln e^{13x}$

69. $e^{\ln 5x^2}$

70. $e^{\ln 7x^2}$

71. $10^{\log \sqrt{x}}$

72. $10^{\log \sqrt[3]{x}}$

Practice PLUS

In Exercises 73–76, write each equation in its equivalent exponential form. Then solve for x.

73. $\log_3(x - 1) = 2$

74. $\log_5(x + 4) = 2$

75. $\log_4 x = -3$

76. $\log_{64} x = \frac{2}{3}$

In Exercises 77–80, evaluate each expression without using a calculator.

77. $\log_3(\log_7 7)$

78. $\log_5(\log_2 32)$

79. $\log_2(\log_3 81)$

80. $\log(\ln e)$

In Exercises 81–86, match each function with its graph. The graphs are labeled (a) through (f), and each graph is displayed in a $[-5, 5, 1]$ by $[-5, 5, 1]$ viewing rectangle.

81. $f(x) = \ln(x + 2)$

82. $f(x) = \ln(x - 2)$

83. $f(x) = \ln x + 2$

84. $f(x) = \ln x - 2$

85. $f(x) = \ln(1 - x)$

86. $f(x) = \ln(2 - x)$

a.

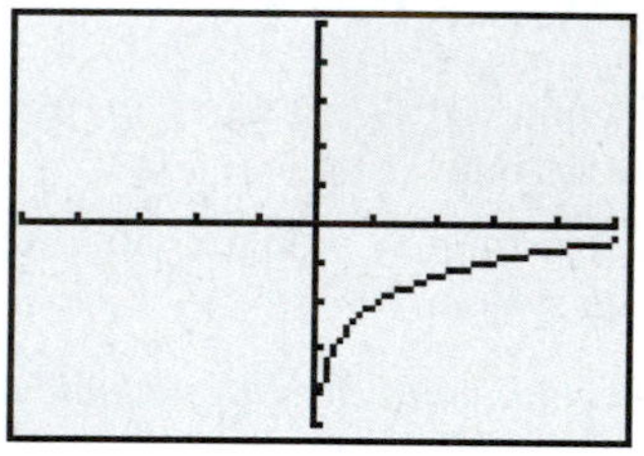

b.

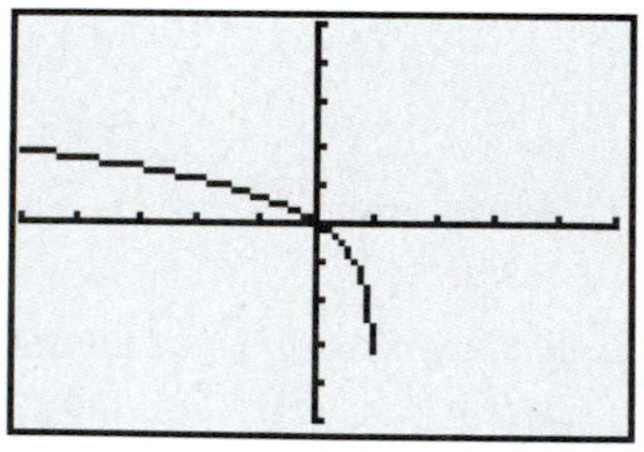

c.

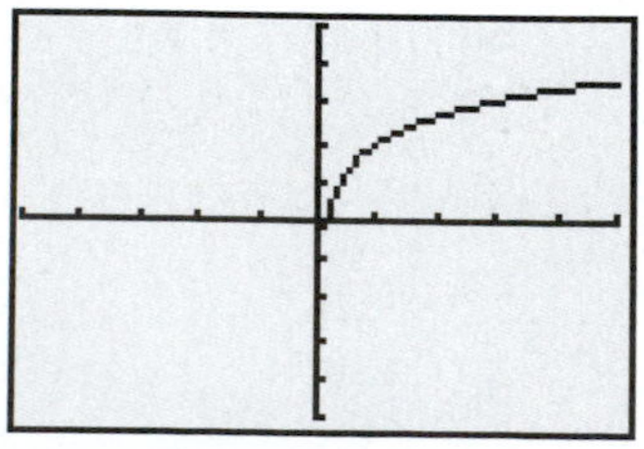

d.

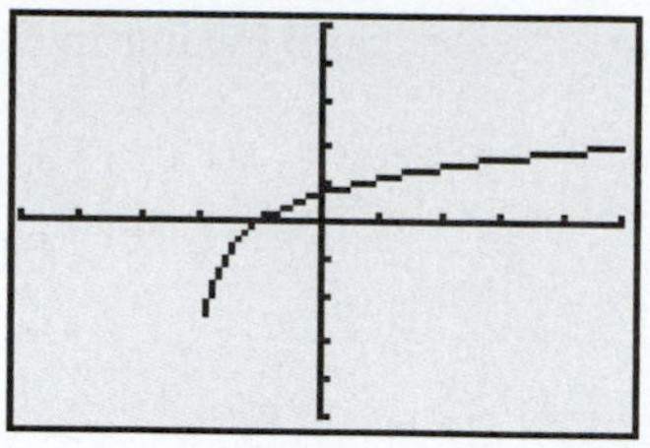

e.

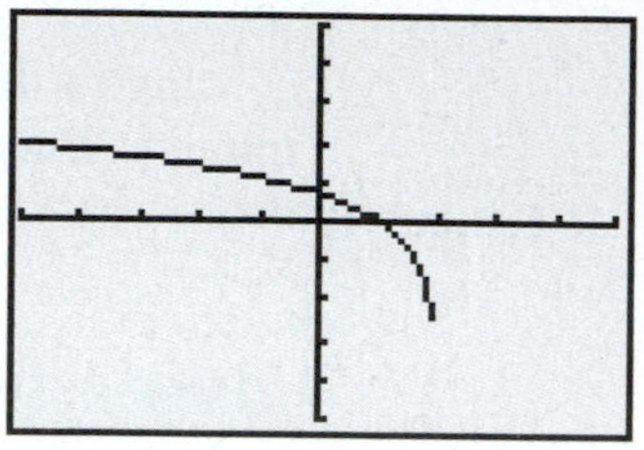

f.

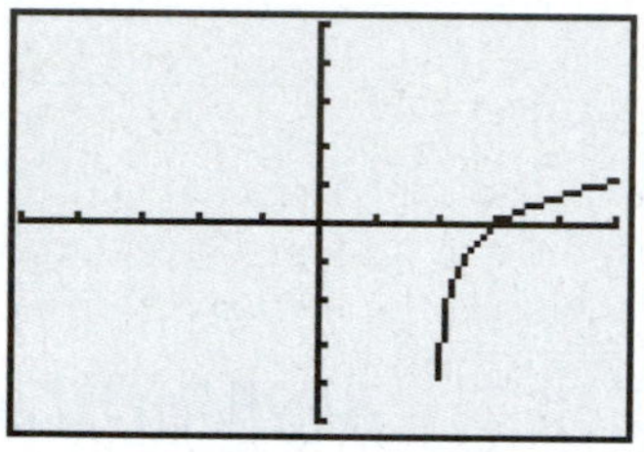

Application Exercises

The percentage of adult height attained by a girl who is x years old can be modeled by

$$f(x) = 62 + 35\log(x - 4),$$

where x represents the girl's age (from 5 to 15) and $f(x)$ represents the percentage of her adult height. Use the formula to solve Exercises 87–88. Round answers to the nearest tenth of a percent.

87. Approximately what percentage of her adult height has a girl attained at age 13?

88. Approximately what percentage of her adult height has a girl attained at age ten?

The bar graph indicates that the percentage of first-year college students expressing antifeminist views declined after 1970. Use this information to solve Exercises 89–90.

Opposition to Feminism among First-Year United States College Students, 1970–2004

Statement: "The activities of married women are best confined to the home and family."

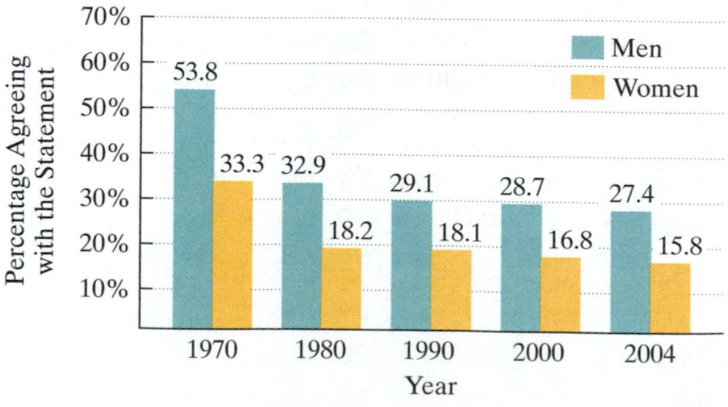

Source: John Macionis, *Sociology, Eleventh Edition,* Prentice Hall, 2007

89. The function

$$f(x) = -7.49 \ln x + 53$$

models the percentage of first-year college men, $f(x)$, expressing antifeminist views (by agreeing with the statement) x years after 1969.

a. Use the function to find the percentage of first-year college men expressing antifeminist views in 2004. Round to one decimal place. Does this function value overestimate or underestimate the percentage displayed by the graph? By how much?

b. Use the function to project the percentage of first-year college men who will express antifeminist views in 2010. Round to one decimal place.

90. The function

$$f(x) = -4.86 \ln x + 32.5$$

models the percentage of first-year college women, $f(x)$, expressing antifeminist views (by agreeing with the statement) x years after 1969.

a. Use the function to find the percentage of first-year college women expressing antifeminist views in 2004. Round to one decimal place. Does this function value overestimate or underestimate the percentage displayed by the graph? By how much?

b. Use the function to project the percentage of first-year college women who will express antifeminist views in 2010. Round to one decimal place.

The loudness level of a sound, D, in decibels, is given by the formula

$$D = 10 \log(10^{12} I),$$

where I is the intensity of the sound, in watts per meter^2. Decibel levels range from 0, a barely audible sound, to 160, a sound resulting in a ruptured eardrum. Use the formula to solve Exercises 91–92.

91. The sound of a blue whale can be heard 500 miles away, reaching an intensity of 6.3×10^6 watts per meter^2. Determine the decibel level of this sound. At close range, can the sound of a blue whale rupture the human eardrum?

92. What is the decibel level of a normal conversation, 3.2×10^{-6} watt per meter^2?

93. Students in a psychology class took a final examination. As part of an experiment to see how much of the course content they remembered over time, they took equivalent forms of the exam in monthly intervals thereafter. The average score for the group, $f(t)$, after t months was modeled by the function

$$f(t) = 88 - 15 \ln(t + 1), \qquad 0 \le t \le 12.$$

a. What was the average score on the original exam?

b. What was the average score, to the nearest tenth, after 2 months? 4 months? 6 months? 8 months? 10 months? one year?

c. Sketch the graph of f (either by hand or with a graphing utility). Describe what the graph indicates in terms of the material retained by the students.

Writing in Mathematics

94. Describe the relationship between an equation in logarithmic form and an equivalent equation in exponential form.

95. What question can be asked to help evaluate $\log_3 81$?

96. Explain why the logarithm of 1 with base b is 0.

97. Describe the following property using words: $\log_b b^x = x$.

98. Explain how to use the graph of $f(x) = 2^x$ to obtain the graph of $g(x) = \log_2 x$.

99. Explain how to find the domain of a logarithmic function.

100. Logarithmic models are well suited to phenomena in which growth is initially rapid, but then begins to level off. Describe something that is changing over time that can be modeled using a logarithmic function.

101. Suppose that a girl is 4 feet 6 inches at age 10. Explain how to use the function in Exercises 87–88 to determine how tall she can expect to be as an adult.

Technology Exercises

In Exercises 102–105, graph f and g in the same viewing rectangle. Then describe the relationship of the graph of g to the graph of f.

102. $f(x) = \ln x, g(x) = \ln(x + 3)$

103. $f(x) = \ln x, g(x) = \ln x + 3$

104. $f(x) = \log x, g(x) = -\log x$

105. $f(x) = \log x, g(x) = \log(x - 2) + 1$

106. Students in a mathematics class took a final examination. They took equivalent forms of the exam in monthly intervals thereafter. The average score, $f(t)$, for the group after t months is modeled by the human memory function $f(t) = 75 - 10 \log(t + 1)$, where $0 \le t \le 12$. Use a graphing utility to graph the function. Then determine how many months will elapse before the average score falls below 65.

107. In parts (a)–(c), graph f and g in the same viewing rectangle.

a. $f(x) = \ln(3x)$, $g(x) = \ln 3 + \ln x$

b. $f(x) = \log(5x^2)$, $g(x) = \log 5 + \log x^2$

c. $f(x) = \ln(2x^3)$, $g(x) = \ln 2 + \ln x^3$

d. Describe what you observe in parts (a)–(c). Generalize this observation by writing an equivalent expression for $\log_b(MN)$, where $M > 0$ and $N > 0$.

e. Complete this statement: The logarithm of a product is equal to ______________________.

108. Graph each of the following functions in the same viewing rectangle and then place the functions in order from the one that increases most slowly to the one that increases most rapidly.

$$y = x, y = \sqrt{x}, y = e^x, y = \ln x, y = x^x, y = x^2$$

Critical Thinking Exercises

Make Sense? *In Exercises 109–112, determine whether each statement "makes sense" or "does not make sense" and explain your reasoning.*

109. I estimate that $\log_8 16$ lies between 1 and 2 because $8^1 = 8$ and $8^2 = 64$.

110. When graphing a logarithmic function, I like to show the graph of its horizontal asymptote.

111. I can evaluate some common logarithms without having to use a calculator.

112. An earthquake of magnitude 8 on the Richter scale is twice as intense as an earthquake of magnitude 4.

In Exercises 113–116, determine whether each statement is true or false. If the statement is false, make the necessary change(s) to produce a true statement.

113. $\dfrac{\log_2 8}{\log_2 4} = \dfrac{8}{4}$

114. $\log(-100) = -2$.

115. The domain of $f(x) = \log_2 x$ is $(-\infty, \infty)$.

116. $\log_b x$ is the exponent to which b must be raised to obtain x.

117. Without using a calculator, find the exact value of

$$\frac{\log_3 81 - \log_\pi 1}{\log_{2\sqrt{2}} 8 - \log 0.001}.$$

118. Without using a calculator, find the exact value of $\log_4[\log_3(\log_2 8)]$.

119. Without using a calculator, determine which is the greater number: $\log_4 60$ or $\log_3 40$.

Review Exercises

120. Solve the system:

$$2x = 11 - 5y$$
$$3x - 2y = -12.$$

(Section 3.1, Example 5)

121. Factor completely:

$$6x^2 - 8xy + 2y^2.$$

(Section 5.4, Example 9)

122. Solve: $x + 3 \le -4$ or $2 - 7x \le 16$.
(Section 4.2, Example 6)

Preview Exercises

Exercises 123–125 will help you prepare for the material covered in the next section. In each exercise, evaluate the indicated logarithmic expressions without using a calculator.

123. a. Evaluate: $\log_2 32$.

b. Evaluate: $\log_2 8 + \log_2 4$.

c. What can you conclude about $\log_2 32$, or $\log_2(8 \cdot 4)$?

124. a. Evaluate: $\log_2 16$.

b. Evaluate: $\log_2 32 - \log_2 2$.

c. What can you conclude about

$$\log_2 16, \text{ or } \log_2\left(\frac{32}{2}\right)?$$

125. a. Evaluate: $\log_3 81$.

b. Evaluate: $2 \log_3 9$.

c. What can you conclude about

$$\log_3 81, \text{ or } \log_3 9^2?$$

SECTION 9.4

Properties of Logarithms

Objectives

1. Use the product rule.
2. Use the quotient rule.
3. Use the power rule.
4. Expand logarithmic expressions.
5. Condense logarithmic expressions.
6. Use the change-of-base property.

We all learn new things in different ways. In this section, we consider important properties of logarithms. What would be the most effective way for you to learn these properties? Would it be helpful to use your graphing utility and discover one of these properties for yourself? To do so, work Exercise 107 in Exercise Set 9.3 before continuing. Would it be helpful to evaluate certain logarithmic expressions that suggest three of the properties? If this is the case, work Preview Exercises 123–125 in Exercise Set 9.3 before continuing. Would the properties become more meaningful if you could see exactly where they come from? If so, you will find details of the proofs of many of these properties in the appendix. The remainder of our work in this chapter will be based on the properties of logarithms that you learn in this section.

1 Use the product rule.

The Product Rule

Properties of exponents correspond to properties of logarithms. For example, when we multiply exponential expressions with the same base, we add exponents:

$$b^m \cdot b^n = b^{m+n}.$$

This property of exponents, coupled with an awareness that a logarithm is an exponent, suggests the following property, called the **product rule**:

Discover for Yourself

We know that $\log 100{,}000 = 5$. Show that you get the same result by writing 100,000 as $1000 \cdot 100$ and then using the product rule. Then verify the product rule by using other numbers whose logarithms are easy to find.

The Product Rule

Let b, M, and N be positive real numbers with $b \neq 1$.

$$\log_b(MN) = \log_b M + \log_b N$$

The logarithm of a product is the sum of the logarithms.

When we use the product rule to write a single logarithm as the sum of two logarithms, we say that we are **expanding a logarithmic expression**. For example, we can use the product rule to expand $\ln(7x)$:

$$\ln(7x) = \ln 7 + \ln x.$$

The logarithm of a product | is | the sum of the logarithms.

EXAMPLE 1 Using the Product Rule

Use the product rule to expand each logarithmic expression:

a. $\log_4(7 \cdot 5)$ b. $\log(10x)$.

Solution

a. $\log_4(7 \cdot 5) = \log_4 7 + \log_4 5$ The logarithm of a product is the sum of the logarithms.

b. $\log(10x) = \log 10 + \log x$ The logarithm of a product is the sum of the logarithms. These are common logarithms with base 10 understood.

$= 1 + \log x$ Because $\log_b b = 1$, then $\log 10 = 1$.

✓ CHECK POINT 1 Use the product rule to expand each logarithmic expression:

a. $\log_6(7 \cdot 11)$ b. $\log(100x)$.

2 Use the quotient rule.

The Quotient Rule

When we divide exponential expressions with the same base, we subtract exponents:

$$\frac{b^m}{b^n} = b^{m-n}.$$

This property suggests the following property of logarithms, called the **quotient rule:**

Discover for Yourself

We know that $\log_2 16 = 4$. Show that you get the same result by writing 16 as $\frac{32}{2}$ and then using the quotient rule. Then verify the quotient rule using other numbers whose logarithms are easy to find.

The Quotient Rule

Let b, M, and N be positive real numbers with $b \neq 1$.

$$\log_b\left(\frac{M}{N}\right) = \log_b M - \log_b N$$

The logarithm of a quotient is the difference of the logarithms.

When we use the quotient rule to write a single logarithm as the difference of two logarithms, we say that we are **expanding a logarithmic expression.** For example, we can use the quotient rule to expand $\log \frac{x}{2}$:

$$\log\left(\frac{x}{2}\right) = \log x - \log 2.$$

The logarithm of a quotient | is | the difference of the logarithms.

EXAMPLE 2 Using the Quotient Rule

Use the quotient rule to expand each logarithmic expression:

a. $\log_7\left(\frac{19}{x}\right)$ b. $\ln\left(\frac{e^3}{7}\right)$.

Solution

a. $\log_7\left(\frac{19}{x}\right) = \log_7 19 - \log_7 x$ — The logarithm of a quotient is the difference of the logarithms.

b. $\ln\left(\frac{e^3}{7}\right) = \ln e^3 - \ln 7$ — The logarithm of a quotient is the difference of the logarithms. These are natural logarithms with base e understood.

$= 3 - \ln 7$ — Because $\ln e^x = x$, then $\ln e^3 = 3$.

✓ CHECK POINT 2 Use the quotient rule to expand each logarithmic expression:

a. $\log_8\left(\frac{23}{x}\right)$ **b.** $\ln\left(\frac{e^5}{11}\right)$.

3 Use the power rule.

The Power Rule

When an exponential expression is raised to a power, we multiply exponents:

$$(b^m)^n = b^{mn}.$$

This property suggests the following property of logarithms, called the **power rule:**

The Power Rule

Let b and M be positive real numbers with $b \neq 1$, and let p be any real number.

$$\log_b M^p = p \log_b M$$

The logarithm of a number with an exponent is the product of the exponent and the logarithm of that number.

When we use the power rule to "pull the exponent to the front," we say that we are **expanding a logarithmic expression**. For example, we can use the power rule to expand $\ln x^2$:

$$\ln x^2 = 2 \ln x.$$

The logarithm of a number with an exponent — is — the product of the exponent and the logarithm of that number.

Figure 9.20 shows the graphs of $y = \ln x^2$ and $y = 2 \ln x$ in $[-5, 5, 1]$ by $[-5, 5, 1]$ viewing rectangles. Are $\ln x^2$ and $2 \ln x$ the same? The graphs illustrate that $y = \ln x^2$ and $y = 2 \ln x$ have different domains. The graphs are only the same if $x > 0$. Thus, we should write

$$\ln x^2 = 2 \ln x \quad \text{for} \quad x > 0.$$

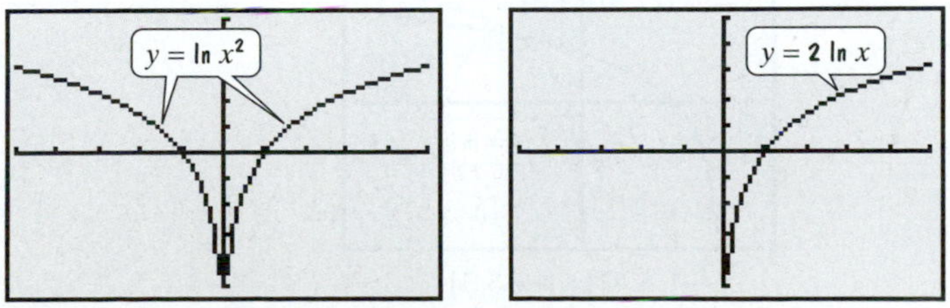

Domain: $(-\infty, 0) \cup (0, \infty)$ Domain: $(0, \infty)$

FIGURE 9.20 $\ln x^2$ and $2 \ln x$ have different domains.

When expanding a logarithmic expression, you might want to determine whether the rewriting has changed the domain of the expression. For the rest of this section, assume that all variables and variable expressions represent positive numbers.

4 Expand logarithmic expressions.

EXAMPLE 3 Using the Power Rule

Use the power rule to expand each logarithmic expression:

a. $\log_5 7^4$ **b.** $\ln \sqrt{x}$ **c.** $\log(4x)^5$.

Solution

a. $\log_5 7^4 = 4 \log_5 7$ — The logarithm of a number with an exponent is the exponent times the logarithm of that number.

b. $\ln \sqrt{x} = \ln x^{\frac{1}{2}}$ — Rewrite the radical using a rational exponent.

$= \frac{1}{2}\ln x$ — Use the power rule to bring the exponent to the front.

c. $\log(4x)^5 = 5 \log(4x)$ — We immediately apply the power rule because the entire variable expression, 4x, is raised to the 5th power.

✓ CHECK POINT 3 Use the power rule to expand each logarithmic expression:

a. $\log_6 8^9$ **b.** $\ln \sqrt[3]{x}$ **c.** $\log(x + 4)^2$.

Expanding Logarithmic Expressions

It is sometimes necessary to use more than one property of logarithms when you expand a logarithmic expression. Properties for expanding logarithmic expressions are as follows:

Properties for Expanding Logarithmic Expressions

For $M > 0$ and $N > 0$:

1. $\log_b(MN) = \log_b M + \log_b N$ — Product rule
2. $\log_b\left(\frac{M}{N}\right) = \log_b M - \log_b N$ — Quotient rule
3. $\log_b M^p = p \log_b M$ — Power rule

Study Tip

The graphs show

$y_1 = \ln(x + 3)$ and $y_2 = \ln x + \ln 3$.

The graphs are not the same:

$$\ln(x + 3) \neq \ln x + \ln 3.$$

In general,

$$\log_b(M + N) \neq \log_b M + \log_b N.$$

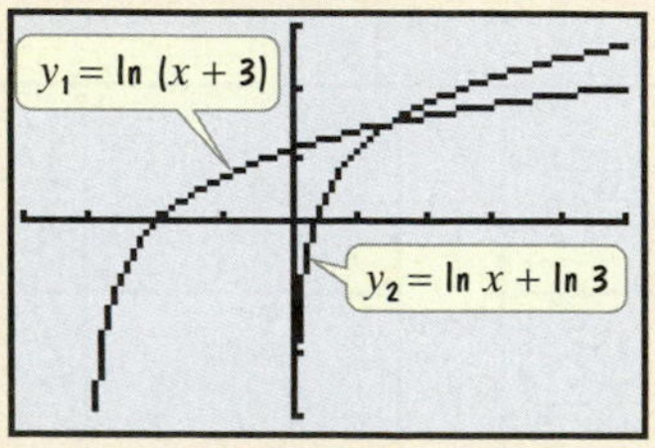

[–4, 5, 1] by [–3, 3, 1]

Try to avoid the following errors:

Incorrect!

$\log_b(M + N) = \log_b M + \log_b N$

$\log_b(M - N) = \log_b M - \log_b N$

$\log_b(M \cdot N) = \log_b M \cdot \log_b N$

$\log_b\left(\frac{M}{N}\right) = \frac{\log_b M}{\log_b N}$

$\frac{\log_b M}{\log_b N} = \log_b M - \log_b N$

$\log_b(MN^p) = p \log_b(MN)$

EXAMPLE 4 Expanding Logarithmic Expressions

Use logarithmic properties to expand each expression as much as possible:

a. $\log_b(x^2\sqrt{y})$ b. $\log_6\left(\frac{\sqrt[3]{x}}{36y^4}\right)$.

Solution We will have to use two or more of the properties for expanding logarithms in each part of this example.

a. $\log_b(x^2\sqrt{y}) = \log_b\left(x^2y^{\frac{1}{2}}\right)$ Use exponential notation.

$= \log_b x^2 + \log_b y^{\frac{1}{2}}$ Use the product rule.

$= 2\log_b x + \frac{1}{2}\log_b y$ Use the power rule.

b. $\log_6\left(\frac{\sqrt[3]{x}}{36y^4}\right) = \log_6\left(\frac{x^{\frac{1}{3}}}{36y^4}\right)$ Use exponential notation.

$= \log_6 x^{\frac{1}{3}} - \log_6(36y^4)$ Use the quotient rule.

$= \log_6 x^{\frac{1}{3}} - (\log_6 36 + \log_6 y^4)$ Use the product rule on $\log_6(36y^4)$.

$= \frac{1}{3}\log_6 x - (\log_6 36 + 4\log_6 y)$ Use the power rule.

$= \frac{1}{3}\log_6 x - \log_6 36 - 4\log_6 y$ Apply the distributive property.

$= \frac{1}{3}\log_6 x - 2 - 4\log_6 y$ $\log_6 36 = 2$ because 2 is the power to which we must raise 6 to get 36. $(6^2 = 36)$

CHECK POINT 4 Use logarithmic properties to expand each expression as much as possible:

a. $\log_b(x^4\sqrt[3]{y})$ b. $\log_5\left(\frac{\sqrt{x}}{25y^3}\right)$.

5 Condense logarithmic expressions.

Condensing Logarithmic Expressions

To **condense a logarithmic expression**, we write the sum or difference of two or more logarithmic expressions as a single logarithmic expression. We use the properties of logarithms to do so:

Properties for Condensing Logarithmic Expressions

For $M > 0$ and $N > 0$:

1. $\log_b M + \log_b N = \log_b(MN)$ Product rule
2. $\log_b M - \log_b N = \log_b\left(\frac{M}{N}\right)$ Quotient rule
3. $p\log_b M = \log_b M^p$ Power rule

Study Tip

These properties are the same as those in the box on page 684. The only difference is that we've reversed the sides in each property from the previous box.

EXAMPLE 5 Condensing Logarithmic Expressions

Write as a single logarithm:

a. $\log_4 2 + \log_4 32$ b. $\log(4x - 3) - \log x$.

Solution

a. $\log_4 2 + \log_4 32 = \log_4(2 \cdot 32)$ — Use the product rule.

$= \log_4 64$ — We now have a single logarithm. However, we can simplify.

$= 3$ — $\log_4 64 = 3$ because $4^3 = 64$.

b. $\log(4x - 3) - \log x = \log\left(\frac{4x - 3}{x}\right)$ — Use the quotient rule.

✓ CHECK POINT 5 Write as a single logarithm:

a. $\log 25 + \log 4$

b. $\log(7x + 6) - \log x$.

Coefficients of logarithms must be 1 before you can condense them using the product and quotient rules. For example, to condense

$$2 \ln x + \ln(x + 1),$$

the coefficient of the first term must be 1. We use the power rule to rewrite the coefficient as an exponent:

1. Use the power rule to make the number in front an exponent.

$$2 \ln x + \ln(x + 1) = \ln x^2 + \ln(x + 1) = \ln[x^2(x + 1)].$$

2. Use the product rule. The sum of logarithms with coefficients of 1 is the logarithm of the product.

EXAMPLE 6 Condensing Logarithmic Expressions

Write as a single logarithm:

a. $\frac{1}{2}\log x + 4\log(x - 1)$

b. $3\ln(x + 7) - \ln x$

c. $4\log_b x - 2\log_b 6 - \frac{1}{2}\log_b y$.

Solution

a. $\frac{1}{2}\log x + 4\log(x - 1)$

$= \log x^{\frac{1}{2}} + \log(x - 1)^4$ — Use the power rule so that all coefficients are 1.

$= \log\left[x^{\frac{1}{2}}(x - 1)^4\right]$ — Use the product rule. The condensed form can be expressed as $\log\left[\sqrt{x}(x - 1)^4\right]$.

b. $3\ln(x + 7) - \ln x$

$= \ln(x + 7)^3 - \ln x$ — Use the power rule so that all coefficients are 1.

$= \ln\left[\frac{(x + 7)^3}{x}\right]$ — Use the quotient rule.

c. $4\log_b x - 2\log_b 6 - \frac{1}{2}\log_b y$

$= \log_b x^4 - \log_b 6^2 - \log_b y^{\frac{1}{2}}$ — Use the power rule so that all coefficients are 1.

$= \log_b x^4 - \left(\log_b 36 + \log_b y^{\frac{1}{2}}\right)$ — Rewrite as a single subtraction.

$= \log_b x^4 - \log_b\left(36y^{\frac{1}{2}}\right)$ — Use the product rule.

$= \log_b\left(\frac{x^4}{36y^{\frac{1}{2}}}\right)$ or $\log_b\left(\frac{x^4}{36\sqrt{y}}\right)$ — Use the quotient rule.

✓ **CHECK POINT 6** Write as a single logarithm:

a. $2 \ln x + \frac{1}{3}\ln(x + 5)$

b. $2 \log(x - 3) - \log x$

c. $\frac{1}{4}\log_b x - 2 \log_b 5 - 10 \log_b y$.

6 Use the change-of-base property.

The Change-of-Base Property

We have seen that calculators give the values of both common logarithms (base 10) and natural logarithms (base e). To find a logarithm with any other base, we can use the following change-of-base property:

The Change-of-Base Property

For any logarithmic bases a and b, and any positive number M,

$$\log_b M = \frac{\log_a M}{\log_a b}.$$

The logarithm of M with base b is equal to the logarithm of M with any new base divided by the logarithm of b with that new base.

In the change-of-base property, base b is the base of the original logarithm. Base a is a new base that we introduce. Thus, the change-of-base property allows us to change from base b to *any* new base a, as long as the newly introduced base is a positive number not equal to 1.

The change-of-base property is used to write a logarithm in terms of quantities that can be evaluated with a calculator. Because calculators contain keys for common (base 10) and natural (base e) logarithms, we will frequently introduce base 10 or base e.

Change-of-Base Property	Introducing Common Logarithms	Introducing Natural Logarithms
$\log_b M = \frac{\log_a M}{\log_a b}$ a is the new introduced base.	$\log_b M = \frac{\log_{10} M}{\log_{10} b}$ 10 is the new introduced base.	$\log_b M = \frac{\log_e M}{\log_e b}$ e is the new introduced base.

Using the notations for common logarithms and natural logarithms, we have the following results:

The Change-of-Base Property: Introducing Common and Natural Logarithms

Introducing Common Logarithms

$$\log_b M = \frac{\log M}{\log b}$$

Introducing Natural Logarithms

$$\log_b M = \frac{\ln M}{\ln b}$$

EXAMPLE 7 Changing Base to Common Logarithms

Use common logarithms to evaluate $\log_5 140$.

Solution Because $\log_b M = \dfrac{\log M}{\log b}$,

$$\log_5 140 = \frac{\log 140}{\log 5} \approx 3.07.$$

Use a calculator: 140 [LOG] [÷] 5 [LOG] [=] or [LOG] 140 [÷] [LOG] 5 [ENTER]. On some calculators, parentheses are needed after 140 and 5.

This means that $\log_5 140 \approx 3.07$.

Discover for Yourself

Find a reasonable estimate of $\log_5 140$ to the nearest whole number: 5 to what power is 140? Compare your estimate to the value obtained in Example 7.

✓ CHECK POINT 7 Use common logarithms to evaluate $\log_7 2506$.

EXAMPLE 8 Changing Base to Natural Logarithms

Use natural logarithms to evaluate $\log_5 140$.

Solution Because $\log_b M = \dfrac{\ln M}{\ln b}$,

$$\log_5 140 = \frac{\ln 140}{\ln 5} \approx 3.07.$$

Use a calculator: 140 [LN] [÷] 5 [LN] [=] or [LN] 140 [÷] [LN] 5 [ENTER]. On some calculators, parentheses are needed after 140 and 5.

We have again shown that $\log_5 140 \approx 3.07$.

✓ CHECK POINT 8 Use natural logarithms to evaluate $\log_7 2506$.

Using Technology

We can use the change-of-base property to graph logarithmic functions with bases other than 10 or e on a graphing utility. For example, **Figure 9.21** shows the graphs of

$$y = \log_2 x \quad \text{and} \quad y = \log_{20} x$$

in a [0, 10, 1] by [−3, 3, 1] viewing rectangle. Because $\log_2 x = \dfrac{\ln x}{\ln 2}$ and $\log_{20} x = \dfrac{\ln x}{\ln 20}$, the functions are entered as

y_1 = [LN] x [÷] [LN] 2

and y_2 = [LN] x [÷] [LN] 20.

On some calculators, parentheses are needed after x, 2, and 20.

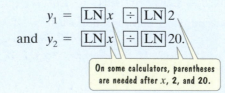

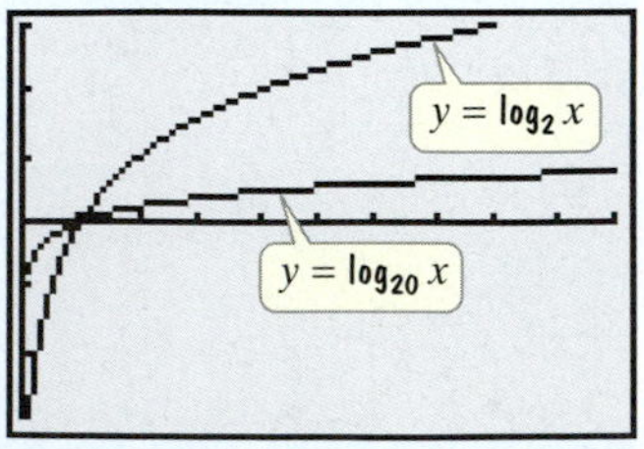

FIGURE 9.21 Using the change-of-base property to graph logarithmic functions

9.4 EXERCISE SET

MyMathLab — MathXL PRACTICE — WATCH — DOWNLOAD — READ — REVIEW

Practice Exercises

In all exercises, assume that all variables and variable expressions represent positive numbers.

In Exercises 1–36, use properties of logarithms to expand each logarithmic expression as much as possible. Where possible, evaluate logarithmic expressions without using a calculator.

1. $\log_5(7 \cdot 3)$
2. $\log_8(13 \cdot 7)$
3. $\log_7(7x)$
4. $\log_9(9x)$
5. $\log(1000x)$
6. $\log(10{,}000x)$
7. $\log_7\left(\frac{7}{x}\right)$
8. $\log_9\left(\frac{9}{x}\right)$
9. $\log\left(\frac{x}{100}\right)$
10. $\log\left(\frac{x}{1000}\right)$
11. $\log_4\left(\frac{64}{y}\right)$
12. $\log_5\left(\frac{125}{y}\right)$
13. $\ln\left(\frac{e^2}{5}\right)$
14. $\ln\left(\frac{e^4}{8}\right)$
15. $\log_b x^3$
16. $\log_b x^7$
17. $\log N^{-6}$
18. $\log M^{-8}$
19. $\ln\sqrt[5]{x}$
20. $\ln\sqrt[7]{x}$
21. $\log_b(x^2y)$
22. $\log_b(xy^3)$
23. $\log_4\left(\frac{\sqrt{x}}{64}\right)$
24. $\log_5\left(\frac{\sqrt{x}}{25}\right)$
25. $\log_6\left(\frac{36}{\sqrt{x+1}}\right)$
26. $\log_8\left(\frac{64}{\sqrt{x+1}}\right)$
27. $\log_b\left(\frac{x^2y}{z^2}\right)$
28. $\log_b\left(\frac{x^3y}{z^2}\right)$
29. $\log\sqrt{100x}$
30. $\ln\sqrt{ex}$
31. $\log\sqrt[3]{\frac{x}{y}}$
32. $\log\sqrt[5]{\frac{x}{y}}$
33. $\log_b\left(\frac{\sqrt{x}y^3}{z^3}\right)$
34. $\log_b\left(\frac{\sqrt[3]{x}y^4}{z^5}\right)$
35. $\log_5\sqrt[3]{\frac{x^2y}{25}}$
36. $\log_2\sqrt[5]{\frac{xy^4}{16}}$

In Exercises 37–60, use properties of logarithms to condense each logarithmic expression. Write the expression as a single logarithm whose coefficient is 1. Where possible, evaluate logarithmic expressions.

37. $\log 5 + \log 2$
38. $\log 250 + \log 4$
39. $\ln x + \ln 7$
40. $\ln x + \ln 3$
41. $\log_2 96 - \log_2 3$
42. $\log_3 405 - \log_3 5$
43. $\log(2x+5) - \log x$
44. $\log(3x+7) - \log x$
45. $\log x + 3\log y$
46. $\log x + 7\log y$
47. $\frac{1}{2}\ln x + \ln y$
48. $\frac{1}{3}\ln x + \ln y$
49. $2\log_b x + 3\log_b y$
50. $5\log_b x + 6\log_b y$
51. $5\ln x - 2\ln y$
52. $7\ln x - 3\ln y$
53. $3\ln x - \frac{1}{3}\ln y$
54. $2\ln x - \frac{1}{2}\ln y$
55. $4\ln(x+6) - 3\ln x$
56. $8\ln(x+9) - 4\ln x$
57. $3\ln x + 5\ln y - 6\ln z$
58. $4\ln x + 7\ln y - 3\ln z$
59. $\frac{1}{2}(\log_5 x + \log_5 y) - 2\log_5(x+1)$
60. $\frac{1}{3}(\log_4 x - \log_4 y) + 2\log_4(x+1)$

In Exercises 61–68, use common logarithms or natural logarithms and a calculator to evaluate to four decimal places.

61. $\log_5 13$
62. $\log_6 17$
63. $\log_{14} 87.5$
64. $\log_{16} 57.2$
65. $\log_{0.1} 17$
66. $\log_{0.3} 19$
67. $\log_\pi 63$
68. $\log_\pi 400$

Practice PLUS

In Exercises 69–74, let $\log_b 2 = A$ and $\log_b 3 = C$. Write each expression in terms of A and C.

69. $\log_b \frac{3}{2}$

70. $\log_b 6$

71. $\log_b 8$

72. $\log_b 81$

73. $\log_b \sqrt{\frac{2}{27}}$

74. $\log_b \sqrt{\frac{3}{16}}$

In Exercises 75–88, determine whether each equation is true or false. Where possible, show work to support your conclusion. If the statement is false, make the necessary change(s) to produce a true statement.

75. $\ln e = 0$

76. $\ln 0 = e$

77. $\log_4(2x^3) = 3\log_4(2x)$

78. $\ln(8x^3) = 3\ln(2x)$

79. $x\log 10^x = x^2$

80. $\ln(x + 1) = \ln x + \ln 1$

81. $\ln(5x) + \ln 1 = \ln(5x)$

82. $\ln x + \ln(2x) = \ln(3x)$

83. $\log(x + 3) - \log(2x) = \dfrac{\log(x + 3)}{\log(2x)}$

84. $\dfrac{\log(x + 2)}{\log(x - 1)} = \log(x + 2) - \log(x - 1)$

85. $\log_6\left(\dfrac{x - 1}{x^2 + 4}\right) = \log_6(x - 1) - \log_6(x^2 + 4)$

86. $\log_6[4(x + 1)] = \log_6 4 + \log_6(x + 1)$

87. $\log_3 7 = \dfrac{1}{\log_7 3}$

88. $e^x = \dfrac{1}{\ln x}$

In Exercises 89–92,

a. *Evaluate the expression in part (a) without using a calculator.*

b. *Use your result from part (a) to write the expression in part (b) as a single logarithm whose coefficient is 1.*

89. a. $\log_3 9$

b. $\log_3 x + 4\log_3 y - 2$

90. a. $\log_2 16$

b. $\log_2 x + 5\log_2 y - 4$

91. a. $\log_{25} 5$

b. $\log_{25} x + \log_{25}(x^2 - 1) - \log_{25}(x + 1) - \dfrac{1}{2}$

92. a. $\log_{36} 6$

b. $\log_{36} x + \log_{36}(x^2 - 4) - \log_{36}(x + 2) - \dfrac{1}{2}$

Application Exercises

93. The loudness level of a sound can be expressed by comparing the sound's intensity to the intensity of a sound barely audible to the human ear. The formula

$$D = 10(\log I - \log I_0)$$

describes the loudness level of a sound, D, in decibels, where I is the intensity of the sound, in watts per meter^2, and I_0 is the intensity of a sound barely audible to the human ear.

a. Express the formula so that the expression in parentheses is written as a single logarithm.

b. Use the form of the formula from part (a) to answer this question. If a sound has an intensity 100 times the intensity of a softer sound, how much larger on the decibel scale is the loudness level of the more intense sound?

94. The formula

$$t = \frac{1}{c}[\ln A - \ln(A - N)]$$

describes the time, t, in weeks, that it takes to achieve mastery of a portion of a task, where A is the maximum learning possible, N is the portion of the learning that is to be achieved, and c is a constant used to measure an individual's learning style.

a. Express the formula so that the expression in brackets is written as a single logarithm.

b. The formula is also used to determine how long it will take chimpanzees and apes to master a task. For example, a typical chimpanzee learning sign language can master a maximum of 65 signs. Use the form of the formula from part (a) to answer this question. How many weeks will it take a chimpanzee to master 30 signs if c for that chimp is 0.03?

Writing in Mathematics

95. Describe the product rule for logarithms and give an example.

96. Describe the quotient rule for logarithms and give an example.

97. Describe the power rule for logarithms and give an example.

98. Without showing the details, explain how to condense $\ln x - 2\ln(x + 1)$.

99. Describe the change-of-base property and give an example.

100. Explain how to use your calculator to find $\log_{14} 283$.

101. You overhear a student talking about a property of logarithms in which division becomes subtraction. Explain what the student means by this.

102. Find ln 2 using a calculator. Then calculate each of the following: $1 - \frac{1}{2}$; $1 - \frac{1}{2} + \frac{1}{3}$; $1 - \frac{1}{2} + \frac{1}{3} - \frac{1}{4}$; $1 - \frac{1}{2} + \frac{1}{3} - \frac{1}{4} + \frac{1}{5}$; .... Describe what you observe.

Technology Exercises

103. **a.** Use a graphing utility (and the change-of-base property) to graph $y = \log_3 x$.

b. Graph $y = 2 + \log_3 x$, $y = \log_3(x + 2)$, and $y = -\log_3 x$ in the same viewing rectangle as $y = \log_3 x$. Then describe the change or changes that need to be made to the graph of $y = \log_3 x$ to obtain each of these three graphs.

104. Graph $y = \log x$, $y = \log(10x)$, and $y = \log(0.1x)$ in the same viewing rectangle. Describe the relationship among the three graphs. What logarithmic property accounts for this relationship?

105. Use a graphing utility and the change-of-base property to graph $y = \log_3 x$, $y = \log_{25} x$, and $y = \log_{100} x$ in the same viewing rectangle.

a. Which graph is on the top in the interval $(0, 1)$? Which is on the bottom?

b. Which graph is on the top in the interval $(1, \infty)$? Which is on the bottom?

c. Generalize by writing a statement about which graph is on top, which is on the bottom, and in which intervals, using $y = \log_b x$ where $b > 1$.

Disprove each statement in Exercises 106–110 by

a. *letting y equal a positive constant of your choice, and*

b. *using a graphing utility to graph the function on each side of the equal sign. The two functions should have different graphs, showing that the equation is not true in general.*

106. $\log(x + y) = \log x + \log y$

107. $\log\frac{x}{y} = \frac{\log x}{\log y}$

108. $\ln(x - y) = \ln x - \ln y$

109. $\ln(xy) = (\ln x)(\ln y)$

110. $\frac{\ln x}{\ln y} = \ln x - \ln y$

Critical Thinking Exercises

Make Sense? *In Exercises 111–114, determine whether each statement "makes sense" or "does not make sense" and explain your reasoning.*

111. Because I cannot simplify the expression $b^m + b^n$ by adding exponents, there is no property for the logarithm of a sum.

112. Because logarithms are exponents, the product, quotient, and power rules remind me of properties for operations with exponents.

113. I can use any positive number other than 1 in the change-of-base property, but the only practical bases are 10 and e because my calculator gives logarithms for these two bases.

114. I expanded $\log_4 \sqrt{\frac{x}{y}}$ by writing the radical using a rational exponent and then applying the quotient rule, obtaining $\frac{1}{2}\log_4 x - \log_4 y$.

In Exercises 115–118, determine whether each statement is true or false. If the statement is false, make the necessary change(s) to produce a true statement.

115. $\ln\sqrt{2} = \frac{\ln 2}{2}$

116. $\frac{\log_7 49}{\log_7 7} = \log_7 49 - \log_7 7$

117. $\log_b(x^3 + y^3) = 3\log_b x + 3\log_b y$

118. $\log_b(xy)^5 = (\log_b x + \log_b y)^5$

119. Use the change-of-base property to prove that

$$\log e = \frac{1}{\ln 10}.$$

120. If $\log 3 = A$ and $\log 7 = B$, find $\log_7 9$ in terms of A and B.

121. Write as a single term that does not contain a logarithm:

$$e^{\ln 8x^5 - \ln 2x^2}.$$

Review Exercises

122. Graph: $5x - 2y > 10$. (Section 4.4, Example 1)

123. Solve: $x - 2(3x - 2) > 2x - 3$. (Section 4.1, Example 3)

124. Divide and simplify: $\frac{\sqrt[3]{40x^2y^6}}{\sqrt[3]{5xy}}$. (Section 7.4, Example 5)

Preview Exercises

Exercises 125–127 will help you prepare for the material covered in the next section.

125. Simplify: $16^{\frac{3}{2}}$.

126. Evaluate $3\ln(2x)$ if $x = \frac{e^4}{2}$.

127. Solve: $\frac{x + 2}{4x + 3} = \frac{1}{x}$.

MID-CHAPTER CHECK POINT Section 9.1–Section 9.4

What You Know: We evaluated and graphed exponential functions $[f(x) = b^x, b > 0$ and $b \neq 1]$, including the natural exponential function $[f(x) = e^x, e \approx 2.718]$. We studied composite and inverse functions, noting that a function has an inverse that is a function if there is no horizontal line that intersects the function's graph more than once. The exponential function passes this horizontal line test and we called the inverse of the exponential function with base b the logarithmic function with base b. We learned that $y = \log_b x$ is equivalent to $b^y = x$. We evaluated and graphed logarithmic functions, including the common logarithmic function $[f(x) = \log_{10} x$ or $f(x) = \log x]$ and the natural logarithmic function $[f(x) = \log_e x$ or $f(x) = \ln x]$. Finally, we used properties of logarithms to expand and condense logarithmic expressions.

In Exercises 1–3, find $(f \circ g)(x)$ and $(g \circ f)(x)$. Are f and g inverses of each other?

1. $f(x) = 3x + 2, g(x) = 4x - 5$
2. $f(x) = \sqrt[3]{7x + 5}, g(x) = \dfrac{x^3 - 5}{7}$
3. $f(x) = \log_5 x, g(x) = 5^x$
4. Let $f(x) = \dfrac{x - 1}{x}$ and $g(x) = \sqrt{x + 3}$.
   a. Find $(f \circ g)(6)$.
   b. Find $(g \circ f)(-1)$.
   c. Find $(f \circ f)(5)$.
   d. Find $(g \circ g)(-2)$.

In Exercises 5–7, find f^{-1}.

5. $f(x) = \dfrac{2x + 5}{4}$
6. $f(x) = 10x^3 - 7$
7. $f = \{(2, 5), (10, -7), (11, -10)\}$

In Exercises 8–10, which graphs represent functions? Among these graphs, which have inverse functions?

8.

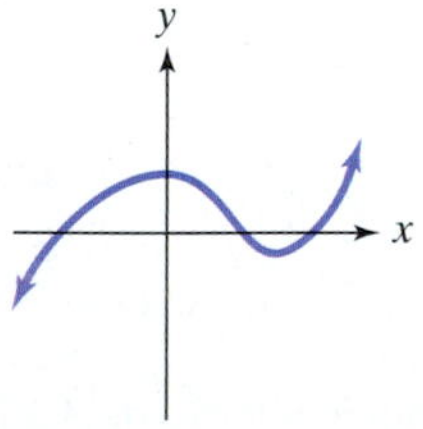

9.

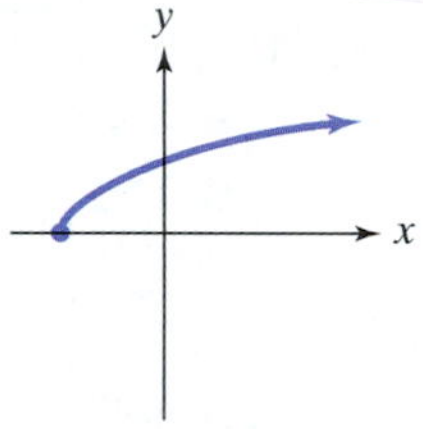

10.

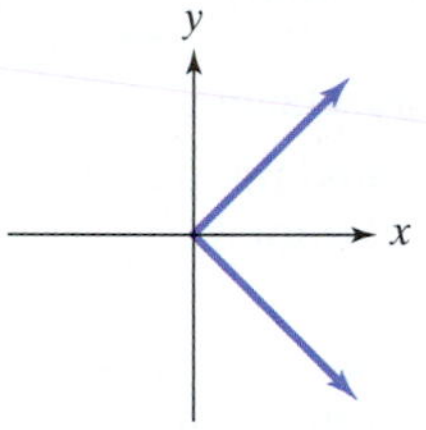

In Exercises 11–14, graph the given function. Give each function's domain and range.

11. $f(x) = 2^x - 3$
12. $f(x) = \left(\frac{1}{3}\right)^x$
13. $f(x) = \log_2 x$
14. $f(x) = \log_2 x + 1$

In Exercises 15–18, find the domain of each function.

15. $f(x) = \log_3(x + 6)$
16. $g(x) = \log_3 x + 6$
17. $h(x) = \log_3(x + 6)^2$
18. $f(x) = 3^{x+6}$

In Exercises 19–29, evaluate each expression without using a calculator. If evaluation is not possible, state the reason.

19. $\log_2 8 + \log_5 25$
20. $\log_3 \frac{1}{9}$
21. $\log_{100} 10$
22. $\log \sqrt[3]{10}$
23. $\log_2(\log_3 81)$
24. $\log_3\left(\log_2 \frac{1}{8}\right)$
25. $6^{\log_6 5}$
26. $\ln e^{\sqrt{7}}$
27. $10^{\log 13}$
28. $\log_{100} 0.1$
29. $\log_\pi \pi^{\sqrt{\pi}}$

In Exercises 30–31, expand and evaluate numerical terms.

30. $\log\left(\dfrac{\sqrt{xy}}{1000}\right)$
31. $\ln(e^{19}x^{20})$

In Exercises 32–34, write each expression as a single logarithm.

32. $8 \log_7 x - \frac{1}{3}\log_7 y$

33. $7 \log_5 x + 2 \log_5 x$

34. $\frac{1}{2}\ln x - 3 \ln y - \ln(z - 2)$

35. Use the formulas

$$A = P\left(1 + \frac{r}{n}\right)^{nt} \quad \text{and} \quad A = Pe^{rt}$$

to solve this exercise. You plan to invest \$8000 for 3 years at an annual rate of 8%. How much more is the return if the interest is compounded continuously than monthly? Round to the nearest dollar.

SECTION 9.5

Exponential and Logarithmic Equations

Objectives

1. Use like bases to solve exponential equations.
2. Use logarithms to solve exponential equations.
3. Use exponential form to solve logarithmic equations.
4. Use the one-to-one property of logarithms to solve logarithmic equations.
5. Solve applied problems involving exponential and logarithmic equations.

At age 20, you inherit \$30,000. You'd like to put aside \$25,000 and eventually have over half a million dollars for early retirement. Is this possible? In this section, you will see how techniques for solving equations with variable exponents provide an answer to this question.

1 Use like bases to solve exponential equations.

Exponential Equations

An **exponential equation** is an equation containing a variable in an exponent. Examples of exponential equations include

$$2^{3x-8} = 16, \quad 4^x = 15, \quad \text{and} \quad 40e^{0.6x} = 240.$$

Some exponential equations can be solved by expressing each side of the equation as a power of the same base. All exponential functions are one-to-one—that is, no two different ordered pairs have the same second component. Thus, if b is a positive number other than 1 and $b^M = b^N$, then $M = N$.

Solving Exponential Equations by Expressing Each Side as a Power of the Same Base

If $b^M = b^N$, then $M = N$.

Express each side as a power of the same base. (points to b^M and b^N)

Set the exponents equal to each other. (points to M and N)

1. Rewrite the equation in the form $b^M = b^N$.
2. Set $M = N$.
3. Solve for the variable.

Using Technology

Graphic Connections

The graphs of

$$y_1 = 2^{3x-8}$$
$$\text{and}\quad y_2 = 16$$

have an intersection point whose x-coordinate is 4. This verifies that $\{4\}$ is the solution set of $2^{3x-8} = 16$.

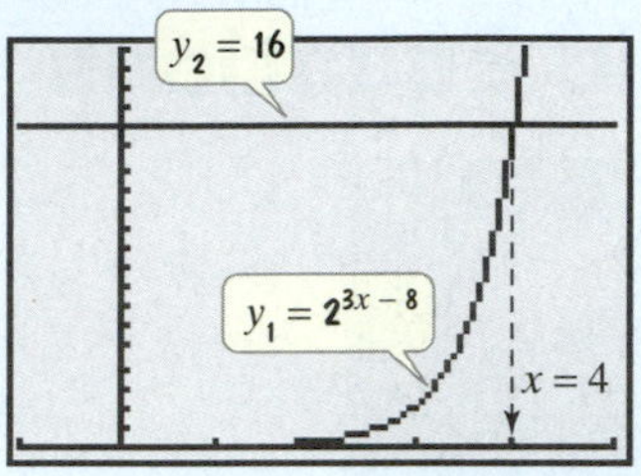

[−1, 5, 1] by [0, 20, 1]

EXAMPLE 1 Solving Exponential Equations

Solve: **a.** $2^{3x-8} = 16$ **b.** $16^x = 64$.

Solution In each equation, express both sides as a power of the same base. Then set the exponents equal to each other.

a. Because 16 is 2^4, we express each side of $2^{3x-8} = 16$ in terms of base 2.

$2^{3x-8} = 16$	This is the given equation.
$2^{3x-8} = 2^4$	Write each side as a power of the same base.
$3x - 8 = 4$	If $b^M = b^N$, $b > 0$ and $b \neq 1$, then $M = N$.
$3x = 12$	Add 8 to both sides.
$x = 4$	Divide both sides by 3.

Check 4:

$$2^{3x-8} = 16$$
$$2^{3\cdot4-8} \stackrel{?}{=} 16$$
$$2^4 \stackrel{?}{=} 16$$
$$16 = 16, \quad \text{true}$$

The solution is 4 and the solution set is $\{4\}$.

b. Because $16 = 4^2$ and $64 = 4^3$, we express each side of $16^x = 64$ in terms of base 4.

$16^x = 64$	This is the given equation.
$(4^2)^x = 4^3$	Write each side as a power of the same base.
$4^{2x} = 4^3$	When an exponential expression is raised to a power, multiply exponents.
$2x = 3$	If two powers of the same base are equal, then the exponents are equal.
$x = \frac{3}{2}$	Divide both sides by 2.

Check $\frac{3}{2}$:

$$16^x = 64$$
$$16^{\frac{3}{2}} \stackrel{?}{=} 64$$
$$(\sqrt{16})^3 \stackrel{?}{=} 64 \qquad b^{\frac{m}{n}} = (\sqrt[n]{b})^m$$
$$4^3 \stackrel{?}{=} 64$$
$$64 = 64, \quad \text{true}$$

The solution is $\frac{3}{2}$ and the solution set is $\{\frac{3}{2}\}$.

Discover for Yourself

The equation $16^x = 64$ can also be solved by writing each side in terms of base 2. Do this. Which solution method do you prefer?

✓ CHECK POINT 1 Solve:

a. $5^{3x-6} = 125$ **b.** $4^x = 32$.

---

2 Use logarithms to solve exponential equations.

Most exponential equations cannot be rewritten so that each side has the same base. Here are two examples:

$$4^x = 15 \qquad\qquad 10^x = 120{,}000.$$

We cannot rewrite both sides in terms of base 2 or base 4.

We cannot rewrite both sides in terms of base 10.

Logarithms are extremely useful in solving these equations. The solution begins with isolating the exponential expression. Notice that the exponential expression is already isolated in both $4^x = 15$ and $10^x = 120{,}000$. Then we take the logarithm on both sides. Why can we do this? All logarithmic relations are functions. Thus, if M and N are positive real numbers and $M = N$, then $\log_b M = \log_b N$.

The base that is used when taking the logarithm on both sides of an equation can be any base at all. If the exponential equation involves base 10, as in $10^x = 120{,}000$, we'll take the common logarithm on both sides. If the exponential equation involves any other base, as in $4^x = 15$, we'll take the natural logarithm on both sides.

Using Logarithms to Solve Exponential Equations

1. Isolate the exponential expression.
2. Take the natural logarithm on both sides of the equation for bases other than 10. Take the common logarithm on both sides of the equation for base 10.
3. Simplify using one of the following properties:

$$\ln b^x = x \ln b \quad \text{or} \quad \ln e^x = x \quad \text{or} \quad \log 10^x = x.$$

4. Solve for the variable.

EXAMPLE 2 Solving Exponential Equations

Solve: **a.** $4^x = 15$ **b.** $10^x = 120{,}000$.

Solution We will use the natural logarithmic function to solve $4^x = 15$ and the common logarithmic function to solve $10^x = 120{,}000$.

a. Because the exponential expression, 4^x, is already isolated on the left side of $4^x = 15$, we begin by taking the natural logarithm on both sides of the equation.

$4^x = 15$ This is the given equation.

$\ln 4^x = \ln 15$ Take the natural logarithm on both sides.

$x \ln 4 = \ln 15$ Use the power rule and bring the variable exponent to the front: $\ln b^x = x \ln b$.

$x = \dfrac{\ln 15}{\ln 4}$ Solve for x by dividing both sides by $\ln 4$.

We now have an exact value for x. We use the exact value for x in the equation's solution set. Thus, the equation's solution is $\frac{\ln 15}{\ln 4}$ and the solution set is $\left\{\frac{\ln 15}{\ln 4}\right\}$. We can obtain a decimal approximation by using a calculator: $x \approx 1.95$. Because $4^2 = 16$, it seems reasonable that the solution to $4^x = 15$ is approximately 1.95.

b. Because the exponential expression, 10^x, is already isolated on the left side of $10^x = 120{,}000$, we begin by taking the common logarithm on both sides of the equation.

$10^x = 120{,}000$ This is the given equation.

$\log 10^x = \log 120{,}000$ Take the common logarithm on both sides.

$x = \log 120{,}000$ Use the inverse property $\log 10^x = x$ on the left.

The equation's solution is $\log 120{,}000$ and the solution set is $\{\log 120{,}000\}$. We can obtain a decimal approximation by using a calculator: $x \approx 5.08$. Because $10^5 = 100{,}000$, it seems reasonable that the solution to $10^x = 120{,}000$ is approximately 5.08.

Discover for Yourself

Keep in mind that the base used when taking the logarithm on both sides of an equation can be any base at all. Solve $4^x = 15$ by taking the common logarithm on both sides. Solve again, this time taking the logarithm with base 4 on both sides. Use the change-of-base property to show that the solutions are the same as the one obtained in Example 2(a).

CHECK POINT 2 Solve: **a.** $5^x = 134$ **b.** $10^x = 8000$.

Find each solution set and then use a calculator to obtain a decimal approximation to two decimal places for the solution.

EXAMPLE 3 Solving an Exponential Equation

Solve: $40e^{0.6x} - 3 = 237$.

Solution We begin by adding 3 to both sides and dividing both sides by 40 to isolate the exponential expression, $e^{0.6x}$. Then we take the natural logarithm on both sides of the equation.

$$40e^{0.6x} - 3 = 237 \quad \text{This is the given equation.}$$
$$40e^{0.6x} = 240 \quad \text{Add 3 to both sides.}$$
$$e^{0.6x} = 6 \quad \text{Isolate the exponential factor by dividing both sides by 40.}$$
$$\ln e^{0.6x} = \ln 6 \quad \text{Take the natural logarithm on both sides.}$$
$$0.6x = \ln 6 \quad \text{Use the inverse property } \ln e^x = x \text{ on the left.}$$
$$x = \frac{\ln 6}{0.6} \approx 2.99 \quad \text{Divide both sides by 0.6 and solve for } x.$$

Thus, the solution of the equation is $\frac{\ln 6}{0.6} \approx 2.99$. Try checking this approximate solution in the original equation to verify that $\left\{\frac{\ln 6}{0.6}\right\}$ is the solution set.

CHECK POINT 3 Solve: $7e^{2x} - 5 = 58$. Find the solution set and then use a calculator to obtain a decimal approximation to two decimal places for the solution.

---

3 Use exponential form to solve logarithmic equations.

Logarithmic Equations

A **logarithmic equation** is an equation containing a variable in a logarithmic expression. Examples of logarithmic equations include

$$\log_4(x + 3) = 2 \quad \text{and} \quad \ln(x + 2) - \ln(4x + 3) = \ln\left(\frac{1}{x}\right).$$

Some logarithmic equations can be expressed in the form $\log_b M = c$. We can solve such equations by rewriting them in exponential form.

Using Exponential Form to Solve Logarithmic Equations

1. Express the equation in the form $\log_b M = c$.
2. Use the definition of a logarithm to rewrite the equation in exponential form:

$$\log_b M = c \quad \text{means} \quad b^c = M.$$

Logarithms are exponents.

3. Solve for the variable.
4. Check proposed solutions in the original equation. Include in the solution set only values for which $M > 0$.

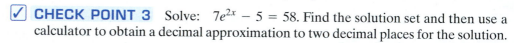

EXAMPLE 4 Solving Logarithmic Equations

Solve: **a.** $\log_4(x+3) = 2$ **b.** $3\ln(2x) = 12$.

Solution The form $\log_b M = c$ involves a single logarithm whose coefficient is 1 on one side and a constant on the other side. Equation (a) is already in this form. We will need to divide both sides of equation (b) by 3 to obtain this form.

a. $\log_4(x+3) = 2$ This is the given equation.

$4^2 = x + 3$ Rewrite in exponential form: $\log_b M = c$ means $b^c = M$.

$16 = x + 3$ Square 4.

$13 = x$ Subtract 3 from both sides.

Check 13:

$\log_4(x+3) = 2$ This is the given logarithmic equation.

$\log_4(13+3) \stackrel{?}{=} 2$ Substitute 13 for x.

$\log_4 16 \stackrel{?}{=} 2$

$2 = 2$, true $\log_4 16 = 2$ because $4^2 = 16$.

This true statement indicates that the solution is 13 and the solution set is $\{13\}$.

b. $3\ln(2x) = 12$ This is the given equation.

$\ln(2x) = 4$ Divide both sides by 3.

$\log_e(2x) = 4$ Rewrite the natural logarithm showing base e. This step is optional.

$e^4 = 2x$ Rewrite in exponential form: $\log_b M = c$ means $b^c = M$.

$\frac{e^4}{2} = x$ Divide both sides by 2.

Check $\frac{e^4}{2}$:

$3\ln(2x) = 12$ This is the given logarithmic equation.

$3\ln\left[2\left(\frac{e^4}{2}\right)\right] \stackrel{?}{=} 12$ Substitute $\frac{e^4}{2}$ for x.

$3\ln e^4 \stackrel{?}{=} 12$ Simplify: $\frac{2}{1}\cdot\frac{e^4}{2} = e^4$.

$3\cdot 4 \stackrel{?}{=} 12$ Because $\ln e^x = x$, we conclude $\ln e^4 = 4$.

$12 = 12$, true

This true statement indicates that the solution is $\frac{e^4}{2}$ and the solution set is $\left\{\frac{e^4}{2}\right\}$.

Using Technology

Graphic Connections

The graphs of

$y_1 = \log_4(x+3)$ and $y_2 = 2$

have an intersection point whose x-coordinate is 13. This verifies that $\{13\}$ is the solution set for $\log_4(x+3) = 2$.

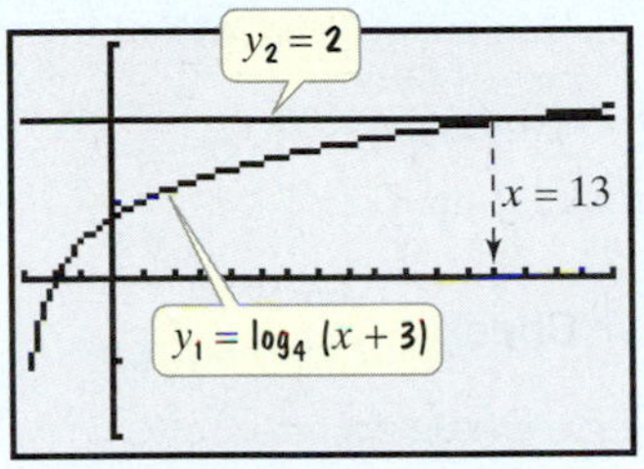

[−3, 17, 1] by [−2, 3, 1]

CHECK POINT 4 Solve:

a. $\log_2(x-4) = 3$ **b.** $4\ln(3x) = 8$.

Logarithmic expressions are defined only for logarithms of positive real numbers. **Always check proposed solutions of a logarithmic equation in the original equation. Exclude from the solution set any proposed solution that produces the logarithm of a negative number or the logarithm of 0.**

To rewrite the logarithmic equation $\log_b M = c$ in the equivalent exponential form $b^c = M$, we need a single logarithm whose coefficient is one. It is sometimes necessary to use properties of logarithms to condense logarithms into a single logarithm. In the next example, we use the product rule for logarithms to obtain a single logarithmic expression on the left side.

EXAMPLE 5 Solving a Logarithmic Equation

Solve: $\log_2 x + \log_2(x-7) = 3$.

Solution

$\log_2 x + \log_2(x-7) = 3$	This is the given equation.
$\log_2[x(x-7)] = 3$	Use the product rule to obtain a single logarithm: $\log_b M + \log_b N = \log_b(MN)$.
$2^3 = x(x-7)$	Rewrite in exponential form.
$8 = x^2 - 7x$	Evaluate 2^3 on the left and apply the distributive property on the right.
$0 = x^2 - 7x - 8$	Set the equation equal to 0.
$0 = (x-8)(x+1)$	Factor.
$x - 8 = 0$ or $x + 1 = 0$	Set each factor equal to 0.
$x = 8$ $\quad x = -1$	Solve for x.

Check 8:

$$\log_2 x + \log_2(x-7) = 3$$
$$\log_2 8 + \log_2(8-7) \stackrel{?}{=} 3$$
$$\log_2 8 + \log_2 1 \stackrel{?}{=} 3$$
$$3 + 0 \stackrel{?}{=} 3$$
$$3 = 3, \quad \text{true}$$

Check −1:

$$\log_2 x + \log_2(x-7) = 3$$
$$\log_2(-1) + \log_2(-1-7) \stackrel{?}{=} 3$$

The number −1 does not check. It produces logarithms of negative numbers. Neither −1 nor −8 are in the domain of a logarithmic function.

The solution is 8 and the solution set is $\{8\}$.

✓ CHECK POINT 5 Solve: $\log x + \log(x-3) = 1$.

4 Use the one-to-one property of logarithms to solve logarithmic equations.

Some logarithmic equations can be expressed in the form $\log_b M = \log_b N$. Because all logarithmic functions are one-to-one, we can conclude that $M = N$.

Using the One-to-One Property of Logarithms to Solve Logarithmic Equations

1. Express the equation in the form $\log_b M = \log_b N$. This form involves a single logarithm whose coefficient is 1 on each side of the equation.
2. Use the one-to-one property to rewrite the equation without logarithms: If $\log_b M = \log_b N$, then $M = N$.
3. Solve for the variable.
4. Check proposed solutions in the original equation. Include in the solution set only values for which $M > 0$ and $N > 0$.

EXAMPLE 6 Solving a Logarithmic Equation

Solve: $\ln(x+2) - \ln(4x+3) = \ln\left(\frac{1}{x}\right)$.

Solution In order to apply the one-to-one property of logarithms, we need a single logarithm whose coefficient is 1 on each side of the equation. The right side is already in this form. We can obtain a single logarithm on the left side by applying the quotient rule.

$$\ln(x+2) - \ln(4x+3) = \ln\left(\frac{1}{x}\right)$$ This is the given equation.

$$\ln\left(\frac{x+2}{4x+3}\right) = \ln\left(\frac{1}{x}\right)$$ Use the quotient rule to obtain a single logarithm on the left side: $\log_b M - \log_b N = \log_b\left(\frac{M}{N}\right)$.

$$\frac{x+2}{4x+3} = \frac{1}{x}$$ Use the one-to-one property: If $\log_b M = \log_b N$, then $M = N$.

$$x(4x+3)\left(\frac{x+2}{4x+3}\right) = x(4x+3)\left(\frac{1}{x}\right)$$ Multiply both sides by $x(4x+3)$, the LCD.

$$x(x+2) = 4x+3$$ Simplify.

$$x^2 + 2x = 4x + 3$$ Apply the distributive property.

$$x^2 - 2x - 3 = 0$$ Subtract $4x + 3$ from both sides and set the equation equal to 0.

$$(x-3)(x+1) = 0$$ Factor.

$$x - 3 = 0 \quad \text{or} \quad x + 1 = 0$$ Set each factor equal to 0.

$$x = 3 \qquad\qquad x = -1$$ Solve for x.

Substituting 3 for x into the original equation produces the true statement $\ln\left(\frac{1}{3}\right) = \ln\left(\frac{1}{3}\right)$. However, substituting -1 produces logarithms of negative numbers. Thus, -1 is not a solution. The solution is 3 and the solution set is $\{3\}$.

Using Technology

Numeric Connections

A graphing utility's TABLE feature can be used to verify that $\{3\}$ is the solution set of

$$\ln(x+2) - \ln(4x+3) = \ln\left(\frac{1}{x}\right).$$

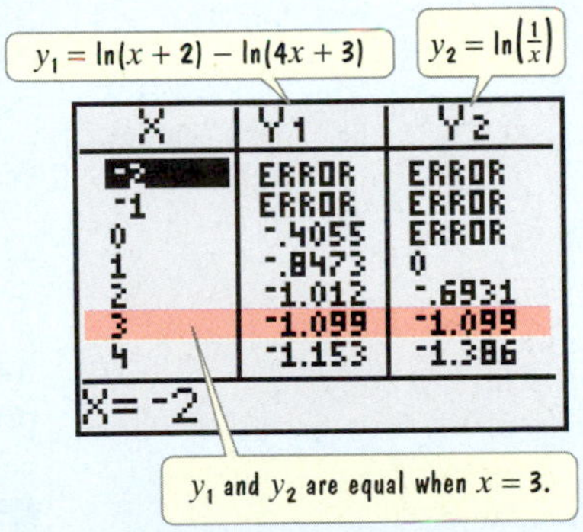

X	Y1	Y2
-2	ERROR	ERROR
-1	ERROR	ERROR
0	-.4055	ERROR
1	-.8473	0
2	-1.012	-.6931
3	-1.099	-1.099
4	-1.153	-1.386

✓ **CHECK POINT 6** Solve: $\ln(x-3) = \ln(7x-23) - \ln(x+1)$.

5 Solve applied problems involving exponential and logarithmic equations.

Applications

Our first applied example provides a mathematical perspective on the old slogan "Alcohol and driving don't mix." In California, where 38% of fatal traffic crashes involve drinking drivers, it is illegal to drive with a blood alcohol concentration of 0.08 or higher. At these levels, drivers may be arrested and charged with driving under the influence.

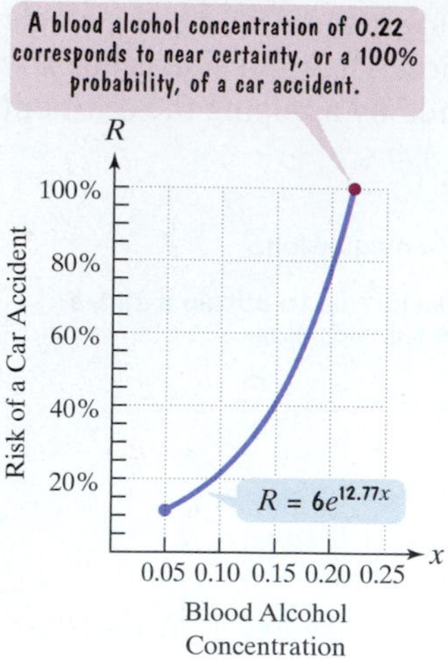

FIGURE 9.22

EXAMPLE 7 Alcohol and Risk of a Car Accident

Medical research indicates that the risk of having a car accident increases exponentially as the concentration of alcohol in the blood increases. The risk is modeled by

$$R = 6e^{12.77x},$$

where x is the blood alcohol concentration and R, given as a percent, is the risk of having a car accident. What blood alcohol concentration corresponds to a 17% risk of a car accident? How is this shown on the graph of R in **Figure 9.22**?

Solution For a risk of 17%, we let $R = 17$ in the equation and solve for x, the blood alcohol concentration.

$$R = 6e^{12.77x}$$ This is the given equation.

$$6e^{12.77x} = 17$$ Substitute 17 for R and (optional) reverse the two sides of the equation.

$$e^{12.77x} = \frac{17}{6}$$ Isolate the exponential factor by dividing both sides by 6.

$$\ln e^{12.77x} = \ln\left(\frac{17}{6}\right)$$ Take the natural logarithm on both sides.

$$12.77x = \ln\left(\frac{17}{6}\right)$$ Use the inverse property $\ln e^x = x$ on the left side.

$$x = \frac{\ln\left(\frac{17}{6}\right)}{12.77} \approx 0.08$$ Divide both sides by 12.77.

For a blood alcohol concentration of 0.08, the risk of a car accident is 17%. This is shown on the graph of R in **Figure 9.22** by the point (0.08, 17) that lies on the blue curve. Take a moment to locate this point on the curve. In many states, it is illegal to drive with a blood alcohol concentration of 0.08.

CHECK POINT 7 Use the formula in Example 7 to solve this problem. What blood alcohol concentration corresponds to a 7% risk of a car accident? (In many states, drivers under the age of 21 can lose their licenses for driving at this level.)

Suppose that you inherit \$30,000 at age 20. Is it possible to invest \$25,000 and have over half a million dollars for early retirement? Our next example illustrates the power of compound interest.

EXAMPLE 8 Revisiting the Formula for Compound Interest

The formula

$$A = P\left(1 + \frac{r}{n}\right)^{nt}$$

describes the accumulated value, A, of a sum of money, P, the principal, after t years at annual percentage rate r (in decimal form) compounded n times a year. How long will it take \$25,000 to grow to \$500,000 at 9% annual interest compounded monthly?

Solution

$$A = P\left(1 + \frac{r}{n}\right)^{nt}$$ This is the given formula.

$$500{,}000 = 25{,}000\left(1 + \frac{0.09}{12}\right)^{12t}$$ A(the desired accumulated value) = 500,000, P(the principal) = 25,000, r(the interest rate) = 9% = 0.09, and $n = 12$ (monthly compounding).

Blitzer Bonus

Playing Doubles: Interest Rates and Doubling Time

One way to calculate what your savings will be worth at some point in the future is to consider doubling time. The following table shows how long it takes for your money to double at different annual interest rates subject to continuous compounding.

Annual Interest Rate	Years to Double
5%	13.9 years
7%	9.9 years
9%	7.7 years
11%	6.3 years

Of course, the first problem is collecting some money to invest. The second problem is finding a reasonably safe investment with a return of 9% or more.

Our goal is to solve the equation for t. Let's reverse the two sides of the equation and then simplify within parentheses.

$$25{,}000\left(1+\frac{0.09}{12}\right)^{12t} = 500{,}000$$ Reverse the two sides of the previous equation.

$$25{,}000(1+0.0075)^{12t} = 500{,}000$$ Divide within parentheses: $\frac{0.09}{12} = 0.0075$.

$$25{,}000(1.0075)^{12t} = 500{,}000$$ Add within parentheses.

$$(1.0075)^{12t} = 20$$ Divide both sides by 25,000.

$$\ln(1.0075)^{12t} = \ln 20$$ Take the natural logarithm on both sides.

$$12t\ln(1.0075) = \ln 20$$ Use the power rule to bring the exponent to the front: $\ln b^x = x\ln b$.

$$t = \frac{\ln 20}{12\ln 1.0075}$$ Solve for t, dividing both sides by $12\ln 1.0075$.

$$\approx 33.4$$ Use a calculator.

After approximately 33.4 years, the \$25,000 will grow to an accumulated value of \$500,000. If you set aside the money at age 20, you can begin enjoying a life of leisure at about age 53.

✓ **CHECK POINT 8** How long, to the nearest tenth of a year, will it take \$1000 to grow to \$3600 at 8% annual interest compounded quarterly?

EXAMPLE 9 The Decline in the Percentage of Young Americans

The bar graph in **Figure 9.23** shows the number of children under 18 as a percentage of the total U.S. population.

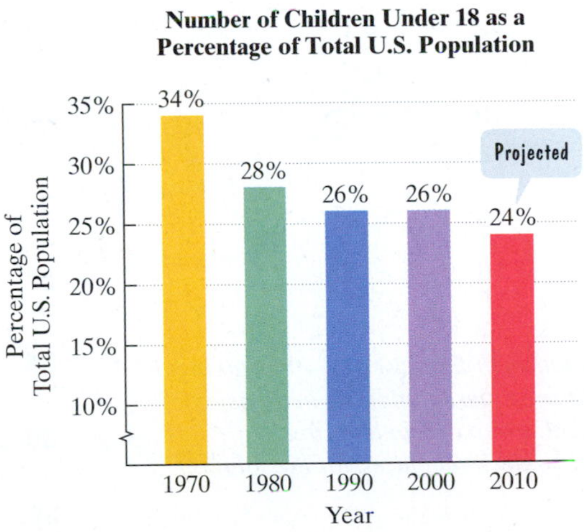

FIGURE 9.23

Source: www.childstats.gov

The function

$$f(x) = 34 - 2.6\ln x$$

models the number of children under 18 as a percentage of the total U.S. population, $f(x)$, x years after 1969. According to the model, when will children under 18 decline to 23% of the total U.S. population? Round to the nearest year.

Solution To find when children under 18 will make up 23% of the U.S. population, we substitute 23 for $f(x)$ and solve for x, the number of years after 1969.

$$f(x) = 34 - 2.6 \ln x \quad \text{This is the given function.}$$

$$23 = 34 - 2.6 \ln x \quad \text{Substitute 23 for } f(x).$$

Our goal is to isolate $\ln x$ and then rewrite the equation in exponential form.

$$-11 = -2.6 \ln x \quad \text{Subtract 34 from both sides.}$$

$$\frac{11}{2.6} = \ln x \quad \text{Divide both sides by } -2.6.$$

$$\frac{11}{2.6} = \log_e x \quad \text{Rewrite the natural logarithm showing base } e. \text{ This step is optional.}$$

$$e^{\frac{11}{2.6}} = x \quad \text{Rewrite in exponential form.}$$

$$69 \approx x \quad \text{Use a calculator.}$$

Approximately 69 years after 1969, in the year 2038, children under 18 will make up 23% of the total U.S. population.

CHECK POINT 9 The function $f(x) = 34.1 \ln x + 117.7$ models the number of U.S. Internet users, $f(x)$, in millions, x years after 1999. By which year, to the nearest year, will there be 200 million Internet users in the United States?

9.5 EXERCISE SET

MyMathLab | MathXL PRACTICE | WATCH | DOWNLOAD | READ | REVIEW

Practice Exercises

Solve each exponential equation in Exercises 1–18 by expressing each side as a power of the same base and then equating exponents.

1. $2^x = 64$
2. $3^x = 81$
3. $5^x = 125$
4. $5^x = 625$
5. $2^{2x-1} = 32$
6. $3^{2x+1} = 27$
7. $4^{2x-1} = 64$
8. $5^{3x-1} = 125$
9. $32^x = 8$
10. $4^x = 32$
11. $9^x = 27$
12. $125^x = 625$
13. $3^{1-x} = \frac{1}{27}$
14. $5^{2-x} = \frac{1}{125}$
15. $6^{\frac{x-3}{4}} = \sqrt{6}$
16. $7^{\frac{x-2}{6}} = \sqrt{7}$
17. $4^x = \dfrac{1}{\sqrt{2}}$
18. $9^x = \dfrac{1}{\sqrt[3]{3}}$

Solve each exponential equation in Exercises 19–40 by taking the logarithm on both sides. Express the solution set in terms of logarithms. Then use a calculator to obtain a decimal approximation, correct to two decimal places, for the solution.

19. $e^x = 5.7$
20. $e^x = 0.83$
21. $10^x = 3.91$
22. $10^x = 8.07$
23. $5^x = 17$
24. $19^x = 143$
25. $5e^x = 25$
26. $9e^x = 99$
27. $3e^{5x} = 1977$
28. $4e^{7x} = 10{,}273$
29. $e^{0.7x} = 13$
30. $e^{0.08x} = 4$
31. $1250e^{0.055x} = 3750$
32. $1250e^{0.065x} = 6250$
33. $30 - (1.4)^x = 0$
34. $135 - (4.7)^x = 0$
35. $e^{1-5x} = 793$
36. $e^{1-8x} = 7957$
37. $7^{x+2} = 410$
38. $5^{x-3} = 137$
39. $2^{x+1} = 5^x$
40. $4^{x+1} = 9^x$

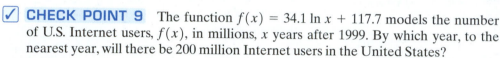

Solve each logarithmic equation in Exercises 41–90. Be sure to reject any value of x that is not in the domain of the original logarithmic expressions. Give the exact answer. Then, where necessary, use a calculator to obtain a decimal approximation, correct to two decimal places, for the solution.

41. $\log_3 x = 4$

42. $\log_5 x = 3$

43. $\log_2 x = -4$

44. $\log_2 x = -5$

45. $\log_9 x = \frac{1}{2}$

46. $\log_{25} x = \frac{1}{2}$

47. $\log x = 2$

48. $\log x = 3$

49. $\log_4(x + 5) = 3$

50. $\log_5(x - 7) = 2$

51. $\log_3(x - 4) = -3$

52. $\log_7(x + 2) = -2$

53. $\log_4(3x + 2) = 3$

54. $\log_2(4x + 1) = 5$

55. $\ln x = 2$

56. $\ln x = 3$

57. $\ln x = -3$

58. $\ln x = -4$

59. $5 \ln(2x) = 20$

60. $6 \ln(2x) = 30$

61. $6 + 2 \ln x = 5$

62. $7 + 3 \ln x = 6$

63. $\ln \sqrt{x + 3} = 1$

64. $\ln \sqrt{x + 4} = 1$

65. $\log_5 x + \log_5(4x - 1) = 1$

66. $\log_6(x + 5) + \log_6 x = 2$

67. $\log_3(x - 5) + \log_3(x + 3) = 2$

68. $\log_2(x - 1) + \log_2(x + 1) = 3$

69. $\log_2(x + 2) - \log_2(x - 5) = 3$

70. $\log_4(x + 2) - \log_4(x - 1) = 1$

71. $\log(3x - 5) - \log(5x) = 2$

72. $\log(2x - 1) - \log x = 2$

73. $\ln(x + 1) - \ln x = 1$

74. $\ln(x + 2) - \ln x = 2$

75. $\log_3(x + 4) = \log_3 7$

76. $\log_2(x - 5) = \log_2 4$

77. $\log(x + 4) = \log x + \log 4$

78. $\log(5x + 1) = \log(2x + 3) + \log 2$

79. $\log(3x - 3) = \log(x + 1) + \log 4$

80. $\log(2x - 1) = \log(x + 3) + \log 3$

81. $2 \log x = \log 25$

82. $3 \log x = \log 125$

83. $\log(x + 4) - \log 2 = \log(5x + 1)$

84. $\log(x + 7) - \log 3 = \log(7x + 1)$

85. $2 \log x - \log 7 = \log 112$

86. $\log(x - 2) + \log 5 = \log 100$

87. $\log x + \log(x + 3) = \log 10$

88. $\log(x + 3) + \log(x - 2) = \log 14$

89. $\ln(x - 4) + \ln(x + 1) = \ln(x - 8)$

90. $\log_2(x - 1) - \log_2(x + 3) = \log_2\left(\frac{1}{x}\right)$

Practice PLUS

In Exercises 91–98, solve each equation.

91. $5^{2x} \cdot 5^{4x} = 125$

92. $3^{x+2} \cdot 3^x = 81$

93. $3^{x^2} = 45$

94. $5^{x^2} = 50$

95. $\log_2(x - 6) + \log_2(x - 4) - \log_2 x = 2$

96. $\log_2(x - 3) + \log_2 x - \log_2(x + 2) = 2$

97. $5^{x^2-12} = 25^{2x}$

98. $3^{x^2-12} = 9^{2x}$

Application Exercises

99. The formula $A = 36.1e^{0.0126t}$ models the population of California, A, in millions, t years after 2005.

a. What was the population of California in 2005?

b. When will the population of California reach 40 million?

100. The formula $A = 22.9e^{0.0183t}$ models the population of Texas, A, in millions, t years after 2005.

a. What was the population of Texas in 2005?

b. When will the population of Texas reach 27 million?

The function $f(x) = 20(0.975)^x$ models the percentage of surface sunlight, $f(x)$, that reaches a depth of x feet beneath the surface of the ocean. The figure shows the graph of this function. Use this information to solve Exercises 101–102.

Intensity of Sunlight Beneath the Ocean's Surface

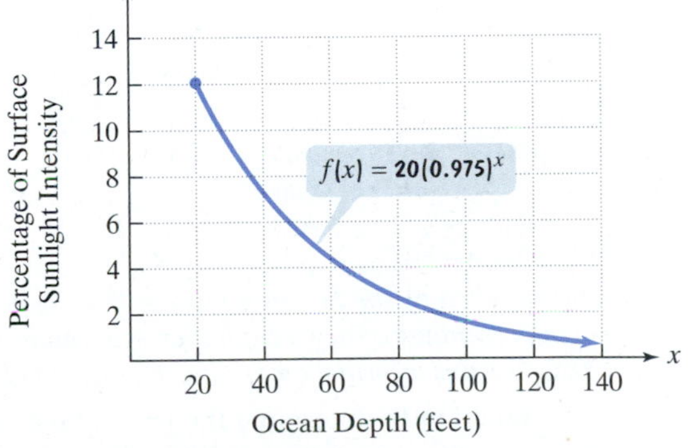

101. Use the function to determine at what depth, to the nearest foot, there is 1% of surface sunlight. How is this shown on the graph of f?

102. Use the function to determine at what depth, to the nearest foot, there is 3% of surface sunlight. How is this shown on the graph of f?

In Exercises 103–106, complete the table for a savings account subject to n compoundings yearly $\left[A = P\left(1 + \frac{r}{n}\right)^{nt}\right]$. *Round answers to one decimal place.*

	Amount Invested	Number of Compounding Periods	Annual Interest Rate	Accumulated Amount	Time t in Years
103.	\$12,500	4	5.75%	\$20,000	
104.	\$7250	12	6.5%	\$15,000	
105.	\$1000	360		\$1400	2
106.	\$5000	360		\$9000	4

In Exercises 107–110, complete the table for a savings account subject to continuous compounding ($A = Pe^{rt}$). *Round answers to one decimal place.*

	Amount Invested	Annual Interest Rate	Accumulated Amount	Time t in Years
107.	\$8000	8%	Double the amount invested	
108.	\$8000		\$12,000	2
109.	\$2350		Triple the amount invested	7
110.	\$17,425	4.25%	\$25,000	

The days when something happens and it's not captured on camera appear to be over. The bar graph shows that by 2007, 68% of cellphones sold in the United States were equipped with cameras.

Percentage of New Cellphones with Cameras

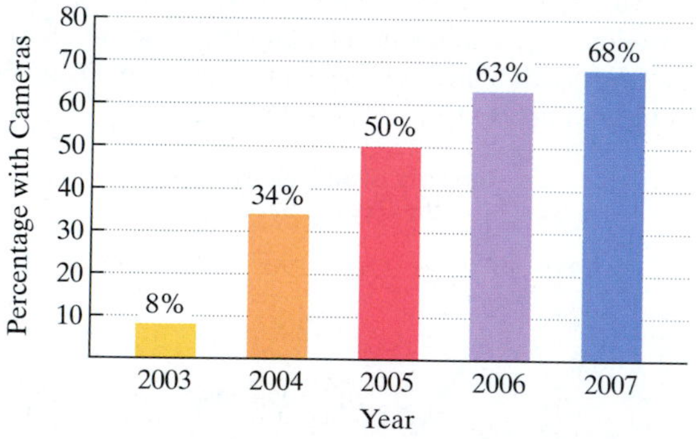

Source: Strategy Analytics Inc.

The data can be modeled by the function $f(x) = 8 + 38 \ln x$, *where* $f(x)$ *is the percentage of new cellphones with cameras x years after 2002. Use this information to solve Exercises 111–112.*

111. a. Use the function to determine the percentage of new cellphones with cameras in 2007. Round to the nearest whole percent. Does this overestimate or underestimate the percent displayed by the graph? By how much?

b. If trends shown from 2003 through 2007 continue, use the function to determine by which year 87% of new cellphones will be equipped with cameras.

112. a. Use the function to determine the percentage of new cellphones with cameras in 2006. Round to the nearest whole percent. Does this overestimate or underestimate the percent displayed by the graph? By how much?

b. If trends shown from 2003 through 2007 continue, use the function to determine by which year all new cellphones will be equipped with cameras.

The function $P(x) = 95 - 30 \log_2 x$ *models the percentage,* $P(x)$, *of students who could recall the important features of a classroom lecture as a function of time, where x represents the number of days that have elapsed since the lecture was given. The figure shows the graph of the function. Use this information to solve Exercises 113–114. Round answers to one decimal place.*

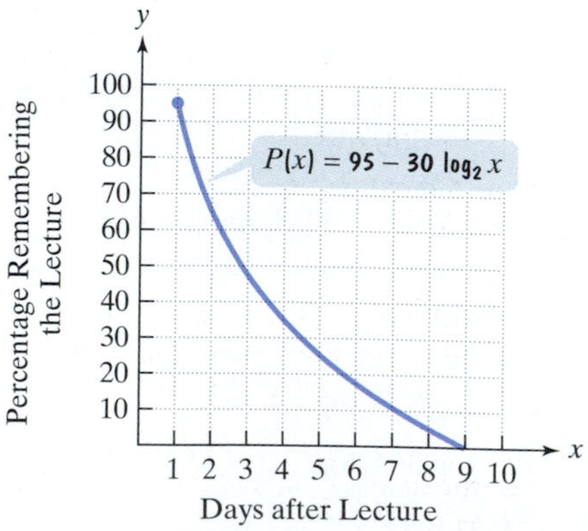

113. After how many days do only half the students recall the important features of the classroom lecture? (Let $P(x) = 50$ and solve for x.) Locate the point on the graph that conveys this information.

114. After how many days have all students forgotten the important features of the classroom lecture? (Let $P(x) = 0$ and solve for x.) Locate the point on the graph that conveys this information.

The pH scale is used to measure the acidity or alkalinity of a solution. The scale ranges from 0 to 14. A neutral solution, such as pure water, has a pH of 7. An acid solution has a pH less than 7 and an alkaline solution has a pH greater than 7. The lower the pH below 7, the more acidic is the solution. Each whole-number decrease in pH represents a tenfold increase in acidity.

The pH Scale

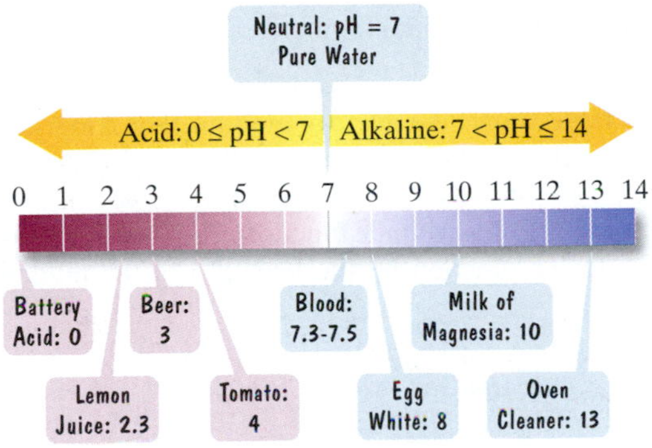

The pH of a solution is given by

$$\text{pH} = -\log x,$$

where x represents the concentration of the hydrogen ions in the solution, in moles per liter. Use the formula to solve Exercises 115–116. Express answers as powers of 10.

115. a. Normal, unpolluted rain has a pH of about 5.6. What is the hydrogen ion concentration?

b. An environmental concern involves the destructive effects of acid rain. The most acidic rainfall ever had a pH of 2.4. What was the hydrogen ion concentration?

c. How many times greater is the hydrogen ion concentration of the acidic rainfall in part (b) than the normal rainfall in part (a)?

116. a. The figure indicates that lemon juice has a pH of 2.3. What is the hydrogen ion concentration?

b. Stomach acid has a pH that ranges from 1 to 3. What is the hydrogen ion concentration of the most acidic stomach?

c. How many times greater is the hydrogen ion concentration of the acidic stomach in part (b) than the lemon juice in part (a)?

Writing in Mathematics

117. What is an exponential equation?

118. Explain how to solve an exponential equation when both sides can be written as a power of the same base.

119. Explain how to solve an exponential equation when both sides cannot be written as a power of the same base. Use $3^x = 140$ in your explanation.

120. What is a logarithmic equation?

121. Explain the differences between solving $\log_3(x - 1) = 4$ and $\log_3(x - 1) = \log_3 4$.

122. In many states, a 17% risk of a car accident with a blood alcohol concentration of 0.08 is the lowest level for charging a motorist with driving under the influence. Do you agree with the 17% risk as a cutoff percentage, or do you feel that the percentage should be lower or higher? Explain your answer. What blood alcohol concentration corresponds to what you believe is an appropriate percentage?

Technology Exercises

In Exercises 123–130, use your graphing utility to graph each side of the equation in the same viewing rectangle. Then use the x-coordinate of the intersection point to find the equation's solution set. Verify this value by direct substitution into the equation.

123. $2^{x+1} = 8$

124. $3^{x+1} = 9$

125. $\log_3(4x - 7) = 2$

126. $\log_3(3x - 2) = 2$

127. $\log(x + 3) + \log x = 1$

128. $\log(x - 15) + \log x = 2$

129. $3^x = 2x + 3$

130. $5^x = 3x + 4$

Hurricanes are one of nature's most destructive forces. These low-pressure areas often have diameters of over 500 miles. The function $f(x) = 0.48\ln(x + 1) + 27$ models the barometric air pressure, $f(x)$, in inches of mercury, at a distance of x miles from the eye of a hurricane. Use this function to solve Exercises 131–132.

131. Graph the function in a [0, 500, 50] by [27, 30, 1] viewing rectangle. What does the shape of the graph indicate about barometric air pressure as the distance from the eye increases?

132. Use an equation to answer this question: How far from the eye of a hurricane is the barometric air pressure 29 inches of mercury? Use the TRACE and ZOOM features or the intersect command of your graphing utility to verify your answer.

133. The function $P(t) = 145e^{-0.092t}$ models a runner's pulse, $P(t)$, in beats per minute, t minutes after a race, where $0 \le t \le 15$. Graph the function using a graphing utility. TRACE along the graph and determine after how many minutes the runner's pulse will be 70 beats per minute. Round to the nearest tenth of a minute. Verify your observation algebraically.

134. The function $W(t) = 2600(1 - 0.51e^{-0.075t})^3$ models the weight, $W(t)$, in kilograms, of a female African elephant at age t years. (1 kilogram $\approx$ 2.2 pounds) Use a graphing utility to graph the function. Then TRACE along the curve to estimate the age of an adult female elephant weighing 1800 kilograms.

Critical Thinking Exercises

Make Sense? *In Exercises 135–138, determine whether each statement "makes sense" or "does not make sense" and explain your reasoning.*

135. Because the equations $2^x = 15$ and $2^x = 16$ are similar, I solved them using the same method.

136. Because the equations

$$\log(3x + 1) = 5 \text{ and } \log(3x + 1) = \log 5$$

are similar, I solved them using the same method.

137. I can solve $4^x = 15$ by writing the equation in logarithmic form.

138. It's important for me to check that the proposed solution of an equation with logarithms gives only logarithms of positive numbers in the original equation.

In Exercises 139–142, determine whether each statement is true or false. If the statement is false, make the necessary change(s) to produce a true statement.

139. If $\log(x + 3) = 2$, then $e^2 = x + 3$.

140. If $\log(7x + 3) - \log(2x + 5) = 4$, then the equation in exponential form is $10^4 = (7x + 3) - (2x + 5)$.

141. If $x = \frac{1}{k}\ln y$, then $y = e^{kx}$.

142. Examples of exponential equations include $10^x = 5.71$, $e^x = 0.72$, and $x^{10} = 5.71$.

143. If \$4000 is deposited into an account paying 3% interest compounded annually and at the same time \$2000 is deposited into an account paying 5% interest compounded annually, after how long will the two accounts have the same balance? Round to the nearest year.

Solve each equation in Exercises 144–146. Check each proposed solution by direct substitution or with a graphing utility.

144. $(\ln x)^2 = \ln x^2$.

145. $(\log x)(2 \log x + 1) = 6$

146. $\ln(\ln x) = 0$

Review Exercises

147. Solve: $\sqrt{x + 4} - \sqrt{x - 1} = 1$. (Section 7.6, Example 4)

148. Solve: $\frac{3}{x + 1} - \frac{5}{x} = \frac{19}{x^2 + x}$. (Section 6.6, Example 5)

149. Simplify: $(-2x^3y^{-2})^{-4}$. (Section 1.6, Example 7)

Preview Exercises

Exercises 150–152 will help you prepare for the material covered in the next section.

150. The formula $A = 10e^{-0.003t}$ models the population of Hungary, A, in millions, t years after 2006.

a. Find Hungary's population, in millions, for 2006, 2007, 2008, and 2009. Round to two decimal places.

b. Is Hungary's population increasing or decreasing?

151. The table shows the average amount that Americans paid for a gallon of gasoline from 2002 through 2006. Create a scatter plot for the data. Based on the shape of the scatter plot, would a logarithmic function, an exponential function, or a linear function be the best choice for modeling the data?

Average Gas Price in the U.S.

Year	Average Price per Gallon
2002	\$1.40
2003	\$1.60
2004	\$1.92
2005	\$2.30
2006	\$2.91

Source: Oil Price Information Service

152. a. Simplify: $e^{\ln 3}$.

b. Use your simplification from part (a) to rewrite 3^x in terms of base e.

SECTION 9.6

Exponential Growth and Decay; Modeling Data

Objectives

1. Model exponential growth and decay.
2. Choose an appropriate model for data.
3. Express an exponential model in base e.

The most casual cruise on the Internet shows how people disagree when it comes to making predictions about the effects of the world's growing population. Some argue that there is a recent slowdown in the growth rate, economies remain robust, and famines in North Korea and Ethiopia are aberrations rather than signs of the future. Others say that the 6.8 billion people on Earth is twice as many as can be supported in middle-class comfort, and the world is running out of arable land and fresh water. Debates about entities that are growing exponentially can be approached mathematically: We can create functions that model data and use these functions to make predictions. In this section, we will show you how this is done.

1 Model exponential growth and decay.

Exponential Growth and Decay

One of algebra's many applications is to predict the behavior of variables. This can be done with *exponential growth* and *decay models*. With exponential growth or decay, quantities grow or decay at a rate directly proportional to their size. Populations that are growing exponentially grow extremely rapidly as they get larger because there are more adults to have offspring. For example, the **growth rate** for world population is approximately 1.2%, or 0.012. This means that each year world population is 1.2% more than what it was in the previous year. In 2007, world population was 6.6 billion. Thus, we compute the world population in 2008 as follows:

$$6.6 \text{ billion} + 1.2\% \text{ of } 6.6 \text{ billion} = 6.6 + (0.012)(6.6) = 6.6792.$$

This computation indicates that 6.6792 billion people populated the world in 2008. The 0.0792 billion represents an increase of 79.2 million people from 2007 to 2008, the equivalent of the population of Germany. Using 1.2% as the annual growth rate, world population for 2009 is found in a similar manner:

$$6.6792 + 1.2\% \text{ of } 6.6792 = 6.6792 + (0.012)(6.6792) \approx 6.759.$$

This computation indicates that approximately 6.759 billion people will populate the world in 2009.

The explosive growth of world population may remind you of the growth of money in an account subject to compound interest. Just as the growth rate for world population is multiplied by the population plus any increase in the population, a compound interest rate is multiplied by your original investment plus any accumulated interest. The balance in an account subject to continuous compounding and world population are special cases of *exponential growth models*.

Study Tip

You have seen the formula for exponential growth before, but with different letters. It is the formula for compound interest with continous compounding.

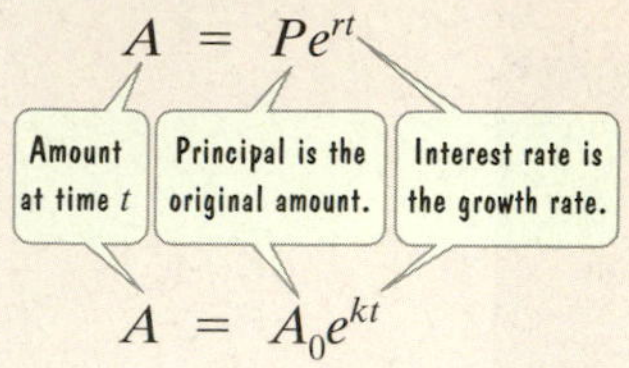

Exponential Growth and Decay Models

The mathematical model for **exponential growth** or **decay** is given by

$$f(t) = A_0e^{kt} \quad \text{or} \quad A = A_0e^{kt}.$$

- **If $k > 0$, the function models the amount, or size, of a *growing* entity.** A_0 is the original amount, or size, of the growing entity at time $t = 0$, A is the amount at time t, and k is a constant representing the growth rate.
- **If $k < 0$, the function models the amount, or size, of a *decaying* entity.** A_0 is the original amount, or size, of the decaying entity at time $t = 0$, A is the amount at time t, and k is a constant representing the decay rate.

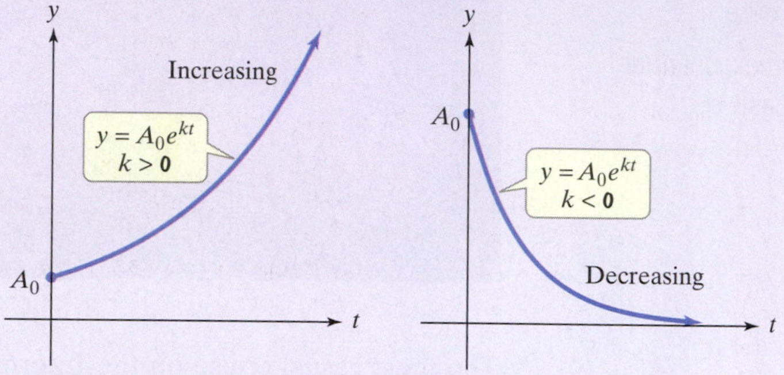

(a) Exponential growth **(b)** Exponential decay

Sometimes we need to use given data to determine k, the rate of growth or decay. After we compute the value of k, we can use the formula $A = A_0e^{kt}$ to make predictions. This idea is illustrated in our first two examples.

EXAMPLE 1 Modeling the Growth of the U.S. Population

The graph in **Figure 9.24** shows the U.S. population, in millions, for five selected years from 1970 through 2007. In 1970, the U.S. population was 203.3 million. By 2007, it had grown to 300.9 million.

a. Find an exponential growth function that models the data for 1970 through 2007.

b. By which year will the U.S. population reach 315 million?

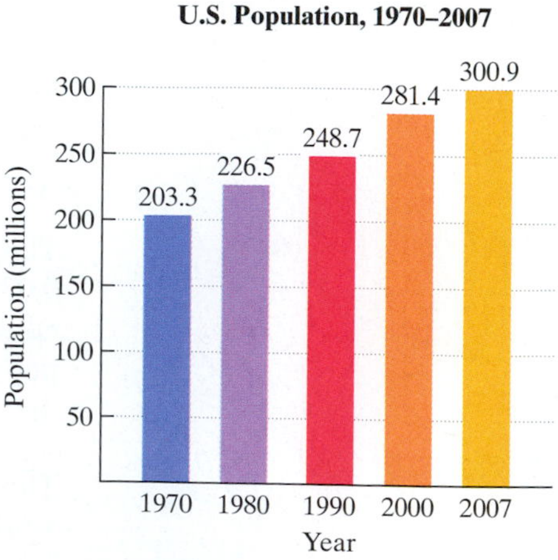

FIGURE 9.24

Source: Bureau of the Census

Solution

a. We use the exponential growth model

$$A = A_0e^{kt}$$

in which t is the number of years after 1970. This means that 1970 corresponds to $t = 0$. At that time the U.S. population was 203.3 million, so we substitute 203.3 for A_0 in the growth model:

$$A = 203.3e^{kt}.$$

We are given that 300.9 million is the population in 2007. Because 2007 is 37 years after 1970, when $t = 37$ the value of A is 300.9. Substituting these numbers into the growth model will enable us to find k, the growth rate. We know that $k > 0$ because the problem involves growth.

$A = 203.3e^{kt}$	Use the growth model with $A_0 = 203.3$.
$300.9 = 203.3e^{k \cdot 37}$	When $t = 37$, $A = 300.9$. Substitute these numbers into the model.
$e^{37k} = \dfrac{300.9}{203.3}$	Isolate the exponential factor by dividing both sides by 203.3. We also reversed the sides.
$\ln e^{37k} = \ln\left(\dfrac{300.9}{203.3}\right)$	Take the natural logarithm on both sides.
$37k = \ln\left(\dfrac{300.9}{203.3}\right)$	Simplify the left side using $\ln e^x = x$.
$k = \dfrac{\ln\left(\dfrac{300.9}{203.3}\right)}{37} \approx 0.011$	Divide both sides by 37 and solve for k. Then use a calculator.

The value of k, approximately 0.011, indicates a growth rate of about 1.1%. We substitute 0.011 for k in the growth model, $A = 203.3e^{kt}$, to obtain an exponential growth function for the U.S. population. It is

$$A = 203.3e^{0.011t},$$

where t is measured in years after 1970.

b. To find the year in which the U.S. population will reach 315 million, substitute 315 for A in the model from part (a) and solve for t.

$A = 203.3e^{0.011t}$	This is the model from part (a).
$315 = 203.3e^{0.011t}$	Substitute 315 for A.
$e^{0.011t} = \dfrac{315}{203.3}$	Divide both sides by 203.3. We also reversed the sides.
$\ln e^{0.011t} = \ln\left(\dfrac{315}{203.3}\right)$	Take the natural logarithm on both sides.
$0.011t = \ln\left(\dfrac{315}{203.3}\right)$	Simplify on the left using $\ln e^x = x$.
$t = \dfrac{\ln\left(\dfrac{315}{203.3}\right)}{0.011} \approx 40$	Divide both sides by 0.011 and solve for t. Then use a calculator.

Because t represents the number of years after 1970, the model indicates that the U.S. population will reach 315 million by 1970 + 40, or in the year 2010.

In Example 1, we used only two data values, the population for 1970 and the population for 2007, to develop a model for U.S. population growth from 1970 through 2007. By not using data for any other years, have we created a model that inaccurately describes both the existing data and future population projections given by the U.S. Census Bureau? Something else to think about: Is an exponential model the best choice for describing U.S. population growth, or might a linear model provide a better description? We return to these issues in Exercises 39–43 in the exercise set.

✓ CHECK POINT 1 In 1990, the population of Africa was 643 million and by 2006 it had grown to 906 million.

a. Use the exponential growth model $A = A_0e^{kt}$, in which t is the number of years after 1990, to find the exponential growth function that models the data.

b. By which year will Africa's population reach 2000 million, or two billion?

Blitzer Bonus

Carbon Dating and Artistic Development

The artistic community was electrified by the discovery in 1995 of spectacular cave paintings in a limestone cavern in France. Carbon dating of the charcoal from the site showed that the images, created by artists of remarkable talent, were 30,000 years old, making them the oldest cave paintings ever found. The artists seemed to have used the cavern's natural contours to heighten a sense of perspective. The quality of the painting suggests that the art of early humans did not mature steadily from primitive to sophisticated in any simple linear fashion.

Our next example involves exponential decay and its use in determining the age of fossils and artifacts. The method is based on considering the percentage of carbon-14 remaining in the fossil or artifact. Carbon-14 decays exponentially with a *half-life* of approximately 5715 years. The **half-life** of a substance is the time required for half of a given sample to disintegrate. Thus, after 5715 years a given amount of carbon-14 will have decayed to half the original amount. Carbon dating is useful for artifacts or fossils up to 80,000 years old. Older objects do not have enough carbon-14 left to determine age accurately.

EXAMPLE 2 Carbon-14 Dating: The Dead Sea Scrolls

a. Use the fact that after 5715 years a given amount of carbon-14 will have decayed to half the original amount to find the exponential decay model for carbon-14.

b. In 1947, earthenware jars containing what are known as the Dead Sea Scrolls were found by an Arab Bedouin herdsman. Analysis indicated that the scroll wrappings contained 76% of their original carbon-14. Estimate the age of the Dead Sea Scrolls.

Solution

a. We begin with the exponential decay model $A = A_0 e^{kt}$. We know that $k < 0$ because the problem involves the decay of carbon-14. After 5715 years ($t = 5715$), the amount of carbon-14 present, A, is half the original amount, A_0. Thus, we can substitute $\frac{A_0}{2}$ for A in the exponential decay model. This will enable us to find k, the decay rate.

$$A = A_0 e^{kt}$$ Begin with the exponential decay model.

$$\frac{A_0}{2} = A_0 e^{k5715}$$ After 5715 years ($t = 5715$), $A = \frac{A_0}{2}$ (because the amount present, A, is half the original amount, A_0).

$$\frac{1}{2} = e^{5715k}$$ Divide both sides of the equation by A_0.

$$\ln\left(\frac{1}{2}\right) = \ln e^{5715k}$$ Take the natural logarithm on both sides.

$$\ln\left(\frac{1}{2}\right) = 5715k$$ Simplify the right side using $\ln e^x = x$.

$$k = \frac{\ln\left(\frac{1}{2}\right)}{5715} \approx -0.000121$$ Divide both sides by 5715 and solve for k.

Substituting for k in the decay model, $A = A_0 e^{kt}$, the model for carbon-14 is

$$A = A_0 e^{-0.000121t}.$$

b. In 1947, the Dead Sea Scrolls contained 76% of their original carbon-14. To find their age in 1947, substitute $0.76A_0$ for A in the model from part (a) and solve for t.

$$A = A_0 e^{-0.000121t}$$ This is the decay model for carbon-14.

$$0.76A_0 = A_0 e^{-0.000121t}$$ A, the amount present, is 76% of the original amount, so $A = 0.76A_0$.

$$0.76 = e^{-0.000121t}$$ Divide both sides of the equation by A_0.

$$\ln 0.76 = \ln e^{-0.000121t}$$ Take the natural logarithm on both sides.

$$\ln 0.76 = -0.000121t$$ Simplify the right side using $\ln e^x = x$.

$$t = \frac{\ln 0.76}{-0.000121} \approx 2268$$ Divide both sides by -0.000121 and solve for t.

The Dead Sea Scrolls are approximately 2268 years old plus the number of years between 1947 and the current year.

✓ **CHECK POINT 2** Strontium-90 is a waste product from nuclear reactors. As a consequence of fallout from atmospheric nuclear tests, we all have a measurable amount of strontium-90 in our bones.

a. Use the fact that after 28 years a given amount of strontium-90 will have decayed to half the original amount to find the exponential decay model for strontium-90.

b. Suppose that a nuclear accident occurs and releases 60 grams of strontium-90 into the atmosphere. How long will it take for strontium-90 to decay to a level of 10 grams.

2 Choose an appropriate model for data.

Modeling Data

Throughout this chapter, we have been working with models that were given. However, we can create functions that model data by observing patterns in scatter plots. **Figure 9.25** shows scatter plots for data that are exponential or logarithmic.

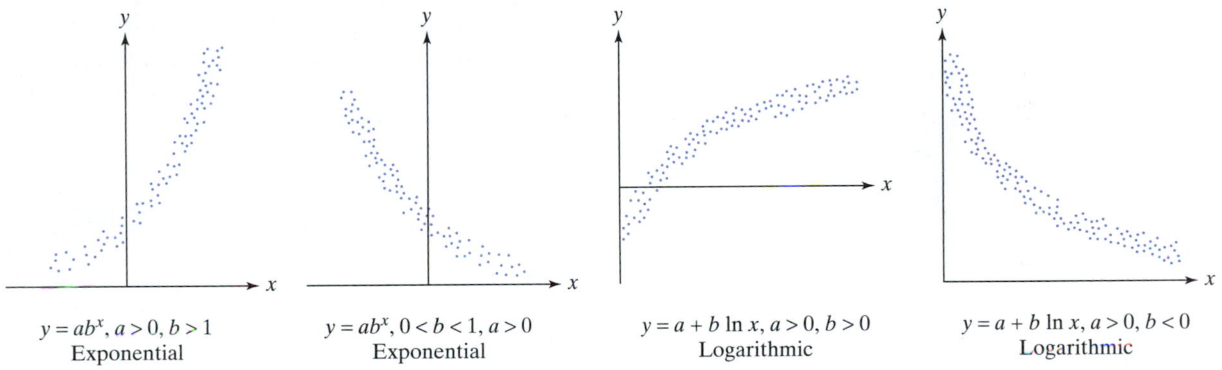

FIGURE 9.25 Scatter plots for exponential or logarithmic models

EXAMPLE 3 Choosing a Model for Data

The bar graph in **Figure 9.26(a)** indicates that for the period from 2002 through 2006, an increasing number of Americans had weight-loss surgery. A scatter plot is shown in **Figure 9.26(b)**. What type of function would be a good choice for modeling the data?

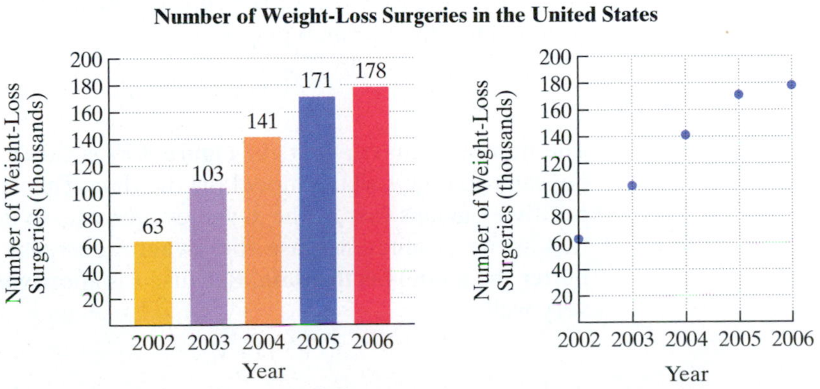

FIGURE 9.26(a)

Source: American Society for Bariatric Surgery

FIGURE 9.26(b)

Solution Because the data in the scatter plot increase rapidly at first and then begin to level off a bit, the shape suggests that a logarithmic function is a good choice for modeling the data.

✓ CHECK POINT 3 **Table 9.2** shows the populations of various cities, in thousands, and the average walking speed, in feet per second, of a person living in the city. Create a scatter plot for the data. Based on the scatter plot, what type of function would be a good choice for modeling the data?

Table 9.2 Population and Walking Speed

Population (thousands)	Walking Speed (feet per second)
5.5	0.6
14	1.0
71	1.6
138	1.9
342	2.2

Source: Mark and Helen Bornstein, "The Pace of Life"

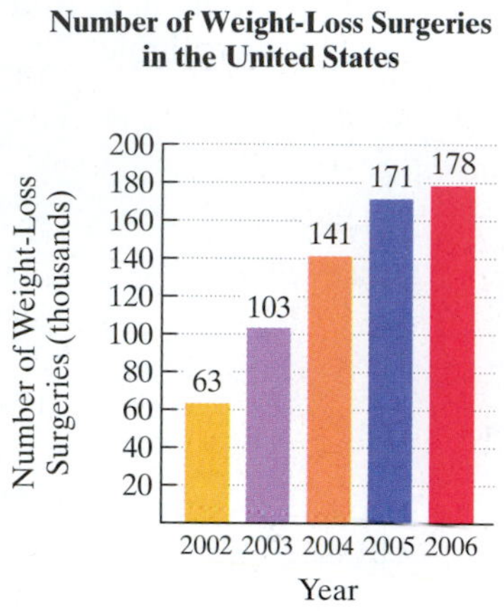

FIGURE 9.26(a) (repeated)

How can we obtain a logarithmic function that models the data for the number of weight-loss surgeries, in thousands, shown in **Figure 9.26(a)**? A graphing utility can be used to obtain a logarithmic model of the form $y = a + b \ln x$. **Because the domain of the logarithmic function is the set of positive numbers, zero must not be a value for x.** What does this mean for the number of weight-loss surgeries that begin in the year 2002? We must start values of x after 0. Thus, we'll assign x to represent the number of years after 2001. This gives us the data shown in **Table 9.3**. Using the logarithmic regression option, we obtain the equation in **Figure 9.27**.

Table 9.3

x, Number of Years after 2001	y, Number of Weight-Loss Surgeries (thousands)
1 (2002)	63
2 (2003)	103
3 (2004)	141
4 (2005)	171
5 (2006)	178

FIGURE 9.27 A logarithmic model for the data in **Table 9.3**

From **Figure 9.27**, we see that the logarithmic model of the data, with numbers rounded to three decimal places, is

$$y = 59.059 + 75.343 \ln x.$$

The number r that appears in **Figure 9.27** is called the **correlation coefficient** and is a measure of how well the model fits the data. The value of r is such that $-1 \le r \le 1$. A positive r means that as the x-values increase, so do the y-values. A negative r means that as the x-values increase, the y-values decrease. **The closer that r is to -1 or 1, the better the model fits the data.** Because r is approximately 0.992, the model fits the data very well.

EXAMPLE 4 Choosing a Model for Data

Figure 9.28(a) at the top of the next page shows world population, in billions, for seven selected years from 1950 through 2006. A scatter plot is shown in **Figure 9.28(b)**. Suggest two types of functions that would be good choices for modeling the data.

World Population, 1950–2006

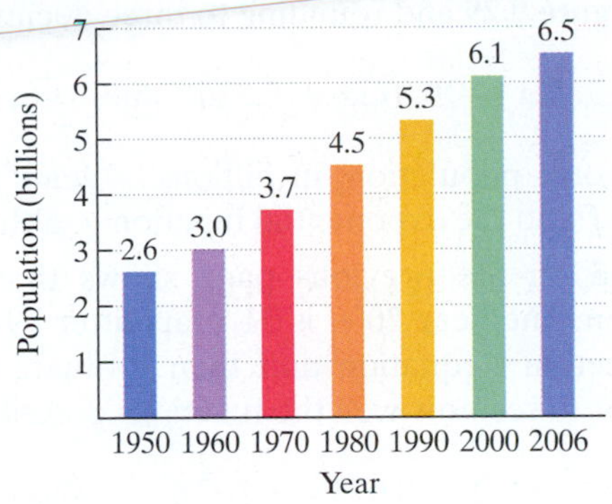

FIGURE 9.28(a)

Source: U.S. Census Bureau, International Database

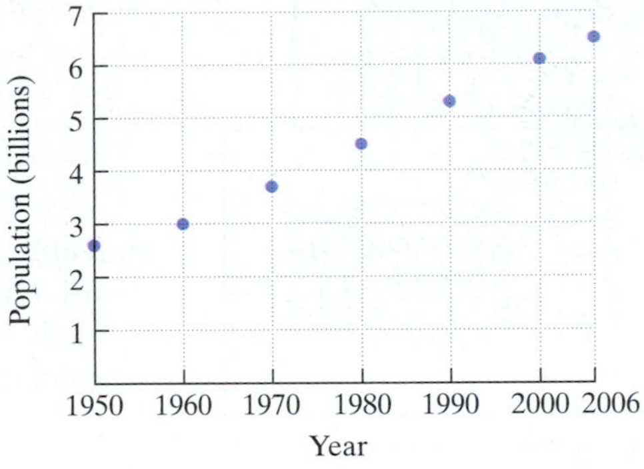

FIGURE 9.28(b)

Solution Because the data in the scatter plot appear to increase more and more rapidly, the shape suggests that an exponential model might be a good choice. Furthermore, we can probably draw a line that passes through or near the seven points. Thus, a linear function would also be a good choice for modeling the data.

CHECK POINT 4 **Table 9.4** shows the percentage of U.S. men who are married or who have been married, by age. Create a scatter plot for the data. Based on the scatter plot, what type of function would be a good choice for modeling the data?

Table 9.4 Percentage of U.S. Men Who Are Married or Who Have Been Married, by Age

Age	Percent
18	2
20	7
25	36
30	61
35	75

Source: National Center for Health Statistics

EXAMPLE 5 Comparing Linear and Exponential Models

The data for world population are shown in **Table 9.5**. Using a graphing utility's linear regression feature and exponential regression feature, we enter the data and obtain the models shown in **Figure 9.29**.

Although the domain of $y = ab^x$ is the set of all real numbers, some graphing utilities only accept positive values for x. That's why we assigned x to represent the number of years after 1949.

Table 9.5

x, Number of Years after 1949	y, World Population (billions)
1 (1950)	2.6
11 (1960)	3.0
21 (1970)	3.7
31 (1980)	4.5
41 (1990)	5.3
51 (2000)	6.1
57 (2006)	6.5

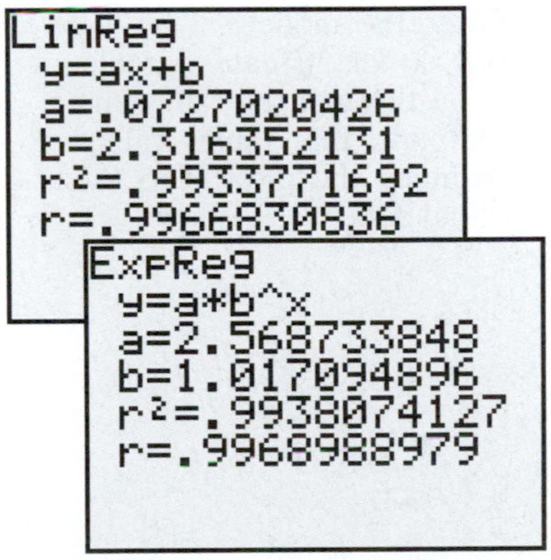

FIGURE 9.29 A linear model and an exponential model for the data in **Table 9.5**

Because r, the correlation coefficient, is close to 1 in each screen in **Figure 9.29**, the models fit the data very well.

a. Use **Figure 9.29** to express each model in function notation, with numbers rounded to three decimal places.

b. How well do the functions model world population in 2000?

c. By one projection, world population is expected to reach 8 billion in the year 2026. Which function serves as a better model for this prediction?

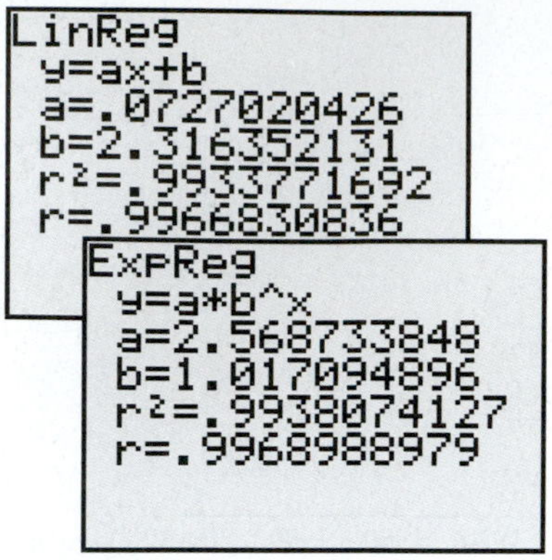

FIGURE 9.29 (repeated)

Solution

a. Using **Figure 9.29** and rounding to three decimal places, the functions

$$f(x) = 0.073x + 2.316 \quad \text{and} \quad g(x) = 2.569(1.017)^x$$

model world population, in billions, x years after 1949. We named the linear function f and the exponential function g, although any letters can be used.

b. **Table 9.5** on the previous page shows that world population in 2000 was 6.1 billion. The year 2000 is 51 years after 1949. Thus, we substitute 51 for x in each function's equation and then evaluate the resulting expressions with a calculator to see how well the functions describe world population in 2000.

$f(x) = 0.073x + 2.316$ — This is the linear model.

$f(51) = 0.073(51) + 2.316$ — Substitute 51 for x.

≈ 6.0 — Use a calculator.

$g(x) = 2.569(1.017)^x$ — This is the exponential model.

$g(51) = 2.569(1.017)^{51}$ — Substitute 51 for x.

≈ 6.1 — Use a calculator: 2.569 [×] 1.017 [y^x] (or [∧]) 51 [=].

Because 6.1 billion was the actual world population in 2000, both functions model world population in 2000 extremely well.

c. Let's see which model comes closest to projecting a world population of 8 billion in the year 2026. Because 2026 is 77 years after 1949 ($2026 - 1949 = 77$), we substitute 77 for x in each function's equation.

$f(x) = 0.073x + 2.316$ — This is the linear model.

$f(77) = 0.073(77) + 2.316$ — Substitute 77 for x.

≈ 7.9 — Use a calculator.

$g(x) = 2.569(1.017)^x$ — This is the exponential model.

$g(77) = 2.569(1.017)^{77}$ — Substitute 77 for x.

≈ 9.4 — Use a calculator: 2.569 [×] 1.017 [y^x] (or [∧]) 77 [=].

The linear function $f(x) = 0.073x + 2.316$ serves as a better model for a projected world population of 8 billion by 2026.

Study Tip

Once you have obtained one or more models for data, you can use a graphing utility's [TABLE] feature to numerically see how well each model describes the data. Enter the models as y_1, y_2, and so on. Create a table, scroll through the table, and compare the table values given by the models to the actual data.

✓ CHECK POINT 5 Use the models in Example 5(a) to solve this problem.

a. World population in 1970 was 3.7 billion. Which function serves as a better model for this year?

b. By one projection, world population is expected to reach 7 billion by 2012. How well do the functions model this projection?

When using a graphing utility to model data, begin with a scatter plot, drawn either by hand or with the graphing utility, to obtain a general picture for the shape of the data. It might be difficult to determine which model best fits the data—linear, logarithmic, exponential, quadratic, or something else. If necessary, use your graphing utility to fit several models to the data. The best model is the one that yields the value r, the correlation coefficient, closest to 1 or -1. Finding a proper fit for data can be almost as much art as it is mathematics. In this era of technology, the process of creating models that best fit data is one that involves more decision making than computation.

3 Express an exponential model in base e.

Expressing $y = ab^x$ in Base e

Graphing utilities display exponential models in the form $y = ab^x$. However, our discussion of exponential growth involved base e. Because of the inverse property $b = e^{\ln b}$, we can rewrite any model in the form $y = ab^x$ in terms of base e.

Expressing an Exponential Model in Base e

$$y = ab^x \quad \text{is equivalent to} \quad y = ae^{(\ln b)\cdot x}.$$

EXAMPLE 6 Rewriting the Model for World Population in Base e

We have seen that the function

$$g(x) = 2.569(1.017)^x$$

models world population, $g(x)$, in billions, x years after 1949. Rewrite the model in terms of base e.

Solution We use the two equivalent equations shown in the voice balloons to rewrite the model in terms of base e.

$y = ab^x$ $\qquad$ $y = ae^{(\ln b)\cdot x}$

$$g(x) = 2.569(1.017)^x \quad \text{is equivalent to} \quad g(x) = 2.569e^{(\ln 1.017)\cdot x}.$$

Using $\ln 1.017 \approx 0.017$, the exponential growth model for world population, $g(x)$, in billions, x years after 1949 is

$$g(x) = 2.569e^{0.017x}.$$

In Example 6, we can replace $g(x)$ with A and x with t so that the model has the same letters as those in the exponential growth model $A = A_0e^{kt}$.

$A = A_o \; e^{kt}$ This is the exponential growth model.

$A = 2.569e^{0.017t}$ This is the model for world population.

The value of k, 0.017, indicates a growth rate of 1.7%. Although this is an excellent model for the data, we must be careful about making projections about world population using this growth function. Why? World population growth rate is now 1.2%, not 1.7%, so our model will overestimate future populations.

✓ CHECK POINT 6 Rewrite $y = 4(7.8)^x$ in terms of base e. Express the answer in terms of a natural logarithm and then round to three decimal places.

9.6 EXERCISE SET

MyMathLab — MathXL PRACTICE · WATCH · DOWNLOAD · READ · REVIEW

Practice Exercises and Application Exercises

The exponential models describe the population of the indicated country, A, in millions, t years after 2006. Use these models to solve Exercises 1–6.

India: $A = 1095.4e^{0.014t}$

Iraq: $A = 26.8e^{0.027t}$

Japan: $A = 127.5e^{0.001t}$

Russia: $A = 142.9e^{-0.004t}$

1. What was the population of Japan in 2006?

2. What was the population of Iraq in 2006?

3. Which country has the greatest growth rate? By what percentage is the population of that country increasing each year?

4. Which country has a decreasing population? By what percentage is the population of that country decreasing each year?

5. When will India's population be 1238 million?

6. When will India's population be 1416 million?

About the size of New Jersey, Israel has seen its population soar to more than 6 million since it was established. With the help of U.S. aid, the country now has a diversified economy rivaling those of other developed Western nations. By contrast, the Palestinians, living under Israeli occupation and a corrupt regime, endure bleak conditions. The graphs show that by 2050, Palestinians in the West Bank, Gaza Strip, and East Jerusalem will outnumber Israelis. Exercises 7–8 involve the projected growth of these two populations.

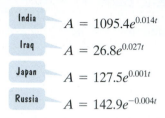

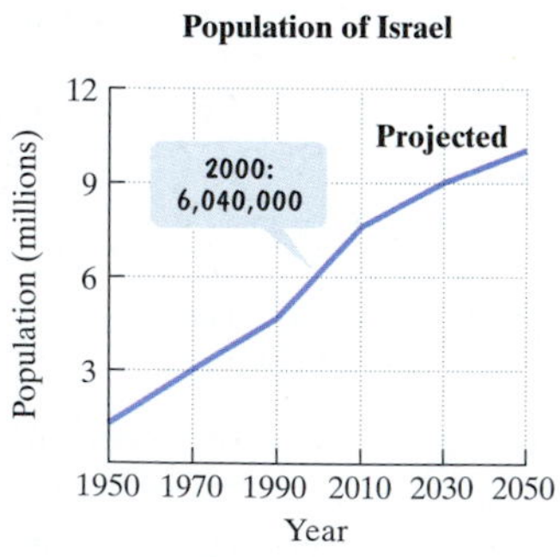

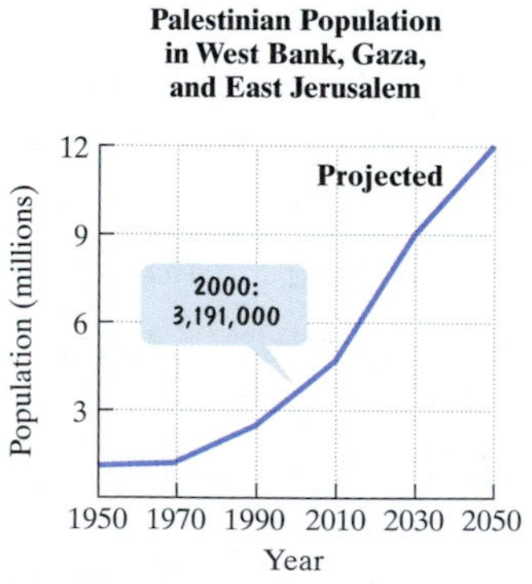

Source: Newsweek

7. a. In 2000, the population of Israel was approximately 6.04 million and by 2050 it is projected to grow to 10 million. Use the exponential growth model $A = A_0e^{kt}$, in which t is the number of years after 2000, to find an exponential growth function that models the data.

b. In which year will Israel's population be 9 million?

8. a. In 2000, the population of the Palestinians in the West Bank, Gaza Strip, and East Jerusalem was approximately 3.2 million and by 2050 it is projected to grow to 12 million. Use the exponential growth model $A = A_0e^{kt}$, in which t is the number of years after 2000, to find an exponential growth function that models the data.

b. In which year will the Palestinian population be 9 million?

An artifact originally had 16 grams of carbon-14 present. The decay model $A = 16e^{-0.000121t}$ describes the amount of carbon-14 present after t years. Use this model to solve Exercises 9–10.

9. How many grams of carbon-14 will be present in 5715 years?

10. How many grams of carbon-14 will be present in 11,430 years?

11. The half-life of the radioactive element krypton-91 is 10 seconds. If 16 grams of krypton-91 are initially present, how many grams are present after 10 seconds? 20 seconds? 30 seconds? 40 seconds? 50 seconds?

12. The half-life of the radioactive element plutonium-239 is 25,000 years. If 16 grams of plutonium-239 are initially present, how many grams are present after 25,000 years? 50,000 years? 75,000 years? 100,000 years? 125,000 years?

Use the exponential decay model for carbon-14, $A = A_0e^{-0.000121t}$, to solve Exercises 13–14.

13. Prehistoric cave paintings were discovered in a cave in France. The paint contained 15% of the original carbon-14. Estimate the age of the paintings.

14. Skeletons were found at a construction site in San Francisco in 1989. The skeletons contained 88% of the expected amount of carbon-14 found in a living person. In 1989, how old were the skeletons?

15. The August 1978 issue of *National Geographic* described the 1964 find of bones of a newly discovered dinosaur weighing 170 pounds, measuring 9 feet, with a 6-inch claw on one toe of each hind foot. The age of the dinosaur was estimated using potassium-40 dating of rocks surrounding the bones.

a. Potassium-40 decays exponentially with a half-life of approximately 1.31 billion years. Use the fact that after 1.31 billion years a given amount of potassium-40 will have decayed to half the original amount to show that the decay model for potassium-40 is given by $A = A_0e^{-0.52912t}$, where t is in billions of years.

b. Analysis of the rocks surrounding the dinosaur bones indicated that 94.5% of the original amount of potassium-40 was still present. Let $A = 0.945A_0$ in the model in part (a) and estimate the age of the bones of the dinosaur.

16. A bird species in danger of extinction has a population that is decreasing exponentially ($A = A_0e^{kt}$). Five years ago the population was at 1400 and today only 1000 of the birds are alive. Once the population drops below 100, the situation will be irreversible. When will this happen?

17. Use the exponential growth model, $A = A_0e^{kt}$, to show that the time it takes a population to double (to grow from A_0 to $2A_0$) is given by $t = \frac{\ln 2}{k}$.

18. Use the exponential growth model, $A = A_0e^{kt}$, to show that the time it takes a population to triple (to grow from A_0 to $3A_0$) is given by $t = \frac{\ln 3}{k}$.

Use the formula $t = \frac{\ln 2}{k}$ that gives the time for a population with a growth rate k to double to solve Exercises 19–20. Express each answer to the nearest whole year.

19. The growth model $A = 4.1e^{0.01t}$ describes New Zealand's population, A, in millions, t years after 2006.

a. What is New Zealand's growth rate?

b. How long will it take New Zealand to double its population?

20. The growth model $A = 107.4e^{0.012t}$ describes Mexico's population, A, in millions, t years after 2003.

a. What is Mexico's growth rate?

b. How long will it take Mexico to double its population?

Exercises 21–26 present data in the form of tables. For each data set shown by the table,

a. *Create a scatter plot for the data.*

b. *Use the scatter plot to determine whether an exponential function, a logarithmic function, or a linear function is the best choice for modeling the data. (If applicable, in Exercise 45 you will use your graphing utility to obtain these functions.)*

21. Percent of Miscarriages, by Age

Woman's Age	Percent of Miscarriages
22	9%
27	10%
32	13%
37	20%
42	38%
47	52%

Source: Time

22. Savings Needed for Health-Care Expenses during Retirement

Age at Death	Savings Needed
80	$219,000
85	$307,000
90	$409,000
95	$524,000
100	$656,000

Source: Employee Benefit Research Institute

23. Intensity and Loudness Level of Various Sounds

Intensity (watts per meter^2)	Loudness Level (decibels)
0.1 (loud thunder)	110
1 (rock concert, 2 yd from speakers)	120
10 (jackhammer)	130
100 (jet takeoff, 40 yd away)	140

24. Temperature Increase in an Enclosed Vehicle

Minutes	Temperature Increase (°F)
10	19°
20	29°
30	34°
40	38°
50	41°
60	43°

25. U.S. Per Capita Consumption of Bottled Water

Year	Per Capita Consumption (gallons)
2001	18.8
2002	20.9
2003	22.4
2004	24.0
2005	26.1
2006	28.3

Source: Beverage Marketing Corporation

26. Percentage of U.S. Consumers Looking for Trans Fats on Food Labels

Year	Percentage of U.S. Consumers
2004	15%
2005	20%
2006	25%
2007	31%

Source: IVPD Group

In Exercises 27–30, rewrite the equation in terms of base e. Express the answer in terms of a natural logarithm and then round to three decimal places.

27. $y = 100(4.6)^x$

28. $y = 1000(7.3)^x$

29. $y = 2.5(0.7)^x$

30. $y = 4.5(0.6)^x$

Writing in Mathematics

31. Nigeria has a growth rate of 0.025 or 2.5%. Describe what this means.

32. How can you tell if an exponential model describes exponential growth or exponential decay?

33. Suppose that a population that is growing exponentially increases from 800,000 people in 1997 to 1,000,000 people in 2000. Without showing the details, describe how to obtain an exponential growth function that models the data.

34. What is the half-life of a substance?

35. Describe the shape of a scatter plot that suggests modeling the data with an exponential function.

36. You take up weightlifting and record the maximum number of pounds you can lift at the end of each week. You start off with rapid growth in terms of the weight you can lift from week to week, but then the growth begins to level off. Describe how to obtain a function that models the number of pounds you can lift at the end of each week. How can you use this function to predict what might happen if you continue the sport?

37. Would you prefer that your salary be modeled exponentially or logarithmically? Explain your answer.

38. One problem with all exponential growth models is that nothing can grow exponentially forever. Describe factors that might limit the size of a population.

Technology Exercises

In Example 1 on page 708, we used two data points and an exponential function to model the population of the United States from 1970 through 2007. The data are shown again in the table. Use all five data points to solve Exercises 39–43.

x, Number of Years after 1969	y, U.S. Population (millions)
1 (1970)	203.3
11 (1980)	226.5
21 (1990)	248.7
31 (2000)	281.4
38 (2007)	300.9

39. a. Use your graphing utility's exponential regression option to obtain a model of the form $y = ab^x$ that fits the data. How well does the correlation coefficient, r, indicate that the model fits the data?

b. Rewrite the model in terms of base e. By what percentage is the population of the United States increasing each year?

40. Use your graphing utility's logarithmic regression option to obtain a model of the form $y = a + b \ln x$ that fits the data. How well does the correlation coefficient, r, indicate that the model fits the data?

41. Use your graphing utility's linear regression option to obtain a model of the form $y = ax + b$ that fits the data. How well does the correlation coefficient, r, indicate that the model fits the data?

42. Use your graphing utility's power regression option to obtain a model of the form $y = ax^b$ that fits the data. How well does the correlation coefficient, r, indicate that the model fits the data?

43. Use the values of r in Exercises 39–42 to select the two models of best fit. Use each of these models to predict by which year the U.S. population will reach 315 million. How do these answers compare to the year we found in Example 1, namely 2010? If you obtained different years, how do you account for this difference?

44. The figure shows the number of people in the United States age 65 and over, with projected figures for the year 2010 and beyond.

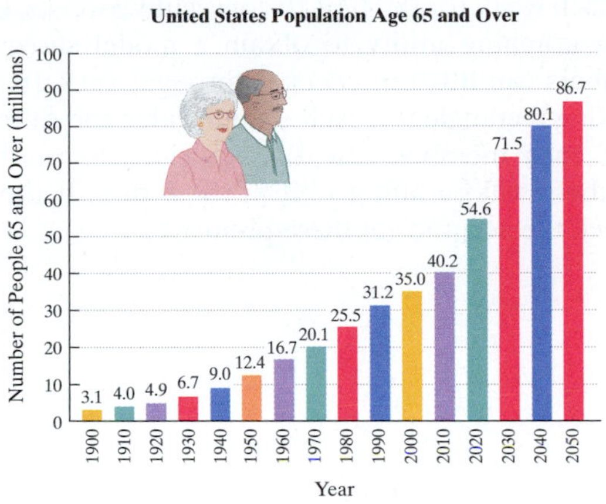

Source: U.S. Bureau of the Census

a. Let x represent the number of years after 1899 and let y represent the U.S. population age 65 and over, in millions. Use your graphing utility to find the model that best fits the data in the bar graph.

b. Rewrite the model in terms of base e. By what percentage is the 65 and over population increasing each year?

45. In Exercises 21–26, you determined the best choice for the kind of function that modeled the data in the table. For each of the exercises that you worked, use a graphing utility to find the actual function that best fits the data. Then use the model to make a reasonable prediction for a value that exceeds those shown in the table's first column.

Critical Thinking Exercises

Make Sense? *In Exercises 46–49, determine whether each statement "makes sense" or "does not make sense" and explain your reasoning.*

46. I used an exponential model with a positive growth rate to describe the depreciation in my car's value over four years.

47. After 100 years, a population whose growth rate is 3% will have three times as many people as a population whose growth rate is 1%.

48. Because carbon-14 decays exponentially, carbon dating can determine the ages of ancient fossils.

49. When I used an exponential function to model Russia's declining population, the growth rate k was negative.

The exponential growth models describe the population of the indicated country, A, in millions, t years after 2006.

Canada $A = 33.1e^{0.009t}$

Uganda $A = 28.2e^{0.034t}$

In Exercises 50–53, use this information to determine whether each statement is true or false. If the statement is false, make the necessary change(s) to produce a true statement.

50. In 2006, Canada's population exceeded Uganda's by 4.9 million.

51. By 2009, the models indicate that Canada's population will exceed Uganda's by approximately 2.8 million.

52. The models indicate that in 2013, Uganda's population will exceed Canada's.

53. Uganda's growth rate is approximately 3.8 times that of Canada's.

54. Over a period of time, a hot object cools to the temperature of the surrounding air. This is described mathematically by Newton's Law of Cooling:

$$T = C + (T_0 - C)e^{-kt},$$

where t is the time it takes for an object to cool from temperature T_0 to temperature T, C is the surrounding air temperature, and k is a positive constant that is associated with the cooling object. A cake removed from the oven has a temperature of 210°F and is left to cool in a room that has a temperature of 70°F. After 30 minutes, the temperature of the cake is 140°F. What is the temperature of the cake after 40 minutes?

Review Exercises

55. Divide:

$$\frac{x^2 - 9}{2x^2 + 7x + 3} \div \frac{x^2 - 3x}{2x^2 + 11x + 5}.$$

(Section 6.1, Example 7)

56. Solve: $x^{\frac{2}{3}} + 2x^{\frac{1}{3}} - 3 = 0.$

(Section 8.4, Example 5)

57. Simplify: $6\sqrt{2} - 2\sqrt{50} + 3\sqrt{98}.$

(Section 7.4, Example 2)

Preview Exercises

Exercises 58–60 will help you prepare for the material covered in the first section of the next chapter.

In Exercises 58–59, let $(x_1, y_1) = (7, 2)$ *and* $(x_2, y_2) = (1, -1)$.

58. Find $\sqrt{(x_2 - x_1)^2 + (y_2 - y_1)^2}$. Express the answer in simplified radical form.

59. Find the point represented by $\left(\frac{x_1 + x_2}{2}, \frac{y_1 + y_2}{2}\right)$.

60. Use a rectangular coordinate system to graph the circle with center $(1, -1)$ and radius 1.

GROUP PROJECT

CHAPTER 9

This activity is intended for three or four people who would like to take up weightlifting. Each person in the group should record the maximum number of pounds that he or she can lift at the end of each week for the first 10 consecutive weeks. Use the logarithmic regression option of a graphing utility to obtain a model showing the amount of weight that group members can lift from week 1 through week 10. Graph each of the models in the same viewing rectangle to observe similarities and differences among weight–growth patterns of each member. Use the functions to predict the amount of weight that group members will be able to lift in the future. If the group continues to work out together, check the accuracy of these predictions.

Chapter 9 Summary

Definitions and Concepts	Examples

Section 9.1 Exponential Functions

Definitions and Concepts

The exponential function with base b is defined by $f(x) = b^x$, where $b > 0$ and $b \neq 1$. The graph contains the point $(0, 1)$. When $b > 1$, the graph rises from left to right. When $0 < b < 1$, the graph falls from left to right. The x-axis is a horizontal asymptote. The domain is $(-\infty, \infty)$; the range is $(0, \infty)$. The natural exponential function is $f(x) = e^x$, where $e \approx 2.71828$.

Examples

Graph $f(x) = 2^x$ and $g(x) = 2^{x-1}$.

x	$f(x) = 2^x$	$g(x) = 2^{x-1}$
-2	$2^{-2} = \frac{1}{4}$	$2^{-3} = \frac{1}{8}$
-1	$2^{-1} = \frac{1}{2}$	$2^{-2} = \frac{1}{4}$
0	$2^0 = 1$	$2^{-1} = \frac{1}{2}$
1	$2^1 = 2$	$2^0 = 1$
2	$2^2 = 4$	$2^1 = 2$

The graph of g is the graph of f shifted one unit to the right.

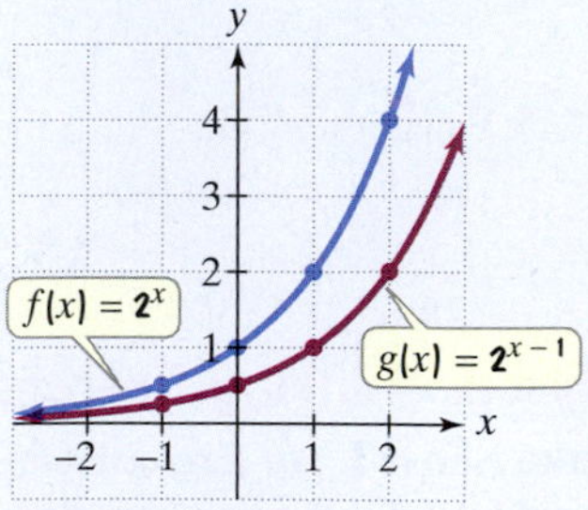

Definitions and Concepts

Formulas for Compound Interest

After t years, the balance, A, in an account with principal P and annual interest rate r is given by the following formulas:

1. For n compoundings per year: $A = P\left(1 + \frac{r}{n}\right)^{nt}$
2. For continuous compounding: $A = Pe^{rt}$

Examples

Select the better investment for \$4000 over 6 years:

- 6% compounded semiannually

$$A = P\left(1 + \frac{r}{n}\right)^{nt}$$

$$= 4000\left(1 + \frac{0.06}{2}\right)^{2 \cdot 6} \approx \$5703$$

- 5.9% compounded continuously

$$A = Pe^{rt} = 4000e^{0.059(6)} \approx \$5699$$

The first investment is better.

Definitions and Concepts	Examples

Section 9.2 Composite and Inverse Functions

Composite Functions

The composite function $f \circ g$ is defined by

$$(f \circ g)(x) = f(g(x)).$$

The composite function $g \circ f$ is defined by

$$(g \circ f)(x) = g(f(x)).$$

Let $f(x) = x^2 + x$ and $g(x) = 2x + 1$.

- $(f \circ g)(x) = f(g(x)) = (g(x))^2 + g(x)$ (Replace x with $g(x)$.)
  $= (2x + 1)^2 + (2x + 1) = 4x^2 + 4x + 1 + 2x + 1$
  $= 4x^2 + 6x + 2$
- $(g \circ f)(x) = g(f(x)) = 2f(x) + 1$ (Replace x with $f(x)$.)
  $= 2(x^2 + x) + 1 = 2x^2 + 2x + 1$

Inverse Functions

If $f(g(x)) = x$ and $g(f(x)) = x$, function g is the inverse of function f, denoted f^{-1} and read "f inverse." The procedure for finding a function's inverse uses a switch-and-solve strategy. Switch x and y, then solve for y.

If $f(x) = 2x - 5$, find $f^{-1}(x)$.

$y = 2x - 5$ Replace f(x) with y.

$x = 2y - 5$ Exchange x and y.

$x + 5 = 2y$ Solve for y.

$\frac{x+5}{2} = y$

$f^{-1}(x) = \frac{x+5}{2}$ Replace y with $f^{-1}(x)$.

The Horizontal Line Test for Inverse Functions

A function, f, has an inverse that is a function, f^{-1}, if there is no horizontal line that intersects the graph of f at more than one point. A one-to-one function is one in which no two different ordered pairs have the same second component. Only one-to-one functions have inverse functions. If the point (a, b) is on the graph of f, then the point (b, a) is on the graph of f^{-1}. The graph of f^{-1} is a reflection of the graph of f about the line $y = x$.

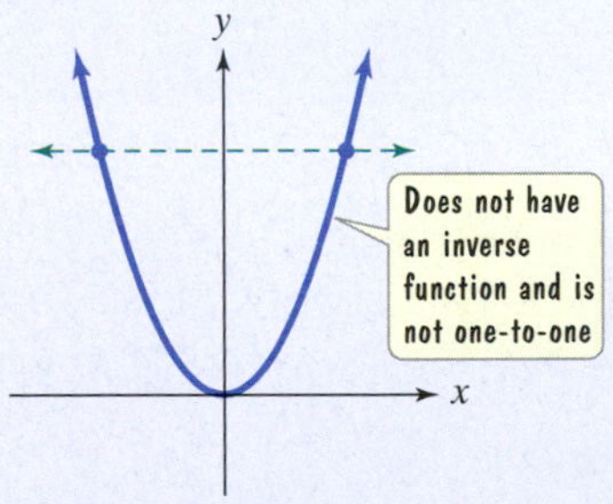
y
x
Does not have an inverse function and is not one-to-one

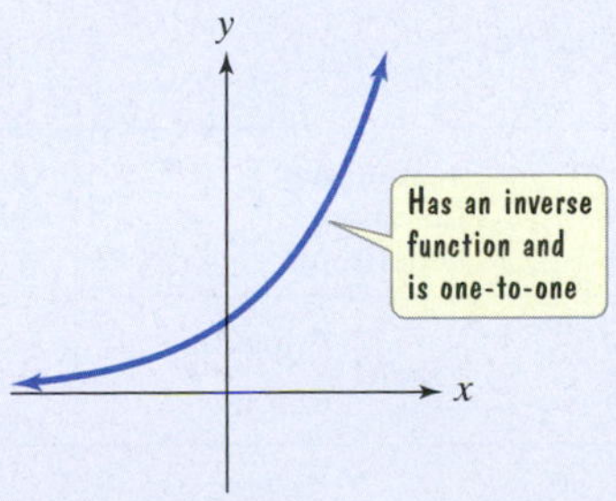
y
x
Has an inverse function and is one-to-one

Definitions and Concepts	Examples

Section 9.3 Logarithmic Functions

Definition of the logarithmic function: For $x > 0$ and $b > 0$, $b \neq 1$, $y = \log_b x$ is equivalent to $b^y = x$. The function $f(x) = \log_b x$ is the logarithmic function with base b. This function is the inverse function of the exponential function with base b.

- Write $\log_2 32 = 5$ in exponential form.

$$2^5 = 32 \quad y = \log_b x \text{ means } b^y = x.$$

- Write $\sqrt{49} = 7$, or $49^{\frac{1}{2}} = 7$, in logarithmic form.

$$\frac{1}{2} = \log_{49} 7 \quad b^y = x \text{ means } y = \log_b x.$$

The graph of $f(x) = \log_b x$ can be obtained from $f(x) = b^x$ by reversing coordinates. The graph of $f(x) = \log_b x$ contains the point $(1, 0)$. If $b > 1$, the graph rises from left to right. If $0 < b < 1$, the graph falls from left to right. The y-axis is a vertical asymptote. The domain is $(0, \infty)$; the range is $(-\infty, \infty)$. $f(x) = \log x$ means $f(x) = \log_{10} x$ and is the common logarithmic function. $f(x) = \ln x$ means $f(x) = \log_e x$ and is the natural logarithmic function. The domain of $f(x) = \log_b g(x)$ consists of all x for which $g(x) > 0$.

- Find the domain: $f(x) = \log_6(4 - x)$.

$$4 - x > 0$$

$$4 > x \quad (\text{or } x < 4)$$

The domain is $\{x|x < 4\}$ or $(-\infty, 4)$.

Basic Logarithmic Properties

Base b ($b > 0$, $b \neq 1$)	Base 10 (Common Logarithms)	Base e (Natural Logarithms)
$\log_b 1 = 0$	$\log 1 = 0$	$\ln 1 = 0$
$\log_b b = 1$	$\log 10 = 1$	$\ln e = 1$
$\log_b b^x = x$	$\log 10^x = x$	$\ln e^x = x$
$b^{\log_b x} = x$	$10^{\log x} = x$	$e^{\ln x} = x$

- $\log_8 1 = 0$ because $\log_b 1 = 0$.
- $\log_4 4 = 1$ because $\log_b b = 1$.
- $\ln e^{8x} = 8x$ because $\ln e^x = x$.
- $e^{\ln \sqrt[3]{x}} = \sqrt[3]{x}$ because $e^{\ln x} = x$.
- $\log_t t^{25} = 25$ because $\log_b b^x = x$.

Section 9.4 Properties of Logarithms

Properties of Logarithms

For $M > 0$ and $N > 0$:

1. *The Product Rule*: $\log_b(MN) = \log_b M + \log_b N$
2. *The Quotient Rule*: $\log_b\left(\frac{M}{N}\right) = \log_b M - \log_b N$
3. *The Power Rule*: $\log_b M^p = p \log_b M$
4. *The Change-of Base Property*:

The General Property	Introducing Common Logarithms	Introducing Natural Logarithms
$\log_b M = \frac{\log_a M}{\log_a b}$	$\log_b M = \frac{\log M}{\log b}$	$\log_b M = \frac{\ln M}{\ln b}$

- Expand: $\log_3(81x^7)$.

$$= \log_3 81 + \log_3 x^7$$

$$= 4 + 7 \log_3 x$$

- Write as a single logarithm: $7 \ln x - 4 \ln y$.

$$= \ln x^7 - \ln y^4 = \ln\left(\frac{x^7}{y^4}\right)$$

- Evaluate: $\log_6 92$.

$$\log_6 92 = \frac{\ln 92}{\ln 6} \approx 2.5237$$

Definitions and Concepts	Examples

Section 9.5 Exponential and Logarithmic Equations

An exponential equation is an equation containing a variable in an exponent. Some exponential equations can be solved by expressing both sides as a power of the same base. Then set the exponents equal to each other:

If $b^M = b^N$, then $M = N$.

Solve: $4^{2x-1} = 64$.

$$\begin{aligned} 4^{2x-1} &= 4^3 \\ 2x - 1 &= 3 \\ 2x &= 4 \\ x &= 2 \end{aligned}$$

The solution is 2 and the solution set is $\{2\}$.

If both sides of an exponential equation cannot be expressed as a power of the same base, isolate the exponential expression. Take the natural logarithm on both sides for bases other than 10 and take the common logarithm on both sides for base 10. Simplify using

$\ln b^x = x \ln b$ or $\ln e^x = x$ or $\log 10^x = x$.

Solve: $7^x = 103$.

$$\begin{aligned} \ln 7^x &= \ln 103 \\ x \ln 7 &= \ln 103 \\ x &= \frac{\ln 103}{\ln 7} \end{aligned}$$

The solution is $\frac{\ln 103}{\ln 7}$ and the solution set is $\left\{\frac{\ln 103}{\ln 7}\right\}$.

A logarithmic equation is an equation containing a variable in a logarithmic expression. Logarithmic equations in the form $\log_b x = c$ can be solved by rewriting in exponential form as $b^c = x$. When checking logarithmic equations, reject proposed solutions that produce the logarithm of a negative number or the logarithm of zero in the original equation.

Solve: $\log_2(3x - 1) = 5$.

$$\begin{aligned} 2^5 &= 3x - 1 \\ 32 &= 3x - 1 \\ 33 &= 3x \\ 11 &= x \end{aligned}$$

The solution is 11 and the solution set is $\{11\}$.

Exponential form

Solve: $3 \ln 2x = 15$.

$$\begin{aligned} \ln 2x &= 5 \\ \log_e 2x &= 5 \\ e^5 &= 2x \\ \frac{e^5}{2} &= x \end{aligned}$$

The solution is $\frac{e^5}{2}$ and the solution set is $\left\{\frac{e^5}{2}\right\}$.

Logarithmic equations in the form $\log_b M = \log_b N$, where $M > 0$ and $N > 0$, can be solved using the one-to-one property of logarithms:

If $\log_b M = \log_b N$, then $M = N$.

Solve: $\log(2x - 1) = \log(4x - 3) - \log x$.

$$\begin{aligned} \log(2x - 1) &= \log\left(\frac{4x - 3}{x}\right) \\ 2x - 1 &= \frac{4x - 3}{x} \\ x(2x - 1) &= 4x - 3 \\ 2x^2 - x &= 4x - 3 \\ 2x^2 - 5x + 3 &= 0 \\ (2x - 3)(x - 1) &= 0 \end{aligned}$$

$2x - 3 = 0$ or $x - 1 = 0$

$x = \frac{3}{2}$ $\qquad$ $x = 1$

Neither number produces the logarithm of 0 or logarithms of negative numbers in the original equation. The solutions are 1 and $\frac{3}{2}$, and the solution set is $\left\{1, \frac{3}{2}\right\}$.

Definitions and Concepts	Examples
Section 9.6 Exponential Growth and Decay; Modeling Data	
Exponential growth and decay models are given by $A = A_0e^{kt}$ in which t represents time, A_0 is the amount present at $t = 0$, and A is the amount present at time t. If $k > 0$, the model describes growth and k is the growth rate. If $k < 0$, the model describes decay and k is the decay rate. Scatter plots for exponential and logarithmic models are shown in **Figure 9.25** on page 711. When using a graphing utility to model data, the closer that the correlation coefficient r is to -1 or 1, the better the model fits the data.	The 1970 population of the Tokyo, Japan, urban area was 16.5 million: in 2000, it was 26.4 million. Write an exponential growth function that describes the population, in millions, t years after 1970. Begin with $A = A_0e^{kt}$. $A = 16.5e^{kt}$ — In 1970 ($t = 0$), the population was 16.5 million. $26.4 = 16.5e^{k \cdot 30}$ — When $t = 30$ (in 2000), $A = 26.4$. $e^{30k} = \frac{26.4}{16.5}$ — Isolate the exponential factor. $\ln e^{30k} = \ln\left(\frac{26.4}{16.5}\right)$ — Take the natural logarithm on both sides. $30k = \ln\left(\frac{26.4}{16.5}\right)$ and $k = \frac{\ln\left(\frac{26.4}{16.5}\right)}{30} \approx 0.016$ The growth function is $A = 16.5e^{0.016t}$. (Growth rate is 0.016 or 1.6%.)
Expressing an Exponential Model in Base *e* $y = ab^x$ is equivalent to $y = ae^{(\ln b)x}$.	Rewrite in terms of base e: $y = 24(7.2)^x$. $y = 24e^{(\ln 7.2)x} \approx 24e^{1.974x}$

CHAPTER 9 REVIEW EXERCISES

9.1 *In Exercises 1–4, set up a table of coordinates for each function. Select integers from −2 to 2, inclusive, for x. Then use the table of coordinates to match the function with its graph. [The graphs are labeled (a) through (d).]*

1. $f(x) = 4^x$
2. $f(x) = 4^{-x}$
3. $f(x) = -4^{-x}$
4. $f(x) = -4^{-x} + 3$

a.

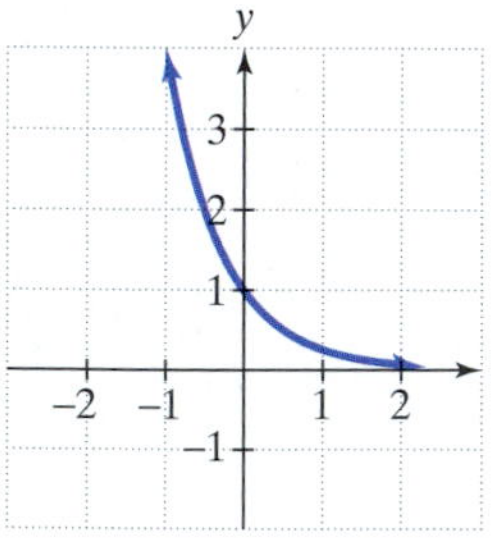

b.

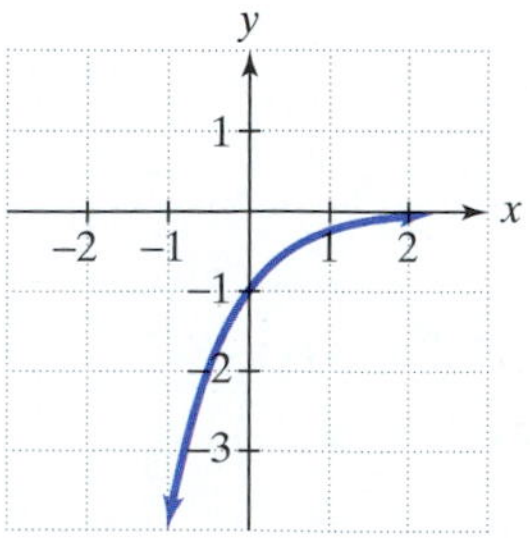

c.

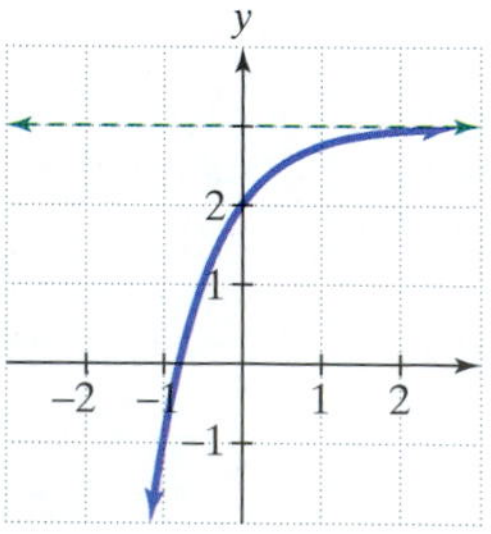

d.

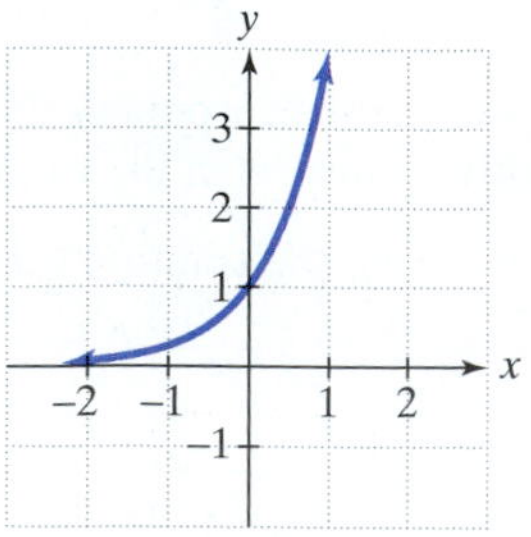

In Exercises 5–8, graph functions f and g in the same rectangular coordinate system. Select integers from −2 to 2, inclusive, for x. Then describe how the graph of g is related to the graph of f. If applicable, use a graphing utility to confirm your hand-drawn graphs.

5. $f(x) = 2^x$ and $g(x) = 2^{x-1}$
6. $f(x) = 2^x$ and $g(x) = \left(\frac{1}{2}\right)^x$
7. $f(x) = 3^x$ and $g(x) = 3^x - 1$
8. $f(x) = 3^x$ and $g(x) = -3^x$

Use the compound interest formulas

$$A = P\left(1 + \frac{r}{n}\right)^{nt} \quad \text{and} \quad A = Pe^{rt}$$

to solve Exercises 9–10.

9. Suppose that you have $5000 to invest. Which investment yields the greater return over 5 years: 5.5% compounded semiannually or 5.25% compounded monthly?

10. Suppose that you have \$14,000 to invest. Which investment yields the greater return over 10 years: 7% compounded monthly or 6.85% compounded continuously?

11. A cup of coffee is taken out of a microwave oven and placed in a room. The temperature, T, in degrees Fahrenheit, of the coffee after t minutes is modeled by the function $T = 70 + 130e^{-0.04855t}$. The graph of the function is shown in the figure.

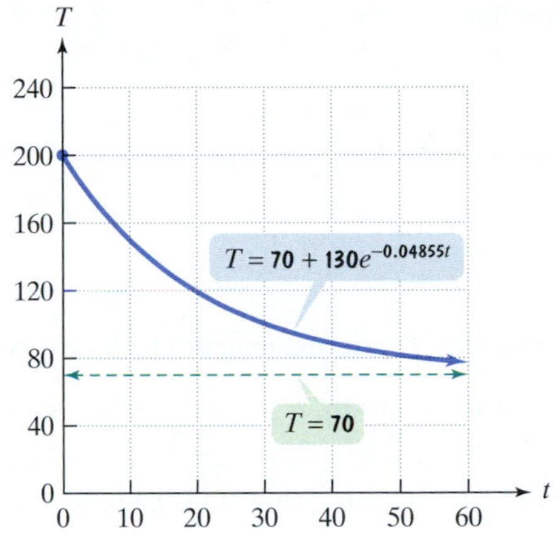

Use the graph to answer each of the following questions.

a. What was the temperature of the coffee when it was first taken out of the microwave?

b. What is a reasonable estimate of the temperature of the coffee after 20 minutes? Use your calculator to verify this estimate.

c. What is the limit of the temperature to which the coffee will cool? What does this tell you about the temperature of the room?

9.2 *In Exercises 12–13, find* **a.** $(f \circ g)(x)$; **b.** $(g \circ f)(x)$; **c.** $(f \circ g)(3)$.

12. $f(x) = x^2 + 3,\ g(x) = 4x - 1$

13. $f(x) = \sqrt{x},\ g(x) = x + 1$

In Exercises 14–15, find $f(g(x))$ and $g(f(x))$ and determine whether each pair of functions f and g are inverses of each other.

14. $f(x) = \frac{3}{5}x + \frac{1}{2}$ and $g(x) = \frac{5}{3}x - 2$

15. $f(x) = 2 - 5x$ and $g(x) = \frac{2 - x}{5}$

The functions in Exercises 16–18 are all one-to-one. For each function,

a. *Find an equation of $f^{-1}(x)$, the inverse function.*

b. *Verify that your equation is correct by showing that $f(f^{-1}(x)) = x$ and $f^{-1}(f(x)) = x$.*

16. $f(x) = 4x - 3$

17. $f(x) = \sqrt{x + 2}$

18. $f(x) = 8x^3 + 1$

Which graphs in Exercises 19–22 represent functions that have inverse functions?

19.

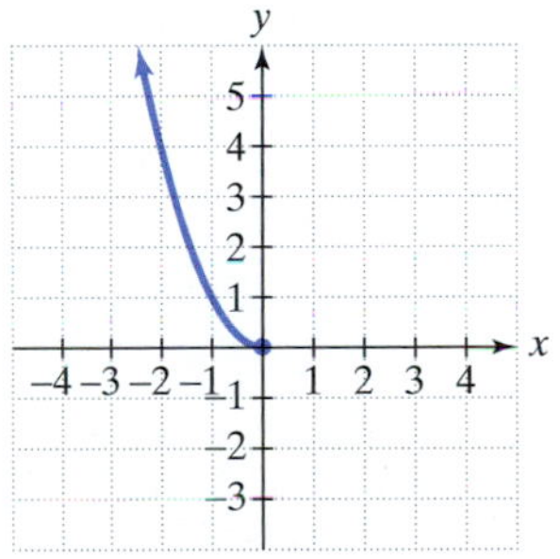

20.

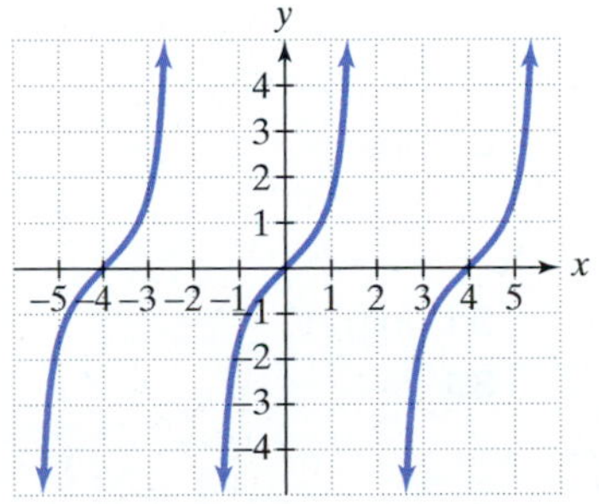

21.

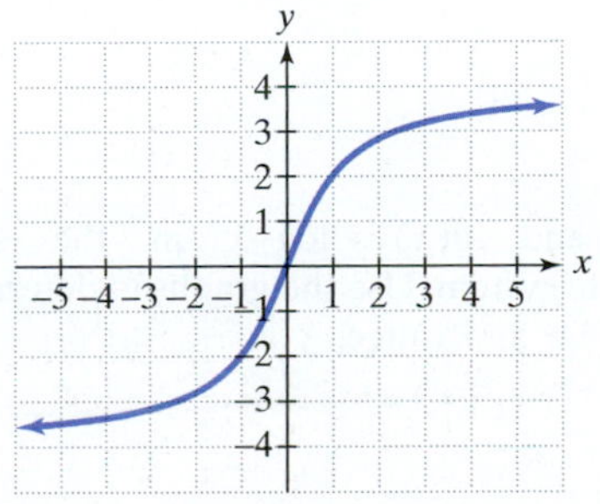

22.

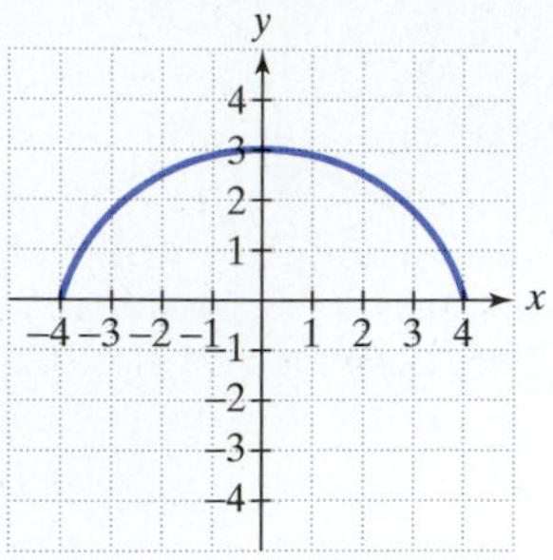

23. Use the graph of f in the figure shown to draw the graph of its inverse function.

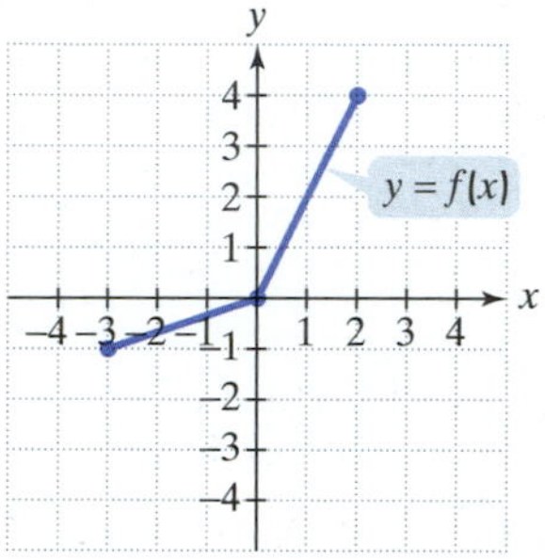

9.3

In Exercises 24–26, write each equation in its equivalent exponential form.

24. $\frac{1}{2} = \log_{49} 7$

25. $3 = \log_4 x$

26. $\log_3 81 = y$

In Exercises 27–29, write each equation in its equivalent logarithmic form.

27. $6^3 = 216$

28. $b^4 = 625$

29. $13^y = 874$

In Exercises 30–40, evaluate each expression without using a calculator. If evaluation is not possible, state the reason.

30. $\log_4 64$

31. $\log_5 \frac{1}{25}$

32. $\log_3(-9)$

33. $\log_{16} 4$

34. $\log_{17} 17$

35. $\log_3 3^8$

36. $\ln e^5$

37. $\log_3 \frac{1}{\sqrt{3}}$

38. $\ln \frac{1}{e^2}$

39. $\log \frac{1}{1000}$

40. $\log_3(\log_8 8)$

41. Graph $f(x) = 2^x$ and $g(x) = \log_2 x$ in the same rectangular coordinate system. Use the graphs to determine each function's domain and range.

42. Graph $f(x) = \left(\frac{1}{3}\right)^x$ and $g(x) = \log_{\frac{1}{3}} x$ in the same rectangular coordinate system. Use the graphs to determine each function's domain and range.

In Exercises 43–45, find the domain of each logarithmic function.

43. $f(x) = \log_8(x + 5)$

44. $f(x) = \log(3 - x)$

45. $f(x) = \ln(x - 1)^2$

In Exercises 46–48, simplify each expression.

46. $\ln e^{6x}$

47. $e^{\ln \sqrt{x}}$

48. $10^{\log 4x^2}$

49. On the Richter scale, the magnitude, R, of an earthquake of intensity I is given by $R = \log \frac{I}{I_0}$, where I_0 is the intensity of a barely felt zero-level earthquake. If the intensity of an earthquake is $1000I_0$, what is its magnitude on the Richter scale?

50. Students in a psychology class took a final examination. As part of an experiment to see how much of the course content they remembered over time, they took equivalent forms of the exam in monthly intervals thereafter. The average score, $f(t)$, for the group after t months is modeled by the function $f(t) = 76 - 18\log(t + 1)$, where $0 \le t \le 12$.

a. What was the average score when the exam was first given?

b. What was the average score, to the nearest tenth, after 2 months? 4 months? 6 months? 8 months? one year?

c. Use the results from parts (a) and (b) to graph f. Describe what the shape of the graph indicates in terms of the material retained by the students.

51. The formula

$$t = \frac{1}{c}\ln\left(\frac{A}{A - N}\right)$$

describes the time, t, in weeks, that it takes to achieve mastery of a portion of a task. In the formula, A represents maximum learning possible, N is the portion of the learning that is to be achieved, and c is a constant used to measure an individual's learning style. A 50-year-old man decides to start running as a way to maintain good health. He feels that the maximum rate he could ever hope to achieve is 12 miles per hour. How many weeks will it take before the man can run 5 miles per hour if $c = 0.06$ for this person?

9.4 *In Exercises 52–55, use properties of logarithms to expand each logarithmic expression as much as possible. Where possible, evaluate logarithmic expressions without using a calculator. Assume that all variables represent positive numbers.*

52. $\log_6(36x^3)$

53. $\log_4\left(\frac{\sqrt{x}}{64}\right)$

54. $\log_2\left(\frac{xy^2}{64}\right)$

55. $\ln \sqrt[3]{\frac{x}{e}}$

In Exercises 56–59, use properties of logarithms to condense each logarithmic expression. Write the expression as a single logarithm whose coefficient is 1.

56. $\log_b 7 + \log_b 3$

57. $\log 3 - 3 \log x$

58. $3 \ln x + 4 \ln y$

59. $\frac{1}{2}\ln x - \ln y$

In Exercises 60–61, use common logarithms or natural logarithms and a calculator to evaluate to four decimal places.

60. $\log_6 72{,}348$

61. $\log_4 0.863$

In Exercises 62–65, determine whether each equation is true or false. Where possible, show work to support your conclusion. If the statement is false, make the necessary change(s) to produce a true statement.

62. $(\ln x)(\ln 1) = 0$

63. $\log(x + 9) - \log(x + 1) = \frac{\log(x + 9)}{\log(x + 1)}$

64. $(\log_2 x)^4 = 4 \log_2 x$

65. $\ln e^x = x \ln e$

9.5 *In Exercises 66–71, solve each exponential equation. Where necessary, express the solution set in terms of natural logarithms and use a calculator to obtain a decimal approximation, correct to two decimal places, for the solution.*

66. $2^{4x-2} = 64$

67. $125^x = 25$

68. $9^x = \frac{1}{27}$

69. $8^x = 12{,}143$

70. $9e^{5x} = 1269$

71. $30e^{0.045x} = 90$

In Exercises 72–81, solve each logarithmic equation.

72. $\log_5 x = -3$

73. $\log x = 2$

74. $\log_4(3x - 5) = 3$

75. $\ln x = -1$

76. $3 + 4\ln(2x) = 15$

77. $\log_2(x + 3) + \log_2(x - 3) = 4$

78. $\log_3(x - 1) - \log_3(x + 2) = 2$

79. $\log_4(3x - 5) = \log_4 3$

80. $\ln(x + 4) - \ln(x + 1) = \ln x$

81. $\log_6(2x + 1) = \log_6(x - 3) + \log_6(x + 5)$

82. The function $P(x) = 14.7e^{-0.21x}$ models the average atmospheric pressure, $P(x)$, in pounds per square inch, at an altitude of x miles above sea level. The atmospheric pressure at the peak of Mt. Everest, the world's highest mountain, is 4.6 pounds per square inch. How many miles above sea level, to the nearest tenth of a mile, is the peak of Mt. Everest?

83. The amount of carbon dioxide in the atmosphere, measured in parts per million, has been increasing as a result of the burning of oil and coal. The buildup of gases and particles traps heat and raises the planet's temperature, a phenomenon called the greenhouse effect. Carbon dioxide accounts for about half of the warming. The function $f(t) = 364(1.005)^t$ projects carbon dioxide concentration, $f(t)$, in parts per million, t years after 2000. Using the projections given by the function, when will the carbon dioxide concentration be double the preindustrial level of 280 parts per million?

84. The function $W(x) = 0.37 \ln x + 0.05$ models the average walking speed, $W(x)$, in feet per second, of residents in a city whose population is x thousand. Visitors to New York City frequently feel they are moving too slowly to keep pace with New Yorkers' average walking speed of 3.38 feet per second. What is the population of New York City? Round to the nearest thousand.

85. Use the compound interest formula

$$A = P\left(1 + \frac{r}{n}\right)^{nt}$$

to solve this problem. How long, to the nearest tenth of a year, will it take \$12,500 to grow to \$20,000 at 6.5% annual interest compounded quarterly?

Use the compound interest formula

$$A = Pe^{rt}$$

to solve Exercises 86–87.

86. How long, to the nearest tenth of a year, will it take \$50,000 to triple in value at 7.5% annual interest compounded continuously?

87. What interest rate is required for an investment subject to continuous compounding to triple in 5 years?

9.6

88. According to the U.S. Bureau of the Census, in 1990 there were 22.4 million residents of Hispanic origin living in the United States. By 2005 the number had increased to 41.9 million. The exponential growth function $A = 22.4e^{kt}$ describes the U.S. Hispanic population, A, in millions, t years after 1990.

a. Find k, correct to three decimal places.

b. Use the resulting model to project the Hispanic resident population in 2010.

c. In which year will the Hispanic resident population reach 60 million?

89. Use the exponential decay model, $A = A_0e^{kt}$, to solve this exercise. The half-life of polonium-210 is 140 days. How long will it take for a sample of this substance to decay to 20% of its original amount?

Exercises 90–91 present data in the form of tables. For each data set shown by the table,

a. *Create a scatter plot for the data.*

b. *Use the scatter plot to determine whether an exponential function or a logarithmic function is the best choice for modeling the data.*

90. Number of U.S. Hotels That Allow Pets

Year	Number
2002	11,327
2003	12,634
2004	13,532
2005	13,884
2006	13,890

Source: *Traveling with Your Pet: The AAA Pet Book*

91. Percentage of U.S. Women Who Are Married or Who Have Been Married, by Age

Age	Percent
18	5%
20	17%
25	51%
30	73%

Source: National Center for Health Statistics

In Exercises 92–93, rewrite the equation in terms of base e. Express the answer in terms of a natural logarithm and then round to three decimal places.

92. $y = 73(2.6)^x$

93. $y = 6.5(0.43)^x$

94. The figure shows world population projections through the year 2150. The data are from the United Nations Family Planning Program and are based on optimistic or pessimistic expectations for successful control of human population growth. Suppose that you are interested in modeling these data using exponential, logarithmic, linear, and quadratic functions. Which function would you use to model each of the projections? Explain your choices. For the choice corresponding to a quadratic model, would your formula involve one with a positive or negative leading coefficient? Explain.

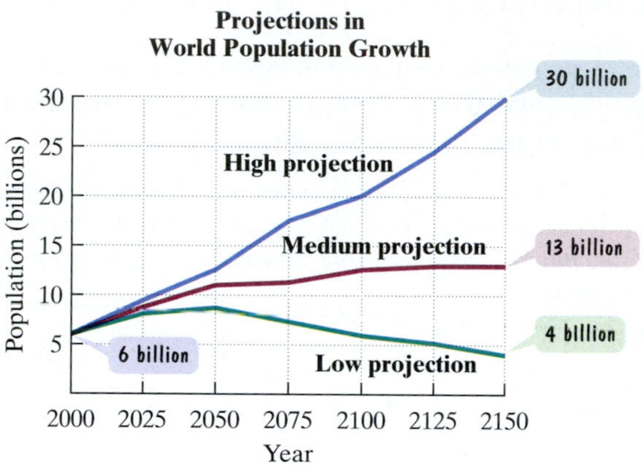

CHAPTER 9 TEST

Remember to use your Chapter Test Prep Video CD to see the worked-out solutions to the test questions you want to review.

1. Graph $f(x) = 2^x$ and $g(x) = 2^{x+1}$ in the same rectangular coordinate system.

2. Use $A = P\left(1 + \frac{r}{n}\right)^{nt}$ and $A = Pe^{rt}$ to solve this problem. Suppose you have \$3000 to invest. Which investment yields the greater return over 10 years: 6.5% compounded semiannually or 6% compounded continuously? How much more (to the nearest dollar) is yielded by the better investment?

3. If $f(x) = x^2 + x$ and $g(x) = 3x - 1$, find $(f \circ g)(x)$ and $(g \circ f)(x)$.

4. If $f(x) = 5x - 7$, find $f^{-1}(x)$.

5. A function f models the amount given to charity as a function of income. The graph of f is shown in the figure.

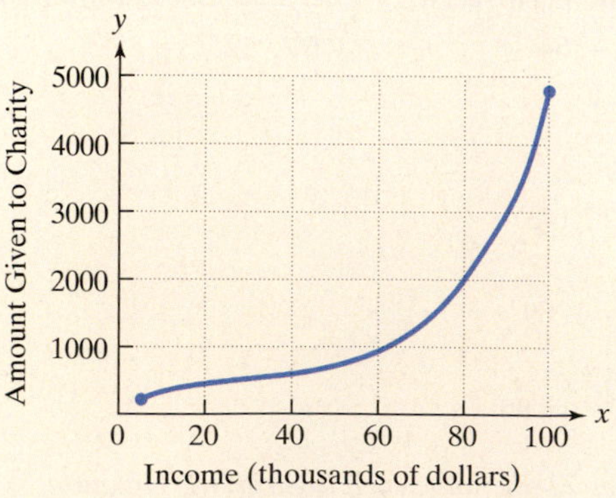

a. Explain why f has an inverse that is a function.

b. Find $f(80)$.

c. Describe in practical terms the meaning of $f^{-1}(2000)$.

6. Write in exponential form: $\log_5 125 = 3$.

7. Write in logarithmic form: $\sqrt{36} = 6$.

8. Graph $f(x) = 3^x$ and $g(x) = \log_3 x$ in the same rectangular coordinate system. Use the graphs to determine each function's domain and range.

In Exercises 9–11, simplify each expression.

9. $\ln e^{5x}$

10. $\log_b b$

11. $\log_6 1$

12. Find the domain: $f(x) = \log_5(x - 7)$.

13. On the decibel scale, the loudness of a sound, in decibels, is given by $D = 10 \log \frac{I}{I_0}$, where I is the intensity of the sound, in watts per meter^2, and I_0 is the intensity of a sound barely audible to the human ear. If the intensity of a sound is $10^{12}I_0$, what is its loudness in decibels? (Such a sound is potentially damaging to the ear.)

In Exercises 14–15, use properties of logarithms to expand each logarithmic expression as much as possible. Where possible, evaluate logarithmic expressions without using a calculator.

14. $\log_4(64x^5)$

15. $\log_3\left(\frac{\sqrt[3]{x}}{81}\right)$

In Exercises 16–17, write each expression as a single logarithm.

16. $6 \log x + 2 \log y$

17. $\ln 7 - 3 \ln x$

18. Use a calculator to evaluate $\log_{15} 71$ to four decimal places.

In Exercises 19–26, solve each equation.

19. $3^{x-2} = 81$

20. $5^x = 1.4$

21. $400e^{0.005x} = 1600$

22. $\log_{25} x = \frac{1}{2}$

23. $\log_6(4x - 1) = 3$

24. $2 \ln(3x) = 8$

25. $\log x + \log(x + 15) = 2$

26. $\ln(x - 4) - \ln(x + 1) = \ln 6$

27. The function

$$A = 82.4e^{-0.002t}$$

models the population of Germany, A, in millions, t years after 2006.

a. What was the population of Germany in 2006?

b. Is the population of Germany increasing or decreasing? Explain.

c. In which year will the population of Germany be 81.4 million?

Use the formulas

$$A = P\left(1 + \frac{r}{n}\right)^{nt} \quad \text{and} \quad A = Pe^{rt}$$

to solve Exercises 28–29.

28. How long, to the nearest tenth of a year, will it take \$4000 to grow to \$8000 at 5% annual interest compounded quarterly?

29. What interest rate is required for an investment subject to continuous compounding to double in 10 years?

30. The 1990 population of Europe was 509 million; in 2000, it was 729 million. Write an exponential growth function that describes the population of Europe, in millions, t years after 1990.

31. Use the exponential decay model for carbon-14, $A = A_0e^{-0.000121t}$, to solve this exercise. Bones of a prehistoric man were discovered and contained 5% of the original amount of carbon-14. How long ago did the man die?

In Exercises 32–35, determine whether the values in each table belong to an exponential function, a logarithmic function, a linear function, or a quadratic function.

32.

x	y
0	3
1	1
2	−1
3	−3
4	−5

33.

x	y
$\frac{1}{3}$	−1
1	0
3	1
9	2
27	3

34.

x	y
0	1
1	5
2	25
3	125
4	625

35.

x	y
0	12
1	3
2	0
3	3
4	12

36. Rewrite $y = 96(0.38)^x$ in terms of base e. Express the answer in terms of a natural logarithm and then round to three decimal places.

CUMULATIVE REVIEW EXERCISES (CHAPTERS 1–9)

In Exercises 1–7, solve each equation, inequality, or system.

1. $8 - (4x - 5) = x - 7$
2. $5x + 4y = 22$
   $3x - 8y = -18$
3. $-3x + 2y + 4z = 6$
   $7x - y + 3z = 23$
   $2x + 3y + z = 7$
4. $|x - 1| > 3$
5. $\sqrt{x + 4} - \sqrt{x - 4} = 2$
6. $x - 4 \geq 0$ and $-3x \leq -6$
7. $2x^2 = 3x - 2$

In Exercises 8–12, graph each function, equation, or inequality in a rectangular coordinate system.

8. $3x = 15 + 5y$
9. $2x - 3y > 6$
10. $f(x) = -\frac{1}{2}x + 1$
11. $f(x) = x^2 + 6x + 8$
12. $f(x) = (x - 3)^2 - 4$
13. Evaluate:

$$\begin{vmatrix} 3 & 1 & 0 \\ 0 & 5 & -6 \\ -2 & -1 & 0 \end{vmatrix}.$$

14. Solve for c: $A = \dfrac{cd}{c + d}$.

In Exercises 15–17, let $f(x) = x^2 + 3x - 15$ and $g(x) = x - 2$. Find each indicated expression.

15. $f(g(x))$
16. $g(f(x))$
17. $g(a + h) - g(a)$
18. If $f(x) = 7x - 3$, find $f^{-1}(x)$.

In Exercises 19–20, find the domain of each function.

19. $f(x) = \dfrac{x - 2}{x^2 - 3x + 2}$
20. $f(x) = \ln(2x - 8)$
21. Write the equation of the linear function whose graph contains the point $(-2, 4)$ and is perpendicular to the line whose equation is $2x + y = 10$.

In Exercises 22–26, perform the indicated operations and simplify, if possible.

22. $\dfrac{-5x^3y^7}{15x^4y^{-2}}$
23. $(4x^2 - 5y)^2$
24. $(5x^3 - 24x^2 + 9) \div (5x + 1)$
25. $\dfrac{\sqrt[3]{32xy^{10}}}{\sqrt[3]{2xy^2}}$
26. $\dfrac{x + 2}{x^2 - 6x + 8} + \dfrac{3x - 8}{x^2 - 5x + 6}$

In Exercises 27–28, factor completely.

27. $x^4 - 4x^3 + 8x - 32$
28. $2x^2 + 12xy + 18y^2$
29. Write as a single logarithm whose coefficient is 1:

$$2 \ln x - \frac{1}{2}\ln y.$$

30. The length of a rectangular carpet is 4 feet greater than twice its width. If the area is 48 square feet, find the carpet's length and width.
31. Working alone, you can mow the lawn in 2 hours and your sister can do it in 3 hours. How long will it take you to do the job if you work together?
32. Your motorboat can travel 15 miles per hour in still water. Traveling with the river's current, the boat can cover 20 miles in the same time it takes to go 10 miles against the current. Find the rate of the current.
33. Use the formula for continuous compounding, $A = Pe^{rt}$, to solve this problem. What interest rate is required for an investment of \$6000 subject to continuous compounding to grow to \$18,000 in 10 years?

CHAPTER 10

Conic Sections and Systems of Nonlinear Equations

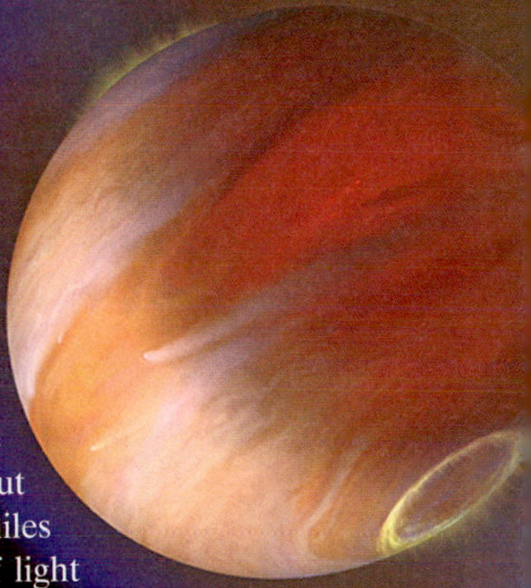

From ripples in water to the path on which humanity journeys through space, certain curves occur naturally throughout the universe. Over two thousand years ago, the ancient Greeks studied these curves, called *conic sections*, without regard to their immediate usefulness simply because studying them elicited ideas that were exciting, challenging, and interesting. The ancient Greeks could not have imagined the applications of these curves in the twenty-first century. They enable the Hubble Space Telescope, a large satellite about the size of a school bus orbiting 375 miles above Earth, to gather distant rays of light and focus them into spectacular images of our evolving universe. They provide doctors with a procedure for dissolving kidney stones painlessly without invasive surgery. In this chapter, we use the rectangular coordinate system to study the conic sections and the mathematics behind their surprising applications.

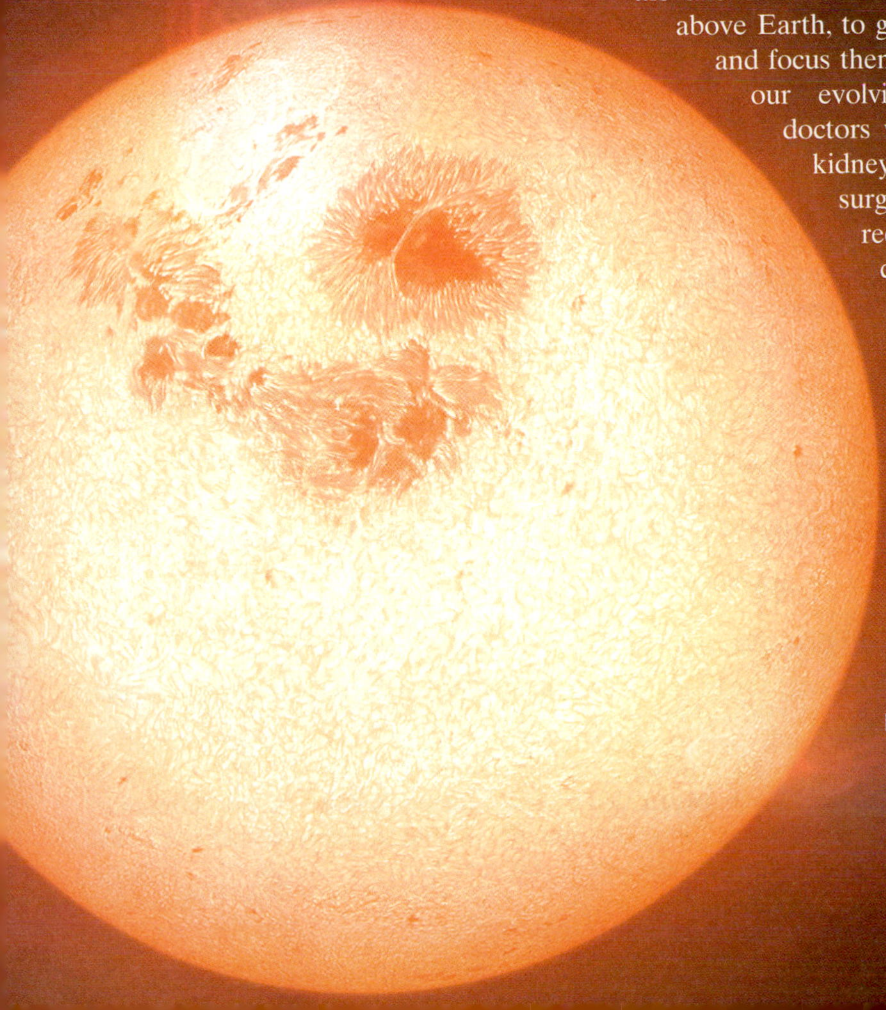

Here's where you'll find applications that move beyond planet Earth:

- Planetary orbits: Section 10.2, page 748; Group Project, page 785
- Halley's Comet: Blitzer Bonus, page 750
- Hubble Space Telescope: Section 10.4, pages 763 and 768.

For a kidney stone here on Earth, see Section 10.2, page 748.

SECTION 10.1

Distance and Midpoint Formulas; Circles

Objectives

1. Find the distance between two points.
2. Find the midpoint of a line segment.
3. Write the standard form of a circle's equation.
4. Give the center and radius of a circle whose equation is in standard form.
5. Convert the general form of a circle's equation to standard form.

It's a good idea to know your way around a circle. Clocks, angles, maps, and compasses are based on circles. Circles occur everywhere in nature: in ripples on water, patterns on a butterfly's wings, and cross sections of trees. Some consider the circle to be the most pleasing of all shapes.

The rectangular coordinate system gives us a unique way of knowing a circle. It enables us to translate a circle's geometric definition into an algebraic equation. To do this, we must first develop a formula for the distance between any two points in rectangular coordinates.

1 Find the distance between two points.

The Distance Formula

Using the Pythagorean Theorem, we can find the distance between the two points $P_1(x_1, y_1)$ and $P_2(x_2, y_2)$ in the rectangular coordinate system. The two points are illustrated in **Figure 10.1**.

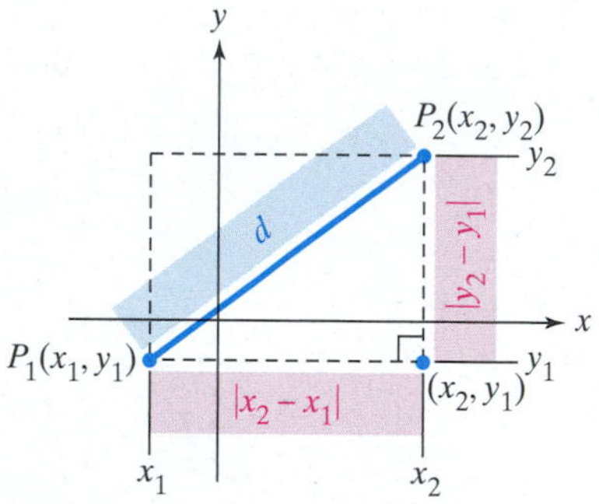

FIGURE 10.1

The distance that we need to find is represented by d and shown in blue. Notice that the distance between two points on the dashed horizontal line is the absolute value of the difference between the x-coordinates of the two points. This distance, $|x_2 - x_1|$, is shown in pink. Similarly, the distance between two points on the dashed vertical line is the absolute value of the difference between the y-coordinates of the two points. This distance, $|y_2 - y_1|$, is also shown in pink.

Because the dashed lines are horizontal and vertical, a right triangle is formed. Thus, we can use the Pythagorean Theorem to find the distance d. Squaring the lengths of the triangle's sides results in positive numbers, so absolute value notation is not necessary.

$$d^2 = (x_2 - x_1)^2 + (y_2 - y_1)^2$$ Apply the Pythagorean Theorem to the right triangle in Figure 10.1.

$$d = \pm\sqrt{(x_2 - x_1)^2 + (y_2 - y_1)^2}$$ Apply the square root property.

$$d = \sqrt{(x_2 - x_1)^2 + (y_2 - y_1)^2}$$ Because distance is nonnegative, write only the principal square root.

This result is called the **distance formula**.

The Distance Formula

The distance, d, between the points (x_1, y_1) and (x_2, y_2) in the rectangular coordinate system is

$$d = \sqrt{(x_2 - x_1)^2 + (y_2 - y_1)^2}.$$

To compute the distance between two points, find the square of the difference between the x-coordinates plus the square of the difference between the y-coordinates. The principal square root of this sum is the distance.

When using the distance formula, it does not matter which point you call (x_1, y_1) and which you call (x_2, y_2).

EXAMPLE 1 Using the Distance Formula

Find the distance between $(-1, 4)$ and $(3, -2)$.

Solution We will let $(x_1, y_1) = (-1, 4)$ and $(x_2, y_2) = (3, -2)$.

$$d = \sqrt{(x_2 - x_1)^2 + (y_2 - y_1)^2}$$ Use the distance formula.

$$= \sqrt{[3 - (-1)]^2 + (-2 - 4)^2}$$ Substitute the given values.

$$= \sqrt{4^2 + (-6)^2}$$ Perform operations inside grouping symbols: $3 - (-1) = 3 + 1 = 4$ and $-2 - 4 = -6$.

$$= \sqrt{16 + 36}$$ Square 4 and -6.

Caution: This does not equal $\sqrt{16} + \sqrt{36}$.

$$= \sqrt{52}$$ Add.

$$= \sqrt{4 \cdot 13} = 2\sqrt{13} \approx 7.21$$ $\sqrt{52} = \sqrt{4 \cdot 13} = \sqrt{4}\sqrt{13} = 2\sqrt{13}$

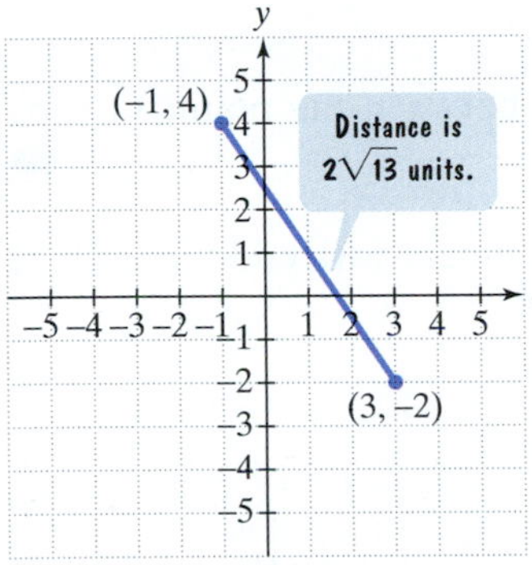

FIGURE 10.2 Finding the distance between two points

The distance between the given points is $2\sqrt{13}$ units, or approximately 7.21 units. The situation is illustrated in **Figure 10.2**.

✓ **CHECK POINT 1** Find the distance between $(-1, -3)$ and $(2, 3)$. Round to two decimal places.

2 Find the midpoint of a line segment.

The Midpoint Formula

The distance formula can be used to derive a formula for finding the midpoint of a line segment between two points. The formula is given as follows:

The Midpoint Formula

Consider a line segment whose endpoints are (x_1, y_1) and (x_2, y_2). The coordinates of the segment's midpoint are

$$\left(\frac{x_1 + x_2}{2}, \frac{y_1 + y_2}{2}\right).$$

To find the midpoint, take the average of the two x-coordinates and the average of the two y-coordinates.

Study Tip

The midpoint formula requires finding the *sum* of coordinates. By contrast, the distance formula requires finding the *difference* of coordinates:

Midpoint: Sum of coordinates

$$\left(\frac{x_1 + x_2}{2}, \frac{y_1 + y_2}{2}\right)$$

Distance: Difference of coordinates

$$\sqrt{(x_2 - x_1)^2 + (y_2 - y_1)^2}$$

It's easy to confuse the two formulas. Be sure to use addition, not subtraction, when applying the midpoint formula.

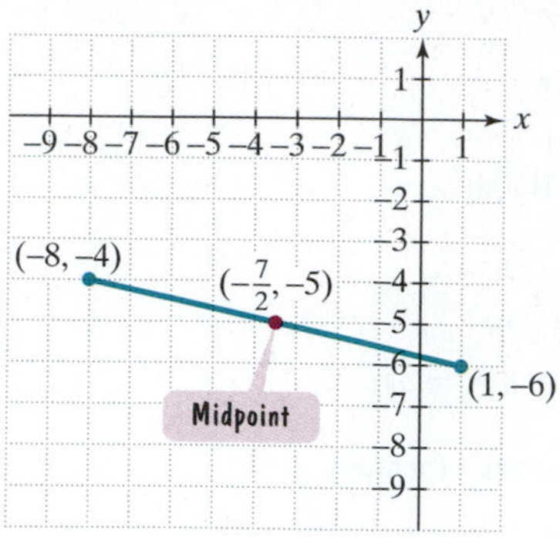

FIGURE 10.3 Finding a line segment's midpoint

EXAMPLE 2 Using the Midpoint Formula

Find the midpoint of the line segment with endpoints $(1, -6)$ and $(-8, -4)$.

Solution To find the coordinates of the midpoint, we average the coordinates of the endpoints.

$$\text{Midpoint} = \left(\frac{1 + (-8)}{2}, \frac{-6 + (-4)}{2}\right) = \left(\frac{-7}{2}, \frac{-10}{2}\right) = \left(-\frac{7}{2}, -5\right)$$

Average the x-coordinates. Average the y-coordinates.

Figure 10.3 illustrates that the point $(-\frac{7}{2}, -5)$ is midway between the points $(1, -6)$ and $(-8, -4)$.

✓ CHECK POINT 2 Find the midpoint of the line segment with endpoints $(1, 2)$ and $(7, -3)$.

Circles

Our goal is to translate a circle's geometric definition into an equation. We begin with this geometric definition.

Definition of a Circle

A **circle** is the set of all points in a plane that are equidistant from a fixed point, called the **center**. The fixed distance from the circle's center to any point on the circle is called the **radius**.

Figure 10.4 is our starting point for obtaining a circle's equation. We've placed the circle into a rectangular coordinate system. The circle's center is (h, k) and its radius is r. We let (x, y) represent the coordinates of any point on the circle.

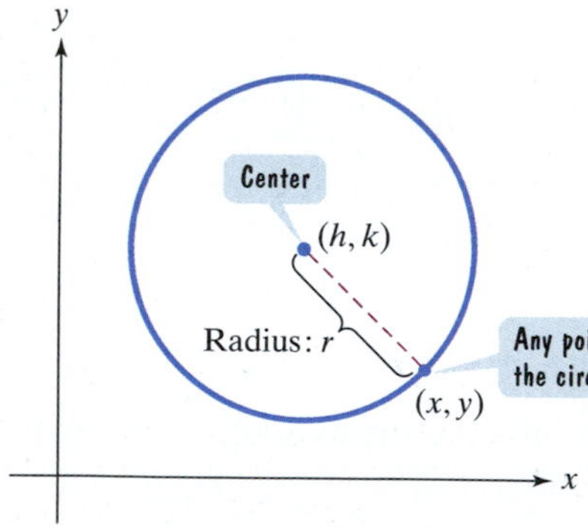

FIGURE 10.4 A circle centered at (h, k) with radius r

What does the geometric definition of a circle tell us about the point (x, y) in **Figure 10.4**? The point is on the circle if and only if its distance from the center is r. We can use the distance formula to express this idea algebraically:

The distance between (x, y) and (h, k) is always r.

$$\sqrt{(x-h)^2 + (y-k)^2} = r.$$

Squaring both sides of $\sqrt{(x-h)^2 + (y-k)^2} = r$ yields the *standard form of the equation of a circle.*

3 Write the standard form of a circle's equation.

The Standard Form of the Equation of a Circle

The **standard form of the equation of a circle** with center (h, k) and radius r is

$$(x - h)^2 + (y - k)^2 = r^2.$$

EXAMPLE 3 Finding the Standard Form of a Circle's Equation

Write the standard form of the equation of the circle with center (0, 0) and radius 2. Graph the circle.

Solution The center is (0, 0). Because the center is represented as (h, k) in the standard form of the equation, $h = 0$ and $k = 0$. The radius is 2, so we will let $r = 2$ in the equation.

$$(x - h)^2 + (y - k)^2 = r^2 \quad \text{This is the standard form of a circle's equation.}$$
$$(x - 0)^2 + (y - 0)^2 = 2^2 \quad \text{Substitute 0 for } h\text{, 0 for } k\text{, and 2 for } r.$$
$$x^2 + y^2 = 4 \quad \text{Simplify.}$$

The standard form of the equation of the circle is $x^2 + y^2 = 4$. **Figure 10.5** shows the graph.

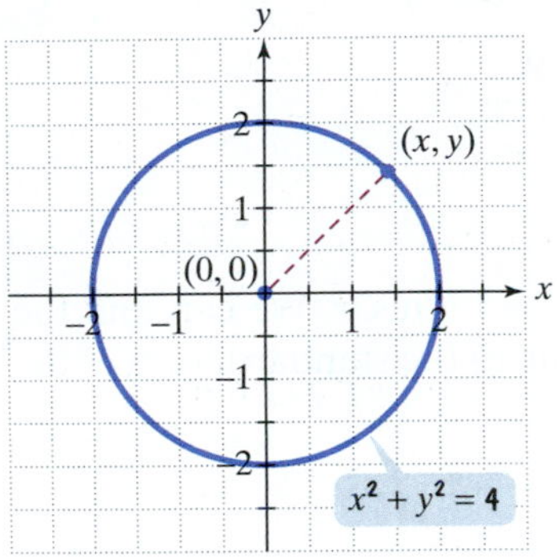

FIGURE 10.5 The graph of $x^2 + y^2 = 4$

✓ **CHECK POINT 3** Write the standard form of the equation of the circle with center (0, 0) and radius 4.

Using Technology

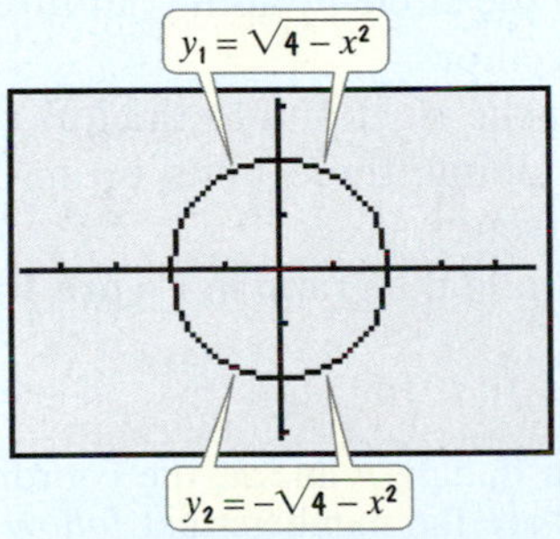

To graph a circle with a graphing utility, first solve the equation for y.

$$x^2 + y^2 = 4$$
$$y^2 = 4 - x^2$$
$$y = \pm\sqrt{4 - x^2} \quad y \text{ is not a function of } x.$$

Graph the two equations

$$y_1 = \sqrt{4 - x^2} \quad \text{and} \quad y_2 = -\sqrt{4 - x^2}$$

in the same viewing rectangle. The graph of $y_1 = \sqrt{4 - x^2}$ is the top semicircle because y is always positive. The graph of $y_2 = -\sqrt{4 - x^2}$ is the bottom semicircle because y is always negative. Use a ZOOM SQUARE setting so that the circle looks like a circle. (Many graphing utilities have problems connecting the two semicircles because the segments directly to the left and to the right of the center become nearly vertical.)

Example 3 and Check Point 3 involved circles centered at the origin. The standard form of the equation of all such circles is $x^2 + y^2 = r^2$, where r is the circle's radius. Now, let's consider a circle whose center is not at the origin.

EXAMPLE 4 Finding the Standard Form of a Circle's Equation

Write the standard form of the equation of the circle with center (−2, 3) and radius 4.

Solution The center is (−2, 3). Because the center is represented as (h, k) in the standard form of the equation, $h = -2$ and $k = 3$. The radius is 4, so we will let $r = 4$ in the equation.

$$(x - h)^2 + (y - k)^2 = r^2 \quad \text{This is the standard form of a circle's equation.}$$
$$[x - (-2)]^2 + (y - 3)^2 = 4^2 \quad \text{Substitute } -2 \text{ for } h\text{, 3 for } k\text{, and 4 for } r.$$
$$(x + 2)^2 + (y - 3)^2 = 16 \quad \text{Simplify.}$$

The standard form of the equation of the circle is $(x + 2)^2 + (y - 3)^2 = 16$.

✓ CHECK POINT 4 Write the standard form of the equation of the circle with center $(5, -6)$ and radius 10.

4 Give the center and radius of a circle whose equation is in standard form.

EXAMPLE 5 Using the Standard Form of a Circle's Equation to Graph the Circle

a. Find the center and radius of the circle whose equation is

$$(x-2)^2 + (y+4)^2 = 9.$$

b. Graph the equation.

Solution

a. We begin by finding the circle's center, (h, k), and its radius, r. We can find the values for h, k, and r by comparing the given equation to the standard form of the equation of a circle, $(x-h)^2 + (y-k)^2 = r^2$.

$$(x-2)^2 + (y+4)^2 = 9$$

$$(x-2)^2 + (y-(-4))^2 = 3^2$$

This is $(x-h)^2$, with $h = 2$. This is $(y-k)^2$, with $k = -4$. This is r^2, with $r = 3$.

We see that $h = 2$, $k = -4$, and $r = 3$. Thus, the circle has center $(h, k) = (2, -4)$ and a radius of 3 units.

b. To graph this circle, first plot the center $(2, -4)$. Because the radius is 3, you can locate at least four points on the circle by going out three units to the right, to the left, up, and down from the center.

The points three units to the right and to the left of $(2, -4)$ are $(5, -4)$ and $(-1, -4)$, respectively. The points three units up and down from $(2, -4)$ are $(2, -1)$ and $(2, -7)$, respectively.

Using these points, we obtain the graph in **Figure 10.6**.

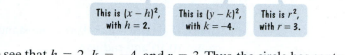

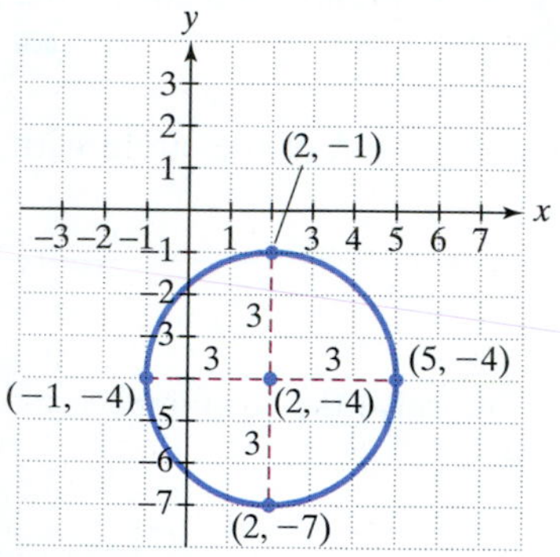

FIGURE 10.6 The graph of $(x-2)^2 + (y+4)^2 = 9$

Study Tip

It's easy to make sign errors when finding h and k, the coordinates of a circle's center, (h, k). Keep in mind that h and k are the numbers that *follow the subtraction signs* in a circle's equation:

$$(x-2)^2 + (y+4)^2 = 9$$

$$(x-2)^2 + (y-(-4))^2 = 9.$$

The number *after* the subtraction is 2: $h = 2$. The number *after* the subtraction is -4: $k = -4$.

✓ CHECK POINT 5 Find the center and radius of the circle whose equation is

$$(x+3)^2 + (y-1)^2 = 4$$

and graph the equation.

If we square $x - 2$ and $y + 4$ in the standard form of the equation of Example 5, we obtain another form for the circle's equation.

$$(x-2)^2 + (y+4)^2 = 9 \quad \text{This is the standard form of the equation from Example 5.}$$

$$x^2 - 4x + 4 + y^2 + 8y + 16 = 9 \quad \text{Square } x - 2 \text{ and } y + 4.$$

$$x^2 + y^2 - 4x + 8y + 20 = 9$$ Combine numerical terms and rearrange terms.

$$x^2 + y^2 - 4x + 8y + 11 = 0$$ Subtract 9 from both sides.

This result suggests that an equation in the form $x^2 + y^2 + Dx + Ey + F = 0$ can represent a circle. This is called the *general form of the equation of a circle.*

The General Form of the Equation of a Circle

The **general form of the equation of a circle** is

$$x^2 + y^2 + Dx + Ey + F = 0.$$

5 Convert the general form of a circle's equation to standard form.

We can convert the general form of the equation of a circle to the standard form $(x - h)^2 + (y - k)^2 = r^2$. We do so by completing the square on x and y. Let's see how this is done.

EXAMPLE 6 Converting the General Form of a Circle's Equation to Standard Form and Graphing the Circle

Write in standard form and graph: $x^2 + y^2 + 4x - 6y - 23 = 0$.

Solution Because we plan to complete the square on both x and y, let's rearrange the terms so that x-terms are arranged in descending order, y-terms are arranged in descending order, and the constant term appears on the right.

$$x^2 + y^2 + 4x - 6y - 23 = 0$$ This is the given equation.

$$(x^2 + 4x \qquad) + (y^2 - 6y \qquad) = 23$$ Rewrite in anticipation of completing the square.

$$(x^2 + 4x + 4) + (y^2 - 6y + 9) = 23 + 4 + 9$$ Complete the square on x: $\frac{1}{2} \cdot 4 = 2$ and $2^2 = 4$, so add 4 to both sides. Complete the square on y: $\frac{1}{2}(-6) = -3$ and $(-3)^2 = 9$, so add 9 to both sides.

Remember that numbers added on the left side must also be added on the right side.

$$(x + 2)^2 + (y - 3)^2 = 36$$ Factor on the left and add on the right.

This last equation is in standard form. We can identify the circle's center and radius by comparing this equation to the standard form of the equation of a circle, $(x - h)^2 + (y - k)^2 = r^2$.

$$(x + 2)^2 + (y - 3)^2 = 36$$

$$(x - (-2))^2 + (y - 3)^2 = 6^2$$

This is $(x - h)^2$, with $h = -2$. This is $(y - k)^2$, with $k = 3$. This is r^2, with $r = 6$.

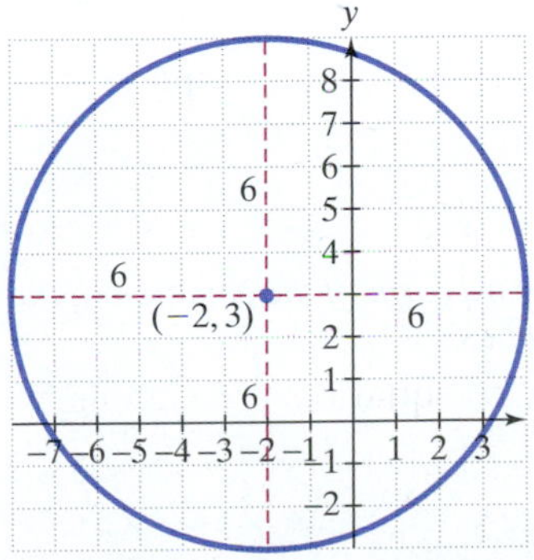

FIGURE 10.7 The graph of $(x + 2)^2 + (y - 3)^2 = 36$

We use the center, $(h, k) = (-2, 3)$, and the radius, $r = 6$, to graph the circle. The graph is shown in **Figure 10.7**.

Using Technology

To graph $x^2 + y^2 + 4x - 6y - 23 = 0$, rewrite the equation as a quadratic equation in y.

$$y^2 - 6y + (x^2 + 4x - 23) = 0$$

Now solve for y using the quadratic formula, with $a = 1$, $b = -6$, and $c = x^2 + 4x - 23$.

$$y = \frac{-b \pm \sqrt{b^2 - 4ac}}{2a} = \frac{-(-6) \pm \sqrt{(-6)^2 - 4 \cdot 1(x^2 + 4x - 23)}}{2 \cdot 1} = \frac{6 \pm \sqrt{36 - 4(x^2 + 4x - 23)}}{2}$$

Because we will enter these equations, there is no need to simplify further. Enter

$$y_1 = \frac{6 + \sqrt{36 - 4(x^2 + 4x - 23)}}{2}$$

and

$$y_2 = \frac{6 - \sqrt{36 - 4(x^2 + 4x - 23)}}{2}.$$

Use a ZOOM SQUARE setting. The graph is shown on the right.

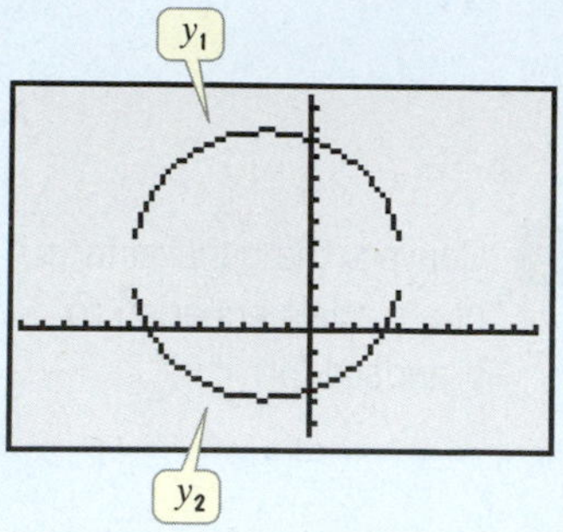

CHECK POINT 6 Write in standard form and graph:
$x^2 + y^2 + 4x - 4y - 1 = 0$.

10.1 EXERCISE SET

MyMathLab

Practice Exercises

In Exercises 1–18, find the distance between each pair of points. If necessary, round answers to two decimal places.

1. $(2, 3)$ and $(14, 8)$
2. $(5, 1)$ and $(8, 5)$
3. $(4, 1)$ and $(6, 3)$
4. $(2, 3)$ and $(3, 5)$
5. $(0, 0)$ and $(-3, 4)$
6. $(0, 0)$ and $(3, -4)$
7. $(-2, -6)$ and $(3, -4)$
8. $(-4, -1)$ and $(2, -3)$
9. $(0, -3)$ and $(4, 1)$
10. $(0, -2)$ and $(4, 3)$
11. $(3.5, 8.2)$ and $(-0.5, 6.2)$
12. $(2.6, 1.3)$ and $(1.6, -5.7)$
13. $(0, -\sqrt{3})$ and $(\sqrt{5}, 0)$
14. $(0, -\sqrt{2})$ and $(\sqrt{7}, 0)$
15. $(3\sqrt{3}, \sqrt{5})$ and $(-\sqrt{3}, 4\sqrt{5})$
16. $(2\sqrt{3}, \sqrt{6})$ and $(-\sqrt{3}, 5\sqrt{6})$
17. $\left(\frac{7}{3}, \frac{1}{5}\right)$ and $\left(\frac{1}{3}, \frac{6}{5}\right)$
18. $\left(-\frac{1}{4}, -\frac{1}{7}\right)$ and $\left(\frac{3}{4}, \frac{6}{7}\right)$

In Exercises 19–30, find the midpoint of the line segment with the given endpoints.

19. $(6, 8)$ and $(2, 4)$
20. $(10, 4)$ and $(2, 6)$
21. $(-2, -8)$ and $(-6, -2)$
22. $(-4, -7)$ and $(-1, -3)$
23. $(-3, -4)$ and $(6, -8)$
24. $(-2, -1)$ and $(-8, 6)$
25. $\left(-\frac{7}{2}, \frac{3}{2}\right)$ and $\left(-\frac{5}{2}, -\frac{11}{2}\right)$
26. $\left(-\frac{2}{5}, \frac{7}{15}\right)$ and $\left(-\frac{2}{5}, -\frac{4}{15}\right)$
27. $(8, 3\sqrt{5})$ and $(-6, 7\sqrt{5})$
28. $(7\sqrt{3}, -6)$ and $(3\sqrt{3}, -2)$
29. $(\sqrt{18}, -4)$ and $(\sqrt{2}, 4)$
30. $(\sqrt{50}, -6)$ and $(\sqrt{2}, 6)$

In Exercises 31–40, write the standard form of the equation of the circle with the given center and radius.

31. Center $(0, 0)$, $r = 7$
32. Center $(0, 0)$, $r = 8$
33. Center $(3, 2)$, $r = 5$
34. Center $(2, -1)$, $r = 4$
35. Center $(-1, 4)$, $r = 2$

36. Center $(-3, 5)$, $r = 3$

37. Center $(-3, -1)$, $r = \sqrt{3}$

38. Center $(-5, -3)$, $r = \sqrt{5}$

39. Center $(-4, 0)$, $r = 10$

40. Center $(-2, 0)$, $r = 6$

In Exercises 41–48, give the center and radius of the circle described by the equation and graph each equation.

41. $x^2 + y^2 = 16$

42. $x^2 + y^2 = 49$

43. $(x - 3)^2 + (y - 1)^2 = 36$

44. $(x - 2)^2 + (y - 3)^2 = 16$

45. $(x + 3)^2 + (y - 2)^2 = 4$

46. $(x + 1)^2 + (y - 4)^2 = 25$

47. $(x + 2)^2 + (y + 2)^2 = 4$

48. $(x + 4)^2 + (y + 5)^2 = 36$

In Exercises 49–56, complete the square and write the equation in standard form. Then give the center and radius of each circle and graph the equation.

49. $x^2 + y^2 + 6x + 2y + 6 = 0$

50. $x^2 + y^2 + 8x + 4y + 16 = 0$

51. $x^2 + y^2 - 10x - 6y - 30 = 0$

52. $x^2 + y^2 - 4x - 12y - 9 = 0$

53. $x^2 + y^2 + 8x - 2y - 8 = 0$

54. $x^2 + y^2 + 12x - 6y - 4 = 0$

55. $x^2 - 2x + y^2 - 15 = 0$

56. $x^2 + y^2 - 6y - 7 = 0$

Practice PLUS

In Exercises 57–60, find the solution set for each system by graphing both of the system's equations in the same rectangular coordinate system and finding all points of intersection. Check all solutions in both equations.

57. $x^2 + y^2 = 16$
$x - y = 4$

58. $x^2 + y^2 = 9$
$x - y = 3$

59. $(x - 2)^2 + (y + 3)^2 = 4$
$y = x - 3$

60. $(x - 3)^2 + (y + 1)^2 = 9$
$y = x - 1$

In Exercises 61–64, write the standard form of the equation of the circle with the given graph.

61.

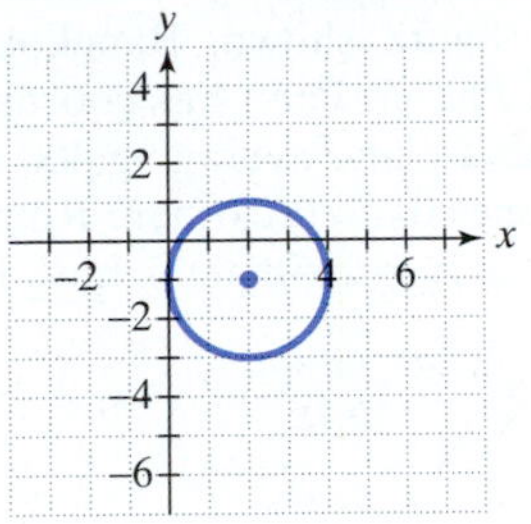

62.

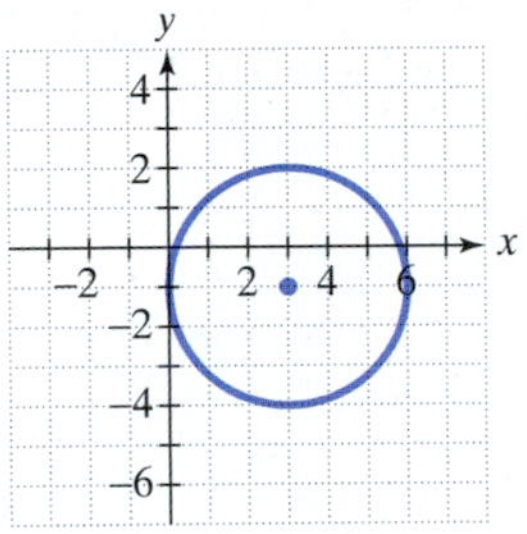

63.

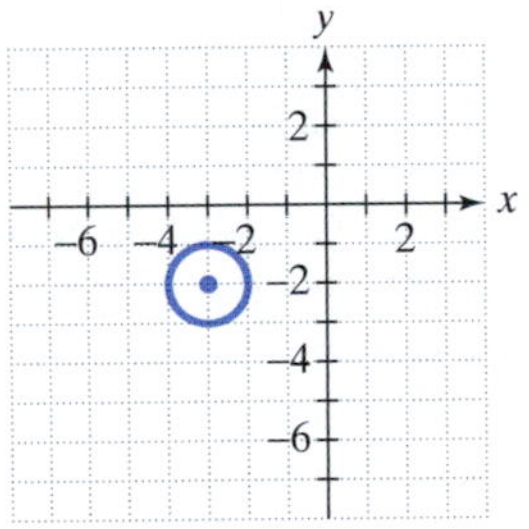

64.

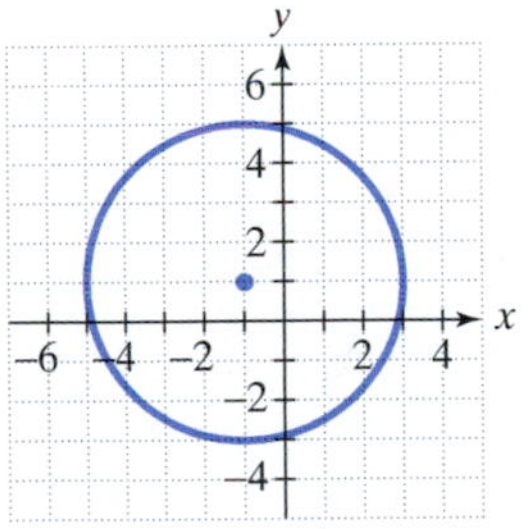

In Exercises 65–66, a line segment through the center of each circle intersects the circle at the points shown.

a. *Find the coordinates of the circle's center.*

b. *Find the radius of the circle.*

c. *Use your answers from parts (a) and (b) to write the standard form of the circle's equation.*

65.

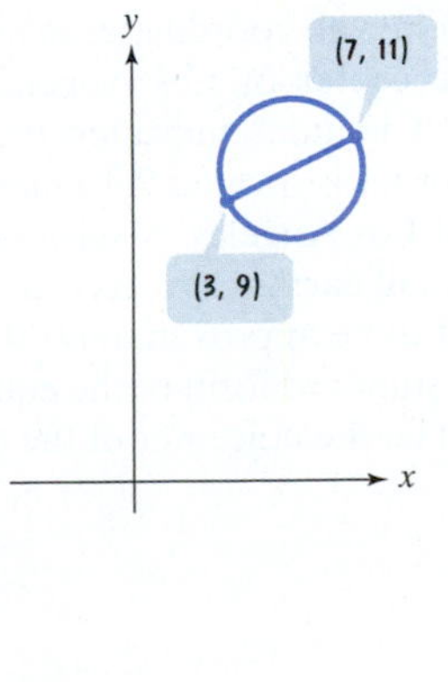

66.

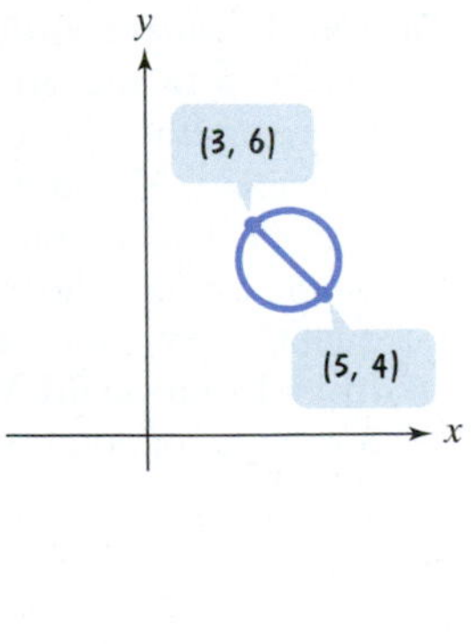

Application Exercises

67. A rectangular coordinate system with coordinates in miles is placed on the map in the figure shown. Bangkok has coordinates $(-115, 170)$ and Phnom Penh has coordinates $(65, 70)$. How long will it take a plane averaging 400 miles per hour to fly directly from one city to the other? Round to the nearest tenth of an hour. Approximately how many minutes is the flight?

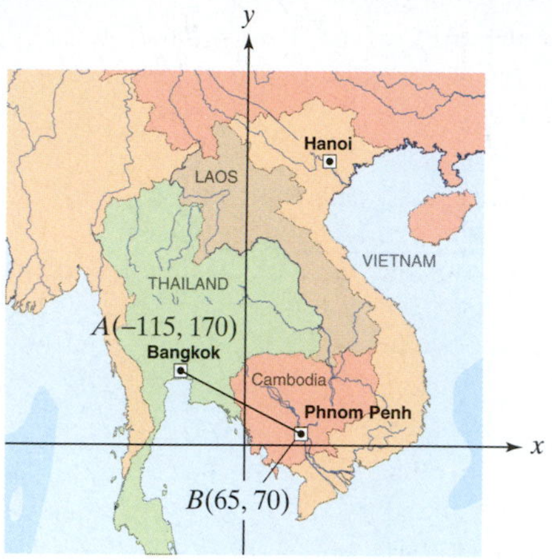

68. The ferris wheel in the figure has a radius of 68 feet. The clearance between the wheel and the ground is 14 feet. The rectangular coordinate system shown has its origin on the ground directly below the center of the wheel. Use the coordinate system to write the equation of the circular wheel.

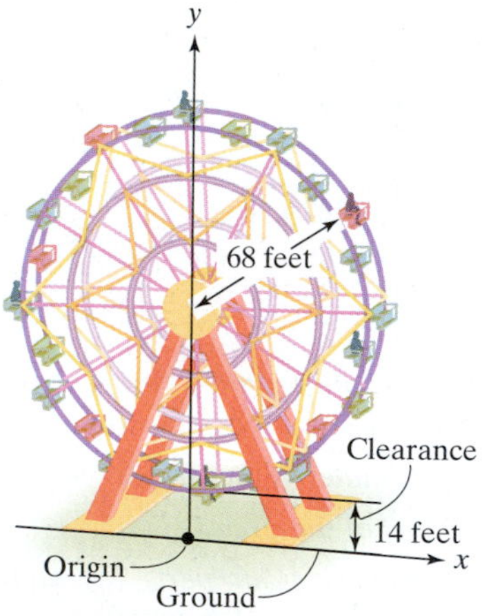

69. A rectangular coordinate system with coordinates in miles is placed with the origin at the center of Los Angeles. The figure at the top of the next column indicates that the University of Southern California is located 2.4 miles west and 2.7 miles south of central Los Angeles. A seismograph on the campus shows that a small earthquake occurred. The quake's epicenter is estimated to be approximately 30 miles from the university. Write the standard form of the equation for the set of points that could be the epicenter of the quake.

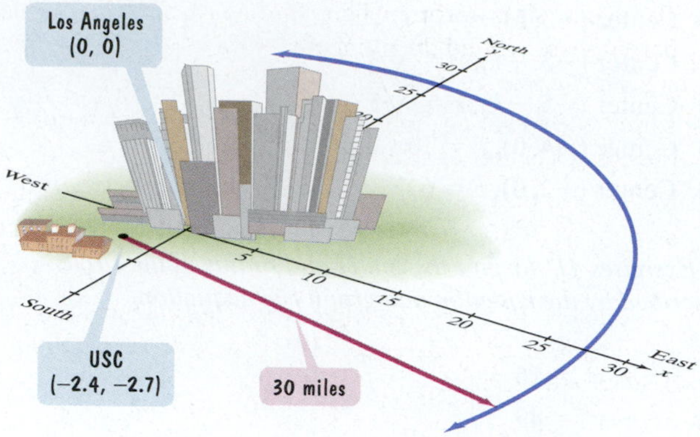

Writing in Mathematics

70. In your own words, describe how to find the distance between two points in the rectangular coordinate system.

71. In your own words, describe how to find the midpoint of a line segment if its endpoints are known.

72. What is a circle? Without using variables, describe how the definition of a circle can be used to obtain a form of its equation.

73. Give an example of a circle's equation in standard form. Describe how to find the center and radius for this circle.

74. How is the standard form of a circle's equation obtained from its general form?

75. Does $(x - 3)^2 + (y - 5)^2 = 0$ represent the equation of a circle? If not, describe the graph of this equation.

76. Does $(x - 3)^2 + (y - 5)^2 = -25$ represent the equation of a circle? What sort of set is the graph of this equation?

Technology Exercises

In Exercises 77–79, use a graphing utility to graph each circle whose equation is given.

77. $x^2 + y^2 = 25$

78. $(y + 1)^2 = 36 - (x - 3)^2$

79. $x^2 + 10x + y^2 - 4y - 20 = 0$

Critical Thinking Exercises

Make Sense? *In Exercises 80–83, determine whether each statement "makes sense" or "does not make sense" and explain your reasoning.*

80. I've noticed that in mathematics, one topic often leads logically to a new topic:

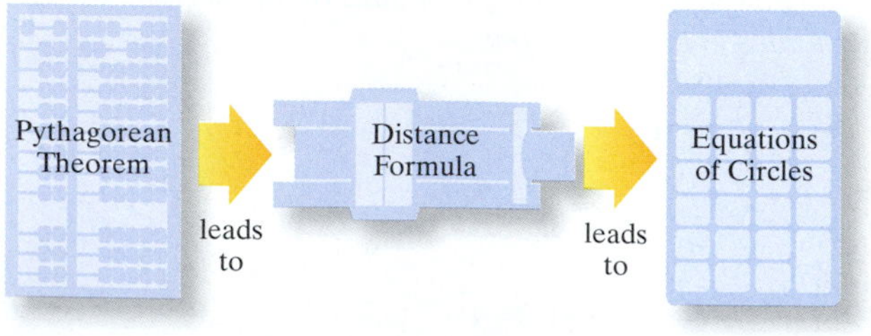

81. To avoid sign errors when finding h and k, I place parentheses around the numbers that follow the subtraction signs in a circle's equation.

82. I used the equation $(x + 1)^2 + (y - 5)^2 = -4$ to identify the circle's center and radius.

83. My graph of $(x - 2)^2 + (y + 1)^2 = 16$ is my graph of $x^2 + y^2 = 16$ translated two units right and one unit down.

In Exercises 84–87, determine whether each statement is true or false. If the statement is false, make the necessary change(s) to produce a true statement.

84. The equation of the circle whose center is at the origin with radius 16 is $x^2 + y^2 = 16$.

85. The graph of $(x - 3)^2 + (y + 5)^2 = 36$ is a circle with radius 6 centered at $(-3, 5)$.

86. The graph of $(x - 4) + (y + 6) = 25$ is a circle with radius 5 centered at $(4, -6)$.

87. The graph of $(x - 3)^2 + (y + 5)^2 = -36$ is a circle with radius 6 centered at $(3, -5)$.

88. Show that the points $A(1, 1 + d)$, $B(3, 3 + d)$, and $C(6, 6 + d)$ are collinear (lie along a straight line) by showing that the distance from A to B plus the distance from B to C equals the distance from A to C.

89. Prove the midpoint formula by using the following procedure.

a. Show that the distance between (x_1, y_1) and $\left(\frac{x_1 + x_2}{2}, \frac{y_1 + y_2}{2}\right)$ is equal to the distance between (x_2, y_2) and $\left(\frac{x_1 + x_2}{2}, \frac{y_1 + y_2}{2}\right)$.

b. Use the procedure from Exercise 88 and the distances from part (a) to show that the points (x_1, y_1), $\left(\frac{x_1 + x_2}{2}, \frac{y_1 + y_2}{2}\right)$, and (x_2, y_2) are collinear.

90. Find all points with y-coordinate 2 so that the distance between $(x, 2)$ and $(2, -1)$ is 5.

91. Find the area of the doughnut-shaped region bounded by the graphs of $(x - 2)^2 + (y + 3)^2 = 25$ and $(x - 2)^2 + (y + 3)^2 = 36$.

92. A **tangent line** to a circle is a line that intersects the circle at exactly one point. The tangent line is perpendicular to the radius of the circle at this point of contact. Write the point-slope form of the equation of a line tangent to the circle whose equation is $x^2 + y^2 = 25$ at the point $(3, -4)$.

Review Exercises

93. If $f(x) = x^2 - 2$ and $g(x) = 3x + 4$, find $f(g(x))$ and $g(f(x))$. (Section 9.2, Example 1)

94. Solve: $2x = \sqrt{7x - 3} + 3$.
(Section 7.6, Example 3)

95. Solve: $|2x - 5| < 10$.
(Section 4.3, Example 4 or Example 6(a))

Preview Exercises

Exercises 96–98 will help you prepare for the material covered in the next section.

96. Set $y = 0$ and find the x-intercepts: $\frac{x^2}{9} + \frac{y^2}{4} = 1$.

97. Set $x = 0$ and find the y-intercepts: $\frac{x^2}{9} + \frac{y^2}{4} = 1$.

98. Divide both sides of $25x^2 + 16y^2 = 400$ by 400 and simplify.

SECTION 10.2

The Ellipse

Objectives

1. Graph ellipses centered at the origin.
2. Graph ellipses not centered at the origin.
3. Solve applied problems involving ellipses.

You took on a summer job driving a truck, delivering books that were ordered online. You're an avid reader, so just being around books sounded appealing. However, now you're feeling a bit shaky driving the truck for the first time. It's 10 feet wide and 9 feet high; compared to your compact car, it feels like you're behind the wheel of a tank. Up ahead you see a sign at the semielliptical entrance to a tunnel: Caution! Tunnel is 10 Feet High at Center Peak. Then you see another sign: Caution! Tunnel is 40 Feet Wide. Will your truck clear the opening of the tunnel's archway?

Mathematics is present in the movements of planets, bridge and tunnel construction, navigational systems used to keep track of a ship's location, manufacture of lenses for telescopes, and even in a procedure for disintegrating kidney stones. The mathematics behind these applications involves conic sections. **Conic sections** are curves that result from the intersection of a right circular cone and a plane. **Figure 10.8** illustrates the four conic sections: the circle, the ellipse, the parabola, and the hyperbola.

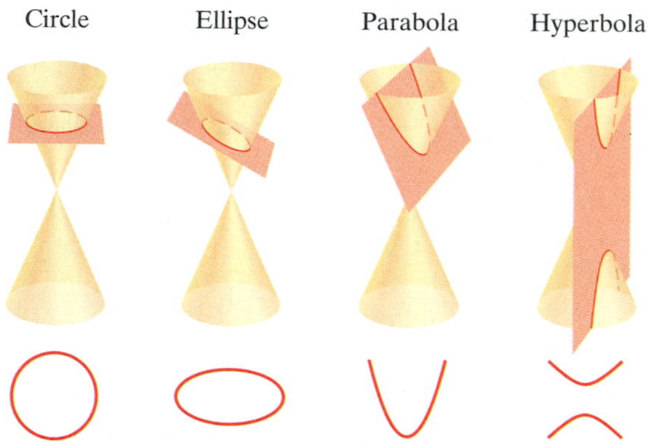

FIGURE 10.8 Obtaining the conic sections by intersecting a plane and a cone

In this section, we study the symmetric oval-shaped curve known as the ellipse. We will use a geometric definition for an ellipse to derive its equation. With this equation, we will determine if your delivery truck will clear the tunnel's entrance.

Definition of an Ellipse

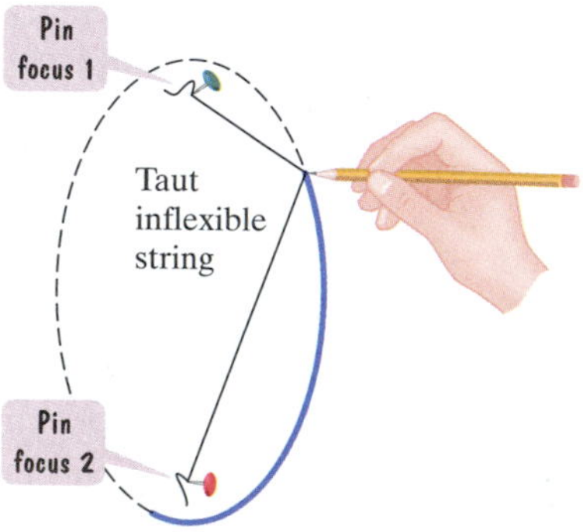

FIGURE 10.9 Drawing an ellipse

Figure 10.9 illustrates how to draw an ellipse. Place pins at two fixed points, each of which is called a focus (plural: foci). If the ends of a fixed length of string are fastened to the pins and we draw the string taut with a pencil, the path traced by the pencil will be an ellipse. Notice that the sum of the distances of the pencil point from the foci remains constant because the length of the string is fixed. This procedure for drawing an ellipse illustrates its geometric definition.

Definition of an Ellipse

An **ellipse** is the set of all points, P, in a plane the sum of whose distances from two fixed points, F_1 and F_2, is constant (see **Figure 10.10**). These two fixed points are called the **foci** (plural of **focus**). The midpoint of the segment connecting the foci is the **center** of the ellipse.

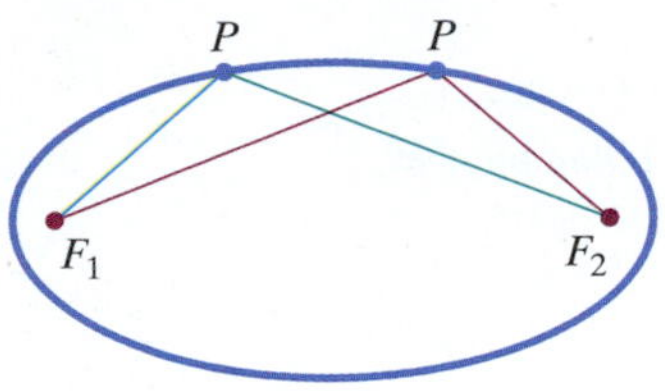

FIGURE 10.10

Figure 10.11 illustrates that an ellipse can be elongated in any direction. We will limit our discussion to ellipses that are elongated horizontally or vertically. The line through the foci intersects the ellipse at two points, called the **vertices** (singular: **vertex**). The line segment that joins the vertices is the **major axis**. Notice that the midpoint of the major axis is the center of the ellipse. The line segment whose endpoints are on the ellipse and that is perpendicular to the major axis at the center is called the **minor axis** of the ellipse.

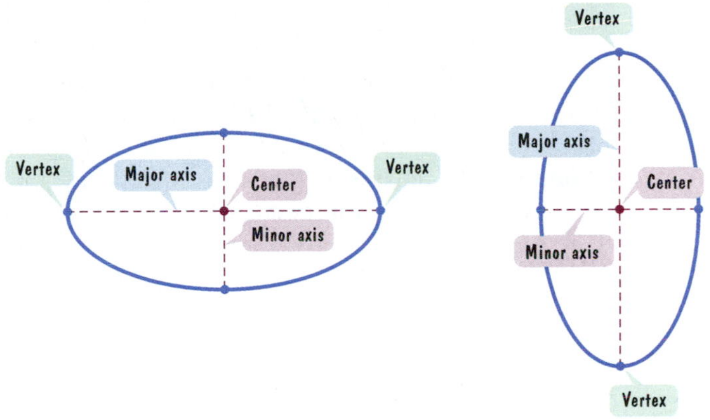

FIGURE 10.11 Horizontal and vertical elongations of an ellipse

Standard Form of the Equation of an Ellipse

The rectangular coordinate system gives us a unique way of describing an ellipse. It enables us to translate an ellipse's geometric definition into an algebraic equation.

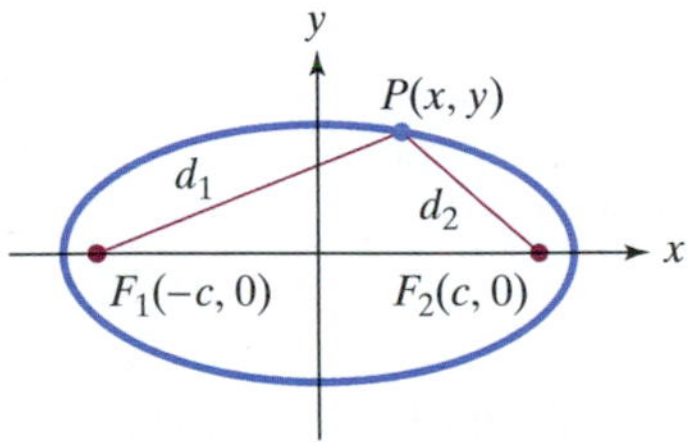

FIGURE 10.12

We start with **Figure 10.12** to obtain an ellipse's equation. We've placed an ellipse that is elongated horizontally into a rectangular coordinate system. The foci are on the x-axis at $(-c, 0)$ and $(c, 0)$, as in **Figure 10.12**. In this way, the center of the ellipse is at the origin. We let (x, y) represent the coordinates of any point, P, on the ellipse.

What does the definition of an ellipse tell us about the point (x, y) in **Figure 10.12**? For any point (x, y) on the ellipse, the sum of the distances to the two foci, $d_1 + d_2$, must be constant. As we shall see, it is convenient to denote this constant by $2a$. Thus, the point (x, y) is on the ellipse if and only if

$$d_1 + d_2 = 2a.$$

$$\sqrt{(x + c)^2 + y^2} + \sqrt{(x - c)^2 + y^2} = 2a \quad \text{Use the distance formula.}$$

After eliminating radicals and simplifying, we obtain

$$(a^2 - c^2)x^2 + a^2y^2 = a^2(a^2 - c^2).$$

Study Tip

The algebraic details behind eliminating the radicals and obtaining the equation shown can be found in the appendix. There you will find a step-by-step derivation of the ellipse's equation.

To simplify $(a^2 - c^2)x^2 + a^2y^2 = a^2(a^2 - c^2)$, let $b^2 = a^2 - c^2$. Substituting b^2 for $a^2 - c^2$, we obtain

$$b^2x^2 + a^2y^2 = a^2b^2.$$

$$\frac{b^2x^2}{a^2b^2} + \frac{a^2y^2}{a^2b^2} = \frac{a^2b^2}{a^2b^2} \quad \text{Divide both sides by } a^2b^2.$$

$$\frac{x^2}{a^2} + \frac{y^2}{b^2} = 1 \quad \text{Simplify.}$$

This last equation is the **standard form of the equation of an ellipse centered at the origin**. There are two such equations, one for a horizontal major axis and one for a vertical major axis.

Standard Forms of the Equations of an Ellipse

The **standard form of the equation of an ellipse** with center at the origin and major and minor axes of lengths $2a$ and $2b$ (where a and b are positive, and $a^2 > b^2$) is

$$\frac{x^2}{a^2} + \frac{y^2}{b^2} = 1 \quad \text{or} \quad \frac{x^2}{b^2} + \frac{y^2}{a^2} = 1.$$

Figure 10.13 illustrates that the vertices are on the major axis, a units from the center. The foci are on the major axis, c units from the center.

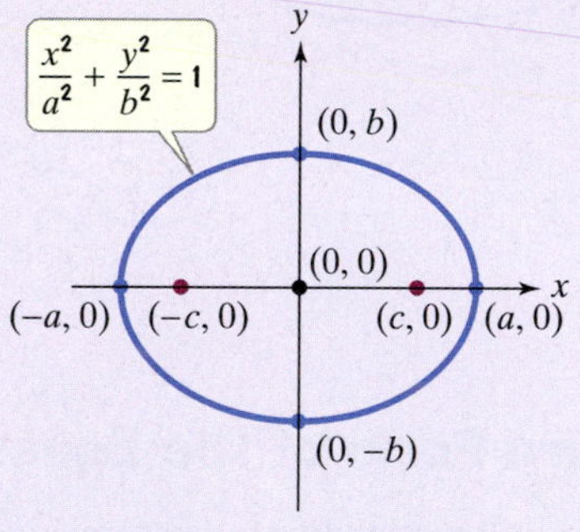

FIGURE 10.13(a) Major axis is horizontal with length $2a$.

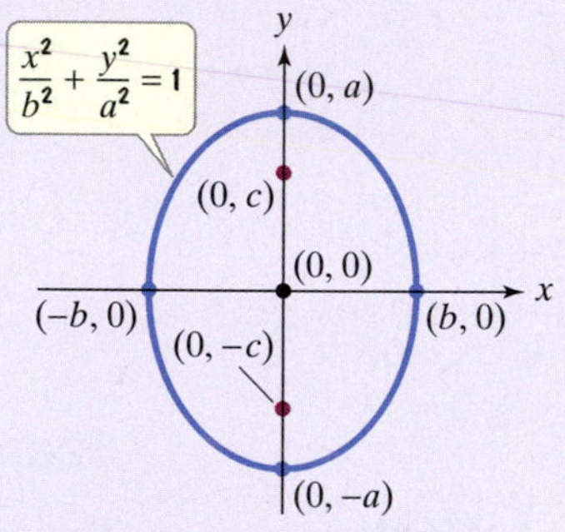

FIGURE 10.13(b) Major axis is vertical with length $2a$.

The intercepts shown in **Figure 10.13(a)** can be obtained algebraically. Let's do this for

$$\frac{x^2}{a^2} + \frac{y^2}{b^2} = 1.$$

x-intercepts: Set $y = 0$.

$$\frac{x^2}{a^2} = 1$$

$$x^2 = a^2$$

$$x = \pm a$$

x-intercepts are $-a$ and a. The graph passes through $(-a, 0)$ and $(a, 0)$, which are the vertices.

y-intercepts: Set $x = 0$.

$$\frac{y^2}{b^2} = 1$$

$$y^2 = b^2$$

$$y = \pm b$$

y-intercepts are $-b$ and b. The graph passes through $(0, -b)$ and $(0, b)$.

1 Graph ellipses centered at the origin.

Using the Standard Form of the Equation of an Ellipse

We can use the standard form of an ellipse's equation to graph the ellipse.

EXAMPLE 1 Graphing an Ellipse Centered at the Origin

Graph the ellipse: $\frac{x^2}{9} + \frac{y^2}{4} = 1$.

Solution The given equation is the standard form of an ellipse's equation with $a^2 = 9$ and $b^2 = 4$.

$$\frac{x^2}{9} + \frac{y^2}{4} = 1$$

$a^2 = 9$. This is the larger of the two denominators.

$b^2 = 4$. This is the smaller of the two denominators.

Because the denominator of the x^2-term is greater than the denominator of the y^2-term, the major axis is horizontal. Based on the standard form of the equation, we know the vertices are $(-a, 0)$ and $(a, 0)$. Because $a^2 = 9$, $a = 3$. Thus, the vertices are $(-3, 0)$ and $(3, 0)$, shown in **Figure 10.14**.

Now let us find the endpoints of the vertical minor axis. According to the standard form of the equation, these endpoints are $(0, -b)$ and $(0, b)$. Because $b^2 = 4$, $b = 2$. Thus, the endpoints of the minor axis are $(0, -2)$ and $(0, 2)$. They are shown in **Figure 10.14**.

You can sketch the ellipse in **Figure 10.14** by locating endpoints on the major and minor axes.

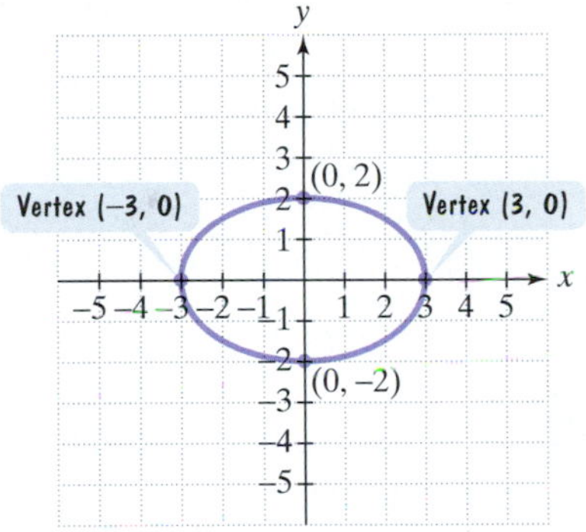

FIGURE 10.14 The graph of $\frac{x^2}{9} + \frac{y^2}{4} = 1$

$$\frac{x^2}{3^2} + \frac{y^2}{2^2} = 1$$

Endpoints of the major axis are 3 units to the right and left of the center.

Endpoints of the minor axis are 2 units up and down from the center.

Using Technology

We graph $\frac{x^2}{9} + \frac{y^2}{4} = 1$ with a graphing utility by solving for y.

$$\frac{y^2}{4} = 1 - \frac{x^2}{9}$$

$$y^2 = 4\left(1 - \frac{x^2}{9}\right)$$

$$y = \pm 2\sqrt{1 - \frac{x^2}{9}}$$

Notice that the square root property requires us to define two functions. Enter

$y_1 = 2$ [√] (1 [−] x [∧] 2 [÷] 9)

and

$$y_2 = -y_1.$$

To see the true shape of the ellipse, use the

[ZOOM SQUARE]

feature so that one unit on the y-axis is the same length as one unit on the x-axis.

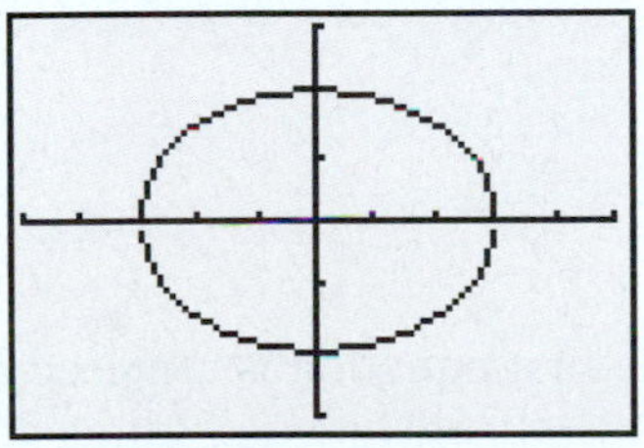

[−5, 5, 1] by [−3, 3, 1]

CHECK POINT 1 Graph the ellipse: $\frac{x^2}{36} + \frac{y^2}{9} = 1$.

EXAMPLE 2 Graphing an Ellipse Centered at the Origin

Graph the ellipse: $25x^2 + 16y^2 = 400$.

Solution We begin by expressing the equation in standard form. Because we want 1 on the right side, we divide both sides by 400.

$$\frac{25x^2}{400} + \frac{16y^2}{400} = \frac{400}{400}$$

$$\frac{x^2}{16} + \frac{y^2}{25} = 1$$

$b^2 = 16$. This is the smaller of the two denominators.

$a^2 = 25$. This is the larger of the two denominators.

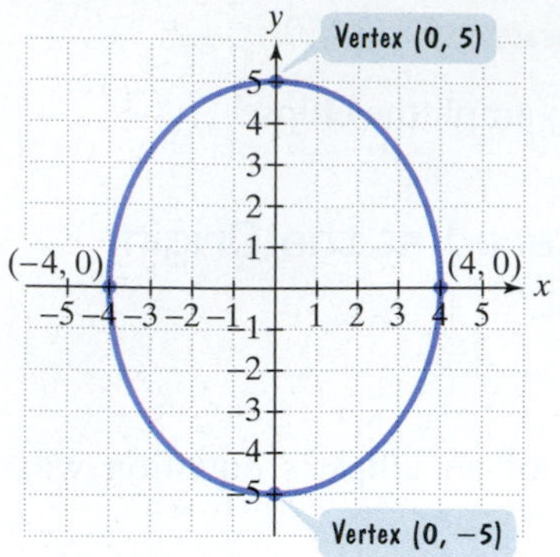

FIGURE 10.15 The graph of $\frac{x^2}{16} + \frac{y^2}{25} = 1$

The equation $\frac{x^2}{16} + \frac{y^2}{25} = 1$ is the standard form of an ellipse's equation with $a^2 = 25$ and $b^2 = 16$. Because the denominator of the y^2-term is greater than the denominator of the x^2-term, the major axis is vertical. Based on the standard form of the equation, we know that the vertices are $(0, -a)$ and $(0, a)$. Because $a^2 = 25, a = 5$. Thus, the vertices are $(0, -5)$ and $(0, 5)$, shown in **Figure 10.15**.

Now let us find the endpoints of the horizontal minor axis. According to the standard form of the equation, these endpoints are $(-b, 0)$ and $(b, 0)$. Because $b^2 = 16, b = 4$. Thus, the endpoints of the minor axis are $(-4, 0)$ and $(4, 0)$. They are shown in **Figure 10.15**.

You can sketch the ellipse in **Figure 10.15** by locating the endpoints on the major and minor axes.

$$\frac{x^2}{4^2} + \frac{y^2}{5^2} = 1$$

Endpoints of the minor axis are 4 units to the right and left of the center.

Endpoints of the major axis are 5 units up and down from the center.

✓ **CHECK POINT 2** Graph the ellipse: $16x^2 + 9y^2 = 144$.

2 Graph ellipses not centered at the origin.

Ellipses Centered at (h, k)

Horizontal and vertical translations can be used to graph ellipses that are not centered at the origin. **Figure 10.16** illustrates that the graphs of

$$\frac{(x-h)^2}{a^2} + \frac{(y-k)^2}{b^2} = 1 \quad \text{and} \quad \frac{x^2}{a^2} + \frac{y^2}{b^2} = 1$$

have the same size and shape. However, the graph of the first equation is centered at (h, k) rather than at the origin.

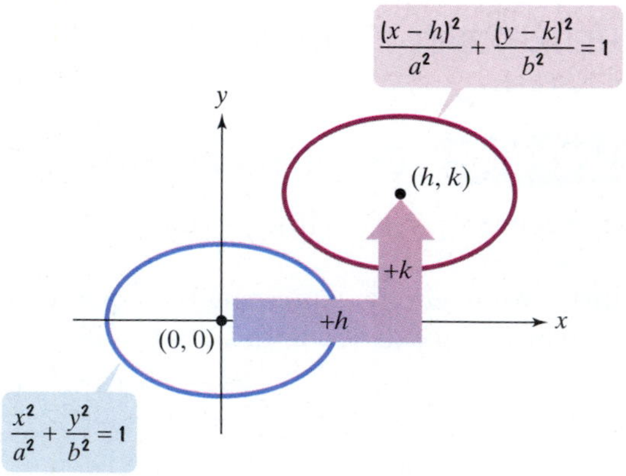

FIGURE 10.16 Translating an ellipse's graph

Table 10.1 on the next page gives the standard forms of equations of ellipses centered at (h, k) and shows their graphs.

Table 10.1 Standard Forms of Equations of Ellipses Centered at (h, k)

Equation	Center	Major Axis	Vertices	Graph
$\frac{(x-h)^2}{a^2}+\frac{(y-k)^2}{b^2}=1$ Endpoints of major axis are a units right and a units left of center. $a^2 > b^2$	(h, k)	Parallel to the x-axis, horizontal	$(h-a, k)$ $(h+a, k)$	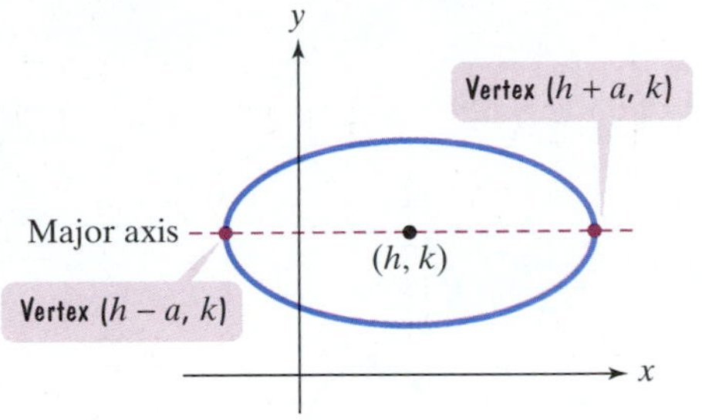
$\frac{(x-h)^2}{b^2}+\frac{(y-k)^2}{a^2}=1$ $a^2 > b^2$ Endpoints of the major axis are a units above and a units below the center.	(h, k)	Parallel to the y-axis, vertical	$(h, k-a)$ $(h, k+a)$	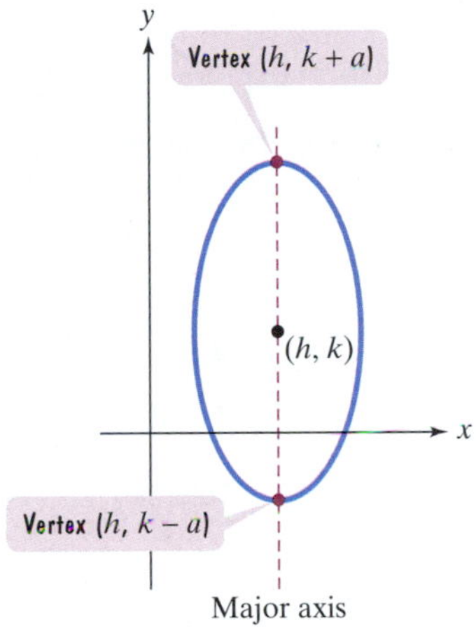

EXAMPLE 3 Graphing an Ellipse Centered at (h, k)

Graph the ellipse: $\frac{(x-1)^2}{4}+\frac{(y+2)^2}{9}=1.$

Solution To graph the ellipse, we need to know its center, (h, k). In the standard forms of equations centered at (h, k), h is the number subtracted from x and k is the number subtracted from y.

This is $(x-h)^2$ with $h = 1$.

This is $(y-k)^2$ with $k = -2$.

$$\frac{(x-1)^2}{4}+\frac{(y-(-2))^2}{9}=1$$

We see that $h = 1$ and $k = -2$. Thus, the center of the ellipse, (h, k), is $(1, -2)$. We can graph the ellipse by locating endpoints on the major and minor axes. To do this, we must identify a^2 and b^2.

$$\frac{(x-1)^2}{4}+\frac{(y+2)^2}{9}=1$$

$b^2 = 4$. This is the smaller of the two denominators.

$a^2 = 9$. This is the larger of the two denominators.

The larger number is under the expression involving y. This means that the major axis is vertical and parallel to the y-axis.

We can sketch the ellipse by locating endpoints on the major and minor axes.

$$\frac{(x-1)^2}{2^2} + \frac{(y+2)^2}{3^2} = 1$$

Endpoints of the minor axis are 2 units to the right and left of the center.

Endpoints of the major axis (the vertices) are 3 units up and down from the center.

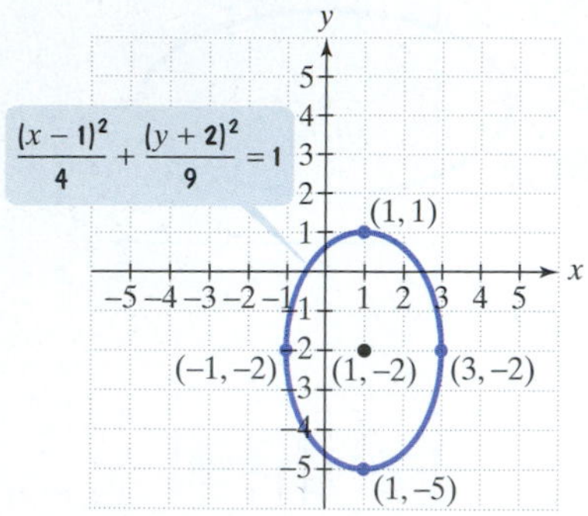

FIGURE 10.17 The graph of an ellipse centered at (1, −2)

We categorize the observations in the voice balloons as follows:

3 units above and below center

For a Vertical Major Axis with Center (1, −2)	
Vertices	**Endpoints of Minor Axis**
(1, −2 + 3) = (1, 1)	(1 + 2, −2) = (3, −2)
(1, −2 − 3) = (1, −5)	(1 − 2, −2) = (−1, −2)

2 units right and left of center

Using the center and these four points, we can sketch the ellipse shown in **Figure 10.17**.

✓ **CHECK POINT 3** Graph the ellipse: $\frac{(x+1)^2}{9} + \frac{(y-2)^2}{4} = 1.$

3 Solve applied problems involving ellipses.

Applications

Ellipses have many applications. German scientist Johannes Kepler (1571–1630) showed that the planets in our solar system move in elliptical orbits, with the sun at a focus. Earth satellites also travel in elliptical orbits, with Earth at a focus.

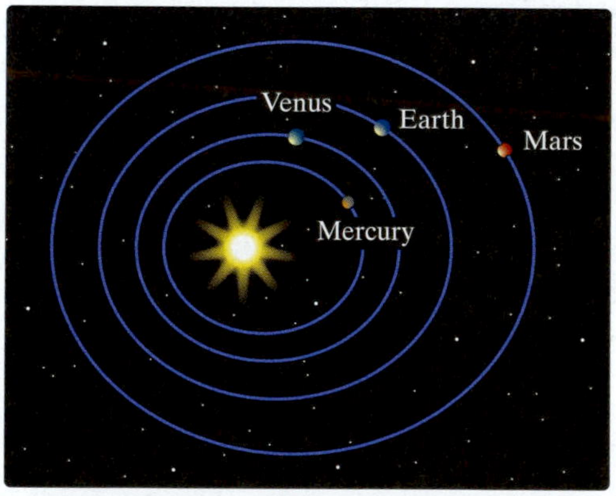

Planets move in elliptical orbits.

One intriguing aspect of the ellipse is that a ray of light or a sound wave emanating from one focus will be reflected from the ellipse to exactly the other focus. A whispering gallery is an elliptical room with an elliptical, dome-shaped ceiling. People standing at the foci can whisper and hear each other quite clearly, while persons in other locations in the room cannot hear them. Statuary Hall in the U.S. Capitol Building is elliptical. President John Quincy Adams, while a member of the House of Representatives, was aware of this acoustical phenomenon. He situated his desk at a focal point of the elliptical ceiling, easily eavesdropping on the private conversations of other House members located near the other focus.

The elliptical reflection principle is used in a procedure for disintegrating kidney stones. The patient is placed within a device that is elliptical in shape. The patient is placed so the kidney is centered at one focus, while ultrasound waves from the other focus hit the walls and are reflected to the kidney stone. The convergence of the ultrasound waves at the kidney stone causes vibrations that shatter it into fragments. The small pieces can then be passed painlessly through the patient's system. The patient recovers in days, as opposed to up to six weeks if surgery is used instead.

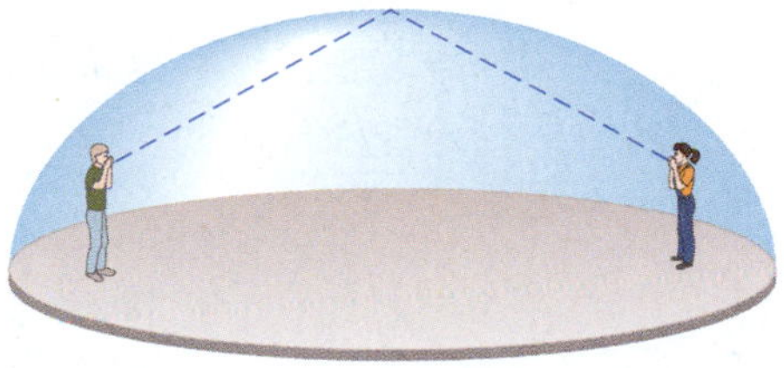

Whispering in an elliptical dome

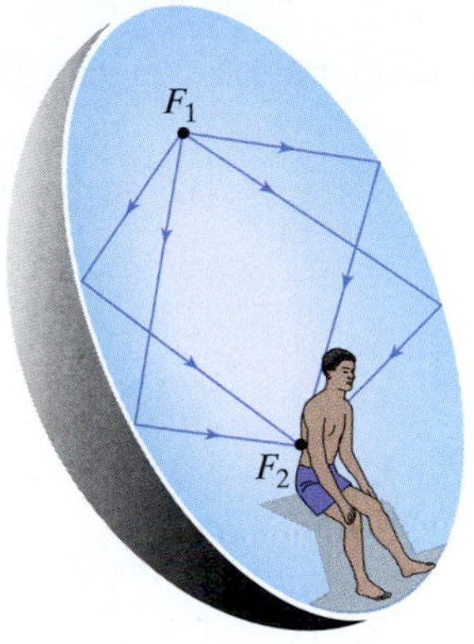

Disintegrating kidney stones

Ellipses are often used for supporting arches of bridges and in tunnel construction. This application forms the basis of our next example.

EXAMPLE 4 An Application Involving an Ellipse

A semielliptical archway over a one-way road has a height of 10 feet and a width of 40 feet (see **Figure 10.18**). Your truck has a width of 10 feet and a height of 9 feet. Will your truck clear the opening of the archway?

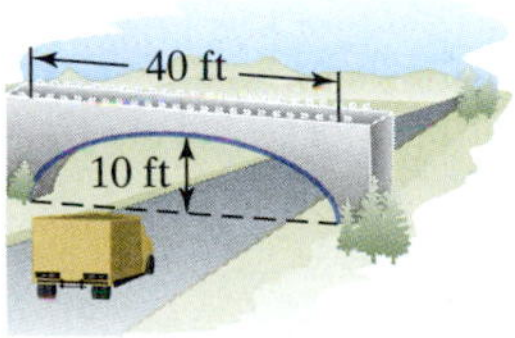

FIGURE 10.18 A semielliptical archway

Solution Because your truck's width is 10 feet, to determine the clearance, we must find the height of the archway 5 feet from the center. If that height is 9 feet or less, the truck will not clear the opening.

In **Figure 10.19**, we've constructed a coordinate system with the x-axis on the ground and the origin at the center of the archway. Also shown is the truck, whose height is 9 feet.

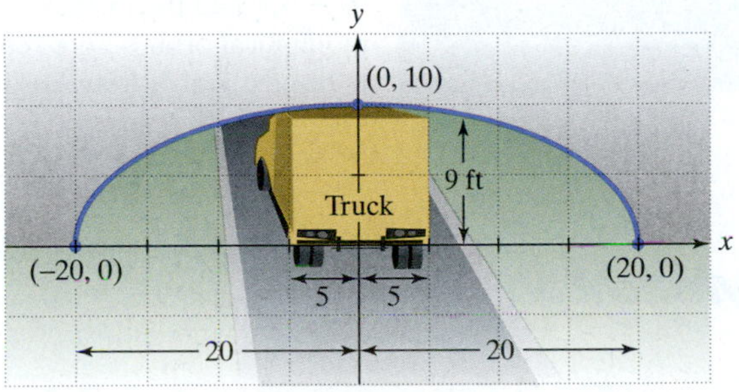

FIGURE 10.19

Using the equation $\frac{x^2}{a^2} + \frac{y^2}{b^2} = 1$, we can express the equation of the blue archway in **Figure 10.19** as $\frac{x^2}{20^2} + \frac{y^2}{10^2} = 1$, or $\frac{x^2}{400} + \frac{y^2}{100} = 1$.

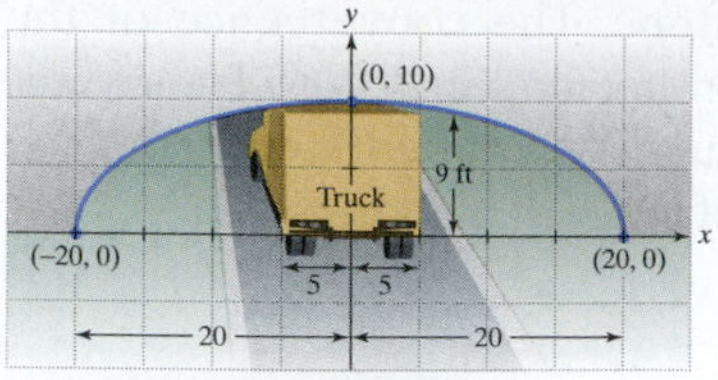

FIGURE 10.19 (repeated)

As shown in **Figure 10.19**, the edge of the 10-foot-wide truck corresponds to $x = 5$. We find the height of the archway 5 feet from the center by substituting 5 for x and solving for y.

$$\frac{5^2}{400} + \frac{y^2}{100} = 1 \qquad \text{Substitute 5 for } x \text{ in } \frac{x^2}{400} + \frac{y^2}{100} = 1.$$

$$\frac{25}{400} + \frac{y^2}{100} = 1 \qquad \text{Square 5.}$$

$$400\left(\frac{25}{400} + \frac{y^2}{100}\right) = 400(1) \qquad \text{Clear fractions by multiplying both sides by 400.}$$

$$25 + 4y^2 = 400 \qquad \text{Use the distributive property and simplify.}$$

$$4y^2 = 375 \qquad \text{Subtract 25 from both sides.}$$

$$y^2 = \frac{375}{4} \qquad \text{Divide both sides by 4.}$$

$$y = \sqrt{\frac{375}{4}} \qquad \text{Take only the positive square root. The archway is above the x-axis, so y is nonnegative.}$$

$$\approx 9.68 \qquad \text{Use a calculator.}$$

Thus, the height of the archway 5 feet from the center is approximately 9.68 feet. Because your truck's height is 9 feet, there is enough room for the truck to clear the archway.

CHECK POINT 4 Will a truck that is 12 feet wide and has a height of 9 feet clear the opening of the archway described in Example 4?

Blitzer Bonus

Halley's Comet

Halley's Comet has an elliptical orbit with the sun at one focus. The comet returns every 76.3 years. The first recorded sighting was in 239 B.C. It was last seen in 1986. At that time, spacecraft went close to the comet, measuring its nucleus to be 7 miles long and 4 miles wide. By 2024, Halley's Comet will have reached the farthest point in its elliptical orbit before returning to be next visible from Earth as it loops around the sun in 2062.

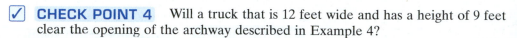

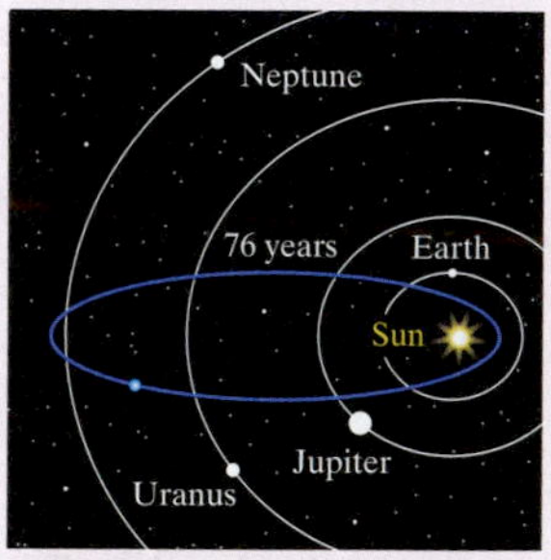

The elliptical orbit of Halley's Comet

10.2 EXERCISE SET

MyMathLab — MathXL PRACTICE · WATCH · DOWNLOAD · READ · REVIEW

Practice Exercises

In Exercises 1–16, graph each ellipse

1. $\frac{x^2}{16} + \frac{y^2}{4} = 1$
2. $\frac{x^2}{25} + \frac{y^2}{16} = 1$
3. $\frac{x^2}{9} + \frac{y^2}{36} = 1$
4. $\frac{x^2}{16} + \frac{y^2}{49} = 1$
5. $\frac{x^2}{25} + \frac{y^2}{64} = 1$
6. $\frac{x^2}{49} + \frac{y^2}{36} = 1$
7. $\frac{x^2}{49} + \frac{y^2}{81} = 1$
8. $\frac{x^2}{64} + \frac{y^2}{100} = 1$
9. $25x^2 + 4y^2 = 100$
10. $9x^2 + 4y^2 = 36$
11. $4x^2 + 16y^2 = 64$
12. $16x^2 + 9y^2 = 144$

13. $25x^2 + 9y^2 = 225$

14. $4x^2 + 25y^2 = 100$

15. $x^2 + 2y^2 = 8$

16. $12x^2 + 4y^2 = 36$

In Exercises 17–20, find the standard form of the equation of each ellipse.

17.

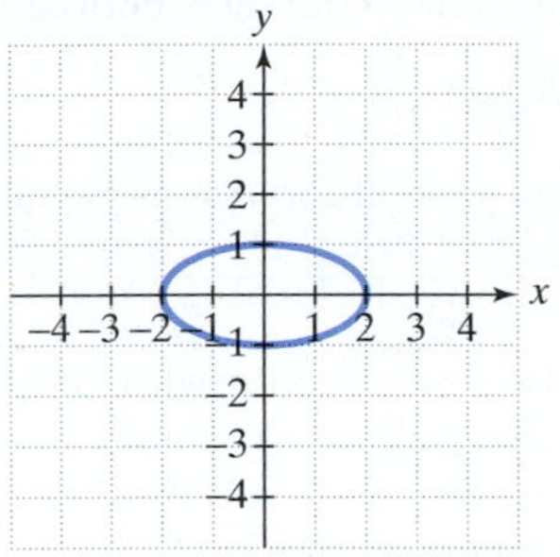

18.

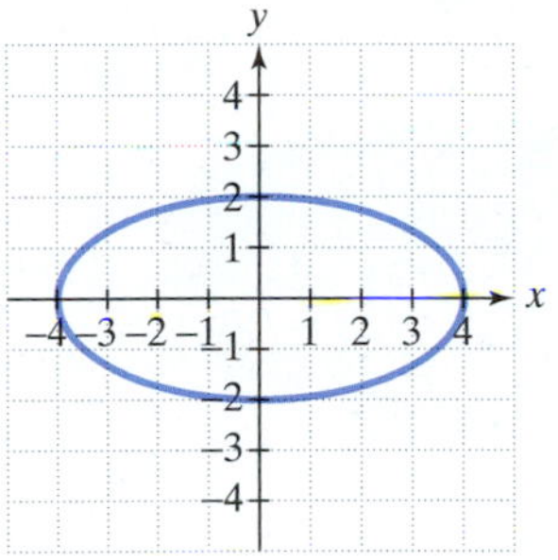

19.

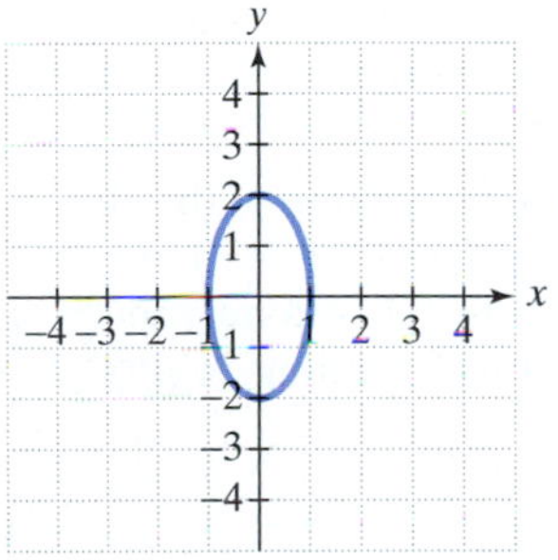

20.

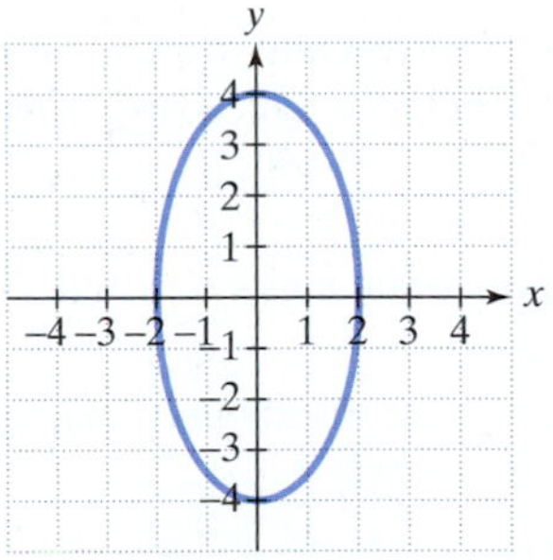

In Exercises 21–32, graph each ellipse.

21. $\frac{(x-2)^2}{9} + \frac{(y-1)^2}{4} = 1$

22. $\frac{(x-1)^2}{16} + \frac{(y+2)^2}{9} = 1$

23. $(x+3)^2 + 4(y-2)^2 = 16$

24. $(x-3)^2 + 9(y+2)^2 = 36$

25. $\frac{(x-4)^2}{9} + \frac{(y+2)^2}{25} = 1$

26. $\frac{(x-3)^2}{9} + \frac{(y+1)^2}{16} = 1$

27. $\frac{x^2}{25} + \frac{(y-2)^2}{36} = 1$

28. $\frac{(x-4)^2}{4} + \frac{y^2}{25} = 1$

29. $\frac{(x+3)^2}{9} + (y-2)^2 = 1$

30. $\frac{(x+2)^2}{16} + (y-3)^2 = 1$

31. $9(x-1)^2 + 4(y+3)^2 = 36$

32. $36(x+4)^2 + (y+3)^2 = 36$

Practice PLUS

In Exercises 33–34, find the standard form of the equation of each ellipse.

33.

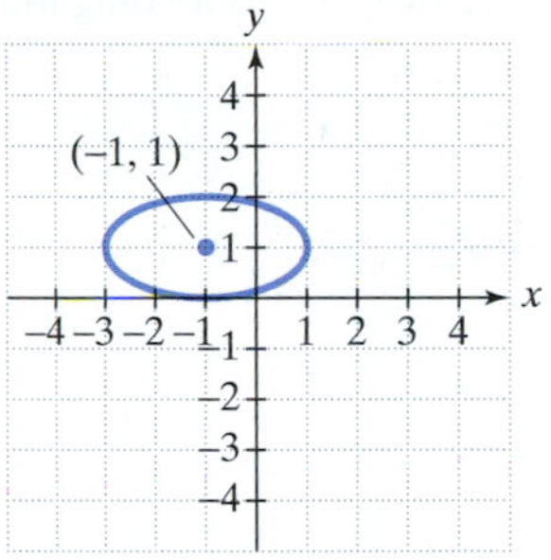

34.

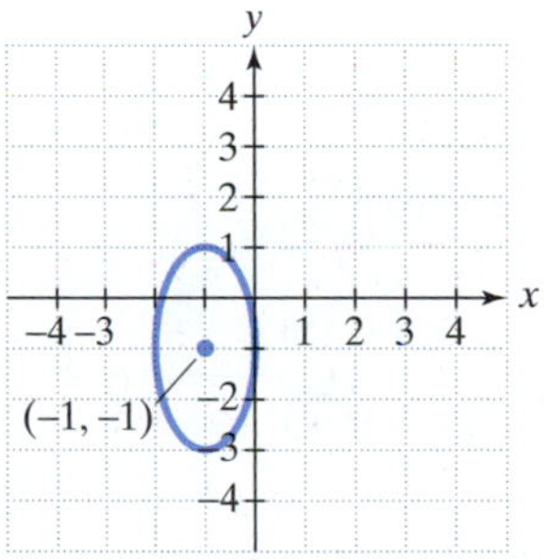

In Exercises 35–40, find the solution set for each system by graphing both of the system's equations in the same rectangular coordinate system and finding all points of intersection. Check all solutions in both equations.

35. $x^2 + y^2 = 1$
$x^2 + 9y^2 = 9$

36. $x^2 + y^2 = 25$
$25x^2 + y^2 = 25$

37. $\frac{x^2}{25} + \frac{y^2}{9} = 1$
$y = 3$

38. $\frac{x^2}{4} + \frac{y^2}{36} = 1$
$x = -2$

39. $4x^2 + y^2 = 4$
$2x - y = 2$

40. $4x^2 + y^2 = 4$
$x + y = 3$

In Exercises 41–42, graph each semiellipse.

41. $y = -\sqrt{16 - 4x^2}$ **42.** $y = -\sqrt{4 - 4x^2}$

Application Exercises

43. Will a truck that is 8 feet wide carrying a load that reaches 7 feet above the ground clear the semielliptical arch on the one-way road that passes under the bridge shown in the figure?

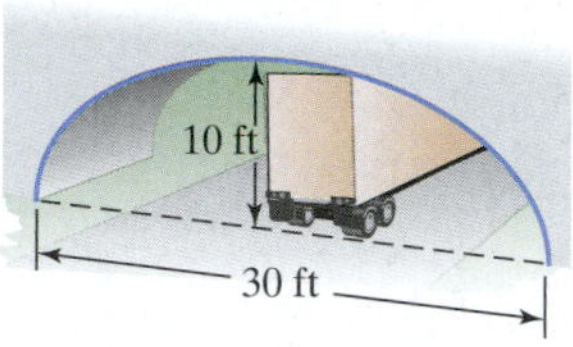

44. A semielliptic archway has a height of 20 feet and a width of 50 feet, as shown in the figure. Can a truck 14 feet high and 10 feet wide drive under the archway without going into the other lane?

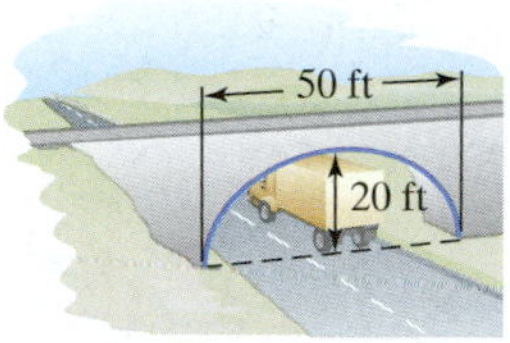

45. The elliptical ceiling in Statuary Hall in the U.S. Capitol Building is 96 feet long and 23 feet tall.

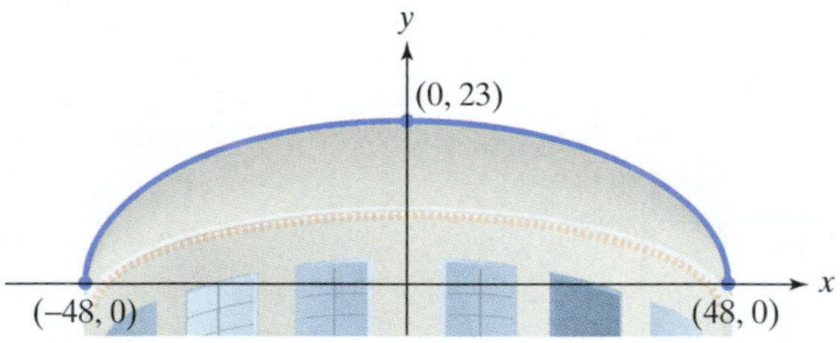

a. Using the rectangular coordinate system in the figure shown, write the standard form of the equation of the elliptical ceiling.

b. John Quincy Adams discovered that he could overhear the conversations of opposing party leaders near the left side of the chamber if he situated his desk at the focus, $(c, 0)$, at the right side of the chamber, where $c^2 = a^2 - b^2$. How far from the center of the ellipse along the major axis did Adams situate his desk? (Round to the nearest foot.)

46. If an elliptical whispering room has a height of 30 feet and a width of 100 feet, where should two people stand if they would like to whisper back and forth and be heard?

Writing in Mathematics

47. What is an ellipse?

48. Describe how to graph $\frac{x^2}{25} + \frac{y^2}{16} = 1$.

49. Describe one similarity and one difference between the graphs of $\frac{x^2}{25} + \frac{y^2}{16} = 1$ and $\frac{x^2}{16} + \frac{y^2}{25} = 1$.

50. Describe one similarity and one difference between the graphs of $\frac{x^2}{25} + \frac{y^2}{16} = 1$ and $\frac{(x-1)^2}{25} + \frac{(y-1)^2}{16} = 1$.

51. An elliptipool is an elliptical pool table with only one pocket. A pool shark places a ball on the table, hits it in what appears to be a random direction, and yet it bounces off the edge, falling directly into the pocket. Explain why this happens.

Technology Exercises

52. Use a graphing utility to graph any five of the ellipses that you graphed by hand in Exercises 1–16.

53. Use a graphing utility to graph any three of the ellipses that you graphed by hand in Exercises 21–32. First solve the given equation for y by using the square root property. Enter each of the two resulting equations to produce each half of the ellipse.

Critical Thinking Exercises

Make Sense? *In Exercises 54–57, determine whether each statement "makes sense" or "does not make sense" and explain your reasoning.*

54. I graphed an ellipse with a horizontal major axis and foci on the y-axis.

55. I graphed an ellipse that was symmetric about its major axis, but not symmetric about its minor axis.

56. You told me that an ellipse centered at the origin has vertices at $(-5, 0)$ and $(5, 0)$, so I was able to graph the ellipse.

57. In a whispering gallery at our science museum, I stood at one focus, my friend stood at the other focus, and we had a clear conversation, very little of which was heard by the 25 museum visitors standing between us.

In Exercises 58–61, determine whether each statement is true or false. If the statement is false, make the necessary change(s) to produce a true statement.

58. The graphs of $x^2 + y^2 = 16$ and $\frac{x^2}{4} + \frac{y^2}{9} = 1$ do not intersect.

59. Some ellipses have equations that define y as a function of x.

60. The graph of $\frac{x^2}{9} + \frac{y^2}{4} = 0$ is an ellipse.

61. The equation $\frac{x^2}{1681} + \frac{y^2}{841} = 1$ models the elliptical opening of a wind tunnel that is 82 feet wide and 58 feet high.

62. Find the standard form of the equation of an ellipse with vertices at $(0, -6)$ and $(0, 6)$, passing through $(2, -4)$.

In Exercises 63–64, convert each equation to standard form by completing the square on x and y. Then graph the ellipse.

63. $9x^2 + 25y^2 - 36x + 50y - 164 = 0$

64. $4x^2 + 9y^2 - 32x + 36y + 64 = 0$

65. An Earth satellite has an elliptical orbit described by

$$\frac{x^2}{(5000)^2} + \frac{y^2}{(4750)^2} = 1.$$

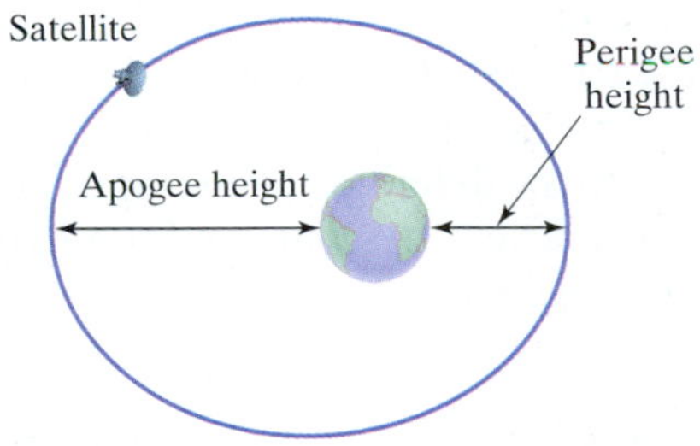

(All units are in miles.) The coordinates of the center of Earth are (16, 0).

a. The perigee of the satellite's orbit is the point that is nearest Earth's center. If the radius of Earth is approximately 4000 miles, find the distance of the perigee above Earth's surface.

b. The apogee of the satellite's orbit is the point that is the greatest distance from Earth's center. Find the distance of the apogee above Earth's surface.

66. The equation of the red ellipse in the figure shown is

$$\frac{x^2}{25} + \frac{y^2}{9} = 1.$$

Write the equation for each circle shown in the figure.

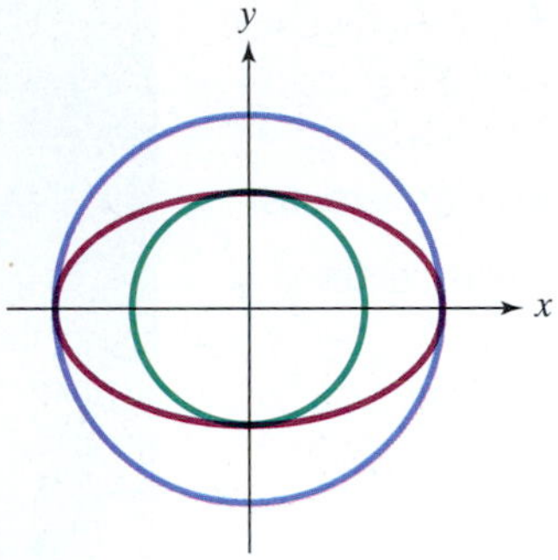

Review Exercises

67. Factor completely: $x^3 + 2x^2 - 4x - 8$. (Section 5.5, Example 4)

68. Simplify: $\sqrt[3]{40x^4y^7}$. (Section 7.3, Example 5)

69. Solve: $\frac{2}{x+2} + \frac{4}{x-2} = \frac{x-1}{x^2-4}$. (Section 6.6, Example 5)

Preview Exercises

Exercises 70–72 will help you prepare for the material covered in the next section.

70. Divide both sides of $4x^2 - 9y^2 = 36$ by 36 and simplify. How does the simplified equation differ from that of an ellipse?

71. Consider the equation $\frac{x^2}{16} - \frac{y^2}{9} = 1$.

a. Find the x-intercepts.

b. Explain why there are no y-intercepts.

72. Consider the equation $\frac{y^2}{9} - \frac{x^2}{16} = 1$.

a. Find the y-intercepts.

b. Explain why there are no x-intercepts.

SECTION 10.3

The Hyperbola

Objectives

1. Locate a hyperbola's vertices.
2. Graph hyperbolas centered at the origin.

St. Mary's Cathedral

Conic sections are often used to create unusual architectural designs. The top of St. Mary's Cathedral in San Francisco is a 2135-cubic-foot dome with walls rising 200 feet above the floor and supported by four massive concrete pylons that extend 94 feet into the ground. Cross sections of the roof are parabolas and hyperbolas. In this section, we study the curve with two parts known as the hyperbola.

FIGURE 10.20 Casting hyperbolic shadows

Definition of a Hyperbola

Figure 10.20 shows a cylindrical lampshade casting two shadows on a wall. These shadows indicate the distinguishing feature of hyperbolas: Their graphs contain two disjoint parts, called **branches**. Although each branch might look like a parabola, its shape is actually quite different.

The definition of a hyperbola is similar to the definition of an ellipse. For an ellipse, the *sum* of the distances from the foci is a constant. By contrast, for a hyperbola, the *difference* of the distances from the foci is a constant.

Definition of a Hyperbola

A **hyperbola** is the set of points in a plane the difference of whose distances from two fixed points, called foci, is constant.

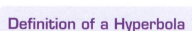

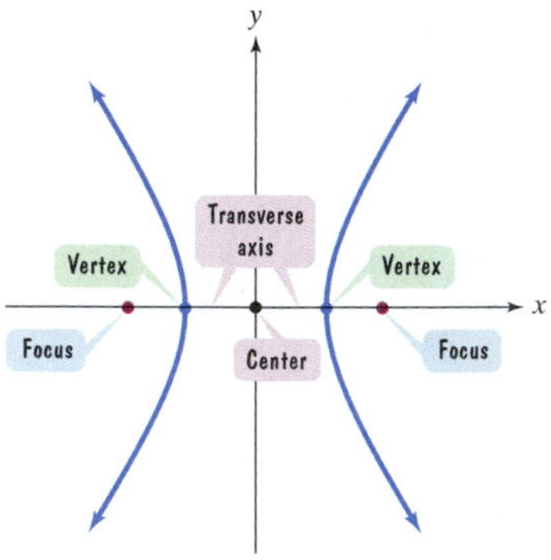

FIGURE 10.21 The two branches of a hyperbola

Figure 10.21 illustrates the two branches of a hyperbola. The line through the foci intersects the hyperbola at two points, called the **vertices**. The line segment that joins the vertices is the **transverse axis**. The midpoint of the transverse axis is the **center** of the hyperbola. Notice that the center lies midway between the vertices, as well as midway between the foci.

Standard Form of the Equation of a Hyperbola

The rectangular coordinate system enables us to translate a hyperbola's geometric definition into an algebraic equation. **Figure 10.22** on the next page is our starting point for obtaining an equation. We place the foci, F_1 and F_2, on the x-axis at the points $(-c, 0)$ and $(c, 0)$. Note that the center of this hyperbola is at the origin. We let (x, y) represent the coordinates of any point, P, on the hyperbola.

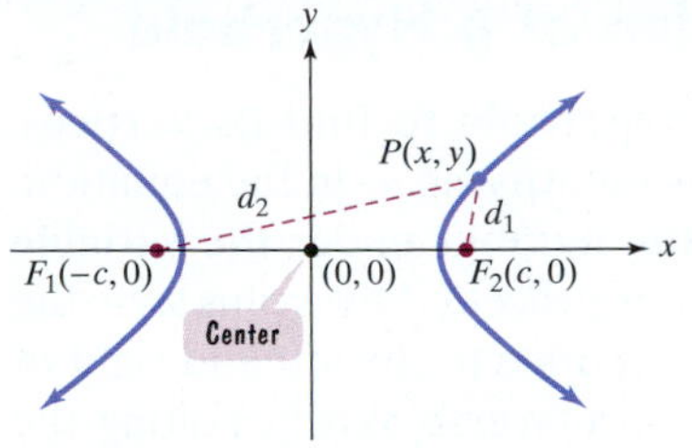

FIGURE 10.22

What does the definition of a hyperbola tell us about the point (x, y) in **Figure 10.22**? For any point (x, y) on the hyperbola, the absolute value of the difference of the distances from the two foci, $|d_2 - d_1|$, must be constant. We denote this constant by $2a$, just as we did for the ellipse. Thus, the point (x, y) is on the hyperbola if and only if

$$|d_2 - d_1| = 2a.$$

$$\left|\sqrt{(x+c)^2 + y^2} - \sqrt{(x-c)^2 + y^2}\right| = 2a \qquad \text{Use the distance formula.}$$

After eliminating radicals and simplifying, we obtain

$$(c^2 - a^2)x^2 - a^2y^2 = a^2(c^2 - a^2).$$

For convenience, let $b^2 = c^2 - a^2$. Substituting b^2 for $c^2 - a^2$ in the preceding equation, we obtain

$$b^2x^2 - a^2y^2 = a^2b^2.$$

$$\frac{b^2x^2}{a^2b^2} - \frac{a^2y^2}{a^2b^2} = \frac{a^2b^2}{a^2b^2} \qquad \text{Divide both sides by } a^2b^2.$$

$$\frac{x^2}{a^2} - \frac{y^2}{b^2} = 1 \qquad \text{Simplify.}$$

This last equation is called the **standard form of the equation of a hyperbola centered at the origin**. There are two such equations. The first is for a hyperbola in which the transverse axis lies on the x-axis. The second is for a hyperbola in which the transverse axis lies on the y-axis.

Standard Forms of the Equations of a Hyperbola

The **standard form of the equation of a hyperbola** with center at the origin is

$$\frac{x^2}{a^2} - \frac{y^2}{b^2} = 1 \quad \text{or} \quad \frac{y^2}{a^2} - \frac{x^2}{b^2} = 1.$$

Figure 10.23(a) illustrates that for the equation on the left, the transverse axis lies on the x-axis. **Figure 10.23(b)** illustrates that for the equation on the right, the transverse axis lies on the y-axis. The vertices are a units from the center and the foci are c units from the center.

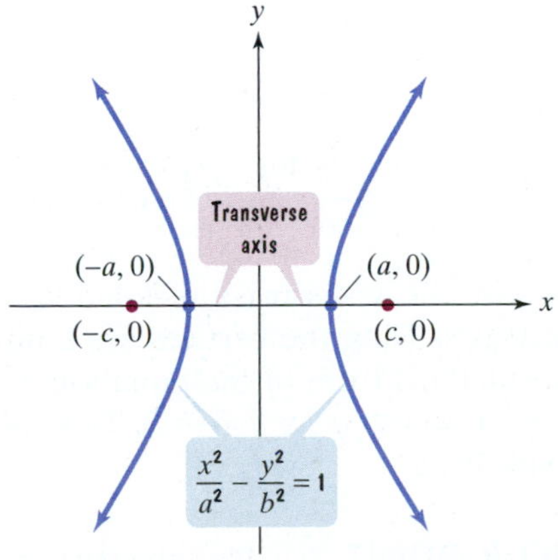

FIGURE 10.23(a) Transverse axis lies on the x-axis.

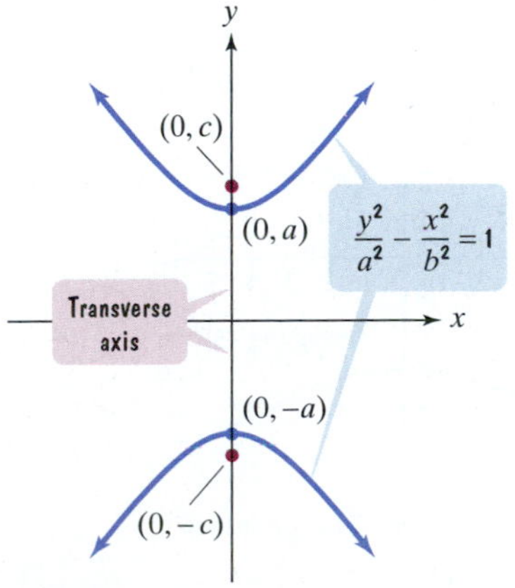

FIGURE 10.23(b) Transverse axis lies on the y-axis.

Study Tip

When the x^2-term is preceded by a plus sign, the transverse axis is horizontal. When the y^2-term is preceded by a plus sign, the transverse axis is vertical.

1 Locate a hyperbola's vertices.

Using the Standard Form of the Equation of a Hyperbola

We can use the standard form of the equation of a hyperbola to find its vertices. Because the vertices are a units from the center, begin by identifying a^2 in the equation. In the standard form of a hyperbola's equation, **a^2 is the number under the variable whose term is preceded by a plus sign (+).** If the x^2-term is preceded by a plus sign, the transverse axis lies along the x-axis. Thus, the vertices are a units to the left and right of the origin. If the y^2-term is preceded by a plus sign, the transverse axis lies along the y-axis. Thus, the vertices are a units above and below the origin.

EXAMPLE 1 Finding Vertices from a Hyperbola's Equation

Find the vertices for each of the following hyperbolas with the given equation:

a. $\dfrac{x^2}{16} - \dfrac{y^2}{9} = 1$ **b.** $\dfrac{y^2}{9} - \dfrac{x^2}{16} = 1.$

Solution Both equations are in standard form. We begin by identifying a^2 and b^2 in each equation.

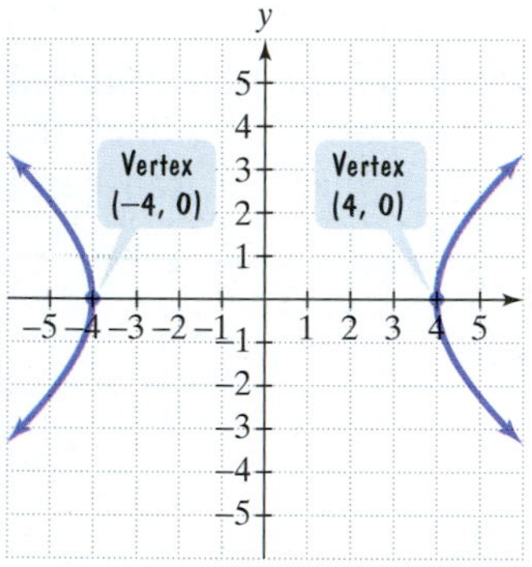

FIGURE 10.24 The graph of $\dfrac{x^2}{16} - \dfrac{y^2}{9} = 1$

a. The first equation is in the form $\dfrac{x^2}{a^2} - \dfrac{y^2}{b^2} = 1.$

$$\frac{x^2}{16} - \frac{y^2}{9} = 1$$

$a^2 = 16$. This is the denominator of the term preceded by a plus sign.

$b^2 = 9$. This is the denominator of the term preceded by a minus sign.

Because the x^2-term is preceded by a plus sign, the transverse axis lies along the x-axis. Thus, the vertices are a units to the *left* and *right* of the origin. Based on the standard form of the equation, we know that the vertices are $(-a, 0)$ and $(a, 0)$. Because $a^2 = 16$, $a = 4$. Thus, the vertices are $(-4, 0)$ and $(4, 0)$, shown in **Figure 10.24**.

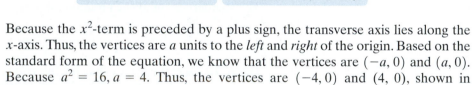

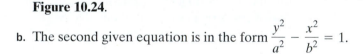

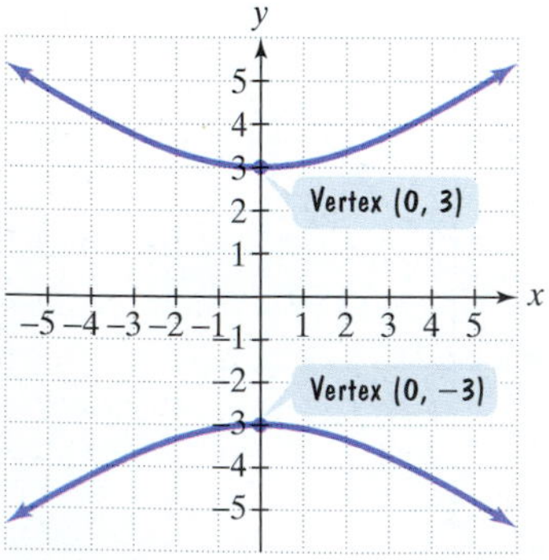

FIGURE 10.25 The graph of $\dfrac{y^2}{9} - \dfrac{x^2}{16} = 1$

b. The second given equation is in the form $\dfrac{y^2}{a^2} - \dfrac{x^2}{b^2} = 1.$

$$\frac{y^2}{9} - \frac{x^2}{16} = 1$$

$a^2 = 9$. This is the denominator of the term preceded by a plus sign.

$b^2 = 16$. This is the denominator of the term preceded by a minus sign.

Because the y^2-term is preceded by a plus sign, the transverse axis lies along the y-axis. Thus, the vertices are a units *above* and *below* the origin. Based on the standard form of the equation, we know that the vertices are $(0, -a)$ and $(0, a)$. Because $a^2 = 9$, $a = 3$. Thus, the vertices are $(0, -3)$ and $(0, 3)$, shown in **Figure 10.25**.

✓ **CHECK POINT 1** Find the vertices for each of the following hyperbolas with the given equation:

a. $\dfrac{x^2}{25} - \dfrac{y^2}{16} = 1$

b. $\dfrac{y^2}{25} - \dfrac{x^2}{16} = 1.$

The Asymptotes of a Hyperbola

As x and y get larger, the two branches of the graph of a hyperbola approach a pair of intersecting lines, called **asymptotes**. The asymptotes pass through the center of the hyperbola and are helpful in graphing hyperbolas.

Figure 10.26 shows the asymptotes for the graphs of hyperbolas centered at the origin. The asymptotes pass through the corners of a rectangle. Note that the dimensions of this rectangle are $2a$ by $2b$.

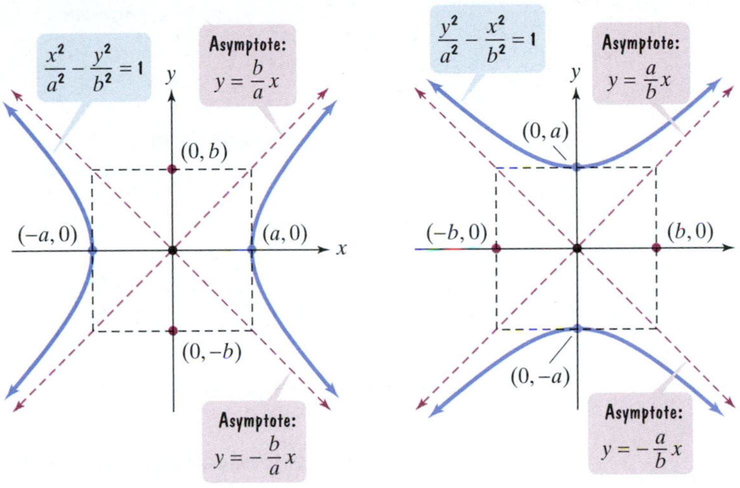

FIGURE 10.26 Asymptotes of a hyperbola

Why are $y = \pm\frac{b}{a}x$ the asymptotes for a hyperbola whose transverse axis is horizontal? The proof can be found in the appendix.

2 Graph hyperbolas centered at the origin.

Graphing Hyperbolas Centered at the Origin

Hyperbolas are graphed using vertices and asymptotes.

Graphing Hyperbolas

1. Locate the vertices.
2. Use dashed lines to draw the rectangle centered at the origin with sides parallel to the axes, crossing one axis at $\pm a$ and the other at $\pm b$.
3. Use dashed lines to draw the diagonals of this rectangle and extend them to obtain the asymptotes.
4. Draw the two branches of the hyperbola by starting at each vertex and approaching the asymptotes.

EXAMPLE 2 Graphing a Hyperbola

Graph the hyperbola: $\frac{x^2}{25} - \frac{y^2}{16} = 1.$

Solution

Step 1. Locate the vertices. The given equation is in the form $\frac{x^2}{a^2} - \frac{y^2}{b^2} = 1$, with $a^2 = 25$ and $b^2 = 16$.

$$\frac{x^2}{25} - \frac{y^2}{16} = 1$$

$a^2 = 25$ $\qquad$ $b^2 = 16$

$$\frac{x^2}{25} - \frac{y^2}{16} = 1$$

$a^2 = 25$ $b^2 = 16$

The given equation (repeated)

Based on the standard form of the equation with the transverse axis on the x-axis, we know that the vertices are $(-a, 0)$ and $(a, 0)$. Because $a^2 = 25$, $a = 5$. Thus, the vertices are $(-5, 0)$ and $(5, 0)$, shown in **Figure 10.27**.

Step 2. Draw a rectangle. Because $a^2 = 25$ and $b^2 = 16$, $a = 5$ and $b = 4$. We construct a rectangle to find the asymptotes, using -5 and 5 on the x-axis (the vertices are located here) and -4 and 4 on the y-axis. The rectangle passes through these four points, shown using dashed lines in **Figure 10.27**.

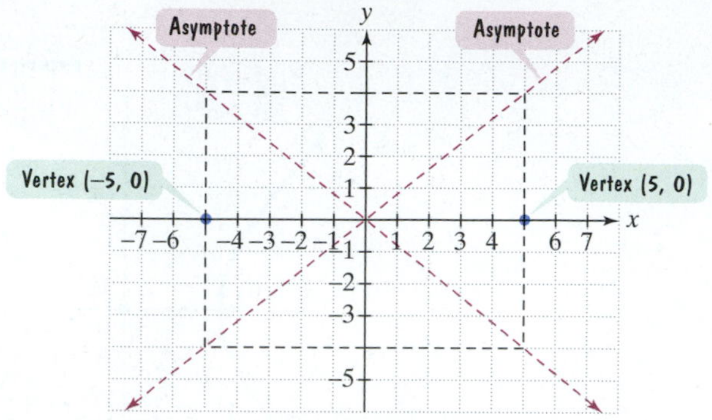

FIGURE 10.27 Preparing to graph $\frac{x^2}{25} - \frac{y^2}{16} = 1$

Step 3. Draw extended diagonals for the rectangle to obtain the asymptotes. We draw dashed lines through the opposite corners of the rectangle, shown in **Figure 10.27**, to obtain the graph of the asymptotes.

Step 4. Draw the two branches of the hyperbola by starting at each vertex and approaching the asymptotes. The hyperbola is shown in **Figure 10.28**.

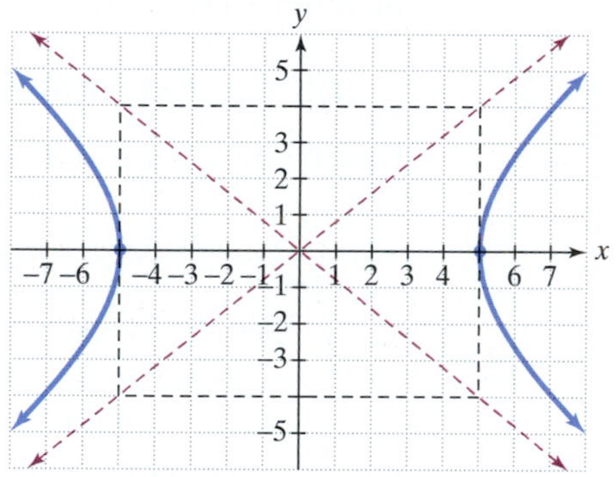

FIGURE 10.28 The graph of $\frac{x^2}{25} - \frac{y^2}{16} = 1$

Using Technology

Graph $\frac{x^2}{25} - \frac{y^2}{16} = 1$ by solving for y:

$$y_1 = \frac{\sqrt{16x^2 - 400}}{5}$$

$$y_2 = -\frac{\sqrt{16x^2 - 400}}{5} = -y_1.$$

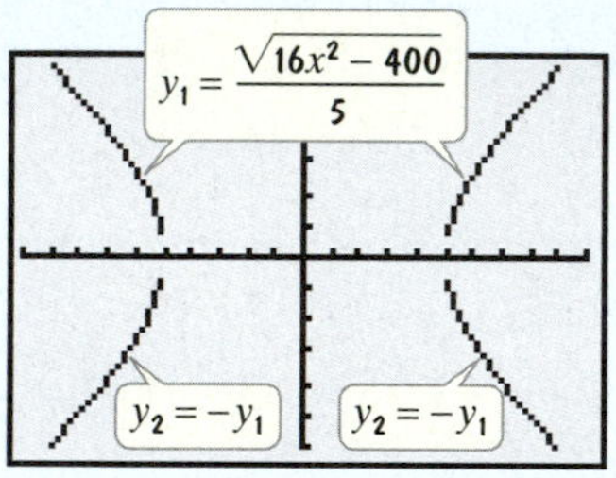

[−10, 10, 1] by [−6, 6, 1]

✓ **CHECK POINT 2** Graph the hyperbola: $\frac{x^2}{36} - \frac{y^2}{9} = 1.$

EXAMPLE 3 Graphing a Hyperbola

Graph the hyperbola: $9y^2 - 4x^2 = 36.$

Solution We begin by writing the equation in standard form. The right side should be 1, so we divide both sides by 36.

$$\frac{9y^2}{36} - \frac{4x^2}{36} = \frac{36}{36}$$

$$\frac{y^2}{4} - \frac{x^2}{9} = 1 \quad \text{Simplify. The right side is now 1.}$$

Now we are ready to use our four-step procedure for graphing hyperbolas.

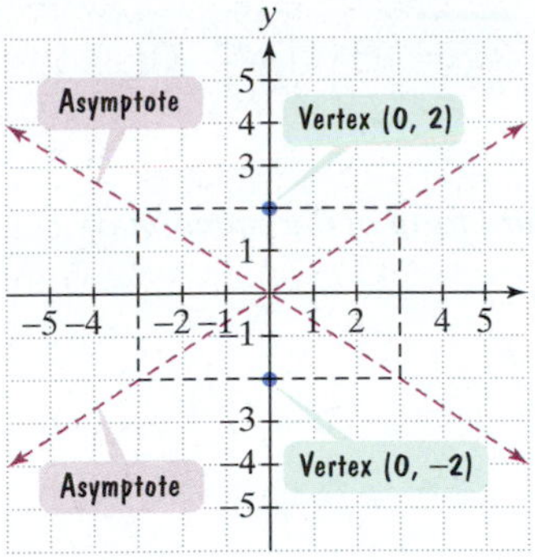

FIGURE 10.29 Preparing to graph $\frac{y^2}{4} - \frac{x^2}{9} = 1$

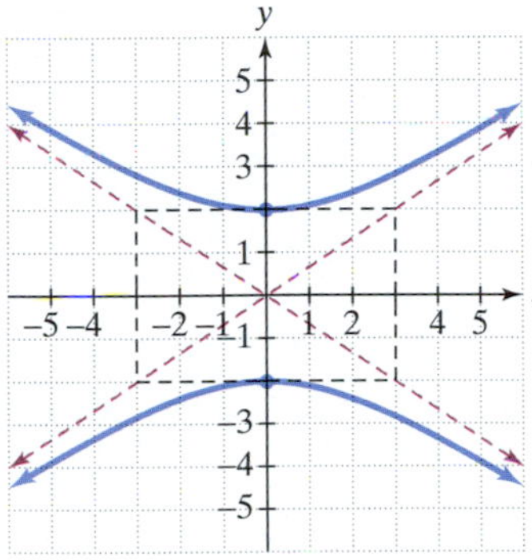

FIGURE 10.30 The graph of $\frac{y^2}{4} - \frac{x^2}{9} = 1$

Step 1. Locate the vertices. The equation that we obtained, $\frac{y^2}{4} - \frac{x^2}{9} = 1$, is in the form $\frac{y^2}{a^2} - \frac{x^2}{b^2} = 1$, with $a^2 = 4$ and $b^2 = 9$.

$$\frac{y^2}{4} - \frac{x^2}{9} = 1$$

$a^2 = 4$ $\quad$ $b^2 = 9$

Based on the standard form of the equation with the transverse axis on the y-axis, we know that the vertices are $(0, -a)$ and $(0, a)$. Because $a^2 = 4$, $a = 2$. Thus, the vertices are $(0, -2)$ and $(0, 2)$, shown in **Figure 10.29**.

Step 2. Draw a rectangle. Because $a^2 = 4$ and $b^2 = 9$, $a = 2$ and $b = 3$. We construct a rectangle to find the asymptotes, using -2 and 2 on the y-axis (the vertices are located here) and -3 and 3 on the x-axis. The rectangle passes through these four points, shown using dashed lines in **Figure 10.29**.

Step 3. Draw extended diagonals of the rectangle to obtain the asymptotes. We draw dashed lines through the opposite corners of the rectangle, shown in **Figure 10.29**, to obtain the graph of the asymptotes.

Step 4. Draw the two branches of the hyperbola by starting at each vertex and approaching the asymptotes. The hyperbola is shown in **Figure 10.30**.

✓ CHECK POINT 3 Graph the hyperbola: $y^2 - 4x^2 = 4$.

Applications

Hyperbolas have many applications. When a jet flies at a speed greater than the speed of sound, the shock wave that is created is heard as a sonic boom. The wave has the shape of a cone. The shape formed as the cone hits the ground is one branch of a hyperbola.

Halley's Comet, a permanent part of our solar system, travels around the sun in an elliptical orbit. Other comets pass through the solar system only once, following a hyperbolic path with the sun as a focus.

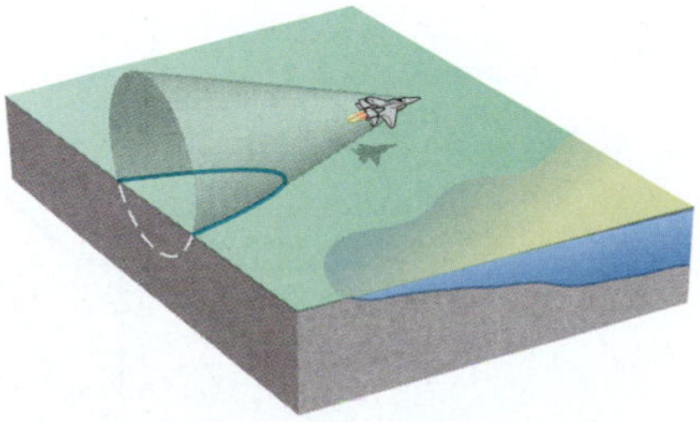

The hyperbolic shape of a sonic boom

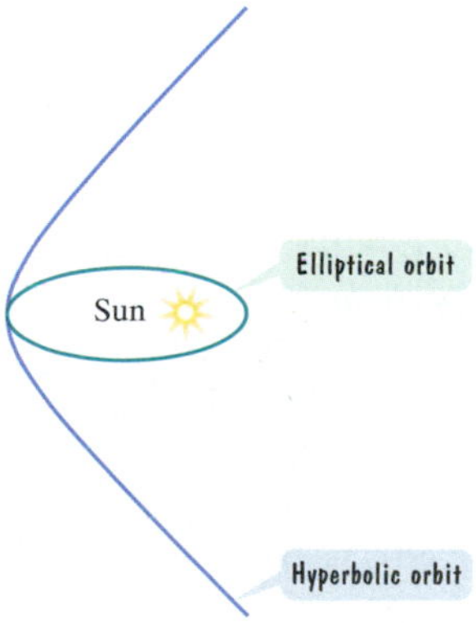

Orbits of comets

Hyperbolas are of practical importance in fields ranging from architecture to navigation. Cooling towers used in the design for nuclear power plants have cross sections that are both ellipses and hyperbolas. Three-dimensional solids whose cross sections are hyperbolas are used in unique architectural creations, including the TWA building at Kennedy Airport in New York City and the St. Louis Science Center Planetarium.

10.3 EXERCISE SET

MyMathLab · MathXL PRACTICE · WATCH · DOWNLOAD · READ · REVIEW

Practice Exercises

In Exercises 1–4, find the vertices of the hyperbola with the given equation. Then match each equation to one of the graphs that are shown and labeled (a)–(d).

1. $\dfrac{x^2}{4} - \dfrac{y^2}{1} = 1$

2. $\dfrac{x^2}{1} - \dfrac{y^2}{4} = 1$

3. $\dfrac{y^2}{4} - \dfrac{x^2}{1} = 1$

4. $\dfrac{y^2}{1} - \dfrac{x^2}{4} = 1$

a.

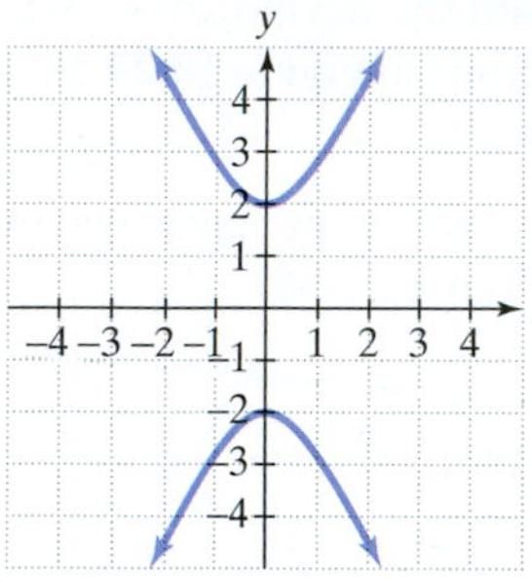

b.

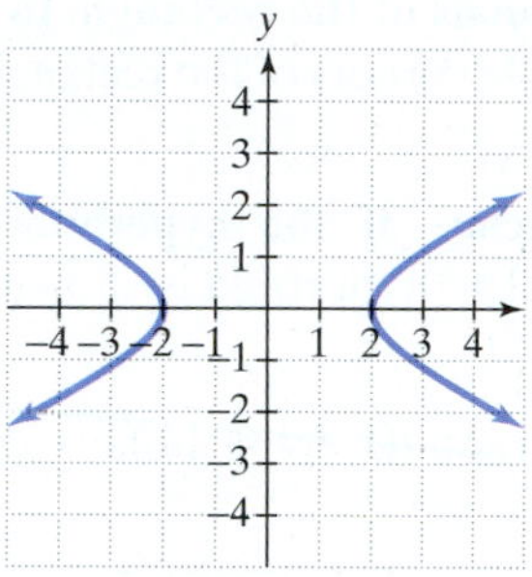

c.

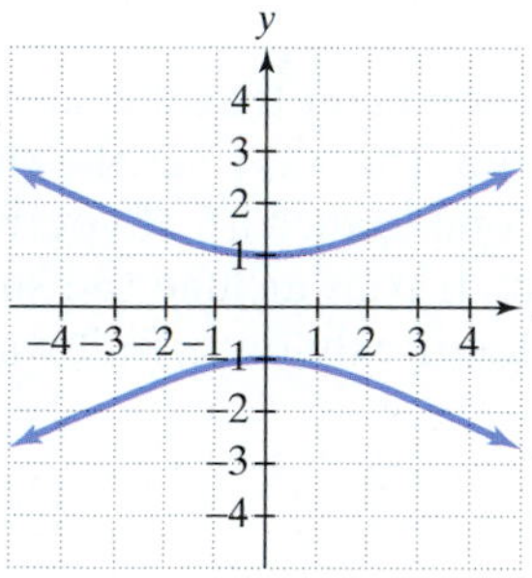

d.

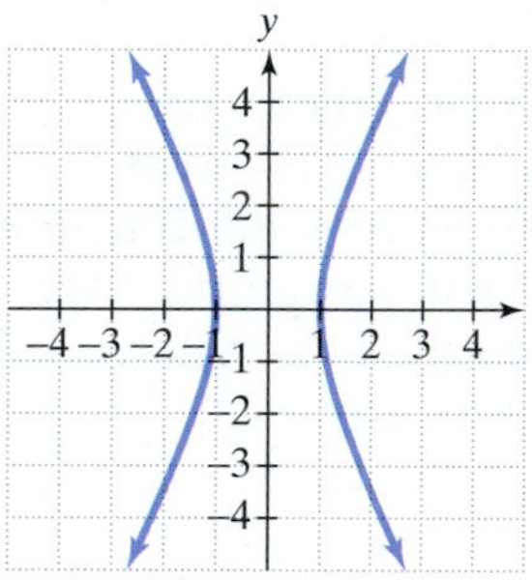

In Exercises 5–18, use vertices and asymptotes to graph each hyperbola.

5. $\dfrac{x^2}{9} - \dfrac{y^2}{25} = 1$

6. $\dfrac{x^2}{16} - \dfrac{y^2}{25} = 1$

7. $\dfrac{x^2}{100} - \dfrac{y^2}{64} = 1$

8. $\dfrac{x^2}{144} - \dfrac{y^2}{81} = 1$

9. $\dfrac{y^2}{16} - \dfrac{x^2}{36} = 1$

10. $\dfrac{y^2}{25} - \dfrac{x^2}{64} = 1$

11. $\dfrac{y^2}{36} - \dfrac{x^2}{25} = 1$

12. $\dfrac{y^2}{100} - \dfrac{x^2}{49} = 1$

13. $9x^2 - 4y^2 = 36$

14. $4x^2 - 25y^2 = 100$

15. $9y^2 - 25x^2 = 225$

16. $16y^2 - 9x^2 = 144$

17. $4x^2 = 4 + y^2$

18. $25y^2 = 225 + 9x^2$

In Exercises 19–22, find the standard form of the equation of each hyperbola.

19.

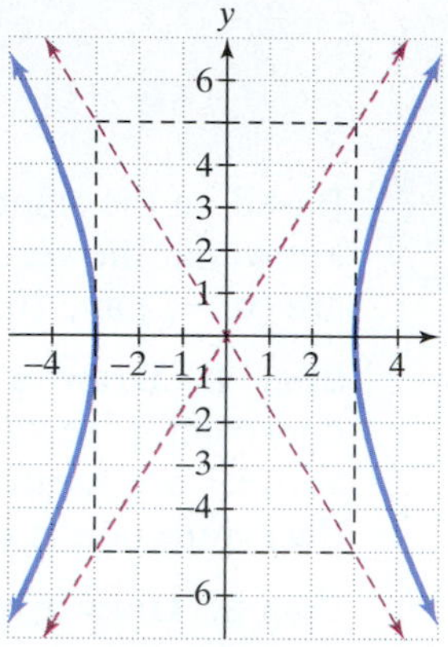

20.

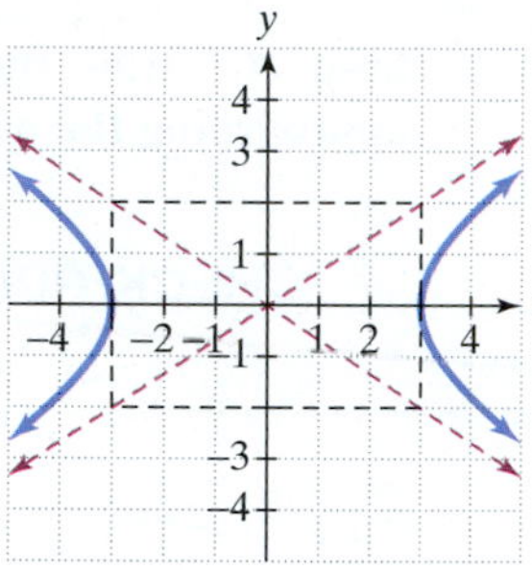

21.

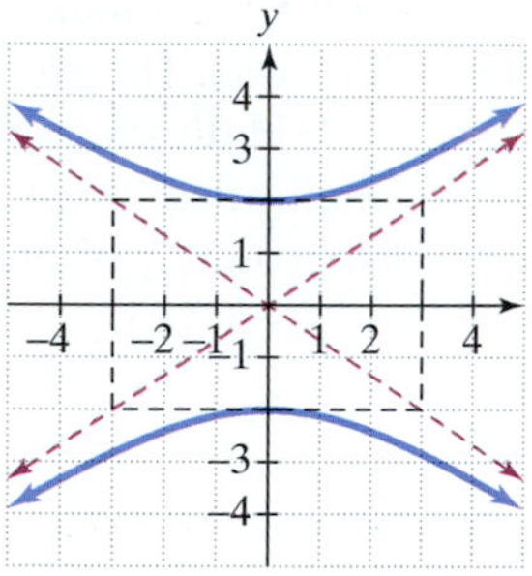

22.

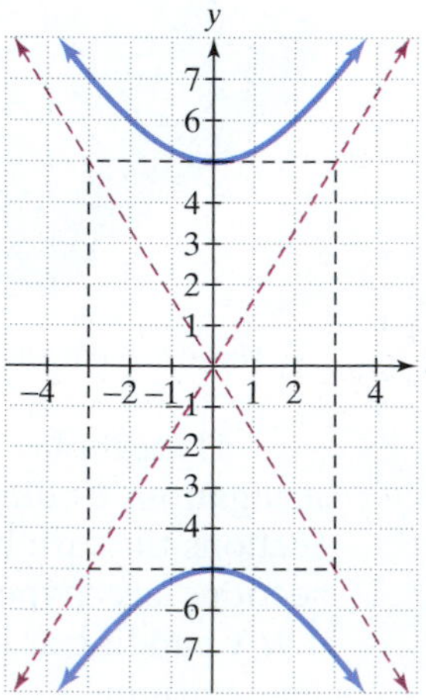

Practice PLUS

In Exercises 23–28, graph each relation. Use the relation's graph to determine its domain and range.

23. $\frac{x^2}{9} - \frac{y^2}{16} = 1$

24. $\frac{x^2}{25} - \frac{y^2}{4} = 1$

25. $\frac{x^2}{9} + \frac{y^2}{16} = 1$

26. $\frac{x^2}{25} + \frac{y^2}{4} = 1$

27. $\frac{y^2}{16} - \frac{x^2}{9} = 1$

28. $\frac{y^2}{4} - \frac{x^2}{25} = 1$

In Exercises 29–32, find the solution set for each system by graphing both of the system's equations in the same rectangular coordinate system and finding points of intersection. Check all solutions in both equations.

29. $x^2 - y^2 = 4$
$x^2 + y^2 = 4$

30. $x^2 - y^2 = 9$
$x^2 + y^2 = 9$

31. $9x^2 + y^2 = 9$
$y^2 - 9x^2 = 9$

32. $4x^2 + y^2 = 4$
$y^2 - 4x^2 = 4$

Application Exercises

33. An architect designs two houses that are shaped and positioned like a part of the branches of the hyperbola whose equation is $625y^2 - 400x^2 = 250{,}000$, where x and y are in yards. How far apart are the houses at their closest point?

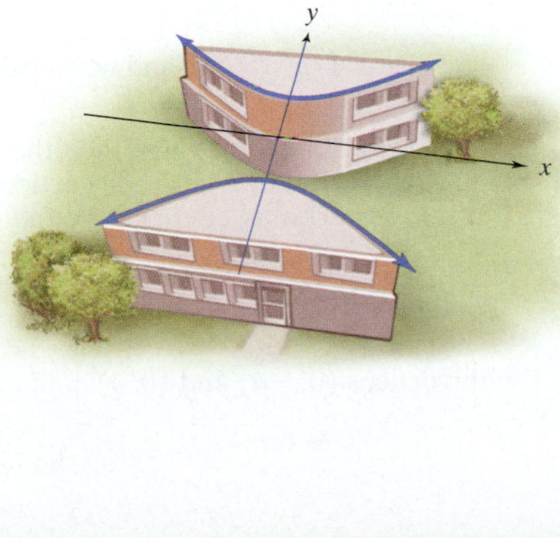

34. Scattering experiments, in which moving particles are deflected by various forces, led to the concept of the nucleus of an atom. In 1911, the physicist Ernest Rutherford (1871–1937) discovered that when alpha particles are directed toward the nuclei of gold atoms, they are eventually deflected along hyperbolic paths, illustrated in the figure. If a particle gets as close as 3 units to the nucleus along a hyperbolic path with an asymptote given by $y = \frac{1}{2}x$, what is the equation of its path?

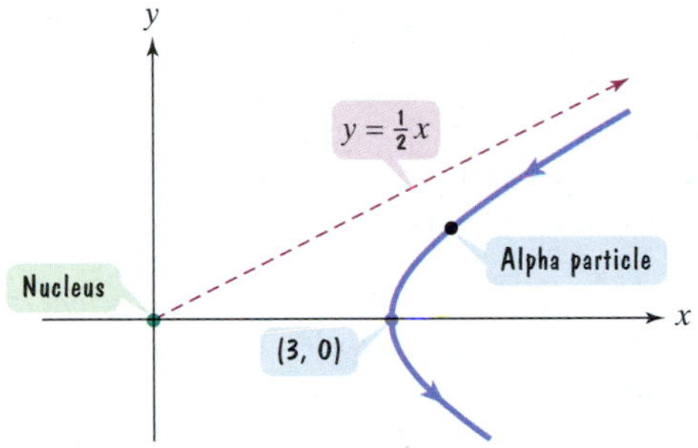

Writing in Mathematics

35. What is a hyperbola?

36. Describe how to graph $\frac{x^2}{9} - \frac{y^2}{1} = 1$.

37. Describe one similarity and one difference between the graphs of $\frac{x^2}{9} - \frac{y^2}{1} = 1$ and $\frac{y^2}{9} - \frac{x^2}{1} = 1$.

38. How can you distinguish an ellipse from a hyperbola by looking at their equations?

39. In 1992, a NASA team began a project called Spaceguard Survey, calling for an international watch for comets that might collide with Earth. Why is it more difficult to detect a possible "doomsday comet" with a hyperbolic orbit than one with an elliptical orbit?

Technology Exercises

40. Use a graphing utility to graph any five of the hyperbolas that you graphed by hand in Exercises 5–18.

41. Use a graphing utility to graph $\frac{x^2}{4} - \frac{y^2}{9} = 0$. Is the graph a hyperbola? In general, what is the graph of $\frac{x^2}{a^2} - \frac{y^2}{b^2} = 0$?

42. Graph $\frac{x^2}{a^2} - \frac{y^2}{b^2} = 1$ and $\frac{x^2}{a^2} - \frac{y^2}{b^2} = -1$ in the same viewing rectangle for values of a^2 and b^2 of your choice. Describe the relationship between the two graphs.

Critical Thinking Exercises

Make Sense? *In Exercises 43–46, determine whether each statement "makes sense" or "does not make sense" and explain your reasoning.*

43. I changed the addition in an ellipse's equation to subtraction and this changed its elongation from horizontal to vertical.
44. I noticed that the definition of a hyperbola closely resembles that of an ellipse in that it depends on the distances between a set of points in a plane to two fixed points, the foci.
45. I graphed a hyperbola centered at the origin that had y-intercepts, but no x-intercepts.
46. I graphed a hyperbola centered at the origin that was symmetric with respect to the x-axis and also symmetric with respect to the y-axis.

In Exercises 47–50, determine whether each statement is true or false. If the statement is false, make the necessary change(s) to produce a true statement.

47. If one branch of a hyperbola is removed from a graph, then the branch that remains must define y as a function of x.
48. All points on the asymptotes of a hyperbola also satisfy the hyperbola's equation.
49. The graph of $\frac{x^2}{9} - \frac{y^2}{4} = 1$ does not intersect the line $y = -\frac{2}{3}x$.
50. Two different hyperbolas can never share the same asymptotes.

The graph of $\frac{(x-h)^2}{a^2} - \frac{(y-k)^2}{b^2} = 1$ *is the same as the graph of* $\frac{x^2}{a^2} - \frac{y^2}{b^2} = 1$*, except the center is at* (h, k) *rather than at the origin. Use this information to graph the hyperbolas in Exercises 51–54.*

51. $\frac{(x-2)^2}{16} - \frac{(y-3)^2}{9} = 1$
52. $\frac{(x+2)^2}{9} - \frac{(y-1)^2}{25} = 1$
53. $(x-3)^2 - 4(y+3)^2 = 4$
54. $x^2 - y^2 - 2x - 4y - 4 = 0$

In Exercises 55–56, find the standard form of the equation of the hyperbola satisfying the given conditions.

55. vertices: $(6, 0)$, $(-6, 0)$; asymptotes: $y = 4x$, $y = -4x$
56. vertices: $(0, 7)$, $(0, -7)$; asymptotes: $y = 5x$, $y = -5x$

Review Exercises

57. Use intercepts and the vertex to graph the quadratic function: $y = -x^2 - 4x + 5$. (Section 8.3, Example 4)
58. Solve: $3x^2 - 11x - 4 \geq 0$. (Section 8.5, Example 1)
59. Solve: $\log_4(3x + 1) = 3$. (Section 9.5, Example 4)

Preview Exercises

Exercises 60–62 will help you prepare for the material covered in the next section.

In Exercises 60–61, graph each parabola with the given equation.

60. $y = x^2 + 4x - 5$
61. $y = -3(x-1)^2 + 2$
62. Set $x = 0$ and find the y-intercepts, correct to one decimal place: $x = -3(y-1)^2 + 2$.

MID-CHAPTER CHECK POINT Section 10.1–Section 10.3

What You Know: We found the length $\left[d = \sqrt{(x_2 - x_1)^2 + (y_2 - y_1)^2}\right]$ and the midpoint $\left[\left(\frac{x_1 + x_2}{2}, \frac{y_1 + y_2}{2}\right)\right]$ of the line segment with endpoints (x_1, y_1) and (x_2, y_2). We learned to graph circles with center (h, k) and radius r $[(x-h)^2 + (y-k)^2 = r^2]$. We graphed ellipses centered at (h, k) $\left[\frac{(x-h)^2}{a^2} + \frac{(y-k)^2}{b^2} = 1 \text{ or } \frac{(x-h)^2}{b^2} + \frac{(y-k)^2}{a^2} = 1, a^2 > b^2\right]$ and we saw that the larger denominator (a^2) determines whether the major axis is horizontal or vertical. We used vertices and asymptotes to graph hyperbolas centered at the origin $\left[\frac{x^2}{a^2} - \frac{y^2}{b^2} = 1 \text{ with vertices } (-a, 0) \text{ and } (a, 0) \text{ or } \frac{y^2}{a^2} - \frac{x^2}{b^2} = 1 \text{ with vertices } (0, -a) \text{ and } (0, a)\right]$.

In Exercises 1–4, graph each circle.

1. $x^2 + y^2 = 9$
2. $(x - 3)^2 + (y + 2)^2 = 25$
3. $x^2 + (y - 1)^2 = 4$
4. $x^2 + y^2 - 4x - 2y - 4 = 0$

In Exercises 5–8, graph each ellipse.

5. $\frac{x^2}{25} + \frac{y^2}{4} = 1$
6. $9x^2 + 4y^2 = 36$
7. $\frac{(x - 2)^2}{16} + \frac{(y + 1)^2}{25} = 1$
8. $\frac{(x + 2)^2}{25} + \frac{(y - 1)^2}{16} = 1$

In Exercises 9–12, graph each hyperbola.

9. $\frac{x^2}{9} - y^2 = 1$
10. $\frac{y^2}{9} - x^2 = 1$
11. $y^2 - 4x^2 = 16$
12. $4x^2 - 49y^2 = 196$

In Exercises 13–18, graph each equation.

13. $x^2 + y^2 = 4$
14. $x + y = 4$
15. $x^2 - y^2 = 4$
16. $x^2 + 4y^2 = 4$
17. $(x + 1)^2 + (y - 1)^2 = 4$
18. $x^2 + 4(y - 1)^2 = 4$

In Exercises 19–20, find the length (in simplified radical form and rounded to two decimal places) and the midpoint of the line segment with the given endpoints.

19. $(2, -2)$ and $(-2, 2)$
20. $(-5, 8)$ and $(-10, 14)$

SECTION 10.4

The Parabola; Identifying Conic Sections

Objectives

1. Graph horizontal parabolas.
2. Identify conic sections by their equations.

At first glance, this image looks like columns of smoke rising from a fire into a starry sky. Those are, indeed, stars in the background, but you are not looking at ordinary smoke columns. These stand almost 6 trillion miles high and are 7000 light-years from Earth—more than 400 million times as far away as the sun.

This NASA photograph is one of a series of stunning images captured from the ends of the universe by the Hubble Space Telescope. The image shows infant star systems the size of our solar system emerging from the gas and dust that shrouded their creation. Using a parabolic mirror that is 94.5 inches in diameter, the Hubble has provided answers to many of the profound mysteries of the cosmos: How big and how old is the universe? How did the galaxies come to exist? Do other Earth-like planets orbit other sun-like stars? In this section, we study parabolas and their applications, including parabolic shapes that gather distant rays of light and focus them into spectacular images.

Definition of a Parabola

In Chapter 8, we studied parabolas, viewing them as graphs of quadratic functions in the form

$$y = a(x - h)^2 + k \quad \text{or} \quad y = ax^2 + bx + c.$$

Study Tip

Here is a summary of what you should already know about graphing parabolas.

Graphing $y = a(x - h)^2 + k$ and $y = ax^2 + bx + c$

1. If $a > 0$, the graph opens upward. If $a < 0$, the graph opens downward.
2. The vertex of $y = a(x - h)^2 + k$ is (h, k).

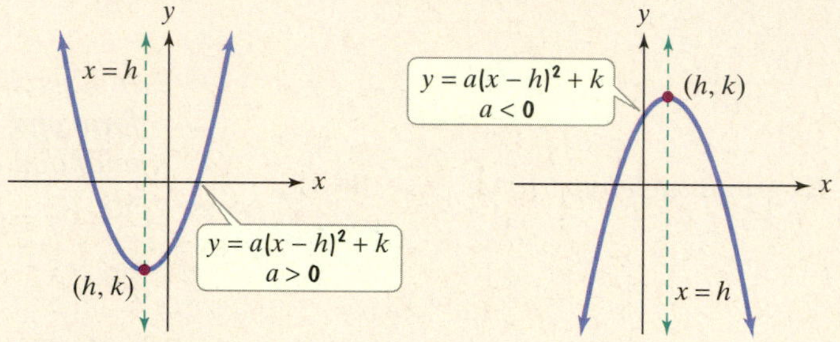

3. The x-coordinate of the vertex of $y = ax^2 + bx + c$ is $x = -\dfrac{b}{2a}$.

Parabolas can be given a geometric definition that enables us to include graphs that open to the left or to the right. The definitions of ellipses and hyperbolas involved two fixed points, the foci. By contrast, the definition of a parabola is based on one point and a line.

Definition of a Parabola

A **parabola** is the set of all points in a plane that are equidistant from a fixed line (the **directrix**) and a fixed point (the **focus**) that is not on the line (see **Figure 10.31**).

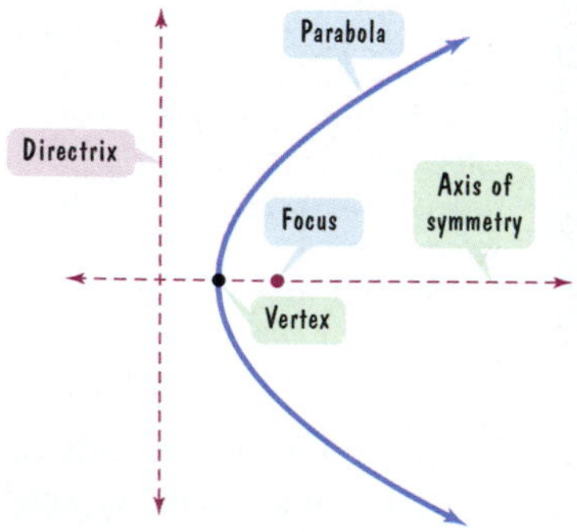

FIGURE 10.31

In **Figure 10.31**, find the line passing through the focus and perpendicular to the directrix. This is the **axis of symmetry** of the parabola. The point of intersection of the parabola with its axis of symmetry is the parabola's **vertex**. Notice that the vertex is midway between the focus and the directrix.

Parabolas can open to the left, right, upward, or downward. **Figure 10.32** shows a basic "family" of four parabolas and their equations. Notice that the red and blue parabolas that open to the left or right are not functions of x because they fail the vertical line test—that is, it is possible to draw vertical lines that intersect these graphs at more than one point.

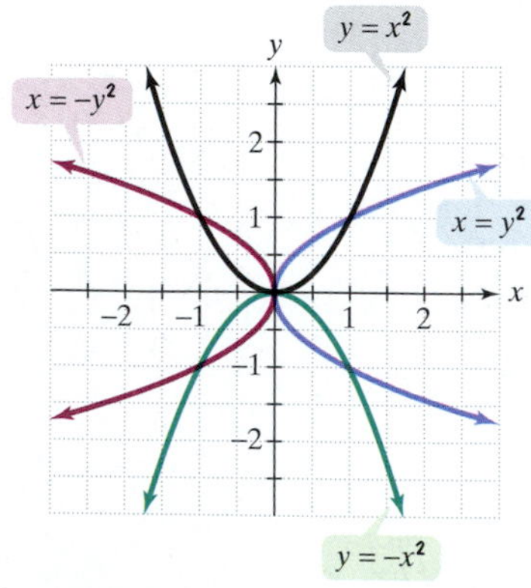

FIGURE 10.32

The equation $x = y^2$ interchanges the variables in the equation $y = x^2$. By interchanging x and y in the two forms of a parabola's equation,

$$y = a(x - h)^2 + k \quad \text{and} \quad y = ax^2 + bx + c,$$

we can obtain the forms of the equations of parabolas that open to the right or to the left.

Parabolas Opening to the Left or to the Right

The graphs of

$$x = a(y - k)^2 + h \quad \text{and} \quad x = ay^2 + by + c$$

are parabolas opening to the left or to the right.

1. If $a > 0$, the graph opens to the right. If $a < 0$, the graph opens to the left.
2. The vertex of $x = a(y - k)^2 + h$ is (h, k).

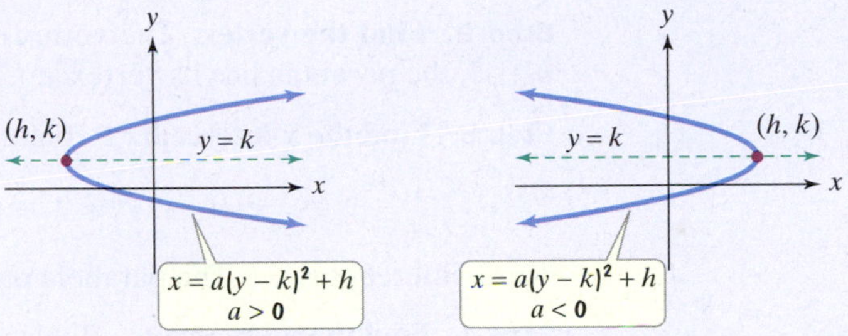

3. The y-coordinate of the vertex of $x = ay^2 + by + c$ is $y = -\frac{b}{2a}$.

1 Graph horizontal parabolas.

Graphing Parabolas Opening to the Left or the Right

Here is a procedure for graphing horizontal parabolas that are not functions. Notice how this procedure is similar to the one that we used in Chapter 8 for graphing vertical parabolas.

Graphing Horizontal Parabolas

To graph $x = a(y - k)^2 + h$ or $x = ay^2 + by + c$,

1. Determine whether the parabola opens to the left or to the right. If $a > 0$, it opens to the right. If $a < 0$, it opens to the left.
2. Determine the vertex of the parabola. The vertex of $x = a(y - k)^2 + h$ is (h, k). The y-coordinate of the vertex of $x = ay^2 + by + c$ is $y = -\frac{b}{2a}$. Substitute this value of y into the equation to find the x-coordinate.
3. Find the x-intercept by replacing y with 0.
4. Find any y-intercepts by replacing x with 0. Solve the resulting quadratic equation for y.
5. Plot the intercepts and the vertex. Connect these points with a smooth curve. If additional points are needed to obtain a more accurate graph, select values for y on each side of the axis of symmetry and compute values for x.

EXAMPLE 1 Graphing a Horizontal Parabola in the Form $x = a(y - k)^2 + h$

Graph: $x = -3(y - 1)^2 + 2$.

Solution We can graph this equation by following the steps in the preceding box. We begin by identifying values for a, k, and h.

$$x = a(y - k)^2 + h$$

$a = -3$ $\quad$ $k = 1$ $\quad$ $h = 2$

$$x = -3(y - 1)^2 + 2$$

Step 1. Determine how the parabola opens. Note that a, the coefficient of y^2, is -3. Thus, $a < 0$; this negative value tells us that the parabola opens to the left.

Step 2. Find the vertex. The vertex of the parabola is at (h, k). Because $k = 1$ and $h = 2$, the parabola has its vertex at $(2, 1)$.

Step 3. Find the x-intercept. Replace y with 0 in $x = -3(y - 1)^2 + 2$.

$$x = -3(0 - 1)^2 + 2 = -3(-1)^2 + 2 = -3(1) + 2 = -1$$

The x-intercept is -1. The parabola passes through $(-1, 0)$.

Step 4. Find the y-intercepts. Replace x with 0 in the given equation.

$$x = -3(y - 1)^2 + 2 \qquad \text{This is the given equation.}$$

$$0 = -3(y - 1)^2 + 2 \qquad \text{Replace } x \text{ with } 0.$$

$$3(y - 1)^2 = 2 \qquad \text{Solve for } y. \text{ Add } 3(y-1)^2 \text{ to both sides of the equation.}$$

$$(y - 1)^2 = \frac{2}{3} \qquad \text{Divide both sides by 3.}$$

$$y - 1 = \sqrt{\frac{2}{3}} \quad \text{or} \quad y - 1 = -\sqrt{\frac{2}{3}} \qquad \text{Apply the square root property.}$$

$$y = 1 + \sqrt{\frac{2}{3}} \qquad y = 1 - \sqrt{\frac{2}{3}} \qquad \text{Add 1 to both sides in each equation.}$$

$$y \approx 1.8 \qquad y \approx 0.2 \qquad \text{Use a calculator.}$$

The y-intercepts are approximately 1.8 and 0.2. The parabola passes through approximately $(0, 1.8)$ and $(0, 0.2)$.

Step 5. Graph the parabola. With a vertex at $(2, 1)$, an x-intercept at -1, and y-intercepts at approximately 1.8 and 0.2, the graph of the parabola is shown in **Figure 10.33**. The axis of symmetry is the horizontal line whose equation is $y = 1$.

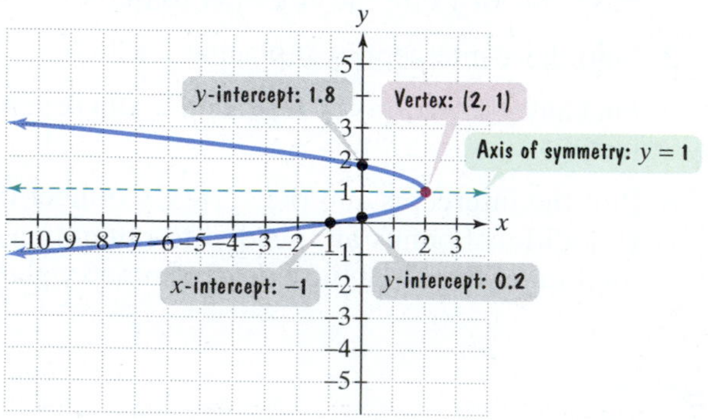

FIGURE 10.33 The graph of $x = -3(y - 1)^2 + 2$

x	y
−1	2
−10	−1

Choose values for y.

Compute values for x using $x = -3(y-1)^2 + 2$:

$x = -3(2-1)^2 + 2 = -3 \cdot 1^2 + 2 = -1$

$x = -3(-1-1)^2 + 2 = -3(-2)^2 + 2 = -10.$

We can possibly improve our graph in **Figure 10.33** by finding additional points on the parabola. Choose values of y on each side of the axis of symmetry, $y = 1$. We use $y = 2$ and $y = -1$. Then we compute values for x. The values in the table of coordinates show that $(-1, 2)$ and $(-10, -1)$ are points on the parabola. Locate each point in **Figure 10.33**.

✓ CHECK POINT 1 Graph: $x = -(y-2)^2 + 1$.

EXAMPLE 2 Graphing a Horizontal Parabola in the Form $x = ay^2 + by + c$

Graph: $x = y^2 + 4y - 5$.

Solution

Step 1. Determine how the parabola opens. Note that a, the coefficient of y^2, is 1. Thus $a > 0$; this positive value tells us that the parabola opens to the right.

Using Technology

To graph $x = y^2 + 4y - 5$ using a graphing utility, rewrite the equation as a quadratic equation in y.

$$y^2 + 4y + (-x - 5) = 0$$

$a = 1$ $b = 4$ $c = -x - 5$

Use the quadratic formula to solve for y and enter the resulting equations.

$$y_1 = \frac{-4 + \sqrt{16 - 4(-x-5)}}{2}$$

$$y_2 = \frac{-4 - \sqrt{16 - 4(-x-5)}}{2}$$

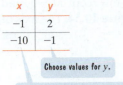

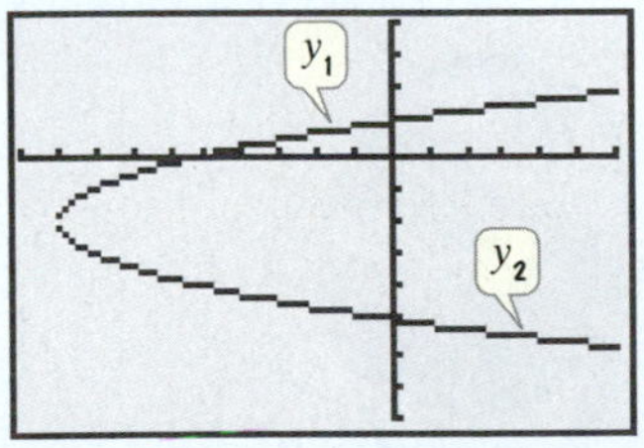

[−10, 6, 1] by [−8, 4, 1]

Step 2. Find the vertex. We know that the y-coordinate of the vertex is $y = -\frac{b}{2a}$. We identify a, b, and c in $x = ay^2 + by + c$.

$$x = y^2 + 4y - 5$$

$a = 1$ $b = 4$ $c = -5$

Substitute the values of a and b into the equation for the y-coordinate:

$$y = -\frac{b}{2a} = -\frac{4}{2 \cdot 1} = -2.$$

The y-coordinate of the vertex is -2. We substitute -2 for y into the parabola's equation, $x = y^2 + 4y - 5$, to find the x-coordinate:

$$x = (-2)^2 + 4(-2) - 5 = 4 - 8 - 5 = -9.$$

The vertex is at $(-9, -2)$.

Step 3. Find the x-intercept. Replace y with 0 in $x = y^2 + 4y - 5$.

$$x = 0^2 + 4 \cdot 0 - 5 = -5$$

The x-intercept is -5. The parabola passes through $(-5, 0)$.

Step 4. Find the y-intercepts. Replace x with 0 in the given equation.

$x = y^2 + 4y - 5$ This is the given equation.

$0 = y^2 + 4y - 5$ Replace x with 0.

$0 = (y-1)(y+5)$ Use factoring to solve the quadratic equation.

$y - 1 = 0$ or $y + 5 = 0$ Set each factor equal to 0.

$y = 1$ $\quad$ $y = -5$ Solve.

The y-intercepts are 1 and -5. The parabola passes through $(0, 1)$ and $(0, -5)$.

Step 5. Graph the parabola. With a vertex at $(-9, -2)$, an x-intercept at -5, and y-intercepts at 1 and -5, the graph of the parabola is shown in **Figure 10.34**. The axis of symmetry is the horizontal line whose equation is $y = -2$.

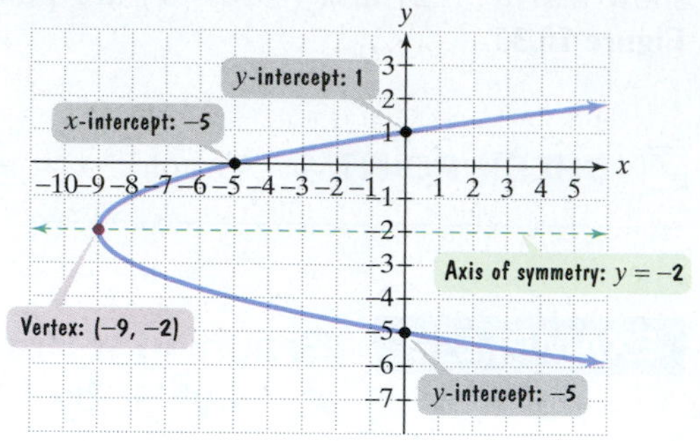

FIGURE 10.34 The graph of $x = y^2 + 4y - 5$

✓ **CHECK POINT 2** Graph: $x = y^2 + 8y + 7$.

Applications

Parabolas have many applications. Cables hung between structures to form suspension bridges form parabolas. Arches constructed of steel and concrete, whose main purpose is strength, are usually parabolic in shape.

We have seen that comets in our solar system travel in orbits that are ellipses and hyperbolas. Some comets also follow parabolic paths. Only comets with elliptical orbits, such as Halley's Comet, return to our part of the galaxy.

If a parabola is rotated about its axis of symmetry, a parabolic surface is formed. **Figure 10.35(a)** shows how a parabolic surface can be used to reflect light. Light originates at the focus. Note how the light is reflected by the parabolic surface, so that the outgoing light is parallel to the axis of symmetry. The reflective properties of parabolic surfaces are used in the design of searchlights [see **Figure 10.35(b)**], automobile headlights, and parabolic microphones.

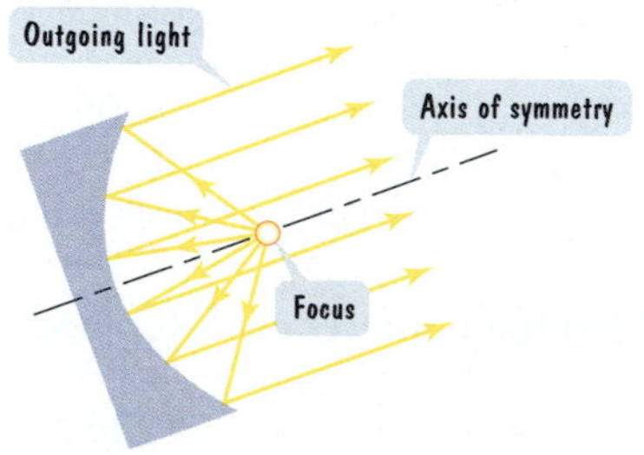

FIGURE 10.35(a) Parabolic surface reflecting light

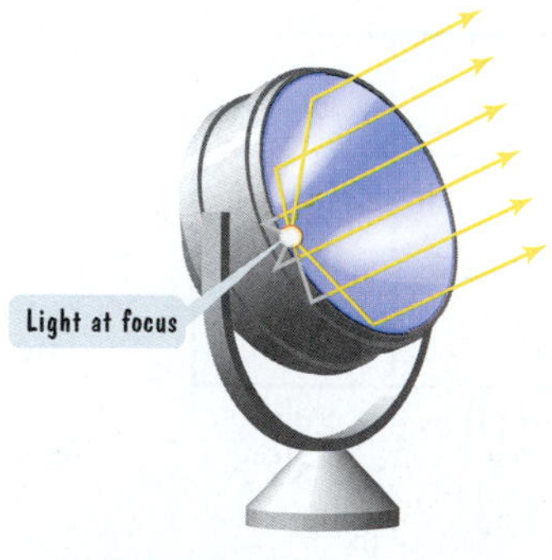

FIGURE 10.35(b) Light from the focus is reflected parallel to the axis of symmetry.

Figure 10.36(a) at the top of next page shows how a parabolic surface can be used to reflect *incoming* light. Note that light rays strike the surface and are reflected *to the focus*. This principle is used in the design of reflecting telescopes, radar, and television satellite dishes. Reflecting telescopes magnify the light from distant stars by reflecting the light from these bodies to the focus of a parabolic mirror [see **Figure 10.36(b)**].

Blitzer Bonus

The Hubble Space Telescope

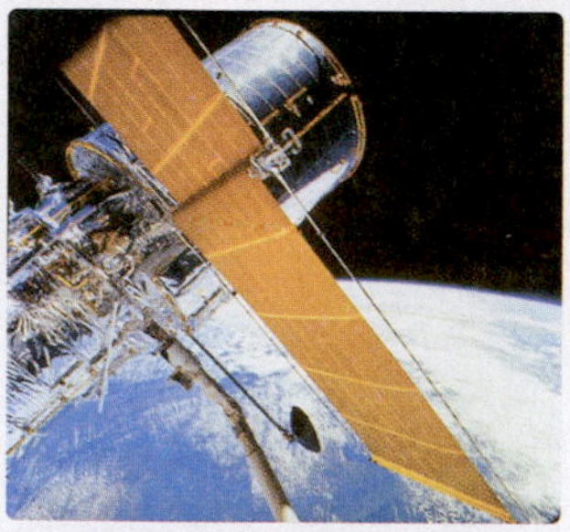

The Hubble Space Telescope

For decades, astronomers hoped to create an observatory above the atmosphere that would provide an unobscured view of the universe. This vision was realized with the 1990 launching of the Hubble Space Telescope. The telescope initially had blurred vision due to problems with its parabolic mirror. The mirror had been ground two millionths of a meter smaller than design specifications. In 1993, astronauts from the Space Shuttle *Endeavor* equipped the telescope with optics to correct the blurred vision. "A small change for a mirror, a giant leap for astronomy," Christopher J. Burrows of the Space Telescope Science Institute said when clear images from the ends of the universe were presented to the public after the repair mission.

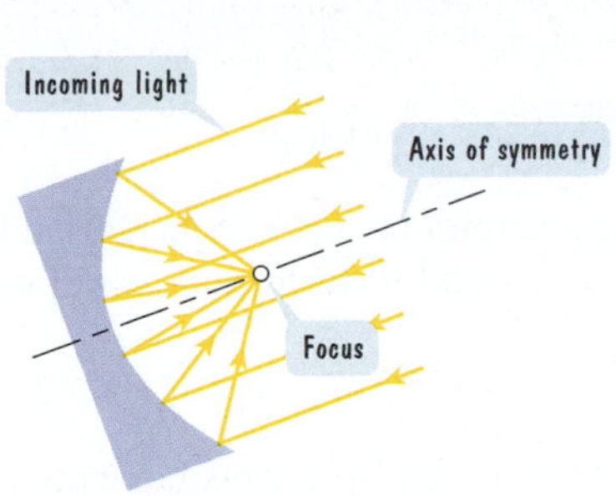

FIGURE 10.36(a) Parabolic surface reflecting incoming light

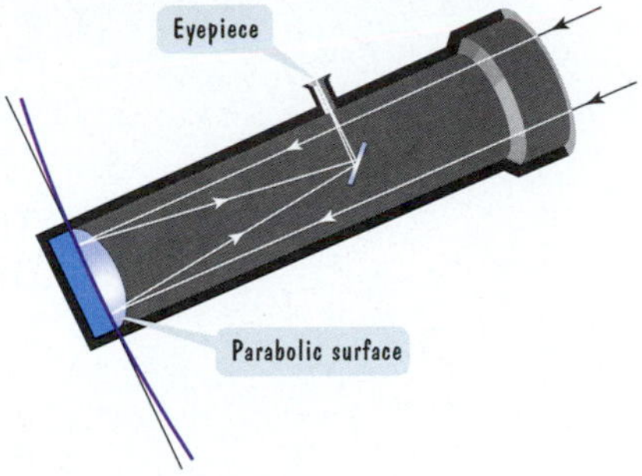

FIGURE 10.36(b) Incoming light rays are reflected to the focus.

2 Identify conic sections by their equations.

Identifying Conic Sections by Their Equations

What do the equations of the conic sections in this chapter have in common? They contain x^2-terms, y^2-terms, or both. Furthermore, they do not contain terms with exponents greater than 2. **Table 10.2** shows how to identify conic sections by their equations.

Table 10.2 Recognizing Conic Sections from Equations

Conic Section	How to Identify the Equation	Example
Circle	When x^2- and y^2-terms are on the same side, they have the same coefficient.	$x^2 + y^2 = 16$
Ellipse	When x^2- and y^2-terms are on the same side, they have different coefficients of the same sign.	$4x^2 + 16y^2 = 64$ or (dividing by 64) $\frac{x^2}{16} + \frac{y^2}{4} = 1$
Hyperbola	When x^2- and y^2-terms are on the same side, they have coefficients with opposite signs.	$9y^2 - 4x^2 = 36$ or (dividing by 36) $\frac{y^2}{4} - \frac{x^2}{9} = 1$
Parabola	Only one of the variables is squared.	$x = y^2 + 4y - 5$

Richard E. Prince "The Cone of Apollonius" (detail), fiberglass, steel, paint, graphite, 51 × 18 × 14 in. Collection: Vancouver Art Gallery, Vancouver, Canada. Photo courtesy of Equinox Gallery, Vancouver, Canada.

EXAMPLE 3 Recognizing Equations of Conic Sections

Indicate whether the graph of each equation is a circle, an ellipse, a hyperbola, or a parabola:

a. $4y^2 = 16 - 4x^2$ **b.** $x^2 = y^2 + 9$ **c.** $x + 7 - 6y = y^2$ **d.** $x^2 = 16 - 16y^2$.

Solution (Throughout the solution, in addition to identifying each equation's graph, we'll also discuss the graph's important features.) If both variables are squared, the graph of the equation is not a parabola. In this case, we collect the x^2- and y^2-terms on the same side of the equation.

a. $4y^2 = 16 - 4x^2$

Both variables, x and y, are squared.

The graph is a circle, an ellipse, or a hyperbola. To see which one it is, we collect the x^2- and y^2-terms on the same side. Add $4x^2$ to both sides. We obtain

$$4x^2 + 4y^2 = 16.$$

Because the coefficients of x^2 and y^2 are the same, namely 4, the equation's graph is a circle. This becomes more obvious if we divide both sides by 4.

$$\frac{4x^2}{4} + \frac{4y^2}{4} = \frac{16}{4}$$ Divide both sides by 4.

$$x^2 + y^2 = 4$$ Simplify.

$$(x - 0)^2 + (y - 0)^2 = 2^2$$ Write in the form $(x - h)^2 + (y - k)^2 = r^2$ with center (h, k) and radius r.

The graph is a circle with center at the origin and radius 2.

b. $x^2 = y^2 + 9$

Both variables, x and y, are squared.

The graph cannot be a parabola. To see if it is a circle, an ellipse, or a hyperbola, we collect the x^2- and y^2-terms on the same side. Subtract y^2 from both sides. We obtain

$$x^2 - y^2 = 9.$$

Because the x^2- and y^2-terms have coefficients with opposite signs, the equation's graph is a hyperbola. This becomes more obvious if we divide both sides by 9 to obtain 1 on the right.

$$\frac{x^2}{9} - \frac{y^2}{9} = 1$$

$a^2 = 9$ $\quad$ $b^2 = 9$

The hyperbola's vertices are $(a, 0)$ and $(-a, 0)$, namely $(3, 0)$ and $(-3, 0)$.

c. $x + 7 - 6y = y^2$

Only one variable, y, is squared.

Because only one variable is squared, the graph of the equation is a parabola. We can express the equation of the horizontal parabola in the form $x = ay^2 + by + c$ by isolating x on the left. We obtain

$$x = y^2 + 6y - 7. \quad \text{Add } 6y - 7 \text{ to both sides.}$$

Because the coefficient of the y^2-term, 1, is positive, the graph of the horizontal parabola opens to the right.

d. $x^2 = 16 - 16y^2$

Both variables, x and y, are squared.

The graph cannot be a parabola. To see if it is a circle, an ellipse, or a hyperbola, we collect the x^2- and y^2-terms on the same side. Add $16y^2$ to both sides. We obtain

$$x^2 + 16y^2 = 16.$$

Because the x^2- and y^2-terms have different coefficients of the same sign, namely 1 and 16, the equation's graph is an ellipse. This becomes more obvious if we divide both sides by 16 to obtain 1 on the right.

$$\frac{x^2}{16} + \frac{y^2}{1} = 1$$

$a^2 = 16$ $\quad$ $b^2 = 1$

An equation in the form $\frac{x^2}{a^2} + \frac{y^2}{b^2} = 1$ is an ellipse. The vertices are $(4, 0)$ and $(-4, 0)$.

✓ CHECK POINT 3 Identify whether the graph of each equation is a circle, an ellipse, a hyperbola, or a parabola:

a. $x^2 = 4y^2 + 16$

b. $x^2 = 16 - 4y^2$

c. $4x^2 = 16 - 4y^2$

d. $x = -4y^2 + 16y.$

10.4 EXERCISE SET

MyMathLab · MathXL PRACTICE · WATCH · DOWNLOAD · READ · REVIEW

Practice Exercises

In Exercises 1–6, the equation of a horizontal parabola is given. For each equation: Determine how the parabola opens. Find the parabola's vertex. Use your results to identify the equation's graph. [The graphs are labeled (a) through (f).]

1. $x = (y - 2)^2 - 1$

2. $x = (y + 2)^2 - 1$

3. $x = (y + 2)^2 + 1$

4. $x = (y - 2)^2 + 1$

5. $x = -(y - 2)^2 + 1$

6. $x = -(y + 2)^2 + 1$

a.

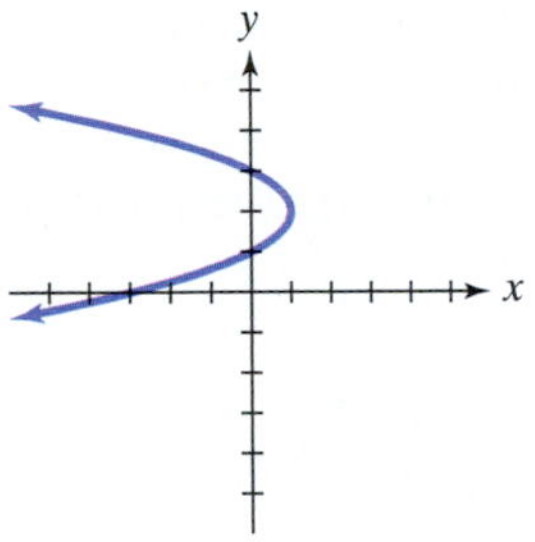

b.

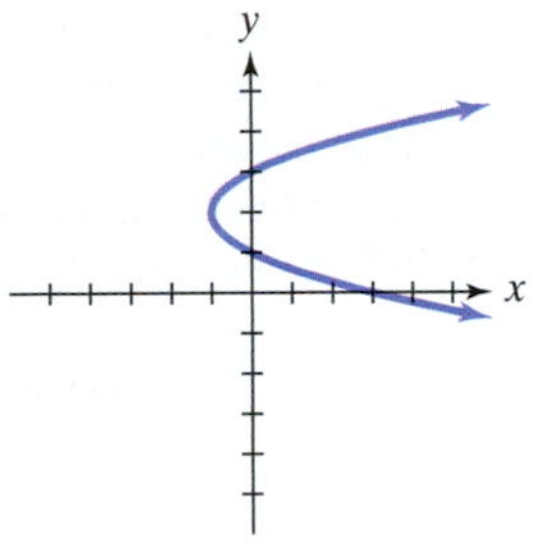

c.

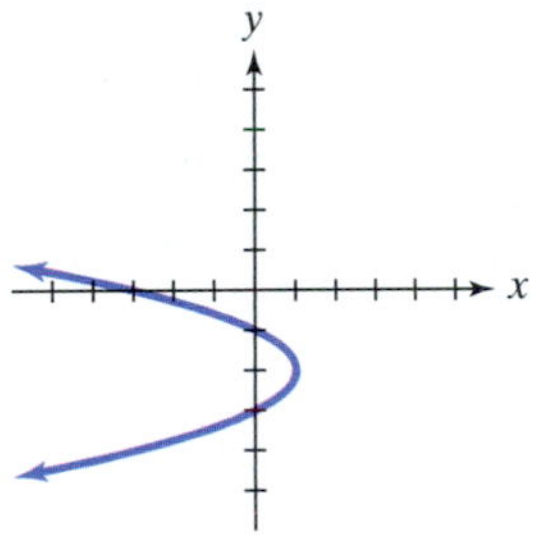

d.

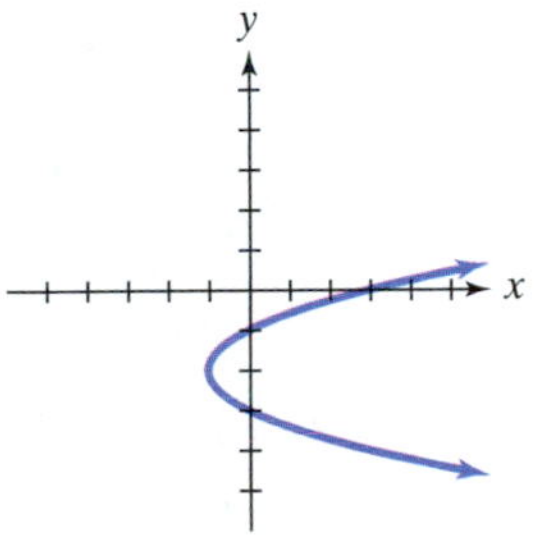

e.

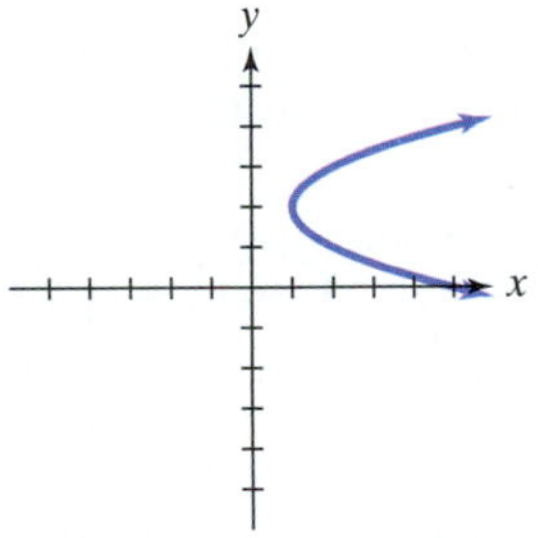

f.

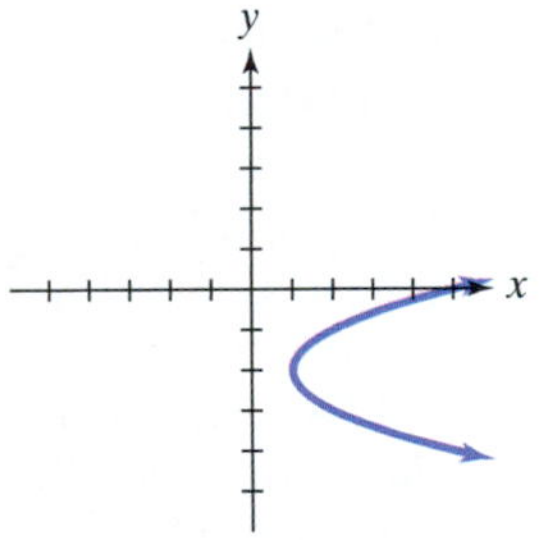

In Exercises 7–18, find the coordinates of the vertex for the horizontal parabola defined by the given equation.

7. $x = 2y^2$

8. $x = 4y^2$

9. $x = (y - 2)^2 + 3$

10. $x = (y - 3)^2 + 4$

11. $x = -4(y + 2)^2 - 1$

12. $x = -2(y + 5)^2 - 1$

13. $x = 2(y - 6)^2$

14. $x = 3(y - 7)^2$

15. $x = y^2 - 6y + 6$

16. $x = y^2 + 6y + 8$

17. $x = 3y^2 + 6y + 7$

18. $x = -2y^2 + 4y + 6$

In Exercises 19–42, use the vertex and intercepts to sketch the graph of each equation. If needed, find additional points on the parabola by choosing values of y on each side of the axis of symmetry.

19. $x = (y - 2)^2 - 4$

20. $x = (y - 3)^2 - 4$

21. $x = (y - 3)^2 - 5$

22. $x = (y + 2)^2 - 3$

23. $x = -(y - 5)^2 + 4$

24. $x = -(y - 3)^2 + 4$

25. $x = (y - 4)^2 + 1$

26. $x = (y - 2)^2 + 3$

27. $x = -3(y - 5)^2 + 3$

28. $x = -2(y + 6)^2 + 2$

29. $x = -2(y + 3)^2 - 1$

30. $x = -3(y + 1)^2 - 2$

31. $x = \frac{1}{2}(y + 2)^2 + 1$

32. $x = \frac{1}{2}(y + 1)^2 + 2$

33. $x = y^2 + 2y - 3$

34. $x = y^2 - 6y + 8$

35. $x = -y^2 - 4y + 5$

36. $x = -y^2 - 6y + 7$

37. $x = y^2 + 6y$

38. $x = y^2 + 4y$

39. $x = -2y^2 - 4y$

40. $x = -3y^2 - 6y$

41. $x = -2y^2 - 4y + 1$

42. $x = -2y^2 + 4y - 3$

In Exercises 43–54, the equation of a parabola is given. Determine:

a. *if the parabola is horizontal or vertical.*

b. *the way the parabola opens.*

c. *the vertex.*

43. $x = 2(y - 1)^2 + 2$

44. $x = 2(y - 3)^2 + 1$

45. $y = 2(x - 1)^2 + 2$

46. $y = 2(x - 3)^2 + 1$

47. $y = -(x + 3)^2 + 4$

48. $y = -(x + 1)^2 + 4$

49. $x = -(y + 3)^2 + 4$

50. $x = -(y + 1)^2 + 4$

51. $y = x^2 - 4x - 1$

52. $y = x^2 + 6x + 10$

53. $x = -y^2 + 4y + 1$

54. $x = -y^2 - 6y - 10$

In Exercises 55–64, indicate whether the graph of each equation is a circle, an ellipse, a hyperbola, or a parabola.

55. $x - 7 - 8y = y^2$

56. $x - 3 - 4y = 6y^2$

57. $4x^2 = 36 - y^2$

58. $4x^2 = 36 + y^2$

59. $x^2 = 36 + 4y^2$

60. $x^2 = 36 - 4y^2$

61. $3x^2 = 12 - 3y^2$

62. $3x^2 = 27 - 3y^2$

63. $3x^2 = 12 + 3y^2$

64. $3x^2 = 27 + 3y^2$

In Exercises 65–74, indicate whether the graph of each equation is a circle, an ellipse, a hyperbola, or a parabola. Then graph the conic section.

65. $x^2 - 4y^2 = 16$

66. $7x^2 - 7y^2 = 28$

67. $4x^2 + 4y^2 = 16$

68. $7x^2 + 7y^2 = 28$

69. $x^2 + 4y^2 = 16$

70. $4x^2 + y^2 = 16$

71. $x = (y - 1)^2 - 4$

72. $x = (y - 4)^2 - 1$

73. $(x - 2)^2 + (y + 1)^2 = 16$

74. $(x - 1)^2 + (y + 2)^2 = 16$

Practice PLUS

In Exercises 75–80, use the vertex and the direction in which the parabola opens to determine the relation's domain and range. Is the relation a function?

75. $x = y^2 + 6y + 5$

76. $x = y^2 - 2y - 5$

77. $y = -x^2 + 4x - 3$

78. $y = -x^2 - 4x + 4$

79. $x = -4(y - 1)^2 + 3$

80. $x = -3(y - 1)^2 - 2$

In Exercises 81–86, find the solution set for each system by graphing both of the system's equations in the same rectangular coordinate system and finding points of intersection. Check all solutions in both equations.

81. $x = (y - 2)^2 - 4$
$y = -\frac{1}{2}x$

82. $x = (y - 3)^2 + 2$
$x + y = 5$

83. $x = y^2 - 3$
$x = y^2 - 3y$

84. $x = y^2 - 5$
$x^2 + y^2 = 25$

85. $x = (y + 2)^2 - 1$
$(x - 2)^2 + (y + 2)^2 = 1$

86. $x = 2y^2 + 4y + 5$
$(x + 1)^2 + (y - 2)^2 = 1$

Application Exercises

87. The George Washington Bridge spans the Hudson River from New York to New Jersey. Its two towers are 3500 feet apart and rise 316 feet above the road. The cable between the towers has the shape of a parabola and the cable just touches the sides of the road midway between the towers. The parabola is positioned in a rectangular coordinate system with its vertex at the origin. The point (1750, 316) lies on the parabola, as shown.

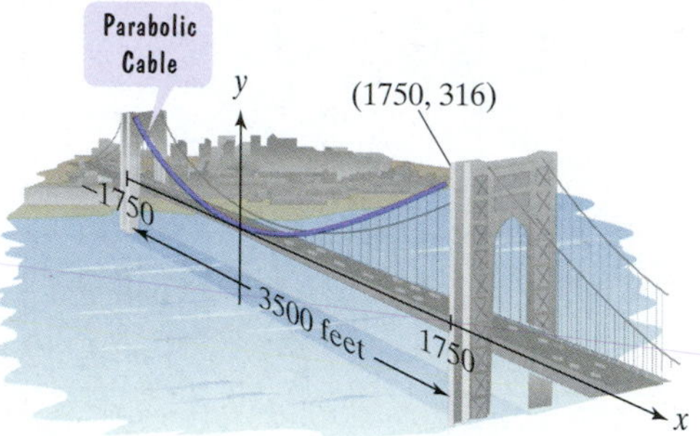

a. Write an equation in the form $y = ax^2$ for the parabolic cable. Do this by substituting 1750 for x and 316 for y and determining the value of a.

b. Use the equation in part (a) to find the height of the cable 1000 feet from a tower. Round to the nearest foot.

88. The towers of the Golden Gate Bridge connecting San Francisco to Marin County are 1280 meters apart and rise 140 meters above the road. The cable between the towers has the shape of a parabola and the cable just touches the sides of the road midway between the towers. The parabola is positioned in a rectangular coordinate system with its vertex at the origin. The point (640, 140) lies on the parabola, as shown.

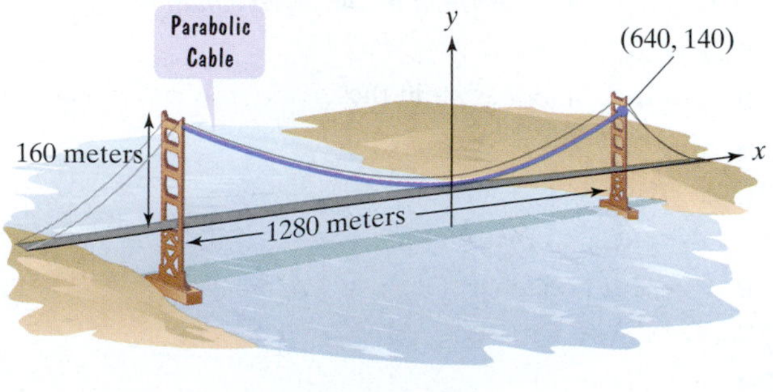

a. Write an equation in the form $y = ax^2$ for the parabolic cable. Do this by substituting 640 for x and 140 for y and determining the value of a.

b. Use the equation in part (a) to find the height of the cable 200 meters from a tower. Round to the nearest meter.

89. A satellite dish is in the shape of a parabolic surface. Signals coming from a satellite strike the surface of the dish and are reflected to the focus, where the receiver is located. The satellite dish shown has a diameter of 12 feet and a depth of 2 feet. The parabola is positioned in a rectangular coordinate system with its vertex at the origin.

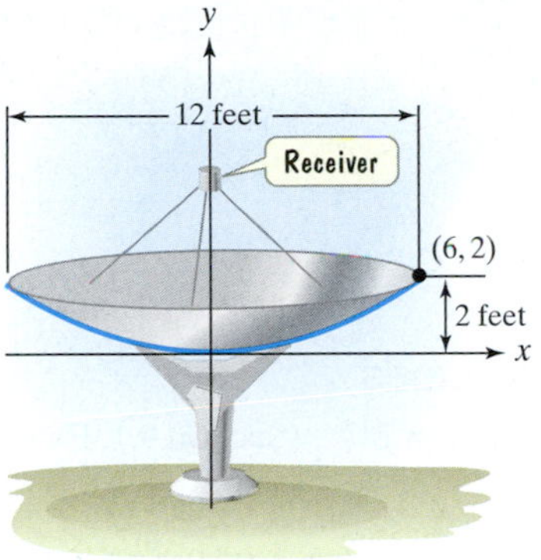

a. Write an equation in the form $y = ax^2$ for the parabola used to shape the dish.

b. The receiver should be placed at the focus $(0, p)$. The value of p is given by the equation $a = \frac{1}{4p}$. How far from the base of the dish should the receiver be placed?

90. An engineer is designing a flashlight using a parabolic reflecting mirror and a light source, as shown. The casting has a diameter of 4 inches and a depth of 2 inches. The parabola is positioned in a rectangular coordinate system with its vertex at the origin.

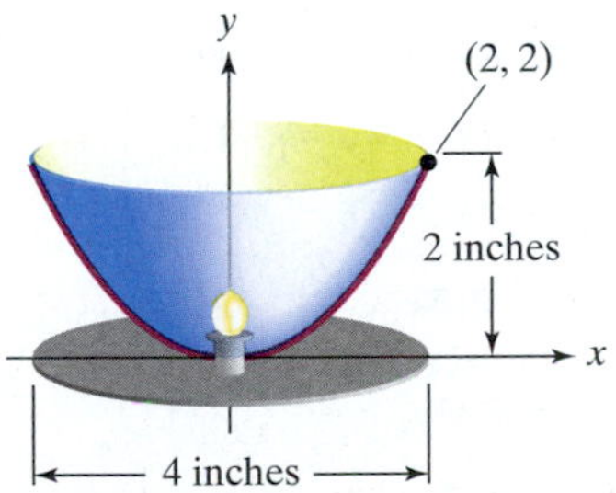

a. Write an equation in the form $y = ax^2$ for the parabola used to shape the mirror.

b. The light source should be placed at the focus $(0, p)$. The value of p is given by the equation $a = \frac{1}{4p}$. Where should the light source be placed relative to the mirror's vertex?

Moiré patterns, such as those shown in Exercises 91–92, can appear when two repetitive patterns overlap to produce a third, sometimes unintended, pattern.

a. *In each exercise, use the name of a conic section to describe the moiré pattern.*

b. *Select one of the following equations that can possibly describe a conic section within the moiré pattern:*

$$x^2 + y^2 = 1; \quad x^2 - y^2 = 1; \quad x^2 + 4y^2 = 4.$$

91.

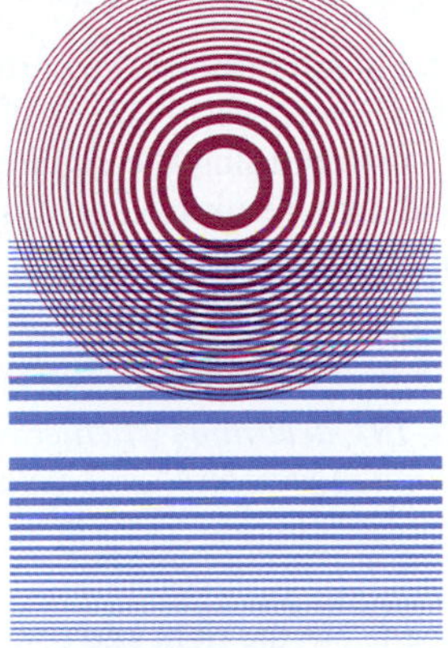

92.

Writing in Mathematics

93. What is a parabola?

94. If you are given an equation of a parabola, explain how to determine if the parabola opens to the right, to the left, upward, or downward.

95. Explain how to use $x = 2(y + 3)^2 - 5$ to find the parabola's vertex.

96. Explain how to use $x = y^2 + 8y + 9$ to find the parabola's vertex.

97. Describe one similarity and one difference between the graphs of $x = 4y^2$ and $x = 4(y - 1)^2 + 2$.

98. How can you distinguish parabolas from other conic sections by looking at their equations?

99. How can you distinguish ellipses from hyperbolas by looking at their equations?

100. How can you distinguish ellipses from circles by looking at their equations?

Technology Exercises

Use a graphing utility to graph the parabolas in Exercises 101–102. Write the given equation as a quadratic equation in y and use the quadratic formula to solve for y. Enter each of the equations to produce the complete graph.

101. $y^2 + 2y - 6x + 13 = 0$

102. $y^2 + 10y - x + 25 = 0$

103. Use a graphing utility to graph any three of the parabolas that you graphed by hand in Exercises 19–42. First solve the given equation for y, possibly using the square root method. Enter each of the two resulting equations to produce the complete graph.

Critical Thinking Exercises

Make Sense? *In Exercises 104–107, determine whether each statement "makes sense" or "does not make sense" and explain your reasoning.*

104. I graphed a parabola that opened to the right and contained a maximum point.

105. I'm graphing an equation that contains neither an x^2-term nor a y^2-term, so the graph cannot be a conic section.

106. Knowing that a parabola opening to the right has a vertex at $(-1, 1)$ gives me enough information to determine its graph.

107. I noticed that depending on the values for A and B, assuming that they are not both zero, the graph of $Ax^2 + By^2 = C$ can represent any of the conic sections other than a parabola.

In Exercises 108–111, determine whether each statement is true or false. If the statement is false, make the necessary change(s) to produce a true statement.

108. The parabola whose equation is $x = 2y - y^2 + 5$ opens to the right.

109. If the parabola whose equation is $x = ay^2 + by + c$ has its vertex at $(3, 2)$ and $a > 0$, then it has no y-intercepts.

110. Some parabolas that open to the right have equations that define y as a function of x.

111. The graph of $x = a(y - k) + h$ is a parabola with vertex at (h, k).

112. Look at the satellite dish shown in Exercise 89. Why must the receiver for a shallow dish be farther from the base of the dish than for a deeper dish of the same diameter?

113. The parabolic arch shown in the figure is 50 feet above the water at the center and 200 feet wide at the base. Will a boat that is 30 feet tall clear the arch 30 feet from the center?

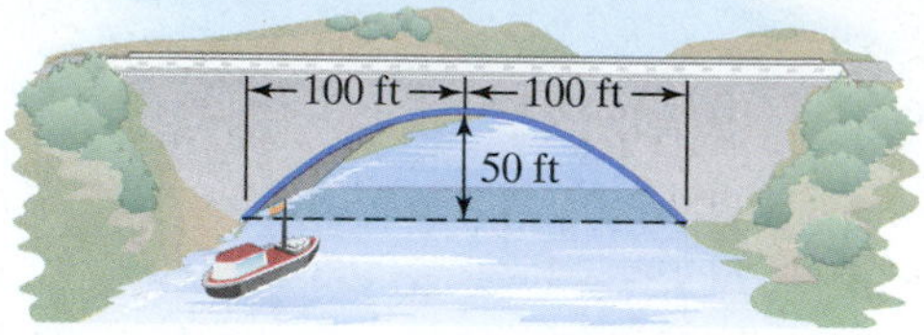

Review Exercises

114. Graph: $f(x) = 2^{1-x}$. (Section 9.1, Example 4)

115. If $f(x) = \frac{1}{3}x - 5$, find $f^{-1}(x)$. (Section 9.2, Example 4)

116. Solve: $(x + 1)^2 + (x + 3)^2 = 4$. (Section 5.7, Example 2)

Preview Exercises

Exercises 117–119 will help you prepare for the material covered in the next section.

117. Solve by the substitution method:

$$4x + 3y = 4$$
$$y = 2x - 7.$$

118. Solve by the addition method:

$$2x + 4y = -4$$
$$3x + 5y = -3.$$

119. Solve: $x^2 = 2(3x - 9) + 10.$

SECTION 10.5

Systems of Nonlinear Equations in Two Variables

Objectives

1. Recognize systems of nonlinear equations in two variables.
2. Solve systems of nonlinear equations by substitution.
3. Solve systems of nonlinear equations by addition.
4. Solve problems using systems of nonlinear equations.

Scientists debate the probability that a "doomsday rock" will collide with Earth. It has been estimated that an asteroid, a tiny planet that revolves around the sun, crashes into Earth about once every 250,000 years, and that such a collision would have disastrous results. In 1908, a small fragment struck Siberia, leveling thousands of acres of trees. One theory about the extinction of dinosaurs 65 million years ago involves Earth's collision with a large asteroid and the resulting drastic changes in Earth's climate.

Understanding the path of Earth and the path of a comet is essential to detecting threatening space debris. Orbits about the sun are not described by linear equations in the form $Ax + By = C$. The ability to solve systems that do not contain linear equations provides NASA scientists watching for troublesome asteroids with a way to locate possible collision points with Earth's orbit.

1 Recognize systems of nonlinear equations in two variables.

Systems of Nonlinear Equations and their Solutions

A **system of** two **nonlinear equations** in two variables, also called a **nonlinear system**, contains at least one equation that cannot be expressed in the form $Ax + By = C$. Here are two examples:

$$x^2 = 2y + 10$$
$$3x - y = 9$$

Not in the form $Ax + By = C$. The term x^2 is not linear.

$$y = x^2 + 3$$
$$x^2 + y^2 = 9.$$

Neither equation is in the form $Ax + By = C$. The terms x^2 and y^2 are not linear.

A **solution** of a nonlinear system in two variables is an ordered pair of real numbers that satisfies all equations in the system. The **solution set** of the system is the set of all such ordered pairs. As with linear systems in two variables, the solution of a nonlinear system (if there is one) corresponds to the intersection point(s) of the graphs of the equations in the system. Unlike linear systems, the graphs can be circles, ellipses, hyperbolas, parabolas, or anything other than two lines. We will solve nonlinear systems using the substitution method and the addition method.

2 Solve systems of nonlinear equations by substitution.

Eliminating a Variable Using the Substitution Method

The substitution method involves converting a nonlinear system into one equation in one variable by an appropriate substitution. The steps in the solution process are exactly the same as those used to solve a linear system by substitution. However, when you obtain an equation in one variable, this equation may not be linear. In our first example, this equation is quadratic.

EXAMPLE 1 Solving a Nonlinear System by the Substitution Method

Solve by the substitution method:

$$x^2 = 2y + 10 \quad \text{(The graph is a parabola.)}$$
$$3x - y = 9. \quad \text{(The graph is a line.)}$$

Solution

Step 1. Solve one of the equations for one variable in terms of the other. We begin by isolating one of the variables raised to the first power in either of the equations. By solving for y in the second equation, which has a coefficient of -1, we can avoid fractions.

$3x - y = 9$	This is the second equation in the given system.
$3x = y + 9$	Add y to both sides.
$3x - 9 = y$	Subtract 9 from both sides.

Step 2. Substitute the expression from step 1 into the other equation. We substitute $3x - 9$ for y in the first equation.

$$y = \boxed{3x - 9} \qquad x^2 = 2\boxed{y} + 10$$

This gives us an equation in one variable, namely

$$x^2 = 2(3x - 9) + 10.$$

The variable y has been eliminated.

Step 3. Solve the resulting equation containing one variable.

$x^2 = 2(3x - 9) + 10$	This is the equation containing one variable.
$x^2 = 6x - 18 + 10$	Use the distributive property.
$x^2 = 6x - 8$	Combine numerical terms on the right.
$x^2 - 6x + 8 = 0$	Move all terms to one side and set the quadratic equation equal to 0.
$(x - 4)(x - 2) = 0$	Factor.
$x - 4 = 0$ or $x - 2 = 0$	Set each factor equal to 0.
$x = 4$ $\quad$ $x = 2$	Solve for x.

Step 4. Back-substitute the obtained values into the equation from step 1. Now that we have the x-coordinates of the solutions, we back-substitute 4 for x and 2 for x into the equation $y = 3x - 9$.

If x is 4, $\quad y = 3(4) - 9 = 3, \quad$ so $(4, 3)$ is a solution.

If x is 2, $\quad y = 3(2) - 9 = -3, \quad$ so $(2, -3)$ is a solution.

Step 5. Check the proposed solutions in both of the system's given equations. We begin by checking $(4, 3)$. Replace x with 4 and y with 3.

$x^2 = 2y + 10$	$3x - y = 9$	These are the given equations.
$4^2 \stackrel{?}{=} 2(3) + 10$	$3(4) - 3 \stackrel{?}{=} 9$	Let $x = 4$ and $y = 3$.
$16 \stackrel{?}{=} 6 + 10$	$12 - 3 \stackrel{?}{=} 9$	Simplify.
$16 = 16$, true	$9 = 9$, true	True statements result.

The ordered pair $(4, 3)$ satisfies both equations. Thus, $(4, 3)$ is a solution of the system.

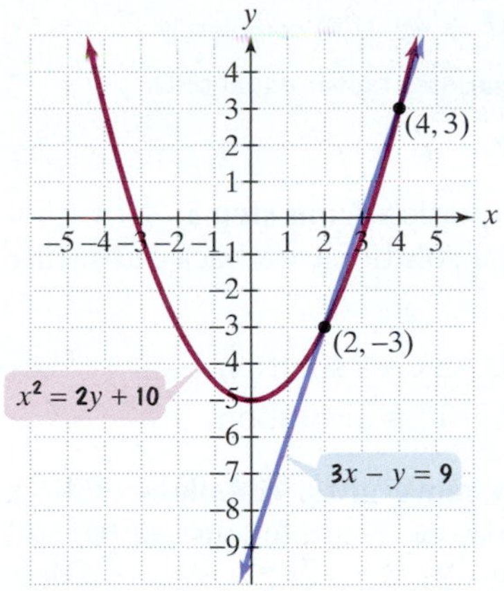

FIGURE 10.37
Points of intersection illustrate the nonlinear system's solutions.

Now let's check $(2, -3)$. Replace x with 2 and y with -3 in both given equations.

$x^2 = 2y + 10$	$3x - y = 9$	These are the given equations.
$2^2 \stackrel{?}{=} 2(-3) + 10$	$3(2) - (-3) \stackrel{?}{=} 9$	Let $x = 2$ and $y = -3$.
$4 \stackrel{?}{=} -6 + 10$	$6 + 3 \stackrel{?}{=} 9$	Simplify.
$4 = 4$, true	$9 = 9$, true	True statements result.

The ordered pair $(2, -3)$ also satisfies both equations and is a solution of the system. The solutions are $(4, 3)$ and $(2, -3)$, and the solution set is $\{(4, 3), (2, -3)\}$.

Figure 10.37 shows the graphs of the equations in the system and the solutions as intersection points.

CHECK POINT 1 Solve by the substitution method:

$$x^2 = y - 1$$
$$4x - y = -1.$$

EXAMPLE 2 Solving a Nonlinear System by the Substitution Method

Solve by the substitution method:

$$x - y = 3 \quad \text{(The graph is a line.)}$$
$$(x - 2)^2 + (y + 3)^2 = 4. \quad \text{(The graph is a circle.)}$$

Solution Graphically, we are finding the intersection of a line and a circle with center $(2, -3)$ and radius 2.

Step 1. Solve one of the equations for one variable in terms of the other. We will solve for x in the linear equation—that is, the first equation. (We could also solve for y.)

$$x - y = 3 \quad \text{This is the first equation in the given system.}$$
$$x = y + 3 \quad \text{Add } y \text{ to both sides.}$$

Step 2. Substitute the expression from step 1 into the other equation. We substitute $y + 3$ for x in the second equation.

$$x = \boxed{y + 3} \qquad (\boxed{x} - 2)^2 + (y + 3)^2 = 4$$

This gives an equation in one variable, namely

$$(y + 3 - 2)^2 + (y + 3)^2 = 4.$$

The variable x has been eliminated.

Step 3. Solve the resulting equation containing one variable.

$(y + 3 - 2)^2 + (y + 3)^2 = 4$	This is the equation containing one variable.
$(y + 1)^2 + (y + 3)^2 = 4$	Combine numerical terms in the first parentheses.
$y^2 + 2y + 1 + y^2 + 6y + 9 = 4$	Use the formula $(A + B)^2 = A^2 + 2AB + B^2$ to square $y + 1$ and $y + 3$.
$2y^2 + 8y + 10 = 4$	Combine like terms on the left.
$2y^2 + 8y + 6 = 0$	Subtract 4 from both sides and set the quadratic equation equal to 0.
$2(y^2 + 4y + 3) = 0$	Factor out 2.

$$2(y+3)(y+1)=0 \qquad \text{Factor } 2(y^2+4y+3) \text{ completely.}$$

$$y+3=0 \quad \text{or} \quad y+1=0 \qquad \text{Set each variable factor equal to 0.}$$

$$y=-3 \qquad\qquad y=-1 \qquad \text{Solve for } y.$$

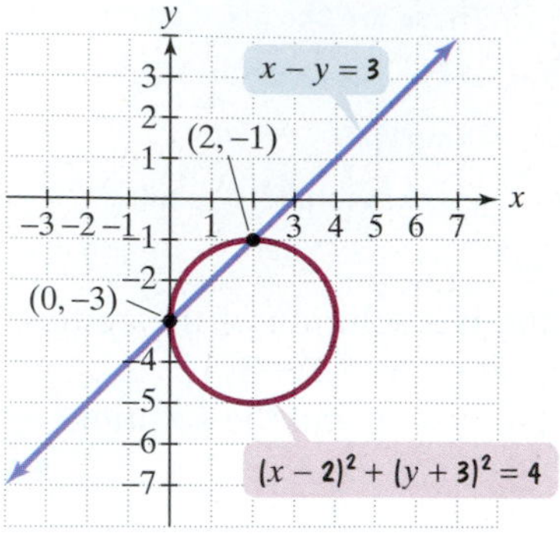

FIGURE 10.38
Points of intersection illustrate the nonlinear system's solutions.

Step 4. **Back-substitute the obtained values into the equation from step 1.** Now that we have $y=-3$ and $y=-1$, or the y-coordinates of the solutions, we back-substitute -3 for y and -1 for y in the equation $x=y+3$.

If $y=-3$: $\quad x=-3+3=0$, $\quad$ so $(0,-3)$ is a solution.
If $y=-1$: $\quad x=-1+3=2$, $\quad$ so $(2,-1)$ is a solution.

Step 5. **Check the proposed solutions in both of the system's given equations.** Take a moment to show that each ordered pair satisfies both equations. The solutions are $(0,-3)$ and $(2,-1)$, and the solution set of the given system is $\{(0,-3),(2,-1)\}$.

Figure 10.38 shows the graphs of the equations in the system and the solutions as intersection points.

✓ **CHECK POINT 2** Solve by the substitution method:

$$x+2y=0$$
$$(x-1)^2+(y-1)^2=5.$$

3 Solve systems of nonlinear equations by addition.

Eliminating a Variable Using the Addition Method

In solving linear systems with two variables, we learned that the addition method works well when each equation is in the form $Ax+By=C$. For nonlinear systems, the addition method can be used when each equation is in the form $Ax^2+By^2=C$. If necessary, we will multiply either equation or both equations by appropriate numbers so that the coefficients of x^2 or y^2 will have a sum of 0. We then add equations. The sum will be an equation in one variable.

EXAMPLE 3 Solving a Nonlinear System by the Addition Method

Solve the system:

$$4x^2+y^2=13 \qquad \text{Equation 1 (The graph is an ellipse.)}$$
$$x^2+y^2=10. \qquad \text{Equation 2 (The graph is a circle.)}$$

Solution We can use the same steps that we did when we solved linear systems by the addition method.

Step 1. **Write both equations in the form $Ax^2+By^2=C$.** Both equations are already in this form, so we can skip this step.

Step 2. **If necessary, multiply either equation or both equations by appropriate numbers so that the sum of the x^2-coefficients or the sum of the y^2-coefficients is 0.** We can eliminate y^2 by multiplying Equation 2 by -1.

$$4x^2+y^2=13 \xrightarrow{\text{No change}} 4x^2+y^2=13$$

$$x^2+y^2=10 \xrightarrow{\text{Multiply by } -1.} -x^2-y^2=-10$$

Steps 3 and 4. Add equations and solve for the remaining variable.

$$\begin{aligned} 4x^2 + y^2 &= 13 \\ -x^2 - y^2 &= -10 \\ \hline 3x^2 &= 3 \end{aligned}$$ Add equations.

$$x^2 = 1$$ Divide both sides by 3.

$$x = \pm 1$$ Use the square root property: If $x^2 = c$, then $x = \pm\sqrt{c}$.

Step 5. Back-substitute and find the values for the other variable. We must back-substitute each value of x into either one of the original equations. Let's use $x^2 + y^2 = 10$, Equation 2. If $x = 1$,

$$1^2 + y^2 = 10$$ Replace x with 1 in Equation 2.

$$y^2 = 9$$ Subtract 1 from both sides.

$$y = \pm 3.$$ Apply the square root property.

$(1, 3)$ and $(1, -3)$ are solutions. If $x = -1$,

$$(-1)^2 + y^2 = 10$$ Replace x with −1 in Equation 2.

$$y^2 = 9$$ The steps are the same as before.

$$y = \pm 3.$$

$(-1, 3)$ and $(-1, -3)$ are solutions.

Step 6. Check. Take a moment to show that each of the four ordered pairs satisfies the given equations, $4x^2 + y^2 = 13$ and $x^2 + y^2 = 10$. The solutions are $(1, 3)$, $(1, -3)$, $(-1, 3)$, and $(-1, -3)$, and the solution set of the given system is $\{(1, 3), (1, -3), (-1, 3), (-1, -3)\}$.

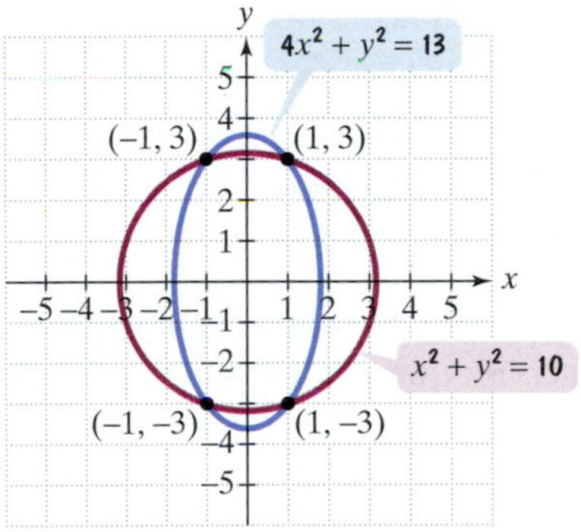

FIGURE 10.39 A system with four solutions

Figure 10.39 shows the graphs of the equations in the system, an ellipse and a circle, and the solutions as intersection points.

✓ **CHECK POINT 3** Solve the system:

$$3x^2 + 2y^2 = 35$$
$$4x^2 + 3y^2 = 48.$$

Study Tip

When solving nonlinear systems, extra solutions may be introduced that do not satisfy both equations in the system. Therefore, you should get into the habit of checking all proposed pairs in each of the system's two equations.

In solving nonlinear systems, we will include only ordered pairs with real numbers in the solution set. We have seen that each of these ordered pairs corresponds to a point of intersection of the system's graphs.

EXAMPLE 4 Solving a Nonlinear System by the Addition Method

Solve the system:

$$y = x^2 + 3$$ Equation 1 (The graph is a parabola.)

$$x^2 + y^2 = 9.$$ Equation 2 (The graph is a circle.)

Solution We could use substitution because Equation 1, $y = x^2 + 3$, has y expressed in terms of x, but substituting $x^2 + 3$ for y in $x^2 + y^2 = 9$ would result in a fourth-degree equation. However, we can rewrite Equation 1 by subtracting x^2 from both sides and adding the equations to eliminate the x^2-terms.

Notice how like terms are arranged in columns.

$$-x^2 + y \quad = 3$$ Subtract x^2 from both sides of Equation 1.

$$x^2 \quad + y^2 = 9$$ This is Equation 2.

$$y + y^2 = 12$$ Add the equations.

We now solve this quadratic equation.

$y + y^2 = 12$	This is the equation containing one variable.
$y^2 + y - 12 = 0$	Subtract 12 from both sides and write the quadratic equation in standard form.
$(y + 4)(y - 3) = 0$	Factor.
$y + 4 = 0$ or $y - 3 = 0$	Set each factor equal to 0.
$y = -4$ $y = 3$	Solve for y.

$y = x^2 + 3$ Equation 1
$x^2 + y^2 = 9$ Equation 2

The given equations (repeated)

To complete the solution, we must back-substitute each value of y into either one of the original equations. We will use $y = x^2 + 3$, Equation 1. First, we substitute -4 for y.

$$-4 = x^2 + 3$$
$$-7 = x^2 \quad \text{Subtract 3 from both sides.}$$

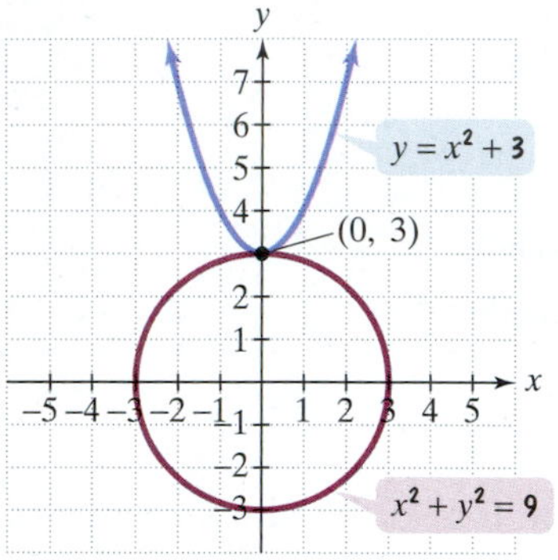

FIGURE 10.40 A system with one real solution.

Because the square of a real number cannot be negative, the equation $x^2 = -7$ does not have real-number solutions. We will not include the imaginary solutions, $x = \pm\sqrt{-7}$, or $i\sqrt{7}$ and $-i\sqrt{7}$, in the ordered pairs that make up the solution set. Thus, we move on to our other value for y, 3, and substitute this value into Equation 1.

$y = x^2 + 3$	This is Equation 1.
$3 = x^2 + 3$	Back-substitute 3 for y.
$0 = x^2$	Subtract 3 from both sides.
$0 = x$	Solve for x.

We showed that if $y = 3$, then $x = 0$. Thus, $(0, 3)$ is the solution with a real ordered pair. Take a moment to show that $(0, 3)$ satisfies the given equations, $y = x^2 + 3$ and $x^2 + y^2 = 9$. The solution set of the system is $\{(0, 3)\}$. **Figure 10.40** shows the system's graphs and the solution as an intersection point.

✓ CHECK POINT 4 Solve the system:

$$y = x^2 + 5$$
$$x^2 + y^2 = 25.$$

4 Solve problems using systems of nonlinear equations.

Applications

Many geometric problems can be modeled and solved by the use of systems of nonlinear equations. We will use our step-by-step strategy for solving problems using mathematical models that are created from verbal conditions.

EXAMPLE 5 An Application of a Nonlinear System

You have 36 yards of fencing to build the enclosure in **Figure 10.41**. Some of this fencing is to be used to build an internal divider. If you'd like to enclose 54 square yards, what are the dimensions of the enclosure?

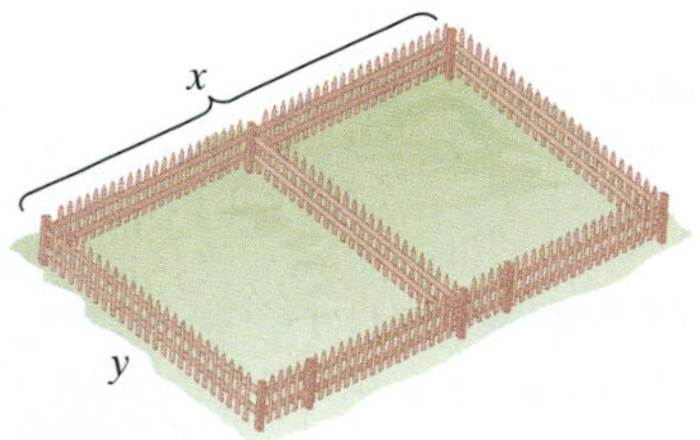

FIGURE 10.41 Building an enclosure

Solution

Step 1. Use variables to represent unknown quantities. Let x = the enclosure's length and y = the enclosure's width. These variables are shown in **Figure 10.41**.

Step 2. Write a system of equations that models the problem's conditions. The first condition is that you have 36 yards of fencing.

Fencing along both lengths	plus	fencing along both widths	plus	fencing for the internal divider	equals	36 yards.
$2x$	$+$	$2y$	$+$	y	$=$	36

Adding like terms, we can express the equation that models the verbal conditions for the fencing as $2x + 3y = 36$.

The second condition is that you'd like to enclose 54 square yards. The rectangle's area, the product of its length and its width, must be 54 square yards.

Length	times	width	is	54 square yards.
x	$\cdot$	y	$=$	54

Step 3. Solve the system and answer the problem's question. We must solve the system

$$2x + 3y = 36 \quad \text{Equation 1}$$
$$xy = 54. \quad \text{Equation 2}$$

We will use substitution. Because Equation 1 has no coefficients of 1 or -1, we will work with Equation 2 and solve for y. Dividing both sides of $xy = 54$ by x, we obtain

$$y = \frac{54}{x}.$$

Now we substitute $\frac{54}{x}$ for y in Equation 1 and solve for x.

$$2x + 3y = 36 \quad \text{This is Equation 1.}$$
$$2x + 3 \cdot \frac{54}{x} = 36 \quad \text{Substitute } \frac{54}{x} \text{ for } y.$$
$$2x + \frac{162}{x} = 36 \quad \text{Multiply.}$$
$$x\left(2x + \frac{162}{x}\right) = 36 \cdot x \quad \text{Clear fractions by multiplying both sides by } x.$$
$$2x^2 + 162 = 36x \quad \text{Use the distributive property on the left side.}$$
$$2x^2 - 36x + 162 = 0 \quad \text{Subtract } 36x \text{ from both sides and write the quadratic equation in standard form.}$$
$$2(x^2 - 18x + 81) = 0 \quad \text{Factor out 2.}$$
$$2(x - 9)^2 = 0 \quad \text{Factor completely using } A^2 - 2AB + B^2 = (A - B)^2.$$
$$x - 9 = 0 \quad \text{Set the repeated factor equal to zero.}$$
$$x = 9 \quad \text{Solve for } x.$$

We back-substitute this value of x into $y = \frac{54}{x}$.

$$\text{If } x = 9, \quad y = \frac{54}{9} = 6.$$

This means that the dimensions of the enclosure in **Figure 10.41** are 9 yards by 6 yards.

Step 4. Check the proposed solution in the original wording of the problem. Take a moment to check that a length of 9 yards and a width of 6 yards results in 36 yards of fencing and an area of 54 square yards.

✓ **CHECK POINT 5** Find the length and width of a rectangle whose perimeter is 20 feet and whose area is 21 square feet.

10.5 EXERCISE SET

MyMathLab | MathXL PRACTICE | WATCH | DOWNLOAD | READ | REVIEW

Practice Exercises

In Exercises 1–18, solve each system by the substitution method.

1. $x + y = 2$
$y = x^2 - 4$

2. $x - y = -1$
$y = x^2 + 1$

3. $x + y = 2$
$y = x^2 - 4x + 4$

4. $2x + y = -5$
$y = x^2 + 6x + 7$

5. $y = x^2 - 4x - 10$
$y = -x^2 - 2x + 14$

6. $y = x^2 + 4x + 5$
$y = x^2 + 2x - 1$

7. $x^2 + y^2 = 25$
$x - y = 1$

8. $x^2 + y^2 = 5$
$3x - y = 5$

9. $xy = 6$
$2x - y = 1$

10. $xy = -12$
$x - 2y + 14 = 0$

11. $y^2 = x^2 - 9$
$2y = x - 3$

12. $x^2 + y = 4$
$2x + y = 1$

13. $xy = 3$
$x^2 + y^2 = 10$

14. $xy = 4$
$x^2 + y^2 = 8$

15. $x + y = 1$
$x^2 + xy - y^2 = -5$

16. $x + y = -3$
$x^2 + 2y^2 = 12y + 18$

17. $x + y = 1$
$(x - 1)^2 + (y + 2)^2 = 10$

18. $2x + y = 4$
$(x + 1)^2 + (y - 2)^2 = 4$

In Exercises 19–28, solve each system by the addition method.

19. $x^2 + y^2 = 13$
$x^2 - y^2 = 5$

20. $4x^2 - y^2 = 4$
$4x^2 + y^2 = 4$

21. $x^2 - 4y^2 = -7$
$3x^2 + y^2 = 31$

22. $3x^2 - 2y^2 = -5$
$2x^2 - y^2 = -2$

23. $3x^2 + 4y^2 - 16 = 0$
$2x^2 - 3y^2 - 5 = 0$

24. $16x^2 - 4y^2 - 72 = 0$
$x^2 - y^2 - 3 = 0$

25. $x^2 + y^2 = 25$
$(x - 8)^2 + y^2 = 41$

26. $x^2 + y^2 = 4$
$x^2 + (y - 3)^2 = 9$

27. $y^2 - x = 4$
$x^2 + y^2 = 4$

28. $x^2 - 2y = 8$
$x^2 + y^2 = 16$

In Exercises 29–42, solve each system by the method of your choice.

29. $3x^2 + 4y^2 = 16$
$2x^2 - 3y^2 = 5$

30. $x + y^2 = 4$
$x^2 + y^2 = 16$

31. $2x^2 + y^2 = 18$
$xy = 4$

32. $x^2 + 4y^2 = 20$
$xy = 4$

33. $x^2 + 4y^2 = 20$
$x + 2y = 6$

34. $3x^2 - 2y^2 = 1$
$4x - y = 3$

35. $x^3 + y = 0$
$x^2 - y = 0$

36. $x^3 + y = 0$
$2x^2 - y = 0$

37. $x^2 + (y - 2)^2 = 4$
$x^2 - 2y = 0$

38. $x^2 - y^2 - 4x + 6y - 4 = 0$
$x^2 + y^2 - 4x - 6y + 12 = 0$

39. $y = (x + 3)^2$
$x + 2y = -2$

40. $(x - 1)^2 + (y + 1)^2 = 5$
$2x - y = 3$

41. $x^2 + y^2 + 3y = 22$
$2x + y = -1$

42. $x - 3y = -5$
$x^2 + y^2 - 25 = 0$

In Exercises 43–46, let x represent one number and let y represent the other number. Use the given conditions to write a system of nonlinear equations. Solve the system and find the numbers.

43. The sum of two numbers is 10 and their product is 24. Find the numbers.

44. The sum of two numbers is 20 and their product is 96. Find the numbers.

45. The difference between the squares of two numbers is 3. Twice the square of the first number increased by the square of the second number is 9. Find the numbers.

46. The difference between the squares of two numbers is 5. Twice the square of the second number subtracted from three times the square of the first number is 19. Find the numbers.

Practice PLUS

In Exercises 47–52, solve each system by the method of your choice.

47. $2x^2 + xy = 6$
$x^2 + 2xy = 0$

48. $4x^2 + xy = 30$
$x^2 + 3xy = -9$

49. $-4x + y = 12$
$y = x^3 + 3x^2$

50. $-9x + y = 45$
$y = x^3 + 5x^2$

51. $\frac{3}{x^2} + \frac{1}{y^2} = 7$
$\frac{5}{x^2} - \frac{2}{y^2} = -3$

52. $\frac{2}{x^2} + \frac{1}{y^2} = 11$
$\frac{4}{x^2} - \frac{2}{y^2} = -14$

In Exercises 53–54, make a rough sketch in a rectangular coordinate system of the graphs representing the equations in each system.

53. The system, whose graphs are a circle and an ellipse, has two solutions. Both solutions can be represented as points on the x-axis.

54. The system, whose graphs are a line with positive slope and a parabola that is not a function, has two solutions.

Application Exercises

55. A planet follows an elliptical path described by $16x^2 + 4y^2 = 64$. A comet follows the parabolic path $y = x^2 - 4$. Where might the comet intersect the orbiting planet?

56. A system for tracking ships indicates that a ship lies on a hyperbolic path described by $2y^2 - x^2 = 1$. The process is repeated and the ship is found to lie on a hyperbolic path described by $2x^2 - y^2 = 1$. If it is known that the ship is located in the first quadrant of the coordinate system, determine its exact location.

57. Find the length and width of a rectangle whose perimeter is 36 feet and whose area is 77 square feet.

58. Find the length and width of a rectangle whose perimeter is 40 feet and whose area is 96 square feet.

Use the formula for the area of a rectangle and the Pythagorean Theorem to solve Exercises 59–60.

59. A small television has a picture with a diagonal measure of 10 inches and a viewing area of 48 square inches. Find the length and width of the screen.

60. The area of a rug is 108 square feet and the length of its diagonal is 15 feet. Find the length and width of the rug.

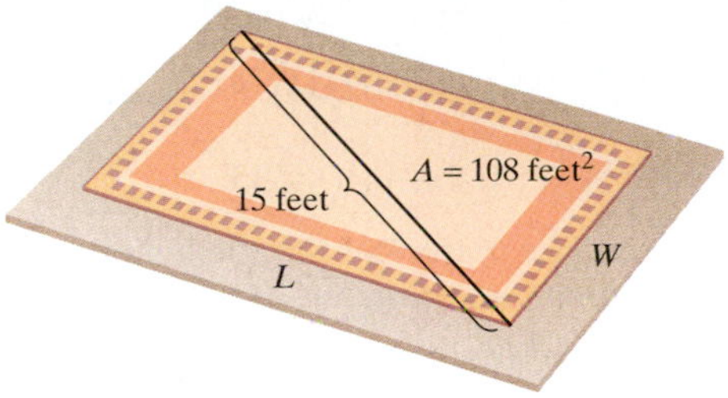

61. The figure shows a square floor plan with a smaller square area that will accommodate a combination fountain and pool. The floor with the fountain-pool area removed has an area of 21 square meters and a perimeter of 24 meters. Find the dimensions of the floor and the dimensions of the square that will accommodate the fountain and pool.

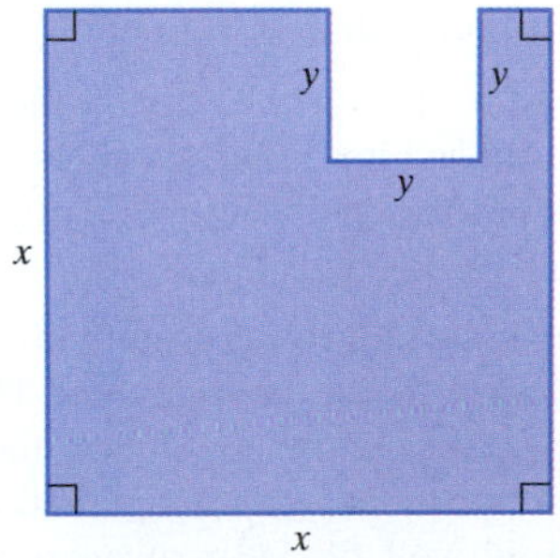

62. The area of the rectangular piece of cardboard shown on the left is 216 square inches. The cardboard is used to make an open box by cutting a 2-inch square from each corner and turning up the sides. If the box is to have a volume of 224 cubic inches, find the length and width of the cardboard that must be used.

63. The bar graph shows that compared to a century ago, work in the United States now involves mostly white-collar service jobs.

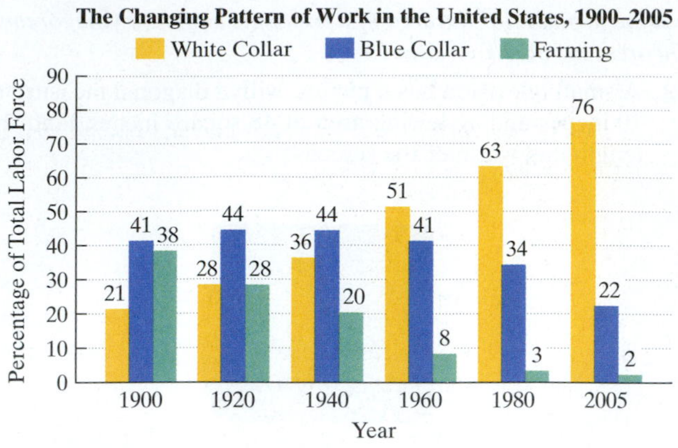

Source: U.S. Department of Labor

The data can be modeled by linear and quadratic functions.

White collar: $0.5x - y = -18$

Blue collar: $y = -0.004x^2 + 0.23x + 41$

Farming: $0.4x + y = 35$

In each function, x represents the number of years after 1900 and y represents the percentage of the total U.S. labor force.

a. Based on the information in the graph, it appears that there was a year when the percentage of white-collar workers in the labor force was the same as the percentage of blue-collar workers in the labor force. According to the graph, between which two decades did this occur?

b. Solve a nonlinear system to determine the year described in part (a). Round to the nearest year. What percentage of the labor force, to the nearest percent, consisted of white-collar workers and what percentage consisted of blue-collar workers?

c. According to the graph, for which year was the percentage of white-collar workers the same as the percentage of farmers? What percentage of U.S. workers were in each of these groups?

d. Solve a linear system to determine the year described in part (c). Round to the nearest year. Use the models to find the percentage of the labor force consisting of white-collar workers and the percentage consisting of farmers. How well do your answers model the actual data specified in part (c)?

Writing in Mathematics

64. What is a system of nonlinear equations? Provide an example with your description.

65. Explain how to solve a nonlinear system using the substitution method. Use $x^2 + y^2 = 9$ and $2x - y = 3$ to illustrate your explanation.

66. Explain how to solve a nonlinear system using the addition method. Use $x^2 - y^2 = 5$ and $3x^2 - 2y^2 = 19$ to illustrate your explanation.

67. The daily demand and supply models for a carrot cake supplied by a bakery to a convenience store are given by the demand model $N = 40 - 3p$ and the supply model $N = \frac{p^2}{10}$, in which p is the price of the cake and N is the number of cakes sold or supplied each day to the convenience store. Explain how to determine the price at which supply and demand are equal. Then describe how to find how many carrot cakes can be supplied and sold each day at this price.

Technology Exercises

68. Verify your solutions to any five exercises from Exercises 1–42 by using a graphing utility to graph the two equations in the system in the same viewing rectangle. Then use the trace or intersection feature to verify the solutions.

69. Write a system of equations, one equation whose graph is a line and the other whose graph is a parabola, that has no ordered pairs that are real numbers in its solution set. Graph the equations using a graphing utility and verify that you are correct.

Critical Thinking Exercises

Make Sense? *In Exercises 70–73, determine whether each statement "makes sense" or "does not make sense" and explain your reasoning.*

70. I use the same steps to solve nonlinear systems as I did to solve linear systems, although I don't obtain linear equations when a variable is eliminated.

71. I graphed a nonlinear system that modeled the elliptical orbits of Earth and Mars, and the graphs indicated the system had a solution with a real ordered pair.

72. Without using any algebra, it's obvious that the nonlinear system consisting of $x^2 + y^2 = 4$ and $x^2 + y^2 = 25$ does not have real-number solutions.

73. I think that the nonlinear system consisting of $x^2 + y^2 = 36$ and $y = (x - 2)^2 - 3$ is easier to solve graphically than by using the substitution method or the addition method.

In Exercises 74–77, determine whether each statement is true or false. If the statement is false, make the necessary change(s) to produce a true statement.

74. A system of two equations in two variables whose graphs are a circle and a line can have four real ordered-pair solutions.

75. A system of two equations in two variables whose graphs are a parabola and a circle can have four real ordered-pair solutions.

76. A system of two equations in two variables whose graphs are two circles must have at least two real ordered-pair solutions.

77. A system of two equations in two variables whose graphs are a parabola and a circle cannot have only one real ordered-pair solution.

78. Find a and b in this figure.

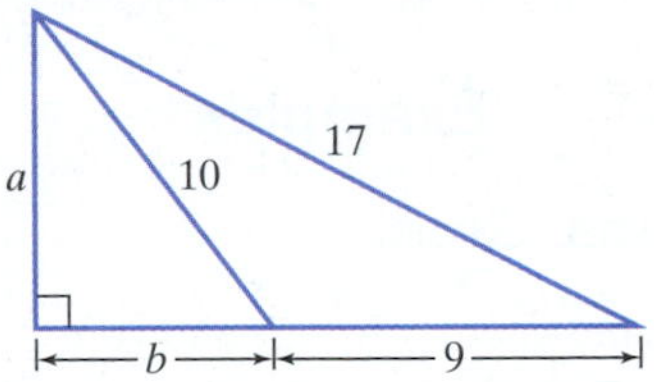

Solve the systems in Exercises 79–80.

79. $\log_y x = 3$
$\log_y(4x) = 5$

80. $\log x^2 = y + 3$
$\log x = y - 1$

Review Exercises

81. Graph: $3x - 2y \le 6$.
(Section 4.4, Example 1)

82. Find the slope of the line passing through $(-2, -3)$ and $(1, 5)$.
(Section 2.4, Example 2)

83. Multiply: $(3x - 2)(2x^2 - 4x + 3)$.
(Section 5.2, Example 3)

Preview Exercises

Exercises 84–86 will help you prepare for the material covered in the first section of the next chapter.

84. Evaluate $\frac{(-1)^n}{3^n - 1}$ for $n = 1, 2, 3,$ and 4.

85. Find the product of all positive integers from n down through 1 for $n = 5$.

86. Evaluate $n^2 + 1$ for all consecutive integers from 1 to 6. Then find the sum of the six evaluations.

GROUP PROJECT CHAPTER 10

■ MODELING PLANETARY MOTION

Polish astronomer Nicolaus Copernicus (1473–1543) was correct in stating that planets in our solar system revolve around the sun and not Earth. However, he incorrectly believed that celestial orbits move in perfect circles, calling his system "the ballet of the planets."

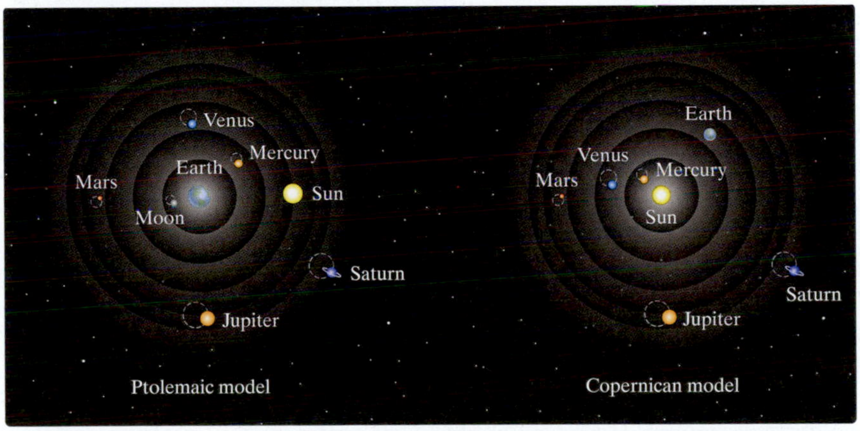

German scientist and mathematician Johannes Kepler (1571–1630) discovered that planets move in elliptical orbits with the sun at one focus. In this exercise, group members will write equations for two of these orbits and use a graphing utility to see what they look like. Use the following information:

Earth's orbit:	Length of major axis: 186 million miles Length of minor axis: 185.8 million miles
Mars's orbit:	Length of major axis: 283.5 million miles Length of minor axis: 278.5 million miles.

a. Group members should write equations in the form $\frac{x^2}{a^2} + \frac{y^2}{b^2} = 1$ for the elliptical orbits of Earth and Mars.

b. Use a graphing utility to graph the two ellipses in the same $[-300, 300, 50]$ by $[-200, 200, 50]$ viewing rectangle. Based on these graphs, explain why early astronomers incorrectly used the Copernican model to describe planetary motion.

c. Describing planetary orbits, Kepler wrote, "The heavenly motions are nothing but a continuous song for several voices, to be perceived by the intellect, not by the ear." Group members should discuss what Kepler meant by this statement.

CHAPTER 10 SUMMARY

Definitions and Concepts	Examples

Section 10.1 Distance and Midpoint Formulas; Circles

The Distance Formula

The distance, d, between the points (x_1, y_1) and (x_2, y_2) is given by

$$d = \sqrt{(x_2 - x_1)^2 + (y_2 - y_1)^2}.$$

Find the distance between $(-3, -5)$ and $(6, -2)$.

$$d = \sqrt{[6 - (-3)]^2 + [-2 - (-5)]^2}$$
$$= \sqrt{9^2 + 3^2} = \sqrt{81 + 9} = \sqrt{90} = 3\sqrt{10} \approx 9.49$$

The Midpoint Formula

The midpoint of the line segment whose endpoints are (x_1, y_1) and (x_2, y_2) is the point with coordinates

$$\left(\frac{x_1 + x_2}{2}, \frac{y_1 + y_2}{2}\right).$$

Find the midpoint of the line segment whose endpoints are $(-3, 6)$ and $(4, 1)$.

$$\text{midpoint} = \left(\frac{-3 + 4}{2}, \frac{6 + 1}{2}\right) = \left(\frac{1}{2}, \frac{7}{2}\right)$$

A circle is the set of all points in a plane that are equidistant from a fixed point, the center. The distance from the center to any point on the circle is the radius.

Standard Form of the Equation of a Circle

The graph of $(x - h)^2 + (y - k)^2 = r^2$ is a circle with center (h, k) and radius r.

General Form of the Equation of a Circle

$$x^2 + y^2 + Dx + Ey + F = 0$$

Convert from the general form to the standard form by completing the square on x and y.

Find the center and radius and graph:

$$(x - 3)^2 + (y + 4)^2 = 16.$$

$$(x - 3)^2 + (y - (-4))^2 = 4^2$$

$h = 3$ $k = -4$ $r = 4$

The center, (h, k), is $(3, -4)$ and the radius, r, is 4.

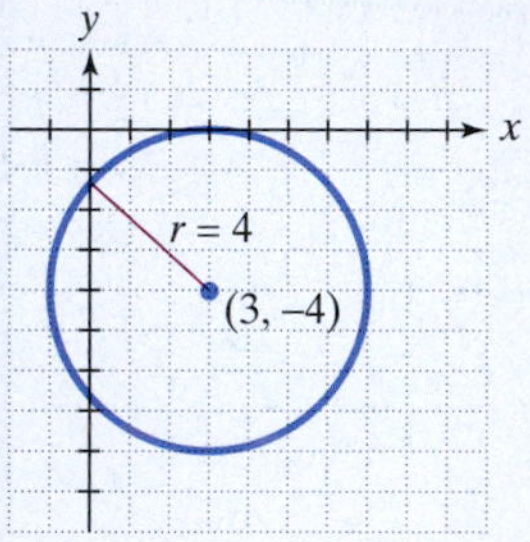

Section 10.2 The Ellipse

An ellipse is the set of all points in a plane the sum of whose distances from two fixed points, the foci, is constant. The midpoint of the segment connecting the foci is the center of the ellipse.

Standard Forms of the Equations of an Ellipse Centered at the Origin ($a^2 > b^2$)

Horizontal with vertices $(a, 0)$ and $(-a, 0)$	Vertical with vertices $(0, a)$ and $(0, -a)$
$\frac{x^2}{a^2} + \frac{y^2}{b^2} = 1$	$\frac{x^2}{b^2} + \frac{y^2}{a^2} = 1$
Endpoints of major axis a units left and right of center; minor axis b units up and down from center.	Endpoints of major axis a units up and down from center; minor axis b units left and right of center.

The equations $\frac{(x - h)^2}{a^2} + \frac{(y - k)^2}{b^2} = 1$ and $\frac{(x - h)^2}{b^2} + \frac{(y - k)^2}{a^2} = 1 (a^2 > b^2)$ represent ellipses centered at (h, k).

Graph: $\frac{(x + 2)^2}{9} + \frac{(y + 4)^2}{25} = 1.$

$$\frac{(x - (-2))^2}{9} + \frac{(y - (-4))^2}{25} = 1$$

$b^2 = 9$ $a^2 = 25$

Center, (h, k), is $(-2, -4)$. With $a^2 = 25$, vertices are 5 units above and below the center. With $b^2 = 9$, endpoints of minor axis are 3 units to the left and right of the center.

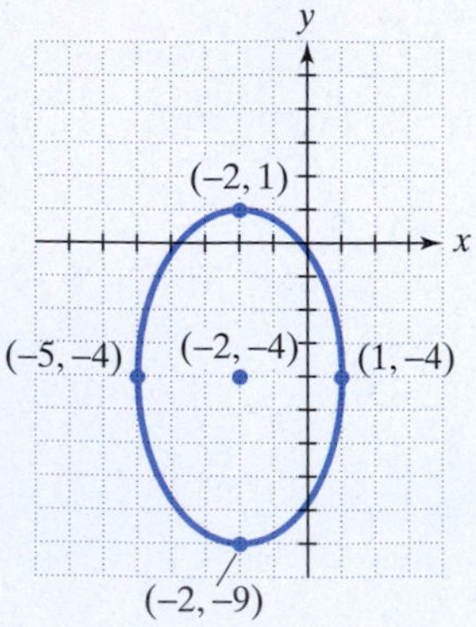

Definitions and Concepts | Examples

Section 10.3 The Hyperbola

A hyperbola is the set of all points in a plane the difference of whose distances from two fixed points, the foci, is constant.

Standard Forms of the Equations of a Hyperbola Centered at the Origin

$$\frac{x^2}{a^2} - \frac{y^2}{b^2} = 1 \qquad \frac{y^2}{a^2} - \frac{x^2}{b^2} = 1$$

Vertices are $(a, 0)$ and $(-a, 0)$. Vertices are $(0, a)$ and $(0, -a)$.

As x and y get larger, the two branches of a hyperbola approach a pair of intersecting lines, called asymptotes. Draw the rectangle centered at the origin with sides parallel to the axes, crossing one axis at $\pm a$ and the other at $\pm b$. Draw the diagonals of this rectangle and extend them to obtain the asymptotes. Draw the two branches of the hyperbola by starting at each vertex and approaching the asymptotes.

Graph: $4x^2 - 9y^2 = 36$.

$$\frac{4x^2}{36} - \frac{9y^2}{36} = \frac{36}{36}$$

$$\frac{x^2}{9} - \frac{y^2}{4} = 1$$

$a^2 = 9$ $\quad$ $b^2 = 4$

Vertices are $(3, 0)$ and $(-3, 0)$. Draw a rectangle using -3 and 3 on the x-axis and -2 and 2 on the y-axis. Its extended diagonals are the asymptotes.

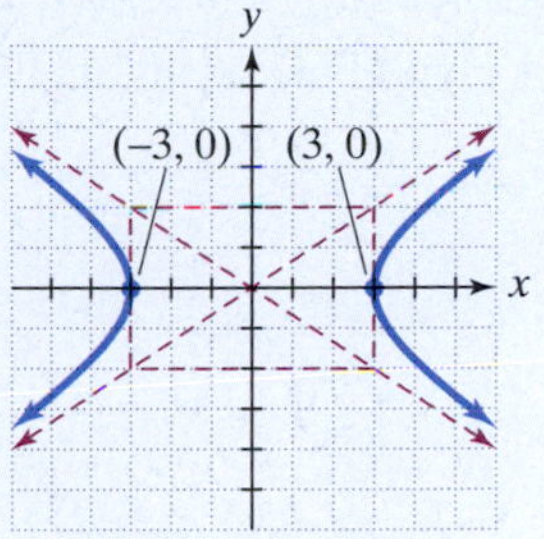

Section 10.4 The Parabola; Identifying Conic Sections

A parabola is the set of all points in a plane that are equidistant from a fixed line, the directrix, and a fixed point, the focus, that is not on the line. The line passing through the focus and perpendicular to the directrix is the axis of symmetry. The point of intersection of the parabola with its axis of symmetry is the vertex.

Equations of Horizontal Parabolas

$x = a(y - k)^2 + h$ $\qquad$ $x = ay^2 + by + c$

Vertex is (h, k). $\qquad$ y-coordinate of vertex is $y = -\dfrac{b}{2a}$.

If $a > 0$, the parabola opens to the right. If $a < 0$, the parabola opens to the left.

Equations of Vertical Parabolas

$y = a(x - h)^2 + k$ $\qquad$ $y = ax^2 + bx + c$

Vertex is (h, k). $\qquad$ x-coordinate of vertex is $x = -\dfrac{b}{2a}$.

If $a > 0$, the parabola opens upward. If $a < 0$, the parabola opens downward.

Find the vertex and graph:

$$x = -(y + 3)^2 + 4.$$

$$x = -1(y - (-3))^2 + 4$$

$a = -1$ $\quad$ $k = -3$ $\quad$ $h = 4$

Parabola opens to the left. Vertex, (h, k), is $(4, -3)$.

x-intercept: Let $y = 0$.

$$x = -(0 + 3)^2 + 4 = -9 + 4 = -5$$

y-intercepts: Let $x = 0$.

$$0 = -(y + 3)^2 + 4$$

$$(y + 3)^2 = 4$$

$$y + 3 = \pm\sqrt{4}$$

$$y = -1 \quad \text{or} \quad y = -5$$

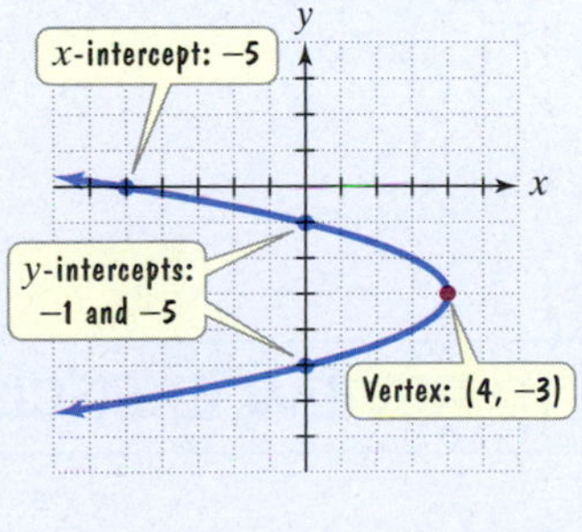

Definitions and Concepts	Examples

Section 10.4 The Parabola; Identifying Conic Sections (continued)

Recognizing Conic Sections from Equations

Conic sections result from the intersection of a cone and a plane. Their equations contain x^2-terms, y^2-terms, or both. For parabolas, only one variable is squared. For circles, ellipses, and hyperbolas, both variables are squared. Collect x^2- and y^2-terms on the same side of the equation to identify the graph:

- Circle: x^2 and y^2 have the same coefficient of the same sign.
- Ellipse: x^2 and y^2 have different coefficients of the same sign.
- Hyperbola: x^2 and y^2 have coefficients with opposite signs.

Examples:

- $9x^2 + 4y^2 = 36$

  Different coefficients of the same sign; ellipse
- $9y^2 = 25x^2 + 225$

  $9y^2 - 25x^2 = 225$

  Coefficients with opposite signs; hyperbola
- $x = -y^2 + 4y$

  Only one variable, y, is squared; parabola
- $\frac{x^2}{9} + \frac{y^2}{9} = 1$

  $x^2 + y^2 = 9$

  Same coefficient; circle

Section 10.5 Systems of Nonlinear Equations in Two Variables

A system of two nonlinear equations in two variables, x and y, contains at least one equation that cannot be expressed in the form $Ax + By = C$. Systems can be solved by the substitution method or the addition method. Each real solution corresponds to a point of intersection of the system's graphs.

Solve: $x^2 + y^2 = 25$

$x - 3y = -5.$

Using substitution: $x = 3y - 5$

$$(3y - 5)^2 + y^2 = 25$$
$$9y^2 - 30y + 25 + y^2 = 25$$
$$10y^2 - 30y = 0$$
$$10y(y - 3) = 0$$
$$y = 0 \quad \text{or} \quad y = 3$$

If $y = 0$: $x = 3y - 5 = 3 \cdot 0 - 5 = -5.$

If $y = 3$: $x = 3y - 5 = 3 \cdot 3 - 5 = 4.$

The solutions are $(-5, 0)$ and $(4, 3)$, and the solution set is $\{(-5, 0), (4, 3)\}$.

CHAPTER 10 REVIEW EXERCISES

10.1 *In Exercises 1–2, find the distance between each pair of points. If necessary, round answers to two decimal places.*

1. $(-2, -3)$ and $(3, 9)$
2. $(-4, 3)$ and $(-2, 5)$

In Exercises 3–4, find the midpoint of the line segment with the given endpoints.

3. $(2, 6)$ and $(-12, 4)$
4. $(4, -6)$ and $(-15, 2)$

In Exercises 5–6, write the standard form of the equation of the circle with the given center and radius.

5. Center $(0, 0)$, $r = 3$
6. Center $(-2, 4)$, $r = 6$

In Exercises 7–10, give the center and radius of each circle and graph its equation.

7. $x^2 + y^2 = 1$
8. $(x + 2)^2 + (y - 3)^2 = 9$

9. $x^2 + y^2 - 4x + 2y - 4 = 0$

10. $x^2 + y^2 - 4y = 0$

10.2 *In Exercises 11–16, graph each ellipse.*

11. $\frac{x^2}{36} + \frac{y^2}{25} = 1$

12. $\frac{x^2}{25} + \frac{y^2}{16} = 1$

13. $4x^2 + y^2 = 16$

14. $4x^2 + 9y^2 = 36$

15. $\frac{(x-1)^2}{16} + \frac{(y+2)^2}{9} = 1$

16. $\frac{(x+1)^2}{9} + \frac{(y-2)^2}{16} = 1$

17. A semielliptic archway has a height of 15 feet at the center and a width of 50 feet, as shown in the figure. The 50-foot width consists of a two-lane road. Can a truck that is 12 feet high and 14 feet wide drive under the archway without going into the other lane?

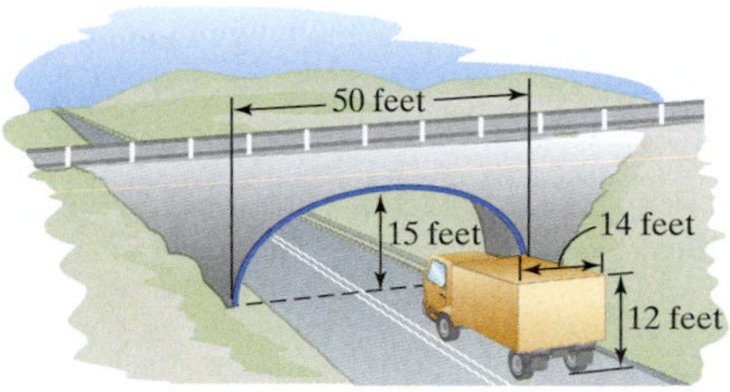

10.3 *In Exercises 18–21, use vertices and asymptotes to graph each hyperbola.*

18. $\frac{x^2}{16} - y^2 = 1$

19. $\frac{y^2}{16} - x^2 = 1$

20. $9x^2 - 16y^2 = 144$

21. $4y^2 - x^2 = 16$

10.4 *In Exercises 22–25, use the vertex and intercepts to sketch the graph of each equation. If needed, find additional points on the parabola by choosing values of y on each side of the axis of symmetry.*

22. $x = (y-3)^2 - 4$

23. $x = -2(y+3)^2 + 2$

24. $x = y^2 - 8y + 12$

25. $x = -y^2 - 4y + 6$

In Exercises 26–32, indicate whether the graph of each equation is a circle, an ellipse, a hyperbola, or a parabola.

26. $x + 8y = y^2 + 10$

27. $16x^2 = 32 - y^2$

28. $x^2 = 25 + 25y^2$

29. $x^2 = 4 - y^2$

30. $36y^2 = 576 + 16x^2$

31. $\frac{(x+3)^2}{9} + \frac{(y-4)^2}{25} = 1$

32. $y = x^2 + 6x + 9$

In Exercises 33–41, indicate whether the graph of each equation is a circle, an ellipse, a hyperbola, or a parabola. Then graph the conic section.

33. $5x^2 + 5y^2 = 180$

34. $4x^2 + 9y^2 = 36$

35. $4x^2 - 9y^2 = 36$

36. $\frac{x^2}{25} + \frac{y^2}{1} = 1$

37. $x + 3 = -y^2 + 2y$

38. $y - 3 = x^2 - 2x$

39. $\frac{(x+2)^2}{16} + \frac{(y-5)^2}{4} = 1$

40. $(x-3)^2 + (y+2)^2 = 4$

41. $x^2 + y^2 + 6x - 2y + 6 = 0$

42. An engineer is designing headlight units for automobiles. The unit has a parabolic surface with a diameter of 12 inches and a depth of 3 inches. The situation is illustrated in the figure, where a coordinate system has been superimposed.

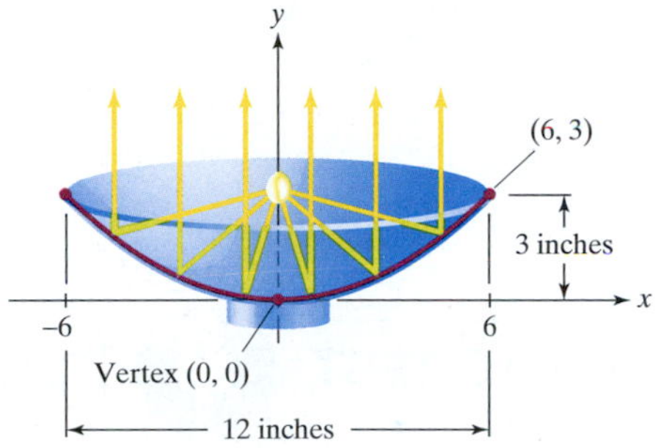

a. Use the point (6, 3) to write an equation in the form $y = ax^2$ for the parabola used to design the headlight.

b. The light source should be placed at the focus $(0, p)$. The value of p is given by the equation $a = \frac{1}{4p}$. Where should the light source be placed? Describe this placement relative to the vertex.

10.5 *In Exercises 43–53, solve each system by the method of your choice.*

43. $5y = x^2 - 1$
$x - y = 1$

44. $y = x^2 + 2x + 1$
$x + y = 1$

45. $x^2 + y^2 = 2$
$x + y = 0$

46. $2x^2 + y^2 = 24$
$x^2 + y^2 = 15$

47. $xy - 4 = 0$
$y - x = 0$

48. $y^2 = 4x$
$x - 2y + 3 = 0$

49. $x^2 + y^2 = 10$
$y = x + 2$

50. $xy = 1$
$y = 2x + 1$

51. $x + y + 1 = 0$
$x^2 + y^2 + 6y - x = -5$

52. $x^2 + y^2 = 13$
$x^2 - y = 7$

53. $2x^2 + 3y^2 = 21$
$3x^2 - 4y^2 = 23$

54. The perimeter of a rectangle is 26 meters and its area is 40 square meters. Find its dimensions.

55. Find the coordinates of all points (x, y) that lie on the line whose equation is $2x + y = 8$, so that the area of the rectangle shown in the figure is 6 square units.

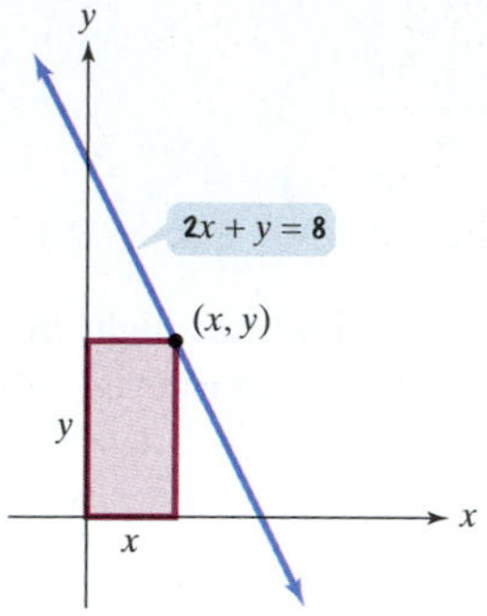

56. Two adjoining square fields with an area of 2900 square feet are to be enclosed with 240 feet of fencing. The situation is represented in the figure. Find the length of each side where a variable appears.

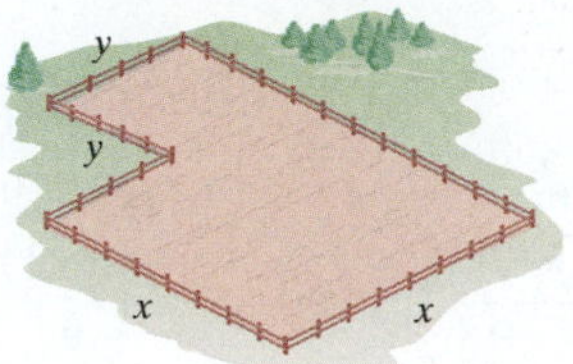

CHAPTER 10 TEST

Remember to use your Chapter Test Prep Video CD to see the worked-out solutions to the test questions you want to review.

1. Find the distance between $(-1, 5)$ and $(2, -3)$. If necessary, round the answer to two decimal places.

2. Find the midpoint of the line segment whose endpoints are $(-5, -2)$ and $(12, -6)$.

3. Write the standard form of the equation of the circle with center $(3, -2)$ and radius 5.

In Exercises 4–5, give the center and radius of each circle.

4. $(x - 5)^2 + (y + 3)^2 = 49$

5. $x^2 + y^2 + 4x - 6y - 3 = 0$

In Exercises 6–7, give the coordinates of the vertex for each parabola.

6. $x = -2(y + 3)^2 + 7$

7. $x = y^2 + 10y + 23$

In Exercises 8–16, indicate whether the graph of each equation is a circle, an ellipse, a hyperbola, or a parabola. Then graph the conic section.

8. $\dfrac{x^2}{4} - \dfrac{y^2}{9} = 1$

9. $4x^2 + 9y^2 = 36$

10. $x = (y + 1)^2 - 4$

11. $16x^2 + y^2 = 16$

12. $25y^2 = 9x^2 + 225$

13. $x = -y^2 + 6y$

14. $\dfrac{(x - 2)^2}{16} + \dfrac{(y + 3)^2}{9} = 1$

15. $(x + 1)^2 + (y + 2)^2 = 9$

16. $\dfrac{x^2}{4} + \dfrac{y^2}{4} = 1$

In Exercises 17–18, solve each system.

17. $x^2 + y^2 = 25$
$x + y = 1$

18. $2x^2 - 5y^2 = -2$
$3x^2 + 2y^2 = 35$

19. The rectangular plot of land shown in the figure is to be fenced along three sides using 39 feet of fencing. No fencing is to be placed along the river's edge. The area of the plot is 180 square feet. What are its dimensions?

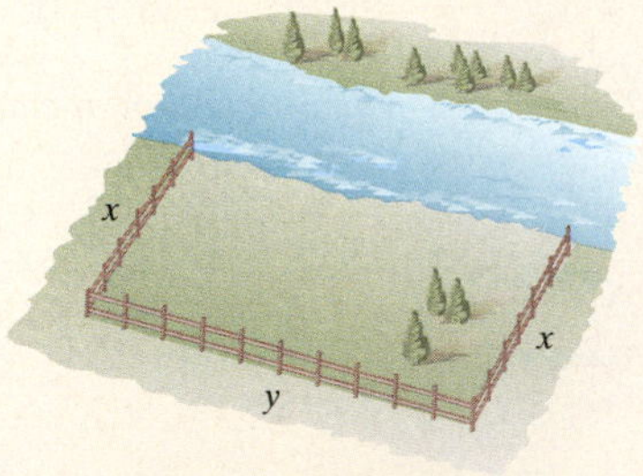

20. A rectangle has a diagonal of 5 feet and a perimeter of 14 feet. Find the rectangle's dimensions.

CUMULATIVE REVIEW EXERCISES (CHAPTERS 1–10)

In Exercises 1–7, solve each equation, inequality, or system.

1. $3x + 7 > 4$ or $6 - x < 1$

2. $x(2x - 7) = 4$

3. $\dfrac{5}{x-3} = 1 + \dfrac{30}{x^2 - 9}$

4. $3x^2 + 8x + 5 < 0$

5. $3^{2x-1} = 81$

6. $30e^{0.7x} = 240$

7. $3x^2 + 4y^2 = 39$
$5x^2 - 2y^2 = -13$

In Exercises 8–11, graph each function, equation, or inequality in a rectangular coordinate system.

8. $f(x) = -\frac{2}{3}x + 4$

9. $3x - y > 6$

10. $x^2 + y^2 + 4x - 6y + 9 = 0$

11. $9x^2 - 4y^2 = 36$

In Exercises 12–15, perform the indicated operations and simplify, if possible.

12. $-2(3^2 - 12)^3 - 45 \div 9 - 3$

13. $(3x^3 - 19x^2 + 17x + 4) \div (3x - 4)$

14. $\sqrt[3]{4x^2y^5} \cdot \sqrt[3]{4xy^2}$

15. $(2 + 3i)(4 - i)$

In Exercises 16–17, factor completely.

16. $12x^3 - 36x^2 + 27x$

17. $x^3 - 2x^2 - 9x + 18$

18. Find the domain: $f(x) = \sqrt{6 - 3x}$.

19. Rationalize the denominator: $\dfrac{1 - \sqrt{x}}{1 + \sqrt{x}}$.

20. Write as a single logarithm: $\frac{1}{3}\ln x + 7\ln y$.

21. Divide using synthetic division:

$(3x^3 - 5x^2 + 2x - 1) \div (x - 2)$.

22. Write a quadratic equation whose solution set is $\{-2\sqrt{3}, 2\sqrt{3}\}$.

23. Two cars leave from the same place at the same time, traveling in opposite directions. The rate of the faster car exceeds that of the slower car by 10 miles per hour. After 2 hours, the cars are 180 miles apart. Find the rate of each car.

24. Rent-a-Truck charges a daily rental rate of \$39 plus \$0.16 per mile. A competing agency, Ace Truck Rentals, charges \$25 a day plus \$0.24 per mile for the same truck. How many miles must be driven in a day to make the daily cost of both agencies the same? What will be the cost?

25. Three apples and two bananas provide 354 calories, and two apples and three bananas provide 381 calories. Find the number of calories in one apple and one banana.

CHAPTER 11

More on Polynomial and Rational Functions

We all recognize symmetry when we see it. The perfect reflections in a lake of a house surrounded by rock formations occurs because the surface of the water is absolutely still. A mild breeze rippling the water's surface would distort this symmetry.

Graphs of some functions have symmetrical qualities. In this chapter, we take a closer look at polynomial and rational functions and the symmetries of their graphs.

Symmetry of the face is one key to physical attractiveness. A relationship between attractive faces and graphs of some polynomial functions opens the discussion of symmetry on page 798 of Section 11.1.

SECTION 11.1

Polynomial Functions and Their Graphs

Objectives

1. Use factoring to find zeros of polynomial functions.
2. Identify zeros and their multiplicities.
3. Understand the relationship between degree and turning points.
4. Identify even or odd functions and recognize their symmetries.
5. Graph polynomial functions.

Have you ever experienced a sudden onset of extreme panic even in the absence of an emergency? Panic attacks are accompanied by various physical symptoms, such as a racing heart and rapid breathing. Within seconds, the heart rate of a person having a panic attack can increase by 50 or more beats per minute. Many people who have a panic attack think that they are having a heart attack. In this section's exercise set, you will see why a polynomial function can be used to model this intense physiological reaction.

Definition of a Polynomial Function

In Chapter 5, we saw that in a polynomial function, the expression that defines the function is a polynomial. This polynomial can be a single term or the sum of two or more terms containing variables with whole-number exponents. We begin this chapter with a more formal definition of a polynomial function. In this definition, the coefficients of the terms are represented by a_n (read "a sub n"), a_{n-1} (read "a sub n minus 1"), a_{n-2}, and so on. The small letters to the lower right of each a are called **subscripts** and are *not exponents*. Subscripts are used to distinguish one constant from another when a large and undetermined number of such constants are needed.

Definition of a Polynomial Function

Let n be a nonnegative integer and let $a_n, a_{n-1}, \ldots, a_2, a_1, a_0$, be real numbers, with $a_n \neq 0$. The function defined by

$$f(x) = a_nx^n + a_{n-1}x^{n-1} + \cdots + a_2x^2 + a_1x + a_0$$

is called a **polynomial function of degree n**. The number a_n, the coefficient of the variable to the highest power, is called the **leading coefficient**.

A constant function $f(x) = c$, where $c \neq 0$, is a polynomial function of degree 0. A linear function $f(x) = mx + b$, where $m \neq 0$, is a polynomial function of degree 1. A quadratic function $f(x) = ax^2 + bx + c$, where $a \neq 0$, is a polynomial function of degree 2. In this section, we focus on polynomial functions of degree 3 or higher.

Study Tip

Here is what you should already know about the graphs of polynomial functions from our work in Chapter 5.

1. Polynomial functions of degree 2 or greater have graphs that are smooth and continuous. These graphs contain only rounded curves with no sharp corners. The graphs have no breaks and can be drawn without lifting your pencil from the rectangular coordinate system.
2. The graph of a polynomial function eventually rises or falls without bound as it moves far to the left or far to the right, called its **end behavior**. The end behavior depends upon the leading term a_nx^n and is given by the **Leading Coefficient Test**.

- If n is odd, the graph has opposite behavior at each end. For $a_n > 0$, the graph falls to the left and rises to the right. For $a_n < 0$, the graph rises to the left and falls to the right.
- If n is even, the graph has the same behavior at each end. For $a_n > 0$, the graph rises to the left and rises to the right. For $a_n < 0$, the graph falls to the left and falls to the right.

To review examples illustrating the Leading Coefficient Test, see Examples 3 through 5 in Section 5.1 on pages 309–310.

1 Use factoring to find zeros of polynomial functions.

Zeros of Polynomial Functions

If f is a polynomial function, then the values of x for which $f(x)$ is equal to 0 are called the **zeros** of f. These values of x are the **roots**, or **solutions**, of the polynomial equation $f(x) = 0$. Each real root of the polynomial equation appears as an x-intercept of the graph of the polynomial function.

EXAMPLE 1 Finding Zeros of a Polynomial Function

Find all zeros of $f(x) = x^3 + 3x^2 - x - 3$.

Solution By definition, the zeros are the values of x for which $f(x)$ is equal to 0. Thus, we set $f(x)$ equal to 0:

$$f(x) = x^3 + 3x^2 - x - 3 = 0.$$

We solve the polynomial equation $x^3 + 3x^2 - x - 3 = 0$ for x as follows:

$x^3 + 3x^2 - x - 3 = 0$ This is the equation needed to find the function's zeros.

$x^2(x + 3) - 1(x + 3) = 0$ Factor x^2 from the first two terms and -1 from the last two terms.

$(x + 3)(x^2 - 1) = 0$ A common factor of $x + 3$ is factored from the expression.

$x + 3 = 0$ or $x^2 - 1 = 0$ Set each factor equal to 0.

$x = -3$ $\quad x^2 = 1$ Solve for x.

$x = \pm 1$ Remember that if $x^2 = d$, then $x = \pm\sqrt{d}$.

The zeros of f are -3, -1, and 1. The graph of f in **Figure 11.1** shows that each zero is an x-intercept. The graph passes through the points $(-3, 0)$, $(-1, 0)$, and $(1, 0)$.

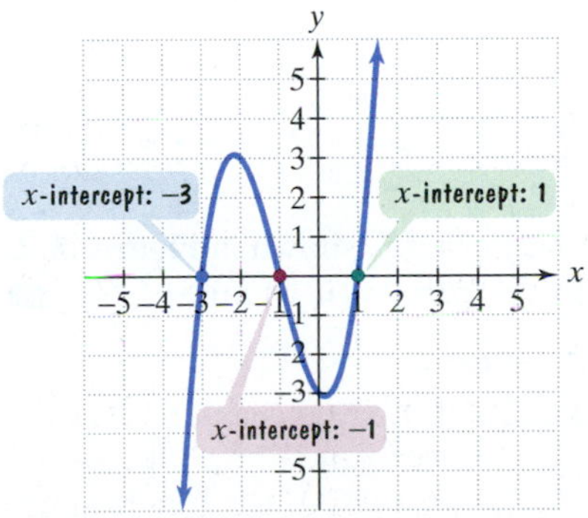

FIGURE 11.1

Using Technology

Graphic and Numeric Connections

A graphing utility can be used to verify that $-3, -1$, and 1 are the three real zeros of $f(x) = x^3 + 3x^2 - x - 3$.

Numeric Check

Display a table for the function.

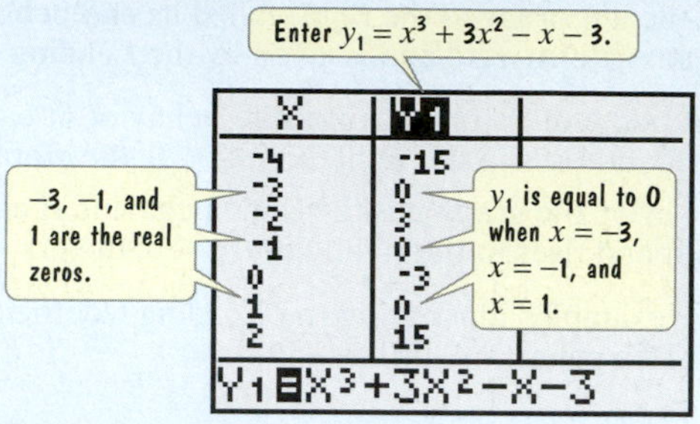

Graphic Check

Display a graph for the function. The x-intercepts indicate that $-3, -1$, and 1 are the real zeros.

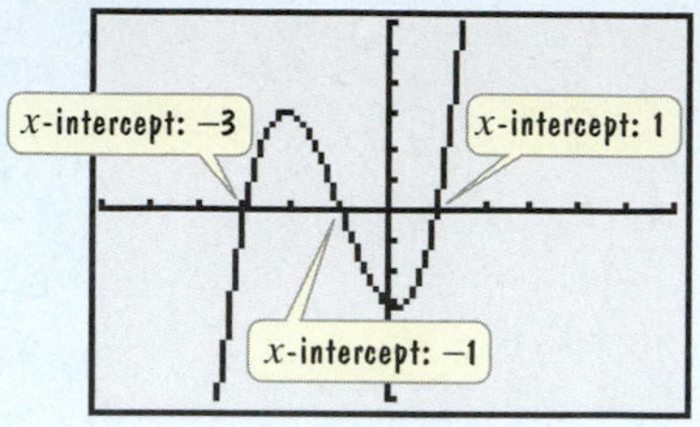

$[-6, 6, 1]$ by $[-6, 6, 1]$

The utility's ZERO feature on the graph of f also verifies that $-3, -1$, and 1 are the function's real zeros.

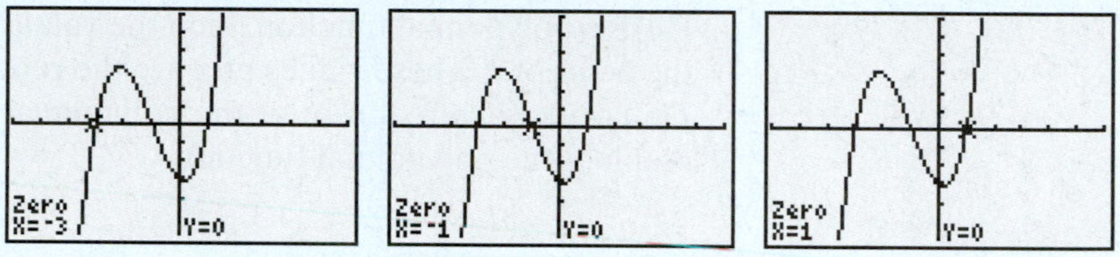

CHECK POINT 1 Find all zeros of $f(x) = x^3 + 2x^2 - 4x - 8$.

EXAMPLE 2 Finding Zeros of a Polynomial Function

Find all zeros of $f(x) = -x^4 + 4x^3 - 4x^2$.

Solution We find the zeros of f by setting $f(x)$ equal to 0 and solving the resulting equation.

$$-x^4 + 4x^3 - 4x^2 = 0 \quad \text{We now have a polynomial equation.}$$

$$x^4 - 4x^3 + 4x^2 = 0 \quad \text{Multiply both sides by } -1. \text{ This step is optional.}$$

$$x^2(x^2 - 4x + 4) = 0 \quad \text{Factor out } x^2.$$

$$x^2(x - 2)^2 = 0 \quad \text{Factor completely.}$$

$$x^2 = 0 \quad \text{or} \quad (x - 2)^2 = 0 \quad \text{Set each factor equal to 0.}$$

$$x = 0 \qquad\qquad x = 2 \quad \text{Solve for } x.$$

The zeros of $f(x) = -x^4 + 4x^3 - 4x^2$ are 0 and 2. The graph of f, shown in **Figure 11.2**, has x-intercepts at 0 and 2. The graph passes through the points $(0, 0)$ and $(2, 0)$. ■

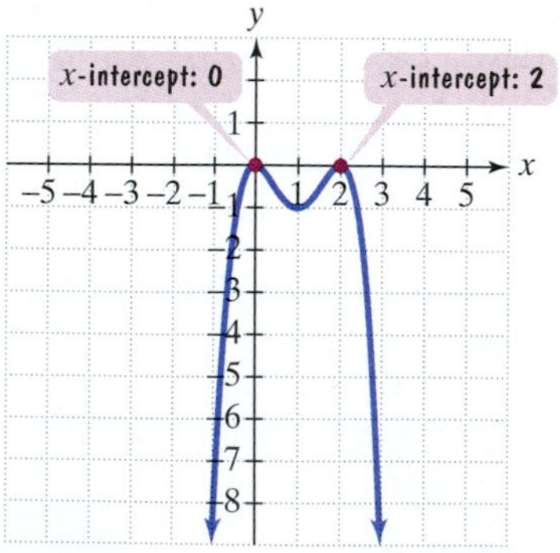

FIGURE 11.2 The zeros of $f(x) = -x^4 + 4x^3 - 4x^2$, namely 0 and 2, are the x-intercepts for the graph of f.

CHECK POINT 2 Find all zeros of $f(x) = x^4 - 4x^2$.

2 Identify zeros and their multiplicities.

Multiplicities of Zeros

We can use the results of factoring to express a polynomial as a product of factors. For instance, in Example 2, we can use our factoring to express the function's equation as follows:

$$f(x) = -x^4 + 4x^3 - 4x^2 = -(x^4 - 4x^3 + 4x^2) = -x^2(x - 2)^2.$$

The factor x occurs twice: $x^2 = x \cdot x$.

The factor $(x - 2)$ occurs twice: $(x - 2)^2 = (x - 2)(x - 2)$.

Notice that each factor occurs twice. In factoring the equation for the polynomial function f, if the same factor $x - r$ occurs k times, but not $k + 1$ times, we call r a **zero with multiplicity k**. For the polynomial function

$$f(x) = -x^2(x - 2)^2,$$

0 and 2 are both zeros with multiplicity 2.

Multiplicity provides another connection between zeros and graphs. The multiplicity of a zero tells us if the graph of a polynomial function touches the x-axis at the zero and turns around, or if the graph crosses the x-axis at the zero. For example, look again at the graph of $f(x) = -x^4 + 4x^3 - 4x^2$ in **Figure 11.2** on the previous page. Each zero, 0 and 2, is a zero with multiplicity 2. The graph of f touches, but does not cross, the x-axis at each of these zeros of even multiplicity. By contrast, a graph crosses the x-axis at zeros of odd multiplicity.

Multiplicity and x-Intercepts

If r is a zero of **even multiplicity**, then the graph **touches** the x-axis **and turns around** at r. If r is a zero of **odd multiplicity**, then the graph **crosses** the x-axis at r. Regardless of whether the multiplicity of a zero is even or odd, graphs tend to flatten out near zeros with multiplicity greater than one.

If a polynomial function's equation is expressed as a product of linear factors, we can quickly identify zeros and their multiplicities.

EXAMPLE 3 Finding Zeros and Their Multiplicities

Find the zeros of $f(x) = (x + 1)(2x - 3)^2$ and give the multiplicity of each zero. State whether the graph crosses the x-axis or touches the x-axis and turns around at each zero.

Solution We find the zeros of f by setting $f(x)$ equal to 0:

$$(x + 1)(2x - 3)^2 = 0.$$

Set each factor equal to 0.

$x + 1 = 0$
$x = -1$

$2x - 3 = 0$
$x = \frac{3}{2}$

$$(x + 1)^1(2x - 3)^2 = 0$$

This exponent is 1. Thus, the multiplicity of -1 is 1.

This exponent is 2. Thus, the multiplicity of $\frac{3}{2}$ is 2.

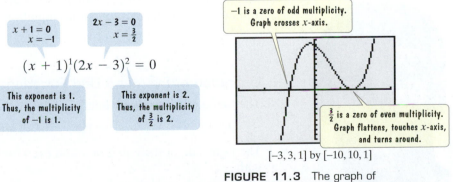

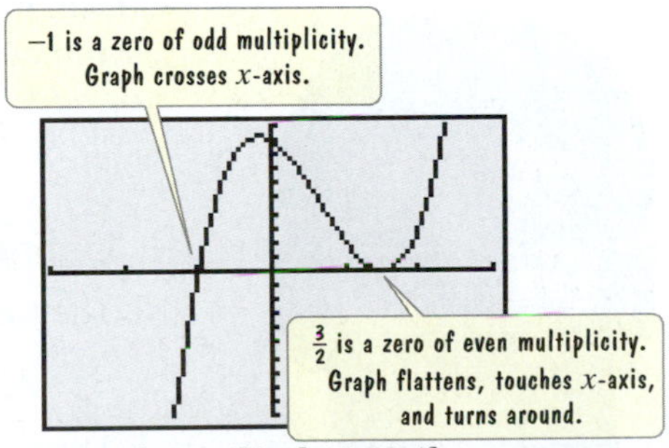

−1 is a zero of odd multiplicity.
Graph crosses x-axis.
3/2 is a zero of even multiplicity.
Graph flattens, touches x-axis, and turns around.

$[-3, 3, 1]$ by $[-10, 10, 1]$

FIGURE 11.3 The graph of $f(x) = (x + 1)(2x - 3)^2$

The zeros of $f(x) = (x + 1)(2x - 3)^2$ are -1, with multiplicity 1, and $\frac{3}{2}$, with multiplicity 2. Because the multiplicity of -1 is odd, the graph crosses the x-axis at this zero. Because the multiplicity of $\frac{3}{2}$ is even, the graph touches the x-axis and turns around at this zero. These relationships are illustrated by the graph of f in **Figure 11.3**.

CHECK POINT 3 Find the zeros of $f(x) = -4\left(x + \frac{1}{2}\right)^2(x - 5)^3$ and give the multiplicity of each zero. State whether the graph crosses the x-axis or touches the x-axis and turns around at each zero.

3 Understand the relationship between degree and turning points.

Turning Points of Polynomial Functions

The graph of $f(x) = x^5 - 6x^3 + 8x + 1$ is shown in **Figure 11.4**. The graph has four smooth **turning points**. At each turning point, the graph changes direction from increasing to decreasing or vice versa. The given equation has 5 as its greatest exponent and is therefore a polynomial function of degree 5. Notice that the graph has four turning points. In general, **if f is a polynomial function of degree n, then the graph of f has at most $n - 1$ turning points.**

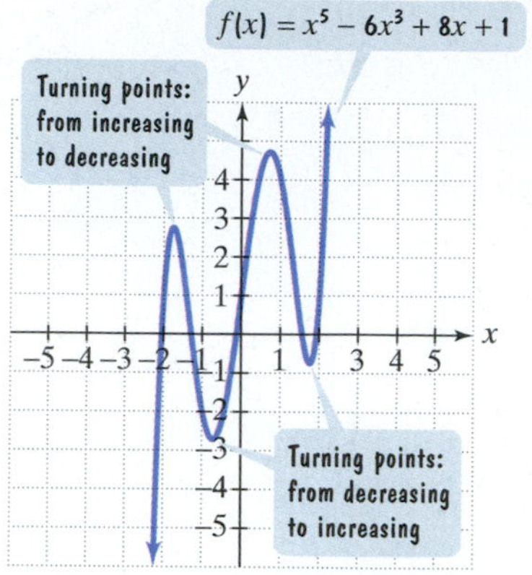

FIGURE 11.4 Graph with four turning points

Figure 11.4 illustrates that each turning point is either a relative high point or a relative low point on the graph of f. Without the aid of a graphing utility or a knowledge of calculus, it is difficult and often impossible to locate turning points of polynomial functions with degrees greater than 2. If necessary, test values can be taken between the x-intercepts to get a general idea of how high the graph rises or how low the graph falls. For the purpose of graphing in this section, a general estimate is sometimes appropriate and necessary.

4 Identify even or odd functions and recognize their symmetries.

Even and Odd Functions and Symmetry

Is beauty in the eye of the beholder? Or are there certain objects (or people) that are so well balanced and proportioned that they are universally pleasing to the eye? What constitutes an attractive human face? In **Figure 11.5**, we've drawn lines between paired features and marked the midpoints. Notice how the features line up almost perfectly. Each half of the face is a mirror image of the other half through the white vertical line.

Did you know that graphs of some polynomial functions exhibit exactly the kind of symmetry shown by the attractive face in **Figure 11.5**? The word *symmetry* comes from the Greek *symmetria*, meaning "the same measure." We can identify graphs with symmetry by looking at a function's equation and determining if the function is *even* or *odd*.

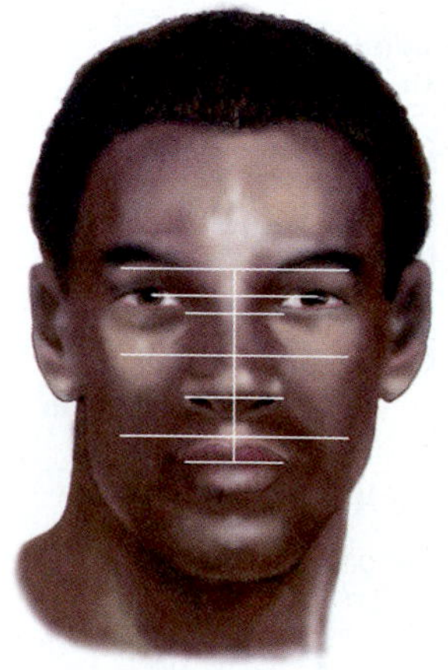
FIGURE 11.5 To most people, an attractive face is one in which each half is an almost perfect mirror image of the other half.

Definition of Even and Odd Functions

The function f is an **even function** if

$$f(-x) = f(x) \quad \text{for all } x \text{ in the domain of } f.$$

The right side of the equation of an even function does not change if x is replaced with $-x$.

The function f is an **odd function** if

$$f(-x) = -f(x) \quad \text{for all } x \text{ in the domain of } f.$$

Every term on the right side of the equation of an odd function changes sign if x is replaced with $-x$.

EXAMPLE 4 Identifying Even or Odd Functions

Identify whether each of the following functions is even, odd, or neither:

a. $f(x) = x^3$ **b.** $g(x) = x^4 - 2x^2$ **c.** $h(x) = x^2 + 2x + 1$.

Solution In each case, replace x with $-x$ and simplify. If the right side of the equation stays the same, the function is even. If every term on the right changes sign, the function is odd.

a. We use the given function's equation, $f(x) = x^3$, to find $f(-x)$.

Use $f(x) = x^3$.

Replace x with $-x$.

$$f(-x) = (-x)^3 = (-x)(-x)(-x) = -x^3$$

There is only one term in the equation $f(x) = x^3$, and the term changed signs when we replaced x with $-x$. Because $f(-x) = -f(x)$, f is an odd function.

b. We use the given function's equation, $g(x) = x^4 - 2x^2$, to find $g(-x)$.

Use $g(x) = x^4 - 2x^2$.

Replace x with $-x$.

$$g(-x) = (-x)^4 - 2(-x)^2 = (-x)(-x)(-x)(-x) - 2(-x)(-x)$$
$$= x^4 - 2x^2$$

The right side of the equation of the given function, $g(x) = x^4 - 2x^2$, did not change when we replaced x with $-x$. Because $g(-x) = g(x)$, g is an even function.

c. We use the given function's equation, $h(x) = x^2 + 2x + 1$, to find $h(-x)$.

Use $h(x) = x^2 + 2x + 1$.

Replace x with $-x$.

$$h(-x) = (-x)^2 + 2(-x) + 1 = x^2 - 2x + 1$$

The right side of the equation of the given function, $h(x) = x^2 + 2x + 1$, changed when we replaced x with $-x$. Thus, $h(-x) \neq h(x)$, so h is not an even function. The sign of *each* of the three terms in the equation for $h(x)$ did not change when we replaced x with $-x$. Only the second term changed signs. Thus, $h(-x) \neq -h(x)$, so h is not an odd function. We conclude that h is neither an even nor an odd function.

CHECK POINT 4 Identify whether each of the following functions is even, odd, or neither:

a. $f(x) = x^2 + 6$ **b.** $g(x) = 7x^3 - x$ **c.** $h(x) = x^5 + 1$.

Now, let's see what knowing whether a function is even or odd tells us about the function's graph. Begin with the even function $f(x) = x^2 - 4$, shown in **Figure 11.6**. The function is even because

$$f(-x) = (-x)^2 - 4 = x^2 - 4 = f(x).$$

Examine the pairs of points shown, such as $(3, 5)$ and $(-3, 5)$. Notice that we obtain the same y-coordinate whenever we evaluate the function at a value of x and the value of its opposite, $-x$. Like the attractive face, each half of the graph is a mirror image of the other half through the y-axis. If we were to fold the paper along the y-axis, the two halves of the graph would coincide. This means that the graph is *symmetric with respect to the y-axis*. A graph is **symmetric with respect to the y-axis** if, for every point (x, y) on the graph, the point $(-x, y)$ is also on the graph. All even functions have graphs with this kind of symmetry.

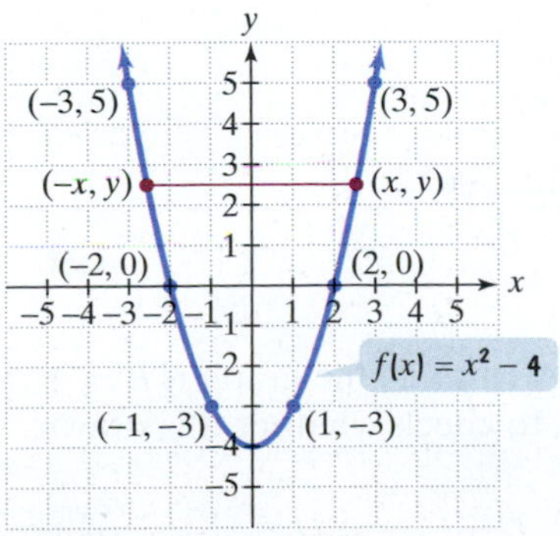

FIGURE 11.6 y-axis symmetry with $f(-x) = f(x)$

Even Functions and y-axis Symmetry

The graph of an even function, in which $f(-x) = f(x)$, is symmetric with respect to the y-axis.

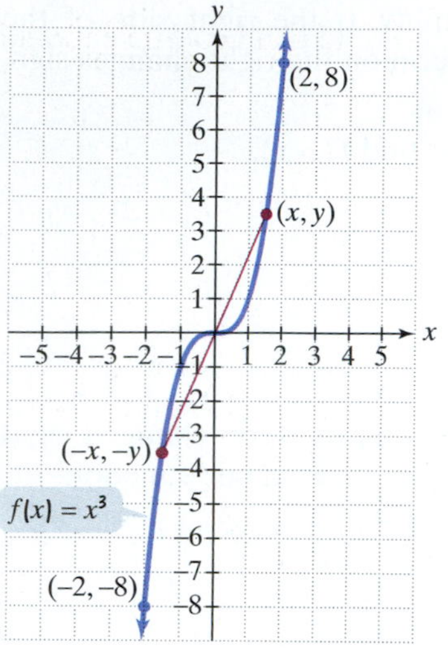

FIGURE 11.7 Origin symmetry with $f(-x) = -f(x)$

Now, consider the graph of the function $f(x) = x^3$. In Example 4, we saw that $f(-x) = -f(x)$, so this is an odd function. Although the graph in **Figure 11.7** is not symmetric with respect to the y-axis, it is symmetric in another way. Look at the pairs of points, such as $(2, 8)$ and $(-2, -8)$. For each point (x, y) on the graph, the point $(-x, -y)$ is also on the graph. The points $(2, 8)$ and $(-2, -8)$ are reflections of one another about the origin. This means that the origin is the midpoint of the line segment connecting the points.

A graph is **symmetric with respect to the origin** if, for every point (x, y) on the graph, the point $(-x, -y)$ is also on the graph. Observe that the first- and third-quadrant portions of $f(x) = x^3$ are reflections of one another with respect to the origin. Notice that $f(x)$ and $f(-x)$ have opposite signs, so that $f(-x) = -f(x)$. All odd functions have graphs with origin symmetry.

Odd Functions and Origin Symmetry

The graph of an odd function in which $f(-x) = -f(x)$ is symmetric with respect to the origin.

5 Graph polynomial functions.

A Strategy for Graphing Polynomial Functions

Here's a general strategy for graphing a polynomial function. A graphing utility is a valuable complement, but not a necessary component, of this strategy. If you are using a graphing utility, some of the steps listed in the following box will help you select a viewing rectangle that shows the important parts of the graph.

Graphing a Polynomial Function

$$f(x) = a_n x^n + a_{n-1}x^{n-1} + \cdots + a_1 x + a_0, \quad a_n \neq 0$$

1. Use the Leading Coefficient Test to determine the graph's end behavior.
2. Find x-intercepts by setting $f(x) = 0$ and solving the resulting polynomial equation. If there is an x-intercept at r as a result of $(x - r)^k$ in the complete factorization of $f(x)$, then
   a. If k is even, the graph touches the x-axis at r and turns around.
   b. If k is odd, the graph crosses the x-axis at r.
   c. If $k > 1$, the graph flattens out near $(r, 0)$.
3. Find the y-intercept by computing $f(0)$.
4. Use symmetry, if applicable, to help draw the graph:
   a. y-axis symmetry: $f(-x) = f(x)$
   b. Origin symmetry: $f(-x) = -f(x)$.
5. Use the fact that the maximum number of turning points of the graph is $n - 1$, where n is the degree of the polynomial function, to check whether it is drawn correctly.

Study Tip

Remember that, without calculus, it is often impossible to give the exact location of turning points. However, you can obtain additional points satisfying the function to estimate how high the graph rises or how low it falls. To obtain these points, use values of x between (and to the left and right of) the x-intercepts.

EXAMPLE 5 Graphing a Polynomial Function

Graph: $f(x) = x^4 - 2x^2 + 1$.

Solution

Step 1. Determine end behavior. Identify the sign of a_n, the leading coefficient, and the degree, n, of the polynomial function.

$$f(x) = x^4 - 2x^2 + 1$$

The leading coefficient, 1, is positive.

The degree of the polynomial function, 4, is even.

Because the degree, 4, is even, the graph has the same behavior at each end. The leading coefficient, 1, is positive. Thus, the graph rises to the left and rises to the right.

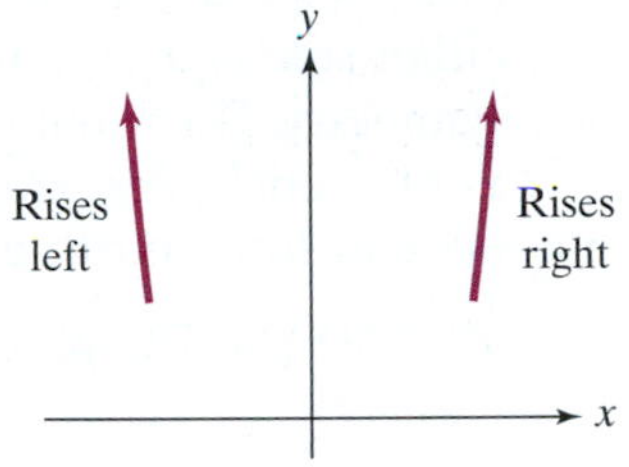

Step 2. Find x-intercepts (zeros of the function) by setting $f(x) = 0$.

$$x^4 - 2x^2 + 1 = 0 \quad \text{Set } f(x) \text{ equal to 0.}$$

$$(x^2 - 1)(x^2 - 1) = 0 \quad \text{Factor.}$$

$$(x + 1)(x - 1)(x + 1)(x - 1) = 0 \quad \text{Factor completely.}$$

$$(x + 1)^2(x - 1)^2 = 0 \quad \text{Express the factoring in a more compact form.}$$

$$(x + 1)^2 = 0 \quad \text{or} \quad (x - 1)^2 = 0 \quad \text{Set each factor equal to 0.}$$

$$x = -1 \qquad\qquad x = 1 \quad \text{Solve for x.}$$

We see that -1 and 1 are both repeated zeros with multiplicity 2. Because of the even multiplicity, the graph touches the x-axis at -1 and 1 and turns around. Furthermore, the graph tends to flatten out near these zeros with multiplicity greater than one.

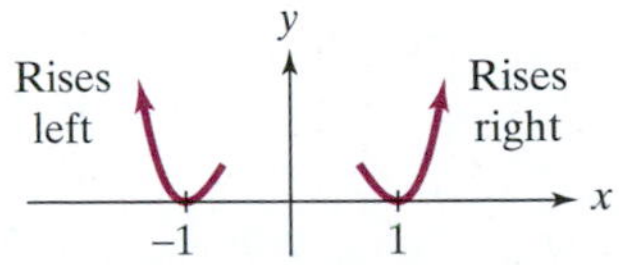

Step 3. Find the y-intercept by computing $f(0)$. We use $f(x) = x^4 - 2x^2 + 1$ and compute $f(0)$.

$$f(0) = 0^4 - 2 \cdot 0^2 + 1 = 1$$

The y-intercept is 1, so the graph passes through $(0, 1)$.

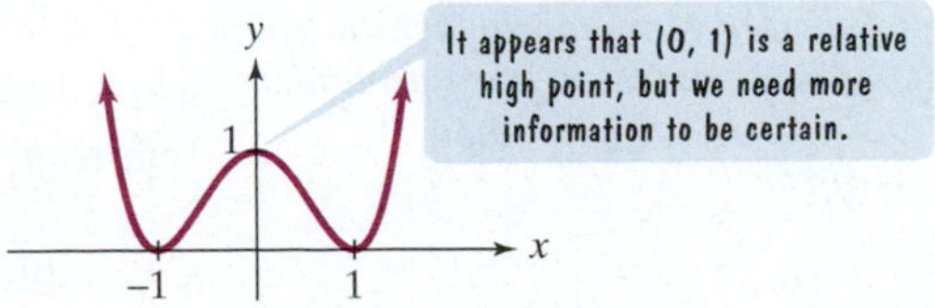

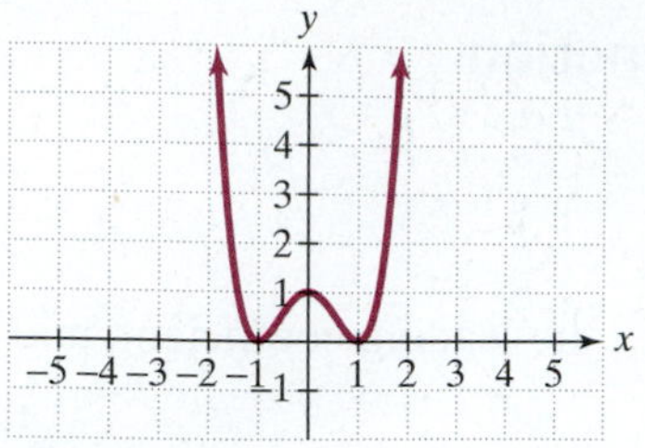

FIGURE 11.8 The graph of $f(x) = x^4 - 2x^2 + 1$

Step 4. **Use possible symmetry to help draw the graph.** Our partial graph at the bottom of the previous page suggests y-axis symmetry. Let's verify this by finding $f(-x)$.

$$f(x) = x^4 - 2x^2 + 1$$

Replace x with $-x$.

$$f(-x) = (-x)^4 - 2(-x)^2 + 1 = x^4 - 2x^2 + 1$$

Because $f(-x) = f(x)$, the graph of f is symmetric with respect to the y-axis. **Figure 11.8** shows the graph of $f(x) = x^4 - 2x^2 + 1$.

Step 5. **Use the fact that the maximum number of turning points of the graph is $n - 1$ to check whether it is drawn correctly.** Because $n = 4$, the maximum number of turning points is $4 - 1$, or 3. Because the graph in **Figure 11.8** has three turning points, we have not violated the maximum number possible. Can you see how this verifies that (0, 1) is a relative high point, as well as a turning point? If the graph rose above 1 on either side of $x = 0$, it would have to rise above 1 on the other side as well because of symmetry. This would require additional turning points to smoothly curve back to the x-intercepts. The graph already has three turning points, which is the maximum number for a fourth-degree polynomial function.

✓ **CHECK POINT 5** Use the five-step strategy to graph $f(x) = x^3 - 3x^2$.

11.1 EXERCISE SET

MyMathLab — MathXL PRACTICE · WATCH · DOWNLOAD · READ · REVIEW

Practice Exercises

In Exercises 1–6, use the Leading Coefficient Test to determine the end behavior of the graph of the polynomial function.

1. $f(x) = 5x^3 + 7x^2 - x + 9$
2. $f(x) = 11x^3 - 6x^2 + x + 3$
3. $f(x) = 5x^4 + 7x^2 - x + 9$
4. $f(x) = 11x^4 - 6x^2 + x + 3$
5. $f(x) = -5x^4 + 7x^2 - x + 9$
6. $f(x) = -11x^4 - 6x^2 + x + 3$

In Exercises 7–14, find the zeros for each polynomial function and give the multiplicity for each zero. State whether the graph crosses the x-axis, or touches the x-axis and turns around, at each zero.

7. $f(x) = 2(x - 5)(x + 4)^2$
8. $f(x) = 3(x + 5)(x + 2)^2$
9. $f(x) = 4(x - 3)(x + 6)^3$
10. $f(x) = -3\left(x + \frac{1}{2}\right)(x - 4)^3$
11. $f(x) = x^3 - 2x^2 + x$
12. $f(x) = x^3 + 4x^2 + 4x$
13. $f(x) = x^3 + 7x^2 - 4x - 28$
14. $f(x) = x^3 + 5x^2 - 9x - 45$

In Exercises 15–26, determine whether each function is even, odd, or neither.

15. $f(x) = x^3 + x$
16. $f(x) = x^3 - x$
17. $g(x) = x^2 + x$
18. $g(x) = x^2 - x$
19. $h(x) = x^2 - x^4$
20. $h(x) = 2x^2 + x^4$
21. $f(x) = x^2 - x^4 + 1$
22. $f(x) = 2x^2 + x^4 + 1$
23. $f(x) = \frac{1}{5}x^6 - 3x^2$
24. $f(x) = 2x^3 - 6x^5$
25. $f(x) = x(1 - x^2)$
26. $f(x) = x^2(1 - x^2)$

In Exercises 27–30,

a. *Identify which graphs are not those of polynomial functions.*

b. *Determine whether each graph is the graph of an even function, an odd function, or a function that is neither even nor odd.*

27.

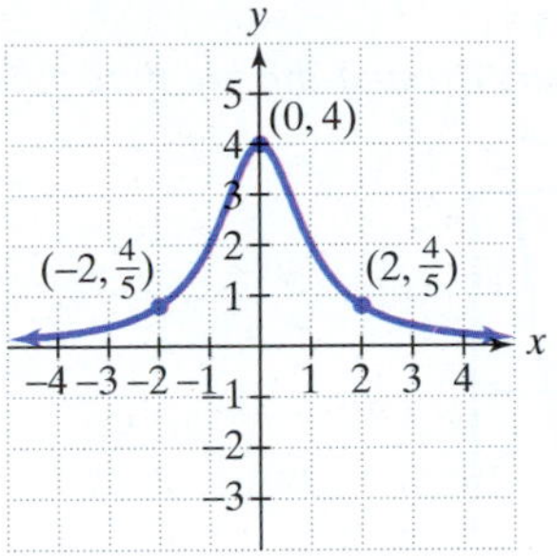

28.

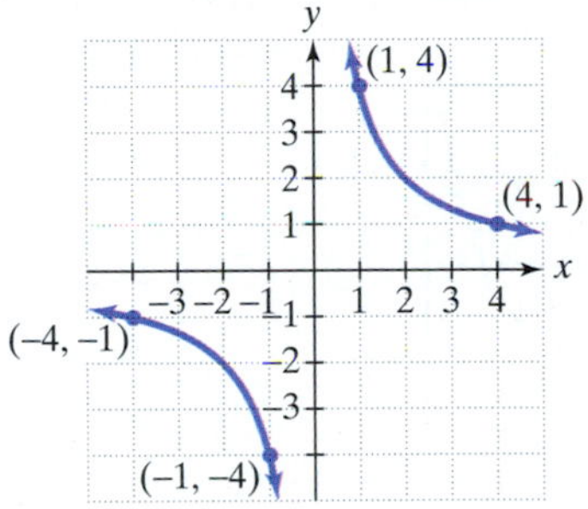

29.

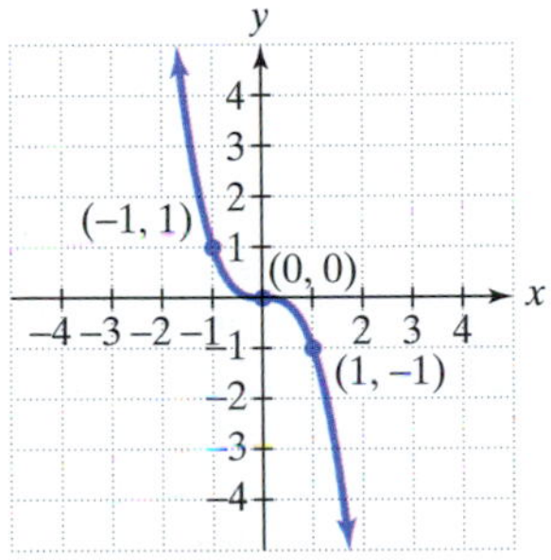

30.

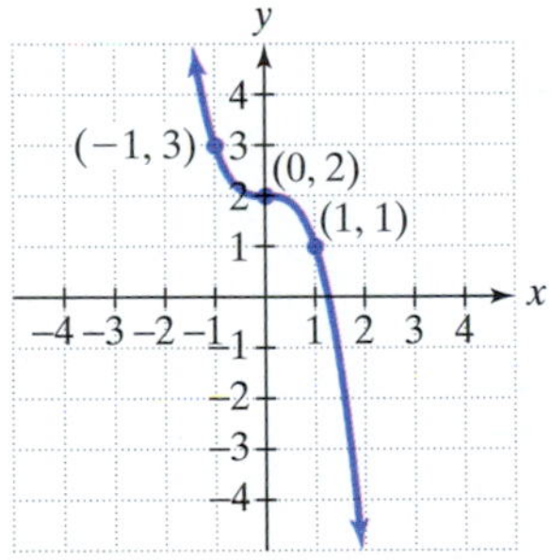

In Exercises 31–54,

a. *Use the Leading Coefficient Test to determine the graph's end behavior.*

b. *Find the x-intercepts. State whether the graph crosses the x-axis, or touches the x-axis and turns around, at each intercept.*

c. *Find the y-intercept.*

d. *Determine whether the graph has y-axis symmetry, origin symmetry, or neither.*

e. *If necessary, find a few additional points and graph the function. Use the maximum number of turning points to check whether it is drawn correctly.*

31. $f(x) = x^3 + 2x^2 - x - 2$

32. $f(x) = x^3 + x^2 - 4x - 4$

33. $f(x) = x^4 - 9x^2$

34. $f(x) = x^4 - x^2$

35. $f(x) = -x^4 + 16x^2$

36. $f(x) = -x^4 + 4x^2$

37. $f(x) = x^4 - 2x^3 + x^2$

38. $f(x) = x^4 - 6x^3 + 9x^2$

39. $f(x) = -2x^4 + 4x^3$

40. $f(x) = -2x^4 + 2x^3$

41. $f(x) = 6x^3 - 9x - x^5$

42. $f(x) = 6x - x^3 - x^5$

43. $f(x) = 3x^2 - x^3$

44. $f(x) = \frac{1}{2} - \frac{1}{2}x^4$

45. $f(x) = -3(x - 1)^2(x^2 - 4)$

46. $f(x) = -2(x - 4)^2(x^2 - 25)$

47. $f(x) = x^2(x - 1)^3(x + 2)$

48. $f(x) = x^3(x + 2)^2(x + 1)$

49. $f(x) = -x^2(x - 1)(x + 3)$

50. $f(x) = -x^2(x + 2)(x - 2)$

51. $f(x) = -2x^3(x - 1)^2(x + 5)$

52. $f(x) = -3x^3(x - 1)^2(x + 3)$

53. $f(x) = (x - 2)^2(x + 4)(x - 1)$

54. $f(x) = (x + 3)(x + 1)^3(x + 4)$

Practice PLUS

In Exercises 55–62, complete graphs of polynomial functions whose zeros are integers are shown. (By "complete graphs," we mean that each graph is continuous and displays the function's end behavior.)

a. *Find the zeros and state whether the multiplicity of each zero is even or odd.*

b. *Write an equation, expressed as the product of factors, of a polynomial function that might have each graph. Use a leading coefficient of 1 or −1, and make the degree of f as small as possible.*

c. *Use both the equation in part (b) and the graph to find the y-intercept.*

55.

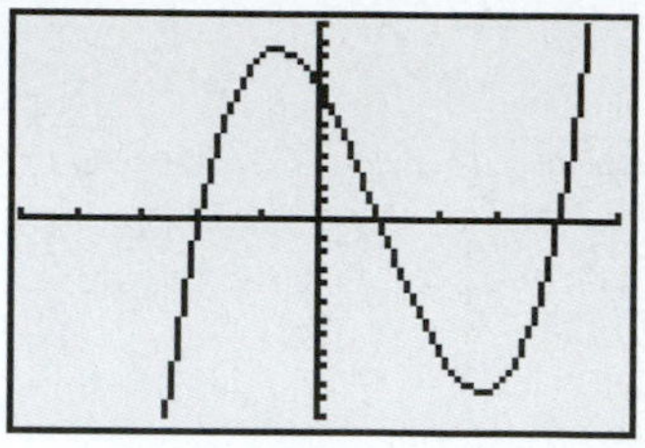

[−5, 5, 1] by [−12, 12, 1]

56.

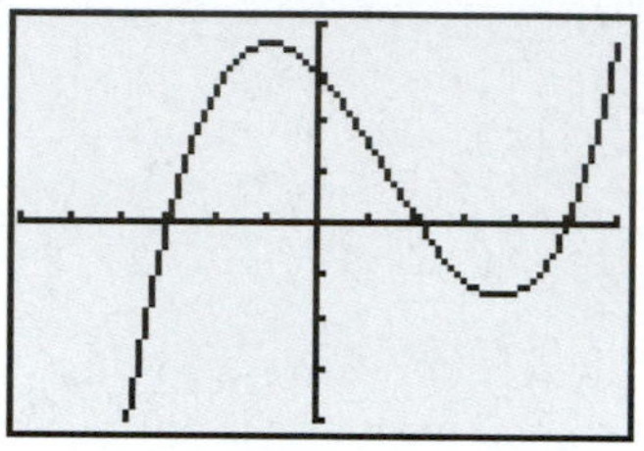

[−6, 6, 1] by [−40, 40, 10]

57.

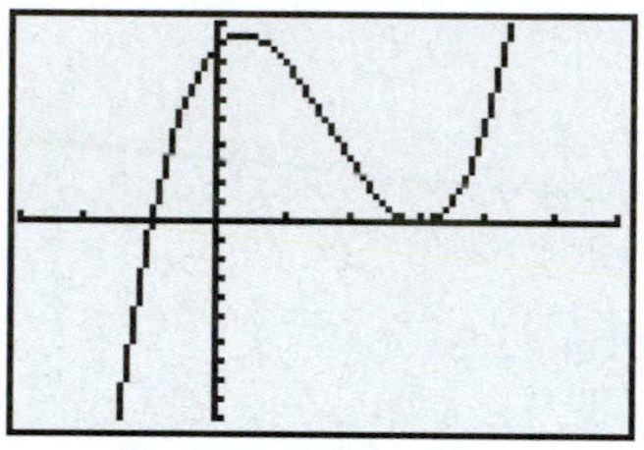

[−3, 6, 1] by [−10, 10, 1]

58.

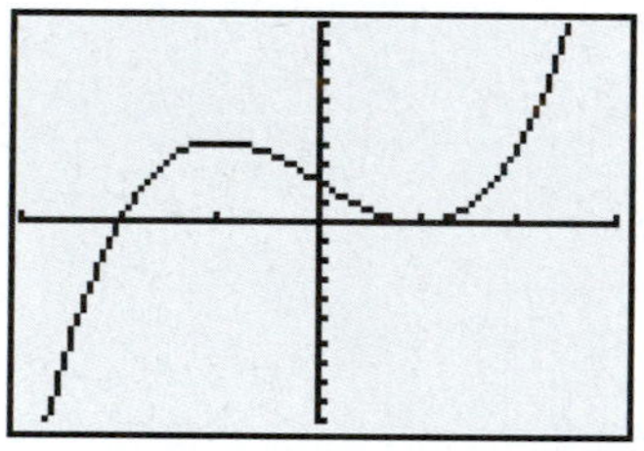

[−3, 3, 1] by [−10, 10, 1]

59.

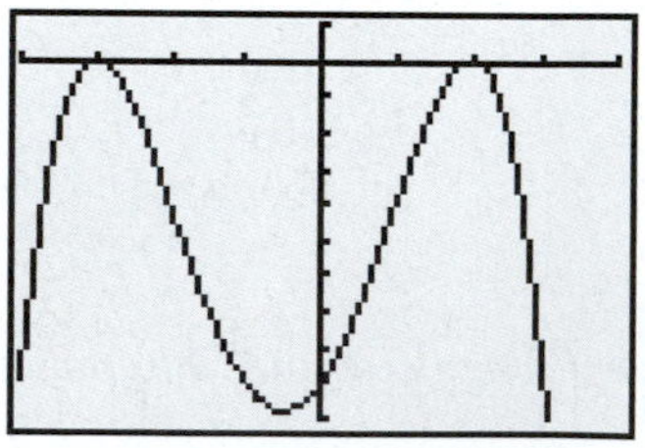

[−4, 4, 1] by [−40, 4, 4]

60.

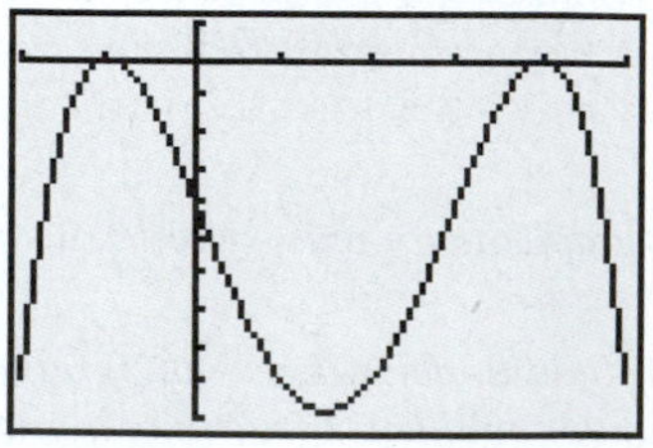

[−2, 5, 1] by [−40, 4, 4]

61. 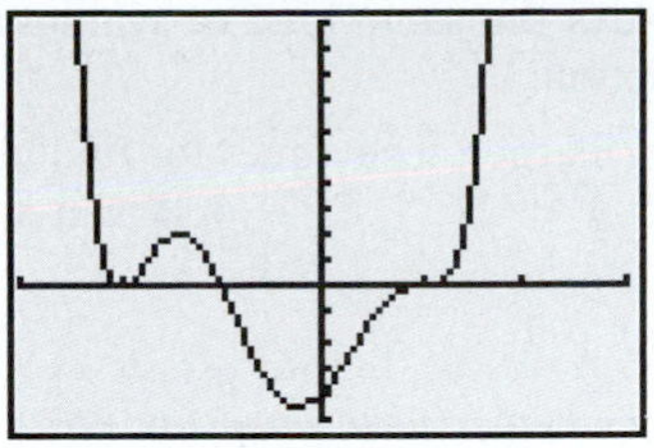

[−3, 3, 1] by [−5, 10, 1]

62. 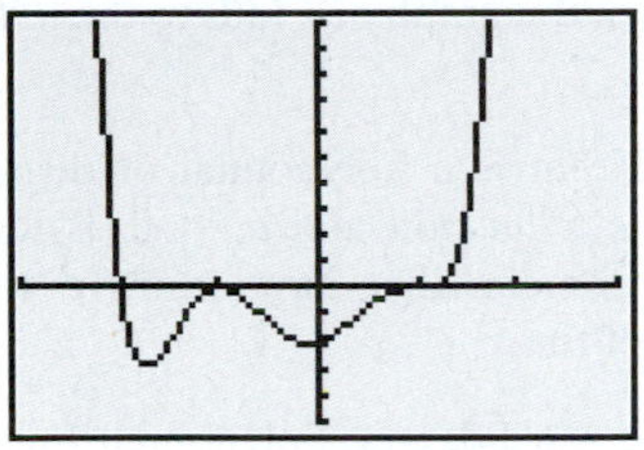

[−3, 3, 1] by [−5, 10, 1]

Application Exercises

63. During a diagnostic evaluation, a 33-year-old woman experienced a panic attack a few minutes after she had been asked to relax her whole body. The graph shows the rapid increase in heart rate during the panic attack.

Heart Rate before and during a Panic Attack

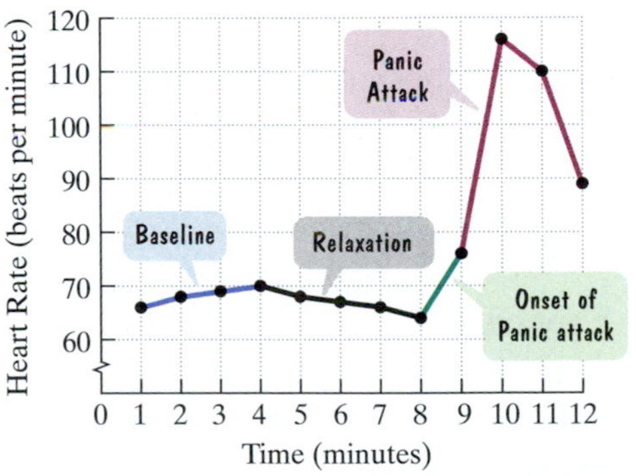

Source: Davis and Palladino, *Psychology, 5th Edition*, Prentice Hall, 2007

a. For which time periods during the diagnostic evaluation was the woman's heart rate increasing?

b. For which time periods during the diagnostic evaluation was the woman's heart rate decreasing?

c. How many turning points (from increasing to decreasing or from decreasing to increasing) occurred for the woman's heart rate during the first 12 minutes of the diagnostic evaluation?

d. Suppose that a polynomial function is used to model the data displayed by the graph using

(time during the evaluation, heart rate).

Use the number of turning points to determine the degree of the polynomial function of best fit.

e. For the model in part (d), should the leading coefficient of the polynomial function be positive or negative? Explain your answer.

f. Use the graph to estimate the woman's maximum heart rate during the first 12 minutes of the diagnostic evaluation. After how many minutes did this occur?

g. Use the graph to estimate the woman's minimum heart rate during the first 12 minutes of the diagnostic evaluation. After how many minutes did this occur?

64. Although it has been more than 50 years since the Supreme Court ruled against school segregation, data from the Civil Rights Project at Harvard University indicate that integration and academic equality remain elusive. The graph shows the percentage of the average African-American student's classmates who were white for the period from 1970 through 2002.

Percentage of the Average African-American Student's Classmates Who Were White

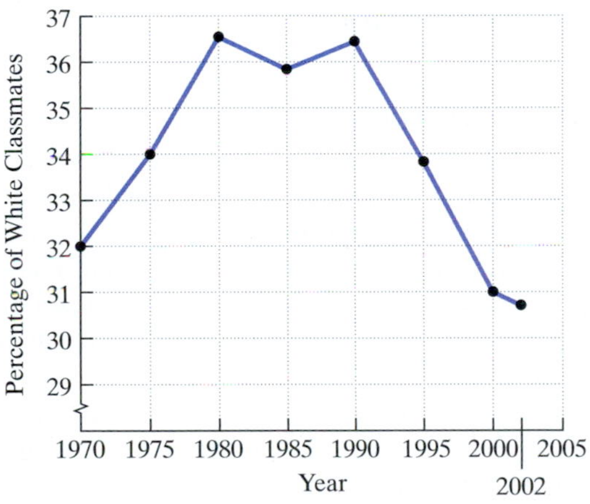

Source: Civil Rights Project, Harvard University

a. For which years was the percentage of white classmates increasing?

b. For which years was the percentage of white classmates decreasing?

c. How many turning points (from increasing to decreasing or from decreasing to increasing) does the graph have for the period shown?

d. Suppose that a polynomial function is used to model the data displayed by the graph using

(number of years after 1970, percentage of the average African-American student's classmates who were white).

Use the number of turning points to determine the degree of the polynomial function of best fit.

e. For the model in part (d), should the leading coefficient of the polynomial function be positive or negative? Explain your answer.

Writing in Mathematics

65. What is a polynomial function of degree n? Why are subscripts used to distinguish the coefficients of the terms?
66. What are the zeros of a polynomial function and how are they found?
67. Explain the relationship between the multiplicity of a zero and whether or not the graph crosses or touches the x-axis at that zero.
68. Explain the relationship between the degree of a polynomial function and the number of turning points on its graph.
69. If you are given a function's equation, how do you determine if the function is even, odd, or neither?
70. If a function is even, describe the symmetry of its graph.
71. If a function is odd, describe the symmetry of its graph.
72. Can the graph of a polynomial function have no x-intercepts? Explain.
73. Can the graph of a polynomial function have no y-intercept? Explain.
74. Describe a strategy for graphing a polynomial function. In your description, mention intercepts, the polynomial's degree, and turning points.

Technology Exercises

75. In Exercises 1–6, you determined end behavior. Use a graphing utility to graph any two functions from these exercises to verify your results.
76. In Exercises 7–14, you found zeros and their multiplicities. Use a graphing utility to graph any two functions from these exercises and use the resulting x-intercepts to verify your results.
77. In Exercises 15–26, you identified even and odd functions. Use a graphing utility to graph any one even function and verify its y-axis symmetry. Then use a graphing utility to graph any one odd function and verify its origin symmetry.

In Exercises 78–79, find the zeros for each polynomial function and give the multiplicity for each zero. Then use a graphing utility with a viewing rectangle large enough to show end behavior to graph each polynomial function and verify the zeros.

78. $f(x) = x^3 - 7x^2 - 4x + 28$

79. $f(x) = 2x^3 + x^2 - 50x - 25$

Critical Thinking Exercises

Make Sense? *In Exercises 80–83, determine whether each statement "makes sense" or "does not make sense" and explain your reasoning.*

80. I graphed $f(x) = (x + 2)^3(x - 4)^2$, and the graph touched the x-axis and turned around at -2.
81. I'm graphing a fourth-degree polynomial function with four turning points.
82. An attractive face displays the same kind of symmetry as the graph of an odd function.
83. Although I have not yet learned techniques for finding the x-intercepts of $f(x) = x^3 + 2x^2 - 5x - 6$, I can easily determine the y-intercept.

In Exercises 84–87, determine whether each statement is true or false. If the statement is false, make the necessary change(s) to produce a true statement.

84. If $f(x) = -x^3 + 4x$, then the graph of f falls to the left and falls to the right.
85. A mathematical model that is a polynomial of degree n whose leading term is $a_n x^n$, n odd and $a_n < 0$, is ideally suited to describe phenomena that have positive values over unlimited periods of time.
86. There is more than one third-degree polynomial function with the same three x-intercepts.
87. The graph of a function with origin symmetry can rise to the left and rise to the right.

Use the descriptions in Exercises 88–89 to write an equation of a polynomial function with the given characteristics. Use a graphing utility to graph your function to see if you are correct. If not, modify the function's equation and repeat this process.

88. Crosses the x-axis at -4, 0, and 3; lies above the x-axis between -4 and 0; lies below the x-axis between 0 and 3
89. Touches the x-axis and turns at 0 and crosses the x-axis at 2; lies below the x-axis between 0 and 2

Review Exercises

90. Solve: $\frac{1}{4x} - \frac{3}{4} = \frac{7}{x}$.
(Section 6.6, Example 1)
91. Subtract: $8\sqrt{45} - 3\sqrt{80}$.
(Section 7.4, Example 2)
92. Solve: $x^2 + 2x = 2$.
(Section 8.2, Example 2)

Preview Exercises

Exercises 93–95 will help you prepare for the material covered in the next section.

93. Use synthetic division to divide the polynomial $f(x) = x^3 + 2x^2 - 5x - 6$ by $x - 2$.
94. Solve: $x^2 + 4x - 1 = 0$.
95. Solve: $x^2 + 4x + 6 = 0$.

SECTION 11.2

Zeros of Polynomial Functions

You stole my formula!

Tartaglia's Secret Formula for One Solution of $x^3 + mx = n$

$$x = \sqrt[3]{\sqrt{\left(\frac{n}{2}\right)^2 + \left(\frac{m}{3}\right)^3} + \frac{n}{2}} - \sqrt[3]{\sqrt{\left(\frac{n}{2}\right)^2 + \left(\frac{m}{3}\right)^3} - \frac{n}{2}}$$

Objectives

1. Use the Factor Theorem to solve a polynomial equation.
2. Use the Rational Zero Theorem to find possible rational zeros.
3. Find zeros of a polynomial function.
4. Solve polynomial equations.
5. Use Descartes's Rule of Signs.

Popularizers of mathematics are sharing bizarre stories that are giving math a secure place in popular culture. One episode, able to compete with the wildest fare served up by television talk shows and the tabloids, involves three Italian mathematicians and, of all things, zeros of polynomial functions.

Tartaglia (1499–1557), poor and starving, has found a formula that gives a root for a third-degree polynomial equation. Cardano (1501–1576) begs Tartaglia to reveal the secret formula, wheedling it from him with the promise he will find the impoverished Tartaglia a patron. Then Cardano publishes his famous work *Ars Magna*, in which he presents Tartaglia's formula as his own. Cardano uses his most talented student, Ferrari (1522–1565), who derived a formula for a root of a fourth-degree polynomial equation, to falsely accuse Tartaglia of plagiarism. The dispute becomes violent and Tartaglia is fortunate to escape alive.

The noise from this "You Stole My Formula" episode is quieted by the work of French mathematician Evariste Galois (1811–1832). Galois proved that there is no general formula for finding roots of polynomial equations of degree 5 or greater. There are, however, methods for finding roots. In this section, we study methods for finding zeros of polynomial functions. Synthetic division plays an important role in this process.

1 Use the Factor Theorem to solve a polynomial equation.

The Factor Theorem

In Chapter 6, we used a shortcut for long division, called **synthetic division**, to divide a polynomial by a binomial of the form $x - c$. For example, we can divide the polynomial $f(x) = x^3 - 4x^2 + 5x + 3$ by $x - 2$ as follows:

Study Tip

If you've forgotten the details behind synthetic division and the Remainder Theorem, see Section 6.5 beginning on page 434.

Coefficients of the dividend, $x^3 - 4x^2 + 5x + 3$

This is c in $x - c$. For $x - 2$, c is 2.

2	1	−4	5	3
	↓	2	−4	2
	1	−2	1	5

Coefficients of the quotient

Remainder

Thus,
$$\frac{f(x)}{x - c} = \frac{x^3 - 4x^2 + 5x + 3}{x - 2} = x^2 - 2x + 1 + \frac{5}{x - 2}.$$

The quotient, $q(x)$, is $x^2 - 2x + 1$.

Do you remember how to check the answer to a long division problem? Find the product of the divisor and the quotient and add the remainder. The result should be the dividend. If the divisor is $x - c$, we can express this idea symbolically:

$$f(x) = (x - c)q(x) + r.$$

Dividend | Divisor | Quotient | The remainder, r, is a constant when dividing by $x - c$.

In Chapter 6, we also studied the **Remainder Theorem**. It states that if the polynomial $f(x)$ is divided by $x - c$, then the remainder is $f(c)$. For example, in the synthetic division shown previously, we divided $f(x) = x^3 - 4x^2 + 5x + 3$ by $x - 2$. The remainder is 5. Thus, $f(2) = 5$. You can verify this by evaluating $f(2)$ directly, substituting 2 for x.

Let's look again at the equation used to check the answer to a long division problem when the divisor is $x - c$.

$$f(x) = (x - c)q(x) + r$$

Dividend | Divisor | Quotient | Constant remainder

By the Remainder Theorem, the remainder r is $f(c)$, so we can substitute $f(c)$ for r:

$$f(x) = (x - c)q(x) + f(c).$$

Notice that if $f(c) = 0$, then

$$f(x) = (x - c)q(x)$$

so that $x - c$ is a factor of $f(x)$. This means that for the polynomial function $f(x)$, if $f(c) = 0$, then $x - c$ is a factor of $f(x)$.

Let's reverse directions and see what happens if $x - c$ is a factor of $f(x)$. This means that

$$f(x) = (x - c)q(x).$$

If we replace x with c, we obtain

$$f(c) = (c - c)q(c) = 0 \cdot q(c) = 0.$$

Thus, if $x - c$ is a factor of $f(x)$, then $f(c) = 0$.

We have proved a result known as the **Factor Theorem**.

The Factor Theorem

Let $f(x)$ be a polynomial.

- **a.** If $f(c) = 0$, then $x - c$ is a factor of $f(x)$.
- **b.** If $x - c$ is a factor of $f(x)$, then $f(c) = 0$.

The example that follows shows how the Factor Theorem can be used to solve a polynomial equation.

EXAMPLE 1 Using the Factor Theorem

Solve the equation $2x^3 - 3x^2 - 11x + 6 = 0$ given that 3 is a zero of $f(x) = 2x^3 - 3x^2 - 11x + 6$.

Solution We are given that 3 is a zero of $f(x) = 2x^3 - 3x^2 - 11x + 6$. This means that $f(3) = 0$. Because $f(3) = 0$, the Factor Theorem tells us that $x - 3$ is a factor of $f(x)$. We'll use synthetic division to divide $f(x)$ by $x - 3$.

3	2	−3	−11	6
		6	9	−6
	2	3	−2	0

$$x - 3\overline{)2x^3 - 3x^2 - 11x + 6} \quad \text{quotient: } 2x^2 + 3x - 2$$

The remainder, 0, verifies that $x - 3$ is a factor of $2x^3 - 3x^2 - 11x + 6$.

Equivalently,

$$2x^3 - 3x^2 - 11x + 6 = (x - 3)(2x^2 + 3x - 2).$$

Now we can solve the polynomial equation.

$2x^3 - 3x^2 - 11x + 6 = 0$ This is the given equation.

$(x - 3)(2x^2 + 3x - 2) = 0$ Factor using the result from the synthetic division.

$(x - 3)(2x - 1)(x + 2) = 0$ Factor the trinomial.

$x - 3 = 0$ or $2x - 1 = 0$ or $x + 2 = 0$ Set each factor equal to 0.

$x = 3$ $\quad x = \frac{1}{2}$ $\quad x = -2$ Solve for x.

The solutions are $-2, \frac{1}{2}$, and 3, and the solution set is $\left\{-2, \frac{1}{2}, 3\right\}$.

Using Technology

Graphic Connections

Because the solution set of

$$2x^3 - 3x^2 - 11x + 6 = 0$$

is $\left\{-2, \frac{1}{2}, 3\right\}$, this implies that the polynomial function

$$f(x) = 2x^3 - 3x^2 - 11x + 6$$

has x-intercepts (or zeros) at $-2, \frac{1}{2}$, and 3. This is verified by the graph of f.

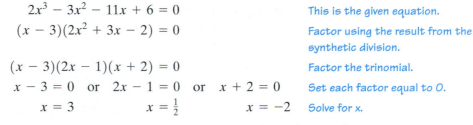

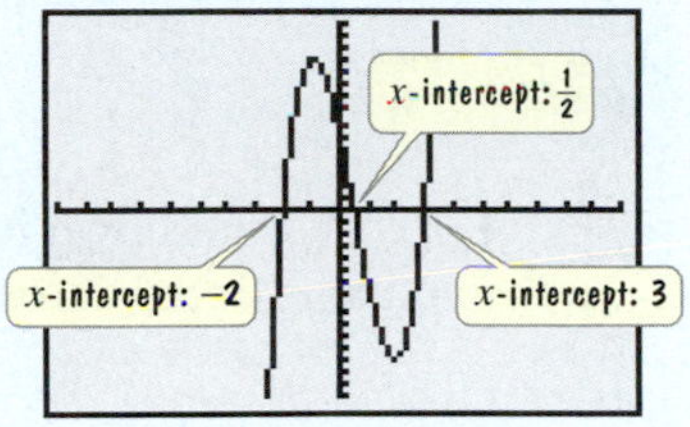

[−10, 10, 1] by [−15, 15, 1]

CHECK POINT 1 Solve the equation $2x^3 - 5x^2 + x + 2 = 0$ given that 2 is a zero of $f(x) = 2x^3 - 5x^2 + x + 2$.

Based on the Factor Theorem, the following statements are useful in solving polynomial equations:

1. If $f(x)$ is divided by $x - c$ and the remainder is zero, then c is a zero of f and c is a root of the polynomial equation $f(x) = 0$.
2. If $f(x)$ is divided by $x - c$ and the remainder is zero, then $x - c$ is a factor of $f(x)$.

2 Use the Rational Zero Theorem to find possible rational zeros.

The Rational Zero Theorem

The Rational Zero Theorem provides us with a tool that we can use to make a list of all possible rational zeros of a polynomial function. Equivalently, the theorem gives all possible rational roots of a polynomial equation. Not every number in the list will be a zero of the function, but every rational zero of the polynomial function will appear somewhere in the list.

The Rational Zero Theorem

If $f(x) = a_n x^n + a_{n-1}x^{n-1} + \cdots + a_1 x + a_0$ has *integer* coefficients and $\frac{p}{q}$ (where $\frac{p}{q}$ is reduced to lowest terms) is a rational zero of f, then p is a factor of the constant term, a_0, and q is a factor of the leading coefficient, a_n.

You can explore the "why" behind the Rational Zero Theorem in Exercise 90 of Exercise Set 11.2. For now, let's see if we can figure out what the theorem tells us about possible rational zeros. To use the theorem, list all the integers that are factors of the constant term, a_0. Then list all the integers that are factors of the leading coefficient, a_n. Finally, list all possible rational zeros:

$$\text{Possible rational zeros} = \frac{\text{Factors of the constant term}}{\text{Factors of the leading coefficient}}.$$

EXAMPLE 2 Using the Rational Zero Theorem

List all possible rational zeros of $f(x) = -x^4 + 3x^2 + 4$.

Solution The constant term is 4. We list all of its factors: ±1, ±2, ±4. The leading coefficient is −1. Its factors are ±1.

Factors of the constant term, 4:	±1, ±2, ±4
Factors of the leading coefficient, −1:	±1

Because

$$\text{Possible rational zeros} = \frac{\text{Factors of the constant term}}{\text{Factors of the leading coefficient}},$$

we must take each number in the first row, ±1, ±2, ±4, and divide by each number in the second row, ±1.

$$\text{Possible rational zeros} = \frac{\text{Factors of } 4}{\text{Factors of } -1} = \frac{\pm1, \pm2, \pm4}{\pm1} = \pm1, \quad \pm2, \quad \pm4$$

Divide ±1 by ±1. Divide ±2 by ±1. Divide ±4 by ±1.

There are six possible rational zeros. The graph of $f(x) = -x^4 + 3x^2 + 4$ is shown in **Figure 11.9**. The x-intercepts are −2 and 2. Thus, −2 and 2 are the actual rational zeros.

Study Tip

Always keep in mind the relationship among zeros, roots, and x-intercepts. The zeros of a function f are the roots, or solutions, of the equation $f(x) = 0$. Furthermore, the real zeros, or real roots, are the x-intercepts of the graph of f.

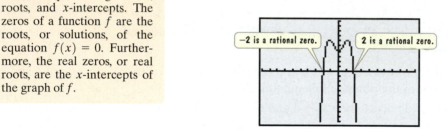

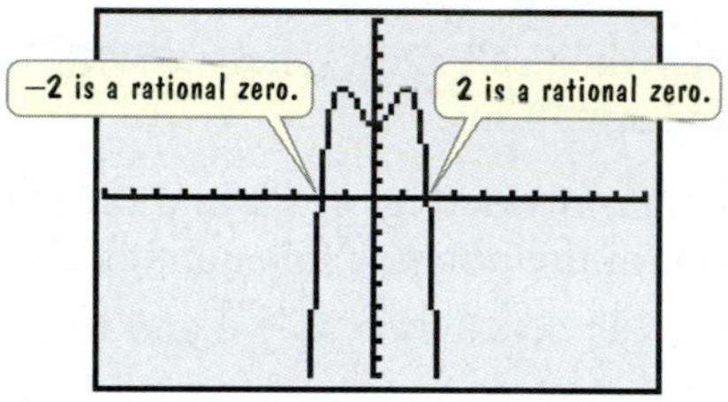

−2 is a rational zero.
2 is a rational zero.

FIGURE 11.9 The graph of $f(x) = -x^4 + 3x^2 + 4$ shows that −2 and 2 are rational zeros.

CHECK POINT 2 List all possible rational zeros of $f(x) = x^3 + 2x^2 - 5x - 6$.

EXAMPLE 3 Using the Rational Zero Theorem

List all possible rational zeros of $f(x) = 15x^3 + 14x^2 - 3x - 2$.

Solution The constant term is −2 and the leading coefficient is 15.

$$\text{Possible rational zeros} = \frac{\text{Factors of the constant term, } -2}{\text{Factors of the leading coefficient, } 15} = \frac{\pm1, \pm2}{\pm1, \pm3, \pm5, \pm15}$$

$$= \pm1, \pm2, \pm\frac{1}{3}, \pm\frac{2}{3}, \pm\frac{1}{5}, \pm\frac{2}{5}, \pm\frac{1}{15}, \pm\frac{2}{15}$$

Divide ±1 and ±2 by ±1. Divide ±1 and ±2 by ±3. Divide ±1 and ±2 by ±5. Divide ±1 and ±2 by ±15.

There are 16 possible rational zeros. The actual solution set of

$$15x^3 + 14x^2 - 3x - 2 = 0$$

is $\left\{-1, -\frac{1}{3}, \frac{2}{5}\right\}$, which contains three of the 16 possible zeros.

☑ **CHECK POINT 3** List all possible rational zeros of $f(x) = 4x^5 + 12x^4 - x - 3$.

---

3 Find zeros of a polynomial function.

How do we determine which (if any) of the possible rational zeros are rational zeros of the polynomial function? To find the first rational zero, we can use a trial-and-error process involving synthetic division: If $f(x)$ is divided by $x - c$ and the remainder is zero, then c is a zero of f. After we identify the first rational zero, we use the result of the synthetic division to factor the original polynomial. Then we set each factor equal to zero to identify any additional rational zeros.

EXAMPLE 4 Finding Zeros of a Polynomial Function

Find all rational zeros of $f(x) = x^3 + 2x^2 - 5x - 6$.

Solution We begin by listing all possible rational zeros.

Possible rational zeros

$$= \frac{\text{Factors of the constant term, } -6}{\text{Factors of the leading coefficient, } 1} = \frac{\pm 1, \pm 2, \pm 3, \pm 6}{\pm 1} = \pm 1, \pm 2, \pm 3, \pm 6$$

Divide the eight numbers in the numerator by ±1.

Now we will use synthetic division to see if we can find a rational zero among the possible rational zeros ±1, ±2, ±3, ±6. Keep in mind that if $f(x)$ is divided by $x - c$ and the remainder is zero, then c is a zero of f. Let's start by testing 1. If 1 is not a rational zero, then we will test other possible rational zeros.

Test 1.

Possible rational zero

Coefficients of $f(x) = x^3 + 2x^2 - 5x - 6$

$$\begin{array}{r|rrrr} 1 & 1 & 2 & -5 & -6 \\ & & 1 & 3 & -2 \\ \hline & 1 & 3 & -2 & -8 \end{array}$$

The nonzero remainder shows that 1 is not a zero.

Test 2.

Possible rational zero

Coefficients of $f(x) = x^3 + 2x^2 - 5x - 6$

$$\begin{array}{r|rrrr} 2 & 1 & 2 & -5 & -6 \\ & & 2 & 8 & 6 \\ \hline & 1 & 4 & 3 & 0 \end{array}$$

The zero remainder shows that 2 is a zero.

The zero remainder tells us that 2 is a zero of the polynomial function $f(x) = x^3 + 2x^2 - 5x - 6$. Equivalently, 2 is a solution, or root, of the polynomial equation $x^3 + 2x^2 - 5x - 6 = 0$. Thus, $x - 2$ is a factor of the polynomial. The first three numbers in the bottom row of the synthetic division on the right, 1, 4, and 3, give the coefficients of the other factor. This factor is $x^2 + 4x + 3$.

$x^3 + 2x^2 - 5x - 6 = 0$ Finding the zeros of $f(x) = x^3 + 2x^2 - 5x - 6$ is the same as finding the roots of this equation.

$(x - 2)(x^2 + 4x + 3) = 0$ Factor using the result from the synthetic division.

$(x - 2)(x + 3)(x + 1) = 0$ Factor completely.

$x - 2 = 0$ or $x + 3 = 0$ or $x + 1 = 0$ Set each factor equal to zero.

$x = 2$ $\quad x = -3$ $\quad x = -1$ Solve for x.

The solutions are -3, -1, and 2, and the solution set is $\{-3, -1, 2\}$. The rational zeros of f are -3, -1, and 2.

☑ **CHECK POINT 4** Find all rational zeros of $f(x) = x^3 + 8x^2 + 11x - 20$.

Our work in Example 4 involved finding zeros of a third-degree polynomial function. The Rational Zero Theorem is a tool that allows us to rewrite such functions as products of two factors, one linear and one quadratic. Zeros of the quadratic factor are found by factoring, the quadratic formula, or the square root property.

EXAMPLE 5 Finding Zeros of a Polynomial Function

Find all zeros of $f(x) = x^3 + 7x^2 + 11x - 3$.

Solution We begin by listing all possible rational zeros.

$$\text{Possible rational zeros} = \frac{\text{Factors of the constant team, } -3}{\text{Factors of the leading coefficient, } 1} = \frac{\pm 1, \pm 3}{\pm 1} = \pm 1, \pm 3$$

Now we will use synthetic division to see if we can find a rational zero among the four possible rational zeros.

Test 1.	Test −1.	Test 3.	Test −3.
$\begin{array}{r\|rrrr} 1 & 1 & 7 & 11 & -3 \\ & & 1 & 8 & 19 \\ \hline & 1 & 8 & 19 & 16 \end{array}$	$\begin{array}{r\|rrrr} -1 & 1 & 7 & 11 & -3 \\ & & -1 & -6 & -5 \\ \hline & 1 & 6 & 5 & -8 \end{array}$	$\begin{array}{r\|rrrr} 3 & 1 & 7 & 11 & -3 \\ & & 3 & 30 & 123 \\ \hline & 1 & 10 & 41 & 120 \end{array}$	$\begin{array}{r\|rrrr} -3 & 1 & 7 & 11 & -3 \\ & & -3 & -12 & 3 \\ \hline & 1 & 4 & -1 & 0 \end{array}$

The zero remainder on the right tells us that −3 is a zero of the polynomial function $f(x) = x^3 + 7x^2 + 11x - 3$. To find all zeros of f, we proceed as follows:

$x^3 + 7x^2 + 11x - 3 = 0$ Finding the zeros of f is the same thing as finding the roots of $f(x) = 0$.

$(x + 3)(x^2 + 4x - 1) = 0$ This result is from the last synthetic division. The first three numbers in the bottom row, 1, 4, and −1, give the coefficients of the second factor.

$x + 3 = 0$ or $x^2 + 4x - 1 = 0$ Set each factor equal to 0.

$x = -3$ Solve the linear equation.

We can use the quadratic formula to solve $x^2 + 4x - 1 = 0$.

$x = \dfrac{-b \pm \sqrt{b^2 - 4ac}}{2a}$ We use the quadratic formula because $x^2 + 4x - 1$ cannot be factored.

$= \dfrac{-4 \pm \sqrt{4^2 - 4(1)(-1)}}{2(1)}$ Let $a = 1$, $b = 4$, and $c = -1$.

$= \dfrac{-4 \pm \sqrt{20}}{2}$ Multiply and subtract under the radical: $4^2 - 4(1)(-1) = 16 - (-4) = 16 + 4 = 20$.

$= \dfrac{-4 \pm 2\sqrt{5}}{2}$ $\sqrt{20} = \sqrt{4 \cdot 5} = 2\sqrt{5}$

$= -2 \pm \sqrt{5}$ Divide the numerator and the denominator by 2.

The solutions are −3 and $-2 \pm \sqrt{5}$, and the solution set is $\{-3, -2 - \sqrt{5}, -2 + \sqrt{5}\}$. The zeros of f are -3, $-2 - \sqrt{5}$, and $-2 + \sqrt{5}$. Among these three real zeros, one zero is rational and two are irrational.

✓ CHECK POINT 5 Find all zeros of $f(x) = x^3 + x^2 - 5x - 2$.

Study Tip

Be sure you are familiar with the various kinds of zeros of polynomial functions. Here's a quick example:

$$f(x) = (x + 3)(2x - 1)(x + \sqrt{2})(x - \sqrt{2})(x - 4 + 5i)(x - 4 - 5i).$$

Zeros: $-3, \quad \frac{1}{2}, \quad -\sqrt{2}, \quad \sqrt{2}, \quad 4 - 5i, \quad 4 + 5i$

Rational zeros: -3, $\frac{1}{2}$
Irrational zeros: $-\sqrt{2}$, $\sqrt{2}$
Complex imaginary zeros: $4 - 5i$, $4 + 5i$
Real zeros: -3, $\frac{1}{2}$, $-\sqrt{2}$, $\sqrt{2}$
Nonreal zeros: $4 - 5i$, $4 + 5i$

If the degree of a polynomial function or equation is 4 or higher, it is often necessary to find more than one linear factor by synthetic division.

One way to speed up the process of finding the first zero is to graph the function. Any x-intercept is a zero.

4 Solve polynomial equations.

EXAMPLE 6 Solving a Polynomial Equation

Solve: $x^4 - 6x^2 - 8x + 24 = 0.$

Solution Recall that we refer to the *zeros* of a polynomial function and the *roots* of a polynomial equation. Because we are given an equation, we will use the word "roots," rather than "zeros," in the solution process. We begin by listing all possible rational roots.

$$\text{Possible rational roots} = \frac{\text{Factors of the constant term, 24}}{\text{Factors of the leading coefficient, 1}}$$

$$= \frac{\pm 1, \pm 2, \pm 3, \pm 4, \pm 6, \pm 8, \pm 12, \pm 24}{\pm 1}$$

$$= \pm 1, \pm 2, \pm 3, \pm 4, \pm 6, \pm 8, \pm 12, \pm 24$$

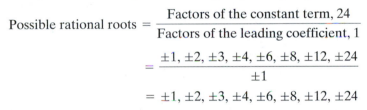

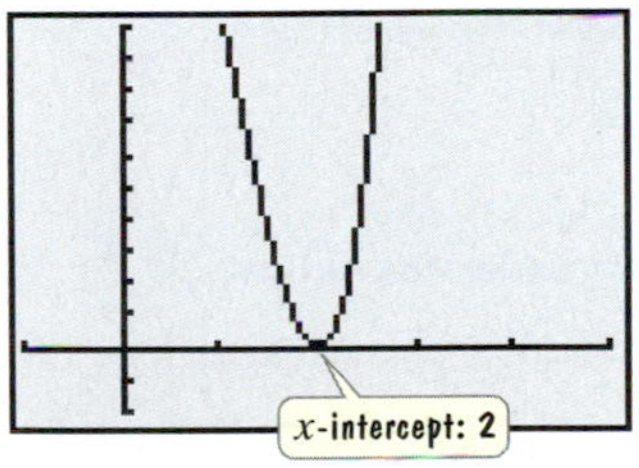

FIGURE 11.10 The graph of $f(x) = x^4 - 6x^2 - 8x + 24$ in a $[-1, 5, 1]$ by $[-2, 10, 1]$ viewing rectangle

Part of the graph of $f(x) = x^4 - 6x^2 - 8x + 24$ is shown in **Figure 11.10**. Because the x-intercept shown in the viewing rectangle is 2, we will test 2 by synthetic division and show that it is a root of the given equation. Without the graph, the procedure would be to start the trial-and-error synthetic division with 1 and proceed until a zero remainder is found, as we did in Example 5.

$$\begin{array}{r|rrrrr} 2 & 1 & 0 & -6 & -8 & 24 \\ & & 2 & 4 & -4 & -24 \\ \hline & 1 & 2 & -2 & -12 & 0 \end{array}$$

Careful! $x^4 - 6x^2 - 8x + 24 = x^4 + 0x^3 - 6x^2 - 8x + 24$

The zero remainder indicates that 2 is a root of $x^4 - 6x^2 - 8x + 24 = 0$.

Now we can rewrite the given equation in factored form.

$x^4 - 6x^2 - 8x + 24 = 0$ This is the given equation.

$(x - 2)(x^3 + 2x^2 - 2x - 12) = 0$ This is the result obtained from the synthetic division. The first four numbers in the bottom row, 1, 2, -2, and -12, give the coefficients of the second factor.

$x - 2 = 0$ or $x^3 + 2x^2 - 2x - 12 = 0$ Set each factor equal to 0.

We can use the same approach to look for rational roots of the polynomial equation $x^3 + 2x^2 - 2x - 12 = 0$, listing all possible rational roots. Without the graph in **Figure 11.10**, the procedure would be to start testing possible rational roots by trial-

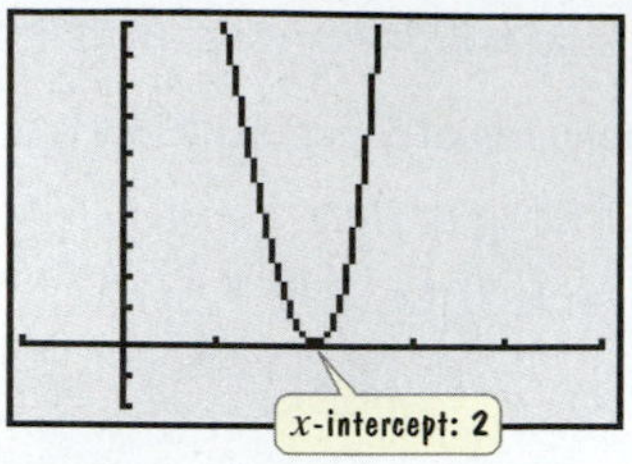

FIGURE 11.10 (repeated)

and-error synthetic division. However, take a second look at the graph in **Figure 11.10**. Because the graph turns around at 2, this means that 2 is a root of even multiplicity. Thus, 2 must also be a root of $x^3 + 2x^2 - 2x - 12 = 0$, confirmed by the following synthetic division.

2	1	2	−2	−12
		2	8	12
	1	4	6	0

These are the coefficients of $x^3 + 2x^2 - 2x - 12$.

The zero remainder indicates that 2 is a root of $x^3 + 2x^2 - 2x - 12 = 0$.

Now we can solve the original equation as follows:

$$x^4 - 6x^2 - 8x + 24 = 0$$ This is the given equation.

$$(x-2)(x^3 + 2x^2 - 2x - 12) = 0$$ This factorization was obtained from the first synthetic division.

$$(x-2)(x-2)(x^2 + 4x + 6) = 0$$ This factorization was obtained from the second synthetic division. The first three numbers in the bottom row, 1, 4, and 6, give the coefficients of the third factor.

$$x - 2 = 0 \quad \text{or} \quad x - 2 = 0 \quad \text{or} \quad x^2 + 4x + 6 = 0$$ Set each factor equal to 0.

$$x = 2 \qquad\qquad x = 2$$ Solve the linear equations.

We can use the quadratic formula to solve $x^2 + 4x + 6 = 0$.

$$x = \frac{-b \pm \sqrt{b^2 - 4ac}}{2a}$$ We use the quadratic formula because $x^2 + 4x + 6$ cannot be factored.

$$= \frac{-4 \pm \sqrt{4^2 - 4(1)(6)}}{2(1)}$$ Let $a = 1$, $b = 4$, and $c = 6$.

$$= \frac{-4 \pm \sqrt{-8}}{2}$$ Multiply and subtract under the radical: $4^2 - 4(1)(6) = 16 - 24 = -8$.

$$= \frac{-4 \pm 2i\sqrt{2}}{2}$$ $\sqrt{-8} = \sqrt{4(2)(-1)} = 2i\sqrt{2}$

$$= -2 \pm i\sqrt{2}$$ Simplify.

The roots of the original equation, $x^4 - 6x^2 - 8x + 24 = 0$, are 2, $-2 - i\sqrt{2}$, and $-2 + i\sqrt{2}$, and the solution set is $\{2, -2 \pm i\sqrt{2}\}$. The graph in **Figure 11.10** shows the only real root, 2, but not the complex imaginary roots, $-2 \pm i\sqrt{2}$.

In Example 6, 2 is a repeated root of the equation with multiplicity 2. The example illustrates two general properties:

Properties of Polynomial Equations

1. If a polynomial equation is of degree n, then counting multiple roots separately, the equation has n roots.
2. If $a + bi$ is a root of a polynomial equation $(b \neq 0)$, then the imaginary number $a - bi$ is also a root. Imaginary roots, if they exist, occur in conjugate pairs.

CHECK POINT 6 Solve: $x^4 - 6x^3 + 22x^2 - 30x + 13 = 0$.

5 Use Descartes's Rule of Signs.

Descartes's Rule of Signs

Because an nth-degree polynomial equation might have roots that are imaginary numbers, we should note that such an equation can have *at most n* real roots. **Descartes's Rule of Signs** provides even more specific information about the number of real zeros that a polynomial can have. The rule is based on considering *variations in sign* between consecutive coefficients. For example, the function $f(x) = 3x^7 - 2x^5 - x^4 + 7x^2 + x - 3$ has three sign changes:

$$f(x) = 3x^7 - 2x^5 - x^4 + 7x^2 + x - 3.$$

sign change — sign change — sign change

"An equation can have as many true [positive] roots as it contains changes of sign, from plus to minus or from minus to plus." René Descartes (1596–1650) in *La Géométrie* (1637)

Descartes's Rule of Signs

Let $f(x) = a_nx^n + a_{n-1}x^{n-1} + \cdots + a_2x^2 + a_1x + a_0$ be a polynomial with real coefficients.

1. The number of *positive real zeros* of f is either
   a. the same as the number of sign changes of $f(x)$
   or
   b. less than the number of sign changes of $f(x)$ by a positive even integer.

   If $f(x)$ has only one variation in sign, then f has exactly one positive real zero.
2. The number of *negative real zeros* of f is either
   a. the same as the number of sign changes of $f(-x)$
   or
   b. less than the number of sign changes of $f(-x)$ by a positive even integer.

   If $f(-x)$ has only one variation in sign, then f has exactly one negative real zero.

Study Tip

The number of real zeros given by Descartes's Rule of Signs includes rational zeros from a list of possible rational zeros, as well as irrational zeros not on the list. It does not include any imaginary zeros.

Table 11.1 illustrates what Descartes's Rule of Signs tells us about the positive real zeros of various polynomial functions.

Table 11.1 Descartes's Rule of Signs and Positive Real Zeros

Polynomial Function	Sign Changes	Conclusion
$f(x) = 3x^7 - 2x^5 - x^4 + 7x^2 + x - 3$ (sign change, sign change, sign change)	3	There are 3 positive real zeros. or There is $3 - 2 = 1$ positive real zero.
$f(x) = 4x^5 + 2x^4 - 3x^2 + x + 5$ (sign change, sign change)	2	There are 2 positive real zeros. or There are $2 - 2 = 0$ positive real zeros.
$f(x) = -7x^6 - 5x^4 + x + 9$ (sign change)	1	There is 1 positive real zero.

EXAMPLE 7 Using Descartes's Rule of Signs

Determine the possible numbers of positive and negative real zeros of $f(x) = x^3 + 2x^2 + 5x + 4$.

Solution

1. To find possibilities for the number of positive real zeros, count the number of sign changes in the equation for $f(x)$. Because all the coefficients are positive, there are no variations in sign. Thus, there are no positive real zeros.
2. To find possibilities for the number of negative real zeros, count the number of sign changes in the equation for $f(-x)$. We obtain this equation by replacing x with $-x$ in the given function.

$$f(x) = x^3 + 2x^2 + 5x + 4$$

Replace x with $-x$.

$$\begin{aligned} f(-x) &= (-x)^3 + 2(-x)^2 + 5(-x) + 4 \\ &= -x^3 + 2x^2 - 5x + 4 \end{aligned}$$

Now count the sign changes.

$$f(x) = -x^3 + 2x^2 - 5x + 4$$

sign change　sign change　sign change

There are three variations in sign. The number of negative real zeros of f is either equal to the number of sign changes, 3, or is less than this number by an even integer. This means that either there are 3 negative real zeros or there is $3 - 2 = 1$ negative real zero.

What do the results of Example 7 mean in terms of solving

$$x^3 + 2x^2 + 5x + 4 = 0?$$

Without using Descartes's Rule of Signs, we list the possible rational roots as follows: Possible rational roots

$$= \frac{\text{Factors of the constant term, 4}}{\text{Factors of the leading coefficient, 1}} = \frac{\pm 1, \pm 2, \pm 4}{\pm 1} = \pm 1, \pm 2, \pm 4.$$

However, Descartes's Rule of Signs informed us that $f(x) = x^3 + 2x^2 + 5x + 4$ has no positive real zeros. Thus, the polynomial equation $x^3 + 2x^2 + 5x + 4 = 0$ has no positive real roots. This means that we can eliminate the positive numbers from our list of possible rational roots. Possible rational roots include only -1, -2, and -4. We can use synthetic division and test the first of these three possible rational roots of $x^3 + 2x^2 + 5x + 4 = 0$ as follows:

Test -1.

$$\begin{array}{r|rrrr} -1 & 1 & 2 & 5 & 4 \\ & & -1 & -1 & -4 \\ \hline & 1 & 1 & 4 & 0 \end{array}$$

The zero remainder shows that -1 is a root.

By solving the equation $x^3 + 2x^2 + 5x + 4 = 0$, you will find that this equation of degree 3 has three roots. One root is -1 and the other two roots are imaginary numbers in a conjugate pair. Verify this by completing the solution process.

✓ CHECK POINT 7 Determine the possible numbers of positive and negative real zeros of $f(x) = x^4 - 14x^3 + 71x^2 - 154x + 120$.

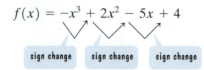

11.2 EXERCISE SET

MyMathLab · MathXL PRACTICE · WATCH · DOWNLOAD · READ · REVIEW

Practice Exercises

1. Solve the equation $x^3 - 9x^2 + 26x - 24 = 0$ given that 3 is a zero of $f(x) = x^3 - 9x^2 + 26x - 24$.
2. Solve the equation $x^3 - 5x^2 + 2x + 8 = 0$ given that 4 is a zero of $f(x) = x^3 - 5x^2 + 2x + 8$.
3. Solve the equation $2x^3 + 3x^2 - 8x + 3 = 0$ given that -3 is a zero of $f(x) = 2x^3 + 3x^2 - 8x + 3$.
4. Solve the equation $15x^3 + 14x^2 - 3x - 2 = 0$ given that -1 is a zero of $f(x) = 15x^3 + 14x^2 - 3x - 2$.
5. Solve the equation $12x^3 + 16x^2 - 5x - 3 = 0$ given that $-\frac{3}{2}$ is a zero of $f(x) = 12x^3 + 16x^2 - 5x - 3$.
6. Solve the equation $3x^3 + 7x^2 - 22x - 8 = 0$ given that $-\frac{1}{3}$ is a zero of $f(x) = 3x^3 + 7x^2 - 22x - 8$.
7. Solve $2x^3 + 3x^2 - 8x - 12 = 0$ given that 2 is a root.
8. Solve $3x^3 - 5x^2 - 4x + 4 = 0$ given that 2 is a root.

In Exercises 9–16, use the Rational Zero Theorem to list all possible rational zeros for each given function.

9. $f(x) = x^3 + x^2 - 4x - 4$
10. $f(x) = x^3 + 3x^2 - 6x - 8$
11. $f(x) = 3x^4 - 11x^3 - x^2 + 19x + 6$
12. $f(x) = 2x^4 + 3x^3 - 11x^2 - 9x + 15$
13. $f(x) = 4x^4 - x^3 + 5x^2 - 2x - 6$
14. $f(x) = 3x^4 - 11x^3 - 3x^2 - 6x + 8$
15. $f(x) = x^5 - x^4 - 7x^3 + 7x^2 - 12x - 12$
16. $f(x) = 4x^5 - 8x^4 - x + 2$

In Exercises 17–22,

a. *List all possible rational zeros.*

b. *Use synthetic division to test the possible rational zeros and find an actual zero.*

c. *Use the zero from part (b) to find all the zeros of the polynomial function.*

17. $f(x) = x^3 + x^2 - 4x - 4$
18. $f(x) = x^3 + 4x^2 + x - 6$
19. $f(x) = 2x^3 - 3x^2 - 11x + 6$
20. $f(x) = 2x^3 - 5x^2 + x + 2$
21. $f(x) = 3x^3 + 7x^2 - 22x - 8$
22. $f(x) = 3x^3 + 8x^2 - 15x + 4$

In Exercises 23–30,

a. *List all possible rational roots.*

b. *Use synthetic division to test the possible rational roots and find an actual root.*

c. *Use the root from part (b) to solve the equation.*

23. $x^3 - 2x^2 - 11x + 12 = 0$
24. $x^3 - 2x^2 - 7x - 4 = 0$
25. $x^3 - 10x - 12 = 0$
26. $x^3 - 5x^2 + 17x - 13 = 0$
27. $6x^3 + 25x^2 - 24x + 5 = 0$
28. $2x^3 - 5x^2 - 6x + 4 = 0$
29. $x^4 - 2x^3 - 5x^2 + 8x + 4 = 0$
30. $x^4 - 2x^2 - 16x - 15 = 0$

In Exercises 31–36, use Descartes's Rule of Signs to determine the possible numbers of positive and negative real zeros for each given function.

31. $f(x) = x^3 + 2x^2 + 5x + 4$
32. $f(x) = x^3 + 7x^2 + x + 7$
33. $f(x) = 5x^3 - 3x^2 + 3x - 1$
34. $f(x) = -2x^3 + x^2 - x + 7$
35. $f(x) = 2x^4 - 5x^3 - x^2 - 6x + 4$
36. $f(x) = 4x^4 - x^3 + 5x^2 - 2x - 6$

In Exercises 37–48, find all zeros of the polynomial function or solve the given polynomial equation. Use the Rational Zero Theorem and Descartes's Rule of Signs as an aid in obtaining the first zero or the first root.

37. $f(x) = x^3 - 4x^2 - 7x + 10$
38. $f(x) = x^3 + 12x^2 + 21x + 10$
39. $2x^3 - x^2 - 9x - 4 = 0$
40. $3x^3 - 8x^2 - 8x + 8 = 0$
41. $x^4 - 3x^3 - 20x^2 - 24x - 8 = 0$
42. $x^4 - x^3 + 2x^2 - 4x - 8 = 0$

43. $f(x) = 3x^4 - 11x^3 - x^2 + 19x + 6$

44. $f(x) = 2x^4 + 3x^3 - 11x^2 - 9x + 15$

45. $4x^4 - x^3 + 5x^2 - 2x - 6 = 0$

46. $3x^4 - 11x^3 - 3x^2 - 6x + 8 = 0$

47. $2x^5 + 7x^4 - 18x^2 - 8x + 8 = 0$

48. $4x^5 + 12x^4 - 41x^3 - 99x^2 + 10x + 24 = 0$

Practice PLUS

Exercises 49–56 show incomplete graphs of given polynomial functions.

a. *Find all the zeros of each function.*

b. *Without using a graphing utility, draw a complete graph of the function.*

49. $f(x) = -x^3 + x^2 + 16x - 16$

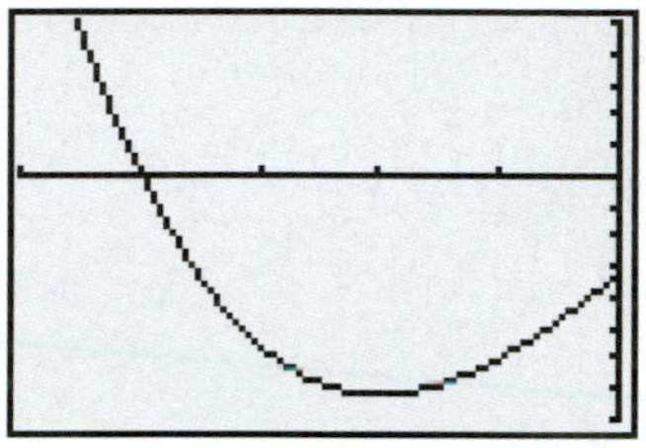

[−5, 0, 1] by [−40, 25, 5]

50. $f(x) = -x^3 + 3x^2 - 4$

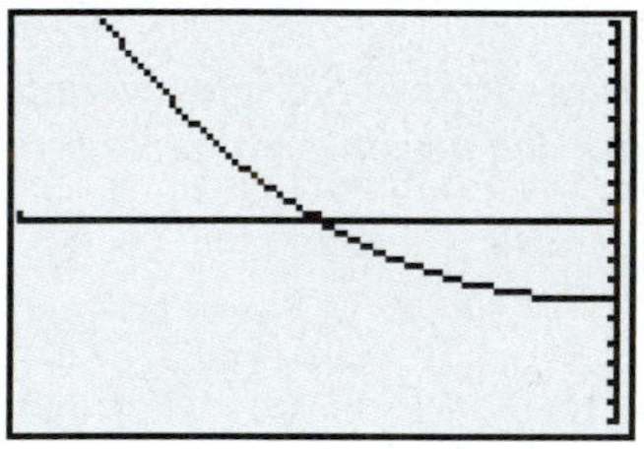

[−2, 0, 1] by [−10, 10, 1]

51. $f(x) = 4x^3 - 8x^2 - 3x + 9$

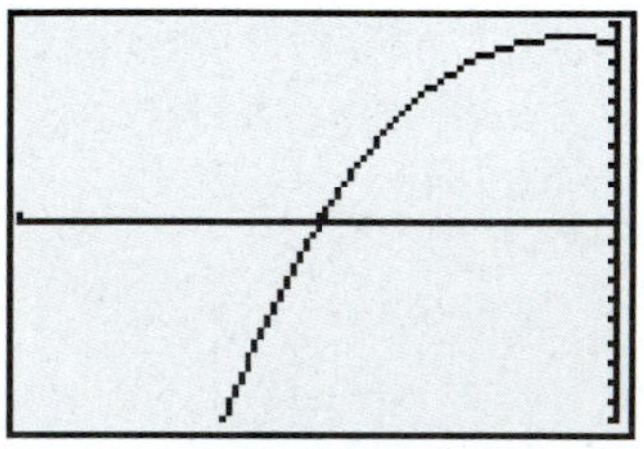

[−2, 0, 1] by [−10, 10, 1]

52. $f(x) = 3x^3 + 2x^2 + 2x - 1$

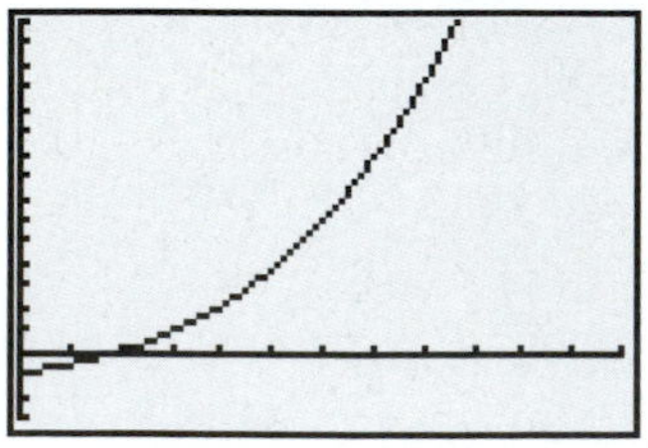

$[0, 2, \frac{1}{6}]$ by [−3, 15, 1]

53. $f(x) = 2x^4 - 3x^3 - 7x^2 - 8x + 6$

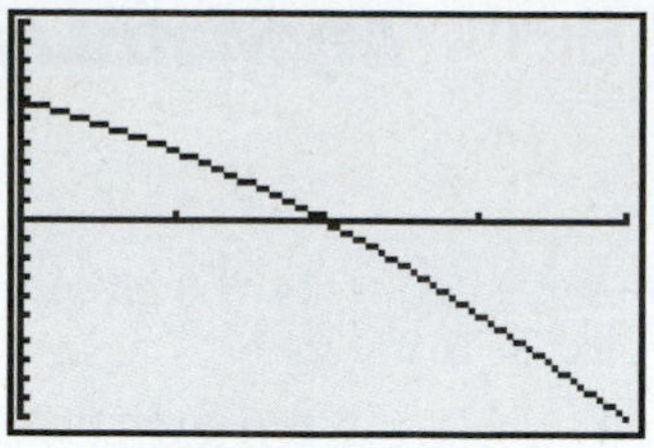

$[0, 1, \frac{1}{4}]$ by [−10, 10, 1]

54. $f(x) = 2x^4 + 2x^3 - 22x^2 - 18x + 36$

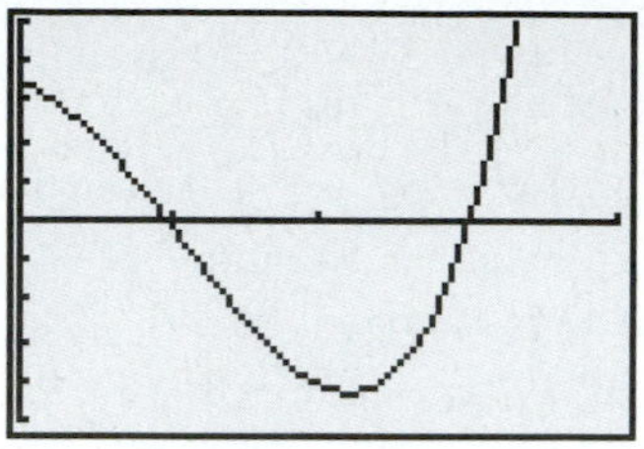

[0, 4, 1] by [−50, 50, 10]

55. $f(x) = 3x^5 + 2x^4 - 15x^3 - 10x^2 + 12x + 8$

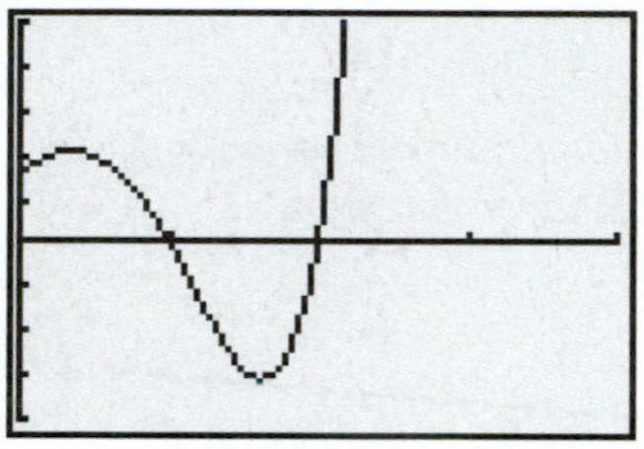

[0, 4, 1] by [−20, 25, 5]

56. $f(x) = -5x^4 + 4x^3 - 19x^2 + 16x + 4$

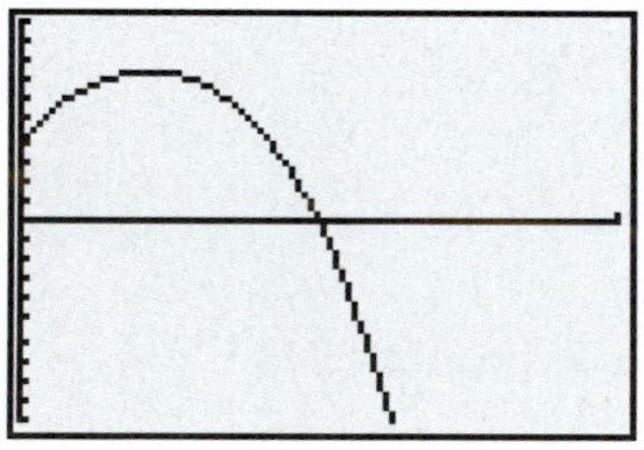

[0, 2, 1] by [−10, 10, 1]

Application Exercises

The graphs are based on a study of the percentage of professional works completed in each age decade of life by 738 people who lived to be at least 79. Use the graphs to solve Exercises 57–58.

Age Trends in Professional Productivity

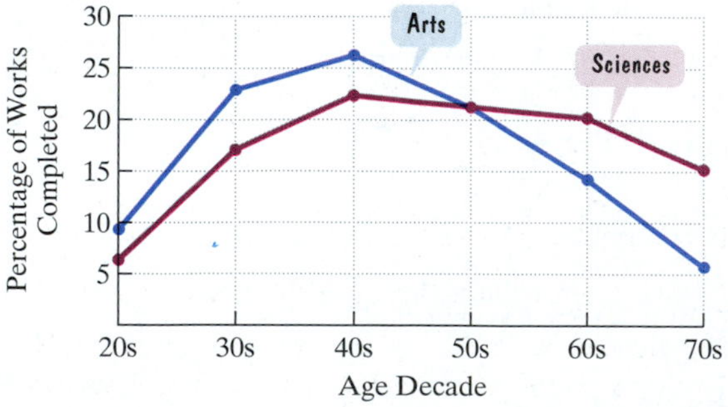

Source: Dennis, W. (1966), Creative productivity between the ages of 28 and 80 years. *Journal of Gerontology*, 21, 1–8.

57. Suppose that a polynomial function f is used to model the data shown in the graph at the bottom of the previous page for the arts using

(age decade, percentage of works completed).

a. Use the graph to solve the polynomial equation $f(x) = 27$. Describe what this means in terms of an age decade and productivity.

b. Describe the degree and the leading coefficient of the function f that can be used to model the data in the graph.

58. Suppose that a polynomial function g is used to model the data shown in the graph at the bottom of the previous page for the sciences using

(age decade, percentage of works completed).

a. Use the graph to solve the polynomial equation $g(x) = 20$. Find only the meaningful value of x and then describe what this means in terms of an age decade and productivity.

b. Describe the degree and the leading coefficient of the function g that can be used to model the data in the graph.

The polynomial function

$$H(x) = -0.001618x^4 + 0.077326x^3 - 1.2367x^2 + 11.460x + 2.914$$

models the age in human years, $H(x)$, of a dog that is x years old, where $x \geq 1$. Although the coefficients make it difficult to solve equations algebraically using this function, a graph of the function makes approximate solutions possible.

Use the graph shown to solve Exercises 59–61. Round all answers to the nearest year.

Dog's Age in Human Years

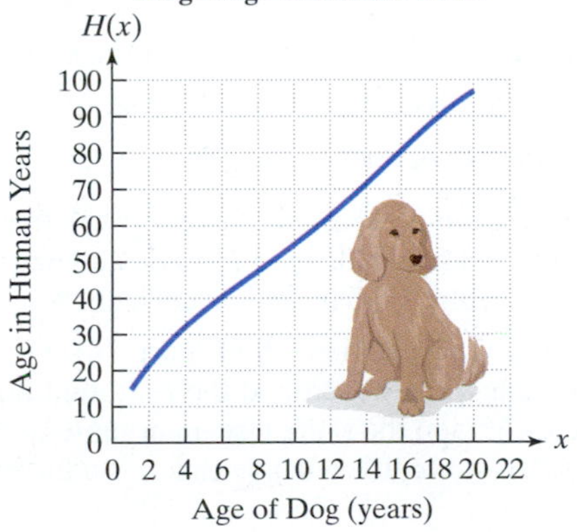

Source: U.C. Davis

59. If you are 25, what is the equivalent age for dogs?

60. If you are 90, what is the equivalent age for dogs?

61. Set up an equation to answer the question in either Exercise 59 or 60. Bring all terms to one side and obtain zero on the other side. What are some of the difficulties involved in solving this equation?

62. A box with an open top is formed by cutting squares out of the corners of a rectangular piece of cardboard 10 inches by 8 inches and then folding up the sides. If x represents the length of the side of the square cut from each corner of the rectangle, what size square must be cut if the volume of the box is to be 48 cubic inches?

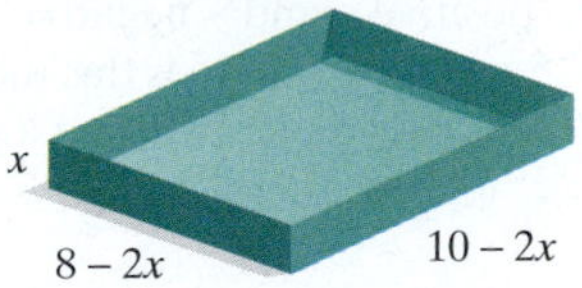

63. The concentration of a drug, in parts per million, in a patient's blood x hours after the drug is administered is given by the function

$$f(x) = -x^4 + 12x^3 - 58x^2 + 132x.$$

How many hours after the drug is administered will it be eliminated from the bloodstream?

Writing in Mathematics

64. How can the Factor Theorem be used to determine if $x - 1$ is a factor of $x^3 - 2x^2 - 11x + 12$?

65. If you know that -2 is a zero of

$$f(x) = x^3 + 7x^2 + 4x - 12,$$

explain how to solve the equation

$$x^3 + 7x^2 + 4x - 12 = 0.$$

66. Describe how to find the possible rational zeros of a polynomial function.

67. Describe how to use Descartes's Rule of signs to determine the possible number of positive real zeros of a polynomial function.

68. Describe how to use Descartes's Rule of Signs to determine the possible number of negative roots of a polynomial equation.

69. Why must every polynomial equation of degree 3 have at least one real root?

70. Explain why the equation $x^4 + 6x^2 + 2 = 0$ has no rational roots.

71. Suppose $\frac{3}{4}$ is a root of a polynomial equation with integer coefficients. What does this tell us about the leading coefficient and the constant term in the equation?

72. Use the graphs for Exercises 57–58 at the bottom of the previous page to describe one similarity and one difference between age trends in professional productivity in the arts and the sciences.

Technology Exercises

The equations in Exercises 73–76 have real roots that are rational. Use the Rational Zero Theorem to list all possible rational roots. Then graph the polynomial function in the given viewing rectangle to determine which possible rational roots are actual roots of the equation.

73. $2x^3 - 15x^2 + 22x + 15 = 0$; $[-1, 6, 1]$ by $[-50, 50, 1]$

74. $6x^3 - 19x^2 + 16x - 4 = 0$; $[0, 2, 1]$ by $[-3, 2, 1]$

75. $2x^4 + 7x^3 - 4x^2 - 27x - 18 = 0$; $[-4, 3, 1]$ by $[-45, 45, 1]$

76. $4x^4 + 4x^3 + 7x^2 - x - 2 = 0$; $[-2, 2, 1]$ by $[-5, 5, 1]$

77. Use Descartes's Rule of Signs to determine the possible numbers of positive and negative real zeros of $f(x) = 3x^4 + 5x^2 + 2$. What does this mean in terms of the graph of f? Verify your result by using a graphing utility to graph f.

78. Use Descartes's Rule of Signs to determine the possible numbers of positive and negative real zeros of $f(x) = x^5 - x^4 + x^3 - x^2 + x - 8$. Verify your result by using a graphing utility to graph f.

79. Write equations for several polynomial functions of odd degree and graph each function. Is it possible for the graph to have no real zeros? Explain. Try doing the same thing for polynomial functions of even degree. Now is it possible to have no real zeros?

Critical Thinking Exercises

Make Sense? *In Exercises 80–83, determine whether each statement "makes sense" or "does not make sense" and explain your reasoning.*

80. By using the quadratic formula, I do not need to bother with synthetic division when solving polynomial equations of degree 3 or higher.

81. I found the zeros of function f, but I still need to find the solutions of the equation $f(x) = 0$.

82. I've noticed that $f(-x)$ is used to explore the number of negative real zeros of a polynomial function, as well as to determine whether a function is even, odd, or neither.

83. The humor in this cartoon is based on the fact that i and π cannot simultaneously be members of a polynomial equation's solution set.

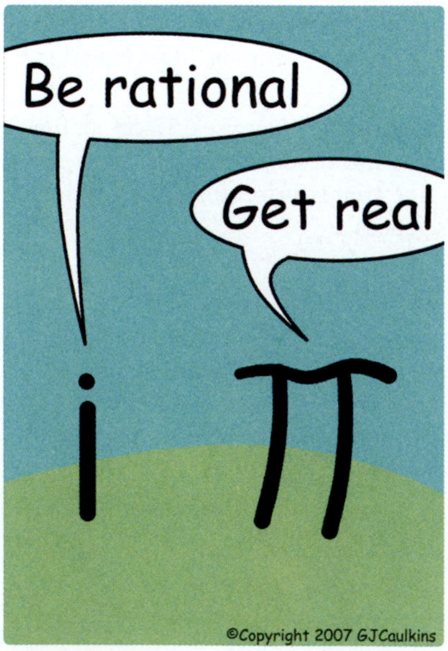

©2007 GJ Caulkins

In Exercises 84–87, determine whether each statement is true or false. If the statement is false, make the necessary change(s) to produce a true statement.

84. The equation $x^3 + 5x^2 + 6x + 1 = 0$ has one positive real root.

85. Descartes's Rule of Signs gives the exact number of positive and negative real roots for a polynomial equation.

86. Every polynomial equation of degree 3 with integer coefficients has at least one rational root.

87. Every polynomial equation of degree n has n distinct solutions.

88. Give an example of a polynomial equation that has no real roots. Describe how you obtained the equation.

89. If the volume of the solid shown in the figure is 208 cubic inches, find the measure of the side designated by x.

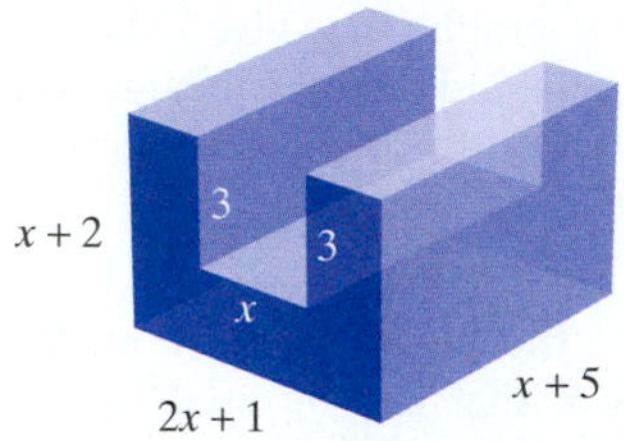

90. In this exercise, we lead you through the steps involved in the proof of the Rational Zero Theorem. Consider the polynomial equation

$$a_n x^n + a_{n-1}x^{n-1} + a_{n-2}x^{n-2} + \cdots + a_1 x + a_0 = 0,$$

and let $\frac{p}{q}$ be a rational root reduced to lowest terms.

a. Substitute $\frac{p}{q}$ for x in the equation and show that the equation can be written as

$$a_n p^n + a_{n-1}p^{n-1}q + a_{n-2}p^{n-2}q^2 + \cdots + a_1 pq^{n-1} = -a_0 q^n.$$

b. Why is p a factor of the left side of the equation?

c. Because p divides the left side, it must also divide the right side. However, because $\frac{p}{q}$ is reduced to lowest terms, p and q have no common factors other than -1 and 1. Because p does divide the right side and has no factors in common with q^n, what can you conclude?

d. Rewrite the equation from part (a) with all terms containing q on the left and the term that does not have a factor of q on the right. Use an argument that parallels parts (b) and (c) to conclude that q is a factor of a_n.

Review Exercises

91. Solve: $x^2 + 2x + 3 > 11$. (Section 8.5, Example 2)

92. Graph: $16x^2 + 9y^2 = 144$. (Section 10.2, Example 2)

93. Factor: $x^3 - 125y^3$. (Section 5.5, Example 9)

Preview Exercises

Exercises 94–96 will help you prepare for the material covered in the next section.

Use the graph of function f to solve each exercise.

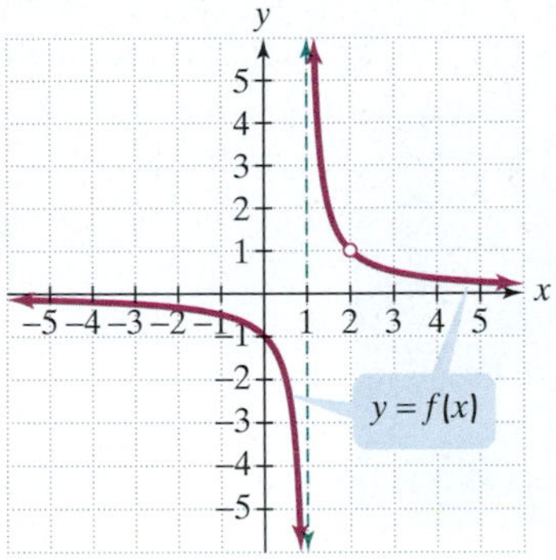

94. For what values of x is the function undefined?

95. Write the equation of the vertical asymptote, or the vertical line that the graph of f approaches but does not touch.

96. Write the equation of the horizontal asymptote, or the horizontal line that the graph of f approaches but does not touch.

MID-CHAPTER CHECK POINT Section 11.1–Section 11.2

What You Know: We learned a number of techniques for finding the zeros of a polynomial function f or, equivalently, finding the roots, or solutions, of the equation $f(x) = 0$. For some functions, the zeros were found by factoring $f(x)$. For other functions, we listed possible rational zeros and used synthetic division and the Factor Theorem to determine the zeros. We saw that graphs cross the x-axis at zeros of odd multiplicity and touch the x-axis and turn around at zeros of even multiplicity. We learned to graph polynomial functions using zeros, the Leading Coefficient Test, intercepts, and symmetry:

$$y\text{-axis symmetry: } f(-x) = f(x);$$
$$\text{origin symmetry: } f(-x) = -f(x).$$

We checked graphs using the fact that a polynomial function of degree n has a graph with at most $n - 1$ turning points.

In Exercises 1–9, find all zeros of each polynomial function. Then graph the function.

1. $f(x) = (x - 2)^2(x + 1)^3$

2. $f(x) = -(x - 2)^2(x + 1)^2$

3. $f(x) = x^3 - x^2 - 4x + 4$

4. $f(x) = x^4 - 5x^2 + 4$

5. $f(x) = -(x + 1)^6$

6. $f(x) = -6x^3 + 7x^2 - 1$

7. $f(x) = 2x^3 - 2x$

8. $f(x) = x^3 - 2x^2 + 26x$

9. $f(x) = -x^3 + 5x^2 - 5x - 3$

In Exercises 10–15, solve each polynomial equation.

10. $x^3 - 3x + 2 = 0$

11. $6x^3 - 11x^2 + 6x - 1 = 0$

12. $(2x + 1)(3x - 2)^3(2x - 7) = 0$

13. $2x^3 + 5x^2 - 200x - 500 = 0$

14. $x^4 - x^3 - 11x^2 = x + 12$

15. $2x^4 + x^3 - 17x^2 - 4x + 6 = 0$

SECTION 11.3

Rational Functions and Their Graphs

Objectives

1. Identify vertical asymptotes.
2. Identify horizontal asymptotes.
3. Graph rational functions.
4. Solve applied problems involving rational functions.

How much did you spend on textbooks this year? Much of what you learn in college takes place directly through your interaction with these books. Their pages ultimately reflect the knowledge you acquire in academia. However, there is still that ghastly moment when your campus store tallies the total cost for the semester's required texts. How does this amount compare with the average amount per college student and what can college students expect in the future? In this section, you will see how rational functions model the cost of these important learning tools.

Rational Functions and the Reciprocal Function

Rational functions were first introduced in Chapter 6. Recall that **rational functions** are quotients of polynomial functions. This means that rational functions can be expressed as

$$f(x) = \frac{p(x)}{q(x)},$$

where p and q are polynomial functions and $q(x) \neq 0$. The **domain** of a rational function is the set of all real numbers except the x-values that make the denominator zero. For example, the domain of the rational function

$$f(x) = \frac{x^2 + 7x + 9}{x(x-2)(x+5)}$$

This is $p(x)$.

This is $q(x)$.

is the set of all real numbers except 0, 2, and −5.

The most basic rational function is the **reciprocal function**, defined by $f(x) = \frac{1}{x}$. The denominator of the reciprocal function is zero only when $x = 0$, so the domain of f is the set of all real numbers except 0.

Let's look at the behavior of f near the excluded value 0. We start by evaluating $f(x)$ to the left of 0—that is, at values of x that are less than 0.

x approaches 0 from the left.

x	−1	−0.5	−0.1	−0.01	−0.001
$f(x) = \frac{1}{x}$	−1	−2	−10	−100	−1000

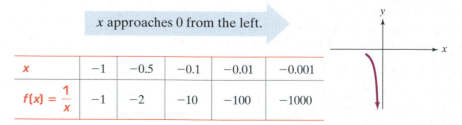

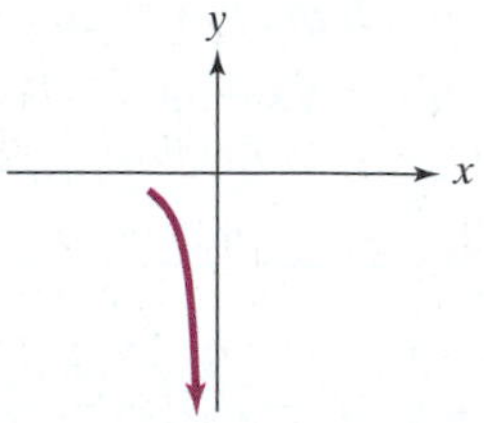

Mathematically, we say that "x approaches 0 from the left." From the table and the accompanying graph, it appears that as x approaches 0 from the left, the function values, $f(x)$, decrease without bound.

Next, we evaluate $f(x)$ to the right of 0—that is, at values of x that are greater than 0.

x approaches 0 from the right.

x	0.001	0.01	0.1	0.5	1
$f(x) = \frac{1}{x}$	1000	100	10	2	1

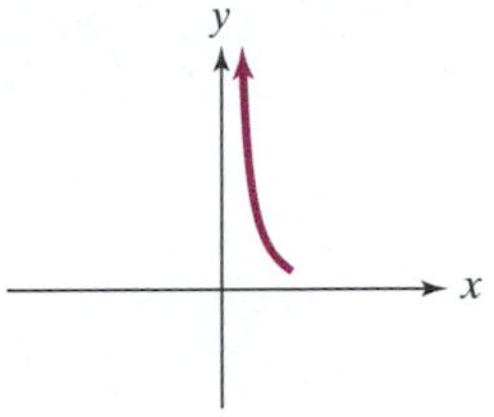

Mathematically, we say that "x approaches 0 from the right." From the table and the accompanying graph, it appears that as x approaches 0 from the right, the function values, $f(x)$, increase without bound.

Now let's see what happens to the function values, $f(x)$, as x gets farther away from the origin. The following tables suggest what happens to $f(x)$ as x increases or decreases without bound.

x increases without bound:

x	1	10	100	1000
$f(x) = \frac{1}{x}$	1	0.1	0.01	0.001

x decreases without bound:

x	-1	-10	-100	-1000
$f(x) = \frac{1}{x}$	-1	-0.1	-0.01	-0.001

It appears that as x increases or decreases without bound, the function values, $f(x)$, are getting progressively closer to 0.

Figure 11.11 illustrates the end behavior of $f(x) = \frac{1}{x}$ as x increases or decreases without bound. The graph shows that the function values, $f(x)$, are approaching 0. This means that as x increases or decreases without bound, the graph of f is approaching the horizontal line $y = 0$ (that is, the x-axis).

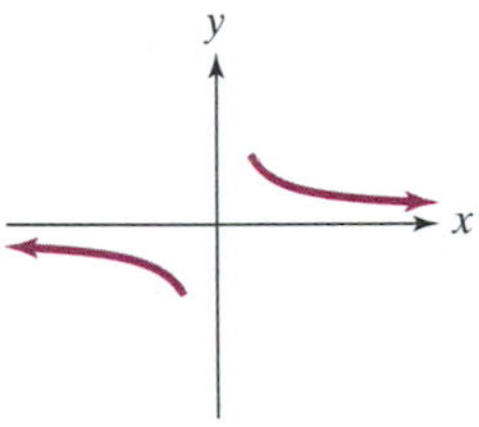

FIGURE 11.11 $f(x)$ approaches 0 as x increases or decreases without bound.

The graph of the reciprocal function $f(x) = \frac{1}{x}$ is shown in **Figure 11.12**. Unlike the graph of a polynomial function, the graph of the reciprocal function has a break and is composed of two distinct branches.

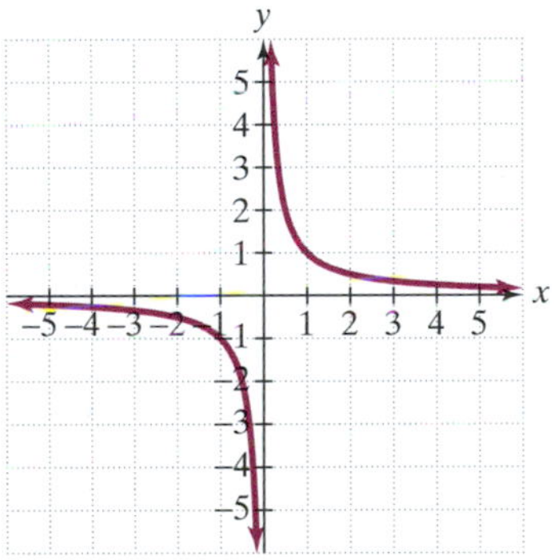

FIGURE 11.12 The graph of the reciprocal function $f(x) = \frac{1}{x}$

Study Tip

If x is far from 0, then $\frac{1}{x}$ is close to 0. By contrast, if x is close to 0, then $\frac{1}{x}$ is far from 0.

Blitzer Bonus

The Reciprocal Function as a Knuckle Tattoo

"I got the tattoo because I like the idea of math not being well behaved. That sounds lame and I really don't mean that in some kind of anarchy-type way. I just think that it's kind of nice that something as perfectly functional as math can kink up around the edges."

Kink up around the edges? We're about to describe the graphic behavior of the reciprocal function using *asymptotes* rather than kink. Asymptotes are lines that graphs approach but never touch. *Asymptote* comes from the Greek word *asymptotos*, meaning "not meeting."

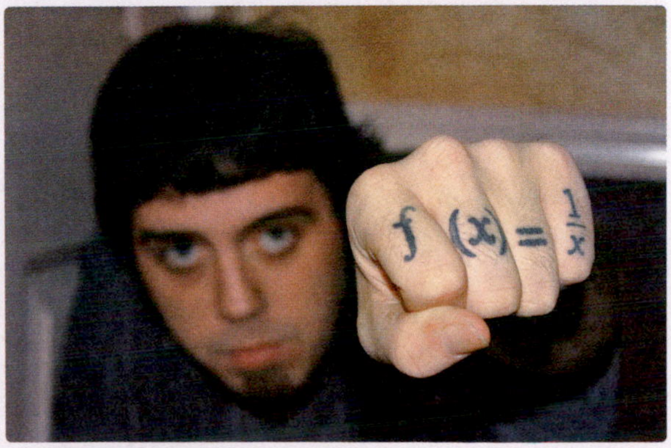

1 Identify vertical asymptotes.

Vertical Asymptotes of Rational Functions

Look again at the graph of $f(x) = \frac{1}{x}$ in **Figure 11.12** on the previous page. The curve approaches, but does not touch, the y-axis. The y-axis, or $x = 0$, is said to be a *vertical asymptote* of the graph. A rational function may have no vertical asymptotes, one vertical asymptote, or several vertical asymptotes. The graph of a rational function never intersects a vertical asymptote. We will use dashed lines to show asymptotes.

Definition of a Vertical Asymptote

The line $x = a$ is a **vertical asymptote** of the graph of a function f if $f(x)$ increases or decreases without bound as x approaches a from either the right or the left.

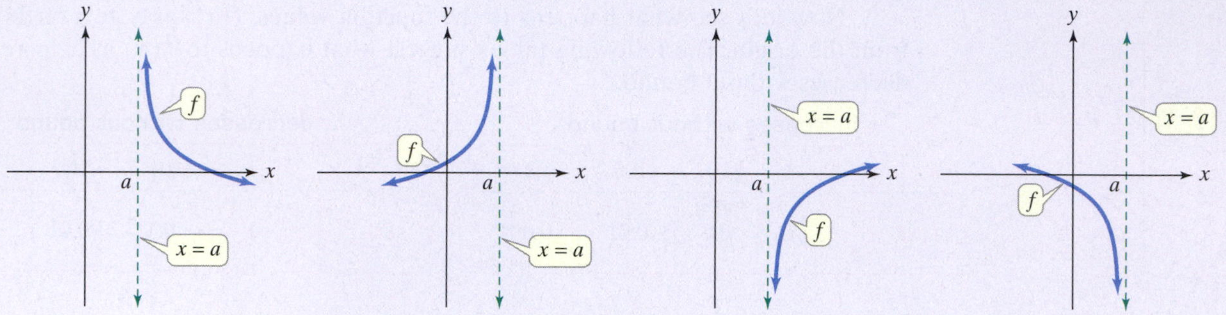

If the graph of a rational function has one or more vertical asymptotes, they can be located using the following theorem:

Locating Vertical Asymptotes

If $f(x) = \frac{p(x)}{q(x)}$ is a rational function in which $p(x)$ and $q(x)$ have no common factors with degree 1 or greater and a is a zero of $q(x)$, the denominator, then $x = a$ is a vertical asymptote of the graph of f.

EXAMPLE 1 Finding the Vertical Asymptotes of a Rational Function

Find the vertical asymptotes, if any, of the graph of each rational function:

a. $f(x) = \frac{x}{x^2 - 9}$ **b.** $g(x) = \frac{x + 3}{x^2 - 9}$ **c.** $h(x) = \frac{x + 3}{x^2 + 9}$.

Solution Factoring is usually helpful in identifying zeros of denominators.

a. $$f(x) = \frac{x}{x^2 - 9} = \frac{x}{(x + 3)(x - 3)}$$

This factor is 0 if $x = -3$.

This factor is 0 if $x = 3$.

There are no common factors in the numerator and the denominator. The zeros of the denominator are -3 and 3. Thus, the lines $x = -3$ and $x = 3$ are the vertical asymptotes for the graph of f. [See **Figure 11.13(a)**.]

b. We will use factoring to see if there are any common factors with degree 1 or greater.

$$g(x) = \frac{x + 3}{x^2 - 9} = \frac{(x + 3)}{(x + 3)(x - 3)} = \frac{1}{x - 3}$$

There is a common factor, $x + 3$, so simplify.

This denominator is 0 if $x = 3$.

The only zero of the denominator of $g(x)$ in simplified form is 3. Thus, the line $x = 3$ is the only vertical asymptote of the graph of g. [See **Figure 11.13(b)**.]

c. We cannot factor the denominator of $h(x)$ over the real numbers.

$$h(x) = \frac{x+3}{x^2+9}$$

No real numbers make this denominator 0.

The denominator has no real zeros. Thus, the graph of h has no vertical asymptotes. [See **Figure 11.13(c)**.]

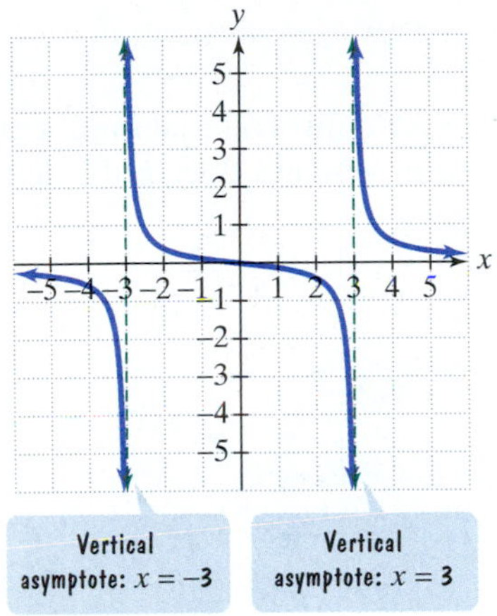
Vertical asymptote: $x = -3$
Vertical asymptote: $x = 3$

FIGURE 11.13(a) The graph of $f(x) = \frac{x}{x^2 - 9}$ has two vertical asymptotes.

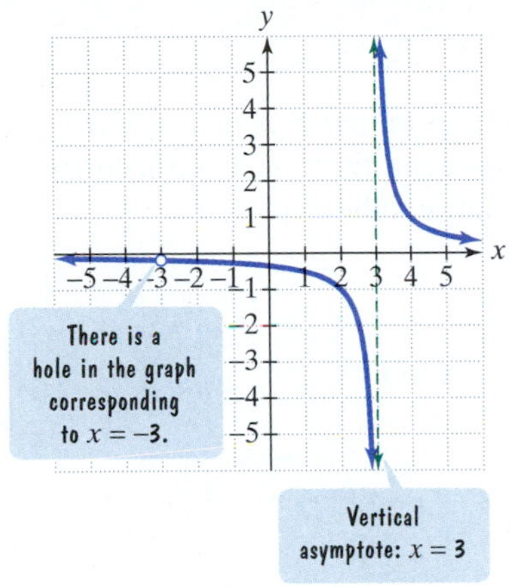
There is a hole in the graph corresponding to $x = -3$.
Vertical asymptote: $x = 3$

FIGURE 11.13(b) The graph of $g(x) = \frac{x+3}{x^2 - 9}$ has one vertical asymptote.

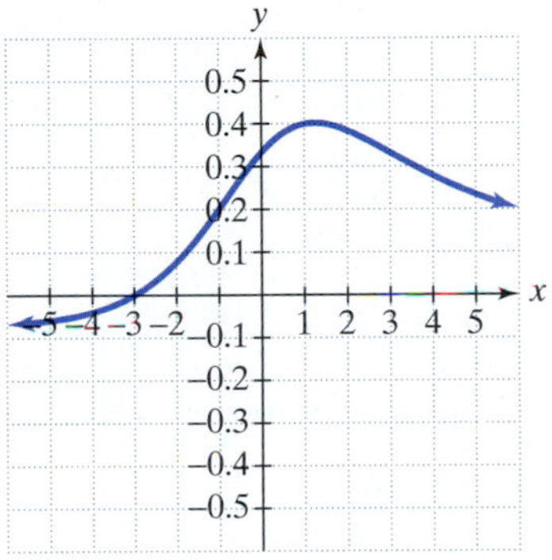

FIGURE 11.13(c) The graph of $h(x) = \frac{x+3}{x^2 + 9}$ has no vertical asymptotes.

CHECK POINT 1 Find the vertical asymptotes, if any, of the graph of each rational function:

a. $f(x) = \frac{x}{x^2 - 1}$ **b.** $g(x) = \frac{x - 1}{x^2 - 1}$ **c.** $h(x) = \frac{x - 1}{x^2 + 1}$.

Using Technology

The graph of the rational function $f(x) = \frac{x}{x^2 - 9}$, drawn by hand in **Figure 11.13(a)**, is graphed below in a $[-5, 5, 1]$ by $[-4, 4, 1]$ viewing rectangle. The graph is shown in connected mode and in dot mode. In connected mode, the graphing utility plots many points and connects the points with curves. In dot mode, the utility plots the same points, but does not connect them.

Connected Mode

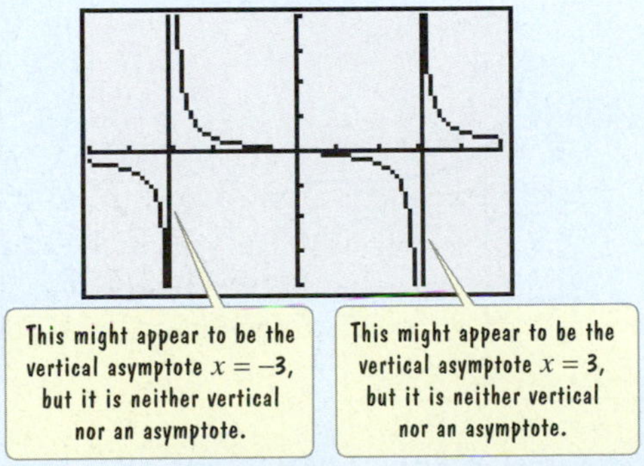
This might appear to be the vertical asymptote $x = -3$, but it is neither vertical nor an asymptote.
This might appear to be the vertical asymptote $x = 3$, but it is neither vertical nor an asymptote.

Dot Mode

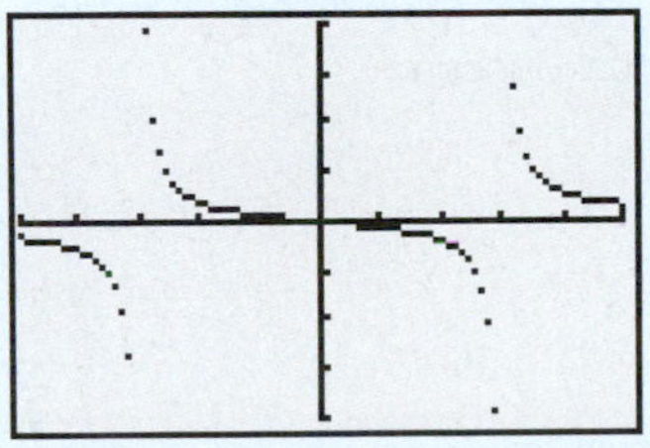

The steep lines in connected mode that are "almost" the vertical asymptotes $x = -3$ and $x = 3$ are not part of the graph and do not represent the vertical asymptotes. The graphing utility has incorrectly connected the last point to the left of $x = -3$ with the first point to the right of $x = -3$. It has also incorrectly connected the last point to the left of $x = 3$ with the first point to the right of $x = 3$. The effect is to create two near-vertical segments that look like asymptotes. This erroneous effect does not appear using dot mode.

A value where the denominator of a function is zero does not necessarily result in a vertical asymptote. There is a hole corresponding to $x = a$, and not a vertical asymptote, in the graph of a function under the following conditions: The value a causes the denominator to be zero, but there is a reduced form of the function's equation in which a does not cause the denominator to be zero.

Consider, for example, the function

$$f(x) = \frac{x^2 - 9}{x - 3}.$$

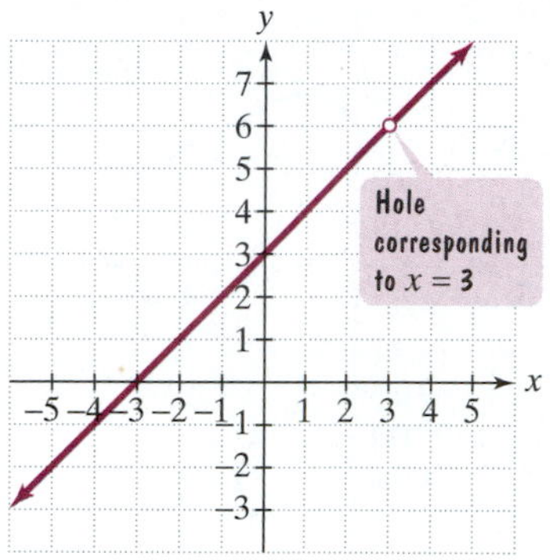

FIGURE 11.14

Because the denominator is zero when $x = 3$, the function's domain is all real numbers except 3. However, there is a reduced form of the equation in which 3 does not cause the denominator to be zero:

$$f(x) = \frac{x^2 - 9}{x - 3} = \frac{(x + 3)\cancel{(x - 3)}}{\cancel{x - 3}} = x + 3, \; x \neq 3.$$

Denominator is zero at $x = 3$.

In this reduced form, 3 does not result in a zero denominator.

Figure 11.14 shows that the graph has a hole corresponding to $x = 3$. Graphing utilities do not show this feature of the graph.

2 Identify horizontal asymptotes.

Horizontal Asymptotes of Rational Functions

Figure 11.12, repeated, shows the graph of the reciprocal function $f(x) = \frac{1}{x}$. As x increases or decreases without bound, the function values are approaching 0. The line $y = 0$ (that is, the x-axis) is a *horizontal asymptote* of the graph. Many, but not all, rational functions have horizontal asymptotes.

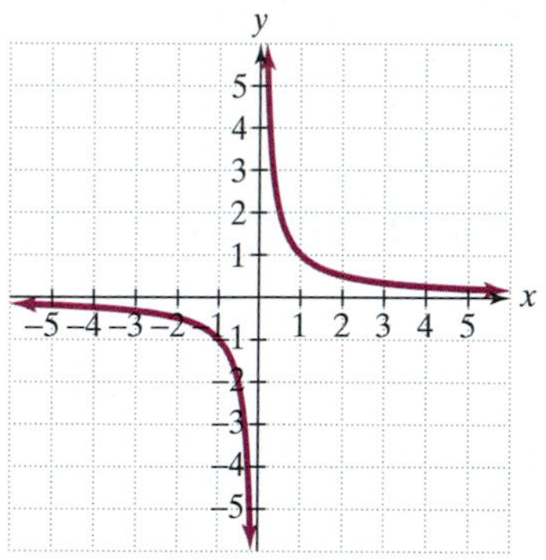

FIGURE 11.12 (repeated) The graph of the reciprocal function $f(x) = \frac{1}{x}$

Definition of a Horizontal Asymptote

The line $y = b$ is a **horizontal asymptote** of the graph of a function f if $f(x)$ approaches b as x increases or decreases without bound.

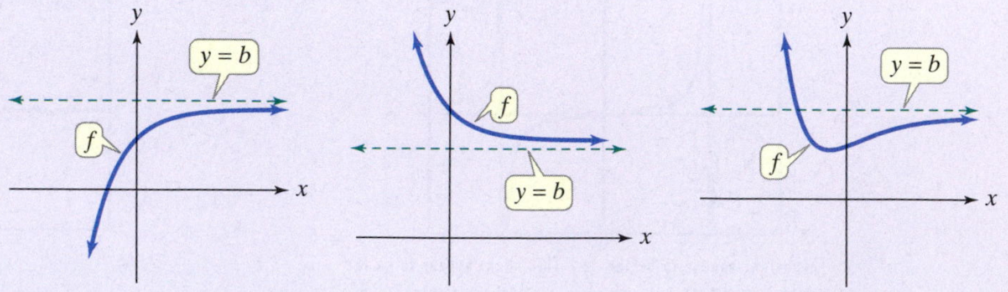

Recall that a rational function may have several vertical asymptotes. By contrast, it can have at most one horizontal asymptote. Although a graph can never intersect a vertical asymptote, it may cross its horizontal asymptote.

If the graph of a rational function has a horizontal asymptote, it can be located by using the following theorem:

Locating Horizontal Asymptotes

Let f be the rational function given by

$$f(x) = \frac{a_n x^n + a_{n-1}x^{n-1} + \cdots + a_1 x + a_0}{b_m x^m + b_{m-1}x^{m-1} + \cdots + b_1 x + b_0}, \quad a_n \neq 0, b_m \neq 0.$$

The degree of the numerator is n. The degree of the denominator is m.

1. If $n < m$, the x-axis, or $y = 0$, is the horizontal asymptote of the graph of f.
2. If $n = m$, the line $y = \frac{a_n}{b_m}$ is the horizontal asymptote of the graph of f.
3. If $n > m$, the graph of f has no horizontal asymptote.

EXAMPLE 2 Finding the Horizontal Asymptote of a Rational Function

Find the horizontal asymptote, if any, of the graph of each rational function:

a. $f(x) = \dfrac{4x}{2x^2 + 1}$ **b.** $g(x) = \dfrac{4x^2}{2x^2 + 1}$ **c.** $h(x) = \dfrac{4x^3}{2x^2 + 1}$.

Solution

a. $f(x) = \dfrac{4x}{2x^2 + 1}$

The degree of the numerator, 1, is less than the degree of the denominator, 2. Thus, the graph of f has the x-axis as a horizontal asymptote [see **Figure 11.15(a)**]. The equation of the horizontal asymptote is $y = 0$.

b. $g(x) = \dfrac{4x^2}{2x^2 + 1}$

The degree of the numerator, 2, is equal to the degree of the denominator, 2. The leading coefficients of the numerator and denominator, 4 and 2, are used to obtain the equation of the horizontal asymptote. The equation of the horizontal asymptote is $y = \frac{4}{2}$ or $y = 2$ [see **Figure 11.15(b)**].

c. $h(x) = \dfrac{4x^3}{2x^2 + 1}$

The degree of the numerator, 3, is greater than the degree of the denominator, 2. Thus, the graph of h has no horizontal asymptote [see **Figure 11.15(c)**].

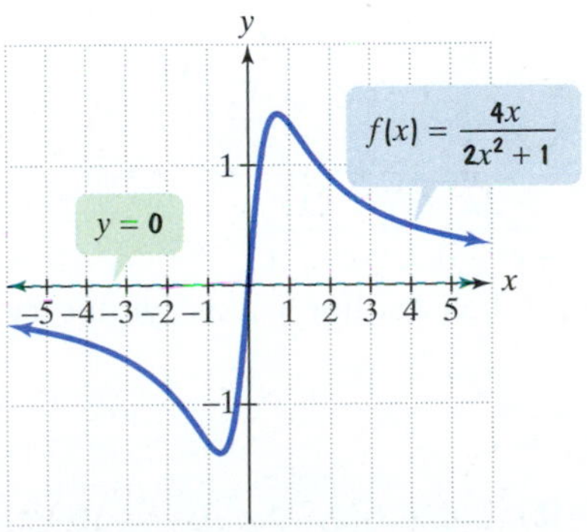

FIGURE 11.15(a) The horizontal asymptote of the graph is $y = 0$.

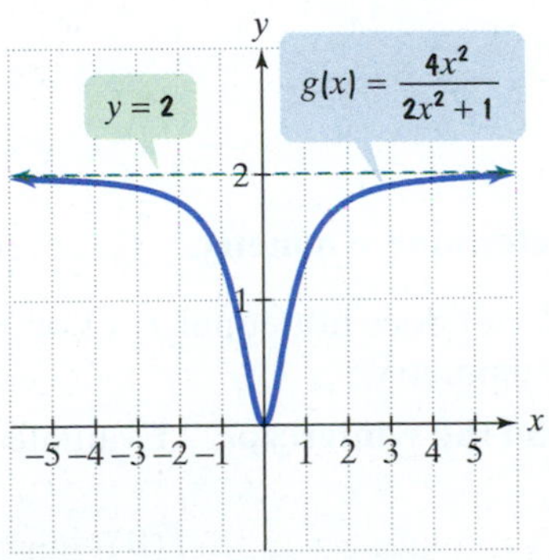

FIGURE 11.15(b) The horizontal asymptote of the graph is $y = 2$.

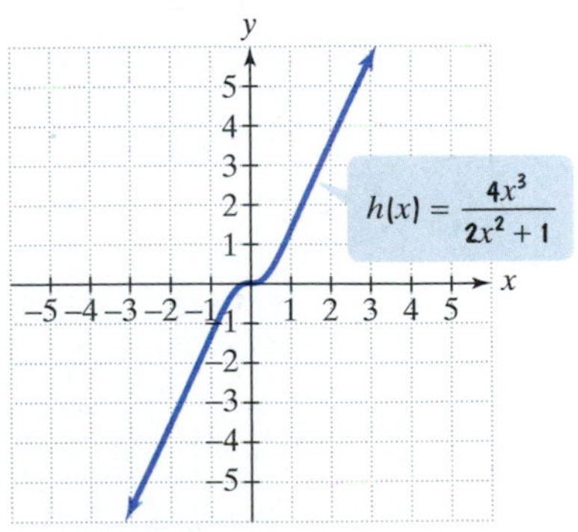

FIGURE 11.15(c) The graph has no horizontal asymptote.

✓ CHECK POINT 2 Find the horizontal asymptote, if any, of the graph of each rational function:

a. $f(x) = \dfrac{9x^2}{3x^2 + 1}$

b. $g(x) = \dfrac{9x}{3x^2 + 1}$

c. $h(x) = \dfrac{9x^3}{3x^2 + 1}$.

---

3 Graph rational functions.

Graphing Rational Functions

Here are some suggestions for graphing rational functions:

Strategy for Graphing a Rational Function

The following strategy can be used to graph

$$f(x) = \frac{p(x)}{q(x)},$$

where p and q are polynomial functions with no common factors with degree 1 or greater.

1. Determine whether the graph of f has symmetry.

   $f(-x) = f(x)$: y-axis symmetry

   $f(-x) = -f(x)$: origin symmetry

2. Find the y-intercept (if there is one) by evaluating $f(0)$.
3. Find the x-intercepts (if there are any) by solving the equation $p(x) = 0$.
4. Find any vertical asymptote(s) by solving the equation $q(x) = 0$.
5. Find the horizontal asymptote (if there is one) using the rules for determining the horizontal asymptote of a rational function.
6. Plot at least one point between and beyond each x-intercept and vertical asymptote.
7. Use the information obtained previously to graph the function between and beyond the vertical asymptotes.

EXAMPLE 3 Graphing a Rational Function

Graph: $f(x) = \dfrac{2x}{x - 1}$.

Solution

Step 1. Determine symmetry. $f(-x) = \dfrac{2(-x)}{-x - 1} = \dfrac{-2x}{-x - 1} = \dfrac{2x}{x + 1}$

Because $f(-x)$ does not equal $f(x)$ or $-f(x)$, the graph has neither y-axis symmetry nor origin symmetry.

Step 2. Find the y-intercept. Evaluate $f(0)$.

$$f(0) = \frac{2 \cdot 0}{0 - 1} = \frac{0}{-1} = 0$$

The y-intercept is 0, so the graph passes through the origin.

Step 3. Find the x-intercept(s). This is done by solving $p(x) = 0$.

$$2x = 0 \quad \text{Set the numerator equal to 0.}$$

$$x = 0 \quad \text{Divide both sides by 2.}$$

There is only one x-intercept. This verifies that the graph passes through the origin.

Step 4. Find the vertical asymptote(s). Solve $q(x) = 0$, thereby finding zeros of the denominator.

$$x - 1 = 0 \quad \text{Set the denominator equal to 0.}$$

$$x = 1 \quad \text{Add 1 to both sides.}$$

The equation of the vertical asymptote is $x = 1$.

Step 5. Find the horizontal asymptote. Because the numerator and denominator of $f(x) = \frac{2x}{x-1}$ have the same degree, 1, the leading coefficients of the numerator and denominator, 2 and 1, respectively, are used to obtain the equation of the horizontal asymptote. The equation is

$$y = \frac{2}{1} = 2.$$

The equation of the horizontal asymptote is $y = 2$.

Step 6. Plot points between and beyond the x-intercept and vertical asymptote. With an x-intercept at 0 and a vertical asymptote at $x = 1$, we evaluate the function at $-2, -1, \frac{1}{2}, 2$, and 4.

x	-2	-1	$\frac{1}{2}$	2	4
$f(x) = \frac{2x}{x-1}$	$\frac{4}{3}$	1	-2	4	$\frac{8}{3}$

Figure 11.16 shows these points, the y-intercept, the x-intercept, and the asymptotes.

Step 7. Graph the function. The graph of $f(x) = \frac{2x}{x-1}$ is shown in **Figure 11.17**.

Using Technology

The graph of $y = \frac{2x}{x-1}$, obtained using the dot mode in a $[-6, 6, 1]$ by $[-6, 6, 1]$ viewing rectangle, verifies that our hand-drawn graph in **Figure 11.17** is correct.

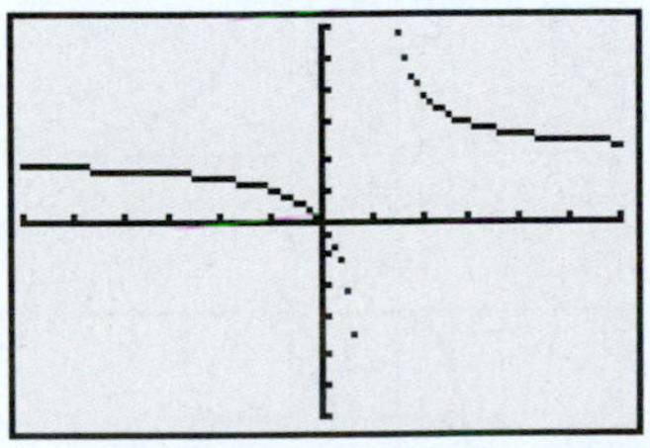

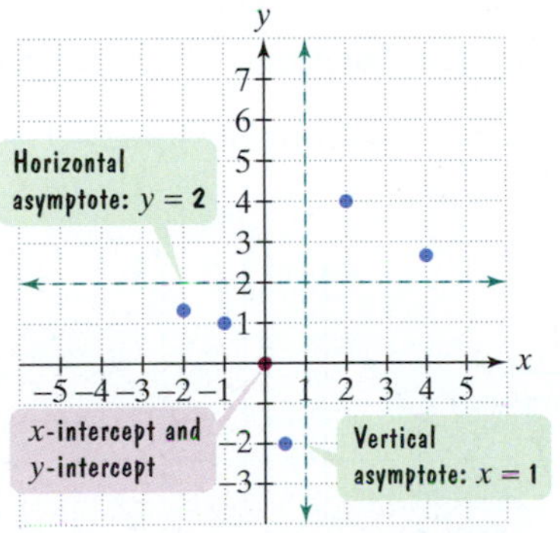

FIGURE 11.16 Preparing to graph the rational function $f(x) = \frac{2x}{x-1}$

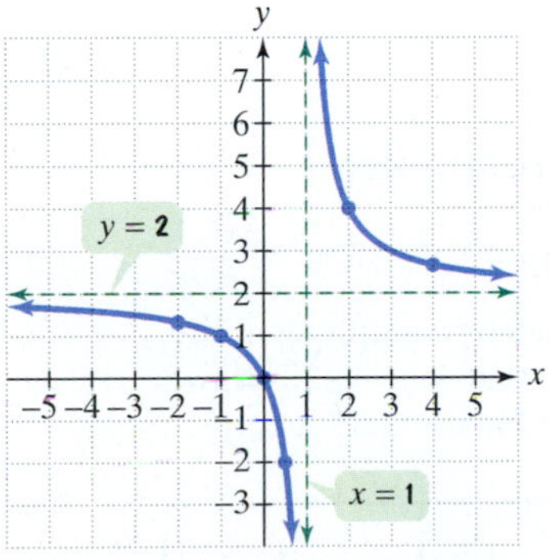

FIGURE 11.17 The graph of $f(x) = \frac{2x}{x-1}$

CHECK POINT 3 Graph: $f(x) = \frac{3x}{x-2}$.

EXAMPLE 4 Graphing a Rational Function

Graph: $f(x) = \dfrac{3x^2}{x^2 - 4}$.

Solution

Step 1. Determine symmetry. $f(-x) = \dfrac{3(-x)^2}{(-x)^2 - 4} = \dfrac{3x^2}{x^2 - 4} = f(x)$

The graph of f is symmetric with respect to the y-axis.

Step 2. Find the y-intercept. $f(0) = \dfrac{3 \cdot 0^2}{0^2 - 4} = \dfrac{0}{-4} = 0$: The y-intercept is 0, so the graph passes through the origin.

Step 3. Find the x-intercept(s). $3x^2 = 0$, so $x = 0$: The x-intercept is 0, verifying that the graph passes through the origin.

Step 4. Find the vertical asymptote(s). Set $q(x) = 0$.

$x^2 - 4 = 0$ Set the denominator equal to 0.

$x^2 = 4$ Add 4 to both sides.

$x = \pm 2$ Use the square root property.

The vertical asymptotes are $x = -2$ and $x = 2$.

Step 5. Find the horizontal asymptote. Because the numerator and denominator of $f(x) = \dfrac{3x^2}{x^2 - 4}$ have the same degree, 2, their leading coefficients, 3 and 1, are used to determine the equation of the horizontal asymptote. The equation is $y = \frac{3}{1} = 3$.

Step 6. Plot points between and beyond the x-intercept and the vertical asymptotes. With an x-intercept at 0 and vertical asymptotes at $x = -2$ and $x = 2$, we evaluate the function at $-3, -1, 1, 3$, and 4.

x	-3	-1	1	3	4
$f(x) = \dfrac{3x^2}{x^2 - 4}$	$\frac{27}{5}$	-1	-1	$\frac{27}{5}$	4

Figure 11.18 shows these points, the y-intercept, the x-intercept, and the asymptotes.

Step 7. Graph the function. The graph of $f(x) = \dfrac{3x^2}{x^2 - 4}$ is shown in **Figure 11.19**. The y-axis symmetry is now obvious.

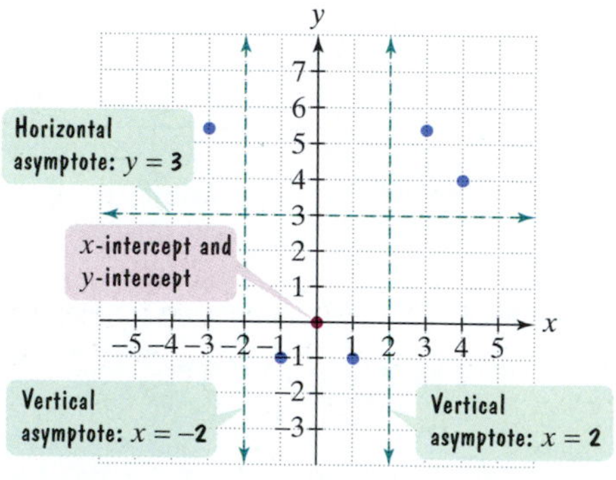

FIGURE 11.18 Preparing to graph $f(x) = \dfrac{3x^2}{x^2 - 4}$

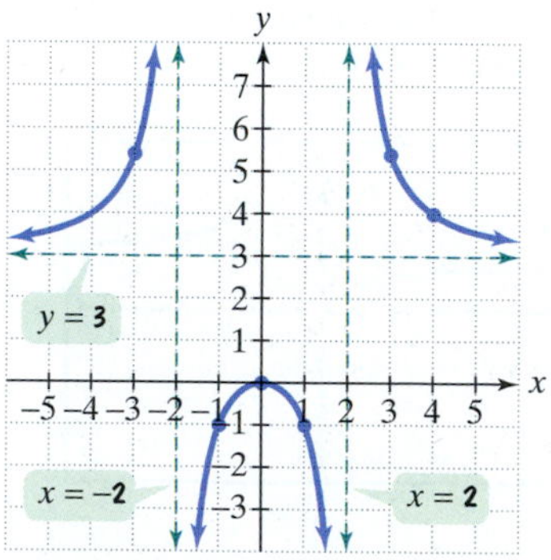

FIGURE 11.19 The graph of $f(x) = \dfrac{3x^2}{x^2 - 4}$

Study Tip

Because the graph has y-axis symmetry, it is not necessary to evaluate the even function at -3 and again at 3.

$$f(-3) = f(3) = \frac{27}{5}$$

This also applies to evaluation at -1 and 1.

Using Technology

The graph of $y = \dfrac{3x^2}{x^2 - 4}$, generated by a graphing utility, verifies that our hand-drawn graph in **Figure 11.19** is correct.

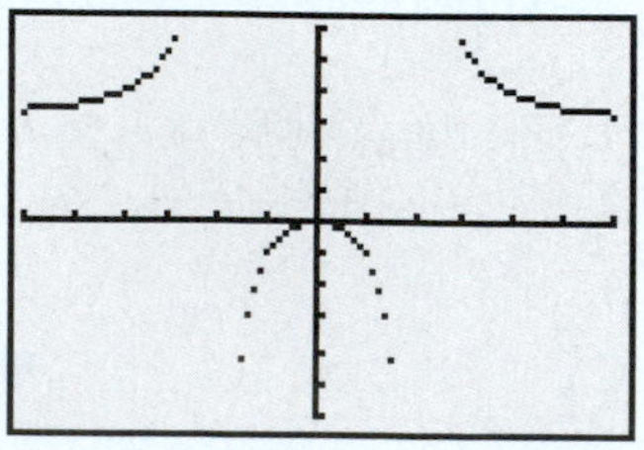

[−6, 6, 1] by [−6, 6, 1]

✓ CHECK POINT 4 Graph: $f(x) = \dfrac{2x^2}{x^2 - 9}$.

Example 5 illustrates that not every rational function has vertical and horizontal asymptotes.

EXAMPLE 5 Graphing a Rational Function

Graph: $f(x) = \dfrac{x^4}{x^2 + 1}$.

Solution

Step 1. Determine symmetry. $f(-x) = \dfrac{(-x)^4}{(-x)^2 + 1} = \dfrac{x^4}{x^2 + 1} = f(x)$

The graph of f is symmetric with respect to the y-axis.

Step 2. Find the y-intercept. $f(0) = \dfrac{0^4}{0^2 + 1} = \dfrac{0}{1} = 0$: The y-intercept is 0.

Step 3. Find the x-intercept(s). $x^4 = 0$, so $x = 0$: The x-intercept is 0.

Step 4. Find the vertical asymptote(s). Set $q(x) = 0$.

$$x^2 + 1 = 0 \quad \text{Set the denominator equal to 0.}$$

$$x^2 = -1 \quad \text{Subtract 1 from both sides.}$$

Although this equation has imaginary roots ($x = \pm i$), there are no real roots. Thus, the graph of f has no vertical asymptotes.

Step 5. Find the horizontal asymptote. Because the degree of the numerator, 4, is greater than the degree of the denominator, 2, there is no horizontal asymptote.

Step 6. Plot points between and beyond the x-intercept and the vertical asymptotes. With an x-intercept at 0 and no vertical asymptotes, let's look at function values at $-2, -1, 1,$ and 2. You can evaluate the function at 1 and 2. Use y-axis symmetry to obtain function values at -1 and -2:

$$f(-1) = f(1) \quad \text{and} \quad f(-2) = f(2).$$

x	-2	-1	1	2
$f(x) = \dfrac{x^4}{x^2+1}$	$\dfrac{16}{5}$	$\dfrac{1}{2}$	$\dfrac{1}{2}$	$\dfrac{16}{5}$

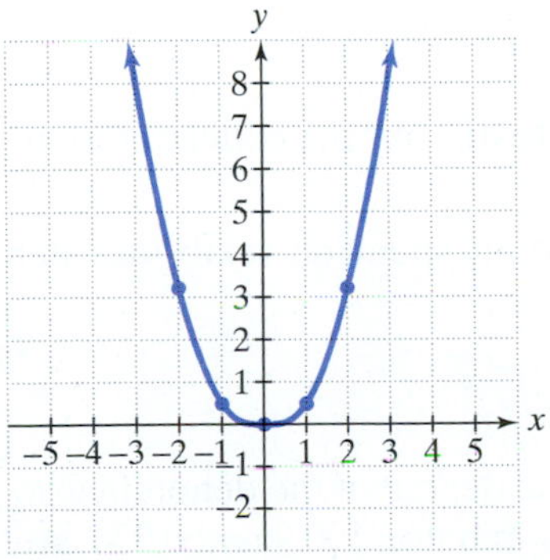

FIGURE 11.20 The graph of $f(x) = \dfrac{x^4}{x^2 + 1}$

Step 7. Graph the function. **Figure 11.20** shows the graph of f using the points obtained from the table and y-axis symmetry. Notice that as x increases or decreases without bound, the function values, $f(x)$, are getting larger without bound.

✓ CHECK POINT 5 Graph: $f(x) = \dfrac{x^4}{x^2 + 2}$.

4 Solve applied problems involving rational functions.

Applications

There are numerous examples of asymptotic behavior in rational functions that describe real-world phenomena.

EXAMPLE 6 Average Amount Spent on Textbooks

Textbook sales at college stores have increased during the past two decades. The function

$$B(x) = 190.9x + 2413.99$$

models textbook sales, $B(x)$, in millions of dollars, x years after 1985. College enrollment has also increased. The function

$$E(x) = 0.234x + 12.54$$

models total college enrollment, $E(x)$, in millions, x years after 1985.

a. Write a function that models the average amount of money spent on textbooks per college student, $M(x)$, in dollars per student, x years after 1985.

b. Predict the average amount of money that will be spent on textbooks per college student in 2010.

c. What is the horizontal asymptote for the function that models the average amount of money spent on textbooks per college student? Describe what this represents in practical terms.

Solution

a. To write a function that models the average amount of money spent on textbooks per college student, divide textbook sales, $B(x)$, in millions of dollars, by college enrollment, $E(x)$, in millions. The rational function

$$M(x) = \frac{190.9x + 2413.99}{0.234x + 12.54}$$

This is $B(x)$.

This is $E(x)$.

models the average amount of money spent on textbooks per college student, $M(x)$, in dollars per student, x years after 1985.

b. We are interested in the average amount that will be spent on textbooks in 2010. Because 2010 is 25 years after 1985, evaluate the function in part (a) at $x = 25$.

$$M(25) = \frac{190.9(25) + 2413.99}{0.234(25) + 12.54} \approx 391$$ Use a calculator.

The average amount of money that will be spent on textbooks per college student in 2010 is approximately \$391.

c. We use the function that models the average amount spent on textbooks per student

$$M(x) = \frac{190.9x + 2413.99}{0.234x + 12.54}$$

in which the degree of the numerator, 1, is equal to the degree of the denominator, 1. The leading coefficients of the numerator and denominator, 190.9 and 0.234, are used to obtain the equation of the horizontal asymptote. The equation of the horizontal asymptote is

$$y = \frac{190.9}{0.234} \quad \text{or} \quad y \approx 816.$$

This means that over an increasing period of time, the average amount of money spent on textbooks will be approaching approximately \$816 per student per year. The horizontal asymptote is illustrated in **Figure 11.21** at the top of the next page.

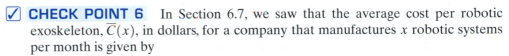

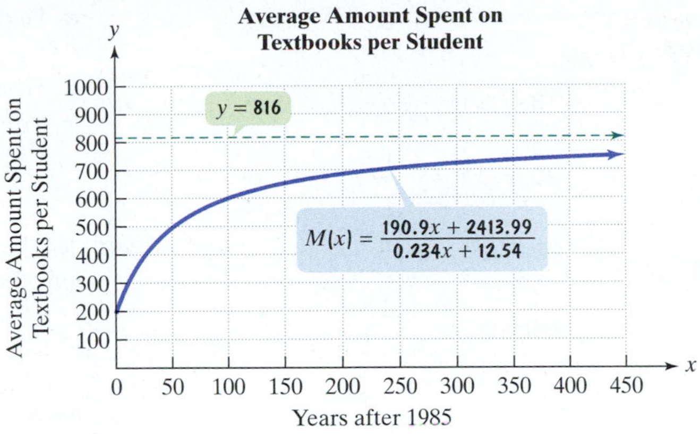

FIGURE 11.21 The average amount of money spent on textbooks is approaching \$816 per student per year.

CHECK POINT 6 In Section 6.7, we saw that the average cost per robotic exoskeleton, $\overline{C}(x)$, in dollars, for a company that manufactures x robotic systems per month is given by

$$\overline{C}(x) = \frac{5000x + 1{,}000{,}000}{x}.$$

What is the horizontal asymptote for this average cost function? Describe what this represents for the company.

11.3 EXERCISE SET

MyMathLab PRACTICE WATCH DOWNLOAD READ REVIEW

Practice Exercises

In Exercises 1–8, find the vertical asymptotes, if any, of the graph of each rational function.

1. $f(x) = \dfrac{x}{x+4}$
2. $f(x) = \dfrac{x}{x-3}$
3. $g(x) = \dfrac{x+3}{x(x+4)}$
4. $g(x) = \dfrac{x+3}{x(x-3)}$
5. $h(x) = \dfrac{x}{x(x+4)}$
6. $h(x) = \dfrac{x}{x(x-3)}$
7. $r(x) = \dfrac{x}{x^2+4}$
8. $r(x) = \dfrac{x}{x^2+3}$

In Exercises 9–16, find the horizontal asymptote, if any, of the graph of each rational function.

9. $f(x) = \dfrac{12x}{3x^2+1}$
10. $f(x) = \dfrac{15x}{3x^2+1}$
11. $g(x) = \dfrac{12x^2}{3x^2+1}$
12. $g(x) = \dfrac{15x^2}{3x^2+1}$
13. $h(x) = \dfrac{12x^3}{3x^2+1}$
14. $h(x) = \dfrac{15x^3}{3x^2+1}$
15. $f(x) = \dfrac{-2x+1}{3x+5}$
16. $f(x) = \dfrac{-3x+7}{5x-2}$

In Exercises 17–38, follow the seven steps on page 828 to graph each rational function.

17. $f(x) = \frac{4x}{x - 2}$

18. $f(x) = \frac{3x}{x - 1}$

19. $f(x) = \frac{2x}{x^2 - 4}$

20. $f(x) = \frac{4x}{x^2 - 1}$

21. $f(x) = \frac{2x^2}{x^2 - 1}$

22. $f(x) = \frac{4x^2}{x^2 - 9}$

23. $f(x) = \frac{-x}{x + 1}$

24. $f(x) = \frac{-3x}{x + 2}$

25. $f(x) = -\frac{1}{x^2 - 4}$

26. $f(x) = -\frac{2}{x^2 - 1}$

27. $f(x) = \frac{2}{x^2 + x - 2}$

28. $f(x) = \frac{-2}{x^2 - x - 2}$

29. $f(x) = \frac{2x^2}{x^2 + 4}$

30. $f(x) = \frac{4x^2}{x^2 + 1}$

31. $f(x) = \frac{x + 2}{x^2 + x - 6}$

32. $f(x) = \frac{x - 4}{x^2 - x - 6}$

33. $f(x) = \frac{x^4}{x^2 + 2}$

34. $f(x) = \frac{2x^4}{x^2 + 1}$

35. $f(x) = \frac{x^2 + x - 12}{x^2 - 4}$

36. $f(x) = \frac{x^2}{x^2 + x - 6}$

37. $f(x) = \frac{3x^2 + x - 4}{2x^2 - 5x}$

38. $f(x) = \frac{x^2 - 4x + 3}{(x + 1)^2}$

Practice PLUS

In Exercises 39–44, the equation for f is given by the simplified expression that results after performing the indicated operation. Write the equation for f and then graph the function.

39. $\frac{5x^2}{x^2 - 4} \cdot \frac{x^2 + 4x + 4}{10x^3}$

40. $\frac{x - 5}{10x - 2} \div \frac{x^2 - 10x + 25}{25x^2 - 1}$

41. $\frac{x}{2x + 6} - \frac{9}{x^2 - 9}$

42. $\frac{2}{x^2 + 3x + 2} - \frac{4}{x^2 + 4x + 3}$

43. $\frac{1 - \frac{3}{x + 2}}{1 + \frac{1}{x - 2}}$

44. $\frac{x - \frac{1}{x}}{x + \frac{1}{x}}$

Application Exercises

45. The local game commission stocks a lake with bass. They introduce 100 bass into the lake. The rational function

$$f(x) = \frac{70x + 100}{0.02x + 1}$$

models the number of bass in the lake, $f(x)$, after x months.

a. Find the number of bass in the lake after 5 months.

b. What is the equation of the horizontal asymptote associated with this function? Describe what this means about the number of bass in the lake over time.

46. The local game commission stocks a lake with bass. They introduce 200 bass into the lake. The rational function

$$f(x) = \frac{140x + 200}{0.02x + 1}$$

models the number of bass in the lake, $f(x)$, after x months.

a. Find the number of bass in the lake after 5 months.

b. What is the equation of the horizontal asymptote associated with this function? Describe what this means about the number of bass in the lake over time.

The bar graph shows the amount, in billions of dollars, that the United States government spent on human resources and total budget outlays for six selected years. (Human resources include education, health, Medicare, Social Security, and veterans benefits and services.)

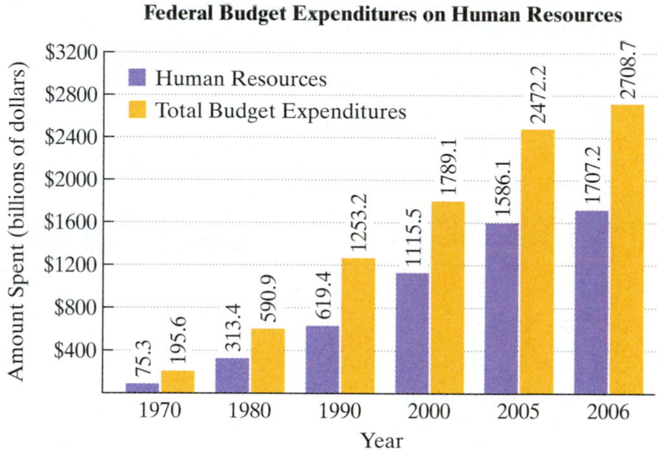

Source: Office of Management and Budget

The function $p(x) = 11x^2 + 40x + 1040$ *models the amount,* $p(x)$*, in billions of dollars, that the United States government spent on human resources x years after 1970. The function* $q(x) = 12x^2 + 230x + 2190$ *models total budget expenditures,* $q(x)$*, in billions of dollars, x years after 1970. Use this information to solve Exercises 47–48.*

47. a. Use p and q to write a rational function that models the fraction of total budget outlays spent on human resources x years after 1970.

b. Use the data displayed by the bar graph to find the percentage of federal expenditures spent on human resources in 2006. Round to the nearest percent.

c. Use the rational function from part (a) to find the percentage of federal expenditures spent on human resources in 2006. Round to the nearest percent. Does this underestimate or overestimate the actual percent that you found in part (b)? By how much?

d. What is the equation of the horizontal asymptote associated with the rational function in part (a)? If trends modeled by the function continue, what percentage of the federal budget will be spent on human resources over time? Round to the nearest percent.

48. a. Use p and q to write a rational function that models the fraction of total budget outlays spent on human resources x years after 1970.

b. Use the data displayed by the bar graph to find the percentage of federal expenditures spent on human resources in 2005. Round to the nearest percent.

c. Use the rational function from part (a) to find the percentage of federal expenditures spent on human resources in 2005. Round to the nearest percent. How does this compare with the actual percent that you found in part (b)?

d. What is the equation of the horizontal asymptote associated with the rational function in part (a)? If trends modeled by the function continue, what percentage of the federal budget will be spent on human resources over time? Round to the nearest percent.

49. A drug is injected into a patient and the concentration of the drug in the bloodstream is monitored. The drug's concentration, $C(t)$, in milligrams per liter, after t hours is modeled by

$$C(t) = \frac{5t}{t^2 + 1}.$$

The graph of this rational function, obtained with a graphing utility, is shown in the figure.

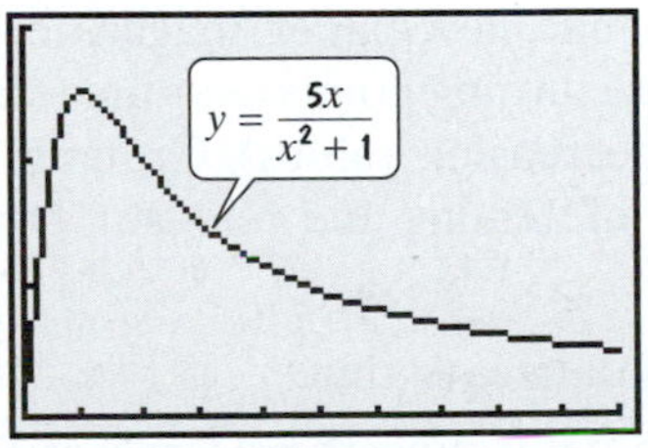

[0, 10, 1] by [0, 3, 1]

a. Use the graph to obtain a reasonable estimate of the drug's concentration after 3 hours.

b. Use the function's equation to determine the drug's concentration after 3 hours.

c. Use the function's equation to find the horizontal asymptote for the graph. Describe what this means about the drug's concentration in the patient's bloodstream as time increases.

50. The temperature, $F(t)$, in degrees Fahrenheit, of a dessert placed in a freezer for t hours is modeled by

$$F(t) = \frac{80}{t^2 + 4t + 1}.$$

a. Find and interpret $F(0)$.

b. Find the temperature of the dessert after 1 hour, 2 hours, 3 hours, 4 hours, and 5 hours.

c. What is the equation of the horizontal asymptote associated with this function? Describe what this means in terms of the dessert's temperature over time.

d. Graph the function.

Writing in Mathematics

51. What is a rational function?

52. What is the reciprocal function? Describe the behavior of its graph.

53. If you are given the equation of a rational function, explain how to find the vertical asymptotes, if any, of the function's graph.

54. If you are given the equation of a rational function, explain how to find the horizontal asymptote, if there is one, of the function's graph.

55. Describe how to graph a rational function.

56. The function $f(x) = \frac{5000x}{x^2 + 36}$ describes the population density, $f(x)$, in people per square mile, in a large city x miles from the city's center. Describe what eventually happens to the population density as the distance from the city's center increases.

57. In Exercises 47(d) and 48(d), you used a rational function to determine the percentage of the federal budget that will be spent on human resources over time. Does the percent that you obtained seem realistic? Why or why not?

Technology Exercises

58. Use a graphing utility to verify any five of your hand-drawn graphs in Exercises 17–38.

59. Use a graphing utility to graph

$$f(x) = \frac{x^2 - 4x + 3}{x - 2} \quad \text{and} \quad g(x) = \frac{x^2 - 5x + 6}{x - 2}.$$

What difference do you observe between the graphs of f and g? How do you account for these differences?

Critical Thinking Exercises

Make Sense? *In Exercises 60–63, determine whether each statement "makes sense" or "does not make sense" and explain your reasoning.*

60. I've graphed a rational function that has two vertical asymptotes and two horizontal asymptotes.

61. My graph of $y = \frac{x-1}{(x-1)(x-2)}$ has vertical asymptotes at $x = 1$ and $x = 2$.

62. The function

$$f(x) = \frac{1.96x + 3.14}{3.04x + 21.79}$$

models the fraction of nonviolent prisoners in New York State prisons x years after 1980. I can conclude from this equation that over time the percentage of nonviolent prisoners will exceed 60%.

63. The function

$$f(x) = \frac{6.5x^2 - 20.4x + 234}{x^2 + 36}$$

models the pH level, $f(x)$, of the human mouth x minutes after a person eats food containing sugar. I can conclude from this equation that over time the pH level drops back to 6.5, which is probably the normal pH level.

In Exercises 64–67, determine whether each statement is true or false. If the statement is false, make the necessary change(s) to produce a true statement.

64. The graph of a rational function cannot have both a vertical asymptote and a horizontal asymptote.

65. It is possible to have a rational function whose graph has no y-intercept.

66. The graph of a rational function can have three vertical asymptotes.

67. The graph of a rational function can never cross a vertical asymptote.

In Exercises 68–69, write the equation of a rational function $f(x) = \frac{p(x)}{q(x)}$ having the indicated properties, in which the degrees of p and q are as small as possible. More than one correct function may be possible. Graph your function using a graphing utility to verify that it has the required properties.

68. f has a vertical asymptote given by $x = 3$, a horizontal asymptote $y = 0$, y-intercept $= -1$, and no x-intercept.

69. f has vertical asymptotes given by $x = -2$ and $x = 2$, a horizontal asymptote $y = 2$, y-intercept $= \frac{9}{2}$, x-intercepts of -3 and 3, and y-axis symmetry.

Review Exercises

70. Solve: $\log_3 x + \log_3(x - 2) = 1$. (Section 9.5, Example 5)

71. Solve: $|4x - 5| = 15$. (Section 4.3, Example 1)

72. Multiply and simplify: $\sqrt{2xy} \cdot \sqrt{10xy^2}$.
(Section 7.3, Example 7)

Preview Exercises

Exercises 73–75 will help you prepare for the material covered in the first section of the next chapter.

73. Evaluate $\frac{(-1)^n}{3^n - 1}$ for $n = 1, 2, 3$, and 4.

74. Find the product of all positive integers from n down through 1 for $n = 5$.

75. Evaluate $i^2 + 1$ for all consecutive integers from 1 to 6 Then find the sum of the six evaluations.

GROUP PROJECT

CHAPTER 11

Each group member should consult an almanac, newspaper, magazine, or the Internet to find data whose graphs contain two or more turning points over time (from decreasing to increasing values or from increasing to decreasing values). Group members should select the three sets of data that are most interesting and relevant. For each data set selected,

a. Create a line graph that illustrates the data over time.
b. Use the number of turning points to determine the degree of the polynomial function of best fit. Should the leading coefficient of the function be positive or negative? Explain your answer.
c. Use the polynomial regression feature of a graphing utility to find the polynomial function that best fits the data.
d. Use the equation of the polynomial function to make a prediction from the data. What circumstances might affect the accuracy of your prediction?

Chapter 11 Summary

Definitions and Concepts	Examples

Section 11.1 Polynomial Functions and Their Graphs

A polynomial function of degree n is defined by

$$f(x) = a_n x^n + a_{n-1}x^{n-1} + \cdots + a_2x^2 + a_1x + a_0,\ a_n \neq 0.$$

The graphs of polynomial functions are smooth and continuous. The end behavior of the graph of a polynomial function depends on the leading term, $a_n x^n$, given by the Leading Coefficient Test.

- If n is odd and $a_n > 0$, the graph falls to the left and rises to the right.
- If n is odd and $a_n < 0$, the graph rises to the left and falls to the right.
- If n is even and $a_n > 0$, the graph rises to the left and rises to the right.
- If n is even and $a_n < 0$, the graph falls to the left and falls to the right.

The values of x for which $f(x)$ is equal to 0 are the zeros of the polynomial function f. These values are the roots of the polynomial equation $f(x) = 0$.

If $x - r$ occurs k times, but not $k + 1$ times, in a polynomial function's factorization, r is a zero with multiplicity k. If $k \geq 2$, r is a repeated zero. If k is even, the graph touches the x-axis and turns at r; if k is odd, it crosses the x-axis at r.

If f is a polynomial of degree n, the graph of f has at most $n - 1$ turning points.

Example

$$f(x) = x^3 - x^2 - 9x + 9$$

- Determine end behavior.

  Leading term is x^3, with $n = 3$ (odd) and leading coefficient is 1, which is greater than 0. The graph falls to the left and rises to the right.

- Find zeros and give multiplicities.

$$x^3 - x^2 - 9x + 9 = 0$$

$$x^2(x - 1) - 9(x - 1) = 0$$

$$(x - 1)(x^2 - 9) = 0$$

$$(x - 1)(x + 3)(x - 3) = 0$$

$$x - 1 = 0 \qquad x + 3 = 0 \qquad x - 3 = 0$$

$$x = 1 \qquad x = -3 \qquad x = 3$$

Each factor occurs once, so each zero has multiplicity 1. The graph crosses the x-axis at each zero, 1, −3, and 3.

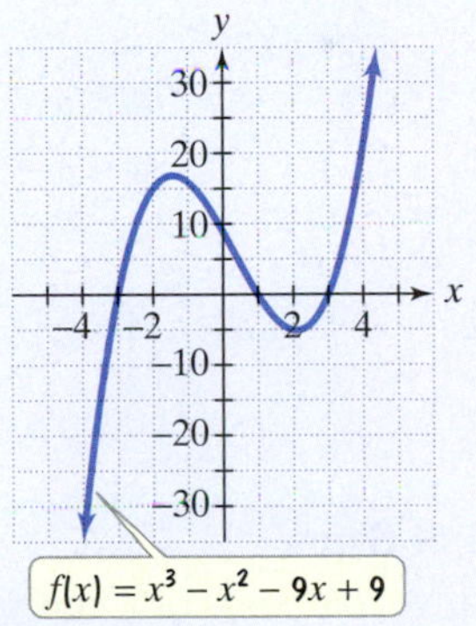

Even and Odd Functions

The graph of an even function in which $f(-x) = f(x)$ is symmetric with respect to the y-axis. The graph of an odd function in which $f(-x) = -f(x)$ is symmetric with respect to the origin.

Example

Is $f(x) = x^3 + 9x$ even, odd, or neither?

$$f(-x) = (-x)^3 + 9(-x)$$

$$= -x^3 - 9x = -f(x)$$

Thus, f is an odd function.

Definitions and Concepts	Examples

Section 11.2 Zeros of Polynomial Functions

The Factor Theorem

If a polynomial function $f(x)$ is divided by $x - c$ and the remainder is zero, c is a zero of f and a root of $f(x) = 0$. If c is a zero of f or a root of $f(x) = 0$, then $x - c$ is a factor of $f(x)$.

The Rational Zero Theorem

Possible rational zeros of a polynomial function are

$$\frac{\text{factors of the constant term}}{\text{factors of the leading coefficient}}.$$

Properties of Polynomial Equations

1. If the equation is of degree n, then counting multiple roots separately, the equation has n roots.
2. If $a + bi$ is a root, then $a - bi$ is also a root.

Solve: $12x^3 - 8x^2 - 3x + 2 = 0$.

Possible rational roots

$$= \frac{\text{Factors of the constant term, 2}}{\text{Factors of the leading coefficient, 12}}$$

$$= \frac{\pm 1, \pm 2}{\pm 1, \pm 2, \pm 3, \pm 4, \pm 6, \pm 12}$$

$$= \pm 1, \pm 2, \pm \tfrac{1}{2}, \pm \tfrac{1}{3}, \pm \tfrac{2}{3}, \pm \tfrac{1}{4}, \pm \tfrac{1}{6}, \pm \tfrac{1}{12}$$

Test $\frac{1}{2}$:

$$\begin{array}{r|rrrr} \frac{1}{2} & 12 & -8 & -3 & 2 \\ & & 6 & -1 & -2 \\ \hline & 12 & -2 & -4 & 0 \end{array}$$

Coefficients of the equation

The zero remainder shows $\frac{1}{2}$ is a root.

Using the result of the synthetic division, rewrite the equation as

$$\left(x - \frac{1}{2}\right)(12x^2 - 2x - 4) = 0.$$

$$\left(x - \frac{1}{2}\right)2(6x^2 - x - 2) = 0$$

$$2\left(x - \frac{1}{2}\right)(3x - 2)(2x + 1) = 0$$

$$x - \frac{1}{2} = 0 \quad \text{or} \quad 3x - 2 = 0 \quad \text{or} \quad 2x + 1 = 0$$

$$x = \frac{1}{2} \qquad x = \frac{2}{3} \qquad x = -\frac{1}{2}$$

The roots are $-\frac{1}{2}, \frac{1}{2}$, and $\frac{2}{3}$, and the solution set is

$$\left\{-\frac{1}{2}, \frac{1}{2}, \frac{2}{3}\right\}.$$

Descartes's Rule of Signs

The number of positive real zeros of f equals the number of sign changes of $f(x)$ or is less than that number by a positive even integer. The number of negative real zeros of f equals the number of sign changes of $f(-x)$ or is less than that number by a positive even integer.

Determine the possible numbers of positive and negative real zeros of

$$f(x) = x^4 - 4x^3 + 7x^2 - 6x - 18.$$

sign change sign change sign change

Three sign changes: Either there are 3 positive real zeros or there is $3 - 2 = 1$ positive real zero.

$$f(-x) = (-x)^4 - 4(-x)^3 + 7(-x)^2 - 6(-x) - 18$$
$$= x^4 + 4x^3 + 7x^2 + 6x - 18$$

sign change

One sign change: There is 1 negative real zero.

Definitions and Concepts	Examples

Section 11.3 Rational Functions and Their Graphs

Rational functions are quotients of polynomial functions. Rational functions can be expressed as:

$$f(x) = \frac{p(x)}{q(x)},$$

where p and q are polynomial functions and $q(x) \neq 0$.

Locating Vertical Asymptotes

If $p(x)$ and $q(x)$ have no common factors with degree 1 or greater and a is a zero of $q(x)$, then $x = a$ is a vertical asymptote of the graph of f.

Locating Horizontal Asymptotes

- If the degree of p is less than the degree of q, the x-axis (that is, $y = 0$) is the horizontal asymptote of the graph of f.
- If the degree of p is equal to the degree of q, the line
$$y = \frac{\text{the leading coefficient of } p}{\text{the leading coefficient of } q}$$
is the horizontal asymptote of the graph of f.
- If the degree of p is greater than the degree of q, the graph of f has no horizontal asymptote.

Graphing Rational Functions

Rational functions can be graphed using symmetry, intercepts, asymptotes, and points between and beyond each x-intercept and vertical asymptote.

Examples

$$f(x) = \frac{x+1}{x^2+2x-3}$$

- Find vertical asymptotes.

$$f(x) = \frac{x+1}{(x+3)(x-1)}$$

This factor is 0 if $x = -3$. This factor is 0 if $x = 1$.

The lines $x = -3$ and $x = 1$ are vertical asymptotes for the graph of f.

- Find the horizontal asymptote.

The degree of the numerator, 1, is less than that of the denominator, 2. The horizontal asymptote is the x-axis, $y = 0$.

- Find intercepts.

$$y\text{-intercept: } f(0) = \frac{0+1}{0^2+2\cdot 0-3} = -\frac{1}{3}$$

$$x\text{-intercept: } x + 1 = 0$$
$$x = -1$$

- Graph the function. Plot points beyond and between vertical asymptotes and the x-intercept. Using $x = -4$, $x = -2$, and $x = 2$, we evaluate the function and obtain

$$\left(-4, -\frac{3}{5}\right), \left(-2, \frac{1}{3}\right), \text{ and } \left(2, \frac{3}{5}\right).$$

Using these points, the asymptotes, and intercepts, the graph is shown as follows.

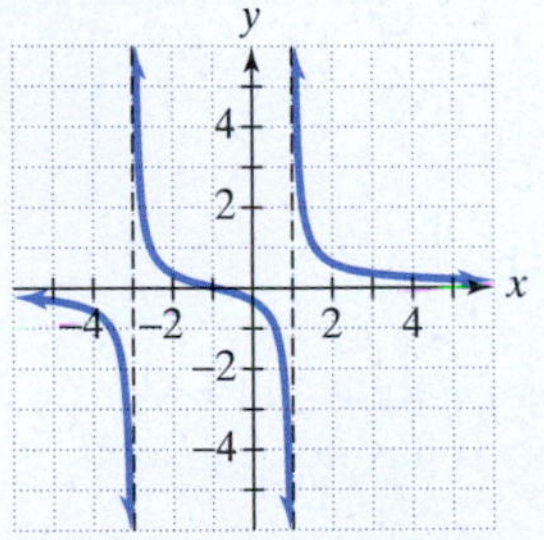

CHAPTER 11 REVIEW EXERCISES

11.1 *In Exercises 1–2, find the zeros for each polynomial function and give the multiplicity of each zero. State whether the graph crosses the x-axis or touches the x-axis and turns around at each zero.*

1. $f(x) = -2(x-1)(x+2)^2(x+5)^3$

2. $f(x) = x^3 - 5x^2 - 25x + 125$

In Exercises 3–5, determine whether each function is even, odd, or neither. State each function's symmetry, if any. If you are using a graphing utility, graph the function and verify its possible symmetry.

3. $f(x) = x^3 - 5x$
4. $f(x) = x^4 - 2x^2 + 1$
5. $f(x) = 2x(1 - x^2)$

In Exercises 6–11,

a. *Use the Leading Coefficient Test to determine the graph's end behavior.*

b. *Determine whether the graph has y-axis symmetry, origin symmetry, or neither.*

c. *Graph the function.*

6. $f(x) = x^3 - x^2 - 9x + 9$

7. $f(x) = 4x - x^3$

8. $f(x) = 2x^3 + 3x^2 - 8x - 12$

9. $f(x) = -x^4 + 25x^2$

10. $f(x) = -x^4 + 6x^3 - 9x^2$

11. $f(x) = 3x^4 - 15x^3$

11.2

12. Solve the equation $x^3 - 17x + 4 = 0$ given that 4 is a zero of $f(x) = x^3 - 17x + 4$.

13. Solve the equation $3x^3 + 10x^2 - x - 12 = 0$ given that -3 is a root.

In Exercises 14–15, use the Rational Zero Theorem to list all possible rational zeros for each given function.

14. $f(x) = x^4 - 6x^3 + 14x^2 - 14x + 5$

15. $f(x) = 3x^5 - 2x^4 - 15x^3 + 10x^2 + 12x - 8$

In Exercises 16–17, use Descartes's Rule of Signs to determine the possible numbers of positive and negative real zeros for each given function.

16. $f(x) = 3x^4 - 2x^3 - 8x + 5$

17. $f(x) = 2x^5 - 3x^3 - 5x^2 + 3x - 1$

18. Use Descartes's Rule of Signs to explain why $2x^4 + 6x^2 + 8 = 0$ has no real roots.

For Exercises 19–24,

a. *List all possible rational roots or rational zeros.*

b. *Use Descartes's Rule of Signs to determine the possible numbers of positive and negative real roots or real zeros.*

c. *Use synthetic division to test the possible rational roots or zeros and find an actual root or zero.*

d. *Use the root or zero from part (c) to find all the zeros or roots.*

19. $f(x) = x^3 + 3x^2 - 4$

20. $f(x) = 6x^3 + x^2 - 4x + 1$

21. $8x^3 - 36x^2 + 46x - 15 = 0$

22. $x^4 - x^3 - 7x^2 + x + 6 = 0$

23. $4x^4 + 7x^2 - 2 = 0$

24. $f(x) = 2x^4 + x^3 - 9x^2 - 4x + 4$

11.3

In Exercises 25–31, find the vertical asymptotes, if any, and the horizontal asymptote, if one exists, of the graph of each rational function. Then graph the rational function.

25. $f(x) = \dfrac{2x}{x^2 - 9}$

26. $f(x) = \dfrac{2x - 4}{x + 3}$

27. $f(x) = \dfrac{4x^2}{x^2 - 1}$

28. $f(x) = \dfrac{x^2}{x^2 + 1}$

29. $f(x) = \dfrac{x^4}{x^2 + 2}$

30. $f(x) = \dfrac{x^2 - 3x - 4}{x^2 - x - 6}$

31. $f(x) = \dfrac{x^2 + 4x + 3}{(x + 2)^2}$

Exercises 32–33 involve rational functions that model the given situation. In each case, find the equation of the horizontal asymptote associated with the function. Then describe what this means in practical terms.

32. $f(x) = \dfrac{150x + 120}{0.05x + 1}$; the number of bass, $f(x)$, after x months in a lake that was stocked with 120 bass

33. $P(x) = \dfrac{72{,}900}{100x^2 + 729}$; the percentage, $P(x)$, of people in the United States with x years of education who are unemployed

CHAPTER 11 TEST

Remember to use your Chapter Test Prep Video CD to see the worked-out solutions to the test questions you want to review.

1. Consider the function $f(x) = x^3 - 5x^2 - 4x + 20$.

a. Use factoring to find all zeros of f.

b. Use the Leading Coefficient Test and the zeros of f to graph the function.

2. Determine whether $f(x) = x^4 - x^2$ is even, odd, or neither. Use your answer to explain why the graph in the figure shown cannot be the graph of f.

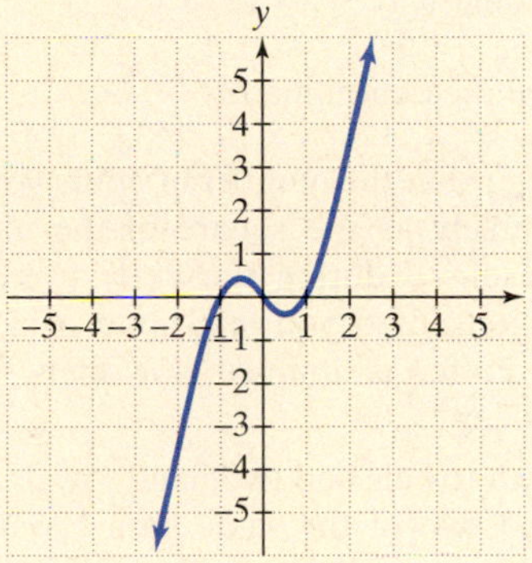

3. The graph of $f(x) = 6x^3 - 19x^2 + 16x - 4$ is shown in the figure.

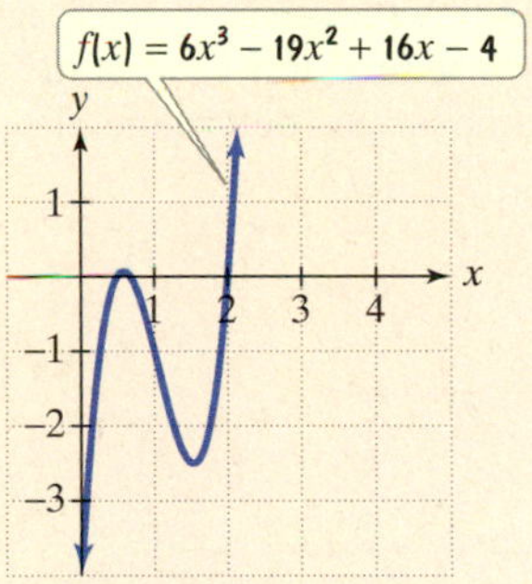

a. Based on the graph of f, find the root of the equation $6x^3 - 19x^2 + 16x - 4 = 0$ that is an integer.

b. Use synthetic division to find the other two roots of $6x^3 - 19x^2 + 16x - 4 = 0$.

4. Use the Rational Zero Theorem to list all possible rational zeros of $f(x) = 2x^3 + 11x^2 - 7x - 6$.

5. Use Descartes's Rule of Signs to determine the possible number of positive and negative real zeros of

$$f(x) = 3x^5 - 2x^4 - 2x^2 + x - 1.$$

6. Solve: $x^3 + 6x^2 - x - 30 = 0$.

7. Consider the function whose equation is given by $f(x) = 2x^4 - x^3 - 13x^2 + 5x + 15$.

a. List all possible rational zeros.

b. Use the graph of f in the figure shown and synthetic division to find all zeros of the function.

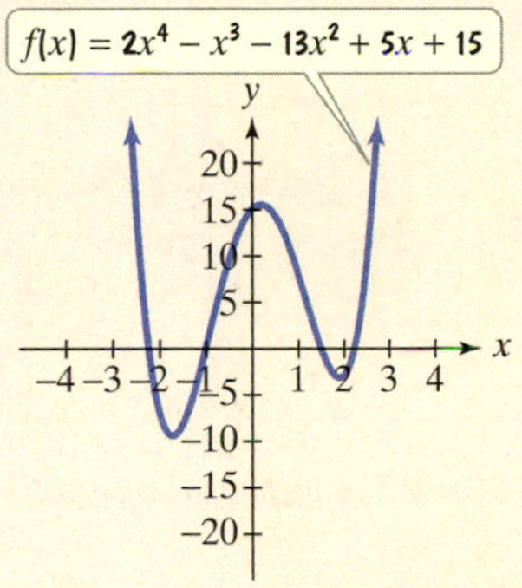

In Exercises 8–10, find the vertical asymptotes, if any, and the horizontal asymptote, if one exists, of the graph of each rational function. Then graph the rational function.

8. $f(x) = \dfrac{4x - 2}{x - 1}$

9. $f(x) = \dfrac{x^2 - 1}{x^2 - 4}$

10. $f(x) = \dfrac{4x^2}{x^2 + 3}$

CUMULATIVE REVIEW EXERCISES (CHAPTERS 1–11)

In Exercises 1–7, solve each equation, inequality, or system.

1. $x^3 - 4x^2 - 10x + 4 = 0$
2. $2x + y = 3$
   $y = 2x^2 - 1$
3. $\sqrt{x} + 6 = x$
4. $x^{\frac{2}{3}} + 3x^{\frac{1}{3}} - 18 = 0$
5. $\log_2 x + \log_2 3 = 1$
6. $x + y - z = -2$
   $3x - 2y - 5z = 7$
   $2x - 3y + 4z = 17$
7. $\dfrac{x-4}{x+6} > 0$

In Exercises 8–10, graph each function, equation, or inequality in a rectangular coordinate system.

8. $\dfrac{x^2}{16} - \dfrac{y^2}{9} = 1$
9. $3x - 2y \leq -6$
10. $f(x) = \dfrac{2x}{x+1}$

In Exercises 11–15, perform the indicated operations and simplify, if possible.

11. $(5x - 7)^2$
12. $(2x - 3)(4x^2 + 6x + 9)$
13. $\dfrac{3}{x^2 - x} - \dfrac{2x}{3x - 3}$
14. $\sqrt{54} - \sqrt{150}$
15. $(1 + 2i)(6 - 3i)$

In Exercises 16–17, factor completely.

16. $3x^3 + 6x^2 - 3x - 6$
17. $25x^2 - 20xy + 4y^2$
18. Write the slope-intercept form of the equation of the line through $(2, -3)$ and parallel to the line whose equation is $4x + y = 7$.
19. Evaluate the determinant: $\begin{vmatrix} 4 & -3 \\ 5 & 7 \end{vmatrix}$.
20. If $f(x) = x^2 - 4x + 7$ and $g(x) = x + 3$, find $f(g(x))$ and simplify.
21. Write as a single logarithm whose coefficient is 1: $\frac{1}{2}\log x + 3\log y$.
22. Rationalize the denominator $\dfrac{9x}{\sqrt[3]{3x^2}}$.
23. You discover that the number of hours you sleep each night varies inversely as the square of the number of cups of coffee consumed during the early evening. If 2 cups of coffee are consumed, you get 8 hours of sleep. If the number of cups of coffee is doubled, how many hours should you expect to sleep?
24. If three bath towels and two hand towels can be purchased for \$31 and two of the same bath towels and five of the same hand towels can be purchased for \$39, determine the cost of each kind of towel.
25. A square is made into a rectangle by increasing two of its sides by 6 centimeters each and decreasing each of its other two sides by 4 centimeters each. If the square and rectangle have the same areas, determine the dimensions of each figure.

CHAPTER 12

Sequences, Induction, and Probability

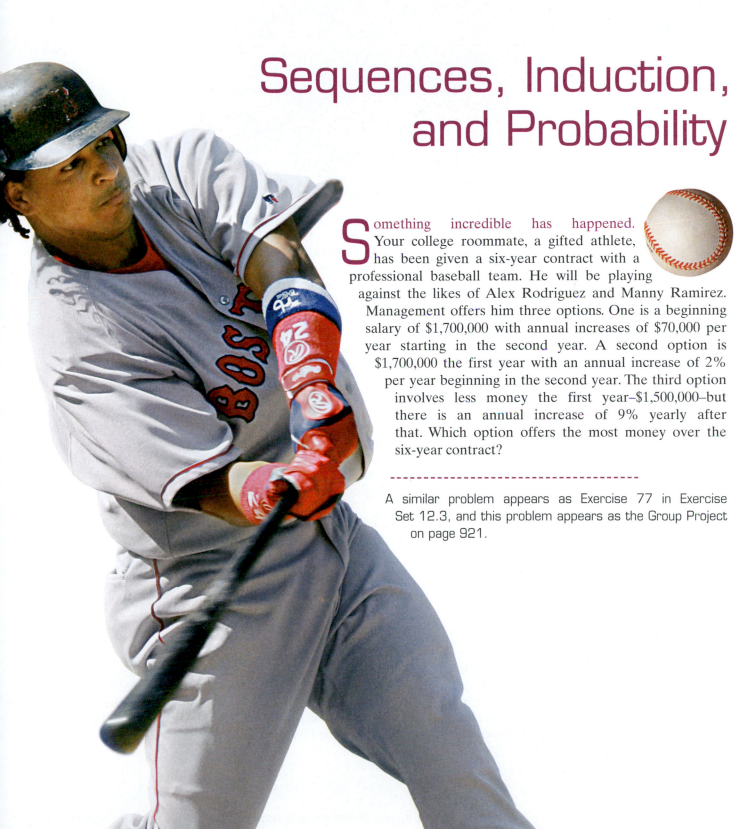

Something incredible has happened. Your college roommate, a gifted athlete, has been given a six-year contract with a professional baseball team. He will be playing against the likes of Alex Rodriguez and Manny Ramirez. Management offers him three options. One is a beginning salary of \$1,700,000 with annual increases of \$70,000 per year starting in the second year. A second option is \$1,700,000 the first year with an annual increase of 2% per year beginning in the second year. The third option involves less money the first year–\$1,500,000–but there is an annual increase of 9% yearly after that. Which option offers the most money over the six-year contract?

A similar problem appears as Exercise 77 in Exercise Set 12.3, and this problem appears as the Group Project on page 921.

SECTION 12.1

Sequences and Summation Notation

Objectives

1. Find particular terms of a sequence from the general term.
2. Use factorial notation.
3. Use summation notation.

Sequences

Many creations in nature involve intricate mathematical designs, including a variety of spirals. For example, the arrangement of the individual florets in the head of a sunflower forms spirals. In some species, there are 21 spirals in the clockwise direction and 34 in the counterclockwise direction. The precise numbers depend on the species of sunflower: 21 and 34, or 34 and 55, or 55 and 89, or even 89 and 144.

This observation becomes more interesting when we consider a sequence of numbers investigated by Leonardo of Pisa, also known as Fibonacci, an Italian mathematician of the thirteenth century. The **Fibonacci sequence** of numbers is an infinite sequence that begins as follows:

$$1, 1, 2, 3, 5, 8, 13, 21, 34, 55, 89, 144, 233, \ldots.$$

The first two terms are 1. Every term thereafter is the sum of the two preceding terms. For example, the third term, 2, is the sum of the first and second terms: $1 + 1 = 2$. The fourth term, 3, is the sum of the second and third terms: $1 + 2 = 3$, and so on. Did you know that the number of spirals in a daisy or a sunflower, 21 and 34, are two Fibonacci numbers? The number of spirals in a pinecone, 8 and 13, and a pineapple, 8 and 13, are also Fibonacci numbers.

We can think of the Fibonacci sequence as a function. The terms of the sequence

$$1, 1, 2, 3, 5, 8, 13, 21, 34, 55, 89, 144, 233, \ldots$$

are the range values for a function f whose domain is the set of positive integers.

Domain:	1,	2,	3,	4,	5,	6,	7,	...
	↓	↓	↓	↓	↓	↓	↓	
Range:	1,	1,	2,	3,	5,	8,	13,	...

Thus, $f(1) = 1, f(2) = 1, f(3) = 2, f(4) = 3, f(5) = 5, f(6) = 8, f(7) = 13$, and so on.

The letter a with a subscript is used to represent function values of a sequence, rather than the usual function notation. The subscripts make up the domain of the sequence and they identify the location of a term. Thus, a_1 represents the first term of the sequence, a_2 represents the second term, a_3 the third term, and so on. This notation is shown for the first six terms of the Fibonacci sequence:

1,	1,	2,	3,	5,	8.
$a_1 = 1$	$a_2 = 1$	$a_3 = 2$	$a_4 = 3$	$a_5 = 5$	$a_6 = 8$

The notation a_n represents the nth term, or **general term**, of a sequence. The entire sequence is represented by $\{a_n\}$.

Blitzer Bonus

Fibonacci Numbers on the Piano Keyboard

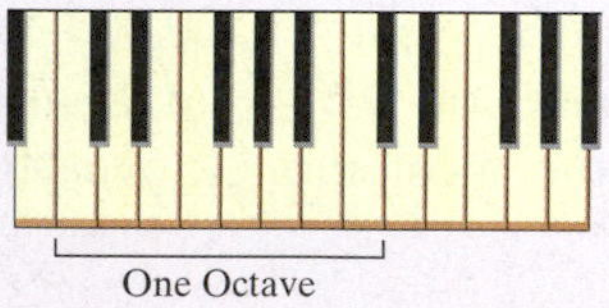

Numbers in the Fibonacci sequence can be found in an octave on the piano keyboard. The octave contains 2 black keys in one cluster and 3 black keys in another cluster, for a total of 5 black keys. It also has 8 white keys, for a total of 13 keys. The numbers 2, 3, 5, 8, and 13 are the third through seventh terms of the Fibonacci sequence.

Definition of a Sequence

An **infinite sequence** $\{a_n\}$ is a function whose domain is the set of positive integers. The function values, or **terms**, of the sequence are represented by

$$a_1, a_2, a_3, a_4, \ldots, a_n, \ldots.$$

Sequences whose domains consist only of the first n positive integers are called **finite sequences**.

1 Find particular terms of a sequence from the general term.

EXAMPLE 1 Writing Terms of a Sequence from the General Term

Write the first four terms of the sequence whose nth term, or general term, is given:

a. $a_n = 3n + 4$ **b.** $a_n = \dfrac{(-1)^n}{3^n - 1}$.

Solution

a. We need to find the first four terms of the sequence whose general term is $a_n = 3n + 4$. To do so, we replace n in the formula with 1, 2, 3, and 4.

a_1, 1st term: $3 \cdot 1 + 4 = 3 + 4 = 7$ a_2, 2nd term: $3 \cdot 2 + 4 = 6 + 4 = 10$

a_3, 3rd term: $3 \cdot 3 + 4 = 9 + 4 = 13$ a_4, 4th term: $3 \cdot 4 + 4 = 12 + 4 = 16$

The first four terms are 7, 10, 13, and 16. The sequence defined by $a_n = 3n + 4$ can be written as

$$7, 10, 13, 16, \ldots, 3n + 4, \ldots.$$

Study Tip

The factor $(-1)^n$ in the general term of a sequence causes the signs of the terms to alternate between positive and negative, depending on whether n is even or odd.

b. We need to find the first four terms of the sequence whose general term is $a_n = \dfrac{(-1)^n}{3^n - 1}$. To do so, we replace each occurrence of n in the formula with 1, 2, 3, and 4.

a_1, 1st term: $\dfrac{(-1)^1}{3^1 - 1} = \dfrac{-1}{3 - 1} = -\dfrac{1}{2}$ a_2, 2nd term: $\dfrac{(-1)^2}{3^2 - 1} = \dfrac{1}{9 - 1} = \dfrac{1}{8}$

a_3, 3rd term: $\dfrac{(-1)^3}{3^3 - 1} = \dfrac{-1}{27 - 1} = -\dfrac{1}{26}$ a_4, 4th term: $\dfrac{(-1)^4}{3^4 - 1} = \dfrac{1}{81 - 1} = \dfrac{1}{80}$

The first four terms are $-\frac{1}{2}, \frac{1}{8}, -\frac{1}{26}$, and $\frac{1}{80}$. The sequence defined by $\dfrac{(-1)^n}{3^n - 1}$ can be written as

$$-\frac{1}{2}, \frac{1}{8}, -\frac{1}{26}, \frac{1}{80}, \ldots, \frac{(-1)^n}{3^n - 1}, \ldots.$$

✓ CHECK POINT 1 Write the first four terms of the sequence whose nth term, or general term, is given:

a. $a_n = 2n + 5$ **b.** $a_n = \dfrac{(-1)^n}{2^n + 1}$.

Although sequences are usually named with the letter a, any lowercase letter can be used. For example, the first four terms of the sequence $\{b_n\} = \left\{\left(\frac{1}{2}\right)^n\right\}$ are $b_1 = \frac{1}{2}$, $b_2 = \frac{1}{4}$, $b_3 = \frac{1}{8}$, and $b_4 = \frac{1}{16}$.

Using Technology

Graphing utilities can write the terms of a sequence and graph them. For example, to find the first six terms of

$\{a_n\} = \left\{\frac{1}{n}\right\}$, enter

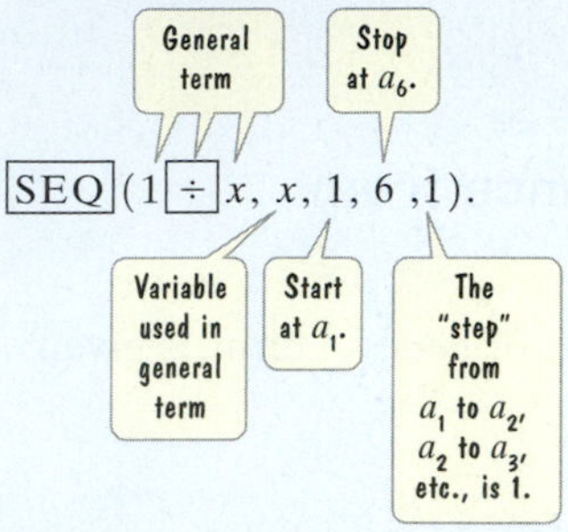

The first few terms of the sequence are shown in the viewing rectangle. By pressing the right arrow key to scroll right, you can see the remaining terms.

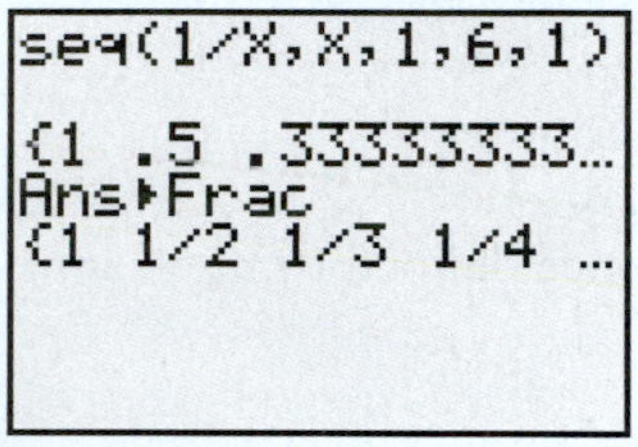

Because a sequence is a function whose domain is the set of positive integers, the **graph of a sequence** is a set of discrete points. For example, consider the sequence whose general term is $a_n = \frac{1}{n}$. How does the graph of this sequence differ from the graph of the rational function $f(x) = \frac{1}{x}$? The graph of $f(x) = \frac{1}{x}$ is shown in **Figure 12.1(a)** for positive values of x. To obtain the graph of the sequence $\{a_n\} = \left\{\frac{1}{n}\right\}$, remove all the points from the graph of f except those whose x-coordinates are positive integers. Thus, we remove all points except $(1, 1)$, $\left(2, \frac{1}{2}\right)$, $\left(3, \frac{1}{3}\right)$, $\left(4, \frac{1}{4}\right)$, and so on. The remaining points are the graph of the sequence $\{a_n\} = \left\{\frac{1}{n}\right\}$, shown in **Figure 12.1(b)**. Notice that the horizontal axis is labeled n and the vertical axis is labeled a_n.

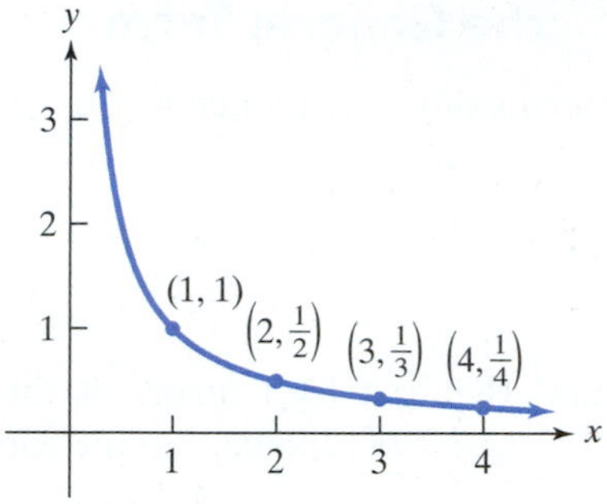

FIGURE 12.1(a) The graph of $f(x) = \frac{1}{x}$, $x > 0$

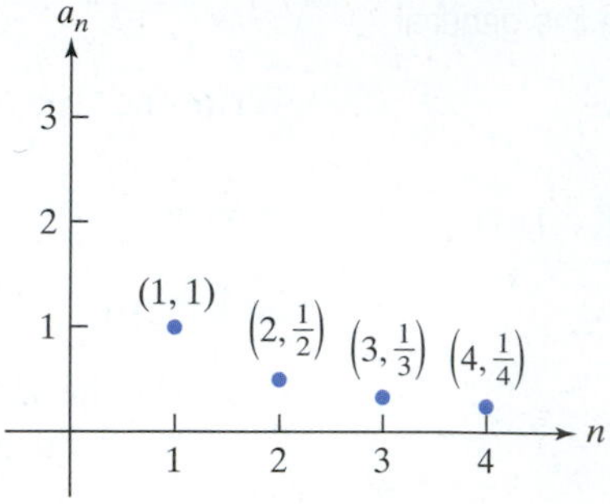

FIGURE 12.1(b) The graph of $\{a_n\} = \left\{\frac{1}{n}\right\}$

Comparing a continuous graph to the graph of a sequence

Factorial Notation

2 Use factorial notation.

Products of consecutive positive integers occur quite often in sequences. These products can be expressed in a special notation, called **factorial notation**.

Factorial Notation

If n is a positive integer, the notation $n!$ (read "n factorial") is the product of all positive integers from n down through 1.

$$n! = n(n - 1)(n - 2)\cdots(3)(2)(1)$$

0! (zero factorial), by definition, is 1.

$$0! = 1$$

The values of $n!$ for the first six positive integers are

$$1! = 1$$
$$2! = 2 \cdot 1 = 2$$
$$3! = 3 \cdot 2 \cdot 1 = 6$$
$$4! = 4 \cdot 3 \cdot 2 \cdot 1 = 24$$
$$5! = 5 \cdot 4 \cdot 3 \cdot 2 \cdot 1 = 120$$
$$6! = 6 \cdot 5 \cdot 4 \cdot 3 \cdot 2 \cdot 1 = 720.$$

Using Technology

Calculators have factorial keys. To find 5 factorial, most calculators use one of the following:

Many Scientific Calculators:

5 [x!]

Many Graphing Calculators:

5 [!] [ENTER]

Because $n!$ becomes quite large as n increases, your calculator will display these larger values in scientific notation.

Factorials affect only the number or variable that they follow unless grouping symbols appear. For example,

$$2 \cdot 3! = 2(3 \cdot 2 \cdot 1) = 2 \cdot 6 = 12$$

whereas

$$(2 \cdot 3)! = 6! = 6 \cdot 5 \cdot 4 \cdot 3 \cdot 2 \cdot 1 = 720.$$

In this sense, factorials are similar to exponents.

EXAMPLE 2 Finding Terms of a Sequence Involving Factorials

Write the first four terms of the sequence whose nth term is

$$a_n = \frac{2^n}{(n-1)!}.$$

Solution We need to find the first four terms of the sequence. To do so, we replace each n in the formula with 1, 2, 3, and 4.

a_1, 1st term: $$\frac{2^1}{(1-1)!} = \frac{2}{0!} = \frac{2}{1} = 2$$

a_2, 2nd term: $$\frac{2^2}{(2-1)!} = \frac{4}{1!} = \frac{4}{1} = 4$$

a_3, 3rd term: $$\frac{2^3}{(3-1)!} = \frac{8}{2!} = \frac{8}{2 \cdot 1} = 4$$

a_4, 4th term: $$\frac{2^4}{(4-1)!} = \frac{16}{3!} = \frac{16}{3 \cdot 2 \cdot 1} = \frac{16}{6} = \frac{8}{3}$$

The first four terms are 2, 4, 4, and $\frac{8}{3}$.

✓ **CHECK POINT 2** Write the first four terms of the sequence whose nth term is

$$a_n = \frac{20}{(n+1)!}.$$

3 Use summation notation.

Summation Notation

It is sometimes useful to find the sum of the first n terms of a sequence. For example, consider the cost of raising a child born in the United States in 2005 to a middle-income ($43,200–$72,600 per year) family, shown in **Table 12.1**.

Table 12.1 The Cost of Raising a Child Born in the U.S. in 2005 to a Middle-Income Family

Year	2005	2006	2007	2008	2009	2010	2011	2012	2013
Average Cost	$10,220	$10,530	$10,850	$11,490	$11,840	$12,220	$12,460	$12,840	$13,230
	Child is under 1.	Child is 1.	Child is 2.	Child is 3.	Child is 4.	Child is 5.	Child is 6.	Child is 7.	Child is 8.

Year	2014	2015	2016	2017	2018	2019	2020	2021	2022
Average Cost	$13,420	$13,830	$14,250	$15,740	$16,220	$16,710	$17,690	$18,230	$18,780
	Child is 9.	Child is 10.	Child is 11.	Child is 12.	Child is 13.	Child is 14.	Child is 15.	Child is 16.	Child is 17.

Source: U.S. Department of Agriculture

We can let a_n represent the cost of raising a child in year n, where $n = 1$ corresponds to 2005, $n = 2$ to 2006, $n = 3$ to 2007, and so on. The terms of the finite sequence in **Table 12.1** are given on the following page.

10,220	10,530	10,850	11,490	11,840	12,220	12,460	12,840	13,230,
a_1	a_2	a_3	a_4	a_5	a_6	a_7	a_8	a_9
13,420	13,830	14,250	15,740	16,220	16,710	17,690	18,230	18,780
a_{10}	a_{11}	a_{12}	a_{13}	a_{14}	a_{15}	a_{16}	a_{17}	a_{18}

Why might we want to add the terms of this sequence? We do this to find the total cost of raising a child born in 2005 from birth through age 17. Thus,

$$\begin{aligned} &a_1 + a_2 + a_3 + a_4 + a_5 + a_6 + a_7 + a_8 + a_9 + a_{10} + a_{11} + a_{12} + a_{13} + a_{14} + a_{15} \\ &\quad + a_{16} + a_{17} + a_{18} \\ &= 10{,}220 + 10{,}530 + 10{,}850 + 11{,}490 + 11{,}840 + 12{,}220 + 12{,}460 + 12{,}840 \\ &\quad + 13{,}230 + 13{,}420 + 13{,}830 + 14{,}250 + 15{,}740 + 16{,}220 + 16{,}710 + 17{,}690 \\ &\quad + 18{,}230 + 18{,}780 \\ &= 250{,}550. \end{aligned}$$

We see that the total cost of raising a child born in 2005 from birth through age 17 is \$250,550.

There is a compact notation for expressing the sum of the first n terms of a sequence. For example, rather than write

$$\begin{aligned} &a_1 + a_2 + a_3 + a_4 + a_5 + a_6 + a_7 + a_8 + a_9 + a_{10} + a_{11} \\ &\quad + a_{12} + a_{13} + a_{14} + a_{15} + a_{16} + a_{17} + a_{18}, \end{aligned}$$

we can use *summation notation* to express the sum as

$$\sum_{i=1}^{18} a_i.$$

We read this expression as "the sum as i goes from 1 to 18 of a_i." The letter i is called the *index of summation* and is not related to the use of i to represent $\sqrt{-1}$.

You can think of the symbol Σ (the uppercase Greek letter sigma) as an instruction to add up the terms of a sequence.

Summation Notation

The sum of the first n terms of a sequence is represented by the **summation notation**

$$\sum_{i=1}^{n} a_i = a_1 + a_2 + a_3 + a_4 + \cdots + a_n,$$

where i is the **index of summation**, n is the **upper limit of summation**, and 1 is the **lower limit of summation.**

Any letter can be used for the index of summation. The letters i, j, and k are used commonly. Furthermore, the lower limit of summation can be an integer other than 1.

When we write out a sum that is given in summation notation, we are **expanding the summation**. Example 3 shows how to do this.

EXAMPLE 3 Using Summation Notation

Expand and evaluate the sum:

a. $\sum_{i=1}^{6} (i^2 + 1)$ **b.** $\sum_{k=4}^{7} [(-2)^k - 5]$ **c.** $\sum_{i=1}^{5} 3.$

Using Technology

Graphing utilities can calculate the sum of a sequence. For example, to find the sum of the sequence in Example 3(a), enter

SUM SEQ $(x^2 + 1, x, 1, 6, 1)$.

Then press ENTER; 97 should be displayed. Use this capability to verify Example 3(b).

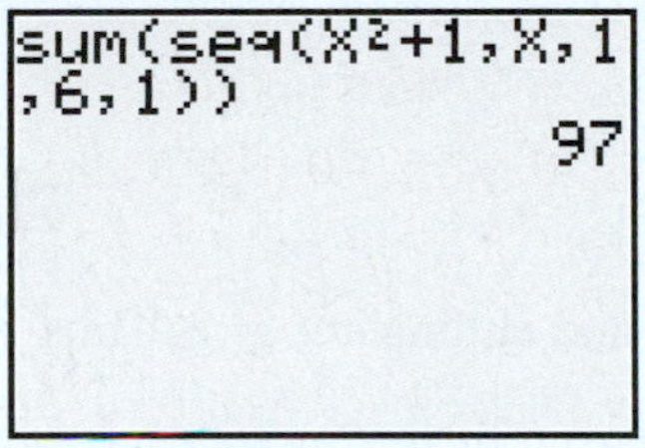

Solution

a. To find $\sum_{i=1}^{6}(i^2 + 1)$, we must replace i in the expression $i^2 + 1$ with all consecutive integers from 1 to 6, inclusive. Then we add.

$$\begin{aligned}\sum_{i=1}^{6}(i^2 + 1) &= (1^2 + 1) + (2^2 + 1) + (3^2 + 1) + (4^2 + 1) \\ &\quad + (5^2 + 1) + (6^2 + 1) \\ &= 2 + 5 + 10 + 17 + 26 + 37 \\ &= 97\end{aligned}$$

b. The index of summation in $\sum_{k=4}^{7}[(-2)^k - 5]$ is k. First we evaluate $(-2)^k - 5$ for all consecutive integers from 4 through 7, inclusive. Then we add.

$$\begin{aligned}\sum_{k=4}^{7}[(-2)^k - 5] &= [(-2)^4 - 5] + [(-2)^5 - 5] \\ &\quad + [(-2)^6 - 5] + [(-2)^7 - 5] \\ &= (16 - 5) + (-32 - 5) + (64 - 5) + (-128 - 5) \\ &= 11 + (-37) + 59 + (-133) \\ &= -100\end{aligned}$$

c. To find $\sum_{i=1}^{5}3$, we observe that every term of the sum is 3. The notation $i = 1$ through 5 indicates that we must add the first five terms of a sequence in which every term is 3.

$$\sum_{i=1}^{5}3 = 3 + 3 + 3 + 3 + 3 = 15$$

CHECK POINT 3 Expand and evaluate the sum:

a. $\sum_{i=1}^{6}2i^2$

b. $\sum_{k=3}^{5}(2^k - 3)$

c. $\sum_{i=1}^{5}4.$

Although the domain of a sequence is the set of positive integers, any integers can be used for the limits of summation. For a given sum, we can vary the upper and lower limits of summation, as well as the letter used for the index of summation. By doing so, we can produce different-looking summation notations for the same sum. For example, the sum of the squares of the first four positive integers, $1^2 + 2^2 + 3^2 + 4^2$, can be expressed in a number of equivalent ways:

$$\sum_{i=1}^{4}i^2 = 1^2 + 2^2 + 3^2 + 4^2 = 30$$

$$\begin{aligned}\sum_{i=0}^{3}(i + 1)^2 &= (0 + 1)^2 + (1 + 1)^2 + (2 + 1)^2 + (3 + 1)^2 \\ &= 1^2 + 2^2 + 3^2 + 4^2 = 30\end{aligned}$$

$$\begin{aligned}\sum_{k=2}^{5}(k - 1)^2 &= (2 - 1)^2 + (3 - 1)^2 + (4 - 1)^2 + (5 - 1)^2 \\ &= 1^2 + 2^2 + 3^2 + 4^2 = 30.\end{aligned}$$

EXAMPLE 4 Writing Sums in Summation Notation

Express each sum using summation notation:

a. $1^3 + 2^3 + 3^3 + \cdots + 7^3$ **b.** $1 + \frac{1}{3} + \frac{1}{9} + \frac{1}{27} + \cdots + \frac{1}{3^{n-1}}$.

Solution In each case, we will use 1 as the lower limit of summation and i for the index of summation.

a. The sum $1^3 + 2^3 + 3^3 + \cdots + 7^3$ has seven terms, each of the form i^3, starting at $i = 1$ and ending at $i = 7$. Thus,

$$1^3 + 2^3 + 3^3 + \cdots + 7^3 = \sum_{i=1}^{7} i^3.$$

b. The sum

$$1 + \frac{1}{3} + \frac{1}{9} + \frac{1}{27} + \cdots + \frac{1}{3^{n-1}}$$

has n terms, each of the form $\frac{1}{3^{i-1}}$, starting at $i = 1$ and ending at $i = n$. Thus,

$$1 + \frac{1}{3} + \frac{1}{9} + \frac{1}{27} + \cdots + \frac{1}{3^{n-1}} = \sum_{i=1}^{n} \frac{1}{3^{i-1}}.$$

CHECK POINT 4 Express each sum using summation notation:

a. $1^2 + 2^2 + 3^2 + \cdots + 9^2$ **b.** $1 + \frac{1}{2} + \frac{1}{4} + \frac{1}{8} + \cdots + \frac{1}{2^{n-1}}$.

12.1 EXERCISE SET

MyMathLab

MathXL PRACTICE WATCH DOWNLOAD READ REVIEW

Practice Exercises

In Exercises 1–16, write the first four terms of each sequence whose general term is given.

1. $a_n = 3n + 2$

2. $a_n = 4n - 1$

3. $a_n = 3^n$

4. $a_n = \left(\frac{1}{3}\right)^n$

5. $a_n = (-3)^n$

6. $a_n = \left(-\frac{1}{3}\right)^n$

7. $a_n = (-1)^n(n + 3)$

8. $a_n = (-1)^{n+1}(n + 4)$

9. $a_n = \frac{2n}{n + 4}$

10. $a_n = \frac{3n}{n + 5}$

11. $a_n = \frac{(-1)^{n+1}}{2^n - 1}$

12. $a_n = \frac{(-1)^{n+1}}{2^n + 1}$

13. $a_n = \frac{n^2}{n!}$

14. $a_n = \frac{(n + 1)!}{n^2}$

15. $a_n = 2(n + 1)!$

16. $a_n = -2(n - 1)!$

In Exercises 17–30, find each indicated sum.

17. $\sum_{i=1}^{6} 5i$

18. $\sum_{i=1}^{6} 7i$

19. $\sum_{i=1}^{4} 2i^2$

20. $\sum_{i=1}^{5} i^3$

21. $\sum_{k=1}^{5} k(k + 4)$

22. $\sum_{k=1}^{4} (k - 3)(k + 2)$

23. $\sum_{i=1}^{4} \left(-\frac{1}{2}\right)^i$

24. $\sum_{i=2}^{4} \left(-\frac{1}{3}\right)^i$

25. $\sum_{i=5}^{9} 11$

26. $\sum_{i=3}^{7} 12$

27. $\sum_{i=0}^{4} \frac{(-1)^i}{i!}$

28. $\sum_{i=0}^{4} \frac{(-1)^{i+1}}{(i + 1)!}$

29. $\sum_{i=1}^{5} \frac{i!}{(i - 1)!}$

30. $\sum_{i=1}^{5} \frac{(i + 2)!}{i!}$

In Exercises 31–42, express each sum using summation notation. Use 1 as the lower limit of summation and i for the index of summation.

31. $1^2 + 2^2 + 3^2 + \cdots + 15^2$

32. $1^4 + 2^4 + 3^4 + \cdots + 12^4$

33. $2 + 2^2 + 2^3 + \cdots + 2^{11}$

34. $5 + 5^2 + 5^3 + \cdots + 5^{12}$

35. $1 + 2 + 3 + \cdots + 30$

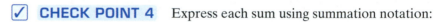

36. $1 + 2 + 3 + \cdots + 40$

37. $\frac{1}{2} + \frac{2}{3} + \frac{3}{4} + \cdots + \frac{14}{14 + 1}$

38. $\frac{1}{3} + \frac{2}{4} + \frac{3}{5} + \cdots + \frac{16}{16 + 2}$

39. $4 + \frac{4^2}{2} + \frac{4^3}{3} + \cdots + \frac{4^n}{n}$

40. $\frac{1}{9} + \frac{2}{9^2} + \frac{3}{9^3} + \cdots + \frac{n}{9^n}$

41. $1 + 3 + 5 + \cdots + (2n - 1)$

42. $a + ar + ar^2 + \cdots + ar^{n-1}$

In Exercises 43–48, express each sum using summation notation. Use a lower limit of summation of your choice and k for the index of summation.

43. $5 + 7 + 9 + 11 + \cdots + 31$

44. $6 + 8 + 10 + 12 + \cdots + 32$

45. $a + ar + ar^2 + \cdots + ar^{12}$

46. $a + ar + ar^2 + \cdots + ar^{14}$

47. $a + (a + d) + (a + 2d) + \cdots + (a + nd)$

48. $(a + d) + (a + d^2) + \cdots + (a + d^n)$

Practice PLUS

In Exercises 49–56, use the graphs of $\{a_n\}$ and $\{b_n\}$ to find each indicated sum.

The Graph of $\{a_n\}$

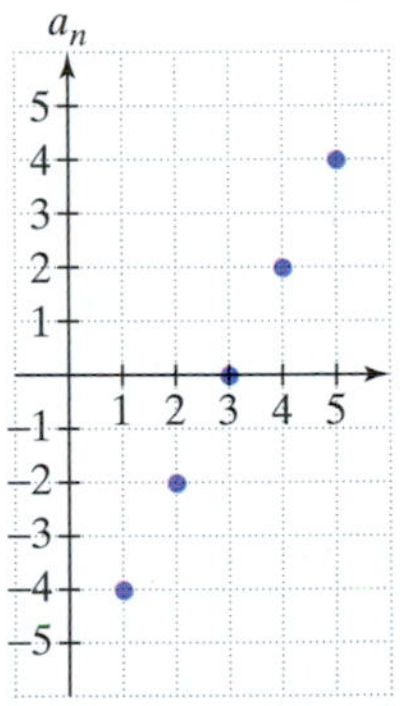

The Graph of $\{b_n\}$

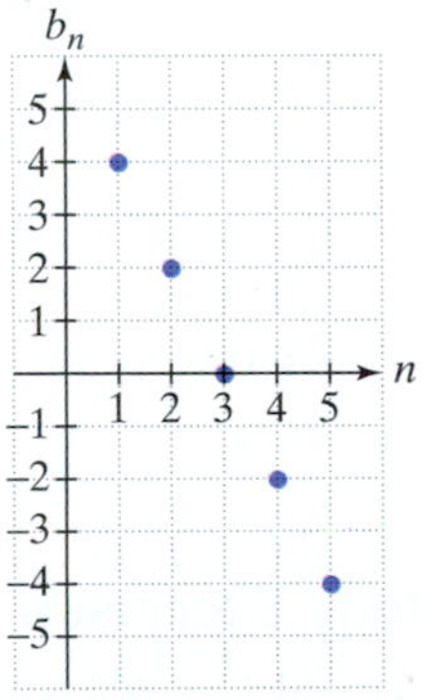

49. $\sum_{i=1}^{5} (a_i^2 + 1)$

50. $\sum_{i=1}^{5} (b_i^2 - 1)$

51. $\sum_{i=1}^{5} (2a_i + b_i)$

52. $\sum_{i=1}^{5} (a_i + 3b_i)$

53. $\sum_{i=4}^{5} \left(\frac{a_i}{b_i}\right)^2$

54. $\sum_{i=4}^{5} \left(\frac{a_i}{b_i}\right)^3$

55. $\sum_{i=1}^{5} a_i^2 + \sum_{i=1}^{5} b_i^2$

56. $\sum_{i=1}^{5} a_i^2 - \sum_{i=3}^{5} b_i^2$

Application Exercises

57. The bar graph shows the number of books published about the September 11, 2001 terrorist attacks from 2001 through 2006.

9/11 Speaks Volumes: Number of Books about the September 11 Terrorist Attacks

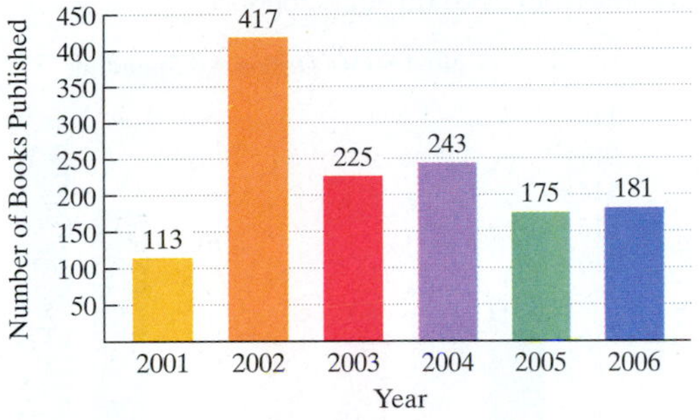

Source: Andrew Grabois, *Books-In-Print*

Let a_n represent the number of books published about the 9/11 attacks in year n, where $n = 1$ corresponds to 2001, $n = 2$ to 2002, and so on.

a. Find $\sum_{i=1}^{6} a_i$. What does this number represent?

b. Find $\frac{\sum_{i=1}^{6} a_i}{6}$. Round to the nearest whole number. What does this number represent?

58. The bar graph shows the U.S. trade deficit in goods and services, in billions of dollars, with China.

United States Trade Deficit with China

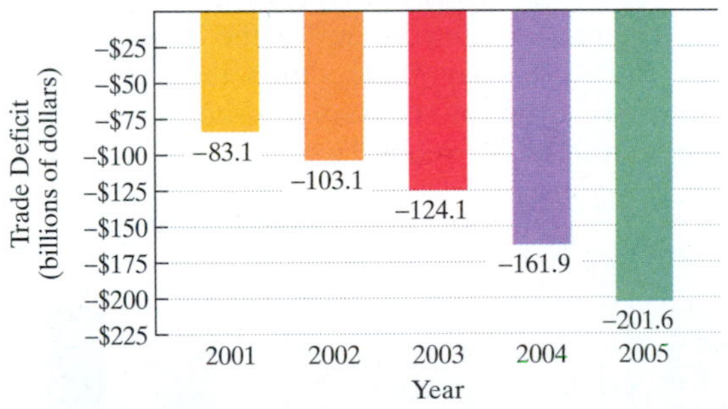

Source: U.S. Census Bureau

Let a_n represent the U.S. trade deficit with China, in billions of dollars, in year n, where $n = 1$ corresponds to 2001, $n = 2$ corresponds to 2002, and so on.

a. Find $\sum_{i=1}^{5} a_i$. What does this number represent?

b. Find $\frac{\sum_{i=1}^{5} a_i}{5}$. What does this number represent?

59. Advertisers don't have to fear that they'll face a sea of "sold out" signs as they rush to the Internet. The growing number of popular sites filled with user-created content, including MySpace.com and YouTube.com, provide plenty of inventory for advertisers who can't find space on top portals such as Yahoo. The bar graph shows U.S. online ad spending, in billions of dollars, from 2000 through 2006.

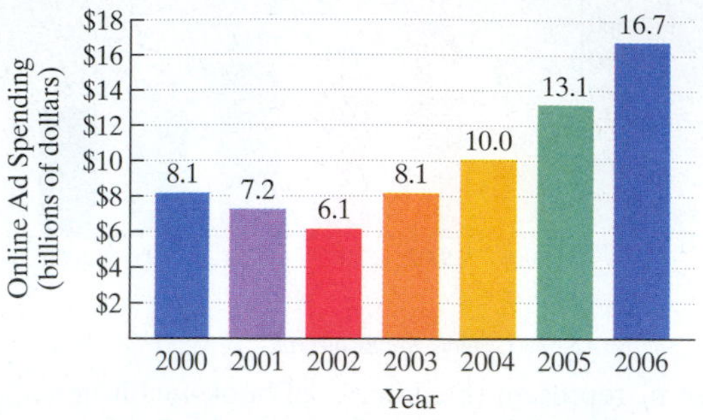

Source: eMarketer

Let a_n represent online ad spending, in billions of dollars, n years after 1999.

a. Use the numbers given in the graph to find and interpret $\frac{1}{7}\sum_{i=1}^{7} a_i$.

b. The finite sequence whose general term is

$$a_n = 0.5n^2 - 1.5n + 8,$$

where $n = 1, 2, 3, \ldots, 7$, models online ad spending, a_n, in billions of dollars, n years after 1999. Use the model to find $\frac{1}{7}\sum_{i=1}^{7} a_i$. Does this underestimate or overestimate the actual sum in part (a)? By how much?

60. More and more television commercial time is devoted to drug companies as hucksters for the benefits and risks of their wares. The bar graph shows the amount that drug companies spent on consumer drug ads, in billions of dollars, from 2002 through 2006.

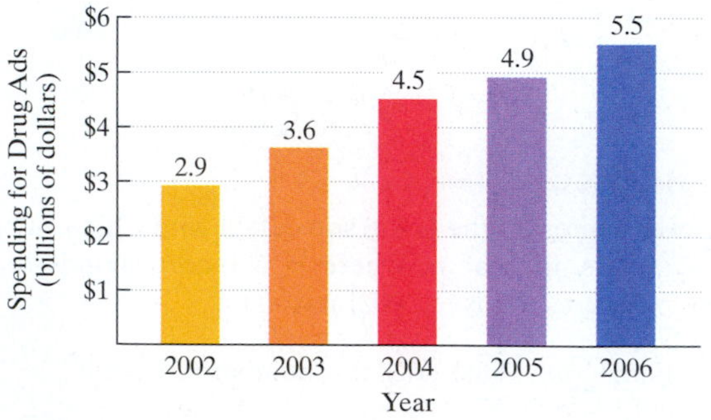

Source: Nielsen Monitor-Plus

Let a_n represent spending for consumer drug ads, in billions of dollars, n years after 2001.

a. Use the numbers given in the graph to find and interpret $\frac{1}{5}\sum_{i=1}^{5} a_i$.

b. The finite sequence whose general term is

$$a_n = 0.65n + 2.3,$$

where $n = 1, 2, 3, 4, 5$, models spending for consumer drug ads, in billions of dollars, n years after 2001. Use the model to find $\frac{1}{5}\sum_{i=1}^{5} a_i$. Does this seem reasonable in terms of the actual sum in part (a), or has model breakdown occurred?

61. A deposit of \$6000 is made in an account that earns 6% interest compounded quarterly. The balance in the account after n quarters is given by the sequence

$$a_n = 6000\left(1 + \frac{0.06}{4}\right)^n, \quad n = 1, 2, 3, \ldots.$$

Find the balance in the account after five years. Round to the nearest cent.

62. A deposit of \$10,000 is made in an account that earns 8% interest compounded quarterly. The balance in the account after n quarters is given by the sequence

$$a_n = 10{,}000\left(1 + \frac{0.08}{4}\right)^n, \quad n = 1, 2, 3, \ldots.$$

Find the balance in the account after six years. Round to the nearest cent.

Writing in Mathematics

63. What is a sequence? Give an example with your description.

64. Explain how to write terms of a sequence if the formula for the general term is given.

65. What does the graph of a sequence look like? How is it obtained?

66. Explain how to find $n!$ if n is a positive integer.

67. What is the meaning of the symbol Σ? Give an example with your description.

68. You buy a new car for \$24,000. At the end of n years, the value of your car is given by the sequence

$$a_n = 24{,}000\left(\frac{3}{4}\right)^n, \quad n = 1, 2, 3, \ldots.$$

Find a_5 and write a sentence explaining what this value represents. Describe the nth term of the sequence in terms of the value of your car at the end of each year.

Technology Exercises

69. Use the SEQ (sequence) capability of a graphing utility to verify the terms of the sequences you obtained for any five sequences from Exercises 1–16.

70. Use the SUM SEQ (sum of the sequence) capability of a graphing utility to verify any five of the sums you obtained in Exercises 17–30.

Many graphing utilities have a sequence-graphing mode that plots the terms of a sequence as points on a rectangular coordinate system. Consult your manual; if your graphing utility has this capability, use it to graph each of the sequences in Exercises 71–74.

What appears to be happening to the terms of each sequence as n gets larger?

71. $a_n = \dfrac{n}{n+1}$; n: [0, 10, 1] by a_n: [0, 1, 0.1]

72. $a_n = \dfrac{100}{n}$; n: [0, 1000, 100] by a_n: [0, 1, 0.1]

73. $a_n = \dfrac{2n^2 + 5n - 7}{n^3}$; n: [0, 10, 1] by a_n: [0, 2, 0.2]

74. $a_n = \dfrac{3n^4 + n - 1}{5n^4 + 2n^2 + 1}$; n: [0, 10, 1] by a_n: [0, 1, 0.1]

Critical Thinking Exercises

Make Sense? *In Exercises 75–78, determine whether each statement "makes sense" or "does not make sense" and explain your reasoning.*

75. Now that I've studied sequences, I realize that the joke in this cartoon is based on the fact that you can't have a negative number of sheep.

WHEN MATHEMATICIANS CAN'T SLEEP

76. By writing $a_1, a_2, a_3, a_4, \ldots, a_n, \ldots$, I can see that the range of a sequence is the set of positive integers.

77. It makes a difference whether or not I use parentheses around the expression following the summation symbol, because the value of $\sum_{i=1}^{8}(i+7)$ is 92, but the value of $\sum_{i=1}^{8} i + 7$ is 43.

78. Without writing out the terms, I can see that $(-1)^{2n}$ in $a_n = \dfrac{(-1)^{2n}}{3n}$ causes the terms to alternate in sign.

In Exercises 79–82, determine whether each statement is true or false. If the statement is false, make the necessary change(s) to produce a true statement.

79. $\sum_{i=1}^{2}(-1)^i 2^i = 0$

80. $\sum_{i=1}^{2} a_i b_i = \sum_{i=1}^{2} a_i \sum_{i=1}^{2} b_i$

81. $\sum_{i=1}^{4} 3i + \sum_{i=1}^{4} 4i = \sum_{i=1}^{4} 7i$

82. $\sum_{i=0}^{6}(-1)^i(i+1)^2 = \sum_{j=1}^{7}(-1)^j j^2$

In Exercises 83–90, find a general term, a_n, for each sequence. More than one answer may be possible.

83. $1, \frac{1}{2}, \frac{1}{3}, \frac{1}{4}, \ldots$

84. $1, 4, 9, 16, \ldots$

85. $-1, 1, -1, 1, \ldots$

86. $1 \cdot 3, 2 \cdot 4, 3 \cdot 5, 4 \cdot 6, \ldots$

87. $\frac{3}{2}, \frac{4}{3}, \frac{5}{4}, \frac{6}{5}, \ldots$

88. $5, 7, 9, 11, \ldots$

89. $\frac{4}{1}, \frac{9}{2}, \frac{16}{3}, \frac{25}{4}, \ldots$

90. $4, -8, 16, -32, \ldots$

91. Evaluate without using a calculator: $\dfrac{600!}{599!}$.

In Exercises 92–93, rewrite each expression as a polynomial in standard form.

92. $\dfrac{(n+4)!}{(n+2)!}$

93. $\dfrac{n!}{(n-3)!}$

In Exercises 94–95, expand and write the answer as a single logarithm with a coefficient of 1.

94. $\sum_{i=1}^{4} \log(2i)$

95. $\sum_{i=2}^{4} 2i \log x$

96. If $a_1 = 7$ and $a_n = a_{n-1} + 5$ for $n \geq 2$, write the first four terms of the sequence.

Review Exercises

97. Simplify: $\sqrt[3]{40x^4y^7}$. (Section 7.3, Example 5)

98. Factor: $27x^3 - 8$. (Section 5.5, Example 9)

99. Solve: $\dfrac{6}{x} + \dfrac{6}{x+2} = \dfrac{5}{2}$.

(Section 6.6, Example 4)

Preview Exercises

Exercises 100–102 will help you prepare for the material covered in the next section.

100. Consider the sequence $8, 3, -2, -7, -12, \ldots$. Find $a_2 - a_1$, $a_3 - a_2$, $a_4 - a_3$, and $a_5 - a_4$. What do you observe?

101. Consider the sequence whose nth term is $a_n = 4n - 3$. Find $a_2 - a_1$, $a_3 - a_2$, $a_4 - a_3$, and $a_5 - a_4$. What do you observe?

102. Use the formula $a_n = 4 + (n-1)(-7)$ to find the eighth term of the sequence $4, -3, -10, \ldots$.

SECTION 12.2

Arithmetic Sequences

Objectives

1. Find the common difference for an arithmetic sequence.
2. Write terms of an arithmetic sequence.
3. Use the formula for the general term of an arithmetic sequence.
4. Use the formula for the sum of the first n terms of an arithmetic sequence.

Your grandmother and her financial counselor are looking at options in case an adult residential facility is needed in the future. The good news is that your grandmother's total assets are \$500,000. The bad news is that yearly adult residential community costs average \$64,130, increasing by \$1800 each year. In this section, we will see how sequences can be used to describe your grandmother's situation and help her to identify realistic options.

1 Find the common difference for an arithmetic sequence.

Arithmetic Sequences

The bar graph in **Figure 12.2** shows annual salaries, rounded to the nearest thousand dollars, of U.S. senators from 2000 to 2005. The graph illustrates that each year salaries increased by \$4 thousand. The sequence of annual salaries

$$142, 146, 150, 154, 158, 162, \ldots$$

shows that each term after the first, 142, differs from the preceding term by a constant amount, namely 4. This sequence is an example of an *arithmetic sequence*.

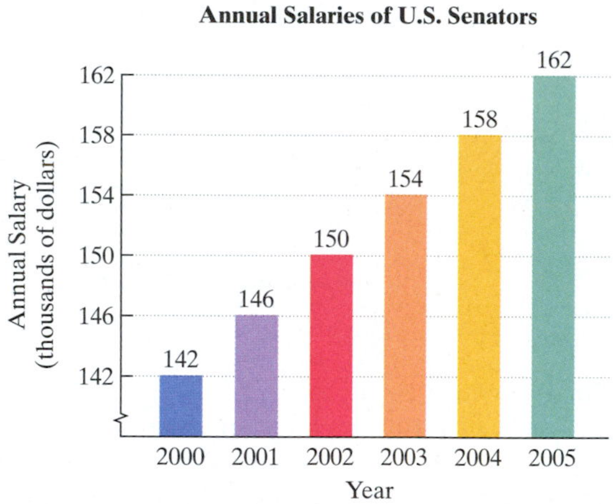

FIGURE 12.2
Source: U.S. Senate

Definition of an Arithmetic Sequence

An **arithmetic sequence** is a sequence in which each term after the first differs from the preceding term by a constant amount. The difference between consecutive terms is called the **common difference** of the sequence.

The common difference, d, is found by subtracting any term from the term that directly follows it. In the following examples, the common difference is found by subtracting the first term from the second term, $a_2 - a_1$.

Arithmetic Sequence	Common Difference
142, 146, 150, 154, 158, . . .	$d = 146 - 142 = 4$
$-5, -2, 1, 4, 7, \ldots$	$d = -2 - (-5) = -2 + 5 = 3$
$8, 3, -2, -7, -12, \ldots$	$d = 3 - 8 = -5$

Figure 12.3 shows the graphs of the last two arithmetic sequences in our list. The common difference for the increasing sequence in **Figure 12.3(a)** is 3. The common difference for the decreasing sequence in **Figure 12.3(b)** is −5.

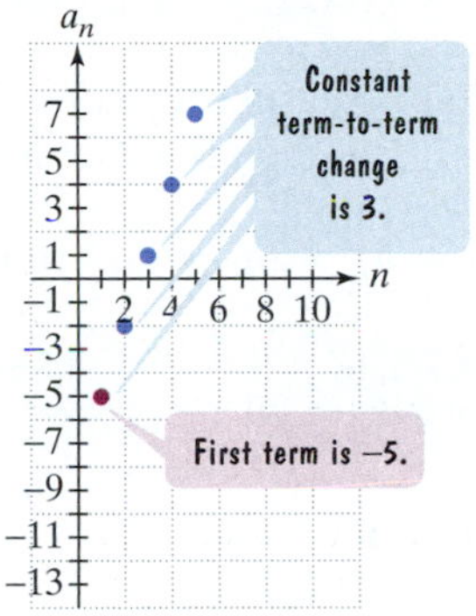

FIGURE 12.3(a) The graph of $\{a_n\} = -5, -2, 1, 4, 7, \ldots$

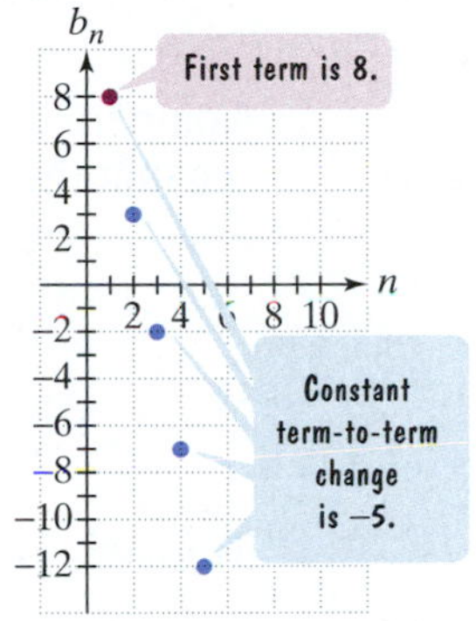

FIGURE 12.3(b) The graph of $\{b_n\} = 8, 3, -2, -7, -12, \ldots$

The graph of each arithmetic sequence in **Figure 12.3** forms a set of discrete points lying on a straight line. This illustrates that **an arithmetic sequence is a linear function whose domain is the set of positive integers**.

If the first term of an arithmetic sequence is a_1, each term after the first is obtained by adding d, the common difference, to the previous term.

2 Write terms of an arithmetic sequence.

EXAMPLE 1 Writing the Terms of an Arithmetic Sequence Using the First Term and the Common Difference

Write the first six terms of the arithmetic sequence with first term 6 and common difference −2.

Solution To find the second term, we add −2 to the first term, 6, giving 4. For the next term, we add −2 to 4, and so on.

$$\begin{aligned}
a_1 \text{ (first term)} &= 6 \\
a_2 \text{ (second term)} &= 6 + (-2) = 4 \\
a_3 \text{ (third term)} &= 4 + (-2) = 2 \\
a_4 \text{ (fourth term)} &= 2 + (-2) = 0 \\
a_5 \text{ (fifth term)} &= 0 + (-2) = -2 \\
a_6 \text{ (sixth term)} &= -2 + (-2) = -4
\end{aligned}$$

The first six terms are

$$6, 4, 2, 0, -2, \text{ and } -4.$$

✓ CHECK POINT 1 Write the first six terms of the arithmetic sequence with first term 100 and common difference −30.

3 Use the formula for the general term of an arithmetic sequence.

The General Term of an Arithmetic Sequence

Consider an arithmetic sequence whose first term is a_1 and whose common difference is d. We are looking for a formula for the general term, a_n. Let's begin by writing the first six terms. The first term is a_1. The second term is $a_1 + d$. The third term is $a_1 + d + d$, or $a_1 + 2d$. Thus, we start with a_1 and add d to each successive term. The first six terms are

$$a_1, \quad a_1 + d, \quad a_1 + 2d, \quad a_1 + 3d, \quad a_1 + 4d, \quad a_1 + 5d.$$

a_1, first term; a_2, second term; a_3, third term; a_4, fourth term; a_5, fifth term; a_6, sixth term

Compare the coefficient of d and the subscript of a denoting the term number. Can you see that the coefficient of d is 1 less than the subscript of a denoting the term number?

$$a_3\text{: third term} = a_1 + 2d \qquad a_4\text{: fourth term} = a_1 + 3d$$

One less than 3, or 2, is the coefficient of d.

One less than 4, or 3, is the coefficient of d.

Thus, the formula for the nth term is

$$a_n\text{: } n\text{th term} = a_1 + (n - 1)d.$$

One less than n, or $n - 1$, is the coefficient of d.

General Term of an Arithmetic Sequence

The nth term (the general term) of an arithmetic sequence with first term a_1 and common difference d is

$$a_n = a_1 + (n - 1)d.$$

EXAMPLE 2 Using the Formula for the General Term of an Arithmetic Sequence

Find the eighth term of the arithmetic sequence whose first term is 4 and whose common difference is −7.

Solution To find the eighth term, a_8, we replace n in the formula with 8, a_1 with 4, and d with −7.

$$a_n = a_1 + (n - 1)d$$

$$a_8 = 4 + (8 - 1)(-7) = 4 + 7(-7) = 4 + (-49) = -45$$

The eighth term is −45. We can check this result by writing the first eight terms of the sequence:

$$4, -3, -10, -17, -24, -31, -38, -45.$$

✓ CHECK POINT 2 Find the ninth term of the arithmetic sequence whose first term is 6 and whose common difference is −5.

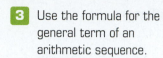

EXAMPLE 3 Using an Arithmetic Sequence to Model Teachers' Earnings

According to the National Education Association, teachers in the United States earned an average of \$30,532 in 1990. This amount has increased by approximately \$1472 per year.

a. Write a formula for the nth term of the arithmetic sequence that models teachers' average earnings n years after 1989.

b. How much will U.S. teachers earn, on average, by the year 2010?

Solution

a. We can express teachers' earnings by the following arithmetic sequence:

$$30{,}532, \quad 32{,}004, \quad 33{,}476, \quad 34{,}948, \ldots$$

a_1: earnings in 1990, 1 year after 1989

a_2: earnings in 1991, 2 years after 1989

a_3: earnings in 1992, 3 years after 1989

a_4: earnings in 1993, 4 years after 1989

In this sequence, a_1, the first term, represents the amount teachers earned in 1990. Each subsequent year this amount increases by \$1472, so $d = 1472$. We use the formula for the general term of an arithmetic sequence to write the nth term of the sequence that describes teachers' earnings n years after 1989.

$$a_n = a_1 + (n-1)d$$ This is the formula for the general term of an arithmetic sequence.

$$a_n = 30{,}532 + (n-1)1472$$ $a_1 = 30{,}532$ and $d = 1472$.

$$a_n = 30{,}532 + 1472n - 1472$$ Distribute 1472 to each term in parentheses.

$$a_n = 1472n + 29{,}060$$ Simplify.

Thus, teachers' earnings n years after 1989 can be modeled by $a_n = 1472n + 29{,}060$.

b. Now we need to find teachers' earnings in 2010. The year 2010 is 21 years after 1989: That is, $2010 - 1989 = 21$. Thus, $n = 21$. We substitute 21 for n in $a_n = 1472n + 29{,}060$.

$$a_{21} = 1472 \cdot 21 + 29{,}060 = 59{,}972$$

The 21st term of the sequence is 59,972. Therefore, U.S. teachers are predicted to earn an average of \$59,972 by the year 2010.

CHECK POINT 3 Thanks to drive-thrus and curbside delivery, Americans are eating more meals behind the wheel. In 2004, we averaged 32 à la car meals, increasing by approximately 0.7 meal per year. (*Source*: *Newsweek*)

a. Write a formula for the nth term of the arithmetic sequence that models the average number of car meals n years after 2003.

b. How many car meals will Americans average by the year 2014?

4 Use the formula for the sum of the first n terms of an arithmetic sequence.

The Sum of the First n Terms of an Arithmetic Sequence

The sum of the first n terms of an arithmetic sequence, denoted by S_n, and called the ***n*th partial sum**, can be found without having to add up all the terms. Let

$$S_n = a_1 + a_2 + a_3 + \cdots + a_n$$

be the sum of the first n terms of an arithmetic sequence. Because d is the common difference between terms, S_n can be written forward and backward, as shown on the next page.

Forward: Start with the first term, a_1. Keep adding d.

Backward: Start with the last term, a_n. Keep subtracting d.

$$
\begin{aligned}
S_n &= a_1 + (a_1 + d) + (a_1 + 2d) + \cdots + a_n \\
S_n &= a_n + (a_n - d) + (a_n - 2d) + \cdots + a_1 \\
\hline
2S_n &= (a_1 + a_n) + (a_1 + a_n) + (a_1 + a_n) + \cdots + (a_1 + a_n)
\end{aligned}
$$

Add the two equations.

Because there are n sums of $(a_1 + a_n)$ on the right side, we can express this side as $n(a_1 + a_n)$. Thus, the last equation can be written as follows:

$$2S_n = n(a_1 + a_n).$$

$$S_n = \frac{n}{2}(a_1 + a_n)$$ Solve for S_n, dividing both sides by 2.

We have proved the following result:

The Sum of the First n Terms of an Arithmetic Sequence

The sum, S_n, of the first n terms of an arithmetic sequence is given by

$$S_n = \frac{n}{2}(a_1 + a_n),$$

in which a_1 is the first term and a_n is the nth term.

To find the sum of the terms of an arithmetic sequence using $S_n = \frac{n}{2}(a_1 + a_n)$, we need to know the first term, a_1, the last term, a_n, and the number of terms, n. The following examples illustrate how to use this formula.

EXAMPLE 4 Finding the Sum of n Terms of an Arithmetic Sequence

Find the sum of the first 100 terms of the arithmetic sequence: 1, 3, 5, 7, ...

Solution By finding the sum of the first 100 terms of 1, 3, 5, 7, . . . , we are finding the sum of the first 100 odd numbers. To find the sum of the first 100 terms, S_{100}, we replace n in the formula with 100.

$$S_n = \frac{n}{2}(a_1 + a_n)$$

$$S_{100} = \frac{100}{2}(a_1 + a_{100})$$

The first term, a_1, is 1.

We must find a_{100}, the 100th term.

We use the formula for the general term of an arithmetic sequence to find a_{100}. The common difference, d, of 1, 3, 5, 7, . . . , is 2.

$a_n = a_1 + (n - 1)d$ — This is the formula for the nth term of an arithmetic sequence. Use it to find the 100th term.

$a_{100} = 1 + (100 - 1) \cdot 2$ — Substitute 100 for n, 2 for d, and 1 (the first term) for a_1.

$= 1 + 99 \cdot 2$ — Perform the subtraction in parentheses.

$= 1 + 198 = 199$ — Multiply $(99 \cdot 2 = 198)$ and then add.

Now we are ready to find the sum of the 100 terms 1, 3, 5, 7, . . . , 199.

$$S_n = \frac{n}{2}(a_1 + a_n)$$ Use the formula for the sum of the first n terms of an arithmetic sequence. Let $n = 100$, $a_1 = 1$, and $a_{100} = 199$.

$$S_{100} = \frac{100}{2}(1 + 199) = 50(200) = 10{,}000$$

The sum of the first 100 odd numbers is 10,000. Equivalently, the 100th partial sum of the sequence 1, 3, 5, 7, . . . is 10,000.

✓ **CHECK POINT 4** Find the sum of the first 15 terms of the arithmetic sequence: 3, 6, 9, 12, ...

Using Technology

To find

$$\sum_{i=1}^{25}(5i - 9)$$

on a graphing utility, enter:

SUM SEQ $(5x - 9, x, 1, 25, 1)$.

Then press ENTER .

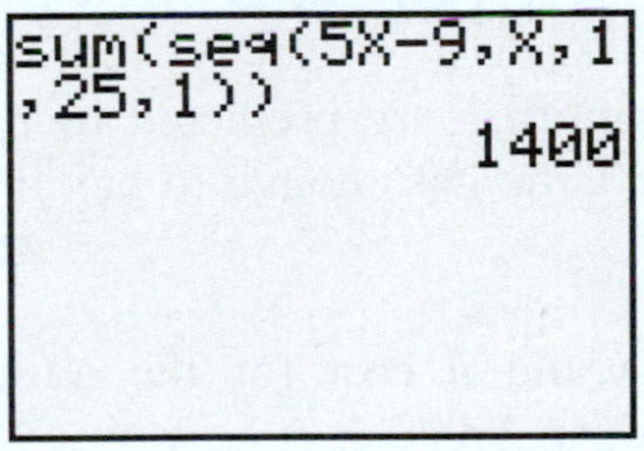

EXAMPLE 5 Using S_n to Evaluate a Summation

Find the following sum: $\sum_{i=1}^{25}(5i - 9)$.

Solution

$$\sum_{i=1}^{25}(5i - 9) = (5 \cdot 1 - 9) + (5 \cdot 2 - 9) + (5 \cdot 3 - 9) + \cdots + (5 \cdot 25 - 9)$$
$$= -4 + 1 + 6 + \cdots + 116$$

By evaluating the first three terms and the last term, we see that $a_1 = -4$, d, the common difference, is $1 - (-4)$ or 5, and a_{25}, the last term, is 116.

$$S_n = \frac{n}{2}(a_1 + a_n)$$ Use the formula for the sum of the first n terms of an arithmetic sequence. Let $n = 25$, $a_1 = -4$, and $a_{25} = 116$.

$$S_{25} = \frac{25}{2}(-4 + 116) = \frac{25}{2}(112) = 1400$$

Thus,

$$\sum_{i=1}^{25}(5i - 9) = 1400.$$

✓ **CHECK POINT 5** Find the following sum: $\sum_{i=1}^{30}(6i - 11)$.

EXAMPLE 6 Modeling Total Nursing Home Costs over a Six-Year Period

Your grandmother has assets of \$500,000. One option that she is considering involves an adult residential community for a six-year period beginning in 2009. The model

$$a_n = 1800n + 64{,}130$$

describes yearly adult residential community costs n years after 2008. Does your grandmother have enough to pay for the facility?

Solution We must find the sum of an arithmetic sequence whose general term is $a_n = 1800n + 64{,}130$. The first term of the sequence corresponds to the facility's costs in the year 2009. The last term corresponds to costs in the year 2014. Because the model describes costs n years after 2008, $n = 1$ describes the year 2009 and $n = 6$ describes the year 2014.

$$a_n = 1800n + 64{,}130$$ This is the given formula for the general term of the sequence.

$$a_1 = 1800 \cdot 1 + 64{,}130 = 65{,}930$$ Find a_1 by replacing n with 1.

$$a_6 = 1800 \cdot 6 + 64{,}130 = 74{,}930$$ Find a_6 by replacing n with 6.

The first year the facility will cost \$65,930. By year six, the facility will cost \$74,930. Now we must find the sum of the costs for all six years. We focus on the sum of the first six terms of the arithmetic sequence

$$65{,}930,\ 67{,}730,\ \ldots,\ 74{,}930.$$

(a_1, a_2, a_6)

We find this sum using the formula for the sum of the first n terms of an arithmetic sequence. We are adding 6 terms: $n = 6$. The first term is 65,930: $a_1 = 65{,}930$. The last term—that is, the sixth term—is 74,930: $a_6 = 74{,}930$.

$$S_n = \frac{n}{2}(a_1 + a_n)$$

$$S_6 = \frac{6}{2}(65{,}930 + 74{,}930) = 3(140{,}860) = 422{,}580$$

Total adult residential community costs for your grandmother are predicted to be \$422,580. Because your grandmother's assets are \$500,000, she has enough to pay for the facility for the six-year period.

CHECK POINT 6 In Example 6, how much would it cost for the adult residential community for a ten-year period beginning in 2009?

12.2 EXERCISE SET

MyMathLab — MathXL PRACTICE — WATCH — DOWNLOAD — READ — REVIEW

Practice Exercises

In Exercises 1–6, find the common difference for each arithmetic sequence.

1. 2, 6, 10, 14, ...
2. 3, 8, 13, 18, ...
3. −7, −2, 3, 8, ...
4. −10, −4, 2, 8, ...
5. 714, 711, 708, 705, ...
6. 611, 606, 601, 596, ...

In Exercises 7–16, write the first six terms of each arithmetic sequence with the given first term, a_1, and common difference, d.

7. $a_1 = 200, d = 20$
8. $a_1 = 300, d = 50$
9. $a_1 = -7, d = 4$
10. $a_1 = -8, d = 5$
11. $a_1 = 300, d = -90$
12. $a_1 = 200, d = -60$
13. $a_1 = \frac{5}{2}, d = -\frac{1}{2}$
14. $a_1 = \frac{3}{4}, d = -\frac{1}{4}$
15. $a_1 = -0.4, d = -1.6$
16. $a_1 = -0.3, d = -1.7$

In Exercises 17–24, use the formula for the general term (the nth term) of an arithmetic sequence to find the indicated term of each sequence with the given first term, a_1, and common difference, d.

17. Find a_6 when $a_1 = 13, d = 4$.
18. Find a_{16} when $a_1 = 9, d = 2$,
19. Find a_{50} when $a_1 = 7, d = 5$,
20. Find a_{60} when $a_1 = 8, d = 6$.
21. Find a_{200} when $a_1 = -40, d = 5$.
22. Find a_{150} when $a_1 = -60, d = 5$.

23. Find a_{60} when $a_1 = 35, d = -3$.

24. Find a_{70} when $a_1 = -32, d = 4$.

In Exercises 25–34, write a formula for the general term (the nth term) of each arithmetic sequence. Then use the formula for a_n to find a_{20}, the 20th term of the sequence.

25. $1, 5, 9, 13, \ldots$

26. $2, 7, 12, 17, \ldots$

27. $7, 3, -1, -5, \ldots$

28. $6, 1, -4, -9, \ldots$

29. $-20, -24, -28, -32, \ldots$

30. $-70, -75, -80, -85, \ldots$

31. $a_1 = -\frac{1}{3}, d = \frac{1}{3}$

32. $a_1 = -\frac{1}{4}, d = \frac{1}{4}$

33. $a_1 = 4, d = -0.3$

34. $a_1 = 5, d = -0.2$

35. Find the sum of the first 20 terms of the arithmetic sequence: $4, 10, 16, 22, \ldots$.

36. Find the sum of the first 25 terms of the arithmetic sequence: $7, 19, 31, 43, \ldots$.

37. Find the sum of the first 50 terms of the arithmetic sequence: $-10, -6, -2, 2, \ldots$.

38. Find the sum of the first 50 terms of the arithmetic sequence: $-15, -9, -3, 3, \ldots$.

39. Find $1 + 2 + 3 + 4 + \cdots + 100$, the sum of the first 100 natural numbers.

40. Find $2 + 4 + 6 + 8 + \cdots + 200$, the sum of the first 100 positive even integers.

41. Find the sum of the first 60 positive even integers.

42. Find the sum of the first 80 positive even integers.

43. Find the sum of the even integers between 21 and 45.

44. Find the sum of the odd integers between 30 and 54.

For Exercises 45–50, write out the first three terms and the last term. Then use the formula for the sum of the first n terms of an arithmetic sequence to find the indicated sum.

45. $\sum_{i=1}^{17} (5i + 3)$

46. $\sum_{i=1}^{20} (6i - 4)$

47. $\sum_{i=1}^{30} (-3i + 5)$

48. $\sum_{i=1}^{40} (-2i + 6)$

49. $\sum_{i=1}^{100} 4i$

50. $\sum_{i=1}^{50} (-4i)$

Practice PLUS

Use the graphs of the arithmetic sequences $\{a_n\}$ and $\{b_n\}$ to solve Exercises 51–58.

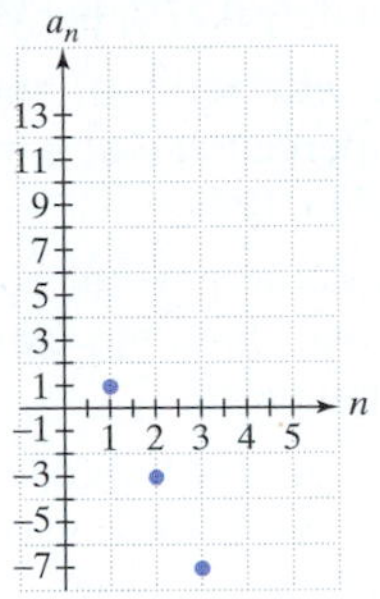

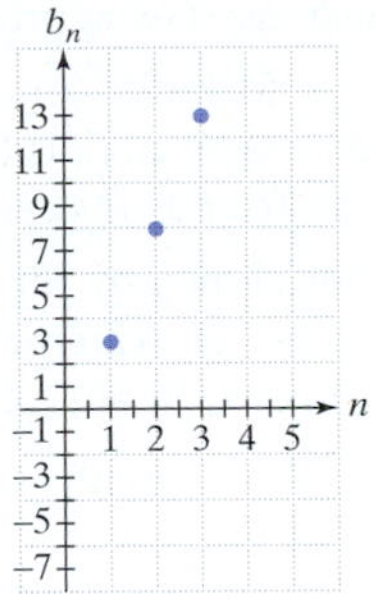

51. Find $a_{14} + b_{12}$.

52. Find $a_{16} + b_{18}$.

53. If $\{a_n\}$ is a finite sequence whose last term is -83, how many terms does $\{a_n\}$ contain?

54. If $\{b_n\}$ is a finite sequence whose last term is 93, how many terms does $\{b_n\}$ contain?

55. Find the difference between the sum of the first 14 terms of $\{b_n\}$ and the sum of the first 14 terms of $\{a_n\}$.

56. Find the difference between the sum of the first 15 terms of $\{b_n\}$ and the sum of the first 15 terms of $\{a_n\}$.

57. Write a linear function $f(x) = mx + b$, whose domain is the set of positive integers, that represents $\{a_n\}$.

58. Write a linear function $g(x) = mx + b$, whose domain is the set of positive integers, that represents $\{b_n\}$.

Use a system of two equations in two variables, a_1 and d, to solve Exercises 59–60.

59. Write a formula for the general term (the nth term) of the arithmetic sequence whose second term, a_2, is 4 and whose sixth term, a_6, is 16.

60. Write a formula for the general term (the nth term) of the arithmetic sequence whose third term, a_3, is 7 and whose eighth term, a_8, is 17.

Application Exercises

The bar graphs show changes that have taken place in the United States over time. Exercises 61–62 involve developing arithmetic sequences that model the data.

Percentage of United States Adults with No Close Friends

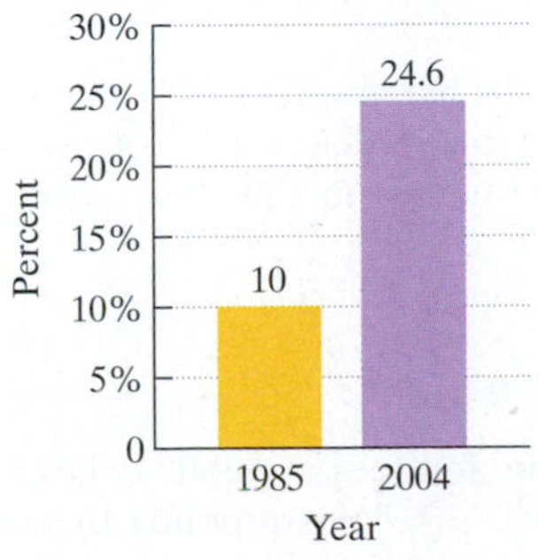

Source: American Sociological Review

Percentage of High School Grades of A+, A, or A− for College Freshmen

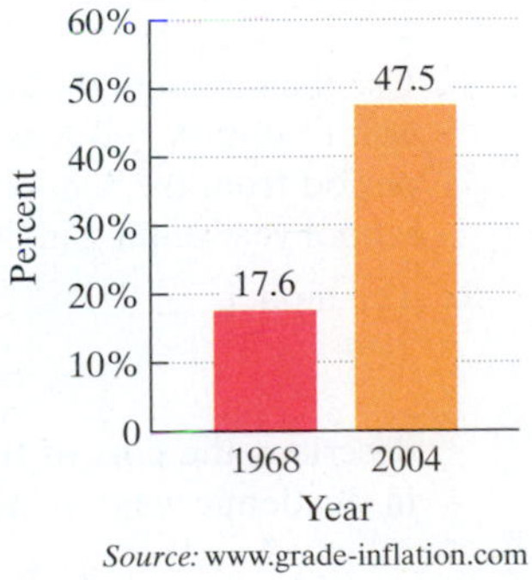

Source: www.grade-inflation.com

(For Exercises 61–62, graphs are shown at the bottom of the previous page.)

61. In 1985, 10% of Americans had no close friends. On average, this has increased by approximately 0.77% per year.

a. Write a formula for the nth term of the arithmetic sequence that models the percentage of Americans with no close friends n years after 1984.

b. If trends shown by the model in part (a) continue, what percentage of Americans will have no close friends in 2011? Round to one decimal place.

62. In 1968, 17.6% of high school grades for college freshmen consisted of A's (A+, A, or A−). On average, this has increased by approximately 0.83% per year.

a. Write a formula for the nth term of the arithmetic sequence that models the percentage of high school grades of A for college freshmen n years after 1967.

b. If trends shown by the model in part (a) continue, what percentage of high school grades for college freshmen will consist of A's in 2018?

63. Company A pays $24,000 yearly with raises of $1600 per year. Company B pays $28,000 yearly with raises of $1000 per year. Which company will pay more in year 10? How much more?

64. Company A pays $23,000 yearly with raises of $1200 per year. Company B pays $26,000 yearly with raises of $800 per year. Which company will pay more in year 10? How much more?

In Exercises 65–66, we revisit the data from Chapter 1 showing the average cost of tuition and fees at public and private four-year U.S. colleges.

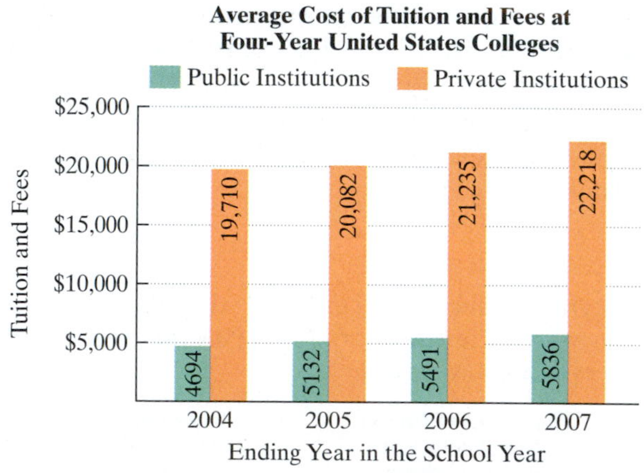

Source: The College Board

65. a. Use the numbers shown in the bar graph to find the total cost of tuition and fees at public colleges for a four-year period from the school year ending in 2004 through the school year ending in 2007.

b. The model

$$a_n = 379n + 4342$$

describes the cost of tuition and fees at public colleges in academic year n, where $n = 1$ corresponds to the school year ending in 2004, $n = 2$ to the school year ending in 2005, and so on. Use this model and the formula for S_n to find the total cost of tuition and fees at public colleges for a four-year period from the school year ending in 2004 through the school year ending in 2007. Does the model understimate or overstimate the actual sum you obtained in part (a)? By how much?

66. a. Use the numbers shown in the bar graph to find the total cost of tuition and fees at private colleges for a four-year period from the school year ending in 2004 through the school year ending in 2007.

b. The model

$$a_n = 868n + 18{,}642$$

describes the cost of tuition and fees at private colleges in academic year n, where $n = 1$ corresponds to the school year ending in 2004, $n = 2$ to the school year ending in 2005, and so on. Use this model and the formula for S_n, to find the total cost of tuition and fees at private colleges for a four-year period from the school year ending in 2004 through the school year ending in 2007. Does the model underestimate or overestimate the actual sum that you obtained in part (a)? By how much?

67. Use one of the models in Exercises 65–66 and the formula for S_n to find the total cost of tuition and fees for your undergraduate education. How well does the model describe your anticipated costs?

68. A company offers a starting yearly salary of $33,000 with raises of $2500 per year. Find the total salary over a ten-year period.

69. You are considering two job offers. Company A will start you at $19,000 a year and guarantee a raise of $2600 per year. Company B will start you at a higher salary, $27,000 a year, but will only guarantee a raise of $1200 per year. Find the total salary that each company will pay over a ten-year period. Which company pays the greater total amount?

70. A theater has 30 seats in the first row, 32 seats in the second row, increasing by 2 seats each row for a total of 26 rows. How many seats are there in the theater?

71. A section in a stadium has 20 seats in the first row, 23 seats in the second row, increasing by 3 seats each row for a total of 38 rows. How many seats are in this section of the stadium?

Writing in Mathematics

72. What is an arithmetic sequence? Give an example with your explanation.

73. What is the common difference in an arithmetic sequence?

74. Explain how to find the general term of an arithmetic sequence.

75. Explain how to find the sum of the first n terms of an arithmetic sequence without having to add up all the terms.

Technology Exercises

76. Use the SEQ (sequence) capability of a graphing utility and the formula you obtained for a_n to verify the value you found for a_{20} in any five exercises from Exercises 25–34.

77. Use the capability of a graphing utility to calculate the sum of a sequence to verify any five of your answers to Exercises 45–50.

Critical Thinking Exercises

Make Sense? *In Exercises 78–81, determine whether each statement "makes sense" or "does not make sense" and explain your reasoning.*

78. Rather than performing the addition, I used the formula $S_n = \frac{n}{2}(a_1 + a_n)$ to find the sum of the first thirty terms of the sequence 2, 4, 8, 16, 32, ...

79. I was able to find the sum of the first fifty terms of an arithmetic sequence even though I did not identify every term.

80. The sequence for the number of seats per row in our movie theater as the rows move toward the back is arithmetic with $d = 1$ so people don't block the view of those in the row behind them.

81. Beginning at 6:45 A.M., a bus stops on my block every 23 minutes, so I used the formula for the nth term of an arithmetic sequence to describe the stopping time for the nth bus of the day.

In Exercises 82–85, determine whether each statement is true or false. If the statement is false, make the necessary change(s) to produce a true statement.

82. The sum of an arithmetic sequence cannot be negative.

83. The common difference for the arithmetic sequence $1, -1, -3, -5, \ldots$ is 2.

84. An arithmetic sequence is a linear function whose domain is the set of natural numbers.

85. The sequence $\log_2 2, \log_2 4, \log_2 8, \log_2 16, \log_2 32, \ldots$ is arithmetic.

86. Give examples of two different arithmetic sequences whose fourth term, a_4, is 10.

87. In the sequence 21,700, 23,172, 24,644, 26,116, . . . , which term is 314,628?

88. A *degree-day* is a unit used to measure the fuel requirements of buildings. By definition, each degree that the average daily temperature is below 65°F is 1 degree-day. For example, an average daily temperature of 42°F constitutes 23 degree-days. If the average temperature on January 1 was 42°F and fell 2°F for each subsequent day up to and including January 10, how many degree-days are included from January 1 to January 10?

89. Show that the sum of the first n positive odd integers,

$$1 + 3 + 5 + \cdots + (2n - 1),$$

is n^2.

Review Exercises

90. Solve: $\log(x^2 - 25) - \log(x + 5) = 3$. (Section 9.5, Example 5)

91. Solve: $x^2 + 3x \le 10$. (Section 8.5, Example 1)

92. Solve for P: $A = \dfrac{Pt}{P + t}$.

(Section 6.7, Example 1)

Preview Exercises

Exercises 93–95 will help you prepare for the material covered in the next section.

93. Consider the sequence $1, -2, 4, -8, 16, \ldots$. Find $\frac{a_2}{a_1}, \frac{a_3}{a_2}, \frac{a_4}{a_3}$, and $\frac{a_5}{a_4}$. What do you observe?

94. Consider the sequence whose nth term is $a_n = 3 \cdot 5^n$. Find $\frac{a_2}{a_1}, \frac{a_3}{a_2}, \frac{a_4}{a_3}$, and $\frac{a_5}{a_4}$. What do you observe?

95. Use the formula $a_n = a_1 3^{n-1}$ to find the 7th term of the sequence 11, 33, 99, 297, ...

SECTION 12.3

Geometric Sequences and Series

Objectives

1. Find the common ratio of a geometric sequence.
2. Write terms of a geometric sequence.
3. Use the formula for the general term of a geometric sequence.
4. Use the formula for the sum of the first n terms of a geometric sequence.
5. Find the value of an annuity.
6. Use the formula for the sum of an infinite geometric series.

Here we are at the closing moments of a job interview. You're shaking hands with the manager. You managed to answer all the tough questions without losing your poise, and now you've been offered a job. As a matter of fact, your qualifications are so terrific that you've been offered two jobs—one just the day before, with a rival company in the same field! One company offers \$30,000 the first year, with increases of 6% per year for four years after that. The other offers \$32,000 the first year, with annual increases of 3% per year after that. Over a five-year period, which is the better offer?

If salary raises amount to a certain percent each year, the yearly salaries over time form a geometric sequence. In this section, we investigate geometric sequences and their properties. After studying the section, you will be in a position to decide which job offer to accept: You will know which company will pay you more over five years.

1 Find the common ratio of a geometric sequence.

Geometric Sequences

Figure 12.4 shows a sequence in which the number of squares is increasing. From left to right, the number of squares is 1, 5, 25, 125, and 625. In this sequence, each term after the first, 1, is obtained by multiplying the preceding term by a constant amount, namely 5. This sequence of increasing numbers of squares is an example of a *geometric sequence.*

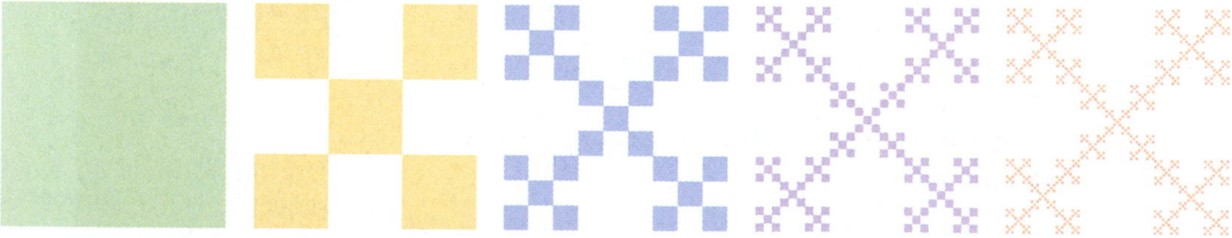

FIGURE 12.4 A geometric sequence of squares

Definition of a Geometric Sequence

A **geometric sequence** is a sequence in which each term after the first is obtained by multiplying the preceding term by a fixed nonzero constant. The amount by which we multiply each time is called the **common ratio** of the sequence.

The common ratio, r, is found by dividing any term after the first term by the term that directly precedes it. In the following examples, the common ratio is found by dividing the second term by the first term, $\frac{a_2}{a_1}$.

Geometric sequence	Common ratio
$1, 5, 25, 125, 625, \ldots$	$r = \frac{5}{1} = 5$
$4, 8, 16, 32, 64, \ldots$	$r = \frac{8}{4} = 2$
$6, -12, 24, -48, 96, \ldots$	$r = \frac{-12}{6} = -2$
$9, -3, 1, -\frac{1}{3}, \frac{1}{9}, \ldots$	$r = \frac{-3}{9} = -\frac{1}{3}$

Study Tip

When the common ratio of a geometric sequence is negative, the signs of the terms alternate.

Figure 12.5 shows a partial graph of the first geometric sequence in our list, 1, 5, 25, 125 .... The graph forms a set of discrete points lying on the exponential function $f(x) = 5^{x-1}$. This illustrates that **a geometric sequence with a positive common ratio other than 1 is an exponential function whose domain is the set of positive integers.**

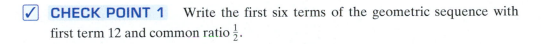

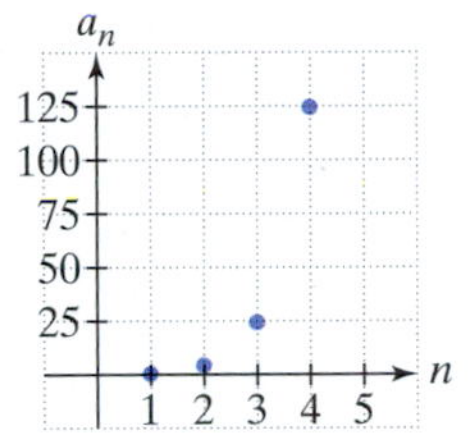

FIGURE 12.5 The graph of $\{a_n\} = 1, 5, 25, 125, \ldots$

2 Write terms of a geometric sequence.

How do we write out the terms of a geometric sequence when the first term and the common ratio are known? We multiply the first term by the common ratio to get the second term, multiply the second term by the common ratio to get the third term, and so on.

EXAMPLE 1 Writing the Terms of a Geometric Sequence

Write the first six terms of the geometric sequence with first term 6 and common ratio $\frac{1}{3}$.

Solution The first term is 6. The second term is $6 \cdot \frac{1}{3}$, or 2. The third term is $2 \cdot \frac{1}{3}$, or $\frac{2}{3}$. The fourth term is $\frac{2}{3} \cdot \frac{1}{3}$, or $\frac{2}{9}$, and so on. The first six terms are

$$6, 2, \frac{2}{3}, \frac{2}{9}, \frac{2}{27}, \text{ and } \frac{2}{81}.$$

✓ **CHECK POINT 1** Write the first six terms of the geometric sequence with first term 12 and common ratio $\frac{1}{2}$.

3 Use the formula for the general term of a geometric sequence.

The General Term of a Geometric Sequence

Consider a geometric sequence whose first term is a_1 and whose common ratio is r. We are looking for a formula for the general term, a_n. Let's begin by writing the first six terms. The first term is a_1. The second term is $a_1 r$. The third term is $a_1 r \cdot r$, or $a_1 r^2$. The fourth term is $a_1 r^2 \cdot r$, or $a_1 r^3$, and so on. Starting with a_1 and multiplying each successive term by r, the first six terms are $a_1, a_1 r, a_1 r^2, a_1 r^3, a_1 r^4$, and $a_1 r^5$.

Compare the exponent on r and the subscript of a denoting the term number:

$$a_1, \quad a_1r, \quad a_1r^2, \quad a_1r^3, \quad a_1r^4, \quad a_1r^5.$$

a_1, first term; a_2, second term; a_3, third term; a_4, fourth term; a_5, fifth term; a_6, sixth term

Can you see that the exponent on r is 1 less than the subscript of a denoting the term number?

$$a_3\text{: third term} = a_1r^2 \qquad a_4\text{: fourth term} = a_1r^3$$

One less than 3, or 2, is the exponent on r.

One less than 4, or 3, is the exponent on r.

Thus, the formula for the nth term is

$$a_n = a_1r^{n-1}.$$

One less than n, or $n - 1$, is the exponent on r.

General Term of a Geometric Sequence

The nth term (the general term) of a geometric sequence with first term a_1 and common ratio r is

$$a_n = a_1r^{n-1}.$$

Study Tip

Be careful with the order of operations when evaluating

$$a_1r^{n-1}.$$

First find r^{n-1}. Then multiply the result by a_1.

EXAMPLE 2 Using the Formula for the General Term of a Geometric Sequence

Find the eighth term of the geometric sequence whose first term is -4 and whose common ratio is -2.

Solution To find the eighth term, a_8, we replace n in the formula with 8, a_1 with -4, and r with -2.

$$a_n = a_1r^{n-1}$$
$$a_8 = -4(-2)^{8-1} = -4(-2)^7 = -4(-128) = 512$$

The eighth term is 512. We can check this result by writing the first eight terms of the sequence:

$$-4, 8, -16, 32, -64, 128, -256, 512.$$

✓ CHECK POINT 2 Find the seventh term of the geometric sequence whose first term is 5 and whose common ratio is -3.

In Chapter 9, we studied exponential functions of the form $f(x) = b^x$ and used an exponential function to model the growth of the U.S. population from 1970 through 2007 (Example 1 on page 708). In our next example, we revisit the country's population growth over a shorter period of time, 2000 through 2006. Because a geometric sequence is an exponential function whose domain is the set of positive integers, geometric and exponential growth mean the same thing.

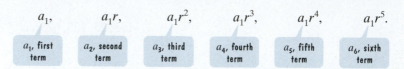

Blitzer Bonus

Geometric Population Growth

Economist Thomas Malthus (1766–1834) predicted that population would increase as a geometric sequence and food production would increase as an arithmetic sequence. He concluded that eventually population would exceed food production. If two sequences, one geometric and one arithmetic, are increasing, the geometric sequence will eventually overtake the arithmetic sequence, regardless of any head start that the arithmetic sequence might initially have.

EXAMPLE 3 Geometric Population Growth

The table shows the population of the United States in 2000, with estimates given by the Census Bureau for 2001 through 2006.

Year	2000	2001	2002	2003	2004	2005	2006
Population (millions)	281.4	284.5	287.6	290.8	294.0	297.2	300.5

a. Show that the population is increasing geometrically.

b. Write the general term for the geometric sequence modeling the population of the United States, in millions, n years after 1999.

c. Project the U.S. population, in millions, for the year 2009.

Solution

a. First, we use the sequence of population growth, 281.4, 284.5, 287.6, 290.8, and so on, to divide the population for each year by the population in the preceding year.

$$\frac{284.5}{281.4} \approx 1.011, \quad \frac{287.6}{284.5} \approx 1.011, \quad \frac{290.8}{287.6} \approx 1.011$$

Continuing in this manner, we will keep getting approximately 1.011. This means that the population is increasing geometrically with $r \approx 1.011$. The population of the United States in any year shown in the sequence is approximately 1.011 times the population the year before.

b. The sequence of the U.S. population growth is

$$281.4, 284.5, 287.6, 290.8, 294.0, 297.2, 300.5, \ldots.$$

Because the population is increasing geometrically, we can find the general term of this sequence using

$$a_n = a_1 r^{n-1}.$$

In this sequence, $a_1 = 281.4$ and [from part (a)] $r \approx 1.011$. We substitute these values into the formula for the general term. This gives the general term for the geometric sequence modeling the U.S. population, in millions, n years after 1999.

$$a_n = 281.4(1.011)^{n-1}$$

c. We can use the formula for the general term, a_n, in part (b) to project the U.S. population for the year 2009. The year 2009 is 10 years after 1999—that is, $2009 - 1999 = 10$. Thus, $n = 10$. We substitute 10 for n in $a_n = 281.4(1.011)^{n-1}$.

$$a_{10} = 281.4(1.011)^{10-1} = 281.4(1.011)^9 \approx 310.5$$

The model projects that the United States will have a population of approximately 310.5 million in the year 2009.

✓ CHECK POINT 3 Write the general term for the geometric sequence

$$3, 6, 12, 24, 48, \ldots.$$

Then use the formula for the general term to find the eighth term.

4 Use the formula for the sum of the first n terms of a geometric sequence.

The Sum of the First n Terms of a Geometric Sequence

The sum of the first n terms of a geometric sequence, denoted by S_n, and called the ***n*th partial sum**, can be found without having to add up all the terms. Recall that the first n terms of a geometric sequence are

$$a_1, a_1 r, a_1 r^2, \ldots, a_1 r^{n-2}, a_1 r^{n-1}.$$

We find the sum of $a_1, a_1r, a_1r^2, \ldots, a_1r^{n-2}, a_1r^{n-1}$ as follows:

$$S_n = a_1 + a_1r + a_1r^2 + \cdots + a_1r^{n-2} + a_1r^{n-1}$$ S_n is the sum of the first n terms of the sequence.

$$rS_n = a_1r + a_1r^2 + a_1r^3 + \cdots + a_1r^{n-1} + a_1r^n$$ Multiply both sides of the equation by r.

$$S_n - rS_n = a_1 - a_1r^n$$ Subtract the second equation from the first equation.

$$S_n(1 - r) = a_1(1 - r^n)$$ Factor out S_n on the left and a_1 on the right.

$$S_n = \frac{a_1(1 - r^n)}{1 - r}.$$ Solve for S_n by dividing both sides by $1 - r$ (assuming that $r \neq 1$).

We have proved the following result:

Study Tip

If the common ratio is 1, the geometric sequence is

$a_1, a_1, a_1, a_1, \ldots.$

The sum of the first n terms of this sequence is na_1:

$$S_n = \underbrace{a_1 + a_1 + a_1 + \cdots + a_1}_{\text{There are } n \text{ terms}} = na_1.$$

The Sum of the First n Terms of a Geometric Sequence

The sum, S_n, of the first n terms of a geometric sequence is given by

$$S_n = \frac{a_1(1 - r^n)}{1 - r}$$

in which a_1 is the first term and r is the common ratio ($r \neq 1$).

To find the sum of the terms of a geometric sequence, we need to know the first term, a_1, the common ratio, r, and the number of terms, n. The following examples illustrate how to use this formula.

EXAMPLE 4 Finding the Sum of the First n Terms of a Geometric Sequence

Find the sum of the first 18 terms of the geometric sequence: $2, -8, 32, -128, \ldots.$

Solution To find the sum of the first 18 terms, S_{18}, we replace n in the formula with 18.

$$S_n = \frac{a_1(1 - r^n)}{1 - r}$$

$$S_{18} = \frac{a_1(1 - r^{18})}{1 - r}$$

The first term, a_1, is 2.

We must find r, the common ratio.

We can find the common ratio by dividing the second term of $2, -8, 32, -128, \ldots$ by the first term.

$$r = \frac{a_2}{a_1} = \frac{-8}{2} = -4$$

Now we are ready to find the sum of the first 18 terms of $2, -8, 32, -128, \ldots.$

$$S_n = \frac{a_1(1 - r^n)}{1 - r}$$ Use the formula for the sum of the first n terms of a geometric sequence.

$$S_{18} = \frac{2[1 - (-4)^{18}]}{1 - (-4)}$$ a_1 (the first term) $= 2$, $r = -4$, and $n = 18$ because we want the sum of the first 18 terms.

$$= -27{,}487{,}790{,}694$$ Use a calculator.

The sum of the first 18 terms is $-27{,}487{,}790{,}694$. Equivalently, this number is the 18th partial sum of the sequence $2, -8, 32, -128, \ldots.$

✓ CHECK POINT 4 Find the sum of the first nine terms of the geometric sequence: $2, -6, 18, -54, \ldots.$

EXAMPLE 5 Using S_n to Evaluate a Summation

Find the following sum: $\sum_{i=1}^{10} 6 \cdot 2^i$.

Solution Let's write out a few terms in the sum.

$$\sum_{i=1}^{10} 6 \cdot 2^i = 6 \cdot 2 + 6 \cdot 2^2 + 6 \cdot 2^3 + \cdots + 6 \cdot 2^{10}$$

Do you see that each term after the first is obtained by multiplying the preceding term by 2? To find the sum of the 10 terms ($n = 10$), we need to know the first term, a_1, and the common ratio, r. The first term is $6 \cdot 2$ or 12: $a_1 = 12$. The common ratio is 2.

$$S_n = \frac{a_1(1 - r^n)}{1 - r}$$ Use the formula for the sum of the first n terms of a geometric sequence.

$$S_{10} = \frac{12(1 - 2^{10})}{1 - 2}$$ a_1 (the first term) = 12, $r = 2$, and $n = 10$ because we are adding ten terms.

$$= 12{,}276$$ Use a calculator.

Thus,

$$\sum_{i=1}^{10} 6 \cdot 2^i = 12{,}276.$$

Using Technology

To find

$$\sum_{i=1}^{10} (6 \cdot 2^i)$$

on a graphing utility, enter:

SUM SEQ $(6 \times 2^x, x, 1, 10, 1)$.

Then press ENTER.

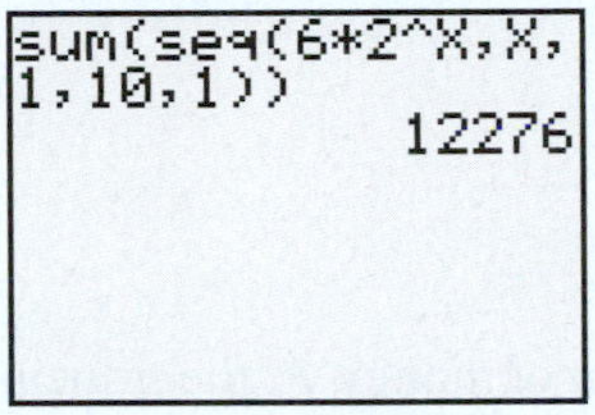

CHECK POINT 5 Find the following sum: $\sum_{i=1}^{8} 2 \cdot 3^i$.

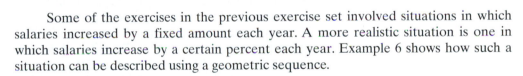

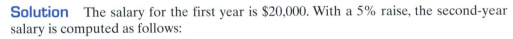

Some of the exercises in the previous exercise set involved situations in which salaries increased by a fixed amount each year. A more realistic situation is one in which salaries increase by a certain percent each year. Example 6 shows how such a situation can be described using a geometric sequence.

EXAMPLE 6 Computing a Lifetime Salary

A union contract specifies that each worker will receive a 5% pay increase each year for the next 30 years. One worker is paid \$20,000 the first year. What is this person's total lifetime salary over a 30-year period?

Solution The salary for the first year is \$20,000. With a 5% raise, the second-year salary is computed as follows:

$$\text{Salary for year 2} = 20{,}000 + 20{,}000(0.05) = 20{,}000(1 + 0.05) = 20{,}000(1.05).$$

Each year, the salary is 1.05 times what it was in the previous year. Thus, the salary for year 3 is 1.05 times 20,000(1.05), or $20{,}000(1.05)^2$. The salaries for the first five years are given in the table.

Yearly Salaries

Year 1	Year 2	Year 3	Year 4	Year 5	. . .
20,000	20,000(1.05)	$20{,}000(1.05)^2$	$20{,}000(1.05)^3$	$20{,}000(1.05)^4$	. . .

The numbers in the bottom row form a geometric sequence with $a_1 = 20{,}000$ and $r = 1.05$. To find the total salary over 30 years, we use the formula for the sum of the first n terms of a geometric sequence, with $n = 30$.

$$S_n = \frac{a_1(1 - r^n)}{1 - r}$$

$$S_{30} = \frac{20{,}000[1 - (1.05)^{30}]}{1 - 1.05}$$ Use $S_n = \frac{a_1(1 - r^n)}{1 - r}$ with $a_1 = 20{,}000$, $r = 1.05$, and $n = 30$.

Total salary over 30 years

$$\approx 1{,}328{,}777$$ Use a calculator.

The total salary over the 30-year period is approximately $1,328,777.

CHECK POINT 6 A job pays a salary of $30,000 the first year. During the next 29 years, the salary increases by 6% each year. What is the total lifetime salary over the 30-year period?

5 Find the value of an annuity.

Annuities

The compound interest formula

$$A = P(1 + r)^t$$

gives the future value, A, after t years, when a fixed amount of money, P, the principal, is deposited in an account that pays an annual interest rate r (in decimal form) compounded once a year. However, money is often invested in small amounts at periodic intervals. For example, to save for retirement, you might decide to place $1000 into an Individual Retirement Account (IRA) at the end of each year until you retire. An **annuity** is a sequence of equal payments made at equal time periods. An IRA is an example of an annuity.

Suppose P dollars is deposited into an account at the end of each year. The account pays an annual interest rate, r, compounded annually. At the end of the first year, the account contains P dollars. At the end of the second year, P dollars is deposited again. At the time of this deposit, the first deposit has received interest earned during the second year. The **value of the annuity** is the sum of all deposits made plus all interest paid. Thus, the value of the annuity after two years is

$$P + P(1 + r).$$

Deposit of P dollars at end of second year

First-year deposit of P dollars with interest earned for a year

The value of the annuity after three years is

$$P + P(1 + r) + P(1 + r)^2.$$

Deposit of P dollars at end of third year

Second-year deposit of P dollars with interest earned for a year

First-year deposit of P dollars with interest earned over two years

The value of the annuity after t years is

$$P + P(1 + r) + P(1 + r)^2 + P(1 + r)^3 + \cdots + P(1 + r)^{t-1}.$$

Deposit of P dollars at end of year t

First-year deposit of P dollars with interest earned over $t - 1$ years

This is the sum of the terms of a geometric sequence with first term P and common ratio $1 + r$. We use the formula

$$S_n = \frac{a_1(1 - r^n)}{1 - r}$$

to find the sum of the terms:

$$S_t = \frac{P[1-(1+r)^t]}{1-(1+r)} = \frac{P[1-(1+r)^t]}{-r} = P\frac{(1+r)^t - 1}{r}.$$

This formula gives the value of an annuity after t years if interest is compounded once a year. We can adjust the formula to find the value of an annuity if equal payments are made at the end of each of n yearly compounding periods.

Value of an Annuity: Interest Compounded n Times Per Year

If P is the deposit made at the end of each compounding period for an annuity at r percent annual interest compounded n times per year, the value, A, of the annuity after t years is

$$A = \frac{P\left[\left(1+\frac{r}{n}\right)^{nt} - 1\right]}{\frac{r}{n}}.$$

EXAMPLE 7 Determining the Value of an Annuity

At age 25, to save for retirement, you decide to deposit \$200 at the end of each month into an IRA that pays 7.5% compounded monthly.

a. How much will you have from the IRA when you retire at age 65?

b. Find the interest.

Solution

a. Because you are 25, the amount that you will have from the IRA when you retire at 65 is its value after 40 years.

$$A = \frac{P\left[\left(1+\frac{r}{n}\right)^{nt} - 1\right]}{\frac{r}{n}}$$

Use the formula for the value of an annuity.

$$A = \frac{200\left[\left(1+\frac{0.075}{12}\right)^{12\cdot 40} - 1\right]}{\frac{0.075}{12}}$$

The annuity involves month-end deposits of \$200: $P = 200$. The interest rate is 7.5%: $r = 0.075$. The interest is compounded monthly: $n = 12$. The number of years is 40: $t = 40$.

$$= \frac{200[(1+0.00625)^{480} - 1]}{0.00625}$$

Using parentheses keys, this can be performed in a single step on a graphing calculator.

$$= \frac{200[(1.00625)^{480} - 1]}{0.00625}$$

$$\approx \frac{200(19.8989 - 1)}{0.00625}$$

Use a calculator to find $(1.00625)^{480}$: 1.00625 [y^x] 480 [=].

$$\approx 604{,}765$$

After 40 years, you will have approximately \$604,765 when retiring at age 65.

Blitzer Bonus

Stashing Cash and Making Taxes Less Taxing

As you prepare for your future career, retirement probably seems very far away. Making regular deposits into an IRA may not be fun, but there is a special incentive from Uncle Sam that makes it far more appealing. Traditional IRAs are **tax-deferred savings plans**. This means that you do not pay taxes on deposits and interest until you begin withdrawals, typically at retirement. Before then, yearly deposits count as adjustments to gross income and are not part of your taxable income. Not only do you get a tax break now, but you ultimately earn more. This is because you do not pay taxes on interest from year to year, allowing earnings to accumulate until you start withdrawals. With a tax code that encourages long-term savings, opening an IRA early in your career is a smart way to gain more control over how you will spend a large part of your life.

b. Interest = Value of the IRA − Total deposits

$$\approx \$604{,}765 - \$200 \cdot 12 \cdot 40$$

$200 per month × 12 months per year × 40 years

$$= \$604{,}765 - \$96{,}000 = \$508{,}765$$

The interest is approximately $508,765, more than five times the amount of your contributions to the IRA.

✓ CHECK POINT 7 At age 30, to save for retirement, you decide to deposit $100 at the end of each month into an IRA that pays 9.5% compounded monthly.

a. How much will you have from the IRA when you retire at age 65?

b. Find the interest.

---

6 Use the formula for the sum of an infinite geometric series.

Geometric Series

An infinite sum of the form

$$a_1 + a_1r + a_1r^2 + a_1r^3 + \cdots + a_1r^{n-1} + \cdots$$

with first term a_1 and common ratio r is called an **infinite geometric series.** How can we determine which infinite geometric series have sums and which do not? We look at what happens to r^n as n gets larger in the formula for the sum of the first n terms of this series, namely

$$S_n = \frac{a_1(1 - r^n)}{1 - r}.$$

If r is any number between −1 and 1, that is, $-1 < r < 1$, the term r^n approaches 0 as n gets larger. For example, consider what happens to r^n for $r = \frac{1}{2}$:

$$\left(\frac{1}{2}\right)^1 = \frac{1}{2} \quad \left(\frac{1}{2}\right)^2 = \frac{1}{4} \quad \left(\frac{1}{2}\right)^3 = \frac{1}{8} \quad \left(\frac{1}{2}\right)^4 = \frac{1}{16} \quad \left(\frac{1}{2}\right)^5 = \frac{1}{32} \quad \left(\frac{1}{2}\right)^6 = \frac{1}{64}.$$

These numbers are approaching 0 as n gets larger.

Take another look at the formula for the sum of the first n terms of a geometric sequence.

$$S_n = \frac{a_1(1 - r^n)}{1 - r}$$

If $-1 < r < 1$, r^n approaches 0 as n gets larger.

Let us replace r^n with 0 in the formula for S_n. This change gives us a formula for the sum of an infinite geometric series with a common ratio between −1 and 1.

The Sum of an Infinite Geometric Series

If $-1 < r < 1$ (equivalently, $|r| < 1$), then the sum of the infinite geometric series

$$a_1 + a_1r + a_1r^2 + a_1r^3 + \cdots$$

in which a_1 is the first term and r is the common ratio is given by

$$S = \frac{a_1}{1 - r}.$$

If $|r| \geq 1$, the infinite series does not have a sum.

To use the formula for the sum of an infinite geometric series, we need to know the first term and the common ratio. For example, consider

$$\frac{1}{2}+\frac{1}{4}+\frac{1}{8}+\frac{1}{16}+\frac{1}{32}+\cdots.$$

First term, a_1, is $\frac{1}{2}$.

Common ratio, r, is $\frac{a_2}{a_1}$. $r=\frac{1}{4}\div\frac{1}{2}=\frac{1}{4}\cdot 2=\frac{1}{2}$

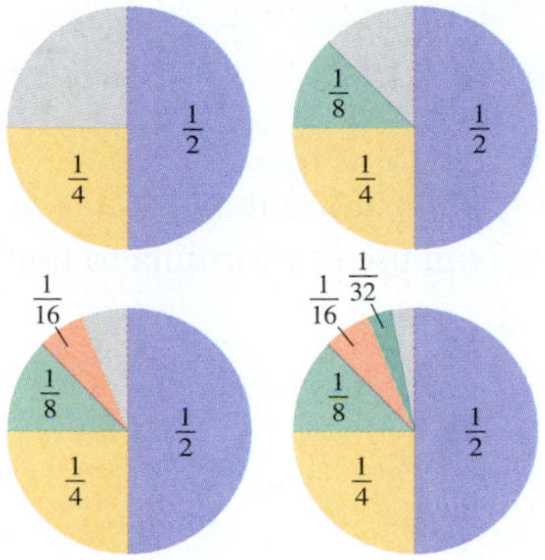

FIGURE 12.6 The sum $\frac{1}{2}+\frac{1}{4}+\frac{1}{8}+\frac{1}{16}+\frac{1}{32}+\cdots$ is approaching 1.

With $r=\frac{1}{2}$, the condition that $|r|<1$ is met, so the infinite geometric series has a sum given by $S=\frac{a_1}{1-r}$. The sum of the series is found as follows:

$$\frac{1}{2}+\frac{1}{4}+\frac{1}{8}+\frac{1}{16}+\frac{1}{32}+\cdots=\frac{a_1}{1-r}=\frac{\frac{1}{2}}{1-\frac{1}{2}}=\frac{\frac{1}{2}}{\frac{1}{2}}=1.$$

Thus, the sum of the infinite geometric series is 1. Notice how this is illustrated in **Figure 12.6**. As more terms are included, the sum is approaching the area of one complete circle.

EXAMPLE 8 Finding the Sum of an Infinite Geometric Series

Find the sum of the infinite geometric series: $\frac{3}{8}-\frac{3}{16}+\frac{3}{32}-\frac{3}{64}+\cdots$.

Solution Before finding the sum, we must find the common ratio.

$$r=\frac{a_2}{a_1}=\frac{-\frac{3}{16}}{\frac{3}{8}}=-\frac{3}{16}\cdot\frac{8}{3}=-\frac{1}{2}$$

Because $r=-\frac{1}{2}$, the condition that $|r|<1$ is met. Thus, the infinite geometric series has a sum.

$$S=\frac{a_1}{1-r}$$

This is the formula for the sum of an infinite geometric series. Let $a_1=\frac{3}{8}$ and $r=-\frac{1}{2}$.

$$=\frac{\frac{3}{8}}{1-\left(-\frac{1}{2}\right)}=\frac{\frac{3}{8}}{\frac{3}{2}}=\frac{3}{8}\cdot\frac{2}{3}=\frac{1}{4}$$

Thus, the sum of $\frac{3}{8}-\frac{3}{16}+\frac{3}{32}-\frac{3}{64}+\cdots$ is $\frac{1}{4}$. Put in an informal way, as we continue to add more and more terms, the sum is approximately $\frac{1}{4}$.

✓ CHECK POINT 8 Find the sum of the infinite geometric series: $3+2+\frac{4}{3}+\frac{8}{9}+\cdots$.

We can use the formula for the sum of an infinite geometric series to express a repeating decimal as a fraction in lowest terms.

EXAMPLE 9 Writing a Repeating Decimal as a Fraction

Express $0.\overline{7}$ as a fraction in lowest terms.

Solution

$$0.\overline{7} = 0.7777\ldots = \frac{7}{10} + \frac{7}{100} + \frac{7}{1000} + \frac{7}{10{,}000} + \cdots$$

Observe that $0.\overline{7}$ is an infinite geometric series with first term $\frac{7}{10}$ and common ratio $\frac{1}{10}$. Because $r = \frac{1}{10}$, the condition that $|r| < 1$ is met. Thus, we can use our formula to find the sum. Therefore,

$$0.\overline{7} = \frac{a_1}{1 - r} = \frac{\frac{7}{10}}{1 - \frac{1}{10}} = \frac{\frac{7}{10}}{\frac{9}{10}} = \frac{7}{10} \cdot \frac{10}{9} = \frac{7}{9}.$$

An equivalent fraction for $0.\overline{7}$ is $\frac{7}{9}$.

✓ CHECK POINT 9 Express $0.\overline{9}$ as a fraction in lowest terms.

Infinite geometric series have many applications, as illustrated in Example 10.

EXAMPLE 10 Tax Rebates and the Multiplier Effect

A tax rebate that returns a certain amount of money to taxpayers can have a total effect on the economy that is many times this amount. In economics, this phenomenon is called the **multiplier effect**. Suppose, for example, that the government reduces taxes so that each consumer has \$2000 more income. The government assumes that each person will spend 70% of this (=\$1400). The individuals and businesses receiving this \$1400 in turn spend 70% of it (=\$980), creating extra income for other people to spend, and so on. Determine the total amount spent on consumer goods from the initial \$2000 tax rebate.

Solution

The total amount spent is given by the infinite geometric series

$$1400 + 980 + 686 + \cdots.$$

(980: 70% of 1400) (686: 70% of 980)

The first term is 1400: $a_1 = 1400$. The common ratio is 70%, or 0.7: $r = 0.7$. Because $r = 0.7$, the condition that $|r| < 1$ is met. Thus, we can use our formula to find the sum. Therefore,

$$1400 + 980 + 686 + \cdots = \frac{a_1}{1 - r} = \frac{1400}{1 - 0.7} \approx 4667.$$

This means that the total amount spent on consumer goods from the initial \$2000 rebate is approximately \$4667.

✓ CHECK POINT 10 Rework Example 10 and determine the total amount spent on consumer goods with a \$1000 tax rebate and 80% spending down the line.

12.3 EXERCISE SET

MyMathLab — MathXL PRACTICE — WATCH — DOWNLOAD — READ — REVIEW

Practice Exercises

In Exercises 1–8, find the common ratio for each geometric sequence.

1. 5, 15, 45, 135, . . .

2. 5, 10, 20, 40, . . .

3. $-15, 30, -60, 120, \ldots$

4. $-2, 6, -18, 54, \ldots$

5. $3, \frac{9}{2}, \frac{27}{4}, \frac{81}{8}, \ldots$

6. $4, \frac{8}{3}, \frac{16}{9}, \frac{32}{27}, \ldots$

7. $4, -0.4, 0.04, -0.004, \ldots$

8. $7, -0.7, 0.07, -0.007, \ldots$

In Exercises 9–16, write the first five terms of each geometric sequence with the given first term, a_1, and common ratio, r.

9. $a_1 = 2, r = 3$

10. $a_1 = 2, r = 4$

11. $a_1 = 20, r = \frac{1}{2}$

12. $a_1 = 24, r = \frac{1}{3}$

13. $a_1 = -4, r = -10$

14. $a_1 = -3, r = -10$

15. $a_1 = -\frac{1}{4}, r = -2$

16. $a_1 = -\frac{1}{16}, r = -4$

In Exercises 17–24, use the formula for the general term (the nth term) of a geometric sequence to find the indicated term of each sequence with the given first term, a_1, and common ratio, r.

17. Find a_8 when $a_1 = 6, r = 2$.

18. Find a_8 when $a_1 = 5, r = 3$.

19. Find a_{12} when $a_1 = 5, r = -2$.

20. Find a_{12} when $a_1 = 4, r = -2$.

21. Find a_6 when $a_1 = 6400, r = -\frac{1}{2}$.

22. Find a_6 when $a_1 = 8000, r = -\frac{1}{2}$.

23. Find a_8 when $a_1 = 1{,}000{,}000, r = 0.1$.

24. Find a_8 when $a_1 = 40{,}000, r = 0.1$.

In Exercises 25–32, write a formula for the general term (the nth term) of each geometric sequence. Then use the formula for a_n to find a_7, the seventh term of the sequence.

25. 3, 12, 48, 192, . . .

26. 3, 15, 75, 375, . . .

27. $18, 6, 2, \frac{2}{3}, \ldots$

28. $12, 6, 3, \frac{3}{2}, \ldots$

29. $1.5, -3, 6, -12, \ldots$

30. $5, -1, \frac{1}{5}, -\frac{1}{25}, \ldots$

31. $0.0004, -0.004, 0.04, -0.4, \ldots$

32. $0.0007, -0.007, 0.07, -0.7, \ldots$

Use the formula for the sum of the first n terms of a geometric sequence to solve Exercises 33–38.

33. Find the sum of the first 12 terms of the geometric sequence: 2, 6, 18, 54, ...

34. Find the sum of the first 12 terms of the geometric sequence: 3, 6, 12, 24, ...

35. Find the sum of the first 11 terms of the geometric sequence: $3, -6, 12, -24, \ldots$.

36. Find the sum of the first 11 terms of the geometric sequence: $4, -12, 36, -108, \ldots$.

37. Find the sum of the first 14 terms of the geometric sequence: $-\frac{3}{2}, 3, -6, 12, \ldots$.

38. Find the sum of the first 14 terms of the geometric sequence: $-\frac{1}{24}, \frac{1}{12}, -\frac{1}{6}, \frac{1}{3}, \ldots$.

In Exercises 39–44, find the indicated sum. Use the formula for the sum of the first n terms of a geometric sequence.

39. $\sum_{i=1}^{8} 3^i$

40. $\sum_{i=1}^{6} 4^i$

41. $\sum_{i=1}^{10} 5 \cdot 2^i$

42. $\sum_{i=1}^{7} 4(-3)^i$

43. $\sum_{i=1}^{6} \left(\frac{1}{2}\right)^{i+1}$

44. $\sum_{i=1}^{6} \left(\frac{1}{3}\right)^{i+1}$

In Exercises 45–52, find the sum of each infinite geometric series.

45. $1 + \frac{1}{3} + \frac{1}{9} + \frac{1}{27} + \cdots$

46. $1 + \frac{1}{4} + \frac{1}{16} + \frac{1}{64} + \cdots$

47. $3 + \frac{3}{4} + \frac{3}{4^2} + \frac{3}{4^3} + \cdots$

48. $5 + \frac{5}{6} + \frac{5}{6^2} + \frac{5}{6^3} + \cdots$

49. $1 - \frac{1}{2} + \frac{1}{4} - \frac{1}{8} + \cdots$

50. $3 - 1 + \frac{1}{3} - \frac{1}{9} + \cdots$

51. $\sum_{i=1}^{\infty} 26(-0.3)^{i-1}$

52. $\sum_{i=1}^{\infty} 51(-0.7)^{i-1}$

In Exercises 53–58, express each repeating decimal as a fraction in lowest terms.

53. $0.\overline{5} = \frac{5}{10} + \frac{5}{100} + \frac{5}{1000} + \frac{5}{10{,}000} + \cdots$

54. $0.\overline{1} = \frac{1}{10} + \frac{1}{100} + \frac{1}{1000} + \frac{1}{10{,}000} + \cdots$

55. $0.\overline{47} = \frac{47}{100} + \frac{47}{10{,}000} + \frac{47}{1{,}000{,}000} + \cdots$

56. $0.\overline{83} = \frac{83}{100} + \frac{83}{10{,}000} + \frac{83}{1{,}000{,}000} + \cdots$

57. $0.\overline{257}$ **58.** $0.\overline{529}$

In Exercises 59–64, the general term of a sequence is given. Determine whether the sequence is arithmetic, geometric, or neither. If the sequence is arithmetic, find the common difference; if it is geometric, find the common ratio.

59. $a_n = n + 5$

60. $a_n = n - 3$

61. $a_n = 2^n$

62. $a_n = \left(\frac{1}{2}\right)^n$

63. $a_n = n^2 + 5$

64. $a_n = n^2 - 3$

Practice PLUS

In Exercises 65–70, let

$$\{a_n\} = -5, 10, -20, 40, \ldots,$$
$$\{b_n\} = 10, -5, -20, -35, \ldots,$$
and $$\{c_n\} = -2, 1, -\tfrac{1}{2}, \tfrac{1}{4}, \ldots$$

65. Find $a_{10} + b_{10}$.

66. Find $a_{11} + b_{11}$.

67. Find the difference between the sum of the first 10 terms of $\{a_n\}$ and the sum of the first 10 terms of $\{b_n\}$.

68. Find the difference between the sum of the first 11 terms of $\{a_n\}$ and the sum of the first 11 terms of $\{b_n\}$.

69. Find the product of the sum of the first 6 terms of $\{a_n\}$ and the sum of the infinite series containing all the terms of $\{c_n\}$.

70. Find the product of the sum of the first 9 terms of $\{a_n\}$ and the sum of the infinite series containing all the terms of $\{c_n\}$.

In Exercises 71–72, find a_2 and a_3 for each geometric sequence.

71. $8, a_2, a_3, 27$

72. $2, a_2, a_3, -54$

In Exercises 73–74, round all answers to the nearest dollar.

73. Here are two ways of investing $30,000 for 20 years.

Lump-Sum Deposit	Rate	Time
$30,000	5% compounded annually	20 years

Periodic Deposits	Rate	Time
$1500 at the end of each year	5% compounded annually	20 years

After 20 years, how much more will you have from the lump-sum investment than from the annuity?

74. Here are two ways of investing $40,000 for 25 years.

Lump-Sum Deposit	Rate	Time
$40,000	6.5% compounded annually	25 years

Periodic Deposits	Rate	Time
$1600 at the end of each year	6.5% compounded annually	25 years

After 25 years, how much more will you have from the lump-sum investment than from the annuity?

Application Exercises

Use the formula for the general term (the nth term) of a geometric sequence to solve Exercises 75–78.

In Exercises 75–76 suppose you save $1 the first day of a month, $2 the second day, $4 the third day, and so on. That is, each day you save twice as much as you did the day before.

75. What will you put aside for savings on the fifteenth day of the month?

76. What will you put aside for savings on the thirtieth day of the month?

77. A professional baseball player signs a contract with a beginning salary of $3,000,000 for the first year and an annual increase of 4% per year beginning in the second year. That is, beginning in year 2, the athlete's salary will be 1.04 times what it was in the previous year. What is the athlete's salary for year 7 of the contract? Round to the nearest dollar.

78. You are offered a job that pays $30,000 for the first year with an annual increase of 5% per year beginning in the second year. That is, beginning in year 2, your salary will be 1.05 times what it was in the previous year. What can you expect to earn in your sixth year on the job? Round to the nearest dollar.

In Exercises 79–80, you will develop geometric sequences that model the population growth for California and Texas, the two most-populated U.S. states.

79. The table shows population estimates for California from 2003 through 2006 from the U.S. Census Bureau.

Year	2003	2004	2005	2006
Population in millions	35.48	35.89	36.13	36.46

a. Divide the population for each year by the population in the preceding year. Round to two decimal places and show that California has a population increase that is approximately geometric.

b. Write the general term of the geometric sequence modeling California's population, in millions, n years after 2002.

c. Use your model from part (b) to project California's population, in millions, for the year 2010. Round to two decimal places.

80. The table shows population estimates for Texas from 2003 through 2006 from the U.S. Census Bureau.

Year	2003	2004	2005	2006
Population in millions	22.12	22.49	22.86	23.41

a. Divide the population for each year by the population in the preceding year. Round to two decimal places and show that Texas has a population increase that is approximately geometric.

b. Write the general term of the geometric sequence modeling Texas's population, in millions, n years after 2002.

c. Use your model from part (b) to project Texas's population, in millions, for the year 2010. Round to two decimal places.

Use the formula for the sum of the first n terms of a geometric sequence to solve Exercises 81–86.

In Exercises 81–82, you save \$1 the first day of a month, \$2 the second day, \$4 the third day, continuing to double your savings each day.

81. What will your total savings be for the first 15 days?

82. What will your total savings be for the first 30 days?

83. A job pays a salary of \$24,000 the first year. During the next 19 years, the salary increases by 5% each year. What is the total lifetime salary over the 20-year period? Round to the nearest dollar.

84. You are investigating two employment opportunities. Company A offers \$30,000 the first year. During the next four years, the salary is guaranteed to increase by 6% per year. Company B offers \$32,000 the first year, with guaranteed annual increases of 3% per year after that. Which company offers the better total salary for a five-year contract? By how much? Round to the nearest dollar.

85. A pendulum swings through an arc of 20 inches. On each successive swing, the length of the arc is 90% of the previous length.

$$20,\quad 0.9(20),\quad 0.9^2(20),\quad 0.9^3(20),\ldots$$

1st swing | 2nd swing | 3rd swing | 4th swing

After 10 swings, what is the total length of the distance the pendulum has swung? Round to the nearest hundredth of an inch.

86. A pendulum swings through an arc of 16 inches. On each successive swing, the length of the arc is 96% of the previous length.

$$16,\quad 0.96(16),\quad (0.96)^2(16),\quad (0.96)^3(16),\ldots$$

1st swing | 2nd swing | 3rd swing | 4th swing

After 10 swings, what is the total length of the distance the pendulum has swung? Round to the nearest hundredth of an inch.

Use the formula for the value of an annuity to solve Exercises 87–92. Round answers to the nearest dollar.

87. To save money for a sabbatical to earn a master's degree, you deposit \$2000 at the end of each year in an annuity that pays 7.5% compounded annually.

a. How much will you have saved at the end of five years?

b. Find the interest.

88. To save money for a sabbatical to earn a master's degree, you deposit \$2500 at the end of each year in an annuity that pays 6.25% compounded annually.

a. How much will you have saved at the end of five years?

b. Find the interest.

89. At age 25, to save for retirement, you decide to deposit \$50 at the end of each month in an IRA that pays 5.5% compounded monthly.

a. How much will you have from the IRA when you retire at age 65?

b. Find the interest.

90. At age 25, to save for retirement, you decide to deposit \$75 at the end of each month in an IRA that pays 6.5% compounded monthly.

a. How much will you have from the IRA when you retire at age 65?

b. Find the interest.

91. To offer scholarship funds to children of employees, a company invests \$10,000 at the end of every three months in an annuity that pays 10.5% compounded quarterly.

a. How much will the company have in scholarship funds at the end of ten years?

b. Find the interest.

92. To offer scholarship funds to children of employees, a company invests \$15,000 at the end of every three months in an annuity that pays 9% compounded quarterly.

a. How much will the company have in scholarship funds at the end of ten years?

b. Find the interest.

Use the formula for the sum of an infinite geometric series to solve Exercises 93–95.

93. A new factory in a small town has an annual payroll of \$6 million. It is expected that 60% of this money will be spent in the town by factory personnel. The people in the town who receive this money are expected to spend 60% of what they receive in the town, and so on. What is the total of all this spending, called the *total economic impact* of the factory, on the town each year?

94. How much additional spending will be generated by a \$10 billion tax rebate if 60% of all income is spent?

95. If the shading process shown in the figure is continued indefinitely, what fractional part of the largest square will eventually be shaded?

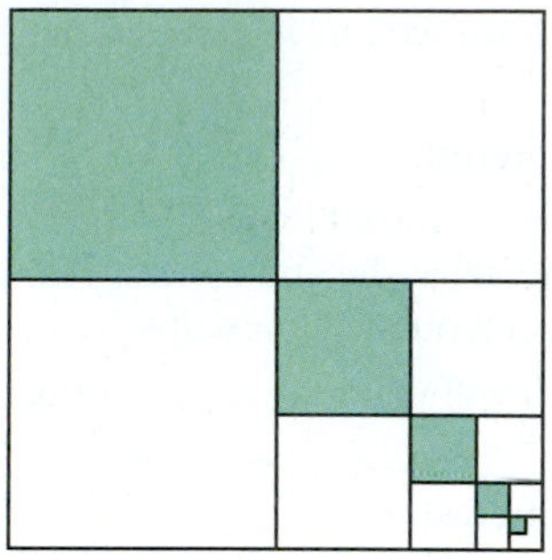

Writing in Mathematics

96. What is a geometric sequence? Give an example with your explanation.

97. What is the common ratio in a geometric sequence?

98. Explain how to find the general term of a geometric sequence.

99. Explain how to find the sum of the first n terms of a geometric sequence without having to add up all the terms.

100. What is an annuity?

101. What is the difference between a geometric sequence and an infinite geometric series?

102. How do you determine if an infinite geometric series has a sum? Explain how to find the sum of such an infinite geometric series.

103. Would you rather have \$10,000,000 and a brand new BMW or 1¢ today, 2¢ tomorrow, 4¢ on day 3, 8¢ on day 4, 16¢ on day 5, and so on, for 30 days? Explain.

104. For the first 30 days of a flu outbreak, the number of students on your campus who become ill is increasing. Which is worse: the number of students with the flu is increasing arithmetically or is increasing geometrically? Explain your answer.

Technology Exercises

105. Use the SEQ (sequence) capability of a graphing utility and the formula you obtained for a_n to verify the value you found for a_7 in any three exercises from Exercises 25–32.

106. Use the capability of a graphing utility to calculate the sum of a sequence to verify any three of your answers to Exercises 39–44.

In Exercises 107–108, use a graphing utility to graph the function. Determine the horizontal asymptote for the graph of f and discuss its relationship to the sum of the given series.

107. **Function**

$$f(x) = \frac{2\left[1 - \left(\frac{1}{3}\right)^x\right]}{1 - \frac{1}{3}}$$

Series

$$2 + 2\left(\frac{1}{3}\right) + 2\left(\frac{1}{3}\right)^2 + 2\left(\frac{1}{3}\right)^3 + \cdots$$

108. **Function**

$$f(x) = \frac{4[1 - (0.6)^x]}{1 - 0.6}$$

Series

$$4 + 4(0.6) + 4(0.6)^2 + 4(0.6)^3 + \cdots$$

Critical Thinking Exercises

Make Sense? *In Exercises 109–112, determine whether each statement "makes sense" or "does not make sense" and explain your reasoning.*

109. There's no end to the number of geometric sequences that I can generate whose first term is 5 if I pick nonzero numbers r and multiply 5 by each value of r repeatedly.

110. I've noticed that the big difference between arithmetic and geometric sequences is that arithmetic sequences are based on addition and geometric sequences are based on multiplication.

111. I modeled California's population growth with a geometric sequence, so my model is an exponential function whose domain is the set of natural numbers.

112. I used a formula to find the sum of the infinite geometric series $3 + 1 + \frac{1}{3} + \frac{1}{9} + \cdots$ and then checked my answer by actually adding all the terms.

In Exercises 113–116, determine whether each statement is true or false. If the statement is false, make the necessary change(s) to produce a true statement.

113. The sequence 2, 6, 24, 120, ... is an example of a geometric sequence.

114. The sum of the geometric series $\frac{1}{2} + \frac{1}{4} + \frac{1}{8} + \cdots + \frac{1}{512}$ can only be estimated without knowing precisely what terms occur between $\frac{1}{8}$ and $\frac{1}{512}$.

115. $10 - 5 + \frac{5}{2} - \frac{5}{4} + \cdots = \dfrac{10}{1 - \frac{1}{2}}$

116. If the nth term of a geometric sequence is $a_n = 3(0.5)^{n-1}$, the common ratio is $\frac{1}{2}$.

117. In a pest-eradication program, sterilized male flies are released into the general population each day. Ninety percent of those flies will survive a given day. How many flies should be released each day if the long-range goal of the program is to keep 20,000 sterilized flies in the population?

118. You are now 25 years old and would like to retire at age 55 with a retirement fund of \$1,000,000. How much should you deposit at the end of each month for the next 30 years in an IRA paying 10% annual interest compounded monthly to achieve your goal? Round to the nearest dollar.

Review Exercises

119. Simplify: $\sqrt{28} - 3\sqrt{7} + \sqrt{63}$. (Section 7.4, Example 2)

120. Solve: $2x^2 = 4 - x$. (Section 8.2, Example 2)

121. Rationalize the denominator: $\frac{6}{\sqrt{3} - \sqrt{5}}$. (Section 7.5, Example 5)

Preview Exercises

Exercises 122–124 will help you prepare for the material covered in the next section.

Each exercise involves observing a pattern in the expanded form of the binomial expression $(a + b)^n$.

$$(a + b)^1 = a + b$$
$$(a + b)^2 = a^2 + 2ab + b^2$$
$$(a + b)^3 = a^3 + 3a^2b + 3ab^2 + b^3$$
$$(a + b)^4 = a^4 + 4a^3b + 6a^2b^2 + 4ab^3 + b^4$$
$$(a + b)^5 = a^5 + 5a^4b + 10a^3b^2 + 10a^2b^3 + 5ab^4 + b^5$$

122. Describe the pattern for the exponents on a.

123. Describe the pattern for the exponents on b.

124. Describe the pattern for the sum of the exponents on the variables in each term.

MID-CHAPTER CHECK POINT Section 12.1– Section 12.3

What You Know: We learned that a sequence is a function whose domain is the set of positive integers. In an arithmetic sequence, each term after the first differs from the preceding term by a constant, the common difference, d. In a geometric sequence, each term after the first is obtained by multiplying the preceding term by a nonzero constant, the common ratio, r. We found the general term of arithmetic sequences $[a_n = a_1 + (n - 1)d]$ and geometric sequences $[a_n = a_1r^{n-1}]$ and used these formulas to find particular terms. We determined the sum of the first n terms of arithmetic sequences $\left[S_n = \frac{n}{2}(a_1 + a_n)\right]$ and geometric sequences $\left[S_n = \frac{a_1(1 - r^n)}{1 - r}\right]$. Finally, we determined the sum of an infinite geometric series, $a_1 + a_1r + a_1r^2 + a_1r^3 + \cdots$, if $-1 < r < 1$ $\left(S = \frac{a_1}{1 - r}\right)$.

In Exercises 1–3, write the first five terms of each sequence. Assume that d represents the common difference of an arithmetic sequence and r represents the common ratio of a geometric sequence.

1. $a_n = (-1)^{n+1}\frac{n}{(n - 1)!}$

2. $a_1 = 5, d = -3$

3. $a_1 = 5, r = -3$

In Exercises 4–6, write a formula for the general term (the nth term) of each sequence. Then use the formula to find the indicated term.

4. $2, 6, 10, 14, \ldots; a_{20}$

5. $3, 6, 12, 24, \ldots; a_{10}$

6. $\frac{3}{2}, 1, \frac{1}{2}, 0, \ldots; a_{30}$

7. Find the sum of the first ten terms of the sequence: $5, 10, 20, 40, \ldots$.

8. Find the sum of the first 50 terms of the sequence: $-2, 0, 2, 4, \ldots$.

9. Find the sum of the first ten terms of the sequence: $-20, 40, -80, 160, \ldots$.

10. Find the sum of the first 100 terms of the sequence: $4, -2, -8, -14, \ldots$.

In Exercises 11–14, find each indicated sum.

11. $\sum_{i=1}^{4}(i + 4)(i - 1)$

12. $\sum_{i=1}^{50}(3i - 2)$

13. $\sum_{i=1}^{6}\left(\frac{3}{2}\right)^i$

14. $\sum_{i=1}^{\infty}\left(-\frac{2}{5}\right)^{i-1}$

15. Express $0.\overline{45}$ as a fraction in lowest terms.

16. Express the sum using summation notation. Use i for the index of summation.

$$\frac{1}{3} + \frac{2}{4} + \frac{3}{5} + \cdots + \frac{18}{20}$$

17. A skydiver falls 16 feet during the first second of a dive, 48 feet during the second second, 80 feet during the third second, 112 feet during the fourth second, and so on. Find the distance that the skydiver falls during the 15th second and the total distance the skydiver falls in 15 seconds.

18. If the average value of a house increases 10% per year, how much will a house costing \$500,000 in year 1 be worth in year 9? Round to the nearest dollar.

SECTION 12.4

The Binomial Theorem

Objectives

1. Evaluate a binomial coefficient.
2. Expand a binomial raised to a power.
3. Find a particular term in a binomial expansion.

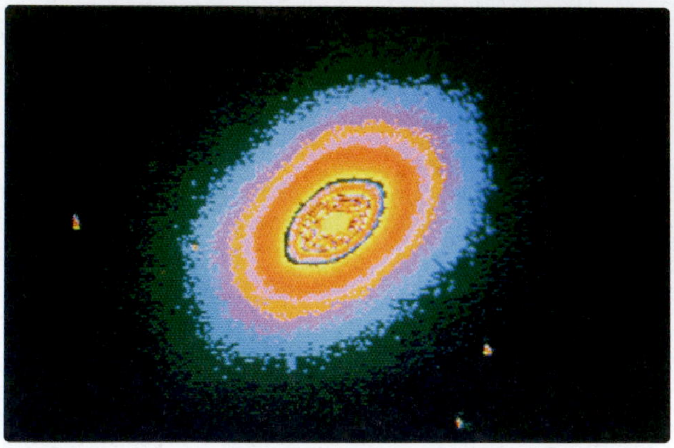

Galaxies are groupings of billions of stars bound together by gravity. Some galaxies, such as the Centaurus galaxy shown here, are elliptical in shape.

Is mathematics discovered or invented? For example, planets revolve in elliptical orbits. Does that mean that the ellipse is out there, waiting for the mind to discover it? Or do people create the definition of an ellipse just as they compose a song? And is it possible for the same mathematics to be discovered/invented by independent researchers separated by time, place, and culture? This is precisely what occurred when mathematicians attempted to find efficient methods for raising binomials to higher and higher powers, such as

$$(x+2)^3, (x+2)^4, (x+2)^5, (x+2)^6,$$

and so on. In this section, we study higher powers of binomials and a method first discovered/invented by great minds in Eastern and in Western cultures working independently.

1 Evaluate a binomial coefficient.

Binomial Coefficients

Before turning to powers of binomials, we introduce a special notation that uses factorials.

Definition of a Binomial Coefficient $\binom{n}{r}$

For nonnegative integers n and r, with $n \geq r$, the expression $\binom{n}{r}$ (read "n above r") is called a **binomial coefficient** and is defined by

$$\binom{n}{r} = \frac{n!}{r!\,(n-r)!}.$$

The symbol ${}_nC_r$ is often used in place of $\binom{n}{r}$ to denote binomial coefficients.

Can you see that the definition of a binomial coefficient involves a fraction with factorials in the numerator and the denominator? When evaluating such an expression, try to reduce the fraction before performing the multiplications. For example, consider $\frac{26!}{21!}$. Rather than writing out 26! as the product of all integers from 26 down to 1, we can express 26! as

$$26! = 26 \cdot 25 \cdot 24 \cdot 23 \cdot 22 \cdot 21!.$$

In this way, we can divide both the numerator and the denominator by the common factor, 21!.

$$\frac{26!}{21!} = \frac{26 \cdot 25 \cdot 24 \cdot 23 \cdot 22 \cdot \cancel{21!}}{\cancel{21!}} = 26 \cdot 25 \cdot 24 \cdot 23 \cdot 22 = 7{,}893{,}600$$

EXAMPLE 1 Evaluating Binomial Coefficients

Evaluate:

a. $\binom{6}{2}$ **b.** $\binom{3}{0}$ **c.** $\binom{9}{3}$ **d.** $\binom{4}{4}$.

Using Technology

Graphing utilities can compute binomial coefficients. For example, to find $\binom{6}{2}$, many utilities require the sequence

6 [nCr] 2 [ENTER].

The graphing utility will display 15. Consult your manual and verify the other evaluations in Example 1.

Solution In each case, we apply the definition of the binomial coefficient.

a. $\binom{6}{2} = \frac{6!}{2!\,(6-2)!} = \frac{6!}{2!\,4!} = \frac{6 \cdot 5 \cdot \cancel{4!}}{2 \cdot 1 \cdot \cancel{4!}} = 15$

b. $\binom{3}{0} = \frac{3!}{0!(3-0)!} = \frac{\cancel{3!}}{0!\cancel{3!}} = \frac{1}{1} = 1$

Remember that $0! = 1$.

c. $\binom{9}{3} = \frac{9!}{3!\,(9-3)!} = \frac{9!}{3!\,6!} = \frac{9 \cdot 8 \cdot 7 \cdot \cancel{6!}}{3 \cdot 2 \cdot 1 \cdot \cancel{6!}} = 84$

d. $\binom{4}{4} = \frac{4!}{4!\,(4-4)!} = \frac{\cancel{4!}}{\cancel{4!}\,0!} = \frac{1}{1} = 1$

CHECK POINT 1 Evaluate:

a. $\binom{6}{3}$ **b.** $\binom{6}{0}$ **c.** $\binom{8}{2}$ **d.** $\binom{3}{3}$.

2 Expand a binomial raised to a power.

The Binomial Theorem

When we write out the *binomial expression* $(a + b)^n$, where n is a positive integer, a number of patterns begin to appear.

$$(a + b)^1 = a + b$$
$$(a + b)^2 = a^2 + 2ab + b^2$$
$$(a + b)^3 = a^3 + 3a^2b + 3ab^2 + b^3$$
$$(a + b)^4 = a^4 + 4a^3b + 6a^2b^2 + 4ab^3 + b^4$$
$$(a + b)^5 = a^5 + 5a^4b + 10a^3b^2 + 10a^2b^3 + 5ab^4 + b^5$$

Each expanded form of the binomial expression is a polynomial. Observe the following patterns:

1. The first term in the expansion of $(a + b)^n$ is a^n. The exponents on a decrease by 1 in each successive term.
2. The exponents on b in the expansion of $(a + b)^n$ increase by 1 in each successive term. In the first term, the exponent on b is 0. (Because $b^0 = 1$, b is not shown in the first term.) The last term is b^n.
3. The sum of the exponents on the variables in any term in the expansion of $(a + b)^n$ is equal to n.
4. The number of terms in the polynomial expansion is one greater than the power of the binomial, n. There are $n + 1$ terms in the expanded form of $(a + b)^n$.

Using these observations, the variable parts of the expansion of $(a + b)^6$ are

$$a^6,\quad a^5b,\quad a^4b^2,\quad a^3b^3,\quad a^2b^4,\quad ab^5,\quad b^6.$$

The first term is a^6, with the exponents on a decreasing by 1 in each successive term. The exponents on b increase from 0 to 6, with the last term being b^6. The sum of the exponents in each term is equal to 6.

We can generalize from these observations to obtain the variable parts of the expansion of $(a + b)^n$. They are

$$a^n,\quad a^{n-1}b,\quad a^{n-2}b^2,\quad a^{n-3}b^3, \ldots,\quad ab^{n-1},\quad b^n.$$

Exponents on a are decreasing by 1. Exponents on b are increasing by 1.

Sum of exponents: $n - 1 + 1 = n$

Sum of exponents: $n - 3 + 3 = n$

Sum of exponents: $1 + n - 1 = n$

If we use binomial coefficients and the pattern for the variable part of each term, a formula called the **Binomial Theorem** can be used to expand any positive integral power of a binomial.

A Formula for Expanding Binomials: The Binomial Theorem

For any positive integer n,

$$(a + b)^n = \binom{n}{0}a^n + \binom{n}{1}a^{n-1}b + \binom{n}{2}a^{n-2}b^2 + \binom{n}{3}a^{n-3}b^3 + \cdots + \binom{n}{n}b^n$$

$$= \sum_{r=0}^{n}\binom{n}{r}a^{n-r}b^r.$$

EXAMPLE 2 Using the Binomial Theorem

Expand: $(x + 2)^4$.

Solution We use the Binomial Theorem

$$(a + b)^n = \binom{n}{0}a^n + \binom{n}{1}a^{n-1}b + \binom{n}{2}a^{n-2}b^2 + \binom{n}{3}a^{n-3}b^3 + \cdots + \binom{n}{n}b^n$$

to expand $(x + 2)^4$. In $(x + 2)^4$, $a = x$, $b = 2$, and $n = 4$. In the expansion, powers of x are in descending order, starting with x^4. Powers of 2 are in ascending order, starting with 2^0. (Because $2^0 = 1$, a 2 is not shown in the first term.) The sum of the exponents on x and 2 in each term is equal to 4, the exponent in the expression $(x + 2)^4$.

$$(x + 2)^4 = \binom{4}{0}x^4 + \binom{4}{1}x^3 \cdot 2 + \binom{4}{2}x^2 \cdot 2^2 + \binom{4}{3}x \cdot 2^3 + \binom{4}{4}2^4$$

These binomial coefficients are evaluated using $\binom{n}{r} = \frac{n!}{r!(n-r)!}$.

$$= \frac{4!}{0!4!}x^4 + \frac{4!}{1!3!}x^3 \cdot 2 + \frac{4!}{2!2!}x^2 \cdot 4 + \frac{4!}{3!1!}x \cdot 8 + \frac{4!}{4!0!} \cdot 16$$

$\frac{4!}{2!2!} = \frac{4 \cdot 3 \cdot 2!}{2! \cdot 2 \cdot 1} = \frac{12}{2} = 6$

Take a few minutes to verify the other factorial evaluations.

$$= 1 \cdot x^4 + 4x^3 \cdot 2 + 6x^2 \cdot 4 + 4x \cdot 8 + 1 \cdot 16$$

$$= x^4 + 8x^3 + 24x^2 + 32x + 16$$

Using Technology

You can use a graphing utility's table feature to find the five binomial coefficients in Example 2.

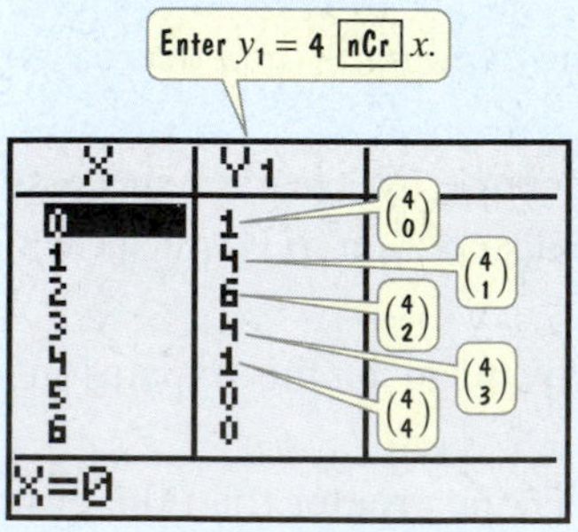

CHECK POINT 2 Expand: $(x + 1)^4$.

EXAMPLE 3 Using the Binomial Theorem

Expand: $(2x - y)^5$.

Solution Because the Binomial Theorem involves the addition of two terms raised to a power, we rewrite $(2x - y)^5$ as $[2x + (-y)]^5$. We use the Binomial Theorem

$$(a + b)^n = \binom{n}{0}a^n + \binom{n}{1}a^{n-1}b + \binom{n}{2}a^{n-2}b^2 + \binom{n}{3}a^{n-3}b^3 + \cdots + \binom{n}{n}b^n$$

to expand $[2x + (-y)]^5$. In $[2x + (-y)]^5$, $a = 2x$, $b = -y$, and $n = 5$. In the expansion, powers of $2x$ are in descending order, starting with $(2x)^5$. Powers of $-y$ are in ascending order, starting with $(-y)^0$. [Because $(-y)^0 = 1$, a $-y$ is not shown in the first term.] The sum of the exponents on $2x$ and $-y$ in each term is equal to 5, the exponent in the expression $(2x - y)^5$.

$$(2x - y)^5 = [2x + (-y)]^5$$

$$= \binom{5}{0}(2x)^5 + \binom{5}{1}(2x)^4(-y) + \binom{5}{2}(2x)^3(-y)^2 + \binom{5}{3}(2x)^2(-y)^3 + \binom{5}{4}(2x)(-y)^4 + \binom{5}{5}(-y)^5$$

Evaluate binomial coefficients using $\binom{n}{r} = \frac{n!}{r!(n-r)!}$.

$$= \frac{5!}{0!5!}(2x)^5 + \frac{5!}{1!4!}(2x)^4(-y) + \frac{5!}{2!3!}(2x)^3(-y)^2 + \frac{5!}{3!2!}(2x)^2(-y)^3 + \frac{5!}{4!1!}(2x)(-y)^4 + \frac{5!}{5!0!}(-y)^5$$

$\frac{5!}{2!3!} = \frac{5 \cdot 4 \cdot 3!}{2 \cdot 1 \cdot 3!} = 10$
Take a few minutes to verify the other factorial evaluations.

$$= 1(2x)^5 + 5(2x)^4(-y) + 10(2x)^3(-y)^2 + 10(2x)^2(-y)^3 + 5(2x)(-y)^4 + 1(-y)^5$$

Raise both factors in these parentheses to the indicated powers.

$$= 1(32x^5) + 5(16x^4)(-y) + 10(8x^3)(-y)^2 + 10(4x^2)(-y)^3 + 5(2x)(-y)^4 + 1(-y)^5$$

Now raise $-y$ to the indicated powers.

$$= 1(32x^5) + 5(16x^4)(-y) + 10(8x^3)y^2 + 10(4x^2)(-y^3) + 5(2x)y^4 + 1(-y^5)$$

Multiplying factors in each of the six terms gives us the desired expansion:

$$(2x - y)^5 = 32x^5 - 80x^4y + 80x^3y^2 - 40x^2y^3 + 10xy^4 - y^5.$$

CHECK POINT 3 Expand: $(x - 2y)^5$.

3 Find a particular term in a binomial expansion.

Finding a Particular Term in a Binomial Expansion

By observing the terms in the formula for expanding binomials, we can find a formula for finding a particular term without writing the entire expansion.

$$\binom{n}{0}a^nb^0 \qquad \binom{n}{1}a^{n-1}b^1 \qquad \binom{n}{2}a^{n-2}b^2$$

1st term, 2nd term, 3rd term

The exponent on b is 1 less than the term number.

Based on the observation in the bottom voice balloon, the $(r + 1)$st term of the expansion of $(a + b)^n$ is the term that contains b^r.

Finding a Particular Term in a Binomial Expansion

The $(r + 1)$st term of the expansion of $(a + b)^n$ is

$$\binom{n}{r}a^{n-r}b^r.$$

EXAMPLE 4 Finding a Single Term of a Binomial Expansion

Find the fourth term in the expansion of $(3x + 2y)^7$.

Solution The fourth term in the expansion of $(3x + 2y)^7$ contains $(2y)^3$. To find the fourth term, first note that $4 = 3 + 1$. Equivalently, the fourth term of $(3x + 2y)^7$ is the $(3 + 1)$st term. Thus, $r = 3$, $a = 3x$, $b = 2y$, and $n = 7$. The fourth term is

$$\binom{7}{3}(3x)^{7-3}(2y)^3 = \binom{7}{3}(3x)^4(2y)^3 = \frac{7!}{3!(7-3)!}(3x)^4(2y)^3.$$

Use the formula for the $(r + 1)$st term of $(a + b)^n$: $\binom{n}{r}a^{n-r}b^r$.

We use $\binom{n}{r} = \frac{n!}{r!(n-r)!}$ to evaluate $\binom{7}{3}$.

Now we need to evaluate the factorial expression and raise $3x$ and $2y$ to the indicated powers. We obtain

$$\frac{7!}{3!\,4!}(81x^4)(8y^3) = \frac{7\cdot6\cdot5\cdot\cancel{4!}}{3\cdot2\cdot1\cdot\cancel{4!}}(81x^4)(8y^3) = 35(81x^4)(8y^3) = 22{,}680x^4y^3.$$

The fourth term of $(3x + 2y)^7$ is $22{,}680x^4y^3$.

✓ **CHECK POINT 4** Find the fifth term in the expansion of $(2x + y)^9$.

Blitzer Bonus

The Universality of Mathematics

Pascal's triangle is an array of numbers showing coefficients of the terms in the expansions of $(a + b)^n$. Although credited to French mathematician Blaise Pascal (1623–1662), the triangular array of numbers appeared in a Chinese document printed in 1303. The Binomial Theorem was known in Eastern cultures prior to its discovery in Europe. The same mathematics is often discovered/invented by independent researchers separated by time, place, and culture.

Binomial Expansions

$(a + b)^0 = 1$

$(a + b)^1 = a + b$

$(a + b)^2 = a^2 + 2ab + b^2$

$(a + b)^3 = a^3 + 3a^2b + 3ab^2 + b^3$

$(a + b)^4 = a^4 + 4a^3b + 6a^2b^2 + 4ab^3 + b^4$

$(a + b)^5 = a^5 + 5a^4b + 10a^3b^2 + 10a^2b^3 + 5ab^4 + b^5$

Pascal's Triangle

Coefficients in the Expansions

1
1 1
1 2 1
1 3 3 1
1 4 6 4 1
1 5 10 10 5 1
1 6 15 20 15 6 1
1 7 21 35 35 21 7 1
1 8 28 56 70 56 28 8 1

Chinese Document: 1303

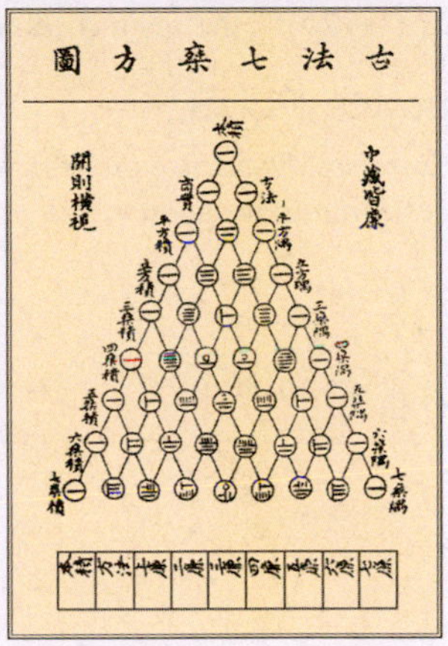

12.4 EXERCISE SET

MyMathLab · MathXL PRACTICE · WATCH · DOWNLOAD · READ · REVIEW

Practice Exercises

In Exercises 1–8, evaluate the given binomial coefficient.

1. $\binom{8}{3}$
2. $\binom{7}{2}$
3. $\binom{12}{1}$
4. $\binom{11}{1}$
5. $\binom{6}{6}$
6. $\binom{15}{2}$
7. $\binom{100}{2}$
8. $\binom{100}{98}$

In Exercises 9–30, use the Binomial Theorem to expand each binomial and express the result in simplified form.

9. $(x + 2)^3$
10. $(x + 4)^3$
11. $(3x + y)^3$
12. $(x + 3y)^3$
13. $(5x - 1)^3$
14. $(4x - 1)^3$
15. $(2x + 1)^4$
16. $(3x + 1)^4$
17. $(x^2 + 2y)^4$
18. $(x^2 + y)^4$
19. $(y - 3)^4$
20. $(y - 4)^4$
21. $(2x^3 - 1)^4$
22. $(2x^5 - 1)^4$
23. $(c + 2)^5$
24. $(c + 3)^5$
25. $(x - 1)^5$
26. $(x - 2)^5$
27. $(3x - y)^5$
28. $(x - 3y)^5$
29. $(2a + b)^6$
30. $(a + 2b)^6$

In Exercises 31–38, write the first three terms in each binomial expansion, expressing the result in simplified form.

31. $(x + 2)^8$
32. $(x + 3)^8$
33. $(x - 2y)^{10}$
34. $(x - 2y)^9$
35. $(x^2 + 1)^{16}$
36. $(x^2 + 1)^{17}$
37. $(y^3 - 1)^{20}$
38. $(y^3 - 1)^{21}$

In Exercises 39–48, find the term indicated in each expansion.

39. $(2x + y)^6$; third term
40. $(x + 2y)^6$; third term

41. $(x - 1)^9$; fifth term

42. $(x - 1)^{10}$; fifth term

43. $(x^2 + y^3)^8$; sixth term

44. $(x^3 + y^2)^8$; sixth term

45. $\left(x - \frac{1}{2}\right)^9$; fourth term

46. $\left(x + \frac{1}{2}\right)^8$; fourth term

47. $(x^2 + y)^{22}$; the term containing y^{14}

48. $(x + 2y)^{10}$; the term containing y^6

Practice PLUS

In Exercises 49–52, use the Binomial Theorem to expand each expression and write the result in simplified form.

49. $(x^3 + x^{-2})^4$

50. $(x^2 + x^{-3})^4$

51. $\left(x^{\frac{1}{3}} - x^{-\frac{1}{3}}\right)^3$

52. $\left(x^{\frac{2}{3}} - \dfrac{1}{\sqrt[3]{x}}\right)^3$

Exercises 53–54 involve expressions containing i, where $i = \sqrt{-1}$. Expand each expression and use powers of i to simplify the result.

53. $(-1 + i\sqrt{3})^3$

54. $(-1 - i\sqrt{3})^3$

In Exercises 55–56, find $\dfrac{f(x + h) - f(x)}{h}$ and simplify.

55. $f(x) = x^4 + 7$

56. $f(x) = x^5 + 8$

57. Find the middle term in the expansion of $\left(\dfrac{3}{x} + \dfrac{x}{3}\right)^{10}$.

58. Find the middle term in the expansion of $\left(\dfrac{1}{x} - x^2\right)^{12}$.

Application Exercises

The graph shows that U.S. smokers have a greater probability of suffering from some ailments than the general adult population. Exercises 59–60 are based on some of the probabilities, expressed as decimals, shown to the right of the bars. In each exercise, use a calculator to determine the probability, correct to four decimal places.

Probability That United States Adults Suffer from Various Ailments

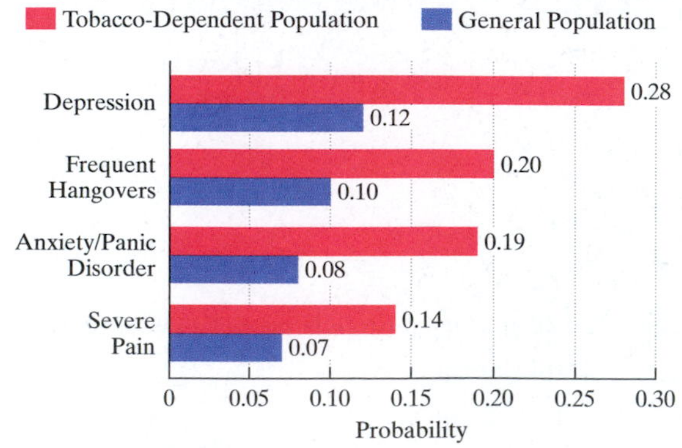

Source: MARS 2005 OTC/DTC

If the probability an event will occur is p and the probability it will not occur is q, then each term in the expansion of $(p + q)^n$ represents a probability.

59. The probability that a smoker suffers from depression is 0.28. If five smokers are randomly selected, the probability that three of them will suffer from depression is the third term of the binomial expansion of

$$(0.28 + 0.72)^5.$$

5 smokers are selected.

Probability a smoker suffers from depression

Probability a smoker does not suffer from depression

What is this probability?

60. The probability that a person in the general population suffers from depression is 0.12. If five people from the general population are randomly selected, the probability that three of them will suffer from depression is the third term of the binomial expansion of

$$(0.12 + 0.88)^5$$

5 people from the general population are selected.

Probability a person in the general population suffers from depression

Probability a person in the general population does not suffer from depression

What is this probability?

Writing in Mathematics

61. Explain how to evaluate $\binom{n}{r}$. Provide an example with your explanation.

62. Describe the pattern in the exponents on a in the expansion of $(a + b)^n$.

63. Describe the pattern in the exponents on b in the expansion of $(a + b)^n$.

64. What is true about the sum of the exponents on a and b in any term in the expansion of $(a + b)^n$?

65. How do you determine how many terms there are in a binomial expansion?

66. Explain how to use the Binomial Theorem to expand a binomial. Provide an example with your explanation.

67. Explain how to find a particular term in a binomial expansion without having to write out the entire expansion.

Technology Exercises

68. Use the [nCr] key on a graphing utility to verify your answers in Exercises 1–8.

In Exercises 69–70, graph each of the functions in the same viewing rectangle. Describe how the graphs illustrate the Binomial Theorem.

69. $f_1(x) = (x + 2)^3$ $\quad f_2(x) = x^3$

$f_3(x) = x^3 + 6x^2$ $\quad f_4(x) = x^3 + 6x^2 + 12x$

$f_5(x) = x^3 + 6x^2 + 12x + 8$

Use a $[-10, 10, 1]$ by $[-30, 30, 10]$ viewing rectangle.

70. $f_1(x) = (x + 1)^4$ $\quad f_2(x) = x^4$
$f_3(x) = x^4 + 4x^3$ $\quad f_4(x) = x^4 + 4x^3 + 6x^2$
$f_5(x) = x^4 + 4x^3 + 6x^2 + 4x$
$f_6(x) = x^4 + 4x^3 + 6x^2 + 4x + 1$
Use a $[-5, 5, 1]$ by $[-30, 30, 10]$ viewing rectangle.

In Exercises 71–73, use the Binomial Theorem to find a polynomial expansion for each function. Then use a graphing utility and an approach similar to the one in Exercises 69 and 70 to verify the expansion.

71. $f_1(x) = (x - 1)^3$

72. $f_1(x) = (x - 2)^4$

73. $f_1(x) = (x + 2)^6$

Critical Thinking Exercises

Make Sense? *In Exercises 74–77, determine whether each statement "makes sense" or "does not make sense" and explain your reasoning.*

74. In order to expand $(x^3 - y^4)^5$, I find it helpful to rewrite the expression inside the parentheses as $x^3 + (-y^4)$.

75. Without writing the expansion of $(x - 1)^6$, I can see that the terms have alternating positive and negative signs.

76. I use binomial coefficients to expand $(a + b)^n$, where $\binom{n}{1}$ is the coefficient of the first term, $\binom{n}{2}$ is the coefficient of the second term, and so on.

77. One of the terms in my binomial expansion is $\binom{7}{5}x^2y^4$.

In Exercises 78–81, determine whether each statement is true or false. If the statement is false, make the necessary change(s) to produce a true statement.

78. The binomial expansion for $(a + b)^n$ contains n terms.

79. The Binomial Theorem can be written in condensed form as
$$(a + b)^n = \sum_{r=0}^{n} \binom{n}{r} a^{n-r}b^r.$$

80. The sum of the binomial coefficients in $(a + b)^n$ cannot be 2^n.

81. There are no values of a and b such that
$$(a + b)^4 = a^4 + b^4.$$

82. Use the Binomial Theorem to expand and then simplify the result: $(x^2 + x + 1)^3$. [*Hint*: Write $x^2 + x + 1$ as $x^2 + (x + 1)$.]

83. Find the term in the expansion of $(x^2 + y^2)^5$ containing x^4 as a factor.

Review Exercises

84. If $f(x) = x^2 + 2x + 3$, find $f(a + 1)$.
(Section 5.2, Example 11)

85. If $f(x) = x^2 + 5x$ and $g(x) = 2x - 3$, find $f(g(x))$ and $g(f(x))$.
(Section 9.2. Example 1)

86. Subtract: $\dfrac{x}{x + 3} - \dfrac{x + 1}{2x^2 - 2x - 24}$.
(Section 6.2, Example 7)

Preview Exercises

Exercises 87–89 will help you prepare for the material covered in the next section.
In Exercises 87–88, show that
$$1 + 2 + 3 + \cdots + n = \frac{n(n + 1)}{2}$$
is true for the given value of n.

87. $n = 3$: Show that $1 + 2 + 3 = \dfrac{3(3 + 1)}{2}$.

88. $n = 5$: Show that $1 + 2 + 3 + 4 + 5 = \dfrac{5(5 + 1)}{2}$.

89. Simplify: $\dfrac{k(k + 1)(2k + 1)}{6} + (k + 1)^2$.

SECTION 12.5

Mathematical Induction

Objectives

1. Understand the principle of mathematical induction.
2. Prove statements using mathematical induction.

After ten years of work, Princeton University's Andrew Wiles proved Fermat's Last Theorem.

Pierre de Fermat (1601–1665) was a lawyer who enjoyed studying mathematics. In a margin of one of his books, he claimed that no positive integers x, y, and z satisfy

$$x^n + y^n = z^n$$

if n is an integer greater than or equal to 3.

If $n = 2$, we can find positive integers satisfying $x^n + y^n = z^n$, or $x^2 + y^2 = z^2$:

$$3^2 + 4^2 = 5^2.$$

However, Fermat claimed that no positive integers satisfy

$$x^3 + y^3 = z^3, \quad x^4 + y^4 = z^4, \quad x^5 + y^5 = z^5,$$

and so on. Fermat claimed to have a proof of his conjecture, but added, "The margin of my book is too narrow to write it down." Some believe that he never had a proof and intended to frustrate his colleagues.

In 1994, 40-year-old Princeton math professor Andrew Wiles proved Fermat's Last Theorem using a principle called *mathematical induction*. In this section, you will learn how to use this powerful method to prove statements about the positive integers.

1 Understand the principle of mathematical induction.

The Principle of Mathematical Induction

How do we prove statements using mathematical induction? Let's consider an example. We will prove a statement that appears to give a correct formula for the sum of the first n positive integers:

$$S_n: 1 + 2 + 3 + \cdots + n = \frac{n(n+1)}{2}.$$

We can verify this statement for, say, the first four positive integers. If $n = 1$, the statement S_1 is

Take the first term on the left. Substitute 1 for n on the right.

$$1 \stackrel{?}{=} \frac{1(1+1)}{2}$$

$$1 \stackrel{?}{=} \frac{1 \cdot 2}{2}$$

$$1 = 1. \checkmark$$

This true statement shows that S_1 is true.

If $n = 2$, the statement S_2 is

Add the first two terms on the left.

$$1 + 2 \stackrel{?}{=} \frac{2(2+1)}{2}$$

Substitute 2 for n on the right.

$$3 \stackrel{?}{=} \frac{2 \cdot 3}{2}$$

$$3 = 3. \checkmark$$

This true statement shows that S_2 is true.

If $n = 3$, the statement S_3 is

Add the first three terms on the left.

$$1 + 2 + 3 \stackrel{?}{=} \frac{3(3+1)}{2}$$

Substitute 3 for n on the right.

$$6 \stackrel{?}{=} \frac{3 \cdot 4}{2}$$

$$6 = 6. \checkmark$$

This true statement shows that S_3 is true.

Finally, if $n = 4$, the statement S_4 is

Add the first four terms on the left.

$$1 + 2 + 3 + 4 \stackrel{?}{=} \frac{4(4+1)}{2}$$

Substitute 4 for n on the right.

$$10 \stackrel{?}{=} \frac{4 \cdot 5}{2}$$

$$10 = 10. \checkmark$$

This true statement shows that S_4 is true.

This approach does *not* prove that the given statement S_n is true for every positive integer n. The fact that the formula produces true statements for $n = 1, 2, 3,$ and 4 does not guarantee that it is valid for all positive integers n. Thus, we need to be able to verify the truth of S_n without verifying the statement for each and every one of the positive integers.

A legitimate proof of the given statement S_n involves a technique called **mathematical induction**.

The Principle of Mathematical Induction

Let S_n be a statement involving the positive integer n. If

1. S_1 is true, and
2. the truth of the statement S_k implies the truth of the statement S_{k+1}, for every positive integer k,

then the statement S_n is true for all positive integers n.

FIGURE 12.7 Falling dominoes illustrate the principle of mathematical induction.

The principle of mathematical induction can be illustrated using an unending line of dominoes, as shown in **Figure 12.7**. If the first domino is pushed over, it knocks down the next, which knocks down the next, and so on, in a chain reaction. To topple all the dominoes in the infinite sequence, two conditions must be satisfied:

1. The first domino must be knocked down.
2. If the domino in position k is knocked down, then the domino in position $k + 1$ must be knocked down.

If the second condition is not satisfied, it does not follow that all the dominoes will topple. For example, suppose the dominoes are spaced far enough apart so that a falling domino does not push over the next domino in the line.

The domino analogy provides the two steps that are required in a proof by mathematical induction.

The Steps in a Proof by Mathematical Induction

Let S_n be a statement involving the positive integer n. To prove that S_n is true for all positive integers n requires two steps.

Step 1. Show that S_1 is true.

Step 2. Show that if S_k is assumed to be true, then S_{k+1} is also true, for every positive integer k.

Notice that to prove S_n, we work only with the statements S_1, S_k, and S_{k+1}. Our first example provides practice in writing these statements.

EXAMPLE 1 Writing S_1, S_k, and S_{k+1}

For the given statement S_n, write the three statements S_1, S_k, and S_{k+1}.

a. $S_n: 1 + 2 + 3 + \cdots + n = \dfrac{n(n+1)}{2}$

b. $S_n: 1^2 + 2^2 + 3^2 + \cdots + n^2 = \dfrac{n(n+1)(2n+1)}{6}$

Solution

a. We begin with

$$S_n: 1 + 2 + 3 + \cdots + n = \frac{n(n+1)}{2}.$$

Write S_1 by taking the first term on the left and replacing n with 1 on the right.

$$S_1: 1 = \frac{1(1+1)}{2}$$

Write S_k by taking the sum of the first k terms on the left and replacing n with k on the right.

$$S_k: 1 + 2 + 3 + \cdots + k = \frac{k(k+1)}{2}$$

Write S_{k+1} by taking the sum of the first $k + 1$ terms on the left and replacing n with $k + 1$ on the right.

$$S_{k+1}: 1 + 2 + 3 + \cdots + (k+1) = \frac{(k+1)[(k+1)+1]}{2}$$

$$S_{k+1}: 1 + 2 + 3 + \cdots + (k+1) = \frac{(k+1)(k+2)}{2}$$ Simplify on the right.

b. We begin with

$$S_n: 1^2 + 2^2 + 3^2 + \cdots + n^2 = \frac{n(n+1)(2n+1)}{6}.$$

Write S_1 by taking the first term on the left and replacing n with 1 on the right.

$$S_1: 1^2 = \frac{1(1+1)(2 \cdot 1 + 1)}{6}$$

Write S_k by taking the sum of the first k terms on the left and replacing n with k on the right.

$$S_k: 1^2 + 2^2 + 3^2 + \cdots + k^2 = \frac{k(k+1)(2k+1)}{6}$$

Write S_{k+1} by taking the sum of the first $k+1$ terms on the left and replacing n with $k+1$ on the right.

$$S_{k+1}: 1^2 + 2^2 + 3^2 + \cdots + (k+1)^2 = \frac{(k+1)[(k+1)+1][2(k+1)+1]}{6}$$

$$S_{k+1}: 1^2 + 2^2 + 3^2 + \cdots + (k+1)^2 = \frac{(k+1)(k+2)(2k+3)}{6}$$ Simplify on the right.

✓ **CHECK POINT 1** For the given statement S_n, write the three statements S_1, S_k, and S_{k+1}.

a. $2 + 4 + 6 + \cdots + 2n = n(n+1)$

b. $1^3 + 2^3 + 3^3 + \cdots + n^3 = \dfrac{n^2(n+1)^2}{4}$

Always simplify S_{k+1} before trying to use mathematical induction to prove that S_n is true. For example, consider

$$S_n: 1^2 + 3^2 + 5^2 + \cdots + (2n-1)^2 = \frac{n(2n-1)(2n+1)}{3}.$$

Begin by writing S_{k+1} as follows:

$$S_{k+1}: 1^2 + 3^2 + 5^2 + \cdots + [2(k+1)-1]^2$$

$$= \frac{(k+1)[2(k+1)-1][2(k+1)+1]}{3}.$$

The sum of the first $k+1$ terms

Replace n with $k+1$ on the right side of S_n.

Now simplify both sides of the equation.

$$S_{k+1}: 1^2 + 3^2 + 5^2 + \cdots + (2k+2-1)^2 = \frac{(k+1)(2k+2-1)(2k+2+1)}{3}$$

$$S_{k+1}: 1^2 + 3^2 + 5^2 + \cdots + (2k+1)^2 = \frac{(k+1)(2k+1)(2k+3)}{3}$$

2 Prove statements using mathematical induction.

Proving Statements about Positive Integers Using Mathematical Induction

Now that we know how to find S_1, S_k, and S_{k+1}, let's see how we can use these statements to carry out the two steps in a proof by mathematical induction. In Examples 2 and 3, we will use the statements S_1, S_k, and S_{k+1} to prove each of the statements S_n that we worked with in Example 1.

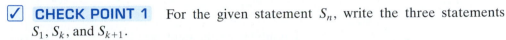

EXAMPLE 2 Proving a Formula by Mathematical Induction

Use mathematical induction to prove that

$$1 + 2 + 3 + \cdots + n = \frac{n(n+1)}{2}$$

for all positive integers n.

Solution

Step 1. Show that S_1 is true. Statement S_1 is

$$1 = \frac{1(1+1)}{2}.$$

Simplifying on the right, we obtain $1 = 1$. This true statement shows that S_1 is true.

Step 2. Show that if S_k is true, then S_{k+1} is true. Using S_k and S_{k+1} from Example 1(a), show that the truth of S_k,

$$1 + 2 + 3 + \cdots + k = \frac{k(k+1)}{2},$$

implies the truth of S_{k+1},

$$1 + 2 + 3 + \cdots + (k+1) = \frac{(k+1)(k+2)}{2}.$$

We will work with S_k. Because we assume that S_k is true, we add the next consecutive integer after k, namely, $k + 1$, to both sides.

$$1 + 2 + 3 + \cdots + k = \frac{k(k+1)}{2}$$ This is S_k, which we assume is true.

$$1 + 2 + 3 + \cdots + k + (k+1) = \frac{k(k+1)}{2} + (k+1)$$ Add $k + 1$ to both sides of the equation.

We do not have to write this k because k is understood to be the integer that precedes $k + 1$.

$$1 + 2 + 3 + \cdots + (k+1) = \frac{k(k+1)}{2} + \frac{2(k+1)}{2}$$ Write the right side with a common denominator of 2.

$$1 + 2 + 3 + \cdots + (k+1) = \frac{(k+1)}{2}(k+2)$$ Factor out the common factor $\frac{k+1}{2}$ on the right.

$$1 + 2 + 3 + \cdots + (k+1) = \frac{(k+1)(k+2)}{2}$$ This final result is the statement S_{k+1}.

We have shown that if we assume that S_k is true and we add $k + 1$ to both sides of S_k, then S_{k+1} is also true. By the principle of mathematical induction, the statement S_n, namely,

$$1 + 2 + 3 + \cdots + n = \frac{n(n+1)}{2},$$

is true for every positive integer n.

✓ CHECK POINT 2 Use mathematical induction to prove that

$$2 + 4 + 6 + \cdots + 2n = n(n+1).$$

for all positive integers n.

EXAMPLE 3 Proving a Formula by Mathematical Induction

Use mathematical induction to prove that

$$1^2 + 2^2 + 3^2 + \cdots + n^2 = \frac{n(n+1)(2n+1)}{6}$$

for all positive integers n.

Solution

Step 1. Show that S_1 is true. Statement S_1 is

$$1^2 = \frac{1(1+1)(2\cdot 1+1)}{6}.$$

Simplifying, we obtain $1 = \frac{1\cdot 2\cdot 3}{6}$. Further simplification on the right gives the statement $1 = 1$. This true statement shows that S_1 is true.

Step 2. Show that if S_k is true, then S_{k+1} is true. Using S_k and S_{k+1} from Example 1(b), show that the truth of

$$S_k: 1^2 + 2^2 + 3^2 + \cdots + k^2 = \frac{k(k+1)(2k+1)}{6}$$

implies the truth of

$$S_{k+1}: 1^2 + 2^2 + 3^2 + \cdots + (k+1)^2 = \frac{(k+1)(k+2)(2k+3)}{6}.$$

We will work with S_k. Because we assume that S_k is true, we add the square of the next consecutive integer after k, namely, $(k+1)^2$, to both sides of the equation.

$$1^2 + 2^2 + 3^2 + \cdots + k^2 = \frac{k(k+1)(2k+1)}{6}$$

This is S_k, assumed to be true. We must work with this and show S_{k+1} is true.

$$1^2 + 2^2 + 3^2 + \cdots + k^2 + (k+1)^2 = \frac{k(k+1)(2k+1)}{6} + (k+1)^2$$

Add $(k+1)^2$ to both sides.

$$1^2 + 2^2 + 3^2 + \cdots + (k+1)^2 = \frac{k(k+1)(2k+1)}{6} + \frac{6(k+1)^2}{6}$$

It is not necessary to write k^2 on the left. Express the right side with the least common denominator, 6.

$$1^2 + 2^2 + 3^2 + \cdots + (k+1)^2 = \frac{(k+1)}{6}[k(2k+1) + 6(k+1)]$$

Factor out the common factor $\frac{k+1}{6}$.

$$1^2 + 2^2 + 3^2 + \cdots + (k+1)^2 = \frac{(k+1)}{6}(2k^2 + 7k + 6)$$

Multiply and combine like terms.

$$1^2 + 2^2 + 3^2 + \cdots + (k+1)^2 = \frac{(k+1)}{6}(k+2)(2k+3)$$

Factor $2k^2 + 7k + 6$.

$$1^2 + 2^2 + 3^2 + \cdots + (k+1)^2 = \frac{(k+1)(k+2)(2k+3)}{6}$$

This final statement is S_{k+1}.

We have shown that if we assume that S_k is true and we add $(k+1)^2$ to both sides of S_k, then S_{k+1} is also true. By the principle of mathematical induction, the statement S_n, namely,

$$1^2 + 2^2 + 3^2 + \cdots + n^2 = \frac{n(n+1)(2n+1)}{6},$$

is true for every positive integer n.

✓ CHECK POINT 3 Use mathematical induction to prove that

$$1^3 + 2^3 + 3^3 + \cdots + n^3 = \frac{n^2(n+1)^2}{4}$$

for all positive integers n.

12.5 EXERCISE SET

MyMathLab MathXL PRACTICE WATCH DOWNLOAD READ REVIEW

Practice Exercises

In Exercises 1–4, a statement S_n about the positive integers is given. Write statements S_1, S_2, and S_3, and show that each of these statements is true.

1. $S_n: 1 + 3 + 5 + \cdots + (2n - 1) = n^2$

2. $S_n: 3 + 4 + 5 + \cdots + (n + 2) = \dfrac{n(n + 5)}{2}$

3. S_n: 2 is a factor of $n^2 - n$.

4. S_n: 3 is a factor of $n^3 - n$.

In Exercises 5–10, a statement S_n about the positive integers is given. Write statements S_k and S_{k+1}, simplifying statement S_{k+1} completely.

5. $S_n: 4 + 8 + 12 + \cdots + 4n = 2n(n + 1)$

6. $S_n: 3 + 4 + 5 + \cdots + (n + 2) = \dfrac{n(n + 5)}{2}$

7. $S_n: 3 + 7 + 11 + \cdots + (4n - 1) = n(2n + 1)$

8. $S_n: 2 + 7 + 12 + \cdots + (5n - 3) = \dfrac{n(5n - 1)}{2}$

9. S_n: 2 is a factor of $n^2 - n + 2$.

10. S_n: 2 is a factor of $n^2 - n$.

In Exercises 11–24, use mathematical induction to prove that each statement is true for every positive integer n.

11. $4 + 8 + 12 + \cdots + 4n = 2n(n + 1)$

12. $3 + 4 + 5 + \cdots + (n + 2) = \dfrac{n(n + 5)}{2}$

13. $1 + 3 + 5 + \cdots + (2n - 1) = n^2$

14. $3 + 6 + 9 + \cdots + 3n = \dfrac{3n(n + 1)}{2}$

15. $3 + 7 + 11 + \cdots + (4n - 1) = n(2n + 1)$

16. $2 + 7 + 12 + \cdots + (5n - 3) = \dfrac{n(5n - 1)}{2}$

17. $1 + 2 + 2^2 + \cdots + 2^{n-1} = 2^n - 1$

18. $1 + 3 + 3^2 + \cdots + 3^{n-1} = \dfrac{3^n - 1}{2}$

19. $2 + 4 + 8 + \cdots + 2^n = 2^{n+1} - 2$

20. $\dfrac{1}{2} + \dfrac{1}{4} + \dfrac{1}{8} + \cdots + \dfrac{1}{2^n} = 1 - \dfrac{1}{2^n}$

21. $1 \cdot 2 + 2 \cdot 3 + 3 \cdot 4 + \cdots + n(n + 1) = \dfrac{n(n + 1)(n + 2)}{3}$

22. $1 \cdot 3 + 2 \cdot 4 + 3 \cdot 5 + \cdots + n(n + 2) = \dfrac{n(n + 1)(2n + 7)}{6}$

23. $\dfrac{1}{1 \cdot 2} + \dfrac{1}{2 \cdot 3} + \dfrac{1}{3 \cdot 4} + \cdots + \dfrac{1}{n(n + 1)} = \dfrac{n}{n + 1}$

24. $\dfrac{1}{2 \cdot 3} + \dfrac{1}{3 \cdot 4} + \dfrac{1}{4 \cdot 5} + \cdots + \dfrac{1}{(n + 1)(n + 2)} = \dfrac{n}{2n + 4}$

Practice PLUS

In Exercises 25–32, use mathematical induction to prove that each statement is true for every positive integer n.

25. $\displaystyle\sum_{i=1}^{n} 5 \cdot 6^i = 6(6^n - 1)$

26. $\displaystyle\sum_{i=1}^{n} 7 \cdot 8^i = 8(8^n - 1)$

27. $n + 2 > n$

28. If $0 < x < 1$, then $0 < x^n < 1$.

29. $(ab)^n = a^n b^n$

30. $\left(\dfrac{a}{b}\right)^n = \dfrac{a^n}{b^n}$

31. $n^2 + n$ is divisible by 2.

32. $n^2 + 3n$ is divisible by 2.

Writing in Mathematics

33. Explain how to use mathematical induction to prove that a statement is true for every positive integer n.

34. Consider the statement S_n given by

$$n^2 - n + 41 \text{ is prime.}$$

Although $S_1, S_2, \ldots, S_{40}$ are true, S_{41} is false. Describe how this is illustrated by the dominoes in the figure. What does this tell you about a pattern, or formula, that seems to work for several values of n?

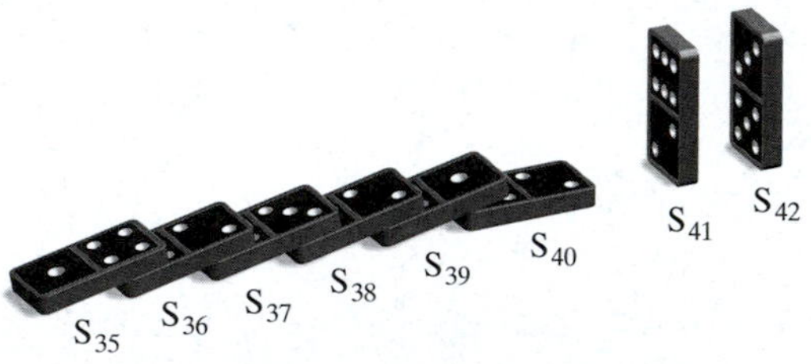

35. Fermat's most notorious theorem, described in the section opener on page 888, baffled the greatest minds for more than three centuries. In 1994, after ten years of work, Princeton University's Andrew Wiles proved Fermat's Last Theorem. *People* magazine put him on its list of "the 25 most intriguing people of the year," the Gap asked him to model jeans, and Barbara Walters chased him for an interview. "Who's Barbara Walters?" asked the bookish Wiles, who had somehow gone through life without a television.

Using the 1993 PBS documentary "Solving Fermat: Andrew Wiles" or information about Andrew Wiles on the Internet, research and write a report on what Wiles did to prove Fermat's Last Theorem, problems along the way, and the role of mathematical induction in the proof.

Critical Thinking Exercises

Make Sense? *In Exercises 36–39, determine whether each statement "makes sense" or "does not make sense" and explain your reasoning.*

36. I use mathematical induction to prove that statements are true for all real numbers n.

37. I begin proofs by mathematical induction by writing S_k and S_{k+1}, both of which I assume to be true.

38. When a line of falling dominoes is used to illustrate the principle of mathematical induction, it is not necessary for all the dominoes to topple.

39. This triangular arrangement of 36 circles illustrates that

$$1 + 2 + 3 + \cdots + n = \frac{n(n+1)}{2}$$

is true for $n = 8$.

In Exercises 40–41, find S_1 through S_5 and then use the pattern to make a conjecture about S_n. Prove the conjectured formula for S_n by mathematical induction.

40. $S_n: \frac{1}{4} + \frac{1}{12} + \frac{1}{24} + \cdots + \frac{1}{2n(n+1)} = ?$

41. $S_n: \left(1 - \frac{1}{2}\right)\left(1 - \frac{1}{3}\right)\left(1 - \frac{1}{4}\right)\cdots\left(1 - \frac{1}{n+1}\right) = ?$

42. Follow the outline below and use mathematical induction to prove the Binomial Theorem:

$$(a+b)^n = \binom{n}{0}a^n + \binom{n}{1}a^{n-1}b + \binom{n}{2}a^{n-2}b^2 + \cdots + \binom{n}{n-1}ab^{n-1} + \binom{n}{n}b^n.$$

a. Verify the formula for $n = 1$.

b. Replace n with k and write the statement that is assumed to be true. Replace n with $k + 1$ and write the statement that must be proved.

c. Multiply both sides of the statement assumed to be true by $a + b$. Add exponents on the left. On the right, distribute a and b, respectively.

d. Collect like terms on the right. At this point, you should have

$$(a+b)^{k+1} = \binom{k}{0}a^{k+1} + \left[\binom{k}{0} + \binom{k}{1}\right]a^k b + \left[\binom{k}{1} + \binom{k}{2}\right]a^{k-1}b^2 + \left[\binom{k}{2} + \binom{k}{3}\right]a^{k-2}b^3 + \cdots + \left[\binom{k}{k-1} + \binom{k}{k}\right]ab^k + \binom{k}{k}b^{k+1}.$$

e. Use the fact that $\binom{n}{r} + \binom{n}{r+1} = \binom{n+1}{r+1}$ to add the binomial sums in brackets. For example, because $\binom{n}{r} + \binom{n}{r+1} = \binom{n+1}{r+1}$, then $\binom{k}{0} + \binom{k}{1} = \binom{k+1}{1}$ and $\binom{k}{1} + \binom{k}{2} = \binom{k+1}{2}$.

f. Because $\binom{k}{0} = \binom{k+1}{0}$ (why?) and $\binom{k}{k} = \binom{k+1}{k+1}$ (why?), substitute these results and the results from part (e) into the equation in part (d). This should give the statement that we were required to prove in the second step of the mathematical induction process.

Review Exercises

43. Solve for t: $V = C(1 - t)$.

(Section 1.5, Example 6)

44. Solve: $x^3 + 2x^2 - 5x - 6 = 0$.

(Section 11.2, Example 6)

45. Give the center and radius. Then graph the equation:

$$x^2 + y^2 - 2x + 4y - 4 = 0.$$

(Section 10.1, Example 6)

Preview Exercises

Exercises 46–48 will help you prepare for the material covered in the next section.

46. Evaluate $\frac{n!}{(n-r)!}$ for $n = 20$ and $r = 3$.

47. Evaluate $\frac{n!}{(n-r)!\,r!}$ for $n = 8$ and $r = 3$.

48. You can choose from two pairs of jeans (one blue, one black) and three T-shirts (one beige, one yellow, and one blue), as shown in the diagram.

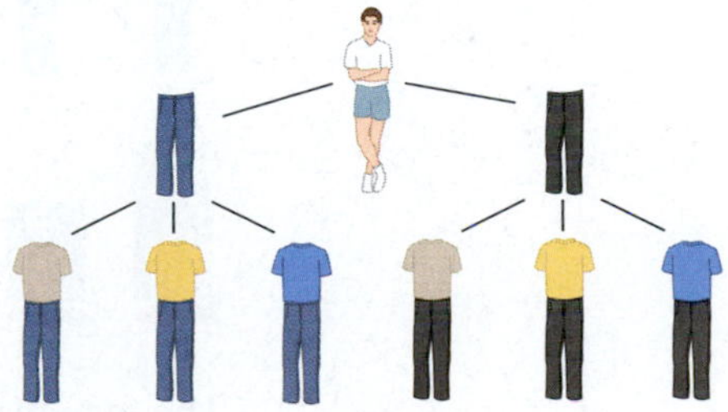

True or false: The diagram shows that you can form 2×3, or 6, different outfits.

SECTION 12.6

Counting Principles, Permutations, and Combinations

Objectives

1. Use the Fundamental Counting Principle.
2. Use the permutations formula.
3. Distinguish between permutation problems and combination problems.
4. Use the combinations formula.

Have you ever imagined what your life would be like if you won the lottery? What changes would you make? Before you fantasize about becoming a person of leisure with a staff of obedient elves, think about this: The probability of winning top prize in the lottery is about the same as the probability of being struck by lightning. There are millions of possible number combinations in lottery games and only one way of winning the grand prize. Determining the probability of winning involves calculating the chance of getting the winning combination from all possible outcomes. In this section, we begin preparing for the surprising world of probability by looking at methods for counting possible outcomes.

1 Use the Fundamental Counting Principle.

The Fundamental Counting Principle

It's early morning, you're groggy, and you have to select something to wear for your 8 A.M. class. (What *were* you thinking of when you signed up for a class at that hour?!) Fortunately, your "lecture wardrobe" is rather limited—just two pairs of jeans to choose from (one blue, one black), three T-shirts to choose from (one beige, one yellow, and one blue), and two pairs of sneakers to select from (one black pair, one red pair). Your possible outfits are shown in **Figure 12.8**.

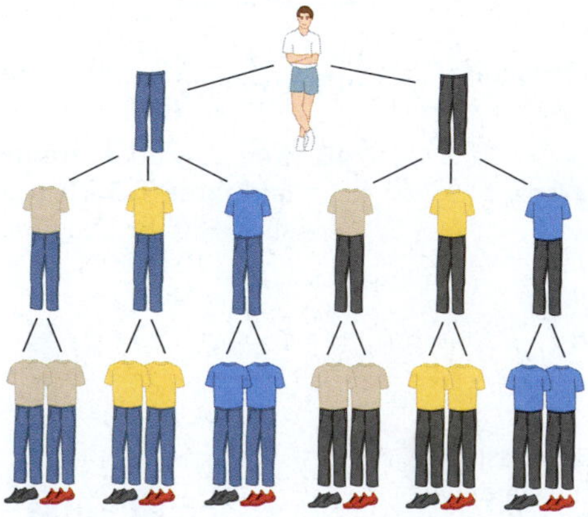

FIGURE 12.8 Selecting a wardrobe

The **tree diagram**, so named because of its branches, shows that you can form twelve outfits from your two pairs of jeans, three T-shirts, and two pairs of sneakers. Notice that the number of outfits can be obtained by multiplying the number of choices for jeans, 2, the number of choices for the T-shirts, 3, and the number of choices for the sneakers, 2:

$$2 \cdot 3 \cdot 2 = 12.$$

We can generalize this idea to any two or more groups of items—not just jeans, T-shirts, and sneakers—with the **Fundamental Counting Principle**:

The Fundamental Counting Principle

The number of ways in which a series of successive things can occur is found by multiplying the number of ways in which each thing can occur.

For example, if you own 30 pairs of jeans, 20 T-shirts, and 12 pairs of sneakers, you have

$$30 \cdot 20 \cdot 12 = 7200$$

choices for your wardrobe!

EXAMPLE 1 Options in Planning a Course Schedule

Next semester you are planning to take three courses—math, English, and humanities. Based on time blocks and highly recommended professors, there are 8 sections of math, 5 of English, and 4 of humanities that you find suitable. Assuming no scheduling conflicts, how many different three-course schedules are possible?

Solution This situation involves making choices with three groups of items.

Math	English	Humanities
8 choices	5 choices	4 choices

We use the Fundamental Counting Principle to find the number of three-course schedules. Multiply the number of choices for each of the three groups:

$$8 \cdot 5 \cdot 4 = 160.$$

Thus, there are 160 different three-course schedules.

✓ **CHECK POINT 1** A pizza can be ordered with three choices of size (small, medium, or large), four choices of crust (thin, thick, crispy, or regular), and six choices of toppings (ground beef, sausage, pepperoni, bacon, mushrooms, or onions). How many different one-topping pizzas can be ordered?

The number of possible ways of playing the first four moves on each side in a game of chess is 318,979,564,000.

EXAMPLE 2 A Multiple-Choice Test

You are taking a multiple-choice test that has ten questions. Each of the questions has four answer choices, with one correct answer per question. If you select one of these four choices for each question and leave nothing blank, in how many ways can you answer the questions?

Solution This situation involves making choices with ten questions.

Question 1	Question 2	Question 3	...	Question 9	Question 10
4 choices	4 choices	4 choices		4 choices	4 choices

We use the Fundamental Counting Principle to determine the number of ways that you can answer the questions on the test. Multiply the number of choices, 4, for each of the ten questions.

$$4 \cdot 4 \cdot 4 \cdot 4 \cdot 4 \cdot 4 \cdot 4 \cdot 4 \cdot 4 \cdot 4 = 4^{10} = 1{,}048{,}576$$

Thus, you can answer the questions in 1,048,576 different ways.

Blitzer Bonus

Running Out of Telephone Numbers

By the year 2020, portable telephones used for business and pleasure will all be videophones. At that time, U.S. population is expected to be 323 million. Faxes, beepers, cell phones, computer phone lines, and business lines may result in certain areas running out of phone numbers. Solution: Add more digits!

With or without extra digits, we expect that the 2020 videophone greeting will still be "hello," a word created by Thomas Edison in 1877. Phone inventor Alexander Graham Bell preferred "ahoy," but "hello" won out, appearing in the *Oxford English Dictionary* in 1883. *(Source: New York Times)*

Are you surprised that there are over one million ways of answering a ten-question multiple-choice test? Of course, there is only one way to answer the test and receive a perfect score. The probability of guessing your way into a perfect score involves calculating the chance of getting a perfect score, just one way, from all 1,048,576 possible outcomes. In short, prepare for the test and do not rely on guessing!

✓ **CHECK POINT 2** You are taking a multiple-choice test that has six questions. Each of the questions has three answer choices, with one correct answer per question. If you select one of these three choices for each question and leave nothing blank, in how many ways can you answer the questions?

EXAMPLE 3 Telephone Numbers in the United States

Telephone numbers in the United States begin with three-digit area codes followed by seven-digit local telephone numbers. Area codes and local telephone numbers cannot begin with 0 or 1. How many different telephone numbers are possible?

Solution This situation involves making choices with ten groups of items.

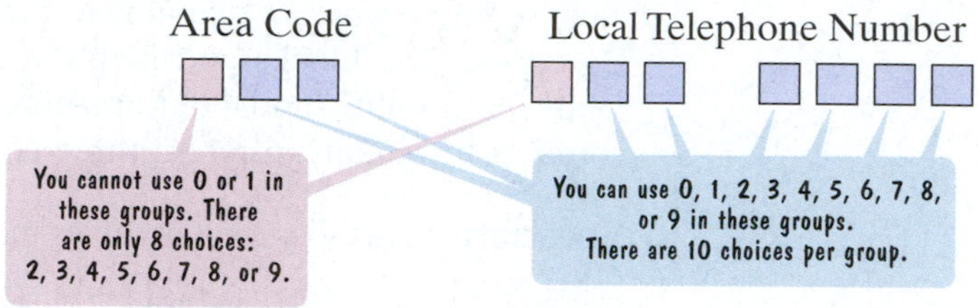

Here are the numbers of choices for each of the ten groups of items:

Area Code			Local Telephone Number						
8	10	10	8	10	10	10	10	10	10

We use the Fundamental Counting Principle to determine the number of different telephone numbers that are possible. The total number of telephone numbers possible is

$$8 \cdot 10 \cdot 10 \cdot 8 \cdot 10 \cdot 10 \cdot 10 \cdot 10 \cdot 10 \cdot 10 = 6{,}400{,}000{,}000.$$

There are six billion four hundred million (6.4×10^9) different telephone numbers that are possible.

✓ **CHECK POINT 3** License plates in a particular state display two letters followed by three numbers, such as AT-887 or BB-013. How many different license plates can be manufactured?

2 Use the permutations formula.

Permutations

You are the coach of a little league baseball team. There are 13 players on the team (and lots of parents hovering in the background, dreaming of stardom for their little "Manny Ramirez"). You need to choose a batting order having 9 players. The order makes a difference, because, for instance, if bases are loaded and "Little Manny" is fourth or fifth at bat, his possible home run will drive in three additional runs. How many batting orders can you form?

You can choose any of 13 players for the first person at bat. Then you will have 12 players from which to choose the second batter, then 11 from which to choose the third batter, and so on. The situation can be shown as follows:

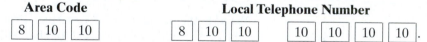

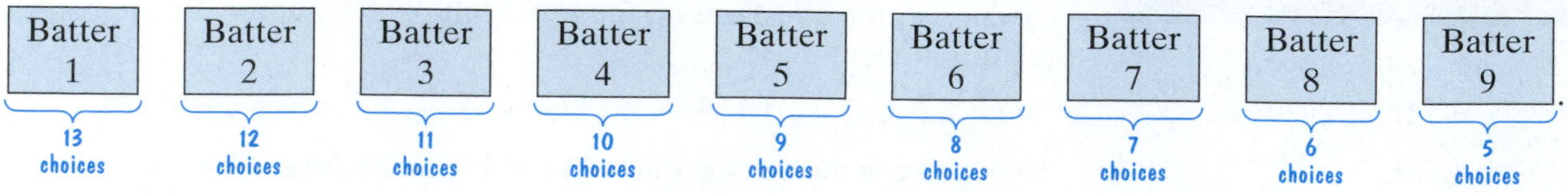

We use the Fundamental Counting Principle to find the number of batting orders. The total number of batting orders is

$$13 \cdot 12 \cdot 11 \cdot 10 \cdot 9 \cdot 8 \cdot 7 \cdot 6 \cdot 5 = 259{,}459{,}200.$$

Nearly 260 million batting orders are possible for your 13-player little league team. Each batting order is called a *permutation* of 13 players taken 9 at a time. The number of permutations of 13 players taken 9 at a time is 259,459,200.

A **permutation** is an ordered arrangement of items that occurs when

- No item is used more than once. (Each of the 9 players in the batting order bats exactly once.)
- The order of arrangement makes a difference.

We can obtain a formula for finding the number of permutations of 13 players taken 9 at a time by rewriting our computation:

$$13 \cdot 12 \cdot 11 \cdot 10 \cdot 9 \cdot 8 \cdot 7 \cdot 6 \cdot 5$$
$$= \frac{13 \cdot 12 \cdot 11 \cdot 10 \cdot 9 \cdot 8 \cdot 7 \cdot 6 \cdot 5 \cdot \boxed{4 \cdot 3 \cdot 2 \cdot 1}}{\boxed{4 \cdot 3 \cdot 2 \cdot 1}} = \frac{13!}{4!} = \frac{13!}{(13-9)!}.$$

Thus, the number of permutations of 13 things taken 9 at a time is $\frac{13!}{(13-9)!}$. The special notation ${}_{13}P_9$ is used to replace the phrase "the number of permutations of 13 things taken 9 at a time." Using this new notation, we can write

$${}_{13}P_9 = \frac{13!}{(13-9)!}.$$

The numerator of this expression is the number of items, 13 team members, expressed as a factorial: 13!. The denominator is also a factorial. It is the factorial of the difference between the number of items, 13, and the number of items in each permutation, 9 batters: $(13 - 9)!$.

The notation ${}_nP_r$ means the **number of permutations of n things taken r at a time.** We can generalize from the situation in which 9 batters were taken from 13 players. By generalizing, we obtain the following formula for the number of permutations if r items are taken from n items.

Permutations of n Things Taken r at a Time

The number of possible permutations if r items are taken from n items is

$${}_nP_r = \frac{n!}{(n-r)!}.$$

Study Tip

Because all permutation problems are also Fundamental Counting problems, they can be solved using the formula for ${}_nP_r$ or using the Fundamental Counting Principle.

Using Technology

Graphing utilities have a menu item for calculating permutations, usually labeled $\boxed{{}_nP_r}$. For example, to find ${}_{20}P_3$, the keystrokes are

20 $\boxed{{}_nP_r}$ 3 $\boxed{\text{ENTER}}$.

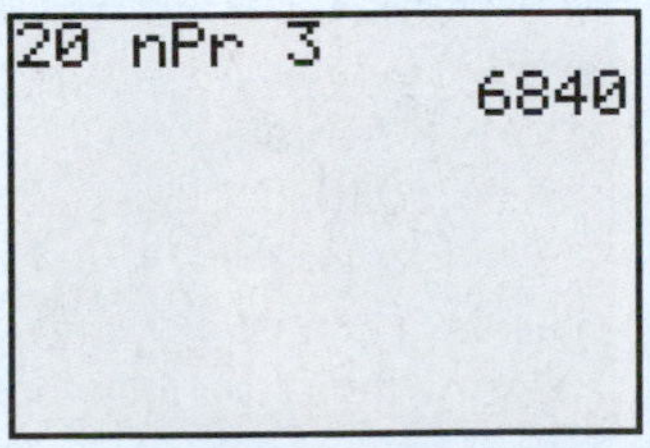

If you are using a scientific calculator, check your manual for the location of the menu item for calculating permutations and the required keystrokes.

EXAMPLE 4 Using the Formula for Permutations

You and 19 of your friends have decided to form an Internet marketing consulting firm. The group needs to choose three officers—a CEO, an operating manager, and a treasurer. In how many ways can those offices be filled?

Solution Your group is choosing $r = 3$ officers from a group of $n = 20$ people (you and 19 friends). The order in which the officers are chosen matters because the CEO, the operating manager, and the treasurer each have different responsibilities. Thus, we are looking for the number of permutations of 20 things taken 3 at a time. We use the formula

$${}_nP_r = \frac{n!}{(n-r)!}$$

with $n = 20$ and $r = 3$.

$${}_{20}P_3 = \frac{20!}{(20-3)!} = \frac{20!}{17!} = \frac{20 \cdot 19 \cdot 18 \cdot 17!}{17!} = \frac{20 \cdot 19 \cdot 18 \cdot \cancel{17!}}{\cancel{17!}} = 20 \cdot 19 \cdot 18 = 6840$$

Thus, there are 6840 different ways of filling the three offices.

✓ **CHECK POINT 4** A corporation has seven members on its board of directors. In how many different ways can it elect a president, vice-president, secretary, and treasurer?

EXAMPLE 5 Using the Formula for Permutations

You need to arrange seven of your favorite books along a small shelf. How many different ways can you arrange the books, assuming that the order of the books makes a difference to you?

Solution Because you are using all seven of your books in every possible arrangement, you are arranging $r = 7$ books from a group of $n = 7$ books. Thus, we are looking for the number of permutations of 7 things taken 7 at a time. We use the formula

$$_nP_r = \frac{n!}{(n-r)!}$$

with $n = 7$ and $r = 7$.

$$_7P_7 = \frac{7!}{(7-7)!} = \frac{7!}{0!} = \frac{7!}{1} = 5040$$

Thus, you can arrange the books in 5040 ways. There are 5040 different possible permutations.

✓ **CHECK POINT 5** In how many ways can 6 books be lined up along a shelf?

Blitzer Bonus

How to Pass the Time for $2\frac{1}{2}$ Million Years

If you were to arrange 15 different books on a shelf and it took you one minute for each permutation, the entire task would take 2,487,965 years.

(Source: Isaac Asimov's *Book of Facts)*

3 Distinguish between permutation problems and combination problems.

Combinations

As the twentieth century drew to a close, *Time* magazine presented a series of special issues on the most influential people of the century. In their issue on heroes and icons (June 14, 1999), they discussed a number of people whose careers became more profitable after their tragic deaths, including Marilyn Monroe, James Dean, Jim Morrison, Kurt Cobain, and Selena.

Imagine that you ask your friends the following question: "Of these five people, which three would you select to be included in a documentary featuring the best of their work?" You are not asking your friends to rank their three favorite artists in any kind of order—they should merely select the three to be included in the documentary.

One friend answers, "Jim Morrison, Kurt Cobain, and Selena." Another responds, "Selena, Kurt Cobain, and Jim Morrison." These two people have the same artists in their group of selections, even if they are named in a different order. We are interested *in which artists are named, not the order in which they are named*, for the documentary. Because the items are taken without regard to order, this is not a permutation problem. No ranking of any sort is involved.

Marilyn Monroe, actress (1927–1962)

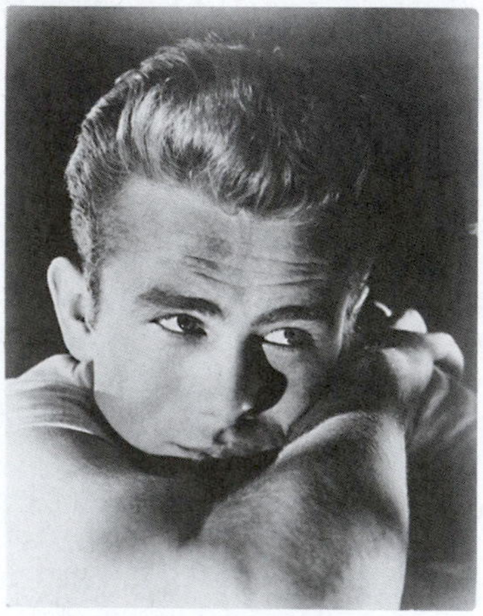

James Dean, actor (1931–1955)

Jim Morrison, musician and lead singer of The Doors (1943–1971)

Kurt Cobain, musician and front man for Nirvana (1967–1994)

Selena, musician of Tejano music (1971–1995)

Later on, you ask your roommate which three artists she would select for the documentary. She names Marilyn Monroe, James Dean, and Selena. Her selection is different from those of your two other friends because different entertainers are cited.

Mathematicians describe the group of artists given by your roommate as a *combination*. A **combination** of items occurs when

- The items are selected from the same group (the five stars who died young and tragically).
- No item is used more than once. (You may adore Selena, but your three selections cannot be Selena, Selena, and Selena.)
- The order of items makes no difference. (Morrison, Cobain, Selena is the same group in the documentary as Selena, Cobain, Morrison.)

Do you see the difference between a permutation and a combination? A permutation is an ordered arrangement of a given group of items. A combination is a group of items taken without regard to their order. **Permutation** problems involve situations in which **order matters**. **Combination** problems involve situations in which the **order** of items **makes no difference**.

EXAMPLE 6 Distinguishing between Permutations and Combinations

For each of the following problems, determine whether the problem is one involving permutations or combinations. (It is not necessary to solve the problem.)

a. Six students are running for student government president, vice-president, and treasurer. The student with the greatest number of votes becomes the president, the second highest vote-getter becomes vice-president, and the student who gets the third largest number of votes will be treasurer. How many different outcomes are possible for these three positions?

b. Six people are on the board of supervisors for your neighborhood park. A three-person committee is needed to study the possibility of expanding the park. How many different committees could be formed from the six people?

c. Baskin-Robbins offers 31 different flavors of ice cream. One of their items is a bowl consisting of three scoops of ice cream, each a different flavor. How many such bowls are possible?

Solution

a. Students are choosing three student government officers from six candidates. The order in which the officers are chosen makes a difference because each of the offices (president, vice-president, treasurer) is different. Order matters. This is a problem involving permutations.

b. A three-person committee is to be formed from the six-person board of supervisors. The order in which the three people are selected does not matter because they are not filling different roles on the committee. Because order makes no difference, this is a problem involving combinations.

c. A three-scoop bowl of three different flavors is to be formed from Baskin-Robbin's 31 flavors. The order in which the three scoops of ice cream are put into the bowl is irrelevant. A bowl with chocolate, vanilla, and strawberry is exactly the same as a bowl with vanilla, strawberry, and chocolate. Different orderings do not change things, and so this is a problem involving combinations.

✓ **CHECK POINT 6** For each of the following problems, explain if the problem is one involving permutations or combinations. (It is not necessary to solve the problem.)

a. How many ways can you select 6 free videos from a list of 200 videos?

b. In a race in which there are 50 runners and no ties, in how many ways can the first three finishers come in?

4 Use the combinations formula.

A Formula for Combinations

We have seen that the notation ${}_nP_r$ means the number of permutations of n things taken r at a time. Similarly, the notation **${}_nC_r$ means the number of combinations of n things taken r at a time.**

We can develop a formula for ${}_nC_r$ by comparing permutations and combinations. Consider the letters A, B, C, and D. The number of permutations of these four letters taken three at a time is

$$ {}_4P_3 = \frac{4!}{(4-3)!} = \frac{4!}{1!} = \frac{4\cdot 3\cdot 2\cdot 1}{1} = 24. $$

Here are the 24 permutations:

ABC,	ABD,	ACD,	BCD,
ACB,	ADB,	ADC,	BDC,
BAC,	BAD,	CAD,	CBD,
BCA,	BDA,	CDA,	CDB,
CAB,	DAB,	DAC,	DBC,
CBA,	DBA,	DCA,	DCB.
This column contains only one combination, ABC.	This column contains only one combination, ABD.	This column contains only one combination, ACD.	This column contains only one combination, BCD.

Because the order of items makes no difference in determining combinations, each column of six permutations represents one combination. There are a total of four combinations:

ABC, ABD, ACD, BCD.

Thus, ${}_4C_3 = 4$: The number of combinations of 4 things taken 3 at a time is 4. With 24 permutations and only four combinations, there are 6, or 3!, times as many permutations as there are combinations.

In general, there are $r!$ times as many permutations of n things taken r at a time as there are combinations of n things taken r at a time. Thus, we find the number of combinations of n things taken r at a time by dividing the number of permutations of n things taken r at a time by $r!$.

$$ {}_nC_r = \frac{{}_nP_r}{r!} = \frac{\frac{n!}{(n-r)!}}{r!} = \frac{n!}{(n-r)!\,r!} $$

Study Tip

The number of combinations if r items are taken from n items cannot be found using the Fundamental Counting Principle and requires the use of the formula shown on the right.

Combinations of *n* Things Taken *r* at a Time

The number of possible combinations if r items are taken from n items is

$$ {}_nC_r = \frac{n!}{(n-r)!\,r!}. $$

Notice that the formula for ${}_nC_r$ is the same as the formula for the binomial coefficient $\binom{n}{r}$.

EXAMPLE 7 Using the Formula for Combinations

A three-person committee is needed to study ways of improving public transportation. How many committees could be formed from the eight people on the board of supervisors?

Using Technology

Graphing utilities have a menu item for calculating combinations, usually labeled ${}_nC_r$. For example, to find ${}_8C_3$, the keystrokes on most graphing utilities are

8 $\boxed{{}_nC_r}$ 3 $\boxed{\text{ENTER}}$.

If you are using a scientific calculator, check your manual to see whether there is a menu item for calculating combinations.

If you use your calculator's factorial key to find $\frac{8!}{5!\,3!}$, be sure to enclose the factorials in the denominator with parentheses

8 $\boxed{!}$ $\boxed{\div}$ $\boxed{(}$ 5 $\boxed{!}$ $\boxed{\times}$ 3 $\boxed{!}$ $\boxed{)}$

pressing $\boxed{=}$ or $\boxed{\text{ENTER}}$ to obtain the answer.

Solution The order in which the three people are selected does not matter. This is a problem of selecting $r = 3$ people from a group of $n = 8$ people. We are looking for the number of combinations of eight things taken three at a time. We use the formula

$$ {}_nC_r = \frac{n!}{(n-r)!\,r!} $$

with $n = 8$ and $r = 3$.

$$ {}_8C_3 = \frac{8!}{(8-3)!\,3!} = \frac{8!}{5!\,3!} = \frac{8\cdot7\cdot6\cdot5!}{5!\cdot3\cdot2\cdot1} = \frac{8\cdot7\cdot6\cdot\cancel{5!}}{\cancel{5!}\cdot3\cdot2\cdot1} = 56 $$

Thus, 56 committees of three people each can be formed from the eight people on the board of supervisors.

☑ **CHECK POINT 7** From a group of 10 physicians, in how many ways can four people be selected to attend a conference on acupuncture?

EXAMPLE 8 Using the Formula for Combinations

In poker, a person is dealt 5 cards from a standard 52-card deck. The order in which you are dealt the 5 cards does not matter. How many different 5-card poker hands are possible?

Solution Because the order in which the 5 cards are dealt does not matter, this is a problem involving combinations. We are looking for the number of combinations of $n = 52$ cards drawn $r = 5$ at a time. We use the formula

$$ {}_nC_r = \frac{n!}{(n-r)!\,r!} $$

with $n = 52$ and $r = 5$.

$$ {}_{52}C_5 = \frac{52!}{(52-5)!\,5!} = \frac{52!}{47!\,5!} = \frac{52\cdot51\cdot50\cdot49\cdot48\cdot\cancel{47!}}{\cancel{47!}\cdot5\cdot4\cdot3\cdot2\cdot1} = 2{,}598{,}960 $$

Thus, there are 2,598,960 different 5-card poker hands possible. It surprises many people that more than 2.5 million 5-card hands can be dealt from a mere 52 cards.

FIGURE 12.9 A royal flush

If you are a poker player, it does not get any better than to be dealt the 5-card hand shown in **Figure 12.9**. This hand is called a *royal flush*. It consists of an ace, king, queen, jack, and 10, all of the same suit: all hearts, all diamonds, all clubs, or all spades. The probability of being dealt a royal flush involves calculating the number of ways of being dealt such a hand: just 4 of all 2,598,960 possible hands. In the next section, we move from counting possibilities to computing probabilities.

☑ **CHECK POINT 8** How many different 4-card hands can be dealt from a deck that has 16 different cards?

12.6 EXERCISE SET

MyMathLab — MathXL PRACTICE, WATCH, DOWNLOAD, READ, REVIEW

Practice Exercises

In Exercises 1–8, use the formula for ${}_nP_r$ to evaluate each expression.

1. ${}_9P_4$
2. ${}_7P_3$
3. ${}_8P_5$
4. ${}_{10}P_4$
5. ${}_6P_6$
6. ${}_9P_9$
7. ${}_8P_0$
8. ${}_6P_0$

In Exercises 9–16, use the formula for ${}_nC_r$ to evaluate each expression.

9. ${}_9C_5$
10. ${}_{10}C_6$
11. ${}_{11}C_4$
12. ${}_{12}C_5$

13. ${}_7C_7$

14. ${}_4C_4$

15. ${}_5C_0$

16. ${}_6C_0$

In Exercises 17–20, does the problem involve permutations or combinations? Explain your answer. (It is not necessary to solve the problem.)

17. A medical researcher needs 6 people to test the effectiveness of an experimental drug. If 13 people have volunteered for the test, in how many ways can 6 people be selected?

18. Fifty people purchase raffle tickets. Three winning tickets are selected at random. If first prize is $1000, second prize is $500, and third prize is $100, in how many different ways can the prizes be awarded?

19. How many different four-letter passwords can be formed from the letters A, B, C, D, E, F, and G if no repetition of letters is allowed?

20. Fifty people purchase raffle tickets. Three winning tickets are selected at random. If each prize is $500, in how many different ways can the prizes be awarded?

Practice PLUS

In Exercises 21–28, evaluate each expression.

21. $\frac{{}_7P_3}{3!} - {}_7C_3$

22. $\frac{{}_{20}P_2}{2!} - {}_{20}C_2$

23. $1 - \frac{{}_3P_2}{{}_4P_3}$

24. $1 - \frac{{}_5P_3}{{}_{10}P_4}$

25. $\frac{{}_7C_3}{{}_5C_4} - \frac{98!}{96!}$

26. $\frac{{}_{10}C_3}{{}_6C_4} - \frac{46!}{44!}$

27. $\frac{{}_4C_2 \cdot {}_6C_1}{{}_{18}C_3}$

28. $\frac{{}_5C_1 \cdot {}_7C_2}{{}_{12}C_3}$

Application Exercises

Use the Fundamental Counting Principle to solve Exercises 29–40.

29. The model of the car you are thinking of buying is available in nine different colors and three different styles (hatchback, sedan, or station wagon). In how many ways can you order the car?

30. A popular brand of pen is available in three colors (red, green, or blue) and four writing tips (bold, medium, fine, or micro). How many different choices of pens do you have with this brand?

31. An ice cream store sells two drinks (sodas or milk shakes), in four sizes (small, medium, large, or jumbo), and five flavors (vanilla, strawberry, chocolate, coffee, or pistachio). In how many ways can a customer order a drink?

32. A restaurant offers the following lunch menu.

Main Course	Vegetables	Beverages	Desserts
Ham	Potatoes	Coffee	Cake
Chicken	Peas	Tea	Pie
Fish	Green beans	Milk	Ice cream
Beef		Soda	

If one item is selected from each of the four groups, in how many ways can a meal be ordered? Describe two such orders.

33. You are taking a multiple-choice test that has five questions. Each of the questions has three answer choices, with one correct answer per question. If you select one of these three choices for each question and leave nothing blank, in how many ways can you answer the questions?

34. You are taking a multiple-choice test that has eight questions. Each of the questions has three answer choices, with one correct answer per question. If you select one of these three choices for each question and leave nothing blank, in how many ways can you answer the questions?

35. In the original plan for area codes in 1945, the first digit could be any number from 2 through 9, the second digit was either 0 or 1, and the third digit could be any number except 0. With this plan, how many different area codes were possible?

36. How many different four-letter radio station call letters can be formed if the first letter must be W or K?

37. Six performers are to present their comedy acts on a weekend evening at a comedy club. One of the performers insists on being the last stand-up comic of the evening. If this performer's request is granted, how many different ways are there to schedule the appearances?

38. Five singers are to perform at a night club. One of the singers insists on being the last performer of the evening. If this singer's request is granted, how many different ways are there to schedule the appearances?

39. In the *Cambridge Encyclopedia of Language* (Cambridge University Press, 1987), author David Crystal presents five sentences that make a reasonable paragraph regardless of their order. The sentences are as follows:

- Mark had told him about the foxes.
- John looked out the window.
- Could it be a fox?
- However, nobody had seen one for months.
- He thought he saw a shape in the bushes.

How many different five-sentence paragraphs can be formed if the paragraph begins with "He thought he saw a shape in the bushes" and ends with "John looked out of the window"?

40. A television programmer is arranging the order that five movies will be seen between the hours of 6 P.M. and 4 A.M. Two of the movies have a G rating and they are to be shown in the first two time blocks. One of the movies is rated NC-17 and it is to be shown in the last of the time blocks, from 2 A.M. until 4 A.M. Given these restrictions, in how many ways can the five movies be arranged during the indicated time blocks?

Use the formula for ${}_nP_r$ to solve Exercises 41–48.

41. A club with ten members is to choose three officers—president, vice-president, and secretary-treasurer. If each office is to be held by one person and no person can hold more than one office, in how many ways can those offices be filled?

42. A corporation has ten members on its board of directors. In how many different ways can it elect a president, vice-president, secretary, and treasurer?

43. For a segment of a radio show, a disc jockey can play 7 songs. If there are 13 songs to select from, in how many ways can the program for this segment be arranged?

44. Suppose you are asked to list, in order of preference, the three best movies you have seen this year. If you saw 20 movies during the year, in how many ways can the three best be chosen and ranked?

45. In a race in which six automobiles are entered and there are no ties, in how many ways can the first three finishers come in?

46. In a production of *West Side Story*, eight actors are considered for the male roles of Tony, Riff, and Bernardo. In how many ways can the director cast the male roles?

47. Nine bands have volunteered to perform at a benefit concert, but there is only enough time for five of the bands to play. How many lineups are possible?

48. How many arrangements can be made using four of the letters of the word COMBINE if no letter is to be used more than once?

Use the formula for ${}_nC_r$ to solve Exercises 49–56.

49. An election ballot asks voters to select three city commissioners from a group of six candidates. In how many ways can this be done?

50. A four-person committee is to be elected from an organization's membership of 11 people. How many different committees are possible?

51. Of 12 possible books, you plan to take 4 with you on vacation. How many different collections of 4 books can you take?

52. There are 14 standbys who hope to get seats on a flight, but only 6 seats are available on the plane. How many different ways can the 6 people be selected?

53. You volunteer to help drive children at a charity event to the zoo, but you can fit only 8 of the 17 children present in your van. How many different groups of 8 children can you drive?

54. Of the 100 people in the U.S. Senate, 18 serve on the Foreign Relations Committee. How many ways are there to select Senate members for this committee (assuming party affiliation is not a factor in selection)?

55. To win at LOTTO in the state of Florida, one must correctly select 6 numbers from a collection of 53 numbers (1 through 53). The order in which the selection is made does not matter. How many different selections are possible?

56. To win in the New York State lottery, one must correctly select 6 numbers from 59 numbers. The order in which the selection is made does not matter. How many different selections are possible?

In Exercises 57–66, solve by the method of your choice.

57. In a race in which six automobiles are entered and there are no ties, in how many ways can the first four finishers come in?

58. A book club offers a choice of 8 books from a list of 40. In how many ways can a member make a selection?

59. A medical researcher needs 6 people to test the effectiveness of an experimental drug. If 13 people have volunteered for the test, in how many ways can 6 people be selected?

60. Fifty people purchase raffle tickets. Three winning tickets are selected at random. If first prize is \$1000, second prize is \$500, and third prize is \$100, in how many different ways can the prizes be awarded?

61. From a club of 20 people, in how many ways can a group of three members be selected to attend a conference?

62. Fifty people purchase raffle tickets. Three winning tickets are selected at random. If each prize is \$500, in how many different ways can the prizes be awarded?

63. How many different four-letter passwords can be formed from the letters A, B, C, D, E, F, and G if no repetition of letters is allowed?

64. Nine comedy acts will perform over two evenings. Five of the acts will perform on the first evening and the order in which the acts perform is important. How many ways can the schedule for the first evening be made?

65. Using 15 flavors of ice cream, how many cones with three different flavors can you create if it is important to you which flavor goes on the top, middle, and bottom?

66. Baskin-Robbins offers 31 different flavors of ice cream. One of their items is a bowl consisting of three scoops of ice cream, each a different flavor. How many such bowls are possible?

Exercises 67–72 are based on the following jokes about books:

- *"Outside of a dog, a book is man's best friend. Inside of a dog, it's too dark to read."—Groucho Marx*
- *"I recently bought a book of free verse. For \$12." —George Carlin*
- *"If a word in the dictionary was misspelled, how would we know?"—Steven Wright*
- *"Encyclopedia is a Latin term. It means 'to paraphrase a term paper.' "—Greg Ray*
- *"A bookstore is one of the only pieces of evidence we have that people are still thinking."—Jerry Seinfeld*
- *"I honestly believe there is absolutely nothing like going to bed with a good book. Or a friend who's read one." —Phyllis Diller*

67. In how many ways can these six jokes be ranked from best to worst?

68. If Phyllis Diller's joke about books is excluded, in how many ways can the remaining five jokes be ranked from best to worst?

69. In how many ways can people select their three favorite jokes from these comments about books?

70. In how many ways can people select their two favorite jokes from these comments about books?

71. If the order in which these jokes are told makes a difference in terms of how they are received, how many ways can they be delivered if George Carlin's joke is delivered first and Jerry Seinfeld's joke is told last?

72. If the order in which these jokes are told makes a difference in terms of how they are received, how many ways can they be delivered if a joke by a man is told first?

Writing in Mathematics

73. Explain the Fundamental Counting Principle.

74. Write an original problem that can be solved using the Fundamental Counting Principle. Then solve the problem.

75. What is a permutation?

76. Describe what $_nP_r$ represents.

77. Write a word problem that can be solved by evaluating $_7P_3$.

78. What is a combination?

79. Explain how to distinguish between permutation and combination problems.

80. Write a word problem that can be solved by evaluating $_7C_3$.

Technology Exercises

81. Use a graphing utility with an $\boxed{_nP_r}$ menu item to verify your answers in Exercises 1–8.

82. Use a graphing utility with an $\boxed{_nC_r}$ menu item to verify your answers in Exercises 9–16.

Critical Thinking Exercises

Make Sense? *In Exercises 83–86, determine whether each statement "makes sense" or "does not make sense" and explain your reasoning.*

83. I used the Fundamental Counting Principle to determine the number of five-digit ZIP codes that are available to the U.S. Postal Service.

84. I used the permutations formula to determine the number of ways the manager of a baseball team can form a 9-player batting order from a team of 25 players.

85. I used the combinations formula to determine how many different four-note sound sequences can be created from the notes C, D, E, F, G, A, and B.

86. I used the permutations formula to determine the number of ways people can select their 9 favorite baseball players from a team of 25 players.

In Exercises 87–90, determine whether each statement is true or false. If the statement is false, make the necessary change(s) to produce a true statement.

87. The number of ways to choose four questions out of ten questions on an essay test is $_{10}P_4$.

88. If $r > 1$, $_nP_r$ is less than $_nC_r$.

89. $_7P_3 = 3!_7C_3$

90. The number of ways to pick a winner and first runner-up in a talent contest with 20 contestants is $_{20}C_2$.

91. Five men and five women line up at a checkout counter in a store. In how many ways can they line up if the first person in line is a woman and the people in line alternate woman, man, woman, man, and so on?

92. How many four-digit odd numbers less than 6000 can be formed using the digits 2, 4, 6, 7, 8, and 9, assuming that digits may be repeated?

93. A mathematics exam consists of 10 multiple-choice questions and 5 open-ended problems in which all work must be shown. If an examinee must answer 8 of the multiple-choice questions and 3 of the open-ended problems, in how many ways can the questions and problems be chosen?

Review Exercises

94. If $f(x) = x^2 + 2x - 5$ and $g(x) = 4x - 1$, find $(f \circ g)(x)$. (Section 9.2. Example 1)

95. Solve: $|2x - 5| > 3$. (Section 4.3, Example 5 or Example 6(b))

96. Graph: $f(x) = (x - 2)^2(x + 1)$. (Section 11.1, Example 5)

Preview Exercises

Exercises 97–99 will help you prepare for the material covered in the next section.

The figure shows that when a die is rolled, there are six equally likely outcomes: 1, 2, 3, 4, 5, or 6. Use this information to solve each exercise.

97. What fraction of the outcomes are less than 5?

98. What fraction of the outcomes are not less than 5?

99. What fraction of the outcomes are even or greater than 3?

SECTION 12.7

Probability

Objectives

1. Compute empirical probability.
2. Compute theoretical probability.
3. Find the probability that an event will not occur.
4. Find the probability of one event or a second event occurring.
5. Find the probability of one event and a second event occurring.

Table 12.2 The Hours of Sleep Americans Get on a Typical Night

Hours of Sleep	Number of Americans, in millions
4 or less	12
5	27
6	75
7	90
8	81
9	9
10 or more	6
	Total: 300

Source: Discovery Health Media

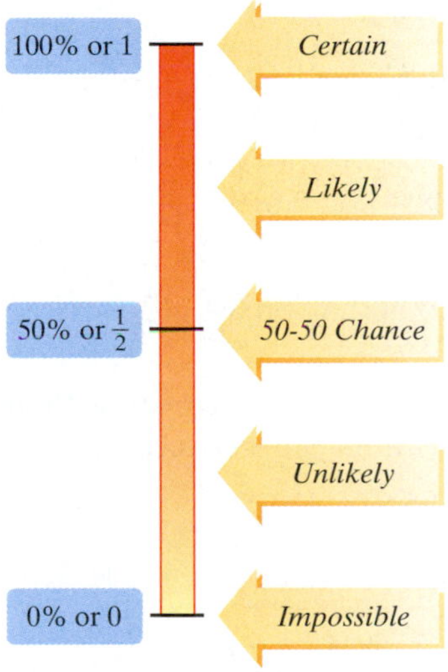

Possible Values for Probabilities

How many hours of sleep do you typically get each night? **Table 12.2** indicates that 75 million out of 300 million Americans are getting six hours of sleep on a typical night. The *probability* of an American getting six hours of sleep on a typical night is $\frac{75}{300}$. This fraction can be reduced to $\frac{1}{4}$, or expressed as 0.25, or 25%. Thus, 25% of Americans get six hours of sleep each night.

We find a probability by dividing one number by another. Probabilities are assigned to an *event*, such as getting six hours of sleep on a typical night. Events that are certain to occur are assigned probabilities of 1, or 100%. For example, the probability that a given individual will eventually die is 1. Although Woody Allen whined, "I don't want to achieve immortality through my work. I want to achieve it through not dying," death (and taxes) are always certain. By contrast, if an event cannot occur, its probability is 0. Regrettably, the probability that Elvis will return and serenade us with one final reprise of "Don't Be Cruel" (and we hope we're not) is 0.

Probabilities of events are expressed as numbers ranging from 0 to 1, or 0% to 100%. The closer the probability of a given event is to 1, the more likely it is that the event will occur. The closer the probability of a given event is to 0, the less likely it is that the event will occur.

Empirical Probability

1 Compute empirical probability.

Empirical probability applies to situations in which we observe how frequently an event occurs. We use the following formula to compute the empirical probability of an event:

Computing Empirical Probability

The **empirical probability** of event E, denoted by $P(E)$, is

$$P(E) = \frac{\text{observed number of times } E \text{ occurs}}{\text{total number of observed occurrences}}.$$

EXAMPLE 1 Empirical Probabilities with Real-World Data

When women turn 40, their gynecologists typically remind them that it is time to undergo mammography screening for breast cancer. The data in **Table 12.3** at the top of the next page are based on 100,000 U.S. women, ages 40 to 50, who participated in mammography screening.

Table 12.3 Mammography Screening on 100,000 U.S. Women, Ages 40 to 50

	Breast Cancer	No Breast Cancer
Positive Mammogram	720	6944
Negative Mammogram	80	92,256

720 + 6944 = 7664 women have positive mammograms.

80 + 92,256 = 92,336 women have negative mammograms.

720 + 80 = 800 women have breast cancer.

6944 + 92,256 = 99,200 women do not have breast cancer.

Source: Gerd Gigerenzer, *Calculated Risks*, Simon and Schuster, 2002

a. Use **Table 12.3** to find the probability that a woman aged 40 to 50 has breast cancer.

b. Among women without breast cancer, find the probability of a positive mammogram.

c. Among women with positive mammograms, find the probability of not having breast cancer.

Solution

a. We begin with the probability that a woman aged 40 to 50 has breast cancer. The probability of having breast cancer is the number of women with breast cancer divided by the total number of women.

$$P(\text{breast cancer}) = \frac{\text{number of women with breast cancer}}{\text{total number of women}}$$

$$= \frac{800}{100{,}000} = \frac{1}{125} = 0.008$$

The empirical probability that a woman aged 40 to 50 has breast cancer is $\frac{1}{125}$, or 0.008.

b. Now, we find the probability of a positive mammogram among women without breast cancer. Thus, we restrict the data to women without breast cancer:

	No Breast Cancer
Positive Mammogram	6944
Negative Mammogram	92,256

Within the restricted data, the probability of a positive mammogram is the number of women with positive mammograms divided by the total number of women.

$$P(\text{positive mammogram}) = \frac{\text{number of women with positive mammograms}}{\text{total number of women in the restricted data}}$$

$$= \frac{6944}{6944 + 92{,}256} = \frac{6944}{99{,}200} = 0.07$$

This is the total number of women without breast cancer.

Among women without breast cancer, the empirical probability of a positive mammogram is $\frac{6944}{99{,}200}$, or 0.07.

c. Now, we find the probability of not having breast cancer among women with positive mammograms. Thus, we restrict the data to women with positive mammograms:

	Breast Cancer	No Breast Cancer
Positive Mammogram	720	6944

Within the restricted data, the probability of not having breast cancer is the number of women with no breast cancer divided by the total number of women.

$$P(\text{no breast cancer}) = \frac{\text{number of women with no breast cancer}}{\text{total number of women in the restricted data}}$$

$$= \frac{6944}{720 + 6944} = \frac{6944}{7664} \approx 0.906$$

This is the total number of women with positive mammograms.

Among women with positive mammograms, the probability of not having breast cancer is $\frac{6944}{7664}$, or approximately 0.906.

✓ **CHECK POINT 1** Use the data in **Table 12.3** to solve this exercise. Express probabilities as fractions and as decimals to three decimal places.

a. Find the probability that a woman aged 40 to 50 has a positive mammogram.

b. Among women with breast cancer, find the probability of a positive mammogram.

c. Among women with positive mammograms, find the probability of having breast cancer.

2 Compute theoretical probability.

Theoretical Probability

You toss a coin. Although it is equally likely to land either heads up, denoted by H, or tails up, denoted by T, the actual outcome is uncertain. Any occurrence for which the outcome is uncertain is called an **experiment**. Thus, tossing a coin is an example of an experiment. The set of all possible outcomes of an experiment is the **sample space** of the experiment, denoted by S. The sample space for the coin-tossing experiment is

$$S = \{H, T\}.$$

Lands heads up   Lands tails up

We can define an event more formally using these concepts. An **event**, denoted by E, is any subcollection, or subset, of a sample space. For example, the subset $E = \{T\}$ is the event of landing tails up when a coin is tossed.

Theoretical probability applies to situations like this, in which the sample space only contains equally likely outcomes, all of which are known. To calculate the theoretical probability of an event, we divide the number of outcomes resulting in the event by the number of outcomes in the sample space.

Computing Theoretical Probability

If an event E has $n(E)$ equally likely outcomes and its sample space S has $n(S)$ equally likely outcomes, the **theoretical probability** of event E, denoted by $P(E)$, is

$$P(E) = \frac{\text{number of outcomes in event } E}{\text{number of outcomes in sample space } S} = \frac{n(E)}{n(S)}.$$

The sum of the theoretical probabilities of all possible outcomes in the sample space is 1.

How can we use this formula to compute the probability of a coin landing tails up? We use the following sets:

$$E = \{T\}. \qquad S = \{H, T\}.$$

This is the event of landing tails up. This is the sample space with all equally likely outcomes.

The probability of a coin landing tails up is

$$P(E) = \frac{\text{number of outcomes that result in tails up}}{\text{total number of possible outcomes}} = \frac{n(E)}{n(S)} = \frac{1}{2}.$$

FIGURE 12.10 Outcomes when a die is rolled

Theoretical probability applies to many games of chance, including dice rolling, lotteries, card games, and roulette. The next example deals with the experiment of rolling a die. **Figure 12.10** illustrates that when a die is rolled, there are six equally likely outcomes. The sample space can be shown as

$$S = \{1, 2, 3, 4, 5, 6\}.$$

EXAMPLE 2 Computing Theoretical Probability

A die is rolled. Find the probability of getting a number less than 5.

Solution The sample space of equally likely outcomes is $S = \{1, 2, 3, 4, 5, 6\}$. There are six outcomes in the sample space, so $n(S) = 6$.

We are interested in the probability of getting a number less than 5. The event of getting a number less than 5 can be represented by

$$E = \{1, 2, 3, 4\}.$$

There are four outcomes in this event, so $n(E) = 4$.

The probability of rolling a number less than 5 is

$$P(E) = \frac{n(E)}{n(S)} = \frac{4}{6} = \frac{2}{3}.$$

✓ **CHECK POINT 2** A die is rolled. Find the probability of getting a number greater than 4.

EXAMPLE 3 Computing Theoretical Probability

Two ordinary six-sided dice are rolled. What is the probability of getting a sum of 8?

Solution Each die has six equally likely outcomes. By the Fundamental Counting Principle, there are $6 \cdot 6$, or 36, equally likely outcomes in the sample space. That is, $n(S) = 36$. The 36 outcomes are shown at the top of the next page as ordered pairs. The five ways of rolling a sum of 8 appear in the green highlighted diagonal.

	Second Die ⚀	⚁	⚂	⚃	⚄	⚅
First Die ⚀	(1, 1)	(1, 2)	(1, 3)	(1, 4)	(1, 5)	(1, 6)
⚁	(2, 1)	(2, 2)	(2, 3)	(2, 4)	(2, 5)	(2, 6)
⚂	(3, 1)	(3, 2)	(3, 3)	(3, 4)	(3, 5)	(3, 6)
⚃	(4, 1)	(4, 2)	(4, 3)	(4, 4)	(4, 5)	(4, 6)
⚄	(5, 1)	(5, 2)	(5, 3)	(5, 4)	(5, 5)	(5, 6)
⚅	(6, 1)	(6, 2)	(6, 3)	(6, 4)	(6, 5)	(6, 6)

$$S = \{(1,1), (1,2), (1,3), (1,4), (1,5), (1,6), (2,1), (2,2), (2,3), (2,4), (2,5), (2,6), (3,1), (3,2), (3,3), (3,4), (3,5), (3,6), (4,1), (4,2), (4,3), (4,4), (4,5), (4,6), (5,1), (5,2), (5,3), (5,4), (5,5), (5,6), (6,1), (6,2), (6,3), (6,4), (6,5), (6,6)\}$$

The phrase "getting a sum of 8" describes the event

$$E = \{(6,2), (5,3), (4,4), (3,5), (2,6)\}.$$

This event has 5 outcomes, so $n(E) = 5$. Thus, the probability of getting a sum of 8 is

$$P(E) = \frac{n(E)}{n(S)} = \frac{5}{36}.$$

✓ **CHECK POINT 3** What is the probability of getting a sum of 5 when two six-sided dice are rolled?

Computing Theoretical Probability without Listing an Event and the Sample Space

In some situations, we can compute theoretical probability without having to write out the elements in each event and in each sample space. For example, suppose you are dealt one card from a standard 52-card deck, illustrated in **Figure 12.11**. The deck has four suits: Hearts and diamonds are red, and clubs and spades are black. Each suit has 13 different face values—A(ace), 2, 3, 4, 5, 6, 7, 8, 9, 10, J(jack), Q(queen), and K(king). Jacks, queens, and kings are called **picture cards** or **face cards**.

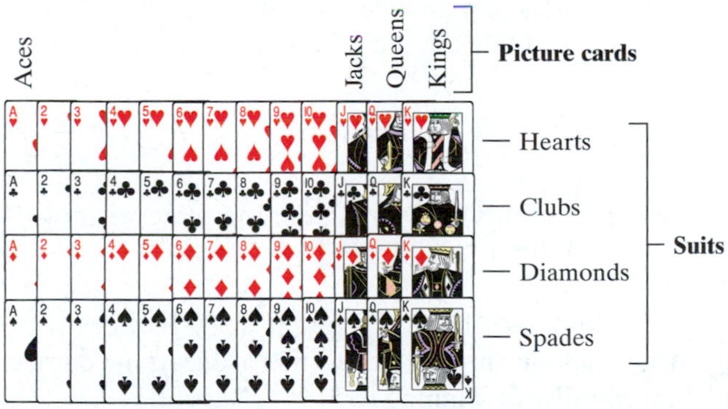

FIGURE 12.11
A standard 52-card deck

EXAMPLE 4 Probability and a Deck of 52 Cards

You are dealt one card from a standard 52-card deck. Find the probability of being dealt a heart.

Solution Let E be the event of being dealt a heart. Because there are 13 hearts in the deck, the event of being dealt a heart can occur in 13 ways. The number of outcomes in event E is 13: $n(E) = 13$. With 52 cards in the deck, the total number of possible ways of being dealt a single card is 52. The number of outcomes in the sample space is 52: $n(S) = 52$. The probability of being dealt a heart is

$$P(E) = \frac{n(E)}{n(S)} = \frac{13}{52} = \frac{1}{4}.$$

✓ **CHECK POINT 4** If you are dealt one card from a standard 52-card deck, find the probability of being dealt a king.

If your state has a lottery drawing each week, the probability that someone will win the top prize is relatively high. If there is no winner this week, it is virtually certain that eventually someone will be graced with millions of dollars. So, why are you so unlucky compared to this undisclosed someone? In Example 5, we provide an answer to this question, using the counting principles discussed in Section 12.6.

EXAMPLE 5 Probability and Combinations: Winning the Lottery

Florida's lottery game, LOTTO, is set up so that each player chooses six different numbers from 1 to 53. If the six numbers chosen match the six numbers drawn randomly, the player wins (or shares) the top cash prize. (As of this writing, the top cash prize has ranged from \$7 million to \$106.5 million.) With one LOTTO ticket, what is the probability of winning this prize?

Solution Because the order of the six numbers does not matter, this is a situation involving combinations. Let E be the event of winning the lottery with one ticket. With one LOTTO ticket, there is only one way of winning. Thus, $n(E) = 1$. The sample space is the set of all possible six-number combinations. We can use the combinations formula

$$_nC_r = \frac{n!}{(n-r)!\,r!}$$

to find the total number of possible combinations. We are selecting $r = 6$ numbers from a collection of $n = 53$ numbers.

$$_{53}C_6 = \frac{53!}{(53-6)!\,6!} = \frac{53!}{47!\,6!} = \frac{53\cdot52\cdot51\cdot50\cdot49\cdot48\cdot\cancel{47!}}{\cancel{47!}\cdot6\cdot5\cdot4\cdot3\cdot2\cdot1} = 22{,}957{,}480$$

There are nearly 23 million number combinations possible in LOTTO. If a person buys one LOTTO ticket, the probability of winning is

$$P(E) = \frac{n(E)}{n(S)} = \frac{1}{22{,}957{,}480} \approx 0.0000000436.$$

The probability of winning the top prize with one LOTTO ticket is $\frac{1}{22{,}957{,}480}$, or about 1 in 23 million.

Suppose that a person buys 5000 different tickets in Florida's LOTTO. Because that person has selected 5000 different combinations of the six numbers, the probability of winning is

$$\frac{5000}{22{,}957{,}480} \approx 0.000218.$$

The chances of winning top prize are about 218 in a million. At \$1 per LOTTO ticket, it is highly probable that our LOTTO player will be \$5000 poorer. Knowing a little probability helps a lotto.

Blitzer Bonus

Comparing the Probability of Dying to the Probability of Winning Florida's LOTTO

As a healthy nonsmoking 30-year-old, your probability of dying this year is approximately 0.001. Divide this probability by the probability of winning LOTTO with one ticket:

$$\frac{0.001}{0.0000000436} \approx 22{,}936.$$

A healthy 30-year-old is nearly 23,000 times more likely to die this year than to win Florida's lottery.

CHECK POINT 5 People lose interest when they do not win at games of chance, including Florida's LOTTO. With drawings twice weekly instead of once, the game described in Example 5 was brought in to bring back lost players and increase ticket sales. The original LOTTO was set up so that each player chose six different numbers from 1 to 49, rather than from 1 to 53, with a lottery drawing only once a week. With one LOTTO ticket, what was the probability of winning the top cash prize in Florida's original LOTTO? Express the answer as a fraction and as a decimal correct to ten places.

3 Find the probability that an event will not occur.

Probability of an Event Not Occurring

If we know $P(E)$, the probability of an event E, we can determine the probability that the event will not occur, denoted by $P(\text{not } E)$. Because the sum of the probabilities of all possible outcomes in any situation is 1,

$$P(E) + P(\text{not } E) = 1.$$

We now solve this equation for $P(\text{not } E)$, the probability that event E will not occur, by subtracting $P(E)$ from both sides. The resulting formula is given in the following box:

The Probability of an Event Not Occurring

The probability that an event E will not occur is equal to 1 minus the probability that it will occur.

$$P(\text{not } E) = 1 - P(E)$$

EXAMPLE 6 The Probability of an Event Not Occurring

The circle graph in **Figure 12.12** shows the distribution, by age group, of the 191 million car drivers in the United States, with all numbers rounded to the nearest million. If one driver is randomly selected from this population, find the probability that the person is not in the 20–29 age group. Express the probability as a simplified fraction.

Number of U.S. Car Drivers, by Age Group

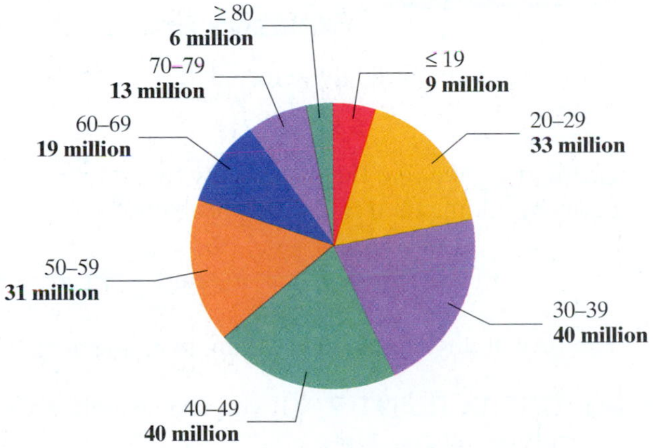

FIGURE 12.12

Source: U.S. Census Bureau

Solution We use the probability that the selected person *is* in the 20–29 age group to find the probability that the selected person is not in this age group.

$$P(\text{not in 20–29 age group})$$
$$= 1 - P(\text{in 20–29 age group})$$
$$= 1 - \frac{33}{191}$$

The graph shows 33 million drivers in the 20–29 age group.

This number, 191 million drivers, was given, but can be obtained by adding the numbers in the eight sectors.

$$= \frac{191}{191} - \frac{33}{191} = \frac{158}{191}$$

The probability that a randomly selected driver is not in the 20–29 age group is $\frac{158}{191}$.

CHECK POINT 6 If one driver is randomly selected from the population represented in **Figure 12.12**, find the probability, expressed as a simplified fraction, that the person is not in the 50–59 age group.

4 Find the probability of one event or a second event occurring.

Or Probabilities with Mutually Exclusive Events

Suppose that you randomly select one card from a deck of 52 cards. Let A be the event of selecting a king and B be the event of selecting a queen. Only one card is selected, so it is impossible to get both a king and a queen. The events of selecting a king and a queen cannot occur simultaneously. They are called *mutually exclusive events*.

Mutually Exclusive Events

If it is impossible for events A and B to occur simultaneously, the events are said to be **mutually exclusive**.

In general, if A and B are mutually exclusive events, the probability that either A or B will occur is determined by adding their individual probabilities.

Or Probabilities with Mutually Exclusive Events

If A and B are mutually exclusive events, then

$$P(A \text{ or } B) = P(A) + P(B).$$

Using set notation, $P(A \cup B) = P(A) + P(B)$.

EXAMPLE 7 The Probability of Either of Two Mutually Exclusive Events Occurring

If one card is randomly selected from a deck of cards, what is the probability of selecting a king or a queen?

Solution We find the probability that either of these mutually exclusive events will occur by adding their individual probabilities.

$$P(\text{king or queen}) = P(\text{king}) + P(\text{queen}) = \frac{4}{52} + \frac{4}{52} = \frac{8}{52} = \frac{2}{13}$$

The probability of selecting a king or a queen is $\frac{2}{13}$.

CHECK POINT 7 If you roll a single, six-sided die, what is the probability of getting either a 4 or a 5?

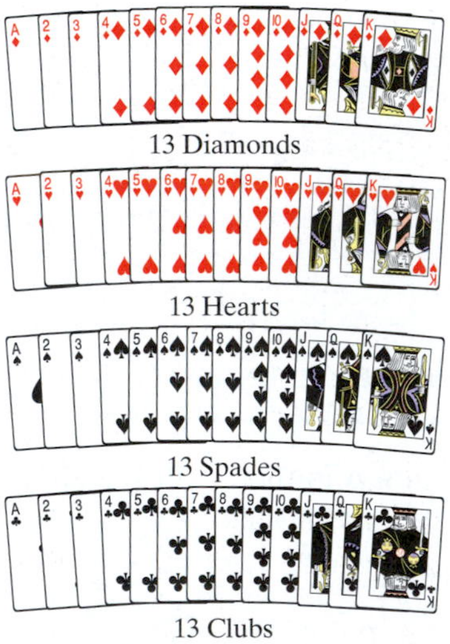

FIGURE 12.13 A deck of 52 cards

Or Probabilities with Events That Are Not Mutually Exclusive

Consider the deck of 52 cards shown in **Figure 12.13**. Suppose that these cards are shuffled and you randomly select one card from the deck. What is the probability of selecting a diamond or a picture card (jack, queen, king)? Begin by adding their individual probabilities.

$$P(\text{diamond}) + P(\text{picture card}) = \frac{13}{52} + \frac{12}{52}$$

There are 13 diamonds in the deck of 52 cards.

There are 12 picture cards in the deck of 52 cards.

FIGURE 12.14 Three diamonds are picture cards.

However, this sum is not the probability of selecting a diamond or a picture card. The problem is that there are three cards that are *simultaneously* diamonds and picture cards, shown in **Figure 12.14**. The events of selecting a diamond and selecting a picture card are not mutually exclusive. It is possible to select a card that is both a diamond and a picture card.

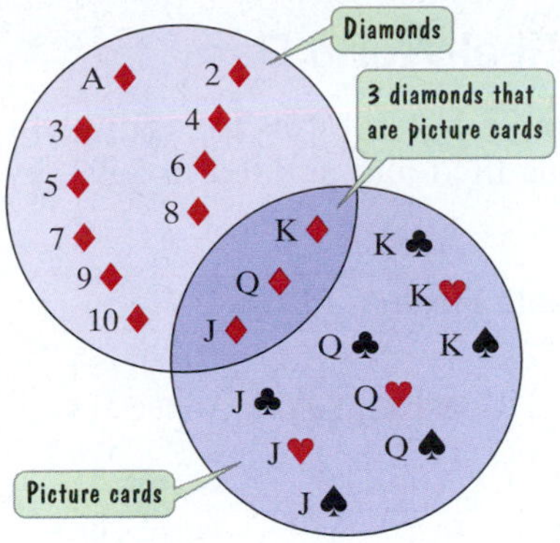

FIGURE 12.15

The situation is illustrated in the diagram in **Figure 12.15**. Why can't we find the probability of selecting a diamond or a picture card by adding their individual probabilities? The diagram shows that three of the cards, the three diamonds that are picture cards, get counted twice when we add the individual probabilities. First the three cards get counted as diamonds and then they get counted as picture cards. In order to avoid the error of counting the three cards twice, we need to subtract the probability of getting a diamond and a picture card, $\frac{3}{52}$, as follows:

$$P(\text{diamond or picture card})$$
$$= P(\text{diamond}) + P(\text{picture card}) - P(\text{diamond and picture card})$$
$$= \frac{13}{52} + \frac{12}{52} - \frac{3}{52} = \frac{13 + 12 - 3}{52} = \frac{22}{52} = \frac{11}{26}.$$

Thus, the probability of selecting a diamond or a picture card is $\frac{11}{26}$.

In general, if A and B are events that are not mutually exclusive, the probability that A or B will occur is determined by adding their individual probabilities and then subtracting the probability that A and B occur simultaneously.

Or Probabilities with Events That Are Not Mutually Exclusive

If A and B are not mutually exclusive events, then

$$P(A \text{ or } B) = P(A) + P(B) - P(A \text{ and } B).$$

Using set notation,

$$P(A \cup B) = P(A) + P(B) - P(A \cap B).$$

EXAMPLE 8 An *Or* Probability with Events That Are Not Mutually Exclusive

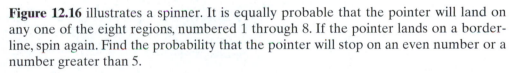

FIGURE 12.16 It is equally probable that the pointer will land on any one of the eight regions.

Figure 12.16 illustrates a spinner. It is equally probable that the pointer will land on any one of the eight regions, numbered 1 through 8. If the pointer lands on a borderline, spin again. Find the probability that the pointer will stop on an even number or a number greater than 5.

Solution It is possible for the pointer to land on a number that is both even and greater than 5. Two of the numbers, 6 and 8, are even and greater than 5. These events are not mutually exclusive. The probability of landing on a number that is even or greater than 5 is calculated as follows:

$$P\begin{pmatrix}\text{even or}\\ \text{greater than 5}\end{pmatrix} = P(\text{even}) + P(\text{greater than 5}) - P\begin{pmatrix}\text{even and}\\ \text{greater than 5}\end{pmatrix}$$

$$= \frac{4}{8} + \frac{3}{8} - \frac{2}{8}$$

Four of the eight numbers, 2, 4, 6, and 8, are even.

Three of the eight numbers, 6, 7, and 8, are greater than 5.

Two of the eight numbers, 6 and 8, are even and greater than 5.

$$= \frac{4 + 3 - 2}{8} = \frac{5}{8}.$$

The probability that the pointer will stop on an even number or a number greater than 5 is $\frac{5}{8}$.

✓ **CHECK POINT 8** Use **Figure 12.16** to find the probability that the pointer will stop on an odd number or a number less than 5.

EXAMPLE 9 An *Or* Probability with Real-World Data

Each year the Internal Revenue Service audits a sample of tax forms to verify their accuracy. **Table 12.4** shows the number of tax returns filed and audited in 2003, by taxable income.

Table 12.4 Tax Returns Filed and Audited, by Taxable Income, 2003

	< \$25,000	\$25,000–\$49,999	\$50,000–\$99,999	≥ \$100,000	Total
Audit	414,357	135,041	113,944	151,969	815,311
No audit	52,792,911	30,964,967	25,502,542	10,775,542	120,035,962
Total	53,207,268	31,100,008	25,616,486	10,927,511	120,851,273

Source: Internal Revenue Service

If one person is randomly selected from the population represented in **Table 12.4**, find the probability that

a. the taxpayer had a taxable income less than \$25,000 or was audited.
b. the taxpayer had a taxable income less than \$25,000 or at least \$100,000.

Express probabilities as decimals rounded to the nearest hundredth.

Solution

a. It is possible to select a taxpayer who both earned less than \$25,000 and was audited. Thus, these events are not mutually exclusive.

$$P(\text{less than } \$25{,}000 \text{ or audited})$$
$$= P(\text{less than } \$25{,}000) + P(\text{audited}) - P(\text{less than } \$25{,}000 \text{ and audited})$$
$$= \frac{53{,}207{,}268}{120{,}851{,}273} + \frac{815{,}311}{120{,}851{,}273} - \frac{414{,}357}{120{,}851{,}273}$$

Of the 120,851,273 taxpayers, 53,207,268 had taxable incomes less than \$25,000.

Of the 120,851,273 taxpayers, 815,311 were audited.

Of the 120,851,273 taxpayers, 414,357 earned less than \$25,000 and were audited.

$$= \frac{53{,}608{,}222}{120{,}851{,}273} \approx 0.44$$

The probability that a taxpayer had a taxable income less than \$25,000 or was audited is approximately 0.44.

b. A taxable income of *at least* \$100,000 means \$100,000 or more. Thus, it is not possible to select a taxpayer with both a taxable income of less than \$25,000 and at least \$100,000. These events are mutually exclusive.

$$P(\text{less than } \$25{,}000 \text{ or at least } \$100{,}000)$$
$$= P(\text{less than } \$25{,}000) + P(\text{at least } \$100{,}000)$$
$$= \frac{53{,}207{,}268}{120{,}851{,}273} + \frac{10{,}927{,}511}{120{,}851{,}273}$$

Of the 120,851,273 taxpayers, 53,207,268 had taxable incomes less than \$25,000.

Of the 120,851,273 taxpayers, 10,927,511 had taxable incomes of \$100,000 or more.

$$= \frac{64{,}134{,}779}{120{,}851{,}273} \approx 0.53$$

The probability that a taxpayer had a taxable income less than \$25,000 or at least \$100,000 is approximately 0.53.

✓ **CHECK POINT 9** If one person is randomly selected from the population represented in **Table 12.4**, find the probability, expressed as a decimal rounded to the nearest hundredth, that

a. the taxpayer had a taxable income of at least \$100,000 or was not audited.

b. the taxpayer had a taxable income less than \$25,000 or between \$50,000 and \$99,999, inclusive.

5 Find the probability of one event and a second event occurring.

And Probabilities with Independent Events

Consider tossing a fair coin two times in succession. The outcome of the first toss, heads or tails, does not affect what happens when you toss the coin a second time. For example, the occurrence of tails on the first toss does not make tails more likely or less likely to occur on the second toss. The repeated toss of a coin produces *independent events* because the outcome of one toss does not influence the outcome of others. Two events are **independent events** if the occurrence of either of them has no effect on the probability of the other.

If two events are independent, we can calculate the probability of the first occurring and the second occurring by multiplying their probabilities.

And Probabilities with Independent Events

If A and B are independent events, then

$$P(A \text{ and } B) = P(A) \cdot P(B).$$

EXAMPLE 10 Independent Events on a Roulette Wheel

FIGURE 12.17
A U.S. roulette wheel

Figure 12.17 shows a U.S. roulette wheel that has 38 numbered slots (1 through 36, 0, and 00). Of the 38 compartments, 18 are black, 18 are red, and two are green. A play has the dealer spin the wheel and a small ball in opposite directions. As the ball slows to a stop, it can land with equal probability on any one of the 38 numbered slots. Find the probability of red occurring on two consecutive plays.

Solution The wheel has 38 equally likely outcomes and 18 are red. Thus, the probability of red occurring on a play is $\frac{18}{38}$, or $\frac{9}{19}$. The result that occurs on each play is independent of all previous results. Thus,

$$P(\text{red and red}) = P(\text{red}) \cdot P(\text{red}) = \frac{9}{19} \cdot \frac{9}{19} = \frac{81}{361} \approx 0.224.$$

The probability of red occurring on two consecutive plays is $\frac{81}{361}$.

Some roulette players incorrectly believe that if red occurs on two consecutive plays, then another color is "due." Because the events are independent, the outcomes of previous spins have no effect on any other spins.

✓ **CHECK POINT 10** Find the probability of green occurring on two consecutive plays on a roulette wheel.

The *and* rule for independent events can be extended to cover three or more events. Thus, if A, B, and C are independent events, then

$$P(A \text{ and } B \text{ and } C) = P(A) \cdot P(B) \cdot P(C).$$

EXAMPLE 11 Independent Events in a Family

The picture in the margin shows a family that has had nine girls in a row. Find the probability of this occurrence.

Solution If two or more events are independent, we can find the probability of them all occurring by multiplying their probabilities. The probability of a baby girl is $\frac{1}{2}$, so the probability of nine girls in a row is $\frac{1}{2}$ used as a factor nine times.

$$P(\text{nine girls in a row}) = \frac{1}{2}\cdot\frac{1}{2}\cdot\frac{1}{2}\cdot\frac{1}{2}\cdot\frac{1}{2}\cdot\frac{1}{2}\cdot\frac{1}{2}\cdot\frac{1}{2}\cdot\frac{1}{2}$$
$$= \left(\frac{1}{2}\right)^9 = \frac{1}{512}$$

The probability of a run of nine girls in a row is $\frac{1}{512}$. (If another child is born into the family, this event is independent of the other nine, and the probability of a girl is still $\frac{1}{2}$.)

✓ **CHECK POINT 11** Find the probability of a family having four boys in a row.

12.7 EXERCISE SET

MyMathLab MathXL PRACTICE WATCH DOWNLOAD READ REVIEW

Practice and Application Exercises

The table shows the distribution, by marital status and gender, of the 212.5 million Americans ages 18 or older. Use the data in the table to solve Exercises 1–10.

Marital Status of the United States Population, Ages 18 or Older, in Millions

	Never Married	Married	Widowed	Divorced	Total
Male	28.6	62.1	2.7	9.0	102.4
Female	23.3	62.8	11.3	12.7	110.1
Total	51.9	124.9	14.0	21.7	212.5

Total male: $28.6 + 62.1 + 2.7 + 9.0 = 102.4$

Total female: $23.3 + 62.8 + 11.3 + 12.7 = 110.1$

Total never married: $28.6 + 23.3 = 51.9$

Total married: $62.1 + 62.8 = 124.9$

Total widowed: $2.7 + 11.3 = 14.0$

Total divorced: $9.0 + 12.7 = 21.7$

Total adult population: $102.4 + 110.1 = 212.5$

Source: U.S. Census Bureau

If one person is randomly selected from the population described in the table, find the probability, to the nearest hundredth, that the person

1. is divorced.
2. has never been married.
3. is female.
4. is male.
5. is a widowed male.
6. is a widowed female.
7. Among those who are divorced, find the probability of selecting a woman.
8. Among those who are divorced, find the probability of selecting a man.
9. Among adult men, find the probability of selecting a married person.
10. Among adult women, find the probability of selecting a married person.

In Exercises 11–16, a die is rolled. Find the probability of getting

11. a 4.

12. a 5.

13. an odd number.

14. a number greater than 3.

15. a number greater than 4.

16. a number greater than 7.

In Exercises 17–20, you are dealt one card from a standard 52-card deck. Find the probability of being dealt

17. a queen.

18. a diamond.

19. a picture card.

20. a card greater than 3 and less than 7.

In Exercises 21–22, a fair coin is tossed two times in succession. The sample space of equally likely outcomes is {HH, HT, TH, TT}. *Find the probability of getting*

21. two heads.

22. the same outcome on each toss.

In Exercises 23–24, you select a family with three children. If M represents a male child and F a female child, the sample space of equally likely outcomes is {MMM, MMF, MFM, MFF, FMM, FMF, FFM, FFF}. *Find the probability of selecting a family with*

23. at least one male child.

24. at least two female children.

In Exercises 25–26, a single die is rolled twice. The 36 equally likely outcomes are shown as follows:

First Roll \ *Second Roll*	⚀	⚁	⚂	⚃	⚄	⚅
⚀	(1, 1)	(1, 2)	(1, 3)	(1, 4)	(1, 5)	(1, 6)
⚁	(2, 1)	(2, 2)	(2, 3)	(2, 4)	(2, 5)	(2, 6)
⚂	(3, 1)	(3, 2)	(3, 3)	(3, 4)	(3, 5)	(3, 6)
⚃	(4, 1)	(4, 2)	(4, 3)	(4, 4)	(4, 5)	(4, 6)
⚄	(5, 1)	(5, 2)	(5, 3)	(5, 4)	(5, 5)	(5, 6)
⚅	(6, 1)	(6, 2)	(6, 3)	(6, 4)	(6, 5)	(6, 6)

Find the probability of getting

25. two numbers whose sum is 4.

26. two numbers whose sum is 6.

27. To play the California lottery, a person has to correctly select 6 out of 51 numbers, paying $1 for each six-number selection. If you pick six numbers that are the same as the ones drawn by the lottery, you win mountains of money. What is the probability that a person with one combination of six numbers will win? What is the probability of winning if 100 different lottery tickets are purchased?

28. A state lottery is designed so that a player chooses six numbers from 1 to 30 on one lottery ticket. What is the probability that a player with one lottery ticket will win? What is the probability of winning if 100 different lottery tickets are purchased?

Exercises 29–30 involve a deck of 52 cards. If necessary, refer to the picture of a deck of cards, **Figure 12.13** *on page 914.*

29. A poker hand consists of five cards.

a. Find the total number of possible five-card poker hands.

b. A diamond flush is a five-card hand consisting of all diamonds. Find the number of possible diamond flushes.

c. Find the probability of being dealt a diamond flush.

30. If you are dealt 3 cards from a shuffled deck of 52 cards, find the probability that all 3 cards are picture cards.

The table shows the educational attainment of the U.S. population, ages 25 and over. Use the data in the table, expressed in millions, to solve Exercises 31–36.

Educational Attainment, in Millions, of the United States Population, Ages 25 and Over

	Less Than 4 Years High School	4 Years High School Only	Some College (Less than 4 years)	4 Years College (or More)	Total
Male	14	25	20	23	82
Female	15	31	24	22	92
Total	29	56	44	45	174

Source: U.S. Census Bureau

Find the probability, expressed as a simplified fraction, that a randomly selected American, aged 25 or over

31. has not completed four years (or more) of college.

32. has not completed four years of high school.

33. has completed four years of high school only or less than four years of college.

34. has completed less than four years of high school or four years of high school only.

35. has completed four years of high school only or is a man.

36. has completed four years of high school only or is a woman.

In Exercises 37–42, you are dealt one card from a 52-card deck. Find the probability that

37. you are not dealt a king.

38. you are not dealt a picture card.

39. you are dealt a 2 or a 3.

40. you are dealt a red 7 or a black 8.

41. you are dealt a 7 or a red card.

42. you are dealt a 5 or a black card.

In Exercises 43–44, it is equally probable that the pointer on the spinner shown will land on any one of the eight regions, numbered 1 through 8. If the pointer lands on a borderline, spin again.

Find the probability that the pointer will stop on

43. an odd number or a number less than 6.

44. an odd number or a number greater than 3.

Use this information to solve Exercises 45–46. The mathematics department of a college has 8 male professors, 11 female professors, 14 male teaching assistants, and 7 female teaching assistants. If a person is selected at random from the group, find the probability that the selected person is

45. a professor or a male.

46. a professor or a female.

In Exercises 47–50, a single die is rolled twice. Find the probability of rolling

47. a 2 the first time and a 3 the second time.

48. a 5 the first time and a 1 the second time.

49. an even number the first time and a number greater than 2 the second time.

50. an odd number the first time and a number less than 3 the second time.

51. If you toss a fair coin six times, what is the probability of getting all heads?

52. If you toss a fair coin seven times, what is the probability of getting all tails?

53. The probability that South Florida will be hit by a major hurricane (category 4 or 5) in any single year is $\frac{1}{16}$. (*Source*: National Hurricane Center)

a. What is the probability that South Florida will be hit by a major hurricane two years in a row?

b. What is the probability that South Florida will be hit by a major hurricane in three consecutive years?

c. What is the probability that South Florida will not be hit by a major hurricane in the next ten years?

d. What is the probability that South Florida will be hit by a major hurricane at least once in the next ten years?

Writing in Mathematics

54. Describe the difference between theoretical probability and empirical probability.

55. Give an example of an event whose probability must be determined empirically rather than theoretically.

56. Write a probability word problem whose answer is one of the following fractions: $\frac{1}{6}$ or $\frac{1}{4}$ or $\frac{1}{3}$.

57. Explain how to find the probability of an event not occurring. Give an example.

58. What are mutually exclusive events? Give an example of two events that are mutually exclusive.

59. Explain how to find *or* probabilities with mutually exclusive events. Give an example.

60. Give an example of two events that are not mutually exclusive.

61. Explain how to find *or* probabilities with events that are not mutually exclusive. Give an example.

62. Explain how to find *and* probabilities with independent events. Give an example.

Critical Thinking Exercises

Make Sense? *In Exercises 63–66, determine whether each statement "makes sense" or "does not make sense" and explain your reasoning.*

63. The probability that Jill will win the election is 0.7 and the probability that she will not win is 0.4.

64. Assuming the next U.S. president will be a Democrat or a Republican, the probability of a Republican president is 0.5.

65. The probability that I will go to graduate school is 1.5.

66. When I toss a coin, the probability of getting heads *or* tails is 1, but the probability of getting heads *and* tails is 0.

67. The target in the figure shown contains four squares. If a dart thrown at random hits the target, find the probability that it will land in a yellow region.

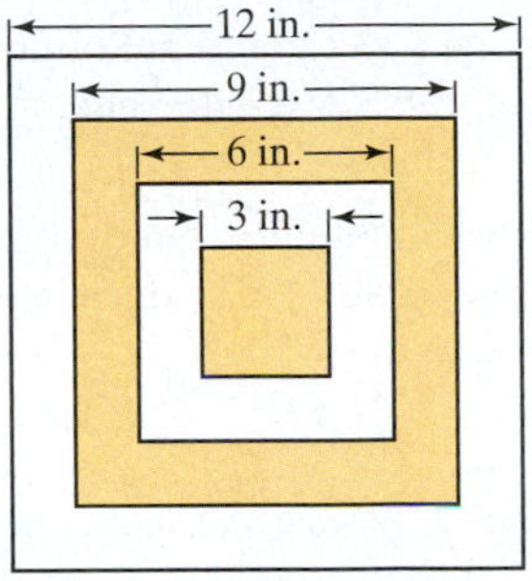

68. Suppose that it is a week in which the cash prize in Florida's LOTTO is promised to exceed \$50 million. If a person purchases 22,957,480 tickets in LOTTO at \$1 per ticket (all possible combinations), isn't this a guarantee of winning the lottery? Because the probability in this situation is 1, what's wrong with doing this?

69. Some three-digit numbers, such as 101 and 313, read the same forward and backward. If you select a number from all three-digit numbers, find the probability that it will read the same forward and backward.

70. In a class of 50 students, 29 are Democrats, 11 are business majors, and 5 of the business majors are Democrats. If one student is randomly selected from the class, find the probability of choosing

a. a Democrat who is not a business major.

b. a student who is neither a Democrat nor a business major.

71. One New Year's Eve, the probability of a person driving while intoxicated or having a driving accident is 0.35. If the probability of driving while intoxicated is 0.32 and the probability of having a driving accident is 0.09, find the probability of a person having a driving accident while intoxicated.

Review Exercises

72. Graph: $f(x) = \dfrac{x^2 - 1}{x^2 - 4}$. (Section 11.3, Example 4)

73. Solve: $\log_2(x + 5) + \log_2(x - 1) = 4$. (Section 9.5, Example 5)

74. Divide $x^3 + 5x^2 + 3x - 10$ by $x + 2$.

(Section 6.5, Example 1)

GROUP PROJECT

CHAPTER 12

Group members serve as a financial team analyzing the three options given to the professional baseball player described in the chapter opener on page 843. As a group, determine which option provides the most amount of money over the six-year contract and which provides the least. Describe one advantage and one disadvantage to each option.

CHAPTER 12 SUMMARY

Definitions and Concepts | **Examples**

Section 12.1 Sequences and Summation Notation

An infinite sequence $\{a_n\}$ is a function whose domain is the set of positive integers. The function values, or terms, are represented by

$$a_1, a_2, a_3, a_4, \ldots, a_n, \ldots.$$

General term: $a_n = \dfrac{(-1)^n}{n^3}$.

$$a_1 = \frac{(-1)^1}{1^3} = -1, \quad a_2 = \frac{(-1)^2}{2^3} = \frac{1}{8},$$

$$a_3 = \frac{(-1)^3}{3^3} = -\frac{1}{27}, \quad a_4 = \frac{(-1)^4}{4^3} = \frac{1}{64}$$

First four terms are $-1, \frac{1}{8}, -\frac{1}{27}$, and $\frac{1}{64}$.

Factorial Notation

$$n! = n(n-1)(n-2)\cdots(3)(2)(1) \quad \text{and} \quad 0! = 1$$

$$6! = 6\cdot5\cdot4\cdot3\cdot2\cdot1 = 720$$

$$3! = 3\cdot2\cdot1 = 6$$

Summation Notation

$$\sum_{i=1}^{n} a_i = a_1 + a_2 + a_3 + a_4 + \cdots + a_n$$

In the summation shown here, i is the index of summation, n is the upper limit of summation, and 1 is the lower limit of summation.

$$\sum_{i=3}^{7}(i^2 - 4)$$

$$= (3^2 - 4) + (4^2 - 4) + (5^2 - 4) + (6^2 - 4) + (7^2 - 4)$$

$$= (9 - 4) + (16 - 4) + (25 - 4) + (36 - 4) + (49 - 4)$$

$$= 5 + 12 + 21 + 32 + 45$$

$$= 115$$

Section 12.2 Arithmetic Sequences

In an arithmetic sequence, each term after the first differs from the preceding term by a constant, the common difference. Subtract any term from the term that directly follows it to find the common difference.

General Term of an Arithmetic Sequence

The nth term (the general term) of an arithmetic sequence with first term a_1 and common difference d is

$$a_n = a_1 + (n-1)d.$$

Find the general term and the tenth term:

$$3, 7, 11, 15, \ldots.$$

$a_1 = 3$ | $d = 7 - 3 = 4$

Using $a_n = a_1 + (n-1)d$,

$$a_n = 3 + (n-1)4 = 3 + 4n - 4 = 4n - 1.$$

The general term is $a_n = 4n - 1$.
The tenth term is $a_{10} = 4\cdot10 - 1 = 39$.

The Sum of the First *n* Terms of an Arithmetic Sequence

The sum, S_n, of the first n terms of an arithmetic sequence is given by

$$S_n = \frac{n}{2}(a_1 + a_n)$$

in which a_1 is the first term and a_n is the nth term.

Find the sum of the first ten terms:

$$2, 5, 8, 11, \ldots.$$

$a_1 = 2$ | $d = 5 - 2 = 3$

First find a_{10}, the 10th term. Using $a_n = a_1 + (n-1)d$,

$$a_{10} = 2 + (10-1)\cdot3 = 2 + 9\cdot3 = 29.$$

Find the sum of the first ten terms using

$$S_n = \frac{n}{2}(a_1 + a_n).$$

$$S_{10} = \frac{10}{2}(a_1 + a_{10}) = 5(2 + 29) = 5(31) = 155$$

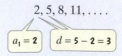

Definitions and Concepts | Examples

Section 12.3 Geometric Sequences and Series

In a geometric sequence, each term after the first is obtained by multiplying the preceding term by a nonzero constant, the common ratio. Divide any term after the first by the term that directly precedes it to find the common ratio.

General Term of a Geometric Sequence

The nth term (the general term) of a geometric sequence with first term a_1 and common ratio r is

$$a_n = a_1 r^{n-1}.$$

Find the general term and the ninth term:

$$12, -6, 3, -\frac{3}{2}, \ldots.$$

$a_1 = 12$ $\quad r = \frac{-6}{12} = -\frac{1}{2}$

Using $a_n = a_1 r^{n-1}$,

$$a_n = 12\left(-\frac{1}{2}\right)^{n-1} \text{ is the general term.}$$

The ninth term is

$$a_9 = 12\left(-\frac{1}{2}\right)^{9-1} = 12\left(-\frac{1}{2}\right)^{8} = \frac{12}{256} = \frac{3}{64}.$$

The Sum of the First n Terms of a Geometric Sequence

The sum, S_n, of the first n terms of a geometric sequence is given by

$$S_n = \frac{a_1(1 - r^n)}{1 - r}$$

in which a_1 is the first term and r is the common ratio $(r \neq 1)$.

Find $\sum_{i=1}^{8} 4 \cdot 3^i$

$$= 4 \cdot 3 + 4 \cdot 3^2 + 4 \cdot 3^3 + \cdots + 4 \cdot 3^8.$$

$a_1 = 12$ $\quad r = \frac{4 \cdot 3^2}{4 \cdot 3} = 3$

Using $S_n = \frac{a_1(1 - r^n)}{1 - r}$,

$$S_8 = \frac{12(1 - 3^8)}{1 - 3} = 39{,}360.$$

The Sum of an Infinite Geometric Series

If $-1 < r < 1$ (equivalently, $|r| < 1$), then the sum of the infinite geometric series

$$a_1 + a_1 r + a_1 r^2 + a_1 r^3 + \cdots$$

in which a_1 is the first term and r is the common ratio is given by

$$S = \frac{a_1}{1 - r}.$$

If $|r| \geq 1$, the infinite series does not have a sum.

Find the sum:

$$6 + \frac{6}{3} + \frac{6}{3^2} + \frac{6}{3^3} + \cdots.$$

$a_1 = 6$ $\quad r = \frac{1}{3}$

Using $S = \frac{a_1}{1 - r}$, the sum is

$$S = \frac{6}{1 - \frac{1}{3}} = \frac{6}{\frac{2}{3}} = 6 \cdot \frac{3}{2} = 9.$$

Section 12.4 The Binomial Theorem

Definition of a Binomial Coefficient

$$\binom{n}{r} = \frac{n!}{r!\,(n - r)!}$$

$$\binom{8}{3} = \frac{8!}{3!\,(8 - 3)!} = \frac{8!}{3!\,5!}$$

$$= \frac{8 \cdot 7 \cdot 6 \cdot \cancel{5!}}{3 \cdot 2 \cdot 1 \cdot \cancel{5!}} = 56$$

Definitions and Concepts	Examples

Section 12.4 The Binomial Theorem (continued)

A Formula for Expanding Binomials: The Binomial Theorem

For any positive integer n,

$$(a+b)^n = \binom{n}{0}a^n + \binom{n}{1}a^{n-1}b + \binom{n}{2}a^{n-2}b^2 + \binom{n}{3}a^{n-3}b^3 + \cdots + \binom{n}{n}b^n.$$

Expand: $(3x-y)^4 = [3x+(-y)]^4$.

$$= \binom{4}{0}(3x)^4 + \binom{4}{1}(3x)^3(-y) + \binom{4}{2}(3x)^2(-y)^2 + \binom{4}{3}(3x)^1(-y)^3 + \binom{4}{4}(-y)^4$$

$$= 1\cdot 81x^4 + 4\cdot 27x^3(-y) + 6\cdot 9x^2y^2 + 4\cdot 3x(-y^3) + 1\cdot y^4$$

$$= 81x^4 - 108x^3y + 54x^2y^2 - 12xy^3 + y^4$$

Finding a Particular Term in a Binomial Expansion

The $(r+1)$st term in the expansion of $(a+b)^n$ is

$$\binom{n}{r}a^{n-r}b^r.$$

The 8th term, or $(7+1)$st term $(r=7)$, of $(x+2y)^{10}$ is

$$\binom{10}{7}x^{10-7}(2y)^7$$

$$= \binom{10}{7}x^3(2y)^7 = 120x^3\cdot 128y^7$$

$$= 15{,}360x^3y^7.$$

Section 12.5 Mathematical Induction

To prove that S_n is true for all positive integers n,

a. Show that S_1 is true.

b. Show that if S_k is assumed to be true, then S_{k+1} is also true, for every positive integer k.

Prove that for every positive integer n,

$$2+4+6+\cdots+2n = n(n+1).$$

a. Statement S_1 is

$$2 = 1(1+1), \text{ or } 2 = 2, \text{ which is true.}$$

b. Show that if S_k,

$$2+4+6+\cdots+2k = k(k+1),$$

is true, then S_{k+1},

$$2+4+6+\cdots+(2k+2) = (k+1)(k+2),$$

is true. Add $2k+2$ to both sides of S_k:

$$2+4+6+\cdots+2k+(2k+2) = k(k+1)+(2k+2)$$

$$2+4+6+\cdots+(2k+2) = k(k+1)+2(k+1)$$

$$2+4+6+\cdots+(2k+2) = (k+1)(k+2)$$

Section 12.6 Counting Principles, Permutations, and Combinations

The Fundamental Counting Principle

The number of ways in which a series of successive things can occur is found by multiplying the number of ways in which each thing can occur.

How many ways can 6 applicants fill three different positions?

$$6\cdot 5\cdot 4 = 120 \text{ ways}$$

Permutations

A permutation from a group of items occurs when no item is used more than once and the order of arrangement makes a difference.

Permutations Formula: The number of possible permutations if r items are taken from n items is ${}_nP_r = \dfrac{n!}{(n-r)!}$.

How many ways can 6 applicants fill three different positions?

$${}_6P_3 = \frac{6!}{(6-3)!} = \frac{6!}{3!} = \frac{6\cdot 5\cdot 4\cdot 3!}{3!}$$

$$= 6\cdot 5\cdot 4 = 120 \text{ ways}$$

Definitions and Concepts	Examples

Section 12.6 Counting Principles, Permutations, and Combinations (continued)

Combinations

A combination from a group of items occurs when no item is used more than once and the order of items makes no difference.

Combinations Formula: The number of possible combinations if r items are taken from n items is ${}_nC_r = \dfrac{n!}{(n-r)!\,r!}$.

How many different sets of 3 books can be selected from 6 books?

$$
\begin{aligned}
{}_6C_3 &= \frac{6!}{(6-3)!\,3!} = \frac{6!}{3!\,3!} \\
&= \frac{6\cdot5\cdot4\cdot3!}{3\cdot2\cdot1\cdot3!} = \frac{6\cdot5\cdot4}{3\cdot2\cdot1} \\
&= 20 \text{ ways}
\end{aligned}
$$

Section 12.7 Probability

Empirical Probability

Empirical probability applies to situations in which we observe the frequency of the occurrence of an event. The empirical probability of event E is

$$P(E) = \frac{\text{observed number of time } E \text{ occurs}}{\text{total number of observed occurrences}}.$$

Teachers in U.S. Catholic High Schools

Religious	Lay	Total
4149	43,581	47,730

Source: National Catholic Education Association

The probability that a randomly selected U.S. Catholic high school teacher belongs to a religious order is

$$\frac{4149}{47{,}730} \approx 0.087.$$

Theoretical Probability

Theoretical probability applies to situations in which the sample space of all equally likely outcomes is known. The theoretical probability of event E is

$$P(E) = \frac{\text{number of outcomes in event } E}{\text{number of outcomes in sample space } S} = \frac{n(E)}{n(S)}.$$

Probability of an event not occurring: $P(\text{not } E) = 1 - P(E)$.

A die is rolled.

$$S = \{1, 2, 3, 4, 5, 6\}$$

Probability of getting a number greater than 4 $(E = \{5, 6\})$ is $\frac{2}{6} = \frac{1}{3}$.

Probability of not getting a number greater than 4 is $1 - \frac{1}{3} = \frac{2}{3}$.

***Or* Probabilities**

If it is impossible for events A and B to occur simultaneously, the events are mutually exclusive.

If A and B are mutually exclusive events, then $P(A \text{ or } B) = P(A) + P(B)$.

If A and B are not mutually exclusive events, then $P(A \text{ or } B) = P(A) + P(B) - P(A \text{ and } B)$.

A die is rolled: $S = \{1, 2, 3, 4, 5, 6\}$.

Probability (2 or 5)

$$= P(2) + P(5) = \frac{1}{6} + \frac{1}{6} = \frac{2}{6} = \frac{1}{3}.$$

Probability (even or greater than 3)

$$= P(\text{even}) + P(>3) - P(\text{even and} > 3)$$

$$= \frac{3}{6} + \frac{3}{6} - \frac{2}{6} = \frac{4}{6} = \frac{2}{3}.$$

This event is {4, 6}.

***And* Probabilities**

Two events are independent if the occurrence of either of them has no effect on the probability of the other.

If A and B are independent events, then

$$P(A \text{ and } B) = P(A) \cdot P(B).$$

The probability of a succession of independent events is the product of each of their probabilities.

A quiz contains six multiple-choice questions. Each question has four answer choices, with one correct answer per question. If you guess at every question, the probability of answering all correctly is

$$\frac{1}{4}\cdot\frac{1}{4}\cdot\frac{1}{4}\cdot\frac{1}{4} = \frac{1}{256}.$$

CHAPTER 12 REVIEW EXERCISES

12.1 *In Exercises 1–4, write the first four terms of each sequence whose general term is given.*

1. $a_n = 7n - 4$

2. $a_n = (-1)^n \frac{n+2}{n+1}$

3. $a_n = \frac{1}{(n-1)!}$

4. $a_n = \frac{(-1)^{n+1}}{2^n}$

In Exercises 5–6, find each indicated sum.

5. $\sum_{i=1}^{5} (2i^2 - 3)$

6. $\sum_{i=0}^{4} (-1)^{i+1} i!$

In Exercises 7–8, express each sum using summation notation. Use i for the index of summation.

7. $\frac{1}{3} + \frac{2}{4} + \frac{3}{5} + \cdots + \frac{15}{17}$

8. $4^3 + 5^3 + 6^3 + \cdots + 13^3$

12.2 *In Exercises 9–11, write the first six terms of each arithmetic sequence.*

9. $a_1 = 7, d = 4$

10. $a_1 = -4, d = -5$

11. $a_1 = \frac{3}{2}, d = -\frac{1}{2}$

In Exercises 12–14, use the formula for the general term (the nth term) of an arithmetic sequence to find the indicated term of each sequence.

12. Find a_6 when $a_1 = 5, d = 3$.

13. Find a_{12} when $a_1 = -8, d = -2$.

14. Find a_{14} when $a_1 = 14, d = -4$.

In Exercises 15–18, write a formula for the general term (the nth term) of each arithmetic sequence. Then use the formula for a_n to find a_{20}, the 20th term of the sequence.

15. $-7, -3, 1, 5, \ldots$

16. $a_1 = 200, d = -20$

17. $a_1 = -12, d = -\frac{1}{2}$

18. $15, 8, 1, -6, \ldots$

19. Find the sum of the first 22 terms of the arithmetic sequence: $5, 12, 19, 26, \ldots$.

20. Find the sum of the first 15 terms of the arithmetic sequence: $-6, -3, 0, 3, \ldots$.

21. Find $3 + 6 + 9 + \cdots + 300$, the sum of the first 100 positive multiples of 3.

In Exercises 22–24, use the formula for the sum of the first n terms of an arithmetic sequence to find the indicated sum.

22. $\sum_{i=1}^{16} (3i + 2)$

23. $\sum_{i=1}^{25} (-2i + 6)$

24. $\sum_{i=1}^{30} (-5i)$

25. The graphic indicates that there are more eyes at school.

Percentage of United States Students Ages 12–18 Seeing Security Cameras at School

Source: Department of Education

In 2001, 39% of students ages 12–18 reported seeing one or more security cameras at their school. On average, this has increased by approximately 4.75% per year since then.

a. Write a formula for the *n*th term of the arithmetic sequence that describes the percentage of students ages 12–18 who reported seeing security cameras at school *n* years after 2000.

b. Use the model to predict the percentage of students ages 12–18 who will report seeing security cameras at school by the year 2013.

26. A company offers a starting salary of $31,500 with raises of $2300 per year. Find the total salary over a ten-year period.

27. A theater has 25 seats in the first row and 35 rows in all. Each successive row contains one additional seat. How many seats are in the theater?

12.3 *In Exercises 28–31, write the first five terms of each geometric sequence.*

28. $a_1 = 3, r = 2$

29. $a_1 = \frac{1}{2}, r = \frac{1}{2}$

30. $a_1 = 16, r = -\frac{1}{4}$

31. $a_1 = -5, r = -1$

In Exercises 32–34, use the formula for the general term (the nth term) of a geometric sequence to find the indicated term of each sequence.

32. Find a_7 when $a_1 = 2, r = 3$.

33. Find a_6 when $a_1 = 16, r = \frac{1}{2}$.

34. Find a_5 when $a_1 = -3, r = 2$.

In Exercises 35–37, write a formula for the general term (the nth term) of each geometric sequence. Then use the formula for a_n to find a_8, the eighth term of the sequence.

35. $1, 2, 4, 8, \ldots$

36. $100, 10, 1, \frac{1}{10}, \ldots$

37. $12, -4, \frac{4}{3}, -\frac{4}{9}, \ldots$

38. Find the sum of the first 15 terms of the geometric sequence: $5, -15, 45, -135, \ldots$.

39. Find the sum of the first 7 terms of the geometric sequence: $8, 4, 2, 1, \ldots$.

In Exercises 40–42, use the formula for the sum of the first n terms of a geometric sequence to find the indicated sum.

40. $\sum_{i=1}^{6} 5^i$

41. $\sum_{i=1}^{7} 3(-2)^i$

42. $\sum_{i=1}^{5} 2\left(\frac{1}{4}\right)^{i-1}$

In Exercises 43–46, find the sum of each infinite geometric series.

43. $9 + 3 + 1 + \frac{1}{3} + \cdots$

44. $2 - 1 + \frac{1}{2} - \frac{1}{4} + \cdots$

45. $-6 + 4 - \frac{8}{3} + \frac{16}{9} - \cdots$

46. $\sum_{i=1}^{\infty} 5(0.8)^i$

In Exercises 47–48, express each repeating decimal as a fraction in lowest terms.

47. $0.\overline{6}$

48. $0.\overline{47}$

49. Projections for the U.S. population, ages 85 and older, are shown in the following table.

Year	2000	2010	2020	2030	2040	2050
Projected Population in millions	4.2	5.9	8.3	11.6	16.2	22.7

Actual 2000 population

Source: U.S. Census Bureau

a. Show that the U.S. population, ages 85 and older, is projected to increase geometrically.

b. Write the general term of the geometric sequence describing the U.S. population, ages 85 and older, in millions, n decades after 1990.

c. Use the model in part (b) to project the U.S. population, ages 85 and older, in 2080.

50. A job pays \$32,000 for the first year with an annual increase of 6% per year beginning in the second year. What is the salary in the sixth year? What is the total salary paid over this six-year period? Round answers to the nearest dollar.

In Exercises 51–52, use the formula for the value of an annuity and round to the nearest dollar.

51. You spend \$10 per week on lottery tickets, averaging \$520 per year. Instead of buying tickets, if you deposited the \$520 at the end of each year in an annuity paying 6% compounded annually,

a. How much would you have after 20 years?

b. Find the interest.

52. To save for retirement, you decide to deposit \$100 at the end of each month in an IRA that pays 5.5% compounded monthly.

a. How much will you have from the IRA after 30 years?

b. Find the interest.

53. A factory in an isolated town has an annual payroll of \$4 million. It is estimated that 70% of this money is spent within the town, that people in the town receiving this money will again spend 70% of what they receive in the town, and so on. What is the total of all this spending in the town each year?

12.4

In Exercises 54–55, evaluate the given binomial coefficient.

54. $\binom{11}{8}$

55. $\binom{90}{2}$

In Exercises 56–59, use the Binomial Theorem to expand each binomial and express the result in simplified form.

56. $(2x + 1)^3$

57. $(x^2 - 1)^4$

58. $(x + 2y)^5$

59. $(x - 2)^6$

In Exercises 60–61, write the first three terms in each binomial expansion, expressing the result in simplified form.

60. $(x^2 + 3)^8$

61. $(x - 3)^9$

In Exercises 62–63, find the term indicated in each expansion.

62. $(x + 2)^5$; fourth term

63. $(2x - 3)^6$; fifth term

12.5

In Exercises 64–67, use mathematical induction to prove that each statement is true for every positive integer n.

64. $5 + 10 + 15 + \cdots + 5n = \frac{5n(n + 1)}{2}$

65. $1 + 4 + 4^2 + \cdots + 4^{n-1} = \frac{4^n - 1}{3}$

66. $2 + 6 + 10 + \cdots + (4n - 2) = 2n^2$

67. $1 \cdot 3 + 2 \cdot 4 + 3 \cdot 5 + \cdots + n(n + 2) = \frac{n(n + 1)(2n + 7)}{6}$

12.6 *In Exercises 68–71, evaluate each expression.*

68. ${}_8P_3$

69. ${}_9P_5$

70. ${}_8C_3$

71. ${}_{13}C_{11}$

In Exercises 72–78, solve by the method of your choice.

72. A popular brand of pen comes in red, green, blue, or black ink. The writing tip can be chosen from extra bold, bold, regular, fine, or micro. How many different choices of pens do you have with this brand?

73. A stock can go up, go down, or stay unchanged. How many possibilities are there if you own five stocks?

74. A club with 15 members is to choose four officers—president, vice-president, secretary, and treasurer. In how many ways can these offices be filled?

75. How many different ways can a director select 4 actors from a group of 20 actors to attend a workshop on performing in rock musicals?

76. From the 20 CDs that you've bought during the past year, you plan to take 3 with you on vacation. How many different sets of three CDs can you take?

77. How many different ways can a director select from 20 male actors and cast the roles of Mark, Roger, Angel, and Collins in the musical *Rent*?

78. In how many ways can five airplanes line up for departure on a runway?

12.7 *Suppose that a survey of 350 college students is taken. Each student is asked the type of college attended (public or private) and the family's income level (low, middle, high). Use the data in the table to solve Exercises 79–84. Express probabilities as simplified fractions.*

	Public	Private	Total
Low	120	20	140
Middle	110	50	160
High	22	28	50
Total	252	98	350

Find the probability that a randomly selected student in the survey

79. attends a public college.

80. is not from a high-income family.

81. is from a middle-income or a high-income family.

82. attends a private college or is from a high-income family.

83. Among people who attend a public college, find the probability that a randomly selected student is from a low-income family.

84. Among people from a middle-income family, find the probability that a randomly selected student attends a private college.

In Exercises 85–86, a die is rolled. Find the probability of

85. getting a number less than 5.

86. getting a number less than 3 or greater than 4.

In Exercises 87–88, you are dealt one card from a 52-card deck. Find the probability of

87. getting an ace or a king.

88. getting a queen or a red card.

In Exercises 89–91, it is equally probable that the pointer on the spinner shown will land on any one of the six regions, numbered 1 through 6, and colored as shown. If the pointer lands on a borderline, spin again. Find the probability of

89. not stopping on yellow.

90. stopping on red or a number greater than 3.

91. stopping on green on the first spin and stopping on a number less than 4 on the second spin.

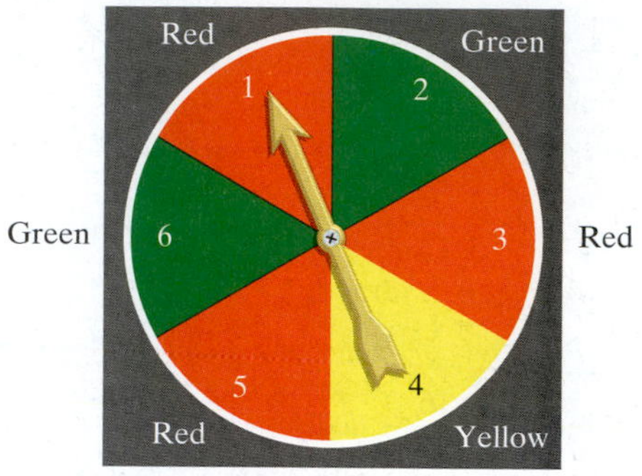

92. A lottery game is set up so that each player chooses five different numbers from 1 to 20. If the five numbers match the five numbers drawn in the lottery, the player wins (or shares) the top cash prize. What is the probability of winning the prize

a. with one lottery ticket?

b. with 100 different lottery tickets?

93. What is the probability of a family having five boys born in a row?

94. The probability of a flood in any given year in a region prone to floods is 0.2.

a. What is the probability of a flood two years in a row?

b. What is the probability of a flood for three consecutive years?

c. What is the probability of no flooding for four consecutive years?

CHAPTER 12 TEST

Remember to use your Chapter Test Prep Video CD to see the worked-out solutions to the test questions you want to review.

1. Write the first five terms of the sequence whose general term is $a_n = \dfrac{(-1)^{n+1}}{n^2}$.

2. Find the indicated sum: $\sum_{i=1}^{5}(i^2 + 10)$.

3. Express the sum using summation notation. Use i for the index of summation.

$$\frac{2}{3} + \frac{3}{4} + \frac{4}{5} + \cdots + \frac{21}{22}$$

In Exercises 4–5, write a formula for the general term (the nth term) of each sequence. Then use the formula to find the twelfth term of the sequence.

4. $4, 9, 14, 19, \ldots$

5. $16, 4, 1, \frac{1}{4}, \ldots$

In Exercises 6–7, use the formula for the sum of the first n terms of an arithmetic sequence.

6. Find the sum of the first ten terms of the arithmetic sequence: $-7, -14, -21, -28, \ldots$.

7. Find $\sum_{i=1}^{20}(3i - 4)$.

In Exercises 8–9, use the formula for the sum of the first n terms of a geometric sequence.

8. Find the sum of the first ten terms of the geometric sequence: $7, -14, 28, -56, \ldots$.

9. Find $\sum_{i=1}^{15}(-2)^i$.

10. Find the sum of the infinite geometric series:

$$4 + \frac{4}{2} + \frac{4}{2^2} + \frac{4}{2^3} + \cdots.$$

11. Express $0.\overline{73}$ in fractional notation.

12. A job pays \$30,000 for the first year with an annual increase of 4% per year beginning in the second year. What is the total salary paid over an eight-year period? Round to the nearest dollar.

13. Evaluate: $\binom{9}{2}$.

14. Use the Binomial Theorem to expand and simplify: $(x^2 - 1)^5$.

15. Use the Binomial Theorem to write the first three terms in the expansion and simplify: $(x + y^2)^8$.

16. Use mathematical induction to prove that for every positive integer n,

$$1 + 4 + 7 + \cdots + (3n - 2) = \frac{n(3n - 1)}{2}.$$

17. A human resource manager has 11 applicants to fill three different positions. Assuming that all applicants are equally qualified for any of the three positions, in how many ways can this be done?

18. From the ten books that you've recently bought but not read, you plan to take four with you on vacation. How many different sets of four books can you take?

19. How many seven-digit local telephone numbers can be formed if the first three digits are 279?

A class is collecting data on eye color and gender. They organize the data they collected into the table shown. Numbers in the table represent the number of students from the class that belong to each of the categories. Use the data to solve Exercises 20–23. Express probabilities as simplified fractions.

	Brown	Blue	Green
Male	22	18	10
Female	18	20	12

Find the probability that a randomly selected student from this class

20. does not have brown eyes.

21. has brown eyes or blue eyes.

22. is female or has green eyes.

23. Among the students with blue eyes, find the probability of selecting a male.

24. A lottery game is set up so that each player chooses six different numbers from 1 to 15. If the six numbers match the six numbers drawn in the lottery, the player wins (or shares) the top cash prize. What is the probability of winning the prize with 50 different lottery tickets?

25. One card is randomly selected from a deck of 52 cards. Find the probability of selecting a black card or a picture card.

26. A group of students consists of 10 male freshmen, 15 female freshmen, 20 male sophomores, and 5 female sophomores. If one person is randomly selected from the group find the probability of selecting a freshman or a female.

27. A quiz consisting of four multiple-choice questions has four answer choices for each question, with one correct answer per question. If a person guesses at every question, what is the probability of answering *all* questions correctly?

28. It is equally probable that the pointer on the spinner shown will land on any one of the eight regions. If the spinner is spun twice, find the probability that the pointer lands on red on the first spin and blue on the second spin.

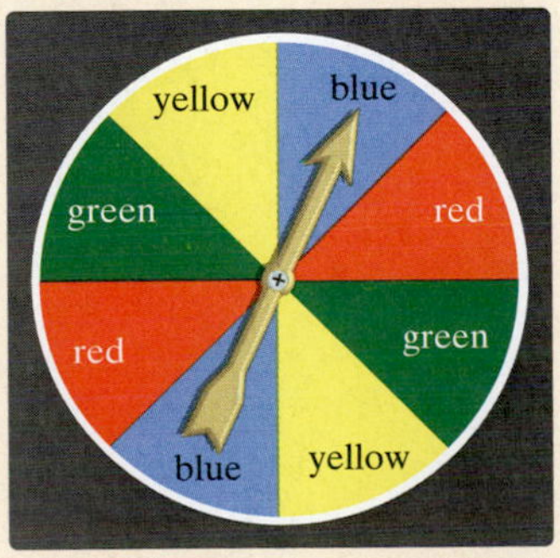

CUMULATIVE REVIEW EXERCISES (CHAPTERS 1–12)

In Exercises 1–11, solve each equation, inequality, or system.

1. $\sqrt{2x+5} - \sqrt{x+3} = 2$
2. $(x-5)^2 = -49$
3. $x^2 + x > 6$
4. $6x - 3(5x+2) = 4(1-x)$
5. $\dfrac{2}{x-3} - \dfrac{3}{x+3} = \dfrac{12}{x^2-9}$
6. $3x + 2 < 4$ and $4 - x > 1$
7. $$\begin{aligned} 3x - 2y + z &= 7 \\ 2x + 3y - z &= 13 \\ x - y + 2z &= -6 \end{aligned}$$
8. $\log_9 x + \log_9(x-8) = 1$
9. $$\begin{aligned} 2x^2 - 3y^2 &= 5 \\ 3x^2 + 4y^2 &= 16 \end{aligned}$$
10. $$\begin{aligned} 2x^2 - y^2 &= -8 \\ x - y &= 6 \end{aligned}$$
11. $3x^3 + 4x^2 - 7x + 2 = 0$

In Exercises 12–17, graph each function, equation, or inequality in a rectangular coordinate system.

12. $f(x) = (x+2)^2 - 4$
13. $y < -3x + 5$
14. $f(x) = 3^{x-2}$
15. $\dfrac{x^2}{16} + \dfrac{y^2}{4} = 1$
16. $x^2 - y^2 = 9$
17. $f(x) = \dfrac{x-1}{x-2}$

In Exercises 18–21, perform the indicated operations and simplify, if possible.

18. $\dfrac{2x+1}{x-5} - \dfrac{4}{x^2-3x-10}$
19. $\dfrac{\dfrac{1}{x-1}+1}{\dfrac{1}{x+1}-1}$
20. $\dfrac{6}{\sqrt{5}-\sqrt{2}}$
21. $8\sqrt{45} + 2\sqrt{5} - 7\sqrt{20}$
22. Rationalize the denominator: $\dfrac{5}{\sqrt[3]{2x^2y}}$.
23. Factor completely: $5ax + 5ay - 4bx - 4by$.
24. Write as a single logarithm: $5 \log x - \dfrac{1}{2}\log y$.
25. Solve for p: $\dfrac{1}{p} + \dfrac{1}{q} = \dfrac{1}{f}$.
26. Find the distance between $(6, -1)$ and $(-3, -4)$. Round to two decimal places.
27. Find the indicated sum: $\displaystyle\sum_{i=2}^{5}(i^3 - 4)$.
28. Find the sum of the first 30 terms of the arithmetic sequence: 2, 6, 10, 14, ...
29. Express $0.\overline{3}$ as a fraction in lowest terms.
30. Use the Binomial Theorem to expand and simplify: $(2x - y^3)^4$.

In Exercises 31–33, find the domain of each function.

31. $f(x) = \dfrac{2}{x^2 + 2x - 15}$
32. $f(x) = \sqrt{2x - 6}$
33. $f(x) = \ln(1 - x)$
34. The length of a rectangular garden is 2 feet more than twice its width. If 22 feet of fencing is needed to enclose the garden, what are its dimensions?
35. With a 6% raise, you will earn $19,610 annually. What is your salary before this raise?
36. The function $F(t) = 1 - k\ln(t+1)$ models the fraction of people, $F(t)$, who remember all the words in a list of nonsense words t hours after memorizing them. After 3 hours, only half the people could remember all the words. Determine the value of k and then predict the fraction of people in the group who will remember all the words after 6 hours. Round to three decimal places and then express the fraction with a denominator of 1000.

APPENDIX

Where Did That Come From? Selected Proofs

Section 9.4 Properties of Logarithms

The Product Rule

Let b, M, and N be positive real numbers with $b \neq 1$.

$$\log_b(MN) = \log_b M + \log_b N$$

Proof. We begin by letting $\log_b M = R$ and $\log_b N = S$.
Now we write each logarithm in exponential form.

$$\log_b M = R \quad \text{means} \quad b^R = M.$$
$$\log_b N = S \quad \text{means} \quad b^S = N.$$

By substituting and using a property of exponents, we see that

$$MN = b^R b^S = b^{R+S}$$

Now we change $MN = b^{R+S}$ to logarithmic form.

$$MN = b^{R+S} \quad \text{means} \quad \log_b(MN) = R + S.$$

Finally, substituting $\log_b M$ for R and $\log_b N$ for S gives us

$$\log_b(MN) = \log_b M + \log_b N,$$

the property that we wanted to prove.
The quotient and power rules for logarithms are proved using similar procedures.

The Change-of-Base Property

For any logarithmic bases a and b, and any positive number M,

$$\log_b M = \frac{\log_a M}{\log_a b}.$$

Proof. To prove the change-of-base property, we let x equal the logarithm on the left side:

$$\log_b M = x.$$

Now we rewrite this logarithm in exponential form.

$$\log_b M = x \quad \text{means} \quad b^x = M.$$

Because b^x and M are equal, the logarithms with base a for each of these expressions must be equal. This means that

$$\log_a b^x = \log_a M$$

$$x \log_a b = \log_a M \quad \text{Apply the power rule for logarithms on the left side.}$$

$$x = \frac{\log_a M}{\log_a b} \quad \text{Solve for } x \text{ by dividing both sides by } \log_a b.$$

In our first step we let x equal $\log_b M$. Replacing x on the left side by $\log_b M$ gives us

$$\log_b M = \frac{\log_a M}{\log_a b},$$

which is the change-of-base property.

Section 10.2 The Ellipse

The Standard Form of the Equation of an Ellipse with a Horizontal Major Axis Centered at the Origin

Proof. Refer to **Figure A.1**.

$$d_1 + d_2 = 2a \quad \text{The sum of the distances from } P \text{ to the foci equals a constant, } 2a.$$

$$\sqrt{(x+c)^2 + y^2} + \sqrt{(x-c)^2 + y^2} = 2a \quad \text{Use the distance formula.}$$

$$\sqrt{(x+c)^2 + y^2} = 2a - \sqrt{(x-c)^2 + y^2} \quad \text{Isolate a radical.}$$

$$(x+c)^2 + y^2 = 4a^2 - 4a\sqrt{(x-c)^2 + y^2} + (x-c)^2 + y^2 \quad \text{Square both sides.}$$

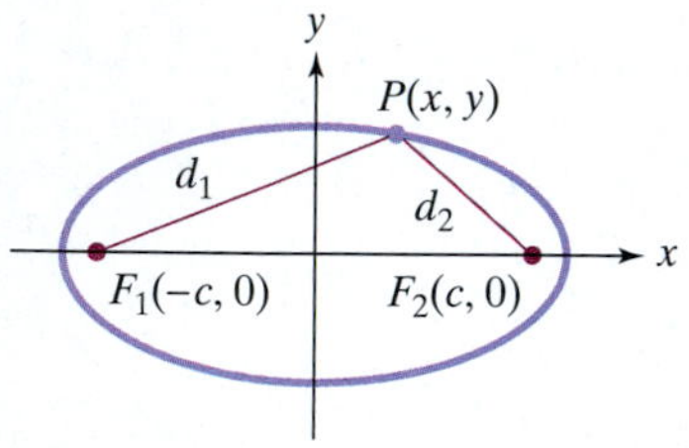

FIGURE A.1

$$x^2 + 2cx + c^2 + y^2 = 4a^2 - 4a\sqrt{(x-c)^2 + y^2} + x^2 - 2cx + c^2 + y^2 \quad \text{Square } x + c \text{ and } x - c.$$

$$4cx - 4a^2 = -4a\sqrt{(x-c)^2 + y^2} \quad \text{Simplify and isolate the radical.}$$

$$cx - a^2 = -a\sqrt{(x-c)^2 + y^2} \quad \text{Divide both sides by 4.}$$

$$(cx - a^2)^2 = a^2[(x-c)^2 + y^2] \quad \text{Square both sides.}$$

$$c^2x^2 - 2a^2cx + a^4 = a^2(x^2 - 2cx + c^2 + y^2) \quad \text{Square } cx - a^2 \text{ and } x - c.$$

$$c^2x^2 - 2a^2cx + a^4 = a^2x^2 - 2a^2cx + a^2c^2 + a^2y^2 \quad \text{Use the distributive property.}$$

$$c^2x^2 + a^4 = a^2x^2 + a^2c^2 + a^2y^2 \quad \text{Add } 2a^2cx \text{ to both sides.}$$

$$c^2x^2 - a^2x^2 - a^2y^2 = a^2c^2 - a^4 \quad \text{Rearrange the terms.}$$

$$(c^2 - a^2)x^2 - a^2y^2 = a^2(c^2 - a^2) \quad \text{Factor out } x^2 \text{ and } a^2 \text{, respectively.}$$

$$(a^2 - c^2)x^2 + a^2y^2 = a^2(a^2 - c^2) \quad \text{Multiply both sides by } -1.$$

Refer to the discussion on page 744 and let $b^2 = a^2 - c^2$ in the preceding equation.

$$b^2x^2 + a^2y^2 = a^2b^2$$

$$\frac{x^2}{a^2} + \frac{y^2}{b^2} = 1 \qquad \text{Divide both sides by } a^2b^2.$$

Section 10.3 The Hyperbola

The Asymptotes of a Hyperbola Centered at the Origin

The hyperbola

$$\frac{x^2}{a^2} - \frac{y^2}{b^2} = 1$$

with a horizontal transverse axis has the two asymptotes

$$y = \frac{b}{a}x \quad \text{and} \quad y = -\frac{b}{a}x.$$

Proof. Begin by solving the hyperbola's equation for y.

$$\frac{x^2}{a^2} - \frac{y^2}{b^2} = 1 \qquad \text{This is the standard form of the equation of a hyperbola.}$$

$$\frac{y^2}{b^2} = \frac{x^2}{a^2} - 1 \qquad \text{We isolate the term involving } y^2 \text{ to solve for } y.$$

$$y^2 = \frac{b^2x^2}{a^2} - b^2 \qquad \text{Multiply both sides by } b^2.$$

$$y^2 = \frac{b^2x^2}{a^2}\left(1 - \frac{a^2}{x^2}\right)$$

Factor out $\frac{b^2x^2}{a^2}$ on the right. Verify that this result is correct by multiplying using the distributive property and obtaining the previous step.

$$y = \pm\sqrt{\frac{b^2x^2}{a^2}\left(1 - \frac{a^2}{x^2}\right)}$$

Solve for y using the square root method: If $u^2 = d$ then $u = \pm\sqrt{d}$.

$$y = \pm\frac{b}{a}x\sqrt{1 - \frac{a^2}{x^2}} \qquad \text{Simplify.}$$

As x increases or decreases without bound, the value of $\frac{a^2}{x^2}$ approaches 0. Consequently, the value of y can be approximated by

$$y = \pm\frac{b}{a}x.$$

This means that the lines whose equations are $y = \frac{b}{a}x$ and $y = -\frac{b}{a}x$ are asymptotes for the graph of the hyperbola.

ANSWERS TO SELECTED EXERCISES

CHAPTER 1

Section 1.1

Check Point Exercises

1. a. $8x + 5$ **b.** $\frac{x}{7} - 2x$ **2.** 21.8; At age 10, the average neurotic level is 21.8. **3.** 608 **4. a.** 33,502 **b.** greater by 612 lobbyists **5. a.** true **b.** true **6. a.** -8 is less than -2.; true **b.** 7 is greater than -3.; true **c.** -1 is less than or equal to -4.; false **d.** 5 is greater than or equal to 5.; true **e.** 2 is greater than or equal to -14.; true

Exercise Set 1.1

1. $x + 5$ **3.** $x - 4$ **5.** $4x$ **7.** $2x + 10$ **9.** $6 - \frac{1}{2}x$ **11.** $\frac{4}{x} - 2$ **13.** $\frac{3}{5 - x}$ **15.** 57 **17.** 10 **19.** $1\frac{1}{9}$ or $\frac{10}{9}$ **21.** 10 **23.** 44 **25.** 46 **27.** $\{1, 2, 3, 4\}$ **29.** $\{-7, -6, -5, -4\}$ **31.** $\{8, 9, 10, \dots\}$ **33.** $\{1, 3, 5, 7, 9\}$ **35.** true **37.** true **39.** false **41.** true **43.** false **45.** true **47.** false **49.** -6 is less than -2.; true **51.** 5 is greater than -7.; true **53.** 0 is less than -4.; false **55.** -4 is less than or equal to 1.; true **57.** -2 is less than or equal to -6.; false **59.** -2 is less than or equal to -2.; true **61.** -2 is greater than or equal to -2.; true **63.** 2 is less than or equal to $-\frac{1}{2}$.; false **65.** true **67.** false **69.** true **71.** false **73.** false **75.** 4.2 **77.** 0.4 **79.** 863; overestimates by 42 **81.** 10°C **83.** 60 ft **103.** does not make sense **105.** does not make sense **107.** false **109.** true **111.** $(2 \cdot 3 + 3) \cdot 5 = 45$ **113.** -6 and -5 **114.** -5 and 5 **115.** 8 **116.** 34; 34

Section 1.2

Check Point Exercises

1. a. 6 **b.** 4.5 **c.** 0 **2. a.** -28 **b.** 0.7 **c.** $-\frac{1}{10}$ **3. a.** 8 **b.** $-\frac{1}{3}$ **4. a.** -3 **b.** 10.5 **c.** $-\frac{3}{5}$ **5. a.** 25 **b.** -25 **c.** -64 **d.** $\frac{81}{625}$ **6. a.** -8 **b.** $\frac{8}{15}$ **7.** 74 **8.** -4 **9.** addition: $9 + 4x$; multiplication: $x \cdot 4 + 9$ **10. a.** $(6 + 12) + x = 18 + x$ **b.** $(-7 \cdot 4)x = -28x$ **11.** $-28x - 8$ **12.** $14x + 15x^2$ or $15x^2 + 14x$ **13.** $12x - 40$ **14.** $42 - 4x$

Exercise Set 1.2

1. 7 **3.** 4 **5.** 7.6 **7.** $\frac{\pi}{2}$ **9.** $\sqrt{2}$ **11.** $-\frac{2}{5}$ **13.** -11 **15.** -4 **17.** -4.5 **19.** $\frac{2}{15}$ **21.** $-\frac{35}{36}$ **23.** -8.2 **25.** -12.4 **27.** 0 **29.** -11 **31.** 5 **33.** 0 **35.** -12 **37.** 18 **39.** -15 **41.** $-\frac{1}{4}$ **43.** 5.5 **45.** $\sqrt{2}$ **47.** -90 **49.** 33 **51.** $-\frac{15}{13}$ **53.** 0 **55.** -8 **57.** 48 **59.** 100 **61.** -100 **63.** -8 **65.** 1 **67.** -1 **69.** $\frac{1}{8}$ **71.** -3 **73.** 45 **75.** 0 **77.** undefined **79.** $\frac{9}{14}$ **81.** -15 **83.** -2 **85.** -24 **87.** 45 **89.** $\frac{1}{121}$ **91.** 14 **93.** $-\frac{8}{3}$ **95.** $-\frac{1}{2}$ **97.** 31 **99.** 37 **101.** addition: $10 + 4x$; multiplication: $x \cdot 4 + 10$ **103.** addition: $-5 + 7x$; multiplication: $x \cdot 7 - 5$ **105.** $(4 + 6) + x = 10 + x$ **107.** $(-7 \cdot 3)x = -21x$ **109.** $\left(-\frac{1}{3} \cdot -3\right)y = y$ **111.** $6x + 15$ **113.** $-14x - 21$ **115.** $-3x + 6$ **117.** $12x$ **119.** $5x^2$ **121.** $10x + 12x^2$ **123.** $18x - 40$ **125.** $8y - 12$ **127.** $16y - 25$ **129.** $12x^2 + 11$ **131.** $x - (x + 4)$; -4 **133.** $6(-5x)$; $-30x$ **135.** $5x - 2x$; $3x$ **137.** $8x - (3x + 6)$; $5x - 6$ **139.** \$389 billion **141.** \$25 billion **143.** $-$\$377 billion; deficit **145.** \$552 billion **147.** 38.55%; overestimates by 3.55% **149. a.** $1200 - 0.07x$ **b.** \$780 **169.** makes sense **171.** does not make sense **173.** false **175.** false **177.** true **179.** $\left(2 \cdot 5 - \frac{1}{2} \cdot 10\right) \cdot 9 = 45$ **181.** $\frac{10}{x} - 4x$ **182.** 42 **183.** true **184.** -5; 0; 3; 4; 3; 0; -5 **185.** -8; -3; 0; 1; 0; -3; -8 **186.** 3; 2; 1; 0; 1; 2; 3

Section 1.3

Check Point Exercises

1.

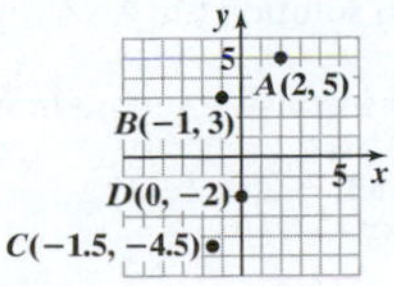

2.

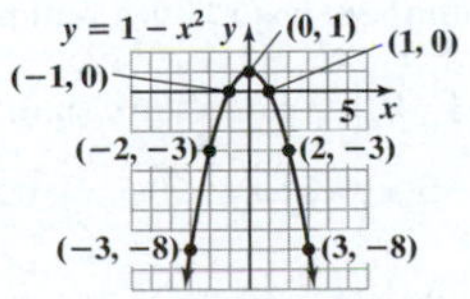

3.

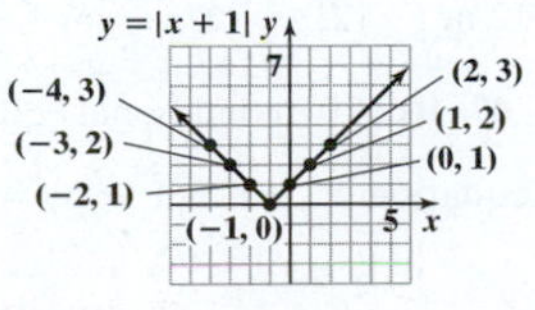

4. a. 0 to 3 hr **b.** 3 to 13 hr **c.** 0.05 mg per 100 ml; after 3 hr **d.** None of the drug is left in the body. **5.** minimum x-value: -100; maximum x-value: 100; distance between tick marks on x-axis: 50; minimum y-value: -100; maximum y-value: 100; distance between tick marks on y-axis: 10

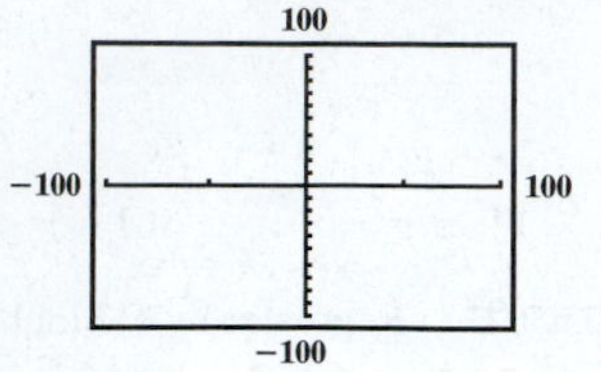

Exercise Set 1.3

1–9.

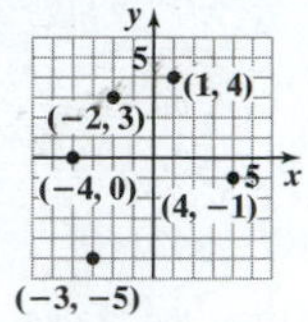

11.

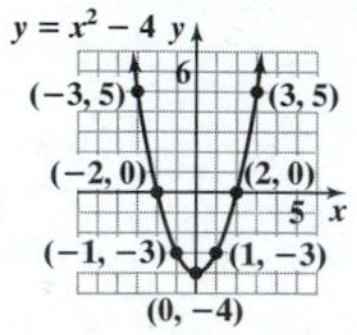

13.

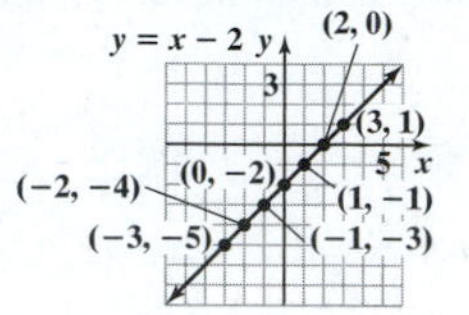

15.

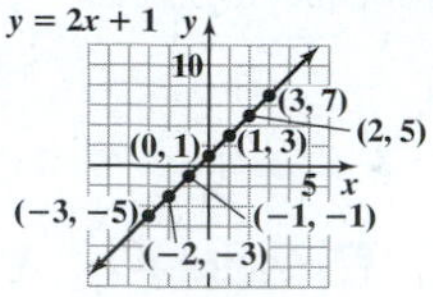

17.

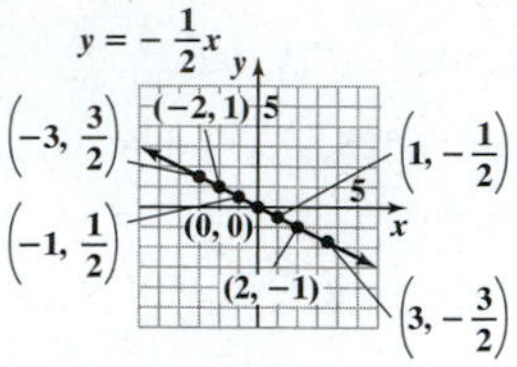

19.

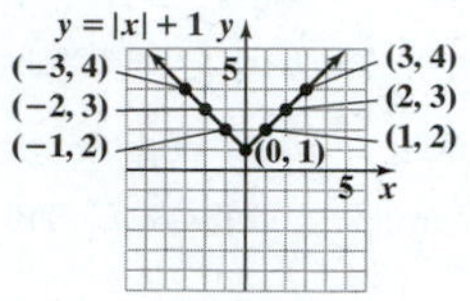

21.

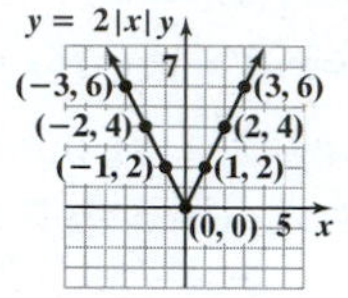

23.

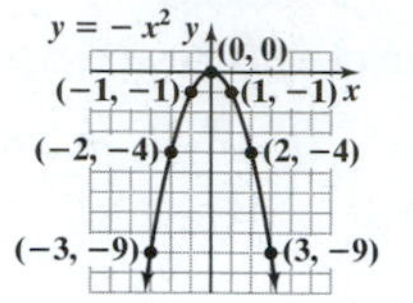

25.

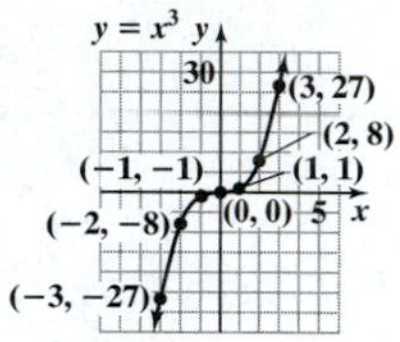

27. c **29.** b **31.** c **33.** no **35.** (2, 0) **37.** $(-2, 4)$ and (1, 1) **39.** $y = 2x + 4$

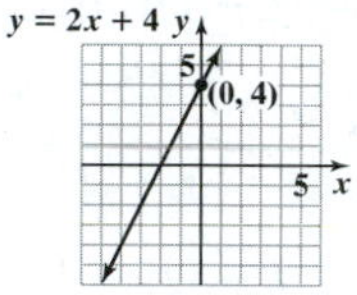

41. $y = 3 - x^2$

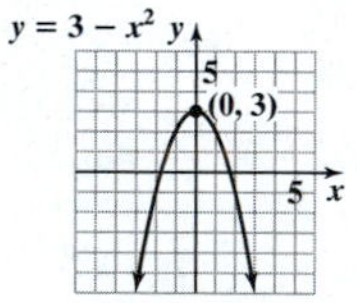

43.

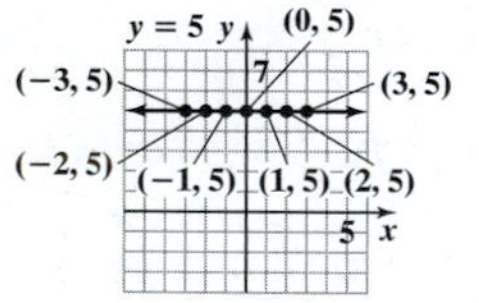

45.

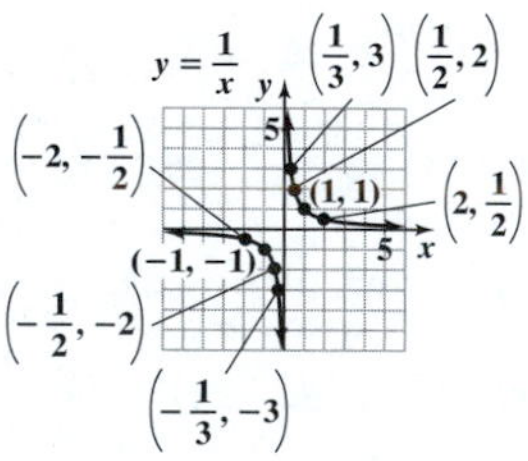

47. 35% **49.** 1945; 94% **51.** 1950–1960; 91% **53.** 8 yr old; 1 awakening **55.** about 1.9 awakenings **57.** a **59.** b **61.** b **63.** c **73.** makes sense **75.** makes sense **77.** false **79.** true **81.** \$15 **83.** 14.3 **84.** 3 **85.** $-14x - 25$ **86.** true **87.** $7 - 3x$ **88.** $15x + 5$

Section 1.4

Check Point Exercises

1. 6 or $\{6\}$ **2.** -1 or $\{-1\}$ **3.** 6 or $\{6\}$ **4.** 1 or $\{1\}$ **5.** no solution or $\varnothing$; inconsistent equation **6.** all real numbers or $\{x \mid x \text{ is a real number}\}$; identity **7.** 2008

Exercise Set 1.4

1. 3 or $\{3\}$ **3.** 11 or $\{11\}$ **5.** 11 or $\{11\}$ **7.** 7 or $\{7\}$ **9.** 13 or $\{13\}$ **11.** 2 or $\{2\}$ **13.** -4 or $\{-4\}$ **15.** 9 or $\{9\}$ **17.** -5 or $\{-5\}$ **19.** 6 or $\{6\}$ **21.** 19 or $\{19\}$ **23.** $\frac{5}{2}$ or $\left\{\frac{5}{2}\right\}$ **25.** 12 or $\{12\}$ **27.** 24 or $\{24\}$ **29.** -15 or $\{-15\}$ **31.** 5 or $\{5\}$ **33.** 13 or $\{13\}$ **35.** -12 or $\{-12\}$ **37.** $\frac{46}{5}$ or $\left\{\frac{46}{5}\right\}$ **39.** all real numbers or $\{x \mid x \text{ is a real number}\}$; identity **41.** no solution or $\varnothing$; inconsistent equation **43.** 0 or $\{0\}$; conditional equation **45.** -10 or $\{-10\}$; conditional equation **47.** no solution or $\varnothing$; inconsistent equation **49.** 0 or $\{0\}$; conditional equation **51.** $3(x - 4) = 3(2 - 2x)$; 2 **53.** $-3(x - 3) = 5(2 - x)$; 0.5 **55.** 2 **57.** -7 **59.** -2 or $\{-2\}$

61. no solution or $\varnothing$ **63.** 10 or $\{10\}$ **65.** -2 or $\{-2\}$ **67. a.** model 1: \$22,228; model 2: \$22,208; Model 1 overestimates the cost by \$10 and model 2 underestimates the cost by \$10. **b.** the school year ending 2012 **69. a.** \$32,000 **b.** \$32,616; \$616 **c.** \$32,597; \$597 **71.** model 1 **73.** 2025 **85.** 3 or $\{3\}$;

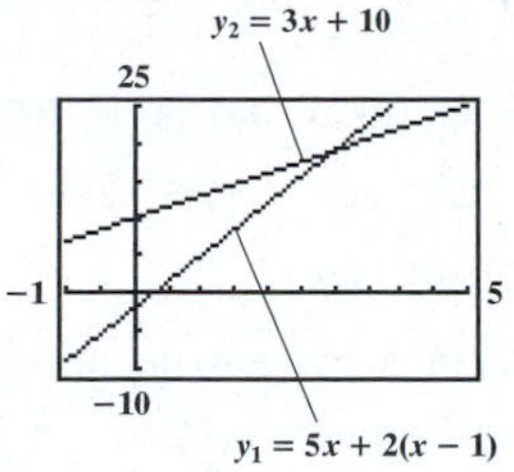

87. -6 or $\{-6\}$;

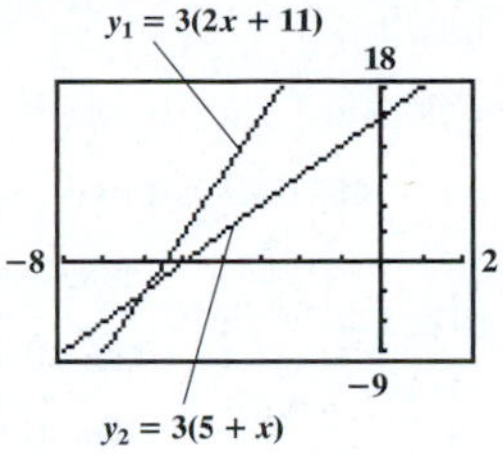

89. makes sense **91.** does not make sense **93.** false **95.** true **97.** $x = \frac{c-b}{a}$ **101.** $\frac{3}{10}$ **102.** -60 **103.**

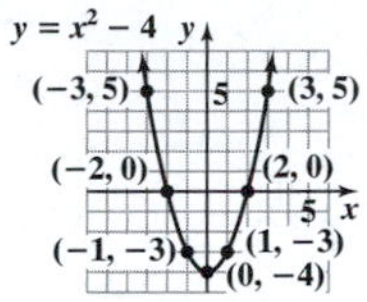

104. a. $3x - 4 = 32$ **b.** 12 **105.** $x + 44$ **106.** $20{,}000 - 2500x$

Mid-Chapter Check Point Exercises

1. $3x + 10$ **2.** 6 or $\{6\}$ **3.** -15 **4.** -7 or $\{-7\}$ **5.** 3 **6.** $13x - 23$ **7.** 0 or $\{0\}$ **8.** $-7x - 34$ **9.** 7 **10.** no solution or $\varnothing$ or inconsistent equation **11.** 3 or $\{3\}$ **12.** -4 **13.** all real numbers or $\{x|x$ is a real number$\}$ or identity **14.** 2 **15.**

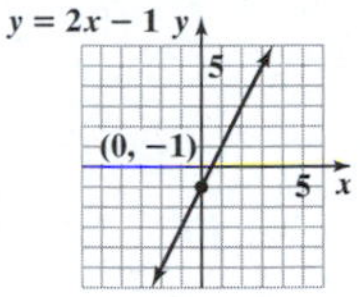

16.

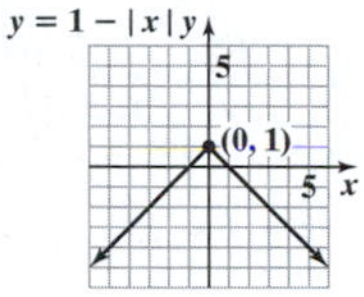

17.

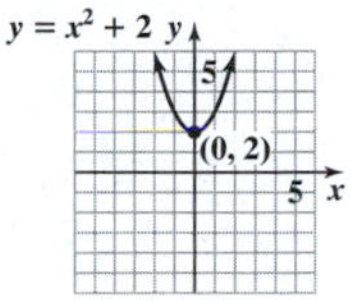

18. true **19.** false **20.** false **21.** true

Section 1.5

Check Point Exercises

1. free: 3; not free: 23; partly free: 20 **2.** 2015 **3.** 300 min **4.** \$1200 **5.** 50 ft by 94 ft **6.** $w = \frac{P - 2l}{2}$ **7.** $h = \frac{V}{lw}$ **8.** $W = 106 + 6H$ or $W = 6H + 106$ **9.** $C = \frac{P}{1 + M}$

Exercise Set 1.5

1. 6 **3.** 25 **5.** 120 **7.** 320 **9.** 19 and 45 **11.** 2 **13.** 8 **15.** all real numbers **17.** sophomore: \$1581; junior: \$2002; senior: \$2846 **19.** 94°, 47°, 39° **21.** 59°, 60°, 61° **23.** 2050 **25.** 2014 **27.** after 5 months; \$165 **29.** 30 times **31. a.** 2014; 22,300 students **b.** $y_1 = 13300 + 1000x$; $y_2 = 26800 - 500x$ **33.** \$420 **35.** \$150 **37.** \$39,000 **39.** \$467.20 **41.** 50 yd by 100 yd **43.** 36 ft by 78 ft **45.** 2 in. **47.** 11 min **49.** $l = \frac{A}{w}$ **51.** $b = \frac{2A}{h}$ **53.** $P = \frac{I}{rt}$ **55.** $p = \frac{T - D}{m}$ **57.** $a = \frac{2A}{h} - b$ or $a = \frac{2A - hb}{h}$ **59.** $h = \frac{3V}{\pi r^2}$ **61.** $m = \frac{y - y_1}{x - x_1}$ **63.** $d_1 = Vt + d_2$ **65.** $x = \frac{C - By}{A}$ **67.** $v = \frac{2s - at^2}{2t}$ **69.** $n = \frac{L - a}{d} + 1$ **71.** $l = \frac{A - 2wh}{2w + 2h}$ **73.** $I = \frac{E}{R + r}$ **83. a.** $y = 19.4 + 0.4x$ **85.** makes sense **87.** does not make sense **89.** true **91.** false **93.** 10 problems **95.** 36 plants **97.** -6 is less than or equal to -6.; true **98.** $\frac{7}{3}$ **99.** -8 or $\{-8\}$ **100. a.** b^7 **b.** b^{10} **c.** Add the exponents. **101. a.** b^4 **b.** b^6 **c.** Subtract the exponents. **102.** -8

Section 1.6

Check Point Exercises

1. a. b^{11} **b.** $40x^5y^{10}$ **2. a.** $(-3)^3$ or -27 **b.** $9x^{11}y^3$ **3. a.** 1 **b.** 1 **c.** -1 **d.** 10 **e.** 1 **4. a.** $\frac{1}{25}$ **b.** $-\frac{1}{27}$ **c.** 16 **d.** $\frac{3y^4}{x^6}$ **5. a.** $\frac{4^3}{7^2} = \frac{64}{49}$ **b.** $\frac{x^2}{5}$ **6. a.** x^{15} **b.** $\frac{1}{y^{14}}$ **c.** b^{12} **7. a.** $16x^4$ **b.** $-27y^6$ **c.** $\frac{y^2}{16x^{10}}$ **8. a.** $\frac{x^{15}}{64}$ **b.** $\frac{16}{x^{12}y^8}$ **c.** $x^{15}y^{20}$ **9. a.** $-12y^9$ **b.** $\frac{4y^{14}}{x^6}$ **c.** $\frac{64}{x^9y^{15}}$

Exercise Set 1.6

1. b^{11} **3.** x^4 **5.** 32 **7.** $6x^6$ **9.** $20y^{12}$ **11.** $100x^{10}y^{12}$ **13.** $21x^5yz^4$ **15.** b^9 **17.** $5x^5$ **19.** x^5y^5 **21.** $10xy^3$ **23.** $-8a^{11}b^8c^4$ **25.** 1 **27.** 1 **29.** -1 **31.** 13 **33.** 1 **35.** $\frac{1}{3^2} = \frac{1}{9}$ **37.** $\frac{1}{(-5)^2} = \frac{1}{25}$ **39.** $-\frac{1}{5^2} = -\frac{1}{25}$ **41.** $\frac{x^2}{y^3}$ **43.** $\frac{8y^3}{x^7}$ **45.** $5^3 = 125$ **47.** $(-3)^4 = 81$ **49.** $\frac{y^5}{x^2}$ **51.** $\frac{b^7c^3}{a^4}$ **53.** x^{60} **55.** $\frac{1}{b^{12}}$ **57.** 7^{20} **59.** $64x^3$ **61.** $9x^{14}$ **63.** $8x^3y^6$ **65.** $9x^4y^{10}$ **67.** $-\frac{x^6}{27}$ **69.** $\frac{y^8}{25x^6}$ **71.** $\frac{x^{20}}{16y^{16}z^8}$ **73.** $\frac{16}{x^4}$ **75.** $\frac{x^6}{25}$ **77.** $\frac{81x^4}{y^4}$ **79.** $\frac{x^{24}}{y^{12}}$ **81.** x^9y^{12} **83.** a^8b^{12} **85.** $\frac{1}{x^6}$

87. $-\frac{4}{x}$ **89.** $\frac{2}{x^7}$ **91.** $\frac{10a^2}{b^5}$ **93.** $\frac{1}{x^9}$ **95.** $-\frac{6}{a^2}$ **97.** $\frac{3y^4z^3}{x^7}$ **99.** $3x^{10}$ **101.** $\frac{1}{x^{10}}$ **103.** $-\frac{5y^8}{x^6}$ **105.** $\frac{8a^9c^{12}}{b}$ **107.** x^{16} **109.** $-\frac{27b^{15}}{a^{18}}$ **111.** 1 **113.** $\frac{81x^{20}}{y^{32}}$ **115.** $\frac{1}{100a^4b^{12}c^8}$ **117.** $10x^2y^4$ **119.** $\frac{8}{3xy^{10}}$ **121.** $\frac{y^5}{8x^{14}}$ **123.** $\frac{1}{128x^7y^{16}}$ **125. a.** 1000 aphids **b.** 16,000 aphids **c.** 125 aphids **127. a.** one person **b.** 10 people **129. a.** (0, 1) **b.** (4, 10) **131.** d **133.** 0.55 astronomical unit **135.** 1.8 astronomical units **147.** makes sense **149.** does not make sense **151.** false **153.** false **155.** false **157.** true **159.** x^{9n} **161.** $x^{3n}y^{6n+3}$ **162.**

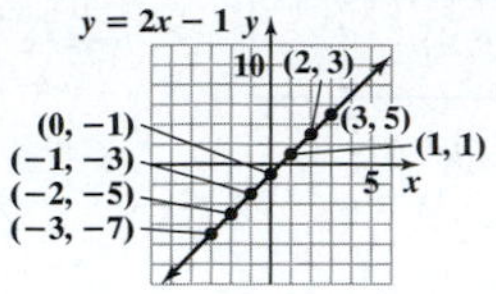

163. $y = \frac{C - Ax}{B}$ **164.** 40 m by 75 m **165.** It moves the decimal point three places to the right. **166.** It moves the decimal point 2 places to the left. **167. a.** 10^5; 100,000 **b.** 10^6; 1,000,000

Section 1.7

Check Point Exercises

1. a. −2,600,000,000 **b.** 0.000003017 **2. a.** 5.21×10^9 **b.** -6.893×10^{-8} **3.** 5.19×10^{11} **4. a.** 3.55×10^{-1} **b.** 4×10^8 **5.** \$6847 **6.** 3.1×10^7 mi or 31 million mi

Exercise Set 1.7

1. 380 **3.** 0.0006 **5.** −7,160,000 **7.** 1.4 **9.** 0.79 **11.** −0.00415 **13.** −60,000,100,000 **15.** 3.2×10^4 **17.** 6.38×10^{17} **19.** -3.17×10^2 **21.** -5.716×10^3 **23.** 2.7×10^{-3} **25.** -5.04×10^{-9} **27.** 7×10^{-3} **29.** 3.14159×10^0 **31.** 6.3×10^7 **33.** 6.4×10^4 **35.** 1.22×10^{-11} **37.** 2.67×10^{13} **39.** 2.1×10^3 **41.** 4×10^5 **43.** 2×10^{-8} **45.** 5×10^3 **47.** 4×10^{15} **49.** 9×10^{-3} **51.** 6×10^{13} **53.** -6.2×10^3 **55.** 1.63×10^{19} **57.** -3.6×10^5 **59.** $\$4.65 \times 10^{10}$ **61.** $\$1 \times 10^8$ **63.** approximately 67 hot dogs per person **65.** $2.5 \times 10^2 = 250$ chickens **67. a.** $\$1.0813 \times 10^4$; \$10,813 **b.** \$901 **69.** Medicare; \$3242 **71.** 1.06×10^{-18} g **73.** 3.1536×10^7 **81.** does not make sense **83.** makes sense **85.** false **87.** false **89.** true **91.** 1.25×10^{-15} **94.** $85x - 26$ **95.** 4 or {4} **96.** $\frac{y^6}{64x^8}$ **97.** set 1 **98.** −170 **99.** $5a + 5h + 7$

Review Exercises

1. $2x - 10$ **2.** $6x + 4$ **3.** $\frac{9}{x} + \frac{1}{2}x$ **4.** 34 **5.** 60 **6.** 15 **7.** {1, 2} **8.** {−3, −2, −1, 0, 1} **9.** false **10.** true **11.** true **12.** −5 is less than 2.; true **13.** −7 is greater than or equal to −3.; false **14.** −7 is less than or equal to −7.; true **15.** 124 ft **16.** 9.7 **17.** 5.003 **18.** 0 **19.** −7.6 **20.** −4.4 **21.** 13 **22.** 60 **23.** $-\frac{1}{10}$ **24.** $-\frac{3}{35}$ **25.** −240 **26.** 16 **27.** −32 **28.** $-\frac{5}{12}$ **29.** 7 **30.** −9.1 **31.** 7 **32.** 9 **33.** −2 **34.** −18 **35.** 55 **36.** 1 **37.** −4 **38.** −13 **39.** $17x - 15$ **40.** $9x^2 + x$ **41.** $5y - 17$ **42.** $10x$ **43.** $-3x - 8$ **44–46.**

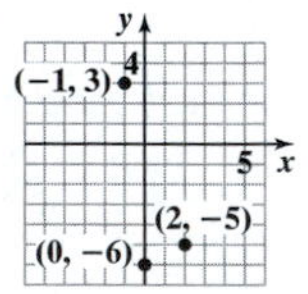

47.

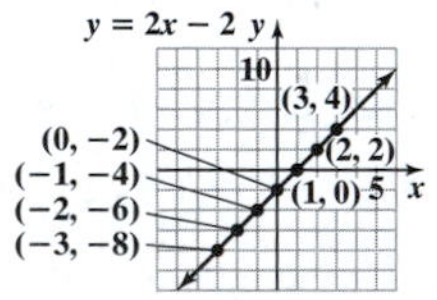

48.

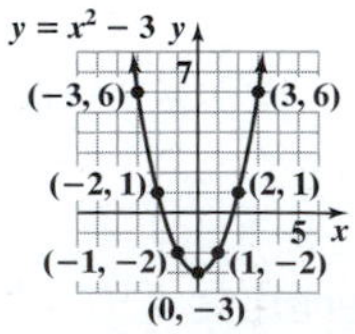

49.

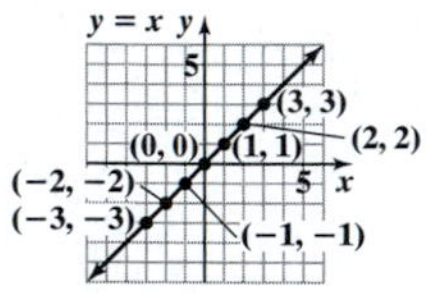

50.

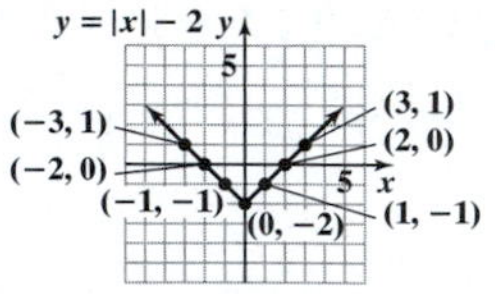

51. minimum x-value: −20; maximum x-value: 40; distance between tick marks on x-axis: 10; minimum y-value: −5; maximum y-value: 5; distance between tick marks on y-axis: 1

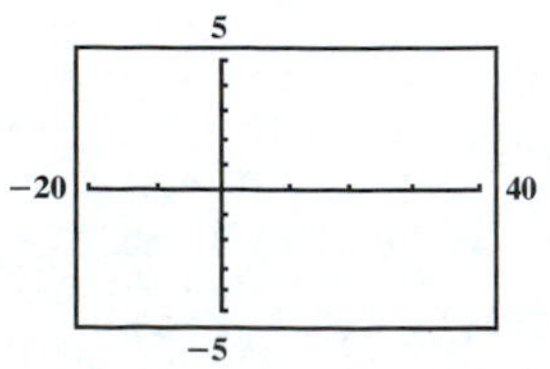

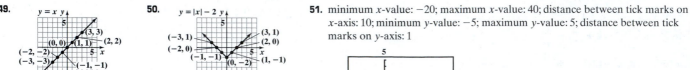

52. 20% **53.** 85 years old **55.** c **56.** 6 or {6} **57.** −10 or {−10} **58.** 5 or {5} **59.** −13 or {−13}

60. -3 or $\{-3\}$ **61.** -1 or $\{-1\}$ **62.** 2 or $\{2\}$ **63.** 2 or $\{2\}$ **64.** $\frac{72}{11}$ or $\left\{\frac{72}{11}\right\}$ **65.** -12 or $\{-12\}$ **66.** $\frac{77}{15}$ or $\left\{\frac{77}{15}\right\}$
67. no solution or $\varnothing$; inconsistent equation **68.** all real numbers or $\{x \mid x \text{ is a real number}\}$; identity **69.** 0 or $\{0\}$; conditional equation
70. $\frac{3}{2}$ or $\left\{\frac{3}{2}\right\}$; conditional equation **71.** no solution or $\varnothing$; inconsistent equation **72. a.** 1997 **b.** overestimates by 2 corporations
73. U2: \$260 million; The Rolling Stones: \$141 million; The Eagles: \$117 million **74.** 25°, 35°, 120° **75. a.** 2018 **b.** \$1177 billion
c. They are shown by the intersection of the graphs at approximately (2018, 1177) **76.** 500 min **77.** \$60 **78.** \$10,000 in sales
79. 44 yd by 126 yd **80. a.** $14{,}100 + 1500x = 41{,}700 - 800x$ **b.** 2017; 32,100 **81.** $h = \frac{3V}{B}$ **82.** $x = \frac{y - y_1}{m} + x_1$
83. $R = \frac{E}{I} - r$ or $R = \frac{E - Ir}{I}$ **84.** $F = \frac{9C + 160}{5}$ or $F = \frac{9}{5}C + 32$ **85.** $g = \frac{s - vt}{t^2}$ **86.** $g = \frac{T}{r + vt}$ **87.** $15x^{13}$ **88.** $\frac{x^2}{y^5}$
89. $\frac{x^4y^7}{9}$ **90.** $\frac{1}{x^{18}}$ **91.** $49x^6y^2$ **92.** $-\frac{8}{y^7}$ **93.** $-\frac{12}{x^7}$ **94.** $3x^{10}$ **95.** $-\frac{a^8}{2b^5}$ **96.** $-24x^7y^4$ **97.** $\frac{3}{4}$ **98.** $\frac{y^{12}}{125x^6}$ **99.** $-\frac{6x^9}{y^5}$
100. $\frac{9x^8y^{14}}{25}$ **101.** $-\frac{x^{21}}{8y^{27}}$ **102.** 7,160,000 **103.** 0.000107 **104.** -4.1×10^{13} **105.** 8.09×10^{-3} **106.** 1.26×10^8 **107.** 2.5×10^1
108. 2.88×10^{13}

Chapter Test

1. $4x - 5$ **2.** 170 **3.** $\{-4, -3, -2, -1\}$ **4.** true **5.** -3 is greater than -1.; false **6.** 259; underestimates by 3 **7.** 17.9
8. -7.6 **9.** $\frac{1}{4}$ **10.** -60 **11.** $\frac{1}{8}$ **12.** 3.1 **13.** -3 **14.** 6 **15.** -4 **16.** $-5x - 18$ **17.** $6y - 27$ **18.** $17x - 22$

19.

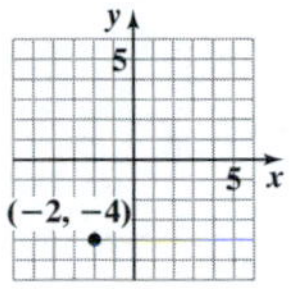

20.

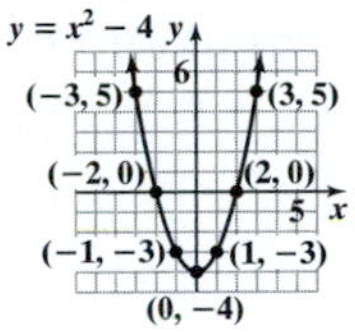

21. 2 or $\{2\}$ **22.** -6 or $\{-6\}$ **23.** no solution or $\varnothing$; inconsistent equation
24. 23 and 49 **25.** 5 yr **26.** 20 prints; \$3.80 **27.** \$50 **28.** 120 yd by 380 yd
29. $h = \frac{3V}{lw}$ **30.** $y = \frac{C - Ax}{B}$ **31.** $-\frac{14}{x^5}$ **32.** $\frac{40}{x^3y^8}$ **33.** $\frac{x^6}{4y^3}$ **34.** $\frac{x^{15}}{64y^6}$
35. $\frac{x^{16}}{9y^{10}}$ **36.** 0.0000038 **37.** 4.07×10^{11} **38.** 5×10^3 **39.** 1.3×10^{10}

CHAPTER 2

Section 2.1

Check Point Exercises

1. domain: $\{0, 10, 20, 30, 34\}$; range: $\{9.1, 6.7, 10.7, 13.2, 15.5\}$ **2. a.** not a function **b.** function **3. a.** 29 **b.** 65 **c.** 46 **d.** $6a + 6h + 9$
4. a. Every element in the domain corresponds to exactly one element in the range. **b.** domain: $\{0, 1, 2, 3, 4\}$; range: $\{3, 0, 1, 2\}$ **c.** 0
d. 2 **e.** $x = 0$ and $x = 4$

Exercise Set 2.1

1. function; domain: $\{1, 3, 5\}$; range: $\{2, 4, 5\}$ **3.** not a function; domain: $\{3, 4\}$; range: $\{4, 5\}$ **5.** function; domain: $\{-3, -2, -1, 0\}$; range: $\{-3, -2, -1, 0\}$ **7.** not a function; domain: $\{1\}$; range: $\{4, 5, 6\}$ **9. a.** 1 **b.** 6 **c.** -7 **d.** $2a + 1$ **e.** $a + 3$ **11. a.** -2
b. -17 **c.** 0 **d.** $12b - 2$ **e.** $3b + 10$ **13. a.** 5 **b.** 8 **c.** 53 **d.** 32 **e.** $48b^2 + 5$ **15. a.** -1 **b.** 26 **c.** 19
d. $2b^2 + 3b - 1$ **e.** $50a^2 + 15a - 1$ **17. a.** $\frac{3}{4}$ **b.** -3 **c.** $\frac{11}{8}$ **d.** $\frac{13}{9}$ **e.** $\frac{2a + 2h - 3}{a + h - 4}$ **f.** Denominator would be zero.
19. a. 6 **b.** 12 **c.** 0 **21. a.** 2 **b.** 1 **c.** -1 and 1 **23.** -2; 10 **25.** -38 **27.** $-2x^3 - 2x$ **29. a.** -1 **b.** 7 **c.** 19
d. 112 **31. a.** {(EL, 1%), (L, 7%), (SL, 11%), (M, 52%), (SC, 13%), (C, 13%), (EC, 3%)} **b.** Yes; Each ideology corresponds to exactly one percentage. **c.** {(1%, EL), (7%, L), (11%, SL), (52%, M), (13%, SC), (13%, C), (3%, EC)} **d.** No; 13% in the domain corresponds to two ideologies in the range, SC and C. **37.** makes sense **39.** makes sense **41.** false **43.** true **45.** true **47.** 3
49. $f(2) = 6$; $f(3) = 9$; $f(4) = 12$; no **50.** 0 **51.** $\frac{y^{10}}{9x^4}$ **52.** $\{-15\}$

53.

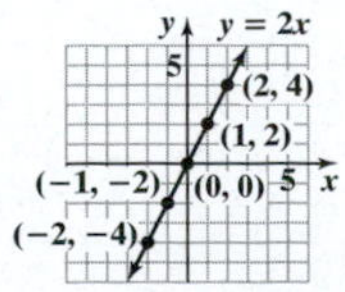

54.

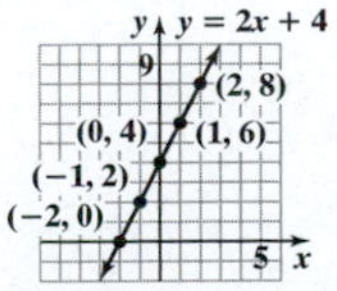

55. a. 3 **b.** -3 and 3 **c.** all real numbers **d.** all real numbers greater than or equal to 1

Section 2.2

Check Point Exercises

1. 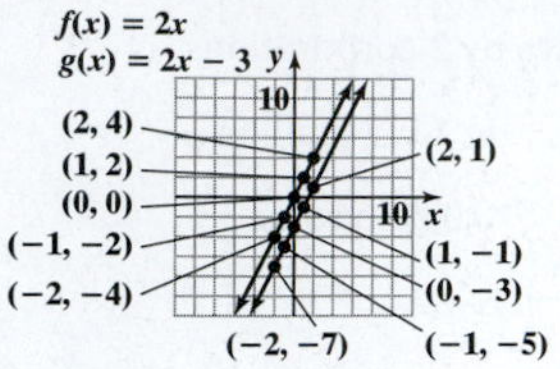

; The graph of g is the graph of f shifted down by 3 units.

2. a. function **b.** function **c.** not a function **3. a.** 400 **b.** 9 **c.** approximately 425
4. a. domain: $\{x|-2 \le x \le 1\}$; range: $\{y|0 \le y \le 3\}$ **b.** domain: $\{x|-2 < x \le 1\}$; range: $\{y|-1 \le y < 2\}$
c. domain: $\{x|-3 \le x < 0\}$; range: $\{y|y = -3, -2, -1\}$

Exercise Set 2.2

1. 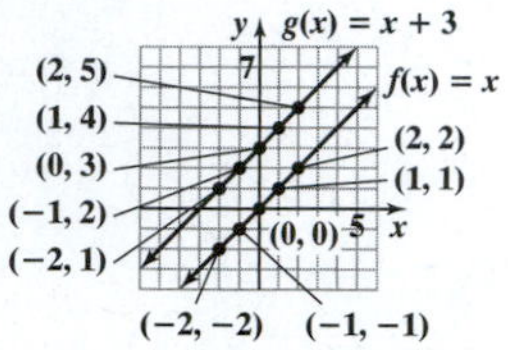

The graph of g is the graph of f shifted up by 3 units.

3. 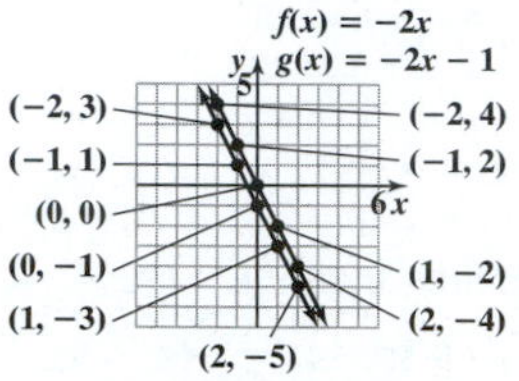

The graph of g is the graph of f shifted down by 1 unit.

5. 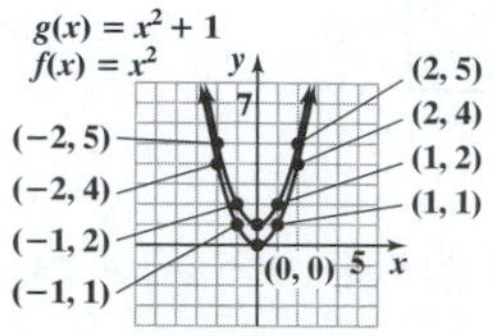

The graph of g is the graph of f shifted up by 1 unit.

7. 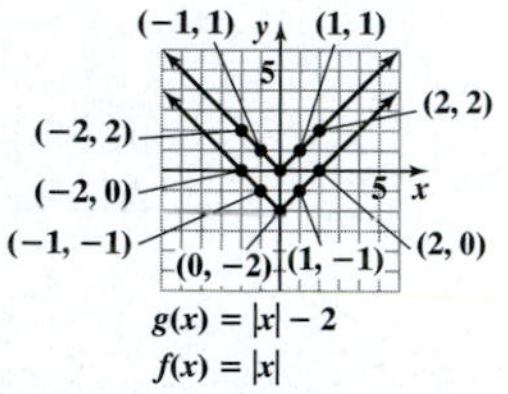

The graph of g is the graph of f shifted down by 2 units.

9. 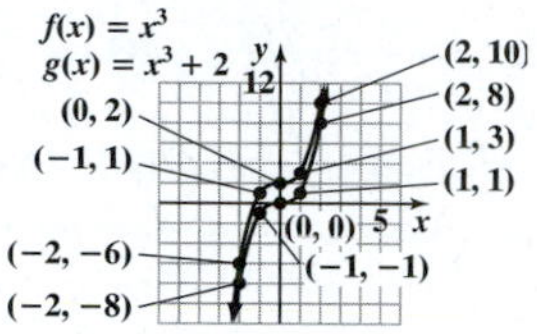

The graph of g is the graph of f shifted up by 2 units.

11. function **13.** not a function **15.** function **17.** not a function **19.** -4 **21.** 4 **23.** 0 **25.** 2 **27.** 2
29. -2 **31.** domain: $\{x|0 \le x < 5\}$; range: $\{y|-1 \le y < 5\}$ **33.** domain: $\{x|x \ge 0\}$; range: $\{y|y \ge 1\}$
35. domain: $\{x|-2 \le x \le 6\}$; range: $\{y|-2 \le y \le 6\}$ **37.** domain: $\{x|x$ is a real number$\}$; range: $\{y|y \le -2\}$
39. domain: $\{x|x = -5, -2, 0, 1, 3\}$; range: $\{y|y = 2\}$ **41. a.** $\{x\,|\,x$ is a real number$\}$ **b.** $\{y\,|\,y \ge -4\}$ **c.** 4 **d.** 2 and 6
e. $(1, 0), (7, 0)$ **f.** $(0, 4)$ **g.** $\{x\,|\,1 < x < 7\}$ **h.** positive **43. a.** 707; Approximately 707,000 bachelor's degrees were awarded to women in 2000.; (20, 707) **b.** overestimates by 2 thousand **45. a.** 68; Approximately 68,000 more bachelor's degrees were awarded to women than to men in 1990.; The points on the graphs with first coordinate 10 are 68 units apart. **b.** overestimates by 13 thousand
47. 440; For 20-year-old drivers, there are 440 accidents per 50 million miles driven.; (20, 440) **49.** $x = 45$; $y = 190$; The minimum number of accidents is 190 per 50 million miles driven and is attributed to 45-year-old drivers. **51.** 3.1; In 1960, Jewish Americans made up about 3.1% of the U.S. population. **53.** 19 and 64; In 1919 and in 1964, Jewish Americans made up about 3% of the U.S. population. **55.** 1940; 3.7%
57. Each year corresponds to only one percentage. **59.** 0.75; It costs \$0.75 to mail a 3-ounce first-class letter. **61.** \$0.58 **69.** makes sense
71. does not make sense **73.** false **75.** true **77.** true **79.** -18 **81.** yes **82.** $\left\{\frac{3}{2}\right\}$ **83.** 76 yd by 236 yd
84. Division by 0 is undefined. **85.** 19 **86.** $10.9x^2 - 35x + 1641$

Section 2.3

Check Point Exercises

1. a. $\{x|x$ is a real number$\}$ **b.** $\{x|x$ is a real number and $x \ne -5\}$ **2. a.** $3x^2 + 6x + 6$ **b.** 78 **3. a.** $\frac{5}{x} - \frac{7}{x-8}$
b. $\{x\,|\,x$ is a real number and $x \ne 0$ and $x \ne 8\}$ **4. a.** 23 **b.** $x^2 - 3x - 3$; 1 **c.** $\frac{x^2 - 2x}{x + 3}; \frac{7}{2}$ **d.** -24
5. a. $(B + D)(x) = 3.9x^2 + 5x + 6451$ **b.** 6573.5 thousand **c.** underestimates by 1.5 thousand

Exercise Set 2.3

1. $\{x|x$ is a real number$\}$ **3.** $\{x|x$ is a real number and $x \ne -4\}$ **5.** $\{x|x$ is a real number and $x \ne 3\}$ **7.** $\{x|x$ is a real number and $x \ne 5\}$
9. $\{x|x$ is a real number and $x \ne -7$ and $x \ne 9\}$ **11. a.** $5x - 5$ **b.** 20 **13. a.** $3x^2 + x - 5$ **b.** 75 **15. a.** $2x^2 - 2$ **b.** 48
17. $\{x|x$ is a real number$\}$ **19.** $\{x|x$ is a real number and $x \ne 5\}$ **21.** $\{x|x$ is a real number and $x \ne 0$ and $x \ne 5\}$

23. $\{x|x \text{ is a real number and } x \neq 2 \text{ and } x \neq -3\}$ **25.** $\{x|x \text{ is a real number and } x \neq 2\}$ **27.** $\{x|x \text{ is a real number}\}$
29. $x^2 + 3x + 2; 20$ **31.** 0 **33.** $x^2 + 5x - 2; 48$ **35.** -8 **37.** -16 **39.** -135 **41.** $\frac{x^2 + 4x}{2 - x}; 5$ **43.** -1
45. $\{x|x \text{ is a real number}\}$ **47.** $\{x|x \text{ is a real number and } x \neq 2\}$ **49.** 5 **51.** -1 **53.** $\{x|-4 \leq x \leq 3\}$
55.

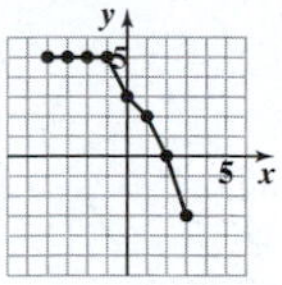

57. -4 **59.** -4 **61. a.** $(M + F)(x) = 3.06x + 235$ **b.** 296.2 million **c.** underestimates by 1.8 million
63. a. $\left(\frac{M}{F}\right)(x) = \frac{1.58x + 114.4}{1.48x + 120.6}$ **b.** 0.967 **c.** overestimates by approximately 0.002

69.

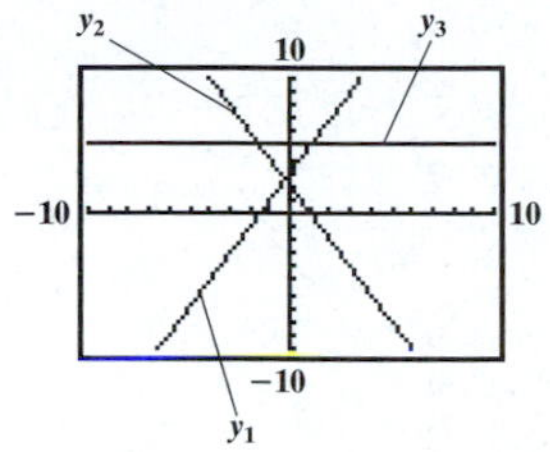

71.

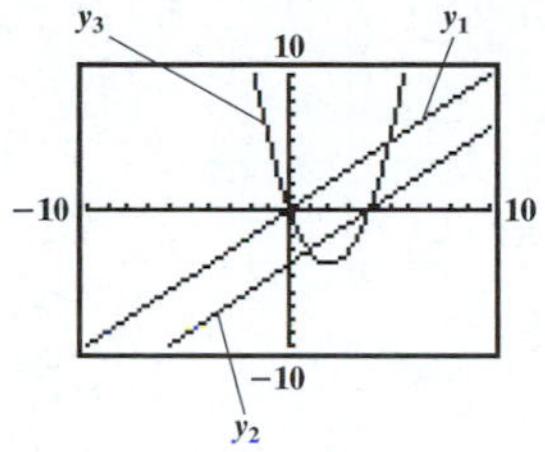

73. No y-value is displayed; y_3 is undefined at $x = 0$. **75.** makes sense **77.** makes sense **79.** true **81.** false
82. $b = \frac{R - 3a}{3}$ or $b = \frac{R}{3} - a$ **83.** $\{7\}$ **84.** $6b + 8$ **85. a.** $\frac{3}{2}$ **b.** -2
86. a.

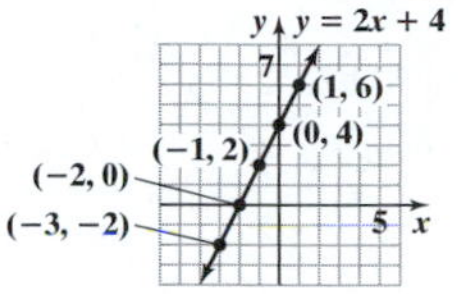

b. $(-2, 0)$ **c.** $(0, 4)$ **87.** $y = -\frac{5}{3}x - 4$

Mid-Chapter Check Point Exercises

1. not a function; domain: $\{1, 2\}$; range: $\{-6, 4, 6\}$ **2.** function; domain: $\{0, 2, 3\}$; range: $\{1, 4\}$ **3.** function; domain: $\{x|-2 \leq x < 2\}$; range: $\{y|0 \leq y \leq 3\}$ **4.** not a function; domain: $\{x|-3 < x \leq 4\}$; range: $\{y|-1 \leq y \leq 2\}$ **5.** not a function; domain: $\{-2, -1, 0, 1, 2\}$; range: $\{-2, -1, 1, 3\}$ **6.** function; domain: $\{x|x \leq 1\}$; range: $\{y|y \geq -1\}$ **7.** No vertical line intersects the graph of f more than once.
8. 3 **9.** -2 **10.** -6 and 2 **11.** $\{x|x \text{ is a real number}\}$ **12.** $\{y|y \leq 4\}$ **13.** $\{x|x \text{ is a real number}\}$
14. $\{x|x \text{ is a real number and } x \neq -2 \text{ and } x \neq 2\}$ **15.** 23 **16.** 23 **17.** $a^2 - 5a - 3$ **18.** $x^2 - 5x + 3; 17$ **19.** $x^2 - x + 13; 33$
20. -36 **21.** $\frac{x^2 - 3x + 8}{-2x - 5}; 12$ **22.** $\left\{x|x \text{ is a real number and } x \neq -\frac{5}{2}\right\}$

Section 2.4

Check Point Exercises

1.

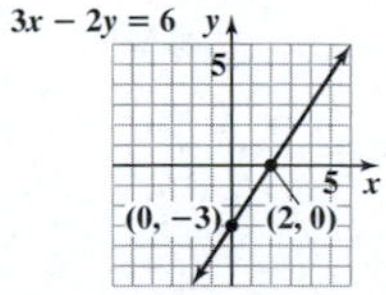

2. a. 6 **b.** $-\frac{7}{5}$ **3.** $m = 2; b = -5$

4.

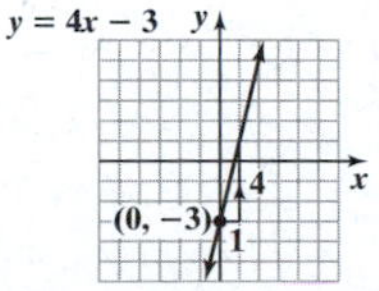

5.

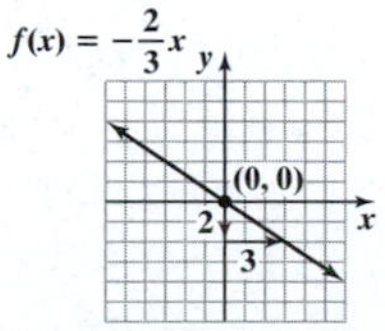

6.

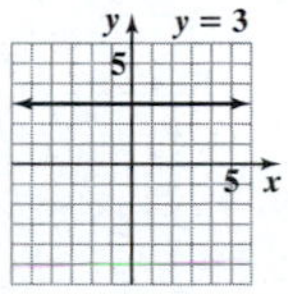

7.

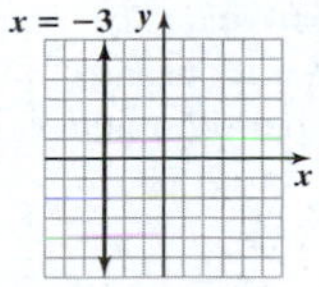

8. 0.57; For the 50–59 age group, the percentage reporting using illegal drugs in the previous month increased by approximately 0.57% each year. **9.** 0.01 mg per 100 mL per hr
10. a. $A(x) = 0.11x + 23.9$ **b.** 28.3 yr old

Exercise Set 2.4

1.

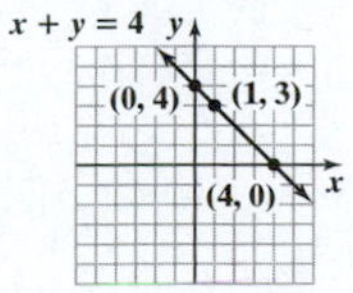

3.

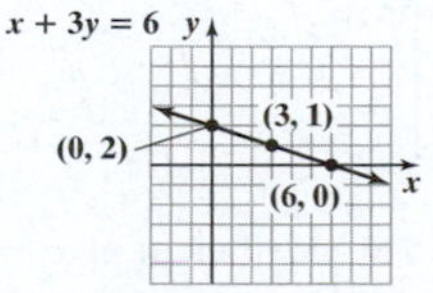

5.

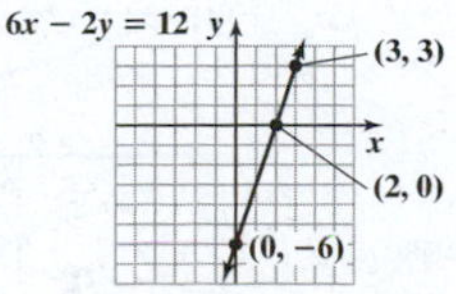

7.

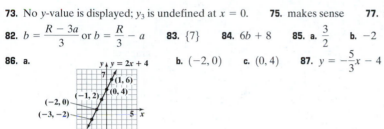

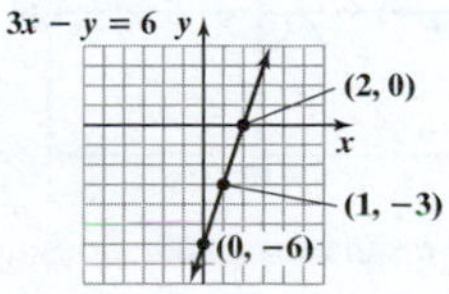

9.
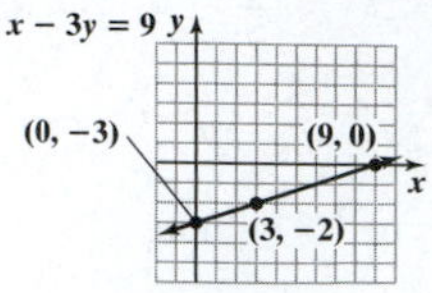

11.
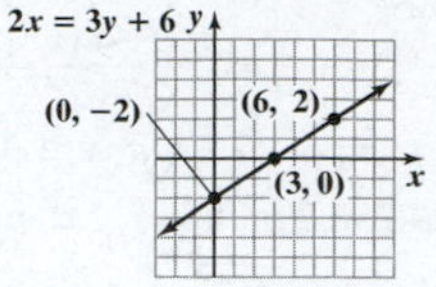

13.
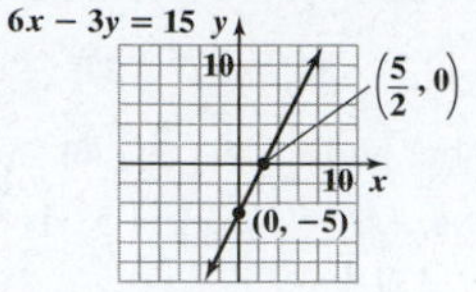

15. $m = 4$; rises **17.** $m = \frac{1}{3}$; rises

19. $m = 0$; horizontal **21.** $m = -\frac{4}{3}$; falls **23.** undefined slope; vertical **25.** L_1: $\frac{2}{3}$; L_2: -2; L_3: $-\frac{1}{2}$

27. $m = 2$; $b = 1$;
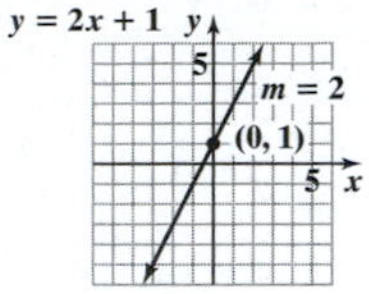

29. $m = -2$; $b = 1$;
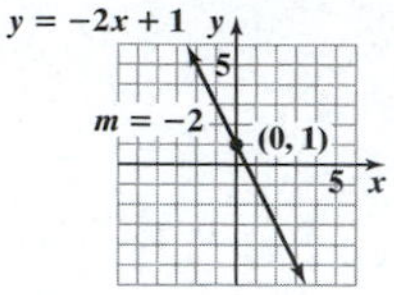

31. $m = \frac{3}{4}$; $b = -2$;
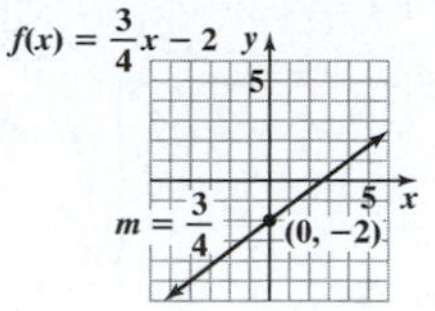

33. $m = -\frac{3}{5}$; $b = 7$;
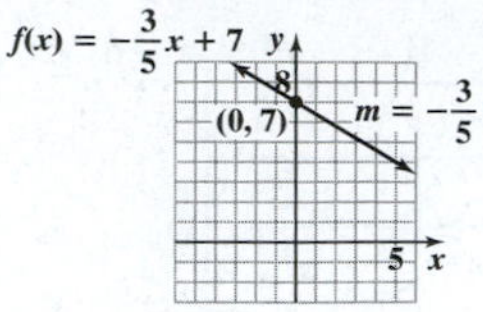

35. $m = -\frac{1}{2}$; $b = 0$;
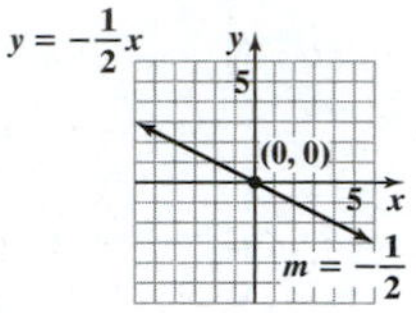

37. $m = 0$; $b = -\frac{1}{2}$;
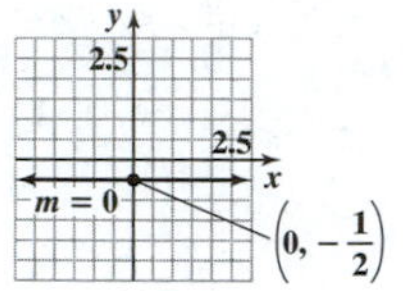

39. **a.** $y = -2x$

b. $m = -2$; $b = 0$

c.
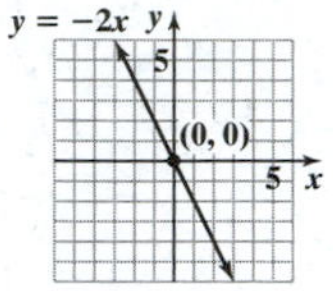

41. **a.** $y = \frac{4}{5}x$

b. $m = \frac{4}{5}$; $b = 0$

c.
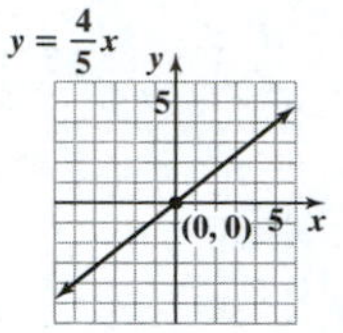

43. **a.** $y = -3x + 2$

b. $m = -3$; $b = 2$

c.
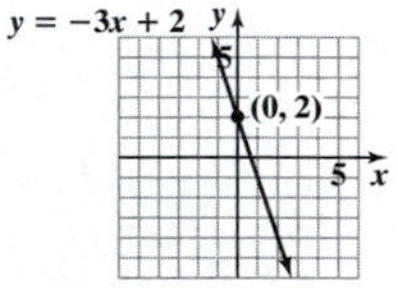

45. **a.** $y = -\frac{5}{3}x + 5$

b. $m = -\frac{5}{3}$; $b = 5$

c.
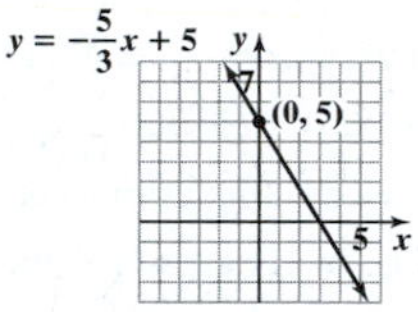

47.
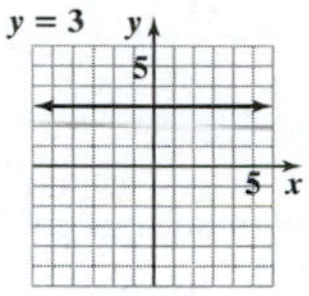

49.
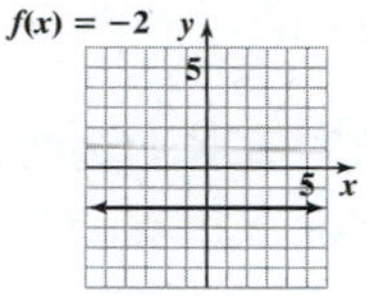

51.
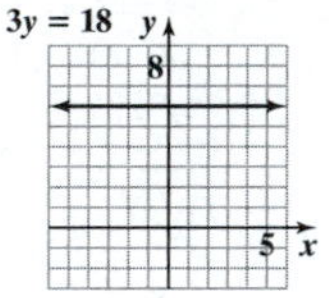

53.
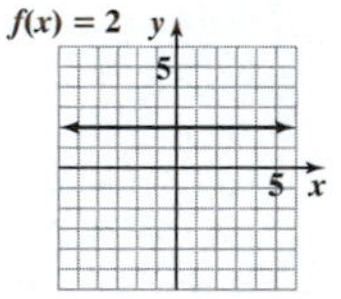

55.
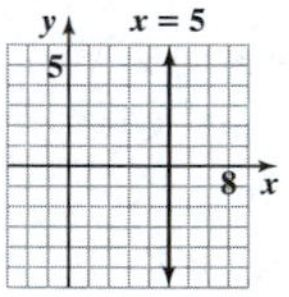

57.
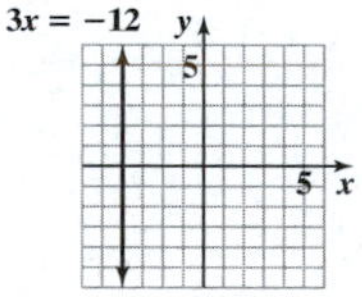

59.
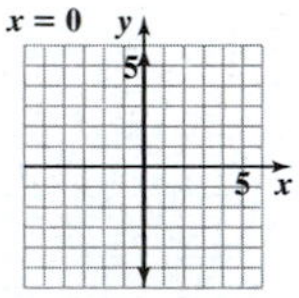

61. $m = -\frac{a}{b}$; falls **63.** undefined slope; vertical **65.** $m = -\frac{A}{B}$; $b = \frac{C}{B}$ **67.** -2

69.
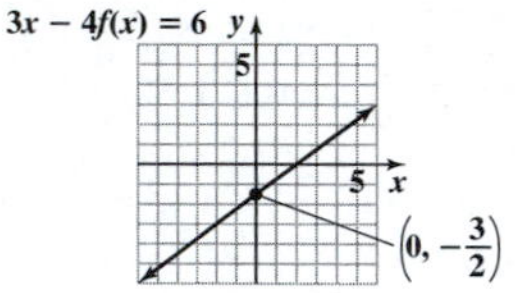

71. 5 **73.** m_1, m_3, m_2, m_4 **75.** $m = 0.01$; The temperature of Earth is increasing by 0.01°F per year.
77. $m = -0.52$; The percentage of U.S. adults who smoke cigarettes is decreasing by 0.52% each year.
79. **a.** 30% **b.** 50% **c.** $m = 4$; average increase of 4% of marriages ending in divorce per year
81. **a.** 254; If no women in a country are literate, the mortality rate of children under 5 is 254 per thousand.
b. -2.4; For each 1% of adult females who are literate, the mortality rate decreases by 2.4 per thousand.
c. $f(x) = -2.4x + 254$ **d.** 134 per thousand **83.** $P(x) = -1.2x + 47$

103. $m = 2$;
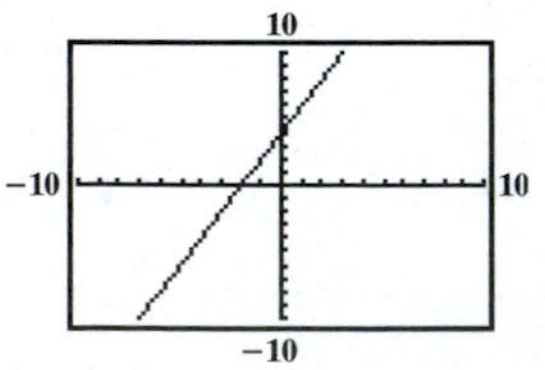

105. $m = -\frac{1}{2}$;
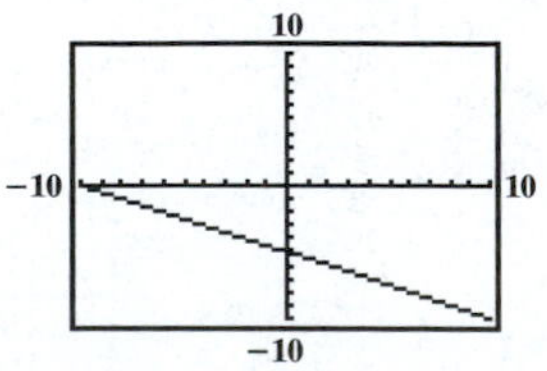

107. does not make sense **109.** does not make sense **111.** false **113.** true **115.** coefficient of x: -6; coefficient of y: 3
117. **a.** $mx_1 + mx_2 + b$ **b.** $mx_1 + mx_2 + 2b$ **c.** no **118.** $16x^4y^6$ **119.** 3.2×10^{-3} **120.** $3x + 17$ **121.** $y = 7x + 33$

122. $y = -\frac{7}{3}x - \frac{2}{3}$ **123. a.** $y = -\frac{1}{4}x + 2; -\frac{1}{4}$ **b.** 4

Section 2.5

Check Point Exercises

1. $y + 3 = -2(x - 4)$; $y = -2x + 5$ or $f(x) = -2x + 5$ **2. a.** $y + 3 = -2(x - 6)$ or $y - 5 = -2(x - 2)$ **b.** $y = -2x + 9$ or $f(x) = -2x + 9$
3. Answers will vary due to rounding.; $f(x) = 0.17x + 72.9$ or $f(x) = 0.17x + 73$; 83.1 yr or 83.2 yr
4. $y - 5 = 3(x + 2)$; $y = 3x + 11$ or $f(x) = 3x + 11$ **5. a.** $m = 3$ **b.** $y + 6 = 3(x + 2)$; $y = 3x$ or $f(x) = 3x$

Exercise Set 2.5

1. $y - 5 = 3(x - 2)$; $f(x) = 3x - 1$ **3.** $y - 6 = 5(x + 2)$; $f(x) = 5x + 16$ **5.** $y + 2 = -4(x + 3)$; $f(x) = -4x - 14$
7. $y - 0 = -5(x + 2)$; $f(x) = -5x - 10$ **9.** $y + \frac{1}{2} = -1(x + 2)$; $f(x) = -x - \frac{5}{2}$ **11.** $y - 0 = \frac{1}{4}(x - 0)$; $f(x) = \frac{1}{4}x$
13. $y + 4 = -\frac{2}{3}(x - 6)$; $f(x) = -\frac{2}{3}x$ **15.** $y - 3 = 1(x - 6)$ or $y - 2 = 1(x - 5)$; $f(x) = x - 3$
17. $y - 0 = 2(x + 2)$ or $y - 4 = 2(x - 0)$; $f(x) = 2x + 4$ **19.** $y - 13 = -2(x + 6)$ or $y - 5 = -2(x + 2)$; $f(x) = -2x + 1$
21. $y - 9 = -\frac{11}{3}(x - 1)$ or $y + 2 = -\frac{11}{3}(x - 4)$; $f(x) = -\frac{11}{3}x + \frac{38}{3}$ **23.** $y + 5 = 0(x + 2)$ or $y + 5 = 0(x - 3)$; $f(x) = -5$
25. $y - 8 = 2(x - 7)$ or $y - 0 = 2(x - 3)$; $f(x) = 2x - 6$ **27.** $y - 0 = \frac{1}{2}(x - 2)$ or $y + 1 = \frac{1}{2}(x - 0)$; $f(x) = \frac{1}{2}x - 1$
29. a. 5 **b.** $-\frac{1}{5}$ **31. a.** -7 **b.** $\frac{1}{7}$ **33. a.** $\frac{1}{2}$ **b.** -2 **35. a.** $-\frac{2}{5}$ **b.** $\frac{5}{2}$ **37. a.** -4 **b.** $\frac{1}{4}$ **39. a.** $-\frac{1}{2}$ **b.** 2
41. a. $\frac{2}{3}$ **b.** $-\frac{3}{2}$ **43. a.** undefined **b.** 0 **45.** $y - 2 = 2(x - 4)$; $y = 2x - 6$ or $f(x) = 2x - 6$
47. $y - 4 = -\frac{1}{2}(x - 2)$; $y = -\frac{1}{2}x + 5$ or $f(x) = -\frac{1}{2}x + 5$ **49.** $y + 10 = -4(x + 8)$; $y = -4x - 42$ or $f(x) = -4x - 42$
51. $y + 3 = -5(x - 2)$; $y = -5x + 7$ or $f(x) = -5x + 7$ **53.** $y - 2 = \frac{2}{3}(x + 2)$; $y = \frac{2}{3}x + \frac{10}{3}$ or $f(x) = \frac{2}{3}x + \frac{10}{3}$
55. $y + 7 = -2(x - 4)$; $y = -2x + 1$ or $f(x) = -2x + 1$ **57.** $f(x) = 5$ **59.** $f(x) = -\frac{1}{2}x + 1$ **61.** $f(x) = -\frac{2}{3}x - 2$
63. $f(x) = 4x - 5$ **65.** $-\frac{A}{B}$ **67. a.** $y - 31.1 = 0.78(x - 10)$ or $y - 38.9 = 0.78(x - 20)$ **b.** $f(x) = 0.78x + 23.3$ **c.** 54.5%
69. a. & b.

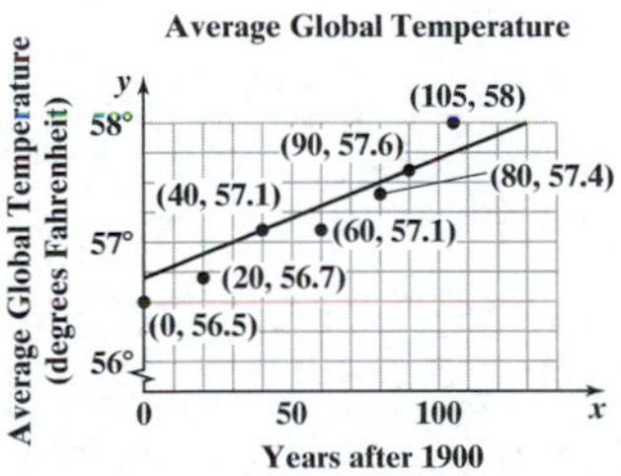

b. $y - 57.1 = 0.01(x - 40)$ or $y - 57.6 = 0.01(x - 90)$; $f(x) = 0.01x + 56.7$ **c.** 58.2°F

71. a. $m \approx 43.1$; The cost of Social Security is projected to increase at a rate of approximately \$43.1 billion per year. **b.** $m \approx 51.4$; The cost of Medicare is projected to increase at a rate of approximately \$51.4 billion per year. **c.** no; The cost of Medicare is projected to increase at a faster rate than the cost of Social Security.

81. a.

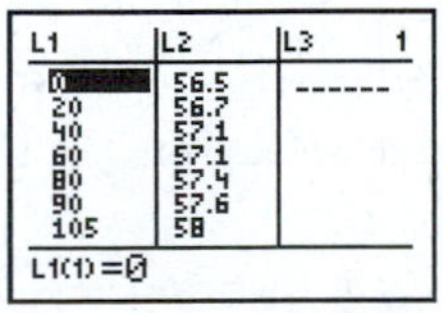

b. & d.

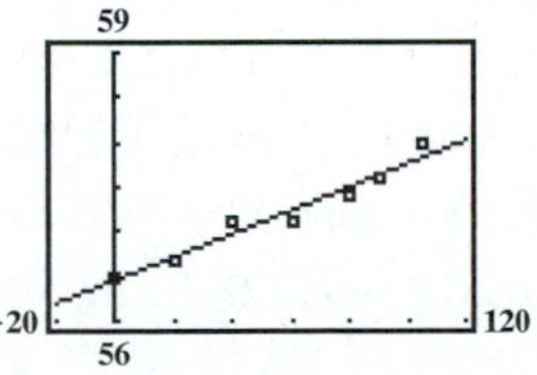

c.

83. makes sense **85.** makes sense **87.** true **89.** true **91.** -4 **93.** $(-40, 74)$ and $(97, -200)$ **95.** 33 **96.** -56
97. 40°, 60°, and 80° **98. a.** yes **b.** yes **99.** $(3, -4)$;

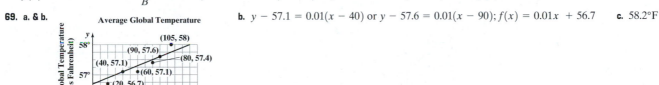

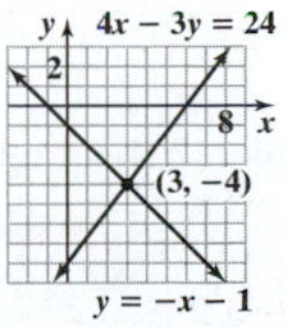

100. $\{1\}$

Review Exercises

1. function; domain: $\{3, 4, 5\}$; range: $\{10\}$ **2.** function; domain: $\{1, 2, 3, 4\}$; range: $\{12, 100, \pi, -6\}$ **3.** not a function; domain: $\{13, 15\}$; range: $\{14, 16, 17\}$ **4. a.** -5 **b.** 16 **c.** -75 **d.** $14a - 5$ **e.** $7a + 9$ **5. a.** 2 **b.** 52 **c.** 70 **d.** $3b^2 - 5b + 2$ **e.** $48a^2 - 20a + 2$

6.

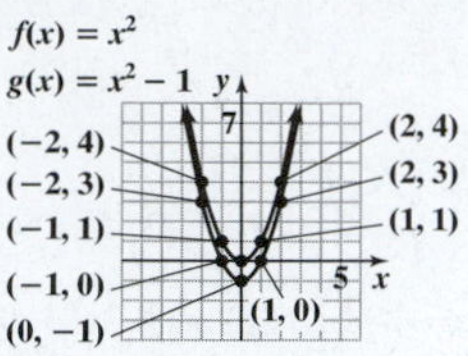

The graph of g is the graph of f shifted down by 1 unit.

7.

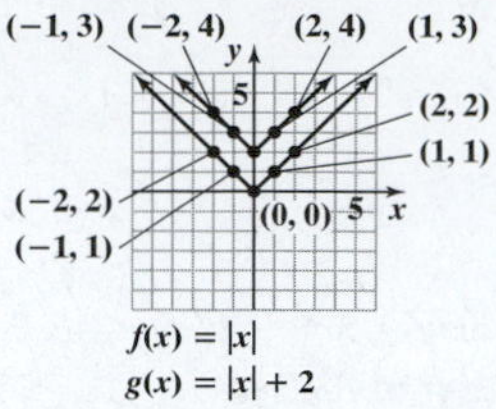

The graph of g is the graph of f shifted up by 2 units.

8. not a function **9.** function **10.** function **11.** not a function **12.** not a function **13.** function **14.** -3 **15.** -2

16. 3 **17.** $\{x|-3 \le x < 5\}$ **18.** $\{y|-5 \le y \le 0\}$

19. a. For each time, there is only one height.
b. 0; The eagle was on the ground after 15 seconds.
c. 45 m
d. 7 and 22; After 7 seconds and after 22 seconds, the eagle's height is 20 meters.
e. Answers will vary.

20. $\{x|x$ is a real number$\}$ **21.** $\{x|x$ is a real number and $x \ne -8\}$ **22.** $\{x|x$ is a real number and $x \ne 5\}$ **23. a.** $6x - 4$ **b.** 14

24. a. $5x^2 + 1$ **b.** 46 **25.** $\{x|x$ is a real number and $x \ne 4\}$ **26.** $\{x|x$ is a real number and $x \ne -6$ and $x \ne -1\}$ **27.** $x^2 - x - 5$; 1

28. 1 **29.** $x^2 - 3x + 5$; 3 **30.** 9 **31.** -120 **32.** $\frac{x^2 - 2x}{x - 5}$; -8 **33.** $\{x|x$ is a real number$\}$ **34.** $\{x|x$ is a real number and $x \ne 5\}$

35.

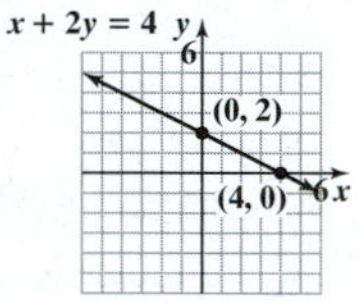

36.

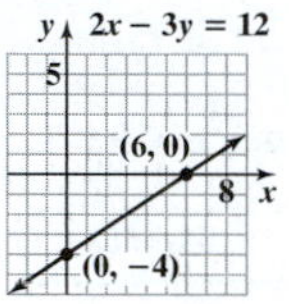

37.

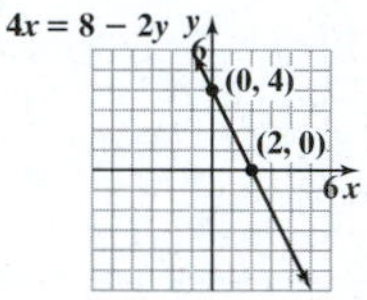

38. 2; rises **39.** $-\frac{2}{3}$; falls

40. undefined; vertical

41. 0; horizontal

42. $m = 2$; $b = -1$;

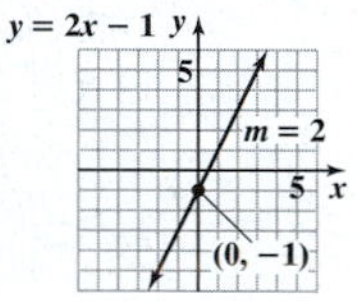

43. $m = -\frac{1}{2}$; $b = 4$;

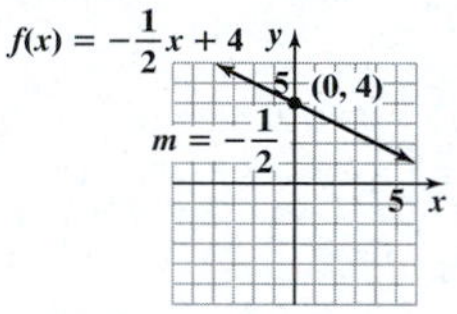

44. $m = \frac{2}{3}$; $b = 0$;

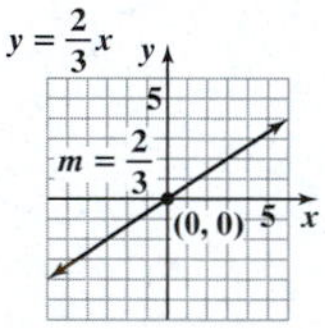

45. $y = -2x + 4$; $m = -2$; $b = 4$ **46.** $y = -\frac{5}{3}x$; $m = -\frac{5}{3}$; $b = 0$ **47.** $y = -\frac{5}{3}x + 2$; $m = -\frac{5}{3}$; $b = 2$

48.

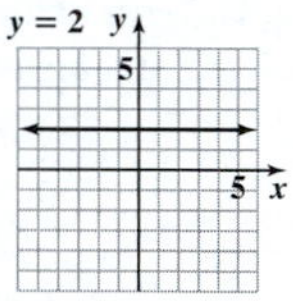

49.

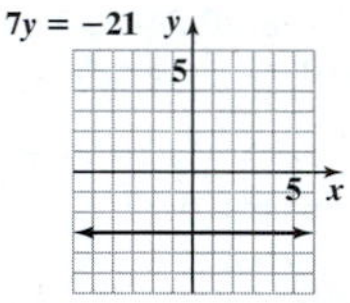

50.

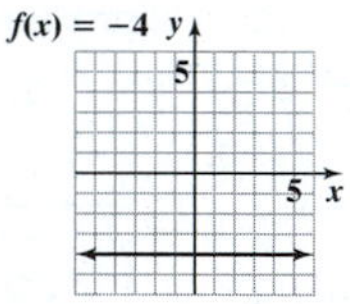

51.

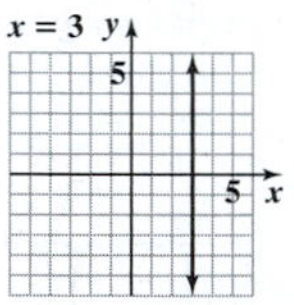

52.

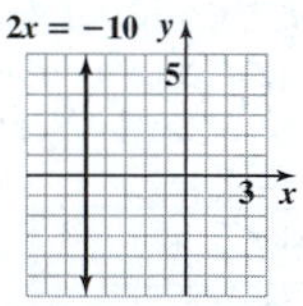

53. -0.27; Record time has been decreasing at a rate of 0.27 second per year since 1900. **54. a.** 137; There was an average increase of approximately 137 discharges per year. **b.** -202; There was an average decrease of approximately 202 discharges per year.

55. a. $F = \frac{9}{5}C + 32$ **b.** 86° **56.** $y - 2 = -6(x + 3)$; $y = -6x - 16$ or $f(x) = -6x - 16$ **57.** $y - 6 = 2(x - 1)$ or $y - 2 = 2(x + 1)$; $y = 2x + 4$ or $f(x) = 2x + 4$ **58.** $y + 7 = -3(x - 4)$; $y = -3x + 5$ or $f(x) = -3x + 5$ **59.** $y - 6 = -3(x + 2)$; $y = -3x$ or $f(x) = -3x$

60. a. $y - 34.6 = 0.8(x - 2)$ or $y - 37.0 = 0.8(x - 5)$ **b.** $f(x) = 0.8x + 33$ **c.** 41 million

Chapter Test

1. function; domain: $\{1, 3, 5, 6\}$; range: $\{2, 4, 6\}$ **2.** not a function; domain: $\{2, 4, 6\}$; range: $\{1, 3, 5, 6\}$ **3.** $3a + 10$ **4.** 28

5.

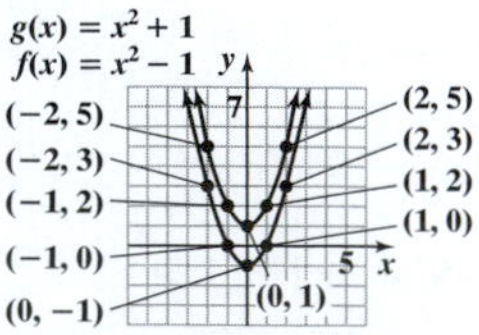

The graph of g is the graph of f shifted up by 2 units.

6. function **7.** not a function **8.** -3 **9.** -2 and 3 **10.** $\{x|x \text{ is a real number}\}$ **11.** $\{y|y \leq 3\}$ **12.** $\{x|x \text{ is a real number and } x \neq 10\}$
13. $x^2 + 5x + 2; 26$ **14.** $x^2 + 3x - 2; -4$ **15.** -15 **16.** $\dfrac{x^2 + 4x}{x + 2}; 3$ **17.** $\{x|x \text{ is a real number and } x \neq -2\}$

18.
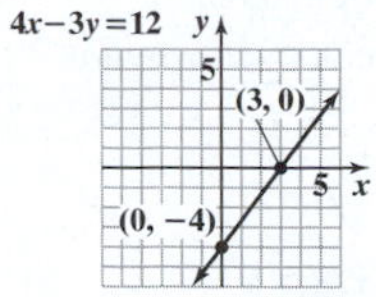

19.
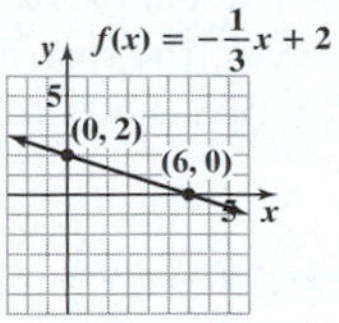

20.
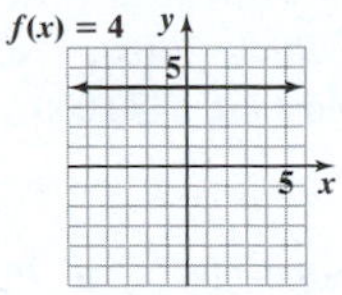

21. $-\frac{1}{2}$; falls **22.** undefined; vertical **23.** 176; In 2005, the number of Super Bowl viewers was 176 million. **24.** 3.6; The number of Super Bowl viewers is increasing at a rate of 3.6 million per year. **25.** $y + 3 = 1(x + 1)$ or $y - 2 = 1(x - 4)$; $y = x - 2$ or $f(x) = x - 2$
26. $y - 3 = 2(x + 2)$; $y = 2x + 7$ or $f(x) = 2x + 7$ **27.** $y + 4 = -\frac{1}{2}(x - 6)$; $y = -\frac{1}{2}x - 1$ or $f(x) = -\frac{1}{2}x - 1$
28. a. $y - 476 = 5(x - 2)$ or $y - 486 = 5(x - 4)$ **b.** $f(x) = 5x + 466$ **c.** 516 per 100,000 residents

Cumulative Review Exercises

1. $\{0, 1, 2, 3\}$ **2.** false **3.** 7 **4.** 15 **5.** $7 + 3x$ or $3x + 7$ **6.** $\{-4\}$ **7.** $\{x|x \text{ is a real number}\}$; identity **8.** $\{-6\}$ **9.** \$2250
10. $t = \dfrac{A - p}{pr}$ **11.** $\dfrac{y^{10}}{9x^8}$ **12.** $\dfrac{9x^{10}}{y^{12}}$ **13.** 2.1×10^{-5} **14.** function; domain: $\{1, 2, 3, 4, 6\}$; range: $\{5\}$

15.
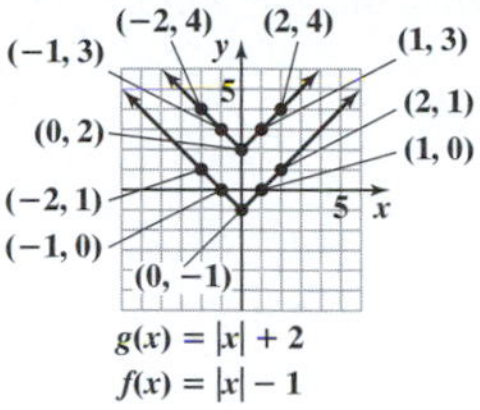

The graph of g is the graph of f shifted up by 3 units.

16. $\{x|x \text{ is a real number and } x \neq 15\}$ **17.** $2x^2 + x + 5; 6$

18.
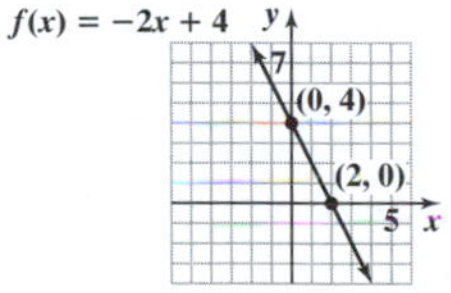

19.
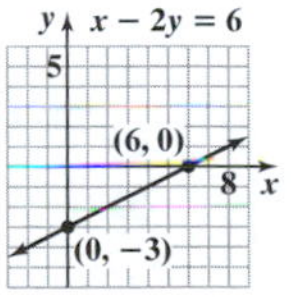

20. $y + 5 = 4(x - 3)$; $y = 4x - 17$ or $f(x) = 4x - 17$

CHAPTER 3

Section 3.1

Check Point Exercises

1. a. not a solution **b.** solution **2.** $(1, 4)$ or $\{(1, 4)\}$ **3.** $(6, 11)$ or $\{(6, 11)\}$ **4.** $(-2, 5)$ or $\{(-2, 5)\}$ **5.** $\left(-\frac{1}{2}, 2\right)$ or $\left\{\left(-\frac{1}{2}, 2\right)\right\}$
6. $(2, -1)$ or $\{(2, -1)\}$ **7.** $\left(\frac{37}{7}, \frac{19}{7}\right)$ or $\left\{\left(\frac{37}{7}, \frac{19}{7}\right)\right\}$ **8.** no solution or $\varnothing$ **9.** $\{(x, y)|x = 4y - 8\}$ or $\{(x, y)|5x - 20y = -40\}$

Exercise Set 3.1

1. solution **3.** not a solution **5.** solution **7.** $\{(3, 1)\}$ **9.** $\left\{\left(\frac{1}{2}, 3\right)\right\}$ **11.** $\{(4, 3)\}$ **13.** $\{(x, y)|2x + 3y = 6\}$ or $\{(x, y)|4x = -6y + 12\}$
15. $\{(1, 0)\}$ **17.** $\varnothing$ **19.** $\{(1, 2)\}$ **21.** $\{(3, 1)\}$ **23.** $\varnothing$ **25.** $\{(2, 4)\}$ **27.** $\{(3, 1)\}$ **29.** $\{(2, 1)\}$ **31.** $\{(2, -3)\}$ **33.** $\varnothing$
35. $\{(3, -2)\}$ **37.** $\varnothing$ **39.** $\{(-5, -1)\}$ **41.** $\left\{(x, y)\middle|y = \frac{2}{5}x - 2\right\}$ or $\{(x, y)|2x - 5y = 10\}$ **43.** $\{(5, 2)\}$ **45.** $\{(2, -3)\}$
47. $\{(-1, 1)\}$ **49.** $\{(-2, -7)\}$ **51.** $\{(7, 2)\}$ **53.** $\{(4, -1)\}$ **55.** $\{(x, y)|2x + 6y = 8\}$ or $\{(x, y)|3x + 9y = 12\}$ **57.** $\left\{\left(\frac{29}{22}, -\frac{5}{11}\right)\right\}$
59. $\{(-2, -1)\}$ **61.** $\{(1, -3)\}$ **63.** $\{(1, -3)\}$ **65.** $\{(4, 3)\}$ **67.** $\{(x, y)|x = 3y - 1\}$ or $\{(x, y)|2x - 6y = -2\}$ **69.** $\varnothing$ **71.** $\{(5, 1)\}$
73. $\left\{\left(\frac{32}{7}, -\frac{20}{7}\right)\right\}$ **75.** $\{(-5, 7)\}$ **77.** $\varnothing$ **79.** $\{(x, y)|x + 2y - 3 = 0\}$ or $\{(x, y)|12 = 8y + 4x\}$ **81.** $\{(0, 0)\}$ **83.** $\{(6, -1)\}$
85. $\left\{\left(\frac{1}{a}, 3\right)\right\}$ **87.** $m = -4, b = 3$ **89.** $y = x - 4$; $y = -\frac{1}{3}x + 4$ **91. a.** approximately (1975, 18); 1975; 18% **b.** 1975; 18%
c. quite well or extremely well **93. a.** $y = 0.04x + 5.48$ **b.** $y = 0.17x + 1.84$ **c.** 2028; 6.6%; Medicare **95. a.** $y = -0.54x + 38$
b. $y = -0.79x + 40$ **c.** 1993; 33.68% **97. a.** 150 sold; 300 supplied **b.** \$100; 250 **109.** makes sense **111.** makes sense
113. false **115.** false **117.** $a = 3, b = 2$ **119.** $\left\{\left(\dfrac{b_2c_1 - b_1c_2}{a_1b_2 - a_2b_1}, \dfrac{a_1c_2 - a_2c_1}{a_1b_2 - a_2b_1}\right)\right\}$ **120.** $\left\{\frac{10}{9}\right\}$ **121.** $-128x^{19}y^8$ **122.** 11
123. $0.15x + 0.07y$ **124.** 15 mL **125.** $80x$

Section 3.2

Check Point Exercises

1. hamburger and fries: 1240; fettuccine Alfredo: 1500 **2.** \$3150 at 9%; \$1850 at 11% **3.** 12% solution: 100 oz; 20% solution: 60 oz **4.** boat: 35 mph; current: 7 mph **5. a.** $C(x) = 300{,}000 + 30x$ **b.** $R(x) = 80x$ **c.** (6000, 480,000); The company will break even when it produces and sells 6000 pairs of shoes. At this level, both revenue and cost are \$480,000. **6.** $P(x) = 50x - 300{,}000$

Exercise Set 3.2

1. 3 and 4 **3.** first number: 2; second number: 5 **5. a.** 1500 units; \$48,000 **b.** $P(x) = 17x - 25{,}500$ **7. a.** 500 units; \$122,500 **b.** $P(x) = 140x - 70{,}000$ **9.** after completing college: 41%; after completing high school: 7% **11.** 22 computers and 14 hard drives **13.** \$2000 at 6% and \$5000 at 8% **15.** first fund: \$8000; second fund: \$6000 **17.** \$17,000 at 12%; \$3000 at a 5% loss **19.** California: 100 gal; French: 100 gal **21.** 18-karat gold: 96 g; 12-karat gold: 204 g **23.** cheaper candy: 30 lb; more expensive candy: 45 lb **25.** 8 nickels and 7 dimes **27.** plane: 130 mph; wind: 30 mph **29.** crew: 6 km/hr; current: 2 km/hr **31.** in still water: 4.5 mph; current: 1.5 mph **33.** 86 and 74 **35.** 80°, 50°, 50° **37.** 70 ft by 40 ft **39.** two-seat tables: 6; four-seat tables: 11 **41.** 500 radios **43.** −6000; When the company produces and sells 200 radios, the loss is \$6000. **45. a.** $P(x) = 20x - 10{,}000$ **b.** \$190,000 **47. a.** $C(x) = 18{,}000 + 20x$ **b.** $R(x) = 80x$ **c.** (300, 24,000); When 300 canoes are produced and sold, both revenue and cost are \$24,000. **49. a.** $C(x) = 30{,}000 + 2500x$ **b.** $R(x) = 3125x$ **c.** (48, 150,000); For 48 sold-out performances, both cost and revenue are \$150,000.

59. (6, 300);

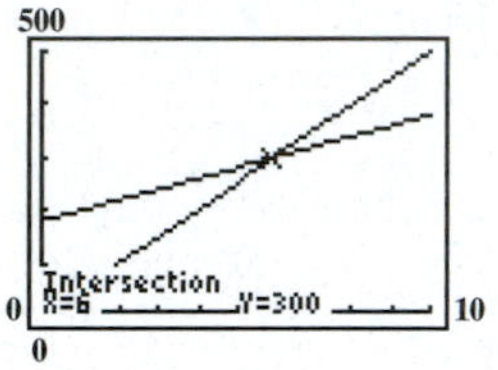

63. does not make sense **65.** makes sense **67.** yes, 8 hexagons and 4 squares **69.** 95 **71.** $y - 5 = -2(x + 2)$ or $y - 13 = -2(x + 6)$; $y = -2x + 1$ or $f(x) = -2x + 1$ **72.** $y - 0 = 1(x + 3)$; $y = x + 3$ or $f(x) = x + 3$ **73.** $\{x | x$ is a real number and $x \neq 3\}$ **74.** yes **75.** $11x + 4y = -3$ **76.** $16a + 4b + c = 1682$

Section 3.3

Check Point Exercises

1. $(-1) - 2(-4) + 3(5) = 22$; $2(-1) - 3(-4) - 5 = 5$; $3(-1) + (-4) - 5(5) = -32$ **2.** $(1, 4, -3)$ or $\{(1, 4, -3)\}$ **3.** $(4, 5, 3)$ or $\{(4, 5, 3)\}$ **4.** $y = 3x^2 - 12x + 13$ or $f(x) = 3x^2 - 12x + 13$

Exercise Set 3.3

1. not a solution **3.** solution **5.** $\{(2, 3, 3)\}$ **7.** $\{(2, -1, 1)\}$ **9.** $\{(1, 2, 3)\}$ **11.** $\{(3, 1, 5)\}$ **13.** $\{(1, 0, -3)\}$ **15.** $\{(1, -5, -6)\}$ **17.** no solution or $\varnothing$ **19.** infinitely many solutions; dependent equations **21.** $\left\{\left(\frac{1}{2}, \frac{1}{3}, -1\right)\right\}$ **23.** $y = 2x^2 - x + 3$ **25.** $y = 2x^2 + x - 5$ **27.** 7, 4, and 5 **29.** $\{(4, 8, 6)\}$ **31.** $y = -\frac{3}{4}x^2 + 6x - 11$ **33.** $\left\{\left(\frac{8}{a}, -\frac{3}{b}, -\frac{5}{c}\right)\right\}$ **35. a.** (0, 1.8), (5, 0.8), (13, 1.9) **b.** $0a + 0b + c = 1.8$; $25a + 5b + c = 0.8$; $169a + 13b + c = 1.9$ **37. a.** $y = -16x^2 + 40x + 200$ **b.** 0; After 5 seconds, the ball hits the ground. **39.** rent: \$9224; cars: \$3020; books: \$447 **41.** \$1200 at 8%; \$2000 at 10%; \$3500 at 12% **43.** 200 \$8 tickets; 150 \$10 tickets; 50 \$12 tickets **45.** 4 oz of food A; 0.5 oz of food B; 1 oz of food C **55.** does not make sense **57.** makes sense **59.** false **61.** true **63.** 60°, 55°, 65° **65.** 30 cm

66.

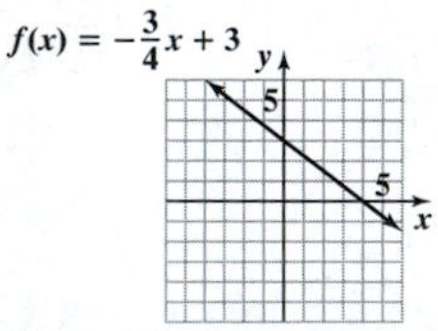

67.

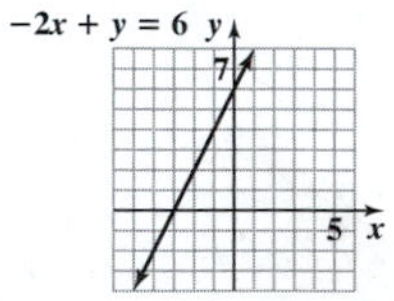

68.

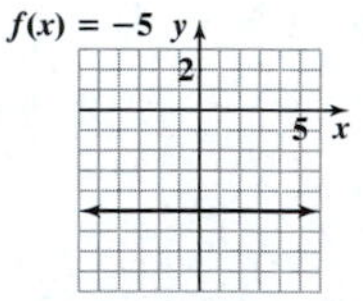

69. $\{(-3, 1)\}$; The value for y is given and the value for x can be found by back-substitution. **70.** $\{(6, 3, 5)\}$; The value for z is given and the values of the other variables can be found by back-substitution. **71.** $\left[\begin{array}{cc|c} 1 & 2 & -1 \\ 0 & -11 & -11 \end{array}\right]$

Mid-Chapter Check Point Exercises

1. $\{(-1, 2)\}$ **2.** $\{(1, -2)\}$ **3.** $\{(6, 10)\}$ **4.** $\{(x, y) | y = 4x - 5\}$ or $\{(x, y) | 8x - 2y = 10\}$ **5.** $\left\{\left(\frac{11}{19}, \frac{7}{19}\right)\right\}$ **6.** $\varnothing$ **7.** $\{(-1, 2, -2)\}$ **8.** $\{(4, -2, 3)\}$ **9.** $\{(3, 2)\}$ **10.** $\{(-2, 1)\}$ **11. a.** $C(x) = 400{,}000 + 20x$ **b.** $R(x) = 100x$ **c.** $P(x) = 80x - 400{,}000$ **d.** (5000, 500,000); The company will break even when it produces and sells 5000 PDAs. At this level, both revenue and cost are \$500,000. **12.** 6 roses and 14 carnations **13.** \$6300 at 5% and \$8700 at 6% **14.** 13% nitrogen: 20 gal; 18% nitrogen: 30 gal **15.** rowing rate in still water: 3 mph; current: 1.5 mph **16.** \$4500 at 2% and \$3500 at 5% **17.** $y = -x^2 + 2x + 3$ **18.** 8 nickels, 6 dimes, 12 quarters

Section 3.4

Check Point Exercises

1. a. $\left[\begin{array}{rrr|r} 1 & 6 & -3 & 7 \\ 4 & 12 & -20 & 8 \\ -3 & -2 & 1 & -9 \end{array}\right]$ **b.** $\left[\begin{array}{rrr|r} 1 & 3 & -5 & 2 \\ 1 & 6 & -3 & 7 \\ -3 & -2 & 1 & -9 \end{array}\right]$ **c.** $\left[\begin{array}{rrr|r} 4 & 12 & -20 & 8 \\ 1 & 6 & -3 & 7 \\ 0 & 16 & -8 & 12 \end{array}\right]$ **2.** $(-1, 2)$ or $\{(-1, 2)\}$ **3.** $(5, 2, 3)$ or $\{(5, 2, 3)\}$

Exercise Set 3.4

1. $\left[\begin{array}{rr|r} 1 & -\frac{3}{2} & 5 \\ 2 & 2 & 5 \end{array}\right]$ **3.** $\left[\begin{array}{rr|r} 1 & -\frac{4}{3} & 2 \\ 3 & 5 & -2 \end{array}\right]$ **5.** $\left[\begin{array}{rr|r} 1 & -3 & 5 \\ 0 & 12 & -6 \end{array}\right]$ **7.** $\left[\begin{array}{rr|r} 1 & -\frac{3}{2} & \frac{7}{2} \\ 0 & \frac{17}{2} & -\frac{17}{2} \end{array}\right]$ **9.** $\left[\begin{array}{rrr|r} 1 & -3 & 2 & 5 \\ 1 & 5 & -5 & 0 \\ 3 & 0 & 4 & 7 \end{array}\right]$

11. $\left[\begin{array}{rrr|r} 1 & -3 & 2 & 0 \\ 0 & 10 & -7 & 7 \\ 2 & -2 & 1 & 3 \end{array}\right]$ **13.** $\left[\begin{array}{rrr|r} 1 & 1 & -1 & 6 \\ 0 & -3 & 3 & -15 \\ 0 & -4 & 2 & -14 \end{array}\right]$ **15.** $\{(4, 2)\}$ **17.** $\{(3, -3)\}$ **19.** $\{(2, -5)\}$ **21.** no solution or $\varnothing$

23. infinitely many solutions; dependent equations **25.** $\{(1, -1, 2)\}$ **27.** $\{(3, -1, -1)\}$ **29.** $\{(1, 2, -1)\}$ **31.** $\{(1, 2, 3)\}$

33. no solution or $\varnothing$ **35.** infinitely many solutions; dependent equations **37.** $\{(-1, 2, -2)\}$

39. $w - x + y + z = 3$; $x - 2y - z = 0$; $y + 6z = 17$; $z = 3$; $\{(2, 1, -1, 3)\}$ **41.** $\left[\begin{array}{rrrr|r} 1 & -1 & 1 & 1 & 3 \\ 0 & 1 & -2 & -1 & 0 \\ 0 & 2 & 1 & 2 & 5 \\ 0 & 6 & -3 & -1 & -9 \end{array}\right]$ **43.** $\{(1, 2, 3, -2)\}$

45. a. $s(t) = -16t^2 + 56t$ **b.** 0; The ball hits the ground 3.5 seconds after it is thrown.; $(3.5, 0)$ **47.** yes: 34%; no: 61%; not sure: 5%

59. makes sense **61.** does not make sense **63.** false **65.** false **67.** $-6a + 13$ **68.** 15 **69.** $-\dfrac{x^{11}}{3y^{36}}$ **70.** 2 **71.** 6 **72.** -31

Section 3.5

Check Point Exercises

1. a. -4 **b.** -17 **2.** $(4, -2)$ or $\{(4, -2)\}$ **3.** 80 **4.** $(2, -3, 4)$ or $\{(2, -3, 4)\}$

Exercise Set 3.5

1. 1 **3.** -29 **5.** 0 **7.** 33 **9.** $-\dfrac{7}{16}$ **11.** $\{(5, 2)\}$ **13.** $\{(2, -3)\}$ **15.** $\{(3, -1)\}$ **17.** $\{(4, 0)\}$ **19.** $\{(4, 2)\}$ **21.** $\{(7, 4)\}$

23. inconsistent; no solution or $\varnothing$ **25.** dependent equations; infinitely many solutions **27.** 72 **29.** -75 **31.** 0 **33.** $\{(-5, -2, 7)\}$

35. $\{(2, -3, 4)\}$ **37.** $\{(3, -1, 2)\}$ **39.** $\{(2, 3, 1)\}$ **41.** -42 **43.** $2x - 4y = 8$; $3x + 5y = -10$ **45.** $\{-11\}$ **47.** $\{4\}$ **49.** 28 sq units

51. yes **53.** $\begin{vmatrix} x & y & 1 \\ 3 & -5 & 1 \\ -2 & 6 & 1 \end{vmatrix} = 0$; $y = -\dfrac{11}{5}x + \dfrac{8}{5}$ **65.** does not make sense **67.** does not make sense **69.** true **71.** false

73. The value is multiplied by -1. **76.** all real numbers or $\{x \mid x \text{ is a real number}\}$ **77.** $y = \dfrac{2x + 7}{3}$ **78.** $\{0\}$ **79.** $\{14\}$

80. $\{-3\}$ **81.** $\{x \mid x \text{ is a real number}\}$

Review Exercises

1. not a solution **2.** solution **3.** $\{(2, 3)\}$ **4.** $\{(x, y) \mid 3x - 2y = 6\}$ or $\{(x, y) \mid 6x - 4y = 12\}$ **5.** $\{(-5, -6)\}$ **6.** $\varnothing$ **7.** $\{(3, 4)\}$

8. $\{(23, -43)\}$ **9.** $\{(-4, 2)\}$ **10.** $\left\{\left(3, \frac{1}{2}\right)\right\}$ **11.** $\{(x, y) \mid y = 4 - x\}$ or $\{(x, y) \mid 3x + 3y = 12\}$ **12.** $\left\{\left(3, \frac{8}{3}\right)\right\}$ **13.** $\varnothing$

14. TV: \$350; stereo: \$370 **15.** \$2500 at 4%; \$6500 at 7% **16.** 10 ml of 34%; 90 ml of 4% **17.** plane: 630 mph; wind: 90 mph

18. 12 ft by 5 ft **19.** loss of \$4500 **20.** $(500, 42{,}500)$; When 500 calculators are produced and sold, both cost and revenue are \$42,500.

21. $P(x) = 45x - 22{,}500$ **22. a.** $C(x) = 60{,}000 + 200x$ **b.** $R(x) = 450x$ **c.** $(240, 108{,}000)$; When 240 desks are produced and sold, both cost and revenue are \$108,000. **23.** no **24.** $\{(0, 1, 2)\}$ **25.** $\{(2, 1, -1)\}$ **26.** infinitely many solutions; dependent equations

27. $y = 3x^2 - 4x + 5$ **28.** 18–29: \$8300; 30–39: \$16,400; 40–49: \$19,500 **29.** $\left[\begin{array}{rr|r} 1 & -8 & 3 \\ 0 & 1 & -2 \end{array}\right]$ **30.** $\left[\begin{array}{rr|r} 1 & -3 & 1 \\ 0 & 7 & -7 \end{array}\right]$ **31.** $\left[\begin{array}{rrr|r} 1 & -1 & \frac{1}{2} & -\frac{1}{2} \\ 1 & 2 & -1 & 2 \\ 6 & 4 & 3 & 5 \end{array}\right]$

32. $\left[\begin{array}{rrr|r} 1 & 2 & 2 & 2 \\ 0 & 1 & -1 & 2 \\ 0 & 0 & 9 & -9 \end{array}\right]$ **33.** $\{(-5, 3)\}$ **34.** no solution or $\varnothing$ **35.** $\{(1, 3, -4)\}$ **36.** $\{(-2, -1, 0)\}$ **37.** 17 **38.** 4

39. -86 **40.** -236 **41.** $\left\{\left(\frac{7}{4}, -\frac{25}{8}\right)\right\}$ **42.** $\{(2, -7)\}$ **43.** $\{(23, -12, 3)\}$ **44.** $\{(-3, 2, 1)\}$ **45.** $y = \frac{5}{8}x^2 - 50x + 1150$; 30-year-old drivers are involved in 212.5 accidents daily and 50-year-old drivers are involved in 212.5 accidents daily.

Chapter Test

1. $\{(2, 4)\}$ **2.** $\{(6, -5)\}$ **3.** $\{(1, -3)\}$ **4.** $\{(x, y)|4x = 2y + 6\}$ or $\{(x, y)|y = 2x - 3\}$ **5.** one-bedroom: 15 units; two-bedroom: 35 units **6.** \$2000 at 6% and \$7000 at 7% **7.** 6% solution: 12 oz; 9% solution: 24 oz **8.** boat: 14 mph; current: 2 mph **9.** $C(x) = 360{,}000 + 850x$ **10.** $R(x) = 1150x$ **11.** (1200, 1,380,000); When 1200 computers are produced and sold, both cost and revenue are \$1,380,000. **12.** $P(x) = 85x - 350{,}000$ **13.** $\{(1, 3, 2)\}$ **14.** $\left[\begin{array}{ccc|c} 1 & 0 & -4 & 5 \\ 0 & -1 & 26 & -20 \\ 2 & -1 & 4 & -3 \end{array}\right]$ **15.** $\{(4, -2)\}$ **16.** $\{(-1, 2, 2)\}$ **17.** 17 **18.** -10 **19.** $\{(-1, -6)\}$ **20.** $\{(4, -3, 3)\}$

Cumulative Review Exercises

1. 1 **2.** $15x - 7$ **3.** $\{-6\}$ **4.** $\{-15\}$ **5.** no solution or $\varnothing$ **6.** \$2000 **7.** $-\dfrac{x^8}{4y^{30}}$ **8.** $-4a - 3$ **9.** $\{x|x$ is a real number and $x \neq -3\}$ **10.** $x^2 - 3x - 1; -1$

11.
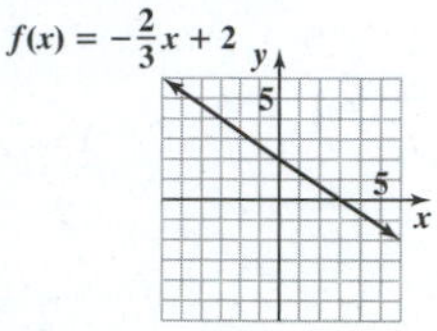

12.
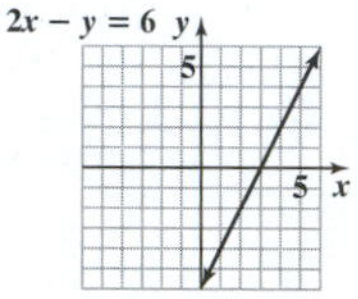

13. $y - 4 = -3(x - 2)$ or $y + 2 = -3(x - 4)$; $y = -3x + 10$ or $f(x) = -3x + 10$ **14.** $y - 0 = -3(x + 1)$; $y = -3x - 3$ or $f(x) = -3x - 3$ **15.** $\left\{\left(7, \frac{1}{3}\right)\right\}$ **16.** $\{(3, 2, 4)\}$ **17.** pad: \$0.80; pen: \$0.20 **18.** 23 **19.** $\{(3, -2, 1)\}$ **20.** $\{(-3, 2)\}$

CHAPTER 4

Section 4.1

Check Point Exercises

1. a. $\{x|-2 \le x < 5\}$

b. $\{x|1 \le x \le 3.5\}$
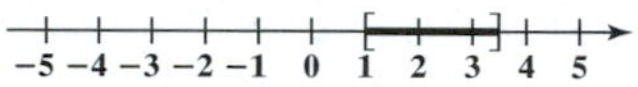

c. $\{x|x < -1\}$
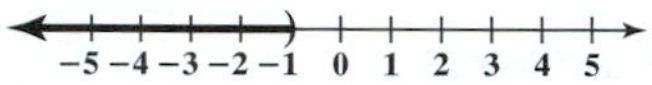

2. $\{x|x > -5\}$ or $(-5, \infty)$
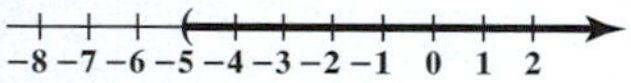

3. $\{x|x < 4\}$ or $(-\infty, 4)$
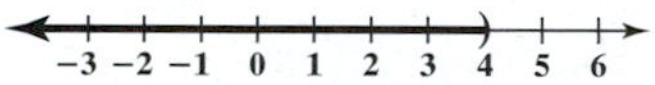

4. $\{x|x \ge 13\}$ or $[13, \infty)$
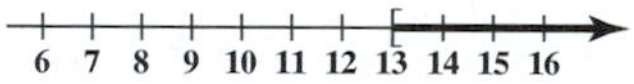

5. a. $\{x|x$ is a real number$\}$ or $\mathbb{R}$ or $(-\infty, \infty)$ **b.** $\varnothing$ **6.** more than 720 miles

Exercise Set 4.1

1. $\{x|1 < x \le 6\}$
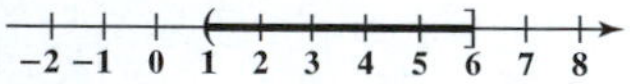

3. $\{x|-5 \le x < 2\}$
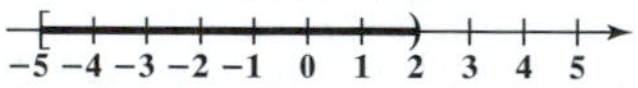

5. $\{x|-3 \le x \le 1\}$
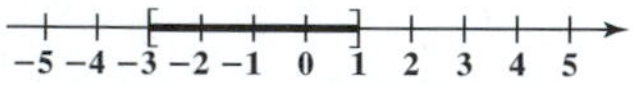

7. $\{x|x > 2\}$
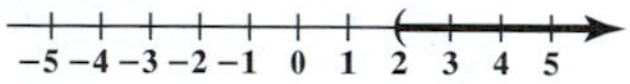

9. $\{x|x \ge -3\}$
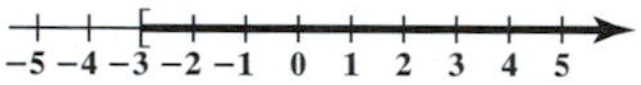

11. $\{x|x < 3\}$

13. $\{x|x < 5.5\}$
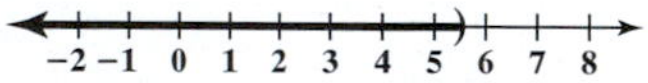

15. $\{x|x < 3\}$ or $(-\infty, 3)$
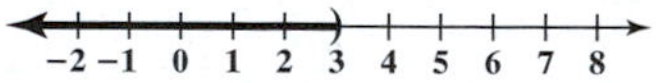

17. $\{x|x \ge 7\}$ or $[7, \infty)$
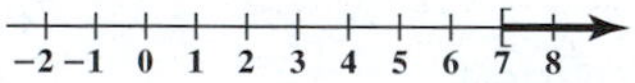

19. $\{x|x \le -4\}$ or $(-\infty, -4]$
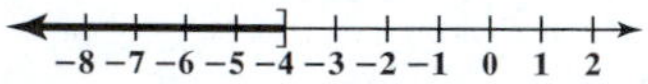

21. $\left\{x \middle| x \le -\frac{2}{5}\right\}$ or $\left(-\infty, -\frac{2}{5}\right]$
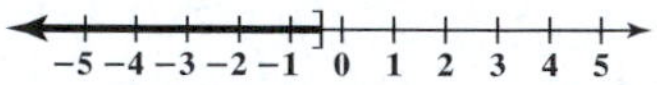

23. $\{x|x \ge 0\}$ or $[0, \infty)$
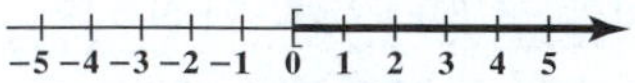

25. $\{x|x < 1\}$ or $(-\infty, 1)$
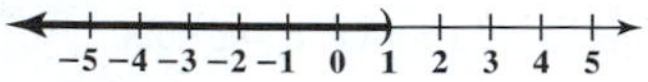

27. $\{x|x \ge 6\}$ or $[6, \infty)$
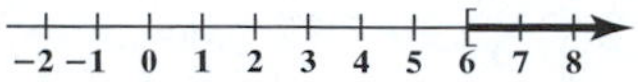

29. $\{x|x \ge -6\}$ or $[-6, \infty)$
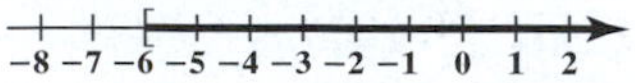

31. $\{x|x < -6\}$ or $(-\infty, -6)$
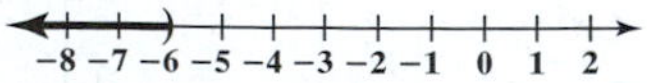

33. $\{x|x \ge 13\}$ or $[13, \infty)$
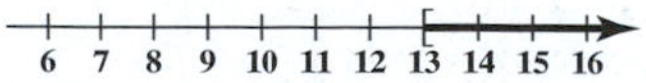

35. $\{x|x$ is a real number$\}$ or $\mathbb{R}$ or $(-\infty, \infty)$
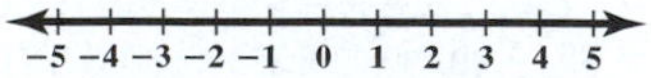

37. $\{x|x$ is a real number$\}$ or $\mathbb{R}$ or $(-\infty, \infty)$

39. $\varnothing$

41. $\{x|x > -6\}$ or $(-6, \infty)$
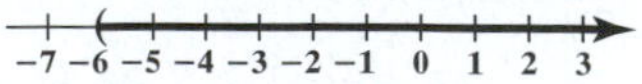

43. $\{x|x \ge -1\}$ or $[-1, \infty)$
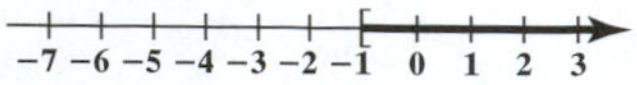

45. $\{x|x < -2\}$ or $(-\infty, -2)$
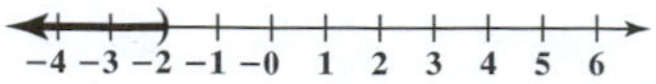

47. $\{x|x < 5\}$ or $(-\infty, 5)$

49. $\left\{x \middle| x \geq -\frac{4}{7}\right\}$ or $\left[-\frac{4}{7}, \infty\right)$ **51.** $\{x|x \geq 6\}$ or $[6, \infty)$ **53.** $\{x|x < 2\}$ or $(-\infty, 2)$

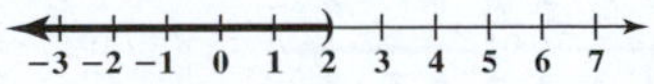

55. $x < \frac{c-b}{a}$ **57.** $\{x|x \leq -3\}$ or $(-\infty, -3]$ **59.** $\{x|x > -1.4\}$ or $(-1.4, \infty)$ **61.** $(0, 4)$ **63.** intimacy ≥ passion or passion ≤ intimacy **65.** commitment > passion or passion < commitment **67.** 9; after 3 years **69.** voting years after 2006 **71.** $\{t|t > 175\}$ or $(175, \infty)$; The women's speed skating times will be less than the men's after the year 2075. **73.** more than 100 miles per day **75.** greater than \$32,000 **77.** more than 6250 tapes **79.** 40 bags or fewer

89.

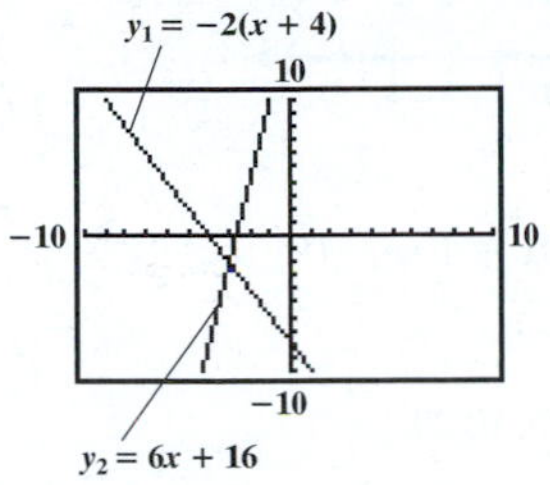

$\{x|x < -3\}$ or $(-\infty, -3)$

91. ∅

93. a. plan A: $4 + 0.10x$; plan B: $2 + 0.15x$

b.

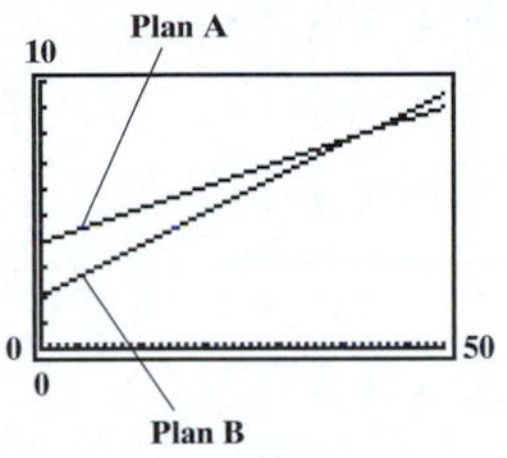

c–d. more than 40 checks per month

95. makes sense **97.** makes sense **99.** false **101.** true **103.** Since $x > y$, $y - x < 0$. Thus, when both sides were multiplied by $y - x$, the sense of the inequality should have been changed. **104.** 29 **105.** $\{(-1, -1, 2)\}$ **106.** $\frac{x^9}{8y^{15}}$ **107. a.** $\{3, 4\}$ **b.** $\{1, 2, 3, 4, 5, 6, 7\}$ **108. a.** $\{x|x < 8\}$ or $(-\infty, 8)$ **b.** $\{x|x < 5\}$ or $(-\infty, 5)$ **c.** any number less than 5 **d.** any number in $[5, 8)$ **109. a.** $\{x|x \geq 1\}$ or $[1, \infty)$ **b.** $\{x|x \geq 3\}$ or $[3, \infty)$ **c.** any number greater than or equal to 3 **d.** any number in $[1, 3)$

Section 4.2

Check Point Exercises

1. $\{3, 7\}$ **2.** $\{x|x < 1\}$ or $(-\infty, 1)$ **3.** ∅ **4.** $\{x|-1 \leq x < 4\}$ or $[-1, 4)$;

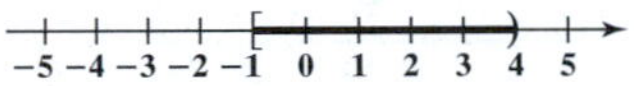

5. $\{3, 4, 5, 6, 7, 8, 9\}$ **6.** $\{x|x \leq 1 \text{ or } x > 3\}$ or $(-\infty, 1] \cup (3, \infty)$ **7.** $\{x|x \text{ is a real number}\}$ or ℝ or $(-\infty, \infty)$

Exercise Set 4.2

1. $\{2, 4\}$ **3.** ∅ **5.** ∅

7. $\{x|x > 6\}$; $(6, \infty)$

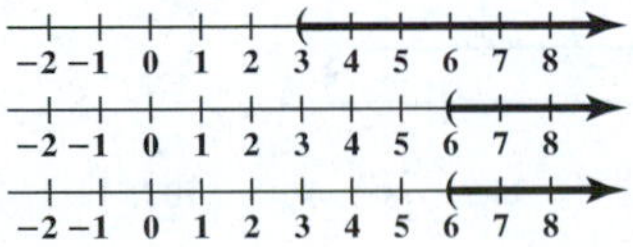

9. $\{x|x \leq 1\}$; $(-\infty, 1]$

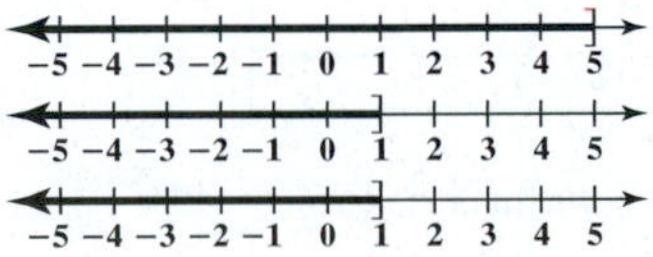

11. $\{x|-1 \leq x < 2\}$; $[-1, 2)$

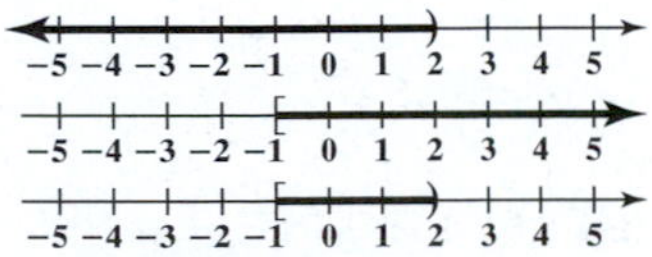

13. ∅

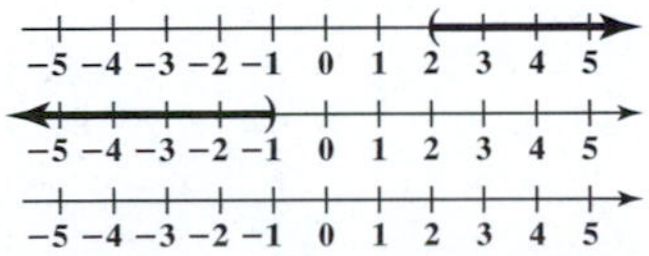

15. $\{x|-6 < x < -4\}$; $(-6, -4)$

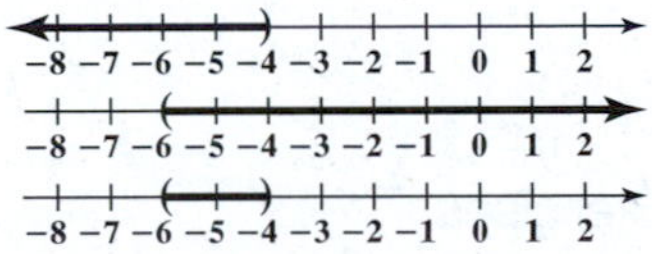

17. $\{x|-3 < x \leq 6\}$; $(-3, 6]$

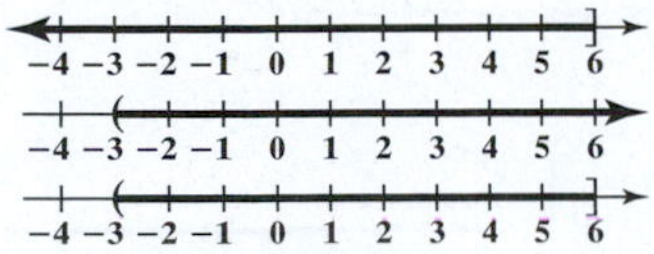

19. $\{x|2 < x < 5\}$; $(2, 5)$

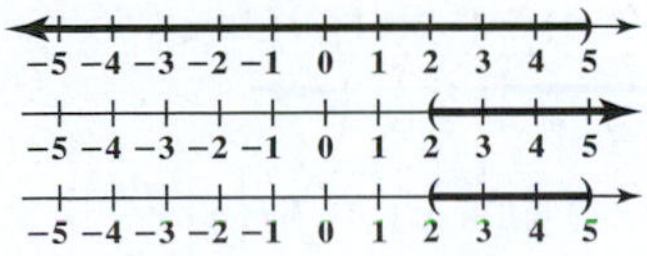

21. ∅

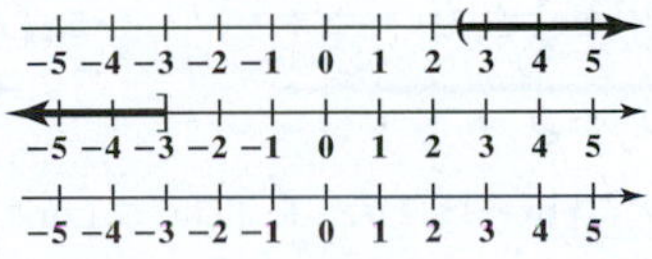

23. $\{x|0 \leq x < 2\}$; $[0, 2)$

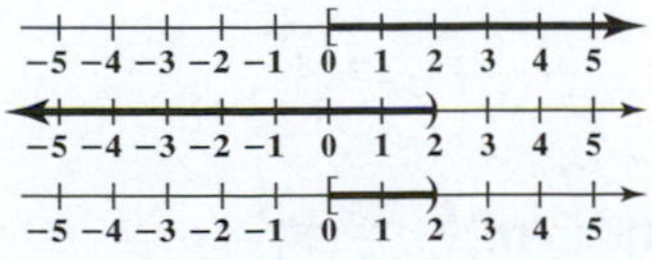

25. $\{x|3 < x < 5\}$; $(3, 5)$

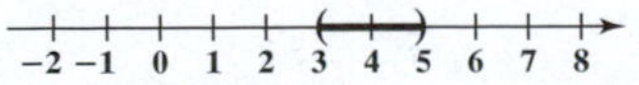

27. $\{x|-1 \leq x < 3\}$; $[-1, 3)$

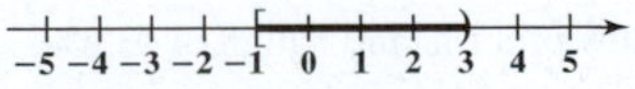

29. $\{x|-5 < x \leq -2\}$; $(-5, -2]$

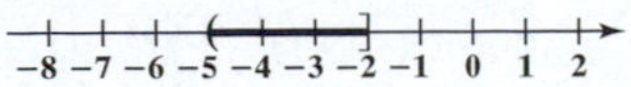

31. $\{x|3 \leq x < 6\}$; $[3, 6)$

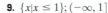

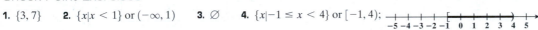

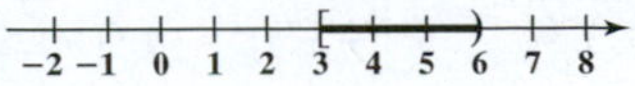

33. $\{1, 2, 3, 4, 5\}$ **35.** $\{1, 2, 3, 4, 5, 6, 7, 8, 10\}$ **37.** $\{a, e, i, o, u\}$

39. $\{x|x > 3\}; (3, \infty)$

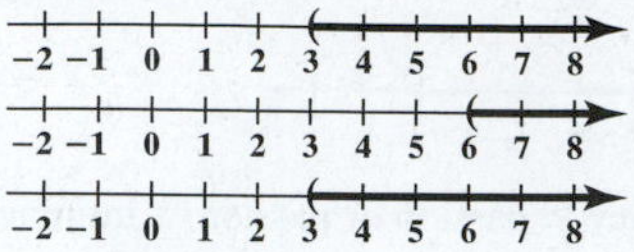

41. $\{x|x \le 5\}; (-\infty, 5]$

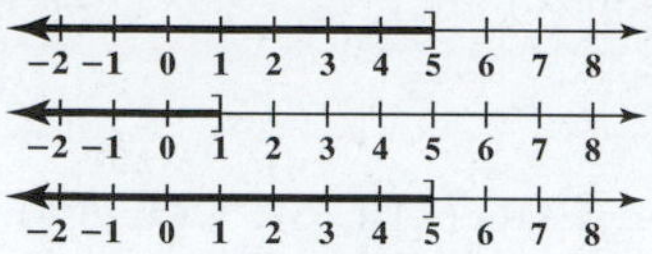

43. $\{x|x$ is a real number$\}; (-\infty, \infty)$

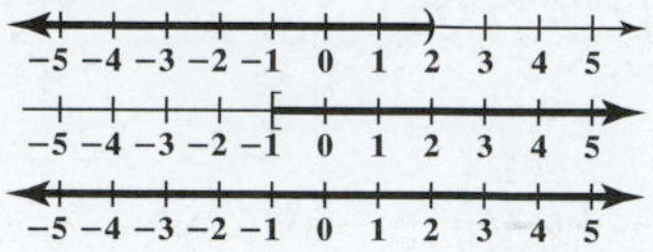

45. $\{x|x < -1 \text{ or } x \ge 2\}; (-\infty, -1) \cup [2, \infty)$

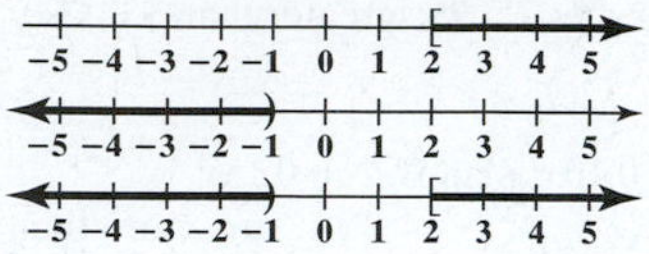

47. $\{x|x < -3 \text{ or } x > 4\}; (-\infty, -3) \cup (4, \infty)$

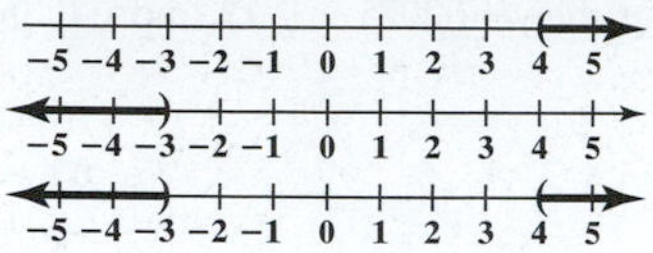

49. $\{x|x \le 1 \text{ or } x \ge 3\}; (-\infty, 1] \cup [3, \infty)$

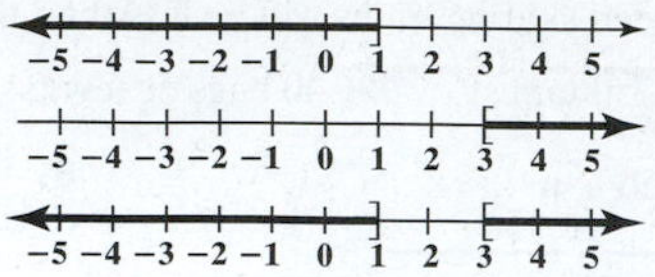

51. $\{x|x$ is a real number$\}; (-\infty, \infty)$

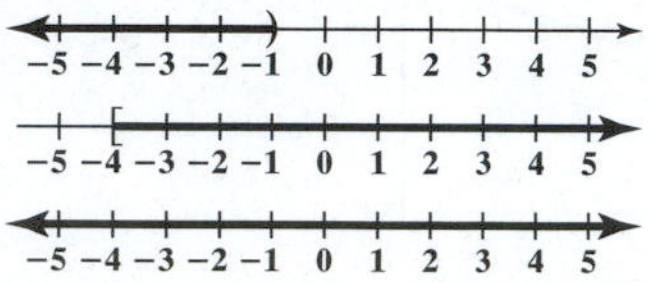

53. $\{x|x < 2\}; (-\infty, 2)$

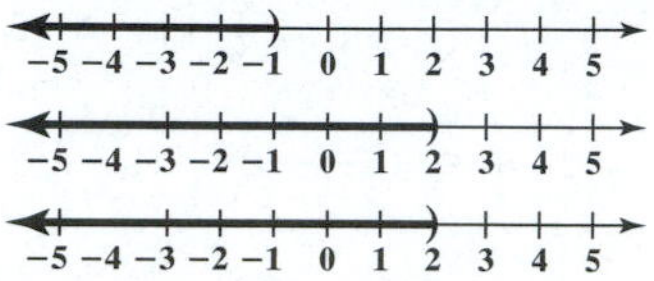

55. $\{x|x > 4\}$ or $(4, \infty)$ **57.** $\{x|x < 0 \text{ or } x > 6\}$ or $(-\infty, 0) \cup (6, \infty)$ **59.** $\frac{b-c}{a} < x < \frac{b+c}{a}$ **61.** $\{x|-1 \le x \le 3\}$ or $[-1, 3]$
63. $\{x|-1 < x < 3\}$ or $(-1, 3)$ **65.** $\{x|-1 \le x < 2\}$ or $[-1, 2)$ **67.** $\{-3, -2, -1\}$ **69.** $\{1974, 1980, 1994\}$
71. $\{1962, 1974, 1980, 1994, 2002\}$ **73.** $\{1962, 2002\}$ **75.** $\varnothing$ **77.** $\{1962, 1974\}$ **79.** between 80 and 110 minutes, inclusive
81. $[76, 126)$; If the highest grade is 100, then $[76, 100]$. **83.** more than 3 and less than 15 crossings per 3-month period
91. $\{x|-2 < x < 6\}$ or $(-2, 6)$;

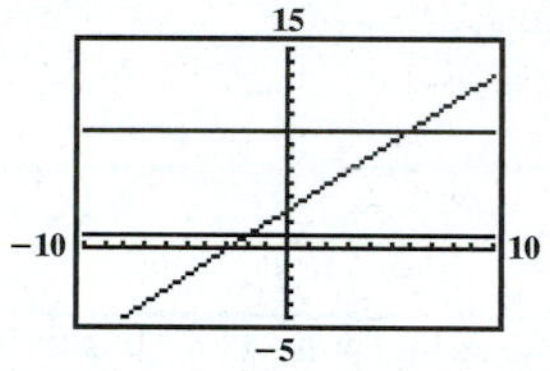

93. $\left\{x \middle| 2 \le x \le \frac{5}{2}\right\}$ or $\left[2, \frac{5}{2}\right]$;

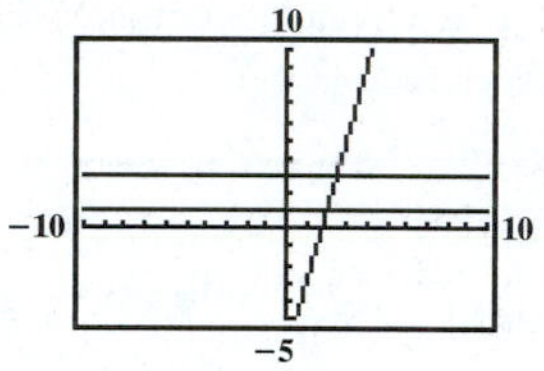

95. Exercise 91:

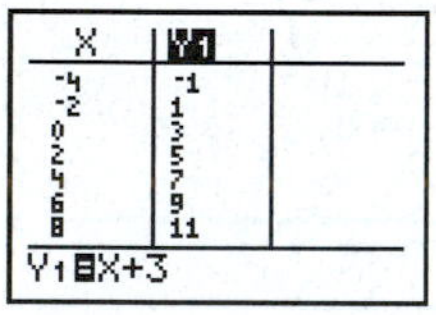

Exercise 93:

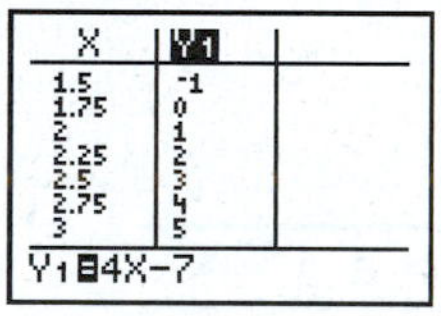

97. makes sense **99.** makes sense **101.** false **103.** false **105.** $(-\infty, 4]$ **107.** $[-1, 4]$ **109.** least: 4 nickels; greatest: 7 nickels
110. $-x^2 + 5x - 9; -15$ **111.** $f(x) = -\frac{1}{2}x + 4$ **112.** $17 - 2x$ **113.** $-\frac{1}{2}$ and 1 **114.** -1 and 3 **115. a.** -5 satisfies the inequality.
b. no

Section 4.3

Check Point Exercises

1. -2 and 3, or $\{-2, 3\}$ **2.** $-\frac{13}{3}$ and 5, or $\left\{-\frac{13}{3}, 5\right\}$ **3.** $\frac{4}{3}$ and 10, or $\left\{\frac{4}{3}, 10\right\}$
4. $\{x|-3 < x < 7\}$ or $(-3, 7)$

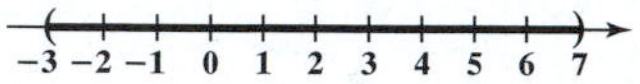

5. $\{x|x \le 1 \text{ or } x \ge 4\}$ or $(-\infty, 1] \cup [4, \infty)$

6. a. $\{x|-3 < x < 7\}$ or $(-3, 7)$ **b.** $\{x|x \le 1 \text{ or } x \ge 4\}$ or $(-\infty, 1] \cup [4, \infty)$ **7.** $\left\{x \middle| -\frac{11}{5} \le x \le 3\right\}$ or $\left[-\frac{11}{5}, 3\right]$;

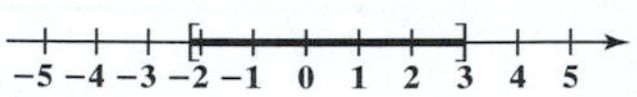

8. $\{x|8.1 \le x \le 13.9\}$ or $[8.1, 13.9]$; The percentage of children in the population who think that not being able to do everything they want is a bad thing is between a low of 8.1% and a high of 13.9%.

Exercise Set 4.3

1. $\{-8, 8\}$ **3.** $\{-5, 9\}$ **5.** $\{-3, 4\}$ **7.** $\{-1, 2\}$ **9.** $\varnothing$ **11.** $\{-3\}$ **13.** $\{-11, -1\}$ **15.** $\{-3, 4\}$ **17.** $\left\{-\frac{13}{3}, 5\right\}$
19. $\left\{-\frac{2}{5}, \frac{2}{5}\right\}$ **21.** $\varnothing$ **23.** $\varnothing$ **25.** $\left\{\frac{1}{2}\right\}$ **27.** $\left\{\frac{3}{4}, 5\right\}$ **29.** $\left\{\frac{5}{3}, 3\right\}$ **31.** $\{0\}$ **33.** $\{4\}$ **35.** $\{4\}$ **37.** $\{-1, 15\}$

39. $\{x|-3 < x < 3\}$ or $(-3, 3)$

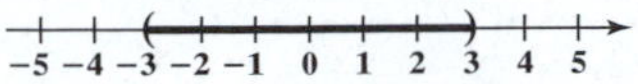

41. $\{x|1 < x < 3\}$ or $(1, 3)$

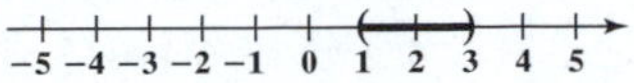

43. $\{x|-3 \le x \le -1\}$ or $[-3, -1]$

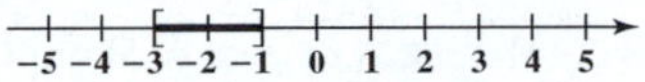

45. $\{x|-1 < x < 7\}$ or $(-1, 7)$

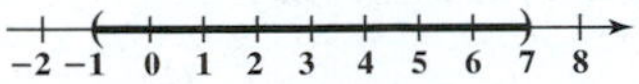

47. $\{x|x < -3 \text{ or } x > 3\}$ or $(-\infty, -3) \cup (3, \infty)$

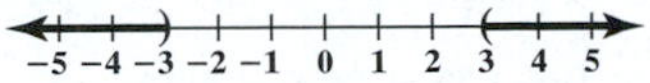

49. $\{x|x < -4 \text{ or } x > -2\}$ or $(-\infty, -4) \cup (-2, \infty)$

51. $\{x|x \le 2 \text{ or } x \ge 6\}$ or $(-\infty, 2] \cup [6, \infty)$

53. $\left\{x \middle| x < \frac{1}{3} \text{ or } x > 5\right\}$ or $\left(-\infty, \frac{1}{3}\right) \cup (5, \infty)$

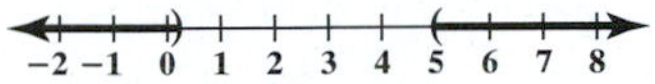

55. $\{x|-5 \le x \le 3\}$ or $[-5, 3]$

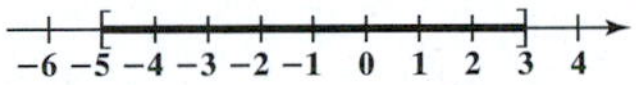

57. $\{x|-6 < x < 0\}$ or $(-6, 0)$

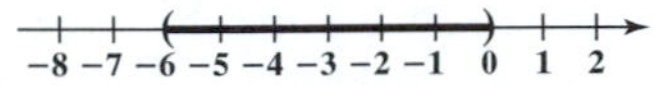

59. $\{x|x \le -5 \text{ or } x \ge 3\}$ or $(-\infty, -5] \cup [3, \infty)$

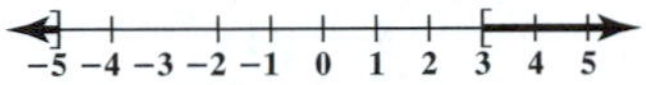

61. $\{x|x < -3 \text{ or } x > 12\}$ or $(-\infty, -3) \cup (12, \infty)$

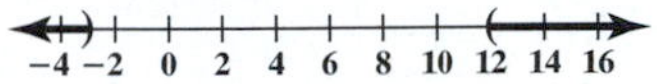

63. $\varnothing$

65. $\{x|x \text{ is a real number}\}$ or $(-\infty, \infty)$

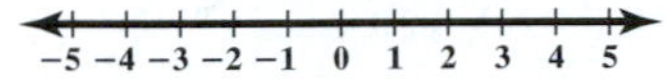

67. $\{x|-9 \le x \le 5\}$ or $[-9, 5]$

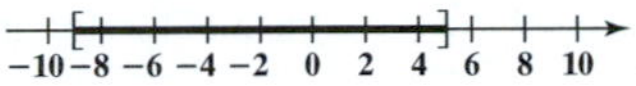

69. $\{x|x < 1 \text{ or } x > 2\}$ or $(-\infty, 1) \cup (2, \infty)$

71. $\{x|x < -3 \text{ or } x > 5\}$ or $(-\infty, -3) \cup (5, \infty)$

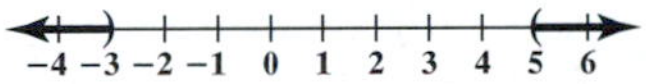

73. $\{x|x \le -1 \text{ or } x \ge 2\}$ or $(-\infty, -1] \cup [2, \infty)$

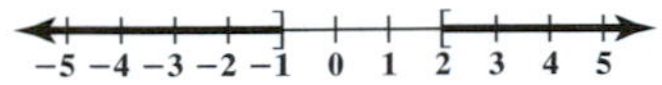

75. $-\frac{3}{2}$ and 4 **77.** -7 and -2 **79.** $\left\{x \middle| -\frac{7}{3} \le x \le 1\right\}$ or $\left[-\frac{7}{3}, 1\right]$ **81.** $\{x|x < -1 \text{ or } x > 4\}$ or $(-\infty, -1) \cup (4, \infty)$

83. $\left(-\infty, -\frac{1}{3}\right] \cup [3, \infty)$ **85.** $\left\{x \middle| \frac{-c-b}{a} < x < \frac{c-b}{a}\right\}$ **87.** $\{3, 5\}$ **89.** $\{x|-2 \le x \le 1\}$ or $[-2, 1]$

91. $\{x|58.6 \le x \le 61.8\}$ or $[58.6, 61.8]$; The percentage of the U.S. population that watched M*A*S*H is between a low of 58.6% and a high of 61.8%.; 1.6% **93.** $\{T|50 \le T \le 64\}$ or $[50, 64]$; The monthly average temperature for San Francisco, CA is between a low of 50°F and a high of 64°F. **95.** $\{x|8.59 \le x \le 8.61\}$ or $[8.59, 8.61]$; A machine part that is supposed to be 8.6 cm is acceptable between a low of 8.59 and a high of 8.61 cm.

97. If the number of outcomes that result in heads is 41 or less or 59 or more, then the coin is unfair.

105. $\{-6, 4\}$;

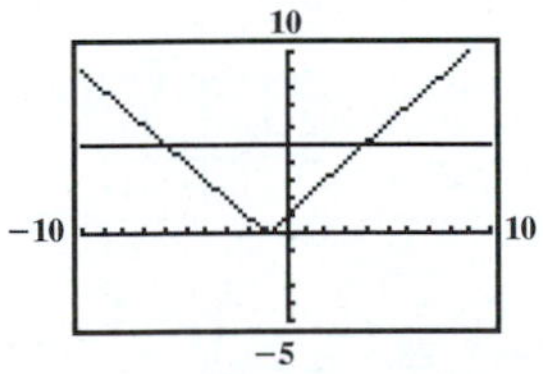

107. $\{2, 3\}$;

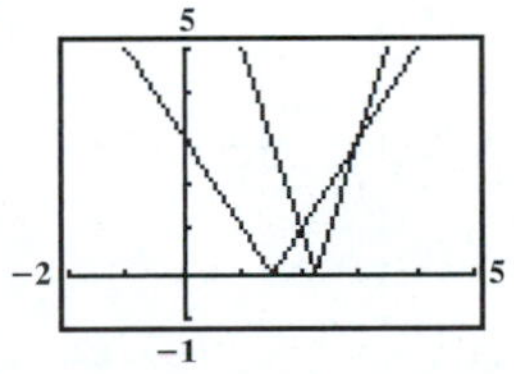

109. $\{x|-2 < x < 3\}$ or $(-2, 3)$;

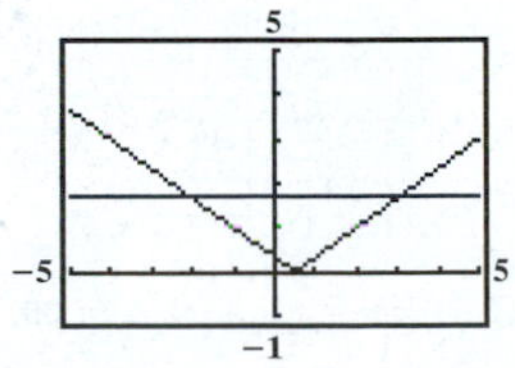

111. $\{x|x < -3 \text{ or } x > 4\}$ or $(-\infty, -3) \cup (4, \infty)$;

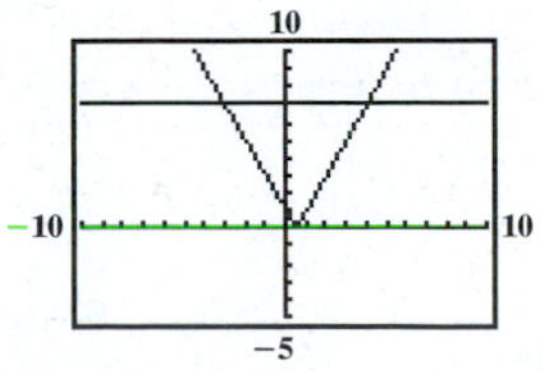

113. $\{x|x \text{ is a real number}\}$ or $(-\infty, \infty)$;

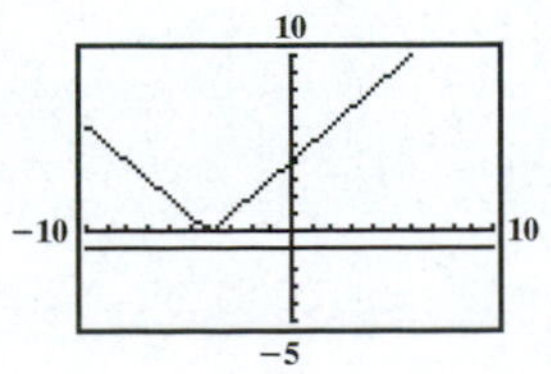

115. makes sense **117.** does not make sense **119.** false **121.** true

123. a. $|x - 4| < 3$ **b.** $|x - 4| \ge 3$ **125.** $\{1\}$

126.

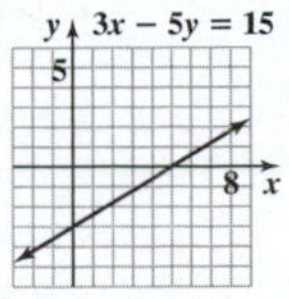

127.

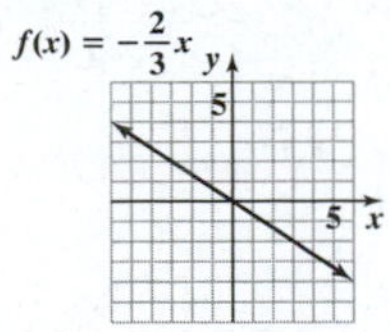

128.

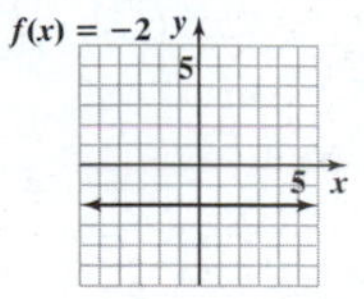

Mid-Chapter Check Point Exercises

1. $\{x|x \le -4\}$ or $(-\infty, -4]$ **2.** $\{x|3 \le x < 5\}$ or $[3, 5)$ **3.** $\left\{\frac{1}{2}, 3\right\}$ **4.** $\{x|x < -1\}$ or $(-\infty, -1)$

5. $\{x|x < -9$ or $x > -5\}$ or $(-\infty, -9) \cup (-5, \infty)$ **6.** $\left\{x \middle| -\frac{2}{3} \le x \le 2\right\}$ or $\left[-\frac{2}{3}, 2\right]$ **7.** $\left\{\frac{1}{2}, \frac{13}{4}\right\}$ **8.** $\{x|-4 < x \le -2\}$ or $(-4, -2]$

9. $\{x|x$ is a real number$\}$ or $(-\infty, \infty)$ **10.** $\{x|x \le -3\}$ or $(-\infty, -3]$ **11.** $\{x|x < -3\}$ or $(-\infty, -3)$

12. $\left\{x \middle| x < -1 \text{ or } x > -\frac{1}{5}\right\}$ or $(-\infty, -1) \cup \left(-\frac{1}{5}, \infty\right)$ **13.** $\{x|x \le -10$ or $x \ge 2\}$ or $(-\infty, -10] \cup [2, \infty)$ **14.** $\varnothing$

15. $\left\{x \middle| x < -\frac{5}{3}\right\}$ or $\left(-\infty, -\frac{5}{3}\right)$ **16.** $\{x|x > 4\}$ or $(4, \infty)$ **17.** $\{-2, 7\}$ **18.** $\varnothing$ **19.** no more than 80 miles per day **20.** $[49\%, 99\%)$

21. at least \$120,000 **22.** at least 750,000 discs each month

Section 4.4

Check Point Exercises

1. $4x - 2y \ge 8$

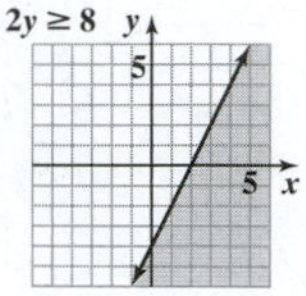

2. $y > -\frac{3}{4}x$

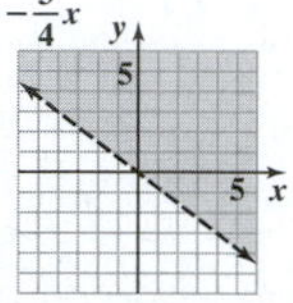

3. a. $y > 1$

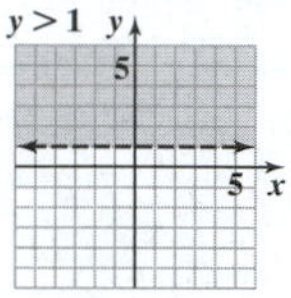

b. $x \le -2$

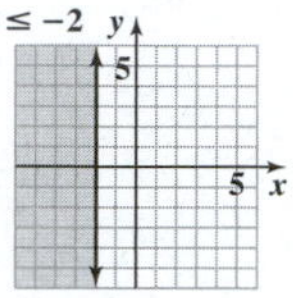

4. $B = (60, 20)$; Using $T = 60$ and $P = 20$, each of the three inequalities for grasslands is true: $60 \ge 35$, true; $5(60) - 7(20) \ge 70$, true; $3(60) - 35(20) \le -140$, true.

5. $x - 3y < 6$, $2x + 3y \ge -6$

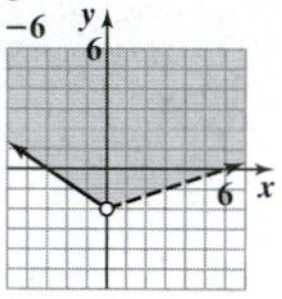

6. $x + y < 2$, $-2 \le x < 1$, $y > -3$

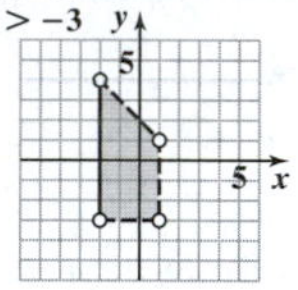

Exercise Set 4.4

1. $x + y \ge 3$

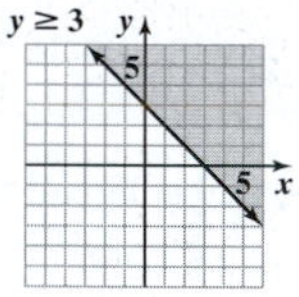

3. $x - y < 5$

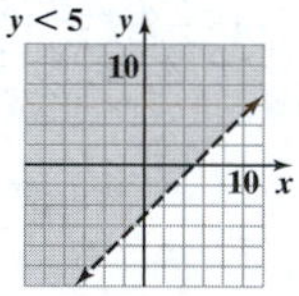

5. $x + 2y > 4$

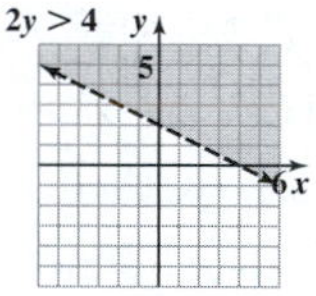

7. $3x - y \le 6$

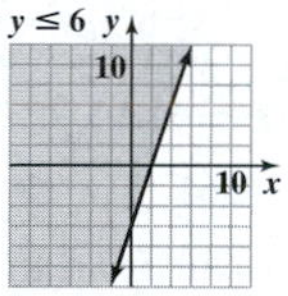

9. $\frac{x}{2} + \frac{y}{3} < 1$

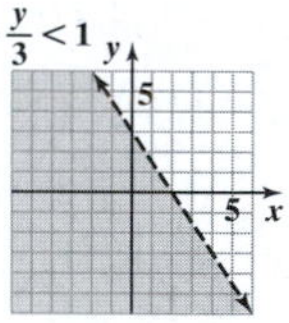

11. $y > \frac{1}{3}x$

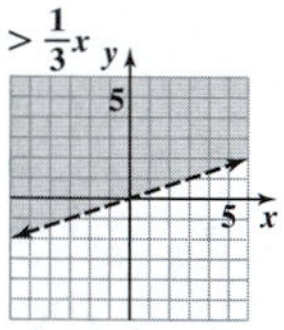

13. $y \le 3x + 2$

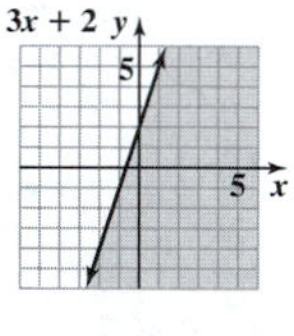

15. $y < -\frac{1}{4}x$

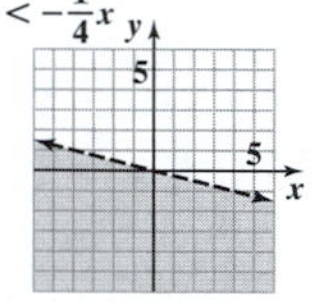

17. $x \le 2$

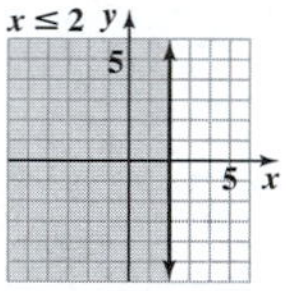

19. $y > -4$

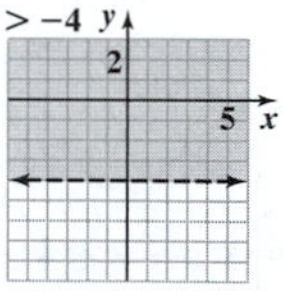

21. $y \ge 0$

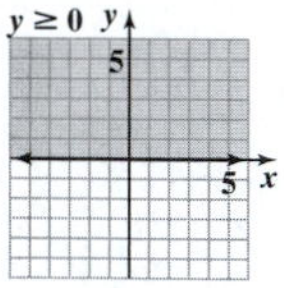

23. $3x + 6y \le 6$, $2x + y \le 8$

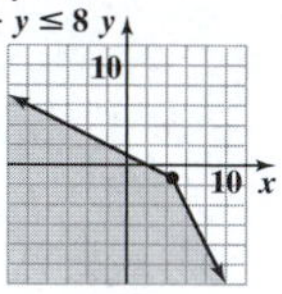

25. $2x - 5y \le 10$, $3x - 2y > 6$

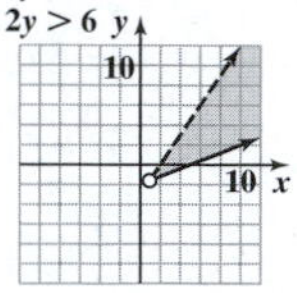

27. $y > 2x - 3$, $y < -x + 6$

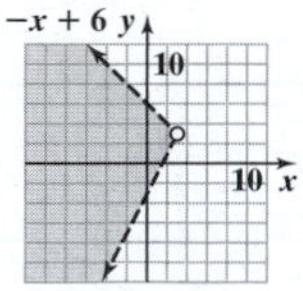

29. $x + 2y \le 4$, $y \ge x - 3$

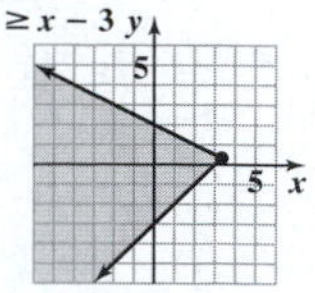

31. $x \le 2$, $y \ge -1$

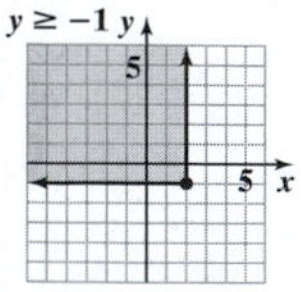

33. $-2 \le x < 5$

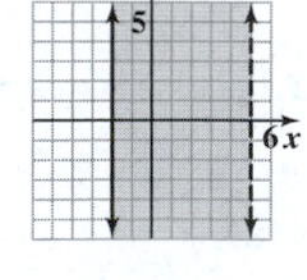

35. $x - y \le 1$, $x \ge 2$

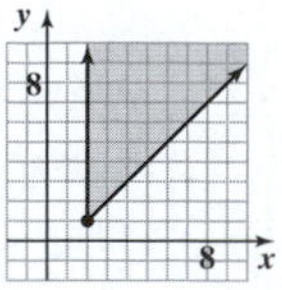

37. $\varnothing$

39. $x + y > 4$, $x + y > -1$

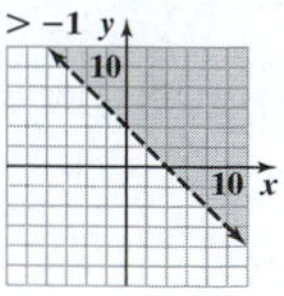

41. 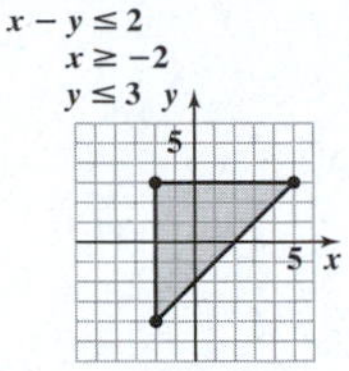

43. 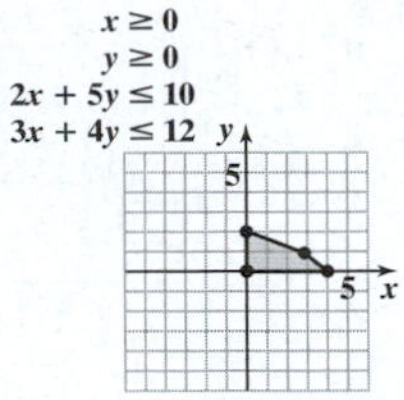

45. 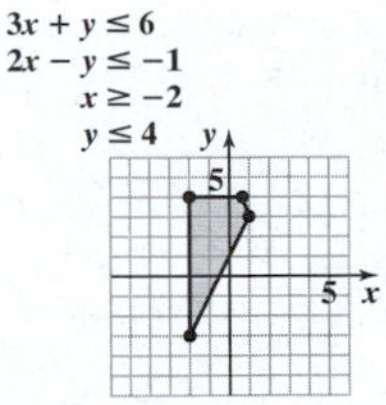

47. $y \geq -2x + 4$

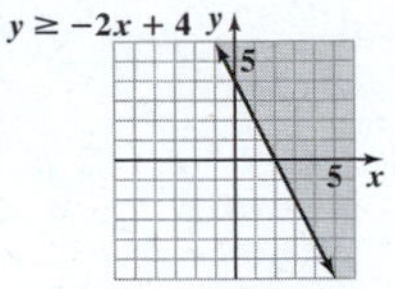

49. $x + y \leq 4$ and $3x + y \leq 6$

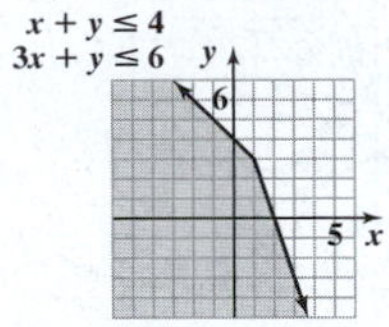

51. $-2 \leq x \leq 2$; $-3 \leq y \leq 3$

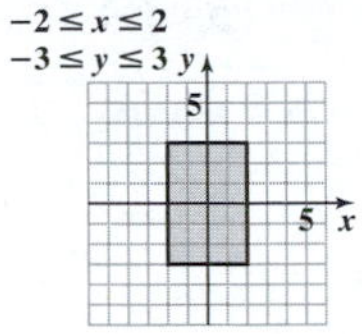

53. 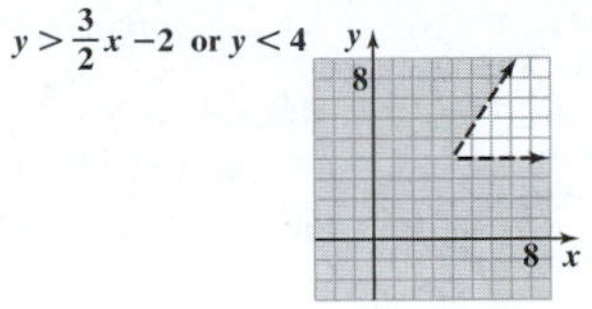

55. no solution **57.** infinitely many solutions **59. a.** $A = (20, 150)$; A 20-year-old with a heart rate of 150 beats per minute is within the target range. **b.** $10 \leq 20 \leq 70$, true; $150 \geq 0.7(220 - 20)$, true; $150 \leq 0.8(220 - 20)$, true **61.** $10 \leq a \leq 70$; $H \geq 0.6(220 - a)$; $H \leq 0.7(220 - a)$

63. a. $y \geq 0$; $x + y \geq 5$; $x \geq 1$; $200x + 100y \leq 700$ **b.** 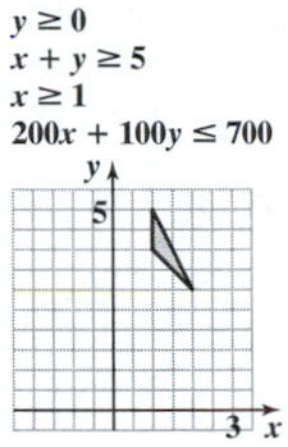

c. 2 nights

75. 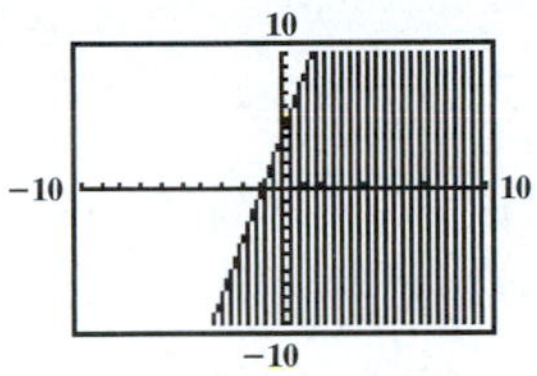

77. 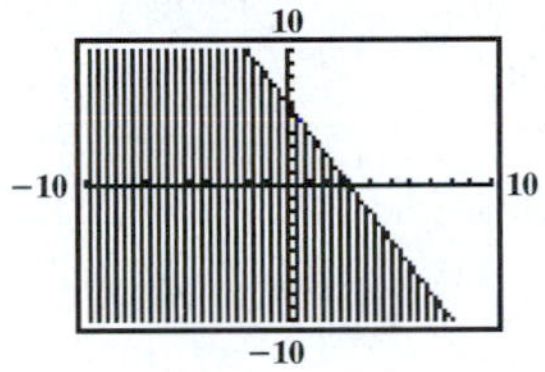

83. makes sense **85.** makes sense **87.** false **89.** true **91.** $x \geq -2$, $y > -1$ **93.** $y \geq 2x - 2$

95. 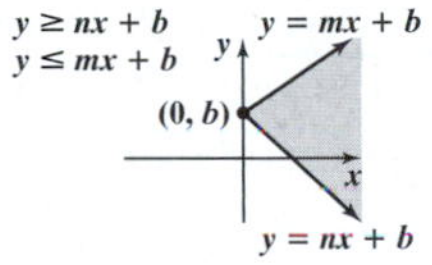

96. $\{(3, 1)\}$ **97.** $\{(2, 4)\}$ **98.** 165

99. a. 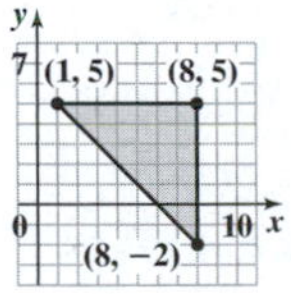

b. $(1, 5), (8, 5), (8, -2)$ **c.** at $(1, 5)$: 13; at $(8, 5)$: 34; at $(8, -2)$: 20

100. a. 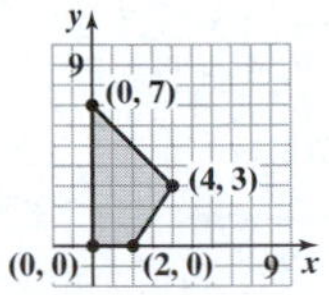

b. $(0, 0), (2, 0), (4, 3), (0, 7)$ **c.** at $(0, 0)$: 0; at $(2, 0)$: 4; at $(4, 3)$: 23; at $(0, 7)$: 35

101. $20x + 10y \leq 80{,}000$

Section 4.5

Check Point Exercises

1. $z = 25x + 55y$ **2.** $x + y \leq 80$ **3.** $30 \leq x \leq 80$, $10 \leq y \leq 30$; $z = 25x + 55y$, $x + y \leq 80$, $30 \leq x \leq 80$, $10 \leq y \leq 30$ **4.** 50 bookshelves and 30 desks; \$2900 **5.** 30

Exercise Set 4.5

1. $(1, 2)$: 17; $(2, 10)$: 70; $(7, 5)$: 65; $(8, 3)$: 58; maximum: 70; minimum: 17 **3.** $(0, 0)$: 0; $(0, 8)$: 400; $(4, 9)$: 610; $(8, 0)$: 320; maximum: 610; minimum: 0

5. a.
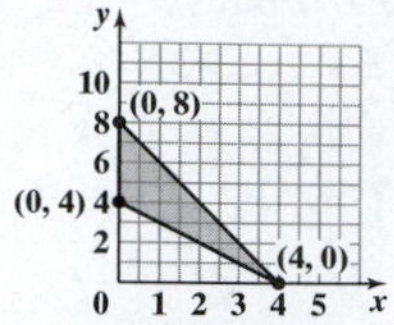

b. (0, 4): 8; (0, 8): 16; (4, 0): 12
c. maximum: 16; at (0, 8)

7. a.
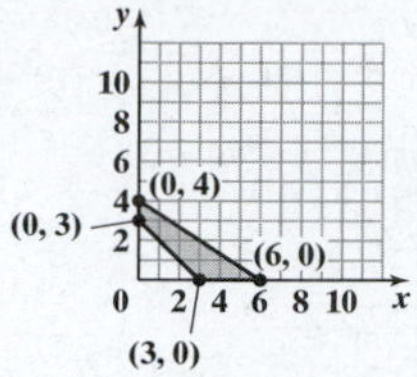

b. (0, 3): 3; (0, 4): 4; (6, 0): 24; (3, 0): 12
c. maximum: 24; at (6, 0)

9. a.
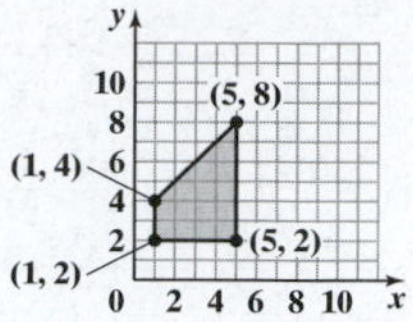

b. (1, 2): −1; (1, 4): −5; (5, 8): −1; (5, 2): 11
c. maximum: 11; at (5, 2)

11. a.
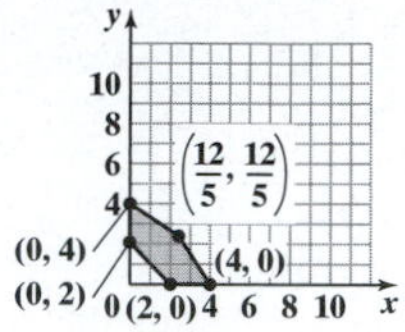

b. (0, 2): 4; (0, 4): 8; $\left(\frac{12}{5}, \frac{12}{5}\right)$: $\frac{72}{5} = 14.4$; (4, 0): 16; (2, 0): 8
c. maximum: 16; at (4, 0)

13. a.
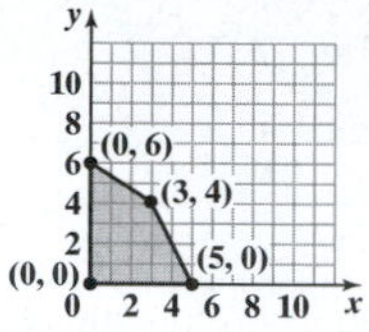

b. (0, 0): 0; (0, 6): 72; (3, 4): 78; (5, 0): 50
c. maximum: 78; at (3, 4)

15. a. $z = 125x + 200y$
b. $x \leq 450$; $y \leq 200$; $600x + 900y \leq 360{,}000$
c.
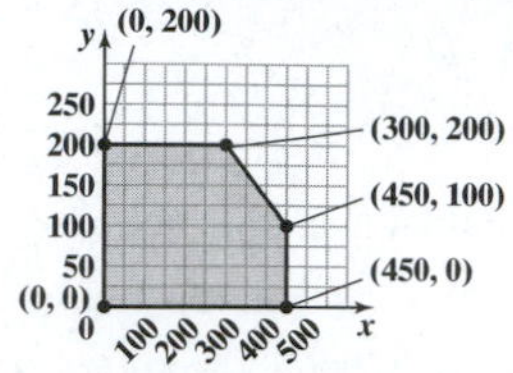

d. 0; 40,000; 77,500; 76,250; 56,250
e. 300; 200; 77,500

17. 40 of model A and 0 of model B **19.** 300 boxes of food and 200 boxes of clothing **21.** 100 parents and 50 students **23.** 10 Boeing 727s and 42 Falcon 20s **29.** does not make sense **31.** makes sense **33.** \$5000 in stocks and \$5000 in bonds **35.** $54x^7y^{15}$ **36.** $L = \frac{12P + W}{2}$ **37.** 10 **38.** $4x^3 + 9x^2 - 13x - 3$ **39.** $5x^3 - 2x^2 + 12x - 15$ **40. a.** g **b.** f

Review Exercises

1. $\{x \mid -2 < x \leq 3\}$
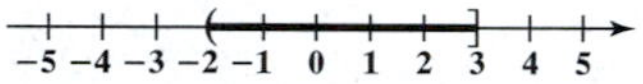

2. $\{x \mid -1.5 \leq x \leq 2\}$

3. $\{x \mid x > -1\}$

4. $\{x \mid x \geq -2\}$; $[-2, \infty)$
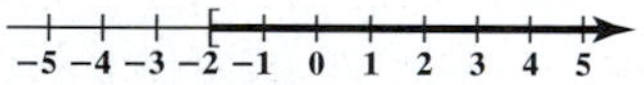

5. $\left\{x \,\middle|\, x \geq \frac{3}{5}\right\}$; $\left[\frac{3}{5}, \infty\right)$
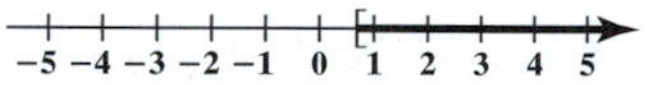

6. $\left\{x \,\middle|\, x < -\frac{21}{2}\right\}$; $\left(-\infty, -\frac{21}{2}\right)$
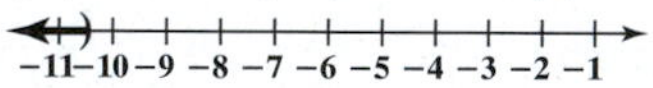

7. $\{x \mid x > -3\}$; $(-3, \infty)$
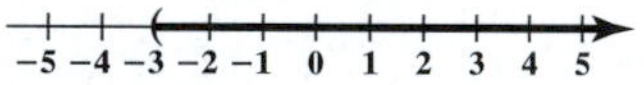

8. $\{x \mid x \leq -2\}$; $(-\infty, -2]$

9. ∅ **10.** more than 50 checks **11.** more than \$13,500 in sales **12.** $\{a, c\}$ **13.** $\{a\}$ **14.** $\{a, b, c, d, e\}$ **15.** $\{a, b, c, d, f, g\}$

16. $\{x \mid x \leq 3\}$; $(-\infty, 3]$

17. $\{x \mid x < 6\}$; $(-\infty, 6)$
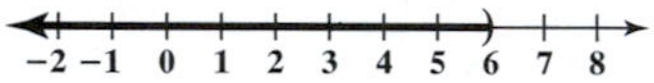

18. $\{x \mid 6 < x < 8\}$; $(6, 8)$
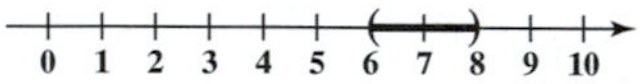

19. $\{x \mid x \leq 1\}$; $(-\infty, 1]$
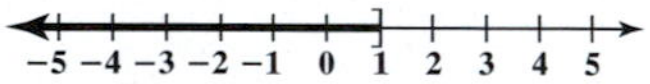

20. ∅

21. $\{x \mid x < 1 \text{ or } x > 2\}$; $(-\infty, 1) \cup (2, \infty)$
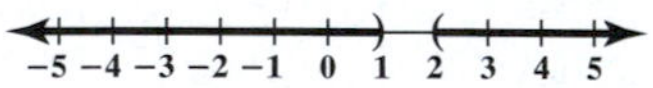

22. $\{x \mid x \leq -4 \text{ or } x > 2\}$; $(-\infty, -4] \cup (2, \infty)$
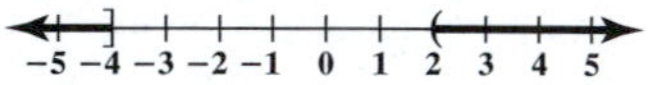

23. $\{x \mid x < -2\}$; $(-\infty, -2)$
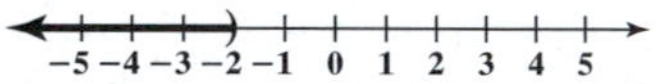

24. $\{x \mid x \text{ is a real number}\}$; $(-\infty, \infty)$

25. $\{x \mid -5 < x \leq 2\}$; $(-5, 2]$
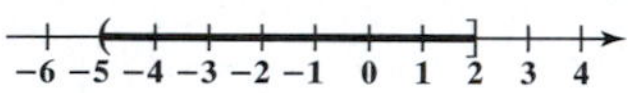

26. $\left\{x \,\middle|\, -\frac{3}{4} \leq x \leq 1\right\}$; $\left[-\frac{3}{4}, 1\right]$
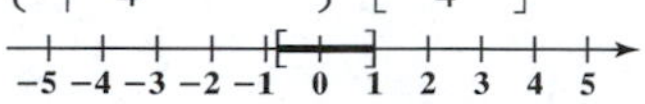

27. [49%, 99%)

28. $\{-4, 3\}$ **29.** ∅ **30.** $\left\{-\frac{11}{2}, \frac{23}{2}\right\}$ **31.** $\left\{-4, -\frac{6}{11}\right\}$

32. $\{x \mid -9 \leq x \leq 6\}$; $[-9, 6]$
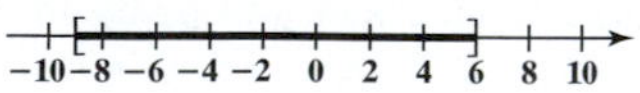

33. $\{x \mid x < -6 \text{ or } x > 0\}$; $(-\infty, -6) \cup (0, \infty)$
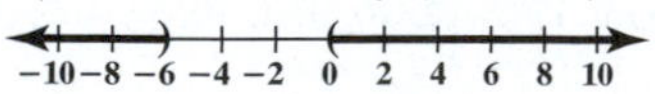

34. $\{x \mid -3 < x < -2\}$; $(-3, -2)$
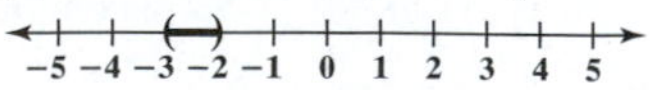

35. $\{x \mid x \leq -5 \text{ or } x \geq 1\}$; $(-\infty, -5] \cup [1, \infty)$;
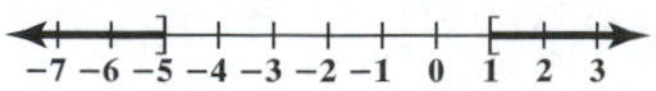

36. ∅ **37.** Approximately 90% of the population sleeps between 5.5 hours and 7.5 hours daily, inclusive.

38. $3x - 4y > 12$

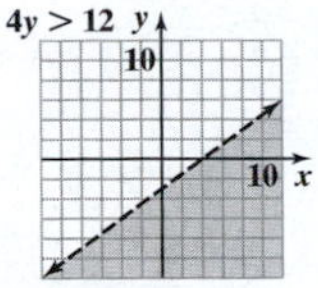

39. $x - 3y \le 6$

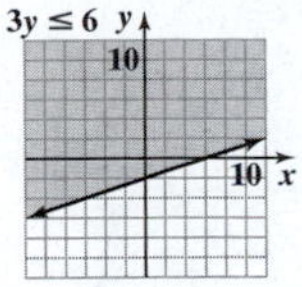

40. $y \le -\frac{1}{2}x + 2$

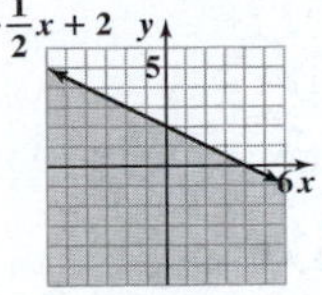

41. $y > \frac{3}{5}x$

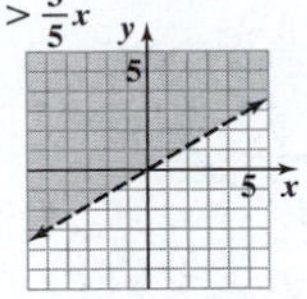

42. $x \le 2$

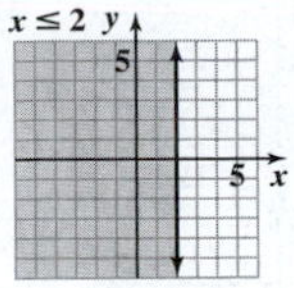

43. $y > -3$

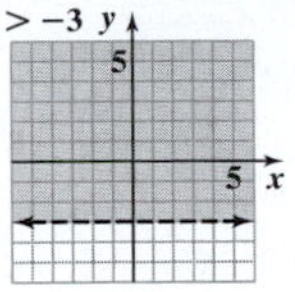

44. $2x - y \le 4$, $x + y \ge 5$

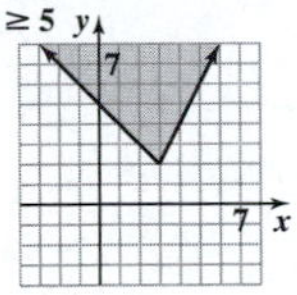

45. $y < -x + 4$, $y > x - 4$

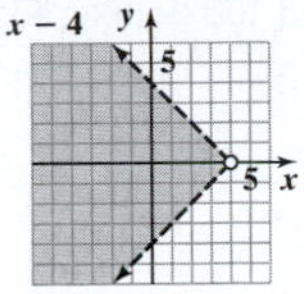

46. $-3 \le x < 5$

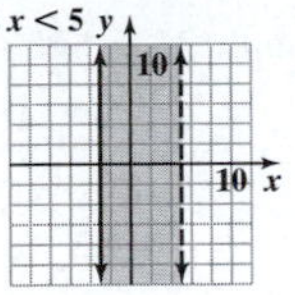

47. $-2 < y \le 6$

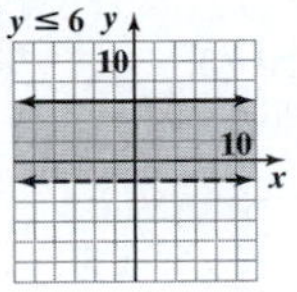

48. $x \ge 3$, $y \le 0$

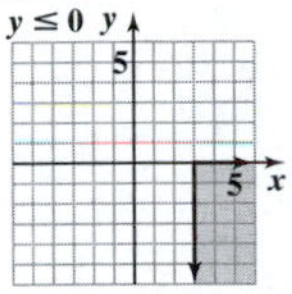

49. $2x - y > -4$, $x \ge 0$

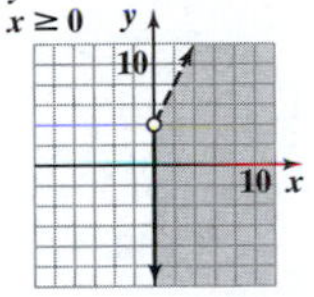

50. $x + y \le 6$, $y \ge 2x - 3$

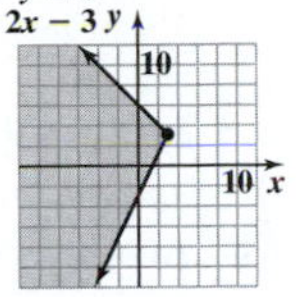

51. $3x + 2y \ge 4$, $x - y \le 3$, $x \ge 0, y \ge 0$

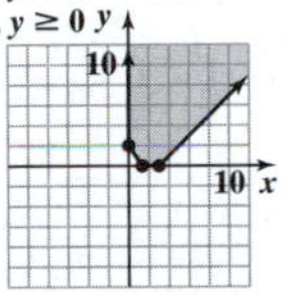

52. $\varnothing$

53. $\left(\frac{1}{2}, \frac{1}{2}\right)$: $\frac{5}{2}$; (2, 2): 10; (4, 0): 8; (1, 0): 2; maximum: 10; minimum: 2

54.

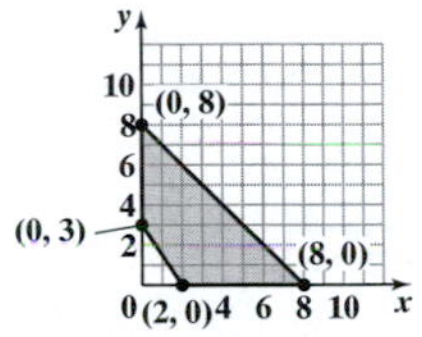

maximum: 24

55.

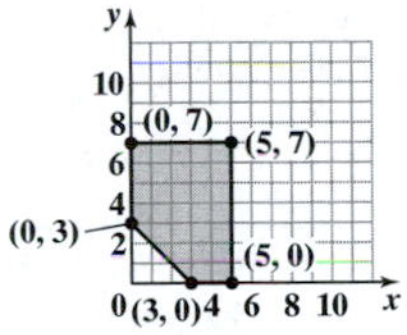

maximum: 33

56.

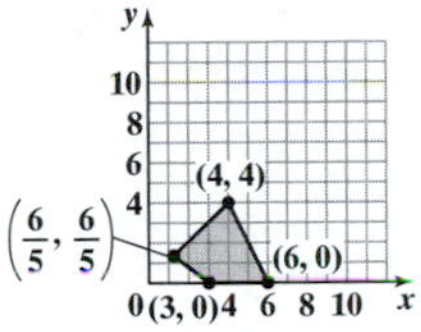

maximum: 44

57. a. $z = 500x + 350y$

c.

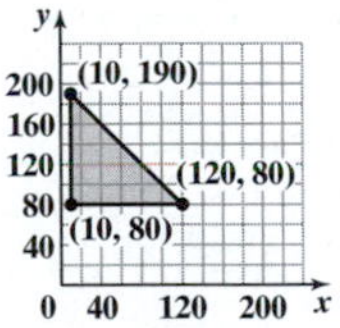

b. $x + y \le 200$; $x \ge 10$; $y \ge 80$

d. (10, 80): 33,000; (10, 190): 71,500; (120, 80): 88,000

e. 120; 80; 88,000

58. 480 of model A and 240 of model B

Chapter Test

1. $\{x \mid -3 \le x < 2\}$

2. $\{x \mid x \le -1\}$

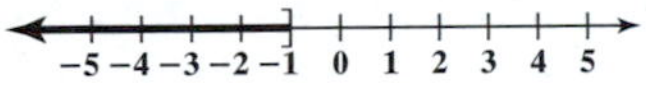

3. $\{x \mid x \le 12\}$; $(-\infty, 12]$

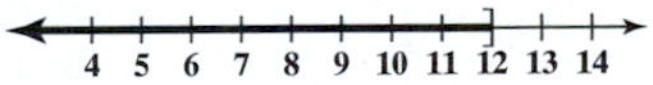

4. $\left\{x \,\middle|\, x \ge \frac{21}{8}\right\}$; $\left[\frac{21}{8}, \infty\right)$

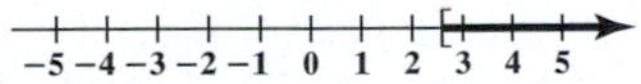

5. more than 200 calls

6. $\{4, 6\}$

7. $\{2, 4, 6, 8, 10, 12, 14\}$

8. $\{x \mid -2 < x < -1\}$; $(-2, -1)$

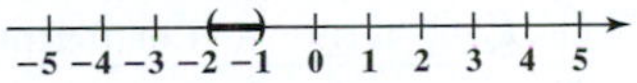

9. $\{x \mid x \ge -2\}$; $[-2, \infty)$

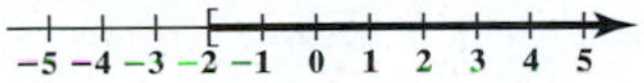

10. $\{x \mid x < 4\}$; $(-\infty, 4)$

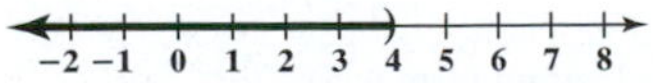

11. $\{x \mid x \le -4 \text{ or } x > 2\}$; $(-\infty, -4] \cup (2, \infty)$

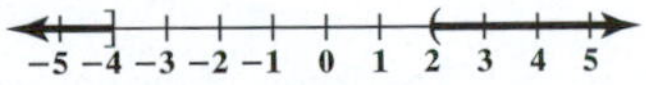

12. $\left\{x \,\middle|\, -7 \le x < \frac{13}{2}\right\}$; $\left[-7, \frac{13}{2}\right)$

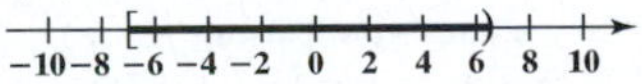

13. $\left\{-2, \frac{4}{5}\right\}$

14. $\left\{-\frac{8}{5}, 7\right\}$

15. $\{x \mid -3 < x < 4\}$; $(-3, 4)$

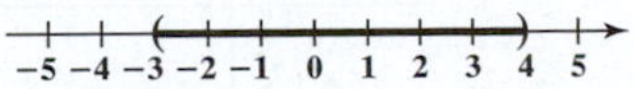

16. $\{x \mid x \le -1 \text{ or } x \ge 4\}$; $(-\infty, -1] \cup [4, \infty)$

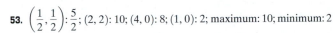

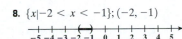

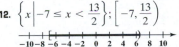

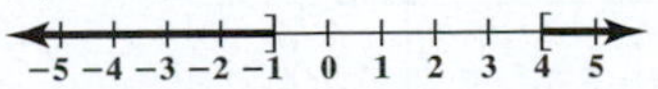

17. $\{b \mid b < 90.6 \text{ or } b > 106.6\}$ or $(-\infty, 90.6) \cup (106.6, \infty)$; Hypothermia: Body temperature below 90.6°F; Hyperthermia: Body temperature above 106.6°F

18. $3x - 2y < 6$

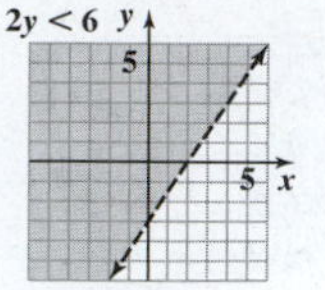

19. $y \geq \frac{1}{2}x - 1$

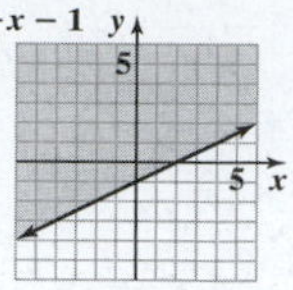

20. $y \leq -1$

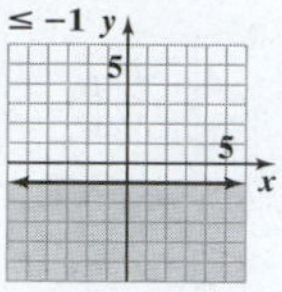

21. $x + y \geq 2$, $x - y \geq 4$

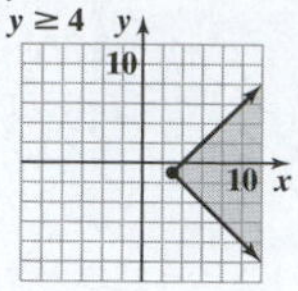

22. $3x + y \leq 9$, $2x + 3y \geq 6$, $x \geq 0$, $y \geq 0$

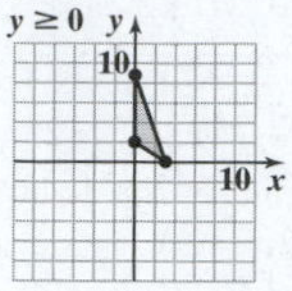

23. $-2 < x \leq 4$

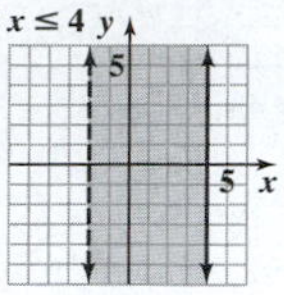

24. maximum: 26 **25.** 50 regular and 100 deluxe; $35,000

Cumulative Review Exercises

1. $\{-1\}$ **2.** $\{8\}$ **3.** $-\frac{2y^7}{3x^5}$ **4.** $22; 4a^2 - 6a + 4$ **5.** $2x^2 + x + 2; 12$ **6.** $f(x) = -\frac{1}{2}x + 4$

7. $f(x) = 2x + 1$

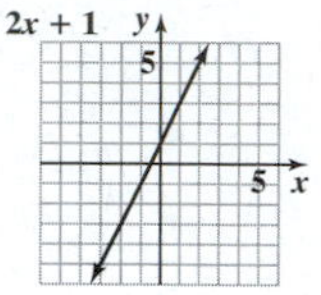

8. $y > 2x$

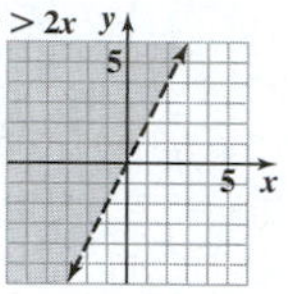

9. $2x - y \geq 6$

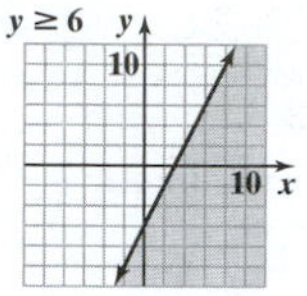

10. $f(x) = -1$

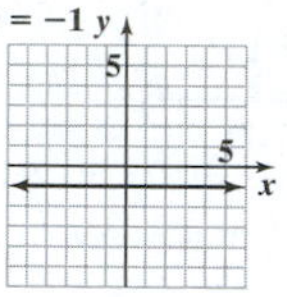

11. $\{(-4, 2, -1)\}$ **12.** $\{(-1, 2)\}$ **13.** -17 **14.** 46 rooms with kitchen facilities and 14 without kitchen facilities **15.** a. and b. are functions.

16. $\{x|x \geq -7\}; [-7, \infty)$

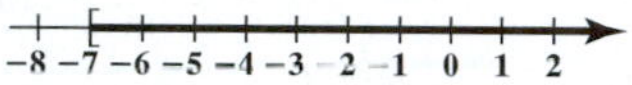

17. $\{x|x < -6\}; (-\infty, -6)$

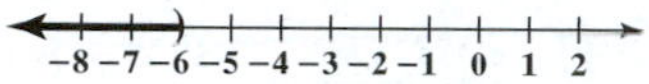

18. $\{x|x \leq 3 \text{ or } x \geq 5\}; (-\infty, 3] \cup [5, \infty)$

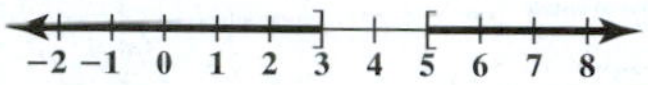

19. $\{x|-10 \leq x \leq 7\}; [-10, 7]$

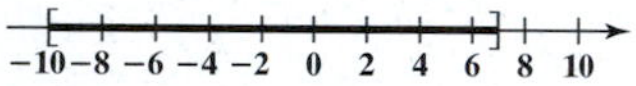

20. $\left\{x \middle| x < \frac{1}{3} \text{ or } x > 5\right\}; \left(-\infty, \frac{1}{3}\right) \cup (5, \infty)$

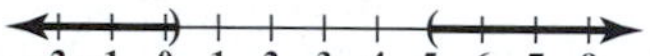

CHAPTER 5

Section 5.1

Check Point Exercises

1.

Term	Coefficient	Degree
$8x^4y^5$	8	9
$-7x^3y^2$	-7	5
$-x^2y$	-1	3
$-5x$	-5	1
11	11	0

The degree of the polynomial is 9, the leading term is $8x^4y^5$, and the leading coefficient is 8.

2. 16 **3.** The graph rises to the left and to the right. **4.** This would not be appropriate over long time periods. Since the graph falls to the right, at some point the ratio would be negative, which is not possible. **5.** The graph does not show the end behavior of the function. The graph should fall to the left. **6.** $-3x^3 + 10x^2 - 10$ **7.** $9xy^3 + 3xy^2 - 15y - 9$ **8.** $10x^3 - 2x^2 + 8x - 10$ **9.** $13x^2y^5 + 2xy^3 - 10$

Exercise Set 5.1

1.

Term	Coefficient	Degree
$-x^4$	-1	4
x^2	1	2

The degree of the polynomial is 4, the leading term is $-x^4$, and the leading coefficient is -1.

3.

Term	Coefficient	Degree
$5x^3$	5	3
$7x^2$	7	2
$-x$	-1	1
9	9	0

The degree of the polynomial is 3, the leading term is $5x^3$, and the leading coefficient is 5.

5.

Term	Coefficient	Degree
$3x^2$	3	2
$-7x^4$	-7	4
$-x$	-1	1
6	6	0

The degree of the polynomial is 4, the leading term is $-7x^4$, and the leading coefficient is -7.

7.

Term	Coefficient	Degree
x^3y^2	1	5
$-5x^2y^7$	-5	9
$6y^2$	6	2
-3	-3	0

The degree of the polynomial is 9, the leading term is $-5x^2y^7$, and the leading coefficient is -5.

9.

Term	Coefficient	Degree
x^5	1	5
$3x^2y^4$	3	6
$7xy$	7	2
$9x$	9	1
-2	-2	0

The degree of the polynomial is 6, the leading term is $3x^2y^4$, and the leading coefficient is 3.

11. 0 **13.** 12 **15.** 56 **17.** -29 **19.** -1 **21-23.** Graph #23 is not that of a polynomial function. **25.** falls to the left and falls to the right; graph (b) **27.** rises to the left and rises to the right; graph (a) **29.** $11x^3 + 7x^2 - 12x - 4$ **31.** $-\frac{2}{5}x^4 + x^3 + \frac{3}{8}x^2$ **33.** $9x^2y - 6xy$ **35.** $2x^2y + 15xy + 15$ **37.** $-9x^4y^2 - 6x^2y^2 - 5x^2y + 2xy$ **39.** $5x^{2n} - 2x^n - 6$ **41.** $12x^3 + 4x^2 + 12x - 14$ **43.** $22y^5 + 9y^4 + 7y^3 - 13y^2 + 3y - 5$ **45.** $-5x^3 + 8xy - 9y^2$ **47.** $x^4y^2 + 8x^3y + y - 6x$ **49.** $y^{2n} + 2y^n - 3$ **51.** $8a^2b^4 + 3ab^2 + 8ab$ **53.** $5x^3 + 3x^2y - xy^2 - 4y^3$ **55.** $5x^4 - x^3 + 5x^2 - 5x + 2$ **57.** $-10x^2y^2 + 4x^2 + 3$ **59.** $-4x^3 - x^2 + 4x + 8$; 7 **61.** $-8x^2 - 2x - 1$; -29 **63.** $-9x^3 - x^2 + 13x + 20$ **65.** 476,398; In 2000, the cumulative number of AIDS deaths in the U.S. was 476,398. **67.** (10, 476,398) **69.** 513,931; 514,071; g **71.** falls to the right; no; Cumulative number of deaths cannot decrease. **73.** no; The graph falls to the right, so eventually there will be a negative number of thefts, which is not possible. **87.** Answers will vary; an example is $f(x) = x^2 - x + 1$. **89.** Answers will vary; an example is $f(x) = -x^3 + x + 1$.

91.

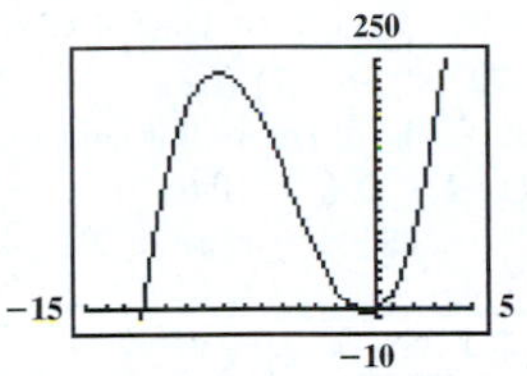

93.

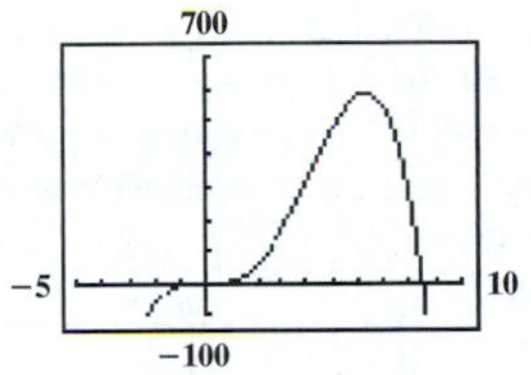

95.

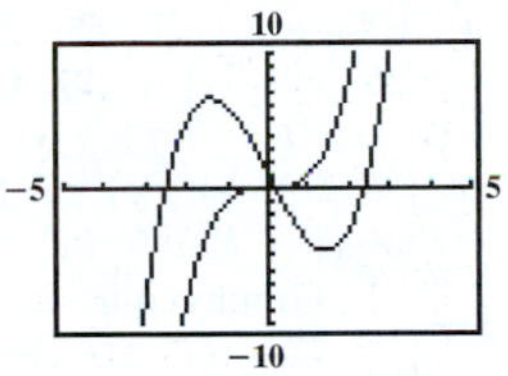

97. makes sense **99.** makes sense **101.** false **103.** false **105.** $3x^{2n} - x^n + 4$ **107.** $9x^2 - 3x + 5$ **108.** $\left\{\frac{2}{3}\right\}$

109.

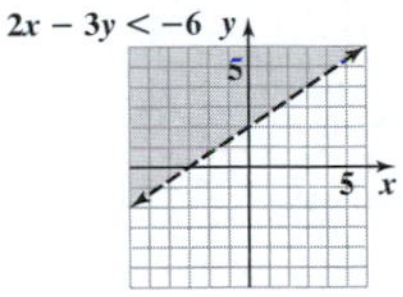

110. $y - 5 = 3(x + 2)$; $y = 3x + 11$ or $f(x) = 3x + 11$ **111.** $10x^7y^9$ **112.** $16x^8 + 6x^5$ **113.** $3x^3 + 19x^2 + 43x + 35$

Section 5.2

Check Point Exercises

1. a. $-18x^7y^{11}$ **b.** $30x^{10}y^6z^8$ **2. a.** $12x^9 - 18x^6 + 24x^4$ **b.** $10x^5y^9 - 8x^7y^7 - 10x^4y^3$ **3.** $6x^3 - 2x^2 - x + 2$ **4.** $12x^2y^6 - 8x^2y^4 + 6xy^5 + 2y^2$ **5. a.** $x^2 + 8x + 15$ **b.** $14x^2 + xy - 4y^2$ **c.** $4x^6 - 12x^4 - 5x^3 + 15x$ **6. a.** $x^2 + 16x + 64$ **b.** $16x^2 + 40xy + 25y^2$ **7. a.** $x^2 - 10x + 25$ **b.** $4x^2 - 24xy^4 + 36y^8$ **8. a.** $x^2 - 9$ **b.** $25x^2 - 49y^2$ **c.** $25a^2b^4 - 16a^2$ **9. a.** $9x^2 + 12x + 4 - 25y^2$ **b.** $4x^2 + 4xy + y^2 + 12x + 6y + 9$ **10. a.** $x^2 - 10x + 21$ **b.** 5 **11. a.** $a^2 + a - 2$ **b.** $2ah + h^2 - 5h$

Exercise Set 5.2

1. $15x^6$ **3.** $15x^3y^{11}$ **5.** $-6x^2y^9z^9$ **7.** $2x^{3n}y^{n-2}$ **9.** $12x^3 + 8x^2$ **11.** $2y^3 - 10y^2$ **13.** $10x^8 - 20x^5 + 45x^3$ **15.** $28x^2y + 12xy^2$ **17.** $18a^3b^5 + 15a^2b^3$ **19.** $-12x^6y^3 + 28x^3y^4 - 24x^2y$ **21.** $-12x^{3n} + 20x^{2n} - 2x^{n+1}$ **23.** $x^3 - x^2 - x - 15$ **25.** $x^3 - 1$ **27.** $a^3 - b^3$ **29.** $x^4 + 5x^3 + x^2 - 11x + 4$ **31.** $x^3 - 4x^2y + 4xy^2 - y^3$ **33.** $x^3y^3 + 8$ **35.** $x^2 + 11x + 28$ **37.** $y^2 - y - 30$ **39.** $10x^2 + 11x + 3$ **41.** $6y^2 - 11y + 4$ **43.** $15x^2 - 22x + 8$ **45.** $2x^2 + xy - 21y^2$ **47.** $14x^2y^2 - 19xy - 3$ **49.** $x^3 - 4x^2 - 5x + 20$ **51.** $8x^5 - 40x^3 + 3x^2 - 15$ **53.** $3x^{2n} + 5x^ny^n - 2y^{2n}$ **55.** $x^2 + 6x + 9$ **57.** $y^2 - 10y + 25$ **59.** $4x^2 + 4xy + y^2$ **61.** $25x^2 - 30xy + 9y^2$ **63.** $4x^4 + 12x^2y + 9y^2$ **65.** $16x^2y^4 - 8x^2y^3 + x^2y^2$ **67.** $a^{2n} + 8a^nb^n + 16b^{2n}$ **69.** $x^2 - 16$ **71.** $25x^2 - 9$ **73.** $16x^2 - 49y^2$ **75.** $y^6 - 4$ **77.** $1 - y^{10}$ **79.** $49x^2y^4 - 100y^2$ **81.** $25a^{2n} - 49$ **83.** $4x^2 + 12x + 9 - 16y^2$ **85.** $x^2 + 2xy + y^2 - 9$ **87.** $25x^2 + 70xy + 49y^2 - 4$ **89.** $25y^2 - 4x^2 - 12x - 9$ **91.** $x^2 + 2xy + y^2 + 2x + 2y + 1$ **93.** $x^4 - 1$ **95. a.** $x^2 + 4x - 12$ **b.** -15 **c.** -12 **97. a.** $x^3 - 27$ **b.** -35 **c.** -27 **99. a.** $a^2 + a + 5$ **b.** $2ah + h^2 - 3h$ **101. a.** $3a^2 + 14a + 15$ **b.** $6ah + 3h^2 + 2h$ **103.** $48xy$ **105.** $-9x^2 + 3x + 9$ **107.** $16x^4 - 625$ **109.** $x^3 - 3x^2 + 3x - 1$ **111.** $(2x - 7)^2 = 4x^2 - 28x + 49$ **113. a.** $x^2 + 6x + 4x + 24$ **b.** $(x + 6)(x + 4) = x^2 + 10x + 24$ **115. a.** $x^2 + 12x + 27$ **b.** $x^2 + 6x + 5$ **c.** $6x + 22$ **117. a.** $4x^2 - 36x + 80$ **b.** $4x^3 - 36x^2 + 80x$ **119. a.** $V(x) = -2x^3 + 10x^2 + 300x$ **b.** rises to the left and falls to the right **c.** no; Because the graph falls to the right, volume will eventually be negative, which is not possible. **d.** 2000; Carry-on luggage with a depth of 10 inches has a volume of 2000 cubic inches. **e.** (10, 2000) **f.** $\{x \mid 0 < x < 15\}$ or (0, 15), although answers may vary.

131.

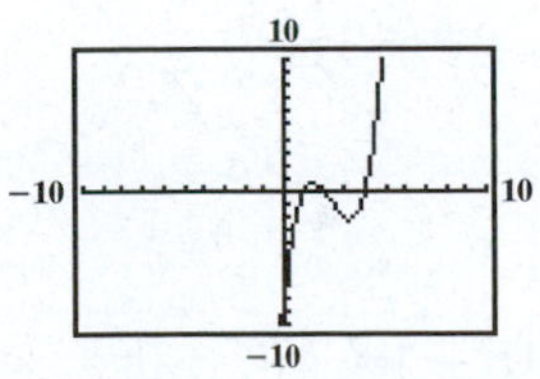

conclusion: $y_1 = y_2$

133.

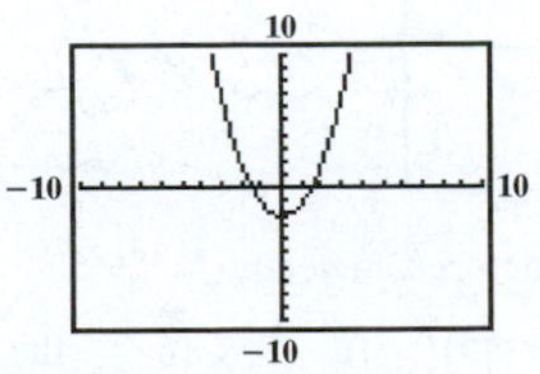

conclusion: $y_1 = y_2$

135. makes sense **137.** makes sense, although answers may vary **139.** false **141.** false **143.** $x^2 + 2x$ **145.** $2x^3 + 12x^2 + 12x + 10$

147. 9 and 11 **148.** $\left\{x \middle| x \le -\frac{14}{3} \text{ or } x \ge 2\right\}$ or $\left(-\infty, -\frac{14}{3}\right] \cup [2, \infty)$ **149.** $\{x|x \ge -3\}$ or $[-3, \infty)$ **150.** 8.034×10^9 **151. a.** $3x^2$ **b.** $6x^2y^2$ **152.** $x^3 - 5x^2 + 3x - 15$ **153.** $3x^3 - 2xy + 12x - 8y$

Section 5.3

Check Point Exercises

1. $10x(2x + 3)$ **2. a.** $3x^2(3x^2 + 7)$ **b.** $5x^3y^2(3 - 5xy)$ **c.** $4x^2y^3(4x^2y^2 - 2xy + 1)$ **3.** $-2x(x^2 - 5x + 3)$ **4. a.** $(x - 4)(3 + 7a)$ **b.** $(a + b)(7x - 1)$ **5.** $(x - 4)(x^2 + 5)$ **6.** $(x + 5)(4x - 3y)$

Exercise Set 5.3

1. $2x(5x + 2)$ **3.** $y(y - 4)$ **5.** $x^2(x + 5)$ **7.** $4x^2(3x^2 - 2)$ **9.** $2x^2(16x^2 + x + 4)$ **11.** $2xy(2xy^2 + 3)$ **13.** $10xy^2(3xy - 1)$ **15.** $2x(6y - 3z + 2w)$ **17.** $3x^2y^4(5xy^2 - 3x^2 + 4y)$ **19.** $5x^2y^4z^2(5xy^2 - 3x^2z^2 + 5yz)$ **21.** $5x^n(3x^n - 5)$ **23.** $-4(x - 3)$ **25.** $-8(x + 6)$ **27.** $-2(x^2 - 3x + 7)$ **29.** $-5(y^2 - 8x)$ **31.** $-4x(x^2 - 8x + 5)$ **33.** $-1(x^2 + 7x - 5)$ **35.** $(x + 3)(4 + a)$ **37.** $(y - 6)(x - 7)$ **39.** $(x + y)(3x - 1)$ **41.** $(3x - 1)(4x^2 + 1)$ **43.** $(x + 3)(2x + 1)$ **45.** $(x + 3)(x + 5)$ **47.** $(x + 7)(x - 4)$ **49.** $(x - 3)(x^2 + 4)$ **51.** $(y - 6)(x + 2)$ **53.** $(y + 1)(x - 7)$ **55.** $(5x - 6y)(2x + 7y)$ **57.** $(4x - 1)(x^2 - 3)$ **59.** $(x - a)(x - b)$ **61.** $(x - 3)(x^2 + 4)$ **63.** $(a - b)(y - x)$ **65.** $(a + 2b)(y^2 - 3x)$ **67.** $(x^n + 1)(y^n + 3)$ **69.** $(a + 1)(b - c)$ **71.** $(x^3 - 5)(1 + 4y)$ **73.** $y^6(3x - 1)^4(6xy - 2y - 7)$ **75.** $(x^2 + 5x - 2)(a + b)$ **77.** $(x + y + z)(a - b + c)$ **79. a.** 16; The ball is 16 feet above the ground after 2 seconds. **b.** 0; The ball is on the ground after 2.5 seconds. **c.** $-8t(2t - 5)$; $f(t) = -8t(2t - 5)$ **d.** 16; 0; yes; no; Answers will vary. **81. a.** $(x - 0.4x)(1 - 0.4) = (0.6x)(0.6) = 0.36x$ **b.** no; 36% **83.** $A = P + Pr + (P + Pr)r = P(1 + r) + Pr(1 + r) = (1 + r)(P + Pr) = P(1 + r)(1 + r) = P(1 + r)^2$ **85.** $A = r(\pi r + 2l)$

93.

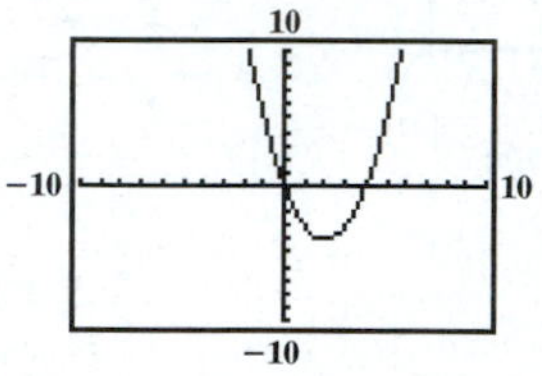

Graphs coincide.; factored correctly

95.

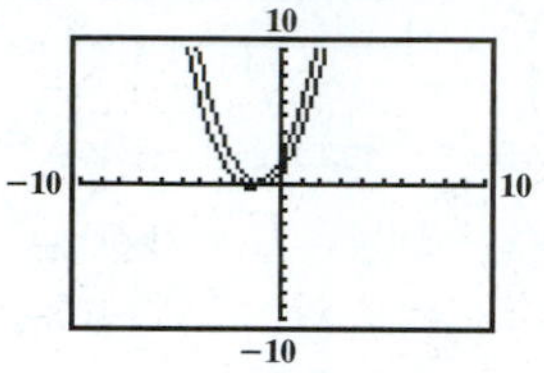

Graphs do not coincide.; $x^2 + 2x + x + 2 = (x + 2)(x + 1)$

97. does not make sense **99.** makes sense **101.** false **103.** true **105.** $x^{2n}(x^{2n} + 1 + x^n)$ **107.** $4y^{2n}(2y^4 + 4y^3 - 3)$ **109.** Answers will vary; an example is $6x^2 - 4x + 9x - 6$. **110.** $\left\{\left(\frac{20}{11}, -\frac{14}{11}\right)\right\}$ **111. a.** function **b.** not a function **112.** length: 8 ft; width: 3 ft **113.** 4 **114.** 2 **115.** 7

Section 5.4

Check Point Exercises

1. $(x + 4)(x + 2)$ or $(x + 2)(x + 4)$ **2.** $(x - 5)(x - 4)$ **3.** $(y + 22)(y - 3)$ **4.** $(x - 3y)(x - 2y)$ **5.** $3x(x - 7)(x + 2)$ **6.** $(x^3 - 5)(x^3 - 2)$ **7.** $(3x - 14)(x - 2)$ **8.** $x^4(3x - 1)(2x + 7)$ **9.** $(2x - y)(x - 3y)$ **10.** $(3y^2 - 2)(y^2 + 4)$ **11.** $(2x - 5)(4x - 1)$

Exercise Set 5.4

1. $(x + 3)(x + 2)$ **3.** $(x + 6)(x + 2)$ **5.** $(x + 5)(x + 4)$ **7.** $(y + 8)(y + 2)$ **9.** $(x - 3)(x - 5)$ **11.** $(y - 2)(y - 10)$ **13.** $(a + 7)(a - 2)$ **15.** $(x + 6)(x - 5)$ **17.** $(x + 4)(x - 7)$ **19.** $(y + 4)(y - 9)$ **21.** prime **23.** $(x - 2y)(x - 7y)$ **25.** $(x + 5y)(x - 6y)$ **27.** prime **29.** $(a - 10b)(a - 8b)$ **31.** $3(x + 3)(x - 2)$ **33.** $2x(x - 3)(x - 4)$ **35.** $3y(y - 2)(y - 3)$ **37.** $2x^2(x + 3)(x - 16)$ **39.** $(x^3 + 2)(x^3 - 3)$ **41.** $(x^2 - 6)(x^2 + 1)$ **43.** $(x + 6)(x + 2)$ **45.** $(3x + 5)(x + 1)$ **47.** $(x + 11)(5x + 1)$ **49.** $(y + 8)(3y - 2)$ **51.** $(y + 2)(4y + 1)$ **53.** $(2x + 3)(5x + 2)$ **55.** $(4x - 3)(2x - 3)$ **57.** $(y - 3)(6y - 5)$ **59.** prime **61.** $(x + y)(3x + y)$ **63.** $(2x + y)(3x - 5y)$ **65.** $(5x - 2y)(3x - 5y)$ **67.** $(3a - 7b)(a + 2b)$ **69.** $5x(3x - 2)(x - 1)$ **71.** $2x^2(3x + 2)(4x - 1)$ **73.** $y^3(5y + 1)(3y - 1)$ **75.** $3(8x + 9y)(x - y)$ **77.** $2b(a + 3)(3a - 10)$ **79.** $2y(2x - y)(3x - 7y)$ **81.** $13x^3y(y + 4)(y - 1)$ **83.** $(2x^2 - 3)(x^2 + 1)$ **85.** $(2x^3 + 5)(x^3 + 3)$ **87.** $(2y^5 + 1)(y^5 + 3)$ **89.** $(5x + 12)(x + 2)$ **91.** $(2x - 13)(x - 2)$ **93.** $(x - 0.3)(x - 0.2)$ **95.** $\left(x + \frac{3}{7}\right)\left(x - \frac{1}{7}\right)$ **97.** $(ax - b)(cx + d)$ **99.** $-x^3y^2(4x - 3y)(x - y)$ **101.** $f(x) = 3x - 13$ and $g(x) = x - 3$, or vice versa **103.** $2x + 1$ by $x + 3$ **105. a.** 32; The diver is 32 feet above the water after 1 second. **b.** 0; The diver hits the water after 2 seconds. **c.** $-16(t - 2)(t + 1)$; $f(t) = -16(t - 2)(t + 1)$ **d.** 32; 0 **107. a.** $x^2 + x + x + x + 1 + 1 = x^2 + 3x + 2$ **b.** $(x + 2)(x + 1)$ **c.** Answers will vary.

117.

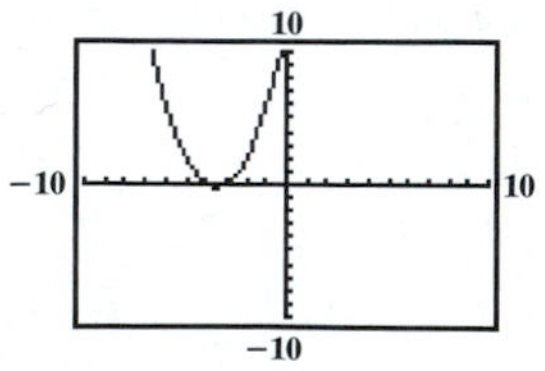

Graphs coincide.; factored correctly

119.

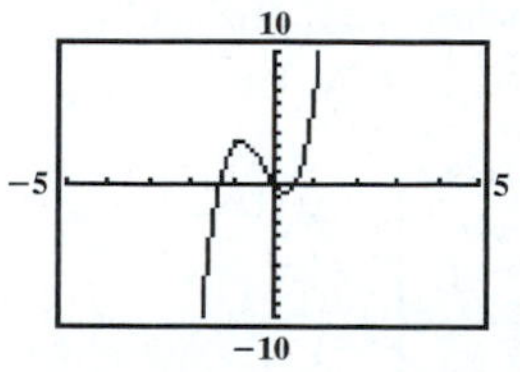

Graphs coincide.; factored correctly

123. makes sense **125.** makes sense **127.** true **129.** false **131.** −16, −8, 8, 16 **133.** $(4x^n - 5)(x^n - 1)$ **135.** $b^2(b^n - 2)(b^n + 5)$ **137.** $d^n(2d - 3)(d - 1)$ **138.** $\{x|x > 5\}$ or $(5, \infty)$ **139.** $\{(2, -1, 3)\}$ **140.** $(x + 2)(4x^2 - 5)$ **141.** $(x + 7)(x + 7)$ or $(x + 7)^2$ **142.** $(x - 4)(x - 4)$ or $(x - 4)^2$ **143.** $(x + 5)(x - 5)$

Mid-Chapter Check Point Exercises

1. $-x^3 + 4x^2 + 6x + 17$ **2.** $-2x^7y^3z^5$ **3.** $30x^5y^3 - 35x^3y^2 - 2x^2y$ **4.** $3x^3 + 4x^2 - 39x + 40$ **5.** $2x^4 - x^3 - 8x^2 + 11x - 4$ **6.** $-x^2 - 5x + 5$ **7.** $-4x^3y - 6x^2y - 7y - 5$ **8.** $8x^2 + 18x - 5$ **9.** $10x^2y^2 - 11xy - 6$ **10.** $9x^2 - 4y^2$ **11.** $6x^3y - 9xy^2 + 2x^2 - 3y$ **12.** $49x^6y^2 - 25x^2$ **13.** $6xh + 3h^2 - 2h$ **14.** $x^4 - 6x^2 + 9$ **15.** $x^5 + 2x^3 + 2x^2 - 15x - 6$ **16.** $4x^2 + 20xy + 25y^2$ **17.** $x^2 + 12x + 36 - 9y^2$ **18.** $x^2 + 2xy + y^2 + 10x + 10y + 25$ **19.** $(x - 8)(x + 3)$ **20.** $(5x + 2)(3y + 1)$ **21.** $(5x - 2)(x + 2)$ **22.** $5(7x^2 + 2x - 10)$ **23.** $9(x - 2)(x + 1)$ **24.** $5x^2y(2xy - 4y + 7)$ **25.** $(3x + 1)(6x + 5)$ **26.** $(4x - 3y)(3x - 4)$ **27.** $(3x - 4)(3x - 1)$ **28.** $(3x^3 + 5)(x^3 + 2)$ **29.** $x(5x - 2)(5x + 7)$ **30.** $(x^3 - 3)(2x - y)$

Section 5.5

Check Point Exercises

1. a. $(4x + 5)(4x - 5)$ **b.** $(10y^3 + 3x^2)(10y^3 - 3x^2)$ **2.** $6y(1 + xy^3)(1 - xy^3)$ **3.** $(4x^2 + 9)(2x + 3)(2x - 3)$ **4.** $(x + 7)(x + 2)(x - 2)$ **5. a.** $(x + 3)^2$ **b.** $(4x + 5y)^2$ **c.** $(2y^2 - 5)^2$ **6.** $(x + 5 + y)(x + 5 - y)$ **7.** $(a + b - 2)(a - b + 2)$ **8. a.** $(x + 3)(x^2 - 3x + 9)$ **b.** $(x^2 + 10y)(x^4 - 10x^2y + 100y^2)$ **9. a.** $(x - 2)(x^2 + 2x + 4)$ **b.** $(1 - 3xy)(1 + 3xy + 9x^2y^2)$

Exercise Set 5.5

1. $(x + 2)(x - 2)$ **3.** $(3x + 5)(3x - 5)$ **5.** $(3 + 5y)(3 - 5y)$ **7.** $(6x + 7y)(6x - 7y)$ **9.** $(xy + 1)(xy - 1)$ **11.** $(3x^2 + 5y^3)(3x^2 - 5y^3)$ **13.** $(x^7 + y^2)(x^7 - y^2)$ **15.** $(x - 3 + y)(x - 3 - y)$ **17.** $(a + b - 2)(a - b + 2)$ **19.** $(x^n + 5)(x^n - 5)$ **21.** $(1 + a^n)(1 - a^n)$ **23.** $2x(x + 2)(x - 2)$ **25.** $2(5 + y)(5 - y)$ **27.** $8(x + y)(x - y)$ **29.** $2xy(x + 3)(x - 3)$ **31.** $a(ab + 7c)(ab - 7c)$ **33.** $5y(1 + xy^3)(1 - xy^3)$ **35.** $8(x^2 + y^2)$ **37.** prime **39.** $(x^2 + 4)(x + 2)(x - 2)$ **41.** $(9x^2 + 1)(3x + 1)(3x - 1)$ **43.** $2x(x^2 + y^2)(x + y)(x - y)$ **45.** $(x + 3)(x + 2)(x - 2)$ **47.** $(x - 7)(x + 1)(x - 1)$ **49.** $(x + 2)^2$ **51.** $(x - 5)^2$ **53.** $(x^2 - 2)^2$ **55.** $(3y + 1)^2$ **57.** $(8y - 1)^2$ **59.** $(x - 6y)^2$ **61.** prime **63.** $(3x + 8y)^2$ **65.** $(x - 3 + y)(x - 3 - y)$ **67.** $(x + 10 + x^2)(x + 10 - x^2)$ **69.** $(3x - 5 + 6y)(3x - 5 - 6y)$ **71.** $(x^2 + x + 1)(x^2 - x - 1)$ **73.** $(z + x - 2y)(z - x + 2y)$ **75.** $(x + 4)(x^2 - 4x + 16)$ **77.** $(x - 3)(x^2 + 3x + 9)$ **79.** $(2y + 1)(4y^2 - 2y + 1)$ **81.** $(5x - 2)(25x^2 + 10x + 4)$ **83.** $(xy + 3)(x^2y^2 - 3xy + 9)$ **85.** $x(4 - x)(16 + 4x + x^2)$ **87.** $(x^2 + 3y)(x^4 - 3x^2y + 9y^2)$ **89.** $(5x^2 - 4y^2)(25x^4 + 20x^2y^2 + 16y^4)$ **91.** $(x + 1)(x^2 - x + 1)(x^6 - x^3 + 1)$ **93.** $(x - 2y)(x^2 - xy + y^2)$ **95.** $(0.2x + 0.3)^2$ or $\frac{1}{100}(2x + 3)^2$ **97.** $x\left(2x - \frac{1}{2}\right)\left(4x^2 + x + \frac{1}{4}\right)$ **99.** $(x - 1)(x^2 + x + 1)(x - 2)(x^2 + 2x + 4)$ **101.** $(x^4 + 1)(x^2 + 4)(x + 2)(x - 2)$ **103.** $(x + 1)(x - 1)(x - 2)(x^2 + 2x + 4)$ **105. a.** $(A + B)^2$ **b.** A^2; AB; AB; B^2 **c.** $A^2 + 2AB + B^2$ **d.** $A^2 + 2AB + B^2 = (A + B)^2$; factoring a perfect square trinomial **107.** $25x^2 - 9 = (5x + 3)(5x - 3)$ **109.** $49x^2 - 36 = (7x + 6)(7x - 6)$ **111.** $3a^3 - 3ab^2 = 3a(a + b)(a - b)$

117.
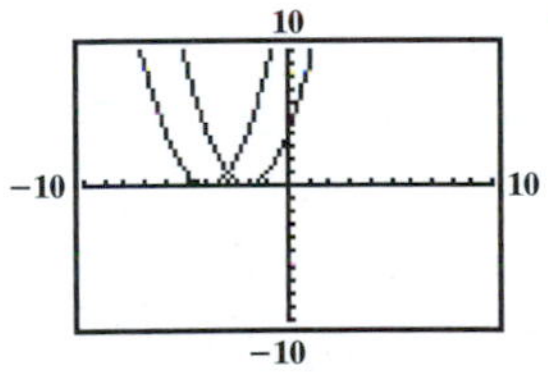

; Graphs do not coincide.; $x^2 + 4x + 4 = (x + 2)^2$

119.
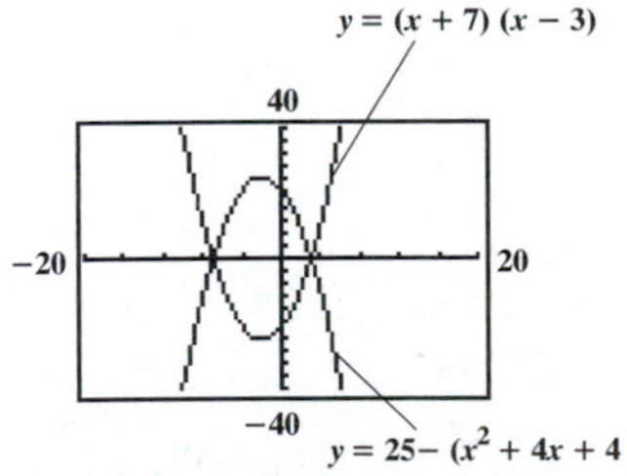

; Graphs do not coincide.; $25 - (x^2 + 4x + 4) = (7 + x)(3 - x)$

121.
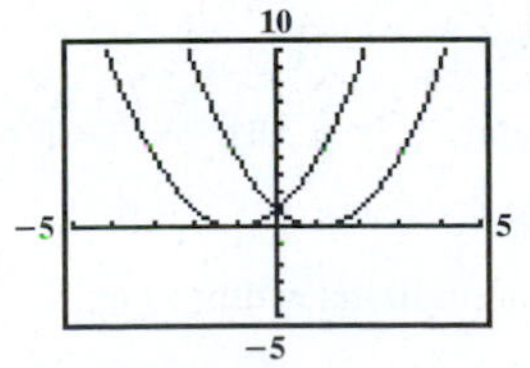

; Graphs do not coincide.; $(x - 3)^2 + 8(x - 3) + 16 = (x + 1)^2$

123.
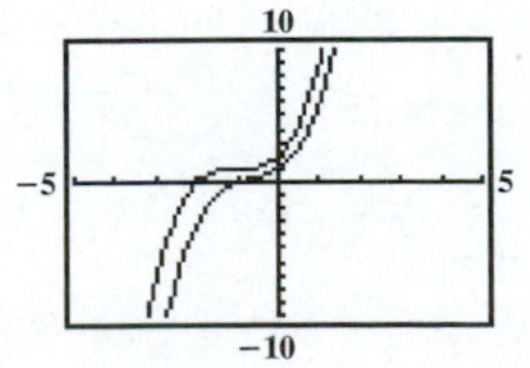

; Graphs do not coincide.; $(x + 1)^3 + 1 = (x + 2)(x^2 + x + 1)$

125. makes sense **127.** does not make sense **129.** false **131.** false **133.** $(y + x)(y^2 - xy + x^2 + 1)$ **135.** $(x^n + y^{4n})(x^{2n} - x^ny^{4n} + y^{8n})$

137. $x^6 - y^6 = (x^3 + y^3)(x^3 - y^3) = (x + y)(x^2 - xy + y^2)(x - y)(x^2 + xy + y^2)$;
$x^6 - y^6 = (x^2 - y^2)(x^4 + x^2y^2 + y^4) = (x + y)(x - y)(x^4 + x^2y^2 + y^4)$; $x^4 + x^2y^2 + y^4 = (x^2 - xy + y^2)(x^2 + xy + y^2)$
139. 1 **140.** $\{x|5 \le x \le 11\}$ or $[5, 11]$ **141.** $\{(-2, 1)\}$ **142.** $(x + 7)(3x - y)$ **143.** $2x(x + 2)^2$ **144.** $5x(x - y)(x - 7y)$
145. $(x + y)(3b + 4)(3b - 4)$

Section 5.6

Check Point Exercises

1. $3x(x - 5)^2$ **2.** $3y(x + 2)(x - 6)$ **3.** $(x + y)(4a + 5)(4a - 5)$
4. $(x + 10 + 6a)(x + 10 - 6a)$ **5.** $x(x + 2)(x^2 - 2x + 4)(x^6 - 8x^3 + 64)$

Exercise Set 5.6

1. $x(x + 4)(x - 4)$ **3.** $3(x + 3)^2$ **5.** $3(3x - 1)(9x^2 + 3x + 1)$ **7.** $(x + 4)(x - 4)(y - 2)$ **9.** $2b(2a + 5)(a - 3)$
11. $(y + 2)(y - 2)(a - 4)$ **13.** $11x(x^2 + y)(x^2 - y)$ **15.** $4x(x^2 + 4)(x + 2)(x - 2)$ **17.** $(x - 4)(x + 3)(x - 3)$
19. $2x^2(x + 3)(x^2 - 3x + 9)$ **21.** $3y(x^2 + 4y^2)(x + 2y)(x - 2y)$ **23.** $3x(2x + 3y)^2$ **25.** $(x - 6 + 7y)(x - 6 - 7y)$ **27.** prime
29. $12xy(x + y)(x - y)$ **31.** $6b(x^2 + y^2)$ **33.** $(x + y)(x - y)(x^2 + xy + y^2)$ **35.** $(x + 6 + 2a)(x + 6 - 2a)$ **37.** $(x^3 - 2)(x^3 + 7)$
39. $(2x - 7)(2 + x^2)$ **41.** $2(3x - 2y)(9x^2 + 6xy + 4y^2)$ **43.** $(x + 5 + y)(x + 5 - y)$ **45.** $(x^4 + y^4)(x^2 + y^2)(x + y)(x - y)$
47. $xy(x + 4y)(x - 4y)$ **49.** $x(1 + 2x)(1 - 2x + 4x^2)$ **51.** $2(4y + 1)(2y - 1)$ **53.** $y(14y^2 + 7y - 10)$ **55.** $3(3x + 2y)^2$
57. $3x(4x^2 + y^2)$ **59.** $x^3y^3(xy - 1)(x^2y^2 + xy + 1)$ **61.** $2(x + 5)(x - 5)$ **63.** $(x - y)(a + 2)(a - 2)$
65. $(c + d - 1)(c + d + 1)[(c + d)^2 + 1]$ **67.** $(p + q)^2(p - q)$ **69.** $(x + 2y)(x - 2y)(x + y)(x - y)$
71. $(x + y + 3)^2$ **73.** $(x - y)^2(x - y + 2)(x - y - 2)$ **75.** $(2x - y^2)(x - 3y^2)$ **77.** $(x - y)(x^2 + xy + y^2 - 1)$
79. $(xy + 1)(x^2y^2 - xy + 1)(x - 2)(x^2 + 2x + 4)$ **81. a.** $x(x + y) - y(x + y)$ **b.** $(x + y)(x - y)$
83. a. $xy + xy + xy + 3x(x) = 3xy + 3x^2$ **b.** $3x(y + x)$ **85. a.** $8x^2 - 2\pi x^2$ **b.** $2x^2(4 - \pi)$

89.

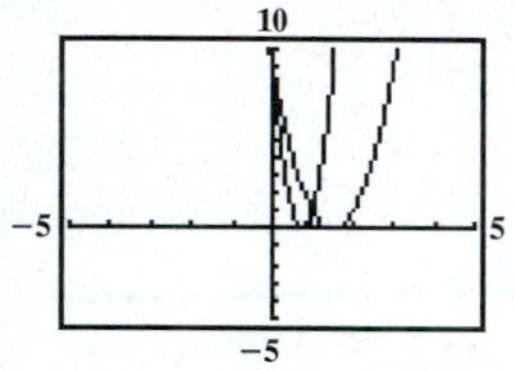

Graphs do not coincide.; $4x^2 - 12x + 9 = (2x - 3)^2$

91.

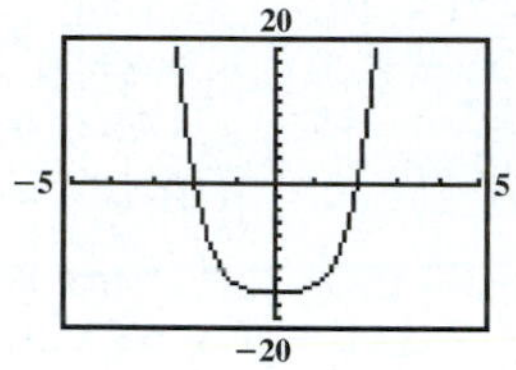

Graphs coincide.; factored correctly

95. makes sense **97.** does not make sense **99.** true **101.** false **103.** $x^n(3x - 1)(x - 4)$ **105.** $(x^2 + 2x + 2)(x^2 - 2x + 2)$
107. $\{-1\}$ **108.** $\frac{8x^9}{y^3}$ **109.** 52 **110.** 0 **111.** 0 **112.** $(x - 5)(x + 3) = 0$

Section 5.7

Check Point Exercises

1. $-\frac{1}{2}$ and 5, or $\left\{-\frac{1}{2}, 5\right\}$ **2. a.** 0 and $\frac{2}{3}$, or $\left\{0, \frac{2}{3}\right\}$ **b.** 5 or $\{5\}$ **c.** -4 and 3, or $\{-4, 3\}$
3. $-2, -\frac{3}{2}$, and 2, or $\left\{-2, -\frac{3}{2}, 2\right\}$ **4.** after 3 seconds; (3, 336) **5.** 2 ft **6.** 5 ft

Exercise Set 5.7

1. $\{-4, 3\}$ **3.** $\{-7, 1\}$ **5.** $\left\{-4, \frac{2}{3}\right\}$ **7.** $\left\{\frac{3}{5}, 1\right\}$ **9.** $\left\{-2, \frac{1}{3}\right\}$ **11.** $\{0, 8\}$ **13.** $\left\{0, \frac{5}{3}\right\}$ **15.** $\{-2\}$ **17.** $\{7\}$ **19.** $\left\{\frac{5}{3}\right\}$
21. $\{-5, 5\}$ **23.** $\left\{-\frac{10}{3}, \frac{10}{3}\right\}$ **25.** $\{-3, 6\}$ **27.** $\{-3, -2\}$ **29.** $\{4\}$ **31.** $\{-2, 8\}$ **33.** $\left\{-1, \frac{2}{5}\right\}$ **35.** $\{-6, -3\}$
37. $\{-5, -4, 5\}$ **39.** $\{-5, 1, 5\}$ **41.** $\{-4, 0, 4\}$ **43.** $\{0, 2\}$ **45.** $\{-5, -3, 0\}$ **47.** 2 and 4; d **49.** -4 and -2; c **51.** $\{-7, -1, 6\}$
53. $\left\{-\frac{4}{3}, 0, \frac{4}{5}, 2\right\}$ **55.** -4 and 8 **57.** $-2, -\frac{1}{2}$, and 2 **59.** -7 and 4 **61.** 2 and 3 **63.** -9 and 5 **65.** 1 second; 0.25, 0.5, 0.75, 1
67. 7 **69.** (7, 21) **71.** length: 9 ft; width: 6 ft **73.** 5 in. **75.** 5 m **77. a.** $4x^2 + 44x$ **b.** 3 ft **79.** length: 10 in.; width: 10 in.
81. length: 12 ft; width: 5 ft **83.** $30\frac{1}{8}$ ft

97.

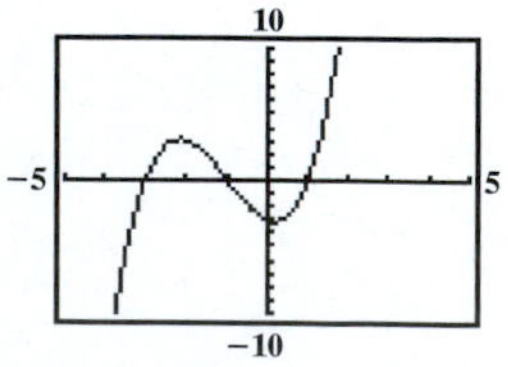

$\{-3, -1, 1\}$

99.

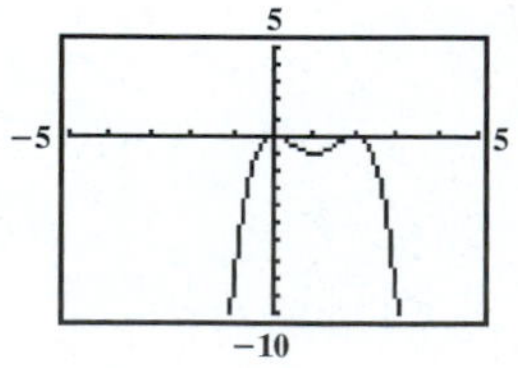

$\{0, 2\}$

101. makes sense **103.** makes sense **105.** false **107.** true **109.** Answers will vary; an example is $x^2 - 4x - 21 = 0$.

111. $\left\{-2, \frac{10}{3}\right\}$ **112.** 12 **113.** \$1700 at 5% and \$1300 at 8% **114.** 30 **115.** $\{x \mid x \text{ is a real number and } x \neq 2\}$
116. $\frac{(x-9)(x+2)}{(2x-1)(x+2)}; \frac{x-9}{2x-1}$

Review Exercises

1.

Term	Coefficient	Degree
$-5x^3$	-5	3
$7x^2$	7	2
$-x$	-1	1
2	2	0

The degree of the polynomial is 3, the leading term is $-5x^3$, and the leading coefficient is -5.

2.

Term	Coefficient	Degree
$8x^4y^2$	8	6
$-7xy^6$	-7	7
$-x^3y$	-1	4

The degree of the polynomial is 7, the leading term is $-7xy^6$, and the leading coefficient is -7.

3. -31 **4. a.** 380,621; There were 380,621 Americans living with AIDS in 2002. **b.** underestimates by 150 **5.** rises to the left and falls to the right; c **6.** rises to the left and rises to the right; b **7.** falls to the left and rises to the right; a **8.** falls to the left and falls to the right; d **9. a.** 20,787; 23,416; Values are the same as those shown for 1999 and 2000. **b.** The graph of this function falls to the right which indicates that the number of newly declared computer majors would eventually be negative, which is not possible. **10.** $x^3 - 6x^2 - x - 9$ **11.** $12x^3y - 2x^2y - 14y - 17$ **12.** $15x^3 - 10x^2 + 11x - 8$ **13.** $-2x^3y^2 - 4x^3y - 8$ **14.** $3x^3 + 5x^2y - xy^2 - 8y^3$ **15.** $-12x^6y^2z^7$ **16.** $2x^8 - 24x^5 - 12x^3$ **17.** $21x^5y^4 - 35x^2y^3 - 7xy^2$ **18.** $6x^3 + 29x^2 + 27x - 20$ **19.** $x^4 + 4x^3 + 4x^2 - x - 2$ **20.** $12x^2 - 23x + 5$ **21.** $15x^2y^2 + 2xy - 8$ **22.** $9x^2 + 42xy + 49y^2$ **23.** $x^4 - 10x^2y + 25y^2$ **24.** $4x^2 - 49y^2$ **25.** $9x^2y^4 - 16x^2$ **26.** $x^2 + 6x + 9 - 25y^2$ **27.** $x^2 + 2xy + y^2 + 8x + 8y + 16$ **28.** $2x^2 - x - 15$; 21 **29. a.** $a^2 - 9a + 10$ **b.** $2ah + h^2 - 7h$ **30.** $8x^2(2x + 3)$ **31.** $2x(1 - 18x)$ **32.** $7xy(3xy - 2y + 1)$ **33.** $9x^2y(2xy - 3)$ **34.** $-4(3x^2 - 2x + 12)$ **35.** $-1(x^2 + 11x - 14)$ **36.** $(x - 1)(x^2 - 2)$ **37.** $(y - 3)(x - 5)$ **38.** $(x - 3y)(5a + 2b)$ **39.** $(x + 5)(x + 3)$ **40.** $(x + 20)(x - 4)$ **41.** $(x + 17y)(x - y)$ **42.** $3x(x - 1)(x - 11)$ **43.** $(x + 7)(3x + 1)$ **44.** $(3x - 2)(2x - 3)$ **45.** $(5x + 4y)(x - 2y)$ **46.** $x(3x + 4)(2x - 1)$ **47.** $(x + 3)(2x + 5)$ **48.** $(x^3 + 6)(x^3 - 5)$ **49.** $(x^2 + 3)(x^2 - 13)$ **50.** $(x + 11)(x + 9)$ **51.** $(5x^3 + 2)(x^3 + 3)$ **52.** $(2x + 5)(2x - 5)$ **53.** $(1 + 9xy)(1 - 9xy)$ **54.** $(x^4 + y^3)(x^4 - y^3)$ **55.** $(x - 1 + y)(x - 1 - y)$ **56.** $(x + 8)^2$ **57.** $(3x - 1)^2$ **58.** $(5x + 2y)^2$ **59.** prime **60.** $(5x - 4y)^2$ **61.** $(x + 9 + y)(x + 9 - y)$ **62.** $(z + 5x - 1)(z - 5x + 1)$ **63.** $(4x + 3)(16x^2 - 12x + 9)$ **64.** $(5x - 2)(25x^2 + 10x + 4)$ **65.** $(xy + 1)(x^2y^2 - xy + 1)$ **66.** $3x(5x + 1)$ **67.** $3x^2(2x + 1)(2x - 1)$ **68.** $4x(5x^3 - 6x^2 + 7x - 3)$ **69.** $x(x - 2)(x - 13)$ **70.** $-2y^2(y - 3)(y - 9)$ **71.** $(3x - 5)^2$ **72.** $5(x + 3)(x - 3)$ **73.** $(2x - 1)(x + 3)(x - 3)$ **74.** $(3x - y)(2x - 7y)$ **75.** $2y(y + 3)^2$ **76.** $(x + 3 + 2a)(x + 3 - 2a)$ **77.** $(2x - 3)(4x^2 + 6x + 9)$ **78.** $x(x^2 + 1)(x + 1)(x - 1)$ **79.** $(x^2 - 3)^2$ **80.** prime **81.** $4(a + 2)(a^2 - 2a + 4)$ **82.** $(x^2 + 9)(x + 3)(x - 3)$ **83.** $(a + 3b)(x - y)$ **84.** $(3x - 5y)(9x^2 + 15xy + 25y^2)$ **85.** $2xy(x + 3)(5x - 4)$ **86.** $(2x^3 + 5)(3x^3 - 1)$ **87.** $(x + 5)(2 + xy)$ **88.** $(y + 2)(y + 5)(y - 5)$ **89.** $(a^4 + 1)(a^2 + 1)(a + 1)(a - 1)$ **90.** $(x - 4)(3 + y)(3 - y)$ **91. a.** $2xy + 2y^2$ **b.** $2y(x + y)$ **92. a.** $x^2 - 4y^2$ **b.** $(x + 2y)(x - 2y)$ **93.** $\{-5, -1\}$ **94.** $\left\{\frac{1}{3}, 7\right\}$ **95.** $\{-8, 7\}$ **96.** $\{0, 4\}$ **97.** $\{-5, -3, 3\}$ **98.** 9 seconds **99. a.** 20 miles per hour **b.** (20, 40) **c.** As a car's speed increases, its stopping distance gets longer at increasingly greater rates. **100.** length: 9 ft; width: 6 ft **101.** 2 in. **102.** 50 yd, 120 yd, and 130 yd

Chapter Test

1. degree: 3; leading coefficient: -6 **2.** degree: 9; leading coefficient: 7 **3.** 6; 4 **4.** falls to the left and falls to the right **5.** falls to the left and rises to the right **6.** $7x^3y - 18x^2y - y - 9$ **7.** $11x^2 - 13x - 6$ **8.** $35x^7y^3$ **9.** $x^3 - 4x^2y + 2xy^2 + y^3$ **10.** $21x^2 - 20xy - 9y^2$ **11.** $4x^2 - 25y^2$ **12.** $16y^2 - 56y + 49$ **13.** $x^2 + 4x + 4 - 9y^2$ **14.** $3x^2 + x - 10$; 60 **15.** $2ah + h^2 - 5h$ **16.** $x^2(14x - 15)$ **17.** $(9y + 5)(9y - 5)$ **18.** $(x + 3)(x + 5)(x - 5)$ **19.** $(5x - 3)^2$ **20.** $(x + 5 + 3y)(x + 5 - 3y)$ **21.** prime **22.** $(y + 2)(y - 18)$ **23.** $(2x + 5)(7x + 3)$ **24.** $5(x - 1)(x^2 + x + 1)$ **25.** $3(2x + y)(2x - y)$ **26.** $2(3x - 1)(2x - 5)$ **27.** $3(x^2 + 1)(x + 1)(x - 1)$ **28.** $(x^4 + y^4)(x^2 + y^2)(x + y)(x - y)$ **29.** $4x^2y(3y^3 + 2xy - 9)$ **30.** $(x^3 + 2)(x^3 - 14)$ **31.** $(x^2 - 6)(x^2 + 4)$ **32.** $3y(4x - 1)(x - 2)$ **33.** $y(y - 3)(y^2 + 2)$ **34.** $\left\{-\frac{1}{3}, 2\right\}$ **35.** $\left\{-1, \frac{6}{5}\right\}$ **36.** $\left\{0, \frac{1}{3}\right\}$ **37.** $\{-1, 1, 4\}$ **38.** 7 seconds **39.** length: 5 yd; width: 3 yd **40.** 12 units, 9 units, and 15 units

Cumulative Review Exercises

1. $\{4\}$ **2.** $\left\{\left(-4, \frac{1}{2}\right)\right\}$ **3.** $\{(2, 1, -1)\}$ **4.** $\{x \mid 2 < x < 3\}$ or $(2, 3)$ **5.** $\{x \mid x \leq -2 \text{ or } x \geq 7\}$ or $(-\infty, -2] \cup [7, \infty)$ **6.** $\left\{1, \frac{5}{2}\right\}$ **7.** $\{-5, 0, 2\}$ **8.** $x = \frac{b}{c - a}$ **9.** $f(x) = 2x + 1$ **10.** winner: 1480 votes; loser: 1320 votes

11.

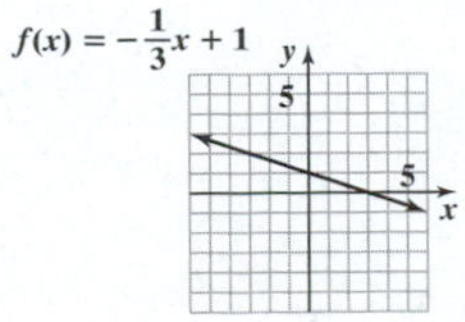

12.

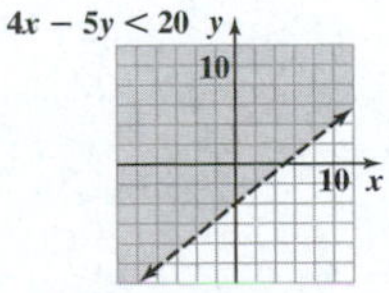

13.

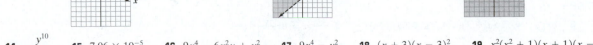

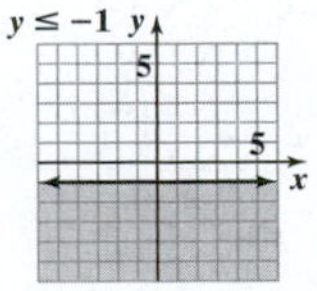

14. $-\frac{y^{10}}{2x^6}$ **15.** 7.06×10^{-5} **16.** $9x^4 - 6x^2y + y^2$ **17.** $9x^4 - y^2$ **18.** $(x + 3)(x - 3)^2$ **19.** $x^2(x^2 + 1)(x + 1)(x - 1)$ **20.** $14x^3y^2(1 - 2x)$

CHAPTER 6

Section 6.1

Check Point Exercises

1. a. 80; The cost to remove 40% of the lake's pollutants is \$80 thousand.; (40, 80) **b.** 180; The cost to remove 60% of the lake's pollutants is \$180 thousand.; (60, 180) **2.** $\left\{x \middle| x \text{ is a real number and } x \neq -3 \text{ and } x \neq \frac{1}{2}\right\}$ or $(-\infty, -3) \cup \left(-3, \frac{1}{2}\right) \cup \left(\frac{1}{2}, \infty\right)$ **3.** $x + 5$
4. a. $\frac{x-5}{3x-1}$ **b.** $\frac{x+4y}{3x(x+y)}$ **5.** $\frac{x+3}{x-4}$ **6.** $\frac{-4(x+2)}{3x(3x-2)}$ or $-\frac{4(x+2)}{3x(3x-2)}$ **7. a.** $9(3x+7)$ **b.** $\frac{x+3}{5}$

Exercise Set 6.1

1. -5; -3; 2 **3.** 0; does not exist; $-\frac{21}{2}$ **5.** $-\frac{7}{2}$; -5; $\frac{11}{5}$ **7.** $\{x|x$ is a real number and $x \neq 5\}$ or $(-\infty, 5) \cup (5, \infty)$
9. $\{x|x$ is a real number and $x \neq -3$ and $x \neq 1\}$ or $(-\infty, -3) \cup (-3, 1) \cup (1, \infty)$ **11.** $\{x|x$ is a real number and $x \neq -5\}$ or $(-\infty, -5) \cup (-5, \infty)$
13. $\{x|x$ is a real number and $x \neq 3$ and $x \neq 5\}$ or $(-\infty, 3) \cup (3, 5) \cup (5, \infty)$ **15.** $\{x|x$ is a real number and $x \neq -\frac{4}{3}$ and $x \neq 2\}$ or $\left(-\infty, -\frac{4}{3}\right) \cup \left(-\frac{4}{3}, 2\right) \cup (2, \infty)$ **17.** 4 **19.** domain: $\{x|x$ is a real number and $x \neq -2$ and $x \neq 2\}$ or $(-\infty, -2) \cup (-2, 2) \cup (2, \infty)$; range: $\{y|y \leq 0$ or $y > 3\}$ or $(-\infty, 0] \cup (3, \infty)$ **21.** As x decreases, the function values are approaching 3.; $y = 3$
23. There is no point on the graph with x-coordinate -2. **25.** The graph is not continuous. Furthermore, it neither rises nor falls without bound to either the left or the right. **27.** $x + 2$ **29.** $\frac{1}{x-3}$ **31.** $\frac{4}{x}$ **33.** $\frac{4}{y+5}$ **35.** $-\frac{1}{3x+5}$ **37.** $\frac{y+7}{y-7}$ **39.** $\frac{x+9}{x-1}$ **41.** cannot be simplified **43.** $-\frac{x+3}{x+4}$ **45.** $\frac{x+5y}{3x-y}$ **47.** $\frac{x^2+2x+4}{x+2}$ **49.** $x^2 - 3$ **51.** $\frac{3}{2}$ **53.** $\frac{x+7}{x}$ **55.** $\frac{x+3}{x-2}$ **57.** $\frac{x+4}{x+2}$
59. $-\frac{2}{y(y+3)}$ **61.** $\frac{y^2+2y+4}{2y}$ **63.** $\frac{x^2+x+1}{x-2}$ **65.** 4 **67.** $\frac{4(x+y)}{3(x-y)}$ **69.** 1 **71.** 1 **73.** $\frac{9}{28}$ **75.** $\frac{7}{10}$ **77.** $\frac{x}{3}$ **79.** $\frac{y-5}{2}$
81. $\frac{(x^2+4)(x-4)}{x-1}$ **83.** $\frac{y-7}{y-5}$ **85.** $(x-1)(2x-1)$ **87.** $\frac{1}{x-2y}$ **89.** $3x^2$ **91.** $\frac{x+1}{x(x-1)}$ **93.** $\frac{-(x-y)}{(x+y)(b+c)}$ **95.** $\frac{a^2+1}{b(a+2)}$
97. $-\frac{a-b}{4c^2}$ or $\frac{b-a}{4c^2}$ **99.** 7 **101.** $2a + h - 5$ **103.** $\left(\frac{f}{g}\right)(x) = -x - 2$; $\left\{x \middle| x \text{ is a real number and } x \neq -2 \text{ and } x \neq \frac{1}{2}\right\}$ or $(-\infty, -2) \cup \left(-2, \frac{1}{2}\right) \cup \left(\frac{1}{2}, \infty\right)$ **105.** 195; The cost to inoculate 60% of the population is \$195 million.; (60, 195)
107. 100; This indicates that we cannot inoculate 100% of the population. **109.** after 6 minutes; about 4.8 **111.** 6.5; Over time the pH level rises back to normal. **113.** 90; Lung cancer has an incidence ratio of 10 between the ages of 55 and 64. 90% of deaths from lung cancer in this group are smoking related.; (10, 90) **115.** $y = 100$; As the incidence ratio increases, the percentage of smoking-related deaths is approaching 100%.

129.

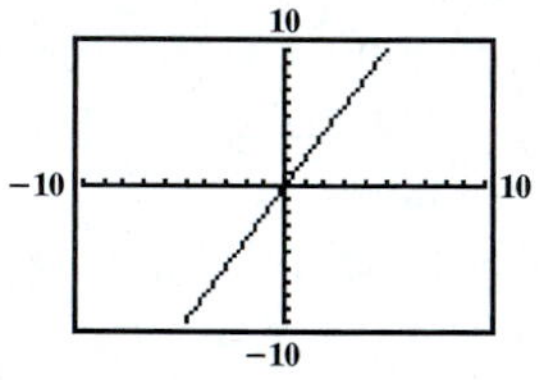

Graphs coincide.; multiplied correctly

131.

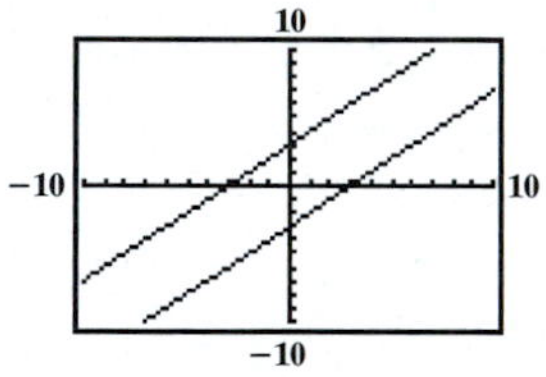

Graphs do not coincide; right side should be $x + 3$.

135. makes sense **137.** does not make sense **139.** false **141.** false

143.

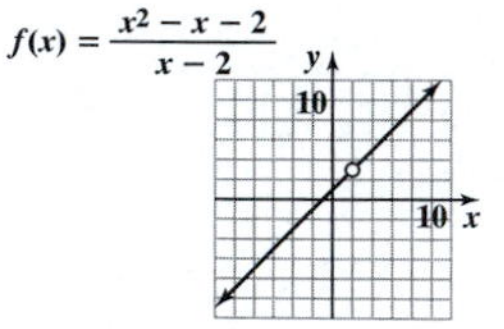

145. $\frac{y^n-1}{y^n+4}$ **147.**

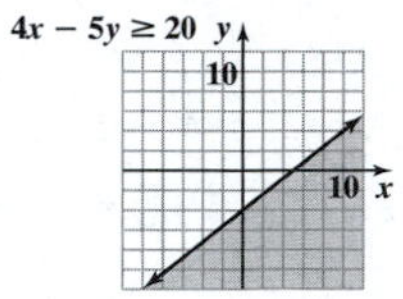

148. $2x^3 - 11x^2 + 3x + 30$ **149.** $16a^8b^{26}c^2$

150. $\frac{2}{5}$ **151.** $\frac{7}{6}$ **152.** $\frac{1}{6}$

Section 6.2

Check Point Exercises

1. $\frac{x-5}{x+3}$ **2.** $\frac{1}{x-y}$ **3.** $18x^2$ **4.** $5x(x+3)(x+3)$ or $5x(x+3)^2$ **5.** $\frac{21+4x}{18x^2}$ or $\frac{4x+21}{18x^2}$
6. $\frac{2x^2-2x+8}{(x-4)(x+4)}$ or $\frac{2(x^2-x+4)}{(x-4)(x+4)}$ **7.** $\frac{x^2-3x+5}{(x-3)(x+1)(x-2)}$ **8.** $\frac{3}{y+2}$ **9.** $\frac{3x-5y}{x-3y}$

Exercise Set 6.2

1. $\frac{2}{3x}$ **3.** $\frac{10x+3}{x-5}$ **5.** $\frac{2x-1}{x+3}$ **7.** $\frac{y-3}{y+3}$ **9.** $\frac{x+1}{4x-3}$ **11.** $\frac{x-3}{x-1}$ **13.** $\frac{4y+3}{2y-1}$ **15.** $\frac{x^2+xy+y^2}{x+y}$ **17.** $175x^2$ **19.** $(x-5)(x+5)$ **21.** $y(y+10)(y-10)$ **23.** $(x+4)(x-4)(x-4)$ **25.** $(y-5)(y-6)(y+1)$ **27.** $(y+2)(2y+3)(y-2)(2y+1)$ **29.** $\frac{3+50x}{5x^2}$ **31.** $\frac{7x-2}{(x-2)(x+1)}$ **33.** $\frac{3x-4}{(x+2)(x-3)}$ **35.** $\frac{2x^2-2x+61}{(x+5)(x-6)}$ **37.** $\frac{20-x}{(x-5)(x+5)}$ **39.** $\frac{13}{(y-3)(y-2)}$ **41.** $\frac{1}{x+3}$ **43.** $\frac{6x^2+14x+10}{(x+1)(x+4)(x+3)}$ **45.** $-\frac{x^2-3x-13}{(x+4)(x-2)(x+1)}$ **47.** $\frac{4x-11}{x-3}$ **49.** $\frac{2y-14}{(y-4)(y+4)}$ **51.** $\frac{x^2+2x-14}{3(x+2)(x-2)}$ **53.** $\frac{16}{x-4}$ **55.** $\frac{12x+7y}{(x-y)(x+y)}$ **57.** $\frac{x^2-18x-30}{(5x+6)(x-2)}$ **59.** $\frac{3x^2+16xy-13y^2}{(x+5y)(x-5y)(x-4y)}$ **61.** $\frac{8x^3-32x^2+23x+6}{(x+2)(x-2)(x-2)(x-1)}$ **63.** $\frac{16a^2-12ab-18b^2}{(3a+4b)(3a-4b)(2a-b)}$ **65.** $\frac{m+6}{(m-1)(m+2)(2m-1)}$ **67.** $\frac{x^2+5x+8}{(x+2)(x+1)}$ **69.** 2 **71.** $-\frac{1}{x(x+h)}$ **73.** $\frac{2d}{a^2+ab+b^2}$ **75.** $(f-g)(x)=\frac{x+2}{x+3}$; $\{x \mid x$ is a real number and $x \neq -5$ and $x \neq -3\}$ or $(-\infty,-5)\cup(-5,-3)\cup(-3,\infty)$ **77.** 11; If you average 0 miles per hour over the speed limits, the total driving time is about 11 hours.; (0, 11) **79.** $\frac{720x+48{,}050}{(x+70)(x+65)}$; 11 **81.** 12 mph; Answers will vary. **83. a.** 307; There are 307 arrests for every 100,000 drivers who are 20 years old.; (20, 307) **b.** $f(x)=\frac{-5x^3+27{,}680x-388{,}150}{x^2+9}$ **c.** 25; 356 **85.** $\frac{2x(2x+15)}{(x+7)(x+8)}$ **91.** Answers will vary.; $\frac{b+a}{ab}$ **93.** makes sense **95.** does not make sense **97.** false **99.** false **101.** $\frac{1}{x^{2n}-1}$ **103.** $\frac{x-y+1}{(x-y)(x-y)}$ **104.** $\frac{y^{10}}{9x^4}$ **105.** $\left\{x \middle| -\frac{13}{3} \le x \le 5\right\}$ or $\left[-\frac{13}{3}, 5\right]$ **106.** $2x(5x+3)(5x-3)$ **107.** xy^2+y^3 **108.** $-h$ **109.** $\frac{x+1}{x-3}$

Section 6.3

Check Point Exercises

1. $\frac{y}{x+y}$ **2.** $-\frac{1}{x(x+7)}$ **3.** $\frac{2x}{x^2+1}$ **4.** $\frac{x+2}{x-5}$

Exercise Set 6.3

1. $\frac{4x+2}{x-3}$ **3.** $\frac{x^2+9}{(x-3)(x+3)}$ **5.** $\frac{y+x}{y-x}$ **7.** $\frac{4-x}{5x-3}$ **9.** $\frac{1}{x-3}$ **11.** $-\frac{1}{x(x+5)}$ **13.** $-x$ **15.** $-\frac{x+1}{x-1}$ **17.** $\frac{(x+y)(x+y)}{xy}$ **19.** $\frac{4x}{x^2+4}$ **21.** $\frac{2y^3+5x^2}{y^3(5-3x^2)}$ **23.** $-\frac{12}{5}$ **25.** $\frac{3ab(b+a)}{(2b+3a)(2b-3a)}$ **27.** $-\frac{x+10}{5x-2}$ **29.** $\frac{2y}{3y+7}$ **31.** $\frac{2b+a}{b-2a}$ **33.** $\frac{4(7x+5)}{69(x+5)}$ **35.** $\frac{3x-2y}{13y+5x}$ **37.** $\frac{m(m+3)}{(m-2)(m+1)}$ **39.** $\frac{3a(a-3)}{(a+4)(3a-5)}$ **41.** $-\frac{4}{(x+2)(x-2)}$ **43.** 6 **45.** $x(x+1)$ **47.** $\frac{x+4}{x+2}$ **49.** $-\frac{3}{a(a+h)}$ **51. a.** $\frac{Pi(1+i)^n}{(1+i)^n-1}$ **b.** \$527 per month **53.** $R=\frac{R_1R_2R_3}{R_2R_3+R_1R_3+R_1R_2}$; about 2.18 ohms

59.

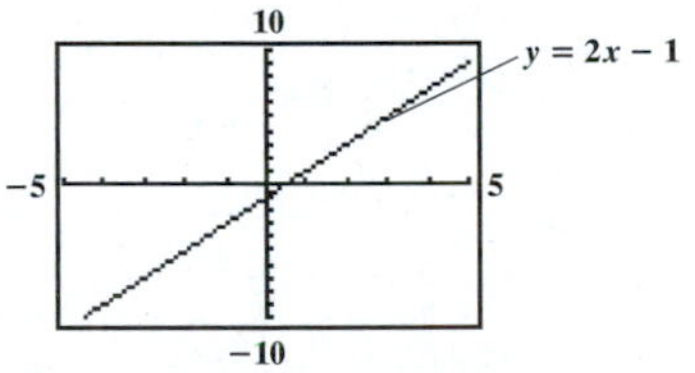

Graphs coincide.; simplified correctly

61.

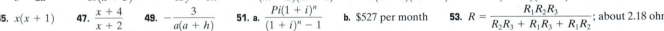

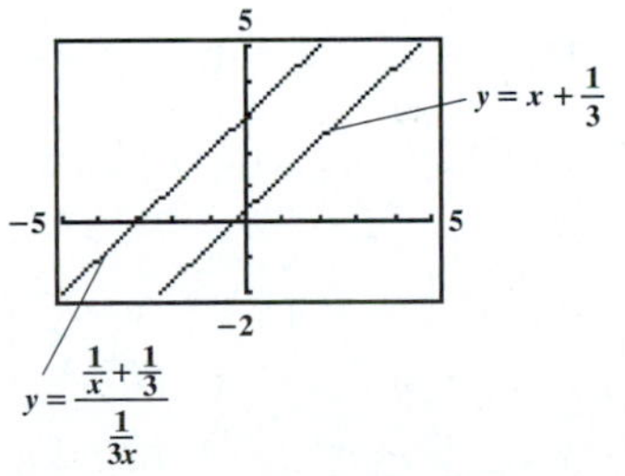

Graphs do not coincide.; $\frac{\frac{1}{x}+\frac{1}{3}}{\frac{1}{3x}}=x+3$

63. does not make sense **65.** does not make sense **67.** $\frac{1}{(x+h+1)(x+1)}$ **69.** $\frac{a+1}{a+2}$ **71.** $\{3, 9\}$ **72.** $16x^4-8x^2y+y^2$ **73.** $\{x \mid 1 < x < 5\}$ or $(1, 5)$ **74.** $2xy^3$ **75.** $35+\frac{2}{21}$ **76.** $7x$

Section 6.4

Check Point Exercises

1. $4x^2-8x+\frac{1}{2}+\frac{1}{x}$ **2.** $3x^2y^2-xy+\frac{2}{y}$ **3.** $3x-8$ **4.** $2x^2-x+3-\frac{6}{2x-1}$ **5.** $2x^2+7x+14+\frac{21x-10}{x^2-2x}$

Exercise Set 6.4

1. $5x^4 - 3x^2 + 2$ **3.** $6x^2 + 2x - 3 - \frac{2}{x}$ **5.** $7x - \frac{7}{4} - \frac{4}{x}$ **7.** $-5x^3 + 10x^2 - \frac{3}{5}x + 8$ **9.** $2a^2b - a - 3b$ **11.** $6x - \frac{3}{y} - \frac{2}{xy^2}$ **13.** $x + 3$ **15.** $x^2 + x - 2$ **17.** $x - 2 + \frac{2}{x - 5}$ **19.** $x + 5 - \frac{10}{2x + 3}$ **21.** $x^2 + 2x + 3 + \frac{1}{x + 1}$ **23.** $2y^2 + 3y - 1$ **25.** $3x^2 - 3x + 1 + \frac{2}{3x + 2}$ **27.** $2x^2 + 4x + 5 + \frac{9}{2x - 4}$ **29.** $2y^2 + y - 2 - \frac{2}{2y - 1}$ **31.** $2y^3 + 3y^2 - 4y + 1$ **33.** $4x^2 + 3x - 8 + \frac{18}{x^2 + 3}$ **35.** $5x^2 + x + 3 + \frac{x + 7}{3x^2 - 1}$ **37.** $\left(\frac{f}{g}\right)(x) = 2x^2 - 9x + 10$ **39.** $\left(\frac{f}{g}\right)(x) = x^3 - x^2 + x - 2$ **41.** $x^3 - x^2y + xy^2 - y^3 + \frac{2y^4}{x + y}$ **43.** $3x^2 + 2x - 1$ **45.** $4x - 7 + \frac{4x + 8}{x^2 + x + 1}$ **47.** $x^3 + x^2 - x - 3 + \frac{-x + 5}{x^2 - x + 2}$ **49.** $4x^2 + 5xy - y^2$ **51.** $\left(\frac{f - g}{h}\right)(x) = 2x^2 - 5x + 8$; $\left\{x \mid x \text{ is a real number and } x \neq -\frac{1}{4}\right\}$ or $\left(-\infty, -\frac{1}{4}\right) \cup \left(-\frac{1}{4}, \infty\right)$ **53.** $x = 3a^2 + 4a - 2$ **55.** 70; At the tax rate of 30%, the government tax revenue is \$70 tens of billions, or \$700 billion.; (30, 70) **57.** $f(x) = 80 + \frac{800}{x - 110}$; 70; yes; Answers will vary.

65.

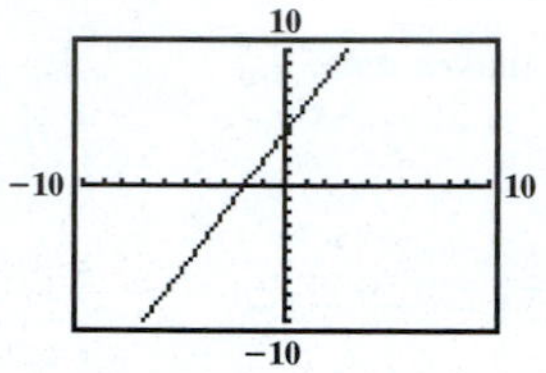

Graphs coincide.; division correct

67.

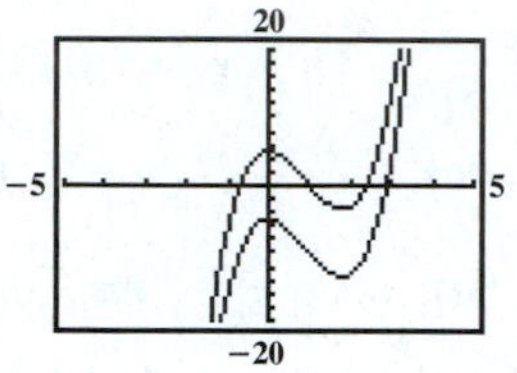

Graphs do not coincide; right side should be $3x^3 - 8x^2 - 5$.

69. makes sense **71.** makes sense **73.** false **75.** false **77.** $x^{2n} + x^n + 3 + \frac{3}{x^n - 5}$ **79.** $x - 2$ **81.** $\left\{x \,\middle|\, x < -\frac{1}{2} \text{ or } x > \frac{7}{2}\right\}$ or $\left(-\infty, -\frac{1}{2}\right) \cup \left(\frac{7}{2}, \infty\right)$ **82.** 4.061×10^7 **83.** $22x + 12$ **84. a.** $5x^2 - 10x + 26 - \frac{44}{x + 2}$ **b.** −10; 26; −44; The numbers are the coefficients of the quotient and the remainder. **85. a.** $3x^2 - 7x + 9 - \frac{10}{x + 1}$ **b.** −7; 9; −10; The numbers are the coefficients of the quotient and the remainder. **86.** $2x^2 + 3x - 2$; $(x - 3)(2x - 1)(x + 2)$

Mid-Chapter Check Point Exercises

1. $\frac{x + 2}{x + 6}$ **2.** $\frac{3(x - 2)}{x - 1}$ **3.** $\frac{x^2 + 3x + 9}{x - 1}$ **4.** $\frac{5x - 3}{x - 2}$ **5.** $\frac{x^2 + 6x - 4}{x^2 + 6x + 1}$ **6.** $2x^3 - 5x^2 - 3x + 6$ **7.** $x + y$ **8.** $4x^6y^2 - 2x^4y + \frac{3}{7}y$ **9.** $\frac{x^2 - 14x - 16}{(x + 6)(x - 2)}$ **10.** $\frac{10}{(x + 2)(x - 2)}$ **11.** $\frac{2(x - 3)}{x - 1}$ **12.** $\frac{x - 5}{x - 7}$ **13.** $2x^2 - x - 3 + \frac{x + 1}{3x^2 - 1}$ **14.** $\frac{5x + 2}{3x - 1}$ **15.** $\frac{5x}{(x - 6)(x - 1)(x + 4)}$ **16.** $\frac{7x + 4}{4(x + 1)}$ **17.** $x + 2$ **18.** $16x^2 - 8x + 4 - \frac{2}{2x + 1}$ **19.** $\frac{3x + 2}{(x + 2)(x - 1)}$ **20.** Domain: $\{x \mid x \text{ is a real number and } x \neq -7 \text{ and } x \neq 2\}$ or $(-\infty, -7) \cup (-7, 2) \cup (2, \infty)$; $f(x) = \frac{5}{x + 7}$

Section 6.5

Check Point Exercises

1. $x^2 - 2x - 3$ **2.** −105 **3.** The remainder is zero.; -1, $-\frac{1}{3}$, and $\frac{2}{5}$, or $\left\{-1, -\frac{1}{3}, \frac{2}{5}\right\}$

Exercise Set 6.5

1. $2x + 5$ **3.** $3x - 8 + \frac{20}{x + 5}$ **5.** $4x^2 + x + 4 + \frac{3}{x - 1}$ **7.** $6x^4 + 12x^3 + 22x^2 + 48x + 93 + \frac{187}{x - 2}$ **9.** $x^3 - 10x^2 + 51x - 260 + \frac{1300}{5 + x}$ **11.** $3x^2 + 3x - 3$ **13.** $x^4 + x^3 + 2x^2 + 2x + 2$ **15.** $x^3 + 4x^2 + 16x + 64$ **17.** $2x^4 - 7x^3 + 15x^2 - 31x + 64 - \frac{129}{x + 2}$ **19.** −25 **21.** −133 **23.** 240 **25.** 1 **27.** The remainder is 0.; $\{-1, 2, 3\}$ **29.** The remainder is 0.; $\left\{-\frac{1}{2}, 1, 2\right\}$ **31.** The remainder is 0.; $\left\{-5, \frac{1}{3}, \frac{1}{2}\right\}$ **33.** 2; The remainder is zero.; −3, −1, and 2, or $\{-3, -1, 2\}$ **35.** 1; The remainder is zero.; $\frac{1}{3}$, $\frac{1}{2}$, and 1, or $\left\{\frac{1}{3}, \frac{1}{2}, 1\right\}$ **37.** $4x^3 - 9x^2 + 7x - 6$ **39.** $0.5x^2 - 0.4x + 0.3$ **41. a.** The remainder is 0. **b.** 3 mm

47. a. For Exercise 27:

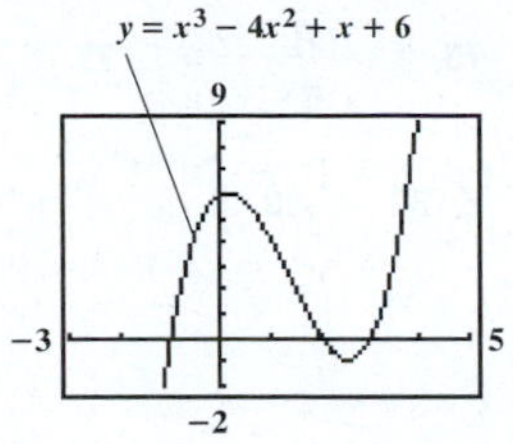

b. For Exercise 29:

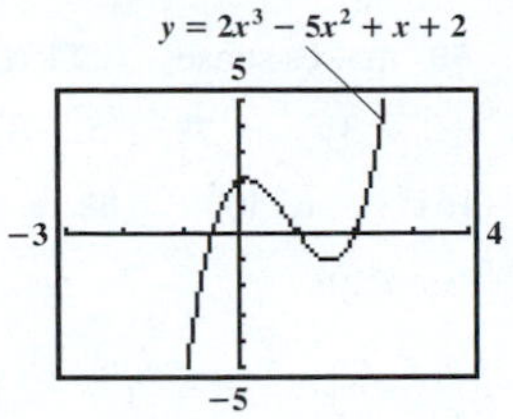

c. For Exercise 31:

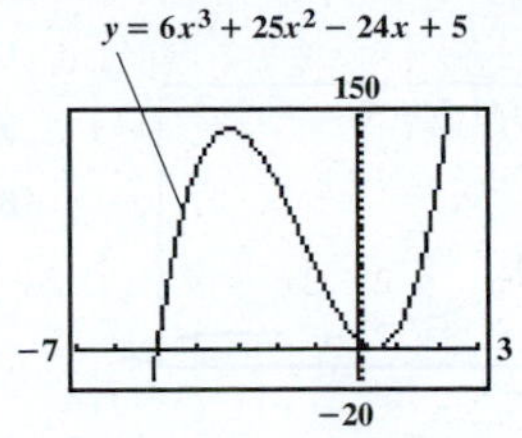

49. makes sense **51.** makes sense **53.** The remainder is 0.; $\{-2, -1, 2, 5\}$ **54.** $\{x|x > -10\}$ or $(-10, \infty)$ **55.** $\{-5, -1\}$ **56.** $\{(5, 3)\}$ **57.** $6x$ **58.** $3x$ **59.** $(x + 3)(x - 3)$

Section 6.6

Check Point Exercises

1. 6 or $\{6\}$ **2.** 4 or $\{4\}$ **3.** no solution or $\varnothing$ **4.** 4 and 6, or $\{4, 6\}$ **5.** 1 and 7, or $\{1, 7\}$ **6.** 50%

Exercise Set 6.6

1. $\{1\}$ **3.** $\{3\}$ **5.** $\{4\}$ **7.** $\{-4\}$ **9.** $\left\{\frac{9}{7}\right\}$ **11.** no solution or $\varnothing$ **13.** $\{2\}$ **15.** $\{2\}$ **17.** $\{5\}$ **19.** $\{-7, -1\}$ **21.** $\{-6, 3\}$ **23.** $\{-1\}$ **25.** no solution or $\varnothing$ **27.** $\{2\}$ **29.** no solution or $\varnothing$ **31.** $\left\{-\frac{4}{3}\right\}$ **33.** $\left\{-6, \frac{1}{2}\right\}$ **35.** 4, 10 **37.** $-3, -\frac{3}{4}$ **39.** $\frac{x - 2}{x(x + 1)}$ **41.** $\{2\}$ **43.** $\{4\}$ **45.** $\frac{-2(x - 4)}{(x - 2)(x^2 + 2x + 4)}$ **47.** 0 **49.** 1, 7 **51.** 60% **53.** 10 days; (10, 8) **55.** $y = 5$; On average, the students remembered 5 words over an extended period of time. **57.** 11 learning trials; (11, 0.95) **59.** As the number of learning trials increases, the proportion of correct responses increases.; Initially, the proportion of correct responses increases rapidly, but slows down as time increases. **61.** 125 liters **63. a.** $f(x) = 0.45x + 7.2$ **b.** $g(x) = -0.38x + 7.4$ **c.** $h(x) = \frac{0.45x + 7.2}{-0.38x + 7.4}$ **d.** approximately 1.5 times as much **e.** approximately 1.5 times as much; Rounded to one decimal place, it gives the exact number. **f.** 2011 **g.** (10, 3.25)

71.

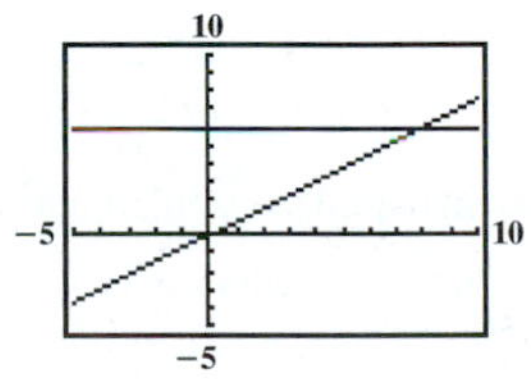

; $\{8\}$

73.

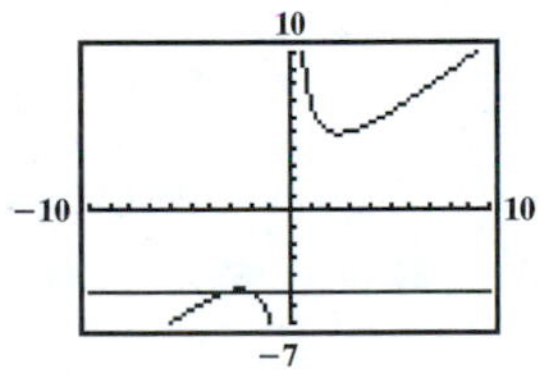

; $\{-3, -2\}$

75.

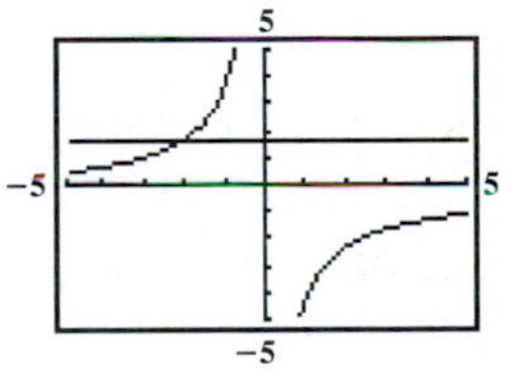

; $\{-2\}$

77. does not make sense **79.** does not make sense **81.** false **83.** true **85.** $\left\{-\frac{19}{5}, 13\right\}$ **87.** $\{-3\}$

88.

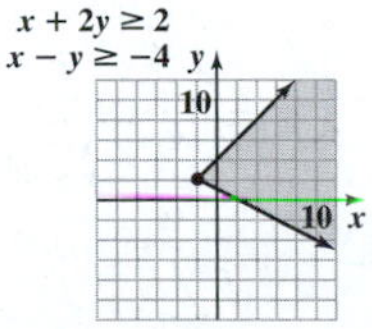

89. $\{15\}$ **90.** $F = \frac{9C + 160}{5}$ **91.** $p = \frac{qf}{q - f}$ **92.** $\{-20, 30\}$ **93.** 40 miles per hour

Section 6.7

Check Point Exercises

1. $x = \frac{yz}{y - z}$ **2. a.** $C(x) = 500{,}000 + 400x$ **b.** $\overline{C}(x) = \frac{500{,}000 + 400x}{x}$ **c.** 10,000 wheelchairs **3.** 12 mph **4.** 6 months **5.** experienced carpenter: 8 hr; apprentice: 24 hr

Exercise Set 6.7

1. $P_1 = \frac{P_2V_2}{V_1}$ **3.** $f = \frac{pq}{q + p}$ **5.** $r = \frac{A - P}{P}$ **7.** $m_1 = \frac{Fd^2}{Gm_2}$ **9.** $x = \overline{x} + zs$ **11.** $R = \frac{E - Ir}{I}$ **13.** $f_1 = \frac{ff_2}{f_2 - f}$ **15.** 50,000 wheelchairs **17.** $y = 400$; Average cost is nearing \$400 as production level increases. **19. a.** $C(x) = 100{,}000 + 100x$ **b.** $\overline{C}(x) = \frac{100{,}000 + 100x}{x}$ **c.** 500 bikes **21.** 5 mph **23.** The time increases as the running rate is close to zero miles per hour. **25.** car: 50 mph; bus: 30 mph **27.** 6 mph **29.** 3 mph **31.** 500 mph **33.** 4.2 ft per sec **35.** 12 miles **37.** 18 min; yes **39.** 4 hr

41. 12 hr **43.** 10 hr **45.** 10 hr **47.** 10 min **49.** 3 **51.** $\frac{1}{3}$ and 3 **53.** 10 **55.** $x = \frac{ab}{a+b}$

65. For Exercise 45:

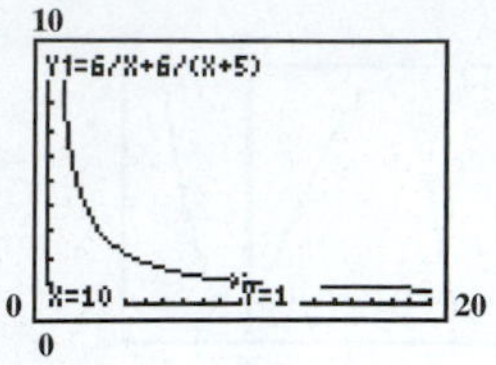

67. makes sense **69.** makes sense **71.** false **73.** true **75.** $f = \frac{p-s}{ps-s}$ **77.** 8 hr **78.** $(x + 2 + 3y)(x + 2 - 3y)$ **79.** $\{(5, -3)\}$ **80.** $\{(3, -1, 2)\}$ **81. a.** 16 **b.** $y = 16x^2$ **c.** 400 **82. a.** 96 **b.** $y = \frac{96}{x}$ **c.** 32 **83.** 8

Section 6.8

Check Point Exercises

1. 66 gal **2.** about 556 ft **3.** 512 cycles per second **4.** 24 min **5.** 96π cubic feet

Exercise Set 6.8

1. 156 **3.** 30 **5.** $\frac{5}{6}$ **7.** 240 **9.** 50 **11.** $x = kyz$; $y = \frac{x}{kz}$ **13.** $x = \frac{kz^3}{y}$; $y = \frac{kz^3}{x}$ **15.** $x = \frac{kyz}{\sqrt{w}}$; $y = \frac{x\sqrt{w}}{kz}$ **17.** $x = kz(y + w)$; $y = \frac{x - kzw}{kz}$ **19.** $x = \frac{kz}{y - w}$; $y = \frac{xw + kz}{x}$ **21.** 5.4 ft **23.** 80 in. **25.** about 607 lb **27.** 32° **29. a.** $L = \frac{1890}{R}$ **b.** an approximate model **c.** 70 yr **31. a.** 90 beats per minute **b.** 95 beats per minute **c.** by the point (63, 30) **33.** 90 milliroentgens per hour **35.** This person has a BMI of 24.4 and is not overweight. **37.** 1800 Btu **39.** $\frac{1}{4}$ of what it was originally **41. a.** $C = \frac{kP_1P_2}{d^2}$ **b.** $k \approx 0.02$; $C = \frac{0.02P_1P_2}{d^2}$ **c.** approximately 39,813 daily phone calls

43. a.

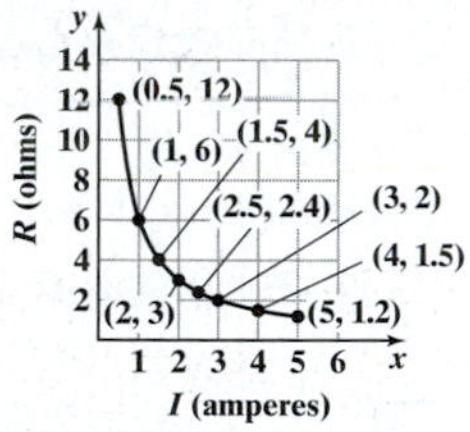

b. Current varies inversely as resistance. **c.** $R = \frac{6}{I}$

49. z varies directly as the square root of x and inversely as the square of y.

53. does not make sense **55.** makes sense **57.** The wind pressure is 4 times more destructive. **59.** Distance is increased by $\sqrt{50}$, or about 7.07, for the space telescope. **60.** -2 **61.** $(x + 3)(x - 3)(y - 3)$ **62.** 4 **63.** 3 **64.** 6 **65.** domain: $[-4, \infty)$; range: $[0, \infty)$

Review Exercises

1. a. $\frac{7}{4}$ **b.** $\frac{3}{4}$ **c.** does not exist **d.** 0 **2.** $\{x|x$ is a real number and $x \neq -4$ and $x \neq 3\}$ or $(-\infty, -4) \cup (-4, 3) \cup (3, \infty)$ **3.** $\{x|x$ is a real number and $x \neq -2$ and $x \neq 1\}$ or $(-\infty, -2) \cup (-2, 1) \cup (1, \infty)$ **4.** $\frac{x^2 - 7}{3x}$ **5.** $\frac{x-1}{x-7}$ **6.** $\frac{3x+2}{x-5}$ **7.** cannot be simplified **8.** $\frac{x^2 + 2x + 4}{x + 2}$ **9.** $\frac{5(x+1)}{3}$ **10.** $\frac{2x-1}{x+1}$ **11.** $\frac{x-7}{x^2}$ **12.** $\frac{1}{3(x+3)}$ **13.** $\frac{1}{2}$ **14.** $\frac{(y+4)(y+2)}{(y-6)(y^2+4y+16)}$ **15.** $\frac{3x^2}{2x-5}$ **16. a.** 50 deer **b.** 150 deer **c.** $y = 225$; The population will approach 225 deer over time. **17.** 4 **18.** $\frac{1}{x+3}$ **19.** $3x + 2$ **20.** $36x^3$ **21.** $(x + 7)(x - 5)(x + 2)$ **22.** $\frac{3x-5}{x(x-5)}$ **23.** $\frac{5x-2}{(x-3)(x+2)(x-2)}$ **24.** $\frac{2x-8}{(x-3)(x-5)}$ or $\frac{2(x-4)}{(x-3)(x-5)}$ **25.** $\frac{4x}{(3x+4)(3x-4)}$ **26.** $\frac{y-3}{(y+1)(y+3)}$ **27.** $\frac{2x^2-9}{(x+3)(x-3)}$ **28.** $3(x + y)$ or $3x + 3y$ **29.** $\frac{3}{8}$ **30.** $\frac{x}{x-5}$ **31.** $\frac{3x+8}{3x+10}$ **32.** $\frac{4(x-2)}{2x+5}$ **33.** $\frac{3x^2+9x}{x^2+8x-33}$ **34.** $\frac{1+x}{1-x}$ **35.** $3x - 6 + \frac{2}{x} - \frac{2}{5x^2}$ **36.** $6x^3y + 2xy - 10x$ **37.** $3x - 7 + \frac{26}{2x+3}$ **38.** $2x^2 - 4x + 1 - \frac{10}{5x-3}$ **39.** $x^5 + 5x^4 + 8x^3 + 16x^2 + 33x + 63 + \frac{128}{x-2}$ **40.** $2x^2 + 3x - 1$ **41.** $4x^2 - 7x + 5 - \frac{4}{x+1}$ **42.** $3x^3 + 6x^2 + 10x + 10$ **43.** $x^3 - 4x^2 + 16x - 64 + \frac{272}{x+4}$ **44.** 3 **45.** 4 **46.** solution **47.** not a solution **48.** The remainder is 0.; $\left\{-1, \frac{1}{3}, \frac{1}{2}\right\}$ **49.** $\{6\}$ **50.** $\{52\}$ **51.** $\{7\}$ **52.** $\left\{-\frac{9}{2}\right\}$ **53.** $\left\{-\frac{1}{2}, 3\right\}$ **54.** $\{-23, 2\}$ **55.** $\{-3, 2\}$ **56.** 80% **57.** $C = R - nP$ **58.** $T_1 = \frac{P_1V_1T_2}{P_2V_2}$ **59.** $P = \frac{A}{rT+1}$ **60.** $R = \frac{R_1R_2}{R_2+R_1}$ **61.** $n = \frac{IR}{E - Ir}$ **62. a.** $C(x) = 50{,}000 + 25x$ **b.** $\overline{C}(x) = \frac{50{,}000 + 25x}{x}$ **c.** 5000 graphing calculators **63.** 12 mph **64.** 9 mph **65.** 2 hr; no

66. faster crew: 36 hr; slower crew: 45 hr **67.** 240 min or 4 hr **68.** \$4935 **69.** 1600 ft **70.** 440 vibrations per second **71.** 112 decibels **72.** 16 hr **73.** 800 cubic feet

Chapter Test

1. $\{x|x$ is a real number and $x \neq 2$ and $x \neq 5\}$ or $(-\infty, 2) \cup (2, 5) \cup (5, \infty)$; $\frac{x}{x-5}$ **2.** $\frac{x}{x-4}$ **3.** $\frac{(x+3)(x-1)}{x+1}$ **4.** $\frac{x-2}{x+5}$ **5.** $\frac{3x-24}{x-3}$ **6.** $\frac{1}{3x+5}$ **7.** $\frac{x^2+2x+15}{(x+3)(x-3)}$ **8.** $\frac{3x-4}{(x+2)(x-3)}$ **9.** $\frac{5x^2-7x+4}{(x+2)(x-1)(x-2)}$ **10.** $\frac{x+3}{x+5}$ **11.** $\frac{x-2}{x-10}$ **12.** $\frac{x-2}{8}$ **13.** $1-x$ **14.** $3x^2y^2+4y^2-\frac{5}{2}y$ **15.** $3x^2-3x+1+\frac{2}{3x+2}$ **16.** $3x^2+2x+3+\frac{9-6x}{x^2-1}$ **17.** $3x^3-4x^2+7$ **18.** 12 **19.** solution **20.** $\{-7, 3\}$ **21.** $\{-1\}$ **22.** 3 years **23.** $a = \frac{Rs}{s-R}$ or $a = -\frac{Rs}{R-s}$ **24. a.** $C(x) = 300{,}000 + 10x$ **b.** $\overline{C}(x) = \frac{300{,}000 + 10x}{x}$ **c.** 20,000 players **25.** 12 hr **26.** 4 mph **27.** 45 foot-candles

Cumulative Review Exercises

1. $\{x|x < -6\}$ or $(-\infty, -6)$ **2.** $\left\{-\frac{1}{2}, 4\right\}$ **3.** $\{(-2, 0, 4)\}$ **4.** $\left\{x \middle| -2 \le x \le \frac{14}{3}\right\}$ or $\left[-2, \frac{14}{3}\right]$ **5.** $\{12\}$ **6.** $s = \frac{2R - Iw}{2I}$ **7.** $\{(3, 2)\}$ **8.** $f(x) = -3x - 2$

9.

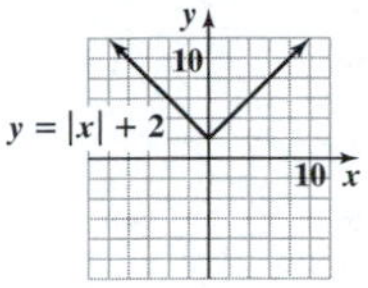

10.

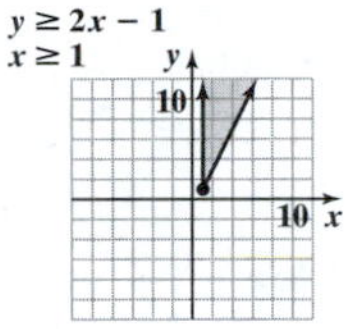

11.

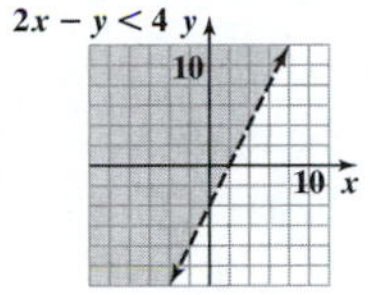

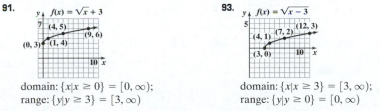

12. $x^2 + 4x + 4 - 9y^2$ **13.** $\frac{x(x+1)}{2x+5}$ **14.** $\frac{x+25}{2(x-4)(x-5)}$ **15.** $3x + 4 + \frac{2}{x+2}$ **16.** $(y-6)(x+2)$ **17.** $2xy(2x+3)(6x-5)$ **18.** 4 sec **19.** basic cable service: \$25; each movie channel: \$10 **20.** 1 ft

CHAPTER 7

Section 7.1

Check Point Exercises

1. a. 8 **b.** -7 **c.** $\frac{4}{5}$ **d.** 0.09 **e.** 5 **f.** 7 **2. a.** 4 **b.** $-\sqrt{24} \approx -4.90$ **3.** $\{x|x \ge 3\}$ or $[3, \infty)$ **4.** approximately 15.1 minutes **5. a.** 7 **b.** $|x+8|$ **c.** $|7x^5|$ or $7|x^5|$ **d.** $|x-3|$ **6. a.** 3 **b.** -2 **7.** $-3x$ **8. a.** 2 **b.** -2 **c.** not a real number **d.** -1 **9. a.** $|x+6|$ **b.** $3x-2$ **c.** 8

Exercise Set 7.1

1. 6 **3.** -6 **5.** not a real number **7.** $\frac{1}{5}$ **9.** $-\frac{3}{4}$ **11.** 0.9 **13.** -0.2 **15.** 3 **17.** 1 **19.** not a real number **21.** 4; 1; 0; not a real number **23.** -5; $-\sqrt{5} \approx -2.24$; -1; not a real number **25.** 4; 2; 1; 6 **27.** $\{x|x \ge 3\}$ or $[3, \infty)$; c **29.** $\{x|x \ge -5\}$ or $[-5, \infty)$; d **31.** $\{x|x \le 3\}$ or $(-\infty, 3]$; e **33.** 5 **35.** 4 **37.** $|x-1|$ **39.** $|6x^2|$ or $6x^2$ **41.** $-|10x^3|$ or $-10|x^3|$ **43.** $|x+6|$ **45.** $-|x-4|$ **47.** 3 **49.** -3 **51.** $\frac{1}{5}$ **53.** $-\frac{3}{10}$ **55.** 3; 2; -1; -4 **57.** -2; 0; 2 **59.** 1 **61.** 2 **63.** -2 **65.** not a real number **67.** -1 **69.** not a real number **71.** -4 **73.** 2 **75.** -2 **77.** x **79.** $|y|$ **81.** $-2x$ **83.** -5 **85.** 5 **87.** $|x+3|$ **89.** $-2(x-1)$

91.

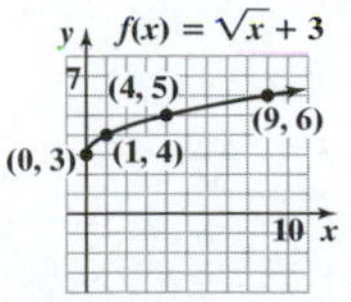

domain: $\{x|x \ge 0\} = [0, \infty)$;
range: $\{y|y \ge 3\} = [3, \infty)$

93.

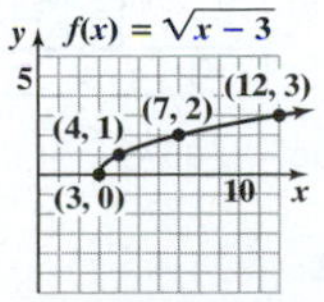

domain: $\{x|x \ge 3\} = [3, \infty)$;
range: $\{y|y \ge 0\} = [0, \infty)$

95. $\{x|x < 15\}$ or $(-\infty, 15)$ **97.** $\{x|1 \le x < 3\}$ or $[1, 3)$ **99.** 3 **101. a.** 40.2 in.; underestimates by 0.6 in. **b.** 0.9 in. per month **c.** 0.2 in. per month; This is a much smaller rate of change.; The graph is not as steep between 50 and 60 as it is between 0 and 10. **103.** 70 mph; The officer should not believe the motorist.; Answers will vary.

115. 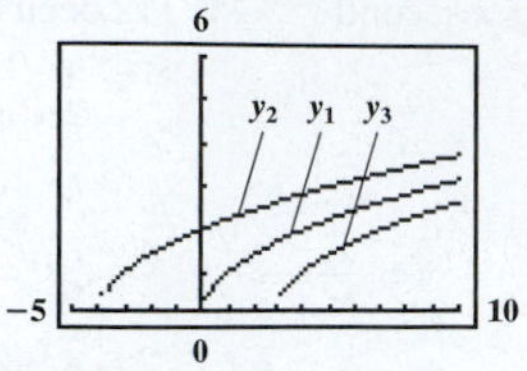

; Answers will vary.

117. 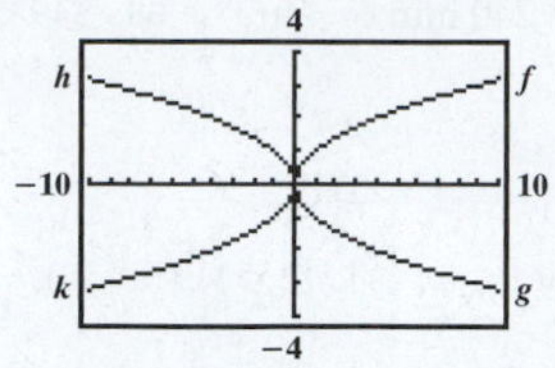

domain of f: $\{x | x \geq 0\}$ or $[0, \infty)$; range of f: $\{y | y \geq 0\}$ or $[0, \infty)$
domain of g: $\{x | x \geq 0\}$ or $[0, \infty)$; range of g: $\{y | y \leq 0\}$ or $(-\infty, 0]$
domain of h: $\{x | x \leq 0\}$ or $(-\infty, 0]$; range of h: $\{y | y \geq 0\}$ or $[0, \infty)$
domain of k: $\{x | x \leq 0\}$ or $(-\infty, 0]$; range of k: $\{y | y \leq 0\}$ or $(-\infty, 0]$

119. does not make sense **121.** makes sense **123.** false **125.** false **127.** Answers will vary; an example is $f(x) = \sqrt{15 - 3x}$.

129. $|(2x + 3)^5|$ **131.** The graph of h is the graph of f shifted left 3 units.

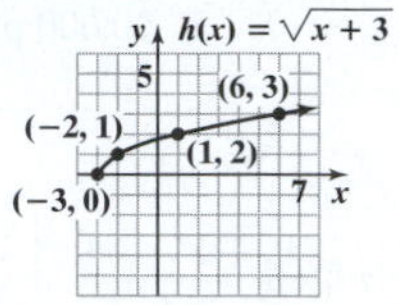

132. $7x + 30$ **133.** $\frac{x^8}{9y^6}$ **134.** $\left\{x \middle| x < -\frac{7}{3} \text{ or } x > 5\right\}$ or $\left(-\infty, -\frac{7}{3}\right) \cup (5, \infty)$ **135.** $\frac{2^7}{x}$ or $\frac{128}{x}$ **136.** $\frac{2}{x^3}$ **137.** $\frac{y^{12}}{x^8}$

Section 7.2

Check Point Exercises

1. a. $\sqrt{25} = 5$ **b.** $\sqrt[3]{-8} = -2$ **c.** $\sqrt[4]{5xy^2}$ **2. a.** $(5xy)^{1/4}$ **b.** $\left(\frac{a^3b}{2}\right)^{1/5}$ **3. a.** $(\sqrt[3]{8})^4 = 16$ **b.** $(\sqrt{25})^3 = 125$ **c.** $-(\sqrt[4]{81})^3 = -27$

4. a. $6^{4/3}$ **b.** $(2xy)^{7/5}$ **5. a.** $\frac{1}{100^{1/2}} = \frac{1}{10}$ **b.** $\frac{1}{8^{1/3}} = \frac{1}{2}$ **c.** $\frac{1}{32^{3/5}} = \frac{1}{8}$ **d.** $\frac{1}{(3xy)^{5/9}}$ **6. a.** $7^{5/6}$ **b.** $\frac{5}{x}$ **c.** $9.1^{3/10}$ **d.** $\frac{y^{1/12}}{x^{1/5}}$

7. a. $\sqrt{x}$ **b.** $2a^4$ **c.** $\sqrt[4]{x^2y}$ **d.** $\sqrt[6]{x}$ **e.** $\sqrt[6]{x}$

Exercise Set 7.2

1. $\sqrt{49} = 7$ **3.** $\sqrt[3]{-27} = -3$ **5.** $-\sqrt[4]{16} = -2$ **7.** $\sqrt[3]{xy}$ **9.** $\sqrt[5]{2xy^3}$ **11.** $(\sqrt{81})^3 = 729$ **13.** $(\sqrt[3]{125})^2 = 25$

15. $(\sqrt[5]{-32})^3 = -8$ **17.** $(\sqrt[3]{27})^2 + (\sqrt[4]{16})^3 = 17$ **19.** $\sqrt[7]{(xy)^4}$ **21.** $7^{1/2}$ **23.** $5^{1/3}$ **25.** $(11x)^{1/5}$ **27.** $x^{3/2}$ **29.** $x^{3/5}$

31. $(x^2y)^{1/5}$ **33.** $(19xy)^{3/2}$ **35.** $(7xy^2)^{5/6}$ **37.** $2xy^{2/3}$ **39.** $\frac{1}{49^{1/2}} = \frac{1}{7}$ **41.** $\frac{1}{27^{1/3}} = \frac{1}{3}$ **43.** $\frac{1}{16^{3/4}} = \frac{1}{8}$ **45.** $\frac{1}{8^{2/3}} = \frac{1}{4}$

47. $\left(\frac{27}{8}\right)^{1/3} = \frac{3}{2}$ **49.** $\frac{1}{(-64)^{2/3}} = \frac{1}{16}$ **51.** $\frac{1}{(2xy)^{7/10}}$ **53.** $\frac{5x}{z^{1/3}}$ **55.** 3 **57.** 4 **59.** $x^{5/6}$ **61.** $x^{3/5}$ **63.** $\frac{1}{x^{5/12}}$ **65.** 25

67. $\frac{1}{y^{1/6}}$ **69.** $32x$ **71.** $5x^2y^3$ **73.** $\frac{x^{1/4}}{y^{3/10}}$ **75.** 3 **77.** $27y^{2/3}$ **79.** $\sqrt[4]{x}$ **81.** $2a^2$ **83.** x^2y^3 **85.** x^6y^6 **87.** $\sqrt[5]{3y}$ **89.** $\sqrt[3]{4a^2}$

91. $\sqrt[3]{x^2y}$ **93.** $\sqrt[6]{2^5}$ or $\sqrt[6]{32}$ **95.** $\sqrt[10]{x^9}$ **97.** $\sqrt[12]{a^{10}b^7}$ **99.** $\sqrt[20]{x}$ **101.** $\sqrt{y}$ **103.** $\sqrt[8]{x}$ **105.** $\sqrt[4]{x^2y}$ **107.** $\sqrt[12]{2x}$

109. x^9y^{15} **111.** $\sqrt[4]{a^3b^3}$ **113.** $x^{2/3} - x$ **115.** $x + 2x^{1/2} - 15$ **117.** $2x^{1/2}(3 + x)$ **119.** $15x^{1/3}(1 - 4x^{2/3})$ **121.** $\frac{x^2}{7y^{3/2}}$ **123.** $\frac{x^3}{y^2}$

125. 58 species of plants **127.** about 1872 calories per day **129. a.** $C = 35.74 + 0.6215t - 35.74v^{4/25} + 0.4275tv^{4/25}$ **b.** 8°F

131. a. $C(v) = 35.74 - 35.74v^{4/25}$ **b.** $C(25) \approx -24$; When the air temperature is 0°F and the wind speed is 25 miles per hour, the windchill temperature is −24°. **c.** the point $(25, -24)$ **133. a.** $L + 1.25S^{1/2} - 9.8D^{1/3} \leq 16.296$ **b.** eligible

145. simplified correctly

X	Y_1	Y_2
1	1	1
2	1.4142	1.4142
3	1.7321	1.7321
4	2	2
5	2.2361	2.2361
6	2.4495	2.4495
7	2.6458	2.6458

X=5

147. Right side should be $x^{1/2}$.

X	Y_1	Y_2
1	1	1
2	1.4142	.70711
3	1.7321	.57735
4	2	.5
5	2.2361	.44721
6	2.4494	.40824
7	2.6457	.37796

X=1

149. does not make sense **151.** does not make sense **153.** false **155.** true **157.** $\frac{1}{4}$ of the cake **159.** $\{x | x \geq 3\}$ or $[3, \infty)$

160. $y = -2x + 11$ or $f(x) = -2x + 11$ **161.**

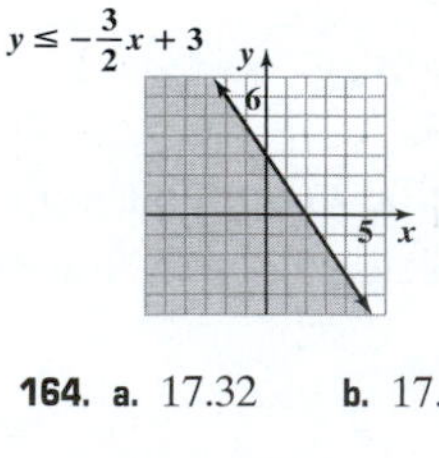

162. $\{(3, 4)\}$

163. a. 8 **b.** 8 **c.** $\sqrt{16} \cdot \sqrt{4} = \sqrt{16 \cdot 4}$ **164. a.** 17.32 **b.** 17.32 **c.** $\sqrt{300} = 10\sqrt{3}$ **165. a.** x^7 **b.** y^4

Section 7.3

Check Point Exercises

1. a. $\sqrt{55}$ **b.** $\sqrt{x^2-16}$ **c.** $\sqrt[3]{60}$ **d.** $\sqrt[7]{12x^4}$ **2. a.** $4\sqrt{5}$ **b.** $2\sqrt[3]{5}$ **c.** $2\sqrt[4]{2}$ **d.** $10|x|\sqrt{2y}$
3. $f(x)=\sqrt{3}|x-2|$ **4.** $x^4y^5z\sqrt{xyz}$ **5.** $2x^3y^4\sqrt[3]{5xy^2}$ **6.** $2x^2z\sqrt[5]{x^2y^2z^3}$ **7. a.** $2\sqrt{3}$ **b.** $100\sqrt[3]{4}$ **c.** $2x^2y\sqrt[4]{2}$

Exercise Set 7.3

1. $\sqrt{15}$ **3.** $\sqrt[3]{18}$ **5.** $\sqrt[4]{33}$ **7.** $\sqrt{33xy}$ **9.** $\sqrt[5]{24x^4}$ **11.** $\sqrt{x^2-9}$ **13.** $\sqrt[6]{(x-4)^5}$ **15.** $\sqrt{x}$ **17.** $\sqrt[4]{\dfrac{3x}{7y}}$ **19.** $\sqrt[7]{77x^5y^3}$
21. $5\sqrt{2}$ **23.** $3\sqrt{5}$ **25.** $5\sqrt{3x}$ **27.** $2\sqrt[3]{2}$ **29.** $3x$ **31.** $-2y\sqrt[3]{2x^2}$ **33.** $6|x+2|$ **35.** $2(x+2)\sqrt[3]{4}$ **37.** $|x-1|\sqrt{3}$
39. $x^3\sqrt{x}$ **41.** $x^4y^4\sqrt{y}$ **43.** $4x\sqrt{3x}$ **45.** $y^2\sqrt[3]{y^2}$ **47.** $x^4y\sqrt[3]{x^2z}$ **49.** $3x^2y^2\sqrt[3]{3x^2}$ **51.** $(x+y)\sqrt[3]{(x+y)^2}$ **53.** $y^3\sqrt[5]{y^2}$
55. $2xy^3\sqrt[3]{2xy^2}$ **57.** $2x^2\sqrt[4]{5x^2}$ **59.** $(x-3)^2\sqrt[4]{(x-3)^2}$ or $(x-3)^2\sqrt{x-3}$ **61.** $2\sqrt{6}$ **63.** $5\sqrt{2xy}$ **65.** $6x$ **67.** $10xy\sqrt{2y}$
69. $60\sqrt{2}$ **71.** $2\sqrt[3]{6}$ **73.** $2x^2\sqrt{10x}$ **75.** $5xy^4\sqrt[3]{x^2y^2}$ **77.** $2xyz\sqrt[4]{x^2z^3}$ **79.** $2xy^2z\sqrt[5]{2y^3z}$ **81.** $(x-y)^2\sqrt[3]{(x-y)^2}$
83. $-6x^3y^3\sqrt[3]{2yz^2}$ **85.** $-6y^2\sqrt[5]{2x^3y}$ **87.** $-6x^3y^3\sqrt{2}$ **89.** $-12x^3y^4\sqrt[4]{x^3}$ **91.** $6\sqrt{3}$ miles; 10.4 miles **93.** $8\sqrt{3}$ ft per sec; 14 ft per sec
95. a. $\dfrac{7.644}{2\sqrt[4]{2}}=\dfrac{3.822}{\sqrt[4]{2}}$ **b.** 3.21 liters of blood per minute per square meter; (32, 3.21)

103. Graphs are the same; simplification is correct.

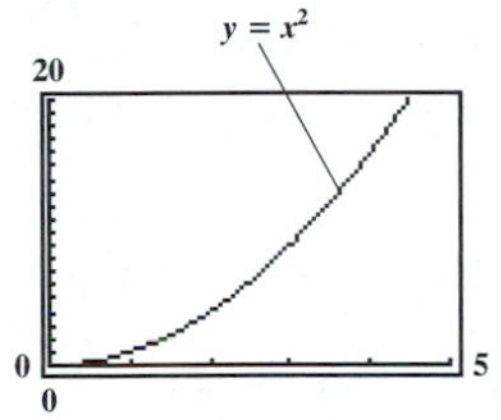

105. Graphs are not the same.; $\sqrt{3x^2-6x+3}=|x-1|\sqrt{3}$

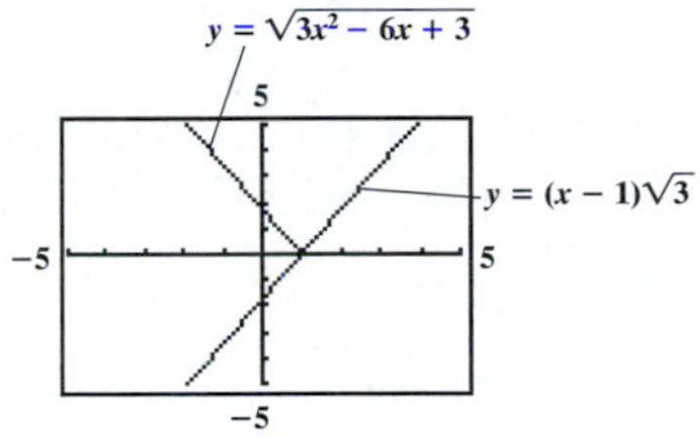

107. makes sense **109.** makes sense **111.** false **113.** false **115.** Its square root is multiplied by $\sqrt{3}$. **117.** $g(x)=\sqrt[3]{4x^2}$
119. $\{x|5\le x\le 11\}$ or $[5,11]$ **120.** $\{(2,-4)\}$ **121.** $(4x-3)(16x^2+12x+9)$ **122. a.** $31x$ **b.** $31\sqrt{2}$ **123. a.** $-8x$
b. $-8\sqrt{2}$ **124.** $\dfrac{y\sqrt[4]{7y}}{x^3}$

Section 7.4

Check Point Exercises

1. a. $10\sqrt{13}$ **b.** $(21-6x)\sqrt[3]{7}$ **c.** $5\sqrt[4]{3x}+2\sqrt[3]{3x}$ **2. a.** $21\sqrt{5}$ **b.** $-12\sqrt{3x}$ **c.** cannot be simplified
3. a. $-9\sqrt[3]{3}$ **b.** $(5+3xy)\sqrt[3]{x^2y}$ **4. a.** $\dfrac{2\sqrt{3}}{5}$ **b.** $\dfrac{3x\sqrt{x}}{y^5}$ **c.** $\dfrac{2y^2\sqrt[3]{y}}{x^4}$ **5. a.** $2x^2\sqrt{5}$ **b.** $\dfrac{5\sqrt{xy}}{2}$ **c.** $2x^2y$

Exercise Set 7.4

1. $11\sqrt{5}$ **3.** $7\sqrt[3]{6}$ **5.** $2\sqrt[5]{2}$ **7.** $\sqrt{13}+2\sqrt{5}$ **9.** $7\sqrt{5}+2\sqrt[3]{x}$ **11.** $4\sqrt{3}$ **13.** $19\sqrt{3}$ **15.** $6\sqrt{2x}$ **17.** $13\sqrt[3]{2}$
19. $(9x+1)\sqrt{5x}$ **21.** $7y\sqrt[3]{2x}$ **23.** $(3x-2)\sqrt[3]{2x}$ **25.** $4\sqrt{x-2}$ **27.** $5x\sqrt[3]{xy^2}$ **29.** $\dfrac{\sqrt{11}}{2}$ **31.** $\dfrac{\sqrt[3]{19}}{3}$ **33.** $\dfrac{x}{6y^4}$ **35.** $\dfrac{2x\sqrt{2x}}{5y^3}$
37. $\dfrac{x\sqrt{x}}{2y}$ **39.** $\dfrac{x^2\sqrt[3]{50x^2}}{3y^4}$ **41.** $\dfrac{y\sqrt[4]{9y^2}}{x^2}$ **43.** $\dfrac{2x^2\sqrt[5]{2x^3}}{y^4}$ **45.** $2\sqrt{2}$ **47.** 2 **49.** $3x$ **51.** x^2y **53.** $2x^2\sqrt{5}$ **55.** $4a^5b^5$
57. $3\sqrt{xy}$ **59.** $2xy$ **61.** $2x^2y^2\sqrt[4]{y^2}$ or $2x^2y^2\sqrt{y}$ **63.** $\sqrt[3]{x+3}$ **65.** $\sqrt[3]{a^2-ab+b^2}$ **67.** $\dfrac{43\sqrt{2}}{35}$ **69.** $-11xy\sqrt{2x}$ **71.** $25x\sqrt{2x}$
73. $7x\sqrt{3xy}$ **75.** $3x\sqrt[3]{5xy}$ **77.** $\left(\dfrac{f}{g}\right)(x)=4x\sqrt{x}$; domain: $\{x|x>0\}$ or $(0,\infty)$ **79.** $\left(\dfrac{f}{g}\right)(x)=2x\sqrt[3]{2x}$; domain: $\{x|x$ is a real number and $x\neq 0\}$ or $(-\infty,0)\cup(0,\infty)$ **81.** $P=18\sqrt{5}$ ft; $A=100$ sq ft **83.** $12\sqrt{5}$ m
85. a. $5\sqrt{10}$; the projected increase in the number of Americans ages 65–84, in millions, from 2020 to 2050 **b.** 15.8; underestimates by 2.7 million
95. Graphs are not the same.; $\sqrt{16x}-\sqrt{9x}=\sqrt{x}$

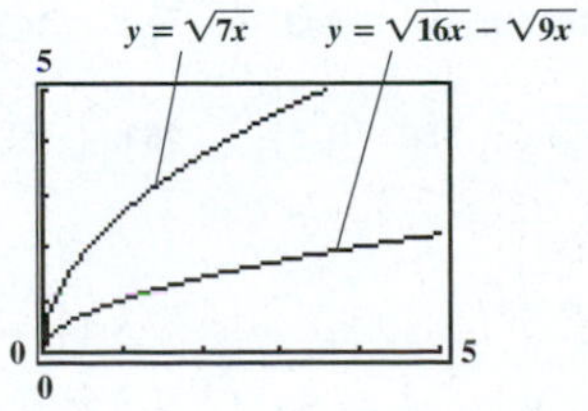

97. makes sense **99.** does not make sense **101.** false **103.** false **105.** $2\sqrt{2}$ **107.** $x^2y^3a^5b\sqrt{b}$ **108.** $\{0\}$
109. $(x-2y)(x-6y)$ **110.** $\frac{3x^2+8x+6}{(x+3)^2(x+2)}$ **111. a.** $7x+35$ **b.** $x\sqrt{7}+\sqrt{35}$ **112. a.** $6x^2+33x+15$ **b.** $27+33\sqrt{2}$
113. $\frac{5\sqrt[5]{8x^2y^4}}{x}$

Mid-Chapter Check Point Exercises

1. 13 **2.** $2x^2y^3\sqrt{2xy}$ **3.** $5\sqrt[3]{4x^2}$ **4.** $12x\sqrt[3]{2x}$ **5.** 1 **6.** $4xy^{1/12}$ **7.** $-\sqrt{3}$ **8.** $\frac{5x\sqrt{5x}}{y^2}$ **9.** $\sqrt[4]{x^3}$ **10.** $3x\sqrt[3]{2x^2}$ **11.** $2\sqrt[3]{10}$
12. $\frac{x^2}{y^4}$ **13.** $x^{7/12}$ **14.** $x\sqrt[3]{y^2}$ **15.** $(x-2)\sqrt[7]{(x-2)^2}$ **16.** $2x^2y^4\sqrt[4]{2x^3y}$ **17.** $14\sqrt[3]{2}$ **18.** $x\sqrt[7]{x^2y^2}$ **19.** $\frac{1}{25}$ **20.** $\sqrt[6]{32}$
21. $\frac{4x}{y^2}$ **22.** $8x^2y^3\sqrt{y}$ **23.** $-10x^3y\sqrt{6y}$ **24.** $\{x|x \le 6\}$ or $(-\infty, 6]$ **25.** $\{x|x$ is a real number$\}$ or $(-\infty, \infty)$

Section 7.5

Check Point Exercises

1. a. $x\sqrt{6}+2\sqrt{15}$ **b.** $y-\sqrt[3]{7y}$ **c.** $36-18\sqrt{10}$ **2. a.** $11+2\sqrt{30}$ **b.** 1 **c.** $a-7$ **3. a.** $\frac{\sqrt{21}}{7}$ **b.** $\frac{\sqrt[3]{6}}{3}$
4. a. $\frac{\sqrt{14xy}}{7y}$ **b.** $\frac{\sqrt[3]{3xy^2}}{3y}$ **c.** $\frac{3\sqrt[5]{4x^3y}}{y}$ **5.** $12\sqrt{3}-18$ **6.** $\frac{3\sqrt{5}+3\sqrt{2}+\sqrt{35}+\sqrt{14}}{3}$ **7.** $\frac{1}{\sqrt{x+3}+\sqrt{x}}$

Exercise Set 7.5

1. $x\sqrt{2}+\sqrt{14}$ **3.** $7\sqrt{6}-6$ **5.** $12\sqrt{2}-6$ **7.** $\sqrt[3]{12}+4\sqrt[3]{10}$ **9.** $2x\sqrt[3]{2}-\sqrt[3]{x^2}$ **11.** $32+11\sqrt{2}$ **13.** $34-15\sqrt{5}$
15. $117-36\sqrt{7}$ **17.** $\sqrt{6}+\sqrt{10}+\sqrt{21}+\sqrt{35}$ **19.** $\sqrt{6}-\sqrt{10}-\sqrt{21}+\sqrt{35}$ **21.** $-48+7\sqrt{6}$ **23.** $8+2\sqrt{15}$
25. $3x-2\sqrt{3xy}+y$ **27.** -44 **29.** -71 **31.** 6 **33.** $6-5\sqrt{x}+x$ **35.** $\sqrt[3]{x^2}+\sqrt[3]{x}-20$ **37.** $2x^2+x\sqrt[3]{y^2}-y\sqrt[3]{y}$ **39.** $\frac{\sqrt{10}}{5}$
41. $\frac{\sqrt{11x}}{x}$ **43.** $\frac{3\sqrt{3y}}{y}$ **45.** $\frac{\sqrt[3]{4}}{2}$ **47.** $3\sqrt[3]{2}$ **49.** $\frac{\sqrt[3]{18}}{3}$ **51.** $\frac{4\sqrt[3]{x^2}}{x}$ **53.** $\frac{\sqrt[3]{2y}}{y}$ **55.** $\frac{7\sqrt[3]{4x}}{2x}$ **57.** $\frac{\sqrt[3]{2x^2y}}{xy}$ **59.** $\frac{3\sqrt[4]{x^3}}{x}$
61. $\frac{3\sqrt[5]{4x^2}}{x}$ **63.** $x\sqrt[5]{8x^3y}$ **65.** $\frac{3\sqrt{3y}}{xy}$ **67.** $-\frac{5a^2\sqrt{3ab}}{b^2}$ **69.** $\frac{\sqrt{2mn}}{2m}$ **71.** $\frac{3\sqrt[4]{x^3y}}{x^2y}$ **73.** $-\frac{6\sqrt[3]{xy}}{x^2y^3}$ **75.** $8\sqrt{5}-16$
77. $\frac{13\sqrt{11}+39}{2}$ **79.** $3\sqrt{5}-3\sqrt{3}$ **81.** $\frac{a+\sqrt{ab}}{a-b}$ **83.** $25\sqrt{2}+15\sqrt{5}$ **85.** $4+\sqrt{15}$ **87.** $\frac{x-2\sqrt{x}-3}{x-9}$ **89.** $\frac{3\sqrt{6}+4}{2}$
91. $\frac{4\sqrt{xy}+4x+y}{y-4x}$ **93.** $\frac{3}{\sqrt{6}}$ **95.** $\frac{2x}{\sqrt[3]{2x^2y}}$ **97.** $\frac{x-9}{x-3\sqrt{x}}$ **99.** $\frac{a-b}{a-2\sqrt{ab}+b}$ **101.** $\frac{1}{\sqrt{x+5}+\sqrt{x}}$ **103.** $\frac{1}{(x+y)(\sqrt{x}-\sqrt{y})}$
105. $\frac{3\sqrt{2}}{2}$ **107.** $-2\sqrt[3]{25}$ **109.** $\frac{7\sqrt{6}}{6}$ **111.** $7\sqrt{3}-7\sqrt{2}$ **113.** 0 **115.** 6 **117.** $\frac{\sqrt{5}+1}{2}$; 1.62 to 1 **119.** $P = 8\sqrt{2}$ in.; $A = 7$ sq in.
121. $2\sqrt{6}$ in.

131.

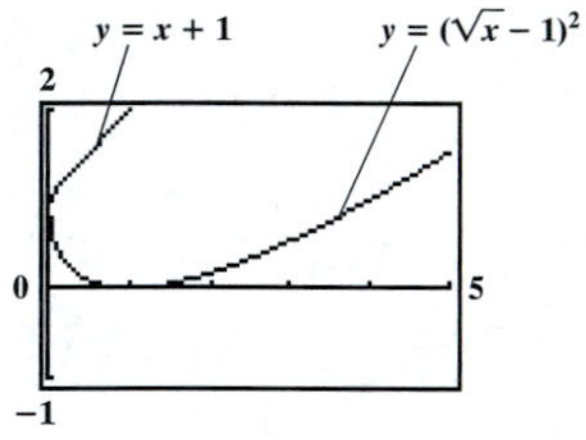

; Graphs are not the same.; $(\sqrt{x}-1)(\sqrt{x}-1) = x-2\sqrt{x}+1$

133.

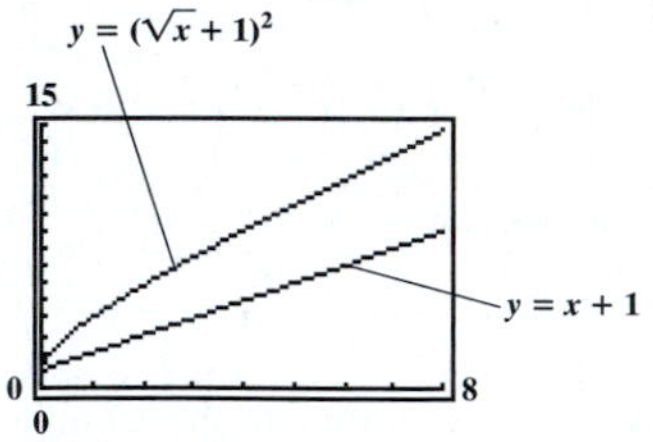

; Graphs are not the same.; $(\sqrt{x}+1)^2 = x+2\sqrt{x}+1$

135. makes sense **137.** does not make sense **139.** false **141.** true **143.** $\left\{\frac{45}{7}\right\}$ **145.** $\frac{5\sqrt{2}+3\sqrt{3}-4\sqrt{6}+2}{23}$
146. $\frac{2x+7}{x^2-4}$ **147.** $\left\{-\frac{7}{2}\right\}$ **148.** 13 **149.** $x+5+2\sqrt{x+4}$ **150.** $\{0, 5\}$ **151.** $\{-5, 2\}$

Section 7.6

Check Point Exercises

1. 20 or $\{20\}$ **2.** no solution or $\varnothing$ **3.** -1 and 3, or $\{-1, 3\}$ **4.** 4 or $\{4\}$ **5.** -12 or $\{-12\}$ **6.** 2060

Exercise Set 7.6

1. $\{6\}$ **3.** $\{17\}$ **5.** no solution or $\varnothing$ **7.** $\{8\}$ **9.** $\{0, 3\}$ **11.** $\{1, 3\}$ **13.** $\{3, 7\}$ **15.** $\{9\}$ **17.** $\{8\}$ **19.** $\{35\}$ **21.** $\{16\}$ **23.** $\{2\}$ **25.** $\{2, 6\}$ **27.** $\left\{\frac{5}{2}\right\}$ **29.** $\{5\}$ **31.** no solution or $\varnothing$ **33.** $\{2, 6\}$ **35.** $\{0, 10\}$ **37.** $\{8\}$ **39.** 4 **41.** -7 **43.** $V = \frac{\pi r^2 h}{3}$ or $V = \frac{1}{3}\pi r^2 h$ **45.** $l = \frac{8t^2}{\pi^2}$ **47.** 8 **49.** 9 **51.** 5.4 ft **53.** by the point (5.4, 1.16) **55. a.** 16; 16% of Americans earning \$25 thousand annually report fair or poor health.; overestimates by 1% **b.** approximately \$30 thousand **57.** 27 sq mi **59.** 149 million km

69.

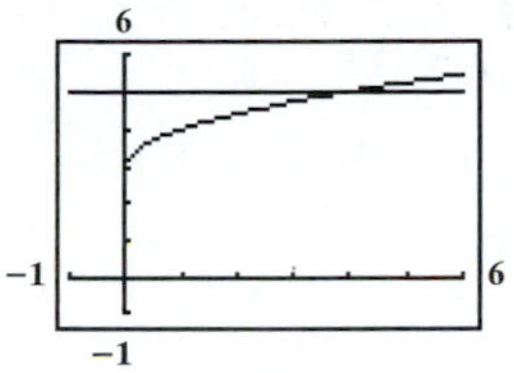

$\{4\}$

71.

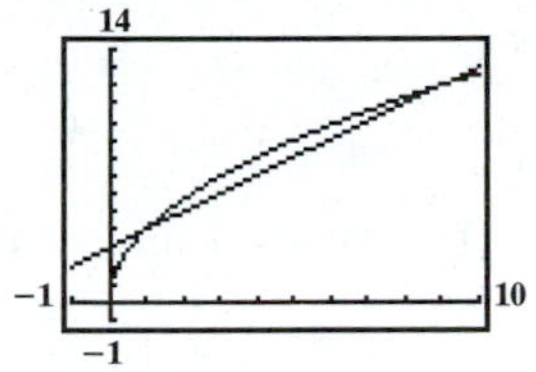

$\{1, 9\}$

73. does not make sense **75.** does not make sense **77.** false **79.** true **81.** $\sqrt{x-7} = 3; \sqrt{x} = 4; 1 + \sqrt{x} = 5$ **83.** $\{16\}$ **85.** $4x^3 - 15x^2 + 47x - 142 + \frac{425}{x+3}$ **86.** $\frac{x+2}{2(x+3)}$ **87.** $(y - 3 + 5x)(y - 3 - 5x)$ **88.** $6 + 13x$ **89.** $15x^2 - 29x - 14$ **90.** $\frac{54 + 43\sqrt{2}}{-46}$

Section 7.7

Check Point Exercises

1. a. $8i$ **b.** $i\sqrt{11}$ **c.** $4i\sqrt{3}$ **2. a.** $8 + i$ **b.** $-10 + 10i$ **3. a.** $63 + 14i$ **b.** $58 - 11i$ **4.** $-\sqrt{35}$ **5.** $\frac{18}{25} + \frac{26}{25}i$ **6.** $-\frac{1}{2} - \frac{3}{4}i$ **7. a.** 1 **b.** i **c.** $-i$

Exercise Set 7.7

1. $10i$ **3.** $i\sqrt{23}$ **5.** $3i\sqrt{2}$ **7.** $3i\sqrt{7}$ **9.** $-6i\sqrt{3}$ **11.** $5 + 6i$ **13.** $15 + i\sqrt{3}$ **15.** $-2 - 3i\sqrt{2}$ **17.** $8 + 3i$ **19.** $8 - 2i$ **21.** $5 + 3i$ **23.** $-1 - 7i$ **25.** $-2 + 9i$ **27.** $8 + 15i$ **29.** $-14 + 17i$ **31.** $9 + 5i\sqrt{3}$ **33.** $-6 + 10i$ **35.** $-21 - 15i$ **37.** $-35 - 14i$ **39.** $7 + 19i$ **41.** $-1 - 31i$ **43.** $3 + 36i$ **45.** 34 **47.** 34 **49.** 11 **51.** $-5 + 12i$ **53.** $21 - 20i$ **55.** $-\sqrt{14}$ **57.** -6 **59.** $-5\sqrt{7}$ **61.** $-2\sqrt{6}$ **63.** $\frac{3}{5} - \frac{1}{5}i$ **65.** $1 + i$ **67.** $\frac{28}{25} + \frac{21}{25}i$ **69.** $-\frac{12}{13} + \frac{18}{13}i$ **71.** $0 + i$ or i **73.** $\frac{3}{10} - \frac{11}{10}i$ **75.** $\frac{11}{13} - \frac{16}{13}i$ **77.** $-\frac{23}{58} + \frac{43}{58}i$ **79.** $0 - \frac{7}{3}i$ or $-\frac{7}{3}i$ **81.** $-\frac{5}{2} - 4i$ **83.** $-\frac{7}{3} + \frac{4}{3}i$ **85.** -1 **87.** $-i$ **89.** -1 **91.** 1 **93.** i **95.** 1 **97.** $-i$ **99.** 0 **101.** $-11 - 5i$ **103.** $-5 + 10i$ **105.** $0 + 47i$ or $47i$ **107.** $1 - i$ **109.** 0 **111.** $10 + 10i$ **113.** $\frac{20}{13} + \frac{30}{13}i$ **115.** $(47 + 13i)$ volts **117.** $(5 + i\sqrt{15}) + (5 - i\sqrt{15}) = 10; (5 + i\sqrt{15})(5 - i\sqrt{15}) = 25 - 15i^2 = 25 + 15 = 40$ **131.** $\sqrt{-9} + \sqrt{-16} = 3i + 4i = 7i$ **133.** does not make sense **135.** does not make sense **137.** false **139.** false **141.** $\frac{14}{25} - \frac{2}{25}i$ **143.** $\frac{8}{5} + \frac{16}{5}i$ **144.** $\frac{x^2}{y^2}$ **145.** $x = \frac{yz}{y - z}$ **146.** $\left\{\frac{21}{11}\right\}$ **147.** $\left\{-4, \frac{1}{2}\right\}$ **148.** $\{-3, 3\}$ **149.** $-\sqrt{6}$ is a solution.

Review Exercises

1. 9 **2.** $-\frac{1}{10}$ **3.** -3 **4.** not a real number **5.** -2 **6.** 5; 1.73; 0; not a real number **7.** 2; -2; -4 **8.** $\{x \mid x \geq 2\}$ or $[2, \infty)$ **9.** $\{x \mid x \leq 25\}$ or $(-\infty, 25]$ **10.** $5|x|$ **11.** $|x + 14|$ **12.** $|x - 4|$ **13.** $4x$ **14.** $2|x|$ **15.** $-2(x + 7)$ **16.** $\sqrt[3]{5xy}$ **17.** $(\sqrt{16})^3 = 64$ **18.** $(\sqrt[5]{32})^4 = 16$ **19.** $(7x)^{1/2}$ **20.** $(19xy)^{5/3}$ **21.** $\frac{1}{8^{2/3}} = \frac{1}{4}$ **22.** $\frac{3x}{a^{4/5}b^{4/5}} = \frac{3x}{\sqrt[5]{a^4b^4}}$ **23.** $x^{7/12}$ **24.** $5^{1/6}$ **25.** $2x^2y$ **26.** $\frac{y^{1/8}}{x^{1/3}}$ **27.** x^3y^4 **28.** $y\sqrt[3]{x}$ **29.** $\sqrt[6]{x^5}$ **30.** $\sqrt[6]{x}$ **31.** $\sqrt[15]{x}$ **32.** \$3150 million **33.** $\sqrt{21xy}$ **34.** $\sqrt[5]{77x^3}$ **35.** $\sqrt[6]{(x-5)^5}$ **36.** $f(x) = \sqrt{7}|x - 1|$ **37.** $2x\sqrt{5x}$ **38.** $3x^2y^2\sqrt[3]{2x^2}$ **39.** $2y^2z\sqrt[4]{2x^3y^3z}$ **40.** $2x^2\sqrt{6x}$ **41.** $2xy\sqrt[3]{2y^2}$ **42.** $xyz^2\sqrt[5]{16y^4z}$ **43.** $\sqrt{x^2 - 1}$ **44.** $8\sqrt[3]{3}$ **45.** $9\sqrt{2}$ **46.** $(3x + y^2)\sqrt[3]{x}$ **47.** $-8\sqrt[3]{6}$ **48.** $\frac{2}{5}\sqrt[3]{2}$ **49.** $\frac{x\sqrt{x}}{10y^2}$ **50.** $\frac{y\sqrt[4]{3y}}{2x^5}$ **51.** $2\sqrt{6}$ **52.** $2\sqrt[3]{2}$ **53.** $2x\sqrt[4]{2x}$ **54.** $10x^2\sqrt{xy}$ **55.** $6\sqrt{2} + 12\sqrt{5}$ **56.** $5\sqrt[3]{2} - \sqrt[3]{10}$ **57.** $-83 + 3\sqrt{35}$ **58.** $\sqrt{xy} - \sqrt{11x} - \sqrt{11y} + 11$ **59.** $13 + 4\sqrt{10}$ **60.** $22 - 4\sqrt{30}$ **61.** -6 **62.** 4 **63.** $\frac{2\sqrt{6}}{3}$ **64.** $\frac{\sqrt{14}}{7}$ **65.** $4\sqrt[3]{3}$ **66.** $\frac{\sqrt{10xy}}{5y}$ **67.** $\frac{7\sqrt[3]{4x}}{x}$ **68.** $\frac{\sqrt[4]{189x^3}}{3x}$ **69.** $\frac{5\sqrt[5]{xy^4}}{2xy}$ **70.** $3\sqrt{3} + 3$ **71.** $\frac{\sqrt{35} - \sqrt{21}}{2}$ **72.** $10\sqrt{5} + 15\sqrt{2}$ **73.** $\frac{x + 8\sqrt{x} + 15}{x - 9}$

74. $\frac{5+\sqrt{21}}{2}$ **75.** $\frac{3\sqrt{2}+2}{7}$ **76.** $\frac{2}{\sqrt{14}}$ **77.** $\frac{3x}{\sqrt[3]{9x^2y}}$ **78.** $\frac{7}{\sqrt{35}+\sqrt{21}}$ **79.** $\frac{2}{5-\sqrt{21}}$ **80.** {16} **81.** no solution or $\varnothing$ **82.** {2} **83.** {8} **84.** {−4, −2} **85. a.** 0.99; The average state cigarette tax per pack was approximately $0.99 six years after 2001, or in 2007.; underestimates by $0.01 **b.** 9 years after 2001, in 2010 **86.** 84 years old **87.** $9i$ **88.** $3i\sqrt{7}$ **89.** $-2i\sqrt{2}$ **90.** $12+2i$ **91.** $-9+4i$ **92.** $-12-8i$ **93.** $29+11i$ **94.** $-7-24i$ **95.** $113+0i$ or 113 **96.** $-2\sqrt{6}+0i$ or $-2\sqrt{6}$ **97.** $\frac{15}{13}-\frac{3}{13}i$ **98.** $\frac{1}{5}+\frac{11}{10}i$ **99.** $\frac{1}{3}-\frac{5}{3}i$ **100.** 1 **101.** $-i$

Chapter Test

1. a. 6 **b.** $\{x \mid x \le 4\}$ or $(-\infty, 4]$ **2.** $\frac{1}{81}$ **3.** $\frac{5y^{1/8}}{x^{1/4}}$ **4.** $\sqrt{x}$ **5.** $\sqrt[20]{x^9}$ **6.** $5|x|\sqrt{3}$ **7.** $|x-5|$ **8.** $2xy^2\sqrt[3]{xy^2}$ **9.** $-\frac{2}{x^2}$ **10.** $\sqrt{50x^2y}$ **11.** $2x\sqrt[4]{2y^3}$ **12.** $-7\sqrt{2}$ **13.** $(2x+y^2)\sqrt[3]{x}$ **14.** $2x\sqrt[3]{x}$ **15.** $12\sqrt{2}-\sqrt{15}$ **16.** $26+6\sqrt{3}$ **17.** $52-14\sqrt{3}$ **18.** $\frac{\sqrt{5x}}{x}$ **19.** $\frac{\sqrt[3]{25x}}{x}$ **20.** $-5+2\sqrt{6}$ **21.** {6} **22.** {16} **23.** {−3} **24.** 49 months **25.** $5i\sqrt{3}$ **26.** $-1+6i$ **27.** $26+7i$ **28.** $-6+0i$ or -6 **29.** $\frac{1}{5}+\frac{7}{5}i$ **30.** $-i$

Cumulative Review Exercises

1. {(−2, −1, −2)} **2.** $\left\{-\frac{1}{3}, 4\right\}$ **3.** $\left\{x \,\middle|\, x > \frac{1}{3}\right\}$ or $\left(\frac{1}{3}, \infty\right)$ **4.** $\left\{\frac{3}{4}\right\}$ **5.** {−1}

6. $x + 2y < 2$
$2y - x > 4$

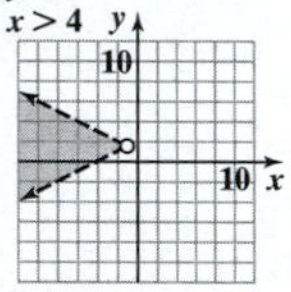

7. $\frac{x^2}{15(x+2)}$ **8.** $\frac{x}{y}$ **9.** $8x^3 - 22x^2 + 11x + 6$ **10.** $\frac{5x-6}{(x-5)(x+3)}$ **11.** −64 **12.** 0 **13.** $-\frac{16-9\sqrt{3}}{13}$ **14.** $2x^2 + x + 5 + \frac{6}{x-2}$ **15.** $-34-3\sqrt{6}$ **16.** $2(3x+2)(4x-1)$ **17.** $(4x^2+1)(2x+1)(2x-1)$ **18.** about 53 lumens **19.** $1500 at 7% and $4500 at 9% **20.** 2650 students

CHAPTER 8

Section 8.1

Check Point Exercises

1. $\pm\sqrt{7}$ or $\{\pm\sqrt{7}\}$ **2.** $\pm\frac{\sqrt{33}}{3}$ or $\left\{\pm\frac{\sqrt{33}}{3}\right\}$ **3.** $\pm\frac{3}{2}i$ or $\left\{\pm\frac{3}{2}i\right\}$ **4.** $3\pm\sqrt{10}$ or $\{3\pm\sqrt{10}\}$ **5. a.** 25; $x^2+10x+25=(x+5)^2$ **b.** $\frac{9}{4}$; $x^2-3x+\frac{9}{4}=\left(x-\frac{3}{2}\right)^2$ **c.** $\frac{9}{64}$; $x^2+\frac{3}{4}x+\frac{9}{64}=\left(x+\frac{3}{8}\right)^2$ **6.** $-2\pm\sqrt{5}$ or $\{-2\pm\sqrt{5}\}$ **7.** $\frac{-3\pm\sqrt{41}}{4}$ or $\left\{\frac{-3\pm\sqrt{41}}{4}\right\}$ **8.** 20% **9.** $10\sqrt{21}$ ft; 45.8 ft

Exercise Set 8.1

1. {±5} **3.** $\{\pm\sqrt{6}\}$ **5.** $\left\{\pm\frac{5}{4}\right\}$ **7.** $\left\{\pm\frac{\sqrt{6}}{3}\right\}$ **9.** $\left\{\pm\frac{4}{5}i\right\}$ **11.** {−10, −4} **13.** $\{3\pm\sqrt{5}\}$ **15.** $\{-2\pm2\sqrt{2}\}$ **17.** $\{5\pm3i\}$ **19.** $\left\{\frac{-3\pm\sqrt{11}}{4}\right\}$ **21.** {−3, 9} **23.** 1; $x^2+2x+1=(x+1)^2$ **25.** 49; $x^2-14x+49=(x-7)^2$ **27.** $\frac{49}{4}$; $x^2+7x+\frac{49}{4}=\left(x+\frac{7}{2}\right)^2$ **29.** $\frac{1}{16}$; $x^2-\frac{1}{2}x+\frac{1}{16}=\left(x-\frac{1}{4}\right)^2$ **31.** $\frac{4}{9}$; $x^2+\frac{4}{3}x+\frac{4}{9}=\left(x+\frac{2}{3}\right)^2$ **33.** $\frac{81}{64}$; $x^2-\frac{9}{4}x+\frac{81}{64}=\left(x-\frac{9}{8}\right)^2$ **35.** {−8, 4} **37.** $\{-3\pm\sqrt{7}\}$ **39.** $\{4\pm\sqrt{15}\}$ **41.** $\{-1\pm i\}$ **43.** $\left\{\frac{-3\pm\sqrt{13}}{2}\right\}$ **45.** $\left\{-\frac{3}{7}, -\frac{1}{7}\right\}$ **47.** $\left\{\frac{-1\pm\sqrt{5}}{2}\right\}$ **49.** $\left\{-\frac{5}{2}, 1\right\}$ **51.** $\left\{\frac{-3\pm\sqrt{6}}{3}\right\}$ **53.** $\left\{\frac{4\pm\sqrt{13}}{3}\right\}$ **55.** $\left\{\frac{1}{4}\pm\frac{1}{4}i\right\}$ **57.** −5, 7 **59.** $-\frac{1}{5}$, 1 **61.** $-2\pm5i$ **63.** $2\pm2i$ **65.** $v=\sqrt{2gh}$ **67.** $r=\frac{\sqrt{AP}}{P}-1$ **69.** $\left\{\frac{-1\pm\sqrt{19}}{6}\right\}$ **71.** {−b, 2b} **73. a.** x^2+8x **b.** 16 **c.** $x^2+8x+16$ **d.** $(x+4)^2$ **75.** 20% **77.** 6.25% **79.** approximately 1990 **81.** $10\sqrt{3}$ sec; 17.3 sec **83.** $3\sqrt{5}$ mi; 6.7 mi **85.** $20\sqrt{2}$ ft; 28.3 ft **87.** $50\sqrt{2}$ ft; 70.7 ft **89.** 10 m

97.

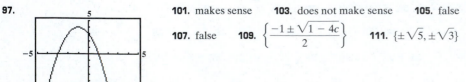

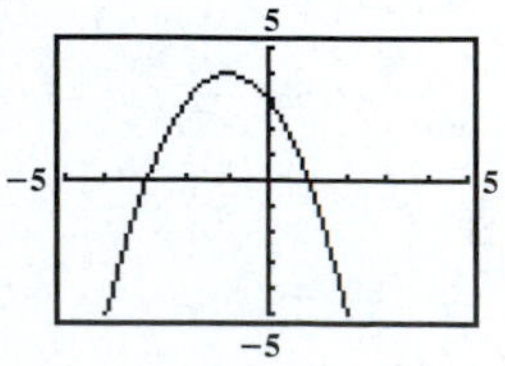

{−3, 1}

101. makes sense **103.** does not make sense **105.** false **107.** false **109.** $\left\{\frac{-1\pm\sqrt{1-4c}}{2}\right\}$ **111.** $\{\pm\sqrt{5}, \pm\sqrt{3}\}$

112. $4 - 2x$ **113.** $(1 - 2x)(1 + 2x + 4x^2)$ **114.** $x^3 - 2x^2 - 4x - 12 - \frac{42}{x - 3}$ **115. a.** $\left\{-\frac{1}{2}, \frac{1}{4}\right\}$ **b.** 36; yes

116. a. $\left\{\frac{1}{3}\right\}$ **b.** 0 **117. a.** $3x^2 + 4x + 2 = 0$ **b.** -8

Section 8.2

Check Point Exercises

1. -5 and $\frac{1}{2}$, or $\left\{-5, \frac{1}{2}\right\}$ **2.** $\frac{3 \pm \sqrt{7}}{2}$ or $\left\{\frac{3 \pm \sqrt{7}}{2}\right\}$ **3.** $-1 \pm i\frac{\sqrt{6}}{3}$ or $\left\{-1 \pm i\frac{\sqrt{6}}{3}\right\}$ **4. a.** 0; one real rational solution **b.** 81; two real rational solutions **c.** -44; two imaginary solutions that are complex conjugates **5. a.** $20x^2 + 7x - 3 = 0$ **b.** $x^2 + 49 = 0$ **6.** 26 years old; The point (26, 115) lies approximately on the blue graph.

Exercise Set 8.2

1. $\{-6, -2\}$ **3.** $\left\{1, \frac{5}{2}\right\}$ **5.** $\left\{\frac{-3 \pm \sqrt{89}}{2}\right\}$ **7.** $\left\{\frac{7 \pm \sqrt{85}}{6}\right\}$ **9.** $\left\{\frac{1 \pm \sqrt{7}}{6}\right\}$ **11.** $\left\{\frac{3}{8} \pm i\frac{\sqrt{87}}{8}\right\}$ **13.** $\{2 \pm 2i\}$ **15.** $\left\{\frac{4}{3} \pm i\frac{\sqrt{5}}{3}\right\}$ **17.** $\left\{-\frac{3}{2}, 4\right\}$ **19.** 52; two real irrational solutions **21.** 4; two real rational solutions **23.** -23; two imaginary solutions **25.** 36; two real rational solutions **27.** -60; two imaginary solutions **29.** 0; one (repeated) real rational solution **31.** $\left\{-\frac{2}{3}, 2\right\}$ **33.** $\{1 \pm \sqrt{2}\}$ **35.** $\left\{\frac{3 \pm \sqrt{65}}{4}\right\}$ **37.** $\left\{0, \frac{8}{3}\right\}$ **39.** $\left\{\frac{-6 \pm 2\sqrt{6}}{3}\right\}$ **41.** $\left\{\frac{2 \pm \sqrt{10}}{3}\right\}$ **43.** $\{2 \pm \sqrt{10}\}$ **45.** $\left\{1, \frac{5}{2}\right\}$ **47.** $x^2 - 2x - 15 = 0$ **49.** $12x^2 + 5x - 2 = 0$ **51.** $x^2 + 36 = 0$ **53.** $x^2 - 2 = 0$ **55.** $x^2 - 20 = 0$ **57.** $x^2 - 2x + 2 = 0$ **59.** $x^2 - 2x - 1 = 0$ **61.** b **63.** a **65.** $1 + \sqrt{7}$ **67.** $\left\{\frac{-1 \pm \sqrt{21}}{2}\right\}$ **69.** $\left\{-2\sqrt{2}, \frac{\sqrt{2}}{2}\right\}$ **71.** $\{-3, 1, -1 \pm i\sqrt{2}\}$ **73.** 33-year-olds and 58-year-olds; The function models the actual data well. **75.** 77.8 ft; (b) **77.** 5.5 m by 1.5 m **79.** 17.6 in. and 18.6 in. **81.** 9.3 in. and 0.7 in. **83.** 7.5 hr and 8.5 hr

93.

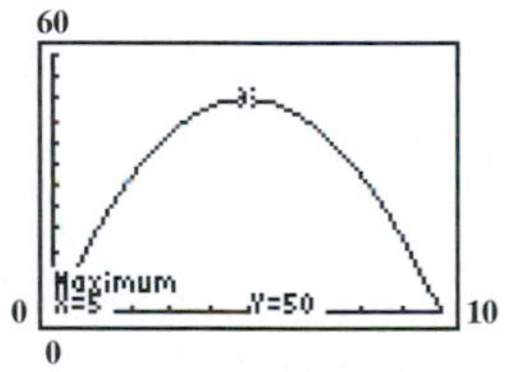

; depth: 5 in.; maximum area: 50 sq in.

95. does not make sense **97.** does not make sense **99.** false **101.** true **103.** 2.4 m; yes **105.** $\left\{-3, \frac{1}{4}\right\}$ **106.** $\{3, 7\}$

107. $\frac{5\sqrt{3} - 5x}{3 - x^2}$ **108.**

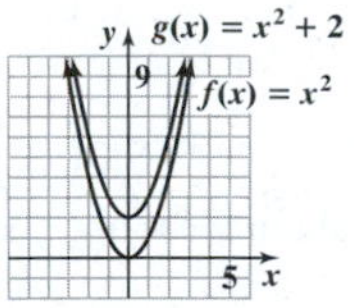

109.

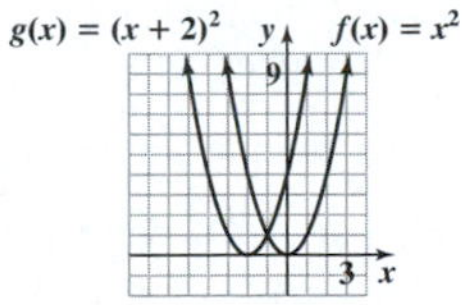

110. 1 and 5

Section 8.3

Check Point Exercises

1.

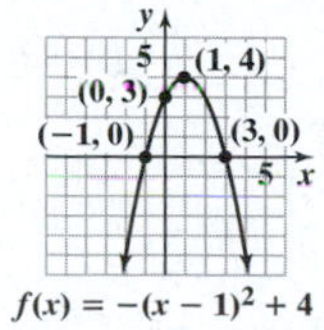

2.

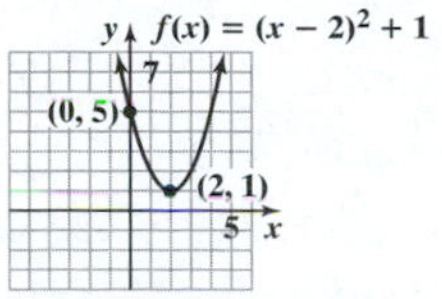

3. $(-2, -9)$

4.

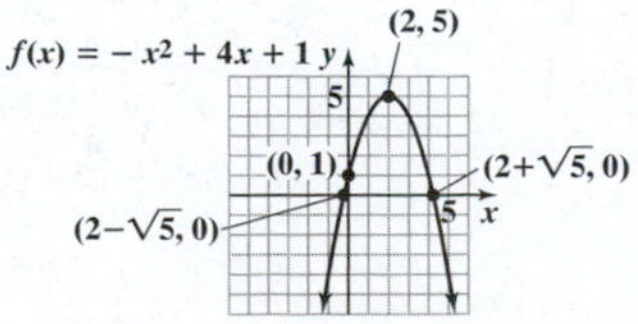

domain: $\{x|x$ is a real number$\}$ or $(-\infty, \infty)$; range: $\{y|y \leq 5\}$ or $(-\infty, 5]$

5. a. minimum **b.** Minimum is 984 at $x = 2$. **c.** domain: $\{x|x$ is a real number$\}$ or $(-\infty, \infty)$; range: $\{y|y \geq 984\}$ or $[984, \infty)$ **6.** 33.7 ft **7.** $4, -4; -16$ **8.** 30 ft by 30 ft; 900 sq ft

Exercise Set 8.3

1. $h(x) = (x - 1)^2 + 1$ **3.** $j(x) = (x - 1)^2 - 1$ **5.** $h(x) = x^2 - 1$ **7.** $g(x) = x^2 - 2x + 1$ **9.** (3, 1) **11.** (−1, 5) **13.** (2, −5) **15.** (−1, 9)

17.

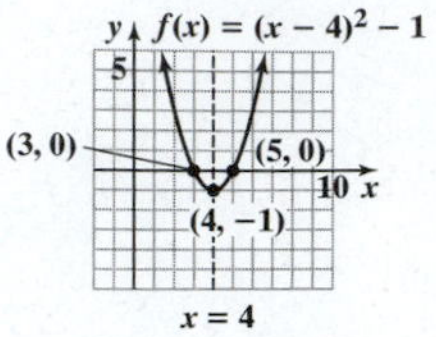

$\{y|y \geq -1\}$ or $[-1, \infty)$

19.

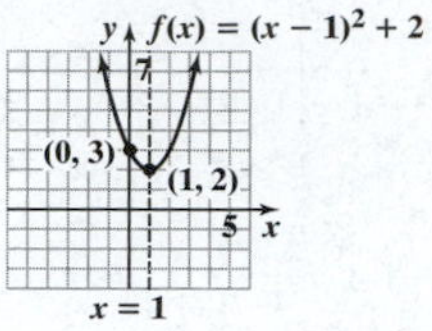

$\{y|y \geq 2\}$ or $[2, \infty)$

21.

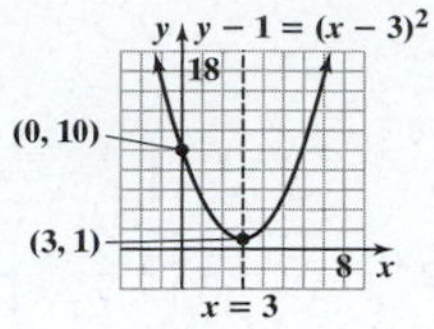

$\{y|y \geq 1\}$ or $[1, \infty)$

23.

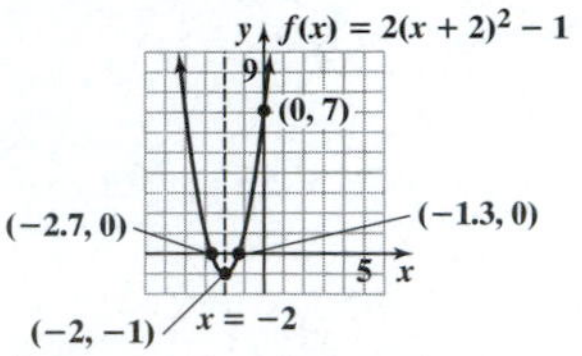

$\{y|y \geq -1\}$ or $[-1, \infty)$

25.

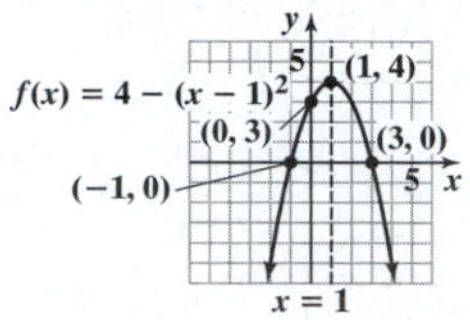

$\{y|y \leq 4\}$ or $(-\infty, 4]$

27.

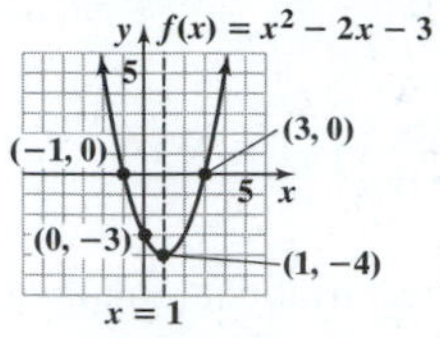

$\{y|y \geq -4\}$ or $[-4, \infty)$

29.

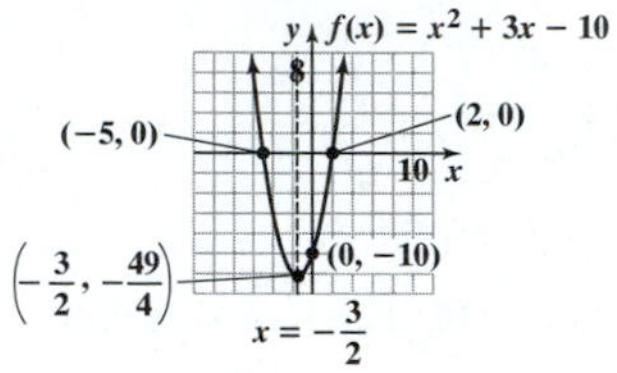

$\left\{y \middle| y \geq -\frac{49}{4}\right\}$ or $\left[-\frac{49}{4}, \infty\right)$

31.

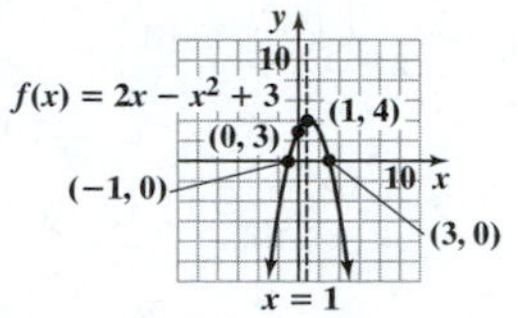

$\{y|y \leq 4\}$ or $(-\infty, 4]$

33.

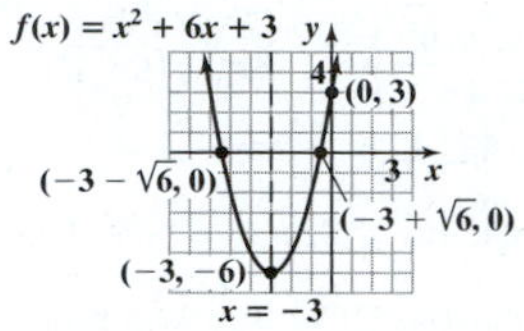

$\{y\,|\,y \geq -6\}$ or $[-6, \infty)$

35.

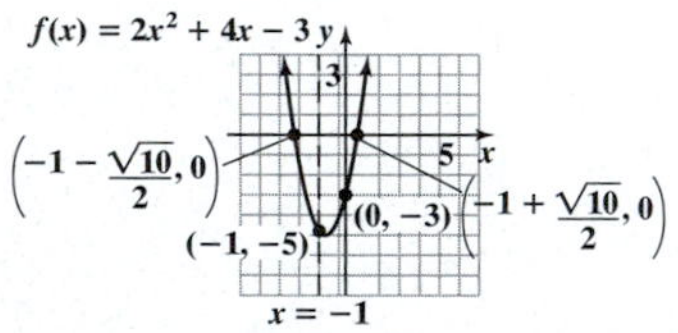

$\{y\,|\,y \geq -5\}$ or $[-5, \infty)$

37.

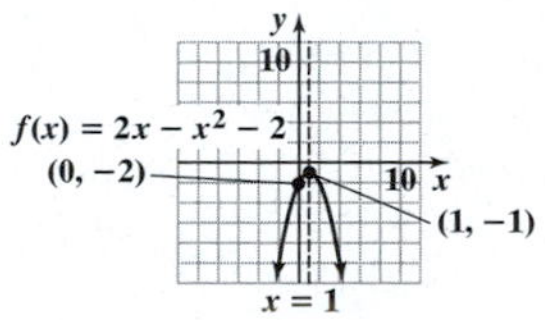

$\{y|y \leq -1\}$ or $(-\infty, -1]$

39. a. minimum **b.** Minimum is −13 at $x = 2$. **c.** domain: $\{x\,|\,x$ is a real number$\}$ or $(-\infty, \infty)$; range: $\{y\,|\,y \geq -13\}$ or $[-13, \infty)$

41. a. maximum **b.** Maximum is 1 at $x = 1$. **c.** domain: $\{x\,|\,x$ is a real number$\}$ or $(-\infty, \infty)$; range: $\{y\,|\,y \leq 1\}$ or $(-\infty, 1]$

43. a. minimum **b.** Minimum is $-\frac{5}{4}$ at $x = \frac{1}{2}$. **c.** domain: $\{x\,|\,x$ is a real number$\}$ or $(-\infty, \infty)$; range: $\left\{y \middle| y \geq -\frac{5}{4}\right\}$ or $\left[-\frac{5}{4}, \infty\right)$

45. domain: $\{x|x$ is a real number$\}$ or $(-\infty, \infty)$; range: $\{y|y \geq -2\}$ or $[-2, \infty)$ **47.** domain: $\{x\,|\,x$ is a real number$\}$ or $(-\infty, \infty)$; range: $\{y\,|\,y \leq -6\}$ or $(-\infty, -6]$ **49.** $f(x) = 2(x - 5)^2 + 3$ **51.** $f(x) = 2(x + 10)^2 - 5$ **53.** $f(x) = -3(x + 2)^2 + 4$

55. $f(x) = 3(x - 11)^2$ **57. a.** 2.75 gallons per person; underestimates by 0.05 gallon **b.** 1992; 2.048 gallons; seems reasonable

59. a. 2 sec; 224 ft **b.** 5.7 sec **c.** 160; 160 feet is the height of the building. **d.**

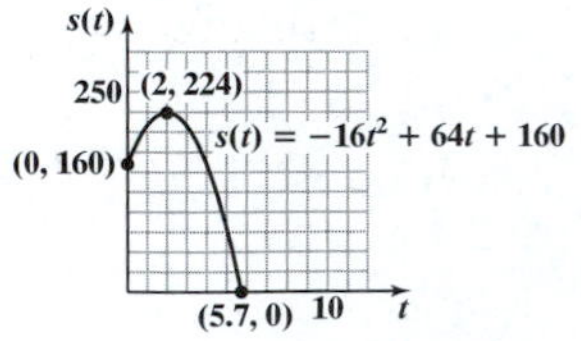

61. 8 and 8; 64 **63.** 8, −8; −64 **65.** length: 300 ft; width: 150 ft; maximum area: 45,000 sq ft **67.** 12.5 yd by 12.5 yd; 156.25 sq yd

69. 5 in.; 50 sq in. **71. a.** $C(x) = 525 + 0.55x$ **b.** $P(x) = -0.001x^2 + 2.45x - 525$ **c.** 1225 sandwiches; $975.63

79. a.

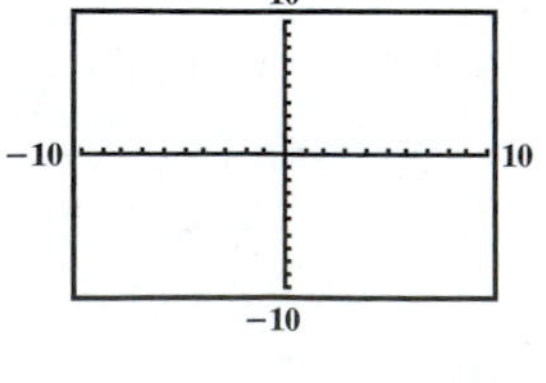

b. (20.5, −120.5) **c.**

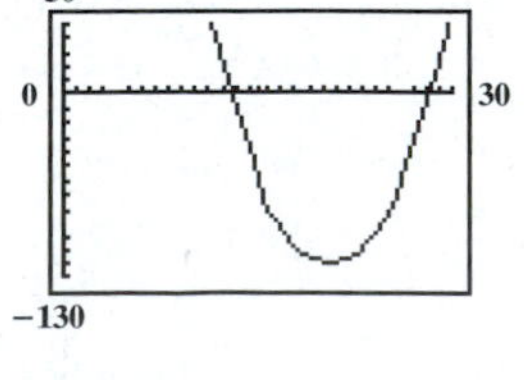

d. Answers will vary.

81. (2.5, 185)

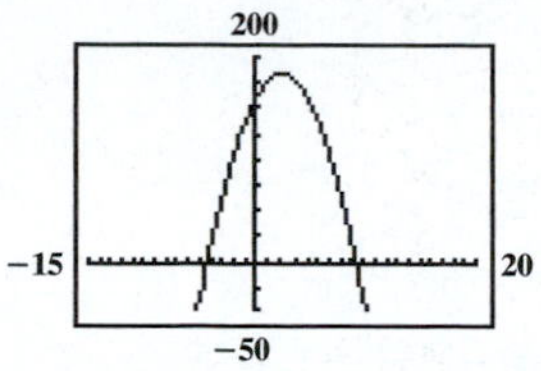

83. (−30, 91)

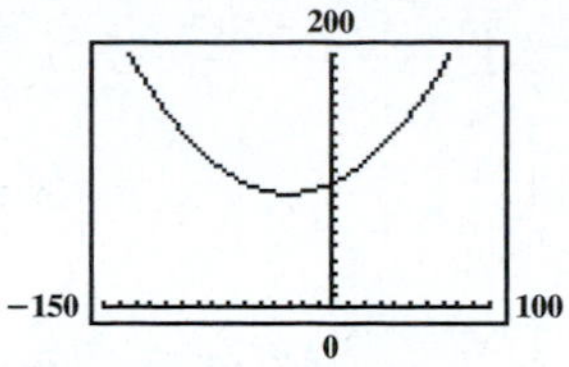

85. makes sense **87.** makes sense **89.** true **91.** false **93.** $x = -2; (-3, -2)$ **95.** $f(x) = \frac{1}{2}(x + 3)^2 - 4$
97. 125 ft by about 166.7 ft; about 20,833 sq ft **99.** {7} **100.** $\frac{x}{x - 2}$ **101.** {(6, −2)}
102. {−1, 9} **103.** $\left\{-2, \frac{5}{2}\right\}$ **104.** $5u^2 + 11u + 2 = 0$

Mid-Chapter Check Point Exercises

1. $\left\{-\frac{1}{3}, \frac{11}{3}\right\}$ **2.** $\left\{-1, \frac{7}{5}\right\}$ **3.** $\left\{\frac{3 \pm \sqrt{15}}{3}\right\}$ **4.** $\{-3 \pm \sqrt{7}\}$ **5.** $\left\{\pm\frac{6\sqrt{5}}{5}\right\}$ **6.** $\left\{\frac{5}{2} \pm i\frac{\sqrt{7}}{2}\right\}$ **7.** $\{\pm i\sqrt{13}\}$ **8.** $\left\{-4, \frac{1}{2}\right\}$
9. $\{-3 \pm 2\sqrt{6}\}$ **10.** $\{2 \pm \sqrt{3}\}$ **11.** $\left\{\frac{3}{4} \pm i\frac{\sqrt{23}}{4}\right\}$ **12.** $\left\{\frac{-3 \pm \sqrt{41}}{4}\right\}$ **13.** {4} **14.** $\{-5 \pm 2\sqrt{7}\}$

15.

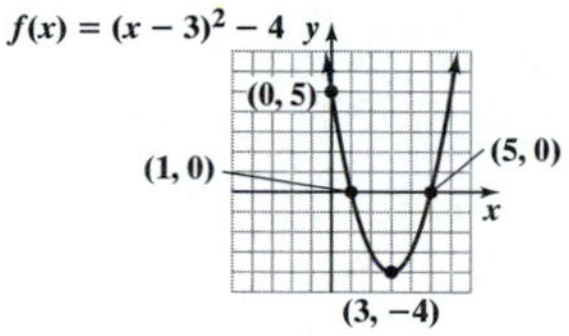

domain: $\{x|x$ is a real number$\}$ or $(-\infty, \infty)$; range: $\{y|y \geq -4\}$ or $[-4, \infty)$

16.

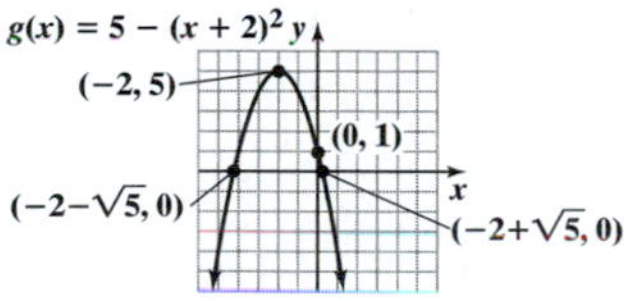

domain: $\{x|x$ is a real number$\}$ or $(-\infty, \infty)$; range: $\{y|y \leq 5\}$ or $(-\infty, 5]$

17.

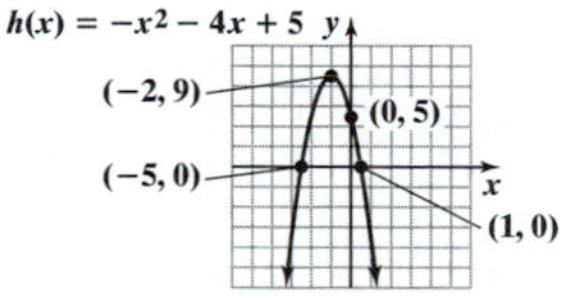

domain: $\{x|x$ is a real number$\}$ or $(-\infty, \infty)$; range: $\{y|y \leq 9\}$ or $(-\infty, 9]$

18.

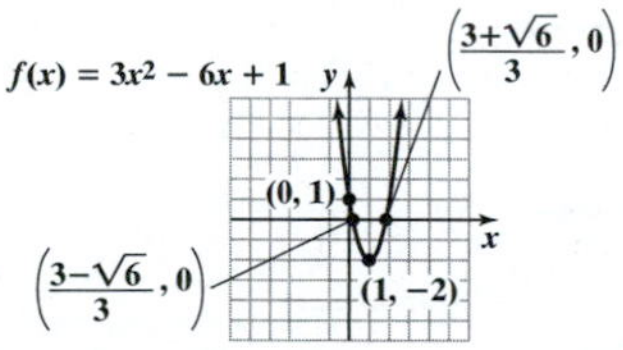

domain: $\{x|x$ is a real number$\}$ or $(-\infty, \infty)$; range: $\{y|y \geq -2\}$ or $[-2, \infty)$

19. two imaginary solutions **20.** two real rational solutions **21.** $8x^2 - 2x - 3 = 0$ **22.** $x^2 - 12 = 0$ **23.** 75 cabinets per day; $1200
24. −9, −9; 81 **25.** 10 in.; 100 sq in.

Section 8.4

Check Point Exercises

1. $-\sqrt{3}, -\sqrt{2}, \sqrt{2}$, and $\sqrt{3}$ or $\{\pm\sqrt{2}, \pm\sqrt{3}\}$ **2.** 16 or {16} **3.** $-\sqrt{6}, -1, 1$, and $\sqrt{6}$, or $\{-\sqrt{6}, -1, 1, \sqrt{6}\}$
4. −1 and 2, or {−1, 2} **5.** $-\frac{1}{27}$ and 64, or $\left\{-\frac{1}{27}, 64\right\}$

Exercise Set 8.4

1. {−2, 2, −1, 1} **3.** $\{-3, 3, -\sqrt{2}, \sqrt{2}\}$ **5.** $\{-2i, 2i, -\sqrt{2}, \sqrt{2}\}$ **7.** {1} **9.** {49} **11.** {25, 64} **13.** {2, 12} **15.** $\{-\sqrt{3}, 0, \sqrt{3}\}$
17. {−5, −2, −1, 2} **19.** $\left\{-\frac{1}{4}, \frac{1}{5}\right\}$ **21.** $\left\{\frac{1}{3}, 2\right\}$ **23.** $\left\{\frac{-2 \pm \sqrt{7}}{3}\right\}$ **25.** {−8, 27} **27.** {−243, 32} **29.** {1} **31.** {−8, −2, 1, 4}
33. −2, 2, −1, and 1; c **35.** 1; e **37.** 2 and 3; f **39.** −5, −4, 1, and 2 **41.** $-\frac{3}{2}$ and $-\frac{1}{3}$ **43.** $\frac{64}{15}$ and $\frac{81}{20}$ **45.** $\frac{5}{2}$ and $\frac{25}{6}$
47. ages 20 and 55; The function models the data well. **53.** {3, 5} **55.** {1} **57.** {−1, 4} **59.** {1, 8} **61.** makes sense
63. does not make sense **65.** true **67.** false **69.** $\left\{\sqrt[3]{-2}, \frac{\sqrt[3]{225}}{5}\right\}$ **71.** $\frac{1}{5x - 1}$ **72.** $\frac{1}{2} + \frac{3}{2}i$ **73.** {(4, −2)}

74. $\left\{-3, \frac{5}{2}\right\}$ **75.** $\{-2, -1, 2\}$ **76.** $\frac{-x-5}{x+3}$

Section 8.5

Check Point Exercises

1. $\{x|x < -4 \text{ or } x > 5\}$ or $(-\infty, -4) \cup (5, \infty)$

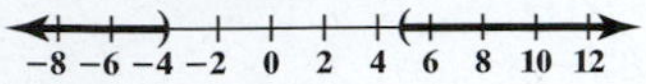

2. $\{x|x \le -3 \text{ or } -1 \le x \le 1\}$ or $(-\infty, -3] \cup [-1, 1]$

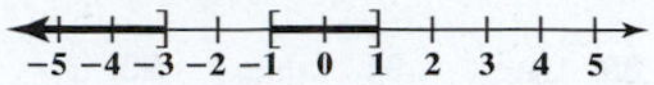

3. $\{x|-2 < x < 5\}$ or $(-2, 5)$

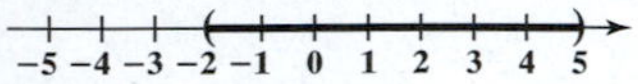

4. $\{x|x < -1 \text{ or } x \ge 1\}$ or $(-\infty, -1) \cup [1, \infty)$

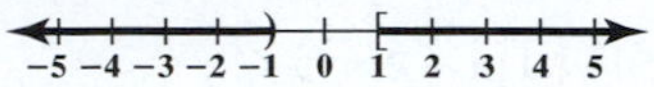

5. between 1 and 4 seconds, excluding $t = 1$ and $t = 4$

Exercise Set 8.5

1. $\{x|x < -2 \text{ or } x > 4\}$ or $(-\infty, -2) \cup (4, \infty)$

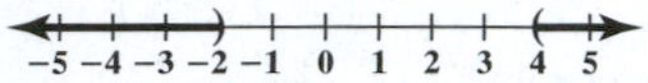

3. $\{x|-3 \le x \le 7\}$ or $[-3, 7]$

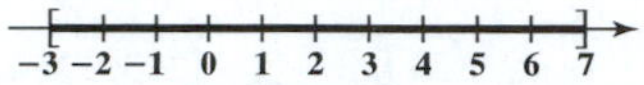

5. $\{x|x < 1 \text{ or } x > 4\}$ or $(-\infty, 1) \cup (4, \infty)$

7. $\{x|x < -4 \text{ or } x > -1\}$ or $(-\infty, -4) \cup (-1, \infty)$

9. $\{x|2 \le x \le 4\}$ or $[2, 4]$

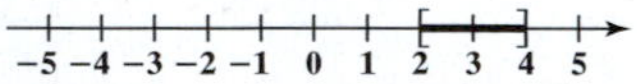

11. $\left\{x \middle| -4 \le x \le \frac{2}{3}\right\}$ or $\left[-4, \frac{2}{3}\right]$

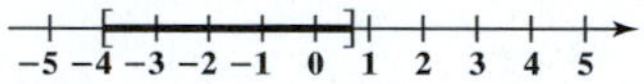

13. $\left\{x \middle| -3 < x < \frac{5}{2}\right\}$ or $\left(-3, \frac{5}{2}\right)$

15. $\left\{x \middle| -1 < x < -\frac{3}{4}\right\}$ or $\left(-1, -\frac{3}{4}\right)$

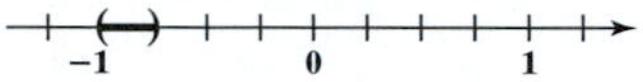

17. $\{x|x \le 0 \text{ or } x \ge 4\}$ or $(-\infty, 0] \cup [4, \infty)$

19. $\left\{x \middle| x < -\frac{3}{2} \text{ or } x > 0\right\}$ or $\left(-\infty, -\frac{3}{2}\right) \cup (0, \infty)$

21. $\{x|0 \le x \le 1\}$ or $[0, 1]$

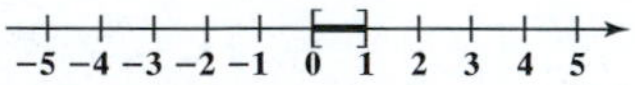

23. $\{x|2 - \sqrt{2} \le x \le 2 + \sqrt{2}\}$ or $[2 - \sqrt{2}, 2 + \sqrt{2}]$

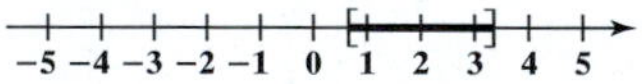

25. $\left\{x \middle| x < \frac{2 - \sqrt{10}}{3} \text{ or } x > \frac{2 + \sqrt{10}}{3}\right\}$ or $\left(-\infty, \frac{2 - \sqrt{10}}{3}\right) \cup \left(\frac{2 + \sqrt{10}}{3}, \infty\right)$;

27. $\left\{x \middle| x \le \frac{5 - \sqrt{33}}{4} \text{ or } x \ge \frac{5 + \sqrt{33}}{4}\right\}$ or $\left(-\infty, \frac{5 - \sqrt{33}}{4}\right] \cup \left[\frac{5 + \sqrt{33}}{4}, \infty\right)$;

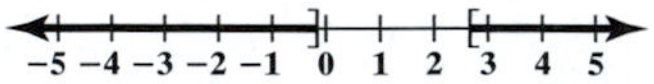

29. no solution or $\varnothing$

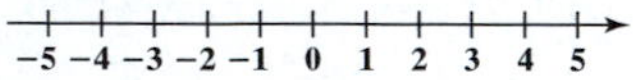

31. $\{x|1 \le x \le 2 \text{ or } x \ge 3\}$ or $[1, 2] \cup [3, \infty)$

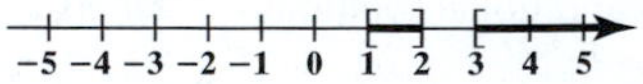

33. $\{x|-2 \le x \le -1 \text{ or } x \ge 1\}$ or $[-2, -1] \cup [1, \infty)$

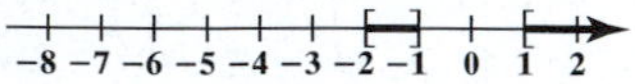

35. $\{x|x < -3\}$ or $(-\infty, -3)$

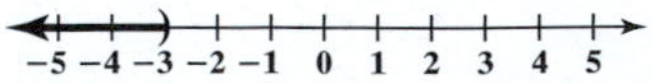

37. $\{x|x > -1\}$ or $(-1, \infty)$

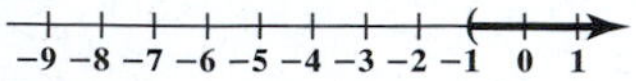

39. $\{x|x = 0 \text{ or } x \ge 9\}$ or $\{0\} \cup [9, \infty)$

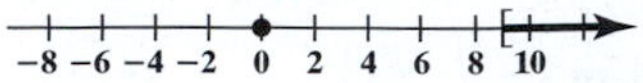

41. $\{x|x < -3 \text{ or } x > 4\}$ or $(-\infty, -3) \cup (4, \infty)$

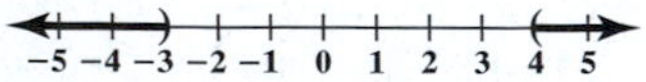

43. $\{x|-4 < x < -3\}$ or $(-4, -3)$

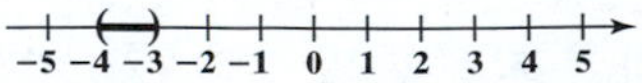

45. $\{x|2 \le x < 4\}$ or $[2, 4)$

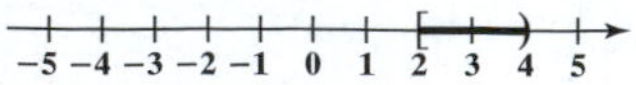

47. $\left\{x \middle| x < -\frac{4}{3} \text{ or } x \ge 2\right\}$ or $\left(-\infty, -\frac{4}{3}\right) \cup [2, \infty)$

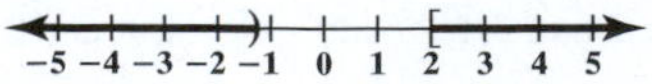

49. $\{x|x < 0 \text{ or } x > 3\}$ or $(-\infty, 0) \cup (3, \infty)$

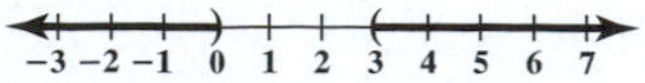

51. $\{x|x < -5 \text{ or } x > -3\}$ or $(-\infty, -5) \cup (-3, \infty)$

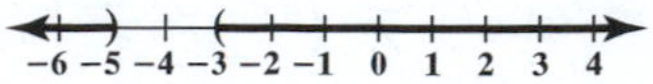

53. $\left\{x \middle| x < \frac{1}{2} \text{ or } x \geq \frac{7}{5}\right\}$ or $\left(-\infty, \frac{1}{2}\right) \cup \left[\frac{7}{5}, \infty\right)$

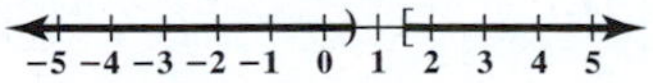

55. $\{x|x \leq -6 \text{ or } x > -2\}$ or $(-\infty, -6] \cup (-2, \infty)$

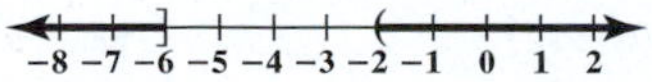

57. $\left\{x \middle| x \leq \frac{1}{2} \text{ or } x \geq 2\right\}$ or $\left(-\infty, \frac{1}{2}\right] \cup [2, \infty)$

59. $\{x|-1 < x < 1\}$ or $(-1, 1)$

61. $\{x|x < -8 \text{ or } -6 < x < 4 \text{ or } x > 6\}$ or $(-\infty, -8) \cup (-6, 4) \cup (6, \infty)$

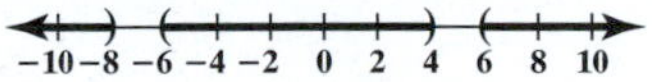

63. $\{x|-3 < x < 2\}$ or $(-3, 2)$

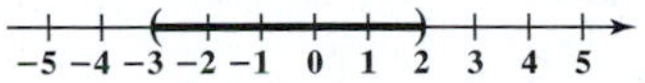

65. $\{x|x < -1 \text{ or } 1 < x < 2 \text{ or } x > 3\}$ or $(-\infty, -1) \cup (1, 2) \cup (3, \infty)$

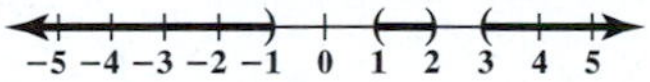

67. $\left\{x \middle| -6 \leq x \leq -\frac{1}{2} \text{ or } x \geq 1\right\}$ or $\left[-6, -\frac{1}{2}\right] \cup [1, \infty)$

69. $\{x|x < -2 \text{ or } -1 \leq x < 2\}$ or $(-\infty, -2) \cup [-1, 2)$

71. between 0 and 3 seconds, excluding $t = 0$ and $t = 3$ **73. a.** dry: 160 ft; wet: 185 ft **b.** dry pavement: graph (b); wet pavement: graph (a) **c.** extremely well; Function values and data are identical. **d.** speeds exceeding 76 miles per hour; points on graph (b) to the right of (76, 540) **75.** The company's production level must be at least 20,000 wheelchairs per month. For values of x greater than or equal to 20,000, the graph lies on or below the line $y = 425$. **77.** The length of the shorter side cannot exceed 6 ft.

83. $\left\{x \middle| -3 \leq x \leq \frac{1}{2}\right\}$ or $\left[-3, \frac{1}{2}\right]$ **85.** $\{x|x < 3 \text{ or } x \geq 8\}$ or $(-\infty, 3) \cup [8, \infty)$ **87.** $\{x|-3 < x < -1 \text{ or } x > 2\}$ or $(-3, -1) \cup (2, \infty)$

89. a. $f(x) = 0.1375x^2 + 0.7x + 37.8$ **b.** speeds exceeding 52 mph **91.** does not make sense **93.** does not make sense **95.** false **97.** true

99. Answers will vary.; example: $\frac{x-3}{x+4} \geq 0$ **101.** $\{2\}$ **103.** $\{x|x < 2 \text{ or } x > 2\}$ or $(-\infty, 2) \cup (2, \infty)$ **105.** $27 - 3x^2 \geq 0$; $\{x|-3 \leq x \leq 3\}$ or $[-3, 3]$ **106.** $\{x|-19 < x < 29\}$ or $(-19, 29)$ **107.** $\frac{2(x-1)}{(x+4)(x-3)}$ **108.** $(x^2 + 4y^2)(x + 2y)(x - 2y)$

109.

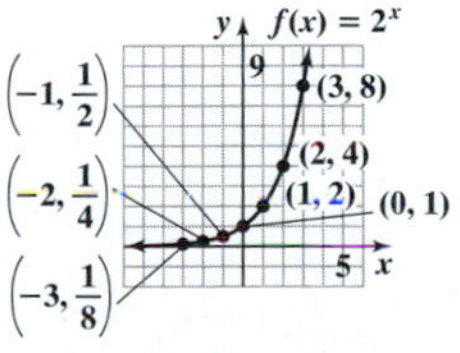

110.

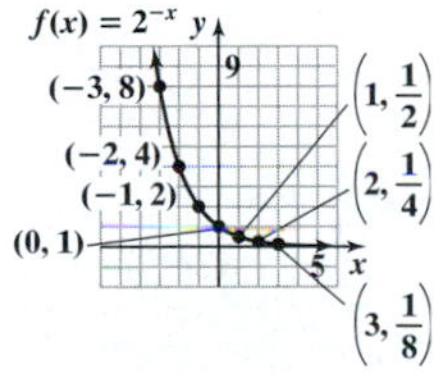

111.

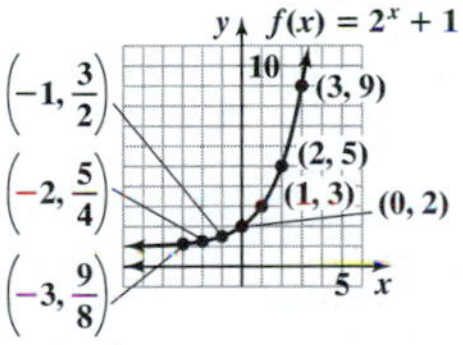

Review Exercises

1. $\{\pm 8\}$ **2.** $\{\pm 5\sqrt{2}\}$ **3.** $\left\{\pm \frac{\sqrt{6}}{3}\right\}$ **4.** $\{4 \pm 3\sqrt{2}\}$ **5.** $\{-7 \pm 6i\}$ **6.** 100; $x^2 + 20x + 100 = (x + 10)^2$ **7.** $\frac{9}{4}$; $x^2 - 3x + \frac{9}{4} = \left(x - \frac{3}{2}\right)^2$

8. $\{3, 9\}$ **9.** $\left\{\frac{7 \pm \sqrt{53}}{2}\right\}$ **10.** $\left\{\frac{-3 \pm \sqrt{41}}{4}\right\}$ **11.** 8% **12.** 14 weeks **13.** $60\sqrt{5}$ m; 134.2 m **14.** $\{1 \pm \sqrt{5}\}$ **15.** $\{1 \pm 3i\sqrt{2}\}$

16. $\left\{\frac{-2 \pm \sqrt{10}}{2}\right\}$ **17.** two imaginary solutions **18.** two real rational solutions **19.** two real irrational solutions **20.** $\left\{-\frac{2}{3}, 4\right\}$

21. $\{-5, 2\}$ **22.** $\left\{\frac{1 \pm \sqrt{21}}{10}\right\}$ **23.** $\{-4, 4\}$ **24.** $\{3 \pm 2\sqrt{2}\}$ **25.** $\left\{\frac{1}{6} \pm i\frac{\sqrt{23}}{6}\right\}$ **26.** $\{4 \pm \sqrt{5}\}$ **27.** $15x^2 - 4x - 3 = 0$

28. $x^2 + 81 = 0$ **29.** $x^2 - 48 = 0$ **30. a.** 261 ft; overestimates by 1 ft **b.** 40 mph **31. a.** by the point (35, 261) **b.** by the point (40, 267) **32.** 8.8 sec

33.

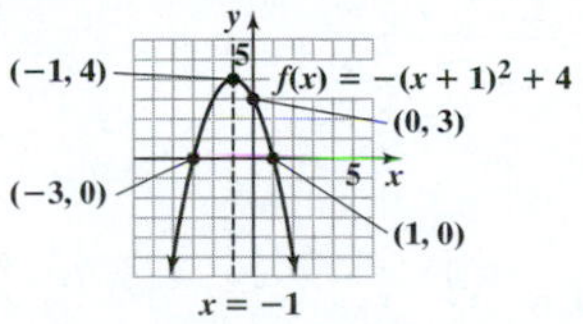

34.

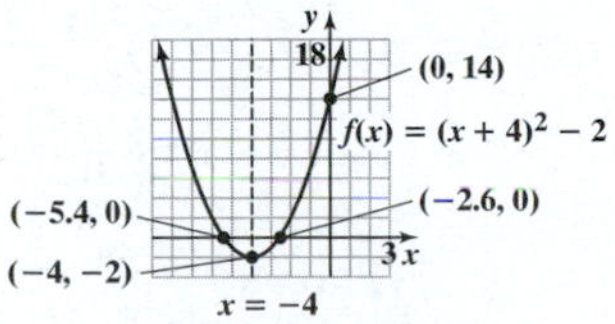

35.

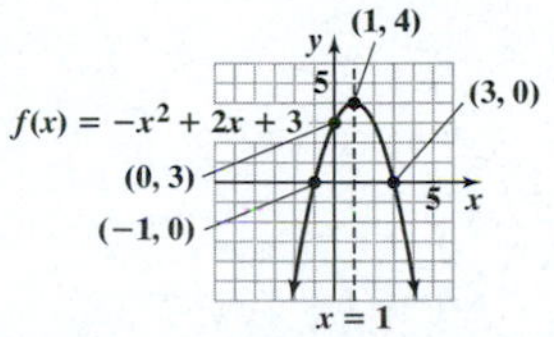

36.

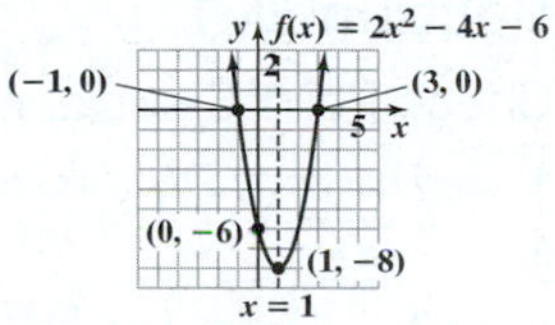

37. 25 in. of rainfall per year; 13.5 in. of growth **38.** 12.5 sec; 2540 feet **39.** 7.2 h; 622 per 100,000 males

40. 250 yd by 500 yd; 125,000 sq yard **41.** −7 and 7; −49 **42.** $\{-\sqrt{2}, \sqrt{2}, -2, 2\}$ **43.** $\{1\}$ **44.** $\{-5, -1, 3\}$ **45.** $\left\{-\frac{1}{8}, \frac{1}{7}\right\}$

46. $\{-27, 64\}$ **47.** $\{16\}$

48. $\left\{x \middle| -3 < x < \frac{1}{2}\right\}$ or $\left(-3, \frac{1}{2}\right)$;

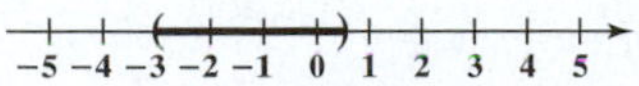

49. $\left\{x \middle| x \leq -4 \text{ or } x \geq -\frac{1}{2}\right\}$ or $(-\infty, -4] \cup \left[-\frac{1}{2}, \infty\right)$;

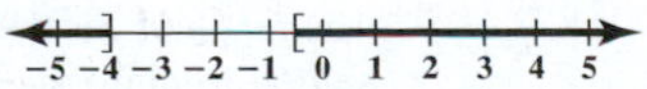

50. $\{x|-3 < x < 0 \text{ or } x > 1\}$ or $(-3, 0) \cup (1, \infty)$;

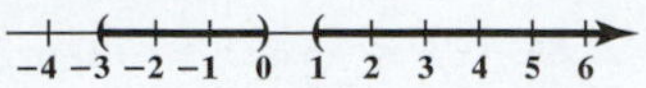

51. $\{x|x < -2 \text{ or } x > 6\}$ or $(-\infty, -2) \cup (6, \infty)$;

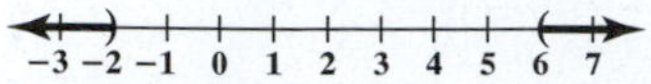

52. $\left\{x \middle| x < 4 \text{ or } x \geq \frac{23}{4}\right\}$ or $(-\infty, 4) \cup \left[\frac{23}{4}, \infty\right)$;

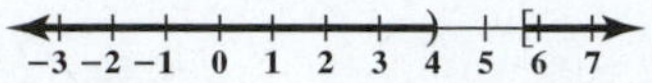

53. between 1 and 2 seconds, excluding $t = 1$ and $t = 2$ **54. a.** 200 beats per minute **b.** between 0 and 4 minutes and more than 12 minutes after the workout; between 0 and 4 minutes; Answers will vary.

Chapter Test

1. $\left\{\pm\frac{\sqrt{10}}{2}\right\}$ **2.** $\{3 \pm 2\sqrt{5}\}$ **3.** $64; x^2 - 16x + 64 = (x - 8)^2$ **4.** $\frac{1}{25}; x^2 + \frac{2}{5}x + \frac{1}{25} = \left(x + \frac{1}{5}\right)^2$ **5.** $\{3 \pm \sqrt{2}\}$ **6.** $50\sqrt{2}$ ft

7. two real irrational solutions **8.** two imaginary solutions **9.** $\left\{-5, \frac{1}{2}\right\}$ **10.** $\{-4 \pm \sqrt{11}\}$ **11.** $\{-2 \pm 5i\}$ **12.** $\left\{\frac{3}{2} \pm \frac{1}{2}i\right\}$

13. $x^2 - 4x - 21 = 0$ **14.** $x^2 + 100 = 0$ **15. a.** 70.2%; overestimates by 0.2% **b.** 2010

16.

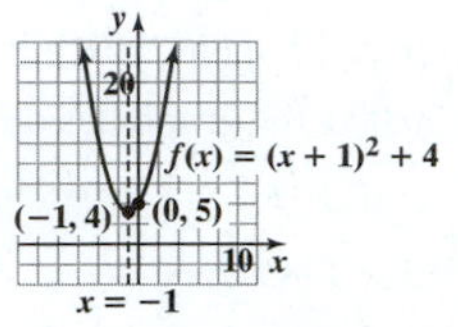

17.

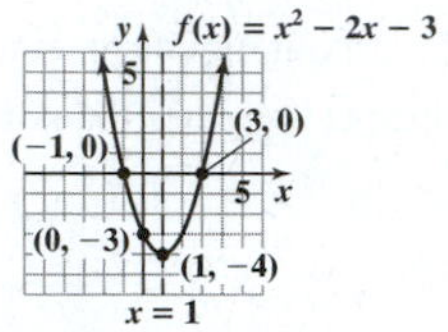

18. after 2 sec; 69 ft **19.** 4.1 sec **20.** 23 computers; \$169 hundreds or \$16,900 **21.** $\{1, 2\}$ **22.** $\{-3, 3, -2, 2\}$ **23.** $\{1, 512\}$

24. $\{x|-3 < x < 4\}$ or $(-3, 4)$;

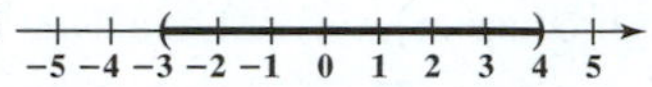

25. $\{x|x < 3 \text{ or } x \geq 10\}$ or $(-\infty, 3) \cup [10, \infty)$;

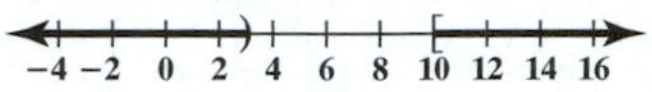

Cumulative Review Exercises

1. $\left\{\frac{2}{3}\right\}$ **2.** $\{(-5, 2)\}$ **3.** $\{(1, 4, -2)\}$ **4.** $\{x|x \geq 10\}$ or $[10, \infty)$ **5.** $\{x|x \leq -4\}$ or $(-\infty, -4]$ **6.** $\{x|x > 2\}$ or $(2, \infty)$

7. $\{x|-2 < x < 3\}$ or $(-2, 3)$ **8.** $\{3, 9\}$ **9.** no solution or $\varnothing$ **10.** $\{12\}$ **11.** $\left\{\frac{-2 \pm \sqrt{14}}{2}\right\}$ **12.** $\{8, 27\}$

13. $\left\{x \middle| -2 \leq x \leq \frac{3}{2}\right\}$ or $\left[-2, \frac{3}{2}\right]$

14.

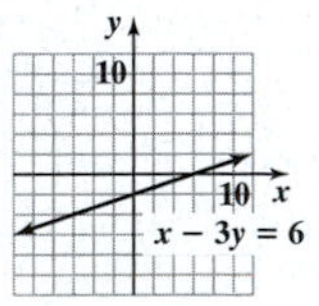

15.

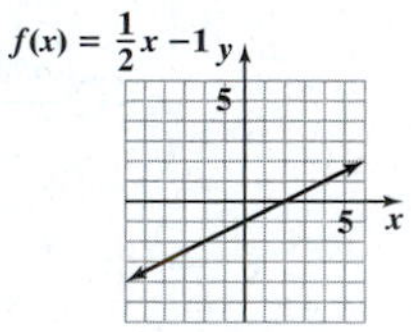

16.

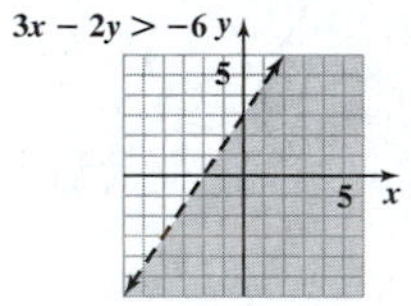

17.

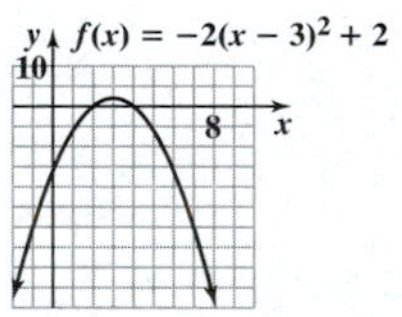

18. $-16x + 24y$ **19.** $-\frac{20x^7}{y^4}$ **20.** $x^2 - 5xy - 16y^2$ **21.** $6x^2 + 13x - 5$ **22.** $9x^4 - 24x^2y + 16y^2$ **23.** $\frac{3x^2 + 6x - 2}{(x + 5)(x + 2)}$

24. $\frac{x - 3}{x}$ **25.** $\frac{x - 4}{3x + 6}$ **26.** $5xy\sqrt{2x}$ **27.** $9\sqrt{2}$ **28.** $44 + 6i$ **29.** $(9x^2 + 1)(3x + 1)(3x - 1)$ **30.** $2x(4x - 1)(3x - 2)$

31. $(x + 3y)(x^2 - 3xy + 9y^2)$ **32.** $x^2 + 2x - 13; 22$ **33.** $x + 5$; $\{x|x$ is a real number and $x \neq 2\}$ or $(-\infty, 2) \cup (2, \infty)$ **34.** $2a + h + 3$

35. $3x^2 - 7x + 18 - \frac{28}{x + 2}$ **36.** $R = -\frac{Ir}{I - 1}$ or $R = \frac{Ir}{1 - I}$ **37.** $y = -3x - 1$ or $f(x) = -3x - 1$ **38.** 6 **39.** \$620 **40.** 13 yd by 4 yd

41. \$2600 at 12% and \$1400 at 14% **42.** 11 amps

CHAPTER 9

Section 9.1

Check Point Exercises

1. approximately \$160; overestimates by \$11

2.

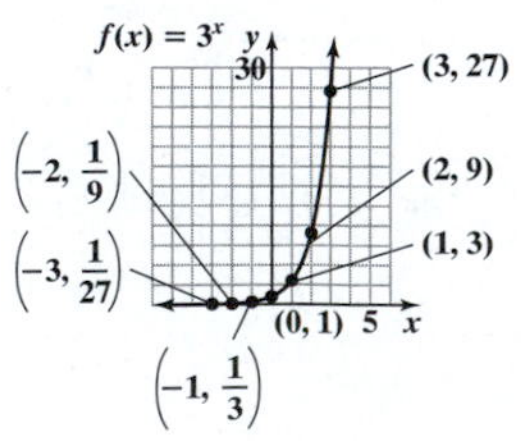

3.

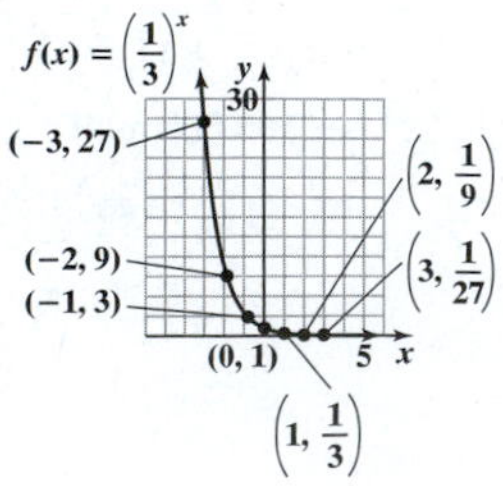

4.

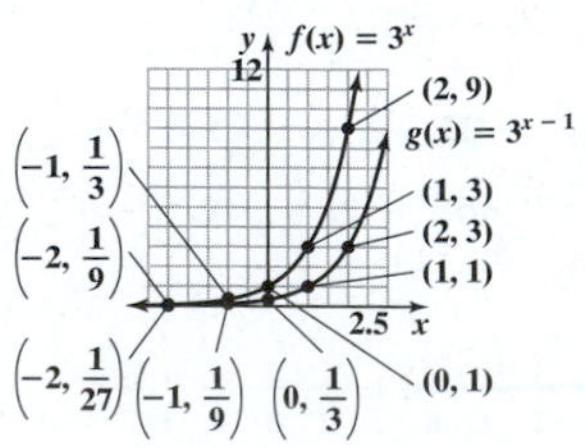

The graph of g is the graph of f shifted 1 unit to the right.

5.

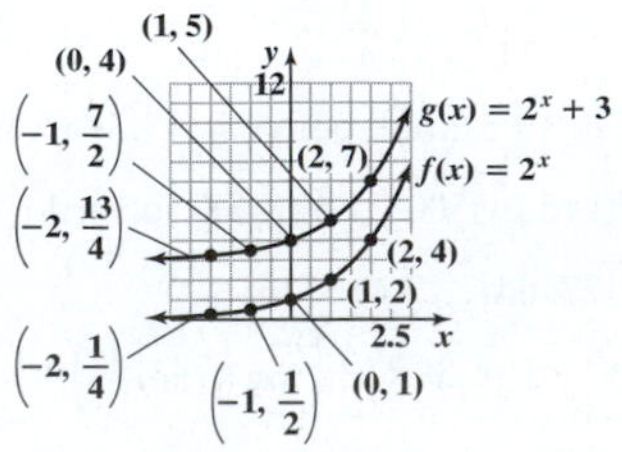

The graph of g is the graph of f shifted up 3 units.

6. approximately 4446 **7. a.** \$14,859.47 **b.** \$14,918.25

Exercise Set 9.1

1. 10.556 **3.** 11.665 **5.** 0.125 **7.** 9.974 **9.** 0.387

11. ; d

x	$f(x)$
-2	$\frac{1}{9}$
-1	$\frac{1}{3}$
0	1
1	3
2	9

13. ; e

x	$f(x)$
-2	$-\frac{8}{9}$
-1	$-\frac{2}{3}$
0	0
1	2
2	8

15. ; f

x	$f(x)$
-2	9
-1	3
0	1
1	$\frac{1}{3}$
2	$\frac{1}{9}$

17.

x	$f(x)$
-2	$\frac{1}{16}$
-1	$\frac{1}{4}$
0	1
1	4
2	16

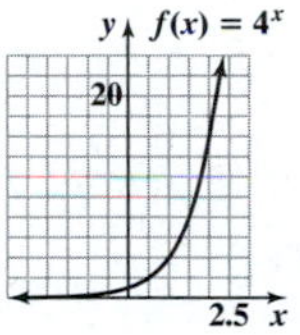

19.

x	$g(x)$
-2	$\frac{4}{9}$
-1	$\frac{2}{3}$
0	1
1	$\frac{3}{2}$
2	$\frac{9}{4}$

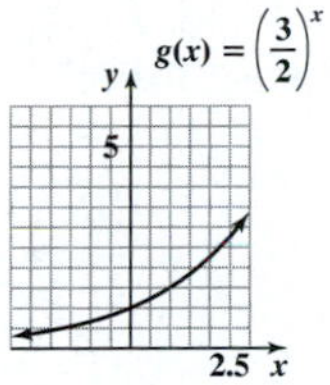

21.

x	$h(x)$
-2	4
-1	2
0	1
1	$\frac{1}{2}$
2	$\frac{1}{4}$

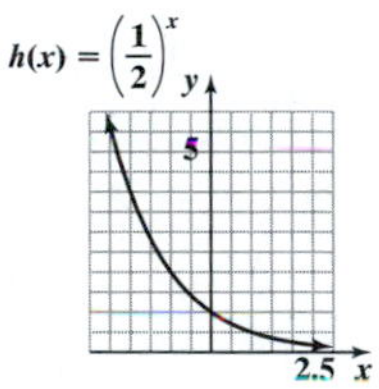

23.

x	$f(x)$
-2	2.78
-1	1.67
0	1
1	0.6
2	0.36

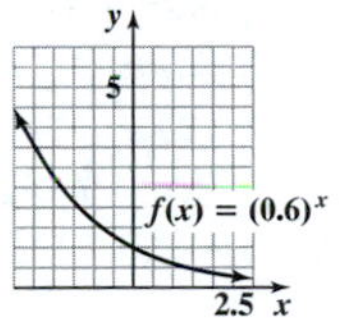

25.

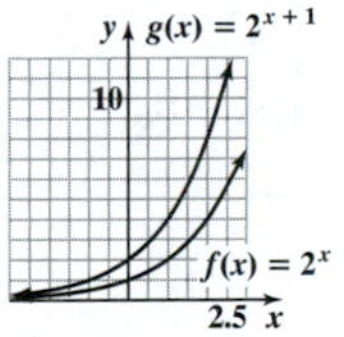

The graph of g is the graph of f shifted 1 unit to the left.

27.

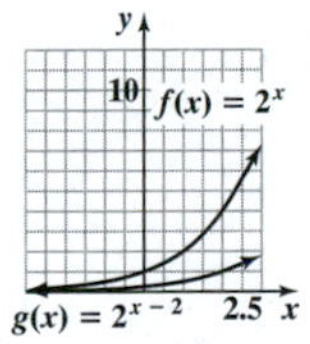

The graph of g is the graph of f shifted 2 units to the right.

29.

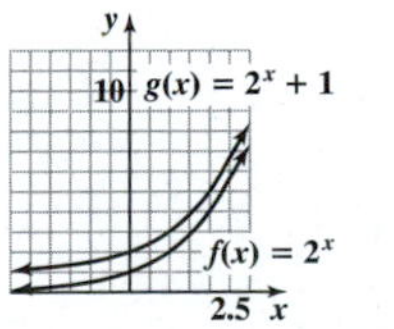

The graph of g is the graph of f shifted up 1 unit.

31.

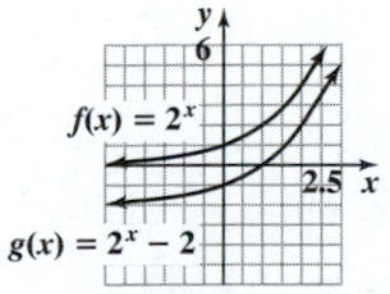

The graph of g is the graph of f shifted down 2 units.

33.

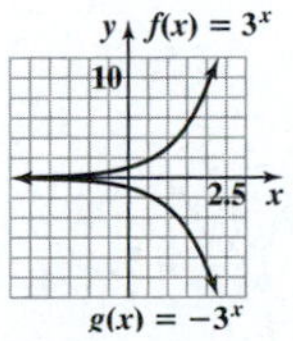

The graph of g is a reflection of the graph of f across the x-axis.

35.

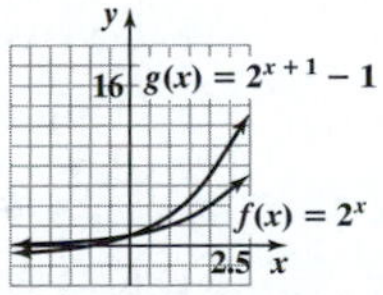

The graph of g is the graph of f shifted 1 unit to the left and 1 unit down.

37.

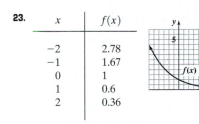

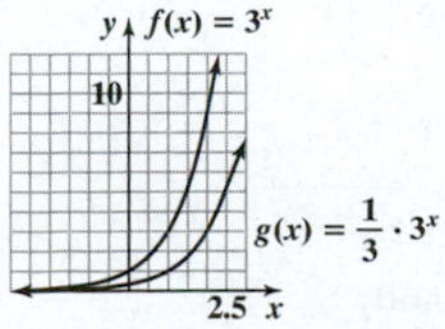

The graph of g is the graph of f stretched vertically by a factor of $\frac{1}{3}$.

39. a. $13,116.51 **b.** $13,157.04 **c.** $13,165.31 **41.** 7% compounded monthly
43. domain: $\{x|x \text{ is a real number}\}$ or $(-\infty, \infty)$; range: $\{y|y > -2\}$ or $(-2, \infty)$
45. domain: $\{x|x \text{ is a real number}\}$ or $(-\infty, \infty)$; range: $\{y|y > 1\}$ or $(1, \infty)$
47. domain: $\{x|x \text{ is a real number}\}$ or $(-\infty, \infty)$; range: $\{y|y > 0\}$ or $(0, \infty)$
49. (0, 1)

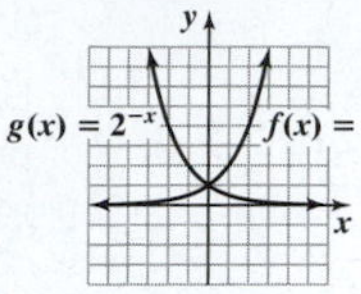

51.

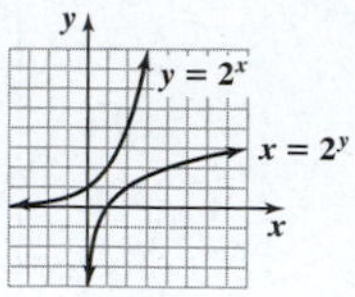

53. a. 574 million **b.** 1148 million **c.** 2295 million **d.** 4590 million **e.** It appears to double. **55.** $832,744 **57. a.** approximately 8.9 million words **b.** approximately 10.4 million words **c.** the linear model **59. a.** 100% **b.** about 68.5% **c.** about 30.8% **d.** about 20% **61.** 11.3; About 11.3% of 30-year-olds have some coronary heart disease.
63. a. about 1429 people **b.** about 24,546 people **c.** The number of ill people cannot exceed the population.; The asymptote indicates that the number of ill people will not exceed 30,000, the population of the town.

69. a. $f(t) = 10{,}000\left(1 + \frac{0.05}{4}\right)^{4t}$; $f(t) = 10{,}000\left(1 + \frac{0.045}{12}\right)^{12t}$

b.

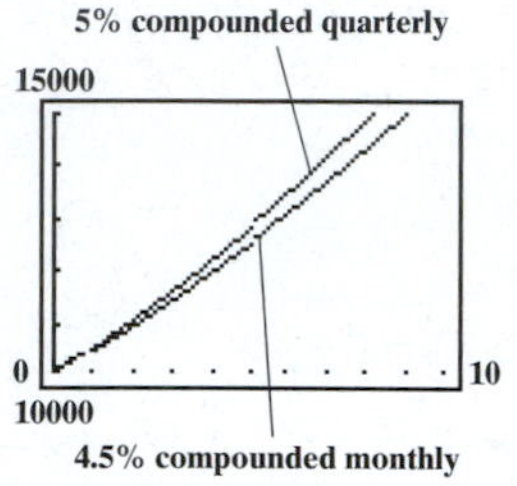

the bank that pays 5% interest compounded quarterly

71. does not make sense **73.** does not make sense **75.** false **77.** false **79. a.** $y = \left(\frac{1}{3}\right)^x$ **b.** $y = \left(\frac{1}{5}\right)^x$ **c.** $y = 5^x$ **d.** $y = 3^x$
81. $b = \frac{Da}{a - D}$ **82.** -1 **83.** $\{-2, 5\}$ **84. a.** 11 **b.** 127 **85.** x **86.** $y = \frac{x + 5}{7}$

Section 9.2

Check Point Exercises

1. a. $5x^2 + 1$ **b.** $25x^2 + 60x + 35$ **2.** $f(g(x)) = 7\left(\frac{x}{7}\right) = x$; $g(f(x)) = \frac{7x}{7} = x$
3. $f(g(x)) = 4\left(\frac{x + 7}{4}\right) - 7 = (x + 7) - 7 = x$; $g(f(x)) = \frac{(4x - 7) + 7}{4} = \frac{4x}{4} = x$ **4.** $f^{-1}(x) = \frac{x - 7}{2}$ **5.** $f^{-1}(x) = \sqrt[3]{\frac{x + 1}{4}}$
6. (b) and (c) **7.**

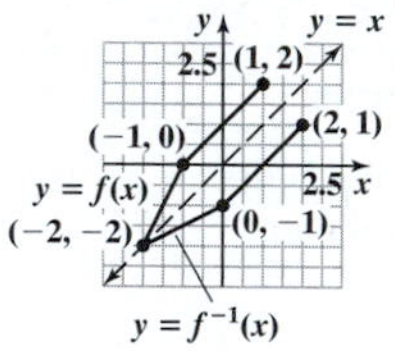

Exercise Set 9.2

1. a. $2x + 14$ **b.** $2x + 7$ **c.** 18 **3. a.** $2x + 5$ **b.** $2x + 9$ **c.** 9 **5. a.** $20x^2 - 11$ **b.** $80x^2 - 120x + 43$ **c.** 69
7. a. $x^4 - 4x^2 + 6$ **b.** $x^4 + 4x^2 + 2$ **c.** 6 **9. a.** $\sqrt{x} - 1$ **b.** $\sqrt{x - 1}$ **c.** 1 **11. a.** x **b.** x **c.** 2 **13. a.** x **b.** x **c.** 2 **15.** $f(g(x)) = x$; $g(f(x)) = x$; inverses **17.** $f(g(x)) = x$; $g(f(x)) = x$; inverses **19.** $f(g(x)) = \frac{5x - 56}{9}$; $g(f(x)) = \frac{5x - 4}{9}$; not inverses
21. $f(g(x)) = x$; $g(f(x)) = x$; inverses **23.** $f(g(x)) = x$; $g(f(x)) = x$; inverses **25. a.** $f^{-1}(x) = x - 3$ **b.** $f(f^{-1}(x)) = (x - 3) + 3 = x$ and $f^{-1}(f(x)) = (x + 3) - 3 = x$ **27. a.** $f^{-1}(x) = \frac{x}{2}$ **b.** $f(f^{-1}(x)) = 2\left(\frac{x}{2}\right) = x$ and $f^{-1}(f(x)) = \frac{2x}{2} = x$ **29. a.** $f^{-1}(x) = \frac{x - 3}{2}$
b. $f(f^{-1}(x)) = 2\left(\frac{x - 3}{2}\right) + 3 = x$ and $f^{-1}(f(x)) = \frac{(2x + 3) - 3}{2} = x$ **31. a.** $f^{-1}(x) = \sqrt[3]{x - 2}$ **b.** $f(f^{-1}(x)) = (\sqrt[3]{x - 2})^3 + 2 = x$ and $f^{-1}(f(x)) = \sqrt[3]{(x^3 + 2) - 2} = x$ **33. a.** $f^{-1}(x) = \sqrt[3]{x} - 2$ **b.** $f(f^{-1}(x)) = ((\sqrt[3]{x} - 2) + 2)^3 = x$ and $f^{-1}(f(x)) = \sqrt[3]{(x + 2)^3} - 2 = x$
35. a. $f^{-1}(x) = \frac{1}{x}$ **b.** $f(f^{-1}(x)) = \frac{1}{\frac{1}{x}} = x$ and $f^{-1}(f(x)) = \frac{1}{\frac{1}{x}} = x$ **37. a.** $f^{-1}(x) = x^2, x \geq 0$ **b.** $f(f^{-1}(x)) = \sqrt{x^2} = x$ and $f^{-1}(f(x)) = (\sqrt{x})^2 = x$ **39. a.** $f^{-1}(x) = \sqrt{x - 1}$ **b.** $f(f^{-1}(x)) = (\sqrt{x - 1})^2 + 1 = x$ and $f^{-1}(f(x)) = \sqrt{(x^2 + 1) - 1} = x$
41. a. $f^{-1}(x) = \frac{3x + 1}{x - 2}$ **b.** $f(f^{-1}(x)) = \frac{2\left(\frac{3x + 1}{x - 2}\right) + 1}{\left(\frac{3x + 1}{x - 2}\right) - 3} = x$ and $f^{-1}(f(x)) = \frac{3\left(\frac{2x + 1}{x - 3}\right) + 1}{\left(\frac{2x + 1}{x - 3}\right) - 2} = x$ **43. a.** $f^{-1}(x) = (x - 3)^3 + 4$

b. $f(f^{-1}(x)) = \sqrt[3]{((x-3)^3+4)-4}+3 = x$ and $f^{-1}(f(x)) = ((\sqrt[3]{x-4}+3)-3)^3+4 = x$ **45.** no inverse **47.** no inverse

49. inverse function **51.**

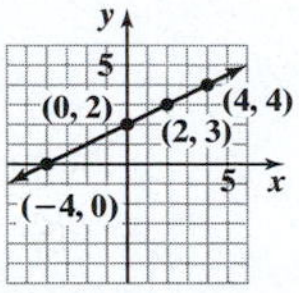

53.

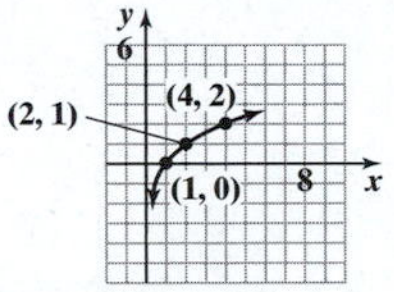

55. 5 **57.** 1 **59.** 2 **61.** 1 **63.** −6 **65.** −7 **67.** 3 **69.** 11

71. a. f represents the price after a \$400 discount, and g represents the price after a 25% discount (75% of the regular price). **b.** $0.75x - 400$; $f \circ g$ represents an additional \$400 discount on a price that has already been reduced by 25%. **c.** $0.75(x - 400) = 0.75x - 300$; $g \circ f$ represents an additional 25% discount on a price that has already been reduced \$400. **d.** $f \circ g$; $0.75x - 400 < 0.75x - 300$, so $f \circ g$ represents the lower price after the two discounts. **e.** $f^{-1}(x) = x + 400$; f^{-1} represents the regular price, since the value of x here is the price after a \$400 discount.
73. a. f: {(Zambia, 4.2), (Colombia, 4.5), (Poland, 3.3), (Italy, 3.3), (U.S., 2.5)} **b.** {(4.2, Zambia), (4.5, Colombia), (3.3, Poland), (3.3, Italy), (2.5, U.S.)}; no; The input 3.3 is associated with two outputs, Poland and Italy. **75. a.** No horizontal line intersects the graph of f in more than one point. **b.** $f^{-1}(0.25)$, or approximately 15, represents the number of people who would have to be in the room so that the probability of two sharing a birthday would be 0.25; $f^{-1}(0.5)$, or approximately 23, represents the number of people so that the probability would be 0.5; $f^{-1}(0.7)$, or approximately 30, represents the number of people so that the probability would be 0.7.
77. $f(g(x)) = \frac{9}{5}\left[\frac{5}{9}(x-32)\right] + 32 = x$ and $g(f(x)) = \frac{5}{9}\left[\left(\frac{9}{5}x+32\right) - 32\right] = x$

85.

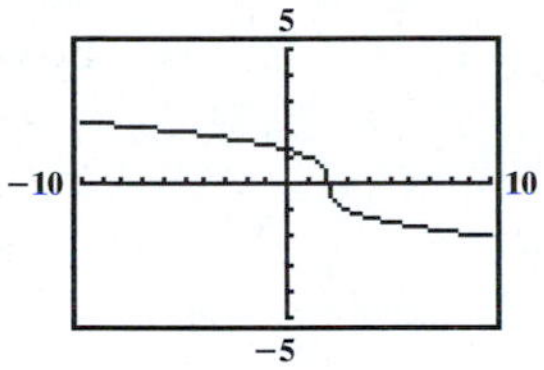

; inverse function

87.

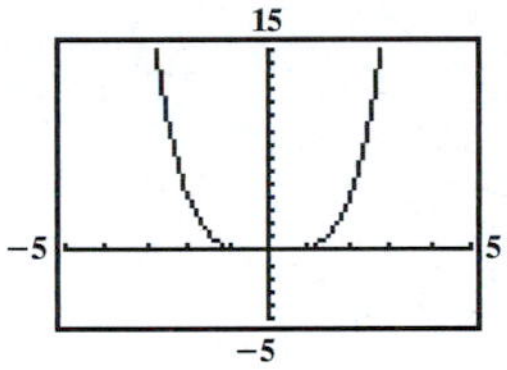

; no inverse function

89.

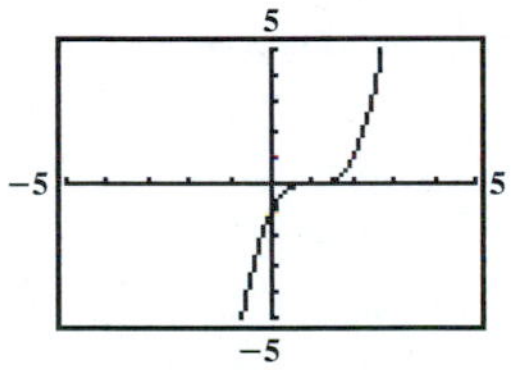

; inverse function

91.

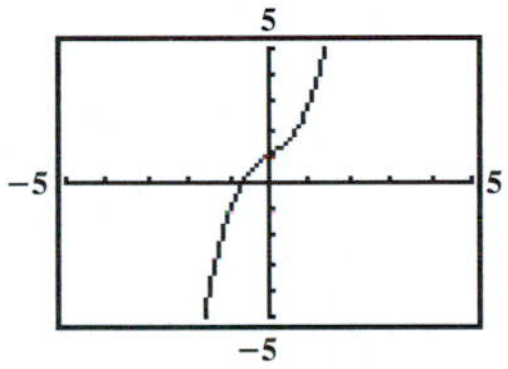

; inverse function

93.

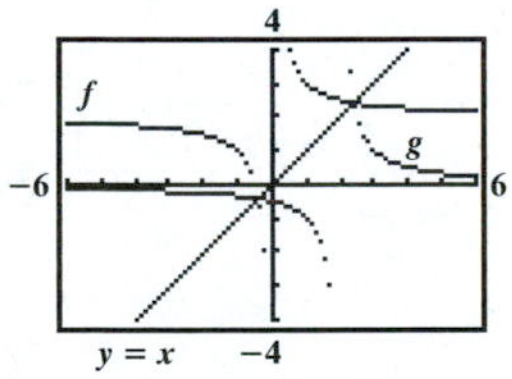

; inverses

95. does not make sense **97.** makes sense **99.** false **101.** true
103. Answers will vary; Examples are $f(x) = \sqrt{x+5}$ and $g(x) = 3x^2$.

105. $f(f(x)) = \dfrac{3\left(\frac{3x-2}{5x-3}\right)-2}{5\left(\frac{3x-2}{5x-3}\right)-3} = \dfrac{3(3x-2)-2(5x-3)}{5(3x-2)-3(5x-3)} = \dfrac{9x-6-10x+6}{15x-10-15x+9} = \dfrac{-x}{-1} = x$

107. 5×10^8 **108.**

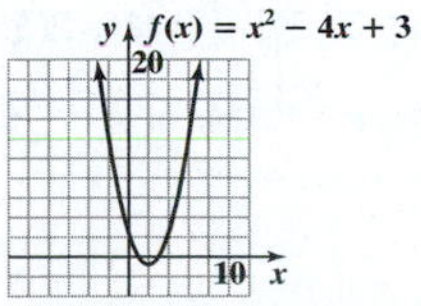

109. {5}

110. There is no method for solving $x = 2^y$ for y. **111.** $\frac{1}{2}$ **112.** $\{x \mid x \neq 3\}$ or $(-\infty, 3) \cup (3, \infty)$

Section 9.3

Check Point Exercises

1. a. $7^3 = x$ **b.** $b^2 = 25$ **c.** $4^y = 26$ **2. a.** $5 = \log_2 x$ **b.** $3 = \log_b 27$ **c.** $y = \log_e 33$ **3. a.** 2 **b.** 1 **c.** $\frac{1}{2}$ **4. a.** 1 **b.** 0

5. a. 8 **b.** 17 **6.**

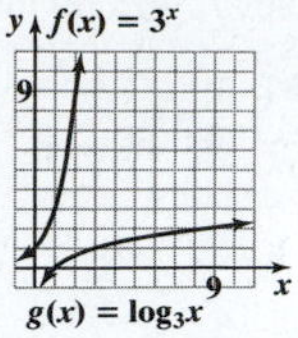

7. $\{x|x > 5\}$ or $(5, \infty)$ **8.** approximately 80% **9.** 4
10. a. $\{x|x < 4\}$ or $(-\infty, 4)$ **b.** $\{x|x \neq 0\}$ or $(-\infty, 0) \cup (0, \infty)$
11. 34°; extremely well

Exercise Set 9.3

1. $2^4 = 16$ **3.** $3^2 = x$ **5.** $b^5 = 32$ **7.** $6^y = 216$ **9.** $\log_2 8 = 3$ **11.** $\log_2 \frac{1}{16} = -4$ **13.** $\log_8 2 = \frac{1}{3}$ **15.** $\log_{13} x = 2$

17. $\log_b 1000 = 3$ **19.** $\log_7 200 = y$ **21.** 2 **23.** 6 **25.** -1 **27.** -3 **29.** $\frac{1}{2}$ **31.** $-\frac{1}{2}$ **33.** $\frac{1}{2}$ **35.** 1 **37.** 0 **39.** 7

41. 19 **43.**

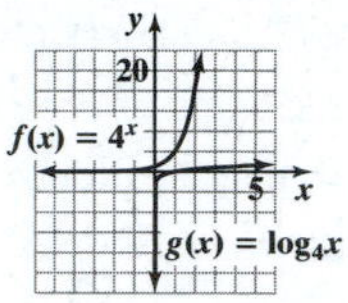

45.

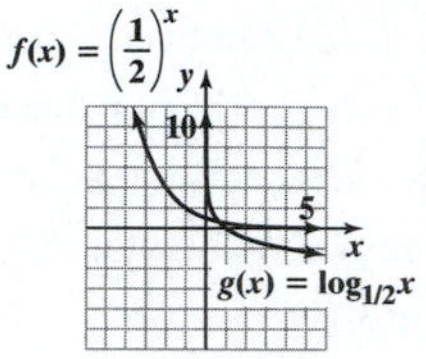

47. $\{x|x > -4\}$ or $(-4, \infty)$ **49.** $\{x|x < 2\}$ or $(-\infty, 2)$
51. $\{x|x \neq 2\}$ or $(-\infty, 2) \cup (2, \infty)$ **53.** 2 **55.** 7 **57.** 33 **59.** 0
61. 6 **63.** -6 **65.** 125 **67.** $9x$ **69.** $5x^2$ **71.** $\sqrt{x}$
73. $3^2 = x - 1$; $\{10\}$ **75.** $4^{-3} = x$; $\left\{\frac{1}{64}\right\}$ **77.** 0 **79.** 2 **81.** d
83. c **85.** b **87.** approximately 95.4%
89. a. 26.4%; underestimates by 1% **b.** 25.2%

91. approximately 188 decibels; yes

93. a. 88
b. 71.5; 63.9; 58.8; 55.0; 52.0; 49.5
c.

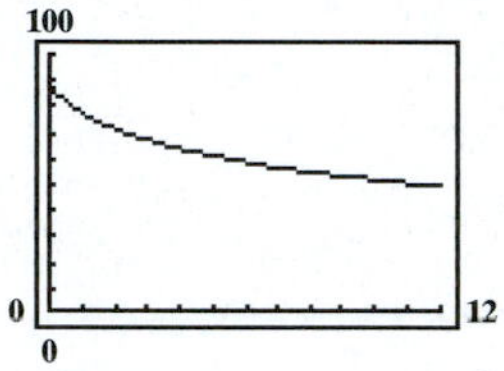

The students remembered less of the material over time.

103.

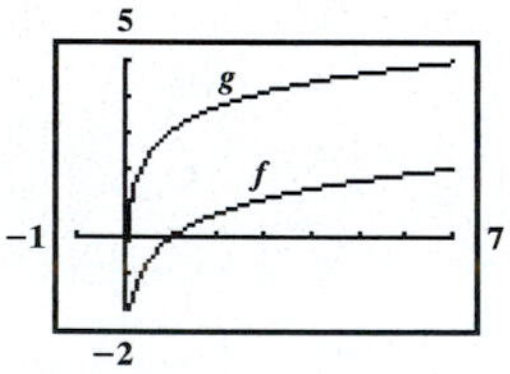

The graph of g is the graph of f shifted up 3 units.

105.

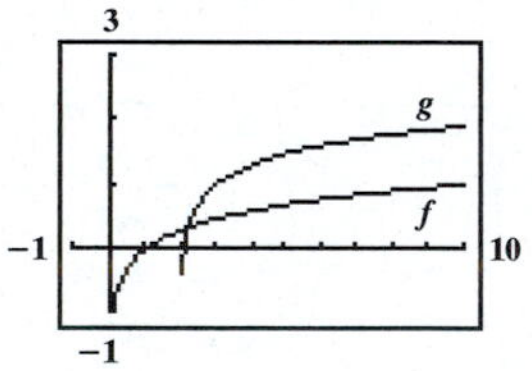

The graph of g is the graph of f shifted 2 units to the right and 1 unit up.

107. a.

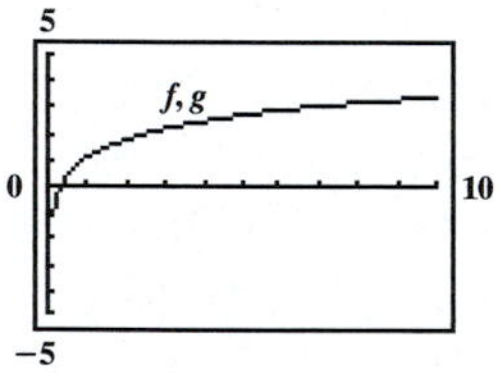

b.

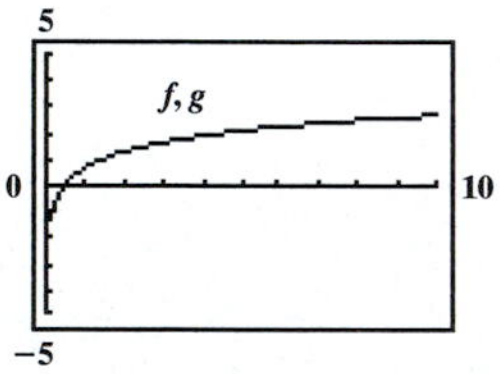

c.

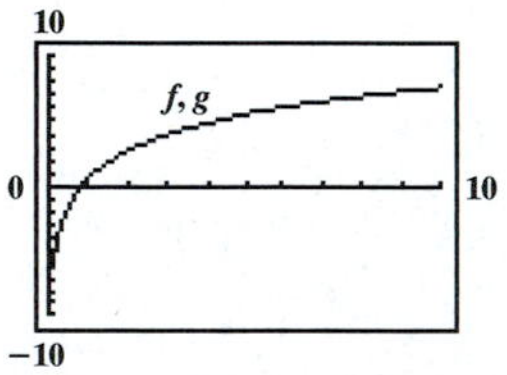

d. In each case, the graphs of f and g are the same.; $\log_b(MN) = \log_b M + \log_b N$
e. the sum of the logarithms of the factors

109. makes sense **111.** makes sense **113.** false **115.** false **117.** $\frac{4}{5}$ **119.** $\log_3 40$ **120.** $\{(-2, 3)\}$ **121.** $2(3x - y)(x - y)$

122. $\{x|x \leq -7 \text{ or } x \geq -2\}$ or $(-\infty, -7] \cup [-2, \infty)$ **123. a.** 5 **b.** 5 **c.** $\log_2(8 \cdot 4) = \log_2 8 + \log_2 4$ **124. a.** 4 **b.** 4
c. $\log_2\left(\frac{32}{2}\right) = \log_2 32 - \log_2 2$ **125. a.** 4 **b.** 4 **c.** $\log_3 9^2 = 2\log_3 9$

Section 9.4

Check Point Exercises

1. a. $\log_6 7 + \log_6 11$ **b.** $2 + \log x$ **2. a.** $\log_8 23 - \log_8 x$ **b.** $5 - \ln 11$ **3. a.** $9 \log_6 8$ **b.** $\frac{1}{3} \ln x$ **c.** $2 \log(x + 4)$

4. a. $4 \log_b x + \frac{1}{3} \log_b y$ **b.** $\frac{1}{2} \log_5 x - 2 - 3 \log_5 y$ **5. a.** $\log 100 = 2$ **b.** $\log\left(\frac{7x + 6}{x}\right)$ **6. a.** $\ln(x^2 \sqrt[3]{x + 5})$ **b.** $\log\left[\frac{(x - 3)^2}{x}\right]$
c. $\log_b\left(\frac{\sqrt[4]{x}}{25y^{10}}\right)$ **7.** $\frac{\log 2506}{\log 7} \approx 4.02$ **8.** $\frac{\ln 2506}{\ln 7} \approx 4.02$

Exercise Set 9.4

1. $\log_5 7 + \log_5 3$ **3.** $1 + \log_7 x$ **5.** $3 + \log x$ **7.** $1 - \log_7 x$ **9.** $\log x - 2$ **11.** $3 - \log_4 y$ **13.** $2 - \ln 5$ **15.** $3 \log_b x$ **17.** $-6 \log N$ **19.** $\frac{1}{5} \ln x$ **21.** $2 \log_b x + \log_b y$ **23.** $\frac{1}{2} \log_4 x - 3$ **25.** $2 - \frac{1}{2} \log_6(x + 1)$ **27.** $2 \log_b x - \log_b y - 2 \log_b z$ **29.** $1 + \frac{1}{2} \log x$ **31.** $\frac{1}{3} \log x - \frac{1}{3} \log y$ **33.** $\frac{1}{2} \log_b x + 3 \log_b y - 3 \log_b z$ **35.** $\frac{2}{3} \log_5 x + \frac{1}{3} \log_5 y - \frac{2}{3}$ **37.** $\log 10 = 1$ **39.** $\ln (7x)$ **41.** $\log_2 32 = 5$ **43.** $\log\left(\frac{2x + 5}{x}\right)$ **45.** $\log (xy^3)$ **47.** $\ln (y\sqrt{x})$ **49.** $\log_b (x^2y^3)$ **51.** $\ln\left(\frac{x^5}{y^2}\right)$ **53.** $\ln\left(\frac{x^3}{\sqrt[3]{y}}\right)$ **55.** $\ln\left[\frac{(x + 6)^4}{x^3}\right]$ **57.** $\ln\left(\frac{x^3y^5}{z^6}\right)$ **59.** $\log_5\left[\frac{\sqrt{xy}}{(x + 1)^2}\right]$ **61.** 1.5937 **63.** 1.6944 **65.** −1.2304 **67.** 3.6193 **69.** $C - A$ **71.** $3A$ **73.** $\frac{1}{2}A - \frac{3}{2}C$ **75.** false; $\ln e = 1$ **77.** false; $\log_4(2x)^3 = 3 \log_4(2x)$ **79.** true **81.** true **83.** false; $\log(x + 3) - \log(2x) = \log\left(\frac{x + 3}{2x}\right)$ **85.** true **87.** true **89. a.** 2 **b.** $\log_3\left(\frac{xy^4}{9}\right)$ **91. a.** $\frac{1}{2}$ **b.** $\log_{25}\left[\frac{x(x^2 - 1)}{5(x + 1)}\right] = \log_{25}\left[\frac{x(x - 1)}{5}\right]$ **93. a.** $D = 10 \log\left(\frac{I}{I_0}\right)$ **b.** 20 decibels

103. a. & b.

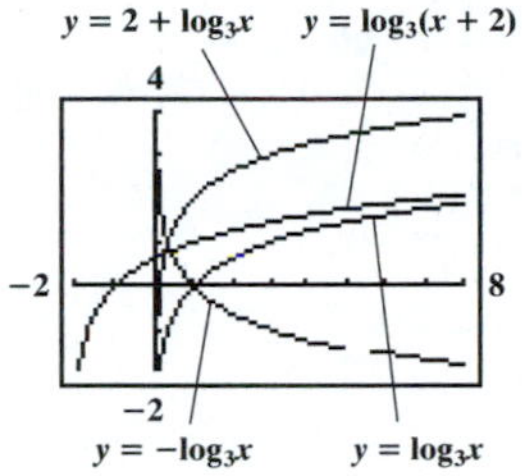

$y = \log_3 x$ is shifted up 2 units to obtain $y = 2 + \log_3 x$,
$y = \log_3 x$ is shifted to the left 2 units to obtain $y = \log_3 (x + 2)$, and
$y = \log_3 x$ is reflected across the x-axis to obtain $y = -\log_3 x$.

105.

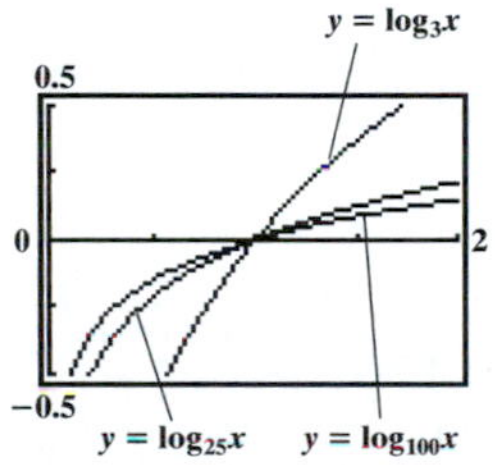

a. $y = \log_{100} x$ is on the top and $y = \log_3 x$ is on the bottom.
b. $y = \log_3 x$ is on the top and $y = \log_{100} x$ is on the bottom.
c. If $y = \log_b x$ is graphed for two different values of b, the graph of the one with the larger base will be on top in the interval $(0, 1)$ and the one with the smaller base will be on top in the interval $(1, \infty)$.

111. makes sense **113.** makes sense **115.** true **117.** false **119.** $\log e = \frac{\ln e}{\ln 10} = \frac{1}{\ln 10}$ **121.** $4x^3$

122.

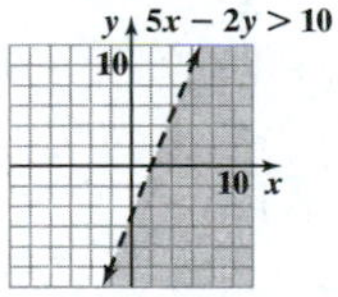

123. $\{x|x < 1\}$ or $(-\infty, 1)$ **124.** $2y\sqrt[3]{xy^2}$ **125.** 64 **126.** 12 **127.** $\{-1, 3\}$

Mid-Chapter Check Point Exercises

1. $(f \circ g)(x) = 12x - 13$; $(g \circ f)(x) = 12x + 3$; no **2.** $(f \circ g)(x) = x$; $(g \circ f)(x) = x$; yes **3.** $(f \circ g)(x) = x$; $(g \circ f)(x) = x$; yes **4. a.** $\frac{2}{3}$ **b.** $\sqrt{5}$ **c.** $-\frac{1}{4}$ **d.** 2 **5.** $f^{-1}(x) = \frac{4x - 5}{2}$ **6.** $f^{-1}(x) = \sqrt[3]{\frac{x + 7}{10}}$ **7.** $f^{-1} = \{(5, 2), (-7, 10), (-10, 11)\}$ **8.** function; no inverse function **9.** function; inverse function **10.** not a function

11.

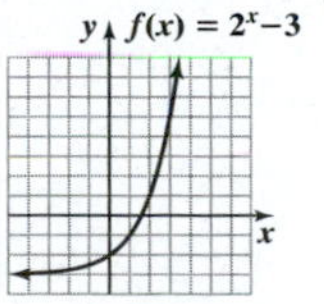

domain: $\{x|x$ is a real number$\}$ or $(-\infty, \infty)$; range: $\{y|y > -3\}$ or $(-3, \infty)$

12.

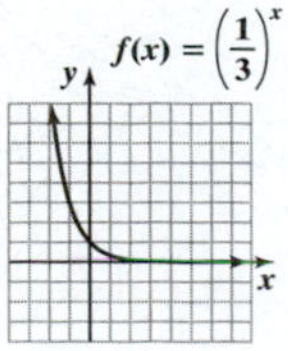

domain: $\{x|x$ is a real number$\}$ or $(-\infty, \infty)$; range: $\{y|y > 0\}$ or $(0, \infty)$

13. 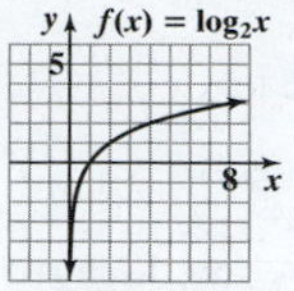

domain: $\{x|x > 0\}$ or $(0, \infty)$; range: $\{y|y$ is a real number$\}$ or $(-\infty, \infty)$

14. 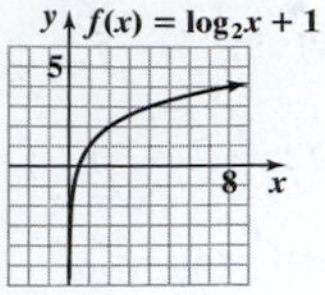

domain: $\{x|x > 0\}$ or $(0, \infty)$; range: $\{y|y$ is a real number$\}$ or $(-\infty, \infty)$

15. $\{x|x > -6\}$ or $(-6, \infty)$ **16.** $\{x|x > 0\}$ or $(0, \infty)$ **17.** $\{x|x \neq -6\}$ or $(-\infty, -6) \cup (-6, \infty)$

18. $\{x|x$ is a real number$\}$ or $(-\infty, \infty)$ **19.** 5 **20.** -2 **21.** $\frac{1}{2}$ **22.** $\frac{1}{3}$ **23.** 2

24. Evaluation is not possible; $\log_2 \frac{1}{8} = -3$ and $\log_3(-3)$ is undefined. **25.** 5 **26.** $\sqrt{7}$ **27.** 13 **28.** $-\frac{1}{2}$ **29.** $\sqrt{\pi}$

30. $\frac{1}{2}\log x + \frac{1}{2}\log y - 3$ **31.** $19 + 20 \ln x$ **32.** $\log_7\left(\frac{x^8}{\sqrt[3]{y}}\right)$ **33.** $\log_5 x^9$ **34.** $\ln\left[\frac{\sqrt{x}}{y^3(z-2)}\right]$ **35.** \$8

Section 9.5

Check Point Exercises

1. a. 3 or $\{3\}$ **b.** $\frac{5}{2}$ or $\left\{\frac{5}{2}\right\}$ **2. a.** $\frac{\ln 134}{\ln 5} \approx 3.04$ or $\left\{\frac{\ln 134}{\ln 5} \approx 3.04\right\}$ **b.** $\log 8000 \approx 3.90$ or $\{\log 8000 \approx 3.90\}$

3. $\frac{\ln 9}{2} = \ln 3 \approx 1.10$ or $\{\ln 3 \approx 1.10\}$ **4. a.** 12 or $\{12\}$ **b.** $\frac{e^2}{3}$ or $\left\{\frac{e^2}{3}\right\}$ **5.** 5 or $\{5\}$ **6.** 4 and 5, or $\{4, 5\}$

7. blood alcohol concentration of 0.01 **8.** 16.2 years **9.** by 2010

Exercise Set 9.5

1. $\{6\}$ **3.** $\{3\}$ **5.** $\{3\}$ **7.** $\{2\}$ **9.** $\left\{\frac{3}{5}\right\}$ **11.** $\left\{\frac{3}{2}\right\}$ **13.** $\{4\}$ **15.** $\{5\}$ **17.** $\left\{-\frac{1}{4}\right\}$ **19.** $\{\ln 5.7 \approx 1.74\}$

21. $\{\log 3.91 \approx 0.59\}$ **23.** $\left\{\frac{\ln 17}{\ln 5} \approx 1.76\right\}$ **25.** $\{\ln 5 \approx 1.61\}$ **27.** $\left\{\frac{\ln 659}{5} \approx 1.30\right\}$ **29.** $\left\{\frac{\ln 13}{0.7} \approx 3.66\right\}$ **31.** $\left\{\frac{\ln 3}{0.055} \approx 19.97\right\}$

33. $\left\{\frac{\ln 30}{\ln 1.4} \approx 10.11\right\}$ **35.** $\left\{\frac{1 - \ln 793}{5} \approx -1.14\right\}$ **37.** $\left\{\frac{\ln 410}{\ln 7} - 2 \approx 1.09\right\}$ **39.** $\left\{\frac{\ln 2}{\ln 5 - \ln 2} \approx 0.76\right\}$ **41.** $\{81\}$ **43.** $\left\{\frac{1}{16}\right\}$

45. $\{3\}$ **47.** $\{100\}$ **49.** $\{59\}$ **51.** $\left\{\frac{109}{27}\right\}$ **53.** $\left\{\frac{62}{3}\right\}$ **55.** $\{e^2 \approx 7.39\}$ **57.** $\{e^{-3} \approx 0.05\}$ **59.** $\left\{\frac{e^4}{2} \approx 27.30\right\}$ **61.** $\{e^{-1/2} \approx 0.61\}$

63. $\{e^2 - 3 \approx 4.39\}$ **65.** $\left\{\frac{5}{4}\right\}$ **67.** $\{6\}$ **69.** $\{6\}$ **71.** no solution or $\varnothing$ **73.** $\left\{\frac{1}{e-1} \approx 0.58\right\}$ **75.** $\{3\}$ **77.** $\left\{\frac{4}{3}\right\}$

79. no solution or $\varnothing$ **81.** $\{5\}$ **83.** $\left\{\frac{2}{9}\right\}$ **85.** $\{28\}$ **87.** $\{2\}$ **89.** no solution or $\varnothing$ **91.** $\left\{\frac{1}{2}\right\}$ **93.** $\left\{\pm\sqrt{\frac{\ln 45}{\ln 3}} \approx \pm 1.86\right\}$

95. $\{12\}$ **97.** $\{-2, 6\}$ **99. a.** 36.1 million **b.** 2013 **101.** 118 ft; by the point (118, 1) **103.** 8.2 **105.** 16.8%

107. 8.7 **109.** 15.7% **111. a.** 69%; overestimates by 1% **b.** 2010 **113.** about 2.8 days; (2.8, 50)

115. a. $10^{-5.6}$ mole per liter **b.** $10^{-2.4}$ mole per liter **c.** $10^{3.2}$ times greater

123. $\{2\}$ 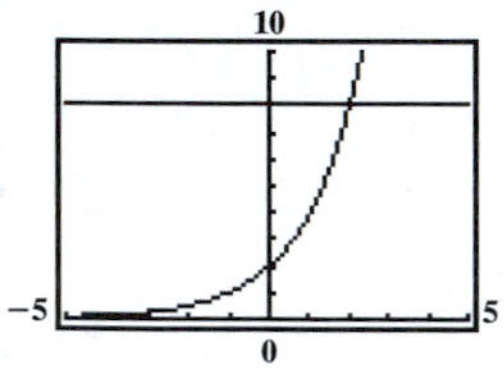

125. $\{4\}$ 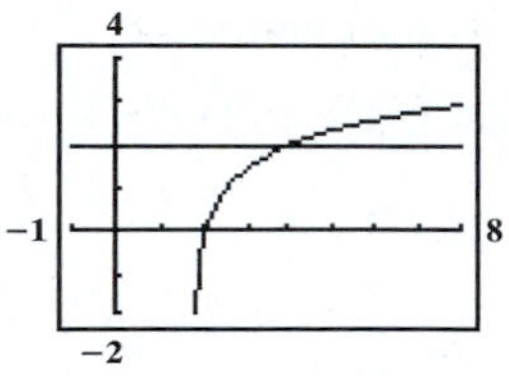

127. $\{2\}$ 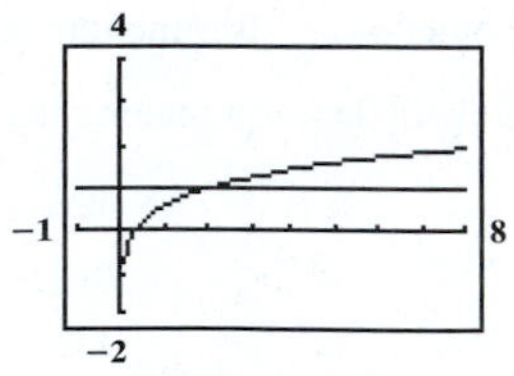

129. $\{-1.39, 1.69\}$ 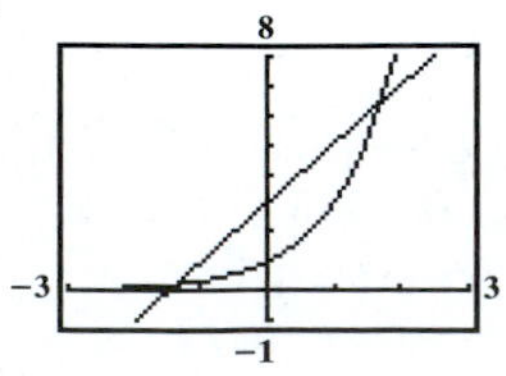

131. 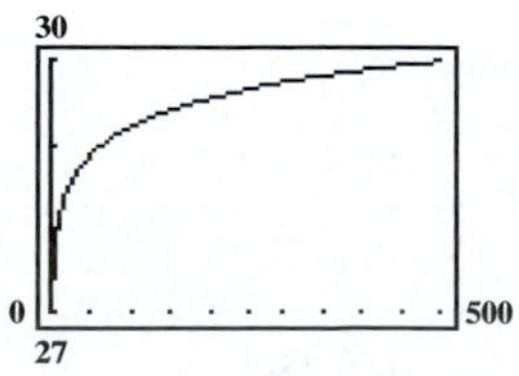

The barometric air pressure increases as the distance from the eye increases.

133. 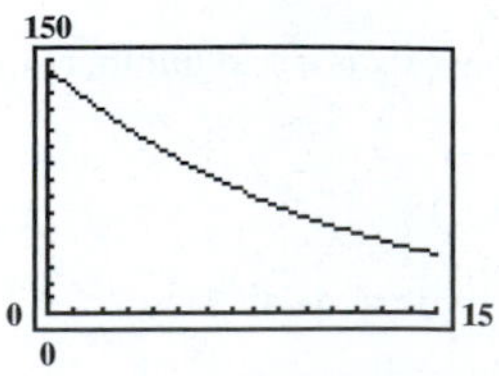

The runner's pulse will be 70 beats per minute after about 7.9 minutes.

135. does not make sense **137.** makes sense **139.** false **141.** true **143.** about 36 yr **145.** $\{10^{-2}, 10^{3/2}\}$

147. $\{5\}$ **148.** $\{-12\}$ **149.** $\frac{y^8}{16x^{12}}$ **150. a.** 10 million; 9.97 million; 9.94 million; 9.91 million **b.** decreasing

151.

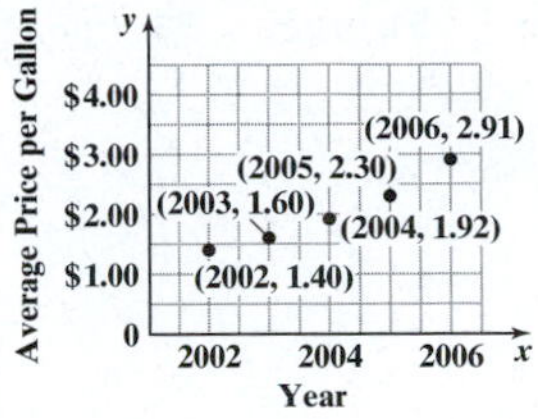

; exponential function **152. a.** 3 **b.** $e^{(\ln 3)x}$

Section 9.6

Check Point Exercises

1. a. $A = 643e^{0.021t}$ **b.** 2044 **2. a.** $A = A_0e^{-0.0248t}$ **b.** about 72.2 years

3.

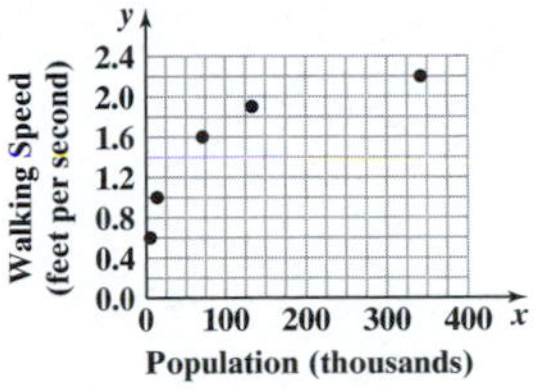

logarithmic function

4.

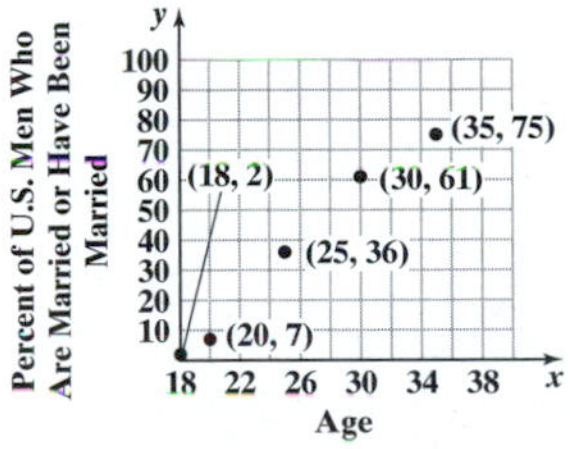

exponential function, although answers may vary

5. a. the exponential function g **b.** f: fairly well; g: not very well **6.** $y = 4e^{(\ln 7.8)x}$; $y = 4e^{2.054x}$

Exercise Set 9.6

1. 127.5 million **3.** Iraq; 2.7% **5.** 2015 **7. a.** $A = 6.04e^{0.01t}$ **b.** 2040 **9.** approximately 8 grams

11. 8 grams after 10 seconds; 4 grams after 20 seconds; 2 grams after 30 seconds; 1 gram after 40 seconds; 0.5 gram after 50 seconds

13. approximately 15,679 years old **15. a.** $\frac{1}{2} = e^{1.31k}$ yields $k = \frac{\ln\left(\frac{1}{2}\right)}{1.31} \approx -0.52912$. **b.** about 0.1069 billion or 106,900,000 years old

17. $2A_0 = A_0e^{kt}$; $2 = e^{kt}$; $\ln 2 = \ln e^{kt}$; $\ln 2 = kt$; $\frac{\ln 2}{k} = t$ **19. a.** 1% **b.** about 69 years

21.

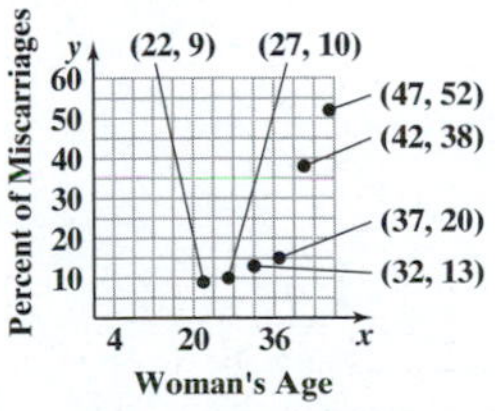

exponential function

23.

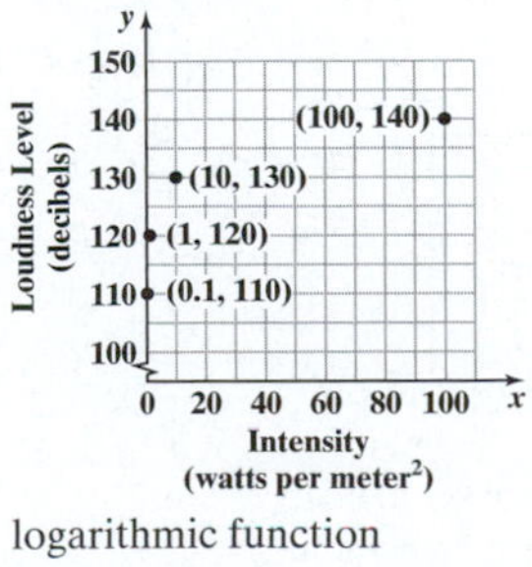

logarithmic function

25.

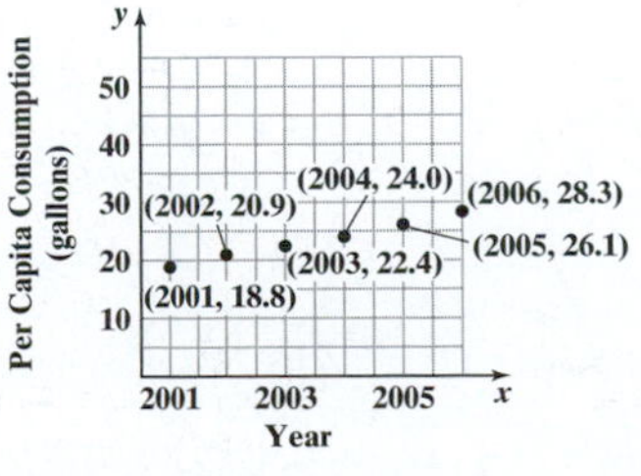

linear function

27. $y = 100e^{(\ln 4.6)x}$; $y = 100e^{1.526x}$ **29.** $y = 2.5e^{(\ln 0.7)x}$; $y = 2.5e^{-0.357x}$

39. a. $y = 200.9(1.011)^x$; $r \approx 0.999$; Since r is close to 1, the model fits the data well. **b.** $y = 200.9e^{\ln(1.011)x}$; $y = 200.9e^{0.0109x}$; by approximately 1% **41.** $y = 2.654x + 198.015$; $r \approx 0.997$; Since r is close to 1, the model fits the data well. **43.** $y = 200.9(1.011)^x$; $y = 2.654x + 198.015$; using exponential, by 2010; using linear, by 2013; Answers will vary. **45.** Models will vary. Examples are given. Predictions will vary. For Exercise 21: $y = 1.402(1.078)^x$; For Exercise 23: $y = 120 + 4.343 \ln x$; For Exercise 25: $y = 1.849x + 16.947$ (x = number of years after 2000) **47.** does not make sense **49.** makes sense **51.** true **53.** true **55.** $\frac{x+5}{x}$ **56.** $\{-27, 1\}$ **57.** $17\sqrt{2}$ **58.** $3\sqrt{5}$ **59.** $\left(4, \frac{1}{2}\right)$ **60.**

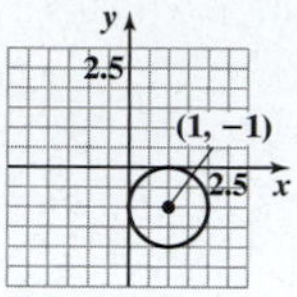

Review Exercises

1.

x	$f(x)$
−2	$\frac{1}{16}$
−1	$\frac{1}{4}$
0	1
1	4
2	16

; d

2.

x	$f(x)$
−2	16
−1	4
0	1
1	$\frac{1}{4}$
2	$\frac{1}{16}$

; a

3.

x	$f(x)$
−2	−16
−1	−4
0	−1
1	$-\frac{1}{4}$
2	$-\frac{1}{16}$

; b

4.

x	$f(x)$
−2	−13
−1	−1
0	2
1	$\frac{11}{4}$
2	$\frac{47}{16}$

; c

5.

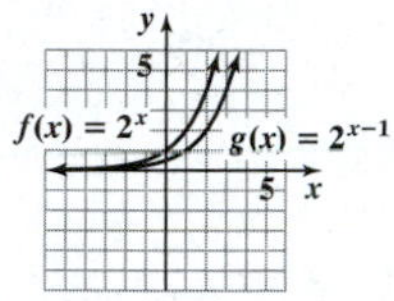

The graph of g is the graph of f shifted to the right 1 unit.

6.

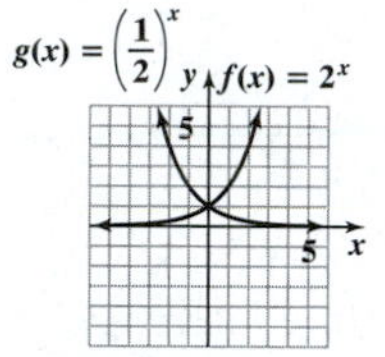

The graph of g is the reflection of f across the y-axis.

7.

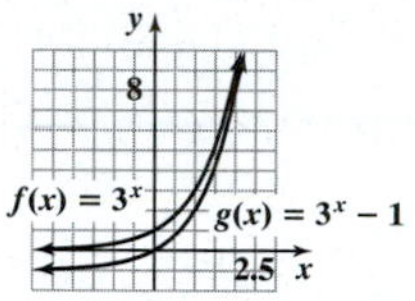

The graph of g is the graph of f shifted down 1 unit.

8.

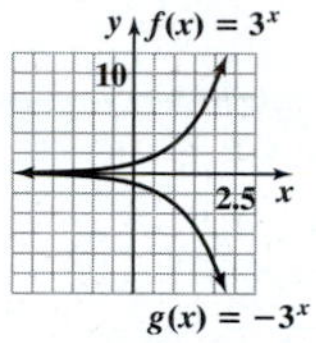

The graph of g is the reflection of f across the x-axis.

9. 5.5% compounded semiannually **10.** 7% compounded monthly **11. a.** 200°F **b.** about 119°F **c.** 70°F; The temperature of the room is 70°F. **12. a.** $16x^2 - 8x + 4$ **b.** $4x^2 + 11$ **c.** 124 **13. a.** $\sqrt{x+1}$ **b.** $\sqrt{x} + 1$ **c.** 2 **14.** $f(g(x)) = x - \frac{7}{10}$; $g(f(x)) = x - \frac{7}{6}$; not inverses **15.** $f(g(x)) = x$; $g(f(x)) = x$; inverses **16. a.** $f^{-1}(x) = \frac{x+3}{4}$ **b.** $f(f^{-1}(x)) = 4\left(\frac{x+3}{4}\right) - 3 = x$ and $f^{-1}(f(x)) = \frac{(4x-3)+3}{4} = x$ **17. a.** $f^{-1}(x) = x^2 - 2, x \geq 0$ **b.** $f(f^{-1}(x)) = \sqrt{(x^2-2)+2} = x$ and $f^{-1}(f(x)) = (\sqrt{x+2})^2 - 2 = x$ **18. a.** $f^{-1}(x) = \frac{\sqrt[3]{x-1}}{2}$ **b.** $f(f^{-1}(x)) = 8\left(\frac{\sqrt[3]{x-1}}{2}\right)^3 + 1 = x$ and $f^{-1}(f(x)) = \frac{\sqrt[3]{(8x^3+1)-1}}{2} = x$ **19.** inverse function **20.** no inverse function **21.** inverse function **22.** no inverse function **23.**

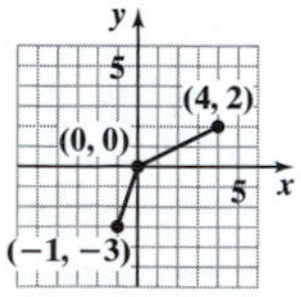

24. $49^{1/2} = 7$ **25.** $4^3 = x$ **26.** $3^y = 81$ **27.** $\log_6 216 = 3$ **28.** $\log_b 625 = 4$ **29.** $\log_{13} 874 = y$ **30.** 3 **31.** −2 **32.** −9 is not in the domain of $y = \log_3 x$. **33.** $\frac{1}{2}$ **34.** 1 **35.** 8 **36.** 5 **37.** $-\frac{1}{2}$ **38.** −2 **39.** −3 **40.** 0

41. domain of f: $\{x|x \text{ is a real number}\}$ or $(-\infty, \infty)$; range of f: $\{y|y > 0\}$ or $(0, \infty)$; domain of g: $\{x|x > 0\}$ or $(0, \infty)$; range of g: $\{y|y \text{ is a real number}\}$ or $(-\infty, \infty)$

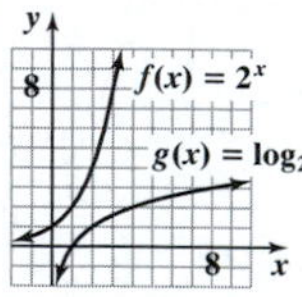

42. domain of f: $\{x|x \text{ is a real number}\}$ or $(-\infty, \infty)$; range of f: $\{y|y > 0\}$ or $(0, \infty)$; domain of g: $\{x|x > 0\}$ or $(0, \infty)$; range of g: $\{y|y \text{ is a real number}\}$ or $(-\infty, \infty)$

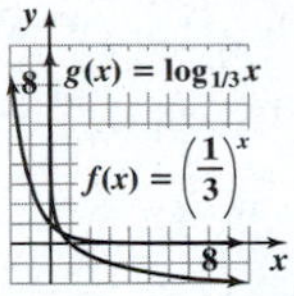

43. $\{x|x > -5\}$ or $(-5, \infty)$ **44.** $\{x|x < 3\}$ or $(-\infty, 3)$ **45.** $\{x|x \neq 1\}$ or $(-\infty, 1) \cup (1, \infty)$ **46.** $6x$ **47.** $\sqrt{x}$ **48.** $4x^2$ **49.** 3

50. a. 76 **b.** 67.4 after 2 months; 63.4 after 4 months; 60.8 after 6 months; 58.8 after 8 months; 55.9 after one year

c.

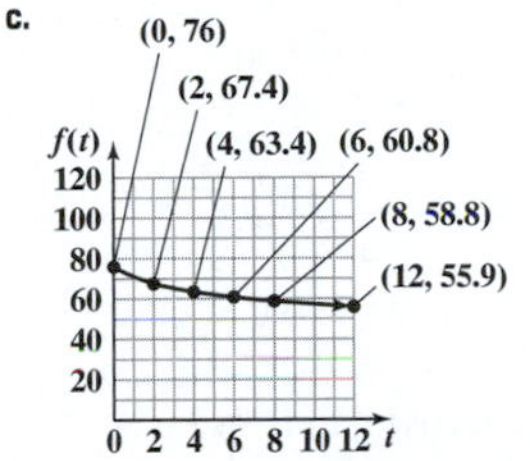

As time increases, the amount of material retained by the students decreases.

51. about 9 weeks **52.** $2 + 3\log_6 x$ **53.** $\frac{1}{2}\log_4 x - 3$ **54.** $\log_2 x + 2\log_2 y - 6$ **55.** $\frac{1}{3}\ln x - \frac{1}{3}$ **56.** $\log_b 21$ **57.** $\log\left(\frac{3}{x^3}\right)$

58. $\ln(x^3y^4)$ **59.** $\ln\left(\frac{\sqrt{x}}{y}\right)$ **60.** 6.2448 **61.** -0.1063 **62.** true **63.** false; $\log(x + 9) - \log(x + 1) = \log\left(\frac{x+9}{x+1}\right)$

64. false; $\log_2 x^4 = 4\log_2 x$ **65.** true **66.** $\{2\}$ **67.** $\left\{\frac{2}{3}\right\}$ **68.** $\left\{-\frac{3}{2}\right\}$ **69.** $\left\{\frac{\ln 12{,}143}{\ln 8} \approx 4.52\right\}$ **70.** $\left\{\frac{\ln 141}{5} \approx 0.99\right\}$

71. $\left\{\frac{\ln 3}{0.045} \approx 24.41\right\}$ **72.** $\left\{\frac{1}{125}\right\}$ **73.** $\{100\}$ **74.** $\{23\}$ **75.** $\left\{\frac{1}{e}\right\}$ **76.** $\left\{\frac{e^3}{2}\right\}$ **77.** $\{5\}$ **78.** no solution or $\varnothing$

79. $\left\{\frac{8}{3}\right\}$ **80.** $\{2\}$ **81.** $\{4\}$ **82.** 5.5 mi **83.** approximately 2086 **84.** approximately 8103 thousand or 8,103,000 **85.** 7.3 years

86. 14.6 years **87.** about 22% **88. a.** 0.042 **b.** about 51.9 million **c.** 2013 **89.** 325 days

90.

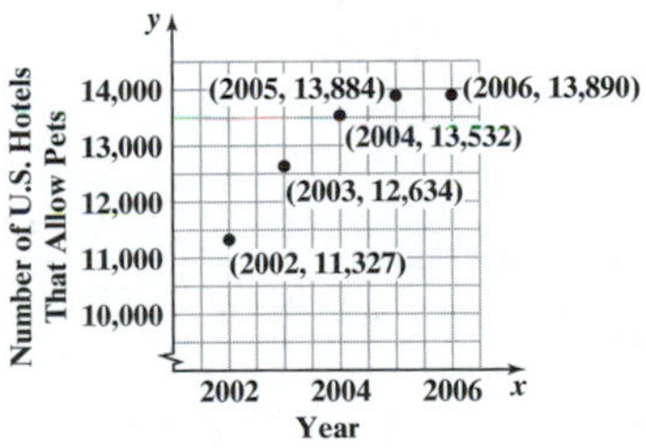

; logarithmic function

91.

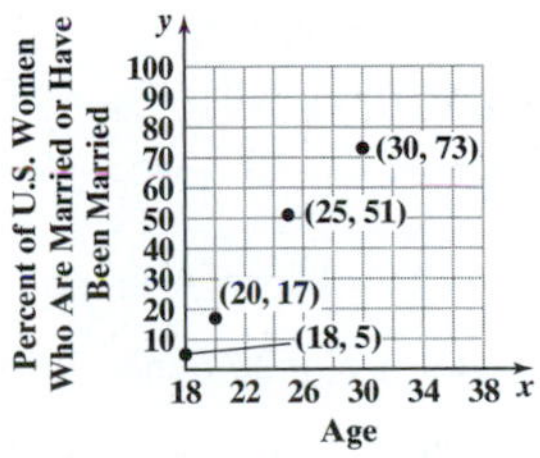

; exponential function, although answers may vary

92. $y = 73e^{(\ln 2.6)x}$; $y = 73e^{0.956x}$ **93.** $y = 6.5e^{(\ln 0.43)x}$; $y = 6.5e^{-0.844x}$ **94.** Answers will vary.

Chapter Test

1.

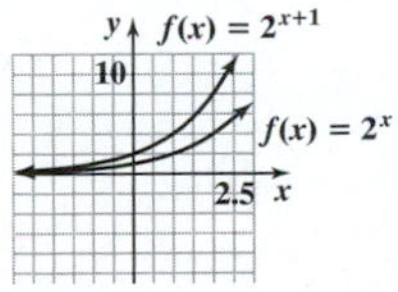

2. 6.5% compounded semiannually; \$221 **3.** $(f \circ g)(x) = 9x^2 - 3x$; $(g \circ f)(x) = 3x^2 + 3x - 1$ **4.** $f^{-1}(x) = \frac{x+7}{5}$

5. a. No horizontal line intersects the graph of f in more than one point. **b.** 2000 **c.** $f^{-1}(2000)$ represents the income, \$80 thousand, of a family that gives \$2000 to charity. **6.** $5^3 = 125$ **7.** $\log_{36} 6 = \frac{1}{2}$

8.

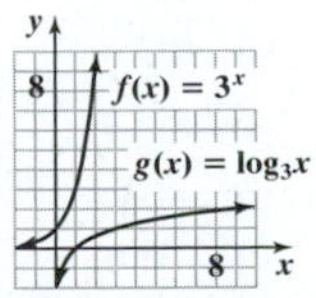

domain of f: $\{x|x \text{ is a real number}\}$ or $(-\infty, \infty)$; range of f: $\{y|y > 0\}$ or $(0, \infty)$; domain of g: $\{x|x > 0\}$ or $(0, \infty)$; range of g: $\{y|y \text{ is a real number}\}$ or $(-\infty, \infty)$

9. $5x$ **10.** 1 **11.** 0 **12.** $\{x|x > 7\}$ or $(7, \infty)$ **13.** 120 decibels **14.** $3 + 5\log_4 x$ **15.** $\frac{1}{3}\log_3 x - 4$ **16.** $\log(x^6y^2)$

17. $\ln\left(\frac{7}{x^3}\right)$ **18.** 1.5741 **19.** $\{6\}$ **20.** $\left\{\frac{\ln 1.4}{\ln 5}\right\}$ **21.** $\left\{\frac{\ln 4}{0.005}\right\}$ **22.** $\{5\}$ **23.** $\left\{\frac{217}{4}\right\}$ **24.** $\left\{\frac{e^4}{3}\right\}$ **25.** $\{5\}$

26. no solution or $\varnothing$ **27. a.** 82.4 million **b.** decreasing; The growth rate, -0.002, is negative. **c.** 2012 **28.** 13.9 years **29.** about 6.9%

30. $A = 509e^{0.036t}$ **31.** about 24,758 years ago **32.** linear **33.** logarithmic **34.** exponential **35.** quadratic

36. $y = 96e^{(\ln 0.38)x}$; $y = 96e^{-0.968x}$

Cumulative Review Exercises

1. $\{4\}$ **2.** $\{(2, 3)\}$ **3.** $\{(2, 0, 3)\}$ **4.** $\{x|x < -2 \text{ or } x > 4\}$ or $(-\infty, -2) \cup (4, \infty)$ **5.** $\{5\}$ **6.** $\{x|x \geq 4\}$ or $[4, \infty)$ **7.** $\left\{\frac{3}{4} \pm i\frac{\sqrt{7}}{4}\right\}$

8.

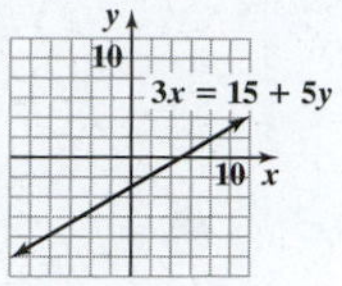

9.

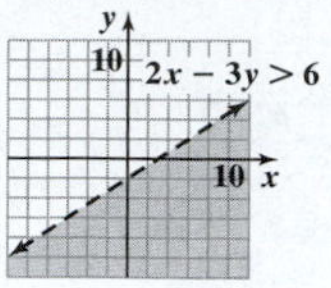

10.

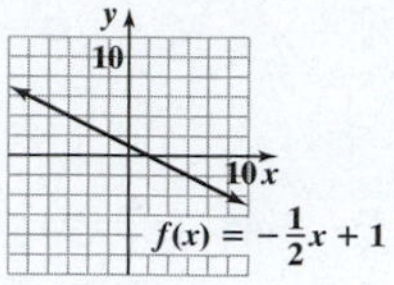

11.

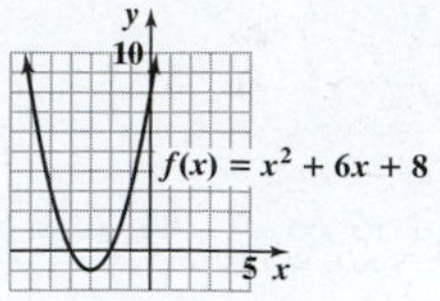

12.

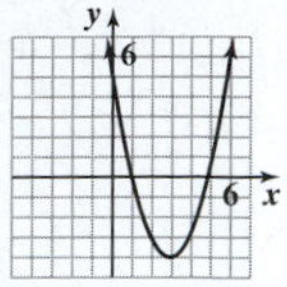

$f(x) = (x - 3)^2 - 4$

13. -6 **14.** $c = \dfrac{Ad}{d - A}$ **15.** $x^2 - x - 17$ **16.** $x^2 + 3x - 17$ **17.** h

18. $f^{-1}(x) = \dfrac{x + 3}{7}$ **19.** $\{x|x \text{ is a real number and } x \neq 1 \text{ and } x \neq 2\}$ or $(-\infty, 1) \cup (1, 2) \cup (2, \infty)$

20. $\{x|x > 4\}$ or $(4, \infty)$ **21.** $y = \frac{1}{2}x + 5$ or $f(x) = \frac{1}{2}x + 5$ **22.** $-\dfrac{y^9}{3x}$ **23.** $16x^4 - 40x^2y + 25y^2$

24. $x^2 - 5x + 1 + \dfrac{8}{5x + 1}$ **25.** $2y^2\sqrt[3]{2y^2}$ **26.** $\dfrac{4x - 13}{(x - 3)(x - 4)}$ **27.** $(x - 4)(x + 2)(x^2 - 2x + 4)$

28. $2(x + 3y)^2$ **29.** $\ln\left(\dfrac{x^2}{\sqrt{y}}\right)$ **30.** length: 12 ft; width: 4 ft **31.** $\frac{6}{5}$ hr or 1 hr and 12 min **32.** 5 mph **33.** approximately 11%

CHAPTER 10

Section 10.1

Check Point Exercises

1. approximately 6.71 units **2.** $\left(4, -\frac{1}{2}\right)$ **3.** $x^2 + y^2 = 16$ **4.** $(x - 5)^2 + (y + 6)^2 = 100$

5. center: $(-3, 1)$; radius: 2 units

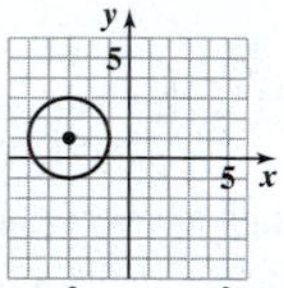

$(x + 3)^2 + (y - 1)^2 = 4$

6. $(x + 2)^2 + (y - 2)^2 = 9$

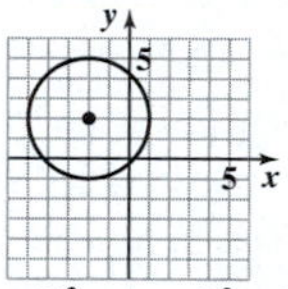

$(x + 2)^2 + (y - 2)^2 = 9$

Exercise Set 10.1

1. 13 units **3.** $2\sqrt{2}$ or 2.83 units **5.** 5 units **7.** $\sqrt{29}$ or 5.39 units **9.** $4\sqrt{2}$ or 5.66 units **11.** $2\sqrt{5}$ or 4.47 units

13. $2\sqrt{2}$ or 2.83 units **15.** $\sqrt{93}$ or 9.64 units **17.** $\sqrt{5}$ or 2.24 units **19.** $(4, 6)$ **21.** $(-4, -5)$ **23.** $\left(\frac{3}{2}, -6\right)$

25. $(-3, -2)$ **27.** $(1, 5\sqrt{5})$ **29.** $(2\sqrt{2}, 0)$ **31.** $x^2 + y^2 = 49$ **33.** $(x - 3)^2 + (y - 2)^2 = 25$ **35.** $(x + 1)^2 + (y - 4)^2 = 4$

37. $(x + 3)^2 + (y + 1)^2 = 3$ **39.** $(x + 4)^2 + y^2 = 100$

41. center: $(0, 0)$; $r = 4$

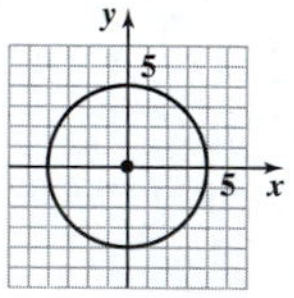

$x^2 + y^2 = 16$

43. center: $(3, 1)$; $r = 6$

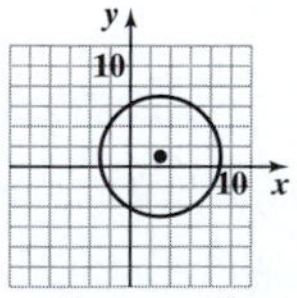

$(x - 3)^2 + (y - 1)^2 = 36$

45. center: $(-3, 2)$; $r = 2$

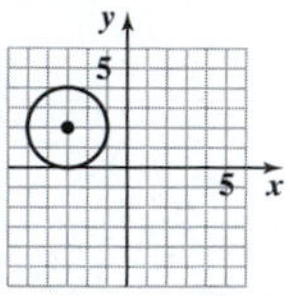

$(x + 3)^2 + (y - 2)^2 = 4$

47. center: $(-2, -2)$; $r = 2$

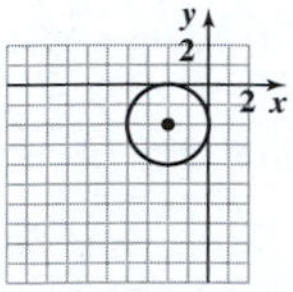

$(x + 2)^2 + (y + 2)^2 = 4$

49. $(x + 3)^2 + (y + 1)^2 = 4$; center: $(-3, -1)$; $r = 2$

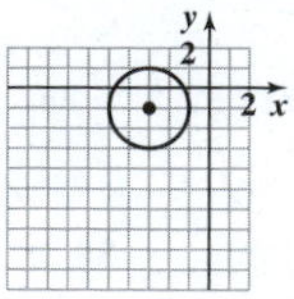

$x^2 + y^2 + 6x + 2y + 6 = 0$

51. $(x - 5)^2 + (y - 3)^2 = 64$; center: $(5, 3)$; $r = 8$

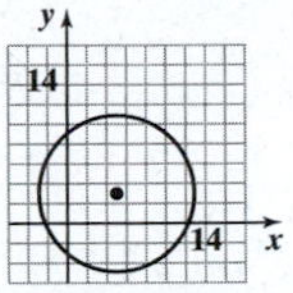

$x^2 + y^2 - 10x - 6y - 30 = 0$

53. $(x + 4)^2 + (y - 1)^2 = 25$; center: $(-4, 1)$; $r = 5$

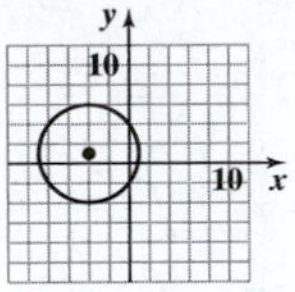

$x^2 + y^2 + 8x - 2y - 8 = 0$

55. $(x - 1)^2 + y^2 = 16$; center: $(1, 0)$; $r = 4$

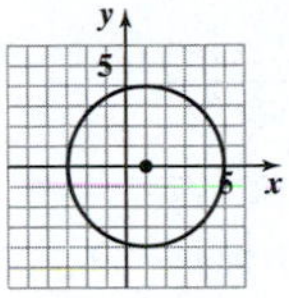

$x^2 - 2x + y^2 - 15 = 0$

57. $x^2 + y^2 = 16$, $x - y = 4$ $\{(0, -4), (4, 0)\}$

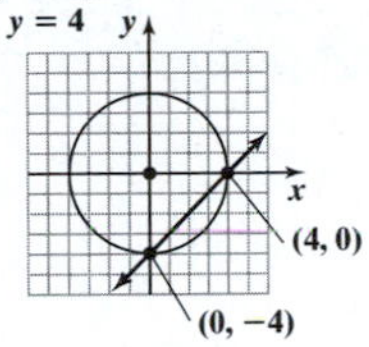

59. $(x - 2)^2 + (y + 3)^2 = 4$, $y = x - 3$ $\{(0, -3), (2, -1)\}$

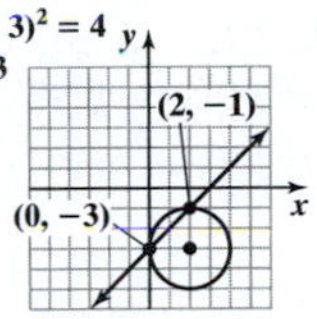

61. $(x - 2)^2 + (y + 1)^2 = 4$ **63.** $(x + 3)^2 + (y + 2)^2 = 1$ **65. a.** $(5, 10)$ **b.** $\sqrt{5}$ **c.** $(x - 5)^2 + (y - 10)^2 = 5$
67. 0.5 hour or 30 minutes **69.** $(x + 2.4)^2 + (y + 2.7)^2 = 900$

77.

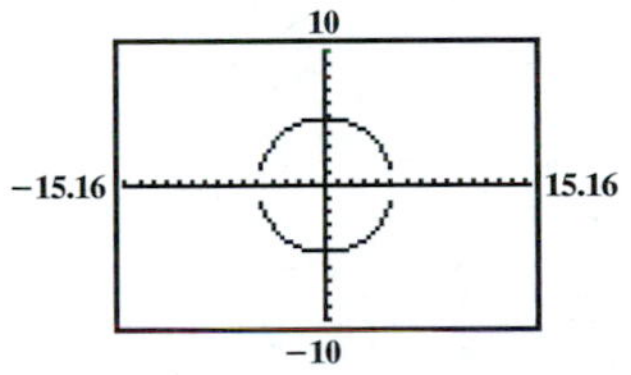

79.

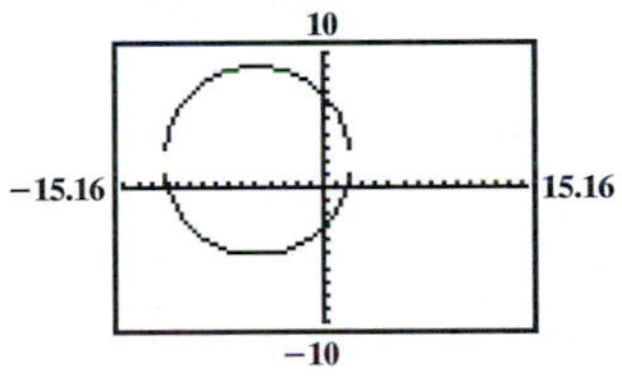

81. makes sense **83.** makes sense **85.** false **87.** false

89. a. $\sqrt{\left(x_1 - \frac{x_1 + x_2}{2}\right)^2 + \left(y_1 - \frac{y_1 + y_2}{2}\right)^2} = \frac{1}{2}\sqrt{(x_1 - x_2)^2 + (y_1 - y_2)^2}$; $\sqrt{\left(x_2 - \frac{x_1 + x_2}{2}\right)^2 + \left(y_2 - \frac{y_1 + y_2}{2}\right)^2} = \frac{1}{2}\sqrt{(x_1 - x_2)^2 + (y_1 - y_2)^2}$

b. The distance from (x_1, y_1) to (x_2, y_2) is $\sqrt{(x_1 - x_2)^2 + (y_1 - y_2)^2}$, which equals the sum of the distances from part (a).

91. 11π sq units or approximately 35 sq units **93.** $f(g(x)) = 9x^2 + 24x + 14$; $g(f(x)) = 3x^2 - 2$ **94.** $\{4\}$

95. $\left\{x \middle| -\frac{5}{2} < x < \frac{15}{2}\right\}$ or $\left(-\frac{5}{2}, \frac{15}{2}\right)$ **96.** -3 and 3 **97.** -2 and 2 **98.** $\frac{x^2}{16} + \frac{y^2}{25} = 1$

Section 10.2

Check Point Exercises

1.

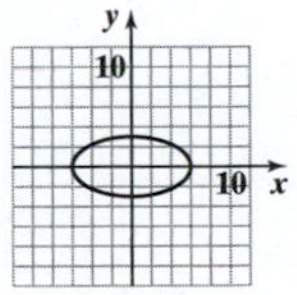

$\frac{x^2}{36} + \frac{y^2}{9} = 1$

2.

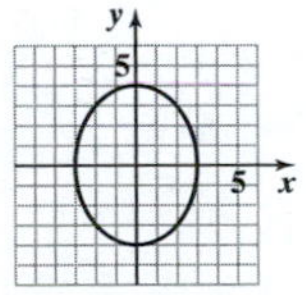

$16x^2 + 9y^2 = 144$

3.

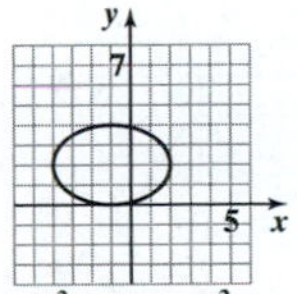

$\frac{(x + 1)^2}{9} + \frac{(y - 2)^2}{4} = 1$

4. Yes, the height of the archway 6 feet from the center is approximately 9.54 feet.

Exercise Set 10.2

1.

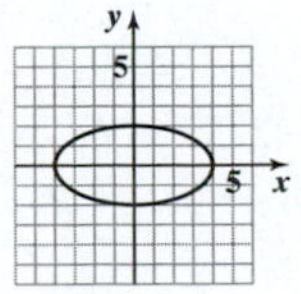

$\frac{x^2}{16} + \frac{y^2}{4} = 1$

3.

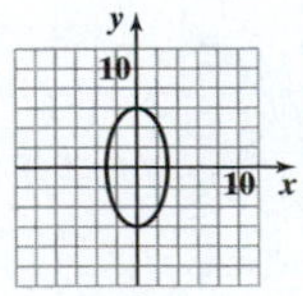

$\frac{x^2}{9} + \frac{y^2}{36} = 1$

5.

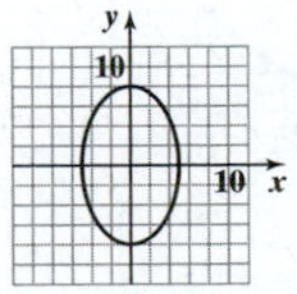

$\frac{x^2}{25} + \frac{y^2}{64} = 1$

7.

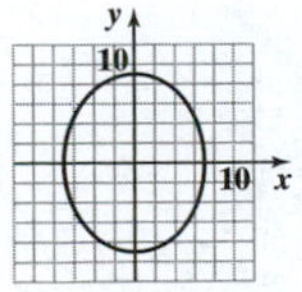

$\frac{x^2}{49} + \frac{y^2}{81} = 1$

9.

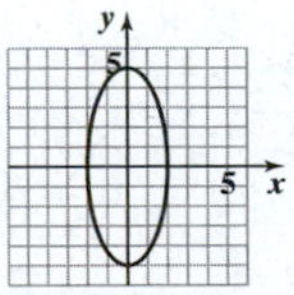

$25x^2 + 4y^2 = 100$

11.

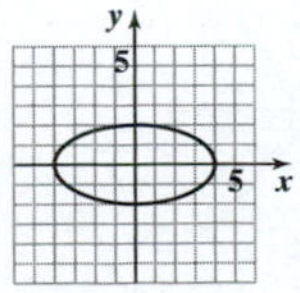

$4x^2 + 16y^2 = 64$

13.

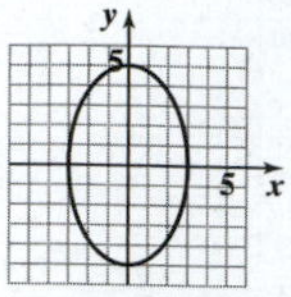

$25x^2 + 9y^2 = 225$

15.

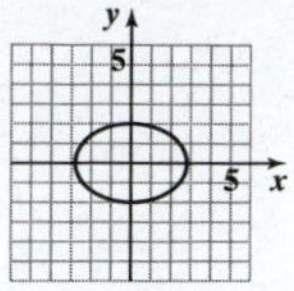

$x^2 + 2y^2 = 8$

17. $\frac{x^2}{4} + \frac{y^2}{1} = 1$ or $\frac{x^2}{4} + y^2 = 1$ **19.** $\frac{x^2}{1} + \frac{y^2}{4} = 1$ or $x^2 + \frac{y^2}{4} = 1$

21.

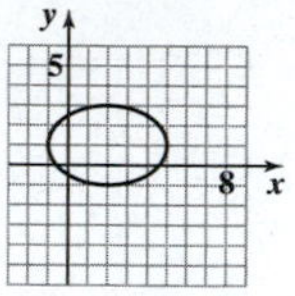

$\frac{(x-2)^2}{9} + \frac{(y-1)^2}{4} = 1$

23.

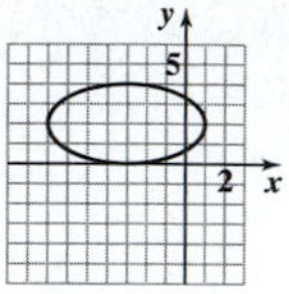

$(x+3)^2 + 4(y-2)^2 = 16$

25.

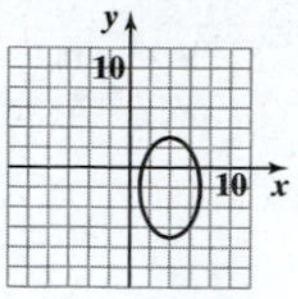

$\frac{(x-4)^2}{9} + \frac{(y+2)^2}{25} = 1$

27.

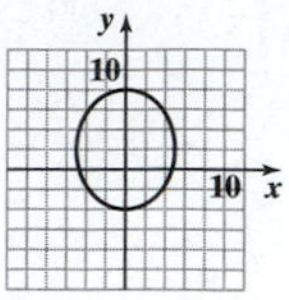

$\frac{x^2}{25} + \frac{(y-2)^2}{36} = 1$

29.

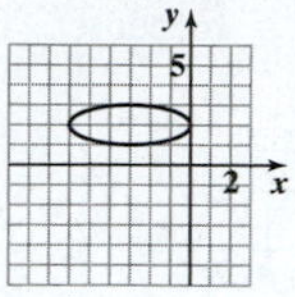

$\frac{(x+3)^2}{9} + (y-2)^2 = 1$

31.

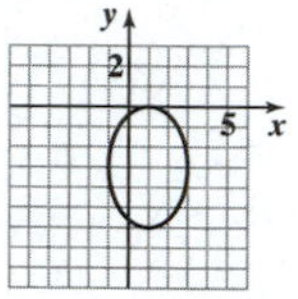

$9(x-1)^2 + 4(y+3)^2 = 36$

33. $\frac{(x+1)^2}{4} + \frac{(y-1)^2}{1} = 1$ **35.** $\{(0, -1), (0, 1)\}$ **37.** $\{(0, 3)\}$ **39.** $\{(0, -2), (1, 0)\}$

41.

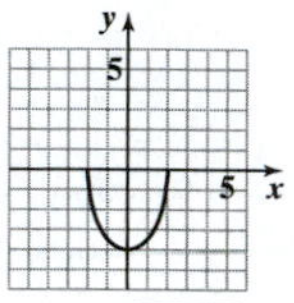

$y = -\sqrt{16 - 4x^2}$

43. Yes, the height of the archway 4 feet from the center is approximately 9.64 feet.

45. a. $\frac{x^2}{48^2} + \frac{y^2}{23^2} = 1$ or $\frac{x^2}{2304} + \frac{y^2}{529} = 1$ **b.** approximately 42 feet

55. does not make sense **57.** makes sense **59.** false **61.** true

63.

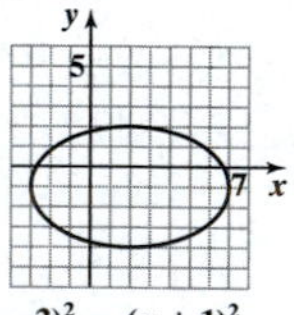

$\frac{(x-2)^2}{25} + \frac{(y+1)^2}{9} = 1$

65. a. 984 mi **b.** 1016 mi **67.** $(x+2)(x+2)(x-2)$ or $(x+2)^2(x-2)$ **68.** $2xy^2\sqrt[3]{5xy}$ **69.** $\{-1\}$ **70.** $\frac{x^2}{9} - \frac{y^2}{4} = 1$; Terms are separated by subtraction rather than by addition. **71. a.** -4 and 4 **b.** The equation $y^2 = -9$ has no real solutions. **72. a.** -3 and 3 **b.** The equation $x^2 = -16$ has no real solutions.

Section 10.3

Check Point Exercises

1. a. $(-5, 0)$ and $(5, 0)$ **b.** $(0, -5)$ and $(0, 5)$

2.

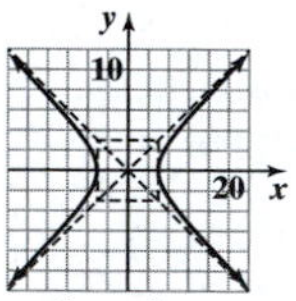

$\frac{x^2}{36} - \frac{y^2}{9} = 1$

3.

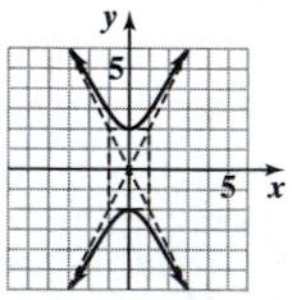

$y^2 - 4x^2 = 4$

Exercise Set 10.3

1. $(-2, 0)$ and $(2, 0)$; b **3.** $(0, -2)$ and $(0, 2)$; a

5.

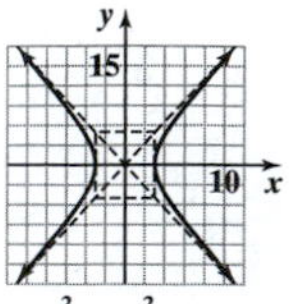

$\frac{x^2}{9} - \frac{y^2}{25} = 1$

7.

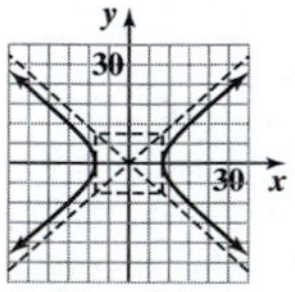

$\frac{x^2}{100} - \frac{y^2}{64} = 1$

9.

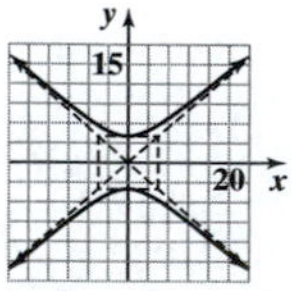

$\frac{y^2}{16} - \frac{x^2}{36} = 1$

11.

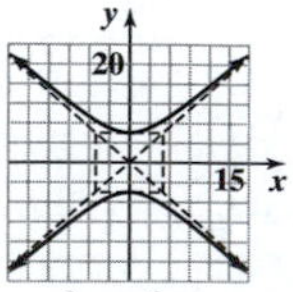

$\frac{y^2}{36} - \frac{x^2}{25} = 1$

13.

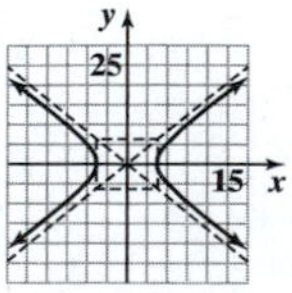

$9x^2 - 4y^2 = 36$

15.

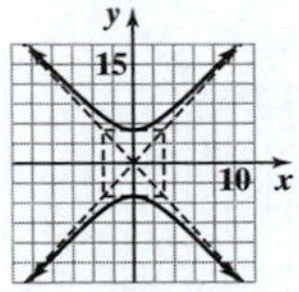

$9y^2 - 25x^2 = 225$

17.

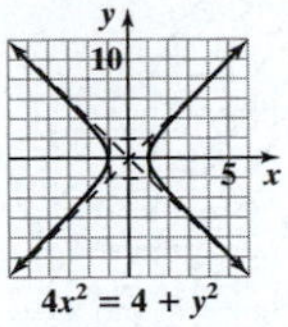

$4x^2 = 4 + y^2$

19. $\frac{x^2}{9} - \frac{y^2}{25} = 1$ **21.** $\frac{y^2}{4} - \frac{x^2}{9} = 1$

23.

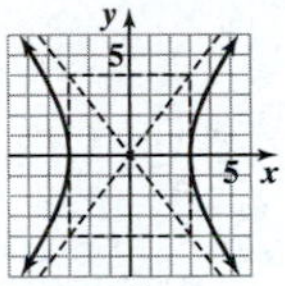

$\frac{x^2}{9} - \frac{y^2}{16} = 1$

domain: $\{x|x \le -3 \text{ or } x \ge 3\}$ or $(-\infty, -3] \cup [3, \infty)$; range: $\{y|y \text{ is a real number}\}$ or $(-\infty, \infty)$

25.

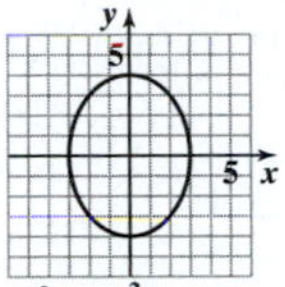

$\frac{x^2}{9} + \frac{y^2}{16} = 1$

domain: $\{x|-3 \le x \le 3\}$ or $[-3, 3]$; range: $\{y|-4 \le y \le 4\}$ or $[-4, 4]$

27.

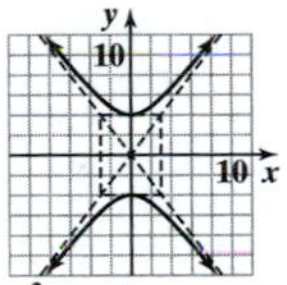

$\frac{y^2}{16} - \frac{x^2}{9} = 1$

domain: $\{x|x \text{ is a real number}\}$ or $(-\infty, \infty)$; range: $\{y|y \le -4 \text{ or } y \ge 4\}$ or $(-\infty, -4] \cup [4, \infty)$

29. $\{(-2, 0), (2, 0)\}$ **31.** $\{(0, -3), (0, 3)\}$ **33.** 40 yd

41.

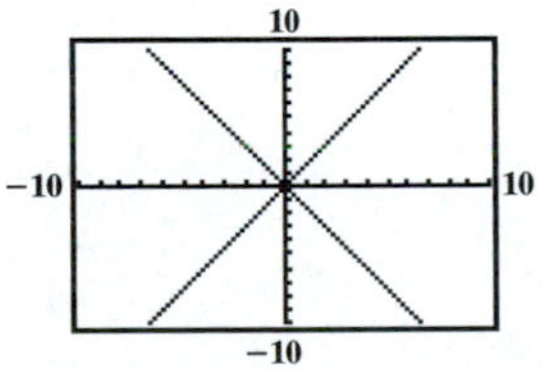

no; two lines; $y = \frac{b}{a}x$ and $y = -\frac{b}{a}x$ **43.** does not make sense **45.** makes sense **47.** false **49.** true

51.

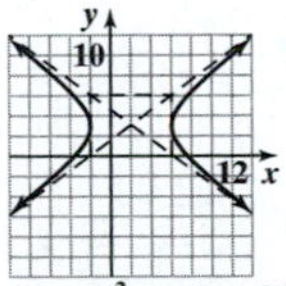

$\frac{(x-2)^2}{16} - \frac{(y-3)^2}{9} = 1$

53.

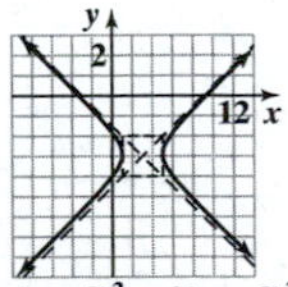

$(x-3)^2 - 4(y+3)^2 = 4$

55. $\frac{x^2}{36} - \frac{y^2}{576} = 1$

57.

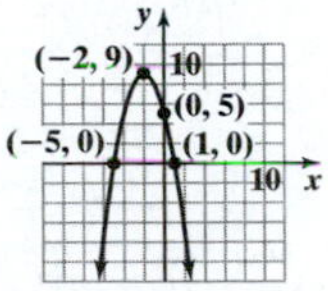

$y = -x^2 - 4x + 5$

58. $\left\{x \middle| x \le -\frac{1}{3} \text{ or } x \ge 4\right\}$ or $\left(-\infty, -\frac{1}{3}\right] \cup [4, \infty)$ **59.** $\{21\}$

60.

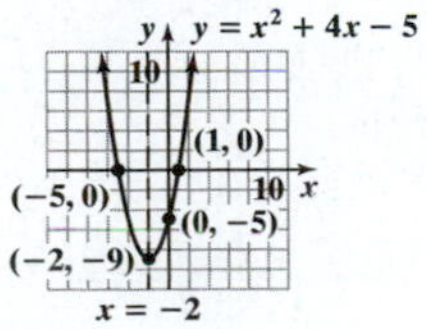

61.

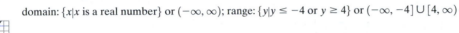

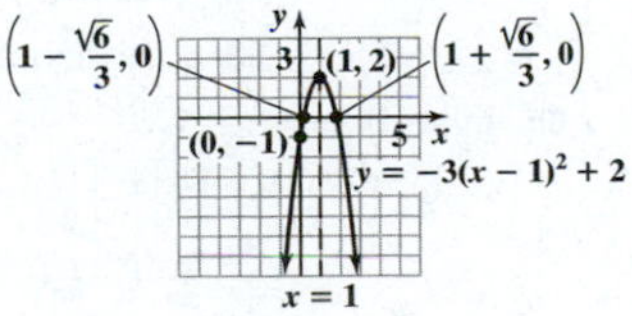

62. 0.2 and 1.8

Mid-Chapter Check Point Exercises

1.
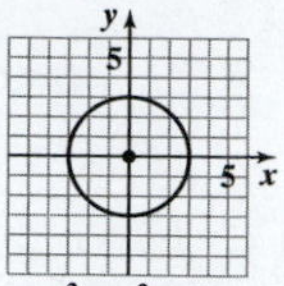
$x^2 + y^2 = 9$

2.
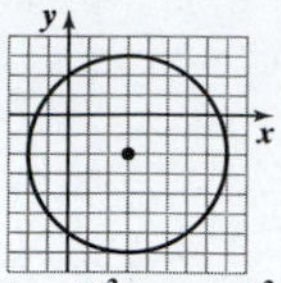
$(x-3)^2 + (y+2)^2 = 25$

3.
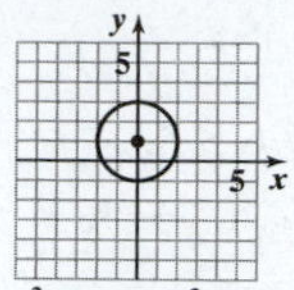
$x^2 + (y-1)^2 = 4$

4.
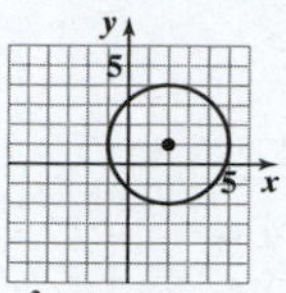
$x^2 + y^2 - 4x - 2y - 4 = 0$

5.
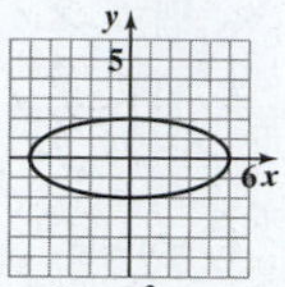
$\frac{x^2}{25} + \frac{y^2}{4} = 1$

6.
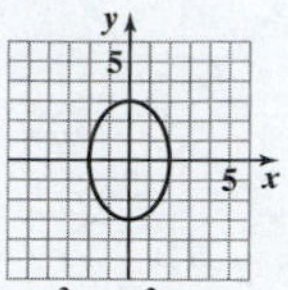
$9x^2 + 4y^2 = 36$

7.
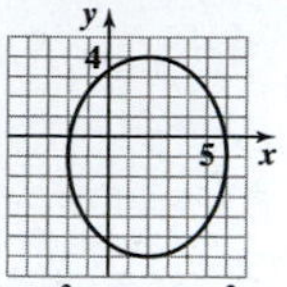
$\frac{(x-2)^2}{16} + \frac{(y+1)^2}{25} = 1$

8.
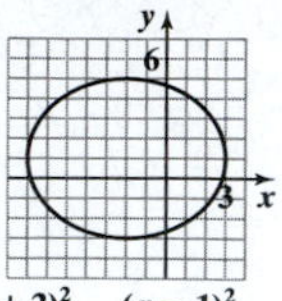
$\frac{(x+2)^2}{25} + \frac{(y-1)^2}{16} = 1$

9.
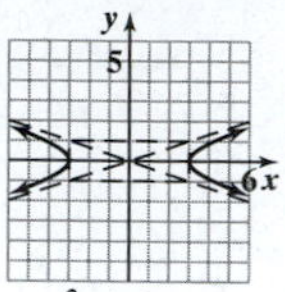
$\frac{x^2}{9} - y^2 = 1$

10.
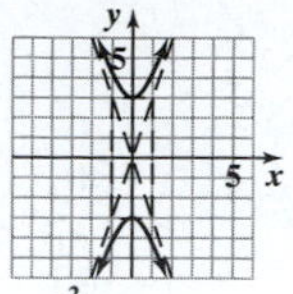
$\frac{y^2}{9} - x^2 = 1$

11.
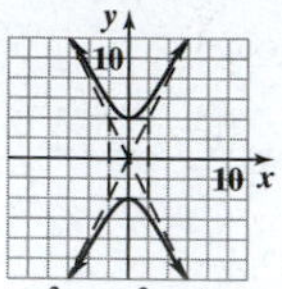
$y^2 - 4x^2 = 16$

12.
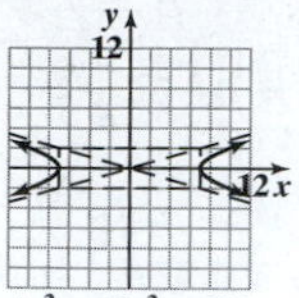
$4x^2 - 49y^2 = 196$

13.
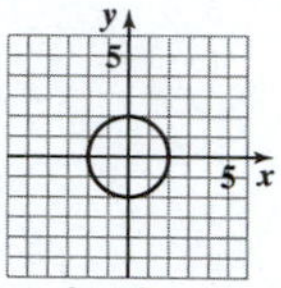
$x^2 + y^2 = 4$

14.
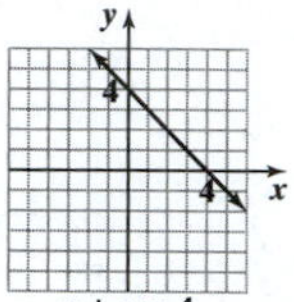
$x + y = 4$

15.
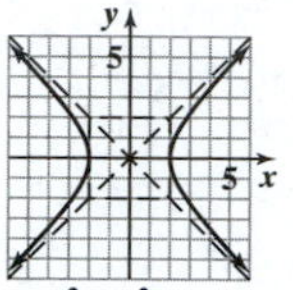
$x^2 - y^2 = 4$

16.
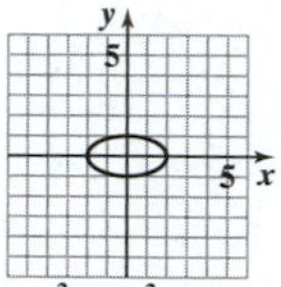
$x^2 + 4y^2 = 4$

17.
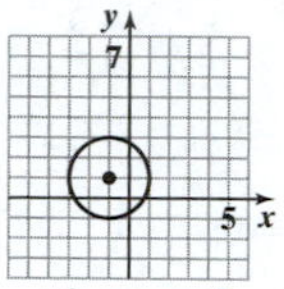
$(x+1)^2 + (y-1)^2 = 4$

18.
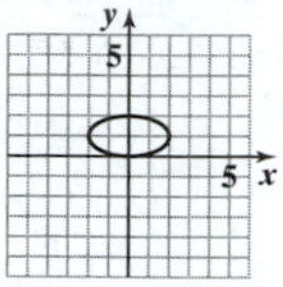
$x^2 + 4(y-1)^2 = 4$

19. $4\sqrt{2} \approx 5.66$ units; $(0, 0)$ **20.** $\sqrt{61} \approx 7.81$ units; $\left(-\frac{15}{2}, 11\right)$

Section 10.4

Check Point Exercises

1.
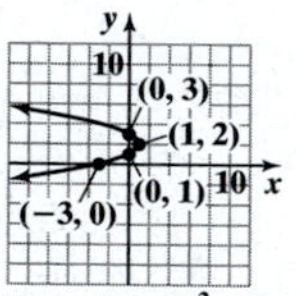

$x = -(y-2)^2 + 1$

2.
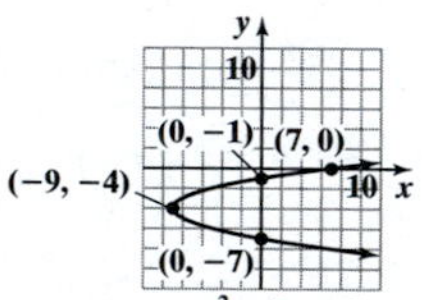

$x = y^2 + 8y + 7$

3. a. hyperbola **b.** ellipse **c.** circle **d.** parabola

Exercise Set 10.4

1. opens to right; $(-1, 2)$; b **3.** opens to right; $(1, -2)$; f **5.** opens to left; $(1, 2)$; a **7.** $(0, 0)$ **9.** $(3, 2)$ **11.** $(-1, -2)$ **13.** $(0, 6)$ **15.** $(-3, 3)$ **17.** $(4, -1)$

19.
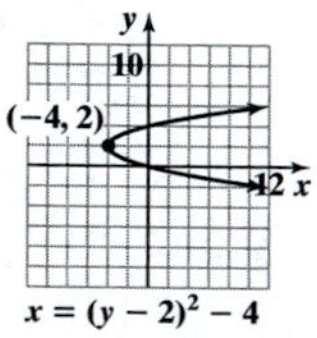

$x = (y-2)^2 - 4$

21.
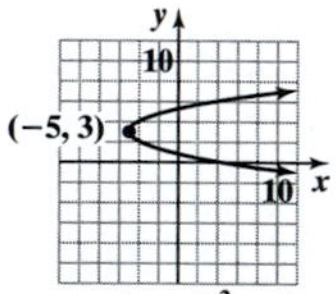

$x = (y-3)^2 - 5$

23.
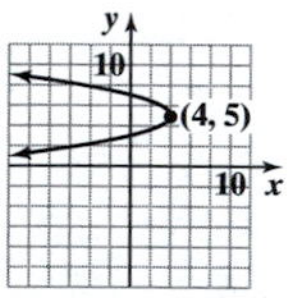

$x = -(y-5)^2 + 4$

25.
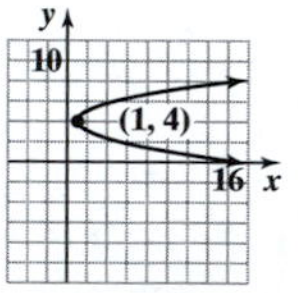

$x = (y-4)^2 + 1$

27.
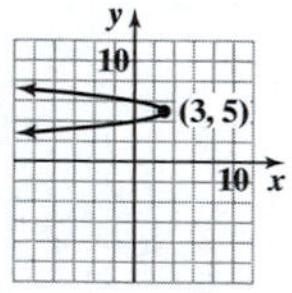

$x = -3(y-5)^2 + 3$

29.
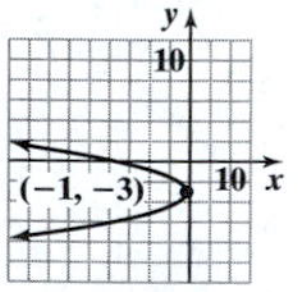

$x = -2(y+3)^2 - 1$

31.
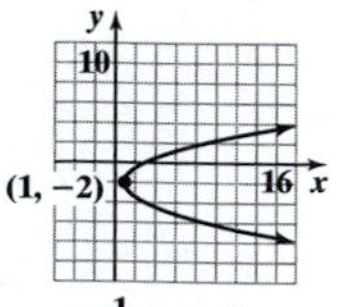

$x = \frac{1}{2}(y+2)^2 + 1$

33.
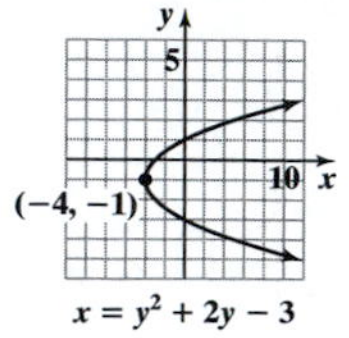

$x = y^2 + 2y - 3$

35.
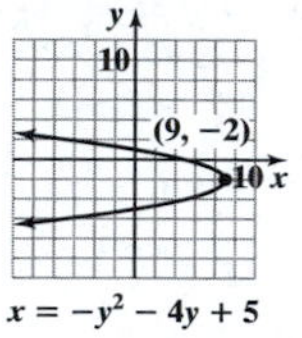

$x = -y^2 - 4y + 5$

37.
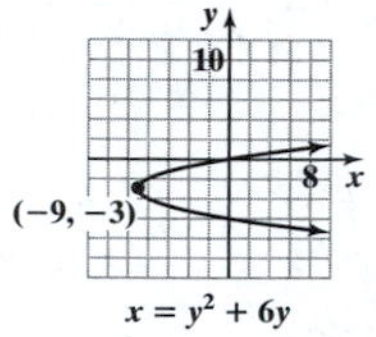

$x = y^2 + 6y$

39.
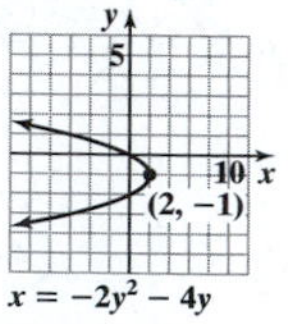

$x = -2y^2 - 4y$

41.
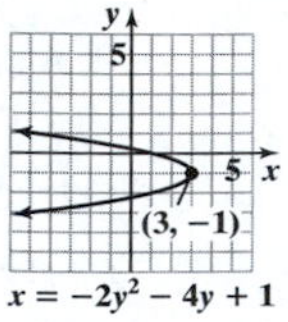

$x = -2y^2 - 4y + 1$

43. a. horizontal **b.** to the right **c.** $(2, 1)$ **45. a.** vertical **b.** upward **c.** $(1, 2)$ **47. a.** vertical **b.** downward **c.** $(-3, 4)$ **49. a.** horizontal **b.** to the left **c.** $(4, -3)$ **51. a.** vertical **b.** upward **c.** $(2, -5)$ **53. a.** horizontal **b.** to the left **c.** $(5, 2)$ **55.** parabola **57.** ellipse **59.** hyperbola **61.** circle **63.** hyperbola **65.** hyperbola

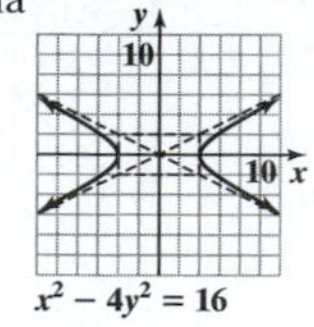
$x^2 - 4y^2 = 16$

67. circle

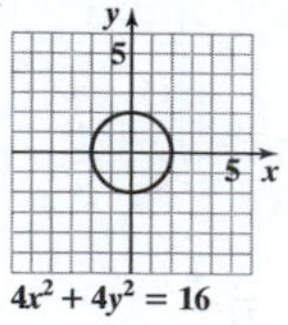
$4x^2 + 4y^2 = 16$

69. ellipse

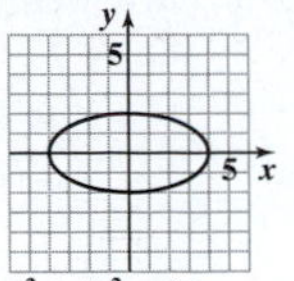

$x^2 + 4y^2 = 16$

71. parabola

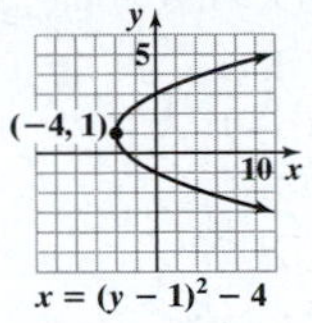

$x = (y - 1)^2 - 4$

73. circle

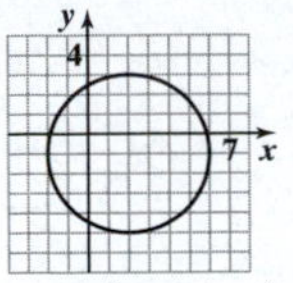

$(x - 2)^2 + (y + 1)^2 = 16$

75. domain: $\{x|x \geq -4\}$ or $[-4, \infty)$; range: $\{y|y$ is a real number$\}$ or $(-\infty, \infty)$; not a function

77. domain: $\{x|x$ is a real number$\}$ or $(-\infty, \infty)$; range: $\{y|y \leq 1\}$ or $(-\infty, 1]$; function

79. domain: $\{x|x \leq 3\}$ or $(-\infty, 3]$; range: $\{y|y$ is a real number$\}$ or $(-\infty, \infty)$; not a function

81. $\{(-4, 2), (0, 0)\}$

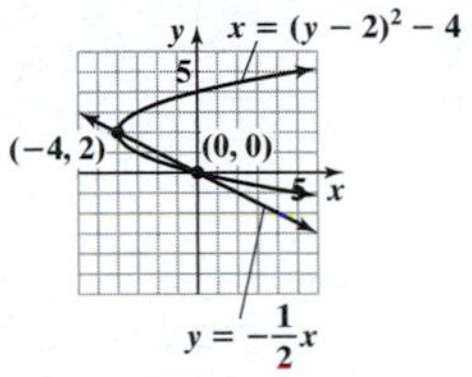

83. $\{(-2, 1)\}$

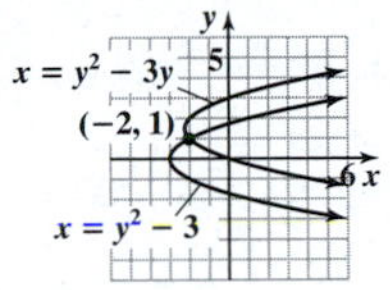

85. $\varnothing$

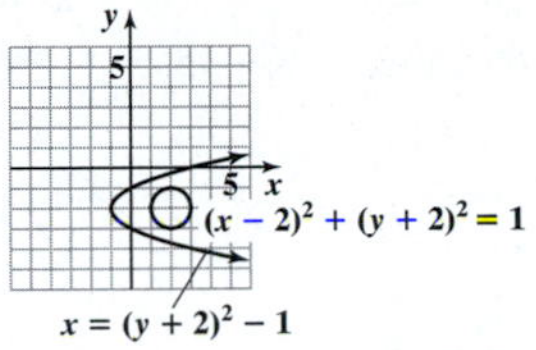

87. a. $y = 0.0001032x^2$ **b.** 58 ft **89. a.** $y = \frac{1}{18}x^2$ **b.** 4.5 ft **91. a.** ellipse **b.** $x^2 + 4y^2 = 4$

101. $y^2 + 2y + (-6x + 13) = 0$;
$y = -1 \pm \sqrt{6x - 12}$

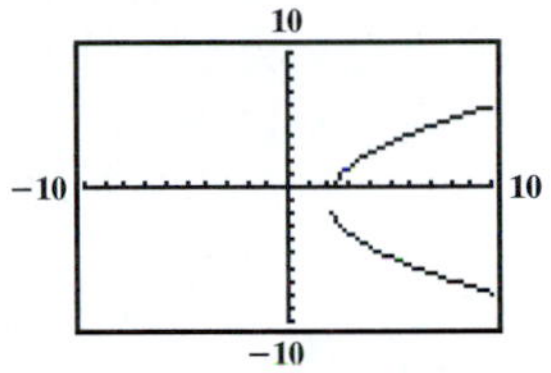

105. makes sense **107.** makes sense **109.** true **111.** false **113.** Yes; the height of the arch 30 feet from the center is 45.5 feet.

114.

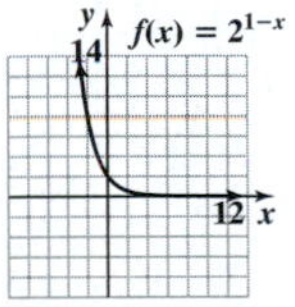

115. $f^{-1}(x) = 3x + 15$ **116.** $\{-3, -1\}$ **117.** $\{(2.5, -2)\}$ **118.** $\{(4, -3)\}$ **119.** $\{2, 4\}$

Section 10.5

Check Point Exercises

1. (0, 1) and (4, 17), or $\{(0, 1), (4, 17)\}$ **2.** $(2, -1)$ and $\left(-\frac{6}{5}, \frac{3}{5}\right)$, or $\left\{(2, -1), \left(-\frac{6}{5}, \frac{3}{5}\right)\right\}$

3. $(3, 2), (-3, 2), (3, -2)$, and $(-3, -2)$, or $\{(3, 2), (-3, 2), (3, -2), (-3, -2)\}$ **4.** (0, 5) or $\{(0, 5)\}$ **5.** length: 7 feet; width: 3 feet

Exercise Set 10.5

1. $\{(2, 0), (-3, 5)\}$ **3.** $\{(2, 0), (1, 1)\}$ **5.** $\{(-3, 11), (4, -10)\}$ **7.** $\{(-3, -4), (4, 3)\}$ **9.** $\left\{(2, 3), \left(-\frac{3}{2}, -4\right)\right\}$

11. $\{(3, 0), (-5, -4)\}$ **13.** $\{(3, 1), (-1, -3), (1, 3), (-3, -1)\}$ **15.** $\{(4, -3), (-1, 2)\}$ **17.** $\{(4, -3), (0, 1)\}$

19. $\{(3, 2), (-3, 2), (3, -2), (-3, -2)\}$ **21.** $\{(3, 2), (-3, 2), (3, -2), (-3, -2)\}$ **23.** $\{(2, 1), (-2, 1), (2, -1), (-2, -1)\}$

25. $\{(3, 4), (3, -4)\}$ **27.** $\{(0, 2), (0, -2), (-1, \sqrt{3}), (-1, -\sqrt{3})\}$ **29.** $\{(2, 1), (-2, 1), (2, -1), (-2, -1)\}$

31. $\{(-1, -4), (1, 4), (2\sqrt{2}, \sqrt{2}), (-2\sqrt{2}, -\sqrt{2})\}$ **33.** $\{(4, 1), (2, 2)\}$ **35.** $\{(0, 0), (-1, 1)\}$ **37.** $\{(0, 0), (2, 2), (-2, 2)\}$

39. $\left\{(-4, 1), \left(-\frac{5}{2}, \frac{1}{4}\right)\right\}$ **41.** $\left\{(-2, 3), \left(\frac{12}{5}, -\frac{29}{5}\right)\right\}$ **43.** $x + y = 10$; $xy = 24$; 6 and 4

45. $x^2 - y^2 = 3$; $2x^2 + y^2 = 9$; 2 and 1, −2 and 1, −2 and −1, or 2 and −1 **47.** $\{(2, -1), (-2, 1)\}$ **49.** $\{(2, 20), (-2, 4), (-3, 0)\}$

51. $\left\{\left(-1, -\frac{1}{2}\right), \left(-1, \frac{1}{2}\right), \left(1, -\frac{1}{2}\right), \left(1, \frac{1}{2}\right)\right\}$

53. Answers will vary.; example:

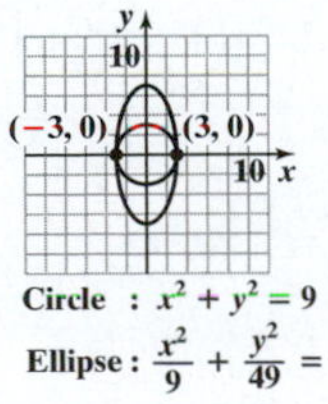

Circle : $x^2 + y^2 = 9$

Ellipse : $\frac{x^2}{9} + \frac{y^2}{49} = 1$

55. $(0, -4), (2, 0), (-2, 0)$

57. length: 11 feet; width: 7 feet

59. length: 8 inches; width: 6 inches

61. large square: 5 meters by 5 meters; small square: 2 meters by 2 meters

63. **a.** between the 1940s and the 1960s **b.** 1949; 43%; 43% **c.** 1920; 28% **d.** 1919; white collar: 27.5%; farmers: 27.4%; fairly well, although answers will vary. **71.** does not make sense **73.** makes sense **75.** true **77.** false

79. $\{(8, 2)\}$ **81.** $3x - 2y \le 6$

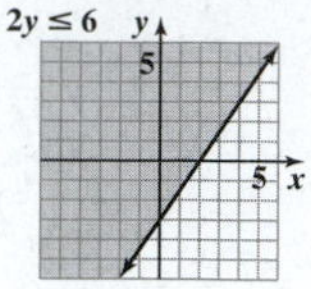

82. $m = \frac{8}{3}$ **83.** $6x^3 - 16x^2 + 17x - 6$

84. rises to the left and rises to the right **85.** $x^2(x - 2)^2$ **86.** $h(-x) = x^2 - 2x + 1$

Review Exercises

1. 13 units **2.** $2\sqrt{2} \approx 2.83$ units **3.** $(-5, 5)$ **4.** $\left(-\frac{11}{2}, -2\right)$ **5.** $x^2 + y^2 = 9$ **6.** $(x + 2)^2 + (y - 4)^2 = 36$

7. center: $(0, 0)$; $r = 1$

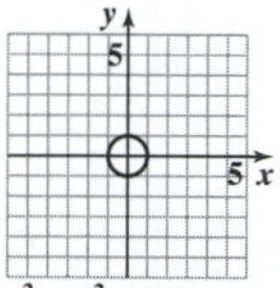

$x^2 + y^2 = 1$

8. center: $(-2, 3)$; $r = 3$

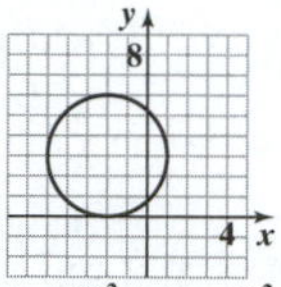

$(x + 2)^2 + (y - 3)^2 = 9$

9. center: $(2, -1)$; $r = 3$

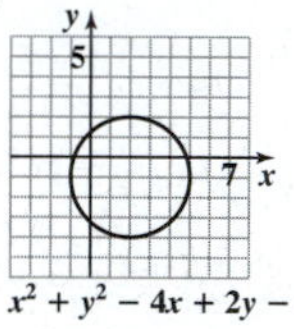

$x^2 + y^2 - 4x + 2y - 4 = 0$

10. center: $(0, 2)$; $r = 2$

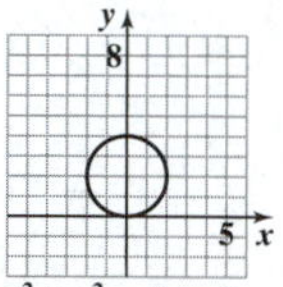

$x^2 + y^2 - 4y = 0$

11.

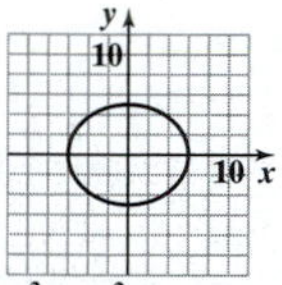

$\frac{x^2}{36} + \frac{y^2}{25} = 1$

12.

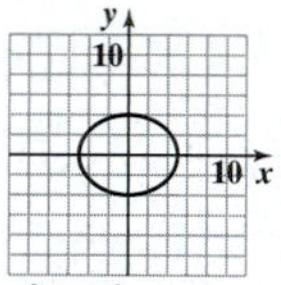

$\frac{x^2}{25} + \frac{y^2}{16} = 1$

13.

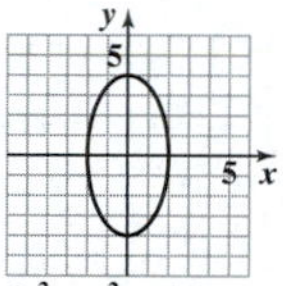

$4x^2 + y^2 = 16$

14.

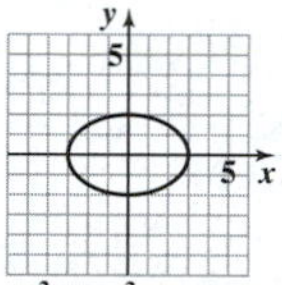

$4x^2 + 9y^2 = 36$

15.

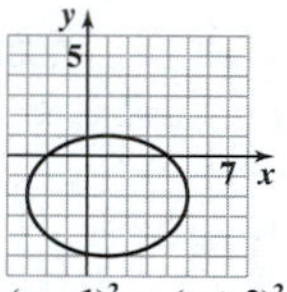

$\frac{(x - 1)^2}{16} + \frac{(y + 2)^2}{9} = 1$

16.

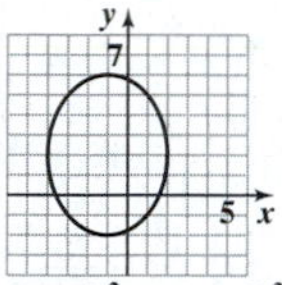

$\frac{(x + 1)^2}{9} + \frac{(y - 2)^2}{16} = 1$

17. Yes, the height of the archway 14 feet from the center is approximately 12.43 feet.

18.

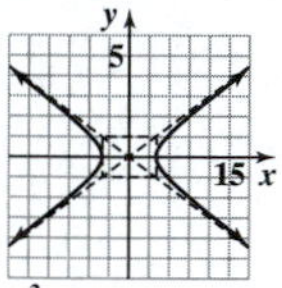

$\frac{x^2}{16} - y^2 = 1$

19.

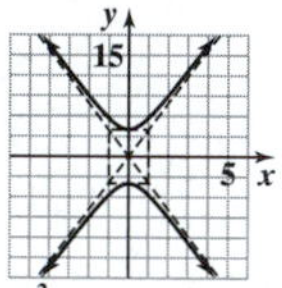

$\frac{y^2}{16} - x^2 = 1$

20.

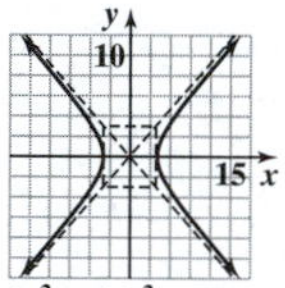

$9x^2 - 16y^2 = 144$

21.

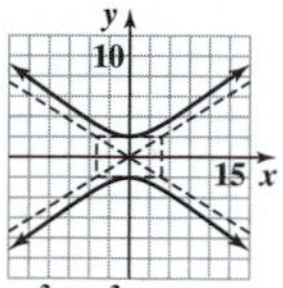

$4y^2 - x^2 = 16$

22.

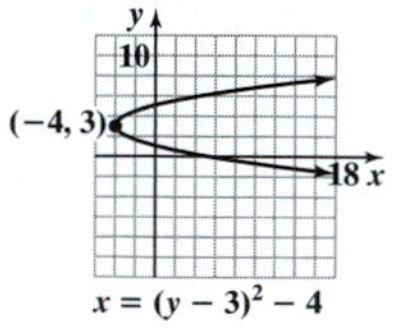

$x = (y - 3)^2 - 4$

23.

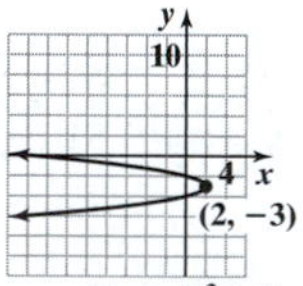

$x = -2(y + 3)^2 + 2$

24.

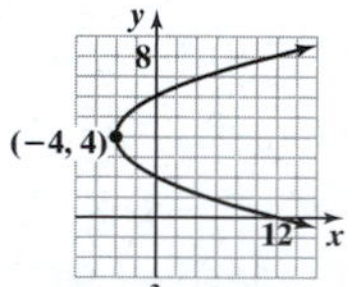

$x = y^2 - 8y + 12$

25.

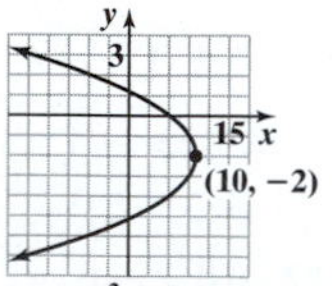

$x = -y^2 - 4y + 6$

26. parabola **27.** ellipse **28.** hyperbola **29.** circle **30.** hyperbola **31.** ellipse **32.** parabola

33. circle

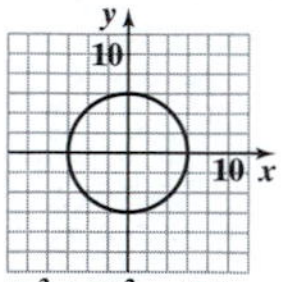

$5x^2 + 5y^2 = 180$

34. ellipse

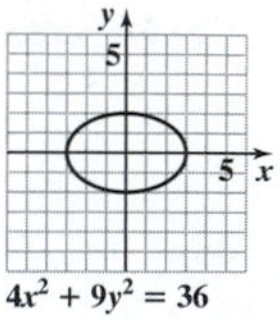

$4x^2 + 9y^2 = 36$

35. hyperbola

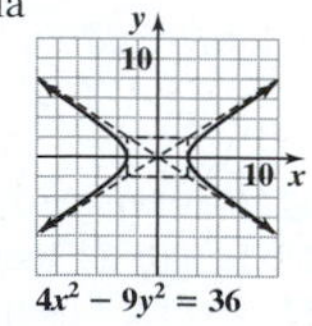

$4x^2 - 9y^2 = 36$

36. ellipse

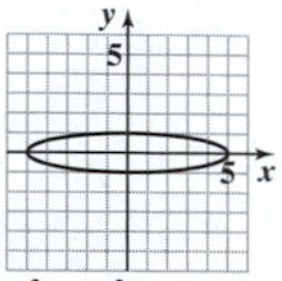

$\frac{x^2}{25} + \frac{y^2}{1} = 1$

37. parabola

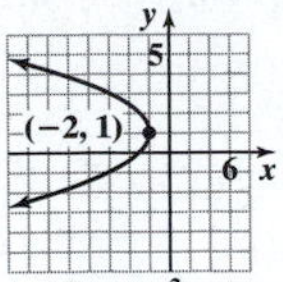

$x + 3 = -y^2 + 2y$

38. parabola

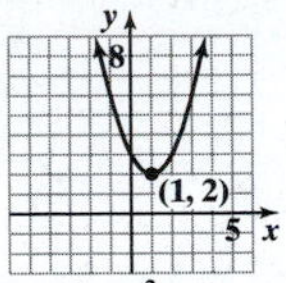

$y - 3 = x^2 - 2x$

39. ellipse

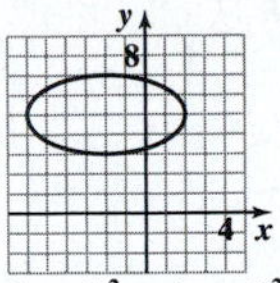

$\frac{(x+2)^2}{16} + \frac{(y-5)^2}{4} = 1$

40. circle

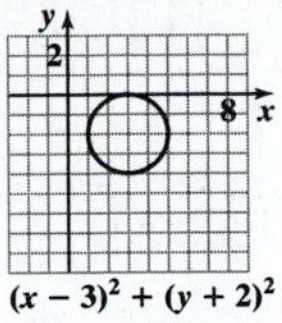

$(x-3)^2 + (y+2)^2 = 4$

41. circle

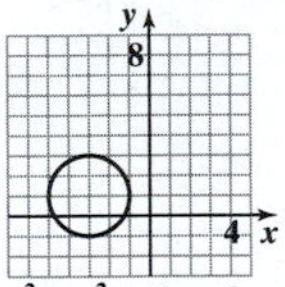

$x^2 + y^2 + 6x - 2y + 6 = 0$

42. a. $y = \frac{1}{12}x^2$ **b.** $(0, 3)$; 3 inches above the vertex **43.** $\{(1, 0), (4, 3)\}$

44. $\{(0, 1), (-3, 4)\}$ **45.** $\{(-1, 1), (1, -1)\}$ **46.** $\{(3, \sqrt{6}), (-3, \sqrt{6}), (3, -\sqrt{6}), (-3, -\sqrt{6})\}$

47. $\{(2, 2), (-2, -2)\}$ **48.** $\{(1, 2), (9, 6)\}$ **49.** $\{(-3, -1), (1, 3)\}$ **50.** $\left\{(-1, -1), \left(\frac{1}{2}, 2\right)\right\}$

51. $\left\{(0, -1), \left(\frac{5}{2}, -\frac{7}{2}\right)\right\}$ **52.** $\{(3, 2), (-3, 2), (2, -3), (-2, -3)\}$

53. $\{(3, 1), (-3, 1), (3, -1)$, and $(-3, -1)\}$ **54.** 8 meters by 5 meters **55.** $(1, 6)$ and $(3, 2)$ **56.** x: 46 ft; y: 28 ft or x: 50 ft; y: 20 ft

Chapter Test

1. $\sqrt{73} \approx 8.54$ units **2.** $\left(\frac{7}{2}, -4\right)$ **3.** $(x-3)^2 + (y+2)^2 = 25$ **4.** center: $(5, -3)$; $r = 7$ **5.** center: $(-2, 3)$; $r = 4$

6. $(7, -3)$ **7.** $(-2, -5)$

8. hyperbola

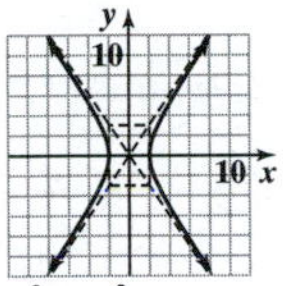

$\frac{x^2}{4} - \frac{y^2}{9} = 1$

9. ellipse

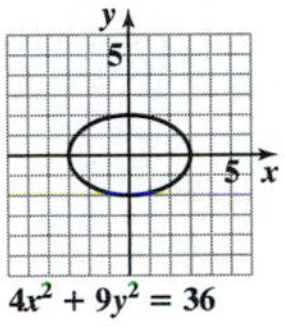

$4x^2 + 9y^2 = 36$

10. parabola

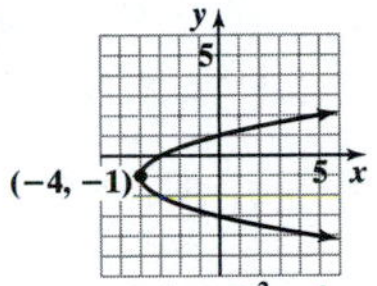

$x = (y+1)^2 - 4$

11. ellipse

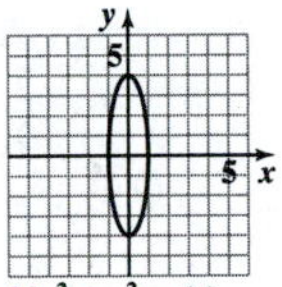

$16x^2 + y^2 = 16$

12. hyperbola

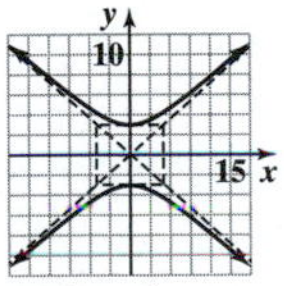

$25y^2 = 9x^2 + 225$

13. parabola

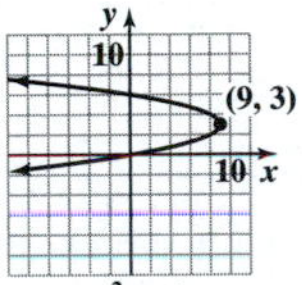

$x = -y^2 + 6y$

14. ellipse

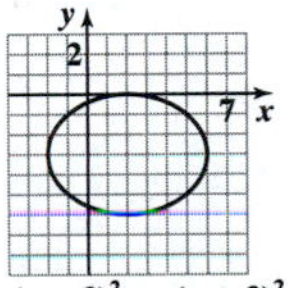

$\frac{(x-2)^2}{16} + \frac{(y+3)^2}{9} = 1$

15. circle

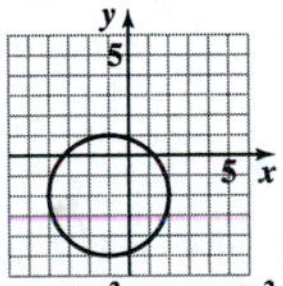

$(x+1)^2 + (y+2)^2 = 9$

16. circle

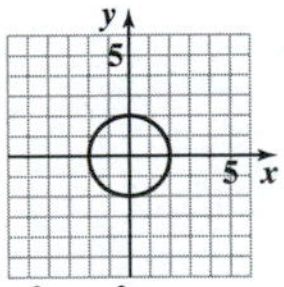

$\frac{x^2}{4} + \frac{y^2}{4} = 1$

17. $\{(4, -3), (-3, 4)\}$ **18.** $\{(3, 2), (-3, 2), (3, -2), (-3, -2)\}$

19. 15 feet by 12 feet or 24 feet by 7.5 feet **20.** 4 feet by 3 feet

Cumulative Review Exercises

1. $\{x|x > -1\}$ or $(-1, \infty)$ **2.** $\left\{-\frac{1}{2}, 4\right\}$ **3.** $\{2\}$ **4.** $\left\{x \middle| -\frac{5}{3} < x < -1\right\}$ or $\left(-\frac{5}{3}, -1\right)$ **5.** $\left\{\frac{5}{2}\right\}$

6. $\left\{\frac{\ln 8}{0.7} \approx 2.97\right\}$ **7.** $\{(1, 3), (-1, 3), (1, -3), (-1, -3)\}$

8.

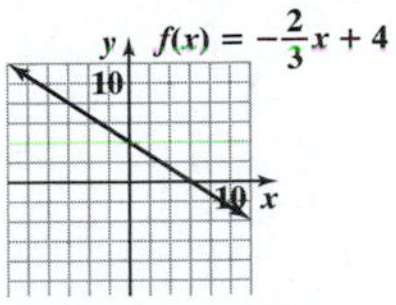

9.

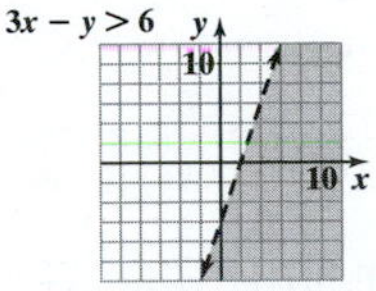

10.

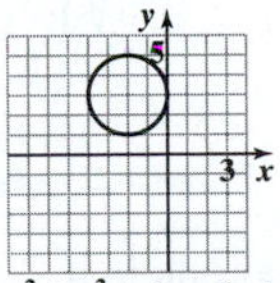

$x^2 + y^2 + 4x - 6y + 9 = 0$

11.

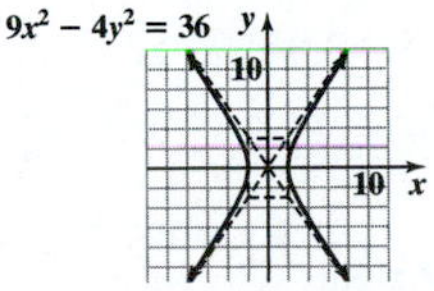

12. 46 **13.** $x^2 - 5x - 1$ **14.** $2xy^2\sqrt[3]{2y}$ **15.** $11 + 10i$ **16.** $3x(2x-3)^2$ **17.** $(x-2)(x+3)(x-3)$

18. $\{x|x \le 2\}$ or $(-\infty, 2]$ **19.** $\frac{(1-\sqrt{x})^2}{1-x}$ or $\frac{1 - 2\sqrt{x} + x}{1-x}$ **20.** $\ln(x^{1/3}y^7)$ **21.** $3x^2 + x + 4 + \frac{7}{x-2}$ **22.** $x^2 - 12 = 0$

23. faster car: 50 mph; slower car: 40 mph **24.** 175 miles; \$67 **25.** apple: 60 calories; banana: 87 calories

CHAPTER 11

Section 11.1

Check Point Exercises

1. -2 and 2 **2.** -2, 0, and 2 **3.** $-\frac{1}{2}$ with multiplicity 2 and 5 with multiplicity 3; touches and turns at $-\frac{1}{2}$ and crosses at 5
4. a. even **b.** odd **c.** neither **5.**

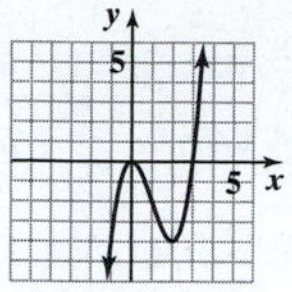

$f(x) = x^3 - 3x^2$

Exercise Set 11.1

1. falls to the left and rises to the right **3.** rises to the left and rises to the right **5.** falls to the left and falls to the right
7. 5 with multiplicity 1 and -4 with multiplicity 2; crosses at 5 and touches and turns at -4
9. 3 with multiplicity 1 and -6 with multiplicity 3; crosses at both 3 and -6 **11.** 0 with multiplicity 1 and 1 with multiplicity 2; crosses at 0 and touches and turns at 1 **13.** -7, -2, and 2 all with multiplicity 1; crosses at -7, -2, and 2 **15.** odd **17.** neither **19.** even
21. even **23.** even **25.** odd **27. a.** not a polynomial function **b.** even **29. a.** could be a polynomial function **b.** odd

31. a. falls to the left and rises to the right
b. -2, -1, and 1; crosses at -2, -1, and 1
c. -2
d. neither
e.

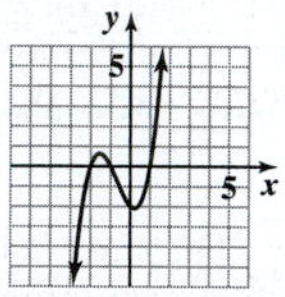

$f(x) = x^3 + 2x^2 - x - 2$

33. a. rises to the left and rises to the right
b. -3, 0, and 3; crosses at -3 and 3, and touches and turns at 0
c. 0
d. even; y-axis symmetry
e.

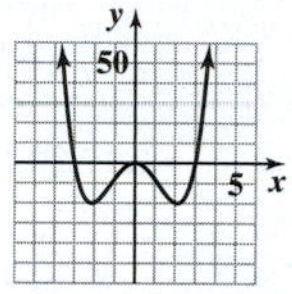

$f(x) = x^4 - 9x^2$

35. a. falls to the left and falls to the right
b. -4, 0, and 4; crosses at -4 and 4, and touches and turns at 0
c. 0
d. even; y-axis symmetry
e.

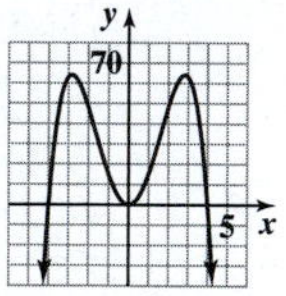

$f(x) = -x^4 + 16x^2$

37. a. rises to the left and rises to the right
b. 0 and 1; touches and turns at 0 and 1
c. 0
d. neither
e.

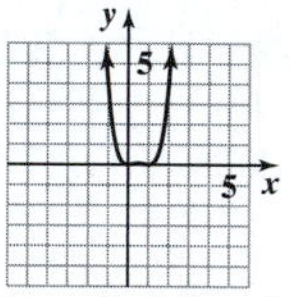

$f(x) = x^4 - 2x^3 + x^2$

39. a. falls to the left and falls to the right
b. 0 and 2; crosses at 0 and 2
c. 0
d. neither
e.

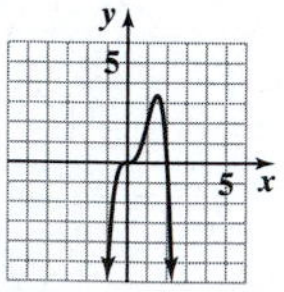

$f(x) = -2x^4 + 4x^3$

41. a. rises to the left and falls to the right
b. $-\sqrt{3}$, 0, and $\sqrt{3}$; crosses at 0, and touches and turns at $-\sqrt{3}$ and $\sqrt{3}$
c. 0
d. odd; origin symmetry
e.

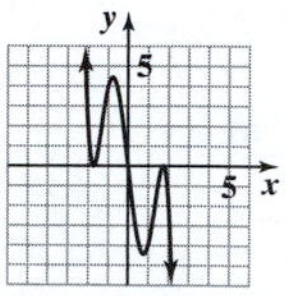

$f(x) = 6x^3 - 9x - x^5$

43. a. rises to the left and falls to the right
b. 0 and 3; crosses at 3, and touches and turns at 0
c. 0
d. neither
e.

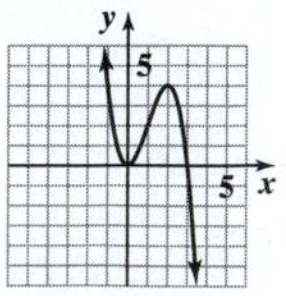

$f(x) = 3x^2 - x^3$

45. a. falls to the left and falls to the right
b. -2, 1, and 2; crosses at -2 and 2, and touches and turns at 1
c. 12
d. neither
e.

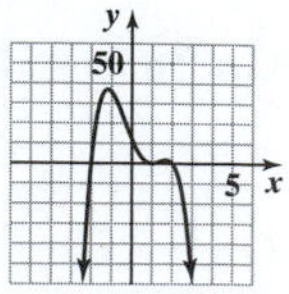

$f(x) = -3(x - 1)^2 (x^2 - 4)$

47. a. rises to the left and rises to the right **b.** -2, 0, and 1; crosses at -2 and 1, and touches and turns at 0 **c.** 0 **d.** neither **e.**

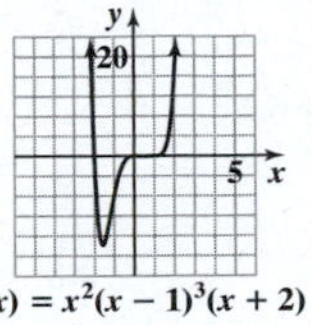

$f(x) = x^2(x-1)^3(x+2)$

49. a. falls to the left and falls to the right **b.** -3, 0, and 1; crosses at -3 and 1, and touches and turns at 0 **c.** 0 **d.** neither **e.**

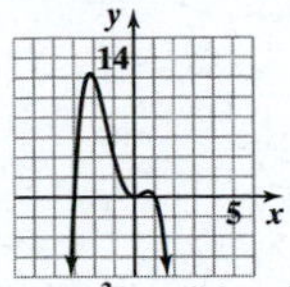

$f(x) = -x^2(x-1)(x+3)$

51. a. falls to the left and falls to the right **b.** -5, 0, and 1; crosses at -5 and 0, and touches and turns at 1 **c.** 0 **d.** neither **e.**

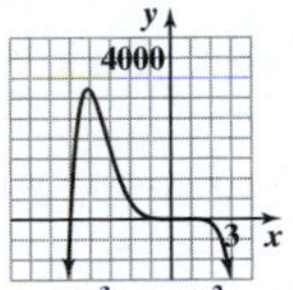

$f(x) = -2x^3(x-1)^2(x+5)$

53. a. rises to the left and rises to the right **b.** -4, 1, and 2; crosses at -4 and 1, and touches and turns at 2 **c.** -16 **d.** neither **e.**

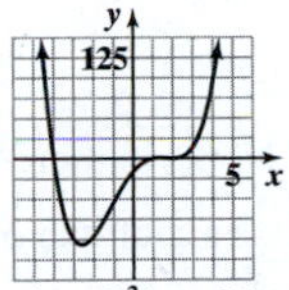

$f(x) = (x-2)^2(x+4)(x-1)$

55. a. -2, odd; 1, odd; 4, odd **b.** $f(x) = (x+2)(x-1)(x-4)$ **c.** 8 **57. a.** -1, odd; 3, even **b.** $f(x) = (x+1)(x-3)^2$ **c.** 9 **59. a.** -3, even; 2, even **b.** $f(x) = -(x+3)^2(x-2)^2$ **c.** -36 **61. a.** -2, even; -1, odd; 1, odd **b.** $f(x) = (x+2)^2(x+1)(x-1)^3$ **c.** -4 **63. a.** from 1 through 4 minutes and from 8 through 10 minutes **b.** from 4 through 8 minutes and from 10 through 12 minutes **c.** 3 **d.** 4 **e.** negative; The graph falls to the left and falls to the right. **f.** 116 ± 1 beats per minute; 10 minutes **g.** 64 ± 1 beats per minute; 8 minutes

79.

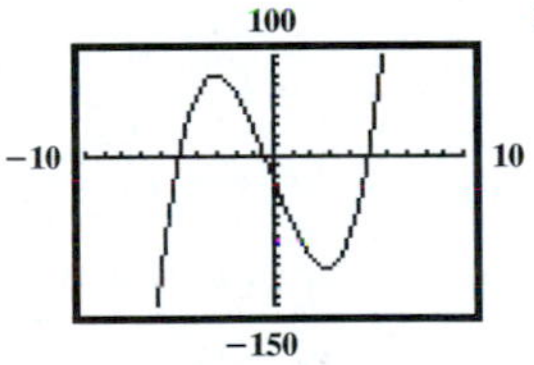

; $-5, -\frac{1}{2}$, and 5, all with multiplicity 1 **81.** does not make sense **83.** makes sense **85.** false **87.** false

89. Answers will vary: An example is $f(x) = x^3 - 2x^2$.

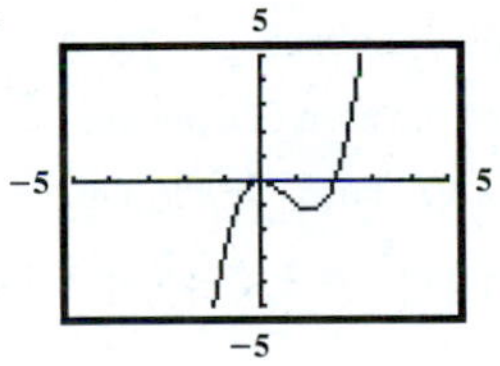

90. $\{-9\}$ **91.** $12\sqrt{5}$ **92.** $\{-1 \pm \sqrt{3}\}$ **93.** $x^2 + 4x + 3$ **94.** $\{-2 \pm \sqrt{5}\}$ **95.** $\{-2 \pm i\sqrt{2}\}$

Section 11.2

Check Point Exercises

1. $-\frac{1}{2}$, 1, and 2, or $\left\{-\frac{1}{2}, 1, 2\right\}$ **2.** $\pm1, \pm2, \pm3, \pm6$ **3.** $\pm1, \pm3, \pm\frac{1}{2}, \pm\frac{3}{2}, \pm\frac{1}{4}, \pm\frac{3}{4}$ **4.** -5, -4, and 1 **5.** 2, $\frac{-3 \pm \sqrt{5}}{2}$ **6.** 1, $2 - 3i$, and $2 + 3i$, or $\{1, 2 \pm 3i\}$ **7.** 4, 2, or 0 positive real zeros and no negative real zeros

Exercise Set 11.2

1. $\{2, 3, 4\}$ **3.** $\left\{-3, \frac{1}{2}, 1\right\}$ **5.** $\left\{-\frac{3}{2}, -\frac{1}{3}, \frac{1}{2}\right\}$ **7.** $\left\{-2, -\frac{3}{2}, 2\right\}$ **9.** $\pm1, \pm2, \pm4$ **11.** $\pm1, \pm2, \pm3, \pm6, \pm\frac{1}{3}, \pm\frac{2}{3}$ **13.** $\pm1, \pm2, \pm3, \pm6, \pm\frac{1}{2}, \pm\frac{3}{2}, \pm\frac{1}{4}, \pm\frac{3}{4}$ **15.** $\pm1, \pm2, \pm3, \pm4, \pm6, \pm12$ **17. a.** $\pm1, \pm2, \pm4$ **b.** -2, -1, or 2 **c.** -2, -1, and 2 **19. a.** $\pm1, \pm2, \pm3, \pm6, \pm\frac{1}{2}, \pm\frac{3}{2}$ **b.** $-2, \frac{1}{2}$, or 3 **c.** $-2, \frac{1}{2}$, and 3 **21. a.** $\pm1, \pm2, \pm4, \pm8, \pm\frac{1}{3}, \pm\frac{2}{3}, \pm\frac{4}{3}, \pm\frac{8}{3}$ **b.** $-4, -\frac{1}{3}$, or 2 **c.** $-4, -\frac{1}{3}$, and 2 **23. a.** $\pm1, \pm2, \pm3, \pm4, \pm6, \pm12$ **b.** -3, 1, or 4 **c.** $\{-3, 1, 4\}$ **25. a.** $\pm1, \pm2, \pm3, \pm4, \pm6, \pm12$ **b.** -2 **c.** $\{-2, 1 \pm \sqrt{7}\}$ **27. a.** $\pm1, \pm5, \pm\frac{1}{2}, \pm\frac{5}{2}, \pm\frac{1}{3}, \pm\frac{5}{3}, \pm\frac{1}{6}, \pm\frac{5}{6}$ **b.** $-5, \frac{1}{3}$, or $\frac{1}{2}$ **c.** $\left\{-5, \frac{1}{3}, \frac{1}{2}\right\}$ **29. a.** $\pm1, \pm2, \pm4$ **b.** -2 or 2 **c.** $\{-2, 2, 1 \pm \sqrt{2}\}$ **31.** no positive real zeros and 3 or 1 negative real zeros **33.** 3 or 1 positive real zeros and no negative real zeros **35.** 2 or 0 positive real zeros and 2 or 0 negative real zeros **37.** -2, 1, and 5 **39.** $\left\{-\frac{1}{2}, \frac{1 \pm \sqrt{17}}{2}\right\}$ **41.** $\{-2, -1, 3 \pm \sqrt{13}\}$

43. $-1, -\frac{1}{3}, 2$, and 3 **45.** $\left\{-\frac{3}{4}, 1, \pm i\sqrt{2}\right\}$ **47.** $\left\{-2, \frac{1}{2}, \pm\sqrt{2}\right\}$

49. a. $-4, 1$, and 4 **b.**

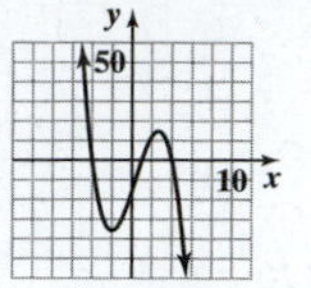

$f(x) = -x^3 + x^2 + 16x - 16$

51. a. -1 and $\frac{3}{2}$ **b.**

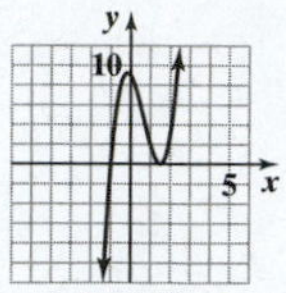

$f(x) = 4x^3 - 8x^2 - 3x + 9$

53. a. $\frac{1}{2}, 3, -1 \pm i$ **b.**

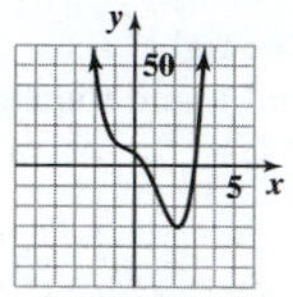

$f(x) = 2x^4 - 3x^3 - 7x^2 - 8x + 6$

55. a. $-2, -1, -\frac{2}{3}, 1$, and 2 **b.**

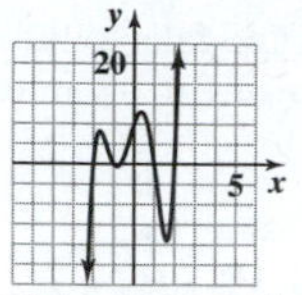

$f(x) = 3x^5 + 2x^4 - 15x^3 - 10x^2 + 12x + 8$

57. a. 40s; Artists completed 27% of their professional works in their 40s. **b.** degree: even; leading coefficient: negative

59. ≈ 3 yr **63.** 6 hr **71.** The leading coefficient is a multiple of 4, and the constant is a multiple of 3.

73. $\pm 1, \pm 3, \pm 5, \pm 15, \pm\frac{1}{2}, \pm\frac{3}{2}, \pm\frac{5}{2}, \pm\frac{15}{2}$;

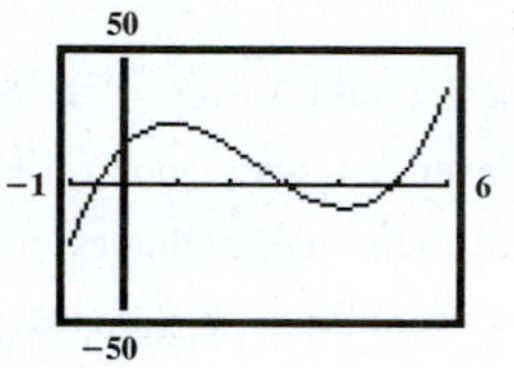

$; -\frac{1}{2}, 3, 5$

75. $\pm 1, \pm 2, \pm 3, \pm 6, \pm 9, \pm 18, \pm\frac{1}{2}, \pm\frac{3}{2}, \pm\frac{9}{2}$;

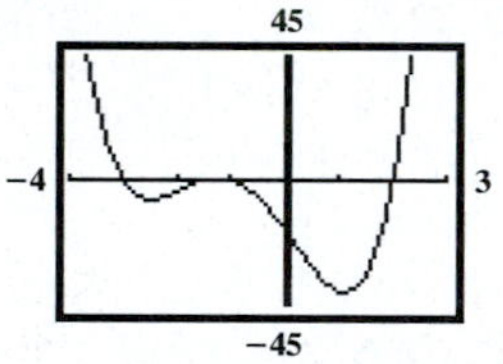

$; -3, -\frac{3}{2}, -1, 2$

77. no positive real zeros and no negative real zeros; The graph has no x-intercepts. **79.** Answers will vary.; A polynomial function of odd degree with integer coefficients must have at least one real zero.; A polynomial function of even degree with integer coefficients may have no real zeros.

81. does not make sense **83.** does not make sense **85.** false **87.** false **89.** 3 in. **91.** $\{x \mid x < -4 \text{ or } x > 2\}$ or $(-\infty, -4) \cup (2, \infty)$

92.

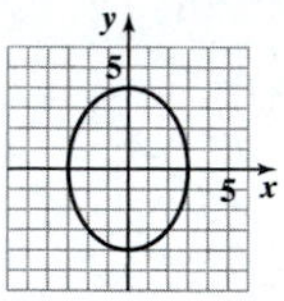

$16x^2 + 9y^2 = 144$

93. $(x - 5y)(x^2 + 5xy + 25y^2)$ **94.** $x = 1$ and $x = 2$ **95.** $x = 1$ **96.** $y = 0$

Mid-Chapter Check Point Exercises

1. -1 and 2

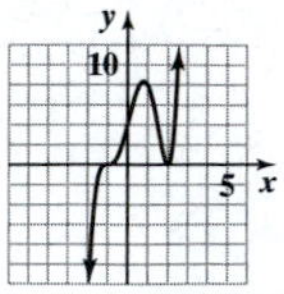

$f(x) = (x - 2)^2(x + 1)^3$

2. -1 and 2

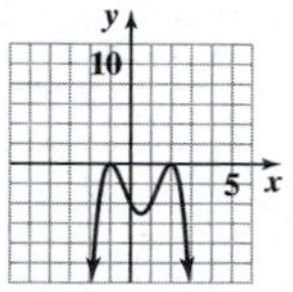

$f(x) = -(x - 2)^2(x + 1)^2$

3. $-2, 1$, and 2

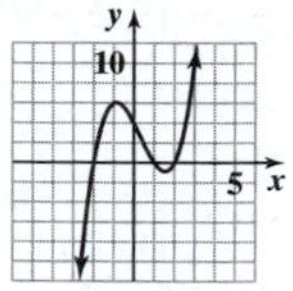

$f(x) = x^3 - x^2 - 4x + 4$

4. $-2, -1, 1$, and 2

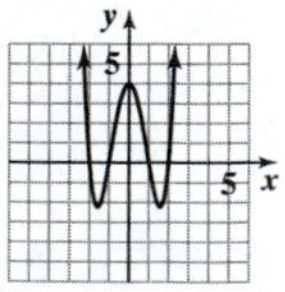

$f(x) = x^4 - 5x^2 + 4$

5. -1

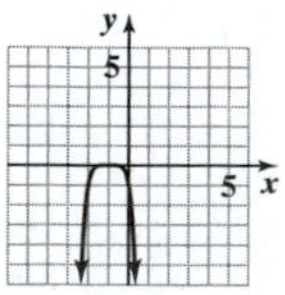

$f(x) = -(x + 1)^6$

6. $-\frac{1}{3}, \frac{1}{2}$, and 1

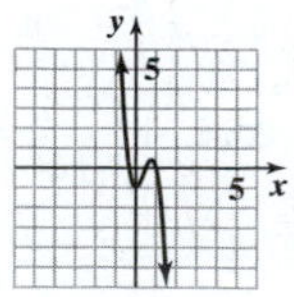

$f(x) = -6x^3 + 7x^2 - 1$

7. $-1, 0$, and 1

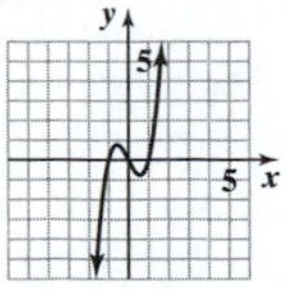

$f(x) = 2x^3 - 2x$

8. $0, 1 \pm 5i$

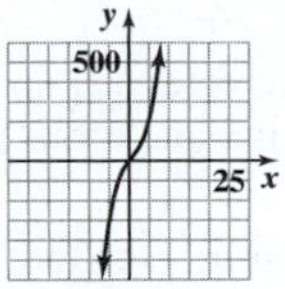

$f(x) = x^3 - 2x^2 + 26x$

9. $3, 1 \pm \sqrt{2}$

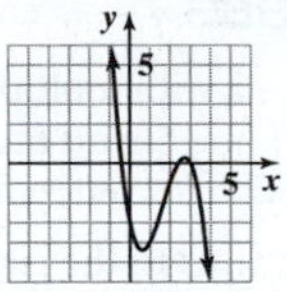

$f(x) = -x^3 + 5x^2 - 5x - 3$

10. $\{-2, 1\}$ **11.** $\{\frac{1}{3}, \frac{1}{2}, 1\}$ **12.** $\{-\frac{1}{2}, \frac{2}{3}, \frac{7}{2}\}$ **13.** $\{-10, -\frac{5}{2}, 10\}$ **14.** $\{-3, 4, \pm i\}$ **15.** $\{-3, \frac{1}{2}, 1 \pm \sqrt{3}\}$

Section 11.3

Check Point Exercises

1. a. $x = -1$ and $x = 1$ **b.** $x = -1$ **c.** no vertical asymptotes **2. a.** $y = 3$ **b.** $y = 0$ **c.** no horizontal asymptote

3.

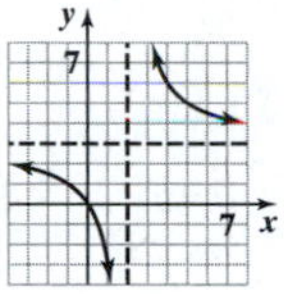

$f(x) = \frac{3x}{x - 2}$

4.

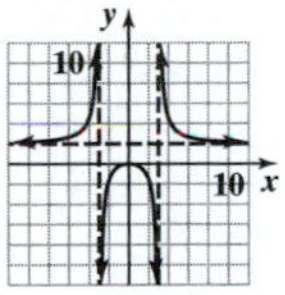

$f(x) = \frac{2x^2}{x^2 - 9}$

5.

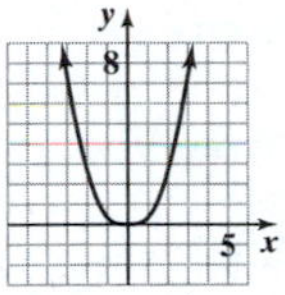

$f(x) = \frac{x^4}{x^2 + 2}$

6. $y = 5000$; As the number of robotic systems manufactured per month increases, the average cost per system approaches \$5000.

Exercise Set 11.3

1. $x = -4$ **3.** $x = -4$ and $x = 0$ **5.** $x = -4$ **7.** no vertical asymptotes **9.** $y = 0$ **11.** $y = 4$ **13.** no horizontal asymptote **15.** $y = -\frac{2}{3}$

17.

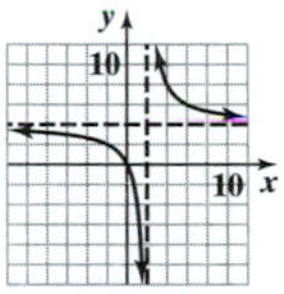

$f(x) = \frac{4x}{x - 2}$

19.

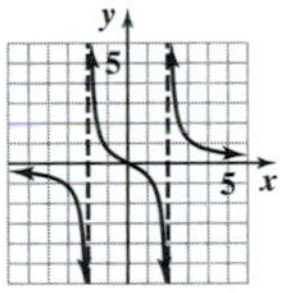

$f(x) = \frac{2x}{x^2 - 4}$

21.

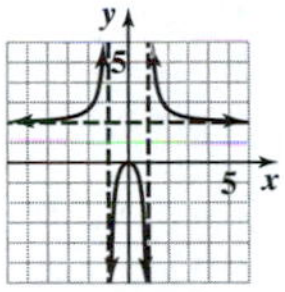

$f(x) = \frac{2x^2}{x^2 - 1}$

23.

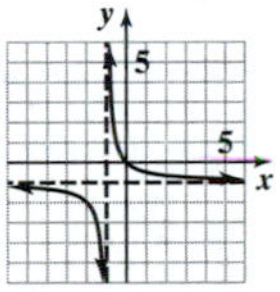

$f(x) = \frac{-x}{x + 1}$

25.

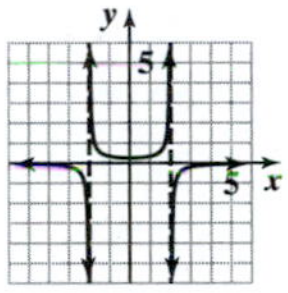

$f(x) = -\frac{1}{x^2 - 4}$

27.

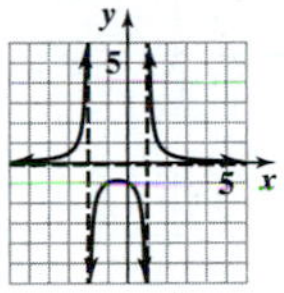

$f(x) = \frac{2}{x^2 + x - 2}$

29.

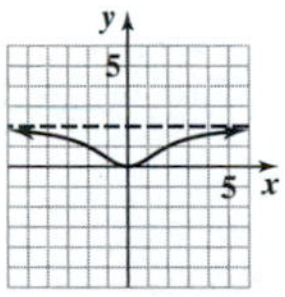

$f(x) = \frac{2x^2}{x^2 + 4}$

31.

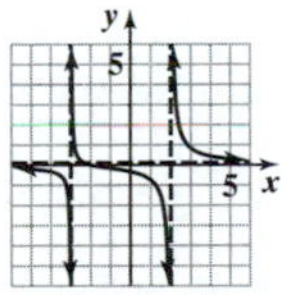

$f(x) = \frac{x + 2}{x^2 + x - 6}$

33.

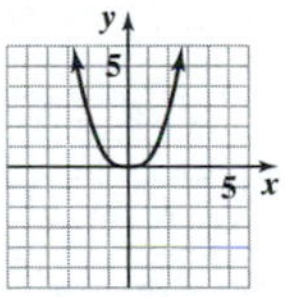

$f(x) = \frac{x^4}{x^2 + 2}$

35.

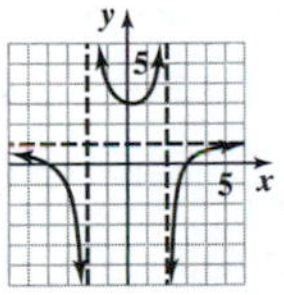

$f(x) = \frac{x^2 + x - 12}{x^2 - 4}$

37.

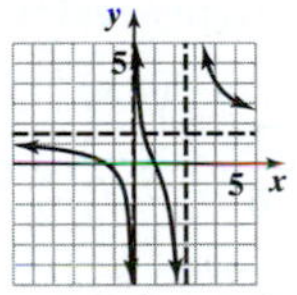

$f(x) = \frac{3x^2 + x - 4}{2x^2 - 5x}$

39.

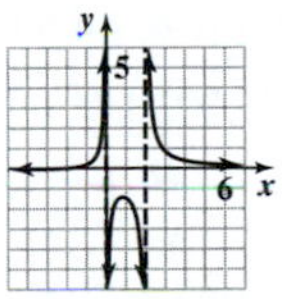

$f(x) = \frac{x + 2}{2x(x - 2)}$

41.

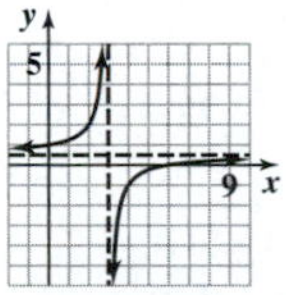

$f(x) = \frac{x - 6}{2(x - 3)}$

43.

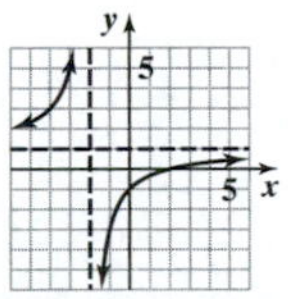

$f(x) = \frac{x - 2}{x + 2}$

45. a. about 409 bass **b.** $y = 3500$; Over time, the number of bass in the lake will approach 3500.

47. a. $f(x) = \frac{11x^2 + 40x + 1040}{12x^2 + 230x + 2190}$ **b.** 63% **c.** 64%; overestimates by 1% **49. a.** approximately 1.5 mg per l **b.** 1.5 mg per l **c.** $y = 0$; Over time, the drug's concentration will approach 0 mg per l.

59.

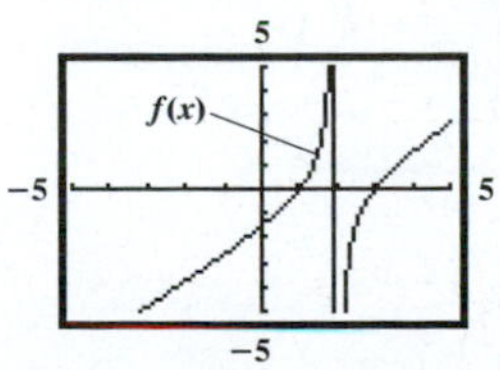

;

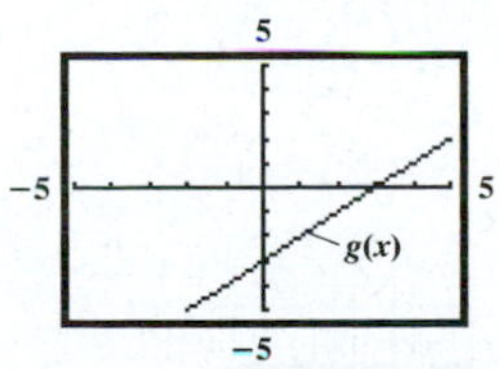

The graph of f has a vertical asymptote, but the graph of g does not.

The equation for g has a common factor in the numerator and the denominator.

61. does not make sense **63.** makes sense **65.** false **67.** true **70.** $\{3\}$ **71.** $\left\{-\frac{5}{2}, 5\right\}$ **72.** $2xy\sqrt{5y}$ **73.** $-\frac{1}{2}; \frac{1}{8}; -\frac{1}{27}; \frac{1}{80}$

74. 120 **75.** 2; 5; 10; 17; 26; 37; Sum is 97.

Review Exercises

1. 1 with multiplicity 1, -2 with multiplicity 2, and -5 with multiplicity 3; crosses at 1 and -5, and touches and turns at -2
2. -5 with multiplicity 1 and 5 with multiplicity 2; crosses at -5, and touches and turns at 5 **3.** odd; origin symmetry
4. even; y-axis symmetry **5.** odd; origin symmetry

6. a. falls to the left and rises to the right
b. neither
c.

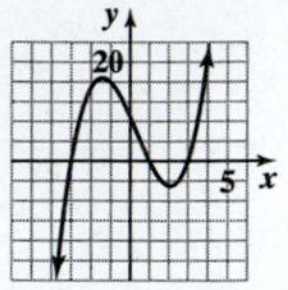

$f(x) = x^3 - x^2 - 9x + 9$

7. a. rises to the left and falls to the right
b. origin symmetry
c.

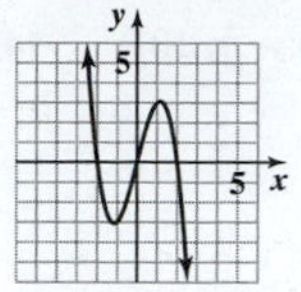

$f(x) = 4x - x^3$

8. a. falls to the left and rises to the right
b. neither
c.

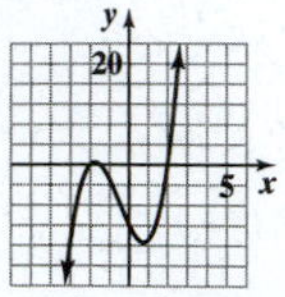

$f(x) = 2x^3 + 3x^2 - 8x - 12$

9. a. falls to the left and falls to the right
b. y-axis symmetry
c.

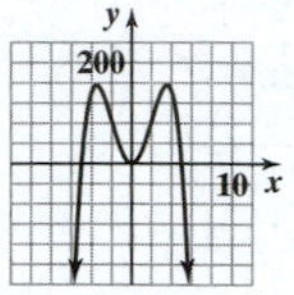

$f(x) = -x^4 + 25x^2$

10. a. falls to the left and falls to the right
b. neither
c.

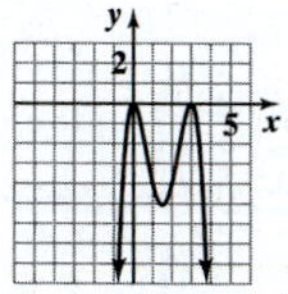

$f(x) = -x^4 + 6x^3 - 9x^2$

11. a. rises to the left and rises to the right
b. neither
c.

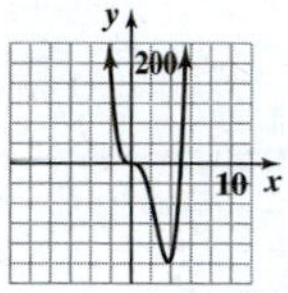

$f(x) = 3x^4 - 15x^3$

12. $\{4, -2 \pm \sqrt{5}\}$ **13.** $\left\{-3, -\frac{4}{3}, 1\right\}$ **14.** $\pm1, \pm5$ **15.** $\pm1, \pm2, \pm4, \pm8, \pm\frac{1}{3}, \pm\frac{2}{3}, \pm\frac{4}{3}, \pm\frac{8}{3}$
16. 2 or 0 positive real zeros and no negative real zeros **17.** 3 or 1 positive real zeros and 2 or 0 negative real zeros
18. There are no sign changes for $f(x)$ or $f(-x)$. **19. a.** $\pm1, \pm2, \pm4$ **b.** 1 positive real zero and 2 or 0 negative real zeros **c.** -2 or 1
d. -2 and 1 **20. a.** $\pm1, \pm\frac{1}{2}, \pm\frac{1}{3}, \pm\frac{1}{6}$ **b.** 2 or 0 positive real zeros and 1 negative real zero **c.** $-1, \frac{1}{3}$, or $\frac{1}{2}$ **d.** $-1, \frac{1}{3}$, and $\frac{1}{2}$
21. a. $\pm1, \pm3, \pm5, \pm15, \pm\frac{1}{2}, \pm\frac{3}{2}, \pm\frac{5}{2}, \pm\frac{15}{2}, \pm\frac{1}{4}, \pm\frac{3}{4}, \pm\frac{5}{4}, \pm\frac{15}{4}, \pm\frac{1}{8}, \pm\frac{3}{8}, \pm\frac{5}{8}, \pm\frac{15}{8}$ **b.** 3 or 1 positive real roots and no negative real roots
c. $\frac{1}{2}, \frac{3}{2}$, or $\frac{5}{2}$ **d.** $\left\{\frac{1}{2}, \frac{3}{2}, \frac{5}{2}\right\}$ **22. a.** $\pm1, \pm2, \pm3, \pm6$ **b.** 2 or 0 positive real roots and 2 or 0 negative real roots **c.** $-2, -1, 1$, or 3
d. $\{-2, -1, 1, 3\}$ **23. a.** $\pm1, \pm2, \pm\frac{1}{2}, \pm\frac{1}{4}$ **b.** 1 positive real root and 1 negative real root **c.** $-\frac{1}{2}$ or $\frac{1}{2}$ **d.** $\left\{-\frac{1}{2}, \frac{1}{2}, \pm i\sqrt{2}\right\}$
24. a. $\pm1, \pm2, \pm4, \pm\frac{1}{2}$ **b.** 2 or 0 positive real roots and 2 or 0 negative real roots **c.** $-2, -1, \frac{1}{2}$, or 2 **d.** $-2, -1, \frac{1}{2}$, and 2

25. $x = -3$ and $x = 3$; $y = 0$

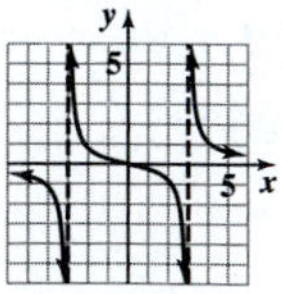

$f(x) = \frac{2x}{x^2 - 9}$

26. $x = -3$; $y = 2$

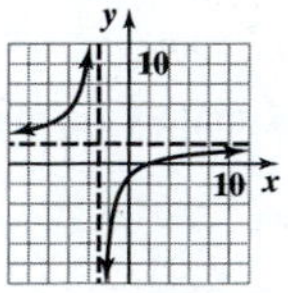

$f(x) = \frac{2x - 4}{x + 3}$

27. $x = -1$ and $x = 1$; $y = 4$

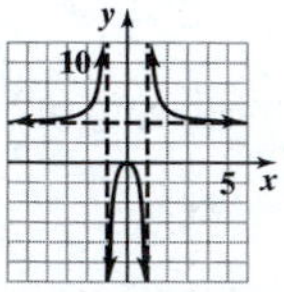

$f(x) = \frac{4x^2}{x^2 - 1}$

28. no vertical asymptotes; $y = 1$

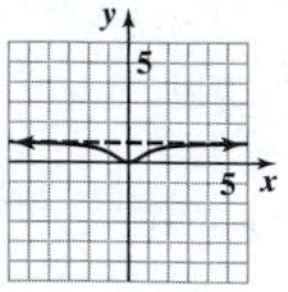

$f(x) = \frac{x^2}{x^2 + 1}$

29. no vertical asymptotes; no horizontal asymptote

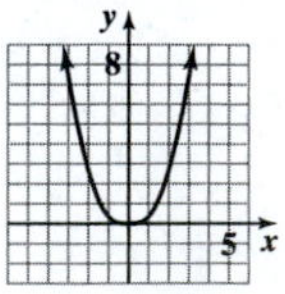

$f(x) = \frac{x^4}{x^2 + 2}$

30. $x = -2$ and $x = 3$; $y = 1$

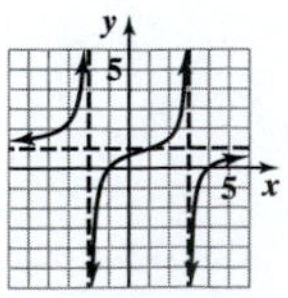

$f(x) = \frac{x^2 - 3x - 4}{x^2 - x - 6}$

31. $x = -2$; $y = 1$

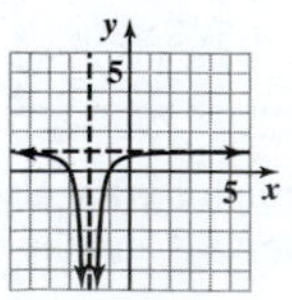

$f(x) = \frac{x^2 + 4x + 3}{(x + 2)^2}$

32. $y = 3000$; Over time, the number of bass will approach 3000.
33. $y = 0$; As the number of years of education increases, the percentage of people unemployed approaches 0.

Chapter Test

1. a. −2, 2, and 5

b.

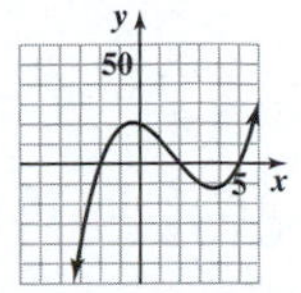

$f(x) = x^3 - 5x^2 - 4x + 20$

2. even; The graph of f should have y-axis symmetry, but the graph in the figure has origin symmetry.

3. a. 2 **b.** $\frac{1}{2}$ and $\frac{2}{3}$

4. $\pm1, \pm2, \pm3, \pm6, \pm\frac{1}{2}, \pm\frac{3}{2}$ **5.** 3 or 1 positive real zeros and no negative real zeros

6. $\{-5, -3, 2\}$ **7. a.** $\pm1, \pm3, \pm5, \pm15, \pm\frac{1}{2}, \pm\frac{3}{2}, \pm\frac{5}{2}, \pm\frac{15}{2}$ **b.** $-\sqrt{5}, -1, \frac{3}{2}$, and $\sqrt{5}$

8. $x = 1$; $y = 4$

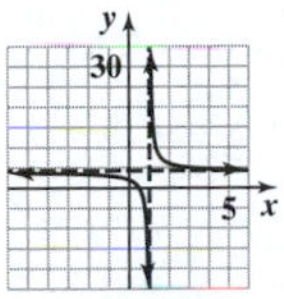

$f(x) = \frac{4x-2}{x-1}$

9. $x = -2$ and $x = 2$; $y = 1$

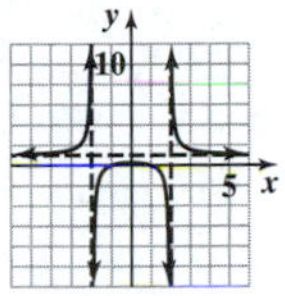

$f(x) = \frac{x^2-1}{x^2-4}$

10. no vertical asymptotes; $y = 4$

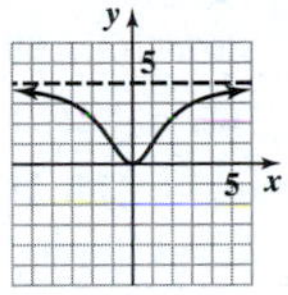

$f(x) = \frac{4x^2}{x^2+3}$

Cumulative Review Exercises

1. $\{-2, 3 \pm \sqrt{7}\}$ **2.** $\{(-2, 7), (1, 1)\}$ **3.** $\{9\}$ **4.** $\{-216, 27\}$ **5.** $\left\{\frac{2}{3}\right\}$ **6.** $\{(2, -3, 1)\}$

7. $\{x \mid x < -6 \text{ or } x > 4\}$ or $(-\infty, -6) \cup (4, \infty)$

8.

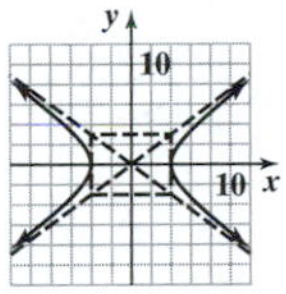

$\frac{x^2}{16} - \frac{y^2}{9} = 1$

9.

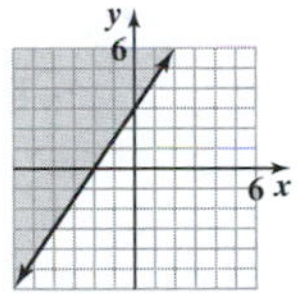

$3x - 2y \le -6$

10.

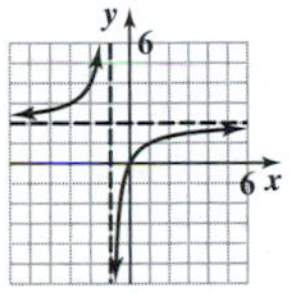

$f(x) = \frac{2x}{x+1}$

11. $25x^2 - 70x + 49$ **12.** $8x^3 - 27$ **13.** $\frac{9-2x^2}{3x(x-1)}$ **14.** $-2\sqrt{6}$ **15.** $12 + 9i$ **16.** $3(x+2)(x+1)(x-1)$ **17.** $(5x - 2y)^2$ **18.** $y = -4x + 5$ or $f(x) = -4x + 5$ **19.** 43 **20.** $x^2 + 2x + 4$ **21.** $\log(\sqrt{x}\,y^3)$ **22.** $3\sqrt[3]{9x}$ **23.** 2 hr **24.** bath towel: \$7; hand towel: \$5 **25.** square: 12 cm by 12 cm; rectangle: 18 cm by 8 cm

CHAPTER 12

Section 12.1

Check Point Exercises

1. a. 7, 9, 11, 13 **b.** $-\frac{1}{3}, \frac{1}{5}, -\frac{1}{9}, \frac{1}{17}$ **2.** $10, \frac{10}{3}, \frac{5}{6}, \frac{1}{6}$ **3. a.** $2(1)^2 + 2(2)^2 + 2(3)^2 + 2(4)^2 + 2(5)^2 + 2(6)^2 = 182$ **b.** $(2^3 - 3) + (2^4 - 3) + (2^5 - 3) = 47$ **c.** $4 + 4 + 4 + 4 + 4 = 20$ **4. a.** $\sum_{i=1}^{9} i^2$ **b.** $\sum_{i=1}^{n} \frac{1}{2^{i-1}}$

Exercise Set 12.1

1. 5, 8, 11, 14 **3.** 3, 9, 27, 81 **5.** −3, 9, −27, 81 **7.** −4, 5, −6, 7 **9.** $\frac{2}{5}, \frac{2}{3}, \frac{6}{7}, 1$ **11.** $1, -\frac{1}{3}, \frac{1}{7}, -\frac{1}{15}$ **13.** $1, 2, \frac{3}{2}, \frac{2}{3}$ **15.** 4, 12, 48, 240 **17.** 105 **19.** 60 **21.** 115 **23.** $-\frac{5}{16}$ **25.** 55 **27.** $\frac{3}{8}$ **29.** 15 **31.** $\sum_{i=1}^{15} i^2$ **33.** $\sum_{i=1}^{11} 2^i$ **35.** $\sum_{i=1}^{30} i$ **37.** $\sum_{i=1}^{14} \frac{i}{i+1}$ **39.** $\sum_{i=1}^{n} \frac{4^i}{i}$ **41.** $\sum_{i=1}^{n} (2i - 1)$ **43.** Answers will vary; examples are: $\sum_{k=1}^{14} (2k + 3)$ or $\sum_{k=2}^{15} (2k + 1)$. **45.** Answers will vary; an example is: $\sum_{k=0}^{12} ar^k$. **47.** Answers will vary; an example is: $\sum_{k=0}^{n} (a + kd)$. **49.** 45 **51.** 0 **53.** 2 **55.** 80 **57. a.** 1354; the total number of books published about the September 11 attacks from 2001 through 2006 **b.** 226; the average number of books published per year about the September 11 attacks from 2001 through 2006 **59. a.** 9.9: Online ad spending averaged \$9.9 billion per year from 2000 through 2006. **b.** 12; overestimates by \$2.1 billion **61.** \$8081.13 **71.** As n gets larger, the terms get closer to 1. **73.** As n gets larger, the terms get closer to 0. **75.** does not make sense **77.** makes sense **79.** false **81.** true **83.** $a_n = \frac{1}{n}$ **85.** $a_n = (-1)^n$ **87.** $a_n = \frac{n+2}{n+1}$ **89.** $a_n = \frac{(n+1)^2}{n}$ **91.** 600 **93.** $n^3 - 3n^2 + 2n$ **95.** $4\log x + 6\log x + 8\log x = \log x^{18}$ **97.** $2xy^2\sqrt[3]{5xy}$ **98.** $(3x - 2)(9x^2 + 6x + 4)$ **99.** $\left\{-\frac{6}{5}, 4\right\}$ **100.** −5; −5; −5; −5; The difference between consecutive terms is always −5. **101.** 4; 4; 4; 4; The difference between consecutive terms is always 4. **102.** −45

Section 12.2

Check Point Exercises

1. 100, 70, 40, 10, −20, −50 **2.** −34 **3. a.** $a_n = 0.7n + 31.3$ **b.** 39 **4.** 360 **5.** 2460 **6.** \$740,300

Exercise Set 12.2

1. 4 **3.** 5 **5.** −3 **7.** 200, 220, 240, 260, 280, 300 **9.** −7, −3, 1, 5, 9, 13 **11.** 300, 210, 120, 30, −60, −150 **13.** $\frac{5}{2}, 2, \frac{3}{2}, 1, \frac{1}{2}, 0$ **15.** −0.4, −2, −3.6, −5.2, −6.8, −8.4 **17.** 33 **19.** 252 **21.** 955 **23.** −142 **25.** $a_n = 4n - 3$; $a_{20} = 77$ **27.** $a_n = 11 - 4n$; $a_{20} = -69$ **29.** $a_n = -4n - 16$; $a_{20} = -96$ **31.** $a_n = \frac{1}{3}n - \frac{2}{3}$; $a_{20} = 6$ **33.** $a_n = 4.3 - 0.3n$; $a_{20} = -1.7$ **35.** 1220 **37.** 4400 **39.** 5050 **41.** 3660 **43.** 396 **45.** $8 + 13 + 18 + \cdots + 88 = 816$ **47.** $2 + (-1) + (-4) + \cdots + (-85) = -1245$ **49.** $4 + 8 + 12 + \cdots + 400 = 20{,}200$ **51.** 7 **53.** 22 **55.** 847 **57.** $f(x) = -4x + 5$ **59.** $a_n = 3n - 2$ **61. a.** $a_n = 0.77n + 9.23$ **b.** 30.0% **63.** company A; \$1400 **65. a.** \$21,153 **b.** \$21,158; overestimates by \$5 **69.** Company A: \$307,000; Company B: \$324,000; Company B pays the greater total amount. **71.** 2869 seats **79.** makes sense **81.** makes sense **83.** false **85.** true **87.** 200th term **89.** $S_n = \frac{n}{2}[1 + (2n - 1)] = \frac{n}{2}(2n) = n^2$ **90.** {1005} **91.** $\{x \mid -5 \le x \le 2\}$ or $[-5, 2]$ **92.** $P = \frac{At}{t - A}$ **93.** −2; −2; −2; −2; The ratio of a term to the term that directly precedes it is always −2. **94.** 5; 5; 5; 5; The ratio of a term to the term that directly precedes it is always 5. **95.** 8019

Section 12.3

Check Point Exercises

1. $12, 6, 3, \frac{3}{2}, \frac{3}{4}, \frac{3}{8}$ **2.** 3645 **3.** $a_n = 3(2)^{n-1}$; $a_8 = 384$ **4.** 9842 **5.** 19,680 **6.** approximately \$2,371,746 **7. a.** \$333,946 **b.** \$291,946 **8.** 9 **9.** $\frac{1}{1}$ or 1 **10.** \$4000

Exercise Set 12.3

1. $r = 3$ **3.** $r = -2$ **5.** $r = \frac{3}{2}$ **7.** $r = -0.1$ **9.** 2, 6, 18, 54, 162 **11.** $20, 10, 5, \frac{5}{2}, \frac{5}{4}$ **13.** −4, 40, −400, 4000, −40,000 **15.** $-\frac{1}{4}, \frac{1}{2}, -1, 2, -4$ **17.** $a_8 = 768$ **19.** $a_{12} = -10{,}240$ **21.** $a_6 = -200$ **23.** $a_8 = 0.1$ **25.** $a_n = 3(4)^{n-1}$; $a_7 = 12{,}288$ **27.** $a_n = 18\left(\frac{1}{3}\right)^{n-1}$; $a_7 = \frac{2}{81}$ **29.** $a_n = 1.5(-2)^{n-1}$; $a_7 = 96$ **31.** $a_n = 0.0004(-10)^{n-1}$; $a_7 = 400$ **33.** 531,440 **35.** 2049 **37.** $\frac{16{,}383}{2}$ or 8191.5 **39.** 9840 **41.** 10,230 **43.** $\frac{63}{128}$ **45.** $\frac{3}{2}$ **47.** 4 **49.** $\frac{2}{3}$ **51.** 20 **53.** $\frac{5}{9}$ **55.** $\frac{47}{99}$ **57.** $\frac{257}{999}$ **59.** arithmetic; $d = 1$ **61.** geometric; $r = 2$ **63.** neither **65.** 2435 **67.** 2280 **69.** −140 **71.** $a_2 = 12$, $a_3 = 18$ **73.** \$30,000 **75.** \$16,384 **77.** approximately \$3,795,957 **79. a.** approximately 1.01 for each division **b.** $a_n = 35.48(1.01)^{n-1}$ **c.** approximately 38.04 million **81.** \$32,767 **83.** approximately \$793,583 **85.** approximately 130.26 in. **87. a.** \$11,617 **b.** \$1617 **89. a.** \$87,052 **b.** \$63,052 **91. a.** \$693,031 **b.** \$293,031 **93.** \$9 million **95.** $\frac{1}{3}$

107.

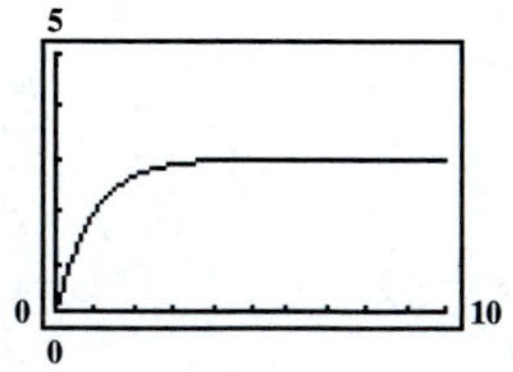

horizontal asymptote: $y = 3$; sum of series: 3

109. makes sense **111.** makes sense **113.** false **115.** false **117.** 2000 flies **119.** $2\sqrt{7}$ **120.** $\left\{\frac{-1 \pm \sqrt{33}}{4}\right\}$ **121.** $-3(\sqrt{3} + \sqrt{5})$ **122.** The exponents on a begin with the exponent on $a + b$ and decrease by 1 in each successive term. **123.** The exponents on b begin with 0, increase by 1 in each successive term, and end with the exponent on $a + b$. **124.** The sum of the exponents is the exponent on $a + b$.

Mid-Chapter Check Point Exercises

1. $1, -2, \frac{3}{2}, -\frac{2}{3}, \frac{5}{24}$ **2.** 5, 2, −1, −4, −7 **3.** 5, −15, 45, −135, 405 **4.** $a_n = 4n - 2$; $a_{20} = 78$ **5.** $a_n = 3(2)^{n-1}$; $a_{10} = 1536$ **6.** $a_n = -\frac{1}{2}n + 2$; $a_{30} = -13$ **7.** 5115 **8.** 2350 **9.** 6820 **10.** −29,300 **11.** 44 **12.** 3725 **13.** $\frac{1995}{64}$ **14.** $\frac{5}{7}$ **15.** $\frac{5}{11}$ **16.** Answers will vary; an example is $\sum_{i=1}^{18} \frac{i}{i+2}$. **17.** 464 ft; 3600 ft **18.** \$1,071,794

Section 12.4

Check Point Exercises

1. a. 20 **b.** 1 **c.** 28 **d.** 1 **2.** $x^4 + 4x^3 + 6x^2 + 4x + 1$ **3.** $x^5 - 10x^4y + 40x^3y^2 - 80x^2y^3 + 80xy^4 - 32y^5$ **4.** $4032x^5y^4$

Exercise Set 12.4

1. 56 **3.** 12 **5.** 1 **7.** 4950 **9.** $x^3 + 6x^2 + 12x + 8$ **11.** $27x^3 + 27x^2y + 9xy^2 + y^3$ **13.** $125x^3 - 75x^2 + 15x - 1$
15. $16x^4 + 32x^3 + 24x^2 + 8x + 1$ **17.** $x^8 + 8x^6y + 24x^4y^2 + 32x^2y^3 + 16y^4$ **19.** $y^4 - 12y^3 + 54y^2 - 108y + 81$
21. $16x^{12} - 32x^9 + 24x^6 - 8x^3 + 1$ **23.** $c^5 + 10c^4 + 40c^3 + 80c^2 + 80c + 32$ **25.** $x^5 - 5x^4 + 10x^3 - 10x^2 + 5x - 1$
27. $243x^5 - 405x^4y + 270x^3y^2 - 90x^2y^3 + 15xy^4 - y^5$ **29.** $64a^6 + 192a^5b + 240a^4b^2 + 160a^3b^3 + 60a^2b^4 + 12ab^5 + b^6$
31. $x^8 + 16x^7 + 112x^6$ **33.** $x^{10} - 20x^9y + 180x^8y^2$ **35.** $x^{32} + 16x^{30} + 120x^{28}$ **37.** $y^{60} - 20y^{57} + 190y^{54}$ **39.** $240x^4y^2$ **41.** $126x^5$
43. $56x^6y^{15}$ **45.** $-\frac{21}{2}x^6$ **47.** $319{,}770x^{16}y^{14}$ **49.** $x^{12} + 4x^7 + 6x^2 + \frac{4}{x^3} + \frac{1}{x^8}$ **51.** $x - 3x^{1/3} + \frac{3}{x^{1/3}} - \frac{1}{x}$ **53.** 8
55. $4x^3 + 6x^2h + 4xh^2 + h^3$ **57.** 252 **59.** 0.1138

69.

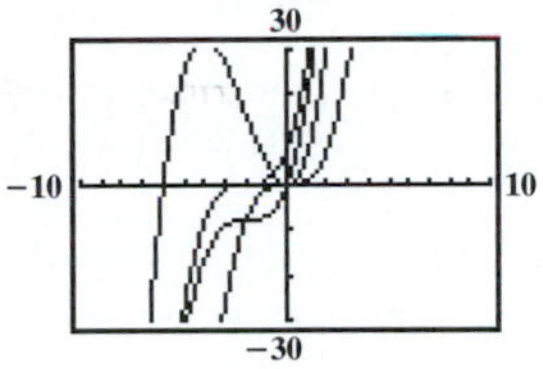

f_2, f_3, and f_4 are approaching $f_1 = f_5$.

71. $x^3 - 3x^2 + 3x - 1$

73. $x^6 + 12x^5 + 60x^4 + 160x^3 + 240x^2 + 192x + 64$

75. makes sense **77.** does not make sense **79.** true **81.** false **83.** $10x^4y^6$ **84.** $a^2 + 4a + 6$
85. $f(g(x)) = 4x^2 - 2x - 6;\ g(f(x)) = 2x^2 + 10x - 3$ **86.** $\frac{2x^2 - 9x - 1}{2(x - 4)(x + 3)}$ **87.** $6 = 6$ **88.** $15 = 15$ **89.** $\frac{(k + 1)(k + 2)(2k + 3)}{6}$

Section 12.5

Check Point Exercises

1. a. $S_1: 2 = 1(1 + 1);\ S_k: 2 + 4 + 6 + \cdots + 2k = k(k + 1);\ S_{k+1}: 2 + 4 + 6 + \cdots + 2(k + 1) = (k + 1)(k + 2)$

b. $S_1: 1^3 = \frac{1^2(1 + 1)^2}{4};\ S_k: 1^3 + 2^3 + 3^3 + \cdots + k^3 = \frac{k^2(k + 1)^2}{4};\ S_{k+1}: 1^3 + 2^3 + 3^3 + \cdots + (k + 1)^3 = \frac{(k + 1)^2(k + 2)^2}{4}$

2. $S_1: 2 = 1(1 + 1);\ S_k: 2 + 4 + 6 + \cdots + 2k = k(k + 1);\ S_{k+1}: 2 + 4 + 6 + \cdots + 2k + 2(k + 1) = (k + 1)(k + 2)$;
S_{k+1} can be obtained by adding $2(k + 1)$ to both sides of S_k.

3. $S_1: 1^3 = \frac{1^2(1 + 1)^2}{4};\ S_k: 1^3 + 2^3 + 3^3 + \cdots + k^3 = \frac{k^2(k + 1)^2}{4};\ S_{k+1}: 1^3 + 2^3 + 3^3 + \cdots + k^3 + (k + 1)^3 = \frac{(k + 1)^2(k + 2)^2}{4}$;
S_{k+1} can be obtained by adding $(k + 1)^3$ to both sides of S_k.

Exercise Set 12.5

1. $S_1: 1 = 1^2;\ S_2: 1 + 3 = 2^2;\ S_3: 1 + 3 + 5 = 3^2$ **3.** S_1: 2 is a factor of $1^2 - 1$; S_2: 2 is a factor of $2^2 - 2$; S_3: 2 is a factor of $3^2 - 3$.

5. $S_k: 4 + 8 + 12 + \cdots + 4k = 2k(k + 1);\ S_{k+1}: 4 + 8 + 12 + \cdots + (4k + 4) = 2(k + 1)(k + 2)$

7. $S_k: 3 + 7 + 11 + \cdots + (4k - 1) = k(2k + 1);\ S_{k+1}: 3 + 7 + 11 + \cdots + (4k + 3) = (k + 1)(2k + 3)$

9. S_k: 2 is a factor of $k^2 - k + 2$; S_{k+1}: 2 is a factor of $k^2 + k + 2$.

11. $S_1: 4 = 2(1)(1 + 1);\ S_k: 4 + 8 + 12 + \cdots + 4k = 2k(k + 1);\ S_{k+1}: 4 + 8 + 12 + \cdots + 4(k + 1) = 2(k + 1)(k + 2)$;
S_{k+1} can be obtained by adding $4(k + 1)$ to both sides of S_k.

13. $S_1: 1 = 1^2;\ S_k: 1 + 3 + 5 + \cdots + (2k - 1) = k^2;\ S_{k+1}: 1 + 3 + 5 + \cdots + (2k + 1) = (k + 1)^2$;
S_{k+1} can be obtained by adding $2k + 1$ to both sides of S_k.

15. $S_1: 3 = 1[2(1) + 1];\ S_k: 3 + 7 + 11 + \cdots + (4k - 1) = k(2k + 1);\ S_{k+1}: 3 + 7 + 11 + \cdots + (4k + 3) = (k + 1)(2k + 3)$;
S_{k+1} can be obtained by adding $4k + 3$ to both sides of S_k.

17. $S_1: 1 = 2^1 - 1;\ S_k: 1 + 2 + 2^2 + \cdots + 2^{k-1} = 2^k - 1;\ S_{k+1}: 1 + 2 + 2^2 + \cdots + 2^k = 2^{k+1} - 1$;
S_{k+1} can be obtained by adding 2^k to both sides of S_k.

19. $S_1: 2 = 2^{1+1} - 2;\ S_k: 2 + 4 + 8 + \cdots + 2^k = 2^{k+1} - 2;\ S_{k+1}: 2 + 4 + 8 + \cdots + 2^{k+1} = 2^{k+2} - 2$;
S_{k+1} can be obtained by adding 2^{k+1} to both sides of S_k.

21. $S_1: 1 \cdot 2 = \frac{1(1 + 1)(1 + 2)}{3};\ S_k: 1 \cdot 2 + 2 \cdot 3 + 3 \cdot 4 + \cdots + k(k + 1) = \frac{k(k + 1)(k + 2)}{3}$;
$S_{k+1}: 1 \cdot 2 + 2 \cdot 3 + 3 \cdot 4 + \cdots + (k + 1)(k + 2) = \frac{(k + 1)(k + 2)(k + 3)}{3}$; S_{k+1} can be obtained by adding $(k + 1)(k + 2)$ to both sides of S_k.

23. $S_1: \frac{1}{1 \cdot 2} = \frac{1}{1+1}$; $S_k: \frac{1}{1 \cdot 2} + \frac{1}{2 \cdot 3} + \frac{1}{3 \cdot 4} + \cdots + \frac{1}{k(k+1)} = \frac{k}{k+1}$; $S_{k+1}: \frac{1}{1 \cdot 2} + \frac{1}{2 \cdot 3} + \frac{1}{3 \cdot 4} + \cdots + \frac{1}{(k+1)(k+2)} = \frac{k+1}{k+2}$;

S_{k+1} can be obtained by adding $\frac{1}{(k+1)(k+2)}$ to both sides of S_k.

25. $S_1: 5 \cdot 6^1 = 6(6^1 - 1)$; $S_k: \sum_{i=1}^{k} 5 \cdot 6^i = 6(6^k - 1)$;

$S_{k+1}: \sum_{i=1}^{k+1} 5 \cdot 6^i = 6(6^{k+1} - 1)$; S_{k+1} can be obtained by adding $5 \cdot 6^{k+1}$ to both sides of S_k.

27. $S_1: 1 + 2 > 1$; $S_k: k + 2 > k$; $S_{k+1}: k + 3 > k + 1$; S_{k+1} can be obtained by adding 1 to both sides of S_k.

29. $S_1: (ab)^1 = a^1b^1$, $S_k: (ab)^k = a^kb^k$; $S_{k+1}: (ab)^{k+1} = a^{k+1}b^{k+1}$; S_{k+1} can be obtained by multiplying both sides of S_k by ab.

31. S_1: 2 is divisible by 2; S_k: $k^2 + k$ is divisible by 2; S_{k+1}: $k^2 + 3k + 2$ is divisible by 2;

S_{k+1} can be obtained from S_k by rewriting $k^2 + 3k + 2$ as $(k^2 + k) + 2(k + 1)$.

37. does not make sense **39.** makes sense

41. $S_1: \frac{1}{2}$; $S_2: \frac{1}{3}$; $S_3: \frac{1}{4}$; $S_4: \frac{1}{5}$; $S_5: \frac{1}{6}$; $S_n: \left(1 - \frac{1}{2}\right)\left(1 - \frac{1}{3}\right)\left(1 - \frac{1}{4}\right) \cdots \left(1 - \frac{1}{n+1}\right) = \frac{1}{n+1}$; $S_1: 1 - \frac{1}{2} = \frac{1}{1+1}$;

$S_k: \left(1 - \frac{1}{2}\right)\left(1 - \frac{1}{3}\right)\left(1 - \frac{1}{4}\right) \cdots \left(1 - \frac{1}{k+1}\right) = \frac{1}{k+1}$; $S_{k+1}: \left(1 - \frac{1}{2}\right)\left(1 - \frac{1}{3}\right)\left(1 - \frac{1}{4}\right) \cdots \left(1 - \frac{1}{k+1}\right)\left(1 - \frac{1}{k+2}\right) = \frac{1}{k+2}$;

S_{k+1} can be obtained by multiplying both sides of S_k by $\left(1 - \frac{1}{k+2}\right)$.

43. $t = -\frac{V - C}{C}$ or $t = \frac{C - V}{C}$ **44.** $\{-3, -1, 2\}$ **45.** center: $(1, -2)$; $r = 3$;

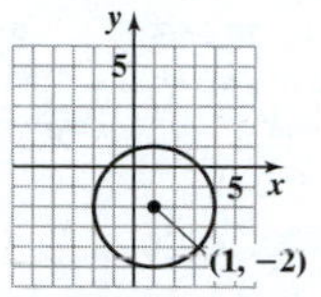

$x^2 + y^2 - 2x + 4y - 4 = 0$

46. 6840 **47.** 56 **48.** true

Section 12.6

Check Point Exercises

1. 72 **2.** 729 **3.** 676,000 **4.** 840 **5.** 720 **6. a.** combinations **b.** permutations **7.** 210 **8.** 1820

Exercise Set 12.6

1. 3024 **3.** 6720 **5.** 720 **7.** 1 **9.** 126 **11.** 330 **13.** 1 **15.** 1 **17.** combinations **19.** permutations **21.** 0 **23.** $\frac{3}{4}$

25. -9499 **27.** $\frac{3}{68}$ **29.** 27 **31.** 40 **33.** 243 **35.** 144 **37.** 120 **39.** 6 **41.** 720 **43.** 8,648,640 **45.** 120 **47.** 15,120

49. 20 **51.** 495 **53.** 24,310 **55.** 22,957,480 **57.** 360 **59.** 1716 **61.** 1140 **63.** 840 **65.** 2730 **67.** 720 **69.** 20 **71.** 24

83. makes sense **85.** does not make sense **87.** false **89.** true **91.** 14,400 **93.** 450

94. $(f \circ g)(x) = 16x^2 - 6$ **95.** $\{x | x < 1 \text{ or } x > 4\}$ or $(-\infty, 1) \cup (4, \infty)$ **96.**

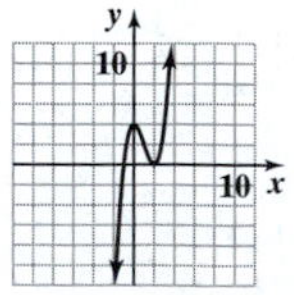

$f(x) = (x - 2)^2 (x + 1)$

97. $\frac{2}{3}$ **98.** $\frac{1}{3}$ **99.** $\frac{2}{3}$

Section 12.7

Check Point Exercises

1. a. $\frac{7664}{100{,}000} \approx 0.077$ **b.** $\frac{720}{800} = 0.9$ **c.** $\frac{720}{7664} \approx 0.094$ **2.** $\frac{1}{3}$ **3.** $\frac{1}{9}$ **4.** $\frac{1}{13}$ **5.** $\frac{1}{13{,}983{,}816} \approx 0.0000000715$ **6.** $\frac{160}{191}$

7. $\frac{1}{3}$ **8.** $\frac{3}{4}$ **9. a.** 0.99 **b.** 0.65 **10.** $\frac{1}{361} \approx 0.00277$ **11.** $\frac{1}{16}$

Exercise Set 12.7

1. 0.10 **3.** 0.52 **5.** 0.01 **7.** 0.59 **9.** 0.61 **11.** $\frac{1}{6}$ **13.** $\frac{1}{2}$ **15.** $\frac{1}{3}$ **17.** $\frac{1}{13}$ **19.** $\frac{3}{13}$ **21.** $\frac{1}{4}$ **23.** $\frac{7}{8}$

25. $\frac{1}{12}$ **27.** $\frac{1}{18{,}009{,}460} \approx 0.0000000555$; $\frac{5}{900{,}473} \approx 0.00000555$ **29. a.** 2,598,960 **b.** 1287 **c.** $\frac{1287}{2{,}598{,}960} \approx 0.000495$

31. $\frac{43}{58}$ **33.** $\frac{50}{87}$ **35.** $\frac{113}{174}$ **37.** $\frac{12}{13}$ **39.** $\frac{2}{13}$ **41.** $\frac{7}{13}$ **43.** $\frac{3}{4}$ **45.** $\frac{33}{40}$ **47.** $\frac{1}{36}$ **49.** $\frac{1}{3}$ **51.** $\frac{1}{64}$

53. a. $\frac{1}{256}$ **b.** $\frac{1}{4096}$ **c.** $\left(\frac{15}{16}\right)^{10} \approx 0.524$ **d.** $1 - \left(\frac{15}{16}\right)^{10} \approx 0.476$ **63.** does not make sense **65.** does not make sense **67.** $\frac{3}{8}$

69. $\frac{1}{10}$ **71.** 0.06 **72.**

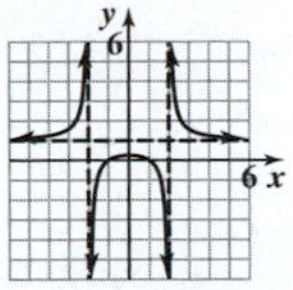

$f(x) = \frac{x^2 - 1}{x^2 - 4}$

73. $\{3\}$ **74.** $x^2 + 3x - 3 - \frac{4}{x + 2}$

Review Exercises

1. 3, 10, 17, 24 **2.** $-\frac{3}{2}, \frac{4}{3}, -\frac{5}{4}, \frac{6}{5}$ **3.** $1, 1, \frac{1}{2}, \frac{1}{6}$ **4.** $\frac{1}{2}, -\frac{1}{4}, \frac{1}{8}, -\frac{1}{16}$ **5.** 95 **6.** -20 **7.** Answers will vary; an example is $\sum_{i=1}^{15} \frac{i}{i + 2}$.

8. Answers will vary; examples are $\sum_{i=4}^{13} i^3$ or $\sum_{i=1}^{10} (i + 3)^3$. **9.** 7, 11, 15, 19, 23, 27 **10.** $-4, -9, -14, -19, -24, -29$ **11.** $\frac{3}{2}, 1, \frac{1}{2}, 0, -\frac{1}{2}, -1$

12. 20 **13.** -30 **14.** -38 **15.** $a_n = 4n - 11; a_{20} = 69$ **16.** $a_n = 220 - 20n; a_{20} = -180$ **17.** $a_n = -\frac{23}{2} - \frac{1}{2}n; a_{20} = -\frac{43}{2}$

18. $a_n = 22 - 7n; a_{20} = -118$ **19.** 1727 **20.** 225 **21.** 15,150 **22.** 440 **23.** -500 **24.** -2325 **25. a.** $a_n = 4.75n + 34.25$ **b.** 96% **26.** \$418,500 **27.** 1470 seats **28.** 3, 6, 12, 24, 48 **29.** $\frac{1}{2}, \frac{1}{4}, \frac{1}{8}, \frac{1}{16}, \frac{1}{32}$ **30.** $16, -4, 1, -\frac{1}{4}, \frac{1}{16}$ **31.** $-5, 5, -5, 5, -5$

32. $a_7 = 1458$ **33.** $a_6 = \frac{1}{2}$ **34.** $a_5 = -48$ **35.** $a_n = 1(2)^{n-1}$ or $a_n = 2^{n-1}; a_8 = 128$ **36.** $a_n = 100\left(\frac{1}{10}\right)^{n-1}; a_8 = \frac{1}{100{,}000} = 0.00001$

37. $a_n = 12\left(-\frac{1}{3}\right)^{n-1}; a_8 = -\frac{4}{729}$ **38.** 17,936,135 **39.** $\frac{127}{8}$ or 15.875 **40.** 19,530 **41.** -258 **42.** $\frac{341}{128}$ **43.** $\frac{27}{2}$ **44.** $\frac{4}{3}$

45. $-\frac{18}{5}$ **46.** 20 **47.** $\frac{2}{3}$ **48.** $\frac{47}{99}$ **49. a.** approximately 1.4 for each division **b.** $a_n = 4.2(1.4)^{n-1}$ **c.** 62.0 million

50. approximately \$42,823; approximately \$223,210 **51. a.** \$19,129 **b.** \$8729 **52. a.** \$91,361 **b.** \$55,361 **53.** $\$9\frac{1}{3}$ million

54. 165 **55.** 4005 **56.** $8x^3 + 12x^2 + 6x + 1$ **57.** $x^8 - 4x^6 + 6x^4 - 4x^2 + 1$ **58.** $x^5 + 10x^4y + 40x^3y^2 + 80x^2y^3 + 80xy^4 + 32y^5$

59. $x^6 - 12x^5 + 60x^4 - 160x^3 + 240x^2 - 192x + 64$ **60.** $x^{16} + 24x^{14} + 252x^{12}$ **61.** $x^9 - 27x^8 + 324x^7$ **62.** $80x^2$ **63.** $4860x^2$

64. $S_1: 5 = \frac{5(1)(1 + 1)}{2}; S_k: 5 + 10 + 15 + \cdots + 5k = \frac{5k(k + 1)}{2}; S_{k+1}: 5 + 10 + 15 + \cdots + 5(k + 1) = \frac{5(k + 1)(k + 2)}{2}$;
S_{k+1} can be obtained by adding $5(k + 1)$ to both sides of S_k.

65. $S_1: 1 = \frac{4^1 - 1}{3}; S_k: 1 + 4 + 4^2 + \cdots + 4^{k-1} = \frac{4^k - 1}{3}; S_{k+1}: 1 + 4 + 4^2 + \cdots + 4^k = \frac{4^{k+1} - 1}{3}$;
S_{k+1} can be obtained by adding 4^k to both sides of S_k.

66. $S_1: 2 = 2(1)^2; S_k: 2 + 6 + 10 + \cdots + (4k - 2) = 2k^2; S_{k+1}: 2 + 6 + 10 + \cdots + (4k + 2) = 2k^2 + 4k + 2$;
S_{k+1} can be obtained by adding $4k + 2$ to both sides of S_k.

67. $S_1: 1 \cdot 3 = \frac{1(1 + 1)[2(1) + 7]}{6}; S_k: 1 \cdot 3 + 2 \cdot 4 + 3 \cdot 5 + \cdots + k(k + 2) = \frac{k(k + 1)(2k + 7)}{6}$;
$S_{k+1}: 1 \cdot 3 + 2 \cdot 4 + 3 \cdot 5 + \cdots + (k + 1)(k + 3) = \frac{(k + 1)(k + 2)(2k + 9)}{6}$;
S_{k+1} can be obtained by adding $(k + 1)(k + 3)$ to both sides of S_k.

68. 336 **69.** 15,120 **70.** 56 **71.** 78 **72.** 20 **73.** 243 **74.** 32,760 **75.** 4845 **76.** 1140 **77.** 116,280 **78.** 120
79. $\frac{18}{25}$ **80.** $\frac{6}{7}$ **81.** $\frac{3}{5}$ **82.** $\frac{12}{35}$ **83.** $\frac{10}{21}$ **84.** $\frac{5}{16}$ **85.** $\frac{2}{3}$ **86.** $\frac{2}{3}$ **87.** $\frac{2}{13}$ **88.** $\frac{7}{13}$ **89.** $\frac{5}{6}$ **90.** $\frac{5}{6}$ **91.** $\frac{1}{6}$

92. a. $\frac{1}{15{,}504} \approx 0.0000645$ **b.** $\frac{25}{3876} \approx 0.00645$ **93.** $\frac{1}{32}$ **94. a.** 0.04 **b.** 0.008 **c.** 0.4096

Chapter Test

1. $1, -\frac{1}{4}, \frac{1}{9}, -\frac{1}{16}, \frac{1}{25}$ **2.** 105 **3.** Answers will vary; examples are $\sum_{i=2}^{21} \frac{i}{i + 1}$ or $\sum_{i=1}^{20} \frac{i + 1}{i + 2}$. **4.** $a_n = 5n - 1; a_{12} = 59$

5. $a_n = 16\left(\frac{1}{4}\right)^{n-1}; a_{12} = \frac{1}{262{,}144}$ **6.** -385 **7.** 550 **8.** -2387 **9.** $-21{,}846$ **10.** 8 **11.** $\frac{73}{99}$ **12.** approximately \$276,427

13. 36 **14.** $x^{10} - 5x^8 + 10x^6 - 10x^4 + 5x^2 - 1$ **15.** $x^8 + 8x^7y^2 + 28x^6y^4$

16. $S_1: 1 = \frac{1[3(1) - 1]}{2}; S_k: 1 + 4 + 7 + \cdots + (3k - 2) = \frac{k(3k - 1)}{2}; S_{k+1}: 1 + 4 + 7 + \cdots + (3k + 1) = \frac{(k + 1)(3k + 2)}{2}$;
S_{k+1} can be obtained by adding $3k + 1$ to both sides of S_k. **17.** 990 **18.** 210 **19.** 10,000 **20.** $\frac{3}{5}$ **21.** $\frac{39}{50}$ **22.** $\frac{3}{5}$

23. $\frac{9}{19}$ **24.** $\frac{10}{1001}$ **25.** $\frac{8}{13}$ **26.** $\frac{3}{5}$ **27.** $\frac{1}{256}$ **28.** $\frac{1}{16}$

Cumulative Review Exercises

1. $\{22\}$ **2.** $\{5 \pm 7i\}$ **3.** $\{x|x < -3 \text{ or } x > 2\}$ or $(-\infty, -3) \cup (2, \infty)$ **4.** $\{-2\}$ **5.** no solution or $\varnothing$ **6.** $\left\{x \middle| x < \frac{2}{3}\right\}$ or $\left(-\infty, \frac{2}{3}\right)$

7. $\{(4, 0, -5)\}$ **8.** $\{9\}$ **9.** $\{(2, 1), (-2, 1), (2, -1), (-2, -1)\}$ **10.** $\{(-14, -20), (2, -4)\}$ **11.** $\left\{\frac{2}{3}, -1 \pm \sqrt{2}\right\}$

12.

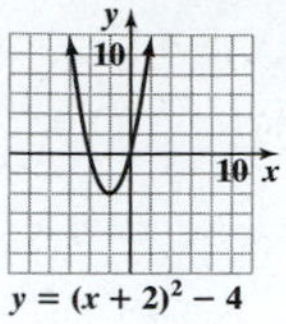

$y = (x + 2)^2 - 4$

13.

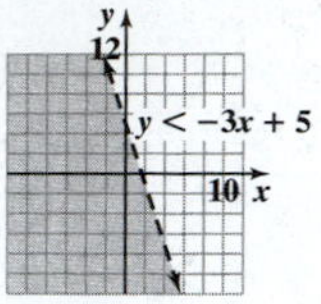

14.

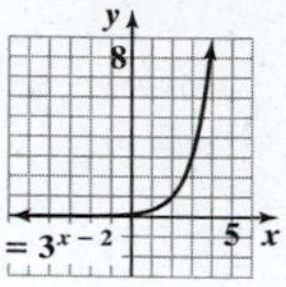

$y = 3^{x-2}$

15.

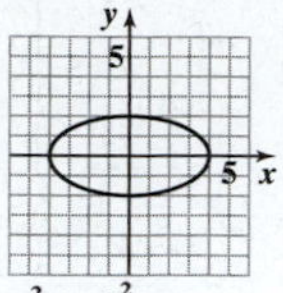

$\frac{x^2}{16} + \frac{y^2}{4} = 1$

16.

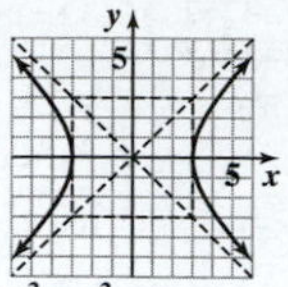

$x^2 - y^2 = 9$

17.

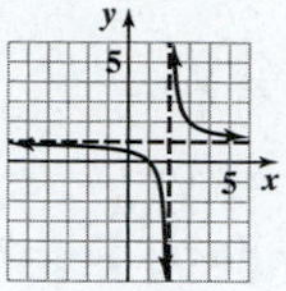

$f(x) = \frac{x-1}{x-2}$

18. $\frac{2x^2 + 5x - 2}{(x - 5)(x + 2)}$ **19.** $-\frac{x + 1}{x - 1}$ **20.** $2\sqrt{5} + 2\sqrt{2}$ **21.** $12\sqrt{5}$ **22.** $\frac{5\sqrt[3]{4xy^2}}{2xy}$ **23.** $(x + y)(5a - 4b)$ **24.** $\log\left(\frac{x^5}{\sqrt{y}}\right)$

25. $p = \frac{qf}{q - f}$ **26.** $3\sqrt{10} \approx 9.49$ units **27.** 208 **28.** 1800 **29.** $\frac{1}{3}$ **30.** $16x^4 - 32x^3y^3 + 24x^2y^6 - 8xy^9 + y^{12}$

31. $\{x|x \text{ is a real number and } x \neq -5 \text{ and } x \neq 3\}$ or $(-\infty, -5) \cup (-5, 3) \cup (3, \infty)$ **32.** $\{x|x \geq 3\}$ or $[3, \infty)$ **33.** $\{x|x < 1\}$ or $(-\infty, 1)$

34. 8 ft by 3 ft **35.** \$18,500 **36.** $k = \frac{1}{2 \ln 4} \approx 0.3607$; about 0.298 or $\frac{298}{1000}$

APPLICATIONS INDEX

A

B

C

D

E

I

J

K

L

P

R

S

T

U

V

W

Y

SUBJECT INDEX

F

G

H

I

Q

PHOTO CREDITS

CHAPTER 1 CO JFPI Studios, Inc./Corbis/Bettmann **p. 2** (left) Kaz Chiba/Image Bank/Getty Images; (right) Digital Vision Ltd./SuperStock, Inc. **p. 13** Mareschal, Tom/Getty Images Inc.-Image Bank **p. 27** Globe Photos, Inc. **p. 37** Gail Albert Halaban/Corbis/Outline **p. 45** Ron Kimball/Kimball Stock **p. 50** Michael Agliolo/Grant Heilman Photography, Inc. **p. 59** U.S. Air Force/AP Wide World Photos **p. 64** Susumu Nishinaga/Photo Researchers, Inc. **p. 76** SuperStock, Inc.

CHAPTER 2 CO Corbis Royalty Free **p. 96** Columbia Pictures/ZUMA/Corbis/Bettmann **p. 105** National Institute for Biological Standards and Control (U.K.)/Science Photo Library/Photo Researchers, Inc. **p. 116** (left) Clay McLachlan/IPN/Aurora & Quanta Productions Inc.; (right) © Robert Llewellyn/CORBIS All Rights Reserved. **p. 126** David Schmidt/Masterfile Corporation **p. 144** © David Stoecklein/CORBIS All Rights Reserved. **p. 155** John Serafin

CHAPTER 3 CO TORU/Getty Images **CO inset** Kyodo/Landov LLC **p. 166** UPPA/Topham/The Image Works **p. 182** (left) Esbin-Anderson/Grant Heilman Photography, Inc.; (right) Frazer Harrison/Getty Images **p. 189** AP/Wide World Photos **p. 193** PEANUTS © United Feature Syndicate, Inc. **p. 196** David W. Hamilton/Getty Images, Inc.-Image Bank **p. 208** © Mike Watson/CORBIS All Rights Reserved. **p. 219** David Parker/Science Museum/Science Photo Library/Photo Researchers, Inc.

CHAPTER 4 CO © Don Mason/CORBIS All Rights Reserved. **p. 240** Phillip & Karen Smith/SuperStock, Inc. **p. 253** Thomas Brummett/Getty Images, Inc.-Photodisc **p. 262** CBS-TV/Picture Desk, Inc./Kobal Collection **p. 276** Frank Clarkson **p. 286** AP Wide World Photos **p. 287** Pascal Parrot/ Corbis/Sygma

CHAPTER 5 CO AFP/Getty Images, Inc.-Agence France Presse **CO inset** Hugh Threlfall/Alamy Images **p. 304** N. Hashimoto/Corbis/Sygma **p. 317** Robert F. Blitzer **p. 326** Steve Lyne/Getty Images **p. 331** Jasper Johns *0 Through 9*, 1961. Oil on canvas, 137 X 105 cm. The Saatchi Collection, Courtesy of the Leo Castelli Gallery. © Jasper Johns/Licensed by VAGA, New York, NY. **p. 339** Bob Watkins/Photofusion **p. 353** © Sandro Vannini/CORBIS **p. 362** Phil Jason/Getty Images, Inc.-Taxi **p. 368** SuperStock, Inc.

CHAPTER 6 CO istockphoto.com **CO inset** istockphoto.com **p. 392** Fritz Hoffmann/The Image Works **p. 406** Bill Pugliano/Liaison Agency/Getty Images, Inc.-Liaison **p. 417** Chip Simons/Chip Simons Photography **p. 425** McAlister, Steve/Getty Images, Inc.-Image Bank **p. 434** Treat Davidson/Photo Researchers, Inc. **p. 441** Sandra Baker/Alamy Images **p. 451** Yuriko Nakao/Corbis/Reuters America LLC **p. 462** © Joseph Sohm/Visions of America/CORBIS All Rights Reserved. **p. 463** (left) A1pix Ltd., UK; (right) A1pix Ltd., UK **p. 465** Gerald French/Corbis/Bettmann **p. 466** SuperStock, Inc. **p. 470** (top and bottom) David Madison/PCN Photography **p. 472** UPI/Corbis/Bettmann

CHAPTER 7 CO PASIEKA/Photo Researchers, Inc. **p. 486** Reuters/Corbis/Reuters America LLC **p. 498** © Fred Bavendam/Peter Arnold, Inc. **p. 508** George Tooker (b. 1920) "Mirror II" 1963, egg tempera on gesso panel, 20 X 20 in., 1968.4. Gift of R. H. Donnelley Erdman (PA 1956). Addison Gallery of American Art, Phillips Academy, Andover, Massachusetts. All Rights Reserved. © George Tooker. **p. 516** Kauko Helavuo/Getty Images, Inc.-Image Bank **p. 524** PEANUTS © United Feature Syndicate, Inc. **p. 533** John G. Ross/Photo Researchers, Inc. **p. 534** Jim West **p. 542** © Steve Chenn/CORBIS **p. 544** © 2005 Roz Chast from Cartoonbank.com. All rights reserved. **p. 551** R. F. Voss "29-Fold M-set Seahorse" computer-generated image. © 1990 R. F. Voss/IBM Research **p. 553** Robotman reprinted by permission of Newspaper Enterprise Association Inc.

CHAPTER 8 CO Photo courtesy of Dr. Andrew Davidhazy **p. 562** Les Stone/The Image Works **p. 575** Mauro Fermariello/Photo Researchers, Inc. **p. 583** Jasper Johns, "Zero."© Jasper Johns/Licensed by VAGA, New York, NY. **p. 589** Joe McBride/Getty Images, Inc.-Stone Allstock **p. 608** Photonica/Getty Images-Photonica Amana America, Inc. **p. 616** © The New Yorker Collection 1995 Warren Miller from cartoonbank.com. All Rights Reserved.

CHAPTER 9 CO (woman sitting) Prentice Hall School Division; **(bundles of money)** MedioImages/Getty Images-Medio Images **p. 640** Jeff Greenberg/PhotoEdit Inc. **p. 646** Tim Davis/Corbis Royalty Free **p. 653** Fred Prouser/Reuters/Landov LLC **p. 668** David Weintraub/Photo Researchers, Inc. **p. 681** Paul Barton/Corbis/Bettmann **p. 693** Purestock/SuperStock, Inc. **p. 707** Mark Leibowitz/Masterfile Corporation **p. 710** Corbis/Sygma

CHAPTER 10 CO ESA/A. Schaller (STScI)/NASA Headquarters **p. 732** Skip Moody/Dembinsky Photo Associates **p. 742** © Kevin Fleming/CORBIS **p. 750** David R. Austen Photography Group **p. 754** Andrea Pistolesi **p. 763** Jeff Hester and Paul Scowen (Arizona State University), and NASA **p. 768** Space Telescope Science Institute **p. 769** Richard E. Prince "The Cone of Apollonius" (detail), fiberglass, steel, paint, graphite, 51 x 18 x 14 in. Collection: Vancouver Art Gallery, Vancouver, Canada. Photo courtesy of Equinox Gallery, Vancouver, Canada. **p. 775** Index Stock Photography

CHAPTER 11 CO Joe Cornish/ Getty Images, Inc.-Stone/Allstock **p. 794** Pierre Perrin/Corbis/Sygma **p. 807** From Hieraymi Candi, "Ni Mediolanensis Medio (Basel, 1554)," Cardano portrait, frontispiece. Smithsonian Institution Libraries. © 2003

Smithsonian Institution. **p. 815** Courtesy of the Library of Congress **p. 820** © 2007 GJ Caulkins **p. 822** Tom Stewart/Corbis/Stock Market **p. 823** Courtesy of knuckletattoos.com

CHAPTER 12 CO Jeff Lewis Photography **CO inset** Darryl Brooks/Brooks Images **p. 844** Dick Morton **p. 853** © Harry Blair **p. 854** David Young-Wolff/PhotoEdit Inc. **p. 864** SuperStock, Inc. **p. 867** Richard Lord/The Image Works **p. 869** (bills) U.S. Bureau of Engraving and Printing **p. 880** Dr. Rudolph Schild/Science Photo Library/Photo Researchers, Inc. **p. 885** From *Science and Civilization in China,* Vol. 3, pg. 135, Fig. "Pascal's triangle as depicted by Chu Shih Chieh," by Joseph Needham, 1959. Reprinted with the permission of Cambridge University Press. **p. 888** Charles Rex Arbogast/AP Wide World Photos **p. 896** AP Wide World Photos **p. 897** SuperStock, Inc. **p. 898** Jonathan Drake/Bloomberg News/Landov LLC **p. 899** Getty Images, Inc.-PhotoDisc **p. 900** (Marilyn Monroe) Picture Desk, Inc./Kobal Collection (James Dean) Imapress/Globe Photos, Inc. (Jim Morrison) Michael Ochs Archives Ltd./Getty Images Inc. (Kurt Cobain) S.I.N./Corbis/Bettmann (Selena) AP Wide World Photos **p. 907** Larry Kolvoord/The Image Works **p. 918** UPI/Corbis/Bettmann

Definitions, Rules, and Formulas

The Real Numbers

Natural Numbers: $\{1, 2, 3, \ldots\}$
Whole Numbers: $\{0, 1, 2, 3, \ldots\}$
Integers: $\{\ldots, -3, -2, -1, 0, 1, 2, 3, \ldots\}$
Rational Numbers: $\{\frac{a}{b} | a \text{ and } b \text{ are integers}, b \neq 0\}$
Irrational Numbers: $\{x | x \text{ is real and not rational}\}$

Basic Rules of Algebra

Commutative: $a + b = b + a$; $ab = ba$
Associative: $(a + b) + c = a + (b + c)$; $(ab)c = a(bc)$
Distributive: $a(b + c) = ab + ac$; $a(b - c) = ab - ac$
Identity: $a + 0 = a$; $a \cdot 1 = a$
Inverse: $a + (-a) = 0$; $a \cdot \frac{1}{a} = 1 (a \neq 0)$
Multiplication Properties: $(-1)a = -a$; $(-1)(-a) = a$; $a \cdot 0 = 0$; $(-a)(b) = (a)(-b) = -ab$; $(-a)(-b) = ab$

Set-Builder Notation, Interval Notation, and Graphs

$(a, b) = \{x | a < x < b\}$
$[a, b) = \{x | a \leq x < b\}$
$(a, b] = \{x | a < x \leq b\}$
$[a, b] = \{x | a \leq x \leq b\}$
$(-\infty, b) = \{x | x < b\}$
$(-\infty, b] = \{x | x \leq b\}$
$(a, \infty) = \{x | x > a\}$
$[a, \infty) = \{x | x \geq a\}$

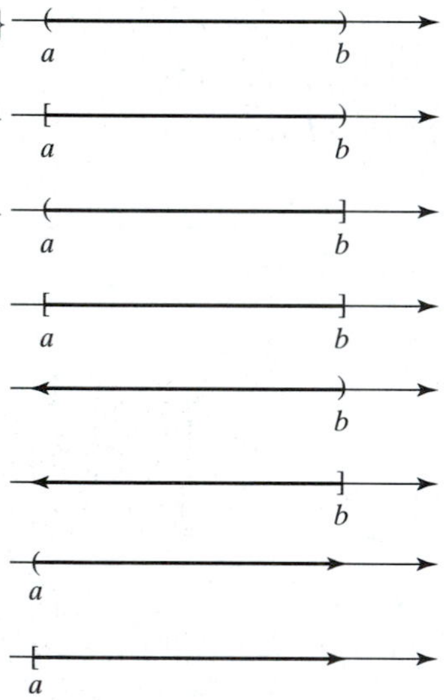

$(-\infty, \infty) = \{x | x \text{ is a real number}\} = \{x | x \in R\}$

Slope Formula

$$\text{slope } (m) = \frac{\text{change in } y}{\text{change in } x} = \frac{y_2 - y_1}{x_2 - x_1} \quad (x_1 \neq x_2)$$

Equations of Lines

1. *Slope-intercept form:* $y = mx + b$
   m is the line's slope and b is its y-intercept.
2. *Standard form:* $Ax + By = C$
3. *Point-slope form:* $y - y_1 = m(x - x_1)$
   m is the line's slope and (x_1, y_1) is a fixed point on the line.
4. *Horizontal line parallel to the x-axis:* $y = b$
5. *Vertical line parallel to the y-axis:* $x = a$

Systems of Equations

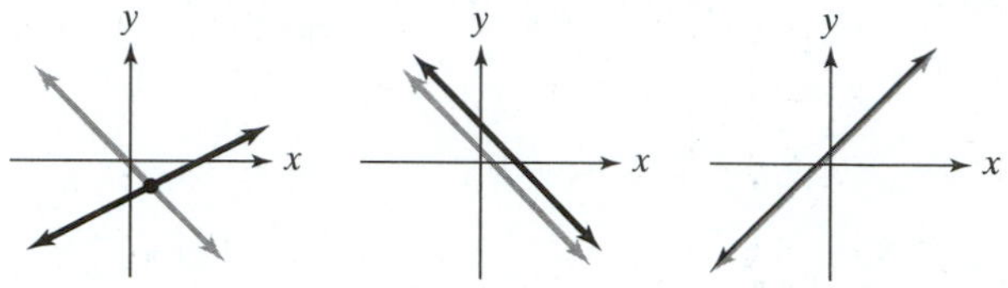

One solution: consistent — No solution: inconsistent — Infinitely many solutions: dependent and consistent

A system of linear equations may be solved: (a) graphically, (b) by the substitution method, (c) by the addition or elimination method, (d) by matrices, or (e) by determinants.

$$\begin{vmatrix} a_1 & b_1 \\ a_2 & b_2 \end{vmatrix} = a_1 b_2 - a_2 b_1$$

Cramer's Rule:

Given a system of a equations of the form

$$a_1 x + b_1 y = c_1, \quad a_2 x + b_2 y = c_2, \quad \text{then } x = \frac{\begin{vmatrix} c_1 & b_1 \\ c_2 & b_2 \end{vmatrix}}{\begin{vmatrix} a_1 & b_1 \\ a_2 & b_2 \end{vmatrix}} \text{ and } y = \frac{\begin{vmatrix} a_1 & c_1 \\ a_2 & c_2 \end{vmatrix}}{\begin{vmatrix} a_1 & b_1 \\ a_2 & b_2 \end{vmatrix}}.$$

Absolute Value

1. $|x| = \begin{cases} x & \text{if } x \geq 0 \\ -x & \text{if } x < 0 \end{cases}$
2. If $|x| = c$, then $x = c$ or $x = -c$. $(c > 0)$
3. If $|x| < c$, then $-c < x < c$. $(c > 0)$
4. If $|x| > c$, then $x < -c$ or $x > c$. $(c > 0)$

Special Factorizations

1. *Difference of two squares:*
   $$A^2 - B^2 = (A + B)(A - B)$$
2. *Perfect square trinomials:*
   $$A^2 + 2AB + B^2 = (A + B)^2$$
   $$A^2 - 2AB + B^2 = (A - B)^2$$

3. *Sum of two cubes:*

$$A^3 + B^3 = (A + B)(A^2 - AB + B^2)$$

4. *Difference of two cubes:*

$$A^3 - B^3 = (A - B)(A^2 + AB + B^2)$$

Variation

English Statement	*Equation*
y varies directly as x.	$y = kx$
y varies directly as x^n.	$y = kx^n$
y varies inversely as x.	$y = \frac{k}{x}$
y varies inversely as x^n.	$y = \frac{k}{x^n}$
y varies jointly as x and z.	$y = kxz$

Exponents

Definitions of Rational Exponents

1. $a^{\frac{1}{n}} = \sqrt[n]{a}$
2. $a^{\frac{m}{n}} = (\sqrt[n]{a})^m$ or $\sqrt[n]{a^m}$
3. $a^{-\frac{m}{n}} = \frac{1}{a^{\frac{m}{n}}}$

Properties of Rational Exponents

If m and n are rational exponents, and a and b are real numbers for which the following expressions are defined, then

1. $b^m \cdot b^n = b^{m+n}$
2. $\frac{b^m}{b^n} = b^{m-n}$
3. $(b^m)^n = b^{mn}$
4. $(ab)^n = a^n b^n$
5. $\left(\frac{a}{b}\right)^n = \frac{a^n}{b^n}$

Radicals

1. If n is even, then $\sqrt[n]{a^n} = |a|$.
2. If n is odd, then $\sqrt[n]{a^n} = a$.
3. The product rule: $\sqrt[n]{a} \cdot \sqrt[n]{b} = \sqrt[n]{ab}$
4. The quotient rule: $\frac{\sqrt[n]{a}}{\sqrt[n]{b}} = \sqrt[n]{\frac{a}{b}}$

Complex Numbers

1. The imaginary unit i is defined as

$$i = \sqrt{-1}, \quad \text{where} \quad i^2 = -1.$$

The set of numbers in the form $a + bi$ is called the set of complex numbers. If $b = 0$, the complex number is a real number. If $b \neq 0$, the complex number is an imaginary number.

2. The complex numbers $a + bi$ and $a - bi$ are conjugates. Conjugates can be multiplied using the formula

$$(A + B)(A - B) = A^2 - B^2.$$

The multiplication of conjugates results in a real number.

3. To simplify powers of i, rewrite the expression in terms of i^2. Then replace i^2 with -1 and simplify.

Quadratic Equations and Functions

1. The solutions of a quadratic equation in standard form

$$ax^2 + bx + c = 0, \quad a \neq 0,$$

are given by the quadratic formula

$$x = \frac{-b \pm \sqrt{b^2 - 4ac}}{2a}.$$

2. The discriminant, $b^2 - 4ac$, of the quadratic equation $ax^2 + bx + c = 0$ determines the number and type of solutions.

Discriminant	**Solutions**
Positive perfect square with a, b, and c rational numbers	2 rational solutions
Positive and not a perfect square	2 irrational solutions
Zero, with a, b, and c rational numbers	1 rational solution
Negative	2 imaginary solutions

3. The graph of the quadratic function

$$f(x) = a(x - h)^2 + k, \quad a \neq 0,$$

is called a parabola. The vertex, or turning point, is (h, k). The graph opens upward if a is positive and downward if a negative. The axis of symmetry is a vertical line passing through the vertex. The graph can be obtained using the vertex, x-intercepts, if any, [set $f(x)$ equal to zero], and the y-intercept (set $x = 0$).

4. A parabola whose equation is in the form

$$f(x) = ax^2 + bx + c, \quad a \neq 0,$$

has its vertex at

$$\left(-\frac{b}{2a}, f\left(-\frac{b}{2a}\right)\right).$$

If $a > 0$, then f has a minimum that occurs at $x = -\frac{b}{2a}$. If $a < 0$, then f has a maximum that occurs at $x = -\frac{b}{2a}$.

Definitions, Rules, and Formulas (continued)

Exponential and Logarithmic Functions

1. Exponential Function: $f(x) = b^x, b > 0, b \neq 1$
Graphs:

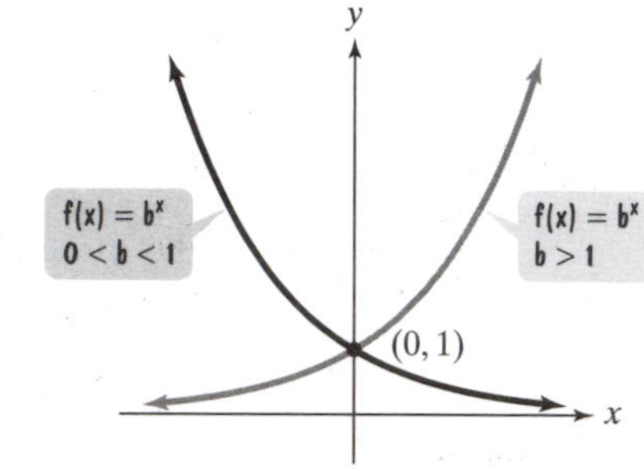

2. Logarithmic Function: $f(x) = \log_b x, b > 0, b \neq 1$
$y = \log_b x$ is equivalent to $x = b^y$.

Graphs:

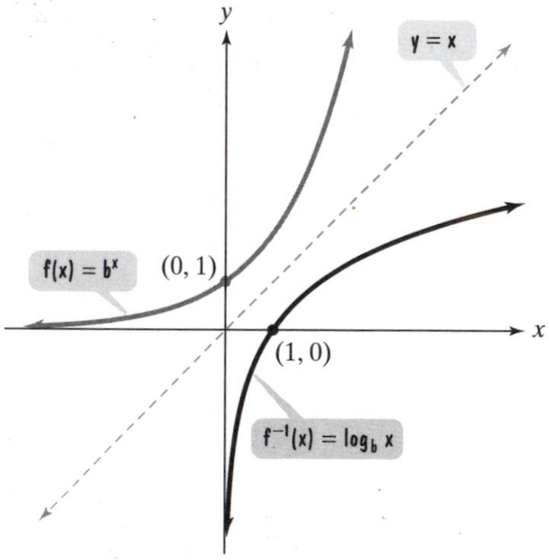

$b > 1$

3. Properties of Logarithms

a. $\log_b(MN) = \log_b M + \log_b N$

b. $\log_b\left(\frac{M}{N}\right) = \log_b M - \log_b N$

c. $\log_b M^p = p \log_b M$

d. $\log_b M = \frac{\log_a M}{\log_a b} = \frac{\ln M}{\ln b} = \frac{\log M}{\log b}$

e. $\log_b b^x = x; \log 10^x = x; \ln e^x = x$

f. $b^{\log_b x} = x; 10^{\log x} = x; e^{\ln x} = x$

Distance and Midpoint Formulas

1. The distance from (x_1, y_1) to (x_2, y_2) is

$$\sqrt{(x_2 - x_1)^2 + (y_2 - y_1)^2}.$$

2. The midpoint of the line segment with endpoints (x_1, y_1) and (x_2, y_2) is

$$\left(\frac{x_1 + x_2}{2}, \frac{y_1 + y_2}{2}\right).$$

Conic Sections

Circle

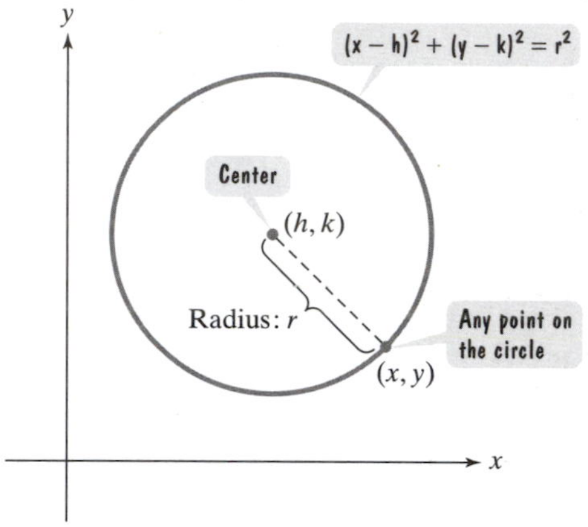

Ellipse

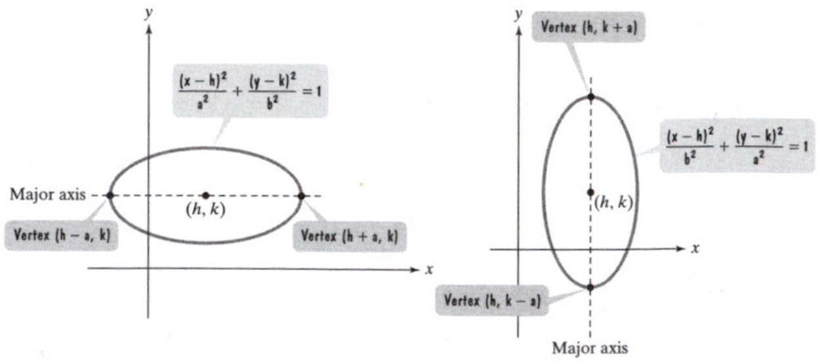

Hyperbola

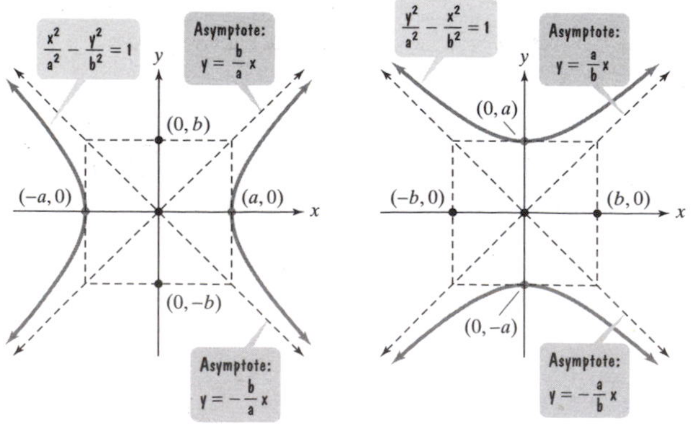